Fundamental Constants

Quantity	Symbol	Value*
Avogadro's number	N_A	$6.022\ 141\ 79 \times 10^{23}\ \text{mol}^{-1}$
Boltzmann's constant	k	$1.380\ 6504 \times 10^{-23}\ \text{J/K}$
Electron charge magnitude	e	$1.602\ 176\ 487 \times 10^{-19}\ \text{C}$
Permeability of free space	μ_0	$4\pi \times 10^{-7}\ \text{T}\cdot\text{m/A}$
Permittivity of free space	ϵ_0	$8.854\ 187\ 817 \times 10^{-12}\ \text{C}^2/(\text{N}\cdot\text{m}^2)$
Planck's constant	h	$6.626\ 068\ 96 \times 10^{-34}\ \text{J}\cdot\text{s}$
Mass of electron	m_e	$9.109\ 382\ 15 \times 10^{-31}\ \text{kg}$
Mass of neutron	m_n	$1.674\ 927\ 211 \times 10^{-27}\ \text{kg}$
Mass of proton	m_p	$1.672\ 621\ 637 \times 10^{-27}\ \text{kg}$
Speed of light in vacuum	c	$2.997\ 924\ 58 \times 10^{8}\ \text{m/s}$
Universal gravitational constant	G	$6.674 \times 10^{-11}\ \text{N}\cdot\text{m}^2/\text{kg}^2$
Universal gas constant	R	$8.314\ 472\ \text{J/(mol}\cdot\text{K)}$

* 2006 CODATA recommended values.

Useful Physical Data

Acceleration due to earth's gravity	$9.80\ \text{m/s}^2 = 32.2\ \text{ft/s}^2$
Atmospheric pressure at sea level	$1.013 \times 10^5\ \text{Pa} = 14.70\ \text{lb/in.}^2$
Density of air (0 °C, 1 atm pressure)	$1.29\ \text{kg/m}^3$
Speed of sound in air (20 °C)	$343\ \text{m/s}$

Water
- Density (4 °C) $1.000 \times 10^3\ \text{kg/m}^3$
- Latent heat of fusion $3.35 \times 10^5\ \text{J/kg}$
- Latent heat of vaporization $2.26 \times 10^6\ \text{J/kg}$
- Specific heat capacity $4186\ \text{J/(kg}\cdot\text{C}^\circ)$

Earth
- Mass $5.98 \times 10^{24}\ \text{kg}$
- Radius (equatorial) $6.38 \times 10^6\ \text{m}$
- Mean distance from sun $1.50 \times 10^{11}\ \text{m}$

Moon
- Mass $7.35 \times 10^{22}\ \text{kg}$
- Radius (mean) $1.74 \times 10^6\ \text{m}$
- Mean distance from earth $3.85 \times 10^8\ \text{m}$

Sun
- Mass $1.99 \times 10^{30}\ \text{kg}$
- Radius (mean) $6.96 \times 10^8\ \text{m}$

Frequently Used Mathematical Symbols

Symbol	Meaning
$=$	is equal to
$\neq$	is not equal to
$\propto$	is proportional to
$>$	is greater than
$<$	is less than
$\approx$	is approximately equal to
$|x|$	absolute value of x (always treated as a positive quantity)
Δ	the difference between two variables (e.g., ΔT is the final temperature minus the initial temperature)
Σ	the sum of two or more variables (e.g., $\sum_{i=1}^{3} x_i = x_1 + x_2 + x_3$)

Conversion Factors

Length

1 in. = 2.54 cm

1 ft = 0.3048 m

1 mi = 5280 ft = 1.609 km

1 m = 3.281 ft

1 km = 0.6214 mi

1 angstrom (Å) = 10^{-10} m

Mass

1 slug = 14.59 kg

1 kg = 1000 grams = 6.852×10^{-2} slug

1 atomic mass unit (u) = 1.6605×10^{-27} kg

(1 kg has a weight of 2.205 lb where the
   acceleration due to gravity is 32.174 ft/s^2)

Time

1 d = 24 h = 1.44×10^3 min = 8.64×10^4 s

1 yr = 365.24 days = 3.156×10^7 s

Speed

1 mi/h = 1.609 km/h = 1.467 ft/s = 0.4470 m/s

1 km/h = 0.6214 mi/h = 0.2778 m/s = 0.9113 ft/s

Force

1 lb = 4.448 N

1 N = 10^5 dynes = 0.2248 lb

Work and Energy

1 J = 0.7376 ft·lb = 10^7 ergs

1 kcal = 4186 J

1 Btu = 1055 J

1 kWh = 3.600×10^6 J

1 eV = 1.602×10^{-19} J

Power

1 hp = 550 ft·lb/s = 745.7 W

1 W = 0.7376 ft·lb/s

Pressure

1 Pa = 1 N/m^2 = 1.450×10^{-4} lb/in.^{2}

1 lb/in.^{2} = 6.895×10^3 Pa

1 atm = 1.013×10^5 Pa = 1.013 bar =
   14.70 lb/in.^{2} = 760 torr

Volume

1 liter = 10^{-3} m^3 = 1000 cm^3 = 0.03531 ft^3

1 ft^3 = 0.02832 m^3 = 7.481 U.S. gallons

1 U.S. gallon = 3.785×10^{-3} m^3 = 0.1337 ft^3

Angle

1 radian = 57.30°

1° = 0.01745 radian

Standard Prefixes Used to Denote Multiples of Ten

Prefix	Symbol	Factor
Tera	T	10^{12}
Giga	G	10^{9}
Mega	M	10^{6}
Kilo	k	10^{3}
Hecto	h	10^{2}
Deka	da	10^{1}
Deci	d	10^{-1}
Centi	c	10^{-2}
Milli	m	10^{-3}
Micro	μ	10^{-6}
Nano	n	10^{-9}
Pico	p	10^{-12}
Femto	f	10^{-15}

Basic Mathematical Formulas

Area of a circle = πr^2

Circumference of a circle = $2\pi r$

Surface area of a sphere = $4\pi r^2$

Volume of a sphere = $\frac{4}{3}\pi r^3$

Pythagorean theorem: $h^2 = h_o^2 + h_a^2$

Sine of an angle: $\sin \theta = h_o/h$

Cosine of an angle: $\cos \theta = h_a/h$

Tangent of an angle: $\tan \theta = h_o/h_a$

Law of cosines: $c^2 = a^2 + b^2 - 2ab \cos \gamma$

Law of sines: $a/\sin \alpha = b/\sin \beta = c/\sin \gamma$

Quadratic formula:

If $ax^2 + bx + c = 0$, then, $x = (-b \pm \sqrt{b^2 - 4ac})/(2a)$

Why WileyPLUS for Physics?

WileyPLUS for Physics is an easy-to-use online learning and assessment system. The problem types and resources are designed to enable and support transferable problem-solving skills and conceptual understanding.

"The problems were set up nicely, and fun to do. There were various difficulties of questions and lots of relevant physics phenomena which were best for understanding the subject. All the additional material [in the online text] is very, very interesting!"

— Student Minh V. Luu, SUNY Geneseo

⊕ *Extensive problem banks include conceptual problems, algorithmic end-of-chapter problems, symbolic notation problems, simulation problems, and tutorial problems.*

⊕ *Instructor-controlled problem-solving assistance within each problem links students to hints, additional examples, videos, and specific sections of the online text.*

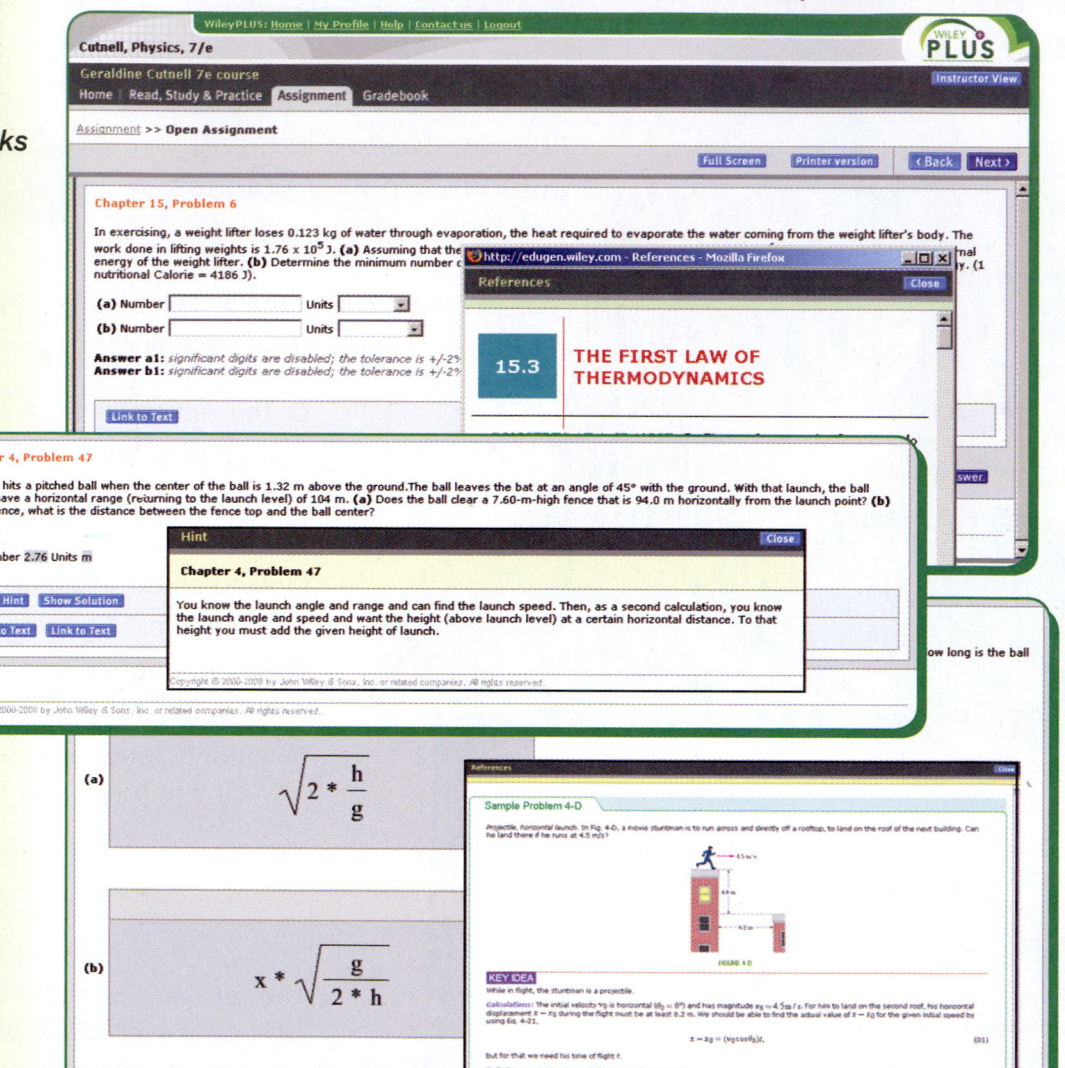

"WileyPLUS complements instructors' lectures, and gives more resources for when I get stuck on homework problems ... well thought out and intuitive, and provides assistance in solving problems with links to the text and useful demonstrations."

— Student Theodore Lee, SUNY Geneseo

See and try WileyPLUS in action!
Details and Demo: www.wileyplus.com

WileyPLUS combines robust course management tools with the complete online text and all of the interactive teaching and learning resources you and your students need in one easy-to-use system.

"WileyPLUS actually got me to do my work on time ... gave a lot of good examples, and got me more prepared for the tests. It really changed the way I studied and learned the material!"
— Student Kelly Gorton, SUNY New Paltz

"The GO Tutorials are great ... very conducive to my understanding of Physics. WileyPLUS gave me another option when I did not know how to solve a problem."
— Justin Ellis, University of Tennessee at Chattanooga

Chapter 4, Problem 41

A car is traveling up a hill that is inclined at an angle θ above the horizontal. Determine the ratio of the magnitude of the normal force to the weight of the car when **(a)** $\theta = 17°$ and **(b)** $\theta = 39°$.

(a) Ratio = [   ]
(b) Ratio = [   ]

[GO Tutorial]

[Link to Text] [Link to Text]

POWERED BY
mapleNET
www.maplesoft.com

Copyright © 2000-2008 by John Wiley & Sons, Inc. o[...]

GO Tutorial   [Close]

This GO Tutorial will provide you with a step-by-step guide on how to approach this problem. When you are finished, go back and try the problem again on your own. To view the original question while you work, you can just drag this screen to the side. **(This GO Tutorial consists of 6 steps).**

Step [1 ▾] : Chapter 4, Problem 41 Solution Step 1

Note: Be aware that the numeric values in this stepped tutorial are different from the numeric values that appear in the question you are attempting to answer.

Concept Question (a) A car is driving up a hill. Is the magnitude of the normal force exerted on the car greater than, equal to, or less than the magnitude of its weight?

○ Equal to.
○ Greater than.
○ It depends on how
○ Less than.

✗ **Incorrect.** The normal force (magnitude = F_N) points perpendicular to the sloping road. In contrast, the weight (magnitude = W) points vertically downward. Hence, it has a component (magnitude = $W_\perp$) that is perpendicular to the hill and a component that is parallel to the slope. Since the car does not move perpendicular to the slope, the normal force and the component of the weight perpendicular to the slope must balance. Thus, $F_N = W_\perp$. Since $W_\perp$ is not the full weight and only a component of it, the magnitude of the normal force cannot be equal to the magnitude of the weight.

Note: Be aware that the numeric values in this stepped tutorial are different from the numeric values that appear in the question you are attempting to answer.

Concept Question (a) A car is driving up a hill. Is the magnitude of the normal force exerted o[...] the car greater than, equal to, or less than the magnitude of its weight?

◉ Equal to.
○ Greater than.

⊕ *Assignable GO™ Tutorial Problems show students how to break down complex problems into simpler steps, and feedback is provided for each step.*

"I loved the GO problems, they broke the problem down and made me realize that I could do this stuff on my own. I have an 'A' because of WileyPLUS."
— Student Nathan Bennett, United States Military Academy

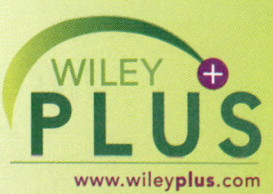

WILEY PLUS
www.wileyplus.com

www.wileyplus.com

Wiley is committed to making your entire *WileyPLUS* experience productive & enjoyable by providing the help, resources, and personal support you & your students need, when you need it. It's all here: www.wileyplus.com –

TECHNICAL SUPPORT:

- A fully searchable knowledge base of FAQs and help documentation, available 24/7
- Live chat with a trained member of our support staff during business hours
- A form to fill out and submit online to ask any question and get a quick response
- **Instructor-only** phone line during business hours: 1.877.586.0192

FACULTY-LED TRAINING THROUGH THE WILEY FACULTY NETWORK:
Register online: www.wherefacultyconnect.com
Connect with your colleagues in a complimentary virtual seminar, with a personal mentor in your field, or at a live workshop to share best practices for teaching with technology.

1ST DAY OF CLASS...AND BEYOND!
Resources You & Your Students Need to Get Started & Use *WileyPLUS* from the first day forward.

- 2-Minute Tutorials on how to set up & maintain your *WileyPLUS* course
- User guides, links to technical support & training options
- *WileyPLUS for Dummies*: Instructors' quick reference guide to using *WileyPLUS*
- Student tutorials & instruction on how to register, buy, and use *WileyPLUS*

YOUR *WileyPLUS* ACCOUNT MANAGER:
Your personal *WileyPLUS* connection for any assistance you need!

SET UP YOUR *WileyPLUS* COURSE IN MINUTES!
Selected *WileyPLUS* courses with QuickStart contain pre-loaded assignments & presentations created by subject matter experts who are also experienced *WileyPLUS* users.

Interested? See and try WileyPLUS in action!
Details and Demo: www.wileyplus.com

8TH EDITION

PHYSICS

ADVANCED EDITION

8TH EDITION

PHYSICS

ADVANCED EDITION

John D. Cutnell & Kenneth W. Johnson

Southern Illinois University at Carbondale

with contributions by

Kent D. Fisher

Columbus State Community College

JOHN WILEY & SONS, INC.

To my wife, Joan Cutnell, a patient friend and my support throughout this project.
To Anne Johnson, my wonderful wife, a caring person, and my best friend.

SPONSORING EDITOR	*Geraldine Osnato*
EXECUTIVE EDITOR	*Stuart Johnson*
SENIOR MEDIA EDITOR	*Thomas Kulesa*
ASSISTANT EDITOR	*Alyson Rentrop*
PRODUCTION EDITOR	*Barbara Russiello*
EXECUTIVE MARKETING MANAGER	*Amanda Wainer*
COVER AND TEXT DESIGNER	*Madelyn Lesure*
SENIOR ILLUSTRATION EDITOR	*Sigmund Malinowski*
ELECTRONIC ILLUSTRATION	*Precision Graphics*
PHOTO RESEARCHER	*Elaine Soares*
PHOTO EDITOR	*Hilary Newman*
COVER PHOTO	*© Paul McKeown/istockphoto*

This book was set in 10/12 Times Roman by Aptara and printed and bound by RR Donnelley. The cover was printed by RR Donnelley.

This book is printed on acid-free paper. ∞

To order books or for customer service, please call 1-800-CALL WILEY (225-5945).

Library of Congress Cataloging in Publication Data:
Cutnell, John D.
  Physics / John D. Cutnell & Kenneth W. Johnson. – 8th ed.

  ISBN 978-0-470-22355-0 (cloth: main) – ISBN 978-0-470-37924-0 (v. 1: pbk.) –
ISBN 978-0-470-37925-7 (v. 2: pbk.)
1. Physics–Textbooks. I. Johnson, Kenneth W. II. Title.
  QC23.2.C87 2008

    2008028063
Main Text (hardcover)	ISBN 978-0-470-22355-0
Volume 1 (paperback)	ISBN 978-0-470-37924-0
Volume 2 (paperback)	ISBN 978-0-470-37925-7
Advanced Edition (hardcover)	ISBN 978-0-470-47544-7
Binder Ready Version	ISBN 978-0-470-40167-5

Printed in the United States of America

10 9 8 7 6 5 4 3 2

BRIEF CONTENTS

CONTENTS

THE PHYSICS OF

The physics of applications of physics principles. To show students that physics has a widespread impact on their lives, we have included a large number of applications of physics principles. Many of these applications are not found in other texts. The most important ones are listed below along with the page number locating the corresponding discussion. They are identified in the margin or in the text of the page on which they occur with the label "The physics of." Biomedical applications are marked with an icon in the shape of a caduceus. The discussions are integrated into the text, so that they occur as a natural part of the physics being presented. It should be noted that the list is not a complete list of all the applications of physics principles to be found in the text. There are many additional applications that are discussed only briefly or that occur in the homework questions and problems.

Continued

PREFACE

We have written this text for students and teachers who are partners in an algebra-based college-level physics course, including the high school course in preparation for the AP Physics exam. The AP exam requires a deep and thorough conceptual understanding of physics, as well as strong problem-solving ability. It is our observation, and also the conclusion of a body of physics education research, that students must internalize physics concepts in order to become expert problem solvers. Thus, this text focuses throughout on the concepts that underpin each new topic. New features in the eighth edition have been designed with both the AP curriculum and the exam in mind. Teachers will find that the goals of the text are aligned directly with the goals of the AP Physics program.

GOALS

CONCEPTUAL UNDERSTANDING Students often regard physics as a collection of equations that can be used blindly to solve problems. However, a good problem-solving technique does not begin with equations. It starts with a firm grasp of physics concepts and how they fit together to provide a coherent description of natural phenomena. Helping students develop a conceptual understanding of physics principles is a primary goal of this text. The features in the text that work toward this goal are:

- *Conceptual Examples*
- *Concepts & Calculations* sections
- *Concepts at a Glance* charts
- *Focus on Concepts* homework material
- *Check Your Understanding* questions
- *Concept Simulations* (an online feature)

REASONING The ability to reason in an organized manner is essential to solving problems, and helping students to improve their reasoning skill is also one of our primary goals. To this end, we have included the following features.

- Explicit reasoning steps in all examples
- *Reasoning Strategies* for solving certain classes of problems
- *Analyzing Multiple-Concept Problems*
- **GO** homework problems (an online feature)
- *Interactive LearningWare* (an online feature)
- *Interactive Solutions* (an online feature)

THE AP PHYSICS B COURSE AND SUCCESS ON THE EXAM The AP Physics course is designed to replicate the typical college-level year-long algebra-based physics course—the college course that often uses this very textbook. The AP exam replicates the level and type of questions that often appear on college physics exams. Thus, the conceptual and reasoning skills developed through the features of this text will help students master the physics and earn college credit and placement by preparing them well for the AP exam.

RELEVANCE Since it is always easier to learn something new if it can be related to day-to-day living, we want to show students that physics principles come into play over and over again in their lives. To emphasize this goal, we have included a wide range of applications of physics principles. Many of these applications are biomedical in nature (for example, transcranial magnetic stimulation). Others deal with modern technology (for example, digital photography). Still others focus on things that we take for granted in our lives (for example, household plumbing). To call attention to the applications, we have used the label *The Physics of* ..., either in the margin or in the text.

ORGANIZATION AND COVERAGE

The text includes 32 chapters and is organized in a fairly standard fashion according to the following sequence: *Mechanics, Thermal Physics, Wave Motion, Electricity and Magnetism, Light and Optics,* and *Modern Physics.* The text is available in a single volume consisting of all 32 chapters. It is also available in two volumes, Volume 1 including Chapters 1–17 (*Mechanics, Thermal Physics,* and *Wave Motion*) and Volume 2 including Chapters 18–32 (*Electricity and Magnetism, Light and Optics,* and *Modern Physics*).

A course that satisfies the College Board's AP Physics B curriculum would include the following sections:

Newtonian Mechanics, chapters 2–7, 10.1–10.4, and 9.1–9.2
Fluid Mechanics, chapter 11
Thermal Physics, chapters 12.1–12.6, 14, and 15
Electricity and Magnetism, chapters 18.1–18.8, 19, 20.1–20.4, 20.6–20.12, 21.1–21.7, 22.1–22.5
Waves and Optics, chapters 16, 17, 24, 25, 26.1–26.9, 27
Atomic and Nuclear Physics, chapters 29.3–29.5, 31

Of course, this text is also designed to provide the necessary resources for *any* algebra-based course. The text is comprehensive, including much more material than will be covered on the AP exam, to the instructor's and students' advantage. Those students who are capable and interested in working at a depth beyond that provided in AP Physics B can read beyond the chapters listed above. Many schools still have a month or more of classes after the mid-May AP exam administration. This text provides numerous options for continued study of physics after the exam. For example, special relativity (chapter 28) often piques student interest but is not part of the AP curriculum. Optional sections, marked with a star (*), should normally be omitted, as they have little impact on the overall development of the material and are unlikely to appear on the AP exam. However, in the weeks after the exam, asterisked sections can provide interesting and useful material.

FEATURES OF THE EIGHTH EDITION

The features in the eighth edition of the text have been fine-tuned to help students in their AP exam preparation. The design of the multiple-choice portion of the AP exam is such that it specifically addresses basic conceptual understandings. For this reason, many of the new features of the eighth edition have incorporated the multiple-choice format. For example, most of our *Conceptual Examples* now appear in a multiple-choice format. Conceptual questions posed in the *Check Your Understanding* sections are available in each section of each chapter. Since the answers are now given at the back of the book, students can immediately check their understanding of the material they have just studied. We include a new feature at the end of each chapter called *Focus on Concepts*, which consists mostly of questions of a conceptual nature, all of which are available for assignment via an online homework management program such as *WileyPLUS* or WebAssign. The appendix on page A–41 describes each of these and other features in detail, showing explicitly how the features can be used as part of an organized effort to prepare students for success on the AP exam.

New! *FOCUS ON CONCEPTS* This new feature is located at the end of every chapter and replaces the *Conceptual Questions* that were part of the seventh edition. It consists primarily of multiple-choice questions that deal with important concepts. Some problems are also included that are designed to avoid mathematical complexity in order to probe basic conceptual understanding. All of the questions and problems are available for assignment via an online homework management program such as *WileyPLUS* or WebAssign.

FOCUS ON CONCEPTS

Note to Instructors: The numbering of the questions shown here reflects the fact that they are only a represent... available online. However, all of the questions are availa...

16. Three identical blocks are being pulled or pushed across a horizontal surface by a force $\vec{F}$, as shown in the drawings. The force $\vec{F}$ in each case has the same magnitude. Rank the kinetic frictional forces that act on the blocks in ascending order (smallest first).

(a) B, C, A   **(b)** C, A, B   **(c)** B, A, C   **(d)** C, B, A   **(e)** A, C, B

A   B   C

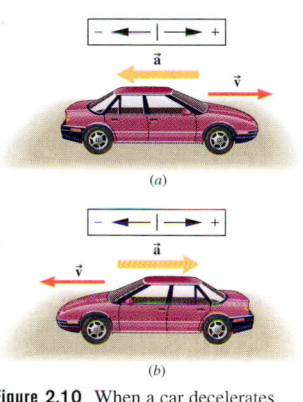

New! **MODIFIED *CONCEPTUAL EXAMPLES*** Conceptual examples appear in every chapter. They are intended as explicit models of how to use physics principles to analyze a situation before attempting to solve a problem numerically that deals with that situation. The *Focus on Concepts* questions provide the homework counterpart to the conceptual examples. Since the majority of the *Focus on Concepts* questions utilize a multiple-choice format, we have modified most of the conceptual examples so they also appear in that format. A small number, however, remain in their original format because they deal with important issues and are not compatible with a multiple-choice presentation.

Conceptual Example 7 Deceleration Versus Negative Acceleration

A car is traveling along a straight road and is decelerating. Which one of the following statements correctly describes the car's acceleration? **(a)** It must be positive. **(b)** It must be negative. **(c)** It could be positive or negative.

Reasoning The term "decelerating" means that the acceleration vector points opposite to the velocity vector and indicates that the car is slowing down. One possibility is that the velocity vector of the car points to the right, in the positive direction, as Figure 2.10a shows. The term "decelerating" implies, then, that the acceleration vector of the car points to the left, which is the negative direction. Another possibility is that the car is traveling to the left, as in Figure 2.10b. Now, since the velocity vector points to the left, the acceleration vector would point opposite, or to the right, which is the positive direction.

Feedback for correct and incorrect answers.

Answers (a) and (b) are incorrect. The term "decelerating" means only that the acceleration vector points opposite to the velocity vector. It is not specified whether the velocity vector of the car points in the positive or negative direction. Therefore, it is not possible to know whether the acceleration is positive or negative.

Most examples are structured so that they lead naturally to homework problems found at the ends of the chapters. These problems contain explicit cross references to the conceptual example.

Answer (c) is correct. As shown in Figure 2.10, the acceleration vector of the car could point in the positive or the negative direction, so that the acceleration could be either positive or negative, depending on the direction in which the car is moving.

Related Homework: *Problems 43, 72*

Figure 2.10 When a car decelerates along a straight road, the acceleration vector points opposite to the velocity vector, as Conceptual Example 7 discusses.

(a)

(b)

New! **EXPANDED *CHECK YOUR UNDERSTANDING*** This feature appears at the ends of selected sections in every chapter and consists of questions in either a multiple-choice or a free-response format. The questions (answers are at the back of the book) are designed to enable students to see if they have understood the concepts discussed in the section. The collection of questions has been substantially increased relative to that present in the seventh edition, and the scope of the questions has been expanded considerably. Teachers who use a classroom response system will also find the questions helpful to use as "clicker" questions.

✓ CHECK YOUR UNDERSTANDING

(The answers are given at the end of the book.)

23. A circus performer hangs stationary from a rope. She then begins to climb upward by pulling herself up, hand over hand. When she starts climbing, is the tension in the rope **(a)** less than, **(b)** equal to, or **(c)** greater than it is when she hangs stationary?

24. A freight train is accelerating on a level track. Other things being equal, would the tension in the coupling between the engine and the first car change if some of the cargo in the last car were transferred to any one of the other cars?

25. Two boxes have masses m_1 and m_2, and m_2 is greater than m_1. The boxes are being pushed across a frictionless horizontal surface. As the drawing shows, there are two possible arrangements, and the pushing force is the same in each. In which arrangement, **(a)** or **(b)**, does the force that the left box applies to the right box have a greater magnitude, or **(c)** is the magnitude the same in both cases?

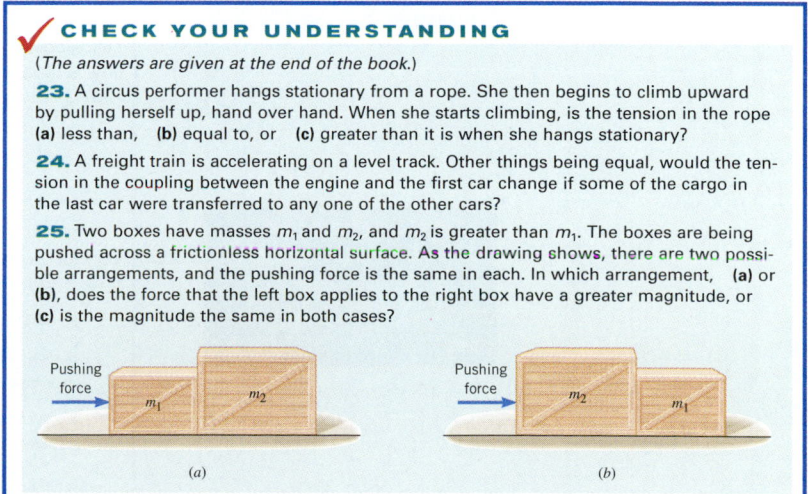

(a)   (b)

EXPANDED AND MODIFIED **GO** *PROBLEMS* Some of the homework problems found in the collection at the end of each chapter are marked with a special **GO** icon. All of these problems are available for assignment via an online homework management program such as *WileyPLUS* or WebAssign. There are 332 **GO** problems, an increase of about 40% over the number present in the seventh edition. In addition, and new to the eighth edition, each of these problems in *WileyPLUS* now includes a guided tutorial option (not graded) that instructors can make available for student access with or without penalty.

Answer input, including direction and units.

Access to the GO tutorial.

Access to a relevant text example.

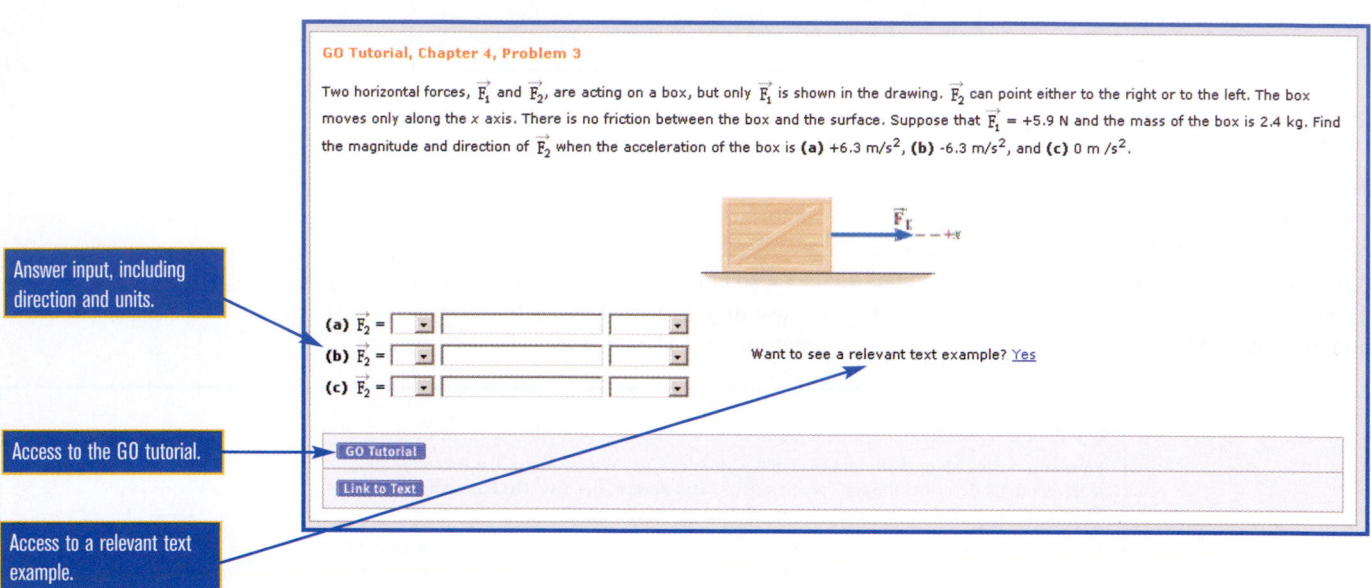

The GO tutorial.

Multiple-choice questions in the GO tutorial include extensive feedback for both correct and incorrect answers.

Multiple-choice questions guide students to the proper conceptual basis for the problem. The GO tutorial also includes calculational steps.

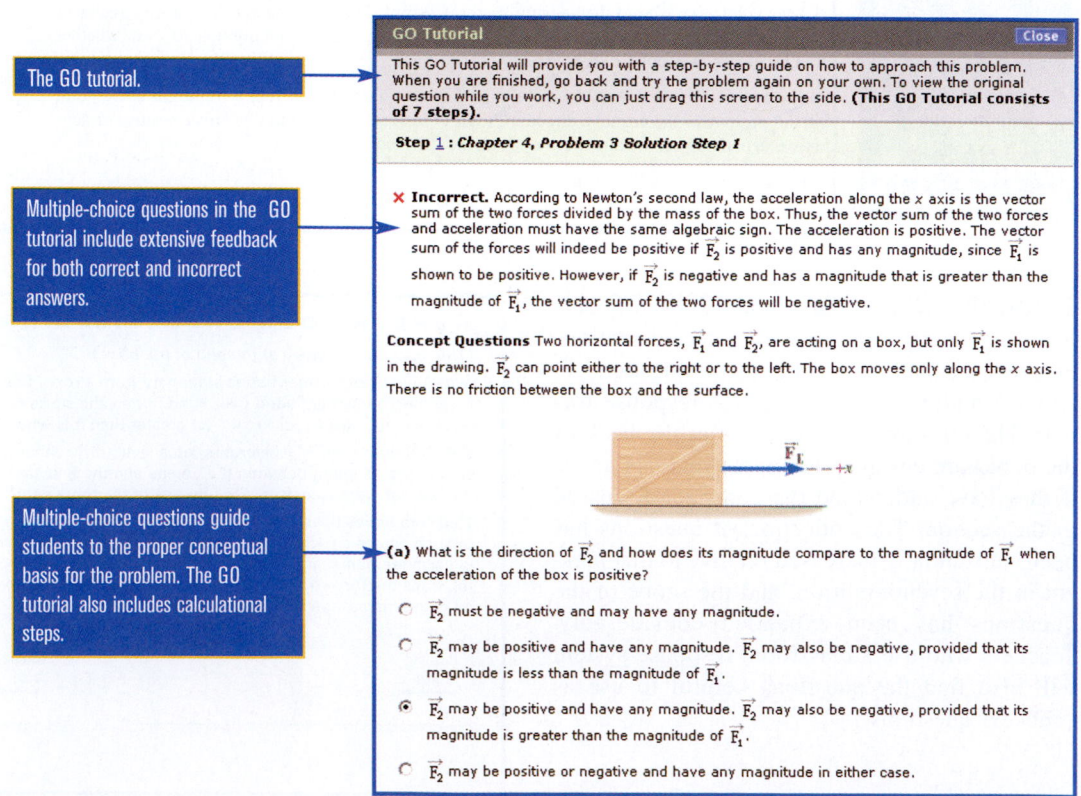

ANALYZING MULTIPLE-CONCEPT PROBLEMS One of the main goals of physics instruction is to help students develop the ability to solve problems that are more thought-provoking than the typical "plug-and-chug" problems. In these more sophisticated or "multiple-concept" problems, students must combine two or more physics concepts before reaching a solution. This is a challenge for them because they must first identify the physics concepts involved in the problem, then associate with each concept an appropriate mathematical equation, and finally assemble the equations to produce a unified algebraic solution. In order to reduce a complex problem into a sum of simpler parts, each Multiple-Concept example consists of four sections: Reasoning, Knowns and Unknowns, Modeling the Problem, and Solution:

> This section discusses the strategy that will be used to solve the problem, and it presents an overview of the physics concepts employed in the solution.

> Each known variable is given a verbal description, an algebraic symbol, and a numerical value. Assigning algebraic symbols is important because the solution is constructed using these symbols. Both explicit data and implicit data are identified because students often focus only on explicitly stated numerical values and overlook data that are present implicitly in the verbal statement of the problem.

> In the left column are the individual steps used in solving the problem. As each step in the left column is presented, the mathematical result of that step is incorporated in the right column into the results from the previous steps, so students can readily see how the individual mathematical equations fit together to produce the desired result.

> This part of the example takes the algebraic equations developed in the modeling section and assembles them into an algebraic solution. Then, the data from the Knowns and Unknowns section are inserted to produce a numerical solution.

> At the end of each Multiple-Concept example, one or more related homework problems are identified, which contain concepts similar to those in the example.

ANALYZING MULTIPLE-CONCEPT PROBLEMS

Example 4 Deep Space 1

The space probe *Deep Space 1* was launched October 24, 1998, and it used a type of engine called an ion propulsion drive. The ion propulsion drive generated only a weak force (or thrust), but could do so for long periods of time using only small amounts of fuel. Suppose the probe, which has a mass of 474 kg, is traveling at an initial speed of 275 m/s. No forces act on it except the 5.60×10^{-2}-N thrust of its engine. This external force $\vec{\mathbf{F}}$ is directed parallel to the displacement $\vec{\mathbf{s}}$, which has a magnitude of 2.42×10^9 m (see Figure 6.6). Determine the final speed of the probe, assuming that its mass remains nearly constant.

The physics of an ion propulsion drive.

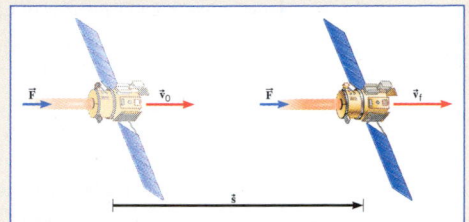

Figure 6.6 The engine of *Deep Space 1* generated a single force $\vec{\mathbf{F}}$ that pointed in the same direction as the displacement $\vec{\mathbf{s}}$. The force performed positive work, causing the space probe to gain kinetic energy.

Reasoning If we can determine the final kinetic energy of the space probe, we can determine its final speed, since kinetic energy is related to mass and speed according to Equation 6.2 and the mass of the probe is known. We will use the work–energy theorem ($W = KE_f - KE_0$), along with the definition of work, to find the final kinetic energy.

Knowns and Unknowns The following list summarizes the data for this problem:

Description	Symbol	Value	Comment
Explicit Data			
Mass	m	474 kg	
Initial speed	v_0	275 m/s	
Magnitude of force	F	5.60×10^{-2} N	
Magnitude of displacement	s	2.42×10^9 m	
Implicit Data			
Angle between force $\vec{\mathbf{F}}$ and displacement $\vec{\mathbf{s}}$	θ	$0°$	The force is parallel to the displacement.
Unknown Variable			
Final speed	v_f	?	

Modeling the Problem

STEP 1 Kinetic Energy An object of mass m and speed v has a kinetic energy KE given by Equation 6.2 as $KE = \frac{1}{2}mv^2$. Using the subscript f to denote the final kinetic energy and the final speed of the probe, we have that

$$KE_f = \frac{1}{2}mv_f^2$$

Solving for v_f gives Equation 1 at the right. The mass m is known but the final kinetic energy KE_f is not, so we will turn to Step 2 to evaluate it.

$$v_f = \sqrt{\frac{2(KE_f)}{m}} \quad (1)$$
$$?$$

STEP 2 The Work–Energy Theorem The work–energy theorem relates the final kinetic energy KE_f of the probe to its initial kinetic energy KE_0 and the work W done by the force of the engine. According to Equation 6.3, this theorem is $W = KE_f - KE_0$. Solving for KE_f shows that

$$KE_f = KE_0 + W$$

The initial kinetic energy can be expressed as $KE_0 = \frac{1}{2}mv_0^2$, so the expression for the final kinetic energy becomes

$$\boxed{KE_f = \frac{1}{2}mv_0^2 + W}$$

This result can be substituted into Equation 1, as indicated at the right. Note from the data table that we know the mass m and the initial speed v_0. The work W is not known and will be evaluated in Step 3.

$$v_f = \sqrt{\frac{2(KE_f)}{m}} \quad (1)$$
$$\boxed{KE_f = \frac{1}{2}mv_0^2 + W} \quad (2)$$
$$?$$

STEP 3 Work The work W is that done by the net external force acting on the space probe. Since there is only the one force $\vec{\mathbf{F}}$ acting on the probe, it is the net force. The work done by this force is given by Equation 6.1 as

$$\boxed{W = (F \cos \theta)s}$$

where F is the magnitude of the force, θ is the angle between the force and the displacement, and s is the magnitude of the displacement. All the variables on the right side of this equation are known, so we can substitute it into Equation 2, as shown in the right column.

$$v_f = \sqrt{\frac{2(KE_f)}{m}} \quad (1)$$
$$\boxed{KE_f = \frac{1}{2}mv_0^2 + W} \quad (2)$$
$$\boxed{W = (F \cos \theta)s}$$

Solution Algebraically combining the results of the three steps, we have

| STEP 1 | STEP 2 | STEP 3 |

$$v_f = \sqrt{\frac{2(KE_f)}{m}} = \sqrt{\frac{2(\frac{1}{2}mv_0^2 + W)}{m}} = \sqrt{\frac{2[\frac{1}{2}mv_0^2 + (F \cos \theta)s]}{m}}$$

The final speed of the space probe is

$$v_f = \sqrt{\frac{2[\frac{1}{2}mv_0^2 + (F \cos \theta)s]}{m}}$$

$$= \sqrt{\frac{2[\frac{1}{2}(474 \text{ kg})(275 \text{ m/s})^2 + (5.60 \times 10^{-2} \text{ N})(\cos 0°)(2.42 \times 10^9 \text{ m})]}{474 \text{ kg}}} = \boxed{805 \text{ m/s}}$$

Related Homework: *Problems 21, 22*

CONCEPTS & CALCULATIONS To emphasize the role of conceptual understanding in solving problems, every chapter includes a *Concepts & Calculations* section. These sections are organized around a special type of example, each of which begins with several conceptual questions that are answered before the quantitative problem is worked out. The purpose of the questions is to focus attention on the concepts with which the problem deals. These examples also provide mini-reviews of material studied earlier in the chapter and in previous chapters.

Concepts & Calculations Example 19   The Buoyant Force

A father (weight $W = 830$ N) and his daughter (weight $W = 340$ N) are spending the day at the lake. They are each sitting on a beach ball that is just submerged beneath the water (see Figure 11.41). Ignoring the weight of the air within the balls and the volumes of their legs that are under water, find the radius of each ball.

Concept Questions and Answers Each beach ball is in equilibrium, being stationary and having no acceleration. Thus, the net force acting on each ball is zero. What balances the downward-acting weight in each case?

Answer   The downward-acting weight is balanced by the upward-acting buoyant force F_B that the water applies to the ball.

In which case is the buoyant force greater?

Answer   The buoyant force acting on the father's beach ball is greater, since it must balance his greater weight.

In the situation described, what determines the magnitude of the buoyant force?

Answer   According to Archimedes' principle, the magnitude of the buoyant force equals the weight of the fluid that the ball displaces. Since the ball is completely submerged, it displaces a volume of water that equals the ball's volume. The weight of this volume of water is the magnitude of the buoyant force.

Which beach ball has the larger radius?

Answer   The father's ball has the larger volume and the larger radius. This follows because a larger buoyant force acts on that ball. For the buoyant force to be larger, that ball must displace a greater volume of water, according to Archimedes' principle. Therefore, the volume of that ball is larger, since the balls are completely submerged.

Solution Since the balls are in equilibrium, the net force acting on each of them must be zero. Therefore, taking upward to be the positive direction, we have

$$\underbrace{\Sigma F}_{\text{Net force}} = F_B - W = 0$$

Figure 11.41 The two bathers are sitting on different-sized beach balls that are just submerged beneath the water.

CONCEPTS AT A GLANCE To provide a coherent picture of how new concepts are built upon previous ones and to reinforce fundamental unifying ideas, the text includes *Concepts at a Glance* flowcharts. Within a chart, new concepts are placed in gold panels, and previously introduced and related concepts are placed in light blue panels. Each chart is discussed in a separate and highlighted paragraph that helps students connect the new concept with previous ones.

▶ **CONCEPTS AT A GLANCE**   The total mechanical energy E is a familiar idea that we originally defined to be the sum of the translational kinetic energy and the gravitational potential energy (see Section 6.5). Then, we included the rotational kinetic energy, as the chart in Figure 9.24 shows. We now expand the total mechanical energy to include the elastic potential energy, as the Concepts-at-a-Glance chart in Figure 10.17 indicates. ◀

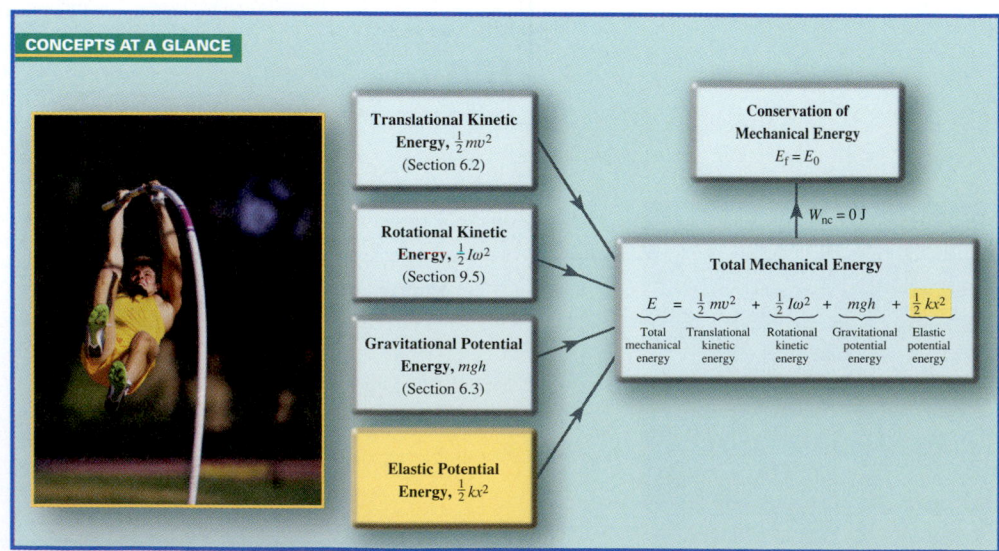

EXPLICIT REASONING STEPS Since reasoning is the cornerstone of problem solving, we have stated the reasoning in all examples. In this step we explain what motivates our procedure for solving the problem before any algebraic or numerical work is done. In the *Concepts & Calculations* examples, the reasoning is presented in a question and answer format.

 Example 6 Ice Skaters

Starting from rest, two skaters push off against each other on smooth level ice, where friction is negligible. As Figure 7.10a shows, one is a woman ($m_1 = 54$ kg), and one is a man ($m_2 = 88$ kg). Part *b* of the drawing shows that the woman moves away with a velocity of $v_{f1} = +2.5$ m/s. Find the "recoil" velocity v_{f2} of the man.

Reasoning For a system consisting of the two skaters on level ice, the sum of the external forces is zero. This is because the weight of each skater is balanced by a corresponding normal force and friction is negligible. The skaters, then, constitute an isolated system, and the principle of conservation of linear momentum applies. We expect the man to have a smaller recoil speed for the following reason: The internal forces that the man and woman exert on each other during pushoff have equal magnitudes but opposite directions, according to Newton's action–reaction law. The man, having the larger mass, experiences a smaller acceleration according to Newton's second law. Hence, he acquires a smaller recoil speed.

Solution The total momentum of the skaters before they push on each other is zero, since they are at rest. Momentum conservation requires that the total momentum remains zero after the skaters have separated, as in part *b* of the drawing:

(a) Before pushoff

(b) After pushoff

REASONING STRATEGY

Applying the Principle of Conservation of Linear Momentum

1. Decide which objects are included in the system.

2. Relative to the system that you have chosen, identify the internal forces and the external forces.

3. Verify that the system is isolated. In other words, verify that the sum of the external forces applied to the system is zero. Only if this sum is zero can the conservation principle be applied. If the sum of the average external forces is not zero, consider a different system for analysis.

4. Set the total final momentum of the isolated system equal to the total initial momentum. Remember that linear momentum is a vector. If necessary, apply the conservation principle separately to the various vector components.

REASONING STRATEGIES A number of the examples in the text deal with well-defined strategies for solving certain types of problems. In such cases, we have included summaries of the steps involved. These summaries, which are titled *Reasoning Strategies*, encourage frequent review of the techniques used and help students focus on the related concepts.

THE PHYSICS OF The text contains about 260 real-world applications that reflect our commitment to showing students how relevant physics is in their lives. Each application is identified in the margin or in the text with the label *The Physics of*, and those that deal with biological material are further marked with an icon in the shape of a caduceus. A complete list of the applications can be found after the Table of Contents.

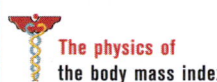 **The physics of the body mass index.**

 Example 3 Body Mass Index

The body mass index (BMI) takes into account your mass in kilograms (kg) and your height in meters (m) and is defined as follows:

$$\text{BMI} = \frac{\text{Mass in kg}}{(\text{Height in m})^2}$$

However, the BMI is often computed using the weight* of a person in pounds (lb) and his or her height in inches (in.). Thus, the expression for the BMI incorporates these quantities, rather than the mass in kilograms and the height in meters. Starting with the definition above, determine the expression for the BMI that uses pounds and inches.

PROBLEM-SOLVING INSIGHTS To reinforce the problem-solving techniques illustrated in the worked-out examples, we have included short statements in the margins or in the text, identified by the label *Problem-solving insight*. These statements help students to develop good problem-solving skills by providing the kind of advice that an instructor might give when explaining a calculation in detail.

Reasoning When the bottom surfaces of the piston and plunger are at the same level, as in Figure 11.14a, Equation 11.5 applies, and we can use it to determine F_1.

Solution According to Equation 11.5, we have

$$F_2 = F_1\left(\frac{A_2}{A_1}\right) \quad \text{or} \quad F_1 = F_2\left(\frac{A_1}{A_2}\right)$$

Problem-solving insight
Note that the relation $F_1 = F_2(A_1/A_2)$, which results from Pascal's principle, applies only when the points 1 and 2 lie at the same depth ($h = 0$ m) in the fluid.

Using $A = \pi r^2$ for the circular areas of the piston and plunger, we find that

$$F_1 = F_2\left(\frac{A_1}{A_2}\right) = F_2\left(\frac{\pi r_1^2}{\pi r_2^2}\right) = (20\ 500\ \text{N})\frac{(0.0120\ \text{m})^2}{(0.150\ \text{m})^2} = \boxed{131\ \text{N}}$$

HOMEWORK MATERIAL The homework material consists of the *Focus on Concepts* questions and the *Problems* found at the end of each chapter. Approximately 700 new questions and problems have been added to this edition. Of course this means there is a greater selection of problems, not that even more problems will be assigned. The problems are labeled according to approximate difficulty. The "no-star" and "one-star" problems are most indicative of the level of reasoning required on the AP exam.

Most of the homework material is available for assignment via an online homework management program such as *WileyPLUS* or WebAssign. In *WileyPLUS* the problems marked with the GO icon are presented in a guided tutorial format that provides enhanced interactivity. The number of such problems in this edition has been increased by about 40%.

12. GO Review Conceptual Example 7 before starting this problem. A uniform plank of length 5.0 m and weight 225 N rests horizontally on two supports, with 1.1 m of the plank hanging over the right support (see the drawing). To what distance x can a person who weighs 450 N walk on the overhanging part of the plank before it just begins to tip?

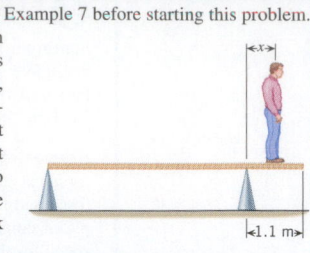

5. ▼ **ssm** The drawing shows a lower leg being exercised. It has a 49-N weight attached to the foot and is extended at an angle θ with respect to the vertical. Consider a rotational axis at the knee. (a) When $\theta = 90.0°$, find the magnitude of the torque that the weight creates. (b) At what angle θ does the magnitude of the torque equal 15 N·m?

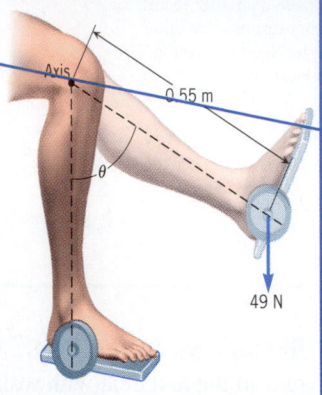

0.55 m

Axis

θ

49 N

In all of the homework material, we have used a variety of real-world situations with realistic data. Those problems marked with a caduceus deal with biomedical situations, and a special effort has been made to increase the amount of this type of homework material.

ADDITIONAL PROBLEMS

85. ssm The main water line enters a house on the first floor. The line has a gauge pressure of 1.90×10^5 Pa. (a) A faucet on the second floor, 6.50 m above the first floor, is turned off. What is the gauge pressure at this faucet? (b) How high could a faucet be before no water would flow from it, even if the faucet were open?

86. Prairie dogs are burrowing rodents. They do not suffocate in their burrows, because the effect of air speed on pressure creates sufficient air circulation. The animals maintain a difference in the shapes of two entrances to the burrow, and because of this difference, the air ($\rho = 1.29$ kg/m³) blows past the openings at different speeds, as the drawing indicates. Assuming that the openings are at the same vertical level, find the difference in air pressure between the openings and indicate which way the air circulates.

Instructors often want to assign homework without identifying a particular section from the text. Such a group of problems is provided under the heading *Additional Problems*.

59. ssm www Two disks are rotating about the same axis. Disk A has a moment of inertia of 3.4 kg·m² and an angular velocity of +7.2 rad/s. Disk B is rotating with an angular velocity of −9.8 rad/s. The two disks are then linked together without the aid of any external torques, so that they rotate as a single unit with an angular velocity of −2.4 rad/s. The axis of rotation for this unit is the same as that for the separate disks. What is the moment of inertia of disk B?

Problems whose solutions appear in the *Student Solutions Manual* are identified with the label **ssm**. Those whose solutions are available online (**www.wiley.com/college/cutnell**) are marked with the label **www**.

CONCEPT SUMMARIES All chapter-ending summaries present an abridged, but complete, version of the material organized section by section and include important equations. The summaries are designed to facilitate student review by including lists of topics and relevant learning aids.

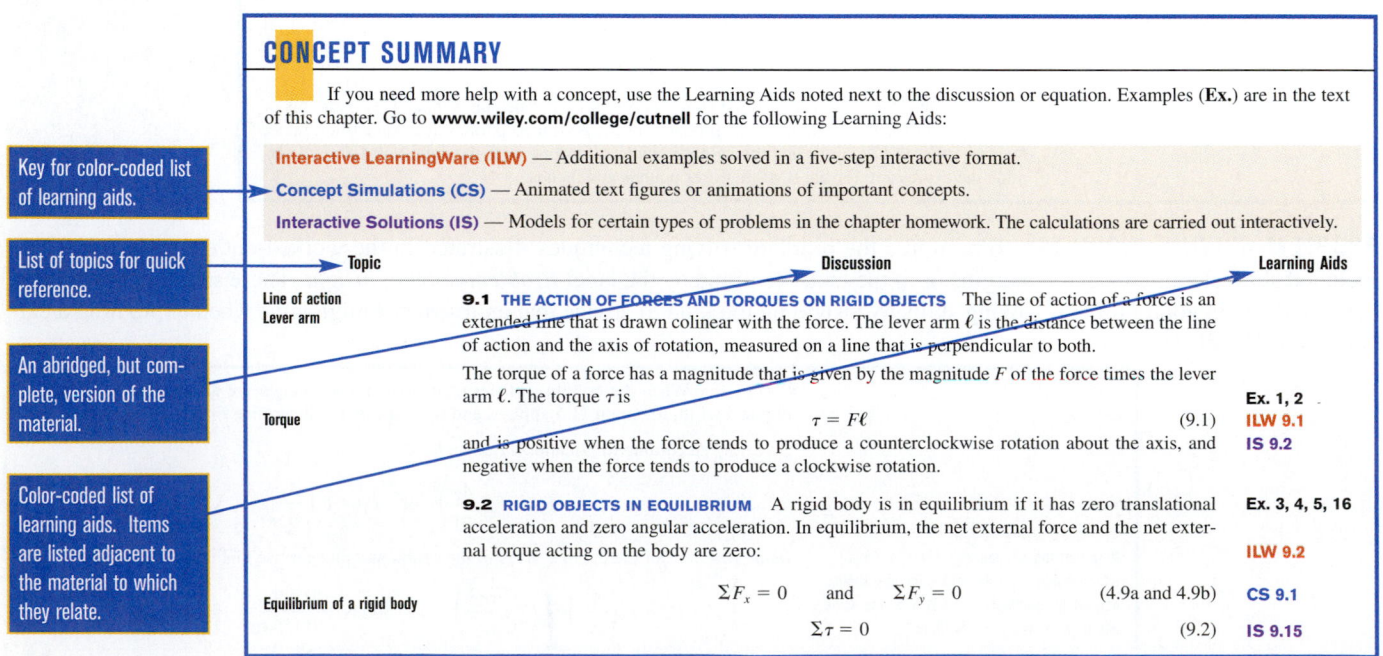

CONCEPT SUMMARY

If you need more help with a concept, use the Learning Aids noted next to the discussion or equation. Examples (**Ex.**) are in the text of this chapter. Go to **www.wiley.com/college/cutnell** for the following Learning Aids:

Interactive LearningWare (ILW) — Additional examples solved in a five-step interactive format.

Concept Simulations (CS) — Animated text figures or animations of important concepts.

Interactive Solutions (IS) — Models for certain types of problems in the chapter homework. The calculations are carried out interactively.

Key for color-coded list of learning aids.

List of topics for quick reference.

An abridged, but complete, version of the material.

Color-coded list of learning aids. Items are listed adjacent to the material to which they relate.

Topic	Discussion	Learning Aids
Line of action Lever arm	**9.1 THE ACTION OF FORCES AND TORQUES ON RIGID OBJECTS** The line of action of a force is an extended line that is drawn colinear with the force. The lever arm ℓ is the distance between the line of action and the axis of rotation, measured on a line that is perpendicular to both.	
Torque	The torque of a force has a magnitude that is given by the magnitude F of the force times the lever arm ℓ. The torque τ is $$\tau = F\ell \qquad (9.1)$$ and is positive when the force tends to produce a counterclockwise rotation about the axis, and negative when the force tends to produce a clockwise rotation.	Ex. 1, 2 ILW 9.1 IS 9.2
Equilibrium of a rigid body	**9.2 RIGID OBJECTS IN EQUILIBRIUM** A rigid body is in equilibrium if it has zero translational acceleration and zero angular acceleration. In equilibrium, the net external force and the net external torque acting on the body are zero: $$\Sigma F_x = 0 \quad \text{and} \quad \Sigma F_y = 0 \qquad \text{(4.9a and 4.9b)}$$ $$\Sigma \tau = 0 \qquad (9.2)$$	Ex. 3, 4, 5, 16 ILW 9.2 CS 9.1 IS 9.15

INTERACTIVE LEARNINGWARE This feature consists of interactive calculational examples available on our Web site (**www.wiley.com/college/cutnell**). Each example is presented in a five-step format designed to improve students' problem-solving skills. The format is similar to that used in the text for the examples in the *Analyzing Multiple-Concept Problems* feature. For student use, the *Interactive LearningWare* examples are referenced in the chapter-ending *Concept Summaries* and in related homework problems.

Animations and drawings help students understand the problem statement.

Multiple-choice questions and answers establish the conceptual basis for the solution.

The pertinent variables are presented interactively in tables that evolve as the example proceeds.

The student models the problem using equations from the text.

The numerical solution is obtained.

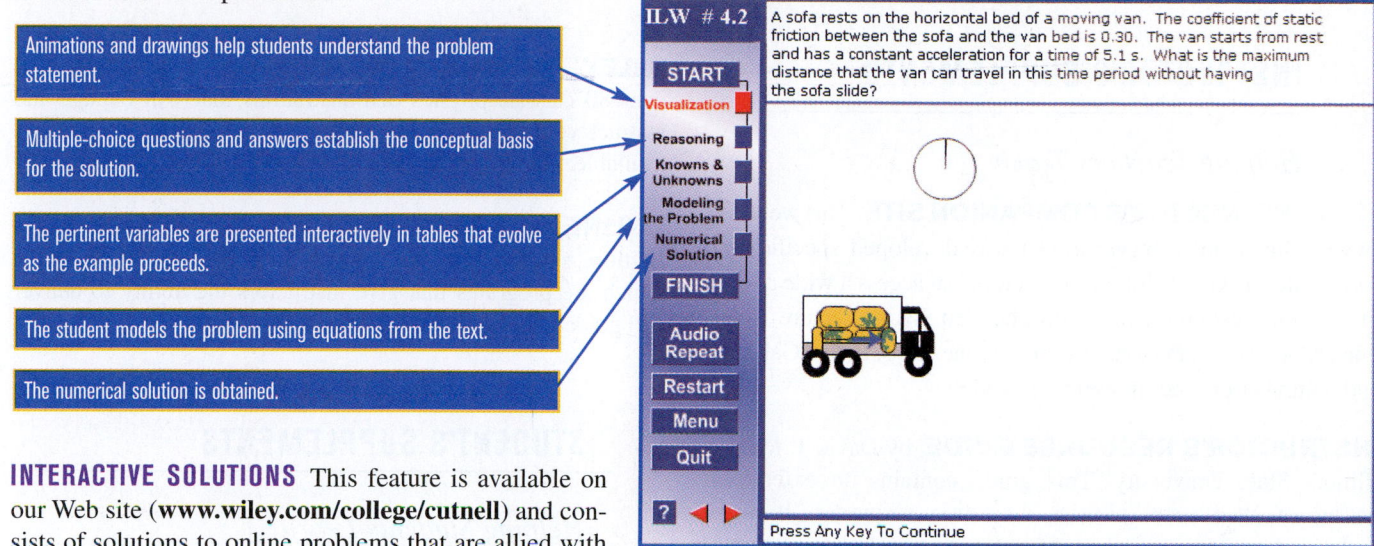

INTERACTIVE SOLUTIONS This feature is available on our Web site (**www.wiley.com/college/cutnell**) and consists of solutions to online problems that are allied with particular homework problems in the text. Each solution is worked out by the student in an interactive manner and is designed to serve as a model for the associated homework problem. The *Interactive Solutions* are referenced in the chapter-ending *Concept Summaries* and in the associated homework problems.

CONCEPT SIMULATIONS This feature is available online at **www.wiley.com/college/cutnell** and consists of simulations that students can use to learn about concepts such as relative velocities, collisions, and ray tracing, since various parameters are under user control. Many of the simulations are directly related to homework material and are specifically referenced in the chapter-ending *Concept Summaries* and in the associated homework problems.

SOLUTIONS The *Solutions* to all of the end-of-chapter problems are available to instructors (and approximately one-half of the solutions to the odd-numbered problems are available to students). In general, the solutions are divided into two parts: *Reasoning* and *Solution*. The *Reasoning* section, like that in the text examples, presents an overview of the physics principles used in solving the problem. The *Solution* section takes the physics principles outlined in the *Reasoning* section and assembles them in a step-by-step manner into an algebraic solution for the problem. The data are then inserted to produce a numerical answer.

Where appropriate, drawings—such as free-body diagrams—are included to aid the student in visualizing the situation. Proper procedures for significant figures are adhered to throughout the solution.

44. *REASONING* The upward normal force (magnitude F_N) that the pilot's seat exerts on him is part of the centripetal force that keeps him on the vertical circular path. However, there is another contribution to the centripetal force, as the free-body diagram at the right shows. This additional contribution is the pilot's downward-acting weight (magnitude W). The centripetal force (magnitude F_c) is the net force that acts on the pilot in the radial direction and must be directed toward the center of the circular path. Noting that the upward direction (toward the center of the circle) is positive in the drawing, the centripetal force is also positive. Thus, $F_c = F_N - W$, or $F_N = F_c + W$. The magnitude of the centripetal force is given by $F_c = mv^2/r$ (Equation 5.3), where m is the mass of the pilot, v is his speed, and r is the radius of the circular path. The magnitude W of the pilot's weight is the product of his mass m and the magnitude g of the acceleration due to gravity, or $W = mg$ (Equation 4.5).

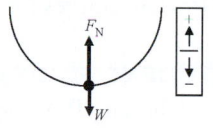

Free-body diagram for the pilot

SOLUTION Dividing the relation $F_N = F_c + W$ by the magnitude of the weight, we can write the ratio F_N/W as

$$\frac{F_N}{W} = \frac{F_c + W}{W} \qquad \text{or} \qquad \frac{F_N}{W} = \frac{F_c}{W} + 1 \tag{1}$$

Substituting $F_c = mv^2/r$ (Equation 5.3) into Equation 1 yields

$$\frac{F_N}{W} = \frac{F_c}{W} + 1 = \frac{mv^2/r}{W} + 1 = \frac{mv^2}{rW} + 1 \tag{2}$$

But $W = mg$ (Equation 4.5), so we substitute this expression into the right hand side of Equation 2. The result is

$$\frac{F_N}{W} = \frac{mv^2}{rW} + 1 = \frac{mv^2}{r(mg)} + 1 = \frac{v^2}{rg} + 1 \tag{3}$$

The ratio F_N/W is

$$\frac{F_N}{W} = \frac{v^2}{rg} + 1 = \frac{(230 \text{ m/s})^2}{(690 \text{ m})(9.80 \text{ m/s}^2)} + 1 = \boxed{8.8}$$

In spite of our best efforts to produce an error-free book, errors no doubt remain. They are solely our responsibility, and we would appreciate hearing of any that you find. We hope that this text makes learning and teaching physics easier and more enjoyable, and we look forward to hearing about your experiences with it. Please feel free to write us care of Physics Editor, Higher Education Division, John Wiley & Sons, Inc., 111 River Street, Hoboken, NJ 07030, or contact us at **www.wiley.com/college/cutnell**

SUPPLEMENTS

INSTRUCTOR'S SUPPLEMENTS

Helping Teachers Teach

INSTRUCTOR'S COMPANION SITE This website (**www.wiley.com/college/cutnell**) was developed specifically for *Physics*, Eighth Edition. Instructors can access a wide range of essential resources, including the Test Bank, Lecture Note PowerPoint slides, Personal Response Questions, Image Gallery, and a number of other important materials.

INSTRUCTOR'S RESOURCE GUIDE by David T. Marx, Illinois State University. This guide contains an extensive listing of Web-based physics education resources. It also includes teaching ideas, lecture notes, demonstration suggestions, and alternative syllabi for courses of different lengths and emphasis. *A Problem Locator Guide* provides an easy way to correlate seventh-edition problem numbers with the corresponding eighth-edition numbers. The guide also contains a chapter on the effective use of Personal Response Systems.

INSTRUCTOR'S SOLUTIONS MANUALS by John D. Cutnell and Kenneth W. Johnson. These manuals provide worked-out solutions for all end-of-chapter *Problems* and answers to the *Focus on Concepts* questions.

TEST BANK by David T. Marx, Illinois State University. This manual includes more than 2200 multiple-choice questions. These items are also available in the *Computerized Test Bank,* which can be found on the *Instructor's Companion Site.*

PERSONAL RESPONSE QUESTIONS by David T. Marx, Illinois State University. This bank of 2200 "clicker" questions is made up of Reading Quiz questions and Interactive Lecture Questions. The Reading Quiz Questions are fairly simple in nature, typically used for attendance taking, for keeping students engaged, and for ensuring that they have completed the assigned reading. Interactive Lecture Questions are more difficult and thought-provoking, intended to promote classroom discussion and to reveal major misconceptions among students.

LECTURE NOTE POWERPOINT SLIDES by Michael Tammaro, University of Rhode Island. These PowerPoint slides contain lecture outlines, figures, and key equations.

WILEY PHYSICS SIMULATIONS CD-ROM This CD contains 50 interactive simulations (Java applets) that can be used for classroom demonstrations.

WILEY PHYSICS DEMONSTRATION DVD contains over 80 classsic physics demonstrations that will engage and instruct your students. An accompanying instructor's guide is available.

ONLINE HOMEWORK AND QUIZZING *Physics,* Eighth Edition, supports WebAssign, LON-CAPA, and *WileyPLUS*, which are programs that give instructors the ability to deliver and grade homework and quizzes online.

STUDENT'S SUPPLEMENTS

Helping Students Learn

STUDENT'S COMPANION SITE This Web site (**www.wiley.com/college/cutnell**) was developed specifically for *Physics,* Eighth Edition, and is designed to assist students further in the study of physics. At this site, students can access the following resources:

- *Interactive Solutions*
- *Concept Simulations*
- *Interactive LearningWare* examples
- Solutions to selected end-of-chapter problems that are identified with a **www** icon in the text.
- Review quizzes for the MCAT exam

STUDENT STUDY GUIDE by John D. Cutnell and Kenneth W. Johnson. This student study guide consists of traditional print materials; with the Student's Companion Site, it provides a rich, interactive environment for review and study.

STUDENT SOLUTIONS MANUAL by John D. Cutnell and Kenneth W. Johnson. This manual provides students with complete worked-out solutions for approximately 600 of the odd-numbered end-of-chapter problems. These problems are indicated in the text with an **ssm** icon.

CLASSROOM ACTIVITY PACK by David T. Marx, Illinois State University. This workbook consists of approximately 60 activities, designed to be done in groups during lecture or discussion sessions. Students work together to answer questions and complete exercises, encouraging deliberation of important concepts. Through these activities, common student misconceptions are highlighted.

ACKNOWLEDGMENTS

When a book has been, as they say, "put to bed," one of the most satisfying moments is sitting back and reflecting on the team that brought it to life. We are indeed fortunate in being associated with such outstanding professionals. Their talent, work ethic, and the pride they take in their work are an inspiration. It is a pleasure to acknowledge their contributions.

Many thanks go to Geraldine Osnato, who became our editor at the onset of this edition, quickly took the helm, and helped us during the many phases of the book's development. We appreciate her efforts, energy, suggestions, and ideas.

We are especially grateful to our former editor Stuart Johnson (now an Executive Editor). Stuart has been with us since the fourth edition, and his impact on the book continues. He is a driving force behind our involvement in *WileyPLUS*, and we are most thankful for his prescient advice.

In our acknowledgments for the last edition, we said that if there were an "Oscar" for the best production editor, we would have given it to Barbara Russiello. After this edition, an Oscar seems inadequate. A Nobel prize would be more appropriate. We owe you "big time," Barbara, for your patient help, talented efforts, and dedication to excellence.

Madelyn Lesure, our cover and text designer, has been with us since the first edition. Once again, she has performed the magic of integrating the book's numerous design elements into a beautiful whole. We simply can't imagine writing a book without Maddy at our side.

It has been a delight to work with Sigmund Malinowski, Senior Illustration Editor. Sigmund has been with us since the third edition, and he coordinated the text and online illustration programs. We are grateful for his knowledge, helpfulness, advice, and willingness to cooperate.

Our proofreader, Gloria Hamilton, continues to amaze us with her phenomenal talents. She is perfection personified. Her instincts for catching the slightest errors and inconsistencies are unparalleled. Athletes and musicians have halls of fame; too bad there isn't one for proofreaders.

Today, a physics text is much more than a collection of paper pages, because online interactive media have become such indispensable educational tools. We are lucky to have Thomas Kulesa, Senior Media Editor, on our team to guide us through the ever-changing landscape of the online world. Thanks, Tom, for always providing expert and thoughtful advice.

We are also thankful to Hilary Newman, Photo Editor, and Elaine Soares, Photo Researcher, for their efforts in selecting the photos in the text.

A book derives enormous benefits from having the insights of someone knowledgeable in marketing strategies, and we are fortunate in having Amanda Wainer, Executive Marketing Manager. With her outstanding knowledge of science textbooks and her keen marketing savvy, Amanda has had a strong influence on the book's success, for which we are very grateful.

Our sincere gratitude goes to Alyson Rentrop, Assistant Editor, for coordinating the extensive supplements package that accompanies the book, to Helen Walden for faithfully copy editing the manuscript, and to Veronica Amour, Editorial Assistant, for acting as liaison between production and media departments.

The sales representatives of John Wiley & Sons, Inc., are a very special group. We deeply appreciate their efforts on our behalf as they are in constant contact with physics departments throughout the country.

Many physicists and their students have generously shared their ideas with us about good pedagogy and helped us by pointing out our errors. For all of their suggestions we are grateful. They have helped us to write more clearly and accurately and have influenced markedly the evolution of this text. To the reviewers of this and previous editions, we especially owe a large debt of gratitude. Specifically, we thank:

Hanadi AbdelSalam, *formerly at Hamilton College*

Edward Adelson, *The Ohio State University*

Alice Hawthorne Allen, *Concord University*

Zaven Altounian, *McGill University*

Joseph Alward, *University of the Pacific*

Joseph Ametepe, *Hollins University*

Chi-Kwan Au, *University of South Carolina*

Santanu Banerjee, *Tougaloo College*

David Bannon, *Oregon State University*

Paul D. Beale, *University of Colorado at Boulder*

Edward E. Beasley, *Gallaudet University*

Rao Bidthanapally, *Oakland University*

Roger Bland, *San Francisco State University*

Michael Bretz, *University of Michigan at Ann Arbor*

Carl Bromberg, *Michigan State University*

Michael Broyles, *Collin County Community College*

Ronald W. Canterna, *University of Wyoming*

Gang Cao, *University of Kentucky*

Neal Cason, *University of Notre Dame*

Kapila Castoldi, *Oakland University*

Anil Chourasia, *Texas A & M University-Commerce*

David Cinabro, *Wayne State University*

Thomas Berry Cobb, *Bowling Green State University*

Lawrence Coleman, *University of California at Davis*

Lattie F. Collins, *East Tennessee State University*

Biman Das, *SUNY Potsdam*

Doyle Davis, *White Mountains Community College*

Steven Davis, *University of Arkansas at Little Rock*

William Dieterle, *California University of Pennsylvania*

Duane Doty, *California State University, Northridge*

Carl Drake, *Jackson State University*

Robert J. Endorf, *University of Cincinnati*

Lewis Ford, *Texas A&M University*

Lyle Ford, *University of Wisconsin-Eau Claire*

Greg Francis, *Montana State University*

C. Sherman Frye, Jr., *Northern Virginia Community College*

John Gagliardi, *Rutgers University*

Silvina Gatica, *Penn State University*

Barry Gilbert, *Rhode Island College*

Joseph Gladden, *University of Mississippi*

Peter Gonthier, *Hope College*

Roy Goodrich, *Louisiana State University*

William Gregg, *Louisiana State University*

David Griffiths, *Oregon State University*

Omar Guerrero, *University of Delaware*

A. J. Haija, *Indiana University of Pennsylvania*

Charles Hakes, *Fort Lewis College*

Dr. Kastro Hamed, *The University of Texas at El Paso*

Parameswar Hari, *California State University, Fresno*

J. Russell Harkay, *Keene State College*

Grant Hart, *Brigham Young University*

Athula Herat, *Slippery Rock University of Pennsylvania*

John Ho, *SUNY Buffalo*

William Hollerman, *University of Louisiana, Lafayette*

Lenore Horner, *Southern Illinois University at Edwardsville*

Doug Ingram, *Texas Christian University*

Shawn Jackson, *University of Arizona*

Darrin Johnson, *University of Minnesota Duluth*

Larry Josbeno, *Corning Community College*

Daniel Kennefick, *University of Arkansas, Fayetteville*

A. Khan, *Jackson State University*

Randy Kobes, *University of Winnipeg*

K. Kothari, *Tuskegee University*

Richard Krantz, *Metropolitan State College, Denver*

Theodore Kruse, *Rutgers University*

Pradeep Kumar, *University of Florida at Gainesville*

Christopher P. Landee, *Clark University*

Alfredo Louro, *University of Calgary*

Donald Luttermoser, *East Tennessee State University*

Sergei Lyuksyutov, *The University of Akron*

Kingshuk Majumdar, *Berea College*

A. John Mallinckrodt, *California State Polytechnic University, Pomona*

Glenn Marsch, *Union University*

Dr. E.L. Mathie, *University of Regina*

John McCullen, *University of Arizona*

B. Wieb van der Meer, *Western Kentucky University*

Donald D. Miller, *Central Missouri State University*

Paul Morris, *Abilene Christian University*

Richard A. Morrow, *University of Maine*

Hermann Nann, *Indiana University, Bloomington*

David Newton, *DeAnza College*

Tom Nelson Oder, *Youngstown State University*

R. Chris Olsen, *Naval Postgraduate School, Monterey, CA*

Michael Ottinger, *Missouri Western State College*

Peter John Polito, *Springfield College*

Jon Pumplin, *Michigan State University*

Oren Quist, *South Dakota State University*

Ricardo Rademacher, *formerly at Northern Kentucky University*

Talat Rahman, *University of Central Florida*

Michael Ram, *SUNY Buffalo*

Jacobo Rapaport, *Ohio University*

Wayne W. Repko, *Michigan State University*

Larry Rowan, *University of North Carolina at Chapel Hill*

Roy S. Rubins, *University of Texas at Arlington*

Dominic Sarsah, *Illinois Valley Community College*

Bartlett Sheinberg, *Houston Community College*

Marc Sher, *College of William & Mary*

James Simmons, *Waynesburg College*

John J. Sinai, *University of Louisville*

Gerald D. Smith, *Huntington College*

Vera Smolyaninova, *Towson University*

Elizabeth Stoddard, *University of Missouri-Kansas City*

Michael Strauss, *University of Oklahoma, Norman*

Virgil Stubblefield, *John A. Logan College*

Daniel Stump, *Michigan State University*

Ronald G. Tabak, *Youngstown State University*

Patrick Tam, *Humboldt State University*

Eddie Tatar, *Idaho State University*

James H. Taylor, *Central Missouri State University*

Francis Tuluri, *Jackson State University*

Robert Tyson, *University of North Carolina at Charlotte*

Timothy Usher, *California State University, San Bernardino*

Mick Veum, *University of Wisconsin-Stevens Point*

James M. Wallace, *formerly at Jackson Community College*

Henry White, *University of Missouri*

Rob Wilson, *formerly at Angelo State University*

Jerry H. Willson, *formerly at Metropolitan State College*

Brian Woodahl, *Indiana University–Purdue University Indianapolis*

C H A P T E R 1

INTRODUCTION AND MATHEMATICAL CONCEPTS

The animation techniques used in the film *Ratatouille* rely on computers and mathematical concepts such as trigonometry and vectors. These mathematical tools will also be useful throughout this book in dealing with the laws of physics. (Buena Vista/Photofest. © 2007 Disney Enterprises, Inc./Pixar)

1.1 THE NATURE OF PHYSICS

The science of physics has developed out of the efforts of men and women to explain our physical environment. These efforts have been so successful that the laws of physics now encompass a remarkable variety of phenomena, including planetary orbits, radio and TV waves, magnetism, and lasers, to name just a few.

The exciting feature of physics is its capacity for predicting how nature will behave in one situation on the basis of experimental data obtained in another situation. Such predictions place physics at the heart of modern technology and, therefore, can have a tremendous impact on our lives. Rocketry and the development of space travel have their roots firmly planted in the physical laws of Galileo Galilei (1564–1642) and Isaac Newton (1642–1727). The transportation industry relies heavily on physics in the development of engines and the design of aerodynamic vehicles. Entire electronics and computer industries owe their existence to the invention of the transistor, which grew directly out of the laws of physics that describe the electrical behavior of solids. The telecommunications industry depends extensively on electromagnetic waves, whose existence was predicted by James Clerk Maxwell (1831–1879) in his theory of electricity and magnetism. The medical profession uses X-ray, ultrasonic, and magnetic resonance methods for obtaining images of the interior of the human body, and physics lies at the core of all these. Perhaps the most widespread impact in modern technology is that due to the laser. Fields ranging from space exploration to medicine benefit from this incredible device, which is a direct application of the principles of atomic physics.

Because physics is so fundamental, it is a required course for students in a wide range of major areas. We welcome you to the study of this fascinating topic. You will learn how to see the world through the "eyes" of physics and to reason as a physicist does. In the process, you will learn how to apply physics principles to a wide range of problems. We hope that you will come to recognize that physics has important things to say about your environment.

1.2 UNITS

Physics experiments involve the measurement of a variety of quantities, and a great deal of effort goes into making these measurements as accurate and reproducible as possible. The first step toward ensuring accuracy and reproducibility is defining the units in which the measurements are made.

In this text, we emphasize the system of units known as *SI units,* which stands for the French phrase "Le **S**ystème **I**nternational d'Unités." By international agreement, this system employs the ***meter*** (m) as the unit of length, the ***kilogram*** (kg) as the unit of mass, and the ***second*** (s) as the unit of time. Two other systems of units are also in use, however. The CGS system utilizes the centimeter (cm), the gram (g), and the second for length, mass, and time, respectively, and the BE or British Engineering system (the gravitational version) uses the foot (ft), the slug (sl), and the second. Table 1.1 summarizes the units used for length, mass, and time in the three systems.

Originally, the meter was defined in terms of the distance measured along the earth's surface between the north pole and the equator. Eventually, a more accurate measurement standard was needed, and by international agreement the meter became the distance between two marks on a bar of platinum–iridium alloy (see Figure 1.1) kept at a temperature of 0 °C. Today, to meet further demands for increased accuracy, the meter is defined as the distance that light travels in a vacuum in a time of 1/299 792 458 second. This definition arises because the speed of light is a universal constant that is defined to be 299 792 458 m/s.

The definition of a kilogram as a unit of mass has also undergone changes over the years. As Chapter 4 discusses, the mass of an object indicates the tendency of the object to continue in motion with a constant velocity. Originally, the kilogram was expressed in terms of a specific amount of water. Today, one kilogram is defined to be the mass of a standard cylinder of platinum–iridium alloy, like the one in Figure 1.2.

As with the units for length and mass, the present definition of the second as a unit of time is different from the original definition. Originally, the second was defined according to the average time for the earth to rotate once about its axis, one day being set equal to 86 400 seconds. The earth's rotational motion was chosen because it is naturally repetitive, occurring over and over again. Today, we still use a naturally occurring repetitive phenomenon to define the second, but of a very different kind. We use the electromagnetic waves emitted by cesium-133 atoms in an atomic clock like that in Figure 1.3. One second is defined as the time needed for 9 192 631 770 wave cycles to occur.*

The units for length, mass, and time, along with a few other units that will arise later, are regarded as ***base*** SI units. The word "base" refers to the fact that these units are used along with various laws to define additional units for other important physical quantities, such as force and energy. The units for such other physical quantities are referred to as ***derived*** units, since they are combinations of the base units. Derived units will be introduced from time to time, as they arise naturally along with the related physical laws.

Figure 1.1 The standard platinum–iridium meter bar. (Courtesy Bureau International des Poids et Mesures, France)

Figure 1.2 The standard platinum–iridium kilogram is kept at the International Bureau of Weights and Measures in Sèvres, France. This copy of the standard kilogram is housed at the National Institute of Standards and Technology. (Courtesy of Sissy Riley, Information Services Division/National Institute of Standards and Technology)

Table 1.1 Units of Measurement

	System		
	SI	CGS	BE
Length	Meter (m)	Centimeter (cm)	Foot (ft)
Mass	Kilogram (kg)	Gram (g)	Slug (sl)
Time	Second (s)	Second (s)	Second (s)

*See Chapter 16 for a discussion of waves in general and Chapter 24 for a discussion of electromagnetic waves in particular.

The value of a quantity in terms of base or derived units is sometimes a very large or very small number. In such cases, it is convenient to introduce larger or smaller units that are related to the normal units by multiples of ten. Table 1.2 summarizes the prefixes that are used to denote multiples of ten. For example, 1000 or 10^3 meters are referred to as 1 kilometer (km), and 0.001 or 10^{-3} meter is called 1 millimeter (mm). Similarly, 1000 grams and 0.001 gram are referred to as 1 kilogram (kg) and 1 milligram (mg), respectively. Appendix A contains a discussion of scientific notation and powers of ten, such as 10^3 and 10^{-3}.

1.3 THE ROLE OF UNITS IN PROBLEM SOLVING

THE CONVERSION OF UNITS

Since any quantity, such as length, can be measured in several different units, it is important to know how to convert from one unit to another. For instance, the foot can be used to express the distance between the two marks on the standard platinum–iridium meter bar. There are 3.281 feet in one meter, and this number can be used to convert from meters to feet, as the following example demonstrates.

 Example 1 The World's Highest Waterfall

The highest waterfall in the world is Angel Falls in Venezuela, with a total drop of 979.0 m (see Figure 1.4). Express this drop in feet.

Reasoning When converting between units, we write down the units explicitly in the calculations and treat them like any algebraic quantity. In particular, we will take advantage of the following algebraic fact: Multiplying or dividing an equation by a factor of 1 does not alter an equation.

Solution Since 3.281 feet = 1 meter, it follows that (3.281 feet)/(1 meter) = 1. Using this factor of 1 to multiply the equation "Length = 979.0 meters," we find that

$$\text{Length} = (979.0 \text{ m})(1) = (979.0 \text{ meters})\left(\frac{3.281 \text{ feet}}{1 \text{ meter}}\right) = \boxed{3212 \text{ feet}}$$

The colored lines emphasize that the units of meters behave like any algebraic quantity and cancel when the multiplication is performed, leaving only the desired unit of feet to describe the answer. In this regard, note that 3.281 feet = 1 meter also implies that (1 meter)/(3.281 feet) = 1. However, we chose not to multiply by a factor of 1 in this form, because the units of meters would not have canceled.

A calculator gives the answer as 3212.099 feet. Standard procedures for significant figures, however, indicate that the answer should be rounded off to four significant figures, since the value of 979.0 meters is accurate to only four significant figures. In this regard, the "1 meter" in the denominator does not limit the significant figures of the answer, because this number is precisely one meter by definition of the conversion factor. Appendix B contains a review of significant figures.

Problem-solving insight. *In any conversion, if the units do not combine algebraically to give the desired result, the conversion has not been carried out properly.* With this in mind, the next example stresses the importance of writing down the units and illustrates a typical situation in which several conversions are required.

 Example 2 Interstate Speed Limit

Express the speed limit of 65 miles/hour in terms of meters/second.

Reasoning As in Example 1, it is important to write down the units explicitly in the calculations and treat them like any algebraic quantity. Here, two well-known relationships come

Figure 1.3 This atomic clock, the NIST-F1, is considered one of the world's most accurate clocks. It keeps time with an uncertainty of about one second in twenty million years. (© Geoffrey Wheeler)

Table 1.2 Standard Prefixes Used to Denote Multiples of Ten

Prefix	Symbol	Factor[a]
tera	T	10^{12}
giga[b]	G	10^{9}
mega	M	10^{6}
kilo	k	10^{3}
hecto	h	10^{2}
deka	da	10^{1}
deci	d	10^{-1}
centi	c	10^{-2}
milli	m	10^{-3}
micro	μ	10^{-6}
nano	n	10^{-9}
pico	p	10^{-12}
femto	f	10^{-15}

[a]Appendix A contains a discussion of powers of ten and scientific notation.
[b]Pronounced jig′a.

into play—namely, 5280 feet = 1 mile and 3600 seconds = 1 hour. As a result, (5280 feet)/(1 mile) = 1 and (3600 seconds)/(1 hour) = 1. In our solution we will use the fact that multiplying and dividing by these factors of unity does not alter an equation.

Solution Multiplying and dividing by factors of unity, we find the speed limit in feet per second as shown below:

$$\text{Speed} = \left(65 \; \frac{\text{miles}}{\text{hour}}\right)(1)(1) = \left(65 \; \frac{\cancel{\text{miles}}}{\cancel{\text{hour}}}\right)\left(\frac{5280 \text{ feet}}{1 \cancel{\text{ mile}}}\right)\left(\frac{1 \cancel{\text{ hour}}}{3600 \text{ s}}\right) = 95 \; \frac{\text{feet}}{\text{second}}$$

To convert feet into meters, we use the fact that (1 meter)/(3.281 feet) = 1:

$$\text{Speed} = \left(95 \; \frac{\text{feet}}{\text{second}}\right)(1) = \left(95 \; \frac{\cancel{\text{feet}}}{\text{second}}\right)\left(\frac{1 \text{ meter}}{3.281 \cancel{\text{ feet}}}\right) = \boxed{29 \; \frac{\text{meters}}{\text{second}}}$$

In addition to their role in guiding the use of conversion factors, units serve a useful purpose in solving problems. They can provide an internal check to eliminate errors, if they are carried along during each step of a calculation and treated like any algebraic factor. In particular, remember that *only quantities with the same units can be added or subtracted* **(Problem-solving insight)**. Thus, at one point in a calculation, if you find yourself adding 12 miles to 32 kilometers, stop and reconsider. Either miles must be converted into kilometers or kilometers must be converted into miles before the addition can be carried out.

A collection of useful conversion factors is given on the page facing the inside of the front cover. The reasoning strategy that we have followed in Examples 1 and 2 for converting between units is outlined as follows:

REASONING STRATEGY

Converting Between Units

1. In all calculations, write down the units explicitly.

2. Treat all units as algebraic quantities. In particular, when identical units are divided, they are eliminated algebraically.

3. Use the conversion factors located on the page facing the inside of the front cover. Be guided by the fact that multiplying or dividing an equation by a factor of 1 does not alter the equation. For instance, the conversion factor of 3.281 feet = 1 meter might be applied in the form (3.281 feet)/(1 meter) = 1. This factor of 1 would be used to multiply an equation such as "Length = 5.00 meters" in order to convert meters to feet.

4. Check to see that your calculations are correct by verifying that the units combine algebraically to give the desired unit for the answer. Only quantities with the same units can be added or subtracted.

Sometimes an equation is expressed in a way that requires specific units to be used for the variables in the equation. In such cases it is important to understand why only certain units can be used in the equation, as the following example illustrates.

Example 3 Body Mass Index

The body mass index (BMI) takes into account your mass in kilograms (kg) and your height in meters (m) and is defined as follows:

$$\text{BMI} = \frac{\text{Mass in kg}}{(\text{Height in m})^2}$$

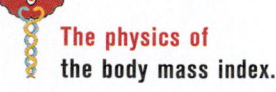

Figure 1.4 Angel Falls in Venezuela is the highest waterfall in the world. (© Kevin Schafer/The Image Bank/Getty Images)

The physics of the body mass index.

However, the BMI is often computed using the weight* of a person in pounds (lb) and his or her height in inches (in.). Thus, the expression for the BMI incorporates these quantities, rather than the mass in kilograms and the height in meters. Starting with the definition above, determine the expression for the BMI that uses pounds and inches.

Reasoning We will begin with the BMI definition and work separately with the numerator and the denominator. We will determine the mass in kilograms that appears in the numerator from the weight in pounds by using the fact that 1 kg corresponds to 2.205 lb. Then, we will determine the height in meters that appears in the denominator from the height in inches with the aid of the facts that 1 m = 3.281 ft and 1 ft = 12 in. These conversion factors are located on the page facing the inside of the front cover of the text.

Solution Since 1 kg corresponds to 2.205 lb, the mass in kilograms can be determined from the weight in pounds in the following way:

$$\text{Mass in kg} = (\text{Weight in lb})\left(\frac{1\ \text{kg}}{2.205\ \text{lb}}\right)$$

Since 1 ft = 12 in. and 1 m = 3.281 ft, we have

$$\text{Height in m} = (\text{Height in in.})\left(\frac{1\ \text{ft}}{12\ \text{in.}}\right)\left(\frac{1\ \text{m}}{3.281\ \text{ft}}\right)$$

Substituting these results into the numerator and denominator of the BMI definition gives

$$\text{BMI} = \frac{\text{Mass in kg}}{(\text{Height in m})^2} = \frac{(\text{Weight in lb})\left(\dfrac{1\ \text{kg}}{2.205\ \text{lb}}\right)}{(\text{Height in in.})^2\left(\dfrac{1\ \text{ft}}{12\ \text{in.}}\right)^2\left(\dfrac{1\ \text{m}}{3.281\ \text{ft}}\right)^2}$$

$$= \left(\frac{1\ \text{kg}}{2.205\ \text{lb}}\right)\left(\frac{12\ \text{in.}}{1\ \text{ft}}\right)^2\left(\frac{3.281\ \text{ft}}{1\ \text{m}}\right)^2\frac{(\text{Weight in lb})}{(\text{Height in in.})^2}$$

$$\boxed{\text{BMI} = \left(703.0\ \frac{\text{kg}\cdot\text{in.}^2}{\text{lb}\cdot\text{m}^2}\right)\frac{(\text{Weight in lb})}{(\text{Height in in.})^2}}$$

For example, if your weight and height are 180 lb and 71 in., your body mass index is 25 kg/m². The BMI can be used to assess approximately whether your weight is normal for your height (see Table 1.3).

Table 1.3 The Body Mass Index

BMI (kg/m²)	Evaluation
Below 18.5	Underweight
18.5–24.9	Normal
25.0–29.9	Overweight
30.0–39.9	Obese
40 and above	Morbidly obese

DIMENSIONAL ANALYSIS

We have seen that many quantities are denoted by specifying both a number and a unit. For example, the distance to the nearest telephone may be 8 meters, or the speed of a car might be 25 meters/second. Each quantity, according to its physical nature, requires a certain *type* of unit. Distance must be measured in a length unit such as meters, feet, or miles, and a time unit will not do. Likewise, the speed of an object must be specified as a length unit divided by a time unit. In physics, the term ***dimension*** is used to refer to the physical nature of a quantity and the type of unit used to specify it. Distance has the dimension of length, which is symbolized as [L], while speed has the dimensions of length [L] divided by time [T], or [L/T]. Many physical quantities can be expressed in terms of a combination of fundamental dimensions such as length [L], time [T], and mass [M]. Later on, we will encounter certain other quantities, such as temperature, which are also fundamental. A fundamental quantity like temperature cannot be expressed as a combination of the dimensions of length, time, mass, or any other fundamental dimension.

Dimensional analysis is used to check mathematical relations for the consistency of their dimensions. As an illustration, consider a car that starts from rest and accelerates to a speed v in a time t. Suppose we wish to calculate the distance x traveled by the car but are not sure whether the correct relation is $x = \frac{1}{2}vt^2$ or $x = \frac{1}{2}vt$. We can decide by checking the quantities on both sides of the equals sign to see whether they have the same

---

*Weight and mass are different concepts, and the relationship between them will be discussed in Section 4.7.

dimensions. If the dimensions are not the same, the relation is incorrect. For $x = \frac{1}{2}vt^2$, we use the dimensions for distance [L], time [T], and speed [L/T] in the following way:

$$x = \tfrac{1}{2}vt^2$$

Dimensions
$$[L] \stackrel{?}{=} \left[\frac{L}{\cancel{T}}\right][T]^2 = [L][T]$$

Problem-solving insight

You can check for errors that may have arisen during algebraic manipulations by performing a dimensional analysis on the final expression.

Dimensions cancel just like algebraic quantities, and pure numerical factors like $\frac{1}{2}$ have no dimensions, so they can be ignored. The dimension on the left of the equals sign does not match those on the right, so the relation $x = \frac{1}{2}vt^2$ cannot be correct. On the other hand, applying dimensional analysis to $x = \frac{1}{2}vt$, we find that

$$x = \tfrac{1}{2}vt$$

Dimensions
$$[L] \stackrel{?}{=} \left[\frac{L}{\cancel{T}}\right][\cancel{T}] = [L]$$

The dimension on the left of the equals sign matches that on the right, so this relation is dimensionally correct. If we know that one of our two choices is the right one, then $x = \frac{1}{2}vt$ is it. In the absence of such knowledge, however, dimensional analysis cannot identify the correct relation. It can only identify which choices *may be* correct, since it does not account for numerical factors like $\frac{1}{2}$ or for the manner in which an equation was derived from physics principles.

✓ CHECK YOUR UNDERSTANDING

(The answers are given at the end of the book.)

1. **(a)** Is it possible for two quantities to have the same dimensions but different units? **(b)** Is it possible for two quantities to have the same units but different dimensions?

2. You can always add two numbers that have the same units (such as 6 meters + 3 meters). Can you always add two numbers that have the same dimensions, such as two numbers that have the dimensions of length [L]?

3. The following table lists four variables, along with their units:

Variable	Units
x	Meters (m)
v	Meters per second (m/s)
t	Seconds (s)
a	Meters per second squared (m/s²)

These variables appear in the following equations, along with a few numbers that have no units. In which of the equations are the units on the left side of the equals sign consistent with the units on the right side?

(a) $x = vt$

(b) $x = vt + \frac{1}{2}at^2$

(c) $v = at$

(d) $v = at + \frac{1}{2}at^3$

(e) $v^3 = 2ax^2$

(f) $t = \sqrt{\dfrac{2x}{a}}$

4. In the equation $y = c^n at^2$ you wish to determine the integer value (1, 2, etc.) of the exponent n. The dimensions of y, a, and t are known. It is also known that c has no dimensions. Can dimensional analysis be used to determine n?

1.4   TRIGONOMETRY

Scientists use mathematics to help them describe how the physical universe works, and trigonometry is an important branch of mathematics. Three trigonometric functions are utilized throughout this text. They are the sine, the cosine, and the tangent of the angle

θ (Greek theta), abbreviated as sin θ, cos θ, and tan θ, respectively. These functions are defined below in terms of the symbols given along with the right triangle in Figure 1.5.

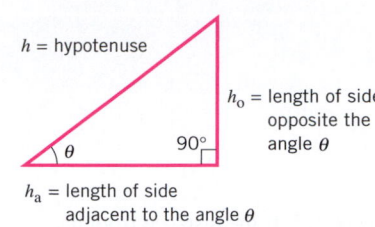

h = hypotenuse

h_o = length of side opposite the angle θ

h_a = length of side adjacent to the angle θ

Figure 1.5 A right triangle.

DEFINITION OF SIN θ, COS θ, AND TAN θ

$$\sin \theta = \frac{h_o}{h} \tag{1.1}$$

$$\cos \theta = \frac{h_a}{h} \tag{1.2}$$

$$\tan \theta = \frac{h_o}{h_a} \tag{1.3}$$

h = length of the **hypotenuse** of a right triangle
h_o = length of the side **opposite** the angle θ
h_a = length of the side **adjacent** to the angle θ

The sine, cosine, and tangent of an angle are numbers without units, because each is the ratio of the lengths of two sides of a right triangle. Example 4 illustrates a typical application of Equation 1.3.

Example 4   Using Trigonometric Functions

On a sunny day, a tall building casts a shadow that is 67.2 m long. The angle between the sun's rays and the ground is $\theta = 50.0°$, as Figure 1.6 shows. Determine the height of the building.

Reasoning We want to find the height of the building. Therefore, we begin with the colored right triangle in Figure 1.6 and identify the height as the length h_o of the side opposite the angle θ. The length of the shadow is the length h_a of the side that is adjacent to the angle θ. The ratio of the length of the opposite side to the length of the adjacent side is the tangent of the angle θ, which can be used to find the height of the building.

Solution We use the tangent function in the following way, with $\theta = 50.0°$ and $h_a = 67.2$ m:

$$\tan \theta = \frac{h_o}{h_a} \tag{1.3}$$

$$h_o = h_a \tan \theta = (67.2 \text{ m})(\tan 50.0°) = (67.2 \text{ m})(1.19) = \boxed{80.0 \text{ m}}$$

The value of tan 50.0° is found by using a calculator.

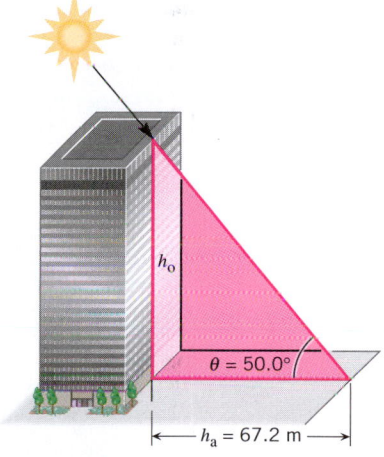

Figure 1.6 From a value for the angle θ and the length h_a of the shadow, the height h_o of the building can be found using trigonometry.

The sine, cosine, or tangent may be used in calculations such as that in Example 4, depending on which side of the triangle has a known value and which side is asked for. However, *the choice of which side of the triangle to label h_o (opposite) and which to label h_a (adjacent) can be made only after the angle θ is identified.*

Problem-solving insight

Often the values for two sides of the right triangle in Figure 1.5 are available, and the value of the angle θ is unknown. The concept of *inverse trigonometric functions* plays an important role in such situations. Equations 1.4–1.6 give the inverse sine, inverse cosine, and inverse tangent in terms of the symbols used in the drawing. For instance, Equation 1.4 is read as "θ equals the angle whose sine is h_o/h."

$$\theta = \sin^{-1}\left(\frac{h_o}{h}\right) \tag{1.4}$$

$$\theta = \cos^{-1}\left(\frac{h_a}{h}\right) \tag{1.5}$$

$$\theta = \tan^{-1}\left(\frac{h_o}{h_a}\right) \tag{1.6}$$

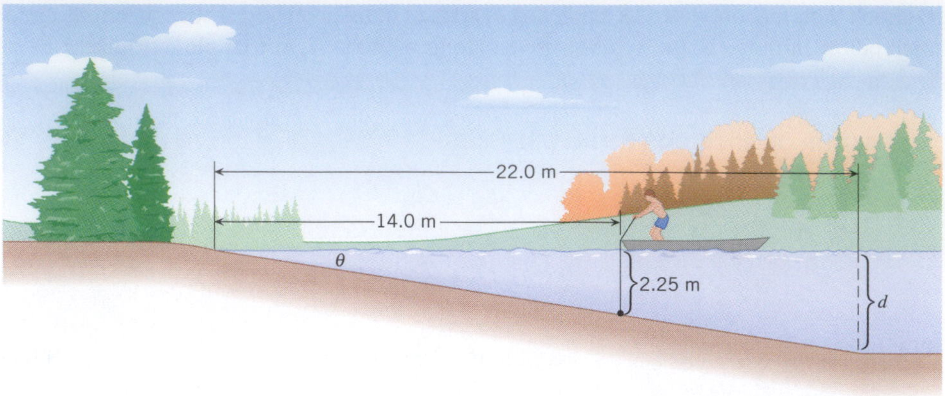

Figure 1.7 If the distance from the shore and the depth of the water at any one point are known, the angle θ can be found with the aid of trigonometry. Knowing the value of θ is useful, because then the depth d at another point can be determined.

The use of -1 as an exponent in Equations 1.4–1.6 *does not mean* "take the reciprocal." For instance, $\tan^{-1}(h_o/h_a)$ does not equal $1/\tan(h_o/h_a)$. Another way to express the inverse trigonometric functions is to use arc sin, arc cos, and arc tan instead of $\sin^{-1}$, $\cos^{-1}$, and $\tan^{-1}$. Example 5 illustrates the use of an inverse trigonometric function.

Example 5 Using Inverse Trigonometric Functions

A lakefront drops off gradually at an angle θ, as Figure 1.7 indicates. For safety reasons, it is necessary to know how deep the lake is at various distances from the shore. To provide some information about the depth, a lifeguard rows straight out from the shore a distance of 14.0 m and drops a weighted fishing line. By measuring the length of the line, the lifeguard determines the depth to be 2.25 m. **(a)** What is the value of θ? **(b)** What would be the depth d of the lake at a distance of 22.0 m from the shore?

Reasoning Near the shore, the lengths of the opposite and adjacent sides of the right triangle in Figure 1.7 are $h_o = 2.25$ m and $h_a = 14.0$ m, relative to the angle θ. Having made this identification, we can use the inverse tangent to find the angle in part (a). For part (b) the opposite and adjacent sides farther from the shore become $h_o = d$ and $h_a = 22.0$ m. With the value for θ obtained in part (a), the tangent function can be used to find the unknown depth. Considering the way in which the lake bottom drops off in Figure 1.7, we expect the unknown depth to be greater than 2.25 m.

Solution **(a)** Using the inverse tangent given in Equation 1.6, we find that

$$\theta = \tan^{-1}\left(\frac{h_o}{h_a}\right) = \tan^{-1}\left(\frac{2.25 \text{ m}}{14.0 \text{ m}}\right) = \boxed{9.13^\circ}$$

(b) With $\theta = 9.13^\circ$, the tangent function given in Equation 1.3 can be used to find the unknown depth farther from the shore, where $h_o = d$ and $h_a = 22.0$ m. Since $\tan\theta = h_o/h_a$, it follows that

$$h_o = h_a \tan\theta$$

$$d = (22.0 \text{ m})(\tan 9.13^\circ) = \boxed{3.54 \text{ m}}$$

which is greater than 2.25 m, as expected.

The right triangle in Figure 1.5 provides the basis for defining the various trigonometric functions according to Equations 1.1–1.3. These functions always involve an angle and two sides of the triangle. There is also a relationship among the lengths of the three sides of a right triangle. This relationship is known as the ***Pythagorean theorem*** and is used often in this text.

PYTHAGOREAN THEOREM

The square of the length of the hypotenuse of a right triangle is equal to the sum of the squares of the lengths of the other two sides:

$$h^2 = h_o^2 + h_a^2 \tag{1.7}$$

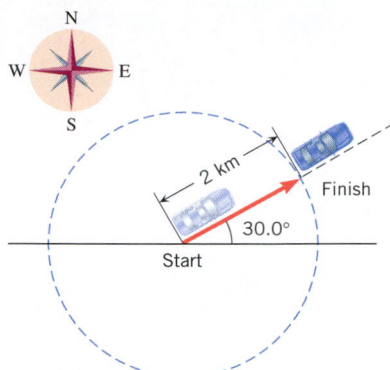

Figure 1.8 A vector quantity has a magnitude and a direction. The colored arrow in this drawing represents a displacement vector.

SCALARS AND VECTORS

The volume of water in a swimming pool might be 50 cubic meters, or the winning time of a race could be 11.3 seconds. In cases like these, only the size of the numbers matters. In other words, *how much* volume or time is there? The 50 specifies the amount of water in units of cubic meters, while the 11.3 specifies the amount of time in seconds. Volume and time are examples of scalar quantities. A ***scalar quantity*** is one that can be described with a single number (including any units) giving its size or magnitude. Some other common scalars are temperature (e.g., 20 °C) and mass (e.g., 85 kg).

While many quantities in physics are scalars, there are also many that are not, and for these quantities the magnitude tells only part of the story. Consider Figure 1.8, which depicts a car that has moved 2 km along a straight line from start to finish. When describing the motion, it is incomplete to say that "the car moved a distance of 2 km." This statement would indicate only that the car ends up somewhere on a circle whose center is at the starting point and whose radius is 2 km. A complete description must include the direction along with the distance, as in the statement "the car moved a distance of 2 km in a direction 30° north of east." A quantity that deals inherently with *both magnitude and direction* is called a ***vector quantity***. Because direction is an important characteristic of vectors, arrows are used to represent them; ***the direction of the arrow gives the direction of the vector.*** The colored arrow in Figure 1.8, for example, is called the *displacement vector,* because it shows how the car is displaced from its starting point. Chapter 2 discusses this particular vector.

The length of the arrow in Figure 1.8 represents the magnitude of the displacement vector. If the car had moved 4 km instead of 2 km from the starting point, the arrow would have been drawn twice as long. ***By convention, the length of a vector arrow is proportional to the magnitude of the vector.***

In physics there are many important kinds of vectors, and the practice of using the length of an arrow to represent the magnitude of a vector applies to each of them. All forces, for instance, are vectors. In common usage a force is a push or a pull, and the direction in which a force acts is just as important as the strength or magnitude of the force. The magnitude of a force is measured in SI units called newtons (N). An arrow representing a force of 20 newtons is drawn twice as long as one representing a force of 10 newtons.

The fundamental distinction between scalars and vectors is the characteristic of direction. Vectors have it, and scalars do not. Conceptual Example 6 helps to clarify this distinction and explains what is meant by the "direction" of a vector.

Conceptual Example 6
Vectors, Scalars, and the Role of Plus and Minus Signs

There are places where the temperature is +20 °C at one time of the year and −20 °C at another time. Do the plus and minus signs that signify positive and negative temperatures imply that temperature is a vector quantity? **(a)** Yes **(b)** No

Reasoning A hallmark of a vector is that there is both a magnitude and a physical direction associated with it, such as 20 meters due east or 20 meters due west.

Answer (a) is incorrect. The plus and minus signs associated with +20 °C and −20°C do not convey a physical direction, such as due east or due west. Therefore, temperature cannot be a vector quantity.

Answer (b) is correct. On a thermometer, the algebraic signs simply mean that the temperature is a number less than or greater than zero on the temperature scale being used and have nothing to do with east, west, or any other physical direction. Temperature, then, is not a vector. It is a scalar, and scalars can sometimes be negative.

Often, for the sake of convenience, quantities such as volume, time, displacement, velocity, and force are represented in physics by symbols. In this text, we write vectors in

Each of these racers in the Tour de France has a velocity as they approach the finish line on the Champs Élysées in Paris. Velocity is an example of a vector quantity, because it has a magnitude (the speed of the racer) and a direction (toward the finish line). (Owen Franken/Corbis Images)

boldface symbols (**this is boldface**) with arrows above them* and write scalars in italic symbols (*this is italic*). Thus, a displacement vector is written as "$\vec{A}$ = 750 m, due east," where the $\vec{A}$ is a boldface symbol. By itself, however, separated from the direction, the magnitude of this vector is a scalar quantity. Therefore, the magnitude is written as "A = 750 m," where the A is an italic symbol without an arrow.

✓ **CHECK YOUR UNDERSTANDING**

(*The answer is given at the end of the book.*)

5. Which of the following statements, if any, involves a vector? **(a)** I walked 2 miles along the beach. **(b)** I walked 2 miles due north along the beach. **(c)** I jumped off a cliff and hit the water traveling at 17 miles per hour. **(d)** I jumped off a cliff and hit the water traveling straight down at a speed of 17 miles per hour. **(e)** My bank account shows a negative balance of −25 dollars.

1.6 VECTOR ADDITION AND SUBTRACTION

ADDITION

Often it is necessary to add one vector to another, and the process of addition must take into account both the magnitude and the direction of the vectors. The simplest situation occurs when the vectors point along the same direction—that is, when they are colinear, as in Figure 1.9. Here, a car first moves along a straight line, with a displacement vector $\vec{A}$ of 275 m, due east. Then, the car moves again in the same direction, with a displacement vector $\vec{B}$ of 125 m, due east. These two vectors add to give the total displacement vector $\vec{R}$, which would apply if the car had moved from start to finish in one step. The symbol $\vec{R}$ is used because the total vector is often called the **resultant vector.** With the tail of the second arrow located at the head of the first arrow, the two lengths simply add to give the length of the total displacement. This kind of vector addition is identical to the familiar addition of two scalar numbers (2 + 3 = 5) *and can be carried out here only because the vectors point along the same direction.* In such cases we add the individual magnitudes to get the magnitude of the total, knowing in advance what the direction must be. Formally, the addition is written as follows:

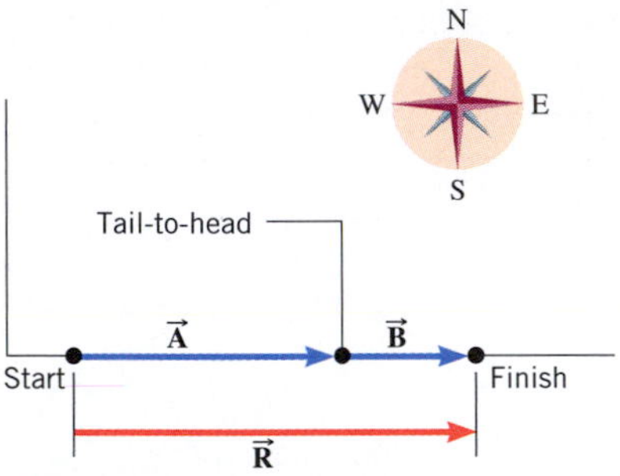

$$\vec{R} = \vec{A} + \vec{B}$$

$$\vec{R} = 275 \text{ m, due east} + 125 \text{ m, due east} = 400 \text{ m, due east}$$

Perpendicular vectors are frequently encountered, and Figure 1.10 indicates how they can be added. This figure applies to a car that first travels with a displacement vector $\vec{A}$ of 275 m, due east, and then with a displacement vector $\vec{B}$ of 125 m, due north. The two vectors add to give a resultant displacement vector $\vec{R}$. Once again, the vectors to be added are arranged in a tail-to-head fashion, and the resultant vector points from the tail of the first to the head of the last vector added. The resultant displacement is given by the vector equation

$$\vec{R} = \vec{A} + \vec{B}$$

The addition in this equation cannot be carried out by writing R = 275 m + 125 m, because the vectors have different directions. Instead, we take advantage of the fact that the triangle in Figure 1.10 is a right triangle and use the Pythagorean theorem (Equation 1.7). According to this theorem, the magnitude of $\vec{R}$ is

$$R = \sqrt{(275 \text{ m})^2 + (125 \text{ m})^2} = 302 \text{ m}$$

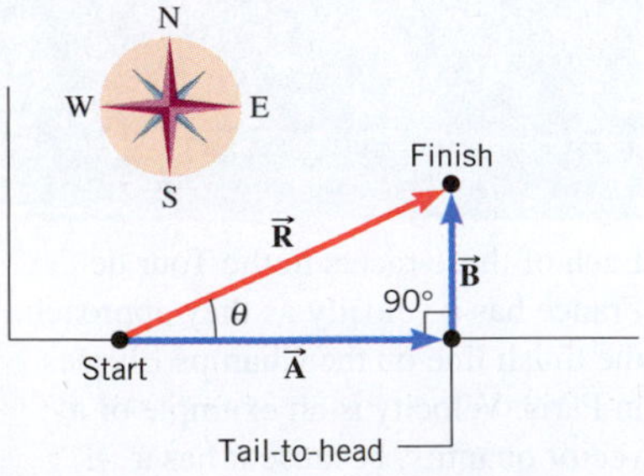

The angle θ in Figure 1.10 gives the direction of the resultant vector. Since the lengths of all three sides of the right triangle are now known, either sin θ, cos θ, or tan θ can be used

Figure 1.9 Two colinear displacement vectors $\vec{A}$ and $\vec{B}$ add to give the resultant displacement vector $\vec{R}$.

Figure 1.10 The addition of two perpendicular displacement vectors $\vec{A}$ and $\vec{B}$ gives the resultant vector $\vec{R}$.

*Vectors are also sometimes written in other texts as boldface symbols without arrows above them.

to determine θ. Noting that $\tan \theta = B/A$ and using the inverse trigonometric function, we find that:

$$\theta = \tan^{-1}\left(\frac{B}{A}\right) = \tan^{-1}\left(\frac{125 \text{ m}}{275 \text{ m}}\right) = 24.4°$$

Thus, the resultant displacement of the car has a magnitude of 302 m and points north of east at an angle of 24.4°. This displacement would bring the car from the start to the finish in Figure 1.10 in a single straight-line step.

When two vectors to be added are not perpendicular, the tail-to-head arrangement does not lead to a right triangle, and the Pythagorean theorem cannot be used. Figure 1.11a illustrates such a case for a car that moves with a displacement $\vec{\mathbf{A}}$ of 275 m, due east, and then with a displacement $\vec{\mathbf{B}}$ of 125 m, in a direction 55.0° north of west. As usual, the resultant displacement vector $\vec{\mathbf{R}}$ is directed from the tail of the first to the head of the last vector added. The vector addition is still given according to

$$\vec{\mathbf{R}} = \vec{\mathbf{A}} + \vec{\mathbf{B}}$$

However, the magnitude of $\vec{\mathbf{R}}$ is not $R = A + B$, because the vectors $\vec{\mathbf{A}}$ and $\vec{\mathbf{B}}$ do not have the same direction, and neither is it $R = \sqrt{A^2 + B^2}$, because the vectors are not perpendicular, so the Pythagorean theorem does not apply. Some other means must be used to find the magnitude and direction of the resultant vector.

One approach uses a graphical technique. In this method, a diagram is constructed in which the arrows are drawn tail to head. The lengths of the vector arrows are drawn to scale, and the angles are drawn accurately (with a protractor, perhaps). Then, the length of the arrow representing the resultant vector is measured with a ruler. This length is converted to the magnitude of the resultant vector by using the scale factor with which the drawing is constructed. In Figure 1.11b, for example, a scale of one centimeter of arrow length for each 10.0 m of displacement is used, and it can be seen that the length of the arrow representing $\vec{\mathbf{R}}$ is 22.8 cm. Since each centimeter corresponds to 10.0 m of displacement, the magnitude of $\vec{\mathbf{R}}$ is 228 m. The angle θ, which gives the direction of $\vec{\mathbf{R}}$, can be measured with a protractor to be $\theta = 26.7°$ north of east.

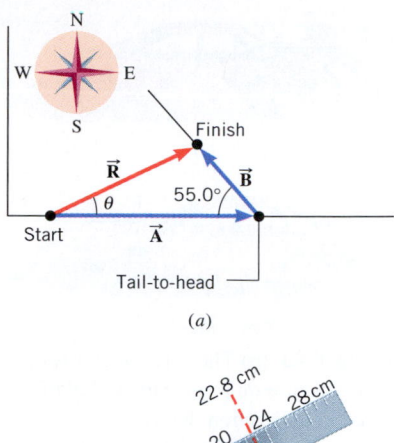

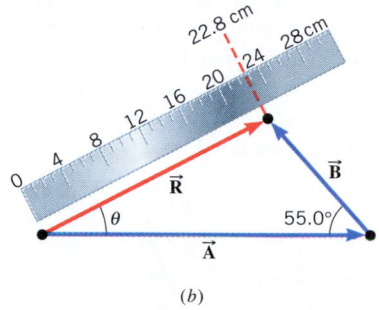

Figure 1.11 (a) The two displacement vectors $\vec{\mathbf{A}}$ and $\vec{\mathbf{B}}$ are neither colinear nor perpendicular but even so they add to give the resultant vector $\vec{\mathbf{R}}$. (b) In one method for adding them together, a graphical technique is used.

SUBTRACTION

The subtraction of one vector from another is carried out in a way that depends on the following fact. *When a vector is multiplied by −1, the magnitude of the vector remains the same, but the direction of the vector is reversed.* Conceptual Example 7 illustrates the meaning of this statement.

 Conceptual Example 7   Multiplying a Vector by −1

Consider two vectors described as follows:

1. A woman climbs 1.2 m up a ladder, so that her displacement vector $\vec{\mathbf{D}}$ is 1.2 m, upward along the ladder, as in Figure 1.12a.

2. A man is pushing with 450 N of force on his stalled car, trying to move it eastward. The force vector $\vec{\mathbf{F}}$ that he applies to the car is 450 N, due east, as in Figure 1.13a.

What are the physical meanings of the vectors $-\vec{\mathbf{D}}$ and $-\vec{\mathbf{F}}$?

(a) $-\vec{\mathbf{D}}$ points upward along the ladder and has a magnitude of −1.2 m; $-\vec{\mathbf{F}}$ points due east and has a magnitude of −450 N. **(b)** $-\vec{\mathbf{D}}$ points downward along the ladder and has a magnitude of −1.2 m; $-\vec{\mathbf{F}}$ points due west and has a magnitude of −450 N. **(c)** $-\vec{\mathbf{D}}$ points downward along the ladder and has a magnitude of 1.2 m; $-\vec{\mathbf{F}}$ points due west and has a magnitude of 450 N.

Reasoning A displacement vector of $-\vec{\mathbf{D}}$ is $(-1)\,\vec{\mathbf{D}}$. The presence of the (-1) factor reverses the direction of the vector, but does not change its magnitude. Similarly, a force vector of $-\vec{\mathbf{F}}$ has the same magnitude as the vector $\vec{\mathbf{F}}$ but has the opposite direction.

Answer (a) and (b) are incorrect. While scalars can sometimes be negative, magnitudes of vectors are never negative.

Figure 1.12 (a) The displacement vector for a woman climbing 1.2 m up a ladder is $\vec{\mathbf{D}}$. (b) The displacement vector for a woman climbing 1.2 m down a ladder is $-\vec{\mathbf{D}}$.

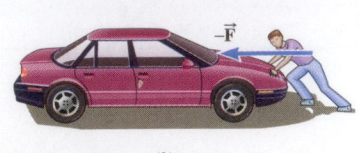

(a)

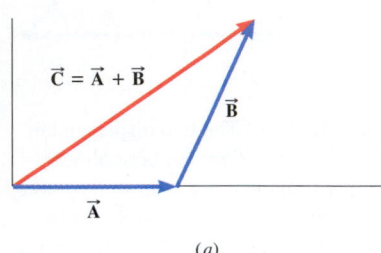

(b)

Figure 1.13 (*a*) The force vector for a man pushing on a car with 450 N of force in a direction due east is $\vec{F}$. (*b*) The force vector for a man pushing on a car with 450 N of force in a direction due west is $-\vec{F}$.

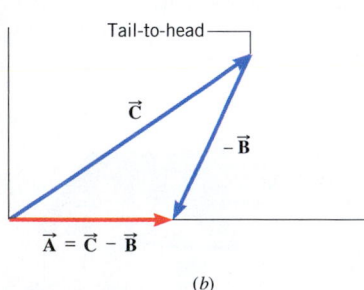

(a)

(b)

Figure 1.14 (*a*) Vector addition according to $\vec{C} = \vec{A} + \vec{B}$. (*b*) Vector subtraction according to $\vec{A} = \vec{C} - \vec{B}$ $= \vec{C} + (-\vec{B})$.

Answer (c) is correct. The vectors $-\vec{D}$ and $-\vec{F}$ have the same magnitudes as $\vec{D}$ and $\vec{F}$, but point in the opposite directions, as indicated in Figures 1.12*b* and 1.13*b*.

Related Homework: *Problem 33*

In practice, vector subtraction is carried out exactly like vector addition, except that one of the vectors added is multiplied by a scalar factor of -1. To see why, look at the two vectors $\vec{A}$ and $\vec{B}$ in Figure 1.14*a*. These vectors add together to give a third vector $\vec{C}$, according to $\vec{C} = \vec{A} + \vec{B}$. Therefore, we can calculate vector $\vec{A}$ as $\vec{A} = \vec{C} - \vec{B}$, which is an example of vector subtraction. However, we can also write this result as $\vec{A} = \vec{C} + (-\vec{B})$ and treat it as vector addition. Figure 1.14*b* shows how to calculate vector $\vec{A}$ by adding the vectors $\vec{C}$ and $-\vec{B}$. Notice that vectors $\vec{C}$ and $-\vec{B}$ are arranged tail to head and that any suitable method of vector addition can be employed to determine $\vec{A}$.

✔ CHECK YOUR UNDERSTANDING

(*The answers are given at the end of the book.*)

6. Two vectors $\vec{A}$ and $\vec{B}$ are added together to give a resultant vector $\vec{R}$: $\vec{R} = \vec{A} + \vec{B}$. The magnitudes of $\vec{A}$ and $\vec{B}$ are 3 m and 8 m, respectively, but the vectors can have any orientation. What is **(a)** the maximum possible value and **(b)** the minimum possible value for the magnitude of $\vec{R}$?

7. Can two nonzero perpendicular vectors be added together so their sum is zero?

8. Can three or more vectors with unequal magnitudes be added together so their sum is zero?

9. In preparation for this question, review Conceptual Example 7. Vectors $\vec{A}$ and $\vec{B}$ satisfy the vector equation $\vec{A} + \vec{B} = 0$. **(a)** How does the magnitude of $\vec{B}$ compare with the magnitude of $\vec{A}$? **(b)** How does the direction of $\vec{B}$ compare with the direction of $\vec{A}$?

10. Vectors $\vec{A}$, $\vec{B}$, and $\vec{C}$ satisfy the vector equation $\vec{A} + \vec{B} = \vec{C}$, and their magnitudes are related by the scalar equation $A^2 + B^2 = C^2$. How is vector $\vec{A}$ oriented with respect to vector $\vec{B}$?

11. Vectors $\vec{A}$, $\vec{B}$, and $\vec{C}$ satisfy the vector equation $\vec{A} + \vec{B} = \vec{C}$, and their magnitudes are related by the scalar equation $A + B = C$. How is vector $\vec{A}$ oriented with respect to vector $\vec{B}$?

THE COMPONENTS OF A VECTOR

1.7

VECTOR COMPONENTS

Suppose a car moves along a straight line from start to finish, as in Figure 1.15, the corresponding displacement vector being $\vec{r}$. The magnitude and direction of the vector $\vec{r}$ give the distance and direction traveled along the straight line. However, the car could also arrive at the finish point by first moving due east, turning through 90°, and then moving due north. This alternative path is shown in the drawing and is associated with the two displacement vectors $\vec{x}$ and $\vec{y}$. The vectors $\vec{x}$ and $\vec{y}$ are called the *x* vector component and the *y* vector component of $\vec{r}$.

Vector components are very important in physics and have two basic features that are apparent in Figure 1.15. One is that the components add together to equal the original vector:

$$\vec{r} = \vec{x} + \vec{y}$$

The components $\vec{x}$ and $\vec{y}$, when added vectorially, convey exactly the same meaning as does the original vector $\vec{r}$: they indicate how the finish point is displaced relative to the starting point. The other feature of vector components that is apparent in Figure 1.15 is that $\vec{x}$ and $\vec{y}$ are not just any two vectors that add together to give the original vector $\vec{r}$: they are perpendicular vectors. This perpendicularity is a valuable characteristic, as we will soon see.

Any type of vector may be expressed in terms of its components, in a way similar to that illustrated for the displacement vector in Figure 1.15. Figure 1.16 shows an arbitrary

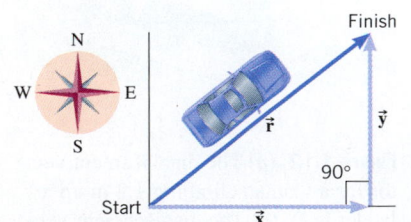

Figure 1.15 The displacement vector $\vec{r}$ and its vector components $\vec{x}$ and $\vec{y}$.

vector $\vec{A}$ and its vector components $\vec{A}_x$ and $\vec{A}_y$. The components are drawn parallel to convenient x and y axes and are perpendicular. They add vectorially to equal the original vector $\vec{A}$:

$$\vec{A} = \vec{A}_x + \vec{A}_y$$

There are times when a drawing such as Figure 1.16 is not the most convenient way to represent vector components, and Figure 1.17 presents an alternative method. The disadvantage of this alternative is that the tail-to-head arrangement of $\vec{A}_x$ and $\vec{A}_y$ is missing, an arrangement that is a nice reminder that $\vec{A}_x$ and $\vec{A}_y$ add together to equal $\vec{A}$.

The definition that follows summarizes the meaning of vector components:

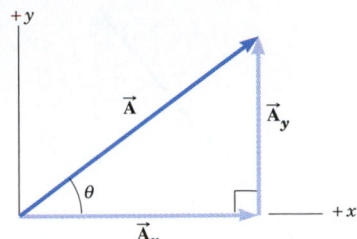

Figure 1.16 An arbitrary vector $\vec{A}$ and its vector components $\vec{A}_x$ and $\vec{A}_y$.

DEFINITION OF VECTOR COMPONENTS

In two dimensions, the vector components of a vector $\vec{A}$ are two perpendicular vectors $\vec{A}_x$ and $\vec{A}_y$ that are parallel to the x and y axes, respectively, and add together vectorially so that $\vec{A} = \vec{A}_x + \vec{A}_y$.

In general, *the components of any vector can be used in place of the vector itself in any calculation where it is convenient to do so* **(Problem-solving insight)**. The values calculated for vector components depend on the orientation of the vector relative to the axes used as a reference. Figure 1.18 illustrates this fact for a vector $\vec{A}$ by showing two sets of axes, one set being rotated clockwise relative to the other. With respect to the black axes, vector $\vec{A}$ has perpendicular vector components $\vec{A}_x$ and $\vec{A}_y$; with respect to the colored rotated axes, vector $\vec{A}$ has different vector components $\vec{A}_x'$ and $\vec{A}_y'$. The choice of which set of components to use is purely a matter of convenience.

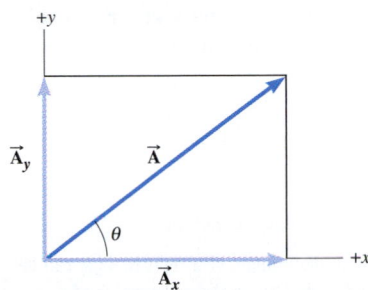

Figure 1.17 This alternative way of drawing the vector $\vec{A}$ and its vector components is completely equivalent to that shown in Figure 1.16.

SCALAR COMPONENTS

It is often easier to work with the *scalar components,* A_x and A_y (note the italic symbols), rather than the vector components $\vec{A}_x$ and $\vec{A}_y$. Scalar components are positive or negative numbers (with units) that are defined as follows: The scalar component A_x has a magnitude equal to that of $\vec{A}_x$ and is given a positive sign if $\vec{A}_x$ points along the $+x$ axis and a negative sign if $\vec{A}_x$ points along the $-x$ axis. The scalar component A_y is defined in a similar manner. The following table shows an example of vector and scalar components:

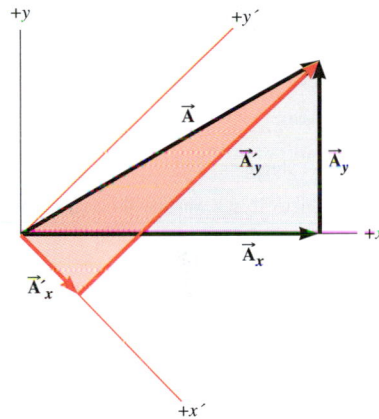

Figure 1.18 The vector components of the vector depend on the orientation of the axes used as a reference.

Vector Components	Scalar Components	Unit Vectors
$\vec{A}_x = 8$ meters, directed along the $+x$ axis	$A_x = +8$ meters	$\vec{A}_x = (+8 \text{ meters}) \, \hat{x}$
$\vec{A}_y = 10$ meters, directed along the $-y$ axis	$A_y = -10$ meters	$\vec{A}_y = (-10 \text{ meters}) \, \hat{y}$

In this text, when we use the term "component," we will be referring to a scalar component, unless otherwise indicated.

Another method of expressing vector components is to use unit vectors. A *unit vector* is a vector that has a magnitude of 1, but no dimensions. We will use a caret (^) to distinguish it from other vectors. Thus,

$\hat{x}$ is a dimensionless unit vector of length 1 that points in the positive x direction, and

$\hat{y}$ is a dimensionless unit vector of length 1 that points in the positive y direction.

These unit vectors are illustrated in Figure 1.19. With the aid of unit vectors, the vector components of an arbitrary vector $\vec{A}$ can be written as $\vec{A}_x = A_x \, \hat{x}$ and $\vec{A}_y = A_y \, \hat{y}$, where A_x and A_y are its scalar components (see the drawing and the third column of the table above). The vector $\vec{A}$ is then written as $\vec{A} = A_x \, \hat{x} + A_y \, \hat{y}$.

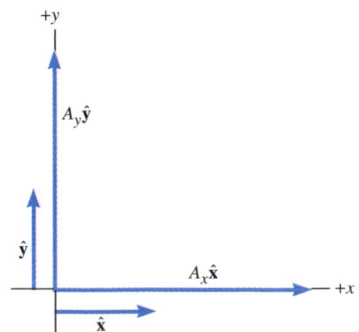

Figure 1.19 The dimensionless unit vectors $\hat{x}$ and $\hat{y}$ have magnitudes equal to 1, and they point in the $+x$ and $+y$ directions, respectively. Expressed in terms of unit vectors, the vector components of the vector $\vec{A}$ are $A_x \, \hat{x}$ and $A_y \, \hat{y}$.

RESOLVING A VECTOR INTO ITS COMPONENTS

If the magnitude and direction of a vector are known, it is possible to find the components of the vector. The process of finding the components is called "resolving the vector into its components." As Example 8 illustrates, this process can be carried out with the

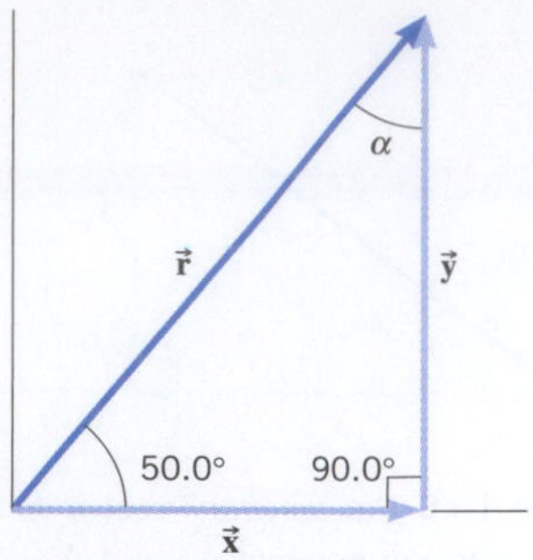

Figure 1.20 The x and y components of the displacement vector $\vec{r}$ can be found using trigonometry.

aid of trigonometry, because the two perpendicular vector components and the original vector form a right triangle.

Example 8 Finding the Components of a Vector

A displacement vector $\vec{r}$ has a magnitude of $r = 175$ m and points at an angle of 50.0° relative to the x axis in Figure 1.20. Find the x and y components of this vector.

Reasoning We will base our solution on the fact that the triangle formed in Figure 1.20 by the vector $\vec{r}$ and its components $\vec{x}$ and $\vec{y}$ is a right triangle. This fact enables us to use the trigonometric sine and cosine functions, as defined in Equations 1.1 and 1.2.

Solution 1 The y component can be obtained using the 50.0° angle and Equation 1.1, $\sin\theta = y/r$:

$$y = r\sin\theta = (175\text{ m})(\sin 50.0°) = \boxed{134\text{ m}}$$

In a similar fashion, the x component can be obtained using the 50.0° angle and Equation 1.2, $\cos\theta = x/r$:

$$x = r\cos\theta = (175\text{ m})(\cos 50.0°) = \boxed{112\text{ m}}$$

Solution 2 The angle α in Figure 1.20 can also be used to find the components. Since $\alpha + 50.0° = 90.0°$, it follows that $\alpha = 40.0°$. The solution using α yields the same answers as in Solution 1:

$$\cos\alpha = \frac{y}{r}$$

$$y = r\cos\alpha = (175\text{ m})(\cos 40.0°) = \boxed{134\text{ m}}$$

$$\sin\alpha = \frac{x}{r}$$

$$x = r\sin\alpha = (175\text{ m})(\sin 40.0°) = \boxed{112\text{ m}}$$

Problem-solving insight

Either acute angle of a right triangle can be used to determine the components of a vector. The choice of angle is a matter of convenience.

Problem-solving insight

You can check to see whether the components of a vector are correct by substituting them into the Pythagorean theorem and verifying that the result is the magnitude of the original vector.

Since the vector components and the original vector form a right triangle, the Pythagorean theorem can be applied to check the validity of calculations such as those in Example 8. Thus, with the components obtained in Example 8, the theorem can be used to verify that the magnitude of the original vector is indeed 175 m, as given initially:

$$r = \sqrt{(112\text{ m})^2 + (134\text{ m})^2} = 175\text{ m}$$

Problem-solving insight

It is possible for one of the components of a vector to be zero. This does not mean that the vector itself is zero, however. ***For a vector to be zero, every vector component must individually be zero.*** Thus, in two dimensions, saying that $\vec{A} = 0$ is equivalent to saying that $\vec{A}_x = 0$ and $\vec{A}_y = 0$. Or, stated in terms of scalar components, if $\vec{A} = 0$, then $A_x = 0$ and $A_y = 0$.

Problem-solving insight

Two vectors are equal if, and only if, they have the same magnitude and direction. Thus, if one displacement vector points east and another points north, they are *not* equal, even if each has the same magnitude of 480 m. In terms of vector components, two vectors $\vec{A}$ and $\vec{B}$ are equal if, and only if, each vector component of one is equal to the corresponding vector component of the other. In two dimensions, if $\vec{A} = \vec{B}$, then $\vec{A}_x = \vec{B}_x$ and $\vec{A}_y = \vec{B}_y$. Alternatively, using scalar components, we write that $A_x = B_x$ and $A_y = B_y$.

✓ CHECK YOUR UNDERSTANDING

(The answers are given at the end of the book.)

12. Which of the following displacement vectors (if any) are equal?

Variable	Magnitude	Direction
$\vec{A}$	100 m	30° north of east
$\vec{B}$	100 m	30° south of west
$\vec{C}$	50 m	30° south of west
$\vec{D}$	100 m	60° east of north

13. Two vectors, $\vec{A}$ and $\vec{B}$, are shown in the drawing.   **(a)** What are the signs (+ or −) of the scalar components of A_x and A_y of the vector $\vec{A}$?   **(b)** What are the signs of the scalar components B_x and B_y of the vector $\vec{B}$?   **(c)** What are the signs of the scalar components R_x and R_y of the vector $\vec{R}$, where $\vec{R} = \vec{A} + \vec{B}$?

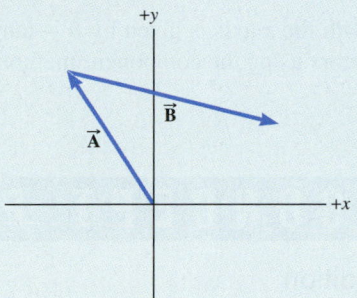

14. Are two vectors with the same magnitude necessarily equal?

15. The magnitude of a vector has doubled, its direction remaining the same. Can you conclude that the magnitude of each component of the vector has doubled?

16. The tail of a vector is fixed to the origin of an x, y axis system. Originally the vector points along the $+x$ axis and has a magnitude of 12 units. As time passes, the vector rotates counterclockwise. What are the sizes of the x and y components of the vector for the following rotational angles?   **(a)** 90°   **(b)** 180°   **(c)** 270°   **(d)** 360°

17. A vector has a component of zero along the x axis of a certain axis system. Does this vector necessarily have a component of zero along the x axis of another (rotated) axis system?

1.8 ADDITION OF VECTORS BY MEANS OF COMPONENTS

The components of a vector provide the most convenient and accurate way of adding (or subtracting) any number of vectors. For example, suppose that vector $\vec{A}$ is added to vector $\vec{B}$. The resultant vector is $\vec{C}$, where $\vec{C} = \vec{A} + \vec{B}$. Figure 1.21$a$ illustrates this vector addition, along with the x and y vector components of $\vec{A}$ and $\vec{B}$. In part b of the drawing, the vectors $\vec{A}$ and $\vec{B}$ have been removed, because we can use the vector components of these vectors in place of them. The vector component $\vec{B}_x$ has been shifted downward and arranged tail to head with vector component $\vec{A}_x$. Similarly, the vector component $\vec{A}_y$ has been shifted to the right and arranged tail to head with the vector component $\vec{B}_y$. The x components are colinear and add together to give the x component of the resultant vector $\vec{C}$. In like fashion, the y components are colinear and add together to give the y component of $\vec{C}$. In terms of scalar components, we can write

$$C_x = A_x + B_x \quad \text{and} \quad C_y = A_y + B_y$$

Figure 1.21 (a) The vectors $\vec{A}$ and $\vec{B}$ add together to give the resultant vector $\vec{C}$. The x and y components of $\vec{A}$ and $\vec{B}$ are also shown. (b) The drawing illustrates that $\vec{C}_x = \vec{A}_x + \vec{B}_x$ and $\vec{C}_y = \vec{A}_y + \vec{B}_y$. ($c$) Vector $\vec{C}$ and its components form a right triangle.

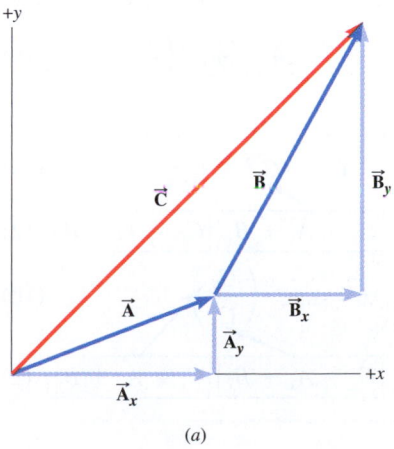

(a)

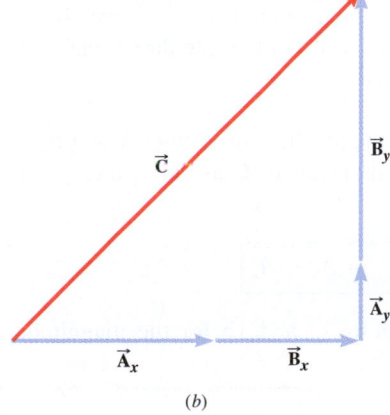

(b)

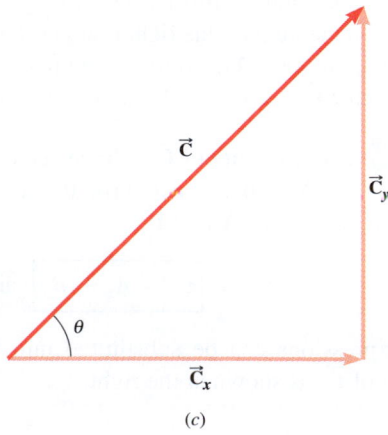

(c)

The vector components $\vec{C}_x$ and $\vec{C}_y$ of the resultant vector form the sides of the right triangle shown in Figure 1.21c. Thus, we can find the magnitude of $\vec{C}$ by using the Pythagorean theorem:

$$C = \sqrt{C_x^2 + C_y^2}$$

The angle θ that $\vec{C}$ makes with the x axis is given by $\theta = \tan^{-1}(C_y/C_x)$. Example 9 illustrates how to add several vectors using the component method.

ANALYZING MULTIPLE-CONCEPT PROBLEMS

Example 9 The Component Method of Vector Addition

A jogger runs 145 m in a direction 20.0° east of north (displacement vector $\vec{A}$) and then 105 m in a direction 35.0° south of east (displacement vector $\vec{B}$). Using components, determine the magnitude and direction of the resultant vector $\vec{C}$ for these two displacements.

Reasoning Figure 1.22 shows the vectors $\vec{A}$ and $\vec{B}$, assuming that the y axis corresponds to the direction due north. The vectors are arranged in a tail-to-head fashion, with the resultant vector $\vec{C}$ drawn from the tail of $\vec{A}$ to the head of $\vec{B}$. The components of the vectors are also shown in the figure. Since $\vec{C}$ and its components form a right triangle (red in the drawing), we will use the Pythagorean theorem and trigonometry to express the magnitude and directional angle θ for $\vec{C}$ in terms of its components. The components of $\vec{C}$ will then be obtained from the components of $\vec{A}$ and $\vec{B}$ and the data given for these two vectors.

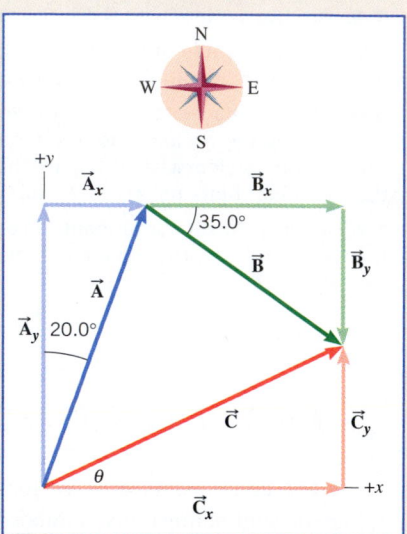

Figure 1.22 The vectors $\vec{A}$ and $\vec{B}$ add together to give the resultant vector $\vec{C}$. The vector components of $\vec{A}$ and $\vec{B}$ are also shown. The resultant vector $\vec{C}$ can be obtained once its components have been found.

Knowns and Unknowns The data for this problem are listed in the table that follows:

Description	Symbol	Value	Comment
Magnitude of vector $\vec{A}$		145 m	
Direction of vector $\vec{A}$		20.0° east of north	See Figure 1.22.
Magnitude of vector $\vec{B}$		105 m	
Direction of vector $\vec{B}$		35.0° south of east	See Figure 1.22.
Unknown Variables			
Magnitude of resultant vector	C	?	
Direction of resultant vector	θ	?	

Modeling the Problem

STEP 1 Magnitude and Direction of $\vec{C}$ In Figure 1.22 the vector $\vec{C}$ and its components $\vec{C}_x$ and $\vec{C}_y$ form a right triangle, as the red arrows show. Applying the Pythagorean theorem to this right triangle shows that the magnitude of $\vec{C}$ is given by Equation 1a at the right. From the red triangle it also follows that the directional angle θ for the vector $\vec{C}$ is given by Equation 1b at the right.

$$C = \sqrt{C_x^2 + C_y^2} \tag{1a}$$

$$\theta = \tan^{-1}\left(\frac{C_y}{C_x}\right) \tag{1b}$$

STEP 2 Components of $\vec{C}$ Since vector $\vec{C}$ is the resultant of vectors $\vec{A}$ and $\vec{B}$, we have $\vec{C} = \vec{A} + \vec{B}$ and can write the scalar components of $\vec{C}$ as the sum of the scalar components of $\vec{A}$ and $\vec{B}$:

$$\boxed{C_x = A_x + B_x} \quad \text{and} \quad \boxed{C_y = A_y + B_y}$$

These expressions can be substituted into Equations 1a and 1b for the magnitude and direction of $\vec{C}$, as shown at the right.

$$C = \sqrt{C_x^2 + C_y^2} \tag{1a}$$

$$\boxed{C_x = A_x + B_x} \quad \boxed{C_y = A_y + B_y} \tag{2}$$

$$\theta = \tan^{-1}\left(\frac{C_y}{C_x}\right) \tag{1b}$$

$$\boxed{C_x = A_x + B_x} \quad \boxed{C_y = A_y + B_y} \tag{2}$$

Solution Algebraically combining the results of each step, we find that

STEP 1   STEP 2

$$C = \sqrt{C_x^2 + C_y^2} = \sqrt{(A_x + B_x)^2 + (A_y + B_y)^2}$$

$$\theta = \tan^{-1}\left(\frac{C_y}{C_x}\right) = \tan^{-1}\left(\frac{A_y + B_y}{A_x + B_x}\right)$$

STEP 1   STEP 2

To use these results we need values for the individual components of $\vec{A}$ and $\vec{B}$.

Referring to Figure 1.22, we find these values to be

$A_x = (145 \text{ m}) \sin 20.0° = 49.6 \text{ m}$ and $A_y = (145 \text{ m}) \cos 20.0° = 136 \text{ m}$

$B_x = (105 \text{ m}) \cos 35.0° = 86.0 \text{ m}$ and $B_y = -(105 \text{ m}) \sin 35.0° = -60.2 \text{ m}$

Note that the component B_y is negative, because $\vec{B}_y$ points downward, in the negative y direction in the drawing. Substituting these values into the results for C and θ gives

$$C = \sqrt{(A_x + B_x)^2 + (A_y + B_y)^2} = \sqrt{(49.6 \text{ m} + 86.0 \text{ m})^2 + (136 \text{ m} - 60.2 \text{ m})^2} = \boxed{155 \text{ m}}$$

$$\theta = \tan^{-1}\left(\frac{A_y + B_y}{A_x + B_x}\right) = \tan^{-1}\left(\frac{136 \text{ m} - 60.2 \text{ m}}{49.6 \text{ m} + 86.0 \text{ m}}\right) = \boxed{29°}$$

Related Homework: *Problems 45, 46, 47, 56, 62*

✓ CHECK YOUR UNDERSTANDING

(The answer is given at the end of the book.)

18. Two vectors, $\vec{A}$ and $\vec{B}$, have vector components that are shown (to the same scale) in the drawing. The resultant vector is labeled $\vec{R}$. Which drawing shows the correct vector sum of $\vec{A} + \vec{B}$? **(a)** 1, **(b)** 2, **(c)** 3, **(d)** 4

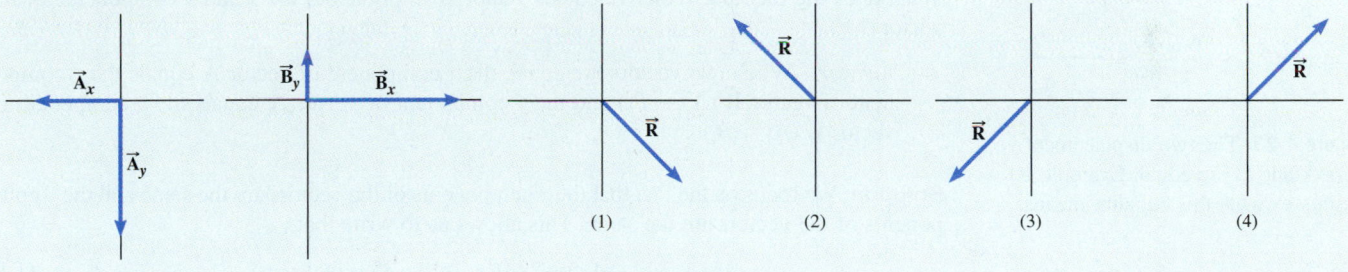

In later chapters we will often use the component method for vector addition. For future reference, the main features of the reasoning strategy used in this technique are summarized below.

REASONING STRATEGY

The Component Method of Vector Addition

1. For each vector to be added, determine the x and y components relative to a conveniently chosen x, y coordinate system. Be sure to take into account the directions of the components by using plus and minus signs to denote whether the components point along the positive or negative axes.

2. Find the algebraic sum of the x components, which is the x component of the resultant vector. Similarly, find the algebraic sum of the y components, which is the y component of the resultant vector.

Continued

3. Use the x and y components of the resultant vector and the Pythagorean theorem to determine the magnitude of the resultant vector.

4. Use either the inverse sine, inverse cosine, or inverse tangent function to find the angle that specifies the direction of the resultant vector.

1.9 CONCEPTS & CALCULATIONS

This chapter has presented an introduction to the mathematics of trigonometry and vectors, which will be used throughout this text. Therefore, in this last section we consider several examples in order to review some of the important features of this mathematics. The three-part format of these examples stresses the role of conceptual understanding in problem solving. First, the problem statement is given. Then, there is a concept question-and-answer section, which is followed by the solution section. The purpose of the concept question-and-answer section is to provide help in understanding the solution and to illustrate how a review of the concepts can help in anticipating some of the characteristics of the numerical answers.

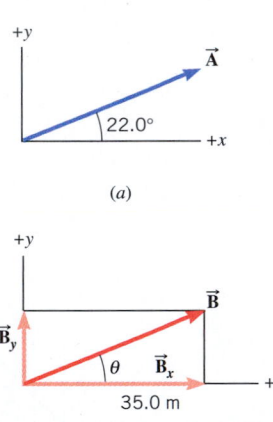

Figure 1.23 The two displacement vectors $\vec{A}$ and $\vec{B}$ are equal. Example 10 discusses what this equality means.

Concepts & Calculations Example 10 Equal Vectors

Figure 1.23 shows two displacement vectors $\vec{A}$ and $\vec{B}$. Vector $\vec{A}$ points at an angle of 22.0° above the x axis and has an unknown magnitude. Vector $\vec{B}$ has an x component $B_x = 35.0$ m and has an unknown y component B_y. These two vectors are equal. Find the magnitude of $\vec{A}$ and the value of B_y.

Concept Questions and Answers What does the fact that vector $\vec{A}$ equals vector $\vec{B}$ imply about the magnitudes and directions of the vectors?

Answer When two vectors are equal, each has the same magnitude and each has the same direction.

What does the fact that vector $\vec{A}$ equals vector $\vec{B}$ imply about the x and y components of the vectors?

Answer When two vectors are equal, the x component of vector $\vec{A}$ equals the x component of vector $\vec{B}$ ($A_x = B_x$) and the y component of vector $\vec{A}$ equals the y component of vector $\vec{B}$ ($A_y = B_y$).

Solution We focus on the fact that the x components of the vectors are the same and the y components of the vectors are the same. This allows us to write that

$$\underbrace{A \cos 22.0°}_{\substack{\text{Component } A_x \\ \text{of vector } \vec{A}}} = \underbrace{35.0 \text{ m}}_{\substack{\text{Component } B_x \\ \text{of vector } \vec{B}}} \qquad (1.8)$$

$$\underbrace{A \sin 22.0°}_{\substack{\text{Component } A_y \\ \text{of vector } \vec{A}}} = \underbrace{B_y}_{\substack{\text{Component } B_y \\ \text{of vector } \vec{B}}} \qquad (1.9)$$

Dividing Equation 1.9 by Equation 1.8 shows that

$$\frac{A \sin 22.0°}{A \cos 22.0°} = \frac{B_y}{35.0 \text{ m}}$$

$$B_y = (35.0 \text{ m}) \frac{\sin 22.0°}{\cos 22.0°} = (35.0 \text{ m}) \tan 22.0° = \boxed{14.1 \text{ m}}$$

Solving Equation 1.8 directly for A gives

$$A = \frac{35.0 \text{ m}}{\cos 22.0°} = \boxed{37.7 \text{ m}}$$

Concepts & Calculations Example 11

⬇ Using Components to Add Vectors

Figure 1.24 shows three displacement vectors $\vec{A}$, $\vec{B}$, and $\vec{C}$. These vectors are arranged in tail-to-head fashion, because they add together to give a resultant displacement $\vec{R}$, which lies along the x axis. Note that the vector $\vec{B}$ is parallel to the x axis. What is the magnitude of the vector $\vec{C}$?

Concept Questions and Answers How is the magnitude of $\vec{C}$ related to its scalar components C_x and C_y?

> *Answer* The magnitude of $\vec{C}$ is given by the Pythagorean theorem in the form $C = \sqrt{C_x^2 + C_y^2}$, since a vector and its components form a right triangle (see Figure 1.16).

Do any of the vectors in Figure 1.24 have a zero value for either their x or y components?

> *Answer* Yes. The vectors $\vec{B}$ and $\vec{R}$ each have a zero value for their y component. This is because these vectors are parallel to the x axis.

What does the fact that $\vec{A}$, $\vec{B}$, and $\vec{C}$ add together to give the resultant $\vec{R}$ tell you about the components of these vectors?

> *Answer* The fact that $\vec{A} + \vec{B} + \vec{C} = \vec{R}$ means that the sum of the x components of $\vec{A}$, $\vec{B}$, and $\vec{C}$ equals the x component of $\vec{R}$ ($A_x + B_x + C_x = R_x$) and the sum of the y components of $\vec{A}$, $\vec{B}$, and $\vec{C}$ equals the y component of $\vec{R}$ ($A_y + B_y + C_y = R_y$).

Solution To begin with, we apply the Pythagorean theorem to relate C, the magnitude of the vector $\vec{C}$, to its scalar components C_x and C_y:

$$C = \sqrt{C_x^2 + C_y^2}$$

To obtain a value for C_x we use the fact that the sum of the x components of $\vec{A}$, $\vec{B}$, and $\vec{C}$ equals the x component of $\vec{R}$:

$$\underbrace{(20.0 \text{ m}) \cos 60.0°}_{A_x} + \underbrace{10.0 \text{ m}}_{B_x} + \underbrace{C_x}_{} = \underbrace{35.0 \text{ m}}_{R_x} \quad \text{or} \quad C_x = 15.0 \text{ m}$$

To obtain a value for C_y we note that the sum of the y components of $\vec{A}$, $\vec{B}$, and $\vec{C}$ equals the y component of $\vec{R}$ (and remember that $\vec{B}$ and $\vec{R}$ have no y components):

$$\underbrace{(20.0 \text{ m}) \sin 60.0°}_{A_y} + \underbrace{0 \text{ m}}_{B_y} + \underbrace{C_y}_{} = \underbrace{0 \text{ m}}_{R_y} \quad \text{or} \quad C_y = -17.3 \text{ m}$$

With these values for C_x and C_y we find that the magnitude of $\vec{C}$ is

$$C = \sqrt{C_x^2 + C_y^2} = \sqrt{(15.0 \text{ m})^2 + (-17.3 \text{ m})^2} = \boxed{22.9 \text{ m}}$$

⬆

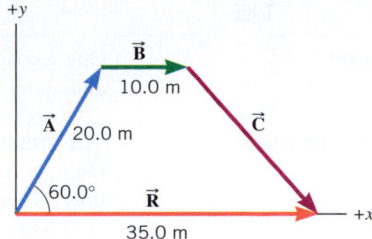

Figure 1.24 The displacement vectors $\vec{A}$, $\vec{B}$, and $\vec{C}$ add together to give the resultant displacement $\vec{R}$, which lies along the x axis. Note that the vector $\vec{B}$ is parallel to the x axis.

CONCEPT SUMMARY

If you need more help with a concept, use the Learning Aids noted next to the discussion or equation. Examples (**Ex.**) are in the text of this chapter. Go to **www.wiley.com/college/cutnell** for the following Learning Aids:

Interactive LearningWare (ILW) — Additional examples solved in a five-step interactive format.

Concept Simulations (CS) — Animated text figures or animations of important concepts.

Interactive Solutions (IS) — Models for certain types of problems in the chapter homework. The calculations are carried out interactively.

Topic	Discussion	Learning Aids
	1.2 UNITS The SI system of units includes the meter (m), the kilogram (kg), and the second (s) as the base units for length, mass, and time, respectively.	
Meter	One meter is the distance that light travels in a vacuum in a time of 1/299 792 458 second.	
Kilogram	One kilogram is the mass of a standard cylinder of platinum–iridium alloy kept at the International Bureau of Weights and Measures.	

Topic	Discussion	Learning Aids
Second	One second is the time for a certain type of electromagnetic wave emitted by cesium-133 atoms to undergo 9 192 631 770 wave cycles.	
Conversion of units	**1.3 THE ROLE OF UNITS IN PROBLEM SOLVING** To convert a number from one unit to another, multiply the number by the ratio of the two units. For instance, to convert 979 meters to feet, multiply 979 meters by the factor (3.281 foot/1 meter).	**Ex. 1, 2, 3**
Dimension	The dimension of a quantity represents its physical nature and the type of unit used to specify it. Three such dimensions are length [L], mass [M], time [T].	
Dimensional analysis	Dimensional analysis is a method for checking mathematical relations for the consistency of their dimensions.	
Sine, cosine, and tangent of an angle θ	**1.4 TRIGONOMETRY** The sine, cosine, and tangent functions of an angle θ are defined in terms of a right triangle that contains θ: $$\sin \theta = \frac{h_\mathrm{o}}{h} \ (1.1) \qquad \cos \theta = \frac{h_\mathrm{a}}{h} \ (1.2) \qquad \tan \theta = \frac{h_\mathrm{o}}{h_\mathrm{a}} \ (1.3)$$ where h_o and h_a are, respectively, the lengths of the sides opposite and adjacent to the angle θ, and h is the length of the hypotenuse.	**Ex. 4** **IS 1.19**
Inverse trigonometric functions	The inverse sine, inverse cosine, and inverse tangent functions are $$\theta = \sin^{-1}\left(\frac{h_\mathrm{o}}{h}\right) \ (1.4) \qquad \theta = \cos^{-1}\left(\frac{h_\mathrm{a}}{h}\right) \ (1.5) \qquad \theta = \tan^{-1}\left(\frac{h_\mathrm{o}}{h_\mathrm{a}}\right) \ (1.6)$$	**Ex. 5**
Pythagorean theorem	The Pythagorean theorem states that the square of the length of the hypotenuse of a right triangle is equal to the sum of the squares of the lengths of the other two sides: $$h^2 = h_\mathrm{o}^2 + h_\mathrm{a}^2 \qquad\qquad (1.7)$$	
Scalars and vectors	**1.5 SCALARS AND VECTORS** A scalar quantity is described completely by its size, which is also called its magnitude. A vector quantity has both a magnitude and a direction. Vectors are often represented by arrows, the length of the arrow being proportional to the magnitude of the vector and the direction of the arrow indicating the direction of the vector.	**Ex. 6**
Graphical method of vector addition and subtraction	**1.6 VECTOR ADDITION AND SUBTRACTION** One procedure for adding vectors utilizes a graphical technique, in which the vectors to be added are arranged in a tail-to-head fashion. The resultant vector is drawn from the tail of the first vector to the head of the last vector.	
	The subtraction of a vector is treated as the addition of a vector that has been multiplied by a scalar factor of -1. Multiplying a vector by -1 reverses the direction of the vector.	**Ex. 7** **IS 1.32**
Vector components	**1.7 THE COMPONENTS OF A VECTOR** In two dimensions, the vector components of a vector $\vec{\mathbf{A}}$ are two perpendicular vectors $\vec{\mathbf{A}}_x$ and $\vec{\mathbf{A}}_y$ that are parallel to the x and y axes, respectively, and that add together vectorially so that $\vec{\mathbf{A}} = \vec{\mathbf{A}}_x + \vec{\mathbf{A}}_y$.	
Scalar components	The scalar component A_x has a magnitude that is equal to that of $\vec{\mathbf{A}}_x$ and is given a positive sign if $\vec{\mathbf{A}}_x$ points along the $+x$ axis and a negative sign if $\vec{\mathbf{A}}_x$ points along the $-x$ axis. The scalar component A_y is defined in a similar manner.	**IS 1.65** **Ex. 8**
Condition for a vector to be zero	A vector is zero if, and only if, each of its vector components is zero.	
Condition for two vectors to be equal	Two vectors are equal if, and only if, they have the same magnitude and direction. Alternatively, two vectors are equal in two dimensions if the x vector components of each are equal and the y vector components of each are equal.	
	1.8 ADDITION OF VECTORS BY MEANS OF COMPONENTS If two vectors $\vec{\mathbf{A}}$ and $\vec{\mathbf{B}}$ are added to give a resultant $\vec{\mathbf{C}}$ such that $\vec{\mathbf{C}} = \vec{\mathbf{A}} + \vec{\mathbf{B}}$, then $$C_x = A_x + B_x \quad \text{and} \quad C_y = A_y + B_y$$ where C_x, A_x, and B_x are the scalar components of the vectors along the x direction, and C_y, A_y, and B_y are the scalar components of the vectors along the y direction.	**Ex. 9** **CS 1.1** **IS 1.55**

FOCUS ON CONCEPTS

Note to Instructors: The numbering of the questions shown here reflects the fact that they are only a representative subset of the total number that are available online. However, all of the questions are available for assignment via an online homework management program such as WileyPLUS or WebAssign.

Section 1.6 Vector Addition and Subtraction

1. During a relay race, runner A runs a certain distance due north and then hands off the baton to runner B, who runs for the same distance in a direction south of east. The two displacement vectors $\vec{A}$ and $\vec{B}$ can be added together to give a resultant vector $\vec{R}$. Which drawing correctly shows the resultant vector? **(a)** 1 **(b)** 2 **(c)** 3 **(d)** 4

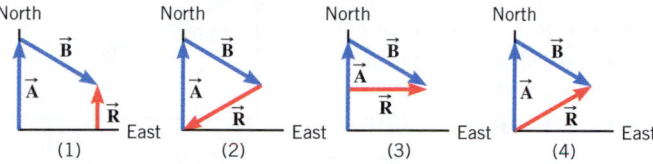

2. How is the magnitude R of the resultant vector $\vec{R}$ in the drawing related to the magnitudes A and B of the vectors $\vec{A}$ and $\vec{B}$? **(a)** The magnitude of the resultant vector $\vec{R}$ is equal to the sum of the magnitudes of $\vec{A}$ and $\vec{B}$, or $R = A + B$. **(b)** The magnitude of the resultant vector $\vec{R}$ is greater than the sum of the magnitudes of $\vec{A}$ and $\vec{B}$, or $R > A + B$. **(c)** The magnitude of the result-ant vector $\vec{R}$ is less than the sum of the magnitudes of $\vec{A}$ and $\vec{B}$, or $R < A + B$.

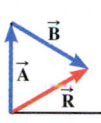

5. The first drawing shows three displacement vectors, $\vec{A}$, $\vec{B}$, and $\vec{C}$, which are added in a tail-to-head fashion. The resultant vector is labeled $\vec{R}$. Which of the following drawings shows the correct resultant vector for $\vec{A} + \vec{B} - \vec{C}$? **(a)** 1 **(b)** 2 **(c)** 3

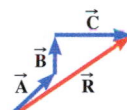 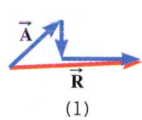 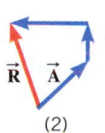 

6. The first drawing shows the sum of three displacement vectors, $\vec{A}$, $\vec{B}$, and $\vec{C}$. The resultant vector is labeled $\vec{R}$. Which of the following drawings shows the correct resultant vector for $\vec{A} - \vec{B} - \vec{C}$? **(a)** 1 **(b)** 2 **(c)** 3

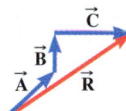 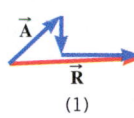 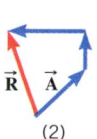 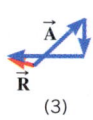

Section 1.7 The Components of a Vector

8. A person is jogging along a straight line, and her displacement is denoted by the vector $\vec{A}$ in the drawings. Which drawing represents the correct vector components, $\vec{A}_x$ and $\vec{A}_y$, for the vector $\vec{A}$? **(a)** 1 **(b)** 2 **(c)** 3 **(d)** 4

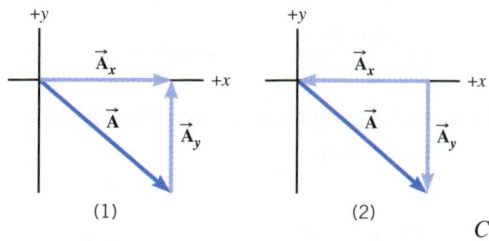

Continued

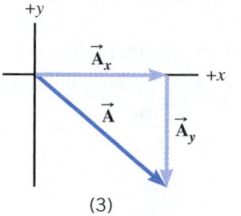 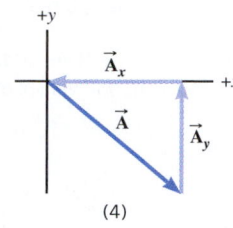

(3) (4)

Question 8

11. A person drives a car for a distance of 450.0 m. The displacement $\vec{A}$ of the car is illustrated in the drawing. What are the *scalar* components of this displacement vector?
(a) $A_x = 0$ m and $A_y = +450.0$ m
(b) $A_x = 0$ m and $A_y = -450.0$ m
(c) $A_x = +450.0$ m and $A_y = +450.0$ m
(d) $A_x = -450.0$ m and $A_y = 0$ m
(e) $A_x = -450.0$ m and $A_y = +450.0$ m

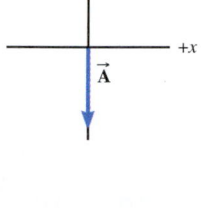

12. Drawing *a* shows a displacement vector $\vec{A}$ (450.0 m along the $-y$ axis). In this x, y coordinate system the scalar components are $A_x = 0$ m and $A_y = -450.0$ m. Suppose that the coordinate system is rotated counterclockwise by 35.0°, but the magnitude (450.0 m) and direction of vector $\vec{A}$ remain unchanged, as in drawing *b*. What are the scalar components, $A_{x'}$ and $A_{y'}$, of the vector $\vec{A}$ in the rotated x', y' coordinate system?
(a) $A_{x'} = -369$ m and $A_{y'} = -258$ m
(b) $A_{x'} = +369$ m and $A_{y'} = -258$ m
(c) $A_{x'} = +258$ m and $A_{y'} = +369$ m
(d) $A_{x'} = +258$ m and $A_{y'} = -369$ m
(e) $A_{x'} = -258$ m and $A_{y'} = -369$ m

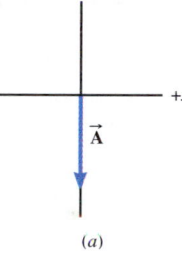

(a)

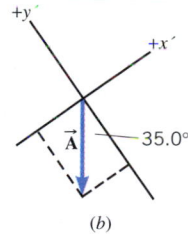

(b)

15. Suppose the vectors $\vec{A}$ and $\vec{B}$ in the drawing have magnitudes of 6.0 m and are directed as shown. What are A_x and B_x, the scalar components of $\vec{A}$ and $\vec{B}$ along the x axis?

	A_x	B_x
(a)	$+(6.0\text{ m})\cos 35° = +4.9$ m	$-(6.0\text{ m})\cos 35° = -4.9$ m
(b)	$+(6.0\text{ m})\sin 35° = +3.4$ m	$-(6.0\text{ m})\cos 35° = -4.9$ m
(c)	$-(6.0\text{ m})\cos 35° = -4.9$ m	$+(6.0\text{ m})\sin 35° = +3.4$ m
(d)	$-(6.0\text{ m})\cos 35° = -4.9$ m	$+(6.0\text{ m})\cos 35° = +4.9$ m
(e)	$-(6.0\text{ m})\sin 35° = -3.4$ m	$+(6.0\text{ m})\sin 35° = +3.4$ m

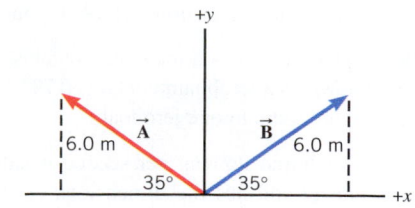

Section 1.8 Addition of Vectors by Means of Components

17. Drawing *a* shows two vectors $\vec{A}$ and $\vec{B}$, and drawing *b* shows their components. The scalar components of these vectors are as follows:

$$A_x = -4.9 \text{ m} \qquad A_y = +3.4 \text{ m}$$
$$B_x = +4.9 \text{ m} \qquad B_y = +3.4 \text{ m}$$

When the vectors $\vec{A}$ and $\vec{B}$ are added, the resultant vector is $\vec{R}$, so that $\vec{R} = \vec{A} + \vec{B}$. What are the values of R_x and R_y, the *x* and *y* components of $\vec{R}$?

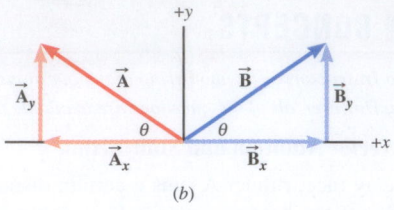

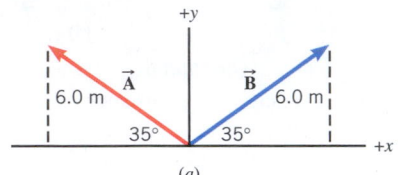

(a)

(b)

Question 17

18. The displacement vectors $\vec{A}$ and $\vec{B}$, when added together, give the resultant vector $\vec{R}$, so that $\vec{R} = \vec{A} + \vec{B}$. Use the data in the drawing to find the magnitude *R* of the resultant vector and the angle θ that it makes with the +*x* axis.

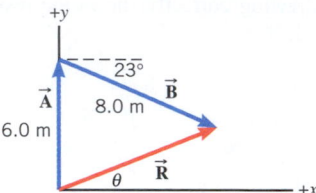

PROBLEMS

Problems that are not marked with a star are considered the easiest to solve. Problems that are marked with a single star (*) are more difficult, while those marked with a double star (**) are the most difficult.

Note to Instructors: Most of the homework problems in this chapter are available for assignment via an online homework management program such as WileyPLUS or WebAssign, and those marked with the icon **GO** *are presented in WileyPLUS using a guided tutorial format that provides enhanced interactivity. See Preface for additional details.*

ssm Solution is in the Student Solutions Manual.
www Solution is available online at www.wiley.com/college/cutnell

 This icon represents a biomedical application.

Section 1.2 Units,
Section 1.3 The Role of Units in Problem Solving

1. **GO** A student sees a newspaper ad for an apartment that has 1330 square feet (ft²) of floor space. How many square meters of area are there?

2. Suppose a man's scalp hair grows at a rate of 0.35 mm per day. What is this growth rate in feet per century?

3. **ssm** Vesna Vulovic survived the longest fall on record without a parachute when her plane exploded and she fell 6 miles, 551 yards. What is this distance in meters?

4. Bicyclists in the Tour de France reach speeds of 34.0 miles per hour (mi/h) on flat sections of the road. What is this speed in **(a)** kilometers per hour (km/h) and **(b)** meters per second (m/s)?

5. Given the quantities $a = 9.7$ m, $b = 4.2$ s, $c = 69$ m/s, what is the value of the quantity $d = a^3/(cb^2)$?

6. The variables *x*, *v*, and *a* have the dimensions of [L], [L]/[T], and [L]/[T]², respectively. These variables are related by an equation that has the form $v^n = 2ax$, where *n* is an integer constant (1, 2, 3, etc.) without dimensions. What must be the value of *n*, so that both sides of the equation have the same dimensions? Explain your reasoning.

7. **ssm** A bottle of wine known as a magnum contains a volume of 1.5 liters. A bottle known as a jeroboam contains 0.792 U.S. gallons. How many magnums are there in one jeroboam?

8. The volume of liquid flowing per second is called the volume flow rate *Q* and has the dimensions of [L]³/[T]. The flow rate of a liquid through a hypodermic needle during an injection can be estimated with the following equation:

$$Q = \frac{\pi R^n (P_2 - P_1)}{8 \eta L}$$

The length and radius of the needle are *L* and *R*, respectively, both of which have the dimension [L]. The pressures at opposite ends of the needle are P_2 and P_1, both of which have the dimensions of [M]/{[L][T]²}. The symbol η represents the viscosity of the liquid and has the dimensions of [M]/{[L][T]}. The symbol π stands for pi and, like the number 8 and the exponent *n*, has no dimensions. Using dimensional analysis, determine the value of *n* in the expression for *Q*.

9. The following are dimensions of various physical parameters that will be discussed later on in the text. Here [L], [T], and [M] denote, respectively, dimensions of length, time, and mass.

	Dimension		Dimension
Distance (*x*)	[L]	Acceleration (*a*)	[L]/[T]²
Time (*t*)	[T]	Force (*F*)	[M][L]/[T]²
Mass (*m*)	[M]	Energy (*E*)	[M][L]²/[T]²
Speed (*v*)	[L]/[T]		

Which of the following equations are dimensionally correct?

(a) $F = ma$ **(d)** $E = max$

(b) $x = \frac{1}{2}at^3$ **(e)** $v = \sqrt{Fx/m}$

(c) $E = \frac{1}{2}mv$

* **10.** A partly full paint can has 0.67 U.S. gallons of paint left in it. **(a)** What is the volume of the paint in cubic meters? **(b)** If all the remaining paint is used to coat a wall evenly (wall area = 13 m²), how thick is the layer of wet paint? Give your answer in meters.

* **11. ssm** A spring is hanging down from the ceiling, and an object of mass *m* is attached to the free end. The object is pulled down, thereby stretching the spring, and then released. The object oscillates up and down, and the time *T* required for one complete up-and-down oscillation is given by the equation $T = 2\pi\sqrt{m/k}$, where *k* is known as the spring constant. What must be the dimension of *k* for this equation to be dimensionally correct?

Section 1.4 Trigonometry

12. A chimpanzee sitting against his favorite tree gets up and walks 51 m due east and 39 m due south to reach a termite mound, where he eats lunch. **(a)** What is the shortest distance between the tree and the termite mound? **(b)** What angle does the shortest distance make with respect to due east?

13. ssm www A highway is to be built between two towns, one of which lies 35.0 km south and 72.0 km west of the other. What is the shortest length of highway that can be built between the two towns, and at what angle would this highway be directed with respect to due west?

14. A monkey is chained to a stake in the ground. The stake is 3.00 m from a vertical pole, and the chain is 3.40 m long. How high can the monkey climb up the pole?

15. GO The corners of a square lie on a circle of diameter *D* = 0.35 m. Each side of the square has a length *L*. Find *L*.

16. GO The drawing shows a person looking at a building on top of which an antenna is mounted. The horizontal distance between the person's eyes and the building is 85.0 m. In part *a* the person is looking at the base of the antenna, and his line of sight makes an

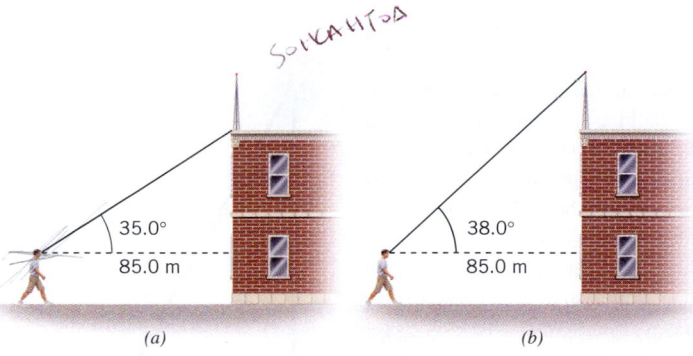

SOKAHTOA

(a)   (b)

angle of 35.0° with the horizontal. In part *b* the person is looking at the top of the antenna, and his line of sight makes an angle of 38.0° with the horizontal. How tall is the antenna?

* **17.** The drawing shows sodium and chloride ions positioned at the corners of a cube that is part of the crystal structure of sodium chloride (common table salt). The edge of the cube is 0.281 nm (1 nm = 1 nanometer = 10⁻⁹ m) in length. Find the distance (in nanometers) between the sodium ion located at one corner of the cube and the chloride ion located on the diagonal at the opposite corner.

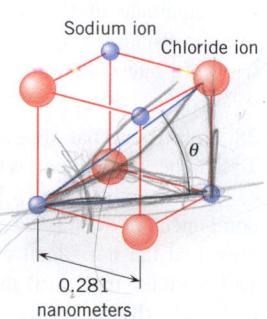

Sodium ion
Chloride ion

θ

0.281 nanometers

18. The two hot-air balloons in the drawing are 48.2 and 61.0 m above the ground. A person in the left balloon observes that the right balloon is 13.3° above the horizontal. What is the horizontal distance *x* between the two balloons?

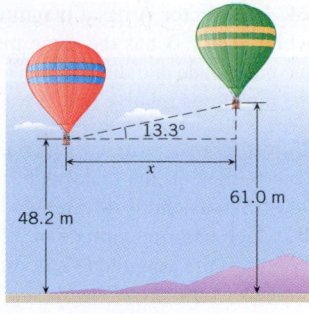

13.3°
x
61.0 m
48.2 m

Problem 18

* **19. Interactive Solution 1.19** at **www.wiley.com/college/cutnell** presents a method for modeling this problem. What is the value of the angle θ in the drawing that accompanies problem 17?

* **20.** You live in the building on the left in the drawing, and a friend lives in the other building. The two of you are having a discussion about the heights of the buildings, and your friend claims that his building is half again as tall as yours. To resolve the issue you climb to the roof of your building and estimate that your line of sight to the top edge of the other building makes an angle of 21° above the horizontal, while your line of sight to the base of the other building makes an angle of 52° below the horizontal. Determine the ratio of the height of the taller building to the height of the shorter building. State whether your friend is right or wrong.

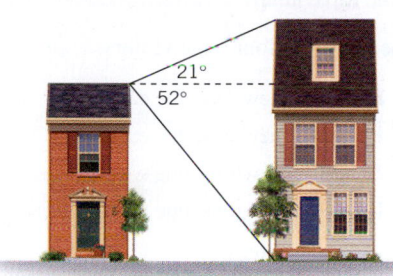

21°
52°

* **21. ssm** Three deer, A, B, and C, are grazing in a field. Deer B is located 62 m from deer A at an angle of 51° north of west. Deer C is located 77° north of east relative to deer A. The distance between deer B and C is 95 m. What is the distance between deer A and C? *(Hint: Consider the law of cosines given in Appendix E.)*

** **22.** An aerialist on a high platform holds on to a trapeze attached to a support by an 8.0-m cord. (See the drawing.) Just before he jumps off the platform, the cord makes an angle of 41° with the vertical. He jumps, swings down, then back up, releasing the trapeze at the instant it is 0.75 m below its initial height. Calculate the angle θ that the trapeze cord makes with the vertical at this instant.

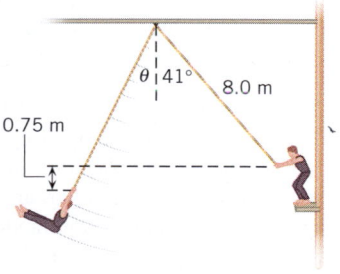

θ 41°
8.0 m
0.75 m

Section 1.6 Vector Addition and Subtraction

23. ssm Two ropes are attached to a heavy box to pull it along the floor. One rope applies a force of 475 newtons in a direction due west; the other applies a force of 315 newtons in a direction due south. As we will see later in the text, force is a vector quantity. **(a)** How much force should be applied by a single rope, and **(b)** in what direction (relative to due west), if it is to accomplish the same effect as the two forces added together?

24. **GO** Vector $\vec{A}$ has a magnitude of 63 units and points due west, while vector $\vec{B}$ has the same magnitude and points due south. Find the magnitude and direction of (a) $\vec{A} + \vec{B}$ and (b) $\vec{A} - \vec{B}$. Specify the directions relative to due west.

25. **ssm www** (a) Two workers are trying to move a heavy crate. One pushes on the crate with a force $\vec{A}$, which has a magnitude of 445 newtons and is directed due west. The other pushes with a force $\vec{B}$, which has a magnitude of 325 newtons and is directed due north. What are the magnitude and direction of the resultant force $\vec{A} + \vec{B}$ applied to the crate? (b) Suppose that the second worker applies a force $-\vec{B}$ instead of $\vec{B}$. What then are the magnitude and direction of the resultant force $\vec{A} - \vec{B}$ applied to the crate? In both cases express the direction relative to due west.

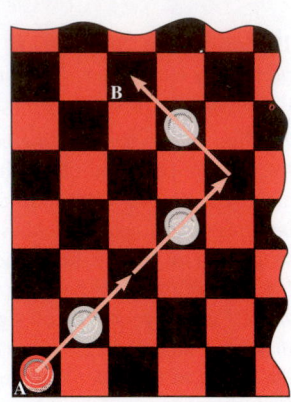

26. **GO** The drawing shows a triple jump on a checkerboard, starting at the center of square A and ending on the center of square B. Each side of a square measures 4.0 cm. What is the magnitude of the displacement of the colored checker during the triple jump?

Problem 26

27. Consider the following four force vectors:

$$\vec{F}_1 = 50.0 \text{ newtons, due east}$$
$$\vec{F}_2 = 10.0 \text{ newtons, due east}$$
$$\vec{F}_3 = 40.0 \text{ newtons, due west}$$
$$\vec{F}_4 = 30.0 \text{ newtons, due west}$$

Which two vectors add together to give a resultant with the smallest magnitude, and which two vectors add to give a resultant with the largest magnitude? In each case specify the magnitude and direction of the resultant.

28. Given the vectors $\vec{P}$ and $\vec{Q}$ shown on the grid, sketch and calculate the magnitudes of the vectors (a) $\vec{M} = \vec{P} + \vec{Q}$ and (b) $\vec{K} = 2\vec{P} - \vec{Q}$. Use the tail-to-head method and express the magnitudes in centimeters with the aid of the grid scale shown in the drawing.

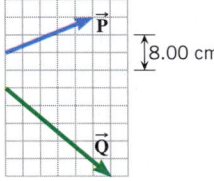

8.00 cm

29. Two bicyclists, starting at the same place, are riding toward the same campground by two different routes. One cyclist rides 1080 m due east and then turns due north and travels another 1430 m before reaching the campground. The second cyclist starts out by heading due north for 1950 m and then turns and heads directly toward the campground. (a) At the turning point, how far is the second cyclist from the campground? (b) What direction (measured relative to due east) must the second cyclist head during the last part of the trip?

* **30.** A jogger travels a route that has two parts. The first is a displacement $\vec{A}$ of 2.50 km due south, and the second involves a displacement $\vec{B}$ that points due east. (a) The resultant displacement $\vec{A} + \vec{B}$ has a magnitude of 3.75 km. What is the magnitude of $\vec{B}$, and what is the direction of $\vec{A} + \vec{B}$ relative to due south? (b) Suppose that $\vec{A} - \vec{B}$ had a magnitude of 3.75 km. What then would be the magnitude of $\vec{B}$, and what is the direction of $\vec{A} - \vec{B}$ relative to due south?

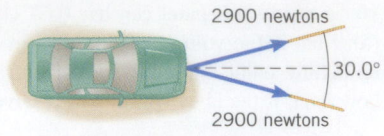

2900 newtons

30.0°

2900 newtons

* **31.** **ssm www** A car is being pulled out of the mud by two forces that are applied by the two ropes shown in the drawing. The dashed line in the drawing bisects the 30.0° angle. The magnitude of the force applied by each rope is 2900 newtons. Arrange the force vectors tail to head and use the graphical technique to answer the following questions. (a) How much force would a single rope need to apply to accomplish the same effect as the two forces added together? (b) How would the single rope be directed relative to the dashed line?

* **32.** Before starting this problem, review **Interactive Solution 1.32** at **www.wiley.com/college/cutnell**. Vector $\vec{A}$ has a magnitude of 12.3 units and points due west. Vector $\vec{B}$ points due north. (a) What is the magnitude of $\vec{B}$ if $\vec{A} + \vec{B}$ has a magnitude of 15.0 units? (b) What is the direction of $\vec{A} + \vec{B}$ relative to due west? (c) What is the magnitude of $\vec{B}$ if $\vec{A} - \vec{B}$ has a magnitude of 15.0 units? (d) What is the direction of $\vec{A} - \vec{B}$ relative to due west?

* **33.** Before starting this problem, review Conceptual Example 7. The force vector $\vec{F}_A$ has a magnitude of 90.0 newtons and points due east. The force vector $\vec{F}_B$ has a magnitude of 135 newtons and points 75° north of east. Use the graphical method and find the magnitude and direction of (a) $\vec{F}_A - \vec{F}_B$ (give the direction with respect to due east) and (b) $\vec{F}_B - \vec{F}_A$ (give the direction with respect to due west).

34. **GO** A force vector has a magnitude of 575 newtons and points at an angle of 36.0° below the positive x axis. What are (a) the x scalar component and (b) the y scalar component of the vector?

35. **ssm** The speed of an object and the direction in which it moves constitute a vector quantity known as the velocity. An ostrich is running at a speed of 17.0 m/s in a direction of 68.0° north of west. What is the magnitude of the ostrich's velocity component that is directed (a) due north and (b) due west?

36. Soccer player #1 is 8.6 m from the goal (see the drawing). If she kicks the ball directly into the net, the ball has a displacement labeled $\vec{A}$. If, on the other hand, she first kicks it to player #2, who then kicks it into the net, the ball undergoes two successive displacements, $\vec{A}_y$ and $\vec{A}_x$. What are the magnitudes and directions of $\vec{A}_x$ and $\vec{A}_y$?

37. You are on a treasure hunt and your map says "Walk due west for 52 paces, then walk 30.0° north of west for 42 paces, and finally walk due north for 25 paces." What is the magnitude of the component of your displacement in the direction (a) due north and (b) due west?

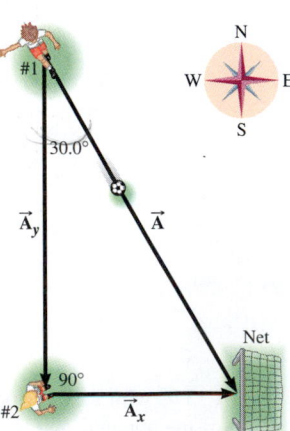

Problem 36

38. Your friend has slipped and fallen. To help her up, you pull with a force $\vec{F}$, as the drawing shows. The vertical component of this force is 130 newtons, and the horizontal component is 150 newtons. Find (a) the magnitude of $\vec{F}$ and (b) the angle θ.

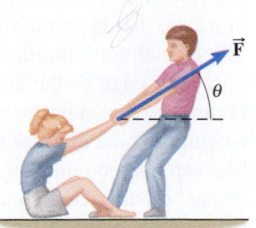

Problem 38

39. ssm The x vector component of a displacement vector $\vec{r}$ has a magnitude of 125 m and points along the negative x axis. The y vector component has a magnitude of 184 m and points along the negative y axis. Find the magnitude and direction of $\vec{r}$. Specify the direction with respect to the negative x axis.

40. GO The components of vector $\vec{A}$ are A_x and A_y (both positive), and the angle that it makes with respect to the positive x axis is θ. Find the angle θ if the components of the displacement vector $\vec{A}$ are **(a)** $A_x = 12$ m and $A_y = 12$ m, **(b)** $A_x = 17$ m and $A_y = 12$ m, and **(c)** $A_x = 12$ m and $A_y = 17$ m.

41. The displacement vector $\vec{A}$ has scalar components of $A_x = 80.0$ m and $A_y = 60.0$ m. The displacement vector $\vec{B}$ has a scalar component of $B_x = 60.0$ m and a magnitude of $B = 75.0$ m. The displacement vector $\vec{C}$ has a magnitude of $C = 100.0$ m and is directed at an angle of $36.9°$ above the $+x$ axis. Two of these vectors are equal. Determine which two, and support your choice with a calculation.

* **42.** Two racing boats set out from the same dock and speed away at the same constant speed of 101 km/h for half an hour (0.500 h), the blue boat headed $25.0°$ south of west, and the green boat headed $37.0°$ south of west. During this half hour **(a)** how much farther west does the blue boat travel, compared to the green boat, and **(b)** how much farther south does the green boat travel, compared to the blue boat? Express your answers in km.

* **43. ssm www** Vector $\vec{A}$ has a magnitude of 6.00 units and points due east. Vector $\vec{B}$ points due north. **(a)** What is the magnitude of $\vec{B}$, if the vector $\vec{A} + \vec{B}$ points $60.0°$ north of east? **(b)** Find the magnitude of $\vec{A} + \vec{B}$.

** **44.** The drawing shows a force vector that has a magnitude of 475 newtons. Find the **(a)** x, **(b)** y, and **(c)** z components of the vector.

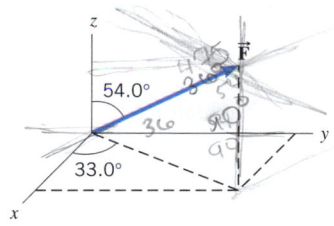

Section 1.8 Addition of Vectors by Means of Components

45. ssm Consult Multiple-Concept Example 9 in preparation for this problem. A golfer, putting on a green, requires three strokes to "hole the ball." During the first putt, the ball rolls 5.0 m due east. For the second putt, the ball travels 2.1 m at an angle of $20.0°$ north of east. The third putt is 0.50 m due north. What displacement (magnitude and direction relative to due east) would have been needed to "hole the ball" on the very first putt?

46. Multiple-Concept Example 9 reviews the concepts that play a role in this problem. As an aid in working this problem, consult **Concept Simulation 1.1** at **www.wiley.com/college/cutnell**. Two forces are applied to a tree stump to pull it out of the ground. Force $\vec{F_A}$ has a magnitude of 2240 newtons and points $34.0°$ south of east, while force $\vec{F_B}$ has a magnitude of 3160 newtons and points due south. Using the component method, find the magnitude and direction of the resultant force $\vec{F_A} + \vec{F_B}$ that is applied to the stump. Specify the direction with respect to due east.

47. Review Multiple-Concept Example 9 before beginning this problem. As an aid in visualizing the concepts in this problem, consult **Concept Simulation 1.1** at **www.wiley.com/college/cutnell**. A football player runs the pattern given in the drawing by the three displacement vectors $\vec{A}$, $\vec{B}$, and $\vec{C}$. The magnitudes of these vectors are $A = 5.00$ m, $B = 15.0$ m, and $C = 18.0$ m. Using the component method, find the magnitude and direction θ of the resultant vector $\vec{A} + \vec{B} + \vec{C}$.

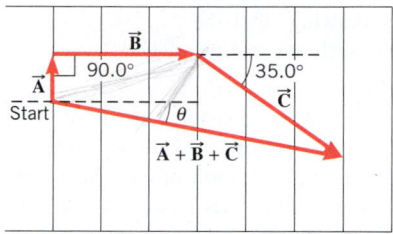

48. GO A baby elephant is stuck in a mud hole. To help pull it out, game keepers use a rope to apply a force $\vec{F_A}$, as part a of the drawing shows. By itself, however, force $\vec{F_A}$ is insufficient. Therefore, two additional forces $\vec{F_B}$ and $\vec{F_C}$ are applied, as in part b of the drawing. Each of these additional forces has the same magnitude F. The magnitude of the resultant force acting on the elephant in part b of the drawing is k times larger than that in part a. Find the ratio F/F_A when $k = 2.00$.

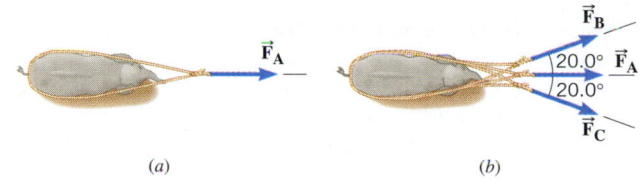

(a)   (b)

49. Displacement vector $\vec{A}$ points due east and has a magnitude of 2.00 km. Displacement vector $\vec{B}$ points due north and has a magnitude of 3.75 km. Displacement vector $\vec{C}$ points due west and has a magnitude of 2.50 km. Displacement vector $\vec{D}$ points due south and has a magnitude of 3.00 km. Find the magnitude and direction (relative to due west) of the resultant vector $\vec{A} + \vec{B} + \vec{C} + \vec{D}$.

50. GO A pilot flies her route in two straight-line segments. The displacement vector $\vec{A}$ for the first segment has a magnitude of 244 km and a direction $30.0°$ north of east. The displacement vector $\vec{B}$ for the second segment has a magnitude of 175 km and a direction due west. The resultant displacement vector is $\vec{R} = \vec{A} + \vec{B}$ and makes an angle θ with the direction due east. Using the component method, find the magnitude of $\vec{R}$ and the directional angle θ.

51. First consult **Concept Simulation 1.1** at **www.wiley.com/college/cutnell**. On a safari, a team of naturalists sets out toward a research station located 4.8 km away in a direction $42°$ north of east. After traveling in a straight line for 2.4 km, they stop and discover that they have been traveling $22°$ north of east, because their guide misread his compass. What are **(a)** the magnitude and **(b)** the direction (relative to due east) of the displacement vector now required to bring the team to the research station?

* **52.** Two geological field teams are working in a remote area. A global positioning system (GPS) tracker at their base camp shows the location of the first team as 38 km away, $19°$ north of west, and the second team as 29 km away, $35°$ east of north. When the first team uses its GPS to check the position of the second team, what does the GPS give for the second team's **(a)** distance from them and **(b)** direction, measured from due east?

*53. **ssm** A sailboat race course consists of four legs, defined by the displacement vectors $\vec{A}$, $\vec{B}$, $\vec{C}$, and $\vec{D}$, as the drawing indicates. The magnitudes of the first three vectors are $A = 3.20$ km, $B = 5.10$ km, and $C = 4.80$ km. The finish line of the course coincides with the starting line. Using the data in the drawing, find the distance of the fourth leg and the angle θ.

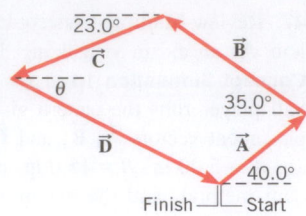

*54. The route followed by a hiker consists of three displacement vectors $\vec{A}$, $\vec{B}$, and $\vec{C}$. Vector $\vec{A}$ is along a measured trail and is 1550 m in a direction 25.0° north of east. Vector $\vec{B}$ is not along a measured trail, but the hiker uses a compass and knows that the direction is 41.0° east of south. Similarly, the direction of vector $\vec{C}$ is 35.0° north of west. The hiker ends up back where she started, so the resultant displacement is zero, or $\vec{A} + \vec{B} + \vec{C} = 0$. Find the magnitudes of (a) vector $\vec{B}$ and (b) vector $\vec{C}$.

*55. **Interactive Solution 1.55** at **www.wiley.com/college/cutnell** presents the solution to a problem that is similar to this one. Vector $\vec{A}$ has a magnitude of 145 units and points 35.0° north of west. Vector $\vec{B}$ points 65.0° east of north. Vector $\vec{C}$ points 15.0° west of south. These three vectors add to give a resultant vector that is zero. Using components, find the magnitudes of (a) vector $\vec{B}$ and (b) vector $\vec{C}$.

*56. Multiple-Concept Example 9 deals with the concepts that are important in this problem. A grasshopper makes four jumps. The displacement vectors are (1) 27.0 cm, due west; (2) 23.0 cm, 35.0° south of west; (3) 28.0 cm, 55.0° south of east; and (4) 35.0 cm, 63.0° north of east. Find the magnitude and direction of the resultant displacement. Express the direction with respect to due west.

ADDITIONAL PROBLEMS

57. You are driving into St. Louis, Missouri, and in the distance you see the famous Gateway-to-the-West arch. This monument rises to a height of 192 m. You estimate your line of sight with the top of the arch to be 2.0° above the horizontal. Approximately how far (in kilometers) are you from the base of the arch?

58. The gondola ski lift at Keystone, Colorado, is 2830 m long. On average, the ski lift rises 14.6° above the horizontal. How high is the top of the ski lift relative to the base?

59. **ssm** Vector $\vec{A}$ points along the $+y$ axis and has a magnitude of 100.0 units. Vector $\vec{B}$ points at an angle of 60.0° above the $+x$ axis and has a magnitude of 200.0 units. Vector $\vec{C}$ points along the $+x$ axis and has a magnitude of 150.0 units. Which vector has (a) the largest x component and (b) the largest y component?

60. Azelastine hydrochloride is an antihistamine nasal spray. A standard-size container holds one fluid ounce (oz) of the liquid. You are searching for this medication in a European drugstore and are asked how many milliliters (mL) there are in one fluid ounce. Using the following conversion factors, determine the number of milliliters in a volume of one fluid ounce: 1 gallon (gal) = 128 oz, 3.785×10^{-3} cubic meters (m^3) = 1 gal, and 1 mL = 10^{-6} m^3.

61. An ocean liner leaves New York City and travels 18.0° north of east for 155 km. How far east and how far north has it gone? In other words, what are the magnitudes of the components of the ship's displacement vector in the directions (a) due east and (b) due north?

62. Multiple-Concept Example 9 provides background pertinent to this problem. The magnitudes of the four displacement vectors shown in the drawing are $A = 16.0$ m, $B = 11.0$ m, $C = 12.0$ m, and $D = 26.0$ m. Determine the magnitude and directional angle for the resultant that occurs when these vectors are added together.

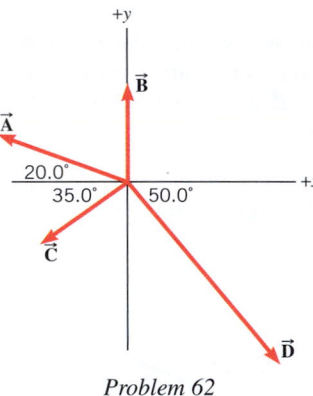

Problem 62

63. **ssm** A circus performer begins his act by walking out along a nearly horizontal high wire. He slips and falls to the safety net, 25.0 ft below. The magnitude of his displacement from the beginning of the walk to the net is 26.7 ft. (a) How far out along the high wire did he walk? (b) Find the angle that his displacement vector makes below the horizontal.

*64. A force vector points at an angle of 52° above the $+x$ axis. It has a y component of $+290$ newtons. Find (a) the magnitude and (b) the x component of the force vector.

*65. **ssm** To review the solution to a similar problem, consult **Interactive Solution 1.65** at **www.wiley.com/college/cutnell**. The magnitude of the force vector $\vec{F}$ is 82.3 newtons. The x component of this vector is directed along the $+x$ axis and has a magnitude of 74.6 newtons. The y component points along the $+y$ axis. (a) Find the direction of $\vec{F}$ relative to the $+x$ axis. (b) Find the component of $\vec{F}$ along the $+y$ axis.

*66. Three forces are applied to an object, as indicated in the drawing. Force $\vec{F}_1$ has a magnitude of 21.0 newtons (21.0 N) and is directed 30.0° to the left of the +y axis. Force $\vec{F}_2$ has a magnitude of 15.0 N and points along the +x axis. What must be the magnitude and direction (specified by the angle θ in the drawing) of the third force $\vec{F}_3$ such that the vector sum of the three forces is 0 N?

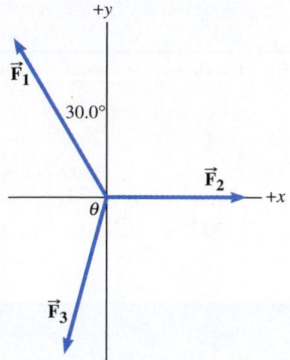

Problem 66

* **67.** At a picnic, there is a contest in which hoses are used to shoot water at a beach ball from three directions. As a result, three forces act on the ball, $\vec{F}_1$, $\vec{F}_2$, and $\vec{F}_3$ (see the drawing). The magnitudes of $\vec{F}_1$ and $\vec{F}_2$ are $F_1 = 50.0$ newtons and $F_2 = 90.0$ newtons. Using a scale drawing and the graphical technique, determine **(a)** the magnitude of $\vec{F}_3$ and **(b)** the angle θ such that the resultant force acting on the ball is zero.

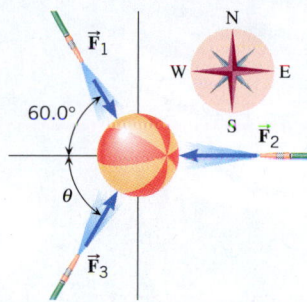

* **68.** A person is standing at the edge of the water and looking out at the ocean (see the drawing). The height of the person's eyes above the water is $h = 1.6$ m, and the radius of the earth is $R = 6.38 \times 10^6$ m. **(a)** How far is it to the horizon? In other words, what is the distance d from the person's eyes to the horizon? *(Note: At the horizon the angle between the line of sight and the radius of the earth is 90°.)* **(b)** Express this distance in miles.

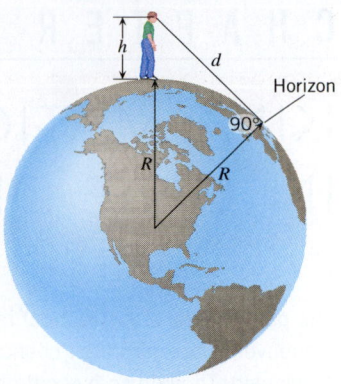

** **69.** What are the x and y components of the vector that must be added to the following three vectors, so that the sum of the four vectors is zero? Due east is the +x direction, and due north is the +y direction.

$$\vec{A} = 113 \text{ units}, 60.0° \text{ south of west}$$
$$\vec{B} = 222 \text{ units}, 35.0° \text{ south of east}$$
$$\vec{C} = 177 \text{ units}, 23.0° \text{ north of east}$$

KINEMATICS
IN ONE DIMENSION

The participants in this tugboat race maneuver to optimize their chances of winning. They do so by controlling the displacement, velocity, and acceleration of their boats. These three concepts and the relationships between them are the focus of this chapter. (© Harald Sund/Getty Images)

DISPLACEMENT

There are two aspects to any motion. In a purely descriptive sense, there is the movement itself. Is it rapid or slow, for instance? Then, there is the issue of what causes the motion or what changes it, which requires that forces be considered. *Kinematics* deals with the concepts that are needed to describe motion, without any reference to forces. The present chapter discusses these concepts as they apply to motion in one dimension, and the next chapter treats two-dimensional motion. *Dynamics* deals with the effect that forces have on motion, a topic that is considered in Chapter 4. Together, kinematics and dynamics form the branch of physics known as *mechanics*. We turn now to the first of the kinematics concepts to be discussed, which is displacement.

To describe the motion of an object, we must be able to specify the location of the object at all times, and Figure 2.1 shows how to do this for one-dimensional motion. In this drawing, the initial position of a car is indicated by the vector labeled $\vec{\mathbf{x}}_0$. The length of $\vec{\mathbf{x}}_0$ is the distance of the car from an arbitrarily chosen origin. At a later time the car has moved to a new position, which is indicated by the vector $\vec{\mathbf{x}}$. The *displacement* of the car $\Delta\vec{\mathbf{x}}$ (read as "delta x" or "the change in x") is a vector drawn from the initial position to the final position. Displacement is a vector quantity in the sense discussed in Section 1.5, for it conveys both a magnitude (the distance between the initial and final positions) and a direction. The displacement can be related to $\vec{\mathbf{x}}_0$ and $\vec{\mathbf{x}}$ by noting from the drawing that

$$\vec{\mathbf{x}}_0 + \Delta\vec{\mathbf{x}} = \vec{\mathbf{x}} \quad \text{or} \quad \Delta\vec{\mathbf{x}} = \vec{\mathbf{x}} - \vec{\mathbf{x}}_0$$

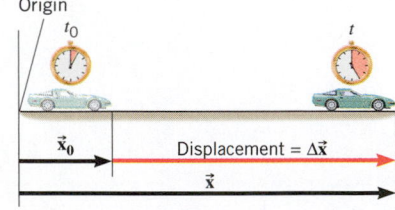

Figure 2.1 The displacement $\Delta\vec{\mathbf{x}}$ is a vector that points from the initial position $\vec{\mathbf{x}}_0$ to the final position $\vec{\mathbf{x}}$.

Thus, the displacement $\Delta\vec{\mathbf{x}}$ is the difference between $\vec{\mathbf{x}}$ and $\vec{\mathbf{x}}_0$, and the Greek letter delta (Δ) is used to signify this difference. It is important to note that the change in any variable is always the final value minus the initial value.

DEFINITION OF DISPLACEMENT

The displacement is a vector that points from an object's initial position to its final position and has a magnitude that equals the shortest distance between the two positions.

SI Unit of Displacement: meter (m)

The SI unit for displacement is the meter (m), but there are other units as well, such as the centimeter and the inch. When converting between centimeters (cm) and inches (in.), remember that 2.54 cm = 1 in.

Often, we will deal with motion along a straight line. In such a case, a displacement in one direction along the line is assigned a positive value, and a displacement in the opposite direction is assigned a negative value. For instance, assume that a car is moving along an east/west direction and that a positive (+) sign is used to denote a direction due east. Then, $\Delta \vec{x} = +500$ m represents a displacement that points to the east and has a magnitude of 500 meters. Conversely, $\Delta \vec{x} = -500$ m is a displacement that has the same magnitude but points in the opposite direction, due west.

The magnitude of the displacement vector is the shortest distance between the initial and final positions of the object. However, this does not mean that displacement and distance are the same physical quantities. In Figure 2.1, for example, the car could reach its final position after going forward and backing up several times. In that case, the total distance traveled by the car would be greater than the magnitude of the displacement vector.

✔ **CHECK YOUR UNDERSTANDING**

(The answer is given at the end of the book.)

1. A honeybee leaves the hive and travels a total distance of 2 km before returning to the hive. What is the magnitude of the displacement vector of the bee?

SPEED AND VELOCITY

2.2

AVERAGE SPEED

One of the most obvious features of an object in motion is how fast it is moving. If a car travels 200 meters in 10 seconds, we say its average speed is 20 meters per second, the *average speed* being the distance traveled divided by the time required to cover the distance:

$$\text{Average speed} = \frac{\text{Distance}}{\text{Elapsed time}} \qquad (2.1)$$

Equation 2.1 indicates that the unit for average speed is the unit for distance divided by the unit for time, or meters per second (m/s) in SI units. Example 1 illustrates how the idea of average speed is used.

Example 1 Distance Run by a Jogger

How far does a jogger run in 1.5 hours (5400 s) if his average speed is 2.22 m/s?

Reasoning The average speed of the jogger is the average distance per second that he travels. Thus, the distance covered by the jogger is equal to the average distance per second (his average speed) multiplied by the number of seconds (the elapsed time) that he runs.

Solution To find the distance run, we rewrite Equation 2.1 as

Distance = (Average speed)(Elapsed time) = (2.22 m/s)(5400 s) = $\boxed{12\ 000\ \text{m}}$

Speed is a useful idea, because it indicates how fast an object is moving. However, speed does not reveal anything about the direction of the motion. To describe both how fast an object moves and the direction of its motion, we need the vector concept of velocity.

AVERAGE VELOCITY

To define the velocity of an object, we will use two concepts that we have already encountered, displacement and time. The building of new concepts from more basic ones is a common theme in physics. In fact the great strength of physics as a science is that it builds a coherent understanding of nature through the development of interrelated concepts.

Suppose that the initial position of the car in Figure 2.1 is $\vec{x}_0$ when the time is t_0. A little later that car arrives at the final position $\vec{x}$ at the time t. The difference between these times is the time required for the car to travel between the two positions. We denote this difference by the shorthand notation Δt (read as "delta t"), where Δt represents the final time t minus the initial time t_0:

$$\underbrace{\Delta t = t - t_0}_{\text{Elapsed time}}$$

Figure 2.2 The velocity of the people riding an up escalator is opposite to the velocity of those riding a down escalator. (API/Alamy Images)

Note that Δt is defined in a manner analogous to $\Delta \vec{x}$, which is the final position minus the initial position ($\Delta \vec{x} = \vec{x} - \vec{x}_0$). Dividing the displacement $\Delta \vec{x}$ of the car by the elapsed time Δt gives the *average velocity* of the car. It is customary to denote the average value of a quantity by placing a horizontal bar above the symbol representing the quantity. The average velocity, then, is written as $\overline{\vec{v}}$, as specified in Equation 2.2:

DEFINITION OF AVERAGE VELOCITY

$$\text{Average velocity} = \frac{\text{Displacement}}{\text{Elapsed time}}$$

$$\overline{\vec{v}} = \frac{\vec{x} - \vec{x}_0}{t - t_0} = \frac{\Delta \vec{x}}{\Delta t} \qquad (2.2)$$

SI Unit of Average Velocity: meter per second (m/s)

Equation 2.2 indicates that the unit for average velocity is the unit for length divided by the unit for time, or meters per second (m/s) in SI units. Velocity can also be expressed in other units, such as kilometers per hour (km/h) or miles per hour (mi/h).

Average velocity is a vector that points in the same direction as the displacement in Equation 2.2. Figure 2.2 illustrates that the velocity of a person confined to move along a line can point either in one direction or in the opposite direction. As with displacement, we will use plus and minus signs to indicate the two possible directions. If the displacement points in the positive direction, the average velocity is positive. Conversely, if the displacement points in the negative direction, the average velocity is negative. Example 2 illustrates these features of average velocity.

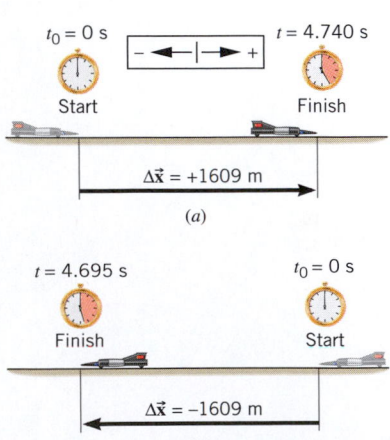

$t_0 = 0$ s $t = 4.740$ s

Start   Finish

$\Delta \vec{x} = +1609$ m

(a)

$t = 4.695$ s $t_0 = 0$ s

Finish   Start

$\Delta \vec{x} = -1609$ m

(b)

Figure 2.3 The arrows in the box at the top of the drawing indicate the positive and negative directions for the displacements of the car, as explained in Example 2.

Example 2 The World's Fastest Jet-Engine Car

Andy Green in the car *ThrustSSC* set a world record of 341.1 m/s (763 mi/h) in 1997. The car was powered by two jet engines, and it was the first one officially to exceed the speed of sound. To establish such a record, the driver makes two runs through the course, one in each direction, to nullify wind effects. Figure 2.3*a* shows that the car first travels from left to right and covers a distance of 1609 m (1 mile) in a time of 4.740 s. Figure 2.3*b* shows that in the reverse direction, the car covers the same distance in 4.695 s. From these data, determine the average velocity for each run.

Reasoning Average velocity is defined as the displacement divided by the elapsed time. In using this definition we recognize that the displacement is not the same as the distance traveled. Displacement takes the direction of the motion into account, and distance does not. During both runs, the car covers the same distance of 1609 m. However, for the first run the displacement is $\Delta \vec{x} = +1609$ m, while for the second it is $\Delta \vec{x} = -1609$ m. The plus and minus signs are essential, because the first run is to the right, which is the positive direction, and the second run is in the opposite or negative direction.

Solution According to Equation 2.2, the average velocities are

Run 1
$$\overline{\vec{v}} = \frac{\Delta \vec{x}}{\Delta t} = \frac{+1609 \text{ m}}{4.740 \text{ s}} = \boxed{+339.5 \text{ m/s}}$$

Run 2
$$\overline{\vec{v}} = \frac{\Delta \vec{x}}{\Delta t} = \frac{-1609 \text{ m}}{4.695 \text{ s}} = \boxed{-342.7 \text{ m/s}}$$

In these answers the algebraic signs convey the directions of the velocity vectors. In particular, for run 2 the minus sign indicates that the average velocity, like the displacement, points to the left in Figure 2.3*b*. The magnitudes of the velocities are 339.5 and 342.7 m/s. The average of these numbers is 341.1 m/s, and this is recorded in the record book.

INSTANTANEOUS VELOCITY

Suppose the magnitude of your average velocity for a long trip was 20 m/s. This value, being an average, does not convey any information about how fast you were moving or the direction of the motion at any instant during the trip. Both can change from one instant to another. Surely there were times when your car traveled faster than 20 m/s and times when it traveled more slowly. The ***instantaneous velocity*** $\vec{v}$ of the car indicates how fast the car moves and the direction of the motion at each instant of time. The magnitude of the instantaneous velocity is called the ***instantaneous speed,*** and it is the number (with units) indicated by the speedometer.

The instantaneous velocity at any point during a trip can be obtained by measuring the time interval Δt for the car to travel a *very small* displacement $\Delta \vec{x}$. We can then compute the average velocity over this interval. If the time Δt is small enough, the instantaneous velocity does not change much during the measurement. Then, the instantaneous velocity $\vec{v}$ at the point of interest is approximately equal to ($\approx$) the average velocity $\overline{\vec{v}}$ computed over the interval, or $\vec{v} \approx \overline{\vec{v}} = \Delta \vec{x} / \Delta t$ (for sufficiently small Δt). In fact, in the limit that Δt becomes infinitesimally small, the instantaneous velocity and the average velocity become equal, so that

$$\vec{v} = \lim_{\Delta t \to 0} \frac{\Delta \vec{x}}{\Delta t} \tag{2.3}$$

The notation $\lim\limits_{\Delta t \to 0} \dfrac{\Delta \vec{x}}{\Delta t}$ means that the ratio $\Delta \vec{x} / \Delta t$ is defined by a limiting process in which smaller and smaller values of Δt are used, so small that they approach zero. As smaller values of Δt are used, $\Delta \vec{x}$ also becomes smaller. However, the ratio $\Delta \vec{x} / \Delta t$ does *not* become zero but, rather, approaches the value of the instantaneous velocity. For brevity, we will use the word *velocity* to mean "instantaneous velocity" and *speed* to mean "instantaneous speed."

✓ CHECK YOUR UNDERSTANDING

(The answers are given at the end of the book).

2. Is the average speed of a vehicle a vector or a scalar quantity?

3. Two buses depart from Chicago, one going to New York and one to San Francisco. Each bus travels at a speed of 30 m/s. Do they have equal velocities?

4. One of the following statements is incorrect.   **(a)** The car traveled around the circular track at a constant velocity.   **(b)** The car traveled around the circular track at a constant speed. Which statement is incorrect?

5. A straight track is 1600 m in length. A runner begins at the starting line, runs due east for the full length of the track, turns around and runs halfway back. The time for this run is five minutes. What is the runner's average velocity, and what is his average speed?

6. The average velocity for a trip has a positive value. Is it possible for the instantaneous velocity at a point during the trip to have a negative value?

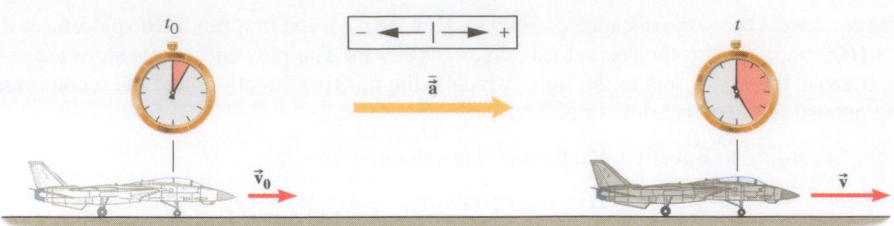

Figure 2.4 During takeoff, the plane accelerates from an initial velocity $\vec{v}_0$ to a final velocity $\vec{v}$ during the time interval $\Delta t = t - t_0$.

A skier accelerates out of the starting gate. (Agence Zoom/Getty Sport/ Getty Images Sport Services)

2.3 ACCELERATION

In a wide range of motions, the velocity changes from moment to moment. To describe the manner in which it changes, the concept of acceleration is needed. The velocity of a moving object may change in a number of ways. For example, it may increase, as it does when the driver of a car steps on the gas pedal to pass the car ahead. Or it may decrease, as it does when the driver applies the brakes to stop at a red light. In either case, the change in velocity may occur over a short or a long time interval. To describe how the velocity of an object changes during a given time interval, we now introduce the new idea of acceleration. This idea depends on two concepts that we have previously encountered, velocity and time. Specifically, the notion of acceleration emerges when the *change* in the velocity is combined with the time during which the change occurs.

The meaning of ***average acceleration*** can be illustrated by considering a plane during takeoff. Figure 2.4 focuses attention on how the plane's velocity changes along the runway. During an elapsed time interval $\Delta t = t - t_0$, the velocity changes from an initial value of $\vec{v}_0$ to a final velocity of $\vec{v}$. The change $\Delta \vec{v}$ in the plane's velocity is its final velocity minus its initial velocity, so that $\Delta \vec{v} = \vec{v} - \vec{v}_0$. The average acceleration $\bar{\vec{a}}$ is defined in the following manner, to provide a measure of how much the velocity changes per unit of elapsed time.

DEFINITION OF AVERAGE ACCELERATION

$$\text{Average acceleration} = \frac{\text{Change in velocity}}{\text{Elapsed time}}$$

$$\bar{\vec{a}} = \frac{\vec{v} - \vec{v}_0}{t - t_0} = \frac{\Delta \vec{v}}{\Delta t} \tag{2.4}$$

SI Unit of Average Acceleration: meter per second squared (m/s^2)

The average acceleration $\bar{\vec{a}}$ is a vector that points in the same direction as $\Delta \vec{v}$, the change in the velocity. Following the usual custom, plus and minus signs indicate the two possible directions for the acceleration vector when the motion is along a straight line.

We are often interested in an object's acceleration at a particular instant of time. The ***instantaneous acceleration*** $\vec{a}$ can be defined by analogy with the procedure used in Section 2.2 for instantaneous velocity:

$$\vec{a} = \lim_{\Delta t \to 0} \frac{\Delta \vec{v}}{\Delta t} \tag{2.5}$$

Equation 2.5 indicates that the instantaneous acceleration is a limiting case of the average acceleration. When the time interval Δt for measuring the acceleration becomes extremely small (approaching zero in the limit), the average acceleration and the instantaneous acceleration become equal. Moreover, in many situations the acceleration is constant, so the acceleration has the same value at any instant of time. In the future, we will use the word *acceleration* to mean "instantaneous acceleration." Example 3 deals with the acceleration of a plane during takeoff.

⬇ **Example 3** Acceleration and Increasing Velocity

Suppose the plane in Figure 2.4 starts from rest ($\vec{v}_0 = 0$ m/s) when $t_0 = 0$ s. The plane accelerates down the runway and at $t = 29$ s attains a velocity of $\vec{v} = +260$ km/h, where the plus sign indicates that the velocity points to the right. Determine the average acceleration of the plane.

Reasoning The average acceleration of the plane is defined as the change in its velocity divided by the elapsed time. The change in the plane's velocity is its final velocity $\vec{v}$ minus its initial velocity $\vec{v}_0$, or $\vec{v} - \vec{v}_0$. The elapsed time is the final time t minus the initial time t_0, or $t - t_0$.

Problem-solving insight

The change in any variable is the final value minus the initial value: for example, the change in velocity is $\Delta \vec{v} = \vec{v} - \vec{v}_0$, and the change in time is ($\Delta t = t - t_0$).

Solution The average acceleration is expressed by Equation 2.4 as

$$\overline{\vec{a}} = \frac{\vec{v} - \vec{v}_0}{t - t_0} = \frac{260 \text{ km/h} - 0 \text{ km/h}}{29 \text{ s} - 0 \text{ s}} = \boxed{+9.0 \; \frac{\text{km/h}}{\text{s}}}$$

The average acceleration calculated in Example 3 is read as "nine kilometers per hour per second." Assuming the acceleration of the plane is constant, a value of $+9.0 \dfrac{\text{km/h}}{\text{s}}$ means the velocity changes by $+9.0$ km/h during each second of the motion. During the first second, the velocity increases from 0 to 9.0 km/h; during the next second, the velocity increases by another 9.0 km/h to 18 km/h, and so on. Figure 2.5 illustrates how the velocity changes during the first two seconds. By the end of the 29th second, the velocity is 260 km/h.

It is customary to express the units for acceleration solely in terms of SI units. One way to obtain SI units for the acceleration in Example 3 is to convert the velocity units from km/h to m/s:

$$\left(260 \; \frac{\text{km}}{\text{h}} \right) \left(\frac{1000 \text{ m}}{1 \text{ km}} \right) \left(\frac{1 \text{ h}}{3600 \text{ s}} \right) = 72 \; \frac{\text{m}}{\text{s}}$$

The average acceleration then becomes

$$\overline{\vec{a}} = \frac{72 \text{ m/s} - 0 \text{ m/s}}{29 \text{ s} - 0 \text{ s}} = +2.5 \text{ m/s}^2$$

where we have used $2.5 \dfrac{\text{m/s}}{\text{s}} = 2.5 \dfrac{\text{m}}{\text{s} \cdot \text{s}} = 2.5 \dfrac{\text{m}}{\text{s}^2}$. An acceleration of $2.5 \dfrac{\text{m}}{\text{s}^2}$ is read as "2.5 meters per second per second" (or "2.5 meters per second squared") and means that the velocity changes by 2.5 m/s during each second of the motion.

Example 4 deals with a case where the motion becomes slower as time passes.

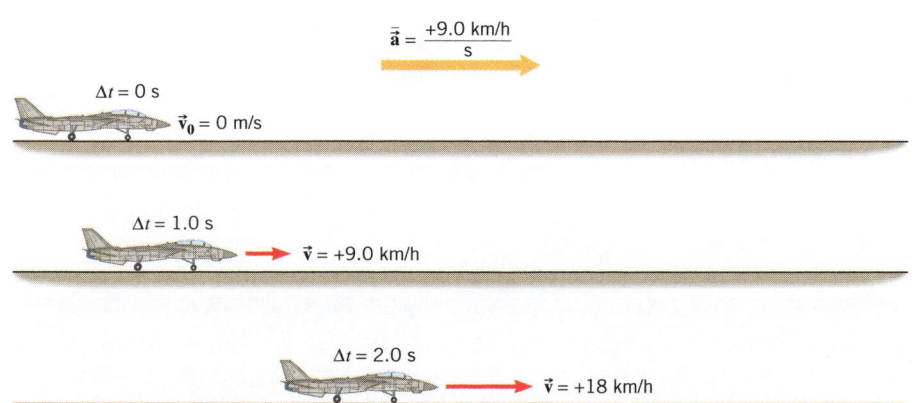

Figure 2.5 An acceleration of $+9.0 \dfrac{\text{km/h}}{\text{s}}$ means that the velocity of the plane changes by $+9.0$ km/h during each second of the motion. The "+" direction for $\overline{\vec{a}}$ and $\vec{v}$ is to the right.

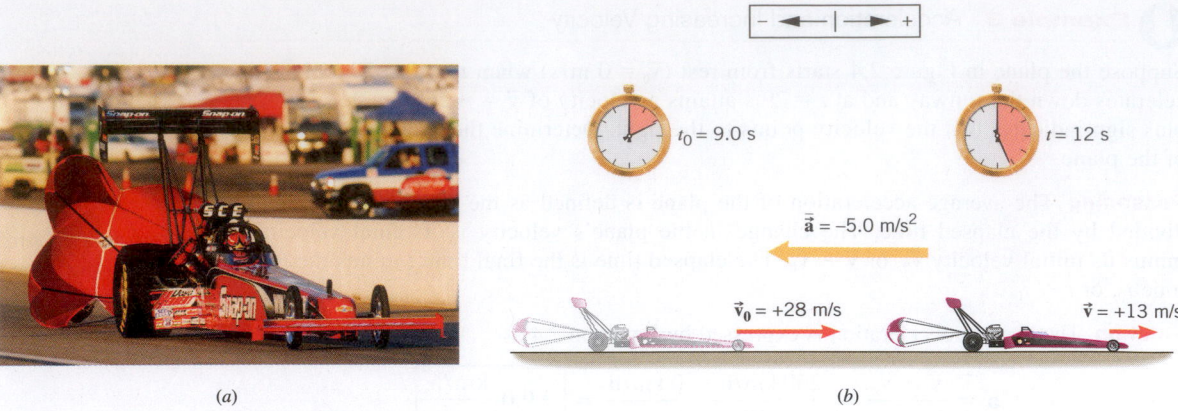

(a)   (b)

Figure 2.6 (*a*) To slow down, a drag racer deploys a parachute and applies the brakes. (*b*) The velocity of the car is decreasing, giving rise to an average acceleration $\vec{\mathbf{a}}$ that points opposite to the velocity. (*a*, Dave Kommel/AutoImagery)

Example 4 Acceleration and Decreasing Velocity

A drag racer crosses the finish line, and the driver deploys a parachute and applies the brakes to slow down, as Figure 2.6 illustrates. The driver begins slowing down when $t_0 = 9.0$ s and the car's velocity is $\vec{\mathbf{v}}_0 = +28$ m/s. When $t = 12.0$ s, the velocity has been reduced to $\vec{\mathbf{v}} = +13$ m/s. What is the average acceleration of the dragster?

Reasoning The average acceleration of an object is always specified as its change in velocity, $\vec{\mathbf{v}} - \vec{\mathbf{v}}_0$, divided by the elapsed time, $t - t_0$. This is true whether the final velocity is less than the initial velocity or greater than the initial velocity.

Solution The average acceleration is, according to Equation 2.4,

$$\vec{\mathbf{a}} = \frac{\vec{\mathbf{v}} - \vec{\mathbf{v}}_0}{t - t_0} = \frac{13 \text{ m/s} - 28 \text{ m/s}}{12.0 \text{ s} - 9.0 \text{ s}} = \boxed{-5.0 \text{ m/s}^2}$$

Figure 2.7 shows how the velocity of the dragster changes during the braking, assuming that the acceleration is constant throughout the motion. The acceleration calculated in Example 4 is negative, indicating that the acceleration points to the left in the drawing. As a result, the acceleration and the velocity point in *opposite* directions. ***Whenever the acceleration and velocity vectors have opposite directions, the object slows down and is said to be "decelerating."*** In contrast, the acceleration and velocity vectors in Figure 2.5 point in the *same* direction, and the object speeds up.

Problem-solving insight

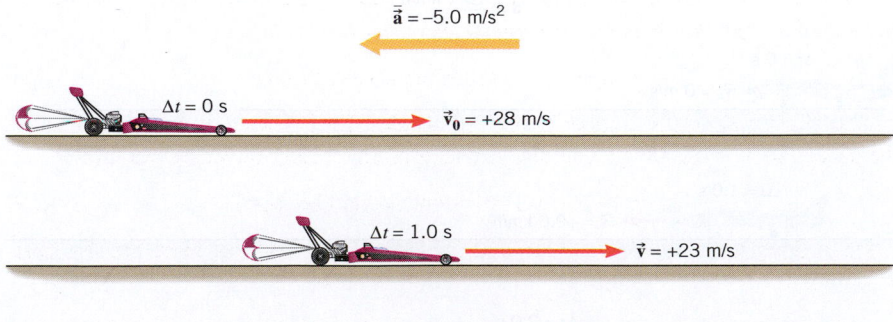

Figure 2.7 Here, an acceleration of -5.0 m/s² means the velocity decreases by 5.0 m/s during each second of elapsed time.

✓ **CHECK YOUR UNDERSTANDING**

(The answers are given at the end of the book.)

7. At one instant of time, a car and a truck are traveling side by side in adjacent lanes of a highway. The car has a greater velocity than the truck has. Does the car necessarily have the greater acceleration?

8. Two cars are moving in the same direction (the positive direction) on a straight road. The acceleration of each car also points in the positive direction. Car 1 has a greater acceleration than car 2 has. Which one of the following statements is true? **(a)** The velocity of car 1 is always greater than the velocity of car 2. **(b)** The velocity of car 2 is always greater than the velocity of car 1. **(c)** In the same time interval, the velocity of car 1 changes by a greater amount than the velocity of car 2 does. **(d)** In the same time interval, the velocity of car 2 changes by a greater amount than the velocity of car 1 does.

9. An object moving with a constant acceleration slows down if the acceleration points in the direction opposite to the direction of the velocity. But can an object ever come to a permanent halt if its acceleration truly remains constant?

10. A runner runs half the remaining distance to the finish line every ten seconds. She runs in a straight line and does not ever reverse her direction. Does her acceleration have a constant magnitude?

As this bald eagle comes in for a landing, its velocity vector points in the same direction as its motion. However, since the eagle is slowing down, its acceleration vector points opposite to the velocity vector. (Tom & Pat Lesson/Photo Researchers, Inc.)

2.4 EQUATIONS OF KINEMATICS FOR CONSTANT ACCELERATION

It is now possible to describe the motion of an object traveling with a constant acceleration along a straight line. To do so, we will use a set of equations known as the equations of kinematics for constant acceleration. These equations entail no new concepts, because they will be obtained by combining the familiar ideas of displacement, velocity, and acceleration. However, they will provide a very convenient way to determine certain aspects of the motion, such as the final position and velocity of a moving object.

In discussing the equations of kinematics, it will be convenient to assume that the object is located at the origin $\vec{x}_0 = 0$ m when $t_0 = 0$ s. With this assumption, the displacement $\Delta\vec{x} = \vec{x} - \vec{x}_0$ becomes $\Delta\vec{x} = \vec{x}$. Furthermore, in the equations that follow, as is customary, we dispense with the use of boldface symbols overdrawn with small arrows for the displacement, velocity, and acceleration vectors. We will, however, continue to convey the directions of these vectors with plus or minus signs.

Consider an object that has an initial velocity of v_0 at time $t_0 = 0$ s and moves for a time t with a constant acceleration a. For a complete description of the motion, it is also necessary to know the final velocity and displacement at time t. The final velocity v can be obtained directly from Equation 2.4:

$$\bar{a} = a = \frac{v - v_0}{t} \quad \text{or} \quad v = v_0 + at \qquad \text{(constant acceleration)} \qquad (2.4)$$

The displacement x at time t can be obtained from Equation 2.2, if a value for the average velocity $\bar{v}$ can be obtained. Considering the assumption that $\vec{x}_0 = 0$ m at $t_0 = 0$ s, we have

$$\bar{v} = \frac{x - x_0}{t - t_0} = \frac{x}{t} \quad \text{or} \quad x = \bar{v}t \qquad (2.2)$$

Because the acceleration is constant, the velocity increases at a constant rate. Thus, the average velocity $\bar{v}$ is midway between the initial and final velocities:

$$\bar{v} = \tfrac{1}{2}(v_0 + v) \qquad \text{(constant acceleration)} \qquad (2.6)$$

Equation 2.6, like Equation 2.4, applies only if the acceleration is constant and cannot be used when the acceleration is changing. The displacement at time t can now be determined as

$$x = \bar{v}t = \tfrac{1}{2}(v_0 + v)t \qquad \text{(constant acceleration)} \qquad (2.7)$$

Notice in Equations 2.4 ($v = v_0 + at$) and 2.7 [$x = \tfrac{1}{2}(v_0 + v)t$] that there are five kinematic variables:

1. x = displacement
2. $a = \bar{a}$ = acceleration (constant)
3. v = final velocity at time t
4. v_0 = initial velocity at time $t_0 = 0$ s
5. t = time elapsed since $t_0 = 0$ s

Each of the two equations contains four of these variables, so if three of them are known, the fourth variable can always be found. Example 5 illustrates how Equations 2.4 and 2.7 are used to describe the motion of an object.

ANALYZING MULTIPLE-CONCEPT PROBLEMS

Example 5 The Displacement of a Speedboat

The speedboat in Figure 2.8 has a constant acceleration of $+2.0$ m/s^2. If the initial velocity of the boat is $+6.0$ m/s, find the boat's displacement after 8.0 seconds.

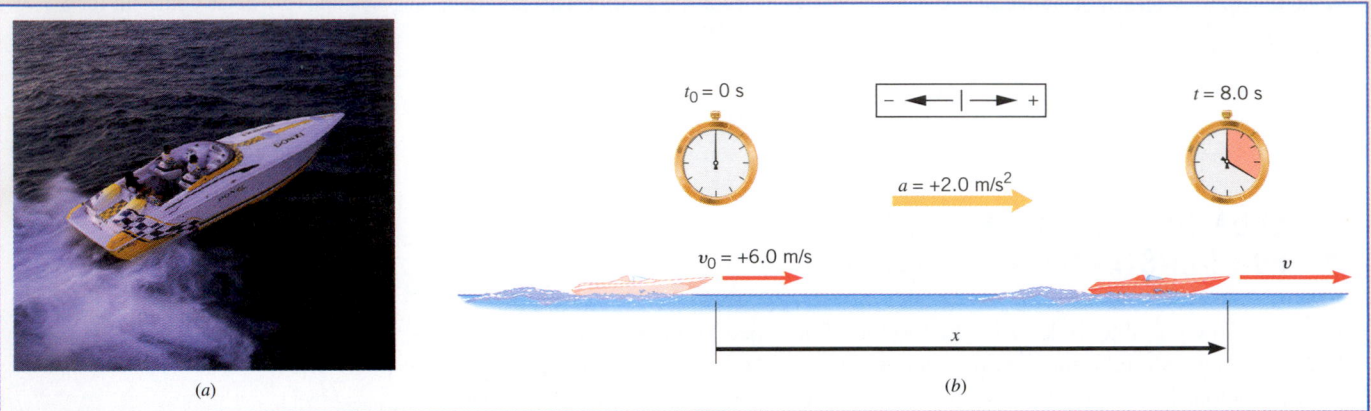

Figure 2.8 (*a*) An accelerating speedboat. (*b*) The boat's displacement x can be determined if the boat's acceleration, initial velocity, and time of travel are known. (*a*, Forest Johnson/Masterfile)

Reasoning As the speedboat accelerates, its velocity is changing. The displacement of the speedboat during a given time interval is equal to the product of its average velocity during that interval and the time. The average velocity, on the other hand, is just one-half the sum of the boat's initial and final velocities. To obtain the final velocity, we will use the definition of constant acceleration, as given in Equation 2.4.

Knowns and Unknowns The numerical values for the three known variables are listed in the table:

Description	Symbol	Value	Comment
Acceleration	a	$+2.0$ m/s^2	Positive, because the boat is accelerating to the right, which is the positive direction. See Figure 2.8*b*.
Initial velocity	v_0	$+6.0$ m/s	Positive, because the boat is moving to the right, which is the positive direction. See Figure 2.8*b*.
Time interval	t	8.0 s	
Unknown Variable			
Displacement of boat	x	?	

Modeling the Problem

STEP 1 Displacement Since the acceleration is constant, the displacement x of the boat is given by Equation 2.7, in which v_0 and v are the initial and final velocities, respectively. In this equation, two of the variables, v_0 and t, are known (see the Knowns and Unknowns table), but the final velocity v is not. However, the final velocity can be determined by employing the definition of acceleration, as Step 2 shows.

$$x = \tfrac{1}{2}(v_0 + v)t \quad (2.7)$$

$$?$$

STEP 2 Acceleration According to Equation 2.4, which is just the definition of constant acceleration, the final velocity v of the boat is

$$v = v_0 + at \quad (2.4)$$

$$x = \tfrac{1}{2}(v_0 + v)t \quad (2.7)$$

$$v = v_0 + at \quad (2.4)$$

All the variables on the right side of the equals sign are known, and we can substitute this relation for v into Equation 2.7, as shown at the right.

Solution Algebraically combining the results of Steps 1 and 2, we find that

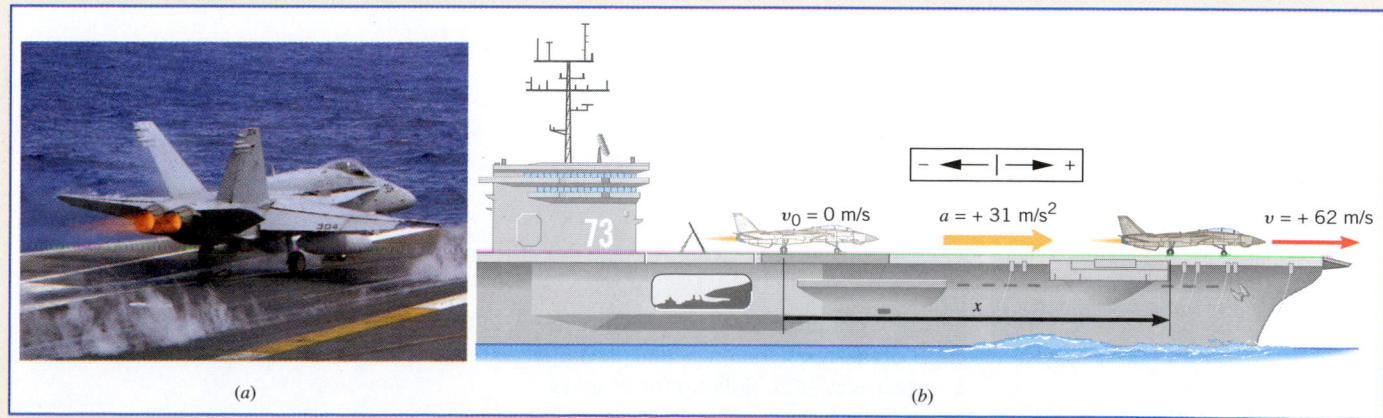

$$x = \tfrac{1}{2}(v_0 + v)t = \tfrac{1}{2}[v_0 + (v_0 + at)]t = v_0t + \tfrac{1}{2}at^2$$

The displacement of the boat after 8.0 s is

$$x = v_0t + \tfrac{1}{2}at^2 = (+6.0 \text{ m/s})(8.0 \text{ s}) + \tfrac{1}{2}(+2.0 \text{ m/s}^2)(8.0 \text{ s})^2 = \boxed{+110 \text{ m}}$$

A calculator would give the answer as 112 m, but this number must be rounded to 110 m, since the data are accurate to only two significant figures.

Related Homework: *Problems 39, 55*

In Example 5, we combined two equations [$x = \tfrac{1}{2}(v_0 + v)t$ and $v = v_0 + at$] into a single equation by algebraically eliminating the final velocity v of the speedboat (which was not known). The result was the following expression for the displacement x of the speedboat:

$$x = v_0t + \tfrac{1}{2}at^2 \qquad \text{(constant acceleration)} \tag{2.8}$$

The first term (v_0t) on the right side of this equation represents the displacement that would result if the acceleration were zero and the velocity remained constant at its initial value of v_0. The second term ($\tfrac{1}{2}at^2$) gives the additional displacement that arises because the velocity changes (a is not zero) to values that are different from its initial value. We now turn to another example of accelerated motion.

ANALYZING MULTIPLE-CONCEPT PROBLEMS

Example 6 Catapulting a Jet

A jet is taking off from the deck of an aircraft carrier, as Figure 2.9 shows. Starting from rest, the jet is catapulted with a constant acceleration of $+31$ m/s^2 along a straight line and reaches a velocity of $+62$ m/s. Find the displacement of the jet.

The physics of catapulting a jet from an aircraft carrier.

Reasoning When the plane is accelerating, its velocity is changing. The displacement of the plane during a given time interval is equal to the product of its average velocity during that interval and the time, the average velocity being equal to one-half the sum of the plane's initial and final velocities. The initial and final velocities are known, but the time is not. However, the time can be determined from a knowledge of the plane's acceleration.

$v_0 = 0$ m/s $\qquad a = +31$ m/s^2 $\qquad v = +62$ m/s

x

(a) (b)

Figure 2.9 (*a*) A plane is being launched from an aircraft carrier. (*b*) During the launch, a catapult accelerates the jet down the flight deck. (*a*, Seth C. Peterson/Department of the Navy)

Continued

Knowns and Unknowns The data for this problem are listed below:

Description	Symbol	Value	Comment
Explicit Data			
Acceleration	a	$+31$ m/s^2	Positive, because the acceleration points in the positive direction. See Figure 2.9b.
Final velocity	v	$+62$ m/s	Positive, because the final velocity points in the positive direction. See Figure 2.9b.
Implicit Data			
Initial velocity	v_0	0 m/s	The plane starts from rest.
Unknown Variable			
Displacement of plane	x	?	

Problem-solving insight
Implicit data are important. For instance, in Example 6 the phrase "starting from rest" means that the initial velocity is zero ($v_0 = 0$ m/s).

Modeling the Problem

STEP 1 **Displacement** The displacement x of the jet is given by Equation 2.7, since the acceleration is constant during launch. Referring to the Knowns and Unknowns table, we see that the initial and final velocities, v_0 and v, in this relation are known. The unknown time t can be determined by using the definition of acceleration (see Step 2).

$$x = \tfrac{1}{2}(v_0 + v)t \qquad (2.7)$$

STEP 2 **Acceleration** The acceleration of an object is defined by Equation 2.4 as the change in its velocity, $v - v_0$, divided by the elapsed time t:

$$a = \frac{v - v_0}{t} \qquad (2.4)$$

Solving for t yields

$$\boxed{t = \frac{v - v_0}{a}}$$

All the variables on the right side of the equals sign are known, and we can substitute this result for t into Equation 2.7, as shown at the right.

$$x = \tfrac{1}{2}(v_0 + v)t \qquad (2.7)$$

$$\boxed{t = \frac{v - v_0}{a}}$$

Solution Algebraically combining the results of Steps 1 and 2, we find that

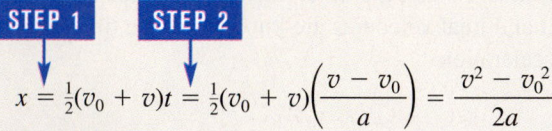

$$\underset{\text{STEP 1}}{\underbrace{x = \tfrac{1}{2}(v_0 + v)t}} = \underset{\text{STEP 2}}{\underbrace{\tfrac{1}{2}(v_0 + v)\left(\frac{v - v_0}{a}\right)}} = \frac{v^2 - v_0^2}{2a}$$

The displacement of the jet is

$$x = \frac{v^2 - v_0^2}{2a} = \frac{(+62 \text{ m/s})^2 - (0 \text{ m/s})^2}{2(+31 \text{ m/s}^2)} = \boxed{+62 \text{ m}}$$

Related Homework: *Problems 27, 51*

In Example 6 we were able to determine the displacement x of the jet from a knowledge of its acceleration a and its initial and final velocities, v_0 and v. The result was $x = (v^2 - v_0^2)/2a$. Solving for v^2 gives

$$v^2 = v_0^2 + 2ax \qquad \text{(constant acceleration)} \qquad (2.9)$$

Equation 2.9 is often used when the time involved in the motion is unknown.

Table 2.1 Equations of Kinematics for Constant Acceleration

Equation Number	Equation	Variables				
		x	a	v	v_0	t
(2.4)	$v = v_0 + at$	—	✓	✓	✓	✓
(2.7)	$x = \frac{1}{2}(v_0 + v)t$	✓	—	✓	✓	✓
(2.8)	$x = v_0 t + \frac{1}{2}at^2$	✓	✓	—	✓	✓
(2.9)	$v^2 = v_0^2 + 2ax$	✓	✓	✓	✓	—

   Table 2.1 presents a summary of the equations that we have been considering. These equations are called the ***equations of kinematics.*** Each equation contains four of the five kinematic variables, as indicated by the check marks (✓) in the table. The next section shows how to apply the equations of kinematics.

✔ **CHECK YOUR UNDERSTANDING**

(The answers are given at the end of the book.)

11. The muzzle velocity of a gun is the velocity of the bullet when it leaves the barrel. The muzzle velocity of one rifle with a short barrel is greater than the muzzle velocity of another rifle that has a longer barrel. In which rifle is the acceleration of the bullet larger?

12. A motorcycle starts from rest and has a constant acceleration. In a time interval *t*, it undergoes a displacement *x* and attains a final velocity *v*. Then *t* is increased so that the displacement is *3x*. In this same increased time interval, what final velocity does the motorcycle attain?

2.5 APPLICATIONS OF THE EQUATIONS OF KINEMATICS

   The equations of kinematics can be applied to any moving object, as long as the acceleration of the object is constant. However, remember that each equation contains four variables. Therefore, numerical values for three of the four must be available if an equation is to be used to calculate the value of the remaining variable. To avoid errors when using these equations, it helps to follow a few sensible guidelines and to be alert for a few situations that can arise during your calculations.

   Problem-solving insight: *Decide at the start which directions are to be called positive (+) and negative (−) relative to a conveniently chosen coordinate origin.* This decision is arbitrary, but important because displacement, velocity, and acceleration are vectors, and their directions must always be taken into account. In the examples that follow, the positive and negative directions will be shown in the drawings that accompany the problems. It does not matter which direction is chosen to be positive. However, once the choice is made, it should not be changed during the course of the calculation.

   Problem-solving insight: *As you reason through a problem before attempting to solve it, be sure to interpret the terms "decelerating" or "deceleration" correctly, should they occur in the problem statement.* These terms are the source of frequent confusion, and Conceptual Example 7 offers help in understanding them.

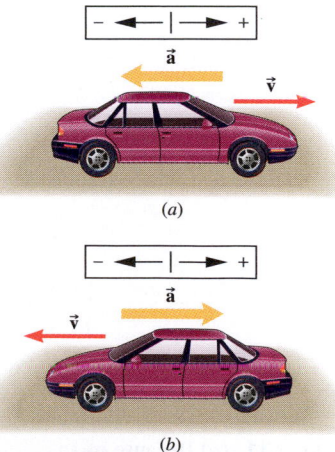

(a)

(b)

Figure 2.10 When a car decelerates along a straight road, the acceleration vector points opposite to the velocity vector, as Conceptual Example 7 discusses.

🔽 **Conceptual Example 7** Deceleration Versus Negative Acceleration

A car is traveling along a straight road and is decelerating. Which one of the following statements correctly describes the car's acceleration? **(a)** It must be positive. **(b)** It must be negative. **(c)** It could be positive or negative.

Reasoning The term "decelerating" means that the acceleration vector points opposite to the velocity vector and indicates that the car is slowing down. One possibility is that the velocity vector of the car points to the right, in the positive direction, as Figure 2.10*a* shows. The term "decelerating" implies, then, that the acceleration vector of the car points to the left, which is the negative direction. Another possibility is that the car is traveling to the left, as in Figure 2.10*b*.

Now, since the velocity vector points to the left, the acceleration vector would point opposite, or to the right, which is the positive direction.

Answers (a) and (b) are incorrect. The term "decelerating" means only that the acceleration vector points opposite to the velocity vector. It is not specified whether the velocity vector of the car points in the positive or negative direction. Therefore, it is not possible to know whether the acceleration is positive or negative.

Answer (c) is correct. As shown in Figure 2.10, the acceleration vector of the car could point in the positive or the negative direction, so that the acceleration could be either positive or negative, depending on the direction in which the car is moving.

Related Homework: *Problems 43, 72*

Problem-solving insight

Sometimes there are two possible answers to a kinematics problem, each answer corresponding to a different situation. Example 8 discusses one such case.

 Example 8 An Accelerating Spacecraft

The physics of the acceleration caused by a retrorocket.

The spacecraft shown in Figure 2.11*a* is traveling with a velocity of $+3250$ m/s. Suddenly the retrorockets are fired, and the spacecraft begins to slow down with an acceleration whose magnitude is 10.0 m/s^2. What is the velocity of the spacecraft when the displacement of the craft is $+215$ km, relative to the point where the retrorockets began firing?

Reasoning Since the spacecraft is slowing down, the acceleration must be opposite to the velocity. The velocity points to the right in the drawing, so the acceleration points to the left, in the negative direction; thus, $a = -10.0$ m/s^2. The three known variables are listed as follows:

Spacecraft Data				
x	a	v	v_0	t
$+215\,000$ m	-10.0 m/s^2	?	$+3250$ m/s	

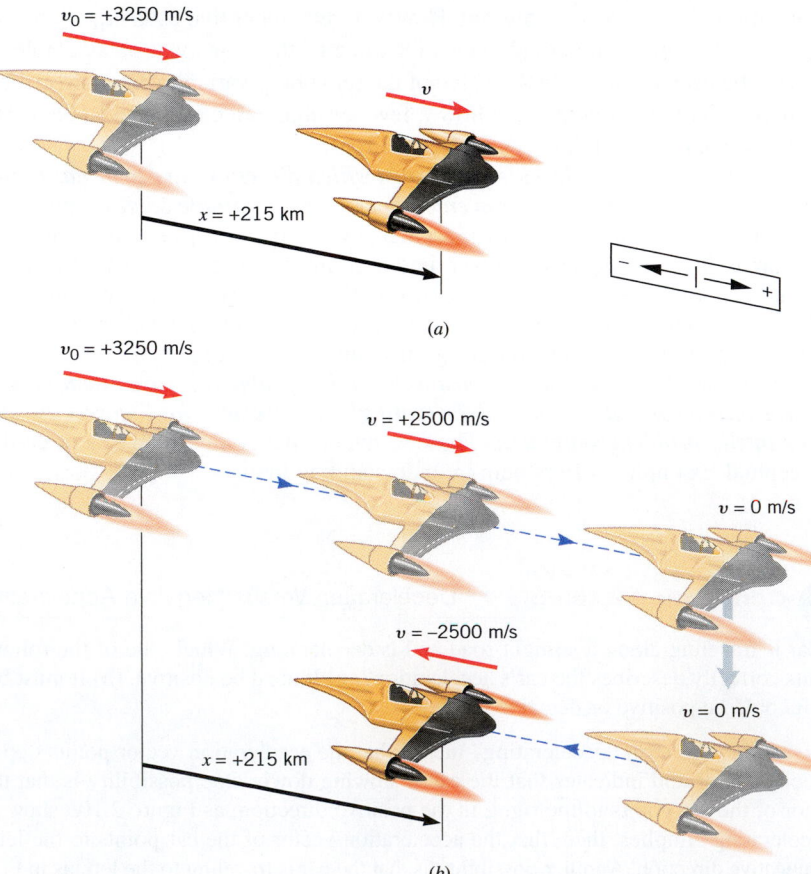

Figure 2.11 (*a*) Because of an acceleration of -10.0 m/s^2, the spacecraft changes its velocity from v_0 to v. (*b*) Continued firing of the retrorockets changes the direction of the craft's motion.

The final velocity v of the spacecraft can be calculated using Equation 2.9, since it contains the four pertinent variables.

Solution From Equation 2.9 ($v^2 = v_0^2 + 2ax$), we find that

$$v = \pm\sqrt{v_0^2 + 2ax} = \pm\sqrt{(3250 \text{ m/s})^2 + 2(-10.0 \text{ m/s}^2)(215\,000 \text{ m})}$$

$$= \boxed{+2500 \text{ m/s}} \quad \text{and} \quad \boxed{-2500 \text{ m/s}}$$

Both of these answers correspond to the *same* displacement ($x = +215$ km), but each arises in a different part of the motion. The answer $v = +2500$ m/s corresponds to the situation in Figure 2.11a, where the spacecraft has slowed to a speed of 2500 m/s, but is still traveling to the right. The answer $v = -2500$ m/s arises because the retrorockets eventually bring the spacecraft to a momentary halt and cause it to reverse its direction. Then it moves to the left, and its speed increases due to the continually firing rockets. After a time, the velocity of the craft becomes $v = -2500$ m/s, giving rise to the situation in Figure 2.11b. In both parts of the drawing the spacecraft has the same displacement, but a greater travel time is required in part b compared to part a.

The motion of two objects may be interrelated, so that they share a common variable. The fact that the motions are interrelated is an important piece of information. In such cases, data for only two variables need be specified for each object. See Interactive LearningWare 2.2 at **www.wiley.com/college/cutnell** for an example that illustrates this.

Often the motion of an object is divided into segments, each with a different acceleration. When solving such problems, it is important to realize that the final velocity for one segment is the initial velocity for the next segment, as Example 9 illustrates.

Problem-solving insight

Problem-solving insight

ANALYZING MULTIPLE-CONCEPT PROBLEMS

Example 9   A Motorcycle Ride

A motorcycle ride consists of two segments, as shown in Figure 2.12a. During segment 1, the motorcycle starts from rest, has an acceleration of $+2.6$ m/s², and has a displacement of $+120$ m. Immediately after segment 1, the motorcycle enters segment 2 and begins slowing down with an acceleration of -1.5 m/s² until its velocity is $+12$ m/s. What is the displacement of the motorcycle during segment 2?

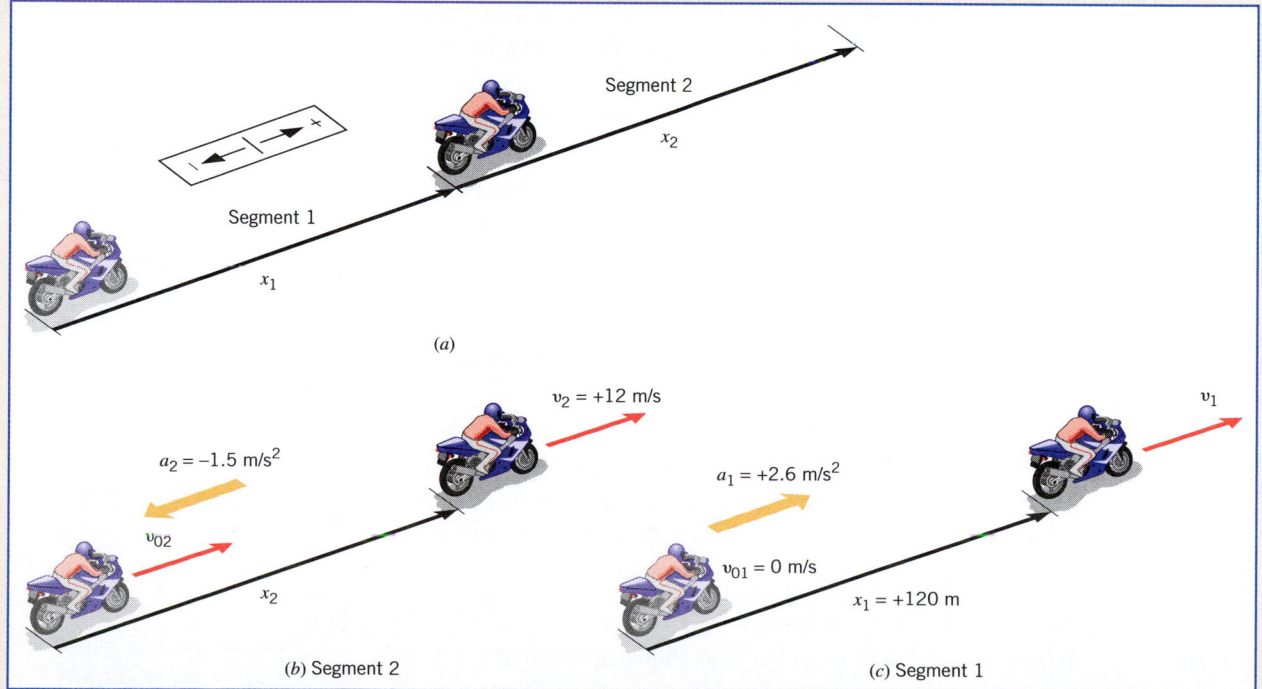

Figure 2.12 (a) This motorcycle ride consists of two segments, each with a different acceleration. (b) The variables for segment 2. (c) The variables for segment 1.

Continued

Reasoning We can use an equation of kinematics from Table 2.1 to find the displacement x_2 for segment 2. To do this, it will be necessary to have values for three of the variables that appear in the equation. Values for the acceleration ($a_2 = -1.5$ m/s^2) and final velocity ($v_2 = +12$ m/s) are given. A value for a third variable, the initial velocity v_{02}, can be obtained by noting that it is also the final velocity of segment 1. The final velocity of segment 1 can be found by using an appropriate equation of kinematics, since three variables (x_1, a_1, and v_{01}) are known for this part of the motion, as the following table reveals.

Knowns and Unknowns The data for this problem are listed in the table:

Description	Symbol	Value	Comment
Explicit Data			
Displacement for segment 1	x_1	+120 m	
Acceleration for segment 1	a_1	+2.6 m/s^2	Positive, because the motorcycle moves in the $+x$ direction and speeds up.
Acceleration for segment 2	a_2	−1.5 m/s^2	Negative, because the motorcycle moves in the $+x$ direction and slows down.
Final velocity for segment 2	v_2	+12 m/s	
Implicit Data			
Initial velocity for segment 1	v_{01}	0 m/s	The motorcycle starts from rest.
Unknown Variable			
Displacement for segment 2	x_2	?	

Modeling the Problem

STEP 1   Displacement During Segment 2   Figure 2.12b shows the situation during segment 2. Two of the variables—the final velocity v_2 and the acceleration a_2—are known, and for convenience we choose Equation 2.9 to find the displacement x_2 of the motorcycle:

$$v_2^2 = v_{02}^2 + 2a_2x_2 \tag{2.9}$$

$$x_2 = \frac{v_2^2 - v_{02}^2}{2a_2} \tag{1}$$

Solving this relation for x_2 yields Equation 1 at the right. Although the initial velocity v_{02} of segment 2 is not known, we will be able to determine its value from a knowledge of the motion during segment 1, as outlined in Steps 2 and 3.

STEP 2   Initial Velocity of Segment 2   Since the motorcycle enters segment 2 immediately after leaving segment 1, the initial velocity v_{02} of segment 2 is equal to the final velocity v_1 of segment 1, or $v_{02} = v_1$. Squaring both sides of this equation gives

$$v_{02}^2 = v_1^2$$

$$x_2 = \frac{v_2^2 - v_{02}^2}{2a_2} \tag{1}$$

$$v_{02}^2 = v_1^2 \tag{2}$$

This result can be substituted into Equation 1 as shown at the right. In the next step v_1 will be determined.

STEP 3   Final Velocity of Segment 1   Figure 2.12c shows the motorcycle during segment 1. Since we know the initial velocity v_{01}, the acceleration a_1, and the displacement x_1, we can employ Equation 2.9 to find the final velocity v_1 at the end of segment 1:

$$v_1^2 = v_{01}^2 + 2a_1x_1$$

$$x_2 = \frac{v_2^2 - v_{02}^2}{2a_2} \tag{1}$$

$$v_{02}^2 = v_1^2 \tag{2}$$

$$v_1^2 = v_{01}^2 + 2a_1x_1 \tag{3}$$

This relation for v_1^2 can be substituted into Equation 2, as shown at the right.

Solution Algebraically combining the results of each step, we find that

$$x_2 = \underset{\text{STEP 1}}{\frac{v_2^2 - v_{02}^2}{2a_2}} = \underset{\text{STEP 2}}{\frac{v_2^2 - v_1^2}{2a_2}} = \underset{\text{STEP 3}}{\frac{v_2^2 - (v_{01}^2 + 2a_1x_1)}{2a_2}}$$

The displacement x_2 of the motorcycle during segment 2 is

$$x_2 = \frac{v_2^2 - (v_{01}^2 + 2a_1x_1)}{2a_2}$$

$$= \frac{(+12 \text{ m/s})^2 - [(0 \text{ m/s})^2 + 2(+2.6 \text{ m/s}^2)(+120 \text{ m})]}{2(-1.5 \text{ m/s}^2)} = \boxed{+160 \text{ m}}$$

Related Homework: *Problems 37, 58, 80*

Now that we have seen how the equations of kinematics are applied to various situations, it's a good idea to summarize the reasoning strategy that has been used. This strategy, which is outlined below, will also be used when we consider freely falling bodies in Section 2.6 and two-dimensional motion in Chapter 3.

REASONING STRATEGY

Applying the Equations of Kinematics

1. Make a drawing to represent the situation being studied. A drawing helps us to see what's happening.

2. Decide which directions are to be called positive ($+$) and negative ($-$) relative to a conveniently chosen coordinate origin. Do not change your decision during the course of a calculation.

3. In an organized way, write down the values (with appropriate plus and minus signs) that are given for any of the five kinematic variables (x, a, v, v_0, and t). Be on the alert for implicit data, such as the phrase "starts from rest," which means that the value of the initial velocity is $v_0 = 0$ m/s. The data summary boxes used in the examples in the text are a good way of keeping track of this information. In addition, identify the variables that you are being asked to determine.

4. Before attempting to solve a problem, verify that the given information contains values for at least three of the five kinematic variables. Once the three known variables are identified along with the desired unknown variable, the appropriate relation from Table 2.1 can be selected. Remember that the motion of two objects may be interrelated, so they may share a common variable. The fact that the motions are interrelated is an important piece of information. In such cases, data for only two variables need be specified for each object.

5. When the motion of an object is divided into segments, as in Example 9, remember that the final velocity of one segment is the initial velocity for the next segment.

6. Keep in mind that there may be two possible answers to a kinematics problem as, for instance, in Example 8. Try to visualize the different physical situations to which the answers correspond.

2.6 FREELY FALLING BODIES

Everyone has observed the effect of gravity as it causes objects to fall downward. In the absence of air resistance, it is found that all bodies at the same location above the earth fall vertically with the same acceleration. Furthermore, if the distance of the fall is small compared to the radius of the earth, the acceleration remains essentially constant

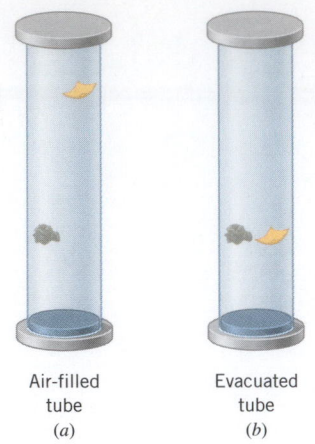

Air-filled   Evacuated
tube   tube
(a)   (b)

Figure 2.13 (a) In the presence of air resistance, the acceleration of the rock is greater than that of the paper. (b) In the absence of air resistance, both the rock and the paper have the same acceleration.

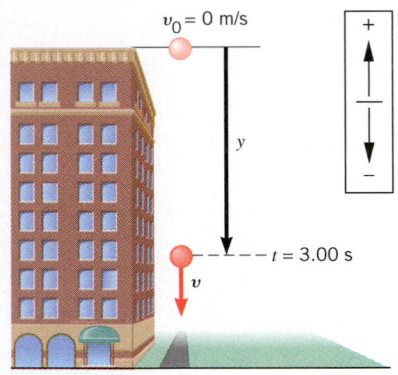

Figure 2.14 The stone, starting with zero velocity at the top of the building, is accelerated downward by gravity.

throughout the descent. This idealized motion, in which air resistance is neglected and the acceleration is nearly constant, is known as *free-fall.* Since the acceleration is constant in free-fall, the equations of kinematics can be used.

The acceleration of a freely falling body is called the ***acceleration due to gravity,*** and its magnitude (without any algebraic sign) is denoted by the symbol g. The acceleration due to gravity is directed downward, toward the center of the earth. Near the earth's surface, g is approximately

$$g = 9.80 \text{ m/s}^2 \quad \text{or} \quad 32.2 \text{ ft/s}^2$$

Unless circumstances warrant otherwise, we will use either of these values for g in subsequent calculations. In reality, however, g decreases with increasing altitude and varies slightly with latitude.

Figure 2.13a shows the well-known phenomenon of a rock falling faster than a sheet of paper. The effect of air resistance is responsible for the slower fall of the paper, for when air is removed from the tube, as in Figure 2.13b, the rock and the paper have exactly the same acceleration due to gravity. In the absence of air, the rock and the paper both exhibit free-fall motion. Free-fall is closely approximated for objects falling near the surface of the moon, where there is no air to retard the motion. A nice demonstration of free-fall was performed on the moon by astronaut David Scott, who dropped a hammer and a feather simultaneously from the same height. Both experienced the same acceleration due to lunar gravity and consequently hit the ground at the same time. The acceleration due to gravity near the surface of the moon is approximately one-sixth as large as that on the earth.

When the equations of kinematics are applied to free-fall motion, it is natural to use the symbol y for the displacement, since the motion occurs in the vertical or y direction. Thus, when using the equations in Table 2.1 for free-fall motion, we will simply replace x with y. There is no significance to this change. The equations have the same algebraic form for either the horizontal or vertical direction, provided that the acceleration remains constant during the motion. We now turn our attention to several examples that illustrate how the equations of kinematics are applied to freely falling bodies.

Example 10   A Falling Stone

A stone is dropped from rest from the top of a tall building, as Figure 2.14 indicates. After 3.00 s of free-fall, what is the displacement y of the stone?

Reasoning The upward direction is chosen as the positive direction. The three known variables are shown in the box below. The initial velocity v_0 of the stone is zero, because the stone is dropped from rest. The acceleration due to gravity is negative, since it points downward in the negative direction.

Stone Data				
y	a	v	v_0	t
?	-9.80 m/s^2		0 m/s	3.00 s

Equation 2.8 contains the appropriate variables and offers a direct solution to the problem. Since the stone moves downward, and upward is the positive direction, we expect the displacement y to have a negative value.

Solution Using Equation 2.8, we find that

$$y = v_0 t + \tfrac{1}{2}at^2 = (0 \text{ m/s})(3.00 \text{ s}) + \tfrac{1}{2}(-9.80 \text{ m/s}^2)(3.00 \text{ s})^2 = \boxed{-44.1 \text{ m}}$$

The answer for y is negative, as expected.

Example 11   The Velocity of a Falling Stone

After 3.00 s of free-fall, what is the velocity v of the stone in Figure 2.14?

Reasoning Because of the acceleration due to gravity, the magnitude of the stone's downward velocity increases by 9.80 m/s during each second of free-fall. The data for the stone are the same as in Example 10, and Equation 2.4 offers a direct solution for the final velocity. Since the stone is moving downward in the negative direction, the value determined for v should be negative.

Solution Using Equation 2.4, we obtain

$$v = v_0 + at = 0 \text{ m/s} + (-9.80 \text{ m/s}^2)(3.00 \text{ s}) = \boxed{-29.4 \text{ m/s}}$$

The velocity is negative, as expected.

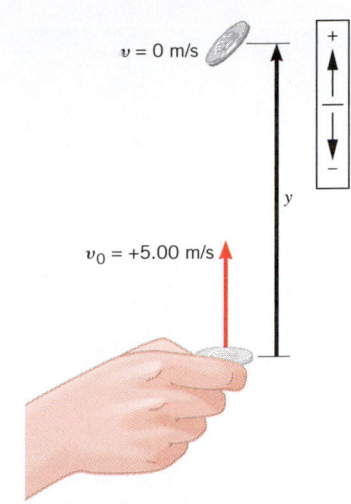

The acceleration due to gravity is always a downward-pointing vector. It describes how the speed increases for an object that is falling freely downward. This same acceleration also describes how the speed decreases for an object moving upward under the influence of gravity alone, in which case the object eventually comes to a momentary halt and then falls back to earth. Examples 12 and 13 show how the equations of kinematics are applied to an object that is moving upward under the influence of gravity.

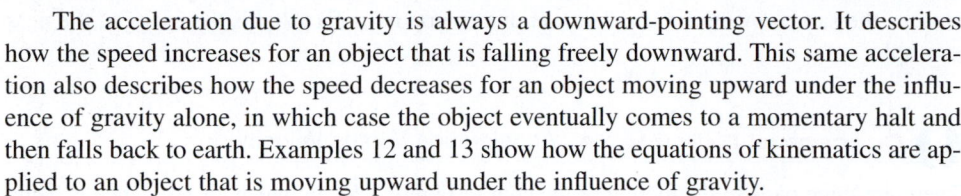 **Example 12** How High Does It Go?

A football game customarily begins with a coin toss to determine who kicks off. The referee tosses the coin up with an initial speed of 5.00 m/s. In the absence of air resistance, how high does the coin go above its point of release?

Reasoning The coin is given an upward initial velocity, as in Figure 2.15. But the acceleration due to gravity points downward. Since the velocity and acceleration point in opposite directions, the coin slows down as it moves upward. Eventually, the velocity of the coin becomes $v = 0$ m/s at the highest point. Assuming that the upward direction is positive, the data can be summarized as shown below:

Figure 2.15 At the start of a football game, a referee tosses a coin upward with an initial velocity of $v_0 = +5.00$ m/s. The velocity of the coin is momentarily zero when the coin reaches its maximum height.

Coin Data

y	a	v	v_0	t
?	-9.80 m/s^2	0 m/s	+5.00 m/s	

With these data, we can use Equation 2.9 ($v^2 = v_0^2 + 2ay$) to find the maximum height y.

Solution Rearranging Equation 2.9, we find that the maximum height of the coin above its release point is

$$y = \frac{v^2 - v_0^2}{2a} = \frac{(0 \text{ m/s})^2 - (5.00 \text{ m/s})^2}{2(-9.80 \text{ m/s}^2)} = \boxed{1.28 \text{ m}}$$

 Example 13 How Long Is It in the Air?

In Figure 2.15, what is the total time the coin is in the air before returning to its release point?

Reasoning During the time the coin travels upward, gravity causes its speed to decrease to zero. On the way down, however, gravity causes the coin to regain the lost speed. Thus, the time for the coin to go up is equal to the time for it to come down. In other words, the total travel time is twice the time for the upward motion. The data for the coin during the upward trip are the same as in Example 12. With these data, we can use Equation 2.4 ($v = v_0 + at$) to find the upward travel time.

Solution Rearranging Equation 2.4, we find that

$$t = \frac{v - v_0}{a} = \frac{0 \text{ m/s} - 5.00 \text{ m/s}}{-9.80 \text{ m/s}^2} = 0.510 \text{ s}$$

The total up-and-down time is twice this value, or $\boxed{1.02 \text{ s}}$.

It is possible to determine the total time by another method. When the coin is tossed upward and returns to its release point, the displacement for the *entire trip* is $y = 0$ m. With this value for the displacement, Equation 2.8 ($y = v_0 t + \frac{1}{2}at^2$) can be used to find the time for the entire trip directly.

Examples 12 and 13 illustrate that the expression "freely falling" does not necessarily mean an object is falling down. A freely falling object is any object moving either upward or downward under the influence of gravity alone. In either case, the object always experiences the same *downward acceleration* due to gravity, a fact that is the focus of the next example.

Conceptual Example 14 | Acceleration Versus Velocity

There are three parts to the motion of the coin in Figure 2.15, in which air resistance is being ignored. On the way up, the coin has an upward-pointing velocity vector with a decreasing magnitude. At the top of its path, the velocity vector of the coin is momentarily zero. On the way down, the coin has a downward-pointing velocity vector with an increasing magnitude. Compare the acceleration vector of the coin with the velocity vector. **(a)** Do the direction and magnitude of the acceleration vector behave in the same fashion as the direction and magnitude of the velocity vector or **(b)** does the acceleration vector have a constant direction and a constant magnitude throughout the motion?

Reasoning Since air resistance is being ignored, the coin is in free-fall motion. This means that the acceleration vector of the coin is the acceleration due to gravity. Acceleration is the rate at which velocity *changes* and is not the same concept as velocity itself.

Answer (a) is incorrect. During the upward and downward parts of the motion, and also at the top of the path, the acceleration due to gravity has a constant downward direction and a constant magnitude of 9.80 m/s². In other words, the acceleration vector of the coin does not behave as the velocity vector does. In particular, the acceleration vector is not zero at the top of the motional path just because the velocity vector is zero there. Acceleration is the rate at which the velocity is changing, and the velocity is changing at the top even though at one instant it is zero.

Answer (b) is correct. The acceleration due to gravity has a constant downward direction and a constant magnitude of 9.80 m/s² at all times during the motion.

The motion of an object that is thrown upward and eventually returns to earth has a symmetry that is useful to keep in mind from the point of view of problem solving. The calculations just completed indicate that a time symmetry exists in free-fall motion, in the sense that the time required for the object to reach maximum height equals the time for it to return to its starting point.

A type of symmetry involving the speed also exists. Figure 2.16 shows the coin considered in Examples 12 and 13. At any displacement y above the point of release, the coin's speed during the upward trip equals the speed at the same point during the downward trip. For instance, when $y = +1.04$ m, Equation 2.9 gives two possible values for the final velocity v, assuming that the initial velocity is $v_0 = +5.00$ m/s:

$$v^2 = v_0{}^2 + 2ay = (5.00 \text{ m/s})^2 + 2(-9.80 \text{ m/s}^2)(1.04 \text{ m}) = 4.62 \text{ m}^2/\text{s}^2$$
$$v = \pm 2.15 \text{ m/s}$$

The value $v = +2.15$ m/s is the velocity of the coin on the upward trip, and $v = -2.15$ m/s is the velocity on the downward trip. The speed in both cases is identical and equals 2.15 m/s. Likewise, the speed just as the coin returns to its point of release is 5.00 m/s, which equals the initial speed. This symmetry involving the speed arises because the coin loses 9.80 m/s in speed each second on the way up and gains back the same amount each second on the way down. In Conceptual Example 15, we use just this kind of symmetry to guide our reasoning as we analyze the motion of a pellet shot from a gun.

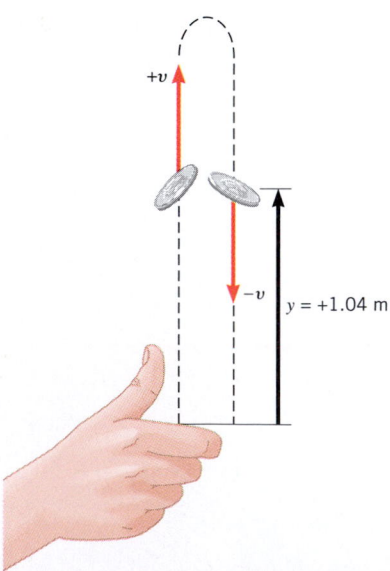

Figure 2.16 For a given displacement along the motional path, the upward speed of the coin is equal to its downward speed, but the two velocities point in opposite directions.

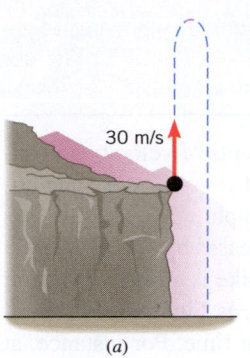 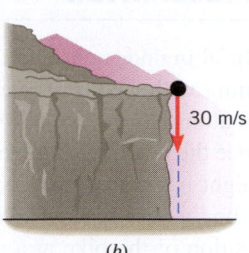 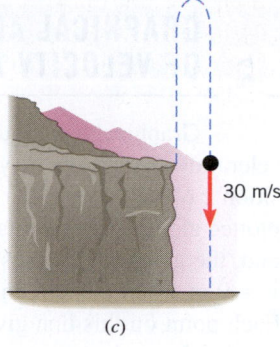

(a) (b) (c)

Figure 2.17 (*a*) From the edge of a cliff, a pellet is fired straight upward from a gun. The pellet's initial speed is 30 m/s. (*b*) The pellet is fired straight downward with an initial speed of 30 m/s. (*c*) In Conceptual Example 15 this drawing plays the central role in reasoning that is based on symmetry.

 Conceptual Example 15 Taking Advantage of Symmetry

Figure 2.17*a* shows a pellet that has been fired straight upward from a gun at the edge of a cliff. The initial speed of the pellet is 30 m/s. It goes up and then falls back down, eventually hitting the ground beneath the cliff. In Figure 2.17*b* the pellet has been fired straight downward at the same initial speed. In the absence of air resistance, would the pellet in Figure 2.17*b* strike the ground with (**a**) a smaller speed than, (**b**) the same speed as, or (**c**) a greater speed than the pellet in Figure 2.17*a*?

Reasoning In the absence of air resistance, the motion is that of free-fall, and the symmetry inherent in free-fall motion offers an immediate answer.

Answers (a) and (c) are incorrect. These answers are incorrect, because they are inconsistent with the symmetry that is discussed next in connection with the correct answer.

Answer (b) is correct. Figure 2.17*c* shows the pellet after it has been fired upward and has fallen back down to its starting point. Symmetry indicates that the speed in Figure 2.17*c* is the same as in Figure 2.17*a*—namely, 30 m/s, as is also the case when the pellet has been actually fired downward. Consequently, whether the pellet is fired as in Figure 2.17*a* or *b*, it begins to move downward from the cliff edge at a speed of 30 m/s. In either case, there is the same acceleration due to gravity and the same displacement from the cliff edge to the ground below. Under these conditions, the pellet reaches the ground with the same speed in both Figures 2.17*a* and *b*.

Related Homework: *Problems 45, 52*

 CHECK YOUR UNDERSTANDING

(The answers are given at the end of the book.)

13. An experimental vehicle slows down and comes to a halt with an acceleration whose magnitude is 9.80 m/s^2. After reversing direction in a negligible amount of time, the vehicle speeds up with an acceleration of 9.80 m/s^2. Except for being horizontal, is this motion (**a**) the same as or   (**b**) different from the motion of a ball that is thrown straight upward, comes to a halt, and falls back to earth? Ignore air resistance.

14. A ball is thrown straight upward with a velocity $\vec{v}_0$ and in a time t reaches the top of its flight path, which is a displacement $\vec{y}$ above the launch point. With a launch velocity of $2\vec{v}_0$, what would be the time required to reach the top of its flight path and what would be the displacement of the top point above the launch point?   (**a**) $4t$ and $2\vec{y}$ (**b**) $2t$ and $4\vec{y}$ (**c**) $2t$ and $2\vec{y}$ (**d**) $4t$ and $4\vec{y}$ (**e**) t and $2\vec{y}$

15. Two objects are thrown vertically upward, first one, and then, a bit later, the other. Is it   (**a**) possible or   (**b**) impossible that both objects reach the same maximum height at the same instant of time?

16. A ball is dropped from rest from the top of a building and strikes the ground with a speed v_f. From ground level, a second ball is thrown straight upward at the same instant that the first ball is dropped. The initial speed of the second ball is $v_0 = v_f$, the same speed with which the first ball eventually strikes the ground. Ignoring air resistance, decide whether the balls cross paths   (**a**) at half the height of the building,   (**b**) above the halfway point, or   (**c**) below the halfway point.

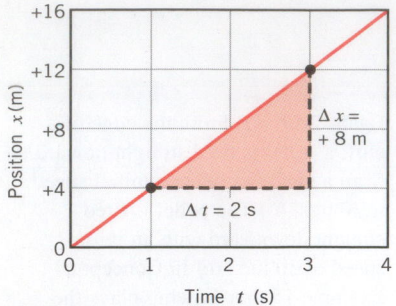

Figure 2.18 A graph of position vs. time for an object moving with a constant velocity of $v = \Delta x/\Delta t = +4$ m/s.

2.7

GRAPHICAL ANALYSIS
OF VELOCITY AND ACCELERATION

Graphical techniques are helpful in understanding the concepts of velocity and acceleration. Suppose a bicyclist is riding with a constant velocity of $v = +4$ m/s. The position x of the bicycle can be plotted along the vertical axis of a graph, while the time t is plotted along the horizontal axis. Since the position of the bike increases by 4 m every second, the graph of x versus t is a straight line. Furthermore, if the bike is assumed to be at $x = 0$ m when $t = 0$ s, the straight line passes through the origin, as Figure 2.18 shows. Each point on this line gives the position of the bike at a particular time. For instance, at $t = 1$ s the position is 4 m, while at $t = 3$ s the position is 12 m.

In constructing the graph in Figure 2.18, we used the fact that the velocity was +4 m/s. Suppose, however, that we were given this graph, but did not have prior knowledge of the velocity. The velocity could be determined by considering what happens to the bike between the times of 1 and 3 s, for instance. The change in time is $\Delta t = 2$ s. During this time interval, the position of the bike changes from +4 to +12 m, and the change in position is $\Delta x = +8$ m. The ratio $\Delta x/\Delta t$ is called the *slope* of the straight line.

$$\text{Slope} = \frac{\Delta x}{\Delta t} = \frac{+8 \text{ m}}{2 \text{ s}} = +4 \text{ m/s}$$

Notice that the slope is equal to the velocity of the bike. This result is no accident, because $\Delta x/\Delta t$ is the definition of average velocity (see Equation 2.2). ***Thus, for an object moving with a constant velocity, the slope of the straight line in a position–time graph gives the velocity.*** Since the position–time graph is a straight line, any time interval Δt can be chosen to calculate the velocity. Choosing a different Δt will yield a different Δx, but the velocity $\Delta x/\Delta t$ will not change. In the real world, objects rarely move with a constant velocity at all times, as the next example illustrates.

⬇ Example 16   A Bicycle Trip

A bicyclist maintains a constant velocity on the outgoing leg of a trip, zero velocity while stopped, and another constant velocity on the way back. Figure 2.19 shows the corresponding position–time graph. Using the time and position intervals indicated in the drawing, obtain the velocities for each segment of the trip.

Reasoning The average velocity $\overline{v}$ is equal to the displacement Δx divided by the elapsed time Δt, $\overline{v} = \Delta x/\Delta t$. The displacement is the final position minus the initial position, which is a positive number for segment 1 and a negative number for segment 3. Note for segment 2 that $\Delta x = 0$ m, since the bicycle is at rest. The drawing shows values for Δt and Δx for each of the three segments.

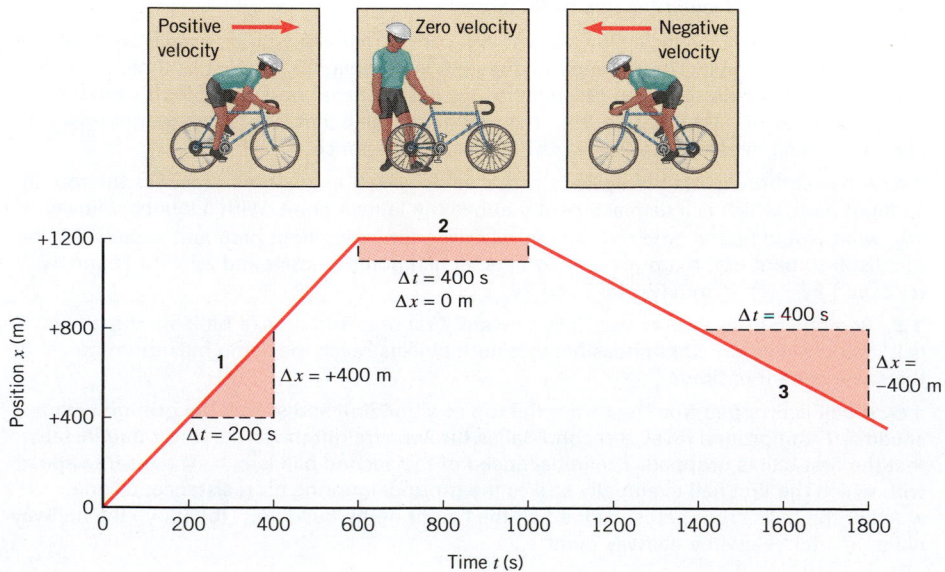

Figure 2.19 This position-vs.-time graph consists of three straight-line segments, each corresponding to a different constant velocity.

Solution The average velocities for the three segments are

Segment 1 $\overline{v} = \dfrac{\Delta x}{\Delta t} = \dfrac{800 \text{ m} - 400 \text{ m}}{400 \text{ s} - 200 \text{ s}} = \dfrac{+400 \text{ m}}{200 \text{ s}} = \boxed{+2 \text{ m/s}}$

Segment 2 $\overline{v} = \dfrac{\Delta x}{\Delta t} = \dfrac{1200 \text{ m} - 1200 \text{ m}}{1000 \text{ s} - 600 \text{ s}} = \dfrac{0 \text{ m}}{400 \text{ s}} = \boxed{0 \text{ m/s}}$

Segment 3 $\overline{v} = \dfrac{\Delta x}{\Delta t} = \dfrac{400 \text{ m} - 800 \text{ m}}{1800 \text{ s} - 1400 \text{ s}} = \dfrac{-400 \text{ m}}{400 \text{ s}} = \boxed{-1 \text{ m/s}}$

In the second segment of the journey the velocity is zero, reflecting the fact that the bike is stationary. Since the position of the bike does not change, segment 2 is a horizontal line that has a zero slope. In the third part of the motion the velocity is negative, because the position of the bike decreases from $x = +800$ m to $x = +400$ m during the 400-s interval shown in the graph. As a result, segment 3 has a negative slope, and the velocity is negative.

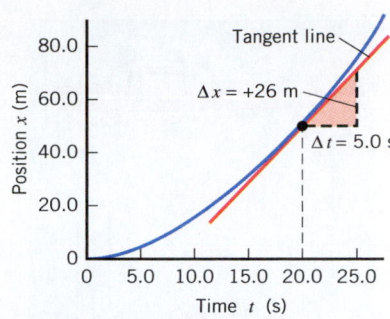

Figure 2.20 When the velocity is changing, the position-vs.-time graph is a curved line. The slope $\Delta x / \Delta t$ of the tangent line drawn to the curve at a given time is the instantaneous velocity at that time.

If the object is accelerating, its velocity is changing. When the velocity is changing, the x-versus-t graph is not a straight line, but is a curve, perhaps like that in Figure 2.20. This curve was drawn using Equation 2.8 ($x = v_0 t + \frac{1}{2} a t^2$), assuming an acceleration of $a = 0.26$ m/s^2 and an initial velocity of $v_0 = 0$ m/s. The velocity at any instant of time can be determined by measuring the slope of the curve at that instant. The slope at any point along the curve is defined to be the slope of the tangent line drawn to the curve at that point. For instance, in Figure 2.20 a tangent line is drawn at $t = 20.0$ s. To determine the slope of the tangent line, a triangle is constructed using an arbitrarily chosen time interval of $\Delta t = 5.0$ s. The change in x associated with this time interval can be read from the tangent line as $\Delta x = +26$ m. Therefore,

$$\text{Slope of tangent line} = \frac{\Delta x}{\Delta t} = \frac{+26 \text{ m}}{5.0 \text{ s}} = +5.2 \text{ m/s}$$

The slope of the tangent line is the instantaneous velocity, which in this case is $v = +5.2$ m/s. This graphical result can be verified by using Equation 2.4 with $v_0 = 0$ m/s: $v = at = (+0.26 \text{ m/s}^2)(20.0 \text{ s}) = +5.2$ m/s.

Insight into the meaning of acceleration can also be gained with the aid of a graphical representation. Consider an object moving with a constant acceleration of $a = +6$ m/s^2. If the object has an initial velocity of $v_0 = +5$ m/s, its velocity at any time is represented by Equation 2.4 as

$$v = v_0 + at = 5 \text{ m/s} + (6 \text{ m/s}^2)t$$

This relation is plotted as the velocity-versus-time graph in Figure 2.21. The graph of v versus t is a straight line that intercepts the vertical axis at $v_0 = 5$ m/s. The slope of this straight line can be calculated from the data shown in the drawing:

$$\text{Slope} = \frac{\Delta v}{\Delta t} = \frac{+12 \text{ m/s}}{2 \text{ s}} = +6 \text{ m/s}^2$$

The ratio $\Delta v / \Delta t$ is, by definition, equal to the average acceleration (Equation 2.4), so ***the slope of the straight line in a velocity–time graph is the average acceleration.***

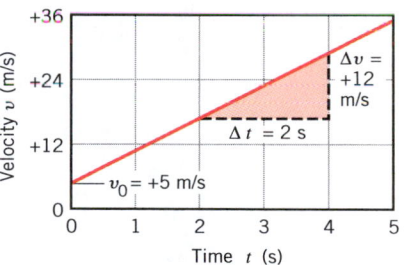

Figure 2.21 A velocity-vs.-time graph that applies to an object with an acceleration of $\Delta v / \Delta t = +6$ m/s^2. The initial velocity is $v_0 = +5$ m/s when $t = 0$ s.

2.8 CONCEPTS & CALCULATIONS

In this chapter we have studied the displacement, velocity, and acceleration vectors. We conclude now by presenting examples that review some of the important features of these concepts. The three-part format of these examples stresses the role of conceptual understanding in problem solving. First, the problem statement is given. Then, there is a concept question-and-answer section, followed by the solution section. The purpose of the concept question-and-answer section is to provide help in understanding the solution and to illustrate how a review of the concepts can help in anticipating some of the characteristics of the numerical answers.

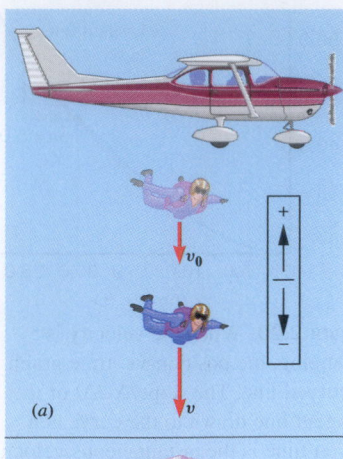

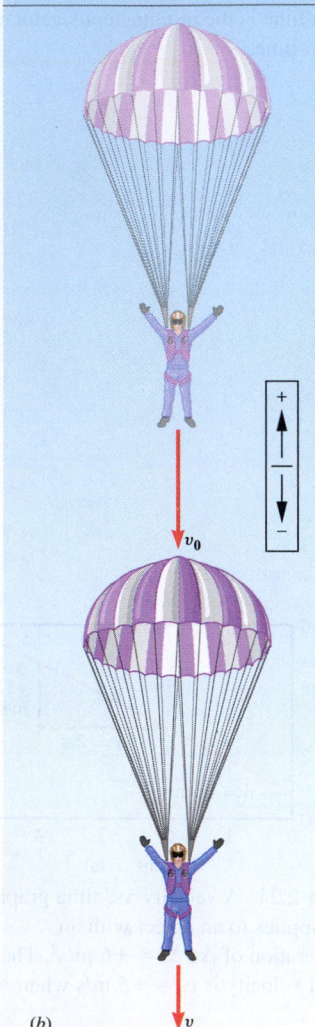

(a)

(b)

Figure 2.22 (a) A skydiver falls initially with her parachute unopened. (b) Later on, she opens her parachute. Her acceleration is different in the two parts of the motion. The initial and final velocities are v_0 and v, respectively.

Concepts & Calculations Example 17

Skydiving

A skydiver is falling straight down, along the negative y direction. **(a)** During the initial part of the fall, her speed increases from 16 to 28 m/s in 1.5 s, as in Figure 2.22a. **(b)** Later, her parachute opens, and her speed decreases from 48 to 26 m/s in 11 s, as in part b of the drawing. In both instances, determine the magnitude and direction of her average acceleration.

Concept Questions and Answers Is her average acceleration positive or negative when her speed is increasing in Figure 2.22a?

Answer Since her speed is increasing, the acceleration vector must point in the same direction as the velocity vector, which points in the negative y direction. Thus, the acceleration is negative.

Is her average acceleration positive or negative when her speed is decreasing in Figure 2.22b?

Answer Since her speed is decreasing, the acceleration vector must point opposite to the velocity vector. Since the velocity vector points in the negative y direction, the acceleration must point in the positive y direction. Thus, the acceleration is positive.

Solution **(a)** Since the skydiver is moving in the negative y direction, her initial velocity is $v_0 = -16$ m/s and her final velocity is $v = -28$ m/s. Her average acceleration $\bar{a}$ is the change in the velocity divided by the elapsed time:

$$\bar{a} = \frac{v - v_0}{t} = \frac{-28 \text{ m/s} - (-16 \text{ m/s})}{1.5 \text{ s}} = \boxed{-8.0 \text{ m/s}^2} \tag{2.4}$$

As expected, her average acceleration is negative. Note that her acceleration is not that due to gravity (-9.8 m/s^2) because of air resistance.

(b) Now the skydiver is slowing down, but still falling along the negative y direction. Her initial and final velocities are $v_0 = -48$ m/s and $v = -26$ m/s, respectively. The average acceleration for this phase of the motion is

$$\bar{a} = \frac{v - v_0}{t} = \frac{-26 \text{ m/s} - (-48 \text{ m/s})}{11 \text{ s}} = \boxed{+2.0 \text{ m/s}^2} \tag{2.4}$$

Now, as anticipated, her average acceleration is positive.

Concepts & Calculations Example 18

A Top-Fuel Dragster

A top-fuel dragster starts from rest and has a constant acceleration of 40.0 m/s^2. What are the **(a)** final velocities and **(b)** displacements of the dragster at the end of 2.0 s and at the end of twice this time, or 4.0 s?

Concept Questions and Answers At a time t the dragster has a certain velocity. When the time doubles to $2t$, does the velocity also double?

Answer Because the dragster has an acceleration of 40.0 m/s^2, its velocity changes by 40.0 m/s during each second of travel. Therefore, since the dragster starts from rest, the velocity is 40.0 m/s at the end of the 1st second, 2 × 40.0 m/s at the end of the 2nd second, 3 × 40.0 m/s at the end of the 3rd second, and so on. Thus, when the time doubles, the velocity also doubles.

When the time doubles to $2t$, does the displacement of the dragster also double?

Answer The displacement of the dragster is equal to its average velocity multiplied by the elapsed time. The average velocity $\bar{v}$ is just one-half the sum of the initial and final velocities, or $\bar{v} = \frac{1}{2}(v_0 + v)$. Since the initial velocity is zero, $v_0 = 0$ m/s and the average velocity is just one-half the final velocity, or $\bar{v} = \frac{1}{2}v$. However, as we have seen, the final velocity is proportional to the elapsed time, since when the time doubles, the final velocity also doubles. Therefore, the displacement, being the product of the average velocity and the time, is proportional to the time squared, or t^2. Consequently, as the time doubles, the displacement does not double, but increases by a factor of four.

Solution (a) According to Equation 2.4, the final velocity v, the initial velocity v_0, the acceleration a, and the elapsed time t are related by $v = v_0 + at$. The final velocities at the two times are

[**t = 2.0 s**] $v = v_0 + at = 0 \text{ m/s} + (40.0 \text{ m/s}^2)(2.0 \text{ s}) = \boxed{80 \text{ m/s}}$

[**t = 4.0 s**] $v = v_0 + at = 0 \text{ m/s} + (40.0 \text{ m/s}^2)(4.0 \text{ s}) = \boxed{160 \text{ m/s}}$

We see that the velocity doubles when the time doubles, as expected.

(**b**) The displacement x is equal to the average velocity multiplied by the time, so

$$x = \underbrace{\tfrac{1}{2}(v_0 + v)t}_{\text{Average velocity}} = \tfrac{1}{2}vt$$

where we have used the fact that $v_0 = 0$ m/s. According to Equation 2.4, the final velocity is related to the acceleration by $v = v_0 + at$, or $v = at$, since $v_0 = 0$ m/s. Therefore, the displacement can be written as $x = \tfrac{1}{2}vt = \tfrac{1}{2}(at)t = \tfrac{1}{2}at^2$. The displacements at the two times are then

[**t = 2.0 s**] $x = \tfrac{1}{2}at^2 = \tfrac{1}{2}(40.0 \text{ m/s}^2)(2.0 \text{ s})^2 = \boxed{80 \text{ m}}$

[**t = 4.0 s**] $x = \tfrac{1}{2}at^2 = \tfrac{1}{2}(40.0 \text{ m/s}^2)(4.0 \text{ s})^2 = \boxed{320 \text{ m}}$

As predicted, the displacement at $t = 4.0$ s is four times the displacement at $t = 2.0$ s.

CONCEPT SUMMARY

If you need more help with a concept, use the Learning Aids noted next to the discussion or equation. Examples (**Ex.**) are in the text of this chapter. Go to **www.wiley.com/college/cutnell** for the following Learning Aids:

Interactive LearningWare (ILW) — Additional examples solved in a five-step interactive format.

Concept Simulations (CS) — Animated text figures or animations of important concepts.

Interactive Solutions (IS) — Models for certain types of problems in the chapter homework. The calculations are carried out interactively.

Topic	Discussion	Learning Aids
Displacement	**2.1 DISPLACEMENT** Displacement is a vector that points from an object's initial position to its final position. The magnitude of the displacement is the shortest distance between the two positions.	
	2.2 SPEED AND VELOCITY The average speed of an object is the distance traveled by the object divided by the time required to cover the distance:	
Average speed	$$\text{Average speed} = \frac{\text{Distance}}{\text{Elapsed time}} \qquad (2.1)$$	**Ex. 1**
	The average velocity $\vec{\mathbf{v}}$ of an object is the object's displacement $\Delta\vec{\mathbf{x}}$ divided by the elapsed time Δt:	
Average velocity	$$\overline{\vec{\mathbf{v}}} = \frac{\Delta\vec{\mathbf{x}}}{\Delta t} \qquad (2.2)$$	**Ex. 2**
	Average velocity is a vector that has the same direction as the displacement. When the elapsed time becomes infinitesimally small, the average velocity becomes equal to the instantaneous velocity $\vec{\mathbf{v}}$, the velocity at an instant of time:	
Instantaneous velocity	$$\vec{\mathbf{v}} = \lim_{\Delta t \to 0} \frac{\Delta\vec{\mathbf{x}}}{\Delta t} \qquad (2.3)$$	
	2.3 ACCELERATION The average acceleration $\vec{\mathbf{a}}$ is a vector. It equals the change $\Delta\vec{\mathbf{v}}$ in the velocity divided by the elapsed time Δt, the change in the velocity being the final minus the initial velocity:	
Average acceleration	$$\overline{\vec{\mathbf{a}}} = \frac{\Delta\vec{\mathbf{v}}}{\Delta t} \qquad (2.4)$$	**Ex. 3, 4, 17**
	When Δt becomes infinitesimally small, the average acceleration becomes equal to the instantaneous acceleration $\vec{\mathbf{a}}$:	**IS 2.22**
Instantaneous acceleration	$$\vec{\mathbf{a}} = \lim_{\Delta t \to 0} \frac{\Delta\vec{\mathbf{v}}}{\Delta t} \qquad (2.5)$$	
	Acceleration is the rate at which the velocity is changing.	

Topic	Discussion	Learning Aids
	2.4 EQUATIONS OF KINEMATICS FOR CONSTANT ACCELERATION 2.5 APPLICATIONS OF THE EQUATIONS OF KINEMATICS The equations of kinematics apply when an object moves with a constant acceleration along a straight line. These equations relate the displacement $x - x_0$, the acceleration a, the final velocity v, the initial velocity v_0, and the elapsed time $t - t_0$. Assuming that $x_0 = 0$ m at $t_0 = 0$ s, the equations of kinematics are	Ex. 5–9, 18
Equations of kinematics	$$v = v_0 + at \qquad (2.4)$$ $$x = \tfrac{1}{2}(v_0 + v)t \qquad (2.7)$$ $$x = v_0 t + \tfrac{1}{2}at^2 \qquad (2.8)$$ $$v^2 = v_0^2 + 2ax \qquad (2.9)$$	CS 2.1, 2.2 ILW 2.1, 2.2 IS 2.36, 2.84
Acceleration due to gravity	**2.6 FREELY FALLING BODIES** In free-fall motion, an object experiences negligible air resistance and a constant acceleration due to gravity. All objects at the same location above the earth have the same acceleration due to gravity. The acceleration due to gravity is directed toward the center of the earth and has a magnitude of approximately 9.80 m/s² near the earth's surface.	Ex. 10–15 CS 2.3 IS 2.54, 2.57
	2.7 GRAPHICAL ANALYSIS OF VELOCITY AND ACCELERATION The slope of a plot of position versus time for a moving object gives the object's velocity. The slope of a plot of velocity versus time gives the object's acceleration.	Ex. 16 CS 2.4

FOCUS ON CONCEPTS

Note to Instructors: The numbering of the questions shown here reflects the fact that they are only a representative subset of the total number that are available online. However, all of the questions are available for assignment via an online homework management program such as WileyPLUS or WebAssign.

Section 2.1   Displacement

1. What is the difference between distance and displacement? **(a)** Distance is a vector, while displacement is not a vector. **(b)** Displacement is a vector, while distance is not a vector. **(c)** There is no difference between the two concepts; they may be used interchangeably.

Section 2.2   Speed and Velocity

3. A jogger runs along a straight and level road for a distance of 8.0 km and then runs back to her starting point. The time for this round-trip is 2.0 h. Which one of the following statements is true? **(a)** Her average speed is 8.0 km/h, but there is not enough information to determine her average velocity. **(b)** Her average speed is 8.0 km/h, and her average velocity is 8.0 km/h. **(c)** Her average speed is 8.0 km/h, and her average velocity is 0 km/h.

Section 2.3   Acceleration

6. The velocity of a train is 80.0 km/h, due west. One and a half hours later its velocity is 65.0 km/h, due west. What is the train's average acceleration? **(a)** 10.0 km/h², due west **(b)** 43.3 km/h², due west **(c)** 10.0 km/h², due east **(d)** 43.3 km/h², due east **(e)** 53.3 km/h², due east

Section 2.4   Equations of Kinematics for Constant Acceleration

10. In which one of the following situations can the equations of kinematics *not* be used? **(a)** When the velocity changes from moment to moment, **(b)** when the velocity remains constant, **(c)** when the acceleration changes from moment to moment, **(d)** when the acceleration remains constant.

13. In a race two horses, Silver Bullet and Shotgun, start from rest and each maintains a constant acceleration. In the same elapsed time Silver Bullet runs 1.20 times farther than Shotgun. According to the equations of kinematics, which one of the following is true concerning the accelerations of the horses? **(a)** $a_{\text{Silver Bullet}} = 1.44\, a_{\text{Shotgun}}$ **(b)** $a_{\text{Silver Bullet}} = a_{\text{Shotgun}}$ **(c)** $a_{\text{Silver Bullet}} = 2.40\, a_{\text{Shotgun}}$ **(d)** $a_{\text{Silver Bullet}} = 1.20\, a_{\text{Shotgun}}$ **(e)** $a_{\text{Silver Bullet}} = 0.72\, a_{\text{Shotgun}}$

Section 2.6   Freely Falling Bodies

19. A rocket is sitting on the launch pad. The engines ignite, and the rocket begins to rise straight upward, picking up speed as it goes. At about 1000 m above the ground the engines shut down, but the rocket continues straight upward, losing speed as it goes. It reaches the top of its flight path and then falls back to earth. Ignoring air resistance, decide which one of the following statements is true. **(a)** All of the rocket's motion, from the moment the engines ignite until just before the rocket lands, is free-fall. **(b)** Only part of the rocket's motion, from just after the engines shut down until just before it lands, is free-fall. **(c)** Only the rocket's motion while the engines are firing is free-fall. **(d)** Only the rocket's motion from the top of its flight path until just before landing is free-fall. **(e)** Only part of the rocket's motion, from just after the engines shut down until it reaches the top of its flight path, is free-fall.

22. The top of a cliff is located a distance H above the ground. At a distance $H/2$ there is a branch that juts out from the side of the cliff, and on this branch a bird's nest is located. Two children throw stones at the nest with the same initial speed, one stone straight downward from the top of the cliff and the other stone straight upward from the ground. In the absence of air resistance, which stone hits the nest in the least amount of time? **(a)** There is insufficient information for an answer. **(b)** Both stones hit the nest in the same amount of time. **(c)** The stone thrown from the ground. **(d)** The stone thrown from the top of the cliff.

Section 2.7   Graphical Analysis of Velocity and Acceleration

24. The graph accompanying this problem shows a three-part motion. For each of the three parts, A, B, and C, identify the direction of the motion. A positive velocity denotes motion to the right. **(a)** A right, B left, C right **(b)** A right, B right, C left **(c)** A right, B left, C left **(d)** A left, B right, C left **(e)** A left, B right, C right

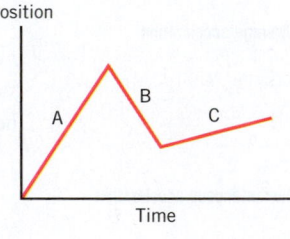

PROBLEMS

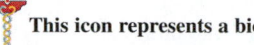

Note to Instructors: Most of the homework problems in this chapter are available for assignment via an online homework management program such as WileyPLUS or WebAssign, and those marked with the icon **GO** *are presented in* WileyPLUS *using a guided tutorial format that provides enhanced interactivity. See Preface for additional details.*

ssm Solution is in the Student Solutions Manual.
www Solution is available online at www.wiley.com/college/cutnell

This icon represents a biomedical application.

Section 2.1 Displacement,
Section 2.2 Speed and Velocity

1. ssm A whale swims due east for a distance of 6.9 km, turns around and goes due west for 1.8 km, and finally turns around again and heads 3.7 km due east. **(a)** What is the total distance traveled by the whale? **(b)** What are the magnitude and direction of the displacement of the whale?

2. **GO** For each of the three pairs of positions listed in the following table, determine the magnitude and direction (positive or negative) of the displacement.

	Initial position x_0	Final position x
(a)	+2.0 m	+6.0 m
(b)	+6.0 m	+2.0 m
(c)	−3.0 m	+7.0 m

3. ssm Due to continental drift, the North American and European continents are drifting apart at an average speed of about 3 cm per year. At this speed, how long (in years) will it take for them to drift apart by another 1500 m (a little less than a mile)?

4. Electrons move through a certain electric circuit at an average speed of 1.1×10^{-2} m/s. How long (in minutes) does it take an electron to traverse 1.5 m of wire in the filament of a light bulb?

5. **GO** The data in the following table describe the initial and final positions of a moving car. The elapsed time for each of the three pairs of positions listed in the table is 0.50 s. Review the concept of average velocity in Section 2.2 and then determine the average velocity (magnitude and direction) for each of the three pairs. Note that the algebraic sign of your answers will convey the direction.

	Initial position x_0	Final position x
(a)	+2.0 m	+6.0 m
(b)	+6.0 m	+2.0 m
(c)	−3.0 m	+7.0 m

6. The Space Shuttle travels at a speed of about 7.6×10^3 m/s. The blink of an astronaut's eye lasts about 110 ms. How many football fields (length = 91.4 m) does the Shuttle cover in the blink of an eye?

7. In 1954 the English runner Roger Bannister broke the four-minute barrier for the mile with a time of 3:59.4 s (3 min and 59.4 s). In 1999 the Moroccan runner Hicham el-Guerrouj set a record of 3:43.13 s for the mile. If these two runners had run in the same race, each running the entire race at the average speed that earned him a place in the record books, el-Guerrouj would have won. By how many meters?

8. An 18-year-old runner can complete a 10.0-km course with an average speed of 4.39 m/s. A 50-year-old runner can cover the same dis-

tance with an average speed of 4.27 m/s. How much later (in seconds) should the younger runner start in order to finish the course *at the same time* as the older runner?

9. A tourist being chased by an angry bear is running in a straight line toward his car at a speed of 4.0 m/s. The car is a distance *d* away. The bear is 26 m behind the tourist and running at 6.0 m/s. The tourist reaches the car safely. What is the maximum possible value for *d*?

* **10.** In reaching her destination, a backpacker walks with an average velocity of 1.34 m/s, due west. This average velocity results because she hikes for 6.44 km with an average velocity of 2.68 m/s, due west, turns around, and hikes with an average velocity of 0.447 m/s, due east. How far east did she walk?

* **11. ssm www** A woman and her dog are out for a morning run to the river, which is located 4.0 km away. The woman runs at 2.5 m/s in a straight line. The dog is unleashed and runs back and forth at 4.5 m/s between his owner and the river, until the woman reaches the river. What is the total distance run by the dog?

* **12.** A car makes a trip due north for three-fourths of the time and due south one-fourth of the time. The average northward velocity has a magnitude of 27 m/s, and the average southward velocity has a magnitude of 17 m/s. What is the average velocity (magnitude and direction) for the entire trip?

** **13.** You are on a train that is traveling at 3.0 m/s along a level straight track. Very near and parallel to the track is a wall that slopes upward at a 12° angle with the horizontal. As you face the window (0.90 m high, 2.0 m wide) in your compartment, the train is moving to the left, as the drawing indicates. The top edge of the wall first appears at window corner A and eventually disappears at window corner B. How much time passes between appearance and disappearance of the upper edge of the wall?

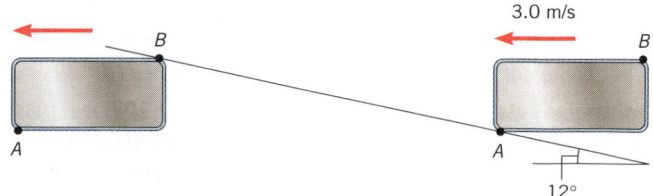

Section 2.3 Acceleration

14. **GO** The data in the following table represent the initial and final velocities for a boat traveling along the *x* axis. The elapsed time for each of the four pairs of velocities in the table is 2.0 s. Review the concept of average acceleration in Section 2.3 and then determine the average acceleration (magnitude and direction) for each of the four pairs. Note that the algebraic sign of your answers will convey the direction.

Continued

	Initial velocity v_0	Final velocity v
(a)	+2.0 m/s	+5.0 m/s
(b)	+5.0 m/s	+2.0 m/s
(c)	−6.0 m/s	−3.0 m/s
(d)	+4.0 m/s	−4.0 m/s

Problem 14

15. **GO** **(a)** Suppose that a NASCAR race car is moving to the right with a constant velocity of +82 m/s. What is the average acceleration of the car? **(b)** Twelve seconds later, the car is halfway around the track and traveling in the opposite direction with the same speed. What is the average acceleration of the car?

16. Over a time interval of 2.16 years, the velocity of a planet orbiting a distant star reverses direction, changing from +20.9 km/s to −18.5 km/s. Find **(a)** the total change in the planet's velocity (in m/s) and **(b)** its average acceleration (in m/s²) during this interval. Include the correct algebraic sign with your answers to convey the directions of the velocity and the acceleration.

17. **ssm** In 1998, NASA launched *Deep Space 1* (DS-1), a spacecraft that successfully flew by the asteroid named 1992 KD (which orbits the sun millions of miles from the earth). The propulsion system of DS-1 worked by ejecting high-speed argon ions out the rear of the engine. The engine slowly increased the velocity of DS-1 by about +9.0 m/s per day. **(a)** How much time (in days) would it take to increase the velocity of DS-1 by +2700 m/s? **(b)** What was the acceleration of DS-1 (in m/s²)?

18. A sprinter explodes out of the starting block with an acceleration of +2.3 m/s², which she sustains for 1.2 s. Then, her acceleration drops to zero for the rest of the race. What is her velocity **(a)** at $t = 1.2$ s and **(b)** at the end of the race?

19. **GO** The initial velocity and acceleration of four moving objects at a given instant in time are given in the following table. Determine the final *speed* of each of the objects, assuming that the time elapsed since $t = 0$ s is 2.0 s.

	Initial velocity v_0	Acceleration a
(a)	+12 m/s	+3.0 m/s²
(b)	+12 m/s	−3.0 m/s²
(c)	−12 m/s	+3.0 m/s²
(d)	−12 m/s	−3.0 m/s²

20. For a standard production car, the highest road-tested acceleration ever reported occurred in 1993, when a Ford RS200 Evolution went from zero to 26.8 m/s (60 mi/h) in 3.275 s. Find the magnitude of the car's acceleration.

21. **ssm** An automobile starts from rest and accelerates to a final velocity in two stages along a straight road. Each stage occupies the same amount of time. In stage 1, the magnitude of the car's acceleration is 3.0 m/s². The magnitude of the car's velocity at the end of stage 2 is 2.5 times greater than it is at the end of stage 1. Find the magnitude of the acceleration in stage 2.

***22.** Consult **Interactive Solution 2.22** at **www.wiley.com/college/cutnell** before beginning this problem. A car is traveling along a straight road at a velocity of +36.0 m/s when its engine cuts out. For the next twelve seconds the car slows down, and its average acceleration is $\bar{a}_1$.

For the next six seconds the car slows down further, and its average acceleration is $\bar{a}_2$. The velocity of the car at the end of the eighteen-second period is +28.0 m/s. The ratio of the average acceleration values is $\bar{a}_1/\bar{a}_2 = 1.50$. Find the velocity of the car at the end of the initial twelve-second interval.

** 23. Two motorcycles are traveling due east with different velocities. However, four seconds later, they have the same velocity. During this four-second interval, cycle A has an average acceleration of 2.0 m/s² due east, while cycle B has an average acceleration of 4.0 m/s² due east. By how much did the speeds *differ* at the beginning of the four-second interval, and which motorcycle was moving faster?

Section 2.4   Equations of Kinematics for Constant Acceleration,
Section 2.5   Applications of the Equations of Kinematics

24. In getting ready to slam-dunk the ball, a basketball player starts from rest and sprints to a speed of 6.0 m/s in 1.5 s. Assuming that the player accelerates uniformly, determine the distance he runs.

25. **ssm** A jetliner, traveling northward, is landing with a speed of 69 m/s. Once the jet touches down, it has 750 m of runway in which to reduce its speed to 6.1 m/s. Compute the average acceleration (magnitude and direction) of the plane during landing.

26. A VW Beetle goes from 0 to 60.0 mi/h with an acceleration of +2.35 m/s². **(a)** How much time does it take for the Beetle to reach this speed? **(b)** A top-fuel dragster can go from 0 to 60.0 mi/h in 0.600 s. Find the acceleration (in m/s²) of the dragster.

27. Before starting this problem, review Multiple-Concept Example 6. The left ventricle of the heart accelerates blood from rest to a velocity of +26 cm/s. **(a)** If the displacement of the blood during the acceleration is +2.0 cm, determine its acceleration (in cm/s²). **(b)** How much time does blood take to reach its final velocity?

28. **(a)** What is the magnitude of the average acceleration of a skier who, starting from rest, reaches a speed of 8.0 m/s when going down a slope for 5.0 s? **(b)** How far does the skier travel in this time?

29. **ssm www** A jogger accelerates from rest to 3.0 m/s in 2.0 s. A car accelerates from 38.0 to 41.0 m/s also in 2.0 s. **(a)** Find the acceleration (magnitude only) of the jogger. **(b)** Determine the acceleration (magnitude only) of the car. **(c)** Does the car travel farther than the jogger during the 2.0 s? If so, how much farther?

30. Consult **Concept Simulation 2.1** at **www.wiley.com/college/cutnell** for help in preparing for this problem. A cheetah is hunting. Its prey runs for 3.0 s at a constant velocity of +9.0 m/s. Starting from rest, what constant acceleration must the cheetah maintain in order to run the same distance as its prey runs in the same time?

31. **ssm www** A speed ramp at an airport is basically a large conveyor belt on which you can stand and be moved along. The belt of one ramp moves at a constant speed such that a person who stands still on it leaves the ramp 64 s after getting on. Clifford is in a real hurry, however, and skips the speed ramp. Starting from rest with an acceleration of 0.37 m/s², he covers the same distance as the ramp does, but in one-fourth the time. What is the speed at which the belt of the ramp is moving?

32. **Concept Simulation 2.2** at **www.wiley.com/college/cutnell** offers a useful review of the concepts that lie at the heart of this problem. Two rockets are flying in the same direction and are side by side at the instant their retrorockets fire. Rocket A has an initial velocity of +5800 m/s, while rocket B has an initial velocity of +8600 m/s. After a time t both rockets are again side by side, the displacement of each being zero. The acceleration of rocket A is −15 m/s². What is the acceleration of rocket B?

a/ -7.9 m/s.
3.2 m

* **33.** [GO] Two cars cover the same distance in a straight line. Car A covers the distance at a constant velocity. Car B starts from rest and maintains a constant acceleration. Both cars cover a distance of 460 m in 210 s. Assume that they are moving in the $+x$ direction. Determine **(a)** the constant velocity of car A, **(b)** the final velocity of car B, and **(c)** the acceleration of car B.

* **34.** [GO] Review **Interactive LearningWare 2.2** at **www.wiley.com/college/cutnell** in preparation for this problem. A race driver has made a pit stop to refuel. After refueling, he starts from rest and leaves the pit area with an acceleration whose magnitude is 6.0 m/s²; after 4.0 s he enters the main speedway. At the same instant, another car on the speedway and traveling at a constant velocity of 70.0 m/s overtakes and passes the entering car. The entering car maintains its acceleration. How much time is required for the entering car to catch the other car?

* **35.** In a historical movie, two knights on horseback start from rest 88.0 m apart and ride directly toward each other to do battle. Sir George's acceleration has a magnitude of 0.300 m/s², while Sir Alfred's has a magnitude of 0.200 m/s². Relative to Sir George's starting point, where do the knights collide?

* **36.** **Interactive Solution 2.36** at **www.wiley.com/college/cutnell** offers help in modeling this problem. A car is traveling at a constant speed of 33 m/s on a highway. At the instant this car passes an entrance ramp, a second car enters the highway from the ramp. The second car starts from rest and has a constant acceleration. What acceleration must it maintain, so that the two cars meet for the first time at the next exit, which is 2.5 km away?

* **37.** Multiple-Concept Example 9 reviews the concepts that are important in this problem. A drag racer, starting from rest, speeds up for 402 m with an acceleration of +17.0 m/s². A parachute then opens, slowing the car down with an acceleration of −6.10 m/s². How fast is the racer moving 3.50×10^2 m after the parachute opens?

* **38.** A speedboat starts from rest and accelerates at +2.01 m/s² for 7.00 s. At the end of this time, the boat continues for an additional 6.00 s with an acceleration of +0.518 m/s². Following this, the boat accelerates at −1.49 m/s² for 8.00 s. **(a)** What is the velocity of the boat at $t = 21.0$ s? **(b)** Find the total displacement of the boat.

* **39.** [GO] Refer to Multiple-Concept Example 5 to review a method by which this problem can be solved. You are driving your car, and the traffic light ahead turns red. You apply the brakes for 3.00 s, and the velocity of the car decreases to +4.50 m/s. The car's deceleration has a magnitude of 2.70 m/s² during this time. What is the car's displacement?

** **40.** A Boeing 747 "Jumbo Jet" has a length of 59.7 m. The runway on which the plane lands intersects another runway. The width of the intersection is 25.0 m. The plane decelerates through the intersection at a rate of 5.70 m/s² and clears it with a final speed of 45.0 m/s. How much time is needed for the plane to clear the intersection?

** **41.** **ssm** A locomotive is accelerating at 1.6 m/s². It passes through a 20.0-m-wide crossing in a time of 2.4 s. After the locomotive leaves the crossing, how much time is required until its speed reaches 32 m/s?

** **42.** In a quarter-mile drag race, two cars start simultaneously from rest, and each accelerates at a constant rate until it either reaches its maximum speed or crosses the finish line. Car A has an acceleration of 11.0 m/s² and a maximum speed of 106 m/s. Car B has an acceleration of 11.6 m/s² and a maximum speed of 92.4 m/s. Which car wins the race, and by how many seconds?

Section 2.6 Freely Falling Bodies

43. **ssm** In preparation for this problem, review Conceptual Example 7. From the top of a cliff, a person uses a slingshot to fire a pebble straight downward, which is the negative direction. The initial speed of the pebble is 9.0 m/s. **(a)** What is the acceleration (magnitude and direction) of the pebble during the downward motion? Is the pebble decelerating? Explain. **(b)** After 0.50 s, how far beneath the cliff top is the pebble?

44. A dynamite blast at a quarry launches a chunk of rock straight upward, and 2.0 s later it is rising at a speed of 15 m/s. Assuming air resistance has no effect on the rock, calculate its speed **(a)** at launch and **(b)** 5.0 s after launch.

45. Review Conceptual Example 15 before attempting this problem. Two identical pellet guns are fired simultaneously from the edge of a cliff. These guns impart an initial speed of 30.0 m/s to each pellet. Gun A is fired straight upward, with the pellet going up and then falling back down, eventually hitting the ground beneath the cliff. Gun B is fired straight downward. In the absence of air resistance, how long after pellet B hits the ground does pellet A hit the ground?

46. [GO] **Concept Simulation 2.3** at **www.wiley.com/college/cutnell** provides some background for this problem. A ball is thrown vertically upward, which is the positive direction. A little later it returns to its point of release. The ball is in the air for a total time of 8.0 s. What is its initial velocity? Neglect air resistance.

47. The drawing shows a device that you can make with a piece of cardboard, which can be used to measure a person's reaction time. Hold the card at the top and suddenly drop it. Ask a friend to try to catch the card between his or her thumb and index finger. Initially, your friend's fingers must be level with the asterisks at the bottom. By noting where your friend catches the card, you can determine his or her reaction time in milliseconds (ms). Calculate the distances d_1, d_2, and d_3.

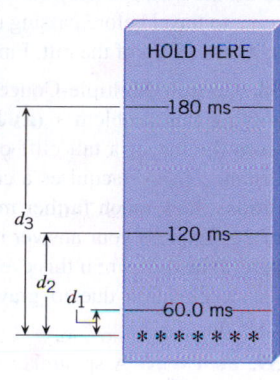

48. Review **Concept Simulation 2.3** at **www.wiley.com/college/cutnell** before attempting this problem. At the beginning of a basketball game, a referee tosses the ball straight up with a speed of 4.6 m/s. A player cannot touch the ball until after it reaches its maximum height and begins to fall down. What is the minimum time that a player must wait before touching the ball?

49. **ssm** A wrecking ball is hanging at rest from a crane when suddenly the cable breaks. The time it takes for the ball to fall halfway to the ground is 1.2 s. Find the time it takes for the ball to fall from rest all the way to the ground.

50. A hot-air balloon is rising upward with a constant speed of 2.50 m/s. When the balloon is 3.00 m above the ground, the balloonist accidentally drops a compass over the side of the balloon. How much time elapses before the compass hits the ground?

51. Multiple-Concept Example 6 reviews the concepts that play a role in this problem. A diver springs upward with an initial speed of 1.8 m/s from a 3.0-m board. **(a)** Find the velocity with which he strikes the water. *[Hint: When the diver reaches the water, his displacement is $y = -3.0$ m (measured from the board), assuming that the downward direction is chosen as the negative direction.]* **(b)** What is the highest point he reaches above the water?

52. Before working this problem, review Conceptual Example 15. A pellet gun is fired straight downward from the edge of a cliff that is 15 m above the ground. The pellet strikes the ground with a speed of 27 m/s. How far above the cliff edge would the pellet have gone had the gun been fired straight upward?

53. ssm From her bedroom window a girl drops a water-filled balloon to the ground, 6.0 m below. If the balloon is released from rest, how long is it in the air?

54. Consult Interactive Solution 2.54 at www.wiley.com/college/cutnell before beginning this problem. A ball is thrown straight upward and rises to a maximum height of 12.0 m above its launch point. At what height above its launch point has the speed of the ball decreased to one-half of its initial value?

55. Consult Multiple-Concept Example 5 in preparation for this problem. The velocity of a diver just before hitting the water is −10.1 m/s, where the minus sign indicates that her motion is directly downward. What is her displacement during the last 1.20 s of the dive?

★56. A golf ball is dropped from rest from a height of 9.50 m. It hits the pavement, then bounces back up, rising just 5.70 m before falling back down again. A boy then catches the ball on the way down when it is 1.20 m above the pavement. Ignoring air resistance, calculate the total amount of time that the ball is in the air, from drop to catch.

★57. Review Interactive Solution 2.57 at www.wiley.com/college/cutnell before beginning this problem. A woman on a bridge 75.0 m high sees a raft floating at a constant speed on the river below. Trying to hit the raft, she drops a stone from rest when the raft has 7.00 m more to travel before passing under the bridge. The stone hits the water 4.00 m in front of the raft. Find the speed of the raft.

★58. Consult Multiple-Concept Example 9 to explore a model for solving this problem. **(a)** Just for fun, a person jumps from rest from the top of a tall cliff overlooking a lake. In falling through a distance H, she acquires a certain speed v. Assuming free-fall conditions, how much farther must she fall in order to acquire a speed of $2v$? Express your answer in terms of H. **(b)** Would the answer to part (a) be different if this event were to occur on another planet where the acceleration due to gravity had a value other than 9.80 m/s^2? Explain.

★59. ssm www A spelunker (cave explorer) drops a stone from rest into a hole. The speed of sound is 343 m/s in air, and the sound of the stone striking the bottom is heard 1.50 s after the stone is dropped. How deep is the hole?

★60. GO Two stones are thrown simultaneously, one straight upward from the base of a cliff and the other straight downward from the top of the cliff. The height of the cliff is 6.00 m. The stones are thrown with the same speed of 9.00 m/s. Find the location (above the base of the cliff) of the point where the stones cross paths.

★61. ssm A cement block accidentally falls from rest from the ledge of a 53.0-m-high building. When the block is 14.0 m above the ground, a man, 2.00 m tall, looks up and notices that the block is directly above him. How much time, at most, does the man have to get out of the way?

★62. A model rocket blasts off from the ground, rising straight upward with a constant acceleration that has a magnitude of 86.0 m/s^2 for 1.70 seconds, at which point its fuel abruptly runs out. Air resistance has no effect on its flight. What maximum altitude (above the ground) will the rocket reach?

★★63. While standing on a bridge 15.0 m above the ground, you drop a stone from rest. When the stone has fallen 3.20 m, you throw a second stone straight down. What initial velocity must you give the second stone if they are both to reach the ground at the same instant? Take the downward direction to be the negative direction.

★★64. Review Interactive LearningWare 2.2 at www.wiley.com/college/cutnell as an aid in solving this problem. A hot air balloon is rising straight up at a constant speed of 7.0 m/s. When the balloon is 12.0 m above the ground, a gun fires a pellet straight up from ground level with an initial speed of 30.0 m/s. Along the paths of the balloon

and the pellet, there are two places where each of them has the same altitude at the same time. How far above ground are these places?

Section 2.7 Graphical Analysis of Velocity and Acceleration

65. **Concept Simulation 2.5** at www.wiley.com/college/cutnell provides a review of the concepts that play a role in this problem. A snowmobile moves according to the velocity–time graph shown in the drawing. What is the snowmobile's average acceleration during each of the segments A, B, and C?

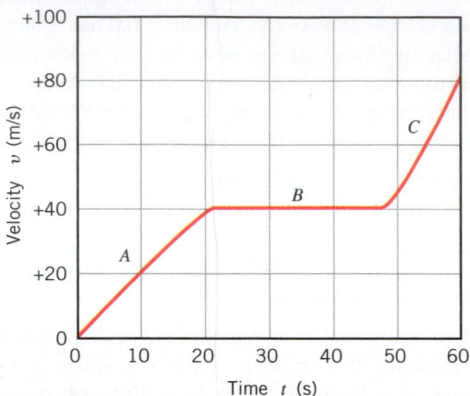

66. Starting at $x = -16$ m at time $t = 0$ s, an object takes 18 s to travel 48 m in the $+x$ direction at a constant velocity. Make a position-versus-time graph of the object's motion and calculate its velocity.

67. **ssm** A person who walks for exercise produces the position–time graph given with this problem. **(a)** Without doing any calculations, decide which segments of the graph (A, B, C, or D) indicate positive, negative, and zero average velocities. **(b)** Calculate the average velocity for each segment to verify your answers to part (a).

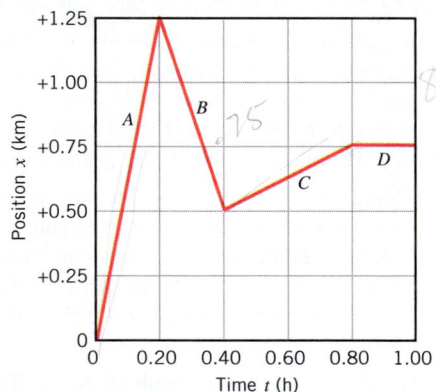

68. A bus makes a trip according to the position–time graph shown in the drawing. What is the average velocity (magnitude and direction) of the bus during each of the segments A, B, and C? Express your answers in km/h.

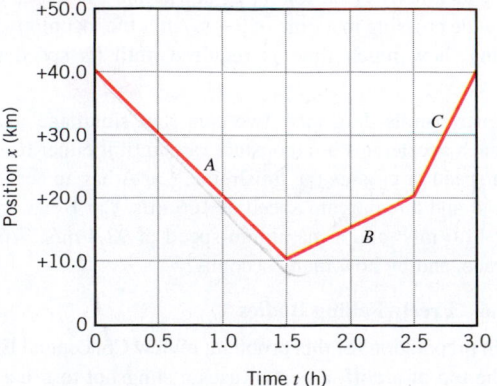

* **69.** **GO** A bus makes a trip according to the position–time graph shown in the illustration. What is the average acceleration (in km/h²) of the bus for the entire 3.5-h period shown in the graph?

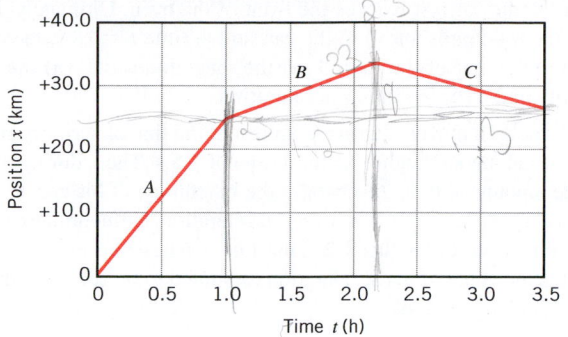

* **70.** A car slows down with an acceleration that has a magnitude of 3.2 m/s². While doing so, it travels 120 m in the $+x$ direction and ends up with a velocity of $+4.0$ m/s. **(a)** What was the car's initial velocity? **(b)** Make a velocity-versus-time graph of the car's motion.

** **71.** **ssm** Two runners start one hundred meters apart and run toward each other. Each runs ten meters during the first second. During each second thereafter, each runner runs ninety percent of the distance he ran in the previous second. Thus, the velocity of each person changes from second to second. However, during any one second, the velocity remains constant. Make a position–time graph for one of the runners. From this graph, determine **(a)** how much time passes before the runners collide and **(b)** the speed with which each is running at the moment of collision.

ADDITIONAL PROBLEMS

72. Review Conceptual Example 7 as background for this problem. A car is traveling to the left, which is the negative direction. The direction of travel remains the same throughout this problem. The car's initial speed is 27.0 m/s, and during a 5.0-s interval, it changes to a final speed of **(a)** 29.0 m/s and **(b)** 23.0 m/s. In each case, find the acceleration (magnitude and algebraic sign) and state whether or not the car is decelerating.

73. **ssm** The greatest height reported for a jump into an airbag is 99.4 m by stuntman Dan Koko. In 1948 he jumped from rest from the top of the Vegas World Hotel and Casino. He struck the airbag at a speed of 39 m/s (88 mi/h). To assess the effects of air resistance, determine how fast he would have been traveling on impact had air resistance been absent.

74. The three-toed sloth is the slowest-moving land mammal. On the ground, the sloth moves at an average speed of 0.037 m/s, considerably slower than the giant tortoise, which walks at 0.076 m/s. After 12 minutes of walking, how much further would the tortoise have gone relative to the sloth?

75. **ssm** Refer to **Concept Simulation 2.4** at **www.wiley.com/college/cutnell** for help in visualizing this problem graphically. A cart is driven by a large propeller or fan, which can accelerate or decelerate the cart. The cart starts out at the position $x = 0$ m, with an initial velocity of $+5.0$ m/s and a constant acceleration due to the fan. The direction to the right is positive. The cart reaches a maximum position of $x = +12.5$ m, where it begins to travel in the negative direction. Find the acceleration of the cart.

76. **Concept Simulation 2.3** at **www.wiley.com/college/cutnell** offers a useful review of the concepts central to this problem. An astronaut on a distant planet wants to determine its acceleration due to gravity. The astronaut throws a rock straight up with a velocity of $+15$ m/s and measures a time of 20.0 s before the rock returns to his hand. What is the acceleration (magnitude and direction) due to gravity on this planet?

77. You step onto a hot beach with your bare feet. A nerve impulse, generated in your foot, travels through your nervous system at an average speed of 110 m/s. How much time does it take for the impulse, which travels a distance of 1.8 m, to reach your brain?

78. A motorcycle has a constant acceleration of 2.5 m/s². Both the velocity and acceleration of the motorcycle point in the same direction. How much time is required for the motorcycle to change its speed from **(a)** 21 to 31 m/s, and **(b)** 51 to 61 m/s?

79. Consult **Concept Simulation 2.1** at **www.wiley.com/college/cutnell** before starting this problem. The Kentucky Derby is held at the Churchill Downs track in Louisville, Kentucky. The track is one and one-quarter miles in length. One of the most famous horses to win this event was Secretariat. In 1973 he set a Derby record that would be hard to beat. His average acceleration during the last four quarter-miles of the race was $+0.0105$ m/s². His velocity at the start of the final mile ($x = +1609$ m) was about $+16.58$ m/s. The acceleration, although small, was very important to his victory. To assess its effect, determine the difference between the time he would have taken to run the final mile at a constant velocity of $+16.58$ m/s and the time he actually took. Although the track is oval in shape, assume it is straight for the purpose of this problem.

* **80.** Multiple-Concept Example 9 illustrates the concepts that are pertinent to this problem. A cab driver picks up a customer and delivers her 2.00 km away, on a straight route. The driver accelerates to the speed limit and, on reaching it, begins to decelerate at once. The magnitude of the deceleration is three times the magnitude of the acceleration. Find the lengths of the acceleration and deceleration phases.

* **81.** **ssm** A bicyclist makes a trip that consists of three parts, each in the same direction (due north) along a straight road. During the first part, she rides for 22 minutes at an average speed of 7.2 m/s. During the second part, she rides for 36 minutes at an average speed of 5.1 m/s. Finally, during the third part, she rides for 8.0 minutes at an average speed of 13 m/s. **(a)** How far has the bicyclist traveled during the entire trip? **(b)** What is her average velocity for the trip?

* **82.** A golfer rides in a golf cart at an average speed of 3.10 m/s for 28.0 s. She then gets out of the cart and starts walking at an average speed of 1.30 m/s. For how long (in seconds) must she walk if her average speed for the entire trip, riding and walking, is 1.80 m/s?

* **83.** **ssm** A log is floating on swiftly moving water. A stone is dropped from rest from a 75-m-high bridge and lands on the log as it passes under the bridge. If the log moves with a constant speed of 5.0 m/s, what is the horizontal distance between the log and the bridge when the stone is released?

* **84.** Review **Interactive Solution 2.84** at **www.wiley.com/college/cutnell** in preparation for this problem. A car is traveling at 20.0 m/s, and the driver sees a traffic light turn red. After 0.530 s (the reaction time), the driver applies the brakes, and the car decelerates at 7.00 m/s². What is the stopping distance of the car, as measured from the point where the driver first sees the red light?

* **85.** Review **Interactive LearningWare 2.2** at www.wiley.com/college/cutnell in preparation for this problem. A police car is traveling at a velocity of 18.0 m/s due north, when a car zooms by at a constant velocity of 42.0 m/s due north. After a reaction time of 0.800 s the policeman begins to pursue the speeder with an acceleration of 5.00 m/s². Including the reaction time, how long does it take for the police car to catch up with the speeder?

** **86.** A ball is dropped from rest from the top of a cliff that is 24 m high. From ground level, a second ball is thrown straight upward at the same instant that the first ball is dropped. The initial speed of the second ball is exactly the same as that with which the first ball eventually hits the ground. In the absence of air resistance, the motions of the balls are just the reverse of each other. Determine how far below the top of the cliff the balls cross paths.

** **87.** **ssm** A train has a length of 92 m and starts from rest with a constant acceleration at time $t = 0$ s. At this instant, a car just reaches the end of the train. The car is moving with a constant velocity. At a time $t = 14$ s, the car just reaches the front of the train. Ultimately, however, the train pulls ahead of the car, and at time $t = 28$ s, the car is again at the rear of the train. Find the magnitudes of **(a)** the car's velocity and **(b)** the train's acceleration.

** **88.** A football player, starting from rest at the line of scrimmage, accelerates along a straight line for a time of 1.5 s. Then, during a negligible amount of time, he changes the magnitude of his acceleration to a value of 1.1 m/s². With this acceleration, he continues in the same direction for another 1.2 s, until he reaches a speed of 3.4 m/s. What is the value of his acceleration (assumed to be constant) during the initial 1.5-s period?

C H A P T E R 3

KINEMATICS IN TWO DIMENSIONS

A child runs through a water fountain outside the Acer Arena (formerly the SuperDome) in Sydney, Australia. The arching water follows a parabolic path whose size depends on the launch velocity and the acceleration due to gravity, assuming that the effects of air resistance can be ignored. (Reuters/© Corbis)

3.1 DISPLACEMENT, VELOCITY, AND ACCELERATION

In Chapter 2 the concepts of displacement, velocity, and acceleration are used to describe an object moving in one dimension. There are also situations in which the motion is along a curved path that lies in a plane. Such two-dimensional motion can be described using the same concepts. In Grand Prix racing, for example, the course follows a curved road, and Figure 3.1 shows a race car at two different positions along it. These positions are identified by the vectors $\vec{r}$ and $\vec{r}_0$, which are drawn from an arbitrary coordinate origin. The *displacement* $\Delta\vec{r}$ of the car is the vector drawn from the initial position $\vec{r}_0$ at time t_0 to the final position $\vec{r}$ at time t. The magnitude of $\Delta\vec{r}$ is the shortest distance between the two positions. In the drawing, the vectors $\vec{r}_0$ and $\Delta\vec{r}$ are drawn tail to head, so it is evident that $\vec{r}$ is the vector sum of $\vec{r}_0$ and $\Delta\vec{r}$. (See Sections 1.5 and 1.6 for a review of vectors and vector addition.) This means that $\vec{r} = \vec{r}_0 + \Delta\vec{r}$, or

$$\text{Displacement} = \Delta\vec{r} = \vec{r} - \vec{r}_0$$

The displacement here is defined as it is in Chapter 2. Now, however, the displacement vector may lie anywhere in a plane, rather than just along a straight line.

The **average velocity** $\vec{\overline{v}}$ of the car between two positions is defined in a manner similar to that in Equation 2.2, as the displacement $\Delta\vec{r} = \vec{r} - \vec{r}_0$ divided by the elapsed time $\Delta t = t - t_0$:

$$\vec{\overline{v}} = \frac{\vec{r} - \vec{r}_0}{t - t_0} = \frac{\Delta\vec{r}}{\Delta t} \tag{3.1}$$

Since both sides of Equation 3.1 must agree in direction, the average velocity vector has the same direction as the displacement $\Delta\vec{r}$. The velocity of the car at an instant of time is

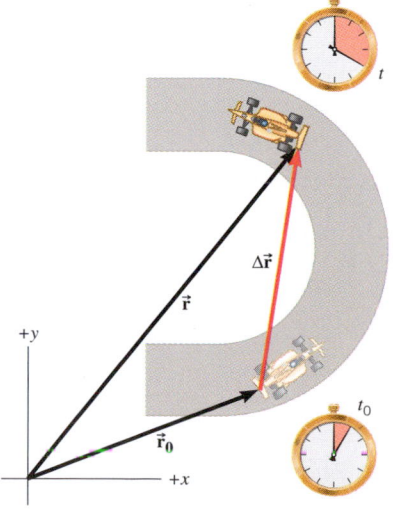

Figure 3.1 The displacement $\Delta\vec{r}$ of the car is a vector that points from the initial position of the car at time t_0 to the final position at time t. The magnitude of $\Delta\vec{r}$ is the shortest distance between the two positions.

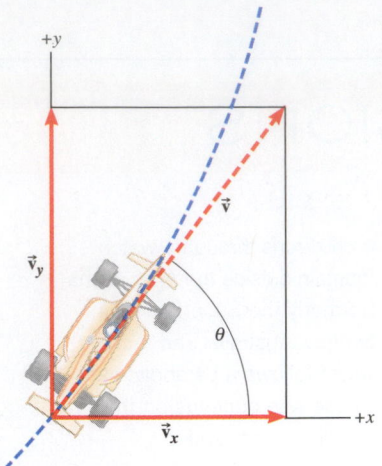

Figure 3.2 The instantaneous velocity $\vec{v}$ and its two vector components $\vec{v}_x$ and $\vec{v}_y$.

its **instantaneous velocity** $\vec{v}$. The average velocity becomes equal to the instantaneous velocity $\vec{v}$ in the limit that Δt becomes infinitesimally small ($\Delta t \rightarrow 0$ s):

$$\vec{v} = \lim_{\Delta t \to 0} \frac{\Delta \vec{r}}{\Delta t}$$

Figure 3.2 illustrates that the instantaneous velocity $\vec{v}$ is tangent to the path of the car. The drawing also shows the vector components $\vec{v}_x$ and $\vec{v}_y$ of the velocity, which are parallel to the x and y axes, respectively.

The **average acceleration** $\vec{\bar{a}}$ is defined just as it is for one-dimensional motion—namely, as the change in velocity, $\Delta\vec{v} = \vec{v} - \vec{v}_0$, divided by the elapsed time Δt:

$$\vec{\bar{a}} = \frac{\vec{v} - \vec{v}_0}{t - t_0} = \frac{\Delta\vec{v}}{\Delta t} \tag{3.2}$$

The average acceleration has the same direction as the change in velocity $\Delta\vec{v}$. In the limit that the elapsed time becomes infinitesimally small, the average acceleration becomes equal to the **instantaneous acceleration** $\vec{a}$:

$$\vec{a} = \lim_{\Delta t \to 0} \frac{\Delta\vec{v}}{\Delta t}$$

The acceleration has a vector component $\vec{a}_x$ along the x direction and a vector component $\vec{a}_y$ along the y direction.

✓ CHECK YOUR UNDERSTANDING

(The answer is given at the end of the book.)

1. Suppose you are driving due east, traveling a distance of 1500 m in 2 minutes. You then turn due north and travel the same distance in the same time. What can be said about the average speeds and the average velocities for the two segments of the trip? **(a)** The average speeds are the same, and the average velocities are the same. **(b)** The average speeds are the same, but the average velocities are different. **(c)** The average speeds are different, but the average velocities are the same.

3.2 EQUATIONS OF KINEMATICS IN TWO DIMENSIONS

To understand how displacement, velocity, and acceleration are applied to two-dimensional motion, consider a spacecraft equipped with two engines that are mounted perpendicular to each other. These engines produce the only forces that the craft experiences, and the spacecraft is assumed to be at the coordinate origin when $t_0 = 0$ s, so that $\vec{r}_0 = 0$ m. At a later time t, the spacecraft's displacement is $\Delta\vec{r} = \vec{r} - \vec{r}_0 = \vec{r}$. Relative to the x and y axes, the displacement $\vec{r}$ has vector components of $\vec{x}$ and $\vec{y}$, respectively.

In Figure 3.3 only the engine oriented along the x direction is firing, and the vehicle accelerates along this direction. It is assumed that the velocity in the y direction is zero, and it remains zero, since the y engine is turned off. The motion of the spacecraft along the x direction is described by the five kinematic variables x, a_x, v_x, v_{0x}, and t. Here the symbol "x" reminds us that we are dealing with the x components of the displacement, velocity, and acceleration vectors. (See Sections 1.7 and 1.8 for a review of vector components.)

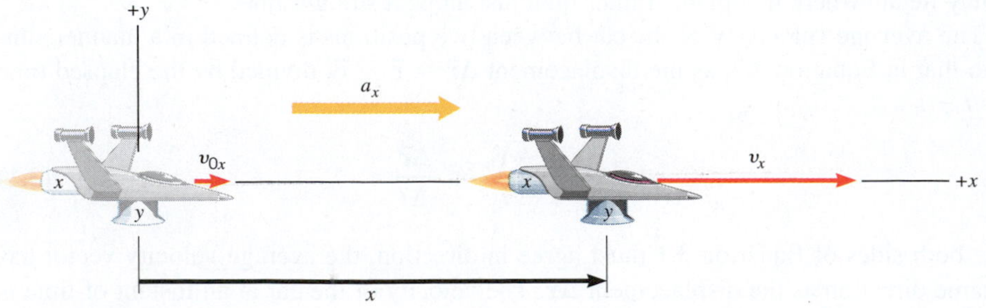

Figure 3.3 The spacecraft is moving with a constant acceleration a_x parallel to the x axis. There is no motion in the y direction, and the y engine is turned off.

Table 3.1 Equations of Kinematics for Constant Acceleration in Two-Dimensional Motion

x Component		Variable	y Component
x		Displacement	y
a_x		Acceleration	a_y
v_x		Final velocity	v_y
v_{0x}		Initial velocity	v_{0y}
t		Elapsed time	t
$v_x = v_{0x} + a_x t$	(3.3a)		$v_y = v_{0y} + a_y t$ (3.3b)
$x = \frac{1}{2}(v_{0x} + v_x)t$	(3.4a)		$y = \frac{1}{2}(v_{0y} + v_y)t$ (3.4b)
$x = v_{0x}t + \frac{1}{2}a_x t^2$	(3.5a)		$y = v_{0y}t + \frac{1}{2}a_y t^2$ (3.5b)
$v_x^2 = v_{0x}^2 + 2a_x x$	(3.6a)		$v_y^2 = v_{0y}^2 + 2a_y y$ (3.6b)

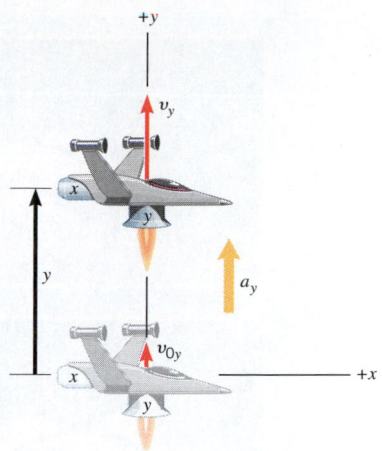

Figure 3.4 The spacecraft is moving with a constant acceleration a_y parallel to the y axis. There is no motion in the x direction, and the x engine is turned off.

The variables x, a_x, v_x, and v_{0x} are scalar components (or "components," for short). As Section 1.7 discusses, these components are positive or negative numbers (with units), depending on whether the associated vector components point in the $+x$ or the $-x$ direction. If the spacecraft has a constant acceleration along the x direction, the motion is exactly like that described in Chapter 2, and the equations of kinematics can be used. For convenience, these equations are written in the left column of Table 3.1.

Figure 3.4 is analogous to Figure 3.3, except that now only the y engine is firing, and the spacecraft accelerates along the y direction. Such a motion can be described in terms of the kinematic variables y, a_y, v_y, v_{0y}, and t. And if the acceleration along the y direction is constant, these variables are related by the equations of kinematics, as written in the right column of Table 3.1. Like their counterparts in the x direction, the scalar components, y, a_y, v_y, and v_{0y}, may be positive ($+$) or negative ($-$) numbers (with units).

If both engines of the spacecraft are firing *at the same time,* the resulting motion takes place in part along the x axis and in part along the y axis, as Figure 3.5 illustrates. The thrust of each engine gives the vehicle a corresponding acceleration component. The x engine accelerates the ship in the x direction and causes a change in the x component of the velocity. Likewise, the y engine causes a change in the y component of the velocity. *It is important to realize that the x part of the motion occurs exactly as it would if the y part did not occur at all. Similarly, the y part of the motion occurs exactly as it would if the x part of the motion did not exist.* In other words, the x and y motions are independent of each other.

Problem-solving insight

▶ **CONCEPTS AT A GLANCE** The independence of the x and y motions lies at the heart of two-dimensional kinematics. It allows us to treat two-dimensional motion as two distinct one-dimensional motions, one for the x direction and one for the y direction. As the Concepts-at-a-Glance chart* in Figure 3.6 illustrates, everything that we have learned in

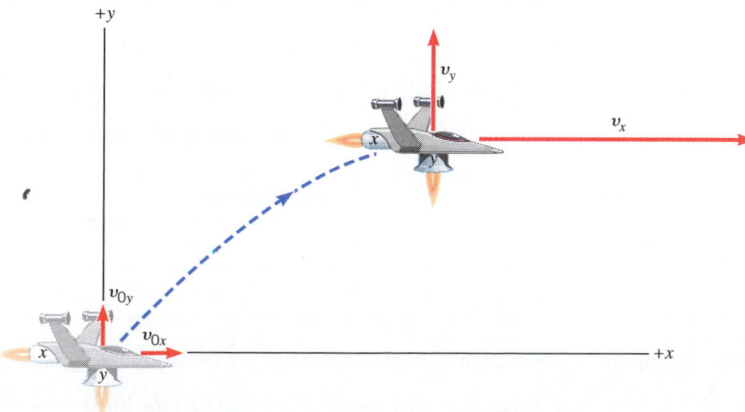

Figure 3.5 The two-dimensional motion of the spacecraft can be viewed as the combination of the separate x and y motions.

*Concepts-at-a-Glance charts occur throughout this text and illustrate diagrammatically how physics builds a coherent understanding of nature through the development of interrelated concepts.

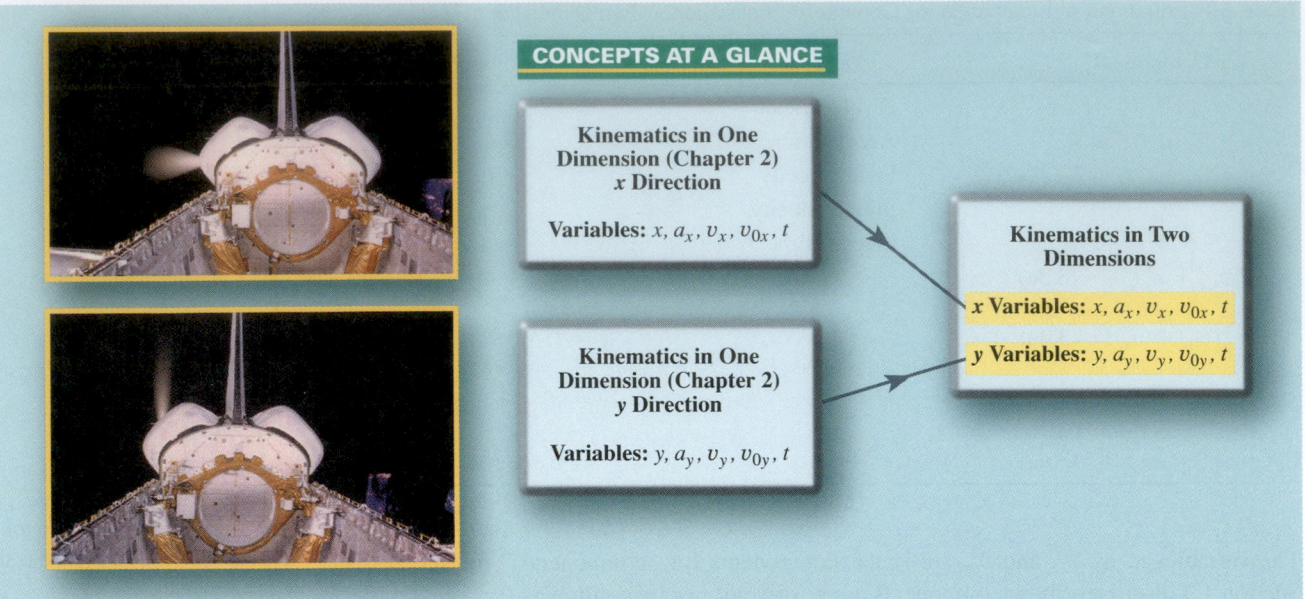

Figure 3.6 **CONCEPTS AT A GLANCE**
In two dimensions, motion along the x direction and motion along the y direction are independent of each other. As a result, each can be analyzed separately according to the procedures for one-dimensional kinematics discussed in Chapter 2. On the space shuttle *Challenger*, motion in perpendicular directions is controlled by thrusters. The photographs show the *Challenger* in orbit with different thrusters activated. (Courtesy NASA)

Chapter 2 about kinematics in one dimension will now be applied separately to each of the two directions. In so doing, we will be able to describe the x and y variables separately and then bring these descriptions together to understand the two-dimensional picture. Examples 1 and 2 take this approach in dealing with a moving spacecraft. ◀

Example 1 The Displacement of a Spacecraft

In Figure 3.5, the directions to the right and upward are the positive directions. In the x direction, the spacecraft has an initial velocity component of $v_{0x} = +22$ m/s and an acceleration component of $a_x = +24$ m/s². In the y direction, the analogous quantities are $v_{0y} = +14$ m/s and $a_y = +12$ m/s². At a time of $t = 7.0$ s, find the x and y components of the spacecraft's displacement.

Reasoning The motion in the x direction and the motion in the y direction can be treated separately, each as a one-dimensional motion subject to the equations of kinematics for constant acceleration (see Table 3.1). By following this procedure we will be able to determine x and y, which specify the spacecraft's location after an elapsed time of 7.0 s.

Solution The data for the motion in the x direction are listed in the following table:

x-Direction Data				
x	a_x	v_x	v_{0x}	t
?	+24 m/s²		+22 m/s	7.0 s

The x component of the craft's displacement can be found by using Equation 3.5a:

$$x = v_{0x}t + \tfrac{1}{2}a_xt^2 = (22 \text{ m/s})(7.0 \text{ s}) + \tfrac{1}{2}(24 \text{ m/s}^2)(7.0 \text{ s})^2 = \boxed{+740 \text{ m}}$$

The data for the motion in the y direction are listed in the following table:

y-Direction Data				
y	a_y	v_y	v_{0y}	t
?	+12 m/s²		+14 m/s	7.0 s

The y component of the craft's displacement can be found by using Equation 3.5b:

$$y = v_{0y}t + \tfrac{1}{2}a_yt^2 = (14 \text{ m/s})(7.0 \text{ s}) + \tfrac{1}{2}(12 \text{ m/s}^2)(7.0 \text{ s})^2 = \boxed{+390 \text{ m}}$$

After 7.0 s, the spacecraft is 740 m to the right and 390 m above the origin.

Problem-solving insight
When the motion is two-dimensional, the time variable t has the same value for both the x and y directions.

ANALYZING MULTIPLE-CONCEPT PROBLEMS

Example 2 The Velocity of a Spacecraft

This example also deals with the spacecraft in Figure 3.5. As in Example 1, the x components of the craft's initial velocity and acceleration are $v_{0x} = +22$ m/s and $a_x = +24$ m/s², respectively. The corresponding y components are $v_{0y} = +14$ m/s and $a_y = +12$ m/s². At a time of $t = 7.0$ s, find the spacecraft's final velocity (magnitude and direction).

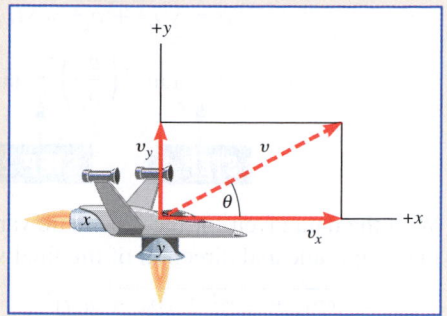

Reasoning Figure 3.7 shows the final velocity vector, which has components v_x and v_y and a magnitude v. The final velocity is directed at an angle θ above the $+x$ axis. The vector and its components form a right triangle, the hypotenuse being the magnitude of the velocity and the components being the other two sides. Thus, we can use the Pythagorean theorem to determine the magnitude v from values for the components v_x and v_y. We can also use trigonometry to determine the directional angle θ.

Figure 3.7 The velocity vector has components v_x and v_y and a magnitude v. The magnitude gives the speed of the spacecraft, and the angle θ gives the direction of travel relative to the positive x direction.

Knowns and Unknowns The data for this problem are listed in the table that follows:

Description	Symbol	Value	Comment
x component of acceleration	a_x	$+24$ m/s²	
x component of initial velocity	v_{0x}	$+22$ m/s	
y component of acceleration	a_y	$+12$ m/s²	
y component of initial velocity	v_{0y}	$+14$ m/s	
Time	t	7.0 s	Same time for x and y directions
Unknown Variables			
Magnitude of final velocity	v	?	
Direction of final velocity	θ	?	

Modeling the Problem

STEP 1 **Final Velocity** In Figure 3.7 the final velocity vector and its components v_x and v_y form a right triangle. Applying the Pythagorean theorem to this right triangle shows that the magnitude v of the final velocity is given in terms of the components by Equation 1a at the right. From the right triangle in Figure 3.7 it also follows that the directional angle θ is given by Equation 1b at the right.

$$v = \sqrt{v_x^2 + v_y^2} \qquad (1a)$$

$$\theta = \tan^{-1}\left(\frac{v_y}{v_x}\right) \qquad (1b)$$

STEP 2 **The Components of the Final Velocity** Values are given for the kinematic variables a_x, v_{0x}, and t in the x direction and for the corresponding variables in the y direction (see the table of knowns and unknowns). For each direction, then, these values allow us to calculate the final velocity components v_x and v_y by using Equation 3.3a and 3.3b from the equations of kinematics.

$$v_x = v_{0x} + a_x t \qquad (3.3a)$$

$$v_y = v_{0y} + a_y t \qquad (3.3b)$$

These expressions can be substituted into Equations 1a and 1b for the magnitude and direction of the final velocity, as shown at the right.

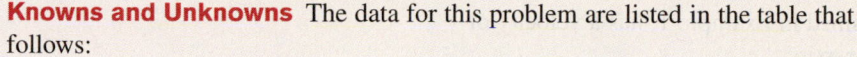

$$v = \sqrt{v_x^2 + v_y^2} \qquad (1a)$$

$$\boxed{v_x = v_{0x} + a_x t} \qquad \boxed{v_y = v_{0y} + a_y t}$$
$$(3.3a,b)$$

$$\theta = \tan^{-1}\left(\frac{v_y}{v_x}\right) \qquad (1b)$$

$$\boxed{v_x = v_{0x} + a_x t} \qquad \boxed{v_y = v_{0y} + a_y t}$$
$$(3.3a,b)$$

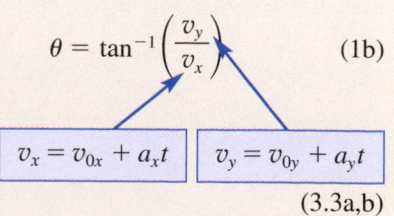

Continued

Solution Algebraically combining the results of each step, we find that

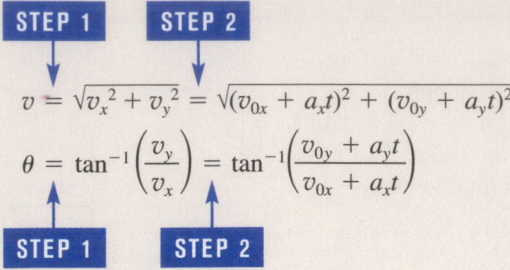

$$v = \sqrt{v_x{}^2 + v_y{}^2} = \sqrt{(v_{0x} + a_x t)^2 + (v_{0y} + a_y t)^2}$$

$$\theta = \tan^{-1}\left(\frac{v_y}{v_x}\right) = \tan^{-1}\left(\frac{v_{0y} + a_y t}{v_{0x} + a_x t}\right)$$

With the data given for the kinematic variables in the x and y directions, we find that the magnitude and direction of the final velocity of the spacecraft are

$$v = \sqrt{(v_{0x} + a_x t)^2 + (v_{0y} + a_y t)^2}$$

$$= \sqrt{[(22 \text{ m/s}) + (24 \text{ m/s}^2)(7.0 \text{ s})]^2 + [(14 \text{ m/s}) + (12 \text{ m/s}^2)(7.0 \text{ s})]^2} = \boxed{210 \text{ m/s}}$$

$$\theta = \tan^{-1}\left(\frac{v_{0y} + a_y t}{v_{0x} + a_x t}\right) = \tan^{-1}\left[\frac{(14 \text{ m/s}) + (12 \text{ m/s}^2)(7.0 \text{ s})}{(22 \text{ m/s}) + (24 \text{ m/s}^2)(7.0 \text{ s})}\right] = \boxed{27°}$$

After 7.0 s, the spacecraft, at the position determined in Example 1, has a velocity of 210 m/s in a direction of 27° above the positive x axis.

Related Homework: *Problem 25*

The following Reasoning Strategy gives an overview of how the equations of kinematics are applied to describe motion in two dimensions, as in Examples 1 and 2.

REASONING STRATEGY

Applying the Equations of Kinematics in Two Dimensions

1. Make a drawing to represent the situation being studied.

2. Decide which directions are to be called positive (+) and negative (−) relative to a conveniently chosen coordinate origin. Do not change your decision during the course of a calculation.

3. Remember that the time variable t has the same value for the part of the motion along the x axis and the part along the y axis.

4. In an organized way, write down the values (with appropriate + and − signs) that are given for any of the five kinematic variables associated with the x direction and the y direction. Be on the alert for implied data, such as the phrase "starts from rest," which means that the values of the initial velocity components are zero: $v_{0x} = 0$ m/s and $v_{0y} = 0$ m/s. The data summary boxes and tables of knowns and unknowns that are used in the examples are a good way of keeping track of this information. In addition, identify the variables that you are being asked to determine.

5. Before attempting to solve a problem, verify that the given information contains values for at least three of the kinematic variables. Do this for the x and the y direction of the motion. Once the three known variables are identified along with the desired unknown variable, the appropriate relations from Table 3.1 can be selected.

6. When the motion is divided into segments, remember that the final velocity for one segment is the initial velocity for the next segment.

7. Keep in mind that a kinematics problem may have two possible answers. Try to visualize the different physical situations to which the answers correspond.

✓ **CHECK YOUR UNDERSTANDING**

(*The answer is given at the end of the book.*)

2. A power boat, starting from rest, maintains a constant acceleration. After a certain time *t*, its displacement and velocity are $\vec{r}$ and $\vec{v}$. At time 2*t*, what would be its displacement and velocity, assuming the acceleration remains the same?   (a) $2\vec{r}$ and $2\vec{v}$ (b) $2\vec{r}$ and $4\vec{v}$ (c) $4\vec{r}$ and $2\vec{v}$ (d) $4\vec{r}$ and $4\vec{v}$

3.3 | PROJECTILE MOTION

The biggest thrill in baseball is a home run. The motion of the ball on its curving path into the stands is a common type of two-dimensional motion called "projectile motion." A good description of such motion can often be obtained with the assumption that air resistance is absent.

Following the approach outlined in Figure 3.6, we consider the horizontal and vertical parts of the motion separately. In the horizontal or *x* direction, the moving object (the projectile) does not slow down in the absence of air resistance. Thus, the *x* component of the velocity remains constant at its initial value or $v_x = v_{0x}$, and the *x* component of the acceleration is $a_x = 0$ m/s². In the vertical or *y* direction, however, the projectile experiences the effect of gravity. As a result, the *y* component of the velocity v_y is not constant, but changes. The *y* component of the acceleration a_y is the downward acceleration due to gravity. If the path or trajectory of the projectile is near the earth's surface, a_y has a magnitude of 9.80 m/s². In this text, then, the phrase "projectile motion" means that $a_x = 0$ m/s² and a_y equals the acceleration due to gravity. Example 3 and other examples in this section illustrate how the equations of kinematics are applied to projectile motion.

⊥ **Example 3**   A Falling Care Package

Figure 3.8 shows an airplane moving horizontally with a constant velocity of $+115$ m/s at an altitude of 1050 m. The directions to the right and upward have been chosen as the positive directions. The plane releases a "care package" that falls to the ground along a curved trajectory. Ignoring air resistance, determine the time required for the package to hit the ground.

Reasoning   The time required for the package to hit the ground is the time it takes for the package to fall through a vertical distance of 1050 m. In falling, it moves to the right, as well as downward, but these two parts of the motion occur independently. Therefore, we can focus solely on the vertical part. We note that the package is moving initially in the horizontal or *x* direction, not in the *y* direction, so that $v_{0y} = 0$ m/s. Furthermore, when the package hits the

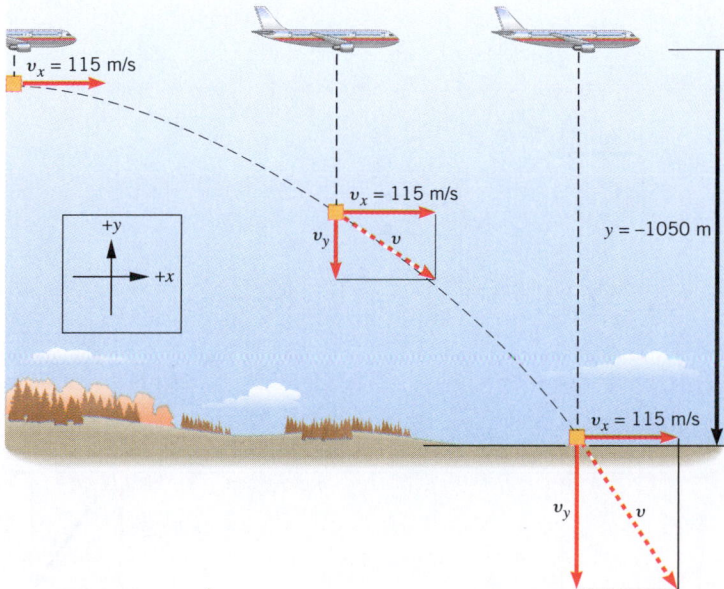

Figure 3.8 The package falling from the plane is an example of projectile motion, as Examples 3 and 4 discuss.

ground, the y component of its displacement is $y = -1050$ m, as the drawing shows. The acceleration is that due to gravity, so $a_y = -9.80$ m/s². These data are summarized as follows:

y-Direction Data				
y	a_y	v_y	v_{0y}	t
-1050 m	-9.80 m/s²		0 m/s	?

With these data, Equation 3.5b ($y = v_{0y}t + \frac{1}{2}a_y t^2$) can be used to find the fall time.

Solution Since $v_{0y} = 0$ m/s, it follows from Equation 3.5b that $y = \frac{1}{2}a_y t^2$ and

$$t = \sqrt{\frac{2y}{a_y}} = \sqrt{\frac{2(-1050 \text{ m})}{-9.80 \text{ m/s}^2}} = \boxed{14.6 \text{ s}}$$

The freely falling package in Example 3 picks up vertical speed on the way downward. The horizontal component of the velocity, however, retains its initial value of $v_{0x} = +115$ m/s throughout the entire descent. Since the plane also travels at a constant horizontal velocity of +115 m/s, it remains directly above the falling package. The pilot always sees the package directly beneath the plane, as the dashed vertical lines in Figure 3.8 show. This result is a direct consequence of the fact that the package has no acceleration in the horizontal direction. In reality, air resistance would slow down the package, and it would not remain directly beneath the plane during the descent.

Figure 3.9 illustrates what happens to two packages that are released simultaneously from the same height, in order to emphasize that the vertical and horizontal parts of the motion in Example 3 occur independently. Package A is dropped from a stationary balloon and falls straight downward toward the ground, since it has no horizontal velocity component ($v_{0x} = 0$ m/s). Package B, on the other hand, is given an initial velocity component of $v_{0x} = +115$ m/s in the horizontal direction, as in Example 3, and follows the path shown in the figure. Both packages hit the ground at the same time. Not only do the packages in Figure 3.9 reach the ground at the same time, but the y components of their velocities are also equal at all points on the way down. However, package B does hit the ground with a greater speed than does package A. Remember, speed is the magnitude of the velocity vector, and the velocity of B has an x component, whereas the velocity of A does not. The magnitude and direction of the velocity vector for package B at the instant just before the package hits the ground is computed in Example 4.

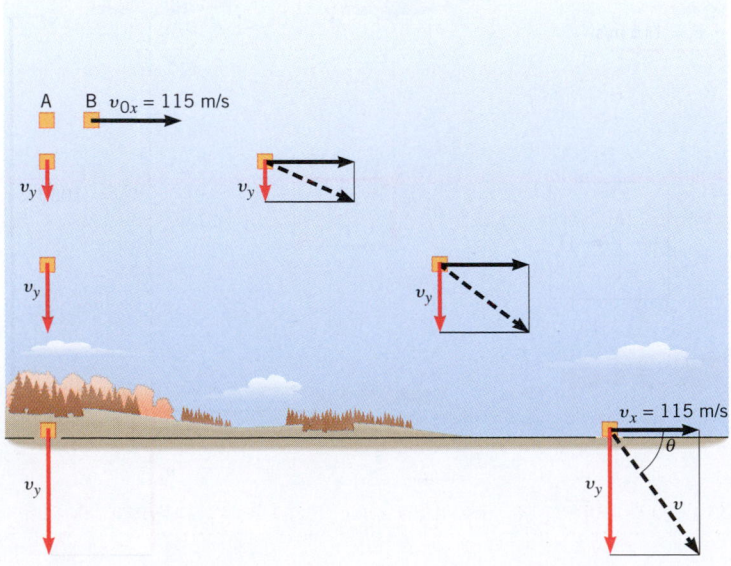

Figure 3.9 Package A and package B are released simultaneously at the same height and strike the ground at the same time because their y variables (y, a_y, and v_{0y}) are the same.

ANALYZING MULTIPLE-CONCEPT PROBLEMS

Example 4 The Velocity of the Care Package

Figure 3.8 shows a care package falling from a plane, and Figure 3.9 shows this package as package B. As in Example 3, the directions to the right and upward are chosen as the positive directions, and the plane is moving horizontally with a constant velocity of +115 m/s at an altitude of 1050 m. Ignoring air resistance, find the magnitude v and the directional angle θ of the final velocity vector that the package has just before it strikes the ground.

Reasoning Figures 3.8 and 3.9 show the final velocity vector, which has components v_x and v_y and a magnitude v. The vector is directed at an angle θ below the horizontal or x direction. We note the right triangle formed by the vector and its components. The hypotenuse of the triangle is the magnitude of the velocity, and the components are the other two sides. As in Example 2, we can use the Pythagorean theorem to express the magnitude or speed v in terms of the components v_x and v_y, and we can use trigonometry to determine the directional angle θ.

Knowns and Unknowns The data for this problem are listed in the table that follows:

Description	Symbol	Value	Comment
Explicit Data, x Direction			
x component of initial velocity	v_{0x}	+ 115 m/s	Package has plane's horizontal velocity at instant of release
Implicit Data, x Direction			
x component of acceleration	a_x	0 m/s^2	No horizontal acceleration, since air resistance is ignored
Explicit Data, y Direction			
y component of displacement	y	− 1050 m	Negative, since upward is positive and package falls downward
Implicit Data, y Direction			
y component of initial velocity	v_{0y}	0 m/s	Package traveling horizontally in x direction at instant of release, not in y direction
y component of acceleration	a_y	− 9.80 m/s^2	Acceleration vector for gravity points downward in the negative direction
Unknown Variables			
Magnitude of final velocity	v	?	
Direction of final velocity	θ	?	

Modeling the Problem

STEP 1 Final Velocity Using the Pythagorean theorem to express the speed v in terms of the components v_x and v_y (see Figure 3.8 or 3.9), we obtain Equation 1a at the right. Furthermore, in a right triangle, the cosine of an angle is the side adjacent to the angle divided by the hypotenuse. With this in mind, we see in Figure 3.8 or Figure 3.9 that the directional angle θ is given by Equation 1b at the right.

$$v = \sqrt{v_x^2 + v_y^2} \tag{1a}$$

$$\theta = \cos^{-1}\left(\frac{v_x}{v}\right) = \cos^{-1}\left(\frac{v_x}{\sqrt{v_x^2 + v_y^2}}\right) \tag{1b}$$

STEP 2 The Components of the Final Velocity Reference to the table of knowns and unknowns shows that, in the x direction, values are available for the kinematic variables v_{0x} and a_x. Since the acceleration a_x is zero, the final velocity component v_x remains unchanged from its initial value of v_{0x}, so we have

$$v_x = v_{0x}$$

In the y direction, values are available for y, v_{0y}, and a_y, so that we can determine the final velocity component v_y using Equation 3.6b from the equations of kinematics:

$$v_y^2 = v_{0y}^2 + 2a_y y \tag{3.6b}$$

These results for v_x and v_y can be substituted into Equations 1a and 1b, as shown at the right.

$$v = \sqrt{v_x^2 + v_y^2} \tag{1a}$$

$$\boxed{v_x = v_{0x}} \quad \boxed{v_y^2 = v_{0y}^2 + 2a_y y} \tag{3.6b}$$

$$\theta = \cos^{-1}\left(\frac{v_x}{\sqrt{v_x^2 + v_y^2}}\right) \tag{1b}$$

$$\boxed{v_x = v_{0x}} \quad \boxed{v_y^2 = v_{0y}^2 + 2a_y y} \tag{3.6b}$$

Continued

Solution Algebraically combining the results of each step, we find that

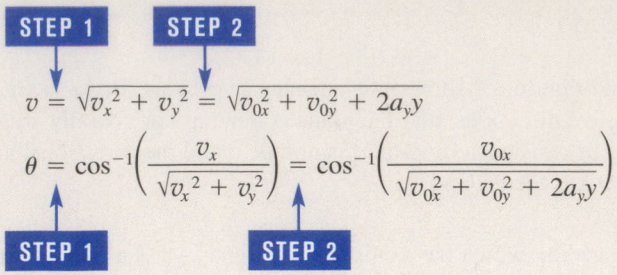

Problem-solving insight

The speed of a projectile at any location along its path is the magnitude v of its velocity at that location: $v = \sqrt{v_x^2 + v_y^2}$. Both the horizontal and vertical velocity components contribute to the speed.

With the data given for the kinematic variables in the x and y directions, we find that the magnitude and direction of the final velocity of the package are

$$v = \sqrt{v_{0x}^2 + v_{0y}^2 + 2a_y y} = \sqrt{(115 \text{ m/s})^2 + (0 \text{ m/s})^2 + 2(-9.80 \text{ m/s}^2)(-1050 \text{ m})} = \boxed{184 \text{ m/s}}$$

$$\theta = \cos^{-1}\left(\frac{v_{0x}}{\sqrt{v_{0x}^2 + v_{0y}^2 + 2a_y y}}\right) = \cos^{-1}\left[\frac{115 \text{ m/s}}{\sqrt{(115 \text{ m/s})^2 + (0 \text{ m/s})^2 + 2(-9.80 \text{ m/s}^2)(-1050 \text{ m})}}\right] = \boxed{51.3°}$$

Related Homework: *Problems 31, 33, 44*

An important feature of projectile motion is that there is no acceleration in the horizontal, or x, direction. Conceptual Example 5 discusses an interesting implication of this feature.

Figure 3.10 The car is moving with a constant velocity to the right, and the rifle is pointed straight up. In the absence of air resistance, a bullet fired from the rifle has no acceleration in the horizontal direction. Example 5 discusses what happens to the bullet.

Conceptual Example 5   I Shot a Bullet into the Air . . .

Suppose you are driving in a convertible with the top down. The car is moving to the right at a constant velocity. As Figure 3.10 illustrates, you point a rifle straight upward and fire it. In the absence of air resistance, would the bullet land **(a)** behind you, **(b)** ahead of you, or **(c)** in the barrel of the rifle?

Reasoning Because there is no air resistance to slow it down, the bullet experiences no horizontal acceleration. Thus, the bullet's horizontal velocity component does not change, and it stays the same as that of the rifle and the car.

Answers (a) and (b) are incorrect. If air resistance were present, it would slow down the bullet and cause it to land behind you, toward the rear of the car. However, air resistance is absent. If the bullet were to land ahead of you, its horizontal velocity component would have to be greater than that of the rifle and the car. This cannot be, since the bullet's horizontal velocity component never changes.

Answer (c) is correct. Since the bullet's horizontal velocity component does not change, it retains its initial value, and remains matched to that of the rifle and the car. As a result, the bullet remains directly above the rifle at all times and would fall directly back into the barrel of the rifle. This situation is analogous to that in Figure 3.8, where the care package, as it falls, remains directly below the plane.

Related Homework: *Problem 41*

Often projectiles, such as footballs and baseballs, are sent into the air at an angle with respect to the ground. From a knowledge of the projectile's initial velocity, a wealth of information can be obtained about the motion. For instance, Example 6 demonstrates how to calculate the maximum height reached by the projectile.

Example 6   The Height of a Kickoff

A placekicker kicks a football at an angle of $\theta = 40.0°$ above the horizontal axis, as Figure 3.11 shows. The initial speed of the ball is $v_0 = 22$ m/s. Ignore air resistance, and find the maximum height H that the ball attains.

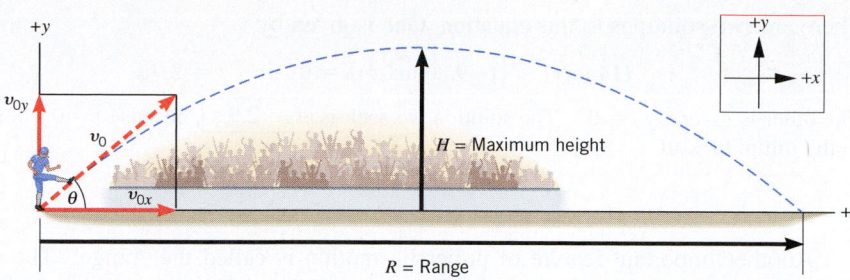

Reasoning The maximum height is a characteristic of the vertical part of the motion, which can be treated separately from the horizontal part. In preparation for making use of this fact, we calculate the vertical component of the initial velocity:

$$v_{0y} = v_0 \sin \theta = +(22 \text{ m/s}) \sin 40.0° = +14 \text{ m/s}$$

The vertical component of the velocity, v_y, decreases as the ball moves upward. Eventually, $v_y = 0$ m/s at the maximum height H. The data below can be used in Equation 3.6b ($v_y^2 = v_{0y}^2 + 2a_y y$) to find the maximum height:

y-Direction Data				
y	a_y	v_y	v_{0y}	t
$H = ?$	-9.80 m/s^2	0 m/s	$+14$ m/s	

Problem-solving insight

When a projectile reaches maximum height, the vertical component of its velocity is momentarily zero ($v_y = 0$ m/s). However, the horizontal component of its velocity is not zero.

Solution From Equation 3.6b, we find that

$$y = H = \frac{v_y^2 - v_{0y}^2}{2a_y} = \frac{(0 \text{ m/s})^2 - (14 \text{ m/s})^2}{2(-9.80 \text{ m/s}^2)} = \boxed{+10 \text{ m}}$$

The height H depends only on the y variables; the same height would have been reached had the ball been thrown *straight up* with an initial velocity of $v_{0y} = +14$ m/s.

It is also possible to find the total time or "hang time" during which the football in Figure 3.11 is in the air. Example 7 shows how to determine this time.

 Example 7 The Time of Flight of a Kickoff

For the motion illustrated in Figure 3.11, ignore air resistance and use the data from Example 6 to determine the time of flight between kickoff and landing.

The physics of
the "hang time" of a football.

Reasoning Given the initial velocity, it is the acceleration due to gravity that determines how long the ball stays in the air. Thus, to find the time of flight we deal with the vertical part of the motion. Since the ball starts at and returns to ground level, the displacement in the y direction is zero. The initial velocity component in the y direction is the same as that in Example 6; that is, $v_{0y} = +14$ m/s. Therefore, we have

y-Direction Data				
y	a_y	v_y	v_{0y}	t
0 m	-9.80 m/s^2		$+14$ m/s	?

The time of flight can be determined from Equation 3.5b ($y = v_{0y}t + \frac{1}{2}a_y t^2$).

Solution Using Equation 3.5b, we find

$$0 \text{ m} = (14 \text{ m/s})t + \tfrac{1}{2}(-9.80 \text{ m/s}^2)t^2 = [(14 \text{ m/s}) + \tfrac{1}{2}(-9.80 \text{ m/s}^2)t]t$$

There are two solutions to this equation. One is given by

$$(14 \text{ m/s}) + \tfrac{1}{2}(-9.80 \text{ m/s}^2)t = 0 \quad \text{or} \quad t = 2.9 \text{ s}$$

The other is given by $t = 0$ s. The solution we seek is $\boxed{t = 2.9 \text{ s}}$, because $t = 0$ s corresponds to the initial kickoff.

Another important feature of projectile motion is called the "range." The range, as Figure 3.11 shows, is the horizontal distance traveled between launching and landing, assuming the projectile returns to the *same vertical level* at which it was fired. Example 8 shows how to obtain the range.

 Example 8 The Range of a Kickoff

For the motion shown in Figure 3.11 and discussed in Examples 6 and 7, ignore air resistance and calculate the range R of the projectile.

Reasoning The range is a characteristic of the horizontal part of the motion. Thus, our starting point is to determine the horizontal component of the initial velocity:

$$v_{0x} = v_0 \cos \theta = +(22 \text{ m/s}) \cos 40.0° = +17 \text{ m/s}$$

Recall from Example 7 that the time of flight is $t = 2.9$ s. Since there is no acceleration in the x direction, v_x remains constant, and the range is simply the product of $v_x = v_{0x}$ and the time.

Solution The range is

$$x = R = v_{0x}t = +(17 \text{ m/s})(2.9 \text{ s}) = \boxed{+49 \text{ m}}$$

The range in the previous example depends on the angle θ at which the projectile is fired above the horizontal. When air resistance is absent, the maximum range results when $\theta = 45°$.

The examples considered thus far have used information about the initial location and velocity of a projectile to determine the final location and velocity. Example 9 deals with the opposite situation and illustrates how the final parameters can be used with the equations of kinematics to determine the initial parameters.

ANALYZING MULTIPLE-CONCEPT PROBLEMS

Example 9 A Home Run

A baseball player hits a home run, and the ball lands in the left-field seats, 7.5 m above the point at which it was hit. It lands with a velocity of 36 m/s at an angle of 28° below the horizontal (see Figure 3.12). The positive directions are upward and to the right in the drawing. Ignoring air resistance, find the magnitude and direction of the initial velocity with which the ball leaves the bat.

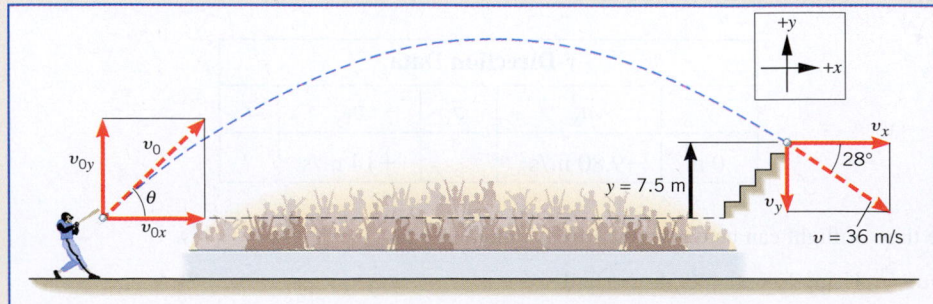

Figure 3.12 The velocity and location of the baseball upon landing can be used to determine its initial velocity, as Example 9 illustrates.

Reasoning Just after the ball is hit, its initial velocity has a magnitude v_0 and components of v_{0x} and v_{0y} and is directed at an angle θ above the horizontal or x direction. Figure 3.12 shows the initial velocity vector and its components. As usual, we will use the Pythagorean theorem to relate v_0 to v_{0x} and v_{0y} and will use trigonometry to determine θ.

Knowns and Unknowns The data for this problem are listed in the table that follows:

Description	Symbol	Value	Comment
Explicit Data			
y component of displacement	y	+7.5 m	Positive, since upward is positive and ball lands above its starting point
Magnitude of final velocity	v	36 m/s	
Direction of final velocity		28°	Below the horizontal (see Figure 3.12)
Implicit Data			
x component of acceleration	a_x	0 m/s	No horizontal acceleration, since air resistances is ignored
y component of acceleration	a_y	−9.80 m/s²	Acceleration vector for gravity points downward in the negative direction
Unknown Variables			
Magnitude of initial velocity	v_0	?	
Direction of initial velocity	θ	?	

Modeling the Problem

STEP 1 **Initial Velocity** The magnitude v_0 of the initial velocity can be related to its components v_{0x} and v_{0y} by using the Pythagorean theorem, since the components are perpendicular to one another. This leads to Equation 1a at the right. Referring to Figure 3.12, we can also use trigonometry to express the directional angle θ in terms of the components v_{0x} and v_{0y}. Thus, we obtain Equation 1b at the right.

$$v_0 = \sqrt{v_{0x}^2 + v_{0y}^2} \tag{1a}$$

$$\theta = \tan^{-1}\left(\frac{v_{0y}}{v_{0x}}\right) \tag{1b}$$

STEP 2 **x Component of the Initial Velocity** To obtain v_{0x}, we note that the acceleration a_x is zero, since air resistance is being ignored. With no acceleration in the x direction, v_{0x} remains unchanged throughout the motion of the ball. Thus, v_{0x} must equal v_x, the x component of the ball's final velocity. We have, then, that

$$\boxed{v_{0x} = v_x = v\cos 28°}$$

This result can be substituted into Equations 1a and 1b, as shown at the right.

$$v_0 = \sqrt{v_{0x}^2 + v_{0y}^2} \tag{1a}$$
$$\boxed{v_{0x} = v\cos 28°}$$
$$\theta = \tan^{-1}\left(\frac{v_{0y}}{v_{0x}}\right) \tag{1b}$$
$$\boxed{v_{0x} = v\cos 28°}$$

STEP 3 **y Component of the Initial Velocity** In contrast to the argument in Step 2, v_{0y} does not equal v_y, the y component of the ball's final velocity, since the ball accelerates in the vertical direction. However, we can use Equation 3.6b from the equations of kinematics to determine v_{0y}:

$$v_y^2 = v_{0y}^2 + 2a_y y \quad \text{or} \quad v_{0y} = +\sqrt{v_y^2 - 2a_y y} \tag{3.6b}$$

The plus sign is chosen for the square root, since the ball's initial velocity component v_{0y} points upward in Figure 3.12. In Equation 3.6b, v_y can be written as

$$v_y = -v\sin 28°$$

where the minus sign is present because v_y points downward in the $-y$ direction in Figure 3.12. We find, then, that

$$\boxed{v_{0y} = +\sqrt{(-v\sin 28°)^2 - 2a_y y}}$$

This result can also be substituted into Equations 1a and 1b, as shown at the right.

$$v_0 = \sqrt{v_{0x}^2 + v_{0y}^2} \tag{1a}$$
$$\boxed{v_{0x} = v\cos 28°}$$
$$\boxed{v_{0y} = +\sqrt{(-v\sin 28°)^2 - 2a_y y}}$$
$$\theta = \tan^{-1}\left(\frac{v_{0y}}{v_{0x}}\right) \tag{1b}$$
$$\boxed{v_{0x} = v\cos 28°}$$
$$\boxed{v_{0y} = +\sqrt{(-v\sin 28°)^2 - 2a_y y}}$$

Continued

Solution Algebraically combining the results of each step, we find that

STEP 1	STEP 2	STEP 3

$$v_0 = \sqrt{v_{0x}^2 + v_{0y}^2} = \sqrt{(v\cos 28°)^2 + v_{0y}^2} = \sqrt{(v\cos 28°)^2 + (-v\sin 28°)^2 - 2a_y y}$$

$$\theta = \tan^{-1}\left(\frac{v_{0y}}{v_{0x}}\right) = \tan^{-1}\left(\frac{v_{0y}}{v\cos 28°}\right) = \tan^{-1}\left(\frac{\sqrt{(-v\sin 28°)^2 - 2a_y y}}{v\cos 28°}\right)$$

STEP 1	STEP 2	STEP 3

With the data given in the table of knowns and unknowns, we find that the magnitude and direction of the ball's initial velocity are

$$v_0 = \sqrt{(v\cos 28°)^2 + (-v\sin 28°)^2 - 2a_y y}$$

$$= \sqrt{[(36 \text{ m/s})\cos 28°]^2 + [-(36 \text{ m/s})\sin 28°]^2 - 2(-9.80 \text{ m/s}^2)(7.5 \text{ m})} = \boxed{38 \text{ m/s}}$$

$$\theta = \tan^{-1}\left(\frac{\sqrt{(-v\sin 28°)^2 - 2a_y y}}{v\cos 28°}\right)$$

$$= \tan^{-1}\left\{\frac{\sqrt{[-(36 \text{ m/s})\sin 28°]^2 - 2(-9.80 \text{ m/s}^2)(7.5 \text{ m})}}{(36 \text{ m/s})\cos 28°}\right\} = \boxed{33°}$$

Related Homework: *Problems 32, 40*

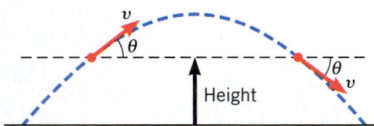

Figure 3.13 The speed v of a projectile at a given height above the ground is the same on the upward and downward parts of the trajectory. The velocities are different, however, since they point in different directions.

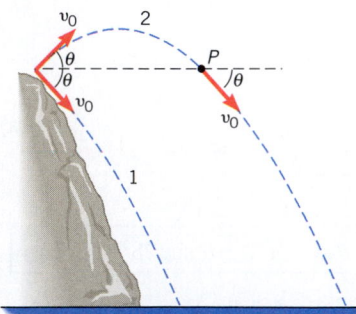

Figure 3.14 Two stones are thrown off the cliff with identical initial speeds v_0, but at equal angles θ that are below and above the horizontal. Conceptual Example 10 compares the velocities with which the stones hit the water below.

In projectile motion, the magnitude of the acceleration due to gravity affects the trajectory in a significant way. For example, a baseball or a golf ball would travel much farther and higher on the moon than on the earth, when launched with the same initial velocity. The reason is that the moon's gravity is only about one-sixth as strong as the earth's.

Section 2.6 points out that certain types of symmetry with respect to time and speed are present for freely falling bodies. These symmetries are also found in projectile motion, since projectiles are falling freely in the vertical direction. In particular, the time required for a projectile to reach its maximum height H is equal to the time spent returning to the ground. In addition, Figure 3.13 shows that the speed v of the object at any height above the ground on the upward part of the trajectory is equal to the speed v at the same height on the downward part. Although the two speeds are the same, the velocities are different, because they point in different directions. Conceptual Example 10 shows how to use this type of symmetry in your reasoning.

Conceptual Example 10 Two Ways to Throw a Stone

From the top of a cliff overlooking a lake, a person throws two stones. The stones have identical initial speeds v_0, but stone 1 is thrown downward at an angle θ below the horizontal, while stone 2 is thrown upward at the same angle above the horizontal, as Figure 3.14 shows. Neglect air resistance and decide which stone, if either, strikes the water with the greater velocity: **(a)** both stones strike the water with the same velocity, **(b)** stone 1 strikes with the greater velocity, **(c)** stone 2 strikes with the greater velocity.

Reasoning Note point P in the drawing, where stone 2 returns to its initial height; here the speed of stone 2 is v_0, (the same as its initial speed), but its velocity is directed at an angle θ below the horizontal. This is exactly the type of projectile symmetry illustrated in Figure 3.13, and this symmetry will lead us to the correct answer.

Answers (b) and (c) are incorrect. You might guess that stone 1, being hurled downward, would strike the water with the greater velocity. Or, you might think that stone 2, having reached a greater height than stone 1, would hit the water with the greater velocity. To understand why neither of these answers is correct, see the response for answer (a) below.

Answer (a) is correct. Let's follow the path of stone 2 as it rises to its maximum height and falls back to earth. When it reaches point P in the drawing, stone 2 has a velocity that is identical to the velocity with which stone 1 is thrown downward from the top of the cliff (see the drawing). From this point on, the velocity of stone 2 changes in exactly the same way as that for stone 1, so both stones strike the water with the same velocity.

Related Homework: *Problems 23, 43*

In all the examples in this section, the projectiles follow a curved trajectory. In general, if the only acceleration is that due to gravity, the shape of the path can be shown to be a *parabola*.

✓ CHECK YOUR UNDERSTANDING

(The answers are given at the end of the book.)

3. A projectile is fired into the air, and it follows the parabolic path shown in the drawing, landing on the right. There is no air resistance. At any instant, the projectile has a velocity $\vec{\mathbf{v}}$ and an acceleration $\vec{\mathbf{a}}$. Which one or more of the drawings could *not* represent the directions for $\vec{\mathbf{v}}$ and $\vec{\mathbf{a}}$ at any point on the trajectory?

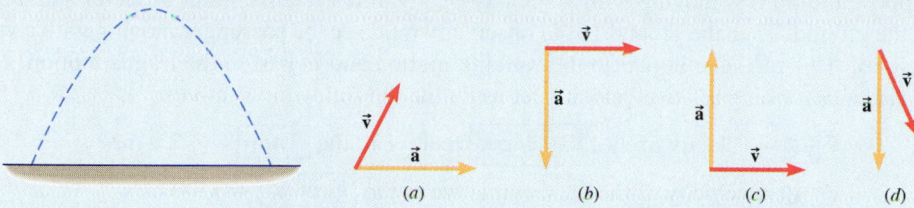

(a)   (b)   (c)   (d)

4. An object is thrown upward at an angle θ above the ground, eventually returning to earth. **(a)** Is there any place along the trajectory where the velocity and acceleration are perpendicular? If so, where? **(b)** Is there any place where the velocity and acceleration are parallel? If so, where?

5. Is the acceleration of a projectile equal to zero when the projectile reaches the top of its trajectory?

6. In baseball, the pitcher's mound is raised to compensate for the fact that the ball falls downward as it travels from the pitcher toward the batter. If baseball were played on the moon, would the pitcher's mound have to be **(a)** higher than, **(b)** lower than, or **(c)** the same height as it is on earth?

7. A tennis ball is hit upward into the air and moves along an arc. Neglecting air resistance, where along the arc is the speed of the ball **(a)** a minimum and **(b)** a maximum?

8. A wrench is accidentally dropped from the top of the mast on a sailboat. Air resistance is negligible. Will the wrench hit at the same place on the deck whether the sailboat is at rest or moving with a constant velocity?

9. A rifle, at a height H above the ground, fires a bullet parallel to the ground. At the same instant and at the same height, a second bullet is dropped from rest. In the absence of air resistance, which bullet, if either, strikes the ground first?

10. A stone is thrown horizontally from the top of a cliff and eventually hits the ground below. A second stone is dropped from rest from the same cliff, falls through the same height, and also hits the ground below. Ignore air resistance. Is each of the following quantities different or the same in the two cases? **(a)** Displacement **(b)** Speed just before impact with the ground **(c)** Time of flight

11. A leopard springs upward at a 45° angle and then falls back to the ground. Air resistance is negligible. Does the leopard, at any point on its trajectory, ever have a speed that is one-half its initial value?

12. Two balls are launched upward from the same spot at different angles with respect to the ground. Both balls rise to the same maximum height. Ball A, however, follows a trajectory that has a greater range than that of ball B. Ignoring air resistance, decide which ball, if either, has the greater launch speed.

Figure 3.15 The velocity of the passenger relative to the ground-based observer is $\vec{v}_{PG}$. It is the vector sum of the velocity $\vec{v}_{PT}$ of the passenger relative to the train and the velocity $\vec{v}_{TG}$ of the train relative to the ground: $\vec{v}_{PG} = \vec{v}_{PT} + \vec{v}_{TG}$.

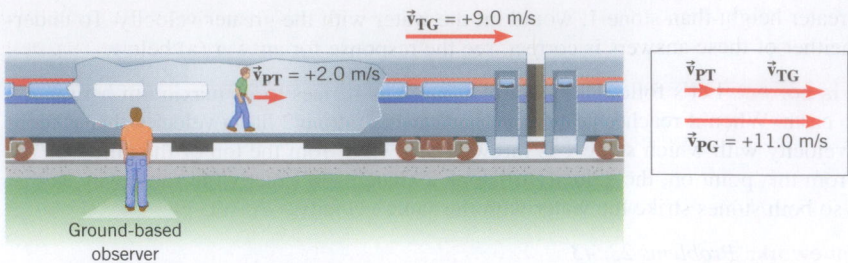

3.4 *RELATIVE VELOCITY

To someone hitchhiking along a highway, two cars speeding by in adjacent lanes seem like a blur. But if the cars have the same velocity, each driver sees the other remaining in place, one lane away. The hitchhiker observes a velocity of perhaps 30 m/s, but each driver observes the other's velocity to be zero. Clearly, the velocity of an object is relative to the observer who is making the measurement.

Figure 3.15 illustrates the concept of relative velocity by showing a passenger walking toward the front of a moving train. The people sitting on the train see the passenger walking with a velocity of +2.0 m/s, where the plus sign denotes a direction to the right. Suppose the train is moving with a velocity of +9.0 m/s relative to an observer standing on the ground. Then the ground-based observer would see the passenger moving with a velocity of +11 m/s, due in part to the walking motion and in part to the train's motion. As an aid in describing relative velocity, let us define the following symbols:

$$\vec{v}\;\boxed{PT} = \text{velocity of the}\;\boxed{\text{Passenger}}\;\text{relative to the}\;\boxed{\text{Train}} = +2.0 \text{ m/s}$$

$$\vec{v}\;\boxed{TG} = \text{velocity of the}\;\boxed{\text{Train}}\;\text{relative to the}\;\boxed{\text{Ground}} = +9.0 \text{ m/s}$$

$$\vec{v}\;\boxed{PG} = \text{velocity of the}\;\boxed{\text{Passenger}}\;\text{relative to the}\;\boxed{\text{Ground}} = +11 \text{ m/s}$$

In terms of these symbols, the situation in Figure 3.15 can be summarized as follows:

$$\vec{v}_{PG} = \vec{v}_{PT} + \vec{v}_{TG} \tag{3.7}$$

or

$$\vec{v}_{PG} = (2.0 \text{ m/s}) + (9.0 \text{ m/s}) = +11 \text{ m/s}$$

According to Equation 3.7*, $\vec{v}_{PG}$ is the vector sum of $\vec{v}_{PT}$ and $\vec{v}_{TG}$, and this sum is shown in the drawing. Had the passenger been walking toward the rear of the train, rather than toward the front, the velocity relative to the ground-based observer would have been $\vec{v}_{PG} = (-2.0 \text{ m/s}) + (9.0 \text{ m/s}) = +7.0 \text{ m/s}$.

Each velocity symbol in Equation 3.7 contains a two-letter subscript. The first letter in the subscript refers to the body that is moving, while the second letter indicates the object relative to which the velocity is measured. For example, $\vec{v}_{TG}$ and $\vec{v}_{PG}$ are the velocities of the **T**rain and **P**assenger measured relative to the **G**round. Similarly, $\vec{v}_{PT}$ is the velocity of the **P**assenger measured by an observer sitting on the **T**rain.

The ordering of the subscript symbols in Equation 3.7 follows a definite pattern. The first subscript (P) on the left side of the equation is also the first subscript on the right side of the equation. Likewise, the last subscript (G) on the left side is also the last subscript on the right side. The third subscript (T) appears only on the right side of the equation as the two "inner" subscripts. The colored boxes below emphasize the pattern of the symbols in the subscripts:

$$\vec{v}\;\boxed{PG} = \vec{v}\;\boxed{P}T + \vec{v}_T\;\boxed{G}$$

In other situations, the subscripts will not necessarily be P, G, and T, but will be compatible with the names of the objects involved in the motion.

The film *Fly Away Home* is based on the remarkable story of how William Lishman and Joe Duff used their ultralight planes to guide geese on a seven-day flight from the Toronto area in Canada to Virginia. The geese matched their velocities to that of the ultralights, with the result that the relative velocity of the geese and the planes was zero. (Courtesy Operation Migration, www.operationmigration.org)

---

*This equation assumes that the train and the ground move in a straight line relative to one another.

Equation 3.7 has been presented in connection with one-dimensional motion, but the result is also valid for two-dimensional motion. Figure 3.16 depicts a common situation that deals with relative velocity in two dimensions. Part *a* of the drawing shows a boat being carried downstream by a river; the engine of the boat is turned off. In part *b*, the engine is turned on, and now the boat moves across the river in a diagonal fashion because of the combined motion produced by the current and the engine. The list below gives the velocities for this type of motion and the objects relative to which they are measured:

$\vec{\mathbf{v}}_{\boxed{BW}}$ = velocity of the $\boxed{Boat}$ relative to the $\boxed{Water}$

$\vec{\mathbf{v}}_{\boxed{WS}}$ = velocity of the $\boxed{Water}$ relative to the $\boxed{Shore}$

$\vec{\mathbf{v}}_{\boxed{BS}}$ = velocity of the $\boxed{Boat}$ relative to the $\boxed{Shore}$

The velocity $\vec{\mathbf{v}}_{BW}$ of the boat relative to the water is the velocity measured by an observer who, for instance, is floating on an inner tube and drifting downstream with the current. When the engine is turned off, the boat also drifts downstream with the current, and $\vec{\mathbf{v}}_{BW}$ is zero. When the engine is turned on, however, the boat can move relative to the water, and $\vec{\mathbf{v}}_{BW}$ is no longer zero. The velocity $\vec{\mathbf{v}}_{WS}$ of the water relative to the shore is the velocity of the current measured by an observer on the shore. The velocity $\vec{\mathbf{v}}_{BS}$ of the boat relative to the shore is due to the combined motion of the boat relative to the water and the motion of the water relative to the shore. In symbols,

$$\vec{\mathbf{v}}_{\boxed{BS}} = \vec{\mathbf{v}}_{\boxed{B}W} + \vec{\mathbf{v}}_{W\boxed{S}}$$

The ordering of the subscripts in this equation is identical to that in Equation 3.7, although the letters have been changed to reflect a different physical situation. Example 11 illustrates the concept of relative velocity in two dimensions.

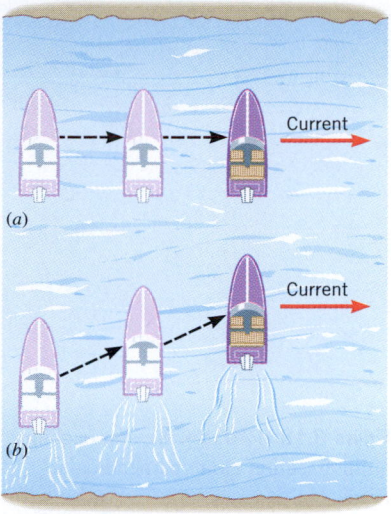

Figure 3.16 (*a*) A boat with its engine turned off is carried along by the current. (*b*) With the engine turned on, the boat moves across the river in a diagonal fashion.

 Example 11   Crossing a River

The engine of a boat drives it across a river that is 1800 m wide. The velocity $\vec{\mathbf{v}}_{BW}$ of the boat relative to the water is 4.0 m/s, directed perpendicular to the current, as in Figure 3.17. The velocity $\vec{\mathbf{v}}_{WS}$ of the water relative to the shore is 2.0 m/s. **(a)** What is the velocity $\vec{\mathbf{v}}_{BS}$ of the boat relative to the shore? **(b)** How long does it take for the boat to cross the river?

Reasoning **(a)** The velocity of the boat relative to the shore is $\vec{\mathbf{v}}_{BS}$. It is the vector sum of the velocity $\vec{\mathbf{v}}_{BW}$ of the boat relative to the water and the velocity $\vec{\mathbf{v}}_{WS}$ of the water relative to the shore: $\vec{\mathbf{v}}_{BS} = \vec{\mathbf{v}}_{BW} + \vec{\mathbf{v}}_{WS}$. Since $\vec{\mathbf{v}}_{BW}$ and $\vec{\mathbf{v}}_{WS}$ are both known, we can use this relation among the velocities, with the aid of trigonometry, to find the magnitude and directional angle of $\vec{\mathbf{v}}_{BS}$.

(b) The component of $\vec{\mathbf{v}}_{BS}$ that is parallel to the width of the river (see Figure 3.17) determines how fast the boat crosses the river; this parallel component is $v_{BS} \sin \theta = v_{BW} = 4.0$ m/s. The time for the boat to cross the river is equal to the width of the river divided by the magnitude of this velocity component.

Solution **(a)** Since the vectors $\vec{\mathbf{v}}_{BW}$ and $\vec{\mathbf{v}}_{WS}$ are perpendicular (see Figure 3.17), the magnitude of $\vec{\mathbf{v}}_{BS}$ can be determined by using the Pythagorean theorem:

$$v_{BS} = \sqrt{(v_{BW})^2 + (v_{WS})^2} = \sqrt{(4.0 \text{ m/s})^2 + (2.0 \text{ m/s})^2} = \boxed{4.5 \text{ m/s}}$$

Thus, the boat moves at a speed of 4.5 m/s with respect to an observer on shore. The direction of the boat relative to the shore is given by the angle θ in the drawing:

$$\tan \theta = \frac{v_{BW}}{v_{WS}} \quad \text{or} \quad \theta = \tan^{-1}\left(\frac{v_{BW}}{v_{WS}}\right) = \tan^{-1}\left(\frac{4.0 \text{ m/s}}{2.0 \text{ m/s}}\right) = \boxed{63°}$$

(b) The time *t* for the boat to cross the river is

$$t = \frac{\text{Width}}{v_{BS} \sin \theta} = \frac{1800 \text{ m}}{4.0 \text{ m/s}} = \boxed{450 \text{ s}}$$

Figure 3.17 The velocity of the boat relative to the shore is $\vec{\mathbf{v}}_{BS}$. It is the vector sum of the velocity $\vec{\mathbf{v}}_{BW}$ of the boat relative to the water and the velocity $\vec{\mathbf{v}}_{WS}$ of the water relative to the shore: $\vec{\mathbf{v}}_{BS} = \vec{\mathbf{v}}_{BW} + \vec{\mathbf{v}}_{WS}$.

Sometimes, situations arise when two vehicles are in relative motion, and it is useful to know the relative velocity of one with respect to the other. Example 12 considers this type of relative motion.

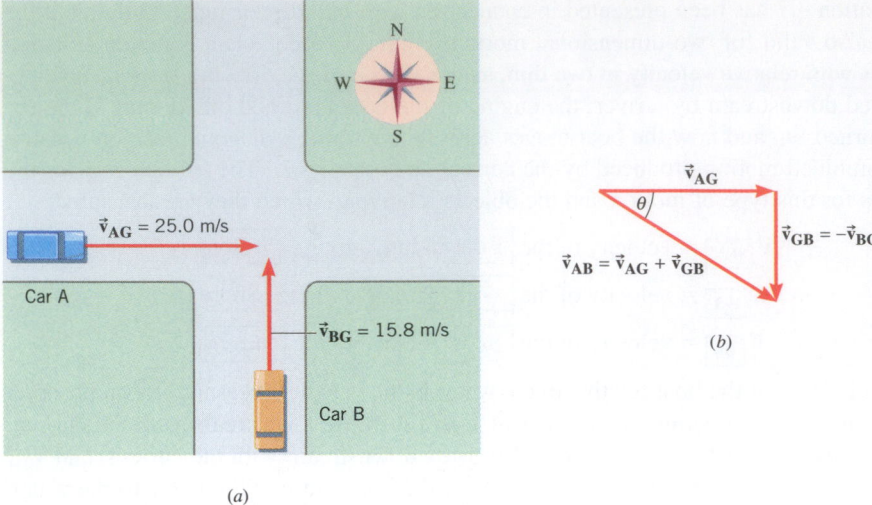

Figure 3.18 Two cars are approaching an intersection along perpendicular roads.

Example 12   Approaching an Intersection

Figure 3.18a shows two cars approaching an intersection along perpendicular roads. The cars have the following velocities:

$$\vec{v}_{\boxed{AG}} = \text{velocity of } \boxed{\text{car A}} \text{ relative to the } \boxed{\text{Ground}} = 25.0 \text{ m/s, eastward}$$

$$\vec{v}_{\boxed{BG}} = \text{velocity of } \boxed{\text{car B}} \text{ relative to the } \boxed{\text{Ground}} = 15.8 \text{ m/s, northward}$$

Find the magnitude and direction of $\vec{v}_{AB}$, where

$$\vec{v}_{\boxed{AB}} = \text{velocity of } \boxed{\text{car A}} \text{ as measured by a passenger in } \boxed{\text{car B}}$$

Reasoning To find $\vec{v}_{AB}$, we use an equation whose subscripts follow the order outlined earlier. Thus,

$$\vec{v}_{\boxed{AB}} = \vec{v}_{\boxed{A}G} + \vec{v}_{G\boxed{B}}$$

In this equation, the term $\vec{v}_{GB}$ is the velocity of the ground relative to a passenger in car B, rather than $\vec{v}_{BG}$, which is given as 15.8 m/s, northward. In other words, the subscripts are reversed. However, $\vec{v}_{GB}$ is related to $\vec{v}_{BG}$ according to

$$\vec{v}_{GB} = -\vec{v}_{BG}$$

This relationship reflects the fact that a passenger in car B, moving northward relative to the ground, looks out the car window and sees the ground moving southward, in the opposite direction. Therefore, the equation $\vec{v}_{AB} = \vec{v}_{AG} + \vec{v}_{GB}$ may be used to find $\vec{v}_{AB}$, provided we recognize $\vec{v}_{GB}$ as a vector that points opposite to the given velocity $\vec{v}_{BG}$. With this in mind, Figure 3.18b illustrates how $\vec{v}_{AG}$ and $\vec{v}_{GB}$ are added vectorially to give $\vec{v}_{AB}$.

Solution From the vector triangle in Figure 3.18b, the magnitude and direction of $\vec{v}_{AB}$ can be calculated as

$$v_{AB} = \sqrt{(v_{AG})^2 + (v_{GB})^2} = \sqrt{(25.0 \text{ m/s})^2 + (-15.8 \text{ m/s})^2} = \boxed{29.6 \text{ m/s}}$$

and

$$\cos \theta = \frac{v_{AG}}{v_{AB}} \quad \text{or} \quad \theta = \cos^{-1}\left(\frac{v_{AG}}{v_{AB}}\right) = \cos^{-1}\left(\frac{25.0 \text{ m/s}}{29.6 \text{ m/s}}\right) = \boxed{32.4°}$$

Problem-solving insight

In general, the velocity of object R relative to object S is always the negative of the velocity of object S relative to R: $\vec{v}_{RS} = -\vec{v}_{SR}$.

The physics of raindrops falling on car windows.

While driving a car, have you ever noticed that the rear window sometimes remains dry, even though rain is falling? This phenomenon is a consequence of relative velocity, as Figure 3.19 helps to explain. Part a shows a car traveling horizontally with a velocity of $\vec{v}_{CG}$ and a raindrop falling vertically with a velocity of $\vec{v}_{RG}$. Both velocities are measured relative to the ground. To determine whether the raindrop hits the window, however, we

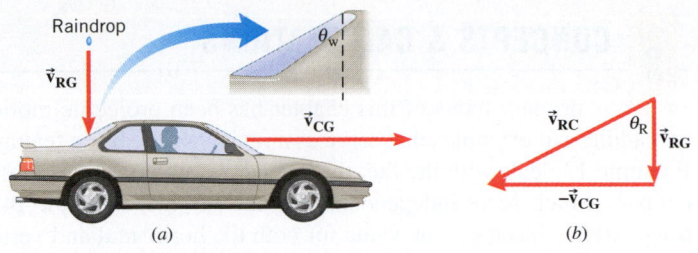

need to consider the velocity of the raindrop relative to the car, not to the ground. This velocity is $\vec{v}_{RC}$, and we know that

$$\vec{v}_{RC} = \vec{v}_{RG} + \vec{v}_{GC} = \vec{v}_{RG} - \vec{v}_{CG}$$

Here, we have used the fact that $\vec{v}_{GC} = -\vec{v}_{CG}$. Part *b* of the drawing shows the tail-to-head arrangement corresponding to this vector subtraction and indicates that the direction of $\vec{v}_{RC}$ is given by the angle θ_R. In comparison, the rear window is inclined at an angle θ_W with respect to the vertical (see the blowup in part *a*). When θ_R is greater than θ_W, the raindrop will miss the window. However, θ_R is determined by the speed v_{RG} of the raindrop and the speed v_{CG} of the car, according to $\theta_R = \tan^{-1}(v_{CG}/v_{RG})$. At higher car speeds, the angle θ_R becomes too large for the drop to hit the window. At a high enough speed, then, the car simply drives out from under each falling drop!

✓ CHECK YOUR UNDERSTANDING

(The answers are given at the end of the book.)

13. Three cars, A, B, and C, are moving along a straight section of a highway. The velocity of A relative to B is $\vec{v}_{AB}$, the velocity of A relative to C is $\vec{v}_{AC}$, and the velocity of C relative to B is $\vec{v}_{CB}$. Fill in the missing velocities in the table.

	$\vec{v}_{AB}$	$\vec{v}_{AC}$	$\vec{v}_{CB}$
(a)	?	+40 m/s	+30 m/s
(b)	?	+50 m/s	−20 m/s
(c)	+60 m/s	+20 m/s	?
(d)	−50 m/s	?	+10 m/s

14. On a riverboat cruise, a plastic bottle is accidentally dropped overboard. A passenger on the boat estimates that the boat pulls ahead of the bottle by 5 meters each second. Is it possible to conclude that the magnitude of the velocity of the boat with respect to the shore is 5 m/s?

15. A plane takes off at St. Louis, flies straight to Denver, and then returns the same way. The plane flies at the same speed with respect to the ground during the entire flight, and there are no head winds or tail winds. Since the earth revolves around its axis once a day, you might expect that the times for the outbound trip and the return trip differ, depending on whether the plane flies against the earth's rotation or with it. Is this true, or are the two times the same?

16. A child is playing on the floor of a recreational vehicle (RV) as it moves along the highway at a constant velocity. He has a toy cannon, which shoots a marble at a fixed angle and speed with respect to the floor. The cannon can be aimed toward the front or the rear of the RV. Is the range toward the front the same as, less than, or greater than the range toward the rear? Answer this question   **(a)** from the child's point of view and **(b)** from the point of view of an observer standing still on the ground.

17. Three swimmers can swim equally fast relative to the water. They have a race to see who can swim across a river in the least time. Swimmer A swims perpendicular to the current and lands on the far shore downstream, because the current has swept him in that direction. Swimmer B swims upstream at an angle to the current and lands on the far shore directly opposite the starting point. Swimmer C swims downstream at an angle to the current in an attempt to take advantage of the current. Who crosses the river in the least time?

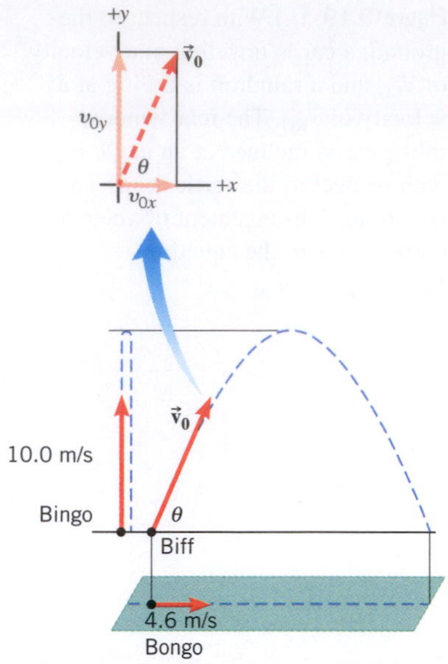

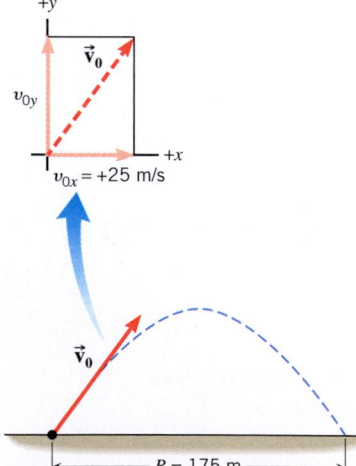

Figure 3.20 Example 13 discusses the projectile motion shown here, in which three circus clowns, Biff, Bongo, and Bingo, are launched simultaneously.

3.5 | CONCEPTS & CALCULATIONS

A primary focus of this chapter has been projectile motion. This section presents two additional examples that serve as a review of the basic features of this type of motion. Example 13 deals with the fact that projectile motion consists of a horizontal and a vertical part, which occur independently of one another. Example 14 stresses the fact that the time variable has the same value for both the horizontal and vertical parts of the motion.

Concepts & Calculations Example 13
Projectile Motion

In a circus act, Biff the clown is fired from a cannon at an initial velocity $\vec{v}_0$ directed at an angle θ above the horizontal, as Figure 3.20 shows. Simultaneously, two other clowns are also launched. Bongo is launched horizontally on roller skates at a speed of 4.6 m/s. He rolls along the ground while Biff flies through the air. When Biff returns to the ground, he lands side by side with his roller-skating friend, who is gliding by just at the instant of landing. The third clown, Bingo, however, is fired straight upward at a speed of 10.0 m/s and reaches the same maximum height at the same instant as Biff. Ignore air resistance, and assume that the roller skates are unimpeded by friction. Find the speed v_0 and the angle θ for Biff.

Concept Questions and Answers Is Bongo's left-to-right motion the same as or different from the horizontal part of Biff's motion along his trajectory?

Answer The horizontal and vertical parts of projectile motion occur independently of one another. Therefore, since Biff lands side by side with Bongo, Bongo's left-to-right motion is identical to the horizontal part of Biff's motion along his trajectory.

Is Biff's motion in the horizontal direction determined by the initial velocity $\vec{v}_0$, just its horizontal component v_{0x}, or just its vertical component v_{0y}?

Answer Just the horizontal component v_{0x} determines Biff's motion in the horizontal direction.

Is Bingo's up-and-down motion the same as or different from the vertical part of Biff's motion along his trajectory?

Answer The horizontal and vertical parts of projectile motion occur independently of one another. Therefore, since they reach the same maximum height at the same instant, Bingo's up-and-down motion is identical to the vertical part of Biff's motion along his trajectory.

Is Biff's initial motion in the vertical direction determined by the initial velocity $\vec{v}_0$, just its horizontal component v_{0x}, or just its vertical component v_{0y}?

Answer Just the vertical component v_{0y} determines Biff's motion in the vertical direction.

Solution Based on the Concept Questions and Answers, we can identify the x and y components of Biff's initial velocity as follows:

$$v_{0x} = 4.6 \text{ m/s} \quad \text{and} \quad v_{0y} = 10.0 \text{ m/s}$$

The initial speed v_0 is the magnitude of the initial velocity, and it, along with the directional angle θ, can be determined from the components:

$$v_0 = \sqrt{(v_{0x})^2 + (v_{0y})^2} = \sqrt{(4.6 \text{ m/s})^2 + (10.0 \text{ m/s})^2} = \boxed{11 \text{ m/s}}$$

$$\theta = \tan^{-1}\left(\frac{v_{0y}}{v_{0x}}\right) = \tan^{-1}\left(\frac{10.0 \text{ m/s}}{4.6 \text{ m/s}}\right) = \boxed{65°}$$

Concepts & Calculations Example 14
Time and Projectile Motion

Figure 3.21 A projectile, launched with a velocity whose horizontal component is $v_{0x} = +25$ m/s, has a range of $R = 175$ m. From these data the vertical component v_{0y} of the initial velocity can be determined.

A projectile is launched from and returns to ground level, as Figure 3.21 shows. Air resistance is absent. The horizontal range of the projectile is $R = 175$ m, and the horizontal component of the launch velocity is $v_{0x} = 25$ m/s. Find the vertical component v_{0y} of the launch velocity.

Concept Questions and Answers What is the final value of the horizontal component v_x of the projectile's velocity?

Answer The final value v_x of the horizontal component of the projectile's velocity is the same as the initial value in the absence of air resistance. In other words, the horizontal motion occurs at a constant velocity of 25 m/s.

Can the time be determined for the horizontal part of the motion?

Answer Yes. In constant-velocity motion, the time is just the horizontal distance (the range) divided by the magnitude of the horizontal component of the projectile's velocity.

Is the time for the horizontal part of the motion the same as the time for the vertical part of the motion?

Answer Yes. The value for the time calculated for the horizontal part of the motion can be used to analyze the vertical part of the motion.

For the vertical part of the motion, what is the displacement of the projectile?

Answer Since the projectile is launched from and returns to ground level, the vertical displacement is zero.

Solution From the constant-velocity horizontal motion, we find that the time is

$$t = \frac{R}{v_{0x}} = \frac{175 \text{ m}}{25 \text{ m/s}} = 7.0 \text{ s}$$

For the vertical part of the motion, we know that the displacement is zero and that the acceleration due to gravity is -9.80 m/s^2, assuming that upward is the positive direction. Therefore, we can use Equation 3.5b to find the initial y component of the velocity:

$$y = v_{0y}t + \tfrac{1}{2}a_y t^2 \quad \text{or} \quad 0 \text{ m} = v_{0y}t + \tfrac{1}{2}a_y t^2$$

$$v_{0y} = -\tfrac{1}{2}a_y t = -\tfrac{1}{2}(-9.80 \text{ m/s}^2)(7.0 \text{ s}) = \boxed{34 \text{ m/s}}$$

CONCEPT SUMMARY

If you need more help with a concept, use the Learning Aids noted next to the discussion or equation. Examples (**Ex.**) are in the text of this chapter. Go to **www.wiley.com/college/cutnell** for the following Learning Aids:

Interactive LearningWare (ILW) — Additional examples solved in a five-step interactive format.

Concept Simulations (CS) — Animated text figures or animations of important concepts.

Interactive Solutions (IS) — Models for certain types of problems in the chapter homework. The calculations are carried out interactively.

Topic	Discussion	Learning Aids
	3.1 DISPLACEMENT, VELOCITY, AND ACCELERATION The position of an object is located with a	
Displacement vector	vector $\vec{r}$ drawn from the coordinate origin to the object. The displacement $\Delta\vec{r}$ of the object is defined as $\Delta\vec{r} = \vec{r} - \vec{r}_0$, where $\vec{r}$ and $\vec{r}_0$ specify its final and initial positions, respectively.	
Average velocity	The average velocity $\bar{\vec{v}}$ of an object moving between two positions is defined as its displacement $\Delta\vec{r} = \vec{r} - \vec{r}_0$ divided by the elapsed time $\Delta t = t - t_0$:	**IS 3.10**

$$\bar{\vec{v}} = \frac{\vec{r} - \vec{r}_0}{t - t_0} = \frac{\Delta\vec{r}}{\Delta t} \tag{3.1}$$

Instantaneous velocity	The instantaneous velocity $\vec{v}$ is the velocity at an instant of time. The average velocity becomes equal to the instantaneous velocity in the limit that Δt becomes infinitesimally small ($\Delta t \to 0$ s):	

$$\vec{v} = \lim_{\Delta t \to 0} \frac{\Delta\vec{r}}{\Delta t}$$

Average acceleration	The average acceleration $\bar{\vec{a}}$ of an object is the change in its velocity, $\Delta\vec{v} = \vec{v} - \vec{v}_0$ divided by the elapsed time $\Delta t = t - t_0$:	**ILW 3.1**

$$\bar{\vec{a}} = \frac{\vec{v} - \vec{v}_0}{t - t_0} = \frac{\Delta\vec{v}}{\Delta t} \tag{3.2}$$

Topic	Discussion	Learning Aids
Instantaneous acceleration	The instantaneous acceleration $\vec{a}$ is the acceleration at an instant of time. The average acceleration becomes equal to the instantaneous acceleration in the limit that the elapsed time Δt becomes infinitesimally small: $$\vec{a} = \lim_{\Delta t \to 0} \frac{\Delta \vec{v}}{\Delta t}$$	

3.2 EQUATIONS OF KINEMATICS IN TWO DIMENSIONS Motion in two dimensions can be described in terms of the time t and the x and y components of four vectors: the displacement, the acceleration, and the initial and final velocities. **Ex. 1, 2**

Independence of the x and y parts of the motion

The x part of the motion occurs exactly as it would if the y part did not occur at all. Similarly, the y part of the motion occurs exactly as it would if the x part of the motion did not exist. The motion can be analyzed by treating the x and y components of the four vectors separately and realizing that the time t is the same for each component.

Equations of kinematics for constant acceleration

When the acceleration is constant, the x components of the displacement, the acceleration, and the initial and final velocities are related by the equations of kinematics, and so are the y components:

x Component		y Component	
$v_x = v_{0x} + a_x t$	(3.3a)	$v_y = v_{0y} + a_y t$	(3.3b)
$x = \frac{1}{2}(v_{0x} + v_x)t$	(3.4a)	$y = \frac{1}{2}(v_{0y} + v_y)t$	(3.4b)
$x = v_{0x}t + \frac{1}{2}a_x t^2$	(3.5a)	$y = v_{0y}t + \frac{1}{2}a_y t^2$	(3.5b)
$v_x^2 = v_{0x}^2 + 2a_x x$	(3.6a)	$v_y^2 = v_{0y}^2 + 2a_y y$	(3.6b)

The directions of these components are conveyed by assigning a plus ($+$) or minus ($-$) sign to each one.

Acceleration in projectile motion

3.3 PROJECTILE MOTION Projectile motion is an idealized kind of motion that occurs when a moving object (the projectile) experiences only the acceleration due to gravity, which acts vertically downward. If the trajectory of the projectile is near the earth's surface, a_y has a magnitude of 9.80 m/s^2. The acceleration has no horizontal component ($a_x = 0 \text{ m/s}^2$), the effects of air resistance being negligible. **Ex. 3–9, 13, 14 / CS 3.1, 3.2 / ILW 3.2 / IS 3.39, 3.73, 3.77**

Symmetries in projectile motion

There are several symmetries in projectile motion: (1) The time to reach maximum height from any point is equal to the time spent returning from the maximum height to that point. (2) The speed of a projectile depends only on its height above its launch point, and not on whether it is moving upward or downward. **Ex. 10**

Adding relative velocities

3.4 RELATIVE VELOCITY The velocity of object A relative to object B is $\vec{v}_{AB}$, and the velocity of object B relative to object C is $\vec{v}_{BC}$. The velocity of A relative to C is (note the ordering of the subscripts) **Ex. 11 / CS 3.3 / IS 3.76**

$$\vec{v}_{AC} = \vec{v}_{AB} + \vec{v}_{BC}$$

While the velocity of object A relative to object B is $\vec{v}_{AB}$, the velocity of B relative to A is $\vec{v}_{BA} = -\vec{v}_{AB}$. **Ex. 12**

FOCUS ON CONCEPTS

Note to Instructors: The numbering of the questions shown here reflects the fact that they are only a representative subset of the total number that are available online. However, all of the questions are available for assignment via an online homework management program such as WileyPLUS or WebAssign.

Section 3.3 Projectile Motion

1. The drawing shows projectile motion at three points along the trajectory. The speeds at the points are v_1, v_2, and v_3. Assume there is no air resistance and rank the speeds, largest to smallest. (Note that the symbol $>$ means "greater than.") **(a)** $v_1 > v_3 > v_2$ **(b)** $v_1 > v_2 > v_3$ **(c)** $v_2 > v_3 > v_1$ **(d)** $v_2 > v_1 > v_3$ **(e)** $v_3 > v_2 > v_1$

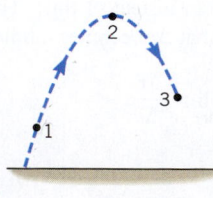

3. Two balls are thrown from the top of a building, as in the drawing. Ball 1 is thrown straight down, and ball 2 is thrown with the same speed, but upward at an angle θ with respect to the horizontal. Consider the motion of the balls after they are released. Which one of the following statements is true? **(a)** The acceleration of ball 1 becomes larger and larger as it falls, because the ball is going faster and faster. **(b)** The acceleration of ball 2 decreases as it rises, becomes zero at the top of the trajectory, and then increases as the ball begins to fall toward the ground. **(c)** Both balls have the same acceleration at all times. **(d)** Ball 2 has an acceleration

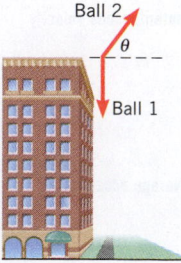

in both the horizontal and vertical directions, but ball 1 has an acceleration only in the vertical direction.

4. Each drawing shows three points along the path of a projectile, one on its way up, one at the top, and one on its way down. The launch point is on the left in each drawing. Which drawing correctly represents the acceleration $\vec{a}$ of the projectile at these three points?
(a) 1 **(b)** 2 **(c)** 3 **(d)** 4 **(e)** 5

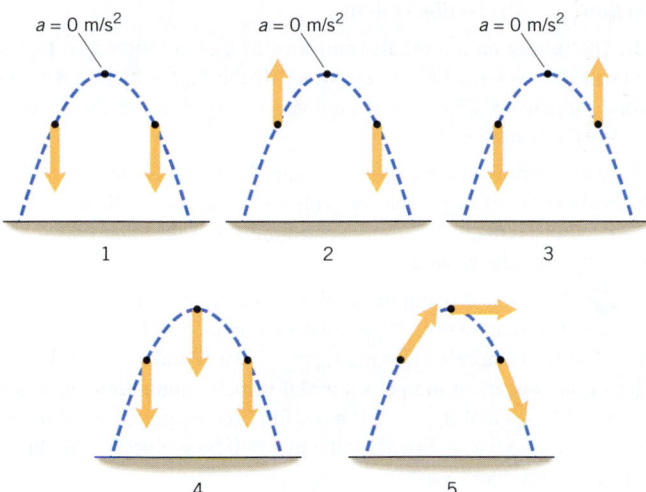

6. Ball 1 is thrown into the air and it follows the trajectory for projectile motion shown in the drawing. At the instant that ball 1 is at the top of its trajectory, ball 2 is dropped from rest from the same height. Which ball reaches the ground first? **(a)** Ball 1 reaches the ground first, since it is moving at the top of the trajectory, while ball 2 is dropped from rest. **(b)** Ball 2 reaches the ground first, because it has the shorter distance to travel. **(c)** Both balls reach the ground at the same time. **(d)** There is not enough information to tell which ball reaches the ground first.

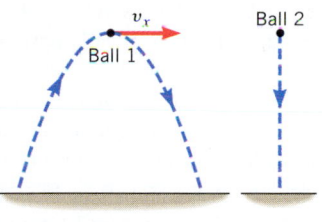

9. Two objects are fired into the air, and the drawing shows the projectile motions. Projectile 1 reaches the greater height, but projectile 2 has the greater range. Which one is in the air for the greater amount of time? **(a)** Projectile 1, because it travels higher than projectile 2. **(b)** Projectile 2, because it has the greater range. **(c)** Both projectiles spend the same amount of time in the air. **(d)** Projectile 2, because it has the smaller initial speed and, therefore, travels more slowly than projectile 1.

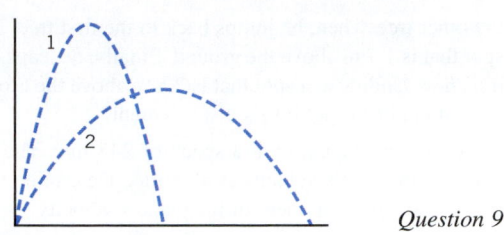

Question 9

Section 3.4 Relative Velocity

14. A slower-moving car is traveling behind a faster-moving bus. The velocities of the two vehicles are as follows:

$\vec{v}_{CG}$ = velocity of the **C**ar relative to the **G**round = +12 m/s

$\vec{v}_{BG}$ = velocity of the **B**us relative to the **G**round = +16 m/s

A passenger on the bus gets up and walks toward the front of the bus with a velocity of $\vec{v}_{PB}$, where $\vec{v}_{PB}$ = velocity of the **P**assenger relative to the **B**us = +2 m/s. What is $\vec{v}_{PC}$, the velocity of the **P**assenger relative to the **C**ar?

(a) +2 m/s + 16 m/s + 12 m/s = +30 m/s
(b) −2 m/s + 16 m/s + 12 m/s = +26 m/s
(c) +2 m/s + 16 m/s − 12 m/s = +6 m/s
(d) −2 m/s + 16 m/s − 12 m/s = +2 m/s

15. Your car is traveling behind a jeep. Both are moving at the same speed, so the velocity of the jeep relative to you is zero. A spare tire is strapped to the back of the jeep. Suddenly the strap breaks, and the tire falls off the jeep. Will your car hit the spare tire before the tire hits the road? Assume that air resistance is absent. **(a)** Yes. As long as the car doesn't slow down, it will hit the tire. **(b)** No. The car will not hit the tire before the tire hits the ground, no matter how close you are to the jeep. **(c)** If the tire falls from a great enough height, the car will hit the tire. **(d)** If the car is far enough behind the jeep, the car will not hit the tire.

16. The drawing shows two cars traveling in different directions with different speeds. Their velocities are:

$\vec{v}_{AG}$ = velocity of car **A** relative to the **G**round = 27.0 m/s, due east

$\vec{v}_{BG}$ = velocity of car **B** relative to the **G**round = 21.0 m/s, due north

The passenger of car B looks out the window and sees car A. What is the velocity (magnitude and direction) of car A as observed by the passenger of car B? In other words, what is the velocity $\vec{v}_{AB}$ of car A relative to car B? Give the directional angle of $\vec{v}_{AB}$ with respect to due east.

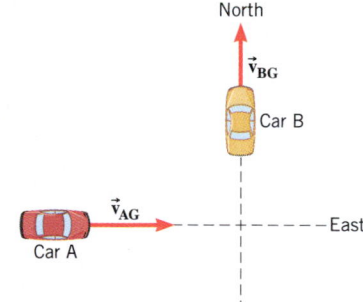

PROBLEMS

Note to Instructors: Most of the homework problems in this chapter are available for assignment via an online homework management program such as WileyPLUS or WebAssign, and those marked with the icon **GO** *are presented in WileyPLUS using a guided tutorial format that provides enhanced interactivity. See Preface for additional details.*

ssm Solution is in the Student Solutions Manual.
www Solution is available online at www.wiley.com/college/cutnell

 This icon represents a biomedical application.

Section 3.1 Displacement, Velocity, and Acceleration

1. **ssm** Two trees have perfectly straight trunks and are both growing perpendicular to the flat horizontal ground beneath them. The

sides of the trunks that face each other are separated by 1.3 m. A frisky squirrel makes three jumps in rapid succession. First, he leaps from the foot of one tree to a spot that is 1.0 m above the ground on the other tree. Then, he jumps back to the first tree, landing on it at a spot that is 1.7 m above the ground. Finally, he leaps back to the other tree, now landing at a spot that is 2.5 m above the ground. What is the magnitude of the squirrel's displacement?

2. A jetliner is moving at a speed of 245 m/s. The vertical component of the plane's velocity is 40.6 m/s. Determine the magnitude of the horizontal component of the plane's velocity.

3. In a football game a kicker attempts a field goal. The ball remains in contact with the kicker's foot for 0.050 s, during which time it experiences an acceleration of 340 m/s². The ball is launched at an angle of 51° above the ground. Determine the horizontal and vertical components of the launch velocity.

4. A meteoroid is speeding through the atmosphere, traveling east at 18.3 km/s while descending at a rate of 11.5 km/s. What is its speed, in km/s?

5. **ssm** A radar antenna is tracking a satellite orbiting the earth. At a certain time, the radar screen shows the satellite to be 162 km away. The radar antenna is pointing upward at an angle of 62.3° from the ground. Find the x and y components (in km) of the position vector of the satellite, relative to the antenna.

6. A mountain-climbing expedition establishes two intermediate camps, labeled A and B in the drawing, above the base camp. What is the magnitude Δr of the displacement between camp A and camp B?

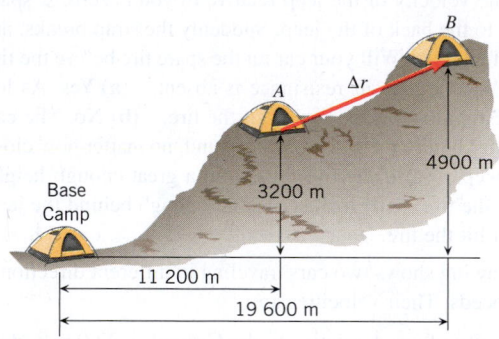

7. **ssm** In diving to a depth of 750 m, an elephant seal also moves 460 m due east of his starting point. What is the magnitude of the seal's displacement?

8. **GO** In a mall, a shopper rides up an escalator between floors. At the top of the escalator, the shopper turns right and walks 9.00 m to a store. The magnitude of the shopper's displacement from the bottom of the escalator to the store is 16.0 m. The vertical distance between the floors is 6.00 m. At what angle is the escalator inclined above the horizontal?

9. **ssm www** A skateboarder, starting from rest, rolls down a 12.0-m ramp. When she arrives at the bottom of the ramp her speed is 7.70 m/s. **(a)** Determine the magnitude of her acceleration, assumed to be constant. **(b)** If the ramp is inclined at 25.0° with respect to the ground, what is the component of her acceleration that is parallel to the ground?

*10. **Interactive Solution 3.10** at **www.wiley.com/college/cutnell** presents a model for solving this problem. The earth moves around the sun in a nearly circular orbit of radius 1.50×10^{11} m. During the three summer months (an elapsed time of 7.89×10^{6} s), the earth moves one-fourth of the distance around the sun. **(a)** What is the average speed of the earth? **(b)** What is the magnitude of the average velocity of the earth during this period?

*11. **Interactive LearningWare 3.1** at **www.wiley.com/college/cutnell** reviews the approach taken in problems such as this one. A bird

watcher meanders through the woods, walking 0.50 km due east, 0.75 km due south, and 2.15 km in a direction 35.0° north of west. The time required for this trip is 2.50 h. Determine the magnitude and direction (relative to due west) of the bird watcher's **(a)** displacement and **(b)** average velocity. Use kilometers and hours for distance and time, respectively.

**Section 3.2   Equations of Kinematics in Two Dimensions,
Section 3.3   Projectile Motion**

12. The punter on a football team tries to kick a football so that it stays in the air for a long "hang time." If the ball is kicked with an initial velocity of 25.0 m/s at an angle of 60.0° above the ground, what is the hang time?

13. **ssm** Suppose that the plane in Example 3 is traveling with twice the horizontal velocity—that is, with a velocity of +230 m/s. If all other factors remain the same, determine the time required for the package to hit the ground.

14. **GO** A puck is moving on an air hockey table. Relative to an x, y coordinate system at time $t = 0$ s, the x components of the puck's initial velocity and acceleration are $v_{0x} = +1.0$ m/s and $a_x = +2.0$ m/s². The y components of the puck's initial velocity and acceleration are $v_{0y} = +2.0$ m/s and $a_y = -2.0$ m/s². Find the magnitude and direction of the puck's velocity at a time of $t = 0.50$ s. Specify the direction relative to the $+x$ axis.

15. A dolphin leaps out of the water at an angle of 35° above the horizontal. The horizontal component of the dolphin's velocity is 7.7 m/s. Find the magnitude of the vertical component of the velocity.

16. A skateboarder shoots off a ramp with a velocity of 6.6 m/s, directed at an angle of 58° above the horizontal. The end of the ramp is 1.2 m above the ground. Let the x axis be parallel to the ground, the $+y$ direction be vertically upward, and take as the origin the point on the ground directly below the top of the ramp. **(a)** How high above the ground is the highest point that the skateboarder reaches? **(b)** When the skateboarder reaches the highest point, how far is this point horizontally from the end of the ramp?

17. **ssm** A hot-air balloon is rising straight up with a speed of 3.0 m/s. A ballast bag is released from rest relative to the balloon when it is 9.5 m above the ground. How much time elapses before the ballast bag hits the ground?

18. **Concept Simulation 3.2** at **www.wiley.com/college/cutnell** reviews the concepts that are important in this problem. A golfer imparts a speed of 30.3 m/s to a ball, and it travels the maximum possible distance before landing on the green. The tee and the green are at the same elevation. **(a)** How much time does the ball spend in the air? **(b)** What is the longest hole in one that the golfer can make, if the ball does not roll when it hits the green?

19. Michael Jordan, formerly of the Chicago Bulls basketball team, had some fanatic fans. They claimed that he was able to jump and remain in the air for two full seconds from launch to landing. Evaluate this claim by calculating the maximum height that such a jump would attain. For comparison, Jordan's maximum jump height has been estimated at about one meter.

20. **GO** A golfer hits a shot to a green that is elevated 3.0 m above the point where the ball is struck. The ball leaves the club at a speed of 14.0 m/s at an angle of 40.0° above the horizontal. It rises to its maximum height and then falls down to the green. Ignoring air resistance, find the speed of the ball just before it lands.

21. **ssm** A golf ball rolls off a horizontal cliff with an initial speed of 11.4 m/s. The ball falls a vertical distance of 15.5 m into a lake below. **(a)** How much time does the ball spend in the air? **(b)** What is the speed v of the ball just before it strikes the water?

22. A space vehicle is coasting at a constant velocity of 21.0 m/s in the $+y$ direction relative to a space station. The pilot of the vehicle fires a RCS (reaction control system) thruster, which causes it to accelerate at 0.320 m/s^2 in the $+x$ direction. After 45.0 s, the pilot shuts off the RCS thruster. After the RCS thruster is turned off, find **(a)** the magnitude and **(b)** the direction of the vehicle's velocity relative to the space station. Express the direction as an angle measured from the $+y$ direction.

23. ssm As preparation for this problem, review Conceptual Example 10. The drawing shows two planes each about to drop an empty fuel tank. At the moment of release each plane has the same speed of 135 m/s, and each tank is at the same height of 2.00 km above the ground. Although the speeds are the same, the velocities are different at the instant of release, because one plane is flying at an angle of 15.0° above the horizontal and the other is flying at an angle of 15.0° below the horizontal. Find the magnitude and direction of the velocity with which the fuel tank hits the ground if it is from **(a)** plane A and **(b)** plane B. In each part, give the directional angles with respect to the horizontal.

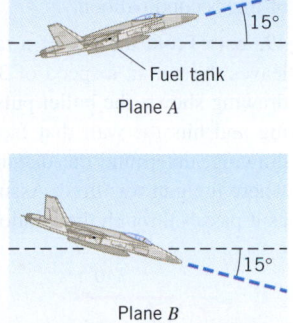

Plane A

Plane B

24. **GO** A criminal is escaping across a rooftop and runs off the roof horizontally at a speed of 5.3 m/s, hoping to land on the roof of an adjacent building. Air resistance is negligible. The horizontal distance between the two buildings is D, and the roof of the adjacent building is 2.0 m below the jumping-off point. Find the maximum value for D.

25. Useful background for this problem can be found in Multiple-Concept Example 2. On a spacecraft two engines fire for a time of 565 s. One gives the craft an acceleration in the x direction of $a_x =$ 5.10 m/s^2, while the other produces an acceleration in the y direction of $a_y = 7.30$ m/s^2. At the end of the firing period, the craft has velocity components of $v_x = 3775$ m/s and $v_y = 4816$ m/s. Find the magnitude and direction of the initial velocity. Express the direction as an angle with respect to the $+x$ axis.

26. Interactive LearningWare 3.2 at **www.wiley.com/college/cutnell** provides a review of the concepts in this problem. On a distant planet, golf is just as popular as it is on earth. A golfer tees off and drives the ball 3.5 times as far as he would have on earth, given the same initial velocities on both planets. The ball is launched at a speed of 45 m/s at an angle of 29° above the horizontal. When the ball lands, it is at the same level as the tee. On the distant planet, what are **(a)** the maximum height and **(b)** the range of the ball?

27. ssm A fire hose ejects a stream of water at an angle of 35.0° above the horizontal. The water leaves the nozzle with a speed of 25.0 m/s. Assuming that the water behaves like a projectile, how far from a building should the fire hose be located to hit the highest possible fire?

28. **GO** A ball is thrown upward at a speed v_0 at an angle of 52° above the horizontal. It reaches a maximum height of 7.5 m. How high would this ball go if it were thrown straight upward at speed v_0?

29. A major-league pitcher can throw a baseball in excess of 41.0 m/s. If a ball is thrown horizontally at this speed, how much will it drop by the time it reaches a catcher who is 17.0 m away from the point of release?

30. A quarterback claims that he can throw the football a horizontal distance of 183 m (200 yd). Furthermore, he claims that he can do this by launching the ball at the relatively low angle of 30.0° above the horizontal. To evaluate this claim, determine the speed with which this quarterback must throw the ball. Assume that the ball is launched and caught at the same vertical level and that air resistance can be ignored. For comparison, a baseball pitcher who can accurately throw a fastball at 45 m/s (100 mph) would be considered exceptional.

31. ssm www Multiple-Concept Example 4 provides useful background for this problem. A diver runs horizontally with a speed of 1.20 m/s off a platform that is 10.0 m above the water. What is his speed just before striking the water?

32. The perspective provided by Multiple-Concept Example 9 is useful here. The highest barrier that a projectile can clear is 13.5 m, when the projectile is launched at an angle of 15.0° above the horizontal. What is the projectile's launch speed?

33. Consult Multiple-Concept Example 4 for background before beginning this problem. Suppose the water at the top of Niagara Falls has a horizontal speed of 2.7 m/s just before it cascades over the edge of the falls. At what vertical distance below the edge does the velocity vector of the water point downward at a 75° angle below the horizontal?

34. **GO** In the absence of air resistance, a projectile is launched from and returns to ground level. It follows a trajectory similar to that in Figure 3.11 and has a range of 23 m. Suppose the launch speed is doubled, and the projectile is fired at the same angle above the ground. What is the new range?

35. A rocket is fired at a speed of 75.0 m/s from ground level, at an angle of 60.0° above the horizontal. The rocket is fired toward an 11.0-m-high wall, which is located 27.0 m away. The rocket attains its launch speed in a negligibly short period of time, after which its engines shut down and the rocket coasts. By how much does the rocket clear the top of the wall?

36. A rifle is used to shoot twice at a target, using identical cartridges. The first time, the rifle is aimed parallel to the ground and directly at the center of the bull's-eye. The bullet strikes the target at a distance of H_A below the center, however. The second time, the rifle is similarly aimed, but from twice the distance from the target. This time the bullet strikes the target at a distance of H_B below the center. Find the ratio H_B/H_A.

* **37. ssm** An airplane with a speed of 97.5 m/s is climbing upward at an angle of 50.0° with respect to the horizontal. When the plane's altitude is 732 m, the pilot releases a package. **(a)** Calculate the distance along the ground, measured from a point directly beneath the point of release, to where the package hits the earth. **(b)** Relative to the ground, determine the angle of the velocity vector of the package just before impact.

* **38.** A child operating a radio-controlled model car on a dock accidentally steers it off the edge. The car's displacement 1.1 s after leaving the dock has a magnitude of 7.0 m. What is the car's speed at the instant it drives off the edge of the dock?

* **39.** As an aid in working this problem, consult **Interactive Solution 3.39** at **www.wiley.com/college/cutnell**. A soccer player kicks the ball toward a goal that is 16.8 m in front of him. The ball leaves his foot at a speed of 16.0 m/s and an angle of 28.0° above the ground. Find the speed of the ball when the goalie catches it in front of the net.

* **40.** See Multiple-Concept Example 9 for the basic idea behind problems such as this. A diver springs upward from a board that is three meters above the water. At the instant she contacts the water her speed is 8.90 m/s and her body makes an angle of 75.0° with respect to the horizontal surface of the water. Determine her initial velocity, both magnitude and direction.

* **41.** Review Conceptual Example 5 and **Concept Simulation 3.1** at www.wiley.com/college/cutnell before beginning this problem. You are traveling in a convertible with the top down. The car is moving at a constant velocity of 25 m/s, due east along flat ground. You throw a tomato straight upward at a speed of 11 m/s. How far has the car moved when you get a chance to catch the tomato?

* **42.** In the javelin throw at a track-and-field event, the javelin is launched at a speed of 29 m/s at an angle of 36° above the horizontal. As the javelin travels upward, its velocity points above the horizontal at an angle that decreases as time passes. How much time is required for the angle to be reduced from 36° at launch to 18°?

* **43.** **ssm www** As preparation for this problem, review Conceptual Example 10. The two stones described there have identical initial speeds of $v_0 = 13.0$ m/s and are thrown at an angle $\theta = 30.0°$, one below the horizontal and one above the horizontal. What is the distance *between* the points where the stones strike the ground?

* **44.** Multiple-Concept Example 4 deals with a situation similar to that presented here. A marble is thrown horizontally with a speed of 15 m/s from the top of a building. When it strikes the ground, the marble has a velocity that makes an angle of 65° with the horizontal. From what height above the ground was the marble thrown?

* **45.** The lob in tennis is an effective tactic when your opponent is near the net. It consists of lofting the ball over his head, forcing him to move quickly away from the net (see the drawing). Suppose that you lob the ball with an initial speed of 15.0 m/s, at an angle of 50.0° above the horizontal. At this instant your opponent is 10.0 m away from the ball. He begins moving away from you 0.30 s later, hoping to reach the ball and hit it back at the moment that it is 2.10 m above its launch point. With what minimum average speed must he move? (Ignore the fact that he can stretch, so that his racket can reach the ball before he does.)

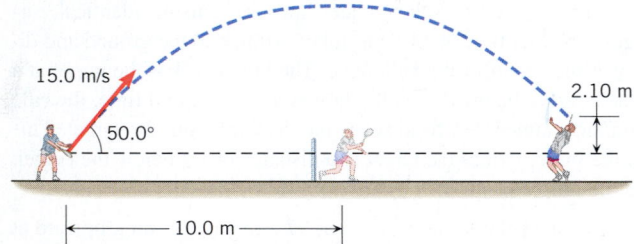

** **46.** **Concept Simulation 3.2** at **www.wiley.com/college/cutnell** reviews principles in this problem. A projectile is launched from ground level at an angle of 12.0° above the horizontal. It returns to ground level. To what value should the launch angle be adjusted, without changing the launch speed, so that the range doubles?

** **47.** **ssm** The drawing shows an exaggerated view of a rifle that has been "sighted in" for a 91.4-meter target. If the muzzle speed of the bullet is $v_0 = 427$ m/s, what are the two possible angles θ_1 and θ_2 between the rifle barrel and the horizontal such that the bullet will hit the target? One of these angles is so large that it is never used in target shooting. (*Hint: The following trigonometric identity may be useful:* $2 \sin \theta \cos \theta = \sin 2\theta$.)

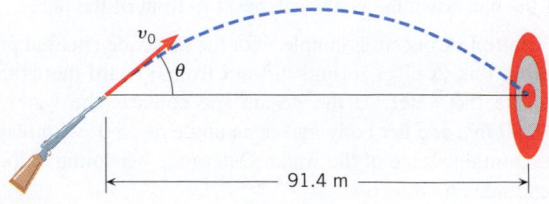

** **48.** In the annual battle of the dorms, students gather on the roofs of Jackson and Walton dorms to launch water balloons at each other with slingshots. The horizontal distance between the buildings is 35.0 m, and the heights of the Jackson and Walton buildings are, respectively, 15.0 m and 22.0 m. Ignore air resistance. **(a)** The first balloon launched by the Jackson team hits Walton dorm 2.0 s after launch, striking it halfway between the ground and the roof. Find the direction of the balloon's initial velocity. Give your answer as an angle measured above the horizontal. **(b)** A second balloon launched at the same angle hits the edge of Walton's roof. Find the initial speed of this second balloon.

** **49.** **ssm** From the top of a tall building, a gun is fired. The bullet leaves the gun at a speed of 340 m/s, parallel to the ground. As the drawing shows, the bullet puts a hole in a window of another building and hits the wall that faces the window. Using the data in the drawing, determine the distances *D* and *H*, which locate the point where the gun was fired. Assume that the bullet does not slow down as it passes through the window.

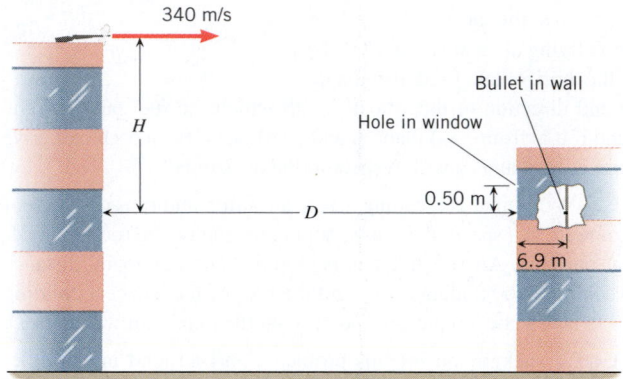

** **50.** A small can is hanging from the ceiling. A rifle is aimed directly at the can, as the figure illustrates. At the instant the gun is fired, the can is released. Ignore air resistance and show that the bullet will always strike the can, regardless of the initial speed of the bullet. Assume that the bullet strikes the can before the can reaches the ground.

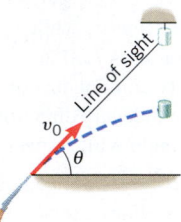

** **51.** A placekicker is about to kick a field goal. The ball is 26.9 m from the goalpost. The ball is kicked with an initial velocity of 19.8 m/s at an angle θ above the ground. Between what two angles, θ_1 and θ_2, will the ball clear the 2.74-m-high crossbar? (*Hint: The following trigonometric identities may be useful:* $\sec \theta = 1/(\cos \theta)$ *and* $\sec^2 \theta = 1 + \tan^2 \theta$.)

Section 3.4 Relative Velocity

52. Two cars, A and B, are traveling in the same direction, although car A is 186 m behind car B. The speed of A is 24.4 m/s, and the speed of B is 18.6 m/s. How much time does it take for A to catch B?

53. **ssm** A swimmer, capable of swimming at a speed of 1.4 m/s in still water (i.e., the swimmer can swim with a speed of 1.4 m/s relative to the water), starts to swim directly across a 2.8-km-wide river. However, the current is 0.91 m/s, and it carries the swimmer downstream. **(a)** How long does it take the swimmer to cross the river? **(b)** How far downstream will the swimmer be upon reaching the other side of the river?

54. **GO** Two friends, Barbara and Neil, are out rollerblading. With respect to the ground, Barbara is skating due south at a speed of 4.0 m/s. Neil is in front of her. With respect to the ground, Neil is skating due west at a speed of 3.2 m/s. Find Neil's velocity (magnitude and direction relative to due west), as seen by Barbara.

55. The escalator that leads down into a subway station has a length of 30.0 m and a speed of 1.8 m/s relative to the ground. A student is coming out of the station by running in the wrong direction on this escalator. The local record time for this trick is 11 s. Relative to the escalator, what speed must the student exceed in order to beat the record?

56. A police officer is driving due north at a constant speed of 29 m/s relative to the ground when she notices a truck on an east–west highway ahead of her, driving west at high speed. She finds that the truck's speed relative to her car is 48 m/s (about 110 mph). **(a)** Sketch the vector triangle that shows how the truck's velocity relative to the ground is related to the police car's velocity relative to the ground and to the truck's velocity relative to the police car. The sketch need not be to scale, but the velocity vectors should be oriented correctly and bear the appropriate labels. **(b)** What is the truck's speed, relative to the ground?

57. **ssm** Two passenger trains are passing each other on adjacent tracks. Train A is moving east with a speed of 13 m/s, and train B is traveling west with a speed of 28 m/s. **(a)** What is the velocity (magnitude and direction) of train A as seen by the passengers in train B? **(b)** What is the velocity (magnitude and direction) of train B as seen by the passengers in train A?

58. On a pleasure cruise a boat is traveling relative to the water at a speed of 5.0 m/s due south. Relative to the boat, a passenger walks toward the back of the boat at a speed of 1.5 m/s. **(a)** What is the magnitude and direction of the passenger's velocity relative to the water? **(b)** How long does it take for the passenger to walk a distance of 27 m on the boat? **(c)** How long does it take for the passenger to cover a distance of 27 m on the water?

59. You are in a hot-air balloon that, relative to the ground, has a velocity of 6.0 m/s in a direction due east. You see a hawk moving directly away from the balloon in a direction due north. The speed of the hawk relative to you is 2.0 m/s. What are the magnitude and direction of the hawk's velocity relative to the ground? Express the directional angle relative to due east.

60. **GO** The captain of a plane wishes to proceed due west. The cruising speed of the plane is 245 m/s relative to the air. A weather report indicates that a 38.0-m/s wind is blowing from the south to the

north. In what direction, measured with respect to due west, should the pilot head the plane?

* **61.** A person looking out the window of a stationary train notices that raindrops are falling vertically down at a speed of 5.0 m/s relative to the ground. When the train moves at a constant velocity, the raindrops make an angle of 25° when they move past the window, as the drawing shows. How fast is the train moving?

* **62.** **GO** Relative to the ground, a car has a velocity of 16.0 m/s, directed due north. Relative to this car, a truck has a velocity of 24.0 m/s, directed 52.0° north of east. What is the magnitude of the truck's velocity relative to the ground?

* **63.** **ssm** Mario, a hockey player, is skating due south at a speed of 7.0 m/s relative to the ice. A teammate passes the puck to him. The puck has a speed of 11.0 m/s and is moving in a direction of 22° west of south, relative to the ice. What are the magnitude and direction (relative to due south) of the puck's velocity, as observed by Mario?

** **64.** A jetliner can fly 6.00 hours on a full load of fuel. Without any wind it flies at a speed of 2.40×10^2 m/s. The plane is to make a round-trip by heading due west for a certain distance, turning around, and then heading due east for the return trip. During the entire flight, however, the plane encounters a 57.8-m/s wind from the jet stream, which blows from west to east. What is the maximum distance that the plane can travel due west and just be able to return home?

** **65.** **ssm** **www** A Coast Guard ship is traveling at a constant velocity of 4.20 m/s, due east, relative to the water. On his radar screen the navigator detects an object that is moving at a constant velocity. The object is located at a distance of 2310 m with respect to the ship, in a direction 32.0° south of east. Six minutes later, he notes that the object's position relative to the ship has changed to 1120 m, 57.0° south of west. What are the magnitude and direction of the velocity of the object relative to the water? Express the direction as an angle with respect to due west.

ADDITIONAL PROBLEMS

66. On a spacecraft, two engines are turned on for 684 s at a moment when the velocity of the craft has x and y components of $v_{0x} = 4370$ m/s and $v_{0y} = 6280$ m/s. While the engines are firing, the craft undergoes a displacement that has components of $x = 4.11 \times 10^6$ m and $y = 6.07 \times 10^6$ m. Find the x and y components of the craft's acceleration.

67. A volleyball is spiked so that it has an initial velocity of 15 m/s directed downward at an angle of 55° below the horizontal. What is the horizontal component of the ball's velocity when the opposing player fields the ball?

68. A spider crawling across a table leaps onto a magazine blocking its path. The initial velocity of the spider is 0.870 m/s at an angle of 35.0° above the table, and it lands on the magazine 0.0770 s after leaving the table. Ignore air resistance. How thick is the magazine? Express your answer in millimeters.

69. Review **Interactive LearningWare 3.2** at **www.wiley.com/college/cutnell** in preparation for this problem. The acceleration due to gravity on the moon has a magnitude of 1.62 m/s². Examples 6–8 deal

with a placekicker kicking a football. Assume that the ball is kicked on the moon instead of on the earth. Find **(a)** the maximum height H and **(b)** the range that the ball would attain on the moon.

70. The 1994 Winter Olympics included the aerials competition in skiing. In this event skiers speed down a ramp that slopes sharply upward at the end. The sharp upward slope launches them into the air, where they perform acrobatic maneuvers. In the women's competition, the end of a typical launch ramp is directed 63° above the horizontal. With this launch angle, a skier attains a height of 13 m above the end of the ramp. What is the skier's launch speed?

71. **ssm** An eagle is flying horizontally at 6.0 m/s with a fish in its claws. It accidentally drops the fish. **(a)** How much time passes before the fish's speed doubles? **(b)** How much additional time would be required for the fish's speed to double again?

72. Baseball player A bunts the ball by hitting it in such a way that it acquires an initial velocity of 1.9 m/s parallel to the ground. Upon contact with the bat the ball is 1.2 m above the ground. Player B wishes to duplicate this bunt, in so far as he also wants to give the ball a velocity parallel to the ground and have his ball travel the same horizontal distance as player A's ball does. However, player B hits the ball when it is 1.5 m above the ground. What is the magnitude of the initial velocity that player B's ball must be given?

* **73.** Before starting this problem consult **Interactive Solution 3.73** at **www.wiley.com/college/cutnell**. A golfer, standing on a fairway, hits a shot to a green that is elevated 5.50 m above the point where she is standing. If the ball leaves her club with a velocity of 46.0 m/s at an angle of 35.0° above the ground, find the time that the ball is in the air before it hits the green.

* **74.** In a stunt being filmed for a movie, a sports car overtakes a truck towing a ramp, drives up and off the ramp, soars into the air, and then lands on top of a flat trailer being towed by a second truck. The tops of the ramp and the flat trailer are the same height above the road, and the ramp is inclined 16° above the horizontal. Both trucks are driving at a constant speed of 11 m/s, and the flat trailer is 15 m from the end of the ramp. Neglect air resistance, and assume that the ramp changes the direction, but not the magnitude, of the car's initial velocity. What is the minimum speed the car must have, *relative to the road*, as it starts up the ramp?

* **75. ssm** An airplane is flying with a velocity of 240 m/s at an angle of 30.0° with the horizontal, as the drawing shows. When the altitude of the plane is 2.4 km, a flare is released from the plane. The flare hits the target on the ground. What is the angle θ?

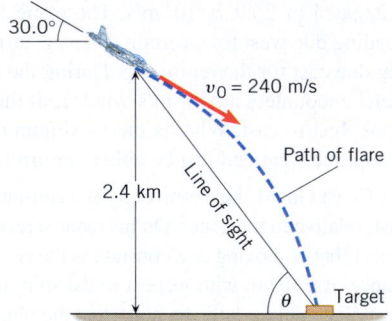

* **76.** As an aid in working this problem, consult **Interactive Solution 3.76** at **www.wiley.com/college/cutnell**. A ferryboat is traveling in a

direction 38.0° north of east with a speed of 5.50 m/s relative to the water. A passenger is walking with a velocity of 2.50 m/s due east relative to the boat. What is the velocity (magnitude and direction) of the passenger with respect to the water? Determine the directional angle relative to due east.

* **77.** First review **Interactive Solution 3.77** at **www.wiley.com/college/ cutnell**. After leaving the end of a ski ramp, a ski jumper lands downhill at a point that is displaced 51.0 m horizontally from the end of the ramp. His velocity, just before landing, is 23.0 m/s and points in a direction 43.0° below the horizontal. Neglecting air resistance and any lift he experiences while airborne, find his initial velocity (magnitude and direction) when he left the end of the ramp. Express the direction as an angle relative to the horizontal.

** **78.** Two boats are heading away from shore. Boat 1 heads due north at a speed of 3.00 m/s relative to the shore. Relative to boat 1, boat 2 is moving 30.0° north of east at a speed of 1.60 m/s. A passenger on boat 2 walks due east across the deck at a speed of 1.20 m/s relative to boat 2. What is the speed of the passenger relative to the shore?

** **79.** Two cannons are mounted as shown in the drawing and rigged to fire simultaneously. They are used in a circus act in which two clowns serve as human cannonballs. The clowns are fired toward each other and collide at a height of 1.00 m above the muzzles of the cannons. Clown A is launched at a 75.0° angle, with a speed of 9.00 m/s. The horizontal separation between the clowns as they leave the cannons is 6.00 m. Find the launch speed v_{0B} and the launch angle θ_B ($>45.0°$) for clown B.

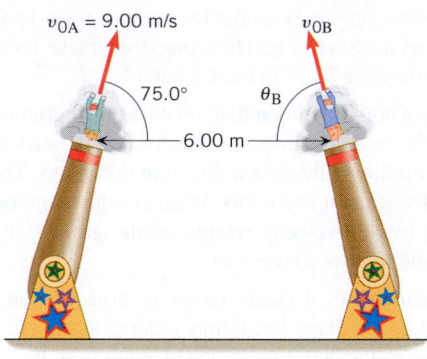

FORCES AND NEWTON'S LAWS OF MOTION

As this downhill skier plows through the snow, her motion is determined by forces due to gravity, friction, and air resistance. This chapter discusses the relationship between the forces acting on an object and the resulting motion. (© John Kelly/The Image Bank/Getty Images)

4.1 THE CONCEPTS OF FORCE AND MASS

In common usage, a *force* is a push or a pull, as the examples in Figure 4.1 illustrate. In basketball, a player launches a shot by pushing on the ball. The tow bar attached to a speeding boat pulls a water skier. Forces such as those that launch the basketball or pull the skier are called *contact forces,* because they arise from the physical contact between two objects. There are circumstances, however, in which two objects exert forces on one another even though they are not touching. Such forces are referred to as *noncontact forces* or *action-at-a-distance forces.* One example of such a noncontact force occurs when a diver is pulled toward the earth because of the force of gravity. The earth exerts this force even when it is not in direct contact with the diver. In Figure 4.1, arrows are used to represent the forces. It is appropriate to use arrows, because a force is a vector quantity and has both a magnitude and a direction. The direction of the arrow gives the direction of the force, and the length is proportional to its strength or magnitude.

The word *mass* is just as familiar as the word *force.* A massive supertanker, for instance, is one that contains an enormous amount of mass. As we will see in the next section, it is difficult to set such a massive object into motion and difficult to bring it to a halt once it is moving. In comparison, a penny does not contain much mass. The emphasis here is on the amount of mass, and the idea of direction is of no concern. Therefore, mass is a scalar quantity.

During the seventeenth century, Isaac Newton, building on the work of Galileo, developed three important laws that deal with force and mass. Collectively they are called "Newton's laws of motion" and provide the basis for understanding the effect that forces have on an object. Because of the importance of these laws, a separate section will be devoted to each one.

(a)

(b)

(c)

Figure 4.1 The arrow labeled $\vec{F}$ represents the force that acts on (*a*) the basketball, (*b*) the water skier, and (*c*) the cliff diver. (*a*, © Nathaniel S. Butler/ NBAE/Getty Images; *b*, © P. Beavis/Corbis; *c*, © Amy and Chuck Wiley/Wales/Index Stock)

4.2 NEWTON'S FIRST LAW OF MOTION

THE FIRST LAW

To gain some insight into Newton's first law, think about the game of ice hockey (Figure 4.2). If a player does not hit a stationary puck, it will remain at rest on the ice. After the puck is struck, however, it coasts on its own across the ice, slowing down only slightly because of friction. Since ice is very slippery, there is only a relatively small amount of friction to slow down the puck. In fact, if it were possible to remove all friction and wind resistance, and if the rink were infinitely large, the puck would coast forever in a straight line at a constant speed. Left on its own, the puck would lose none of the velocity imparted to it at the time it was struck. This is the essence of Newton's first law of motion:

> **NEWTON'S FIRST LAW OF MOTION**
>
> An object continues in a state of rest or in a state of motion at a constant velocity (constant speed in a constant direction), unless compelled to change that state by a net force.

In the first law the phrase "net force" is crucial. Often, several forces act simultaneously on a body, and *the net force is the vector sum of all of them.* Individual forces matter only to the extent that they contribute to the total. For instance, if friction and other opposing forces were absent, a car could travel forever at 30 m/s in a straight line, without using any gas after it had come up to speed. In reality gas is needed, but only so that the engine can produce the necessary force to cancel opposing forces such as friction. This cancellation ensures that there is no net force to change the state of motion of the car.

When an object moves at a constant speed in a constant direction, its velocity is constant. Newton's first law indicates that a state of rest (zero velocity) and a state of constant velocity are completely equivalent, in the sense that neither one requires the application of a net force to sustain it. *The purpose served when a net force acts on an object is not to sustain the object's velocity, but, rather, to change it.*

INERTIA AND MASS

A greater net force is required to change the velocity of some objects than of others. For instance, a net force that is just enough to cause a bicycle to pick up speed will cause only an imperceptible change in the motion of a freight train. In comparison to the bicycle, the train has a much greater tendency to remain at rest. Accordingly, we say that the train has more *inertia* than the bicycle. Quantitatively, the inertia of an object is measured by its *mass.* The following definition of inertia and mass indicates why Newton's first law is sometimes called the law of inertia:

Figure 4.2 The game of ice hockey can give some insight into Newton's laws of motion. (© Corbis)

DEFINITION OF INERTIA AND MASS

Inertia is the natural tendency of an object to remain at rest or in motion at a constant velocity. The mass of an object is a quantitative measure of inertia.

SI Unit of Inertia and Mass: kilogram (kg)

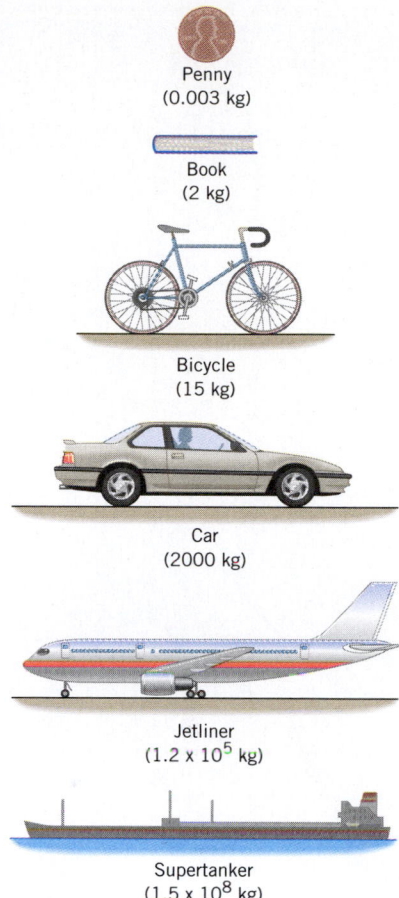

Penny
(0.003 kg)

Book
(2 kg)

Bicycle
(15 kg)

Car
(2000 kg)

Jetliner
$(1.2 \times 10^5 \text{ kg})$

Supertanker
$(1.5 \times 10^8 \text{ kg})$

Figure 4.3 The masses of various objects.

The SI unit for mass is the kilogram (kg), whereas the units in the CGS system and the BE system are the gram (g) and the slug (sl), respectively. Conversion factors between these units are given on the page facing the inside of the front cover. Figure 4.3 gives the masses of various objects, ranging from a penny to a supertanker. The larger the mass, the greater is the inertia. Often the words "mass" and "weight" are used interchangeably, but this is incorrect. Mass and weight are different concepts, and Section 4.7 will discuss the distinction between them.

The physics of seat belts. Figure 4.4 shows a useful application of inertia. Automobile seat belts unwind freely when pulled gently, so they can be buckled. But in an accident, they hold you safely in place. One seat-belt mechanism consists of a ratchet wheel, a locking bar, and a pendulum. The belt is wound around a spool mounted on the ratchet wheel. While the car is at rest or moving at a constant velocity, the pendulum hangs straight down, and the locking bar rests horizontally, as the gray part of the drawing shows. Consequently, nothing prevents the ratchet wheel from turning, and the seat belt can be pulled out easily. When the car suddenly slows down in an accident, however, the relatively massive lower part of the pendulum keeps moving forward because of its inertia. The pendulum swings on its pivot into the position shown in color and causes the locking bar to block the rotation of the ratchet wheel, thus preventing the seat belt from unwinding.

AN INERTIAL REFERENCE FRAME

Newton's first law (and also the second law) can appear to be invalid to certain observers. Suppose, for instance, that you are a passenger riding in a friend's car. While the car moves at a constant speed along a straight line, you do not feel the seat pushing against your back to any unusual extent. This experience is consistent with the first law, which indicates that in the absence of a net force you should move with a constant velocity. Suddenly the driver floors the gas pedal. Immediately you feel the seat pressing against your back as the car accelerates. Therefore, you sense that a force is being applied to you. The first law leads you to believe that your motion should change, and, relative to the ground outside, your motion does change. But *relative to the car,* you can see that your motion does *not* change, because you remain stationary with respect to the car. Clearly, Newton's first law does not hold for observers who use the accelerating car as a frame of reference. As a result, such a reference frame is said to be *noninertial*. All accelerating reference frames are noninertial. In contrast, observers for whom the law of inertia is valid are said to be using *inertial reference frames* for their observations, as defined below:

DEFINITION OF AN INERTIAL REFERENCE FRAME

An inertial reference frame is one in which Newton's law of inertia is valid.

The acceleration of an inertial reference frame is zero, so it moves with a constant velocity. All of Newton's laws of motion are valid in inertial reference frames, and when we apply these laws, we will be assuming such a reference frame. In particular, the earth itself is a good approximation of an inertial reference frame.

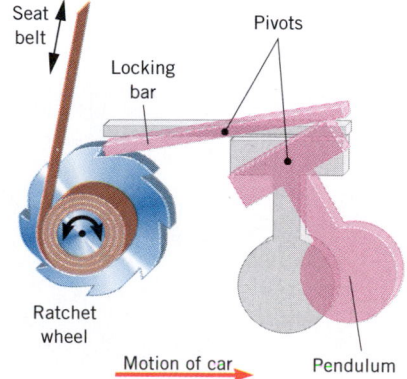

Seat belt

Pivots

Locking bar

Ratchet wheel

Motion of car

Pendulum

Figure 4.4 Inertia plays a central role in one seat-belt mechanism. The gray part of the drawing applies when the car is at rest or moving at a constant velocity. The colored parts show what happens when the car suddenly slows down, as in an accident.

✓ CHECK YOUR UNDERSTANDING

(The answer is given at the end of the book.)

1. Which of the following statements can be explained by Newton's first law? (A); When your car suddenly comes to a halt, you lunge forward. (B); When your car rapidly accelerates, you are pressed backward against the seat.   **(a)** Neither A nor B   **(b)** Both A and B   **(c)** A but not B   **(d)** B but not A

4.3 NEWTON'S SECOND LAW OF MOTION

Newton's first law indicates that if no net force acts on an object, then the velocity of the object remains unchanged. The second law deals with what happens when a net force does act. Consider a hockey puck once again. When a player strikes a stationary puck, he causes the velocity of the puck to change. In other words, he makes the puck accelerate. The cause of the acceleration is the force that the hockey stick applies. As long as this force acts, the velocity increases, and the puck accelerates. Now, suppose another player strikes the puck and applies twice as much force as the first player does. The greater force produces a greater acceleration. In fact, if the friction between the puck and the ice is negligible, and if there is no wind resistance, the acceleration of the puck is directly proportional to the force. Twice the force produces twice the acceleration. Moreover, the acceleration is a vector quantity, just as the force is, and points in the same direction as the force.

Often, several forces act on an object simultaneously. Friction and wind resistance, for instance, do have some effect on a hockey puck. In such cases, it is the net force, or the vector sum of all the forces acting, that is important. Mathematically, the net force is written as $\Sigma\vec{\mathbf{F}}$, where the Greek capital letter Σ (sigma) denotes the vector sum. Newton's second law states that the acceleration is proportional to the net force acting on the object.

In Newton's second law, the net force is only one of two factors that determine the acceleration. The other is the inertia or mass of the object. After all, the same net force that imparts an appreciable acceleration to a hockey puck (small mass) will impart very little acceleration to a semitrailer truck (large mass). Newton's second law states that for a given net force, the magnitude of the acceleration is inversely proportional to the mass. Twice the mass means one-half the acceleration, if the same net force acts on both objects. Thus, the second law shows how the acceleration depends on both the net force and the mass, as given in Equation 4.1.

NEWTON'S SECOND LAW OF MOTION

When a net external force $\Sigma\vec{\mathbf{F}}$ acts on an object of mass m, the acceleration $\vec{\mathbf{a}}$ that results is directly proportional to the net force and has a magnitude that is inversely proportional to the mass. The direction of the acceleration is the same as the direction of the net force.

$$\vec{\mathbf{a}} = \frac{\Sigma\vec{\mathbf{F}}}{m} \quad \text{or} \quad \Sigma\vec{\mathbf{F}} = m\vec{\mathbf{a}} \tag{4.1}$$

SI Unit of Force: $\text{kg} \cdot \text{m/s}^2 = \text{newton (N)}$

Note that the net force in Equation 4.1 includes only the forces that the environment exerts on the object of interest. Such forces are called *external forces.* In contrast, *internal forces* are forces that one part of an object exerts on another part of the object and are not included in Equation 4.1.

According to Equation 4.1, the SI unit for force is the unit for mass (kg) times the unit for acceleration (m/s^2), or

$$\text{SI unit for force} = (\text{kg})\left(\frac{\text{m}}{\text{s}^2}\right) = \frac{\text{kg} \cdot \text{m}}{\text{s}^2}$$

The combination of $\text{kg} \cdot \text{m/s}^2$ is called a *newton* (N) and is a derived SI unit, not a base unit; 1 newton = 1 N = 1 $\text{kg} \cdot \text{m/s}^2$.

In the CGS system, the procedure for establishing the unit of force is the same as with SI units, except that mass is expressed in grams (g) and acceleration in cm/s^2. The resulting unit for force is the *dyne;* 1 dyne = 1 $\text{g} \cdot \text{cm/s}^2$.

Table 4.1 Units for Mass, Acceleration, and Force

System	Mass	Acceleration	Force
SI	kilogram (kg)	meter/second^2 (m/s^2)	newton (N)
CGS	gram (g)	centimeter/second^2 (cm/s^2)	dyne (dyn)
BE	slug (sl)	foot/second^2 (ft/s^2)	pound (lb)

In the BE system, the unit for force is defined to be the pound (lb),* and the unit for acceleration is ft/s^2. With this procedure, Newton's second law can then be used to obtain the BE unit for mass:

$$\text{BE unit for force} = \text{lb} = (\text{unit for mass})\left(\frac{\text{ft}}{\text{s}^2}\right)$$

$$\text{Unit for mass} = \frac{\text{lb} \cdot \text{s}^2}{\text{ft}}$$

The combination of lb·s^2/ft is the unit for mass in the BE system and is called the *slug* (sl); 1 slug = 1 sl = 1 lb·s^2/ft.

Table 4.1 summarizes the various units for mass, acceleration, and force. Conversion factors between force units from different systems are provided on the page facing the inside of the front cover.

When using the second law to calculate the acceleration, it is necessary to determine the net force that acts on the object. In this determination a *free-body diagram* helps enormously. A free-body diagram is a diagram that represents the object and the forces that act on it. Only the forces that *act on the object* appear in a free-body diagram. Forces that the object exerts on its environment are not included. Example 1 illustrates the use of a free-body diagram.

 Example 1   Pushing a Stalled Car

Two people are pushing a stalled car, as Figure 4.5a indicates. The mass of the car is 1850 kg. One person applies a force of 275 N to the car, while the other applies a force of 395 N. Both forces act in the same direction. A third force of 560 N also acts on the car, but in a direction opposite to that in which the people are pushing. This force arises because of friction and the extent to which the pavement opposes the motion of the tires. Find the acceleration of the car.

Reasoning   According to Newton's second law, the acceleration is the net force divided by the mass of the car. To determine the net force, we use the free-body diagram in Figure 4.5b. In this diagram, the car is represented as a dot, and its motion is along the +x axis. The diagram makes it clear that the forces all act along one direction. Therefore, they can be added as colinear vectors to obtain the net force.

Problem-solving insight

A free-body diagram is very helpful when applying Newton's second law. Always start a problem by drawing the free-body diagram.

Solution   From Equation 4.1, the acceleration is $a = (\Sigma F)/m$. The net force is

$$\Sigma F = +275 \text{ N} + 395 \text{ N} - 560 \text{ N} = +110 \text{ N}$$

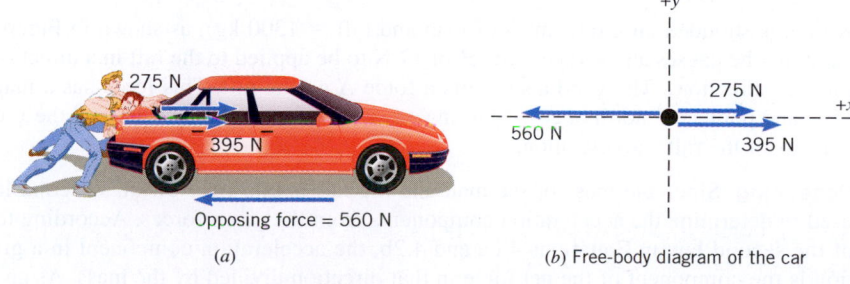

Opposing force = 560 N

(a)

(b) Free-body diagram of the car

Figure 4.5 (*a*) Two people push a stalled car. A force created by friction and the pavement opposes their efforts. (*b*) A free-body diagram that shows the horizontal forces acting on the car.

*We refer here to the gravitational version of the BE system, in which a force of one pound is defined to be the pull of the earth on a certain standard body at a location where the acceleration due to gravity is 32.174 ft/s^2.

The acceleration can now be obtained:

$$a = \frac{\Sigma F}{m} = \frac{+110 \text{ N}}{1850 \text{ kg}} = \boxed{+0.059 \text{ m/s}^2} \tag{4.1}$$

Problem-solving insight

The direction of the acceleration is always the same as the direction of the net force.

The plus sign indicates that the acceleration points along the $+x$ axis, in the same direction as the net force.

✓ CHECK YOUR UNDERSTANDING

(*The answers are given at the end of the book.*)

2. The net external force acting on an object is zero. Which one of the following statements is true?　**(a)** The object can only be stationary.　**(b)** The object can only be traveling with a constant velocity.　**(c)** The object can be either stationary or traveling with a constant velocity.　**(d)** The object can only be traveling with a velocity that is changing.

3. In Case A an object is moving straight downward with a constant speed of 9.80 m/s, while in Case B an object is moving straight downward with a constant acceleration of magnitude 9.80 m/s². Which one of the following is true?　**(a)** A nonzero net external force acts on the object in both cases.　**(b)** A nonzero net external force acts on the object in neither case.　**(c)** A nonzero net external force acts on the object in Case A only.　**(d)** A nonzero net external force acts on the object in Case B only.

4.4 THE VECTOR NATURE OF NEWTON'S SECOND LAW OF MOTION

When a football player throws a pass, the direction of the force he applies to the ball is important. Both the force and the resulting acceleration of the ball are vector quantities, as are all forces and accelerations. The directions of these vectors can be taken into account in two dimensions by using x and y components. The net force $\Sigma\vec{F}$ in Newton's second law has components ΣF_x and ΣF_y, while the acceleration $\vec{a}$ has components a_x and a_y. Consequently, Newton's second law, as expressed in Equation 4.1, can be written in an equivalent form as two equations, one for the x components and one for the y components:

$$\Sigma F_x = ma_x \tag{4.2a}$$
$$\Sigma F_y = ma_y \tag{4.2b}$$

This procedure is similar to that employed in Chapter 3 for the equations of two-dimensional kinematics (see Table 3.1). The components in Equations 4.2a and 4.2b are scalar components and will be either positive or negative numbers, depending on whether they point along the positive or negative x or y axis. The remainder of this section deals with examples that show how these equations are used.

 Example 2　Applying Newton's Second Law Using Components

A man is stranded on a raft (mass of man and raft = 1300 kg), as shown in Figure 4.6a. By paddling, he causes an average force $\vec{P}$ of 17 N to be applied to the raft in a direction due east (the $+x$ direction). The wind also exerts a force $\vec{A}$ on the raft. This force has a magnitude of 15 N and points 67° north of east. Ignoring any resistance from the water, find the x and y components of the raft's acceleration.

Problem-solving insight

Applications of Newton's second law always involve the net external force, which is the vector sum of all the external forces that act on an object. Each component of the net force leads to a corresponding component of the acceleration.

Reasoning　Since the mass of the man and the raft is known, Newton's second law can be used to determine the acceleration components from the given forces. According to the form of the second law in Equations 4.2a and 4.2b, the acceleration component in a given direction is the component of the net force in that direction divided by the mass. As an aid in determining the components ΣF_x and ΣF_y of the net force, we use the free-body diagram in Figure 4.6b. In this diagram, the directions due east and due north are the $+x$ and $+y$ directions, respectively.

Solution Figure 4.6b shows the force components:

Force	x Component	y Component
$\vec{P}$	$+17\,N$	$0\,N$
$\vec{A}$	$+(15\,N)\cos 67° = +6\,N$	$+(15\,N)\sin 67° = +14\,N$
	$\Sigma F_x = +17\,N + 6\,N = +23\,N$	$\Sigma F_y = +14\,N$

The plus signs indicate that ΣF_x points in the direction of the $+x$ axis and ΣF_y points in the direction of the $+y$ axis. The x and y components of the acceleration point in the directions of ΣF_x and ΣF_y, respectively, and can now be calculated:

$$a_x = \frac{\Sigma F_x}{m} = \frac{+23\,N}{1300\,kg} = \boxed{+0.018\,m/s^2} \tag{4.2a}$$

$$a_y = \frac{\Sigma F_y}{m} = \frac{+14\,N}{1300\,kg} = \boxed{+0.011\,m/s^2} \tag{4.2b}$$

These acceleration components are shown in Figure 4.6c.

Example 3 The Displacement of a Raft

At the moment that the forces $\vec{P}$ and $\vec{A}$ begin acting on the raft in Example 2, the velocity of the raft is 0.15 m/s, in a direction due east (the $+x$ direction). Assuming that the forces are maintained for 65 s, find the x and y components of the raft's displacement during this time interval.

Reasoning Once the net force acting on an object and the object's mass have been used in Newton's second law to determine the acceleration, it becomes possible to use the equations of kinematics to describe the resulting motion. We know from Example 2 that the acceleration components are $a_x = +0.018\,m/s^2$ and $a_y = +0.011\,m/s^2$, and it is given here that the initial velocity components are $v_{0x} = +0.15\,m/s$ and $v_{0y} = 0\,m/s$. Thus, Equation 3.5a $(x = v_{0x}t + \frac{1}{2}a_x t^2)$ and Equation 3.5b $(y = v_{0y}t + \frac{1}{2}a_y t^2)$ can be used with $t = 65\,s$ to determine the x and y components of the raft's displacement.

Solution According to Equations 3.5a and 3.5b, the x and y components of the displacement are

$$x = v_{0x}t + \tfrac{1}{2}a_x t^2 = (0.15\,m/s)(65\,s) + \tfrac{1}{2}(0.018\,m/s^2)(65\,s)^2 = \boxed{48\,m}$$

$$y = v_{0y}t + \tfrac{1}{2}a_y t^2 = (0\,m/s)(65\,s) + \tfrac{1}{2}(0.011\,m/s^2)(65\,s)^2 = \boxed{23\,m}$$

Figure 4.6d shows the final location of the raft.

✔ **CHECK YOUR UNDERSTANDING**

(The answers are given at the end of the book.)

4. Newton's second law indicates that when a net force acts on an object, it must accelerate. Does this mean that when two or more forces are applied to an object simultaneously, it must accelerate?

5. All of the following, except one, cause the acceleration of an object to double. Which one is the exception? **(a)** All forces acting on the object double. **(b)** The net force acting on the object doubles. **(c)** Both the net force acting on the object and the mass of the object double. **(d)** The net force acting on the object remains the same, while the mass of the object is reduced by a factor of two.

4.5 NEWTON'S THIRD LAW OF MOTION

Imagine you are in a football game. You line up facing your opponent, the ball is snapped, and the two of you crash together. No doubt, you feel a force. But think about your opponent. He too feels something, for while he is applying a force to you, you are

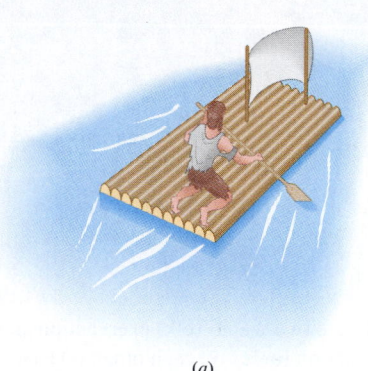

(a)

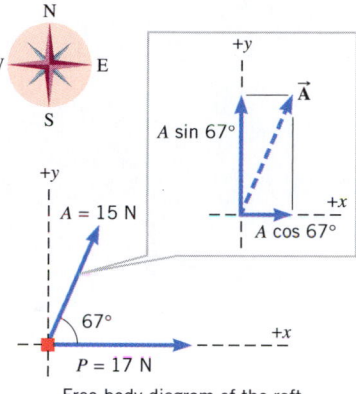

Free-body diagram of the raft

(b)

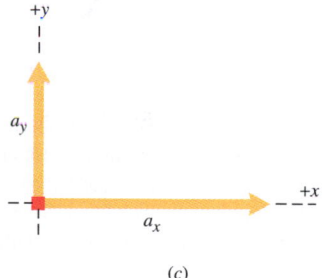

(c)

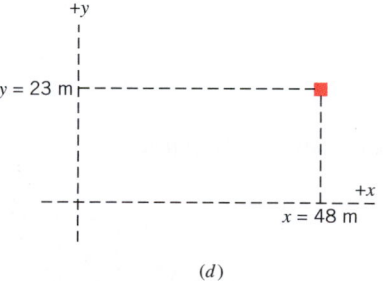

(d)

Figure 4.6 (a) A man is paddling a raft, as in Examples 2 and 3. (b) The free-body diagram shows the forces $\vec{P}$ and $\vec{A}$ that act on the raft. Forces acting on the raft in a direction perpendicular to the surface of the water play no role in the examples and are omitted for clarity. (c) The raft's acceleration components a_x and a_y. (d) In 65 s, the components of the raft's displacement are $x = 48\,m$ and $y = 23\,m$.

These two wapiti (elk) exert action and reaction forces on each other. (© First Light/Corbis Images)

applying a force to him. In other words, there isn't just one force on the line of scrimmage; there is a pair of forces. Newton was the first to realize that all forces occur in pairs and there is no such thing as an isolated force, existing all by itself. His third law of motion deals with this fundamental characteristic of forces.

NEWTON'S THIRD LAW OF MOTION

Whenever one body exerts a force on a second body, the second body exerts an oppositely directed force of equal magnitude on the first body.

The third law is often called the "action–reaction" law, because it is sometimes quoted as follows: "For every action (force) there is an equal, but opposite, reaction."

Figure 4.7 illustrates how the third law applies to an astronaut who is drifting just outside a spacecraft and who pushes on the spacecraft with a force $\vec{\mathbf{P}}$. According to the third law, the spacecraft pushes back on the astronaut with a force $-\vec{\mathbf{P}}$ that is equal in magnitude but opposite in direction. In Example 4, we examine the accelerations produced by each of these forces.

Example 4   The Accelerations Produced by Action and Reaction Forces

Suppose that the mass of the spacecraft in Figure 4.7 is m_S = 11 000 kg and that the mass of the astronaut is m_A = 92 kg. In addition, assume that the astronaut exerts a force of $\vec{\mathbf{P}}$ = +36 N on the spacecraft. Find the accelerations of the spacecraft and the astronaut.

Reasoning According to Newton's third law, when the astronaut applies the force $\vec{\mathbf{P}}$ = +36 N to the spacecraft, the spacecraft applies a reaction force $-\vec{\mathbf{P}}$ = −36 N to the astronaut. As a result, the spacecraft and the astronaut accelerate in opposite directions. Although the action and reaction forces have the same magnitude, they do not create accelerations of the same magnitude, because the spacecraft and the astronaut have different masses. According to Newton's second law, the astronaut, having a much smaller mass, will experience a much larger acceleration. In applying the second law, we note that the net force acting on the spacecraft is $\Sigma\vec{\mathbf{F}} = \vec{\mathbf{P}}$, while the net force acting on the astronaut is $\Sigma\vec{\mathbf{F}} = -\vec{\mathbf{P}}$.

Solution Using the second law, we find that the acceleration of the spacecraft is

$$\vec{\mathbf{a}}_S = \frac{\vec{\mathbf{P}}}{m_S} = \frac{+36 \text{ N}}{11\ 000 \text{ kg}} = \boxed{+0.0033 \text{ m/s}^2}$$

The acceleration of the astronaut is

$$\vec{\mathbf{a}}_A = \frac{-\vec{\mathbf{P}}}{m_A} = \frac{-36 \text{ N}}{92 \text{ kg}} = \boxed{-0.39 \text{ m/s}^2}$$

Problem-solving insight

Even though the magnitudes of the action and reaction forces are always equal, these forces do not necessarily produce accelerations that have equal magnitudes, since each force acts on a different object that may have a different mass.

The physics of
automatic trailer brakes.

There is a clever application of Newton's third law in some rental trailers. As Figure 4.8 illustrates, the tow bar connecting the trailer to the rear bumper of a car contains a mechanism that can automatically actuate brakes on the trailer wheels. This mechanism works without the need for electrical connections between the car and the trailer. When the driver applies the car brakes, the car slows down. Because of inertia, however, the trailer continues to roll forward and begins pushing against the bumper. In reaction, the bumper pushes back on the tow bar. The reaction force is used by the mechanism in the tow bar to "push the brake pedal" for the trailer.

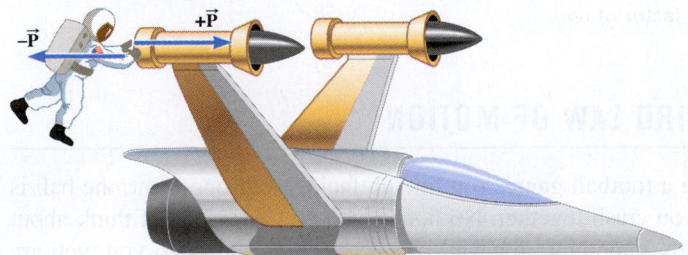

Figure 4.7 The astronaut pushes on the spacecraft with a force $+\vec{\mathbf{P}}$. According to Newton's third law, the spacecraft simultaneously pushes back on the astronaut with a force $-\vec{\mathbf{P}}$.

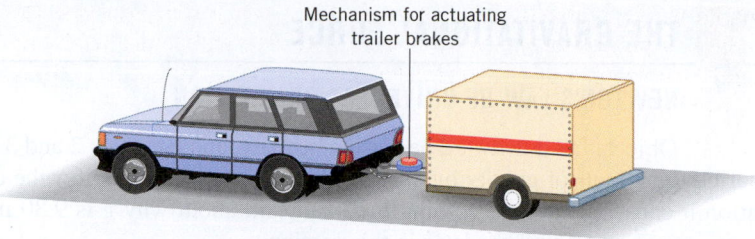

Mechanism for actuating
trailer brakes

Figure 4.8 Some rental trailers include an automatic brake-actuating mechanism.

✔ **CHECK YOUR UNDERSTANDING**

(*The answer is given at the end of the book.*)

6. A father and his seven-year-old daughter are facing each other on ice skates. With their hands, they push off against one another. Which one or more of the following statements is (are) true?   **(a)** Each experiences an acceleration that has a different magnitude.   **(b)** Each experiences an acceleration of the same magnitude.   **(c)** Each experiences a pushing force that has a different magnitude.   **(d)** Each experiences a pushing force that has the same magnitude.

4.6 TYPES OF FORCES: AN OVERVIEW

Newton's three laws of motion make it clear that forces play a central role in determining the motion of an object. In the next four sections some common forces will be discussed: the gravitational force (Section 4.7), the normal force (Section 4.8), frictional forces (Section 4.9), and the tension force (Section 4.10). In later chapters, we will encounter still others, such as electric and magnetic forces. It is important to realize that Newton's second law is always valid, regardless of which of these forces may act on an object. One does not have a different law for every type of common force. Thus, we need only to determine what forces are acting on an object, add them together to form the net force, and then use Newton's second law to determine the object's acceleration.

In nature there are two general types of forces, fundamental and nonfundamental. Fundamental forces are the ones that are truly unique, in the sense that all other forces can be explained in terms of them. Only three fundamental forces have been discovered:

1. Gravitational force
2. Strong nuclear force
3. Electroweak force

The gravitational force is discussed in the next section. The strong nuclear force plays a primary role in the stability of the nucleus of the atom (see Section 31.2). The electroweak force is a single force that manifests itself in two ways (see Section 32.6). One manifestation is the electromagnetic force that electrically charged particles exert on one another (see Sections 18.5, 21.2, and 21.8). The other manifestation is the so-called weak nuclear force that plays a role in the radioactive disintegration of certain nuclei (see Section 31.5).

Except for the gravitational force, all of the forces discussed in this chapter are nonfundamental, because they are related to the electromagnetic force. They arise from the interactions between the electrically charged particles that comprise atoms and molecules. Our understanding of which forces are fundamental, however, is continually evolving. For instance, in the 1860s and 1870s James Clerk Maxwell showed that the electric force and the magnetic force could be explained as manifestations of a single electromagnetic force. Then, in the 1970s, Sheldon Glashow (1932–), Abdus Salam (1926–1996), and Steven Weinberg (1933–) presented the theory that explains how the electromagnetic force and the weak nuclear force are related to the electroweak force. They received a Nobel prize in 1979 for their achievement. Today, efforts continue that have the goal of further reducing the number of fundamental forces.

THE GRAVITATIONAL FORCE

4.7

NEWTON'S LAW OF UNIVERSAL GRAVITATION

Objects fall downward because of gravity, and Chapters 2 and 3 discuss how to describe the effects of gravity by using a value of $g = 9.80$ m/s^2 for the downward acceleration it causes. However, nothing has been said about why g is 9.80 m/s^2. The reason is fascinating, as we will see later in this section.

The acceleration due to gravity is like any other acceleration, and Newton's second law indicates that it must be caused by a net force. In addition to his famous three laws of motion, Newton also provided a coherent understanding of the **gravitational force.** His "law of universal gravitation" is stated as follows:

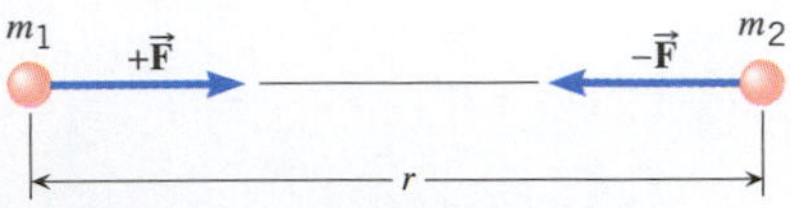

NEWTON'S LAW OF UNIVERSAL GRAVITATION

Every particle in the universe exerts an attractive force on every other particle. A particle is a piece of matter, small enough in size to be regarded as a mathematical point. For two particles that have masses m_1 and m_2 and are separated by a distance r, the force that each exerts on the other is directed along the line joining the particles (see Figure 4.9) and has a magnitude given by

$$F = G \frac{m_1 m_2}{r^2} \tag{4.3}$$

The symbol G denotes the universal gravitational constant, whose value is found experimentally to be

$$G = 6.674 \times 10^{-11} \text{ N} \cdot \text{m}^2/\text{kg}^2$$

Figure 4.9 The two particles, whose masses are m_1 and m_2, are attracted by gravitational forces $+\vec{\mathbf{F}}$ and $-\vec{\mathbf{F}}$.

The constant G that appears in Equation 4.3 is called the **universal gravitational constant,** because it has the same value for all pairs of particles anywhere in the universe, no matter what their separation. The value for G was first measured in an experiment by the English scientist Henry Cavendish (1731–1810), more than a century after Newton proposed his law of universal gravitation.

To see the main features of Newton's law of universal gravitation, look at the two particles in Figure 4.9. They have masses m_1 and m_2 and are separated by a distance r. In the picture, it is assumed that a force pointing to the right is positive. The gravitational forces point along the line joining the particles and are

$+\vec{\mathbf{F}}$, the gravitational force exerted on particle 1 by particle 2
$-\vec{\mathbf{F}}$, the gravitational force exerted on particle 2 by particle 1

These two forces have equal magnitudes and opposite directions. They act on different bodies, causing them to be mutually attracted. In fact, these forces are an action–reaction pair, as required by Newton's third law. Example 5 shows that the magnitude of the gravitational force is extremely small for ordinary values of the masses and the distance between them.

 Example 5   Gravitational Attraction

What is the magnitude of the gravitational force that acts on each particle in Figure 4.9, assuming $m_1 = 12$ kg (approximately the mass of a bicycle), $m_2 = 25$ kg, and $r = 1.2$ m?

Reasoning and Solution   The magnitude of the gravitational force can be found using Equation 4.3:

$$F = G \frac{m_1 m_2}{r^2} = (6.67 \times 10^{-11} \text{ N} \cdot \text{m}^2/\text{kg}^2) \frac{(12 \text{ kg})(25 \text{ kg})}{(1.2 \text{ m})^2} = \boxed{1.4 \times 10^{-8} \text{ N}}$$

For comparison, you exert a force of about 1 N when pushing a doorbell, so that the gravitational force is exceedingly small in circumstances such as those here. This result is due to the fact that G itself is very small. However, if one of the bodies has a large mass, like that of the earth (5.98×10^{24} kg), the gravitational force can be large.

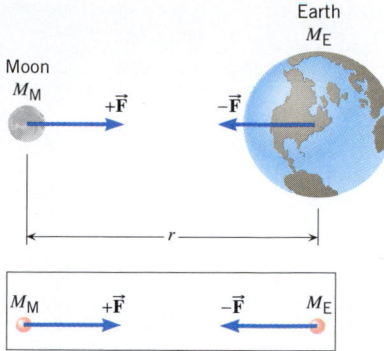

As expressed by Equation 4.3, Newton's law of gravitation applies only to particles. However, most familiar objects are too large to be considered particles. Nevertheless, the law of universal gravitation can be applied to such objects with the aid of calculus. Newton was able to prove that an object of finite size can be considered to be a particle for purposes of using the gravitation law, provided the mass of the object is distributed with spherical symmetry about its center. Thus, Equation 4.3 can be applied when each object is a sphere whose mass is spread uniformly over its entire volume. Figure 4.10 shows this kind of application, assuming that the earth and the moon are such uniform spheres of matter. In this case, r is the distance *between the centers of the spheres* and not the distance between the outer surfaces. The gravitational forces that the spheres exert on each other are the same as if the entire mass of each were concentrated at its center. Even if the objects are not uniform spheres, Equation 4.3 can be used to a good degree of approximation if the sizes of the objects are small relative to the distance of separation r.

Figure 4.10 The gravitational force that each uniform sphere of matter exerts on the other is the same as if each sphere were a particle with its mass concentrated at its center. The earth (mass M_E) and the moon (mass M_M) approximate such uniform spheres.

WEIGHT

The weight of an object exists because of the gravitational pull of the earth, according to the following definition:

DEFINITION OF WEIGHT

The weight of an object on or above the earth is the gravitational force that the earth exerts on the object. The weight always acts downward, toward the center of the earth. On or above another astronomical body, the weight is the gravitational force exerted on the object by that body.

SI Unit of Weight: newton (N)

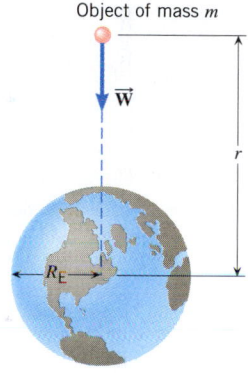

Figure 4.11 On or above the earth, the weight $\vec{W}$ of an object is the gravitational force exerted on the object by the earth.

Using W for the magnitude of the weight,* m for the mass of the object, and M_E for the mass of the earth, it follows from Equation 4.3 that

$$W = G \frac{M_E m}{r^2} \qquad (4.4)$$

Equation 4.4 and Figure 4.11 both emphasize that an object has weight whether or not it is resting on the earth's surface, because the gravitational force is acting even when the distance r is not equal to the radius R_E of the earth. However, the gravitational force becomes weaker as r increases, since r is in the denominator of Equation 4.4. Figure 4.12, for example, shows how the weight of the Hubble Space Telescope becomes smaller as the distance r from the center of the earth increases. In Example 6 the telescope's weight is determined when it is on earth and in orbit.

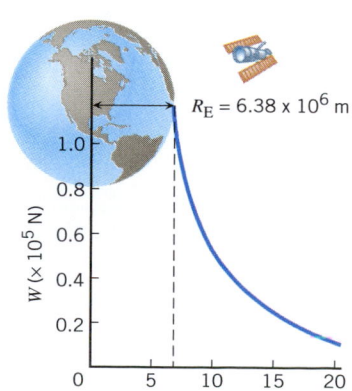

Figure 4.12 The weight of the Hubble Space Telescope decreases as the telescope gets farther from the earth. The distance from the center of the earth to the telescope is r.

Example 6 The Hubble Space Telescope

The mass of the Hubble Space Telescope is 11 600 kg. Determine the weight of the telescope **(a)** when it was resting on the earth and **(b)** as it is in its orbit 598 km above the earth's surface.

---

*Often, the word "weight" and the phrase "magnitude of the weight" are used interchangeably, even though weight is a vector. Generally, the context makes it clear when the direction of the weight vector must be taken into account.

Reasoning The weight of the Hubble Space Telescope is the gravitational force exerted on it by the earth. According to Equation 4.4, the weight varies inversely as the square of the radial distance r. Thus, we expect the telescope's weight on the earth's surface (r smaller) to be greater than its weight in orbit (r larger).

Solution **(a)** On the earth's surface, the weight is given by Equation 4.4 with $r = 6.38 \times 10^6$ m (the earth's radius):

$$W = G\frac{M_{\text{E}}m}{r^2} = \frac{(6.67 \times 10^{-11}\ \text{N} \cdot \text{m}^2/\text{kg}^2)(5.98 \times 10^{24}\ \text{kg})(11\ 600\ \text{kg})}{(6.38 \times 10^6\ \text{m})^2}$$

$$\boxed{W = 1.14 \times 10^5\ \text{N}}$$

(b) When the telescope is 598 km above the surface, its distance from the center of the earth is

$$r = 6.38 \times 10^6\ \text{m} + 598 \times 10^3\ \text{m} = 6.98 \times 10^6\ \text{m}$$

The weight now can be calculated as in part (a), except that the new value of r must be used: $\boxed{W = 0.950 \times 10^5\ \text{N}}$. As expected, the weight is less in orbit.

> **Problem-solving insight**
>
> When applying Newton's gravitation law to uniform spheres of matter, remember that the distance r is between the centers of the spheres, not between the surfaces.

The space age has forced us to broaden our ideas about weight. For instance, an astronaut weighs only about one-sixth as much on the moon as on the earth. To obtain his weight on the moon from Equation 4.4, it is only necessary to replace M_{E} by M_{M} (the mass of the moon) and let $r = R_{\text{M}}$ (the radius of the moon).

RELATION BETWEEN MASS AND WEIGHT

> **Problem-solving insight**
>
> Mass and weight are different quantities. They cannot be interchanged when solving problems.

Although massive objects weigh a lot on the earth, mass and weight are not the same quantity. As Section 4.2 discusses, mass is a quantitative measure of inertia. As such, mass is an intrinsic property of matter and does not change as an object is moved from one location to another. Weight, in contrast, is the gravitational force acting on the object and can vary, depending on how far the object is above the earth's surface or whether it is located near another body such as the moon.

The relation between weight W and mass m can be written in two ways:

$$W = \boxed{G\frac{M_{\text{E}}}{r^2}}\ m \tag{4.4}$$

$$W = m\ \boxed{g} \tag{4.5}$$

Equation 4.4 is Newton's law of universal gravitation, and Equation 4.5 is Newton's second law (net force equals mass times acceleration) incorporating the acceleration g due to gravity. These expressions make the distinction between mass and weight stand out. The weight of an object whose mass is m depends on the values for the universal gravitational constant G, the mass M_{E} of the earth, and the distance r. These three parameters together determine the acceleration g due to gravity. The specific value of $g = 9.80$ m/s^2 applies only when r equals the radius R_{E} of the earth. For larger values of r, as would be the case on top of a mountain, the effective value of g is less than 9.80 m/s^2. The fact that g decreases as the distance r increases means that the weight likewise decreases. The mass of the object, however, does not depend on these effects and does not change. Conceptual Example 7 further explores the difference between mass and weight.

 Conceptual Example 7 Mass Versus Weight

A vehicle designed for exploring the moon's surface is being tested on earth, where it weighs roughly six times more than it will on the moon. The acceleration of the vehicle along the ground is measured. To achieve the same acceleration on the moon, will the required net force be **(a)** the same as, **(b)** greater than, or **(c)** less than that on earth?

Reasoning Do not be misled by the fact that the vehicle weighs more on earth. The greater weight occurs only because the mass and radius of the earth are different than the mass and radius of the moon. In any event, in Newton's second law the net force is proportional to the vehicle's mass, not its weight.

Answers (b) and (c) are incorrect. According to Newton's second law, for a given acceleration, the net force depends only on the mass. If the required net force were greater or smaller on the moon than it is on the earth, the implication would be that the vehicle's mass is different on the moon than it is on earth, which is contrary to fact.

Answer (a) is correct. The net force $\Sigma\vec{F}$ required to accelerate the vehicle is specified by Newton's second law as $\Sigma\vec{F} = m\vec{a}$, where m is the vehicle's mass and $\vec{a}$ is the acceleration. For a given acceleration, the net force depends only on the mass, which is the same on the moon as it is on the earth. Therefore, the required net force is the same on the moon as it is on the earth.

Related Homework: *Problems 21, 26*

The Lunar Roving Vehicle that astronaut Eugene Cernan is driving on the moon and the Lunar Excursion Module (behind the Roving Vehicle) have the same mass that they have on the earth. However, their weight is different on the moon than on the earth, as Conceptual Example 7 discusses. (NASA/Johnson Space Center)

✓ CHECK YOUR UNDERSTANDING

(The answers are given at the end of the book.)

7. When a body is moved from sea level to the top of a mountain, what changes, **(a)** the body's mass, **(b)** its weight, or **(c)** both its mass and its weight?

8. Object A weighs twice as much as object B at the same spot on the earth. Would the same be true at a given spot on Mars?

9. Three particles have identical masses. Each particle experiences only the gravitational forces due to the other two particles. How should the particles be arranged so each one experiences a net gravitational force that has the same magnitude? **(a)** On the corners of an equilateral triangle **(b)** On three of the four corners of a square **(c)** On the corners of a right triangle

10. Two objects with masses m_1 and m_2 are separated by a distance $2d$. Mass m_2 is greater than mass m_1. A third object has a mass m_3. All three objects are located on the same straight line. The net gravitational force acting on the third object is zero. Which one of the drawings correctly represents the locations of the objects?

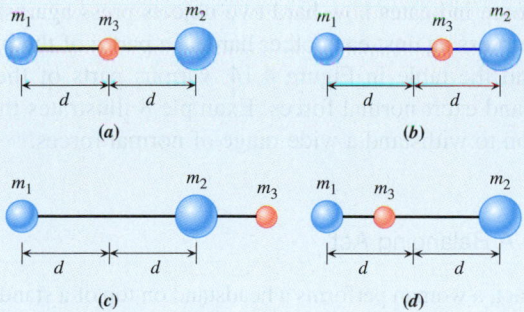

THE NORMAL FORCE

4.8

THE DEFINITION AND INTERPRETATION OF THE NORMAL FORCE

In many situations, an object is in contact with a surface, such as a tabletop. Because of the contact, there is a force acting on the object. The present section discusses only one component of this force, the component that acts perpendicular to the surface. The next section discusses the component that acts parallel to the surface. The perpendicular component is called the ***normal force.***

DEFINITION OF THE NORMAL FORCE

The normal force $\vec{F}_N$ is one component of the force that a surface exerts on an object with which it is in contact—namely, the component that is perpendicular to the surface.

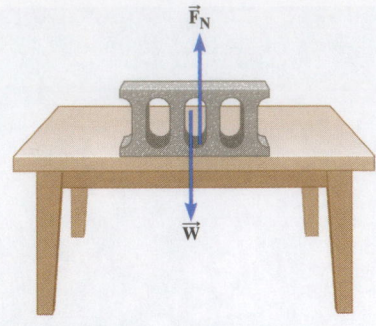

Figure 4.13 Two forces act on the block, its weight $\vec{\mathbf{W}}$ and the normal force $\vec{\mathbf{F}}_N$ exerted by the surface of the table.

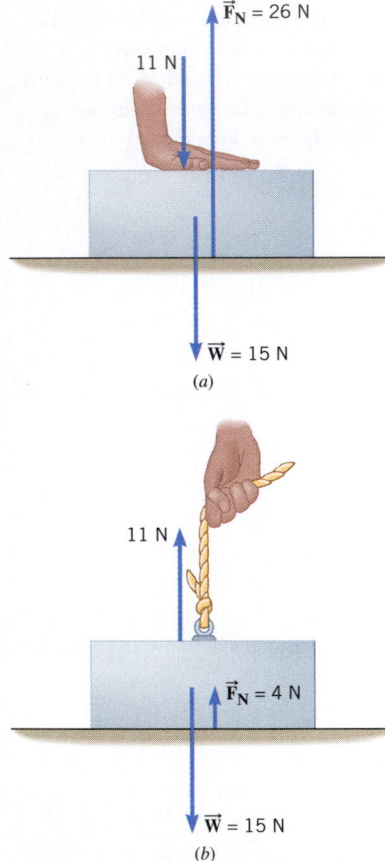

Figure 4.14 (*a*) The normal force $\vec{\mathbf{F}}_N$ is greater than the weight of the box, because the box is being pressed downward with an 11-N force. (*b*) The normal force is smaller than the weight, because the rope supplies an upward force of 11 N that partially supports the box.

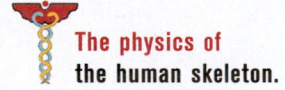

The physics of the human skeleton.

Figure 4.13 shows a block resting on a horizontal table and identifies the two forces that act on the block, the weight $\vec{\mathbf{W}}$ and the normal force $\vec{\mathbf{F}}_N$. To understand how an inanimate object, such as a tabletop, can exert a normal force, think about what happens when you sit on a mattress. Your weight causes the springs in the mattress to compress. As a result, the compressed springs exert an upward force (the normal force) on you. In a similar manner, the weight of the block causes invisible "atomic springs" in the surface of the table to compress, thus producing a normal force on the block.

Newton's third law plays an important role in connection with the normal force. In Figure 4.13, for instance, the block exerts a force on the table by pressing down on it. Consistent with the third law, the table exerts an oppositely directed force of equal magnitude on the block. This reaction force is the normal force. The magnitude of the normal force indicates how hard the two objects press against each other.

If an object is resting on a horizontal surface and there are no vertically acting forces except the object's weight and the normal force, the magnitudes of these two forces are equal; that is, $F_N = W$. This is the situation in Figure 4.13. The weight must be balanced by the normal force for the object to remain at rest on the table. If the magnitudes of these forces were not equal, there would be a net force acting on the block, and the block would accelerate either upward or downward, in accord with Newton's second law.

If other forces in addition to $\vec{\mathbf{W}}$ and $\vec{\mathbf{F}}_N$ act in the vertical direction, the magnitudes of the normal force and the weight are no longer equal. In Figure 4.14*a*, for instance, a box whose weight is 15 N is being pushed downward against a table. The pushing force has a magnitude of 11 N. Thus, the total downward force exerted on the box is 26 N, and this must be balanced by the upward-acting normal force if the box is to remain at rest. In this situation, then, the normal force is 26 N, which is considerably larger than the weight of the box.

Figure 4.14*b* illustrates a different situation. Here, the box is being pulled upward by a rope that applies a force of 11 N. The net force acting on the box due to its weight and the rope is only 4 N, downward. To balance this force, the normal force needs to be only 4 N. It is not hard to imagine what would happen if the force applied by the rope were increased to 15 N—exactly equal to the weight of the box. In this situation, the normal force would become zero. In fact, the table could be removed, since the block would be supported entirely by the rope. The situations in Figure 4.14 are consistent with the idea that the magnitude of the normal force indicates how hard two objects press against each other. Clearly, the box and the table press against each other harder in part *a* of the picture than in part *b*.

Like the box and the table in Figure 4.14, various parts of the human body press against one another and exert normal forces. Example 8 illustrates the remarkable ability of the human skeleton to withstand a wide range of normal forces.

Example 8   A Balancing Act

In a circus balancing act, a woman performs a headstand on top of a standing performer's head, as Figure 4.15*a* illustrates. The woman weighs 490 N, and the standing performer's head and neck weigh 50 N. It is primarily the seventh cervical vertebra in the spine that supports all the weight above the shoulders. What is the normal force that this vertebra exerts on the neck and head of the standing performer **(a)** before the act and **(b)** during the act?

Reasoning To begin, we draw a free-body diagram for the neck and head of the standing performer. Before the act, there are only two forces, the weight of the standing performer's head and neck, and the normal force. During the act, an additional force is present due to the woman's weight. In both cases, the upward and downward forces must balance for the head and neck to remain at rest. This condition of balance will lead us to values for the normal force.

Solution **(a)** Figure 4.15*b* shows the free-body diagram for the standing performer's head and neck before the act. The only forces acting are the normal force $\vec{\mathbf{F}}_N$ and the 50-N weight. These two forces must balance for the standing performer's head and neck to remain at rest. Therefore, the seventh cervical vertebra exerts a normal force of $\boxed{F_N = 50 \text{ N}}$.

(b) Figure 4.15*c* shows the free-body diagram that applies during the act. Now, the total downward force exerted on the standing performer's head and neck is 50 N + 490 N = 540 N, which must be balanced by the upward normal force, so that $\boxed{F_N = 540 \text{ N}}$.

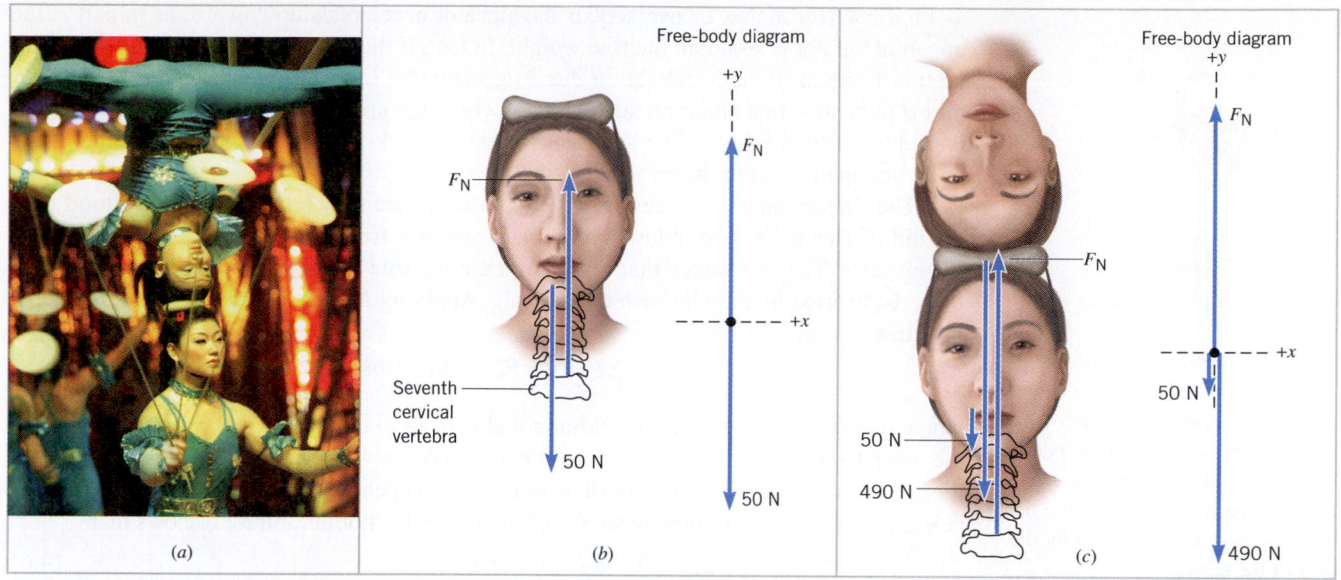

Figure 4.15 (*a*) A young woman keeps her balance during a performance by China's Sincuan Acrobatic group. A free-body diagram is shown (above the shoulders) for the standing performer (*b*) before the act and (*c*) during the act. For convenience, the scales used for the vectors in parts *b* and *c* are different. (*a*, SUPRI/Reuters/Landov LLC)

In summary, the normal force does not necessarily have the same magnitude as the weight of the object. The value of the normal force depends on what other forces are present. It also depends on whether the objects in contact are accelerating. In one situation that involves accelerating objects, the magnitude of the normal force can be regarded as a kind of "apparent weight," as we will now see.

APPARENT WEIGHT

Usually, the weight of an object can be determined with the aid of a scale. However, even though a scale is working properly, there are situations in which it does not give the correct weight. In such situations, the reading on the scale gives only the "apparent" weight, rather than the gravitational force or "true" weight. The apparent weight is the force that the object exerts on the scale with which it is in contact.

To see the discrepancies that can arise between true weight and apparent weight, consider the scale in the elevator in Figure 4.16. The reasons for the discrepancies will be explained shortly. A person whose true weight is 700 N steps on the scale. If the elevator is at rest or moving with a constant velocity (either upward or downward), the scale registers the true weight, as Figure 4.16*a* illustrates.

If the elevator is accelerating, the apparent weight and the true weight are not equal. When the elevator accelerates upward, the apparent weight is greater than the true weight,

Figure 4.16 (*a*) When the elevator is not accelerating, the scale registers the true weight (*W* = 700 N) of the person. (*b*) When the elevator accelerates upward, the apparent weight (1000 N) exceeds the true weight. (*c*) When the elevator accelerates downward, the apparent weight (400 N) is less than the true weight. (*d*) The apparent weight is zero if the elevator falls freely—that is, if it falls with the acceleration due to gravity.

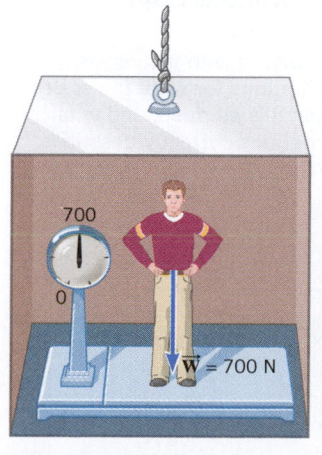

(*a*) No acceleration ($\vec{\mathbf{v}}$ = constant)

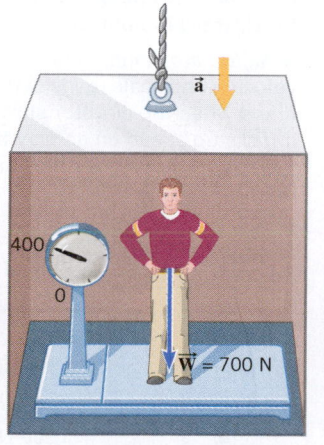

(*b*) Upward acceleration

(*c*) Downward acceleration

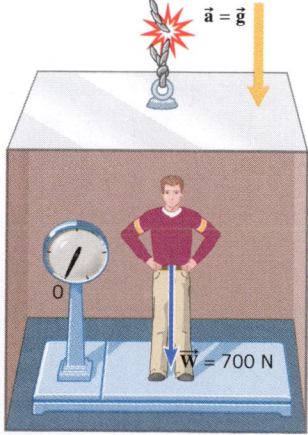

(*d*) Free-fall

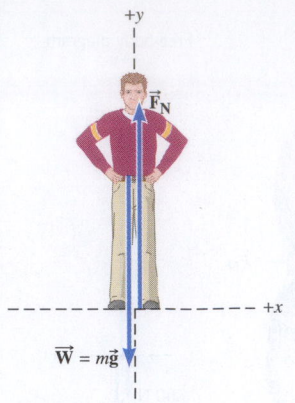

Figure 4.17 A free-body diagram showing the forces acting on the person riding in the elevator of Figure 4.16. $\vec{\mathbf{W}}$ is the true weight, and $\vec{\mathbf{F}}_N$ is the normal force exerted on the person by the platform of the scale.

as Figure 4.16*b* shows. Conversely, if the elevator accelerates downward, as in part *c*, the apparent weight is less than the true weight. In fact, if the elevator falls freely, so its acceleration is equal to the acceleration due to gravity, the apparent weight becomes zero, as part *d* indicates. In a situation such as this, where the apparent weight is zero, the person is said to be "weightless." The apparent weight, then, does not equal the true weight if the scale and the person on it are accelerating.

The discrepancies between true weight and apparent weight can be understood with the aid of Newton's second law. Figure 4.17 shows a free-body diagram of the person in the elevator. The two forces that act on him are the true weight $\vec{\mathbf{W}} = m\vec{\mathbf{g}}$ and the normal force $\vec{\mathbf{F}}_N$ exerted by the platform of the scale. Applying Newton's second law in the vertical direction gives

$$\Sigma F_y = +F_N - mg = ma$$

where a is the acceleration of the elevator and person. In this result, the symbol g stands for the magnitude of the acceleration due to gravity and can never be a negative quantity. However, the acceleration a may be either positive or negative, depending on whether the elevator is accelerating upward (+) or downward (−). Solving for the normal force F_N shows that

$$F_N = \underbrace{mg}_{\substack{\text{Apparent} \\ \text{weight}}} + \underbrace{ma}_{\substack{\text{True} \\ \text{weight}}} \qquad (4.6)$$

In Equation 4.6, F_N is the magnitude of the normal force exerted on the person by the scale. But in accord with Newton's third law, F_N is also the magnitude of the downward force that the person exerts on the scale—namely, the apparent weight.

Equation 4.6 contains all the features shown in Figure 4.16. If the elevator is not accelerating, $a = 0 \text{ m/s}^2$, and the apparent weight equals the true weight. If the elevator accelerates upward, a is positive, and the equation shows that the apparent weight is greater than the true weight. If the elevator accelerates downward, a is negative, and the apparent weight is less than the true weight. If the elevator falls freely, $a = -g$, and the apparent weight is zero. The apparent weight is zero because when both the person and the scale fall freely, they cannot push against one another. In this text, when the weight is given, it is assumed to be the true weight, unless stated otherwise.

✓ **CHECK YOUR UNDERSTANDING**

(*The answers are given at the end of the book.*)

11. A stack of books whose true weight is 165 N is placed on a scale in an elevator. The scale reads 165 N. From this information alone, can you tell whether the elevator is moving with a constant velocity of 2 m/s upward, is moving with a constant velocity of 2 m/s downward, or is at rest?

12. A 10-kg suitcase is placed on a scale that is in an elevator. In which direction is the elevator accelerating when the scale reads 75 N and when it reads 120 N? **(a)** Downward when it reads 75 N and upward when it reads 120 N **(b)** Upward when it reads 75 N and downward when it reads 120 N **(c)** Downward in both cases **(d)** Upward in both cases

13. You are standing on a scale in an elevator that is moving upward with a constant velocity. The scale reads 600 N. The following table shows five options for what the scale reads when the elevator slows down as it comes to a stop, when it is stopped, and when it picks up speed on its way back down. Which one of the five options correctly describes the scale's readings? Note that the symbol < means "less than" and > means "greater than."

Option	Elevator slows down as it comes to a halt	Elevator is stopped	Elevator picks up speed on its way back down
(a)	> 600 N	> 600 N	> 600 N
(b)	< 600 N	600 N	< 600 N
(c)	> 600 N	600 N	< 600 N
(d)	< 600 N	< 600 N	< 600 N
(e)	< 600 N	600 N	> 600 N

Figure 4.18 This photo, shot from underneath a transparent surface, shows a tire rolling under wet conditions. The channels in the tire collect and divert water away from the regions where the tire contacts the surface, thus providing better traction. (Courtesy Goodyear Tire & Rubber Co.)

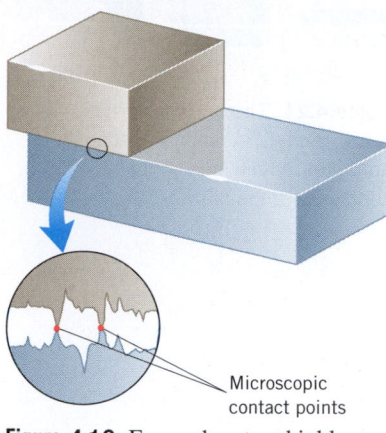

Figure 4.19 Even when two highly polished surfaces are in contact, they touch only at relatively few points.

4.9 STATIC AND KINETIC FRICTIONAL FORCES

When an object is in contact with a surface, there is a force acting on the object. The previous section discusses the component of this force that is perpendicular to the surface, which is called the normal force. When the object moves or attempts to move along the surface, there is also a component of the force that is parallel to the surface. This parallel force component is called the *frictional force,* or simply *friction.*

In many situations considerable engineering effort is expended trying to reduce friction. For example, oil is used to reduce the friction that causes wear and tear in the pistons and cylinder walls of an automobile engine. Sometimes, however, friction is absolutely necessary. Without friction, car tires could not provide the traction needed to move the car. In fact, the raised tread on a tire is designed to maintain friction. On a wet road, the spaces in the tread pattern (see Figure 4.18) provide channels for the water to collect and be diverted away. Thus, these channels largely prevent the water from coming between the tire surface and the road surface, where it would reduce friction and allow the tire to skid.

Surfaces that appear to be highly polished can actually look quite rough when examined under a microscope. Such an examination reveals that two surfaces in contact touch only at relatively few spots, as Figure 4.19 illustrates. The microscopic area of contact for these spots is substantially less than the apparent macroscopic area of contact between the surfaces—perhaps thousands of times less. At these contact points the molecules of the different bodies are close enough together to exert strong attractive intermolecular forces on one another, leading to what are known as "cold welds." Frictional forces are associated with these welded spots, but the exact details of how frictional forces arise are not well understood. However, some empirical relations have been developed that make it possible to account for the effects of friction.

Figure 4.20 helps to explain the main features of the type of friction known as *static friction.* The block in this drawing is initially at rest on a table, and as long as there is no attempt to move the block, there is no static frictional force. Then, a horizontal force $\vec{F}$ is applied to the block by means of a rope. If $\vec{F}$ is small, as in part *a,* experience tells us that the block still does not move. Why? It does not move because the static frictional force $\vec{f}_s$ exactly cancels the effect of the applied force. The direction of $\vec{f}_s$ is opposite to that of $\vec{F}$, and the magnitude of $\vec{f}_s$ equals the magnitude of the applied force, $f_s = F$. Increasing the applied force in Figure 4.20 by a small amount still does not cause the block to move. There is no movement because the static frictional force also increases by an amount that cancels out the increase in the applied force (see part *b* of the drawing). If the applied force continues to increase, however, there comes a point when the block finally "breaks away"

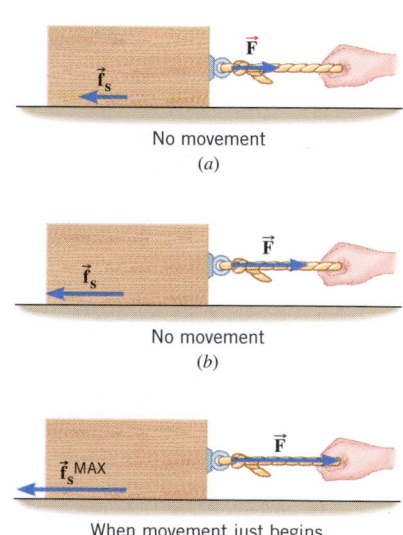

No movement
(a)

No movement
(b)

When movement just begins
(c)

Figure 4.20 Applying a small force $\vec{F}$ to the block, as in parts *a* and *b*, produces no movement, because the static frictional force $\vec{f}_s$ exactly balances the applied force. (*c*) The block just begins to move when the applied force is slightly greater than the maximum static frictional force $\vec{f}_s^{\text{MAX}}$.

Figure 4.21 The maximum static frictional force $\vec{f}_s^{\text{MAX}}$ would be the same, no matter which side of the block is in contact with the table.

and begins to slide. The force just before breakaway represents the *maximum static frictional force* $\vec{f}_s^{\text{MAX}}$ that the table can exert on the block (see part *c* of the drawing). Any applied force that is greater than $\vec{f}_s^{\text{MAX}}$ cannot be balanced by static friction, and the resulting net force accelerates the block to the right.

Experimental evidence shows that, to a good degree of approximation, the maximum static frictional force between a pair of dry, unlubricated surfaces has two main characteristics. It is independent of the apparent macroscopic area of contact between the objects, provided that the surfaces are hard or nondeformable. For instance, in Figure 4.21 the maximum static frictional force that the surface of the table can exert on a block is the same, whether the block is resting on its largest or its smallest side. The other main characteristic of $\vec{f}_s^{\text{MAX}}$ is that its magnitude is proportional to the magnitude of the normal force $\vec{F}_N$. As Section 4.8 points out, the magnitude of the normal force indicates how hard two surfaces are being pressed together. The harder they are pressed, the larger is f_s^{MAX}, presumably because the number of "cold-welded," microscopic contact points is increased. Equation 4.7 expresses the proportionality between f_s^{MAX} and F_N with the aid of a proportionality constant μ_s, which is called the ***coefficient of static friction.***

STATIC FRICTIONAL FORCE

The magnitude f_s of the static frictional force can have any value from zero up to a maximum value of f_s^{MAX}, depending on the applied force. In other words, $f_s \leq f_s^{\text{MAX}}$, where the symbol $\leq$ is read as "less than or equal to." The equality holds only when f_s attains its maximum value, which is

$$f_s^{\text{MAX}} = \mu_s F_N \tag{4.7}$$

In Equation 4.7, μ_s is the coefficient of static friction, and F_N is the magnitude of the normal force.

It should be emphasized that Equation 4.7 relates only the *magnitudes* of $\vec{f}_s^{\text{MAX}}$ and $\vec{F}_N$, *not the vectors themselves.* This equation does not imply that the directions of the vectors are the same. In fact, $\vec{f}_s^{\text{MAX}}$ is parallel to the surface, while $\vec{F}_N$ is perpendicular to it.

The coefficient of static friction, being the ratio of the magnitudes of two forces $(\mu_s = f_s^{\text{MAX}}/F_N)$, has no units. Also, it depends on the type of material from which each surface is made (steel on wood, rubber on concrete, etc.), the condition of the surfaces (polished, rough, etc.), and other variables such as temperature. Table 4.2 gives some typical values of μ_s for various surfaces. Example 9 illustrates the use of Equation 4.7 for determining the maximum static frictional force.

Table 4.2 Approximate Values of the Coefficients of Friction for Various Surfaces[a]

Materials	Coefficient of Static Friction, μ_s	Coefficient of Kinetic Friction, μ_k
Glass on glass (dry)	0.94	0.4
Ice on ice (clean, 0 °C)	0.1	0.02
Rubber on dry concrete	1.0	0.8
Rubber on wet concrete	0.7	0.5
Steel on ice	0.1	0.05
Steel on steel (dry hard steel)	0.78	0.42
Teflon on Teflon	0.04	0.04
Wood on wood	0.35	0.3

[a]The last column gives the coefficients of kinetic friction, a concept that will be discussed shortly.

ANALYZING MULTIPLE-CONCEPT PROBLEMS

Example 9 The Force Needed To Start a Skier Moving

A skier is standing motionless on a horizontal patch of snow. She is holding onto a horizontal tow rope, which is about to pull her forward (see Figure 4.22a). The skier's mass is 59 kg, and the coefficient of static friction between the skis and snow is 0.14. What is the magnitude of the maximum force that the tow rope can apply to the skier without causing her to move?

Reasoning When the rope applies a relatively small force, the skier does not accelerate. The reason is that the static frictional force opposes the applied force and the two forces have the same magnitude. We can apply Newton's second law in the horizontal direction to this situation. In order for the rope to pull the skier forward, it must exert a force large enough to overcome the *maximum* static frictional force acting on the skis. The magnitude of the maximum static frictional force depends on the coefficient of static friction (which is known) and on the magnitude of the normal force. We can determine the magnitude of the normal force by using Newton's second law, along with the fact that the skier does not accelerate in the vertical direction.

Knowns and Unknowns The data for this problem are as follows:

Description	Symbol	Value
Mass of skier	m	59 kg
Coefficient of static friction	μ_s	0.14
Unknown Variable		
Magnitude of maximum horizontal force that tow rope can apply	F	?

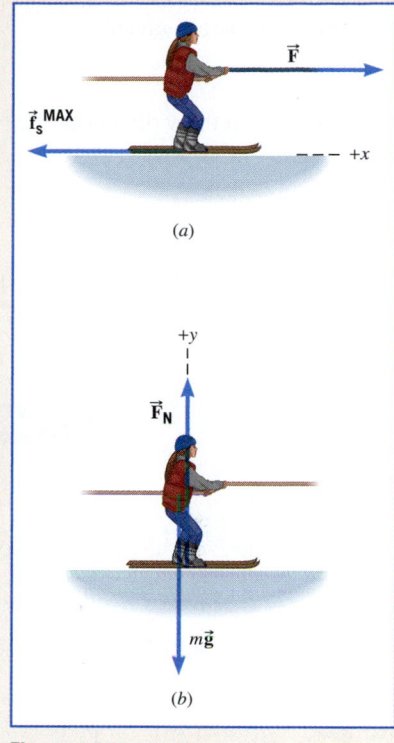

Figure 4.22 (a) Two horizontal forces act on the skier in the horizontal direction just before she begins to move. (b) Two vertical forces act on the skier.

Modeling the Problem

STEP 1 Newton's Second Law (Horizontal Direction) Figure 4.22a shows the two horizontal forces that act on the skier just before she begins to move: the force $\vec{F}$ applied by the tow rope and the maximum static frictional force $\vec{f}_s^{\text{MAX}}$. Since the skier is standing motionless, she is not accelerating in the horizontal or x direction, so $a_x = 0 \text{ m/s}^2$. Applying Newton's second law (Equation 4.2a) to this situation, we have

$$\Sigma F_x = ma_x = 0$$

Since the net force ΣF_x in the x direction is $\Sigma F_x = +F - f_s^{\text{MAX}}$, Newton's second law can be written as $+F - f_s^{\text{MAX}} = 0$. Thus,

$$F = f_s^{\text{MAX}}$$

We do not know f_s^{MAX}, but its value will be determined in Steps 2 and 3.

$$F = f_s^{\text{MAX}} \qquad (1)$$

$?$

STEP 2 The Maximum Static Frictional Force The magnitude f_s^{MAX} of the maximum static frictional force is related to the coefficient of static friction μ_s and the magnitude F_N of the normal force by Equation 4.7:

$$\boxed{f_s^{\text{MAX}} = \mu_s F_N} \qquad (4.7)$$

We now substitute this result into Equation 1, as indicated in the right column. The coefficient of static friction is known, but F_N is not. An expression for F_N will be obtained in the next step.

$$F = f_s^{\text{MAX}} \qquad (1)$$

$$\boxed{f_s^{\text{MAX}} = \mu_s F_N} \qquad (4.7)$$

$?$

Continued

STEP 3 **Newton's Second Law (Vertical Direction)** We can find the magnitude F_N of the normal force by noting that the skier does not accelerate in the vertical or y direction, so $a_y = 0$ m/s^2. Figure 4.22b shows the two vertical forces that act on the skier: the normal force $\vec{F}_N$ and her weight $m\vec{g}$. Applying Newton's second law (Equation 4.2b) to the vertical direction gives

$$\Sigma F_y = ma_y = 0$$

The net force in the y direction is $\Sigma F_y = +F_N - mg$, so Newton's second law becomes $+F_N - mg = 0$. Thus,

$$\boxed{F_N = mg}$$

We now substitute this result into Equation 4.7, as shown at the right.

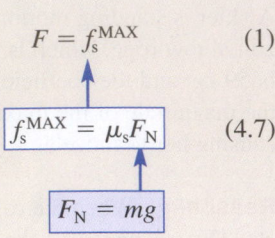

$$F = f_s^{MAX} \qquad (1)$$

$$\boxed{f_s^{MAX} = \mu_s F_N} \qquad (4.7)$$

$$\boxed{F_N = mg}$$

Solution Algebraically combining the results of the three steps, we have

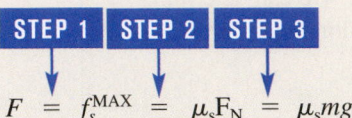

| STEP 1 | STEP 2 | STEP 3 |

$$F = f_s^{MAX} = \mu_s F_N = \mu_s mg$$

The magnitude F of the maximum force is

$$F = \mu_s mg = (0.14)(59 \text{ kg})(9.80 \text{ m/s}^2) = \boxed{81 \text{ N}}$$

If the force exerted by the tow rope exceeds this value, the skier will begin to accelerate forward.

Related Homework: *Problems 44, 108, 118*

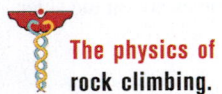

The physics of
rock climbing.

Static friction is often essential, as it is to the rock climber in Figure 4.23, for instance. She presses outward against the walls of the rock formation with her hands and feet to create sufficiently large normal forces, so that the static frictional forces can support her weight.

Once two surfaces begin sliding over one another, the static frictional force is no longer of any concern. Instead, a type of friction known as **kinetic* friction** comes into play. The kinetic frictional force opposes the relative sliding motion. If you have ever pushed an object across a floor, you may have noticed that it takes less force to keep the object sliding than it takes to get it going in the first place. In other words, the kinetic frictional force is usually less than the static frictional force.

Experimental evidence indicates that the kinetic frictional force $\vec{f}_k$ has three main characteristics, to a good degree of approximation. It is independent of the apparent area of contact between the surfaces (see Figure 4.21). It is independent of the speed of the sliding motion, if the speed is small. And lastly, the magnitude of the kinetic frictional force is proportional to the magnitude of the normal force. Equation 4.8 expresses this proportionality with the aid of a proportionality constant μ_k, which is called the **coefficient of kinetic friction.**

*The word "kinetic" is derived from the Greek word *kinetikos,* meaning "pertaining to motion."

Figure 4.23 In maneuvering her way up El Matador at Devil's Tower National Monument in Wyoming, this rock climber uses the static frictional forces between her hands and feet and the vertical rock walls to support her weight. (© Corey Rich/Aurora Photos)

KINETIC FRICTIONAL FORCE

The magnitude f_k of the kinetic frictional force is given by

$$f_k = \mu_k F_N \qquad (4.8)$$

In Equation 4.8, μ_k is the coefficient of kinetic friction, and F_N is the magnitude of the normal force.

Equation 4.8, like Equation 4.7, is a relationship between only the magnitudes of the frictional and normal forces. The directions of these forces are perpendicular to each other. Moreover, like the coefficient of static friction, the coefficient of kinetic friction is a number without units and depends on the type and condition of the two surfaces that are in contact. As indicated in Table 4.2, values for μ_k are typically less than those for μ_s, reflecting the fact that kinetic friction is generally less than static friction. The next example illustrates the effect of kinetic friction.

ANALYZING MULTIPLE-CONCEPT PROBLEMS

Example 10 Sled Riding

A sled and its rider are moving at a speed of 4.0 m/s along a horizontal stretch of snow, as Figure 4.24a illustrates. The snow exerts a kinetic frictional force on the runners of the sled, so the sled slows down and eventually comes to a stop. The coefficient of kinetic friction is 0.050. What is the displacement x of the sled?

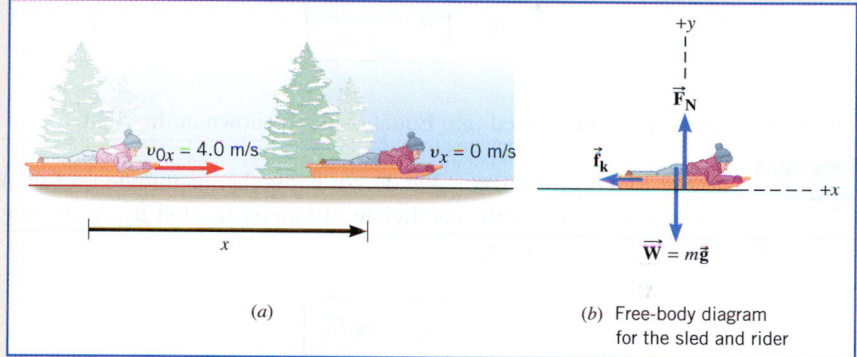

(a)

(b) Free-body diagram for the sled and rider

Reasoning As the sled slows down, its velocity is decreasing. As our discussions in Chapters 2 and 3 indicate, the changing velocity is described by an acceleration (which in this case is a deceleration since the sled is slowing down). Assuming that the acceleration is constant, we can use one of the equations of kinematics from Chapter 3 to relate the displacement to the initial and final velocities and to the acceleration. The acceleration of the sled is not given directly. However, we can determine it by using Newton's second law of motion, which relates the acceleration to the net force (which is the kinetic frictional force in this case) acting on the sled and its mass.

Figure 4.24 (a) The moving sled decelerates because of the kinetic frictional force. (b) Three forces act on the moving sled, the weight $\vec{W}$ of the sled and its rider, the normal force $\vec{F}_N$, and the kinetic frictional force $\vec{f}_k$. The free-body diagram for the sled shows these forces.

Knowns and Unknowns The data for this problem are listed in the table:

Description	Symbol	Value	Comment
Explicit Data			
Initial velocity	v_{0x}	+4.0 m/s	Positive, because the velocity points in the +x direction. See drawing.
Coefficient of kinetic friction	μ_k	0.050	
Implicit Data			
Final velocity	v_x	0 m/s	The sled comes to a stop.
Unknown Variable			
Displacement	x	?	

Continued

Modeling the Problem

STEP 1 **Displacement** To obtain the displacement x of the sled we will use Equation 3.6a from the equations of kinematics:

$$v_x^2 = v_{0x}^2 + 2a_x x$$

Solving for the displacement x gives the result shown at the right. This equation is useful because two of the variables, v_{0x} and v_x, are known and the acceleration a_x can be found by applying Newton's second law to the accelerating sled (see Step 2).

$$x = \frac{v_x^2 - v_{0x}^2}{2a_x} \qquad (1)$$

$?$

STEP 2 **Newton's Second Law** Newton's second law, as given in Equation 4.2a, states that the acceleration a_x is equal to the net force ΣF_x divided by the mass m:

$$a_x = \frac{\Sigma F_x}{m}$$

The free-body diagram in Figure 4.24b shows that the only force acting on the sled in the horizontal or x direction is the kinetic frictional force $\vec{f}_k$. We can write this force as $-f_k$, where f_k is the magnitude of the force and the minus sign indicates that it points in the $-x$ direction. Since the net force is $\Sigma F_x = -f_k$, Equation 4.2a becomes

$$\boxed{a_x = \frac{-f_k}{m}}$$

This result can now be substituted into Equation 1, as shown at the right.

$$x = \frac{v_x^2 - v_{0x}^2}{2a_x} \qquad (1)$$

$$\boxed{a_x = \frac{-f_k}{m}} \qquad (2)$$

$?$

STEP 3 **Kinetic Frictional Force** We do not know the magnitude f_k of the kinetic frictional force, but we do know the coefficient of kinetic friction μ_k. According to Equation 4.8, the two are related by

$$\boxed{f_k = \mu_k F_N} \qquad (4.8)$$

where F_N is the magnitude of the normal force. This relation can be substituted into Equation 2, as shown at the right. An expression for F_N will be obtained in the next step.

$$x = \frac{v_x^2 - v_{0x}^2}{2a_x} \qquad (1)$$

$$\boxed{a_x = \frac{-f_k}{m}} \qquad (2)$$

$$\boxed{f_k = \mu_k F_N} \qquad (4.8)$$

$?$

STEP 4 **Normal Force** The magnitude F_N of the normal force can be found by noting that the sled does not accelerate in the vertical or y direction ($a_y = 0 \text{ m/s}^2$). Thus, Newton's second law, as given in Equation 4.2b becomes

$$\Sigma F_y = ma_y = 0$$

There are two forces acting on the sled in the y direction, the normal force $\vec{F}_N$ and its weight $\vec{W}$ [see part (b) of the drawing]. Therefore, the net force in the y direction is

$$\Sigma F_y = +F_N - W$$

where $W = mg$ (Equation 4.5). Thus, Newton's second law becomes

$$+F_N - mg = 0 \quad \text{or} \quad \boxed{F_N = mg}$$

This result for F_N can be substituted into Equation 4.8, as shown at the right.

$$x = \frac{v_x^2 - v_{0x}^2}{2a_x} \qquad (1)$$

$$\boxed{a_x = \frac{-f_k}{m}} \qquad (2)$$

$$\boxed{f_k = \mu_k F_N} \qquad (4.8)$$

$$\boxed{F_N = mg}$$

Solution Algebraically combining the results of each step, we find that

STEP 1   STEP 2   STEP 3   STEP 4

$$x = \frac{v_x^2 - v_{0x}^2}{2a_x} = \frac{v_x^2 - v_{0x}^2}{2\left(\dfrac{-f_k}{m}\right)} = \frac{v_x^2 - v_{0x}^2}{2\left(\dfrac{-\mu_k F_N}{m}\right)} = \frac{v_x^2 - v_{0x}^2}{2\left(\dfrac{-\mu_k mg}{m}\right)} = \frac{v_x^2 - v_{0x}^2}{2(-\mu_k g)}$$

Note that the mass m of the sled and rider is algebraically eliminated from the final result. Thus, the displacement of the sled is

$$x = \frac{v_x^2 - v_{0x}^2}{2(-\mu_k g)} = \frac{(0 \text{ m/s})^2 - (+4.0 \text{ m/s})^2}{2[-(0.050)(9.80 \text{ m/s}^2)]} = \boxed{+16 \text{ m}}$$

Related Homework: *Problems 48, 49, 85*

Static friction opposes the impending relative motion between two objects, while kinetic friction opposes the relative sliding motion that actually does occur. In either case, *relative motion* is opposed. However, this opposition to relative motion does not mean that friction prevents or works against the motion of *all* objects. For instance, the foot of a person walking exerts a force on the earth, and the earth exerts a reaction force on the foot. This reaction force is a static frictional force, and it opposes the impending backward motion of the foot, propelling the person forward in the process. Kinetic friction can also cause an object to move, all the while opposing relative motion, as it does in Example 10. In this example the kinetic frictional force acts on the sled and opposes the relative motion of the sled and the earth. Newton's third law indicates, however, that since the earth exerts the kinetic frictional force on the sled, the sled must exert a reaction force on the earth. In response, the earth accelerates, but because of the earth's huge mass, the motion is too slight to be noticed.

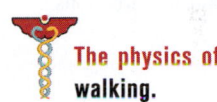

The physics of walking.

✓ **CHECK YOUR UNDERSTANDING**

(The answers are given at the end of the book.)

14. Suppose that the coefficients of static and kinetic friction have values such that $\mu_s = 1.4\,\mu_k$ for a crate in contact with a cement floor. Which one of the following statements is true?   **(a)** The magnitude of the static frictional force is always 1.4 times the magnitude of the kinetic frictional force.   **(b)** The magnitude of the kinetic frictional force is always 1.4 times the magnitude of the static frictional force.   **(c)** The magnitude of the maximum static frictional force is 1.4 times the magnitude of the kinetic frictional force.

15. A person has a choice of either pushing or pulling a sled at a constant velocity, as the drawing illustrates. Friction is present. If the angle θ is the same in both cases, does it require less force to push or to pull the sled?

16. A box has a weight of 150 N and is being pulled across a horizontal floor by a force that has a magnitude of 110 N. The pulling force can point horizontally, or it can point above the horizontal at an angle θ. When the pulling force points horizontally, the kinetic frictional force acting on the box is twice as large as when the pulling force points at the angle θ. Find θ.

17. A box rests on the floor of an elevator. Because of static friction, a force is required to start the box sliding across the floor when the elevator is   **(a)** stationary,   **(b)** accelerating upward, and   **(c)** accelerating downward. Rank the forces required in these three situations in ascending order—that is, smallest first.

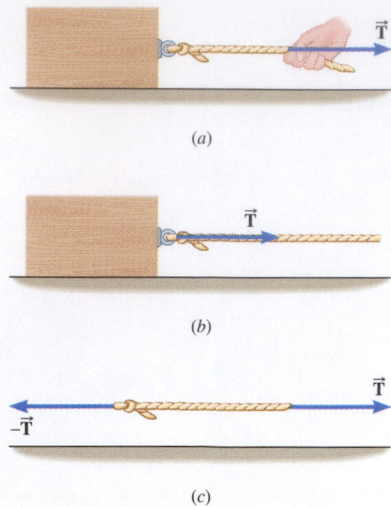

(a)

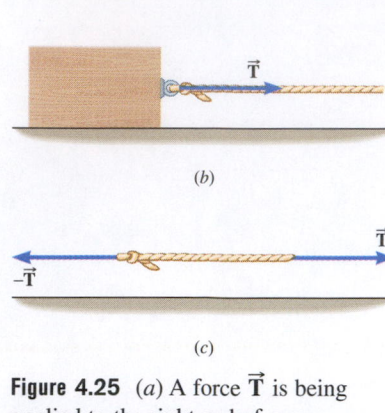

(b)

(c)

Figure 4.25 (a) A force $\vec{T}$ is being applied to the right end of a rope. (b) The force is transmitted to the box. (c) Forces are applied to both ends of the rope. These forces have equal magnitudes and opposite directions.

4.10 THE TENSION FORCE

Forces are often applied by means of cables or ropes that are used to pull an object. For instance, Figure 4.25a shows a force $\vec{T}$ being applied to the right end of a rope attached to a box. Each particle in the rope in turn applies a force to its neighbor. As a result, the force is applied to the box, as part b of the drawing shows.

In situations such as that in Figure 4.25, we say that the force $\vec{T}$ is applied to the box because of the tension in the rope, meaning that the tension and the force applied to the box have the same magnitude. However, the word "tension" is commonly used to mean the tendency of the rope to be pulled apart. To see the relationship between these two uses of the word "tension," consider the left end of the rope, which applies the force $\vec{T}$ to the box. In accordance with Newton's third law, the box applies a reaction force to the rope. The reaction force has the same magnitude as $\vec{T}$ but is oppositely directed. In other words, a force $-\vec{T}$ acts on the left end of the rope. Thus, forces of equal magnitude act on opposite ends of the rope, as in Figure 4.25c, and tend to pull it apart.

In the previous discussion, we have used the concept of a "massless" rope ($m = 0$ kg) without saying so. In reality, a massless rope does not exist, but it is useful as an idealization when applying Newton's second law. According to the second law, a net force is required to accelerate an object that has mass. In contrast, no net force is needed to accelerate a massless rope, since $\Sigma\vec{F} = m\vec{a}$ and $m = 0$ kg. Thus, when a force $\vec{T}$ is applied to one end of a massless rope, none of the force is needed to accelerate the rope. As a result, the force $\vec{T}$ is also applied undiminished to the object attached at the other end, as we assumed in Figure 4.25.* If the rope had mass, however, some of the force $\vec{T}$ would have to be used to accelerate the rope. The force applied to the box would then be less than $\vec{T}$, and the tension would be different at different locations along the rope. In this text we will assume that a rope connecting one object to another is massless, unless stated otherwise. The ability of a massless rope to transmit tension undiminished from one end to the other is not affected when the rope passes around objects such as the pulley in Figure 4.26 (provided the pulley itself is massless and frictionless).

✓ CHECK YOUR UNDERSTANDING

(The answer is given at the end of the book.)

18. A rope is used in a tug-of-war between two teams of five people each. Both teams are equally strong, so neither team wins. An identical rope is tied to a tree, and the same ten people pull just as hard on the loose end as they did in the contest. In both cases, the people pull steadily with no jerking. Which rope sustains the greater tension, **(a)** the rope tied to the tree or **(b)** the rope in the tug-of-war, or **(c)** do the ropes sustain the same tension?

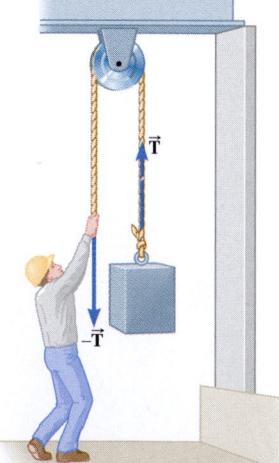

Figure 4.26 The force $\vec{T}$ applied at one end of a massless rope is transmitted undiminished to the other end, even when the rope bends around a pulley, provided the pulley is also massless and friction is absent.

4.11 EQUILIBRIUM APPLICATIONS OF NEWTON'S LAWS OF MOTION

Have you ever been so upset that it took days to recover your "equilibrium?" In this context, the word "equilibrium" refers to a balanced state of mind, one that is not changing wildly. In physics, the word "equilibrium" also refers to a lack of change, but in the sense that the velocity of an object isn't changing. If its velocity doesn't change, an object is not accelerating. Our definition of equilibrium, then, is as follows:

DEFINITION OF EQUILIBRIUM†

An object is in equilibrium when it has zero acceleration.

*If a rope is not accelerating, $\vec{a}$ is zero in the second law, and $\Sigma\vec{F} = m\vec{a} = 0$, regardless of the mass of the rope. Then, the rope can be ignored, no matter what mass it has.

†In this discussion of equilibrium we ignore rotational motion, which is discussed in Chapters 8 and 9. In Section 9.2 a more complete treatment of the equilibrium of a rigid object is presented and takes into account the concept of torque and the fact that objects can rotate.

CONCEPTS AT A GLANCE

External Forces

1. **Gravitational Force** (Section 4.7)
2. **Normal Force** (Section 4.8)
3. **Frictional Forces** (Section 4.9)
4. **Tension Force** (Section 4.10)

Equilibrium (Section 4.11)

$\vec{\mathbf{a}} = 0 \text{ m/s}^2$

Newton's Second Law

$\Sigma\vec{\mathbf{F}} = m\vec{\mathbf{a}}$

$\vec{\mathbf{a}} \neq 0 \text{ m/s}^2$

Nonequilibrium (Section 4.12)

Equilibrium   Nonequilibrium

Figure 4.27 CONCEPTS AT A GLANCE Both equilibrium and nonequilibrium problems can be solved with the aid of Newton's second law. For equilibrium situations, such as the gymnast holding the "Iron Cross" position in the left photograph, the acceleration is zero ($\vec{\mathbf{a}} = 0 \text{ m/s}^2$). For nonequilibrium situations, such as the freely falling gymnast in the right photograph, the acceleration is not zero ($\vec{\mathbf{a}} \neq 0 \text{ m/s}^2$). (Both, Mike Powell/Allsport/Getty Images)

► **CONCEPTS AT A GLANCE** The concept of equilibrium arises directly from Newton's second law. The Concepts-at-a-Glance chart in Figure 4.27 illustrates this important point. When the acceleration of an object is zero ($\vec{\mathbf{a}} = 0 \text{ m/s}^2$), the object is in equilibrium, as the upper-right part of the chart indicates. The present section includes several examples involving equilibrium situations. On the other hand, when the acceleration is not zero ($\vec{\mathbf{a}} \neq 0 \text{ m/s}^2$), we have a nonequilibrium situation, as the lower-right part of the chart suggests. Section 4.12 deals with nonequilibrium applications of Newton's second law. ◄

Since the acceleration is zero for an object in equilibrium, all of the acceleration components are also zero. In two dimensions, this means that $a_x = 0 \text{ m/s}^2$ and $a_y = 0 \text{ m/s}^2$. Substituting these values into the second law ($\Sigma F_x = ma_x$ and $\Sigma F_y = ma_y$) shows that the x component and the y component of the net force must each be zero. In other words, the forces acting on an object in equilibrium must balance. Thus, in two dimensions, the equilibrium condition is expressed by two equations:

$$\Sigma F_x = 0 \tag{4.9a}$$
$$\Sigma F_y = 0 \tag{4.9b}$$

In using Equations 4.9a and 4.9b to solve equilibrium problems, we will use the following five-step reasoning strategy:

REASONING STRATEGY

Analyzing Equilibrium Situations

1. Select the object (often called the "system") to which Equations 4.9a and 4.9b are to be applied. It may be that two or more objects are connected by means of a rope or a cable. If so, it may be necessary to treat each object separately according to the following steps.

2. Draw a free-body diagram for each object chosen above. Be sure to include only forces that act on the object. *Do not include forces that the object exerts on its environment.*

3. Choose a set of x, y axes for each object and resolve all forces in the free-body diagram into components that point along these axes. Select the axes so that as many forces as possible point along one or the other of the two axes. Such a choice minimizes the calculations needed to determine the force components.

Continued

4. Apply Equations 4.9a and 4.9b by setting the sum of the *x* components and the sum of the *y* components of the forces each equal to zero.

5. Solve the two equations obtained in Step 4 for the desired unknown quantities, remembering that two equations can yield answers for only two unknowns at most.

Example 11 illustrates how these steps are followed. It deals with a traction device in which three forces act together to bring about the equilibrium.

 Example 11   Traction for the Foot

 The physics of traction for a foot injury.

Figure 4.28*a* shows a traction device used with a foot injury. The weight of the 2.2-kg object creates a tension in the rope that passes around the pulleys. Therefore, tension forces $\vec{T}_1$ and $\vec{T}_2$ are applied to the pulley on the foot. (It may seem surprising that the rope applies a force to either side of the foot pulley. A similar effect occurs when you place a finger inside a rubber band and push downward. You can feel each side of the rubber band pulling upward on the finger.) The foot pulley is kept in equilibrium because the foot also applies a force $\vec{F}$ to it. This force arises in reaction (Newton's third law) to the pulling effect of the forces $\vec{T}_1$ and $\vec{T}_2$. Ignoring the weight of the foot, find the magnitude of $\vec{F}$.

Reasoning The forces $\vec{T}_1$, $\vec{T}_2$, and $\vec{F}$ keep the pulley on the foot at rest. The pulley, therefore, has no acceleration and is in equilibrium. As a result, the sum of the *x* components and the sum of the *y* components of the three forces must each be zero. Figure 4.28*b* shows the free-body diagram of the pulley on the foot. The *x* axis is chosen to be along the direction of force $\vec{F}$, and the components of the forces $\vec{T}_1$ and $\vec{T}_2$ are indicated in the drawing. (See Section 1.7 for a review of vector components.)

Solution Since the sum of the *y* components of the forces is zero, it follows that

$$\Sigma F_y = +T_1 \sin 35° - T_2 \sin 35° = 0 \tag{4.9b}$$

or $T_1 = T_2$. In other words, the magnitudes of the tension forces are equal. In addition, the sum of the *x* components of the forces is zero, so we have that

$$\Sigma F_x = +T_1 \cos 35° + T_2 \cos 35° - F = 0 \tag{4.9a}$$

Solving for *F* and letting $T_1 = T_2 = T$, we find that $F = 2T \cos 35°$. However, the tension *T* in the rope is determined by the weight of the 2.2-kg object: $T = mg$, where *m* is its mass and *g* is the acceleration due to gravity. Therefore, the magnitude of $\vec{F}$ is

$$F = 2T \cos 35° = 2mg \cos 35° = 2(2.2 \text{ kg})(9.80 \text{ m/s}^2) \cos 35° = \boxed{35 \text{ N}}$$

Example 12 presents another situation in which three forces are responsible for the equilibrium of an object. However, in this example all the forces have different magnitudes.

 Example 12   Replacing an Engine

An automobile engine has a weight $\vec{W}$, whose magnitude is $W = 3150$ N. This engine is being positioned above an engine compartment, as Figure 4.29*a* illustrates. To position the engine, a

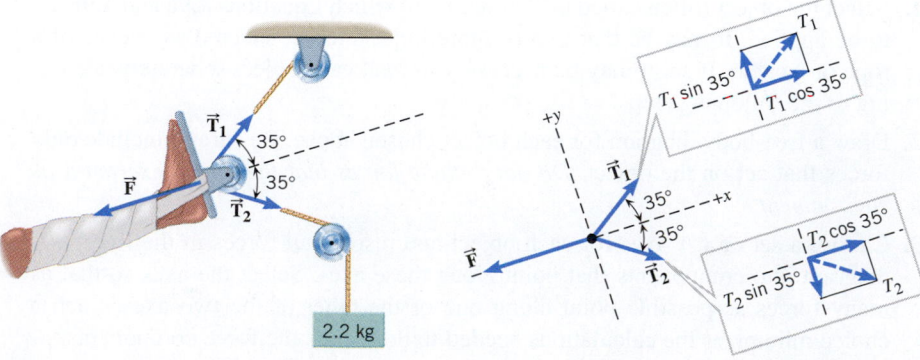

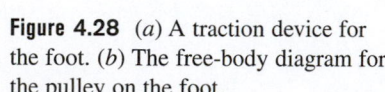

Figure 4.28 (*a*) A traction device for the foot. (*b*) The free-body diagram for the pulley on the foot.

(*a*)

(*b*) Free-body diagram for the foot pulley

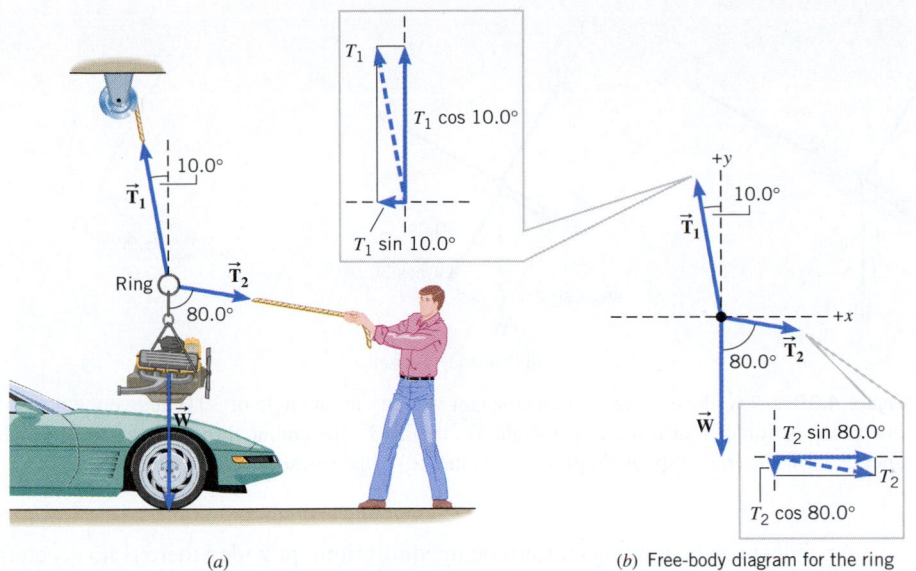

Figure 4.29 (*a*) The ring is in equilibrium because of the three forces $\vec{T}_1$ (the tension force in the supporting cable), $\vec{T}_2$ (the tension force in the positioning rope), and $\vec{W}$ (the weight of the engine). (*b*) The free-body diagram for the ring.

(*b*) Free-body diagram for the ring

worker is using a rope. Find the tension $\vec{T}_1$ in the supporting cable and the tension $\vec{T}_2$ in the positioning rope.

Reasoning Under the influence of the forces $\vec{W}$, $\vec{T}_1$, and $\vec{T}_2$ the ring in Figure 4.29*a* is at rest and, therefore, in equilibrium. Consequently, the sum of the *x* components and the sum of the *y* components of these forces must each be zero; $\Sigma F_x = 0$ and $\Sigma F_y = 0$. By using these relations, we can find T_1 and T_2. Figure 4.29*b* shows the free-body diagram of the ring and the force components for a suitable *x*, *y* axis system.

Solution The free-body diagram shows the components for each of the three forces, and the components are listed in the following table:

Force	*x* Component	*y* Component
$\vec{T}_1$	$-T_1 \sin 10.0°$	$+T_1 \cos 10.0°$
$\vec{T}_2$	$+T_2 \sin 80.0°$	$-T_2 \cos 80.0°$
$\vec{W}$	0	$-W$

The plus signs in the table denote components that point along the positive axes, and the minus signs denote components that point along the negative axes. Setting the sum of the *x* components and the sum of the *y* components equal to zero leads to the following equations:

$$\Sigma F_x = -T_1 \sin 10.0° + T_2 \sin 80.0° = 0 \tag{4.9a}$$

$$\Sigma F_y = +T_1 \cos 10.0° - T_2 \cos 80.0° - W = 0 \tag{4.9b}$$

Solving the first of these equations for T_1 shows that

$$T_1 = \left(\frac{\sin 80.0°}{\sin 10.0°} \right) T_2$$

Substituting this expression for T_1 into the second equation gives

$$\left(\frac{\sin 80.0°}{\sin 10.0°} \right) T_2 \cos 10.0° - T_2 \cos 80.0° - W = 0$$

$$T_2 = \frac{W}{\left(\dfrac{\sin 80.0°}{\sin 10.0°} \right) \cos 10.0° - \cos 80.0°}$$

Setting $W = 3150$ N in this result yields $\boxed{T_2 = 582 \text{ N}}$.

Since $T_1 = \left(\dfrac{\sin 80.0°}{\sin 10.0°} \right) T_2$ and $T_2 = 582$ N, it follows that $\boxed{T_1 = 3.30 \times 10^3 \text{ N}}$.

Problem-solving insight

When an object is in equilibrium, as here in Example 12, the net force is zero, $\Sigma \vec{F} = 0$. This does not mean that each individual force is zero. It means that the vector sum of all the forces is zero.

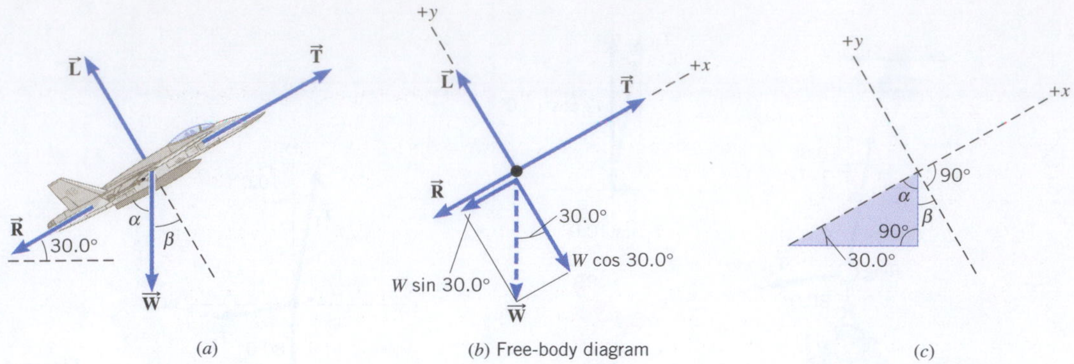

Figure 4.30 (*a*) A plane moves with a constant velocity at an angle of 30.0° above the horizontal due to the action of four forces, the weight $\vec{\mathbf{W}}$, the lift $\vec{\mathbf{L}}$, the engine thrust $\vec{\mathbf{T}}$, and the air resistance $\vec{\mathbf{R}}$. (*b*) The free-body diagram for the plane. (*c*) This geometry occurs often in physics.

An object can be moving and still be in equilibrium, provided there is no acceleration. Example 13 illustrates such a case, and the solution is again obtained using the five-step reasoning strategy summarized at the beginning of the section.

Example 13 Equilibrium at Constant Velocity

A jet plane is flying with a constant speed along a straight line, at an angle of 30.0° above the horizontal, as Figure 4.30*a* indicates. The plane has a weight $\vec{\mathbf{W}}$ whose magnitude is $W = 86\,500$ N, and its engines provide a forward thrust $\vec{\mathbf{T}}$ of magnitude $T = 103\,000$ N. In addition, the lift force $\vec{\mathbf{L}}$ (directed perpendicular to the wings) and the force $\vec{\mathbf{R}}$ of air resistance (directed opposite to the motion) act on the plane. Find $\vec{\mathbf{L}}$ and $\vec{\mathbf{R}}$.

Problem-solving insight

A moving object is in equilibrium if it moves with a constant velocity; then its acceleration is zero. A zero acceleration is the fundamental characteristic of an object in equilibrium.

Reasoning Figure 4.30*b* shows the free-body diagram of the plane, including the forces $\vec{\mathbf{W}}$, $\vec{\mathbf{L}}$, $\vec{\mathbf{T}}$, and $\vec{\mathbf{R}}$. Since the plane is not accelerating, it is in equilibrium, and the sum of the *x* components and the sum of the *y* components of these forces must be zero. The lift force $\vec{\mathbf{L}}$ and the force $\vec{\mathbf{R}}$ of air resistance can be obtained from these equilibrium conditions. To calculate the components, we have chosen axes in the free-body diagram that are rotated by 30.0° from their usual horizontal–vertical positions. This has been done purely for convenience, since the weight $\vec{\mathbf{W}}$ is then the only force that does not lie along either axis.

Solution When determining the components of the weight, it is necessary to realize that the angle β in Figure 4.30*a* is 30.0°. Part *c* of the drawing focuses attention on the geometry that is responsible for this fact. There it can be seen that $\alpha + \beta = 90°$ and $\alpha + 30.0° = 90°$, with the result that $\beta = 30.0°$. The table below lists the components of the forces that act on the jet.

Force	*x* Component	*y* Component
$\vec{\mathbf{W}}$	$-W \sin 30.0°$	$-W \cos 30.0°$
$\vec{\mathbf{L}}$	0	$+L$
$\vec{\mathbf{T}}$	$+T$	0
$\vec{\mathbf{R}}$	$-R$	0

Setting the sum of the *x* component of the forces to zero gives

$$\Sigma F_x = -W \sin 30.0° + T - R = 0 \tag{4.9a}$$

$$R = T - W \sin 30.0° = 103\,000 \text{ N} - (86\,500 \text{ N}) \sin 30.0° = \boxed{59\,800 \text{ N}}$$

Setting the sum of the *y* component of the forces to zero gives

$$\Sigma F_y = -W \cos 30.0° + L = 0 \tag{4.9b}$$

$$L = W \cos 30.0° = (86\,500 \text{ N}) \cos 30.0° = \boxed{74\,900 \text{ N}}$$

✓ CHECK YOUR UNDERSTANDING

(The answers are given at the end of the book.)

19. In which one of the following situations could an object possibly be in equilibrium? **(a)** Three forces act on the object; the forces all point along the same line but may have different directions. **(b)** Two perpendicular forces act on the object. **(c)** A single force acts on the object. **(d)** In none of the situations described in (a), (b), and (c) could the object possibly be in equilibrium.

20. A stone is thrown from the top of a cliff. Air resistance is negligible. As the stone falls, is it **(a)** in equilibrium or **(b)** not in equilibrium?

21. During the final stages of descent, a sky diver with an open parachute approaches the ground with a constant velocity. There is no wind to blow him from side to side. Which one of the following statements is true? **(a)** The sky diver is not in equilibrium. **(b)** The force of gravity is the only force acting on the sky diver, so that he is in equilibrium. **(c)** The sky diver is in equilibrium because no forces are acting on him. **(d)** The sky diver is in equilibrium because two forces act on him, the downward-acting force of gravity and the upward-acting force of the parachute.

22. A crate hangs from a ring at the middle of a rope, as the drawing illustrates. A person is pulling on the right end of the rope to keep the crate in equilibrium. Can the rope ever be made to be perfectly horizontal?

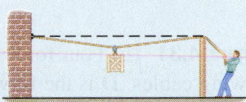

NONEQUILIBRIUM APPLICATIONS OF NEWTON'S LAWS OF MOTION

When an object is accelerating, it is not in equilibrium, as indicated in Figure 4.27. The forces acting on it are not balanced, so the net force is not zero in Newton's second law. However, with one exception, the reasoning strategy followed in solving nonequilibrium problems is identical to that used in equilibrium situations. The exception occurs in Step 4 of the five steps outlined at the beginning of the last section. Since the object is now accelerating, the representation of Newton's second law in Equations 4.2a and 4.2b applies instead of Equations 4.9a and 4.9b:

$$\Sigma F_x = ma_x \quad (4.2a) \qquad \text{and} \qquad \Sigma F_y = ma_y \quad (4.2b)$$

Example 14 uses these equations in a situation where the forces are applied in directions similar to those in Example 11, except that now an acceleration is present.

Example 14 Towing a Supertanker

A supertanker of mass $m = 1.50 \times 10^8$ kg is being towed by two tugboats, as in Figure 4.31*a*. The tensions in the towing cables apply the forces $\vec{T}_1$ and $\vec{T}_2$ at equal angles of 30.0° with respect to the tanker's axis. In addition, the tanker's engines produce a forward drive force $\vec{D}$, whose magnitude is $D = 75.0 \times 10^3$ N. Moreover, the water applies an opposing force $\vec{R}$, whose magnitude is $R = 40.0 \times 10^3$ N. The tanker moves forward with an acceleration that points along the tanker's axis and has a magnitude of 2.00×10^{-3} m/s². Find the magnitudes of the tensions $\vec{T}_1$ and $\vec{T}_2$.

Reasoning The unknown forces $\vec{T}_1$ and $\vec{T}_2$ contribute to the net force that accelerates the tanker. To determine $\vec{T}_1$ and $\vec{T}_2$, therefore, we analyze the net force, which we will do using components. The various force components can be found by referring to the free-body diagram for the tanker in Figure 4.31*b*, where the ship's axis is chosen as the *x* axis. We will then use Newton's second law in its component form, $\Sigma F_x = ma_x$ and $\Sigma F_y = ma_y$, to obtain the magnitudes of $\vec{T}_1$ and $\vec{T}_2$.

Solution The individual force components are summarized as follows:

Force	*x* Component	*y* Component
$\vec{T}_1$	$+T_1 \cos 30.0°$	$+T_1 \sin 30.0°$
$\vec{T}_2$	$+T_2 \cos 30.0°$	$-T_2 \sin 30.0°$
$\vec{D}$	$+D$	0
$\vec{R}$	$-R$	0

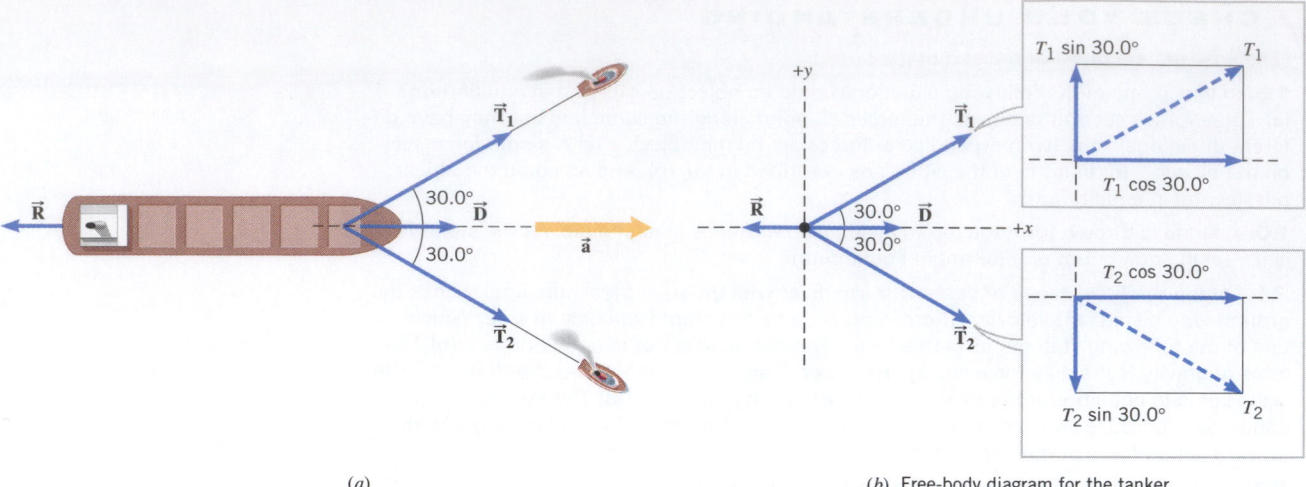

(a)   (b) Free-body diagram for the tanker

Figure 4.31 (a) Four forces act on a supertanker: $\vec{T}_1$ and $\vec{T}_2$ are the tension forces due to the towing cables, $\vec{D}$ is the forward drive force produced by the tanker's engines, and $\vec{R}$ is the force with which the water opposes the tanker's motion. (b) The free-body diagram for the tanker.

Since the acceleration points along the x axis, there is no y component of the acceleration ($a_y = 0$ m/s²). Consequently, the sum of the y components of the forces must be zero:

$$\Sigma F_y = +T_1 \sin 30.0° - T_2 \sin 30.0° = 0$$

This result shows that the magnitudes of the tensions in the cables are equal, $T_1 = T_2$. Since the ship accelerates along the x direction, the sum of the x components of the forces is not zero. The second law indicates that

$$\Sigma F_x = T_1 \cos 30.0° + T_2 \cos 30.0° + D - R = ma_x$$

Since $T_1 = T_2$, we can replace the two separate tension symbols by a single symbol T, the magnitude of the tension. Solving for T gives

$$T = \frac{ma_x + R - D}{2 \cos 30.0°}$$

$$= \frac{(1.50 \times 10^8 \text{ kg})(2.00 \times 10^{-3} \text{ m/s}^2) + 40.0 \times 10^3 \text{ N} - 75.0 \times 10^3 \text{ N}}{2 \cos 30.0°}$$

$$= \boxed{1.53 \times 10^5 \text{ N}}$$

It often happens that two objects are connected somehow, perhaps by a drawbar like that used when a truck pulls a trailer. If the tension in the connecting device is of no interest, the objects can be treated as a single composite object when applying Newton's second law. However, if it is necessary to find the tension, as in the next example, then the second law must be applied separately to at least one of the objects.

Example 15   Hauling a Trailer

A truck is hauling a trailer along a level road, as Figure 4.32a illustrates. The mass of the truck is $m_1 = 8500$ kg and that of the trailer is $m_2 = 27\ 000$ kg. The two move along the x axis with an acceleration of $a_x = 0.78$ m/s². Ignoring the retarding forces of friction and air resistance, determine **(a)** the tension $\vec{T}$ in the horizontal drawbar between the trailer and the truck and **(b)** the force $\vec{D}$ that propels the truck forward.

Reasoning Since the truck and the trailer accelerate along the horizontal direction and friction is being ignored, only forces that have components in the horizontal direction are of

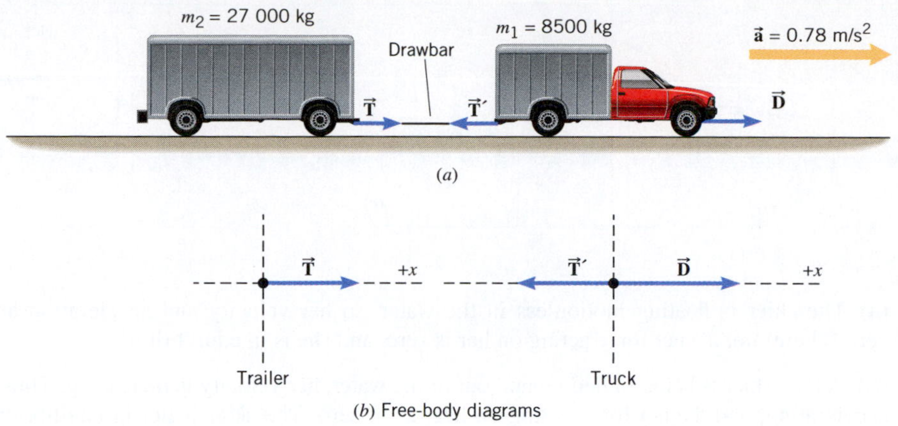

$m_2 = 27\ 000$ kg

Drawbar

$m_1 = 8500$ kg

$\vec{a} = 0.78$ m/s^2

$\vec{T}$ $\vec{T}'$ $\vec{D}$

(a)

$\vec{T}$ $+x$ $\vec{T}'$ $\vec{D}$ $+x$

Trailer

Truck

(b) Free-body diagrams

Figure 4.32 (a) The force $\vec{D}$ acts on the truck and propels it forward. The drawbar exerts the tension force $\vec{T}'$ on the truck and the tension force $\vec{T}$ on the trailer. (b) The free-body diagrams for the trailer and the truck, ignoring the vertical forces.

interest. Therefore, Figure 4.32 omits the weight and the normal force, which act vertically. To determine the tension force $\vec{T}$ in the drawbar, we draw the free-body diagram for the trailer and apply Newton's second law, $\Sigma F_x = ma_x$. Similarly, we can determine the propulsion force $\vec{D}$ by drawing the free-body diagram for the truck and applying Newton's second law.

Solution (a) The free-body diagram for the trailer is shown in Figure 4.32b. There is only one horizontal force acting on the trailer, the tension force $\vec{T}$ due to the drawbar. Therefore, it is straightforward to obtain the tension from $\Sigma F_x = m_2 a_x$, since the mass of the trailer and the acceleration are known:

$$\Sigma F_x = T = m_2 a_x = (27\ 000\ \text{kg})(0.78\ \text{m/s}^2) = \boxed{21\ 000\ \text{N}}$$

(b) Two horizontal forces act on the truck, as the free-body diagram in Figure 4.32b shows. One is the desired force $\vec{D}$. The other is the force $\vec{T}'$. According to Newton's third law, $\vec{T}'$ is the force with which the trailer pulls back on the truck, in reaction to the truck pulling forward. If the drawbar has negligible mass, the magnitude of $\vec{T}'$ is equal to the magnitude of $\vec{T}$—namely, 21 000 N. Since the magnitude of $\vec{T}'$, the mass of the truck, and the acceleration are known, $\Sigma F_x = m_1 a_x$ can be used to determine the drive force:

$$\Sigma F_x = +D - T' = m_1 a_x$$

$$D = m_1 a_x + T' = (8500\ \text{kg})(0.78\ \text{m/s}^2) + 21\ 000\ \text{N} = \boxed{28\ 000\ \text{N}}$$

In Section 4.11 we examined situations where the net force acting on an object is zero, and in this section we have considered two examples where the net force is not zero. Conceptual Example 16 illustrates a common situation where the net force is zero at certain times but is not zero at other times.

 Conceptual Example 16 The Motion of a Water Skier

Figure 4.33 shows a water skier at four different moments:

(a) The skier is floating motionless in the water.

(b) The skier is being pulled out of the water and up onto the skis.

(c) The skier is moving at a constant speed along a straight line.

(d) The skier has let go of the tow rope and is slowing down.

For each moment, explain whether the net force acting on the skier is zero.

Reasoning and Solution According to Newton's second law, if an object has zero acceleration, the net force acting on it is zero. In such a case, the object is in equilibrium. In contrast, if the object has an acceleration, the net force acting on it is not zero. Such an object is not in equilibrium. We will consider the acceleration in each of the four phases of the motion to decide whether the net force is zero.

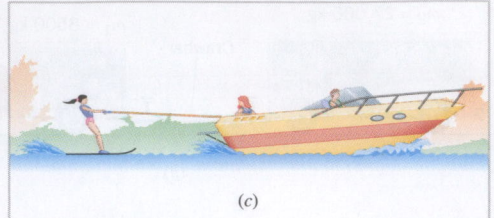

Figure 4.33 A water skier (*a*) floating in water, (*b*) being pulled up by the boat, (*c*) moving at a constant velocity, and (*d*) slowing down.

(a) The skier is floating motionless in the water, so her velocity and acceleration are both zero. Therefore, the net force acting on her is zero, and she is in equilibrium.

(b) As the skier is being pulled up and out of the water, her velocity is increasing. Thus, she is accelerating, and the net force acting on her is not zero. The skier is not in equilibrium. The direction of the net force is shown in Figure 4.33*b*.

(c) The skier is now moving at a constant speed along a straight line (Figure 4.33*c*), so her velocity is constant. Since her velocity is constant, her acceleration is zero. Thus, the net force acting on her is zero, and she is again in equilibrium, even though she is moving.

(d) After the skier lets go of the tow rope, her speed decreases, so she is decelerating. Thus, the net force acting on her is not zero, and she is not in equilibrium. The direction of the net force is shown in Figure 4.33*d*.

Related Homework: *Problem 78*

The force of gravity is often present among the forces that affect the acceleration of an object. Examples 17–19 deal with typical situations.

ANALYZING MULTIPLE-CONCEPT PROBLEMS

Example 17   Hauling a Crate

A flatbed truck is carrying a crate up a 10.0° hill, as Figure 4.34*a* illustrates. The coefficient of static friction between the truck bed and the crate is 0.350. Find the maximum acceleration that the truck can attain before the crate begins to slip backward relative to the truck.

Reasoning The crate will not slip as long as it has the same acceleration as the truck. Therefore, a net force must act on the crate to accelerate it, and the static frictional force $\vec{f}_s$ contributes to this net force. Since the crate tends to slip backward, the static frictional force is directed forward, up the hill.

As the acceleration of the truck increases, $\vec{f}_s$ must also increase to produce a corresponding increase in the acceleration of the crate. However, the static frictional force can increase only until its maximum value $\vec{f}_s^{\text{MAX}}$ is reached, at which point the crate and truck have the maximum acceleration $\vec{a}^{\text{MAX}}$. If the acceleration increases even more, the crate will slip.

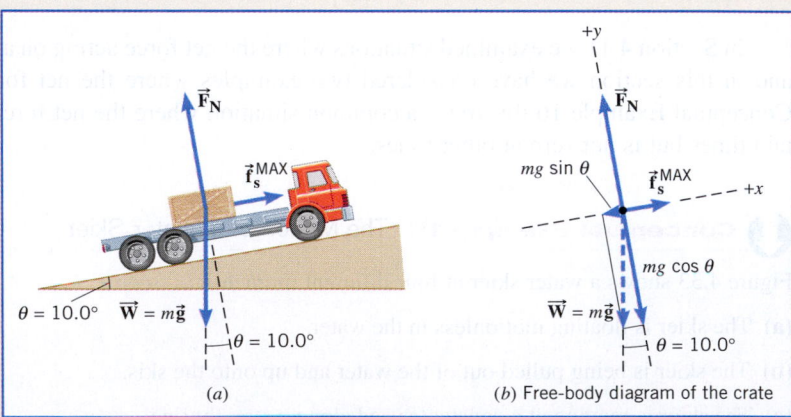

Figure 4.34 (*a*) A crate on a truck is kept from slipping by the static frictional force $\vec{f}_s^{\text{MAX}}$. The other forces that act on the crate are its weight $\vec{W}$ and the normal force $\vec{F}_N$. (*b*) The free-body diagram of the crate.

To find $\vec{a}^{\text{MAX}}$, we will employ Newton's second law, the definition of weight, and the relationship between the maximum static frictional force and the normal force.

Knowns and Unknowns   The data for this problem are as follows:

Description	Symbol	Value
Angle of hill	θ	10.0°
Coefficient of static friction	μ_s	0.350
Unknown Variable		
Maximum acceleration before crate slips	a^{MAX}	?

Modeling the Problem

STEP 1   Newton's Second Law (*x* direction)   With the *x* direction chosen to be parallel to the acceleration of the truck, Newton's second law for this direction can be written as (see Equation 4.2a) $a_x^{\text{MAX}} = (\Sigma F_x)/m$, where ΣF_x is the net force acting on the crate in the *x* direction and *m* is the crate's mass. Using the *x* components of the forces shown in Figure 4.34*b*, we find that the net force is $\Sigma F_x = -mg \sin \theta + f_s^{\text{MAX}}$. Substituting this expression into Newton's second law gives

$$a_x^{\text{MAX}} = \frac{\Sigma F_x}{m} = \frac{-mg \sin \theta + f_s^{\text{MAX}}}{m}$$

$$a_x^{\text{MAX}} = \frac{-mg \sin \theta + f_s^{\text{MAX}}}{m} \quad (1)$$

The acceleration due to gravity g and the angle θ are known, but m and f_s^{MAX} are not. We will now turn our attention to finding f_s^{MAX}.

STEP 2   The Maximum Static Frictional Force   The magnitude f_s^{MAX} of the maximum static frictional force is related to the coefficient of static friction μ_s and the magnitude F_N of the normal force by Equation 4.7:

$$f_s^{\text{MAX}} = \mu_s F_N \quad (4.7)$$

$$a_x^{\text{MAX}} = \frac{-mg \sin \theta + f_s^{\text{MAX}}}{m} \quad (1)$$

$$f_s^{\text{MAX}} = \mu_s F_N \quad (4.7)$$

This result can be substituted into Equation 1, as shown at the right. Although μ_s is known, F_N is not known. An expression for F_N will be found in Step 3, however.

STEP 3   Newton's Second Law (*y* direction)   We can determine the magnitude F_N of the normal force by noting that the crate does not accelerate in the *y* direction ($a_y = 0$ m/s²). Thus, Newton's second law as given in Equation 4.2b becomes

$$\Sigma F_y = ma_y = 0$$

There are two forces acting on the crate in the *y* direction (see Figure 4.34*b*): the normal force $+F_N$ and the *y* component of the weight $-mg \cos \theta$ (The minus sign is included because this component points along the negative *y* direction.) Thus, the net force is $\Sigma F_y = +F_N - mg \cos \theta$. Newton's second law becomes

$$+F_N - mg \cos \theta = 0 \quad \text{or} \quad \boxed{F_N = mg \cos \theta}$$

$$a_x^{\text{MAX}} = \frac{-mg \sin \theta + f_s^{\text{MAX}}}{m} \quad (1)$$

$$f_s^{\text{MAX}} = \mu_s F_N \quad (4.7)$$

$$\boxed{F_N = mg \cos \theta}$$

This result for F_N can be substituted into Equation 4.7, as indicated at the right.

Continued

Solution Algebraically combining the results of the three steps, we find that

| STEP 1 | STEP 2 | STEP 3 |

$$a_x^{\text{MAX}} = \frac{-mg\sin\theta + f_s^{\text{MAX}}}{m} = \frac{-mg\sin\theta + \mu_s F_N}{m} = \frac{-\cancel{m}g\sin\theta + \mu_s \cancel{m}g\cos\theta}{\cancel{m}}$$

$$= -g\sin\theta + \mu_s g\cos\theta$$

Note that the mass m of the crate is algebraically eliminated from the final result. Thus, the maximum acceleration is

$$a_x^{\text{MAX}} = -g\sin\theta + \mu_s g\cos\theta$$

$$= -(9.80 \text{ m/s}^2)\sin 10.0° + (0.350)(9.80 \text{ m/s}^2)\cos 10.0° = \boxed{1.68 \text{ m/s}^2}$$

Related Homework: *Problems 50, 81, 86*

⬇ Example 18   Accelerating Blocks

Block 1 (mass $m_1 = 8.00$ kg) is moving on a frictionless 30.0° incline. This block is connected to block 2 (mass $m_2 = 22.0$ kg) by a massless cord that passes over a massless and frictionless pulley (see Figure 4.35a). Find the acceleration of each block and the tension in the cord.

Reasoning Since both blocks accelerate, there must be a net force acting on each one. The key to this problem is to realize that Newton's second law can be used separately for each block to relate the net force and the acceleration. Note also that both blocks have accelerations of the same magnitude a, since they move as a unit. We assume that block 1 accelerates up the incline and choose this direction to be the $+x$ axis. If block 1 in reality accelerates down the incline, then the value obtained for the acceleration will be a negative number.

Solution Three forces act on block 1: (1) $\vec{W}_1$ is its weight $[W_1 = m_1 g = (8.00 \text{ kg}) \times (9.80 \text{ m/s}^2) = 78.4 \text{ N}]$, (2) $\vec{T}$ is the force applied because of the tension in the cord, and (3) $\vec{F}_N$ is the normal force that the incline exerts. Figure 4.35b shows the free-body diagram for block 1. The weight is the only force that does not point along the x, y axes, and its x and y components are given in the diagram. Applying Newton's second law ($\Sigma F_x = m_1 a_x$) to block 1 shows that

$$\Sigma F_x = -W_1 \sin 30.0° + T = m_1 a$$

where we have set $a_x = a$. This equation cannot be solved as it stands, since both T and a are unknown quantities. To complete the solution, we next consider block 2.

Two forces act on block 2, as the free-body diagram in Figure 4.35b indicates: (1) $\vec{W}_2$ is its weight $[W_2 = m_2 g = (22.0 \text{ kg})(9.80 \text{ m/s}^2) = 216 \text{ N}]$ and (2) $\vec{T}'$ is exerted as a result of

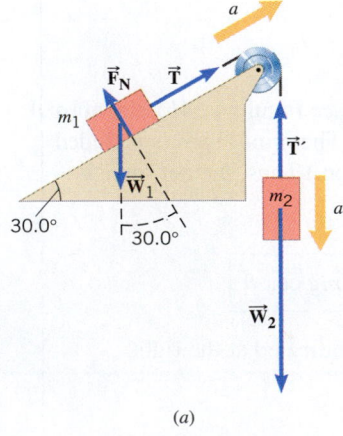

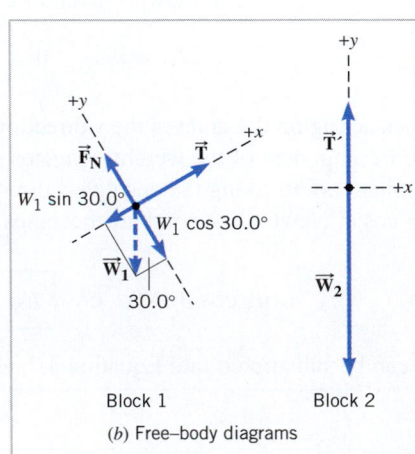

Figure 4.35 (a) Three forces act on block 1: its weight $\vec{W}_1$, the normal force $\vec{F}_N$, and the force $\vec{T}$ due to the tension in the cord. Two forces act on block 2: its weight $\vec{W}_2$ and the force $\vec{T}'$ due to the tension. The acceleration is labeled according to its magnitude a. (b) Free-body diagrams for the two blocks.

Block 1

Block 2

(b) Free–body diagrams

block 1 pulling back on the connecting cord. Since the cord and the frictionless pulley are massless, the magnitudes of $\vec{T}'$ and $\vec{T}$ are the same: $T' = T$. Applying Newton's second law ($\Sigma F_y = m_2 a_y$) to block 2 reveals that

$$\Sigma F_y = T - W_2 = m_2(-a)$$

The acceleration a_y has been set equal to $-a$ since block 2 moves downward along the $-y$ axis in the free-body diagram, consistent with the assumption that block 1 moves up the incline. Now there are two equations in two unknowns, and they may be solved simultaneously (see Appendix C) to give T and a:

$$\boxed{T = 86.3 \text{ N}} \quad \text{and} \quad \boxed{a = 5.89 \text{ m/s}^2}$$

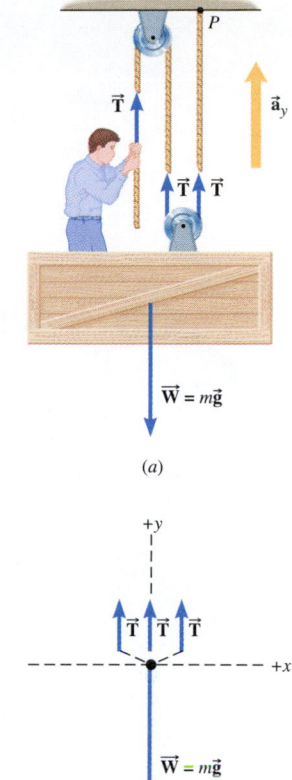

(a)

$\vec{W} = m\vec{g}$

 Example 19 Hoisting a Scaffold

A window washer on a scaffold is hoisting the scaffold up the side of a building by pulling downward on a rope, as in Figure 4.36a. The magnitude of the pulling force is 540 N, and the combined mass of the worker and the scaffold is 155 kg. Find the upward acceleration of the unit.

Reasoning The worker and the scaffold form a single unit, on which the rope exerts a force in three places. The left end of the rope exerts an upward force $\vec{T}$ on the worker's hands. This force arises because he pulls downward with a 540-N force, and the rope exerts an oppositely directed force of equal magnitude on him, in accord with Newton's third law. Thus, the magnitude T of the upward force is $T = 540$ N and is the magnitude of the tension in the rope. If the masses of the rope and each pulley are negligible and if the pulleys are friction-free, the tension is transmitted undiminished along the rope. Then, a 540-N tension force $\vec{T}$ acts upward on the left side of the scaffold pulley (see part a of the drawing). A tension force is also applied to the point P, where the rope attaches to the roof. The roof pulls back on the rope in accord with the third law, and this pull leads to the 540-N tension force $\vec{T}$ that acts on the right side of the scaffold pulley. In addition to the three upward forces, the weight of the unit must be taken into account [$W = mg = (155 \text{ kg})(9.80 \text{ m/s}^2) = 1520$ N]. Part b of the drawing shows the free-body diagram.

(b) Free-body diagram of the unit

Figure 4.36 (a) A window washer pulls down on the rope to hoist the scaffold up the side of a building. The force $\vec{T}$ results from the effort of the window washer and acts on him and the scaffold in three places, as discussed in Example 19. (b) The free-body diagram of the unit comprising the man and the scaffold.

Solution Newton's second law ($\Sigma F_y = ma_y$) can be applied to calculate the acceleration a_y:

$$\Sigma F_y = +T + T + T - W = ma_y$$

$$a_y = \frac{3T - W}{m} = \frac{3(540 \text{ N}) - 1520 \text{ N}}{155 \text{ kg}} = \boxed{0.65 \text{ m/s}^2}$$

 CHECK YOUR UNDERSTANDING

(*The answers are given at the end of the book.*)

23. A circus performer hangs stationary from a rope. She then begins to climb upward by pulling herself up, hand over hand. When she starts climbing, is the tension in the rope **(a)** less than, **(b)** equal to, or **(c)** greater than it is when she hangs stationary?

24. A freight train is accelerating on a level track. Other things being equal, would the tension in the coupling between the engine and the first car change if some of the cargo in the last car were transferred to any one of the other cars?

25. Two boxes have masses m_1 and m_2, and m_2 is greater than m_1. The boxes are being pushed across a frictionless horizontal surface. As the drawing shows, there are two possible arrangements, and the pushing force is the same in each. In which arrangement, **(a)** or **(b)**, does the force that the left box applies to the right box have a greater magnitude, or **(c)** is the magnitude the same in both cases?

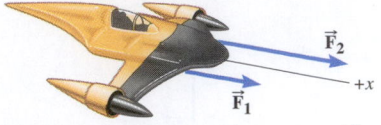

Figure 4.37 Two horizontal forces, $\vec{F}_1$ and $\vec{F}_2$, act on the spacecraft. A third force $\vec{F}_3$ also acts but is not shown.

4.13

CONCEPTS & CALCULATIONS

Newton's three laws of motion provide the basis for understanding the effect of forces on the motion of an object, as we have seen. The second law is especially important, because it provides the quantitative relationship between net force and acceleration. The examples in this section serve as a review of the essential features of this relationship.

Concepts & Calculations Example 20
Velocity, Acceleration, and Newton's Second Law of Motion

Figure 4.37 shows two forces, $\vec{F}_1 = +3000$ N and $\vec{F}_2 = +5000$ N, acting on a spacecraft; the plus signs indicate that the forces are directed along the $+x$ axis. A third force $\vec{F}_3$ also acts on the spacecraft but is not shown in the drawing. The craft is moving with a constant velocity of $+850$ m/s. Find the magnitude and direction of $\vec{F}_3$.

Concept Questions and Answers Suppose the spacecraft were stationary. What would be the direction of $\vec{F}_3$?

Answer If the spacecraft were stationary, its acceleration would be zero. According to Newton's second law, the acceleration of an object is proportional to the net force acting on it. Thus, the net force must also be zero. But the net force is the vector sum of the three forces in this case. Therefore, the force $\vec{F}_3$ must have a direction such that it balances to zero the forces $\vec{F}_1$ and $\vec{F}_2$. Since $\vec{F}_1$ and $\vec{F}_2$ point along the $+x$ axis in Figure 4.37, $\vec{F}_3$ must then point along the $-x$ axis.

When the spacecraft is moving at a constant velocity of $+850$ m/s, what is the direction of $\vec{F}_3$?

Answer Since the velocity is constant, the acceleration is still zero. As a result, everything we said in the stationary case applies again here. The net force is zero, and the force $\vec{F}_3$ must point along the $-x$ axis in Figure 4.37.

Solution Since the velocity is constant, the acceleration is zero. The net force must also be zero, so that

$$\Sigma F_x = F_1 + F_2 + F_3 = 0$$

Solving for F_3 yields

$$F_3 = -(F_1 + F_2) = -(3000 \text{ N} + 5000 \text{ N}) = \boxed{-8000 \text{ N}}$$

The minus sign in the answer means that $\vec{F}_3$ points opposite to the sum of $\vec{F}_1$ and $\vec{F}_2$, or along the $-x$ axis in Figure 4.37. The force $\vec{F}_3$ has a magnitude of 8000 N, which is the magnitude of the sum of the forces $\vec{F}_1$ and $\vec{F}_2$. The answer is independent of the velocity of the spacecraft, as long as that velocity remains constant.

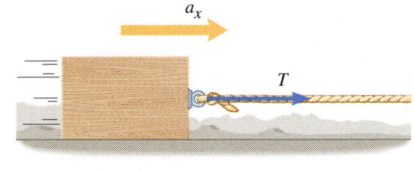

(a)

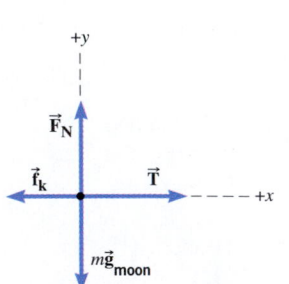

(b) Free-body diagram for the block

Figure 4.38 (a) A block is sliding on a horizontal surface on the moon. The tension in the rope is $\vec{T}$. (b) The free-body diagram for the block, including a kinetic frictional force $\vec{f}_k$.

Concepts & Calculations Example 21
The Importance of Mass

On earth a block has a weight of 88 N. This block is sliding on a horizontal surface on the moon, where the acceleration due to gravity is 1.60 m/s². As Figure 4.38a shows, the block is being pulled by a horizontal rope in which the tension is $T = 24$ N. The coefficient of kinetic friction between the block and the surface is $\mu_k = 0.20$. Determine the acceleration of the block.

Concept Questions and Answers Which of Newton's laws of motion provides a way to determine the acceleration of the block?

Answer Newton's second law allows us to calculate the acceleration as $a_x = \Sigma F_x / m$, where ΣF_x is the net force acting in the horizontal direction and m is the mass of the block.

This problem deals with a situation on the moon, but the block's mass on the moon is not given. Instead, the block's earth-weight is given. Why can the earth-weight be used to obtain a value for the block's mass that applies on the moon?

Answer Since the block's earth-weight W_{earth} is related to the block's mass according to $W_{earth} = mg_{earth}$, we can use $W_{earth} = 88$ N and $g_{earth} = 9.80$ m/s² to obtain m. But mass is an intrinsic property of the block and does not depend on whether the block is on the earth or on the moon. Therefore, the value obtained for m applies on the moon as well as on the earth.

Does the net force ΣF_x equal the tension T?

Answer No. The net force ΣF_x is the vector sum of all the external forces acting in the horizontal direction. It includes the kinetic frictional force f_k as well as the tension T.

Solution Figure 4.38*b* shows the free-body diagram for the block. The net force along the x axis is $\Sigma F_x = +T - f_k$, where T is the magnitude of the tension in the rope and f_k is the magnitude of the kinetic frictional force. According to Equation 4.8, f_k is related to the magnitude F_N of the normal force by $f_k = \mu_k F_N$, where μ_k is the coefficient of kinetic friction. The acceleration a_x of the block is given by Newton's second law as

$$a_x = \frac{\Sigma F_x}{m} = \frac{+T - \mu_k F_N}{m}$$

We can obtain an expression for F_N by noting that the block does not move in the y direction, so $a_y = 0 \text{ m/s}^2$. Therefore, the net force ΣF_y along the y direction must also be zero. An examination of the free-body diagram reveals that $\Sigma F_y = +F_N - mg_{\text{moon}} = 0$, so that $F_N = mg_{\text{moon}}$. The acceleration in the x direction becomes

$$a_x = \frac{+T - \mu_k mg_{\text{moon}}}{m}$$

Using the earth-weight of the block to determine its mass, we find

$$W_{\text{earth}} = mg_{\text{earth}} \quad \text{or} \quad m = \frac{W_{\text{earth}}}{g_{\text{earth}}} = \frac{88 \text{ N}}{9.80 \text{ m/s}^2} = 9.0 \text{ kg}$$

The acceleration of the block is, then,

$$a_x = \frac{+T - \mu_k mg_{\text{moon}}}{m} = \frac{24 \text{ N} - (0.20)(9.0 \text{ kg})(1.60 \text{ m/s}^2)}{9.0 \text{ kg}} = \boxed{+2.3 \text{ m/s}^2}$$

CONCEPT SUMMARY

If you need more help with a concept, use the Learning Aids noted next to the discussion or equation. Examples (**Ex.**) are in the text of this chapter. Go to **www.wiley.com/college/cutnell** for the following Learning Aids:

Interactive LearningWare (ILW) — Additional examples solved in a five-step interactive format.

Concept Simulations (CS) — Animated text figures or animations of important concepts.

Interactive Solutions (IS) — Models for certain types of problems in the chapter homework. The calculations are carried out interactively.

Topic	Discussion	Learning Aids
Contact and noncontact forces	**4.1 THE CONCEPTS OF FORCE AND MASS** A force is a push or a pull and is a vector quantity. Contact forces arise from the physical contact between two objects. Noncontact forces are also called action-at-a-distance forces, because they arise without physical contact between two objects.	
Mass	Mass is a property of matter that determines how difficult it is to accelerate or decelerate an object. Mass is a scalar quantity.	
Newton's first law	**4.2 NEWTON'S FIRST LAW OF MOTION** Newton's first law of motion, sometimes called the law of inertia, states that an object continues in a state of rest or in a state of motion at a constant velocity unless compelled to change that state by a net force.	
Inertia Mass Inertial reference frame	Inertia is the natural tendency of an object to remain at rest or in motion at a constant velocity. The mass of a body is a quantitative measure of inertia and is measured in an SI unit called the kilogram (kg). An inertial reference frame is one in which Newton's law of inertia is valid.	

Topic	Discussion	Learning Aids

4.3 NEWTON'S SECOND LAW OF MOTION   4.4 THE VECTOR NATURE OF NEWTON'S SECOND LAW OF MOTION   Newton's second law of motion states that when a net force $\Sigma\vec{F}$ acts on an object of mass m, the acceleration $\vec{a}$ of the object can be obtained from the following equation:

Newton's second law (vector form)

$$\Sigma\vec{F} = m\vec{a} \qquad (4.1)$$ **CS 4.1**

This is a vector equation and, for motion in two dimensions, is equivalent to the following two equations:

Newton's second law (component form)

$$\Sigma F_x = ma_x \qquad (4.2a)$$
$$\Sigma F_y = ma_y \qquad (4.2b)$$

Ex. 1–3, 20, 21
IS 4.11
ILW 4.11

In these equations the x and y subscripts refer to the scalar components of the force and acceleration vectors. The SI unit of force is the Newton (N).

Free-body diagram

When determining the net force, a free-body diagram is helpful. A free-body diagram is a diagram that represents the object and the forces acting on it.

Newton's third law of motion

4.5 NEWTON'S THIRD LAW OF MOTION   Newton's third law of motion, often called the action–reaction law, states that whenever one object exerts a force on a second object, the second object exerts an oppositely directed force of equal magnitude on the first object.   **Ex. 4**

Fundamental forces

4.6 TYPES OF FORCES: AN OVERVIEW   Only three fundamental forces have been discovered: the gravitational force, the strong nuclear force, and the electroweak force. The electroweak force manifests itself as either the electromagnetic force or the weak nuclear force.

4.7 THE GRAVITATIONAL FORCE   Newton's law of universal gravitation states that every particle in the universe exerts an attractive force on every other particle. For two particles that are separated by a distance r and have masses m_1 and m_2, the law states that the magnitude of this attractive force is

Newton's law of universal gravitation

$$F = G\frac{m_1 m_2}{r^2} \qquad (4.3)$$ **Ex. 5, 6**

The direction of this force lies along the line between the particles. The constant G has a value of $G = 6.674 \times 10^{-11}\ \text{N·m}^2/\text{kg}^2$ and is called the universal gravitational constant.

The weight W of an object on or above the earth is the gravitational force that the earth exerts on the object and can be calculated from the mass m of the object and the magnitude g of the acceleration due to the earth's gravity according to

Weight and mass

$$W = mg \qquad (4.5)$$ **Ex. 7**

Normal force

4.8 THE NORMAL FORCE   The normal force $\vec{F}_N$ is one component of the force that a surface exerts on an object with which it is in contact—namely, the component that is perpendicular to the surface.   **Ex. 8**

The apparent weight is the force that an object exerts on the platform of a scale and may be larger or smaller than the true weight mg if the object and the scale have an acceleration a (+ if upward, − if downward). The apparent weight is

Apparent weight

$$\text{Apparent weight} = mg + ma \qquad (4.6)$$

4.9 STATIC AND KINETIC FRICTIONAL FORCES   A surface exerts a force on an object with which it is in contact. The component of the force perpendicular to the surface is called the normal force.

Friction

The component parallel to the surface is called friction.

The force of static friction between two surfaces opposes any impending relative motion of the surfaces. The magnitude of the static frictional force depends on the magnitude of the applied force and can assume any value up to a maximum of

Maximum static frictional force

$$f_s^{\text{MAX}} = \mu_s F_N \qquad (4.7)$$

Ex. 9
CS 4.2, 4.4
ILW 4.2

where μ_s is the coefficient of static friction and F_N is the magnitude of the normal force.

The force of kinetic friction between two surfaces sliding against one another opposes the relative motion of the surfaces. This force has a magnitude given by

Kinetic frictional force

$$f_k = \mu_k F_N \qquad (4.8)$$

Ex. 10
CS 4.3
IS 4.47, 4.65

where μ_k is the coefficient of kinetic friction.

Topic	Discussion	Learning Aids
Tension	**4.10 THE TENSION FORCE** The word "tension" is commonly used to mean the tendency of a rope to be pulled apart due to forces that are applied at each end. Because of tension, a rope transmits a force from one end to the other. When a rope is accelerating, the force is transmitted undiminished only if the rope is massless.	
Definition of equilibrium	**4.11 EQUILIBRIUM APPLICATIONS OF NEWTON'S LAWS OF MOTION** An object is in equilibrium when the object has zero acceleration, or, in other words, when it moves at a constant velocity (which may be zero). The sum of the forces that act on an object in equilibrium is zero. Under equilibrium conditions in two dimensions, the separate sums of the force components in the x direction and in the y direction must each be zero:	
The equilibrium condition	$$\Sigma F_x = 0 \qquad (4.9a)$$ $$\Sigma F_y = 0 \qquad (4.9b)$$	**Ex. 11, 12, 13** **ILW 4.3**
	4.12 NONEQUILIBRIUM APPLICATIONS OF NEWTON'S LAWS OF MOTION If an object is not in equilibrium, then Newton's second law must be used to account for the acceleration:	
	$$\Sigma F_x = ma_x \qquad (4.2a)$$ $$\Sigma F_y = ma_y \qquad (4.2b)$$	**Ex. 14–19** **ILW 4.4** **IS 4.83**

FOCUS ON CONCEPTS

Note to Instructors: The numbering of the questions shown here reflects the fact that they are only a representative subset of the total number that are available online. However, all of the questions are available for assignment via an online homework management program such as WileyPLUS or WebAssign.

Section 4.2   Newton's First Law of Motion

1. An object is moving at a constant velocity. All but one of the following statements could be true. Which one cannot be true? **(a)** No forces act on the object. **(b)** A single force acts on the object. **(c)** Two forces act simultaneously on the object. **(d)** Three forces act simultaneously on the object.

3. A cup of coffee is sitting on a table in a recreational vehicle (RV). The cup slides toward the rear of the RV. According to Newton's first law, which one or more of the following statements could describe the motion of the RV? (A) The RV is at rest, and the driver suddenly accelerates. (B) The RV is moving forward, and the driver suddenly accelerates. (C) The RV is moving backward, and the driver suddenly hits the brakes. **(a)** A **(b)** B **(c)** C **(d)** A and B **(e)** A, B, and C

Section 4.4   The Vector Nature of Newton's Second Law of Motion

5. Two forces act on a moving object that has a mass of 27 kg. One force has a magnitude of 12 N and points due south, while the other force has a magnitude of 17 N and points due west. What is the acceleration of the object? **(a)** 0.63 m/s² directed 55° south of west **(b)** 0.44 m/s² directed 24° south of west **(c)** 0.77 m/s² directed 35° south of west **(d)** 0.77 m/s² directed 55° south of west **(e)** 1.1 m/s² directed 35° south of west

Section 4.5   Newton's Third Law of Motion

7. Which one of the following is true, according to Newton's laws of motion? Ignore friction. **(a)** A sports utility vehicle (SUV) hits a stationary motorcycle. Since it is stationary, the motorcycle sustains a greater force than the SUV does. **(b)** A semitrailer truck crashes all the way through a wall. Since the wall collapses, the wall sustains a greater force than the truck does. **(c)** Sam (18 years old) and his sister (9 years old) go ice skating. They push off against each other and fly apart. Sam flies off with the greater acceleration. **(d)** Two astronauts on a space walk are throwing a ball back and forth between each other. In this game of catch the distance between them remains constant. **(e)** None of the above is true, according to the third law.

8. Two ice skaters, Paul and Tom, are each holding on to opposite ends of the same rope. Each pulls the other toward him. The magnitude of Paul's acceleration is 1.25 times greater than the magnitude of Tom's acceleration. What is the ratio of Paul's mass to Tom's mass? **(a)** 0.67 **(b)** 0.80 **(c)** 0.25 **(d)** 1.25 **(e)** 0.50

Section 4.7   The Gravitational Force

9. In another solar system a planet has twice the earth's mass and three times the earth's radius. Your weight on this planet is _____ times your earth-weight. Assume that the masses of the earth and of the other planet are uniformly distributed. **(a)** 0.667 **(b)** 2.000 **(c)** 0.111 **(d)** 0.444 **(e)** 0.222

11. What is the mass on Mercury of an object that weighs 784 N on the earth's surface? **(a)** 80.0 kg **(b)** 48.0 kg **(c)** 118 kg **(d)** 26.0 kg **(e)** There is not enough information to calculate the mass.

Section 4.8   The Normal Force

12. The apparent weight of a passenger in an elevator is greater than his true weight. Which one of the following is true? **(a)** The elevator is either moving upward with an increasing speed or moving upward with a decreasing speed. **(b)** The elevator is either moving upward with an increasing speed or moving downward with an increasing speed. **(c)** The elevator is either moving upward with a decreasing speed or moving downward with a decreasing speed. **(d)** The elevator is either moving upward with an increasing speed or moving downward with a decreasing speed. **(e)** The elevator is either moving upward with a decreasing speed or moving downward with an increasing speed.

13. The drawings show three examples of the force with which someone pushes against a vertical wall. In each case the magnitude of the pushing force is the same. Rank the normal forces that the wall applies to the pusher in ascending order (smallest first). **(a)** C, B, A **(b)** B, A, C **(c)** A, C, B **(d)** B, C, A **(e)** C, A, B

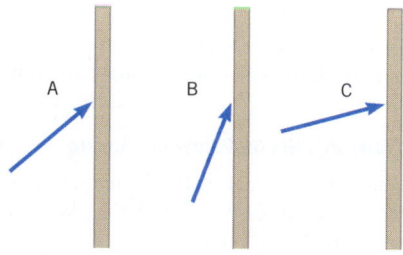

Section 4.9   Static and Kinetic Frictional Forces

15. The drawing shows three blocks, each with the same mass, stacked one upon the other. The bottom block rests on a frictionless horizontal surface and is being pulled by a force $\vec{F}$ that is parallel to this surface. The surfaces where the blocks touch each other have identical coefficients of static friction. Which one of the following correctly describes the magnitude of the *net* force of static friction f_s that acts on each block?

(a) $f_{s,A} = f_{s,B} = f_{s,C}$ (b) $f_{s,A} = f_{s,B} = \frac{1}{2}f_{s,C}$

(c) $f_{s,A} = 0$ and $f_{s,B} = \frac{1}{2}f_{s,C}$ (d) $f_{s,C} = 0$ and $f_{s,A} = \frac{1}{2}f_{s,B}$

(e) $f_{s,A} = f_{s,C} = \frac{1}{2}f_{s,B}$

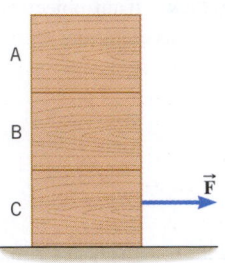

16. Three identical blocks are being pulled or pushed across a horizontal surface by a force $\vec{F}$, as shown in the drawings. The force $\vec{F}$ in each case has the same magnitude. Rank the kinetic frictional forces that act on the blocks in ascending order (smallest first).

(a) B, C, A   (b) C, A, B   (c) B, A, C   (d) C, B, A   (e) A, C, B

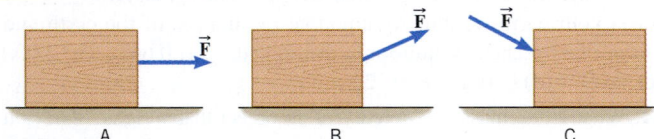

Section 4.10   The Tension Force

18. A heavy block is suspended from a ceiling using pulleys in three different ways, as shown in the drawings. Rank the tension in the rope that passes over the pulleys in ascending order (smallest first).

(a) B, A, C   (b) C, B, A   (c) A, B, C   (d) C, A, B   (e) B, C, A

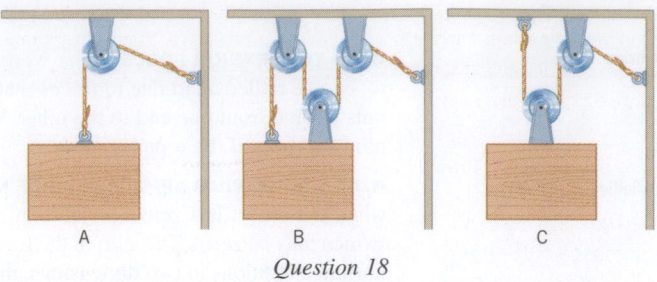

Section 4.11   Equilibrium Applications of Newton's Laws of Motion

20. A certain object is in equilibrium. Which one of the following statements is *not* true?   (a) The object must be at rest.   (b) The object has a constant velocity.   (c) The object has no acceleration.   (d) No net force acts on the object.

23. Two identical boxes are being pulled across a horizontal floor at a constant velocity by a horizontal pulling force of 176 N that is applied to one of the boxes, as the drawing shows. There is kinetic friction between each box and the floor. Find the tension in the rope between the boxes.   (a) 176 N   (b) 88.0 N   (c) 132 N   (d) 44.0 N   (e) There is not enough information to calculate the tension.

Section 4.12   Nonequilibrium Applications of Newton's Laws of Motion

25. A man is standing on a platform that is connected to a pulley arrangement, as the drawing shows. By pulling upward on the rope with a force $\vec{P}$ the man can raise the platform and himself. The total mass of the man plus the platform is 94.0 kg. What pulling force should the man apply to create an upward acceleration of 1.20 m/s^2?

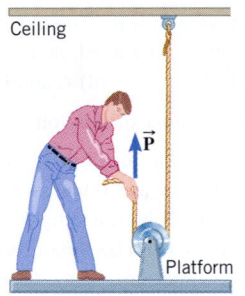

PROBLEMS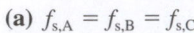

Note to Instructors: *Most of the homework problems in this chapter are available for assignment via an online homework management program such as* WileyPLUS *or* WebAssign, *and those marked with the icon* *are presented in* WileyPLUS *using a guided tutorial format that provides enhanced interactivity. See Preface for additional details.*

ssm   Solution is in the Student Solutions Manual.
www   Solution is available online at www.wiley.com/college/cutnell

This icon represents a biomedical application.

Section 4.3   Newton's Second Law of Motion

1. An airplane has a mass of 3.1×10^4 kg and takes off under the influence of a constant net force of 3.7×10^4 N. What is the net force that acts on the plane's 78-kg pilot?

2. **Concept Simulation 4.1** at www.wiley.com/college/cutnell reviews the concepts that are important in this problem. The speed of a bobsled is increasing because it has an acceleration of 2.4 m/s^2. At a given instant in time, the forces resisting the motion, including kinetic friction and air resistance, total 450 N. The combined mass of the bobsled and its riders is 270 kg.   **(a)** What is the magnitude of the force propelling the bobsled forward?   **(b)** What is the magnitude of the net force that acts on the bobsled?

3. **GO** Two horizontal forces, $\vec{F}_1$ and $\vec{F}_2$, are acting on a box, but only $\vec{F}_1$ is shown in the drawing. $\vec{F}_2$ can point either to the right or to the left. The box moves only along the x axis. There is no friction between the box and the surface. Suppose that $\vec{F}_1 = +9.0$ N and the mass of the box is 3.0 kg. Find the magnitude and direction of $\vec{F}_2$ when the acceleration of the box is   **(a)** $+5.0$ m/s², **(b)** -5.0 m/s², and   **(c)** 0 m/s².

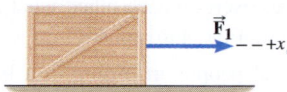

4. In the amusement park ride known as Magic Mountain Superman, powerful magnets accelerate a car and its riders from rest to 45 m/s (about 100 mi/h) in a time of 7.0 s. The combined mass of the car and riders is 5.5×10^3 kg. Find the average net force exerted on the car and riders by the magnets.

5. **ssm** When a 58-g tennis ball is served, it accelerates from rest to a speed of 45 m/s. The impact with the racket gives the ball a constant acceleration over a distance of 44 cm. What is the magnitude of the net force acting on the ball?

6. Review **Interactive LearningWare 4.1** at **www.wiley.com/college/ cutnell** in preparation for this problem. During a circus performance, a 72-kg human cannonball is shot out of an 18-m-long cannon. If the human cannonball spends 0.95 s in the cannon, determine the average net force exerted on him in the barrel of the cannon.

7. **ssm Interactive LearningWare 4.1** at **www.wiley.com/college/ cutnell** reviews the approach taken in problems such as this one. A 1580-kg car is traveling with a speed of 15.0 m/s. What is the magnitude of the horizontal net force that is required to bring the car to a halt in a distance of 50.0 m?

8. **GO** A person with a black belt in karate has a fist that has a mass of 0.70 kg. Starting from rest, this fist attains a velocity of 8.0 m/s in 0.15 s. What is the magnitude of the average net force applied to the fist to achieve this level of performance?

* 9. **ssm www** Two forces $\vec{F}_A$ and $\vec{F}_B$ are applied to an object whose mass is 8.0 kg. The larger force is $\vec{F}_A$. When both forces point due east, the object's acceleration has a magnitude of 0.50 m/s². However, when $\vec{F}_A$ points due east and $\vec{F}_B$ points due west, the acceleration is 0.40 m/s², due east. Find   **(a)** the magnitude of $\vec{F}_A$ and   **(b)** the magnitude of $\vec{F}_B$.

* 10. An electron is a subatomic particle ($m = 9.11 \times 10^{-31}$ kg) that is subject to electric forces. An electron moving in the $+x$ direction accelerates from an initial velocity of $+5.40 \times 10^5$ m/s to a final velocity of $+2.10 \times 10^6$ m/s while traveling a distance of 0.038 m. The electron's acceleration is due to two electric forces parallel to the x axis: $\vec{F}_1 = +7.50 \times 10^{-17}$ N, and $\vec{F}_2$, which points in the $-x$ direction. Find the magnitudes of   **(a)** the net force acting on the electron and **(b)** the electric force $\vec{F}_2$.

Section 4.4   The Vector Nature of Newton's Second Law of Motion,   Section 4.5   Newton's Third Law of Motion

11. Review **Interactive Solution 4.11** at **www.wiley.com/college/ cutnell** before starting this problem. Two forces, $\vec{F}_1$ and $\vec{F}_2$, act on the 7.00-kg block shown in the drawing. The magnitudes of the forces are $F_1 = 59.0$ N and $F_2 = 33.0$ N. What is the horizontal acceleration (magnitude and direction) of the block?

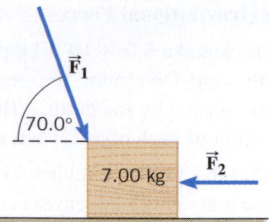

Problem 11

12. At an instant when a soccer ball is in contact with the foot of a player kicking it, the horizontal or x component of the ball's acceleration is 810 m/s² and the vertical or y component of its acceleration is 1100 m/s². The ball's mass is 0.43 kg. What is the magnitude of the net force acting on the soccer ball at this instant?

13. **ssm Interactive LearningWare 4.4** at **www.wiley.com/college/ cutnell** provides a review of the concepts in this problem. A rocket of mass 4.50×10^5 kg is in flight. Its thrust is directed at an angle of 55.0° above the horizontal and has a magnitude of 7.50×10^6 N. Find the magnitude and direction of the rocket's acceleration. Give the direction as an angle above the horizontal.

14. When a parachute opens, the air exerts a large drag force on it. This upward force is initially greater than the weight of the sky diver and, thus, slows him down. Suppose the weight of the sky diver is 915 N and the drag force has a magnitude of 1027 N. The mass of the sky diver is 93.4 kg. What are the magnitude and direction of his acceleration?

15. Airplane flight recorders must be able to survive catastrophic crashes. Therefore, they are typically encased in crash-resistant steel or titanium boxes that are subjected to rigorous testing. One of the tests is an impact shock test, in which the box must survive being thrown at high speeds against a barrier. A 41-kg box is thrown at a speed of 220 m/s and is brought to a halt in a collision that lasts for a time of 6.5 ms. What is the magnitude of the average net force that acts on the box during the collision?

16. **GO** Two skaters, a man and a woman, are standing on ice. Neglect any friction between the skate blades and the ice. The mass of the man is 82 kg, and the mass of the woman is 48 kg. The woman pushes on the man with a force of 45 N due east. Determine the acceleration (magnitude and direction) of   **(a)** the man and **(b)** the woman.

* 17. A duck has a mass of 2.5 kg. As the duck paddles, a force of 0.10 N acts on it in a direction due east. In addition, the current of the water exerts a force of 0.20 N in a direction of 52° south of east. When these forces begin to act, the velocity of the duck is 0.11 m/s in a direction due east. Find the magnitude and direction (relative to due east) of the displacement that the duck undergoes in 3.0 s while the forces are acting.

** 18. At a time when mining asteroids has become feasible, astronauts have connected a line between their 3500-kg space tug and a 6200-kg asteroid. Using their tug's engine, they pull on the asteroid with a force of 490 N. Initially the tug and the asteroid are at rest, 450 m apart. How much time does it take for the tug and the asteroid to meet?

** 19. **ssm www** A 325-kg boat is sailing 15.0° north of east at a speed of 2.00 m/s. Thirty seconds later, it is sailing 35.0° north of east at a speed of 4.00 m/s. During this time, three forces act on the boat: a 31.0-N force directed 15.0° north of east (due to an auxiliary engine), a 23.0-N force directed 15.0° south of west (resistance due to the water), and $\vec{F}_W$ (due to the wind). Find the magnitude and direction of the force $\vec{F}_W$. Express the direction as an angle with respect to due east.

Section 4.7   The Gravitational Force

20. GO A 5.0-kg rock and a 3.0×10^{-4}-kg pebble are held near the surface of the earth.   (a) Determine the magnitude of the gravitational force exerted on each by the earth.   (b) Calculate the magnitude of the acceleration of each object when released.

21. In preparation for this problem, review Conceptual Example 7. A space traveler whose mass is 115 kg leaves earth. What are his weight and mass   (a) on earth and   (b) in interplanetary space where there are no nearby planetary objects?

22. A 55-kg bungee jumper has fallen far enough that her bungee cord is beginning to stretch and resist her downward motion. Find the force (magnitude and direction) exerted on her by the bungee cord at an instant when her downward acceleration has a magnitude of 7.6 m/s². Ignore the effects of air resistance.

23. GO A raindrop has a mass of 5.2×10^{-7} kg and is falling near the surface of the earth. Calculate the magnitude of the gravitational force exerted   (a) on the raindrop by the earth and   (b) on the earth by the raindrop.

24. A bowling ball (mass = 7.2 kg, radius = 0.11 m) and a billiard ball (mass = 0.38 kg, radius = 0.028 m) may each be treated as uniform spheres. What is the magnitude of the maximum gravitational force that each can exert on the other?

25. ssm Saturn has an equatorial radius of 6.00×10^7 m and a mass of 5.67×10^{26} kg.   (a) Compute the acceleration of gravity at the equator of Saturn.   (b) What is the ratio of a person's weight on Saturn to that on earth?

26. Review Conceptual Example 7 in preparation for this problem. In tests on earth a lunar surface exploration vehicle (mass = 5.90×10^3 kg) achieves a forward acceleration of 0.220 m/s². To achieve this same acceleration on the moon, the vehicle's engines must produce a drive force of 1.43×10^3 N. What is the magnitude of the frictional force that acts on the vehicle on the moon?

27. ssm Synchronous communications satellites are placed in a circular orbit that is 3.59×10^7 m above the surface of the earth. What is the magnitude of the acceleration due to gravity at this distance?

28. The drawing (not to scale) shows one alignment of the sun, earth, and moon. The gravitational force $\vec{F}_{SM}$ that the sun exerts on the moon is perpendicular to the force $\vec{F}_{EM}$ that the earth exerts on the moon. The masses are: mass of sun = 1.99×10^{30} kg, mass of earth = 5.98×10^{24} kg, mass of moon = 7.35×10^{22} kg. The distances shown in the drawing are $r_{SM} = 1.50 \times 10^{11}$ m and $r_{EM} = 3.85 \times 10^8$ m. Determine the magnitude of the net gravitational force on the moon.

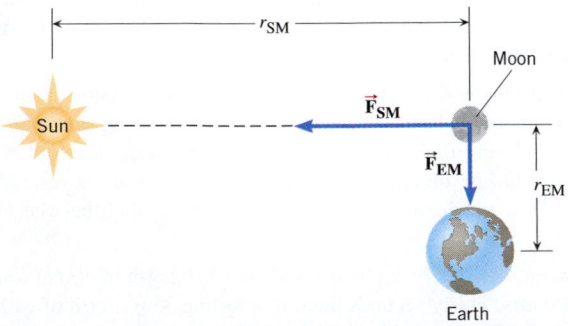

29. (a) Calculate the magnitude of the gravitational force exerted on a 425-kg satellite that is a distance of two earth radii from the center of the earth.   (b) What is the magnitude of the gravitational force exerted on the earth by the satellite?   (c) Determine the magnitude of the satellite's acceleration.   (d) What is the magnitude of the earth's acceleration?

30. A space traveler weighs 540.0 N on earth. What will the traveler weigh on another planet whose radius is twice that of earth and whose mass is three times that of earth?

31. ssm The mass of a robot is 5450 kg. This robot weighs 3620 N more on planet A than it does on planet B. Both planets have the same radius of 1.33×10^7 m. What is the difference $M_A - M_B$ in the masses of these planets?

* 32. Three uniform spheres are located at the corners of an equilateral triangle. Each side of the triangle has a length of 1.20 m. Two of the spheres have a mass of 2.80 kg each. The third sphere (mass unknown) is released from rest. Considering only the gravitational forces that the spheres exert on each other, what is the magnitude of the initial acceleration of the third sphere?

* 33. ssm www Several people are riding in a hot-air balloon. The combined mass of the people and balloon is 310 kg. The balloon is motionless in the air, because the downward-acting weight of the people and balloon is balanced by an upward-acting "buoyant" force. If the buoyant force remains constant, how much mass should be dropped overboard so the balloon acquires an upward acceleration of 0.15 m/s²?

* 34. Jupiter is the largest planet in our solar system, having a mass and radius that are, respectively, 318 and 11.2 times that of earth. Suppose that an object falls from rest near the surface of each planet and that the acceleration due to gravity remains constant during the fall. Each object falls the same distance before striking the ground. Determine the ratio of the time of fall on Jupiter to that on earth.

* 35. The sun is more massive than the moon, but the sun is farther from the earth. Which one exerts a greater gravitational force on a person standing on the earth? Give your answer by determining the ratio F_{sun}/F_{moon} of the magnitudes of the gravitational forces. Use the data on the inside of the front cover.

* 36. As a moon follows its orbit around a planet, the maximum gravitational force exerted on the moon by the planet exceeds the minimum gravitational force by 11%. Find the ratio r_{max}/r_{min}, where r_{max} is the moon's maximum distance from the center of the planet and r_{min} is the minimum distance.

** 37. Two particles are located on the x axis. Particle 1 has a mass m and is at the origin. Particle 2 has a mass $2m$ and is at $x = +L$. A third particle is placed between particles 1 and 2. Where on the x axis should the third particle be located so that the magnitude of the gravitational force on *both* particle 1 and particle 2 doubles? Express your answer in terms of L.

Section 4.8   The Normal Force,
Section 4.9   Static and Kinetic Frictional Forces

38. A 35-kg crate rests on a horizontal floor, and a 65-kg person is standing on the crate. Determine the magnitude of the normal force that   (a) the floor exerts on the crate and   (b) the crate exerts on the person.

39. ssm A student presses a book between his hands, as the drawing indicates. The forces that he exerts on the front and back covers of the book are perpendicular to the book and are horizontal. The book weighs 31 N. The coefficient of static friction between his hands and the book is 0.40. To keep the book from falling, what is the magnitude of the minimum pressing force that each hand must exert?

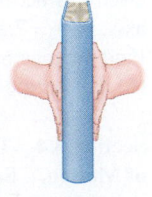

40. A 95.0-kg person stands on a scale in an elevator. What is the apparent weight when the elevator is   (a) accelerating upward with an acceleration of 1.80 m/s²,   (b) moving upward at a constant speed, and   (c) accelerating downward with an acceleration of 1.30 m/s²?

41. **GO** A car is traveling up a hill that is inclined at an angle θ above the horizontal. Determine the ratio of the magnitude of the normal force to the weight of the car when **(a)** $\theta = 15°$ and **(b)** $\theta = 35°$.

42. In a European country a bathroom scale displays its reading in kilograms. When a man stands on this scale, it reads 92.6 kg. When he pulls down on a chin-up bar installed over the scale, the reading decreases to 75.1 kg. What is the magnitude of the force he exerts on the chin-up bar?

43. A Mercedes-Benz 300SL ($m = 1700$ kg) is parked on a road that rises 15° above the horizontal. What are the magnitudes of **(a)** the normal force and **(b)** the static frictional force that the ground exerts on the tires?

44. **GO** Consult Multiple-Concept Example 9 to explore a model for solving this problem. A person pushes on a 57-kg refrigerator with a horizontal force of -267 N; the minus sign indicates that the force points in the $-x$ direction. The coefficient of static friction is 0.65. **(a)** If the refrigerator does not move, what are the magnitude and direction of the static frictional force that the floor exerts on the refrigerator? **(b)** What is the magnitude of the largest pushing force that can be applied to the refrigerator before it just begins to move?

45. **ssm** A 20.0-kg sled is being pulled across a horizontal surface at a constant velocity. The pulling force has a magnitude of 80.0 N and is directed at an angle of 30.0° above the horizontal. Determine the coefficient of kinetic friction.

46. A 6.00-kg box is sliding across the horizontal floor of an elevator. The coefficient of kinetic friction between the box and the floor is 0.360. Determine the kinetic frictional force that acts on the box when the elevator is **(a)** stationary, **(b)** accelerating upward with an acceleration whose magnitude is 1.20 m/s², and **(c)** accelerating downward with an acceleration whose magnitude is 1.20 m/s².

47. Review **Interactive Solution 4.47** at **www.wiley.com/college/ cutnell** in preparation for this problem. An 81-kg baseball player slides into second base. The coefficient of kinetic friction between the player and the ground is 0.49. **(a)** What is the magnitude of the frictional force? **(b)** If the player comes to rest after 1.6 s, what was his initial velocity?

* **48.** Consult Multiple-Concept Example 10 in preparation for this problem. Traveling at a speed of 16.1 m/s, the driver of an automobile suddenly locks the wheels by slamming on the brakes. The coefficient of kinetic friction between the tires and the road is 0.720. What is the speed of the automobile after 1.30 s have elapsed? Ignore the effects of air resistance.

* **49.** **ssm** Refer to Multiple-Concept Example 10 for help in solving problems like this one. An ice skater is gliding horizontally across the ice with an initial velocity of $+6.3$ m/s. The coefficient of kinetic friction between the ice and the skate blades is 0.081, and air resistance is negligible. How much time elapses before her velocity is reduced to $+2.8$ m/s?

* **50.** Multiple-Concept Example 17 reviews the basic concepts involved in this problem. Air rushing over the wings of high-performance race cars generates unwanted horizontal air resistance but also causes a vertical *downforce*, which helps the cars hug the track more securely. The coefficient of static friction between the track and the tires of a 690-kg race car is 0.87. What is the magnitude of the maximum acceleration at which the car can speed up without its tires slipping when a 4060-N downforce and an 1190-N horizontal-air-resistance force act on it?

** **51.** The drawing shows a 25.0-kg crate that is initially at rest. Note that the view is one looking down on the top of the crate. Two forces,

$\vec{F}_1$ and $\vec{F}_2$, are applied to the crate, and it begins to move. The coefficient of kinetic friction between the crate and the floor is $\mu_k = 0.350$. Determine the magnitude and direction (relative to the x axis) of the acceleration of the crate.

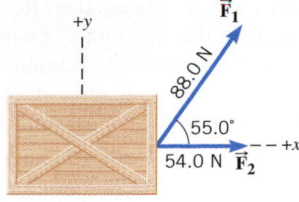

Top view

Section 4.10 The Tension Force, Section 4.11 Equilibrium Applications of Newton's Laws of Motion

52. Review **Interactive LearningWare 4.3** at **www.wiley.com/college/cutnell** in preparation for this problem. The helicopter in the drawing is moving horizontally to the right at a constant velocity. The weight of the helicopter is $W = 53\,800$ N. The lift force $\vec{L}$ generated by the rotating blade makes an angle of 21.0° with respect to the vertical. **(a)** What is the magnitude of the lift force? **(b)** Determine the magnitude of the air resistance $\vec{R}$ that opposes the motion.

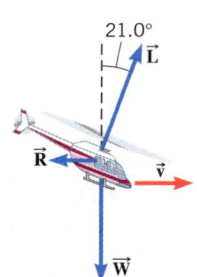

53. **ssm** Three forces act on a moving object. One force has a magnitude of 80.0 N and is directed due north. Another has a magnitude of 60.0 N and is directed due west. What must be the magnitude and direction of the third force, such that the object continues to move with a constant velocity?

54. A supertanker (mass = 1.70×10^8 kg) is moving with a constant velocity. Its engines generate a forward thrust of 7.40×10^5 N. Determine **(a)** the magnitude of the resistive force exerted on the tanker by the water and **(b)** the magnitude of the upward buoyant force exerted on the tanker by the water.

55. The drawing shows a wire tooth brace used by orthodontists. The topmost tooth is protruding slightly, and the tension in the wire exerts two forces $\vec{T}$ and $\vec{T}'$ on this tooth in order to bring it back into alignment. If the forces have the same magnitude of 21.0 N, what is the magnitude of the net force exerted on the tooth by these forces?

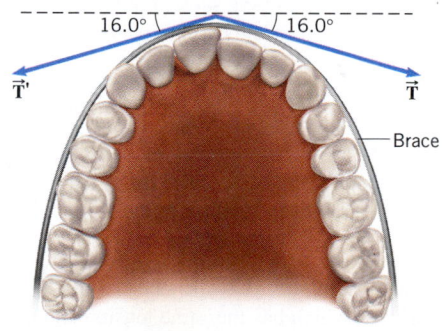

56. Part *a* of the drawing shows a bucket of water suspended from the pulley of a well; the tension in the rope is 92.0 N. Part *b* shows the same bucket of water being pulled up from the well at a constant velocity. What is the tension in the rope in part *b*?

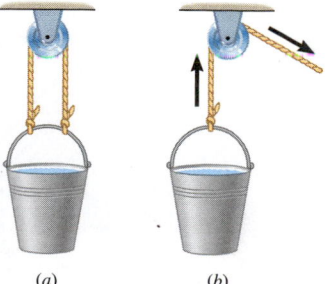

(*a*)   (*b*)

57. ssm As preparation for this problem, review Example 13. Suppose that the pilot in Example 13 suddenly jettisons 2800 N of fuel. If the plane is to continue moving with the same velocity under the influence of the same air resistance $\vec{R}$, by how much does the pilot have to reduce **(a)** the thrust and **(b)** the lift?

58. A worker stands still on a roof sloped at an angle of 36° above the horizontal. He is prevented from slipping by a static frictional force of 390 N. Find the mass of the worker.

59. **GO** A 1.40-kg bottle of vintage wine is lying horizontally in a rack, as shown in the drawing. The two surfaces on which the bottle rests are 90.0° apart, and the right surface makes an angle of 45.0° with respect to the horizontal. Each surface exerts a force on the bottle that is perpendicular to the surface. Both forces have the same magnitude F. Find the value of F.

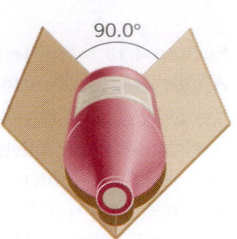

60. The drawing shows a circus clown who weighs 890 N. The coefficient of static friction between the clown's feet and the ground is 0.53. He pulls vertically downward on a rope that passes around three pulleys and is tied around his feet. What is the minimum pulling force that the clown must exert to yank his feet out from under himself?

* 61. The drawing shows box 1 resting on a table, with box 2 resting on top of box 1. A massless rope passes over a massless, frictionless pulley. One end of the rope is connected to box 2, and the other end is connected to box 3. The weights of the three boxes are $W_1 = 55$ N, $W_2 = 35$ N, and $W_3 = 28$ N. Determine the magnitude of the normal force that the table exerts on box 1.

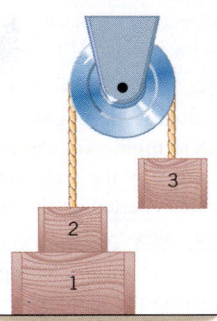

Problem 61

* 62. A mountain climber, in the process of crossing between two cliffs by a rope, pauses to rest. She weighs 535 N. As the drawing shows, she is closer to the left cliff than to the right cliff, with the result that the tensions in the left and right sides of the rope are not the same. Find the tensions in the rope to the left and to the right of the mountain climber.

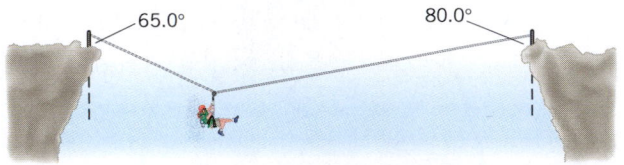

* 63. ssm A 44-kg chandelier is suspended 1.5 m below a ceiling by three wires, each of which has the same tension and the same length of 2.0 m (see the drawing). Find the tension in each wire.

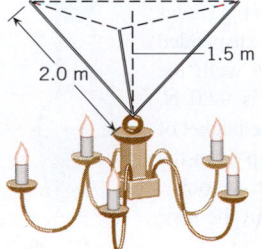

* 64. The person in the drawing is standing on crutches. Assume that the force exerted on each crutch by the ground is directed along the crutch, as the force vectors in the drawing indicate. If the coefficient of static friction between a crutch and the ground is 0.90, determine the largest angle θ^{MAX} that the crutch can have just before it begins to slip on the floor.

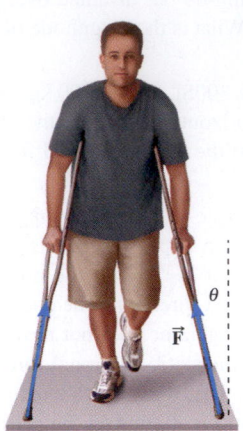

* 65. Consult **Interactive Solution 4.65** at **www.wiley.com/college/ cutnell** before beginning this problem. A toboggan slides down a hill and has a constant velocity. The angle of the hill is 8.00° with respect to the horizontal. What is the coefficient of kinetic friction between the surface of the hill and the toboggan?

* 66. **GO** A block is pressed against a vertical wall by a force $\vec{P}$, as the drawing shows. This force can either push the block upward at a constant velocity or allow it to slide downward at a constant velocity. The magnitude of the force is different in the two cases, while the directional angle θ is the same. Kinetic friction exists between the block and the wall, and the coefficient of kinetic friction is 0.250. The weight of the block is 39.0 N, and the directional angle for the force $\vec{P}$ is $\theta = 30.0°$. Determine the magnitude of $\vec{P}$ when the block slides **(a)** up the wall and **(b)** down the wall.

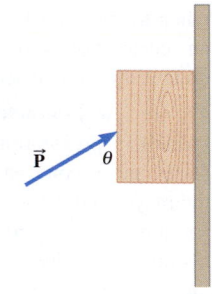

** 67. ssm A bicyclist is coasting straight down a hill at a constant speed. The combined mass of the rider and bicycle is 80.0 kg, and the hill is inclined at 15.0° with respect to the horizontal. Air resistance opposes the motion of the cyclist. Later, the bicyclist climbs the same hill at the same constant speed. How much force (directed parallel to the hill) must be applied to the bicycle in order for the bicyclist to climb the hill?

** 68. Review **Interactive LearningWare 4.3** at **www.wiley.com/college/ cutnell** in preparation for this problem. A kite is hovering over the ground at the end of a straight 43-m line. The tension in the line has a magnitude of 16 N. Wind blowing on the kite exerts a force of 19 N, directed 56° above the horizontal. Note that the line attached to the kite is *not* oriented at an angle of 56° above the horizontal. Find the height of the kite, relative to the person holding the line.

** 69. A damp washcloth is hung over the edge of a table to dry. Thus, part (mass = m_{on}) of the washcloth rests on the table and part (mass = m_{off}) does not. The coefficient of static friction between the table and the washcloth is 0.40. Determine the maximum fraction $[m_{off}/(m_{on} + m_{off})]$ that can hang over the edge without causing the whole washcloth to slide off the table.

Section 4.12 Nonequilibrium Applications of Newton's Laws of Motion

70. A 1450-kg submarine rises straight up toward the surface. Seawater exerts both an upward buoyant force of 16 140 N on the submarine and a downward resistive force of 1030 N. What is the submarine's acceleration?

71. Only two forces act on an object (mass = 4.00 kg), as in the drawing. Find the magnitude and direction (relative to the x axis) of the acceleration of the object.

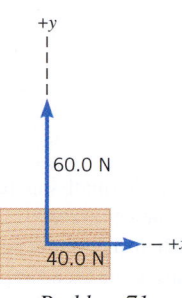

Problem 71

72. A fisherman is fishing from a bridge and is using a "45-N test line." In other words, the line will sustain a maximum force of 45 N without breaking. What is the weight of the heaviest fish that can be pulled up vertically when the line is reeled in **(a)** at a constant speed and **(b)** with an acceleration whose magnitude is 2.0 m/s²?

73. **ssm** A 1380-kg car is moving due east with an initial speed of 27.0 m/s. After 8.00 s the car has slowed down to 17.0 m/s. Find the magnitude and direction of the net force that produces the deceleration.

74. A helicopter flies over the arctic ice pack at a constant altitude, towing an airborne 129-kg laser sensor that measures the thickness of the ice (see the drawing). The helicopter and the sensor both move only in the horizontal direction and have a horizontal acceleration of magnitude 2.84 m/s². Ignoring air resistance, find the tension in the cable towing the sensor.

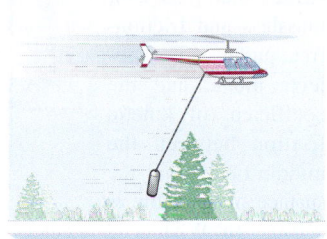

75. **ssm** In a supermarket parking lot, an employee is pushing ten empty shopping carts, lined up in a straight line. The acceleration of the carts is 0.050 m/s². The ground is level, and each cart has a mass of 26 kg. **(a)** What is the net force acting on any one of the carts? **(b)** Assuming friction is negligible, what is the force exerted by the fifth cart on the sixth cart?

76. In the drawing, the weight of the block on the table is 422 N and that of the hanging block is 185 N. Ignoring all frictional effects and assuming the pulley to be massless, find **(a)** the acceleration of the two blocks and **(b)** the tension in the cord.

77. **ssm www** A student is skateboarding down a ramp that is 6.0 m long and inclined at 18° with respect to the horizontal. The initial speed of the skateboarder at the top of the ramp is 2.6 m/s. Neglect friction and find the speed at the bottom of the ramp.

78. Review Conceptual Example 16 as background for this problem. The water skier there has a mass of 73 kg. Find the magnitude of the net force acting on the skier when **(a)** she is accelerated from rest to a speed of 11 m/s in 8.0 s and **(b)** she lets go of the tow rope and glides to a halt in 21 s.

79. A car is towing a boat on a trailer. The driver starts from rest and accelerates to a velocity of +11 m/s in a time of 28 s. The combined mass of the boat and trailer is 410 kg. The frictional force acting on the trailer can be ignored. What is the tension in the hitch that connects the trailer to the car?

80. A man seeking to set a world record wants to tow a 109 000-kg airplane along a runway by pulling horizontally on a cable attached to the airplane. The mass of the man is 85 kg, and the coefficient of static friction between his shoes and the runway is 0.77. What is the greatest acceleration the man can give the airplane? Assume that the airplane is on wheels that turn without any frictional resistance.

* **81.** The principles used to solve this problem are similar to those in Multiple-Concept Example 17. A 205-kg log is pulled up a ramp by means of a rope that is parallel to the surface of the ramp. The ramp is inclined at 30.0° with respect to the horizontal. The coefficient of kinetic friction between the log and the ramp is 0.900, and the log has an acceleration of 0.800 m/s². Find the tension in the rope.

* **82.** **GO** To hoist himself into a tree, a 72.0-kg man ties one end of a nylon rope around his waist and throws the other end over a branch of the tree. He then pulls downward on the free end of the rope with a force of 358 N. Neglect any friction between the rope and the branch, and determine the man's upward acceleration.

* **83.** **ssm** Review **Interactive Solution 4.83** at **www.wiley.com/ college/cutnell** before starting this problem. The drawing shows Robin Hood (mass = 77.0 kg) about to escape from a dangerous situation. With one hand, he is gripping the rope that holds up a chandelier (mass = 195 kg). When he cuts the rope where it is tied to the floor, the chandelier will fall, and he will be pulled up toward a balcony above. Ignore the friction between the rope and the beams over which it slides, and find **(a)** the acceleration with which Robin is pulled upward and **(b)** the tension in the rope while Robin escapes.

* **84.** A train consists of 50 cars, each of which has a mass of 6.8 × 10³ kg. The train has an acceleration of +8.0 × 10⁻² m/s². Ignore friction and determine the tension in the coupling **(a)** between the 30th and 31st cars and **(b)** between the 49th and 50th cars.

* **85.** Consult Multiple-Concept Example 10 for insight into solving this type of problem. A box is sliding up an incline that makes an angle of 15.0° with respect to the horizontal. The coefficient of kinetic friction between the box and the surface of the incline is 0.180. The initial speed of the box at the bottom of the incline is 1.50 m/s. How far does the box travel along the incline before coming to rest?

* **86.** This problem uses the same concepts as Multiple-Concept Example 17. In Problem 80, an 85-kg man plans to tow a 109 000-kg airplane along a runway by pulling horizontally on a cable attached to it. Suppose that he instead attempts the feat by pulling the cable at an angle of 9.0° above the horizontal. The coefficient of static friction between his shoes and the runway is 0.77. What is the greatest acceleration the man can give the airplane? Assume that the airplane is on wheels that turn without any frictional resistance.

* **87.** The alarm at a fire station rings and an 86-kg fireman, starting from rest, slides down a pole to the floor below (a distance of 4.0 m). Just before landing, his speed is 1.4 m/s. What is the magnitude of the kinetic frictional force exerted on the fireman as he slides down the pole?

* **88.** GO Two blocks are sliding to the right across a horizontal surface, as the drawing shows. In Case A the mass of each block is 3.0 kg. In Case B the mass of block 1 (the block behind) is 6.0 kg, and the mass of block 2 is 3.0 kg. No frictional force acts on block 1 in either Case A or Case B. However, a kinetic frictional force of 5.8 N does act on block 2 in both cases and opposes the motion. For both Case A and Case B determine   (a) the magnitude of the forces with which the blocks push against each other and   (b) the magnitude of the acceleration of the blocks.

Case A   Case B

* **89.** ssm At an airport, luggage is unloaded from a plane into the three cars of a luggage carrier, as the drawing shows. The acceleration of the carrier is $0.12 \ \text{m/s}^2$, and friction is negligible. The coupling bars have negligible mass. By how much would the tension in *each* of the coupling bars A, B, and C change if 39 kg of luggage were removed from car 2 and placed in   (a) car 1 and   (b) car 3? If the tension changes, specify whether it increases or decreases.

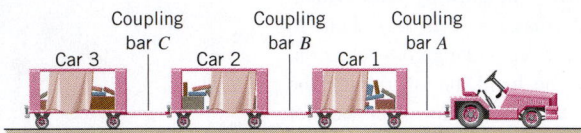

* **90.** Consult **Interactive LearningWare 4.2** at **www.wiley.com/college/cutnell** before beginning this problem. A truck is traveling at a speed of 25.0 m/s along a level road. A crate is resting on the bed of the truck, and the coefficient of static friction between the crate and the truck bed is 0.650. Determine the shortest distance in which the truck can come to a halt without causing the crate to slip forward relative to the truck.

** **91.** In the drawing, the rope and the pulleys are massless, and there is no friction. Find   (a) the tension in the rope and   (b) the acceleration of the 10.0-kg block. *(Hint: The larger mass moves twice as far as the smaller mass.)*

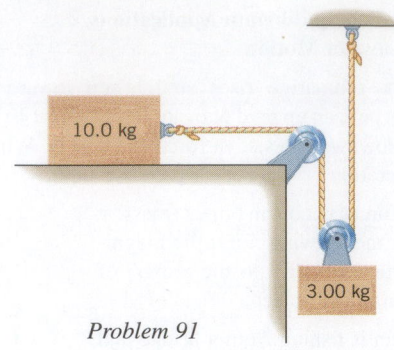

Problem 91

** **92.** A small sphere is hung by a string from the ceiling of a van. When the van is stationary, the sphere hangs vertically. However, when the van accelerates, the sphere swings backward so that the string makes an angle of θ with respect to the vertical.   (a) Derive an expression for the magnitude a of the acceleration of the van in terms of the angle θ and the magnitude g of the acceleration due to gravity.   (b) Find the acceleration of the van when $\theta = 10.0°$.   (c) What is the angle θ when the van moves with a constant velocity?

** **93.** ssm The drawing shows three objects. They are connected by strings that pass over massless and friction-free pulleys. The objects move, and the coefficient of kinetic friction between the middle object and the surface of the table is 0.100.   (a) What is the acceleration of the three objects?   (b) Find the tension in each of the two strings.

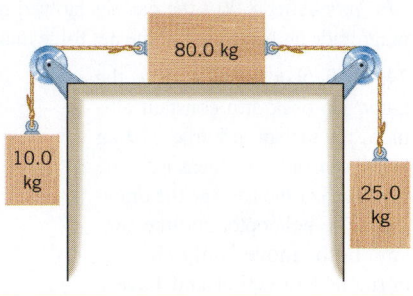

** **94.** A 5.00-kg block is placed on top of a 12.0-kg block that rests on a frictionless table. The coefficient of static friction between the two blocks is 0.600. What is the maximum horizontal force that can be applied before the 5.00-kg block begins to slip relative to the 12.0-kg block, if the force is applied to   (a) the more massive block and   (b) the less massive block?

ADDITIONAL PROBLEMS

95. ssm On earth, two parts of a space probe weigh 11 000 N and 3400 N. These parts are separated by a center-to-center distance of 12 m and may be treated as uniform spherical objects. Find the magnitude of the gravitational force that each part exerts on the other out in space, far from any other objects.

96. The space probe *Deep Space 1* was launched on October 24, 1998. Its mass was 474 kg. The goal of the mission was to test a new kind of engine called an ion propulsion drive. This engine generated only a weak thrust, but it could do so over long periods of time with the consumption of only small amounts of fuel. The mission was spectacularly successful. At a thrust of 56 mN how many days were required for the probe to attain a velocity of 805 m/s (1800 mi/h), assuming that the probe started from rest and that the mass remained nearly constant?

97. ssm A rocket blasts off from rest and attains a speed of 45 m/s in 15 s. An astronaut has a mass of 57 kg. What is the astronaut's apparent weight during takeoff?

98. Only two forces act on an object (mass = 3.00 kg), as in the drawing. Find the magnitude and direction (relative to the x axis) of the acceleration of the object.

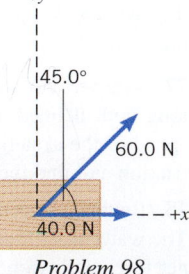

Problem 98

99. ssm A 60.0-kg crate rests on a level floor at a shipping dock. The coefficients of static and kinetic friction are 0.760 and 0.410, respectively. What horizontal pushing force is required to   (a) just start the crate moving and   (b) slide the crate across the dock at a constant speed?

100. **Concept Simulation 4.1** at **www.wiley.com/college/cutnell** reviews the central idea in this problem. A boat has a mass of 6800 kg. Its engines generate a drive force of 4100 N due west, while the wind exerts a force of 800 N due east and the water exerts a resistive force of 1200 N due east. What is the magnitude and direction of the boat's acceleration?

101. A 1.14×10^4-kg lunar landing craft is about to touch down on the surface of the moon, where the acceleration due to gravity is 1.60 m/s². At an altitude of 165 m the craft's downward velocity is 18.0 m/s. To slow down the craft, a retrorocket is firing to provide an upward thrust. Assuming the descent is vertical, find the magnitude of the thrust needed to reduce the velocity to zero at the instant when the craft touches the lunar surface.

102. A woman stands on a scale in a moving elevator. Her mass is 60.0 kg, and the combined mass of the elevator and scale is an additional 815 kg. Starting from rest, the elevator accelerates upward. During the acceleration, the hoisting cable applies a force of 9410 N. What does the scale read during the acceleration?

103. **ssm** A 15-g bullet is fired from a rifle. It takes 2.50×10^{-3} s for the bullet to travel the length of the barrel, and it exits the barrel with a speed of 715 m/s. Assuming that the acceleration of the bullet is constant, find the average net force exerted on the bullet.

104. A person in a kayak starts paddling, and it accelerates from 0 to 0.60 m/s in a distance of 0.41 m. If the combined mass of the person and the kayak is 73 kg, what is the magnitude of the net force acting on the kayak?

105. Mars has a mass of 6.46×10^{23} kg and a radius of 3.39×10^6 m. **(a)** What is the acceleration due to gravity on Mars? **(b)** How much would a 65-kg person weigh on this planet?

106. The drawing shows three particles far away from any other objects and located on a straight line. The masses of these particles are $m_A = 363$ kg, $m_B = 517$ kg, and $m_C = 154$ kg. Find the magnitude and direction of the net gravitational force acting on **(a)** particle A, **(b)** particle B, and **(c)** particle C.

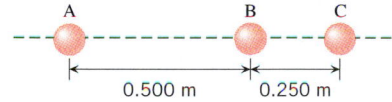

* 107. **ssm** Two objects (45.0 and 21.0 kg) are connected by a massless string that passes over a massless, frictionless pulley. The pulley hangs from the ceiling. Find **(a)** the acceleration of the objects and **(b)** the tension in the string.

* 108. The central ideas in this problem are reviewed in Multiple-Concept Example 9. One block rests upon a horizontal surface. A second identical block rests upon the first one. The coefficient of static friction between the blocks is the same as the coefficient of static friction between the lower block and the horizontal surface. A horizontal force is applied to the upper block, and the magnitude of the force is slowly increased. When the force reaches 47.0 N, the upper block just begins to slide. The force is then removed from the upper block, and the blocks are returned to their original configuration. What is the magnitude of the horizontal force that should be applied to the lower block so that it just begins to slide out from under the upper block?

* 109. A skier is pulled up a slope at a constant velocity by a tow bar. The slope is inclined at 25.0° with respect to the horizontal. The force applied to the skier by the tow bar is parallel to the slope. The skier's mass is 55.0 kg, and the coefficient of kinetic friction between the skis and the snow is 0.120. Find the magnitude of the force that the tow bar exerts on the skier.

* 110. **Interactive LearningWare 4.3** at **www.wiley.com/college/cutnell** reviews the principles that play a role in this problem. During a storm, a tree limb breaks off and comes to rest across a barbed wire fence at a point that is not in the middle between two fence posts. The limb exerts a downward force of 151 N on the wire. The left section of the wire makes an angle of 14.0° relative to the horizontal and sustains

a tension of 447 N. Find the magnitude and direction of the tension that the right section of the wire sustains.

* 111. **ssm** A person whose weight is 5.20×10^2 N is being pulled up vertically by a rope from the bottom of a cave that is 35.1 m deep. The maximum tension that the rope can withstand without breaking is 569 N. What is the shortest time, starting from rest, in which the person can be brought out of the cave?

* 112. Refer to **Concept Simulation 4.4** at **www.wiley.com/college/cutnell** for background relating to this problem. The drawing shows a large cube (mass = 25 kg) being accelerated across a horizontal frictionless surface by a horizontal force $\vec{P}$.

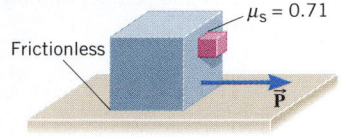

A small cube (mass = 4.0 kg) is in contact with the front surface of the large cube and will slide downward unless $\vec{P}$ is sufficiently large. The coefficient of static friction between the cubes is 0.71. What is the smallest magnitude that $\vec{P}$ can have in order to keep the small cube from sliding downward?

* 113. A space probe has two engines. Each generates the same amount of force when fired, and the directions of these forces can be independently adjusted. When the engines are fired simultaneously and each applies its force in the same direction, the probe, starting from rest, takes 28 s to travel a certain distance. How long does it take to travel the same distance, again starting from rest, if the engines are fired simultaneously and the forces that they apply to the probe are perpendicular?

* 114. A spacecraft is on a journey to the moon. At what point, as measured from the center of the earth, does the gravitational force exerted on the spacecraft by the earth balance that exerted by the moon? This point lies on a line between the centers of the earth and the moon. The distance between the earth and the moon is 3.85×10^8 m, and the mass of the earth is 81.4 times as great as that of the moon.

* 115. **ssm** A person is trying to judge whether a picture (mass = 1.10 kg) is properly positioned by temporarily pressing it against a wall. The pressing force is perpendicular to the wall. The coefficient of static friction between the picture and the wall is 0.660. What is the minimum amount of pressing force that must be used?

** 116. As part *a* of the drawing shows, two blocks are connected by a rope that passes over a set of pulleys. One block has a weight of 412 N, and the other has a weight of 908 N. The rope and the pulleys are massless and there is no friction. **(a)** What is the acceleration of the lighter block? **(b)** Suppose that the heavier block is removed, and a downward force of 908 N is provided by someone pulling on the rope, as part *b* of the drawing shows. Find the acceleration of the remaining block. **(c)** Explain why the answers in (a) and (b) are different.

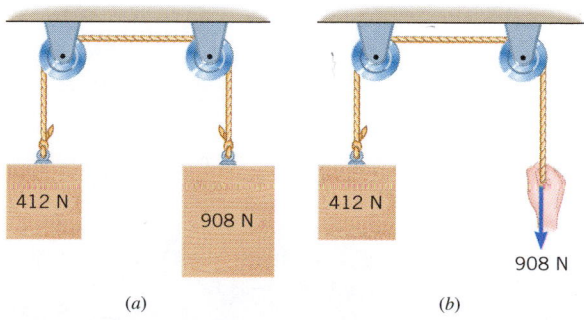

** 117. **ssm** A penguin slides at a constant velocity of 1.4 m/s down an icy incline. The incline slopes above the horizontal at an angle of 6.9°. At the bottom of the incline, the penguin slides onto a horizontal patch of ice. The coefficient of kinetic friction between the penguin and the ice is the same for the incline as for the horizontal patch. How much time is required for the penguin to slide to a halt after entering the horizontal patch of ice?

** 118. The basic concepts in this problem are presented in Multiple-Concept Example 9. A 225-kg crate rests on a surface that is inclined above the horizontal at an angle of 20.0°. A horizontal force (magnitude = 535 N and parallel to the ground, not the incline) is required

to start the crate moving down the incline. What is the coefficient of static friction between the crate and the incline?

** 119. While moving in, a new homeowner is pushing a box across the floor at a constant velocity. The coefficient of kinetic friction between the box and the floor is 0.41. The pushing force is directed downward at an angle θ below the horizontal. When θ is greater than a certain value, it is not possible to move the box, no matter how large the pushing force is. Find that value of θ.

DYNAMICS
OF UNIFORM CIRCULAR MOTION

When traveling around a circular turn, each of these race cars experiences a net force and an acceleration that point toward the center of the circle. (© Jared Tilton/ASP/Cal Sport Media/NewsCom)

5.1   UNIFORM CIRCULAR MOTION

There are many examples of motion on a circular path. Of the many possibilities, we single out those that satisfy the following definition:

DEFINITION OF UNIFORM CIRCULAR MOTION

Uniform circular motion is the motion of an object traveling at a constant (uniform) speed on a circular path.

As an example of uniform circular motion, Figure 5.1 shows a model airplane on a guide-line. The speed of the plane is the magnitude of the velocity vector $\vec{v}$, and since the speed is constant, the vectors in the drawing have the same magnitude at all points on the circle.

Sometimes it is more convenient to describe uniform circular motion by specifying the period of the motion, rather than the speed. The **period T** is the time required to travel once around the circle—that is, to make one complete revolution. There is a relationship between period and speed, since speed v is the distance traveled (here, the circumference of the circle $= 2\pi r$) divided by the time T:

$$v = \frac{2\pi r}{T} \tag{5.1}$$

If the radius is known, as in Example 1, the speed can be calculated from the period, or vice versa.

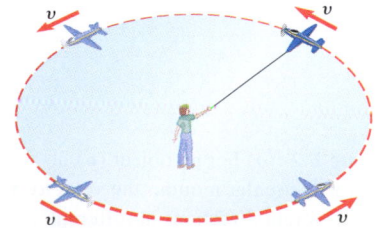

Figure 5.1 The motion of a model airplane flying at a constant speed on a horizontal circular path is an example of uniform circular motion.

Example 1 A Tire-Balancing Machine

The wheel of a car has a radius of $r = 0.29$ m and is being rotated at 830 revolutions per minute (rpm) on a tire-balancing machine. Determine the speed (in m/s) at which the outer edge of the wheel is moving.

Reasoning The speed v can be obtained directly from $v = 2\pi r/T$, but first the period T is needed. The period is the time for one revolution, and it must be expressed in seconds, because the problem asks for the speed in meters per second.

Solution Since the tire makes 830 revolutions in one minute, the number of minutes required for a single revolution is

$$\frac{1}{830 \text{ revolutions/min}} = 1.2 \times 10^{-3} \text{ min/revolution}$$

Therefore, the period is $T = 1.2 \times 10^{-3}$ min, which corresponds to 0.072 s. Equation 5.1 can now be used to find the speed:

$$v = \frac{2\pi r}{T} = \frac{2\pi(0.29 \text{ m})}{0.072 \text{ s}} = \boxed{25 \text{ m/s}}$$

The definition of uniform circular motion emphasizes that the speed, or the magnitude of the velocity vector, is constant. It is equally significant that the direction of the vector is *not constant*. In Figure 5.1, for instance, the velocity vector changes direction as the plane moves around the circle. Any change in the velocity vector, even if it is only a change in direction, means that an acceleration is occurring. This particular acceleration is called "centripetal acceleration," because it points toward the center of the circle, as the next section explains.

5.2 CENTRIPETAL ACCELERATION

In this section we determine how the magnitude a_c of the centripetal acceleration depends on the speed v of the object and the radius r of the circular path. We will see that $a_c = v^2/r$.

In Figure 5.2a an object (symbolized by a dot ●) is in uniform circular motion. At time t_0 the velocity is tangent to the circle at point O, and at a later time t the velocity is tangent at point P. As the object moves from O to P, the radius traces out the angle θ, and the velocity vector changes direction. To emphasize the change, part b of the picture shows the velocity vector removed from point P, shifted parallel to itself, and redrawn with its tail at point O. The angle β between the two vectors indicates the change in direction. Since the radii CO and CP are perpendicular to the tangents at points O and P, respectively, it follows that $\alpha + \beta = 90°$ and $\alpha + \theta = 90°$. Therefore, angle β and angle θ are equal.

As always, acceleration is the change $\Delta\vec{v}$ in velocity divided by the elapsed time Δt, or $\vec{a} = \Delta\vec{v}/\Delta t$. Figure 5.3$a$ shows the two velocity vectors oriented at the angle θ with

Figure 5.2 (*a*) For an object (●) in uniform circular motion, the velocity $\vec{v}$ has different directions at different places on the circle. (*b*) The velocity vector has been removed from point P, shifted parallel to itself, and redrawn with its tail at point O.

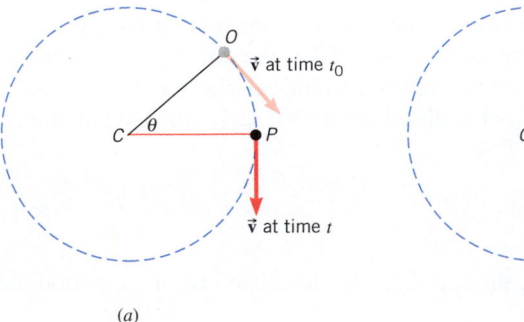

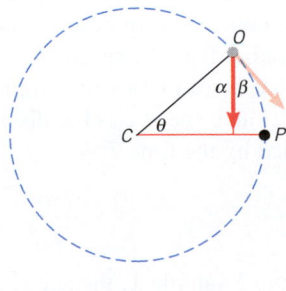

(*a*)

(*b*)

respect to one another, together with the vector $\Delta\vec{v}$ that represents the change in velocity. The change $\Delta\vec{v}$ is the increment that must be added to the velocity at time t_0, so that the resultant velocity has the new direction after an elapsed time $\Delta t = t - t_0$. Figure 5.3b shows the sector of the circle *COP*. In the limit that Δt is very small, the arc length *OP* is approximately a straight line whose length is the distance $v\Delta t$ traveled by the object. In this limit, *COP* is an isosceles triangle, as is the triangle in part *a* of the drawing. Since both triangles have equal apex angles θ, they are similar, so that

$$\frac{\Delta v}{v} = \frac{v\,\Delta t}{r}$$

This equation can be solved for $\Delta v/\Delta t$, to show that the magnitude a_c of the centripetal acceleration is given by $a_c = v^2/r$.

Centripetal acceleration is a vector quantity and, therefore, has a direction as well as a magnitude. The direction is toward the center of the circle, and Conceptual Example 2 helps us to set the stage for explaining this important fact.

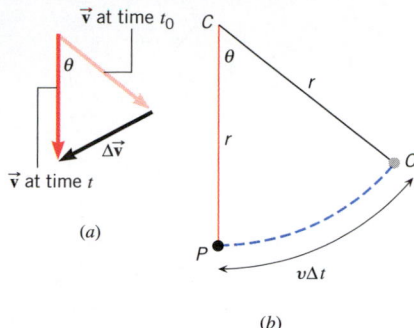

 Conceptual Example 2   Which Way Will the Object Go?

In Figure 5.4 an object, such as a model airplane on a guideline, is in uniform circular motion. The object is symbolized by a dot ($\bullet$), and at point *O* it is released suddenly from its circular path. For instance, suppose that the guideline for the model plane is cut suddenly. Does the object move **(a)** along the straight tangent line between points *O* and *A* or **(b)** along the circular arc between points *O* and *P*?

Reasoning Newton's first law of motion (see Section 4.2) guides our reasoning. This law states that an object continues in a state of rest or in a state of motion at a constant velocity (i.e., at a constant speed along a straight line) unless compelled to change that state by a net force. When an object is suddenly released from its circular path, there is no longer a net force being applied to the object. In the case of the model airplane, the guideline cannot apply a force, since it is cut. Gravity certainly acts on the plane, but the wings provide a lift force that balances the weight of the plane.

Answer (b) is incorrect. An object such as a model airplane will remain on a circular path only if a net force keeps it there. Since there is no net force, it cannot travel on the circular arc.

Answer (a) is correct. In the absence of a net force, the plane or any object would continue to move at a constant speed along a straight line in the direction it had at the time of release, consistent with Newton's first law. This speed and direction are given in Figure 5.4 by the velocity vector $\vec{v}$.

Related Homework: *Problems 3, 4*

Figure 5.3 (*a*) The directions of the velocity vector at times *t* and t_0 differ by the angle θ. (*b*) When the object moves along the circle from *O* to *P*, the radius *r* traces out the same angle θ. Here, the sector *COP* has been rotated clockwise by 90° relative to its orientation in Figure 5.2.

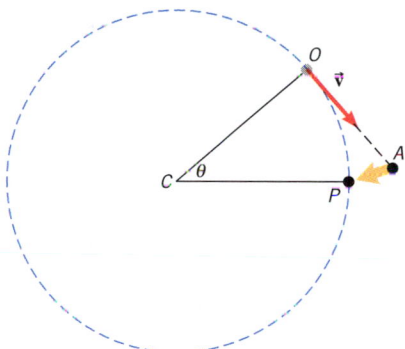

Figure 5.4 If an object ($\bullet$) moving on a circular path were released from its path at point *O*, it would move along the straight tangent line *OA* in the absence of a net force.

As Example 2 discusses, the object in Figure 5.4 would travel on a tangent line if it were released from its circular path suddenly at point *O*. It would move in a straight line to point *A* in the time it would have taken to travel on the circle to point *P*. It is as if, in the process of remaining on the circle, the object drops through the distance *AP*, and *AP* is directed toward the center of the circle in the limit that the angle θ is small. Thus, the object in uniform circular motion accelerates toward the center of the circle at every moment. Since the word "centripetal" means "moving toward a center", the acceleration is called ***centripetal acceleration.***

CENTRIPETAL ACCELERATION

Magnitude: The centripetal acceleration of an object moving with a speed v on a circular path of radius r has a magnitude a_c given by

$$a_c = \frac{v^2}{r} \tag{5.2}$$

Direction: The centripetal acceleration vector always points toward the center of the circle and continually changes direction as the object moves.

The following example illustrates the effect of the radius r on the centripetal acceleration.

Example 3
 The Effect of Radius on Centripetal Acceleration

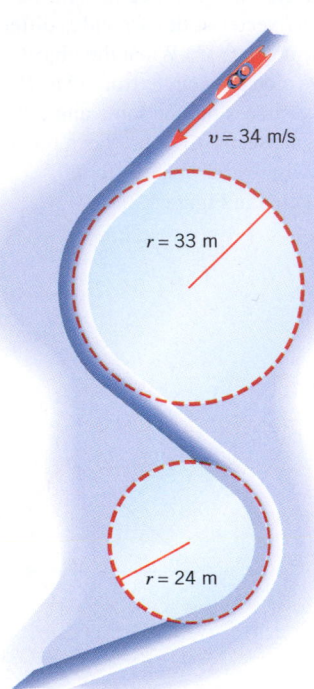

The physics of a bobsled track.

The bobsled track at the 1994 Olympics in Lillehammer, Norway, contained turns with radii of 33 m and 24 m, as Figure 5.5 illustrates. Find the centripetal acceleration at each turn for a speed of 34 m/s, a speed that was achieved in the two-man event. Express the answers as multiples of $g = 9.8$ m/s^2.

Reasoning In each case, the magnitude of the centripetal acceleration can be obtained from $a_c = v^2/r$. Since the radius r is in the denominator on the right side of this expression, we expect the acceleration to be smaller when r is larger.

Solution From $a_c = v^2/r$ it follows that

Radius = 33 m $\qquad\qquad a_c = \dfrac{(34\ \text{m/s})^2}{33\ \text{m}} = 35\ \text{m/s}^2 = \boxed{3.6\ g}$

Radius = 24 m $\qquad\qquad a_c = \dfrac{(34\ \text{m/s})^2}{24\ \text{m}} = 48\ \text{m/s}^2 = \boxed{4.9\ g}$

The centripetal acceleration is indeed smaller when the radius is larger. In fact, with r in the denominator on the right of $a_c = v^2/r$, the acceleration approaches zero when the radius becomes very large. Uniform circular motion along the arc of an infinitely large circle entails no acceleration, because it is just like motion at a constant speed along a straight line.

Figure 5.5 This bobsled travels at the same speed around two curves with different radii. For the turn with the larger radius, the sled has a smaller centripetal acceleration.

In Section 4.11 we learned that an object is in equilibrium when it has zero acceleration. Conceptual Example 4 discusses whether an object undergoing uniform circular motion can ever be at equilibrium.

Conceptual Example 4
 Uniform Circular Motion and Equilibrium

A car moves at a constant speed along a straight line as it approaches a circular turn. In which of the following parts of the motion is the car in equilibrium? **(a)** As it moves along the straight line toward the circular turn, **(b)** as it is going around the turn, **(c)** as it moves away from the turn along a straight line.

Reasoning An object is in equilibrium when it has no acceleration, according to the definition given in Section 4.11. If the object's velocity remains constant, both in magnitude and direction, its acceleration is zero.

Answer (b) is incorrect. As the car goes around the turn, the direction of travel changes, so the car has a centripetal acceleration that is characteristic of uniform circular motion. Because of this acceleration, the car is not in equilibrium during the turn.

Answers (a) and (c) are correct. As the car either approaches the turn or moves away from the turn it is traveling along a straight line, and both the speed and direction of the motion are constant. Thus, the velocity vector does not change, and there is no acceleration. Consequently, for these parts of the motion, the car is in equilibrium.

Related Homework: *Problem 49*

We have seen that going around tight turns (smaller r) and gentle turns (larger r) at the same speed entails different centripetal accelerations. And most drivers know that such turns "feel" different. This feeling is associated with the force that is present in uniform circular motion, and we turn to this topic in the next section.

✓ CHECK YOUR UNDERSTANDING

(The answers are given at the end of the book.)

1. The car in the drawing is moving clockwise around a circular section of road at a constant speed. What are the directions of its velocity and acceleration at   **(a)** position 1 and   **(b)** position 2? Specify your responses as north, east, south, or west.

2. The speedometer of your car shows that you are traveling at a constant speed of 35 m/s. Is it possible that your car is accelerating?

3. Consider two people, one on the earth's surface at the equator and the other at the north pole. Which has the larger centripetal acceleration?

4. Which of the following statements about centripetal acceleration is true?   **(a)** An object moving at a constant velocity cannot have a centripetal acceleration.   **(b)** An object moving at a constant speed may have a centripetal acceleration.

5. A car is traveling at a constant speed along the road *ABCDE* shown in the drawing. Sections *AB* and *DE* are straight. Rank the accelerations in each of the four sections according to magnitude, listing the smallest first.

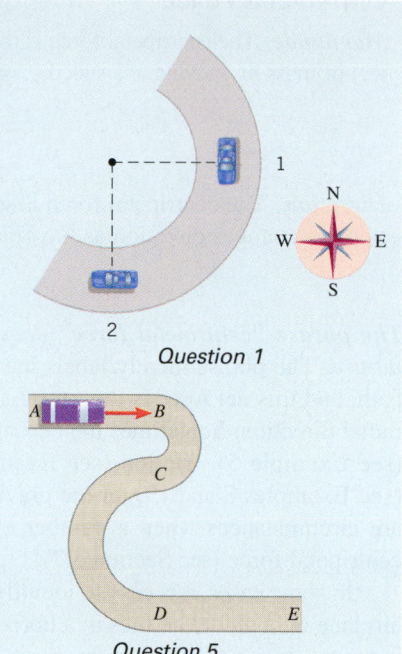

Question 1

Question 5

CENTRIPETAL FORCE

▶ **CONCEPTS AT A GLANCE**   Newton's second law indicates that whenever an object accelerates, there must be a net force to create the acceleration. Thus, in uniform circular motion there must be a net force to produce the centripetal acceleration. As the Concepts-at-a-Glance chart in Figure 5.6 indicates, the second law gives this net force as the product of the object's mass m and its acceleration v^2/r. The net force causing the centripetal acceleration is called the ***centripetal force*** $\vec{F}_c$ and points in the same direction as the acceleration—that is, toward the center of the circle. ◀

CONCEPTS AT A GLANCE

Figure 5.6   CONCEPTS AT A GLANCE Uniform circular motion entails a centripetal acceleration $a_c = v^2/r$. The net force required to produce this acceleration is called the centripetal force and is given by Newton's second law as $F_c = mv^2/r$. The photograph shows a fairground ride in which a rider spins around on a vertical circle. He experiences a centripetal force that comes in part from gravity and in part from the push of his seat against his back. (Bernd Mellmann/Alamy Images)

CENTRIPETAL FORCE

Magnitude: The centripetal force is the name given to the net force required to keep an object of mass m, moving at a speed v, on a circular path of radius r, and it has a magnitude of

$$F_c = \frac{mv^2}{r} \tag{5.3}$$

Direction: The centripetal force always points toward the center of the circle and continually changes direction as the object moves.

The phrase "centripetal force" does not denote a new and separate force created by nature. The phrase merely labels the net force pointing toward the center of the circular path, and this net force is the vector sum of all the force components that point along the radial direction. Sometimes the centripetal force consists of a single force such as tension (see Example 5), friction (see Example 7), the normal force or a component thereof (see Examples 8 and 13), or the gravitational force (see Examples 9–11). However, there are circumstances when a number of different forces contribute simultaneously to the centripetal force (see Section 5.7).

In some cases, it is easy to identify the source of the centripetal force, as when a model airplane on a guideline flies in a horizontal circle. The only force pulling the plane inward is the tension in the line, so this force alone (or a component of it) is the centripetal force. Example 5 illustrates the fact that higher speeds require greater tensions.

 Example 5   The Effect of Speed on Centripetal Force

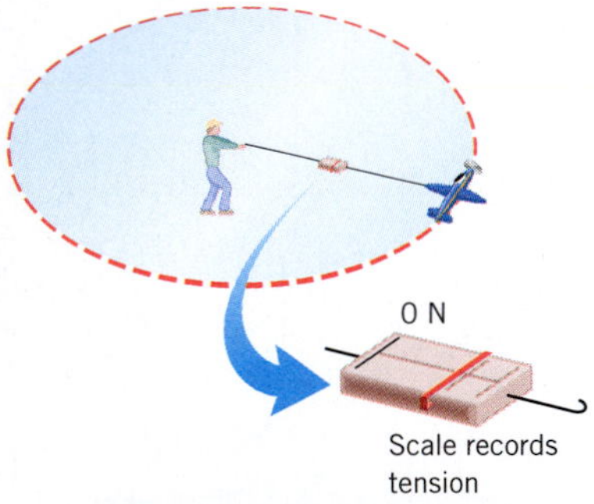

The model airplane in Figure 5.7 has a mass of 0.90 kg and moves at a constant speed on a circle that is parallel to the ground. The path of the airplane and its guideline lie in the same horizontal plane, because the weight of the plane is balanced by the lift generated by its wings. Find the tension T in the guideline (length = 17 m) for speeds of 19 and 38 m/s.

Reasoning Since the plane flies on a circular path, it experiences a centripetal acceleration that is directed toward the center of the circle. According to Newton's second law of motion, this acceleration is produced by a net force that acts on the plane, and this net force is called the centripetal force. The centripetal force is also directed toward the center of the circle. Since the tension T in the guideline is the only force pulling the plane inward, it must be the centripetal force.

Solution Equation 5.3 gives the tension directly: $F_c = T = mv^2/r$.

Speed = 19 m/s
$$T = \frac{(0.90 \text{ kg})(19 \text{ m/s})^2}{17 \text{ m}} = \boxed{19 \text{ N}}$$

Speed = 38 m/s
$$T = \frac{(0.90 \text{ kg})(38 \text{ m/s})^2}{17 \text{ m}} = \boxed{76 \text{ N}}$$

Figure 5.7 The scale records the tension in the guideline. See Example 5.

Conceptual Example 6 deals with another case where it is easy to identify the source of the centripetal force.

 Conceptual Example 6   A Trapeze Act

In a circus, a man hangs upside down from a trapeze, legs bent over the bar and arms downward, holding his partner (see Figure 5.8). Is it harder for the man to hold his partner **(a)** when the partner hangs straight down and is stationary or **(b)** when the partner is swinging through the straight-down position?

Figure 5.8 See Example 6 for a discussion of the role of centripetal force in this trapeze act. (Ludwig Goppenhammer/Ludwig's Flying Trapeze)

Reasoning Whenever an object moves on a circular path, it experiences a centripetal acceleration that is directed toward the center of the path. A net force, known as the centripetal force, is required to produce this acceleration.

Answer (a) is incorrect. When the man and his partner are stationary, the man's arms must support only his partner's weight. When the two are swinging, however, the man's arms must provide the additional force required to produce the partner's centripetal acceleration. Thus, it is easier, not harder, for the man to hold his partner when the partner hangs straight down and is stationary.

Answer (b) is correct. When the two are swinging, the partner is moving on a circular arc and, therefore, has a centripetal acceleration. The man's arms must support the partner's weight and must simultaneously exert an additional pull to provide the centripetal force that produces this acceleration. Because of the additional pull, it is harder for the man to hold his partner while swinging.

Related Homework: *Problems 18, 19*

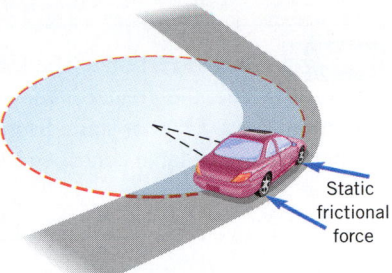

Figure 5.9 When the car moves without skidding around a curve, static friction between the road and the tires provides the centripetal force to keep the car on the road.

When a car moves at a steady speed around an unbanked curve, the centripetal force keeping the car on the curve comes from the static friction between the road and the tires, as Figure 5.9 indicates. It is static, rather than kinetic friction, because the tires are not slipping with respect to the radial direction. If the static frictional force is insufficient, given the speed and the radius of the turn, the car will skid off the road. Example 7 shows how an icy road can limit safe driving.

ANALYZING MULTIPLE-CONCEPT PROBLEMS

Example 7 Centripetal Force and Safe Driving

At what maximum speed can a car safely negotiate a horizontal unbanked turn (radius = 51 m) in dry weather (coefficient of static friction = 0.95) and icy weather (coefficient of static friction = 0.10)?

Reasoning The speed v at which the car of mass m can negotiate a turn of radius r is related to the centripetal force F_c that is available, according to $F_c = mv^2/r$ (Equation 5.3). Static friction provides the centripetal force and can provide a maximum force f_s^{MAX} given by $f_s^{MAX} = \mu_s F_N$ (Equation 4.7), in which μ_s is the coefficient of static friction and F_N is the normal force. Thus, we will use the fact that $F_c = f_s^{MAX}$ to obtain the maximum speed from Equation 5.3. It will also be necessary to evaluate the normal force, which we will do by considering that it must balance the weight of the car. Experience indicates that the maximum speed should be greater for the dry road than for the icy road.

Knowns and Unknowns The data for this problem are as follows:

Description	Symbol	Value	Comment
Radius of turn	r	51 m	
Coefficient of static friction	μ_s	0.95	Dry conditions
Coefficient of static friction	μ_s	0.10	Icy conditions
Unknown Variable			
Speed of car	v	?	

Modeling the Problem

STEP 1 **Speed and Centripetal Force** According to Equation 5.3, we have

$$F_c = \frac{mv^2}{r}$$

Solving for the speed v gives Equation 1 at the right. To use this result, it is necessary to consider the centripetal force F_c, which we do in Step 2.

$$v = \sqrt{\frac{rF_c}{m}} \qquad (1)$$

?

Continued

STEP 2   Static Friction   The force of static friction is the centripetal force, and the greater its value, the greater is the speed at which the car can negotiate the turn. The maximum force f_s^{MAX} of static friction is given by $f_s^{MAX} = \mu_s F_N$ (Equation 4.7). Thus, the maximum available centripetal force is

$$F_c = f_s^{MAX} = \mu_s F_N$$

which we can substitute into Equation 1, as shown at the right. The next step considers the normal force F_N.

$$v = \sqrt{\frac{rF_c}{m}} \quad (1)$$

$$\boxed{F_c = \mu_s F_N} \quad (2)$$

$$?$$

STEP 3   The Normal Force   The fact that the car does not accelerate in the vertical direction tells us that the normal force balances the car's weight mg, or

$$F_N = mg$$

This result for the normal force can now be substituted into Equation 2, as indicated at the right.

$$v = \sqrt{\frac{rF_c}{m}} \quad (1)$$

$$\boxed{F_c = \mu_s F_N} \quad (2)$$

$$F_N = mg$$

Solution   Algebraically combining the results of each step, we find that

| STEP 1 | STEP 2 | STEP 3 |

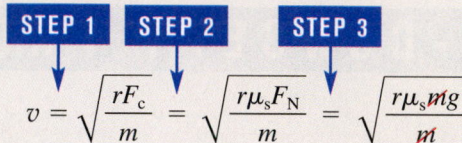

$$v = \sqrt{\frac{rF_c}{m}} = \sqrt{\frac{r\mu_s F_N}{m}} = \sqrt{\frac{r\mu_s mg}{m}}$$

The mass m of the car is eliminated algebraically from this result, and we find that the maximum speeds are

Dry road ($\mu_s = 0.95$) $v = \sqrt{r\mu_s g} = \sqrt{(51 \text{ m})(0.95)(9.8 \text{ m/s}^2)} = \boxed{22 \text{ m/s}}$

Icy road ($\mu_s = 0.10$) $v = \sqrt{r\mu_s g} = \sqrt{(51 \text{ m})(0.10)(9.8 \text{ m/s}^2)} = \boxed{7.1 \text{ m/s}}$

As expected, the dry road allows a greater maximum speed.

Related Homework: *Problems 16, 17, 20, 27*

Problem-solving insight
When using an equation to obtain a numerical answer, algebraically solve for the unknown variable in terms of the known variables. Then substitute in the numbers for the known variables, as this example shows.

A passenger in Figure 5.9 must also experience a centripetal force to remain on the circular path. However, if the upholstery is very slippery, there may not be enough static friction to keep him in place as the car makes a tight turn at high speed. Then, when viewed from inside the car, he appears to be thrown toward the outside of the curve. What really happens is that the passenger slides off on a tangent to the circle, until he encounters a source of centripetal force to keep him in place while the car turns. This occurs when the passenger bumps into the side of the car, which pushes on him with the necessary force.

The physics of flying an airplane in a banked turn.

Sometimes the source of the centripetal force is not obvious. A pilot making a turn, for instance, banks or tilts the plane at an angle to create the centripetal force. As a plane flies, the air pushes upward on the wing surfaces with a net lifting force $\vec{L}$ that is perpendicular to the wing surfaces, as Figure 5.10a shows. Part b of the drawing illustrates that when the plane is banked at an angle θ, a component $L \sin \theta$ of the lifting force is directed toward the center of the turn. It is this component that provides the centripetal force. Greater speeds and/or tighter turns require greater centripetal forces. In such situations, the pilot must bank the plane at a larger angle, so that a larger component of the lift points toward the center of the turn. The technique of banking into a turn also has an application in the construction of high-speed roadways, where the road itself is banked to achieve a similar effect, as the next section discusses.

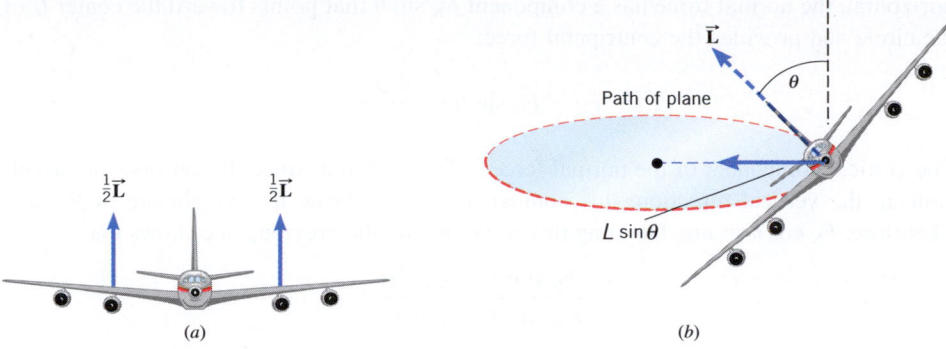

(a)

(b)

Figure 5.10 (a) The air exerts an upward lifting force $\frac{1}{2}\vec{L}$ on each wing. (b) When a plane executes a circular turn, the plane banks at an angle θ. The lift component $L \sin \theta$ is directed toward the center of the circle and provides the centripetal force.

✓ CHECK YOUR UNDERSTANDING

(The answers are given at the end of the book.)

6. A car is traveling in uniform circular motion on a section of road whose radius is r (see the drawing). The road is slippery, and the car is just on the verge of sliding. **(a)** If the car's speed were doubled, what would have to be the smallest radius in order that the car does not slide? Express your answer in terms of r. **(b)** What would be your answer to part (a) if the car were replaced by one that weighed twice as much, the car's speed still being doubled?

7. Other things being equal, would it be easier to drive at high speed around an unbanked horizontal curve on the moon than to drive around the same curve on the earth?

8. What is the chance of a light car safely rounding an unbanked curve on an icy road as compared to that of a heavy car: worse, the same, or better? Assume that both cars have the same speed and are equipped with identical tires.

9. A penny is placed on a rotating turntable. Where on the turntable does the penny require the largest centripetal force to remain in place, at the center of the turntable or at the edge of the turntable?

$$\mu_s mg = m \frac{v_1^2}{r}$$

$$\boxed{v_1 = \sqrt{\mu_s g r_1}}$$

$$\mu_s mg = m \frac{(2v_1)^2}{r_2}$$

$$r_2 = \frac{4v_1^2}{\mu_s g} = \frac{4(\mu_s g r_1)}{\mu_s g}$$

$$\boxed{r_2 = 4r_1}$$

BANKED CURVES

When a car travels without skidding around an unbanked curve, the static frictional force between the tires and the road provides the centripetal force. The reliance on friction can be eliminated completely for a given speed, however, if the curve is banked at an angle relative to the horizontal, much in the same way that a plane is banked while making a turn.

Figure 5.11a shows a car going around a friction-free banked curve. The radius of the curve is r, where r is measured parallel to the horizontal and not to the slanted surface. Part b shows the normal force $\vec{F}_N$ that the road applies to the car, the normal force being perpendicular to the road. Because the roadbed makes an angle θ with respect to the

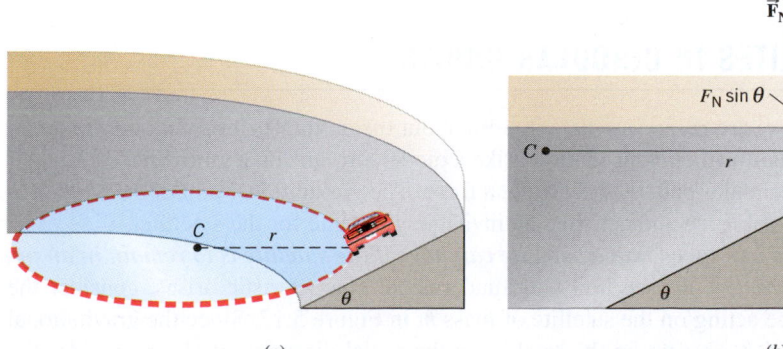

(a)

(b)

Figure 5.11 (a) A car travels on a circle of radius r on a frictionless banked road. The banking angle is θ, and the center of the circle is at C. (b) The forces acting on the car are its weight $m\vec{g}$ and the normal force $\vec{F}_N$. A component $F_N \sin \theta$ of the normal force provides the centripetal force.

horizontal, the normal force has a component $F_N \sin \theta$ that points toward the center C of the circle and provides the centripetal force:

$$F_c = F_N \sin \theta = \frac{mv^2}{r}$$

The vertical component of the normal force is $F_N \cos \theta$ and, since the car does not accelerate in the vertical direction, this component must balance the weight mg of the car. Therefore, $F_N \cos \theta = mg$. Dividing this equation into the previous one shows that

$$\frac{F_N \sin \theta}{F_N \cos \theta} = \frac{mv^2/r}{mg}$$

$$\tan \theta = \frac{v^2}{rg} \tag{5.4}$$

Equation 5.4 indicates that, for a given speed v, the centripetal force needed for a turn of radius r can be obtained from the normal force by banking the turn at an angle θ, independent of the mass of the vehicle. Greater speeds and smaller radii require more steeply banked curves—that is, larger values of θ. At a speed that is too small for a given θ, a car would slide down a frictionless banked curve; at a speed that is too large, a car would slide off the top. The next example deals with a famous banked curve.

 Example 8 The Daytona 500

The physics of the Daytona International Speedway.

The Daytona 500 is the major event of the NASCAR (National Association for Stock Car Auto Racing) season. It is held at the Daytona International Speedway in Daytona, Florida. The turns in this oval track have a maximum radius (at the top) of $r = 316$ m and are banked steeply, with $\theta = 31°$ (see Figure 5.11). Suppose these maximum-radius turns were frictionless. At what speed would the cars have to travel around them?

Reasoning In the absence of friction, the horizontal component of the normal force that the track exerts on the car must provide the centripetal force. Therefore, the speed of the car is given by Equation 5.4.

Solution From Equation 5.4, it follows that

$$v = \sqrt{rg \tan \theta} = \sqrt{(316 \text{ m})(9.80 \text{ m/s}^2) \tan 31°} = \boxed{43 \text{ m/s (96 mph)}}$$

Drivers actually negotiate the turns at speeds up to 195 mph, however, which requires a greater centripetal force than that implied by Equation 5.4 for frictionless turns. Static friction provides the additional force.

✓ **CHECK YOUR UNDERSTANDING**

(The answer is given at the end of the book.)

10. Go to **Concept Simulation 5.2** at **www.wiley.com/college/cutnell** to review the concepts involved in this question. Two cars are identical, except for the type of tread design on their tires. The cars are driven at the same speed and enter the same unbanked horizontal turn. Car A cannot negotiate the turn, but car B can. Which tread design, the one on car A or the one on car B, yields a larger coefficient of static friction between the tires and the road?

5.5 SATELLITES IN CIRCULAR ORBITS

Today there are many satellites in orbit about the earth. The ones in circular orbits are examples of uniform circular motion. Like a model airplane on a guideline, each satellite is kept on its circular path by a centripetal force. The gravitational pull of the earth provides the centripetal force and acts like an invisible guideline for the satellite.

There is only one speed that a satellite can have if the satellite is to remain in an orbit with a fixed radius. To see how this fundamental characteristic arises, consider the gravitational force acting on the satellite of mass m in Figure 5.12. Since the gravitational force is the only force acting on the satellite in the radial direction, it alone provides the centripetal force. Therefore, using Newton's law of gravitation (Equation 4.3), we have

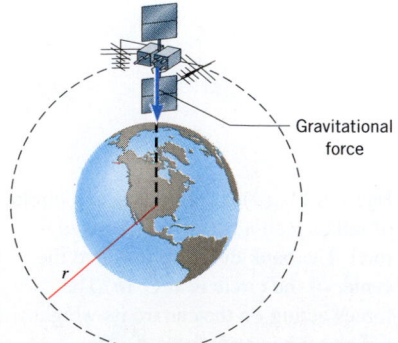

Gravitational force

Figure 5.12 For a satellite in circular orbit around the earth, the gravitational force provides the centripetal force.

$$F_c = G\frac{mM_E}{r^2} = \frac{mv^2}{r}$$

where G is the universal gravitational constant, M_E is the mass of the earth, and r is the distance from the center of the earth to the satellite. Solving for the speed v of the satellite gives

$$v = \sqrt{\frac{GM_E}{r}} \qquad (5.5)$$

If the satellite is to remain in an orbit of radius r, the speed must have precisely this value. Note that the radius r of the orbit is in the denominator in Equation 5.5. This means that the closer the satellite is to the earth, the smaller is the value for r and the greater the orbital speed must be.

The mass m of the satellite does not appear in Equation 5.5, having been eliminated algebraically. ***Consequently, for a given orbit, a satellite with a large mass has exactly the same orbital speed as a satellite with a small mass.*** However, more effort is certainly required to lift the larger-mass satellite into orbit. The orbital speed of one famous artificial satellite is determined in the following example.

Figure 5.13 The Hubble Space Telescope orbiting the earth. (Courtesy NASA)

 Example 9   Orbital Speed of the Hubble Space Telescope

Determine the speed of the Hubble Space Telescope (see Figure 5.13) orbiting at a height of 598 km above the earth's surface.

Reasoning Before Equation 5.5 can be applied, the orbital radius r must be determined *relative to the center of the earth*. Since the radius of the earth is approximately 6.38×10^6 m, and the height of the telescope above the earth's surface is 0.598×10^6 m, the orbital radius is $r = 6.98 \times 10^6$ m.

Solution The orbital speed is

$$v = \sqrt{\frac{GM_E}{r}} = \sqrt{\frac{(6.67 \times 10^{-11}\,\text{N}\cdot\text{m}^2/\text{kg}^2)(5.98 \times 10^{24}\,\text{kg})}{6.98 \times 10^6\,\text{m}}}$$

$$v = \boxed{7.56 \times 10^3\,\text{m/s}\ (16\,900\ \text{mi/h})}$$

The physics of the Hubble Space Telescope.

Problem-solving insight

The orbital radius r that appears in the relation $v = \sqrt{GM_E/r}$ is the distance from the satellite to the center of the earth (not to the surface of the earth).

Many applications of satellite technology affect our lives. An increasingly important application is the network of 24 satellites called the Global Positioning System (GPS), which can be used to determine the position of an object to within 15 m or less. Figure 5.14 illustrates how the system works. Each GPS satellite carries a highly accurate atomic clock, whose time is transmitted to the ground continually by means of radio waves. In the drawing, a car carries a computerized GPS receiver that can detect the waves

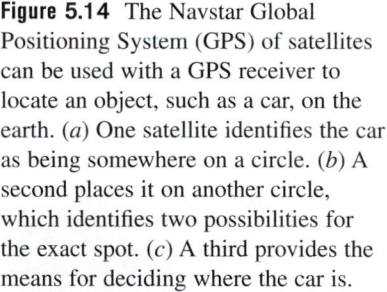

Figure 5.14 The Navstar Global Positioning System (GPS) of satellites can be used with a GPS receiver to locate an object, such as a car, on the earth. (*a*) One satellite identifies the car as being somewhere on a circle. (*b*) A second places it on another circle, which identifies two possibilities for the exact spot. (*c*) A third provides the means for deciding where the car is.

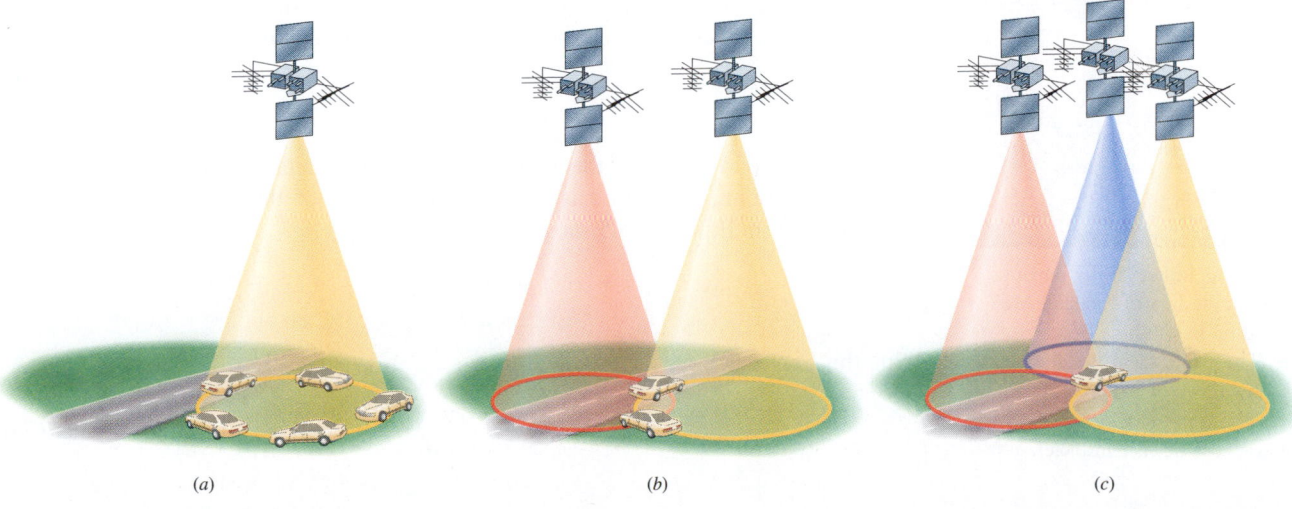

(*a*)   (*b*)   (*c*)

The physics of the Global Positioning System.

and is synchronized to the satellite clock. The receiver can therefore determine the distance between the car and a satellite from a knowledge of the travel time of the waves and the speed at which they move. This speed, as we will see in Chapter 24, is the speed of light and is known with great precision. A measurement using a single satellite locates the car somewhere on a circle, as Figure 5.14a shows, while a measurement using a second satellite locates the car on another circle. The intersection of the circles reveals two possible positions for the car, as in Figure 5.14b. With the aid of a third satellite, a third circle can be established, which intersects the other two and identifies the car's exact position, as in Figure 5.14c. The use of ground-based radio beacons to provide additional reference points leads to a system called Differential GPS, which can locate objects even more accurately than the satellite-based system alone. Navigational systems for automobiles and portable systems that tell hikers and people with visual impairments where they are located are two of the many uses for the GPS technique. GPS applications are so numerous that they have developed into a multibillion dollar industry.

Equation 5.5 applies to man-made earth satellites or to natural satellites like the moon. It also applies to circular orbits about any astronomical object, provided M_{E} is replaced by the mass of the object on which the orbit is centered. Example 10, for instance, shows how scientists have applied this equation to conclude that a supermassive black hole is probably located at the center of the galaxy known as M87. This galaxy is located at a distance of about 50 million light-years away from the earth. (One light-year is the distance that light travels in a year, or 9.5×10^{15} m.)

 Example 10 A Supermassive Black Hole

The physics of locating a black hole.

The Hubble telescope has detected the light being emitted from different regions of galaxy M87, which is shown in Figure 5.15. The black circle identifies the center of the galaxy. From the characteristics of this light, astronomers have determined that the orbiting speed is 7.5×10^5 m/s for matter located at a distance of 5.7×10^{17} m from the center. Find the mass M of the object located at the galactic center.

Reasoning and Solution Replacing M_{E} in Equation 5.5 with M gives $v = \sqrt{GM/r}$, which can be solved to show that

$$M = \frac{v^2 r}{G} = \frac{(7.5 \times 10^5 \text{ m/s})^2 (5.7 \times 10^{17} \text{ m})}{6.67 \times 10^{-11} \text{ N} \cdot \text{m}^2/\text{kg}^2}$$

$$= \boxed{4.8 \times 10^{39} \text{ kg}}$$

The ratio of this incredibly large mass to the mass of our sun is $(4.8 \times 10^{39} \text{ kg})/(2.0 \times 10^{30} \text{ kg}) = 2.4 \times 10^9$. Thus, matter equivalent to 2.4 billion suns is located at the center of galaxy M87. Considering that the volume of space in which this matter is located contains relatively few visible stars, researchers have concluded that the data provide strong evidence for the existence of a supermassive black hole. The term "black hole" is used because the tremendous mass prevents even light from escaping. The light that forms the image in Figure 5.15 comes not from the black hole itself, but from matter that surrounds it.

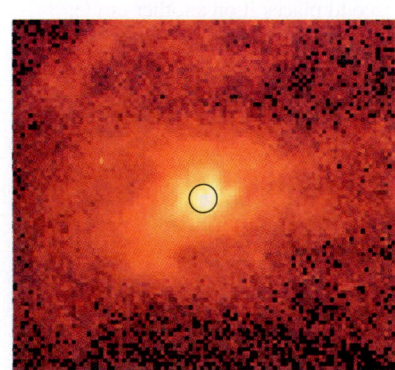

Figure 5.15 This image of the ionized gas (yellow) at the heart of galaxy M87 was obtained by the Hubble Space Telescope. The circle identifies the center of the galaxy, at which a black hole is thought to exist. (Courtesy NASA and Space Telescope Science Institute)

The period T of a satellite is the time required for one orbital revolution. As in any uniform circular motion, the period is related to the speed of the motion by $v = 2\pi r/T$. Substituting v from Equation 5.5 shows that

$$\sqrt{\frac{GM_{\mathrm{E}}}{r}} = \frac{2\pi r}{T}$$

Solving this expression for the period T gives

$$T = \frac{2\pi r^{3/2}}{\sqrt{GM_{\mathrm{E}}}} \tag{5.6}$$

Although derived for earth orbits, Equation 5.6 can also be used for calculating the periods of those planets in nearly circular orbits about the sun, if M_E is replaced by the mass M_S of the sun and r is interpreted as the distance between the center of the planet and the center of the sun. The fact that the period is proportional to the three-halves power of the orbital radius is known as Kepler's third law, and it is one of the laws discovered by Johannes Kepler (1571–1630) during his studies of planetary motion. Kepler's third law also holds for elliptical orbits, which will be discussed in Chapter 9.

An important application of Equation 5.6 occurs in the field of communications, where "synchronous satellites" are put into a circular orbit that is in the plane of the equator, as Figure 5.16 shows. The orbital period is chosen to be one day, which is also the time it takes for the earth to turn once about its axis. Therefore, these satellites move around their orbits in a way that is synchronized with the rotation of the earth. For earth-based observers, synchronous satellites have the useful characteristic of appearing in fixed positions in the sky and can serve as "stationary" relay stations for communication signals sent up from the earth's surface. **The physics of digital satellite system TV.** This is exactly what is done in the digital satellite systems that are a popular alternative to cable TV. As the blowup in Figure 5.16 indicates, a small "dish" antenna on your house picks up the digital TV signals relayed back to earth by the satellite. After being decoded, these signals are delivered to your TV set. All synchronous satellites are in orbit at the same height above the earth's surface, as Example 11 shows.

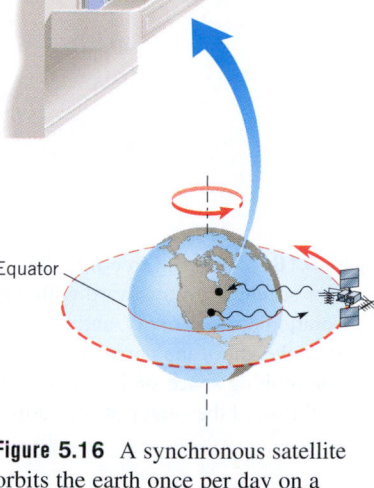

Figure 5.16 A synchronous satellite orbits the earth once per day on a circular path in the plane of the equator. Digital satellite system television uses such satellites as relay stations. TV signals are sent up from the earth's surface and then rebroadcast down to your own small dish antenna.

 Example 11 The Orbital Radius for Synchronous Satellites

What is the height H above the earth's surface at which all synchronous satellites (regardless of mass) must be placed in orbit?

Reasoning The period T of a synchronous satellite is one day, so we can use Equation 5.6 to find the distance r from the satellite to the center of the earth. To find the height H of the satellite above the earth's surface we will have to take into account the fact that the earth itself has a radius of 6.38×10^6 m.

Solution A period of one day* corresponds to $T = 8.64 \times 10^4$ s. In using this value it is convenient to rearrange the equation $T = 2\pi r^{3/2}/\sqrt{GM_E}$ as follows:

$$r^{3/2} = \frac{T\sqrt{GM_E}}{2\pi} = \frac{(8.64 \times 10^4 \text{ s})\sqrt{(6.67 \times 10^{-11}\text{ N} \cdot \text{m}^2/\text{kg}^2)(5.98 \times 10^{24}\text{ kg})}}{2\pi}$$

By squaring and then taking the cube root, we find that $r = 4.23 \times 10^7$ m. Since the radius of the earth is approximately 6.38×10^6 m, the height of the satellite above the earth's surface is

$$H = 4.23 \times 10^7 \text{ m} - 0.64 \times 10^7 \text{ m} = \boxed{3.59 \times 10^7 \text{ m (22 300 mi)}}$$

✓ CHECK YOUR UNDERSTANDING

(The answer is given at the end of the book.)

11. Two satellites are placed in orbit, one about Mars and the other about Jupiter, such that the orbital speeds are the same. Mars has the smaller mass. Is the radius of the satellite in orbit about Mars less than, greater than, or equal to the radius of the satellite orbiting Jupiter?

5.6 APPARENT WEIGHTLESSNESS AND ARTIFICIAL GRAVITY

The idea of life on board an orbiting satellite conjures up visions of astronauts floating around in a state of "weightlessness," as in Figure 5.17. Actually, this state should be called "apparent weightlessness," because it is similar to the condition of zero apparent

Figure 5.17 In a state of apparent weightlessness, astronaut and pilot Susan L. Still floats in the orbiting Spacelab Module. (Shuttle Mission Imagery/NASA Media Services)

---

*Successive appearances of the sun define the solar day of 24 h or 8.64×10^4 s. The sun moves against the background of the stars, however, and the time required for the earth to turn once on its axis relative to the fixed stars is 23 h 56 min, which is called the sidereal day. The sidereal day should be used in Example 11, but the neglect of this effect introduces an error of less than 0.4% in the answer.

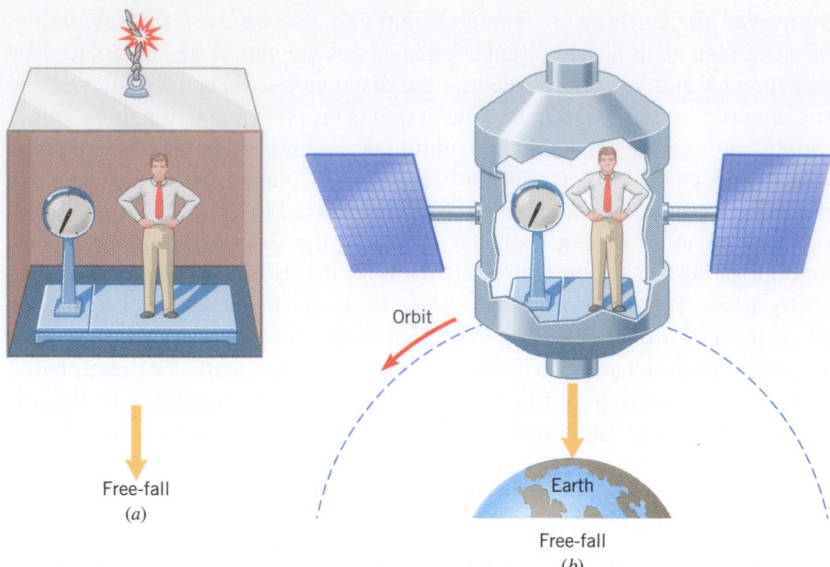

Figure 5.18 (*a*) During free-fall, the elevator accelerates downward with the acceleration due to gravity, and the apparent weight of the person is zero. (*b*) The orbiting space station is also in free-fall toward the center of the earth.

weight that occurs in an elevator during free-fall. Conceptual Example 12 explores this similarity.

Conceptual Example 12 Apparent Weightlessness and Free-Fall

Figure 5.18 shows a person on a scale in a freely falling elevator and in a satellite in a circular orbit. Assume that when the person is standing stationary on the earth, his weight is 800 N (180 lbs). In each case, what apparent weight is recorded by the scale? **(a)** The scale in the elevator records 800 N while that in the satellite records 0 N. **(b)** The scale in the elevator records 0 N while that in the satellite records 800 N. **(c)** Both scales record 0 N.

The physics of apparent weightlessness.

Reasoning As Section 4.8 discusses, apparent weight is the force that an object exerts on the platform of a scale. This force depends on whether or not the object and the platform are accelerating together.

Answer (a) is incorrect. The scale and the person in the free-falling elevator are accelerating toward the earth at the same rate. Therefore, they cannot push against one another, and so the scale does not record an apparent weight of 800 N.

Answer (b) is incorrect. The scale and the person in the satellite are accelerating toward the center of the earth at the same rate (they have the same centripetal acceleration). Therefore, they cannot push against one another, and so the scale does not record an apparent weight of 800 N.

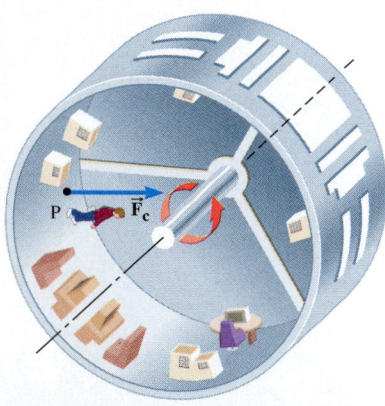

Figure 5.19 The surface of the rotating space station pushes on an object with which it is in contact and thereby provides the centripetal force that keeps the object moving on a circular path.

Answer (c) is correct. The scale and the person in the elevator fall together and, therefore, they cannot push against one another. Therefore, the scale records an apparent weight of 0 N. In the orbiting satellite in Figure 5.18*b*, both the person and the scale are in uniform circular motion. Objects in uniform circular motion continually accelerate or "fall" toward the center of the circle in order to remain on the circular path. Consequently, both the person and the scale "fall" with the same acceleration toward the center of the earth and cannot push against one another. Thus, the apparent weight in the satellite is zero, just as it is in the freely falling elevator.

The physics of artificial gravity.

The physiological effects of prolonged apparent weightlessness are only partially known. To minimize such effects, it is likely that artificial gravity will be provided in large space stations of the future. To help explain artificial gravity, Figure 5.19 shows a space station rotating about an axis. Because of the rotational motion, any object located at a point P on the interior surface of the station experiences a centripetal force directed toward the axis. The surface of the station provides this force by pushing on the feet of an astronaut, for instance. The centripetal force can be adjusted to match the astronaut's earth-weight by properly selecting the rotational speed of the space station, as Examples 13 and 14 illustrate.

ANALYZING MULTIPLE-CONCEPT PROBLEMS

Example 13 Artificial Gravity

At what speed must the interior surface of the space station ($r = 1700$ m) move in Figure 5.19, so that the astronaut at point P experiences a push on his feet that equals his weight on earth?

Reasoning The floor of the rotating space station exerts a normal force on the feet of the astronaut. This is the centripetal force ($F_c = mv^2/r$) that keeps the astronaut moving on a circular path. Since the magnitude of the normal force equals the astronaut's weight on earth, we can determine the speed v of the space station's floor.

Knowns and Unknowns The data for this problem are given in the following table:

Description	Symbol	Value
Radius of space station	r	1700 m
Unknown Variable		
Speed of space station's floor	v	?

Modeling the Problem

STEP 1 **Speed and Centripetal Force**   The centripetal force acting on the astronaut is given by Equation 5.3 as

$$F_c = \frac{mv^2}{r}$$

$$v = \sqrt{\frac{rF_c}{m}} \qquad (1)$$

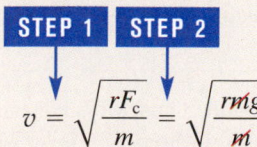

Solving for the speed v gives Equation 1 at the right. Step 2 considers the centripetal force F_c that appears in this result.

STEP 2 **Magnitude of the Centripetal Force**   The normal force applied to the astronaut's feet by the floor is the centripetal force and has a magnitude equal to the astronaut's earth-weight. This earth-weight is given by Equation 4.5 as the astronaut's mass m times the magnitude g of the acceleration due to the earth's gravity. Thus, we have for the centripetal force that

$$F_c = mg$$

$$v = \sqrt{\frac{rF_c}{m}} \qquad (1)$$

$$F_c = mg$$

which can be substituted into Equation 1, as indicated at the right.

Solution Algebraically combining the results of each step, we find that

STEP 1	STEP 2

$$v = \sqrt{\frac{rF_c}{m}} = \sqrt{\frac{rm g}{m}}$$

The astronaut's mass m is eliminated algebraically from this result, so the speed of the space station floor is

$$v = \sqrt{rg} = \sqrt{(1700 \text{ m})(9.80 \text{ m/s}^2)} = \boxed{130 \text{ m/s}}$$

Related Homework: *Problem 34*

ANALYZING MULTIPLE-CONCEPT PROBLEMS

Example 14   A Rotating Space Laboratory

A space laboratory is rotating to create artificial gravity, as Figure 5.20 indicates. Its period of rotation is chosen so the outer ring ($r_O = 2150$ m) simulates the acceleration due to gravity on earth (9.80 m/s^2). What should be the radius r_I of the inner ring, so it simulates the acceleration due to gravity on the surface of Mars (3.72 m/s^2)?

Reasoning   The value given for either acceleration corresponds to the centripetal acceleration $a_c = v^2/r$ (Equation 5.2) in the corresponding ring. This expression can be solved for the radius r. However, we are given no direct information about the speed v. Instead, it is stated that the period of rotation T is chosen so that the outer ring simulates gravity on earth. Although no value is given for T, we know that the laboratory is a rigid structure, so that all points on it make one revolution in the same time. This is an important observation, because it means that both rings have the same value for T. Thus, we will be able to use the fact that $v = 2\pi r/T$ (Equation 5.1) together with $a_c = v^2/r$ in order to find the radius of the inner ring.

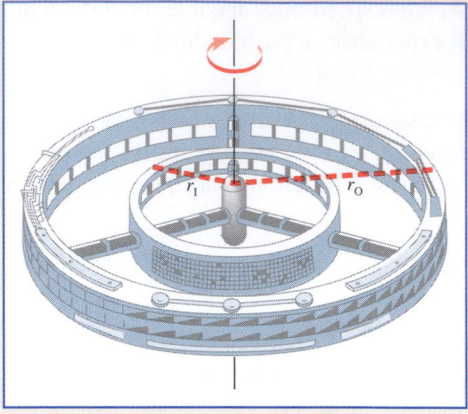

Figure 5.20   The outer ring (radius = r_O) of this rotating space laboratory simulates gravity on earth, while the inner ring (radius = r_I) simulates gravity on Mars.

Knowns and Unknowns   The data for this problem are shown in the following table:

Description	Symbol	Value	Comment
Radius of outer ring	r_O	2150 m	
Centripetal acceleration of outer ring	$a_{c,O}$	9.80 m/s^2	Acceleration due to gravity on earth
Centripetal acceleration of inner ring	$a_{c,I}$	3.72 m/s^2	Acceleration due to gravity on Mars
Unknown Variable			
Radius of inner ring	r_I	?	

Modeling the Problem

STEP 1   **Inner Ring**   According to Equation 5.2, the centripetal acceleration of the inner ring is $a_{c,I} = v_I^2/r_I$, which can be solved for the radius r_I:

$$r_I = \frac{v_I^2}{a_{c,I}}$$

The speed v_I is given as $v_I = 2\pi r_I/T$ (see Equation 5.1). With this substitution, the expression for the inner radius becomes

$$r_I = \frac{(2\pi r_I/T)^2}{a_{c,I}}$$

$$r_I = \frac{a_{c,I}T^2}{4\pi^2} \qquad (1)$$

Solving for the radius r_I gives Equation 1 at the right. The term T^2 in Equation 1 is unknown, but it can be determined, as Step 2 discusses.

STEP 2   **Outer Ring**   The inner and outer rings both have the same period. Thus, we can obtain the period by considering the outer ring. The same approach used in Step 1 for the inner ring can be applied to the outer ring, with the result that the radius r_O is given by an expression analogous to Equation 1:

$$r_O = \frac{a_{c,O}T^2}{4\pi^2}$$

Note that, except for T, all the variables in this result are known (see data table). Solving for T^2 gives

Problem-solving insight

All points on a rotating rigid body have the same value for the period T of the motion.

$$T^2 = \frac{4\pi^2 r_O}{a_{c,\,O}}$$

$$r_I = \frac{a_{c,\,I} T^2}{4\pi^2} \qquad (1)$$

$$T^2 = \frac{4\pi^2 r_O}{a_{c,\,O}}$$

As shown at the right, this expression can be substituted for T^2 in Equation 1.

Solution Algebraically combining the results of each step, we find that

$$r_I = \overset{\text{STEP 1}}{\frac{a_{c,\,I} T^2}{4\pi^2}} = \overset{\text{STEP 2}}{\frac{a_{c,\,I}(4\pi^2 r_O/a_{c,\,O})}{4\pi^2}}$$

$$r_I = \frac{a_{c,\,I} r_O}{a_{c,\,O}} = \frac{(3.72 \text{ m/s}^2)(2150 \text{ m})}{9.80 \text{ m/s}^2} = \boxed{816 \text{ m}}$$

Related Homework: *Problems 40, 50*

✓ CHECK YOUR UNDERSTANDING

(The answer is given at the end of the book.)

12. The acceleration due to gravity on the moon is one-sixth that on earth.   **(a)** Is the true weight of a person on the moon less than, greater than, or equal to the true weight of the same person on earth?   **(b)** Is the apparent weight of a person in orbit about the moon less than, greater than, or equal to the apparent weight of the same person in orbit about the earth?

---

*VERTICAL CIRCULAR MOTION

Motorcycle stunt drivers perform a feat in which they drive their cycles around a vertical circular track, as in Figure 5.21a. Usually, the speed varies in this stunt. When the speed of travel on a circular path changes from moment to moment, the motion is said to be nonuniform. Nonetheless, we can use the concepts that apply to uniform circular motion to gain considerable insight into the motion that occurs on a vertical circle.

There are four points on a vertical circle where the centripetal force can be identified easily, as Figure 5.21b indicates. As you look at Figure 5.21b, keep in mind that the centripetal force is not a new and separate force of nature. Instead, at each point the centripetal force is the net sum of all the force components oriented along the radial direction, and it points toward the center of the circle. The drawing shows only the weight of the cycle plus rider (magnitude = mg) and the normal force pushing on the cycle (magnitude = F_N). The propulsion and braking forces are omitted for simplicity, because they do not act in the radial direction. The magnitude of the centripetal force at each of the four points is given as follows in terms of mg and F_N:

(1) $\underbrace{F_{N1} - mg}_{= F_{c1}} = \dfrac{m v_1^2}{r}$ (3) $\underbrace{F_{N3} + mg}_{= F_{c3}} = \dfrac{m v_3^2}{r}$

(2) $\underbrace{F_{N2}}_{= F_{c2}} = \dfrac{m v_2^2}{r}$ (4) $\underbrace{F_{N4}}_{= F_{c4}} = \dfrac{m v_4^2}{r}$

As the cycle goes around, the magnitude of the normal force changes. It changes because the speed changes and because the weight does not have the same effect at every point. At the bottom, the normal force and the weight oppose one another, giving a centripetal force of magnitude $F_{N1} - mg$. At the top, the normal force and the weight reinforce each other to provide a centripetal force whose magnitude is $F_{N3} + mg$. At points 2 and 4

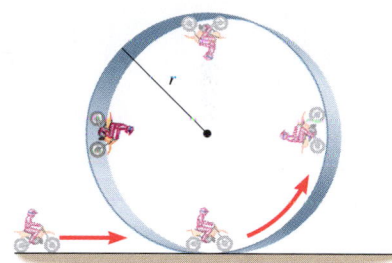

(a)

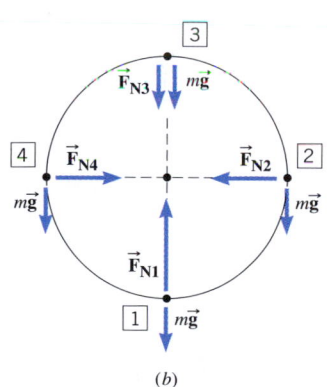

(b)

Figure 5.21 (a) A vertical loop-the-loop motorcycle stunt. (b) The normal force $\vec{F}_N$ and the weight $m\vec{g}$ of the cycle and the rider are shown here at four locations.

Problem-solving insight

Centripetal force $\vec{F}_c$ is the name given to the net force that points toward the center of a circular path. As shown here, there may be more than one force that contributes to this net force.

**The physics of
the loop-the-loop
motorcycle stunt.**

on either side, only F_{N2} and F_{N4} provide the centripetal force. The weight is tangent to the circle at points 2 and 4 and has no component pointing toward the center. If the speed at each of the four places is known, along with the mass and the radius, the normal forces can be determined.

Riders who perform the loop-the-loop trick must have at least a minimum speed at the top of the circle to remain on the track. This speed can be determined by considering the centripetal force at point 3. The speed v_3 in the equation $F_{N3} + mg = mv_3{}^2/r$ is a minimum when F_{N3} is zero. Then, the speed is given by $v_3 = \sqrt{rg}$. At this speed, the track does not exert a normal force to keep the cycle on the circle, because the weight mg provides all the centripetal force. Under these conditions, the rider experiences an apparent weightlessness like that discussed in Section 5.6, because for an instant the rider and the cycle are falling freely toward the center of the circle.

✓ **CHECK YOUR UNDERSTANDING**

(The answers are given at the end of the book.)

13. Would a change in the earth's mass affect   **(a)** the banking of airplanes as they turn, **(b)** the banking of roadbeds,   **(c)** the speeds with which satellites are put into circular orbits, and   **(d)** the performance of the loop-the-loop motorcycle stunt?

14. A stone is tied to a string and whirled around in a circle at a constant speed. Is the string more likely to break when the circle is horizontal or when it is vertical? Assume that the constant speed is the same in each case.

5.8 CONCEPTS & CALCULATIONS

In uniform circular motion the concepts of acceleration and force play central roles. Example 15 deals with acceleration, which we first discussed in Chapter 2, particularly in connection with the equations of kinematics. This example emphasizes the difference between the acceleration that is used in the equations of kinematics and the acceleration that arises in uniform circular motion.

 Concepts & Calculations Example 15   Acceleration

At time $t = 0$ s, automobile A is traveling at a speed of 18 m/s along a straight road and is picking up speed with an acceleration that has a magnitude of 3.5 m/s² (Figure 5.22a). At time $t = 0$ s, automobile B is traveling at a speed of 18 m/s in uniform circular motion as it negotiates a turn (Figure 5.22b). It has a centripetal acceleration whose magnitude is also 3.5 m/s². Determine the speed of each automobile when $t = 2.0$ s.

Concept Questions and Answers   Which automobile has a constant acceleration?

Answer   Acceleration is a vector, and for it to be constant, both its magnitude and direction must be constant. Automobile A has a constant acceleration, because its acceleration has a constant magnitude of 3.5 m/s² and its direction always points forward along the straight road. Automobile B has an acceleration with a constant magnitude of 3.5 m/s², but the acceleration does not have a constant direction. Automobile B is in uniform circular motion, so it has a centripetal acceleration, which points toward the center of the circle at every instant. Therefore, the direction of the acceleration vector continually changes, and the vector is not constant.

For which automobile do the equations of kinematics apply?

Answer   The equations of kinematics apply for automobile A, because it has a constant acceleration, which must be the case when you use these equations. They do not apply to automobile B, because it does not have a constant acceleration.

Solution   To determine the speed of automobile A at $t = 2.0$ s, we use Equation 2.4 from the equations of kinematics:

$$v = v_0 + at = 18 \text{ m/s} + (3.5 \text{ m/s}^2)(2.0 \text{ s}) = \boxed{25 \text{ m/s}}$$

To determine the speed of automobile B we note that this car is in uniform circular motion and, therefore, has a constant speed as it goes around the turn. At a time of $t = 2.0$ s its speed is the same as it was at $t = 0$ s—namely, $v = \boxed{18 \text{ m/s}}$.

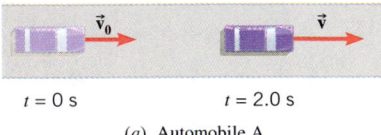

$t = 0$ s $t = 2.0$ s

(a) Automobile A

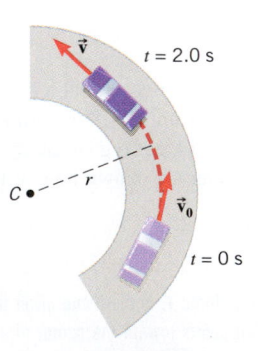

(b) Automobile B

Figure 5.22 *(a)* Automobile A accelerates to the right along a straight road. *(b)* Automobile B travels at a constant speed along a circular turn. It is also accelerating, but the acceleration is a centripetal acceleration.

The next example deals with force, and it stresses that the centripetal force is the net force along the radial direction. As we will see, this means that all forces along the radial direction must be taken into account when identifying the centripetal force. In addition, the example is a good review of the tension force, which we first encountered in Chapter 4.

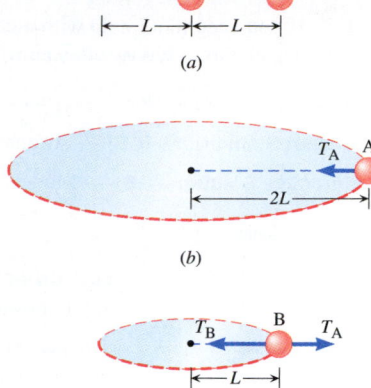

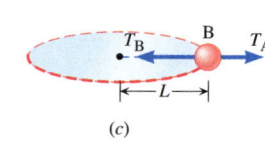

Figure 5.23 Two identical balls attached to a rigid massless rod are whirled around on horizontal circles, as Example 16 explains.

⬇ Concepts & Calculations Example 16   Centripetal Force

Ball A is attached to one end of a rigid massless rod, while an identical ball B is attached to the center of the rod, as Figure 5.23a illustrates. Each ball has a mass of $m = 0.50$ kg, and the length of each half of the rod is $L = 0.40$ m. This arrangement is held by the empty end and is whirled around in a horizontal circle at a constant rate, so each ball is in uniform circular motion. Ball A travels at a constant speed of $v_A = 5.0$ m/s. Find the tension in each half of the rod.

Concept Questions and Answers How many tension forces contribute to the centripetal force that acts on ball A?

Answer As Figure 5.23b illustrates, only a single tension force of magnitude T_A acts on ball A. It points to the left in the drawing and is due to the tension in the rod between the two balls. This force alone provides the centripetal force keeping ball A on its circular path of radius 2L.

How many tension forces contribute to the centripetal force that acts on ball B?

Answer As Figure 5.23c shows, two tension forces act on ball B. One has a magnitude T_B and points to the left in the drawing, which is the positive direction. It is due to the tension in the left half of the rod. The other has a magnitude T_A and points to the right. It is due to the tension in the right half of the rod. The centripetal force acting on ball B points toward the center of the circle and is the vector sum of these two forces, or $T_B - T_A$.

Is the speed of ball B the same as the speed of ball A?

Answer No, it is not. The reason is that ball A travels farther than ball B in the same time. Consider a time of one period, which is the same for either ball, since the arrangement is a rigid unit. It is the time for either ball to travel once around its circular path. In this time ball A travels a distance equal to the circumference of its path, which is $2\pi(2L)$. In contrast, the circumference for ball B is only $2\pi L$. Therefore, the speed of ball B is one-half the speed of ball A, or $v_B = 2.5$ m/s.

Solution Applying Equation 5.3 to each ball, we have

Ball A
$$\underbrace{T_A}_{\text{Centripetal force, } F_c} = \frac{mv_A^2}{2L}$$

Ball B
$$\underbrace{T_B - T_A}_{\text{Centripetal force, } F_c} = \frac{mv_B^2}{L}$$

The tension in the right half of the rod follows directly from the first of these equations:

$$T_A = \frac{mv_A^2}{2L} = \frac{(0.50 \text{ kg})(5.0 \text{ m/s})^2}{2(0.40 \text{ m})} = \boxed{16 \text{ N}}$$

Adding the equations for ball A and ball B gives the following result for the tension in the left half of the rod:

$$T_B = \frac{mv_B^2}{L} + \frac{mv_A^2}{2L} = \frac{(0.50 \text{ kg})(2.5 \text{ m/s})^2}{0.40 \text{ m}} + \frac{(0.50 \text{ kg})(5.0 \text{ m/s})^2}{2(0.40 \text{ m})} = \boxed{23 \text{ N}}$$

CONCEPT SUMMARY

If you need more help with a concept, use the Learning Aids noted next to the discussion or equation. Examples (**Ex.**) are in the text of this chapter. Go to **www.wiley.com/college/cutnell** for the following Learning Aids:

Interactive LearningWare (ILW) — Additional examples solved in a five-step interactive format.

Concept Simulations (CS) — Animated text figures or animations of important concepts.

Interactive Solutions (IS) — Models for certain types of problems in the chapter homework. The calculations are carried out interactively.

Topic	Discussion	Learning Aids
	5.1 UNIFORM CIRCULAR MOTION Uniform circular motion is the motion of an object traveling at a constant (uniform) speed on a circular path.	
Period of the motion	The period T is the time required for the object to travel once around the circle. The speed v of the object is related to the period and the radius r of the circle by	**Ex. 1**

$$v = \frac{2\pi r}{T} \tag{5.1}$$

| | **5.2 CENTRIPETAL ACCELERATION** An object in uniform circular motion experiences an acceleration, known as centripetal acceleration. The magnitude a_c of the centripetal acceleration is | **Ex. 2, 3, 4, 15** **ILW 5.1** |
| Centripetal acceleration (magnitude) | | **IS 5.51** |

$$a_c = \frac{v^2}{r} \tag{5.2}$$

Centripetal acceleration (direction)	where v is the speed of the object and r is the radius of the circle. The direction of the centripetal acceleration vector always points toward the center of the circle and continually changes as the object moves.	
	5.3 CENTRIPETAL FORCE To produce a centripetal acceleration, a net force pointing toward the center of the circle is required. This net force is called the centripetal force, and its magnitude F_c is	**Ex. 5, 6, 7, 16** **ILW 5.2**
Centripetal force (magnitude)		**CS 5.1, 5.2**

$$F_c = \frac{mv^2}{r} \tag{5.3}$$

Centripetal force (direction)	where m and v are the mass and speed of the object, and r is the radius of the circle. The direction of the centripetal force vector, like that of the centripetal acceleration vector, always points toward the center of the circle.	**IS 5.23**
	5.4 BANKED CURVES A vehicle can negotiate a circular turn without relying on static friction to provide the centripetal force, provided the turn is banked at an angle relative to the horizontal. The angle θ at which a friction-free curve must be banked is related to the speed v of the vehicle, the radius r of the curve, and the magnitude g of the acceleration due to gravity by	**Ex. 8** **IS 5.29**
Angle of a banked turn		

$$\tan \theta = \frac{v^2}{rg} \tag{5.4}$$

| | **5.5 SATELLITES IN CIRCULAR ORBITS** When a satellite orbits the earth, the gravitational force provides the centripetal force that keeps the satellite moving in a circular orbit. The speed v and period T of a satellite depend on the mass M_E of the earth and the radius r of the orbit according to | |
| Orbital speed | | **Ex. 9, 10, 11** |

$$v = \sqrt{\frac{GM_E}{r}} \tag{5.5}$$

| Orbital period | | **IS 5.35** |

$$T = \frac{2\pi r^{3/2}}{\sqrt{GM_E}} \tag{5.6}$$

where G is the universal gravitational constant.

| | **5.6 APPARENT WEIGHTLESSNESS AND ARTIFICIAL GRAVITY** The apparent weight of an object is the force that it exerts on a scale with which it is in contact. All objects, including people, on board an orbiting satellite are in free-fall, since they experience negligible air resistance and they have an acceleration that is equal to the acceleration due to gravity. When a person is in free-fall, his or her apparent weight is zero, because both the person and the scale fall freely and cannot push against one another. | **Ex. 12, 13, 14** |
| | **5.7 VERTICAL CIRCULAR MOTION** Vertical circular motion occurs when an object, such as a motorcycle, moves on a vertical circular path. The speed of the object often varies from moment to moment, and so do the magnitudes of the centripetal acceleration and centripetal force. | **IS 5.47** |

FOCUS ON CONCEPTS

Note to Instructors: The numbering of the questions shown here reflects the fact that they are only a representative subset of the total number that are available online. However, all of the questions are available for assignment via an online homework management program such as WileyPLUS or WebAssign.

Section 5.2 Centripetal Acceleration

1. Two cars are traveling at the same constant speed v. Car A is moving along a straight section of the road, while B is rounding a circular turn. Which statement is true about the acceleration of the cars? **(a)** The acceleration of both cars is zero, since they are traveling at a constant speed. **(b)** Car A is accelerating, but car B is not accelerating. **(c)** Car A is not accelerating, but car B is accelerating. **(d)** Both cars are accelerating.

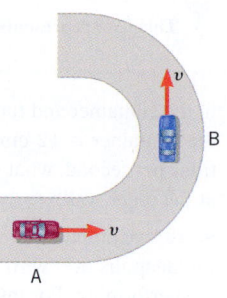

3. Two cars are driving at the same constant speed v around a racetrack. However, they are traveling through turns that have different radii, as shown in the drawing. Which statement is true about the magnitude of the centripetal acceleration of each car? **(a)** The magnitude of the centripetal acceleration of each car is the same, since the cars are moving at the same speed. **(b)** The magnitude of the centripetal acceleration of the car at A is greater than that of the car at B, since the radius of the circular track is smaller at A. **(c)** The magnitude of the centripetal acceleration of the car at A is greater than that of the car at B, since the radius of the circular track is greater at A. **(d)** The magnitude of the centripetal acceleration of the car at A is less than that of the car at B, since the radius of the circular track is smaller at A.

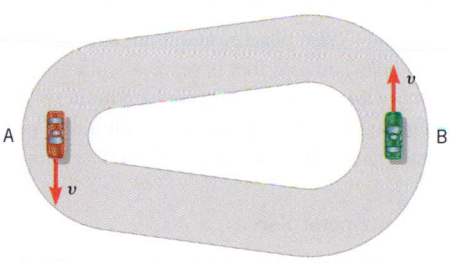

Section 5.3 Centripetal Force

7. The drawing shows two identical stones attached to cords that are being whirled on a tabletop at the same speed. The radius of the larger circle is twice that of the smaller circle. How is the tension T_1 in the longer cord related to the tension T_2 in the shorter cord? **(a)** $T_1 = T_2$ **(b)** $T_1 = 2T_2$ **(c)** $T_1 = 4T_2$ **(d)** $T_1 = \frac{1}{2}T_2$ **(e)** $T_1 = \frac{1}{4}T_2$

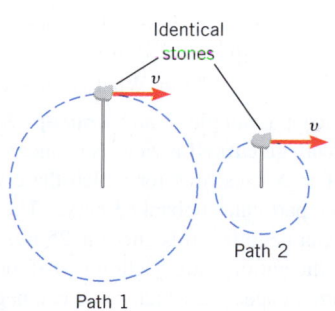

Identical stones

Path 2

Path 1

8. Three particles have the following masses (in multiples of m_0) and move on three different circles with the following speeds (in multiples of v_0) and radii (in multiples of r_0):

Particle	Mass	Speed	Radius
1	m_0	$2v_0$	r_0
2	m_0	$3v_0$	$3r_0$
3	$2m_0$	$2v_0$	$4r_0$

Rank the particles according to the magnitude of the centripetal force that acts on them, largest first. **(a)** 1, 2, 3 **(b)** 1, 3, 2 **(c)** 2, 1, 3 **(d)** 2, 3, 1 **(e)** 3, 2, 1

Section 5.4 Banked Curves

10. Two identical cars, one on the moon and one on the earth, have the same speed and are rounding banked turns that have the same radius r. There are two forces acting on each car, its weight mg and the normal force F_N exerted by the road. Recall that the weight of an object on the moon is about one-sixth of its weight on earth. How does the centripetal force on the moon compare with that on the earth? **(a)** The centripetal forces are the same. **(b)** The centripetal force on the moon is less than that on the earth. **(c)** The centripetal force on the moon is greater than that on the earth.

Section 5.5 Satellites in Circular Orbits

11. Two identical satellites are in orbit about the earth. One orbit has a radius r and the other $2r$. The centripetal force on the satellite in the larger orbit is _____ as that on the satellite in the smaller orbit. **(a)** the same **(b)** twice as great **(c)** four times as great **(d)** half as great **(e)** one-fourth as great

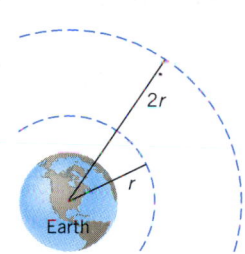

Section 5.7 Vertical Circular Motion

15. The drawing shows an extreme skier at the bottom of a ski jump. At this point the track is circular with a radius r. Two forces act on the skier, her weight mg and the normal force F_N. Which relation describes how the net force acting on her is related to her mass m and speed v and to the radius r? Assume that "up" is the positive direction.

(a) $F_N + mg = \dfrac{mv^2}{r}$ **(b)** $F_N - mg = \dfrac{mv^2}{r}$

(c) $F_N = \dfrac{mv^2}{r}$ **(d)** $-mg = \dfrac{mv^2}{r}$

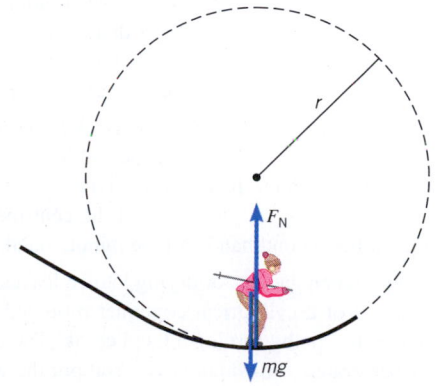

PROBLEMS

Note to Instructors: Most of the homework problems in this chapter are available for assignment via an online homework management program such as WileyPLUS or WebAssign, and those marked with the icon GO *are presented in WileyPLUS using a guided tutorial format that provides enhanced interactivity. See Preface for additional details.*

ssm Solution is in the Student Solutions Manual.
www Solution is available online at www.wiley.com/college/cutnell

 This icon represents a biomedical application.

Section 5.1 Uniform Circular Motion,
Section 5.2 Centripetal Acceleration

1. **ssm** How long does it take a plane, traveling at a constant speed of 110 m/s, to fly once around a circle whose radius is 2850 m?

2. GO The following table lists data for the speed and radius of three examples of uniform circular motion. Find the magnitude of the centripetal acceleration for each example.

	Radius	Speed
Example 1	0.50 m	12 m/s
Example 2	Infinitely large	35 m/s
Example 3	1.8 m	2.3 m/s

3. Review Conceptual Example 2 in preparation for this problem. In Figure 5.4, an object, after being released from its circular path, travels the distance OA in the same time it would have moved from O to P on the circle. The speed of the object on and off the circle remains constant at the same value. Suppose that the radius of the circle in Figure 5.4 is 3.6 m and the angle θ is 25°. What is the distance OA?

4. Review Conceptual Example 2 as background for this problem. One kind of slingshot consists of a pocket that holds a pebble and is whirled on a circle of radius r. The pebble is released from the circle at the angle θ, so that it will hit the target. The distance to the target from the center of the circle is d. (See the drawing, which is not to scale.) The circular path is parallel to the ground, and the target lies in the plane of the circle. The distance d is ten times the radius r. Ignore the effect of gravity in pulling the stone downward after it is released and find the angle θ.

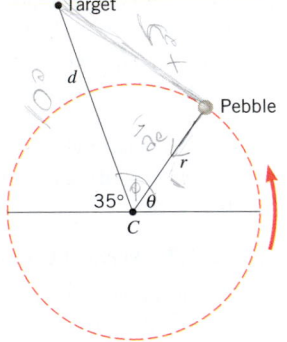

5. **ssm** A car travels at a constant speed around a circular track whose radius is 2.6 km. The car goes once around the track in 360 s. What is the magnitude of the centripetal acceleration of the car?

6. The aorta is a major artery, rising upward from the left ventricle of the heart and curving down to carry blood to the abdomen and lower half of the body. The curved artery can be approximated as a semicircular arch whose diameter is 5.0 cm. If blood flows through the aortic arch at a speed of 0.32 m/s, what is the magnitude (in m/s^2) of the blood's centripetal acceleration?

7. GO **Interactive LearningWare 5.1** at www.wiley.com/college/cutnell explores how to solve similar concept problems to this one. The second hand and the minute hand on one type of clock are the same length. Find the ratio ($a_{c, \text{second}}/a_{c, \text{minute}}$) of the centripetal accelerations of the tips of the second hand and the minute hand.

8. There is a clever kitchen gadget for drying lettuce leaves after you wash them. It consists of a cylindrical container mounted so that it can be rotated about its axis by turning a hand crank. The outer wall of the cylinder is perforated with small holes. You put the wet leaves in the container and turn the crank to spin off the water. The radius of the container is 12 cm. When the cylinder is rotating at 2.0 revolutions per second, what is the magnitude of the centripetal acceleration at the outer wall?

9. **ssm** Computer-controlled display screens provide drivers in the Indianapolis 500 with a variety of information about how their cars are performing. For instance, as a car is going through a turn, a speed of 221 mi/h (98.8 m/s) and centripetal acceleration of 3.00 g (three times the acceleration due to gravity) are displayed. Determine the radius of the turn (in meters).

* 10. The large blade of a helicopter is rotating in a horizontal circle. The length of the blade is 6.7 m, measured from its tip to the center of the circle. Find the ratio of the centripetal acceleration at the end of the blade to that which exists at a point located 3.0 m from the center of the circle.

* 11. A centrifuge is a device in which a small container of material is rotated at a high speed on a circular path. Such a device is used in medical laboratories, for instance, to cause the more dense red blood cells to settle through the less dense blood serum and collect at the bottom of the container. Suppose the centripetal acceleration of the sample is 6.25×10^3 times as large as the acceleration due to gravity. How many revolutions per minute is the sample making, if it is located at a radius of 5.00 cm from the axis of rotation?

* 12. The earth rotates once per day about an axis passing through the north and south poles, an axis that is perpendicular to the plane of the equator. Assuming the earth is a sphere with a radius of 6.38×10^6 m, determine the speed and centripetal acceleration of a person situated **(a)** at the equator and **(b)** at a latitude of 30.0° north of the equator.

Section 5.3 Centripetal Force

13. In a skating stunt known as crack-the-whip, a number of skaters hold hands and form a straight line. They try to skate so that the line rotates about the skater at one end, who acts as the pivot. The skater farthest out has a mass of 80.0 kg and is 6.10 m from the pivot. He is skating at a speed of 6.80 m/s. Determine the magnitude of the centripetal force that acts on him.

14. GO **Interactive LearningWare 5.2** at www.wiley.com/college/cutnell illustrates good problem-solving techniques for this type of problem. Two cars are traveling at the same speed of 27 m/s on a curve that has a radius of 120 m. Car A has a mass of 1100 kg, and car B has a mass of 1600 kg. Find the magnitude of the centripetal acceleration and the magnitude of the centripetal force for each car.

15. **ssm** A 0.015-kg ball is shot from the plunger of a pinball machine. Because of a centripetal force of 0.028 N, the ball follows a circular arc whose radius is 0.25 m. What is the speed of the ball?

16. Multiple-Concept Example 7 and **Concept Simulation 5.2** at www.wiley.com/college/cutnell review the concepts that play a role in this problem. Car A uses tires for which the coefficient of static friction is 1.1 on a particular unbanked curve. The maximum speed at which the car can negotiate this curve is 25 m/s. Car B uses tires for which the coefficient of static friction is 0.85 on the same curve. What is the maximum speed at which car B can negotiate the curve?

17. Review Multiple-Concept Example 7 to see the basic ideas underlying this problem. A stone has a mass of 6.0×10^{-3} kg and is wedged into the tread of an automobile tire, as the drawing shows. The coefficient of static friction between the stone and each side of the tread channel is 0.90. When the tire surface is rotating at a maximum speed of 13 m/s, the stone flies out of the tread. The magnitude F_N of the normal force that each side of the tread channel exerts on the stone is 1.8 N. Assume that only static friction supplies the centripetal force, and determine the radius r of the tire.

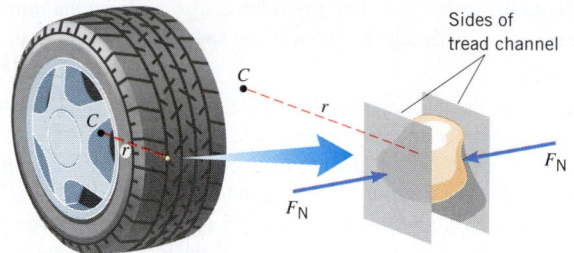

18. For background pertinent to this problem, review Conceptual Example 6. In Figure 5.8 the man hanging upside down is holding a partner who weighs 475 N. Assume that the partner moves on a circle that has a radius of 6.50 m. At a swinging speed of 4.00 m/s, what force must the man apply to his partner in the straight-down position?

19. See Conceptual Example 6 to review the concepts involved in this problem. A 9.5-kg monkey is hanging by one arm from a branch and is swinging on a vertical circle. As an approximation, assume a radial distance of 85 cm between the branch and the point where the monkey's mass is located. As the monkey swings through the lowest point on the circle, it has a speed of 2.8 m/s. Find **(a)** the magnitude of the centripetal force acting on the monkey and **(b)** the magnitude of the tension in the monkey's arm.

20. GO Multiple-Concept Example 7 deals with the concepts that are important in this problem. A penny is placed at the outer edge of a disk (radius = 0.150 m) that rotates about an axis perpendicular to the plane of the disk at its center. The period of the rotation is 1.80 s. Find the minimum coefficient of friction necessary to allow the penny to rotate along with the disk.

*21. A block is hung by a string from the inside roof of a van. When the van goes straight ahead at a speed of 28 m/s, the block hangs vertically down. But when the van maintains this same speed around an unbanked curve (radius = 150 m), the block swings toward the outside of the curve. Then the string makes an angle θ with the vertical. Find θ.

*22. An 830-kg race car can drive around an unbanked turn at a maximum speed of 58 m/s without slipping. The turn has a radius of curvature of 160 m. Air flowing over the car's wing exerts a downward-pointing force (called the *downforce*) of 11 000 N on the car. **(a)** What is the coefficient of static friction between the track and the car's tires? **(b)** What would be the maximum speed if no downforce acted on the car?

*23. Interactive Solution 5.23 at **www.wiley.com/college/ cutnell** illustrates a method for modeling this problem. A "swing" ride at a carnival consists of chairs that are swung in a circle by 15.0-m cables attached to a vertical rotating pole, as the drawing shows. Suppose the total mass of a chair and its occupant is 179 kg. **(a)** Determine the tension in the cable attached to the chair. **(b)** Find the speed of the chair.

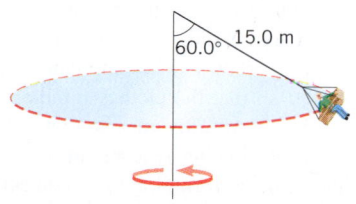

Section 5.4 Banked Curves

24. A woman is riding a Jet Ski at a speed of 26 m/s and notices a seawall straight ahead. The farthest she can lean the craft in order to make a turn is 22°. This situation is like that of a car on a curve that is banked at an angle of 22°. If she tries to make the turn without slowing down, what is the minimum distance from the seawall that she can being making her turn and still avoid a crash?

25. On a banked race track, the smallest circular path on which cars can move has a radius of 112 m, while the largest has a radius of 165 m, as the drawing illustrates. The height of the outer wall is 18 m. Find **(a)** the smallest and **(b)** the largest speed at which cars can move on this track without relying on friction.

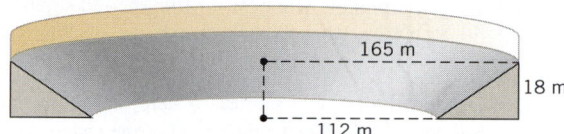

26. Two banked curves have the same radius. Curve A is banked at an angle of 13°, and curve B is banked at an angle of 19°. A car can travel around curve A without relying on friction at a speed of 18 m/s. At what speed can this car travel around curve B without relying on friction?

27. Before attempting this problem, review Examples 7 and 8. Two curves on a highway have the same radii. However, one is unbanked and the other is banked at an angle θ. A car can safely travel along the unbanked curve at a maximum speed v_0 under conditions when the coefficient of static friction between the tires and the road is $\mu_s = 0.81$. The banked curve is frictionless, and the car can negotiate it at the same maximum speed v_0. Find the angle θ of the banked curve.

*28. GO A jet flying at 123 m/s banks to make a horizontal circular turn. The radius of the turn is 3810 m, and the mass of the jet is 2.00×10^5 kg. Calculate the magnitude of the necessary lifting force.

*29. As an aid for this problem, consult **Interactive Solution 5.29** at **www.wiley.com/ college/cutnell**. A racetrack has the shape of an inverted cone, as the drawing shows. On this surface the cars race in circles that are parallel to the ground. For a speed of 34.0 m/s, at what value of the distance d should a driver locate his car if he wishes to stay on a circular path without depending on friction?

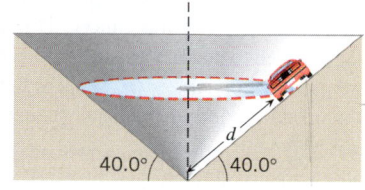

**30. The drawing shows a baggage carousel at an airport. Your suitcase has not slid all the way down the slope and is going around at a constant speed on a circle ($r = 11.0$ m) as the carousel turns. The coefficient of static friction between the suitcase and the carousel is 0.760, and the angle θ in the drawing is 36.0°. How much time is required for your suitcase to go around once?

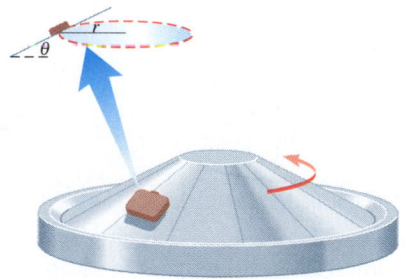

Section 5.5   Satellites in Circular Orbits,
Section 5.6   Apparent Weightlessness and Artificial Gravity

31. **GO** Two satellites are in circular orbits around the earth. The orbit for satellite A is at a height of 360 km above the earth's surface, while that for satellite B is at a height of 720 km. Find the orbital speed for each satellite.

32. A satellite is placed in orbit 6.00×10^5 m above the surface of the planet Jupiter. Jupiter has a mass of 1.90×10^{27} kg and a radius of 7.14×10^7 m. Find the orbital speed of the satellite.

33. **ssm www** A satellite is in a circular orbit around an unknown planet. The satellite has a speed of 1.70×10^4 m/s, and the radius of the orbit is 5.25×10^6 m. A second satellite also has a circular orbit around this same planet. The orbit of this second satellite has a radius of 8.60×10^6 m. What is the orbital speed of the second satellite?

34. Multiple-Concept Example 13 offers a helpful perspective for this problem. Suppose the surface (radius = r) of the space station in Figure 5.19 is rotating at 35.8 m/s. What must be the value of r for the astronauts to weigh one-half of their earth-weight?

35. Review **Interactive Solution 5.35** at **www.wiley.com/college/cutnell** before beginning this problem. Two satellites, A and B, are in different circular orbits about the earth. The orbital speed of satellite A is three times that of satellite B. Find the ratio (T_A/T_B) of the periods of the satellites.

36. The moon orbits the earth at a distance of 3.85×10^8 m. Assume that this distance is between the centers of the earth and the moon and that the mass of the earth is 5.98×10^{24} kg. Find the period for the moon's motion around the earth. Express the answer in days and compare it to the length of a month.

* **37.** **ssm** A satellite moves on a circular earth orbit that has a radius of 6.7×10^6 m. A model airplane is flying on a 15-m guideline in a horizontal circle. The guideline is parallel to the ground. Find the speed of the plane such that the plane and the satellite have the same centripetal acceleration.

* **38.** Two newly discovered planets follow circular orbits around a star in a distant part of the galaxy. The orbital speeds of the planets are determined to be 43.3 km/s and 58.6 km/s. The slower planet's orbital period is 7.60 years. **(a)** What is the mass of the star? **(b)** What is the orbital period of the faster planet, in years?

* **39.** **GO** A satellite has a mass of 5850 kg and is in a circular orbit 4.1×10^5 m above the surface of a planet. The period of the orbit is 2.00 hours. The radius of the planet is 4.15×10^6 m. What would be the true weight of the satellite if it were at rest on the planet's surface?

** **40.** Multiple-Concept Example 14 deals with the issues upon which this problem focuses. To create artificial gravity, the space station shown in the drawing is rotating at a rate of 1.00 rpm. The radii of the cylindrically shaped chambers have the ratio $r_A/r_B =$ 4.00. Each chamber A simulates an acceleration due to gravity of 10.0 m/s². Find values for **(a)** r_A, **(b)** r_B, and **(c)** the acceleration due to gravity that is simulated in chamber B.

Chamber A

Chamber B

Chamber A

Section 5.7   Vertical Circular Motion

41. **ssm** A roller coaster at an amusement park has a dip that bottoms out in a vertical circle of radius r. A passenger feels the seat of the car pushing upward on her with a force equal to twice her weight as she goes through the dip. If $r = 20.0$ m, how fast is the roller coaster traveling at the bottom of the dip?

42. A special electronic sensor is embedded in the seat of a car that takes riders around a circular loop-the-loop ride at an amusement park. The sensor measures the magnitude of the normal force that the seat exerts on a rider. The loop-the-loop ride is in the vertical plane and its radius is 21 m. Sitting on the seat before the ride starts, a rider is level and stationary, and the electronic sensor reads 770 N. At the top of the loop, the rider is upside down and moving, and the sensor reads 350 N. What is the speed of the rider at the top of the loop?

43. **ssm** For the normal force in Figure 5.21 to have the same magnitude at all points on the vertical track, the stunt driver must adjust the speed to be different at different points. Suppose, for example, that the track has a radius of 3.0 m and that the driver goes past point 1 at the bottom with a speed of 15 m/s. What speed must she have at point 3, so that the normal force at the top has the same magnitude as it did at the bottom?

44. Pilots of high-performance fighter planes can be subjected to large centripetal accelerations during high-speed turns. Because of these accelerations, the pilots are subjected to forces that can be much greater than their body weight, leading to an accumulation of blood in the abdomen and legs. As a result, the brain becomes starved for blood, and the pilot can lose consciousness ("black out"). The pilots wear "anti-G suits" to help keep the blood from draining out of the brain. To appreciate the forces that a fighter pilot must endure, consider the magnitude F_N of the normal force that the pilot's seat exerts on him at the bottom of a dive. The magnitude of the pilot's weight is W. The plane is traveling at 230 m/s on a vertical circle of radius 690 m. Determine the ratio F_N/W. For comparison, note that blackout can occur for values of F_N/W as small as 2 if the pilot is not wearing an anti-G suit.

45. **GO** A 0.20-kg ball on a stick is whirled on a vertical circle at a constant speed. When the ball is at the three o'clock position, the stick tension is 16 N. Find the tensions in the stick when the ball is at the twelve o'clock and at the six o'clock positions.

* **46.** **GO** A motorcycle is traveling up one side of a hill and down the other side. The crest of the hill is a circular arc with a radius of 45.0 m. Determine the maximum speed that the cycle can have while moving over the crest without losing contact with the road.

* **47.** Reviewing **Interactive Solution 5.47** at **www.wiley.com/college/cutnell** will help in solving this problem. A stone is tied to a string (length = 1.10 m) and whirled in a circle at the same constant speed in two different ways. First, the circle is horizontal and the string is nearly parallel to the ground. Next, the circle is vertical. In the vertical case the maximum tension in the string is 15.0% larger than the tension that exists when the circle is horizontal. Determine the speed of the stone.

** **48.** In an automatic clothes dryer, a hollow cylinder moves the clothes on a vertical circle (radius $r = 0.32$ m), as the drawing shows. The appliance is designed so that the clothes tumble gently as they dry. This means that when a piece of clothing reaches an angle of θ above the horizontal, it loses contact with the wall of the cylinder and falls onto the clothes below. How many revolutions per second should the cylinder make in order that the clothes lose contact with the wall when $\theta = 70.0°$?

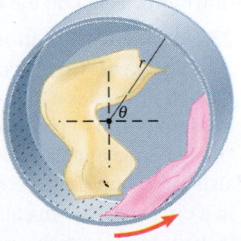

49. ssm Review Example 3, which deals with the bobsled in Figure 5.5. Also review Conceptual Example 4. The mass of the sled and its two riders in Figure 5.5 is 350 kg. Find the magnitude of the centripetal force that acts on the sled during the turn with a radius of **(a)** 33 m and **(b)** 24 m.

50. Consult Multiple-Concept Example 14 for background pertinent to this problem. In designing rotating space stations to provide for artificial-gravity environments, one of the constraints that must be considered is motion sickness. Studies have shown that the negative effects of motion sickness begin to appear when the rotational motion is faster than two revolutions per minute. On the other hand, the magnitude of the centripetal acceleration at the astronauts' feet should equal the magnitude of the acceleration due to gravity on earth. Thus, to eliminate the difficulties with motion sickness, designers must choose the distance between the astronauts' feet and the axis about which the space station rotates to be greater than a certain minimum value. What is this minimum value?

51. Interactive Solution 5.51 at **www.wiley.com/college/cutnell** shows a method for modeling this problem. The blade of a windshield wiper moves through an angle of 90.0° in 0.40 s. The tip of the blade moves on the arc of a circle that has a radius of 0.45 m. What is the magnitude of the centripetal acceleration of the tip of the blade?

52. Concept Simulation 5.1 at **www.wiley.com/college/cutnell** reviews the concepts that are involved in this problem. A child is twirling a 0.0120-kg plastic ball on a string in a horizontal circle whose radius is 0.100 m. The ball travels once around the circle in 0.500 s. **(a)** Determine the centripetal force acting on the ball. **(b)** If the speed is doubled, does the centripetal force double? If not, by what factor does the centripetal force increase?

53. The National Aeronautics and Space Administration (NASA) studies the physiological effects of large accelerations on astronauts. Some of these studies use a machine known as a centrifuge. This machine consists of a long arm, to one end of which is attached a chamber in which the astronaut sits. The other end of the arm is connected to an axis about which the arm and chamber can be rotated. The astronaut moves on a circular path, much like a model airplane flying in a circle on a guideline. The chamber is located 15 m from the center of the circle. At what speed must the chamber move so that an astronaut is subjected to 7.5 times the acceleration due to gravity?

54. At an amusement park there is a ride in which cylindrically shaped chambers spin around a central axis. People sit in seats facing the axis, their backs against the outer wall. At one instant the outer wall moves at a speed of 3.2 m/s, and an 83-kg person feels a 560-N force pressing against his back. What is the radius of a chamber?

55. ssm A motorcycle has a constant speed of 25.0 m/s as it passes over the top of a hill whose radius of curvature is 126 m. The mass of the motorcycle and driver is 342 kg. Find the magnitude of **(a)** the centripetal force and **(b)** the normal force that acts on the cycle.

56. A satellite is in a circular orbit about the earth ($M_E = 5.98 \times 10^{24}$ kg). The period of the satellite is 1.20×10^4 s. What is the speed at which the satellite travels?

★ 57. ssm Each of the space shuttle's main engines is fed liquid hydrogen by a high-pressure pump. Turbine blades inside the pump rotate at 617 rev/s. A point on one of the blades traces out a circle with a radius of 0.020 m as the blade rotates. **(a)** What is the magnitude of the centripetal acceleration that the blade must sustain at this point? **(b)** Express this acceleration as a multiple of $g = 9.80$ m/s^2.

★ 58. A computer is reading data from a rotating CD-ROM. At a point that is 0.030 m from the center of the disc, the centripetal acceleration is 120 m/s^2. What is the centripetal acceleration at a point that is 0.050 m from the center of the disc?

★ 59. ssm www The hammer throw is a track-and-field event in which a 7.3-kg ball (the "hammer") is whirled around in a circle several times and released. It then moves upward on the familiar curving path of projectile motion and eventually returns to earth some distance away. The world record for this distance is 86.75 m, achieved in 1986 by Yuriy Sedykh. Ignore air resistance and the fact that the ball is released above the ground rather than at ground level. Furthermore, assume that the ball is whirled on a circle that has a radius of 1.8 m and that its velocity at the instant of release is directed 41° above the horizontal. Find the magnitude of the centripetal force acting on the ball just prior to the moment of release.

★★ 60. At amusement parks, there is a popular ride where the floor of a rotating cylindrical room falls away, leaving the backs of the riders "plastered" against the wall. Suppose the radius of the room is 3.30 m and the speed of the wall is 10.0 m/s when the floor falls away. **(a)** What is the source of the centripetal force acting on the riders? **(b)** How much centripetal force acts on a 55.0-kg rider? **(c)** What is the minimum coefficient of static friction that must exist between a rider's back and the wall, if the rider is to remain in place when the floor drops away?

★★ 61. ssm www Redo Example 5, assuming that there is no upward lift on the plane generated by its wings. Without such lift, the guideline slopes downward due to the weight of the plane. For purposes of significant figures, use 0.900 kg for the mass of the plane, 17.0 m for the length of the guideline, and 19.0 and 38.0 m/s for the speeds.

WORK AND ENERGY

This rapidly moving surfer has kinetic energy, or energy of motion, because the wave, in accelerating him to his present speed, has done work on him. This chapter discusses the relationship between work and kinetic energy. (© Sport the library/Sports Chrome, Inc.)

6.1 WORK DONE BY A CONSTANT FORCE

Work is a familiar concept. For example, it takes work to push a stalled car. In fact, more work is done when the pushing force is greater or when the displacement of the car is greater. Force and displacement are, in fact, the two essential elements of work, as Figure 6.1 illustrates. The drawing shows a constant pushing force $\vec{\mathbf{F}}$ that points in the same direction as the resulting displacement $\vec{\mathbf{s}}$.* In such a case, the work W is defined as the magnitude F of the force times the magnitude s of the displacement: $W = Fs$. The work done to push a car is the same whether the car is moved north to south or east to west, provided that the amount of force used and the distance moved are the same. Since work does not convey directional information, it is a scalar quantity.

The equation $W = Fs$ indicates that the unit of work is the unit of force times the unit of distance, or the newton·meter in SI units. One newton·meter is referred to as a *joule* (J) (rhymes with "cool"), in honor of James Joule (1818–1889) and his research into the nature of work, energy, and heat. Table 6.1 summarizes the units for work in several systems of measurement.

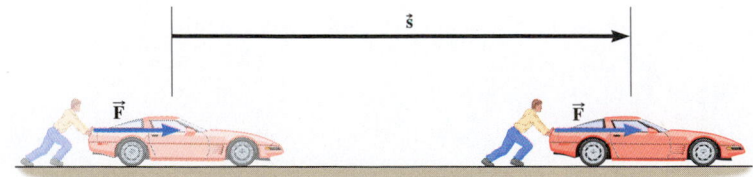

Figure 6.1 Work is done when a force $\vec{\mathbf{F}}$ pushes a car through a displacement $\vec{\mathbf{s}}$.

*When discussing work, it is customary to use the symbol $\vec{\mathbf{s}}$ for the displacement, rather than $\vec{\mathbf{x}}$ or $\vec{\mathbf{y}}$.

Table 6.1 Units of Measurement for Work

System	Force	×	Distance	=	Work
SI	newton (N)		meter (m)		joule (J)
CGS	dyne (dyn)		centimeter (cm)		erg
BE	pound (lb)		foot (ft)		foot · pound (ft · 1b)

The definition of work as $W = Fs$ does have one surprising feature: If the distance s is zero, the work is zero, even if a force is applied. Pushing on an immovable object, such as a brick wall, may tire your muscles, but there is no work done of the type we are discussing. In physics, the idea of work is intimately tied up with the idea of motion. If the object does not move, the force acting on the object does no work.

Often, the force and displacement do not point in the same direction. For instance, Figure 6.2a shows a suitcase-on-wheels being pulled to the right by a force that is applied along the handle. The force is directed at an angle θ relative to the displacement. In such a case, only the component of the force along the displacement is used in defining work. As Figure 6.2b shows, this component is $F \cos \theta$, and it appears in the general definition below:

DEFINITION OF WORK DONE BY A CONSTANT* FORCE

The work done on an object by a constant force $\vec{\mathbf{F}}$ is

$$W = (F \cos \theta)s \tag{6.1}$$

where F is the magnitude of the force, s is the magnitude of the displacement, and θ is the angle between the force and the displacement.

SI Unit of Work: newton · meter = joule (J)

When the force points in the same direction as the displacement, then $\theta = 0°$, and Equation 6.1 reduces to $W = Fs$. Example 1 shows how Equation 6.1 is used to calculate work.

 Example 1 Pulling a Suitcase-on-Wheels

Find the work done by a 45.0-N force in pulling the suitcase in Figure 6.2a at an angle $\theta = 50.0°$ for a distance $s = 75.0$ m.

Reasoning The pulling force causes the suitcase to move a distance of 75.0 m and does work. However, the force makes an angle of 50.0° with the displacement, and we must take this angle into account by using the definition of work given by Equation 6.1.

Solution The work done by the 45.0-N force is

$$W = (F \cos \theta)s = [(45.0 \text{ N}) \cos 50.0°](75.0 \text{ m}) = \boxed{2170 \text{ J}}$$

The answer is expressed in newton · meters or joules (J).

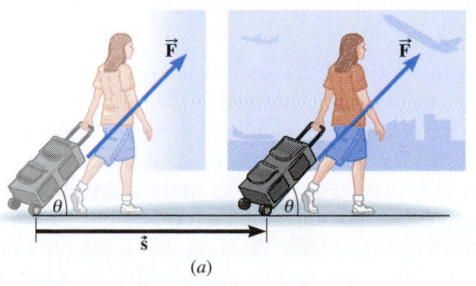

(a)

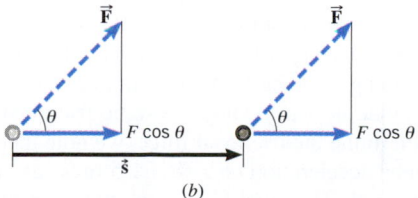

(b)

Figure 6.2 (a) Work can be done by a force $\vec{\mathbf{F}}$ that points at an angle θ relative to the displacement $\vec{\mathbf{s}}$. (b) The force component that points along the displacement is $F \cos \theta$.

---

*Section 6.9 considers the work done by a variable force.

The definition of work in Equation 6.1 takes into account only the component of the force in the direction of the displacement. The force component perpendicular to the displacement does no work. To do work, there must be a force *and* a displacement, and since there is no displacement in the perpendicular direction, there is no work done by the perpendicular component of the force. If the entire force is perpendicular to the displacement, the angle θ in Equation 6.1 is 90°, and the force does no work at all.

Work can be either positive or negative, depending on whether a component of the force points in the same direction as the displacement or in the opposite direction. Example 2 illustrates how positive and negative work arise.

The physics of positive and negative "reps" in weight lifting.

Example 2   Bench-Pressing

The weight lifter in Figure 6.3a is bench-pressing a barbell whose weight is 710 N. In part *b* of the figure, he raises the barbell a distance of 0.65 m above his chest, and in part *c* he lowers it the same distance. The weight is raised and lowered at a constant velocity. Determine the work done on the barbell by the weight lifter during **(a)** the lifting phase and **(b)** the lowering phase.

Reasoning To calculate the work, it is necessary to know the force exerted by the weight lifter. The barbell is raised and lowered at a constant velocity and, therefore, is in equilibrium. Consequently, the force $\vec{F}$ exerted by the weight lifter must balance the weight of the barbell, so $F = 710$ N. During the lifting phase, the force $\vec{F}$ and displacement $\vec{s}$ are in the same direction, as Figure 6.3b shows. The angle between them is θ = 0°. When the barbell is lowered, however, the force and displacement are in opposite directions, as in Figure 6.3c. The angle between the force and the displacement is now θ = 180°. With these observations, we can find the work.

Solution **(a)** During the lifting phase, the work done by the force $\vec{F}$ is given by Equation 6.1 as

$$W = (F \cos \theta)s = [(710 \text{ N}) \cos 0°](0.65 \text{ m}) = \boxed{460 \text{ J}}$$

(b) The work done during the lowering phase is

$$W = (F \cos \theta)s$$
$$= [(710 \text{ N}) \cos 180°](0.65 \text{ m}) = \boxed{-460 \text{ J}}$$

since cos 180° = −1. The work is negative, because the force is opposite to the displacement. Weight lifters call each complete up-and-down movement of the barbell a repetition, or "rep." The lifting of the weight is referred to as the positive part of the rep, and the lowering is known as the negative part.

Example 3 deals with the work done by a static frictional force when it acts on a crate that is resting on the bed of an accelerating truck.

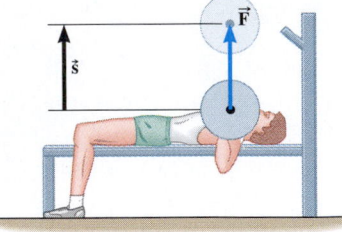

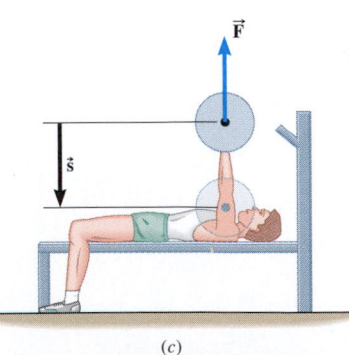

Example 3   Accelerating a Crate

Figure 6.4a shows a 120-kg crate on the flatbed of a truck that is moving with an acceleration of $a = +1.5$ m/s² along the positive *x* axis. The crate does not slip with respect to the truck as the truck undergoes a displacement whose magnitude is $s = 65$ m. What is the total work done on the crate by all of the forces acting on it?

Reasoning The free-body diagram in Figure 6.4b shows the forces that act on the crate: (1) the weight $\vec{W}$ of the crate, (2) the normal force $\vec{F}_N$ exerted by the flatbed, and (3) the static frictional force $\vec{f}_s$, which is exerted by the flatbed in the forward direction and keeps the crate from slipping backward. The weight and the normal force are perpendicular to the displacement, so they do no work. Only the static frictional force does work, since it acts in the *x* direction. To determine the frictional force, we note that the crate does not slip and, therefore, must have the same acceleration of $a = +1.5$ m/s² as does the truck. The force creating this acceleration is the static frictional force, and, knowing the mass of the crate and its acceleration, we can use Newton's second law to obtain its magnitude. Then, knowing the frictional force and the displacement, we can determine the total work done on the crate.

Figure 6.3 (*a*) In the bench press, work is done during both the lifting and lowering phases. (*b*) During the lifting phase, the force $\vec{F}$ does positive work. (*c*) During the lowering phase, the force does negative work. (*a*, Rachel Epstein/ The Image Works)

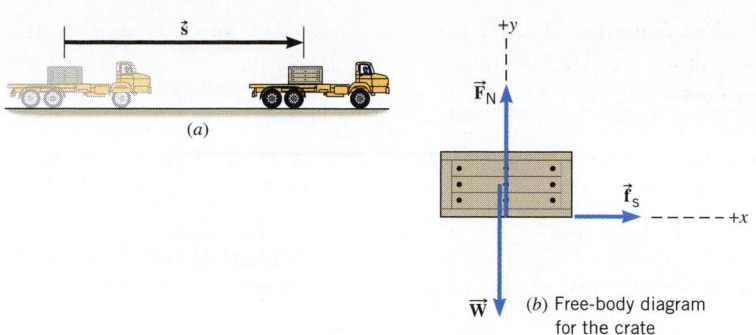

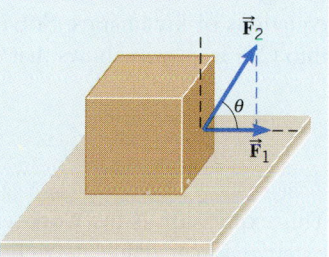

(b) Free-body diagram for the crate

Figure 6.4 (*a*) The truck and crate are accelerating to the right for a distance of $s = 65$ m. (*b*) The free-body diagram for the crate.

Solution From Newton's second law, we find that the magnitude f_s of the static frictional force is

$$f_s = ma = (120 \text{ kg})(1.5 \text{ m/s}^2) = 180 \text{ N}$$

The total work is that done by the static frictional force and is

$$W = (f_s \cos \theta)s = (180 \text{ N})(\cos 0°)(65 \text{ m}) = \boxed{1.2 \times 10^4 \text{ J}} \qquad (6.1)$$

The work is positive, because the frictional force is in the same direction as the displacement ($\theta = 0°$).

✓ CHECK YOUR UNDERSTANDING

(*The answers are given at the end of the book.*)

1. Two forces $\vec{F}_1$ and $\vec{F}_2$ are acting on the box shown in the drawing, causing the box to move across the floor. The two force vectors are drawn to scale. Which one of the following statements is correct? **(a)** $\vec{F}_2$ does more work than $\vec{F}_1$ does. **(b)** $\vec{F}_1$ does more work than $\vec{F}_2$ does. **(c)** Both forces do the same amount of work. **(d)** Neither force does any work.

2. A box is being moved with a velocity $\vec{v}$ by a force $\vec{P}$ (in the same direction as $\vec{v}$) along a level horizontal floor. The normal force is $\vec{F}_N$, the kinetic frictional force is $\vec{f}_k$, and the weight is $m\vec{g}$. Which one of the following statements is correct? **(a)** $\vec{P}$ does positive work, $\vec{F}_N$ and $\vec{f}_k$ do zero work, and $m\vec{g}$ does negative work. **(b)** $\vec{F}_N$ does positive work, $\vec{P}$ and $\vec{f}_k$ do zero work, and $m\vec{g}$ does negative work. **(c)** $\vec{f}_k$ does positive work, $\vec{F}_N$ and $m\vec{g}$ do zero work, and $\vec{P}$ does negative work. **(d)** $\vec{P}$ does positive work, $\vec{F}_N$ and $m\vec{g}$ do zero work, and $\vec{f}_k$ does negative work.

Question 1

3. A force does positive work on a particle that has a displacement pointing in the $+x$ direction. This same force does negative work on a particle that has a displacement pointing in the $+y$ direction. In which quadrant of the x, y coordinate system does the force lie? **(a)** First **(b)** Second **(c)** Third **(d)** Fourth

4. A suitcase is hanging straight down from your hand as you ride an escalator. Your hand exerts a force on the suitcase, and this force does work. This work is **(a)** positive when you ride up and negative when you ride down, **(b)** negative when you ride up and positive when you ride down, **(c)** positive when you ride up or down, **(d)** negative when you ride up or down.

6.2 THE WORK–ENERGY THEOREM AND KINETIC ENERGY

Most people expect that if you do work, you get something as a result. In physics, when a net force performs work on an object, there is always a result from the effort. The result is a change in the *kinetic energy* of the object. As we will now see, the relationship that relates work to the change in kinetic energy is known as the *work–energy theorem.* This theorem is obtained by bringing together three basic concepts that we've already learned about. First we'll apply Newton's second law of motion, $\Sigma F = ma$, which relates the net force ΣF to the acceleration a of an object. Then, we'll determine the work done by the net force when the object moves through a certain distance. Finally, we'll use Equation 2.9, one of the equations of kinematics, to relate the distance and acceleration to the initial and final speeds of the object. The result of this approach will be the work–energy theorem.

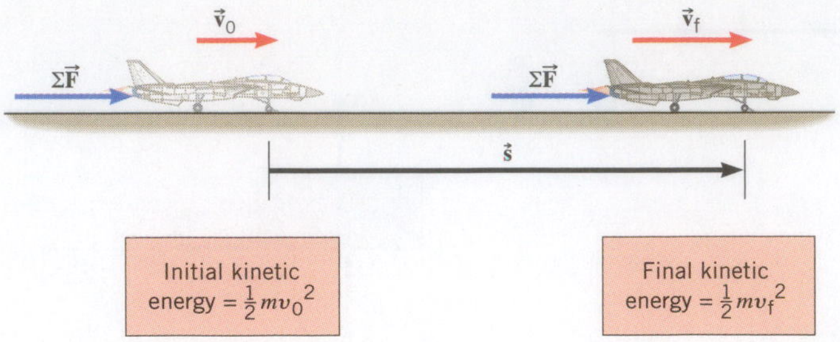

Figure 6.5 A constant net external force $\Sigma\vec{F}$ acts over a displacement $\vec{s}$ and does work on the plane. As a result of the work done, the plane's kinetic energy changes.

To gain some insight into the idea of kinetic energy and the work–energy theorem, look at Figure 6.5, where a constant net external force $\Sigma\vec{F}$ acts on an airplane of mass m. This net force is the vector sum of all the external forces acting on the plane, and, for simplicity, it is assumed to have the same direction as the displacement $\vec{s}$. According to Newton's second law, the net force produces an acceleration a, given by $a = \Sigma F/m$. Consequently, the speed of the plane changes from an initial value of v_0 to a final value of v_f.* Multiplying both sides of $\Sigma F = ma$ by the distance s gives

$$\underbrace{(\Sigma F)s}_{\substack{\text{Work done by} \\ \text{net ext. force}}} = mas$$

The left side of this equation is the work done by the net external force. The term as on the right side can be related to v_0 and v_f by using $v_f^2 = v_0^2 + 2as$ (Equation 2.9) from the equations of kinematics. Solving this equation to give $as = \frac{1}{2}(v_f^2 - v_0^2)$ and substituting into $(\Sigma F)s = mas$ shows that

$$\underbrace{(\Sigma F)s}_{\substack{\text{Work done by} \\ \text{net ext. force}}} = \underbrace{\tfrac{1}{2}mv_f^2}_{\substack{\text{Final} \\ \text{KE}}} - \underbrace{\tfrac{1}{2}mv_0^2}_{\substack{\text{Initial} \\ \text{KE}}}$$

This expression is the work–energy theorem. Its left side is the work W done by the net external force, while its right side involves the difference between two terms, each of which has the form $\frac{1}{2}(\text{mass})(\text{speed})^2$. The quantity $\frac{1}{2}(\text{mass})(\text{speed})^2$ is called kinetic energy (KE) and plays a significant role in physics, as we will soon see.

DEFINITION OF KINETIC ENERGY

The kinetic energy KE of an object with mass m and speed v is given by

$$\text{KE} = \tfrac{1}{2}mv^2 \tag{6.2}$$

SI Unit of Kinetic Energy: joule (J)

The SI unit of kinetic energy is the same as the unit for work, the joule. Kinetic energy, like work, is a scalar quantity. These are not surprising observations, for work and kinetic energy are closely related, as is clear from the following statement of the work–energy theorem.

THE WORK–ENERGY THEOREM

When a net external force does work W on an object, the kinetic energy of the object changes from its initial value of KE_0 to a final value of KE_f, the difference between the two values being equal to the work:

$$W = KE_f - KE_0 = \tfrac{1}{2}mv_f^2 - \tfrac{1}{2}mv_0^2 \tag{6.3}$$

The work–energy theorem may be derived for any direction of the force relative to the displacement, not just the situation in Figure 6.5. In fact, the force may even vary from point to point along a path that is curved rather than straight, and the theorem remains valid. According to the work–energy theorem, a moving object has kinetic energy,

*For extra emphasis, the final speed is now represented by the symbol v_f, rather than v.

because work was done to accelerate the object from rest to a speed v_f.* Conversely, an object with kinetic energy can perform work, if it is allowed to push or pull on another object. Example 4 illustrates the work–energy theorem and considers a single force that does work to change the kinetic energy of a space probe.

ANALYZING MULTIPLE-CONCEPT PROBLEMS

Example 4 Deep Space 1

The space probe *Deep Space 1* was launched October 24, 1998, and it used a type of engine called an ion propulsion drive. The ion propulsion drive generated only a weak force (or thrust), but could do so for long periods of time using only small amounts of fuel. Suppose the probe, which has a mass of 474 kg, is traveling at an initial speed of 275 m/s. No forces act on it except the 5.60×10^{-2}-N thrust of its engine. This external force $\vec{F}$ is directed parallel to the displacement $\vec{s}$, which has a magnitude of 2.42×10^9 m (see Figure 6.6). Determine the final speed of the probe, assuming that its mass remains nearly constant.

The physics of an ion propulsion drive.

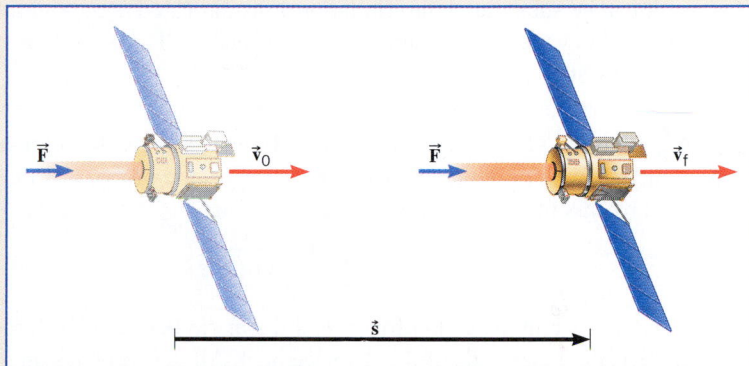

Figure 6.6 The engine of *Deep Space 1* generated a single force $\vec{F}$ that pointed in the same direction as the displacement $\vec{s}$. The force performed positive work, causing the space probe to gain kinetic energy.

Reasoning If we can determine the final kinetic energy of the space probe, we can determine its final speed, since kinetic energy is related to mass and speed according to Equation 6.2 and the mass of the probe is known. We will use the work–energy theorem ($W = KE_f - KE_0$), along with the definition of work, to find the final kinetic energy.

Knowns and Unknowns The following list summarizes the data for this problem:

Description	Symbol	Value	Comment
Explicit Data			
Mass	m	474 kg	
Initial speed	v_0	275 m/s	
Magnitude of force	F	5.60×10^{-2} N	
Magnitude of displacement	s	2.42×10^9 m	
Implicit Data			
Angle between force $\vec{F}$ and displacement $\vec{s}$	θ	0°	The force is parallel to the displacement.
Unknown Variable			
Final speed	v_f	?	

Modeling the Problem

STEP 1 Kinetic Energy An object of mass m and speed v has a kinetic energy KE given by Equation 6.2 as $KE = \frac{1}{2}mv^2$. Using the subscript f to denote the final kinetic energy and the final speed of the probe, we have that

$$KE_f = \tfrac{1}{2}mv_f^2$$

Solving for v_f gives Equation 1 at the right. The mass m is known but the final kinetic energy KE_f is not, so we will turn to Step 2 to evaluate it.

$$v_f = \sqrt{\frac{2(KE_f)}{m}} \qquad (1)$$

*Strictly speaking, the work–energy theorem, as given by Equation 6.3, applies only to a single particle, which occupies a mathematical point in space. A macroscopic object, however, is a collection or system of particles and is spread out over a region of space. Therefore, when a force is applied to a macroscopic object, the point of application of the force may be anywhere on the object. To take into account this and other factors, a discussion of work and energy is required that is beyond the scope of this text. The interested reader may refer to A. B. Arons, *The Physics Teacher,* October 1989, p. 506.

Continued

STEP 2 **The Work–Energy Theorem** The work–energy theorem relates the final kinetic energy KE_f of the probe to its initial kinetic energy KE_0 and the work W done by the force of the engine. According to Equation 6.3, this theorem is $W = KE_f - KE_0$. Solving for KE_f shows that

$$KE_f = KE_0 + W$$

The initial kinetic energy can be expressed as $KE_0 = \frac{1}{2}mv_0^2$, so the expression for the final kinetic energy becomes

$$\boxed{KE_f = \tfrac{1}{2}mv_0^2 + W}$$

This result can be substituted into Equation 1, as indicated at the right. Note from the data table that we know the mass m and the initial speed v_0. The work W is not known and will be evaluated in Step 3.

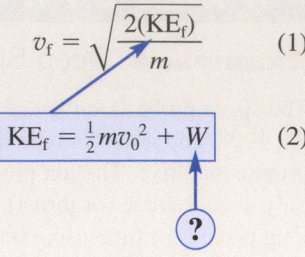

$$v_f = \sqrt{\frac{2(KE_f)}{m}} \qquad (1)$$

$$\boxed{KE_f = \tfrac{1}{2}mv_0^2 + W} \qquad (2)$$

$$?$$

STEP 3 **Work** The work W is that done by the net external force acting on the space probe. Since there is only the one force $\vec{F}$ acting on the probe, it is the net force. The work done by this force is given by Equation 6.1 as

$$\boxed{W = (F \cos \theta)s}$$

where F is the magnitude of the force, θ is the angle between the force and the displacement, and s is the magnitude of the displacement. All the variables on the right side of this equation are known, so we can substitute it into Equation 2, as shown in the right column.

$$v_f = \sqrt{\frac{2(KE_f)}{m}} \qquad (1)$$

$$KE_f = \tfrac{1}{2}mv_0^2 + W \qquad (2)$$

$$\boxed{W = (F \cos \theta)s}$$

Solution Algebraically combining the results of the three steps, we have

STEP 1 **STEP 2** **STEP 3**

$$v_f = \sqrt{\frac{2(KE_f)}{m}} = \sqrt{\frac{2(\frac{1}{2}mv_0^2 + W)}{m}} = \sqrt{\frac{2[\frac{1}{2}mv_0^2 + (F \cos \theta)s]}{m}}$$

The final speed of the space probe is

$$v_f = \sqrt{\frac{2[\frac{1}{2}mv_0^2 + (F \cos \theta)s]}{m}}$$

$$= \sqrt{\frac{2[\frac{1}{2}(474 \text{ kg})(275 \text{ m/s})^2 + (5.60 \times 10^{-2} \text{ N})(\cos 0°)(2.42 \times 10^9 \text{ m})]}{474 \text{ kg}}} = \boxed{805 \text{ m/s}}$$

Related Homework: *Problems 21, 22*

In Example 4 only the force of the engine does work. If several external forces act on an object, they must be added together vectorially to give the net force. The work done by the net force can then be related to the change in the object's kinetic energy by using the work–energy theorem, as in the next example.

ANALYZING MULTIPLE-CONCEPT PROBLEMS

Example 5 Downhill Skiing

A 58-kg skier is coasting down a 25° slope, as Figure 6.7a shows. Near the top of the slope, her speed is 3.6 m/s. She accelerates down the slope because of the gravitational force, even though a kinetic frictional force of magnitude 71 N opposes her motion. Ignoring air resistance, determine the speed at a point that is displaced 57 m downhill.

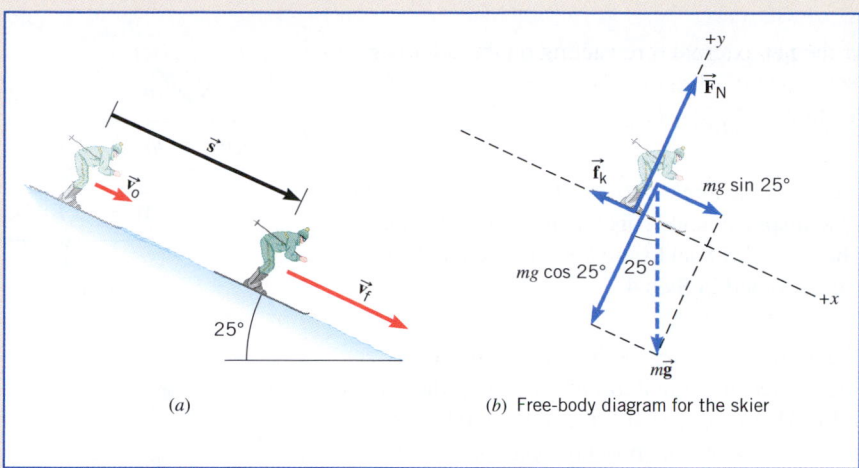

Figure 6.7 (*a*) A skier coasting downhill. (*b*) The free-body diagram for the skier.

(*a*)

(*b*) Free-body diagram for the skier

Reasoning The skier's speed at a point 57 m downhill (her final speed) depends on her final kinetic energy. According to the work–energy theorem, her final kinetic energy, in turn, is related to her initial kinetic energy (which we can calculate directly) and the work done by the net external force that acts on her. The work can be evaluated directly from its definition.

Knowns and Unknowns The data for this problem are listed below:

Description	Symbol	Value	Comment
Explicit Data			
Mass	m	58 kg	
Initial speed	v_0	3.6 m/s	
Magnitude of kinetic frictional force	f_k	71 N	
Magnitude of skier's displacement	s	57 m	
Angle of slope above horizontal		25°	
Implicit Data			
Angle between net force acting on skier and her displacement	θ	0°	The skier is accelerating down the slope, so the direction of the net force is parallel to her displacement.
Unknown Variable			
Final speed	v_f	?	

Modeling the Problem

STEP 1   Kinetic Energy   The final speed v_f of the skier is related to her final kinetic energy KE_f and mass m by Equation 6.2:

$$KE_f = \tfrac{1}{2}mv_f^2$$

Solving for v_f gives Equation 1 at the right. Her mass is known, but her final kinetic energy is not, so we turn to Step 2 to evaluate it.

$$v_f = \sqrt{\frac{2(KE_f)}{m}} \qquad (1)$$

STEP 2   The Work–Energy Theorem   The final kinetic energy KE_f of the skier is related to her initial kinetic energy KE_0 and the work W done by the net external force acting on her by the work–energy theorem (Equation 6.3): $KE_f = KE_0 + W$. The initial kinetic energy can be written as $KE_0 = \tfrac{1}{2}mv_0^2$, so the expression for the final kinetic energy becomes

$$\boxed{KE_f = \tfrac{1}{2}mv_0^2 + W}$$

$$v_f = \sqrt{\frac{2(KE_f)}{m}} \qquad (1)$$

$$\boxed{KE_f = \tfrac{1}{2}mv_0^2 + W} \qquad (2)$$

This result can be substituted into Equation 1, as indicated at the right. The mass m and the initial speed v_0 are known. The work W is not known, and will be evaluated in Steps 3 and 4.

Continued

STEP 3 **Work** The work W done by the net external force acting on the skier is given by Equation 6.1 as

$$W = (\Sigma F \cos \theta)s$$

where ΣF is the magnitude of the net force, θ is the angle between the net force and the displacement, and s is the magnitude of the displacement. This result can be substituted into Equation 2, as shown at the right. The variables θ and s are known (see the data table), and the net external force will be determined in Step 4.

$$v_f = \sqrt{\frac{2(KE_f)}{m}} \qquad (1)$$

$$KE_f = \tfrac{1}{2}mv_0^2 + W \qquad (2)$$

$$W = (\Sigma F \cos \theta)s \qquad (3)$$

$?$

STEP 4 **The Net External Force** Figure 6.7b is a free-body diagram for the skier and shows the three external forces acting on her: the gravitational force $m\vec{g}$, the kinetic frictional force $\vec{f}_k$, and the normal force $\vec{F}_N$. The net external force along the y axis is zero, because there is no acceleration in that direction (the normal force balances the component $mg \cos 25°$ of the weight perpendicular to the slope). Using the data from the table of knowns and unknowns, we find that the net external force along the x axis is

$$\Sigma F = mg \sin 25° - f_k$$
$$= (58 \text{ kg})(9.80 \text{ m/s}^2) \sin 25° - 71 \text{ N} = 170 \text{ N}$$

$$v_f = \sqrt{\frac{2(KE_f)}{m}} \qquad (1)$$

$$KE_f = \tfrac{1}{2}mv_0^2 + W \qquad (2)$$

$$W = (\Sigma F \cos \theta)s \qquad (3)$$

$$\Sigma F = mg \sin 25° - f_k = 170 \text{ N}$$

All the variables on the right side of this equation are known, so we substitute this expression into Equation 3 for the net external force (see the right column).

Solution Algebraically combining the four steps, we arrive at the following expression for the final speed of the skier:

STEP 1   **STEP 2**   **STEPS 3 AND 4**

$$v_f = \sqrt{\frac{2(KE_f)}{m}} = \sqrt{\frac{2(\tfrac{1}{2}mv_0^2 + W)}{m}} = \sqrt{\frac{2[\tfrac{1}{2}mv_0^2 + (170 \text{ N})(\cos \theta)s]}{m}}$$

The final speed of the skier is

$$v_f = \sqrt{\frac{2[\tfrac{1}{2}mv_0^2 + (170 \text{ N})(\cos \theta)s]}{m}}$$

$$= \sqrt{\frac{2[\tfrac{1}{2}(58 \text{ kg})(3.6 \text{ m/s})^2 + (170 \text{ N})(\cos 0°)(57 \text{ m})]}{58 \text{ kg}}} = \boxed{19 \text{ m/s}}$$

Related Homework: *Problems 24, 28*

Problem-solving insight

Example 5 emphasizes that *the work–energy theorem deals with the work done by the net external force. The work–energy theorem does not apply to the work done by an individual force,* unless that force happens to be the only one present, in which case it is the net force. If the work done by the net force is *positive,* as in Example 5, the kinetic energy of the object *increases.* If the work done is *negative,* the kinetic energy *decreases.* If the work is zero, the kinetic energy remains the same. Conceptual Example 6 explores these ideas further.

 Conceptual Example 6 Work and Kinetic Energy

Figure 6.8 illustrates a satellite moving about the earth in a circular orbit and in an elliptical orbit. The only external force that acts on the satellite is the gravitational force. In which orbit does the kinetic energy of the satellite change, **(a)** the circular orbit or **(b)** the elliptical orbit?

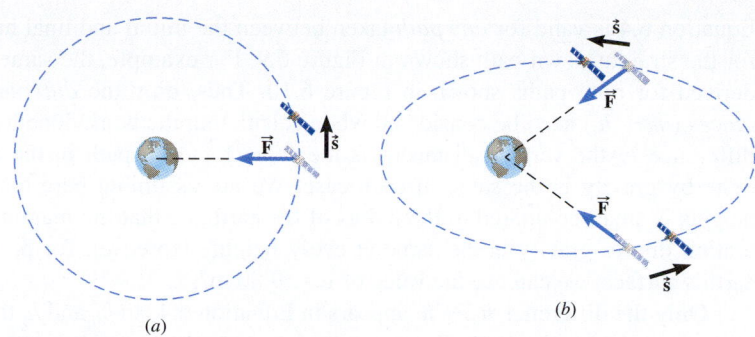

(a)   (b)

Figure 6.8 (a) In a circular orbit, the gravitational force $\vec{F}$ is always perpendicular to the displacement $\vec{s}$ of the satellite and does no work. (b) In an elliptical orbit, there can be a component of the force along the displacement, and, consequently, work can be done.

Reasoning The gravitational force is the only force acting on the satellite, so it is the net force. With this fact in mind, we will apply the work–energy theorem, which states that the work done by the net force equals the change in the kinetic energy.

Answer (a) is incorrect. For the circular orbit in Figure 6.8a the gravitational force $\vec{F}$ does no work on the satellite, since the force is perpendicular to the instantaneous displacement $\vec{s}$ at all times. Thus, the work done by the net force is zero, and according to the work–energy theorem (Equation 6.3), the kinetic energy of the satellite (and, hence, its speed) remains the same everywhere on the orbit.

Answer (b) is correct. For the elliptical orbit in Figure 6.8b the gravitational force $\vec{F}$ does do work. For example, as the satellite moves toward the earth in the top part of Figure 6.8b, there is a component of $\vec{F}$ that points in the same direction as the displacement. Consequently, $\vec{F}$ does positive work during this part of the orbit, and the kinetic energy of the satellite increases. When the satellite moves away from the earth, as in the lower part of Figure 6.8b, $\vec{F}$ has a component that points opposite to the displacement and, therefore, does negative work. As a result, the kinetic energy of the satellite decreases.

Related Homework: *Problem 18*

✓ CHECK YOUR UNDERSTANDING

(*The answers are given at the end of the book.*)

5. A sailboat is moving at a constant velocity. Is work being done by a net external force acting on the boat?

6. A ball has a speed of 15 m/s. Only one external force acts on the ball. After this force acts, the speed of the ball is 7 m/s. Has the force done **(a)** positive, **(b)** zero, or **(c)** negative work on the ball?

7. A rocket is at rest on the launch pad. When the rocket is launched, its kinetic energy increases. Consider all of the forces acting on the rocket during the launch, and decide whether the following statement is true or false: The amount by which the kinetic energy of the rocket increases during the launch is equal to the work done by the force generated by the rocket's engine.

8. A net external force acts on a particle that is moving along a straight line. This net force is not zero. Which one of the following statements is correct? **(a)** The velocity, but not the kinetic energy, of the particle is changing. **(b)** The kinetic energy, but not the velocity, of the particle is changing. **(c)** Both the velocity and the kinetic energy of the particle are changing.

6.3 GRAVITATIONAL POTENTIAL ENERGY

WORK DONE BY THE FORCE OF GRAVITY

The gravitational force is a well-known force that can do positive or negative work, and Figure 6.9 helps to show how the work can be determined. This drawing depicts a basketball of mass m moving vertically downward, the force of gravity $m\vec{g}$ being the only force acting on the ball. The initial height of the ball is h_0, and the final height is h_f, both distances measured from the earth's surface. The displacement $\vec{s}$ is downward and has a magnitude of $s = h_0 - h_f$. To calculate the work W_{gravity} done on the ball by the force of gravity, we use $W = (F \cos \theta)s$ with $F = mg$ and $\theta = 0°$, since the force and displacement are in the same direction:

$$W_{\text{gravity}} = (mg \cos 0°)(h_0 - h_f) = mg(h_0 - h_f) \qquad (6.4)$$

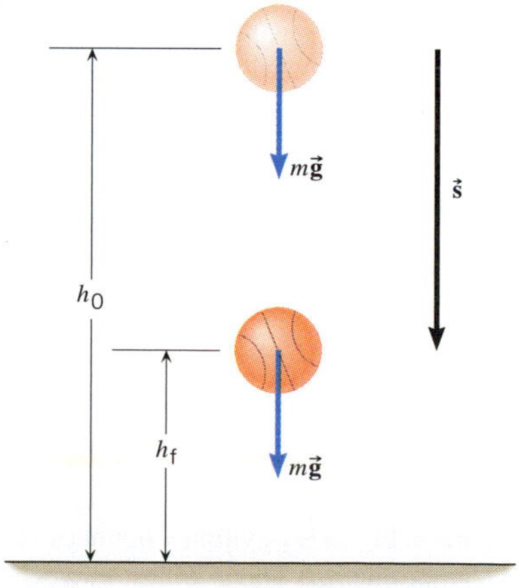

Figure 6.9 Gravity exerts a force $m\vec{g}$ on the basketball. Work is done by the gravitational force as the basketball falls from a height of h_0 to a height of h_f.

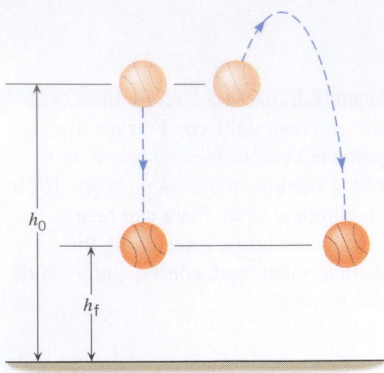

Figure 6.10 An object can move along different paths in going from an initial height of h_0 to a final height of h_f. In each case, the work done by the gravitational force is the same [$W_{\text{gravity}} = mg(h_0 - h_f)$], since the change in vertical distance ($h_0 - h_f$) is the same.

Equation 6.4 is valid for *any path* taken between the initial and final heights, and not just for the straight-down path shown in Figure 6.9. For example, the same expression can be derived for both paths shown in Figure 6.10. Thus, only the *difference in vertical distances* ($h_0 - h_f$) need be considered when calculating the work done by gravity. Since the difference in the vertical distances is the same for each path in the drawing, the work done by gravity is the same in each case. We are assuming here that the difference in heights is small compared to the radius of the earth, so that the magnitude g of the acceleration due to gravity is the same at every height. Moreover, for positions close to the earth's surface, we can use the value of $g = 9.80 \text{ m/s}^2$.

Only the difference $h_0 - h_f$ appears in Equation 6.4, so h_0 and h_f themselves need not be measured from the earth. For instance, they could be measured relative to a level that is one meter above the ground, and $h_0 - h_f$ would still have the same value. Example 7 illustrates how the work done by gravity is used with the work–energy theorem.

Example 7 A Gymnast on a Trampoline

A gymnast springs vertically upward from a trampoline as in Figure 6.11a. The gymnast leaves the trampoline at a height of 1.20 m and reaches a maximum height of 4.80 m before falling back down. All heights are measured with respect to the ground. Ignoring air resistance, determine the initial speed v_0 with which the gymnast leaves the trampoline.

Reasoning We can find the initial speed of the gymnast (mass = m) by using the work–energy theorem, provided the work done by the net external force can be determined. Since only the gravitational force acts on the gymnast in the air, it is the net force, and we can evaluate the work by using the relation $W_{\text{gravity}} = mg(h_0 - h_f)$.

Solution Figure 6.11b shows the gymnast moving upward. The initial and final heights are $h_0 = 1.20 \text{ m}$ and $h_f = 4.80 \text{ m}$, respectively. The initial speed is v_0 and the final speed is $v_f = 0 \text{ m/s}$, since the gymnast comes to a momentary halt at the highest point. Since $v_f = 0 \text{ m/s}$, the final kinetic energy is $\text{KE}_f = 0 \text{ J}$, and the work–energy theorem becomes $W = \text{KE}_f - \text{KE}_0 = -\text{KE}_0$. The work W is that due to gravity, so this theorem reduces to $W_{\text{gravity}} = mg(h_0 - h_f) = -\frac{1}{2}mv_0^2$. Solving for v_0 gives

$$v_0 = \sqrt{-2g(h_0 - h_f)} = \sqrt{-2(9.80 \text{ m/s}^2)(1.20 \text{ m} - 4.80 \text{ m})} = \boxed{8.40 \text{ m/s}}$$

GRAVITATIONAL POTENTIAL ENERGY

An object in motion has kinetic energy. There are also other types of energy. For example, an object may possess energy by virtue of its position relative to the earth and is said to have gravitational potential energy. A pile driver, for instance, is used to pound "piles," or structural support beams, into the ground. The device contains a massive hammer that is raised to a height h and dropped (see Figure 6.12), so the hammer has the potential to do

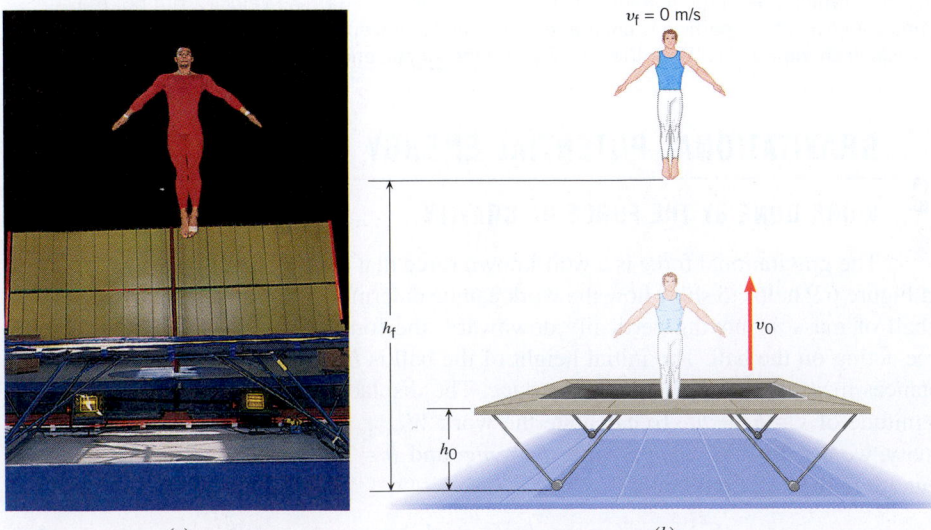

Figure 6.11 (a) A gymnast bounces on a trampoline. (b) The gymnast moves upward with an initial speed v_0 and reaches maximum height with a final speed of zero. (a, James D. Wilson/Trust of James D. Wilson)

$v_f = 0 \text{ m/s}$

h_f

h_0

v_0

(a)

(b)

the work of driving the pile into the ground. The greater the height h, the greater is the potential for doing work, and the greater is the gravitational potential energy.

Now, let's obtain an expression for the gravitational potential energy. Our starting point is Equation 6.4 for the work done by the gravitational force as an object moves from an initial height h_0 to a final height h_f:

$$W_{\text{gravity}} = \underbrace{mgh_0}_{\substack{\text{Initial} \\ \text{gravitational} \\ \text{potential energy} \\ \text{PE}_0}} - \underbrace{mgh_f}_{\substack{\text{Final} \\ \text{gravitational} \\ \text{potential energy} \\ \text{PE}_f}} \qquad (6.4)$$

This equation indicates that the work done by the gravitational force is equal to the difference between the initial and final values of the quantity mgh. The value of mgh is larger when the height is larger and smaller when the height is smaller. We are led, then, to identify the quantity mgh as the **gravitational potential energy.** The concept of potential energy is associated only with a type of force known as a "conservative" force, as we will discuss in Section 6.4.

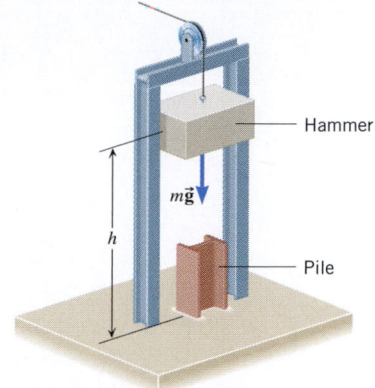

Figure 6.12 In a pile driver, the gravitational potential energy of the hammer relative to the ground is PE = mgh.

DEFINITION OF GRAVITATIONAL POTENTIAL ENERGY

The gravitational potential energy PE is the energy that an object of mass m has by virtue of its position relative to the surface of the earth. That position is measured by the height h of the object relative to an arbitrary zero level:

$$\text{PE} = mgh \qquad (6.5)$$

SI Unit of Gravitational Potential Energy: joule (J)

Gravitational potential energy, like work and kinetic energy, is a scalar quantity and has the same SI unit as they do, the joule. It is the *difference* between two potential energies that is related by Equation 6.4 to the work done by the force of gravity. Therefore, the zero level for the heights can be taken anywhere, as long as both h_0 and h_f are measured relative to the same zero level. The gravitational potential energy depends on both the object and the earth (m and g, respectively), as well as the height h. Therefore, the gravitational potential energy belongs to the object and the earth as a system, although one often speaks of the object alone as possessing the gravitational potential energy.

✓ **CHECK YOUR UNDERSTANDING**

(*The answer is given at the end of the book.*)

9. In a simulation on earth, an astronaut in his space suit climbs up a vertical ladder. On the moon, the same astronaut makes the exact same climb. Which one of the following statements correctly describes how the gravitational potential energy of the astronaut changes during the climb?   **(a)** It changes by a greater amount on the earth.   **(b)** It changes by a greater amount on the moon.   **(c)** The change is the same in both cases.

6.4 CONSERVATIVE VERSUS NONCONSERVATIVE FORCES

The gravitational force has an interesting property that when an object is moved from one place to another, the work done by the gravitational force does not depend on the choice of path. In Figure 6.10, for instance, an object moves from an initial height h_0 to a final height h_f along two different paths. As Section 6.3 discusses, the work done by gravity depends only on the initial and final heights, and not on the path between these heights. For this reason, the gravitational force is called a **conservative force,** according to version 1 of the following definition:

DEFINITION OF A CONSERVATIVE FORCE

Version 1 A force is conservative when the work it does on a moving object is independent of the path between the object's initial and final positions.
Version 2 A force is conservative when it does no net work on an object moving around a closed path, starting and finishing at the same point.

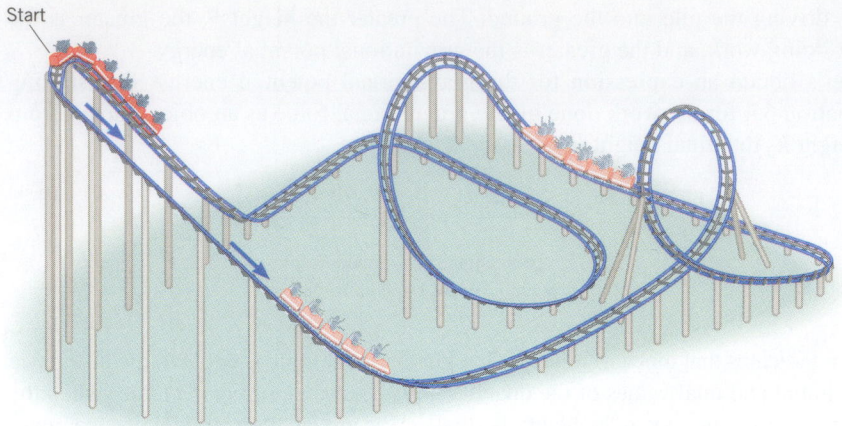

Start

Figure 6.13 A roller coaster track is an example of a closed path.

Figure 6.13 helps us to illustrate version 2 of the definition of a conservative force. The picture shows a roller coaster car racing through dips and double dips, ultimately returning to its starting point. This kind of path, which begins and ends at the same place, is called a *closed* path. Gravity provides the only force that does work on the car, assuming that there is no friction or air resistance. Of course, the track exerts a normal force, but this force is always directed perpendicular to the motion and does no work. On the downward parts of the trip, the gravitational force does positive work, increasing the car's kinetic energy. Conversely, on the upward parts of the motion, the gravitational force does negative work, decreasing the car's kinetic energy. Over the entire trip, the gravitational force does as much positive work as negative work, so the net work is zero, and the car returns to its starting point with the same kinetic energy it had at the start. Therefore, consistent with version 2 of the definition of a conservative force, $W_{gravity} = 0$ J for a closed path.

The gravitational force is our first example of a conservative force. Later, we will encounter others, such as the elastic force of a spring and the electrical force of electrically charged particles. With each conservative force we will associate a potential energy, as we have already done in the gravitational case (see Equation 6.5). For other conservative forces, however, the algebraic form of the potential energy will differ from that in Equation 6.5.

Not all forces are conservative. A force is nonconservative if the work it does on an object moving between two points depends on the path of the motion between the points. The kinetic frictional force is one example of a nonconservative force. When an object slides over a surface and kinetic friction is present, the frictional force points opposite to the sliding motion and does negative work. Between any two points, greater amounts of work are done over longer paths between the points, so that the work depends on the choice of path. Thus, the kinetic frictional force is nonconservative. Air resistance is another nonconservative force. ***The concept of potential energy is not defined for a nonconservative force.***

For a closed path, the total work done by a nonconservative force is not zero as it is for a conservative force. In Figure 6.13, for instance, a frictional force would oppose the motion and slow down the car. Unlike gravity, friction would do negative work on the car throughout the entire trip, on *both* the up and down parts of the motion. Assuming that the car makes it back to the starting point, the car would have *less* kinetic energy than it had originally. Table 6.2 gives some examples of conservative and nonconservative forces.

In normal situations, conservative forces (such as gravity) and nonconservative forces (such as friction and air resistance) act simultaneously on an object. Therefore, we write the work W done by the net external force as $W = W_c + W_{nc}$, where W_c is the work done by the conservative forces and W_{nc} is the work done by the nonconservative forces. According to the work–energy theorem, the work done by the net external force is equal to the change in the object's kinetic energy, or $W_c + W_{nc} = \frac{1}{2}mv_f^2 - \frac{1}{2}mv_0^2$. If the only conservative force acting is the gravitational force, then $W_c = W_{gravity} = mg(h_0 - h_f)$, and the work–energy theorem becomes

$$mg(h_0 - h_f) + W_{nc} = \tfrac{1}{2}mv_f^2 - \tfrac{1}{2}mv_0^2$$

Table 6.2 Some Conservative and Nonconservative Forces

Conservative Forces

Gravitational force (Ch. 4)

Elastic spring force (Ch. 10)

Electric force (Ch. 18, 19)

Nonconservative Forces

Static and kinetic frictional forces

Air resistance

Tension

Normal force

Propulsion force of a rocket

The work done by the gravitational force can be moved to the right side of this equation, with the result that

$$W_{nc} = (\tfrac{1}{2}mv_f^2 - \tfrac{1}{2}mv_0^2) + (mgh_f - mgh_0) \tag{6.6}$$

In terms of kinetic and potential energies, we find that

$$W_{nc} = \underbrace{(KE_f - KE_0)}_{} + \underbrace{(PE_f - PE_0)}_{} \tag{6.7a}$$

Net work done by nonconservative forces   Change in kinetic energy   Change in gravitational potential energy

Equation 6.7a states that the net work W_{nc} done by all the external nonconservative forces equals the change in the object's kinetic energy plus the change in its gravitational potential energy. It is customary to use the delta symbol (Δ) to denote such changes; thus, $\Delta KE = (KE_f - KE_0)$ and $\Delta PE = (PE_f - PE_0)$. With the delta notation, the work–energy theorem takes the form

$$W_{nc} = \Delta KE + \Delta PE \tag{6.7b}$$

In the next two sections, we will show why the form of the work–energy theorem expressed by Equations 6.7a and 6.7b is useful.

Problem-solving insight

The change in both the kinetic and the potential energy is always the final value minus the initial value: $\Delta KE = KE_f - KE_0$ and $\Delta PE = PE_f - PE_0$.

6.5 THE CONSERVATION OF MECHANICAL ENERGY

The concept of work and the work–energy theorem have led us to the conclusion that an object can possess two kinds of energy: kinetic energy, KE, and gravitational potential energy, PE. The sum of these two energies is called the ***total mechanical energy*** E, so that $E = KE + PE$. The concept of total mechanical energy will be extremely useful in describing the motion of objects in this and other chapters. Later on, in a number of places, we will update the definition of total mechanical energy to include other types of potential energies in addition to the gravitational form.

By rearranging the terms on the right side of Equation 6.7a, the work–energy theorem can be expressed in terms of the total mechanical energy:

$$W_{nc} = (KE_f - KE_0) + (PE_f - PE_0) \tag{6.7a}$$
$$= \underbrace{(KE_f + PE_f)}_{E_f} - \underbrace{(KE_0 + PE_0)}_{E_0}$$

or

$$W_{nc} = E_f - E_0 \tag{6.8}$$

Remember: Equation 6.8 is just another form of the work–energy theorem. It states that W_{nc}, the net work done by external nonconservative forces, changes the total mechanical energy from an initial value of E_0 to a final value of E_f.

The conciseness of the work–energy theorem in the form $W_{nc} = E_f - E_0$ allows an important basic principle of physics to stand out. This principle is known as the conservation of mechanical energy. Suppose that the net work W_{nc} done by external nonconservative forces is zero, so $W_{nc} = 0$ J. Then, Equation 6.8 reduces to

$$E_f = E_0 \tag{6.9a}$$

$$\underbrace{\tfrac{1}{2}mv_f^2 + mgh_f}_{E_f} = \underbrace{\tfrac{1}{2}mv_0^2 + mgh_0}_{E_0} \tag{6.9b}$$

Equation 6.9a indicates that the final mechanical energy is equal to the initial mechanical energy. Consequently, the total mechanical energy *remains constant all along the path between the initial and final points,* never varying from the initial value of E_0. A quantity that remains constant throughout the motion is said to be "conserved." The fact that the total mechanical energy is conserved when $W_{nc} = 0$ J is called the ***principle of conservation of mechanical energy.***

CONCEPTS AT A GLANCE

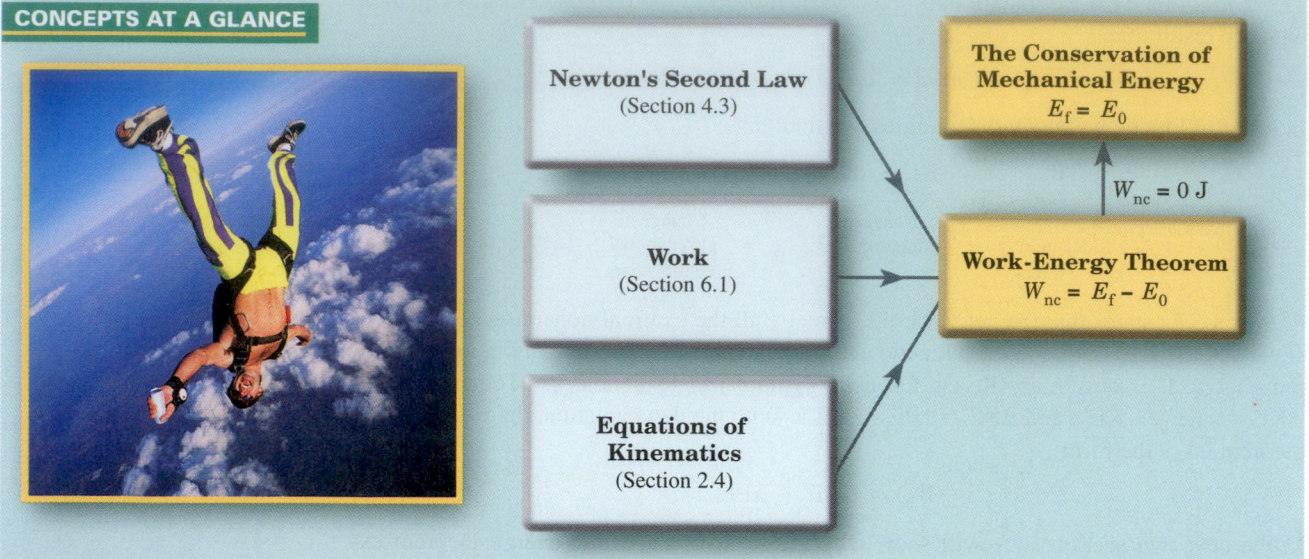

Newton's Second Law (Section 4.3)

Work (Section 6.1)

Equations of Kinematics (Section 2.4)

Work-Energy Theorem
$$W_{nc} = E_f - E_0$$

$W_{nc} = 0$ J

The Conservation of Mechanical Energy
$$E_f = E_0$$

Figure 6.14 CONCEPTS AT A GLANCE The work–energy theorem leads to the principle of conservation of mechanical energy under circumstances in which $W_{nc} = 0$ J, where W_{nc} is the net work done by external nonconservative forces. To the extent that air resistance, a nonconservative force, can be ignored, the total mechanical energy of this skydiver is conserved as he falls toward the earth. (Digital Vision/Fotosearch, LLC)

THE PRINCIPLE OF CONSERVATION OF MECHANICAL ENERGY

The total mechanical energy ($E = KE + PE$) of an object remains constant as the object moves, provided that the net work done by external nonconservative forces is zero, $W_{nc} = 0$ J.

▶ **CONCEPTS AT A GLANCE** The Concepts-at-a-Glance chart in Figure 6.14 outlines how the conservation of mechanical energy arises naturally from concepts that we have encountered earlier. In the middle are the three concepts that we used to develop the work–energy theorem: Newton's second law, work, and the equations of kinematics. On the right side of the chart is the work–energy theorem, stated in the form $W_{nc} = E_f - E_0$. If, as the chart shows, $W_{nc} = 0$ J, then the final and initial total mechanical energies are equal. In other words, $E_f = E_0$, which is the mathematical statement of the conservation of mechanical energy. ◀

The principle of conservation of mechanical energy offers keen insight into the way in which the physical universe operates. While the sum of the kinetic and potential energies at any point is conserved, the two forms may be interconverted or transformed into one another. Kinetic energy of motion is converted into potential energy of position, for instance, when a moving object coasts up a hill. Conversely, potential energy is converted into kinetic energy when an object is allowed to fall. Figure 6.15 illustrates such transformations of energy for a bobsled run, assuming that nonconservative forces, such as friction and wind resistance, can be ignored. The normal force, being directed perpendicular to the path, does no work. Only the force of gravity does work, so the total mechanical

KE	PE	$E = KE + PE$	
0 J	600 000 J	600 000 J	$v_0 = 0$ m/s
200 000 J	400 000 J	600 000 J	
400 000 J	200 000 J	600 000 J	
600 000 J	0 J	600 000 J	

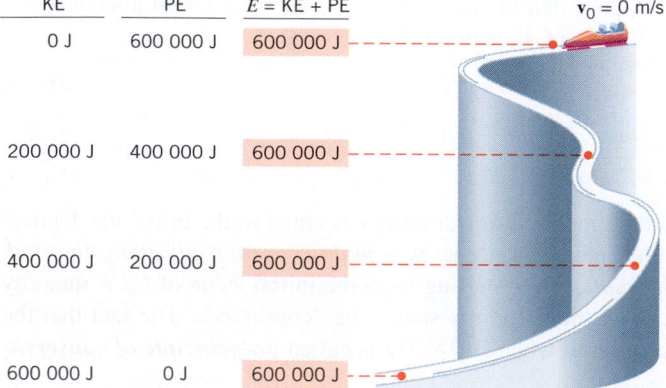

Figure 6.15 If friction and wind resistance are ignored, a bobsled run illustrates how kinetic and potential energy can be interconverted, while the total mechanical energy remains constant. The total mechanical energy is 600 000 J, being all potential energy at the top and all kinetic energy at the bottom.

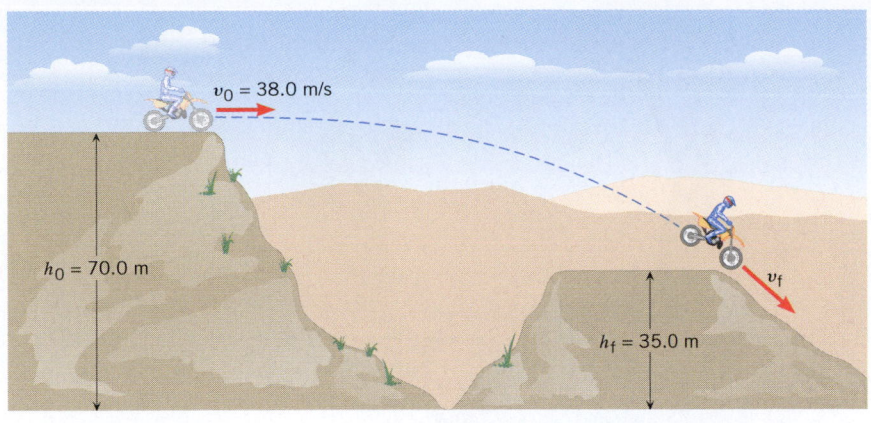

Figure 6.16 A daredevil jumping a canyon.

energy E remains constant at all points along the run. The conservation principle is well known for the ease with which it can be applied, as in the following example.

 Example 8   A Daredevil Motorcyclist

A motorcyclist is trying to leap across the canyon shown in Figure 6.16 by driving horizontally off the cliff at a speed of 38.0 m/s. Ignoring air resistance, find the speed with which the cycle strikes the ground on the other side.

Reasoning Once the cycle leaves the cliff, no forces other than gravity act on the cycle, since air resistance is being ignored. Thus, the work done by external nonconservative forces is zero, $W_{nc} = 0$ J. Accordingly, the principle of conservation of mechanical energy holds, so the total mechanical energy is the same at the final and initial positions of the motorcycle. We will use this important observation to determine the final speed of the cyclist.

Solution The principle of conservation of mechanical energy is written as

$$\underbrace{\tfrac{1}{2}mv_f^2 + mgh_f}_{E_f} = \underbrace{\tfrac{1}{2}mv_0^2 + mgh_0}_{E_0} \tag{6.9b}$$

The mass m of the rider and cycle can be eliminated algebraically from this equation, since m appears as a factor in every term. Solving for v_f gives

$$v_f = \sqrt{v_0^2 + 2g(h_0 - h_f)}$$

$$v_f = \sqrt{(38.0 \text{ m/s})^2 + 2(9.80 \text{ m/s}^2)(70.0 \text{ m} - 35.0 \text{ m})} = \boxed{46.2 \text{ m/s}}$$

Problem-solving insight

Be on the alert for factors, such as the mass m here in Example 8, that sometimes can be eliminated algebraically when using the conservation of mechanical energy.

Examples 9 and 10 emphasize that the principle of conservation of mechanical energy can be applied even when forces act perpendicular to the path of a moving object.

 Conceptual Example 9   The Favorite Swimming Hole

A rope is tied to a tree limb. It is used by a swimmer who, starting from rest, swings down toward the water below, as in Figure 6.17. Only two forces act on him during his descent, the nonconservative force $\vec{T}$, which is due to the tension in the rope, and his weight, which is due to the conservative gravitational force. There is no air resistance. His initial height h_0 and final height h_f are known. Can we use the principle of conservation of mechanical energy to find his speed v_f at the point where he lets go of the rope, even though a nonconservative external force is present? **(a)** Yes. **(b)** No.

Reasoning The principle of conservation of mechanical energy can be used even in the presence of nonconservative forces, provided that the net work W_{nc} done by the nonconservative forces is zero, $W_{nc} = 0$ J.

Answer (b) is incorrect. The mere presence of nonconservative forces does not prevent us from applying the principle of conservation of mechanical energy. The deciding factor is

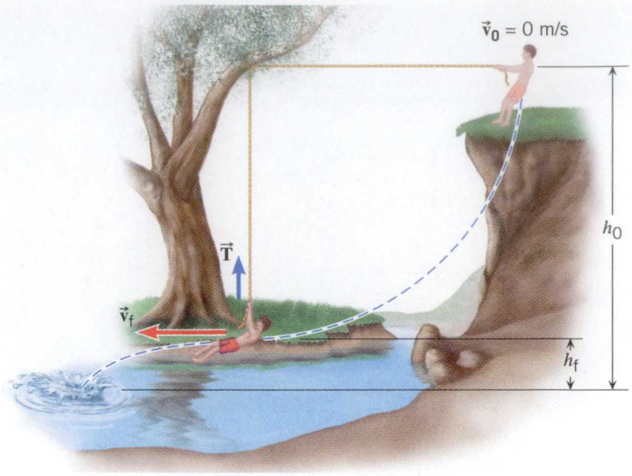

$\vec{v}_0 = 0$ m/s

$\vec{T}$

$\vec{v}_f$

h_0

h_f

Figure 6.17 During the downward swing, the tension $\vec{T}$ in the rope acts perpendicular to the circular arc and, hence, does no work on the person.

Problem-solving insight

When nonconservative forces are perpendicular to the motion, we can still use the principle of conservation of mechanical energy, because such "perpendicular" forces do no work.

whether the net work done by the nonconservative forces is zero. In this case, the tension force $\vec{T}$ is always perpendicular to the circle (see the drawing). Thus, the angle θ between $\vec{T}$ and the instantaneous displacement of the person is always 90°. According to Equation 6.1, the work is proportional to the cosine of this angle, or cos 90°, which is zero, with the result that the work is also zero. As a result, the tension here does not prevent us from using the conservation principle.

Answer (a) is correct. As the person swings downward, he follows a circular path, as shown in Figure 6.17. Since the tension force is the only nonconservative force acting and is always perpendicular to this path, the corresponding net work is $W_{nc} = 0$ J, and we can apply the principle of conservation of mechanical energy to find the speed v_f.

Related Homework: *Problems 42, 86*

The next example illustrates how the conservation of mechanical energy is applied to the breathtaking drop of a roller coaster.

**The physics of
a giant roller coaster.**

Figure 6.18 The Kingda Ka roller coaster in Six Flags Great Adventure, located in Jackson Township, New Jersey, is a giant. It includes a vertical drop of 127 m. (© Splash News and Pictures/NewsCom)

Example 10 The Kingda Ka

The Kingda Ka is a giant among roller coasters (see Figure 6.18). Located in Jackson Township, New Jersey, the ride includes a vertical drop of 127 m. Suppose that the coaster has a speed of 6.0 m/s at the top of the drop. Neglect friction and air resistance and find the speed of the riders at the bottom.

Reasoning Since we are neglecting friction and air resistance, we may set the work done by these forces equal to zero. A normal force from the seat acts on each rider, but this force is perpendicular to the motion, so it does not do any work. Thus, the work done by external nonconservative forces is zero, $W_{nc} = 0$ J, and we may use the principle of conservation of mechanical energy to find the speed of the riders at the bottom.

Solution The principle of conservation of mechanical energy states that

$$\underbrace{\tfrac{1}{2}mv_f^2 + mgh_f}_{E_f} = \underbrace{\tfrac{1}{2}mv_0^2 + mgh_0}_{E_0} \qquad (6.9b)$$

The mass m of the rider appears as a factor in every term in this equation and can be eliminated algebraically. Solving for the final speed gives

$$v_f = \sqrt{v_0^2 + 2g(h_0 - h_f)}$$
$$v_f = \sqrt{(6.0 \text{ m/s})^2 + 2(9.80 \text{ m/s}^2)(127 \text{ m})} = \boxed{50.3 \text{ m/s } (112 \text{ mi/h})}$$

where the vertical drop is $h_0 - h_f = 127$ m.

When applying the principle of conservation of mechanical energy to solving problems, we have been using the following reasoning strategy:

REASONING STRATEGY

Applying the Principle of Conservation of Mechanical Energy

1. Identify the external conservative and nonconservative forces that act on the object. For this principle to apply, the total work done by nonconservative forces must be zero, $W_{nc} = 0$ J. A nonconservative force that is perpendicular to the displacement of the object does no work, for example.

2. Choose the location where the gravitational potential energy is taken to be zero. This location is arbitrary but must not be changed during the course of solving a problem.

3. Set the final total mechanical energy of the object equal to the initial total mechanical energy, as in Equations 6.9a and 6.9b. The total mechanical energy is the sum of the kinetic and potential energies.

✓ CHECK YOUR UNDERSTANDING

(The answers are given at the end of the book.)

10. Suppose the total mechanical energy of an object is conserved. Which one or more of the following statements is/are true?   **(a)** If the kinetic energy decreases, the gravitational potential energy increases.   **(b)** If the gravitational potential energy decreases, the kinetic energy increases.   **(c)** If the kinetic energy does not change, the gravitational potential energy also does not change.

11. In Example 10 the Kingda Ka roller coaster starts with a speed of 6.0 m/s at the top of the drop and attains a speed of 50.3 m/s when it reaches the bottom. If the roller coaster were to then start up an identical hill, what speed would it attain when it reached the top? Assume that friction and air resistance are absent.   **(a)** Greater than 6.0 m/s   **(b)** Exactly 6.0 m/s   **(c)** Between 0 m/s and 6.0m/s   **(d)** 0 m/s

12. The drawing shows an empty fuel tank about to be released by three different jet planes. At the moment of release, each plane has the same speed, and each tank is at the same height above the ground. However, the directions of the velocities of the planes are different.

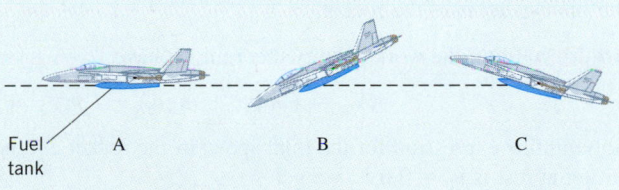

Fuel tank   A   B   C

Which tank has the largest speed upon hitting the ground? Ignore friction and air resistance.   **(a)** A   **(b)** B   **(c)** C   **(d)** Each tank hits the ground with the same speed.

13. In which one or more of the following situations is the principle of conservation of mechanical energy obeyed?   **(a)** An object moves uphill with an increasing speed.   **(b)** An object moves uphill with a decreasing speed.   **(c)** An object moves uphill with a constant speed.   **(d)** An object moves downhill with an increasing speed.   **(e)** An object moves downhill with a decreasing speed.   **(f)** An object moves downhill with a constant speed.

6.6 NONCONSERVATIVE FORCES AND THE WORK-ENERGY THEOREM

Most moving objects experience nonconservative forces, such as friction, air resistance, and propulsive forces, and the work W_{nc} done by the net external nonconservative force is not zero. In these situations, the difference between the final and initial total mechanical energies is equal to W_{nc}, according to $W_{nc} = E_f - E_0$ (Equation 6.8). Consequently, the total mechanical energy is not conserved. The next two examples illustrate how Equation 6.8 is used when nonconservative forces are present and do work.

Example 11   The Kingda Ka Revisited

In Example 10, we ignored nonconservative forces, such as friction. In reality, however, such forces are present when the roller coaster descends. The actual speed of the riders at the bottom is 45.0 m/s, which is less than that determined in Example 10. Assuming again that the coaster has a speed of 6.0 m/s at the top, find the work done by nonconservative forces on a 55.0-kg rider during the descent from a height h_0 to a height h_f, where $h_0 - h_f = 127$ m.

Reasoning Since the speed at the top, the final speed, and the vertical drop are given, we can determine the initial and final total mechanical energies of the rider. The work–energy theorem, $W_{nc} = E_f - E_0$, can then be used to determine the work W_{nc} done by the nonconservative forces.

Solution The work–energy theorem is

$$W_{nc} = \underbrace{(\tfrac{1}{2}mv_f^2 + mgh_f)}_{E_f} - \underbrace{(\tfrac{1}{2}mv_0^2 + mgh_0)}_{E_0} \qquad (6.8)$$

Rearranging the terms on the right side of this equation gives

$$W_{nc} = \tfrac{1}{2}m(v_f^2 - v_0^2) - mg(h_0 - h_f)$$

$$W_{nc} = \tfrac{1}{2}(55.0\,\text{kg})[(45.0\,\text{m/s})^2 - (6.0\,\text{m/s})^2] - (55.0\,\text{kg})(9.80\,\text{m/s}^2)(127\,\text{m}) = \boxed{-1.4 \times 10^4\,\text{J}}$$

Problem-solving insight

As illustrated here and in Example 3, a nonconservative force such as friction can do negative or positive work. It does negative work when it has a component opposite to the displacement and slows down the object. It does positive work when it has a component in the direction of the displacement and speeds up the object.

Example 12   Fireworks

A 0.20-kg rocket in a fireworks display is launched from rest and follows an erratic flight path to reach the point P, as Figure 6.19 shows. Point P is 29 m above the starting point. In the process, 425 J of work is done on the rocket by the nonconservative force generated by the burning propellant. Ignoring air resistance and the mass lost due to the burning propellant, find the speed v_f of the rocket at the point P.

Reasoning The only nonconservative force acting on the rocket is the force generated by the burning propellant, and the work done by this force is $W_{nc} = 425$ J. Because work is done by a nonconservative force, we use the work–energy theorem in the form $W_{nc} = E_f - E_0$ to find the final speed v_f of the rocket.

Solution From the work–energy theorem we have

$$W_{nc} = (\tfrac{1}{2}mv_f^2 + mgh_f) - (\tfrac{1}{2}mv_0^2 + mgh_0) \qquad (6.8)$$

Solving this expression for the final speed of the rocket and noting that the initial speed of the rocket at rest is $v_0 = 0$ m/s, we get

$$v_f = \sqrt{\frac{2[W_{nc} + \tfrac{1}{2}mv_0^2 - mg(h_f - h_0)]}{m}}$$

$$v_f = \sqrt{\frac{2[425\,\text{J} + \tfrac{1}{2}(0.20\,\text{kg})(0\,\text{m/s})^2 - (0.20\,\text{kg})(9.80\,\text{m/s}^2)(29\,\text{m})]}{0.20\,\text{kg}}} = \boxed{61\,\text{m/s}}$$

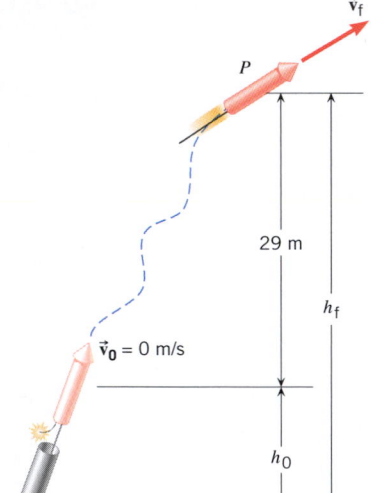

Figure 6.19 A fireworks rocket, moving along an erratic flight path, reaches a point P that is 29 m above the launch point.

✓ CHECK YOUR UNDERSTANDING

(*The answers are given at the end of the book.*)

14. A net external nonconservative force does positive work on a particle. Based solely on this information, you are justified in reaching only one of the following conclusions. Which one is it?   **(a)** The kinetic and potential energies of the particle both decrease.   **(b)** The kinetic and potential energies of the particle both increase.   **(c)** Neither the kinetic nor the potential energy of the particle changes.   **(d)** The total mechanical energy of the particle decreases.   **(e)** The total mechanical energy of the particle increases.

15. In one case, a sports car, its engine running, is driven up a hill at a constant speed. In another case, a truck approaches a hill, and its driver turns off the engine at the bottom of the hill. The truck then coasts up the hill. Which vehicle is obeying the principle of conservation of mechanical energy? Ignore friction and air resistance.   **(a)** Both the sports car and the truck   **(b)** Only the sports car   **(c)** Only the truck

6.7 POWER

In many situations, the time it takes to do work is just as important as the amount of work that is done. Consider two automobiles that are identical in all respects (e.g., both have the same mass), except that one has a "souped-up" engine. The car with the souped-up engine can go from 0 to 27 m/s (60 mph) in 4 seconds, while the other car requires 8 seconds to achieve the same speed. Each engine does work in accelerating its car, but one does it more quickly. Where cars are concerned, we associate the quicker performance with an engine that has a larger horsepower rating. A large horsepower rating means that the engine can do a large amount of work in a short time. In physics, the horsepower rating is just one way to measure an engine's ability to generate power. The idea of *power* incorporates both the concepts of work and time, for power is work done per unit time.

DEFINITION OF AVERAGE POWER

Average power $\overline{P}$ is the average rate at which work W is done, and it is obtained by dividing W by the time t required to perform the work:

$$\overline{P} = \frac{\text{Work}}{\text{Time}} = \frac{W}{t} \qquad (6.10a)$$

SI Unit of Power: joule/s = watt (W)

The definition of average power presented in Equation 6.10a involves work. However, the work–energy theorem relates the work done by a net external force to the change in the energy of the object (see, for example, Equations 6.3 and 6.8). Therefore, we can also define average power as the rate at which the energy is changing, or as the change in energy divided by the time during which the change occurs:

$$\overline{P} = \frac{\text{Change in energy}}{\text{Time}} \qquad (6.10b)$$

Since work, energy, and time are scalar quantities, power is also a scalar quantity. The unit in which power is expressed is that of work divided by time, or a joule per second in SI units. One joule per second is called a watt (W), in honor of James Watt (1736–1819), the developer of the steam engine. The unit of power in the BE system is the foot·pound per second (ft·lb/s), although the familiar horsepower (hp) unit is used frequently for specifying the power generated by electric motors and internal combustion engines:

1 horsepower = 550 foot·pounds/second = 745.7 watts

Table 6.3 summarizes the units for power in the various systems of measurement.

Equation 6.10b provides the basis for understanding the production of power in the human body. In this context the "Change in energy" on the right-hand side of the equation refers to the energy produced by metabolic processes, which, in turn, is derived from the food we eat. Table 6.4 gives typical metabolic rates of energy production needed to sustain various activities. Running at 15 km/h (9.3 mi/h), for example, requires metabolic power sufficient to operate eighteen 75-watt light bulbs, and the metabolic power used in sleeping would operate a single 75-watt bulb.

 **The physics of human metabolism.**

Table 6.3 Units of Measurement for Power

System	Work	÷	Time	=	Power
SI	joule (J)		second (s)		watt (W)
CGS	erg		second (s)		erg per second (erg/s)
BE	foot·pound (ft·lb)		second (s)		foot·pound per second (ft·lb/s)

Table 6.4 Human Metabolic Rates[a]

Activity	Rate (watts)
Running (15 km/h)	1340 W
Skiing	1050 W
Biking	530 W
Walking (5 km/h)	280 W
Sleeping	77 W

[a]For a young 70-kg male.

An alternative expression for power can be obtained from Equation 6.1, which indicates that the work W done when a constant net force of magnitude F points in the same direction as the displacement is $W = (F \cos 0°)s = Fs$. Dividing both sides of this equation by the time t it takes for the force to move the object through the distance s, we obtain

$$\frac{W}{t} = \frac{Fs}{t}$$

But W/t is the average power $\overline{P}$, and s/t is the average speed $\overline{v}$, so that

$$\overline{P} = F\overline{v} \tag{6.11}$$

The next example illustrates the use of Equation 6.11.

ANALYZING MULTIPLE-CONCEPT PROBLEMS

Example 13   The Power to Accelerate a Car

A car, starting from rest, accelerates in the $+x$ direction (see Figure 6.20). It has a mass of 1.10×10^3 kg and maintains an acceleration of $+4.60$ m/s^2 for 5.00 s. Assume that a single horizontal force (not shown) accelerates the vehicle. Determine the average power generated by this force.

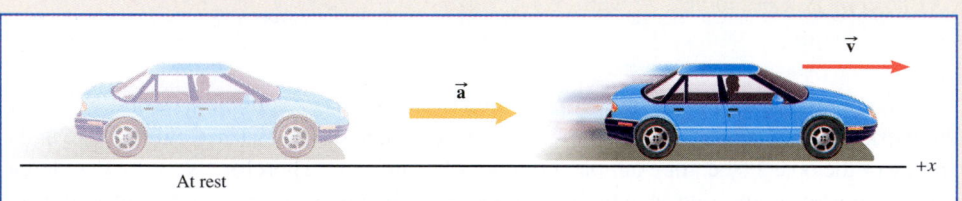

Figure 6.20 A car, starting from rest, accelerates in the $+x$ direction for 5.00 s.

Reasoning The average power $\overline{P}$ can be found by using the relation $\overline{P} = F\overline{v}$ (Equation 6.11), provided that the magnitude F of the horizontal force and the average speed $\overline{v}$ of the car can be determined. The force can be obtained from Newton's second law as the product of the car's mass and acceleration. The average speed can be determined by using the equations of kinematics.

Knowns and Unknowns The data for this problem are listed below:

Description	Symbol	Value	Comment
Explicit Data			
Acceleration	a	$+4.60$ m/s^2	
Mass	m	1.10×10^3 kg	
Time	t	5.00 s	
Implicit Data			
Initial velocity	v_0	0 m/s	Car starts from rest.
Unknown Variable			
Average power	$\overline{P}$	?	

Modeling the Problem

STEP 1   Average Power   The average power $\overline{P}$ is the product of the magnitude F of the horizontal force acting on the car and the car's average speed $\overline{v}$, as given at the right. At this point we do not have numerical values for either F or $\overline{v}$, but in Steps 2 and 3 we will show how to obtain them.

$$\overline{P} = F\overline{v} \tag{6.11}$$

?

STEP 2   **Newton's Second Law**   According to Newton's second law, Equation 4.1, the net force is equal to the mass m of the car times its acceleration a. Since there is only one horizontal force acting on the car, it is the net force. Thus,

$$F = ma$$

This expression also represents the magnitude F of the force, since a is a positive number. Values are known for both m and a, so we substitute this expression for F into Equation 6.11, as indicated at the right. In Step 3 an expression for $\bar{v}$ will be found.

$$\overline{P} = F\overline{v} \qquad (6.11)$$

$$\boxed{F = ma} \qquad \bigcirc{?}$$

STEP 3   **Equations of Kinematics**   Since the acceleration is constant, the equations of kinematics apply, and the average velocity $\bar{v}$ of the car is one-half the sum of its initial velocity v_0 and its final velocity v (Equation 2.6): $\bar{v} = \frac{1}{2}(v_0 + v)$. The final velocity is related to the initial velocity, the acceleration a, and the time t (all of which are known) by Equation 2.4: $v = v_0 + at$. Substituting this expression for v into the equation above for $\bar{v}$ yields

$$\bar{v} = \tfrac{1}{2}(v_0 + v) = \tfrac{1}{2}[v_0 + (v_0 + at)] \quad \text{or} \quad \boxed{\bar{v} = v_0 + \tfrac{1}{2}at}$$

This result, which represents the average velocity, is also equal to the average speed (which is the magnitude of the average velocity), since v_0 is zero and a is positive. As indicated at the right, this expression for $\bar{v}$ can be substituted into Equation 6.11.

$$\overline{P} = F\overline{v} \qquad (6.11)$$

$$\boxed{F = ma} \qquad \boxed{\bar{v} = v_0 + \tfrac{1}{2}at}$$

Solution   Algebraically combining the results of the three steps, we arrive at the following expression for the average power:

| STEP 1 | | STEP 2 | | STEP 3 |

$$\overline{P} \quad = \quad F\overline{v} \quad = \quad (ma)\,\overline{v} \quad = \quad (ma)(v_0 + \tfrac{1}{2}at)$$

The average power generated by the net force that accelerates the car is

$$\overline{P} = (ma)(v_0 + \tfrac{1}{2}at)$$
$$= (1.10 \times 10^3 \text{ kg})(4.60 \text{ m/s}^2)[0 \text{ m/s} + \tfrac{1}{2}(4.60 \text{ m/s}^2)(5.00 \text{ s})] = \boxed{5.82 \times 10^4 \text{ W (78 hp)}}$$

Related Homework: *Problem 67*

✓ **CHECK YOUR UNDERSTANDING**

(The answers are given at the end of the book.)

16. Engine A has a greater power rating than engine B. Which one of the following statements correctly describes the abilities of these engines to do work?   **(a)** Engines A and B can do the same amount of work, but engine A can do it more quickly.   **(b)** Engines A and B can do the same amount of work in the same amount of time.   **(c)** In the same amount of time, engine B can do more work than engine A.

17. Is it correct to conclude that one engine is doing twice the work that another is doing just because it is generating twice the power?   **(a)** Yes   **(b)** No

6.8 OTHER FORMS OF ENERGY AND THE CONSERVATION OF ENERGY

Up to now, we have considered only two types of energy, kinetic energy and gravitational potential energy. There are many other types, however. Electrical energy is used to run electrical appliances. Energy in the form of heat is utilized in cooking food. Moreover, the work done by the kinetic frictional force often appears as heat, as you can experience by rubbing your hands back and forth. When gasoline is burned, some of the stored chemical energy is released and does the work of moving cars, airplanes, and boats. The chemical energy stored in food provides the energy needed for metabolic processes.

The research of many scientists, most notably Albert Einstein, led to the discovery that mass itself is one manifestation of energy. Einstein's famous equation, $E_0 = mc^2$, describes how mass m and energy E_0 are related, where c is the speed of light in a vacuum and has a value of 3.00×10^8 m/s. Because the speed of light is so large, this equation implies that very small masses are equivalent to large amounts of energy. The relationship between mass and energy will be discussed further in Chapter 28.

We have seen that kinetic energy can be converted into gravitational potential energy and vice versa. In general, energy of all types can be converted from one form to another. Part of the chemical energy stored in food, for example, is transformed into gravitational potential energy when a hiker climbs a mountain. Suppose a 65-kg hiker eats a 250-Calorie* snack, which contains 1.0×10^6 J of chemical energy. If this were 100% converted into potential energy $mg(h_f - h_0)$, the change in height would be

$$h_f - h_0 = \frac{1.0 \times 10^6 \text{ J}}{(65 \text{ kg})(9.8 \text{ m/s}^2)} = 1600 \text{ m}$$

The physics of transforming chemical energy in food into mechanical energy.

At a more realistic conversion efficiency of 50%, the change in height would be 800 m. Similarly, in a moving car the chemical energy of gasoline is converted into kinetic energy, as well as into electrical energy and heat.

Whenever energy is transformed from one form to another, it is found that no energy is gained or lost in the process; the total of all the energies before the process is equal to the total of the energies after the process. This observation leads to the following important principle:

THE PRINCIPLE OF CONSERVATION OF ENERGY

Energy can neither be created nor destroyed, but can only be converted from one form to another.

Learning how to convert energy from one form to another more efficiently is one of the main goals of modern science and technology.

6.9 WORK DONE BY A VARIABLE FORCE

The work W done by a constant force (constant in both magnitude and direction) is given by Equation 6.1 as $W = (F \cos \theta)s$. Quite often, situations arise in which the force is not constant but changes with the displacement of the object. For instance, Figure 6.21a shows an archer using a high-tech compound bow. This type of bow consists of a series of pulleys and strings that produce a force-versus-displacement graph like that in Figure 6.21b. One of the key features of the compound bow is that the force rises to a maximum as the string is drawn back, and then falls to 60% of this maximum value when the string is fully drawn. The reduced force at $s = 0.500$ m makes it much easier for the archer to hold the fully drawn bow while aiming the arrow.

When the force varies with the displacement, as in Figure 6.21b, we cannot use the relation $W = (F \cos \theta)s$ to find the work, because this equation is valid only when the force is constant. However, we can use a graphical method. In this method we divide the total displacement into very small segments, Δs_1, Δs_2, and so on (see Figure 6.22a). For each segment, the *average value* of the force component is indicated by a short horizontal line. For example, the short horizontal line for segment Δs_1 is labeled $(F \cos \theta)_1$ in Figure 6.22a. We can then use this average value as the constant-force component in Equation 6.1 and determine an approximate value for the work ΔW_1 done during the first segment: $\Delta W_1 = (F \cos \theta)_1 \Delta s_1$. But this work is just the area of the colored rectangle in the drawing.

(a)

Figure 6.21 (a) A compound bow. (b) A plot of $F \cos \theta$ versus s as the bowstring is drawn back. (a, Ray Laskowitz/Brand X-X Image/Image State

*Energy content in food is typically given in units called Calories, which we will discuss in Section 12.7.

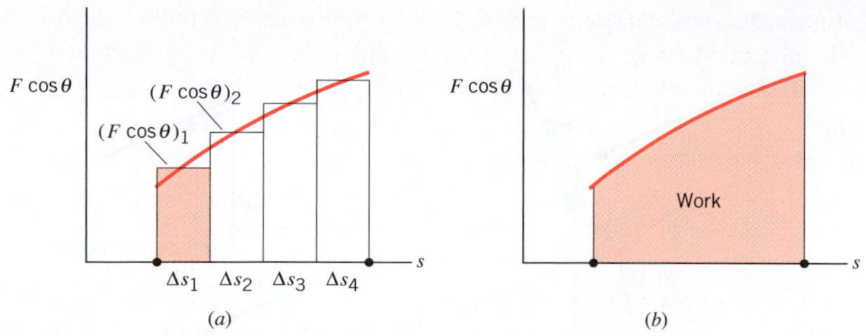

Figure 6.22 (a) The work done by the average-force component $(F \cos \theta)_1$ during the small displacement Δs_1 is $(F \cos \theta)_1 \Delta s_1$, which is the area of the colored rectangle. (b) The work done by a variable force is equal to the colored area under the $F \cos \theta$-versus-s curve.

The word "area" here refers to the area of a rectangle that has a width of Δs_1 and a height of $(F \cos \theta)_1$; it does not mean an area in square meters, such as the area of a parcel of land. In a like manner, we can calculate an approximate value for the work for each segment. Then we add the results for the segments to get, approximately, the work W done by the variable force:

$$W \approx (F \cos \theta)_1 \, \Delta s_1 + (F \cos \theta)_2 \Delta s_2 + \cdots$$

The symbol $\approx$ means "approximately equal to." The right side of this equation is the sum of all the rectangular areas in Figure 6.22a and is an approximate value for the area shaded in color under the graph in Figure 6.22b. If the rectangles are made narrower and narrower by decreasing each Δs, the right side of this equation eventually becomes equal to the area under the graph. Thus, we define the work done by a variable force as follows: *The work done by a variable force in moving an object is equal to the area under the graph of $F \cos \theta$ versus s.* Example 14 illustrates how to use this graphical method to determine the approximate work done when a high-tech compound bow is drawn.

Problem-solving insight

 Example 14 Work and the Compound Bow

The physics of the compound bow.

Find the work that the archer must do in drawing back the string of the compound bow in Figure 6.21 from 0 to 0.500 m.

Reasoning The work is equal to the colored area under the curved line in Figure 6.21b. For convenience, this area is divided into a number of small squares, each having an area of $(9.00 \text{ N})(2.78 \times 10^{-2} \text{ m}) = 0.250 \text{ J}$. The area can be found by counting the number of squares under the curve and multiplying by the area per square.

Solution We estimate that there are 242 colored squares in the drawing. Since each square represents 0.250 J of work, the total work done is

$$W = (242 \text{ squares}) \left(0.250 \, \frac{\text{J}}{\text{square}} \right) = \boxed{60.5 \text{ J}}$$

When the arrow is fired, part of this work is imparted to it as kinetic energy.

6.10 CONCEPTS & CALCULATIONS

This section contains examples that discuss one or more conceptual questions, followed by a related quantitative problem. Example 15 reviews the important concept of work and illustrates how forces can give rise to positive, negative, and zero work. Example 16 examines the all-important conservation of mechanical energy and the work–energy theorem.

 Concepts & Calculations Example 15 Skateboarding and Work

The skateboarder in Figure 6.23a is coasting down a ramp, and there are three forces acting on her: her weight $\vec{W}$ (magnitude = 675 N), a frictional force $\vec{f}$ (magnitude = 125 N) that opposes her motion, and a normal force $\vec{F}_N$ (magnitude = 612 N). Determine the net work done by the three forces when she coasts for a distance of 9.2 m.

Figure 6.23 (*a*) Three forces act on the skateboarder: $\vec{\mathbf{W}}$ = weight, $\vec{\mathbf{f}}$ = frictional force, $\vec{\mathbf{F}}_N$ = normal force. For clarity, the frictional force is not drawn to scale. (*b*) The orientation of the three forces relative to the displacement $\vec{\mathbf{s}}$ of the skateboarder.

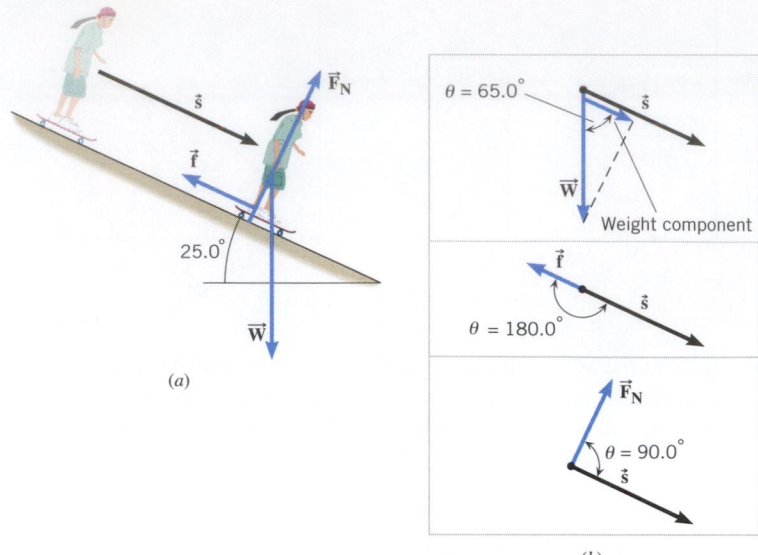

(*a*)

(*b*)

Concept Questions and Answers Figure 6.23*b* shows each force, along with the displacement $\vec{\mathbf{s}}$ of the skateboarder. By examining these diagrams and without doing any numerical calculations, determine whether the work done by each force is positive, negative, or zero. Provide a reason for each answer.

Answer The work done by a force is positive if the force has a component that points in the same direction as the displacement. The work is negative if there is a force component pointing opposite to the displacement. The work done by the weight $\vec{\mathbf{W}}$ is positive, because the weight has a component that points in the same direction as the displacement. The top drawing in Figure 6.23*b* shows this weight component. The work done by the frictional force $\vec{\mathbf{f}}$ is negative, because it points opposite to the direction of the displacement ($\theta = 180.0°$ in the middle drawing). The work done by the normal force $\vec{\mathbf{F}}_N$ is zero, because the normal force is perpendicular to the displacement ($\theta = 90.0°$ in the bottom drawing) and does not have a component along the displacement.

Solution The work W done by a force is given by Equation 6.1 as $W = (F\cos\theta)s$, where F is the magnitude of the force, s is the magnitude of the displacement, and θ is the angle between the force and the displacement. Figure 6.23*b* shows this angle for each of the three forces. The work done by each force is computed in the following table. The net work is the algebraic sum of the three values.

Force	Magnitude F of the Force	Angle, θ	Magnitude s of the Displacement	Work Done by the Force $W = (F\cos\theta)s$
$\vec{\mathbf{W}}$	675 N	65.0°	9.2 m	$W = (675\text{ N})(\cos 65.0°)$ $\times (9.2\text{ m}) = +2620\text{ J}$
$\vec{\mathbf{f}}$	125 N	180.0°	9.2 m	$W = (125\text{ N})(\cos 180.0°)$ $\times (9.2\text{ m}) = -1150\text{ J}$
$\vec{\mathbf{F}}_N$	612 N	90.0°	9.2 m	$W = (612\text{ N})(\cos 90.0°)$ $\times (9.2\text{ m}) = 0\text{ J}$

The net work done by the three forces is

$$+2620\text{ J} + (-1150\text{ J}) + 0\text{ J} = \boxed{+1470\text{ J}}$$

Concepts & Calculations Example 16

 Conservation of Mechanical Energy and the Work–Energy Theorem

Figure 6.24 shows a 0.41-kg block sliding from A to B along a frictionless surface. When the block reaches B, it continues to slide along the horizontal surface BC where a kinetic frictional force acts. As a result, the block slows down, coming to rest at C. The kinetic energy of the block at A is 37 J, and the heights of A and B are 12.0 and 7.0 m above the ground, respectively.

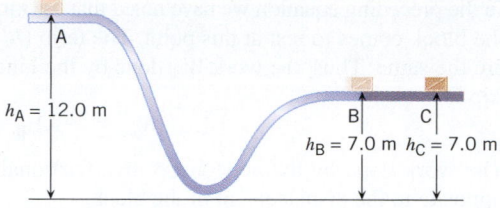

Figure 6.24 The block slides on a frictionless surface from A to B. From B to C, a kinetic frictional force slows down the block until it comes to rest at C.

(a) What is the kinetic energy of the block when it reaches B? **(b)** How much work does the kinetic frictional force do during the BC segment of the trip?

Concept Questions and Answers Is the total mechanical energy of the block conserved as the block goes from A to B? Why or why not?

Answer The total mechanical energy of the block is conserved if the net work done by the nonconservative forces is zero, or $W_{nc} = 0$ J (see Section 6.5). Only two forces act on the block during its trip from A to B: its weight and the normal force. The block's weight, as we have seen in Section 6.4, is a conservative force. The normal force, on the other hand, is a nonconservative force. However, it is always perpendicular to the displacement of the block, so it does no work. Thus, we conclude that $W_{nc} = 0$ J, with the result that the total mechanical energy is conserved during the AB part of the trip.

When the block reaches point B, has its kinetic energy increased, decreased, or remained the same relative to what it had at A? Provide a reason for your answer.

Answer As we have seen, the total mechanical energy is the sum of the kinetic and gravitational potential energies and remains constant during the trip. Therefore, as one type of energy decreases, the other must increase for the sum to remain constant. Since B is lower than A, the gravitational potential energy at B is less than that at A. As a result, the kinetic energy at B must be greater than that at A.

Is the total mechanical energy of the block conserved as the block goes from B to C? Justify your answer.

Answer During this part of the trip, a kinetic frictional force acts on the block. This force is nonconservative, and it does work on the block, just as the frictional force does work on the skateboarder in Example 15. Consequently, the net work W_{nc} done by nonconservative forces is not zero ($W_{nc} \neq 0$ J), so the total mechanical energy is not conserved during this part of the trip.

Solution **(a)** Since the total mechanical energy is conserved during the AB segment, we can set the total mechanical energy at B equal to that at A.

$$\underbrace{KE_B + mgh_B}_{\substack{\text{Total mechanical} \\ \text{energy at B}}} = \underbrace{KE_A + mgh_A}_{\substack{\text{Total mechanical} \\ \text{energy at A}}}$$

Solving this equation for the kinetic energy at B gives

$$KE_B = KE_A + mg(h_A - h_B)$$
$$= 37 \text{ J} + (0.41 \text{ kg})(9.80 \text{ m/s}^2)(12.0 \text{ m} - 7.0 \text{ m}) = \boxed{57 \text{ J}}$$

As expected, the kinetic energy at B is greater than that at A.

(b) During the BC part of the trip, the total mechanical energy is not conserved because a kinetic frictional force is present. The work W_{nc} done by this nonconservative force is given by the work–energy theorem (see Equation 6.8) as the final total mechanical energy minus the initial total mechanical energy:

$$W_{nc} = \underbrace{KE_C + mgh_C}_{\substack{\text{Total mechanical} \\ \text{energy at C}}} - \underbrace{(KE_B + mgh_B)}_{\substack{\text{Total mechanical} \\ \text{energy at B}}}$$

Rearranging this equation gives

$$W_{nc} = \underbrace{KE_C}_{=0 \text{ J}} - KE_B + mg\underbrace{(h_C - h_B)}_{=0 \text{ m}}$$

In the preceding equation we have noted that the kinetic energy KE_C at C is equal to zero, because the block comes to rest at this point. The term $(h_C - h_B)$ is also zero, because the two heights are the same. Thus, the work W_{nc} done by the kinetic frictional force during the BC part of the trip is

$$W_{nc} = -KE_B = \boxed{-57 \text{ J}}$$

The work done by the nonconservative frictional force is negative, because this force points opposite to the displacement of the block.

CONCEPT SUMMARY

If you need more help with a concept, use the Learning Aids noted next to the discussion or equation. Examples (**Ex.**) are in the text of this chapter. Go to **www.wiley.com/college/cutnell** for the following Learning Aids:

Interactive LearningWare (ILW) — Additional examples solved in a five-step interactive format.

Concept Simulations (CS) — Animated text figures or animations of important concepts.

Interactive Solutions (IS) — Models for certain types of problems in the chapter homework. The calculations are carried out interactively.

Topic	Discussion	Learning Aids
	6.1 WORK DONE BY A CONSTANT FORCE The work W done by a constant force acting on an object is	
Work done by constant force	$$W = (F \cos \theta)s \qquad (6.1)$$	**Ex. 1**
	where F is the magnitude of the force, s is the magnitude of the displacement, and θ is the angle between the force and the displacement vectors. Work is a scalar quantity and can be positive or negative, depending on whether the force has a component that points, respectively, in the same direction as the displacement or in the opposite direction. The work is zero if the force is perpendicular ($\theta = 90°$) to the displacement.	**Ex. 2, 3** **ILW 6.1** **Ex. 15**
	6.2 THE WORK–ENERGY THEOREM AND KINETIC ENERGY The kinetic energy KE of an object of mass m and speed v is	
Kinetic energy	$$KE = \tfrac{1}{2}mv^2 \qquad (6.2)$$	
	The work–energy theorem states that the work W done by the net external force acting on an object equals the difference between the object's final kinetic energy KE_f and initial kinetic energy KE_0:	**Ex. 4, 5, 6** **CS 6.1**
Work–energy theorem	$$W = KE_f - KE_0 \qquad (6.3)$$	**ILW 6.2**
	The kinetic energy increases when the net force does positive work and decreases when the net force does negative work.	
	6.3 GRAVITATIONAL POTENTIAL ENERGY The work done by the force of gravity on an object of mass m is	
Work done by force of gravity	$$W_{\text{gravity}} = mg(h_0 - h_f) \qquad (6.4)$$	**Ex. 7**
	where h_0 and h_f are the initial and final heights of the object, respectively.	
	Gravitational potential energy PE is the energy that an object has by virtue of its position. For an object near the surface of the earth, the gravitational potential energy is given by	
Gravitational potential energy	$$PE = mgh \qquad (6.5)$$	
	where h is the height of the object relative to an arbitrary zero level.	
Conservative force	**6.4 CONSERVATIVE VERSUS NONCONSERVATIVE FORCES** A conservative force is one that does the same work in moving an object between two points, independent of the path taken between the points. Alternatively, a force is conservative if the work it does in moving an object around any closed path is zero. A force is nonconservative if the work it does on an object moving between two points depends on the path of the motion between the points.	
Nonconservative force		
	6.5 THE CONSERVATION OF MECHANICAL ENERGY The total mechanical energy E is the sum of the kinetic energy and potential energy:	
Total mechanical energy	$$E = KE + PE$$	

Topic	Discussion	Learning Aids
Alternate form for work-energy theorem	The work–energy theorem can be expressed in an alternate form as $$W_{nc} = E_f - E_0 \qquad (6.8)$$ where W_{nc} is the net work done by the external nonconservative forces, and E_f and E_0 are the final and initial total mechanical energies, respectively.	
Principle of conservation of mechanical energy	The principle of conservation of mechanical energy states that the total mechanical energy E remains constant along the path of an object, provided that the net work done by external nonconservative forces is zero. Whereas E is constant, KE and PE may be transformed into one another.	**Ex. 8, 9, 10** **ILW 6.3** **IS 6.39, 6.43**
	6.6 NONCONSERVATIVE FORCES AND THE WORK-ENERGY THEOREM, **6.7 POWER** Average power $\bar{P}$ is the work done per unit time or the rate at which work is done:	**Ex. 11, 12, 16** **IS 6.59, 6.83**
Average power	$$\bar{P} = \frac{\text{Work}}{\text{Time}} \qquad (6.10a)$$ It is also the rate at which energy changes: $$\bar{P} = \frac{\text{Change in energy}}{\text{Time}} \qquad (6.10b)$$	**IS 6.65**
	When a force of magnitude F acts on an object moving with an average speed $\bar{v}$, the average power is given by $$\bar{P} = F\bar{v} \qquad (6.11)$$	**Ex. 13**
Principle of conservation of energy	**6.8 OTHER FORMS OF ENERGY AND THE CONSERVATION OF ENERGY** The principle of conservation of energy states that energy cannot be created or destroyed but can only be transformed from one form to another.	
Work done by a variable force	**6.9 WORK DONE BY A VARIABLE FORCE** The work done by a variable force of magnitude F in moving an object through a displacement of magnitude s is equal to the area under the graph of $F \cos \theta$ versus s. The angle θ is the angle between the force and displacement vectors.	**Ex. 14**

FOCUS ON CONCEPTS

Note to Instructors: The numbering of the questions shown here reflects the fact that they are only a representative subset of the total number that are available online. However, all of the questions are available for assignment via an online homework management program such as WileyPLUS or WebAssign.

Section 6.1   Work Done by a Constant Force

1. The same force $\vec{F}$ pushes in three different ways on a box moving with a velocity $\vec{v}$, as the drawings show. Rank the work done by the force $\vec{F}$ in ascending order (smallest first): **(a)** A, B, C   **(b)** A, C, B   **(c)** B, A, C   **(d)** C, B, A   **(e)** C, A, B

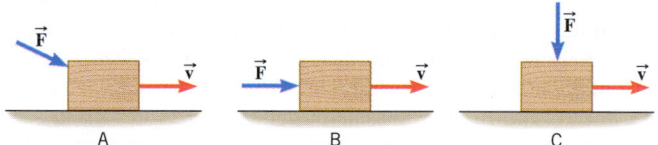

| A | B | C |

2. Consider the force $\vec{F}$ shown in the drawing. This force acts on an object that can move along the positive or negative x axis, or along the positive or negative y axis. The work done by this force is positive when the displacement of the object is along the _____ axis or along the _____ axis: **(a)** $-x$, $-y$ **(b)** $-x$, $+y$ **(c)** $+x$, $+y$ **(d)** $+x$, $-y$

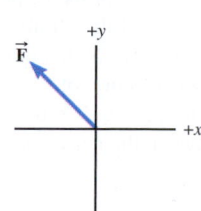

Section 6.2   The Work–Energy Theorem and Kinetic Energy

8. Two forces $\vec{F}_1$ and $\vec{F}_2$ act on a particle. As a result the speed of the particle increases. Which one of the following is NOT possible? **(a)** The work done by $\vec{F}_1$ is positive, and the work done by $\vec{F}_2$ is zero.

(b) The work done by $\vec{F}_1$ is zero, and the work done by $\vec{F}_2$ is positive. **(c)** The work done by each force is positive.   **(d)** The work done by each force is negative.   **(e)** The work done by $\vec{F}_1$ is positive, and the work done by $\vec{F}_2$ is negative.

9. Force $\vec{F}_1$ acts on a particle and does work W_1. Force $\vec{F}_2$ acts simultaneously on the particle and does work W_2. The speed of the particle does not change. Which one of the following must be true?   **(a)** W_1 is zero and W_2 is positive   **(b)** $W_1 = -W_2$ **(c)** W_1 is positive and W_2 is zero   **(d)** W_1 is positive and W_2 is positive

Section 6.4   Conservative Versus Nonconservative Forces

11. A person is riding on a Ferris wheel. When the wheel makes one complete turn, the net work done on the person by the gravitational force _____. **(a)** is positive **(b)** is negative **(c)** is zero **(d)** depends on how fast the wheel moves **(e)** depends on the diameter of the wheel

Section 6.5   The Conservation of Mechanical Energy

13. In which one of the following circumstances could mechanical energy not possibly be conserved, even if friction and air resistance are absent?   **(a)** A car moves up a hill, its velocity continually decreasing along the way.   **(b)** A car moves down a hill, its velocity continually increasing along the way.   **(c)** A car moves along level ground at a constant velocity.   **(d)** A car moves up a hill at a constant velocity.

14. A ball is fixed to the end of a string, which is attached to the ceiling at point P. As the drawing shows, the ball is projected downward at A with the launch speed v_0. Traveling on a circular path, the ball comes to a halt at point B. What enables the ball to reach point B, which is above point A? Ignore friction and air resistance. **(a)** The work done by the tension in the string **(b)** The ball's initial gravitational potential energy **(c)** The ball's initial kinetic energy **(d)** The work done by the gravitational force

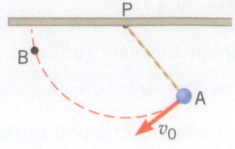

Section 6.6 Nonconservative Forces and the Work–Energy Theorem

21. In which one of the following circumstances does the principle of conservation of mechanical energy apply, even though a nonconservative force acts on the moving object? **(a)** The nonconservative force points in the same direction as the displacement of the object. **(b)** The nonconservative force is perpendicular to the displacement of the object. **(c)** The nonconservative force has a direction that is opposite to the displacement of the object. **(d)** The nonconservative force has a component that points in the same direction as the displacement of the object. **(e)** The nonconservative force has a component that points opposite to the displacement of the object.

22. A 92.0-kg skydiver with an open parachute falls straight downward through a vertical height of 325 m. The skydiver's velocity remains constant. What is the work done by the nonconservative force of air resistance, which is the only nonconservative force acting? **(a)** -2.93×10^5 J **(b)** 0 J **(c)** $+2.93 \times 10^5$ J **(d)** The answer is not obtainable, because insufficient information about the skydiver's speed is given.

Section 6.7 Power

25. The power needed to accelerate a projectile from rest to its launch speed v in a time t is 43.0 W. How much power is needed to accelerate the same projectile from rest to a launch speed of $2v$ in a time of $\frac{1}{2}t$?

PROBLEMS

*Note to Instructors: Most of the homework problems in this chapter are available for assignment via an online homework management program such as WileyPLUS or WebAssign, and those marked with the icon **GO** are presented in WileyPLUS using a guided tutorial format that provides enhanced interactivity. See Preface for additional details.*

ssm   Solution is in the Student Solutions Manual.
www   Solution is available online at www.wiley.com/college/cutnell

  This icon represents a biomedical application.

Section 6.1 Work Done by a Constant Force

1. ssm A water-skier, moving at a speed of 9.30 m/s, is being pulled by a tow rope that makes an angle of 37.0° with respect to the velocity of the boat (see the drawing). The tow rope is parallel to the water. The skier is moving in the same direction as the boat. If the tension in the tow rope is 135 N, determine the work that it does in 12.0 s.

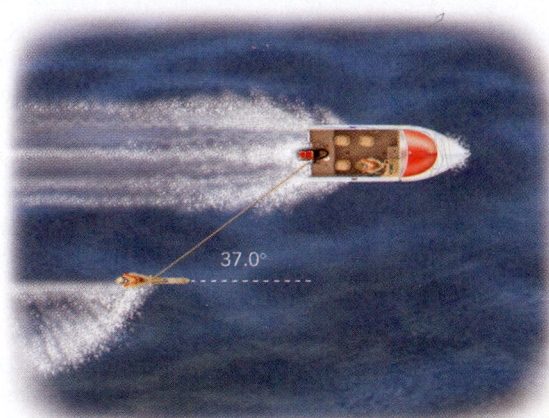

2. GO You are moving into an apartment and take the elevator to the 6th floor. Suppose your weight is 685 N and that of your belongings is 915 N. **(a)** Determine the work done by the elevator in lifting you and your belongings up to the 6th floor (15.2 m) at a constant velocity. **(b)** How much work does the elevator do on you alone (without belongings) on the downward trip, which is also made at a constant velocity?

3. A cable lifts a 1200-kg elevator at a constant velocity for a distance of 35 m. What is the work done by **(a)** the tension in the cable and **(b)** the elevator's weight?

4. A 75.0-kg man is riding an escalator in a shopping mall. The escalator moves the man at a constant velocity from ground level to the floor above, a vertical height of 4.60 m. What is the work done on the man by **(a)** the gravitational force and **(b)** the escalator?

5. ssm Suppose in Figure 6.2 that $+1.10 \times 10^3$ J of work is done by the force $\vec{F}$ (magnitude = 30.0 N) in moving the suitcase a distance of 50.0 m. At what angle θ is the force oriented with respect to the ground?

6. A person pulls a toboggan for a distance of 35.0 m along the snow with a rope directed 25.0° above the snow. The tension in the rope is 94.0 N. **(a)** How much work is done on the toboggan by the tension force? **(b)** How much work is done if the same tension is directed parallel to the snow?

7. See **Interactive LearningWare 6.1** at **www.wiley.com/college/ cutnell** for background on this problem. The drawing shows a plane diving toward the ground and then climbing back upward. During each of these motions, the lift force $\vec{L}$ acts perpendicular to the displacement $\vec{s}$, which has the same magnitude, 1.7×10^3 m, in each case. The engines of the plane exert a thrust $\vec{T}$, which points in the direction of the displacement and has the same magnitude during the dive and the climb. The weight $\vec{W}$ of the plane has a magnitude of 5.9×10^4 N. In both motions, net work is performed due to the combined action of the forces $\vec{L}$, $\vec{T}$, and $\vec{W}$. **(a)** Is more net work done during the dive or the climb? Explain. **(b)** Find the difference between the net work done during the dive and the climb.

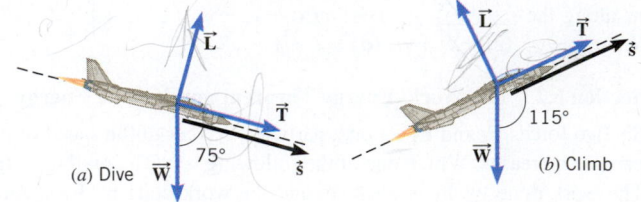

8. A person pushes a 16.0-kg shopping cart at a constant velocity for a distance of 22.0 m. She pushes in a direction 29.0° below the horizontal. A 48.0-N frictional force opposes the motion of the cart. **(a)** What is the magnitude of the force that the shopper exerts? Determine the work done by **(b)** the pushing force, **(c)** the frictional force, and **(d)** the gravitational force.

*9. ssm A 2.40×10^2-N force is pulling an 85.0-kg refrigerator across a horizontal surface. The force acts at an angle of 20.0° above the surface. The coefficient of kinetic friction is 0.200, and the refrigerator moves a distance of 8.00 m. Find **(a)** the work done by the pulling force, and **(b)** the work done by the kinetic frictional force.

*10. As a sailboat sails 52 m due north, a breeze exerts a constant force $\vec{F_1}$ on the boat's sails. This force is directed at an angle west of due north. A force $\vec{F_2}$ of the same magnitude directed due north would do the same amount of work on the sailboat over a distance of just 47 m. What is the angle between the direction of the force $\vec{F_1}$ and due north?

*11. GO A husband and wife take turns pulling their child in a wagon along a horizontal sidewalk. Each exerts a constant force and pulls the wagon through the same displacement. They do the same amount of work, but the husband's pulling force is directed 58° above the horizontal, and the wife's pulling force is directed 38° above the horizontal. The husband pulls with a force whose magnitude is 67 N. What is the magnitude of the pulling force exerted by his wife?

**12. A 1200-kg car is being driven up a 5.0° hill. The frictional force is directed opposite to the motion of the car and has a magnitude of $f = 524$ N. A force $\vec{F}$ is applied to the car by the road and propels the car forward. In addition to these two forces, two other forces act on the car: its weight $\vec{W}$ and the normal force $\vec{F_N}$ directed perpendicular to the road surface. The length of the road up the hill is 290 m. What should be the magnitude of $\vec{F}$, so that the net work done by all the forces acting on the car is +150 kJ?

Section 6.2 The Work–Energy Theorem and Kinetic Energy

13. GO A water-skier is being pulled by a tow rope attached to a boat. As the driver pushes the throttle forward, the skier accelerates. A 70.3-kg water-skier has an initial speed of 6.10 m/s. Later, the speed increases to 11.3 m/s. Determine the work done by the net external force acting on the skier.

14. It takes 185 kJ of work to accelerate a car from 23.0 m/s to 28.0 m/s. What is the car's mass?

15. ssm www Refer to **Concept Simulation 6.1** at **www.wiley.com/college/cutnell** for a review of the concepts with which this problem deals. A 0.075-kg arrow is fired horizontally. The bowstring exerts an average force of 65 N on the arrow over a distance of 0.90 m. With what speed does the arrow leave the bow?

16. Starting from rest, a 1.9×10^{-4}-kg flea springs straight upward. While the flea is pushing off from the ground, the ground exerts an average upward force of 0.38 N on it. This force does $+2.4 \times 10^{-4}$ J of work on the flea. **(a)** What is the flea's speed when it leaves the ground? **(b)** How far upward does the flea move while it is pushing off? Ignore both air resistance and the flea's weight.

17. A fighter jet is launched from an aircraft carrier with the aid of its own engines and a steam-powered catapult. The thrust of its engines is 2.3×10^5 N. In being launched from rest it moves through a distance of 87 m and has a kinetic energy of 4.5×10^7 J at lift-off. What is the work done on the jet by the catapult?

18. As background for this problem, review Conceptual Example 6. A 7420-kg satellite has an elliptical orbit, as in Figure 6.8*b*. The point on the orbit that is farthest from the earth is called the *apogee* and is at the far right side of the drawing. The point on the orbit that is closest to the earth is called the *perigee* and is at the left side of the drawing.

Suppose that the speed of the satellite is 2820 m/s at the apogee and 8450 m/s at the perigee. Find the work done by the gravitational force when the satellite moves from **(a)** the apogee to the perigee and **(b)** the perigee to the apogee.

19. ssm The hammer throw is a track-and-field event in which a 7.3-kg ball (the "hammer"), starting from rest, is whirled around in a circle several times and released. It then moves upward on the familiar curving path of projectile motion. In one throw, the hammer is given a speed of 29 m/s. For comparison, a .22 caliber bullet has a mass of 2.6 g and, starting from rest, exits the barrel of a gun with a speed of 410 m/s. Determine the work done to launch the motion of **(a)** the hammer and **(b)** the bullet.

20. GO An asteroid is moving along a straight line. A force acts along the displacement of the asteroid and slows it down. The asteroid has a mass of 4.5×10^4 kg, and the force causes its speed to change from 7100 to 5500 m/s. **(a)** What is the work done by the force? **(b)** If the asteroid slows down over a distance of 1.8×10^6 m, determine the magnitude of the force.

21. Multiple-Concept Example 4 and **Interactive LearningWare 6.2** at **www.wiley.com/college/cutnell** review the concepts that are important in this problem. A 5.0×10^4-kg space probe is traveling at a speed of 11 000 m/s through deep space. Retrorockets are fired along the line of motion to reduce the probe's speed. The retrorockets generate a force of 4.0×10^5 N over a distance of 2500 km. What is the final speed of the probe?

*22. The concepts in this problem are similar to those in Multiple-Concept Example 4, except that the force doing the work in this problem is the tension in the cable. A rescue helicopter lifts a 79-kg person straight up by means of a cable. The person has an upward acceleration of 0.70 m/s² and is lifted from rest through a distance of 11 m. **(a)** What is the tension in the cable? How much work is done by **(b)** the tension in the cable and **(c)** the person's weight? **(d)** Use the work–energy theorem and find the final speed of the person.

*23. ssm A sled is being pulled across a horizontal patch of snow. Friction is negligible. The pulling force points in the same direction as the sled's displacement, which is along the +x axis. As a result, the kinetic energy of the sled increases by 38%. By what percentage would the sled's kinetic energy have increased if this force had pointed 62° above the +x axis?

*24. Consult Multiple-Concept Example 5 for insight into solving this problem. A skier slides horizontally along the snow for a distance of 21 m before coming to rest. The coefficient of kinetic friction between the skier and the snow is $\mu_k = 0.050$. Initially, how fast was the skier going?

*25. ssm www A 6200-kg satellite is in a circular earth orbit that has a radius of 3.3×10^7 m. A net external force must act on the satellite to make it change to a circular orbit that has a radius of 7.0×10^6 m. What work must the net external force do?

*26. GO Under the influence of its drive force, a snowmobile is moving at a constant velocity along a horizontal patch of snow. When the drive force is shut off, the snowmobile coasts to a halt. The snowmobile and its rider have a mass of 136 kg. Under the influence of a drive force of 205 N, it is moving at a constant velocity whose magnitude is 5.50 m/s. The drive force is then shut off. Find **(a)** the distance in which the snowmobile coasts to a halt and **(b)** the time required to do so.

**27. The model airplane in Figure 5.7 is flying at a speed of 22 m/s on a horizontal circle of radius 16 m. The mass of the plane is 0.90 kg. The person holding the guideline pulls it in until the radius becomes 14 m. The plane speeds up, and the tension in the guideline becomes four times greater. What is the net work done on the plane?

**28. Multiple-Concept Example 5 reviews many of the concepts that play a role in this problem. An extreme skier, starting from rest, coasts down a mountain slope that makes an angle of 25.0° with the horizontal. The coefficient of kinetic friction between her skis and the snow is 0.200. She coasts down a distance of 10.4 m before coming to the edge of a cliff. Without slowing down, she skis off the cliff and lands downhill at a point whose vertical distance is 3.50 m below the edge. How fast is she going just before she lands?

Section 6.3 Gravitational Potential Energy,
Section 6.4 Conservative versus Nonconservative Forces

29. When an 81.0-kg adult uses a spiral staircase to climb to the second floor of his house, his gravitational potential energy increases by 2.00×10^3 J. By how much does the potential energy of an 18.0-kg child increase when the child climbs a normal staircase to the second floor?

30. Juggles and Bangles are clowns. Juggles stands on one end of a teeter-totter at rest on the ground. Bangles jumps off a platform 2.5 m above the ground and lands on the other end of the teeter-totter, launching Juggles into the air. Juggles rises to a height of 3.3 m above the ground, at which point he has the same amount of gravitational potential energy as Bangles had before he jumped, assuming both potential energies are measured using the ground as the reference level. Bangles' mass is 86 kg. What is Juggles' mass?

31. ssm A bicyclist rides 5.0 km due east, while the resistive force from the air has a magnitude of 3.0 N and points due west. The rider then turns around and rides 5.0 km due west, back to her starting point. The resistive force from the air on the return trip has a magnitude of 3.0 N and points due east. **(a)** Find the work done by the resistive force during the round trip. **(b)** Based on your answer to part (a), is the resistive force a conservative force? Explain.

32. A shot-putter puts a shot (weight = 71.1 N) that leaves his hand at a distance of 1.52 m above the ground. **(a)** Find the work done by the gravitational force when the shot has risen to a height of 2.13 m above the ground. Include the correct sign for the work. **(b)** Determine the change ($\Delta PE = PE_f - PE_0$) in the gravitational potential energy of the shot.

33. A 0.60-kg basketball is dropped out of a window that is 6.1 m above the ground. The ball is caught by a person whose hands are 1.5 m above the ground. **(a)** How much work is done on the ball by its weight? What is the gravitational potential energy of the basketball, relative to the ground, when it is **(b)** released and **(c)** caught? **(d)** How is the change ($PE_f - PE_0$) in the ball's gravitational potential energy related to the work done by its weight?

34. **GO** "Rocket Man" has a propulsion unit strapped to his back. He starts from rest on the ground, fires the unit, and accelerates straight upward. At a height of 16 m, his speed is 5.0 m/s. His mass, including the propulsion unit, has the approximately constant value of 136 kg. Find the work done by the force generated by the propulsion unit.

35. ssm A 55.0-kg skateboarder starts out with a speed of 1.80 m/s. He does +80.0 J of work on himself by pushing with his feet against the ground. In addition, friction does −265 J of work on him. In both cases, the forces doing the work are nonconservative. The final speed of the skateboarder is 6.00 m/s. **(a)** Calculate the change ($\Delta PE = PE_f - PE_0$) in the gravitational potential energy. **(b)** How much has the vertical height of the skater changed, and is the skater above or below the starting point?

Section 6.5 The Conservation of Mechanical Energy

36. A 35-kg girl is bouncing on a trampoline. During a certain interval after she leaves the surface of the trampoline, her kinetic energy decreases to 210 J from 440 J. How high does she rise during this interval? Neglect air resistance.

37. ssm A 2.00-kg rock is released from rest at a height of 20.0 m. Ignore air resistance and determine the kinetic energy, gravitational potential energy, and total mechanical energy at each of the following heights: 20.0, 10.0, and 0 m.

38. **GO** The drawing shows two boxes resting on frictionless ramps. One box is relatively light and sits on a steep ramp. The other box is heavier and rests on a ramp that is less steep. The boxes are released from rest at A and allowed to slide down the ramps. The two boxes have masses of 11 and 44 kg. If A and B are 4.5 and 1.5 m, respectively, above the ground, determine the speed of **(a)** the lighter box and **(b)** the heavier box when each reaches B. **(c)** What is the ratio of the kinetic energy of the heavier box to that of the lighter box at B?

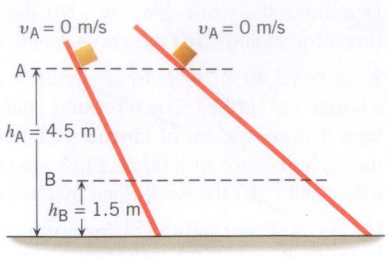

39. Interactive Solution 6.39 at **www.wiley.com/college/cutnell** presents a model for solving this problem. A slingshot fires a pebble from the top of a building at a speed of 14.0 m/s. The building is 31.0 m tall. Ignoring air resistance, find the speed with which the pebble strikes the ground when the pebble is fired **(a)** horizontally, **(b)** vertically straight up, and **(c)** vertically straight down.

40. The skateboarder in the drawing starts down the left side of the ramp with an initial speed of 5.4 m/s. If nonconservative forces, such as kinetic friction and air resistance, are negligible, what would be the height h of the highest point reached by the skateboarder on the right side of the ramp?

41. A 47.0-g golf ball is driven from the tee with an initial speed of 52.0 m/s and rises to a height of 24.6 m. **(a)** Neglect air resistance and determine the kinetic energy of the ball at its highest point. **(b)** What is its speed when it is 8.0 m below its highest point?

42. Consult Conceptual Example 9 in preparation for this problem. **Interactive LearningWare 6.3** at **www.wiley.com/college/cutnell**

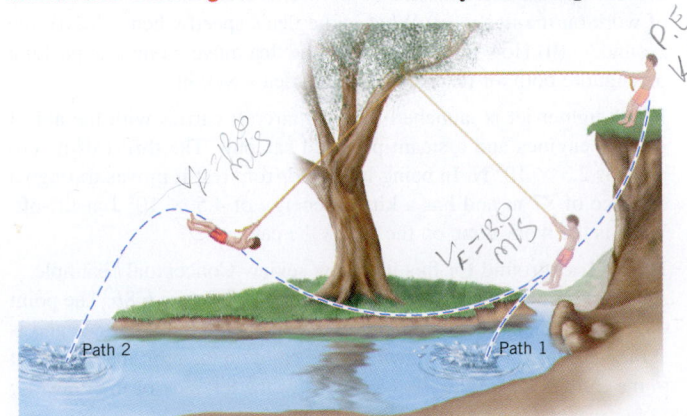

also provides useful background. The drawing shows a person who, starting from rest at the top of a cliff, swings down at the end of a rope, releases it, and falls into the water below. There are two paths by which the person can enter the water. Suppose he enters the water at a speed of 13.0 m/s via path 1. How fast is he moving on path 2 when he releases the rope at a height of 5.20 m above the water? Ignore the effects of air resistance.

*43. Review **Interactive Solution 6.43** at **www.wiley.com/college/cutnell** for background on this problem. A wrecking ball swings at the end of a 12.0-m cable on a vertical circular arc. The crane operator manages to give the ball a speed of 5.00 m/s as the ball passes through the lowest point of its swing, and then gives the ball no further assistance. Friction and air resistance are negligible. What speed v_f does the ball have when the cable makes an angle of 20.0° with respect to the vertical?

*44. The drawing shows a skateboarder moving at 5.4 m/s along a horizontal section of a track that is slanted upward by 48° above the horizontal at its end, which is 0.40 m above the ground. When she leaves the track, she follows the characteristic path of projectile motion. Ignoring friction and air resistance, find the maximum height H to which she rises above the end of the track.

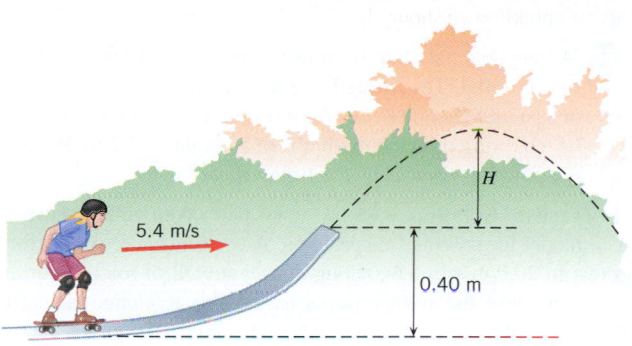

*45. **ssm** A water slide is constructed so that swimmers, starting from rest at the top of the slide, leave the end of the slide traveling horizontally. As the drawing shows, one person hits the water 5.00 m from the end of the slide in a time of 0.500 s after leaving the slide. Ignoring friction and air resistance, find the height H in the drawing.

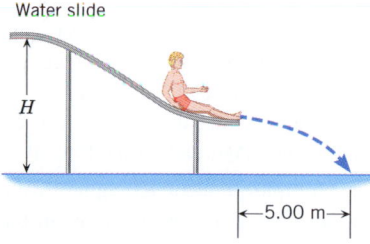

Water slide

*46. A semitrailer is coasting downhill along a mountain highway when its brakes fail. The driver pulls onto a runaway-truck ramp that is inclined at an angle of 14.0° above the horizontal. The semitrailer coasts to a stop after traveling 154 m along the ramp. What was the truck's initial speed? Neglect air resistance and friction.

*47. A skier starts from rest at the top of a hill. The skier coasts down the hill and up a second hill, as the drawing illustrates. The crest of the second hill is circular, with a radius of $r = 36$ m. Neglect friction and air resistance. What must be the height h of the first hill so that the skier just loses contact with the snow at the crest of the second hill?

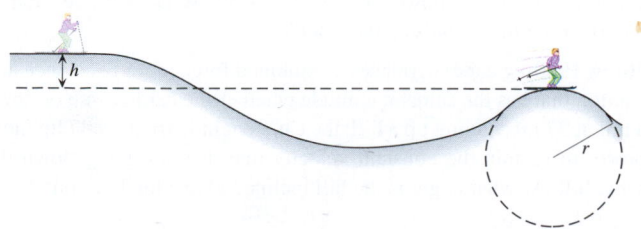

*48. **GO** The drawing shows two frictionless inclines that begin at ground level ($h = 0$ m) and slope upward at the same angle θ. One track is longer than the other, however. Identical blocks are projected up each track with the same initial speed v_0. One the longer track the block slides upward until it reaches a maximum height H above the ground. On the shorter track the block slides upward, flies off the end of the track at a height H_1 above the ground, and then follows the familiar parabolic trajectory of projectile motion. At the highest point of this trajectory, the block is a height H_2 above the end of the track. The initial total mechanical energy of each block is the same and is all kinetic energy. The initial speed of each block is $v_0 = 7.00$ m/s, and each incline slopes upward at an angle of $\theta = 50.0°$. The block on the shorter track leaves the track at a height of $H_1 = 1.25$ m above the ground. Find (a) the height H for the block on the longer track and (b) the total height $H_1 + H_2$ for the block on the shorter track.

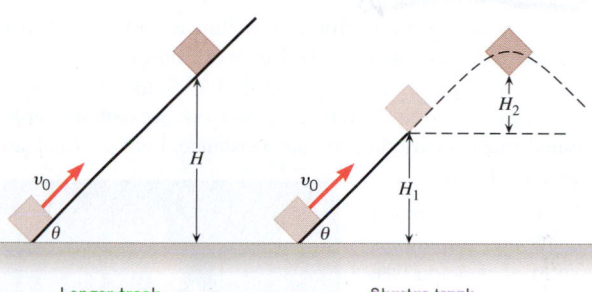

Longer track   Shorter track

**49. The drawing shows a version of the loop-the-loop trick for a small car. If the car is given an initial speed of 4.0 m/s, what is the largest value that the radius r can have if the car is to remain in contact with the circular track at all times?

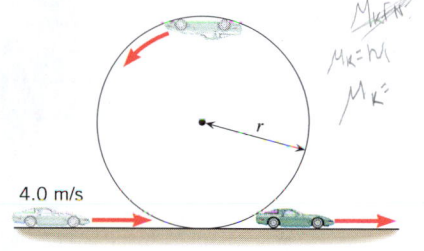

**50. A person starts from rest at the top of a large frictionless spherical surface, and slides into the water below (see the drawing). At what angle θ does the person leave the surface? (*Hint: When the person leaves the surface, the normal force is zero.*)

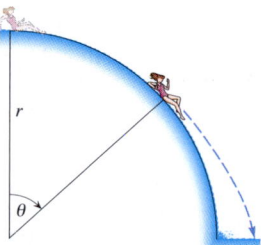

Section 6.6 Nonconservative Forces and the Work–Energy Theorem

51. The surfer in the photo is catching a wave. Suppose she starts at the top of the wave with a speed of 1.4 m/s and moves down the wave until her speed increases to 9.5 m/s. The drop in her vertical height is 2.7 m. If her mass is 59 kg, how much work is done by the (nonconservative) force of the wave?

(Pierre Tostee/Allsport/Getty Images)

52. Starting from rest, a 93-kg firefighter slides down a fire pole. The average frictional force exerted on him by the pole has a magnitude of 810 N, and his speed at the bottom of the pole is 3.4 m/s. How far did he slide down the pole?

53. ssm A roller coaster car (375 kg) moves from A (5.00 m above the ground) to B (20.0 m above the ground). Two nonconservative forces are present: friction does -2.00×10^4 J of work on the car, and a chain mechanism does $+3.00 \times 10^4$ J of work to help the car up a long climb. What is the change in the car's kinetic energy, $\Delta KE = KE_f - KE_0$, from A to B?

54. GO A student, starting from rest, slides down a water slide. On the way down, a kinetic frictional force (a nonconservative force) acts on her. The student has a mass of 83.0 kg, and the height of the water slide is 11.8 m. If the kinetic frictional force does -6.50×10^3 J of work, how fast is the student going at the bottom of the slide?

55. ssm The (nonconservative) force propelling a 1.50×10^3-kg car up a mountain road does 4.70×10^6 J of work on the car. The car starts from rest at sea level and has a speed of 27.0 m/s at an altitude of 2.00×10^2 m above sea level. Obtain the work done on the car by the combined forces of friction and air resistance, both of which are nonconservative forces.

56. One of the new events in the 2002 Winter Olympics was the sport of skeleton (see the photo). Starting at the top of a steep, icy track, a rider jumps onto a sled (known as a skeleton) and proceeds— belly down and head first—to slide down the track. The track has fifteen turns and drops 104 m in elevation from top to bottom. **(a)** In the absence of nonconservative forces, such as friction and air resistance, what would be the speed of a rider at the bottom of the track? Assume that the

(Jacques Demarthon/Getty Images)

speed of the rider at the beginning of the run is relatively small and can be ignored. **(b)** In reality, the best riders reach the bottom with a speed of 35.8 m/s (about 80 mi/h). How much work is done on an 86.0-kg rider and skeleton by nonconservative forces?

*57. A 63-kg skier coasts up a snow-covered hill that makes an angle of 25° with the horizontal. The initial speed of the skier is 6.6 m/s. After coasting 1.9 m up the slope, the skier has a speed of 4.4 m/s. **(a)** Find the work done by the kinetic frictional force that acts on the skis. **(b)** What is the magnitude of the kinetic frictional force?

*58. At a carnival, you can try to ring a bell by striking a target with a 9.00-kg hammer. In response, a 0.400-kg metal piece is sent upward toward the bell, which is 5.00 m above. Suppose that 25.0% of the hammer's kinetic energy is used to do the work of sending the metal piece upward. How fast must the hammer be moving when it strikes the target so that the bell just barely rings?

*59. Refer to **Interactive Solution 6.59** at **www.wiley.com/college/cutnell** for a review of the approach taken in this type of problem. A 67.0-kg person jumps from rest off a 3.00-m-high tower straight down into the water. Neglect air resistance. She comes to rest 1.10 m under the surface of the water. Determine the magnitude of the average force that the water exerts on the diver. This force is nonconservative.

*60. A pitcher throws a 0.140-kg baseball, and it approaches the bat at a speed of 40.0 m/s. The bat does $W_{nc} = 70.0$ J of work on the ball in

hitting it. Ignoring air resistance, determine the speed of the ball after the ball leaves the bat and is 25.0 m above the point of impact.

**61. ssm www A truck is traveling at 11.1 m/s down a hill when the brakes on all four wheels lock. The hill makes an angle of 15.0° with respect to the horizontal. The coefficient of kinetic friction between the tires and the road is 0.750. How far does the truck skid before coming to a stop?

Section 6.7 Power

62. Bicyclists in the Tour de France do enormous amounts of work during a race. For example, the average power per kilogram generated by Lance Armstrong ($m = 75.0$ kg) is 6.50 W per kilogram of his body mass. **(a)** How much work does he do during a 135-km race in which his average speed is 12.0 m/s? **(b)** Often, the work done is expressed in nutritional Calories rather than in joules. Express the work done in part (a) in terms of nutritional Calories, noting that 1 joule = 2.389×10^{-4} nutritional Calories.

63. ssm One kilowatt·hour (kWh) is the amount of work or energy generated when one kilowatt of power is supplied for a time of one hour. A kilowatt·hour is the unit of energy used by power companies when figuring your electric bill. Determine the number of joules of energy in one kilowatt·hour.

64. GO A helicopter, starting from rest, accelerates straight up from the roof of a hospital. The lifting force does work in raising the helicopter. An 810-kg helicopter rises from rest to a speed of 7.0 m/s in a time of 3.5 s. During this time it climbs to a height of 8.2 m. What is the average power generated by the lifting force?

65. Interactive Solution 6.65 at **www.wiley.com/college/cutnell** offers a model for solving this problem. A car accelerates uniformly from rest to 20.0 m/s in 5.6 s along a level stretch of road. Ignoring friction, determine the average power required to accelerate the car if **(a)** the weight of the car is 9.0×10^3 N and **(b)** the weight of the car is 1.4×10^4 N.

66. You are working out on a rowing machine. Each time you pull the rowing bar (which simulates the oars) toward you, it moves a distance of 1.2 m in a time of 1.5 s. The readout on the display indicates that the average power you are producing is 82 W. What is the magnitude of the force that you exert on the handle?

*67. Multiple-Concept Example 13 presents useful background for this problem. The cheetah is one of the fastest-accelerating animals, because it can go from rest to 27 m/s (about 60 mi/h) in 4.0 s. If its mass is 110 kg, determine the average power developed by the cheetah during the acceleration phase of its motion. Express your answer in **(a)** watts and **(b)** horsepower.

*68. Some gliders are launched from the ground by means of a winch, which rapidly reels in a towing cable attached to the glider. What average power must the winch supply in order to accelerate a 184-kg ultralight glider from rest to 26.0 m/s over a horizontal distance of 48.0 m? Assume that friction and air resistance are negligible, and that the tension in the winch cable is constant.

**69. ssm The motor of a ski boat generates an average power of 7.50×10^4 W when the boat is moving at a constant speed of 12 m/s. When the boat is pulling a skier at the same speed, the engine must generate an average power of 8.30×10^4 W. What is the tension in the tow rope that is pulling the skier?

**70. A 1900-kg car experiences a combined force of air resistance and friction that has the same magnitude whether the car goes up or down a hill at 27 m/s. Going up a hill, the car's engine produces 47 hp more power to sustain the constant velocity than it does going down the same hill. At what angle is the hill inclined above the horizontal?

Section 6.9   Work Done by a Variable Force

71. ssm The drawing shows the force-versus-displacement graph for two different bows. These graphs give the force that an archer must apply to draw the bowstring. **(a)** For which bow is more work required to draw the bow fully from $s = 0$ to $s = 0.50$ m? Give your reasoning. **(b)** Estimate the additional work required for the bow identified in part (a) compared to the other bow.

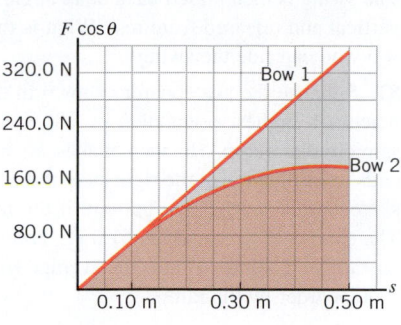

72. The graph shows how the force component $F \cos \theta$ along the displacement varies with the magnitude s of the displacement. Find the work done by the force. *(Hint: Recall how the area of a triangle is related to the triangle's base and height.)*

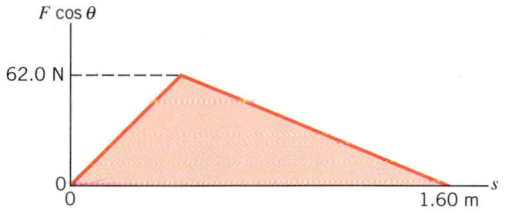

73. The graph shows the net external force component $F \cos \theta$ along the displacement as a function of the magnitude s of the displacement. The graph applies to a 65-kg ice skater. How much work does the net force component do on the skater from **(a)** 0 to 3.0 m and **(b)** 3.0 m to 6.0 m? **(c)** If the initial speed of the skater is 1.5 m/s when $s = 0$ m, what is the speed when $s = 6.0$ m?

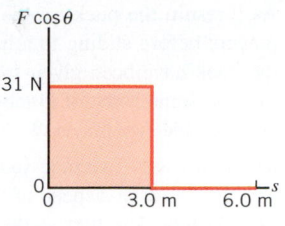

74. Review Example 14, in which the work done in drawing the bowstring in Figure 6.21 from $s = 0$ to $s = 0.500$ m is determined. In part b of the figure, the force component $F \cos \theta$ reaches a maximum at $s = 0.306$ m. Find the percentage of the total work that is done when the bowstring is moved **(a)** from $s = 0$ to 0.306 m and **(b)** from $s = 0.306$ to 0.500 m.

75. GO A net external force is applied to a 6.00-kg object that is initially at rest. The net force component along the displacement of the object varies with the magnitude of the displacement as shown in the drawing. What is the speed of the object at $s = 20.0$ m?

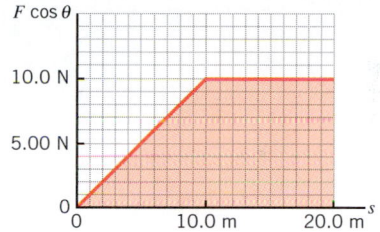

ADDITIONAL PROBLEMS

76. The brakes of a truck cause it to slow down by applying a retarding force of 3.0×10^3 N to the truck over a distance of 850 m. What is the work done by this force on the truck? Is the work positive or negative? Why?

77. ssm Consult **Interactive LearningWare 6.3** at **www.wiley.com/college/cutnell** for a review of the concepts on which this problem is based. A gymnast is swinging on a high bar. The distance between his waist and the bar is 1.1 m, as the drawing shows. At the top of the swing his speed is momentarily zero. Ignoring friction and treating the gymnast as if all of his mass is located at his waist, find his speed at the bottom of the swing.

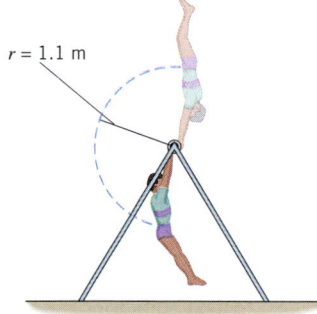

78. A projectile of mass 0.750 kg is shot straight up with an initial speed of 18.0 m/s. **(a)** How high would it go if there were no air friction? **(b)** If the projectile rises to a maximum height of only 11.8 m, determine the magnitude of the average force due to air resistance.

79. ssm During a tug-of-war, team A pulls on team B by applying a force of 1100 N to the rope between them. How much work does team A do if they pull team B toward them a distance of 2.0 m?

80. A 75.0-kg skier rides a 2830-m-long lift to the top of a mountain. The lift makes an angle of 14.6° with the horizontal. What is the change in the skier's gravitational potential energy?

***81.** A 1.00×10^2-kg crate is being pushed across a horizontal floor by a force $\vec{P}$ that makes an angle of 30.0° below the horizontal. The coefficient of kinetic friction is 0.200. What should be the magnitude of $\vec{P}$, so that the net work done by it and the kinetic frictional force is zero?

***82.** In 2.0 minutes, a ski lift raises four skiers at constant speed to a height of 140 m. The average mass of each skier is 65 kg. What is the average power provided by the tension in the cable pulling the lift?

***83. ssm Interactive Solution 6.83** at **www.wiley.com/college/cutnell** offers help for this problem. A basketball of mass 0.60 kg is dropped from rest from a height of 1.05 m. It rebounds to a height of 0.57 m. **(a)** How much mechanical energy was lost during the collision with the floor? **(b)** A basketball player dribbles the ball from a height of 1.05 m by exerting a constant downward force on it for a distance of 0.080 m. In dribbling, the player compensates for the mechanical energy lost during each bounce. If the ball now returns to a height of 1.05 m, what is the magnitude of the force?

***84.** In attempting to pass the puck to a teammate, a hockey player gives it an initial speed of 1.7 m/s. However, this speed is inadequate to compensate for the kinetic friction between the puck and the ice.

As a result, the puck travels only one-half the distance between the players before sliding to a halt. What minimum initial speed should the puck have been given so that it reached the teammate, assuming that the same force of kinetic friction acted on the puck everywhere between the two players?

*85. **ssm www** Two pole-vaulters just clear the bar at the same height. The first lands at a speed of 8.90 m/s, and the second lands at a speed of 9.00 m/s. The first vaulter clears the bar at a speed of 1.00 m/s. Ignore air resistance and friction and determine the speed at which the second vaulter clears the bar.

**86. Conceptual Example 9 provides background for this problem. A swing is made from a rope that will tolerate a maximum tension of 8.00×10^2 N without breaking. Initially, the swing hangs vertically. The swing is then pulled back at an angle of 60.0° with respect to the vertical and released from rest. What is the mass of the heaviest person who can ride the swing?

**87. Suppose the skateboarder shown in the drawing for problem 40 reaches a highest point of $h = 1.80$ m above the right side of the semicircular ramp. He then makes an incomplete midair turn and ends up sliding down the right side of the ramp on his back. When the skateboarder reaches the bottom of the ramp, his speed is 6.40 m/s. The skateboarder's mass is 61.0 kg, and the radius of the semicircular ramp is 2.70 m. What is the average frictional force exerted on the skateboarder by the ramp?

CHAPTER 7

IMPULSE AND MOMENTUM

The chairman of Pan American Silver Corporation is hit in the face with a pie during fundraising for BC Children's Hospital in Vancouver, British Columbia. Part of the pie sticks to his face, in what is known as a completely inelastic collision. Such collisions are discussed in this chapter. (©Andy Clark/Reuters/Corbis)

7.1 THE IMPULSE–MOMENTUM THEOREM

There are many situations in which the force acting on an object is not constant, but varies with time. For instance, Figure 7.1a shows a baseball being hit, and part b of the figure illustrates approximately how the force applied to the ball by the bat changes during the time of contact. The magnitude of the force is zero at the instant t_0 just before the bat touches the ball. During contact, the force rises to a maximum and then returns to zero at the time t_f when the ball leaves the bat. The time interval $\Delta t = t_f - t_0$ during which the bat and ball are in contact is quite short, being only a few thousandths of a second, although the maximum force can be very large, often exceeding thousands of newtons. For comparison, the graph also shows the magnitude $\overline{F}$ of the average force exerted on the ball during the time of contact. Figure 7.2 depicts other situations in which a time-varying force is applied to a ball.

(a)

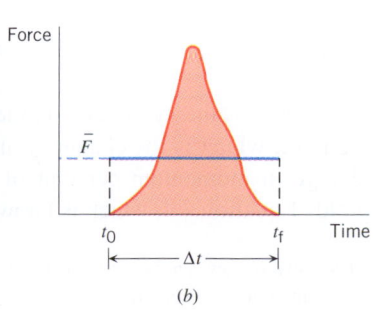

(b)

Figure 7.1 (a) The collision time between a bat and a ball is very short, often less than a millisecond, but the force can be quite large. (b) When the bat strikes the ball, the magnitude of the force exerted on the ball rises to a maximum value and then returns to zero when the ball leaves the bat. The time interval during which the force acts is Δt, and the magnitude of the average force is $\overline{F}$. (a, Chuck Savage/Corbis Images)

To describe how a time-varying force affects the motion of an object, we will introduce two new ideas: the impulse of a force and the linear momentum of an object. These ideas will be used with Newton's second law of motion to produce an important result known as the impulse–momentum theorem. This theorem plays a central role in describing collisions, such as that between a ball and a bat. Later on, we will see also that the theorem leads in a natural way to one of the most fundamental laws in physics, the conservation of linear momentum.

If a baseball is to be hit well, both the magnitude of the force and the time of contact are important. When a large average force acts on the ball for a long enough time, the ball is hit solidly. To describe such situations, we bring together the average force and the time of contact, calling the product of the two the *impulse* of the force.

DEFINITION OF IMPULSE

The impulse $\vec{\mathbf{J}}$ of a force is the product of the average force $\overline{\vec{\mathbf{F}}}$ and the time interval Δt during which the force acts:

$$\vec{\mathbf{J}} = \overline{\vec{\mathbf{F}}}\,\Delta t \qquad (7.1)$$

Impulse is a vector quantity and has the same direction as the average force.

SI Unit of Impulse: newton·second (N·s)

When a ball is hit, it responds to the value of the impulse. A large impulse produces a large response; that is, the ball departs from the bat with a large velocity. However, we know from experience that the more massive the ball, the less velocity it has after leaving the bat. Both mass and velocity play a role in how an object responds to a given impulse, and the effect of each of them is included in the concept of *linear momentum*, which is defined as follows:

DEFINITION OF LINEAR MOMENTUM

The linear momentum $\vec{\mathbf{p}}$ of an object is the product of the object's mass m and velocity $\vec{\mathbf{v}}$:

$$\vec{\mathbf{p}} = m\vec{\mathbf{v}} \qquad (7.2)$$

Linear momentum is a vector quantity that points in the same direction as the velocity.

SI Unit of Linear Momentum: kilogram·meter/second (kg·m/s)

Figure 7.2 In each of these situations, the force applied to the ball varies with time. The time of contact is small, but the maximum force can be large. (*top,* Jonathan Daniel/MLS/Getty Images; *bottom,* Robert Laberge/Getty Images Sport Services)

Newton's second law of motion can now be used to reveal a relationship between impulse and momentum. Figure 7.3 shows a ball with an initial velocity $\vec{\mathbf{v}}_0$ approaching a bat, being struck by the bat, and then departing with a final velocity $\vec{\mathbf{v}}_f$. When the velocity of an object changes from $\vec{\mathbf{v}}_0$ to $\vec{\mathbf{v}}_f$ during a time interval Δt, the average acceleration $\overline{\vec{\mathbf{a}}}$ is given by Equation 2.4 as

$$\overline{\vec{\mathbf{a}}} = \frac{\vec{\mathbf{v}}_f - \vec{\mathbf{v}}_0}{\Delta t}$$

According to Newton's second law, $\Sigma\overline{\vec{\mathbf{F}}} = m\overline{\vec{\mathbf{a}}}$, the average acceleration is produced by the net average force $\Sigma\overline{\vec{\mathbf{F}}}$. Here $\Sigma\overline{\vec{\mathbf{F}}}$ represents the vector sum of all the average forces that act on the object. Thus,

$$\Sigma\overline{\vec{\mathbf{F}}} = m\left(\frac{\vec{\mathbf{v}}_f - \vec{\mathbf{v}}_0}{\Delta t}\right) = \frac{m\vec{\mathbf{v}}_f - m\vec{\mathbf{v}}_0}{\Delta t} \qquad (7.3)$$

In this result, the numerator on the far right is the final momentum minus the initial momentum, which is the change in momentum. Thus, the net average force is given by the change in momentum per unit of time.* Multiplying both sides of Equation 7.3 by Δt yields Equation 7.4, which is known as the *impulse–momentum theorem.*

*The equality between the net force and the change in momentum per unit time is the version of the second law of motion presented originally by Newton.

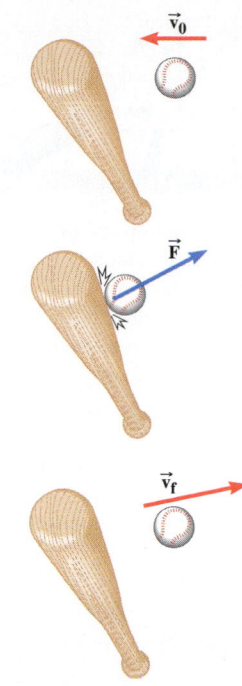

IMPULSE–MOMENTUM THEOREM

When a net average force $\Sigma\vec{F}$ acts on an object during a time interval Δt, the impulse of this force is equal to the change in momentum of the object:

$$\underbrace{(\Sigma\vec{F})\,\Delta t}_{\text{Impulse}} = \underbrace{m\vec{v}_f}_{\substack{\text{Final} \\ \text{momentum}}} - \underbrace{m\vec{v}_0}_{\substack{\text{Initial} \\ \text{momentum}}} \qquad (7.4)$$

Impulse = Change in momentum

During a collision, it is often difficult to measure the net average force $\Sigma\vec{F}$, so it is not easy to determine the impulse $(\Sigma\vec{F})\Delta t$ directly. On the other hand, it is usually straightforward to measure the mass and velocity of an object, so that its momentum $m\vec{v}_f$ just after the collision and $m\vec{v}_0$ just before the collision can be found. Thus, the impulse–momentum theorem allows us to gain information about the impulse indirectly by measuring the change in momentum that the impulse causes. Then, armed with a knowledge of the contact time Δt, we can evaluate the net average force. Examples 1 and 2 illustrate how the theorem is used in this way.

 Example 1   A Well-Hit Ball

A baseball ($m = 0.14$ kg) has an initial velocity of $\vec{v}_0 = -38$ m/s as it approaches a bat. We have chosen the direction of approach as the negative direction. The bat applies an average force $\vec{F}$ that is much larger than the weight of the ball, and the ball departs from the bat with a final velocity of $\vec{v}_f = +58$ m/s. **(a)** Determine the impulse applied to the ball by the bat. **(b)** Assuming that the time of contact is $\Delta t = 1.6 \times 10^{-3}$ s, find the average force exerted on the ball by the bat.

Reasoning Two forces act on the ball during impact, and together they constitute the net average force: the average force $\vec{F}$ exerted by the bat, and the weight of the ball. Since $\vec{F}$ is much greater than the weight of the ball, we neglect the weight. Thus, the net average force is equal to $\vec{F}$, or $\Sigma\vec{F} = \vec{F}$. In hitting the ball, the bat imparts an impulse to it. We cannot use Equation 7.1 ($\vec{J} = \vec{F}\,\Delta t$) to determine the impulse $\vec{J}$ directly, since $\vec{F}$ is not known. We can find the impulse indirectly, however, by turning to the impulse–momentum theorem, which states that the impulse is equal to the ball's final momentum minus its initial momentum. With values for the impulse and the time of contact, Equation 7.1 can be used to determine the average force applied by the bat to the ball.

Solution **(a)** According to the impulse–momentum theorem, the impulse $\vec{J}$ applied to the ball is

$$\vec{J} = m\vec{v}_f - m\vec{v}_0$$

$$= \underbrace{(0.14\text{ kg})(+58\text{ m/s})}_{\text{Final momentum}} - \underbrace{(0.14\text{ kg})(-38\text{ m/s})}_{\text{Initial momentum}}$$

$$= \boxed{+13.4\text{ kg}\cdot\text{m/s}}$$

Problem-solving insight: Momentum is a vector and, as such, has a magnitude and a direction. For motion in one dimension, be sure to indicate the direction by assigning a plus or minus sign to it, as in this example.

(b) Now that the impulse is known, the contact time can be used in Equation 7.1 to find the average force $\vec{F}$ exerted by the bat on the ball:

$$\vec{F} = \frac{\vec{J}}{\Delta t} = \frac{+13.4\text{ kg}\cdot\text{m/s}}{1.6 \times 10^{-3}\text{ s}}$$

$$= \boxed{+8400\text{ N}}$$

The force is positive, indicating that it points opposite to the velocity of the approaching ball. A force of 8400 N corresponds to 1900 lb, such a large value being necessary to change the ball's momentum during the brief contact time.

Figure 7.3 When a bat hits a ball, an average force $\vec{F}$ is applied to the ball by the bat. As a result, the ball's velocity changes from an initial value of $\vec{v}_0$ (top drawing) to a final value of $\vec{v}_f$ (bottom drawing).

This karate expert applies an impulse to the stack of concrete bricks. As a result, the momentum of the stack changes as the bricks break. (Photodisc/Fotosearch, LLC)

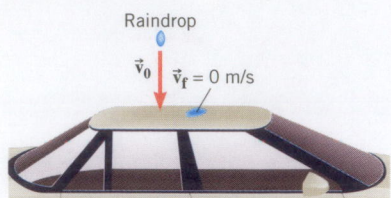

Figure 7.4 A raindrop falling on a car roof has an initial velocity of $\vec{v}_0$ just before striking the roof. The final velocity of the raindrop is $\vec{v}_f = 0$ m/s, because it comes to rest on the roof.

 Example 2   A Rainstorm

During a storm, rain comes straight down with a velocity of $\vec{v}_0 = -15$ m/s and hits the roof of a car perpendicularly (see Figure 7.4). The mass of rain per second that strikes the car roof is 0.060 kg/s. Assuming that the rain comes to rest upon striking the car ($\vec{v}_f = 0$ m/s), find the average force exerted by the rain on the roof.

Reasoning This example differs from Example 1 in an important way. Example 1 gives information about the ball and asks for the force applied to the ball. In contrast, the present example gives information about the rain but asks for the force acting on the roof. However, the force exerted on the roof by the rain and the force exerted on the rain by the roof have equal magnitudes and opposite directions, according to Newton's law of action and reaction (see Section 4.5). Thus, we will find the force exerted on the rain and then apply the law of action and reaction to obtain the force on the roof. Two forces act on the rain while it impacts with the roof: the average force $\vec{\vec{F}}$ exerted by the roof and the weight of the rain. These two forces constitute the net average force. By comparison, however, the force $\vec{\vec{F}}$ is much greater than the weight, so we may neglect the weight. Thus, the net average force becomes equal to $\vec{\vec{F}}$, or $\Sigma\vec{\vec{F}} = \vec{\vec{F}}$. The value of $\vec{\vec{F}}$ can be obtained by applying the impulse–momentum theorem to the rain.

Solution The average force $\vec{\vec{F}}$ needed to reduce the rain's velocity from $\vec{v}_0 = -15$ m/s to $\vec{v}_f = 0$ m/s is given by Equation 7.4 as

$$\vec{\vec{F}} = \frac{m\vec{v}_f - m\vec{v}_0}{\Delta t} = -\left(\frac{m}{\Delta t}\right)\vec{v}_0$$

The term $m/\Delta t$ is the mass of rain per second that strikes the roof, so that $m/\Delta t = 0.060$ kg/s. Thus, the average force exerted on the rain by the roof is

$$\vec{\vec{F}} = -(0.060 \text{ kg/s})(-15 \text{ m/s}) = +0.90 \text{ N}$$

This force is in the positive or upward direction, which is reasonable since the roof exerts an upward force on each falling drop in order to bring it to rest. According to the action–reaction law, the force exerted on the roof by the rain also has a magnitude of 0.90 N but points downward: Force on roof = $\boxed{-0.90 \text{ N}}$.

As you reason through problems such as those in Examples 1 and 2, take advantage of the impulse–momentum theorem. It is a powerful statement that can lead to significant insights. The following Conceptual Example further illustrates its use.

 Conceptual Example 3   Hailstones Versus Raindrops

In Example 2 rain is falling on the roof of a car and exerts a force on it. Instead of rain, suppose hail is falling. The hail comes straight down at a mass rate of $m/\Delta t = 0.060$ kg/s and an initial velocity of $\vec{v}_0 = -15$ m/s and strikes the roof perpendicularly, just as the rain does in Example 2. However, unlike rain, hail usually does not come to rest after striking a surface. Instead, the hailstones bounce off the roof of the car. If hail fell instead of rain, would the force on the roof be (a) smaller than, (b) equal to, or (c) greater than that calculated in Example 2?

Reasoning The raindrops and the hailstones fall in the same manner. That is, both fall with the same initial velocity and mass rate, and both strike the roof perpendicularly. However, there is an important difference: the raindrops come to rest (see Figure 7.4) after striking the roof, whereas the hailstones bounce upward (see Figure 7.5). According to the impulse–momentum theorem (Equation 7.4), the impulse that acts on an object is equal to the change in the object's momentum. This change is $m\vec{v}_f - m\vec{v}_0 = m\Delta\vec{v}$ and is proportional to the change $\Delta\vec{v}$ in the velocity. For a raindrop, the change in velocity is from $\vec{v}_0$ (downward) to zero. For a hailstone, the change is from $\vec{v}_0$ (downward) to $\vec{v}_f$ (upward). Thus, a raindrop and a hailstone experience different changes in velocity, and, hence, different changes in momentum and different impulses.

Answers (a) and (b) are incorrect. The change $\Delta\vec{v}$ in the velocity of a raindrop is smaller than that of a hailstone, since a raindrop does not rebound after striking the roof. According to the impulse–momentum theorem, a smaller impulse acts on a raindrop. But impulse is the product of the average force and the time interval Δt. Since the same amount of mass falls in the same time interval in either case, Δt is the same for both a raindrop and a hailstone. The smaller impulse

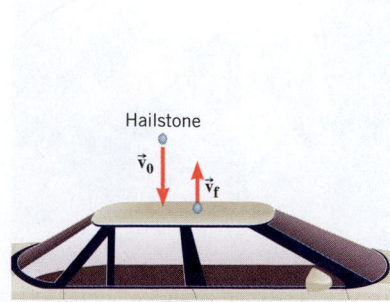

Figure 7.5 Hailstones have a downward velocity of $\vec{v}_0$ just before striking this car roof. They rebound with an upward velocity of $\vec{v}_f$.

acting on a raindrop, then, means that the car roof exerts a smaller force on it. By Newton's third law, this implies that a raindrop, not a hailstone, exerts a smaller force on the roof.

Answer (c) is correct. A hailstone experiences a larger change in momentum than does a raindrop, since a hailstone rebounds after striking the roof. Therefore, according to the impulse–momentum theorem, a greater impulse acts on a hailstone. An impulse is the product of the average force and the time interval Δt. Since the same amount of mass falls in the same time interval in either case, Δt is the same for a hailstone as for a raindrop. The greater impulse acting on a hailstone means that the car roof exerts a greater force on a hailstone than on a raindrop. According to Newton's action–reaction law (see Section 4.5) then, the car roof experiences a greater force from the hail than from the rain.

Related Homework: *Problems 3, 9*

✓ CHECK YOUR UNDERSTANDING

(The answers are given at the end of the book.)

1. Two identical automobiles have the same speed, one traveling east and one traveling west. Do these cars have the same momentum?

2. In Times Square in New York City, people celebrate on New Year's Eve. Some just stand around, but many move about randomly. Consider a group consisting of all of these people. Approximately, what is the total linear momentum of this group at any given instant?

3. Two objects have the same momentum. Do the velocities of these objects necessarily have **(a)** the same directions and **(b)** the same magnitudes?

4. (a) Can a single object have a kinetic energy but no momentum? **(b)** Can a group of two or more objects have a total kinetic energy that is not zero but a total momentum that is zero?

5. Suppose you are standing on the edge of a dock and jump straight down. If you land on sand your stopping time is much shorter than if you land on water. Using the impulse–momentum theorem as a guide, determine which one of the following statements is correct. **(a)** In bringing you to a halt, the sand exerts a greater impulse on you than does the water. **(b)** In bringing you to a halt, the sand and the water exert the same impulse on you, but the sand exerts a greater average force. **(c)** In bringing you to a halt, the sand and the water exert the same impulse on you, but the sand exerts a smaller average force.

6. An airplane is flying horizontally with a constant momentum during a time interval Δt. **(a)** Is there a *net* impulse acting on the plane during this time? Use the impulse–momentum theorem to guide your thinking. **(b)** In the horizontal direction, both the thrust generated by the engines and air resistance act on the plane. Considering your answer to part **(a)**, how is the impulse of the thrust related (in magnitude and direction) to the impulse of the force due to the air resistance?

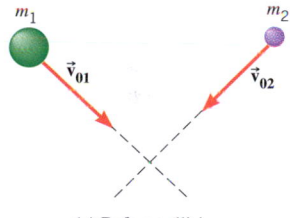

(a) Before collision

7.2 THE PRINCIPLE OF CONSERVATION OF LINEAR MOMENTUM

It is worthwhile to compare the impulse–momentum theorem to the work–energy theorem discussed in Chapter 6. The impulse–momentum theorem states that the impulse produced by a net force is equal to the change in the object's momentum, while the work–energy theorem states that the work done by a net force is equal to the change in the object's kinetic energy. The work–energy theorem leads directly to the principle of conservation of mechanical energy (see Figure 6.14), and, as we will shortly see, the impulse–momentum theorem also leads to a conservation principle, known as the conservation of linear momentum.

We begin by applying the impulse–momentum theorem to a midair collision between two objects. The two objects (masses m_1 and m_2) are approaching each other with initial velocities $\vec{v}_{01}$ and $\vec{v}_{02}$, as Figure 7.6*a* shows. The collection of objects being studied is referred to as the "system." In this case, the system contains only the two objects. They interact during the collision in part *b* and then depart with the final velocities $\vec{v}_{f1}$ and $\vec{v}_{f2}$ shown in part *c*. Because of the collision, the initial and final velocities are not the same.

Two types of forces act on the system:

1. ***Internal forces:*** Forces that the objects within the system exert on each other.

2. ***External forces:*** Forces exerted on the objects by agents external to the system.

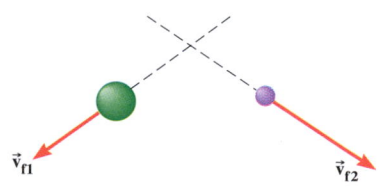

(b) During collision

(c) After collision

Figure 7.6 (*a*) The velocities of the two objects before the collision are $\vec{v}_{01}$ and $\vec{v}_{02}$. (*b*) During the collision, each object exerts a force on the other. These forces are $\vec{F}_{12}$ and $\vec{F}_{21}$. (*c*) The velocities after the collision are $\vec{v}_{f1}$ and $\vec{v}_{f2}$.

During the collision in Figure 7.6b, $\vec{\mathbf{F}}_{12}$ is the force exerted on object 1 by object 2, while $\vec{\mathbf{F}}_{21}$ is the force exerted on object 2 by object 1. These forces are action–reaction forces that are equal in magnitude but opposite in direction, so $\vec{\mathbf{F}}_{12} = -\vec{\mathbf{F}}_{21}$. They are internal forces, since they are forces that the two objects within the system exert on each other. The force of gravity also acts on the objects, their weights being $\vec{\mathbf{W}}_1$ and $\vec{\mathbf{W}}_2$. These weights, however, are external forces, because they are applied by the earth, which is outside the system. Friction and air resistance would also be considered external forces, although these forces are ignored here for the sake of simplicity. The impulse–momentum theorem, as applied to each object, gives the following results:

Object 1

$$(\underbrace{\vec{\mathbf{W}}_1}_{\substack{\text{External} \\ \text{force}}} + \underbrace{\vec{\mathbf{F}}_{12}}_{\substack{\text{Internal} \\ \text{force}}})\,\Delta t = m_1\vec{\mathbf{v}}_{f1} - m_1\vec{\mathbf{v}}_{01}$$

Object 2

$$(\underbrace{\vec{\mathbf{W}}_2}_{\substack{\text{External} \\ \text{force}}} + \underbrace{\vec{\mathbf{F}}_{21}}_{\substack{\text{Internal} \\ \text{force}}})\,\Delta t = m_2\vec{\mathbf{v}}_{f2} - m_2\vec{\mathbf{v}}_{02}$$

Adding these equations produces a single result for the system as a whole:

$$(\underbrace{\vec{\mathbf{W}}_1 + \vec{\mathbf{W}}_2}_{\substack{\text{External} \\ \text{forces}}} + \underbrace{\vec{\mathbf{F}}_{12} + \vec{\mathbf{F}}_{21}}_{\substack{\text{Internal} \\ \text{forces}}})\,\Delta t = \underbrace{(m_1\vec{\mathbf{v}}_{f1} + m_2\vec{\mathbf{v}}_{f2})}_{\substack{\text{Total final} \\ \text{momentum } \vec{\mathbf{P}}_f}} - \underbrace{(m_1\vec{\mathbf{v}}_{01} + m_2\vec{\mathbf{v}}_{02})}_{\substack{\text{Total initial} \\ \text{momentum } \vec{\mathbf{P}}_0}}$$

On the right side of this equation, the quantity $m_1\vec{\mathbf{v}}_{f1} + m_2\vec{\mathbf{v}}_{f2}$ is the vector sum of the final momenta for each object, or the total final momentum $\vec{\mathbf{P}}_f$ of the system. Likewise, the quantity $m_1\vec{\mathbf{v}}_{01} + m_2\vec{\mathbf{v}}_{02}$ is the total initial momentum $\vec{\mathbf{P}}_0$. Therefore, the result above becomes

$$\left(\begin{array}{c} \textbf{Sum of average} \\ \textbf{external forces} \end{array} + \begin{array}{c} \textbf{Sum of average} \\ \textbf{internal forces} \end{array}\right)\Delta t = \vec{\mathbf{P}}_f - \vec{\mathbf{P}}_0 \qquad (7.5)$$

The advantage of the internal/external force classification is that the internal forces always add together to give zero, as a consequence of Newton's law of action–reaction; $\vec{\mathbf{F}}_{12} = -\vec{\mathbf{F}}_{21}$ so that $\vec{\mathbf{F}}_{12} + \vec{\mathbf{F}}_{21} = 0$. Cancellation of the internal forces occurs no matter how many parts there are to the system and allows us to ignore the internal forces, as Equation 7.6 indicates:

$$\textbf{(Sum of average external forces)}\ \Delta t = \vec{\mathbf{P}}_f - \vec{\mathbf{P}}_0 \qquad (7.6)$$

We developed this result with gravity as the only external force. But, in general, the sum of the external forces on the left includes *all* external forces.

With the aid of Equation 7.6, it is possible to see how the conservation of linear momentum arises. Suppose that the sum of the external forces is zero. A system for which this is true is called an ***isolated system***. Then Equation 7.6 indicates that

$$0 = \vec{\mathbf{P}}_f - \vec{\mathbf{P}}_0 \qquad \text{or} \qquad \vec{\mathbf{P}}_f = \vec{\mathbf{P}}_0 \qquad (7.7a)$$

In other words, the final total momentum of the isolated system after the objects in Figure 7.6 collide is the same as the initial total momentum.* Explicitly writing out the final and initial momenta for the two-body collision, we obtain for Equation 7.7a that

$$\underbrace{m_1\vec{\mathbf{v}}_{f1} + m_2\vec{\mathbf{v}}_{f2}}_{\vec{\mathbf{P}}_f} = \underbrace{m_1\vec{\mathbf{v}}_{01} + m_2\vec{\mathbf{v}}_{02}}_{\vec{\mathbf{P}}_0} \qquad (7.7b)$$

---

*Technically, the initial and final momenta are equal when the impulse of the sum of the external forces is zero—that is, when the left-hand side of Equation 7.6 is zero. Somtimes, however, the initial and final momenta are very nearly equal even when the sum of the external forces is not zero. This occurs when the time Δt during which the forces act is so short that it is effectively zero. Then, the left-hand side of Equation 7.6 is approximately zero.

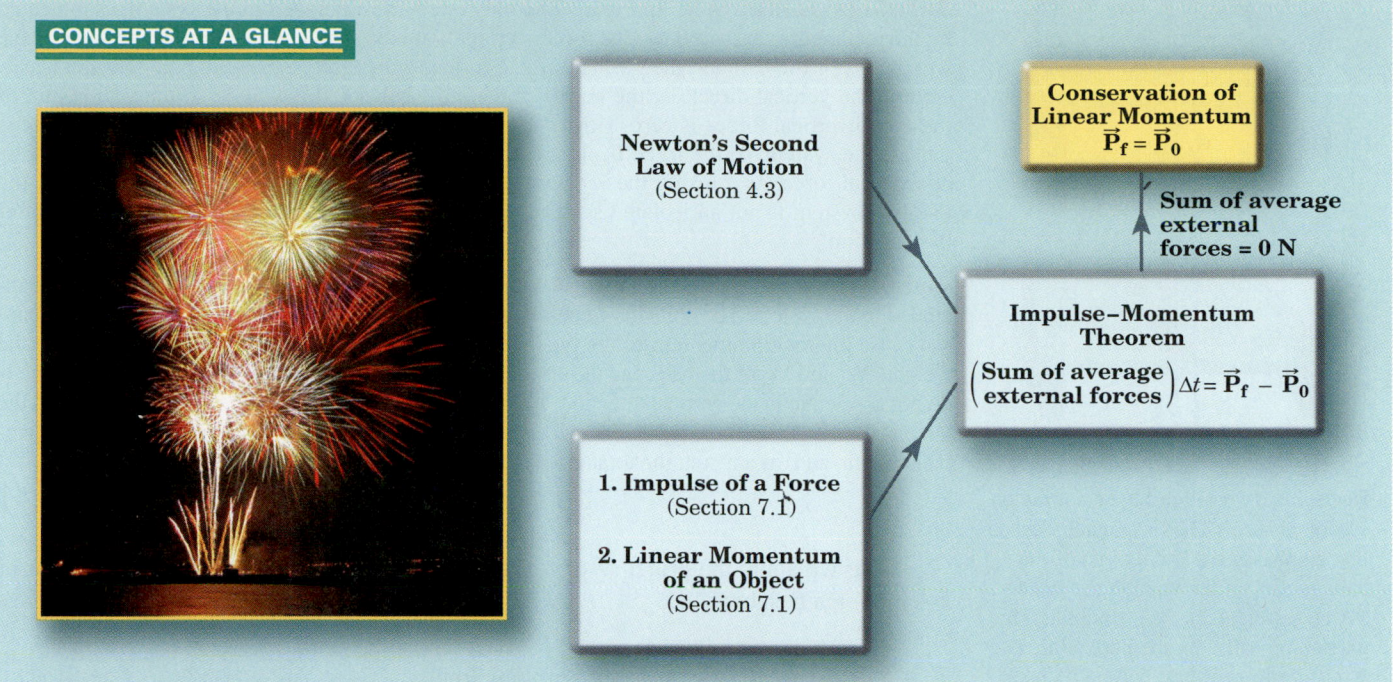

CONCEPTS AT A GLANCE

Newton's Second
Law of Motion
(Section 4.3)

Conservation of
Linear Momentum
$\vec{P}_f = \vec{P}_0$

Sum of average
external
forces = 0 N

Impulse–Momentum
Theorem

$\left(\begin{array}{c}\text{Sum of average}\\\text{external forces}\end{array}\right)\Delta t = \vec{P}_f - \vec{P}_0$

1. Impulse of a Force
(Section 7.1)

2. Linear Momentum
of an Object
(Section 7.1)

Figure 7.7 CONCEPTS AT A GLANCE
The impulse–momentum theorem leads
to the principle of conservation of
linear momentum when the sum of the
average external forces acting on a
system is zero. To the extent that this
principle applies to the fireworks
explosions in the photograph, it follows
that the total linear momentum of a
rocket before it explodes is equal to the
total linear momentum of the fragments
that exist after it explodes. (Norbert
Schwerin/The Image Works)

This result is an example of a general principle known as the **principle of conservation of linear momentum.**

PRINCIPLE OF CONSERVATION OF LINEAR MOMENTUM

The total linear momentum of an isolated system remains constant (is conserved). An isolated system is one for which the vector sum of the average external forces acting on the system is zero.

▶ **CONCEPTS AT A GLANCE**   The Concepts-at-a-Glance chart in Figure 7.7 presents an overview of how concepts that we have studied lead to the conservation of linear momentum. In the middle, the chart shows that the origin of the impulse–momentum theorem lies in Newton's second law and the ideas of the impulse of a force and the linear momentum of an object. The theorem is stated at the right of the chart in the form (**Sum of average external forces**) $\Delta t = \vec{P}_f - \vec{P}_0$. If the sum of the average external forces is zero, then the final and initial total momenta are equal, or $\vec{P}_f = \vec{P}_0$, which is the mathematical statement of the conservation of linear momentum. ◀

The conservation-of-momentum principle applies to a system containing any number of objects, regardless of the internal forces, provided the system is isolated. Whether the system is isolated depends on whether the vector sum of the external forces is zero. Judging whether a force is internal or external depends on which objects are included in the system, as Conceptual Example 4 illustrates.

 Conceptual Example 4   Is the Total Momentum Conserved?

Imagine two balls colliding on a billiard table that is friction-free. Using the momentum-conservation principle as a guide, decide which statement is correct: **(a)** The total momentum of the system that contains only one of the two balls is the same before and after the collision. **(b)** The total momentum of the system that contains both of the two balls is the same before and after the collision.

Reasoning   The total momentum of an isolated system is the same before and after the collision; in such a situation the total momentum is said to be conserved. An isolated system is one for which the vector sum of the average external forces acting on the system is zero. To decide whether statement (a) or (b) is correct, we need to examine the one-ball and two-ball systems and see if they are, in fact, isolated.

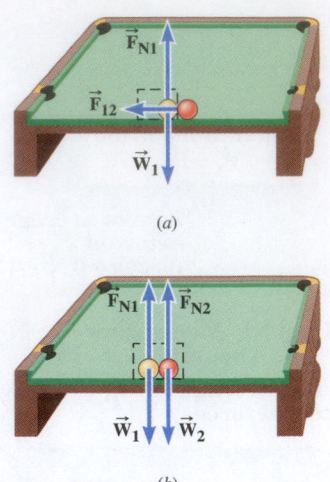

(a)

(b)

Figure 7.8 Two billiard balls collide on a pool table. (a) The rectangular dashed box emphasizes that only one of the balls (ball 1) is included in the system; $\vec{W}_1$ is its weight, $\vec{F}_{N1}$ is the normal force exerted on ball 1 by the pool table, and $\vec{F}_{12}$ is the force exerted on ball 1 by ball 2. (b) Now both balls are included in the system; $\vec{W}_2$ is the weight of ball 2, and $\vec{F}_{N2}$ is the nromal force acting on it.

Problem-solving insight

The conservation of linear momentum is applicable only when the net external force acting on the system is zero. Therefore, the first step in applying momentum conservation is to be sure that the net external force is zero.

Answer (a) is incorrect. In Figure 7.8a only one ball is included in the system, as indicated by the rectangular dashed box. The forces acting on this system are all external and include the weight $\vec{W}_1$ of the ball and the normal force $\vec{F}_{N1}$ due to the table. Since the ball does not accelerate in the vertical direction, the normal force must balance the weight, so the vector sum of these two vertical forces is zero. However, there is a third external force to consider. Ball 2 is outside the system, so the force $\vec{F}_{12}$ that it applies to the system (ball 1) during the collision is an external force. As a result, the vector sum of the three external forces is not zero. Thus, the one-ball system is not an isolated system, and its momentum is not the same before and after the collision.

Answer (b) is correct. The rectangular dashed box in Figure 7.8b shows that both balls are included in the system. The collision forces are not shown, because they are internal forces and cannot cause the total momentum of the two-ball system to change. The external forces include the weights $\vec{W}_1$ and $\vec{W}_2$ of the balls and the upward-pointing normal forces $\vec{F}_{N1}$ and $\vec{F}_{N2}$. Since the balls do not accelerate in the vertical direction, $\vec{F}_{N1}$ balances $\vec{W}_1$, and $\vec{F}_{N2}$ balances $\vec{W}_2$. Furthermore, the table is friction-free. Thus, there is no net external force to change the total momentum of the two-ball system, and, as a result, the total momentum is the same before and after the collision.

Next, we apply the principle of conservation of linear momentum to the problem of assembling a freight train.

Example 5 Assembling a Freight Train

A freight train is being assembled in a switching yard, and Figure 7.9 shows two boxcars. Car 1 has a mass of $m_1 = 65 \times 10^3$ kg and moves at a velocity of $v_{01} = +0.80$ m/s. Car 2, with a mass of $m_2 = 92 \times 10^3$ kg and a velocity of $v_{02} = +1.3$ m/s, overtakes car 1 and couples to it. Neglecting friction, find the common velocity v_f of the cars after they become coupled.

Reasoning The two boxcars constitute the system. The sum of the external forces acting on the system is zero, because the weight of each car is balanced by a corresponding normal force, and friction is being neglected. Thus, the system is isolated, and the principle of conservation of linear momentum applies. The coupling forces that each car exerts on the other are internal forces and do not affect the applicability of this principle.

Solution Momentum conservation indicates that

$$\underbrace{(m_1 + m_2)v_f}_{\substack{\text{Total momentum} \\ \text{after collision}}} = \underbrace{m_1v_{01} + m_2v_{02}}_{\substack{\text{Total momentum} \\ \text{before collision}}}$$

This equation can be solved for v_f, the common velocity of the two cars after the collision:

$$v_f = \frac{m_1v_{01} + m_2v_{02}}{m_1 + m_2}$$

$$= \frac{(65 \times 10^3 \text{ kg})(0.80 \text{ m/s}) + (92 \times 10^3 \text{ kg})(1.3 \text{ m/s})}{(65 \times 10^3 \text{ kg} + 92 \times 10^3 \text{ kg})} = \boxed{+1.1 \text{ m/s}}$$

In the previous example it can be seen that the velocity of car 1 increases, while the velocity of car 2 decreases as a result of the collision. The acceleration and deceleration arise at the moment the cars become coupled, because the cars exert internal forces on each other. The powerful feature of the momentum-conservation principle is that it allows us to determine the changes in velocity without knowing what the internal forces are. Example 6 further illustrates this feature.

Figure 7.9 (a) The boxcar on the left eventually catches up with the other boxcar and (b) couples to it. The coupled cars move together with a common velocity after the collision.

(a) Before coupling

(b) After coupling

 **Example 6**   Ice Skaters

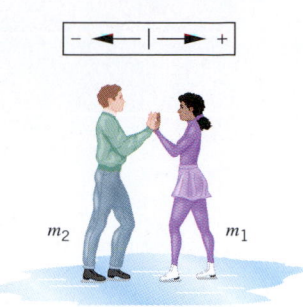

Starting from rest, two skaters push off against each other on smooth level ice, where friction is negligible. As Figure 7.10*a* shows, one is a woman (m_1 = 54 kg), and one is a man (m_2 = 88 kg). Part *b* of the drawing shows that the woman moves away with a velocity of v_{f1} = +2.5 m/s. Find the "recoil" velocity v_{f2} of the man.

(*a*) Before pushoff

Reasoning   For a system consisting of the two skaters on level ice, the sum of the external forces is zero. This is because the weight of each skater is balanced by a corresponding normal force and friction is negligible. The skaters, then, constitute an isolated system, and the principle of conservation of linear momentum applies. We expect the man to have a smaller recoil speed for the following reason: The internal forces that the man and woman exert on each other during pushoff have equal magnitudes but opposite directions, according to Newton's action–reaction law. The man, having the larger mass, experiences a smaller acceleration according to Newton's second law. Hence, he acquires a smaller recoil speed.

Solution   The total momentum of the skaters before they push on each other is zero, since they are at rest. Momentum conservation requires that the total momentum remains zero after the skaters have separated, as in part *b* of the drawing:

$$\underbrace{m_1 v_{f1} + m_2 v_{f2}}_{\substack{\text{Total momentum} \\ \text{after pushing}}} = \underbrace{0}_{\substack{\text{Total momentum} \\ \text{before pushing}}}$$

(*b*) After pushoff

Figure 7.10   (*a*) In the absence of friction, two skaters pushing on each other constitute an isolated system. (*b*) As the skaters move away, the total linear momentum of the system remains zero, which is what it was initially.

Solving for the recoil velocity of the man gives

$$v_{f2} = \frac{-m_1 v_{f1}}{m_2} = \frac{-(54 \text{ kg})(+2.5 \text{ m/s})}{88 \text{ kg}} = \boxed{-1.5 \text{ m/s}}$$

The minus sign indicates that the man moves to the left in the drawing. After the skaters separate, the total momentum of the system remains zero, because momentum is a vector quantity, and the momenta of the man and the woman have equal magnitudes but opposite directions.

It is important to realize that the total linear momentum may be conserved even when the kinetic energies of the individual parts of a system change. In Example 6, for instance, the initial kinetic energy is zero, since the skaters are stationary. But after they push off, the skaters are moving, so each has kinetic energy. The kinetic energy changes because work is done by the internal force that each skater exerts on the other. However, internal forces cannot change the total linear momentum of a system, since the total linear momentum of an isolated system is conserved in the presence of such forces.

Problem-solving insight

When applying the principle of conservation of linear momentum, we have been following a definite reasoning strategy that is summarized as follows:

REASONING STRATEGY

Applying the Principle of Conservation of Linear Momentum

1. Decide which objects are included in the system.

2. Relative to the system that you have chosen, identify the internal forces and the external forces.

3. Verify that the system is isolated. In other words, verify that the sum of the external forces applied to the system is zero. Only if this sum is zero can the conservation principle be applied. If the sum of the average external forces is not zero, consider a different system for analysis.

4. Set the total final momentum of the isolated system equal to the total initial momentum. Remember that linear momentum is a vector. If necessary, apply the conservation principle separately to the various vector components.

✓ **CHECK YOUR UNDERSTANDING**

(*The answers are given at the end of the book.*)

7. An object slides along the surface of the earth and slows down because of kinetic friction. If the object alone is considered as the system, the kinetic frictional force must be identified as an external force that, according to Equation 7.4, decreases the momentum of the system. **(a)** If *both* the object and the earth are considered to be the system, is the force of kinetic friction still an external force? **(b)** Can the frictional force change the total linear momentum of the two-body system?

8. A satellite explodes in outer space, far from any other body, sending thousands of pieces in all directions. Is the linear momentum of the satellite before the explosion less than, equal to, or greater than, the total linear momentum of all the pieces after the explosion?

9. On a distant asteroid, a large catapult is used to throw chunks of stone into space. Could such a device be used as a propulsion system to move the asteroid closer to the earth?

10. A canoe with two people aboard is coasting with an initial momentum of $+110$ kg · m/s. Then, one of the people (person 1) dives off the back of the canoe. During this time, the net average external force acting on the system (the canoe and the two people) is zero. The table lists four possibilities for the final momentum of person 1 and the final momentum of person 2 plus the canoe, immediately after person 1 dives off. Only one possibility could be correct. Which one is it?

Final Momenta

	Person 1	Person 2 and Canoe
(a)	-60 kg · m/s	$+170$ kg · m/s
(b)	-30 kg · m/s	$+110$ kg · m/s
(c)	-40 kg · m/s	-70 kg · m/s
(d)	$+80$ kg · m/s	-30 kg · m/s

11. You are a passenger on a jetliner that is flying at a constant velocity. You get up from your seat and walk toward the front of the plane. Because of this action, your forward momentum increases. Does the forward momentum of the plane itself decrease, remain the same, or increase?

12. An ice boat is coasting on a frozen lake. Friction between the ice and the boat is negligible, and so is air resistance. Nothing is propelling the boat. From a bridge someone jumps straight down into the boat, which continues to coast straight ahead. **(a)** Does the total horizontal momentum of the boat plus the jumper change? **(b)** Does the speed of the boat itself increase, decrease, or remain the same?

13. Concept Simulation 7.1 at **www.wiley.com/college/cutnell** reviews the concepts that are pertinent in this question. In movies, Superman hovers in midair, grabs a villain by the neck, and throws him forward. Superman, however, remains stationary. This is not possible, because it violates which one or more of the following? **(a)** The law of conservation of energy **(b)** Newton's second law **(c)** Newton's third law **(d)** The principle of conservation of linear momentum

14. The energy released by the exploding gunpowder in a cannon propels the cannonball forward. Simultaneously, the cannon recoils. The mass of the cannonball is less than that of the cannon. Which has the greater kinetic energy, the launched cannonball or the recoiling cannon? Assume that momentum conservation applies.

7.3 COLLISIONS IN ONE DIMENSION

As discussed in the last section, the total linear momentum is conserved when two objects collide, provided they constitute an isolated system. When the objects are atoms or subatomic particles, the total kinetic energy of the system is often conserved also. In other words, the total kinetic energy of the particles before the collision equals the total kinetic energy of the particles after the collision, so that kinetic energy gained by one particle is lost by another.

In contrast, when two macroscopic objects collide, such as two cars, the total kinetic energy after the collision is generally less than that before the collision. Kinetic energy is lost mainly in two ways: (1) It can be converted into heat because of friction, and (2) it is spent in creating permanent distortion or damage, as in an automobile collision. With very hard objects, such as a solid steel ball and a marble floor, the permanent distortion suffered upon collision is much smaller than with softer objects and, consequently, less kinetic energy is lost.

Collisions are often classified according to whether the total kinetic energy changes during the collision:

1. **Elastic collision:** One in which the total kinetic energy of the system after the collision is equal to the total kinetic energy before the collision.

2. **Inelastic collision:** One in which the total kinetic energy of the system is *not* the same before and after the collision; if the objects stick together after colliding, the collision is said to be *completely inelastic*.

The boxcars coupling together in Figure 7.9 provide an example of a completely inelastic collision. When a collision is completely inelastic, the greatest amount of kinetic energy is lost. Example 7 shows how one particular elastic collision is described using the conservation of linear momentum and the fact that no kinetic energy is lost.

Problem-solving insight

As long as the net external force is zero, the conservation of linear momentum applies to any type of collision. This is true whether the collision is elastic or inelastic.

ANALYZING MULTIPLE-CONCEPT PROBLEMS

Example 7 A Head-On Collision

Figure 7.11 illustrates an elastic head-on collision between two balls. One ball has a mass of $m_1 = 0.250$ kg and an initial velocity of $+5.00$ m/s. The other has a mass of $m_2 = 0.800$ kg and is initially at rest. No external forces act on the balls. What are the velocities of the balls after the collision?

Reasoning Three facts will guide our solution. The first is that the collision is elastic, so that the kinetic energy of the two-ball system is the same before and after the balls collide. The second fact is that the collision occurs head-on. This means that the velocities before and after the balls collide all point along the same line. In other words, the collision occurs in one dimension. Lastly, no external forces act on the balls, with the result that the two-ball system is isolated and its total linear momentum is conserved. We expect that ball 1, having the smaller mass, will rebound to the left after striking ball 2, which is more massive. Ball 2 will be driven to the right in the process.

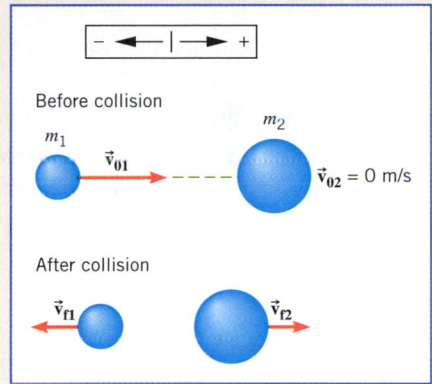

Figure 7.11 A 0.250-kg ball, traveling with an initial velocity of $v_{01} = +5.00$ m/s, undergoes an elastic collision with a 0.800-kg ball that is initially at rest.

Knowns and Unknowns The data for this problem are given in the following table:

Description	Symbol	Value	Comment
Explicit Data			
Mass of ball 1	m_1	0.250 kg	
Initial velocity of ball 1	v_{01}	+5.00 m/s	Before collision occurs.
Mass of ball 2	m_2	0.800 kg	
Implicit Data			
Initial velocity of ball 2	v_{02}	0 m/s	Before collision occurs, ball is initially at rest.
Unknown Variables			
Final velocity of ball 1	v_{f1}	?	After collision occurs.
Final velocity of ball 2	v_{f2}	?	After collision occurs.

Modeling the Problem

STEP 1 Elastic Collision Since the collision is elastic, the kinetic energy of the two-ball system is the same before and after the balls collide:

$$\underbrace{\tfrac{1}{2}m_1 v_{f1}^2 + \tfrac{1}{2}m_2 v_{f2}^2}_{\substack{\text{Total kinetic energy}\\ \text{after collision}}} = \underbrace{\tfrac{1}{2}m_1 v_{01}^2 + 0}_{\substack{\text{Total kinetic energy}\\ \text{before collision}}}$$

Continued

Here we have utilized the fact that ball 2 is at rest before the collision. Thus, its initial velocity v_{02} is zero and so is its initial kinetic energy. Solving this equation for v_{f1}^2, the square of the velocity of ball 1 after the collision, gives Equation 1 at the right. To use this result, we need a value for v_{f2}, which we will obtain in Step 2.

$$v_{f1}^2 = v_{01}^2 - \frac{m_2}{m_1}v_{f2}^2 \qquad (1)$$

STEP 2 **Conservation of Linear Momentum** The total linear momentum of the two-ball system is conserved, because no external forces act on the system; momentum conservation does not depend on whether or not the collision is elastic. Conservation of linear momentum indicates that

$$\underbrace{m_1 v_{f1} + m_2 v_{f2}}_{\substack{\text{Total momentum} \\ \text{after collision}}} = \underbrace{m_1 v_{01} + 0}_{\substack{\text{Total momentum} \\ \text{before collision}}}$$

We have again utilized the fact that ball 2 is at rest before the collision; since its initial velocity v_{02} is zero, so is its initial momentum. Solving the expression above for v_{f2} gives

$$\boxed{v_{f2} = \frac{m_1}{m_2}(v_{01} - v_{f1})}$$

$$v_{f1}^2 = v_{01}^2 - \frac{m_2}{m_1}v_{f2}^2 \qquad (1)$$

$$\boxed{v_{f2} = \frac{m_1}{m_2}(v_{01} - v_{f1})} \qquad (2)$$

which can be substituted into Equation 1, as shown at the right.

Solution Algebraically combining the results of each step, we find that

$$\overset{\text{STEP 1}}{\underset{\downarrow}{v_{f1}^2}} = v_{01}^2 - \frac{m_2}{m_1}v_{f2}^2 \overset{\text{STEP 2}}{\underset{\downarrow}{=}} v_{01}^2 - \frac{m_2}{m_1}\left[\frac{m_1}{m_2}(v_{01} - v_{f1})\right]^2$$

Solving for v_{f1} shows that

$$v_{f1} = \left(\frac{m_1 - m_2}{m_1 + m_2}\right)v_{01} = \left(\frac{0.250 \text{ kg} - 0.800 \text{ kg}}{0.250 \text{ kg} + 0.800 \text{ kg}}\right)(+5.00 \text{ m/s}) = \boxed{-2.62 \text{ m/s}} \qquad (7.8a)$$

Substituting the expression for v_{f1} into Equation 2 gives

$$v_{f2} = \left(\frac{2m_1}{m_1 + m_2}\right)v_{01} = \left[\frac{2(0.250 \text{ kg})}{0.250 \text{ kg} + 0.800 \text{ kg}}\right](+5.00 \text{ m/s}) = \boxed{+2.38 \text{ m/s}} \qquad (7.8b)$$

The negative value for v_{f1} indicates that ball 1 rebounds to the left after the collision in Figure 7.11, while the positive value for v_{f2} indicates that ball 2 moves to the right, as expected.

Related Homework: *Problems 29, 37, 46*

We can get a feel for an elastic collision by dropping a steel ball onto a hard surface, such as a marble floor. If the collision were elastic, the ball would rebound to its original height, as Figure 7.12a illustrates. In contrast, a partially deflated basketball exhibits little rebound from a relatively soft asphalt surface, as in part b, indicating that a fraction of the ball's kinetic energy is dissipated during the inelastic collision. The very deflated basketball in part c has no bounce at all, and a maximum amount of kinetic energy is lost during the completely inelastic collision.

Figure 7.12 (a) A hard steel ball would rebound to its original height after striking a hard marble surface if the collision were elastic. (b) A partially deflated basketball has little bounce on a soft asphalt surface. (c) A very deflated basketball has no bounce at all.

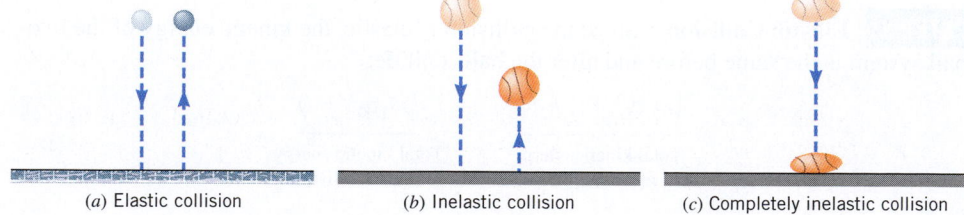

(a) Elastic collision   (b) Inelastic collision   (c) Completely inelastic collision

The next example illustrates a completely inelastic collision in a device called a "ballistic pendulum." This device can be used to measure the speed of a bullet.

ANALYZING MULTIPLE-CONCEPT PROBLEMS

Example 8 A Ballistic Pendulum

The physics of measuring the speed of a bullet.

A ballistic pendulum can be used to measure the speed of a projectile, such as a bullet. The ballistic pendulum shown in Figure 7.13a consists of a stationary 2.50-kg block of wood suspended by a wire of negligible mass. A 0.0100-kg bullet is fired into the block, and the block (with the bullet in it) swings to a maximum height of 0.650 m above the initial position (see part *b* of the drawing). Find the speed with which the bullet is fired, assuming that air resistance is negligible.

Reasoning The physics of the ballistic pendulum can be divided into two parts. The first is the completely inelastic collision between the bullet and the block. The total linear momentum of the system (block plus bullet) is conserved during the collision, because the suspension wire supports the system's weight, which means that the sum of the external forces acting on the system is nearly zero. The second part of the physics is the resulting motion of the block and bullet as they swing upward after the collision. As the system swings upward, the principle of conservation of mechanical energy applies, since nonconservative forces do no work (see Section 6.5). The tension force in the wire does no work because it acts perpendicular to the motion. Since air resistance is negligible, we can ignore the work it does. The conservation principles for linear momentum and mechanical energy provide the basis for our solution.

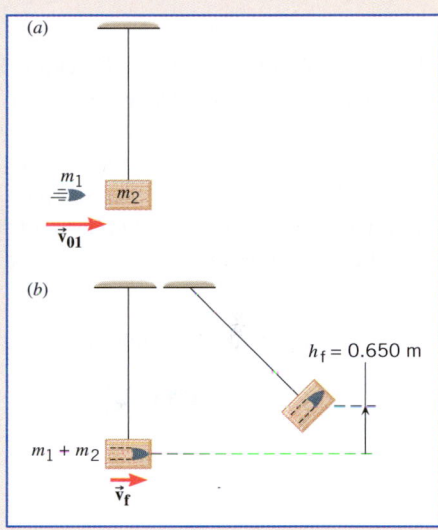

Figure 7.13 (*a*) A bullet approaches a ballistic pendulum. (*b*) The block and bullet swing upward after the collision.

Knowns and Unknowns The data for this problem are as follows:

Description	Symbol	Value	Comment
Explicit Data			
Mass of bullet	m_1	0.0100 kg	
Mass of block	m_2	2.50 kg	
Height to which block plus bullet swings	h_f	0.650 m	Maximum height of swing.
Implicit Data			
Initial velocity of block	v_{02}	0 m/s	Before collision, block is stationary.
Unknown Variable			
Speed with which bullet is fired	v_{01}	?	Before collision with block.

Modeling the Problem

STEP 1 Completely Inelastic Collision Just after the bullet collides with it, the block (with the bullet in it) has a speed v_f. Since linear momentum is conserved, the total momentum of the block–bullet system after the collision is the same as it is before the collision:

$$\underbrace{(m_1 + m_2)v_f}_{\substack{\text{Total momentum} \\ \text{after collision}}} = \underbrace{m_1 v_{01} + 0}_{\substack{\text{Total momentum} \\ \text{before collision}}}$$

Note that the block is at rest before the collision ($v_{02} = 0$ m/s), so its initial momentum is zero. Solving this equation for v_{01} gives Equation 1 at the right. To find a value for the speed v_f in this equation, we turn to Step 2.

$$v_{01} = \left(\frac{m_1 + m_2}{m_1}\right)v_f \qquad (1)$$

?

Continued

STEP 2 **Conservation of Mechanical Energy** The speed v_f immediately after the collision can be obtained from the maximum height h_f to which the system swings, by using the principle of conservation of mechanical energy:

$$\underbrace{(m_1 + m_2)gh_f}_{\substack{\text{Total mechanical energy} \\ \text{at top of swing, all} \\ \text{potential}}} = \underbrace{\tfrac{1}{2}(m_1 + m_2)v_f^2}_{\substack{\text{Total mechanical energy} \\ \text{at bottom of swing,} \\ \text{all kinetic}}}$$

This result can be solved for v_f to show that

$$v_f = \sqrt{2gh_f}$$

which can be substituted into Equation 1, as shown at the right. In applying the energy-conservation principle, it is tempting say that the total potential energy at the top of the swing is equal to the total kinetic energy of the bullet before it strikes the block $[(m_1 + m_2)gh_f = \tfrac{1}{2}m_1 v_{01}^2]$ and solve directly for v_{01}. This is incorrect, however, because the collision between the bullet and the block is inelastic, so that some of the bullet's initial kinetic energy is dissipated during the collision (due to friction and structural damage to the block and bullet).

$$v_{01} = \left(\frac{m_1 + m_2}{m_1}\right)v_f \qquad (1)$$

$$v_f = \sqrt{2gh_f}$$

Solution Algebraically combining the results of each step, we find that

$$\overset{\text{STEP 1}}{v_{01} =} \left(\frac{m_1 + m_2}{m_1}\right) \overset{\text{STEP 2}}{v_f =} \left(\frac{m_1 + m_2}{m_1}\right)\sqrt{2gh_f}$$

$$v_{01} = \left(\frac{0.0100\ \text{kg} + 2.50\ \text{kg}}{0.0100\ \text{kg}}\right)\sqrt{2(9.80\ \text{m/s}^2)(0.650\ \text{m})} = \boxed{+896\ \text{m/s}}$$

Related Homework: *Problem 55*

✓ CHECK YOUR UNDERSTANDING

(The answers are given at the end of the book.)

15. Two balls collide in a one-dimensional elastic collision. The two balls constitute a system, and the net external force acting on them is zero. The table shows four possible sets of values for the initial and final momenta of the two balls, as well as their initial and final kinetic energies. Only one set of values could be correct. Which set is it?

		Initial (Before Collision)		Final (After Collision)	
		Momentum	Kinetic Energy	Momentum	Kinetic Energy
(a)	Ball 1:	+4 kg · m/s	12 J	−5 kg · m/s	10 J
	Ball 2:	−3 kg · m/s	5 J	−1 kg · m/s	7 J
(b)	Ball 1:	+7 kg · m/s	22 J	+5 kg · m/s	18 J
	Ball 2:	+2 kg · m/s	8 J	+4 kg · m/s	15 J
(c)	Ball 1:	−5 kg · m/s	12 J	−6 kg · m/s	15 J
	Ball 2:	−8 kg · m/s	31 J	−9 kg · m/s	25 J
(d)	Ball 1:	+9 kg · m/s	25 J	+6 kg · m/s	18 J
	Ball 2:	+4 kg · m/s	15 J	+7 kg · m/s	22 J

16. In an elastic collision, is the kinetic energy of *each* object the same before and after the collision?

17. Concept Simulation 7.2 at **www.wiley.com/college/cutnell** illustrates the concepts that are involved in this question. Also review Multiple-Concept Example 7. Suppose two objects collide head on, as in Example 7, where initially object 1 (mass = m_1) is moving and object 2 (mass = m_2) is stationary. Now assume that they have the same mass, so $m_1 = m_2$. Which one of the following statements is true? **(a)** Both objects have the same velocity (magnitude and direction) after the collision. **(b)** Object 1 rebounds with one-half its initial speed, while object 2 moves to the right, as in Figure 7.11, with one-half the speed that object 1 had before the collision. **(c)** Object 1 stops completely, while object 2 acquires the same velocity (magnitude and direction) that object 1 had before the collision.

7.4 COLLISIONS IN TWO DIMENSIONS

The collisions discussed so far have been head-on, or one-dimensional, because the velocities of the objects all point along a single line before and after contact. Collisions often occur, however, in two or three dimensions. Figure 7.14 shows a two-dimensional case in which two balls collide on a horizontal frictionless table.

For the system consisting of the two balls, the external forces include the weights of the balls and the corresponding normal forces produced by the table. Since each weight is balanced by a normal force, the sum of the external forces is zero, and the total momentum of the system is conserved, as Equation 7.7b indicates. Momentum is a vector quantity, however, and in two dimensions the x and y components of the total momentum are conserved separately. In other words, Equation 7.7b is equivalent to the following two equations:

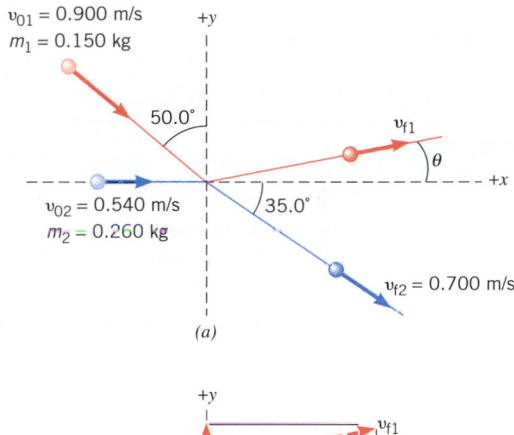

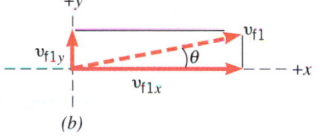

Figure 7.14 (*a*) Top view of two balls colliding on a horizontal frictionless table. (*b*) This part of the drawing shows the *x* and *y* components of the velocity of ball 1 after the collision.

x Component
$$\underbrace{m_1 v_{f1x} + m_2 v_{f2x}}_{P_{fx}} = \underbrace{m_1 v_{01x} + m_2 v_{02x}}_{P_{0x}} \qquad (7.9a)$$

y Component
$$\underbrace{m_1 v_{f1y} + m_2 v_{f2y}}_{P_{fy}} = \underbrace{m_1 v_{01y} + m_2 v_{02y}}_{P_{0y}} \qquad (7.9b)$$

Problem-solving insight: Momentum, being a vector quantity, has a magnitude and a direction. In two dimensions, take into account the direction by using vector components and assigning a plus or minus sign to each component.

These equations are written for a system that contains two objects. If a system contains more than two objects, a mass-times-velocity term must be included for each additional object on either side of Equations 7.9a and 7.9b. Example 9 shows how to deal with a two-dimensional collision when the total linear momentum of the system is conserved.

ANALYZING MULTIPLE-CONCEPT PROBLEMS

Example 9 A Collision in Two Dimensions

For the situation in Figure 7.14*a*, use momentum conservation to determine the magnitude and direction of the final velocity of ball 1 after the collision.

Reasoning We can obtain the magnitude and direction of the final velocity of ball 1 from the *x* and *y* components of this velocity. These components can be determined by using the momentum-conservation principle in its component form, as expressed in Equations 7.9a and 7.9b.

Continued

Knowns and Unknowns The following data are given in Figure 7.14a:

Description	Symbol	Value	Comment
Mass of ball 1	m_1	0.150 kg	
Magnitude of initial velocity of ball 1	v_{01}	0.900 m/s	Before collision occurs.
Directional angle of initial velocity of ball 1		50.0°	See Figure 7.14a.
Mass of ball 2	m_2	0.260 kg	
Magnitude of initial velocity of ball 2	v_{02}	0.540 m/s	Before collision occurs.
Magnitude of final velocity of ball 2	v_{f2}	0.700 m/s	After collision occurs.
Directional angle of final velocity of ball 2		35.0°	See Figure 7.14a.
Unknown Variables			
Magnitude of final velocity of ball 1	v_{f1}	?	After collision occurs.
Directional angle of final velocity of ball 1	θ	?	After collision occurs. See Figure 7.14a.

Modeling the Problem

STEP 1 **Vector Components** Figure 7.14b shows the final velocity v_{f1} of ball 1 oriented at an angle θ above the $+x$ axis. Also shown are the components v_{f1x} and v_{f1y}. The components are perpendicular, so that we can use the Pythagorean theorem and trigonometry and write Equations 1a and 1b at the right.

$$v_{f1} = \sqrt{v_{f1x}^2 + v_{f1y}^2} \qquad (1a)$$

$$\theta = \tan^{-1}\left(\frac{v_{f1y}}{v_{f1x}}\right) \qquad (1b)$$

STEP 2 **Conservation of Linear Momentum** Momentum is a vector quantity, so the conservation principle applies separately to the components of the momentum. Applying momentum conservation (Equation 7.9a) to the x direction we find that

x Component

$$\underbrace{(0.150 \text{ kg})v_{f1x}}_{\text{Ball 1, after}} + \underbrace{(0.260 \text{ kg})(0.700 \text{ m/s}) \cos 35.0°}_{\text{Ball 2, after}}$$

$$= \underbrace{(0.150 \text{ kg})(0.900 \text{ m/s}) \sin 50.0°}_{\text{Ball 1, before}} + \underbrace{(0.260 \text{ kg})(0.540 \text{ m/s})}_{\text{Ball 2, before}}$$

Applying momentum conservation (Equation 7.9b) to the y direction we find that

y Component

$$\underbrace{(0.150 \text{ kg})v_{f1y}}_{\text{Ball 1, after}} + \underbrace{(0.260 \text{ kg})[-(0.700 \text{ m/s}) \sin 35.0°]}_{\text{Ball 2, after}}$$

$$= \underbrace{(0.150 \text{ kg})[-(0.900 \text{ m/s}) \cos 50.0°]}_{\text{Ball 1, before}} + \underbrace{0}_{\text{Ball 2, before}}$$

$$v_{f1} = \sqrt{v_{f1x}^2 + v_{f1y}^2} \qquad (1a)$$

$$\boxed{v_{f1x} = +0.63 \text{ m/s}}$$

$$\boxed{v_{f1y} = +0.12 \text{ m/s}}$$

$$\theta = \tan^{-1}\left(\frac{v_{f1y}}{v_{f1x}}\right) \qquad (1b)$$

$$\boxed{v_{f1x} = +0.63 \text{ m/s}}$$

$$\boxed{v_{f1y} = +0.12 \text{ m/s}}$$

These equations can be solved to obtain values for the components v_{f1x} and v_{f1y}:

$$\boxed{v_{f1x} = +0.63 \text{ m/s}} \quad \text{and} \quad \boxed{v_{f1y} = +0.12 \text{ m/s}}$$

The substitution of these values into Equations 1a and 1b is shown at the right.

Solution Algebraically combining the results of each step, we find that

STEP 1   STEP 2

$$v_{f1} = \sqrt{v_{f1x}^2 + v_{f1y}^2} = \sqrt{(0.63 \text{ m/s})^2 + (0.12 \text{ m/s})^2} = \boxed{0.64 \text{ m/s}}$$

$$\theta = \tan^{-1}\left(\frac{v_{f1y}}{v_{f1x}}\right) = \tan^{-1}\left(\frac{0.12 \text{ m/s}}{0.63 \text{ m/s}}\right) = \boxed{11°}$$

STEP 1   STEP 2

Related Homework: *Problems 34, 44*

7.5 CENTER OF MASS

In previous sections, we have encountered situations in which objects interact with one another, such as the two skaters pushing off in Example 6. In these situations, the mass of the system is located in several places, and the various objects move relative to each other before, after, and even during the interaction. It is possible, however, to speak of a kind of average location for the total mass by introducing a concept known as the **center of mass** (abbreviated as "cm"). With the aid of this concept, we will be able to gain additional insight into the principle of conservation of linear momentum.

The center of mass is a point that represents the average location for the total mass of a system. Figure 7.15, for example, shows two particles of mass m_1 and m_2 that are located on the x axis at the positions x_1 and x_2, respectively. The position x_{cm} of the center-of-mass point from the origin is defined to be

Center of mass

$$x_{cm} = \frac{m_1 x_1 + m_2 x_2}{m_1 + m_2} \qquad (7.10)$$

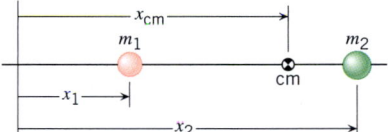

Figure 7.15 The center of mass cm of the two particles is located on a line between them and lies closer to the more massive particle.

Each term in the numerator of this equation is the product of a particle's mass and position, while the denominator is the total mass of the system. If the two masses are equal, we expect the average location of the total mass to be midway between the particles. With $m_1 = m_2 = m$, Equation 7.10 becomes $x_{cm} = (mx_1 + mx_2)/(m + m) = \frac{1}{2}(x_1 + x_2)$, which indeed corresponds to the point midway between the particles. Alternatively, suppose that $m_1 = 5.0$ kg and $x_1 = 2.0$ m, while $m_2 = 12$ kg and $x_2 = 6.0$ m. Then we expect the average location of the total mass to be located closer to particle 2, since it is more massive. Equation 7.10 is also consistent with this expectation, for it gives

$$x_{cm} = \frac{(5.0 \text{ kg})(2.0 \text{ m}) + (12 \text{ kg})(6.0 \text{ m})}{5.0 \text{ kg} + 12 \text{ kg}} = 4.8 \text{ m}$$

If a system contains more than two particles, the center-of-mass point can be determined by generalizing Equation 7.10. For three particles, for instance, the numerator would contain a third term $m_3 x_3$, and the total mass in the denominator would be $m_1 + m_2 + m_3$. For a macroscopic object, which contains many, many particles, the center-of-mass point is located at the geometric center of the object, provided that the mass is distributed symmetrically about the center. Such would be the case for a billiard ball. For objects such as a golf club, the mass is not distributed symmetrically, and the center-of-mass point is not located at the geometric center of the club. The driver used to launch a golf ball from the tee, for instance, has more mass in the club head than in the handle, so the center-of-mass point is closer to the head than to the handle.

Equation 7.10 (and its generalization to more than two particles) deals with particles that lie along a straight line. A system of particles, however, may include particles that do not all lie along a single straight line. For particles lying in a plane, an equation like Equation 7.10 uses the x coordinates of each particle to give the x coordinate of the center of mass. A similar equation uses the y coordinates of each particle to give the y coordinate of

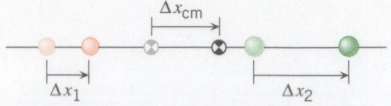

Figure 7.16 During a time interval Δt, the displacements of the particles are Δx_1 and Δx_2, while the displacement of the center of mass is Δx_{cm}.

the center of mass. If a system contains rigid objects, each of them may be treated as if it were a particle with all of the mass located at the object's own center of mass. In this way, a collection of rigid objects becomes a collection of particles, and the x and y coordinates of the center of mass of the collection can be determined as described earlier in this paragraph.

To see how the center-of-mass concept is related to momentum conservation, suppose that the two particles in a system are moving, as they would be during a collision. With the aid of Equation 7.10, we can determine the velocity v_{cm} of the center-of-mass point. During a time interval Δt, the particles experience displacements of Δx_1 and Δx_2, as Figure 7.16 shows. They have different displacements during this time because they have different velocities. Equation 7.10 can be used to find the displacement Δx_{cm} of the center of mass by replacing x_{cm} by Δx_{cm}, x_1 by Δx_1, and x_2 by Δx_2:

$$\Delta x_{cm} = \frac{m_1 \Delta x_1 + m_2 \Delta x_2}{m_1 + m_2}$$

Now we divide both sides of this equation by the time interval Δt. In the limit as Δt becomes infinitesimally small, the ratio $\Delta x_{cm}/\Delta t$ becomes equal to the instantaneous velocity v_{cm} of the center of mass. (See Section 2.2 for a review of instantaneous velocity.) Likewise, the ratios $\Delta x_1/\Delta t$ and $\Delta x_2/\Delta t$ become equal to the instantaneous velocities v_1 and v_2, respectively. Thus, we have

Velocity of center of mass $$v_{cm} = \frac{m_1 v_1 + m_2 v_2}{m_1 + m_2} \qquad (7.11)$$

The numerator $(m_1 v_1 + m_2 v_2)$ on the right-hand side in Equation 7.11 is the momentum of particle 1 $(m_1 v_1)$ plus the momentum of particle 2 $(m_2 v_2)$, which is the total linear momentum of the system. In an isolated system, the total linear momentum does not change because of an interaction such as a collision. Therefore, Equation 7.11 indicates that the velocity v_{cm} of the center of mass does not change either. To emphasize this important point, consider the collision discussed in Example 7. With the data from that example, we can apply Equation 7.11 to determine the velocity of the center of mass before and after the collision:

Before collision $$v_{cm} = \frac{(0.250 \text{ kg})(+5.00 \text{ m/s}) + (0.800 \text{ kg})(0 \text{ m/s})}{0.250 \text{ kg} + 0.800 \text{ kg}} = +1.19 \text{ m/s}$$

After collision $$v_{cm} = \frac{(0.250 \text{ kg})(-2.62 \text{ m/s}) + (0.800 \text{ kg})(+2.38 \text{ m/s})}{0.250 \text{ kg} + 0.800 \text{ kg}} = +1.19 \text{ m/s}$$

Thus, the velocity of the center of mass is the same before and after the objects interact during a collision in which the total linear momentum is conserved.

In a two-dimensional collision such as that discussed in Example 9, the velocity of the center of mass is also the same before and after the collision, provided that the total linear momentum is conserved. Figure 7.17 illustrates this fact. The drawing reproduces Figure 7.14a, except that the initial and final center-of-mass locations are indicated, along with the velocity vector for the center of mass at these two places (see the vectors labeled v_{cm}). As the drawing shows, this velocity vector is the same before and after the balls collide. Thus, the center of mass moves along a straight line as the two balls approach the collision point and continues along the same straight line following the collision.

Figure 7.17 This drawing shows the two balls in Figure 7.14a and the path followed by the center-of-mass point as the balls approach the collision point and then move away from it. Because of momentum conservation, the velocity of the center of mass of the balls is the same before and after the collision (see the vectors labeled v_{cm}). As a result, the center of mass moves along the same straight-line path before and after the collision.

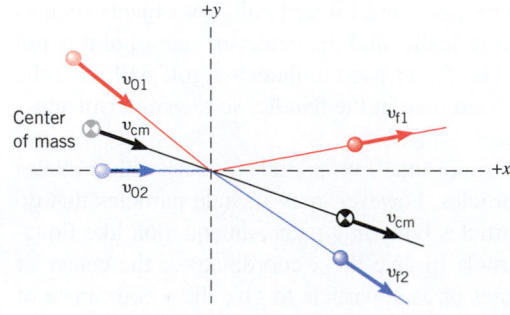

✓ **CHECK YOUR UNDERSTANDING**

(The answers are given at the end of the book.)

18. Would you expect the center of mass of a baseball bat to be located halfway between the ends of the bat, nearer the lighter end, or nearer the heavier end?

19. A sunbather is lying on a floating raft that is stationary. She then gets up and walks to one end of the raft. Consider the sunbather and raft as an isolated system.   **(a)** What is the velocity of the center of mass of this system while she is walking?   **(b)** Does the raft itself move while she is walking? If so, what is the direction of the raft's velocity relative to that of the sunbather?

20. Water, dripping at a constant rate from a faucet, falls to the ground. At any instant there are many drops in the air between the faucet and the ground. Where does the center of mass of the drops lie relative to the halfway point between the faucet and the ground?
(a) Above it   **(b)** Below it   **(c)** Exactly at the halfway point

7.6 CONCEPTS & CALCULATIONS

Momentum and energy are two of the most fundamental concepts in physics. As we have seen in this chapter, momentum is a vector and, like all vectors, has a magnitude and a direction. In contrast, energy is a scalar quantity, as Chapter 6 discusses, and does not have a direction associated with it. Example 10 provides the opportunity to review how the vector nature of momentum and the scalar nature of kinetic energy influence calculations using these quantities.

 Concepts & Calculations Example 10   A Scalar and a Vector

Two joggers, Jim and Tom, are both running at a speed of 4.00 m/s. Jim has a mass of 90.0 kg, and Tom has a mass of 55.0 kg. Find the kinetic energy and momentum of the two-jogger system when **(a)** Jim and Tom are both running due north (Figure 7.18a) and **(b)** Jim is running due north and Tom is running due south (Figure 7.18b).

Concept Questions and Answers   Does the total kinetic energy have a smaller value in case (a) or (b), or is it the same in both cases?

Answer   Everything is the same in both cases, except that both joggers are running due north in (a), but one is running due north and one is running due south in (b). Kinetic energy is a scalar quantity and does not depend on directional differences such as these. Therefore, the kinetic energy of the two-jogger system is the same in both cases.

Does the total momentum have a smaller magnitude in case (a) or (b), or is it the same for each?

Answer   Momentum is a vector. As a result, when we add Jim's momentum to Tom's momentum to get the total for the two-jogger system, we must take the directions into account. For example, we can use positive to denote north and negative to denote south. In case (a) two positive momentum vectors are added, while in case (b) one positive and one negative vector are added. Due to the partial cancellation in case (b), the resulting total momentum will have a smaller magnitude than in case (a).

Solution   **(a)** The kinetic energy of the two-jogger system can be obtained by applying the definition of kinetic energy as $\frac{1}{2}mv^2$ (see Equation 6.2) to each jogger:

$$KE_{system} = \tfrac{1}{2}m_{Jim}v_{Jim}^2 + \tfrac{1}{2}m_{Tom}v_{Tom}^2$$

$$= \tfrac{1}{2}(90.0 \text{ kg})(4.00 \text{ m/s})^2 + \tfrac{1}{2}(55.0 \text{ kg})(4.00 \text{ m/s})^2 = \boxed{1160 \text{ J}}$$

The momentum of each jogger is his mass times his velocity, according to the definition given in Equation 7.2. Using P_{system} to denote the momentum of the two-jogger system and denoting due north as the positive direction, we have

$$P_{system} = m_{Jim}v_{Jim} + m_{Tom}v_{Tom}$$

$$= (90.0 \text{ kg})(+4.00 \text{ m/s}) + (55.0 \text{ kg})(+4.00 \text{ m/s}) = \boxed{580 \text{ kg} \cdot \text{m/s}}$$

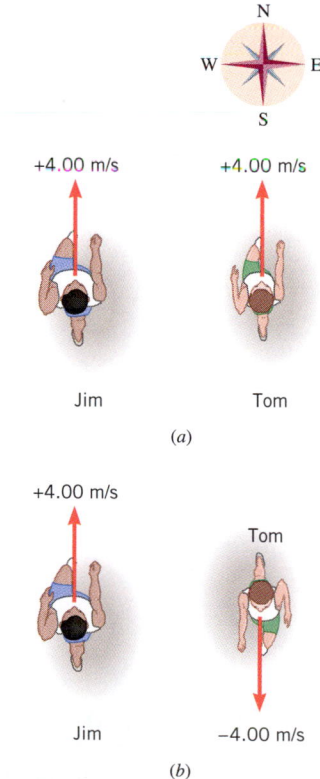

Figure 7.18 (a) Two joggers are running due north with the same velocities. (b) One jogger is running due north and the other due south, with velocities that have the same magnitudes but different directions.

(b) In these calculations, we note that Tom's velocity is in the due south direction, so it is -4.00 m/s, not $+4.00$ m/s. Proceeding as in part (a), we find

$$KE_{system} = \tfrac{1}{2}m_{Jim}v_{Jim}^2 + \tfrac{1}{2}m_{Tom}v_{Tom}^2$$

$$= \tfrac{1}{2}(90.0 \text{ kg})(4.00 \text{ m/s})^2 + \tfrac{1}{2}(55.0 \text{ kg})(-4.00 \text{ m/s})^2 = \boxed{1160 \text{ J}}$$

$$P_{system} = m_{Jim}v_{Jim} + m_{Tom}v_{Tom}$$

$$= (90.0 \text{ kg})(+4.00 \text{ m/s}) + (55.0 \text{ kg})(-4.00 \text{ m/s}) = \boxed{140 \text{ kg} \cdot \text{m/s}}$$

Momentum and kinetic energy are not the same concepts, and the next example explores further some of the differences between them.

Concepts & Calculations Example 11 Momentum and Kinetic Energy

The following table gives mass and speed data for the two objects in Figure 7.19. Find the magnitude of the momentum and the kinetic energy for each object.

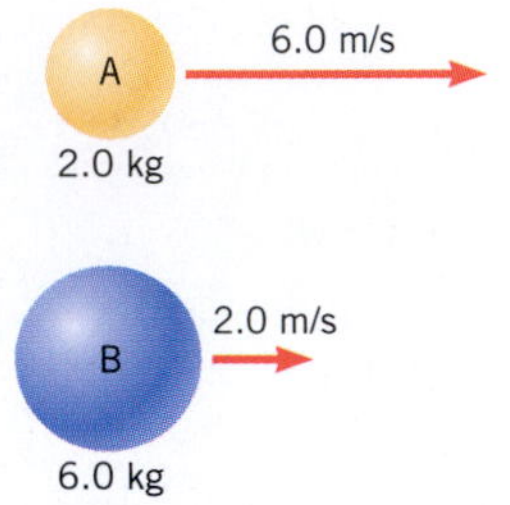

Figure 7.19 The two objects have different masses and speeds. Are their momenta and kinetic energies the same?

	Mass	Speed
Object A	2.0 kg	6.0 m/s
Object B	6.0 kg	2.0 m/s

Concept Questions and Answers Is it possible for two objects to have different speeds when their momenta have the same magnitude?

Answer Yes. The magnitude of an object's momentum is the product of its mass and its speed, according to Equation 7.2. Both mass and speed play a role. The speed can be reduced, for instance, but if the mass is increased proportionally, the magnitude of the momentum will be the same.

If two objects have the same momentum, do they necessarily have the same kinetic energy?

Answer No. Kinetic energy is $KE = \tfrac{1}{2}mv^2$. Momentum is $p = mv$, which means that $v = p/m$. Substituting this relation into the kinetic energy expression shows that $KE = \tfrac{1}{2}m(p/m)^2 = p^2/(2m)$. Thus, two objects can have the same momentum p, but if each has different mass m, their kinetic energies will be different.

Solution Using Equation 7.2 for the momentum and Equation 6.2 for the kinetic energy, we find that

Momenta

$$p_A = m_A v_A = (2.0 \text{ kg})(6.0 \text{ m/s}) = \boxed{12 \text{ kg} \cdot \text{m/s}}$$

$$p_B = m_B v_B = (6.0 \text{ kg})(2.0 \text{ m/s}) = \boxed{12 \text{ kg} \cdot \text{m/s}}$$

Kinetic energies

$$KE_A = \tfrac{1}{2}m_A v_A^2 = \tfrac{1}{2}(2.0 \text{ kg})(6.0 \text{ m/s})^2 = \boxed{36 \text{ J}}$$

$$KE_B = \tfrac{1}{2}m_B v_B^2 = \tfrac{1}{2}(6.0 \text{ kg})(2.0 \text{ m/s})^2 = \boxed{12 \text{ J}}$$

CONCEPT SUMMARY

If you need more help with a concept, use the Learning Aids noted next to the discussion or equation. Examples (**Ex.**) are in the text of this chapter. Go to **www.wiley.com/college/cutnell** for the following Learning Aids:

Interactive LearningWare (ILW) — Additional examples solved in a five-step interactive format.

Concept Simulations (CS) — Animated text figures or animations of important concepts.

Interactive Solutions (IS) — Models for certain types of problems in the chapter homework. The calculations are carried out interactively.

Topic	Discussion	Learning Aids

7.1 THE IMPULSE–MOMENTUM THEOREM The impulse $\vec{\mathbf{J}}$ of a force is the product of the average force $\vec{\mathbf{F}}$ and the time interval Δt during which the force acts:

Impulse

$$\vec{\mathbf{J}} = \vec{\mathbf{F}}\,\Delta t \qquad (7.1)$$

Impulse is a vector that points in the same direction as the average force.

The linear momentum $\vec{\mathbf{p}}$ of an object is the product of the object's mass m and velocity $\vec{\mathbf{v}}$:   **Ex. 10, 11**

Linear momentum

$$\vec{\mathbf{p}} = m\vec{\mathbf{v}} \qquad (7.2)$$

Linear momentum is a vector that points in the same direction as the velocity. The total linear momentum of a system of objects is the vector sum of the momenta of the individual objects.

The impulse–momentum theorem states that when a net force $\Sigma\vec{\mathbf{F}}$ acts on an object, the impulse   **Ex. 1, 2, 3**
of the net force is equal to the change in momentum of the object: **ILW 7.1**
  IS 7.10

Impulse–momentum theorem

$$(\Sigma\vec{\mathbf{F}})\Delta t = m\vec{\mathbf{v}}_{\mathbf{f}} - m\vec{\mathbf{v}}_{\mathbf{0}} \qquad (7.4)$$

7.2 THE PRINCIPLE OF CONSERVATION OF LINEAR MOMENTUM External forces are those forces
Isolated system that agents external to the system exert on objects within the system. An isolated system is one for
which the vector sum of the average external forces acting on the system is zero.

The principle of conservation of linear momentum states that the total linear momentum of an isolated system remains constant. For a two-body collision, the conservation of linear momentum can be written as

Conservation of linear
momentum

$$\underbrace{m_1\vec{\mathbf{v}}_{\mathbf{f1}} + m_2\vec{\mathbf{v}}_{\mathbf{f2}}}_{\substack{\text{Final total linear}\\\text{momentum}}} = \underbrace{m_1\vec{\mathbf{v}}_{\mathbf{01}} + m_2\vec{\mathbf{v}}_{\mathbf{02}}}_{\substack{\text{Initial total linear}\\\text{momentum}}} \qquad (7.7\text{b})$$

Ex. 4, 5, 6
CS 7.1
ILW 7.2
IS 7.22

where m_1 and m_2 are the masses, $\vec{\mathbf{v}}_{\mathbf{f1}}$ and $\vec{\mathbf{v}}_{\mathbf{f2}}$ are the final velocities, and $\vec{\mathbf{v}}_{\mathbf{01}}$ and $\vec{\mathbf{v}}_{\mathbf{02}}$ are the initial velocities of the objects.

Elastic collision   **7.3 COLLISIONS IN ONE DIMENSION** An elastic collision is one in which the total kinetic energy   **Ex. 7, 8**
of the system after the collision is equal to the total kinetic energy of the system before the collision.   **CS 7.2**

Inelastic collision An inelastic collision is one in which the total kinetic energy of the system is not the same   **IS 7.24, 7.39**
before and after the collision. If the objects stick together after the collision, the collision is said
to be completely inelastic.

7.4 COLLISIONS IN TWO DIMENSIONS When the total linear momentum is conserved in a
two-dimensional collision, the x and y components of the total linear momentum are conserved
separately. For a collision between two objects, the conservation of total linear momentum can be
written as

$$\underbrace{m_1 v_{\mathrm{f1}x} + m_2 v_{\mathrm{f2}x}}_{\substack{x\text{ component of final}\\\text{total linear momentum}}} = \underbrace{m_1 v_{01x} + m_2 v_{02x}}_{\substack{x\text{ component of initial}\\\text{total linear momentum}}} \qquad (7.9\text{a})$$

Ex. 9
ILW 7.3

Conservation of linear
momentum

$$\underbrace{m_1 v_{\mathrm{f1}y} + m_2 v_{\mathrm{f2}y}}_{\substack{y\text{ component of final}\\\text{total linear momentum}}} = \underbrace{m_1 v_{01y} + m_2 v_{02y}}_{\substack{y\text{ component of initial}\\\text{total linear momentum}}} \qquad (7.9\text{b})$$

7.5 CENTER OF MASS The location of the center of mass of two particles lying on the x axis is
given by

Location of center of mass

$$x_{\mathrm{cm}} = \frac{m_1 x_1 + m_2 x_2}{m_1 + m_2} \qquad (7.10)$$ **IS 7.60**

where m_1 and m_2 are the masses of the particles and x_1 and x_2 are their positions relative to the coordinate origin. If the particles move with velocities v_1 and v_2, the velocity v_{cm} of the center of mass is

Velocity of center of mass

$$v_{\mathrm{cm}} = \frac{m_1 v_1 + m_2 v_2}{m_1 + m_2} \qquad (7.11)$$

If the total linear momentum of a system of particles remains constant during an interaction such
as a collision, the velocity of the center of mass also remains constant.

FOCUS ON CONCEPTS

Note to Instructors: *The numbering of the questions shown here reflects the fact that they are only a representative subset of the total number that are available online. However, all of the questions are available for assignment via an online homework management program such as WileyPLUS or WebAssign.*

Section 7.1 The Impulse–Momentum Theorem

1. Two identical cars are traveling at the same speed. One is heading due east and the other due north, as the drawing shows. Which statement is true regarding the kinetic energies and momenta of the cars? (a) They both have the same kinetic energies and the same momenta. (b) They have the same kinetic energies, but different momenta. (c) They have different kinetic energies, but the same momentum. (d) They have different kinetic energies and different momentum.

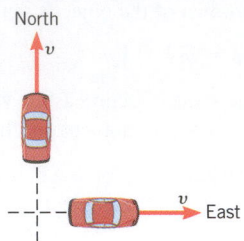

2. Six runners have the mass (in multiples of m_0), speed (in multiples of v_0), and direction of travel that are indicated in the table. Which two runners have identical momenta? (a) B and C (b) A and C (c) C and D (d) A and E (e) D and F

Runner	Mass	Speed	Direction of Travel
A	$\frac{1}{2}m_0$	v_0	Due north
B	m_0	v_0	Due east
C	m_0	$2v_0$	Due south
D	$2m_0$	v_0	Due west
E	m_0	$\frac{1}{2}v_0$	Due north
F	$2m_0$	$2v_0$	Due west

6. A particle is moving along the $+x$ axis, and the graph shows its momentum p as a function of time t. Use the impulse–momentum theorem to rank (largest to smallest) the three regions according to the *magnitude* of the impulse applied to the particle. (a) A, B, C (b) A, C, B (c) A and C (a tie), B (d) C, A, B (e) B, A, C

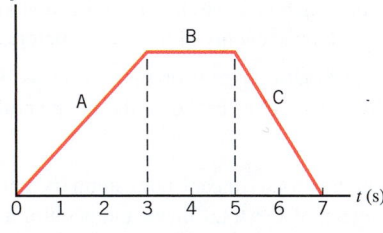

7. A particle moves along the $+x$ axis, and the graph shows its momentum p as a function of time t. In each of the four regions a force, which may or may not be nearly zero, is applied to the particle. In which region is the *magnitude* of the force largest and in which region is it smallest? (a) B largest, D smallest (b) C largest, B smallest (c) A largest, D smallest (d) C largest, A smallest (e) A largest, C smallest

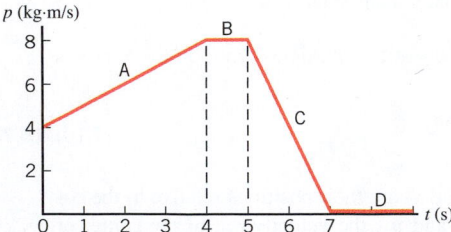

Section 7.2 The Principle of Conservation of Linear Momentum

10. As the text discusses, the conservation of linear momentum is applicable only when the system of objects is an isolated system. Which of the systems listed below are isolated systems?

1. A ball is dropped from the top of a building. The system is the ball.

2. A ball is dropped from the top of a building. The system is the ball and the earth.

3. A billiard ball collides with a stationary billiard ball on a frictionless pool table. The system is the moving ball.

4. A car slides to a halt in an emergency. The system is the car.

5. A space probe is moving in deep space where gravitational and other forces are negligible. The system is the space probe.

(a) Only 2 and 5 are isolated systems. (b) Only 1 and 3 are isolated systems. (c) Only 3 and 5 are isolated systems. (d) Only 4 and 5 are isolated systems. (e) Only 5 is an isolated system.

Section 7.3 Collisions in One Dimension

13. Two objects are involved in a completely inelastic one-dimensional collision. The net external force acting on them is zero. The table lists four possible sets of the initial and final momenta and kinetic energies of the two objects. Which is the only set that could occur?

		Initial (Before Collision)		Final (After Collision)	
		Momentum	Kinetic Energy	Momentum	Kinetic Energy
a.	Object 1:	+6 kg·m/s	15 J	+8 kg·m/s	9 J
	Object 2:	0 kg·m/s	0 J		
b.	Object 1:	+8 kg·m/s	5 J	+6 kg·m/s	12 J
	Object 2:	−2 kg·m/s	7 J		
c.	Object 1:	−3 kg·m/s	1 J	+1 kg·m/s	4 J
	Object 2:	+4 kg·m/s	6 J		
d.	Object 1:	0 kg·m/s	3 J	−8 kg·m/s	11 J
	Object 2:	−8 kg·m/s	8 J		

Section 7.4 Collisions in Two Dimensions

15. Object 1 is moving along the x axis with an initial momentum of $+16$ kg·m/s, where the $+$ sign indicates that it is moving to the right. As the drawing shows, object 1 collides with a second object that is initially at rest. The collision is not head-on, so the objects move off in different directions after the collision. The net external force acting on the two-object system is zero. After the collision, object 1 has a momentum whose y component is -5 kg·m/s. What is the y component of the momentum of object 2 after the collision? (a) 0 kg·m/s (b) $+16$ kg·m/s (c) $+5$ kg·m/s (d) -16 kg·m/s (e) The y component of the momentum of object 2 cannot be determined.

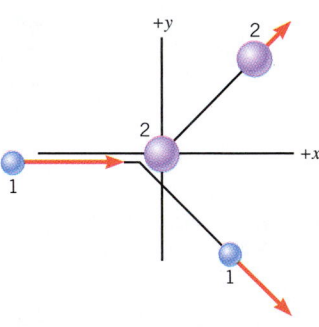

Section 7.5 Center of Mass

17. The drawing shows three particles that are moving with different velocities. Two of the particles have mass m, and the third has a mass $2m$. At the instant shown, the center of mass (cm) of the three particles is at the coordinate origin. What is the velocity v_{cm} (magnitude and direction) of the center of mass?

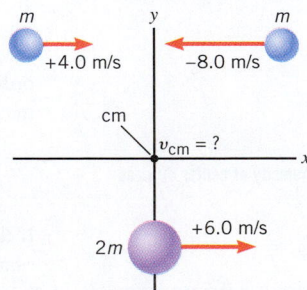

PROBLEMS

Note to Instructors: *Most of the homework problems in this chapter are available for assignment via an online homework management program such as* WileyPLUS *or* WebAssign, *and those marked with the icon* *are presented in* WileyPLUS *using a guided tutorial format that provides enhanced interactivity. See Preface for additional details.*

ssm Solution is in the Student Solutions Manual.
www Solution is available online at www.wiley.com/college/cutnell

💊 **This icon represents a biomedical application.**

Section 7.1 The Impulse–Momentum Theorem

1. ssm One average force $\vec{\mathbf{F}}_1$ has a magnitude that is three times as large as that of another average force $\vec{\mathbf{F}}_2$. Both forces produce the same impulse. The average force $\vec{\mathbf{F}}_1$ acts for a time interval of 3.2 ms. For what time interval does the average force $\vec{\mathbf{F}}_2$ act?

2. GO In a performance test, each of two cars takes 9.0 s to accelerate from rest to 27 m/s. Car A has a mass of 1400 kg, and car B has a mass of 1900 kg. Find the net average force that acts on each car during the test.

3. Before starting this problem, review Conceptual Example 3. Suppose that the hail described there bounces off the roof of the car with a velocity of +15 m/s. Ignoring the weight of the hailstones, calculate the force exerted by the hail on the roof. Compare your answer to that obtained in Example 2 for the rain, and verify that your answer is consistent with the conclusion reached in Conceptual Example 3.

4. A model rocket is constructed with a motor that can provide a total impulse of 29.0 N·s. The mass of the rocket is 0.175 kg. What is the speed that this rocket achieves when launched from rest? Neglect the effects of gravity and air resistance.

5. ssm A volleyball is spiked so that its incoming velocity of +4.0 m/s is changed to an outgoing velocity of −21 m/s. The mass of the volleyball is 0.35 kg. What impulse does the player apply to the ball?

6. A baseball ($m = 149$ g) approaches a bat horizontally at a speed of 40.2 m/s (90 mi/h) and is hit straight back at a speed of 45.6 m/s (102 mi/h). If the ball is in contact with the bat for a time of 1.10 ms, what is the average force exerted on the ball by the bat? Neglect the weight of the ball, since it is so much less than the force of the bat. Choose the direction of the incoming ball as the positive direction.

7. A space probe is traveling in outer space with a momentum that has a magnitude of 7.5×10^7 kg · m/s. A retrorocket is fired to slow down the probe. It applies a force to the probe that has a magnitude of 2.0×10^6 N and a direction opposite to the probe's motion. It fires for a period of 12 s. Determine the momentum of the probe after the retrorocket ceases to fire.

8. 💊 When jumping straight down, you can be seriously injured if you land stiff-legged. One way to avoid injury is to bend your knees upon landing to reduce the force of the impact. A 75-kg man just before contact with the ground has a speed of 6.4 m/s. **(a)** In a stiff-legged landing he comes to a halt in 2.0 ms. Find the average net force that acts on him during this time. **(b)** When he bends his knees, he comes to a halt in 0.10 s. Find the average net force now. **(c)** During the landing, the force of the ground on the man points upward, while the force due to gravity points downward. The average net force acting on the man includes both of these forces. Taking into account the directions of the forces, find the force of the ground on the man in parts (a) and (b).

9. Refer to Conceptual Example 3 as an aid in understanding this problem. A hockey goalie is standing on ice. Another player fires a puck ($m = 0.17$ kg) at the goalie with a velocity of +65 m/s. **(a)** If the goalie catches the puck with his glove in a time of 5.0×10^{-3} s, what is the average force (magnitude and direction) exerted on the goalie by the puck? **(b)** Instead of catching the puck, the goalie slaps it with his stick and returns the puck straight back to the player with a velocity of −65 m/s. The puck and stick are in contact for a time of 5.0×10^{-3} s. Now what is the average force exerted on the goalie by the puck? Verify that your answers to parts (a) and (b) are consistent with the conclusion of Conceptual Example 3.

* **10.** Consult **Interactive Solution 7.10** at **www.wiley.com/college/cutnell** for a review of problem-solving skills that are involved in this problem. A stream of water strikes a stationary turbine blade horizontally, as the drawing illustrates. The incident water stream has a velocity of +16.0 m/s, while the exiting water stream has a velocity of −16.0 m/s. The mass of water per second that strikes the blade is 30.0 kg/s. Find the magnitude of the average force exerted on the water by the blade.

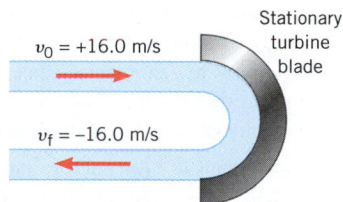

$v_0 = +16.0$ m/s

$v_f = -16.0$ m/s

Stationary turbine blade

* **11. GO** A student ($m = 63$ kg) falls freely from rest and strikes the ground. During the collision with the ground, he comes to rest in a time of 0.040 s. The average force exerted on him by the ground is +18 000 N, where the upward direction is taken to be the positive direction. From what height did the student fall? Assume that the only force acting on him during the collision is that due to the ground.

* **12.** A 0.500-kg ball is dropped from rest at a point 1.20 m above the floor. The ball rebounds straight upward to a height of 0.700 m. What are the magnitude and direction of the impulse of the net force applied to the ball during the collision with the floor?

* **13. ssm www** A golf ball strikes a hard, smooth floor at an angle of 30.0° and, as the drawing shows, rebounds at the same angle. The mass of the ball is 0.047 kg, and its speed is 45 m/s just before and after striking the floor. What is the magnitude of the impulse applied to the golf ball by the floor? (*Hint: Note that only the vertical component of the ball's momentum changes during impact with the floor, and ignore the weight of the ball.*)

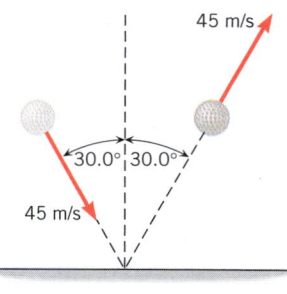

45 m/s

30.0° 30.0°

45 m/s

* **14.** An 85-kg jogger is heading due east at a speed of 2.0 m/s. A 55-kg jogger is heading 32° north of east at a speed of 3.0 m/s. Find the magnitude and direction of the sum of the momenta of the two joggers.

** **15.** A dump truck is being filled with sand. The sand falls straight downward from rest from a height of 2.00 m above the truck bed, and the mass of sand that hits the truck per second is 55.0 kg/s. The truck is parked on the platform of a weight scale. By how much does the scale reading exceed the weight of the truck and sand?

Section 7.2 The Principle of Conservation of Linear Momentum

16. A 2.3-kg cart is rolling across a frictionless, horizontal track toward a 1.5-kg cart that is held initially at rest. The carts are loaded with strong magnets that cause them to attract one another. Thus, the speed of each cart increases. At a certain instant before the carts collide, the first cart's velocity is $+4.5$ m/s, and the second cart's velocity is -1.9 m/s. **(a)** What is the total momentum of the system of the two carts at this instant? **(b)** What was the velocity of the first cart when the second cart was still at rest?

17. In a football game, a receiver is standing still, having just caught a pass. Before he can move, a tackler, running at a velocity of $+4.5$ m/s, grabs him. The tackler holds onto the receiver, and the two move off together with a velocity of $+2.6$ m/s. The mass of the tackler is 115 kg. Assuming that momentum is conserved, find the mass of the receiver.

18. **GO** As the drawing illustrates, two disks with masses m_1 and m_2 are

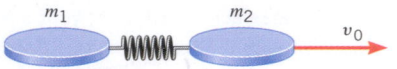

moving horizontally to the right at a speed of v_0. They are on an air-hockey table, which supports them with an essentially frictionless cushion of air. They move as a unit, with a compressed spring between them, which has a negligible mass. Then the spring is released and allowed to push the disks outward. Consider the situation where disk 1 comes to a momentary halt shortly after the spring is released. Assuming that $m_1 = 1.2$ kg, $m_2 = 2.4$ kg, and $v_0 = +5.0$ m/s, find the velocity of disk 2 at that moment.

19. **ssm** Consult **Concept Simulation 7.1** at **www.wiley.com/college/cutnell** in preparation for this problem. Two friends, Al and Jo, have a combined mass of 168 kg. At an ice skating rink they stand close together on skates, at rest and facing each other, with a compressed spring between them. The spring is kept from pushing them apart because they are holding each other. When they release their arms, Al moves off in one direction at a speed of 0.90 m/s, while Jo moves off in the opposite direction at a speed of 1.2 m/s. Assuming that friction is negligible, find Al's mass.

20. **GO** A wagon is rolling forward on level ground. Friction is negligible. The person sitting in the wagon is holding a rock. The total mass of the wagon, rider, and rock is 95.0 kg. The mass of the rock is 0.300 kg. Initially the wagon is rolling forward at a speed of 0.500 m/s. Then the person throws the rock with a speed of 16.0 m/s. Both speeds are relative to the ground. Find the speed of the wagon after the rock is thrown directly forward in one case and directly backward in another.

21. **ssm** A two-stage rocket moves in space at a constant velocity of 4900 m/s. The two stages are then separated by a small explosive charge placed between them. Immediately after the explosion the velocity of the 1200-kg upper stage is 5700 m/s in the same direction as before the explosion. What is the velocity (magnitude and direction) of the 2400-kg lower stage after the explosion?

* **22.** To view an interactive solution to a problem that is very similar to this one, go to **www.wiley.com/college/cutnell** and select **Interactive Solution 7.22.** A fireworks rocket is moving at a speed of 45.0 m/s. The rocket suddenly breaks into two pieces of equal mass, which fly off with velocities $\vec{v}_1$ and $\vec{v}_2$, as shown in the drawing. What is the magnitude of **(a)** $\vec{v}_1$ and **(b)** $\vec{v}_2$?

* **23.** **ssm** By accident, a large plate is dropped and breaks into three pieces. The pieces fly apart parallel to the floor. As the plate

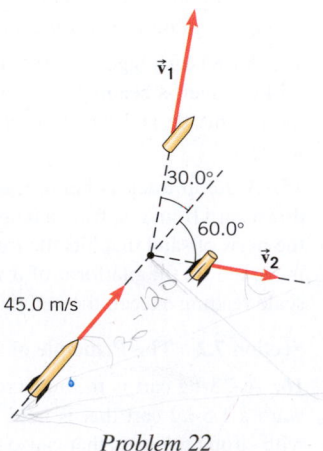

Problem 22

falls, its momentum has only a vertical component, and no component parallel to the floor. After the collision, the component of the total momentum parallel to the floor must remain zero, since the net external force acting on the plate has no component parallel to the floor. Using the data shown in the drawing, find the masses of pieces 1 and 2.

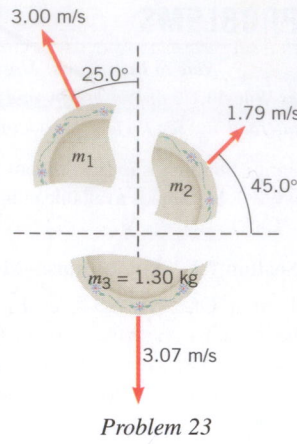

Problem 23

* **24.** To view an interactive solution to a problem that is similar to this one, go to **www.wiley.com/college/cutnell** and select **Interactive Solution 7.24.** A 0.015-kg bullet is fired straight up at a falling wooden block that has a mass of 1.8 kg. The bullet has a speed of 810 m/s when it strikes the block. The block originally was dropped from rest from the top of a building and had been falling for a time t when the collision with the bullet occurred. As a result of the collision, the block (with the bullet in it) reverses direction, rises, and comes to a momentary halt at the top of the building. Find the time t.

* **25.** Two ice skaters have masses m_1 and m_2 and are initially stationary. Their skates are identical. They push against one another, as in Figure 7.10, and move in opposite directions with different speeds. While they are pushing against each other, any kinetic frictional forces acting on their skates can be ignored. However, once the skaters separate, kinetic frictional forces eventually bring them to a halt. As they glide to a halt, the magnitudes of their accelerations are equal, and skater 1 glides twice as far as skater 2. What is the ratio m_1/m_2 of their masses?

** **26.** A wagon is coasting at a speed v_A along a straight and level road. When ten percent of the wagon's mass is thrown off the wagon, parallel to the ground and in the forward direction, the wagon is brought to a halt. If the direction in which this mass is thrown is exactly reversed, but the speed of this mass relative to the wagon remains the same, the wagon accelerates to a new speed v_B. Calculate the ratio v_B/v_A.

** **27.** **ssm www** A cannon of mass 5.80×10^3 kg is rigidly bolted to the earth so it can recoil only by a negligible amount. The cannon fires an 85.0-kg shell horizontally with an initial velocity of $+551$ m/s. Suppose the cannon is then unbolted from the earth, and no external force hinders its recoil. What would be the velocity of a shell fired by this loose cannon? *(Hint: In both cases assume that the burning gunpowder imparts the same kinetic energy to the system.)*

**Section 7.3   Collisions in One Dimension,
Section 7.4   Collisions in Two Dimensions**

28. After sliding down a snow-covered hill on an inner tube, Ashley is coasting across a level snowfield at a constant velocity of $+2.7$ m/s. Miranda runs after her at a velocity of $+4.5$ m/s and hops on the inner tube. How fast do the two of them slide across the snow together on the inner tube? Ashley's mass is 71 kg and Miranda's is 58 kg. Ignore the mass of the inner tube and any friction between the inner tube and the snow.

29. Multiple-Concept Example 7 presents a model for solving problems such as this one. A 1055-kg van, stopped at a traffic light, is hit directly in the rear by a 715-kg car traveling with a velocity of $+2.25$ m/s. Assume that the transmission of the van is in neutral, the brakes are not being applied, and the collision is elastic. What is the final velocity of **(a)** the car and **(b)** the van?

30. **GO** One object is at rest, and another is moving. The two collide in a one-dimensional, completely inelastic collision. In other words, they stick together after the collision and move off with a common velocity. Momentum is conserved. The speed of the object that is moving initially is 25 m/s. The masses of the two objects are 3.0 and 8.0 kg. Determine the final speed of the two-object system after the collision for the case when the large-mass object is the one moving initially and the case when the small-mass object is the one moving initially.

31. **ssm** A projectile (mass = 0.20 kg) is fired at and embeds itself in a target (mass = 2.50 kg). The target (with the projectile in it) flies off after being struck. What percentage of the projectile's incident kinetic energy does the target (with the projectile in it) carry off after being struck?

32. A car (mass = 1100 kg) is traveling at 32 m/s when it collides head-on with a sport utility vehicle (mass = 2500 kg) traveling in the opposite direction. In the collision, the two vehicles come to a halt. At what speed was the sport utility vehicle traveling?

33. **ssm** **www** A 5.00-kg ball, moving to the right at a velocity of +2.00 m/s on a frictionless table, collides head-on with a stationary 7.50-kg ball. Find the final velocities of the balls if the collision is **(a)** elastic and **(b)** completely inelastic.

34. Multiple-Concept Example 9 reviews the concepts that play roles in this problem. The drawing shows a collision between two pucks on an air-hockey table. Puck A has a mass of 0.025 kg and is moving along the x axis with a velocity of +5.5 m/s. It makes a collision with puck B, which has a mass of 0.050 kg and is initially at rest. The collision is not head-on. After the collision, the two pucks fly apart with the angles shown in the drawing. Find the final speed of **(a)** puck A and **(b)** puck B.

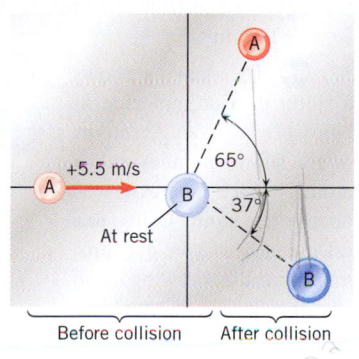

35. **ssm** Batman (mass = 91 kg) jumps straight down from a bridge into a boat (mass = 510 kg) in which a criminal is fleeing. The velocity of the boat is initially +11 m/s. What is the velocity of the boat after Batman lands in it?

36. **GO** Object A is moving due east, while object B is moving due north. They collide and stick together in a completely inelastic collision. Momentum is conserved. Object A has a mass of m_A = 17.0 kg and an initial velocity of $\vec{v}_{0A}$ = 8.00 m/s, due east. Object B, however, has a mass of m_B = 29.0 kg and an initial velocity of $\vec{v}_{0B}$ = 5.00 m/s, due north. Find the magnitude and direction of the total momentum of the two-object system after the collision.

37. Multiple-Concept Example 7 deals with some of the concepts that are used to solve this problem. A cue ball (mass = 0.165 kg) is at rest on a frictionless pool table. The ball is hit dead center by a pool stick, which applies an impulse of +1.50 N·s to the ball. The ball then slides along the table and makes an elastic head-on collision with a second ball of equal mass that is initially at rest. Find the velocity of the second ball just after it is struck.

****38.** **GO** A ball is attached to one end of a wire, the other end being fastened to the ceiling. The wire is held horizontal, and the ball is released from rest (see the drawing). It swings downward and strikes a block initially at rest on a horizontal frictionless surface. Air resistance is negligible, and the collision is elastic. The masses

of the ball and block are, respectively, 1.60 kg and 2.40 kg, and the length of the wire is 1.20 m. Find the velocity (magnitude and direction) of the ball **(a)** just before the collision, and **(b)** just after the collision.

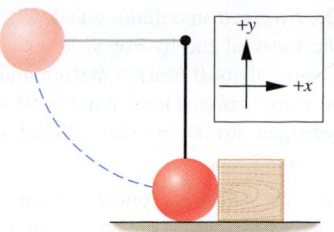

Problem 38

****39.** **Interactive Solution 7.39** at **www.wiley.com/college/ cutnell** shows modeling for a similar problem. An automobile has a mass of 2100 kg and a velocity of +17 m/s. It makes a rear-end collision with a stationary car whose mass is 1900 kg. The cars lock bumpers and skid off together with the wheels locked. **(a)** What is the velocity of the two cars just after the collision? **(b)** Find the impulse (magnitude and direction) that acts on the skidding cars from just after the collision until they come to a halt. **(c)** If the coefficient of kinetic friction between the wheels of the cars and the pavement is μ_k = 0.68, determine how far the cars skid before coming to rest.

****40.** A girl is skipping stones across a lake. One of the stones accidentally ricochets off a toy boat that is initially at rest in the water (see the drawing). The 0.072-kg stone strikes the boat at a velocity of 13 m/s, 15° below due east, and ricochets off at a velocity of 11 m/s, 12° above due east. After being struck by the stone, the boat's velocity is 2.1 m/s, due east. What is the mass of the boat? Assume the water offers no resistance to the boat's motion.

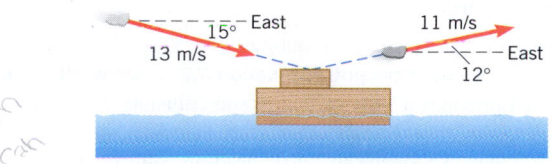

****41.** **ssm** A 50.0-kg skater is traveling due east at a speed of 3.00 m/s. A 70.0-kg skater is moving due south at a speed of 7.00 m/s. They collide and hold on to each other after the collision, managing to move off at an angle θ south of east, with a speed of v_f. Find **(a)** the angle θ and **(b)** the speed v_f, assuming that friction can be ignored.

****42.** **GO** A 4.00-g bullet is moving horizontally with a velocity of +355 m/s, where the + sign indicates that it is moving to the right (see part *a* of the drawing). The bullet is approaching two blocks resting on a horizontal frictionless surface. Air resistance is negligible. The bullet passes completely through the first block (an inelastic collision) and embeds itself in the second one, as indicated in part *b*. Note that both blocks are moving after the collision with the bullet. The mass of the first block is 1150 g, and its velocity is +0.550 m/s after the bullet passes through it. The mass of the second block is 1530 g. **(a)** What is the velocity of the second block after the bullet imbeds itself? **(b)** Find the ratio of the total kinetic energy after the collision to that before the collision.

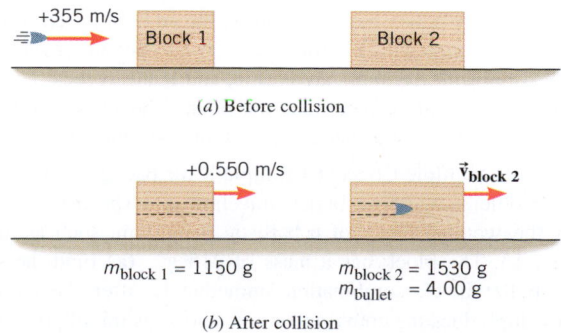

(*a*) Before collision

(*b*) After collision

* **43.** An electron collides elastically with a stationary hydrogen atom. The mass of the hydrogen atom is 1837 times that of the electron. Assume that all motion, before and after the collision, occurs along the same straight line. What is the ratio of the kinetic energy of the hydrogen atom after the collision to that of the electron before the collision?

* **44.** See Multiple-Concept Example 9 to review the approach used in problems of this type. Three guns are aimed at the center of a circle, and each fires a bullet simultaneously. The directions in which they fire are 120° apart. Two of the bullets have the same mass of 4.50×10^{-3} kg and the same speed of 324 m/s. The other bullet has an unknown mass and a speed of 575 m/s. The bullets collide at the center and mash into a stationary lump. What is the unknown mass?

** **45. ssm** Starting with an initial speed of 5.00 m/s at a height of 0.300 m, a 1.50-kg ball swings downward and strikes a 4.60-kg ball that is at rest, as the drawing shows. **(a)** Using the principle of conservation of mechanical energy, find the speed of the 1.50-kg ball just before impact. **(b)** Assuming that the collision is elastic, find the velocities (magnitude and direction) of both balls just after the collision. **(c)** How high does each ball swing after the collision, ignoring air resistance?

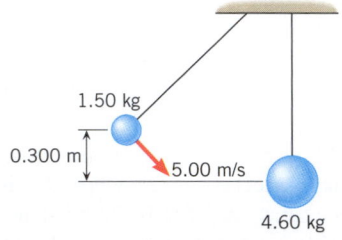

** **46.** Multiple-Concept Example 7 outlines the general approach to problems like this one. **Concept Simulation 7.2** at **www.wiley.com/college/cutnell** provides a view of this elastic collision. Two identical balls are traveling toward each other with velocities of −4.0 and +7.0 m/s, and they experience an elastic head-on collision. Obtain the velocities (magnitude and direction) of each ball after the collision.

** **47.** A ball is dropped from rest from the top of a 6.10-m-tall building, falls straight downward, collides inelastically with the ground, and bounces back. The ball loses 10.0% of its kinetic energy every time it collides with the ground. How many bounces can the ball make and still reach a windowsill that is 2.44 m above the ground?

Section 7.5 Center of Mass

48. Two stars in a binary system orbit around their center of mass. The centers of the two stars are 7.17×10^{11} m apart. The larger of the two stars has a mass of 3.70×10^{30} kg, and its center is 2.08×10^{11} m from the system's center of mass. What is the mass of the smaller star?

49. ssm Consider the two moving boxcars in Example 5. Determine the velocity of their center of mass **(a)** before and **(b)** after the collision. **(c)** Should your answer in part (b) be less than, greater than, or equal to the common velocity v_f of the two coupled cars after the collision? Justify your answer.

50. GO John's mass is 86 kg, and Barbara's is 55 kg. He is standing on the x axis at $x_J = +9.0$ m, while she is standing on the x axis at $x_B = +2.0$ m. They switch positions. How far and in which direction does their center of mass move as a result of the switch?

51. Two balls are approaching each other head-on. Their velocities are +9.70 and −11.8 m/s. Determine the velocity of the center of mass of the two balls **(a)** if they have the same mass and **(b)** if the mass of one ball ($v = 9.70$ m/s) is twice the mass of the other ball ($v = -11.8$ m/s).

* **52.** GO The drawing shows a sulfur dioxide molecule. It consists of two oxygen atoms and a sulfur atom. A sulfur atom is twice as massive as an oxygen atom. Using this information and the data provided in the drawing, find **(a)** the x coordinate and **(b)** the y coordinate of the center of mass of the sulfur dioxide molecule. Express your answers in nanometers (1 nm = 10^{-9} m).

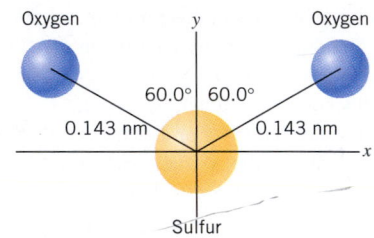

53. ssm www A lumberjack (mass = 98 kg) is standing at rest on one end of a floating log (mass = 230 kg) that is also at rest. The lumberjack runs to the other end of the log, attaining a velocity of +3.6 m/s relative to the shore, and then hops onto an identical floating log that is initially at rest. Neglect any friction and resistance between the logs and the water. **(a)** What is the velocity of the first log just before the lumberjack jumps off? **(b)** Determine the velocity of the second log if the lumberjack comes to rest on it.

54. A golf ball bounces down a flight of steel stairs, striking several steps on the way down, but never hitting the edge of a step. The ball starts at the top step with a vertical velocity component of zero. If all the collisions with the stairs are elastic, and if the vertical height of the staircase is 3.00 m, determine the bounce height when the ball reaches the bottom of the stairs. Neglect air resistance.

55. Consult Multiple-Concept Example 8 for background pertinent to this problem. A 2.50-g bullet, traveling at a speed of 425 m/s, strikes the wooden block of a ballistic pendulum, such as that in Figure 7.13. The block has a mass of 215 g. **(a)** Find the speed of the bullet–block combination immediately after the collision. **(b)** How high does the combination rise above its initial position?

56. Interactive LearningWare 7.1 at **www.wiley.com/college/cutnell** provides a review of the concepts that are involved in this problem. A 62.0-kg person, standing on a diving board, dives straight down into the water. Just before striking the water, her speed is 5.50 m/s. At a time of 1.65 s after she enters the water, her speed is reduced to 1.10 m/s. What is the net average force (magnitude and direction) that acts on her when she is in the water?

57. Interactive LearningWare 7.2 at **www.wiley.com/college/cutnell** provides a review of the concepts in this problem. For tests using a *ballistocardiograph*, a patient lies on a horizontal platform that is supported on jets of air. Because of the air jets, the friction impeding the horizontal motion of the platform is negligible. Each time the heart beats, blood is pushed out of the heart in a direction that is nearly parallel to the platform. Since momentum must be conserved, the body and the platform recoil, and this recoil can be detected to provide information about the heart. For each beat, suppose that 0.050 kg of blood is pushed out of the heart with a velocity of +0.25 m/s and that the mass of the patient and platform is 85 kg. Assuming that the patient does not slip with respect to the platform, and that the patient and platform start from rest, determine the recoil velocity.

58. A 46-kg skater is standing still in front of a wall. By pushing against the wall she propels herself backward with a velocity of -1.2 m/s. Her hands are in contact with the wall for 0.80 s. Ignore friction and wind resistance. Find the magnitude and direction of the average force she exerts on the wall (which has the same magnitude as, but opposite direction to, the force that the wall applies to her).

59. **GO** **Concept Simulation 7.1** at **www.wiley.com/college/cutnell** illustrates the physics principles in this problem. An astronaut in his space suit and with a propulsion unit (empty of its gas propellant) strapped to his back has a mass of 146 kg. The astronaut begins a space walk at rest, with a completely filled propulsion unit. During the space walk, the unit ejects some gas with a velocity of $+32$ m/s. As a result, the astronaut recoils with a velocity of -0.39 m/s. After the gas is ejected, the mass of the astronaut (now wearing a partially empty propulsion unit) is 165 kg. What percentage of the gas propellant was ejected from the completely filled propulsion unit?

60. **Interactive Solution 7.60** at **www.wiley.com/college/cutnell** presents a method for modeling this problem. The carbon monoxide molecule (CO) consists of a carbon atom and an oxygen atom separated by a distance of 1.13×10^{-10} m. The mass m_C of the carbon atom is 0.750 times the mass m_O of the oxygen atom, or $m_C = 0.750\,m_O$. Determine the location of the center of mass of this molecule relative to the carbon atom.

* **61.** **ssm** During July 1994 the comet Shoemaker–Levy 9 smashed into Jupiter in a spectacular fashion. The comet actually consisted of 21 distinct pieces, the largest of which had a mass of approximately 4.0×10^{12} kg and a speed of 6.0×10^4 m/s. Jupiter, the largest planet in the solar system, has a mass of 1.9×10^{27} kg and an orbital speed of 1.3×10^4 m/s. If this piece of the comet had hit Jupiter head-on, what would have been the *change* (magnitude only) in Jupiter's orbital speed (not its final speed)?

* **62.** A 40.0-kg boy, riding a 2.50-kg skateboard at a velocity of $+5.30$ m/s across a level sidewalk, jumps forward to leap over a wall. Just after leaving contact with the board, the boy's velocity relative to the sidewalk is 6.00 m/s, 9.50° above the horizontal. Ignore any friction between the skateboard and the sidewalk. What is the skateboard's velocity relative to the sidewalk at this instant? Be sure to include the correct algebraic sign with your answer.

* **63.** The lead female character in the movie *Diamonds Are Forever* is standing at the edge of an offshore oil rig. As she fires a gun, she is driven back over the edge and into the sea. Suppose the mass of a bullet is 0.010 kg and its velocity is $+720$ m/s. Her mass (including the gun) is 51 kg. **(a)** What recoil velocity does she acquire in response to a single shot from a stationary position, assuming that no external force keeps her in place? **(b)** Under the same assumption, what would be her recoil velocity if, instead, she shoots a blank cartridge that ejects a mass of 5.0×10^{-4} kg at a velocity of $+720$ m/s?

* **64.** The drawing shows a human figure in a sitting position. For purposes of this problem, there are three parts to the figure, and the center of mass of each one is shown in the drawing. These parts are: (1) the torso, neck, and head (total mass = 41 kg) with a center of mass located on the y axis at a point 0.39 m above the origin, (2) the upper legs (mass = 17 kg) with a center of mass located on the x axis at a point 0.17 m to the right of the origin, and (3) the lower legs and feet (total mass = 9.9 kg) with a center of mass located 0.43 m to the right of and 0.26 m below the origin. Find the x and y coordinates of the center of mass of the human figure. Note that the mass of the arms and hands (approximately 12% of the whole-body mass) has been ignored to simplify the drawing.

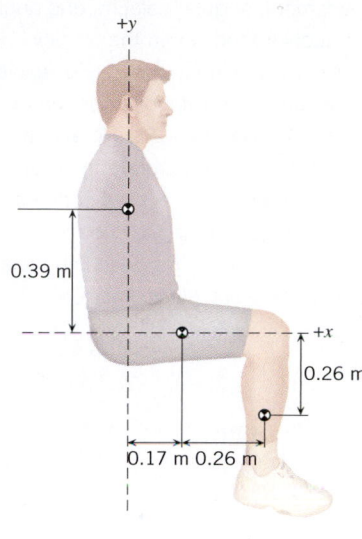

** **65.** **ssm** Two people are standing on a 2.0-m-long platform, one at each end. The platform floats parallel to the ground on a cushion of air, like a hovercraft. One person throws a 6.0-kg ball to the other, who catches it. The ball travels nearly horizontally. Excluding the ball, the total mass of the platform and people is 118 kg. Because of the throw, this 118-kg mass recoils. How far does it move before coming to rest again?

CHAPTER 8

ROTATIONAL KINEMATICS

By using the concepts of angular displacement, angular velocity, and angular acceleration within the framework of rotational kinematics, it is possible to describe the rotational motion of these Peruvian dancers. Rotational kinematics is the central theme of this chapter, and the three concepts are analogous to those used in Chapters 2 and 3 for describing linear motion. (© Hugh Sitton/Zefa/Corbis)

8.1 ROTATIONAL MOTION AND ANGULAR DISPLACEMENT

In the simplest kind of rotation, points on a rigid object move on circular paths. In Figure 8.1, for example, we see the circular paths for points A, B, and C on a spinning skater. The centers of all such circular paths define a line, called the ***axis of rotation***.

The angle through which a rigid object rotates about a fixed axis is called the ***angular displacement***. Figure 8.2 shows how the angular displacement is measured for a rotating compact disc (CD). Here, the axis of rotation passes through the center of the disc and is perpendicular to its surface. On the surface of the CD we draw a radial line, which is a line that intersects the axis of rotation perpendicularly. As the CD turns, we observe the angle through which this line moves relative to a convenient reference line that does not rotate. The radial line moves from its initial orientation at angle θ_0 to a final orientation at angle θ (Greek letter theta). In the process, the line sweeps out the angle $\theta - \theta_0$. As with other differences that we have encountered ($\Delta x = x - x_0$, $\Delta v = v - v_0$, $\Delta t = t - t_0$), it is customary to denote the difference between the final and initial angles by the notation $\Delta\theta$ (read as "delta theta"): $\Delta\theta = \theta - \theta_0$. The angle $\Delta\theta$ is the angular displacement. A rotating object may rotate either counterclockwise or clockwise, and standard convention calls a counterclockwise displacement positive and a clockwise displacement negative.

DEFINITION OF ANGULAR DISPLACEMENT

When a rigid body rotates about a fixed axis, the angular displacement is the angle $\Delta\theta$ swept out by a line passing through any point on the body and intersecting the axis of rotation perpendicularly. By convention, the angular displacement is positive if it is counterclockwise and negative if it is clockwise.

SI Unit of Angular Displacement: radian (rad)*

*The radian is neither a base SI unit nor a derived one. It is regarded as a supplementary SI unit.

Figure 8.1 When an object rotates, points on the object, such as A, B, or C, move on circular paths. The centers of the circles form a line that is the axis of rotation.

Angular displacement is often expressed in one of three units. The first is the familiar *degree*, and it is well known that there are 360 degrees in a circle. The second unit is the *revolution (rev)*, one revolution representing one complete turn of 360°. The most useful unit from a scientific viewpoint, however, is the SI unit called the *radian (rad)*. Figure 8.3 shows how the radian is defined, again using a CD as an example. The picture focuses attention on a point P on the disc. This point starts out on the stationary reference line, so that $\theta_0 = 0$ rad, and the angular displacement is $\Delta\theta = \theta - \theta_0 = \theta$. As the disc rotates, the point traces out an arc of length s, which is measured along a circle of radius r. Equation 8.1 defines the angle θ in radians:

$$\theta \text{ (in radians)} = \frac{\text{Arc length}}{\text{Radius}} = \frac{s}{r} \qquad (8.1)$$

According to this definition, an angle in radians is the ratio of two lengths; for example, meters/meters. In calculations, therefore, the radian is treated as a number without units and has no effect on other units that it multiplies or divides.

To convert between degrees and radians, it is only necessary to remember that the arc length of an entire circle of radius r is the circumference $2\pi r$. Therefore, according to Equation 8.1, *the number of radians that corresponds to 360°, or one revolution, is*

$$\theta = \frac{2\pi r}{r} = 2\pi \text{ rad}$$

Since 2π rad corresponds to 360°, the number of degrees in one radian is

$$1 \text{ rad} = \frac{360°}{2\pi} = 57.3°$$

It is useful to express an angle θ in radians, because then the arc length s subtended at any radius r can be calculated by multiplying θ by r. Example 1 illustrates this point and also shows how to convert between degrees and radians.

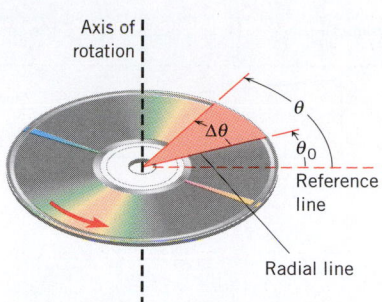

Figure 8.2 The angular displacement of a CD is the angle $\Delta\theta$ swept out by a radial line as the disc turns about its axis of rotation.

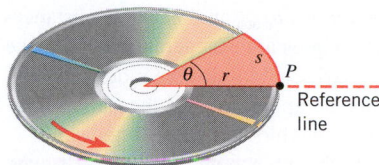

Figure 8.3 In radian measure, the angle θ is defined to be the arc length s divided by the radius r.

 Example 1 Adjacent Synchronous Satellites

Synchronous or "stationary" communications satellites are put into an orbit whose radius is $r = 4.23 \times 10^7$ m. The orbit is in the plane of the equator, and two adjacent satellites have an angular separation of $\theta = 2.00°$, as Figure 8.4 illustrates. Find the arc length s (see the drawing) that separates the satellites.

The physics of communications satellites.

Reasoning Since the radius r and the angle θ are known, we may find the arc length s by using the relation θ (in radians) $= s/r$. But first, the angle must be converted to radians from degrees.

Solution To convert 2.00° into radians, we use the fact that 2π radians is equivalent to 360°:

$$2.00° = (2.00 \text{ degrees})\left(\frac{2\pi \text{ radians}}{360 \text{ degrees}}\right) = 0.0349 \text{ radians}$$

From Equation 8.1, it follows that the arc length between the satellites is

$$s = r\theta = (4.23 \times 10^7 \text{ m})(0.0349 \text{ rad}) = \boxed{1.48 \times 10^6 \text{ m (920 miles)}}$$

The radian, being a unitless quantity, is dropped from the final result, leaving the answer expressed in meters.

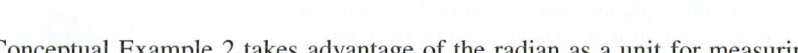

Conceptual Example 2 takes advantage of the radian as a unit for measuring angles and explains the spectacular phenomenon of a total solar eclipse.

 Conceptual Example 2 A Total Eclipse of the Sun

The diameter of the sun is about 400 times greater than that of the moon. By coincidence, the sun is also about 400 times farther from the earth than is the moon. For an observer on earth, compare the angles subtended by the sun and the moon. **(a)** The angle subtended by the sun is

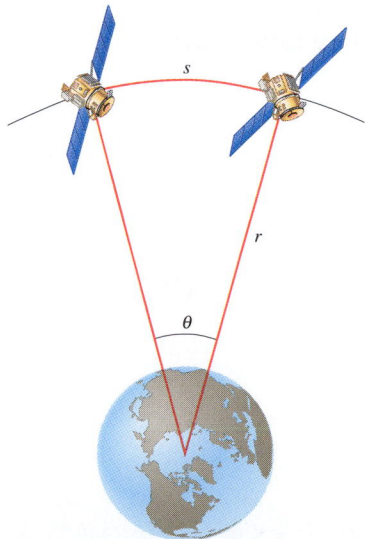

Figure 8.4 Two adjacent synchronous satellites have an angular separation of $\theta = 2.00°$. The distances and angles have been exaggerated for clarity.

s_{sun}

s_{moon}

r_{moon}

θ_{moon}

r_{sun}

θ_{sun}

(a)

(b)

(c)

Figure 8.5 (a) The angles subtended by the moon and sun at the eyes of the observer are θ_{moon} and θ_{sun}. (The distances and angles are exaggerated for the sake of clarity.) (b) Since the moon and sun subtend approximately the same angle, the moon blocks nearly all the sun's light from reaching the observer's eyes. (c) The result is a total solar eclipse. (c, Roger Ressmeyer/Corbis Images)

much greater than that subtended by the moon. **(b)** The angle subtended by the sun is much smaller than that subtended by the moon. **(c)** The angles subtended by the sun and the moon are approximately equal.

Reasoning Equation 8.1 ($\theta = s/r$), which is the definition of an angle in radians, will guide us. The distance r between either the sun and the earth or the moon and the earth is great enough that the arc length s is very nearly equal to the diameter of the sun or the moon.

Answers (a) and (b) are incorrect. Figure 8.5a shows a person on earth viewing the sun and the moon. For the diameters s of the sun and the moon, we know that $s_{sun} \approx 400\, s_{moon}$ (where the symbol $\approx$ means "approximately equal to"). For the distances r between the sun and the earth or the moon and the earth, we know that $r_{sun} \approx 400\, r_{moon}$. Because of these facts, the ratio s/r is approximately the same for the sun and for the moon. Therefore, the subtended angle ($\theta = s/r$) is approximately the same for the sun and the moon.

Answer (c) is correct. Applying Equation 8.1 to the case of the sun and the moon gives $\theta_{sun} = s_{sun}/r_{sun}$ and $\theta_{moon} = s_{moon}/r_{moon}$. We know, however, that $s_{sun} \approx 400\, s_{moon}$ and $r_{sun} \approx 400\, r_{moon}$. Substitution into the expression for θ_{sun} gives

$$\theta_{sun} = \frac{s_{sun}}{r_{sun}} \approx \frac{400\, s_{moon}}{400\, r_{moon}} = \theta_{moon}$$

The physics of a total solar eclipse.

Figure 8.5b shows what happens when the moon comes between the sun and the earth. Since the angles subtended by the sun and the moon are nearly equal, the moon blocks most of the sun's light from reaching the observer's eyes, and a total solar eclipse like that in Figure 8.5c occurs.

Related Homework: *Problems 6, 18*

✓ CHECK YOUR UNDERSTANDING

(The answers are given at the end of the book.)

1. In the drawing, the flat triangular sheet ABC is lying in the plane of the paper. This sheet is going to rotate about first one axis and then another axis. Both of these axes lie in the plane of the paper and pass through point A. For each of the axes the points B and C move on separate circular paths that have the same radii. Identify these two axes.

C

A

B

2. Three objects are visible in the night sky. They have the following diameters (in multiples of d) and subtend the following angles (in multiples of θ_0) at the eye of the observer. Object A has a diameter of $4d$ and subtends an angle of $2\theta_0$. Object B has a diameter of $3d$ and subtends an angle of $\theta_0/2$. Object C has a diameter of $d/2$ and subtends an angle of $\theta_0/8$. Rank them in descending order (greatest first) according to their distance from the observer.

8.2 ANGULAR VELOCITY AND ANGULAR ACCELERATION

ANGULAR VELOCITY

In Section 2.2 we introduced the idea of linear velocity to describe how fast an object moves and the direction of its motion. The average linear velocity $\bar{\vec{v}}$ was defined as the linear displacement $\Delta\vec{x}$ of the object divided by the time Δt required for the displacement to occur, or $\bar{\vec{v}} = \Delta\vec{x}/\Delta t$ (see Equation 2.2). We now introduce the analogous idea of angular velocity to describe the motion of a rigid object rotating about a fixed axis. The **average angular velocity** $\bar{\omega}$ (Greek letter omega) is defined as the angular displacement $\Delta\theta = \theta - \theta_0$ divided by the elapsed time Δt during which the displacement occurs.

DEFINITION OF AVERAGE ANGULAR VELOCITY

$$\frac{\text{Average angular}}{\text{velocity}} = \frac{\text{Angular displacement}}{\text{Elapsed time}}$$

$$\bar{\omega} = \frac{\theta - \theta_0}{t - t_0} = \frac{\Delta\theta}{\Delta t} \tag{8.2}$$

SI Unit of Angular Velocity: radian per second (rad/s)

The SI unit for angular velocity is the radian per second (rad/s), although other units such as revolutions per minute (rev/min or rpm) are also used. In agreement with the sign convention adopted for angular displacement, angular velocity is positive when the rotation is counterclockwise and negative when it is clockwise. Example 3 shows how the concept of average angular velocity is applied to a gymnast.

 Example 3   Gymnast on a High Bar

A gymnast on a high bar swings through two revolutions in a time of 1.90 s, as Figure 8.6 suggests. Find the average angular velocity (in rad/s) of the gymnast.

Reasoning The average angular velocity of the gymnast in rad/s is the angular displacement in radians divided by the elapsed time. However, the angular displacement is given as two revolutions, so we begin by converting this value into radian measure.

Solution The angular displacement (in radians) of the gymnast is

$$\Delta\theta = -2.00 \text{ revolutions} \left(\frac{2\pi \text{ radians}}{1 \text{ revolution}} \right) = -12.6 \text{ radians}$$

where the minus sign denotes that the gymnast rotates clockwise (see the drawing). The average angular velocity is

$$\bar{\omega} = \frac{\Delta\theta}{\Delta t} = \frac{-12.6 \text{ rad}}{1.90 \text{ s}} = \boxed{-6.63 \text{ rad/s}} \tag{8.2}$$

Figure 8.6 Swinging on a high bar.

The *instantaneous angular velocity* ω is the angular velocity that exists at any given instant. To measure it, we follow the same procedure used in Chapter 2 for the instantaneous linear velocity. In this procedure, a small angular displacement $\Delta\theta$ occurs during a small time interval Δt. The time interval is so small that it approaches zero ($\Delta t \rightarrow 0$), and in this limit, the measured average angular velocity, $\bar{\omega} = \Delta\theta/\Delta t$, becomes the instantaneous angular velocity ω:

$$\omega = \lim_{\Delta t \to 0} \bar{\omega} = \lim_{\Delta t \to 0} \frac{\Delta\theta}{\Delta t} \tag{8.3}$$

The magnitude of the instantaneous angular velocity, without reference to whether it is a positive or negative quantity, is called the **instantaneous angular speed**. If a rotating object has a constant angular velocity, the instantaneous value and the average value are the same.

ANGULAR ACCELERATION

In linear motion, a changing velocity means that an acceleration is occurring. Such is also the case in rotational motion; a changing angular velocity means that an **angular acceleration** is occurring. There are many examples of angular acceleration. For instance, as a compact disc recording is played, the disc turns with an angular velocity that is continually decreasing. And when the push buttons of an electric blender are changed from a lower setting to a higher setting, the angular velocity of the blades increases. We will define the average angular acceleration in a fashion analogous to that used for the average linear acceleration. Recall that the average linear acceleration $\vec{\mathbf{a}}$ is equal to the change $\Delta\vec{\mathbf{v}}$ in the linear velocity of an object divided by the elapsed time Δt: $\vec{\mathbf{a}} = \Delta\vec{\mathbf{v}}/\Delta t$ (Equation 2.4). When the angular velocity changes from an initial value of ω_0 at time t_0 to a final value of ω at time t, the average angular acceleration $\overline{\alpha}$ (Greek letter alpha) is defined similarly:

DEFINITION OF AVERAGE ANGULAR ACCELERATION

$$\text{Average angular acceleration} = \frac{\text{Change in angular velocity}}{\text{Elapsed time}}$$

$$\overline{\alpha} = \frac{\omega - \omega_0}{t - t_0} = \frac{\Delta\omega}{\Delta t} \tag{8.4}$$

SI Unit of Average Angular Acceleration: radian per second squared (rad/s²)

The SI unit for average angular acceleration is the unit for angular velocity divided by the unit for time, or (rad/s)/s = rad/s². An angular acceleration of +5 rad/s², for example, means that the angular velocity of the rotating object increases by +5 radians per second during each second of acceleration.

The **instantaneous angular acceleration** α is the angular acceleration at a given instant. In discussing linear motion, we assumed a condition of constant acceleration, so that the average and instantaneous accelerations were identical ($\vec{\mathbf{a}} = \vec{\mathbf{a}}$). Similarly, we assume that the angular acceleration is constant, so that the instantaneous angular acceleration α and the average angular acceleration $\overline{\alpha}$ are the same ($\overline{\alpha} = \alpha$). The next example illustrates the concept of angular acceleration.

 Example 4 A Jet Revving Its Engines

A jet awaiting clearance for takeoff is momentarily stopped on the runway. As seen from the front of one engine, the fan blades are rotating with an angular velocity of −110 rad/s, where the negative sign indicates a clockwise rotation (see Figure 8.7). As the plane takes off, the angular velocity of the blades reaches −330 rad/s in a time of 14 s. Find the angular acceleration, assuming it to be constant.

Reasoning Since the angular acceleration is constant, it is equal to the average angular acceleration. The average acceleration is the change in the angular velocity, $\omega - \omega_0$, divided by the elapsed time, $t - t_0$.

Solution Applying the definition of average angular acceleration given in Equation 8.4, we find that

$$\overline{\alpha} = \frac{\omega - \omega_0}{t - t_0} = \frac{(-330 \text{ rad/s}) - (-110 \text{ rad/s})}{14 \text{ s}} = \boxed{-16 \text{ rad/s}^2}$$

Thus, the magnitude of the angular velocity increases by 16 rad/s during each second that the blades are accelerating. The negative sign in the answer indicates that the direction of the angular acceleration is also in the clockwise direction.

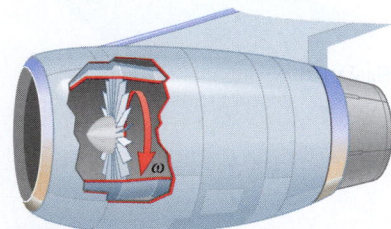

Figure 8.7 The fan blades of a jet engine have an angular velocity ω in a clockwise direction.

3. A pair of scissors is being used to cut a piece of paper in half. Does each blade of the scissors have the same angular velocity (both magnitude and direction) at a given instant?

4. An electric clock is hanging on a wall. As you are watching the second hand rotate, the clock's battery stops functioning, and the second hand comes to a halt over a brief period of time. Which one of the following statements correctly describes the angular velocity ω and angular acceleration α of the second hand as it slows down?   **(a)** ω and α are both negative. **(b)** ω is positive and α is negative.   **(c)** ω is negative and α is positive.   **(d)** ω and α are both positive.

8.3 THE EQUATIONS OF ROTATIONAL KINEMATICS

In Chapters 2 and 3 the concepts of displacement, velocity, and acceleration were introduced. We then combined these concepts and developed a set of equations called the equations of kinematics for constant acceleration (see Tables 2.1 and 3.1). These equations are a great aid in solving problems involving linear motion in one and two dimensions.

We now take a similar approach for rotational motion by bringing together the ideas of angular displacement, angular velocity, and angular acceleration to produce a set of equations called the equations of rotational kinematics for constant angular acceleration. These equations, like those developed in Chapters 2 and 3 for linear motion, will prove very useful in solving problems that involve rotational motion.

A complete description of rotational motion requires values for the angular displacement $\Delta\theta$, the angular acceleration α, the final angular velocity ω, the initial angular velocity ω_0, and the elapsed time Δt. In Example 4, for instance, only the angular displacement of the fan blades during the 14-s interval is missing. Such missing information can be calculated, however. For convenience in the calculations, we assume that the orientation of the rotating object is given by $\theta_0 = 0$ rad at time $t_0 = 0$ s. Then, the angular displacement becomes $\Delta\theta = \theta - \theta_0 = \theta$, and the time interval becomes $\Delta t = t - t_0 = t$.

In Example 4, the angular velocity of the fan blades changes at a constant rate from an initial value of $\omega_0 = -110$ rad/s to a final value of $\omega = -330$ rad/s. Therefore, the average angular velocity is midway between the initial and final values:

$$\overline{\omega} = \tfrac{1}{2}[(-110 \text{ rad/s}) + (-330 \text{ rad/s})] = -220 \text{ rad/s}$$

In other words, when the angular acceleration is constant, the average angular velocity is given by

$$\overline{\omega} = \tfrac{1}{2}(\omega_0 + \omega) \tag{8.5}$$

With a value for the average angular velocity, Equation 8.2 can be used to obtain the angular displacement of the fan blades:

$$\theta = \overline{\omega}t = (-220 \text{ rad/s})(14 \text{ s}) = -3100 \text{ rad}$$

In general, when the angular acceleration is constant, the angular displacement can be obtained from

$$\theta = \overline{\omega}t = \tfrac{1}{2}(\omega_0 + \omega)t \tag{8.6}$$

This equation and Equation 8.4 provide a complete description of rotational motion under the condition of constant angular acceleration. Equation 8.4 (with $t_0 = 0$ s) and Equation 8.6 are compared with the analogous results for linear motion in the first two rows of Table 8.1.

Table 8.1 The Equations of Kinematics for Rotational and Linear Motion

Rotational Motion (α = constant)		Linear Motion (a = constant)	
$\omega = \omega_0 + \alpha t$	(8.4)	$v = v_0 + at$	(2.4)
$\theta = \tfrac{1}{2}(\omega_0 + \omega)t$	(8.6)	$x = \tfrac{1}{2}(v_0 + v)t$	(2.7)
$\theta = \omega_0 t + \tfrac{1}{2}\alpha t^2$	(8.7)	$x = v_0 t + \tfrac{1}{2}at^2$	(2.8)
$\omega^2 = \omega_0^2 + 2\alpha\theta$	(8.8)	$v^2 = v_0^2 + 2ax$	(2.9)

Table 8.2 Symbols Used in Rotational and Linear Kinematics

Rotational Motion	Quantity	Linear Motion
θ	Displacement	x
ω_0	Initial velocity	v_0
ω	Final velocity	v
α	Acceleration	a
t	Time	t

The purpose of this comparison is to emphasize that the mathematical forms of Equations 8.4 and 2.4 are identical, as are the forms of Equations 8.6 and 2.7. Of course, the symbols used for the rotational variables are different from those used for the linear variables, as Table 8.2 indicates.

In Chapter 2, Equations 2.4 and 2.7 are used to derive the remaining two equations of kinematics (Equations 2.8 and 2.9). These additional equations convey no new information but are convenient to have when solving problems. Similar derivations can be carried out here. The results are listed as Equations 8.7 and 8.8 below and in Table 8.1; they can be inferred directly from their counterparts in linear motion by making the substitution of symbols indicated in Table 8.2:

$$\theta = \omega_0 t + \tfrac{1}{2}\alpha t^2 \qquad (8.7)$$

$$\omega^2 = \omega_0^2 + 2\alpha\theta \qquad (8.8)$$

The four equations in the left column of Table 8.1 are called the **equations of rotational kinematics for constant angular acceleration.** The following example illustrates that they are used in the same fashion as the equations of linear kinematics.

 Example 5   Blending with a Blender

The blades of an electric blender are whirling with an angular velocity of $+375$ rad/s while the "puree" button is pushed in, as Figure 8.8 shows. When the "blend" button is pressed, the blades accelerate and reach a greater angular velocity after the blades have rotated through an angular displacement of $+44.0$ rad. The angular acceleration has a constant value of $+1740$ rad/s^2. Find the final angular velocity of the blades.

Reasoning The three known variables are listed in the table below, along with a question mark indicating that a value for the final angular velocity ω is being sought.

θ	α	ω	ω_0	t
$+44.0$ rad	$+1740$ rad/s^2	?	$+375$ rad/s	

We can use Equation 8.8, because it relates the angular variables θ, α, ω, and ω_0.

Solution From Equation 8.8 ($\omega^2 = \omega_0^2 + 2\alpha\theta$) it follows that

$$\omega = +\sqrt{\omega_0^2 + 2\alpha\theta} = +\sqrt{(375 \text{ rad/s})^2 + 2(1740 \text{ rad/s}^2)(44.0 \text{ rad})} = \boxed{+542 \text{ rad/s}}$$

The negative root is disregarded, since the blades do not reverse their direction of rotation.

Problem-solving insight

Each equation of rotational kinematics contains four of the five kinematic variables, θ, α, ω, ω_0, and t. Therefore, it is necessary to have values for three of these variables if one of the equations is to be used to determine a value for an unknown variable.

Problem-solving insight: *The equations of rotational kinematics can be used with any self-consistent set of units for θ, α, ω, ω_0, and t.* Radians are used in Example 5 only because data are given in terms of radians. Had the data for θ, α, and ω_0 been provided in rev, rev/s^2, and rev/s, respectively, then Equation 8.8 could have been used to determine the answer for ω directly in rev/s. In any case, the reasoning strategy for applying the kinematics equations is summarized as follows.

Figure 8.8 The angular velocity of the blades in an electric blender changes each time a different push button is chosen.

REASONING STRATEGY

Applying the Equations of Rotational Kinematics

1. Make a drawing to represent the situation being studied, showing the direction of rotation.

2. Decide which direction of rotation is to be called positive ($+$) and which direction is to be called negative ($-$). In this text we choose the counterclockwise direction to be positive and the clockwise direction to be negative, but this is arbitrary. However, do not change your decision during the course of a calculation.

3. In an organized way, write down the values (with appropriate + and − signs) that are given for any of the five rotational kinematic variables (θ, α, ω, ω_0, and t). Be on the alert for implied data, such as the phrase "starts from rest," which means that the value of the initial angular velocity is $\omega_0 = 0$ rad/s. The data box in Example 5 is a good way to keep track of this information. In addition, identify the variable(s) that you are being asked to determine.

4. Before attempting to solve a problem, verify that the given information contains values for at least three of the five kinematic variables. Once the three variables are identified for which values are known, the appropriate relation from Table 8.1 can be selected.

5. When the rotational motion is divided into segments, the final angular velocity of one segment is the initial angular velocity for the next segment.

6. Keep in mind that there may be two possible answers to a kinematics problem. Try to visualize the different physical situations to which the answers correspond.

✓ **CHECK YOUR UNDERSTANDING**

(The answers are given at the end of the book.)

5. The blades of a ceiling fan start from rest and, after two revolutions, have an angular speed of 0.50 rev/s. The angular acceleration of the blades is constant. What is the angular speed after eight revolutions?

6. Equation 8.7 ($\theta = \omega_0 t + \frac{1}{2}\alpha t^2$) is being used to solve a problem in rotational kinematics. Which one of the following sets of values for the variables ω_0, α, and t *cannot* be substituted directly into this equation to calculate a value for θ?   **(a)** $\omega_0 = 1.0$ rad/s, $\alpha = 1.8$ rad/s^2, and $t = 3.8$ s   **(b)** $\omega_0 = 0.16$ rev/s, $\alpha = 1.8$ rad/s^2, and $t = 3.8$ s   **(c)** $\omega_0 = 0.16$ rev/s, $\alpha = 0.29$ rev/s^2, and $t = 3.8$ s

8.4 ANGULAR VARIABLES AND TANGENTIAL VARIABLES

In the familiar ice-skating stunt known as crack-the-whip, a number of skaters attempt to maintain a straight line as they skate around the one person (the pivot) who remains in place. Figure 8.9 shows each skater moving on a circular arc and includes the corresponding velocity vector at the instant portrayed in the picture. For every individual skater, the vector is drawn tangent to the appropriate circle and, therefore, is called the ***tangential velocity*** $\vec{v}_T$. The magnitude of the tangential velocity is referred to as the ***tangential speed***.

Of all the skaters involved in the stunt, the one farthest from the pivot has the hardest job. Why? Because, in keeping the line straight, this skater covers more distance than anyone else. To accomplish this, he must skate faster than anyone else and, thus, must have the largest tangential speed. In fact, the line remains straight only if each person skates with the correct tangential speed. The skaters closer to the pivot must move with smaller tangential speeds than those farther out, as indicated by the magnitudes of the tangential velocity vectors in Figure 8.9.

With the aid of Figure 8.10, it is possible to show that the tangential speed of any skater is directly proportional to his distance r from the pivot, assuming a given angular speed for the rotating line. When the line rotates as a rigid unit for a time t, it sweeps out the angle θ shown in the drawing. The distance s through which a skater moves along a circular arc can be calculated from $s = r\theta$ (Equation 8.1), provided θ is measured in radians. Dividing both sides of this equation by t gives $s/t = r(\theta/t)$. The term s/t is the tangential speed v_T (e.g., in meters/second) of the skater, while θ/t is the angular speed ω (in radians/second) of the line:

$$v_T = r\omega \qquad (\omega \text{ in rad/s}) \qquad (8.9)$$

In this expression, the terms v_T and ω refer to the magnitudes of the tangential and angular velocities, respectively, and are numbers without algebraic signs.

It is important to emphasize that the angular speed ω in Equation 8.9 must be expressed in radian measure (e.g., in rad/s); no other units, such as revolutions per second,

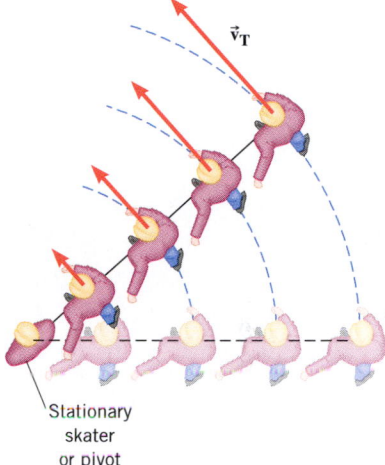

Figure 8.9 When doing the stunt known as crack-the-whip, each skater along the radial line moves on a circular arc. The tangential velocity $\vec{v}_T$ of each skater is represented by an arrow that is tangent to each arc.

The physics of "crack-the-whip."

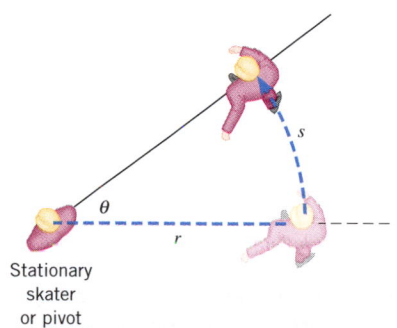

Figure 8.10 During a time t, the line of skaters sweeps through an angle θ. An individual skater, located at a distance r from the stationary skater, moves through a distance s on a circular arc.

are acceptable. This restriction arises because the equation was derived by using the definition of radian measure, $s = r\theta$.

The real challenge for the crack-the-whip skaters is to keep the line straight while making it pick up angular speed—that is, while giving it an angular acceleration. To make the angular speed of the line increase, each skater must increase his tangential speed, since the two speeds are related according to $v_T = r\omega$. Of course, the fact that a skater must skate faster and faster means that he must accelerate, and his tangential acceleration a_T can be related to the angular acceleration α of the line. If the time is measured relative to $t_0 = 0$ s, the definition of linear acceleration is given by Equation 2.4 as $a_T = (v_T - v_{T0})/t$, where v_T and v_{T0} are the final and initial tangential speeds, respectively. Substituting $v_T = r\omega$ for the tangential speed shows that

$$a_T = \frac{v_T - v_{T0}}{t} = \frac{(r\omega) - (r\omega_0)}{t} = r\left(\frac{\omega - \omega_0}{t}\right)$$

Since $\alpha = (\omega - \omega_0)/t$ according to Equation 8.4, it follows that

$$a_T = r\alpha \qquad (\alpha \text{ in rad/s}^2) \tag{8.10}$$

This result shows that, for a given value of α, the tangential acceleration a_T is proportional to the radius r, so the skater farthest from the pivot must have the largest tangential acceleration. In this expression, the terms a_T and α refer to the magnitudes of the numbers involved, without reference to any algebraic sign. Moreover, as is the case for ω in $v_T = r\omega$, only radian measure can be used for α in Equation 8.10.

There is an advantage to using the angular velocity ω and the angular acceleration α to describe the rotational motion of a rigid object. The advantage is that these angular quantities describe the motion of the *entire object*. In contrast, the tangential quantities v_T and a_T describe only the motion of a single point on the object, and Equations 8.9 and 8.10 indicate that different points located at different distances r have different tangential velocities and accelerations. Example 6 stresses this advantage.

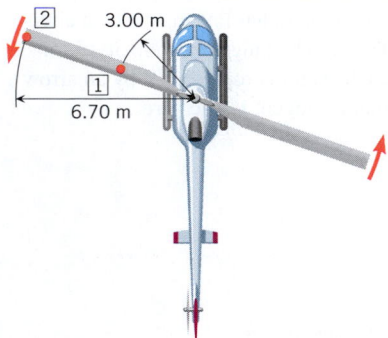

Figure 8.11 Points 1 and 2 on the rotating blade of the helicopter have the same angular speed and acceleration, but they have *different* tangential speeds and accelerations.

Example 6 A Helicopter Blade

A helicopter blade has an angular speed of $\omega = 6.50$ rev/s and has an angular acceleration of $\alpha = 1.30$ rev/s². For points 1 and 2 on the blade in Figure 8.11, find the magnitudes of **(a)** the tangential speeds and **(b)** the tangential accelerations.

Reasoning Since the radius r for each point and the angular speed ω of the helicopter blade are known, we can find the tangential speed v_T for each point by using the relation $v_T = r\omega$. However, since this equation can be used only with radian measure, the angular speed ω must be converted to rad/s from rev/s. In a similar manner, the tangential acceleration a_T for points 1 and 2 can be found using $a_T = r\alpha$, provided the angular acceleration α is expressed in rad/s² rather than in rev/s².

Solution **(a)** Converting the angular speed ω to rad/s from rev/s, we obtain

$$\omega = \left(6.50 \, \frac{\text{rev}}{\text{s}}\right)\left(\frac{2\pi \, \text{rad}}{1 \, \text{rev}}\right) = 40.8 \, \frac{\text{rad}}{\text{s}}$$

The tangential speed for each point is

Point 1 $\qquad v_T = r\omega = (3.00 \text{ m})(40.8 \text{ rad/s}) = \boxed{122 \text{ m/s (273 mph)}}$ $\qquad$ (8.9)

Point 2 $\qquad v_T = r\omega = (6.70 \text{ m})(40.8 \text{ rad/s}) = \boxed{273 \text{ m/s (611 mph)}}$ $\qquad$ (8.9)

The rad unit, being dimensionless, does not appear in the final answers.

(b) Converting the angular acceleration α to rad/s² from rev/s², we find

$$\alpha = \left(1.30 \, \frac{\text{rev}}{\text{s}^2}\right)\left(\frac{2\pi \, \text{rad}}{1 \, \text{rev}}\right) = 8.17 \, \frac{\text{rad}}{\text{s}^2}$$

The tangential accelerations can now be determined:

Point 1 $\qquad a_T = r\alpha = (3.00 \text{ m})(8.17 \text{ rad/s}^2) = \boxed{24.5 \text{ m/s}^2}$ $\qquad$ (8.10)

Point 2 $\qquad a_T = r\alpha = (6.70 \text{ m})(8.17 \text{ rad/s}^2) = \boxed{54.7 \text{ m/s}^2}$ $\qquad$ (8.10)

Problem-solving insight

When using Equations 8.9 and 8.10 to relate tangential and angular quantities, remember that the angular quantity is always expressed in radian measure. These equations are not valid if angles are expressed in degrees or revolutions.

✓ **CHECK YOUR UNDERSTANDING**

(*The answers are given at the end of the book.*)

7. A thin rod rotates at a constant angular speed. In case A the axis of rotation is perpendicular to the rod at its center. In case B the axis is perpendicular to the rod at one end. In which case, if either, are there points on the rod that have the same tangential speeds?

8. It is possible to build a clock in which the tips of the hour hand and the second hand move with the same tangential speed. This is normally never done, however. Why? **(a)** The length of the hour hand would be 3600 times greater than the length of the second hand. **(b)** The hour hand and the second hand would have the same length. **(c)** The length of the hour hand would be 3600 times smaller than the length of the second hand.

9. The earth rotates once per day about its axis, which is perpendicular to the plane of the equator and passes through the north geographic pole. Where on the earth's surface should you stand in order to have the smallest possible tangential speed?

10. A building is located on the earth's equator. As the earth rotates about its axis, which floor of the building has the greater tangential speed? **(a)** The first floor **(b)** The tenth floor **(c)** The twentieth floor

11. The blade of a lawn mower is rotating at an angular speed of 17 rev/s. The tangential speed of the outer edge of the blade is 32 m/s. What is the radius of the blade?

8.5 CENTRIPETAL ACCELERATION AND TANGENTIAL ACCELERATION

When an object picks up speed as it moves around a circle, it has a tangential acceleration, as discussed in the last section. In addition, the object also has a centripetal acceleration, as emphasized in Chapter 5. That chapter deals with *uniform circular motion,* in which a particle moves at a constant tangential speed on a circular path. The tangential speed v_T is the magnitude of the tangential velocity vector. Even when the magnitude of the tangential velocity is constant, an acceleration is present, since the direction of the velocity changes continually. Because the resulting acceleration points toward the center of the circle, it is called the centripetal acceleration. Figure 8.12*a* shows the centripetal acceleration $\vec{a}_c$ for a model airplane flying in uniform circular motion on a guide wire. The magnitude of $\vec{a}_c$ is

$$a_c = \frac{v_T^2}{r} \qquad (5.2)$$

The subscript "T" has now been included in this equation as a reminder that it is the tangential speed that appears in the numerator.

The centripetal acceleration can be expressed in terms of the angular speed ω by using $v_T = r\omega$ (Equation 8.9):

$$a_c = \frac{v_T^2}{r} = \frac{(r\omega)^2}{r} = r\omega^2 \qquad (\omega \text{ in rad/s}) \qquad (8.11)$$

Only radian measure (rad/s) can be used for ω in this result, since the relation $v_T = r\omega$ presumes radian measure.

While considering uniform circular motion in Chapter 5, we ignored the details of how the motion is established in the first place. In Figure 8.12*b*, for instance, the engine of the plane produces a thrust in the tangential direction, and this force leads to a tangential acceleration. In response, the tangential speed of the plane increases from moment to moment, until the situation shown in the drawing results. While the tangential speed is changing, the motion is called *nonuniform circular motion.*

Figure 8.12*b* illustrates an important feature of nonuniform circular motion. Since the direction and the magnitude of the tangential velocity are both changing, the airplane experiences two acceleration components simultaneously. The changing direction means that there is a centripetal acceleration $\vec{a}_c$. The magnitude of $\vec{a}_c$ at any moment can be calculated using the value of the instantaneous angular speed and the radius: $a_c = r\omega^2$. The fact that the magnitude of the tangential velocity is changing means that there is also a tangential acceleration $\vec{a}_T$. The magnitude of $\vec{a}_T$ can be determined from the angular acceleration α according to $a_T = r\alpha$, as the previous section explains. If the magnitude F_T of the net

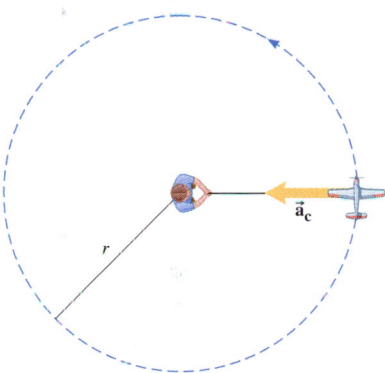

(*a*) Uniform circular motion

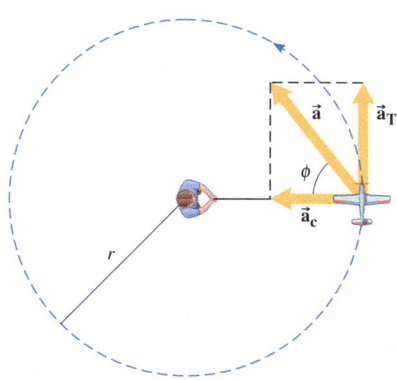

(*b*) Nonuniform circular motion

Figure 8.12 (*a*) If a model airplane flying on a guide wire has a constant tangential speed, the motion is uniform circular motion, and the plane experiences only a centripetal acceleration $\vec{a}_c$. (*b*) Nonuniform circular motion occurs when the tangential speed changes. Then there is a tangential acceleration $\vec{a}_T$ in addition to the centripetal acceleration.

tangential force and the mass m are known, a_T also can be calculated using Newton's second law, $F_T = ma_T$. Figure 8.12b shows the two acceleration components. The total acceleration is given by the vector sum of $\vec{a}_c$ and $\vec{a}_T$. Since $\vec{a}_c$ and $\vec{a}_T$ are perpendicular, the magnitude of the total acceleration $\vec{a}$ can be obtained from the Pythagorean theorem as $a = \sqrt{a_c^2 + a_T^2}$, while the angle ϕ in the drawing can be determined from $\tan \phi = a_T/a_c$. The next example applies these concepts to a discus thrower.

ANALYZING MULTIPLE-CONCEPT PROBLEMS

Example 7 A Discus Thrower

Discus throwers often warm up by throwing the discus with a twisting motion of their bodies. Figure 8.13a illustrates a top view of such a warm-up throw. Starting from rest, the thrower accelerates the discus to a final angular velocity of +15.0 rad/s in a time of 0.270 s before releasing it. During the acceleration, the discus moves on a circular arc of radius 0.810 m. Find the magnitude a of the total acceleration of the discus just before the discus is released.

Reasoning Since the tangential speed of the discus increases as the thrower turns, the discus simultaneously experiences a tangential acceleration $\vec{a}_T$ and a centripetal acceleration $\vec{a}_c$ that are oriented at right angles to each other (see the drawing). The magnitude a of the total acceleration is $a = \sqrt{a_c^2 + a_T^2}$, where a_c and a_T are the magnitudes of the centripetal and tangential accelerations. The magnitude of the centripetal acceleration can be evaluated from Equation 8.11 ($a_c = r\omega^2$), and the magnitude of the tangential acceleration follows from Equation 8.10 ($a_T = r\alpha$). The angular acceleration α can be found from its definition in Equation 8.4.

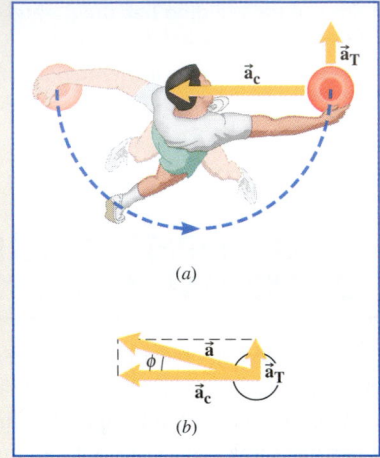

(a)

(b)

Figure 8.13 (a) A discus thrower and the centripetal acceleration $\vec{a}_c$ and tangential acceleration $\vec{a}_T$ of the discus. (b) The total acceleration $\vec{a}$ of the discus just before the discus is released is the vector sum of $\vec{a}_c$ and $\vec{a}_T$.

Knowns and Unknowns The data for this problem are:

Description	Symbol	Value	Comment
Explicit Data			
Final angular velocity	ω	+15.0 rad/s	Positive, because discus moves counterclockwise (see drawing).
Time	t	0.270 s	
Radius of circular arc	r	0.810 m	
Implicit Data			
Initial angular velocity	ω_0	0 rad/s	Discus starts from rest.
Unknown Variable			
Magnitude of total acceleration	a	?	

Modeling the Problem

STEP 1 **Total Acceleration** Figure 8.13b shows the two perpendicular components of the acceleration of the discus at the moment of its release. The centripetal acceleration $\vec{a}_c$ arises because the discus is traveling on a circular path; this acceleration always points toward the center of the circle (see Section 5.2). The tangential acceleration $\vec{a}_T$ arises because the tangential velocity of the discus is increasing; this acceleration is tangent to the circle (see Section 8.4). Since $\vec{a}_c$ and $\vec{a}_T$ are perpendicular, we can use the Pythagorean theorem to find the magnitude a of the total acceleration, as given by Equation 1 at the right. Values for a_c and a_T will be obtained in the next two steps.

$$a = \sqrt{a_c^2 + a_T^2} \qquad (1)$$

? ?

STEP 2 **Centripetal Acceleration** The magnitude a_c of the centripetal acceleration can be evaluated from Equation 8.11 as

$$a_c = r\omega^2$$

where r is the radius of the circle and ω is the angular speed of the discus at the moment of its release. Both r and ω are known, so we can substitute this expression for a_c into Equation 1, as indicated at the right. In Step 3, we will evaluate the magnitude a_T of the tangential acceleration.

$$a = \sqrt{a_c^2 + a_T^2} \qquad (1)$$

$$a_c = r\omega^2 \qquad \boxed{?}$$

STEP 3 **Tangential Acceleration** According to Equation 8.10, the magnitude a_T of the tangential acceleration is $a_T = r\alpha$, where r is the radius of the path and α is the magnitude of the angular acceleration. The angular acceleration is defined (see Equation 8.4) as the change $\omega - \omega_0$ in the angular velocity divided by the time t, or $\alpha = (\omega - \omega_0)/t$. Thus, the tangential acceleration can be written as

$$a_T = r\alpha = r\left(\frac{\omega - \omega_0}{t}\right)$$

All the variables on the right side of this equation are known, so we substitute this expression for a_T into Equation 1 (see the right column).

$$a = \sqrt{a_c^2 + a_T^2} \qquad (1)$$

$$a_c = r\omega^2 \qquad a_T = r\left(\frac{\omega - \omega_0}{t}\right)$$

Solution Algebraically combining the results of the three steps, we have:

STEP 1   STEP 2   STEP 3

$$a = \sqrt{a_c^2 + a_T^2} = \sqrt{r^2\omega^4 + a_T^2} = \sqrt{r^2\omega^4 + r^2\left(\frac{\omega - \omega_0}{t}\right)^2}$$

The magnitude of the total acceleration of the discus is

$$a = \sqrt{r^2\omega^4 + r^2\left(\frac{\omega - \omega_0}{t}\right)^2}$$

$$= \sqrt{(0.810 \text{ m})^2 (15.0 \text{ rad/s})^4 + (0.810 \text{ m})^2\left(\frac{15.0 \text{ rad/s} - 0 \text{ rad/s}}{0.270 \text{ s}}\right)^2} = \boxed{188 \text{ m/s}^2}$$

Note that we can also determine the angle ϕ between the total acceleration of the discus and its centripetal acceleration (see Figure 8.13b). From trigonometry, we have that $\tan \phi = a_T/a_c$, so

$$\phi = \tan^{-1}\left(\frac{a_T}{a_c}\right) = \tan^{-1}\left[\frac{r\left(\dfrac{\omega - \omega_0}{t}\right)}{r\omega^2}\right] = \tan^{-1}\left[\frac{\left(\dfrac{15.0 \text{ rad/s} - 0 \text{ rad/s}}{0.270 \text{ s}}\right)}{(15.0 \text{ rad/s})^2}\right] = 13.9°$$

Related Homework: *Problems 47, 68, 70*

✓ **CHECK YOUR UNDERSTANDING**

(The answers are given at the end of the book.)

12. A car is up on a hydraulic lift at a garage. The wheels are free to rotate, and the drive wheels are rotating with a constant angular velocity. Which one of the following statements is true? **(a)** A point on the rim has no tangential and no centripetal acceleration. **(b)** A point on the rim has both a nonzero tangential acceleration and a nonzero centripetal acceleration. **(c)** A point on the rim has a nonzero tangential acceleration but no centripetal acceleration. **(d)** A point on the rim has no tangential acceleration but does have a nonzero centripetal acceleration.

Continued

13. Section 5.6 discusses how the uniform circular motion of a space station can be used to create artificial gravity. This can be done by adjusting the angular speed of the space station, so the centripetal acceleration at an astronaut's feet equals g, the magnitude of the acceleration due to the earth's gravity (See Figure 5.19). If such an adjustment is made, will the acceleration at the astronaut's head due to the artificial gravity be **(a)** greater than, **(b)** equal to, or **(c)** less than g?

14. A bicycle is turned upside down, the front wheel is spinning (see the drawing), and there is an angular acceleration. At the instant shown, there are six points on the wheel that have arrows associated with them. Which one of the following quantities could the arrows *not* represent? **(a)** Tangential velocity **(b)** Centripetal acceleration **(c)** Tangential acceleration

15. A rotating object starts from rest and has a constant angular acceleration. Three seconds later the centripetal acceleration of a point on the object has a magnitude of 2.0 m/s². What is the magnitude of the centripetal acceleration of this point six seconds after the motion begins?

8.6 ROLLING MOTION

Rolling motion is a familiar situation that involves rotation, as Figure 8.14 illustrates for the case of an automobile tire. The essence of rolling motion is that there is *no slipping* at the point of contact where the tire touches the ground. To a good approximation, the tires on a normally moving automobile roll and do not slip. In contrast, the squealing tires that accompany the start of a drag race are rotating, but they are not rolling while they rapidly spin and slip against the ground.

When the tires in Figure 8.14 roll, there is a relationship between the angular speed at which the tires rotate and the linear speed (assumed constant) at which the car moves forward. In part *b* of the drawing, consider the points labeled *A* and *B* on the left tire. Between these points we apply a coat of red paint to the tread of the tire; the length of this circular arc of paint is *s*. The tire then rolls to the right until point *B* comes in contact with the ground. As the tire rolls, all the paint comes off the tire and sticks to the ground, leaving behind the horizontal red line shown in the drawing. The axle of the wheel moves through a linear distance *d*, which is equal to the length of the horizontal strip of paint. Since the tire does not slip, the distance *d* must be equal to the circular arc length *s*, measured along the outer edge of the tire: $d = s$. Dividing both sides of this equation by the elapsed time *t* shows that $d/t = s/t$. The term d/t is the speed at which the axle moves parallel to the ground—namely, the linear speed *v* of the car. The term s/t is the tangential speed v_T at which a point on the outer edge of the tire moves relative to the axle. In addition, v_T is related to the angular speed ω about the axle according to $v_T = r\omega$ (Equation 8.9). Therefore, it follows that

$$\underbrace{v}_{\substack{\text{Linear} \\ \text{speed}}} = \underbrace{v_T = r\omega}_{\substack{\text{Tangential} \\ \text{speed}}} \qquad (\omega \text{ in rad/s}) \qquad (8.12)$$

If the car in Figure 8.14 has a linear acceleration $\vec{a}$ parallel to the ground, a point on the tire's outer edge experiences a tangential acceleration $\vec{a}_T$ relative to the axle. The same kind of reasoning as that used in the last paragraph reveals that the magnitudes of these accelerations are the same and that they are related to the angular acceleration α of the wheel relative to the axle:

$$\underbrace{a}_{\substack{\text{Linear} \\ \text{acceleration}}} = \underbrace{a_T = r\alpha}_{\substack{\text{Tangential} \\ \text{acceleration}}} \qquad (\alpha \text{ in rad/s}^2) \qquad (8.13)$$

Equations 8.12 and 8.13 may be applied to any rolling motion, because the object does not slip against the surface on which it is rolling. Example 8 illustrates the basic features of rolling motion.

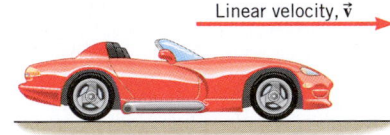

Linear velocity, $\vec{v}$

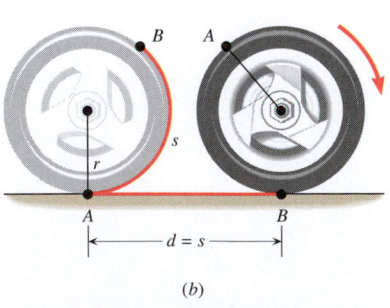

(a)

(b)

Figure 8.14 (*a*) An automobile moves with a linear speed *v*. (*b*) If the tires roll and do not slip, the distance *d* through which an axle moves equals the circular arc length *s* along the outer edge of a tire.

ANALYZING MULTIPLE-CONCEPT PROBLEMS

Example 8 An Accelerating Car

An automobile, starting from rest, has a linear acceleration to the right whose magnitude is 0.800 m/s^2 (see
Figure 8.15). During the next 20.0 s, the tires roll
and do not slip. The radius of each wheel is 0.330 m.
At the end of this time, what is the angle through which
each wheel has rotated?

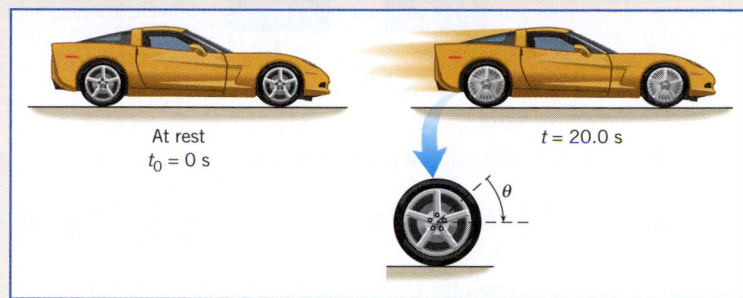

At rest
$t_0 = 0$ s

$t = 20.0$ s

θ

Reasoning Since the magnitude a of the car's linear
acceleration is constant and the tires roll (no slipping),
the relation $a = r\alpha$ (Equation 8.13) reveals that the mag-

Figure 8.15 As the car accelerates to the right, each wheel rotates through
an angle θ.

nitude α of the angular acceleration is also constant. Therefore, the equations of kinematics for rotational motion apply, and we
will use $\theta = \omega_0 t + \frac{1}{2}\alpha t^2$ (Equation 8.7) to find the angle θ through which each tire rotates. Two kinematic variables are known
(the initial angular velocity ω_0 and the time t). However, to use such an equation we also need to know the angular acceleration
of each tire. To find the angular acceleration, we will use Equation 8.13.

Knowns and Unknowns The data for this problem are:

Description	Symbol	Value	Comment
Explicit Data			
Time	t	20.0 s	
Magnitude of linear acceleration	a	0.800 m/s^2	
Tire radius	r	0.330 m	
Implicit Data			
Initial angular velocity	ω_0	0 rad/s	Car starts from rest.
Unknown Variable			
Angle through which each wheel rotates	θ	?	

Modeling the Problem

STEP 1 **Angular Displacement** Since the initial angular velocity ω_0 and the time t
are known, we begin by selecting Equation 8.7 from the equations of kinematics for rotational motion. This equation relates the angular displacement θ to these variables as
shown at the right. Equation 8.6 could also be used, but it provides a less convenient starting point. At this point we do not know the angular acceleration α, so it will be evaluated
in the next step.

$$\theta = \omega_0 t + \tfrac{1}{2}\alpha t^2 \qquad (8.7)$$

?

STEP 2 **Angular Acceleration** Angular acceleration, being a vector, has a magnitude and a direction. Because each wheel rolls (no slipping), the magnitude of its angular
acceleration is equal to a/r (Equation 8.13), where a is the magnitude of the car's linear
acceleration and r is the radius of the wheel.

As the car accelerates to the right, the tires rotate faster and faster in the clockwise, or
negative, direction (see Figure 8.15), so the angular acceleration is also negative.
Inserting a minus sign into the expression a/r to denote the direction of the angular
acceleration gives

$$\alpha = -\frac{a}{r}$$

$$\theta = \omega_0 t + \tfrac{1}{2}\alpha t^2 \qquad (8.7)$$

$$\alpha = -\frac{a}{r}$$

All the variables on the right side of this equation are known, and we substitute it into
Equation 8.7 (see the right column).

Continued

Solution Algebraically combining the results of the steps above, we have:

STEP 1   STEP 2

$$\theta = \omega_0 t + \tfrac{1}{2}\alpha t^2 = \omega_0 t + \tfrac{1}{2}\left(-\dfrac{a}{r}\right)t^2$$

The angle through which each wheel rotates is

$$\theta = \omega_0 t + \tfrac{1}{2}\left(-\dfrac{a}{r}\right)t^2 = (0 \text{ rad/s})(20.0 \text{ s}) + \tfrac{1}{2}\left(-\dfrac{0.800 \text{ m/s}^2}{0.330 \text{ m}}\right)(20.0 \text{ s})^2 = \boxed{-485 \text{ rad}}$$

The minus sign in this result means that the wheel has rotated in the clockwise direction.

Related Homework: *Problem 57*

✔ **CHECK YOUR UNDERSTANDING**

(The answers are given at the end of the book.)

16. The speedometer of a truck is set to read the linear speed of the truck, but uses a device that actually measures the angular speed of the rolling tires that came with the truck. However, the owner replaces the tires with larger-diameter versions. Does the reading on the speedometer after the replacement give a speed that is **(a)** less than, **(b)** equal to, or **(c)** greater than the true linear speed of the truck?

17. Rolling motion is an example that involves rotation about an axis that is not fixed. Give three other examples of rotational motion about an axis that is not fixed.

8.7 *THE VECTOR NATURE OF ANGULAR VARIABLES

We have presented angular velocity and angular acceleration by taking advantage of the analogy between angular variables and linear variables. Like the linear velocity and the linear acceleration, the angular quantities are also vectors and have a direction as well as a magnitude. As yet, however, we have not discussed the directions of these vectors.

When a rigid object rotates about a fixed axis, it is the axis that identifies the motion, and the angular velocity vector points along this axis. Figure 8.16 shows how to determine the direction using a *right-hand rule:*

> ***Right-Hand Rule*** Grasp the axis of rotation with your right hand, so that your fingers circle the axis in the same sense as the rotation. Your extended thumb points along the axis in the direction of the angular velocity vector.

No part of the rotating object moves in the direction of the angular velocity vector.

Angular acceleration arises when the angular velocity changes, and the acceleration vector also points along the axis of rotation. The acceleration vector has the same direction as the *change* in the angular velocity. That is, when the magnitude of the angular velocity (which is the angular speed) is increasing, the angular acceleration vector points in the same direction as the angular velocity. Conversely, when the magnitude of the angular velocity is decreasing, the angular acceleration vector points in the direction opposite to the angular velocity.

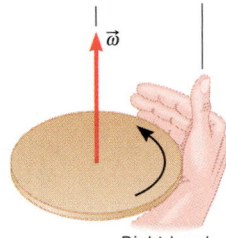

Right hand

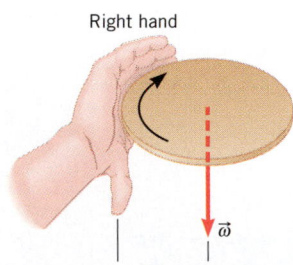

Right hand

Figure 8.16 The angular velocity vector $\vec{\omega}$ of a rotating object points along the axis of rotation. The direction along the axis depends on the sense of the rotation and can be determined with the aid of a right-hand rule (see text).

8.8 CONCEPTS & CALCULATIONS

In this chapter we have studied the concepts of angular displacement, angular velocity, and angular acceleration. We conclude with some examples that review important features of these ideas. Example 9 illustrates that the angular acceleration and the angular velocity can have the same or the opposite direction, depending on whether the angular speed is increasing or decreasing.

Concepts & Calculations Example 9

 Riding a Mountain Bike

A rider on a mountain bike is traveling to the left in Figure 8.17. Each wheel has an angular velocity of +21.7 rad/s, where, as usual, the plus sign indicates that the wheel is rotating in the counterclockwise direction. **(a)** To pass another cyclist, the rider pumps harder, and the angular velocity of the wheels increases from +21.7 to +28.5 rad/s in a time of 3.50 s. **(b)** After passing the cyclist, the rider begins to coast, and the angular velocity of the wheels decreases from +28.5 to +15.3 rad/s in a time of 10.7 s. In both instances, determine the magnitude and direction of the angular acceleration (assumed constant) of the wheels.

Concept Questions and Answers Is the angular acceleration positive or negative when the rider is passing the cyclist and the angular speed of the wheels is increasing?

  Answer Since the angular speed is increasing, the angular acceleration has the same direction as the angular velocity, which is the counterclockwise, or positive, direction (see Figure 8.17a).

Is the angular acceleration positive or negative when the rider is coasting and the angular speed of the wheels is decreasing?

  Answer Since the angular speed is decreasing during the coasting phase, the direction of the angular acceleration is opposite to that of the angular velocity. The angular velocity is in the counterclockwise (positive) direction, so the angular acceleration must be in the clockwise (negative) direction (see Figure 8.17b).

Solution **(a)** The angular acceleration α is the change in the angular velocity, $\omega - \omega_0$, divided by the elapsed time t:

$$\alpha = \frac{\omega - \omega_0}{t} = \frac{+28.5 \text{ rad/s} - (+21.7 \text{ rad/s})}{3.50 \text{ s}} = \boxed{+1.9 \text{ rad/s}^2} \qquad (8.4)$$

As expected, the angular acceleration is positive (counterclockwise).

(b) Now the wheels are slowing down, but still rotating in the positive (counterclockwise) direction. The angular acceleration for this part of the motion is

$$\alpha = \frac{\omega - \omega_0}{t} = \frac{+15.3 \text{ rad/s} - (+28.5 \text{ rad/s})}{10.7 \text{ s}} = \boxed{-1.23 \text{ rad/s}^2} \qquad (8.4)$$

Now, as anticipated, the angular acceleration is negative (clockwise).

(a) Angular speed increasing   (b) Angular speed decreasing

Figure 8.17 (*a*) When the angular speed of the wheel is increasing, the angular velocity ω and the angular acceleration α point in the same direction (counterclockwise in this drawing). (*b*) When the angular speed is decreasing, the angular velocity and the angular acceleration point in opposite directions.

Example 10 reviews the two different types of acceleration, centripetal and tangential, that a car can have when it travels on a circular road.

Concepts & Calculations Example 10

 A Circular Roadway and the Acceleration of Your Car

Suppose you are driving a car in a counterclockwise direction on a circular road whose radius is $r = 390$ m (see Figure 8.18). You look at the speedometer and it reads a steady 32 m/s (about 72 mi/h). **(a)** What is the angular speed of the car? **(b)** Determine the acceleration (magnitude and direction) of the car. **(c)** To avoid a rear-end collision with a vehicle ahead, you apply the

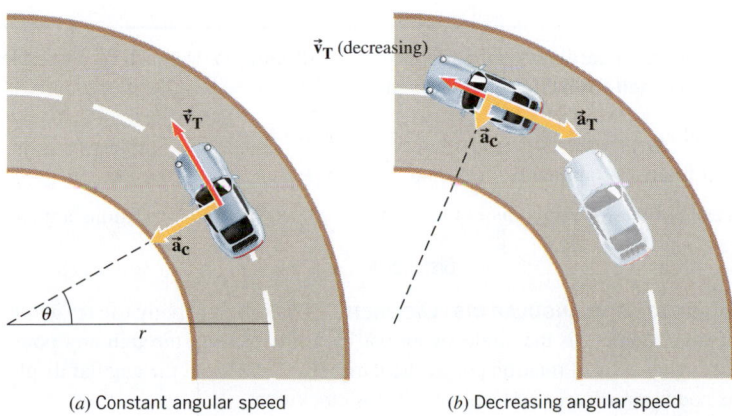

(a) Constant angular speed   (b) Decreasing angular speed

Figure 8.18 (*a*) A car is traveling in a counterclockwise direction at a constant angular speed. The tangential velocity of the car is $\vec{v}_T$ and its centripetal acceleration is $\vec{a}_c$. (*b*) The car is slowing down, so the tangential acceleration $\vec{a}_T$ points opposite to the tangential velocity $\vec{v}_T$.

brakes and reduce your angular speed to 4.9×10^{-2} rad/s in a time of 4.0 s. What is the tangential acceleration (magnitude and direction) of the car?

Concept Questions and Answers Does an object traveling at a constant tangential speed (for example, $v_T = 32$ m/s) along a circular path have an acceleration?

Answer Yes. Recall that an object has an acceleration if its velocity is changing in time. The velocity has two attributes, a magnitude (or speed) and a direction. In this instance the speed is not changing, since it is steady at 32 m/s. However, the direction of the velocity is changing continually as the car moves on the circular road. As Sections 5.2 and 8.5 discuss, this change in direction gives rise to an acceleration, called the centripetal acceleration $\vec{a}_c$. The centripetal acceleration is directed toward the center of the circle (see Figure 8.18a).

Is there a tangential acceleration $\vec{a}_T$ when the angular speed of an object changes (e.g., when the car's angular speed decreases to 4.9×10^{-2} rad/s)?

Answer Yes. When the car's angular speed ω decreases, for example, its tangential speed v_T also decreases. This is because they are related by by $v_T = r\omega$ (Equation 8.9), where r is the radius of the circular road. A decreasing tangential speed v_T, in turn, means that the car has a tangential acceleration $\vec{a}_T$. The direction of the tangential acceleration $\vec{a}_T$ must be *opposite* to that of the tangential velocity $\vec{v}_T$, because the tangential speed is decreasing. (If the two vectors were in the same direction, the tangential speed would be increasing.) Figure 8.18b shows these two vectors.

Solution **(a)** The angular speed ω of the car is equal to its tangential speed v_T (the speed indicated by the speedometer) divided by the radius of the circular road:

$$\omega = \frac{v_T}{r} = \frac{32 \text{ m/s}}{390 \text{ m}} = \boxed{8.2 \times 10^{-2} \text{ rad/s}} \tag{8.9}$$

(b) The acceleration is the centripetal acceleration and arises because the tangential velocity is changing direction as the car travels around the circular path. The magnitude of the centripetal acceleration is

$$a_c = r\omega^2 = (390 \text{ m})(8.2 \times 10^{-2} \text{ rad/s})^2 = \boxed{2.6 \text{ m/s}^2} \tag{8.11}$$

As always, the centripetal acceleration is directed toward the center of the circle.

(c) The tangential acceleration arises because the tangential speed of the car is changing as time passes. The magnitude of the tangential acceleration is related to the magnitude of the angular acceleration by Equation 8.10. The angular acceleration is given by Equation 8.4 as

$$\alpha = \frac{\omega - \omega_0}{t} = \frac{4.9 \times 10^{-2} \text{ rad/s} - 8.2 \times 10^{-2} \text{ rad/s}}{4.0 \text{ s}} = -8.3 \times 10^{-3} \text{ rad/s}^2$$

The magnitude of the angular acceleration is 8.3×10^{-3} rad/s². Therefore, the magnitude of the tangential acceleration is

$$a_T = r\alpha = (390 \text{ m})(8.3 \times 10^{-3} \text{ rad/s}^2) = \boxed{3.2 \text{ m/s}^2} \tag{8.10}$$

Since the car is slowing down, the tangential acceleration is directed opposite to the direction of the tangential velocity, as shown in Figure 8.18b.

CONCEPT SUMMARY

If you need more help with a concept, use the Learning Aids noted next to the discussion or equation. Examples (**Ex.**) are in the text of this chapter. Go to **www.wiley.com/college/cutnell** for the following Learning Aids:

Interactive LearningWare (ILW) — Additional examples solved in a five-step interactive format.

Concept Simulations (CS) — Animated text figures or animations of important concepts.

Interactive Solutions (IS) — Models for certain types of problems in the chapter homework. The calculations are carried out interactively.

Topic	Discussion	Learning Aids
Angular displacement	**8.1 ROTATIONAL MOTION AND ANGULAR DISPLACEMENT** When a rigid body rotates about a fixed axis, the angular displacement is the angle swept out by a line passing through any point on the body and intersecting the axis of rotation perpendicularly. By convention, the angular displacement is positive if it is counterclockwise and negative if it is clockwise.	

Topic	Discussion	Learning Aids
Radian	The radian (rad) is the SI unit of angular displacement. In radians, the angle θ is defined as the circular arc of length s traveled by a point on the rotating body divided by the radial distance r of the point from the axis:	
	$$\theta \text{ (in radians)} = \frac{s}{r} \qquad (8.1)$$	**Ex. 1, 2**

8.2 ANGULAR VELOCITY AND ANGULAR ACCELERATION The average angular velocity $\overline{\omega}$ is the angular displacement $\Delta\theta$ divided by the elapsed time Δt:

Average angular velocity	$$\overline{\omega} = \frac{\Delta\theta}{\Delta t} \qquad (8.2)$$	**Ex. 3**
Instantaneous angular velocity	As Δt approaches zero, the average angular velocity becomes equal to the instantaneous angular velocity ω. The magnitude of the instantaneous angular velocity is called the instantaneous angular speed.	

The average angular acceleration $\overline{\alpha}$ is the change $\Delta\omega$ in the angular velocity divided by the elapsed time Δt:

Average angular acceleration	$$\overline{\alpha} = \frac{\Delta\omega}{\Delta t} \qquad (8.4)$$	**Ex. 4**
Instantaneous angular acceleration	As Δt approaches zero, the average angular acceleration becomes equal to the instantaneous angular acceleration α.	

8.3 THE EQUATIONS OF ROTATIONAL KINEMATICS The equations of rotational kinematics apply when a rigid body rotates with a constant angular acceleration about a fixed axis. These equations relate the angular displacement $\theta - \theta_0$, the angular acceleration α, the final angular velocity ω, the initial angular velocity ω_0, and the elapsed time $t - t_0$. Assuming that $\theta_0 = 0$ rad at $t_0 = 0$ s, the equations of rotational kinematics are

	$$\omega = \omega_0 + \alpha t \qquad (8.4)$$	**Ex. 5, 9**
	$$\theta = \tfrac{1}{2}(\omega + \omega_0)t \qquad (8.6)$$	**ILW 8.1**
Equations of rotational kinematics	$$\theta = \omega_0 t + \tfrac{1}{2}\alpha t^2 \qquad (8.7)$$	
	$$\omega^2 = \omega_0{}^2 + 2\alpha\theta \qquad (8.8)$$	**IS 8.29**

These equations may be used with any self-consistent set of units and are not restricted to radian measure.

8.4 ANGULAR VARIABLES AND TANGENTIAL VARIABLES When a rigid body rotates through an angle θ about a fixed axis, any point on the body moves on a circular arc of length s and radius r. Such a point has a tangential velocity (magnitude $= v_\text{T}$) and, possibly, a tangential acceleration (magnitude $= a_\text{T}$). The angular and tangential variables are related by the following equations:

	$$s = r\theta \qquad (\theta \text{ in rad}) \qquad (8.1)$$	**Ex. 6**
Relations between angular and tangential variables	$$v_\text{T} = r\omega \qquad (\omega \text{ in rad/s}) \qquad (8.9)$$	
	$$a_\text{T} = r\alpha \qquad (\alpha \text{ in rad/s}^2) \qquad (8.10)$$	**IS 8.39, 8.69**

These equations refer to the magnitudes of the variables involved, without reference to positive or negative signs, and only radian measure can be used when applying them.

8.5 CENTRIPETAL ACCELERATION AND TANGENTIAL ACCELERATION The magnitude a_c of the centripetal acceleration of a point on an object rotating with uniform or nonuniform circular motion can be expressed in terms of the radial distance r of the point from the axis and the angular speed ω:

Centripetal acceleration	$$a_\text{c} = r\omega^2 \qquad (\omega \text{ in rad/s}) \qquad (8.11)$$	**Ex. 7, 10** **ILW 8.2**
Total Acceleration	This point experiences a total acceleration $\vec{a}$ that is the vector sum of two perpendicular acceleration components, the centripetal acceleration $\vec{a}_\text{c}$ and the tangential acceleration $\vec{a}_\text{T}$; $\vec{a} = \vec{a}_\text{c} + \vec{a}_\text{T}$.	

8.6 ROLLING MOTION The essence of rolling motion is that there is no slipping at the point where the object touches the surface upon which it is rolling. As a result, the tangential speed v_T of a point on the outer edge of a rolling object, measured relative to the axis through the center of the object, is equal to the linear speed v with which the object moves parallel to the surface. In other words, we have

$$v = v_\text{T} = r\omega \qquad (\omega \text{ in rad/s}) \qquad (8.12)$$

CS 8.1

Topic	Discussion	Learning Aids
	The magnitudes of the tangential acceleration a_T and the linear acceleration a of a rolling object are similarly related:	**IS 8.55**
	$$a = a_T = r\alpha \qquad (\alpha \text{ in rad/s}^2) \qquad (8.13)$$	**Ex. 8**
Right-hand rule	**8.7 THE VECTOR NATURE OF ANGULAR VARIABLES** The direction of the angular velocity vector is given by a right-hand rule. Grasp the axis of rotation with your right hand, so that your fingers circle the axis in the same sense as the rotation. Your extended thumb points along the axis in the direction of the angular velocity vector.	
	The angular acceleration vector has the same direction as the change in the angular velocity.	

FOCUS ON CONCEPTS

Note to Instructors: The numbering of the questions shown here reflects the fact that they are only a representative subset of the total number that are available online. However, all of the questions are available for assignment via an online homework management program such as WileyPLUS or WebAssign.

Section 8.1   Rotational Motion and Angular Displacement

1. The moon is 3.85×10^8 m from the earth and has a diameter of 3.48×10^6 m. You have a pea (diameter = 0.50 cm) and a dime (diameter = 1.8 cm). You close one eye and hold each object at arm's length (71 cm) between your open eye and the moon. Which objects, if any, completely cover your view of the moon? Assume that the moon and both objects are sufficiently far from your eye that the given diameters are equal to arc lengths when calculating angles. **(a)** Both **(b)** Neither **(c)** Pea **(d)** Dime

Section 8.2   Angular Velocity and Angular Acceleration

3. The radius of the circle traced out by the second hand on a clock is 6.00 cm. In a time t the tip of the second hand moves through an arc length of 24.0 cm. Determine the value of t in seconds.

4. A rotating object has an angular acceleration of $\alpha = 0$ rad/s^2. Which one or more of the following three statements is consistent with a zero angular acceleration?   A. The angular velocity is $\omega = 0$ rad/s at all times.   B. The angular velocity is $\omega = 10$ rad/s at all times.   C. The angular displacement θ has the same value at all times.   **(a)** A, B, and C   **(b)** A and B, but not C   **(c)** A only   **(d)** B only   **(e)** C only

Section 8.3   The Equations of Rotational Kinematics

6. A rotating wheel has a constant angular acceleration. It has an angular velocity of 5.0 rad/s at time $t = 0$ s, and 3.0 s later has an angular velocity of 9.0 rad/s. What is the angular displacement of the wheel during the 3.0-s interval?   **(a)** 15 rad   **(b)** 21 rad   **(c)** 27 rad   **(d)** There is not enough information given to determine the angular displacement.

7. A rotating object starts from rest at $t = 0$ s and has a constant angular acceleration. At a time of $t = 7.0$ s the object has an angular velocity of $\omega = 16$ rad/s. What is its angular velocity at a time of $t = 14$ s?

Section 8.4   Angular Variables and Tangential Variables

10. A merry-go-round at a playground is a circular platform that is mounted parallel to the ground and can rotate about an axis that is perpendicular to the platform at its center. The angular speed of the merry-go-round is constant, and a child at a distance of 1.4 m from the axis has a tangential speed of 2.2 m/s. What is the tangential speed of another child, who is located at a distance of 2.1 m from the axis? **(a)** 1.5 m/s **(b)** 3.3 m/s **(c)** 2.2 m/s **(d)** 5.0 m/s **(e)** 0.98 m/s

11. A small fan has blades that have a radius of 0.0600 m. When the fan is turned on, the tips of the blades have a tangential acceleration of 22.0 m/s^2 as the fan comes up to speed. What is the angular acceleration α of the blades?

Section 8.5   Centripetal Acceleration and Tangential Acceleration

13. A wheel rotates with a constant angular speed ω. Which one of the following is true concerning the angular acceleration α of the wheel, the tangential acceleration a_T of a point on the rim of the wheel, and the centripetal acceleration a_c of a point on the rim?

(a) $\alpha = 0$ rad/s^2, $a_T = 0$ m/s^2, and $a_c = 0$ m/s^2

(b) $\alpha = 0$ rad/s^2, $a_T \neq 0$ m/s^2, and $a_c = 0$ m/s^2

(c) $\alpha \neq 0$ rad/s^2, $a_T = 0$ m/s^2, and $a_c = 0$ m/s^2

(d) $\alpha = 0$ rad/s^2, $a_T = 0$ m/s^2, and $a_c \neq 0$ m/s^2

(e) $\alpha \neq 0$ rad/s^2, $a_T \neq 0$ m/s^2, and $a_c \neq 0$ m/s^2

14. A platform is rotating with an angular speed of 3.00 rad/s and an angular acceleration of 11.0 rad/s^2. At a point on the platform that is 1.25 m from the axis of rotation, what is the magnitude of the *total* acceleration a?

Section 8.6   Rolling Motion

15. The radius of each wheel on a bicycle is 0.400 m. The bicycle travels a distance of 3.0 km. Assuming that the wheels do not slip, how many revolutions does each wheel make?

(a) 1.2×10^3 revolutions

(b) 2.4×10^2 revolutions

(c) 6.0×10^3 revolutions

(d) 8.4×10^{-4} revolutions

(e) Since the time of travel is not given, there is not enough information for a solution.

PROBLEMS

Note to Instructors: *Most of the homework problems in this chapter are available for assignment via an online homework management program such as* WileyPLUS *or* WebAssign, *and those marked with the icon* **GO** *are presented in* WileyPLUS *using a guided tutorial format that provides enhanced interactivity. See Preface for additional details.*

ssm Solution is in the Student Solutions Manual.
www Solution is available online at www.wiley.com/college/cutnell

 This icon represents a biomedical application.

Section 8.1 Rotational Motion and Angular Displacement,

Section 8.2 Angular Velocity and Angular Acceleration

1. ssm In Europe, surveyors often measure angles in *grads*. There are 100 grads in one-quarter of a circle. How many grads are there in one radian?

2. GO The table that follows lists four pairs of initial and final angles of a wheel on a moving car. The elapsed time for each pair of angles is 2.0 s. For each of the four pairs, determine the average angular velocity (magnitude and direction as given by the algebraic sign of your answer).

	Initial angle θ_0	Final angle θ
(a)	0.45 rad	0.75 rad
(b)	0.94 rad	0.54 rad
(c)	5.4 rad	4.2 rad
(d)	3.0 rad	3.8 rad

3. The earth spins on its axis once a day and orbits the sun once a year ($365\frac{1}{4}$ days). Determine the average angular velocity (in rad/s) of the earth as it **(a)** spins on its axis and **(b)** orbits the sun. In each case, take the positive direction for the angular displacement to be the direction of the earth's motion.

4. The trap-jaw ant can snap its mandibles shut in as little as 1.3×10^{-4} s. In order to shut, both mandibles rotate through a 90° angle. What is the average angular velocity of one of the mandibles of the trap-jaw ant when the mandibles snap shut?

5. ssm A pitcher throws a curveball that reaches the catcher in 0.60 s. The ball curves because it is spinning at an average angular velocity of 330 rev/min (assumed constant) on its way to the catcher's mitt. What is the angular displacement of the baseball (in radians) as it travels from the pitcher to the catcher?

6. Conceptual Example 2 provides some relevant background for this problem. A jet is circling an airport control tower at a distance of 18.0 km. An observer in the tower watches the jet cross in front of the moon. As seen from the tower, the moon subtends an angle of 9.04×10^{-3} radians. Find the distance traveled (in meters) by the jet as the observer watches the nose of the jet cross from one side of the moon to the other.

7. GO The table that follows lists four pairs of initial and final angular velocities for a rotating fan blade. The elapsed time for each of the four pairs of angular velocities is 4.0 s. For each of the four pairs, find the average angular acceleration (magnitude and direction as given by the algebraic sign of your answer).

Problem 7

	Initial angular velocity ω_0	Final angular velocity ω
(a)	+2.0 rad/s	+5.0 rad/s
(b)	+5.0 rad/s	+2.0 rad/s
(c)	−7.0 rad/s	−3.0 rad/s
(d)	+4.0 rad/s	−4.0 rad/s

8. GO The initial angular velocity and the angular acceleration of four rotating objects at the same instant in time are listed in the table that follows. For each of the objects (a), (b), (c), and (d), determine the final angular *speed* after an elapsed time of 2.0 s.

	Initial angular velocity ω_0	Angular acceleration α
(a)	+12 rad/s	+3.0 rad/s²
(b)	+12 rad/s	−3.0 rad/s²
(c)	−12 rad/s	+3.0 rad/s²
(d)	−12 rad/s	−3.0 rad/s²

9. ssm A Ferris wheel rotates at an angular velocity of 0.24 rad/s. Starting from rest, it reaches its operating speed with an average angular acceleration of 0.030 rad/s². How long does it take the wheel to come up to operating speed?

10. A CD has a playing time of 74 minutes. When the music starts, the CD is rotating at an angular speed of 480 revolutions per minute (rpm). At the end of the music, the CD is rotating at 210 rpm. Find the magnitude of the average angular acceleration of the CD. Express your answer in rad/s².

*11. The sun appears to move across the sky, because the earth spins on its axis. To a person standing on the earth, the sun subtends an angle of $\theta_{\text{sun}} = 9.28 \times 10^{-3}$ rad (see Conceptual Example 2). How much time (in seconds) does it take for the sun to move a distance equal to its own diameter?

*12. A space station consists of two donut-shaped living chambers, A and B, that have the radii shown in the drawing. As the station rotates, an astronaut in chamber A is moved 2.40×10^2 m along a circular arc. How far along a circular arc is an astronaut in chamber B moved during the same time?

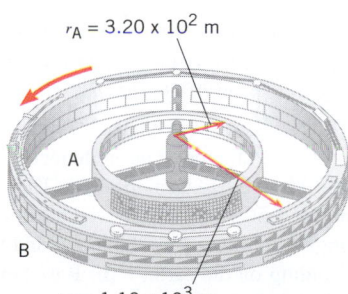

$r_A = 3.20 \times 10^2$ m

$r_B = 1.10 \times 10^3$ m

*13. Two people start at the same place and walk around a circular lake in opposite directions. One walks with an angular speed of 1.7×10^{-3} rad/s, while the other has an angular speed of 3.4×10^{-3} rad/s. How long will it be before they meet?

*14. GO A propeller is rotating about an axis perpendicular to its center, as the drawing shows. The axis is parallel to the ground. An arrow is fired at the propeller, travels parallel to the axis, and passes through one of the open spaces between the propeller blades. The angular open spaces between the three propeller blades are each $\pi/3$ rad (60.0°). The vertical drop of the arrow may be ignored. There is a maximum value ω for the angular speed of the propeller, beyond which the arrow cannot pass through an open space without being struck by one of the blades. Find this maximum value when the arrow has the lengths L and speeds v shown in the following table.

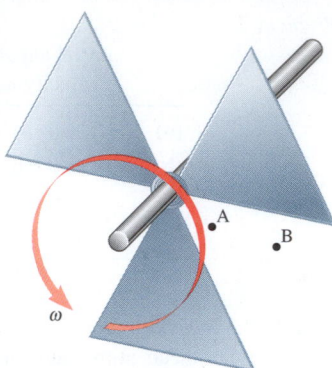

	L	v
(a)	0.71 m	75.0 m/s
(b)	0.71 m	91.0 m/s
(c)	0.81 m	91.0 m/s

*15. A baton twirler throws a spinning baton directly upward. As it goes up and returns to the twirler's hand, the baton turns through four revolutions. Ignoring air resistance and assuming that the average angular speed of the baton is 1.80 rev/s, determine the height to which the center of the baton travels above the point of release.

*16. The drawing shows a device that can be used to measure the speed of a bullet. The device consists of two rotating disks, separated by a distance of $d = 0.850$ m, and rotating with an angular speed of 95.0 rad/s. The bullet first passes through the left disk and then through the right disk. It is found that the angular displacement between the two bullet holes is $\theta = 0.240$ rad. From these data, determine the speed of the bullet.

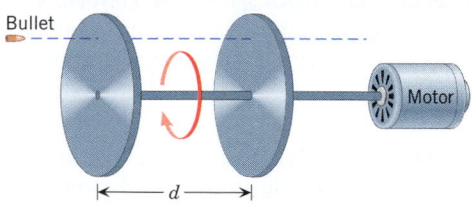

*17. ssm A stroboscope is a light that flashes on and off at a constant rate. It can be used to illuminate a rotating object, and if the flashing rate is adjusted properly, the object can be made to appear stationary. (a) What is the shortest time between flashes of light that will make a three-bladed propeller appear stationary when it is rotating with an angular speed of 16.7 rev/s? (b) What is the next shortest time?

**18. Review Conceptual Example 2 before attempting to work this problem. The moon has a diameter of 3.48×10^6 m and is a distance of 3.85×10^8 m from the earth. The sun has a diameter of 1.39×10^9 m and is 1.50×10^{11} m from the earth. (a) Determine (in radians) the angles subtended by the moon and the sun, as measured by a person standing on the earth. (b) Based on your answers to part (a), decide whether a total eclipse of the sun is really "total." Give your reasoning. (c) Determine the ratio, expressed as a percentage, of the apparent circular area of the moon to the apparent circular area of the sun.

**19. A quarterback throws a pass that is a perfect spiral. In other words, the football does not wobble, but spins smoothly about an axis passing through each end of the ball. Suppose the ball spins at 7.7 rev/s. In addition, the ball is thrown with a linear speed of 19 m/s at an angle of 55° with respect to the ground. If the ball is caught at the same height at which it left the quarterback's hand, how many revolutions has the ball made while in the air?

Section 8.3   The Equations of Rotational Kinematics

20. A gymnast is performing a floor routine. In a tumbling run she spins through the air, increasing her angular velocity from 3.00 to 5.00 rev/s while rotating through one-half of a revolution. How much time does this maneuver take?

21. ssm The drill bit of a variable-speed electric drill has a constant angular acceleration of 2.50 rad/s². The initial angular speed of the bit is 5.00 rad/s. After 4.00 s, (a) what angle has the bit turned through and (b) what is the bit's angular speed?

22. A wind turbine is initially spinning at a constant angular speed. As the wind's strength gradually increases, the turbine experiences a constant angular acceleration of 0.140 rad/s². After making 2870 revolutions, its angular speed is 137 rad/s. (a) What is the initial angular velocity of the turbine? (b) How much time elapses while the turbine is speeding up?

23. ssm www The shaft of a pump starts from rest and has an angular acceleration of 3.00 rad/s² for 18.0 s. At the end of this interval, what is (a) the shaft's angular speed and (b) the angle through which the shaft has turned?

24. GO A car is traveling along a road, and its engine is turning over with an angular velocity of +220 rad/s. The driver steps on the accelerator, and in a time of 10.0 s the angular velocity increases to +280 rad/s. (a) What would have been the angular displacement of the engine if its angular velocity had remained constant at the initial value of +220 rad/s during the entire 10.0-s interval? (b) What would have been the angular displacement if the angular velocity had been equal to its final value of +280 rad/s during the entire 10.0-s interval? (c) Determine the actual value of the angular displacement during the 10.0-s interval.

25. The wheels of a bicycle have an angular velocity of +20.0 rad/s. Then, the brakes are applied. In coming to rest, each wheel makes an angular displacement of +15.92 revolutions. (a) How much time does it take for the bike to come to rest? (b) What is the angular acceleration (in rad/s²) of each wheel?

*26. A top is a toy that is made to spin on its pointed end by pulling on a string wrapped around the body of the top. The string has a length of 64 cm and is wound around the top at a spot where its radius is 2.0 cm. The thickness of the string is negligible. The top is initially at rest. Someone pulls the free end of the string, thereby unwinding it and giving the top an angular acceleration of +12 rad/s². What is the final angular velocity of the top when the string is completely unwound?

*27. ssm Interactive LearningWare 8.1 at www.wiley.com/college/cutnell reviews how to solve problems such as this one. A motorcyclist is traveling along a road and accelerates for 4.50 s to pass another cyclist. The angular acceleration of each wheel is +6.70 rad/s², and, just after passing, the angular velocity of each wheel is +74.5 rad/s, where the plus signs indicate counterclockwise directions. What is the angular displacement of each wheel during this time?

*28. A dentist causes the bit of a high-speed drill to accelerate from an angular speed of 1.05×10^4 rad/s to an angular speed of 3.14×10^4 rad/s. In the process, the bit turns through 1.88×10^4 rad. Assuming a constant angular acceleration, how long would it take the bit to reach its maximum speed of 7.85×10^4 rad/s, starting from rest?

*29. **Interactive Solution 8.29** at **www.wiley.com/college/cutnell** offers a model for this problem. The drive propeller of a ship starts from rest and accelerates at 2.90×10^{-3} rad/s² for 2.10×10^3 s. For the next 1.40×10^3 s the propeller rotates at a constant angular speed. Then it decelerates at 2.30×10^{-3} rad/s² until it slows (without reversing direction) to an angular speed of 4.00 rad/s. Find the total angular displacement of the propeller.

*30. After 10.0 s, a spinning roulette wheel at a casino has slowed down to an angular velocity of +1.88 rad/s. During this time, the wheel has an angular acceleration of −5.04 rad/s². Determine the angular displacement of the wheel.

*31. At the local swimming hole, a favorite trick is to run horizontally off a cliff that is 8.3 m above the water. One diver runs off the edge of the cliff, tucks into a "ball," and rotates on the way down with an average angular speed of 1.6 rev/s. Ignore air resistance and determine the number of revolutions she makes while on the way down.

*32. **GO** At a county fair there is a betting game that involves a spinning wheel. As the drawing shows, the wheel is set into rotational motion with the beginning of the angular section labeled "1" at the marker at the top of the wheel. The wheel

Problem 32

then decelerates and eventually comes to a halt on one of the numbered sections. The wheel in the drawing is divided into twelve sections, each of which is an angle of 30.0°. Determine the numbered section on which the wheel comes to a halt when the deceleration of the wheel has a magnitude of 0.200 rev/s² and the initial angular velocity is **(a)** +1.20 rev/s and **(b)** +1.47 rev/s.

33. ssm www A child, hunting for his favorite wooden horse, is running on the ground around the edge of a stationary merry-go-round. The angular speed of the child has a constant value of 0.250 rad/s. At the instant the child spots the horse, one-quarter of a turn away, the merry-go-round begins to move (in the direction the child is running) with a constant angular acceleration of 0.0100 rad/s². What is the shortest time it takes for the child to catch up with the horse?

Section 8.4   Angular Variables and Tangential Variables

34. **GO** A fan blade is rotating with a constant angular acceleration of +12.0 rad/s². At what point on the blade, as measured from the axis of rotation, does the magnitude of the tangential acceleration equal that of the acceleration due to gravity?

35. On a reel-to-reel tape deck, the tape is pulled past the playback head at a constant linear speed of 0.381 m/s. **(a)** Using the data in part *a* of the drawing, find the angular speed of the take-up reel. **(b)** After 2.40×10^3 s, the take-up reel is almost full, as part *b* of the drawing indicates. Find the average angular acceleration of the reel and specify whether the acceleration indicates an increasing or decreasing angular velocity.

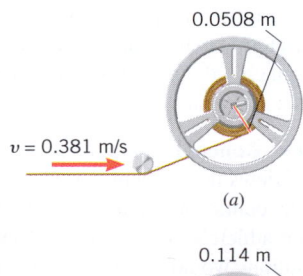

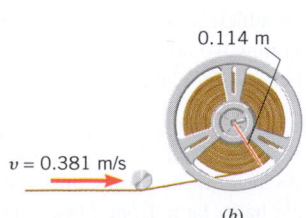

36. Some bacteria are propelled by biological motors that spin hair-like flagella. A typical bacterial motor turning at a constant angular velocity has a radius of 1.5×10^{-8} m, and a tangential speed at the rim of 2.3×10^{-5} m/s. **(a)** What is the angular speed (the

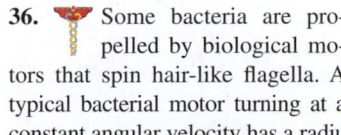

Problem 35

magnitude of the angular velocity) of this bacterial motor?   **(b)** How long does it take the motor to make one revolution?

37. **ssm** A string trimmer is a tool for cutting grass and weeds; it utilizes a length of nylon "string" that rotates about an axis perpendicular to one end of the string. The string rotates at an angular speed of 47 rev/s, and its tip has a tangential speed of 54 m/s. What is the length of the rotating string?

38. In 9.5 s a fisherman winds 2.6 m of fishing line onto a reel whose radius is 3.0 cm (assumed to be constant as an approximation). The line is being reeled in at a constant speed. Determine the angular speed of the reel.

39. **Interactive Solution 8.39** at **www.wiley.com/college/cutnell** offers one approach to solving this problem. The drawing shows a chain-saw blade. The rotating sprocket tip at the end of

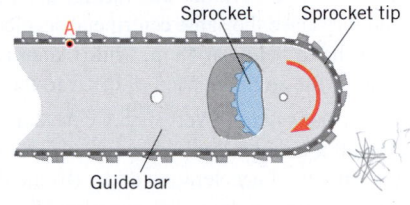

the guide bar has a radius of 4.0×10^{-2} m. The linear speed of a chain link at point A is 5.6 m/s. Find the angular speed of the sprocket tip in rev/s.

40. The earth has a radius of 6.38×10^6 m and turns on its axis once every 23.9 h.   **(a)** What is the tangential speed (in m/s) of a person living in Ecuador, a country that lies on the equator?   **(b)** At what latitude (i.e., the angle θ in the drawing) is the tangential speed one-third that of a person living in Ecuador?

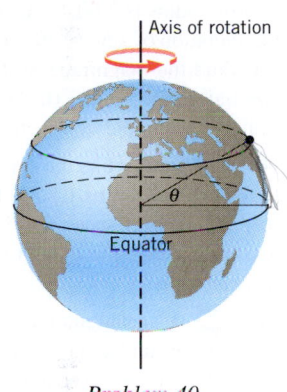

Problem 40

*41. **ssm** A compact disc (CD) contains music on a spiral track. Music is put onto a CD with the assumption that, during playback, the music will be detected at a *constant tangential speed* at any point. Since $v_T = r\omega$, a CD rotates at a smaller angular speed for music near the outer edge and a larger angular speed for music near the inner part of the disc. For music at the outer edge ($r = 0.0568$ m), the angular speed is 3.50 rev/s. Find **(a)** the constant tangential speed at which music is detected and **(b)** the angular speed (in rev/s) for music at a distance of 0.0249 m from the center of a CD.

*42. **GO** A person lowers a bucket into a well by turning the hand crank, as the drawing illustrates. The crank handle moves with a constant tangential speed of 1.20 m/s on its circular path. The rope holding the bucket unwinds without slipping on the barrel of the crank. Find the linear speed with which the bucket moves down the well.

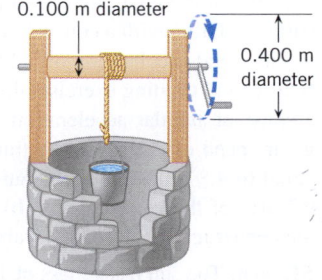

*43. A thin rod (length = 1.50 m) is oriented vertically, with its bottom end attached to the floor by means of a frictionless hinge. The mass of the rod may be ignored, compared to the mass of an object fixed to the top of the rod. The rod, starting from rest, tips over and rotates downward.   **(a)** What is the angular speed of the rod just before it strikes the floor? (*Hint: Consider using the principle of conservation of mechanical energy.*)   **(b)** What is the magnitude of the angular acceleration of the rod just before it strikes the floor?

**44. One type of slingshot can be made from a length of rope and a leather pocket for holding the stone. The stone can be thrown by whirling it rapidly in a horizontal circle and releasing it at the right moment. Such a slingshot is used to throw a stone from the edge of a cliff, the point of release being 20.0 m above the base of the cliff. The stone lands on the ground below the cliff at a point X. The horizontal distance of point X from the base of the cliff (directly beneath the point of release) is thirty times the radius of the circle on which the stone is whirled. Determine the angular speed of the stone at the moment of release.

Section 8.5 Centripetal Acceleration and Tangential Acceleration

45. ssm Two Formula One racing cars are negotiating a circular turn, and they have the same centripetal acceleration. However, the path of car A has a radius of 48 m, while that of car B is 36 m. Determine the ratio of the angular speed of car A to the angular speed of car B.

46. A racing car travels with a constant tangential speed of 75.0 m/s around a circular track of radius 625 m. Find **(a)** the magnitude of the car's total acceleration and **(b)** the direction of its total acceleration relative to the radial direction.

47. ssm Review Multiple-Concept Example 7 in this chapter and **Interactive LearningWare 8.2** at **www.wiley.com/college/cutnell** as aids in solving this problem. A train is rounding a circular curve whose radius is 2.00×10^2 m. At one instant, the train has an angular acceleration of 1.50×10^{-3} rad/s² and an angular speed of 0.0500 rad/s. **(a)** Find the magnitude of the total acceleration (centripetal plus tangential) of the train. **(b)** Determine the angle of the total acceleration relative to the radial direction.

48. The earth orbits the sun once a year $(3.16 \times 10^7$ s) in a nearly circular orbit of radius 1.50×10^{11} m. With respect to the sun, determine **(a)** the angular speed of the earth, **(b)** the tangential speed of the earth, and **(c)** the magnitude and direction of the earth's centripetal acceleration.

*49. GO A rectangular plate is rotating with a constant angular speed about an axis that passes perpendicularly through one corner, as the drawing shows. The centripetal acceleration measured at corner A is n times as great as that measured at corner B. What is the ratio L_1/L_2 of the lengths of the sides of the rectangle when $n = 2.00$?

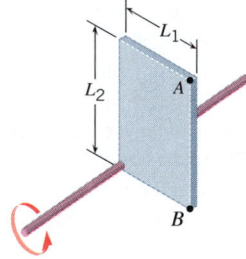

*50. In a large centrifuge used for training pilots and astronauts, a small chamber is fixed at the end of a rigid arm that rotates in a horizontal circle. A trainee riding in the chamber of a centrifuge rotating with a constant angular speed of 2.5 rad/s experiences a centripetal acceleration of 3.2 times the acceleration due to gravity. In a second training exercise, the centrifuge speeds up from rest with a constant angular acceleration. When the centrifuge reaches an angular speed of 2.5 rad/s, the trainee experiences a total acceleration equal to 4.8 times the acceleration due to gravity. **(a)** How long is the arm of the centrifuge? **(b)** What is the angular acceleration of the centrifuge in the second training exercise?

*51. ssm The sun has a mass of 1.99×10^{30} kg and is moving in a circular orbit about the center of our galaxy, the Milky Way. The radius of the orbit is 2.3×10^4 light-years (1 light-year $= 9.5 \times 10^{15}$ m), and the angular speed of the sun is 1.1×10^{-15} rad/s. **(a)** Determine the tangential speed of the sun. **(b)** What is the magnitude of the net force that acts on the sun to keep it moving around the center of the Milky Way?

**52. An electric drill starts from rest and rotates with a constant angular acceleration. After the drill has rotated through a certain angle, the magnitude of the centripetal acceleration of a point on the drill is twice the magnitude of the tangential acceleration. What is the angle?

Section 8.6 Rolling Motion

Note: All problems in this section assume that there is no slipping of the surfaces in contact during the rolling motion.

53. ssm www A motorcycle accelerates uniformly from rest and reaches a linear speed of 22.0 m/s in a time of 9.00 s. The radius of each tire is 0.280 m. What is the magnitude of the angular acceleration of each tire?

54. Suppose you are riding a stationary exercise bicycle, and the electronic meter indicates that the wheel is rotating at 9.1 rad/s. The wheel has a radius of 0.45 m. If you ride the bike for 35 min, how far would you have gone if the bike could move?

55. Refer to **Interactive Solution 8.55** at **www.wiley.com/college/cutnell** in preparation for this problem. A car is traveling with a speed of 20.0 m/s along a straight horizontal road. The wheels have a radius of 0.300 m. If the car speeds up with a linear acceleration of 1.50 m/s² for 8.00 s, find the angular displacement of each wheel during this period.

56. An automobile tire has a radius of 0.330 m, and its center moves forward with a linear speed of $v = 15.0$ m/s. **(a)** Determine the angular speed of the wheel. **(b)** Relative to the axle, what is the tangential speed of a point located 0.175 m from the axle?

*57. Multiple-Concept Example 8 provides useful background for part b of this problem. A motorcycle, which has an initial linear speed of 6.6 m/s, decelerates to a speed of 2.1 m/s in 5.0 s. Each wheel has a radius of 0.65 m and is rotating in a counterclockwise (positive) direction. What is **(a)** the constant angular acceleration (in rad/s²) and **(b)** the angular displacement (in rad) of each wheel?

*58. GO A dragster starts from rest and accelerates down a track. Each tire has a radius of 0.320 m and rolls without slipping. At a distance of 384 m, the angular speed of the wheels is 288 rad/s. Determine **(a)** the linear speed of the dragster and **(b)** the magnitude of the angular acceleration of its wheels.

*59. The two-gear combination shown in the drawing is being used to lift the load L with a constant upward speed of 2.50 m/s. The rope that is attached to the load is being wound onto a cylinder behind the big gear. The depth of the teeth of the gears is negligible compared to the radii. Determine the angular velocity (magnitude and direction) of **(a)** the larger gear and **(b)** the smaller gear.

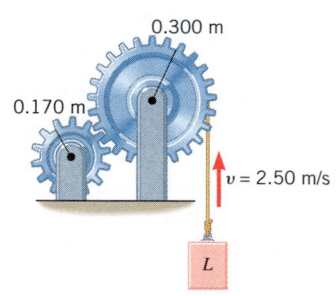

*60. Concept Simulation 8.1 at **www.wiley.com/college/cutnell** reviews the concept that plays the central role in this problem. Over the course of a multi-stage 4520-km bicycle race, the front wheel of an athlete's bicycle makes 2.18×10^6 revolutions. How many revolutions would the wheel have made during the race if its radius had been 1.2 cm larger?

*61. The penny-farthing is a bicycle that was popular between 1870 and 1890. As the drawing shows, this type of bicycle has a large front wheel and a small rear wheel. On a Sunday ride in the park the front wheel (radius $= 1.20$ m) makes 276 revolutions. How many revolutions does the rear wheel (radius $= 0.340$ m) make?

*62. **GO** A ball of radius 0.200 m rolls with a constant linear speed of 3.60 m/s along a horizontal table. The ball rolls off the edge and falls a vertical distance of 2.10 m before hitting the floor. What is the angular displacement of the ball while the ball is in the air?

63. ssm Consult **Concept Simulation 8.1** at **www.wiley.com/college/cutnell** for help in understanding the concepts that are important in this problem. Take two quarters and lay them on a table. Press down on one quarter so it cannot move. Then, starting at the 12:00 position, roll the other quarter along the edge of the stationary quarter, as the drawing suggests. How many revolutions does the rolling quarter make when it travels once around the circumference of the stationary quarter? Surprisingly, the answer is *not* one revolution. *(Hint: Review the paragraph just before Equation 8.12 that discusses how the distance traveled by the axle of a wheel is related to the circular arc length along the outer edge of the wheel.)*

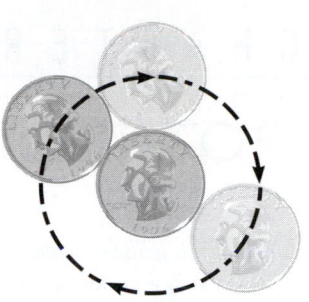

ADDITIONAL PROBLEMS

64. Our sun rotates in a circular orbit about the center of the Milky Way galaxy. The radius of the orbit is 2.2×10^{20} m, and the angular speed of the sun is 1.1×10^{-15} rad/s. How long (in years) does it take for the sun to make one revolution around the center?

65. **ssm** A 220-kg speedboat is negotiating a circular turn (radius = 32 m) around a buoy. During the turn, the engine causes a net tangential force of magnitude 550 N to be applied to the boat. The initial tangential speed of the boat going into the turn is 5.0 m/s. **(a)** Find the tangential acceleration. **(b)** After the boat is 2.0 s into the turn, find the centripetal acceleration.

66. A basketball player is balancing a spinning basketball on the tip of his finger. The angular velocity of the ball slows down from 18.5 to 14.1 rad/s. During the slow-down, the angular displacement is 85.1 rad. Determine the time it takes for the ball to slow down.

67. **ssm** An electric fan is running on HIGH. After the LOW button is pressed, the angular speed of the fan decreases to 83.8 rad/s in 1.75 s. The deceleration is 42.0 rad/s^2. Determine the initial angular speed of the fan.

68. Refer to Multiple-Concept Example 7 for insight into this problem. During a tennis serve, a racket is given an angular acceleration of magnitude 160 rad/s^2. At the top of the serve, the racket has an angular speed of 14 rad/s. If the distance between the top of the racket and the shoulder is 1.5 m, find the magnitude of the total acceleration of the top of the racket.

69. Refer to **Interactive Solution 8.69** at **www.wiley.com/college/cutnell** to review a model for solving this problem. The take-up reel of a cassette tape has an average radius of 1.4 cm. Find the length of tape (in meters) that passes around the reel in 13 s when the reel rotates at an average angular speed of 3.4 rad/s.

*70. Multiple-Concept Example 7 explores the approach taken in problems such as this one. The blades of a ceiling fan have a radius of 0.380 m and are rotating about a fixed axis with an angular velocity of +1.50 rad/s. When the switch on the fan is turned to a higher speed, the blades acquire an angular acceleration of +2.00 rad/s^2. After 0.500 s have elapsed since the switch was reset, what is **(a)** the total acceleration (in m/s^2) of a point on the tip of a blade and **(b)** the angle ϕ between the total acceleration $\vec{a}$ and the centripetal acceleration $\vec{a}_c$? (See Figure 8.12b.)

*71. **ssm** A baseball pitcher throws a baseball horizontally at a linear speed of 42.5 m/s (about 95 mi/h). Before being caught, the baseball travels a horizontal distance of 16.5 m and rotates through an angle of 49.0 rad. The baseball has a radius of 3.67 cm and is rotating about an axis as it travels, much like the earth does. What is the tangential speed of a point on the "equator" of the baseball?

*72. The drawing shows a graph of the angular velocity of a rotating wheel as a function of time. Although not shown in the graph, the angular velocity continues to increase at the same rate until $t = 8.0$ s. What is the angular displacement of the wheel from 0 to 8.0 s?

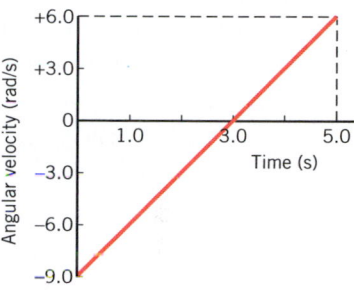

*73. A spinning wheel on a fireworks display is initially rotating in a counterclockwise direction. The wheel has an angular acceleration of -4.00 rad/s^2. Because of this acceleration, the angular velocity of the wheel changes from its initial value to a final value of -25.0 rad/s. While this change occurs, the angular displacement of the wheel is zero. (Note the similarity to that of a ball being thrown vertically upward, coming to a momentary halt, and then falling downward to its initial position.) Find the time required for the change in the angular velocity to occur

*74. **Interactive LearningWare 8.2** at **www.wiley.com/college/cutnell** provides a review of the concepts that are important in this problem. A racing car, starting from rest, travels around a circular turn of radius 23.5 m. At a certain instant, the car is still accelerating, and its angular speed is 0.571 rad/s. At this time, the total acceleration (centripetal plus tangential) makes an angle of 35.0° with respect to the radius. (The situation is similar to that in Figure 8.12b.) What is the magnitude of the total acceleration?

**75. The drawing shows a golf ball passing through a windmill at a miniature golf course. The windmill has 8 blades and rotates at an angular speed of 1.25 rad/s. The opening between successive blades is equal to the width of a blade. A golf ball (diameter 4.50×10^{-2} m) is just passing by one of the rotating blades. What must be the *minimum* speed of the ball so that it will not be hit by the next blade?

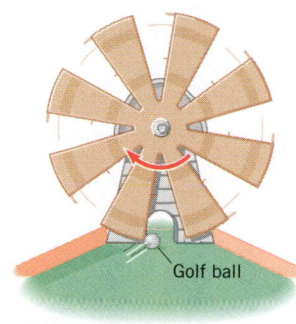
Golf ball

**76 The differential gear of a car axle allows the wheel on the left side of a car to rotate at a different angular speed than the wheel on the right side. A car is driving at a constant speed around a circular track on level ground, completing each lap in 19.5 s. The distance between the tires on the left and right sides of the car is 1.60 m, and the radius of each wheel is 0.350 m. What is the difference between the angular speeds of the wheels on the left and right sides of the car?

CHAPTER 9

ROTATIONAL DYNAMICS

Ten Chinese acrobats balance on a bike during a performance in Shanghai, China. The ensemble is in equilibrium as the bike moves across the stage at a constant velocity. The weight of each acrobat and the bicycle, as well as the location of each weight, determines a quantity called *torque*. We will see that the net torque acting on an object in equilibrium is zero. (© John Slater/Corbis)

9.1
THE ACTION OF FORCES AND TORQUES ON RIGID OBJECTS

The mass of most rigid objects, such as a propeller or a wheel, is spread out and not concentrated at a single point. These objects can move in a number of ways. Figure 9.1*a* illustrates one possibility called translational motion, in which all points on the body travel on parallel paths (not necessarily straight lines). In pure translation there is no rotation of any line in the body. Because translational motion can occur along a curved line, it is often called curvilinear motion or linear motion. Another possibility is rotational motion, which may occur in combination with translational motion, as is the case for the somersaulting gymnast in Figure 9.1*b*.

We have seen many examples of how a net force affects linear motion by causing an object to accelerate. We now need to take into account the possibility that a rigid object can also have an angular acceleration. A net external force causes linear motion to change, but what causes rotational motion to change? For example, something causes the rotational velocity of a speedboat's propeller to change when the boat accelerates. Is it simply the net force? As it turns out, it is not the net external force, but rather the net external torque that causes the rotational velocity to change. Just as greater net forces cause greater linear accelerations, greater net torques cause greater rotational or angular accelerations.

Figure 9.2 helps to explain the idea of torque. When you push on a door with a force $\vec{\mathbf{F}}$, as in part *a*, the door opens more quickly when the force is larger. Other things being equal, a larger force generates a larger torque. However, the door does not open as quickly if you apply the same force at a point closer to the hinge, as in part *b*, because the force now produces less torque. Furthermore, if your push is directed nearly at the hinge, as in part *c*, you will have a hard time opening the door at all, because the torque is nearly zero. In summary, the torque depends on the magnitude of the force, on the point where the force is applied relative to the axis of rotation (the hinge in Figure 9.2), and on the direction of the force.

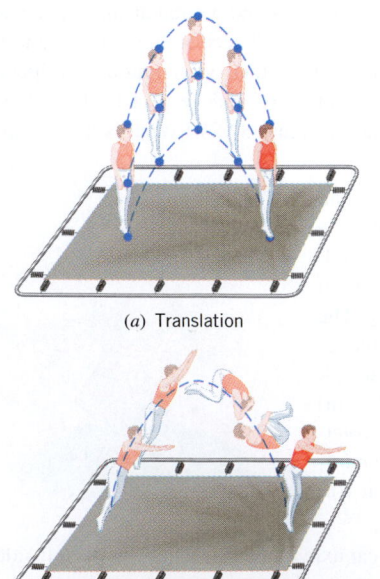

(a) Translation

(b) Combined translation and rotation

Figure 9.1 Examples of (*a*) translational motion and (*b*) combined translational and rotational motions.

For simplicity, we deal with situations in which the force lies in a plane that is perpendicular to the axis of rotation. In Figure 9.3, for instance, the axis is perpendicular to the page and the force lies in the plane of the paper. The drawing shows the line of action and the lever arm of the force, two concepts that are important in the definition of torque. The *line of action* is an extended line drawn colinear with the force. The *lever arm* is the distance ℓ between the line of action and the axis of rotation, measured on a line that is perpendicular to both. The torque is represented by the symbol τ (Greek letter *tau*), and its magnitude is defined as the magnitude of the force times the lever arm:

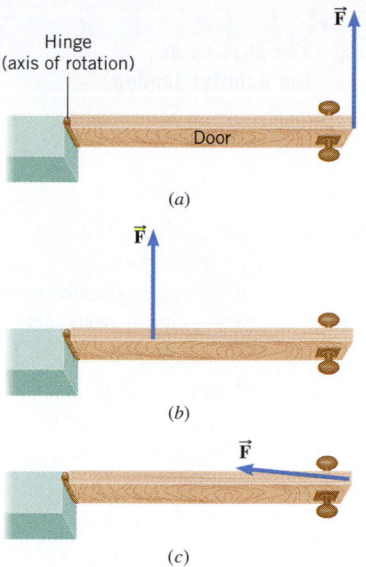

DEFINITION OF TORQUE

   Magnitude of torque = (Magnitude of the force) × (Lever arm) = $F\ell$ (9.1)

Direction: The torque τ is positive when the force tends to produce a counterclockwise rotation about the axis, and negative when the force tends to produce a clockwise rotation.

SI Unit of Torque: newton · meter (N · m)

Equation 9.1 indicates that forces of the same magnitude can produce *different* torques, depending on the value of the lever arm, and Example 1 illustrates this important feature.

Figure 9.2 With a force of a given magnitude, a door is easier to open by (*a*) pushing at the outer edge than by (*b*) pushing closer to the axis of rotation (the hinge). (*c*) Pushing into the hinge makes it difficult to open the door.

 Example 1 Different Lever Arms, Different Torques

In Figure 9.3 a force (magnitude = 55 N) is applied to a door. However, the lever arms are different in the three parts of the drawing: (*a*) $\ell = 0.80$ m, (*b*) $\ell = 0.60$ m, and (*c*) $\ell = 0$ m. Find the torque in each case.

Reasoning In each case the lever arm is the perpendicular distance between the axis of rotation and the line of action of the force. In part *a* this perpendicular distance is equal to the width of the door. In parts *b* and *c*, however, the lever arm is less than the width. Because the lever arm is different in each case, the torque is different, even though the magnitude of the applied force is the same.

Solution Using Equation 9.1, we find the following values for the torques:

(a) $\tau = +F\ell = +(55 \text{ N})(0.80 \text{ m}) = \boxed{+44 \text{ N} \cdot \text{m}}$

(b) $\tau = +F\ell = +(55 \text{ N})(0.60 \text{ m}) = \boxed{+33 \text{ N} \cdot \text{m}}$

(c) $\tau = +F\ell = +(55 \text{ N})(0 \text{ m}) = \boxed{0 \text{ N} \cdot \text{m}}$

In parts *a* and *b* the torques are positive, since the forces tend to produce a counterclockwise rotation of the door. In part *c* the line of action of *F* passes through the axis of rotation (the hinge). Hence, the lever arm is zero, and the torque is zero.

In our bodies, muscles and tendons produce torques about various joints. Example 2 illustrates how the Achilles tendon produces a torque about the ankle joint.

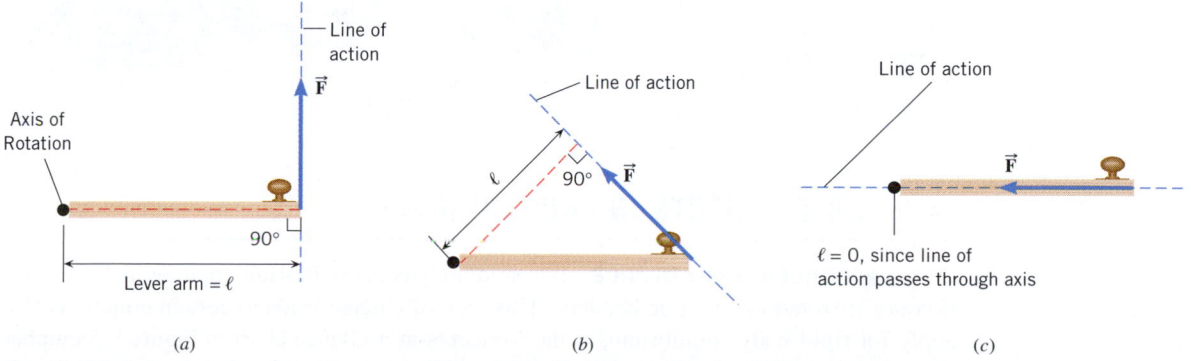

Figure 9.3 In this top view, the hinges of a door appear as a black dot (●) and define the axis of rotation. The line of action and lever arm ℓ are illustrated for a force applied to the door (*a*) perpendicularly and (*b*) at an angle. (*c*) The lever arm is zero because the line of action passes through the axis of rotation.

The physics of the Achilles tendon.

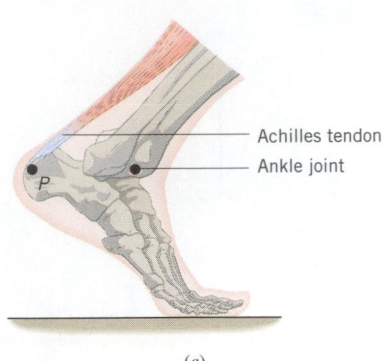

Achilles tendon
Ankle joint

P

(a)

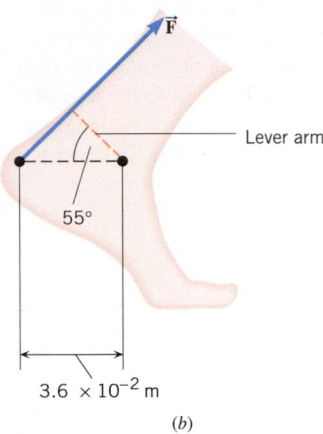

$\vec{F}$

Lever arm

55°

3.6 × 10⁻² m

(b)

Figure 9.4 The force $\vec{F}$ generated by the Achilles tendon produces a clockwise (negative) torque about the ankle joint.

Example 2 The Achilles Tendon

Figure 9.4*a* shows the ankle joint and the Achilles tendon attached to the heel at point *P*. The tendon exerts a force $\vec{F}$ (magnitude = 720 N), as Figure 9.4*b* indicates. Determine the torque (magnitude and direction) of this force about the ankle joint, which is located 3.6×10^{-2} m away from point *P*.

Reasoning To calculate the magnitude of the torque, it is necessary to have a value for the lever arm ℓ. However, the lever arm is not the given distance of 3.6×10^{-2} m. Instead, the lever arm is the perpendicular distance between the axis of rotation at the ankle joint and the line of action of the force $\vec{F}$. In Figure 9.4*b* this distance is indicated by the dashed red line.

Solution From the drawing, it can be seen that the lever arm is $\ell = (3.6 \times 10^{-2}$ m$)\cos 55°$. The magnitude of the torque is

$$F\ell = (720 \text{ N})(3.6 \times 10^{-2} \text{ m})\cos 55° = 15 \text{ N}\cdot\text{m} \qquad (9.1)$$

The force $\vec{F}$ tends to produce a clockwise rotation about the ankle joint, so the torque is negative: $\boxed{\tau = -15 \text{ N}\cdot\text{m}}$.

✓ CHECK YOUR UNDERSTANDING

(The answers are given at the end of the book.)

1. The drawing shows an overhead view of a horizontal bar that is free to rotate about an axis perpendicular to the page. Two forces act on the bar, and they have the same magnitude. However, one force is perpendicular to the bar, and the other makes an angle ϕ with respect to it. The angle ϕ can be 90°, 45°, or 0°. Rank the values of ϕ according to the magnitude of the net torque (the sum of the torques) that the two forces produce, largest net torque first.

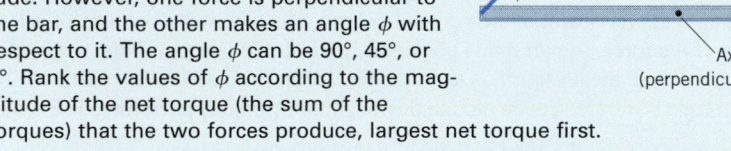

$\vec{F}$
$\vec{F}$
ϕ
Bar (overhead view)
Axis (perpendicular to page)

2. Sometimes, even with a wrench, one cannot loosen a nut that is frozen tightly to a bolt. It is often possible to loosen the nut by slipping one end of a long pipe over the wrench handle and pushing at the other end of the pipe. With the aid of the pipe, does the applied force produce a smaller torque, a greater torque, or the same torque on the nut?

3. Is it possible **(a)** for a large force to produce a small, or even zero, torque and **(b)** for a small force to produce a large torque?

4. The photograph shows a workman struggling to keep a stack of boxes balanced on a dolly. The man's right foot is on the axle of the dolly. Assuming that the boxes are identical, which one creates the greatest torque with respect to the axle?

(© Paul Miller Photography)

9.2 RIGID OBJECTS IN EQUILIBRIUM

▶ **CONCEPTS AT A GLANCE** If a rigid body is in equilibrium, neither its linear motion nor its rotational motion changes. This lack of change leads to certain equations that apply for rigid-body equilibrium, as the Concepts-at-a-Glance chart in Figure 9.5 emphasizes. For instance, an object whose linear motion is not changing has no acceleration $\vec{a}$. Therefore, the net force $\Sigma\vec{F}$ applied to the object must be zero, since $\Sigma\vec{F} = m\vec{a}$ and $\vec{a} = 0$. For two-dimensional motion the x and y components of the net force are separately zero: $\Sigma F_x = 0$ and $\Sigma F_y = 0$ (Equations 4.9a and 4.9b). In calculating the net force, we include

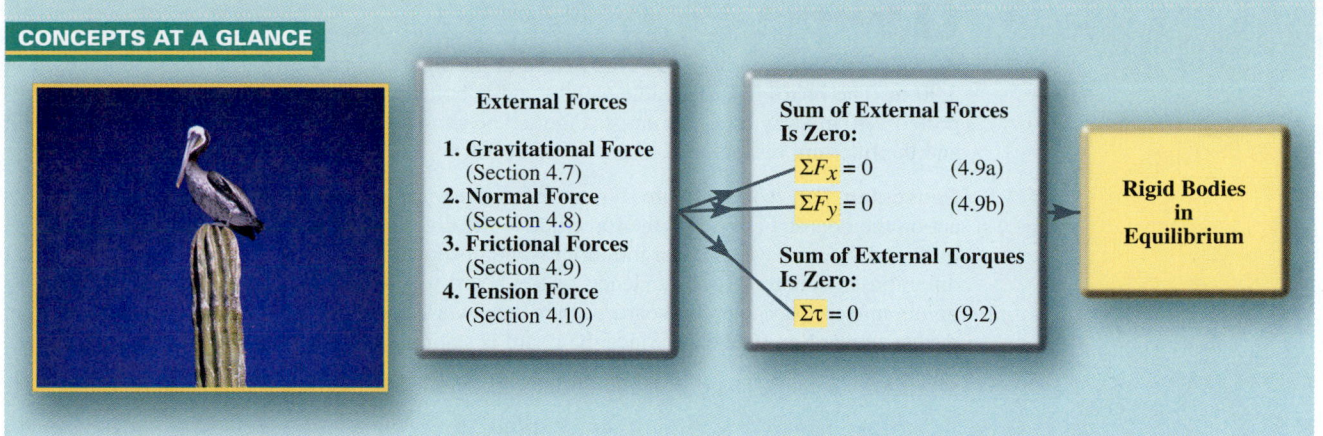

CONCEPTS AT A GLANCE

External Forces	Sum of External Forces Is Zero:	Rigid Bodies in Equilibrium
1. **Gravitational Force** (Section 4.7) 2. **Normal Force** (Section 4.8) 3. **Frictional Forces** (Section 4.9) 4. **Tension Force** (Section 4.10)	$\Sigma F_x = 0$ (4.9a) $\Sigma F_y = 0$ (4.9b) **Sum of External Torques Is Zero:** $\Sigma \tau = 0$ (9.2)	

only forces from external agents, or *external forces.** In addition to linear motion, we must consider rotational motion, which also does not change under equilibrium conditions. This means that the net external torque acting on the object must be zero, because torque is what causes rotational motion to change. Using the symbol $\Sigma \tau$ to represent the net external torque (the sum of all positive and negative torques), we have

$$\Sigma \tau = 0 \qquad (9.2)$$

We define rigid-body equilibrium, then, in the following way. ◄

Figure 9.5 CONCEPTS AT A GLANCE A body is in equilibrium when the sum of the external forces is zero and when the sum of the external torques is zero. These conditions apply to the pelican, since it is nicely balanced on top of the cactus and is in equilibrium. (Macduff Everton/The Image Bank/Getty Images)

EQUILIBRIUM OF A RIGID BODY

A rigid body is in equilibrium if it has zero translational acceleration and zero angular acceleration. In equilibrium, the sum of the externally applied forces is zero, and the sum of the externally applied torques is zero:

$$\Sigma F_x = 0 \quad \text{and} \quad \Sigma F_y = 0 \qquad \text{(4.9a and 4.9b)}$$

$$\Sigma \tau = 0 \qquad (9.2)$$

The reasoning strategy for analyzing the forces and torques acting on a body in equilibrium is given below. The first four steps of the strategy are essentially the same as those outlined in Section 4.11, where only forces are considered. Steps 5 and 6 have been added to account for any external torques that may be present. Example 3 illustrates how this reasoning strategy is applied to a diving board.

REASONING STRATEGY

Applying the Conditions of Equilibrium to a Rigid Body

1. Select the object to which the equations for equilibrium are to be applied.

2. Draw a free-body diagram that shows all the external forces acting on the object.

3. Choose a convenient set of x, y axes and resolve all forces into components that lie along these axes.

4. Apply the equations that specify the balance of forces at equilibrium: $\Sigma F_x = 0$ and $\Sigma F_y = 0$.

5. Select a convenient axis of rotation. The choice is arbitrary. Identify the point where each external force acts on the object, and calculate the torque produced by each force about the chosen axis. Set the sum of the torques equal to zero: $\Sigma \tau = 0$.

6. Solve the equations in Steps 4 and 5 for the desired unknown quantities.

*We ignore internal forces that one part of an object exerts on another part, because they occur in action–reaction pairs, each of which consists of oppositely directed forces of equal magnitude. The effect of one force cancels the effect of the other, as far as the acceleration of the entire object is concerned.

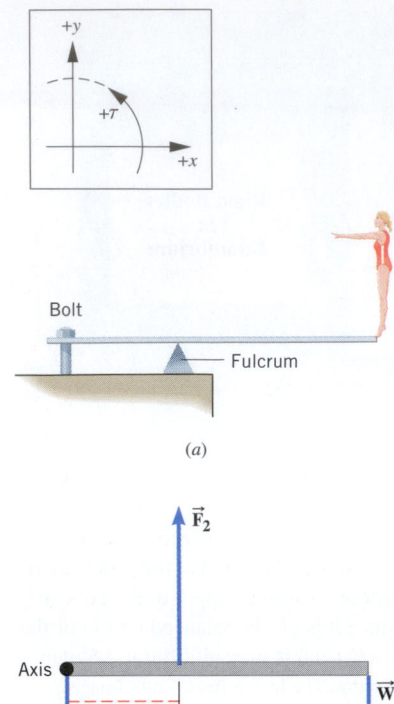

Bolt

Fulcrum

(a)

Axis

$\ell_2 = 1.40 \text{ m}$

$\ell_w = 3.90 \text{ m}$

$\vec{F_2}$

$\vec{F_1}$

$\vec{W}$

(b) Free-body diagram of the diving board

Figure 9.6 (a) A diver stands at the end of a diving board. (b) The free-body diagram for the diving board. The box at the upper left shows the positive x and y directions for the forces, as well as the positive direction (counterclockwise) for the torques.

Example 3 A Diving Board

A woman whose weight is 530 N is poised at the right end of a diving board with a length of 3.90 m. The board has negligible weight and is bolted down at the left end, while being supported 1.40 m away by a fulcrum, as Figure 9.6a shows. Find the forces $\vec{F}_1$ and $\vec{F}_2$ that the bolt and the fulcrum, respectively, exert on the board.

Reasoning Part b of the figure shows the free-body diagram of the diving board. Three forces act on the board: $\vec{F}_1$, $\vec{F}_2$, and the force due to the diver's weight $\vec{W}$. In choosing the directions of $\vec{F}_1$ and $\vec{F}_2$, we have used our intuition: $\vec{F}_1$ points downward because the bolt must pull in that direction to counteract the tendency of the board to rotate clockwise about the fulcrum; $\vec{F}_2$ points upward, because the board pushes downward against the fulcrum, which, in reaction, pushes upward on the board. Since the board is stationary, it is in equilibrium.

Solution Since the board is in equilibrium, the sum of the vertical forces must be zero:

$$\Sigma F_y = -F_1 + F_2 - W = 0 \qquad (4.9b)$$

Similarly, the sum of the torques must be zero, $\Sigma \tau = 0$. For calculating torques, we select an axis that passes through the left end of the board and is perpendicular to the page. (We will see shortly that this choice is arbitrary.) The force $\vec{F}_1$ produces no torque since it passes through the axis and, therefore, has a zero lever arm, while $\vec{F}_2$ creates a counterclockwise (positive) torque, and $\vec{W}$ produces a clockwise (negative) torque. The free-body diagram shows the lever arms for the torques:

$$\Sigma \tau = +F_2 \ell_2 - W \ell_w = 0 \qquad (9.2)$$

Solving this equation for F_2 yields

$$F_2 = \frac{W \ell_w}{\ell_2} = \frac{(530 \text{ N})(3.90 \text{ m})}{1.40 \text{ m}} = \boxed{1480 \text{ N}}$$

This value for F_2, along with $W = 530$ N, can be substituted into Equation 4.9b to show that $\boxed{F_1 = 950 \text{ N}}$.

In Example 3 the sum of the external torques is calculated using an axis that passes through the left end of the diving board. **Problem-solving insight:** *However, the choice of the axis is completely arbitrary, because if an object is in equilibrium, it is in equilibrium with respect to any axis whatsoever.* Thus, the sum of the external torques is zero, no matter where the axis is placed. (See Interactive LearningWare 9.2 at **www.wiley.com/college/cutnell** for a second version of Example 3 in which the axis is chosen differently.) One usually chooses the location so that the lines of action of one or more of the unknown forces pass through the axis. Such a choice simplifies the torque equation, because the torques produced by these forces are zero. For instance, in Example 3 the torque due to the force $\vec{F}_1$ does not appear in Equation 9.2, because the lever arm of this force is zero.

In a calculation of torque, the lever arm of the force must be determined relative to the axis of rotation. In Example 3 the lever arms are obvious, but sometimes a little care is needed in determining them, as in the next example.

Example 4 Fighting a Fire

In Figure 9.7a an 8.00-m ladder of weight $W_L = 355$ N leans against a smooth vertical wall. The term "smooth" means that the wall can exert only a normal force directed perpendicular to the wall and cannot exert a frictional force parallel to it. A firefighter, whose weight is $W_F = 875$ N, stands 6.30 m up from the bottom of the ladder. Assume that the ladder's weight acts at the ladder's center and neglect the hose's weight. Find the forces that the wall and the ground exert on the ladder.

Reasoning Part b of the figure shows the free-body diagram of the ladder. The following forces act on the ladder:

1. Its weight $\vec{W}_L$

2. A force due to the weight $\vec{W}_F$ of the firefighter

3. The force $\vec{P}$ applied to the top of the ladder by the wall and directed perpendicular to the wall

4. The forces $\vec{G}_x$ and $\vec{G}_y$, which are the horizontal and vertical components of the force exerted by the ground on the bottom of the ladder

The ground, unlike the wall, is not smooth, so that the force $\vec{G}_x$ is produced by static friction and prevents the ladder from slipping. The force $\vec{G}_y$ is the normal force applied to the ladder by the ground. The ladder is in equilibrium, so the sum of these forces and the sum of the torques produced by them must be zero.

Solution Since the net force acting on the ladder is zero, we have

$$\Sigma F_x = G_x - P = 0 \tag{4.9a}$$

$$\Sigma F_y = G_y - W_L - W_F = 0 \tag{4.9b}$$

$$\text{or} \quad G_y = W_L + W_F$$

$$= 355 \text{ N} + 875 \text{ N} = \boxed{1230 \text{ N}}$$

Equation 4.9a cannot be solved as it stands, because it contains two unknown variables. However, another equation can be obtained from the fact that the net torque acting on an object in equilibrium is zero. In calculating torques, we use an axis at the left end of the ladder, directed perpendicular to the page, as Figure 9.7c indicates. This axis is convenient, because $\vec{G}_x$ and $\vec{G}_y$ produce no torques about it, their lever arms being zero. Consequently, these forces will not appear in the equation representing the balance of torques. The lever arms for the remaining forces are shown in part c as red dashed lines. The following list summarizes these forces, their lever arms, and the torques:

Force	Lever Arm	Torque
$W_L = 355 \text{ N}$	$\ell_L = (4.00 \text{ m}) \cos 50.0°$	$-W_L \ell_L$
$W_F = 875 \text{ N}$	$\ell_F = (6.30 \text{ m}) \cos 50.0°$	$-W_F \ell_F$
P	$\ell_P = (8.00 \text{ m}) \sin 50.0°$	$+P \ell_P$

Problem-solving insight: Sometimes it is necessary to use trigonometry to determine the lever arms from the distances given in a problem, as here in Example 4.

Setting the sum of the torques equal to zero gives

$$\Sigma \tau = -W_L \ell_L - W_F \ell_F + P \ell_P = 0 \tag{9.2}$$

Solving this equation for P gives

$$P = \frac{W_L \ell_L + W_F \ell_F}{\ell_P}$$

$$= \frac{(355 \text{ N})(4.00 \text{ m}) \cos 50.0° + (875 \text{ N})(6.30 \text{ m}) \cos 50.0°}{(8.00 \text{ m}) \sin 50.0°} = \boxed{727 \text{ N}}$$

Substituting $P = 727 \text{ N}$ into Equation 4.9a indicates that $G_x = P = \boxed{727 \text{ N}}$.

To a large extent the directions of the forces acting on an object in equilibrium can be deduced using intuition. Sometimes, however, the direction of an unknown force is not obvious, and it is inadvertently drawn reversed in the free-body diagram. This kind of mistake causes no difficulty. **Problem-solving insight: *Choosing the direction of an unknown force backward in the free-body diagram simply means that the value determined for the force will be a negative number***, as the next example illustrates.

 **Example 5** Bodybuilding

(a)

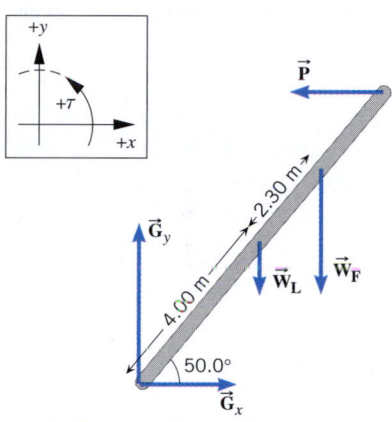

(b) Free-body diagram of the ladder

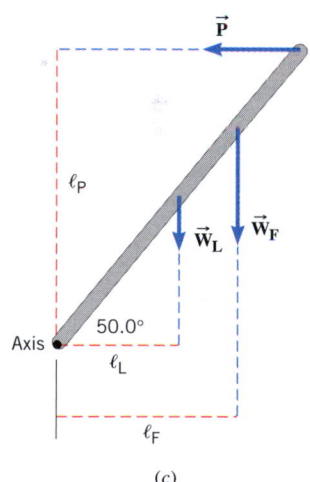

(c)

Figure 9.7 (a) A ladder leaning against a smooth wall. (b) The free-body diagram for the ladder. (c) Three of the forces that act on the ladder and their lever arms. The axis of rotation is at the lower end of the ladder and is perpendicular to the page.

The physics of bodybuilding. A bodybuilder holds a dumbbell of weight $\vec{W}_d$ as in Figure 9.8a. His arm is horizontal and weighs $W_a = 31.0 \text{ N}$. The deltoid muscle is assumed to be the only muscle acting and is attached to the arm as shown. The maximum force $\vec{M}$ that the

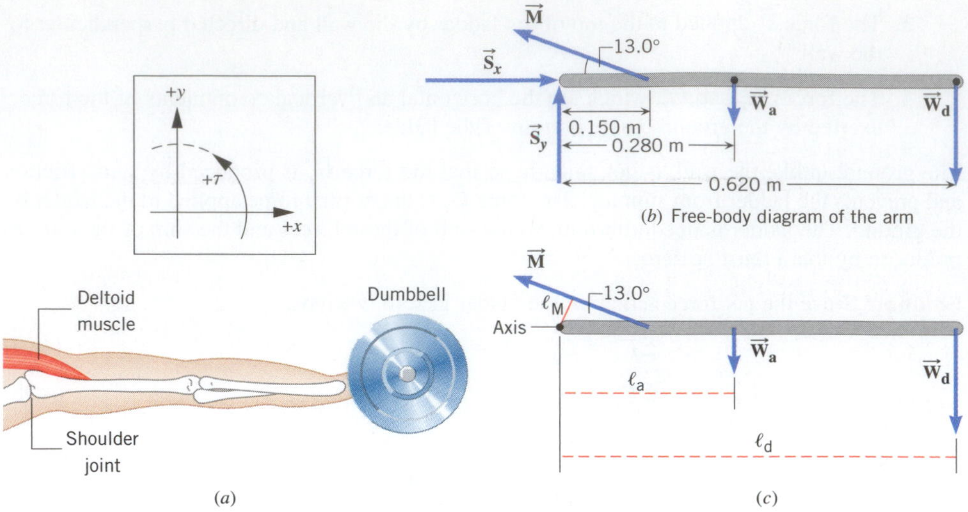

(b) Free-body diagram of the arm

Figure 9.8 (a) The fully extended, horizontal arm of a bodybuilder supports a dumbbell. (b) The free-body diagram for the arm. (c) Three of the forces that act on the arm and their lever arms. The axis of rotation at the left end of the arm is perpendicular to the page. Force vectors are not to scale.

(a)

(c)

deltoid muscle can supply has a magnitude of 1840 N. Figure 9.8b shows the distances that locate where the various forces act on the arm. What is the weight of the heaviest dumbbell that can be held, and what are the horizontal and vertical force components, $\vec{S}_x$ and $\vec{S}_y$, that the shoulder joint applies to the left end of the arm?

Reasoning Figure 9.8b is the free-body diagram for the arm. Note that $\vec{S}_x$ is directed to the right, because the deltoid muscle pulls the arm in toward the shoulder joint, and the joint pushes back in accordance with Newton's third law. The direction of the force $\vec{S}_y$, however, is less obvious, and we are alert for the possibility that the direction chosen in the free-body diagram is backward. If so, the value obtained for $\vec{S}_y$ will be negative.

Solution The arm is in equilibrium, so the net force acting on it is zero:

$$\Sigma F_x = S_x - M \cos 13.0° = 0 \tag{4.9a}$$

$$\text{or}\quad S_x = M \cos 13.0° = (1840 \text{ N}) \cos 13.0° = \boxed{1790 \text{ N}}$$

$$\Sigma F_y = S_y + M \sin 13.0° - W_a - W_d = 0 \tag{4.9b}$$

Equation 4.9b cannot be solved at this point, because it contains two unknowns, S_y and W_d. However, since the arm is in equilibrium, the torques acting on the arm must balance, and this fact provides another equation. To calculate torques, we choose an axis through the left end of the arm and perpendicular to the page. With this axis, the torques due to $\vec{S}_x$ and $\vec{S}_y$ are zero, because the line of action of each force passes through the axis and the lever arm of each force is zero. The list below summarizes the remaining forces, their lever arms (see Figure 9.8c), and the torques.

Force	Lever Arm	Torque
$W_a = 31.0$ N	$\ell_a = 0.280$ m	$-W_a\ell_a$
W_d	$\ell_d = 0.620$ m	$-W_d\ell_d$
$M = 1840$ N	$\ell_M = (0.150 \text{ m}) \sin 13.0°$	$+M\ell_M$

The condition specifying a zero net torque is

$$\Sigma \tau = -W_a\ell_a - W_d\ell_d + M\ell_M = 0 \tag{9.2}$$

Solving this equation for W_d yields

$$W_d = \frac{-W_a\ell_a + M\ell_M}{\ell_d}$$

$$= \frac{-(31.0 \text{ N})(0.280 \text{ m}) + (1840 \text{ N})(0.150 \text{ m}) \sin 13.0°}{0.620 \text{ m}} = \boxed{86.1 \text{ N}}$$

Problem-solving insight

When a force is negative, such as the vertical force $S_y = -297$ N in this example, it means that the direction of the force is opposite to that chosen originally.

Substituting this value for W_d into Equation 4.9b and solving for S_y gives $\boxed{S_y = -297 \text{ N}}$. The minus sign indicates that the choice of direction for S_y in the free-body diagram is wrong. In reality, S_y has a magnitude of 297 N but is directed downward, not upward.

✓ CHECK YOUR UNDERSTANDING

(The answers are given at the end of the book.)

5. Three forces (magnitudes either *F* or 2*F*) act on each of the thin, square sheets shown in the drawing. In parts A and B of the drawing, the force labeled $2\vec{F}$ acts at the center of the sheet. When considering angular acceleration, use an axis of rotation that is perpendicular to the plane of a sheet at its center. Determine in which drawing **(a)** the translational acceleration is equal to zero, but the angular acceleration is not equal to zero; **(b)** the translational acceleration is not equal to zero, but the angular acceleration is equal to zero; and **(c)** both the translational and angular accelerations are zero.

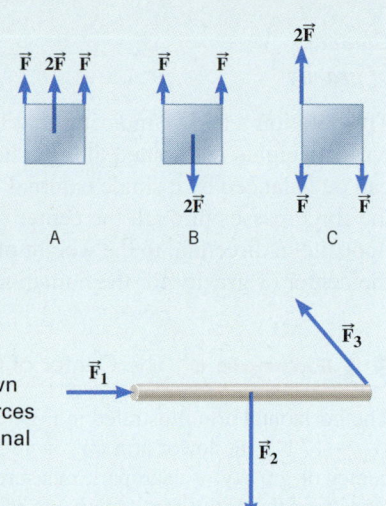

6. The free-body diagram in the drawing shows the forces that act on a thin rod. The three forces are drawn to scale and lie in the plane of the paper. Are these forces sufficient to keep the rod in equilibrium, or are additional forces necessary?

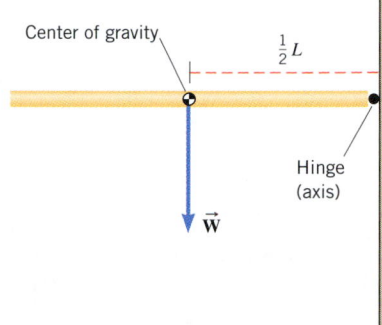

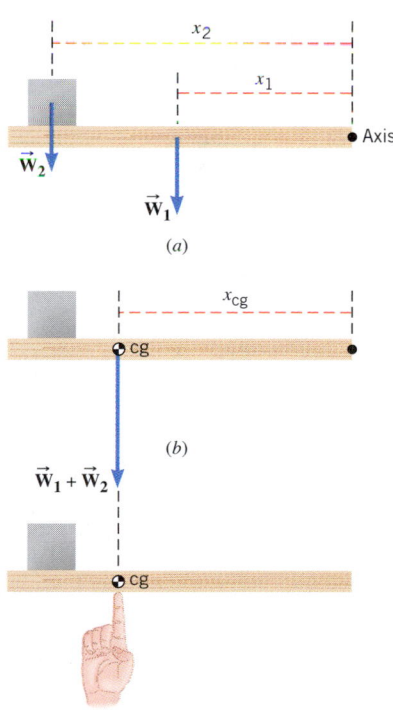

Figure 9.9 A thin, uniform, horizontal rod of length *L* is attached to a vertical wall by a hinge. The center of gravity of the rod is at its geometrical center.

9.3 CENTER OF GRAVITY

Often, it is important to know the torque produced by the weight of an *extended* body. In Examples 4 and 5, for instance, it is necessary to determine the torques caused by the weight of the ladder and the arm, respectively. In both cases the weight is considered to act at a definite point for the purpose of calculating the torque. This point is called the **center of gravity** (abbreviated "cg").

> **DEFINITION OF CENTER OF GRAVITY**
>
> The center of gravity of a rigid body is the point at which its weight can be considered to act when the torque due to the weight is being calculated.

When an object has a symmetrical shape and its weight is distributed uniformly, the center of gravity lies at its geometrical center. For instance, Figure 9.9 shows a thin, uniform, horizontal rod of length *L* attached to a vertical wall by a hinge. The center of gravity of the rod is located at the geometrical center. The lever arm for the weight $\vec{W}$ is *L*/2, and the magnitude of the torque is *W*(*L*/2). In a similar fashion, the center of gravity of any symmetrically shaped and uniform object, such as a sphere, disk, cube, or cylinder, is located at its geometrical center. However, this does not mean that the center of gravity must lie within the object itself. The center of gravity of a compact disc recording, for instance, lies at the center of the hole in the disc and is, therefore, "outside" the object.

Suppose we have a group of objects, with known weights and centers of gravity, and it is necessary to know the center of gravity for the group as a whole. As an example, Figure 9.10*a* shows a group composed of two parts: a horizontal uniform board (weight $\vec{W}_1$) and a uniform box (weight $\vec{W}_2$) near the left end of the board. The center of gravity can be determined by calculating the net torque created by the board and box about an axis that is picked arbitrarily to be at the right end of the board. Part *a* of the figure shows the weights $\vec{W}_1$ and $\vec{W}_2$ and their corresponding lever arms x_1 and x_2. The net torque is $\Sigma\tau = W_1 x_1 + W_2 x_2$. It is also possible to calculate the net torque by treating the total weight $\vec{W}_1 + \vec{W}_2$ as if it were located at the center of gravity and had the lever arm x_{cg}, as part *b* of the drawing indicates: $\Sigma\tau = (W_1 + W_2)x_{\text{cg}}$. The two values for the net torque must be the same, so that

$$W_1 x_1 + W_2 x_2 = (W_1 + W_2)x_{\text{cg}}$$

Figure 9.10 (*a*) A box rests near the left end of a horizontal board. (*b*) The total weight ($\vec{W}_1 + \vec{W}_2$) acts at the center of gravity of the group. (*c*) The group can be balanced by applying an external force (due to the index finger) at the center of gravity.

This expression can be solved for x_{cg}, which locates the center of gravity relative to the axis:

***Center
of gravity***

$$x_{cg} = \frac{W_1 x_1 + W_2 x_2 + \cdots}{W_1 + W_2 + \cdots}$$

(9.3)

The notation "$+ \cdots$" indicates that Equation 9.3 can be extended to account for any number of weights distributed along a horizontal line. Figure 9.10*c* illustrates that the group can be balanced by a single external force (due to the index finger), if the line of action of the force passes through the center of gravity, and if the force is equal in magnitude, but opposite in direction, to the weight of the group. Example 6 demonstrates how to calculate the center of gravity for the human arm.

Example 6 The Center of Gravity of an Arm

The horizontal arm illustrated in Figure 9.11 is composed of three parts: the upper arm (weight $W_1 = 17$ N), the lower arm ($W_2 = 11$ N), and the hand ($W_3 = 4.2$ N). The drawing shows the center of gravity of each part, measured with respect to the shoulder joint. Find the center of gravity of the entire arm, relative to the shoulder joint.

Reasoning and Solution The coordinate x_{cg} of the center of gravity is given by

$$x_{cg} = \frac{W_1 x_1 + W_2 x_2 + W_3 x_3}{W_1 + W_2 + W_3}$$

(9.3)

$$= \frac{(17\text{ N})(0.13\text{ m}) + (11\text{ N})(0.38\text{ m}) + (4.2\text{ N})(0.61\text{ m})}{17\text{ N} + 11\text{ N} + 4.2\text{ N}} = \boxed{0.28\text{ m}}$$

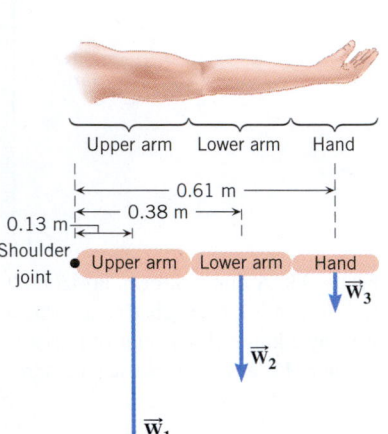

Figure 9.11 The three parts of a human arm, and the weight and center of gravity for each.

The center of gravity plays an important role in determining whether a group of objects remains in equilibrium as the weight distribution within the group changes. A change in the weight distribution causes a change in the position of the center of gravity, and if the change is too great, the group will not remain in equilibrium. Conceptual Example 7 discusses a shift in the center of gravity that led to an embarrassing result.

Conceptual Example 7 Overloading a Cargo Plane

Figure 9.12*a* shows a stationary cargo plane with its front landing gear 9 meters off the ground. This accident occurred because the plane was overloaded toward the rear. How did a shift in the center of gravity of the loaded plane cause the accident?

Reasoning and Solution Figure 9.12*b* shows a drawing of a correctly loaded plane, with the center of gravity located between the front and the rear landing gears. The weight $\vec{W}$ of the plane and cargo acts downward at the center of gravity, and the normal forces $\vec{F}_{N1}$ and $\vec{F}_{N2}$ act upward at the front and at the rear landing gear, respectively. With respect to an axis at the rear landing gear, the counterclockwise torque due to $\vec{F}_{N1}$ balances the clockwise torque due to $\vec{W}$, and the plane remains in equilibrium. Figure 9.12*c* shows the plane with too much cargo loaded toward the rear, just after the plane has begun to rotate counterclockwise. Because of the overloading, the center of gravity has shifted behind the rear landing gear. The torque due to $\vec{W}$ is now counterclockwise and is not balanced by any clockwise torque. Due to the unbalanced counterclockwise torque, the plane rotates until its tail hits the ground, which applies an upward force to the tail. The clockwise torque due to this upward force balances the counterclockwise torque due to $\vec{W}$, and the plane comes again into an equilibrium state, this time with the front landing gear 9 meters off the ground.

Related Homework: *Problems 12, 16, 20*

As we have seen in Example 7, the center of gravity plays an important role in the equilibrium orientation of airplanes. It also plays a similar role in the design of vehicles that can be safely driven with minimal risk to the passengers. Example 8 discusses this role in the context of sport utility vehicles.

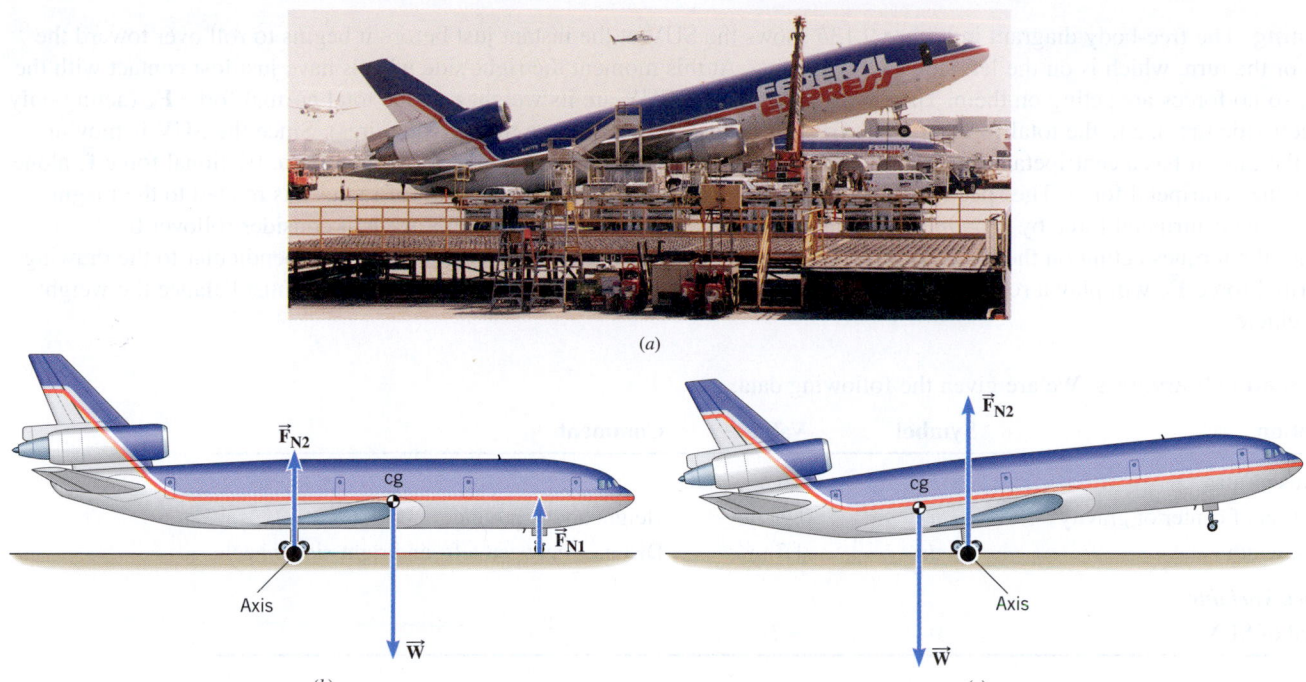

Figure 9.12 (*a*) This stationary cargo plane is sitting on its tail at Los Angeles International Airport, after being overloaded toward the rear. (*b*) In a correctly loaded plane, the center of gravity is between the front and the rear landing gears. (*c*) When the plane is overloaded toward the rear, the center of gravity shifts behind the rear landing gear, and the accident in part *a* occurs. (*a*, © AP/World Wide Photos)

ANALYZING MULTIPLE-CONCEPT PROBLEMS

Example 8   The Static Stability Factor and Rollover

The physics of
the static stability factor and rollover.

Figure 9.13*a* shows a sport utility vehicle (SUV) that is moving away from you and negotiating a horizontal turn. The radius of the turn is 16 m, and its center is on the right in the drawing. The center of gravity of the vehicle is 0.94 m above the ground and, as an approximation, is assumed to be located midway between the wheels on the left and right sides. The separation between these wheels is the track width and is 1.7 m. What is the greatest speed at which the SUV can negotiate the turn without rolling over?

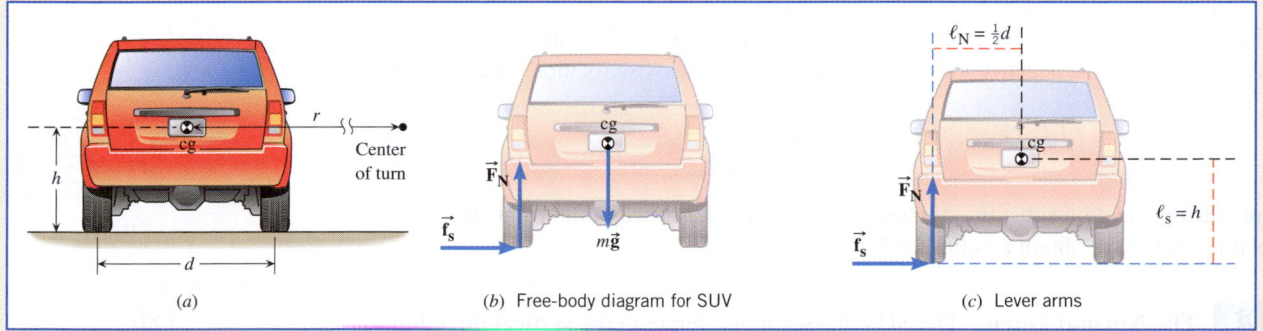

Figure 9.13 (*a*) A sport utility vehicle (SUV) is shown moving away from you, following a road curving to the right. The radius of the turn is *r*. The SUV's center of gravity (⊙) is located at a height *h* above the ground, and its wheels are separated by a distance *d*. (*b*) This free-body diagram shows the SUV at the instant it begins to roll over toward the outside of the turn. The forces acting on it are its weight $m\vec{g}$, the total normal force $\vec{F}_N$ (acting only on the left-side tires), and the total force of static friction $\vec{f}_s$ (also acting only on the left-side tires). (*c*) The lever arms ℓ_N (for $\vec{F}_N$) and ℓ_s (for $\vec{f}_s$) are for an axis passing through the center of gravity and perpendicular to the page.

Continued

Reasoning The free-body diagram in Figure 9.13b shows the SUV at the instant just before it begins to roll over toward the outside of the turn, which is on the left side of the drawing. At this moment the right-side wheels have just lost contact with the ground, so no forces are acting on them. The forces acting on the SUV are its weight $m\vec{g}$, the total normal force $\vec{F}_N$ (acting only on the left-side tires), and the total force of static friction $\vec{f}_s$ (also acting only on the left-side tires). Since the SUV is moving around the turn, it has a centripetal acceleration and must be experiencing a centripetal force. The static frictional force $\vec{f}_s$ alone provides the centripetal force. The speed v at which the SUV (mass m) negotiates the turn of radius r is related to the magnitude F_c of the centripetal force by $F_c = mv^2/r$ (Equation 5.3). After applying this relation, we will consider rollover by analyzing the torques acting on the SUV with respect to an axis through the center of gravity and perpendicular to the drawing. The normal force $\vec{F}_N$ will play a role in this analysis, and it will be evaluated by using the fact that it must balance the weight of the vehicle.

Knowns and Unknowns We are given the following data:

Description	Symbol	Value	Comment
Radius of turn	r	16 m	
Location of center of gravity	h	0.94 m	Height above ground.
Track width	d	1.7 m	Distance between left- and right-side wheels.
Unknown Variable			
Speed of SUV	v	?	

Modeling the Problem

STEP 1 Speed and Centripetal Force According to Equation 5.3, the magnitude of the centripetal force (provided solely by the static frictional force of magnitude f_s) is related to the speed v of the car by

$$F_c = f_s = \frac{mv^2}{r}$$

Solving for the speed gives Equation 1 at the right. To use this result, we must have a value for f_s, which we obtain in Step 2.

$$v = \sqrt{\frac{rf_s}{m}} \qquad (1)$$

$?$

STEP 2 Torques Figure 9.13b shows the free-body diagram for the SUV at the instant just before rollover begins, when the total normal force $\vec{F}_N$ and the total static frictional force $\vec{f}_s$ are acting only on the left-side wheels, the right-side wheels having just lost contact with the ground. At this moment, the sum of the torques due to these two forces is zero for an axis through the center of gravity and perpendicular to the page. Each torque has a magnitude given by Equation 9.1 as the magnitude of the force times the lever arm for the force, and part c of the drawing shows the lever arms. No lever arm is shown for the weight, because it passes through the axis and, therefore, contributes no torque. The force $\vec{f}_s$ produces a positive torque, since it causes a counterclockwise rotation about the chosen axis, while the force $\vec{F}_N$ produces a negative torque, since it causes a clockwise rotation. Thus, we have

$$f_s\ell_s - F_N\ell_N = 0 \quad \text{or} \quad \boxed{f_s = \frac{F_N\ell_N}{\ell_s} = \frac{F_N d}{2h}}$$

This result for f_s can be substituted into Equation 1, as illustrated at the right. We now proceed to Step 3, to obtain a value for F_N.

$$v = \sqrt{\frac{rf_s}{m}} \qquad (1)$$

$$f_s = \frac{F_N d}{2h} \qquad (2)$$

$?$

STEP 3 The Normal Force The SUV does not accelerate in the vertical direction, so the normal force must balance the car's weight mg, or

$$\boxed{F_N = mg}$$

The substitution of this result into Equation 2 is shown at the right.

$$v = \sqrt{\frac{rf_s}{m}} \qquad (1)$$

$$f_s = \frac{F_N d}{2h} \qquad (2)$$

$$F_N = mg$$

Solution The results of each step can be combined algebraically to show that

$$\underbrace{v}_{} \overset{\text{STEP 1}}{=} \sqrt{\frac{rf_s}{m}} \overset{\text{STEP 2}}{=} \sqrt{\frac{rF_N d}{2hm}} \overset{\text{STEP 3}}{=} \sqrt{\frac{r \cancel{m} g d}{2h \cancel{m}}}$$

The final expression is independent of the mass of the SUV, since m has been eliminated algebraically. The greatest speed at which the SUV can negotiate the turn without rolling over is

$$v = \sqrt{rg\underbrace{\left(\frac{d}{2h}\right)}_{\text{SSF}}} = \sqrt{(16 \text{ m})(9.80 \text{ m/s}^2)\left[\frac{1.7 \text{ m}}{2(0.94 \text{ m})}\right]} = \boxed{12 \text{ m/s}}$$

At this speed, the SUV will negotiate the turn just on the verge of rolling over. In the final result and beneath the term $\dfrac{d}{2h}$, we have included the label SSF, which stands for *static stability factor*. The SSF provides one measure of how susceptible a vehicle is to rollover. Higher values of the SSF are better, because they lead to larger values for v, the greatest speed at which the vehicle can negotiate a turn without rolling over. Note that greater values for the track width d (more widely separated wheels) and smaller values for the height h (center of gravity closer to the ground) lead to higher values of the SSF. A Formula One racing car, for example, with its low center of gravity and large track width, is much less prone to rolling over than is an SUV, when both are driven at the same speed around the same curve.

Related Homework: *Problem 17*

The center of gravity of an object with an irregular shape and a nonuniform weight distribution can be found by suspending the object from two different points P_1 and P_2, one at a time. Figure 9.14a shows the object at the moment of release, when its weight $\vec{W}$, acting at the center of gravity, has a nonzero lever arm ℓ relative to the axis shown in the drawing. At this instant the weight produces a torque about the axis. The tension force $\vec{T}$ applied to the object by the suspension cord produces no torque because its line of action passes through the axis. Hence, in part a there is a net torque applied to the object, and the object begins to rotate. Friction eventually brings the object to rest as in part b, where the center of gravity lies directly below the point of suspension. In such an orientation, the line of action of the weight passes through the axis, so there is no longer any net torque. In the absence of a net torque the object remains at rest. By suspending the object from a second point P_2 (see Figure 9.14c), a second

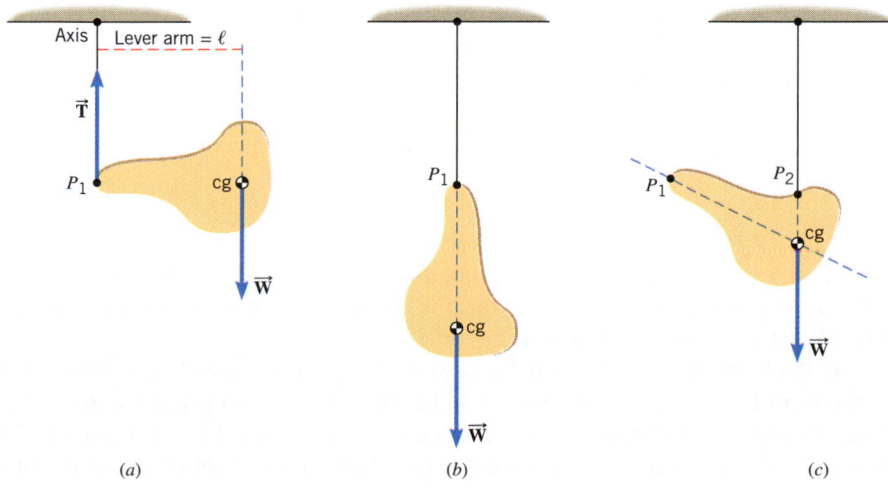

Figure 9.14 The center of gravity (cg) of an object can be located by suspending the object from two different points, P_1 and P_2, one at a time.

line through the object can be established, along which the center of gravity must also lie. The center of gravity, then, must be at the intersection of the two lines.

The center of gravity is closely related to the center-of-mass concept discussed in Section 7.5. To see why they are related, let's replace each occurrence of the weight in Equation 9.3 by $W = mg$, where m is the mass of a given object and g is the magnitude of the acceleration due to gravity at the location of the object. Suppose that g has the same value everywhere the objects are located. Then it can be algebraically eliminated from each term on the right side of Equation 9.3. The resulting equation, which contains only masses and distances, is the same as Equation 7.10, which defines the location of the center of mass. Thus, the two points are identical. For ordinary-sized objects, like cars and boats, the center of gravity coincides with the center of mass.

✓ CHECK YOUR UNDERSTANDING

(The answers are given at the end of the book.)

7. Starting in the spring, fruit begins to grow on the outer end of a branch on a pear tree. As the fruit grows, does the center of gravity of the pear-growing branch **(a)** move toward the pears at the end of the branch, **(b)** move away from the pears, or **(c)** not move at all?

8. The drawing shows a wine rack for a single bottle of wine that seems to defy common sense as it balances on a tabletop. Where is the center of gravity of the combined wine rack and bottle of wine located? **(a)** At the neck of the bottle where it passes through the wine rack **(b)** Directly above the point where the wine rack touches the tabletop **(c)** At a location to the right of where the wine rack touches the tabletop

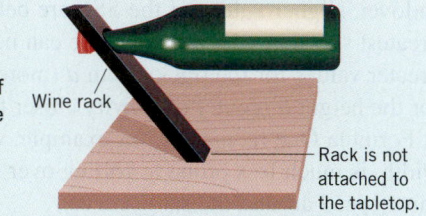

Wine rack

Rack is not attached to the tabletop.

9. Bob and Bill have the same weight and wear identical shoes. When they both keep their feet flat on the floor and their bodies straight, Bob can lean forward farther than Bill can before falling. Other things being equal, whose center of gravity is closer to the ground when both are standing erect?

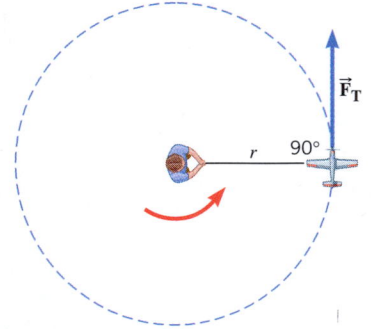

Figure 9.15 A model airplane on a guideline has a mass m and is flying on a circle of radius r (top view). A net tangential force $\vec{F}_T$ acts on the plane.

9.4 NEWTON'S SECOND LAW FOR ROTATIONAL MOTION ABOUT A FIXED AXIS

The goal of this section is to put Newton's second law into a form suitable for describing the rotational motion of a rigid object about a fixed axis. We begin by considering a particle moving on a circular path. Figure 9.15 presents a good approximation of this situation by using a small model plane on a guideline of negligible mass. The plane's engine produces a net external tangential force F_T that gives the plane a tangential acceleration a_T. In accord with Newton's second law, it follows that $F_T = ma_T$. The torque τ produced by this force is $\tau = F_T r$, where the radius r of the circular path is also the lever arm. As a result, the torque is $\tau = ma_T r$. However, the tangential acceleration is related to the angular acceleration α according to $a_T = r\alpha$ (Equation 8.10), where α must be expressed in rad/s². With this substitution for a_T, the torque becomes

$$\tau = (mr^2)\alpha \qquad (9.4)$$

$$\underbrace{}_{\substack{\text{Moment} \\ \text{of inertia } I}}$$

Equation 9.4 is the form of Newton's second law we have been seeking. It indicates that the net external torque τ is directly proportional to the angular acceleration α. The constant of proportionality is $I = mr^2$, which is called the *moment of inertia of the particle*. The SI unit for moment of inertia is kg·m².

If all objects were single particles, it would be just as convenient to use the second law in the form $F_T = ma_T$ as in the form $\tau = I\alpha$. The advantage in using $\tau = I\alpha$ is that it can be applied to any rigid body rotating about a fixed axis, and not just to a particle. To illustrate how this advantage arises, Figure 9.16a shows a flat sheet of material that rotates

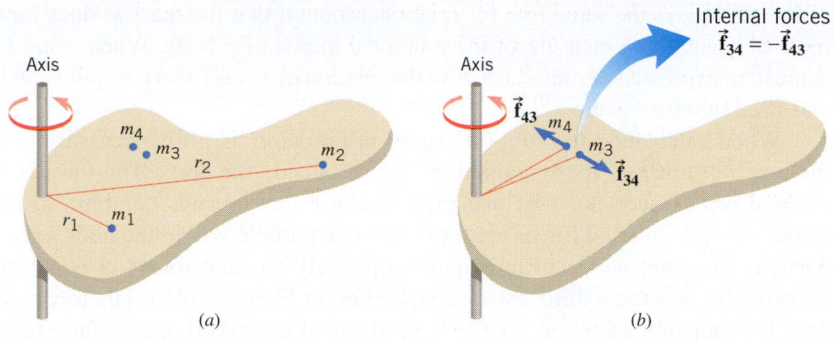

Figure 9.16 (*a*) A rigid body consists of a large number of particles, four of which are shown. (*b*) The internal forces that particles 3 and 4 exert on each other obey Newton's law of action and reaction.

about an axis perpendicular to the sheet. The sheet is composed of a number of mass particles, $m_1, m_2, \ldots, m_N$, where N is very large. Only four particles are shown for the sake of clarity. Each particle behaves in the same way as the model airplane in Figure 9.15 and obeys the relation $\tau = (mr^2)\alpha$:

$$\tau_1 = (m_1 r_1^2)\alpha$$
$$\tau_2 = (m_2 r_2^2)\alpha$$
$$\vdots$$
$$\tau_N = (m_N r_N^2)\alpha$$

In these equations each particle has the same angular acceleration α, since the rotating object is assumed to be rigid. Adding together the N equations and factoring out the common value of α, we find that

$$\underbrace{\Sigma\tau}_{\substack{\text{Net} \\ \text{external torque}}} = \underbrace{(\Sigma mr^2)}_{\substack{\text{Moment} \\ \text{of inertia}}}\alpha \qquad (9.5)$$

where the expression $\Sigma\tau = \tau_1 + \tau_2 + \cdots + \tau_N$ is the sum of the external torques, and $\Sigma mr^2 = m_1 r_1^2 + m_2 r_2^2 + \cdots + m_N r_N^2$ represents the sum of the individual moments of inertia. The latter quantity is the **moment of inertia *I* of the body:**

Moment of inertia
of a body
$$I = m_1 r_1^2 + m_2 r_2^2 + \cdots + m_N r_N^2 = \Sigma mr^2 \qquad (9.6)$$

In this equation, r is the perpendicular radial distance of each particle from the axis of rotation. Combining Equation 9.6 with Equation 9.5 gives the following result:

ROTATIONAL ANALOG OF NEWTON'S SECOND LAW
FOR A RIGID BODY ROTATING ABOUT A FIXED AXIS

$$\text{Net external torque} = \left(\begin{array}{c}\text{Moment of} \\ \text{inertia}\end{array}\right) \times \left(\begin{array}{c}\text{Angular} \\ \text{acceleration}\end{array}\right)$$

$$\Sigma\tau = I\alpha \qquad (9.7)$$

Requirement: α must be expressed in rad/s^2.

The version of Newton's second law given in Equation 9.7 applies only for rigid bodies. The word "rigid" means that the distances r_1, r_2, r_3, etc. that locate each particle m_1, m_2, m_3, etc. (see Figure 9.16*a*) do not change during the rotational motion. In other words, a rigid body is one that does not change its shape while undergoing an angular acceleration in response to an applied net external torque.

The form of the second law for rotational motion, $\Sigma\tau = I\alpha$, is similar to the equation for translational (linear) motion, $\Sigma F = ma$, and is valid only in an inertial frame. The moment

Table 9.1 Moments of Inertia I for Various Rigid Objects of Mass M

Thin-walled hollow cylinder or hoop

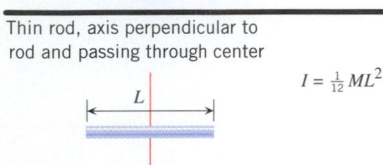

$I = MR^2$

Solid cylinder or disk

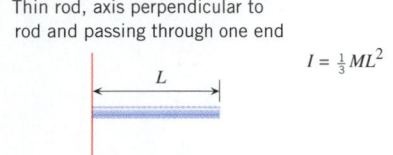

$I = \frac{1}{2}MR^2$

Thin rod, axis perpendicular to rod and passing through center

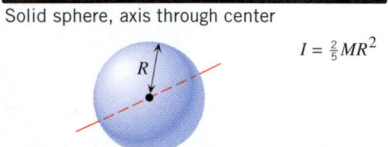

$I = \frac{1}{12}ML^2$

Thin rod, axis perpendicular to rod and passing through one end

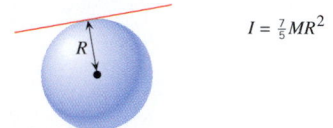

$I = \frac{1}{3}ML^2$

Solid sphere, axis through center

$I = \frac{2}{5}MR^2$

Solid sphere, axis tangent to surface

$I = \frac{7}{5}MR^2$

Thin-walled spherical shell, axis through center

$I = \frac{2}{3}MR^2$

Thin rectangular sheet, axis parallel to one edge and passing through center of other edge

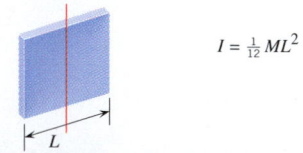

$I = \frac{1}{12}ML^2$

Thin rectangular sheet, axis along one edge

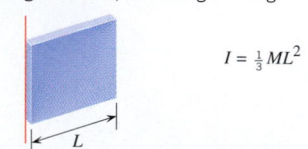

$I = \frac{1}{3}ML^2$

of inertia I plays the same role for rotational motion that the mass m does for translational motion. Thus, I is a measure of the rotational inertia of a body. When using Equation 9.7, α must be expressed in rad/s², because the relation $a_T = r\alpha$ (which requires radian measure) was used in the derivation.

When calculating the sum of torques in Equation 9.7, it is necessary to include only the *external torques,* those applied by agents outside the body. The torques produced by internal forces need not be considered, because they always combine to produce a net torque of zero. Internal forces are those that one particle within the body exerts on another particle. They always occur in pairs of oppositely directed forces of equal magnitude, in accord with Newton's third law (see m_3 and m_4 in Figure 9.16b). The forces in such a pair have the same line of action, so they have identical lever arms and produce torques of equal magnitudes. One torque is counterclockwise, while the other is clockwise, the net torque from the pair being zero.

It can be seen from Equation 9.6 that the moment of inertia depends on both the mass of each particle and its distance from the axis of rotation. The farther a particle is from the axis, the greater is its contribution to the moment of inertia. Therefore, although a rigid object possesses a unique total mass, it does not have a unique moment of inertia, as indicated by Example 9. This example shows how the moment of inertia can change when the axis of rotation changes. The procedure illustrated in Example 9 can be extended using integral calculus to evaluate the moment of inertia of a rigid object with a continuous mass distribution, and Table 9.1 gives some typical results. These results depend on the total mass of the object, its shape, and the location and orientation of the axis.

Problem-solving insight: The moment of inertia depends on the location and orientation of the axis relative to the particles that make up the object.

Example 9 The Moment of Inertia Depends on Where the Axis Is

Two particles each have a mass M and are fixed to the ends of a thin rigid rod, whose mass can be ignored. The length of the rod is L. Find the moment of inertia when this object rotates relative to an axis that is perpendicular to the rod at **(a)** one end and **(b)** the center. (See Figure 9.17.)

Reasoning When the axis of rotation changes, the distance r between the axis and each particle changes. In determining the moment of inertia using $I = \Sigma mr^2$, we must be careful to use the distances that apply for each axis.

Solution **(a)** Particle 1 lies on the axis, as part a of the drawing shows, and has a zero radial distance: $r_1 = 0$. In contrast, particle 2 moves on a circle whose radius is $r_2 = L$. Noting that $m_1 = m_2 = M$, we find that the moment of inertia is

$$I = \Sigma mr^2 = m_1 r_1^2 + m_2 r_2^2 = M(0)^2 + M(L)^2 = \boxed{ML^2} \quad (9.6)$$

(b) Part b of the drawing shows that particle 1 no longer lies on the axis but now moves on a circle of radius $r_1 = L/2$. Particle 2 moves on a circle with the same radius, $r_2 = L/2$. Therefore,

$$I = \Sigma mr^2 = m_1 r_1^2 + m_2 r_2^2 = M(L/2)^2 + M(L/2)^2 = \boxed{\tfrac{1}{2}ML^2}$$

This value differs from the value in part (a) because the axis of rotation is different.

Figure 9.17 Two particles, masses m_1 and m_2, are attached to the ends of a massless rigid rod. The moment of inertia of this object is different, depending on whether the rod rotates about an axis through (a) the end or (b) the center of the rod.

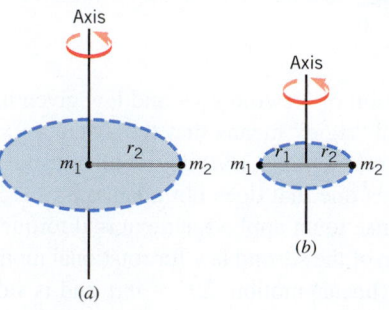

CONCEPTS AT A GLANCE

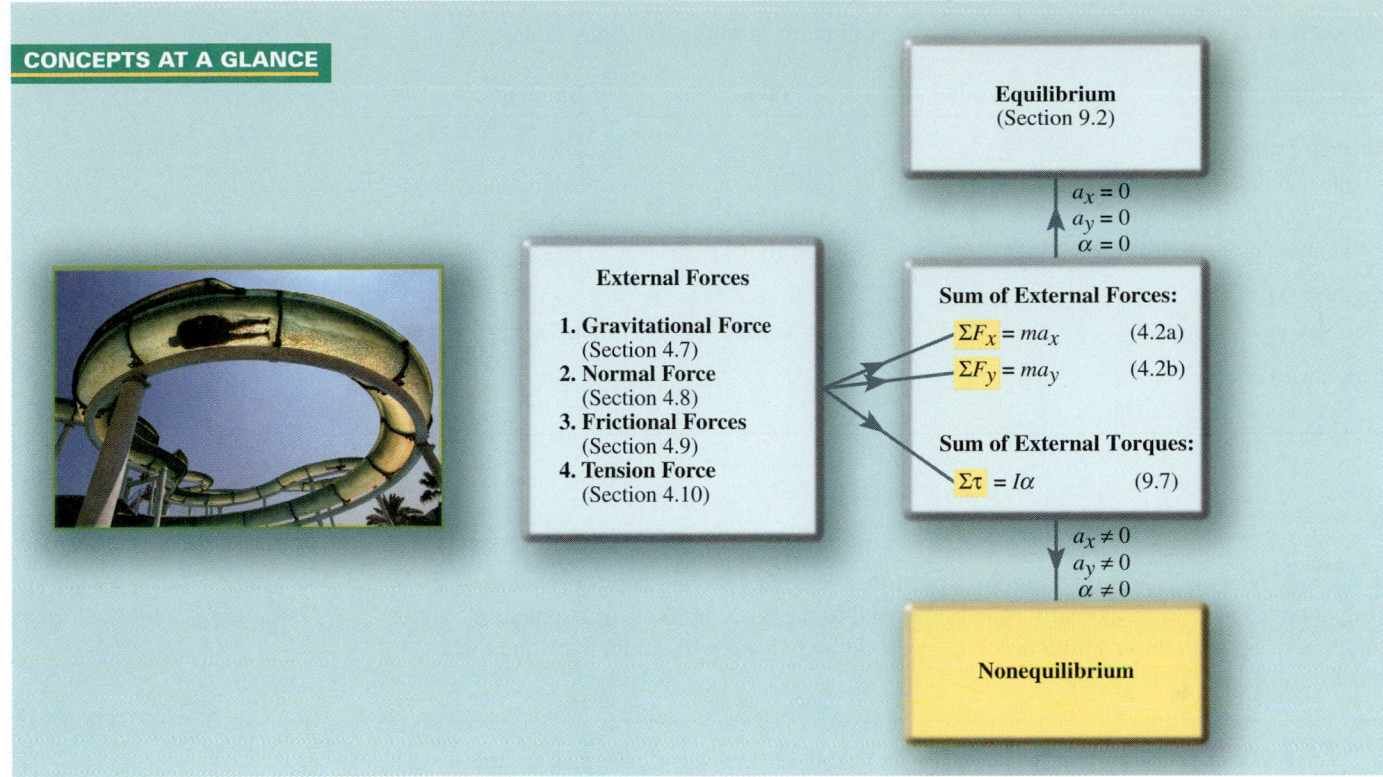

Figure 9.18 **CONCEPTS AT A GLANCE** An object is in equilibrium when its translational acceleration components, a_x and a_y, and its angular acceleration α are zero. If a_x, a_y, or α is not zero, the object has an acceleration and is not in equilibrium. As this vacationer spirals around and down the water slide, he has translational and angular accelerations and is not in equilibrium. (John Berry/ The Image Works)

▶ **CONCEPTS AT A GLANCE** When forces act on a rigid object, they can affect its motion in two ways. They can produce a translational acceleration a (components a_x and a_y). The forces can also produce torques, which can produce an angular acceleration α. In general, we can deal with the resulting combined motion by using Newton's second law. For the translational motion, we use the law in the form $\Sigma F = ma$. For the rotational motion, we use the law in the form $\Sigma \tau = I\alpha$. The Concepts-at-a-Glance chart in Figure 9.18 illustrates the essence of this joint usage of Newton's second law. When a (both components) and α are zero, there is no acceleration of any kind, and the object is in equilibrium. This is the situation already discussed in Section 9.2. If any component of a or α is nonzero, we have accelerated motion, and the object is not in equilibrium. Examples 10, 11, and 12 deal with this type of situation. ◀

ANALYZING MULTIPLE-CONCEPT PROBLEMS

Example 10 The Torque of an Electric Saw Motor

The motor in an electric saw brings the circular blade from rest up to the rated angular velocity of 80.0 rev/s in 240.0 rev. One type of blade has a moment of inertia of $1.41 \times 10^{-3} \ \text{kg} \cdot \text{m}^2$. What net torque (assumed constant) must the motor apply to the blade?

Reasoning Newton's second law for rotational motion, $\Sigma \tau = I\alpha$ (Equation 9.7), can be used to find the net torque $\Sigma \tau$. However, when using the second law, we will need a value for the angular acceleration α, which can be obtained by using one of the equations of rotational kinematics. In addition, we must remember that the value for α must be expressed in rad/s², not rev/s², because Equation 9.7 requires radian measure.

Continued

Knowns and Unknowns The given data are summarized in the following table:

Description	Symbol	Value	Comment
Explicit Data			
Final angular velocity	ω	80.0 rev/s	Must be converted to rad/s.
Angular displacement	θ	240.0 rev	Must be converted to rad.
Moment of inertia	I	$1.41 \times 10^{-3}\ \text{kg} \cdot \text{m}^2$	
Implicit Data			
Initial angular velocity	ω_0	0 rad/s	Blade starts from rest.
Unknown Variable			
Net torque applied to blade	$\Sigma\tau$	?	

Modeling the Problem

STEP 1 **Newton's Second Law for Rotation** Newton's second law for rotation (Equation 9.7) specifies the net torque $\Sigma\tau$ applied to the blade in terms of the blade's moment of inertia I and angular acceleration α. In Step 2, we will obtain the value for α that is needed in Equation 9.7.

$$\Sigma\tau = I\alpha \qquad (9.7)$$

STEP 2 **Rotational Kinematics** As the data table indicates, we have data for the angular displacement θ, the final angular velocity ω, and the initial angular velocity ω_0. With these data, Equation 8.8 from the equations of rotational kinematics can be used to determine the angular acceleration α:

$$\omega^2 = \omega_0^2 + 2\alpha\theta \qquad (8.8)$$

Solving for α gives

$$\alpha = \frac{\omega^2 - \omega_0^2}{2\theta}$$

which can be substituted into Equation 9.7, as shown at the right.

$$\Sigma\tau = I\alpha \qquad (9.7)$$

$$\alpha = \frac{\omega^2 - \omega_0^2}{2\theta}$$

Solution The results of each step can be combined algebraically to show that

$$\underset{\text{STEP 1}}{\Sigma\tau} = \underset{\text{STEP 2}}{I\alpha} = I\left(\frac{\omega^2 - \omega_0^2}{2\theta}\right)$$

In this result for $\Sigma\tau$, we must use radian measure for the variables ω, ω_0, and θ. To convert from revolutions (rev) to radians (rad), we will use the fact that 1 rev = 2π rad. Thus, the net torque applied by the motor to the blade is

$$\Sigma\tau = I\left(\frac{\omega^2 - \omega_0^2}{2\theta}\right)$$

$$= (1.41 \times 10^{-3}\,\text{kg} \cdot \text{m}^2)\left\{\frac{\left[\left(80.0\,\frac{\text{rev}}{\text{s}}\right)\left(\frac{2\pi\,\text{rad}}{1\,\text{rev}}\right)\right]^2 - (0\,\text{rad/s})^2}{2(240.0\,\text{rev})\left(\frac{2\pi\,\text{rad}}{1\,\text{rev}}\right)}\right\} = \boxed{0.118\,\text{N} \cdot \text{m}}$$

Related Homework: *Problems 31, 35, 42, 73*

Figure 9.19 A rider applies a force $\vec{F}$ to the circular handrail. The torque produced by this force is the product of its magnitude and the lever arm ℓ about the axis of rotation. (Dan Coffee/The Image Bank/Getty Images)

To accelerate a wheelchair, the rider applies a force to a handrail attached to each wheel. The torque generated by this force is the product of the magnitude of the force and the lever arm. As Figure 9.19 illustrates, the lever arm is just the radius of the circular rail, which is designed to be as large as possible. Thus, a relatively large torque can be generated for a given force, allowing the rider to accelerate quickly.

The physics of wheelchairs.

Example 10 shows how Newton's second law for rotational motion is used when design considerations demand an adequately large angular acceleration. There are also situations in which it is desirable to have as little angular acceleration as possible, and Conceptual Example 11 deals with one of them.

 Conceptual Example 11 Archery and Bow Stabilizers

Archers can shoot with amazing accuracy, especially using modern bows such as the one in Figure 9.20. Notice the bow stabilizer, a long, thin rod that extends from the front of the bow and has a relatively massive cylinder at the tip. Advertisements claim that the stabilizer helps to steady the archer's aim. Which of the following explains why this is true? The addition of the stabilizer (**a**) decreases the bow's moment of inertia, making it easier for the archer to hold the bow steady; (**b**) has nothing to do with the bow's moment of inertia; (**c**) increases the bow's moment of inertia, making it easier for the archer to hold the bow steady.

The physics of archery and bow stabilizers.

Reasoning An axis of rotation (the black dot) has been added to Figure 9.20. This axis passes through the archer's left shoulder and is perpendicular to the plane of the paper. Any angular acceleration of the archer's body about this axis will lead to a rotation of the bow and, thus, will degrade the archer's aim. The angular acceleration will depend on any unbalanced torques that occur while the archer's tensed muscles try to hold the drawn bow, as well as the bow's moment of inertia.

Answers (a) and (b) are incorrect. Adding a stabilizer to the bow increases its mass. According to the definition of the moment of inertia of a body (Equation 9.6), the increase in mass leads to an increase in the bow's moment of inertia.

Answer (c) is correct. Newton's second law for rotational motion states that the angular acceleration α of the bow is given by $\alpha = (\Sigma \tau)/I$ (Equation 9.7), where $\Sigma \tau$ is the net torque acting on the bow and I is its moment of inertia. The stabilizer increases I, especially the relatively massive cylinder at the tip, since it is so far from the axis. Note that the moment of inertia is in the denominator on the right side of this equation. Therefore, to the extent that I is larger, a given net torque $\Sigma \tau$ will create a smaller angular acceleration and, hence, less disturbance of the aim.

Figure 9.20 The long, thin rod extending from the front of the bow is a stabilizer. The stabilizer helps to steady the archer's aim, as Conceptual Example 11 discusses. (Rob Griffith/©AP/Wide World Photos)

Rotational motion and translational motion sometimes occur together. The next example deals with an interesting situation in which both angular acceleration and translational acceleration must be considered.

ANALYZING MULTIPLE-CONCEPT PROBLEMS

Example 12 Hoisting a Crate

A crate of mass 451 kg is being lifted by the mechanism shown in Figure 9.21a. The two cables are wrapped around their respective pulleys, which have radii of 0.600 and 0.200 m. The pulleys are fastened together to form a dual pulley and turn as a single unit about the center axle, relative to which the combined moment of inertia is 46.0 kg·m². The cables roll on the dual pulley without slipping. A tension of magnitude 2150 N is maintained in the cable attached to the motor. Find the angular acceleration of the dual pulley and the tension in the cable connected to the crate.

Reasoning To determine the angular acceleration of the dual pulley and the tension in the cable attached to the crate, we will apply Newton's second law to the pulley and the crate separately. Four external forces act on the dual pulley, as its free-body diagram in Figure 9.21b shows. These are (1) the tension $\vec{T}_1$ in the cable connected to the motor, (2) the tension $\vec{T}_2$ in the cable attached to the crate, (3) the pulley's weight $\vec{W}_p$, and (4) the reaction force $\vec{P}$ exerted on the dual pulley by the axle. The force $\vec{P}$ arises because the two cables and the pulley's weight pull the pulley down and to the left into the axle, and the axle pushes back, thus keeping the pulley in place. The net torque that

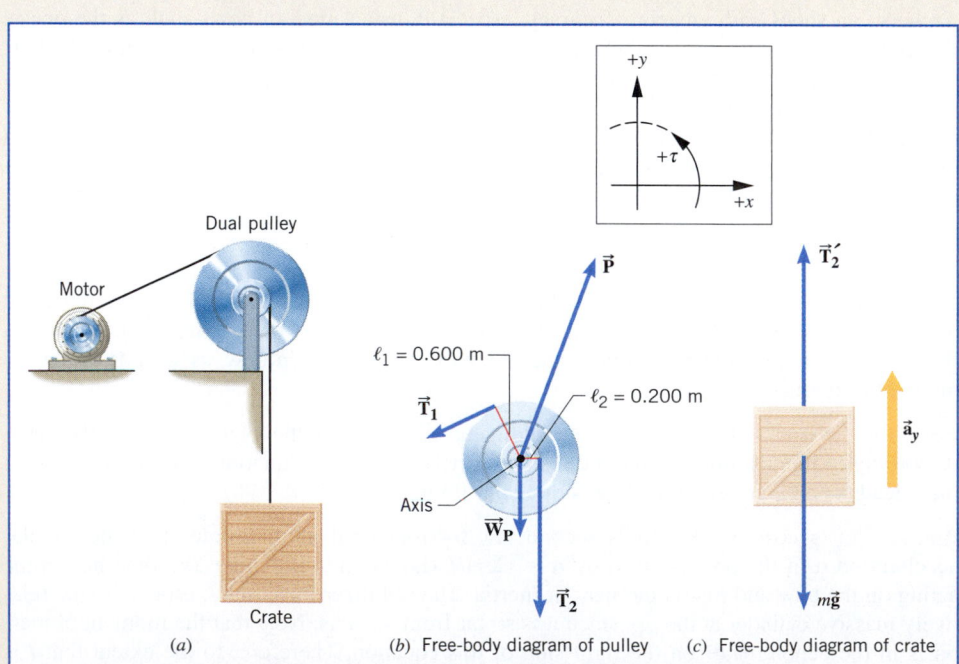

Figure 9.21 (a) The crate is lifted upward by the motor and pulley arrangement. The free-body diagram is shown for (b) the dual pulley and (c) the crate.

results from these forces obeys Newton's second law for rotational motion (Equation 9.7). Two external forces act on the crate, as its free-body diagram in Figure 9.21c indicates. These are (1) the cable tension $\vec{T}_2'$ and (2) the weight $m\vec{g}$ of the crate. The net force that results from these forces obeys Newton's second law for translational motion (Equation 4.2b).

Knowns and Unknowns The following table summarizes the given information:

Description	Symbol	Value	Comment
Mass of crate	m	451 kg	
Radius of outer part of dual pulley	ℓ_1	0.600 m	See Figure 9.21b.
Radius of inner part of dual pulley	ℓ_2	0.200 m	See Figure 9.21b.
Moment of inertia of dual pulley	I	46.0 kg·m²	
Magnitude of tension in cable attached to motor	T_1	2150 N	
Unknown Variables			
Angular acceleration of dual pulley	α	?	
Magnitude of tension in cable attached to crate	T_2	?	

Modeling the Problem

STEP 1 **Newton's Second Law for Rotation**   Using the lever arms ℓ_1 and ℓ_2 shown in Figure 9.21b, we can apply the second law to the rotational motion of the dual pulley.

$$\Sigma\tau = T_1\ell_1 - T_2\ell_2 = I\alpha \qquad (9.7)$$

Note that the forces $\vec{P}$ and $\vec{W}_\mathbf{p}$ have zero lever arms, since their lines of action pass through the axle. Thus, these forces contribute nothing to the net torque. Minor rearrangement of Equation 9.7 gives Equation 1 at the right. In this result, all of the variables are known, except the angular acceleration α and the tension T_2. Therefore, before Equation 1 can be used to determine α, a value for T_2 is required. To obtain this value, we turn to Step 2.

$$I\alpha = T_1\ell_1 - T_2\ell_2 \qquad (1)$$
?

STEP 2 **Newton's Second Law for Translation**   Applying Newton's second law to the upward translational motion of the crate gives (see part c of the drawing)

$$\Sigma F_y = T_2' - mg = ma_y \qquad (4.2b)$$

Note that the magnitude of the tension in the cable between the crate and the pulley is $T_2' = T_2$, so that we can solve Equation 4.2b for the tension and obtain

$$\boxed{T_2' = T_2 = mg + ma_y}$$

This result for T_2 can be substituted into Equation 1, as shown at the right. The mass m of the crate is known, but the linear acceleration a_y of the crate is not, so we proceed to Step 3 to determine its value.

$$I\alpha = T_1\ell_1 - T_2\ell_2 \qquad (1)$$
$$\boxed{T_2 = mg + ma_y} \qquad (2)$$
?

STEP 3 **Rolling Motion**   Because the cable attached to the crate rolls on the pulley without slipping, the linear acceleration a_y of the crate is related to the angular acceleration α of the pulley via $a_y = r\alpha$ (Equation 8.13), where $r = \ell_2$. Thus, we have

$$\boxed{a_y = \ell_2\alpha}$$

We complete our modeling by substituting this result for a_y into Equation 2, as indicated at the right.

$$I\alpha = T_1\ell_1 - T_2\ell_2 \qquad (1)$$
$$\boxed{T_2 = mg + ma_y} \qquad (2)$$
$$\boxed{a_y = \ell_2\alpha}$$

Solution Combining the results of each step algebraically, we have

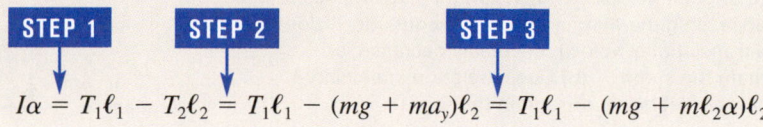

$$\overset{\text{STEP 1}}{I\alpha} = \overset{\text{STEP 2}}{T_1\ell_1 - T_2\ell_2} = T_1\ell_1 - \overset{\text{STEP 3}}{(mg + ma_y)\ell_2} = T_1\ell_1 - (mg + m\ell_2\alpha)\ell_2$$

Continued

Solving for α gives

$$I\alpha = T_1\ell_1 - mg\ell_2 - m\ell_2^2\alpha \quad\text{or}\quad (I + m\ell_2^2)\alpha = T_1\ell_1 - mg\ell_2$$

$$\alpha = \frac{T_1\ell_1 - mg\ell_2}{I + m\ell_2^2} = \frac{(2150\text{ N})(0.600\text{ m}) - (451\text{ kg})(9.80\text{ m/s}^2)(0.200\text{ m})}{46.0\text{ kg}\cdot\text{m}^2 + (451\text{ kg})(0.200\text{ m})^2} = \boxed{6.3\text{ rad/s}^2}$$

Equation 1 can be solved for T_2 to show that

$$T_2 = \frac{T_1\ell_1 - I\alpha}{\ell_2} = \frac{(2150\text{ N})(0.600\text{ m}) - (46.0\text{ kg}\cdot\text{m}^2)(6.3\text{ rad/s}^2)}{0.200\text{ m}} = \boxed{5.00\times10^3\text{ N}}$$

Related Homework: *Problems 44, 47, 80*

We have seen that Newton's second law for rotational motion, $\Sigma\tau = I\alpha$, has the same form as the law for translational motion, $\Sigma F = ma$, so each rotational variable has a translational analog: torque τ and force F are analogous quantities, as are moment of inertia I and mass m, and angular acceleration α and linear acceleration a. The other physical concepts developed for studying translational motion, such as kinetic energy and momentum, also have rotational analogs. For future reference, Table 9.2 itemizes these concepts and their rotational analogs.

✓ **CHECK YOUR UNDERSTANDING**

(The answers are given at the end of the book.)

10. Three massless rods (A, B, and C) are free to rotate about an axis at their left end (see the drawing). The same force $\vec{F}$ is applied to the right end of each rod. Objects with different masses are attached to the rods, but the total mass ($3m$) of the objects is the same for each rod. Rank the angular acceleration of the rods, largest to smallest.

11. A flat triangular sheet of uniform material is shown in the drawing. There are three possible axes of rotation, each perpendicular to the sheet and passing through one corner, A, B, or C. For which axis is the greatest net external torque required to bring the triangle up to an angular speed of 10.0 rad/s in 10.0 s, starting from rest? Assume that the net torque is kept constant while it is being applied.

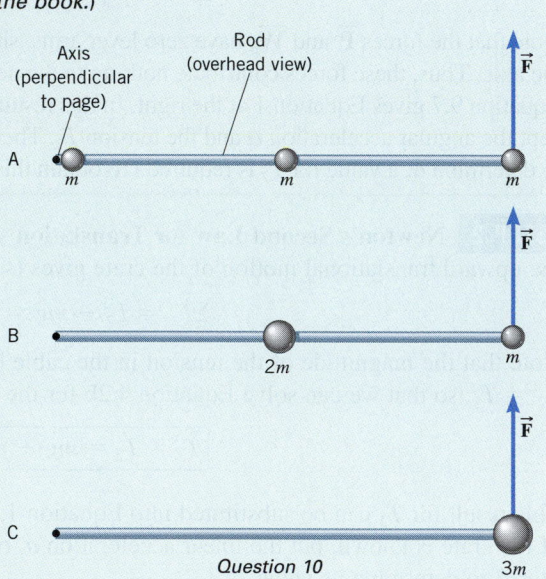

Question 10

12. At a given instant an object has an angular velocity. It also has an angular acceleration due to torques that are present. Therefore, the angular velocity is changing. Does the angular velocity at this instant increase, decrease, or remain the same **(a)** if additional torques are applied so as to make the net torque suddenly equal to zero and **(b)** if all the torques are suddenly removed?

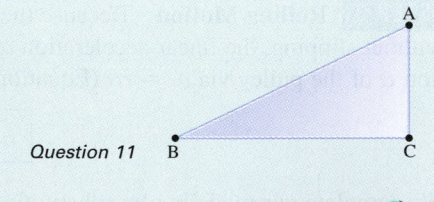

Question 11

13. The space probe in the drawing is initially moving with a constant translational velocity and zero angular velocity. **(a)** When the two engines are fired, each generating a thrust of magnitude T, does the translational velocity increase, decrease, or remain the same? **(b)** Does the angular velocity increase, decrease, or remain the same?

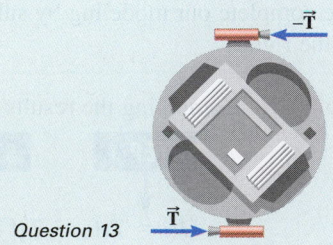

Question 13

Table 9.2 Analogies Between Rotational and Translational Concepts

Physical Concept	Rotational	Translational
Displacement	θ	s
Velocity	ω	v
Acceleration	α	a
The cause of acceleration	Torque τ	Force F
Inertia	Moment of inertia I	Mass m
Newton's second law	$\Sigma\tau = I\alpha$	$\Sigma F = ma$
Work	$\tau\theta$	Fs
Kinetic energy	$\frac{1}{2}I\omega^2$	$\frac{1}{2}mv^2$
Momentum	$L = I\omega$	$p = mv$

9.5 ROTATIONAL WORK AND ENERGY

Work and energy are among the most fundamental and useful concepts in physics. Chapter 6 discusses their application to translational motion. These concepts are equally useful for rotational motion, provided they are expressed in terms of angular variables.

The work W done by a constant force that points in the same direction as the displacement is $W = Fs$ (Equation 6.1), where F and s are the magnitudes of the force and the displacement, respectively. To see how this expression can be rewritten using angular variables, consider Figure 9.22. Here a rope is wrapped around a wheel and is under a constant tension F. If the rope is pulled out a distance s, the wheel rotates through an angle $\theta = s/r$ (Equation 8.1), where r is the radius of the wheel and θ is in radians. Thus, $s = r\theta$, and the work done by the tension force in turning the wheel is $W = Fs = Fr\theta$. However, Fr is the torque τ applied to the wheel by the tension, so the rotational work can be written as follows:

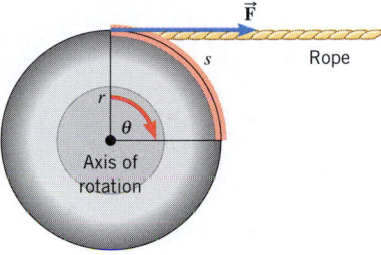

Figure 9.22 The force $\vec{F}$ does work in rotating the wheel through the angle θ.

DEFINITION OF ROTATIONAL WORK

The rotational work W_R done by a constant torque τ in turning an object through an angle θ is

$$W_R = \tau\theta \tag{9.8}$$

Requirement: θ must be expressed in radians.

SI Unit of Rotational Work: joule (J)

Section 6.2 discusses the work–energy theorem and kinetic energy. There we saw that the work done on an object by a net external force causes the translational kinetic energy ($\frac{1}{2}mv^2$) of the object to change. In an analogous manner, the rotational work done by a net external torque causes the rotational kinetic energy to change. A rotating body possesses kinetic energy, because its constituent particles are moving. If the body is rotating with an angular speed ω, the tangential speed v_T of a particle at a distance r from the axis is $v_T = r\omega$ (Equation 8.9). Figure 9.23 shows two such particles. If a particle's mass is m, its kinetic energy is $\frac{1}{2}mv_T^2 = \frac{1}{2}mr^2\omega^2$. The kinetic energy of the entire rotating body, then, is the sum of the kinetic energies of the particles:

$$\text{Rotational KE} = \Sigma(\tfrac{1}{2}mr^2\omega^2) = \tfrac{1}{2}\underbrace{(\Sigma mr^2)}_{\substack{\text{Moment of} \\ \text{inertia, } I}}\omega^2$$

In this result, the angular speed ω is the same for all particles in a rigid body and, therefore, has been factored outside the summation. According to Equation 9.6, the term in

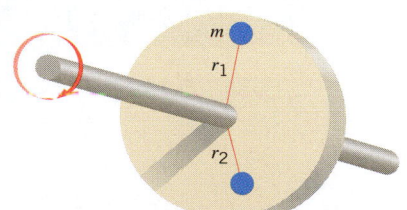

Figure 9.23 The rotating wheel is composed of many particles, two of which are shown.

parentheses is the moment of inertia, $I = \Sigma mr^2$, so the rotational kinetic energy takes the following form:

DEFINITION OF ROTATIONAL KINETIC ENERGY

The rotational kinetic energy KE_R of a rigid object rotating with an angular speed ω about a fixed axis and having a moment of inertia I is

$$KE_R = \tfrac{1}{2}I\omega^2 \qquad (9.9)$$

Requirement: ω must be expressed in rad/s.

SI Unit of Rotational Kinetic Energy: joule (**J**)

► **CONCEPTS AT A GLANCE** Kinetic energy is one part of an object's total mechanical energy. The total mechanical energy is the sum of the kinetic and potential energies and obeys the principle of conservation of mechanical energy (see Section 6.5). The Concepts-at-a-Glance chart in Figure 9.24 shows how rotational kinetic energy is incorporated into this principle. Specifically, we need to remember that translational and rotational motion can occur simultaneously. When a bicycle coasts down a hill, for instance, its tires are both translating and rotating. An object such as a rolling bicycle tire has both translational and rotational kinetic energies, so that the total mechanical energy is

$$\underbrace{E}_{\substack{\text{Total} \\ \text{mechanical} \\ \text{energy}}} = \underbrace{\tfrac{1}{2}mv^2}_{\substack{\text{Translational} \\ \text{kinetic energy}}} + \underbrace{\tfrac{1}{2}I\omega^2}_{\substack{\text{Rotational} \\ \text{kinetic energy}}} + \underbrace{mgh}_{\substack{\text{Gravitational} \\ \text{potential energy}}}$$

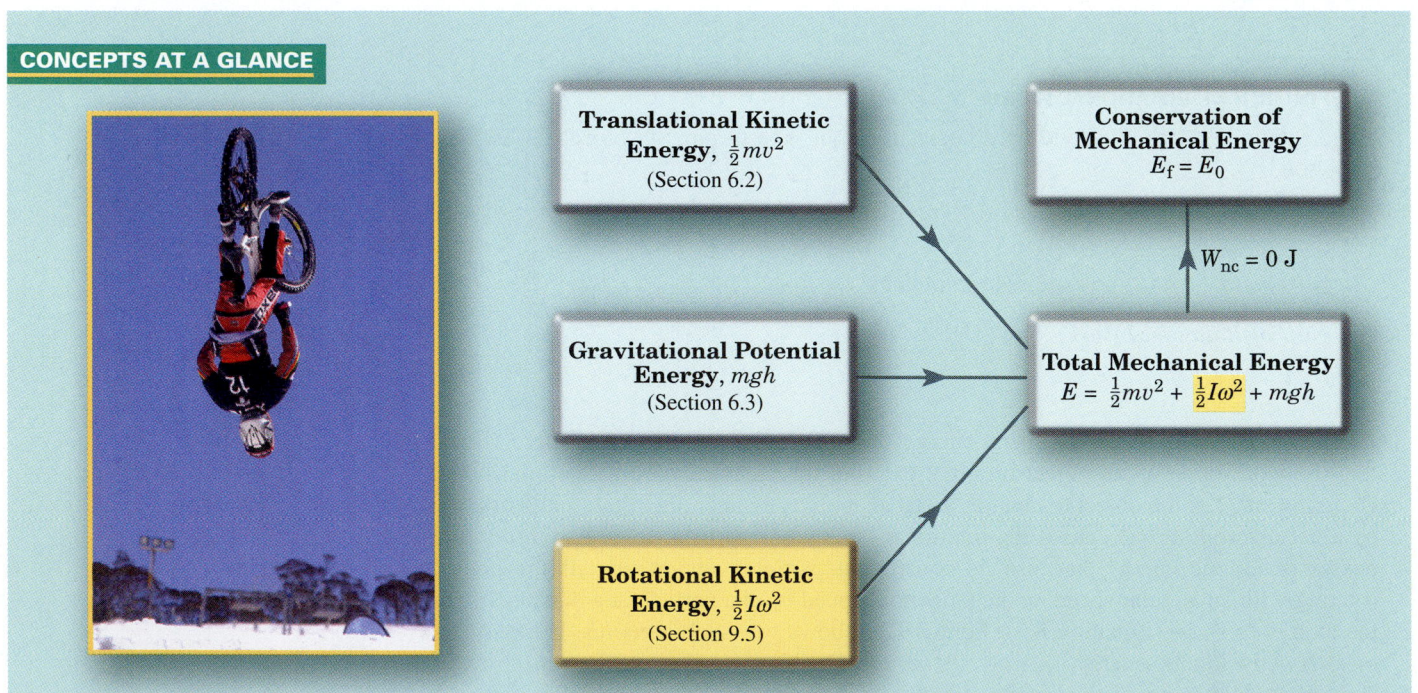

CONCEPTS AT A GLANCE

Translational Kinetic Energy, $\tfrac{1}{2}mv^2$ (Section 6.2)

Gravitational Potential Energy, mgh (Section 6.3)

Rotational Kinetic Energy, $\tfrac{1}{2}I\omega^2$ (Section 9.5)

Total Mechanical Energy $E = \tfrac{1}{2}mv^2 + \tfrac{1}{2}I\omega^2 + mgh$

Conservation of Mechanical Energy $E_f = E_0$

$W_{nc} = 0$ J

Figure 9.24 **CONCEPTS AT A GLANCE** The principle of conservation of mechanical energy can be applied to a rigid object that has both translational and rotational motion, provided the rotational kinetic energy is included in the total mechanical energy and provided that W_{nc}, the net work done by external nonconservative forces and torques, is zero. The principle can be applied to the motion of this mountain biker as he performs a flip high in the air, to the extent that the shapes of his bike and his body are constant and air resistance is negligible. (Chris McGrath/Getty Images Sport Services)

Here m is the mass of the object, v is the translational speed of its center of mass, I is its moment of inertia about an axis through the center of mass, ω is its angular speed, and h is the height of the object's center of mass relative to an arbitrary zero level. Mechanical energy is conserved if W_{nc}, the net work done by external nonconservative forces and external torques, is zero. If the total mechanical energy is conserved as an object moves, its final total mechanical energy E_f equals its initial total mechanical energy E_0: $E_f = E_0$. ◄

Example 13 illustrates the effect of combined translational and rotational motion in the context of how the total mechanical energy of a cylinder is conserved as it rolls down an incline.

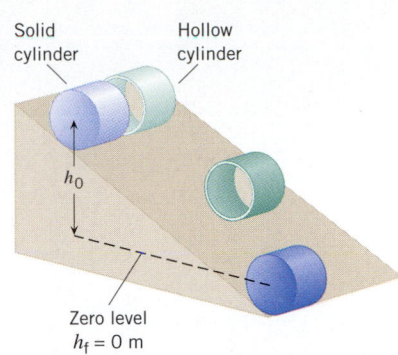

Figure 9.25 A hollow cylinder and a solid cylinder start from rest and roll down the incline plane. The conservation of mechanical energy can be used to show that the solid cylinder, having the greater translational speed, reaches the bottom first.

⬇ Example 13 Rolling Cylinders

A thin-walled hollow cylinder (mass $= m_h$, radius $= r_h$) and a solid cylinder (mass $= m_s$, radius $= r_s$) start from rest at the top of an incline (Figure 9.25). Both cylinders start at the same vertical height h_0 and roll down the incline without slipping. All heights are measured relative to an arbitrarily chosen zero level that passes through the center of mass of a cylinder when it is at the bottom of the incline (see the drawing). Ignoring energy losses due to retarding forces, determine which cylinder has the greatest translational speed on reaching the bottom.

Reasoning Only the conservative force of gravity does work on the cylinders, so the total mechanical energy is conserved as they roll down. The total mechanical energy E at any height h above the zero level is the sum of the translational kinetic energy ($\frac{1}{2}mv^2$), the rotational kinetic energy ($\frac{1}{2}I\omega^2$), and the gravitational potential energy (mgh):

$$E = \tfrac{1}{2}mv^2 + \tfrac{1}{2}I\omega^2 + mgh$$

As the cylinders roll down, potential energy is converted into kinetic energy, but the kinetic energy is shared between the translational form ($\frac{1}{2}mv^2$) and the rotational form ($\frac{1}{2}I\omega^2$). The object with more of its kinetic energy in the translational form will have the greater translational speed at the bottom of the incline. We expect the solid cylinder to have the greater translational speed, because more of its mass is located near the rotational axis and, thus, possesses less rotational kinetic energy.

Solution The total mechanical energy E_f at the bottom ($h_f = 0$ m) is the same as the total mechanical energy E_0 at the top ($h = h_0$, $v_0 = 0$ m/s, $\omega_0 = 0$ rad/s):

$$\tfrac{1}{2}mv_f^2 + \tfrac{1}{2}I\omega_f^2 + mgh_f = \tfrac{1}{2}mv_0^2 + \tfrac{1}{2}I\omega_0^2 + mgh_0$$

$$\tfrac{1}{2}mv_f^2 + \tfrac{1}{2}I\omega_f^2 = mgh_0$$

Since each cylinder rolls without slipping, the final rotational speed ω_f and the final translational speed v_f of its center of mass are related according to Equation 8.12, $\omega_f = v_f/r$, where r is the radius of the cylinder. Substituting this expression for ω_f into the energy-conservation equation and solving for v_f yields

$$v_f = \sqrt{\frac{2mgh_0}{m + I/r^2}}$$

Setting $m = m_h$, $r = r_h$, and $I = mr_h^2$ for the hollow cylinder and then setting $m = m_s$, $r = r_s$, and $I = \frac{1}{2}mr_s^2$ for the solid cylinder (see Table 9.1), we find that the two cylinders have the following translational speeds at the bottom of the incline:

Hollow cylinder $$v_f = \sqrt{gh_0}$$

Solid cylinder $$v_f = \sqrt{\frac{4gh_0}{3}} = 1.15\sqrt{gh_0}$$

The solid cylinder, having the greater translational speed, arrives at the bottom first.

14. Two solid balls are placed side by side at the top of an incline plane and, starting from rest, are allowed to roll down the incline. Which ball, if either, has the greater translational speed at the bottom if **(a)** they have the same radii, but one is more massive than the other; and **(b)** they have the same mass, but one has a larger radius?

15. A thin sheet of plastic is uniform and has the shape of an equilateral triangle. Consider two axes for rotation. Both are perpendicular to the plane of the triangle, axis A passing through the center of the triangle and axis B passing through one corner. If the angular speed ω about each axis is the same, for which axis does the triangle have the greater rotational kinetic energy?

16. A hoop, a solid cylinder, a spherical shell, and a solid sphere are placed at rest at the top of an incline. All the objects have the same radius. They are then released at the same time. What is the order in which they reach the bottom (fastest first)?

9.6 ANGULAR MOMENTUM

In Chapter 7 the linear momentum p of an object is defined as the product of its mass m and linear velocity v; that is, $p = mv$. For rotational motion the analogous concept is called the ***angular momentum*** L. The mathematical form of angular momentum is analogous to that of linear momentum, with the mass m and the linear velocity v being replaced with their rotational counterparts, the moment of inertia I and the angular velocity ω.

DEFINITION OF ANGULAR MOMENTUM

The angular momentum L of a body rotating about a fixed axis is the product of the body's moment of inertia I and its angular velocity ω with respect to that axis:

$$L = I\omega \tag{9.10}$$

Requirement: ω must be expressed in rad/s.

SI Unit of Angular Momentum: $\text{kg} \cdot \text{m}^2/\text{s}$

▶ **CONCEPTS AT A GLANCE** Linear momentum is an important concept in physics because the total linear momentum of a system is conserved when the sum of the average external forces acting on the system is zero. Then, the final total linear momentum P_f and the initial total linear momentum P_0 are the same: $P_f = P_0$. Figure 7.7 outlines the conceptual development of the conservation of linear momentum. The Concepts-at-a-Glance chart in Figure 9.26 outlines a similar development for angular momentum. In constructing this chart, we recall that each translational variable is analogous to a corresponding rotational variable, as Table 9.2 indicates. Therefore, if we replace "force" by "torque" and "linear momentum" by "angular momentum" in Figure 7.7, we obtain Figure 9.26. This chart indicates that when the sum of the average external torques is zero, the final and initial angular momenta are the same: $L_f = L_0$, which is the ***principle of conservation of angular momentum.*** ◀

PRINCIPLE OF CONSERVATION OF ANGULAR MOMENTUM

The total angular momentum of a system remains constant (is conserved) if the net average external torque acting on the system is zero.

Example 14 illustrates an interesting consequence of the conservation of angular momentum.

CONCEPTS AT A GLANCE

Figure 9.26 **CONCEPTS AT A GLANCE** The impulse–momentum theorem for rotational motion leads to the principle of conservation of angular momentum when the sum of the external torques is zero. This concept diagram is similar to that in Figure 7.7, which shows how the conservation of linear momentum arises from the impulse–momentum theorem. The angular momentum of each of these spinning skaters is conserved while they are in the air, assuming that air resistance is negligible. (Paolo Cocco/AFP/Getty Images Sport Services)

 Conceptual Example 14 A Spinning Skater

In Figure 9.27a an ice skater is spinning with both arms and a leg outstretched. In Figure 9.27b she pulls her arms and leg inward. As a result of this maneuver, her angular velocity ω increases dramatically. Why? Neglect any air resistance and assume that friction between her skates and the ice is negligible. **(a)** A net external torque acts on the skater, causing ω to increase. **(b)** No net external torque acts on her; she is simply obeying the conservation of angular momentum. **(c)** Due to the movements of her arms and legs, a net internal torque acts on the skater, causing her angular momentum and ω to increase.

**The physics of
a spinning ice skater.**

Reasoning Considering the skater as the system, we will use the conservation of angular momentum as a guide; it indicates that only a net external torque can cause the total angular momentum of a system to change.

Answer (a) is incorrect. There is no net external torque acting on the skater, because air resistance and the friction between her skates and the ice are negligible.

Answer (c) is incorrect. The movements of her arms and legs do produce internal torques. However, only *external* torques, not internal torques, can change the angular momentum of a system.

Answer (b) is correct. Since any air resistance and friction are negligible, the net external torque acting on the skater is zero, and the skater's angular momentum is conserved as she pulls her arms and leg inward. However, angular momentum is the product of the moment of inertia I and the angular velocity ω (see Equation 9.10). By moving the mass of her arms and leg inward, the skater decreases the distance r of the mass from the axis of rotation and, consequently, decreases her moment of inertia I ($I = \Sigma mr^2$). Since the product of I and ω is constant, then ω must increase as I decreases. Thus, as she pulls her arms and leg inward, she spins with a larger angular velocity.

Related Homework: *Problem 64*

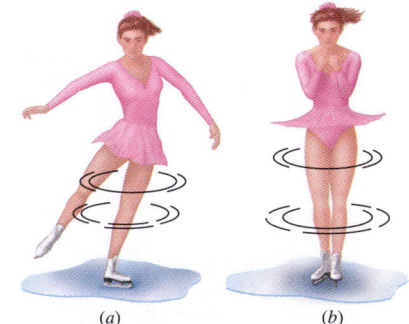

The next example involves a satellite and illustrates another application of the principle of conservation of angular momentum.

Figure 9.27 (*a*) A skater spins slowly on one skate, with both arms and one leg outstretched. (*b*) As she pulls her arms and leg in toward the rotational axis, her angular velocity ω increases.

The physics of
a satellite in orbit about the earth.

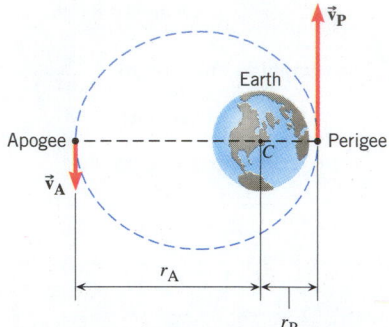

Figure 9.28 A satellite is moving in an elliptical orbit about the earth. The gravitational force exerts no torque on the satellite, so the angular momentum of the satellite is conserved.

 Example 15 A Satellite in an Elliptical Orbit

An artificial satellite is placed into an elliptical orbit about the earth, as illustrated in Figure 9.28. Telemetry data indicate that its point of closest approach (called the *perigee*) is $r_P = 8.37 \times 10^6$ m from the center of the earth, and its point of greatest distance (called the *apogee*) is $r_A = 25.1 \times 10^6$ m from the center of the earth. The speed of the satellite at the perigee is $v_P = 8450$ m/s. Find its speed v_A at the apogee.

Reasoning The only force of any significance that acts on the satellite is the gravitational force of the earth. However, at any instant, this force is directed toward the center of the earth and passes through the axis about which the satellite instantaneously rotates. Therefore, the gravitational force exerts *no torque* on the satellite (the lever arm is zero). Consequently, the angular momentum of the satellite remains constant at all times.

Solution Since the angular momentum is the same at the apogee (A) and the perigee (P), it follows that $I_A\omega_A = I_P\omega_P$. Furthermore, the orbiting satellite can be considered a point mass, so its moment of inertia is $I = mr^2$ (see Equation 9.4). In addition, the angular speed ω of the satellite is related to its tangential speed v_T by $\omega = v_T/r$ (Equation 8.9). If these relations are used at the apogee and perigee, the conservation of angular momentum gives the following result:

$$I_A\omega_A = I_P\omega_P \quad \text{or} \quad (mr_A{}^2)\left(\frac{v_A}{r_A}\right) = (mr_P{}^2)\left(\frac{v_P}{r_P}\right)$$

$$v_A = \frac{r_P v_P}{r_A} = \frac{(8.37 \times 10^6 \text{ m})(8450 \text{ m/s})}{25.1 \times 10^6 \text{ m}} = \boxed{2820 \text{ m/s}}$$

The answer is independent of the mass of the satellite. The satellite behaves just like the skater in Figure 9.27, because its speed is greater at the perigee, where the moment of inertia is smaller.

The result in Example 15 indicates that a satellite does not have a constant speed in an elliptical orbit. The speed changes from a maximum at the perigee to a minimum at the apogee; the closer the satellite comes to the earth, the faster it travels. Planets moving around the sun in elliptical orbits exhibit the same kind of behavior, and Johannes Kepler (1571–1630) formulated his famous second law based on observations of such characteristics of planetary motion. Kepler's second law states that, in a given amount of time, a line joining any planet to the sun sweeps out the same amount of area no matter where the planet is on its elliptical orbit, as Figure 9.29 illustrates. The conservation of angular momentum can be used to show why the law is valid, by means of a calculation similar to that in Example 15.

✓ **CHECK YOUR UNDERSTANDING**

(The answers are given at the end of the book.)

17. A woman is sitting on the spinning seat of a piano stool with her arms folded. Ignore any friction in the spinning stool. What happens to her **(a)** angular velocity and **(b)** angular momentum when she extends her arms outward?

18. Review Conceptual Example 14 as an aid in answering this question. Suppose the ice cap at the South Pole melted and the water was distributed uniformly over the earth's oceans. Would the earth's angular velocity increase, decrease, or remain the same?

19. Conceptual Example 14 provides background for this question. A cloud of interstellar gas is rotating. Because the gravitational force pulls the gas particles together, the cloud shrinks, and, under the right conditions, a star may ultimately be formed. Would the angular velocity of the star be less than, equal to, or greater than the angular velocity of the rotating gas?

20. A person is hanging motionless from a vertical rope over a swimming pool. She lets go of the rope and drops straight down. After letting go, is it possible for her to curl into a ball and start spinning?

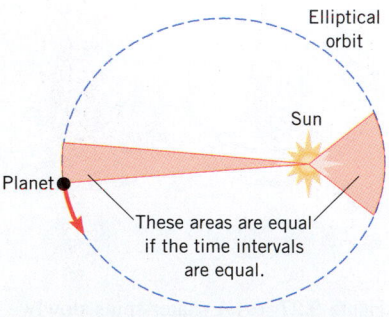

Figure 9.29 Kepler's second law of planetary motion states that a line joining a planet to the sun sweeps out equal areas in equal time intervals.

9.7

CONCEPTS & CALCULATIONS

In this chapter we have seen that a rotational or angular acceleration results when a net external torque acts on an object. In contrast, when a net external force acts on an object, it leads to a translational or linear acceleration, as Chapter 4 discusses. Torque and force, then, are fundamentally different concepts, and Example 16 focuses on this fact.

Concepts & Calculations Example 16
 Torque and Force

Figure 9.30a shows a uniform crate resting on a horizontal surface. The crate has a square cross section and a weight of $W = 580$ N, which is uniformly distributed. At the bottom right edge of the surface is a small obstruction that prevents the crate from sliding when a horizontal pushing force $\vec{P}$ is applied to the left side. However, if this force is great enough, the crate will begin to tip or rotate over the obstruction. Determine the minimum pushing force that leads to tipping.

Concept Questions and Answers What causes tipping—the force $\vec{P}$ or the torque that it creates?

Answer Tipping is a rotational or angular motion. Since the crate starts from rest, an angular acceleration is needed, which can only be created by a net external torque. Thus, the torque created by the force $\vec{P}$ causes the tipping.

A given force can create a variety of torques, depending on the lever arm of the force with respect to the rotational axis. In this case, the rotational axis is located at the small obstruction in Figure 9.30a and is perpendicular to the page. For this axis, the lever arm of the force $\vec{P}$ is ℓ_P, as the drawing shows. Where should $\vec{P}$ be applied so that a minimum force will give the necessary torque? In other words, should the lever arm be a minimum or a maximum?

Answer The magnitude of the torque is the product of the magnitude of the force and the lever arm. Thus, for a minimum force, the lever arm should be a maximum, and the force $\vec{P}$ should be applied at the upper left corner of the crate, as in Figure 9.30b.

Consider the crate just at the instant *before* it begins to rotate. At this instant, the crate is in equilibrium. What must be true about the sum of the external torques acting on the crate?

Answer Because the crate is in equilibrium, the sum of the external torques must be zero.

Solution Only the pushing force $\vec{P}$ and the weight $\vec{W}$ of the crate produce external torques with respect to a rotational axis through the lower right corner of the crate. The obstruction also applies a force to the crate, but it creates no torque, since its line of action passes through the axis. Since the sum of the external torques is zero, we refer to Figure 9.30b for the lever arms and write

$$\Sigma \tau = -P\ell_P + W\ell_W = 0$$

The lever arm for the force $\vec{P}$ is $\ell_P = L$, where L is the length of the side of the crate. The lever arm for the weight is $\ell_W = L/2$, because the crate is uniform, and the center of gravity is at the center of the crate. Substituting these lever arms in the torque equation, we obtain

$$-PL + W(\tfrac{1}{2}L) = 0 \quad \text{or} \quad P = \tfrac{1}{2}W = \tfrac{1}{2}(580 \text{ N}) = \boxed{290 \text{ N}}$$

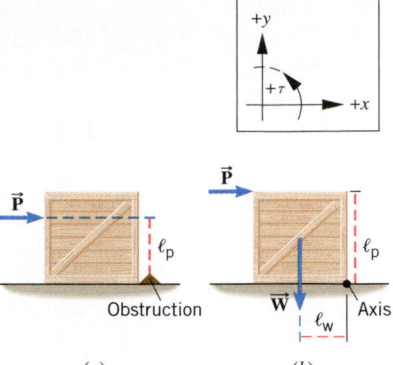

Figure 9.30 (*a*) A horizontal pushing force $\vec{P}$ is applied to a uniform crate, which has a square cross section and a weight $\vec{W}$. The crate rests on the ground, up against a small obstruction. (*b*) Some of the forces acting on the crate and their lever arms.

For rotational motion, the moment of inertia plays a key role. We have seen in this chapter that the moment of inertia and the net external torque determine the angular acceleration of a rotating object, according to Newton's second law. The angular acceleration, in turn, can be used in the equations of rotational kinematics, provided that it remains constant, as Chapter 8 discusses. When applied together in this way, Newton's second law and the equations of rotational kinematics are particularly useful in accounting for a wide variety of rotational motion. The following example reviews this approach in a situation where a rotating object is slowing down.

Concepts & Calculations Example 17
 Which Sphere Takes Longer to Stop?

Two spheres are each rotating at an angular speed of 24 rad/s about axes that pass through their centers. Each has a radius of 0.20 m and a mass of 1.5 kg. However, as Figure 9.31 shows, one is solid and the other is a thin-walled spherical shell. Suddenly, a net external torque due to friction (magnitude = 0.12 N·m) begins to act on each sphere and slows the motion down. How long does it take each sphere to come to a halt?

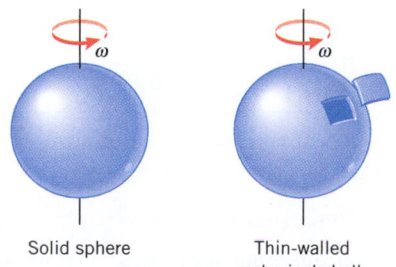

Figure 9.31 The two spheres have identical radii and masses; initially, they also have the same angular velocities. Which one comes to rest first as they slow down due to the same frictional torque?

Concept Questions and Answers Which sphere has the greater moment of inertia and why?

Answer   Referring to Table 9.1, we see that the solid sphere has a moment of inertia of $\frac{2}{5}MR^2$, while the shell has a moment of inertia of $\frac{2}{3}MR^2$. Since the masses and radii of the spheres are the same, it follows that the shell has the greater moment of inertia. The reason is that more of the mass of the shell is located farther from the rotational axis than is the case for the solid sphere. In the solid sphere, some of the mass is located close to the axis and, therefore, does not contribute as much to the moment of inertia.

Which sphere has the angular acceleration (a deceleration) with the smaller magnitude?

Answer   Newton's second law for rotation (Equation 9.7) specifies that the angular acceleration is $\alpha = (\Sigma\tau)/I$, where $\Sigma\tau$ is the net external torque and I is the moment of inertia. Since the moment of inertia is in the denominator, the angular acceleration is smaller when I is greater. Because it has the greater moment of inertia, the shell has the angular acceleration with the smaller magnitude.

Which sphere takes the longer time to come to a halt?

Answer   Since the angular acceleration of the shell has the smaller magnitude, the shell requires a longer time for the deceleration to reduce its angular velocity to zero.

Solution   According to Equation 8.4 from the equations of rotational kinematics, the time is given by $t = (\omega - \omega_0)/\alpha$ where ω and ω_0 are, respectively, the final and initial angular velocities. From Newton's second law as given in Equation 9.7 we know that the angular acceleration is $\alpha = (\Sigma\tau)/I$. Substituting this into the expression for the time gives

$$t = \frac{\omega - \omega_0}{(\Sigma\tau)/I} = \frac{I(\omega - \omega_0)}{\Sigma\tau}$$

In applying this result, we arbitrarily choose the direction of the initial rotation to be positive. With this choice, the torque must be negative, since it causes a deceleration. Using the proper moments of inertia, we find the following times for the spheres to come to a halt:

Solid sphere $t = \dfrac{I(\omega - \omega_0)}{\Sigma\tau} = \dfrac{\frac{2}{5}MR^2(\omega - \omega_0)}{\Sigma\tau}$

$$t = \frac{\frac{2}{5}(1.5 \text{ kg})(0.20 \text{ m})^2[(0 \text{ rad/s}) - (24 \text{ rad/s})]}{-0.12 \text{ N}\cdot\text{m}} = \boxed{4.8 \text{ s}}$$

Spherical shell $t = \dfrac{I(\omega - \omega_0)}{\Sigma\tau} = \dfrac{\frac{2}{3}MR^2(\omega - \omega_0)}{\Sigma\tau}$

$$t = \frac{\frac{2}{3}(1.5 \text{ kg})(0.20 \text{ m})^2[(0 \text{ rad/s}) - (24 \text{ rad/s})]}{-0.12 \text{ N}\cdot\text{m}} = \boxed{8.0 \text{ s}}$$

As expected, the shell requires a longer time to come to a halt.

CONCEPT SUMMARY

If you need more help with a concept, use the Learning Aids noted next to the discussion or equation. Examples (**Ex.**) are in the text of this chapter. Go to **www.wiley.com/college/cutnell** for the following Learning Aids:

Interactive LearningWare (ILW) — Additional examples solved in a five-step interactive format.

Concept Simulations (CS) — Animated text figures or animations of important concepts.

Interactive Solutions (IS) — Models for certain types of problems in the chapter homework. The calculations are carried out interactively.

Topic	Discussion	Learning Aids
Line of action Lever arm	**9.1 THE ACTION OF FORCES AND TORQUES ON RIGID OBJECTS** The line of action of a force is an extended line that is drawn colinear with the force. The lever arm ℓ is the distance between the line of action and the axis of rotation, measured on a line that is perpendicular to both.	

Topic	Discussion	Learning Aids
Torque	The torque of a force has a magnitude that is given by the magnitude F of the force times the lever arm ℓ. The torque τ is $$\tau = F\ell \qquad (9.1)$$ and is positive when the force tends to produce a counterclockwise rotation about the axis, and negative when the force tends to produce a clockwise rotation.	Ex. 1, 2 ILW 9.1 IS 9.2
	9.2 RIGID OBJECTS IN EQUILIBRIUM A rigid body is in equilibrium if it has zero translational acceleration and zero angular acceleration. In equilibrium, the net external force and the net external torque acting on the body are zero:	Ex. 3, 4, 5, 16 ILW 9.2
Equilibrium of a rigid body	$$\Sigma F_x = 0 \quad \text{and} \quad \Sigma F_y = 0 \qquad \text{(4.9a and 4.9b)}$$ $$\Sigma\tau = 0 \qquad (9.2)$$	CS 9.1 IS 9.15
	9.3 CENTER OF GRAVITY The center of gravity of a rigid object is the point where its entire weight can be considered to act when calculating the torque due to the weight. For a symmetrical body with uniformly distributed weight, the center of gravity is at the geometrical center of the body. When a number of objects whose weights are W_1, W_2, . . . are distributed along the x axis at locations x_1, x_2, . . . , the center of gravity x_{cg} is located at	
Definition of center of gravity	$$x_{cg} = \frac{W_1 x_1 + W_2 x_2 + \cdots}{W_1 + W_2 + \cdots} \qquad (9.3)$$	Ex. 6, 7, 8 CS 9.2
	The center of gravity is identical to the center of mass, provided the acceleration due to gravity does not vary over the physical extent of the objects.	
	9.4 NEWTON'S SECOND LAW FOR ROTATIONAL MOTION ABOUT A FIXED AXIS The moment of inertia I of a body composed of N particles is	
Moment of inertia	$$I = m_1 r_1^2 + m_2 r_2^2 + \cdots + m_N r_N^2 = \Sigma m r^2 \qquad (9.6)$$ where m is the mass of a particle and r is the perpendicular distance of the particle from the axis of rotation.	Ex. 9
Newton's second law for rotational motion	For a rigid body rotating about a fixed axis, Newton's second law for rotational motion is $$\Sigma\tau = I\alpha \qquad (\alpha \text{ in rad/s}^2) \qquad (9.7)$$ where $\Sigma\tau$ is the net external torque applied to the body, I is the moment of inertia of the body, and α is its angular acceleration.	Ex. 10, 11, 12, 17 CS 9.3, 9.4 IS 9.73
Rotational work	**9.5 ROTATIONAL WORK AND ENERGY** The rotational work W_R done by a constant torque τ in turning a rigid body through an angle θ is $$W_R = \tau\theta \qquad (\theta \text{ in radians}) \qquad (9.8)$$	
Rotational kinetic energy	The rotational kinetic energy KE_R of a rigid object rotating with an angular speed ω about a fixed axis and having a moment of inertia I is $$KE_R = \tfrac{1}{2}I\omega^2 \qquad (\omega \text{ in rad/s}) \qquad (9.9)$$	
Total mechanical energy	The total mechanical energy E of a rigid body is the sum of its translational kinetic energy ($\tfrac{1}{2}mv^2$), its rotational kinetic energy ($\tfrac{1}{2}I\omega^2$), and its gravitational potential energy (mgh): $$E = \tfrac{1}{2}mv^2 + \tfrac{1}{2}I\omega^2 + mgh$$ where m is the mass of the object, v is the translational speed of its center of mass, I is its moment of inertia about an axis through the center of mass, ω is its angular speed, and h is the height of the object's center of mass relative to an arbitrary zero level.	ILW 9.3 IS 9.53
Conservation of total mechanical energy	The total mechanical energy is conserved if the net work done by external nonconservative forces and external torques is zero. When the total mechanical energy is conserved, the final total mechanical energy E_f equals the initial total mechanical energy E_0: $E_f = E_0$.	Ex. 13
	9.6 ANGULAR MOMENTUM The angular momentum of a rigid body rotating with an angular velocity ω about a fixed axis and having a moment of inertia I with respect to that axis is	
Angular momentum	$$L = I\omega \qquad (\omega \text{ in rad/s}) \qquad (9.10)$$	
Conservation of angular momentum	The principle of conservation of angular momentum states that the total angular momentum of a system remains constant (is conserved) if the net average external torque acting on the system is zero. When the total angular momentum is conserved, the final angular momentum L_f equals the initial angular momentum L_0: $L_f = L_0$.	Ex. 14, 15 IS 9.63

FOCUS ON CONCEPTS

Note to Instructors: The numbering of the questions shown here reflects the fact that they are only a representative subset of the total number that are available online. However, all of the questions are available for assignment via an online homework management program such as WileyPLUS or WebAssign.

Section 9.1 The Action of Forces and Torques on Rigid Objects

1. The wheels on a moving bicycle have both translational (or linear) and rotational motions. What is meant by the phrase "a rigid body, such as a bicycle wheel, is in equilibrium"? **(a)** The body cannot have translational or rotational motion of any kind. **(b)** The body can have translational motion, but it cannot have rotational motion. **(c)** The body cannot have translational motion, but it can have rotational motion. **(d)** The body can have translational and rotational motions, as long as its translational acceleration and angular acceleration are zero.

Section 9.2 Rigid Objects in Equilibrium

3. The drawing illustrates an overhead view of a door and its axis of rotation. The axis is perpendicular to the page. There are four forces acting on the door, and they have the same magnitude. Rank the torque τ that each force produces, largest to smallest. **(a)** $\tau_4, \tau_3, \tau_2, \tau_1$ **(b)** τ_3, τ_2, τ_1 and τ_4 (a two-way tie) **(c)** $\tau_2, \tau_4, \tau_3, \tau_1$ **(d)** $\tau_1, \tau_4, \tau_3, \tau_2$ **(e)** τ_2, τ_3 and τ_4 (a two-way tie), τ_1

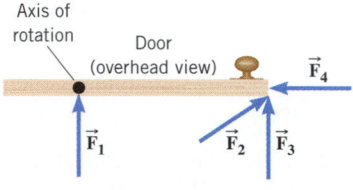

6. Five hockey pucks are sliding across frictionless ice. The drawing shows a top view of the pucks and the three forces that act on each one. As shown, the forces have different magnitudes (F, $2F$, or $3F$), and are applied at different points on the pucks. Only one of the five pucks can be in equilibrium. Which puck is it? **(a)** 1 **(b)** 2 **(c)** 3 **(d)** 4 **(e)** 5

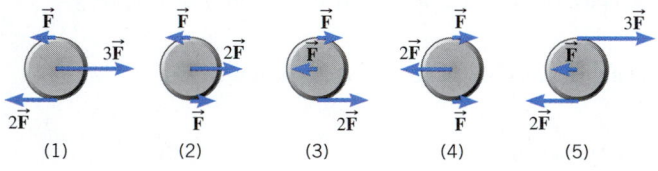

8. The drawing shows a top view of a square box lying on a frictionless floor. Three forces, which are drawn to scale, act on the box. Consider an angular acceleration with respect to an axis through the center of the box (perpendicular to the page). Which one of the following statements is correct? **(a)** The box will have a translational acceleration but not an angular acceleration. **(b)** The box will have both a translational and an angular acceleration. **(c)** The box will have an angular acceleration but not a translational acceleration. **(d)** The box will have neither a translational nor an angular acceleration. **(e)** It is not possible to determine whether the box will have a translational or an angular acceleration.

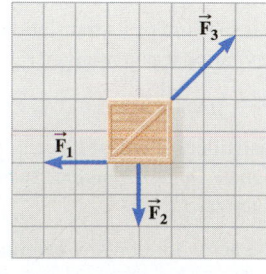

Section 9.4 Newton's Second Law for Rotational Motion About a Fixed Axis

10. The drawing shows three objects rotating about a vertical axis. The mass of each object is given in terms of m_0, and its perpendicular distance from the axis is specified in terms of r_0. Rank the three objects according to their moments of inertia, largest to smallest. **(a)** A, B, C **(b)** A, C, B **(c)** B, A, C **(d)** B, C, A **(e)** C, A, B

12. Two blocks are placed at the ends of a horizontal massless board, as in the drawing. The board is kept from rotating and rests on a support that serves as an axis of rotation. The moment of inertia of this system relative to the axis is 12 kg·m². Determine the magnitude of the angular acceleration when the system is allowed to rotate.

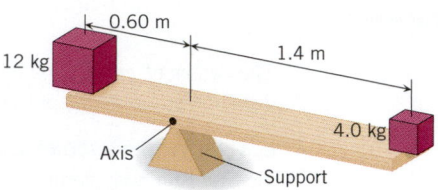

Problem 10

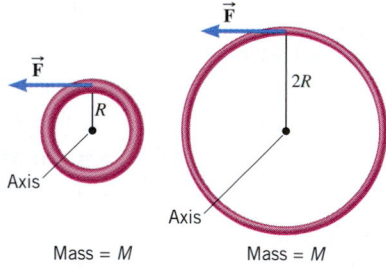

13. The same force $\vec{F}$ is applied to the edge of two hoops (see the drawing). The hoops have the same mass, although the radius of the larger hoop is twice the radius of the smaller one. The entire mass of each hoop is concentrated at its rim, so the moment of inertia is $I = Mr^2$, where M is the mass and r is the radius. Which hoop has the greater angular acceleration, and how many times as great is it compared to the angular acceleration of the other hoop? **(a)** The smaller hoop; two times as great **(b)** The smaller hoop; four times as great **(c)** The larger hoop; two times as great **(d)** The larger hoop; four times as great **(e)** Both have the same angular acceleration.

Section 9.5 Rotational Work and Energy

16. Two hoops, starting from rest, roll down identical inclined planes. The work done by nonconservative forces, such as air resistance, is zero ($W_{nc} = 0$ J). Both have the same mass M, but, as the drawing shows, one hoop has twice the radius of the other. The moment of inertia for each hoop is $I = Mr^2$, where r is its radius. Which hoop, if either, has the greater total kinetic energy (translational

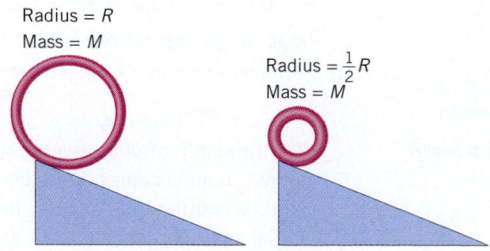

plus rotational) at the bottom of the incline? **(a)** The larger hoop **(b)** The smaller hoop **(c)** Both have the same total kinetic energy.

Section 9.6   Angular Momentum

17. Under what condition(s) is the angular momentum of a rotating body, such as a spinning ice skater, conserved? **(a)** Each external force acting on the body must be zero. **(b)** Each external force and each external torque acting on the body must be zero. **(c)** Each external force may be nonzero, but the sum of the forces must be zero.

(d) Each external torque may be nonzero, but the sum of the torques must be zero.

18. An ice skater is spinning on frictionless ice with her arms extended outward. She then pulls her arms in toward her body, reducing her moment of inertia. Her angular momentum is conserved, so as she reduces her moment of inertia, her angular velocity increases and she spins faster. Compared to her initial rotational kinetic energy, her final rotational kinetic energy is **(a)** the same **(b)** larger, because her angular speed is larger **(c)** smaller, because her moment of inertia is smaller.

PROBLEMS

Note to Instructors: *Most of the homework problems in this chapter are available for assignment via an online homework management program such as* WileyPLUS *or* WebAssign, *and those marked with the icon* **GO** *are presented in* WileyPLUS *using a guided tutorial format that provides enhanced interactivity. See Preface for additional details.*

ssm   Solution is in the Student Solutions Manual.
www   Solution is available online at www.wiley.com/college/cutnell

This icon represents a biomedical application.

Section 9.1   The Action of Forces and Torques on Rigid Objects

1. In San Francisco a very simple technique is used to turn around a cable car when it reaches the end of its route. The car rolls onto a turntable, which can rotate about a vertical axis through its center. Then, two people push perpendicularly on the car, one at each end, as shown in the drawing. The turntable is rotated one-half of a revolution to turn the car around. If the length of the car is 9.20 m and each person pushes with a 185-N force, what is the magnitude of the net torque applied to the car?

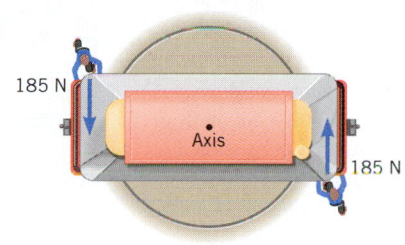

2. Interactive Solution 9.2 at www.wiley.com/college/cutnell presents a model for solving this problem. The wheel of a car has a radius of 0.350 m. The engine of the car applies a torque of 295 N·m to this wheel, which does not slip against the road surface. Since the wheel does not slip, the road must be applying a force of static friction to the wheel that produces a countertorque. Moreover, the car has a constant velocity, so this countertorque balances the applied torque. What is the magnitude of the static frictional force?

3. ssm You are installing a new spark plug in your car, and the manual specifies that it be tightened to a torque that has a magnitude of 45 N·m. Using the data in the drawing, determine the magnitude F of the force that you must exert on the wrench.

4. Two children hang by their hands from the same tree branch. The branch is straight, and grows out from the tree trunk at an angle of 27.0° above the horizontal. One child, with a mass of 44.0 kg, is hanging 1.30 m along the branch from the tree trunk. The other child, with a mass of 35.0 kg, is hanging 2.10 m from the tree trunk. What is the magnitude of the net torque exerted on the branch by the children? Assume that the axis is

located where the branch joins the tree trunk and is perpendicular to the plane formed by the branch and the trunk.

5. ssm The drawing shows a lower leg being exercised. It has a 49-N weight attached to the foot and is extended at an angle θ with respect to the vertical. Consider a rotational axis at the knee. **(a)** When $\theta = 90.0°$, find the magnitude of the torque that the weight creates. **(b)** At what angle θ does the magnitude of the torque equal 15 N·m?

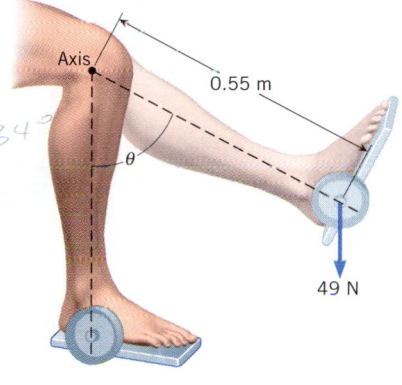

6. A square, 0.40 m on a side, is mounted so that it can rotate about an axis that passes through the center of the square. The axis is perpendicular to the plane of the square. A force of 15 N lies in this plane and is applied to the square. What is the magnitude of the maximum torque that such a force could produce?

***7. ssm Interactive LearningWare 9.1** at **www.wiley.com/college/cutnell** reviews the concepts that are important in this problem. A rod is lying on the top of a table. One end of the rod is hinged to the table so that the rod can rotate freely on the tabletop. Two forces, both parallel to the tabletop, act on the rod at the same place. One force is directed perpendicular to the rod and has a magnitude of 38.0 N. The second force has a magnitude of 55.0 N and is directed at an angle θ with respect to the rod. If the sum of the torques due to the two forces is zero, what must be the angle θ?

***8. GO** One end of a meter stick is pinned to a table, so the stick can rotate freely in a plane parallel to the tabletop. Two forces, both parallel to the tabletop, are applied to the stick in such a way that the net torque is zero. The first force has a magnitude of 2.00 N and is applied perpendicular to the length of the stick at the free end. The second force has a magnitude of 6.00 N and acts at a 30.0° angle with respect to the length of the stick. Where along the stick is the 6.00-N force applied? Express this distance with respect to the end of the stick that is pinned.

*9. A pair of forces with equal magnitudes, opposite directions, and different lines of action is called a "couple." When a couple acts on a rigid object, the couple produces a torque that does *not* depend on the location of the axis. The drawing shows a couple acting on a tire wrench, each force being perpendicular to the wrench. Determine an expression for the torque produced by the couple when the axis is perpendicular to the tire and passes through (a) point A, (b) point B, and (c) point C. Express your answers in terms of the magnitude F of the force and the length L of the wrench.

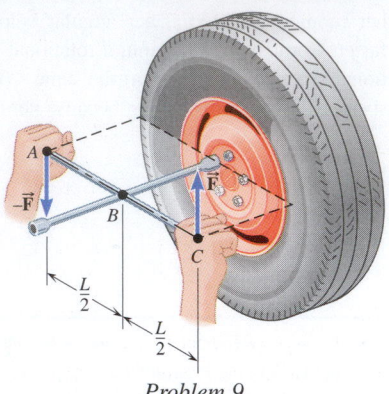

Problem 9

**10. A rotational axis is directed perpendicular to the plane of a square and is located as shown in the drawing. Two forces, $\vec{F}_1$ and $\vec{F}_2$, are applied to diagonally opposite corners, and act along the sides of the square, first as shown in part *a* and then as shown in part *b* of the drawing. In each case the net torque produced by the forces is zero. The square is one meter on a side, and the magnitude of $\vec{F}_2$ is three times that of $\vec{F}_1$. Find the distances a and b that locate the axis.

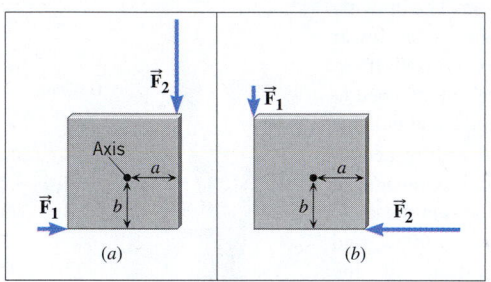

Section 9.2 Rigid Objects in Equilibrium,
Section 9.3 Center of Gravity

11. A person is standing on a level floor. His head, upper torso, arms, and hands together weigh 438 N and have a center of gravity that is 1.28 m above the floor. His upper legs weigh 144 N and have a center of gravity that is 0.760 m above the floor. Finally, his lower legs and feet together weigh 87 N and have a center of gravity that is 0.250 m above the floor. Relative to the floor, find the location of the center of gravity for his entire body.

12. **GO** Review Conceptual Example 7 before starting this problem. A uniform plank of length 5.0 m and weight 225 N rests horizontally on two supports, with 1.1 m of the plank hanging over the right support (see the drawing). To what distance x can a person who weighs 450 N walk on the overhanging part of the plank before it just begins to tip?

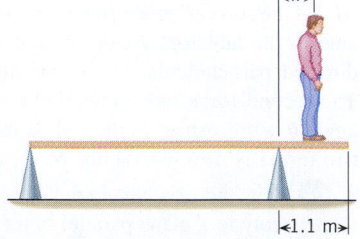

13. **ssm Concept Simulation 9.1** at **www.wiley.com/college/cutnell** illustrates how the forces can vary in problems of this type. A hiker, who weighs 985 N, is strolling through the woods and crosses a small horizontal bridge. The bridge is uniform, weighs 3610 N, and rests on two concrete supports, one at each end. He stops one-fifth of the way along the bridge. What is the magnitude of the force that a concrete support exerts on the bridge (a) at the near end and (b) at the far end?

14. **GO** The drawing shows a rectangular piece of wood. The forces applied to corners B and D have the same magnitude of 12 N and are directed parallel to the long and short sides of the rectangle. The long side of the rectangle is twice as long as the short side. An axis of rotation is shown perpendicular to the plane of the rectangle at its center. A third force (not shown in the drawing) is applied to corner A, directed along the short side of the rectangle (either toward B or away from B), such that the piece of wood is at equilibrium. Find the magnitude and direction of the force applied to corner A.

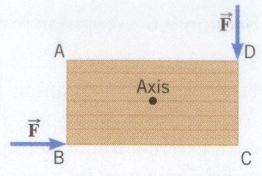

15. **Interactive Solution 9.15** at **www.wiley.com/college/cutnell** illustrates how to model this type of problem. A person exerts a horizontal force of 190 N in the test apparatus shown in the drawing. Find the horizontal force $\vec{M}$ (magnitude and direction) that his flexor muscle exerts on his forearm.

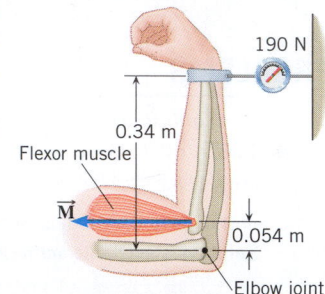

16. Conceptual Example 7 provides useful background for this problem. Workers have loaded a delivery truck in such a way that its center of gravity is only slightly forward of the rear axle, as shown in the drawing. The mass of the truck and its contents is 7460 kg. Find the magnitudes of the forces exerted by the ground on (a) the front wheels and (b) the rear wheels of the truck.

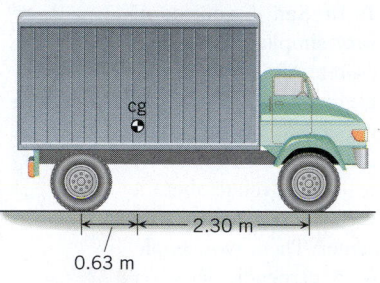

17. Review Multiple-Concept Example 8 before beginning this problem. A sport utility vehicle (SUV) and a sports car travel around the same horizontal curve. The SUV has a static stability factor of 0.80 and can negotiate the curve at a maximum speed of 18 m/s without rolling over. The sports car has a static stability factor of 1.4. At what maximum speed can the sports car negotiate the curve without rolling over?

18. **GO** The wheels, axle, and handles of a wheelbarrow weigh 60.0 N. The load chamber and its contents weigh 525 N. The drawing shows these two forces in two different wheelbarrow designs. To support the wheelbarrow in equilibrium, the man's hands apply a force $\vec{F}$ to the handles that is directed vertically upward. Consider a rotational axis

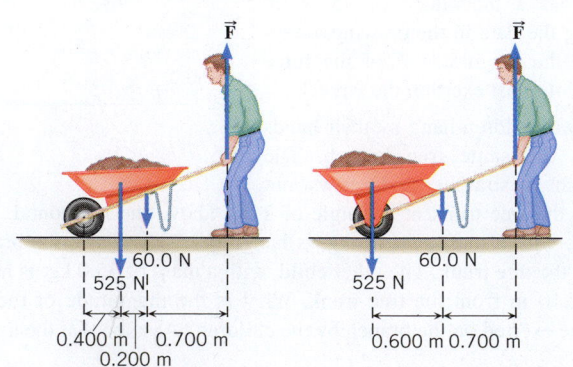

at the point where the tire contacts the ground, directed perpendicular to the plane of the paper. Find the magnitude of the man's force for both designs.

19. **ssm** A lunch tray is being held in one hand, as the drawing illustrates. The mass of the tray itself is 0.200 kg, and its center of gravity is located at its geometrical center. On the tray is a 1.00-kg plate of food and a 0.250-kg cup of coffee. Obtain the force $\vec{T}$ exerted by the thumb and the force $\vec{F}$ exerted by the four fingers. Both forces act perpendicular to the tray, which is being held parallel to the ground.

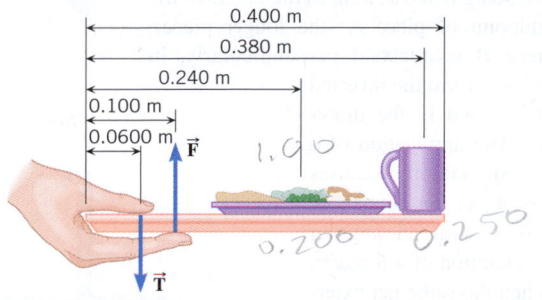

20. Review **Concept Simulation 9.2** at www.wiley.com/college/cutnell and Conceptual Example 7 as background material for this problem. A jet transport has a weight of 1.00×10^6 N and is at rest on the runway. The two rear wheels are 15.0 m behind the front wheel, and the plane's center of gravity is 12.6 m behind the front wheel. Determine the normal force exerted by the ground on **(a)** the front wheel and on **(b)** *each* of the two rear wheels.

21. **GO** The drawing shows a uniform horizontal beam attached to a vertical wall by a frictionless hinge and supported from below at an angle $\theta = 39°$ by a brace that is attached to a pin. The beam has a weight of 340 N. Three additional forces keep the beam in equilibrium. The brace applies a force $\vec{P}$ to the right end of the beam that is directed upward at the angle θ with respect to the horizontal. The hinge applies a force to the left end of the beam that has a horizontal component $\vec{H}$ and a vertical component $\vec{V}$. Find the magnitudes of these three forces.

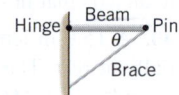

22. In an isometric exercise a person places a hand on a scale and pushes vertically downward, keeping the forearm horizontal. This is possible because the triceps muscle applies an upward force $\vec{M}$ perpendicular to the arm, as the drawing indicates. The forearm weighs 22.0 N and has a center of gravity as indicated. The scale registers 111 N. Determine the magnitude of $\vec{M}$.

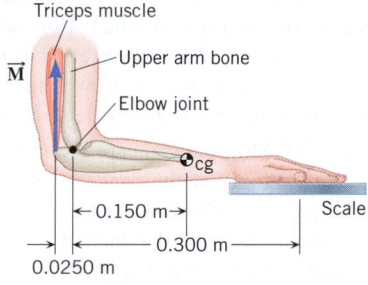

*23. **ssm** A uniform board is leaning against a smooth vertical wall. The board is at an angle θ above the horizontal ground. The coefficient of static friction between the ground and the lower end of the board is 0.650. Find the smallest value for the angle θ, such that the lower end of the board does not slide along the ground.

*24. The drawing shows a bicycle wheel resting against a small step whose height is $h = 0.120$ m. The weight and radius of the wheel are $W = 25.0$ N and $r = 0.340$ m,

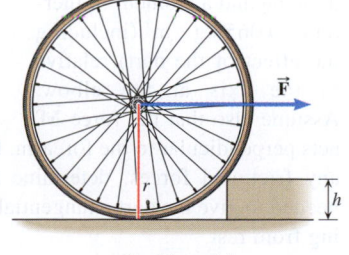

Problem 24

respectively. A horizontal force $\vec{F}$ is applied to the axle of the wheel. As the magnitude of $\vec{F}$ increases, there comes a time when the wheel just begins to rise up and loses contact with the ground. What is the magnitude of the force when this happens?

*25. A 1220-N uniform beam is attached to a vertical wall at one end and is supported by a cable at the other end. A 1960-N crate hangs from the far end of the beam. Using the data shown in the drawing, find **(a)** the magnitude of the tension in the wire and **(b)** the magnitude of the horizontal and vertical components of the force that the wall exerts on the left end of the beam.

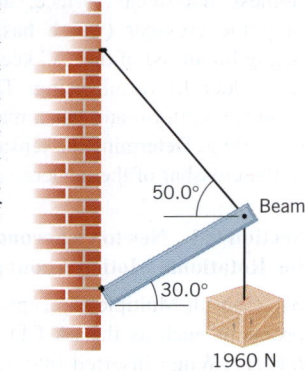

*26. A wrecking ball (weight = 4800 N) is supported by a boom, which may be assumed to be uniform and has a weight of 3600 N. As the drawing shows, a support cable runs from the top of the boom to the tractor. The angle between the support cable and the horizontal is 32°, and the angle between the boom and the horizontal is 48°. Find **(a)** the tension in the support cable and **(b)** the magnitude of the force exerted on the lower end of the boom by the hinge at point P.

*27. **ssm** A man holds a 178-N ball in his hand, with the forearm horizontal (see the drawing). He can support the ball in this position because of the flexor muscle force $\vec{M}$, which is applied perpendicular to the forearm. The forearm weighs 22.0 N and has a center of gravity as indicated. Find **(a)** the magnitude of $\vec{M}$ and **(b)** the magnitude and direction of the force applied by the upper arm bone to the forearm at the elbow joint.

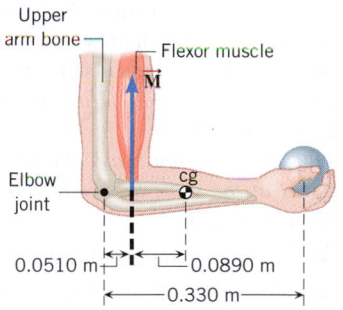

**28. A man drags a 72-kg crate across the floor at a constant velocity by pulling on a strap attached to the bottom of the crate. The crate is tilted 25° above the horizontal, and the strap is inclined 61° above the horizontal. The center of gravity of the crate coincides with its geometrical center, as indicated in the drawing. Find the magnitude of the tension in the strap.

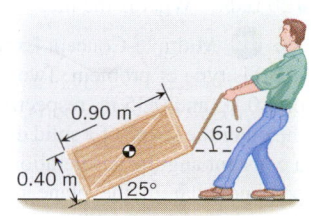

29. **ssm www An inverted "V" is made of uniform boards and weighs 356 N. Each side has the same length and makes a 30.0° angle with the vertical, as the drawing shows. Find the magnitude of the static frictional force that acts on the lower end of each leg of the "V."

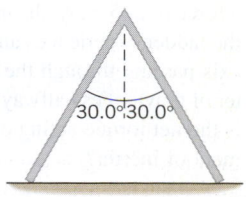

**30. The drawing shows an A-shaped stepladder. Both sides of the ladder are equal in length. This ladder is standing on a frictionless horizontal surface, and only the crossbar (which has a negligible mass) of the "A" keeps the ladder from collapsing. The ladder is uniform and has a mass of 20.0 kg. Determine the tension in the crossbar of the ladder.

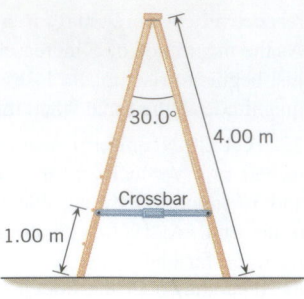

Section 9.4 Newton's Second Law for Rotational Motion About a Fixed Axis

31. Consult Multiple-Concept Example 10 to review an approach to problems such as this. A CD has a mass of 17 g and a radius of 6.0 cm. When inserted into a player, the CD starts from rest and accelerates to an angular velocity of 21 rad/s in 0.80 s. Assuming the CD is a uniform solid disk, determine the net torque acting on it.

32. A uniform solid disk with a mass of 24.3 kg and a radius of 0.314 m is free to rotate about a frictionless axle. Forces of 90.0 and 125 N are applied to the disk, as the drawing illustrates. What is **(a)** the net torque produced by the two forces and **(b)** the angular acceleration of the disk?

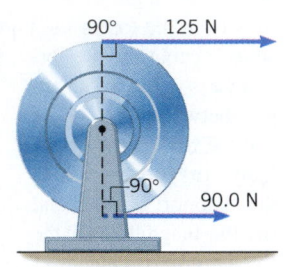

33. A solid circular disk has a mass of 1.2 kg and a radius of 0.16 m. Each of three identical thin rods has a mass of 0.15 kg. The rods are attached perpendicularly to the plane of the disk at its outer edge to form a three-legged stool (see the drawing). Find the moment of inertia of the stool with respect to an axis that is perpendicular to the plane of the disk at its center. *(Hint: When considering the moment of inertia of each rod, note that all of the mass of each rod is located at the same perpendicular distance from the axis.)*

34. A solid cylindrical disk has a radius of 0.15 m. It is mounted to an axle that is perpendicular to the circular end of the disk at its center. When a 45-N force is applied tangentially to the disk, perpendicular to the radius, the disk acquires an angular acceleration of 120 rad/s². What is the mass of the disk?

35. **GO** Multiple-Concept Example 10 provides one model for solving this type of problem. Two wheels have the same mass and radius of 4.0 kg and 0.35 m, respectively. One has the shape of a hoop and the other the shape of a solid disk. The wheels start from rest and have a constant angular acceleration with respect to a rotational axis that is perpendicular to the plane of the wheel at its center. Each turns through an angle of 13 rad in 8.0 s. Find the net external torque that acts on each wheel.

36. A 9.75-m ladder with a mass of 23.2 kg lies flat on the ground. A painter grabs the top end of the ladder and pulls straight upward with a force of 245 N. At the instant the top of the ladder leaves the ground, the ladder experiences an angular acceleration of 1.80 rad/s² about an axis passing through the bottom end of the ladder. The ladder's center of gravity lies halfway between the top and bottom ends. **(a)** What is the net torque acting on the ladder? **(b)** What is the ladder's moment of inertia?

37. ssm A rotating door is made from four rectangular sections, as indicated in the drawing. The mass of each section is 85 kg. A person pushes on the outer edge of one section with a force of $F = 68$ N that is directed perpendicular to the section. Determine the magnitude of the door's angular acceleration.

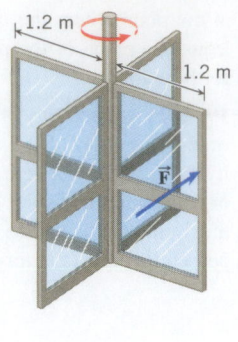

38. A long, thin rod is cut into two pieces, one being twice as long as the other. To the midpoint of piece A (the longer piece), piece B is attached perpendicularly, in order to form the inverted "T" shown in the drawing. The application of a net external torque causes this object to rotate about axis 1 with an angular acceleration of 4.6 rad/s². When the same net external torque is used to cause the object to rotate about axis 2, what is the angular acceleration?

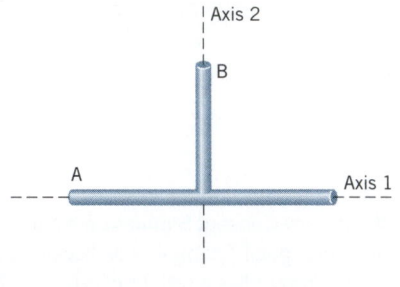

39. ssm A particle is located at each corner of an imaginary cube. Each edge of the cube is 0.25 m long, and each particle has a mass of 0.12 kg. What is the moment of inertia of these particles with respect to an axis that lies along one edge of the cube?

*40. A 15.0-m length of hose is wound around a reel, which is initially at rest. The moment of inertia of the reel is 0.44 kg·m², and its radius is 0.160 m. When the reel is turning, friction at the axle exerts a torque of magnitude 3.40 N·m on the reel. If the hose is pulled so that the tension in it remains a constant 25.0 N, how long does it take to completely unwind the hose from the reel? Neglect the mass of the hose, and assume that the hose unwinds without slipping.

*41. ssm Two thin rectangular sheets (0.20 m × 0.40 m) are identical. In the first sheet the axis of rotation lies along the 0.20-m side, and in the second it lies along the 0.40-m side. The same torque is applied to each sheet. The first sheet, starting from rest, reaches its final angular velocity in 8.0 s. How long does it take for the second sheet, starting from rest, to reach the same angular velocity?

*42. Multiple-Concept Example 10 reviews the approach and some of the concepts that are pertinent to this problem. The drawing shows a model for the motion of the human forearm in throwing a dart. Because of the force $\vec{M}$ applied by the triceps muscle, the forearm can rotate about an axis at the elbow joint. Assume that the forearm has the dimensions shown in the drawing and a moment of inertia of 0.065 kg · m² (including the effect of the dart) relative to the axis at the elbow. Assume also that the force $\vec{M}$ acts perpendicular to the forearm. Ignoring the effect of gravity and any frictional forces, determine the magnitude of the force $\vec{M}$ needed to give the dart a tangential speed of 5.0 m/s in 0.10 s, starting from rest.

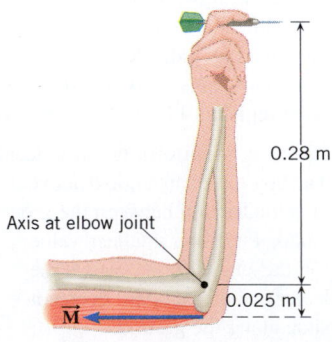

*43. GO The drawing shows two identical systems of objects; each consists of the same three small balls connected by massless rods. In both systems the axis is perpendicular to the page, but it is located at a different place, as shown. The same force of magnitude F is applied to the same ball in each system (see the drawing). The masses of the balls are $m_1 = 9.00$ kg, $m_2 = 6.00$ kg, and $m_3 = 7.00$ kg. The magnitude of the force is $F = 424$ N. **(a)** For each of the two systems, determine the moment of inertia about the given axis of rotation. **(b)** Calculate the torque (magnitude and direction) acting on each system. **(c)** Both systems start from rest, and the direction of the force moves with the system and always points along the 4.00-m rod. What is the angular velocity of each system after 5.00 s?

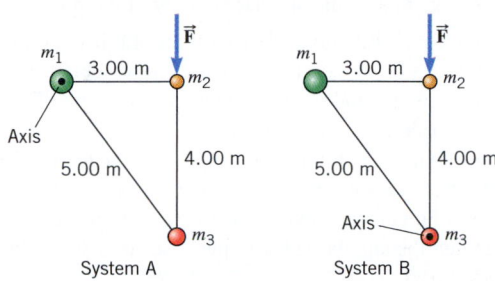

System A   System B

*44. Multiple-Concept Example 12 reviews the concepts that play roles in this problem. A block (mass = 2.0 kg) is hanging from a massless cord that is wrapped around a pulley (moment of inertia $= 1.1 \times 10^{-3}$ kg · m²), as the drawing shows. Initially the pulley is prevented from rotating and the block is stationary. Then, the pulley is allowed to rotate as the block falls. The cord does not slip relative to the pulley as the block falls. Assume that the radius of the cord around the pulley remains constant at a value of 0.040 m during the block's descent. Find the angular acceleration of the pulley and the tension in the cord.

*45. **ssm** A stationary bicycle is raised off the ground, and its front wheel ($m = 1.3$ kg) is rotating at an angular velocity of 13.1 rad/s (see the drawing). The front brake is then applied for 3.0 s, and the wheel slows down to 3.7 rad/s. Assume that all the mass of the wheel is concentrated in the rim, the radius of which is 0.33 m. The coefficient of kinetic friction between each brake pad and the rim is $\mu_k = 0.85$. What is the magnitude of the normal force that *each* brake pad applies to the rim?

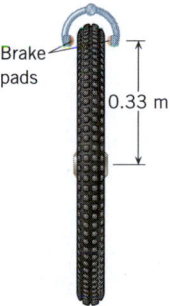

Brake pads

0.33 m

*46. The *parallel axis theorem* provides a useful way to calculate the moment of inertia I about an arbitrary axis. The theorem states that $I = I_{cm} + Mh^2$, where I_{cm} is the moment of inertia of the object relative to an axis that passes through the center of mass and is parallel to the axis of interest, M is the total mass of the object, and h is the perpendicular distance between the two axes. Use this theorem and information to determine an expression for the moment of inertia of a solid cylinder of radius R relative to an axis that lies on the surface of the cylinder and is perpendicular to the circular ends.

47. Multiple-Concept Example 12 deals with a situation that has similarities to this one and uses some of the same concepts that are needed here. See **Concept Simulation 9.4 at **www.wiley.com/college/cutnell** to review the principles involved in this problem. By means of a rope whose mass is negligible, two blocks are suspended over a pulley, as the drawing shows. The pulley can be treated as a uniform solid cylindrical disk. The downward acceleration of the 44.0-kg block is observed to be exactly one-half the acceleration due to gravity. Noting that the tension in the rope is not the same on each side of the pulley, find the mass of the pulley.

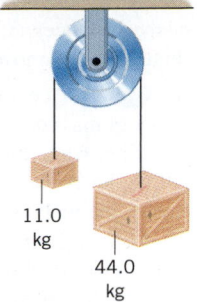

11.0 kg

44.0 kg

Section 9.5 Rotational Work and Energy

48. Calculate the kinetic energy that the earth has because of **(a)** its rotation about its own axis and **(b)** its motion around the sun. Assume that the earth is a uniform sphere and that its path around the sun is circular. For comparison, the total energy used in the United States in one year is about 1.1×10^{20} J.

49. **ssm** A flywheel is a solid disk that rotates about an axis that is perpendicular to the disk at its center. Rotating flywheels provide a means for storing energy in the form of rotational kinetic energy and are being considered as a possible alternative to batteries in electric cars. The gasoline burned in a 300-mile trip in a typical midsize car produces about 1.2×10^9 J of energy. How fast would a 13-kg flywheel with a radius of 0.30 m have to rotate to store this much energy? Give your answer in rev/min.

50. GO Two thin rods of length L are rotating with the same angular speed ω (in rad/s) about axes that pass perpendicularly through one end. Rod A is massless but has a particle of mass 0.66 kg attached to its free end. Rod B has a mass of 0.66 kg, which is distributed uniformly along its length. The length of each rod is 0.75 m, and the angular speed is 4.2 rad/s. Find the kinetic energies of rod A with its attached particle and of rod B.

51. **ssm** Three objects lie in the x, y plane. Each rotates about the z axis with an angular speed of 6.00 rad/s. The mass m of each object and its perpendicular distance r from the z axis are as follows: (1) $m_1 = 6.00$ kg and $r_1 = 2.00$ m, (2) $m_2 = 4.00$ kg and $r_2 = 1.50$ m, (3) $m_3 = 3.00$ kg and $r_3 = 3.00$ m. **(a)** Find the tangential speed of each object. **(b)** Determine the total kinetic energy of this system using the expression $KE = \frac{1}{2}m_1v_1^2 + \frac{1}{2}m_2v_2^2 + \frac{1}{2}m_3v_3^2$. **(c)** Obtain the moment of inertia of the system. **(d)** Find the rotational kinetic energy of the system using the relation $KE_R = \frac{1}{2}I\omega^2$ to verify that the answer is the same as the answer to (b).

52. A helicopter has two blades (see Figure 8.12); each blade has a mass of 240 kg and can be approximated as a thin rod of length 6.7 m. The blades are rotating at an angular speed of 44 rad/s. **(a)** What is the total moment of inertia of the two blades about the axis of rotation? **(b)** Determine the rotational kinetic energy of the spinning blades.

*53. **Interactive Solution 9.53** at **www.wiley.com/college/cutnell** offers a model for solving problems of this type. A solid sphere is rolling on a surface. What fraction of its total kinetic energy is in the form of rotational kinetic energy about the center of mass?

*54. Starting from rest, a basketball rolls from the top of a hill to the bottom, reaching a translational speed of 6.6 m/s. Ignore frictional losses. **(a)** What is the height of the hill? **(b)** Released from rest at the same height, a can of frozen juice rolls to the bottom of the same hill. What is the translational speed of the frozen juice can when it reaches the bottom?

*55. **ssm** A solid cylinder and a thin-walled hollow cylinder (see Table 9.1) have the same mass and radius. They are rolling horizontally along the ground toward the bottom of an incline. The center of mass of each cylinder has the same translational speed. The cylinders roll up the incline and reach their highest points. Calculate the ratio of the distances ($s_{\text{solid}}/s_{\text{hollow}}$) along the incline through which each center of mass moves.

*56. GO Two identical wheels are moving on horizontal surfaces. The center of mass of each has the same linear speed. However, one wheel is rolling, while the other is sliding on a frictionless surface without rolling. Each wheel then encounters an incline plane. One continues to roll up the incline, while the other continues to slide up. Eventually they come to a momentary halt, because the gravitational force slows them down. Each wheel is a disk of mass 2.0 kg. On the horizontal surfaces the center of mass of each wheel moves with a linear speed of 6.0 m/s. **(a)** What is the total kinetic energy of each wheel? **(b)** Determine the maximum height reached by each wheel as it moves up the incline.

*57. A bowling ball encounters a 0.760-m vertical rise on the way back to the ball rack, as the drawing illustrates. Ignore frictional losses and assume that the mass of the ball is distributed uniformly. The translational speed of the ball is 3.50 m/s at the bottom of the rise. Find the translational speed at the top.

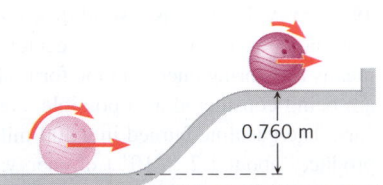

0.760 m

**58. A tennis ball, starting from rest, rolls down the hill in the drawing. At the end of the hill the ball becomes airborne, leaving at an angle of 35° with respect to the ground. Treat the ball as a thin-walled spherical shell, and determine the range x.

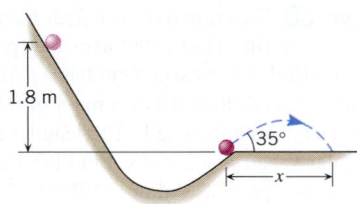

1.8 m

35°

x

Section 9.6 Angular Momentum

59. ssm www Two disks are rotating about the same axis. Disk A has a moment of inertia of 3.4 kg · m² and an angular velocity of +7.2 rad/s. Disk B is rotating with an angular velocity of −9.8 rad/s. The two disks are then linked together without the aid of any external torques, so that they rotate as a single unit with an angular velocity of −2.4 rad/s. The axis of rotation for this unit is the same as that for the separate disks. What is the moment of inertia of disk B?

60. A solid disk rotates in the horizontal plane at an angular velocity of 0.067 rad/s with respect to an axis perpendicular to the disk at its center. The moment of inertia of the disk is 0.10 kg · m². From above, sand is dropped straight down onto this rotating disk, so that a thin uniform ring of sand is formed at a distance of 0.40 m from the axis. The sand in the ring has a mass of 0.50 kg. After all the sand is in place, what is the angular velocity of the disk?

61. GO As seen from above, a playground carousel is rotating counterclockwise about its center on frictionless bearings. A person standing still on the ground grabs onto one of the bars on the carousel very close to its outer edge and climbs aboard. Thus, this person begins with an angular speed of zero and ends up with a nonzero angular speed, which means that he underwent a counterclockwise angular acceleration. The carousel has a radius of 1.50 m, an initial angular speed of 3.14 rad/s, and a moment of inertia of 125 kg · m². The mass of the person is 40.0 kg. Find the final angular speed of the carousel after the person climbs aboard.

62. Just after a motorcycle rides off the end of a ramp and launches into the air, its engine is turning counterclockwise at 7700 rev/min. The motorcycle rider forgets to throttle back, so the engine's angular speed increases to 12 500 rev/min. As a result, the rest of the motorcycle (including the rider) begins to rotate clockwise about the engine at 3.8 rev/min. Calculate the ratio I_E/I_M of the moment of inertia of the engine to the moment of inertia of the rest of the motorcycle (and the rider). Ignore torques due to gravity and air resistance.

63. **Interactive Solution 9.63** at www.wiley.com/college/cutnell illustrates one way of solving a problem similar to this one. A thin rod has a length of 0.25 m and rotates in a circle on a frictionless tabletop. The axis is perpendicular to the length of the rod at one of its ends. The rod has an angular velocity of 0.32 rad/s and a moment of inertia of 1.1×10^{-3} kg·m². A bug standing on the axis decides to crawl out to the other end of the rod. When the bug (mass = 4.2×10^{-3} kg) gets where it's going, what is the angular velocity of the rod?

64. GO Conceptual Example 14 provides useful background for this problem. A playground carousel is free to rotate about its center on frictionless bearings, and air resistance is negligible. The carousel itself (without riders) has a moment of inertia of 125 kg·m². When one person is standing on the carousel at a distance of 1.50 m from the center, the carousel has an angular velocity of 0.600 rad/s. However, as this person moves inward to a point located 0.750 m from the center, the angular velocity increases to 0.800 rad/s. What is the person's mass?

*65. A cylindrically shaped space station is rotating about the axis of the cylinder to create artificial gravity. The radius of the cylinder is 82.5 m. The moment of inertia of the station without people is 3.00×10^9 kg·m². Suppose that 500 people, with an average mass of 70.0 kg each, live on this station. As they move radially from the outer surface of the cylinder toward the axis, the angular speed of the station changes. What is the maximum possible percentage change in the station's angular speed due to the radial movement of the people?

*66. A thin, uniform rod is rotating at an angular velocity of 7.0 rad/s about an axis that is perpendicular to the rod at its center. As the drawing indicates, the rod is hinged at two places, one-quarter of the length from each end. Without the aid of external torques, the rod suddenly assumes a "u" shape, with the arms of the "u" parallel to the rotation axis. What is the angular velocity of the rotating "u"?

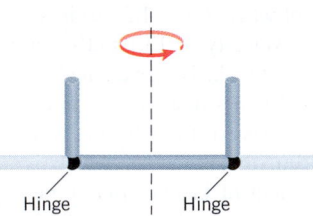

Hinge   Hinge

**67. A small 0.500-kg object moves on a frictionless horizontal table in a circular path of radius 1.00 m. The angular speed is 6.28 rad/s. The object is attached to a string of negligible mass that passes through a small hole in the table at the center of the circle. Someone under the table begins to pull the string downward to make the circle smaller. If the string will tolerate a tension of no more than 105 N, what is the radius of the smallest possible circle on which the object can move?

**68. A platform is rotating at an angular speed of 2.2 rad/s. A block is resting on this platform at a distance of 0.30 m from the axis. The coefficient of static friction between the block and the platform is 0.75. Without any external torque acting on the system, the block is moved toward the axis. Ignore the moment of inertia of the platform and determine the smallest distance from the axis at which the block can be relocated and still remain in place as the platform rotates.

ADDITIONAL PROBLEMS

69. The drawing shows a jet engine suspended beneath the wing of an airplane. The weight $\vec{W}$ of the engine is 10 200 N and acts as shown in the drawing. In flight the engine produces a thrust $\vec{T}$ of 62 300 N that is parallel to the ground. The rotational axis in the drawing is perpendicular to the plane of the paper. With respect to this axis, find the magnitude of the torque due to **(a)** the weight and **(b)** the thrust.

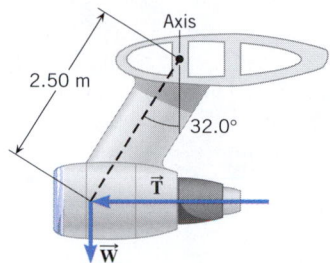

70. When some stars use up their fuel, they undergo a catastrophic explosion called a *supernova*. This explosion blows much or all of the star's mass outward, in the form of a rapidly expanding spherical shell. As a simple model of the supernova process, assume that the star is a solid sphere of radius R that is initially rotating at 2.0 revolutions per day. After the star explodes, find the angular velocity, in revolutions per day, of the expanding supernova shell when its radius is $4.0R$. Assume that all of the star's original mass is contained in the shell.

71. A ceiling fan is turned on and a net torque of 1.8 N·m is applied to the blades. The blades have a total moment of inertia of 0.22 kg·m². What is the angular acceleration of the blades?

72. The drawing shows a person (weight, $W = 584$ N) doing push-ups. Find the normal force exerted by the floor on *each* hand and *each* foot, assuming that the person holds this position.

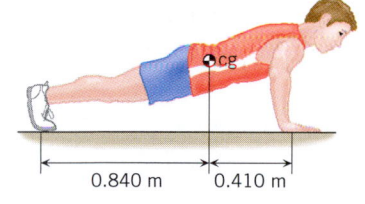

73. Interactive Solution 9.73 at **www.wiley.com/college/cutnell** presents a method for modeling this problem. Multiple-Concept Example 10 offers useful background for problems like this. A cylinder is rotating about an axis that passes through the center of each circular end piece. The cylinder has a radius of 0.0830 m, an angular speed of 76.0 rad/s, and a moment of inertia of 0.615 kg·m². A brake shoe presses against the surface of the cylinder and applies a tangential frictional force to it. The frictional force reduces the angular speed of the cylinder by a factor of two during a time of 6.40 s. **(a)** Find the magnitude of the angular deceleration of the cylinder. **(b)** Find the magnitude of the force of friction applied by the brake shoe.

*74. The drawing shows an outstretched arm (0.61 m in length) that is parallel to the floor. The arm is pulling downward against the ring attached to the pulley system, in order to hold the 98-N weight stationary. To pull the arm downward, the latissimus dorsi muscle applies the force $\vec{M}$ in the drawing, at a point that is 0.069 m from the shoulder joint and oriented at an angle of 29°. The arm has a weight of 47 N and a center of gravity (cg) that is located 0.28 m from the shoulder joint. Find the magnitude of $\vec{M}$.

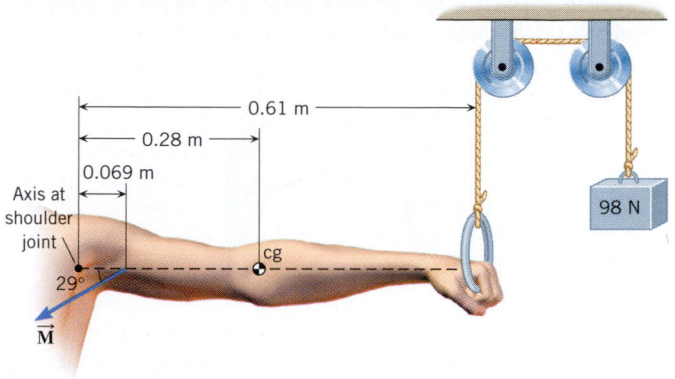

Problem 74

*75. ssm www A massless, rigid board is placed across two bathroom scales that are separated by a distance of 2.00 m. A person lies on the board. The scale under his head reads 425 N, and the scale under his feet reads 315 N. **(a)** Find the weight of the person. **(b)** Locate the center of gravity of the person relative to the scale beneath his head.

*76. A woman who weighs 5.00×10^2 N is leaning against a smooth vertical wall, as the drawing shows. Find **(a)** the force $\vec{F}_N$ (directed perpendicular to the wall) exerted on her shoulders by the wall and the **(b)** horizontal and **(c)** vertical components of the force exerted on her shoes by the ground.

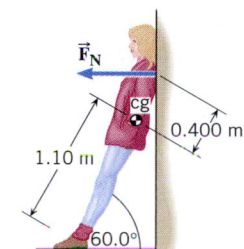

*77. ssm In outer space two identical space modules are joined together by a massless cable. These modules are rotating about their center of mass, which is at the center of the cable because the modules are identical (see the drawing). In each module, the cable is connected to a motor, so that the modules can pull each other together. The initial tangential speed of each module is $v_0 = 17$ m/s. Then they pull together until the distance between them is reduced by a factor of two. Determine the final tangential speed v_f for each module.

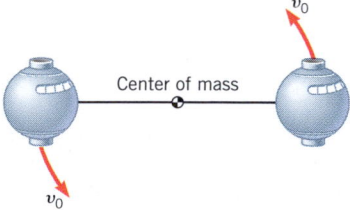

*78. One end of a thin rod is attached to a pivot, about which it can rotate without friction. Air resistance is absent. The rod has a length of 0.80 m and is uniform. It is hanging vertically straight downward. The end of the rod nearest the floor is given a linear speed v_0, so that the rod begins to rotate upward about the pivot. What must be the value of v_0, such that the rod comes to a momentary halt in a straight-up orientation, exactly opposite to its initial orientation?

**79. Two vertical walls are separated by a distance of 1.5 m, as the drawing shows. Wall 1 is smooth, while wall 2 is not smooth. A uniform board is propped between them. The coefficient of static friction between the board and wall 2 is 0.98. What is the length of the longest board that can be propped between the walls?

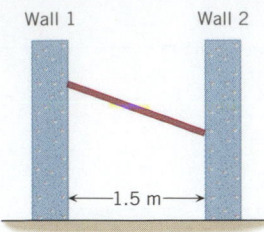

Wall 1   Wall 2

←—1.5 m—→

**80. See Multiple-Concept Example 12 to review some of the concepts that come into play here. The crane shown in the drawing is lifting a 180-kg crate upward with an acceleration of 1.2 m/s². The cable from the crate passes over a solid cylindrical pulley at the top of the boom. The pulley has a mass of 130 kg. The cable is then wound onto a hollow cylindrical drum that is mounted on the deck of the crane. The mass of the drum is 150 kg, and its radius is 0.76 m. The engine applies a counterclockwise torque to the drum in order to wind up the cable. What is the magnitude of this torque? Ignore the mass of the cable.

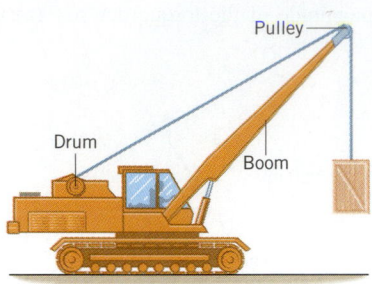

Pulley

Drum

Boom

CHAPTER 10

SIMPLE HARMONIC MOTION AND ELASTICITY

Bungee jumping has a certain attraction for thrill-seekers. It depends critically on the elastic properties of the bungee cord and its ability to store elastic potential energy. The up-and-down motion of the jumper on the end of the cord near the conclusion of his trip is related to simple harmonic motion. This chapter discusses simple harmonic motion and elastic potential energy. (© Mark Potts/SUPERSTOCK)

10.1 THE IDEAL SPRING AND SIMPLE HARMONIC MOTION

Springs are familiar objects that have many applications, ranging from push-button switches on electronic components, to automobile suspension systems, to mattresses. In use, they can be stretched or compressed. For example, the top drawing in Figure 10.1 shows a spring being stretched. Here a hand applies a pulling force F_x^{Applied} to the spring. The subscript x reminds us that F_x^{Applied} lies along the x axis (not shown in the drawing), which is parallel to the length of the spring. In response, the spring stretches and undergoes a displacement of x from its original, or "unstrained," length. The bottom drawing in Figure 10.1 illustrates the spring being compressed. Now the hand applies a pushing force to the spring, and it again undergoes a displacement from its unstrained length.

Experiment reveals that for relatively small displacements, the force F_x^{Applied} required to stretch or compress a spring is directly proportional to the displacement x, or $F_x^{\text{Applied}} \propto x$. As is customary, this proportionality may be converted into an equation by introducing a proportionality constant k:

$$F_x^{\text{Applied}} = kx \qquad (10.1)$$

The constant k is called the **spring constant,** and Equation 10.1 shows that it has the dimensions of force per unit length (N/m). A spring that behaves according to $F_x^{\text{Applied}} = kx$ is said to be an **ideal spring.** Example 1 illustrates one application of such a spring.

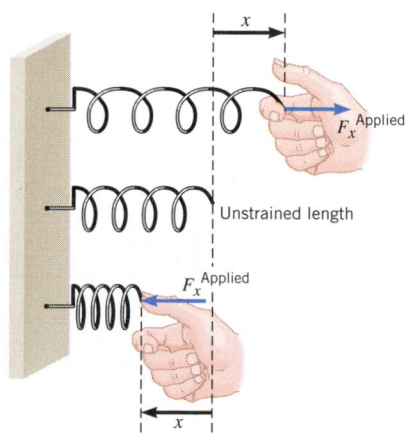

Figure 10.1 An ideal spring obeys the equation $F_x^{\text{Applied}} = kx$, where F_x^{Applied} is the force applied to the spring, x is the displacement of the spring from its unstrained length, and k is the spring constant.

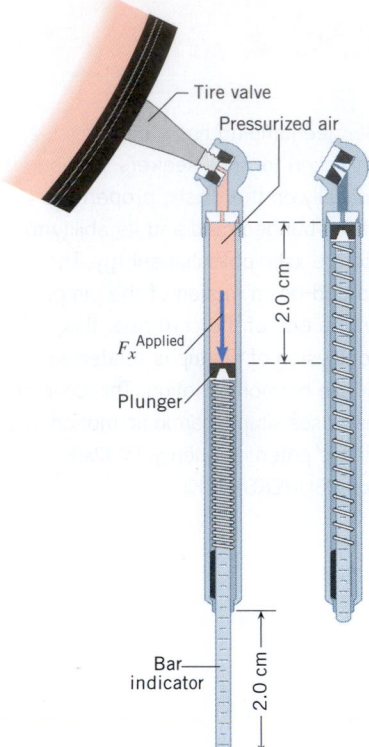

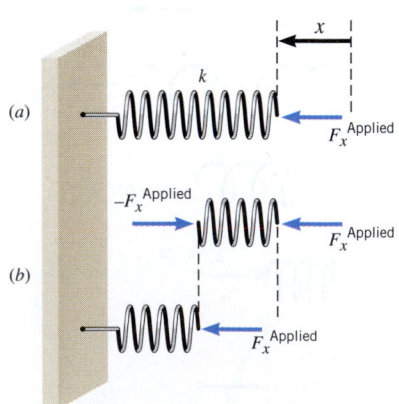

Figure 10.2 In a tire pressure gauge, the pressurized air from the tire exerts a force F_x^{Applied} that compresses a spring.

Figure 10.3 (a) The 10-coil spring has a spring constant k. The applied force is F_x^{Applied}, and the displacement of the spring from its unstrained length is x. (b) The spring in part a is divided in half, so that the forces acting on the two 5-coil springs can be analyzed.

Example 1 A Tire Pressure Gauge

When a tire pressure gauge is pressed against a tire valve, as in Figure 10.2, the air in the tire pushes against a plunger attached to a spring. Suppose the spring constant of the spring is $k = 320$ N/m and the bar indicator of the gauge extends 2.0 cm when the gauge is pressed against the tire valve. What force does the air in the tire apply to the spring?

Reasoning We assume that the spring is an ideal spring, so that the relation $F_x^{\text{Applied}} = kx$ is obeyed. The spring constant k is known, as is the displacement x. Therefore, we can determine the force applied to the spring.

Solution The force needed to compress the spring is given by Equation 10.1 as

$$F_x^{\text{Applied}} = kx = (320 \text{ N/m})(0.020 \text{ m}) = \boxed{6.4 \text{ N}}$$

Thus, the exposed length of the bar indicator indicates the force that the air pressure in the tire exerts on the spring. We will see later that pressure is force per unit area, so force is pressure times area. Since the area of the plunger surface is fixed, the bar indicator can be marked in units of pressure.

Sometimes the spring constant k is referred to as the **stiffness** of the spring, because a large value for k means the spring is "stiff," in the sense that a large force is required to stretch or compress it. Conceptual Example 2 examines what happens to the stiffness of a spring when the spring is cut into two shorter pieces.

Conceptual Example 2 Are Shorter Springs Stiffer Springs?

Figure 10.3a shows a 10-coil spring that has a spring constant k. When this spring is cut in half, so there are two 5-coil springs, is the spring constant of each of the shorter springs (**a**) $\frac{1}{2}k$ or (**b**) $2k$?

Reasoning When the length of a spring is increased or decreased, the change in length is distributed over the entire spring. Greater forces are required to cause changes that are a greater fraction of the spring's initial length, since such changes distort the atomic structure of the spring material to a greater extent.

Answer (a) is incorrect. When a force F_x^{Applied} is applied to the 10-coil spring, as in Figure 10.3a, the displacement of the spring from its unstrained length is x. If this same force were applied to a 5-coil spring that had a spring constant of $\frac{1}{2}k$, the displacement would be $2x$ because the spring is only half as stiff. Since this displacement would be a larger fraction of the length of the 5-coil spring, it would require a force greater than F_x^{Applied}, not equal to F_x^{Applied}. Therefore, the spring constant cannot be $\frac{1}{2}k$.

Answer (b) is correct. As indicated in Figure 10.3a, the displacement of the spring from its unstrained length is x when a force F_x^{Applied} is applied to the 10-coil spring. Figure 10.3b shows the spring divided in half between the fifth and sixth coils (counting from the right). The spring is in equilibrium, so the net force acting on the right half (coils 1–5) must be zero. Thus, as part b shows, a force of $-F_x^{\text{Applied}}$ must act on coil 5 in order to balance the force F_x^{Applied} that acts on coil 1. It is the adjacent coil 6 that exerts the force $-F_x^{\text{Applied}}$, and Newton's action–reaction law now comes into play. It tells us that coil 5, in response, exerts an oppositely directed force of equal magnitude on coil 6. In other words, the force F_x^{Applied} is also exerted on the left half of the spring, as part b also indicates. As a result, the left half compresses by an amount that is one-half the displacement x experienced by the 10-coil spring. We conclude, then, that the 5-coil spring must be twice as stiff as the 10-coil spring. In general, the spring constant is inversely proportional to the number of coils in the spring, so shorter springs are stiffer springs, other things being equal.

Related Homework: *Problems 12, 75*

To stretch or compress a spring, a force must be applied to it. In accord with Newton's third law, the spring exerts an oppositely directed force of equal magnitude. This reaction force is applied by the spring to the agent that does the pulling or pushing. In other words, the reaction force is applied to the object attached to the spring. The reaction force is also called a "restoring force," for a reason that will be clarified shortly. The restoring force of an ideal spring is obtained from the relation $F_x^{\text{Applied}} = kx$ by including the minus sign required by Newton's action–reaction law, as indicated in Equation 10.2.

HOOKE'S LAW* RESTORING FORCE OF AN IDEAL SPRING

The restoring force of an ideal spring is

$$F_x = -kx \tag{10.2}$$

where k is the spring constant and x is the displacement of the spring from its unstrained length. The minus sign indicates that the restoring force always points in a direction opposite to the displacement of the spring from its unstrained length.

In Chapter 4 we encountered four types of forces: the gravitational force, the normal force, frictional forces, and the tension force. These forces can contribute to the net external force, which Newton's second law relates to the mass and acceleration of an object. The restoring force of a spring can also contribute to the net external force. Once again, we see the unifying theme of Newton's second law, in that individual forces contribute to the net force, which, in turn, is responsible for the acceleration. Newton's second law plays a central role in describing the motion of objects attached to springs.

Figure 10.4 helps to explain why the phrase "restoring force" is used. In the picture, an object of mass m is attached to a spring on a frictionless table. In part A, the spring has been stretched to the right, so it exerts the leftward-pointing force F_x. When the object is released, this force pulls it to the left, restoring it toward its equilibrium position. However, consistent with Newton's first law, the moving object has inertia and coasts beyond the equilibrium position, compressing the spring as in part B. The force exerted by the spring now points to the right and, after bringing the object to a momentary halt, acts to restore the object to its equilibrium position. But the object's inertia again carries it beyond the equilibrium position, this time stretching the spring and leading to the restoring force F_x shown in part C. The back-and-forth motion illustrated in the drawing then repeats itself, continuing forever, since no friction acts on the object or the spring.

When the restoring force has the mathematical form given by $F_x = -kx$, the type of friction-free motion illustrated in Figure 10.4 is designated as "simple harmonic motion." By attaching a pen to the object and moving a strip of paper past it at a steady rate,

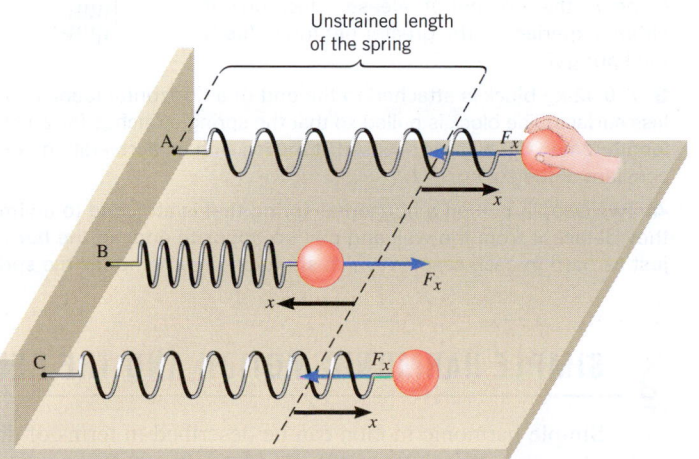

Unstrained length of the spring

Figure 10.4 The restoring force F_x (see blue arrows) produced by an ideal spring always points opposite to the displacement x (see black arrows) of the spring and leads to a back-and-forth motion of the object.

*As we will see in Section 10.8, Equation 10.2 is similar to a relationship first discovered by Robert Hooke (1635–1703).

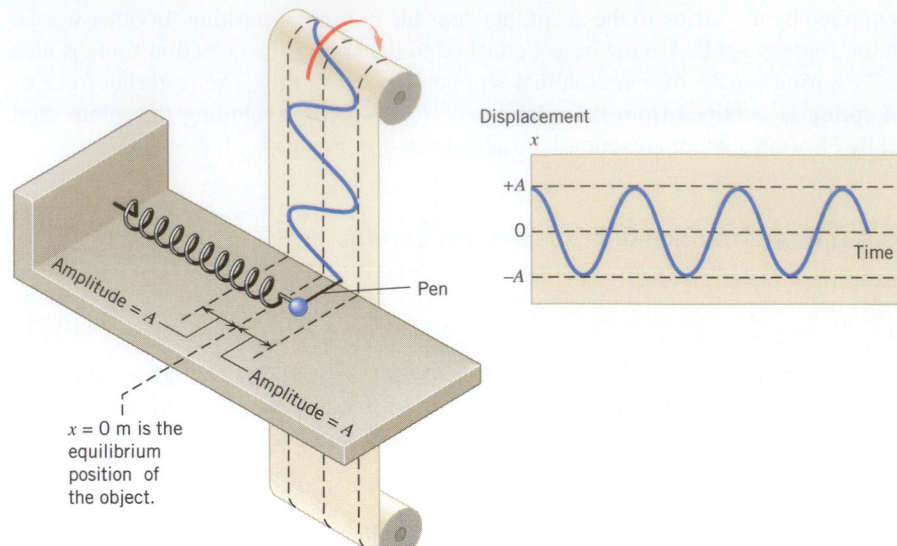

Figure 10.5 When an object moves in simple harmonic motion, a graph of its position as a function of time has a sinusoidal shape with an amplitude A. A pen attached to the object records the graph.

we can record the position of the vibrating object as time passes. Figure 10.5 illustrates the resulting graphical record of simple harmonic motion. The maximum excursion from equilibrium is the ***amplitude*** A of the motion. The shape of this graph is characteristic of simple harmonic motion and is called "sinusoidal," because it has the shape of a trigonometric sine or cosine function.

The restoring force also leads to simple harmonic motion when the object is attached to a vertical spring, just as it does when the spring is horizontal. When the spring is vertical, however, the weight of the object causes the spring to stretch, and the motion occurs with respect to the equilibrium position of the object on the stretched spring, as Figure 10.6 indicates. The amount of initial stretching d_0 due to the weight can be calculated by equating the weight to the magnitude of the restoring force that supports it; thus, $mg = kd_0$, which gives $d_0 = mg/k$.

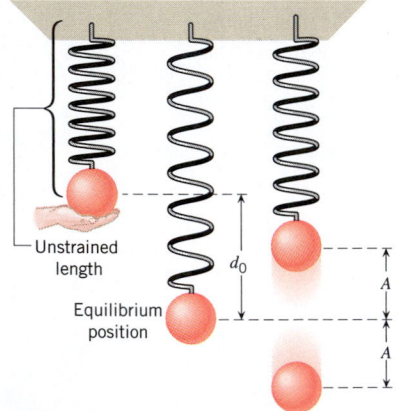

Figure 10.6 The weight of an object on a vertical spring stretches the spring by an amount d_0. Simple harmonic motion of amplitude A occurs with respect to the equilibrium position of the object on the stretched spring.

✓ **CHECK YOUR UNDERSTANDING**

(*The answers are given at the end of the book.*)

1. A steel ball is dropped onto a very hard floor. Over and over again, the ball rebounds to its original height (assuming that no energy is lost during the collision with the floor). Is the motion of the ball simple harmonic motion?

2. The drawing shows identical springs that are attached to a box in two different ways. Initially, the springs are unstrained. The box is then pulled to the right and released. In each case the initial displacement of the box is the same. At the moment of release, which box, if either, experiences the greater net force due to the springs?

3. A 0.42-kg block is attached to the end of a horizontal ideal spring and rests on a frictionless surface. The block is pulled so that the spring stretches for 2.1 cm relative to its unstrained length. When the block is released, it moves with an acceleration of 9.0 m/s². What is the spring constant of the spring?

4. Two people pull on a horizontal spring that is attached to an immovable wall. Then, they detach it from the wall and pull on opposite ends of the horizontal spring. They pull just as hard in each case. In which situation, if either, does the spring stretch more?

10.2 **SIMPLE HARMONIC MOTION AND THE REFERENCE CIRCLE**

Simple harmonic motion can be described in terms of displacement, velocity, and acceleration, and the model in Figure 10.7 is helpful in explaining these characteristics. The model consists of a small ball attached to the top of a rotating turntable. The ball is moving in uniform circular motion (see Section 5.1) on a path known as the ***reference circle.*** As the ball moves, its shadow falls on a strip of film, which is moving upward at a steady rate and recording where the shadow is. A comparison of the film with the paper in Figure 10.5 reveals the same kind of patterns, suggesting that the model is useful.

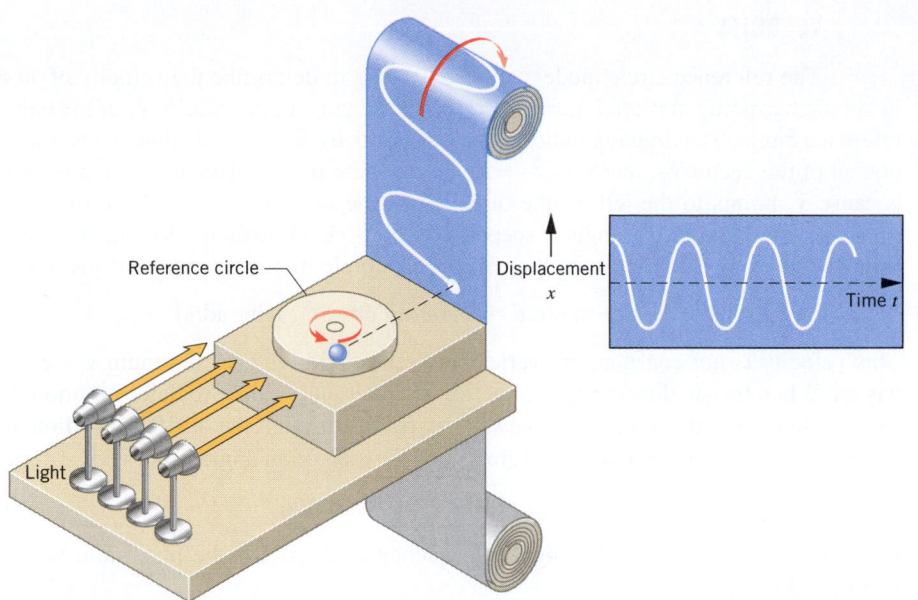

Figure 10.7 The ball mounted on the turntable moves in uniform circular motion, and its shadow, projected on a moving strip of film, executes simple harmonic motion.

DISPLACEMENT

Figure 10.8 shows the reference circle (radius = A) and indicates how to determine the displacement of the shadow on the film. The ball starts on the x axis at $x = +A$ and moves through the angle θ in a time t. The circular motion is uniform, so the ball moves with a constant angular speed ω (in rad/s), and the angle has a value (in rad) of $\theta = \omega t$. The displacement x of the shadow is just the projection of the radius A onto the x axis:

$$x = A \cos \theta = A \cos \omega t \tag{10.3}$$

Figure 10.9 shows a graph of this equation. As time passes, the shadow of the ball oscillates between the values of $x = +A$ and $x = -A$, corresponding to the limiting values of $+1$ and -1 for the cosine of an angle. The radius A of the reference circle, then, is the amplitude of the simple harmonic motion.

As the ball moves one revolution or cycle around the reference circle, its shadow executes one cycle of back-and-forth motion. For any object in simple harmonic motion, the time required to complete one cycle is the ***period T,*** as Figure 10.9 indicates. The value of T depends on the angular speed ω of the ball because the greater the angular speed, the shorter the time it takes to complete one revolution. We can obtain the relationship between ω and T by recalling that $\omega = \Delta\theta/\Delta t$ (Equation 8.2), where $\Delta\theta$ is the angular displacement of the ball and Δt is the time interval. For one cycle, $\Delta\theta = 2\pi$ rad and $\Delta t = T$, so that

$$\omega = \frac{2\pi}{T} \qquad (\omega \text{ in rad/s}) \tag{10.4}$$

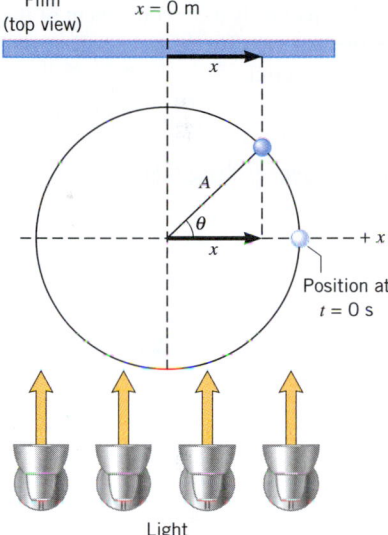

Figure 10.8 A top view of a ball on a turntable. The ball's shadow on the film has a displacement x that depends on the angle θ through which the ball has moved on the reference circle.

Instead of the period, we often speak of the ***frequency f*** of the motion, the frequency being the number of cycles of the motion per second. For example, if an object on a spring completes 10 cycles in one second, the frequency is $f = 10$ cycles/s. The period T, or the time for one cycle, would be $\frac{1}{10}$ s. Thus, frequency and period are related according to

$$f = \frac{1}{T} \tag{10.5}$$

Usually, one cycle per second is referred to as one hertz (Hz), the unit being named after Heinrich Hertz (1857–1894). One thousand cycles per second is called one kilohertz (kHz). Thus, five thousand cycles per second, for instance, can be written as 5 kHz.

Using the relationships $\omega = 2\pi/T$ and $f = 1/T$, we can relate the angular speed ω (in rad/s) to the frequency f (in cycles/s or Hz):

$$\omega = \frac{2\pi}{T} = 2\pi f \qquad (\omega \text{ in rad/s}) \tag{10.6}$$

Because ω is directly proportional to the frequency f, ω is often called the ***angular frequency.***

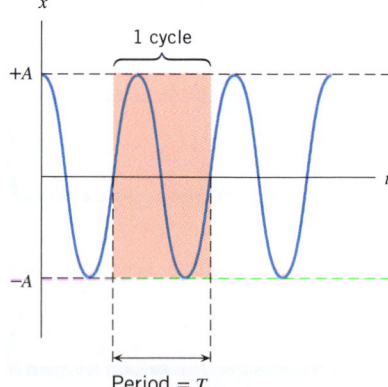

Figure 10.9 For simple harmonic motion, the graph of displacement x versus time t is a sinusoidal curve. The period T is the time required for one complete motional cycle.

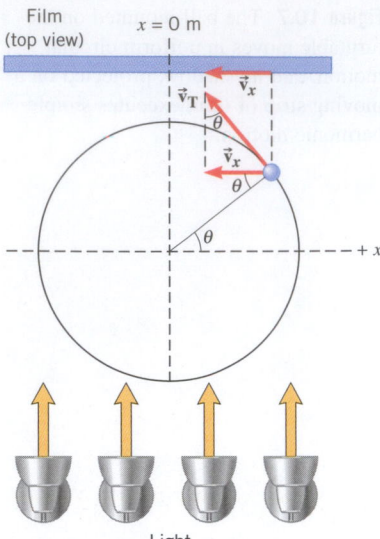

Figure 10.10 The velocity $\vec{v}_x$ of the ball's shadow is the x component of the tangential velocity $\vec{v}_T$ of the ball on the reference circle.

The physics of a loudspeaker diaphragm.

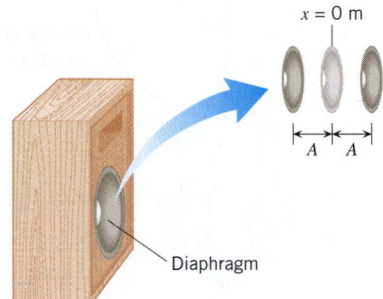

Figure 10.11 The diaphragm of a loudspeaker generates a sound by moving back and forth in simple harmonic motion.

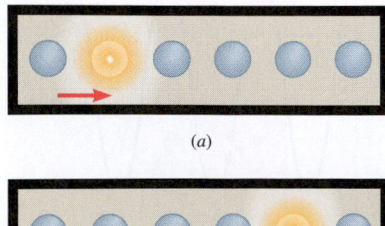

(a)

(b)

Figure 10.12 The motion of a lighted bulb is from (a) left to right and then from (b) right to left.

VELOCITY

The reference circle model can also be used to determine the velocity of an object in simple harmonic motion. Figure 10.10 shows the tangential velocity $\vec{v}_T$ of the ball on the reference circle. The drawing indicates that the velocity $\vec{v}_x$ of the shadow is just the x component of the vector $\vec{v}_T$; that is, $v_x = -v_T \sin\theta$, where $\theta = \omega t$. The minus sign is necessary because $\vec{v}_x$ points to the left, in the direction of the negative x axis. Since the tangential speed v_T is related to the angular speed ω by $v_T = r\omega$ (Equation 8.9) and since $r = A$, it follows that $v_T = A\omega$. Therefore, the velocity in simple harmonic motion is given by

$$v_x = -v_T \sin\theta = -A\omega \sin\omega t \qquad (\omega \text{ in rad/s}) \qquad (10.7)$$

This velocity is *not* constant, but varies between maximum and minimum values as time passes. When the shadow changes direction at either end of the oscillatory motion, the velocity is momentarily zero. When the shadow passes through the $x = 0$ m position, the velocity has a maximum magnitude of $A\omega$, since the sine of an angle is between $+1$ and -1:

$$v_{max} = A\omega \qquad (\omega \text{ in rad/s}) \qquad (10.8)$$

Both the amplitude A and the angular frequency ω determine the maximum velocity, as Example 3 emphasizes.

⬇ Example 3 The Maximum Speed of a Loudspeaker Diaphragm

The diaphragm of a loudspeaker moves back and forth in simple harmonic motion to create sound, as in Figure 10.11. The frequency of the motion is $f = 1.0$ kHz and the amplitude is $A = 0.20$ mm. **(a)** What is the maximum speed of the diaphragm? **(b)** Where in the motion does this maximum speed occur?

Reasoning The maximum speed of an object vibrating in simple harmonic motion is $v_{max} = A\omega$ (ω in rad/s), according to Equation 10.8. The angular frequency ω is related to the frequency f by $\omega = 2\pi f$, according to Equation 10.6.

Solution **(a)** Using Equations 10.8 and 10.6, we find that the maximum speed of the vibrating diaphragm is

$$v_{max} = A\omega = A(2\pi f) = (0.20 \times 10^{-3}\text{ m})(2\pi)(1.0 \times 10^3\text{ Hz}) = \boxed{1.3\text{ m/s}}$$

(b) The speed of the diaphragm is zero when the diaphragm momentarily comes to rest at either end of its motion: $x = +A$ and $x = -A$. Its maximum speed occurs midway between these two positions, or at $x = 0$ m.

Simple harmonic motion is not just any kind of vibratory motion. It is a very specific kind and, among other things, must have the velocity given by Equation 10.7. For instance, advertising signs often use a "moving light" display to grab your attention. Conceptual Example 4 examines the back-and-forth motion in one such display, to see whether it is simple harmonic motion.

⬇ Conceptual Example 4 Moving Lights

Over the entrance to a restaurant is mounted a strip of equally spaced light bulbs, as Figure 10.12a illustrates. Starting at the left end, each bulb turns on in sequence for one-half second. Thus, a lighted bulb appears to move from left to right. After the last bulb on the right turns on, the apparent motion reverses. The lighted bulb then appears to move to the left, as part b of the drawing indicates. As a result, the lighted bulb appears to oscillate back and forth. Is the apparent motion simple harmonic motion? **(a)** No **(b)** Yes

Reasoning In simple harmonic motion the velocity of the moving object must have the velocity specified by $v = -A\omega \sin\omega t$ (Equation 10.7). This velocity is not constant as time passes.

Answer (b) is incorrect. Since the bulbs are equally spaced and each one remains lit for the same amount of time, the apparent motion in Figure 10.12a (or in Figure 10.12b) occurs at a constant velocity and, therefore, cannot be simple harmonic motion.

Answer (a) is correct. The apparent motion is not simple harmonic motion. If it were, the speed would be zero at each end of the sign and would increase to a maximum speed at the center, consistent with Equation 10.7. However, the speed is constant because the bulbs are equally spaced and each remains on for the same amount of time.

ACCELERATION

In simple harmonic motion, the velocity is not constant; consequently, there must be an acceleration. This acceleration can also be determined with the aid of the reference-circle model. As Figure 10.13 shows, the ball on the reference circle moves in uniform circular motion, and, therefore, has a centripetal acceleration $\vec{a}_c$ that points toward the center of the circle. The acceleration $\vec{a}_x$ of the shadow is the x component of the centripetal acceleration; $a_x = -a_c \cos \theta$. The minus sign is needed because the acceleration of the shadow points to the left. Recalling that the centripetal acceleration is related to the angular speed ω by $a_c = r\omega^2$ (Equation 8.11) and using $r = A$, we find that $a_c = A\omega^2$. With this substitution and the fact that $\theta = \omega t$, the acceleration in simple harmonic motion becomes

$$a_x = -a_c \cos \theta = -A\omega^2 \cos \omega t \qquad (\omega \text{ in rad/s}) \qquad (10.9)$$

The acceleration, like the velocity, does *not* have a constant value as time passes. The maximum magnitude of the acceleration is

$$a_{max} = A\omega^2 \qquad (\omega \text{ in rad/s}) \qquad (10.10)$$

Although both the amplitude A and the angular frequency ω determine the maximum value, the frequency has a particularly strong effect, because it is squared. Example 5 shows that the acceleration can be remarkably large in a practical situation.

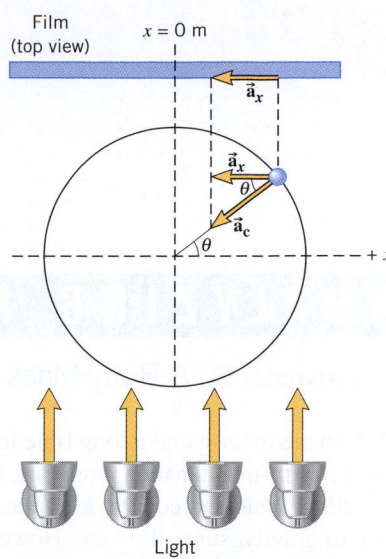

Figure 10.13 The acceleration $\vec{a}_x$ of the ball's shadow is the x component of the centripetal acceleration $\vec{a}_c$ of the ball on the reference circle.

 Example 5   The Loudspeaker Revisited—The Maximum Acceleration

The loudspeaker diaphragm in Figure 10.11 is vibrating at a frequency of $f = 1.0$ kHz, and the amplitude of the motion is $A = 0.20$ mm. **(a)** What is the maximum acceleration of the diaphragm, and **(b)** where does this maximum acceleration occur?

Reasoning The maximum acceleration of an object vibrating in simple harmonic motion is $a_{max} = A\omega^2$ (ω in rad/s), according to Equation 10.10. Equation 10.6 shows that the angular frequency ω is related to the frequency f by $\omega = 2\pi f$.

Solution **(a)** Using Equations 10.10 and 10.6, we find that the maximum acceleration of the vibrating diaphragm is

$$a_{max} = A\omega^2 = A(2\pi f)^2 = (0.20 \times 10^{-3} \text{ m})[2\pi(1.0 \times 10^3 \text{ Hz})]^2$$
$$= \boxed{7.9 \times 10^3 \text{ m/s}^2}$$

This is an incredible acceleration, being more than 800 times the acceleration due to gravity, and the diaphragm must be built to withstand it.

(b) The maximum acceleration occurs when the force acting on the diaphragm is a maximum. The maximum force arises when the diaphragm is at the ends of its path, where the displacement is greatest. Thus, the maximum acceleration occurs at $x = +A$ and $x = -A$ in Figure 10.11.

Problem-solving insight

Do not confuse the vibrational frequency f with the angular frequency ω. The value for f is in hertz (cycles per second). The value for ω is in radians per second. The two are related by $\omega = 2\pi f$.

FREQUENCY OF VIBRATION

With the aid of Newton's second law ($\Sigma F_x = ma_x$), it is possible to determine the frequency at which an object of mass m vibrates on a spring. We assume that the mass of the spring itself is negligible and that the only force acting on the object in the horizontal direction is due to the spring—that is, the Hooke's law restoring force. Thus, the net force is $\Sigma F_x = -kx$, and Newton's second law becomes $-kx = ma_x$, where a_x is the acceleration of the object. The displacement and acceleration of an oscillating spring are, respectively, $x = A \cos \omega t$ (Equation 10.3) and $a_x = -A\omega^2 \cos \omega t$ (Equation 10.9). Substituting these expressions for x and a_x into the relation $-kx = ma_x$, we find that

$$-k(A \cos \omega t) = m(-A\omega^2 \cos \omega t)$$

which yields

$$\omega = \sqrt{\frac{k}{m}} \qquad (\omega \text{ in rad/s}) \qquad (10.11)$$

In this expression, the angular frequency ω must be in radians per second. Larger spring constants k and smaller masses m result in larger frequencies. Example 6 illustrates an application of Equation 10.11.

ANALYZING MULTIPLE-CONCEPT PROBLEMS

Example 6 A Body-Mass Measurement Device

Astronauts who spend a long time in orbit measure their body masses as part of their health-maintenance programs. On earth, it is simple to measure body weight W with a scale and convert it to mass m using the magnitude g of the acceleration due to gravity, since $W = mg$. However, this procedure does not work in orbit, because both the scale and the astronaut are in free fall and cannot press against each other (see Conceptual Example 12 in Chapter 5). Instead, astronauts use a body-mass measurement device, as Figure 10.14 illustrates. The device consists of a spring-mounted chair in which the astronaut sits. The chair is then started oscillating in simple harmonic motion. The period of the motion is measured electronically and is automatically converted into a value of the astronaut's mass, after the mass of the chair is taken into account. The spring used in one such device has a spring constant of 606 N/m, and the mass of the chair is 12.0 kg. The measured oscillation period is 2.41 s. Find the mass of the astronaut.

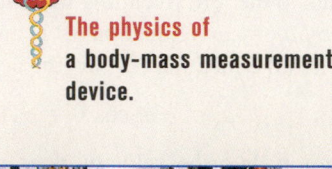

The physics of a body-mass measurement device.

Reasoning Since the astronaut and chair are oscillating in simple harmonic motion, the total mass m_{total} of the two is related to the angular frequency ω of the motion by $\omega = \sqrt{k/m_{\text{total}}}$, or $m_{\text{total}} = k/\omega^2$. The angular frequency, in turn, is related to the period of the motion by $\omega = 2\pi/T$. These two relations will enable us to find the total mass of the astronaut and chair. From this result, we will subtract the mass of the chair to obtain the mass of the astronaut.

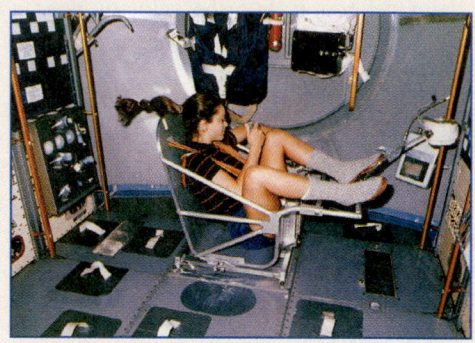

Figure 10.14 Astronaut Tamara Jernigan uses a body-mass measurement device to measure her mass while in orbit. (Courtesy NASA)

Knowns and Unknowns The data for this problem are:

Description	Symbol	Value
Spring constant	k	606 N/m
Mass of chair	m_{chair}	12.0 kg
Period of oscillation	T	2.41 s
Unknown Variable		
Mass of astronaut	m_{astro}	?

Modeling the Problem

STEP 1 **Angular Frequency of Vibration** The mass m_{astro} of the astronaut is equal to the total mass m_{total} of the astronaut and chair minus the mass m_{chair} of the chair:

$$m_{\text{astro}} = m_{\text{total}} - m_{\text{chair}} \qquad (1)$$

The mass of the chair is known. Since the astronaut and chair oscillate in simple harmonic motion, the angular frequency ω of the oscillation is related to the spring constant k of the spring and the total mass m_{total} by Equation 10.11:

$$\omega = \sqrt{\frac{k}{m_{\text{total}}}} \quad \text{or} \quad m_{\text{total}} = \frac{k}{\omega^2}$$

$$m_{\text{astro}} = \frac{k}{\omega^2} - m_{\text{chair}} \qquad (2)$$

Substituting this result for m_{total} into Equation 1 gives Equation 2 at the right. The spring constant and mass of the chair are known, but the angular frequency ω is not; we will evaluate it in Step 2.

STEP 2 **Angular Frequency and Period** The angular frequency ω of the oscillating motion is inversely related to its period T by

$$\omega = \frac{2\pi}{T} \qquad (10.4)$$

All the variables on the right side of this equation are known, and we substitute it into Equation 2, as indicated in the right column.

$$m_{astro} = \frac{k}{\omega^2} - m_{chair} \qquad (2)$$

$$\omega = \frac{2\pi}{T} \qquad (10.4)$$

Solution Algebraically combining the results of the steps above, we have:

STEP 1   STEP 2

$$m_{astro} = \frac{k}{\omega^2} - m_{chair} = \frac{k}{\left(\dfrac{2\pi}{T}\right)^2} - m_{chair} = \frac{kT^2}{4\pi^2} - m_{chair}$$

$$= \frac{(606 \text{ N/m})(2.41 \text{ s})^2}{4\pi^2} - 12.0 \text{ kg} = \boxed{77.2 \text{ kg}}$$

Related Homework: *Problems 23, 77*

Example 6 indicates that the mass of the vibrating object influences the frequency of simple harmonic motion. Electronic sensors are being developed that take advantage of this effect in detecting and measuring small amounts of chemicals. These sensors utilize tiny quartz crystals that vibrate when an electric current passes through them. If a crystal is coated with a substance that absorbs a particular chemical, then the mass of the coated crystal increases as the chemical is absorbed and, according to the relation $f = \frac{1}{2\pi}\sqrt{k/m}$ (Equations 10.6 and 10.11), the frequency of the simple harmonic motion decreases. The change in frequency is detected electronically, and the sensor is calibrated to give the mass of the absorbed chemical.

The physics of detecting and measuring small amounts of chemicals.

✓ **CHECK YOUR UNDERSTANDING**

(*The answers are given at the end of the book.*)

5. The drawing shows plots of the displacement x versus the time t for three objects undergoing simple harmonic motion. Which object—I, II, or III—has the greatest maximum velocity?

6. In Figure 10.13 the shadow moves in simple harmonic motion. Where on the path of the shadow is the acceleration equal to zero?

7. A particle is oscillating in simple harmonic motion. The time required for the particle to travel through one complete cycle is equal to the period of the motion, no matter what the amplitude is. But how can this be, since larger amplitudes mean that the particle travels farther?

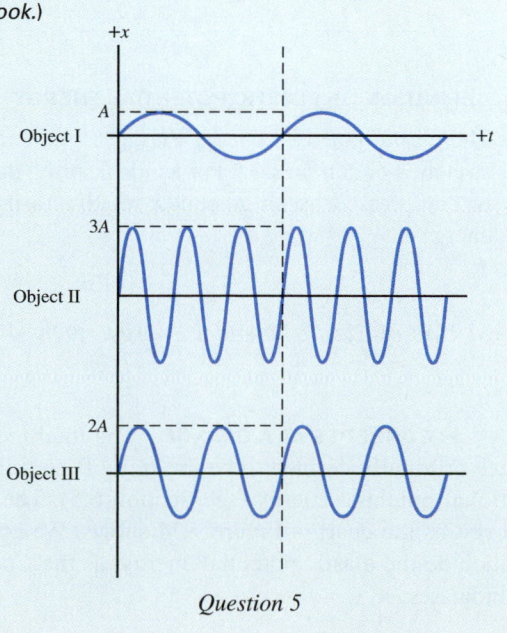

Question 5

10.3 ## ENERGY AND SIMPLE HARMONIC MOTION

We saw in Chapter 6 that an object above the surface of the earth has gravitational potential energy. Therefore, when the object is allowed to fall, like the hammer of the pile driver in Figure 6.12, it can do work. A spring also has potential energy when the spring is stretched or compressed, which we refer to as *elastic potential energy.* Because of elastic potential energy, a stretched or compressed spring can do work on an object that is attached to the spring. For instance, Figure 10.15 shows a door-closing unit that is often found on screen doors. When the door is opened, a spring inside the unit is compressed and has elastic potential energy. When the door is released, the compressed spring expands and does the work of closing the door.

To find an expression for the elastic potential energy, we will determine the work done by the spring force on an object. Figure 10.16 shows an object attached to one end of a stretched spring. When the object is released, the spring contracts and pulls the object from its initial position x_0 to its final position x_f. The work W done by a constant force is given by Equation 6.1 as $W = (F \cos \theta)s$, where F is the magnitude of the force, s is the magnitude of the displacement ($s = x_0 - x_f$), and θ is the angle between the force and the displacement. The magnitude of the spring force is not constant, however. Equation 10.2 gives the spring force as $F_x = -kx$, and as the spring contracts, the magnitude of this force changes from kx_0 to kx_f. In using Equation 6.1 to determine the work, we can account for the changing magnitude by using an average magnitude $\overline{F}_x$ in place of the constant magnitude F_x. Because the dependence of the spring force on x is linear, the magnitude of the average force is just one-half the sum of the initial and final values, or $\overline{F}_x = \frac{1}{2}(kx_0 + kx_f)$. The work $W_{elastic}$ done by the average spring force is, then,

$$W_{elastic} = (\overline{F}_x \cos \theta)s = \tfrac{1}{2}(kx_0 + kx_f)(\cos 0°)(x_0 - x_f)$$

$$W_{elastic} = \underbrace{\tfrac{1}{2}kx_0^2}_{\substack{\text{Initial elastic} \\ \text{potential energy}}} - \underbrace{\tfrac{1}{2}kx_f^2}_{\substack{\text{Final elastic} \\ \text{potential energy}}} \tag{10.12}$$

In the calculation above, θ is $0°$, since the spring force has the same direction as the displacement. Equation 10.12 indicates that the work done by the spring force is equal to the difference between the initial and final values of the quantity $\frac{1}{2}kx^2$. The quantity $\frac{1}{2}kx^2$ is analogous to the quantity mgh, which we identified in Section 6.3 as the gravitational potential energy. Here, we identify the quantity $\frac{1}{2}kx^2$ as the elastic potential energy. Equation 10.13 indicates that the elastic potential energy is a maximum for a fully stretched or compressed spring and zero for a spring that is neither stretched nor compressed ($x = 0$ m).

The physics of a door-closing unit.

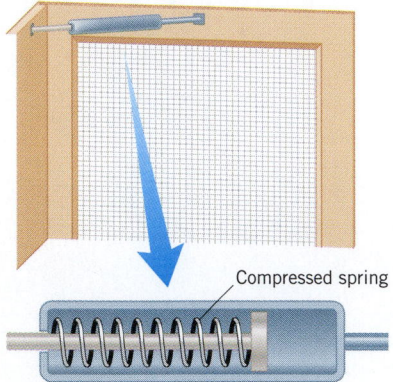

Figure 10.15 A door-closing unit. The elastic potential energy stored in the compressed spring is used to close the door.

Compressed spring

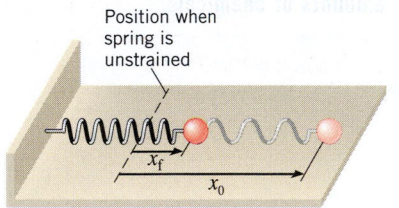

Position when spring is unstrained

x_f

x_0

Figure 10.16 When the object is released, its displacement changes from an initial value of x_0 to a final value of x_f.

DEFINITION OF ELASTIC POTENTIAL ENERGY

The elastic potential energy $PE_{elastic}$ is the energy that a spring has by virtue of being stretched or compressed. For an ideal spring that has a spring constant k and is stretched or compressed by an amount x relative to its unstrained length, the elastic potential energy is

$$PE_{elastic} = \tfrac{1}{2}kx^2 \tag{10.13}$$

SI Unit of Elastic Potential Energy: joule (J)

► **CONCEPTS AT A GLANCE** The total mechanical energy E is a familiar idea that we originally defined to be the sum of the translational kinetic energy and the gravitational potential energy (see Section 6.5). Then, we included the rotational kinetic energy, as the chart in Figure 9.24 shows. We now expand the total mechanical energy to include the elastic potential energy, as the Concepts-at-a-Glance chart in Figure 10.17 indicates. ◄

CONCEPTS AT A GLANCE

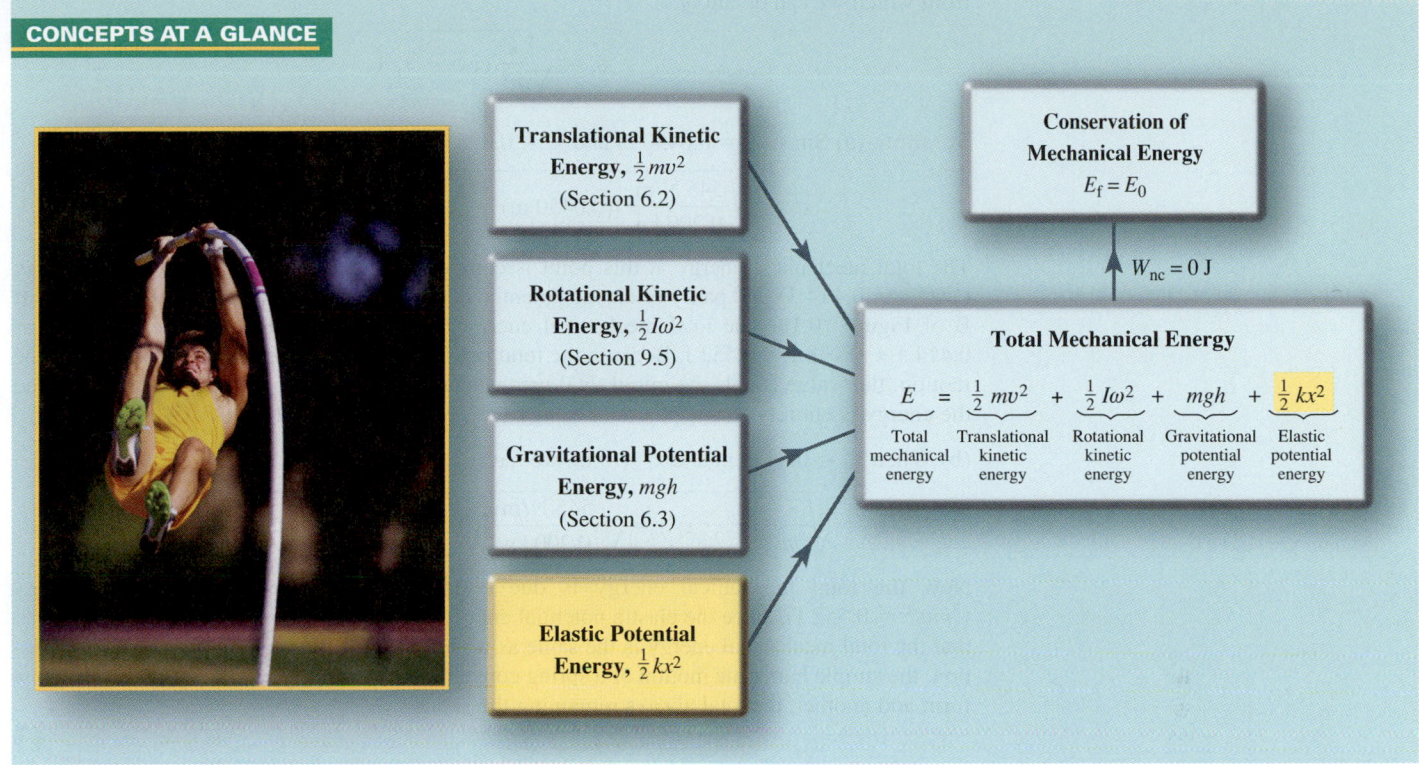

Figure 10.17 CONCEPTS AT A GLANCE The elastic potential energy is added to other energies to give the total mechanical energy, which is conserved if W_{nc}, the net work done by external nonconservative forces, is zero. As this pole-vaulter swings upward, the total mechanical energy remains constant, to the extent that the nonconservative forces, such as those due to friction and air resistance, are negligible. (Mike Powell/Getty Images)

In Equation 10.14 the elastic potential energy is included as part of the total mechanical energy:

$$\underbrace{E}_{\substack{\text{Total}\\\text{mechanical}\\\text{energy}}} = \underbrace{\tfrac{1}{2}mv^2}_{\substack{\text{Translational}\\\text{kinetic}\\\text{energy}}} + \underbrace{\tfrac{1}{2}I\omega^2}_{\substack{\text{Rotational}\\\text{kinetic}\\\text{energy}}} + \underbrace{mgh}_{\substack{\text{Gravitational}\\\text{potential}\\\text{energy}}} + \underbrace{\tfrac{1}{2}kx^2}_{\substack{\text{Elastic}\\\text{potential}\\\text{energy}}} \qquad (10.14)$$

As Section 6.5 discusses, the total mechanical energy is conserved when external nonconservative forces (such as friction) do no net work; that is, when $W_{nc} = 0$ J. Then, the final and initial values of E are the same: $E_f = E_0$. The principle of conservation of total mechanical energy is the subject of the next example.

⬇ Example 7   An Object on a Horizontal Spring

Figure 10.18 shows an object of mass $m = 0.200$ kg that is vibrating on a horizontal frictionless table. The spring has a spring constant of $k = 545$ N/m. The spring is stretched initially to $x_0 = 4.50$ cm and is then released from rest (see part A of the drawing). Determine the final translational speed v_f of the object when the final displacement of the spring is (a) $x_f = 2.25$ cm and (b) $x_f = 0$ cm.

Reasoning The conservation of mechanical energy indicates that, in the absence of friction (a nonconservative force), the final and initial total mechanical energies are the same (see Figure 10.17):

$$E_f = E_0$$

$$\tfrac{1}{2}mv_f^2 + \tfrac{1}{2}I\omega_f^2 + mgh_f + \tfrac{1}{2}kx_f^2 = \tfrac{1}{2}mv_0^2 + \tfrac{1}{2}I\omega_0^2 + mgh_0 + \tfrac{1}{2}kx_0^2$$

Since the object is moving on a horizontal table, the final and initial heights are the same: $h_f = h_0$. The object is not rotating, so its angular speed is zero: $\omega_f = \omega_0 = 0$ rad/s. Also, as the problem states, the initial translational speed of the object is zero, $v_0 = 0$ m/s. With these substitutions, the conservation-of-energy equation becomes

$$\tfrac{1}{2}mv_f^2 + \tfrac{1}{2}kx_f^2 = \tfrac{1}{2}kx_0^2$$

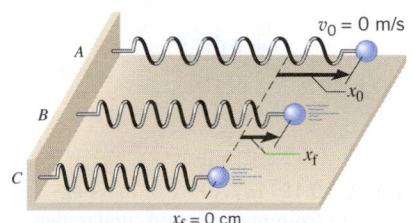

Figure 10.18 The total mechanical energy of this system is entirely elastic potential energy (A), partly elastic potential energy and partly kinetic energy (B), and entirely kinetic energy (C).

from which we can obtain v_f:

$$v_f = \sqrt{\frac{k}{m}(x_0^2 - x_f^2)}$$

Solution (a) Since $x_0 = 0.0450$ m and $x_f = 0.0225$ m, the final translational speed is

$$v_f = \sqrt{\frac{545 \text{ N/m}}{0.200 \text{ kg}}[(0.0450 \text{ m})^2 - (0.0225 \text{ m})^2]} = \boxed{2.03 \text{ m/s}}$$

The total mechanical energy at this point is composed partly of translational kinetic energy ($\frac{1}{2}mv_f^2 = 0.414$ J) and partly of elastic potential energy ($\frac{1}{2}kx_f^2 = 0.138$ J), as indicated in part B of Figure 10.18. The total mechanical energy E is the sum of these two energies: $E = 0.414$ J $+ 0.138$ J $= 0.552$ J. Because the total mechanical energy remains constant during the motion, this value equals the initial total mechanical energy when the object is stationary and the energy is entirely elastic potential energy ($E_0 = \frac{1}{2}kx_0^2 = 0.552$ J).

(b) When $x_0 = 0.0450$ m and $x_f = 0$ m, we have

$$v_f = \sqrt{\frac{k}{m}(x_0^2 - x_f^2)} = \sqrt{\frac{545 \text{ N/m}}{0.200 \text{ kg}}[(0.0450 \text{ m})^2 - (0 \text{ m})^2]} = \boxed{2.35 \text{ m/s}}$$

Now the total mechanical energy is due entirely to the translational kinetic energy ($\frac{1}{2}mv_f^2 = 0.552$ J), since the elastic potential energy is zero (see part C of Figure 10.18). Note that the total mechanical energy is the same as it is in Solution part (a). In the absence of friction, the simple harmonic motion of a spring converts the different types of energy between one form and another, the total always remaining the same.

Conceptual Example 8 takes advantage of energy conservation to illustrate what happens to the maximum speed, amplitude, and angular frequency of a simple harmonic oscillator when its mass is changed suddenly at a certain point in the motion.

Conceptual Example 8
Changing the Mass of a Simple Harmonic Oscillator

Figure 10.19*a* shows a box of mass m attached to a spring that has a force constant k. The box rests on a horizontal, frictionless surface. The spring is initially stretched to $x = A$ and then released from rest. The box executes simple harmonic motion that is characterized by a maximum speed v_x^{max}, an amplitude A, and an angular frequency ω. When the box is passing through the point where the spring is unstrained ($x = 0$ m), a second box of the same mass m and speed v_x^{max} is attached to it, as in part *b* of the drawing. Discuss what happens to **(a)** the maximum speed, **(b)** the amplitude, and **(c)** the angular frequency of the subsequent simple harmonic motion.

Reasoning and Solution **(a)** The maximum speed of an object in simple harmonic motion occurs when the object is passing through the point where the spring is unstrained ($x = 0$ m), as in Figure 10.19*b*. Since the second box is attached at this point with the same speed, *the maximum speed of the two-box system remains the same as that of the one-box system.*

(b) At the same speed, the maximum kinetic energy of the two boxes is twice that of a single box, since the mass is twice as much. Subsequently, when the two boxes move to the left and compress the spring, their kinetic energy is converted into elastic potential energy. Since the two boxes have twice as much kinetic energy as one box alone, the two will have twice as much elastic potential energy when they come to a halt at the extreme left. Here, we are using the principle of conservation of mechanical energy, which applies since friction is absent. But the elastic potential energy is proportional to the amplitude squared (A^2) of the motion, *so the amplitude of the two-box system is $\sqrt{2}$ times as great as that of the one-box system.*

(c) The angular frequency ω of a simple harmonic oscillator is $\omega = \sqrt{k/m}$ (Equation 10.11). Since the mass of the two-box system is twice the mass of the one-box system, *the angular frequency of the two-box system is $\sqrt{2}$ times as small as that of the one-box system.*

Related Homework: *Problem 81*

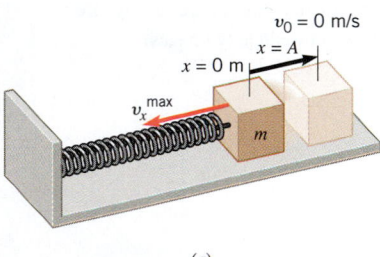

(a)

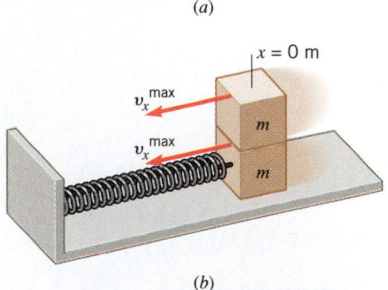

(b)

Figure 10.19 (*a*) A box of mass m, starting from rest at $x = A$, undergoes simple harmonic motion about $x = 0$ m. (*b*) When $x = 0$ m, a second box, with the same mass and speed, is attached.

In the previous two examples, gravitational potential energy plays no role because the spring is horizontal. The next example illustrates that gravitational potential energy must be taken into account when a spring is oriented vertically.

 Example 9 A Falling Ball on a Vertical Spring

A 0.20-kg ball is attached to a vertical spring, as in Figure 10.20. The spring constant of the spring is 28 N/m. The ball, supported initially so that the spring is neither stretched nor compressed, is released from rest. In the absence of air resistance, how far does the ball fall before being brought to a momentary stop by the spring?

Reasoning Since air resistance is absent, only the conservative forces of gravity and the spring act on the ball. Therefore, the principle of conservation of mechanical energy applies:

$$E_f = E_0$$

$$\tfrac{1}{2}mv_f^2 + \tfrac{1}{2}I\omega_f^2 + mgh_f + \tfrac{1}{2}ky_f^2 = \tfrac{1}{2}mv_0^2 + \tfrac{1}{2}I\omega_0^2 + mgh_0 + \tfrac{1}{2}ky_0^2$$

Note that we have replaced x with y in the elastic-potential-energy terms ($\tfrac{1}{2}ky_f^2$ and $\tfrac{1}{2}ky_0^2$), in recognition of the fact that the spring is moving in the vertical direction. The problem states that the final and initial translational speeds of the ball are zero: $v_f = v_0 = 0$ m/s. The ball and spring do not rotate; therefore, the final and initial angular speeds are also zero: $\omega_f = \omega_0 = 0$ rad/s. As Figure 10.20 indicates, the initial height of the ball is h_0, and the final height is $h_f = 0$ m. In addition, the spring is unstrained ($y_0 = 0$ m) to begin with, and so it has no elastic potential energy initially. With these substitutions, the conservation-of-mechanical-energy equation reduces to

$$\tfrac{1}{2}ky_f^2 = mgh_0$$

This result shows that the initial gravitational potential energy (mgh_0) is converted into elastic potential energy ($\tfrac{1}{2}ky_f^2$). When the ball falls to its lowest point, its displacement is $y_f = -h_0$, where the minus sign indicates that the displacement is downward. Substituting this result into the equation above and solving for h_0 yields $h_0 = 2mg/k$.

Solution The distance that the ball falls before coming to a momentary halt is

$$h_0 = \frac{2mg}{k} = \frac{2(0.20 \text{ kg})(9.8 \text{ m/s}^2)}{28 \text{ N/m}} = \boxed{0.14 \text{ m}}$$

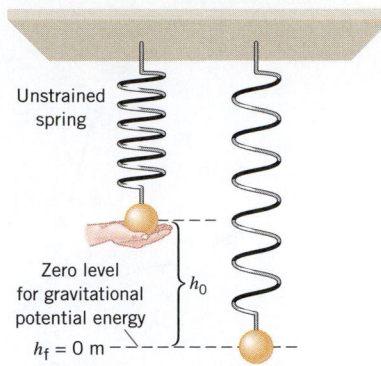

Figure 10.20 The ball is supported initially so that the spring is unstrained. After being released from rest, the ball falls through the distance h_0 before being momentarily stopped by the spring.

Problem-solving insight

When evaluating the total mechanical energy E, always include a potential energy term for every conservative force acting on the system. In Example 9 there are two such terms, gravitational and elastic.

✓ **CHECK YOUR UNDERSTANDING**

(*The answers are given at the end of the book.*)

8. Is more elastic potential energy stored in a spring when the spring is compressed by one centimeter than when it is stretched by the same amount?

9. A block is attached to the end of a horizontal ideal spring and rests on a frictionless surface. The block is pulled so that the spring stretches relative to its unstrained length. In each of the following three cases, the spring is stretched initially by the same amount. Rank the amplitudes of the resulting simple harmonic motion in decreasing order (largest first). **(a)** The block is released from rest. **(b)** The block is given an initial speed v_0. **(c)** The block is given an initial speed $\tfrac{1}{2}v_0$.

10. A block is attached to a horizontal spring and slides back and forth on a frictionless horizontal surface. A second identical block is suddenly attached to the first block. The attachment is accomplished by joining the blocks at one extreme end of the oscillation cycle. The velocities of the blocks are exactly matched at the instant of joining. How do the **(a)** amplitude, **(b)** frequency, and **(c)** maximum speed of the simple harmonic motion change?

10.4 THE PENDULUM

A **simple pendulum** consists of a particle of mass m, attached to a frictionless pivot P by a cable of length L and negligible mass. When the particle is pulled away from its

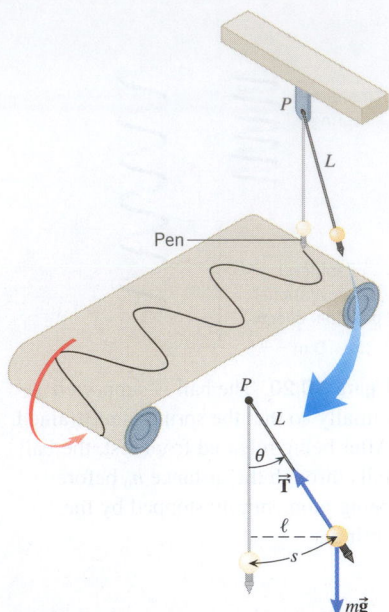

Figure 10.21 A simple pendulum swinging back and forth about the pivot P. If the angle θ is small (about 10° or less), the swinging is approximately simple harmonic motion.

equilibrium position by an angle θ and released, it swings back and forth as Figure 10.21 shows. By attaching a pen to the bottom of the swinging particle and moving a strip of paper beneath it at a steady rate, we can record the position of the particle as time passes. The graphical record reveals a pattern that is similar (but not identical) to the sinusoidal pattern for simple harmonic motion.

Gravity causes the back-and-forth rotation about the axis at P. The rotation speeds up as the particle approaches the lowest point and slows down on the upward part of the swing. Eventually the angular speed is reduced to zero, and the particle swings back. As Section 9.4 discusses, a net torque is required to change the angular speed. The gravitational force $m\vec{g}$ produces this torque. (The tension $\vec{T}$ in the cable creates no torque, because it points directly at the pivot P and, therefore, has a zero lever arm.) According to Equation 9.1, the magnitude of the torque τ is the product of the magnitude mg of the gravitational force and the lever arm ℓ, so that $\tau = -(mg)\ell$. The minus sign is included since the torque is a restoring torque; it acts to reduce the angle θ [the angle θ is positive (counterclockwise), while the torque is negative (clockwise)]. The lever arm ℓ is the perpendicular distance between the line of action of $m\vec{g}$ and the pivot P. In Figure 10.21, ℓ is very nearly equal to the arc length s of the circular path when the angle θ is small (about 10° or less). Furthermore, if θ is expressed in radians, the arc length and the radius L of the circular path are related, according to $s = L\theta$ (Equation 8.1). It follows, then, that $\ell \approx s = L\theta$, and the gravitational torque is

$$\tau \approx -\underbrace{mgL}_{k'}\,\theta$$

In the equation above, the term mgL has a constant value k', independent of θ. *For small angles*, then, the torque that restores the pendulum to its vertical equilibrium position is proportional to the angular displacement θ. The expression $\tau = -k'\theta$ has the same form as the Hooke's law restoring force for an ideal spring, $F = -kx$. Therefore, we expect the frequency of the back-and-forth movement of the pendulum to be given by an equation analogous to Equation 10.11 ($\omega = 2\pi f = \sqrt{k/m}$). In place of the spring constant k, the constant $k' = mgL$ will appear, and, as usual in rotational motion, in place of the mass m, the moment of inertia I will appear:

$$\omega = 2\pi f = \sqrt{\frac{mgL}{I}} \quad \text{(small angles only)} \quad (10.15)$$

The moment of inertia of a particle of mass m, rotating at a radius $r = L$ about an axis, is given by $I = mL^2$ (Equation 9.6). Substituting this expression for I into Equation 10.15 reveals that for a simple pendulum

Simple pendulum $\qquad \omega = 2\pi f = \sqrt{\dfrac{g}{L}} \quad \text{(small angles only)} \quad (10.16)$

The mass of the particle has been eliminated algebraically from this expression, so only the length L and the magnitude g of the acceleration due to gravity determine the frequency of a simple pendulum. If the angle of oscillation is large, the pendulum does not exhibit simple harmonic motion, and Equation 10.16 does not apply. Equation 10.16 provides the basis for using a pendulum to keep time, as Example 10 demonstrates.

 Example 10 Keeping Time

Figure 10.22 shows a clock that uses a pendulum to keep time. Determine the length of a simple pendulum that will swing back and forth in simple harmonic motion with a period of 1.00 s.

Reasoning When a simple pendulum is swinging back and forth in simple harmonic motion, its frequency f is given by Equation 10.16 as $f = \frac{1}{2\pi}\sqrt{g/L}$, where g is the magnitude of the acceleration due to gravity and L is the length of the pendulum. We also know from Equation 10.5 that the frequency is the reciprocal of the period T, so $f = 1/T$. Thus, the equation above becomes $1/T = \frac{1}{2\pi}\sqrt{g/L}$. We can solve this equation for the length L.

Solution The length of the pendulum is

Figure 10.22 This pendulum clock keeps time as the pendulum swings back and forth. (Richard Megna/Fundamental Photographs)

$$L = \frac{T^2 g}{4\pi^2} = \frac{(1.00\ \text{s})^2(9.80\ \text{m/s}^2)}{4\pi^2} = \boxed{0.248\ \text{m}}$$

It is not necessary that the object in Figure 10.21 be a particle at the end of a cable. It may be a rigid extended object, in which case the pendulum is called a *physical pendulum*. For small oscillations, Equation 10.15 still applies, but the moment of inertia I is no longer mL^2. The proper value for the rigid object must be used. (See Section 9.4 for a discussion of moment of inertia.) In addition, the length L for a physical pendulum is the distance between the axis at P and the center of gravity of the object. The next example deals with an important type of physical pendulum.

ANALYZING MULTIPLE-CONCEPT PROBLEMS

Example 11 Pendulum Motion and Walking

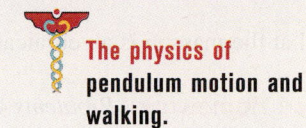

The physics of pendulum motion and walking.

When we walk, our legs alternately swing forward about the hip joint as a pivot. In this motion the leg is acting approximately as a physical pendulum. Treating the leg as a thin uniform rod of length 0.80 m, find the time it takes for the leg to swing forward.

Reasoning The time it takes for the leg to swing forward is one-half of the period T of the pendulum motion, which is related to the frequency f of the motion by $T = 1/f$ (Equation 10.5). The frequency of a physical pendulum is given by $f = \frac{1}{2\pi}\sqrt{mgL/I}$ (Equation 10.15), where m and I are, respectively, its mass and moment of inertia, and L is the distance between the pivot at the hip joint and the center of gravity of the leg. By combining these two relations we will be able to find the time it takes for the leg to swing forward.

Knowns and Unknowns The data for this problem are:

Description	Symbol	Value
Length of leg	L_{leg}	0.80 m
Unknown Variable		
Time for leg to swing forward	t	?

Modeling the Problem

STEP 1 **Period and Frequency** The time t it takes for the leg to swing forward is one-half the period T, or

$$t = \tfrac{1}{2}T$$

The period, in turn, is related to the frequency f by $T = 1/f$ (see Equation 10.5). Substituting this expression for T into the equation above for t gives Equation 1 at the right. The frequency of oscillation will be evaluated in Step 2.

$$t = \tfrac{1}{2}\left(\frac{1}{f}\right) \quad (1)$$

$\uparrow$

$?$

STEP 2 **The Frequency of a Physical Pendulum** According to Equation 10.15, the frequency of a physical pendulum is given by

$$f = \frac{1}{2\pi}\sqrt{\frac{mgL}{I}} \quad (10.15)$$

Since the leg is being approximated as a thin uniform rod, the distance L between the pivot and the center of gravity is one-half the length L_{leg} of the leg, so $L = \frac{1}{2}L_{\text{leg}}$. Using this relation in Equation 10.15 gives Equation 2 at the right, which we then substitute into Equation 1, as shown. The moment of inertia I will be evaluated in Step 3.

$$t = \tfrac{1}{2}\left(\frac{1}{f}\right) \quad (1)$$

$\uparrow$

$$f = \frac{1}{2\pi}\sqrt{\frac{mg\left(\tfrac{1}{2}L_{\text{leg}}\right)}{I}} \quad (2)$$

$\uparrow$

$?$

STEP 3 **Moment of Inertia** The leg is being approximated as a thin uniform rod of length L_{leg} that rotates about an axis that is perpendicular to one end. The moment of inertia I for such an object is given in Table 9.1 as

$$I = \tfrac{1}{3}mL_{\text{leg}}^2$$

We substitute this expression for I into Equation 2, as shown in the right column.

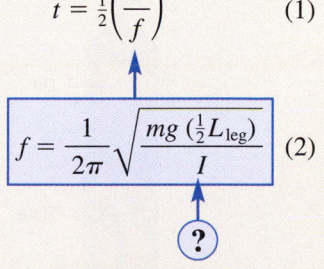

$$t = \tfrac{1}{2}\left(\frac{1}{f}\right) \quad (1)$$

$\uparrow$

$$f = \frac{1}{2\pi}\sqrt{\frac{mg\left(\tfrac{1}{2}L_{\text{leg}}\right)}{I}} \quad (2)$$

$\uparrow$

$$I = \tfrac{1}{3}mL_{\text{leg}}^2$$

Continued

Solution Algebraically combining the results of the steps above, we have:

STEP 1 | STEP 2 | STEP 3

$$t = \frac{1}{2}\left(\frac{1}{f}\right) = \frac{1}{2}\left[\frac{1}{\frac{1}{2\pi}\sqrt{\frac{mg\left(\frac{1}{2}L_{leg}\right)}{I}}}\right] = \frac{1}{2}\left[\frac{1}{\frac{1}{2\pi}\sqrt{\frac{mg\left(\frac{1}{2}L_{leg}\right)}{\frac{1}{3}mL_{leg}^2}}}\right]$$

$$= \pi\sqrt{\frac{2L_{leg}}{3g}} = \pi\sqrt{\frac{2(0.80\text{ m})}{3(9.80\text{ m/s}^2)}} = \boxed{0.73\text{ s}}$$

Note that the mass m is algebraically eliminated from the final result.

Related Homework: *Problems 48, 49*

✓ CHECK YOUR UNDERSTANDING

(The answers are given at the end of the book.)

11. Suppose that a grandfather clock (a simple pendulum) is running slowly; that is, the time it takes to complete each cycle is longer than it should be. Should you **(a)** shorten or **(b)** lengthen the pendulum to make the clock keep correct time?

12. Consult **Concept Simulation 10.2** at **www.wiley.com/college/cutnell** to review the concept that is important here. In principle, the motions of a simple pendulum and an object on an ideal spring can both be used to provide the basic time interval or period used in a clock. Which of the two kinds of clocks becomes more inaccurate when carried to the top of a high mountain?

13. Concept Simulation 10.2 at **www.wiley.com/college/cutnell** deals with the concept on which this question is based. Suppose you were kidnapped and held prisoner by space invaders in a completely isolated room, with nothing but a watch and a pair of shoes (including shoe laces of known length). How could you determine whether you were on earth or on the moon?

14. Two people are sitting on playground swings. One is pulled back 4° from the vertical and the other is pulled back 8°. They are both released at the same instant. Will they both come back to their starting points at the same time?

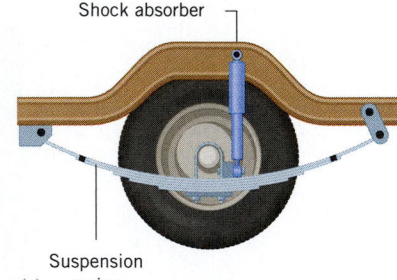

Shock absorber

Suspension
(a) spring

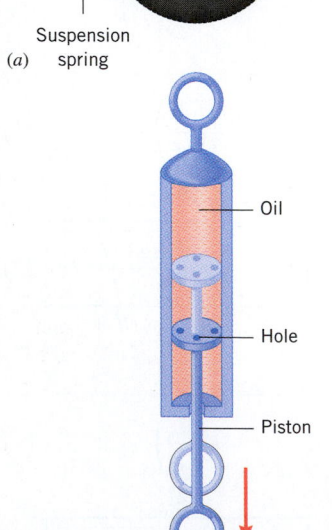

Oil

Hole

Piston

(b)

Figure 10.23 (a) A shock absorber mounted in the suspension system of an automobile and (b) a simplified, cutaway view of the shock absorber.

10.5 | DAMPED HARMONIC MOTION

In simple harmonic motion, an object oscillates with a constant amplitude, because there is no mechanism for dissipating energy. In reality, however, friction or some other energy-dissipating mechanism is always present. In the presence of energy dissipation, the oscillation amplitude decreases as time passes, and the motion is no longer simple harmonic motion. Instead, it is referred to as ***damped harmonic motion,*** the decrease in amplitude being called "damping."

The physics of a shock absorber. One widely used application of damped harmonic motion is in the suspension system of an automobile. Figure 10.23a shows a shock absorber attached to a main suspension spring of a car. A shock absorber is designed to introduce damping forces, which reduce the vibrations associated with a bumpy ride. As part *b* of the drawing shows, a shock absorber consists of a piston in a reservoir of oil. When the piston moves in response to a bump in the road, holes in the piston head permit the piston to pass through the oil. Viscous forces that arise during this movement cause the damping.

Figure 10.24 illustrates the different degrees of damping that can exist. As applied to the example of a car's suspension system, these graphs show the vertical position of the chassis after it has been pulled upward by an amount A_0 at time $t_0 = 0$ s and then released. Part *a* of the figure compares undamped or simple harmonic motion in curve 1 (red) to slightly damped motion in curve 2 (green). In damped harmonic motion, the chassis oscillates with decreasing amplitude and eventually comes to rest. As the degree of damping is increased from curve 2 to curve 3 (gold), the car makes fewer oscillations before coming

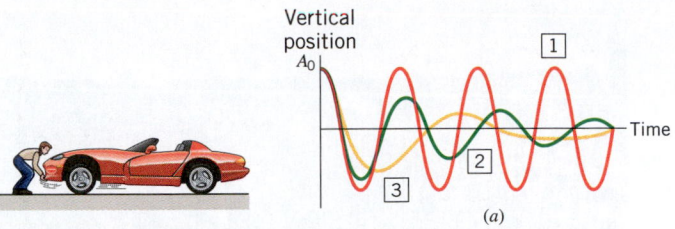

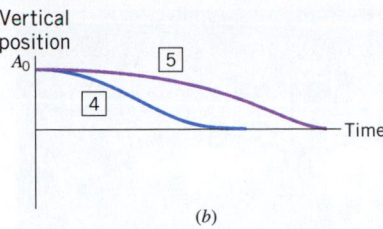

Figure 10.24 Damped harmonic motion. The degree of damping increases from curve 1 to curve 5. Curve 1 (red) represents undamped or simple harmonic motion. Curves 2 (green) and 3 (gold) show under-damped motion. Curve 4 (blue) represents critically damped harmonic motion. Curve 5 (purple) shows overdamped motion.

to a halt. Part *b* of the drawing shows that as the degree of damping is increased further, there comes a point when the car does not oscillate at all after it is released but, rather, settles directly back to its equilibrium position, as in curve 4 (blue). The smallest degree of damping that completely eliminates the oscillations is termed "critical damping," and the motion is said to be *critically damped.*

Figure 10.24*b* also shows that the car takes the longest time to return to its equilibrium position in curve 5 (purple), where the degree of damping is above the value for critical damping. When the damping exceeds the critical value, the motion is said to be *overdamped.* In contrast, when the damping is less than the critical level, the motion is said to be *underdamped* (curves 2 and 3). Typical automobile shock absorbers are designed to produce underdamped motion somewhat like that in curve 3.

✓ **CHECK YOUR UNDERSTANDING**

(The answer is given at the end of the book.)

15. The shock absorbers on a car are badly in need of replacement and introduce very little damping. Does the number of occupants in the car affect the vibration frequency of the car's suspension system?

10.6 **DRIVEN HARMONIC MOTION AND RESONANCE**

In damped harmonic motion, a mechanism such as friction dissipates or reduces the energy of an oscillating system, with the result that the amplitude of the motion decreases in time. This section discusses the opposite effect—namely, the increase in amplitude that results when energy is continually added to an oscillating system.

To set an object on an ideal spring into simple harmonic motion, some agent must apply a force that stretches or compresses the spring initially. Suppose that this force is applied at all times, not just for a brief initial moment. The force could be provided, for example, by a person who simply pushes and pulls the object back and forth. The resulting motion is known as **driven harmonic motion,** because the additional force drives or controls the behavior of the object to a large extent. The additional force is identified as the **driving force.**

Figure 10.25 illustrates one particularly important example of driven harmonic motion. Here, the driving force has the same frequency as the spring system and always points in the direction of the object's velocity. The frequency of the spring system is $f = (1/2\pi)\sqrt{k/m}$ and is called a **natural frequency** because it is the frequency at which the spring system naturally oscillates. Since the driving force and the velocity always have the same direction, positive work is done on the object at all times, and the total mechanical energy of the system increases. As a result, the amplitude of the vibration becomes larger and will increase without limit if there is no damping force to dissipate the energy being added by the driving force. The situation depicted in Figure 10.25 is known as **resonance.**

Figure 10.25 Resonance occurs when the frequency of the driving force (blue arrows) matches a frequency at which the object naturally vibrates. The red arrows represent the velocity of the object.

RESONANCE

Resonance is the condition in which a time-dependent force can transmit large amounts of energy to an oscillating object, leading to a large-amplitude motion. In the absence of damping, resonance occurs when the frequency of the force matches a natural frequency at which the object will oscillate.

(a)

(b)

Figure 10.26 The Bay of Fundy at (a) high tide and (b) low tide. In some places the water level changes by almost 15 m. (Francois Gohier/Photo Researchers, Inc.)

The physics of
high tides at the Bay of Fundy.

The role played by the frequency of a driving force is a critical one. The matching of this frequency with a natural frequency of vibration allows even a relatively weak force to produce a large-amplitude vibration, because the effect of each push–pull cycle is cumulative.

Resonance can occur with any object that can oscillate, and springs need not be involved. The greatest tides in the world occur in the Bay of Fundy, which lies between the Canadian provinces of New Brunswick and Nova Scotia. Figure 10.26 shows the enormous difference between the water level at high and low tides, a difference that in some locations averages about 15 m. This phenomenon is partly due to resonance. The time, or period, that it takes for the tide to flow into and ebb out of a bay depends on the size of the bay, the topology of the bottom, and the configuration of the shoreline. The ebb and flow of the water in the Bay of Fundy has a period of 12.5 hours, which is very close to the lunar tidal period of 12.42 hours. The tide then "drives" water into and out of the Bay of Fundy at a frequency (once per 12.42 hours) that nearly matches the natural frequency of the bay (once per 12.5 hours). The result is the extraordinary high tide in the bay. (You can create a similar effect in a bathtub full of water by moving back and forth in synchronism with the waves you're causing.)

✓ **CHECK YOUR UNDERSTANDING**

(*The answer is given at the end of the book.*)

16. A car travels at a constant speed over a road that contains a series of equally spaced bumps. The spacing between bumps is *d*. The mass of the car is *m*, and the spring constant of the car's suspension springs is *k*. Because of resonance, a particularly jarring ride results. Ignoring the effect of the car's shock absorbers, derive an expression for the car's speed *v* in terms of *d*, *m*, and *k*, as well as some numerical constants.

10.7 ELASTIC DEFORMATION

STRETCHING, COMPRESSION, AND YOUNG'S MODULUS

We have seen that a spring returns to its original shape when the force compressing or stretching it is removed. In fact, all materials become distorted in some way when they are squeezed or stretched, and many of them, such as rubber, return to their original shape when the squeezing or stretching is removed. Such materials are said to be "elastic." From an atomic viewpoint, elastic behavior has its origin in the forces that atoms exert on each other, and Figure 10.27 symbolizes these forces with the aid of springs. It is because of these atomic-level "springs" that a material tends to return to its initial shape once the forces that cause the deformation are removed.

The interatomic forces that hold the atoms of a solid together are particularly strong, so considerable force must be applied to stretch or compress a solid object. Experiments

Figure 10.27 The forces between atoms act like springs. The atoms are represented by red spheres, and the springs between some atoms have been omitted for clarity.

have shown that the magnitude of the force can be expressed by the following relation, provided that the amount of stretch or compression is small compared to the original length of the object:

$$F = Y\left(\frac{\Delta L}{L_0}\right)A \qquad (10.17)$$

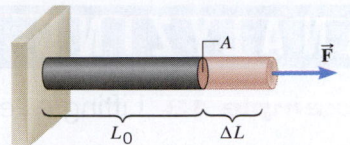

Figure 10.28 In this diagram, $\vec{F}$ denotes the stretching force, A the cross-sectional area, L_0 the original length of the rod, and ΔL the amount of stretch.

As Figure 10.28 shows, F denotes the magnitude of the stretching force applied perpendicularly to the surface at the end, A is the cross-sectional area of the rod, ΔL is the increase in length, and L_0 is the original length. An analogous picture applies in the case of a force that causes compression. The term Y is a proportionality constant called *Young's modulus*, after Thomas Young (1773–1829). Solving Equation 10.17 for Y shows that Young's modulus has units of force per unit area (N/m²). *It should be noted that the magnitude of the force in Equation 10.17 is proportional to the fractional increase (or decrease) in length* $\Delta L/L_0$, *rather than the absolute change* ΔL. The magnitude of the force is also proportional to the cross-sectional area A, which need not be circular, but can have any shape (e.g., rectangular).

 The physics of surgical implants. Table 10.1 reveals that the value of Young's modulus depends on the nature of the material. The values for metals are much larger than those for bone, for example. Equation 10.17 indicates that, for a given force, the material with the greater value of Y undergoes the smaller change in length. This difference between the changes in length is the reason why surgical implants (e.g., artificial hip joints), which are often made from stainless steel or titanium alloys, can lead to chronic deterioration of the bone that is in contact with the implanted prosthesis.

 The physics of bone structure. Forces that are applied as in Figure 10.28 and cause stretching are called "tensile" forces, because they create a tension in the material, much like the tension in a rope. Equation 10.17 also applies when the force compresses the material along its length. In this situation, the force is applied in a direction opposite to that shown in Figure 10.28, and ΔL stands for the amount by which the original length L_0 decreases. Table 10.1 indicates, for example, that bone has different values of Young's modulus for compression and tension, the value for tension being greater. Such differences are related to the structure of the material. The solid part of bone consists of collagen fibers (a protein material) distributed throughout hydroxyapatite (a mineral). The collagen acts like the steel rods in reinforced concrete and increases the value of Y for tension relative to the value of Y for compression.

Most solids have Young's moduli that are rather large, reflecting the fact that a large force is needed to change the length of a solid object by even a small amount, as Example 12 illustrates.

Table 10.1 Values for the Young's Modulus of Solid Materials

Material	Young's Modulus Y (N/m²)
Aluminum	6.9×10^{10}
Bone	
Compression	9.4×10^{9}
Tension	1.6×10^{10}
Brass	9.0×10^{10}
Brick	1.4×10^{10}
Copper	1.1×10^{11}
Mohair	2.9×10^{9}
Nylon	3.7×10^{9}
Pyrex glass	6.2×10^{10}
Steel	2.0×10^{11}
Teflon	3.7×10^{8}
Titanium	1.2×10^{11}
Tungsten	3.6×10^{11}

Example 12 Bone Compression

 The physics of bone compression. A circus performer supports the combined weight (1080 N) of a number of colleagues (see Figure 10.29). Each thighbone (femur) of this performer has a length of 0.55 m and an effective cross-sectional area of 7.7×10^{-4} m². Determine the amount by which each thighbone compresses under the extra weight.

Reasoning The additional weight supported by each thighbone is $F = \frac{1}{2}(1080 \text{ N}) = 540 \text{ N}$, and Table 10.1 indicates that Young's modulus for bone compression is 9.4×10^{9} N/m². Since the length and cross-sectional area of the thighbone are also known, we can use Equation 10.17 to find the amount by which the additional weight compresses the thighbone.

Solution The amount of compression ΔL of each thighbone is

$$\Delta L = \frac{FL_0}{YA} = \frac{(540 \text{ N})(0.55 \text{ m})}{(9.4 \times 10^{9} \text{ N/m}^2)(7.7 \times 10^{-4} \text{ m}^2)} = \boxed{4.1 \times 10^{-5} \text{ m}}$$

This is a very small change, the fractional decrease being $\Delta L/L_0 = 0.000\,075$.

Figure 10.29 The entire weight of the balanced group is supported by the legs of the performer who is lying on her back. (Thierry Orban/Corbis Sygma)

In any situation where a cable is used to apply a force to an object, the cable stretches, as Example 13 illustrates.

ANALYZING MULTIPLE-CONCEPT PROBLEMS

Example 13 Lifting a Jeep

A helicopter is using a steel cable to lift a 2100-kg jeep. The unstretched length of the cable is 16 m, and its radius is 5.0×10^{-3} m. By what amount does the cable stretch when the jeep is hoisted straight upward with an acceleration of $+1.5$ m/s²?

Reasoning Figure 10.30a shows the force $-\vec{F}$ that the jeep exerts on the lower end of the cable, thereby stretching it. We can find the amount ΔL that the cable stretches by using Equation 10.17, $\Delta L = FL_0/(YA)$, where F is the magnitude of the force, L_0 and A are the unstretched length and cross-sectional area of the cable, respectively, and Y is Young's modulus for steel. All the quantities, except F, in this equation are known or can be readily determined. We will employ Newton's second law to find F, since the upward acceleration and mass of the jeep are known.

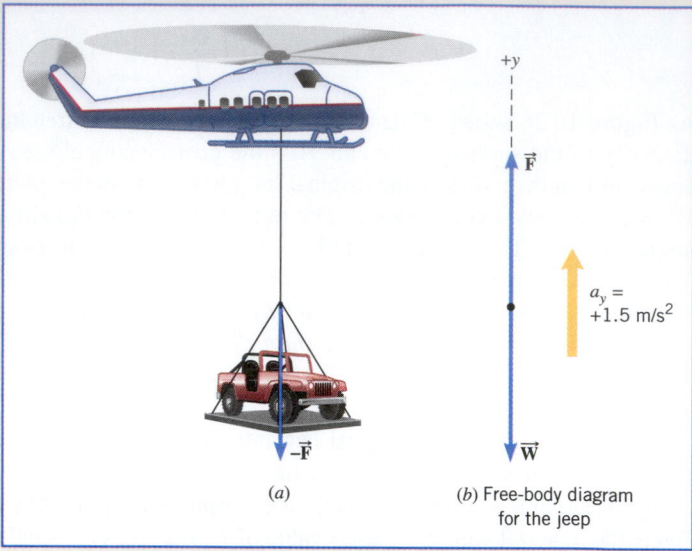

Figure 10.30 (a) The jeep applies a force $-\vec{F}$ to the lower end of the cable, thereby stretching it. (b) The free-body diagram for the jeep, showing the two forces that act on it.

Knowns and Unknowns The data for this problem are listed as follows:

Description	Symbol	Value
Mass of jeep	m	2100 kg
Acceleration of jeep	a_y	$+1.5$ m/s²
Unstretched length of cable	L_0	16 m
Radius of cable	r	5.0×10^{-3} m
Unknown Variable		
Amount that cable stretches	ΔL	?

Modeling the Problem

STEP 1 Elastic Deformation The amount ΔL that the cable stretches is related to the magnitude F of the stretching force by Equation 10.17 as

$$F = Y\left(\frac{\Delta L}{L_0}\right)A$$

Since the cross section of the cable is circular, its area is $A = \pi r^2$, where r is the radius. Solving Equation 10.17 for the change ΔL in the cable's length, and substituting πr^2 for the area, we have

$$\Delta L = \frac{FL_0}{YA} = \frac{FL_0}{Y(\pi r^2)}$$

$$\Delta L = \frac{FL_0}{Y(\pi r^2)} \qquad (1)$$

All the variables except F on the right side of this equation are known, and we will obtain an expression for it in Step 2.

STEP 2 Newton's Second Law To evaluate the magnitude F of the stretching force, we turn to Newton's second law. Figure 10.30a shows that the force exerted on the lower end of the cable by the jeep is $-\vec{F}$. According to Newton's third law, the action–reaction law, the force exerted on the jeep by the cable is $+\vec{F}$. Figure 10.30b shows the free-body diagram for the jeep and the two forces that act on the jeep: the force $+\vec{F}$ pulling it upward and its downward-acting weight $\vec{W} = m\vec{g}$ (see Equation 4.5). According to Newton's second law (Equation 4.2b), the net force ΣF_y acting on the jeep in the y direction is equal to its mass m times its acceleration a_y, or $\Sigma F_y = ma_y$. Taking "up" as the positive direction, we write the net force as $\Sigma F_y = +F - mg$. Thus,

$$\underbrace{+F - mg}_{\Sigma F_y} = ma_y$$

Solving this equation for F gives

$$\boxed{F = ma_y + mg = m(a_y + g)}$$

$$\Delta L = \frac{FL_0}{Y(\pi r^2)} \qquad (1)$$

$$\boxed{F = m(a_y + g)}$$

All the quantities on the right side of this equation are known, so we substitute this result into Equation 1, as indicated in the right column.

Solution Algebraically combining the results of the two steps, we have

STEP 1   **STEP 2**

$$\Delta L = \frac{FL_0}{Y(\pi r^2)} = \frac{m(a_y + g)L_0}{Y(\pi r^2)}$$

Noting that Young's modulus for steel is $Y = 2.0 \times 10^{11}$ N/m^2 (see Table 10.1), we find that the amount by which the cable stretches is

$$\Delta L = \frac{m(a_y + g)L_0}{Y(\pi r^2)} = \frac{(2100 \text{ kg})(1.5 \text{ m/s}^2 + 9.8 \text{ m/s}^2)(16 \text{ m})}{(2.0 \times 10^{11} \text{ N/m}^2)[\pi(5.0 \times 10^{-3} \text{ m})^2]} = \boxed{2.4 \times 10^{-2} \text{ m}}$$

Related Homework: *Problems 55, 71*

SHEAR DEFORMATION AND THE SHEAR MODULUS

It is possible to deform a solid object in a way other than by stretching or compressing it. For instance, place a book on a rough table and push on the top cover, as in Figure 10.31a. Notice that the top cover, and the pages below it, become shifted relative to the stationary bottom cover. The resulting deformation is called a *shear deformation* and occurs because of the combined effect of the force $\vec{F}$ applied (by the hand) to the top of the book and the force $-\vec{F}$ applied (by the table) to the bottom of the book. In general, shearing forces cause a solid object to change its shape. In Figure 10.31 the directions of the forces are parallel to the covers of the book, each of which has an area A, as illustrated in part b of the drawing. These two forces have equal magnitudes, but opposite directions, so the book remains in equilibrium. Equation 10.18 gives the magnitude F of the force needed to produce an amount of shear ΔX for an object with thickness L_0:

$$F = S\left(\frac{\Delta X}{L_0}\right)A \qquad (10.18)$$

This equation is very similar to Equation 10.17. The constant of proportionality S is called the *shear modulus* and, like Young's modulus, has units of force per unit area (N/m^2). The value of S depends on the nature of the material, and Table 10.2 gives some representative values. Example 14 illustrates how to determine the shear modulus of a favorite dessert.

Table 10.2 Values for the Shear Modulus of Solid Materials

Material	Shear Modulus S (N/m^2)
Aluminum	2.4×10^{10}
Bone	1.2×10^{10}
Brass	3.5×10^{10}
Copper	4.2×10^{10}
Lead	5.4×10^9
Nickel	7.3×10^{10}
Steel	8.1×10^{10}
Tungsten	1.5×10^{11}

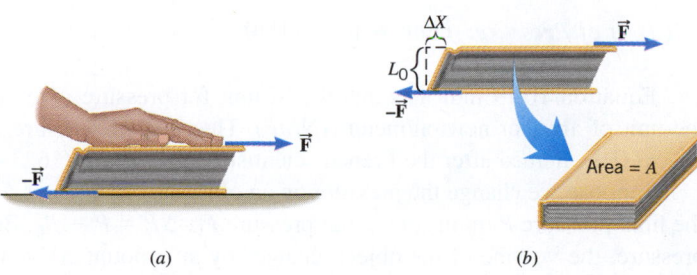

(a)   (b)

Figure 10.31 (*a*) An example of a shear deformation. The shearing forces $\vec{F}$ and $-\vec{F}$ are applied parallel to the top and bottom covers of the book. (*b*) The shear deformation is ΔX. The area of each cover is A, and the thickness of the book is L_0.

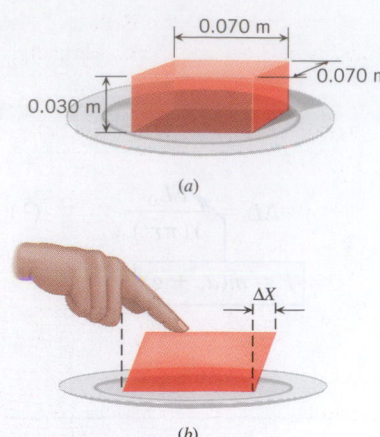

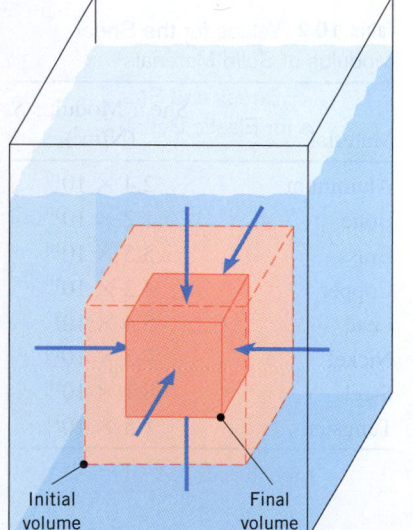

Figure 10.32 (*a*) A block of Jell-O and (*b*) a shearing force applied to it.

Example 14 J-E-L-L-O

A block of Jell-O is resting on a plate. Figure 10.32*a* gives the dimensions of the block. You are bored, impatiently waiting for dinner, and push tangentially across the top surface with a force of $F = 0.45$ N, as in part *b* of the drawing. The top surface moves a distance $\Delta X = 6.0 \times 10^{-3}$ m relative to the bottom surface. Use this idle gesture to measure the shear modulus of Jell-O.

Reasoning The finger applies a force that is parallel to the top surface of the Jell-O block. The shape of the block changes, because the top surface moves a distance ΔX relative to the bottom surface. The magnitude of the force required to produce this change in shape is given by Equation 10.18 as $F = S(\Delta X / L_0)A$. We know the values for all the variables in this relation except S, which, therefore, can be determined.

Solution Solving Equation 10.18 for the shear modulus S, we find that $S = FL_0/(A\,\Delta X)$, where $A = (0.070 \text{ m})(0.070 \text{ m})$ is the area of the top surface, and $L_0 = 0.030$ m is the thickness of the block:

$$S = \frac{FL_0}{A\,\Delta X} = \frac{(0.45 \text{ N})(0.030 \text{ m})}{(0.070 \text{ m})(0.070 \text{ m})(6.0 \times 10^{-3} \text{ m})} = \boxed{460 \text{ N/m}^2}$$

Jell-O can be deformed easily, so its shear modulus is significantly less than that of a more rigid material like steel (see Table 10.2).

Although Equations 10.17 and 10.18 are algebraically similar, they refer to different kinds of deformations. The tensile force in Figure 10.28 is perpendicular to the surface whose area is A, whereas the shearing force in Figure 10.31 is parallel to that surface. Furthermore, the ratio $\Delta L/L_0$ in Equation 10.17 is different from the ratio $\Delta X/L_0$ in Equation 10.18. The distances ΔL and L_0 are parallel, whereas ΔX and L_0 are perpendicular. Young's modulus refers to a *change in length* of one dimension of a solid object as a result of tensile or compressive forces. The shear modulus refers to a *change in shape* of a solid object as a result of shearing forces.

VOLUME DEFORMATION AND THE BULK MODULUS

When a compressive force is applied along one dimension of a solid, the length of that dimension decreases. It is also possible to apply compressive forces so that the size of every dimension (length, width, and depth) decreases, leading to a decrease in volume, as Figure 10.33 illustrates. This kind of overall compression occurs, for example, when an object is submerged in a liquid, and the liquid presses inward everywhere on the object. The forces acting in such situations are applied perpendicular to every surface, and it is more convenient to speak of the perpendicular force per unit area, rather than the amount of any one force in particular. The magnitude of the perpendicular force per unit area is called the *pressure P*.

DEFINITION OF PRESSURE

The pressure P is the magnitude F of the force acting perpendicular to a surface divided by the area A over which the force acts:

$$P = \frac{F}{A} \tag{10.19}$$

Pressure is a scalar, not a vector, quantity.

SI Unit of Pressure: $\text{N/m}^2 = \text{pascal (Pa)}$

Figure 10.33 The arrows denote the forces that push perpendicularly on every surface of an object immersed in a liquid. The magnitude of the force per unit area is the pressure. When the pressure increases, the volume of the object decreases.

Equation 10.19 indicates that the SI unit for pressure is the unit of force divided by the unit of area, or newton/meter^2 (N/m^2). This unit of pressure is often referred to as a *pascal* (Pa), named after the French scientist Blaise Pascal (1623–1662).

Suppose we change the pressure on an object by an amount ΔP, where ΔP represents the final pressure P minus the initial pressure P_0: $\Delta P = P - P_0$. Because of this change in pressure, the volume of the object changes by an amount $\Delta V = V - V_0$, where V and V_0

are the final and initial volumes, respectively. Such a pressure change occurs, for example, when a swimmer dives deeper into the water. Experiment reveals that the change ΔP in pressure needed to change the volume by an amount ΔV is directly proportional to the fractional change $\Delta V/V_0$ in the volume:

$$\Delta P = -B\left(\frac{\Delta V}{V_0}\right) \tag{10.20}$$

This relation is analogous to Equations 10.17 and 10.18, except that the area A in those equations does not appear here explicitly; the area is already taken into account by the concept of pressure (magnitude of the force per unit area). The proportionality constant B is known as the **bulk modulus**. The minus sign occurs because an increase in pressure (ΔP positive) always creates a decrease in volume (ΔV negative), and B is given as a positive quantity. Like Young's modulus and the shear modulus, the bulk modulus has units of force per unit area (N/m^2), and its value depends on the nature of the material. Table 10.3 gives representative values of the bulk modulus.

✓ **CHECK YOUR UNDERSTANDING**

(The answers are given at the end of the book.)

17. Young's modulus for steel is greater than that for a particular unknown material. What does this mean about how these materials compress when used in construction?   **(a)** Steel compresses much more easily than the unknown material does.   **(b)** The unknown material compresses more easily than steel does.   **(c)** Young's modulus has nothing to do with compression, so not enough information is given for an answer.

18. Two rods are made from the same material. One has a circular cross section, and the other has a square cross section. The circle just fits within the square. When the same force is applied to stretch these rods, they each stretch by the same amount. Which rod, if either, is longer?

19. A trash compactor crushes empty aluminum cans, thereby reducing the total volume, so that $\Delta V/V_0 = -0.75$ in Equation 10.20. Can the value given in Table 10.3 for the bulk modulus of aluminum be used to calculate the change ΔP in pressure generated in the trash compactor?

20. Both sides of the relation $F = S(\Delta X/L_0)A$ (Equation 10.18) can be divided by the area A to give F/A on the left side. Can this F/A term be called a pressure, such as the pressure that appears in $\Delta P = -B(\Delta V/V_0)$ (Equation 10.20)?

Table 10.3 Values for the Bulk Modulus of Solid and Liquid Materials

Material	Bulk Modulus B [N/m^2 (=Pa)]
Solids	
Aluminum	7.1×10^{10}
Brass	6.7×10^{10}
Copper	1.3×10^{11}
Diamond	4.43×10^{11}
Lead	4.2×10^{10}
Nylon	6.1×10^{9}
Osmium	4.62×10^{11}
Pyrex glass	2.6×10^{10}
Steel	1.4×10^{11}
Liquids	
Ethanol	8.9×10^{8}
Oil	1.7×10^{9}
Water	2.2×10^{9}

10.8 STRESS, STRAIN, AND HOOKE'S LAW

Equations 10.17, 10.18, and 10.20 specify the amount of force needed for a given amount of elastic deformation, and they are repeated in Table 10.4 to emphasize their common features. The left side of each equation is the magnitude of the force per unit area required to cause an elastic deformation. In general, the ratio of the magnitude of the force to the area is called the **stress**. The right side of each equation involves the change in a quantity (ΔL, ΔX, or ΔV) divided by a quantity (L_0 or V_0) relative to which the change is compared. The terms $\Delta L/L_0$, $\Delta X/L_0$, and $\Delta V/V_0$ are unitless ratios, and each is referred to as the **strain** that results from the applied stress. In the case of stretch and compression, the strain is the fractional change in length, whereas in volume deformation it is the fractional change in volume. In shear deformation the strain refers to a change in shape of the object. Experiments show that these three equations, with constant values for Young's modulus, the shear modulus, and the bulk modulus, apply to a wide range of materials. Therefore, stress and strain are directly proportional to one another, a relationship first discovered by Robert Hooke (1635–1703) and now referred to as **Hooke's law**.

Table 10.4 Stress and Strain Relations for Elastic Behavior

$\dfrac{F}{A}$	$= Y$	$\left(\dfrac{\Delta L}{L_0}\right)$	(10.17)
$\dfrac{F}{A}$	$= S$	$\left(\dfrac{\Delta X}{L_0}\right)$	(10.18)
ΔP	$= B$	$\left(\dfrac{-\Delta V}{V_0}\right)$	(10.20)

Stress proportional Strain
to

HOOKE'S LAW FOR STRESS AND STRAIN

Stress is directly proportional to strain.

SI Unit of Stress: newton per square meter (N/m^2) = pascal (Pa)

SI Unit of Strain: Strain is a unitless quantity.

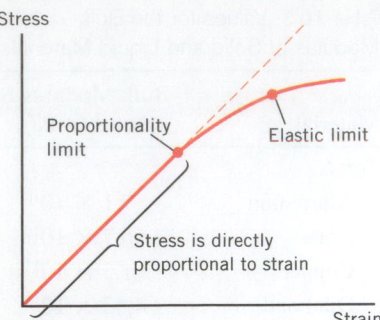

Figure 10.34 Hooke's law (stress is directly proportional to strain) is valid only up to the proportionality limit of a material. Beyond this limit, Hooke's law no longer applies. Beyond the elastic limit, the material remains deformed even when the stress is removed.

In reality, materials obey Hooke's law only up to a certain limit, as Figure 10.34 shows. As long as stress remains proportional to strain, a plot of stress versus strain is a straight line. The point on the graph where the material begins to deviate from straight-line behavior is called the "proportionality limit." Beyond the proportionality limit stress and strain are no longer directly proportional. However, if the stress does not exceed the "elastic limit" of the material, the object will return to its original size and shape once the stress is removed. The "elastic limit" is the point beyond which the object no longer returns to its original size and shape when the stress is removed; the object remains permanently deformed.

✓ **CHECK YOUR UNDERSTANDING**

(The answer is given at the end of the book.)

21. The block in the drawing rests on the ground. Which face—A, B, or C—experiences the largest stress and which face experiences the smallest stress when the block is resting on it?

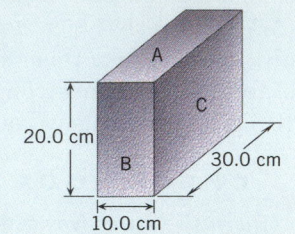

CONCEPTS & CALCULATIONS

10.9

This chapter has examined an important kind of vibratory motion known as simple harmonic motion. Specifically, it has discussed how the motion's displacement, velocity, and acceleration vary with time and explained what determines the frequency of the motion. In addition, we saw that the elastic force is conservative, so that the total mechanical energy is conserved if nonconservative forces, such as friction and air resistance, are absent. We conclude now by presenting some examples that review important features of simple harmonic motion.

Concepts & Calculations Example 15
A Diver Vibrating in Simple Harmonic Motion

A 75-kg diver is standing at the end of a diving board while it is vibrating up and down in simple harmonic motion, as indicated in Figure 10.35. The diving board has an effective spring constant of $k = 4100$ N/m, and the vertical distance between the highest and lowest points in the motion is 0.30 m. **(a)** What is the amplitude of the motion? **(b)** Starting when the diver is at the highest point, what is his speed one-quarter of a period later? **(c)** If the vertical distance between his highest and lowest points were doubled to 0.60 m, what would be the time required for the diver to make one complete motional cycle?

Concept Questions and Answers How is the amplitude A related to the vertical distance between the highest and lowest points of the diver's motion?

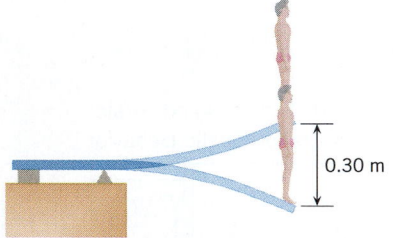

Figure 10.35 A diver at the end of a diving board is bouncing up and down in simple harmonic motion.

Answer The amplitude is the distance from the midpoint of the motion to either the highest or the lowest point. Thus, the amplitude is one-half the vertical distance between the highest and lowest points in the motion.

Starting from the top, where is the diver located one-quarter of a period later, and what can be said about his speed at this point?

Answer The time for the diver to complete one motional cycle is defined as the period. In one cycle, the diver moves downward from the highest point to the lowest point and then moves upward and returns to the highest point. In a time equal to one-quarter of a period, the diver completes one-quarter of this cycle and, therefore, is halfway between the highest and lowest points. His speed is momentarily zero at the highest and lowest points and is a maximum at the halfway point.

If the amplitude of the motion were to double, would the period also double?

Answer No. The period is the time to complete one cycle, and it is equal to the distance traveled during one cycle divided by the average speed. If the amplitude doubles, the distance also doubles. However, the average speed also doubles. We can verify this by examining

Equation 10.7, which gives the diver's velocity as $v_y = -A\omega \sin \omega t$. The speed is the magnitude of this value, or $A\omega \sin \omega t$. Since the speed is proportional to the amplitude A, the speed at every point in the cycle also doubles when the amplitude doubles. Thus, the average speed doubles. However, the period, being the distance divided by the average speed, does not change.

Solution (a) Since the amplitude A is one-half the vertical distance between the highest and lowest points in the motion, $A = \frac{1}{2}(0.30 \text{ m}) = \boxed{0.15 \text{ m}}$.

(b) When the diver is halfway between the highest and lowest points, his speed is a maximum. The maximum speed of an object vibrating in simple harmonic motion is given by Equation 10.8 as $v_{max} = A\omega$, where A is the amplitude of the motion and ω is the angular frequency. The angular frequency can be determined from Equation 10.11 as $\omega = \sqrt{k/m}$, where k is the effective spring constant of the diving board and m is the mass of the diver. The maximum speed is

$$v_{max} = A\omega = A\sqrt{\frac{k}{m}} = (0.15 \text{ m})\sqrt{\frac{4100 \text{ N/m}}{75 \text{ kg}}} = \boxed{1.1 \text{ m/s}}$$

(c) The period is the same, regardless of the amplitude of the motion. From Equation 10.4 we know that the period T and the angular speed ω are related by $T = 2\pi/\omega$, where $\omega = \sqrt{k/m}$. Thus, the period can be written as

$$T = \frac{2\pi}{\omega} = \frac{2\pi}{\sqrt{k/m}} = 2\pi\sqrt{\frac{m}{k}} = 2\pi\sqrt{\frac{75 \text{ kg}}{4100 \text{ N/m}}} = \boxed{0.85 \text{ s}}$$

Concepts & Calculations Example 16
Bungee Jumping and the Conservation of Mechanical Energy

A 68.0-kg bungee jumper is standing on a tall platform ($h_0 = 46.0$ m), as indicated in Figure 10.36. The bungee cord has an unstrained length of $L_0 = 9.00$ m and, when stretched, behaves like an ideal spring with a spring constant of $k = 66.0$ N/m. The jumper falls from rest, and it is assumed that the only forces acting on him are his weight and, for the latter part of the descent, the elastic force of the bungee cord. What is his speed (it is not zero) when he is at the following heights above the water (see the drawing): (a) $h_A = 37.0$ m and (b) $h_B = 15.0$ m?

The physics of bungee jumping.

Concept Questions and Answers Can we use the conservation of mechanical energy to find his speed at any point during the descent?

Answer Yes. His weight and the elastic force of the bungee cord are the only forces acting on him and are conservative forces. Therefore, the total mechanical energy remains constant (is conserved) during his descent.

What types of energy does he have when he is standing on the platform?

Answer Since he's at rest, he has neither translational nor rotational kinetic energy. The bungee cord is not stretched, so there is no elastic potential energy. Relative to the water, however, he does have gravitational potential energy, since he is 46.0 m above it.

What types of energy does he have at point A?

Answer Since he's moving downward, he possesses translational kinetic energy. He is not rotating, though, so his rotational kinetic energy is zero. Because the bungee cord is still not stretched at this point, there is no elastic potential energy. However, he still has gravitational potential energy relative to the water, because he is 37.0 m above it.

What types of energy does he have at point B?

Answer He has translational kinetic energy, because he's still moving. He's not rotating, so his rotational kinetic energy is zero. The bungee cord is stretched at this point, so there is elastic potential energy. He also has gravitational potential energy, because he is still 15.0 m above the water.

Solution (a) The total mechanical energy is the sum of the kinetic and potential energies, as expressed in Equation 10.14:

$$\underbrace{E}_{\substack{\text{Total} \\ \text{mechanical} \\ \text{energy}}} = \underbrace{\tfrac{1}{2}mv^2}_{\substack{\text{Translational} \\ \text{kinetic} \\ \text{energy}}} + \underbrace{\tfrac{1}{2}I\omega^2}_{\substack{\text{Rotational} \\ \text{kinetic} \\ \text{energy}}} + \underbrace{mgh}_{\substack{\text{Gravitational} \\ \text{potential} \\ \text{energy}}} + \underbrace{\tfrac{1}{2}ky^2}_{\substack{\text{Elastic} \\ \text{potential} \\ \text{energy}}}$$

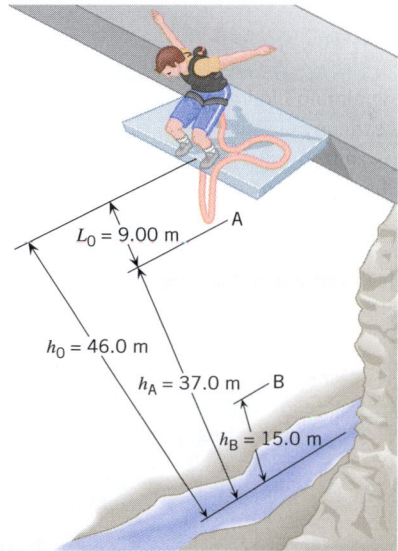

Figure 10.36 A bungee jumper jumps from a height of $h_0 = 46.0$ m. The length of the unstrained bungee cord is $L_0 = 9.00$ m.

As in Example 9, we have replaced x by y in the elastic potential energy $(\frac{1}{2}ky^2)$, since the bungee cord stretches in the vertical direction. The conservation of mechanical energy states that the total mechanical energy at point A is equal to that at the platform:

$$\underbrace{\tfrac{1}{2}mv_A^2 + \tfrac{1}{2}I\omega_A^2 + mgh_A + \tfrac{1}{2}ky_A^2}_{\text{Total mechanical energy at A}} = \underbrace{\tfrac{1}{2}mv_0^2 + \tfrac{1}{2}I\omega_0^2 + mgh_0 + \tfrac{1}{2}ky_0^2}_{\text{Total mechanical energy at the platform}}$$

While standing on the platform, the jumper is at rest, so $v_0 = 0$ m/s and $\omega_0 = 0$ rad/s. The bungee cord is not stretched, so $y_0 = 0$ m. At point A, the jumper is not rotating, $\omega_A = 0$ rad/s, and the bungee cord is still not stretched, $y_A = 0$ m. With these substitutions, the conservation of mechanical energy becomes

$$\tfrac{1}{2}mv_A^2 + mgh_A = mgh_0$$

Solving for the speed at point A yields

$$v_A = \sqrt{2g(h_0 - h_A)} = \sqrt{2(9.80 \text{ m/s}^2)(46.0 \text{ m} - 37.0 \text{ m})} = \boxed{13 \text{ m/s}}$$

(b) At point B the total mechanical energy is the same as it was at the platform, so

$$\underbrace{\tfrac{1}{2}mv_B^2 + \tfrac{1}{2}I\omega_B^2 + mgh_B + \tfrac{1}{2}ky_B^2}_{\text{Total mechanical energy at B}} = \underbrace{mgh_0}_{\substack{\text{Total mechanical} \\ \text{energy at} \\ \text{the platform}}}$$

We set $\omega_B = 0$ rad/s, since there is no rotational motion. Furthermore, the bungee cord stretches by an amount $y_B = h_0 - L_0 - h_B$ (see the drawing). Therefore, we have

$$v_B = \sqrt{2g(h_0 - h_B) - \left(\frac{k}{m}\right)(h_0 - L_0 - h_B)^2}$$

$$= \sqrt{2(9.80 \text{ m/s}^2)(46.0 \text{ m} - 15.0 \text{ m}) - \left(\frac{66.0 \text{ N/m}}{68.0 \text{ kg}}\right)(46.0 \text{ m} - 9.00 \text{ m} - 15.0 \text{ m})^2}$$

$$= \boxed{11.7 \text{ m/s}}$$

CONCEPT SUMMARY

If you need more help with a concept, use the Learning Aids noted next to the discussion or equation. Examples (**Ex.**) are in the text of this chapter. Go to **www.wiley.com/college/cutnell** for the following Learning Aids:

Interactive LearningWare (ILW) — Additional examples solved in a five-step interactive format.

Concept Simulations (CS) — Animated text figures or animations of important concepts.

Interactive Solutions (IS) — Models for certain types of problems in the chapter homework. The calculations are carried out interactively.

Topic	Discussion	Learning Aids
	10.1 THE IDEAL SPRING AND SIMPLE HARMONIC MOTION The force that must be applied to stretch or compress an ideal spring is	
Force applied to an ideal spring	$F_x^{\text{Applied}} = kx \qquad (10.1)$	**Ex. 1, 2** **IS 10.10**
	where k is the spring constant and x is the displacement of the spring from its unstrained length.	
	A spring exerts a restoring force on an object attached to the spring. The restoring force F_x produced by an ideal spring is	
Restoring force of an ideal spring	$F_x = -kx \qquad (10.2)$	
	where the minus sign indicates that the restoring force points opposite to the displacement of the spring.	
Simple harmonic motion	Simple harmonic motion is the oscillatory motion that occurs when a restoring force of the form $F_x = -kx$ acts on an object. A graphical record of position versus time for an object in simple har-	
Amplitude	monic motion is sinusoidal. The amplitude A of the motion is the maximum distance that the object moves away from its equilibrium position.	

Topic	Discussion	Learning Aids
	10.2 SIMPLE HARMONIC MOTION AND THE REFERENCE CIRCLE The period T of simple harmonic motion is the time required to complete one cycle of the motion, and the frequency f is the number of cycles per second that occurs. Frequency and period are related according to	
Period and frequency	$$f = \frac{1}{T} \qquad (10.5)$$	
	The frequency f (in Hz) is related to the angular frequency ω (in rad/s) according to	
Angular frequency	$$\omega = 2\pi f \qquad (\omega \text{ in rad/s}) \qquad (10.6)$$	
	The maximum speed of an object in simple harmonic motion is	
Maximum speed	$$v_{\text{max}} = A\omega \qquad (\omega \text{ in rad/s}) \qquad (10.8)$$	**Ex. 3, 4**
	where A is the amplitude of the motion.	
	The maximum acceleration of an object in simple harmonic motion is	**Ex. 5**
Maximum acceleration	$$a_{\text{max}} = A\omega^2 \qquad (\omega \text{ in rad/s}) \qquad (10.10)$$	**IS 10.83**
	The angular frequency of simple harmonic motion is	**Ex. 6, 15**
Angular frequency of simple harmonic motion	$$\omega = \sqrt{\frac{k}{m}} \qquad (\omega \text{ in rad/s}) \qquad (10.11)$$	**ILW 10.1**
	10.3 ENERGY AND SIMPLE HARMONIC MOTION The elastic potential energy of an object attached to an ideal spring is	
Elastic potential energy	$$PE_{\text{elastic}} = \tfrac{1}{2}kx^2 \qquad (10.13)$$	
	The total mechanical energy E of such a system is the sum of its translational and rotational kinetic energies, gravitational potential energy, and elastic potential energy:	**Ex. 7, 8, 9, 16**
Total mechanical energy	$$E = \tfrac{1}{2}mv^2 + \tfrac{1}{2}I\omega^2 + mgh + \tfrac{1}{2}kx^2 \qquad (10.14)$$	**ILW 10.2**
	If external nonconservative forces like friction do no net work, the total mechanical energy of the system is conserved:	**CS 10.1**
Conservation of mechanical energy	$$E_{\text{f}} = E_0$$	**IS 10.31, 10.37, 10.85**
	10.4 THE PENDULUM A simple pendulum is a particle of mass m attached to a frictionless pivot by a cable whose length is L and whose mass is negligible. The small-angle ($\leq 10°$) back-and-forth swinging of a simple pendulum is simple harmonic motion, but large-angle movement is not. The frequency f of the motion is given by	**CS 10.2**
Frequency of a simple pendulum	$$2\pi f = \sqrt{\frac{g}{L}} \qquad (\text{small angles only}) \qquad (10.16)$$	**Ex. 10**
	A physical pendulum consists of a rigid object, with moment of inertia I and mass m, suspended from a frictionless pivot. For small-angle displacements, the frequency f of simple harmonic motion for a physical pendulum is given by	
Frequency of a physical pendulum	$$2\pi f = \sqrt{\frac{mgL}{I}} \qquad (\text{small angles only}) \qquad (10.15)$$	**Ex. 11**
	where L is the distance between the axis of rotation and the center of gravity of the rigid object.	
Damped harmonic motion Critical damping	**10.5 DAMPED HARMONIC MOTION** Damped harmonic motion is motion in which the amplitude of oscillation decreases as time passes. Critical damping is the minimum degree of damping that eliminates any oscillations in the motion as the object returns to its equilibrium position.	**CS 10.3**
Driven harmonic motion Resonance	**10.6 DRIVEN HARMONIC MOTION AND RESONANCE** Driven harmonic motion occurs when a driving force acts on an object along with the restoring force. Resonance is the condition under which the driving force can transmit large amounts of energy to an oscillating object, leading to large-amplitude motion. In the absence of damping, resonance occurs when the frequency of the driving force matches a natural frequency at which the object oscillates.	
	10.7 ELASTIC DEFORMATION One type of elastic deformation is stretch and compression. The magnitude F of the force required to stretch or compress an object of length L_0 and cross-sectional area A by an amount ΔL is (see Figure 10.28)	
	$$F = Y\left(\frac{\Delta L}{L_0}\right)A \qquad (10.17)$$	**Ex. 12, 13**
Young's modulus	where Y is a constant called Young's modulus.	

Topic	Discussion	Learning Aids
	Another type of elastic deformation is shear. The magnitude F of the shearing force required to create an amount of shear ΔX for an object of thickness L_0 and cross-sectional area A is (see Figure 10.31)	

$$F = S\left(\frac{\Delta X}{L_0}\right)A \qquad (10.18) \quad \textbf{Ex. 14}$$

| Shear modulus | where S is a constant called the shear modulus. | |
| | A third type of elastic deformation is volume deformation, which has to do with pressure. The pressure P is the magnitude F of the force acting perpendicular to a surface divided by the area A over which the force acts: | |

| Pressure | | |

$$P = \frac{F}{A} \qquad (10.19)$$

| | The SI unit for pressure is N/m², a unit known as a pascal (Pa): 1 Pa = 1 N/m². | |
| | The change ΔP in pressure needed to change the volume V_0 of an object by an amount ΔV is (see Figure 10.33) | |

$$\Delta P = -B\left(\frac{\Delta V}{V_0}\right) \qquad (10.20)$$

Bulk modulus	where B is a constant known as the bulk modulus.	
Stress and strain	**10.8 STRESS, STRAIN, AND HOOKE'S LAW** Stress is the magnitude of the force per unit area applied to an object and causes strain. For stretch/compression, the strain is the fractional change $\Delta L/L_0$ in length. For shear, the strain reflects the change in shape of the object and is given by $\Delta X/L_0$ (see Figure 10.31). For volume deformation, the strain is the fractional change in volume $\Delta V/V_0$. Hooke's law states that stress is directly proportional to strain.	
Hooke's law		

FOCUS ON CONCEPTS

Note to Instructors: The numbering of the questions shown here reflects the fact that they are only a representative subset of the total number that are available online. However, all of the questions are available for assignment via an online homework management program such as WileyPLUS or WebAssign.

Section 10.1 The Ideal Spring and Simple Harmonic Motion

2. Which one of the following graphs correctly represents the restoring force F of an ideal spring as a function of the displacement x of the spring from its unstrained length?

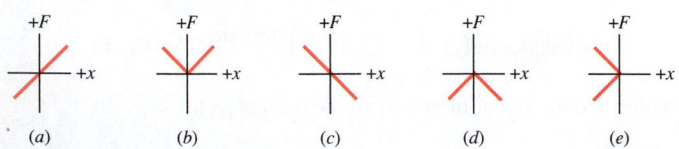

Section 10.2 Simple Harmonic Motion and the Reference Circle

3. You have two springs. One has a greater spring constant than the other. You also have two objects, one with a greater mass than the other. Which object should be attached to which spring, so that the resulting spring–object system has the greatest possible period of oscillation? **(a)** The object with the greater mass should be attached to the spring with the greater spring constant. **(b)** The object with the greater mass should be attached to the spring with the smaller spring constant. **(c)** The object with the smaller mass should be attached to the spring with the smaller spring constant. **(d)** The object with the smaller mass should be attached to the spring with the greater spring constant.

4. An object is oscillating in simple harmonic motion with an amplitude A and an angular frequency ω. What should you do to increase the maximum speed of the motion? **(a)** Reduce both A and ω by 10%. **(b)** Increase A by 10% and reduce ω by 10%. **(c)** Reduce A by 10% and increase ω by 10%. **(d)** Increase both A and ω by 10%.

Section 10.3 Energy and Simple Harmonic Motion

11. The kinetic energy of an object attached to a horizontal ideal spring is denoted by KE and the elastic potential energy by PE. For the simple harmonic motion of this object the maximum kinetic energy and the maximum elastic potential energy during an oscillation cycle are KE_{max} and PE_{max}, respectively. In the absence of friction, air resistance, and any other nonconservative forces, which of the following equations applies to the object–spring system?

A. KE + PE = constant

B. KE_{max} = PE_{max}

(a) A, but not B   **(b)** B, but not A   **(c)** A and B   **(d)** Neither A nor B

13. A block is attached to a horizontal spring. On top of this block rests another block. The two-block system slides back and forth in simple harmonic motion on a frictionless horizontal surface. At one extreme end of the oscillation cycle, where the blocks come to a momentary halt before reversing the direction of their motion, the top block is suddenly lifted vertically upward, without changing the zero velocity of the bottom block. The simple harmonic motion then continues. What happens to the amplitude and the angular frequency of the ensuing motion? **(a)** The amplitude remains the same, and the angular frequency increases. **(b)** The amplitude increases, and the angular frequency remains the same. **(c)** Both the amplitude and the angular frequency remain the same. **(d)** Both the amplitude and the angular frequency decrease.   **(e)** Both the amplitude and the angular frequency increase.

Section 10.4   The Pendulum

14. Five simple pendulums are shown in the drawings. The lengths of the pendulums are drawn to scale, and the masses are either m or $2m$, as shown. Which pendulum has the smallest angular frequency of oscillation?   **(a)** A   **(b)** B   **(c)** C   **(d)** D   **(e)** E

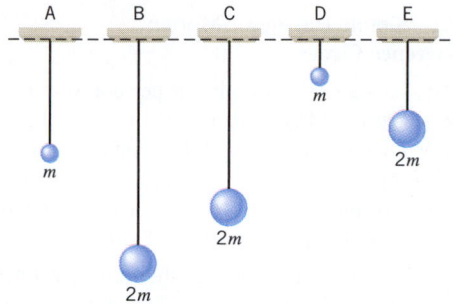

Section 10.5   Damped Harmonic Motion

16. An object on a spring is oscillating in simple harmonic motion. Suddenly, friction appears and causes the energy of the system to be dissipated. The system now exhibits _____.   **(a)** driven harmonic motion   **(b)** Hooke's-law type of motion   **(c)** damped harmonic motion

Section 10.6   Driven Harmonic Motion and Resonance

17. An external force (in addition to the spring force) is continually applied to an object of mass m attached to a spring that has a spring constant k. The frequency of this external force is such that resonance occurs. Then the frequency of this external force is doubled, and the force is applied to one of the spring systems shown in the drawing.

With which system would resonance occur?   **(a)** A   **(b)** B   **(c)** C   **(d)** D   **(e)** E

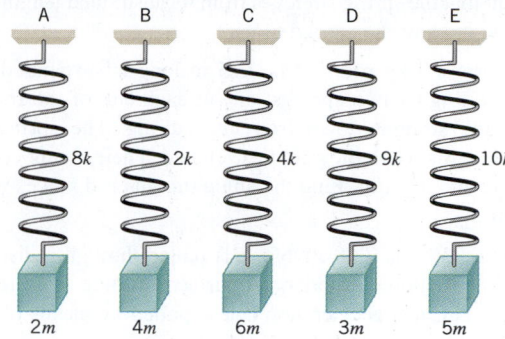

Section 10.7   Elastic Deformation

18. Drawings A and B show two cylinders that are identical in all respects, except that one is hollow. Identical forces are applied to each cylinder in order to stretch them. Which cylinder, if either, stretches more?   **(a)** A and B both stretch by the same amount.   **(b)** A stretches more than B.   **(c)** B stretches more than A.   **(d)** Insufficient information is given for an answer.

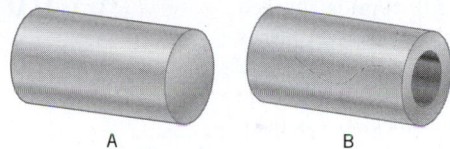

Section 10.8   Stress, Strain, and Hooke's Law

20. A material has a shear modulus of 5.0×10^9 N/m^2. A shear stress of 8.5×10^6 N/m^2 is applied to a piece of the material. What is the resulting shear strain.

PROBLEMS

Note to Instructors: Most of the homework problems in this chapter are available for assignment via an online homework management program such as WileyPLUS *or* WebAssign, *and those marked with the icon* **GO** *are presented in* WileyPLUS *using a guided tutorial format that provides enhanced interactivity. See Preface for additional details.*

Note: Unless otherwise indicated, the values for Young's modulus Y, the shear modulus S, and the bulk modulus B are given, respectively, in Table 10.1, Table 10.2, and Table 10.3.

ssm   Solution is in the Student Solutions Manual.
www   Solution is available online at www.wiley.com/college/cutnell

 This icon represents a biomedical application.

Section 10.1   The Ideal Spring and Simple Harmonic Motion

1. **ssm** A hand exerciser utilizes a coiled spring. A force of 89.0 N is required to compress the spring by 0.0191 m. Determine the force needed to compress the spring by 0.0508 m.

2. The graph shows the force F_x that an archer applies to the string of a long bow versus the string's displacement x. Drawing back this bow is analogous to stretching a spring. From the data in the graph determine the effective spring constant of the bow.

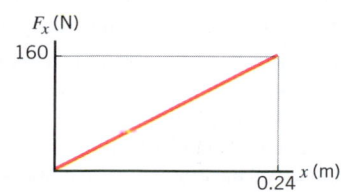

3. In a room that is 2.44 m high, a spring (unstrained length $= 0.30$ m) hangs from the ceiling. A board whose length is 1.98 m is attached to the free end of the spring. The board hangs straight down, so that its 1.98-m length is perpendicular to the floor. The weight of the board (104 N) stretches the spring so that the lower end of the board just

extends to, but does not touch, the floor. What is the spring constant of the spring?

4. A person who weighs 670 N steps onto a spring scale in the bathroom, and the spring compresses by 0.79 cm.   **(a)** What is the spring constant?   **(b)** What is the weight of another person who compresses the spring by 0.34 cm?

5. **ssm** A car is hauling a 92-kg trailer, to which it is connected by a spring. The spring constant is 2300 N/m. The car accelerates with an acceleration of 0.30 m/s^2. By how much does the spring stretch?

6. **GO** A spring lies on a horizontal table, and the left end of the spring is attached to a wall. The other end is connected to a box. The box is pulled to the right, stretching the spring. Static friction exists between the box and the table, so when the spring is stretched only by a small amount and the box is released, the box does not move. The mass of the box is 0.80 kg, and the spring has a spring constant of 59 N/m. The coefficient of static friction between the box and the table on which it rests is $\mu_s = 0.74$. How far can the spring be stretched from its unstrained position without the box moving when it is released?

7. A 0.70-kg block is hung from and stretches a spring that is attached to the ceiling. A second block is attached to the first one, and the amount that the spring stretches from its unstrained length triples. What is the mass of the second block?

*8. A uniform 1.4-kg rod that is 0.75 m long is suspended at rest from the ceiling by two springs, one at each end of the rod. Both springs hang straight down from the ceiling. The springs have identical lengths when they are unstretched. Their spring constants are 59 N/m and 33 N/m. Find the angle that the rod makes with the horizontal.

*9. **ssm** In 0.750 s, a 7.00-kg block is pulled through a distance of 4.00 m on a frictionless horizontal surface, starting from rest. The block has a constant acceleration and is pulled by means of a horizontal spring that is attached to the block. The spring constant of the spring is 415 N/m. By how much does the spring stretch?

*10. **Interactive Solution 10.10** at **www.wiley.com/college/cutnell** discusses a method used to solve this problem. To measure the static friction coefficient between a 1.6-kg block and a vertical wall, the setup shown in the drawing is used. A spring (spring constant = 510 N/m) is attached to the block. Someone pushes on the end of the spring in a direction perpendicular to the wall until the block does not slip downward. The spring is compressed by 0.039 m. What is the coefficient of static friction?

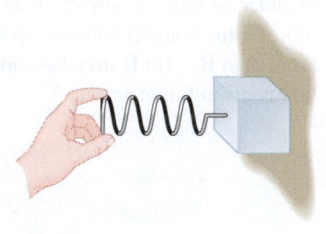

*11. **ssm** A small ball is attached to one end of a spring that has an unstrained length of 0.200 m. The spring is held by the other end, and the ball is whirled around in a horizontal circle at a speed of 3.00 m/s. The spring remains nearly parallel to the ground during the motion and is observed to stretch by 0.010 m. By how much would the spring stretch if it were attached to the ceiling and the ball allowed to hang straight down, motionless?

*12. Review Conceptual Example 2 as an aid in solving this problem. An object is attached to the lower end of a 100-coil spring that is hanging from the ceiling. The spring stretches by 0.160 m. The spring is then cut into two identical springs of 50 coils each. As the drawing shows, each spring is attached between the ceiling and the object. By how much does each spring stretch?

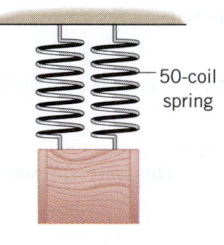

50-coil spring

**13. A 30.0-kg block is resting on a flat horizontal table. On top of this block is resting a 15.0-kg block, to which a horizontal spring is attached, as the drawing illustrates. The spring constant of the spring is 325 N/m. The coefficient of kinetic friction between the lower block and the table is 0.600, and the coefficient of static friction between the two blocks is 0.900. A horizontal force $\vec{F}$ is applied to the lower block as shown. This force is increasing in such a way as to keep the blocks moving at a *constant speed*. At the point where the upper block begins to slip on the lower block, determine (a) the amount by which the spring is compressed and (b) the magnitude of the force $\vec{F}$.

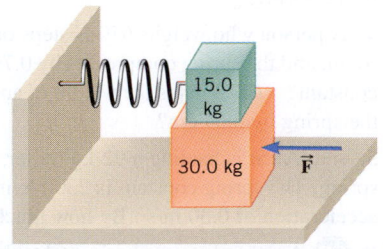

**14. A 15.0-kg block rests on a horizontal table and is attached to one end of a massless, horizontal spring. By pulling horizontally on the other end of the spring, someone causes the block to accelerate uniformly and reach a speed of 5.00 m/s in 0.500 s. In the process, the spring is stretched by 0.200 m. The block is then pulled at a *constant speed* of 5.00 m/s, during which time the spring is stretched by only 0.0500 m. Find (a) the spring constant of the spring and (b) the coefficient of kinetic friction between the block and the table.

Section 10.2 Simple Harmonic Motion and the Reference Circle

15. ssm The shock absorbers in the suspension system of a car are in such bad shape that they have no effect on the behavior of the springs attached to the axles. Each of the identical springs attached to the front axle supports 320 kg. A person pushes down on the middle of the front end of the car and notices that it vibrates through five cycles in 3.0 s. Find the spring constant of either spring.

16. When responding to sound, the human eardrum vibrates about its equilibrium position. Suppose an eardrum is vibrating with an amplitude of 6.3×10^{-7} m and a maximum speed of 2.9×10^{-3} m/s. (a) What is the frequency (in Hz) of the eardrum's vibration? (b) What is the maximum acceleration of the eardrum?

17. In **Concept Simulation 10.3** at **www.wiley.com/college/cutnell** you can explore the concepts in this problem. A block of mass $m = 0.750$ kg is fastened to an unstrained horizontal spring whose spring constant is $k = 82.0$ N/m. The block is given a displacement of $+0.120$ m, where the $+$ sign indicates that the displacement is along the $+x$ axis, and then released from rest. (a) What is the force (magnitude and direction) that the spring exerts on the block just before the block is released? (b) Find the angular frequency ω of the resulting oscillatory motion. (c) What is the maximum speed of the block? (d) Determine the magnitude of the maximum acceleration of the block.

18. Concept Simulation 10.3 at **www.wiley.com/college/cutnell** illustrates the concepts pertinent to this problem. An 0.80-kg object is attached to one end of a spring, as in Figure 10.5, and the system is set into simple harmonic motion. The displacement x of the object as a function of time is shown in the drawing. With the aid of these data, determine (a) the amplitude A of the motion, (b) the angular frequency ω, (c) the spring constant k, (d) the speed of the object at $t = 1.0$ s, and (e) the magnitude of the object's acceleration at $t = 1.0$ s.

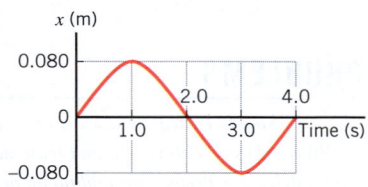

19. A person bounces up and down on a trampoline, while always staying in contact with it. The motion is simple harmonic motion, and it takes 1.90 s to complete one cycle. The height of each bounce above the equilibrium position is 45.0 cm. Determine (a) the amplitude and (b) the angular frequency of the motion. (c) What is the maximum speed attained by the person?

*20. **GO** Objects of equal mass are oscillating up and down in simple harmonic motion on two different vertical springs. The spring constant of spring 1 is 174 N/m. The motion of the object on spring 1 has twice the amplitude as the motion of the object on spring 2. The magnitude of the maximum velocity is the same in each case. Find the spring constant of spring 2.

*21. **ssm Interactive LearningWare 10.1** at **www.wiley.com/college/cutnell** reviews the concepts involved in this problem. A spring stretches by 0.018 m when a 2.8-kg object is suspended from its end. How much mass should be attached to this spring so that its frequency of vibration is $f = 3.0$ Hz?

* **22.** An object attached to a horizontal spring is oscillating back and forth along a frictionless surface. The maximum speed of the object is 1.25 m/s, and its maximum acceleration is 6.89 m/s². How much time elapses between an instant when the object's speed is at a maximum and the next instant when its acceleration is at a maximum?

* **23.** Multiple-Concept Example 6 reviews the principles that play a role in this problem. A bungee jumper, whose mass is 82 kg, jumps from a tall platform. After reaching his lowest point, he continues to oscillate up and down, reaching the low point two more times in 9.6 s. Ignoring air resistance and assuming that the bungee cord is an ideal spring, determine its spring constant.

** **24.** A tray is moved horizontally back and forth in simple harmonic motion at a frequency of $f = 2.00$ Hz. On this tray is an empty cup. Obtain the coefficient of static friction between the tray and the cup, given that the cup begins slipping when the amplitude of the motion is 5.00×10^{-2} m.

Section 10.3   Energy and Simple Harmonic Motion

25. GO The drawing shows three situations in which a block is attached to a spring. The position labeled "0 m" represents the unstrained position of the spring. The block is moved from an initial position x_0 to a final position x_f, the magnitude of the displacement being denoted by the symbol s. Suppose the spring has a spring constant of $k = 46.0$ N/m. Using the data provided in the drawing, determine the total work done by the restoring force of the spring for each situation.

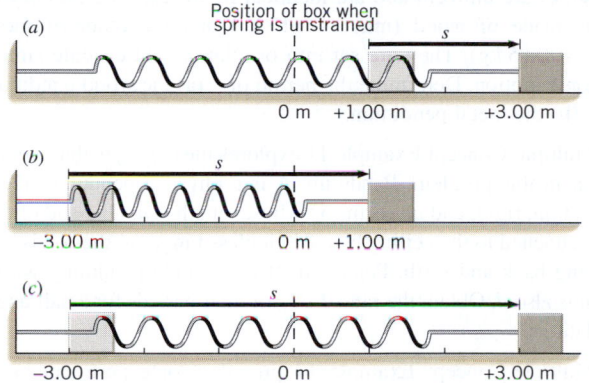

26. A pen contains a spring with a spring constant of 250 N/m. When the tip of the pen is in its retracted position, the spring is compressed 5.0 mm from its unstrained length. In order to push the tip out and lock it into its writing position, the spring must be compressed an additional 6.0 mm. How much work is done by the spring force to ready the pen for writing? Be sure to include the proper algebraic sign with your answer.

27. A spring is hung from the ceiling. A 0.450-kg block is then attached to the free end of the spring. When released from rest, the block drops 0.150 m before momentarily coming to rest, after which it moves back upward. (a) What is the spring constant of the spring? (b) Find the angular frequency of the block's vibrations.

28. A 3.2-kg block is hanging stationary from the end of a vertical spring that is attached to the ceiling. The elastic potential energy of this spring/mass system is 1.8 J. What is the elastic potential energy of the system when the 3.2-kg block is replaced by a 5.0-kg block?

29. ssm **Concept Simulation 10.1** at **www.wiley.com/college/cutnell** allows you to explore the concepts in this problem. A 2.00-kg object is hanging from the end of a vertical spring. The spring constant is 50.0 N/m. The object is pulled 0.200 m downward and released from rest. Complete the following table by calculating the translational kinetic energy, the gravitational potential energy, the

elastic potential energy, and the total mechanical energy E for each vertical position listed. The vertical positions h indicate distances above the point of release, where $h = 0$ m.

h (meters)	KE	PE (gravity)	PE (elastic)	E
0				
0.200				
0.400				

30. A vertical spring with a spring constant of 450 N/m is mounted on the floor. From directly above the spring, which is unstrained, a 0.30-kg block is dropped from rest. It collides with and sticks to the spring, which is compressed by 2.5 cm in bringing the block to a momentary halt. Assuming air resistance is negligible, from what height (in cm) above the compressed spring was the block dropped?

31. Refer to **Interactive Solution 10.31** at **www.wiley.com/college/cutnell** for help in solving this problem. A heavy-duty stapling gun uses a 0.140-kg metal rod that rams against the staple to eject it. The rod is attached to and pushed by a stiff spring called a "ram spring" ($k = 32\,000$ N/m). The mass of this spring may be ignored. The ram spring is compressed by 3.0×10^{-2} m from its unstrained length and then released from rest. Assuming that the ram spring is oriented vertically and is still compressed by 0.8×10^{-2} m when the downward-moving ram hits the staple, find the speed of the ram at the instant of contact.

32. A rifle fires a 2.10×10^{-2}-kg pellet straight upward, because the pellet rests on a compressed spring that is released when the trigger is pulled. The spring has a negligible mass and is compressed by 9.10×10^{-2} m from its unstrained length. The pellet rises to a maximum height of 6.10 m above its position on the compressed spring. Ignoring air resistance, determine the spring constant.

33. ssm A 1.00×10^{-2}-kg block is resting on a horizontal frictionless surface and is attached to a horizontal spring whose spring constant is 124 N/m. The block is shoved parallel to the spring axis and is given an initial speed of 8.00 m/s, while the spring is initially unstrained. What is the amplitude of the resulting simple harmonic motion?

* **34.** Using the data given in Concepts & Calculations Example 16, determine how far the bungee jumper is from the water when he reaches the lowest point in his fall.

* **35.** A horizontal spring is lying on a frictionless surface. One end of the spring is attached to a wall, and the other end is connected to a movable object. The spring and object are compressed by 0.065 m, released from rest, and subsequently oscillate back and forth with an angular frequency of 11.3 rad/s. What is the speed of the object at the instant when the spring is *stretched* by 0.048 m relative to its unstrained length?

* **36.** GO A vertical ideal spring is mounted on the floor and has a spring constant of 170 N/m. A 0.64-kg block is placed on the spring in two different ways. (a) In one case, the block is placed on the spring and not released until it rests stationary on the spring in its equilibrium position. Determine the amount (magnitude only) by which the spring is compressed. (b) In a second situation, the block is released from rest immediately after being placed on the spring and falls downward until it comes to a momentary halt. Determine the amount (magnitude only) by which the spring is now compressed.

* **37.** ssm Consult **Interactive Solution 10.37** at **www.wiley.com/college/cutnell** to explore a model for solving this problem. A spring is compressed by 0.0620 m and is used to launch an object horizontally with a speed of 1.50 m/s. If the object were attached to the spring, at what angular frequency (in rad/s) would it oscillate?

* **38.** A 0.60-kg metal sphere oscillates at the end of a vertical spring. As the spring stretches from 0.12 to 0.23 m (relative to its unstrained length), the speed of the sphere decreases from 5.70 to 4.80 m/s. What is the spring constant of the spring?

* **39.** A block rests on a frictionless horizontal surface and is attached to a spring. When set into simple harmonic motion, the block oscillates back and forth with an angular frequency of 7.0 rad/s. The drawing indicates the position of the block when the spring is unstrained. This position is labeled "$x = 0$ m." The drawing also shows a small bottle located 0.080 m to the right of this position. The block is pulled to the right, stretching the spring by 0.050 m, and is then thrown to the left. In order for the block to knock over the bottle, it must be thrown with a speed exceeding v_0. Ignoring the width of the block, find v_0.

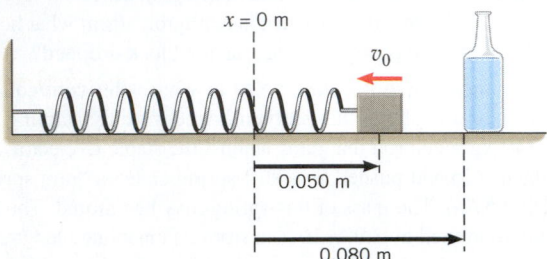

** **40.** **Interactive LearningWare 10.2** at **www.wiley.com/college/cutnell** explores the approach taken in problems such as this one. A spring is mounted vertically on the floor. The mass of the spring is negligible. A certain object is placed on the spring to compress it. When the object is pushed further down by just a bit and then released, one up/down oscillation cycle occurs in 0.250 s. However, when the object is pushed down by 5.00×10^{-2} m to point P and then released, the object flies entirely off the spring. To what height above point P does the object rise in the absence of air resistance?

** **41.** **ssm** A 70.0-kg circus performer is fired from a cannon that is elevated at an angle of 40.0° above the horizontal. The cannon uses strong elastic bands to propel the performer, much in the same way that a slingshot fires a stone. Setting up for this stunt involves stretching the bands by 3.00 m from their unstrained length. At the point where the performer flies free of the bands, his height above the floor is the same as the height of the net into which he is shot. He takes 2.14 s to travel the horizontal distance of 26.8 m between this point and the net. Ignore friction and air resistance and determine the effective spring constant of the firing mechanism.

** **42.** A 0.200-m uniform bar has a mass of 0.750 kg and is released from rest in the vertical position, as the drawing indicates. The spring is initially unstrained and has a spring constant of $k = 25.0$ N/m. Find the tangential speed with which end A strikes the horizontal surface.

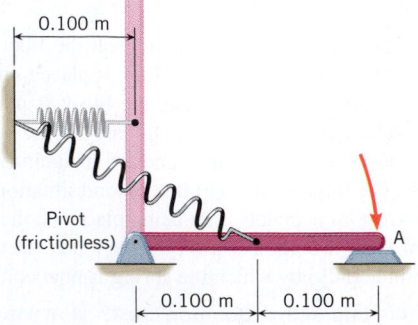

Section 10.4 The Pendulum

43. A simple pendulum is made from a 0.65-m-long string and a small ball attached to its free end. The ball is pulled to one side

through a small angle and then released from rest. After the ball is released, how much time elapses before it attains its greatest speed?

44. **GO** **Concept Simulation 10.2** at **www.wiley.com/college/cutnell** allows you to explore the effect of the acceleration due to gravity on pendulum motion, which is the focus of this problem. Astronauts on a distant planet set up a simple pendulum of length 1.2 m. The pendulum executes simple harmonic motion and makes 100 complete vibrations in 280 s. What is the magnitude of the acceleration due to gravity on this planet?

45. The length of a simple pendulum is 0.79 m and the mass of the particle (the "bob") at the end of the cable is 0.24 kg. The pendulum is pulled away from its equilibrium position by an angle of 8.50° and released from rest. Assume that friction can be neglected and that the resulting oscillatory motion is simple harmonic motion. **(a)** What is the angular frequency of the motion? **(b)** Using the position of the bob at its lowest point as the reference level, determine the total mechanical energy of the pendulum as it swings back and forth. **(c)** What is the bob's speed as it passes through the lowest point of the swing?

46. A simple pendulum is swinging back and forth through a small angle, its motion repeating every 1.25 s. How much longer should the pendulum be made in order to increase its period by 0.20 s?

47. **GO** Two physical pendulums (not simple pendulums) are made from meter sticks that are suspended from the ceiling at one end. The sticks are uniform and are identical in all respects, except that one is made of wood (mass = 0.17 kg) and the other of metal (mass = 0.85 kg). They are set into oscillation and execute simple harmonic motion. Determine the period of **(a)** the wood pendulum and **(b)** the metal pendulum.

* **48.** Multiple-Concept Example 11 explores the concepts that are important in this problem. Pendulum A is a physical pendulum made from a thin, rigid, and uniform rod whose length is d. One end of this rod is attached to the ceiling by a frictionless hinge, so the rod is free to swing back and forth. Pendulum B is a simple pendulum whose length is also d. Obtain the ratio T_A/T_B of their periods for small-angle oscillations.

* **49.** Multiple-Concept Example 11 provides some pertinent background for this problem. A pendulum is constructed from a thin, rigid, and uniform rod with a small sphere attached to the end opposite the pivot. This arrangement is a good approximation to a simple pendulum (period = 0.66 s), because the mass of the sphere (lead) is much greater than the mass of the rod (aluminum). When the sphere is removed, the pendulum is no longer a simple pendulum, but is then a physical pendulum. What is the period of the physical pendulum?

** **50.** A point on the surface of a solid sphere (radius = R) is attached directly to a pivot on the ceiling. The sphere swings back and forth as a physical pendulum with a small amplitude. What is the length of a simple pendulum that has the same period as this physical pendulum? Give your answer in terms of R.

Section 10.7 Elastic Deformation,
Section 10.8 Stress, Strain, and Hooke's Law

51. **ssm** A tow truck is pulling a car out of a ditch by means of a steel cable that is 9.1 m long and has a radius of 0.50 cm. When the car just begins to move, the tension in the cable is 890 N. How much has the cable stretched?

52. Two stretched cables both experience the same stress. The first cable has a radius of 3.5×10^{-3} m and is subject to a stretching force of 270 N. The radius of the second cable is 5.1×10^{-3} m. Determine the stretching force acting on the second cable.

53. ssm The pressure increases by 1.0×10^4 N/m² for every meter of depth beneath the surface of the ocean. At what depth does the volume of a Pyrex glass cube, 1.0×10^{-2} m on an edge at the ocean's surface, decrease by 1.0×10^{-10} m³?

54. The femur is a bone in the leg whose minimum cross-sectional area is about 4.0×10^{-4} m². A compressional force in excess of 6.8×10^4 N will fracture this bone. **(a)** Find the maximum stress that this bone can withstand. **(b)** What is the strain that exists under a maximum-stress condition?

55. Multiple-Concept Example 13 presents a model for solving this type of problem. A 59-kg water skier is being pulled by a nylon tow rope that is attached to a boat. The unstretched length of the rope is 12 m and its cross-sectional area is 2.0×10^{-5} m². As the skier moves, a resistive force (due to the water) of magnitude 130 N acts on her; this force is directed opposite to her motion. What is the change in the length of the rope when the skier has an acceleration whose magnitude is 0.85 m/s²?

56. The drawing shows a 160-kg crate hanging from the end of a steel bar. The length of the bar is 0.10 m, and its cross-sectional area is 3.2×10^{-4} m². Neglect the weight of the bar itself and determine **(a)** the shear stress on the bar and **(b)** the vertical deflection ΔY of the right end of the bar.

57. A 3500-kg statue is placed on top of a cylindrical concrete $(Y = 2.3 \times 10^{10}$ N/m²) stand. The stand has a cross-sectional area of 7.3×10^{-2} m² and a height of 1.8 m. By how much does the statue compress the stand?

58. A copper cube, 0.30 m on a side, is subjected to two shearing forces, each of which has a magnitude $F = 6.0 \times 10^6$ N (see the drawing). Find the angle θ (in degrees), which is one measure of how the shape of the block has been altered by shear deformation.

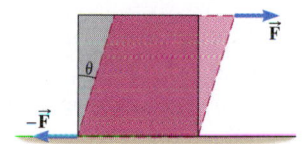

59. ssm Two metal beams are joined together by four rivets, as the drawing indicates. Each rivet has a radius of 5.0×10^{-3} m and is to be exposed to a shearing stress of no more than 5.0×10^8 Pa. What is the maximum tension $\vec{T}$ that can be applied to each beam, assuming that each rivet carries one-fourth of the total load?

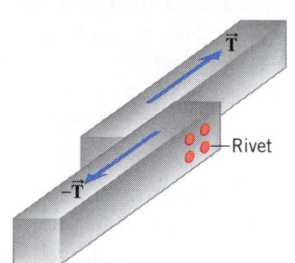

60. A copper cylinder and a brass cylinder are stacked end to end, as in the drawing. Each cylinder has a radius of 0.25 cm. A compressive force of $F = 6500$ N is applied to the right end of the brass cylinder. Find the amount by which the length of the stack decreases.

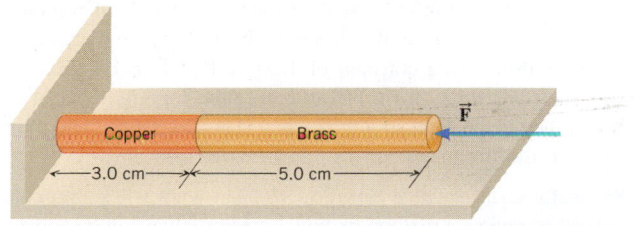

61. Between each pair of vertebrae in the spinal column is a cylindrical disc of cartilage. Typically, this disc has a radius of about 3.0×10^{-2} m and a thickness of about 7.0×10^{-3} m. The shear modulus of cartilage is 1.2×10^7 N/m². Suppose that a shear-

ing force of magnitude 11 N is applied parallel to the top surface of the disc while the bottom surface remains fixed in place. How far does the top surface move relative to the bottom surface?

62. GO One end of a piano wire is wrapped around a cylindrical tuning peg and the other end is fixed in place. The tuning peg is turned so as to stretch the wire. The piano wire is made from steel $(Y = 2.0 \times 10^{11}$ N/m²). It has a radius of 0.80 mm and an unstrained length of 0.76 m. The radius of the tuning peg is 1.8 mm. Initially, there is no tension in the wire, but when the tuning peg is turned, tension develops. Find the tension in the wire when the tuning peg is turned through two revolutions.

* 63. A block of copper is securely fastened to the floor. A force of 1800 N is applied to the top surface of the block, as the drawing shows. Find **(a)** the amount by which the height of the block is changed and **(b)** the shear deformation of the block.

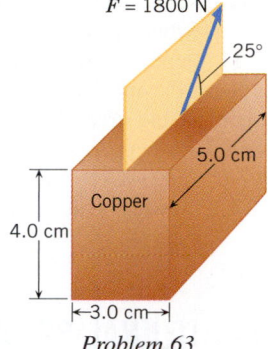

Problem 63

* 64. GO A piece of mohair taken from an Angora goat has a radius of 31×10^{-6} m. What is the least number of identical pieces of mohair needed to suspend a 75-kg person, so the strain experienced by each piece is less than 0.010? Assume that the tension is the same in all the pieces.

* 65. ssm Two rods are identical in all respects except one: one rod is made from aluminum and the other from tungsten. The rods are joined end to end, in order to make a single rod that is twice as long as either the aluminum or tungsten rod. What is the effective value of Young's modulus for this composite rod? That is, what value $Y_{\text{Composite}}$ of Young's modulus should be used in Equation 10.17 when applied to the composite rod? Note that the change $\Delta L_{\text{Composite}}$ in the length of the composite rod is the sum of the changes in length of the aluminum and tungsten rods.

* 66. A die is designed to punch holes with a radius of 1.00×10^{-2} m in a metal sheet that is 3.0×10^{-3} m thick, as the drawing illustrates. To punch through the sheet, the die must exert a shearing stress of 3.5×10^8 Pa. What force $\vec{F}$ must be applied to the die?

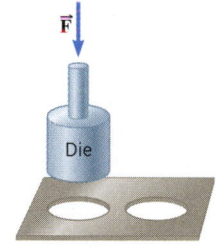

Problem 66

* 67. A gymnast does a one-arm handstand. The humerus, which is the upper arm bone (between the elbow and the shoulder joint), may be approximated as a 0.30-m-long cylinder with an outer radius of 1.00×10^{-2} m and a hollow inner core with a radius of 4.0×10^{-3} m. Excluding the arm, the mass of the gymnast is 63 kg. **(a)** What is the compressional strain of the humerus? **(b)** By how much is the humerus compressed?

* 68. GO Depending on how you fall, you can break a bone easily. The severity of the break depends on how much energy the bone absorbs in the accident, and to evaluate this let us treat the bone as an ideal spring. The maximum applied force of compression that one man's thighbone can endure without breaking is 7.0×10^4 N. The minimum effective cross-sectional area of the bone is 4.0×10^{-4} m², its length is 0.55 m, and Young's modulus is $Y = 9.4 \times 10^9$ N/m². The mass of the man is 65 kg. He falls straight down without rotating, strikes the ground stiff-legged on one foot, and comes to a halt without rotating. To see that it is easy to break a thighbone when falling in this fashion, find the maximum distance through which his center of gravity can fall without his breaking a bone.

*   **69. ssm www** An 8.0-kg stone at the end of a steel wire is being whirled in a circle at a constant tangential speed of 12 m/s. The stone is moving on the surface of a frictionless horizontal table. The wire is 4.0 m long and has a radius of 1.0×10^{-3} m. Find the strain in the wire.

*   **70.** The dimensions of a rectangular block of brass are 0.010 m, 0.020 m, and 0.040 m. The block is to be glued to a table and subjected to a horizontal force of 770 N, as in Figure 10.31. Note that there are three possibilities for the surface of the block that is in contact with the table. What is the maximum possible distance the top surface can move, relative to the bottom surface?

*   **71.** Consult Multiple-Concept Example 13 to review a model for solving this type of problem. A 61-kg snow skier is being pulled up a 12° slope by a steel cable. The cable has a cross-sectional area of 7.8×10^{-5} m². The cable applies a force to the skier, and, in doing so, the cable stretches by 2.0×10^{-4} m. A frictional force of magnitude 68 N acts on the skis and is directed opposite to the skier's motion. If the skier's acceleration up the slope has a magnitude of 1.1 m/s², what is the original (unstretched) length of the cable?

**   **72.** A 6.8-kg bowling ball is attached to the end of a nylon cord with a cross-sectional area of 3.4×10^{-5} m². The other end of the cord is fixed to the ceiling. When the bowling ball is pulled to one side and released from rest, it swings downward in a circular arc. At the instant it reaches its lowest point, the bowling ball is 1.4 m lower than the point it was released from, and the cord is stretched 2.7×10^{-3} m from its unstrained length. What is the unstrained length of the cord? *Hint: When calculating any quantity other than the strain, ignore the increase in the length of the cord.*

**   **73. ssm** The drawing shows two crates that are connected by a steel wire that passes over a pulley. The unstretched length of the wire is 1.5 m, and its cross-sectional area is 1.3×10^{-5} m². The pulley is frictionless and massless. When the crates are accelerating, determine the change in length of the wire. Ignore the mass of the wire.

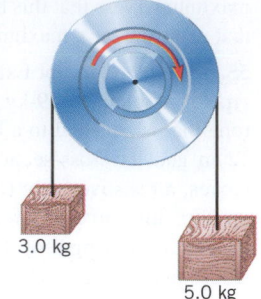

3.0 kg

5.0 kg

Problem 73

ADDITIONAL PROBLEMS

74. A loudspeaker diaphragm is producing a sound for 2.5 s by moving back and forth in simple harmonic motion. The angular frequency of the motion is 7.54×10^4 rad/s. How many times does the diaphragm move back and forth?

75. Refer to Conceptual Example 2 as an aid in solving this problem. A 100-coil spring has a spring constant of 420 N/m. It is cut into four shorter springs, each of which has 25 coils. One end of a 25-coil spring is attached to a wall. An object of mass 46 kg is attached to the other end of the spring, and the system is set into horizontal oscillation. What is the angular frequency of the motion?

76. A spiral staircase winds up to the top of a tower in an old castle. To measure the height of the tower, a rope is attached to the top of the tower and hung down the center of the staircase. However, nothing is available with which to measure the length of the rope. Therefore, at the bottom of the rope a small object is attached so as to form a simple pendulum that just clears the floor. The period of the pendulum is measured to be 9.2 s. What is the height of the tower?

77. ssm www Multiple-Concept Example 6 presents a model for solving this problem. As far as vertical oscillations are concerned, a certain automobile can be considered to be mounted on four identical springs, each having a spring constant of 1.30×10^5 N/m. Four identical passengers sit down inside the car, and it is set into a vertical oscillation that has a period of 0.370 s. If the mass of the empty car is 1560 kg, determine the mass of *each* passenger. Assume that the mass of the car and its passengers is distributed evenly over the springs.

78. When subjected to a force of compression, the length of a bone decreases by 2.7×10^{-5} m. When this same bone is subjected to a tensile force of the same magnitude, by how much does it stretch?

79. An archer, about to shoot an arrow, is applying a force of +240 N to a drawn bowstring. The bow behaves like an ideal spring whose spring constant is 480 N/m. What is the displacement of the bowstring?

80. A piece of aluminum is surrounded by air at a pressure of 1.01×10^5 Pa. The aluminum is placed in a vacuum chamber where the pressure is reduced to zero. Determine the fractional change $\Delta V/V_0$ in the volume of the aluminum.

*   **81.** Review Conceptual Example 8 before starting this problem. A block is attached to a horizontal spring and oscillates back and forth on a frictionless horizontal surface at a frequency of 3.00 Hz. The amplitude of the motion is 5.08×10^{-2} m. At the point where the block has its maximum speed, it suddenly splits into two identical parts, only one part remaining attached to the spring. **(a)** What is the amplitude and the frequency of the simple harmonic motion that exists after the block splits? **(b)** Repeat part (a), assuming that the block splits when it is at one of its extreme positions.

*   **82. GO** A spring is resting vertically on a table. A small box is dropped onto the top of the spring and compresses it. Suppose the spring has a spring constant of 450 N/m and the box has a mass of 1.5 kg. The speed of the box just before it makes contact with the spring is 0.49 m/s. **(a)** Determine the magnitude of the spring's displacement at an instant when the acceleration of the box is zero. **(b)** What is the magnitude of the spring's displacement when the spring is fully compressed?

*   **83. Interactive Solution 10.83** at **www.wiley.com/college/cutnell** presents a model for solving this problem. A vertical spring (spring constant = 112 N/m) is mounted on the floor. A 0.400-kg block is placed on top of the spring and pushed down to start it oscillating in simple harmonic motion. The block is not attached to the spring. **(a)** Obtain the frequency (in Hz) of the motion. **(b)** Determine the amplitude at which the block will lose contact with the spring.

*   **84.** An 86.0-kg climber is scaling the vertical wall of a mountain. His safety rope is made of nylon that, when stretched, behaves like a spring with a spring constant of 1.20×10^3 N/m. He accidentally slips and falls freely for 0.750 m before the rope runs out of slack. How much is the rope stretched when it breaks his fall and momentarily brings him to rest?

*   **85.** Refer to **Interactive Solution 10.85** at **www.wiley.com/college/cutnell** to review a method by which this problem can be solved. An 11.2-kg block and a 21.7-kg block are resting on a horizontal frictionless surface. Between the two is squeezed a spring (spring constant = 1330 N/m). The spring is compressed by 0.141 m from its unstrained length and is not attached to either block. With what speed does each

block move away after the mechanism keeping the spring squeezed is released and the spring falls away?

* **86.** A 1.0×10^{-3}-kg spider is hanging vertically by a thread that has a Young's modulus of 4.5×10^9 N/m^2 and a radius of 13×10^{-6} m. Suppose that a 95-kg person is hanging vertically on an aluminum wire. What is the radius of the wire that would exhibit the same strain as the spider's thread, when the thread is stressed by the full weight of the spider?

* **87. ssm** The front spring of a car's suspension system has a spring constant of 1.50×10^6 N/m and supports a mass of 215 kg. The wheel has a radius of 0.400 m. The car is traveling on a bumpy road, on which the distance between the bumps is equal to the circumference of the wheel. Due to resonance, the wheel starts to vibrate strongly when the car is traveling at a certain minimum linear speed. What is this speed?

** **88.** A 1.00×10^{-2}-kg bullet is fired horizontally into a 2.50-kg wooden block attached to one end of a massless horizontal spring ($k = 845$ N/m). The other end of the spring is fixed in place, and the spring is unstrained initially. The block rests on a horizontal, frictionless surface. The bullet strikes the block perpendicularly and quickly comes to a halt within it. As a result of this completely inelastic collision, the spring is compressed along its axis and causes the block/bullet to oscillate with an amplitude of 0.200 m. What is the speed of the bullet?

** **89. ssm www** A solid brass sphere is subjected to a pressure of 1.0×10^5 Pa due to the earth's atmosphere. On Venus the pressure due to the atmosphere is 9.0×10^6 Pa. By what fraction $\Delta r / r_0$ (including the algebraic sign) does the radius of the sphere change when it is exposed to the Venusian atmosphere? Assume that the change in radius is very small relative to the initial radius.

** **90.** A copper rod (length $= 2.0$ m, radius $= 3.0 \times 10^{-3}$ m) hangs down from the ceiling. A 9.0-kg object is attached to the lower end of the rod. The rod acts as a "spring," and the object oscillates vertically with a small amplitude. Ignoring the rod's mass, find the frequency f of the simple harmonic motion.

** **91. ssm** The drawing shows a top view of a frictionless horizontal surface, where there are two springs with particles of mass m_1 and m_2 attached to them. Each spring has a spring constant of 120 N/m. The particles are pulled to the right and then released from the positions shown in the drawing. How much time passes before the particles are side by side for the first time at $x = 0$ m if **(a)** $m_1 = m_2 = 3.0$ kg and **(b)** $m_1 = 3.0$ kg and $m_2 = 27$ kg?

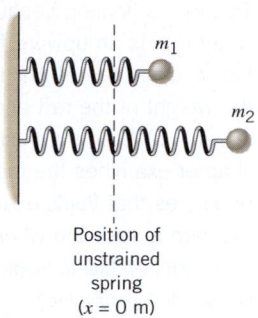

Position of unstrained spring ($x = 0$ m)

** **92.** A cylindrically shaped piece of collagen (a substance found in the body in connective tissue) is being stretched by a force that increases from 0 to 3.0×10^{-2} N. The length and radius of the collagen are, respectively, 2.5 and 0.091 cm, and Young's modulus is 3.1×10^6 N/m^2. **(a)** If the stretching obeys Hooke's law, what is the spring constant k for collagen? **(b)** How much work is done by the variable force that stretches the collagen? (See Section 6.9 for a discussion of the work done by a variable force.)

CHAPTER 11

FLUIDS

This raft is floating because the water exerts an upward force, called the "buoyant force," that balances the weight of the raft and everyone aboard it. Water is a fluid, and this chapter examines the forces and pressures that fluids exert when they are at rest and when they are in motion. (© Martin Siepmann/Age Fotostock America, Inc.)

11.1 MASS DENSITY

Fluids are materials that can flow, and they include both gases and liquids. Air is the most common gas, and flows from place to place as wind. Water is the most familiar liquid, and flowing water has many uses, from generating hydroelectric power to white-water rafting. The *mass density* of a liquid or gas is one of the important factors that determine its behavior as a fluid. As indicated below, the mass density is the mass per unit volume and is denoted by the Greek letter rho (ρ).

DEFINITION OF MASS DENSITY

The mass density ρ is the mass m of a substance divided by its volume V:

$$\rho = \frac{m}{V} \tag{11.1}$$

SI Unit of Mass Density: kg/m^3

Equal volumes of different substances generally have different masses, so the density depends on the nature of the material, as Table 11.1 indicates. Gases have the smallest densities because gas molecules are relatively far apart and a gas contains a large fraction of empty space. In contrast, the molecules are much more tightly packed in liquids and solids, and the tighter packing leads to larger densities. The densities of gases are very sensitive to changes in temperature and pressure. However, for the range of temperatures and pressures encountered in this text, the densities of liquids and solids do not differ much from the values in Table 11.1.

It is the mass of a substance, not its weight, that enters into the definition of density. In situations where weight is needed, it can be calculated from the mass density, the volume, and the acceleration due to gravity, as Example 1 illustrates.

Example 1 Blood as a Fraction of Body Weight

The body of a man whose weight is 690 N contains about 5.2×10^{-3} m³ (5.5 qt) of blood. **(a)** Find the blood's weight and **(b)** express it as a percentage of the body weight.

Reasoning To find the weight W of the blood, we need the mass m, since $W = mg$ (Equation 4.5), where g is the magnitude of the acceleration due to gravity. According to Table 11.1, the density of blood is 1060 kg/m³, so the mass of the blood can be found by using the given volume of 5.2×10^{-3} m³ in Equation 11.1.

Solution **(a)** According to Equation 4.5, the blood's weight is $W = mg$. Equation 11.1 can be solved for m to show that the mass is $m = \rho V$. Substituting this result into Equation 4.5 gives

$$W = mg = (\rho V)g = (1060 \text{ kg/m}^3)(5.2 \times 10^{-3} \text{ m}^3)(9.80 \text{ m/s}^2) = \boxed{54 \text{ N}}$$

(b) The percentage of body weight contributed by the blood is

$$\text{Percentage} = \frac{54 \text{ N}}{690 \text{ N}} \times 100 = \boxed{7.8\%}$$

A convenient way to compare densities is to use the concept of *specific gravity.* The specific gravity of a substance is its density divided by the density of a standard reference material, usually chosen to be water at 4 °C.

$$\text{Specific gravity} = \frac{\text{Density of substance}}{\text{Density of water at 4 °C}} = \frac{\text{Density of substance}}{1.000 \times 10^3 \text{ kg/m}^3} \qquad (11.2)$$

Being the ratio of two densities, specific gravity has no units. For example, Table 11.1 reveals that diamond has a specific gravity of 3.52, since the density of diamond is 3.52 times greater than the density of water at 4 °C.

The next two sections deal with the important concept of pressure. We will see that the density of a fluid is one factor determining the pressure that a fluid exerts.

Table 11.1 Mass Densities[a] of Common Substances

Substance	Mass Density ρ (kg/m³)
Solids	
Aluminum	2700
Brass	8470
Concrete	2200
Copper	8890
Diamond	3520
Gold	19 300
Ice	917
Iron (steel)	7860
Lead	11 300
Quartz	2660
Silver	10 500
Wood (yellow pine)	550
Liquids	
Blood (whole, 37 °C)	1060
Ethyl alcohol	806
Mercury	13 600
Oil (hydraulic)	800
Water (4 °C)	1.000×10^3
Gases	
Air	1.29
Carbon dioxide	1.98
Helium	0.179
Hydrogen	0.0899
Nitrogen	1.25
Oxygen	1.43

[a]Unless otherwise noted, densities are given at 0 °C and 1 atm pressure.

11.2 PRESSURE

People who have fixed a flat tire know something about pressure. The final step in the job is to reinflate the tire to the proper pressure. The underinflated tire is soft because it contains an insufficient number of air molecules to push outward against the rubber and give the tire that solid feel. When air is added from a pump, the number of molecules and the collective force they exert are increased. The air molecules within a tire are free to wander throughout its entire volume, and in the course of their wandering they collide with one another and with the inner walls of the tire. The collisions with the walls allow the air to exert a force against every part of the wall surface, as Figure 11.1 shows. The pressure P exerted by a fluid is defined in Section 10.7 (Equation 10.19) as the magnitude F of the force acting perpendicular to a surface divided by the area A over which the force acts:

$$P = \frac{F}{A} \qquad (11.3)$$

The SI unit for pressure is a newton/meter² (N/m²), a combination that is referred to as a *pascal (Pa)*. A pressure of 1 Pa is a very small amount. Many common situations involve pressures of approximately 10^5 Pa, an amount referred to as one *bar* of pressure. Alternatively, force can be measured in pounds and area in square inches, so another unit for pressure is pounds per square inch (lb/in.²), often abbreviated as "psi."

Because of its pressure, the air in a tire applies a force to any surface with which the air is in contact. Suppose, for instance, that a small cube is inserted inside the tire. As Figure 11.1 shows, the air pressure causes a force to act perpendicularly on each face of the cube. In a similar fashion, a liquid such as water also exerts pressure. A swimmer, for

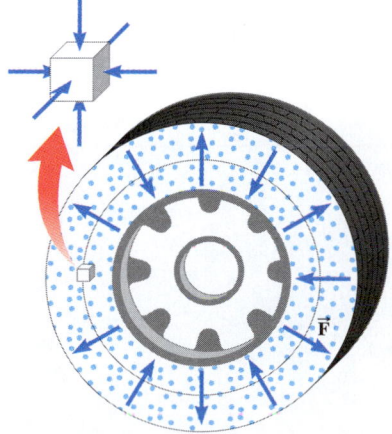

Figure 11.1 In colliding with the inner walls of the tire, the air molecules (blue dots) exert a force on every part of the wall surface. If a small cube were inserted inside the tire, the cube would experience forces (blue arrows) acting perpendicular to each of its six faces.

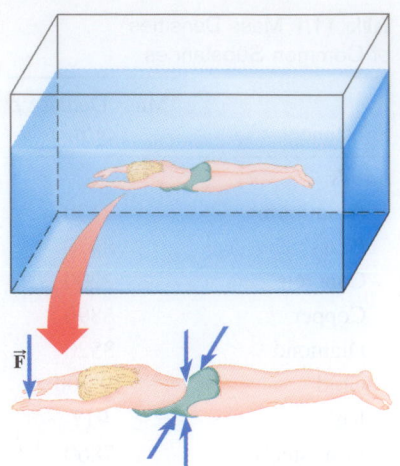

Figure 11.2 Water applies a force perpendicular to each surface within the water, including the walls and bottom of the swimming pool, and all parts of the swimmer's body.

Problem-solving insight

Force is a vector, but pressure is not.

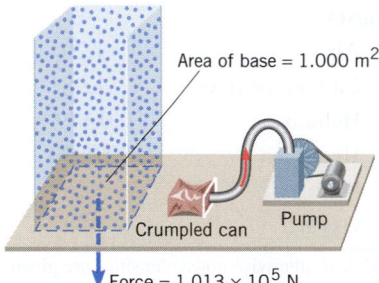

Figure 11.3 Atmospheric pressure at sea level is 1.013×10^5 Pa, which is sufficient to crumple a can if the inside air is pumped out.

Figure 11.4 Lynx have large paws that act as natural snowshoes. (Peter Weimann/ Animals Animals/Earth Scenes)

example, feels the water pushing perpendicularly inward everywhere on her body, as Figure 11.2 illustrates. In general, a static fluid cannot produce a force parallel to a surface, for if it did, the surface would apply a reaction force to the fluid, consistent with Newton's action–reaction law. In response, the fluid would flow and would not then be static.

While fluid pressure can generate a force, which is a vector quantity, *pressure itself is not a vector.* In the definition of pressure, $P = F/A$, the symbol F refers only to the magnitude of the force, so that pressure has no directional characteristic. The force generated by the pressure of a static fluid is always perpendicular to the surface that the fluid contacts, as Example 2 illustrates.

Example 2 The Force on a Swimmer

Suppose that the pressure acting on the back of a swimmer's hand is 1.2×10^5 Pa, a realistic value near the bottom of the diving end of a pool. The surface area of the back of the hand is 8.4×10^{-3} m^2. **(a)** Determine the magnitude of the force that acts on it. **(b)** Discuss the direction of the force.

Reasoning From the definition of pressure in Equation 11.3, we can see that the magnitude of the force is the pressure times the area. The direction of the force is always perpendicular to the surface that the water contacts.

Solution **(a)** A pressure of 1.2×10^5 Pa is 1.2×10^5 N/m^2. From Equation 11.3, we find

$$F = PA = (1.2 \times 10^5 \text{ N/m}^2)(8.4 \times 10^{-3} \text{ m}^2) = \boxed{1.0 \times 10^3 \text{ N}}$$

This is a rather large force, about 230 lb.

(b) In Figure 11.2, the hand (palm downward) is oriented parallel to the bottom of the pool. Since the water pushes perpendicularly against the back of the hand, the force $\vec{F}$ is directed downward in the drawing. This downward-acting force is balanced by an upward-acting force on the palm, so that the hand is in equilibrium. If the hand were rotated by 90°, the directions of these forces would also be rotated by 90°, always being perpendicular to the hand.

A person need not be under water to experience the effects of pressure. Walking about on land, we are at the bottom of the earth's atmosphere, which is a fluid and pushes inward on our bodies just like the water in a swimming pool. As Figure 11.3 indicates, there is enough air above the surface of the earth to create the following pressure at sea level:

Atmospheric pressure at sea level 1.013×10^5 Pa = 1 atmosphere

This pressure corresponds to 14.70 lb/in.^{2} and is referred to as one *atmosphere (atm),* a significant amount of pressure. Look, for instance, in Figure 11.3 at the results of pumping out the air from within a gasoline can. With no internal air to push outward, the inward push of the external air is unbalanced and is strong enough to crumple the can.

The physics of lynx paws. In contrast to the situation in Figure 11.3, reducing the pressure is sometimes beneficial. Lynx, for example, are well suited for hunting on snow because of their oversize paws (see Figure 11.4). The large paws act as snowshoes and distribute the weight over a large area. Thus, they reduce the weight per unit area, or the pressure that the cat applies to the surface, which helps to keep it from sinking into the snow.

✓ **CHECK YOUR UNDERSTANDING**

(The answers are given at the end of the book.)

1. As you climb a mountain, your ears "pop" because of the changes in atmospheric pressure. In which direction, outward or inward, does your eardrum move **(a)** as you climb up and **(b)** as you climb down?

2. A bottle of juice is sealed under partial vacuum, with a lid on which a red dot or "button" is painted. Around the button the following phrase is printed: "Button pops up when seal is broken." Why does the button remain pushed in when the seal is intact? **(a)** The pressure inside the bottle is greater than the pressure outside the bottle. **(b)** The pressure inside the bottle is less than the pressure outside the bottle. **(c)** There is a greater force acting on the interior surface of the seal than acts on the exterior surface.

3. A method for resealing a partially full bottle of wine under a vacuum uses a specially designed rubber stopper to close the bottle. A simple pump is attached to the stopper, and to remove air from the bottle, the plunger of the pump is pulled up and then released. After about 15 pull-and-release cycles the wine is under a partial vacuum. On the fifteenth pull-and-release cycle, does it require **(a)** more force, **(b)** less force, or **(c)** the same force to pull the plunger up than it did on the first cycle?

11.3 PRESSURE AND DEPTH IN A STATIC FLUID

The deeper an underwater swimmer goes, the more strongly the water pushes on his body and the greater is the pressure that he experiences. To determine the relation between pressure and depth, we turn to Newton's second law ($\Sigma\vec{F} = m\vec{a}$). In using the second law, we will focus on two external forces that act on the fluid. One is the gravitational force—that is, the weight of the fluid. The other is the collisional force that is responsible for fluid pressure, as Section 11.2 discusses. Since the fluid is at rest, its acceleration is zero ($\vec{a} = 0$ m/s^2), and it is in equilibrium. By applying the second law in the form $\Sigma\vec{F} = 0$, we will derive a relation between pressure and depth. This relation is especially important because it leads to Pascal's principle (Section 11.5) and Archimedes' principle (Section 11.6), both of which are essential in describing the properties of static fluids.

Figure 11.5 shows a container of fluid and focuses attention on one column of the fluid. The free-body diagram in the figure shows all the vertical forces acting on the column. On the top face (area $= A$), the fluid pressure P_1 generates a downward force whose magnitude is P_1A. Similarly, on the bottom face, the pressure P_2 generates an upward force of magnitude P_2A. The pressure P_2 is greater than the pressure P_1 because the bottom face supports the weight of more fluid than the upper one does. In fact, the excess weight supported by the bottom face is exactly the weight of the fluid within the column. As the free-body diagram indicates, this weight is mg, where m is the mass of the fluid and g is the magnitude of the acceleration due to gravity. Since the column is in equilibrium, we can set the sum of the vertical forces equal to zero and find that

$$\Sigma F_y = P_2A - P_1A - mg = 0 \quad \text{or} \quad P_2A = P_1A + mg$$

The mass m is related to the density ρ and the volume V of the column by $m = \rho V$ (Equation 11.1). Since the volume is the cross-sectional area A times the vertical dimension h, we have $m = \rho A h$. With this substitution, the condition for equilibrium becomes $P_2A = P_1A + \rho A h g$. The area A can be eliminated algebraically from this expression, with the result that

$$P_2 = P_1 + \rho g h \tag{11.4}$$

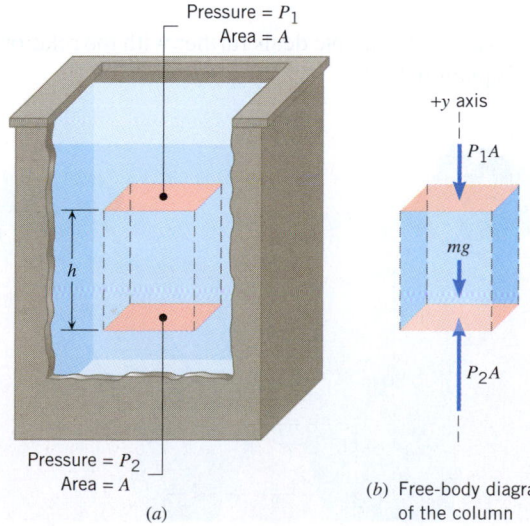

Pressure = P_1
Area = A

+y axis

P_1A

mg

P_2A

Pressure = P_2
Area = A

(a)

(b) Free-body diagram of the column

Figure 11.5 (a) A container of fluid in which one column of the fluid is outlined. The fluid is at rest. (b) The free-body diagram, showing the vertical forces acting on the column.

(a)

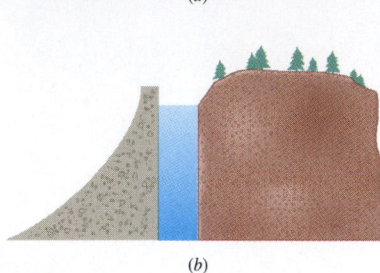

(b)

Figure 11.6 (a) The Hoover Dam in Nevada and Lake Mead behind it. (b) A hypothetical reservoir formed by removing most of the water from Lake Mead. Conceptual Example 3 compares the dam needed for this hypothetical reservoir with the Hoover Dam. (Adam G. Sylvester/Photo Researchers, Inc.)

Equation 11.4 indicates that if the pressure P_1 is known at a higher level, the larger pressure P_2 at a deeper level can be calculated by adding the increment ρgh. In determining the pressure increment ρgh, we assumed that the density ρ is the same at any vertical distance h or, in other words, the fluid is incompressible. The assumption is reasonable for liquids, since the bottom layers can support the upper layers with little compression. In a gas, however, the lower layers are compressed markedly by the weight of the upper layers, with the result that the density varies with vertical distance. For example, the density of our atmosphere is larger near the earth's surface than it is at higher altitudes. When applied to gases, the relation $P_2 = P_1 + \rho gh$ can be used only when h is small enough that any variation in ρ is negligible.

A significant feature of Equation 11.4 is that the pressure increment ρgh is affected by the vertical distance h, but not by any horizontal distance within the fluid. Conceptual Example 3 helps to clarify this feature.

Conceptual Example 3 The Hoover Dam

Lake Mead is the largest wholly artificial reservoir in the United States and was formed after the completion of the Hoover Dam in 1936. As Figure 11.6a suggests, the water in the reservoir backs up behind the dam for a considerable distance (about 200 km or 120 miles). Suppose that all the water were removed, except for a relatively narrow vertical column in contact with the dam. Figure 11.6b shows a side view of this hypothetical situation, in which the water against the dam has the same depth as in Figure 11.6a. How would the dam needed to contain the water in this hypothetical reservoir compare with the Hoover Dam? Would it need to be (a) less massive or (b) equally massive?

Reasoning Imagine a small square in the inner face of the dam, located beneath the water. The magnitude of the force on this square is the product of its area and the pressure of the water, according to Equation 11.3. However, the relation $P_2 = P_1 + \rho gh$ (Equation 11.4) indicates that the pressure at a given point depends on the vertical distance h that the small square is below the water. Thus, the force exerted by the water at any given location depends on the depth at that location.

Answer (a) is incorrect. Since our hypothetical reservoir contains much less water than Lake Mead, it is tempting to say that a less massive structure would be required. Note from the Reasoning section that the force exerted on a given section of the dam depends on how far below the surface the section is located. The horizontal distance of the water backed up behind the dam does not appear in Equation 11.4, and, therefore, has no effect on the pressure and, hence, on the force.

Answer (b) is correct. The force exerted on a given section of the dam depends only on how far that section is located vertically below the surface (see the Reasoning section). Certainly, as one goes deeper and deeper, the water pressure and force become greater. But no matter how deep one goes, the force that the water applies on a given section of the dam does not depend on the amount of water backed up behind the dam. Thus, the dam for our imaginary reservoir would sustain the same forces that the Hoover Dam sustains and, therefore, would need to be equally massive.

The next example deals further with the relationship between pressure and depth given by Equation 11.4.

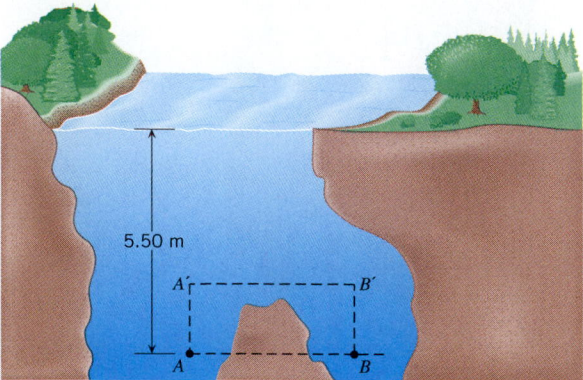

Figure 11.7 The pressures at points A and B are the same, since both points are located at the same vertical distance of 5.50 m beneath the surface of the water.

Example 4 The Swimming Hole

Figure 11.7 shows the cross section of a swimming hole. Points *A* and *B* are both located at a distance of *h* = 5.50 m below the surface of the water. Find the pressure at each of these two points.

Reasoning The pressure at point *B* is the same as that at point *A*, since both are located at the *same vertical distance* beneath the surface and only the vertical distance *h* affects the pressure increment ρgh in Equation 11.4. To understand this important feature more clearly, consider the path *AA′B′B* in Figure 11.7. The pressure decreases on the way up along the vertical segment *AA′* and increases by the same amount on the way back down along segment *B′B*. Since no change in pressure occurs along the horizontal segment *A′B′*, the pressure is the same at *A* and *B*.

Solution The pressure acting on the surface of the water is the atmospheric pressure of 1.01×10^5 Pa. Using this value as P_1 in Equation 11.4, we can determine a value for the pressure P_2 at either point *A* or *B*, both of which are located 5.50 m under the water. Table 11.1 gives the density of water as 1.000×10^3 kg/m³.

$$P_2 = P_1 + \rho gh$$

$$P_2 = 1.01 \times 10^5 \text{ Pa} + (1.000 \times 10^3 \text{ kg/m}^3)(9.80 \text{ m/s}^2)(5.50 \text{ m}) = \boxed{1.55 \times 10^5 \text{ Pa}}$$

Figure 11.8 shows an irregularly shaped container of liquid. Reasoning similar to that used in Example 4 leads to the conclusion that the pressure is the same at points *A*, *B*, *C*, and *D*, since each is at the same vertical distance *h* beneath the surface. In effect, the arteries in our bodies are also an irregularly shaped "container" for the blood. The next example examines the blood pressure at different places in this "container."

 Example 5 Blood Pressure

Blood in the arteries is flowing, but as a first approximation, the effects of this flow can be ignored and the blood treated as a static fluid. Estimate the amount by which the blood pressure P_2 in the anterior tibial artery at the foot exceeds the blood pressure P_1 in the aorta at the heart when a person is **(a)** reclining horizontally as in Figure 11.9*a* and **(b)** standing as in Figure 11.9*b*.

Reasoning and Solution **(a)** When the body is horizontal, there is little or no vertical separation between the feet and the heart. Since *h* = 0 m,

$$P_2 - P_1 = \rho gh = \boxed{0 \text{ Pa}} \qquad (11.4)$$

(b) When an adult is standing up, the vertical separation between the feet and the heart is about 1.35 m, as Figure 11.9*b* indicates. Table 11.1 gives the density of blood as 1060 kg/m³, so that

$$P_2 - P_1 = \rho gh = (1060 \text{ kg/m}^3)(9.80 \text{ m/s}^2)(1.35 \text{ m}) = \boxed{1.40 \times 10^4 \text{ Pa}}$$

Sometimes fluid pressure places limits on how a job can be done. Conceptual Example 6 illustrates how fluid pressure restricts the height to which water can be pumped.

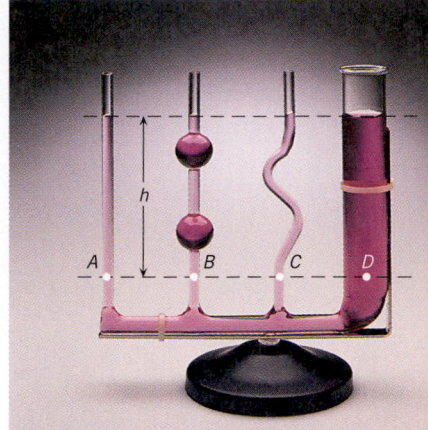

Figure 11.8 Since points *A*, *B*, *C*, and D are at the same distance *h* beneath the liquid surface, the pressure at each of them is the same. (Richard Megna/ Fundamental Photographs)

Problem-solving insight

The pressure at any point in a fluid depends on the vertical distance *h* of the point beneath the surface. However, for a given vertical distance, the pressure is the same, no matter where the point is located horizontally in the fluid.

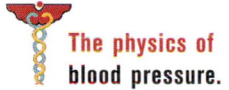 **The physics of blood pressure.**

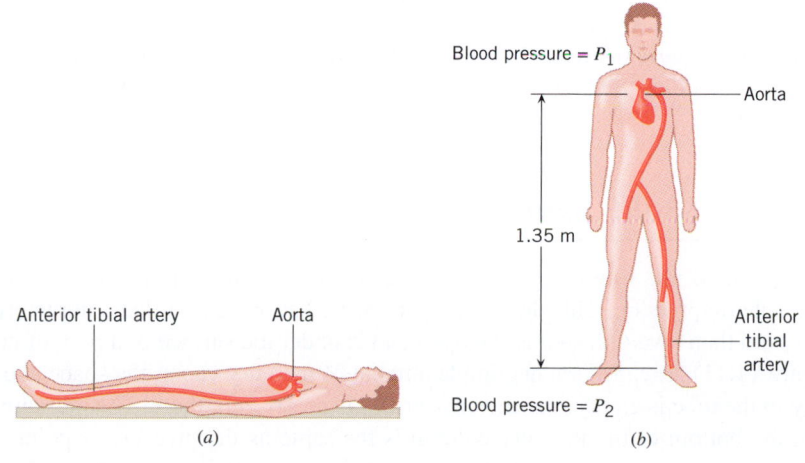

Figure 11.9 The blood pressure in the feet can exceed that in the heart, depending on whether a person is (*a*) reclining horizontally or (*b*) standing.

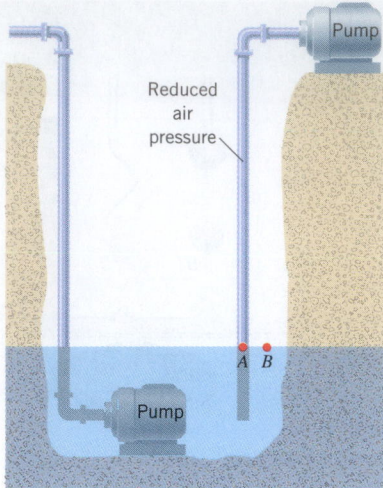

Figure 11.10 A water pump can be placed at the bottom of a well or at ground level. Conceptual Example 6 discusses the two placements.

Conceptual Example 6 Pumping Water

The physics of pumping water from a well. Figure 11.10 shows two methods for pumping water from a well. In one method, the pump is submerged in the water at the bottom of the well, while in the other, it is located at ground level. If the well is shallow, either technique can be used. However, if the well is very deep, only one of the methods works. Which pumping method works, **(a)** the submerged pump or **(b)** the pump located at ground level?

Reasoning To answer this question, we need to examine the nature of the job done by the pump in each place. The pump at the bottom of the well pushes water up the pipe, while the pump at ground level does not push water at all. Instead, the ground-level pump removes air from the pipe, creating a partial vacuum within it. (It's acting just like you do when drinking through a straw. You draw some of the air out of the straw, and the external air pressure pushes the liquid up into it.)

Answer (b) is incorrect. As the pump at ground level removes air from the pipe, the pressure above the water within the pipe is reduced (see point *A* in the drawing). The greater air pressure outside the pipe (see point *B*) pushes water up the pipe. However, even the strongest pump can only remove *all* of the air. Once the air is completely removed, an increase in pump strength does not increase the height to which the water is pushed by the external air pressure. Thus, the ground-level pump can only cause water to rise to a certain maximum height and cannot be used for very deep wells.

Answer (a) is correct. For a very deep well, the column of water becomes very tall, and the pressure at the bottom of the pipe becomes large, due to the pressure increment ρgh in the relation $P_2 = P_1 + \rho gh$ (Equation 11.4). However, as long as the pump can push with sufficient strength to overcome the large pressure, it can shove the next increment of water into the pipe, so the method can be used for very deep wells.

Related Homework: *Problems 21, 93*

✓ CHECK YOUR UNDERSTANDING

(The answers are given at the end of the book.)

4. A scuba diver is swimming under water, and the graph shows a plot of the water pressure acting on the diver as a function of time. In each of the three regions, **(a)** A → B, **(b)** B → C, and **(c)** C → D, does the depth of the diver increase, decrease, or remain constant?

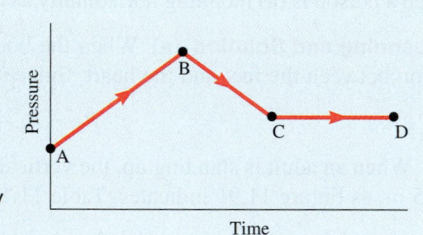

5. A 15-meter-high tank is closed and completely filled with water. A valve is then opened at the bottom of the tank and water begins to flow out. When the water stops flowing, will the tank be completely empty, or will there still be a noticeable amount of water in it?

6. Could you use a straw to sip a drink on the moon? **(a)** Yes. It would be no different than drinking with a straw on earth. **(b)** No, because there is no air on the moon and, therefore, no air pressure to push the liquid up the straw. **(c)** Yes, and it is easier on the moon because the acceleration due to gravity on the moon is only $\frac{1}{6}$ of that on the earth.

7. A scuba diver is below the surface of the water when a storm approaches, dropping the air pressure above the water. Would a sufficiently sensitive pressure gauge attached to his wrist register this drop in air pressure? Assume that the diver's wrist does not move as the storm approaches.

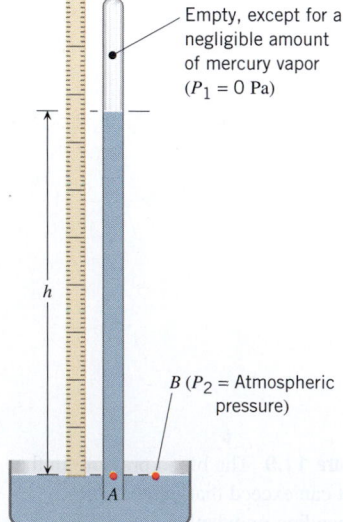

Figure 11.11 A mercury barometer.

Empty, except for a negligible amount of mercury vapor ($P_1 = 0$ Pa)

h

B (P_2 = Atmospheric pressure)

A

11.4 PRESSURE GAUGES

One of the simplest pressure gauges is the mercury barometer used for measuring atmospheric pressure. This device is a tube sealed at one end, filled completely with mercury, and then inverted, so that the open end is under the surface of a pool of mercury (see Figure 11.11). Except for a negligible amount of mercury vapor, the space above the mercury in the tube is empty, and the pressure P_1 is nearly zero there. The pressure P_2 at point *A* at the bottom of the mercury column is the same as the pressure at point *B*—namely, atmospheric pressure—because these two points are at the same level. With $P_1 = 0$ Pa and $P_2 = P_{atm}$, it follows from Equation 11.4 that $P_{atm} = 0$ Pa $+ \rho g h$. Thus, the atmospheric pressure can be determined from the height h of the mercury in the tube, the density ρ of mercury, and the acceleration due to gravity. Usually, weather forecasters report the pressure in terms of the height h, expressing it in millimeters or inches of mercury. For instance, using $P_{atm} = 1.013 \times 10^5$ Pa and $\rho = 13.6 \times 10^3$ kg/m^3 for the density of mercury, we find that $h = P_{atm}/(\rho g) = 760$ mm (29.9 inches).* Slight variations from this value occur, depending on weather conditions and altitude.

Figure 11.12 shows another kind of pressure gauge, the open-tube manometer. The phrase "open-tube" refers to the fact that one side of the U-tube is open to atmospheric pressure. The tube contains a liquid, often mercury, and its other side is connected to the container whose pressure P_2 is to be measured. When the pressure in the container is equal to the atmospheric pressure, the liquid levels in both sides of the U-tube are the same. When the pressure in the container is greater than atmospheric pressure, as in Figure 11.12, the liquid in the tube is pushed downward on the left side and upward on the right side. The relation $P_2 = P_1 + \rho g h$ can be used to determine the container pressure. Atmospheric pressure exists at the top of the right column, so that $P_1 = P_{atm}$. The pressure P_2 is the same at points A and B, so we find that $P_2 = P_{atm} + \rho g h$, or

$$P_2 - P_{atm} = \rho g h$$

The height h is proportional to $P_2 - P_{atm}$, which is called the **gauge pressure.** The gauge pressure is the amount by which the container pressure differs from atmospheric pressure. The actual value for P_2 is called the **absolute pressure.**

The sphygmomanometer is a familiar device for measuring blood pressure. As Figure 11.13 illustrates, a squeeze bulb can be used to inflate the cuff with air, which cuts off the flow of blood through the artery below the cuff. When the release valve is opened, the cuff pressure drops. Blood begins to flow again when the pressure created by the heart at the peak of its beating cycle exceeds the cuff pressure. Using a stethoscope to listen for the initial flow, the operator can measure the corresponding cuff gauge pressure with, for example, an open-tube manometer. This cuff gauge pressure is called the *systolic* pressure. Eventually, there comes a point when even the pressure created by the heart at the low point of its beating cycle is sufficient to cause blood to flow. Identifying this point with the stethoscope, the operator can measure the corresponding cuff gauge pressure, which is referred to as the *diastolic* pressure. The systolic and diastolic pressures are reported in millimeters of mercury, and values of less than 120 and 80, respectively, are typical of a young, healthy heart.

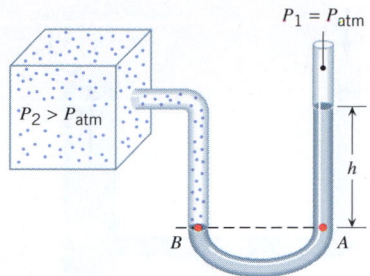

$P_1 = P_{atm}$

$P_2 > P_{atm}$

B A

Figure 11.12 The U-shaped tube is called an open-tube manometer and can be used to measure the pressure P_2 in a container.

Problem-solving insight

When solving problems that deal with pressure, be sure to note the distinction between gauge pressure and absolute pressure.

The physics of a sphygmomanometer.

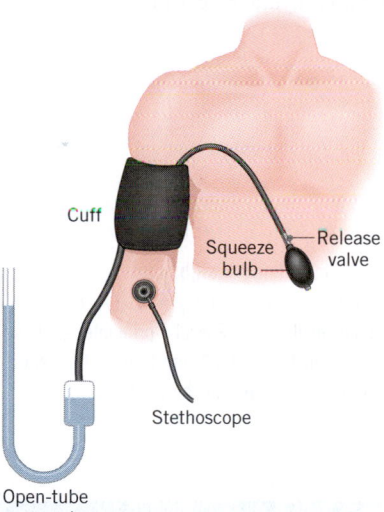

Cuff

Squeeze bulb

Release valve

Stethoscope

Open-tube manometer

Figure 11.13 A sphygmomanometer is used to measure blood pressure.

11.5 PASCAL'S PRINCIPLE

As we have seen, the pressure in a fluid increases with depth, due to the weight of the fluid above the point of interest. A completely enclosed fluid may be subjected to an additional pressure by the application of an external force. For example, Figure 11.14a shows two interconnected cylindrical chambers. The chambers have different diameters and, together with the connecting tube, are completely filled with a liquid. The larger chamber is sealed at the top with a cap, while the smaller one is fitted with a movable piston. Consider the pressure P_1 at a point immediately beneath the piston. According to the definition of pressure, it is the magnitude F_1 of the external force divided by the area A_1 of the piston: $P_1 = F_1/A_1$. If it is necessary to know the pressure P_2 at any deeper place in the liquid, we just add to the value of P_1 the increment $\rho g h$, which takes into account the depth h below the piston: $P_2 = P_1 + \rho g h$. The important feature here is this: The pressure P_1 adds to the pressure $\rho g h$ due to the depth of the liquid at any point, whether that point is in the smaller chamber, the connecting tube, or the larger chamber. Therefore, if the applied pressure P_1 is increased or decreased, the pressure at any other point within the confined liquid changes correspondingly. This behavior is described by **Pascal's principle.**

*A pressure of one millimeter of mercury is sometimes referred to as one *torr*, to honor the inventor of the barometer, Evangelista Torricelli (1608–1647). Thus, one atmosphere of pressure is 760 torr.

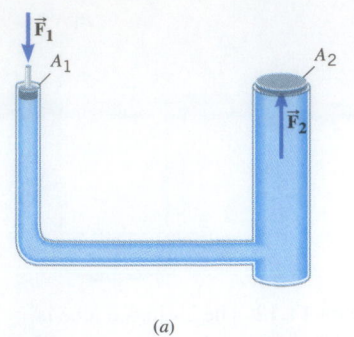

(a)

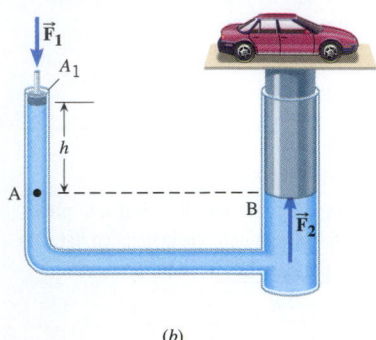

(b)

Figure 11.14 (a) An external force $\vec{F}_1$ is applied to the piston on the left. As a result, a force $\vec{F}_2$ is exerted on the cap on the chamber on the right. (b) The familiar hydraulic car lift.

PASCAL'S PRINCIPLE

Any change in the pressure applied to a completely enclosed fluid is transmitted undiminished to all parts of the fluid and the enclosing walls.

The usefulness of the arrangement in Figure 11.14a becomes apparent when we calculate the force F_2 applied by the liquid to the cap on the right side. The area of the cap is A_2 and the pressure there is P_2. As long as the tops of the left and right chambers are at the same level, the pressure increment $\rho g h$ is zero, so that the relation $P_2 = P_1 + \rho g h$ becomes $P_2 = P_1$. Consequently, $F_2/A_2 = F_1/A_1$, and

$$F_2 = F_1\left(\frac{A_2}{A_1}\right) \tag{11.5}$$

If area A_2 is larger than area A_1, a large force $\vec{F}_2$ can be applied to the cap on the right chamber, starting with a smaller force $\vec{F}_1$ on the left. Depending on the ratio of the areas A_2/A_1, the force $\vec{F}_2$ can be large indeed, as in the familiar hydraulic car lift shown in Figure 11.14b. In this device the force $\vec{F}_2$ is not applied to a cap that seals the larger chamber, but, rather, to a movable plunger that lifts a car. Examples 7 and 8 deal with a hydraulic car lift.

Example 7 A Car Lift

In the hydraulic car lift shown in Figure 11.14b, the input piston on the left has a radius of $r_1 = 0.0120$ m and a negligible weight. The output plunger on the right has a radius of $r_2 = 0.150$ m. The combined weight of the car and the plunger is 20 500 N. Since the output force has a magnitude of $F_2 = 20\,500$ N, it supports the car. Suppose that the bottom surfaces of the piston and plunger are at the same level, so that $h = 0$ m in Figure 11.14b. What is the magnitude F_1 of the input force that is needed so that $F_2 = 20\,500$ N?

Reasoning When the bottom surfaces of the piston and plunger are at the same level, as in Figure 11.14a, Equation 11.5 applies, and we can use it to determine F_1.

Solution According to Equation 11.5, we have

$$F_2 = F_1\left(\frac{A_2}{A_1}\right) \quad \text{or} \quad F_1 = F_2\left(\frac{A_1}{A_2}\right)$$

Problem-solving insight
Note that the relation $F_1 = F_2(A_1/A_2)$, which results from Pascal's principle, applies only when the points 1 and 2 lie at the same depth ($h = 0$ m) in the fluid.

Using $A = \pi r^2$ for the circular areas of the piston and plunger, we find that

$$F_1 = F_2\left(\frac{A_1}{A_2}\right) = F_2\left(\frac{\pi r_1^2}{\pi r_2^2}\right) = (20\,500\text{ N})\frac{(0.0120\text{ m})^2}{(0.150\text{ m})^2} = \boxed{131\text{ N}}$$

ANALYZING MULTIPLE-CONCEPT PROBLEMS

Example 8 A Car Lift Revisited

The data are the same as in Example 7. Suppose now, however, that the bottom surfaces of the piston and plunger are at *different* levels, such that $h = 1.10$ m in Figure 11.14b. The car lift uses hydraulic oil that has a density of 8.00×10^2 kg/m³. What is the magnitude F_1 of the input force that is now needed to produce an output force having a magnitude of $F_2 = 20\,500$ N?

The physics of
a hydraulic car lift.

Reasoning Since the bottom surfaces of the piston and plunger are at different levels, Equation 11.5 no longer applies. Our approach will be based on the definition of pressure (Equation 11.3) and the relation between pressure and depth given in Equation 11.4. It is via Equation 11.4 that we will take into account the fact that the bottom of the output plunger is $h = 1.10$ m below the bottom of the input piston.

Knowns and Unknowns   We have the following data:

Description	Symbol	Value	Comment
Radius of input piston	r_1	0.0120 m	
Radius of output piston	r_2	0.150 m	
Magnitude of output force	F_2	20 500 N	
Density of hydraulic oil	ρ	8.00×10^2 kg/m³	
Level difference between input piston and output plunger	h	1.10 m	See Figure 11.14b.
Unknown Variable			
Magnitude of input force	F_1	?	

Modeling the Problem

STEP 1   Force and Pressure   The input force F_1 is determined by the pressure P_1 acting on the bottom surface of the input piston, which has an area A_1. According to Equation 11.3, the input force is $F_1 = P_1 A_1$, where $A_1 = \pi r_1^2$ because the piston has a circular cross section with a radius r_1. With this expression for the area A_1, the input force can be written as in Equation 1 at the right. We have a value for r_1, but P_1 is unknown, so we turn to Step 2 in order to obtain it.

$$F_1 = P_1 \pi r_1^2 \qquad (1)$$

STEP 2   Pressure and Depth in a Static Fluid   In Figure 11.14b, the bottom surface of the plunger at point B is at the same level as point A, which is at a depth h beneath the input piston. Equation 11.4 applies, so that

$$P_2 = P_1 + \rho gh \quad \text{or} \quad \boxed{P_1 = P_2 - \rho gh}$$

This result for P_1 can be substituted into Equation 1, as shown at the right. Now it is necessary to have a value for P_2, the pressure at the bottom surface of the output plunger, which we obtain in Step 3.

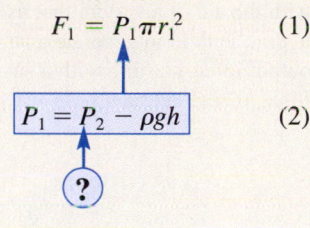

$$F_1 = P_1 \pi r_1^2 \qquad (1)$$
$$\boxed{P_1 = P_2 - \rho gh} \qquad (2)$$

STEP 3   Pressure and Force   By using Equation 11.3, we can relate the unknown pressure P_2 to the given output force F_2 and the area A_2 of the bottom surface of the output plunger: $P_2 = F_2/A_2$. The area A_2 is a circle with a radius r_2, so $A_2 = \pi r_2^2$. Thus, the pressure P_2 is

$$\boxed{P_2 = \frac{F_2}{\pi r_2^2}}$$

At the right, this result is substituted into Equation 2.

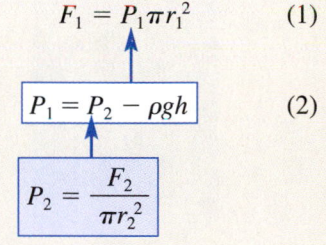

$$F_1 = P_1 \pi r_1^2 \qquad (1)$$
$$P_1 = P_2 - \rho gh \qquad (2)$$
$$\boxed{P_2 = \frac{F_2}{\pi r_2^2}}$$

Solution   Combining the results of each step algebraically, we find that

$$\overset{\text{STEP 1}}{\underset{}{}} \quad \overset{\text{STEP 2}}{\underset{}{}} \qquad \overset{\text{STEP 3}}{\underset{}{}}$$
$$F_1 = P_1 \pi r_1^2 = (P_2 - \rho gh)\pi r_1^2 = \left(\frac{F_2}{\pi r_2^2} - \rho gh \right)\pi r_1^2$$

Thus, we find that the necessary input force is

$$F_1 = \frac{F_2 r_1^2}{r_2^2} - \rho gh \pi r_1^2$$

$$= \frac{(20\,500 \text{ N})(0.0120 \text{ m})^2}{(0.150 \text{ m})^2} - (8.00 \times 10^2 \text{ kg/m}^3)(9.80 \text{ m/s}^2)(1.10 \text{ m})\pi (0.0120 \text{ m})^2$$

$$= \boxed{127 \text{ N}}$$

The answer here is less than the answer in Example 7 because the weight of the 1.10-m column of hydraulic oil provides some of the input force to support the car and plunger.

Related Homework: *Problem 35*

Figure 11.15 At this energy plant, trucks carrying wood waste are emptied with the aid of a system that uses a hydraulic fluid to generate a large output force, starting with a small input force. (Creatas/PictureQuest)

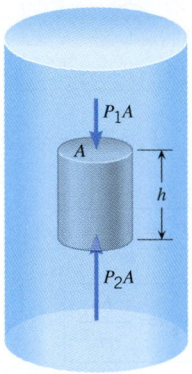

Figure 11.16 The fluid applies a downward force P_1A to the top face of the submerged cylinder and an upward force P_2A to the bottom face.

In a device such as a hydraulic car lift, the same amount of work is done by both the input and output forces in the absence of friction. The larger output force $\vec{\mathbf{F}}_2$ moves through a smaller distance, while the smaller input force $\vec{\mathbf{F}}_1$ moves through a larger distance. The work, being the product of the magnitude of the force and the distance, is the same in either case since mechanical energy is conserved (assuming that friction is negligible).

An enormous variety of clever devices use hydraulic fluids, just as the car lift does. In Figure 11.15, for instance, the fluids multiply a small input force into the large output force required to rotate a truck into a position where it can be easily emptied.

11.6 ARCHIMEDES' PRINCIPLE

Anyone who has tried to push a beach ball under the water has felt how the water pushes back with a strong upward force. This upward force is called the ***buoyant force,*** and all fluids apply such a force to objects that are immersed in them. The buoyant force exists because fluid pressure is larger at greater depths.

In Figure 11.16 a cylinder of height h is being held under the surface of a liquid. The pressure P_1 on the top face generates the downward force P_1A, where A is the area of the face. Similarly, the pressure P_2 on the bottom face generates the upward force P_2A. Since the pressure is greater at greater depths, the upward force exceeds the downward force. Consequently, the liquid applies to the cylinder a net upward force, or buoyant force, whose magnitude F_B is

$$F_B = P_2A - P_1A = (P_2 - P_1)A = \rho g h A$$

We have substituted $P_2 - P_1 = \rho g h$ from Equation 11.4 into this result. In so doing, we find that the buoyant force equals $\rho g h A$. The quantity hA is the volume of liquid that the cylinder moves aside or displaces in being submerged, and ρ denotes the density of the liquid, not the density of the material from which the cylinder is made. Therefore, $\rho h A$ gives the mass m of the displaced fluid, so that the buoyant force equals mg, the weight of the displaced fluid. The phrase "weight of the displaced fluid" refers to the weight of the fluid that would spill out if the container were filled to the brim before the cylinder is inserted into the liquid. The buoyant force is not a new type of force. It is just the name given to the net upward force exerted by the fluid on the object.

The shape of the object in Figure 11.16 is not important. No matter what its shape, the buoyant force pushes it upward in accord with ***Archimedes' principle.*** It was an impressive accomplishment that the Greek scientist Archimedes (ca. 287–212 BC) discovered the essence of this principle so long ago.

ARCHIMEDES' PRINCIPLE

Any fluid applies a buoyant force to an object that is partially or completely immersed in it; the magnitude of the buoyant force equals the weight of the fluid that the object displaces:

$$\underbrace{F_B}_{\substack{\text{Magnitude of} \\ \text{buoyant force}}} = \underbrace{W_{\text{fluid}}}_{\substack{\text{Weight of} \\ \text{displaced fluid}}} \qquad (11.6)$$

The effect that the buoyant force has depends on its strength compared with the strengths of the other forces that are acting. For example, if the buoyant force is strong enough to balance the force of gravity, an object will float in a fluid. Figure 11.17 explores this possibility. In part *a*, a block that weighs 100 N displaces some liquid, and the liquid applies a buoyant force F_B to the block, according to Archimedes' principle. Nevertheless, if the block were released, it would fall further into the liquid because the buoyant force is not sufficiently strong to balance the weight of the block. In part *b*, however, enough of the block is submerged to provide a buoyant force that can balance the 100-N weight, so the block is in equilibrium and floats when released. If the buoyant force were not large enough to balance the weight, even with the block completely submerged, the block would sink. Even if an object sinks, there is still a buoyant force acting on it; it's just that the buoyant force is not large enough to balance the weight. Example 9 provides additional insight into what determines whether an object floats or sinks in a fluid.

Example 9 A Swimming Raft

A solid, square pinewood raft measures 4.0 m on a side and is 0.30 m thick. **(a)** Determine whether the raft floats in water, and **(b)** if so, how much of the raft is beneath the surface (see the distance h in Figure 11.18).

Reasoning To determine whether the raft floats, we will compare the weight of the raft to the maximum possible buoyant force and see whether there could be enough buoyant force to balance the weight. If so, then the value of the distance h can be obtained by utilizing the fact that the floating raft is in equilibrium, with the magnitude of the buoyant force equaling the raft's weight.

Solution **(a)** The weight of the raft is $W_{raft} = m_{pine}g$ (Equation 4.5), where m_{pine} is the mass of the raft and can be calculated as $m_{pine} = \rho_{pine}V_{pine}$ (Equation 11.1). The pinewood's density is $\rho_{pine} = 550$ kg/m^3 (Table 11.1), and its volume is $V_{pine} = 4.0$ m $\times$ 4.0 m $\times$ 0.30 m $= 4.8$ m^3. Thus, we find the weight of the raft to be

$$W_{raft} = m_{pine}g = (\rho_{pine}V_{pine})g = (550 \text{ kg/m}^3)(4.8 \text{ m}^3)(9.80 \text{ m/s}^2) = 26\,000 \text{ N}$$

The maximum possible buoyant force occurs when the entire raft is under the surface, displacing a volume of water that is $V_{water} = V_{pine} = 4.8$ m^3. According to Archimedes' principle, the weight of this volume of water is the maximum buoyant force F_B^{MAX}. It can be obtained using the density of water:

$$F_B^{MAX} = \rho_{water}V_{water}\,g = (1.000 \times 10^3 \text{ kg/m}^3)(4.8 \text{ m}^3)(9.80 \text{ m/s}^2) = 47\,000 \text{ N}$$

Since the maximum possible buoyant force exceeds the 26 000-N weight of the raft, the raft will float only partially submerged at a distance h beneath the water.

Problem-solving insight: When using Archimedes' principle to find the buoyant force F_B that acts on an object, be sure to use the density of the displaced fluid, not the density of the object.

(b) We now find the value of h. The buoyant force balances the raft's weight, so $F_B = 26\,000$ N. However, according to Equation 11.6, the magnitude of the buoyant force is also the weight of the displaced water, so $F_B = 26\,000$ N $= W_{water}$. Using the density of water, we can also express the weight of the displaced water as $W_{water} = \rho_{water}V_{water}g$, where the water volume is $V_{water} = 4.0$ m $\times$ 4.0 m $\times h$. As a result,

$$26\,000 \text{ N} = W_{water} = \rho_{water}(4.0 \text{ m} \times 4.0 \text{ m} \times h)g$$

$$h = \frac{26\,000 \text{ N}}{\rho_{water}(4.0 \text{ m} \times 4.0 \text{ m})g}$$

$$= \frac{26\,000 \text{ N}}{(1.000 \times 10^3 \text{ kg/m}^3)(4.0 \text{ m} \times 4.0 \text{ m})(9.80 \text{ m/s}^2)} = \boxed{0.17 \text{ m}}$$

In order to decide whether the raft will float in part (a) of Example 9, we compared the weight of the raft ($\rho_{pine}V_{pine}$)g to the maximum possible buoyant force ($\rho_{water}V_{water}$)$g = (\rho_{water}V_{pine})g$. The comparison depends only on the densities ρ_{pine} and ρ_{water}. The take-home message is this: Any object that is *solid throughout* will float in a liquid if the density of the object is less than or equal to the density of the liquid. For instance, at 0 °C ice has a density of 917 kg/m^3, whereas water has a density of 1000 kg/m^3. Therefore, ice floats in water.

Although a solid piece of a high-density material like steel will sink in water, such materials can, nonetheless, be used to make floating objects. A supertanker, for example, floats because it is *not* solid metal. It contains enormous amounts of empty space and, because of its shape, displaces enough water to balance its own large weight. Conceptual Example 10 focuses on an interesting aspect of a floating ship.

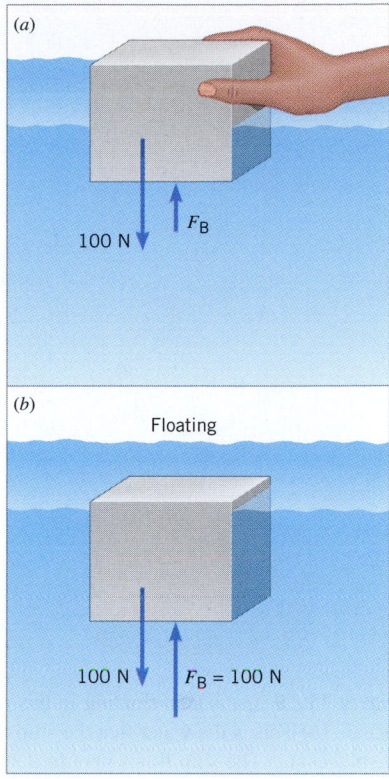

Figure 11.17 (*a*) An object of weight 100 N is being immersed in a liquid. The deeper the object is, the more liquid it displaces, and the stronger the buoyant force is. (*b*) The buoyant force matches the 100-N weight, so the object floats.

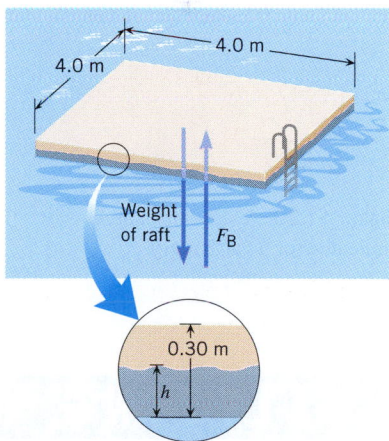

Figure 11.18 A raft floating with a distance h beneath the water.

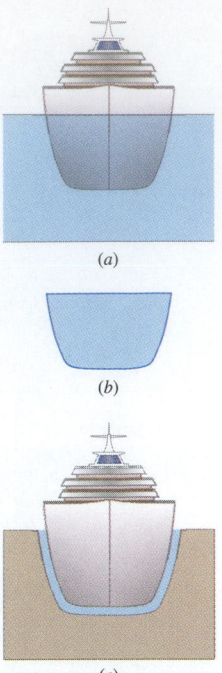

Figure 11.19 (*a*) A ship floating in the ocean. (*b*) This is the water that the ship displaces. (*c*) The ship floats here in a canal that has a cross section similar in shape to that in part *b*.

⬇ Conceptual Example 10 How Much Water Is Needed to Float a Ship?

A ship floating in the ocean is a familiar sight. But is all that water really necessary? Can an ocean vessel float in the amount of water that a swimming pool contains, for instance?

Reasoning and Solution In principle, a ship can float in much less than the amount of water in a swimming pool. To see why, look at Figure 11.19. Part *a* shows the ship floating in the ocean because it contains empty space within its hull and displaces enough water to balance its own weight. Part *b* shows the water that the ship displaces, which, according to Archimedes' principle, has a weight that equals the ship's weight. Although this wedge-shaped portion of water represents the water *displaced* by the ship, it is not the amount that must be present to float the ship, as part *c* illustrates. This part of the drawing shows a canal, the cross section of which matches the shape in part *b*. *All that is needed, in principle, is a thin section of water that separates the hull of the floating ship from the sides of the canal.* This thin section of water could have a very small volume indeed.

The physics of a state-of-charge battery indicator. Archimedes' principle is used in some car batteries to alert the owner that recharging is necessary, via a state-of-charge indicator, such as the one illustrated in Figure 11.20. The battery includes a viewing port that looks down through a plastic rod, which extends into the battery acid. Attached to the end of this rod is a "cage" containing a green ball. The cage has holes in it that allow the acid to enter. When the battery is charged, the density of the acid is great enough that its buoyant force makes the ball rise to the top of the cage, to just beneath the plastic rod. The viewing port shows a green dot. As the battery discharges, the density of the acid decreases. Since the buoyant force is the weight of the acid displaced by the ball, the buoyant force also decreases. As a result, the ball sinks into one of the two chambers oriented at an angle beneath the plastic rod (see Figure 11.20). With the ball no longer visible, the viewing port shows a dark or black dot, warning that the battery charge is low.

Archimedes' principle has allowed us to determine how an object can float in a liquid. This principle also applies to gases, as the next example illustrates.

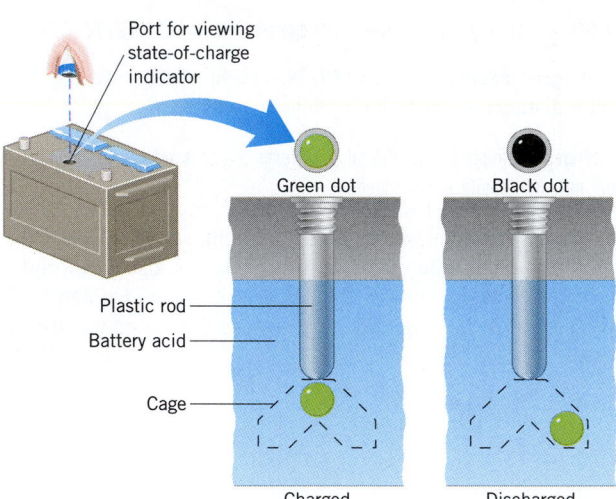

Port for viewing state-of-charge indicator

Green dot — Black dot

Plastic rod

Battery acid

Cage

Charged — Discharged

Figure 11.20 A state-of-charge indicator for a car battery.

ANALYZING MULTIPLE-CONCEPT PROBLEMS

Example 11 A Goodyear Airship

Normally, a Goodyear airship, such as the one in Figure 11.21, contains about 5.40×10^3 m^3 of helium (He), whose density is 0.179 kg/m^3. Find the weight W_L of the load that the airship can carry in equilibrium at an altitude where the density of air is 1.20 kg/m^3.

The physics of a Goodyear airship.

Reasoning The airship and its load are in equilibrium. Thus, the buoyant force F_B applied to the airship by the surrounding air balances the weight W_{He} of the helium and the weight W_L of the load, including the solid parts of the airship. The free-body diagram in Figure 11.21*b* shows these forces. The buoyant force is the weight of the air that the ship displaces, according to Archimedes' principle. The weight of air or helium is given as $W = mg$ (Equation 4.5). Using the density ρ and volume V, we can express the mass m as $m = \rho V$ (Equation 11.1).

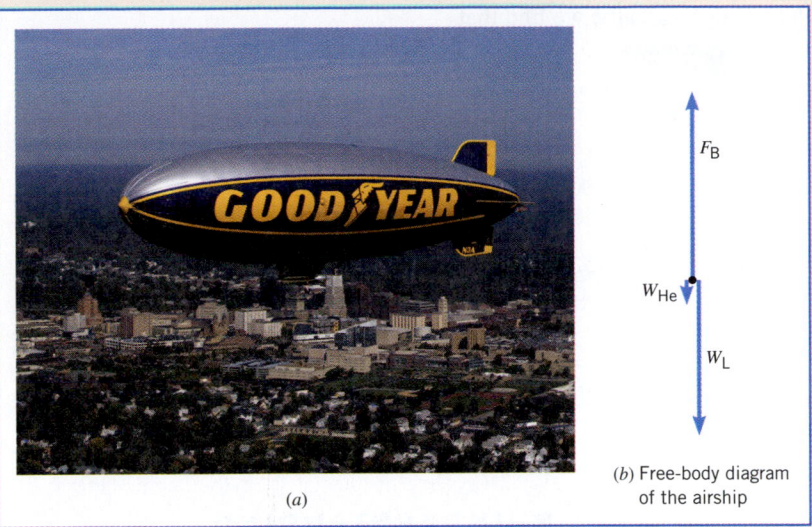

(a)

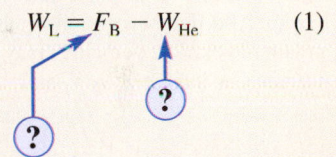

(b) Free-body diagram
of the airship

Figure 11.21 (a) A helium-filled Goodyear airship. (b) The free-body diagram of the airship, including the load weight W_L, the weight W_{He} of the helium, and the buoyant force F_B. (a, Aaron Vandersommers/Goodyear)

Knowns and Unknowns The following table summarizes the data:

Description	Symbol	Value
Volume of helium in airship	V_{He}	5.40×10^3 m³
Density of helium	ρ_{He}	0.179 kg/m³
Density of air	ρ_{air}	1.20 kg/m³
Unknown Variable		
Weight of load	W_L	?

Modeling the Problem

STEP 1 Equilibrium   Because the forces in the free-body diagram (see Figure 11.21b) balance at equilibrium, we have

$$W_{He} + W_L = F_B$$

Rearranging this result gives Equation 1 at the right. Neither the buoyant force F_B nor the weight W_{He} of the helium is known. We will determine them in Steps 2 and 3.

$$W_L = F_B - W_{He} \qquad (1)$$

STEP 2 Weight and Density   According to Equation 4.5, the weight is given by $W = mg$. On the other hand, the density ρ is defined as the mass m divided by the volume V (see Equation 11.1), so we know that $m = \rho V$. Thus, the weight of the helium in the airship is

$$W_{He} = m_{He} g = \rho_{He} V_{He} g$$

The result can be substituted into Equation 1, as shown at the right. We turn now to Step 3 to evaluate the buoyant force.

$$W_L = F_B - W_{He} \qquad (1)$$
$$W_{He} = \rho_{He} V_{He} g$$

STEP 3 Archimedes' Principle   The buoyant force is given by Archimedes' principle as the weight of the air displaced by the airship. Thus, following the approach in Step 2, we can write the buoyant force as follows:

$$F_B = W_{air} = \rho_{air} V_{air} g$$

In this result, V_{air} is very nearly the same as V_{He}, since the volume occupied by the materials of the ship's outer structure is negligible compared to V_{He}. Assuming that $V_{air} = V_{He}$, we see that the expression for the buoyant force becomes

$$F_B = \rho_{air} V_{He} g$$

Substitution of this value for the buoyant force into Equation 1 is shown at the right.

$$W_L = F_B - W_{He} \qquad (1)$$
$$W_{He} = \rho_{He} V_{He} g$$
$$F_B = \rho_{air} V_{He} g$$

Continued

Solution Combining the results of each step algebraically, we find that

STEP 1 STEP 2 STEP 3

$$W_L = F_B - W_{He} = F_B - \rho_{He}V_{He}g = \rho_{air}V_{He}g - \rho_{He}V_{He}g$$

The weight of the load that the airship can carry at an altitude where $\rho_{air} = 1.20$ kg/m^3 is, then,

$$W_L = (\rho_{air} - \rho_{He})V_{He}g = (1.20 \text{ kg/m}^3 - 0.179 \text{ kg/m}^3)(5.40 \times 10^3 \text{ m}^3)(9.80 \text{ m/s}^2)$$

$$= \boxed{5.40 \times 10^4 \text{ N}}$$

Related Homework: *Problems 41, 50*

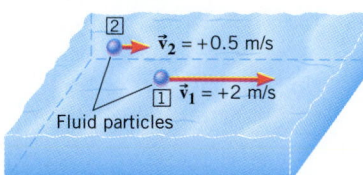

Figure 11.22 Two fluid particles in a stream. At different locations in the stream the particle velocities may be different, as indicated by $\vec{v}_1$ and $\vec{v}_2$.

Figure 11.23 The flow of water in white-water rapids is an example of turbulent flow. (Gary M Prior/Allsport/ Getty Images Sport Services)

✓ **CHECK YOUR UNDERSTANDING**

(The answers are given at the end of the book.)

8. A glass is filled to the brim with water and has an ice cube floating in it. When the ice cube melts, what happens? **(a)** Water spills out of the glass. **(b)** The water level in the glass drops. **(c)** The water level in the glass does not change.

9. A steel beam is suspended completely under water by a cable that is attached to one end of the beam, so it hangs vertically. Another identical beam is also suspended completely under water, but by a cable that is attached to the center of the beam, so it hangs horizontally. Which beam, if either, experiences the greater buoyant force? Neglect any change in water density with depth.

10. A glass beaker, filled to the brim with water, is resting on a scale. A solid block is then placed in the water, causing some of it to spill over. The water that spills is wiped away, and the beaker is still filled to the brim. How do the initial and final readings on the scale compare if the block is made from **(a)** wood (whose density is less than that of water) and **(b)** iron (whose density is greater than that of water)?

11. On a distant planet the acceleration due to gravity is less than it is on earth. Would you float more easily in water on this planet than on earth?

12. As a person dives toward the bottom of a swimming pool, the pressure increases noticeably. Does the buoyant force acting on her also increase? Neglect any change in water density with depth.

13. A pot is partially filled with water, in which a plastic cup is floating. Inside the floating cup is a small block of lead. When the lead block is removed from the cup and placed into the water, it sinks to the bottom. When this happens, does the water level in the pot **(a)** rise, **(b)** fall, or **(c)** remain the same?

11.7 FLUIDS IN MOTION

Fluids can move or flow in many ways. Water may flow smoothly and slowly in a quiet stream or violently over a waterfall. The air may form a gentle breeze or a raging tornado. To deal with such diversity, it helps to identify some of the basic types of fluid flow.

Fluid flow can be steady or unsteady. In *steady flow* the velocity of the fluid particles at any point is constant as time passes. For instance, in Figure 11.22 a fluid particle flows with a velocity of $\vec{v}_1 = +2$ m/s past point 1. In steady flow every particle passing through this point has this same velocity. At another location the velocity may be different, as in a river, which usually flows fastest near its center and slowest near its banks. Thus, at point 2 in the figure, the fluid velocity is $\vec{v}_2 = +0.5$ m/s, and if the flow is steady, all particles passing through this point have a velocity of $+0.5$ m/s. *Unsteady flow* exists whenever the velocity at a point in the fluid changes as time passes. *Turbulent flow* is an extreme kind of unsteady flow and occurs when there are sharp obstacles or bends in the path of a fast-moving fluid, as in the rapids in Figure 11.23. In turbulent flow, the velocity at a point changes erratically from moment to moment, both in magnitude and in direction.

Fluid flow can be compressible or incompressible. Most liquids are nearly incompressible; that is, the density of a liquid remains almost constant as the pressure changes. To a good approximation, then, liquids flow in an incompressible manner. In contrast, gases are highly compressible. However, there are situations in which the density of a flowing gas remains constant enough that the flow can be considered incompressible.

Fluid flow can be viscous or nonviscous. A viscous fluid, such as honey, does not flow readily and is said to have a large viscosity.* In contrast, water is less viscous and flows more readily; water has a smaller viscosity than honey. The flow of a viscous fluid is an energy-dissipating process. The viscosity hinders neighboring layers of fluid from sliding freely past one another. A fluid with zero viscosity flows in an unhindered manner with no dissipation of energy. Although no real fluid has zero viscosity at normal temperatures, some fluids have negligibly small viscosities. An incompressible, nonviscous fluid is called an *ideal fluid.*

When the flow is steady, *streamlines* are often used to represent the trajectories of the fluid particles. A streamline is a line drawn in the fluid such that a tangent to the streamline at any point is parallel to the fluid velocity at that point. Figure 11.24 shows the velocity vectors at three points along a streamline. The fluid velocity can vary (in both magnitude and direction) from point to point along a streamline, but at any given point, the velocity is constant in time, as required by the condition of steady flow. In fact, steady flow is often called *streamline flow.*

Figure 11.25a illustrates a method for making streamlines visible by using small tubes to release a colored dye into the moving liquid. The dye does not immediately mix with the liquid and is carried along a streamline. In the case of a flowing gas, such as that in a wind tunnel, streamlines are often revealed by smoke streamers, as part b of the figure shows.

In steady flow, the pattern of streamlines is steady in time, and, as Figure 11.25a indicates, no two streamlines cross one another. If they did cross, every particle arriving at the crossing point could go one way or the other. This would mean that the velocity at the crossing point would change from moment to moment, a condition that does not exist in steady flow.

✔ **CHECK YOUR UNDERSTANDING**

(The answer is given at the end of the book.)

14. In steady flow, the velocity $\vec{v}$ of a fluid particle at any point is constant in time. On the other hand, the fluid in a pipe accelerates when it moves from a larger-diameter section of the pipe into a smaller-diameter section, so the velocity is increasing during the transition. Does the condition of steady flow rule out such an acceleration?

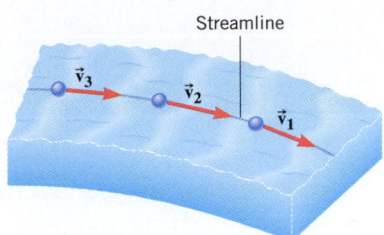

Figure 11.24 At any point along a streamline, the velocity vector of the fluid particle at that point is tangent to the streamline.

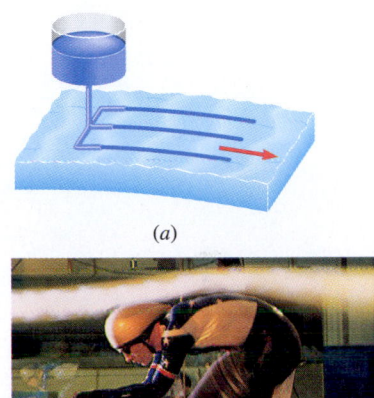

(a)

(b)

Figure 11.25 (*a*) In the steady flow of a liquid, a colored dye reveals the streamlines. (*b*) A smoke streamer reveals a streamline pattern for the air flowing around this pursuit cyclist, as he tests his bike for wind resistance in a wind tunnel. (*b*, Nicholas Pinturas/ Getty Images)

11.8 THE EQUATION OF CONTINUITY

Have you ever used your thumb to control the water flowing from the end of a hose, as in Figure 11.26? If so, you have seen that the water velocity increases when your thumb reduces the cross-sectional area of the hose opening. This kind of fluid behavior is described by the *equation of continuity.* This equation expresses the following simple idea: If a fluid enters one end of a pipe at a certain rate (e.g., 5 kilograms per second), then fluid must also leave at the same rate, assuming that there are no places between the entry and exit points to add or remove fluid. The mass of fluid per second (e.g., 5 kg/s) that flows through a tube is called the *mass flow rate.*

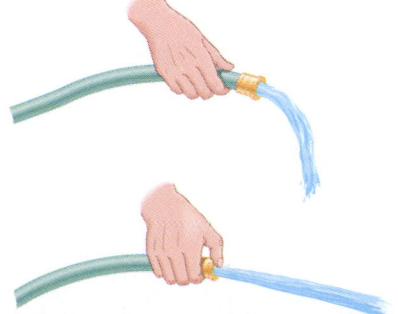

Figure 11.26 When the end of a hose is partially closed off, thus reducing its cross-sectional area, the fluid velocity increases.

*See Section 11.11 for a discussion of viscosity.

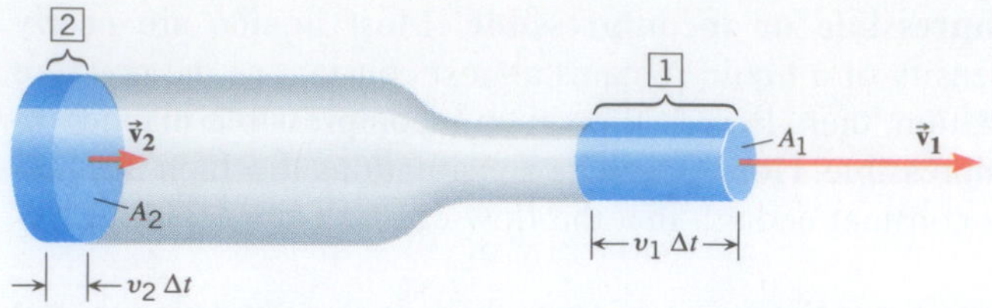

Figure 11.27 In general, a fluid flowing in a tube that has different cross-sectional areas A_1 and A_2 at positions 1 and 2 also has different velocities $\vec{v}_1$ and $\vec{v}_2$ at these positions.

Figure 11.27 shows a small mass of fluid or fluid element (dark blue) moving along a tube. Upstream at position 2, where the tube has a cross-sectional area A_2, the fluid has a speed v_2 and a density ρ_2. Downstream at location 1, the corresponding quantities are A_1, v_1, and ρ_1. During a small time interval Δt, the fluid at point 2 moves a distance of $v_2 \Delta t$, as the drawing shows. The volume of fluid that has flowed past this point is the cross-sectional area times this distance, or $A_2 v_2 \Delta t$. The mass Δm_2 of this fluid element is the product of the density and volume: $\Delta m_2 = \rho_2 A_2 v_2 \Delta t$. Dividing Δm_2 by Δt gives the mass flow rate (the mass per second):

$$\frac{\text{Mass flow rate}}{\text{at position 2}} = \frac{\Delta m_2}{\Delta t} = \rho_2 A_2 v_2 \tag{11.7a}$$

Similar reasoning leads to the mass flow rate at position 1:

$$\frac{\text{Mass flow rate}}{\text{at position 1}} = \frac{\Delta m_1}{\Delta t} = \rho_1 A_1 v_1 \tag{11.7b}$$

Since no fluid can cross the sidewalls of the tube, the mass flow rates at positions 1 and 2 must be equal. However, these positions were selected arbitrarily, so the mass flow rate has the same value everywhere in the tube, an important result known as the *equation of continuity*. The equation of continuity is an expression of the fact that mass is conserved (i.e., neither created nor destroyed) as the fluid flows.

EQUATION OF CONTINUITY

The mass flow rate ($\rho A v$) has the same value at every position along a tube that has a single entry and a single exit point for fluid flow. For two positions along such a tube

$$\rho_1 A_1 v_1 = \rho_2 A_2 v_2 \tag{11.8}$$

where ρ = fluid density (kg/m^3)
   A = cross-sectional area of tube (m^2)
   v = fluid speed (m/s)

SI Unit of Mass Flow Rate: kg/s

The density of an incompressible fluid does not change during flow, so that $\rho_1 = \rho_2$, and the equation of continuity reduces to

Incompressible fluid $\qquad\qquad\qquad A_1 v_1 = A_2 v_2 \tag{11.9}$

The quantity Av represents the volume of fluid per second (measured in m^3/s, for instance) that passes through the tube and is referred to as the *volume flow rate Q:*

$$Q = \text{Volume flow rate} = Av \tag{11.10}$$

Equation 11.9 shows that where the tube's cross-sectional area is large, the fluid speed is small, and, conversely, where the tube's cross-sectional area is small, the speed is large. Example 12 explores this behavior in more detail for the hose in Figure 11.26.

Example 12 A Garden Hose

A garden hose has an unobstructed opening with a cross-sectional area of $2.85 \times 10^{-4} \text{ m}^2$, from which water fills a bucket in 30.0 s. The volume of the bucket is $8.00 \times 10^{-3} \text{ m}^3$ (about two gallons). Find the speed of the water that leaves the hose through **(a)** the unobstructed opening and **(b)** an obstructed opening with half as much area.

Reasoning If we can determine the volume flow rate Q, we can obtain the speed of the water from Equation 11.10 as $v = Q/A$, since the area A is given. We can find the volume flow rate from the volume of the bucket and its fill time.

Solution **(a)** The volume flow rate Q is equal to the volume of the bucket divided by the fill time. Therefore, the speed of the water is

$$v = \frac{Q}{A} = \frac{(8.00 \times 10^{-3} \text{ m}^3)/(30.0 \text{ s})}{2.85 \times 10^{-4} \text{ m}^2} = \boxed{0.936 \text{ m/s}} \qquad (11.10)$$

(b) Water can be considered incompressible, so the equation of continuity can be applied in the form $A_1 v_1 = A_2 v_2$. When the opening of the hose is unobstructed, its area is A_1, and the speed of the water is that determined in part (a)—namely, $v_1 = 0.936$ m/s. Since $A_2 = \frac{1}{2} A_1$, we find that

$$v_2 = \left(\frac{A_1}{A_2}\right) v_1 = \left(\frac{A_1}{\frac{1}{2} A_1}\right)(0.936 \text{ m/s}) = \boxed{1.87 \text{ m/s}} \qquad (11.9)$$

The next example applies the equation of continuity to the flow of blood.

 Example 13 A Clogged Artery

In the condition known as atherosclerosis, a deposit, or atheroma, forms on the arterial wall and reduces the opening through which blood can flow. In the carotid artery in the neck, blood flows three times faster through a partially blocked region than it does through an unobstructed region. Determine the ratio of the effective radii of the artery at the two places.

Reasoning Blood, like most liquids, is incompressible, and the equation of continuity in the form of $A_1 v_1 = A_2 v_2$ (Equation 11.9) can be applied. In applying this equation, we use the fact that the area of a circle is πr^2.

Solution From Equation 11.9, it follows that

$$\underbrace{(\pi r_{\text{U}}^2) v_{\text{U}}}_{\substack{\text{Unobstructed} \\ \text{volume flow rate}}} = \underbrace{(\pi r_{\text{O}}^2) v_{\text{O}}}_{\substack{\text{Obstructed} \\ \text{volume flow rate}}}$$

The ratio of the radii is

$$\frac{r_{\text{U}}}{r_{\text{O}}} = \sqrt{\frac{v_{\text{O}}}{v_{\text{U}}}} = \sqrt{3} = \boxed{1.7}$$

 The physics of a clogged artery.

Problem-solving insight

The equation of continuity in the form $A_1 v_1 = A_2 v_2$ applies only when the density of the fluid is constant. If the density is not constant, the equation of continuity is $\rho_1 A_1 v_1 = \rho_2 A_2 v_2$

✓ **CHECK YOUR UNDERSTANDING**

(The answers are given at the end of the book.)

15. Water flows from left to right through the five sections (A, B, C, D, E) of the pipe shown in the drawing. In which section(s) does the water speed increase, decrease, and remain constant? Treat the water as an incompressible fluid.

	Speed Increases	Speed Decreases	Speed Is Constant
(a)	A, B	D, E	C
(b)	D	B	A, C, E
(c)	D, E	A, B	C
(d)	B	D	A, C, E
(e)	A, B	C, D	E

Continued

16. See **Concept Simulation 11.1** at **www.wiley.com/college/cutnell**. In case A, water falls downward from a faucet. In case B, water shoots upward, as in a fountain. In each case, as the water moves downward or upward, it has a cross-sectional area that **(a)** does not change, **(b)** becomes larger in A and smaller in B, **(c)** becomes smaller in A and larger in B, **(d)** becomes larger in both cases, **(e)** becomes smaller in both cases.

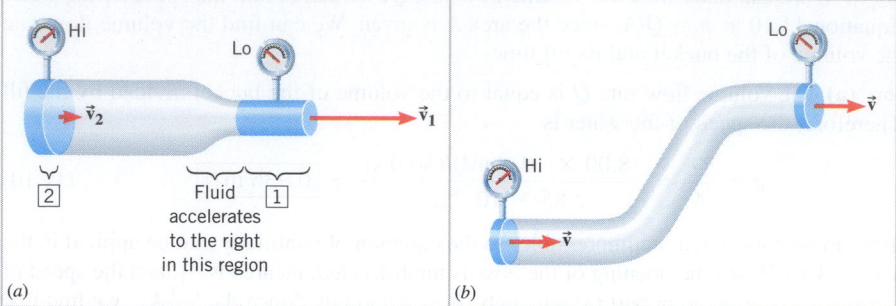

Figure 11.28 (*a*) In this horizontal pipe, the pressure in region 2 is greater than that in region 1. The difference in pressures leads to the net force that accelerates the fluid to the right. (*b*) When the fluid changes elevation, the pressure at the bottom is greater than the pressure at the top, assuming that the cross-sectional area of the pipe is constant.

11.9 BERNOULLI'S EQUATION

For *steady flow*, the speed, pressure, and elevation of an *incompressible and nonviscous* fluid are related by an equation discovered by Daniel Bernoulli (1700–1782). To derive **Bernoulli's equation**, we will use the work–energy theorem. This theorem, which is introduced in Chapter 6, states that the net work W_{nc} done on an object by external nonconservative forces is equal to the change in the total mechanical energy of the object (see Equation 6.8). As mentioned earlier, the pressure within a fluid is caused by collisional forces, which are nonconservative. Therefore, when a fluid is accelerated because of a difference in pressures, work is being done by nonconservative forces ($W_{nc} \neq 0$ J), and this work changes the total mechanical energy of the fluid from an initial value of E_0 to a final value of E_f. The total mechanical energy is not conserved. We will now see how the work–energy theorem leads directly to Bernoulli's equation.

To begin with, let us make two observations about a moving fluid. First, whenever a fluid is flowing in a horizontal pipe and encounters a region of reduced cross-sectional area, the pressure of the fluid drops, as the pressure gauges in Figure 11.28*a* indicate. The reason for this follows from Newton's second law. When moving from the wider region 2 to the narrower region 1, the fluid speeds up or accelerates, consistent with the conservation of mass (as expressed by the equation of continuity). According to the second law, the accelerating fluid must be subjected to an unbalanced force. However, there can be an unbalanced force only if the pressure in region 2 exceeds the pressure in region 1. We will see that the difference in pressures is given by Bernoulli's equation. The second observation is that if the fluid moves to a higher elevation, the pressure at the lower level is greater than the pressure at the higher level, as in Figure 11.28*b*. The basis for this observation is our previous study of static fluids, and Bernoulli's equation will confirm it, provided that the cross-sectional area of the pipe does not change.

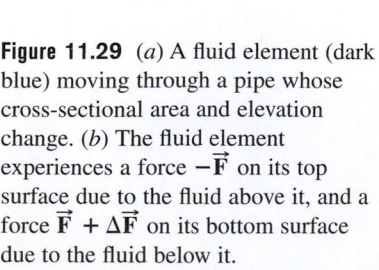

Figure 11.29 (*a*) A fluid element (dark blue) moving through a pipe whose cross-sectional area and elevation change. (*b*) The fluid element experiences a force $-\vec{F}$ on its top surface due to the fluid above it, and a force $\vec{F} + \Delta\vec{F}$ on its bottom surface due to the fluid below it.

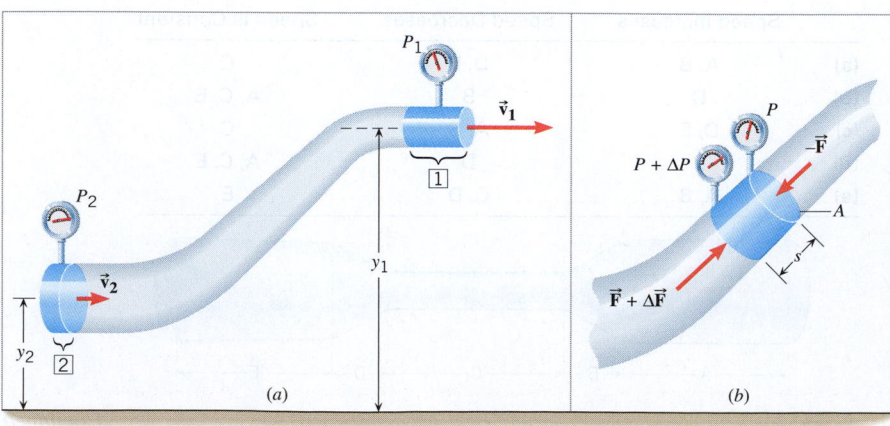

To derive Bernoulli's equation, consider Figure 11.29a. This drawing shows a fluid element of mass m, upstream in region 2 of a pipe. Both the cross-sectional area and the elevation are different at different places along the pipe. The speed, pressure, and elevation in this region are v_2, P_2, and y_2, respectively. Downstream in region 1 these variables have the values v_1, P_1, and y_1. As Chapter 6 discusses, an object moving under the influence of gravity has a total mechanical energy E that is the sum of the kinetic energy KE and the gravitational potential energy PE: $E = \text{KE} + \text{PE} = \frac{1}{2}mv^2 + mgy$. When work W_{nc} is done on the fluid element by external nonconservative forces, the total mechanical energy changes. According to the work–energy theorem, the work equals the change in the total mechanical energy:

$$W_{nc} = E_1 - E_2 = \underbrace{(\tfrac{1}{2}mv_1^2 + mgy_1)}_{\substack{\text{Total mechanical}\\\text{energy in region 1}}} - \underbrace{(\tfrac{1}{2}mv_2^2 + mgy_2)}_{\substack{\text{Total mechanical}\\\text{energy in region 2}}} \tag{6.8}$$

Figure 11.29b helps us understand how the work W_{nc} arises. On the top surface of the fluid element, the surrounding fluid exerts a pressure P. This pressure gives rise to a force of magnitude $F = PA$, where A is the cross-sectional area. On the bottom surface, the surrounding fluid exerts a slightly greater pressure, $P + \Delta P$, where ΔP is the pressure difference between the ends of the element. As a result, the force on the bottom surface has a magnitude of $F + \Delta F = (P + \Delta P)A$. The magnitude of the *net* force pushing the fluid element up the pipe is $\Delta F = (\Delta P)A$. When the fluid element moves through its own length s, the work done is the product of the magnitude of the net force and the distance: Work $= (\Delta F)s = (\Delta P)As$. The quantity As is the volume V of the element, so the work is $(\Delta P)V$. The total work done on the fluid element in moving it from region 2 to region 1 is the sum of the small increments of work $(\Delta P)V$ done as the element moves along the pipe. This sum amounts to $W_{nc} = (P_2 - P_1)V$, where $P_2 - P_1$ is the pressure difference between the two regions. With this expression for W_{nc}, the work–energy theorem becomes

$$W_{nc} = (P_2 - P_1)V = (\tfrac{1}{2}mv_1^2 + mgy_1) - (\tfrac{1}{2}mv_2^2 + mgy_2)$$

By dividing both sides of this result by the volume V, recognizing that m/V is the density ρ of the fluid, and rearranging terms, we obtain Bernoulli's equation.

BERNOULLI'S EQUATION

In the steady flow of a nonviscous, incompressible fluid of density ρ, the pressure P, the fluid speed v, and the elevation y at any two points (1 and 2) are related by

$$P_1 + \tfrac{1}{2}\rho v_1^2 + \rho g y_1 = P_2 + \tfrac{1}{2}\rho v_2^2 + \rho g y_2 \tag{11.11}$$

Since the points 1 and 2 were selected arbitrarily, the term $P + \frac{1}{2}\rho v^2 + \rho g y$ has a constant value at all positions in the flow. For this reason, Bernoulli's equation is sometimes expressed as $P + \frac{1}{2}\rho v^2 + \rho g y = \text{constant}$.

Equation 11.11 can be regarded as an extension of the earlier result that specifies how the pressure varies with depth in a static fluid ($P_2 = P_1 + \rho g h$), the terms $\frac{1}{2}\rho v_1^2$ and $\frac{1}{2}\rho v_2^2$ accounting for the effects of fluid speed. Bernoulli's equation reduces to the result for static fluids when the speed of the fluid is the same everywhere ($v_1 = v_2$), as it is when the cross-sectional area remains constant. Under such conditions, Bernoulli's equation is $P_1 + \rho g y_1 = P_2 + \rho g y_2$. After rearrangement, this result becomes

$$P_2 = P_1 + \rho g(y_1 - y_2) = P_1 + \rho g h$$

which is the result (Equation 11.4) for static fluids.

11.10 APPLICATIONS OF BERNOULLI'S EQUATION

When a moving fluid is contained in a horizontal pipe, all parts of it have the same elevation ($y_1 = y_2$), and Bernoulli's equation simplifies to

$$P_1 + \tfrac{1}{2}\rho v_1^2 = P_2 + \tfrac{1}{2}\rho v_2^2 \tag{11.12}$$

Thus, the quantity $P + \frac{1}{2}\rho v^2$ remains constant throughout a horizontal pipe; if v increases, P decreases, and vice versa. This is the result that we deduced qualitatively from Newton's second law (Section 11.9). Conceptual Example 14 illustrates it.

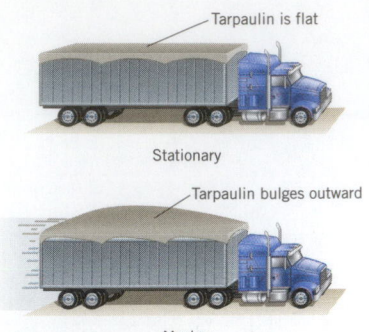

Tarpaulin is flat

Stationary

Tarpaulin bulges outward

Moving

Figure 11.30 The tarpaulin that covers the cargo is flat when the truck is stationary but bulges outward when the truck is moving.

 Conceptual Example 14 Tarpaulins and Bernoulli's Equation

A tarpaulin is a piece of canvas that is used to cover a cargo, like that pulled by the truck in Figure 11.30. Whenever the truck stops, the tarpaulin lies flat. Why does it bulge outward whenever the truck is speeding down the highway? (**a**) The air rushing over the outside surface of the canvas creates a higher pressure than does the stationary air inside the cargo area. (**b**) The air rushing over the outside surface of the canvas creates a lower pressure than does the stationary air inside the cargo area. (**c**) The air inside the cargo area heats up, thus increasing the pressure on the tarp and pushing it outward.

Reasoning When the truck is stationary, the air outside and inside the cargo area is stationary, so the pressure is the same in both places. This pressure applies the same force to the outer and inner surfaces of the canvas, with the result that the tarpaulin lies flat. When the truck is moving, the outside air rushes over the top surface of the canvas, and the pressure generated by the moving air is different than the pressure of the stationary air.

Answer (a) is incorrect. A higher pressure outside and a lower pressure in the cargo area would cause the tarpaulin to sink inward, not bulge outward.

Answer (c) is incorrect. A heating effect would not disappear every time the truck stops and reappear only when the truck is moving.

Answer (b) is correct. In accord with Bernoulli's equation (Equation 11.12), the moving air outside the canvas has a lower pressure than does the stationary air inside the cargo area. The greater inside pressure generates a greater force on the inner surface of the canvas, and the tarpaulin bulges outward.

Related Homework: *Problem 61*

Example 15 applies Equation 11.12 to a dangerous physiological condition known as an aneurysm.

ANALYZING MULTIPLE-CONCEPT PROBLEMS

Example 15 An Enlarged Blood Vessel

An aneurysm is an abnormal enlargement of a blood vessel such as the aorta. Because of an aneurysm, the cross-sectional area A_1 of the aorta increases to a value of $A_2 = 1.7\,A_1$. The speed of the blood ($\rho = 1060\ \text{kg/m}^3$) through a normal portion of the aorta is $v_1 = 0.40$ m/s. Assuming that the aorta is horizontal (the person is lying down), determine the amount by which the pressure P_2 in the enlarged region exceeds the pressure P_1 in the normal region.

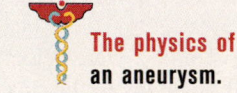 **The physics of an aneurysm.**

Reasoning Bernoulli's equation (Equation 11.12) may be used to find the pressure difference between two points in a fluid moving horizontally. However, in order to use this relation we need to know the speed of the blood in the enlarged region of the artery, as well as the speed in the normal section. We can obtain the speed in the enlarged region by using the equation of continuity (Equation 11.9), which relates it to the speed in the normal region and the cross-sectional areas of the two parts.

Knowns and Unknowns The given data are summarized as follows:

Description	Symbol	Value	Comment
Normal cross-sectional area of aorta	A_1		No value given.
Enlarged cross-sectional area of aorta	A_2	$1.7\,A_1$	
Density of blood	ρ	1060 kg/m³	
Speed of blood in normal portion of aorta	v_1	0.40 m/s	
Unknown Variable			
Amount by which pressure P_2 exceeds pressure P_1	$P_2 - P_1$	?	P_2 refers to enlarged region and P_1 to normal region.

Modeling the Problem

STEP 1 **Bernoulli's Equation** In the form pertinent to horizontal flow, Bernoulli's equation is given by Equation 11.12:

$$P_1 + \tfrac{1}{2}\rho v_1^2 = P_2 + \tfrac{1}{2}\rho v_2^2$$

Rearranging this expression gives Equation 1 at the right, which we can use to determine $P_2 - P_1$. In this result, the density ρ and the speed v_1 are given. However, the speed v_2 is unknown, and we will obtain a value for it in Step 2.

$$P_2 - P_1 = \tfrac{1}{2}\rho(v_1^2 - v_2^2) \qquad (1)$$

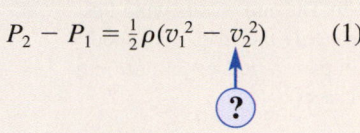

STEP 2 **The Equation of Continuity** For an incompressible fluid like blood, the equation of continuity is given by Equation 11.9 as

$$A_1 v_1 = A_2 v_2 \quad \text{or} \quad \boxed{v_2 = \left(\frac{A_1}{A_2}\right) v_1}$$

This expression for v_2 can be substituted into Equation 1, as shown at the right.

$$P_2 - P_1 = \tfrac{1}{2}\rho(v_1^2 - v_2^2) \qquad (1)$$

$$\boxed{v_2 = \left(\frac{A_1}{A_2}\right) v_1}$$

Solution Combining the results of each step algebraically, we find that

$$\overset{\text{STEP 1}}{} \quad \overset{\text{STEP 2}}{}$$

$$P_2 - P_1 = \tfrac{1}{2}\rho(v_1^2 - v_2^2) = \tfrac{1}{2}\rho\left\{ v_1^2 - \left[\left(\frac{A_1}{A_2}\right) v_1\right]^2 \right\}$$

Since it is given that $A_2 = 1.7\,A_1$, this result for $P_2 - P_1$ reveals that

$$P_2 - P_1 = \tfrac{1}{2}\rho\left\{ v_1^2 - \left[\left(\frac{A_1}{1.7\,A_1}\right) v_1\right]^2 \right\} = \tfrac{1}{2}\rho v_1^2 \left(1 - \frac{1}{1.7^2}\right)$$

$$= \tfrac{1}{2}(1060 \text{ kg/m}^3)(0.40 \text{ m/s})^2 \left(1 - \frac{1}{1.7^2}\right) = \boxed{55 \text{ Pa}}$$

This positive answer indicates that P_2 is greater than P_1. The excess pressure puts added stress on the already weakened tissue of the arterial wall at the aneurysm.

Related Homework: *Problems 64, 65, 74*

The physics of household plumbing. The impact of fluid flow on pressure is widespread. Figure 11.31, for instance, illustrates how household plumbing takes into account the implications of Bernoulli's equation. The U-shaped section of pipe beneath the sink is called a "trap," because it traps water, which serves as a barrier to prevent sewer gas from leaking into the house. Part *a* of the drawing shows poor plumbing. When water from the

Figure 11.31 In a household plumbing system, a vent is necessary to equalize the pressures at points A and B, thus preventing the trap from being emptied. An empty trap allows sewer gas to enter the house.

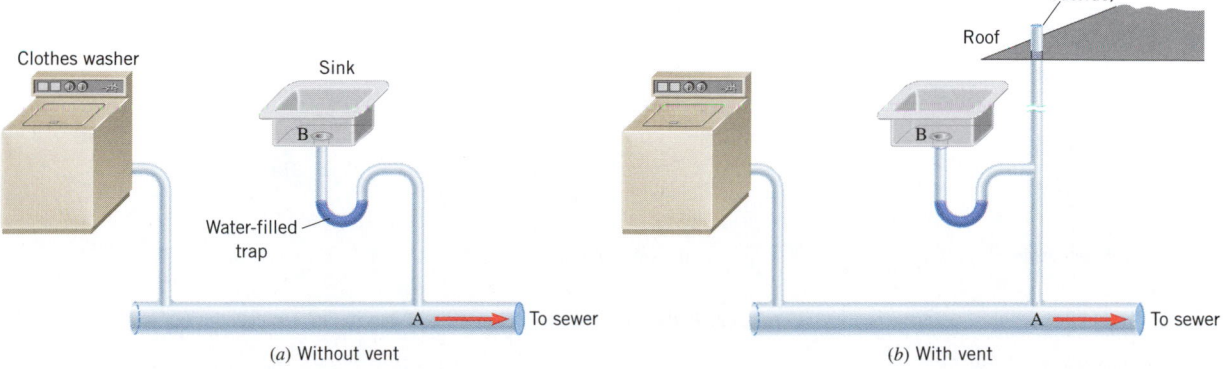

(a) Without vent

(b) With vent

Figure 11.32 (*a*) Air flowing
around an airplane wing. The
wing is moving to the right.
(*b*) The end of this wing has
roughly the shape indicated
in part *a*. (*b*, Joe McBride/Getty
Images)

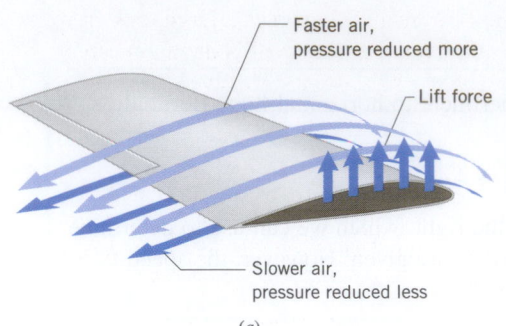

(*a*)

(*b*)

clothes washer rushes through the sewer pipe, the high-speed flow causes the pressure at
point A to drop. The pressure at point B in the sink, however, remains at the higher atmospheric pressure. As a result of this pressure difference, the water is pushed out of the trap
and into the sewer line, leaving no protection against sewer gas. A correctly designed system is vented to the outside of the house, as in Figure 11.31*b*. The vent ensures that the
pressure at A remains the same as that at B (atmospheric pressure), even when water from
the clothes washer is rushing through the pipe. Thus, the purpose of the vent is to prevent
the trap from being emptied, not to provide an escape route for sewer gas.

**The physics of
airplane wings.**

 One of the most spectacular examples of how fluid flow affects pressure is the dy-
namic lift on airplane wings. Figure 11.32*a* shows a wing moving to the right, with the
air flowing leftward past the wing. Hence, according to Bernoulli's equation, the pressures on the top of and beneath the wing are both lower than atmospheric pressure.
However, the pressure above the wing is reduced relative to the pressure under the wing.
This is due to the wing's shape, which causes the air to travel faster (more reduction in
pressure) over the curved top surface and more slowly (less reduction in pressure) over
the flatter bottom. Thus, the wing is lifted upward. Part *b* of the figure shows the wing of
an airplane.

**The physics of
a curveball.**

 The curveball, one of a baseball pitcher's main weapons, is another illustration of the
effects of fluid flow. Figure 11.33*a* shows a baseball moving to the right with no spin. The
view is from above, looking down toward the ground. Here, air flows with the same speed
around both sides of the ball, and the pressure is reduced on both sides by the same amount
relative to atmospheric pressure. No net force exists to make the ball curve to either side.
However, when the ball is given a spin, the air close to its surface is dragged around with
it, and the situation changes. The speed of the air on one-half of the ball is increased, and
the pressure there is even more reduced relative to atmospheric pressure. On the other half
of the ball, the speed of the air is decreased, leading to a lesser reduction in pressure than
occurs without spin. Part *b* of the picture shows the effects of a counterclockwise spin. The
baseball experiences a net deflection force and curves on its way from the pitcher's mound
to the plate, as part *c* shows.*

Figure 11.33 These views of a baseball
are from above, looking down toward
the ground, with the ball moving to the
right. (*a*) Without spin, the ball does
not curve to either side. (*b*) A spinning
ball curves in the direction of the
deflection force. (*c*) The spin in part *b*
causes the ball to curve as shown here.

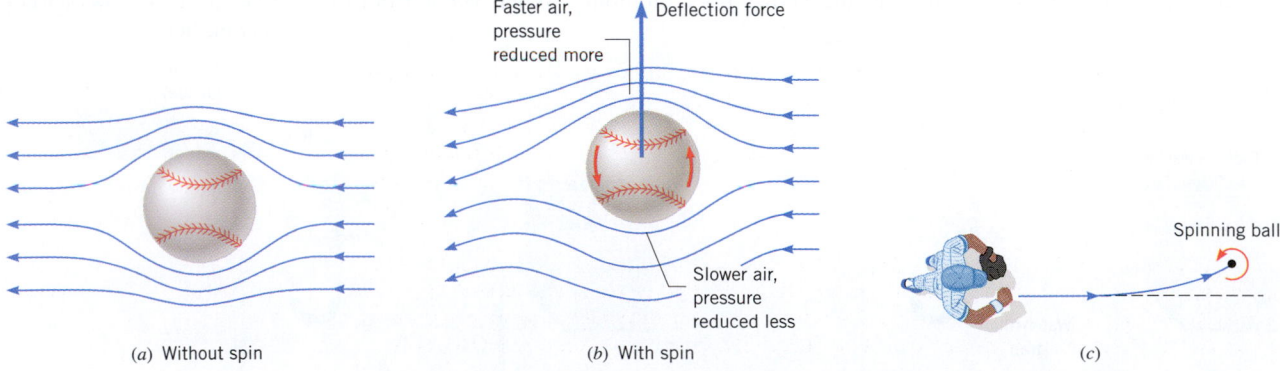

(*a*) Without spin

(*b*) With spin

(*c*)

*In the jargon used in baseball, the pitch shown in Figure 11.33 parts *b* and *c* is called a "slider."

As a final application of Bernoulli's equation, Figure 11.34a shows a large tank from which water is emerging through a small pipe near the bottom. Bernoulli's equation can be used to determine the speed (called the efflux speed) at which the water leaves the pipe, as the next example shows.

 Example 16   Efflux Speed

The tank in Figure 11.34a is open to the atmosphere at the top. Find an expression for the speed of the liquid leaving the pipe at the bottom.

Reasoning We assume that the liquid behaves as an ideal fluid. Therefore, we can apply Bernoulli's equation, and in preparation for doing so, we locate two points in the liquid in Figure 11.34a. Point 1 is just outside the efflux pipe, and point 2 is at the top surface of the liquid. The pressure at each of these points is equal to the atmospheric pressure, a fact that will be used to simplify Bernoulli's equation.

Solution Since the pressures at points 1 and 2 are the same, we have $P_1 = P_2$, and Bernoulli's equation becomes $\frac{1}{2}\rho v_1^2 + \rho g y_1 = \frac{1}{2}\rho v_2^2 + \rho g y_2$. The density ρ can be eliminated algebraically from this result, which can then be solved for the square of the efflux speed v_1:

$$v_1^2 = v_2^2 + 2g(y_2 - y_1) = v_2^2 + 2gh$$

We have substituted $h = y_2 - y_1$ for the height of the liquid above the efflux tube. If the tank is very large, the liquid level changes only slowly, and the speed at point 2 can be set equal to zero, so that $\boxed{v_1 = \sqrt{2gh}}$.

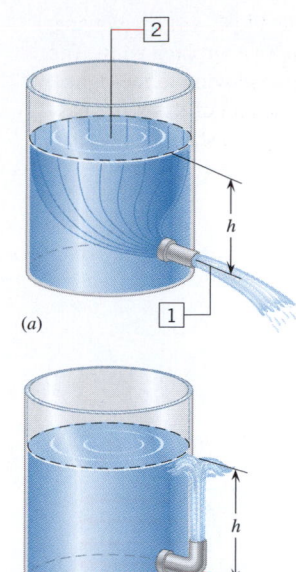

Figure 11.34 (a) Bernoulli's equation can be used to determine the speed of the liquid leaving the small pipe. (b) An ideal fluid (no viscosity) will rise to the fluid level in the tank after leaving a vertical outlet nozzle.

In Example 16 the liquid is assumed to be an ideal fluid, and the speed with which it leaves the pipe is the same as if the liquid had freely fallen through a height h (see Equation 2.9 with $x = h$ and $a = g$). This result is known as **Torricelli's theorem.** If the outlet pipe were pointed directly upward, as in Figure 11.34b, the liquid would rise to a height h equal to the fluid level above the pipe. However, if the liquid is not an ideal fluid, its viscosity cannot be neglected. Then, the efflux speed would be less than that given by Bernoulli's equation, and the liquid would rise to a height less than h.

✓ **CHECK YOUR UNDERSTANDING**

(*The answers are given at the end of the book.*)

17. Fluid is flowing from left to right through a pipe (see the drawing). Points A and B are at the same elevation, but the cross-sectional areas of the pipe are different. Points B and C are at different elevations, but the cross-sectional areas are the same. Rank the pressures at the three points, highest to lowest:
(a) A and B (a tie), C   **(b)** C, A and B (a tie)   **(c)** B, C, A   **(d)** C, B, A   **(e)** A, B, C

18. Have you ever had a large truck pass you from the opposite direction on a narrow two-lane road? You probably noticed that your car was pulled toward the truck as it passed. How does the speed of the air (and, hence, its pressure) between your car and the truck compare to the speed of the air on the opposite side of your car? The speed of the air between the two vehicles is   **(a)** less and produces a smaller air pressure   **(b)** less and produces a greater air pressure   **(c)** greater and produces a smaller air pressure   **(d)** greater and produces a greater air pressure.

19. Hold two sheets of paper by adjacent corners, so that they hang downward. The sheets should be parallel and slightly separated, so that you can see the floor through the gap between them. Blow air strongly down through the gap. What happens to the sheets?
(a) Nothing   **(b)** The sheets move further apart.   **(c)** The sheets come closer together.

20. Suppose that you are a right-handed batter in a baseball game, so the ball is moving from your left to your right. You are caught unprepared, looking directly at the ball as it

Continued

passes by for a strike. If the ball curves upward on its way to the plate, which way is it spinning? **(a)** Clockwise **(b)** Counterclockwise

21. You are sitting on a stationary train next to an *open* window, and the pressure of the air in your inner ear is equal to the pressure outside your ear. The train starts up, and as it accelerates to a high speed, your ears "pop." Your eardrums respond to a decrease or increase in the air pressure by "popping" outward or inward, respectively. Assume that the air pressure in your inner ear has not had time to change, so it remains the same as when the train was stationary. Do your eardrums "pop" **(a)** outward or **(b)** inward?

22. Sometimes the weather conditions at an airport give rise to air that has an unusually low density. What effect does such a low-density condition have on a plane's ability to generate the required lift force for takeoff? **(a)** It has no effect. **(b)** It makes it easier. **(c)** It makes it harder.

11.11 *VISCOUS FLOW

In an ideal fluid there is no viscosity to hinder the fluid layers as they slide past one another. Within a pipe of uniform cross section, every layer of an ideal fluid moves with the same velocity, even the layer next to the wall, as Figure 11.35a shows. When viscosity is present, the fluid layers have different velocities, as part *b* of the drawing illustrates. The fluid at the center of the pipe has the greatest velocity. In contrast, the fluid layer next to the wall surface does not move at all, because it is held tightly by intermolecular forces. So strong are these forces that if a solid surface moves, the adjacent fluid layer moves along with it and remains at rest *relative* to the moving surface.

To help introduce viscosity in a quantitative fashion, Figure 11.36a shows a viscous fluid between two parallel plates. The top plate is free to move, while the bottom one is stationary. If the top plate is to move with a velocity $\vec{v}$ relative to the bottom plate, a force $\vec{F}$ is required. For a highly viscous fluid, like thick honey, a large force is needed; for a less viscous fluid, like water, a smaller one will do. As part *b* of the drawing suggests, we may imagine the fluid to be composed of many thin horizontal layers. When the top plate moves, the intermediate fluid layers slide over each other. The velocity of each layer is different, changing uniformly from $\vec{v}$ at the top plate to zero at the bottom plate. The resulting flow is called *laminar flow,* since a thin layer is often referred to as a *lamina.* As each layer moves, it is subjected to viscous forces from its neighbors. The purpose of the force $\vec{F}$ is to compensate for the effect of these forces, so that any layer can move with a constant velocity.

The amount of force required in Figure 11.36a depends on several factors. Larger areas *A*, being in contact with more fluid, require larger forces, so that the force is proportional to the contact area ($F \propto A$). For a given area, greater speeds require larger forces, with the result that the force is proportional to the speed ($F \propto v$). The force is also inversely proportional to the perpendicular distance *y* between the top and bottom plates ($F \propto 1/y$). The larger the distance *y*, the smaller is the force required to achieve a given speed with a given contact area. These three proportionalities can be expressed simultaneously in the following way: $F \propto Av/y$. Equation 11.13 expresses this relationship with the aid of a proportionality constant η (Greek letter *eta*), which is called the **coefficient of viscosity** or simply the **viscosity.**

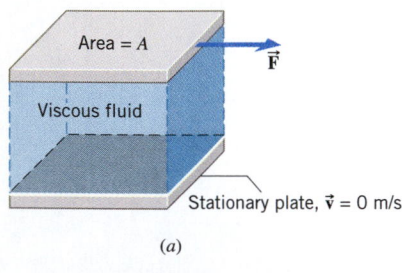

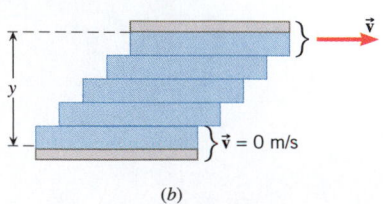

Figure 11.35 (*a*) In ideal (nonviscous) fluid flow, all fluid particles across the pipe have the same velocity. (*b*) In viscous flow, the speed of the fluid is zero at the surface of the pipe and increases to a maximum along the center axis.

Figure 11.36 (*a*) A force $\vec{F}$ is applied to the top plate, which is in contact with a viscous fluid. (*b*) Because of the force $\vec{F}$, the top plate and the adjacent layer of fluid move with a constant velocity $\vec{v}$.

FORCE NEEDED TO MOVE A LAYER OF VISCOUS FLUID WITH A CONSTANT VELOCITY

The magnitude of the tangential force $\vec{F}$ required to move a fluid layer at a constant speed *v*, when the layer has an area *A* and is located a perpendicular distance *y* from an immobile surface, is given by

$$F = \frac{\eta A v}{y} \tag{11.13}$$

where η is the coefficient of viscosity.

SI Unit of Viscosity: Pa·s

Common or CGS Unit of Viscosity: poise (P)

By solving Equation 11.13 for the viscosity, $\eta = Fy/(vA)$, it can be seen that the SI unit for viscosity is $N \cdot m/[(m/s) \cdot m^2] = Pa \cdot s$. Another common unit for viscosity is the *poise* (P), which is used in the CGS system of units and is named after the French physician Jean Poiseuille (1799–1869; pronounced, approximately, as Pwah-zoy′). The following relation exists between the two units:

$$1 \text{ poise (P)} = 0.1 \text{ Pa} \cdot \text{s}$$

Values of viscosity depend on the nature of the fluid. Under ordinary conditions, the viscosities of liquids are significantly *larger* than those of gases. Moreover, the viscosities of either liquids or gases depend markedly on temperature. Usually, the viscosities of liquids decrease as the temperature is increased. Anyone who has heated honey or oil, for example, knows that these fluids flow much more freely at an elevated temperature. In contrast, the viscosities of gases increase as the temperature is raised. An ideal fluid has $\eta = 0$ P.

Viscous flow occurs in a wide variety of situations, such as oil moving through a pipeline or a liquid being forced through the needle of a hypodermic syringe. Figure 11.37 identifies the factors that determine the volume flow rate Q (in m³/s) of the fluid. First, a difference in pressures $P_2 - P_1$ must be maintained between any two locations along the pipe for the fluid to flow. In fact, Q is proportional to $P_2 - P_1$, a greater pressure difference leading to a larger flow rate. Second, a long pipe offers greater resistance to the flow than a short pipe does, and Q is inversely proportional to the length L. [Because of this fact, long pipelines, such as the Alaskan pipeline, have pumping stations at various places along the line to compensate for a drop in pressure (see Figure 11.38).] Third, high-viscosity fluids flow less readily than low-viscosity fluids, and Q is inversely proportional to the viscosity η. Finally, the volume flow rate is larger in a pipe of larger radius, other things being equal. The dependence on the radius R is a surprising one, Q being proportional to the fourth power of the radius, or R^4. If, for instance, the pipe radius is reduced to one-half of its original value, the volume flow rate is reduced to one-sixteenth of its original value, assuming the other variables remain constant. The mathematical relation for Q in terms of these parameters was discovered by Poiseuille and is known as *Poiseuille's law.*

POISEUILLE'S LAW

A fluid whose viscosity is η, flowing through a pipe of radius R and length L, has a volume flow rate Q given by

$$Q = \frac{\pi R^4 (P_2 - P_1)}{8 \eta L} \tag{11.14}$$

where P_1 and P_2 are the pressures at the ends of the pipe.

Example 17 illustrates the use of Poiseuille's law.

 Example 17 Giving an Injection

A hypodermic syringe is filled with a solution whose viscosity is 1.5×10^{-3} Pa·s. As Figure 11.39 shows, the plunger area of the syringe is 8.0×10^{-5} m², and the length of the needle is 0.025 m. The internal radius of the needle is 4.0×10^{-4} m. The gauge pressure in a vein is 1900 Pa (14 mm of mercury). What force must be applied to the plunger, so that 1.0×10^{-6} m³ of solution can be injected in 3.0 s?

Reasoning The necessary force is the pressure applied to the plunger times the area of the plunger. Since viscous flow is occurring, the pressure is different at different points along the syringe. However, the barrel of the syringe is so wide that little pressure difference is required to sustain the flow up to point 2, where the fluid encounters the narrow needle. Consequently, the pressure applied to the plunger is nearly equal to the pressure P_2 at point 2. To find this pressure, we apply Poiseuille's law to the needle. Poiseuille's law indicates that $P_2 - P_1 = 8 \eta L Q / (\pi R^4)$. We note that the pressure P_1 is given as a gauge pressure, which, in this case, is the amount of pressure in excess of atmospheric pressure. This causes no difficulty, because we need to find the amount of force in excess of the force applied to the plunger by the atmosphere. The volume flow rate Q can be obtained from the time needed to inject the known volume of solution.

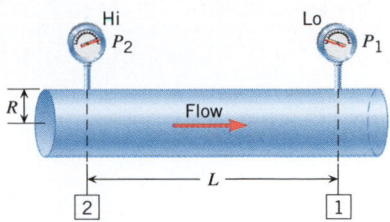

Figure 11.37 For viscous flow, the difference in pressures $P_2 - P_1$, the radius R and length L of the tube, and the viscosity η of the fluid influence the volume flow rate.

The physics of pipeline pumping stations.

Figure 11.38 As oil flows along the Alaskan pipeline, the pressure drops because oil is a viscous fluid. Pumping stations are located along the pipeline to compensate for the drop in pressure. (Ken Graham/Getty Images)

The physics of a hypodermic syringe.

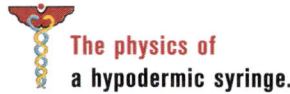

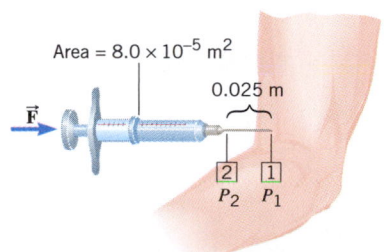

Figure 11.39 The difference in pressure $P_2 - P_1$ required to sustain the fluid flow through a hypodermic needle can be found with the aid of Poiseuille's law.

Solution The volume flow rate is $Q = (1.0 \times 10^{-6} \text{ m}^3)/(3.0 \text{ s}) = 3.3 \times 10^{-7} \text{ m}^3/\text{s}$. According to Poiseuille's law (Equation 11.14), the required pressure difference is

$$P_2 - P_1 = \frac{8\eta LQ}{\pi R^4} = \frac{8(1.5 \times 10^{-3} \text{ Pa} \cdot \text{s})(0.025 \text{ m})(3.3 \times 10^{-7} \text{ m}^3/\text{s})}{\pi (4.0 \times 10^{-4} \text{ m})^4} = 1200 \text{ Pa}$$

Since $P_1 = 1900$ Pa, the pressure P_2 must be $P_2 = 1200$ Pa + 1900 Pa = 3100 Pa. The force that must be applied to the plunger is this pressure times the plunger area:

$$F = (3100 \text{ Pa})(8.0 \times 10^{-5} \text{ m}^2) = \boxed{0.25 \text{ N}}$$

11.12 CONCEPTS & CALCULATIONS

Pressure plays an important role in the behavior of fluids. As we have seen in this chapter, pressure is the magnitude of the force acting perpendicular to a surface divided by the area of the surface. Pressure should not be confused, however, with the force itself. The idea of a force is introduced in Chapter 4. The next example serves to emphasize that pressure and force are different concepts. In addition, it reviews the techniques that Chapters 4 and 9 discuss for analyzing objects that are in equilibrium.

Concepts & Calculations Example 18 Pressure and Force

Figure 11.40*a* shows a rear view of a loaded two-wheeled wheelbarrow on a horizontal surface. It has balloon tires and a weight of $W = 684$ N, which is uniformly distributed. The left tire has a contact area with the ground of $A_L = 6.6 \times 10^{-4}$ m^2, whereas the right tire is underinflated and has a contact area of $A_R = 9.9 \times 10^{-4}$ m^2. Find the force and the pressure that each tire applies to the ground.

Concept Questions and Answers Force is a vector. Therefore, both a direction and a magnitude are needed to specify it. Are both a direction and a magnitude needed to specify a pressure?

> *Answer* No. Only a magnitude is needed to specify a pressure because pressure is a scalar quantity, not a vector quantity. The definition of pressure is the magnitude of the force acting perpendicular to a surface divided by the area of the surface. This definition does not include directional information.

The problem asks for the force that each tire applies to the ground. How is this force related to the force that the ground applies to each tire?

> *Answer* As Newton's law of action–reaction indicates, the force that a tire applies to the ground has the same magnitude but opposite direction as the force that the ground applies to the tire.

Do the left and right tires apply the same force to the ground?

> *Answer* Yes. Since the weight is uniformly distributed, each tire supports half the load. Therefore, each tire applies the same force to the ground.

Do the left and right tires apply the same pressure to the ground?

> *Answer* No. Pressure involves the contact area as well as the magnitude of the force. Since the force magnitudes are the same but the contact areas are different, the pressures are different, too.

Solution Since the wheelbarrow is at rest, it is in equilibrium and the net force acting on it must be zero, as Chapters 4 and 9 discuss. Referring to the free-body diagram in Figure 11.40*b* and taking upward as the positive direction, we have

$$\underbrace{\Sigma F_y = F_L + F_R - W = 0}_{\text{Net force}}$$

where F_L and F_R are, respectively, the forces that the ground applies to the left and right tires. At equilibrium the net torque acting on the wheelbarrow must also be zero, as Chapter 9 discusses. In the free-body diagram the weight acts at the center of gravity, which is at the center of the wheelbarrow. Using this point as the axis of rotation for torques, we can see that the

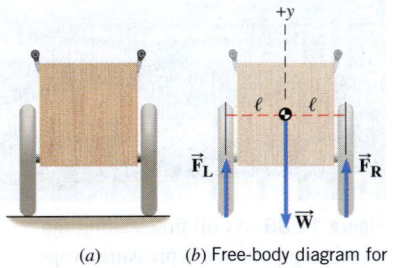

(a)   (b) Free-body diagram for the wheelbarrow

Figure 11.40 (*a*) Rear view of a wheelbarrow with balloon tires. (*b*) Free-body diagram of the wheelbarrow, showing its weight $\vec{W}$ and the forces $\vec{F}_L$ and $\vec{F}_R$ that the ground applies, respectively, to the left and right tires.

forces F_L and F_R have the same lever arm ℓ, which is half the width of the wheelbarrow. For this axis, the weight W has a zero lever arm and, hence, a zero torque. Therefore, it does not contribute to the net torque. Remembering that counterclockwise torques are positive and setting the net torque equal to zero, we have

$$\underbrace{\Sigma \tau}_{\text{Net torque}} = F_R\ell - F_L\ell = 0 \quad \text{or} \quad F_R = F_L$$

Substituting this result into the force equation gives

$$F_L + F_L - W = 0$$
$$F_L = F_R = \tfrac{1}{2}W = \tfrac{1}{2}(684 \text{ N}) = 342 \text{ N}$$

The value of 342 N is the magnitude of the upward force that the ground applies to each tire. Using Newton's action–reaction law, we conclude that each tire applies to the ground a $\boxed{\text{downward force of 342 N}}$. Using this value and the given contact areas, we find that the left and right tires apply the following pressures P_L and P_R to the ground:

$$P_L = \frac{F}{A_L} = \frac{342 \text{ N}}{6.6 \times 10^{-4} \text{ m}^2} = \boxed{5.2 \times 10^5 \text{ Pa}}$$

$$P_R = \frac{F}{A_R} = \frac{342 \text{ N}}{9.9 \times 10^{-4} \text{ m}^2} = \boxed{3.5 \times 10^5 \text{ Pa}}$$

Figure 11.41 The two bathers are sitting on different-sized beach balls that are just submerged beneath the water.

One manifestation of fluid pressure is the buoyant force that is applied to any object immersed in a fluid. In Section 11.6 we looked at this force, as well as Archimedes' principle, which specifies a convenient way to determine it. Example 19 focuses on the essence of this principle.

 Concepts & Calculations Example 19 The Buoyant Force

A father (weight $W = 830$ N) and his daughter (weight $W = 340$ N) are spending the day at the lake. They are each sitting on a beach ball that is just submerged beneath the water (see Figure 11.41). Ignoring the weight of the air within the balls and the volumes of their legs that are under water, find the radius of each ball.

Concept Questions and Answers Each beach ball is in equilibrium, being stationary and having no acceleration. Thus, the net force acting on each ball is zero. What balances the downward-acting weight in each case?

Answer The downward-acting weight is balanced by the upward-acting buoyant force F_B that the water applies to the ball.

In which case is the buoyant force greater?

Answer The buoyant force acting on the father's beach ball is greater, since it must balance his greater weight.

In the situation described, what determines the magnitude of the buoyant force?

Answer According to Archimedes' principle, the magnitude of the buoyant force equals the weight of the fluid that the ball displaces. Since the ball is completely submerged, it displaces a volume of water that equals the ball's volume. The weight of this volume of water is the magnitude of the buoyant force.

Which beach ball has the larger radius?

Answer The father's ball has the larger volume and the larger radius. This follows because a larger buoyant force acts on that ball. For the buoyant force to be larger, that ball must displace a greater volume of water, according to Archimedes' principle. Therefore, the volume of that ball is larger, since the balls are completely submerged.

Solution Since the balls are in equilibrium, the net force acting on each of them must be zero. Therefore, taking upward to be the positive direction, we have

$$\underbrace{\Sigma F}_{\text{Net force}} = F_B - W = 0$$

Archimedes' principle specifies that the magnitude of the buoyant force is the weight of the water displaced by the ball. Using the definition of density given in Equation 11.1, the mass of the displaced water is $m = \rho V$, where $\rho = 1.00 \times 10^3$ kg/m^3 is the density of water (see Table 11.1) and V is the volume displaced. Since all of the ball is submerged, $V = \frac{4}{3}\pi r^3$, assuming that the ball remains spherical. The weight of the displaced water is $mg = \rho(\frac{4}{3}\pi r^3)g$. With this value for the buoyant force, the force equation becomes

$$F_B - W = \rho(\tfrac{4}{3}\pi r^3)g - W = 0$$

Solving for the radius, we find that

Father $\qquad r = \sqrt[3]{\dfrac{3W}{4\pi\rho g}} = \sqrt[3]{\dfrac{3(830 \text{ N})}{4\pi(1.00 \times 10^3 \text{ kg/m}^3)(9.80 \text{ m/s}^2)}} = \boxed{0.27 \text{ m}}$

Daughter $\qquad r = \sqrt[3]{\dfrac{3W}{4\pi\rho g}} = \sqrt[3]{\dfrac{3(340 \text{ N})}{4\pi(1.00 \times 10^3 \text{ kg/m}^3)(9.80 \text{ m/s}^2)}} = \boxed{0.20 \text{ m}}$

As expected, the radius of the father's beach ball is greater.

CONCEPT SUMMARY

If you need more help with a concept, use the Learning Aids noted next to the discussion or equation. Examples (**Ex.**) are in the text of this chapter. Go to **www.wiley.com/college/cutnell** for the following Learning Aids:

Interactive LearningWare (ILW) — Additional examples solved in a five-step interactive format.

Concept Simulations (CS) — Animated text figures or animations of important concepts.

Interactive Solutions (IS) — Models for certain types of problems in the chapter homework. The calculations are carried out interactively.

Topic	Discussion	Learning Aids
	11.1 MASS DENSITY Fluids are materials that can flow, and they include gases and liquids.	
	The mass density ρ of a substance is its mass m divided by its volume V:	
Mass density	$$\rho = \frac{m}{V}$$	(11.1) **Ex. 1**
	The specific gravity of a substance is its mass density divided by the density of water at 4 °C $(1.000 \times 10^3$ kg/m$^3)$:	
Specific gravity	$$\text{Specific gravity} = \frac{\text{Density of substance}}{1.000 \times 10^3 \text{ kg/m}^3}$$	(11.2)
	11.2 PRESSURE The pressure P exerted by a fluid is the magnitude F of the force acting perpendicular to a surface embedded in the fluid divided by the area A over which the force acts:	
Pressure	$$P = \frac{F}{A}$$	(11.3) **Ex. 2, 18** **IS 11.13**
	The SI unit for measuring pressure is the pascal (Pa); 1 Pa = 1 N/m^2.	
Atmospheric pressure	One atmosphere of pressure is 1.013×10^5 Pa or 14.7 lb/in.^{2}	
	11.3 PRESSURE AND DEPTH IN A STATIC FLUID In the presence of gravity, the upper layers of a fluid push downward on the layers beneath, with the result that fluid pressure is related to depth. In an incompressible static fluid whose density is ρ, the relation is	
	$$P_2 = P_1 + \rho g h$$	(11.4) **Ex. 3–6**
	where P_1 is the pressure at one level, P_2 is the pressure at a level that is h meters deeper, and g is the magnitude of the acceleration due to gravity.	
	11.4 PRESSURE GAUGES Two basic types of pressure gauges are the mercury barometer and the open-tube manometer.	
Gauge pressure and absolute pressure	The gauge pressure is the amount by which a pressure P differs from atmospheric pressure. The absolute pressure is the actual value for P.	
Pascal's principle	**11.5 PASCAL'S PRINCIPLE** Pascal's principle states that any change in the pressure applied to a completely enclosed fluid is transmitted undiminished to all parts of the fluid and the enclosing walls.	**Ex. 7, 8** **IS 11.35**

Topic	Discussion	Learning Aids
Buoyant force	**11.6 ARCHIMEDES' PRINCIPLE** The buoyant force is the upward force that a fluid applies to an object that is partially or completely immersed in it.	
Archimedes' principle	Archimedes' principle states that the magnitude of the buoyant force equals the weight of the fluid that the partially or completely immersed object displaces: $$\underbrace{F_B}_{\substack{\text{Magnitude of}\\\text{buoyant force}}} = \underbrace{W_{\text{fluid}}}_{\substack{\text{Weight of}\\\text{displaced fluid}}} \qquad (11.6)$$	**Ex. 9, 10, 11, 19** ILW 11.1
Steady flow	**11.7 FLUIDS IN MOTION, 11.8 THE EQUATION OF CONTINUITY** In steady flow, the velocity of the fluid particles at any point is constant as time passes.	
Ideal fluid	An incompressible, nonviscous fluid is known as an ideal fluid.	
Mass flow rate	The mass flow rate of a fluid with a density ρ, flowing with a speed v in a pipe of cross-sectional area A, is the mass per second (kg/s) flowing past a point and is given by $$\text{Mass flow rate} = \rho A v \qquad (11.7)$$	
Equation of continuity	The equation of continuity expresses the fact that mass is conserved: what flows into one end of a pipe flows out the other end, assuming there are no additional entry or exit points in between. Expressed in terms of the mass flow rate, the equation of continuity is $$\rho_1 A_1 v_1 = \rho_2 A_2 v_2 \qquad (11.8)$$ where the subscripts 1 and 2 denote two points along the pipe.	
Incompressible fluid	If a fluid is incompressible, the density at any two points is the same, $\rho_1 = \rho_2$. For an incompressible fluid, the equation of continuity becomes $$A_1 v_1 = A_2 v_2 \qquad (11.9)$$ The product Av is known as the volume flow rate Q (in m³/s):	**Ex. 12, 13** CS 11.1
Volume flow rate	$$Q = \text{Volume flow rate} = Av \qquad (11.10)$$	
Bernoulli's equation	**11.9 BERNOULLI'S EQUATION, 11.10 APPLICATIONS OF BERNOULLI'S EQUATION** In the steady flow of an ideal fluid whose density is ρ, the pressure P, the fluid speed v, and the elevation y at any two points (1 and 2) in the fluid are related by Bernoulli's equation: $$P_1 + \tfrac{1}{2}\rho v_1^2 + \rho g y_1 = P_2 + \tfrac{1}{2}\rho v_2^2 + \rho g y_2 \qquad (11.11)$$ When the flow is horizontal ($y_1 = y_2$), Bernoulli's equation indicates that higher fluid speeds are associated with lower fluid pressures.	**Ex. 14, 15, 16** ILW 11.2 IS 11.73
Force needed to move a layer of viscous fluid with a constant speed	**11.11 VISCOUS FLOW** The magnitude F of the tangential force required to move a fluid layer at a constant speed v, when the layer has an area A and is located a perpendicular distance y from an immobile surface, is given by $$F = \frac{\eta A v}{y} \qquad (11.13)$$ where η is the coefficient of viscosity.	
Poiseuille's law	A fluid whose viscosity is η, flowing through a pipe of radius R and length L, has a volume flow rate Q given by $$Q = \frac{\pi R^4 (P_2 - P_1)}{8 \eta L} \qquad (11.14)$$ where P_1 and P_2 are the pressures at the ends of the pipe.	**Ex. 17** IS 11.81

FOCUS ON CONCEPTS

Note to Instructors: The numbering of the questions shown here reflects the fact that they are only a representative subset of the total number that are available online. However, all of the questions are available for assignment via an online homework management program such as WileyPLUS or WebAssign.

Section 11.3 Pressure and Depth in a Static Fluid

1. The drawing shows three containers filled to the same height with the same fluid. In which container, if any, is the pressure at the bottom greatest? **(a)** Container A, because its bottom has the greatest surface area. **(b)** All three containers have the same pressure at the bottom. **(c)** Container A, because it has the greatest volume of fluid. **(d)** Container B, because it has the least volume of fluid. **(e)** Container C, because its bottom has the least surface area.

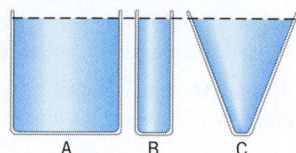

4. Two liquids, 1 and 2, are in equilibrium in a U-tube that is open at both ends, as in the drawing. The liquids do not mix, and liquid 1 rests on top of liquid 2. How is the density ρ_1 of liquid 1 related to the density ρ_2 of liquid 2? **(a)** ρ_1 is equal to ρ_2 because the liquids are in equilibrium. **(b)** ρ_1 is greater than ρ_2. **(c)** ρ_1 is less than ρ_2. **(d)** There is not enough information to tell which liquid has the greater density.

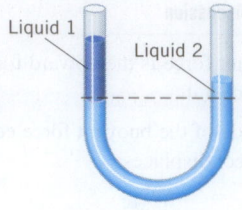

Liquid 1
Liquid 2

Section 11.6 Archimedes' Principle

9. A beaker is filled to the brim with water. A solid object of mass 3.00 kg is lowered into the beaker so that the object is fully submerged in the water (see the drawing). During this process, 2.00 kg of water flows over the rim and out of the beaker. What is the buoyant force that acts on the submerged object, and, when released, does the object rise, sink, or remain in place? **(a)** 29.4 N; the object rises. **(b)** 29.4 N; the object sinks. **(c)** 19.6 N; the object rises. **(d)** 19.6 N; the object sinks. **(e)** 19.6 N; the object remains in place.

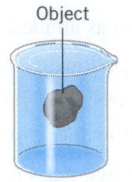

Object

10. Three solid objects are floating in a liquid, as in the drawing. They have different weights and volumes, but have the same thickness (the dimension perpendicular to the page). Rank the objects according to their density, largest first. **(a)** A, B, C **(b)** A, C, B **(c)** B, A, C **(d)** B, C, A **(e)** C, A, B

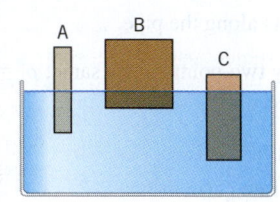

A   B   C

Section 11.8 The Equation of Continuity

12. A hollow pipe is submerged in a stream of water so that the length of the pipe is parallel to the velocity of the water. If the water speed doubles and the cross-sectional area of the pipe triples, what happens to the volume flow rate of the water passing through the pipe? **(a)** The volume flow rate does not change. **(b)** The volume flow rate increases by a factor of 2. **(c)** The volume flow rate increases by a factor of 3. **(d)** The volume flow rate increases by a factor of 4. **(e)** The volume flow rate increases by a factor of 6.

13. In the drawing, water flows from a wide section of a pipe to a narrow section. In which part of the pipe is the volume flow rate the greatest? **(a)** The wide section **(b)** The narrow section **(c)** The volume flow rate is the same in both sections of the pipe.

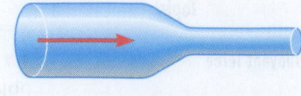

Section 11.9 Bernoulli's Equation

16. Blood flows through a section of a horizontal artery that is partially blocked by a deposit along the artery wall. A hemoglobin molecule moves from the narrow region into the wider region. What happens to the pressure acting on the molecule? **(a)** The pressure increases. **(b)** The pressure decreases. **(c)** There is no change in the pressure.

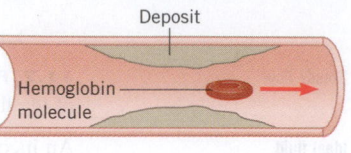

Deposit
Hemoglobin molecule

18. Water is flowing down through the pipe shown in the drawing. Point A is at a higher elevation than B and C are. The cross-sectional areas are the same at A and B but are wider at C. Rank the pressures at the three points, largest first. **(a)** P_A, P_B, P_C **(b)** P_C, P_B, P_A **(c)** P_B, P_C, P_A

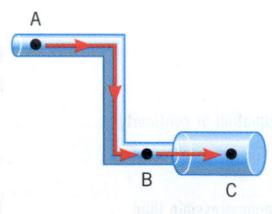

A
B   C

Section 11.11 Viscous Flow

20. A *viscous* fluid is flowing through two horizontal pipes. The pressure difference $P_2 - P_1$ between the ends of each pipe is the same. The pipes have the same radius, although one is twice as long as the other. How does the volume flow rate Q_B in the longer pipe compare with the rate Q_A in the shorter pipe? **(a)** Q_B is the same as Q_A. **(b)** Q_B is twice as large as Q_A. **(c)** Q_B is four times as large as Q_A. **(d)** Q_B is one-half as large as Q_A. **(e)** Q_B is one-fourth as large as Q_A.

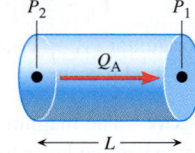
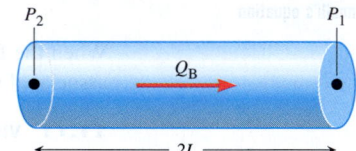

P_2 Q_A P_1 P_2 Q_B P_1

L $2L$

PROBLEMS

WILEY PLUS

Note to Instructors: *Most of the homework problems in this chapter are available for assignment via an online homework management program such as WileyPLUS or WebAssign, and those marked with the icon*  GO *are presented in WileyPLUS using a guided tutorial format that provides enhanced interactivity. See Preface for additional details.*

Section 11.1 Mass Density

1. ssm A water bed for sale has dimensions of 1.83 m × 2.13 m × 0.229 m. The floor of the bedroom will tolerate an additional weight of no more than 6660 N. Find the weight of the water in the bed and determine whether the bed should be purchased.

2. A cylindrical storage tank has a radius of 1.22 m. When filled to a height of 3.71 m, it holds 14 300 kg of a liquid industrial solvent. What is the density of the solvent?

3. ssm Accomplished silver workers in India can pound silver into incredibly thin sheets, as thin as 3.00×10^{-7} m (about one-hundredth of the thickness of this sheet of paper). Find the area of such a sheet that can be formed from 1.00 kg of silver.

4. Neutron stars consist only of neutrons and have unbelievably high densities. A typical mass and radius for a neutron star might be 2.7×10^{28} kg and 1.2×10^3 m. **(a)** Find the density of such a star. **(b)** If a dime ($V = 2.0 \times 10^{-7}$ m³) were made from this material, how much would it weigh (in pounds)?

5. The *karat* is a dimensionless unit that is used to indicate the proportion of gold in a gold-containing alloy. An alloy that is one karat gold contains a weight of pure gold that is one part in twenty-four.

What is the volume of gold in a 14.0-karat gold necklace whose weight is 1.27 N?

*6. An irregularly shaped chunk of concrete has a hollow spherical cavity inside. The mass of the chunk is 33 kg, and the volume enclosed by the outside surface of the chunk is 0.025 m³. What is the radius of the spherical cavity?

*7. A bar of gold measures 0.15 m × 0.050 m × 0.050 m. How many gallons of water have the same mass as this bar?

*8. **GO** A gold prospector finds a solid rock that is composed solely of quartz and gold. The mass and volume of the rock are, respectively, 12.0 kg and 4.00 × 10⁻³ m³. Find the mass of the gold in the rock.

*9. **ssm www** A hypothetical spherical planet consists entirely of iron. What is the period of a satellite that orbits this planet just above its surface? Consult Table 11.1 as necessary.

**10. An antifreeze solution is made by mixing ethylene glycol ($\rho = 1116$ kg/m³) with water. Suppose that the specific gravity of such a solution is 1.0730. Assuming that the total volume of the solution is the sum of its parts, determine the volume percentage of ethylene glycol in the solution.

Section 11.2 Pressure

11. **ssm** An airtight box has a removable lid of area 1.3×10^{-2} m² and negligible weight. The box is taken up a mountain where the air pressure outside the box is 0.85×10^5 Pa. The inside of the box is completely evacuated. What is the magnitude of the force required to pull the lid off the box?

12. United States currency is printed using intaglio presses that generate a printing pressure of 8.0×10^4 lb/in.² A $20 bill is 6.1 in. by 2.6 in. Calculate the magnitude of the force that the printing press applies to one side of the bill.

13. **Interactive Solution 11.13** at **www.wiley.com/college/cutnell** is a model for this problem. A solid concrete block weighs 169 N and is resting on the ground. Its dimensions are 0.400 m × 0.200 m × 0.100 m. A number of identical blocks are stacked on top of this one. What is the smallest number of whole blocks (including the one on the ground) that can be stacked so that their weight creates a pressure of at least two atmospheres on the ground beneath the first block?

14. A person who weighs 625 N is riding a 98-N mountain bike. Suppose that the entire weight of the rider and bike is supported equally by the two tires. If the gauge pressure in each tire is 7.60×10^5 Pa, what is the area of contact between each tire and the ground?

15. A glass bottle of soda is sealed with a screw cap. The absolute pressure of the carbon dioxide inside the bottle is 1.80×10^5 Pa. Assuming that the top and bottom surfaces of the cap each have an area of 4.10×10^{-4} m², obtain the magnitude of the force that the screw thread exerts on the cap in order to keep it on the bottle. The air pressure outside the bottle is one atmosphere.

*16. A log splitter uses a pump with hydraulic oil to push a piston, which is attached to a chisel. The pump can generate a pressure of 2.0×10^7 Pa in the hydraulic oil, and the piston has a radius of 0.050 m. In a stroke lasting 25 s, the piston moves 0.60 m. What is the power needed to operate the log splitter's pump?

*17. **GO** A suitcase (mass $m = 16$ kg) is resting on the floor of an elevator. The part of the suitcase in contact with the floor measures 0.50 m × 0.15 m. The elevator is moving upward with an acceleration of magnitude 1.5 m/s². What pressure (in excess of atmospheric pressure) is applied to the floor beneath the suitcase?

*18. ⚕ As the initially empty urinary bladder fills with urine and expands, its internal pressure increases by 3300 Pa, which triggers the micturition reflex (the feeling of the need to urinate). The drawing shows a horizontal, square section of the bladder wall with an edge length of 0.010 m. Because the bladder is stretched, four tension forces of equal magnitude T act on the square section, one at each edge, and each force is directed at an angle θ below the horizontal. What is the magnitude T of the tension force acting on one edge of the section when the internal bladder pressure is 3300 Pa and each of the four tension forces is directed 5.0° below the horizontal?

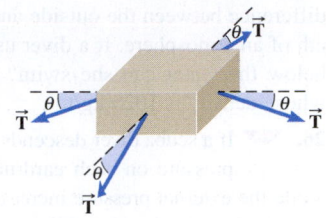

19. **ssm A cylinder (with circular ends) and a hemisphere are solid throughout and made from the same material. They are resting on the ground, the cylinder on one of its ends and the hemisphere on its flat side. The weight of each causes the same pressure to act on the ground. The cylinder is 0.500 m high. What is the radius of the hemisphere?

Section 11.3 Pressure and Depth in a Static Fluid,
Section 11.4 Pressure Gauges

20. The Mariana trench is located in the floor of the Pacific Ocean at a depth of about 11 000 m below the surface of the water. The density of seawater is 1025 kg/m³. **(a)** If an underwater vehicle were to explore such a depth, what force would the water exert on the vehicle's observation window (radius = 0.10 m)? **(b)** For comparison, determine the weight of a jetliner whose mass is 1.2×10^5 kg.

21. As background for this problem, review Conceptual Example 6. A submersible pump is put under the water at the bottom of a well and is used to push water up through a pipe. What minimum output gauge pressure must the pump generate to make the water reach the nozzle at ground level, 71 m above the pump?

22. **GO** A meat baster consists of a squeeze bulb attached to a plastic tube. When the bulb is squeezed and released, with the open end of the tube under the surface of the basting sauce, the sauce rises in the tube to a distance h, as the drawing shows. Using 1.013×10^5 Pa for the atmospheric pressure and 1200 kg/m³ for the density of the sauce, find the absolute pressure in the bulb when the distance h is **(a)** 0.15 m and **(b)** 0.10 m.

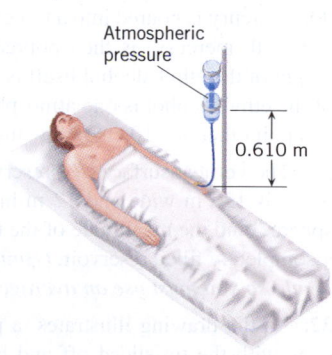

23. **ssm** Measured along the surface of the water, a rectangular swimming pool has a length of 15 m. Along this length, the flat bottom of the pool slopes downward at an angle of 11° below the horizontal, from one end to the other. By how much does the pressure at the bottom of the deep end exceed the pressure at the bottom of the shallow end?

24. ⚕ The drawing shows an intravenous feeding. With the distance shown, nutrient solution ($\rho = 1030$ kg/m³) can just barely enter the blood in the vein. What is the gauge pressure of the venous blood? Express your answer in millimeters of mercury.

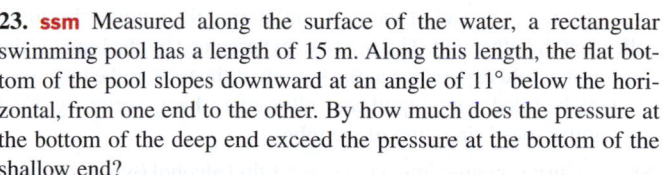

Problem 24

25. ⚕ The human lungs can function satisfactorily up to a limit where the pressure

difference between the outside and inside of the lungs is one-twentieth of an atmosphere. If a diver uses a snorkel for breathing, how far below the water can she swim? Assume the diver is in salt water whose density is 1025 kg/m³.

26. ☤ If a scuba diver descends too quickly into the sea, the internal pressure on each eardrum remains at atmospheric pressure, while the external pressure increases due to the increased water depth. At sufficient depths, the difference between the external and internal pressures can rupture an eardrum. Eardrums can rupture when the pressure difference is as little as 35 kPa. What is the depth at which this pressure difference could occur? The density of seawater is 1025 kg/m³.

27. ssm A water tower is a familiar sight in many towns. The purpose of such a tower is to provide storage capacity and to provide sufficient pressure in the pipes that deliver the water to customers. The drawing shows a spherical reservoir that contains 5.25×10^5 kg of water when full. The reservoir is vented to the atmosphere at the top. For a full reservoir, find the gauge pressure that the water has at the faucet in **(a)** house A and **(b)** house B. Ignore the diameter of the delivery pipes.

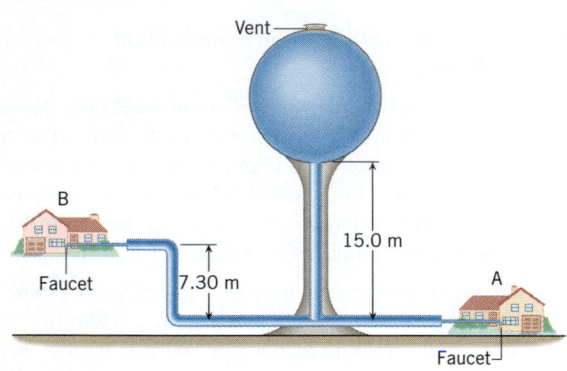

*28. Figure 11.11 shows a mercury barometer. Consider two barometers, one using mercury and another using an unknown liquid. Suppose that the pressure above the liquid in each tube is maintained at the same value P, between zero and atmospheric pressure. The height of the unknown liquid is 16 times greater than the height of the mercury. Find the density of the unknown liquid.

*29. ssm A tube is sealed at both ends and contains a 0.0100-m-long portion of liquid. The length of the tube is large compared to 0.0100 m. There is no air in the tube, and the vapor in the space above the liquid may be ignored. The tube is whirled around in a horizontal circle at a constant angular speed. The axis of rotation passes through one end of the tube, and during the motion, the liquid collects at the other end. The pressure experienced by the liquid is the same as it would experience at the bottom of the tube, if the tube were completely filled with liquid and allowed to hang vertically. Find the angular speed (in rad/s) of the tube.

*30. Mercury is poured into a tall glass. Ethyl alcohol (which does not mix with mercury) is then poured on top of the mercury until the height of the ethyl alcohol itself is 110 cm. The air pressure at the top of the ethyl alcohol is one atmosphere. What is the absolute pressure at a point that is 7.10 cm below the ethyl alcohol–mercury interface?

*31. The vertical surface of a reservoir dam that is in contact with the water is 120 m wide and 12 m high. The air pressure is one atmosphere. Find the magnitude of the total force acting on this surface in a completely filled reservoir. (*Hint: The pressure varies linearly with depth, so you must use an average pressure.*)

**32. As the drawing illustrates, a pond has the shape of an inverted cone with the tip sliced off and has a depth of 5.00 m. The atmospheric pressure above the pond is 1.01×10^5 Pa. The circular top

surface (radius $= R_2$) and circular bottom surface (radius $= R_1$) of the pond are both parallel to the ground. The magnitude of the force acting on the top surface is the same as the magnitude of the force acting on the bottom surface. Obtain **(a)** R_2 and **(b)** R_1.

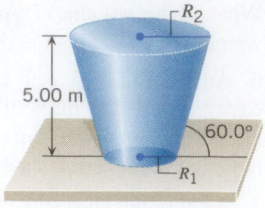

Section 11.5 Pascal's Principle

33. ssm The atmospheric pressure above a swimming pool changes from 755 to 765 mm of mercury. The bottom of the pool is a rectangle (12 m $\times$ 24 m). By how much does the force on the bottom of the pool increase?

34. A barber's chair with a person in it weighs 2100 N. The output plunger of a hydraulic system begins to lift the chair when the barber's foot applies a force of 55 N to the input piston. Neglect any height difference between the plunger and the piston. What is the ratio of the radius of the plunger to the radius of the piston?

35. Interactive Solution 11.35 at **www.wiley.com/college/cutnell** is a model for this problem. Multiple-Concept Example 8 also presents an approach to problems of this kind. The hydraulic oil in a car lift has a density of 8.30×10^2 kg/m³. The weight of the input piston is negligible. The radii of the input piston and output plunger are 7.70×10^{-3} m and 0.125 m, respectively. What input force F is needed to support the 24 500-N combined weight of a car and the output plunger, when **(a)** the bottom surfaces of the piston and plunger are at the same level, and **(b)** the bottom surface of the output plunger is 1.30 m *above* that of the input piston?

36. A hydraulic cylinder with a radius of 0.0281 m has a piston on the left end and a safety valve at the right end. The safety valve is a circular opening with a radius of 0.006 50 m, sealed with a disk. The disk is held in place by a spring (spring constant $= 885$ N/m) that is compressed 0.0100 m from its unstrained length. What is the magnitude of the minimum force that must be exerted on the piston in order to open the safety valve?

*37. The drawing shows a hydraulic system used with disc brakes. The force $\vec{F}$ is applied perpendicularly to the brake pedal. The pedal rotates about the axis shown in the drawing and causes a force to be applied perpendicularly to the input piston (radius $= 9.50 \times 10^{-3}$ m) in the master cylinder. The resulting pressure is transmitted by the brake fluid to the output plungers (radii $= 1.90 \times 10^{-2}$ m), which are covered with the brake linings. The linings are pressed against both sides of a disc attached to the rotating wheel. Suppose that the magnitude of $\vec{F}$ is 9.00 N. Assume that the input piston and the output plungers are at the same vertical level, and find the force applied to each side of the rotating disc.

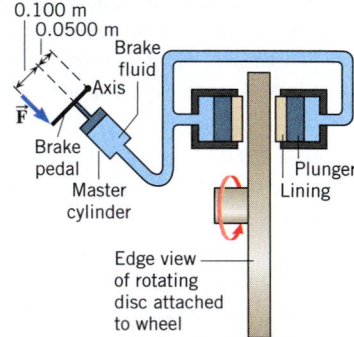

*38. 🟢 **GO** The drawing shows a hydraulic chamber with a spring (spring constant $= 1600$ N/m) attached to the input piston and a rock of mass 40.0 kg resting on the output plunger. The piston and plunger are nearly at the same height, and each has a negligible mass. By how much is the spring compressed from its unstrained position?

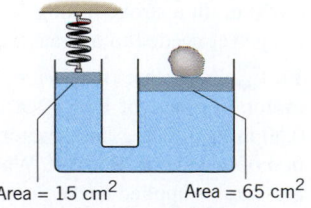

*39. ssm A dump truck uses a hydraulic cylinder, as the drawing illustrates. When activated by the operator, a pump injects hydraulic oil into the cylinder at an absolute pressure of 3.54×10^6 Pa and drives the output plunger, which has a radius of 0.150 m. Assuming that the plunger remains perpendicular to the floor of the load bed, find the torque that the plunger creates about the axis identified in the drawing.

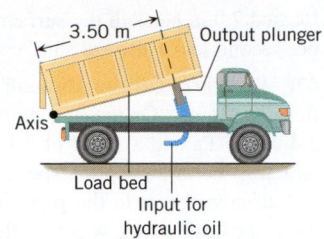

Section 11.6 Archimedes' Principle

40. The density of ice is 917 kg/m³, and the density of sea water is 1025 kg/m³. A swimming polar bear climbs onto a piece of floating ice that has a volume of 5.2 m³. What is the weight of the heaviest bear that the ice can support without sinking completely beneath the water?

41. ssm Multiple-Concept Example 11 reviews the concepts that are important in this problem. What is the radius of a hydrogen-filled balloon that would carry a load of 5750 N (in addition to the weight of the hydrogen) when the density of air is 1.29 kg/m³?

42. GO A hydrometer is a device used to measure the density of a liquid. It is a cylindrical tube weighted at one end, so that it floats with the heavier end downward. The tube is contained inside a large "medicine dropper," into which the liquid is drawn using the squeeze bulb (see the drawing). For use with your car, marks are put on the tube so that the level at which it floats indicates whether the liquid is battery acid (more dense) or antifreeze (less dense). The hydrometer has a weight of $W = 5.88 \times 10^{-2}$ N and a cross-sectional area of $A = 7.85 \times 10^{-5}$ m². How far from the bottom of the tube should the mark be put that denotes **(a)** battery acid ($\rho = 1280$ kg/m³) and **(b)** antifreeze ($\rho = 1073$ kg/m³)?

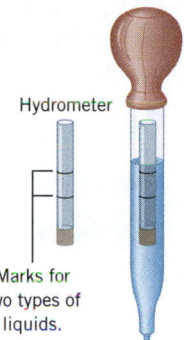

Hydrometer

Marks for two types of liquids.

43. A duck is floating on a lake with 25% of its volume beneath the water. What is the average density of the duck?

44. **(a)** The mass and the radius of the sun are 1.99×10^{30} kg and 6.96×10^8 m. What is its density? **(b)** If a solid object is made from a material that has the same density as the sun, would it sink or float in water? Why? **(c)** Would a solid object sink or float in water if it were made from a material whose density was the same as that of the planet Saturn (mass $= 5.7 \times 10^{26}$ kg, radius $= 6.0 \times 10^7$ m)? Provide a reason for your answer.

45. ssm www An 81-kg person puts on a life jacket, jumps into the water, and floats. The jacket has a volume of 3.1×10^{-2} m³ and is completely submerged under the water. The volume of the person's body that is under water is 6.2×10^{-2} m³. What is the density of the life jacket?

46. A lost shipping container is found resting on the ocean floor and completely submerged. The container is 6.1 m long, 2.4 m wide, and 2.6 m high. Salvage experts attach a spherical balloon to the top of the container and inflate it with air pumped down from the surface. When the balloon's radius is 1.5 m, the shipping container just begins to rise toward the surface. What is the mass of the container? Ignore the mass of the balloon and the air within it. Do *not* neglect the buoyant force exerted on the shipping container by the water. The density of seawater is 1025 kg/m³.

*47. An object is solid throughout. When the object is completely submerged in ethyl alcohol, its apparent weight is 15.2 N. When completely submerged in water, its apparent weight is 13.7 N. What is the volume of the object?

*48. GO A hot-air balloon is accelerating upward under the influence of two forces, its weight and the buoyant force. For simplicity, consider the weight to be only that of the hot air within the balloon, thus ignoring the balloon fabric and the basket. The hot air inside the balloon has a density of $\rho_{\text{hot air}} = 0.93$ kg/m³, and the density of the cool air outside is $\rho_{\text{cool air}} = 1.29$ kg/m³. What is the acceleration of the rising balloon?

*49. ssm A Kennedy half-dollar has a mass that is 1.150×10^{-2} kg. The coin is a mixture of silver and copper, and in water weighs 0.1011 N. Determine the mass of silver in the coin.

*50. Refer to Multiple-Concept Example 11 to see a problem similar to this one. What is the smallest number of whole logs ($\rho = 725$ kg/m³, radius $= 0.0800$ m, length $= 3.00$ m) that can be used to build a raft that will carry four people, each of whom has a mass of 80.0 kg?

**51. ssm A solid cylinder (radius $= 0.150$ m, height $= 0.120$ m) has a mass of 7.00 kg. This cylinder is floating in water. Then oil ($\rho = 725$ kg/m³) is poured on top of the water until the situation shown in the drawing results. How much of the height of the cylinder is in the oil?

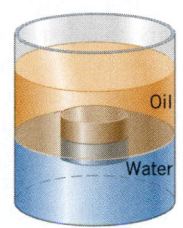

Oil

Water

**52. Interactive LearningWare 11.1 at www. wiley.com/college/cutnell provides a review of the concepts that are important in this problem. A spring is attached to the bottom of an empty swimming pool, with the axis of the spring oriented vertically. An 8.00-kg block of wood ($\rho = 840$ kg/m³) is fixed to the top of the spring and compresses it. Then the pool is filled with water, completely covering the block. The spring is now observed to be stretched twice as much as it had been compressed. Determine the percentage of the block's total volume that is hollow. Ignore any air in the hollow space.

**53. A lighter-than-air balloon and its load of passengers and ballast are floating stationary above the earth. Ballast is weight (of negligible volume) that can be dropped overboard to make the balloon rise. The radius of this balloon is 6.25 m. Assuming a constant value of 1.29 kg/m³ for the density of air, determine how much weight must be dropped overboard to make the balloon rise 105 m in 15.0 s.

Section 11.8 The Equation of Continuity

54. A fuel pump sends gasoline from a car's fuel tank to the engine at a rate of 5.88×10^{-2} kg/s. The density of the gasoline is 735 kg/m³, and the radius of the fuel line is 3.18×10^{-3} m. What is the speed at which the gasoline moves through the fuel line?

55. ssm A patient recovering from surgery is being given fluid intravenously. The fluid has a density of 1030 kg/m³, and 9.5×10^{-4} m³ of it flows into the patient every six hours. Find the mass flow rate in kg/s.

56. GO Water flows straight down from an open faucet. The cross-sectional area of the faucet is 1.8×10^{-4} m², and the speed of the water is 0.85 m/s as it leaves the faucet. Ignoring air resistance, find the cross-sectional area of the water stream at a point 0.10 m below the faucet.

57. A room has a volume of 120 m³. An air-conditioning system is to replace the air in this room every twenty minutes, using ducts that have a square cross section. Assuming that air can be treated as an incompressible fluid, find the length of a side of the square if the air speed within the ducts is **(a)** 3.0 m/s and **(b)** 5.0 m/s.

58. Concept Simulation 11.1 at www.wiley.com/college/cutnell reviews a central concept in this problem. **(a)** The

volume flow rate in an artery supplying the brain is 3.6×10^{-6} m³/s. If the radius of the artery is 5.2 mm, determine the average blood speed. **(b)** Find the average blood speed at a constriction in the artery if the constriction reduces the radius by a factor of 3. Assume that the volume flow rate is the same as that in part (a).

*59. **Concept Simulation 11.1** at **www.wiley.com/college/cutnell** reviews the central idea in this problem. In an adjustable nozzle for a garden hose, a cylindrical plug is aligned along the axis of the hose and can be inserted into the hose opening. The purpose of the plug is to change the speed of the water leaving the hose. The speed of the water passing around the plug is to be three times as large as the speed of the water before it encounters the plug. Find the ratio of the plug radius to the inside hose radius.

*60. ▼ The aorta carries blood away from the heart at a speed of about 40 cm/s and has a radius of approximately 1.1 cm. The aorta branches eventually into a large number of tiny capillaries that distribute the blood to the various body organs. In a capillary, the blood speed is approximately 0.07 cm/s, and the radius is about 6×10^{-4} cm. Treat the blood as an incompressible fluid, and use these data to determine the approximate number of capillaries in the human body.

Section 11.9 Bernoulli's Equation,
Section 11.10 Applications of Bernoulli's Equation

61. **ssm** Review Conceptual Example 14 as an aid in understanding this problem. Suppose that a 15-m/s wind is blowing across the roof of your house. The density of air is 1.29 kg/m³. **(a)** Determine the reduction in pressure (below atmospheric pressure of stationary air) that accompanies this wind. **(b)** Explain why some roofs are "blown outward" during high winds.

62. ▼ One way to administer an inoculation is with a "gun" that shoots the vaccine through a narrow opening. No needle is necessary, for the vaccine emerges with sufficient speed to pass directly into the tissue beneath the skin. The speed is high, because the vaccine ($\rho = 1100$ kg/m³) is held in a reservoir where a high pressure pushes it out. The pressure on the surface of the vaccine in one gun is 4.1×10^{6} Pa above the atmospheric pressure outside the narrow opening. The dosage is small enough that the vaccine's surface in the reservoir is nearly stationary during an inoculation. The vertical height between the vaccine's surface in the reservoir and the opening can be ignored. Find the speed at which the vaccine emerges.

63. **ssm** An airplane wing is designed so that the speed of the air across the top of the wing is 251 m/s when the speed of the air below the wing is 225 m/s. The density of the air is 1.29 kg/m³. What is the lifting force on a wing of area 24.0 m²?

64. **GO** Consult Multiple-Concept Example 15 to review the concepts on which this problem depends. Water flowing out of a horizontal pipe emerges through a nozzle. The radius of the pipe is 1.9 cm, and the radius of the nozzle is 0.48 cm. The speed of the water in the pipe is 0.62 m/s. Treat the water as an ideal fluid, and determine the absolute pressure of the water in the pipe.

65. ▼ See Multiple-Concept Example 15 to review the concepts that are pertinent to this problem. The blood speed in a normal segment of a horizontal artery is 0.11 m/s. An abnormal segment of the artery is narrowed down by an arteriosclerotic plaque to one-fourth the normal cross-sectional area. What is the difference in blood pressures between the normal and constricted segments of the artery?

66. **GO** A ship is floating on a lake. Its hold is the interior space beneath its deck; the hold is empty and is open to the atmosphere. The hull has a hole in it, which is below the water line, so water leaks into the hold. The effective area of the hole is 8.0×10^{-3} m² and is located 2.0 m beneath the surface of the lake. What volume of water per second leaks into the ship?

67. Water is circulating through a closed system of pipes in a two-floor apartment. On the first floor, the water has a gauge pressure of 3.4×10^{5} Pa and a speed of 2.1 m/s. However, on the second floor, which is 4.0 m higher, the speed of the water is 3.7 m/s. The speeds are different because the pipe diameters are different. What is the gauge pressure of the water on the second floor?

68. **Interactive LearningWare 11.2** at **www.wiley.com/college/cutnell** reviews the approach taken in problems such as this one. A small crack occurs at the base of a 15.0-m-high dam. The effective crack area through which water leaves is 1.30×10^{-3} m². **(a)** Ignoring viscous losses, what is the speed of water flowing through the crack? **(b)** How many cubic meters of water per second leave the dam?

*69. **ssm** A Venturi meter is a device that is used for measuring the speed of a fluid within a pipe. The drawing shows a gas flowing at speed v_2 through a horizontal section of pipe whose cross-sectional area is $A_2 = 0.0700$ m². The gas has a density of $\rho = 1.30$ kg/m³. The Venturi meter has a cross-sectional area of $A_1 = 0.0500$ m² and has been substituted for a section of the larger pipe. The pressure difference between the two sections is $P_2 - P_1 = 120$ Pa. Find **(a)** the speed v_2 of the gas in the larger, original pipe and **(b)** the volume flow rate Q of the gas.

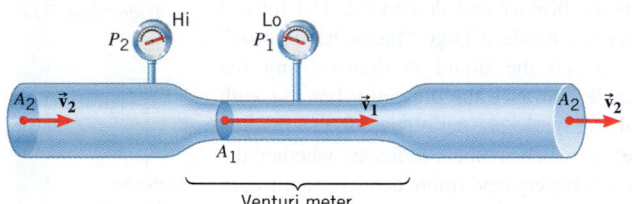

Venturi meter

*70. A hand-pumped water gun is held level at a height of 0.75 m above the ground and fired. The water stream from the gun hits the ground a horizontal distance of 7.3 m from the muzzle. Find the gauge pressure of the water gun's reservoir at the instant when the gun is fired. Assume that the speed of the water in the reservoir is zero and that the water flow is steady. Ignore both air resistance and the height difference between the reservoir and the muzzle.

*71. In a very large closed tank, the absolute pressure of the air above the water is 6.01×10^{5} Pa. The water leaves the bottom of the tank through a nozzle that is directed straight upward. The opening of the nozzle is 4.00 m below the surface of the water. **(a)** Find the speed at which the water leaves the nozzle. **(b)** Ignoring air resistance and viscous effects, determine the height to which the water rises.

*72. An airplane has an effective wing surface area of 16 m² that is generating the lift force. In level flight the air speed over the top of the wings is 62.0 m/s, while the air speed beneath the wings is 54.0 m/s. What is the weight of the plane?

*73. **Interactive Solution 11.73** at **www.wiley.com/college/cutnell** presents one method for modeling this problem. The construction of a flat rectangular roof (5.0 m × 6.3 m) allows it to withstand a maximum net outward force that is 22 000 N. The density of the air is 1.29 kg/m³. At what wind speed will this roof blow outward?

*74. The concepts that play roles in this problem are similar to those in Multiple-Concept Example 15, except that the fluid here moves upward rather than remaining horizontal. A liquid is flowing through a horizontal pipe whose radius is 0.0200 m. The pipe bends straight upward through a height of 10.0 m and joins another horizontal pipe whose radius is 0.0400 m. What volume flow rate will keep the pressures in the two horizontal pipes the same?

**75. ssm A uniform rectangular plate is hanging vertically downward from a hinge that passes along its left edge. By blowing air at 11.0 m/s *over the top of the plate only,* it is possible to keep the plate in a horizontal position, as illustrated in part *a* of the drawing. To what value should the air speed be reduced so that the plate is kept at a 30.0° angle with respect to the vertical, as in part *b* of the drawing? (*Hint: Apply Bernoulli's equation in the form of Equation 11.12.*)

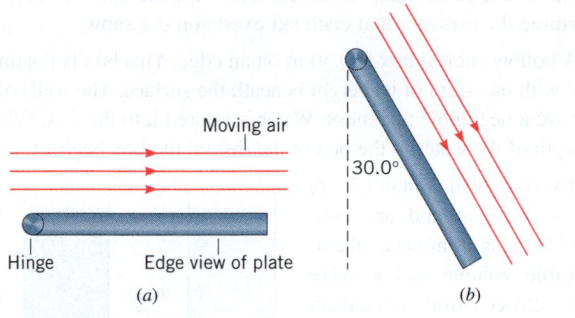

Moving air

30.0°

Hinge Edge view of plate

(a) (b)

76. A siphon tube is useful for removing liquid from a tank. The siphon tube is first filled with liquid, and then one end is inserted into the tank. Liquid then drains out the other end, as the drawing illustrates. **(a) Using reasoning similar to that employed in obtaining Torricelli's theorem, derive an expression for the speed v of the fluid emerging from the tube. This expression should give v in terms of the vertical height y and the acceleration due to gravity g. (Note that this speed does not depend on the depth d of the tube below the surface of the liquid.) **(b)** At what value of the vertical distance y will the siphon stop working? **(c)** Derive an expression for the absolute pressure at the highest point in the siphon (point A) in terms of the atmospheric pressure P_0, the fluid density ρ, g, and the heights h and y. (Note that the fluid speed at point A is the same as the speed of the fluid emerging from the tube, because the cross-sectional area of the tube is the same everywhere.)

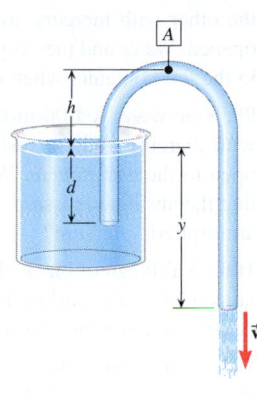

Section 11.11 Viscous Flow

77. ssm www A 1.3-m length of horizontal pipe has a radius of 6.4×10^{-3} m. Water within the pipe flows with a volume flow rate of 9.0×10^{-3} m³/s out of the right end of the pipe and into the air. What is the pressure in the flowing water at the left end of the pipe if the water behaves as **(a)** an ideal fluid and **(b)** a viscous fluid ($\eta = 1.00 \times 10^{-3}$ Pa·s)?

78. Poiseuille's law remains valid as long as the fluid flow is laminar. For sufficiently high speed, however, the flow becomes turbulent, even if the fluid is moving through a smooth pipe with no restrictions. It is found experimentally that the flow is laminar as long as the *Reynolds number* Re is less than about 2000: Re $= 2\bar{v}\rho R/\eta$. Here $\bar{v}$, ρ, and η are, respectively, the average speed, density, and viscosity of the fluid, and R is the radius of the pipe. Calculate the highest average speed that blood ($\rho = 1060$ kg/m³, $\eta = 4.0 \times 10^{-3}$ Pa·s) could have and still remain in laminar flow when it flows through the aorta ($R = 8.0 \times 10^{-3}$ m).

79. In the human body, blood vessels can dilate, or increase their radii, in response to various stimuli, so that the volume flow rate of the blood increases. Assume that the pressure at either end of a blood vessel, the length of the vessel, and the viscosity of the blood remain the same, and determine the factor $R_{dilated}/R_{normal}$ by which the radius of a vessel must change in order to double the volume flow rate of the blood through the vessel.

80. GO A blood transfusion is being set up in an emergency room for an accident victim. Blood has a density of 1060 kg/m³ and a viscosity of 4.0×10^{-3} Pa·s. The needle being used has a length of 3.0 cm and an inner radius of 0.25 mm. The doctor wishes to use a volume flow rate through the needle of 4.5×10^{-8} m³/s. What is the distance h above the victim's arm where the level of the blood in the transfusion bottle should be located? As an approximation, assume that the level of the blood in the transfusion bottle and the point where the needle enters the vein in the arm have the same pressure of one atmosphere. (In reality, the pressure in the vein is slightly above atmospheric pressure.)

81. ssm Interactive Solution 11.81 at **www.wiley.com/college/ cutnell** illustrates a model for solving this problem. A pressure difference of 1.8×10^3 Pa is needed to drive water ($\eta = 1.0 \times 10^{-3}$ Pa·s) through a pipe whose radius is 5.1×10^{-3} m. The volume flow rate of the water is 2.8×10^{-4} m³/s. What is the length of the pipe?

82. A volume of 7.2 m³ of glycerol ($\eta = 0.934$ Pa·s) is pumped through a 15-m length of pipe in 55 minutes. The pressure at the input end of the pipe is 8.6×10^5 Pa, and the pressure at the output end is atmospheric pressure. What is the pipe's radius?

*83. GO Two hoses are connected to the same outlet using a Y-connector, as the drawing shows. The hoses A and B have the same length, but hose B has the larger radius. Each is open to the atmosphere at the end where the water exits. Water flows through both hoses as a viscous fluid, and Poiseuille's law $[Q = \pi R^4(P_2 - P_1)/8\eta L]$ applies to each. In this law, P_2 is the pressure upstream, P_1 is the pressure downstream, and Q is the volume flow rate. The ratio of the radius of hose B to the radius of hose A is $R_B/R_A = 1.50$. Find the ratio of the speed of the water in hose B to the speed in hose A.

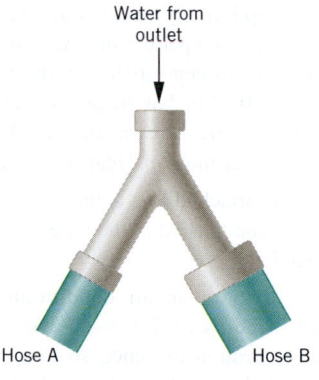

Water from outlet

Hose A Hose B

*84. When an object moves through a fluid, the fluid exerts a viscous force $\vec{F}$ on the object that tends to slow it down. For a small sphere of radius R, moving slowly with a speed v, the magnitude of the viscous force is given by Stokes' law, $F = 6\pi\eta Rv$, where η is the viscosity of the fluid. **(a)** What is the viscous force on a sphere of radius $R = 5.0 \times 10^{-4}$ m that is falling through water ($\eta = 1.00 \times 10^{-3}$ Pa·s) when the sphere has a speed of 3.0 m/s? **(b)** The speed of the falling sphere increases until the viscous force balances the weight of the sphere. Thereafter, no net force acts on the sphere, and it falls with a constant speed called the "terminal speed." If the sphere has a mass of 1.0×10^{-5} kg, what is its terminal speed?

ADDITIONAL PROBLEMS

85. ssm The main water line enters a house on the first floor. The line has a gauge pressure of 1.90×10^5 Pa. **(a)** A faucet on the second floor, 6.50 m above the first floor, is turned off. What is the gauge pressure at this faucet? **(b)** How high could a faucet be before no water would flow from it, even if the faucet were open?

86. Prairie dogs are burrowing rodents. They do not suffocate in their burrows, because the effect of air speed on pressure creates sufficient air circulation. The animals maintain a difference in the shapes of two entrances to the burrow, and because of this difference, the air ($\rho = 1.29$ kg/m³) blows past the openings at different speeds, as the drawing indicates. Assuming that the openings are at the same vertical level, find the difference in air pressure between the openings and indicate which way the air circulates.

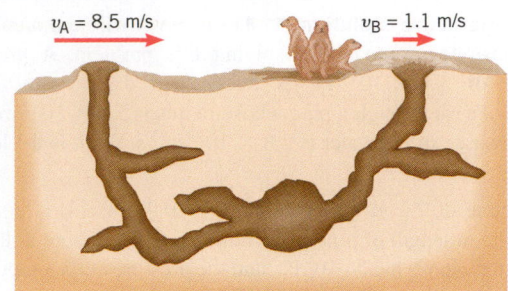

$v_A = 8.5$ m/s $v_B = 1.1$ m/s

87. ssm A 0.10-m × 0.20-m × 0.30-m block is suspended from a wire and is completely under water. What buoyant force acts on the block?

88. At a given instant, the blood pressure in the heart is 1.6×10^4 Pa. If an artery in the brain is 0.45 m above the heart, what is the pressure in the artery? Ignore any pressure changes due to blood flow.

89. One of the concrete pillars that support a house is 2.2 m tall and has a radius of 0.50 m. The density of concrete is about 2.2×10^3 kg/m³. Find the weight of this pillar in pounds (1 N = 0.2248 lb).

90. An underground pump initially forces water through a horizontal pipe at a flow rate of 740 gallons per minute. After several years of operation, corrosion and mineral deposits have reduced the inner radius of the pipe to 0.19 m from 0.24 m, but the pressure difference between the ends of the pipe is the same as it was initially. Find the final flow rate in the pipe in gallons per minute. Treat water as a viscous fluid.

91. One end of a wire is attached to a ceiling, and a solid brass ball is tied to the lower end. The tension in the wire is 120 N. What is the radius of the brass ball?

92. A cylindrical air duct in an air conditioning system has a length of 5.5 m and a radius of 7.2×10^{-2} m. A fan forces air ($\eta = 1.8 \times 10^{-5}$ Pa·s) through the duct, so that the air in a room (volume = 280 m³) is replenished every ten minutes. Determine the difference in pressure between the ends of the air duct.

93. Review Conceptual Example 6 as an aid in understanding this problem. Consider the pump on the right side of Figure 11.10, which acts to reduce the air pressure in the pipe. The air pressure outside the pipe is one atmosphere. Find the maximum depth from which this pump can extract water from the well.

94. A paperweight, when weighed in air, has a weight of W = 6.9 N. When completely immersed in water, however, it has a weight of $W_{\text{in water}} = 4.3$ N. Find the volume of the paperweight.

95. ssm A water line with an internal radius of 6.5×10^{-3} m is connected to a shower head that has 12 holes. The speed of the water in the line is 1.2 m/s. **(a)** What is the volume flow rate in the line? **(b)** At what speed does the water leave one of the holes (effective hole radius = 4.6×10^{-4} m) in the head?

96. A 58-kg skier is going down a slope oriented 35° above the horizontal. The area of each ski in contact with the snow is 0.13 m². Determine the pressure that each ski exerts on the snow.

97. A hollow cubical box is 0.30 m on an edge. This box is floating in a lake with one-third of its height beneath the surface. The walls of the box have a negligible thickness. Water is poured into the box. What is the depth of the water in the box at the instant the box begins to sink?

98. Two identical containers are open at the top and are connected at the bottom via a tube of negligible volume and a valve that is closed. Both containers are filled initially to the same height of 1.00 m, one with water, the other with mercury, as the drawing indicates. The valve is then opened. Water and mercury are immiscible. Determine the fluid level in the left container when equilibrium is reestablished.

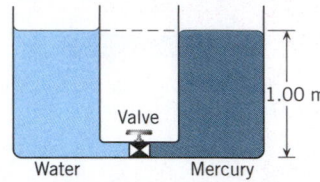

99. ssm www A 1.00-m-tall container is filled to the brim, partway with mercury and the rest of the way with water. The container is open to the atmosphere. What must be the depth of the mercury so that the absolute pressure on the bottom of the container is twice the atmospheric pressure?

100. A full can of black cherry soda has a mass of 0.416 kg. It contains 3.54×10^{-4} m³ of liquid. Assuming that the soda has the same density as water, find the volume of aluminum used to make the can.

101. A fountain sends a stream of water straight up into the air to a maximum height of 5.00 m. The effective cross-sectional area of the pipe feeding the fountain is 5.00×10^{-4} m². Neglecting air resistance and any viscous effects, determine how many gallons per minute are being used by the fountain. (Note: 1 gal = 3.79×10^{-3} m³.)

102. One kilogram of glass ($\rho = 2.60 \times 10^3$ kg/m³) is shaped into a hollow spherical shell that just barely floats in water. What are the inner and outer radii of the shell? Do not assume that the shell is thin.

103. A house has a roof with the dimensions shown in the drawing. Determine the magnitude and direction of the net force that the atmosphere applies to the roof when the outside pressure rises suddenly by 10.0 mm of mercury, before the pressure in the attic can adjust.

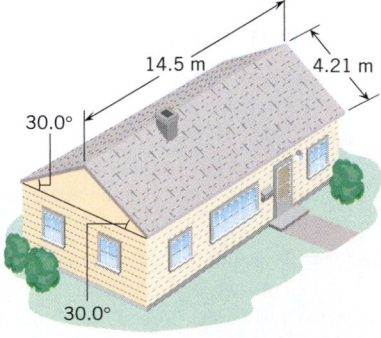

Problem 103

104. Two circular holes, one larger than the other, are cut in the side of a large water tank whose top is open to the atmosphere. The center of one of these holes is located twice as far beneath the surface of the water as the other. The volume flow rate of the water coming out of the holes is the same. **(a)** Decide which hole is located nearest the surface of the water. **(b)** Calculate the ratio of the radius of the larger hole to the radius of the smaller hole.

CHAPTER **12**

TEMPERATURE AND HEAT

It is winter, and these Macaque monkeys are enjoying a dip in the Jigokudani hot springs in Japan. The water in the photograph exists in three forms or phases, solid (snow), liquid, and gas (water vapor). Water can change from one phase to another, and heat plays a role in the change, as we will see in this chapter. (© Shusuke Sezai/Corbis)

12.1 COMMON TEMPERATURE SCALES

To measure temperature we use a thermometer. Many thermometers make use of the fact that materials usually expand with increasing temperature. For example, Figure 12.1 shows the common mercury-in-glass thermometer, which consists of a mercury-filled glass bulb connected to a capillary tube. When the mercury is heated, it expands into the capillary tube, the amount of expansion being proportional to the change in temperature. The outside of the glass is marked with an appropriate scale for reading the temperature.

A number of different temperature scales have been devised, two popular choices being the *Celsius* (formerly, centigrade) and *Fahrenheit scales.* Figure 12.1 illustrates these scales. Historically,* both scales were defined by assigning two temperature points on the scale and then dividing the distance between them into a number of equally spaced intervals. One point was chosen to be the temperature at which ice melts under one atmosphere of pressure (the "ice point"), and the other was the temperature at which water boils under one atmosphere of pressure (the "steam point"). On the Celsius scale, an ice point of 0 °C (0 degrees Celsius) and a steam point of 100 °C were selected. On the Fahrenheit scale, an ice point of 32 °F (32 degrees Fahrenheit) and a steam point of 212 °F were chosen. The Celsius scale is used worldwide, while the Fahrenheit scale is used mostly in the United States, often in home medical thermometers.

There is a subtle difference in the way the temperature of an object is reported, as compared to a *change* in its temperature. For example, the temperature of the human body is about 37 °C, where the symbol °C stands for "degrees Celsius." However, the *change* between two temperatures is specified in "Celsius degrees" (C°)—not in "degrees Celsius." Thus, if the

*Today, the Celsius and Fahrenheit scales are defined in terms of the Kelvin temperature scale; Section 12.2 discusses the Kelvin scale.

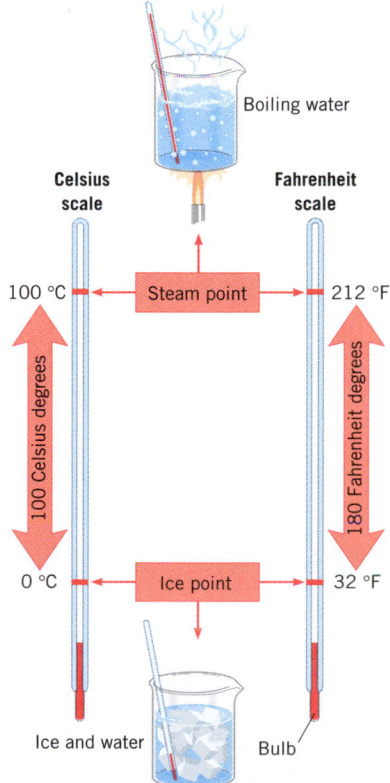

Figure 12.1 The Celsius and Fahrenheit temperature scales.

body temperature rises to 39 °C, the change in temperature is 2 Celsius degrees or 2 C°, not 2 °C.

As Figure 12.1 indicates, the separation between the ice and steam points on the Celsius scale is divided into 100 Celsius degrees, while on the Fahrenheit scale the separation is divided into 180 Fahrenheit degrees. Therefore, the size of the Celsius degree is larger than that of the Fahrenheit degree by a factor of $\frac{180}{100}$, or $\frac{9}{5}$. Examples 1 and 2 illustrate how to convert between the Celsius and Fahrenheit scales using this factor.

 Example 1 Converting from a Fahrenheit to a Celsius Temperature

A healthy person has an oral temperature of 98.6 °F. What would this reading be on the Celsius scale?

Reasoning and Solution A temperature of 98.6 °F is 66.6 Fahrenheit degrees above the ice point of 32.0 °F. Since 1 C° = $\frac{9}{5}$ F°, the difference of 66.6 F° is equivalent to

$$(66.6 \text{ F}°)\left(\frac{1 \text{ C}°}{\frac{9}{5} \text{ F}°}\right) = 37.0 \text{ C}°$$

Thus, the person's temperature is 37.0 Celsius degrees above the ice point. Adding 37.0 Celsius degrees to the ice point of 0 °C on the Celsius scale gives a Celsius temperature of $\boxed{37.0\,°\text{C}}$.

 Example 2 Converting from a Celsius to a Fahrenheit Temperature

A time and temperature sign on a bank indicates that the outdoor temperature is −20.0 °C. Find the corresponding temperature on the Fahrenheit scale.

Reasoning and Solution The temperature of −20.0 °C is 20.0 Celsius degrees *below* the ice point of 0 °C. This number of Celsius degrees corresponds to

$$(20.0 \text{ C}°)\left(\frac{\frac{9}{5} \text{ F}°}{1 \text{ C}°}\right) = 36.0 \text{ F}°$$

The temperature, then, is 36.0 Fahrenheit degrees below the ice point. Subtracting 36.0 Fahrenheit degrees from the ice point of 32.0 °F on the Fahrenheit scale gives a Fahrenheit temperature of $\boxed{-4.0\,°\text{F}}$.

The reasoning strategy used in Examples 1 and 2 for converting between different temperature scales is summarized below.

REASONING STRATEGY

Converting Between Different Temperature Scales

1. Determine the magnitude of the difference between the stated temperature and the ice point on the initial scale.

2. Convert this number of degrees from one scale to the other scale by using the appropriate conversion factor. For conversion between the Celsius and Fahrenheit scales, the factor is based on the fact that 1 C° = $\frac{9}{5}$ F°.

3. Add or subtract the number of degrees on the new scale to or from the ice point on the new scale.

✓ **CHECK YOUR UNDERSTANDING**

(*The answer is given at the end of the book.*)

1. On a new temperature scale the steam point is 348 °X, and the ice point is 112 °X. What is the temperature on this scale that corresponds to 28.0 °C?

12.2 THE KELVIN TEMPERATURE SCALE

Although the Celsius and Fahrenheit scales are widely used, the *Kelvin temperature scale* has greater scientific significance. It was introduced by the Scottish physicist William Thompson (Lord Kelvin, 1824–1907), and in his honor each degree on the scale is called a kelvin (K). By international agreement, the symbol K is not written with a degree sign (°), nor is the word "degrees" used when quoting temperatures. For example, a temperature of 300 K (not 300 °K) is read as "three hundred kelvins," not "three hundred degrees kelvin." The kelvin is the SI base unit for temperature.

Figure 12.2 compares the Kelvin and Celsius scales. The size of one kelvin is identical to the size of one Celsius degree because there are one hundred divisions between the ice and steam points on both scales. As we will discuss shortly, experiments have shown that there exists a lowest possible temperature, below which no substance can be cooled. This lowest temperature is defined to be the zero point on the Kelvin scale and is referred to as *absolute zero*.

The ice point (0 °C) occurs at 273.15 K on the Kelvin scale. Thus, the Kelvin temperature T and the Celsius temperature T_c are related by

$$T = T_c + 273.15 \qquad (12.1)$$

The number 273.15 in Equation 12.1 is an experimental result, obtained in studies that utilize a gas-based thermometer.

When a gas confined to a fixed volume is heated, its pressure increases. Conversely, when the gas is cooled, its pressure decreases. For example, the air pressure in automobile tires can rise by as much as 20% after the car has been driven and the tires have become warm. The change in gas pressure with temperature is the basis for the *constant-volume gas thermometer.*

A constant-volume gas thermometer consists of a gas-filled bulb to which a pressure gauge is attached, as in Figure 12.3. The gas is often hydrogen or helium at a low density, and the pressure gauge can be a U-tube manometer filled with mercury. The bulb is placed in thermal contact with the substance whose temperature is being measured. The volume of the gas is held constant by raising or lowering the *right* column of the U-tube manometer in order to keep the mercury level in the *left* column at the same reference level. The absolute pressure of the gas is proportional to the height h of the mercury on the right. As the temperature changes, the pressure changes and can be used to indicate the temperature, once the constant-volume gas thermometer has been calibrated.

Suppose that the absolute pressure of the gas in Figure 12.3 is measured at different temperatures. If the results are plotted on a pressure-versus-temperature graph, a straight line is obtained, as in Figure 12.4. If the straight line is extended or extrapolated to lower and lower temperatures, the line crosses the temperature axis at −273.15 °C. In reality, no gas can be cooled to this temperature, because all gases liquify before reaching it. However, helium and hydrogen liquify at such low temperatures that they are often used in the thermometer. This kind of graph can be obtained for different amounts and types of low-density gases. In all cases, a straight line is found that extrapolates to −273.15° C on the temperature axis, which suggests that the value of −273.15° C has fundamental significance. The significance of this number is that it is the *absolute zero point* for temperature measurement. The phrase "absolute zero" means that temperatures lower than −273.15 °C cannot be reached by continually cooling a gas or any other substance. If lower temperatures could be reached, then further extrapolation of the straight line in Figure 12.4 would suggest that negative absolute gas pressures could exist. Such a situation would be impossible, because a negative absolute gas pressure has no meaning. Thus, the Kelvin scale is chosen so that its zero temperature point is the lowest temperature attainable.

12.3 THERMOMETERS

All thermometers make use of the change in some physical property with temperature. A property that changes with temperature is called a *thermometric property.* For example, the thermometric property of the mercury thermometer is the length of the mercury

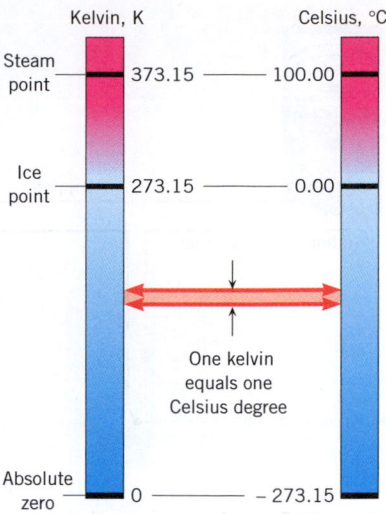

Figure 12.2 A comparison of the Kelvin and Celsius temperature scales.

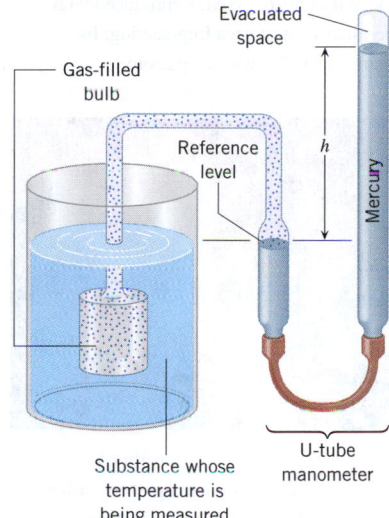

Figure 12.3 A constant-volume gas thermometer.

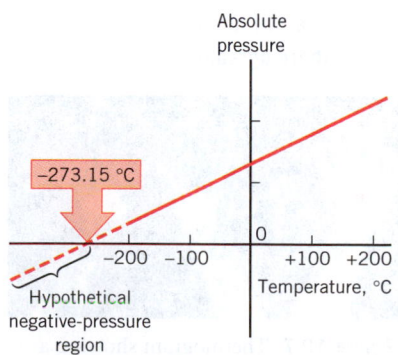

Figure 12.4 A plot of absolute pressure versus temperature for a low-density gas at constant volume. The graph is a straight line and, when extrapolated (dashed line), crosses the temperature axis at −273.15 °C.

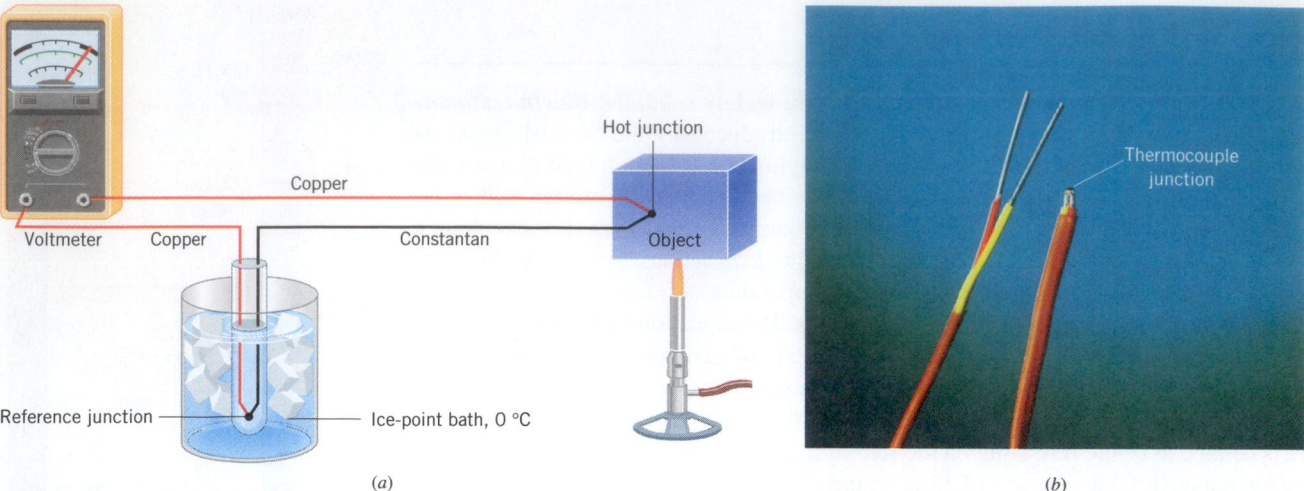

(a)

(b)

Figure 12.5 (*a*) A thermocouple is made from two different types of wires, copper and constantan in this case. (*b*) A thermocouple junction between two different wires. (*b*, © Omega Engineering, Inc. All rights reserved. Reproduced with permission of Omega Engineering, Inc., Stamford, CT, www.omega.com.)

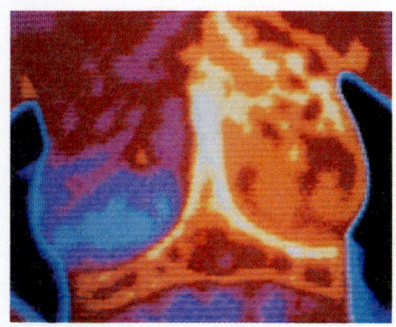

Figure 12.6 Invasive carcinoma (cancer) of the breast registers colors from red to yellow/white in this thermogram, indicating markedly elevated temperatures. (Science Photo Library/ Photo Researchers, Inc.)

The physics of thermography.

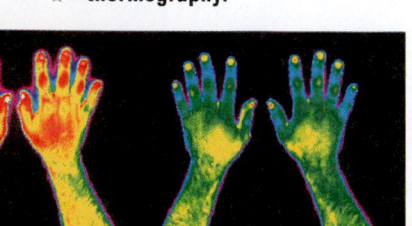

Figure 12.7 Thermogram showing a smoker's forearms before (*left*) and 5 minutes after (*right*) he has smoked a cigarette. Temperatures range from over 34 °C (white) to about 28 °C (blue). (Dr. Arthur Tucker/Science Photo Library/Photo Researchers, Inc.)

column, while in the constant-volume gas thermometer it is the pressure of the gas. Several other important thermometers and their thermometric properties will now be discussed.

The *thermocouple* is a thermometer used extensively in scientific laboratories. It consists of thin wires of different metals, welded together at the ends to form two junctions, as Figure 12.5 illustrates. Often the metals are copper and constantan (a copper–nickel alloy). One of the junctions, called the "hot" junction, is placed in thermal contact with the object whose temperature is being measured. The other junction, termed the "reference" junction, is kept at a known constant temperature (usually an ice–water mixture at 0 °C). The thermocouple generates a voltage that depends on the *difference in temperature* between the two junctions. This voltage is the thermometric property and is measured by a voltmeter, as the drawing indicates. With the aid of calibration tables, the temperature of the hot junction can be obtained from the voltage. Thermocouples are used to measure temperatures as high as 2300 °C or as low as −270 °C.

Most substances offer resistance to the flow of electricity, and this resistance changes with temperature. As a result, electrical resistance provides another thermometric property. *Electrical resistance thermometers* are often made from platinum wire, because platinum has excellent mechanical and electrical properties in the temperature range from −270 °C to +700 °C. The resistance of platinum wire is known as a function of temperature. Thus, the temperature of a substance can be determined by placing the resistance thermometer in thermal contact with the substance and measuring the resistance of the wire.

Radiation emitted by an object can also be used to indicate temperature. At low to moderate temperatures, the predominant radiation emitted is infrared. As the temperature is raised, the intensity of the radiation increases substantially. In one interesting application, an infrared camera registers the intensity of the infrared radiation produced at different locations on the human body. The camera is connected to a color monitor that displays the different infrared intensities as different colors. This "thermal painting" is called a *thermograph* or *thermogram*. Thermography is an important diagnostic tool in medicine. For example, breast cancer is indicated in the thermogram in Figure 12.6 by the elevated temperatures associated with malignant tissue. In another application, Figure 12.7 shows thermographic images of a smoker's forearms before (left) and 5 minutes after (right) he has smoked a cigarette. After smoking, the forearms are cooler due to the effect of nicotine, which causes vasoconstriction (narrowing of the blood vessels) and reduces blood flow, a result that can lead to a higher risk from blood clotting. Temperatures in these images range from over 34 °C to about 28 °C and are indicated in decreasing order by the colors white, red, yellow, green, and blue.

Oceanographers and meteorologists also use thermograms extensively, to map the temperature distribution on the surface of the earth. For example, Figure 12.8 shows a satellite image of the sea-surface temperature of the Pacific Ocean. The region depicted in red is the 1997/98 El Niño, a large area of the ocean, approximately twice the width of the United States, where temperatures reached abnormally high values. This El Niño caused major weather changes in certain regions of the earth.

LINEAR THERMAL EXPANSION

12.4

NORMAL SOLIDS

Have you ever found the metal lid on a glass jar too tight to open? One solution is to run hot water over the lid, which loosens it because the metal expands more than the glass does. To varying extents, most materials expand when heated and contract when cooled. The increase in any one dimension of a solid is called *linear expansion,* linear in the sense that the expansion occurs along a line. Figure 12.9 illustrates the linear expansion of a rod whose length is L_0 when the temperature is T_0. When the temperature of the rod increases to $T_0 + \Delta T$, the length becomes $L_0 + \Delta L$, where ΔT and ΔL are the changes in temperature and length, respectively. Conversely, when the temperature decreases to $T_0 - \Delta T$, the length decreases to $L_0 - \Delta L$.

For modest temperature changes, experiments show that the change in length is directly proportional to the change in temperature ($\Delta L \propto \Delta T$). In addition, the change in length is proportional to the initial length of the rod, a fact that can be understood with the aid of Figure 12.10. Part *a* of the drawing shows two identical rods. Each rod has a length L_0 and expands by ΔL when the temperature increases by ΔT. Part *b* shows the two heated rods combined into a single rod, for which the total expansion is the sum of the expansions of each part—namely, $\Delta L + \Delta L = 2\Delta L$. Clearly, the amount of expansion doubles if the rod is twice as long to begin with. In other words, the change in length is directly proportional to the original length ($\Delta L \propto L_0$). Equation 12.2 expresses the fact that ΔL is proportional to both L_0 and ΔT ($\Delta L \propto L_0 \Delta T$) by using a proportionality constant α, which is called the *coefficient of linear expansion.*

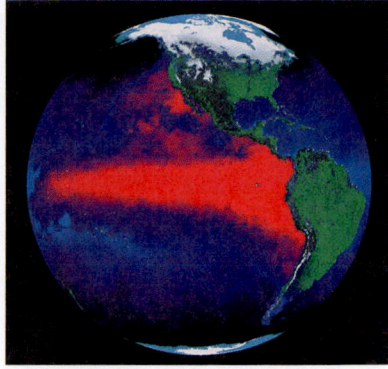

Figure 12.8 A thermogram of the 1997/98 El Niño (red), a large region of abnormally high temperatures in the Pacific Ocean. (Courtesy NOAA)

LINEAR THERMAL EXPANSION OF A SOLID

The length L_0 of an object changes by an amount ΔL when its temperature changes by an amount ΔT:

$$\Delta L = \alpha L_0 \Delta T \tag{12.2}$$

where α is the coefficient of linear expansion.

Common Unit for the Coefficient of Linear Expansion: $\dfrac{1}{C^\circ} = (C^\circ)^{-1}$

Solving Equation 12.2 for α shows that $\alpha = \Delta L/(L_0 \Delta T)$. Since the length units of ΔL and L_0 algebraically cancel, the coefficient of linear expansion α has the unit of $(C^\circ)^{-1}$ when the temperature difference ΔT is expressed in Celsius degrees (C°). Different materials with the same initial length expand and contract by different amounts as the

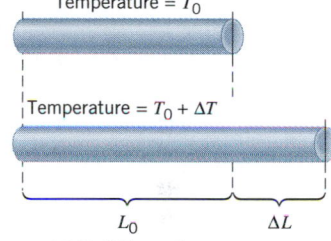

Figure 12.9 When the temperature of a rod is raised by ΔT, the length of the rod increases by ΔL.

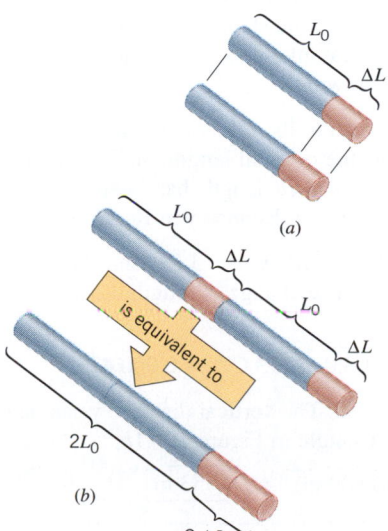

Figure 12.10 (*a*) Each of two identical rods expands by ΔL when heated. (*b*) When the rods are combined into a single rod of length $2L_0$, the "combined" rod expands by $2 \Delta L$.

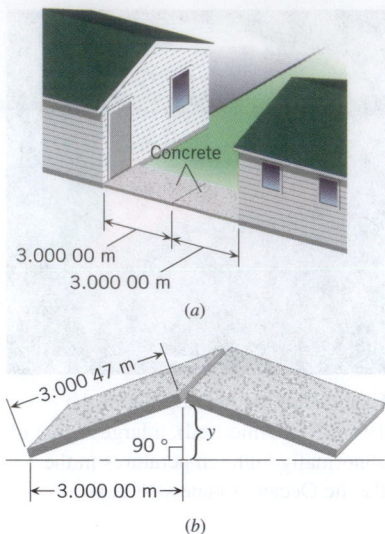

3.000 00 m

3.000 00 m

(a)

3.000 47 m

90° } y

3.000 00 m

(b)

Figure 12.11 (a) Two concrete slabs completely fill the space between the buildings. (b) When the temperature increases, each slab expands, causing the sidewalk to buckle.

Table 12.1 Coefficients of Thermal Expansion for Solids and Liquids[a]

Substance	Coefficient of Thermal Expansion (C°)⁻¹	
	Linear (α)	Volume (β)
Solids		
Aluminum	23×10^{-6}	69×10^{-6}
Brass	19×10^{-6}	57×10^{-6}
Concrete	12×10^{-6}	36×10^{-6}
Copper	17×10^{-6}	51×10^{-6}
Glass (common)	8.5×10^{-6}	26×10^{-6}
Glass (Pyrex)	3.3×10^{-6}	9.9×10^{-6}
Gold	14×10^{-6}	42×10^{-6}
Iron or steel	12×10^{-6}	36×10^{-6}
Lead	29×10^{-6}	87×10^{-6}
Nickel	13×10^{-6}	39×10^{-6}
Quartz (fused)	0.50×10^{-6}	1.5×10^{-6}
Silver	19×10^{-6}	57×10^{-6}
Liquids[b]		
Benzene	—	1240×10^{-6}
Carbon tetrachloride	—	1240×10^{-6}
Ethyl alcohol	—	1120×10^{-6}
Gasoline	—	950×10^{-6}
Mercury	—	182×10^{-6}
Methyl alcohol	—	1200×10^{-6}
Water	—	207×10^{-6}

[a]The values for α and β pertain to a temperature near 20 °C.
[b]Since liquids do not have fixed shapes, the coefficient of linear expansion is not defined for them.

temperature changes, so the value of α depends on the nature of the material. Table 12.1 shows some typical values. Coefficients of linear expansion also vary somewhat depending on the range of temperatures involved, but the values in Table 12.1 are adequate approximations. Example 3 deals with a situation in which a dramatic effect due to thermal expansion can be observed, even though the change in temperature is small.

Example 3 Buckling of a Sidewalk

A concrete sidewalk is constructed between two buildings on a day when the temperature is 25 °C. The sidewalk consists of two slabs, each three meters in length and of negligible thickness (Figure 12.11a). As the temperature rises to 38 °C, the slabs expand, but no space is provided for thermal expansion. The buildings do not move, so the slabs buckle upward. Determine the vertical distance y in part b of the drawing.

Reasoning The expanded length of each slab is equal to its original length plus the change in length ΔL due to the rise in temperature. We know the original length, and Equation 12.2 can be used to find the change in length. Once the expanded length has been determined, the Pythagorean theorem can be employed to find the vertical distance y in Figure 12.11b.

Solution The change in temperature is $\Delta T = 38\ °C - 25\ °C = 13\ C°$, and the coefficient of linear expansion for concrete is given in Table 12.1. The change in length of each slab associated with this temperature change is

$$\Delta L = \alpha L_0 \Delta T = [12 \times 10^{-6}\ (C°)^{-1}](3.0\ m)(13\ C°) = 0.000\ 47\ m \quad (12.2)$$

The expanded length of each slab is, thus, 3.000 47 m. The vertical distance y can be obtained by applying the Pythagorean theorem to the right triangle in Figure 12.11b:

$$y = \sqrt{(3.000\ 47\ m)^2 - (3.000\ 00\ m)^2} = \boxed{0.053\ m}$$

Figure 12.12 An expansion joint in a bridge. (Richard Choy/Peter Arnold, Inc.)

The buckling of a sidewalk is one consequence of not providing sufficient room for thermal expansion. To eliminate such problems, engineers incorporate expansion joints or spaces at intervals along bridge roadbeds, as Figure 12.12 shows.

The physics of an antiscalding device. Although Example 3 shows how thermal expansion can cause problems, there are also times when it can be useful. For instance, each year thousands of children are taken to emergency rooms suffering from burns caused by scalding tap water. Such accidents can be reduced with the aid of the antiscalding device shown in Figure 12.13. This device screws onto the end of a faucet and quickly shuts off the flow of water when it becomes too hot. As the water temperature rises, the actuator spring expands and pushes the plunger forward, shutting off the flow. When the water cools, the spring contracts and the water flow resumes.

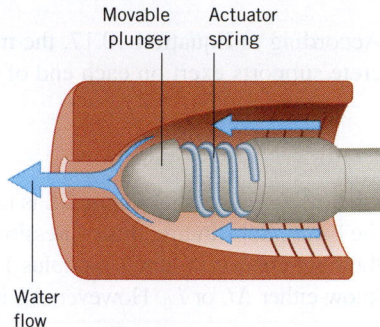

Figure 12.13 An antiscalding device.

THERMAL STRESS

If the concrete slabs in Figure 12.11 had not buckled upward, they would have been subjected to immense forces from the buildings. The forces needed to prevent a solid object from expanding must be strong enough to counteract any change in length that would occur due to a change in temperature. Although the change in temperature may be small, the forces—and hence the stresses—can be enormous. They can, in fact, lead to serious structural damage. Example 4 illustrates just how large the stresses can be.

ANALYZING MULTIPLE-CONCEPT PROBLEMS

Example 4 The Stress on a Steel Beam

A steel beam is used in the roadbed of a bridge. The beam is mounted between two concrete supports when the temperature is 23 °C, with no room provided for thermal expansion (see Figure 12.14). What compressional stress must the concrete supports apply to each end of the beam, if they are to keep the beam from expanding when the temperature rises to 42 °C? Assume that the distance between the concrete supports does not change as the temperature rises.

Reasoning When the temperature rises by an amount ΔT, the natural tendency of the beam is to expand. If the beam were free to expand, it would lengthen by an amount $\Delta L = \alpha L_0 \Delta T$ (Equation 12.2). However, the concrete supports prevent this expansion from occurring by exerting a compressional force on each end of the beam. The magnitude F of this force depends on ΔL through the relation $F = YA(\Delta L/L_0)$ (Equation 10.17), where Y is Young's modulus for steel and A is the cross-sectional area of the beam. According to the discussion in Section 10.8, the compressional stress is equal to the magnitude of the compressional force divided by the cross-sectional area, or Stress = F/A.

The physics of thermal stress.

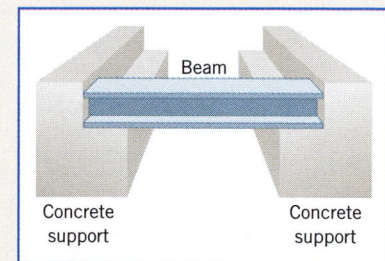

Figure 12.14 A steel beam is mounted between concrete supports with no room provided for thermal expansion.

Knowns and Unknowns The data for this problem are listed in the table:

Description	Symbol	Value
Initial temperature	T_0	23 °C
Final temperature	T	42 °C
Unknown Variable		
Stress	–	?

Modeling the Problem

STEP 1 Stress and Force The compressional stress is defined as the magnitude F of the compressional force divided by the cross-sectional area A of the beam (see Section 10.8), or

$$\text{Stress} = \frac{F}{A}$$

Continued

According to Equation 10.17, the magnitude of the compressional force that the concrete supports exert on each end of the steel beam is given by

$$F = YA\left(\frac{\Delta L}{L_0}\right)$$

where Y is Young's modulus, ΔL is the change in length, and L_0 is the original length of the beam. Substituting this expression for F into the definition of stress gives Equation 1 in the right column. Young's modulus Y for steel is known (see Table 10.1), but we do not know either ΔL or L_0. However, ΔL is proportional to L_0, so we will focus on ΔL in Step 2.

$$\text{Stress} = Y\frac{\Delta L}{L_0} \qquad (1)$$

STEP 2 **Linear Thermal Expansion** If it were free to do so, the beam would have expanded by an amount $\Delta L = \alpha L_0 \Delta T$ (Equation 12.2), where α is the coefficient of linear expansion. The change in temperature is the final temperature T minus the initial temperature T_0, or $\Delta T = T - T_0$. Thus, the beam would have expanded by an amount

$$\Delta L = \alpha L_0(T - T_0)$$

In this expression the variables T and T_0 are known, and α is available in Table 12.1. We substitute this relation for ΔL into Equation 1, as indicated in the right column.

$$\text{Stress} = Y\frac{\Delta L}{L_0} \qquad (1)$$

$$\Delta L = \alpha L_0(T - T_0)$$

Solution Algebraically combining the results of each step, we have

STEP 1 **STEP 2**

$$\text{Stress} = Y\frac{\Delta L}{L_0} = Y\frac{\alpha L_0(T - T_0)}{L_0} = Y\alpha(T - T_0)$$

Note that the original length L_0 of the beam is eliminated algebraically from this result. Taking the value of $Y = 2.0 \times 10^{11}$ N/m^2 from Table 10.1 and $\alpha = 12 \times 10^{-6}$ (C°)$^{-1}$ from Table 12.1, we find that

$$\text{Stress} = Y\alpha(T - T_0)$$

$$= (2.0 \times 10^{11} \text{ N/m}^2)[12 \times 10^{-6} \text{ (C°)}^{-1}](42 \text{ °C} - 23 \text{ °C}) = \boxed{4.6 \times 10^7 \text{ N/m}^2}$$

This is a large stress, equivalent to nearly one million pounds per square foot.

Related Homework: *Problems 20, 27, 98*

THE BIMETALLIC STRIP

A **bimetallic strip** is made from two thin strips of metal that have *different* coefficients of linear expansion, as Figure 12.15a shows. Often brass [$\alpha = 19 \times 10^{-6}$ (C°)$^{-1}$] and steel [$\alpha = 12 \times 10^{-6}$ (C°)$^{-1}$] are selected. The two pieces are welded or riveted together. When the bimetallic strip is heated, the brass, having the larger value of α, expands more than the steel. Since the two metals are bonded together, the bimetallic strip bends into an arc as in part b, with the longer brass piece having a larger radius than the steel piece. When the strip is cooled, the bimetallic strip bends in the opposite direction, as in part c.

The physics of an automatic coffee maker. Bimetallic strips are frequently used as adjustable automatic switches in electrical appliances. Figure 12.16 shows an automatic coffee maker that turns off when the coffee is brewed to the selected strength. In part a, while the brewing cycle is on, electricity passes through the heating coil that heats the water. The electricity can flow because the contact mounted on the bimetallic strip touches the contact mounted on the "strength" adjustment knob, thus providing a continuous path for the electricity. When the bimetallic strip gets hot enough to bend away, as in part b of the drawing, the contacts separate. The electricity stops because it no longer has a continuous path along which to flow, and the brewing cycle is shut off. Turning the "strength" knob adjusts the brewing time by adjusting the distance through which the bimetallic strip must bend for the contact points to separate.

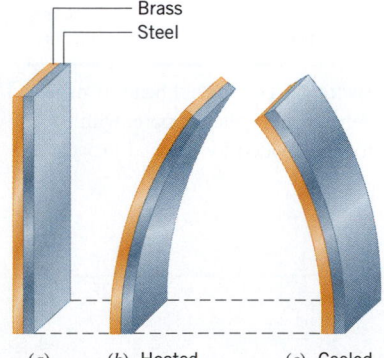

Brass
Steel

(a) (b) Heated (c) Cooled

Figure 12.15 (a) A bimetallic strip and how it behaves when (b) heated and (c) cooled.

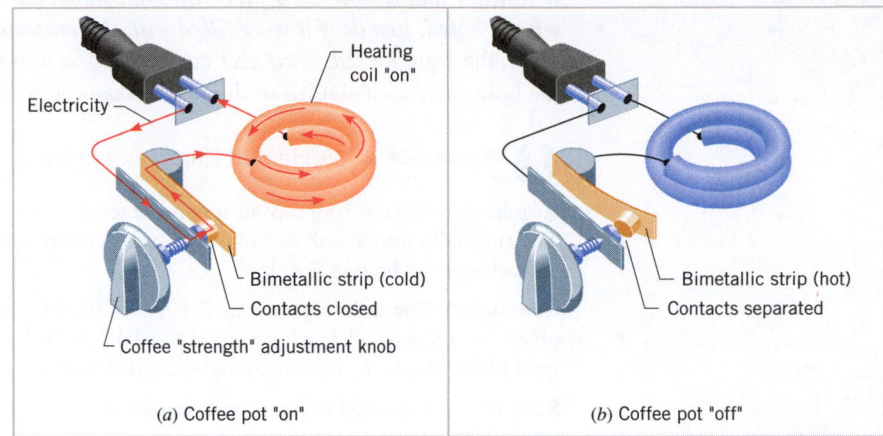

(a) Coffee pot "on"   (b) Coffee pot "off"

Heating coil

Figure 12.16 A bimetallic strip controls whether this coffee pot is (a) "on" (strip cold, straight) or (b) "off" (strip hot, bent).

THE EXPANSION OF HOLES

An interesting example of linear expansion occurs when there is a hole in a piece of solid material. We know that the material itself expands when heated, but what about the hole? Does it expand, contract, or remain the same? Conceptual Example 5 provides some insight into the answer to this question.

Conceptual Example 5
Do Holes Expand or Contract When the Temperature Increases?

Figure 12.17a shows eight square tiles that are attached together and arranged to form a square pattern with a hole in the center. If the tiles are heated, does the size of the hole **(a)** decrease or **(b)** increase?

Reasoning We can analyze this problem by disassembling the pattern into separate tiles, heating them, and then reassembling the pattern. What happens to each of the individual tiles can be explained using what we know about linear expansion.

Answer (a) is incorrect. When a tile is heated both its length and width expand. It is tempting to think, therefore, that the hole in the pattern decreases as the surrounding tiles expand into it. However, this is not correct, because any one tile is prevented from expanding into the hole by the expansion of the tiles next to it.

Answer (b) is correct. Since each tile expands upon heating, the pattern also expands, and the hole along with it, as shown in Figure 12.17b. In fact, if we had a ninth tile that was identical to the others and heated it to the same extent, it would fit exactly into the hole, as Figure 12.7c indicates. Thus, not only does the hole expand, it does so exactly as each of the tiles does. Since the ninth tile is made of the same material as the others, we see that the hole expands just as if it were made of the material of the surrounding tiles. The thermal expansion of the hole and the surrounding material is analogous to a photographic enlargement: everything is enlarged, including holes.

Related Homework: *Problems 13, 23*

Instead of the separate tiles in Example 5, we could have used a square plate with a square hole in the center. The hole in the plate would have expanded just like the hole in the pattern of tiles. Furthermore, the same conclusion applies to a hole of any shape. Thus,

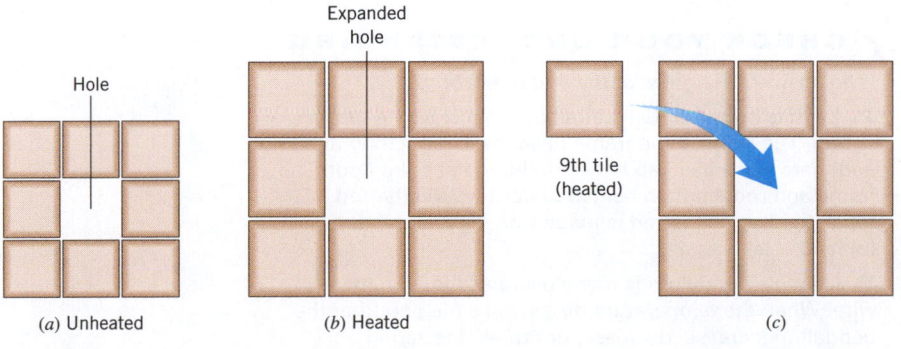

Expanded hole

Hole

9th tile (heated)

(a) Unheated   (b) Heated   (c)

Figure 12.17 (a) The tiles are arranged to form a square pattern with a hole in the center. (b) When the tiles are heated, the pattern, including the hole in the center, expands. (c) The expanded hole is the same size as a heated tile.

Problem-solving insight

it follows that *a hole in a piece of solid material expands when heated and contracts when cooled, just as if it were filled with the material that surrounds it.* If the hole is circular, the equation $\Delta L = \alpha L_0 \Delta T$ can be used to find the change in any linear dimension of the hole, such as its radius or diameter. Example 6 illustrates this type of linear expansion.

 Example 6 A Heated Engagement Ring

A gold engagement ring has an inner diameter of 1.5×10^{-2} m and a temperature of 27 °C. The ring falls into a sink of hot water whose temperature is 49 °C. What is the change in the diameter of the hole in the ring?

Reasoning The hole expands as if it were filled with gold, so the change in the diameter is given by $\Delta L = \alpha L_0 \Delta T$, where $\alpha = 14 \times 10^{-6}$ (C°)$^{-1}$ is the coefficient of linear expansion for gold (Table 12.1), L_0 is the original diameter, and ΔT is the change in temperature.

Solution The change in the ring's diameter is

$$\Delta L = \alpha L_0 \Delta T = [14 \times 10^{-6} \, (\text{C}°)^{-1}](1.5 \times 10^{-2} \text{ m})(49\,°\text{C} - 27\,°\text{C}) = \boxed{4.6 \times 10^{-6} \text{ m}}$$

The previous two examples illustrate that holes expand like the surrounding material when heated. Therefore, holes in materials with larger coefficients of linear expansion expand more than those in materials with smaller coefficients of linear expansion. Conceptual Example 7 explores this aspect of thermal expansion.

 Conceptual Example 7 Expanding Cylinders

Figure 12.18 shows a cross-sectional view of three cylinders, A, B, and C. One is made from lead, one from brass, and one from steel. All three have the same temperature, and they barely fit inside each other. As the cylinders are heated to the same higher temperature, C falls off, while A becomes tightly wedged against B. Which cylinder is made from which material? **(a)** A is brass, B is lead, C is steel **(b)** A is lead, B is brass, C is steel **(c)** A is lead, B is steel, C is brass **(d)** A is brass, B is steel, C is lead **(e)** A is steel, B is brass, C is lead **(f)** A is steel, B is lead, C is brass

Reasoning We will consider how the outer and inner diameters of each cylinder change as the temperature is raised. In particular, with regard to the inner diameter we note that a hole expands as if it were filled with the surrounding material. According to Table 12.1, lead has the greatest coefficient of linear expansion, followed by brass, and then by steel. Thus, the outer and inner diameters of the lead cylinder change the most, while those of the steel cylinder change the least.

Answers (a), (b), (e), and (f) are incorrect. Since the steel cylinder expands the least, it cannot be the outer one, for if it were, the greater expansion of the middle cylinder would prevent the steel cylinder from falling off, as outer cylinder C actually does. The steel cylinder also cannot be the inner one, because then the greater expansion of the middle cylinder would allow the steel cylinder to fall out, contrary to what is observed for inner cylinder A.

Answers (c) and (d) are correct. Since the steel cylinder cannot be on the outside or on the inside, it must be the middle cylinder B. Figure 12.18a shows the lead cylinder as the outer cylinder C. It will fall off as the temperature is raised, since lead expands more than steel. The brass inner cylinder A expands more than the steel cylinder that surrounds it and becomes tightly wedged, as observed. Similar reasoning applies also to Figure 12.18b, which shows the brass cylinder as the outer cylinder and the lead cylinder as the inner one, since both brass and lead expand more than steel.

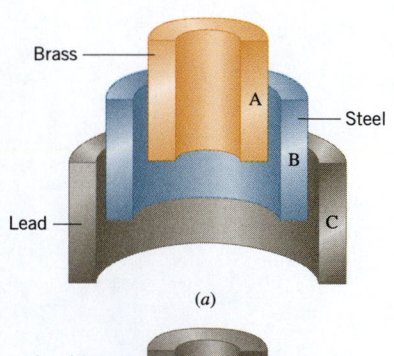

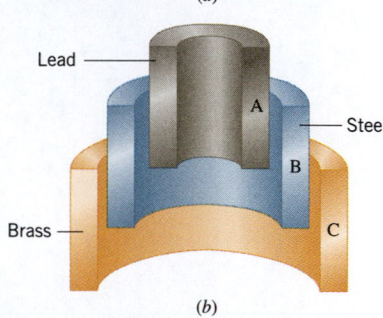

Figure 12.18 Conceptual Example 7 discusses the arrangements of the three cylinders shown in cutaway views in parts *a* and *b*.

✓ **CHECK YOUR UNDERSTANDING**

(*The answers are given at the end of the book.*)

2. A rod is hung from an aluminum frame, as the drawing shows. The rod and the frame have the same temperature, and there is a small gap between the rod and the floor. The frame and rod are then heated uniformly. Will the rod ever touch the floor if the rod is made from **(a)** aluminum, **(b)** lead, **(c)** brass?

3. A simple pendulum is made using a long, thin metal wire. When the temperature drops, does the period of the pendulum increase, decrease, or remain the same?

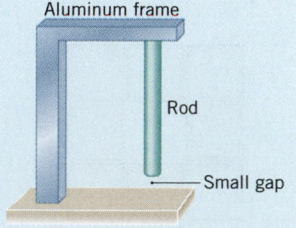

4. For added strength, many highways and buildings are constructed with reinforced concrete (concrete that is reinforced with embedded steel rods). Table 12.1 shows that the coefficient of linear expansion α for concrete is the same as that for steel. Why is it important that these two coefficients be the same?

5. One type of cooking pot is made from stainless steel and has a copper coating over the outside of the bottom. At room temperature the bottom of this pot is flat, but when heated the bottom is not flat. When the bottom of this pot is heated, is it bowed outward or inward?

6. A metal ball has a diameter that is slightly *greater* than the diameter of a hole that has been cut into a metal plate. The coefficient of linear expansion for the metal from which the ball is made is *greater* than that for the metal of the plate. Which one or more of the following procedures can be used to make the ball pass through the hole? **(a)** Raise the temperatures of the ball and the plate by the same amount. **(b)** Lower the temperatures of the ball and the plate by the same amount. **(c)** Heat the ball and cool the plate. **(d)** Cool the ball and heat the plate.

7. A hole is cut through an aluminum plate. A brass ball has a diameter that is slightly *smaller* than the diameter of the hole. The plate and the ball have the same temperature at all times. Should the plate and ball both be heated or both be cooled to *prevent* the ball from falling through the hole?

12.5 VOLUME THERMAL EXPANSION

The volume of a normal material increases as the temperature increases. Most solids and liquids behave in this fashion. By analogy with linear thermal expansion, the change in volume ΔV is proportional to the change in temperature ΔT and to the initial volume V_0, provided the change in temperature is not too large. These two proportionalities can be converted into Equation 12.3 with the aid of a proportionality constant β, known as the *coefficient of volume expansion.* The algebraic form of this equation is similar to that for linear expansion, $\Delta L = \alpha L_0 \Delta T$.

VOLUME THERMAL EXPANSION

The volume V_0 of an object changes by an amount ΔV when its temperature changes by an amount ΔT:

$$\Delta V = \beta V_0 \Delta T \qquad (12.3)$$

where β is the coefficient of volume expansion.

Common Unit for the Coefficient of Volume Expansion: $(C°)^{-1}$

The unit for β, like that for α, is $(C°)^{-1}$. Values for β depend on the nature of the material, and Table 12.1 lists some examples measured near 20 °C. The values of β for liquids are substantially larger than those for solids, because liquids typically expand more than solids, given the same initial volumes and temperature changes. Table 12.1 also shows that, for most solids, the coefficient of volume expansion is three times as much as the coefficient of linear expansion: $\beta = 3\alpha$.

If a cavity exists within a solid object, the volume of the cavity increases when the object expands, just as if the cavity were filled with the surrounding material. The expansion of the cavity is analogous to the expansion of a hole in a sheet of material. Accordingly, the change in volume of a cavity can be found using the relation $\Delta V = \beta V_0 \Delta T$, where β is the coefficient of volume expansion of the material that surrounds the cavity. Example 8 illustrates this point.

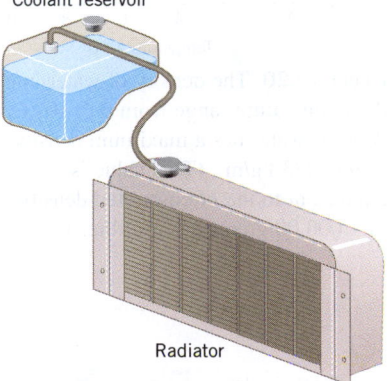

Coolant reservoir

Radiator

Figure 12.19 An automobile radiator and a coolant reservoir for catching the overflow from the radiator.

Example 8 An Automobile Radiator

A small plastic container, called the coolant reservoir, catches the radiator fluid that overflows when an automobile engine becomes hot (see Figure 12.19). The radiator is made of copper, and the coolant has a coefficient of volume expansion of $\beta = 4.10 \times 10^{-4}\ (C°)^{-1}$. If the radiator is filled to its 15-quart capacity when the engine is cold (6.0 °C), how much overflow will spill into the reservoir when the coolant reaches its operating temperature of 92 °C?

The physics of the overflow of an automobile radiator.

Reasoning When the temperature increases, both the coolant and the radiator expand. If they were to expand by the same amount, there would be no overflow. However, the liquid coolant expands more than the radiator, and the overflow volume is the amount of coolant expansion *minus* the amount of the radiator cavity expansion.

Solution When the temperature increases by 86 C°, the coolant expands by an amount

$$\Delta V = \beta V_0 \Delta T = [4.10 \times 10^{-4}\ (\text{C}°)^{-1}](15\ \text{quarts})(86\ \text{C}°) = 0.53\ \text{quarts} \quad (12.3)$$

The radiator cavity expands as if it were filled with copper $[\beta = 51 \times 10^{-6}\ (\text{C}°)^{-1}$; see Table 12.1]. The expansion of the radiator cavity is

$$\Delta V = \beta V_0 \Delta T = [51 \times 10^{-6}\ (\text{C}°)^{-1}](15\ \text{quarts})(86\ \text{C}°) = 0.066\ \text{quarts}$$

The overflow volume is 0.53 quarts − 0.066 quarts = $\boxed{0.46\ \text{quarts}}$.

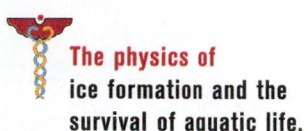

The physics of ice formation and the survival of aquatic life.

Although most substances expand when heated, a few do not. For instance, if water at 0 °C is heated, its volume *decreases* until the temperature reaches 4 °C. Above 4 °C water behaves normally, and its volume increases as the temperature increases. Because a given mass of water has a minimum volume at 4°C, the density (mass per unit volume) of water is greatest at 4 °C, as Figure 12.20 shows.

The fact that water has its greatest density at 4 °C, rather than at 0 °C, has important consequences for the way in which a lake freezes. When the air temperature drops, the surface layer of water is chilled. As the temperature of the surface layer drops toward 4 °C, this layer becomes more dense than the warmer water below. The denser water sinks and pushes up the deeper and warmer water, which in turn is chilled at the surface. This process continues until the temperature of the entire lake reaches 4 °C. Further cooling of the surface water below 4 °C makes it *less dense* than the deeper layers; consequently, the surface layer does not sink but stays on top. Continued cooling of the top layer to 0 °C leads to the formation of ice that floats on the water, because ice has a smaller density than water at any temperature. Below the ice, however, the water temperature remains above 0 °C. The sheet of ice acts as an insulator that reduces the loss of heat from the lake, especially if the ice is covered with a blanket of snow, which is also an insulator. As a result, lakes usually do not freeze solid, even during prolonged cold spells, so fish and other aquatic life can survive.

The physics of bursting water pipes. The fact that the density of ice is smaller than the density of water has an important consequence for homeowners, who have to contend with the possibility of bursting water pipes during severe winters. Water often freezes in a section of pipe exposed to unusually cold temperatures. The ice can form an immovable plug that prevents the subsequent flow of water, as Figure 12.21 illustrates. When water (larger density) turns to ice (smaller density), its volume expands by 8.3%. Therefore, when more water freezes at the left side of the plug, the expanding ice pushes liquid back into the pipe leading to the street connection, and no damage is done. However, when ice forms on the right side of the plug, the expanding ice pushes liquid to the right—but it has nowhere to go if the faucet is closed. As ice continues to form and expand, the water pressure between the plug and faucet rises. Even a small increase in the amount of ice produces a large increase in the pressure. This situation is analogous to the thermal stress discussed in Multi-Concept Example 4, where a small change in the length of the steel beam produces a large stress on the concrete supports. The entire section of pipe to the right of the blockage experiences the same elevated pressure, according to Pascal's principle (Section 11.5). Therefore, the pipe can burst at any point where it is structurally weak, even within the heated space of the building. If you should lose heat during the winter, there is a simple way to prevent pipes from bursting. Simply open the faucet so it drips a little. The excessive pressure will be relieved.

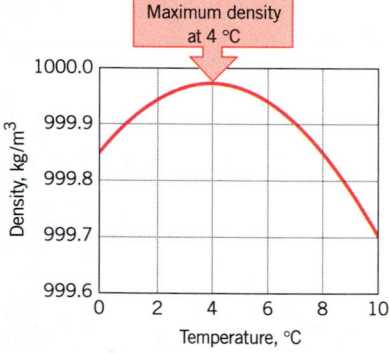

Figure 12.20 The density of water in the temperature range from 0 to 10 °C. At 4 °C water has a maximum density of 999.973 kg/m³. (This value is equivalent to the often-quoted density of 1.000 00 grams per milliliter.)

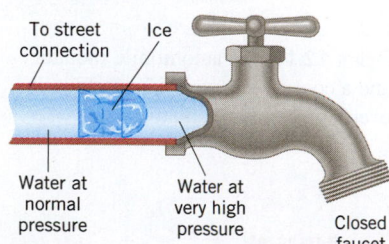

Figure 12.21 As water freezes and expands, enormous pressure is applied to the liquid water between the ice and the closed faucet.

✓ CHECK YOUR UNDERSTANDING

(*The answers are given at the end of the book.*)

8. Suppose that liquid mercury and glass both had the same coefficient of volume expansion. Would a mercury-in-glass thermometer still work?

9. Is the buoyant force provided by warmer water (above 4 °C) greater than, less than, or equal to the buoyant force provided by cooler water (also above 4 °C)?

12.6

HEAT AND INTERNAL ENERGY

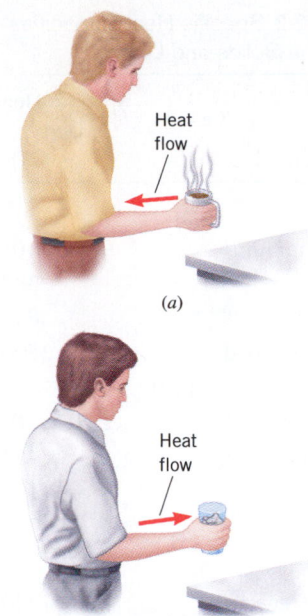

An object with a high temperature is said to be hot, and the word "hot" brings to mind the word "heat." *Heat* flows from a hotter object to a cooler object when the two are placed in contact. It is for this reason that a cup of hot coffee feels hot to the touch, while a glass of ice water feels cold. When the person in Figure 12.22a touches the coffee cup, heat flows from the hotter cup into the cooler hand. When the person touches the glass in part b of the drawing, heat again flows from hot to cold, in this case from the warmer hand into the colder glass. The response of the nerves in the hand to the arrival or departure of heat prompts the brain to identify the coffee cup as being hot and the glass as being cold.

What exactly is heat? As the following definition indicates, heat is a form of energy, energy in transit from hot to cold.

DEFINITION OF HEAT

Heat is energy that flows from a higher-temperature object to a lower-temperature object because of the difference in temperatures.

SI Unit of Heat: joule (J)

Being a kind of energy, heat is measured in the same units used for work, kinetic energy, and potential energy. Thus, the SI unit for heat is the joule.

The heat that flows from hot to cold in Figure 12.22 originates in the *internal energy* of the hot substance. The internal energy of a substance is the sum of the molecular kinetic energy (due to random motion of the molecules), the molecular potential energy (due to forces that act between the atoms of a molecule and between molecules), and other kinds of molecular energy. When heat flows in circumstances where no work is done, the internal energy of the hot substance decreases and the internal energy of the cold substance increases. Although heat may originate in the internal energy supply of a substance, *it is not correct to say that a substance contains heat*. The substance has internal energy, not heat. The word "heat" only refers to the energy actually in transit from hot to cold.

The next two sections consider some effects of heat. For instance, when preparing spaghetti, the first thing that a cook does is to heat the water. Heat from the stove causes the internal energy of the water to increase. Associated with this increase is a rise in temperature. After a while, the temperature reaches 100 °C, and the water begins to boil. During boiling, the added heat causes the water to change from a liquid to a vapor phase (steam). The next section investigates how the addition (or removal) of heat causes the temperature of a substance to change. Then, Section 12.8 discusses the relationship between heat and phase change, such as that which occurs when water boils.

Figure 12.22 Heat is energy in transit from hot to cold. (*a*) Heat flows from the hotter coffee cup to the colder hand. (*b*) Heat flows from the warmer hand to the colder glass of ice water.

12.7

HEAT AND TEMPERATURE CHANGE: SPECIFIC HEAT CAPACITY

SOLIDS AND LIQUIDS

Greater amounts of heat are needed to raise the temperature of solids or liquids to higher values. A greater amount of heat is also required to raise the temperature of a greater mass of material. Similar comments apply when the temperature is lowered, except that heat must be removed. For limited temperature ranges, experiment shows that the heat Q is directly proportional to the change in temperature ΔT and to the mass m. These two proportionalities are expressed below in Equation 12.4, with the help of a proportionality constant c that is referred to as the *specific heat capacity* of the material.

HEAT SUPPLIED OR REMOVED IN CHANGING THE TEMPERATURE OF A SUBSTANCE

The heat Q that must be supplied or removed to change the temperature of a substance of mass m by an amount ΔT is

$$Q = cm\,\Delta T \qquad\qquad (12.4)$$

where c is the specific heat capacity of the substance.

Common Unit for Specific Heat Capacity: $J/(kg \cdot C°)$

Table 12.2 Specific Heat Capacities[a] of Some Solids and Liquids

Substance	Specific Heat Capacity, c $J/(kg \cdot C°)$
Solids	
Aluminum	9.00×10^2
Copper	387
Glass	840
Human body (37 °C, average)	3500
Ice (-15 °C)	2.00×10^3
Iron or steel	452
Lead	128
Silver	235
Liquids	
Benzene	1740
Ethyl alcohol	2450
Glycerin	2410
Mercury	139
Water (15 °C)	4186

[a]Except as noted, the values are for 25 °C and 1 atm of pressure.

Figure 12.23 Cats, such as this Bengal tiger, often pant to get rid of excess heat. (PhotoDisc Blue/PhotoDisc, Inc./ Getty Images)

Solving Equation 12.4 for the specific heat capacity shows that $c = Q/(m \, \Delta T)$, so the unit for specific heat capacity is $J/(kg \cdot C°)$. Table 12.2 reveals that the value of the specific heat capacity depends on the nature of the material. Examples 9, 10, and 11 illustrate the use of Equation 12.4.

 Example 9 A Hot Jogger

In a half hour, a 65-kg jogger can generate 8.0×10^5 J of heat. This heat is removed from the jogger's body by a variety of means, including the body's own temperature-regulating mechanisms. If the heat were not removed, how much would the jogger's body temperature increase?

Reasoning The increase in body temperature depends on the amount of heat Q generated by the jogger, her mass m, and the specific heat capacity c of the human body. Since numerical values are known for these three variables, we can determine the potential rise in temperature by using Equation 12.4.

Solution Table 12.2 gives the average specific heat capacity of the human body as 3500 J/(kg·C°). With this value, Equation 12.4 shows that

$$\Delta T = \frac{Q}{cm} = \frac{8.0 \times 10^5 \text{ J}}{[3500 \text{ J/(kg} \cdot \text{C°)}](65 \text{ kg})} = \boxed{3.5 \text{ C°}}$$

An increase in body temperature of 3.5 °C could be life-threatening. One way in which the jogger's body prevents it from occurring is to remove excess heat by perspiring. In contrast, cats, such as the one in Figure 12.23, do not perspire but often pant to remove excess heat.

 Example 10 Taking a Hot Shower

Cold water at a temperature of 15 °C enters a heater, and the resulting hot water has a temperature of 61 °C. A person uses 120 kg of hot water in taking a shower. **(a)** Find the energy needed to heat the water. **(b)** Assuming that the utility company charges $0.10 per kilowatt·hour for electrical energy, determine the cost of heating the water.

Reasoning The amount Q of heat needed to raise the water temperature can be found from the relation $Q = cm \, \Delta T$, since the specific heat capacity, mass, and temperature change of the water are known. To determine the cost of this energy, we multiply the cost per unit of energy ($0.10 per kilowatt·hour) by the amount of energy used, expressed in energy units of kilowatt·hours.

Solution **(a)** The amount of heat needed to heat the water is

$$Q = cm \, \Delta T = [4186 \text{ J/(kg} \cdot \text{C°)}](120 \text{ kg})(61 \text{ °C} - 15 \text{ °C}) = \boxed{2.3 \times 10^7 \text{ J}} \quad (12.4)$$

(b) The kilowatt·hour (kWh) is the unit of energy that utility companies use in your electric bill. To calculate the cost, we need to determine the number of joules in one kilowatt·hour. Recall that 1 kilowatt is 1000 watts (1 Kw = 1000 W), 1 watt is 1 joule per second (1 W = 1 J/s; see Table 6.3), and 1 hour is equal to 3600 seconds (1 h = 3600 s). Thus,

$$1 \text{ kWh} = (1 \text{ kWh}) \left(\frac{1000 \text{ W}}{1 \text{ kW}} \right) \left(\frac{1 \text{ J/s}}{1 \text{ W}} \right) \left(\frac{3600 \text{ s}}{1 \text{ h}} \right) = 3.60 \times 10^6 \text{ J}$$

The number of kilowatt·hours of energy used to heat the water is

$$(2.3 \times 10^7 \text{ J}) \left(\frac{1 \text{ kWh}}{3.60 \times 10^6 \text{ J}} \right) = 6.4 \text{ kWh}$$

At a cost of $0.10 per kWh, the bill for the heat is $\boxed{\$0.64}$ or 64 cents.

ANALYZING MULTIPLE-CONCEPT PROBLEMS

Example 11 Heating a Swimming Pool

Figure 12.24 shows a swimming pool on a sunny day. If the water absorbs 2.00×10^9 J of heat from the sun, what is the change in the volume of the water?

Reasoning As the water heats up, its volume increases. According to the relation $\Delta V = \beta V_0 \Delta T$ (Equation 12.3), the change ΔV in volume depends on the change ΔT in temperature. The change in temperature, in turn, depends on the amount of heat Q absorbed by the water and on the mass m of water being heated, since $Q = cm\,\Delta T$ (Equation 12.4). To evaluate the mass, we recognize that it depends on the density ρ and the initial volume V_0, since $\rho = m/V_0$ (Equation 11.1). These three relations will be used to determine the change in volume of the water.

Figure 12.24 When water absorbs heat Q from the sun, the water expands. (David Wall/Danita Delimont)

Knowns and Unknowns The data for this problem are:

Description	Symbol	Value
Heat absorbed by water	Q	2.00×10^9 J
Unknown Variable		
Change in volume of water	ΔV	?

Modeling the Problem

STEP 1 **Volume Thermal Expansion** When the temperature of the water changes by an amount ΔT, the volume of the water changes by an amount ΔV, as given by Equation 12.3: $\Delta V = \beta V_0 \Delta T$, where V_0 is the initial volume and β is the coefficient of volume expansion for water. Both V_0 and ΔT are unknown, however, so we will deal with ΔT in Step 2 and V_0 in Step 3.

$$\Delta V = \beta V_0 \Delta T \qquad (12.3)$$

STEP 2 **Heat and Change in Temperature** When heat Q is supplied to the water, the water temperature changes by an amount ΔT. The relation between Q and ΔT is given by Equation 12.4 as $Q = cm\,\Delta T$, where c is the specific heat capacity and m is the mass of the water. Solving this equation for ΔT yields

$$\Delta T = \frac{Q}{cm}$$

which can be substituted into Equation 12.3, as shown at the right. We know the heat Q absorbed by the water, and we will consider the mass m in the next step.

$$\Delta V = \beta V_0 \Delta T \qquad (12.3)$$

$$\Delta T = \frac{Q}{cm} \qquad (1)$$

STEP 3 **Mass Density** The mass m and initial volume V_0 of the water are related to the mass density ρ by Equation 11.1 as $\rho = m/V_0$. Solving this equation for the mass yields

$$m = \rho V_0$$

This expression for m can be substituted into Equation 1, as indicated in the right column.

$$\Delta V = \beta V_0 \Delta T \qquad (12.3)$$

$$\Delta T = \frac{Q}{cm} \qquad (1)$$

$$m = \rho V_0$$

Solution Algebraically combining the results of the three steps, we have

$$\boxed{\text{STEP 1}}\quad\boxed{\text{STEP 2}}\quad\boxed{\text{STEP 3}}$$

$$\Delta V = \beta V_0 \Delta T = \beta V_0\left(\frac{Q}{cm}\right) = \beta V_0\left[\frac{Q}{c(\rho V_0)}\right] = \frac{\beta Q}{c\rho}$$

Note that the initial volume V_0 of the water is eliminated algebraically, so it does not appear in the final expression for ΔV. Taking values of $\beta = 207 \times 10^{-6}$ (C°)$^{-1}$ from Table 12.1, $c = 4186$ J/(kg·C°) from Table 12.2, and $\rho = 1.000 \times 10^3$ kg/m³ from Table 11.1, we have that

$$\Delta V = \frac{\beta Q}{c\rho} = \frac{[207 \times 10^{-6}\,(\text{C°})^{-1}](2.00 \times 10^9\,\text{J})}{[4186\,\text{J/(kg·C°)}](1.000 \times 10^3\,\text{kg/m}^3)} = \boxed{9.89 \times 10^{-2}\,\text{m}^3}$$

Related Homework: *Problems 53, 55*

GASES

As we will see in Section 15.6, the value of the specific heat capacity depends on whether the pressure or volume is held constant while energy in the form of heat is added to or removed from a substance. The distinction between constant pressure and constant volume is usually not important for solids and liquids but is significant for gases. As we will see in Section 15.6, a greater value for the specific heat capacity is obtained for a gas at constant pressure than for a gas at constant volume.

HEAT UNITS OTHER THAN THE JOULE

There are three heat units other than the joule in common use. One kilocalorie (1 kcal) was defined historically as the amount of heat needed to raise the temperature of one kilogram of water by one Celsius degree.* With $Q = 1.00$ kcal, $m = 1.00$ kg, and $\Delta T = 1.00$ C°, the equation $Q = cm\,\Delta T$ shows that such a definition is equivalent to a specific heat capacity for water of $c = 1.00$ kcal/(kg·C°). Similarly, one calorie (1 cal) was defined as the amount of heat needed to raise the temperature of one gram of water by one Celsius degree, which yields a value of $c = 1.00$ cal/(g·C°). (Nutritionists use the word "Calorie," with a capital C, to specify the energy content of foods; this use is unfortunate, since 1 Calorie = 1000 calories = 1 kcal.) The British thermal unit (Btu) is the other commonly used heat unit and was defined historically as the amount of heat needed to raise the temperature of one pound of water by one Fahrenheit degree.

It was not until the time of James Joule (1818–1889) that the relationship between energy in the form of work (in units of joules) and energy in the form of heat (in units of kilocalories) was firmly established. Joule's experiments revealed that the performance of mechanical work, like rubbing your hands together, can make the temperature of a substance rise, just as the absorption of heat can. His experiments and those of later workers have shown that

$$1 \text{ kcal} = 4186 \text{ joules} \quad \text{or} \quad 1 \text{ cal} = 4.186 \text{ joules}$$

Because of its historical significance, this conversion factor is known as the **mechanical equivalent of heat**.

CALORIMETRY

In Section 6.8 we encountered the principle of conservation of energy, which states that energy can be neither created nor destroyed, but can only be converted from one form to another. There we dealt with kinetic and potential energies. In this chapter we have expanded our concept of energy to include heat, which is energy that flows from a higher-temperature object to a lower-temperature object because of the difference in temperature. No matter what its form, whether kinetic energy, potential energy, or heat, energy can be neither created nor destroyed. This fact governs the way objects at different temperatures come to an equilibrium temperature when they are placed in contact. If there is no heat loss to the external surroundings, the heat lost by the hotter objects equals the heat gained by the cooler ones, a process that is consistent with the conservation of energy. Just this kind of process occurs within a thermos. A perfect thermos would prevent any heat from leaking out or in. However, energy in the form of heat can flow *between* materials inside the thermos to the extent that they have different temperatures; for example, between ice cubes and warm tea. The transfer of energy continues until a common temperature is reached at thermal equilibrium.

The kind of heat transfer that occurs within a thermos of iced tea also occurs within a calorimeter, which is the experimental apparatus used in a technique known as *calorimetry*. Figure 12.25 shows that, like a thermos, a calorimeter is essentially an insulated container. It can be used to determine the specific heat capacity of a substance, as the next example illustrates.

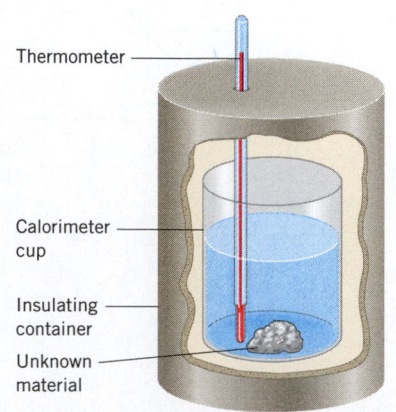

Figure 12.25 A calorimeter can be used to measure the specific heat capacity of an unknown material.

Thermometer

Calorimeter cup

Insulating container

Unknown material

*From 14.5 to 15.5 °C.

Example 12   Measuring the Specific Heat Capacity

The calorimeter cup in Figure 12.25 is made from 0.15 kg of aluminum and contains 0.20 kg of water. Initially, the water and the cup have a common temperature of 18.0 °C. A 0.040-kg mass of unknown material is heated to a temperature of 97.0 °C and then added to the water. The temperature of the water, the cup, and the unknown material is 22.0 °C after thermal equilibrium is reestablished. Ignoring the small amount of heat gained by the thermometer, find the specific heat capacity of the unknown material.

Reasoning Since energy is conserved and there is negligible heat flow between the calorimeter and the outside surroundings, the heat gained by the cold water and the aluminum cup as they warm up is equal to the heat lost by the unknown material as it cools down. Each quantity of heat can be calculated using the relation $Q = cm\,\Delta T$, where, in calorimetry, the change in temperature ΔT is always the higher temperature minus the lower temperature. The equation "Heat gained = Heat lost" contains a single unknown quantity, the desired specific heat capacity.

Solution

$$\underbrace{(cm\,\Delta T)_{\text{Al}} + (cm\,\Delta T)_{\text{water}}}_{\substack{\text{Heat gained by}\\ \text{aluminum and water}}} = \underbrace{(cm\,\Delta T)_{\text{unknown}}}_{\substack{\text{Heat lost by}\\ \text{unknown material}}}$$

$$c_{\text{unknown}} = \frac{c_{\text{Al}} m_{\text{Al}}\,\Delta T_{\text{Al}} + c_{\text{water}} m_{\text{water}}\,\Delta T_{\text{water}}}{m_{\text{unknown}}\,\Delta T_{\text{unknown}}}$$

The changes in temperature for the three substances are $\Delta T_{\text{Al}} = \Delta T_{\text{water}} = 22.0\ °\text{C} - 18.0\ °\text{C} = 4.0\ \text{C}°$, and $\Delta T_{\text{unknown}} = 97.0\ °\text{C} - 22.0\ °\text{C} = 75.0\ \text{C}°$. Table 12.2 contains values for the specific heat capacities of aluminum and water. Substituting these data into the equation above, we find that

$$c_{\text{unknown}} = \frac{[9.00 \times 10^2\ \text{J/(kg}\cdot\text{C}°)](0.15\ \text{kg})(4.0\ \text{C}°) + [4186\ \text{J/(kg}\cdot\text{C}°)](0.20\ \text{kg})(4.0\ \text{C}°)}{(0.040\ \text{kg})(75.0\ \text{C}°)}$$

$$= \boxed{1300\ \text{J/(kg}\cdot\text{C}°)}$$

Problem-solving insight

In the equation "Heat gained = Heat lost," both sides must have the same algebraic sign. Therefore, when calculating heat contributions, always write any temperature changes as the higher minus the lower temperature.

✓ CHECK YOUR UNDERSTANDING

(The answers are given at the end of the book.)

10. Two different objects are supplied with equal amounts of heat. Which one or more of the following statements explain why their temperature changes would not necessarily be the same?   **(a)** The objects have the same mass but are made from materials that have different specific heat capacities.   **(b)** The objects are made from the same material but have different masses.   **(c)** The objects have the same mass and are made from the same material.

11. Two objects are made from the same material but have different masses. The two are placed in contact, and neither one loses any heat to the environment. Which object experiences the temperature change with the greater magnitude, or does each object experience a temperature change of the same magnitude?

12. Consider an object of mass m that experiences a change ΔT in its temperature. Various possibilities for these variables are listed in the table. Rank these possibilities in descending order (largest first) according to how much heat is needed to bring about the change in temperature.

	m (kg)	ΔT (C°)
(a)	2.0	15
(b)	1.5	40
(c)	3.0	25
(d)	2.5	20

12.8 HEAT AND PHASE CHANGE: LATENT HEAT

Surprisingly, there are situations in which the addition or removal of heat does not cause a temperature change. Consider a well-stirred glass of iced tea that has come to thermal equilibrium. Even though heat enters the glass from the warmer room, the temperature of the tea does not rise above 0 °C as long as ice cubes are present. Apparently the heat is being used for some purpose other than raising the temperature. In fact, the heat is being used to melt the ice, and only when all of it is melted will the temperature of the liquid begin to rise.

Figure 12.26 The three phases of water: solid ice is floating in liquid water, and water vapor (invisible) is present in the air. (Klein/Peter Arnold, Inc.)

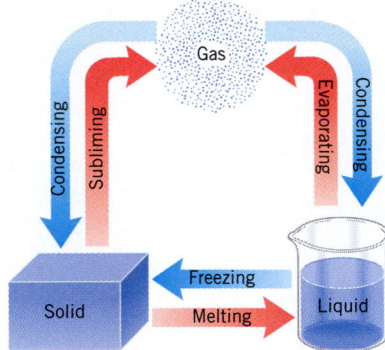

Figure 12.27 Three familiar phases of matter—solid, liquid, and gas—and the phase changes that can occur between any two of them.

An important point illustrated by the iced tea example is that there is more than one type or phase of matter. For instance, some of the water in the glass is in the solid phase (ice) and some is in the liquid phase. The gas or vapor phase is the third familiar phase of matter. In the gas phase, water is referred to as water vapor or steam. All three phases of water are present in the scene depicted in Figure 12.26, although the water vapor is not visible in the photograph.

Matter can change from one phase to another, and heat plays a role in the change. Figure 12.27 summarizes the various possibilities. A solid can *melt* or *fuse* into a liquid if heat is added, while the liquid can *freeze* into a solid if heat is removed. Similarly, a liquid can *evaporate* into a gas if heat is supplied, while the gas can *condense* into a liquid if heat is taken away. Rapid evaporation, with the formation of vapor bubbles within the liquid, is called *boiling*. Finally, a solid can sometimes change directly into a gas if heat is provided. We say that the solid *sublimes* into a gas. Examples of sublimation are (1) solid carbon dioxide, CO_2 (dry ice), turning into gaseous CO_2 and (2) solid naphthalene (moth balls) turning into naphthalene fumes. Conversely, if heat is removed under the right conditions, the gas will condense directly into a solid.

Figure 12.28 displays a graph that indicates what typically happens when heat is added to a material that changes phase. The graph records temperature versus heat added and refers to water at the normal atmospheric pressure of 1.01×10^5 Pa. The water starts off as ice at the subfreezing temperature of $-30\ °C$. As heat is added, the temperature of the ice increases, in accord with the specific heat capacity of ice [2000 J/(kg·C°)]. Not until the temperature reaches the normal melting/freezing point of 0 °C does the water begin to change phase. Then, when heat is added, the solid changes into the liquid, the temperature staying at 0 °C until *all the ice has melted*. Once all the material is in the liquid phase, additional heat causes the temperature to increase again, now in accord with the specific heat capacity of liquid water [4186 J/(kg·C°)]. When the temperature reaches the normal boiling/condensing point of 100 °C, the water begins to change from the liquid to the gas phase and continues to do so as long as heat is added. The temperature remains at 100 °C *until all liquid is gone*. When all of the material is in the gas phase, additional heat once again causes the temperature to rise, this time according to the specific heat capacity of water vapor at constant atmospheric pressure [2020 J/(kg·C°)]. Conceptual Example 13 applies the information in Figure 12.28 to a familiar situation.

Conceptual Example 13 Saving Energy

Suppose you are cooking spaghetti, and the instructions say "boil the pasta in water for ten minutes." To cook spaghetti in an open pot using the least amount of energy, should you (a) turn up the burner to its fullest so the water vigorously boils or (b) turn down the burner so the water barely boils?

Reasoning The spaghetti needs to cook at the temperature of boiling water for ten minutes. In an open pot the pressure is atmospheric pressure, and water boils at 100 °C, regardless of whether it is vigorously boiling or just barely boiling. To convert water into steam requires energy in the form of heat from the burner, and the greater the amount of water converted, the greater the amount of energy needed.

Answer (a) is incorrect. Causing the water to boil vigorously just wastes energy unnecessarily. All it accomplishes is to convert more water into steam.

Answer (b) is correct. Keeping the water just barely boiling uses the least amount of energy to keep the spaghetti at 100 °C, because it minimizes the amount of water converted into steam.

Figure 12.28 The graph shows the way the temperature of water changes as heat is added, starting with ice at $-30\ °C$. The pressure is atmospheric pressure.

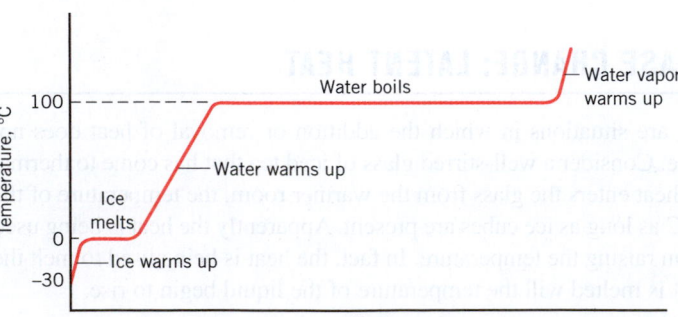

When a substance changes from one phase to another, the amount of heat that must be added or removed depends on the type of material and the nature of the phase change. The heat per kilogram associated with a phase change is referred to as *latent heat:*

HEAT SUPPLIED OR REMOVED IN CHANGING THE PHASE OF A SUBSTANCE

The heat Q that must be supplied or removed to change the phase of a mass m of a substance is

$$Q = mL \qquad (12.5)$$

where L is the latent heat of the substance.

SI Unit of Latent Heat: J/kg

The ***latent heat of fusion*** L_f refers to the change between solid and liquid phases, the ***latent heat of vaporization*** L_v applies to the change between liquid and gas phases, and the ***latent heat of sublimation*** L_s refers to the change between solid and gas phases.

Table 12.3 gives some typical values of latent heats of fusion and vaporization. For instance, the latent heat of fusion for water is $L_f = 3.35 \times 10^5$ J/kg. Thus, 3.35×10^5 J of heat must be supplied to melt one kilogram of ice at 0 °C into liquid water at 0 °C; conversely, this amount of heat must be removed from one kilogram of liquid water at 0 °C to freeze the liquid into ice at 0 °C. In comparison, the latent heat of vaporization for water has the much larger value of $L_v = 22.6 \times 10^5$ J/kg. When water boils at 100 °C, 22.6×10^5 J of heat must be supplied for each kilogram of liquid turned into steam. And when steam condenses at 100 °C, this amount of heat is released from each kilogram of steam that changes back into liquid. Liquid water at 100 °C is hot enough by itself to cause a bad burn, and the additional effect of the large latent heat can cause severe tissue damage if condensation occurs on the skin.

The physics of steam burns.

The physics of high-tech clothing.

By taking advantage of the latent heat of fusion, designers can now engineer clothing that can absorb or release heat to help maintain a comfortable and approximately constant temperature close to your body. As the photograph in Figure 12.29 shows, the fabric in this type of clothing is coated with microscopic balls of heat-resistant plastic that contain a substance known as a "phase-change material" (PCM). When you are enjoying your favorite winter sport, for example, it is easy to become overheated. The PCM prevents this by melting, absorbing excess body heat in the process. When you are taking a break and cooling down, however, the PCM freezes and releases heat to keep you warm. The temperature range over which the PCM can maintain a comfort zone is related to its melting/freezing temperature, which is determined by its chemical composition.

Examples 14 and 15 illustrate how to take into account the effect of latent heat when using the conservation-of-energy principle.

Table 12.3 Latent Heats[a] of Fusion and Vaporization

Substance	Melting Point (°C)	Latent Heat of Fusion, L_f (J/kg)	Boiling Point (°C)	Latent Heat of Vaporization, L_v (J/kg)
Ammonia	−77.8	33.2×10^4	−33.4	13.7×10^5
Benzene	5.5	12.6×10^4	80.1	3.94×10^5
Copper	1083	20.7×10^4	2566	47.3×10^5
Ethyl alcohol	−114.4	10.8×10^4	78.3	8.55×10^5
Gold	1063	6.28×10^4	2808	17.2×10^5
Lead	327.3	2.32×10^4	1750	8.59×10^5
Mercury	−38.9	1.14×10^4	356.6	2.96×10^5
Nitrogen	−210.0	2.57×10^4	−195.8	2.00×10^5
Oxygen	−218.8	1.39×10^4	−183.0	2.13×10^5
Water	0.0	33.5×10^4	100.0	22.6×10^5

[a]The values pertain to 1 atm pressure.

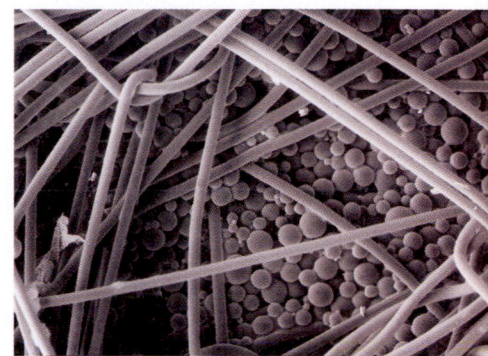

Figure 12.29 This highly magnified image shows a fabric that has been coated with microscopic balls of heat-resistant plastic. The balls contain a substance known as a "phase-change material," the melting and freezing of which absorbs and releases heat. Such fabrics automatically adjust in reaction to your body heat to help maintain a constant temperature next to your skin. (Courtesy Outlast Technologies, Boulder, CO)

Example 14 Ice-Cold Lemonade

Ice at 0 °C is placed in a Styrofoam cup containing 0.32 kg of lemonade at 27 °C. The specific heat capacity of lemonade is virtually the same as that of water; that is, $c = 4186$ J/(kg·C°). After the ice and lemonade reach an equilibrium temperature, some ice still remains. The latent heat of fusion for water is $L_f = 3.35 \times 10^5$ J/kg. Assume that the mass of the cup is so small that it absorbs a negligible amount of heat, and ignore any heat lost to the surroundings. Determine the mass of ice that has melted.

Reasoning According to the principle of energy conservation, the heat gained by the melting ice equals the heat lost by the cooling lemonade. According to Equation 12.5, the heat gained by the melting ice is $Q = mL_f$, where m is the mass of the melted ice, and L_f is the latent heat of fusion for water. The heat lost by the lemonade is given by $Q = cm\,\Delta T$, where ΔT is the higher temperature of 27 °C minus the lower equilibrium temperature. The equilibrium temperature is 0 °C, because there is some ice remaining, and ice is in equilibrium with liquid water when the temperature is 0 °C.

Solution

$$\underbrace{(mL_f)_{\text{ice}}}_{\substack{\text{Heat gained} \\ \text{by ice}}} = \underbrace{(cm\,\Delta T)_{\text{lemonade}}}_{\substack{\text{Heat lost} \\ \text{by lemonade}}}$$

The mass m_{ice} of ice that has melted is

$$m_{\text{ice}} = \frac{(cm\,\Delta T)_{\text{lemonade}}}{L_f} = \frac{[4186\ \text{J/(kg·C°)}](0.32\ \text{kg})(27\ \text{°C} - 0\ \text{°C})}{3.35 \times 10^5\ \text{J/kg}} = \boxed{0.11\ \text{kg}}$$

Example 15 Getting Ready for a Party

A 7.00-kg glass bowl [$c = 840$ J/(kg·C°)] contains 16.0 kg of punch at 25.0 °C. Two-and-a-half kilograms of ice [$c = 2.00 \times 10^3$ J/(kg·C°)] are added to the punch. The ice has an initial temperature of −20.0 °C, having been kept in a very cold freezer. The punch may be treated as if it were water [$c = 4186$ J/(kg·C°)], and it may be assumed that there is no heat flow between the punch bowl and the external environment. The latent heat of fusion for water is 3.35×10^5 J/kg. When thermal equilibrium is reached, all the ice has melted, and the final temperature of the mixture is above 0 °C. Determine this temperature.

Reasoning The final temperature can be found by using the conservation of energy principle: the heat gained is equal to the heat lost. Heat is gained (a) by the ice in warming up to the melting point, (b) by the ice in changing phase from a solid to a liquid, and (c) by the liquid that results from the ice warming up to the final temperature; heat is lost (d) by the punch and (e) by the bowl in cooling down. The heat gained or lost by each component in changing temperature can be determined from the relation $Q = cm\,\Delta T$, where ΔT is the higher temperature minus the lower temperature. The heat gained when water changes phase from a solid to a liquid at 0 °C is $Q = mL_f$, where m is the mass of water and L_f is the latent heat of fusion.

Solution The heat gained or lost by each component is listed as follows:

(a) Heat gained when ice warms to 0.0 °C $\quad = [2.00 \times 10^3\ \text{J/(kg·C°)}](2.50\ \text{kg})[0.0\ \text{°C} - (-20.0\ \text{°C})]$

(b) Heat gained when ice melts at 0.0 °C $\quad = (2.50\ \text{kg})(3.35 \times 10^5\ \text{J/kg})$

(c) Heat gained when melted ice (liquid) warms to temperature T $\quad = [4186\ \text{J/(kg·C°)}](2.50\ \text{kg})(T - 0.0\ \text{°C})$

(d) Heat lost when punch cools to temperature T $\quad = [4186\ \text{J/(kg·C°)}](16.0\ \text{kg})(25.0\ \text{°C} - T)$

(e) Heat lost when bowl cools to temperature T $\quad = [840\ \text{J/(kg·C°)}](7.00\ \text{kg})(25.0\ \text{°C} - T)$

Setting the heat gained equal to the heat lost gives:

$$\underbrace{(a) + (b) + (c)}_{\text{Heat gained}} = \underbrace{(d) + (e)}_{\text{Heat lost}}$$

This equation can be solved to show that $\boxed{T = 11\ °C}$.

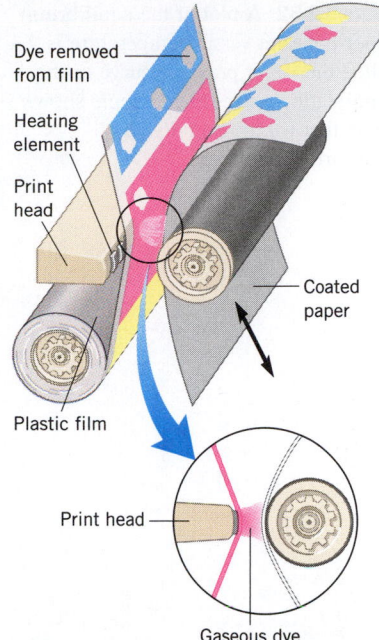

The physics of a dye-sublimation color printer. An interesting application of the phase change between a solid and a gas is found in one kind of color printer used with computers. A dye-sublimation printer uses a thin plastic film coated with separate panels of cyan (blue), yellow, and magenta pigment or dye. A full spectrum of colors is produced by using combinations of tiny spots of these dyes. As Figure 12.30 shows, the coated film passes in front of a print head that extends across the width of the paper and contains 2400 heating elements. When a heating element is turned on, the dye in front of it absorbs heat and goes from a solid to a gas—it sublimes—with no liquid phase in between. A coating on the paper absorbs the gaseous dye on contact, producing a small spot of color. The intensity of the spot is controlled by the heating element, since each element can produce 256 different temperatures; the hotter the element, the greater the amount of dye transferred to the paper. The paper makes three separate passes across the print head, once for each of the dyes. The final result is an image of near-photographic quality. Some printers also employ a fourth pass, in which a clear plastic coating is deposited over the photograph. This coating makes the print waterproof and also helps to prevent premature fading.

Figure 12.30 A dye-sublimation printer. As the plastic film passes in front of the print head, the heat from a given heating element causes one of three pigments or dyes on the film to sublime from a solid to a gas. The gaseous dye is absorbed onto the coated paper as a dot of color. The size of the dots on the paper has been exaggerated for clarity.

✓ **CHECK YOUR UNDERSTANDING**

(The answers are given at the end of the book.)

13. Fruit blossoms are permanently damaged at temperatures of about −4 °C (a hard freeze). Orchard owners sometimes spray a film of water over the blossoms to protect them when a hard freeze is expected. Why does this technique offer protection?

14. When ice cubes are used to cool a drink, both their mass and temperature are important in how effective they are. The table lists several possibilities for the mass and temperature of the ice cubes used to cool one particular drink. Rank the possibilities in descending order (best first) according to their cooling effectiveness. Note that the latent heat of phase change and the specific heat capacity must be considered.

	Mass of ice cubes	Temperature of ice cubes
(a)	m	−6.0 °C
(b)	$\frac{1}{2}m$	−12 °C
(c)	$2m$	−3.0 °C

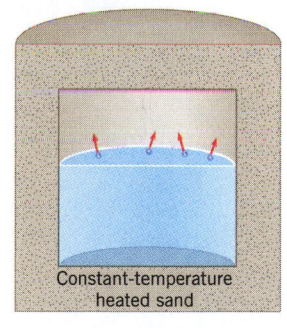

Constant-temperature heated sand

(a)

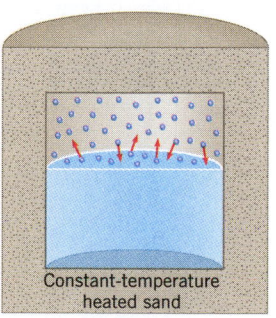

Constant-temperature heated sand

(b)

Figure 12.31 (*a*) Some of the molecules begin entering the vapor phase in the evacuated space above the liquid. (*b*) Equilibrium is reached when the number of molecules entering the vapor phase equals the number returning to the liquid.

12.9 **EQUILIBRIUM BETWEEN PHASES OF MATTER*

Under specific conditions of temperature and pressure, a substance can exist at equilibrium in more than one phase at the same time. Consider Figure 12.31, which shows a container kept at a constant temperature by a large reservoir of heated sand. Initially the container is evacuated. Part *a* shows it just after it has been partially filled with a liquid and a few fast-moving molecules are escaping the liquid and forming a vapor phase. These molecules pick up the required energy (the latent heat of vaporization) during collisions with neighboring molecules in the liquid. However, the reservoir of heated sand replenishes the energy carried away, thus maintaining the constant temperature. At first, the movement of molecules is predominantly from liquid to vapor, although some molecules in the vapor phase do reenter the liquid. As the molecules accumulate in the vapor, the number reentering the liquid eventually equals the number entering the vapor, at which point equilibrium is established, as in part *b*. From this point on, the concentration of molecules in the vapor phase does not change, and the vapor pressure remains constant. The pressure of the vapor that coexists in equilibrium with the liquid is called the *equilibrium vapor pressure* of the liquid.

Figure 12.32 A plot of the equilibrium vapor pressure versus temperature is called the vapor pressure curve or the vaporization curve, the example shown being that for the liquid/vapor equilibrium of water.

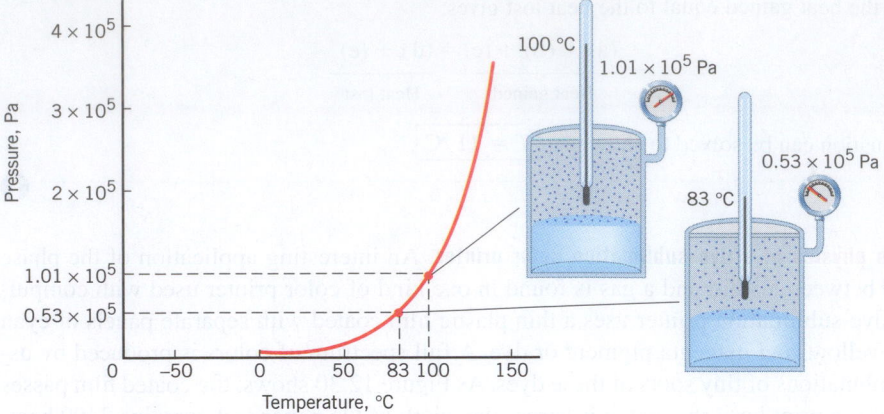

The equilibrium vapor pressure does not depend on the volume of space above the liquid. If more space were provided, more liquid would vaporize, until equilibrium was reestablished at the same vapor pressure, assuming the same temperature is maintained. In fact, the equilibrium vapor pressure depends only on the temperature of the liquid; a higher temperature causes a higher pressure, as the graph in Figure 12.32 indicates for the specific case of water. Only when the temperature and vapor pressure correspond to a point on the curved line, which is called the *vapor pressure curve* or the *vaporization curve,* do liquid and vapor phases coexist at equilibrium.

To illustrate the use of a vaporization curve, let's consider what happens when water boils in a pot that is *open to the air.* Assume that the air pressure acting on the water is 1.01×10^5 Pa (one atmosphere). When boiling occurs, bubbles of water vapor form throughout the liquid, rise to the surface, and break. For these bubbles to form and rise, the pressure of the vapor inside them must at least equal the air pressure acting on the surface of the water. According to Figure 12.32, a value of 1.01×10^5 Pa corresponds to a temperature of 100 °C. Consequently, water boils at 100 °C at one atmosphere of pressure. In general, a *liquid boils at the temperature for which its vapor pressure equals the external pressure.* Water will not boil, then, at sea level if the temperature is only 83 °C, because at this temperature the vapor pressure of water is only 0.53×10^5 Pa (see Figure 12.32), a value less than the external pressure of 1.01×10^5 Pa. However, water does boil at 83 °C on a mountain at an altitude of just under five kilometers, because the atmospheric pressure there is 0.53×10^5 Pa.

The fact that water can boil at a temperature less than 100 °C leads to an interesting phenomenon that Conceptual Example 16 discusses.

(a) Water boiling

Water
Cork

(b) Water boiling again

Figure 12.33 (a) Water is boiling at a temperature of 100 °C and a pressure of one atmosphere. (b) The water boils at a temperature that is less than 100 °C, because the cool water reduces the pressure above the water in the flask.

⬇ **Conceptual Example 16** How to Boil Water That Is Cooling Down

Figure 12.33a shows water boiling in an open flask. Shortly after the flask is removed from the burner, the boiling stops. A cork is then placed in the neck of the flask to seal it, and water is poured over the neck of the flask, as in part *b* of the drawing. To restart the boiling, should the water poured over the neck be **(a)** cold or **(b)** hot—but not boiling?

Reasoning When the open flask is removed from the burner, the water begins to cool and the pressure above its surface is one atmosphere (1.01×10^5 Pa). Boiling quickly stops, because water cannot boil when its temperature is less than 100 °C and the pressure above its surface is one atmosphere. To restart the boiling, it is necessary either to reheat the water to 100 °C or reduce the pressure above the water in the corked flask to something less than one atmosphere so that boiling can occur at a temperature less than 100 °C.

Answer (b) is incorrect. Certainly, pouring hot water over the corked flask will reheat the water. However, since the water being poured is not boiling, its temperature must be less than 100 °C. Therefore, it cannot reheat the water within the flask to 100 °C and restart the boiling.

Answer (a) is correct. When cold water is poured over the corked flask, it causes some of the water vapor inside to condense. Consequently, the pressure above the liquid in the flask drops. When it drops to the value of the vapor pressure of the water in the flask at its current temperature (which is now less than 100 °C), the boiling restarts.

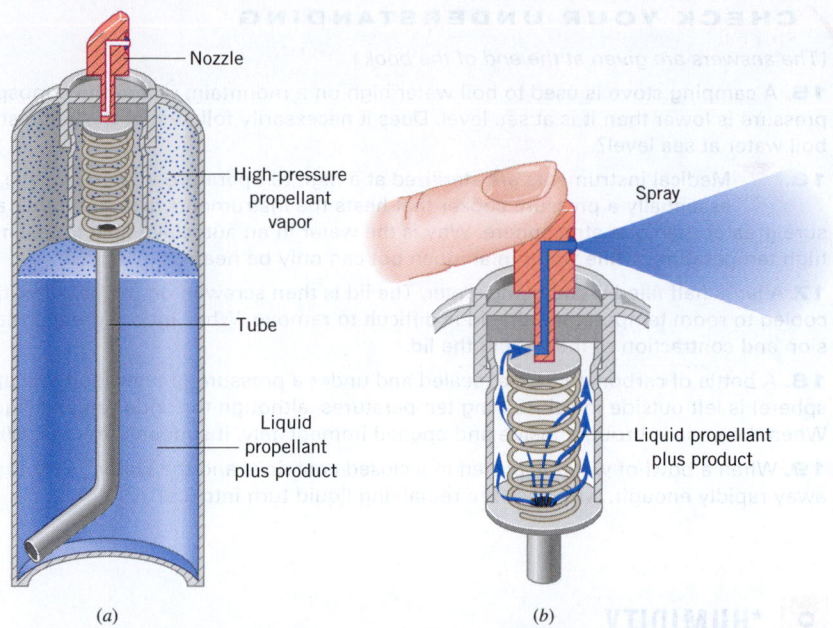

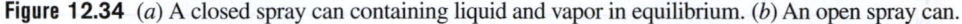

(a)   (b)

Figure 12.34 (a) A closed spray can containing liquid and vapor in equilibrium. (b) An open spray can.

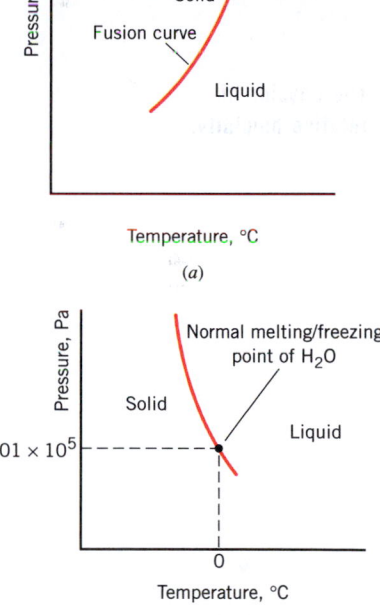

The physics of spray cans. The operation of spray cans is based on the equilibrium between a liquid and its vapor. Figure 12.34a shows that a spray can contains a liquid propellant that is mixed with the product (such as hair spray). Inside the can, propellant vapor forms over the liquid. A propellant is chosen that has an equilibrium vapor pressure that is greater than atmospheric pressure at room temperature. Consequently, when the nozzle of the can is pressed, as in part b of the drawing, the vapor pressure forces the liquid propellant and product up the tube in the can and out the nozzle as a spray. When the nozzle is released, the coiled spring reseals the can and the propellant vapor builds up once again to its equilibrium value.

As is the case for liquid/vapor equilibrium, a solid can be in equilibrium with its liquid phase only at specific conditions of temperature and pressure. For each temperature, there is a single pressure at which the two phases can coexist in equilibrium. A plot of the equilibrium pressure versus equilibrium temperature is referred to as the *fusion curve,* and Figure 12.35a shows a typical curve for a normal substance. A normal substance expands on melting (e.g., carbon dioxide and sulfur). Since higher pressures make it more difficult for such materials to expand, a higher melting temperature is needed for a higher pressure, and the fusion curve slopes upward to the right. Part b of the picture illustrates the fusion curve for water, one of the few substances that contract when they melt. Higher pressures make it easier for such substances to melt. Consequently, a lower melting temperature is associated with a higher pressure, and the fusion curve slopes downward to the right.

It should be noted that just because two phases can coexist in equilibrium does not necessarily mean that they will. Other factors may prevent it. For example, water in an *open* bowl may never come into equilibrium with water vapor if air currents are present. Under such conditions the liquid, perhaps at a temperature of 25 °C, attempts to establish the corresponding equilibrium vapor pressure of 3.2×10^3 Pa. If air currents continually blow the water vapor away, however, equilibrium will never be established, and eventually the water will evaporate completely. Each kilogram of water that goes into the vapor phase takes along the latent heat of vaporization. Because of this heat loss, the remaining liquid would become cooler, except for the fact that the surroundings replenish the loss.

In the case of the human body, water is exuded by the sweat glands and evaporates from a much larger area than the surface of a typical bowl of water. The removal of heat along with the water vapor is called evaporative cooling and is one mechanism that the body uses to maintain its constant temperature.

Figure 12.35 (a) The fusion curve for a normal substance that expands on melting. (b) The fusion curve for water, one of the few substances that contract on melting.

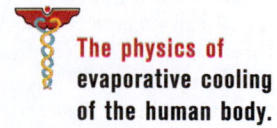

The physics of evaporative cooling of the human body.

It takes less heat to boil water high on a mountain, because the boiling point becomes less than 100 °C as the air pressure decreases at higher elevations. (Gregg Adams/Stone/Getty Images)

The physics of relative humidity.

12.10 *HUMIDITY

Air is a mixture of gases, including nitrogen, oxygen, and water vapor. The total pressure of the mixture is the sum of the partial pressures of the component gases. The **partial pressure** of a gas is the pressure it would exert if it alone occupied the entire volume at the same temperature as the mixture. The partial pressure of water vapor in air depends on weather conditions. It can be as low as zero or as high as the equilibrium vapor pressure of water at the given temperature.

To provide an indication of how much water vapor is in the air, weather forecasters usually give the **relative humidity.** If the relative humidity is too low, the air contains such a small amount of water vapor that skin and mucous membranes tend to dry out. If the relative humidity is too high, especially on a hot day, we become very uncomfortable and our skin feels "sticky." Under such conditions, the air holds so much water vapor that the water exuded by sweat glands cannot evaporate efficiently. The relative humidity is defined as the ratio (expressed as a percentage) of the partial pressure of water vapor in the air to the equilibrium vapor pressure at a given temperature.

$$\begin{array}{c}\text{Percent}\\\text{relative}\\\text{humidity}\end{array} = \dfrac{\begin{array}{c}\text{Partial pressure}\\\text{of water vapor}\end{array}}{\begin{array}{c}\text{Equilibrium vapor pressure of}\\\text{water at the existing temperature}\end{array}} \times 100 \qquad (12.6)$$

The term in the denominator on the right of Equation 12.6 is given by the vaporization curve of water and is the pressure of the water vapor in equilibrium with the liquid. At a given temperature, the partial pressure of the water vapor in the air cannot exceed this value. If it did, the vapor would not be in equilibrium with the liquid and would condense as dew or rain to reestablish equilibrium.

When the partial pressure of the water vapor equals the equilibrium vapor pressure of water at a given temperature, the relative humidity is 100%. In such a situation, the vapor is said to be *saturated* because it is present in the maximum amount, as it would be above a pool of liquid at equilibrium in a closed container. If the relative humidity is less than 100%, the water vapor is said to be *unsaturated.* Example 17 demonstrates how to find the relative humidity.

◖ **Example 17** Relative Humidities

One day, the partial pressure of water vapor in the air is 2.0×10^3 Pa. Using the vaporization curve for water in Figure 12.36, determine the relative humidity if the temperature is **(a)** 32 °C and **(b)** 21 °C.

Reasoning and Solution (a) According to Figure 12.36, the equilibrium vapor pressure of water at 32 °C is 4.8×10^3 Pa. Equation 12.6 reveals that the relative humidity is

$$\text{Relative humidity at 32 °C} = \frac{2.0 \times 10^3 \text{ Pa}}{4.8 \times 10^3 \text{ Pa}} \times 100 = \boxed{42\%}$$

(b) A similar calculation shows that

$$\text{Relative humidity at 21 °C} = \frac{2.0 \times 10^3 \text{ Pa}}{2.5 \times 10^3 \text{ Pa}} \times 100 = \boxed{80\%}$$

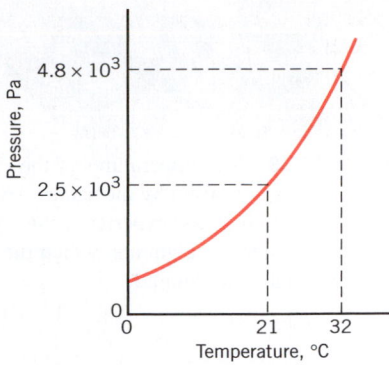

Figure 12.36 The vaporization curve of water.

When air containing a given amount of water vapor is cooled, a temperature is reached in which the partial pressure of the vapor equals the equilibrium vapor pressure. This temperature is known as the *dew point.* For instance, Figure 12.37 shows that if the partial pressure of water vapor is 3.2×10^3 Pa, the dew point is 25 °C. This partial pressure would correspond to a relative humidity of 100%, if the ambient temperature were equal to the dew-point temperature. **The physics of fog formation.** Hence, the dew point is the temperature below which water vapor in the air condenses in the form of liquid drops (dew or fog). The closer the actual temperature is to the dew point, the closer the relative humidity is to 100%. Thus, for fog to form, the air temperature must drop below the dew point. Similarly, water condenses on the outside of a cold glass when the temperature of the air next to the glass falls below the dew point. **The physics of a home dehumidifier.** The cold coils in a home dehumidifier (see Figure 12.38) function very much in the same way that the cold glass does. The coils are kept cold by a circulating refrigerant. When the air blown across them by the fan cools below the dew point, water vapor condenses in the form of droplets, which collect in a receptacle.

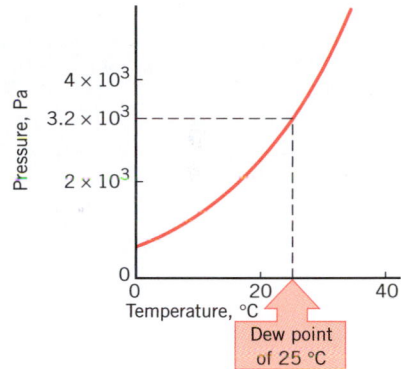

Figure 12.37 On the vaporization curve of water, the dew point is the temperature that corresponds to the actual partial pressure of water vapor in the air.

✓ **CHECK YOUR UNDERSTANDING**

(The answers are given at the end of the book.)

20. A bowl of water is covered tightly and allowed to sit at a constant temperature of 23 °C for a long time. What is the relative humidity in the space between the surface of the water and the cover?

21. Is it possible for dew to form on Tuesday night and not on Monday night, even though Monday night is the cooler night?

22. Two rooms in a mansion have the same temperature. One of these rooms contains an indoor swimming pool. On a cold day the windows of one of the two rooms are "steamed up." Which room is it likely to be?

CONCEPTS & CALCULATIONS

This section contains examples that discuss one or more conceptual questions, followed by a related quantitative problem. Example 18 provides insight on the variables involved when the length and volume of an object change due to a temperature change. Example 19 discusses how different factors affect the temperature change of an object to which heat is being added.

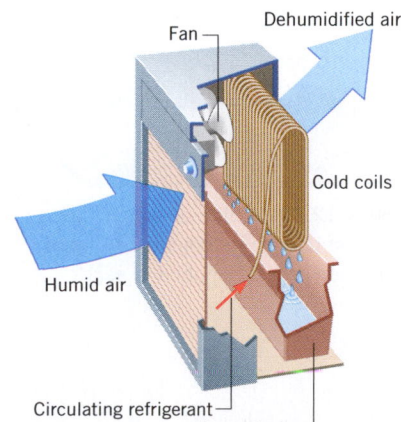

Figure 12.38 The cold coils of a dehumidifier cool the air blowing across them to below the dew point, and water vapor condenses out of the air.

Concepts & Calculations Example 18
 Linear and Volume Thermal Expansion

Figure 12.39 shows three rectangular blocks made from the same material. The initial dimensions of each are expressed as multiples of D, where $D = 2.00$ cm. The blocks are heated and their temperatures increase by 35.0 C°. The coefficients of linear and volume expansion are $\alpha = 1.50 \times 10^{-5}$ (C°)$^{-1}$ and $\beta = 4.50 \times 10^{-5}$ (C°)$^{-1}$, respectively. Determine the change in the **(a)** vertical heights and **(b)** volumes of the blocks.

Concept Questions and Answers Does the change in the vertical height of a block depend only on its height, or does it also depend on its width and depth? Without doing any calculations, rank the blocks according to their change in height, largest first.

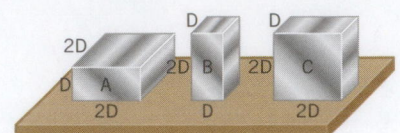

Figure 12.39 The temperatures of the three blocks are raised by the same amount. Which one(s) experience the greatest change in height and which the greatest change in volume?

Answer According to Equation 12.2, $\Delta L = \alpha L_0 \Delta T$, the change ΔL in the height of an object depends on its original height L_0, the change in temperature ΔT, and the coefficient of linear expansion, α. It does not depend on the width or depth of the object. Since blocks B and C have twice the height of A, their heights will increase by twice as much as that of A. The heights of B and C, however, will increase by the same amount, even though C is twice as wide.

Does the change in the volume of a block depend only on its height, or does it also depend on its width and depth? Without doing any calculations, rank the blocks according to their greatest change in volume, largest first.

Answer According to Equation 12.3, $\Delta V = \beta V_0 \Delta T$, the change ΔV in the volume of an object depends on its original volume V_0, the change in temperature ΔT, and the coefficient of volume expansion, β. Thus, the change in volume depends on height, width, and depth, because the original volume is the product of these three dimensions. The initial volumes of A, B, and C, are, respectively, $D \times 2D \times 2D = 4D^3$, $2D \times D \times D = 2D^3$, and $2D \times 2D \times D = 4D^3$. Blocks A and C have equal volumes, which are greater than that of B. Thus, we expect A and C to exhibit the greatest increase in volume, while B exhibits the smallest increase.

Solution (a) The change in the height of each block is given by Equation 12.2 as

$$\Delta L_A = \alpha D \Delta T = [1.50 \times 10^{-5}\,(\text{C}°)^{-1}](2.00\text{ cm})(35.0\text{ C}°) = \boxed{1.05 \times 10^{-3}\text{ cm}}$$

$$\Delta L_B = \alpha(2D)\Delta T = [1.50 \times 10^{-5}\,(\text{C}°)^{-1}](2 \times 2.00\text{ cm})(35.0\text{ C}°) = \boxed{2.10 \times 10^{-3}\text{ cm}}$$

$$\Delta L_C = \alpha(2D)\Delta T = [1.50 \times 10^{-5}\,(\text{C}°)^{-1}](2 \times 2.00\text{ cm})(35.0\text{ C}°) = \boxed{2.10 \times 10^{-3}\text{ cm}}$$

As expected, the heights of B and C increase more than the height of A.

(b) The change in the volume of each block is given by Equation 12.3 as

$$\Delta V_A = \beta(D \times 2D \times 2D)\Delta T$$
$$= [4.50 \times 10^{-5}\,(\text{C}°)^{-1}][4(2.00\text{ cm})^3](35.0\text{ C}°) = \boxed{5.04 \times 10^{-2}\text{ cm}^3}$$

$$\Delta V_B = \beta(2D \times D \times D)\Delta T$$
$$= [4.50 \times 10^{-5}\,(\text{C}°)^{-1}][2(2.00\text{ cm})^3](35.0\text{ C}°) = \boxed{2.52 \times 10^{-2}\text{ cm}^3}$$

$$\Delta V_C = \beta(2D \times 2D \times D)\Delta T$$
$$= [4.50 \times 10^{-5}\,(\text{C}°)^{-1}][4(2.00\text{ cm})^3](35.0\text{ C}°) = \boxed{5.04 \times 10^{-2}\text{ cm}^3}$$

As discussed earlier, the greatest change in volume occurs with A and C followed by B.

Concepts & Calculations Example 19

Heat and Temperature Changes

Objects A and B in Figure 12.40 are made from copper, but the mass of B is three times the mass of A. Object C is made from glass and has the same mass as B. The *same amount* of heat Q is supplied to each one: $Q = 14$ J. Determine the rise in temperature for each.

Concept Questions and Answers Which object, A or B, experiences the greater rise in temperature?

Answer Consider an extreme example. Suppose a cup and a swimming pool are filled with water. For the same heat input, would you intuitively expect the temperature of the cup to rise more than the temperature of the pool? Yes, because the cup has less mass. Another way to arrive at this conclusion is to solve Equation 12.4 for the change in temperature: $\Delta T = Q/(cm)$. Since the heat Q and the specific heat capacity c are the same for A and B, ΔT is inversely proportional to the mass m. So the object with the smaller mass experiences the larger temperature change. Therefore, A has a greater temperature change than B.

Which object, B or C, experiences the greater rise in temperature?

Answer Objects B and C are made from different materials. To see how the rise in temperature depends on the type of material, let's again use Equation 12.4: $\Delta T = Q/(cm)$. Since the heat Q and the mass m are the same for B and C, ΔT is inversely proportional to the specific heat capacity c. So the object with the smaller specific heat capacity experiences the larger temperature change. Table 12.2 indicates that the specific heat capacities

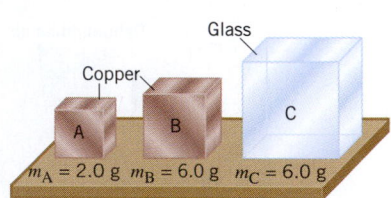

Figure 12.40 Which block has the greatest change in temperature when the same heat is supplied to each?

of copper and glass are 387 and 840 J/(kg·C°), respectively. Since B is made from copper, which has the smaller specific heat capacity, it has the greater temperature change.

Solution According to Equation 12.4, the temperature change for each object is

$$\Delta T_A = \frac{Q}{c_A m_A} = \frac{14\ \text{J}}{[387\ \text{J/(kg·C°)}](2.0 \times 10^{-3}\ \text{kg})} = \boxed{18\ \text{C}°}$$

$$\Delta T_B = \frac{Q}{c_B m_B} = \frac{14\ \text{J}}{[387\ \text{J/(kg·C°)}](6.0 \times 10^{-3}\ \text{kg})} = \boxed{6.0\ \text{C}°}$$

$$\Delta T_C = \frac{Q}{c_C m_C} = \frac{14\ \text{J}}{[840\ \text{J/(kg·C°)}](6.0 \times 10^{-3}\ \text{kg})} = \boxed{2.8\ \text{C}°}$$

As anticipated, ΔT_A is greater than ΔT_B, and ΔT_B is greater than ΔT_C.

CONCEPT SUMMARY

If you need more help with a concept, use the Learning Aids noted next to the discussion or equation. Examples (**Ex.**) are in the text of this chapter. Go to **www.wiley.com/college/cutnell** for the following Learning Aids:

Interactive LearningWare (ILW) — Additional examples solved in a five-step interactive format.

Concept Simulations (CS) — Animated text figures or animations of important concepts.

Interactive Solutions (IS) — Models for certain types of problems in the chapter homework. The calculations are carried out interactively.

Topic	Discussion	Learning Aids
Celsius temperature scale Fahrenheit temperature scale	**12.1 COMMON TEMPERATURE SCALES** On the Celsius temperature scale, there are 100 equal divisions between the ice point (0 °C) and the steam point (100 °C). On the Fahrenheit temperature scale, there are 180 equal divisions between the ice point (32 °F) and the steam point (212 °F).	**Ex. 1, 2**
Kelvin temperature scale	**12.2 THE KELVIN TEMPERATURE SCALE** For scientific work, the Kelvin temperature scale is the scale of choice. One kelvin (K) is equal in size to one Celsius degree. However, the temperature T on the Kelvin scale differs from the temperature T_c on the Celsius scale by an additive constant of 273.15: $$T = T_c + 273.15 \qquad (12.1)$$	
Absolute zero	The lower limit of temperature is called absolute zero and is designated as 0 K on the Kelvin scale.	
Thermometric property	**12.3 THERMOMETERS** The operation of any thermometer is based on the change in some physical property with temperature; this physical property is called a thermometric property. Examples of thermometric properties are the length of a column of mercury, electrical voltage, and electrical resistance.	
Linear thermal expansion	**12.4 LINEAR THERMAL EXPANSION** Most substances expand when heated. For linear expansion, an object of length L_0 experiences a change ΔL in length when the temperature changes by ΔT: $$\Delta L = \alpha L_0 \Delta T \qquad (12.2)$$ where α is the coefficient of linear expansion.	**Ex. 3, 18**
Thermal stress	For an object held rigidly in place, a thermal stress can occur when the object attempts to expand or contract. The stress can be large, even for small temperature changes.	**Ex. 4**
How a hole in a plate expands or contracts	When the temperature changes, a hole in a plate of solid material expands or contracts as if the hole were filled with the surrounding material.	**Ex. 5, 6, 7**
Volume thermal expansion	**12.5 VOLUME THERMAL EXPANSION** For volume expansion, the change ΔV in the volume of an object of volume V_0 is given by $$\Delta V = \beta V_0 \Delta T \qquad (12.3)$$ where β is the coefficient of volume expansion.	**Ex. 8, 18** **ILW 12.1**
How a cavity expands or contracts	When the temperature changes, a cavity in a piece of solid material expands or contracts as if the cavity were filled with the surrounding material.	**IS 12.29**

Topic	Discussion	Learning Aids
Internal energy Heat	**12.6 HEAT AND INTERNAL ENERGY** The internal energy of a substance is the sum of the kinetic, potential, and other kinds of energy that the molecules of the substance have. Heat is energy that flows from a higher-temperature object to a lower-temperature object because of the difference in temperatures. The SI unit for heat is the joule (J).	
	12.7 HEAT AND TEMPERATURE CHANGE: SPECIFIC HEAT CAPACITY The heat Q that must be supplied or removed to change the temperature of a substance of mass m by an amount ΔT is	
Heat needed to change the temperature	$$Q = cm\,\Delta T \qquad (12.4)$$ where c is a constant known as the specific heat capacity.	Ex. 9, 10, 11, 12, 19
Energy conservation and heat	When materials are placed in thermal contact within a perfectly insulated container, the principle of energy conservation requires that heat lost by warmer materials equals heat gained by cooler materials.	IS 12.47, 12.51
	Heat is sometimes measured with a unit called the kilocalorie (kcal). The conversion factor between kilocalories and joules is known as the mechanical equivalent of heat:	
Mechanical equivalent of heat	$$1\ \text{kcal} = 4186\ \text{joules}$$	ILW 12.2
	12.8 HEAT AND PHASE CHANGE: LATENT HEAT Heat must be supplied or removed to make a material change from one phase to another. The heat Q that must be supplied or removed to change the phase of a mass m of a substance is	
Heat needed to change the phase	$$Q = mL \qquad (12.5)$$ where L is the latent heat of the substance and has SI units of J/kg. The latent heats of fusion, vaporization, and sublimation refer, respectively, to the solid/liquid, the liquid/vapor, and the solid/vapor phase changes.	Ex. 13, 14, 15 IS 12.95
	12.9 EQUILIBRIUM BETWEEN PHASES OF MATTER The equilibrium vapor pressure of a substance is the pressure of the vapor phase that is in equilibrium with the liquid phase. For a given substance, vapor pressure depends only on temperature. For a liquid, a plot of the equilibrium vapor pressure versus temperature is called the vapor pressure curve or vaporization curve.	Ex. 16
Vapor pressure curve		
Fusion curve	The fusion curve gives the combinations of temperature and pressure for equilibrium between solid and liquid phases.	
	12.10 HUMIDITY The relative humidity is defined as follows:	
Relative humidity	$$\begin{array}{c}\text{Percent}\\ \text{relative}\\ \text{humidity}\end{array} = \dfrac{\begin{array}{c}\text{Partial pressure}\\ \text{of water vapor}\end{array}}{\begin{array}{c}\text{Equilibrium vapor pressure of}\\ \text{water at the existing temperature}\end{array}} \times 100 \qquad (12.6)$$	Ex. 17
Dew point	The dew point is the temperature below which the water vapor in the air condenses. On the vaporization curve of water, the dew point is the temperature that corresponds to the actual pressure of water vapor in the air.	

FOCUS ON CONCEPTS

Note to Instructors: The numbering of the questions shown here reflects the fact that they are only a representative subset of the total number that are available online. However, all of the questions are available for assignment via an online homework management program such as WileyPLUS or WebAssign.

Section 12.2 The Kelvin Temperature Scale

1. Which one of the following statements correctly describes the Celsius and the Kelvin temperature scales? **(a)** The size of the degree on the Celsius scale is larger than that on the Kelvin scale by a factor of 9/5. **(b)** Both scales assign the same temperature to the ice point, but they assign different temperatures to the steam point. **(c)** Both scales assign the same temperature to the steam point, but they assign different temperatures to the ice point. **(d)** The Celsius scale assigns the same values to the ice and the steam points that the Kelvin scale assigns. **(e)** The size of the degree on each scale is the same.

Section 12.4 Linear Thermal Expansion

2. The drawing shows two thin rods, one made from aluminum $[\alpha = 23 \times 10^{-6}\ (\text{C}°)^{-1}]$ and the other from steel $[\alpha = 12 \times 10^{-6}\ (\text{C}°)^{-1}]$. Each rod has the same length and the same initial temperature and is attached at one end to an immovable wall, as shown. The temperatures of the rods are increased, both by the same amount, until the gap between the rods is closed. Where do the rods meet when the

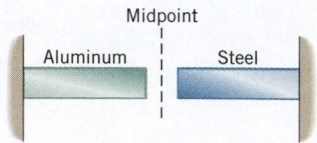

gap is closed? **(a)** The rods meet exactly at the midpoint. **(b)** The rods meet to the right of the midpoint. **(c)** The rods meet to the left of the midpoint.

4. A ball is slightly too large to fit through a hole in a flat plate. The drawing shows two arrangements of this situation. In Arrangement I the ball is made from metal A and the plate from metal B. When both the ball and the plate are cooled by the same number of Celsius degrees, the ball passes through the hole. In Arrangement II the ball is also made from metal A, but the plate is made from metal C. Here, the ball passes through the hole when both the ball and the plate are heated by the same number of Celsius degrees. Rank the coefficients of linear thermal expansion of metals A, B, and C in descending order (largest first): **(a)** α_B, α_A, α_C **(b)** α_B, α_C, α_A **(c)** α_C, α_B, α_A **(d)** α_C, α_A, α_B **(e)** α_A, α_B, α_C

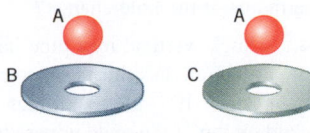

Arrangement I   Arrangement II

Section 12.5 Volume Thermal Expansion

6. A solid sphere and a solid cube are made from the same material. The sphere would just fit within the cube, if it could. Both begin at the same temperature, and both are heated to the same temperature. Which object, if either, has the greater change in volume? **(a)** The sphere. **(b)** The cube. **(c)** Both have the same change in volume. **(d)** Insufficient information is given for an answer.

7. A container can be made from steel [$\beta = 36 \times 10^{-6}$ (C°)$^{-1}$] or lead [$\beta = 87 \times 10^{-6}$ (C°)$^{-1}$]. A liquid is poured into the container, filling it to the brim. The liquid is either water [$\beta = 207 \times 10^{-6}$ (C°)$^{-1}$] or ethyl alcohol [$\beta = 1120 \times 10^{-6}$ (C°)$^{-1}$]. When the full container is heated, some liquid spills out. To keep the overflow to a minimum, from what material should the container be made and what should the liquid be? **(a)** Lead, water **(b)** Steel, water **(c)** Lead, ethyl alcohol **(d)** Steel, ethyl alcohol

Section 12.7 Heat and Temperature Change: Specific Heat Capacity

9. Which of the following cases (if any) requires the greatest amount of heat? In each case the material is the same. **(a)** 1.5 kg of the material is to be heated by 7.0 C°. **(b)** 3.0 kg of the material is to be heated by 3.5 C°. **(c)** 0.50 kg of the material is to be heated by 21 C°. **(d)** 0.75 kg of the material is to be heated by 14 C°. **(e)** The amount of heat required is the same in each of the four previous cases.

10. The following three hot samples have the same temperature. The same amount of heat is removed from each sample. Which one experiences the smallest drop in temperature, and which one experiences the largest drop?

> Sample A. 4.0 kg of water [$c = 4186$ J/(kg·C°)]
> Sample B. 2.0 kg of oil [$c = 2700$ J/(kg·C°)]
> Sample C. 9.0 kg of dirt [$c = 1050$ J/(kg·C°)]

(a) C smallest and A largest **(b)** B smallest and C largest **(c)** A smallest and B largest **(d)** C smallest and B largest **(e)** B smallest and A largest

Section 12.8 Heat and Phase Change: Latent Heat

13. The latent heat of fusion for water is 33.5×10^4 J/kg, while the latent heat of vaporization is 22.6×10^5 J/kg. What mass m of water must be frozen in order to release the amount of heat that 1.00 kg of steam releases when it condenses?

Section 12.9 Equilibrium Between Phases of Matter

15. Which one or more of the following techniques can be used to freeze water?

> A. Cooling the water below its normal freezing point of 0 °C at the normal atmospheric pressure of 1.01×10^5 Pa
>
> B. Cooling the water below its freezing point of -1 °C at a pressure greater than 1.01×10^5 Pa
>
> C. Rapidly pumping away the water vapor above the liquid in an insulated container (The insulation prevents heat flowing from the surroundings into the remaining liquid.)

(a) Only A **(b)** Only B **(c)** Only A and B **(d)** A, B, and C **(e)** Only C

Section 12.10 Humidity

17. Which of the following three statements concerning relative humidity values of 30% and 40% are true? Note that when the relative humidity is 30%, the air temperature may be different than it is when the relative humidity is 40%.

> A. It is possible that at a relative humidity of 30% there is a smaller partial pressure of water vapor in the air than there is at a relative humidity of 40%.
>
> B. It is possible that there is the same partial pressure of water vapor in the air at 30% and at 40% relative humidity.
>
> C. It is possible that at a relative humidity of 30% there is a greater partial pressure of water vapor in the air than there is at a relative humidity of 40%.

(a) A, B, and C **(b)** Only A and B **(c)** Only A and C **(d)** Only B and C **(e)** Only A

PROBLEMS

Note to Instructors: Most of the homework problems in this chapter are available for assignment via an online homework management program such as WileyPLUS or WebAssign, and those marked with the icon **GO** *are presented in WileyPLUS using a guided tutorial format that provides enhanced interactivity. See Preface for additional details.*

Note: For problems in this set, use the values of α and β in Table 12.1, and the values of c, L_f, and L_v in Tables 12.2 and 12.3, unless stated otherwise.

ssm Solution is in the Student Solutions Manual.
www Solution is available online at www.wiley.com/college/cutnell

 This icon represents a biomedical application.

Section 12.1 Common Temperature Scales,
Section 12.2 The Kelvin Temperature Scale,
Section 12.3 Thermometers

1. ssm A temperature of absolute zero occurs at -273.15 °C. What is this temperature on the Fahrenheit scale?

2. Suppose you are hiking down the Grand Canyon. At the top, the temperature early in the morning is a cool 3 °C. By late afternoon, the temperature at the bottom of the canyon has warmed to a sweltering 34 °C. What is the *difference* between the higher and lower temperatures in **(a)** Fahrenheit degrees and **(b)** kelvins?

3. On the moon the surface temperature ranges from 375 K during the day to 1.00×10^2 K at night. What are these temperatures on the **(a)** Celsius and **(b)** Fahrenheit scales?

4. **GO** The drawing shows two thermometers, A and B, whose temperatures are measured in °A and °B. The ice and boiling points of water are also indicated. **(a)** Using the data in the drawing, determine the number of B degrees on the B scale that correspond to 1 A° on the A scale. **(b)** If the temperature of a substance reads +40.0 °A on the A scale, what would that temperature read on the B scale?

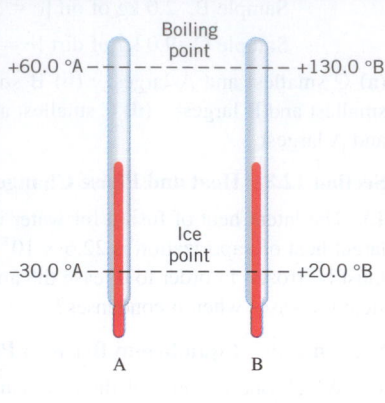

5. **ssm** Dermatologists often remove small precancerous skin lesions by freezing them quickly with liquid nitrogen, which has a temperature of 77 K. What is this temperature on the **(a)** Celsius and **(b)** Fahrenheit scales?

6. What's your normal body temperature? It may not be 98.6 °F, the oft-quoted average that was determined in the nineteenth century. A more recent study has reported an average temperature of 98.2 °F. What is the *difference* between these averages, expressed in Celsius degrees?

* **7.** **GO** A constant-volume gas thermometer (see Figures 12.3 and 12.4) has a pressure of 5.00×10^3 Pa when the gas temperature is 0.00 °C. What is the temperature (in °C) when the pressure is 2.00×10^3 Pa?

* **8.** If a nonhuman civilization were to develop on Saturn's largest moon, Titan, its scientists might well devise a temperature scale based on the properties of methane, which is much more abundant on the surface than water is. Methane freezes at −182.6 °C on Titan, and boils at −155.2 °C. Taking the boiling point of methane as 100.0 °M (degrees Methane) and its freezing point as 0 °M, what temperature on the Methane scale corresponds to the absolute zero point of the Kelvin scale?

* **9.** **ssm** On the Rankine temperature scale, which is sometimes used in engineering applications, the ice point is at 491.67 °R and the steam point is at 671.67 °R. Determine a relationship (analogous to Equation 12.1) between the Rankine and Fahrenheit temperature scales.

Section 12.4 Linear Thermal Expansion

10. An aluminum baseball bat has a length of 0.86 m at a temperature of 17 °C. When the temperature of the bat is raised, the bat lengthens by 0.000 16 m. Determine the final temperature of the bat.

11. **ssm** **www** Find the approximate length of the Golden Gate Bridge if it is known that the steel in the roadbed expands by 0.53 m when the temperature changes from +2 to +32 °C.

12. A steel aircraft carrier is 370 m long when moving through the icy North Atlantic at a temperature of 2.0 °C. By how much does the carrier lengthen when it is traveling in the warm Mediterranean Sea at a temperature of 21 °C?

13. Conceptual Example 5 provides background for this problem. A hole is drilled through a copper plate whose temperature is 11 °C.

(a) When the temperature of the plate is increased, will the radius of the hole be larger or smaller than the radius at 11 °C? Why? **(b)** When the plate is heated to 110 °C, by what fraction $\Delta r/r_0$ will the radius of the hole change?

14. A thick, vertical iron pipe has an inner diameter of 0.065 m. A thin aluminum disk, heated to a temperature of 85 °C, has a diameter that is 3.9×10^{-5} m greater than the pipe's inner diameter. The disk is laid on top of the open upper end of the pipe, perfectly centered on it, and allowed to cool. What is the temperature of the aluminum disk when the disk falls into the pipe? Ignore the temperature change of the pipe.

15. **ssm** When the temperature of a coin is raised by 75 C°, the coin's diameter increases by 2.3×10^{-5} m. If the original diameter of the coin is 1.8×10^{-2} m, find the coefficient of linear expansion.

16. One January morning in 1943, a warm chinook wind rapidly raised the temperature in Spearfish, South Dakota, from below freezing to +12.0 °C. As the chinook died away, the temperature fell to −20.0 °C in 27.0 minutes. Suppose that a 19-m aluminum flagpole were subjected to this temperature change. Find the average speed at which its height would decrease, assuming the flagpole responded instantaneously to the changing temperature.

17. **GO** One rod is made from lead and another from quartz. The rods are heated and experience the same change in temperature. The change in length of each rod is the same. If the initial length of the lead rod is 0.10 m, what is the initial length of the quartz rod?

* **18.** Concrete sidewalks are always laid in sections, with gaps between each section. For example, the drawing shows three identical 2.4-m sections, the outer two of which are against immovable walls. The two identical gaps between the sections are provided so that thermal expansion will not create the thermal stress that could lead to cracks. What is the minimum gap width necessary to account for an increase in temperature of 32 C°?

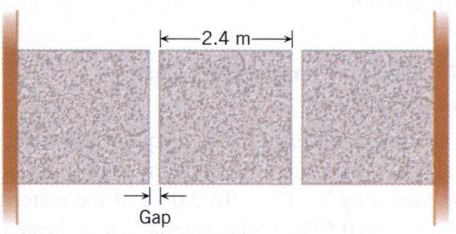

* **19.** The brass bar and the aluminum bar in the drawing are each attached to an immovable wall. At 28 °C the air gap between the rods is 1.3×10^{-3} m. At what temperature will the gap be closed?

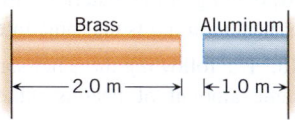

* **20.** Multiple-Concept Example 4 illustrates the concepts that are pertinent to this problem. A cylindrical brass rod (cross-sectional area = 1.3×10^{-5} m²) hangs vertically straight down from a ceiling. When an 860-N block is hung from the lower end of the rod, the rod stretches. The rod is then cooled such that it contracts to its original length. By how many degrees must the temperature be lowered?

* **21.** **ssm** **www** A simple pendulum consists of a ball connected to one end of a thin brass wire. The period of the pendulum is 2.0000 s. The temperature rises by 140 C°, and the length of the wire increases. Determine the period of the heated pendulum.

* **22.** As the drawing shows, two thin strips of metal are bolted together at one end; both have the

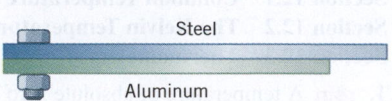

same temperature. One is steel, and the other is aluminum. The steel strip is 0.10% longer than the aluminum strip. By how much should the temperature of the strips be increased, so that the strips have the same length?

*23. Consult Conceptual Example 5 for background pertinent to this problem. A lead sphere has a diameter that is 0.050% larger than the inner diameter of a steel ring when each has a temperature of 70.0 °C. Thus, the ring will not slip over the sphere. At what common temperature will the ring just slip over the sphere?

*24. GO A ball and a thin plate are made from different materials and have the same initial temperature. The ball does not fit through a hole in the plate, because the diameter of the ball is slightly larger than the diameter of the hole. However, the ball will pass through the hole when the ball and the plate are both heated to a common higher temperature. In each of the arrangements in the drawing the diameter of the ball is 1.0×10^{-5} m larger than the diameter of the hole in the thin plate, which has a diameter of 0.10 m. The initial temperature of each arrangement is 25.0 °C. At what temperature will the ball fall through the hole in each arrangement?

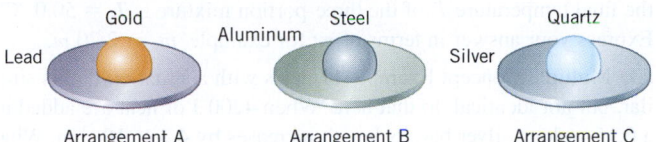

25. ssm A steel ruler is calibrated to read true at 20.0 °C. A draftsman uses the ruler at 40.0 °C to draw a line on a 40.0 °C copper plate. As indicated on the warm ruler, the length of the line is 0.50 m. To what temperature should the plate be cooled, such that the length of the line truly becomes 0.50 m?

26. A steel bicycle wheel (without the rubber tire) is rotating freely with an angular speed of 18.00 rad/s. The temperature of the wheel changes from −100.0 to +300.0 °C. No net external torque acts on the wheel, and the mass of the spokes is negligible. **(a) Does the angular speed increase or decrease as the wheel heats up? Why? **(b)** What is the angular speed at the higher temperature?

**27. Consult Multiple-Concept Example 4 for insight into solving this problem. An aluminum wire of radius 3.0×10^{-4} m is stretched between the ends of a concrete

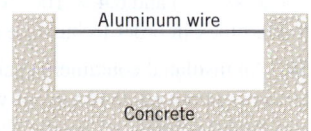

block, as the drawing illustrates. When the system (wire and concrete) is at 35 °C, the tension in the wire is 50.0 N. What is the tension in the wire when the system is heated to 185 °C?

Section 12.5 Volume Thermal Expansion

28. A flask is filled with 1.500 L (L = liter) of a liquid at 97.1 °C. When the liquid is cooled to 15.0 °C, its volume is only 1.383 L, however. Neglect the contraction of the flask and use Table 12.1 to identify the liquid.

29. Interactive Solution 12.29 at **www.wiley.com/college/cutnell** presents a model for solving problems of this type. A thin spherical shell of silver has an inner radius of 2.0×10^{-2} m when the temperature is 18 °C. The shell is heated to 147 °C. Find the change in the interior volume of the shell.

30. A test tube contains 2.54×10^{-4} m³ of liquid carbon tetrachloride at a temperature of 75.0 °C. The test tube and the carbon tetrachloride are cooled to a temperature of −13.0 °C, which is above the freezing point of carbon tetrachloride. Find the volume of carbon tetrachloride in the test tube at −13.0 °C.

31. ssm A lead object and a quartz object each have the same initial volume. The volume of each increases by the same amount, because

the temperature increases. If the temperature of the lead object increases by 4.0 C°, by how much does the temperature of the quartz object increase?

32. At a temperature of 0 °C, the mass and volume of a fluid are 825 kg and 1.17 m³. The coefficient of volume expansion is 1.26×10^{-3} (C°)$^{-1}$. **(a)** What is the density of the fluid at this temperature? **(b)** What is the density of the fluid when the temperature has risen to 20.0 °C?

33. Consult **Interactive LearningWare 12.1** at **www.wiley.com/college/cutnell** for help in solving this problem. During an all-night cram session, a student heats up a one-half liter (0.50×10^{-3} m³) glass (Pyrex) beaker of cold coffee. Initially, the temperature is 18 °C, and the beaker is filled to the brim. A short time later when the student returns, the temperature has risen to 92 °C. The coefficient of volume expansion of coffee is the same as that of water. How much coffee (in cubic meters) has spilled out of the beaker?

34. GO An aluminum can is filled to the brim with a liquid. The can and the liquid are heated so their temperatures change by the same amount. The can's initial volume at 5 °C is 3.5×10^{-4} m³. The coefficient of volume expansion for aluminum is 69×10^{-6} (C°)$^{-1}$. When the can and the liquid are heated to 78 °C, 3.6×10^{-6} m³ of liquid spills over. What is the coefficient of volume expansion of the liquid?

35. ssm Suppose that the steel gas tank in your car is completely filled when the temperature is 17 °C. How many gallons will spill out of the twenty-gallon tank when the temperature rises to 35 °C?

36. Interactive LearningWare 12.1 at **www.wiley.com/college/cutnell** provides some useful background for this problem. Many hot-water heating systems have a reservoir tank connected directly to the pipeline, to allow for expansion when the water becomes hot. The heating system of a house has 76 m of copper pipe whose inside radius is 9.5×10^{-3} m. When the water and pipe are heated from 24 to 78 °C, what must be the minimum volume of the reservoir tank to hold the overflow of water?

*37. A solid aluminum sphere has a radius of 0.50 m and a temperature of 75 °C. The sphere is then completely immersed in a pool of water whose temperature is 25 °C. The sphere cools, while the water temperature remains nearly at 25 °C, because the pool is very large. The sphere is weighed in the water immediately after being submerged (before it begins to cool) and then again after cooling to 25 °C. **(a)** Which weight is larger? Why? **(b)** Use Archimedes' principle to find the magnitude of the *difference* between the weights.

*38. GO At the bottom of an old mercury-in-glass thermometer is a 45-mm³ reservoir filled with mercury. When the thermometer was placed under your tongue, the warmed mercury would expand into a very narrow cylindrical channel, called a capillary, whose radius was 1.7×10^{-2} mm. Marks were placed along the capillary that indicated the temperature. Ignore the thermal expansion of the glass and determine how far (in mm) the mercury would expand into the capillary when the temperature changed by 1.0 C°.

*39. ssm The bulk modulus of water is $B = 2.2 \times 10^{9}$ N/m². What change in pressure ΔP (in atmospheres) is required to keep water from expanding when it is heated from 15 to 25 °C?

*40. A spherical brass shell has an interior volume of 1.60×10^{-3} m³. Within this interior volume is a solid steel ball that has a volume of 0.70×10^{-3} m³. The space between the steel ball and the inner surface of the brass shell is filled completely with mercury. A small hole is drilled through the brass, and the temperature of the arrangement is increased by 12 C°. What is the volume of the mercury that spills out of the hole?

**41.   ssm www Two identical thermometers made of Pyrex glass contain, respectively, identical volumes of mercury and methyl alcohol. If the expansion of the glass is taken into account, how many times greater is the distance between the degree marks on the methyl alcohol thermometer than the distance on the mercury thermometer?

**42. The column of mercury in a barometer (see Figure 11.11) has a height of 0.760 m when the pressure is one atmosphere and the temperature is 0.0 °C. Ignoring any change in the glass containing the mercury, what will be the height of the mercury column for the same one atmosphere of pressure when the temperature rises to 38.0 °C on a hot day? *(Hint: The pressure in the barometer is given by Pressure = ρgh, and the density ρ of the mercury changes when the temperature changes.)*

Section 12.6   Heat and Internal Energy,
Section 12.7   Heat and Temperature Change:
Specific Heat Capacity

43.   ssm Ideally, when a thermometer is used to measure the temperature of an object, the temperature of the object itself should not change. However, if a significant amount of heat flows from the object to the thermometer, the temperature will change. A thermometer has a mass of 31.0 g, a specific heat capacity of $c = 815$ J/(kg·C°), and a temperature of 12.0 °C. It is immersed in 119 g of water, and the final temperature of the water and thermometer is 41.5 °C. What was the temperature of the water before the insertion of the thermometer?

44. If the price of electrical energy is $0.10 per kilowatt·hour, what is the cost of using electrical energy to heat the water in a swimming pool (12.0 m × 9.00 m × 1.5 m) from 15 to 27 °C?

45. An ice chest at a beach party contains 12 cans of soda at 5.0 °C. Each can of soda has a mass of 0.35 kg and a specific heat capacity of 3800 J/(kg·C°). Someone adds a 6.5-kg watermelon at 27 °C to the chest. The specific heat capacity of watermelon is nearly the same as that of water. Ignore the specific heat capacity of the chest and determine the final temperature T of the soda and watermelon.

46. When you drink cold water, your body must expend metabolic energy in order to maintain normal body temperature (37 °C) by warming up the water in your stomach. Could drinking ice water, then, substitute for exercise as a way to "burn calories?" Suppose you expend 430 kilocalories during a brisk hour-long walk. How many liters of ice water (0 °C) would you have to drink in order to use up 430 kilocalories of metabolic energy? For comparison, the stomach can hold about 1 liter.

47. Review **Interactive Solution 12.47** at **www.wiley.com/college/cutnell** for help in approaching this problem. When resting, a person has a metabolic rate of about 3.0×10^5 joules per hour. The person is submerged neck-deep into a tub containing 1.2×10^3 kg of water at 21.00 °C. If the heat from the person goes only into the water, find the water temperature after half an hour.

48.   GO Two bars of identical mass are at 25 °C. One is made from glass and the other from another substance. The specific heat capacity of glass is 840 J/(kg·C°). When identical amounts of heat are supplied to each, the glass bar reaches a temperature of 88 °C, while the other bar reaches 250.0 °C. What is the specific heat capacity of the other substance?

49.   ssm At a fabrication plant, a hot metal forging has a mass of 75 kg and a specific heat capacity of 430 J/(kg·C°). To harden it, the forging is immersed in 710 kg of oil that has a temperature of 32 °C and a specific heat capacity of 2700 J/(kg·C°). The final temperature of the oil and forging at thermal equilibrium is 47 °C. Assuming that heat flows only between the forging and the oil, determine the initial temperature of the forging.

50. A piece of glass has a temperature of 83.0 °C. Liquid that has a temperature of 43.0 °C is poured over the glass, completely covering it, and the temperature at equilibrium is 53.0 °C. The mass of the glass and the liquid is the same. Ignoring the container that holds the glass and liquid and assuming that the heat lost to or gained from the surroundings is negligible, determine the specific heat capacity of the liquid.

*51.   Interactive Solution 12.51 at **www.wiley.com/college/cutnell** deals with one approach to solving problems such as this. A 0.35-kg coffee mug is made from a material that has a specific heat capacity of 920 J/(kg·C°) and contains 0.25 kg of water. The cup and water are at 15 °C. To make a cup of coffeee, a small electric heater is immersed in the water and brings it to a boil in three minutes. Assume that the cup and water always have the same temperature and determine the minimum power rating of this heater.

*52.   GO Three portions of the same liquid are mixed in a container that prevents the exchange of heat with the environment. Portion A has a mass m and a temperature of 94.0 °C, portion B also has a mass m but a temperature of 78.0 °C, and portion C has a mass m_C and a temperature of 34.0 °C. What must be the mass of portion C so that the final temperature T_f of the three-portion mixture is $T_f = 50.0$ °C? Express your answer in terms of m; for example, $m_c = 2.20\ m$.

*53. Multiple-Concept Example 11 deals with a situation that is similar, but not identical, to that here. When 4200 J of heat are added to a 0.15-m-long silver bar, its length increases by 4.3×10^{-3} m. What is the mass of the bar?

*54. The heating element of a water heater in an apartment building has a maximum power output of 28 kW. Four residents of the building take showers at the same time, and each receives heated water at a volume flow rate of 14×10^{-5} m³/s. If the water going into the heater has a temperature of 11 °C, what is the maximum possible temperature of the hot water that each showering resident receives?

*55.   ssm Multiple-Concept Example 11 uses the same physics principles as those employed in this problem. A block of material has a mass of 130 kg and a volume of 4.6×10^{-2} m³. The material has a specific heat capacity and coefficient of volume expansion, respectively, of 750 J/(kg·C°) and 6.4×10^{-5} (C°)$^{-1}$. How much heat must be added to the block in order to increase its volume by 1.2×10^{-5} m³?

**56. An insulated container is partly filled with oil. The lid of the container is removed, 0.125 kg of water heated to 90.0 °C is poured in, and the lid is replaced. As the water and the oil reach equilibrium, the volume of the oil increases by 1.20×10^{-5} m³. The density of the oil is 924 kg/m³, its specific heat capacity is 1970 J/(kg·C°), and its coefficient of volume expansion is 721×10^{-6} (C°)$^{-1}$. What is the temperature when the oil and the water reach equilibrium?

Section 12.8   Heat and Phase Change: Latent Heat

57.   ssm How much heat must be added to 0.45 kg of aluminum to change it from a solid at 130 °C to a liquid at 660 °C (its melting point)? The latent heat of fusion for aluminum is 4.0×10^5 J/kg.

58. To help prevent frost damage, fruit growers sometimes protect their crop by spraying it with water when overnight temperatures are expected to go below freezing. When the water turns to ice during the night, heat is released into the plants, thereby giving a measure of protection against the cold. Suppose a grower sprays 7.2 kg of water at 0 °C onto a fruit tree. **(a)** How much heat is released by the water when it freezes? **(b)** How much would the temperature of a 180-kg tree rise if it absorbed the heat released in part (a)? Assume that the specific heat capacity of the tree is 2.5×10^3 J/(kg·C°) and that no phase change occurs within the tree itself.

59. Assume that the pressure is one atmosphere and determine the heat required to produce 2.00 kg of water vapor at 100.0 °C, starting with   **(a)** 2.00 kg of water at 100.0 °C and   **(b)** 2.00 kg of liquid water at 0.0 °C.

60. **GO** **(a)** Objects A and B have the same mass of 3.0 kg. They melt when 3.0×10^4 J of heat is added to A and when 9.0×10^4 J is added to B. Determine the latent heat of fusion for the substance from which each object is made.   **(b)** Find the heat required to melt object A when its mass is 6.0 kg.

61. **ssm** Find the mass of water that vaporizes when 2.10 kg of mercury at 205 °C is added to 0.110 kg of water at 80.0 °C.

62. A mass $m = 0.054$ kg of benzene vapor at its boiling point of 80.1 °C is to be condensed by mixing the vapor with water at 41 °C. What is the minimum mass of water required to condense all of the benzene vapor? Assume that the mixing and condensation take place in a perfectly insulating container.

63. A person eats a container of strawberry yogurt. The Nutritional Facts label states that it contains 240 Calories (1 Calorie = 4186 J). What mass of perspiration would one have to lose to get rid of this energy? At body temperature, the latent heat of vaporization of water is 2.42×10^6 J/kg.

64. A woman finds the front windshield of her car covered with ice at −12.0 °C. The ice has a thickness of 4.50×10^{-4} m, and the windshield has an area of 1.25 m². The density of ice is 917 kg/m³. How much heat is required to melt the ice?

65. A thermos contains 150 cm³ of coffee at 85 °C. To cool the coffee, you drop two 11-g ice cubes into the thermos. The ice cubes are initially at 0 °C and melt completely. What is the final temperature of the coffee? Treat the coffee as if it were water.

*66. A snow maker at a resort pumps 130 kg of lake water per minute and sprays it into the air above a ski run. The water droplets freeze in the air and fall to the ground, forming a layer of snow. If all the water pumped into the air turns to snow, and the snow cools to the ambient air temperature of −7.0 °C, how much heat does the snow-making process release each minute? Assume that the temperature of the lake water is 12.0 °C, and use 2.00×10^3 J/(kg·C°) for the specific heat capacity of snow.

*67. **ssm** Ice at −10.0 °C and steam at 130 °C are brought together at atmospheric pressure in a perfectly insulated container. After thermal equilibrium is reached, the liquid phase at 50.0 °C is present. Ignoring the container and the equilibrium vapor pressure of the liquid at 50.0 °C, find the ratio of the mass of steam to the mass of ice. The specific heat capacity of steam is 2020 J/(kg·C°).

*68. **GO** Water at 23.0 °C is sprayed onto 0.180 kg of molten gold at 1063 °C (its melting point). The water boils away, forming steam at 100.0 °C and leaving solid gold at 1063 °C. What is the minimum mass of water that must be used?

*69. **ssm www** An unknown material has a normal melting/freezing point of −25.0 °C, and the liquid phase has a specific heat capacity of 160 J/(kg·C°). One-tenth of a kilogram of the solid at −25.0 °C is put into a 0.150-kg aluminum calorimeter cup that contains 0.100 kg of glycerin. The temperature of the cup and the glycerin is initially 27.0 °C. All the unknown material melts, and the final temperature at equilibrium is 20.0 °C. The calorimeter neither loses energy to nor gains energy from the external environment. What is the latent heat of fusion of the unknown material?

*70. Occasionally, huge icebergs are found floating on the ocean's currents. Suppose one such iceberg is 120 km long, 35 km wide, and 230 m thick.   **(a)** How much heat would be required to melt this iceberg (assumed to be at 0 °C) into liquid water at 0 °C? The density of ice is 917 kg/m³.   **(b)** The annual energy consumption by the United States is about 1.1×10^{20} J. If this energy were delivered to the

iceberg every year, how many years would it take before the ice melted?

*71. **ssm** It is claimed that if a lead bullet goes fast enough, it can melt completely when it comes to a halt suddenly, and all its kinetic energy is converted into heat via friction. Find the minimum speed of a lead bullet (initial temperature = 30.0 °C) for such an event to happen.

*72. Two grams of liquid water are at 0 °C, and another two grams are at 100 °C. Heat is removed from the water at 0 °C, completely freezing it at 0 °C. This heat is then used to vaporize some of the water at 100 °C. What is the mass (in grams) of the liquid water that remains?

**73. A locomotive wheel is 1.00 m in diameter. A 25.0-kg steel band has a temperature of 20.0 °C and a diameter that is 6.00×10^{-4} m less than that of the wheel. What is the smallest mass of water vapor at 100 °C that can be condensed on the steel band to heat it, so that it will fit onto the wheel? Do not ignore the water that results from the condensation.

Section 12.9   Equilibrium Between Phases of Matter, Section 12.10   Humidity

74. Use the vapor pressure curve that accompanies this problem to determine the temperature at which liquid carbon dioxide exists in equilibrium with its vapor phase when the vapor pressure is 3.5×10^6 Pa.

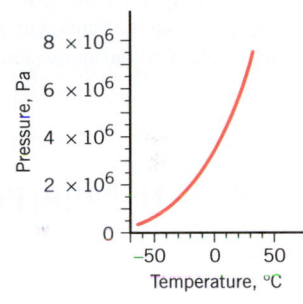
Problem 74

75. **ssm** At a temperature of 10 °C the percent relative humidity is R_{10}, and at 40 °C it is R_{40}. At each of these temperatures the partial pressure of water vapor in the air is the same. Using the vapor pressure curve for water that accompanies this problem, determine the ratio R_{10}/R_{40} of the two humidity values.

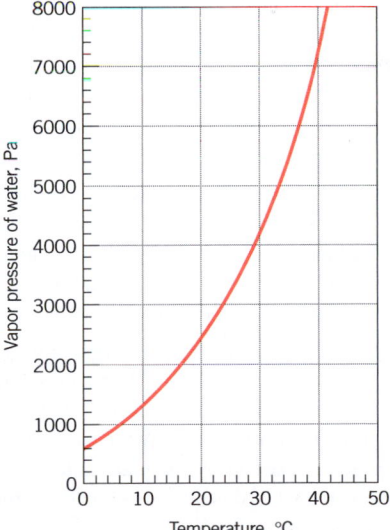

76. On a certain evening, the dew point is 14 °C and the relative humidity is 50.0%. How many Celsius degrees must the temperature fall in order for the relative humidity to increase to 69%? Use the vapor pressure curve for water that accompanies Problem 75 as needed. Assume that the dew point does not change as the temperature falls.

77. Suppose that air in the human lungs has a temperature of 37 °C, and the partial pressure of water vapor has a value of 5.5×10^3 Pa. What is the relative humidity in the lungs? Consult the vapor pressure curve for water that accompanies Problem 75.

78. GO The vapor pressure of water at 10 °C is 1300 Pa. **(a)** What percentage of atmospheric pressure is this? Take atmospheric pressure to be 1.013×10^5 Pa. **(b)** What percentage of the total air pressure at 10 °C is due to water vapor when the relative humidity is 100%? **(c)** The vapor pressure of water at 35 °C is 5500 Pa. What is the relative humidity at this temperature if the partial pressure of water in the air has not changed from what it was at 10 °C when the relative humidity was 100%?

79. The temperature of 2.0 kg of water is 100.0 °C, but the water is not boiling, because the external pressure acting on the water surface is 3.0×10^5 Pa. Using the vapor pressure curve for water given in Figure 12.32, determine the amount of heat that must be added to the water to bring it to the point where it just begins to boil.

80. A container is fitted with a movable piston of negligible mass and radius $r = 0.061$ m. Inside the container is liquid water in equilibrium with its vapor, as the drawing shows. The piston remains stationary with a 120-kg block on top of it. The air pressure acting on the top of the piston is one atmosphere. By using the vaporization curve for water in Figure 12.32, find the temperature of the water.

81. A woman has been outdoors where the temperature is 10 °C. She walks into a 25 °C house, and her glasses "steam up." Using the vapor pressure curve for water that accompanies Problem 75, find the smallest possible value for the relative humidity of the room.

82. The temperature of the air in a room is 36 °C. A person turns on a dehumidifier and notices that when the cooling coils reach 30 °C, water begins to condense on them. What is the relative humidity in the room? Use the vapor pressure curve that accompanies Problem 75.

83. ssm At a picnic, a glass contains 0.300 kg of tea at 30.0 °C, which is the air temperature. To make iced tea, someone adds 0.0670 kg of ice at 0.0 °C and stirs the mixture. When all the ice melts and the final temperature is reached, the glass begins to fog up, because water vapor condenses on the outer glass surface. Using the vapor pressure curve for water that accompanies Problem 75, ignoring the specific heat capacity of the glass, and treating the tea as if it were water, estimate the relative humidity.

84. A tall column of water is open to the atmosphere. At a depth of 10.3 m below the surface, the water is boiling. What is the temperature at this depth? Use the vaporization curve for water in Figure 12.32, as needed.

ADDITIONAL PROBLEMS

85. A steel section of the Alaskan pipeline had a length of 65 m and a temperature of 18 °C when it was installed. What is its change in length when the temperature drops to a frigid −45 °C?

86. The latent heat of vaporization of H_2O at body temperature (37.0 °C) is 2.42×10^6 J/kg. To cool the body of a 75-kg jogger [average specific heat capacity = 3500 J/(kg·C°)] by 1.5 C°, how many kilograms of water in the form of sweat have to be evaporated?

87. ssm Liquid nitrogen boils at a chilly −195.8 °C when the pressure is one atmosphere. A silver coin of mass 1.5×10^{-2} kg and temperature 25 °C is dropped into the boiling liquid. What mass of nitrogen boils off as the coin cools to −195.8 °C?

88. A 0.200-kg piece of aluminum that has a temperature of −155 °C is added to 1.5 kg of water that has a temperature of 3.0 °C. At equilibrium the temperature is 0.0 °C. Ignoring the container and assuming that the heat exchanged with the surroundings is negligible, determine the mass of water that has been frozen into ice.

89. A commonly used method of fastening one part to another part is called "shrink fitting." A steel rod has a diameter of 2.0026 cm, and a flat plate contains a hole whose diameter is 2.0000 cm. The rod is cooled so that it just fits into the hole. When the rod warms up, the enormous thermal stress exerted by the plate holds the rod securely to the plate. By how many Celsius degrees should the rod be cooled?

90. A thin rod consists of two parts joined together. One-third of it is silver and two-thirds is gold. The temperature decreases by 26 C°. Determine the fractional decrease $\dfrac{\Delta L}{L_{0,\,\text{Silver}} + L_{0,\,\text{Gold}}}$ in the rod's length, where $L_{0,\,\text{Silver}}$ and $L_{0,\,\text{Gold}}$ are the initial lengths of the silver and gold rods.

91. ssm A copper kettle contains water at 24 °C. When the water is heated to its boiling point, the volume of the kettle expands by 1.2×10^{-5} m³. Determine the volume of the kettle at 24 °C.

92. When you take a bath, how many kilograms of hot water (49.0 °C) must you mix with cold water (13.0 °C) so that the temperature of the bath is 36.0 °C? The total mass of water (hot plus cold) is 191 kg. Ignore any heat flow between the water and its external surroundings.

93. ssm What is the relative humidity on a day when the temperature is 30 °C and the dew point is 10 °C? Use the vapor pressure curve that accompanies Problem 75.

94. A copper–constantan thermocouple generates a voltage of 4.75×10^{-3} volts when the temperature of the hot junction is 110.0 °C and the reference junction is kept at a temperature of 0.0 °C. If the voltage is proportional to the difference in temperature between the junctions, what is the temperature of the hot junction when the voltage is 1.90×10^{-3} volts?

95. Interactive Solution 12.95 at **www.wiley.com/college/cutnell** provides a model for solving problems such as this. A 42-kg block of ice at 0 °C is sliding on a horizontal surface. The initial speed of the ice is 7.3 m/s and the final speed is 3.5 m/s. Assume that the part of the block that melts has a very small mass and that all the heat generated by kinetic friction goes into the block of ice. Determine the mass of ice that melts into water at 0 °C.

96. Equal masses of two different liquids have the same temperature of 25.0 °C. Liquid A has a freezing point of −68.0 °C and a specific heat capacity of 1850 J/(kg·C°). Liquid B has a freezing point of −96.0 °C and a specific heat capacity of 2670 J/(kg·C°). The same amount of heat must be removed from each liquid in order to freeze it into a solid at its respective freezing point. Determine the difference $L_{f,\,A} - L_{f,\,B}$ between the latent heats of fusion for these liquids.

97. Refer to **Interactive LearningWare 12.2** at **www.wiley.com/college/cutnell** for a review of the concepts that play roles in this problem. The box of a well-known breakfast cereal states that one ounce of the cereal contains 110 Calories (1 food Calorie = 4186 J). If 2.0% of this energy could be converted by a weight lifter's body into work done in lifting a barbell, what is the heaviest barbell that could be lifted a distance of 2.1 m?

*98. Multiple-Concept Example 4 reviews the concepts that are involved in this problem. A ruler is accurate when the temperature is 25 °C. When the temperature drops to −14 °C, the ruler shrinks and no longer measures distances accurately. However, the ruler can be made to read correctly if a force of magnitude 1.2×10^3 N is applied to each end so as to stretch it back to its original length. The ruler has a cross-sectional area of 1.6×10^{-5} m^2, and it is made from a material whose coefficient of linear expansion is 2.5×10^{-5} (C°)$^{-1}$. What is Young's modulus for the material from which the ruler is made?

*99. **ssm** A rock of mass 0.20 kg falls from rest from a height of 15 m into a pail containing 0.35 kg of water. The rock and water have the same initial temperature. The specific heat capacity of the rock is 1840 J/(kg·C°). Ignore the heat absorbed by the pail itself, and determine the rise in the temperature of the rock and water.

**100. A wire is made by attaching two segments together, end to end. One segment is made of aluminum and the other is steel. The effective coefficient of linear expansion of the two-segment wire is 19×10^{-6} (C°)$^{-1}$. What fraction of the length is aluminum?

**101. A steel rod ($\rho = 7860$ kg/m^3) has a length of 2.0 m. It is bolted at both ends between immobile supports. Initially there is no tension in the rod, because the rod just fits between the supports. Find the tension that develops when the rod loses 3300 J of heat.

**102. An 85.0-N backpack is hung from the middle of an aluminum wire, as the drawing shows. The temperature of the wire then drops by 20.0 C°. Find the tension in the wire at the lower temperature. Assume that the distance between the supports does not change, and ignore any thermal stress.

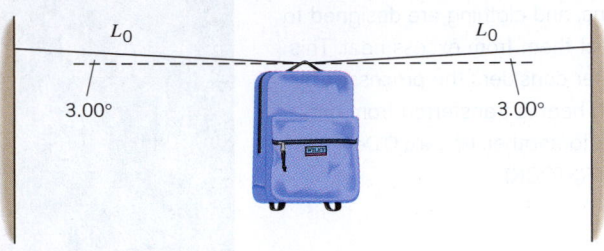

CHAPTER 13

THE TRANSFER OF HEAT

These workers are pouring a (very) hot liquid into forms. Their helmets, aprons, and clothing are designed to protect them from excess heat. This chapter considers the processes by which heat is transferred from one place to another. (© Dale O'Dell/ SUPERSTOCK)

13.1 **CONVECTION**

When heat is transferred to or from a substance, the internal energy of the substance can change, as we saw in Chapter 12. This change in internal energy is accompanied by a change in temperature or a change in phase. The transfer of heat affects us in many ways. For instance, within our homes furnaces distribute heat on cold days, and air conditioners remove it on hot days. Our bodies constantly transfer heat in one direction or another, to prevent the adverse effects of hypo- and hyperthermia. And virtually all our energy originates in the sun and is transferred to us over a distance of 150 million kilometers through the void of space. Today's sunlight provides the energy to drive photosynthesis in the plants that provide our food and, hence, metabolic energy. Ancient sunlight nurtured the organic matter that became the fossil fuels of oil, natural gas, and coal. This chapter examines the three processes by which heat is transferred: convection, conduction, and radiation.

When part of a fluid is warmed, such as the air above a fire, the volume of that part of the fluid expands, and the density decreases. According to Archimedes' principle (see Section 11.6), the surrounding cooler and denser fluid exerts a buoyant force on the warmer fluid and pushes it upward. As warmer fluid rises, the surrounding cooler fluid replaces it. This cooler fluid, in turn, is warmed and pushed upward. Thus, a continuous flow is established, which carries along heat. Whenever heat is transferred by the bulk movement of a gas or a liquid, the heat is said to be transferred by *convection*. The fluid flow itself is called a *convection current*.

CONVECTION

Convection is the process in which heat is carried from place to place by the bulk movement of a fluid.

The smoke rising from the industrial fire in Figure 13.1 is one visible result of convection. Figure 13.2 shows the less visible example of convection currents in a pot of water being heated on a gas burner. The currents distribute the heat from the burning gas to all parts of the water. Conceptual Example 1 deals with some of the important roles that convection plays in the home.

Conceptual Example 1   Hot Water Baseboard Heating and Refrigerators

Hot water baseboard heating units are frequently used in homes, and a cooling coil is a major component of a refrigerator. The locations of these heating and cooling devices are different because each is designed to maximize the production of convection currents. Where should the heating unit and the cooling coil be located? **(a)** Heating unit near the floor of the room and cooling coil near the top of the refrigerator **(b)** Heating unit near the ceiling of the room and cooling coil near the bottom of the refrigerator

Reasoning An important goal for the heating system is to distribute heat throughout a room. The analogous goal for the cooling coil is to remove heat from all of the space within a refrigerator. In each case, the heating or cooling device must be positioned so that convection makes the goal achievable.

Answer (b) is incorrect. If the heating unit were placed near the ceiling of the room, warm air from the unit would remain there, because warm air does not fall (it rises). Thus, there would be very little natural movement (or convection) of air to distribute the heat throughout the room. If the cooling coil were located near the bottom of the refrigerator, the cool air would remain there, because cool air does not rise (it sinks). There would be very little convection to carry the heat from other parts of the refrigerator to the coil for removal.

Answer (a) is correct. The air above the baseboard unit is heated, like the air above a fire. Buoyant forces from the surrounding cooler air push the warm air upward. Cooler air near the ceiling is displaced downward and then warmed by the baseboard heating unit, causing the convection current illustrated in Figure 13.3a. Within the refrigerator, air in contact with the top-mounted coil is cooled, its volume decreases, and its density increases. The surrounding warmer and less dense air cannot provide sufficient buoyant force to support the cooler air, which sinks downward. In the process, warmer air near the bottom of the refrigerator is displaced upward and is then cooled by the coil, establishing the convection current shown in Figure 13.3b.

Figure 13.1 The plumes of thick, gray smoke from this industrial fire rise hundreds of meters into the air because of convection. (Scott Shaw/The Plain Dealer/© AP/Wide World Photos)

The physics of heating and cooling by convection.

Figure 13.2 Convection currents are set up when a pot of water is heated.

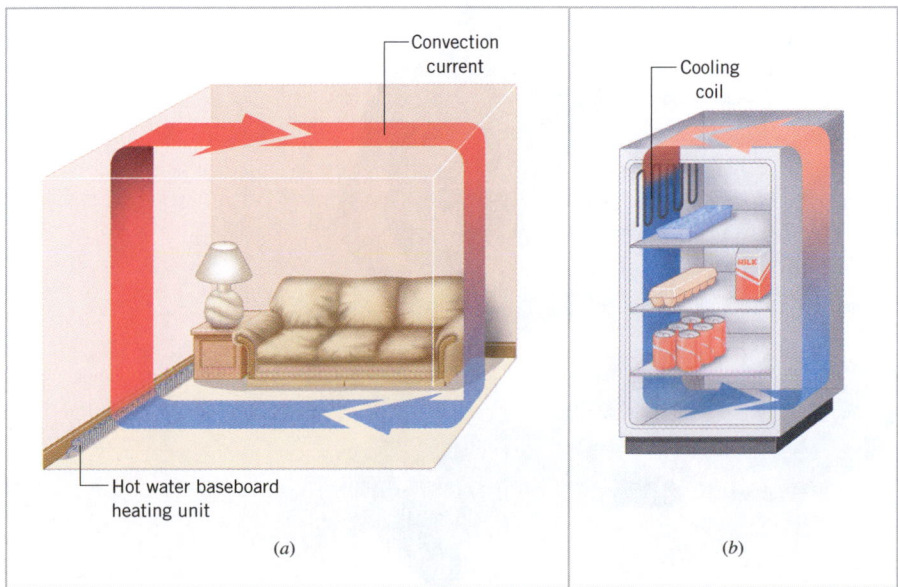

Convection current

Hot water baseboard heating unit

(a)

Cooling coil

(b)

Figure 13.3 (*a*) Air warmed by the baseboard heating unit is pushed to the top of the room by the cooler and denser air. (*b*) Air cooled by the cooling coil sinks to the bottom of the refrigerator. In both (*a*) and (*b*) a convection current is established.

Another example of convection occurs when the ground, heated by the sun's rays, warms the neighboring air. Surrounding cooler and denser air pushes the heated air upward. The resulting updraft or "thermal" can be quite strong, depending on the amount of heat that the ground can supply. As Figure 13.4 illustrates, these thermals can be used by glider pilots to gain considerable altitude. Birds such as eagles utilize thermals in a similar fashion.

It is usual for air temperature to decrease with increasing altitude, and the resulting upward convection currents are important for dispersing pollutants from industrial sources and automobile exhaust systems. Sometimes, however, meteorological conditions cause a layer to form in the atmosphere where the temperature increases with increasing altitude. Such a layer is called an *inversion layer* because its temperature profile is inverted compared to the usual situation. An inversion layer arrests the normal upward convection currents, causing a stagnant-air condition in which the concentration of pollutants increases substantially. This condition leads to a smog layer that can often be seen hovering over large cities.

We have been discussing *natural convection,* in which a temperature difference causes the density at one place in a fluid to be different from the density at another. Sometimes, natural convection is inadequate to transfer sufficient amounts of heat. In such cases *forced convection* is often used, and an external device such as a pump or a fan mixes the warmer and cooler portions of the fluid. Figure 13.5 shows an application of forced convection that is revolutionizing the way in which the effects of overheating are being treated. Athletes, for example, are especially prone to overheating, and the device illustrated in Figure 13.5 is appearing more and more frequently at athletic events. The technique is known as rapid thermal exchange and takes advantage of specialized blood vessels called arteriovenous anastomoses (AVAs) that are found in the palms of the hands (and soles of the feet). These blood vessels are used to help dissipate unwanted heat from the body. The device in the drawing consists of a small chamber containing a curved metal plate, through which cool water is circulated from a refrigerated supply. The overheated athlete inserts his hand into the chamber and places his palm on the plate. The chamber seals around the wrist and is evacuated slightly to reduce the air pressure and thereby promote circulation of blood through the hand. Forced convection plays two roles in this treatment. It causes the water to circulate through the metal plate and remove heat from the blood in the AVAs. Also, the cooled blood returns through veins to the heart, which pumps it throughout the body, thus lowering the body temperature and relieving the effects of overheating.

The physics of
"thermals."

The physics of
an inversion layer.

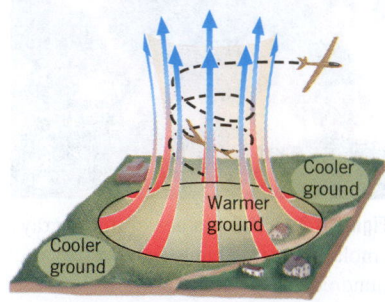

Figure 13.4 Updrafts, or thermals, are caused by the convective movement of air that the ground has warmed.

The physics of
rapid thermal exchange.

Figure 13.5 An overheated athlete uses a rapid-thermal-exchange device to cool down. He places the palm of his hand on a curved metal plate within a slightly evacuated chamber. Forced convection circulates cool water through the plate, which cools the blood flowing through the hand. The cooled blood returns through veins to the heart, which circulates it throughout the body.

Figure 13.6 shows the application of forced convection in an automobile engine. As in the previous application, forced convection occurs in two ways. First, a pump circulates radiator fluid (water and antifreeze) through the engine to remove excess heat from the combustion process. Second, a radiator fan draws air through the radiator. Heat is transferred from the hotter radiator fluid to the cooler air, thereby cooling the fluid.

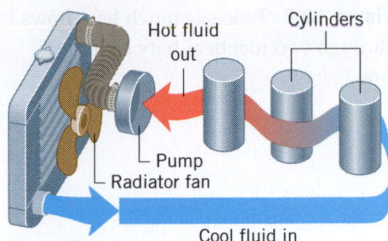

Figure 13.6 The forced convection generated by a pump circulates radiator fluid through an automobile engine to remove excess heat.

✓ **CHECK YOUR UNDERSTANDING**

(*The answer is given at the end of the book.*)

1. The transfer of heat by convection is *smallest* in   **(a)** solids,   **(b)** liquids,   **(c)** gases.

13.2 CONDUCTION

Anyone who has fried a hamburger in an all-metal skillet knows that the metal handle becomes hot. Somehow, heat is transferred from the burner to the handle. Clearly, heat is not being transferred by the bulk movement of the metal or the surrounding air, so convection can be ruled out. Instead, heat is transferred directly through the metal by a process called *conduction.*

CONDUCTION

Conduction is the process whereby heat is transferred directly through a material, with any bulk motion of the material playing no role in the transfer.

One mechanism for conduction occurs when the atoms or molecules in a hotter part of the material vibrate or move with greater energy than those in a cooler part. By means of collisions, the more energetic molecules pass on some of their energy to their less energetic neighbors. For example, imagine a gas filling the space between two walls that face each other and are maintained at different temperatures. Molecules strike the hotter wall, absorb energy from it, and rebound with a greater kinetic energy than when they arrived. As these more energetic molecules collide with their less energetic neighbors, they transfer some of their energy to them. Eventually, this energy is passed on until it reaches the molecules next to the cooler wall. These molecules, in turn, collide with the wall, giving up some of their energy to it in the process. Through such molecular collisions, heat is conducted from the hotter to the cooler wall.

A similar mechanism for the conduction of heat occurs in metals. Metals are different from most substances in having a pool of electrons that are more or less free to wander throughout the metal. These free electrons can transport energy and allow metals to transfer heat very well. The free electrons are also responsible for the excellent electrical conductivity that metals have.

Those materials that conduct heat well are called *thermal conductors,* and those that conduct heat poorly are known as *thermal insulators.* Most metals are excellent thermal conductors; wood, glass, and most plastics are common thermal insulators. Thermal insulators have many important applications. Virtually all new housing construction incorporates thermal insulation in attics and walls to reduce heating and cooling costs. And the wooden or plastic handles on many pots and pans reduce the flow of heat to the cook's hand.

To illustrate the factors that influence the conduction of heat, Figure 13.7 displays a rectangular bar. The ends of the bar are in thermal contact with two bodies, one of which

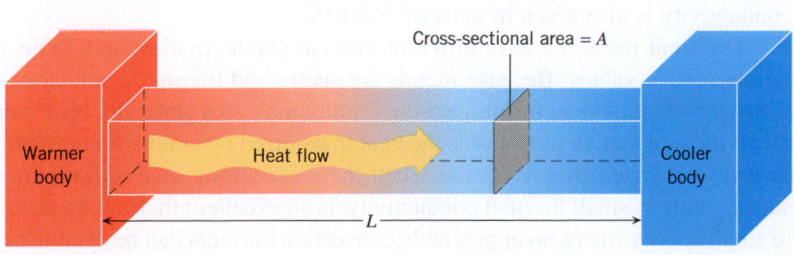

Figure 13.7 Heat is conducted through the bar when the ends of the bar are maintained at different temperatures. The heat flows from the warmer to the cooler end.

Figure 13.8 Twice as much heat flows through two identical bars as through one.

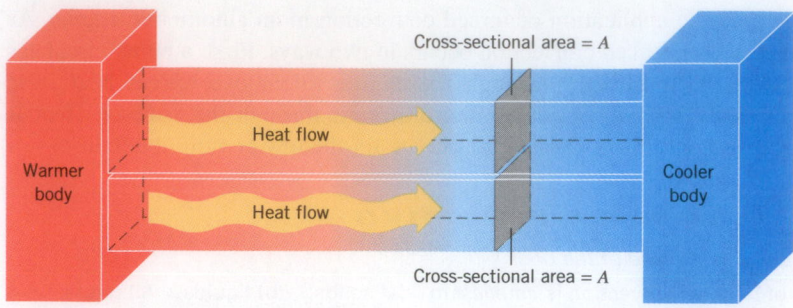

is kept at a constant higher temperature, while the other is kept at a constant lower temperature. Although not shown for the sake of clarity, the sides of the bar are insulated, so the heat lost through them is negligible. The amount of heat Q conducted through the bar from the warmer end to the cooler end depends on a number of factors:

1. Q is proportional to the time t during which conduction takes place ($Q \propto t$). More heat flows in longer time periods.

2. Q is proportional to the temperature difference ΔT between the ends of the bar ($Q \propto \Delta T$). A larger difference causes more heat to flow. No heat flows when both ends have the same temperature and $\Delta T = 0$ C°.

3. Q is proportional to the cross-sectional area A of the bar ($Q \propto A$). Figure 13.8 helps to explain this fact by showing two identical bars (insulated sides not shown) placed between the warmer and cooler bodies. Clearly, twice as much heat flows through two bars as through one, because the cross-sectional area has been doubled.

4. Q is inversely proportional to the length L of the bar ($Q \propto 1/L$). Greater lengths of material conduct less heat. To experience this effect, put two insulated mittens (the pot holders that cooks keep near the stove) on the *same hand*. Then, touch a hot pot and notice that it feels cooler than when you wear only one mitten, signifying that less heat passes through the greater thickness ("length") of material.

These proportionalities can be stated together as $Q \propto (A\ \Delta T)t/L$. Equation 13.1 expresses this result with the aid of a proportionality constant k, which is called the ***thermal conductivity.***

Table 13.1 Thermal Conductivities[a] of Selected Materials

Substance	Thermal Conductivity, k [J/(s·m·C°)]
Metals	
Aluminum	240
Brass	110
Copper	390
Iron	79
Lead	35
Silver	420
Steel (stainless)	14
Gases	
Air	0.0256
Hydrogen (H_2)	0.180
Nitrogen (N_2)	0.0258
Oxygen (O_2)	0.0265
Other Materials	
Asbestos	0.090
Body fat	0.20
Concrete	1.1
Diamond	2450
Glass	0.80
Goose down	0.025
Ice (0 °C)	2.2
Styrofoam	0.010
Water	0.60
Wood (oak)	0.15
Wool	0.040

[a]Except as noted, the values pertain to temperatures near 20 °C.

CONDUCTION OF HEAT THROUGH A MATERIAL

The heat Q conducted during a time t through a bar of length L and cross-sectional area A is

$$Q = \frac{(kA\ \Delta T)t}{L} \qquad (13.1)$$

where ΔT is the temperature difference between the ends of the bar (the higher temperature minus the lower temperature) and k is the thermal conductivity of the material.

SI Unit of Thermal Conductivity: J/(s·m·C°)

Since $k = QL/(tA\ \Delta T)$, the SI unit for thermal conductivity is J·m/(s·m²·C°) or J/(s·m·C°). The SI unit of power is the joule per second (J/s), or watt (W), so the thermal conductivity is also given in units of W/(m·C°).

Different materials have different thermal conductivities, and Table 13.1 gives some representative values. Because metals are such good thermal conductors, they have large thermal conductivities. In comparison, liquids and gases generally have small thermal conductivities. In fact, in most fluids the heat transferred by conduction is negligible compared to that transferred by convection when there are strong convection currents. Air, for instance, with its small thermal conductivity, is an excellent thermal insulator when confined to small spaces where no appreciable convection currents can be established. Goose down,

Styrofoam, and wool derive their fine insulating properties in part from the small dead-air spaces within them, as Figure 13.9 illustrates. **The physics of dressing warmly.** We also take advantage of dead-air spaces when we dress "in layers" during very cold weather and put on several layers of relatively thin clothing rather than one thick layer. The air trapped between the layers acts as an excellent insulator.

Example 2 deals with the role that conduction through body fat plays in regulating body temperature.

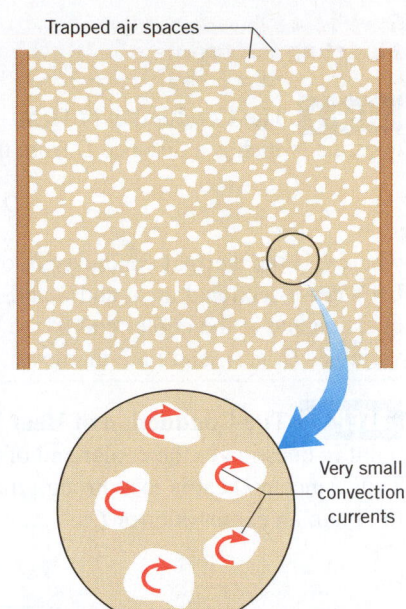

Trapped air spaces

Very small convection currents

Figure 13.9 Styrofoam is an excellent thermal insulator because it contains many small, dead-air spaces. These small spaces inhibit heat transfer by convection currents, and air itself has a very low thermal conductivity.

 Example 2 Heat Transfer in the Human Body

The physics of heat transfer in the human body. When excessive heat is produced within the body, it must be transferred to the skin and dispersed if the temperature at the body interior is to be maintained at the normal value of 37.0 °C. One possible mechanism for transfer is conduction through body fat. Suppose that heat travels through 0.030 m of fat in reaching the skin, which has a total surface area of 1.7 m² and a temperature of 34.0 °C. Find the amount of heat that reaches the skin in half an hour (1800 s).

Reasoning and Solution In Table 13.1 the thermal conductivity of body fat is given as $k = 0.20 \text{ J/(s·m·C°)}$. According to Equation 13.1,

$$Q = \frac{(kA \, \Delta T)t}{L}$$

$$Q = \frac{[0.20 \text{ J/(s·m·C°)}](1.7 \text{ m}^2)(37.0 \text{ °C} - 34.0 \text{ °C})(1800 \text{ s})}{0.030 \text{ m}} = \boxed{6.1 \times 10^4 \text{ J}}.$$

For comparison, a jogger can generate over ten times this amount of heat in a half hour. Thus, conduction through body fat is not a particularly effective way of removing excess heat. Heat transfer via blood flow to the skin is more effective and has the added advantage that the body can vary the blood flow as needed (see Problem 7).

Example 3 uses Equation 13.1 to determine what the temperature is at a point between the warmer and cooler ends of the bar in Figure 13.7.

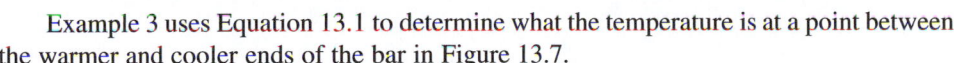

ANALYZING MULTIPLE-CONCEPT PROBLEMS

Example 3 The Temperature at a Point Between the Ends of a Bar

In Figure 13.7 the temperatures at the ends of the bar are 85.0 °C at the warmer end and 27.0 °C at the cooler end. The bar has a length of 0.680 m. What is the temperature at a point that is 0.220 m from the cooler end of the bar?

Reasoning The point in question is closer to the cooler end than to the warmer end of the bar. It might be expected, therefore, that the temperature at this point is less than halfway between 27.0 °C and 85.0 °C. We will demonstrate that this is, in fact, the case, by applying Equation 13.1. This expression applies because no heat escapes through the insulated sides of the bar, and we will use it twice to determine the desired temperature.

Knowns and Unknowns The available data are as follows:

Description	Symbol	Value
Temperature at warmer end	T_W	85.0 °C
Temperature at cooler end	T_C	27.0 °C
Length of bar	L	0.680 m
Distance from cooler end	D	0.220 m
Unknown Variable		
Temperature at distance D from cooler end	T	?

Continued

Modeling the Problem

STEP 1 **The Conduction of Heat** The heat Q conducted in a time t past the point in question (which is a distance D from the cooler end of the bar) is given by Equation 13.1 as

$$Q = \frac{kA(T - T_C)t}{D}$$

where k is the thermal conductivity of the material from which the bar is made, A is the bar's cross-sectional area, and T and T_C are, respectively, the temperature at the point in question and at the cooler end of the bar. Solving for T gives Equation 1 at the right. The variables Q, k, A, and t are unknown, so we proceed to Step 2 to deal with them.

$$T = T_C + \frac{QD}{kAt} \qquad (1)$$

STEP 2 **The Conduction of Heat Revisited** The heat Q that is conducted from the point in question to the cooler end of the bar originates at the warmer end of the bar. Thus, since no heat is lost through the sides, we may apply Equation 13.1 a second time to obtain an expression for Q:

$$Q = \frac{kA(T_W - T_C)t}{L}$$

where T_W and T_C are, respectively, the temperatures at the warmer and cooler ends of the bar, which has a length L. This expression for Q can be substituted into Equation 1, as indicated at the right. The terms k, A, and t remain to be dealt with. Fortunately, however, values for them are unnecessary, because they can be eliminated algebraically from the final calculation.

$$T = T_C + \frac{QD}{kAt} \qquad (1)$$

$$Q = \frac{kA(T_W - T_C)t}{L}$$

Solution Combining the results of each step algebraically, we find that

$$\overset{\text{STEP 1}}{} \qquad \overset{\text{STEP 2}}{}$$

$$T = T_C + \frac{QD}{kAt} = T_C + \frac{\left[\dfrac{kA(T_W - T_C)t}{L}\right]D}{kAt}$$

Simplifying this result gives

$$T = T_C + \frac{\dfrac{kA(T_W - T_C)t}{L}D}{kAt} = T_C + \frac{(T_W - T_C)D}{L}$$

$$= 27.0\,°C + \frac{(85.0\,°C - 27.0\,°C)(0.220\ m)}{0.680\ m} = \boxed{45.8\,°C}$$

As expected, this temperature is less than halfway between 27.0 °C and 85.0 °C.

Related Homework: *Problem 16*

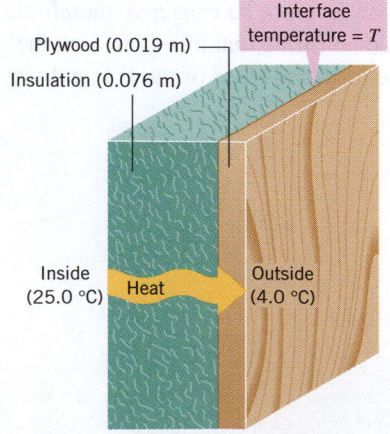

Figure 13.10 Heat flows through the insulation and plywood from the warmer inside to the cooler outside.

Virtually all homes contain insulation in the walls to reduce heat loss. Example 4 illustrates how to determine this loss with and without insulation.

Example 4 Layered Insulation

The physics of layered insulation. One wall of a house consists of 0.019-m-thick plywood backed by 0.076-m-thick insulation, as Figure 13.10 shows. The temperature at the inside surface is 25.0 °C, while the temperature at the outside surface is 4.0 °C, both being constant. The thermal conductivities of the insulation and the plywood are, respectively, 0.030 and 0.080 J/(s·m·C°), and the area of the wall is 35 m². Find the heat conducted through the wall in one hour **(a)** with the insulation and **(b)** without the insulation.

Reasoning The temperature T at the insulation–plywood interface (see Figure 13.10) must be determined before the heat conducted through the wall can be obtained. In calculating this temperature, we use the fact that no heat is accumulating in the wall because the inner and outer temperatures are constant. Therefore, the heat conducted through the insulation must equal the heat conducted through the plywood during the same time; that is, $Q_{\text{insulation}} = Q_{\text{plywood}}$. Each of the Q values can be expressed as $Q = (kA\,\Delta T)t/L$, according to Equation 13.1, leading to an expression that can be solved for the interface temperature. Once a value for T is available, Equation 13.1 can be used to obtain the heat conducted through the wall.

Problem-solving insight: When heat is conducted through a multi-layered material and the high and low temperatures are constant, the heat conducted through each layer is the same.

Solution (a) Using Equation 13.1 and the fact that $Q_{\text{insulation}} = Q_{\text{plywood}}$, we find that

$$\left[\frac{(kA\,\Delta T)t}{L}\right]_{\text{insulation}} = \left[\frac{(kA\,\Delta T)t}{L}\right]_{\text{plywood}}$$

$$\frac{[0.030\ \text{J/(s}\cdot\text{m}\cdot\text{C}°)]\,A(25.0\ °\text{C} - T)t}{0.076\ \text{m}} = \frac{[0.080\ \text{J/(s}\cdot\text{m}\cdot\text{C}°)]\,A(T - 4.0\ °\text{C})t}{0.019\ \text{m}}$$

Note that on each side of the equals sign we have written ΔT as the higher temperature minus the lower temperature. Eliminating the area A and time t algebraically and solving this equation for T reveals that the temperature at the insulation–plywood interface is $T = 5.8\ °\text{C}$.

The heat conducted through the wall is either $Q_{\text{insulation}}$ or Q_{plywood}, since the two quantities are equal. Choosing $Q_{\text{insulation}}$ and using $T = 5.8\ °\text{C}$ in Equation 13.1, we find that

$$Q_{\text{insulation}} = \frac{[0.030\ \text{J/(s}\cdot\text{m}\cdot\text{C}°)](35\ \text{m}^2)(25.0\ °\text{C} - 5.8\ °\text{C})(3600\ \text{s})}{0.076\ \text{m}} = \boxed{9.5 \times 10^5\ \text{J}}$$

(b) It is straightforward to use Equation 13.1 to calculate the amount of heat that would flow through the plywood in one hour if the insulation were absent:

$$Q_{\text{plywood}} = \frac{[0.080\ \text{J/(s}\cdot\text{m}\cdot\text{C}°)](35\ \text{m}^2)(25.0\ °\text{C} - 4.0\ °\text{C})(3600\ \text{s})}{0.019\ \text{m}} = \boxed{110 \times 10^5\ \text{J}}$$

Without insulation, the heat loss is increased by a factor of about 12.

Figure 13.11 During the night of January 25, 2005, temperatures in some areas of Florida dipped below freezing. Farmers sprayed strawberry plants with water to put a coat of ice on them and insulate them against the subfreezing temperatures. (Chris O'Meara/©AP/Wide World Photos)

The physics of protecting fruit plants from freezing. Fruit growers sometimes protect their crops by spraying them with water when overnight temperatures are expected to drop below freezing. Some fruit crops, like the strawberries in Figure 13.11, can withstand temperatures down to freezing (0 °C), but not below freezing. When water is sprayed on the plants, it can freeze and release heat (see Section 12.8), some of which goes into warming the plant. In addition, both water and ice have relatively small thermal conductivities, as Table 13.1 indicates. Thus, they also protect the crop by acting as thermal insulators that reduce heat loss from the plants.

Although a layer of ice may be beneficial to strawberries, it is not so desirable inside a refrigerator, as Conceptual Example 5 discusses.

 Conceptual Example 5 An Iced-up Refrigerator

In a refrigerator, heat is removed by a cold refrigerant fluid that circulates within a tubular space embedded inside a metal plate, as Figure 13.12 illustrates. A good refrigerator cools food as quickly as possible. Which arrangement works best: (a) an aluminum plate coated with ice, (b) an aluminum plate without ice, (c) a stainless steel plate coated with ice, or (d) a stainless steel plate without ice?

Reasoning Figure 13.12 (see the blow-ups) shows the metal cooling plate with and without a layer of ice. Without ice, heat passes by conduction through the metal plate to the refrigerant fluid within. For a given temperature difference across the thickness of the metal, the rate of heat transfer depends on the thermal conductivity of the metal. When the plate becomes coated with ice, any heat that is removed by the refrigerant fluid must first be transferred by conduction through the ice before it encounters the metal plate.

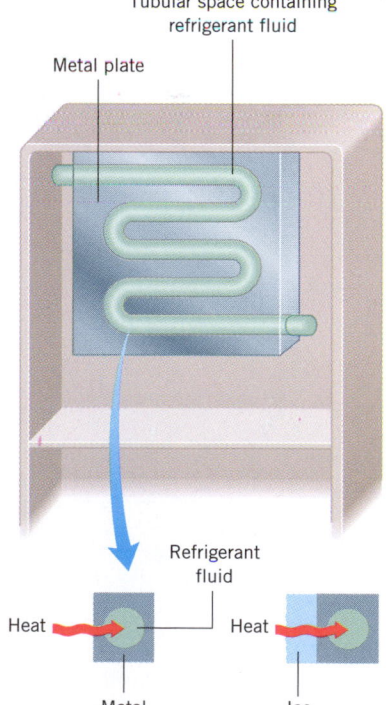

Figure 13.12 In a refrigerator, cooling is accomplished by a cold refrigerant fluid that circulates through a tubular space embedded within a metal plate. Sometimes the plate becomes coated with a layer of ice.

Answers (a), (c), and (d) are incorrect. For answers (a) and (c), the relation $Q = \dfrac{(kA\,\Delta T)t}{L}$ (Equation 13.1) indicates that the heat conducted per unit time (Q/t) is inversely proportional to the thickness L of the ice. As ice builds up, the heat removed per unit time by the cooling plate decreases. Thus, when covered with ice, the cooling plate—regardless of whether it's made from aluminum or stainless steel—does not work as well as a plate that is ice-free. Answer (d)—the stainless steel plate without ice—is incorrect, because heat is transferred more readily through a plate that has a greater thermal conductivity, and stainless steel has a smaller thermal conductivity than does aluminum (see Table 13.1).

Answer (b) is correct. The relation $Q = \dfrac{(kA\,\Delta T)t}{L}$ (Equation 13.1) shows that the heat conducted per unit time (Q/t) is directly proportional to the thermal conductivity k of the metal plate. Since the thermal conductivity of aluminum is more than 17 times greater than the thermal conductivity of stainless steel (see Table 13.1), aluminum is the preferred plate. The aluminum plate arrangement works best without an ice buildup. When ice builds up, the heat removed per unit time decreases because of the increased thickness of material through which the heat must pass.

✓ CHECK YOUR UNDERSTANDING

(The answers are given at the end of the book.)

2. A poker used in a fireplace is held at one end, while the other end is in the fire. In terms of being cooler to the touch, should a poker be made from **(a)** a high-thermal-conductivity material, **(b)** a low-thermal-conductivity material, or **(c)** can either type be used?

3. Several days after a snowstorm, the outdoor temperature remains below freezing. The roof on one house is uniformly covered with snow. On a neighboring house, however, the snow on the roof has completely melted. Which house is better insulated?

4. Concrete walls often contain steel reinforcement bars. Does the steel **(a)** enhance, **(b)** degrade, or **(c)** have no effect on the insulating value of the concrete wall? (Consult Table 13.1.)

5. To keep your hands as warm as possible during skiing, should you wear mittens or gloves? (Mittens, except for the thumb, do not have individual finger compartments.) Assume that the mittens and gloves are the same size and are made of the same material. You should wear: **(a)** gloves, because the individual finger compartments mean that the gloves have a smaller thermal conductivity; **(b)** gloves, because the individual finger compartments mean that the gloves have a larger thermal conductivity; **(c)** mittens, because they have less surface area exposed to the cold air.

6. A water pipe is buried slightly beneath the ground. The ground is covered with a thick layer of snow, which contains a lot of small dead-air spaces within it. The air temperature suddenly drops to well below freezing. The accumulation of snow **(a)** has no effect on whether the water in the pipe freezes, **(b)** causes the water in the pipe to freeze more quickly than if the snow were not there, **(c)** helps prevent the water in the pipe from freezing.

7. Some animals have hair strands that are hollow, air-filled tubes. Others have hair strands that are solid. Which kind, if either, would be more likely to give an animal an advantage for surviving in very cold climates?

8. Two bars are placed between plates whose temperatures are T_{hot} and T_{cold} (see the drawing). The thermal conductivity of bar 1 is six times that of bar 2 ($k_1 = 6k_2$), but bar 1 has only one-third the cross-sectional area ($A_1 = \frac{1}{3}A_2$). Ignore any heat loss through the sides of the bars. What can you conclude about the amounts of heat Q_1 and Q_2, respectively, that bar 1 and bar 2 conduct in a given amount of time? **(a)** $Q_1 = \frac{1}{4}Q_2$ **(b)** $Q_1 = \frac{1}{8}Q_2$ **(c)** $Q_1 = 2Q_2$ **(d)** $Q_1 = 4Q_2$ **(e)** $Q_1 = Q_2$

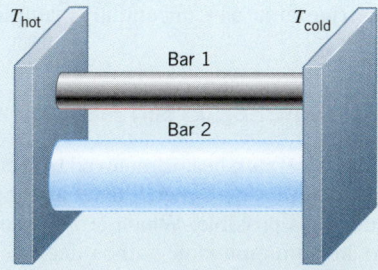

9. A piece of Styrofoam and a piece of wood are joined together to form a layered slab. The two pieces have the same thickness and cross-sectional area, but the Styrofoam has the smaller thermal conductivity. The temperature of the exposed Styrofoam surface is greater than the temperature of the exposed wood surface, both temperatures being constant. Is the temperature of the Styrofoam–wood interface **(a)** closer to the higher temperature of the exposed Styrofoam surface, **(b)** closer to the lower temperature of the exposed wood surface, or **(c)** halfway between the two temperatures?

RADIATION

Energy from the sun is brought to earth by large amounts of visible light waves, as well as by substantial amounts of infrared and ultraviolet waves. These waves are known as electromagnetic waves, a class that also includes the microwaves used for cooking and the radio waves used for AM and FM broadcasts. The sunbather in Figure 13.13 feels hot because her body absorbs energy from the sun's electromagnetic waves. Anyone who has stood by a roaring fire or put a hand near an incandescent light bulb has experienced a similar effect. Thus, fires and light bulbs also emit electromagnetic waves, and when the energy of such waves is absorbed, it can have the same effect as heat.

The process of transferring energy via electromagnetic waves is called *radiation,* and, unlike convection or conduction, it does not require a material medium. Electromagnetic waves from the sun, for example, travel through the void of space during their journey to earth.

Figure 13.13 Suntans are produced by ultraviolet rays. (Ron Chapple/Thinkstock/Alamy Images)

RADIATION

Radiation is the process in which energy is transferred by means of electromagnetic waves.

All bodies continuously radiate energy in the form of electromagnetic waves. Even an ice cube radiates energy, although so little of it is in the form of visible light that an ice cube cannot be seen in the dark. Likewise, the human body emits insufficient visible light to be seen in the dark. However, as Figures 12.6 and 12.7 illustrate, the infrared waves radiating from the body can be detected in the dark by electronic cameras. Generally, an object does not emit much visible light until the temperature of the object exceeds about 1000K. Then a characteristic red glow appears, like that of a heating coil on an electric stove. When its temperature reaches about 1700K, an object begins to glow white-hot, like the tungsten filament in an incandescent light bulb.

In the transfer of energy by radiation, the absorption of electromagnetic waves is just as important as their emission. The surface of an object plays a significant role in determining how much radiant energy the object will absorb or emit. The two blocks in sunlight in Figure 13.14, for example, are identical, except that one has a rough surface coated with lampblack (a fine black soot), while the other has a highly polished silver surface. As the thermometers indicate, the temperature of the black block rises at a much faster rate than that of the silvery block. This is because lampblack absorbs about 97% of the incident radiant energy, while the silvery surface absorbs only about 10%. The remaining part of the incident energy is reflected in each case. We see the lampblack as black in color because it reflects so little of the light falling on it, while the silvery surface looks like a mirror because it reflects so much light. Since the color black is associated with nearly complete absorption of visible light, the term *perfect blackbody* or, simply, *blackbody* is used when referring to an object that absorbs *all* the electromagnetic waves falling on it.

All objects emit and absorb electromagnetic waves simultaneously. When a body has the same constant temperature as its surroundings, the amount of radiant energy being absorbed must balance the amount being emitted in a given interval of time. The block coated with lampblack absorbs and emits the same amount of radiant energy, and the silvery block does too. In either case, if absorption were greater than emission, the block would experience a net gain in energy. As a result, the temperature of the block would rise and not be constant. Similarly, if emission were greater than absorption, the temperature would fall. Since absorption and emission are balanced, *a material that is a good absorber, like lampblack, is also a good emitter, and a material that is a poor absorber, like polished silver, is also a poor emitter.* A perfect blackbody, being a perfect absorber, is also a perfect emitter.

The fact that a black surface is both a good absorber and a good emitter is the reason people are uncomfortable wearing dark clothes during the summer. Dark clothes absorb a large fraction of the sun's radiation and then reemit it in all directions. About one-half of the emitted radiation is directed inward toward the body and creates the sensation of warmth. Light-colored clothes, in contrast, are cooler to wear, because they absorb and reemit relatively little of the incident radiation.

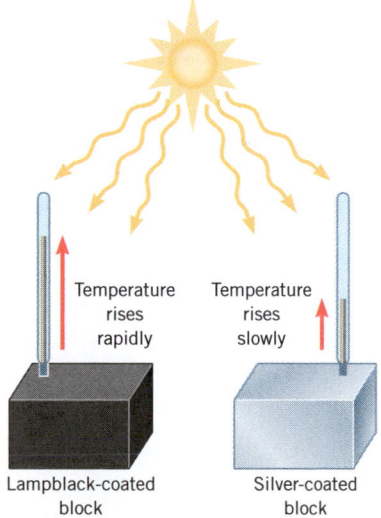

Temperature rises rapidly

Temperature rises slowly

Lampblack-coated block

Silver-coated block

Figure 13.14 The temperature of the block coated with lampblack rises faster than the temperature of the block coated with silver because the black surface absorbs radiant energy from the sun at the greater rate.

The physics of summer clothing.

The physics of
a white sifaka lemur warming up.

(a)

(b)

Figure 13.15 (a) Most lemurs, like this one, are nocturnal and have dark fur. (Wolfgang Kaehler/Corbis-Images) (b) The species of lemur called the white sifaka, however, is active during the day and has white fur. (Nigel Dennis/ Wildlife Pictures/Peter Arnold, Inc.)

The use of light colors for comfort also occurs in nature. Most lemurs, for instance, are nocturnal and have dark fur like the lemur shown in Figure 13.15a. Since they are active at night, the dark fur poses no disadvantage in absorbing excessive sunlight. Figure 13.15b shows a species of lemur called the white sifaka, which lives in semiarid regions where there is little shade. The white color of the fur may help in thermoregulation, by reflecting sunlight, but during the cool mornings, reflection of sunlight would hinder warming up. However, these lemurs have black skin and only sparse fur on their bellies, and to warm up in the morning, they turn their dark bellies toward the sun. The dark color enhances the absorption of sunlight.

The amount of radiant energy Q emitted by a perfect blackbody is proportional to the radiation time interval t ($Q \propto t$). The longer the time, the greater is the amount of energy radiated. Experiment shows that Q is also proportional to the surface area A ($Q \propto A$). An object with a large surface area radiates more energy than one with a small surface area, other things being equal. Finally, experiment reveals that Q is proportional to the *fourth power of the Kelvin temperature T* ($Q \propto T^4$), so the emitted energy increases markedly with increasing temperature. If, for example, the Kelvin temperature of an object doubles, the object emits 2^4 or 16 times more energy. Combining these factors into a single proportionality, we see that $Q \propto T^4 At$. This proportionality is converted into an equation by inserting a proportionality constant σ, known as the *Stefan–Boltzmann constant*. It has been found experimentally that $\sigma = 5.67 \times 10^{-8}$ J/(s·m²·K⁴):

$$Q = \sigma T^4 At$$

The relationship above holds only for a perfect emitter. Most objects are not perfect emitters, however. Suppose that an object radiates only about 80% of the visible light energy that a perfect emitter would radiate, so Q (for the object) $= (0.80)\sigma T^4 At$. The factor such as the 0.80 in this equation is called the *emissivity e* and is a dimensionless number between zero and one. The emissivity is the ratio of the energy an object actually radiates to the energy the object would radiate if it were a perfect emitter. For visible light, the value of e for the human body, for instance, varies between about 0.65 and 0.80, the smaller values pertaining to lighter skin colors. For infrared radiation, e is nearly one for all skin colors. For a perfect blackbody emitter, $e = 1$. Including the factor e on the right side of the expression $Q = \sigma T^4 At$ leads to the *Stefan–Boltzmann law of radiation*.

THE STEFAN–BOLTZMANN LAW OF RADIATION

The radiant energy Q, emitted in a time t by an object that has a Kelvin temperature T, a surface area A, and an emissivity e, is given by

$$Q = e\sigma T^4 At \qquad (13.2)$$

where $\sigma = 5.67 \times 10^{-8}$ J/(s·m²·K⁴) is the Stefan–Boltzmann constant.

In Equation 13.2, the Stefan–Boltzmann constant σ is a universal constant in the sense that its value is the same for all bodies, regardless of the nature of their surfaces. The emissivity e, however, depends on the condition of the surface.

Example 6 shows how the Stefan–Boltzmann law can be used to determine the size of a star.

 Example 6 A Supergiant Star

The supergiant star Betelgeuse has a surface temperature of about 2900 K (about one-half that of our sun) and emits a radiant power (in joules per second, or watts) of approximately 4×10^{30} W (about 10 000 times as great as that of our sun). Assuming that Betelgeuse is a perfect emitter (emissivity $e = 1$) and spherical, find its radius.

Reasoning According to the Stefan–Boltzmann law, the power emitted is $Q/t = e\sigma T^4 A$. A star with a relatively small temperature T can have a relatively large radiant power Q/t only if the area A is large. As we will see, Betelgeuse has a very large surface area, so its radius is enormous.

Solution Solving the Stefan–Boltzmann law for the area, we find

$$A = \frac{Q/t}{e\sigma T^4}$$

But the surface area of a sphere is $A = 4\pi r^2$, so $r = \sqrt{A/4\pi}$. Therefore, we have

$$r = \sqrt{\frac{Q/t}{4\pi e\sigma T^4}} = \sqrt{\frac{4 \times 10^{30}\text{ W}}{4\pi(1)[5.67 \times 10^{-8}\text{ J/(s}\cdot\text{m}^2\cdot\text{K}^4)](2900\text{ K})^4}} = \boxed{3 \times 10^{11}\text{ m}}$$

For comparison, Mars orbits the sun at a distance of 2.28×10^{11} m. Betelgeuse is certainly a "supergiant."

Problem-solving insight

First solve an equation for the unknown in terms of the known variables. Then substitute numbers for the known variables, as this example shows.

The next example explains how to apply the Stefan–Boltzmann law when an object, such as a wood stove, simultaneously emits and absorbs radiant energy.

Example 7   An Unused Wood-Burning Stove

An unused wood-burning stove has a constant temperature of 18 °C (291 K), which is also the temperature of the room in which the stove stands. The stove has an emissivity of 0.900 and a surface area of 3.50 m². What is the *net* radiant power generated by the stove?

Reasoning Power is the change in energy per unit time (Equation 6.10b), or Q/t, which, according to the Stefan–Boltzmann law, is $Q/t = e\sigma T^4 A$ (Equation 13.2). In this problem, however, we need to find the *net* power produced by the stove. The net power is the power the stove emits minus the power the stove absorbs. The power the stove absorbs comes from the walls, ceiling, and floor of the room, all of which emit radiation.

Solution Remembering that temperature must be expressed in kelvins when using the Stefan–Boltzmann law, we find that

$$\begin{array}{l}\text{Power emitted}\\\text{by unheated}\\\text{stove at 18 °C}\end{array} = \frac{Q}{t} = e\sigma T^4 A \qquad (13.2)$$

Problem-solving insight

In the Stefan–Boltzmann law of radiation, the temperature T must be expressed in kelvins, not in degrees Celsius or degrees Fahrenheit.

$$= (0.900)[5.67 \times 10^{-8}\text{ J/(s}\cdot\text{m}^2\cdot\text{K}^4)](291\text{ K})^4(3.50\text{ m}^2) = 1280\text{ W}$$

The fact that the unheated stove emits 1280 W of power and yet maintains a constant temperature means that the stove also absorbs 1280 W of radiant power from its surroundings. Thus, the *net* power generated by the unheated stove is zero:

$$\begin{array}{l}\text{Net power}\\\text{generated by}\\\text{stove at 18 °C}\end{array} = \underbrace{1280\text{ W}}_{\substack{\text{Power emitted}\\\text{by stove at}\\\text{18 °C}}} - \underbrace{1280\text{ W}}_{\substack{\text{Power emitted by}\\\text{room at 18 °C and}\\\text{absorbed by stove}}} = \boxed{0\text{ W}}$$

ANALYZING MULTIPLE-CONCEPT PROBLEMS

Example 8   A Heated Wood-Burning Stove

The physics of
a wood-burning stove.

The wood-burning stove in Example 7 (emissivity = 0.900 and surface area = 3.50 m²) is being used to heat a room. The fire keeps the stove surface at a constant 198 °C (471 K) and the room at a constant 29 °C (302 K). Determine the *net* radiant power generated by the stove.

Reasoning Power is the change in energy per unit time, and according to the Stefan–Boltzmann law, it is $Q/t = e\sigma T^4 A$ (Equation 13.2). As in Example 7, however, we seek the *net* power, which is the power the stove emits minus the power the stove absorbs from its environment. Since the stove has a higher temperature than its environment, the stove emits more power than it absorbs. Thus, we will find that the net radiant power emitted by the stove is no longer zero, as it is in Example 7. In fact, it is the net power that the stove radiates that has warmed the room to its temperature of 29 °C and sustains that temperature.

Continued

Knowns and Unknowns The following data are available:

Description	Symbol	Value	Comment
Emissivity of stove	e	0.900	
Surface area of stove	A	3.50 m²	
Temperature of stove surface	T	198 °C (471 K)	Temperature in kelvins must be used.
Temperature of room	T_0	29 °C (302 K)	Temperature in kelvins must be used.
Unknown Variable			
Net power generated by stove	P_{net}	?	

Modeling the Problem

STEP 1 Net Power The net power P_{net} generated by the stove is the power $P_{emitted}$ that the stove emits minus the power $P_{absorbed}$ that the stove absorbs from its environment, as expressed in Equation 1 at the right. In Steps 2 and 3 we evaluate the emitted and absorbed powers.

$$P_{net} = P_{emitted} - P_{absorbed} \quad (1)$$

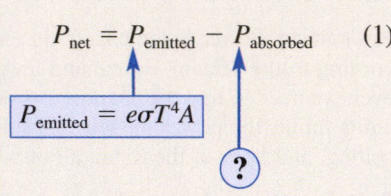

STEP 2 Emitted Power According to the Stefan–Boltzmann law (Equation 13.2), the heated stove (temperature T) emits a power that is

$$P_{emitted} = \frac{Q_{emitted}}{t} = e\sigma T^4 A$$

This expression can be substituted into Equation 1, as indicated at the right. In Step 3, we discuss the absorbed power.

$$P_{net} = P_{emitted} - P_{absorbed} \quad (1)$$

$$P_{emitted} = e\sigma T^4 A$$

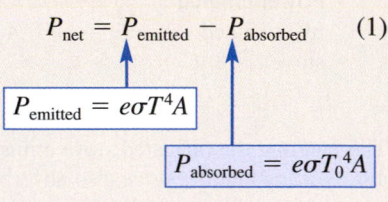

STEP 3 Absorbed Power The radiant power that the stove absorbs from the room is identical to the power that the stove would emit at the constant room temperature of 29 °C (302 K). The reasoning here is exactly like that in Example 7. With T_0 representing the temperature (in Kelvins) of the room, the Stefan–Boltzmann law indicates that

$$P_{absorbed} = \frac{Q_{absorbed}}{t} = e\sigma T_0^4 A$$

We can also substitute this result into Equation 1, as shown in the right column.

$$P_{net} = P_{emitted} - P_{absorbed} \quad (1)$$

$$P_{emitted} = e\sigma T^4 A$$

$$P_{absorbed} = e\sigma T_0^4 A$$

Solution The results of each step can be combined algebraically to show that

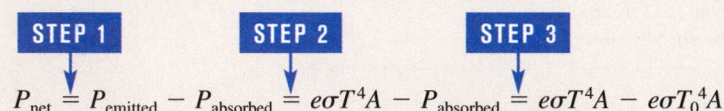

$$P_{net} \overset{\text{STEP 1}}{=} P_{emitted} - P_{absorbed} \overset{\text{STEP 2}}{=} e\sigma T^4 A - P_{absorbed} \overset{\text{STEP 3}}{=} e\sigma T^4 A - e\sigma T_0^4 A$$

Thus, the *net* radiant power the stove produces from the fuel it burns is

$$P_{net} = e\sigma A(T^4 - T_0^4)$$

$$= 0.900[5.67 \times 10^{-8} \text{ J/(s} \cdot \text{m}^2 \cdot \text{K}^4)](3.50 \text{ m}^2)[(471 \text{ K})^4 - (302 \text{ K})^4] = \boxed{7.30 \times 10^3 \text{ W}}$$

Related Homework: *Problems 25, 27, 30, 38*

Example 8 illustrates that when an object has a higher temperature than its surroundings, the object emits a net radiant power $P_{net} = (Q/t)_{net}$. The net power is the power the object emits minus the power it absorbs. Applying the Stefan–Boltzmann law as in Example 8 leads to the following expression for P_{net} when the temperature of the object is T and the temperature of the environment is T_0:

$$P_{net} = e\sigma A(T^4 - T_0^4) \quad (13.3)$$

✓ CHECK YOUR UNDERSTANDING

(The answers are given at the end of the book.)

10. One way that heat is transferred from place to place inside the human body is by the flow of blood. Which one of the following heat transfer processes—forced convection, conduction, or radiation—best describes this action of the blood?

11. Two strips of material, A and B, are identical, except that they have emissivities of 0.4 and 0.7, respectively. The strips are heated to the same temperature and have a red glow. A brighter glow signifies that more energy per second is being radiated. Which strip has the brighter glow?

12. One day during the winter the sun has been shining all day. Toward sunset a light snow begins to fall. It collects without melting on a cement playground, but melts instantly on contact with a black asphalt road adjacent to the playground. Why the difference? **(a)** Being black, asphalt has a higher emissivity than cement, so the asphalt absorbs more radiant energy from the sun during the day and, consequently, warms above the freezing point. **(b)** Being black, asphalt has a lower emissivity than cement, so it absorbs more radiant energy from the sun during the day and, consequently, warms above the freezing point.

13. If you were stranded in the mountains in cold weather, it would help to minimize energy losses from your body if you curled up into the tightest ball possible. Which factor in the relation $Q = e\sigma T^4 At$ (Equation 13.2) are you using to the best advantage by curling into a ball?   **(a)** e **(b)** σ **(c)** T **(d)** A **(e)** t

14. Two identical cubes have the same temperature. One of them, however, is cut in two and the pieces are separated (see the drawing). The radiant energy emitted by the cube cut into two pieces is $Q_{two\ pieces}$ and that emitted by the uncut cube is Q_{cube}. What is true about the radiant energy emitted in a given time?   **(a)** $Q_{two\ pieces} = 2Q_{cube}$ **(b)** $Q_{two\ pieces} = \frac{4}{3}Q_{cube}$ **(c)** $Q_{two\ pieces} = Q_{cube}$ **(d)** $Q_{two\ pieces} = \frac{1}{2}Q_{cube}$ **(e)** $Q_{two\ pieces} = \frac{1}{3}Q_{cube}$

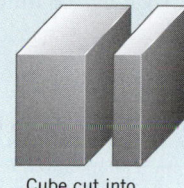
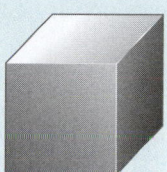

Cube cut into two pieces   Uncut cube

15. Two objects have the same size and shape. Object A has an emissivity of 0.3, and object B has an emissivity of 0.6. Each radiates the same power. How is the Kelvin temperature T_A of A related to the Kelvin temperature T_B of B?   **(a)** $T_A = T_B$ **(b)** $T_A = 2T_B$ **(c)** $T_A = \frac{1}{2}T_B$ **(d)** $T_A = \sqrt{2}\,T_B$ **(e)** $T_A = \sqrt[4]{2}\,T_B$

13.4 APPLICATIONS

To keep heating and air conditioning bills to a minimum, it pays to use good thermal insulation in your home. Insulation inhibits convection between inner and outer walls and minimizes heat transfer by conduction. With respect to conduction, the logic behind home insulation ratings comes directly from Equation 13.1. According to this equation, the heat per unit time Q/t flowing through a thickness of material is $Q/t = kA\,\Delta T/L$. Keeping the value for Q/t to a minimum means using materials that have small thermal conductivities k and large thicknesses L. Construction engineers, however, prefer to use Equation 13.1 in the slightly different form shown below:

$$\frac{Q}{t} = \frac{A\,\Delta T}{L/k}$$

The term L/k in the denominator is called the *R value*. An *R* value expresses in a single number the combined effects of thermal conductivity and thickness. Larger *R* values reduce the heat per unit time flowing through the material and, therefore, mean better insulation. It is also convenient to use *R* values to describe layered slabs formed by sandwiching together a number of materials with different thermal conductivities and different thicknesses. The *R* values for the individual layers can be added to give a single *R* value for the entire slab. It should be noted, however, that *R* values are expressed using units of feet, hours, F°, and BTU for thickness, time, temperature, and heat, respectively.

The physics of rating thermal insulation by *R* values.

When it is in the earth's shadow, an orbiting satellite is shielded from the intense electromagnetic waves emitted by the sun. But when it moves out of the earth's shadow, the satellite experiences the full effect of these waves. As a result, the temperature within a

The physics of regulating the temperature of an orbiting satellite.

Figure 13.16 Highly reflective metal foil covering this satellite (the Hubble Space Telescope) minimizes temperature changes. (Courtesy NASA)

satellite would decrease and increase sharply during an orbital period and sensitive electronic circuitry would suffer, unless precautions are taken. To minimize temperature fluctuations, satellites are often covered with a highly reflecting and, hence, poorly absorbing metal foil, as Figure 13.16 shows. By reflecting much of the sunlight, the foil minimizes temperature rises. Being a poor absorber, the foil is also a poor emitter and reduces radiant energy losses. Reducing these losses keeps the temperature from falling excessively.

The physics of a thermos bottle. A thermos bottle, sometimes referred to as a Dewar flask, reduces the rate at which hot liquids cool down or cold liquids warm up. A thermos usually consists of a double-walled glass vessel with silvered inner walls (see Figure 13.17) and minimizes heat transfer via convection, conduction, and radiation. The space between the walls is evacuated to reduce energy losses due to conduction and convection. The silvered surfaces reflect most of the radiant energy that would otherwise enter or leave the liquid in the thermos. Finally, little heat is lost through the glass or the rubberlike gaskets and stopper, because these materials have relatively small thermal conductivities.

The physics of a halogen cooktop stove. Halogen cooktops use radiant energy to heat pots and pans. A halogen cooktop uses several quartz–iodine lamps, like the ones in ultrabright automobile headlights. These lamps are electrically powered and mounted below a ceramic top. (See Figure 13.18.) The electromagnetic energy they radiate passes through the ceramic top and is absorbed directly by the bottom of the pot. Consequently, the pot heats up very quickly, rivaling the time of a pot on an open gas burner.

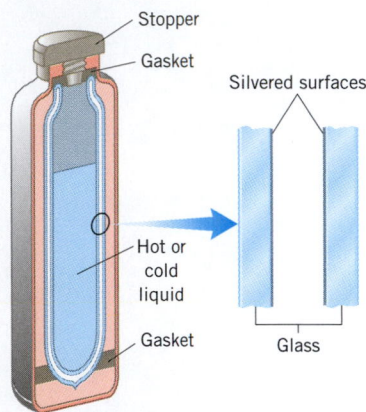

Figure 13.17 A thermos bottle minimizes energy transfer due to convection, conduction, and radiation.

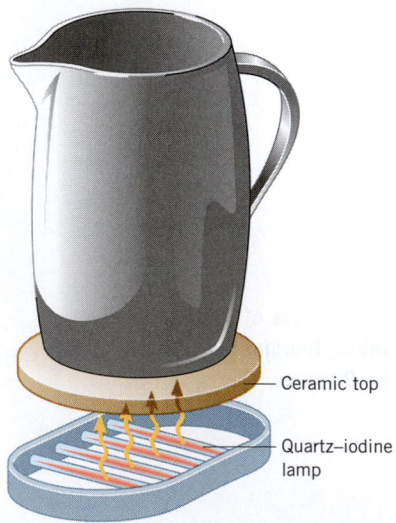

Figure 13.18 In a halogen cooktop, quartz–iodine lamps emit a large amount of electromagnetic energy that is absorbed directly by a pot or pan.

13.5 CONCEPTS & CALCULATIONS

Heat transfer by conduction is governed by Equation 13.1, as we have seen. The next example illustrates a familiar application of this relation in the kitchen. It also gives us the opportunity to review the idea of latent heat of vaporization, which Section 12.8 discusses.

 Concepts & Calculations Example 9 Boiling Water

Two pots are identical, except that in one case the flat bottom is aluminum and in the other it is copper. Each pot contains the same amount of boiling water and sits on a heating element that has a temperature of 155 °C. In the aluminum-bottom pot, the water boils away completely in 360 s. How long does it take the water in the copper-bottom pot to boil away completely?

Concept Questions and Answers Is the heat needed to boil away the water in the aluminum-bottom pot less than, greater than, or the same as the heat needed in the copper-bottom pot?

Answer The heat Q needed is the same in each case. When water boils, it changes from the liquid phase to the vapor phase. The heat that is required to make the water boil away is $Q = mL_v$, according to Equation 12.5, where m is the mass of the water and L_v is the latent heat of vaporization for water. Since the amount of water in each pot is the same, the mass of water is the same in each case. Moreover, the latent heat is a characteristic of water and, therefore, is also the same in each case.

One of the factors in Equation 13.1 that influences the amount of heat conducted through the bottom of each pot is the temperature difference ΔT between the upper and lower surfaces of the pot's bottom. Is this temperature difference for the aluminum-bottom pot less than, greater than, or the same as that for the copper-bottom pot?

Answer For each pot the temperature difference is the same. At the upper surface of each pot bottom the temperature is 100.0 °C, because water boils at 100.0 °C under normal conditions of atmospheric pressure. The temperature remains at 100.0 °C until all the water is gone. For each pot the temperature at the lower surface of the pot bottom is 155 °C, the temperature of the heating element. Therefore, $\Delta T = 155\ °C - 100.0\ °C = 55\ C°$ for each pot.

Is the time required to boil away the water completely in the copper-bottom pot less than, greater than, or the same as that required for the aluminum-bottom pot?

Answer The time is less for the copper-bottom pot. The factors that influence the amount of heat conducted in a given time are the thermal conductivity, the area of the bottom, the temperature difference across the bottom, and the thickness of the bottom. All of these factors are the same for each pot except for the thermal conductivity, which is greater for copper (Cu) than for aluminum (Al) (see Table 13.1). The greater the thermal conductivity, the greater the heat that is conducted in a given time, other things being equal. Therefore, less time is required to boil away the water using the copper-bottom pot.

Solution Applying Equation 13.1 to the conduction of heat into both pots and using Equation 12.5 to express the heat needed to boil away the water, we have

$$Q_{Al} = \frac{(k_{Al}A\,\Delta T)t_{Al}}{L} = mL_v$$

$$Q_{Cu} = \frac{(k_{Cu}A\,\Delta T)t_{Cu}}{L} = mL_v$$

In these two equations the area A, the temperature difference ΔT, the thickness L of the pot bottom, the mass m of the water, and the latent heat of vaporization of water L_v have the same values. Therefore, we can set the two heats equal and obtain

$$\frac{(k_{Al}A\,\Delta T)t_{Al}}{L} = \frac{(k_{Cu}A\,\Delta T)t_{Cu}}{L} \quad \text{or} \quad k_{Al}t_{Al} = k_{Cu}t_{Cu}$$

Solving for t_{Cu} and taking values for the thermal conductivities from Table 13.1, we find

$$t_{Cu} = \frac{k_{Al}t_{Al}}{k_{Cu}} = \frac{[240\ \text{J/(s}\cdot\text{m}\cdot\text{C}°)](360\ \text{s})}{390\ \text{J/(s}\cdot\text{m}\cdot\text{C}°)} = \boxed{220\ \text{s}}$$

As expected, the boil-away time is less for the copper-bottom pot.

Heat transfer by conduction is only one way in which heat gets from place to place. Heat transfer by radiation is another way, and it is governed by the Stefan–Boltzmann law of radiation, as Section 13.3 discusses. Example 10 deals with a case in which heat loss by radiation leads to freezing of water. It stresses the importance of the area from which the radiation occurs and also provides a review of the idea of latent heat of fusion, which Section 12.8 discusses.

 Concepts & Calculations Example 10 Freezing Water

One half of a kilogram of liquid water at 273 K (0 °C) is placed outside on a day when the temperature is 261 K (−12 °C). Assume that heat is lost from the water only by means of radiation and that the emissivity of the radiating surface is 0.60. How long does it take for the water to freeze into ice at 0 °C when the surface area from which the radiation occurs is **(a)** 0.035 m² (as it could be in a cup) and **(b)** 1.5 m² (as it could be if the water were spilled out to form a thin sheet)?

Concept Questions and Answers In case (a) is the heat that must be removed to freeze the water less than, greater than, or the same as in case (b)?

Answer The heat that must be removed is the same in both cases. When water freezes, it changes from the liquid phase to the solid phase. The heat that must be removed to make the water freeze is $Q = mL_f$, according to Equation 12.5, where m is the mass of the water and L_f is the latent heat of fusion for water. The mass is the same in both cases and so is L_f, since it is a characteristic of the water.

The loss of heat by radiation depends on the temperature of the radiating object. Does the temperature of the water change as the water freezes?

Answer No. The temperature of the water does not change as the freezing process takes place. The heat removed serves only to change the water from the liquid to the solid phase, as Section 12.8 discusses. Only after all the water has frozen does the temperature of the ice begin to fall below 0 °C.

The water both loses and gains heat by radiation. How, then, can heat transfer by radiation lead to freezing of the water?

Answer The water freezes because it loses more heat by radiation than it gains. The gain occurs because the environment radiates heat and the water absorbs it. However, the temperature

T_0 of the environment is less than the temperature T of the water. As a result, the environmental radiation cannot offset completely the loss of heat due to radiation from the water.

Will it take longer for the water to freeze in case (a) when the area is smaller or in case (b) when the area is larger?

Answer It will take longer when the area is smaller. This is because the amount of energy radiated in a given time is proportional to the area from which the radiation occurs. A smaller area means that less energy is radiated per second, so more time will be required to freeze the water by removing heat via radiation.

Solution We use Equation 13.3 to take into account that the water both gains and loses heat via radiation. This expression gives the net power lost, the net power being the net heat divided by the time. Thus, we have

$$\frac{Q}{t} = e\sigma A(T^4 - T_0^4) \quad \text{or} \quad t = \frac{Q}{e\sigma A(T^4 - T_0^4)}$$

Using Equation 12.5 to express the heat Q as $Q = mL_f$ and taking the latent heat of fusion for water from Table 12.3 ($L_f = 33.5 \times 10^4$ J/kg), we find

(a) Smaller area

$$t = \frac{mL_f}{e\sigma A(T^4 - T_0^4)}$$

$$= \frac{(0.50 \text{ kg})(33.5 \times 10^4 \text{ J/kg})}{0.60[5.67 \times 10^{-8} \text{ J/(s·m}^2\text{·K}^4)](0.035 \text{ m}^2)[(273 \text{ K})^4 - (261 \text{ K})^4]} = \boxed{1.5 \times 10^5 \text{ s } (42 \text{ h})}$$

(b) Larger area

$$t = \frac{mL_f}{e\sigma A(T^4 - T_0^4)}$$

$$= \frac{(0.50 \text{ kg})(33.5 \times 10^4 \text{ J/kg})}{0.60[5.67 \times 10^{-8} \text{ J/(s·m}^2\text{·K}^4)](1.5 \text{ m}^2)[(273 \text{ K})^4 - (261 \text{ K})^4]} = \boxed{3.6 \times 10^3 \text{ s } (1.0 \text{ h})}$$

As expected, the freezing time is longer when the area is smaller.

CONCEPT SUMMARY

If you need more help with a concept, use the Learning Aids noted next to the discussion or equation. Examples (**Ex.**) are in the text of this chapter. Go to **www.wiley.com/college/cutnell** for the following Learning Aids:

Interactive LearningWare (ILW) — Additional examples solved in a five-step interactive format.

Concept Simulations (CS) — Animated text figures or animations of important concepts.

Interactive Solutions (IS) — Models for certain types of problems in the chapter homework. The calculations are carried out interactively.

Topic	Discussion	Learning Aids
	13.1 CONVECTION Convection is the process in which heat is carried from place to place by the bulk movement of a fluid.	
Natural convection	During natural convection, the warmer, less dense part of a fluid is pushed upward by the buoyant force provided by the surrounding cooler and denser part.	**Ex. 1**
Forced convection	Forced convection occurs when an external device, such as a fan or a pump, causes the fluid to move.	
	13.2 CONDUCTION Conduction is the process whereby heat is transferred directly through a material, with any bulk motion of the material playing no role in the transfer.	
Thermal conductors and thermal insulators	Materials that conduct heat well, such as most metals, are known as thermal conductors. Materials that conduct heat poorly, such as wood, glass, and most plastics, are referred to as thermal insulators.	
	The heat Q conducted during a time t through a bar of length L and cross-sectional area A is	**Ex. 2, 3, 4, 5, 9**
Conduction of heat through a material	$$Q = \frac{(kA\,\Delta T)t}{L} \qquad (13.1)$$	**CS 13.1** **ILW 13.1** **IS 13.9, 13.17**
	where ΔT is the temperature difference between the ends of the bar and k is the thermal conductivity of the material.	

Topic	Discussion	Learning Aids
	13.3 RADIATION Radiation is the process in which energy is transferred by means of electromagnetic waves.	
Absorbers and emitters	All objects, regardless of their temperature, simultaneously absorb and emit electromagnetic waves. Objects that are good absorbers of radiant energy are also good emitters, and objects that are poor absorbers are also poor emitters.	
A perfect blackbody	An object that absorbs all the radiation incident upon it is called a perfect blackbody. A perfect blackbody, being a perfect absorber, is also a perfect emitter.	
	The radiant energy Q emitted during a time t by an object whose surface area is A and whose Kelvin temperature is T is given by the Stefan–Boltzmann law of radiation:	
Stefan–Boltzmann law of radiation	$$Q = e\sigma T^4 At \qquad (13.2)$$	**Ex. 6**
Emissivity	where $\sigma = 5.67 \times 10^{-8}$ J/(s·m²·K⁴) is the Stefan–Boltzmann constant and e is the emissivity, a dimensionless number characterizing the surface of the object. The emissivity lies between 0 and 1, being zero for a nonemitting surface and one for a perfect blackbody.	
	The net radiant power is the power an object emits minus the power it absorbs. The net radiant power P_{net} emitted by an object of temperature T located in an environment of temperature T_0 is	**Ex. 7, 8, 10**
Net radiant power	$$P_{\text{net}} = e\sigma A(T^4 - T_0^4) \qquad (13.3)$$	**ILW 13.2**

FOCUS ON CONCEPTS

Note to Instructors: *The numbering of the questions shown here reflects the fact that they are only a representative subset of the total number that are available online. However, all of the questions are available for assignment via an online homework management program such as* WileyPLUS *or* WebAssign.

Section 13.2 Conduction

1. The heat conducted through a bar depends on which of the following?

  A. The coefficient of linear expansion

  B. The thermal conductivity

  C. The specific heat capacity

  D. The length of the bar

  E. The cross-sectional area of the bar

(a) A, B, and D **(b)** A, C, and D **(c)** B, C, D, and E **(d)** B, D, and E **(e)** C, D, and E

2. Two bars are conducting heat from a region of higher temperature to a region of lower temperature. The bars have identical lengths and cross-sectional areas, but are made from different materials. In the drawing they are placed "in parallel" between the two temperature regions in arrangement A, whereas they are placed end to end in arrangement B. In which arrangement is the heat that is conducted the greatest? **(a)** The heat conducted is the same in both arrangements. **(b)** Arrangement A **(c)** Arrangement B **(d)** It is not possible to determine which arrangement conducts more heat.

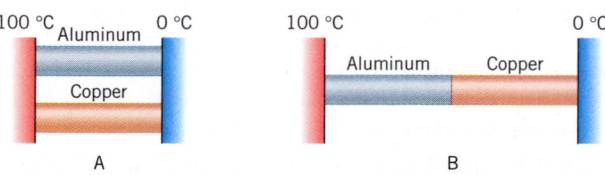

4. The drawing shows a composite slab consisting of three materials through which heat is conducted from left to right. The materials have identical thicknesses and cross-sectional areas. Rank the materials according to their thermal conductivities, largest first.
(a) k_1, k_2, k_3 **(b)** k_1, k_3, k_2 **(c)** k_2, k_1, k_3 **(d)** k_2, k_3, k_1 **(e)** k_3, k_2, k_1

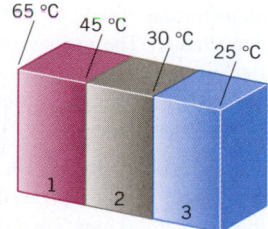

6. The long single bar on the left in the drawing has a thermal conductivity of 240 J/(s·m·C°). The ends of the bar are at temperatures of 400 and 200 °C, and the temperature of its midpoint is halfway between these two temperatures, or 300 °C. The two bars on the right are half as long as the bar on the left, and the thermal conductivities of these bars are different (see the drawing). All of the bars have the same cross-sectional area. What can be said about the temperature at the point where the two bars on the right are joined together?
(a) The temperature at the point where the two bars are joined together is 300 °C. **(b)** The temperature at the point where the two bars are joined together is greater than 300 °C. **(c)** The temperature at the point where the two bars are joined together is less than 300 °C.

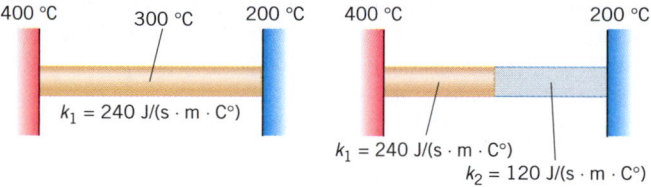

Section 13.3 Radiation

8. Three cubes are made from the same material. As the drawing indicates, they have different sizes and temperatures. Rank the cubes according to the radiant energy they emit per second, largest first.
(a) A, B, C **(b)** A, C, B **(c)** B, A, C **(d)** B, C, A **(e)** C, B, A

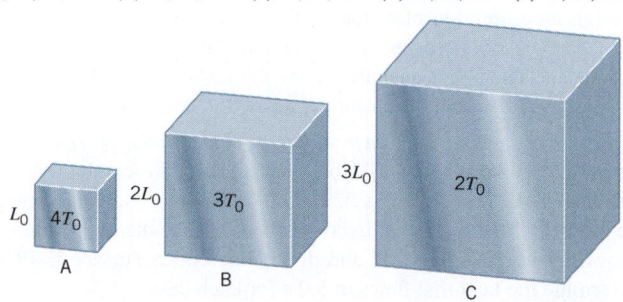

10. An astronaut in the space shuttle has two objects that are identical in all respects, except that one is painted black and the other is painted silver. Initially, they are at the same temperature. When taken from inside the space shuttle and placed in outer space, which object, if either, cools down at a faster rate?   **(a)** The object painted black   **(b)** The object painted silver   **(c)** Both objects cool down at the same rate.   **(d)** It is not possible to determine which object cools down at the faster rate.

11. The emissivity e of object B is $\frac{1}{16}$ that of object A, although both objects are identical in size and shape. If the objects radiate the same energy per second, what is the ratio T_B/T_A of their Kelvin temperatures?   **(a)** $\frac{1}{16}$ **(b)** $\frac{1}{4}$ **(c)** $\frac{1}{2}$ **(d)** 2   **(e)** 4

PROBLEMS

Note to Instructors: Most of the homework problems in this chapter are available for assignment via an online homework management program such as WileyPLUS or WebAssign, and those marked with the icon **GO** *are presented in WileyPLUS using a guided tutorial format that provides enhanced interactivity. See Preface for additional details.*

Note: For problems in this set, use the values for thermal conductivities given in Table 13.1 unless stated otherwise.

ssm   Solution is in the Student Solutions Manual.
www   Solution is available online at www.wiley.com/college/cutnell

This icon represents a biomedical application.

Section 13.2   Conduction

1. The amount of heat per second conducted from the blood capillaries beneath the skin to the surface is 240 J/s. The energy is transferred a distance of 2.0×10^{-3} m through a body whose surface area is 1.6 m². Assuming that the thermal conductivity is that of body fat, determine the temperature difference between the capillaries and the surface of the skin.

2. The temperature in an electric oven is 160 °C. The temperature at the outer surface in the kitchen is 50 °C. The oven (surface area = 1.6 m²) is insulated with material that has a thickness of 0.020 m and a thermal conductivity of 0.045 J/(s·m·C°).   **(a)** How much energy is used to operate the oven for six hours?   **(b)** At a price of $0.10 per kilowatt·hour for electrical energy, what is the cost of operating the oven?

3. ssm **Concept Simulation 13.1** at www.wiley.com/college/cutnell illustrates the concepts pertinent to this problem. A person's body is covered with 1.6 m² of wool clothing. The thickness of the wool is 2.0×10^{-3} m. The temperature at the outside surface of the wool is 11 °C, and the skin temperature is 36 °C. How much heat per second does the person lose due to conduction?

4. **GO** Two objects are maintained at constant temperatures, one hot and one cold. Two identical bars can be attached end to end, as in part a of the drawing, or one on top of the other, as in part b. When either of these arrangements is placed between the hot and the cold objects for the same amount of time, heat Q flows from left to right. Find the ratio Q_a/Q_b.

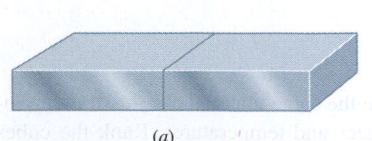

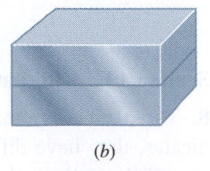

(a)   (b)

5. ssm Due to a temperature difference ΔT, heat is conducted through an aluminum plate that is 0.035 m thick. The plate is then replaced by a stainless steel plate that has the same temperature difference and cross-sectional area. How thick should the steel plate be so that the same amount of heat per second is conducted through it?

6. **GO** The block in the drawing has dimensions $L_0 \times 2L_0 \times 3L_0$, where $L_0 = 0.30$ m. The block has a thermal conductivity of 250 J/(s·m·C°). In drawings A, B, and C, heat is conducted through the block in three different directions; in each case the temperature of the warmer surface is 35 °C and that of the cooler surface is 19 °C. Determine the heat that flows in 5.0 s for each case.

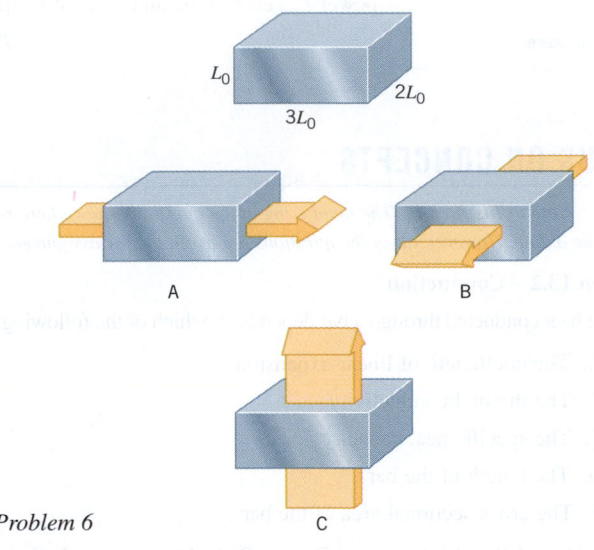

Problem 6

7. **ssm** **www** In the conduction equation $Q = (kA\ \Delta T)t/L$, the combination of factors kA/L is called the *conductance*. The human body has the ability to vary the conductance of the tissue beneath the skin by means of vasoconstriction and vasodilation, in which the flow of blood to the veins and capillaries underlying the skin is decreased and increased, respectively. The conductance can be adjusted over a range such that the tissue beneath the skin is equivalent to a thickness of 0.080 mm of Styrofoam or 3.5 mm of air. By what factor can the body adjust the conductance?

8. A copper pipe with an outer radius of 0.013 m runs from an outdoor wall faucet into the interior of a house. The temperature of the faucet is 4.0 °C, and the temperature of the pipe, at 3.0 m from the faucet, is 25 °C. In fifteen minutes, the pipe conducts a total of 270 J of heat to the outdoor faucet from the house interior. Find the inner radius of the pipe. Ignore any water inside the pipe.

9. Consult **Interactive Solution 13.9** at www.wiley.com/college/cutnell to explore a model for solving this problem. One end of a brass bar is maintained at 306 °C, while the other end is kept at a constant, but lower, temperature. The cross-sectional area of the bar is 2.6×10^{-4} m². Because of insulation, there is negligible heat loss through the sides of the bar. Heat flows through the bar, however, at the rate of 3.6 J/s. What is the temperature of the bar at a point 0.15 m from the hot end?

10. **GO** A wall in a house contains a single window. The window consists of a single pane of glass whose area is 0.16 m² and whose thickness is 2.0 mm. Treat the wall as a slab of the insulating material

Styrofoam whose area and thickness are 18 m² and 0.10 m, respectively. Heat is lost via conduction through the wall and the window. The temperature difference between the inside and outside is the same for the wall and the window. Of the total heat lost by the wall and the window, what is the percentage lost by the window?

11. Interactive LearningWare 13.1 at www.wiley.com/college/cutnell explores the approach taken in problems such as this one. A composite rod is made from stainless steel and iron and has a length of 0.50 m. The cross section of this composite rod is shown in the drawing and consists of a square within a circle. The square cross section of the steel is 1.0 cm on a side. The temperature at one end of the rod is 78 °C, while it is 18 °C at the other end. Assuming that no heat exits through the cylindrical outer surface, find the total amount of heat conducted through the rod in two minutes.

Iron

Stainless steel

*12. A cubical piece of heat-shield tile from the space shuttle measures 0.10 m on a side and has a thermal conductivity of 0.065 J/(s·m·C°). The outer surface of the tile is heated to a temperature of 1150 °C, while the inner surface is maintained at a temperature of 20.0 °C. **(a)** How much heat flows from the outer to the inner surface of the tile in five minutes?   **(b)** If this amount of heat were transferred to two liters (2.0 kg) of liquid water, by how many Celsius degrees would the temperature of the water rise?

*13. **ssm** Three building materials, plasterboard [$k = 0.30$ J/(s·m·C°)], brick [$k = 0.60$ J/(s·m·C°)], and wood [$k = 0.10$ J/(s·m·C°)], are sandwiched together as the drawing illustrates. The temperatures at the inside and outside surfaces are 27 °C and 0 °C, respectively. Each material has the same thickness and cross-sectional area. Find the temperature **(a)** at the plasterboard–brick interface and **(b)** at the brick–wood interface.

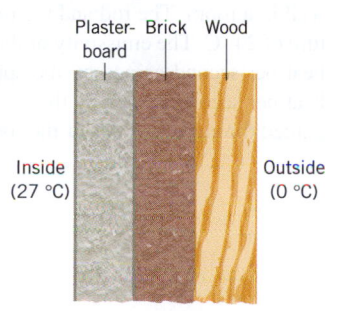

Plaster-board   Brick   Wood

Inside (27 °C)

Outside (0 °C)

*14. **GO** A copper rod has a length of 1.5 m and a cross-sectional area of 4.0×10^{-4} m². One end of the rod is in contact with boiling water and the other with a mixture of ice and water. What is the mass of ice per second that melts? Assume that no heat is lost through the side surface of the rod.

*15. **GO** A pot of water is boiling under one atmosphere of pressure. Assume that heat enters the pot only through its bottom, which is copper and rests on a heating element. In two minutes, the mass of water boiled away is $m = 0.45$ kg. The radius of the pot bottom is $R = 6.5$ cm, and the thickness is $L = 2.0$ mm. What is the temperature T_E of the heating element in contact with the pot?

*16. Multiple-Concept Example 3 discusses an approach to problems such as this. The ends of a thin bar are maintained at different temperatures. The temperature of the cooler end is 11 °C, while the temperature at a point 0.13 m from the cooler end is 23 °C and the temperature of the warmer end is 48 °C. Assuming that heat flows only along the length of the bar (the sides are insulated), find the length of the bar.

*17. Refer to **Interactive Solution 13.17** at www.wiley.com/college/cutnell for help in solving this problem. In an aluminum pot, 0.15 kg of water at 100 °C boils away in four minutes. The bottom of the pot is 3.1×10^{-3} m thick and has a surface area of 0.015 m². To prevent the water from boiling too rapidly, a stainless steel plate has been placed between the pot and the heating element. The plate is 1.4×10^{-3} m

thick, and its area matches that of the pot. Assuming that heat is conducted into the water only through the bottom of the pot, find the temperature at **(a)** the aluminum–steel interface and **(b)** the steel surface in contact with the heating element.

**18. The drawing shows a solid cylindrical rod made from a center cylinder of lead and an outer concentric jacket of copper. Except for its ends, the rod is insulated (not shown), so that the loss of heat from the curved surface is negligible. When a temperature difference is maintained between its ends, this rod conducts one-half the amount of heat that it would conduct if it were solid copper. Determine the ratio of the radii r_1/r_2.

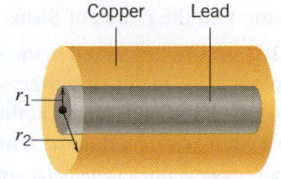

Copper   Lead

r_1

r_2

19. **ssm www Two cylindrical rods have the same mass. One is made of silver (density = 10 500 kg/m³), and one is made of iron (density = 7860 kg/m³). Both rods conduct the same amount of heat per second when the same temperature difference is maintained across their ends. What is the ratio (silver-to-iron) of **(a)** the lengths and **(b)** the radii of these rods?

Section 13.3   Radiation

20. **GO** Light bulb 1 operates with a filament temperature of 2700 K, whereas light bulb 2 has a filament temperature of 2100 K. Both filaments have the same emissivity, and both bulbs radiate the same power. Find the ratio A_1/A_2 of the filament areas of the bulbs.

21. **ssm www** How many days does it take for a perfect blackbody cube (0.0100 m on a side, 30.0 °C) to radiate the same amount of energy that a one-hundred-watt light bulb uses in one hour?

22. In an old house, the heating system uses radiators, which are hollow metal devices through which hot water or steam circulates. In one room the radiator has a dark color (emissivity = 0.75). It has a temperature of 62 °C. The new owner of the house paints the radiator a lighter color (emissivity = 0.50). Assuming that it emits the same radiant power as it did before being painted, what is the temperature (in degrees Celsius) of the newly painted radiator?

23. A person is standing outdoors in the shade where the temperature is 28 °C.   **(a)** What is the radiant energy absorbed per second by his head when it is covered with hair? The surface area of the hair (assumed to be flat) is 160 cm² and its emissivity is 0.85.   **(b)** What would be the radiant energy absorbed per second by the same person if he were bald and the emissivity of his head were 0.65?

24. A baking dish is removed from a hot oven and placed on a cooling rack. As the dish cools down to 35 °C from 175 °C, its net radiant power decreases to 12.0 W. What was the net radiant power of the baking dish when it was first removed from the oven? Assume that the temperature in the kitchen remains at 22 °C as the dish cools down.

25. Multiple-Concept Example 8 reviews the approach that is used in problems such as this. A person eats a dessert that contains 260 Calories. (This "Calorie" unit, with a capital C, is the one used by nutritionists; 1 Calorie = 4186 J. See Section 12.7.) The skin temperature of this individual is 36 °C and that of her environment is 21 °C. The emissivity of her skin is 0.75 and its surface area is 1.3 m². How much time would it take for her to emit a *net* radiant energy from her body that is equal to the energy contained in this dessert?

26. **GO** An object is inside a room that has a constant temperature of 293 K. Via radiation, the object emits three times as much power as it absorbs from the room. What is the temperature (in kelvins) of the object? Assume that the temperature of the object remains constant.

27. ssm Review Multiple-Concept Example 8 before attempting this problem. Suppose the stove in that example had a surface area of only 2.00 m². What would its temperature (in kelvins) have to be so that it still generated a net power of 7300 W?

28. Sirius B is a white star that has a surface temperature (in kelvins) that is four times that of our sun. Sirius B radiates only 0.040 times the power radiated by the sun. Our sun has a radius of 6.96×10^8 m. Assuming that Sirius B has the same emissivity as the sun, find the radius of Sirius B.

29. ssm A car parked in the sun absorbs energy at a rate of 560 watts per square meter of surface area. The car reaches a temperature at which it radiates energy at this same rate. Treating the car as a perfect radiator ($e = 1$), find the temperature.

30. Multiple-Concept Example 8 discusses the ideas on which this problem depends. **Interactive LearningWare 13.2** at **www.wiley.com/college/cutnell** reviews the concepts that are involved in this problem. Suppose the skin temperature of a naked person is 34 °C when the person is standing inside a room whose temperature is 25 °C. The skin area of the individual is 1.5 m². **(a)** Assuming the emissivity is 0.80, find the net loss of radiant power from the body. **(b)** Determine the number of food Calories of energy (1 food Calorie = 4186 J) that are lost in one hour due to the net loss rate obtained in part (a). Metabolic conversion of food into energy replaces this loss.

*31. Liquid helium is stored at its boiling-point temperature of 4.2 K in a spherical container ($r = 0.30$ m). The container is a perfect blackbody radiator. The container is surrounded by a spherical shield whose temperature is 77 K. A vacuum exists in the space between the container and the shield. The latent heat of vaporization for helium is 2.1×10^4 J/kg. What mass of liquid helium boils away through a venting valve in one hour?

*32. Part (a) of the drawing shows a rectangular bar whose dimensions are $L_0 \times 2L_0 \times 3L_0$. The bar is at the same constant temperature as the room (not shown) in which it is located. The bar is then cut, lengthwise, into two identical pieces, as shown in part (b) of the drawing. The temperature of each piece is the same as that of the original bar. **(a)** What is the ratio of the power absorbed by the two bars in part (b) of the drawing to the single bar in part (a)? **(b)** Suppose that the temperature of the single bar in part (a) is 450.0 K. What would the temperature (in kelvins) of the room and the two bars in part (b) have to be so that the two bars absorb the same power as the single bar in part (a)?

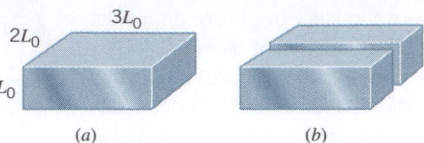

(a)   (b)

**33. ssm A solid cylinder is radiating power. It has a length that is ten times its radius. It is cut into a number of smaller cylinders, each of which has the same length. Each small cylinder has the same temperature as the original cylinder. The total radiant power emitted by the pieces is twice that emitted by the original cylinder. How many smaller cylinders are there?

**34. One end of a 0.25-m copper rod with a cross-sectional area of 1.2×10^{-4} m² is driven into the center of a sphere of ice at 0 °C (radius = 0.15 m). The portion of the rod that is embedded in the ice is also at 0 °C. The rod is horizontal and its other end is fastened to a wall in a room. The rod and the room are kept at a constant temperature of 24 °C. The emissivity of the ice is 0.90. What is the ratio of the heat per second gained by the sphere through conduction to the net heat per second gained by the ice due to radiation? Neglect any heat gained through the sides of the rod.

ADDITIONAL PROBLEMS

35. ssm One end of an iron poker is placed in a fire where the temperature is 502 °C, and the other end is kept at a temperature of 26 °C. The poker is 1.2 m long and has a radius of 5.0×10^{-3} m. Ignoring the heat lost along the length of the poker, find the amount of heat conducted from one end of the poker to the other in 5.0 s.

36. Concept Simulation 13.1 at **www.wiley.com/college/cutnell** illustrates the concepts pertinent to this problem. A refrigerator has a surface area of 5.3 m². It is lined with 0.075-m-thick insulation whose thermal conductivity is 0.030 J/(s·m·C°). The interior temperature is kept at 5 °C, while the temperature at the outside surface is 25 °C. How much heat per second is being removed from the unit?

37. The amount of radiant power produced by the sun is approximately 3.9×10^{26} W. Assuming the sun to be a perfect blackbody sphere with a radius of 6.96×10^8 m, find its surface temperature (in kelvins).

38. Consult Multiple-Concept Example 8 to see the concepts that are pertinent here. A person's body is producing energy internally due to metabolic processes. If the body loses more energy than metabolic processes are generating, its temperature will drop. If the drop is severe, it can be life-threatening. Suppose that a person is unclothed and energy is being lost via radiation from a body surface area of 1.40 m², which has a temperature of 34 °C and an emissivity

of 0.700. Also suppose that metabolic processes are producing energy at a rate of 115 J/s. What is the temperature of the coldest room in which this person could stand and not experience a drop in body temperature?

39. The concrete wall of a building is 0.10 m thick. The temperature inside the building is 20.0 °C, while the temperature outside is 0.0 °C. Heat is conducted through the wall. When the building is unheated, the inside temperature falls to 0.0 °C, and heat conduction ceases. However, the wall does emit radiant energy when its temperature is 0.0 °C. The radiant energy emitted per second per square meter is the same as the heat lost per second per square meter due to conduction. What is the emissivity of the wall?

*40. A solid sphere has a temperature of 773 K. The sphere is melted down and recast into a cube that has the same emissivity and emits the same radiant power as the sphere. What is the cube's temperature?

*41. In a house the temperature at the surface of a window is 25 °C. The temperature outside at the window surface is 5.0 °C. Heat is lost through the window via conduction, and the heat lost per second has a certain value. The temperature outside begins to fall, while the conditions inside the house remain the same. As a result, the heat lost per second increases. What is the temperature at the outside window surface when the heat lost per second doubles?

*42. **GO** Two pots are identical except that the flat bottom of one is aluminum, whereas that of the other is copper. Water in these pots is boiling away at 100.0 °C at the same rate. The temperature of the heating element on which the aluminum bottom is sitting is 155.0 °C. Assume that heat enters the water only through the bottoms of the pots and find the temperature of the heating element on which the copper bottom rests.

43. ssm www Two cylindrical rods are identical, except that one has a thermal conductivity k_1 and the other has a thermal conductivity k_2. As the drawing shows, they are placed between two walls that are maintained at different temperatures T_W (warmer) and T_C (cooler). When the rods are arranged as in part a of the drawing, a total heat Q' flows from the warmer to the cooler wall, but when the rods are arranged as in part b, the total heat flow is Q. Assuming that the conductivity k_2 is twice as great as k_1 and that heat flows only along the lengths of the rods, determine the ratio Q'/Q.

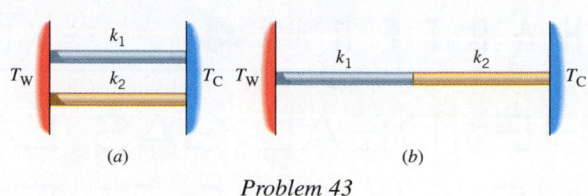

Problem 43

**44. A small sphere (emissivity = 0.90, radius = r_1) is located at the center of a spherical asbestos shell (thickness = 1.0 cm, outer radius = r_2). The thickness of the shell is small compared to the inner and outer radii of the shell. The temperature of the small sphere is 800.0 °C, while the temperature of the inner surface of the shell is 600.0 °C, both temperatures remaining constant. Assuming that $r_2/r_1 = 10.0$ and ignoring any air inside the shell, find the temperature of the outer surface of the shell.

CHAPTER 14

THE IDEAL GAS LAW AND KINETIC THEORY

To the extent that the air in these hot-air balloons behaves like an ideal gas, its pressure, volume, and temperature are related by the ideal gas law, which is one of the central topics of this chapter. (© Christophe Karaba/Reuters/Corbis)

14.1 MOLECULAR MASS, THE MOLE, AND AVOGADRO'S NUMBER

Often, we wish to compare the mass of one atom with another. To facilitate the comparison, a mass scale known as the *atomic mass scale* has been established. To set up this scale, a reference value (along with a unit) is chosen for one of the elements. The unit is called the *atomic mass unit* (symbol: u). By international agreement, the reference element is chosen to be the most abundant type or isotope* of carbon, which is called carbon-12. Its atomic mass† is defined to be exactly twelve atomic mass units, or 12 u. The relationship between the atomic mass unit and the kilogram is

$$1 \text{ u} = 1.6605 \times 10^{-27} \text{ kg}$$

The atomic masses of all the elements are listed in the periodic table, part of which is shown in Figure 14.1. The complete periodic table is given on the inside of the back cover. In general, the masses listed are average values and take into account the various isotopes of an element that exist naturally. For brevity, the unit "u" is often omitted from the table. For example, a magnesium atom (Mg) has an average atomic mass of 24.305 u, while that for the lithium atom (Li) is 6.941 u; thus, atomic magnesium is more massive than atomic lithium by a factor of (24.305 u)/(6.941 u) = 3.502. In the periodic table, the atomic mass of carbon (C) is given as 12.011 u, rather than exactly 12 u, because a small amount (about 1%) of the naturally occurring material is an isotope called carbon-13. The value of 12.011 u is an average that reflects the small contribution of carbon-13.

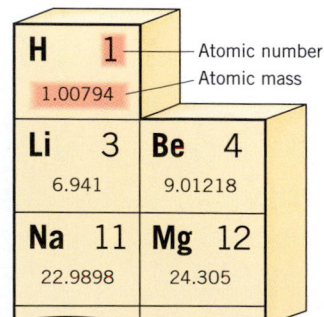

Figure 14.1 A portion of the periodic table showing the atomic number and atomic mass of each element. In the periodic table it is customary to omit the symbol "u" denoting the atomic mass unit.

*Isotopes are discussed in Section 31.1.
†In chemistry the expression "atomic weight" is frequently used in place of "atomic mass."

The molecular mass of a molecule is the sum of the atomic masses of its atoms. For instance, the elements hydrogen and oxygen have atomic masses of 1.007 94 u and 15.9994 u, respectively, so the molecular mass of a water molecule (H_2O) is, therefore, 2(1.007 94 u) + 15.9994 u = 18.0153 u.

Macroscopic amounts of materials contain large numbers of atoms or molecules. Even in a small volume of gas, 1 cm³, for example, the number is enormous. It is convenient to express such large numbers in terms of a single unit, the **gram-mole,** or simply the **mole** (symbol: *mol*). **One gram-mole of a substance contains as many particles (atoms or molecules) as there are atoms in 12 grams of the isotope carbon-12.** Experiment shows that 12 grams of carbon-12 contain 6.022×10^{23} atoms. The number of atoms per mole is known as **Avogadro's number N_A,** after the Italian scientist Amedeo Avogadro (1776–1856):

$$N_A = 6.022 \times 10^{23} \text{ mol}^{-1}$$

Thus, the number of moles n contained in any sample is the number of particles N in the sample divided by the number of particles per mole N_A (Avogadro's number):

$$n = \frac{N}{N_A}$$

Although defined in terms of carbon atoms, the concept of a mole can be applied to any collection of objects by noting that one mole contains Avogadro's number of objects. Thus, one mole of atomic sulfur contains 6.022×10^{23} sulfur atoms, one mole of water contains 6.022×10^{23} H_2O molecules, and one mole of golf balls contains 6.022×10^{23} golf balls. The mole is the SI base unit for expressing "the amount of a substance."

The number n of moles contained in a sample can also be found from its mass. To see how, multiply and divide the right-hand side of the previous equation by the mass $m_{particle}$ of a single particle, expressed in grams:

$$n = \frac{m_{particle}N}{m_{particle}N_A} = \frac{m}{\text{Mass per mole}}$$

The numerator $m_{particle}N$ is the mass of a particle times the number of particles in the sample, which is the mass m of the sample expressed in grams. The denominator $m_{particle}N_A$ is the mass of a particle times the number of particles per mole, which is the mass per mole, expressed in grams per mole. **The mass per mole (in g/mol) of any substance has the same numerical value as the atomic or molecular mass of the substance (in atomic mass units).** To understand this fact, consider the carbon-12 and sodium atoms as examples. The mass per mole of carbon-12 is 12 g/mol, since, by definition, 12 grams of carbon-12 contain one mole of atoms. On the other hand, the mass per mole of sodium (Na) is 22.9898 g/mol for the following reason: as indicated in Figure 14.1, a sodium atom is more massive than a carbon-12 atom by the ratio of their atomic masses, (22.9898 u)/(12 u) = 1.915 82. Therefore, the mass per mole of sodium is 1.915 82 times as great as that of carbon-12, which means equivalently that (1.915 82)(12 g/mol) = 22.9898 g/mol. Thus, the numerical value of the mass per mole of sodium (22.9898) is the same as the numerical value of its atomic mass.

Since one gram-mole of a substance contains Avogadro's number of particles (atoms or molecules), the mass $m_{particle}$ of a particle (in grams) can be obtained by dividing the mass per mole (in g/mol) by Avogadro's number:

$$m_{particle} = \frac{\text{Mass per mole}}{N_A}$$

Example 1 illustrates how to use the concepts of the mole, atomic mass, and Avogadro's number to determine the number of atoms and molecules present in two famous gemstones.

Problem-solving insight

 Example 1 The Hope Diamond and the Rosser Reeves Ruby

Figure 14.2*a* shows the Hope diamond (44.5 carats), which is almost pure carbon. Figure 14.2*b* shows the Rosser Reeves ruby (138 carats), which is primarily aluminum oxide (Al_2O_3). One carat is equivalent to a mass of 0.200 g. Determine **(a)** the number of carbon atoms in the diamond and **(b)** the number of Al_2O_3 molecules in the ruby.

The physics of gemstones.

(a)

(b)

Figure 14.2 (a) The Hope diamond surrounded by 16 smaller diamonds. (Dane A. Penland/Gem & Mineral Collection/Smithsonian Institution) (b) The Rosser Reeves ruby. (Chip Clark/Smithsonian Institution) Both gems are on display at the Smithsonian Institution in Washington, D.C.

Reasoning The number N of atoms (or molecules) in a sample is the number of moles n times the number of atoms per mole N_A (Avogadro's number); $N = nN_A$. We can determine the number of moles by dividing the mass of the sample m by the mass per mole of the substance.

Solution (a) The Hope diamond's mass is $m = (44.5 \text{ carats})[(0.200 \text{ g})/(1 \text{ carat})] = 8.90 \text{ g}$. Since the average atomic mass of naturally occurring carbon is 12.011 u (see the periodic table on the inside of the back cover), the mass per mole of this substance is 12.011 g/mol. The number of moles of carbon in the Hope diamond is

$$n = \frac{m}{\text{Mass per mole}} = \frac{8.90 \text{ g}}{12.011 \text{ g/mol}} = 0.741 \text{ mol}$$

The number of carbon atoms in the Hope diamond is

$$N = nN_A = (0.741 \text{ mol})(6.022 \times 10^{23} \text{ atoms/mol}) = \boxed{4.46 \times 10^{23} \text{ atoms}}$$

(b) The mass of the Rosser Reeves ruby is $m = (138 \text{ carats})[(0.200 \text{ g})/(1 \text{ carat})] = 27.6 \text{ g}$. The molecular mass of an aluminum oxide molecule (Al_2O_3) is the sum of the atomic masses of its atoms, which are 26.9815 u for aluminum and 15.9994 u for oxygen (see the periodic table on the inside of the back cover):

$$\text{Molecular mass} = \underbrace{2(26.9815 \text{ u})}_{\substack{\text{Mass of 2} \\ \text{aluminum atoms}}} + \underbrace{3(15.9994 \text{ u})}_{\substack{\text{Mass of 3} \\ \text{oxygen atoms}}} = 101.9612 \text{ u}$$

Thus, the mass per mole of Al_2O_3 is 101.9612 g/mol. Calculations like those in part (a) reveal that the Rosser Reeves ruby contains 0.271 mol or $\boxed{1.63 \times 10^{23} \text{ molecules of } Al_2O_3}$.

✓ CHECK YOUR UNDERSTANDING

(*The answers are given at the end of the book.*)

1. Consider one mole of hydrogen (H_2) and one mole of oxygen (O_2). Which, if either, has the greater number of molecules and which, if either, has the greater mass?

2. The molecules of substances A and B are composed of different atoms. However, the two substances have the same mass densities. Consider the possibilities for the molecular masses of the two types of molecules and decide whether 1 m³ of substance A contains the same number of molecules as 1 m³ of substance B.

3. A gas mixture contains equal masses of the monatomic gases argon (atomic mass = 39.948 u) and neon (atomic mass = 20.179 u). These two are the only gases present. Of the total number of atoms in the mixture, what percentage is neon?

14.2 THE IDEAL GAS LAW

An *ideal gas* is an idealized model for real gases that have sufficiently low densities. The condition of low density means that the molecules of the gas are so far apart that they do not interact (except during collisions that are effectively elastic). The ideal gas law expresses the relationship between the absolute pressure, the Kelvin temperature, the volume, and the number of moles of the gas.

In discussing the constant-volume gas thermometer, Section 12.2 has already explained the relationship between the absolute pressure and Kelvin temperature of a low-density gas. This thermometer utilizes a small amount of gas (e.g., hydrogen or helium) placed inside a bulb and kept at a constant volume. Since the density is low, the gas behaves as an ideal gas. Experiment reveals that a plot of gas pressure versus temperature is a straight line, as in Figure 12.4. This plot is redrawn in Figure 14.3, with the change that the temperature axis is now labeled in kelvins rather than in degrees Celsius. The graph indicates that the absolute pressure P is directly proportional to the Kelvin temperature T ($P \propto T$), for a fixed volume and a fixed number of molecules.

The relation between absolute pressure and the number of molecules of an ideal gas is simple. Experience indicates that it is possible to increase the pressure of a gas by adding more molecules; this is exactly what happens when a tire is pumped up. When the volume and temperature of a low-density gas are kept constant, doubling the number of molecules doubles the pressure. Thus, the absolute pressure of an ideal gas at constant temperature and constant volume is proportional to the number of molecules or, equivalently, to the number of moles n of the gas ($P \propto n$).

To see how the absolute pressure of a gas depends on the volume of the gas, look at the partially filled balloon in Figure 14.4a. This balloon is "soft," because the pressure of the air is low. However, if all the air in the balloon is squeezed into a smaller "bubble," as in part b of the figure, the "bubble" has a tighter feel. This tightness indicates that the pressure in the smaller volume is high enough to stretch the rubber substantially. Thus, it is possible to increase the pressure of a gas by reducing its volume, and if the number of molecules and the temperature are kept constant, the absolute pressure of an ideal gas is inversely proportional to its volume V ($P \propto 1/V$).

The three relations just discussed for the absolute pressure of an ideal gas can be expressed as a single proportionality, $P \propto nT/V$. This proportionality can be written as an equation by inserting a proportionality constant R, called the **universal gas constant.** Experiments have shown that $R = 8.31$ J/(mol·K) for any real gas with a density sufficiently low to ensure ideal gas behavior. The resulting equation is called the **ideal gas law.**

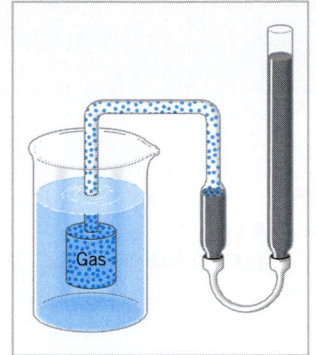

IDEAL GAS LAW

The absolute pressure P of an ideal gas is directly proportional to the Kelvin temperature T and the number of moles n of the gas and is inversely proportional to the volume V of the gas: $P = R(nT/V)$. In other words,

$$PV = nRT \qquad (14.1)$$

where R is the universal gas constant and has the value of 8.31 J/(mol·K).

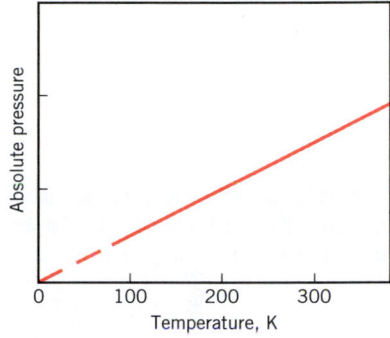

Figure 14.3 The pressure inside a constant-volume gas thermometer is directly proportional to the Kelvin temperature, a proportionality that is characteristic of an ideal gas.

Sometimes, it is convenient to express the ideal gas law in terms of the total number of particles N, instead of the number of moles n. To obtain such an expression, we multiply and divide by Avogadro's number $N_A = 6.022 \times 10^{23}$ particles/mol* on the right in Equation 14.1 and recognize that the product nN_A is equal to the total number N of particles:

$$PV = nRT = nN_A \left(\frac{R}{N_A} \right) T = N \left(\frac{R}{N_A} \right) T$$

The constant term R/N_A is referred to as **Boltzmann's constant,** in honor of the Austrian physicist Ludwig Boltzmann (1844–1906), and is represented by the symbol k:

$$k = \frac{R}{N_A} = \frac{8.31 \text{ J/(mol·K)}}{6.022 \times 10^{23} \text{ mol}^{-1}} = 1.38 \times 10^{-23} \text{ J/K}$$

(a)

(b)

Figure 14.4 (a) The air pressure in the partially filled balloon can be increased by decreasing the volume of the balloon, as illustrated in (b). (Andy Washnik)

*"Particles" is not an SI unit and is often omitted. Then, particles/mol = 1/mol = mol^{-1}.

With this substitution, the ideal gas law becomes

$$PV = NkT \qquad (14.2)$$

Example 2 presents an application of the ideal gas law.

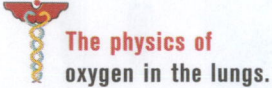

The physics of oxygen in the lungs.

⬇ **Example 2** Oxygen in the Lungs

In the lungs, a thin respiratory membrane separates tiny sacs of air (absolute pressure = 1.00×10^5 Pa) from the blood in the capillaries. These sacs are called alveoli, and it is from them that oxygen enters the blood. The average radius of the alveoli is 0.125 mm, and the air inside contains 14% oxygen. Assuming that the air behaves as an ideal gas at body temperature (310 K), find the number of oxygen molecules in one of the sacs.

Reasoning The pressure and temperature of the air inside an alveolus are known, and its volume can be determined since we know the radius. Thus, the ideal gas law in the form $PV = NkT$ can be used directly to find the number N of air particles inside one of the sacs. The number of oxygen molecules is 14% of the number of air particles.

Solution The volume of a spherical sac is $V = \frac{4}{3}\pi r^3$. Solving Equation 14.2 for the number of air particles, we have

$$N = \frac{PV}{kT} = \frac{(1.00 \times 10^5 \text{ Pa})[\frac{4}{3}\pi(0.125 \times 10^{-3} \text{ m})^3]}{(1.38 \times 10^{-23} \text{ J/K})(310 \text{ K})} = 1.9 \times 10^{14}$$

The number of oxygen molecules is 14% of this value, or $0.14N = \boxed{2.7 \times 10^{13}}$.

With the aid of the ideal gas law, it can be shown that one mole of an ideal gas occupies a volume of 22.4 liters at a temperature of 273 K (0 °C) and a pressure of one atmosphere (1.013×10^5 Pa). These conditions of temperature and pressure are known as *standard temperature and pressure (STP)*. Conceptual Example 3 discusses another interesting application of the ideal gas law.

The physics of rising beer bubbles.

⬇ **Conceptual Example 3** Beer Bubbles on the Rise

If you look carefully at the bubbles rising in a glass of beer (see Figure 14.5), you'll see them grow in size as they move upward, often doubling in volume by the time they reach the surface. Beer bubbles contain mostly carbon dioxide (CO_2), a gas that is dissolved in the beer because of the fermentation process. Which variable describing the gas is responsible for the growth of the rising bubbles? **(a)** The Kelvin temperature T **(b)** The absolute pressure P **(c)** The number of moles n

Reasoning The variables T, P, and n are related to the volume V of a bubble by the ideal gas law ($V = nRT/P$). We assume that this law applies and use it to guide our thinking. According to this law, an increase in temperature, a decrease in pressure, or an increase in the number of moles could account for the growth in size of the upward-moving bubbles.

Answers (a) and (b) are incorrect. Temperature can be eliminated immediately, since it is constant throughout the beer. Pressure cannot be dismissed so easily. As a bubble rises, its depth decreases, and so does the fluid pressure that a bubble experiences. Since volume is inversely proportional to pressure according to the ideal gas law, at least part of the bubble growth is due to the decreasing pressure of the surrounding beer. However, some bubbles *double in volume* on the way up. To account for the doubling, there would need to be two atmospheres of pressure at the bottom of the glass, compared to the one atmosphere at the top. The pressure increment due to depth is ρgh (see Equation 11.4), so an extra pressure of one atmosphere at the bottom would mean 1.01×10^5 Pa $= \rho gh$. Solving for h with ρ equal to the density of water reveals that $h = 10.3$ m. Since most beer glasses are only about 0.2 m tall, we can rule out a change in pressure as the major cause of the change in volume.

Answer (c) is correct. The process of elimination brings us to the conclusion that the number of moles of CO_2 in a bubble must somehow be increasing on the way up. This is, in fact, the case. Each bubble acts as a nucleation site for CO_2 molecules dissolved in the surrounding beer, so as a bubble moves upward, it accumulates carbon dioxide and grows larger.

Related Homework: *Problem 28*

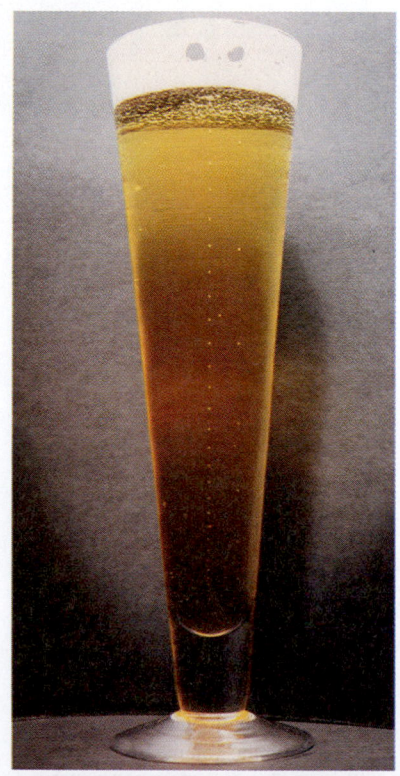

Figure 14.5 The bubbles in a glass of beer grow larger as they move upward. (Courtesy Richard Zare, Stanford University)

Historically, the work of several investigators led to the formulation of the ideal gas law. The Irish scientist Robert Boyle (1627–1691) discovered that at a constant temperature, the absolute pressure of a fixed mass (fixed number of moles) of a low-density gas is inversely proportional to its volume ($P \propto 1/V$). This fact is often called Boyle's law and can be derived from the ideal gas law by noting that $P = nRT/V = $ constant$/V$ when n and T are constants. Alternatively, if an ideal gas changes from an initial pressure and volume (P_i, V_i) to a final pressure and volume (P_f, V_f), it is possible to write $P_iV_i = nRT$ and $P_fV_f = nRT$. Since the right sides of these equations are equal, we may equate the left sides to give the following concise way of expressing **Boyle's law:**

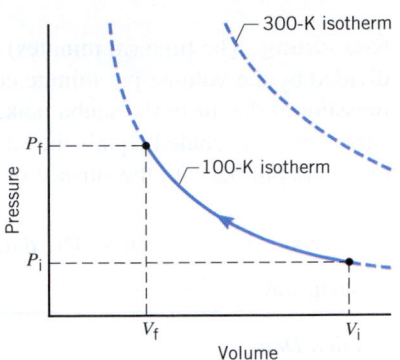

Constant T, constant n
$$P_iV_i = P_fV_f \qquad (14.3)$$

Figure 14.6 illustrates how pressure and volume change according to Boyle's law for a fixed number of moles of an ideal gas at a constant temperature of 100 K. The gas begins with an initial pressure and volume of P_i and V_i and is compressed. The pressure increases as the volume decreases, according to $P = nRT/V$, until the final pressure and volume of P_f and V_f are reached. The curve that passes through the initial and final points is called an **isotherm,** meaning "same temperature." If the temperature had been 300 K, rather than 100 K, the compression would have occurred along the 300-K isotherm. Different isotherms do not intersect. Example 4 deals with an application of Boyle's law to scuba diving.

Figure 14.6 A pressure-versus-volume plot for a gas at a constant temperature is called an isotherm. For an ideal gas, each isotherm is a plot of the equation $P = nRT/V = $ constant$/V$.

ANALYZING MULTIPLE-CONCEPT PROBLEMS

Example 4 Scuba* Diving

The physics of scuba diving.

When a scuba diver descends to greater depths, the water pressure increases. The air pressure inside the body cavities (e.g., lungs, sinuses) must be maintained at the same pressure as that of the surrounding water; otherwise the cavities would collapse. A special valve automatically adjusts the pressure of the air coming from the scuba tank to ensure that the air pressure equals the water pressure at all times. The scuba gear in Figure 14.7a consists of a 0.0150-m³ tank filled with compressed air at an absolute pressure of 2.02×10^7 Pa. Assume that the diver consumes air at the rate of 0.0300 m³ per minute and that the temperature of the air does not change as the diver goes deeper into the water. How long (in minutes) can a diver stay under water at a depth of 10.0 m? Take the density of sea water to be 1025 kg/m³.

(a)   (b)

Figure 14.7 (a) The air pressure inside the body cavities of a scuba diver must be maintained at the same level as the pressure of the surrounding water. (Shirley Vanderbilt/Index Stock) (b) The pressure P_2 at a depth h is greater than the pressure P_1 at the surface.

*The word is an acronym for *self-contained underwater breathing apparatus.*

Continued

Reasoning The time (in minutes) that a scuba diver can remain under water is equal to the volume of air that is available divided by the volume per minute consumed by the diver. The volume of air available to the diver depends on the volume and pressure of the air in the scuba tank, as well as the pressure of the air inhaled by the diver, according to Boyle's law. The pressure of the air inhaled equals the water pressure that acts on the diver. This pressure can be found from a knowledge of the diver's depth beneath the surface of the water.

Knowns and Unknowns The data for this problem are listed below:

Description	Symbol	Value	Comment
Explicit Data			
Volume of air in tank	V_i	0.0150 m³	
Pressure of air in tank	P_i	2.02×10^7 Pa	
Rate of air consumption	C	0.0300 m³/min	
Mass density of sea water	ρ	1025 kg/m³	
Depth of diver	h	10.0 m	
Implicit Data			
Air pressure at surface of water	P_1	1.01×10^5 Pa	Atmospheric pressure at sea level (see Section 11.2).
Unknown Variable			
Time that diver can remain at 10.0-m depth	t	?	

Modeling the Problem

STEP 1 Duration of the Dive The air inside the scuba tank has an initial pressure of P_i and a volume of V_i (the volume of the tank). A scuba diver does *not* breathe the air *directly* from the tank, because the tank pressure of 2.02×10^7 Pa is nearly 200 times atmospheric pressure and would cause his lungs to explode. Instead, a valve on the tank adjusts the pressure of the air being sent to the diver so it equals the surrounding water pressure P_f. The time t (in minutes) that the diver can remain under water is equal to the total volume of air consumed by the diver divided by the rate C (in cubic meters per minute) at which the air is consumed:

$$t = \frac{\text{Total volume of air consumed}}{C}$$

The total volume of air consumed is the volume V_f available at the breathing pressure P_f minus the volume V_i of the scuba tank, because this amount of air always remains behind in the tank. Thus, we have Equation 1 at the right. The volume V_i and the rate C are known, but the final volume V_f is not, so we turn to Step 2 to evaluate it.

$$t = \frac{V_f - V_i}{C} \qquad (1)$$

?

STEP 2 Boyle's Law Since the temperature of the air remains constant, the air volume V_f available to the diver at the pressure P_f is related to the initial pressure P_i and volume V_i of air in the tank by Boyle's law $P_iV_i = P_fV_f$ (Equation 14.3). Solving for V_f yields

$$\boxed{V_f = \frac{P_iV_i}{P_f}}$$

$$t = \frac{V_f - V_i}{C} \qquad (1)$$

$$\boxed{V_f = \frac{P_iV_i}{P_f}} \qquad (2)$$

?

This expression for V_f can be substituted into Equation 1, as indicated at the right. The initial pressure P_i and volume V_i are given. However, we still need to determine the pressure P_f of the air inhaled by the diver, and we will evaluate it in the next step.

STEP 3 Pressure and Depth in a Static Fluid Figure 14.7*b* shows the diver at a depth h below the surface of the water. The absolute pressure P_2 at this depth is related to the pressure P_1 at the surface of the water by Equation 11.4: $P_2 = P_1 + \rho gh$, where ρ

is the mass density of sea water and g is the magnitude of the acceleration due to gravity. Since P_1 is the air pressure at the surface of the water, it is atmospheric pressure. Recall that the valve on the scuba tank adjusts the pressure P_f of the air inhaled by the diver to be equal to the pressure P_2 of the surrounding water. Thus, $P_2 = P_f$, and Equation 11.4 becomes

$$P_f = P_1 + \rho gh$$

We now substitute this expression for P_f into Equation 2, as indicated in the right column.

$$t = \frac{V_f - V_i}{C} \tag{1}$$

$$V_f = \frac{P_i V_i}{P_f} \tag{2}$$

$$\boxed{P_f = P_1 + \rho gh}$$

Solution Algebraically combining the results of the three steps, we have

STEP 1	STEP 2	STEP 3

$$t = \frac{V_f - V_i}{C} = \frac{\dfrac{P_i V_i}{P_f} - V_i}{C} = \frac{\dfrac{P_i V_i}{P_1 + \rho gh} - V_i}{C}$$

The time that the diver can remain at a depth of 10.0 m is

$$t = \frac{\dfrac{P_i V_i}{P_1 + \rho gh} - V_i}{C}$$

$$= \frac{\dfrac{(2.02 \times 10^7 \text{ Pa})(0.0150 \text{ m}^3)}{1.01 \times 10^5 \text{ Pa} + (1025 \text{ kg/m}^3)(9.80 \text{ m/s}^2)(10.0 \text{ m})} - 0.0150 \text{ m}^3}{0.0300 \text{ m}^3/\text{min}} = \boxed{49.6 \text{ min}}$$

Note that at a fixed consumption rate C, greater values for h lead to smaller values for t. In other words, a deeper dive must have a shorter duration.

Related Homework: *Problems 24, 28*

Problem-solving insight

When using the ideal gas law, either directly or in the form of Boyle's law, remember that the pressure P must be the absolute pressure, not the gauge pressure.

Another investigator whose work contributed to the formulation of the ideal gas law was the Frenchman Jacques Charles (1746–1823). He discovered that at a constant pressure, the volume of a fixed mass (fixed number of moles) of a low-density gas is directly proportional to the Kelvin temperature ($V \propto T$). This relationship is known as Charles' law and can be obtained from the ideal gas law by noting that $V = nRT/P = (\text{constant})T$, if n and P are constant. Equivalently, when an ideal gas changes from an initial volume and temperature (V_i, T_i) to a final volume and temperature (V_f, T_f), it is possible to write $V_i/T_i = nR/P$ and $V_f/T_f = nR/P$. Thus, one way of stating **Charles' law** is

Constant P, constant n

$$\frac{V_i}{T_i} = \frac{V_f}{T_f} \tag{14.4}$$

✓ **CHECK YOUR UNDERSTANDING**

(The answers are given at the end of the book.)

4. A tightly sealed house has a large ceiling fan that blows air out of the house and into the attic. The owners turn the fan on and forget to open any windows or doors. What happens to the air pressure in the house after the fan has been on for a while, and does it become easier or harder for the fan to do its job?

5. Above the liquid in a can of hair spray is a gas at a relatively high pressure. The label on the can includes the warning "DO NOT STORE AT HIGH TEMPERATURES." Why is the warning given?

6. What happens to the pressure in a tightly sealed house when the electric furnace turns on and runs for a while?

Continued

7. When you climb a mountain, your eardrums "pop" outward as the air pressure decreases. When you come down, they pop inward as the pressure increases. At the sea coast, you swim through a completely submerged passage and emerge into a pocket of air trapped within a cave. As the tide comes in, the water level in the cave rises, and your eardrums pop. Is this popping analogous to what happens as you climb up or climb down a mountain?

8. Atmospheric pressure decreases with increasing altitude. Given this fact, explain why helium-filled weather balloons are underinflated when they are launched from the ground. Assume that the temperature does not change much as the balloon rises.

9. A slippery cork is being pressed into an almost full (but not 100% full) bottle of wine. When released, the cork slowly slides back out. However, if half the wine is removed from the bottle before the cork is inserted, the cork does not slide out. Explain.

10. Consider equal masses of three monatomic gases: argon (atomic mass = 39.948 u), krypton (atomic mass = 83.80 u), and xenon (atomic mass = 131.29 u). The pressure and volume of each gas is the same. Which gas has the greatest and which the smallest temperature?

14.3 KINETIC THEORY OF GASES

As useful as it is, the ideal gas law provides no insight as to how pressure and temperature are related to properties of the molecules themselves, such as their masses and speeds. To show how such microscopic properties are related to the pressure and temperature of an ideal gas, this section examines the dynamics of molecular motion. The pressure that a gas exerts on the walls of a container is due to the force exerted by the gas molecules when they collide with the walls. Therefore, we will begin by combining the notion of collisional forces exerted by a fluid (Section 11.2) with Newton's second and third laws of motion (Sections 4.3 and 4.5). These concepts will allow us to obtain an expression for the pressure in terms of microscopic properties. We will then combine this with the ideal gas law to show that the average translational kinetic energy $\overline{KE}$ of a particle in an ideal gas is $\overline{KE} = \frac{3}{2}kT$, where k is Boltzmann's constant and T is the Kelvin temperature. In the process, we will also see that the internal energy U of a monatomic ideal gas is $U = \frac{3}{2}nRT$, where n is the number of moles and R is the universal gas constant.

THE DISTRIBUTION OF MOLECULAR SPEEDS

A macroscopic container filled with a gas at standard temperature and pressure contains a large number of particles (atoms or molecules). These particles are in constant, random motion, colliding with each other and with the walls of the container. In the course of one second, a particle undergoes many collisions, and each one changes the particle's speed and direction of motion. As a result, the atoms or molecules have different speeds. It is possible, however, to speak about an average particle speed. At any given instant, some particles have speeds less than, some near, and some greater than the average. For conditions of low gas density, the distribution of speeds within a large collection of molecules at a constant temperature was calculated by the Scottish physicist James Clerk Maxwell (1831–1879). Figure 14.8 displays the Maxwell speed distribution curves for O_2 gas at two different temperatures.

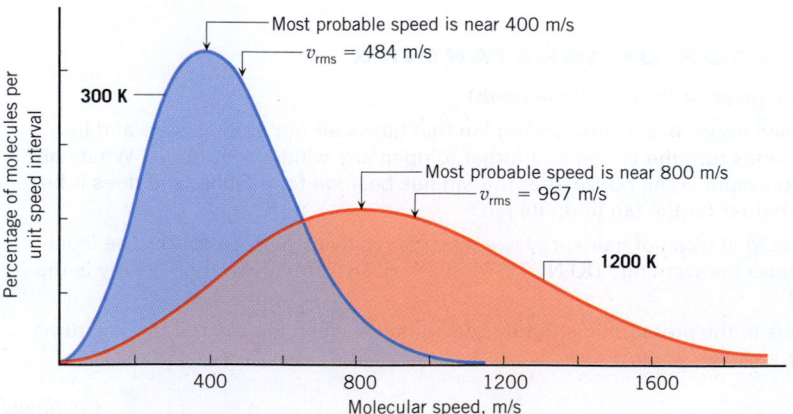

Figure 14.8 The Maxwell distribution curves for molecular speeds in oxygen gas at temperatures of 300 and 1200 K.

When the temperature is 300 K, the maximum in the curve indicates that the most probable speed is about 400 m/s. At a temperature of 1200 K, the distribution curve shifts to the right, and the most probable speed increases to about 800 m/s. One particularly useful type of average speed, known as the rms speed and written as v_{rms}, is also shown in the drawing. When the temperature of the oxygen gas is 300 K the rms speed is 484 m/s, and it increases to 967 m/s when the temperature rises to 1200 K. The meaning of the rms speed and the reason why it is so important will be discussed shortly.

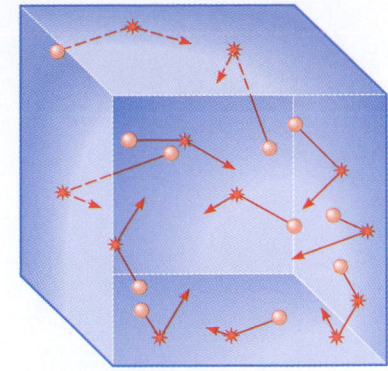

Figure 14.9 The pressure that a gas exerts is caused by the collisions of its molecules with the walls of the container.

KINETIC THEORY

If a ball is thrown against a wall, it exerts a force on the wall. As Figure 14.9 suggests, gas particles do the same thing, except that their masses are smaller and their speeds are greater. The number of particles is so great and they strike the wall so often that the effect of their individual impacts appears as a continuous force. Dividing the magnitude of this force by the area of the wall gives the pressure exerted by the gas.

To calculate the force, consider an ideal gas composed of N identical particles in a cubical container whose sides have length L. Except for elastic* collisions, these particles do not interact. Figure 14.10 focuses attention on one particle of mass m as it strikes the right wall perpendicularly and rebounds elastically. While approaching the wall, the particle has a velocity $+v$ and linear momentum $+mv$ (see Section 7.1 for a review of linear momentum). The particle rebounds with a velocity $-v$ and momentum $-mv$, travels to the left wall, rebounds again, and heads back toward the right. The time t between collisions with the right wall is the round-trip distance $2L$ divided by the speed of the particle; that is, $t = 2L/v$. According to Newton's second law of motion, in the form of the impulse–momentum theorem, the average force exerted on the particle by the wall is given by the change in the particle's momentum per unit time:

$$\text{Average force} = \frac{\text{Final momentum} - \text{Initial momentum}}{\text{Time between successive collisions}} \tag{7.4}$$

$$= \frac{(-mv) - (+mv)}{2L/v} = \frac{-mv^2}{L}$$

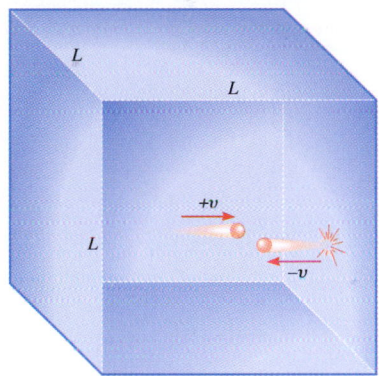

Figure 14.10 A gas particle is shown colliding elastically with the right wall of the container and rebounding from it.

According to Newton's law of action–reaction, the force applied to the wall by the particle is equal in magnitude to this value, but oppositely directed (i.e., $+mv^2/L$). The magnitude F of the *total* force exerted on the right wall is equal to the number of particles that collide with the wall during the time t multiplied by the average force exerted by each particle. Since the N particles move randomly in three dimensions, one-third of them on the average strike the right wall during the time t. Therefore, the total force is

$$F = \left(\frac{N}{3}\right)\left(\frac{m\overline{v^2}}{L}\right)$$

In this result v^2 has been replaced by $\overline{v^2}$, the *average value* of the squared speed. The collection of particles possesses a Maxwell distribution of speeds, so an average value for v^2 must be used, rather than a value for any individual particle. The square root of the quantity $\overline{v^2}$ is called the **root-mean-square speed,** or, for short, the *rms speed;* $v_{rms} = \sqrt{\overline{v^2}}$. With this substitution, the total force becomes

$$F = \left(\frac{N}{3}\right)\left(\frac{mv_{rms}^2}{L}\right)$$

Pressure is force per unit area, so the pressure P acting on a wall of area L^2 is

$$P = \frac{F}{L^2} = \left(\frac{N}{3}\right)\left(\frac{mv_{rms}^2}{L^3}\right)$$

Since the volume of the box is $V = L^3$, the equation above can be written as

$$PV = \tfrac{2}{3}N(\tfrac{1}{2}mv_{rms}^2) \tag{14.5}$$

*The term "elastic" is used here to mean that *on the average*, in a large number of particles, there is no gain or loss of translational kinetic energy because of collisions.

Equation 14.5 relates the macroscopic properties of the gas—its pressure and volume—to the microscopic properties of the constituent particles—their mass and speed. Since the term $\frac{1}{2}mv_{rms}^2$ is the average translational kinetic energy $\overline{KE}$ of an individual particle, it follows that

$$PV = \frac{2}{3}N(\overline{KE})$$

This result is similar to the ideal gas law, $PV = NkT$ (Equation 14.2). Both equations have identical terms on the left, so the terms on the right must be equal: $\frac{2}{3}N(\overline{KE}) = NkT$. Therefore,

$$\overline{KE} = \frac{1}{2}mv_{rms}^2 = \frac{3}{2}kT \qquad (14.6)$$

Equation 14.6 is significant, because it allows us to interpret temperature in terms of the motion of gas particles. This equation indicates that the Kelvin temperature is directly proportional to the average translational kinetic energy per particle in an ideal gas, no matter what the pressure and volume are. On the average, the particles have greater kinetic energies when the gas is hotter than when it is cooler. Conceptual Example 5 discusses a common misconception about the relation between kinetic energy and temperature.

 Conceptual Example 5 Does a Single Particle Have a Temperature?

Each particle in a gas has kinetic energy. Furthermore, the equation $\frac{1}{2}mv_{rms}^2 = \frac{3}{2}kT$ establishes the relationship between the average kinetic energy per particle and the temperature of an ideal gas. Is it valid, then, to conclude that a single particle has a temperature?

Reasoning and Solution We know that a gas contains an enormous number of particles that are traveling with a distribution of speeds, such as those indicated by the graphs in Figure 14.8. Therefore, the particles do not all have the same kinetic energy, but possess a distribution of kinetic energies ranging from very nearly zero to extremely large values. If each particle had a temperature that was associated with its kinetic energy, there would be a whole range of different temperatures within the gas. This is not so, for a gas at thermal equilibrium has only one temperature (see Section 15.2), a temperature that would be registered by a thermometer placed in the gas. Thus, temperature is a property that characterizes the gas as a whole, a fact that is inherent in the relation $\frac{1}{2}mv_{rms}^2 = \frac{3}{2}kT$. The term v_{rms} is a kind of *average particle speed*. Therefore, $\frac{1}{2}mv_{rms}^2$ is the *average kinetic energy* per particle and is characteristic of the gas as a whole. Since the Kelvin temperature is proportional to $\frac{1}{2}mv_{rms}^2$, it is also a characteristic of the gas as a whole and cannot be ascribed to each gas particle individually. Thus, *a single gas particle does not have a temperature.*

If two ideal gases have the same temperature, the relation $\frac{1}{2}mv_{rms}^2 = \frac{3}{2}kT$ indicates that the average kinetic energy of each kind of gas particle is the same. In general, however, the rms speeds of the different particles are not the same, because the masses may be different. The next example illustrates these facts and shows how rapidly gas particles move at normal temperatures.

ANALYZING MULTIPLE-CONCEPT PROBLEMS

Example 6 The Speed of Molecules in Air

Air is primarily a mixture of nitrogen N_2 (molecular mass = 28.0 u) and oxygen O_2 (molecular mass = 32.0 u). Assume that each behaves like an ideal gas and determine the rms speed of the nitrogen and oxygen molecules when the air temperature is 293 K.

Reasoning As Figure 14.8 illustrates, the same type of molecules (e.g., O_2 molecules) within a gas have different speeds, even though the gas itself has a constant temperature. The rms speed v_{rms} of the molecules is a kind of *average speed* and is related to the kinetic energy (also an average) according to Equation 6.2. Thus, the average translational kinetic energy $\overline{KE}$ of a molecule is $\overline{KE} = \frac{1}{2}mv_{rms}^2$, where m is the mass of a molecule. We know that the average kinetic energy depends on the Kelvin temperature T of the gas through the relation $\overline{KE} = \frac{3}{2}kT$ (Equation 14.6). We can find the rms speed of each type of molecule, then, from a knowledge of its mass and the temperature.

Knowns and Unknowns The data for this problem are:

Description	Symbol	Value
Molecular mass of nitrogen (N_2)	—	28.0 u
Molecular mass of oxygen (O_2)	—	32.0 u
Air temperature	T	293 K
Unknown Variable		
Rms speed of nitrogen and oxygen molecules	v_{rms}	?

Modeling the Problem

STEP 1 Rms Speed and Average Kinetic Energy The rms speed v_{rms} of a molecule in a gas is related to the average kinetic energy $\overline{KE}$ by

$$\overline{KE} = \tfrac{1}{2}mv_{rms}^2 \qquad (6.2)$$

where m is its mass. Solving for v_{rms} gives Equation 1 at the right. We will deal with the average kinetic energy $\overline{KE}$ in Step 2.

$$v_{rms} = \sqrt{\frac{2\,\overline{KE}}{m}} \qquad (1)$$

$$\boxed{?}$$

STEP 2 Average Kinetic Energy and Temperature Since the gas is assumed to be an ideal gas, the average kinetic energy of a molecule is directly proportional to the Kelvin temperature T of the gas according to Equation 14.6:

$$\boxed{\overline{KE} = \tfrac{3}{2}kT}$$

where k is Boltzmann's constant. By substituting this relation into Equation 1 at the right, we can obtain an expression for the rms speed.

$$v_{rms} = \sqrt{\frac{2\,\overline{KE}}{m}} \qquad (1)$$

$$\boxed{\overline{KE} = \tfrac{3}{2}kT}$$

Solution Algebraically combining the results of the two steps, we have

STEP 1   **STEP 2**

$$v_{rms} = \sqrt{\frac{2\,\overline{KE}}{m}} = \sqrt{\frac{2(\tfrac{3}{2}kT)}{m}}$$

In this result, m is the mass of a gas molecule (in kilograms). Since the molecular masses of nitrogen and oxygen are given in atomic mass units (28.0 u and 32.0 u, respectively), we must convert them to kilograms by using the conversion factor $1\ u = 1.6605 \times 10^{-27}$ kg (see Section 14.1). Thus,

Nitrogen $\qquad m_{N_2} = (28.0\ u)\left(\dfrac{1.6605 \times 10^{-27}\ kg}{1\ u}\right) = 4.65 \times 10^{-26}\ kg$

Oxygen $\qquad m_{O_2} = (32.0\ u)\left(\dfrac{1.6605 \times 10^{-27}\ kg}{1\ u}\right) = 5.31 \times 10^{-26}\ kg$

The rms speed for each type of molecule is

Nitrogen $\qquad v_{rms} = \sqrt{\dfrac{2(\tfrac{3}{2}kT)}{m_{N_2}}} = \sqrt{\dfrac{2\left[\tfrac{3}{2}(1.38 \times 10^{-23}\ J/K)(293\ K)\right]}{4.65 \times 10^{-26}\ kg}} = \boxed{511\ m/s}$

Oxygen $\qquad v_{rms} = \sqrt{\dfrac{2(\tfrac{3}{2}kT)}{m_{O_2}}} = \sqrt{\dfrac{2\left[\tfrac{3}{2}(1.38 \times 10^{-23}\ J/K)(293\ K)\right]}{5.31 \times 10^{-26}\ kg}} = \boxed{478\ m/s}$

Problem-solving insight

The average translational kinetic energy is the same for all ideal-gas molecules at the same temperature, regardless of their masses. The rms translational speed of the molecules is not the same, however, because it depends on the mass.

Note that the nitrogen and oxygen molecules have the same average kinetic energy, since the temperature is the same for both. The fact that nitrogen has the greater rms speed is due to its smaller mass.

Related Homework: *Problems 33, 58*

The equation $\overline{KE} = \frac{3}{2}kT$ has also been applied to particles much larger than atoms or molecules. The English botanist Robert Brown (1773–1858) observed through a microscope that pollen grains suspended in water move on very irregular, zigzag paths. This Brownian motion can also be observed with other particle suspensions, such as fine smoke particles in air. In 1905, Albert Einstein (1879–1955) showed that Brownian motion could be explained as a response of the large suspended particles to impacts from the moving molecules of the fluid medium (e.g., water or air). As a result of the impacts, the suspended particles have the same average translational kinetic energy as the fluid molecules— namely, $\overline{KE} = \frac{3}{2}kT$. Unlike the molecules, however, the particles are large enough to be seen through a microscope and, because of their relatively large mass, have a comparatively small average speed.

THE INTERNAL ENERGY OF A MONATOMIC IDEAL GAS

Chapter 15 deals with the science of thermodynamics, in which the concept of internal energy plays an important role. Using the results just developed for the average translational kinetic energy, we conclude this section by expressing the internal energy of a monatomic ideal gas in a form that is suitable for use later on.

The internal energy of a substance is the sum of the various kinds of energy that the atoms or molecules of the substance possess. A monatomic ideal gas is composed of single atoms. These atoms are assumed to be so small that the mass is concentrated at a point, with the result that the moment of inertia I about the center of mass is negligible. Thus, the rotational kinetic energy $\frac{1}{2}I\omega^2$ is also negligible. Vibrational kinetic and potential energies are absent, because the atoms are not connected by chemical bonds and, except for elastic collisions, do not interact. As a result, the internal energy U is the total translational kinetic energy of the N atoms that constitute the gas: $U = N(\frac{1}{2}mv_{rms}^2)$. Since $\frac{1}{2}mv_{rms}^2 = \frac{3}{2}kT$ according to Equation 14.6, the internal energy can be written in terms of the Kelvin temperature as

$$U = N(\tfrac{3}{2}kT)$$

Usually, U is expressed in terms of the number of moles n, rather than the number of atoms N. Using the fact that Boltzmann's constant is $k = R/N_A$, where R is the universal gas constant and N_A is Avogadro's number, and realizing that $N/N_A = n$, we find that

Monatomic ideal gas $\qquad\qquad\qquad U = \tfrac{3}{2}nRT \qquad\qquad\qquad$ (14.7)

Thus, the internal energy depends on the number of moles and the Kelvin temperature of the gas. In fact, it can be shown that the internal energy is proportional to the Kelvin temperature for *any type* of ideal gas (e.g., monatomic, diatomic, etc.). For example, when hot-air balloonists turn on the burner, they increase the temperature, and hence the internal energy per mole, of the air inside the balloon (see Figure 14.11).

Figure 14.11 Since air behaves approximately as an ideal gas, the internal energy per mole inside a hot-air balloon increases as the temperature rises. (Erik Lam/Alamy Images)

✓ CHECK YOUR UNDERSTANDING

(The answers are given at the end of the book.)

11. The kinetic theory of gases assumes that, for a given collision time, a gas molecule rebounds with the same speed after colliding with the wall of a container. If the speed after the collision were less than the speed before the collision, the duration of the collision remaining the same, would the pressure of the gas be greater than, equal to, or less than the pressure predicted by kinetic theory?

12. If the temperature of an ideal gas were doubled from 50 to 100 °C, would the average translational kinetic energy of the gas particles also double?

13. The pressure of a monatomic ideal gas doubles, while the volume decreases to one-half its initial value. Does the internal energy of the gas increase, decrease, or remain unchanged?

14. The atoms in a container of helium (He) have the same translational rms speed as the atoms in a container of argon (Ar). Treating each gas as an ideal gas, decide which, if either, has the greater temperature.

15. The pressure of a monatomic ideal gas is doubled, while its volume is reduced by a factor of four. What is the ratio of the new rms speed of the atoms to the initial rms speed?

*DIFFUSION

You can smell the fragrance of a perfume at some distance from an open bottle because perfume molecules leave the space above the liquid in the bottle, where they are relatively concentrated, and spread out into the air, where they are less concentrated. During their journey, they collide with other molecules, so their paths resemble the zigzag paths characteristic of Brownian motion. The process in which molecules move from a region of higher concentration to one of lower concentration is called *diffusion*. Diffusion also occurs in liquids and solids, and Figure 14.12 illustrates ink diffusing through water. However, compared to the rate of diffusion in gases, the rate is generally smaller in liquids and even smaller in solids. The host medium, such as the air or water in the examples above, is referred to as the *solvent,* while the diffusing substance, like the perfume molecules or the ink in Figure 14.12, is known as the *solute.* Relatively speaking, diffusion is a slow process, even in a gas. Conceptual Example 7 illustrates why.

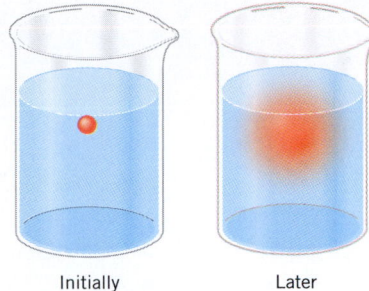

Initially   Later

Figure 14.12 A drop of ink placed in water eventually becomes completely dispersed because of diffusion.

⬇ Conceptual Example 7 Why Diffusion Is Relatively Slow

The fragrance from an open bottle of perfume takes several seconds or sometimes even minutes to reach the other side of a room by the process of diffusion. Which of the following accounts for the fact that diffusion is relatively slow? **(a)** The nature of Brownian motion **(b)** The relatively slow translational rms speeds that characterize gas molecules at room temperature

Reasoning The important characteristic of the paths followed by objects in Brownian motion is their zigzag shapes. We have calculated typical translational rms speeds for gas molecules near room temperature in Example 6, and those results will guide our reasoning here.

Answer (b) is incorrect. In Example 6 we have seen that a gas molecule near room temperature has a translational rms speed of hundreds of meters per second. Such speeds are not slow. It would take a molecule traveling at such a speed just a fraction of a second to cross an ordinary room.

Answer (a) is correct. When a perfume molecule diffuses through air, it makes millions of collisions each second with air molecules. As Figure 14.13 illustrates, the velocity of the molecule changes abruptly because of each collision, but between collisions, it moves in a straight line. Although it does move very fast between collisions, a perfume molecule wanders only slowly away from the bottle because of the zigzag path. It would take a long time indeed to diffuse in this manner across a room. Usually, however, convection currents are present and carry the fragrance to the other side of the room in a matter of seconds or minutes.

Related Homework: *Problems 49, 51*

⬆

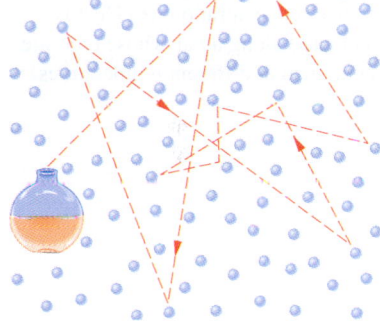

Figure 14.13 A perfume molecule collides with millions of air molecules during its journey, so the path has a zigzag shape. Although the air molecules are shown as stationary, they are also moving.

Diffusion is the basis for drug delivery systems that bypass the need to administer medication orally or via injections. Figure 14.14 shows one such system, the transdermal patch. The word "transdermal" means "across the skin." Such patches, for example, are

The physics of drug delivery systems.

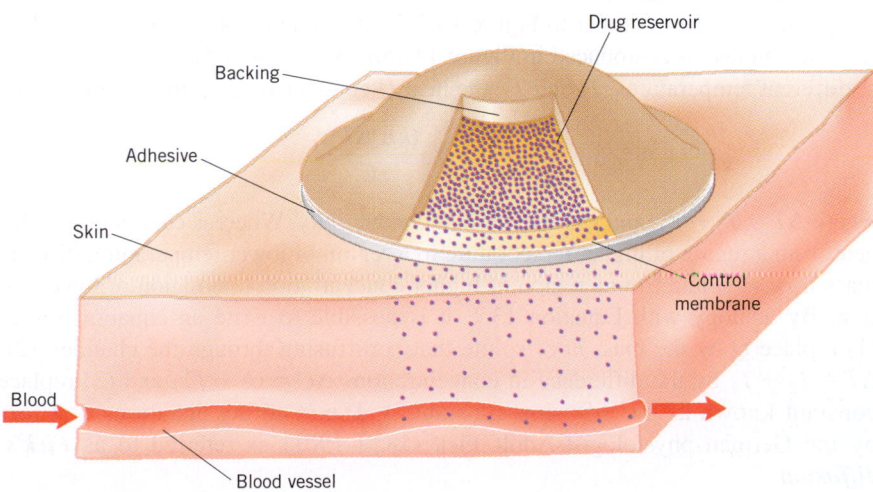

Drug reservoir

Backing

Adhesive

Skin

Control membrane

Blood

Blood vessel

Figure 14.14 Using diffusion, a transdermal patch delivers a drug directly into the skin, where it enters blood vessels. The backing contains the drug within the reservoir, and the control membrane limits the rate of diffusion into the skin.

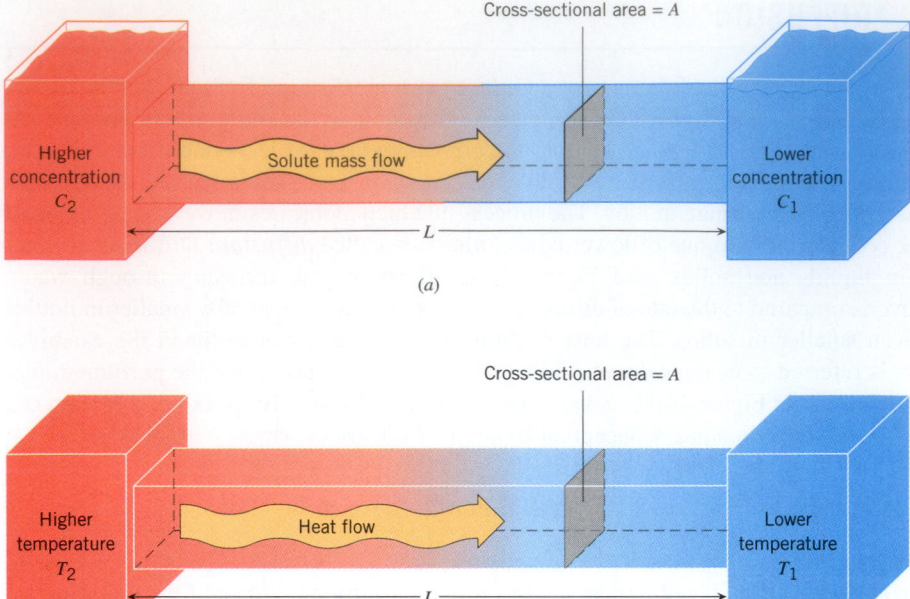

Figure 14.15 (*a*) Solute diffuses through the channel from the region of higher concentration to the region of lower concentration. (*b*) Heat is conducted along a bar whose ends are maintained at different temperatures.

used to deliver nicotine in programs designed to help you stop smoking. The patch is attached to the skin using an adhesive, and the backing of the patch contains the drug within a reservoir. The concentration of the drug in the reservoir is relatively high, just like the concentration of perfume molecules above the liquid in a bottle. The drug diffuses slowly through a control membrane and directly into the skin, where its concentration is relatively low. Diffusion carries it into the blood vessels present in the skin. The purpose of the control membrane is to limit the rate of diffusion, which can also be adjusted in the reservoir by dissolving the drug in a neutral material to lower its initial concentration. Another diffusion-controlled drug delivery system utilizes capsules that are inserted surgically beneath the skin. Contraceptives are administered in this fashion, for instance. The drug in the capsule diffuses slowly into the bloodstream over extended periods that can be as long as a year.

The diffusion process can be described in terms of the arrangement in Figure 14.15*a*. A hollow channel of length L and cross-sectional area A is filled with a fluid. The left end of the channel is connected to a container in which the solute concentration C_2 is relatively high, while the right end is connected to a container in which the solute concentration C_1 is lower. These concentrations are defined as the total mass of the solute molecules divided by the volume of the solution (e.g., 0.1 kg/m^3). Because of the difference in concentration between the ends of the channel, $\Delta C = C_2 - C_1$, there is a net diffusion of the solute from the left end to the right end.

Figure 14.15*a* is similar to Figure 13.7 for the conduction of heat along a bar, which, for convenience, is reproduced in Figure 14.15*b*. When the ends of the bar are maintained at different temperatures, T_2 and T_1, the heat Q conducted along the bar in a time t is

$$Q = \frac{(kA\,\Delta T)t}{L} \qquad (13.1)$$

where $\Delta T = T_2 - T_1$, and k is the thermal conductivity. Whereas conduction is the flow of heat from a region of higher temperature to a region of lower temperature, diffusion is the mass flow of solute from a region of higher concentration to a region of lower concentration. By analogy with Equation 13.1, it is possible to write an equation for diffusion: (1) replace Q by the mass m of solute that is diffusing through the channel, (2) replace $\Delta T = T_2 - T_1$ by the difference in concentrations $\Delta C = C_2 - C_1$, and (3) replace k by a constant known as the diffusion constant D. The resulting equation, first formulated by the German physiologist Adolf Fick (1829–1901), is referred to as ***Fick's law of diffusion.***

FICK'S LAW OF DIFFUSION

The mass m of solute that diffuses in a time t through a solvent contained in a channel of length L and cross-sectional area A is*

$$m = \frac{(DA\,\Delta C)t}{L} \qquad (14.8)$$

where ΔC is the concentration difference between the ends of the channel and D is the diffusion constant.

SI Unit for the Diffusion Constant: m^2/s

It can be verified from Equation 14.8 that the diffusion constant has units of m^2/s, the exact value depending on the nature of the solute and the solvent. For example, the diffusion constant for ink in water is different from that for ink in benzene. Example 8 illustrates an important application of Fick's law.

 Example 8 Water Given Off by Plant Leaves

Large amounts of water can be given off by plants. It has been estimated, for instance, that a single sunflower plant can lose up to a pint of water a day during the growing season. Figure 14.16 shows a cross-sectional view of a leaf. Inside the leaf, water passes from the liquid phase to the vapor phase at the walls of the mesophyll cells. The water vapor then diffuses through the intercellular air spaces and eventually exits the leaf through small openings, called stomatal pores. The diffusion constant for water vapor in air is $D = 2.4 \times 10^{-5}\ \text{m}^2/\text{s}$. A stomatal pore has a cross-sectional area of about $A = 8.0 \times 10^{-11}\ \text{m}^2$ and a length of about $L = 2.5 \times 10^{-5}\ \text{m}$. The concentration of water vapor on the interior side of a pore is roughly $C_2 = 0.022\ \text{kg/m}^3$, while the concentration on the outside is approximately $C_1 = 0.011\ \text{kg/m}^3$. Determine the mass of water vapor that passes through a stomatal pore in one hour.

Reasoning and Solution Fick's law of diffusion shows that

$$m = \frac{(DA\,\Delta C)t}{L} \qquad (14.8)$$

$$m = \frac{(2.4 \times 10^{-5}\ \text{m}^2/\text{s})(8.0 \times 10^{-11}\ \text{m}^2)(0.022\ \text{kg/m}^3 - 0.011\ \text{kg/m}^3)(3600\ \text{s})}{2.5 \times 10^{-5}\ \text{m}}$$

$$= \boxed{3.0 \times 10^{-9}\ \text{kg}}$$

This amount of water may not seem significant. However, a single leaf may have as many as a million stomatal pores, so the water lost by an entire plant can be substantial.

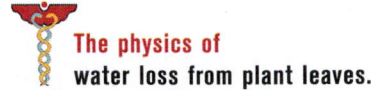
The physics of water loss from plant leaves.

✓ CHECK YOUR UNDERSTANDING

(The answers are given at the end of the book.)

16. In the lungs, oxygen in very small sacs called alveoli diffuses into the blood. The diffusion occurs directly through the walls of the sacs, which have a thickness L. The total effective area A across which diffusion occurs is the sum of the individual areas (each quite small) of the various sac walls. Considering the fact that the mass m of oxygen that enters the blood per second needs to be large and referring to Fick's law of diffusion, what can you deduce about L and about the total number of sacs present in the lungs?

17. The same solute is diffusing through the same solvent in each of three cases. For each case, the table gives the length and cross-sectional area of the diffusion channel. The concentration difference between the ends of the diffusion channel is the same in each case. Rank the diffusion rates (in kg/s) in descending order (largest first).

Case	Length	Cross-Sectional Area
(a)	$\frac{1}{2}L$	A
(b)	L	$\frac{1}{4}A$
(c)	$\frac{1}{3}L$	$2A$

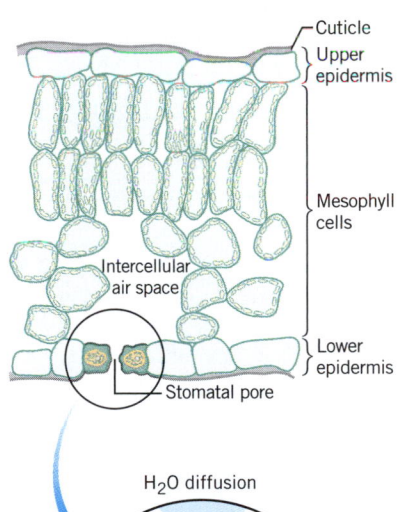

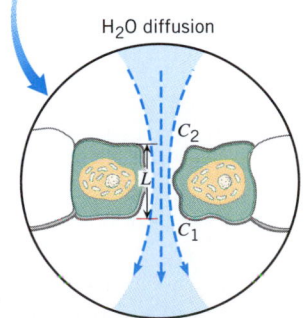

Figure 14.16 A cross-sectional view of a leaf. Water vapor diffuses out of the leaf through a stomatal pore.

*Fick's law assumes that the temperature of the solvent is constant throughout the channel. Experiments indicate that the diffusion constant depends strongly on the temperature.

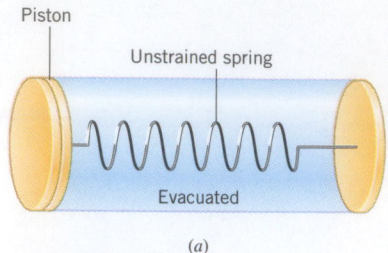

Piston

Unstrained spring

Evacuated

(a)

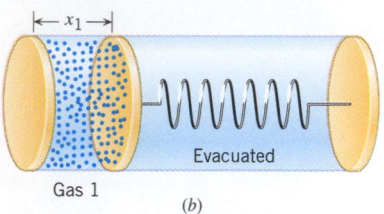

$\leftarrow x_1 \rightarrow$

Evacuated

Gas 1

(b)

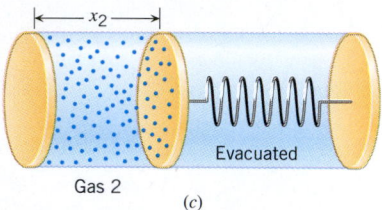

$\leftarrow x_2 \rightarrow$

Evacuated

Gas 2

(c)

Figure 14.17 Two ideal gases at different temperatures compress the spring by different amounts.

14.5 CONCEPTS & CALCULATIONS

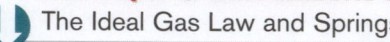

This chapter introduces the ideal gas law, which is a relation between the pressure, volume, temperature, and number of moles of an ideal gas. In Section 10.1, we examined how the compression of a spring depends on the force applied to it. Example 9 reviews how a gas produces a force and why an ideal gas at different temperatures causes a spring to compress by different amounts.

Concepts & Calculations Example 9
The Ideal Gas Law and Springs

Figure 14.17 shows three identical chambers containing a piston and a spring whose spring constant is $k = 5.8 \times 10^4$ N/m. The chamber in part a is completely evacuated, and the piston just touches its left end. In this position, the spring is unstrained. In part b of the drawing, 0.75 mol of ideal gas 1 is introduced into the chamber, and the spring compresses by $x_1 = 15$ cm. In part c, 0.75 mol of ideal gas 2 is introduced into the chamber, and the spring compresses by $x_2 = 24$ cm. Find the temperature of each gas.

Concept Questions and Answers Which gas exerts the greater force on the piston?

Answer We know that a greater force is required to compress a spring by a greater amount. Therefore, gas 2 exerts the greater force.

How is the force required to compress a spring related to the displacement of the spring from its unstrained position?

Answer According to Equation 10.1, the applied force F_x^{Applied} required to compress a spring is directly proportional to the displacement x of the spring from its unstrained position; $F_x^{\text{Applied}} = kx$, where k is the spring constant of the spring.

Which gas exerts the greater pressure on the piston?

Answer According to Equation 11.3, pressure is defined as the magnitude F of the force acting perpendicular to the surface of the piston divided by the area A of the piston; $P = F/A$. Since gas 2 exerts the greater force, and the area of the piston is the same for both gases, gas 2 exerts the greater pressure.

Which gas has the greater temperature?

Answer According to the ideal gas law, $T = PV/(nR)$, gas 2 has the greater temperature. Both gases contain the same number n of moles. However, gas 2 has both a greater pressure P and greater volume V. Thus, it has the greater temperature.

Solution We can use the ideal gas law in the form $T = PV/(nR)$ to determine the temperature T of each gas. First, however, we must find the pressure. According to Equation 11.3, the pressure is the magnitude F of the force that the gas exerts on the piston divided by the area A of the piston, so $P = F/A$. Recall that the force F_x^{Applied} applied to the piston to compress the spring is related to the displacement x of the spring by $F_x^{\text{Applied}} = kx$ (Equation 10.1). Thus, $F = F_x^{\text{Applied}} = kx$, and the pressure can be written as $P = kx/A$. Using this expression for the pressure in the ideal gas law gives

$$T = \frac{PV}{nR} = \frac{\left(\dfrac{kx}{A}\right)V}{nR}$$

However, the cylindrical volume V of the gas is equal to the product of the distance x and the area A, so $V = xA$. With this substitution, the expression for the temperature becomes

$$T = \frac{\left(\dfrac{kx}{A}\right)V}{nR} = \frac{\left(\dfrac{kx}{A}\right)(xA)}{nR} = \frac{kx^2}{nR}$$

The temperatures of the gases are

Gas 1
$$T_1 = \frac{kx_1^2}{nR} = \frac{(5.8 \times 10^4 \text{ N/m})(15 \times 10^{-2} \text{ m})^2}{(0.75 \text{ mol})[8.31 \text{ J/(mol}\cdot\text{K)}]} = \boxed{210 \text{ K}}$$

Gas 2 $T_2 = \dfrac{kx_2^2}{nR} = \dfrac{(5.8 \times 10^4 \text{ N/m})(24 \times 10^{-2} \text{ m})^2}{(0.75 \text{ mol})[8.31 \text{ J/(mol} \cdot \text{K)}]} = \boxed{540 \text{ K}}$

As anticipated, gas 2, which compresses the spring more, has the higher temperature.

The kinetic theory of gases is important because it allows us to understand the relation between the macroscopic properties of a gas, such as pressure and temperature, and the microscopic properties of its particles, such as speed and mass. The following example reviews the essential features of this theory.

Concepts & Calculations Example 10
Hydrogen Atoms in Outer Space

In outer space the density of matter is extremely low, about one atom per cm³. The matter is mainly hydrogen atoms ($m = 1.67 \times 10^{-27}$ kg) whose rms speed is 260 m/s. A cubical box, 2.0 m on a side, is placed in outer space, and the hydrogen atoms are allowed to enter. **(a)** What is the magnitude of the force that the atoms exert on one wall of the box? **(b)** Determine the pressure that the atoms exert. **(c)** Does outer space have a temperature, and, if so, what is it?

Concept Questions and Answers Why do hydrogen atoms exert a force on the walls of the box?

Answer   Every time an atom collides with a wall and rebounds, the atom exerts a force on the wall. Imagine opening your hand so it is flat and having someone throw a ball straight at it. Your hand is like the wall, and as the ball rebounds, you can feel the force. Intuitively, you would expect the force to become greater as the speed and mass of the ball become greater. This is indeed the case.

Do the atoms generate a pressure on the walls of the box?

Answer   Yes. Pressure is defined as the magnitude of the force exerted perpendicularly on a wall divided by the area of the wall. Since the atoms exert a force, they also produce a pressure.

Do hydrogen atoms in outer space have a temperature? If so, how is the temperature related to the microscopic properties of the atoms?

Answer   Yes. The Kelvin temperature is proportional to the average kinetic energy of an atom. The average kinetic energy, in turn, is proportional to the mass of an atom and the square of the rms speed. Since we know both these quantities, we can determine the temperature of the gas.

Solution **(a)** The magnitude F of the force exerted on a wall is given by (see Section 14.3)

$$F = \left(\frac{N}{3}\right)\left(\frac{mv_{\text{rms}}^2}{L}\right)$$

where N is the number of atoms in the box, m is the mass of a single atom, v_{rms} is the rms speed of the atoms, and L is the length of one side of the box. The volume of the cubical box is $(2.0 \times 10^2 \text{ cm})^3 = 8.0 \times 10^6 \text{ cm}^3$. The number N of atoms is equal to the number of atoms per cubic centimeter times the volume of the box in cubic centimeters:

$$N = \left(\frac{1 \text{ atom}}{\text{cm}^3}\right)(8.0 \times 10^6 \text{ cm}^3) = 8.0 \times 10^6$$

The magnitude of the force acting on one wall is

$$F = \left(\frac{N}{3}\right)\left(\frac{mv_{\text{rms}}^2}{L}\right) = \left(\frac{8.0 \times 10^6}{3}\right)\left[\frac{(1.67 \times 10^{-27} \text{ kg})(260 \text{ m/s})^2}{2.0 \text{ m}}\right]$$

$$= \boxed{1.5 \times 10^{-16} \text{ N}}$$

(b) The pressure is the magnitude of the force divided by the area A of a wall:

$$P = \frac{F}{A} = \frac{1.5 \times 10^{-16} \text{ N}}{(2.0 \text{ m})^2} = \boxed{3.8 \times 10^{-17} \text{ Pa}}$$

(c) According to Equation 14.6, the Kelvin temperature T of the hydrogen atoms is related to the average kinetic energy of an atom by $\frac{3}{2}kT = \frac{1}{2}mv_{rms}^2$, where k is Boltzmann's constant. Solving this equation for the temperature gives

$$T = \frac{mv_{rms}^2}{3k} = \frac{(1.67 \times 10^{-27} \text{ kg})(260 \text{ m/s})^2}{3(1.38 \times 10^{-23} \text{ J/K})} = \boxed{2.7 \text{ K}}$$

This is a frigid 2.7 kelvins above absolute zero.

CONCEPT SUMMARY

If you need more help with a concept, use the Learning Aids noted next to the discussion or equation. Examples (**Ex.**) are in the text of this chapter. Go to **www.wiley.com/college/cutnell** for the following Learning Aids:

Interactive LearningWare (ILW) — Additional examples solved in a five-step interactive format.

Concept Simulations (CS) — Animated text figures or animations of important concepts.

Interactive Solutions (IS) — Models for certain types of problems in the chapter homework. The calculations are carried out interactively.

Topic	Discussion	Learning Aids
Atomic mass unit Molecular mass	**14.1 MOLECULAR MASS, THE MOLE, AND AVOGADRO'S NUMBER** Each element in the periodic table is assigned an atomic mass. One atomic mass unit (u) is exactly one-twelfth the mass of an atom of carbon-12. The molecular mass of a molecule is the sum of the atomic masses of its atoms.	
	The number of moles n contained in a sample is equal to the number of particles N (atoms or molecules) in the sample divided by the number of particles per mole N_A,	
Number of moles	$$n = \frac{N}{N_A}$$	
Avogadro's number	where N_A is called Avogadro's number and has a value of $N_A = 6.022 \times 10^{23}$ particles per mole. The number of moles is also equal to the mass m of the sample (expressed in grams) divided by the mass per mole (expressed in grams per mole):	
Number of moles	$$n = \frac{m}{\text{Mass per mole}}$$	Ex. 1
Mass per mole	The mass per mole (in g/mol) of a substance has the same numerical value as the atomic or molecular mass of one of its particles (in atomic mass units).	
	The mass $m_{particle}$ of a particle (in grams) can be obtained by dividing the mass per mole (in g/mol) by Avogadro's number:	
Mass of a particle	$$m_{particle} = \frac{\text{Mass per mole}}{N_A}$$	
	14.2 THE IDEAL GAS LAW The ideal gas law relates the absolute pressure P, the volume V, the number n of moles, and the Kelvin temperature T of an ideal gas according to	
Ideal gas law	$$PV = nRT \qquad (14.1)$$	
Universal gas constant	where $R = 8.31$ J/(mol·K) is the universal gas constant. An alternative form of the ideal gas law is	
Ideal gas law	$$PV = NkT \qquad (14.2)$$	**Ex. 2, 3, 9**
Boltzmann's constant	where N is the number of particles and $k = \dfrac{R}{N_A}$ is Boltzmann's constant. A real gas behaves as an ideal gas when its density is low enough that its particles do not interact, except via elastic collisions.	**IS 14.25, 14.29**
	A form of the ideal gas law that applies when the number of moles and the temperature are constant is known as Boyle's law. Using the subscripts "i" and "f" to denote, respectively, initial and final conditions, we can write Boyle's law as	
Boyle's law	$$P_i V_i = P_f V_f \qquad (14.3)$$	**Ex. 4**
	A form of the ideal gas law that applies when the number of moles and the pressure are constant is called Charles' law:	
Charles' law	$$\frac{V_i}{T_i} = \frac{V_f}{T_f} \qquad (14.4)$$	
Maxwell speed distribution	**14.3 KINETIC THEORY OF GASES** The distribution of particle speeds in an ideal gas at constant temperature is the Maxwell speed distribution (see Figure 14.8). The kinetic theory of gases	

Topic	Discussion	Learning Aids
Average translational kinetic energy **Root-mean-square speed**	indicates that the Kelvin temperature T of an ideal gas is related to the average translational kinetic energy $\overline{KE}$ of a particle according to $$\overline{KE} = \tfrac{1}{2}mv_{rms}^2 = \tfrac{3}{2}kT \qquad (14.6)$$ where v_{rms} is the root-mean-square speed of the particles.	**Ex. 5, 6, 10** **IS 14.39, 14.61**
Internal energy	The internal energy U of n moles of a monatomic ideal gas is $$U = \tfrac{3}{2}nRT \qquad (14.7)$$ The internal energy of any type of ideal gas (e.g., monatomic, diatomic) is proportional to its Kelvin temperature.	
Fick's law of diffusion	**14.4 DIFFUSION** Diffusion is the process whereby solute molecules move through a solvent from a region of higher solute concentration to a region of lower solute concentration. Fick's law of diffusion states that the mass m of solute that diffuses in a time t through the solvent in a channel of length L and cross-sectional area A is $$m = \frac{(DA\,\Delta C)t}{L} \qquad (14.8)$$ where ΔC is the solute concentration difference between the ends of the channel and D is the diffusion constant.	**Ex. 7, 8**

FOCUS ON CONCEPTS

Note to Instructors: All of the questions shown here are available for assignment via an online homework management program such as WileyPLUS or WebAssign.

Section 14.1   Molecular Mass, the Mole, and Avogadro's Number

1. All but one of the following statements are true. Which one is not true?   **(a)** A mass (in grams) equal to the molecular mass (in atomic mass units) of a pure substance contains the same number of molecules, no matter what the substance is.   **(b)** One mole of any pure substance contains the same number of molecules.   **(c)** Ten grams of a pure substance contains twice as many molecules as five grams of the substance.   **(d)** Ten grams of a pure substance contains the same number of molecules, no matter what the substance is.   **(e)** Avogadro's number of molecules of a pure substance and one mole of the substance have the same mass.

2. A mixture of ethyl alcohol (molecular mass = 46.1 u) and water (molecular mass = 18.0 u) contains one mole of molecules. The mixture contains 20.0 g of ethyl alcohol. What mass m of water does it contain?

Section 14.2   The Ideal Gas Law

3. For an ideal gas, each of the following unquestionably leads to an increase in the pressure of the gas, except one. Which one is it?   **(a)** Increasing the temperature and decreasing the volume, while keeping the number of moles of the gas constant   **(b)** Increasing the temperature, the volume, and the number of moles of the gas   **(c)** Increasing the temperature, while keeping the volume and the number of moles of the gas constant   **(d)** Increasing the number of moles of the gas, while keeping the temperature and the volume constant   **(e)** Decreasing the volume, while keeping the temperature and the number of moles of the gas constant

4. The cylinder in the drawing contains 3.00 mol of an ideal gas. By moving the piston, the volume of the gas is reduced to one-fourth its initial value, while the temperature is held constant. How many moles Δn of the gas must be allowed to escape through the valve, so that the pressure of the gas does not change?

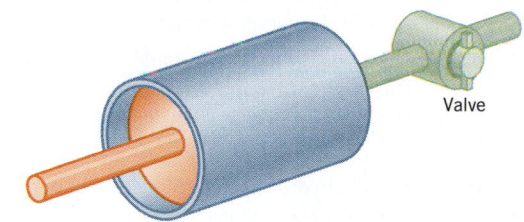

Valve

5. Carbon monoxide is a gas at 0 °C and a pressure of 1.01×10^5 Pa. It is a diatomic gas, each of its molecules consisting of one carbon atom (atomic mass = 12.0 u) and one oxygen atom (atomic mass = 16.0 u). Assuming that carbon monoxide is an ideal gas, calculate its density ρ.

Section 14.3   Kinetic Theory of Gases

6. If the speed of every atom in a monatomic ideal gas were doubled, by what factor would the Kelvin temperature of the gas be multiplied?   **(a)** 4   **(b)** 2   **(c)** 1   **(d)** $\tfrac{1}{2}$ **(e)** $\tfrac{1}{4}$

7. The atomic mass of a nitrogen atom (N) is 14.0 u, while that of an oxygen atom (O) is 16.0 u. Three diatomic gases have the same temperature: nitrogen (N_2), oxygen (O_2), and nitric oxide (NO). Rank these gases in ascending order (smallest first), according to the values of their translational rms speeds: **(a)** O_2, N_2, NO   **(b)** NO, N_2, O_2 **(c)** N_2, NO, O_2 **(d)** O_2, NO, N_2 **(e)** N_2, O_2, NO

8. The pressure of a monatomic ideal gas is doubled, while the volume is cut in half. By what factor is the internal energy of the gas multiplied?   **(a)** $\tfrac{1}{4}$ **(b)** $\tfrac{1}{2}$ **(c)** 1   **(d)** 2   **(e)** 4

Section 14.4 Diffusion

9. The following statements concern how to increase the rate of diffusion (in kg/s). All but one statement are always true. Which one is not necessarily true? **(a)** Increase the cross-sectional area of the diffusion channel, keeping constant its length and the difference in solute concentrations between its ends. **(b)** Increase the difference in solute concentrations between the ends of the diffusion channel, keeping constant its cross-sectional area and its length. **(c)** Decrease the length of the diffusion channel, keeping constant its cross-sectional area and the difference in solute concentrations between its ends.

(d) Increase the cross-sectional area of the diffusion channel, and decrease its length, keeping constant the difference in solute concentrations between its ends. **(e)** Increase the cross-sectional area of the diffusion channel, increase the difference in solute concentrations between its ends, and increase its length.

10. The diffusion rate for a solute is 4.0×10^{-11} kg/s in a solvent-filled channel that has a cross-sectional area of 0.50 cm² and a length of 0.25 cm. What would be the diffusion rate m/t in a channel with a cross-sectional area of 0.30 cm² and a length of 0.10 cm?

PROBLEMS

Note to Instructors: *Most of the homework problems in this chapter are available for assignment via an online homework management program such as WileyPLUS or WebAssign, and those marked with the icon* **GO** *are presented in* WileyPLUS *using a guided tutorial format that provides enhanced interactivity. See Preface for additional details.*

Note: The pressures referred to in these problems are absolute pressures, unless indicated otherwise.

ssm Solution is in the Student Solutions Manual.
www Solution is available online at www.wiley.com/college/cutnell

 This icon represents a biomedical application.

Section 14.1 Molecular Mass, the Mole, and Avogadro's Number

1. ssm Hemoglobin has a molecular mass of 64 500 u. Find the mass (in kg) of one molecule of hemoglobin.

2. Manufacturers of headache remedies routinely claim that their own brands are more potent pain relievers than the competing brands. Their way of making the comparison is to compare the number of molecules in the standard dosage. Tylenol uses 325 mg of acetaminophen ($C_8H_9NO_2$) as the standard dose, while Advil uses 2.00×10^2 mg of ibuprofen ($C_{13}H_{18}O_2$). Find the number of molecules of pain reliever in the standard doses of **(a)** Tylenol and **(b)** Advil.

3. The artificial sweetener NutraSweet is a chemical called aspartame ($C_{14}H_{18}N_2O_5$). What is **(a)** its molecular mass (in atomic mass units) and **(b)** the mass (in kg) of an aspartame molecule?

4. GO A certain element has a mass per mole of 196.967 g/mol. What is the mass of a single atom in **(a)** atomic mass units and **(b)** kilograms? **(c)** How many moles of atoms are in a 285-g sample?

5. ssm The active ingredient in the allergy medication Claritin contains carbon (C), hydrogen (H), chlorine (Cl), nitrogen (N), and oxygen (O). Its molecular formula is $C_{22}H_{23}ClN_2O_2$. The standard adult dosage utilizes 1.572×10^{19} molecules of this species. Determine the mass (in grams) of the active ingredient in the standard dosage.

6. A mass of 135 g of a certain element is known to contain 30.1×10^{23} atoms. What is the element?

***7.** A runner weighs 580 N (about 130 lb), and 71% of this weight is water. **(a)** How many moles of water are in the runner's body? **(b)** How many water molecules (H_2O) are there?

***8. GO** Consider a mixture of three different gases: 1.20 g of argon (molecular mass = 39.948 g/mol), 2.60 g of neon (molecular mass = 20.180 g/mol), and 3.20 g of helium (molecular mass = 4.0026 g/mol). For this mixture, determine the percentage of the total number of atoms that corresponds to each of the components.

***9. ssm** A cylindrical glass of water (H_2O) has a radius of 4.50 cm and a height of 12.0 cm. The density of water is 1.00 g/cm³. How many moles of water molecules are contained in the glass?

***10.** The preparation of homeopathic "remedies" involves the repeated dilution of solutions containing an active ingredient such as arsenic trioxide (As_2O_3). Suppose one begins with 18.0 g of arsenic trioxide dissolved in water, and repeatedly dilutes the solution with pure water, each dilution reducing the amount of arsenic trioxide

remaining in the solution by a factor of 100. Assuming perfect mixing at each dilution, what is the maximum number of dilutions one may perform so that at least one molecule of arsenic trioxide remains in the diluted solution? For comparison, homeopathic "remedies" are commonly diluted 15 or even 30 times.

Section 14.2 The Ideal Gas Law

11. ssm It takes 0.16 g of helium (He) to fill a balloon. How many grams of nitrogen (N_2) would be required to fill the balloon to the same pressure, volume, and temperature?

12. An empty oven with a volume of 0.150 m³ and a temperature of 296 K is turned on. The oven is vented so that the air pressure inside it is always the same as the air pressure of the environment. Initially, the air pressure is 1.00×10^5 Pa, but after the oven has warmed up to a final temperature of 453 K, the atmospheric air pressure has decreased to 9.50×10^4 Pa due to a change in weather conditions. How many moles of air leave the oven while it is heating up?

13. An ideal gas at 15.5 °C and a pressure of 1.72×10^5 Pa occupies a volume of 2.81 m³. **(a)** How many moles of gas are present? **(b)** If the volume is raised to 4.16 m³ and the temperature raised to 28.2 °C, what will be the pressure of the gas?

14. GO Four closed tanks, A, B, C, and D, each contain an ideal gas. The table gives the absolute pressure and volume of the gas in each tank. In each case, there is 0.10 mol of gas. Using this number and the data in the table, compute the temperature of the gas in each tank.

	A	B	C	D
Absolute pressure (Pa)	25.0	30.0	20.0	2.0
Volume (m³)	4.0	5.0	5.0	75

15. Oxygen for hospital patients is kept in special tanks, where the oxygen has a pressure of 65.0 atmospheres and a temperature of 288 K. The tanks are stored in a separate room, and the oxygen is pumped to the patient's room, where it is administered at a pressure of 1.00 atmosphere and a temperature of 297 K. What volume does 1.00 m³ of oxygen in the tanks occupy at the conditions in the patient's room?

16. A Goodyear blimp typically contains 5400 m³ of helium (He) at an absolute pressure of 1.1×10^5 Pa. The temperature of the helium is 280 K. What is the mass (in kg) of the helium in the blimp?

17. A clown at a birthday party has brought along a helium cylinder, with which he intends to fill balloons. When full, each balloon contains 0.034 m³ of helium at an absolute pressure of 1.2×10^5 Pa. The cylinder contains helium at an absolute pressure of 1.6×10^7 Pa and has a volume of 0.0031 m³. The temperature of the helium in the tank and in the balloons is the same and remains constant. What is the maximum number of balloons that can be filled?

18. **GO** The volume of an ideal gas is held constant. Determine the ratio P_2/P_1 of the final pressure to the initial pressure when the temperature of the gas rises **(a)** from 35.0 to 70.0 K and **(b)** from 35.0 to 70.0 °C.

19. **ssm** In a diesel engine, the piston compresses air at 305 K to a volume that is one-sixteenth of the original volume and a pressure that is 48.5 times the original pressure. What is the temperature of the air after the compression?

20. A 0.030-m³ container is initially evacuated. Then, 4.0 g of water is placed in the container, and, after some time, all the water evaporates. If the temperature of the water vapor is 388 K, what is its pressure?

21. On the sunlit surface of Venus, the atmospheric pressure is 9.0×10^6 Pa, and the temperature is 740 K. On the earth's surface the atmospheric pressure is 1.0×10^5 Pa, while the surface temperature can reach 320 K. These data imply that Venus has a "thicker" atmosphere at its surface than does the earth, which means that the number of molecules per unit volume (N/V) is greater on the surface of Venus than on the earth. Find the ratio $(N/V)_{\text{Venus}}/(N/V)_{\text{Earth}}$.

22. When you push down on the handle of a bicycle pump, a piston in the pump cylinder compresses the air inside the cylinder. When the pressure in the cylinder is greater than the pressure inside the inner tube to which the pump is attached, air begins to flow from the pump to the inner tube. As a biker slowly begins to push down the handle of a bicycle pump, the pressure inside the cylinder is 1.0×10^5 Pa, and the piston in the pump is 0.55 m above the bottom of the cylinder. The pressure inside the inner tube is 2.4×10^5 Pa. How far down must the biker push the handle before air begins to flow from the pump to the inner tube? Ignore the air in the hose connecting the pump to the inner tube, and assume that the temperature of the air in the pump cylinder does not change.

*23. **ssm www** The drawing shows two thermally insulated tanks. They are connected by a valve that is initially closed. Each tank contains neon gas at the pressure, temperature, and volume indicated in the drawing. When the valve is opened, the contents of the two tanks mix, and the pressure becomes constant throughout. **(a)** What is the final temperature? Ignore any change in temperature of the tanks themselves. (*Hint: The heat gained by the gas in one tank is equal to that lost by the other.*) **(b)** What is the final pressure?

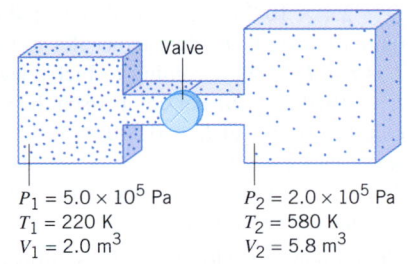

Valve

$P_1 = 5.0 \times 10^5$ Pa $P_2 = 2.0 \times 10^5$ Pa
$T_1 = 220$ K $T_2 = 580$ K
$V_1 = 2.0$ m³ $V_2 = 5.8$ m³

*24. Multiple-Concept Example 4 reviews the principles that play roles in this problem. A primitive diving bell consists of a cylindrical tank with one end open and one end closed. The tank is lowered into a fresh-water lake, open end downward. Water rises into the tank, compressing the trapped air, whose temperature remains constant during the descent. The tank is brought to a halt when the distance between the surface of the water in the tank and the surface of the lake is 40.0 m. Atmospheric pressure at the surface of the lake is 1.01×10^5 Pa. Find the fraction of the tank's volume that is filled with air.

*25. Refer to **Interactive Solution 14.25** at **www.wiley.com/college/ cutnell** for help with problems like this one. An apartment has a room whose dimensions are 2.5 m × 4.0 m × 5.0 m. Assume that the air in the room is composed of 79% nitrogen (N_2) and 21% oxygen (O_2). At a temperature of 22 °C and a pressure of 1.01×10^5 Pa, what is the mass (in grams) of the air?

*26. **GO** The drawing shows an ideal gas confined to a cylinder by a massless piston that is attached to an ideal spring. Outside the cylinder is a vacuum. The cross-sectional area of the piston is $A = 2.50 \times 10^{-3}$ m². The initial pressure, volume, and temperature of the gas are, respectively, P_0, $V_0 = 6.00 \times 10^{-4}$ m³, and $T_0 = 273$ K, and the spring is initially stretched by an amount $x_0 = 0.0800$ m with respect to its unstrained length. The gas is heated, so that its final pressure, volume, and temperature are P_f, V_f, and T_f, and the spring is stretched by an amount $x_f = 0.1000$ m with respect to its unstrained length. What is the final temperature of the gas?

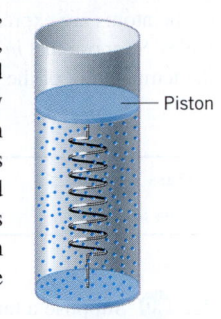

Piston

*27. **ssm** The relative humidity is 55% on a day when the temperature is 30.0 °C. Using the graph that accompanies Problem 75 in Chapter 12, determine the number of moles of water vapor per cubic meter of air.

*28. Multiple-Concept Example 4 and Conceptual Example 3 are pertinent to this problem. A bubble, located 0.200 m beneath the surface in a glass of beer, rises to the top. The air pressure at the top is 1.01×10^5 Pa. Assume that the density of beer is the same as that of fresh water. If the temperature and number of moles of CO_2 in the bubble remain constant as the bubble rises, find the ratio of the bubble's volume at the top to its volume at the bottom.

*29. **Interactive Solution 14.29** at **www.wiley.com/college/cutnell** offers one approach to this problem. One assumption of the ideal gas law is that the atoms or molecules themselves occupy a negligible volume. Verify that this assumption is reasonable by considering gaseous xenon (Xe). Xenon has an atomic radius of 2.0×10^{-10} m. For STP conditions, calculate the percentage of the total volume occupied by the atoms.

**30. A spherical balloon is made from a material whose mass is 3.00 kg. The thickness of the material is negligible compared to the 1.50-m radius of the balloon. The balloon is filled with helium (He) at a temperature of 305 K and just floats in air, neither rising nor falling. The density of the surrounding air is 1.19 kg/m³. Find the absolute pressure of the helium gas.

31. **ssm A cylindrical glass beaker of height 1.520 m rests on a table. The bottom half of the beaker is filled with a gas, and the top half is filled with liquid mercury that is exposed to the atmosphere. The gas and mercury do not mix because they are separated by a frictionless movable piston of negligible mass and thickness. The initial temperature is 273 K. The temperature is increased until a value is reached when one-half of the mercury has spilled out. Ignore the thermal expansion of the glass and the mercury, and find this temperature.

**32. A gas fills the right portion of a horizontal cylinder whose radius is 5.00 cm. The initial pressure of the gas is 1.01×10^5 Pa. A frictionless movable piston separates the gas from the left portion of the cylinder, which is evacuated and contains an ideal spring, as the drawing shows. The piston is initially held in place by a pin. The spring is initially unstrained, and the length of the gas-filled portion is 20.0 cm. When the pin is removed and the gas is allowed to expand, the length of the gas-filled chamber doubles. The initial and final temperatures are equal. Determine the spring constant of the spring.

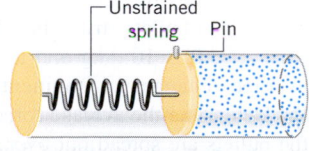

Unstrained spring Pin

Section 14.3   Kinetic Theory of Gases

33. Consult Multiple-Concept Example 6 to review the principles involved in this problem. Near the surface of Venus, the rms speed of carbon dioxide molecules (CO_2) is 650 m/s. What is the temperature (in kelvins) of the atmosphere at that point?

34. **GO** Four tanks A, B, C, and D are filled with monatomic ideal gases. For each tank, the mass of an individual atom and the rms speed of the atoms are expressed in terms of m and v_{rms}, respectively (see the table). Suppose that $m = 3.32 \times 10^{-26}$ kg, and $v_{rms} = 1223$ m/s. Find the temperature of the gas in each tank.

	A	B	C	D
Mass	m	m	$2m$	$2m$
Rms speed	v_{rms}	$2v_{rms}$	v_{rms}	$2v_{rms}$

35. **GO** Suppose a tank contains 680 m^3 of neon (Ne) at an absolute pressure of 1.01×10^5 Pa. The temperature is changed from 293.2 to 294.3 K. What is the increase in the internal energy of the neon?

36. If the translational rms speed of the water vapor molecules (H_2O) in air is 648 m/s, what is the translational rms speed of the carbon dioxide molecules (CO_2) in the same air? Both gases are at the same temperature.

37. **ssm** Initially, the translational rms speed of a molecule of an ideal gas is 463 m/s. The pressure and volume of this gas are kept constant, while the number of molecules is doubled. What is the final translational rms speed of the molecules?

38. Two gas cylinders are identical. One contains the monatomic gas argon (Ar), and the other contains an equal mass of the monatomic gas krypton (Kr). The pressures in the cylinders are the same, but the temperatures are different. Determine the ratio $\dfrac{\overline{KE}_{Krypton}}{\overline{KE}_{Argon}}$ of the average kinetic energy of a krypton atom to the average kinetic energy of an argon atom.

39. Consult **Interactive Solution 14.39** at **www.wiley.com/college/cutnell** to see how this problem can be solved. Very fine smoke particles are suspended in air. The translational rms speed of a smoke particle is 2.8×10^{-3} m/s, and the temperature is 301 K. Find the mass of a particle.

40. A container holds 2.0 mol of gas. The total average kinetic energy of the gas molecules in the container is equal to the kinetic energy of an 8.0×10^{-3}-kg bullet with a speed of 770 m/s. What is the Kelvin temperature of the gas?

*41. **ssm www** The pressure of sulfur dioxide (SO_2) is 2.12×10^4 Pa. There are 421 moles of this gas in a volume of 50.0 m^3. Find the translational rms speed of the sulfur dioxide molecules.

*42. **GO** Helium (He), a monatomic gas, fills a 0.010-m^3 container. The pressure of the gas is 6.2×10^5 Pa. How long would a 0.25-hp engine have to run (1 hp = 746 W) to produce an amount of energy equal to the internal energy of this gas?

43. **ssm www In 10.0 s, 200 bullets strike and embed themselves in a wall. The bullets strike the wall perpendicularly. Each bullet has a mass of 5.0×10^{-3} kg and a speed of 1200 m/s. **(a)** What is the average change in momentum per second for the bullets? **(b)** Determine the average force exerted on the wall. **(c)** Assuming the bullets are spread out over an area of 3.0×10^{-4} m^2, obtain the average pressure they exert on this region of the wall.

**44. A cubical box with each side of length 0.300 m contains 1.000 moles of neon gas at room temperature (293 K). What is the

average rate (in atoms/s) at which neon atoms collide with one side of the container? The mass of a single neon atom is 3.35×10^{-26} kg.

Section 14.4   Diffusion

45. Insects do not have lungs as we do, nor do they breathe through their mouths. Instead, they have a system of tiny tubes, called tracheae, through which oxygen diffuses into their bodies. The tracheae begin at the surface of an insect's body and penetrate into the interior. Suppose that a trachea is 1.9 mm long with a cross-sectional area of 2.1×10^{-9} m^2. The concentration of oxygen in the air outside the insect is 0.28 kg/m^3, and the diffusion constant is 1.1×10^{-5} m^2/s. If the mass per second of oxygen diffusing through a trachea is 1.7×10^{-12} kg/s, find the oxygen concentration at the interior end of the tube.

46. A large tank is filled with methane gas at a concentration of 0.650 kg/m^3. The valve of a 1.50-m pipe connecting the tank to the atmosphere is inadvertently left open for twelve hours. During this time, 9.00×10^{-4} kg of methane diffuses out of the tank, leaving concentration of methane in the tank essentially unchanged. The diffusion constant for methane in air is 2.10×10^{-5} m^2/s. What is the cross-sectional area of the pipe? Assume that the concentration of methane in the atmosphere is zero.

47. **ssm** The diffusion constant for the alcohol ethanol in water is 12.4×10^{-10} m^2/s. A cylinder has a cross-sectional area of 4.00 cm^2 and a length of 2.00 cm. A difference in ethanol concentration of 1.50 kg/m^3 is maintained between the ends of the cylinder. In one hour, what mass of ethanol diffuses through the cylinder?

48. **GO** When a gas is diffusing through air in a diffusion channel, the diffusion rate is the number of gas atoms per second diffusing from one end of the channel to the other end. The faster the atoms move, the greater is the diffusion rate, so the diffusion rate is proportional to the rms speed of the atoms. The atomic mass of ideal gas A is 1.0 u, and that of ideal gas B is 2.0 u. For diffusion through the same channel under the same conditions, find the ratio of the diffusion rate of gas A to the diffusion rate of gas B.

49. Review Conceptual Example 7 before working this problem. For water vapor in air at 293 K, the diffusion constant is $D = 2.4 \times 10^{-5}$ m^2/s. As outlined in Problem 51(a), the time required for the first solute molecules to traverse a channel of length L is $t = L^2/(2D)$, according to Fick's law. **(a)** Find the time t for $L = 0.010$ m. **(b)** For comparison, how long would a water molecule take to travel $L = 0.010$ m at the translational rms speed of water molecules (assumed to be an ideal gas) at a temperature of 293 K? **(c)** Explain why the answer to part (a) is so much longer than the answer to part (b).

*50. Carbon tetrachloride (CCl_4) is diffusing through benzene (C_6H_6), as the drawing illustrates. The concentration of CCl_4 at the left end of the tube is maintained at 1.00×10^{-2} kg/m^3, and the diffusion constant is 20.0×10^{-10} m^2/s. The CCl_4 enters the tube at a mass rate of 5.00×10^{-13} kg/s. Using these data and those shown in the drawing, find **(a)** the mass of CCl_4 per second that passes point A and **(b)** the concentration of CCl_4 at point A.

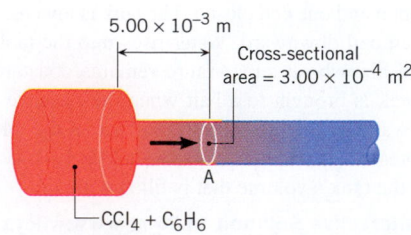

*51. **ssm www** Review Conceptual Example 7 as background for this problem. It is possible to convert Fick's law into a form that is useful when the concentration is zero at one end of the diffusion channel ($C_1 = 0$ in Figure 14.15a). **(a)** Noting that AL is the volume V of the channel and that m/V is the average concentration of solute in the channel, show that Fick's law becomes $t = L^2/(2D)$. This form of Fick's law can be used to estimate the time required for the first solute molecules to traverse the channel. **(b)** A bottle of perfume is opened in a room where convection currents are absent. Assuming that the diffusion constant for perfume in air is 1.0×10^{-5} m²/s, estimate the minimum time required for the perfume to be smelled 2.5 cm away.

**52. The drawing shows a container that is partially filled with 2.0 grams of water. The temperature is maintained at a constant 20 °C. The space above the liquid contains air that is completely saturated with water vapor. A tube of length 0.15 m and cross-sectional area

3.0×10^{-4} m² connects the water vapor at one end to air that remains completely dry at the other end. The diffusion constant for water vapor in air is 2.4×10^{-5} m²/s. How long does it take for the water in the container to evaporate completely? *(Hint: Refer to Problem 75 in Chapter 12 to find the pressure of the water vapor.)*

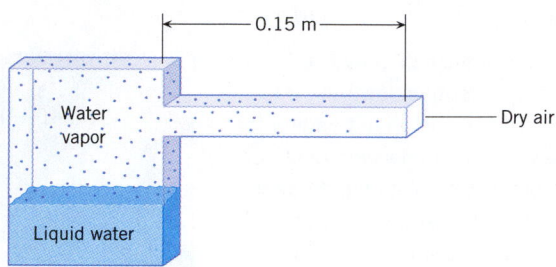

0.15 m — Water vapor — Dry air — Liquid water

ADDITIONAL PROBLEMS

53. At the start of a trip, a driver adjusts the absolute pressure in her tires to be 2.81×10^5 Pa when the outdoor temperature is 284 K. At the end of the trip she measures the pressure to be 3.01×10^5 Pa. Ignoring the expansion of the tires, find the air temperature inside the tires at the end of the trip.

54. A young male adult takes in about 5.0×10^{-4} m³ of fresh air during a normal breath. Fresh air contains approximately 21% oxygen. Assuming that the pressure in the lungs is 1.0×10^5 Pa and that air is an ideal gas at a temperature of 310 K, find the number of oxygen molecules in a normal breath.

55. **ssm** What is the density (in kg/m³) of nitrogen gas (molecular mass = 28 u) at a pressure of 2.0 atmospheres and a temperature of 310 K?

56. **GO** The diffusion constant for the amino acid glycine in water has a value of 1.06×10^{-9} m²/s. In a 2.0-cm-long tube with a cross-sectional area of 1.5×10^{-4} m², the mass rate of diffusion is $m/t = 4.2 \times 10^{-14}$ kg/s, because the glycine concentration is maintained at a value of 8.3×10^{-3} kg/m³ at one end of the tube and at a lower value at the other end. What is the lower concentration?

57. **ssm** The average value of the squared speed $\overline{v^2}$ does not equal the square of the average speed $(\overline{v})^2$. To verify this fact, consider three particles with the following speeds: $v_1 = 3.0$ m/s, $v_2 = 7.0$ m/s, and $v_3 = 9.0$ m/s. Calculate **(a)** $\overline{v^2} = \frac{1}{3}(v_1^2 + v_2^2 + v_3^2)$ and **(b)** $(\overline{v})^2 = [\frac{1}{3}(v_1 + v_2 + v_3)]^2$.

58. Refer to Multiple-Concept Example 6 for insight into the concepts used in this problem. An oxygen molecule is moving near the earth's surface. Another oxygen molecule is moving in the ionosphere (the uppermost part of the earth's atmosphere) where the Kelvin temperature is three times greater. Determine the ratio of the translational rms speed in the ionosphere to the translational rms speed near the earth's surface.

59. **ssm** A tank contains 0.85 mol of molecular nitrogen (N_2). Determine the mass (in grams) of nitrogen that must be *removed* from the tank in order to lower the pressure from 38 to 25 atm. Assume that the volume and temperature of the nitrogen in the tank do not change.

*60. Estimate the spacing between the centers of neighboring atoms in a piece of solid aluminum, based on a knowledge of the density (2700 kg/m³) and atomic mass (26.9815 u) of aluminum. *(Hint: Assume that the volume of the solid is filled with many small cubes, with one atom at the center of each.)*

*61. **Interactive Solution 14.61** at **www.wiley.com/college/cutnell** provides a model for problems of this type. The temperature near the surface of the earth is 291 K. A xenon atom (atomic mass = 131.29 u) has a kinetic energy equal to the average translational kinetic energy and is moving straight up. If the atom does not collide with any other atoms or molecules, how high up will it go before coming to rest? Assume that the acceleration due to gravity is constant throughout the ascent.

*62. Compressed air can be pumped underground into huge caverns as a form of energy storage. The volume of a cavern is 5.6×10^5 m³, and the pressure of the air in it is 7.7×10^6 Pa. Assume that air is a diatomic ideal gas whose internal energy U is given by $U = \frac{5}{2}nRT$. If one home uses 30.0 kW·h of energy per day, how many homes could this internal energy serve for one day?

**63. When perspiration on the human body absorbs heat, some of the perspiration turns into water vapor. The latent heat of vaporization at body temperature (37 °C) is 2.42×10^6 J/kg. The heat absorbed is approximately equal to the average energy $\overline{E}$ given to a single water molecule (H_2O) times the number of water molecules that are vaporized. What is $\overline{E}$?

**64. The mass of a hot-air balloon and its occupants is 320 kg (excluding the hot air inside the balloon). The air outside the balloon has a pressure of 1.01×10^5 Pa and a density of 1.29 kg/m³. To lift off, the air inside the balloon is heated. The volume of the heated balloon is 650 m³. The pressure of the heated air remains the same as the pressure of the outside air. To what temperature (in kelvins) must the air be heated so that the balloon just lifts off? The molecular mass of air is 29 u.

THERMODYNAMICS

A high-performance Funny Car dragster accelerates down the track. Such dragsters use a heat engine for propulsion that can develop 8000 or more horsepower. Note the "header flames" from the exhaust system. The laws that govern the use of heat and work form the basis of thermodynamics, the subject of this chapter. (© David Allio/Icon SMI/Corbis)

Figure 15.1 The air in this colorful hot-air balloon is one example of a thermodynamic system.
(© William Panzer/Alamy Images)

15.1 THERMODYNAMIC SYSTEMS AND THEIR SURROUNDINGS

We have studied heat (Chapter 12) and work (Chapter 6) as separate topics. Often, however, they occur simultaneously. In an automobile engine, for instance, fuel is burned at a relatively high temperature, some of its internal energy is used for doing the work of driving the pistons up and down, and the excess heat is removed by the cooling system to prevent overheating. *Thermodynamics* is the branch of physics that is built upon the fundamental laws that heat and work obey.

In thermodynamics the collection of objects on which attention is being focused is called the *system,* while everything else in the environment is called the *surroundings.* For example, the system in an automobile engine could be the burning gasoline, while the surroundings would then include the pistons, the exhaust system, the radiator, and the outside air. The system and its surroundings are separated by walls of some kind. Walls that permit heat to flow through them, such as those of the engine block, are called *diathermal walls.* Perfectly insulating walls that do not permit heat to flow between the system and its surroundings are known as *adiabatic walls.*

To understand what the laws of thermodynamics have to say about the relationship between heat and work, it is necessary to describe the physical condition or *state of a system.* We might be interested, for instance, in the hot air within the balloon in Figure 15.1. The hot air itself would be the system, and the skin of the balloon provides the walls that separate this system from the surrounding cooler air. The state of the system would be specified by giving values for the pressure, volume, temperature, and mass of the hot air.

As this chapter discusses, there are four laws of thermodynamics. We begin with the one known as the zeroth law and then consider the remaining three.

15.2 THE ZEROTH LAW OF THERMODYNAMICS

The zeroth law of thermodynamics deals with the concept of *thermal equilibrium.* Two systems are said to be in thermal equilibrium if there is no net flow of heat between them when they are brought into thermal contact. For instance, you are definitely *not* in thermal equilibrium with the water in Lake Michigan in January. Just dive into it, and you will find out how quickly your body loses heat to the frigid water. To help explain the central idea of the zeroth law of thermodynamics, Figure 15.2a shows two systems labeled A and B. Each is within a container whose adiabatic walls are made from insulation that prevents the flow of heat, and each has the same temperature, as indicated by a thermometer. In part b, one wall of each container is replaced by a thin silver sheet, and the two sheets are touched together. Silver has a large thermal conductivity, so heat flows through it readily and the silver sheets behave as diathermal walls. Even though the diathermal walls would permit it, no net flow of heat occurs in part b, indicating that the two systems are in thermal equilibrium. There is no net flow of heat because the two systems have the same temperature. We see, then, that *temperature is the indicator of thermal equilibrium in the sense that there is no net flow of heat between two systems in thermal contact that have the same temperature.*

In Figure 15.2 the thermometer plays an important role. System A is in equilibrium with the thermometer, and so is system B. In each case, the thermometer registers the same temperature, thereby indicating that the two systems are equally hot. Consequently, systems A and B are found to be in thermal equilibrium with each other. In effect, the thermometer is a third system. The fact that system A and system B are each in thermal equilibrium with this third system at the same temperature means that they are in thermal equilibrium with each other. This finding is an example of the *zeroth law of thermodynamics.*

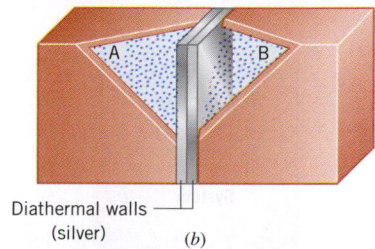

Figure 15.2 (*a*) Systems A and B are surrounded by adiabatic walls and register the same temperature on a thermometer. (*b*) When A is put into thermal contact with B through diathermal walls, no net flow of heat occurs between the systems.

THE ZEROTH LAW OF THERMODYNAMICS

Two systems individually in thermal equilibrium with a third system* are in thermal equilibrium with each other.

The zeroth law establishes temperature as the indicator of thermal equilibrium and implies that all parts of a system must be in thermal equilibrium if the system is to have a definable single temperature. In other words, there can be no flow of heat within a system that is in thermal equilibrium.

15.3 THE FIRST LAW OF THERMODYNAMICS

The atoms and molecules of a substance have kinetic and potential energy. These and other kinds of molecular energy constitute the internal energy of a substance. When a substance participates in a process involving energy in the form of work and heat, the internal energy of the substance can change. The relationship between work, heat, and changes in the internal energy is known as the first law of thermodynamics. We will now see that the first law of thermodynamics is an expression of the conservation of energy.

Suppose that a system gains heat Q and that this is the only effect occurring. Consistent with the law of conservation of energy, the internal energy of the system increases from an initial value of U_i to a final value of U_f, the change being $\Delta U = U_f - U_i = Q$. In writing this equation, we use the convention that *heat Q is positive when the system gains heat and negative when the system loses heat.* The internal energy of a system can also change because of work. If a system does work W on its surroundings and there is no heat flow, energy conservation indicates that the internal energy of the system decreases from U_i to U_f, the change now being $\Delta U = U_f - U_i = -W$. The minus sign is included because we follow the convention that *work is positive when it is done by the system and negative when it is*

Problem-solving insight

Problem-solving insight

---

*The state of the third system is the same when it is in thermal equilibrium with either of the two systems. In Figure 15.2, for example, the mercury level is the same in the thermometer in either system.

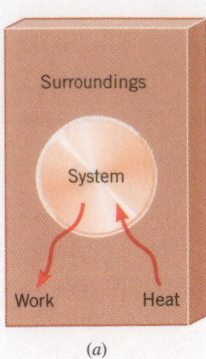

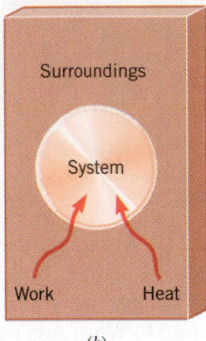

Figure 15.3 (*a*) The system gains energy in the form of heat but loses energy because work is done by the system. (*b*) The system gains energy in the form of heat and also gains energy because work is done on the system.

Problem-solving insight

When using the first law of thermodynamics, as expressed by Equation 15.1, be careful to follow the proper sign conventions for the heat Q and the work W.

done on the system. A system can gain or lose energy simultaneously in the form of heat Q and work W. The change in internal energy due to both factors is given by Equation 15.1. Thus, the *first law of thermodynamics* is just the conservation-of-energy principle applied to heat, work, and the change in the internal energy.

THE FIRST LAW OF THERMODYNAMICS

The internal energy of a system changes from an initial value U_i to a final value of U_f due to heat Q and work W:

$$\Delta U = U_f - U_i = Q - W \tag{15.1}$$

Q is positive when the system gains heat and negative when it loses heat. W is positive when work is done by the system and negative when work is done on the system.

Example 1 illustrates the use of Equation 15.1 and the sign conventions for Q and W.

 Example 1   Positive and Negative Work

Figure 15.3 illustrates a system and its surroundings. In part *a*, the system gains 1500 J of heat from its surroundings, and 2200 J of work is done *by* the system on the surroundings. In part *b*, the system also gains 1500 J of heat, but 2200 J of work is done *on* the system by the surroundings. In each case, determine the change in the internal energy of the system.

Reasoning In Figure 15.3*a* the system loses more energy in doing work than it gains in the form of heat, so the internal energy of the system decreases. Thus, we expect the change in the internal energy, $\Delta U = U_f - U_i$, to be negative. In part *b* of the drawing, the system gains energy in the form of both heat and work. The internal energy of the system increases, and we expect ΔU to be positive.

Solution (**a**) The heat is positive, $Q = +1500$ J, since it is gained by the system. The work is positive, $W = +2200$ J, since it is done *by* the system. According to the first law of thermodynamics

$$\Delta U = Q - W = (+1500 \text{ J}) - (+2200 \text{ J}) = \boxed{-700 \text{ J}} \tag{15.1}$$

The minus sign for ΔU indicates that the internal energy has decreased, as expected.

(**b**) The heat is positive, $Q = +1500$ J, since it is gained by the system. But the work is negative, $W = -2200$ J, since it is done *on* the system. Thus,

$$\Delta U = Q - W = (+1500 \text{ J}) - (-2200 \text{ J}) = \boxed{+3700 \text{ J}} \tag{15.1}$$

The plus sign for ΔU indicates that the internal energy has increased, as expected.

In the first law of thermodynamics, the internal energy U, heat Q, and work W are energy quantities, and each is expressed in energy units such as joules. However, there is a fundamental difference between U, on the one hand, and Q and W on the other. The next example sets the stage for explaining this difference.

 Example 2   An Ideal Gas

The temperature of three moles of a monatomic ideal gas is reduced from $T_i = 540$ K to $T_f = 350$ K by two different methods. In the first method 5500 J of heat flows into the gas, whereas in the second, 1500 J of heat flows into it. In each case find (**a**) the change in the internal energy and (**b**) the work done by the gas.

Reasoning Since the internal energy of a monatomic ideal gas is $U = \frac{3}{2}nRT$ (Equation 14.7) and since the number of moles n is fixed, only a change in temperature T can alter the internal energy. Because the change in T is the same in both methods, the change in U is also the same. From the given temperatures, the change ΔU in internal energy can be determined. Then, the first law of thermodynamics can be used with ΔU and the given heat values to calculate the work for each of the methods.

Solution (a) Using Equation 14.7 for the internal energy of a monatomic ideal gas, we find for each method of adding heat that

$$\Delta U = \tfrac{3}{2}nR(T_f - T_i) = \tfrac{3}{2}(3.0 \text{ mol})[8.31 \text{ J/(mol} \cdot \text{K)}](350 \text{ K} - 540 \text{ K}) = \boxed{-7100 \text{ J}}$$

(b) Since ΔU is now known and the heat is given in each method, Equation 15.1 can be used to determine the work:

1st method $W = Q - \Delta U = 5500 \text{ J} - (-7100 \text{ J}) = \boxed{12\ 600 \text{ J}}$

2nd method $W = Q - \Delta U = 1500 \text{ J} - (-7100 \text{ J}) = \boxed{8600 \text{ J}}$

In each method the gas does work, but it does more in the first method.

To understand the difference between U and either Q or W, consider the value for ΔU in Example 2. In both methods ΔU is the same. Its value is determined once the initial and final temperatures are specified because the internal energy of an ideal gas depends only on the temperature. Temperature is one of the variables (along with pressure and volume) that define the state of a system. ***The internal energy depends only on the state of a system, not on the method by which the system arrives at a given state.*** In recognition of this characteristic, internal energy is referred to as a ***function of state.*** In contrast, heat and work are not functions of state because they have different values for each different method used to make the system change from one state to another, as in Example 2.

Problem-solving insight

✓ CHECK YOUR UNDERSTANDING

(The answer is given at the end of the book.)

1. A gas is enclosed within a chamber that is fitted with a frictionless piston. The piston is then pushed in, thereby compressing the gas. Which statement below regarding this process is consistent with the first law of thermodynamics?   **(a)** The internal energy of the gas will increase.   **(b)** The internal energy of the gas will decrease.   **(c)** The internal energy of the gas will not change.   **(d)** The internal energy of the gas may increase, decrease, or remain the same, depending on the amount of heat that the gas gains or loses.

15.4 THERMAL PROCESSES

A system can interact with its surroundings in many ways, and the heat and work that come into play always obey the first law of thermodynamics. This section introduces four common thermal processes. In each case, the process is assumed to be ***quasi-static,*** which means that it occurs slowly enough that a uniform pressure and temperature exist throughout all regions of the system at all times.

An isobaric process is one that occurs at constant pressure. For instance, Figure 15.4 shows a substance (solid, liquid, or gas) contained in a chamber fitted with a frictionless piston. The pressure P experienced by the substance is always the same, because it is determined by the external atmosphere and the weight of the piston and the block resting on it. Heating the substance makes it expand and do work W in lifting the piston and block through the displacement $\vec{s}$. The work can be calculated from $W = Fs$ (Equation 6.1), where F is the magnitude of the force and s is the magnitude of the displacement. The force is generated by the pressure P acting on the bottom surface of the piston (area = A), according to $F = PA$ (Equation 10.19). With this substitution for F, the work becomes $W = (PA)s$. But the product $A \cdot s$ is the change in volume of the material, $\Delta V = V_f - V_i$, where V_f and V_i are the final and initial volumes, respectively. Thus, the relation is

Isobaric process $W = P\,\Delta V = P(V_f - V_i)$ (15.2)

Consistent with our sign convention, this result predicts a positive value for the work done *by a system* when it expands isobarically (V_f exceeds V_i). Equation 15.2 also applies to an

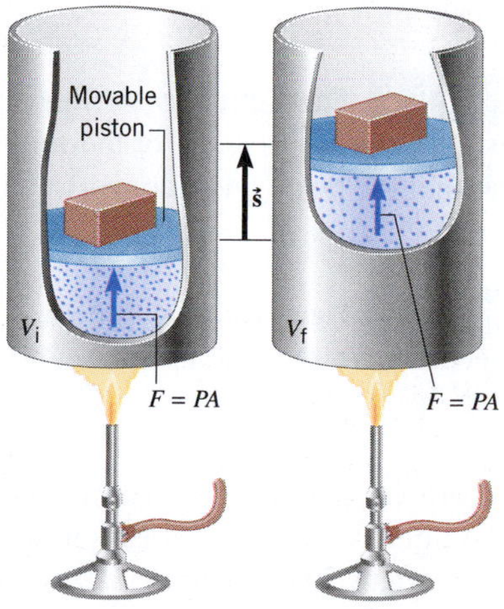

Figure 15.4 The substance in the chamber is expanding isobarically because the pressure is held constant by the external atmosphere and the weight of the piston and the block.

*The fact that an ideal gas is used in Example 2 does not restrict our conclusion. Had a real (nonideal) gas or other material been used, the only difference would have been that the expression for the internal energy would have been more complicated. It might have involved the volume V, as well as the temperature T, for instance.

isobaric compression (V_f less than V_i). Then, the work is negative, since work must be done *on the system* to compress it. Example 3 emphasizes that $W = P \, \Delta V$ applies to any system, solid, liquid, or gas, as long as the pressure remains constant while the volume changes.

ANALYZING MULTIPLE-CONCEPT PROBLEMS

Example 3   Isobaric Expansion of Water

One gram of water is placed in the cylinder in Figure 15.4, and the pressure is maintained at 2.0×10^5 Pa. The temperature of the water is raised by 31 C°. In one case, the water is in the liquid phase, expands by the small amount of 1.0×10^{-8} m³, and has a specific heat capacity of 4186 J/(kg·C°). In another case, the water is in the gas phase, expands by the much greater amount of 7.1×10^{-5} m³, and has a specific heat capacity of 2020 J/(kg·C°). Determine the change in the internal energy of the water in each case.

Reasoning The change ΔU in the internal energy is given by the first law of thermodynamics as $\Delta U = Q - W$ (Equation 15.1). The heat Q may be evaluated as $Q = cm \, \Delta T$ (Equation 12.4). Finally, since the process occurs at a constant pressure (isobaric), the work W may be found using $W = P \, \Delta V$ (Equation 15.2).

Knowns and Unknowns The following table summarizes the given data:

Description	Symbol	Value	Comment
Mass of water	m	1.0 g	0.0010 kg
Pressure on water	P	2.0×10^5 Pa	Pressure is constant.
Increase in temperature	ΔT	31 C°	
Increase in volume of liquid	ΔV_{liquid}	1.0×10^{-8} m³	Expansion occurs.
Specific heat capacity of liquid	c_{liquid}	4186 J/(kg·C°)	
Increase in volume of gas	ΔV_{gas}	7.1×10^{-5} m³	Expansion occurs.
Specific heat capacity of gas	c_{gas}	2020 J/(kg·C°)	
Unknown Variables			
Change in internal energy of liquid	ΔU_{liquid}	?	
Change in internal energy of gas	ΔU_{gas}	?	

Modeling the Problem

STEP 1 The First Law of Thermodynamics   The change ΔU in the internal energy is given by the first law of thermodynamics, as shown at the right. In Equation 15.1, neither the heat Q nor the work W is known, so we turn to Steps 2 and 3 to evaluate them.

$$\Delta U = Q - W \qquad (15.1)$$

STEP 2 Heat and Specific Heat Capacity   According to Equation 12.4, the heat Q needed to raise the temperature of a mass m of material by an amount ΔT is

$$Q = cm \, \Delta T$$

where c is the material's specific heat capacity. Data are available for all of the terms on the right side of this expression, which can be substituted into Equation 15.1, as shown at the right. The remaining unknown variable in Equation 15.1 is the work W, and we evaluate it in Step 3.

$$\Delta U = Q - W \qquad (15.1)$$
$$Q = cm \, \Delta T \qquad (12.4)$$

STEP 3 Work Done at Constant Pressure   Under constant-pressure, or isobaric, conditions, the work W done is given by Equation 15.2 as

$$W = P \, \Delta V$$

where P is the pressure acting on the material and ΔV is the change in the volume of the material. Substitution of this expression into Equation 15.1 is shown at the right.

$$\Delta U = Q - W \qquad (15.1)$$
$$Q = cm \, \Delta T \qquad (12.4)$$
$$W = P \, \Delta V \qquad (15.2)$$

Solution Combining the results of each step algebraically, we find that

[STEP 1]   [STEP 2]   [STEP 3]

$$\Delta U = Q - W = cm\,\Delta T - W = cm\,\Delta T - P\,\Delta V$$

Applying this result to the liquid and to the gaseous water gives

$$\Delta U_{\text{liquid}} = c_{\text{liquid}}m\,\Delta T - P\,\Delta V_{\text{liquid}}$$

$$= [4186\ \text{J/(kg·C°)}](0.0010\ \text{kg})(31\ \text{C°}) - (2.0 \times 10^5\ \text{Pa})(1.0 \times 10^{-8}\ \text{m}^3)$$

$$= 130\ \text{J} - 0.0020\ \text{J} = \boxed{130\ \text{J}}$$

$$\Delta U_{\text{gas}} = c_{\text{gas}}m\,\Delta T - P\,\Delta V_{\text{gas}}$$

$$= [2020\ \text{J/(kg·C°)}](0.0010\ \text{kg})(31\ \text{C°}) - (2.0 \times 10^5\ \text{Pa})(7.1 \times 10^{-5}\ \text{m}^3)$$

$$= 63\ \text{J} - 14\ \text{J} = \boxed{49\ \text{J}}$$

For the liquid, virtually all the 130 J of heat serves to change the internal energy, since the volume change and the corresponding work of expansion are so small. In contrast, a significant fraction of the 63 J of heat added to the gas causes work of expansion to be done, so that only 49 J is left to change the internal energy.

Related Homework: *Problem 14*

It is often convenient to display thermal processes graphically. For instance, Figure 15.5 shows a plot of pressure versus volume for an isobaric expansion. Since the pressure is constant, the graph is a horizontal straight line, beginning at the initial volume V_i and ending at the final volume V_f. In terms of such a plot, the work $W = P(V_f - V_i)$ is the area under the graph, which is the shaded rectangle of height P and width $V_f - V_i$.

Another common thermal process is an **isochoric process, one that occurs at constant volume.** Figure 15.6a illustrates an isochoric process in which a substance (solid, liquid, or gas) is heated. The substance would expand if it could, but the rigid container keeps the volume constant, so the pressure–volume plot shown in Figure 15.6b is a vertical straight line. Because the volume is constant, the pressure inside rises, and the substance exerts more and more force on the walls. Although enormous forces can be generated in the closed container, no work is done ($W = 0$ J), since the walls do not move. Consistent with zero work being done, the area under the vertical straight line in Figure 15.6b is zero. Since no work is done, the first law of thermodynamics indicates that the heat in an isochoric process serves only to change the internal energy: $\Delta U = Q - W = Q$.

A third important thermal process is an **isothermal process, one that takes place at constant temperature.** The next section illustrates the important features of an isothermal process when the system is an ideal gas.

Last, there is the **adiabatic process, one that occurs without the transfer of heat.** Since there is no heat transfer, Q equals zero, and the first law indicates that $\Delta U = Q - W = -W$. Thus, when work is done by a system adiabatically, W is positive and the internal energy decreases by exactly the amount of the work done. When work is done on a system adiabatically, W is negative and the internal energy increases correspondingly. The next section discusses an adiabatic process for an ideal gas.

A process may be complex enough that it is not recognizable as one of the four just discussed. For instance, Figure 15.7 shows a process for a gas in which the pressure, volume, and temperature are changed along the straight line from X to Y. With the aid of integral calculus, the following can be proved. **Problem-solving insight:** *The area under a pressure–volume graph is the work for any kind of process.* Thus, the area representing the work has been colored in Figure 15.7. The volume increases, so that work is done by the gas. This work is positive by convention, as is the area. In contrast, if a process reduces the volume, work is done on the gas, and this work is negative by convention. Correspondingly, the area under the pressure–volume graph would be assigned a negative value. In Example 4, we determine the work for the case shown in Figure 15.7.

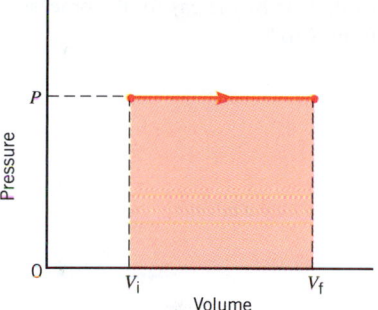

Figure 15.5 For an isobaric process, a pressure-versus-volume plot is a horizontal straight line, and the work done [$W = P(V_f - V_i)$] is the colored rectangular area under the graph.

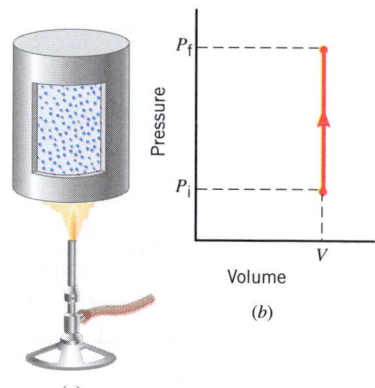

Figure 15.6 (a) The substance in the chamber is being heated isochorically because the rigid chamber keeps the volume constant. (b) The pressure–volume plot for an isochoric process is a vertical straight line. The area under the graph is zero, indicating that no work is done.

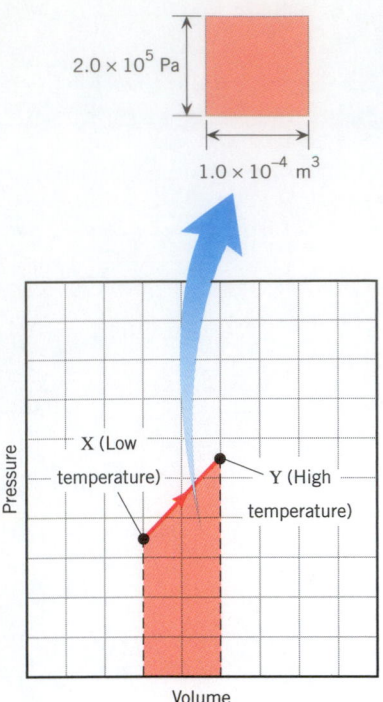

Figure 15.7 The colored area gives the work done by the gas for the process from X to Y.

Example 4 Work and the Area Under a Pressure–Volume Graph

Determine the work for the process in which the pressure, volume, and temperature of a gas are changed along the straight line from X to Y in Figure 15.7.

Reasoning The work is given by the area (in color) under the straight line between X and Y. Since the volume increases, work is done by the gas on the surroundings, so the work is positive. The area can be found by counting squares in Figure 15.7 and multiplying by the area per square.

Solution We estimate that there are 8.9 colored squares in the drawing. The area of one square is $(2.0 \times 10^5 \text{ Pa})(1.0 \times 10^{-4} \text{ m}^3) = 2.0 \times 10^1 \text{ J}$, so the work is

$$W = +(8.9 \text{ squares})(2.0 \times 10^1 \text{ J/square}) = \boxed{+180 \text{ J}}$$

✓ CHECK YOUR UNDERSTANDING

(The answers are given at the end of the book.)

2. The drawing shows a pressure-versus-volume plot for a three-step process: A → B, B → C, and C → A. For each step, the work can be positive, negative, or zero. Which answer in the table correctly describes the work for the three steps?

	Work Done by the System		
	A → B	B → C	C → A
(a)	Positive	Negative	Negative
(b)	Positive	Positive	Negative
(c)	Negative	Negative	Positive
(d)	Positive	Negative	Zero
(e)	Negative	Positive	Zero

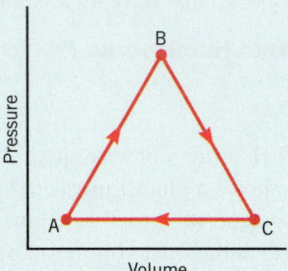

3. Is it possible for the temperature of a substance to rise without heat flowing into the substance?   **(a)** Yes, provided that the volume of the substance does not change.   **(b)** Yes, provided that the substance expands and does positive work.   **(c)** Yes, provided that work is done on the substance and it contracts.

4. The drawing shows a pressure-versus-volume graph in which a gas expands at constant pressure from *A* to *B*, and then goes from *B* to *C* at constant volume. Complete the table by deciding whether each of the four unspecified quantities is positive (+), negative (−), or zero (0).

	ΔU	Q	W
A → B	+	?	?
B → C	?	+	?

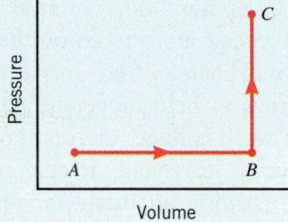

5. When a solid melts at constant pressure, the volume of the resulting liquid does not differ much from the volume of the solid. According to the first law of thermodynamics, how does the internal energy of the liquid compare to the internal energy of the solid? The internal energy of the liquid is   **(a)** greater than,   **(b)** the same as,   **(c)** less than the internal energy of the solid.

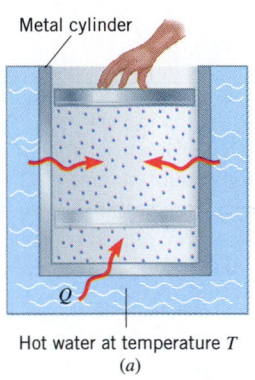

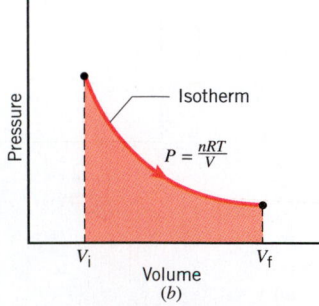

Figure 15.8 (*a*) The ideal gas in the cylinder is expanding isothermally at temperature *T*. The force holding the piston in place is reduced slowly, so the expansion occurs quasi-statically. (*b*) The work done by the gas is given by the colored area.

15.5 THERMAL PROCESSES USING AN IDEAL GAS

ISOTHERMAL EXPANSION OR COMPRESSION

When a system performs work isothermally, the temperature remains constant. In Figure 15.8*a*, for instance, a metal cylinder contains *n* moles of an ideal gas, and the large mass of hot water maintains the cylinder and gas at a constant Kelvin temperature *T*. The piston is held in place initially so the volume of the gas is V_i. As the external force applied to the piston is reduced quasi-statically, the pressure decreases as the gas expands to the final volume V_f. Figure 15.8*b* gives a plot of pressure ($P = nRT/V$) versus volume for the process. The solid red line in the graph is called an isotherm (meaning "constant temperature") and represents the relation between pressure and volume when the temperature is held constant. The work W done by the gas is *not* given by $W = P\,\Delta V = P(V_f - V_i)$ because the pressure is not constant. Nevertheless, the work is equal to the area under the graph. The techniques of integral calculus lead to the following result* for W:

Isothermal expansion or compression of an ideal gas

$$W = nRT \ln\left(\frac{V_f}{V_i}\right) \qquad (15.3)$$

Where does the energy for this work originate? Since the internal energy of any ideal gas is proportional to the Kelvin temperature ($U = \frac{3}{2}nRT$ for a monatomic ideal gas, for example), the internal energy remains constant throughout an isothermal process, and the change in internal energy is zero. As a result, the first law of thermodynamics becomes $\Delta U = 0 = Q - W$. In other words, $Q = W$, and the energy for the work originates in the hot water. Heat flows into the gas from the water, as Figure 15.8a illustrates. If the gas is compressed isothermally, Equation 15.3 still applies, and heat flows out of the gas into the water. Example 5 deals with the isothermal expansion of an ideal gas.

 Example 5 Isothermal Expansion of an Ideal Gas

Two moles of the monatomic gas argon expand isothermally at 298 K, from an initial volume of $V_i = 0.025$ m³ to a final volume of $V_f = 0.050$ m³. Assuming that argon is an ideal gas, find (a) the work done by the gas, (b) the change in the internal energy of the gas, and (c) the heat supplied to the gas.

Reasoning and Solution (a) The work done by the gas can be found from Equation 15.3:

$$W = nRT \ln\left(\frac{V_f}{V_i}\right) = (2.0 \text{ mol})[8.31 \text{ J/(mol·K)}](298 \text{ K}) \ln\left(\frac{0.050 \text{ m}^3}{0.025 \text{ m}^3}\right) = \boxed{+3400 \text{ J}}$$

(b) The internal energy of a monatomic ideal gas is $U = \frac{3}{2}nRT$ (Equation 14.7) and does not change when the temperature is constant. Therefore, $\boxed{\Delta U = 0 \text{ J}}$.

(c) The heat Q supplied can be determined from the first law of thermodynamics:

$$Q = \Delta U + W = 0 \text{ J} + 3400 \text{ J} = \boxed{+3400 \text{ J}} \qquad (15.1)$$

ADIABATIC EXPANSION OR COMPRESSION

When a system performs work adiabatically, no heat flows into or out of the system. Figure 15.9a shows an arrangement in which n moles of an ideal gas do work under adiabatic conditions, expanding quasi-statically from an initial volume V_i to a final volume V_f. The arrangement is similar to that in Figure 15.8 for isothermal expansion. However, a different amount of work is done here, because the cylinder is now surrounded by insulating material that prevents the flow of heat from occurring, so $Q = 0$ J. According to the first law of thermodynamics, the change in internal energy is $\Delta U = Q - W = -W$. Since the internal energy of an ideal monatomic gas is $U = \frac{3}{2}nRT$ (Equation 14.7), it follows directly that $\Delta U = U_f - U_i = \frac{3}{2}nR(T_f - T_i)$, where T_i and T_f are the initial and final Kelvin temperatures. With this substitution, the relation $\Delta U = -W$ becomes

Adiabatic expansion or compression of a monatomic ideal gas

$$W = \frac{3}{2}nR(T_i - T_f) \qquad (15.4)$$

When an ideal gas expands adiabatically, it does positive work, so W is positive in Equation 15.4. Therefore, the term $T_i - T_f$ is also positive, so the final temperature of the gas must be less than the initial temperature. The internal energy of the gas is reduced to

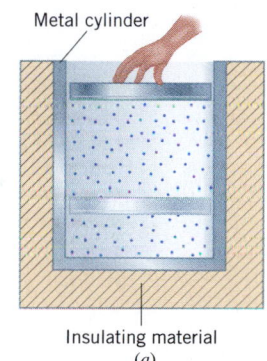

Metal cylinder

Insulating material
(a)

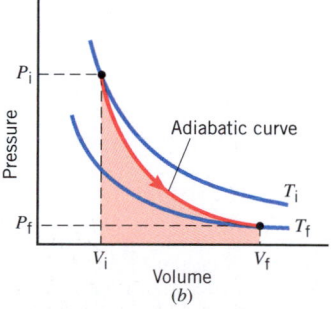

Pressure vs. Volume

(b)

Figure 15.9 (a) The ideal gas in the cylinder is expanding adiabatically. The force holding the piston in place is reduced slowly, so the expansion occurs quasi-statically. (b) A plot of pressure versus volume yields the adiabatic curve shown in red, which intersects the isotherms (blue) at the initial temperature T_i and the final temperature T_f. The work done by the gas is given by the colored area.

*In this result, "ln" denotes the natural logarithm to the base $e = 2.71828$. The natural logarithm is related to the common logarithm to the base ten by $\ln(V_f/V_i) = 2.303 \log(V_f/V_i)$.

Table 15.1 Summary of Thermal Processes

Type of Thermal Process	Work Done	First Law of Thermodynamics $(\Delta U = Q - W)$
Isobaric (constant pressure)	$W = P(V_f - V_i)$	$\Delta U = Q - \underbrace{P(V_f - V_i)}_{W}$
Isochoric (constant volume)	$W = 0 \text{ J}$	$\Delta U = Q - \underbrace{0 \text{ J}}_{W}$
Isothermal (constant temperature)	$W = nRT \ln\left(\dfrac{V_f}{V_i}\right)$ (for an ideal gas)	$\underbrace{0 \text{ J}}_{\substack{\Delta U \text{ for an} \\ \text{ideal gas}}} = Q - \underbrace{nRT \ln\left(\dfrac{V_f}{V_i}\right)}_{W}$
Adiabatic (no heat flow)	$W = \tfrac{3}{2}nR(T_i - T_f)$ (for a monatomic ideal gas)	$\Delta U = \underbrace{0 \text{ J}}_{Q} - \underbrace{\tfrac{3}{2}nR(T_i - T_f)}_{W}$

provide the necessary energy to do the work, and because the internal energy is proportional to the Kelvin temperature, the temperature decreases. Figure 15.9b shows a plot of pressure versus volume for an adiabatic process. The adiabatic curve (red) intersects the isotherms (blue) at the higher initial temperature $[T_i = P_iV_i/(nR)]$ and also at the lower final temperature $[T_f = P_fV_f/(nR)]$. The colored area under the adiabatic curve represents the work done.

The reverse of an adiabatic expansion is an adiabatic compression (W is negative), and Equation 15.4 indicates that the final temperature exceeds the initial temperature. The energy provided by the agent doing the work increases the internal energy of the gas. As a result, the gas becomes hotter.

The equation that gives the adiabatic curve (red) between the initial pressure and volume (P_i, V_i) and the final pressure and volume (P_f, V_f) in Figure 15.9b can be derived using integral calculus. The result is

Adiabatic expansion or compression of an ideal gas

$$P_iV_i^{\gamma} = P_fV_f^{\gamma} \tag{15.5}$$

where the exponent γ (Greek gamma) is the ratio of the specific heat capacities at constant pressure and constant volume, $\gamma = c_P/c_V$. Equation 15.5 applies in conjunction with the ideal gas law, because *each point* on the adiabatic curve satisfies the relation $PV = nRT$.

Table 15.1 summarizes the work done in the four types of thermal processes that we have been considering. For each process it also shows how the first law of thermodynamics depends on the work and other variables.

✓ CHECK YOUR UNDERSTANDING

(The answers are given at the end of the book.)

6. One hundred joules of heat is added to a gas, and the gas expands at constant pressure. Is it possible that the internal energy increases by 100 J?   **(a)** Yes   **(b)** No; the increase in the internal energy is less than 100 J, since work is done by the gas.   **(c)** No; the increase in the internal energy is greater than 100 J, since work is done by the gas.

7. A gas is compressed isothermally, and its internal energy increases. Is the gas an ideal gas?   **(a)** No, because if the temperature of an ideal gas remains constant, its internal energy must also remain constant.   **(b)** No, because if the temperature of an ideal gas remains constant, its internal energy must decrease.   **(c)** Yes, because if the temperature of an ideal gas remains constant, its internal energy must increase.

8. A material undergoes an isochoric process that is also adiabatic. Is the internal energy of the material at the end of the process   **(a)** greater than,   **(b)** less than, or   **(c)** the same as it was at the start?

9. The drawing shows an arrangement for an adiabatic free expansion or "throttling" process. The process is adiabatic because the entire arrangement is contained within perfectly insulating walls. The gas in chamber A rushes suddenly into chamber B through a hole in the partition. Chamber B is initially evacuated, so the gas expands there under zero external pressure and the work ($W = P\,\Delta V$) it does is zero. Assume that the gas is an ideal gas. How does the final temperature of the gas after expansion compare to its initial temperature? The final temperature is   **(a)** greater than,   **(b)** less than,   **(c)** the same as the initial temperature.

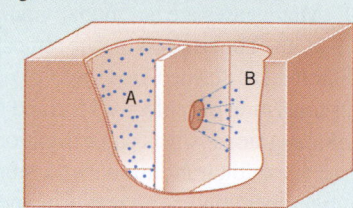

15.6 SPECIFIC HEAT CAPACITIES

In this section the first law of thermodynamics is used to gain an understanding of the factors that determine the specific heat capacity of a material. Remember, when the temperature of a substance changes as a result of heat flow, the change in temperature ΔT and the amount of heat Q are related according to $Q = cm\,\Delta T$ (Equation 12.4). In this expression c denotes the specific heat capacity in units of J/(kg·C°), and m is the mass in kilograms. Now, however, it is more convenient to express the amount of material as the number of moles n, rather than the number of kilograms. Therefore, we replace the expression $Q = cm\,\Delta T$ with the following analogous expression:

$$Q = Cn\,\Delta T \tag{15.6}$$

where the capital letter C (as opposed to the lowercase c) refers to the **molar specific heat capacity** in units of J/(mol·K). In addition, the unit for measuring the temperature change ΔT is the kelvin (K) rather than the Celsius degree (C°), and $\Delta T = T_f - T_i$, where T_f and T_i are the final and initial temperatures. For gases it is necessary to distinguish between the molar specific heat capacities C_P and C_V, which apply, respectively, to conditions of constant pressure and constant volume. With the help of the first law of thermodynamics and an ideal gas as an example, it is possible to see why C_P and C_V differ.

To determine the molar specific heat capacities, we must first calculate the heat Q needed to raise the temperature of an ideal gas from T_i to T_f. According to the first law, $Q = \Delta U + W$. We also know that the internal energy of a monatomic ideal gas is $U = \frac{3}{2}nRT$ (Equation 14.7). As a result, $\Delta U = U_f - U_i = \frac{3}{2}nR(T_f - T_i)$. When the heating process occurs at constant pressure, the work done is given by Equation 15.2: $W = P\,\Delta V = P(V_f - V_i)$. For an ideal gas, $PV = nRT$, so the work becomes $W = nR(T_f - T_i)$. On the other hand, when the volume is constant, $\Delta V = 0$ m³, and the work done is zero. The calculation of the heat is summarized below:

$$Q = \Delta U + W$$

$$Q_{\text{constant pressure}} = \tfrac{3}{2}nR(T_f - T_i) + nR(T_f - T_i) = \tfrac{5}{2}nR(T_f - T_i)$$

$$Q_{\text{constant volume}} = \tfrac{3}{2}nR(T_f - T_i) + 0$$

The molar specific heat capacities can now be determined, since Equation 15.6 indicates that $C = Q/[n(T_f - T_i)]$:

Constant pressure
for a monatomic
ideal gas

$$C_P = \frac{Q_{\text{constant pressure}}}{n(T_f - T_i)} = \tfrac{5}{2}R \tag{15.7}$$

Constant volume
for a monatomic
ideal gas

$$C_V = \frac{Q_{\text{constant volume}}}{n(T_f - T_i)} = \tfrac{3}{2}R \tag{15.8}$$

The ratio γ of the specific heats is

Monatomic
ideal gas

$$\gamma = \frac{C_P}{C_V} = \frac{\tfrac{5}{2}R}{\tfrac{3}{2}R} = \frac{5}{3} \tag{15.9}$$

For real monatomic gases near room temperature, experimental values of C_P and C_V give ratios very close to the theoretical value of $\frac{5}{3}$.

Many gases are not monatomic. Instead, they consist of molecules formed from more than one atom. The oxygen in our atmosphere, for example, is a diatomic gas, because it consists of molecules formed from two oxygen atoms. Similarly, atmospheric nitrogen is a diatomic gas consisting of molecules formed from two nitrogen atoms. Whereas the individual atoms in a monatomic ideal gas can exhibit only translational motion, the molecules in a diatomic ideal gas can exhibit translational and rotational motion, as well as vibrational motion at sufficiently high temperatures. The result of such additional motions is that Equations 15.7–15.9 do not apply to a diatomic ideal gas. Instead, if the temperature is sufficiently moderate that the diatomic molecules do not vibrate, the molar specific heat capacities of a diatomic ideal gas are $C_P = \frac{7}{2}R$ and $C_V = \frac{5}{2}R$, with the result that $\gamma = \frac{C_P}{C_V} = \frac{7}{5}$.

The difference between C_P and C_V arises because work is done when the gas expands in response to the addition of heat under conditions of constant pressure, whereas no work is done under conditions of constant volume. For a monatomic ideal gas, C_P exceeds C_V by an amount equal to R, the ideal gas constant:

$$C_P - C_V = R \qquad (15.10)$$

In fact, it can be shown that Equation 15.10 applies to any kind of ideal gas—monatomic, diatomic, etc.

✓ CHECK YOUR UNDERSTANDING

(The answers are given at the end of the book.)

10. Suppose that a material contracts when it is heated. Following the same line of reasoning used in the text to reach Equations 15.7 and 15.8, deduce the relationship between the specific heat capacity at constant pressure (C_P) and the specific heat capacity at constant volume (C_V). Which of the following describes the relationship? **(a)** $C_P = C_V$ **(b)** C_P is greater than C_V **(c)** C_P is less than C_V

11. You want to heat a gas so that its temperature will be as high as possible. Should you heat the gas under conditions of **(a)** constant pressure or **(b)** constant volume? **(c)** It does not matter what the conditions are.

15.7 THE SECOND LAW OF THERMODYNAMICS

Ice cream melts when left out on a warm day. A cold can of soda warms up on a hot day at a picnic. Ice cream and soda never become colder when left in a hot environment, for heat always flows spontaneously from hot to cold, and never from cold to hot. The spontaneous flow of heat is the focus of one of the most profound laws in all of science, the *second law of thermodynamics.*

THE SECOND LAW OF THERMODYNAMICS: THE HEAT FLOW STATEMENT

Heat flows spontaneously from a substance at a higher temperature to a substance at a lower temperature and does not flow spontaneously in the reverse direction.

It is important to realize that the second law of thermodynamics deals with a different aspect of nature than does the first law of thermodynamics. The second law is a statement about the natural tendency of heat to flow from hot to cold, whereas the first law deals with energy conservation and focuses on both heat and work. A number of important devices depend on heat and work in their operation, and to understand such devices both laws are needed. For instance, an automobile engine is a type of heat engine because it uses heat to produce work. In discussing heat engines, Sections 15.8 and 15.9 will bring together the first and second laws to analyze engine efficiency. Then, in Section 15.10 we will see that refrigerators, air conditioners, and heat pumps also utilize heat and work and are closely related to heat engines. The way in which these three appliances operate also depends on both the first and second laws of thermodynamics.

15.8 HEAT ENGINES

A *heat engine* is any device that uses heat to perform work. It has three essential features:

1. Heat is supplied to the engine at a relatively high input temperature from a place called the *hot reservoir.*

2. Part of the input heat is used to perform work by the *working substance* of the engine, which is the material within the engine that actually does the work (e.g., the gasoline–air mixture in an automobile engine).

3. The remainder of the input heat is rejected to a place called the *cold reservoir,* which has a temperature lower than the input temperature.

Figure 15.10 illustrates these features. The symbol Q_H refers to the input heat, and the subscript H indicates the hot reservoir. Similarly, the symbol Q_C stands for the rejected heat, and the subscript C denotes the cold reservoir. The symbol W refers to the work done. The vertical bars enclosing each of these three symbols in the drawing are included to emphasize that we are concerned here with the absolute values, or magnitudes, of the symbols. Thus, $|Q_H|$ indicates the magnitude of the input heat, $|Q_C|$ denotes the magnitude of the rejected heat, and $|W|$ stands for the magnitude of the work done. **Problem-solving insight:** *Since $|Q_H|$, $|Q_C|$, and $|W|$ refer to magnitudes only, they never have negative values assigned to them when they appear in equations.*

To be highly efficient, a heat engine must produce a relatively large amount of work from as little input heat as possible. Therefore, the *efficiency e* of a heat engine is defined as the ratio of the magnitude of the work $|W|$ done by the engine to the magnitude of the input heat $|Q_H|$:

$$e = \frac{|W|}{|Q_H|} \tag{15.11}$$

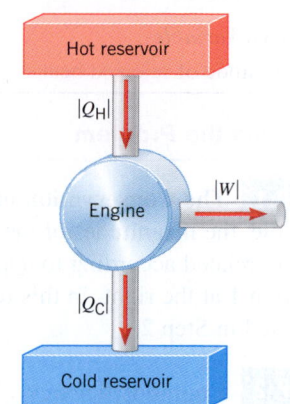

Figure 15.10 This schematic representation of a heat engine shows the input heat (magnitude $= |Q_H|$) that originates from the hot reservoir, the work (magnitude $= |W|$) that the engine does, and the heat (magnitude $= |Q_C|$) that the engine rejects to the cold reservoir.

If the input heat were converted entirely into work, the engine would have an efficiency of 1.00, since $|W| = |Q_H|$; such an engine would be 100% efficient. *Efficiencies are often quoted as percentages obtained by multiplying the ratio $|W|/|Q_H|$ by a factor of 100.* Thus, an efficiency of 68% would mean that a value of 0.68 is used for the efficiency in Equation 15.11.

An engine, like any device, must obey the principle of conservation of energy. Some of the engine's input heat $|Q_H|$ is converted into work $|W|$, and the remainder $|Q_C|$ is rejected to the cold reservoir. If there are no other losses in the engine, the principle of energy conservation requires that

$$|Q_H| = |W| + |Q_C| \tag{15.12}$$

Solving this equation for $|W|$ and substituting the result into Equation 15.11 leads to the following alternative expression for the efficiency e of a heat engine:

$$e = \frac{|Q_H| - |Q_C|}{|Q_H|} = 1 - \frac{|Q_C|}{|Q_H|} \tag{15.13}$$

Example 6 illustrates how the concepts of efficiency and energy conservation are applied to a heat engine.

ANALYZING MULTIPLE-CONCEPT PROBLEMS

Example 6   An Automobile Engine

An automobile engine has an efficiency of 22.0% and produces 2510 J of work. How much heat is rejected by the engine?

Reasoning Energy conservation indicates that the amount of heat rejected to the cold reservoir is the part of the input heat that is *not* converted into work. The work is given, and the input heat can be obtained since the efficiency of the engine is also given.

Continued

Knowns and Unknowns The following data are available:

Description	Symbol	Value
Efficiency of engine	e	22.0% (0.220)
Magnitude of work	$|W|$	2510 J
Unknown Variable		
Magnitude of rejected heat	$|Q_C|$	?

Modeling the Problem

STEP 1 **The Conservation of Energy** According to the energy-conservation principle, the magnitudes of the input heat $|Q_H|$, the work done $|W|$, and the rejected heat $|Q_C|$ are related according to $|Q_H| = |W| + |Q_C|$ (Equation 15.12). Solving for $|Q_C|$ gives Equation 1 at the right. In this result, $|W|$ is known, but $|Q_H|$ is not, although it will be evaluated in Step 2.

$$|Q_C| = |Q_H| - |W| \quad (1)$$
$$\circlearrowleft ?$$

STEP 2 **Engine Efficiency** Equation 15.11 gives the engine efficiency as $e = |W|/|Q_H|$. Solving for $|Q_H|$, we find that

$$|Q_H| = \frac{|W|}{e}$$

which can be substituted into Equation 1 as shown in the right column.

$$|Q_C| = |Q_H| - |W| \quad (1)$$
$$|Q_H| = \frac{|W|}{e}$$

Solution Combining the results of each step algebraically, we find that

| STEP 1 | STEP 2 |

$$|Q_C| = |Q_H| - |W| = \frac{|W|}{e} - |W|$$

The magnitude of the rejected heat, then, is

$$|Q_C| = |W|\left(\frac{1}{e} - 1\right) = (2510 \text{ J})\left(\frac{1}{0.220} - 1\right) = \boxed{8900 \text{ J}}$$

Problem-solving insight

When efficiency is stated as a percentage (e.g., 22.0%), it must be converted to a decimal fraction (e.g., 0.220) before being used in an equation.

Related Homework: *Problems 45, 84*

In Example 6, less than one-quarter of the input heat is converted into work because the efficiency of the automobile engine is only 22.0%. If the engine were 100% efficient, all the input heat would be converted into work. Unfortunately, nature does not permit 100%-efficient heat engines to exist, as the next section discusses.

15.9 CARNOT'S PRINCIPLE AND THE CARNOT ENGINE

What is it that allows a heat engine to operate with maximum efficiency? The French engineer Sadi Carnot (1796–1832) proposed that a heat engine has maximum efficiency when the processes within the engine are reversible. *A reversible process is one in which both the system and its environment can be returned to exactly the states they were in before the process occurred.*

In a reversible process, *both* the system and its environment can be returned to their initial states. Therefore, a process that involves an energy-dissipating mechanism, such as friction, cannot be reversible because the energy wasted due to friction would alter the system or the environment or both. There are also reasons other than friction why a process may not be reversible. For instance, the spontaneous flow of heat from a hot substance to a cold substance is irreversible, even though friction is not present. For heat to flow in the reverse direction, work must be done, as we will see in Section 15.10. The agent doing such work must be located in the environment of the hot and cold substances, and, therefore, the environment must change while the heat is moved back from cold to hot. Since the system and the environment cannot *both* be returned to their initial states, the process of spontaneous heat flow is irreversible. In fact, all spontaneous processes are irreversible, such as the explosion of an unstable chemical or the bursting of a bubble. When the word "reversible" is used in connection with engines, it does not just mean a gear that allows the engine to operate a device in reverse. All cars have a reverse gear, for instance, but no automobile engine is thermodynamically reversible, since friction exists no matter which way the car moves.

Today, the idea that the efficiency of a heat engine is a maximum when the engine operates reversibly is referred to as *Carnot's principle.*

CARNOT'S PRINCIPLE: AN ALTERNATIVE STATEMENT OF THE SECOND LAW OF THERMODYNAMICS

No irreversible engine operating between two reservoirs at constant temperatures can have a greater efficiency than a reversible engine operating between the same temperatures. Furthermore, all reversible engines operating between the same temperatures have the same efficiency.

Carnot's principle is quite remarkable, for no mention is made of the working substance of the engine. It does not matter whether the working substance is a gas, a liquid, or a solid. As long as the process is reversible, the efficiency of the engine is a maximum. However, Carnot's principle does *not* state, or even imply, that a reversible engine has an efficiency of 100%.

It can be shown that if Carnot's principle were not valid, it would be possible for heat to flow spontaneously from a cold substance to a hot substance, in violation of the second law of thermodynamics. In effect, then, Carnot's principle is another way of expressing the second law.

No real engine operates reversibly. Nonetheless, the idea of a reversible engine provides a useful standard for judging the performance of real engines. Figure 15.11 shows a reversible engine, called a *Carnot engine,* that is particularly useful as an idealized model. An important feature of a Carnot engine is that all input heat (magnitude $= |Q_H|$) originates from a hot reservoir *at a single temperature* T_H and all rejected heat (magnitude $= |Q_C|$) goes into a cold reservoir *at a single temperature* T_C. This important feature is emphasized in Problem 61, which focuses on a pressure-versus-volume plot for a Carnot engine that utilizes an ideal gas as its working substance.

Carnot's principle implies that the efficiency of a reversible engine is independent of the working substance of the engine, and therefore can depend only on the temperatures of the hot and cold reservoirs. Since efficiency is $e = 1 - |Q_C|/|Q_H|$ according to Equation 15.13, the ratio $|Q_C|/|Q_H|$ can depend only on the reservoir temperatures. This observation led Lord Kelvin to propose a *thermodynamic temperature scale.* He proposed that the thermodynamic temperatures of the cold and hot reservoirs be defined such that their ratio is equal to $|Q_C|/|Q_H|$. Thus, the thermodynamic temperature scale is related to the heats absorbed and rejected by a Carnot engine, and is independent of the working substance. If a reference temperature is properly chosen, it can be shown that the thermodynamic temperature scale is identical to the Kelvin scale introduced in Section 12.2 and used in the ideal gas law. As a result, the ratio of the magnitude of the rejected heat $|Q_C|$ to the magnitude of the input heat $|Q_H|$ is

$$\frac{|Q_C|}{|Q_H|} = \frac{T_C}{T_H} \qquad (15.14)$$

where the temperatures T_C and T_H *must be expressed in kelvins.*

The efficiency e_{Carnot} of a Carnot engine can be written in a particularly useful way by substituting Equation 15.14 into Equation 15.13 for the efficiency, $e = 1 - |Q_C|/|Q_H|$:

Efficiency of a Carnot engine

$$e_{\text{Carnot}} = 1 - \frac{T_C}{T_H} \qquad (15.15)$$

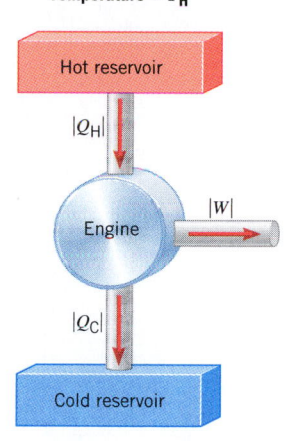

Temperature $= T_H$

Temperature $= T_C$

Figure 15.11 A Carnot engine is a reversible engine in which all input heat $|Q_H|$ originates from a hot reservoir at a single temperature T_H, and all rejected heat $|Q_C|$ goes into a cold reservoir at a single temperature T_C. The work done by the engine is $|W|$.

This relation gives the *maximum possible efficiency* for a heat engine operating between two Kelvin temperatures T_C and T_H, and the next example illustrates its application.

The physics of
extracting work from a warm ocean.

 Example 7 A Tropical Ocean as a Heat Engine

Water near the surface of a tropical ocean has a temperature of 298.2 K (25.0 °C), whereas water 700 m beneath the surface has a temperature of 280.2 K (7.0 °C). It has been proposed that the warm water be used as the hot reservoir and the cool water as the cold reservoir of a heat engine. Find the maximum possible efficiency for such an engine.

Reasoning The maximum possible efficiency is the efficiency that a Carnot engine would have (Equation 15.15) operating between temperatures of $T_H = 298.2$ K and $T_C = 280.2$ K.

Solution Using $T_H = 298.2$ K and $T_C = 280.2$ K in Equation 15.15, we find that

$$e_{Carnot} = 1 - \frac{T_C}{T_H} = 1 - \frac{280.2 \text{ K}}{298.2 \text{ K}} = \boxed{0.060 \ (6.0\%)}$$

Problem-solving insight

When determining the efficiency of a Carnot engine, be sure the temperatures T_C and T_H of the cold and hot reservoirs are expressed in kelvins; degrees Celsius or degrees Fahrenheit will not do.

In Example 7 the maximum possible efficiency is only 6.0%. The small efficiency arises because the Kelvin temperatures of the hot and cold reservoirs are so close. A greater efficiency is possible only when there is a greater difference between the reservoir temperatures. However, there are limits on how large the efficiency of a heat engine can be, as Conceptual Example 8 discusses.

 Conceptual Example 8 Natural Limits
on the Efficiency of a Heat Engine

Consider a hypothetical engine that receives 1000 J of heat as input from a hot reservoir and delivers 1000 J of work, rejecting no heat to a cold reservoir whose temperature is above 0 K. Which law of thermodynamics does this engine violate? **(a)** The first law **(b)** The second law **(c)** Both the first and second laws

Reasoning The first law of thermodynamics is an expression of energy conservation. The second law states that no irreversible engine operating between two reservoirs at constant temperatures can have a greater efficiency than a reversible engine operating between the same temperatures. The efficiency of such a reversible engine is e_{Carnot}, the efficiency of a Carnot engine.

Answers (a) and (c) are incorrect. From the point of view of energy conservation, nothing is wrong with an engine that converts 1000 J of heat into 1000 J of work. Energy has been neither created nor destroyed; it has only been transformed from one form (heat) into another form (work). Therefore, this engine does not violate the first law of thermodynamics.

Answer (b) is correct. Since all of the input heat is converted into work, the efficiency of the engine is 1, or 100%. But Equation 15.15, which is based on the second law of thermodynamics, indicates that the maximum possible efficiency is $e_{Carnot} = 1 - T_C/T_H$, where T_C and T_H are the temperatures of the cold and hot reservoirs, respectively. Since we are told that T_C is above 0 K, it is clear that the ratio T_C/T_H is greater than zero, so the maximum possible efficiency is less than 1 (or less than 100%). The engine, therefore, violates the second law of thermodynamics, which limits the efficiencies of heat engines to values less than 100%.

Example 8 has emphasized that *even a perfect heat engine has an efficiency that is less than 1.0 or 100%.* In this regard, we note that the maximum possible efficiency, as given by Equation 15.15, approaches 1.0 when T_C approaches absolute zero (0 K). However, experiments have shown that it is not possible to cool a substance to absolute zero (see Section 15.12), so nature does not allow a 100%-efficient heat engine to exist. As a result, there will always be heat rejected to a cold reservoir whenever a heat engine is used to do work, even if friction and other irreversible processes are eliminated completely. This rejected heat is a form of thermal pollution. The second law of thermodynamics requires that at least some thermal pollution be generated whenever heat engines are used to perform work. This kind of thermal pollution can be reduced only if society reduces its dependence on heat engines to do work.

The physics of
thermal pollution.

✓ **CHECK YOUR UNDERSTANDING**

(The answers are given at the end of the book.)

12. The second law of thermodynamics, in the form of Carnot's principle, indicates that the most efficient heat engine operating between two temperatures is a reversible one. Does this mean that a reversible engine operating between the temperatures of 600 and 400 K must be more efficient than an *irreversible* engine operating between 700 and 300 K?

13. Concept Simulation 15.1 at **www.wiley.com/college/cutnell** allows you to explore the concepts that relate to this question. Three reversible engines, A, B, and C, use the same cold reservoir for their exhaust heats. However, they use different hot reservoirs that have the following temperatures: (A) 1000 K, (B) 1100 K, and (C) 900 K. Rank these engines in order of increasing efficiency (smallest efficiency first).   **(a)** A, C, B   **(b)** C, B, A   **(c)** B, A, C   **(d)** C, A, B

14. In **Concept Simulation 15.1** at **www.wiley.com/college/cutnell** you can explore the concepts that are important in this question. Suppose that you wish to improve the efficiency of a Carnot engine. Which answer describes the best way?   **(a)** Lower the Kelvin temperature of the cold reservoir by a factor of four.   **(b)** Raise the Kelvin temperature of the hot reservoir by a factor of four.   **(c)** Cut the Kelvin temperature of the cold reservoir in half and double the Kelvin temperature of the hot reservoir.   **(d)** All three choices give the same improvement in efficiency.

15. Consider a hypothetical device that takes 10 000 J of heat from a hot reservoir and 5000 J of heat from a cold reservoir (whose temperature is greater than 0 K) and produces 15 000 J of work. What can be said about this device?   **(a)** It violates the first law of thermodynamics but not the second law.   **(b)** It violates the second law of thermodynamics but not the first law.   **(c)** It violates both the first and second laws of thermodynamics.   **(d)** It does not violate either the first or the second law of thermodynamics.

15.10 REFRIGERATORS, AIR CONDITIONERS, AND HEAT PUMPS

The natural tendency of heat is to flow from hot to cold, as indicated by the second law of thermodynamics. However, if work is used, heat can be *made* to flow from cold to hot, against its natural tendency. Refrigerators, air conditioners, and heat pumps are, in fact, devices that do just that. As Figure 15.12 illustrates, these devices use work (magnitude $= |W|$) to extract heat (magnitude $= |Q_C|$) from the cold reservoir and deposit heat (magnitude $= |Q_H|$) into the hot reservoir. Generally speaking, such a process is called a *refrigeration process.* A comparison of the left and right sides of this drawing shows that the directions of the arrows symbolizing heat and work in a refrigeration process are opposite to those in an engine process. Nonetheless, energy is conserved during a refrigeration process, just as it is in an engine process, so $|Q_H| = |W| + |Q_C|$. Moreover, if the process occurs reversibly, we have ideal devices that are called Carnot refrigerators, Carnot air conditioners, and Carnot heat pumps. For these ideal devices, the relation $|Q_C|/|Q_H| = T_C/T_H$ (Equation 15.14) applies, just as it does for a Carnot engine.

The physics of refrigerators. In a *refrigerator,* the interior of the unit is the cold reservoir, while the warmer exterior is the hot reservoir. As Figure 15.13 illustrates, the refrigerator takes heat from the food inside and deposits it into the kitchen, along with the energy needed to do the work of making the heat flow from cold to hot. For this reason, the

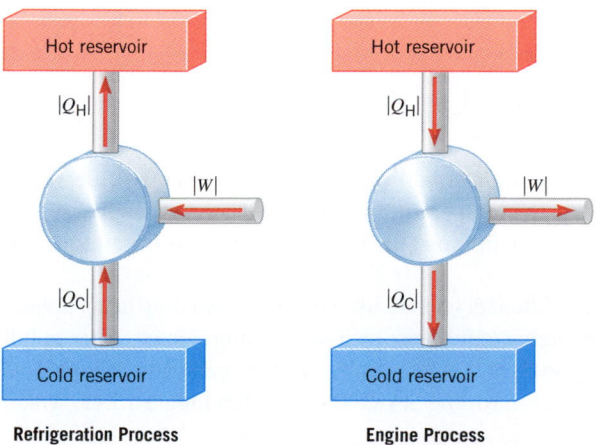

Refrigeration Process   **Engine Process**

Figure 15.12 In the refrigeration process on the left, work $|W|$ is used to remove heat $|Q_C|$ from the cold reservoir and deposit heat $|Q_H|$ into the hot reservoir. Compare this with the engine process on the right.

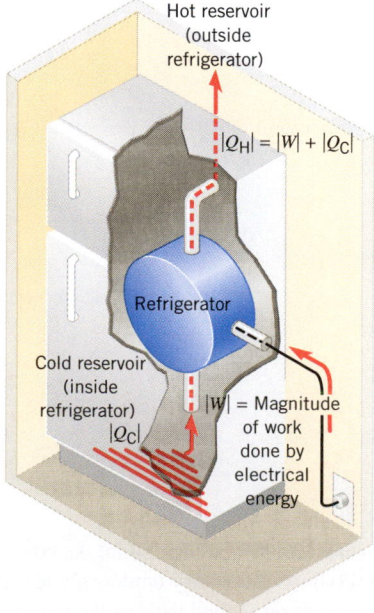

Figure 15.13 A refrigerator.

The physics of air conditioners.

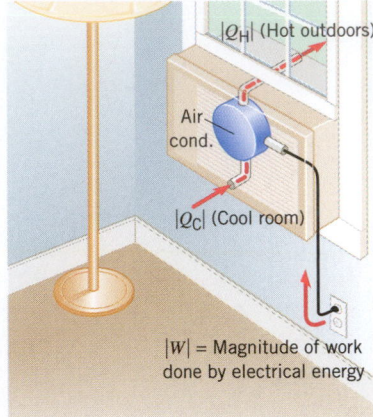

Figure 15.14 A window air conditioner removes heat from a room, which is the cold reservoir, and deposits heat outdoors, which is the hot reservoir.

outside surfaces (usually the sides and back) of most refrigerators are warm to the touch while the units operate.

An *air conditioner* is like a refrigerator, except that the room itself is the cold reservoir and the outdoors is the hot reservoir. Figure 15.14 shows a window unit, which cools a room by removing heat and depositing it outside, along with the work used to make the heat flow from cold to hot. Conceptual Example 9 considers a common misconception about refrigerators and air conditioners.

Conceptual Example 9
You Can't Beat the Second Law of Thermodynamics

Is it possible (A) to cool your kitchen by leaving the refrigerator door open or (B) to cool your bedroom by putting a window air conditioner on the floor by the bed? (**a**) Only A is possible. (**b**) Only B is possible. (**c**) Both are possible. (**d**) Neither is possible.

Reasoning During a refrigeration process (be it in a refrigerator or in an air conditioner), heat (magnitude $= |Q_C|$) is removed from a cold reservoir and heat (magnitude $= |Q_H|$) is deposited into a hot reservoir. Moreover, according to the second law of thermodynamics, work (magnitude $= |W|$) is required to move this heat from the cold reservoir to the hot reservoir. The principle of conservation of energy states that $|Q_H| = |W| + |Q_C|$ (Equation 15.12), and we will use this as a guide in assessing the possibilities.

Answers (a), (b), and (c) are incorrect. If you wanted to cool your kitchen by leaving the refrigerator door open, the refrigerator would have to take heat from directly in front of the open door and pump *less* heat out the back of the unit and into the kitchen (since the refrigerator is supposed to be cooling the entire kitchen). Likewise, if you tried to cool your entire bedroom by placing the air conditioner on the floor by the bed, the air conditioner would have to take heat (magnitude $= |Q_C|$) from directly in front of the unit and deposit *less* heat (magnitude $= |Q_H|$) out the back. According to the second law of thermodynamics this cannot happen, since $|Q_H| = |W| + |Q_C|$; that is, $|Q_H|$ is greater than (not less than) $|Q_C|$ because $|W|$ is greater than zero.

Answer (d) is correct. The heat (magnitude $= |Q_C|$) removed from the air directly in front of the open refrigerator is deposited back into the kitchen at the rear of the unit. Moreover, according to the second law of thermodynamics, work (magnitude $= |W|$) is needed to move that heat from cold to hot, and the energy from this work is also deposited into the kitchen as additional heat. Thus, the open refrigerator puts into the kitchen an amount of heat $|Q_H| = |W| + |Q_C|$, which is *more* than it removes from in front of the open refrigerator. Thus, rather than cooling the kitchen, the open refrigerator warms it up. Putting an air conditioner on the floor to cool your bedroom is similarly a no-win game. The heat pumped out the back of the air conditioner and into the bedroom is greater than the heat pulled into the front of the unit. Consequently, the air conditioner actually warms the bedroom.

Related Homework: *Problem 71*

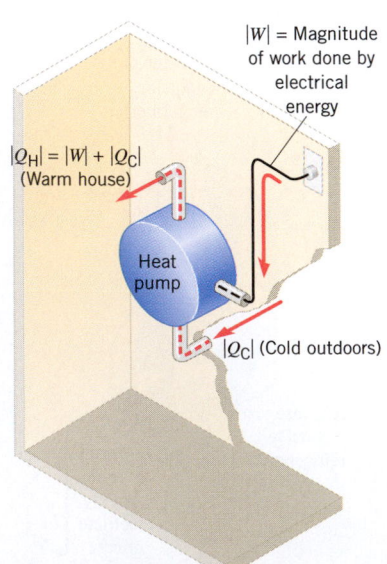

Figure 15.15 In a heat pump the cold reservoir is the wintry outdoors, and the hot reservoir is the inside of the house.

The quality of a refrigerator or air conditioner is rated according to its coefficient of performance. Such appliances perform well when they remove a relatively large amount of heat (magnitude $= |Q_C|$) from a cold reservoir by using as small an amount of work (magnitude $= |W|$) as possible. Therefore, the coefficient of performance is defined as the ratio of $|Q_C|$ to $|W|$, and the greater this ratio is, the better the performance is:

Refrigerator or air conditioner	Coefficient of performance $= \dfrac{|Q_C|}{|W|}$	(15.16)

Commercially available refrigerators and air conditioners have coefficients of performance in the range 2 to 6, depending on the temperatures involved. The coefficients of performance for these real devices are less than those for ideal, or Carnot, refrigerators and air conditioners.

In a sense, refrigerators and air conditioners operate like pumps. They pump heat "uphill" from a lower temperature to a higher temperature, just as a water pump forces water uphill from a lower elevation to a higher elevation. It would be appropriate to call them heat pumps. However, the name "heat pump" is reserved for the device illustrated in Figure 15.15, which is a home heating appliance. The **heat pump** uses work $|W|$ to make heat $|Q_C|$ from the wintry outdoors (the cold reservoir) flow up the temperature "hill" into a warm house (the hot reservoir). According to the conservation of energy, the heat pump deposits inside the house an amount of heat $|Q_H| = |W| + |Q_C|$. The air conditioner and the heat pump do closely related jobs. The air conditioner refrigerates the inside of the house and heats up the outdoors, while the heat pump refrigerates the outdoors and heats up the inside. These jobs are so closely related that most heat pump systems serve in a dual capacity, being equipped with a switch that converts them from heaters in the winter into air conditioners in the summer.

The physics of **heat pumps.**

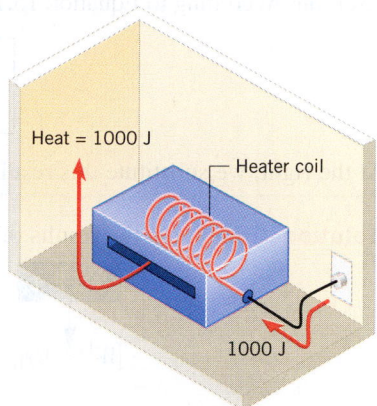

Heat pumps are popular for home heating in today's energy-conscious world, and it is easy to understand why. Suppose that 1000 J of energy is available for home heating. Figure 15.16 shows that a conventional electric heating system uses this 1000 J to heat a coil of wire, just as in a toaster. A fan blows air across the hot coil, and forced convection carries the 1000 J of heat into the house. In contrast, the heat pump in Figure 15.15 does not use the 1000 J directly as heat. Instead, it uses the 1000 J to do the work (magnitude = $|W|$) of pumping heat (magnitude = $|Q_C|$) from the cooler outdoors into the warmer house and, in so doing, delivers an amount of energy $|Q_H| = |W| + |Q_C|$. With $|W| = 1000$ J, this becomes $|Q_H| = 1000$ J $+ |Q_C|$, so that the heat pump delivers more than 1000 J of heat, whereas the conventional electric heating system delivers only 1000 J. The next example shows how the basic relations $|Q_H| = |W| + |Q_C|$ and $|Q_C|/|Q_H| = T_C/T_H$ are used with heat pumps.

Figure 15.16 This conventional electric heating system is delivering 1000 J of heat to the living room.

ANALYZING MULTIPLE-CONCEPT PROBLEMS

Example 10   A Heat Pump

An ideal, or Carnot, heat pump is used to heat a house to a temperature of 294 K (21 °C). How much work must the pump do to deliver 3350 J of heat into the house on a day when the outdoor temperature is 273 K (0 °C) and on another day when the outdoor temperature is 252 K (−21 °C)?

Reasoning The conservation of energy dictates that the heat delivered into the house (the hot reservoir) equals the energy from the work done by the heat pump plus the energy in the form of heat taken from the cold outdoors (the cold reservoir). The heat delivered into the house is given, so that we can use energy conservation to determine the work, provided that we can obtain a value for the heat taken from the outdoors. Since we are dealing with an ideal heat pump, we can obtain this value by using Equation 15.14, which relates the ratio of the magnitudes of the heats for the cold and hot reservoirs to the ratio of the reservoir temperatures (in kelvins).

Knowns and Unknowns The following data are available:

Description	Symbol	Value	Comment
Temperature of hot reservoir (interior of house)	T_H	294 K	Temperature in kelvins must be used.
Temperature of cold reservoir (outdoors)	T_C	273 K or 252 K	Temperature in kelvins must be used.
Magnitude of heat delivered into house	$|Q_H|$	3350 J	
Unknown Variable			
Magnitude of work done by pump	$|W|$	?	

Modeling the Problem

STEP 1   The Conservation of Energy   The energy-conservation principle requires that $|Q_H| = |W| + |Q_C|$ (Equation 15.12), where $|Q_H|$, $|W|$, and $|Q_C|$ are, respectively, the magnitudes of the heat delivered into the house (the hot reservoir), the work done by the heat pump, and the heat taken from the cold outdoors (the cold reservoir). Solving for $|W|$ gives Equation 1 at the right. In this result, we have a value for $|Q_H|$ but not for $|Q_C|$. Step 2 deals with this missing information.

$$|W| = |Q_H| - |Q_C| \qquad (1)$$

Continued

STEP 2 **Heat and Thermodynamic Temperature** To obtain a value for $|Q_C|$, we will use the information given about the temperatures T_H for the hot reservoir and T_C for the cold reservoir. According to Equation 15.14, $|Q_C|/|Q_H| = T_C/T_H$, which can be solved for $|Q_C|$:

$$|Q_C| = |Q_H|\left(\frac{T_C}{T_H}\right)$$

At the right, we substitute this result into Equation 1.

$$|W| = |Q_H| - |Q_C| \qquad (1)$$

$$|Q_C| = |Q_H|\left(\frac{T_C}{T_H}\right)$$

Solution Combining the results of each step algebraically, we find that

STEP 1   **STEP 2**

$$|W| = |Q_H| - |Q_C| = |Q_H| - |Q_H|\left(\frac{T_C}{T_H}\right)$$

It follows that the magnitude of the work for the two given outdoor temperatures is

Outdoor temperature of 273 K
$$|W| = |Q_H|\left(1 - \frac{T_C}{T_H}\right) = (3350 \text{ J})\left(1 - \frac{273 \text{ K}}{294 \text{ K}}\right) = \boxed{240 \text{ J}}$$

Outdoor temperature of 252 K
$$|W| = |Q_H|\left(1 - \frac{T_C}{T_H}\right) = (3350 \text{ J})\left(1 - \frac{252 \text{ K}}{294 \text{ K}}\right) = \boxed{479 \text{ J}}$$

More work must be done when the outdoor temperature is lower, because the heat is pumped up a greater temperature "hill."

Related Homework: *Problem 88*

Problem-solving insight

When applying Equation 15.14 ($|Q_C|/|Q_H| = T_C/T_H$) to heat pumps, refrigerators, or air conditioners, be sure the temperatures T_C and T_H are expressed in kelvins; degrees Celsius or degrees Fahrenheit will not do.

It is also possible to specify a coefficient of performance for heat pumps. However, unlike refrigerators and air conditioners, the job of a heat pump is to heat, not to cool. As a result, the coefficient of performance of a heat pump is the ratio of the magnitude of the heat $|Q_H|$ delivered into the house to the magnitude of the work $|W|$ required to deliver it:

Heat pump $$\text{Coefficient of performance} = \frac{|Q_H|}{|W|} \qquad (15.17)$$

The coefficient of performance depends on the indoor and outdoor temperatures. Commercial units have coefficients of about 3 to 4 under favorable conditions.

✓ **CHECK YOUR UNDERSTANDING**

(The answers are given at the end of the book.)

16. Each drawing represents a hypothetical heat engine or a hypothetical heat pump and shows the corresponding heats and work. Only one of these hypothetical situations is allowed in nature. Which is it?

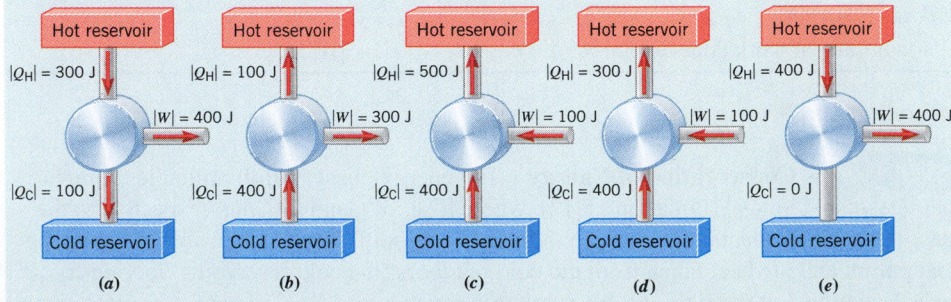

(a)   (b)   (c)   (d)   (e)

17. A refrigerator is kept in a garage that is not heated in the cold winter or air-conditioned in the hot summer. Does it cost more for this refrigerator to make a kilogram of ice cubes in the winter or in the summer? **(a)** In the summer **(b)** In the winter **(c)** It costs the same in both seasons.

18. The coefficient of performance of a heat pump that is removing heat from the cold outdoors **(a)** must always be less than one, **(b)** can be either less than or greater than one, **(c)** must always be greater than one.

19. A kitchen air conditioner and a refrigerator both remove heat from a cold reservoir and deposit it in a hot reservoir. However, the air conditioner _____ the kitchen, while the refrigerator _____ the kitchen. **(a)** cools, cools **(b)** cools, warms **(c)** warms, warms **(d)** warms, cools

20. On a summer day a window air conditioner cycles on and off, according to how the temperature within the room changes. When are you more likely to be able to fry an egg on the outside part of the unit? **(a)** When the unit is on **(b)** When the unit is off **(c)** It does not matter whether the unit is on or off.

15.11 ENTROPY

A Carnot engine has the maximum possible efficiency for its operating conditions because the processes occurring within it are reversible. Irreversible processes, such as friction, cause real engines to operate at less than maximum efficiency, for they reduce our ability to use heat to perform work. As an extreme example, imagine that a hot object is placed in thermal contact with a cold object, so heat flows spontaneously, and hence irreversibly, from hot to cold. Eventually both objects reach the same temperature, and $T_C = T_H$. A Carnot engine using these two objects as heat reservoirs is unable to do work, because the efficiency of the engine is zero $[e_{Carnot} = 1 - (T_C/T_H) = 0]$. In general, irreversible processes cause us to lose some, but not necessarily all, of the ability to perform work. This partial loss can be expressed in terms of a concept called **entropy.**

To introduce the idea of entropy we recall the relation $|Q_C|/|Q_H| = T_C/T_H$ (Equation 15.14) that applies to a Carnot engine. It is possible to rearrange this equation as $|Q_C|/T_C = |Q_H|/T_H$, which focuses attention on the heat Q divided by the Kelvin temperature T. The quantity Q/T is called the change in the entropy ΔS:

$$\Delta S = \left(\frac{Q}{T}\right)_R \tag{15.18}$$

In this expression the temperature T must be in kelvins, and the subscript R refers to the word "reversible." It can be shown that Equation 15.18 applies to any process in which heat enters (Q is positive) or leaves (Q is negative) a system reversibly at a constant temperature. Such is the case for the heat that flows into and out of the reservoirs of a Carnot engine. Equation 15.18 indicates that the SI unit for entropy is a joule per kelvin (J/K).

Entropy, like internal energy, is a function of the state or condition of the system. Only the state of a system determines the entropy S that a system has. Therefore, the change in entropy ΔS is equal to the entropy of the final state of the system minus the entropy of the initial state.

We can now describe what happens to the entropy of a Carnot engine. As the engine operates, the entropy of the hot reservoir decreases, since heat of magnitude $|Q_H|$ departs reversibly at a Kelvin temperature T_H. The corresponding change in the entropy is $\Delta S_H = -|Q_H|/T_H$, where the minus sign is needed to indicate a decrease, since the symbol $|Q_H|$ denotes only the magnitude of the heat. In contrast, the entropy of the cold reservoir increases by an amount $\Delta S_C = +|Q_C|/T_C$, for the rejected heat reversibly enters the cold reservoir at a Kelvin temperature T_C. The total change in entropy is

$$\Delta S_C + \Delta S_H = +\frac{|Q_C|}{T_C} - \frac{|Q_H|}{T_H} = 0$$

because $|Q_C|/T_C = |Q_H|/T_H$ according to Equation 15.14.

The fact that the total change in entropy is zero for a Carnot engine is a specific illustration of a general result. It can be proved that when *any* reversible process occurs, the change in the entropy of the universe is zero; $\Delta S_{universe} = 0$ J/K for a reversible process. The word "universe" means that $\Delta S_{universe}$ takes into account the entropy changes of all parts of

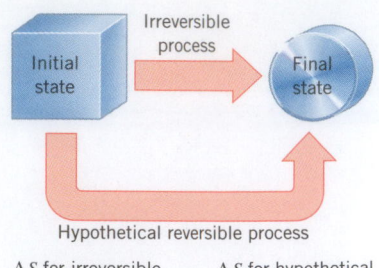

Hypothetical reversible process

ΔS for irreversible process = ΔS for hypothetical reversible process

Figure 15.17 Although the relation $\Delta S = (Q/T)_R$ applies to reversible processes, it can be used as part of an indirect procedure to find the entropy change for an irreversible process. This drawing illustrates the procedure discussed in the text.

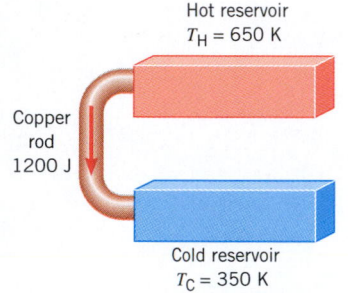

Figure 15.18 Heat flows spontaneously from a hot reservoir to a cold reservoir.

the system and all parts of the environment. *Reversible processes do not alter the total entropy of the universe.* To be sure, the entropy of one part of the universe may change because of a reversible process, but if so, the entropy of another part changes in the opposite way by the same amount.

What happens to the entropy of the universe when an *irreversible* process occurs is more complex, because the expression $\Delta S = (Q/T)_R$ does not apply directly. However, if a system changes irreversibly from an initial state to a final state, this expression can be used to calculate ΔS indirectly, as Figure 15.17 indicates. We imagine a hypothetical reversible process that causes the system to change between *the same initial and final states* and then find ΔS for this reversible process. The value obtained for ΔS also applies to the irreversible process that actually occurs, since only the nature of the initial and final states, and not the path between them, determines ΔS. Example 11 illustrates this indirect method and shows that spontaneous (irreversible) processes increase the entropy of the universe.

 Example 11   The Entropy of the Universe Increases

Figure 15.18 shows 1200 J of heat flowing spontaneously through a copper rod from a hot reservoir at 650 K to a cold reservoir at 350 K. Determine the amount by which this irreversible process changes the entropy of the universe, assuming that no other changes occur.

Reasoning   The hot-to-cold heat flow is irreversible, so the relation $\Delta S = (Q/T)_R$ is applied to a hypothetical process whereby the 1200 J of heat is taken reversibly from the hot reservoir and added reversibly to the cold reservoir.

Solution   The total entropy change of the universe is the algebraic sum of the entropy changes for each reservoir:

$$\Delta S_{universe} = \underbrace{-\frac{1200\ J}{650\ K}}_{\substack{\text{Entropy lost} \\ \text{by hot reservoir}}} + \underbrace{\frac{1200\ J}{350\ K}}_{\substack{\text{Entropy gained} \\ \text{by cold reservoir}}} = \boxed{+1.6\ J/K}$$

The irreversible process causes the entropy of the universe to increase by 1.6 J/K.

Example 11 is a specific illustration of a general result: *Any irreversible process increases the entropy of the universe.* In other words, $\Delta S_{universe} > 0$ J/K for an irreversible process. Reversible processes do not alter the entropy of the universe, whereas irreversible processes cause the entropy to increase. Therefore, the entropy of the universe continually increases, like time itself, and entropy is sometimes called "time's arrow." It can be shown that this behavior of the entropy of the universe provides a completely general statement of the second law of thermodynamics, which applies not only to heat flow but also to all kinds of other processes.

THE SECOND LAW OF THERMODYNAMICS STATED IN TERMS OF ENTROPY

The total entropy of the universe does not change when a reversible process occurs ($\Delta S_{universe} = 0$ J/K) and increases when an irreversible process occurs ($\Delta S_{universe} > 0$ J/K).

When an irreversible process occurs and the entropy of the universe increases, the energy available for doing work decreases, as the next example illustrates.

 Example 12   Energy Unavailable for Doing Work

Suppose that 1200 J of heat is used as input for an engine under two different conditions. In Figure 15.19a the heat is supplied by a hot reservoir whose temperature is 650 K. In part *b* of the drawing, the heat flows irreversibly through a copper rod into a second reservoir whose temperature is 350 K and then enters the engine. In either case, a 150-K reservoir is used as the cold reservoir. For each case, determine the maximum amount of work that can be obtained from the 1200 J of heat.

Reasoning According to Equation 15.11, the work (magnitude $= |W|$) obtained from the engine is the product of its efficiency e and the input heat (magnitude $= |Q_H|$), or $|W| = e|Q_H|$. For a given input heat, the maximum amount of work is obtained when the efficiency is a maximum—that is, when the engine is a Carnot engine. The efficiency of a Carnot engine is given by Equation 15.15 as $e_{Carnot} = 1 - T_C/T_H$. Therefore, the efficiency may be determined from the Kelvin temperatures of the hot and cold reservoirs.

Solution

Before irreversible heat flow

$$e_{Carnot} = 1 - \frac{T_C}{T_H} = 1 - \frac{150\ K}{650\ K} = 0.77$$

$$|W| = (e_{Carnot})(1200\ J) = (0.77)(1200\ J) = \boxed{920\ J}$$

After irreversible heat flow

$$e_{Carnot} = 1 - \frac{T_C}{T_H} = 1 - \frac{150\ K}{350\ K} = 0.57$$

$$|W| = (e_{Carnot})(1200\ J) = (0.57)(1200\ J) = \boxed{680\ J}$$

When the 1200 J of input heat is taken from the 350-K reservoir instead of the 650-K reservoir, the efficiency of the Carnot engine is smaller. As a result, less work (680 J versus 920 J) can be extracted from the input heat.

Example 12 shows that 240 J less work (920 J − 680 J) can be performed when the input heat is obtained from the hot reservoir with the lower temperature. In other words, the irreversible process of heat flow through the copper rod causes energy to become unavailable for doing work in the amount of $W_{unavailable} = 240$ J. Example 11 shows that this irreversible process simultaneously causes the entropy of the universe to increase by an amount $\Delta S_{universe} = +1.6$ J/K. These values for $W_{unavailable}$ and $\Delta S_{universe}$ are in fact related. If you multiply $\Delta S_{universe}$ by 150 K, which is the lowest Kelvin temperature in Example 12, you obtain $W_{unavailable} = (150\ K) \times (1.6\ J/K) = 240$ J. This illustrates the following general result:

$$W_{unavailable} = T_0 \Delta S_{universe} \tag{15.19}$$

where T_0 is the Kelvin temperature of the coldest heat reservoir. Since irreversible processes cause the entropy of the universe to increase, they cause energy to be degraded, in the sense that part of the energy becomes unavailable for the performance of work. In contrast, there is no penalty when reversible processes occur, because for them $\Delta S_{universe} = 0$ J/K, and there is no loss of work.

Entropy can also be interpreted in terms of order and disorder. As an example, consider a block of ice (Figure 15.20) with each of its H_2O molecules fixed rigidly in place in a highly structured and ordered arrangement. In comparison, the puddle of water into which the ice melts is disordered and unorganized, because the molecules in a liquid are free to move from place to place. Heat is required to melt the ice and produce the disorder. Moreover, heat flow into a system increases the entropy of the system, according to $\Delta S = (Q/T)_R$. We associate an increase in entropy, then, with an increase in disorder. Conversely, we associate a decrease in entropy with a decrease in disorder or a greater degree of order. Example 13 illustrates an order-to-disorder change and the increase of entropy that accompanies it.

Figure 15.19 Heat in the amount of $|Q_H| = 1200$ J is used as input for an engine under two different conditions in parts a and b.

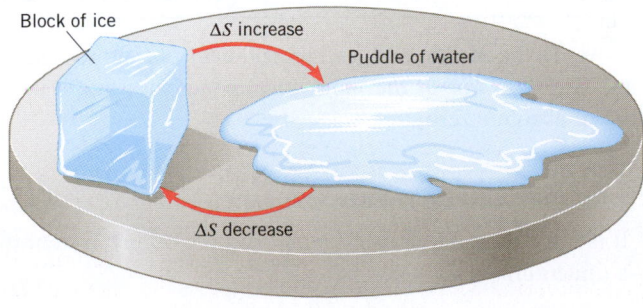
Block of ice — ΔS increase — Puddle of water — ΔS decrease

Figure 15.20 A block of ice is an example of an ordered system relative to a puddle of water.

Figure 15.21 With the aid of explosives, demolition experts caused the Kingdome in Seattle, Washington, to go from the ordered state (lower entropy) shown in the top photograph to the disordered state (higher entropy) shown in the bottom photograph. (Anthony Bolante/Reuters/Landov LLC)

 Example 13 Order to Disorder

Find the change in entropy that results when a 2.3-kg block of ice melts slowly (reversibly) at 273 K (0 °C).

Reasoning Since the phase change occurs reversibly at a constant temperature, the change in entropy can be found by using Equation 15.18, $\Delta S = (Q/T)_R$, where Q is the heat absorbed by the melting ice. This heat can be determined by using the relation $Q = mL_f$ (Equation 12.5), where m is the mass and $L_f = 3.35 \times 10^5$ J/kg is the latent heat of fusion of water (see Table 12.3).

Solution Using Equation 15.18 and Equation 12.5, we find that the change in entropy is

$$\Delta S = \left(\frac{Q}{T}\right)_R = \frac{mL_f}{T} = \frac{(2.3 \text{ kg})(3.35 \times 10^5 \text{ J/kg})}{273 \text{ K}} = \boxed{+2.8 \times 10^3 \text{ J/K}}$$

a result that is positive, since the ice absorbs heat as it melts.

Figure 15.21 shows another order-to-disorder change that can be described in terms of entropy.

✓ CHECK YOUR UNDERSTANDING

(The answers are given at the end of the book.)

21. Two equal amounts of water are mixed together in an insulated container, and no work is done in the process. The initial temperatures of the water are different, but the mixture reaches a uniform temperature. Do the internal energy and entropy of the water increase, decrease, or remain constant as a result of the mixing process?

	Internal Energy of the Water	Entropy of the Water
(a)	Increases	Increases
(b)	Decreases	Decreases
(c)	Remains constant	Decreases
(d)	Remains constant	Increases
(e)	Remains constant	Remains constant

22. An event happens somewhere in the universe and, as a result, the entropy of an object changes by −5 J/K. Consistent with the second law of thermodynamics, which one (or more) of the following is a possible value for the entropy change for the rest of the universe? **(a)** −5 J/K **(b)** 0 J/K **(c)** +5 J/K **(d)** +10 J/K

23. In each of the following cases, which has the greater entropy, a handful of popcorn kernels or the popcorn that results from them; a salad before or after it has been tossed; and a messy apartment or a neat apartment?

24. A glass of water contains a teaspoon of dissolved sugar. After a while, the water evaporates, leaving behind sugar crystals. The entropy of the sugar crystals is less than the entropy of the dissolved sugar because the sugar crystals are in a more ordered state. Why doesn't this process violate the second law of thermodynamics? **(a)** Because, considering what happens to the water, the total entropy of the universe also decreases. **(b)** Because, considering what happens to the water, the total entropy of the universe increases. **(c)** Because the second law does not apply to this situation.

25. A builder uses lumber to construct a building, which is unfortunately destroyed in a fire. Thus, the lumber existed at one time or another in three different states: (A) as unused building material, (B) as a building, and (C) as a burned-out shell of a building. Rank these three states in order of decreasing entropy (largest first). **(a)** C, B, A **(b)** A, B, C **(c)** C, A, B **(d)** A, C, B **(e)** B, A, C

15.12 THE THIRD LAW OF THERMODYNAMICS

To the zeroth, first, and second laws of thermodynamics we add the third (and last) law. The ***third law of thermodynamics*** indicates that it is impossible to reach a temperature of absolute zero.

THE THIRD LAW OF THERMODYNAMICS

It is not possible to lower the temperature of any system to absolute zero ($T = 0$ K) in a finite number of steps.

This law, like the second law, can be expressed in a number of ways, but a discussion of them is beyond the scope of this text. The third law is needed to explain a number of experimental observations that cannot be explained by the other laws of thermodynamics.

15.13 CONCEPTS & CALCULATIONS

The first law of thermodynamics is basically a restatement of the conservation-of-energy principle in terms of heat and work. Example 14 emphasizes this important fact by showing that the conservation principle and the first law provide the same approach to a problem. In addition, the example reviews the concept of latent heat of sublimation (see Section 12.8) and the ideal gas law (see Section 14.2).

Concepts & Calculations Example 14
The Sublimation of Zinc

The sublimation of zinc (mass per mole $= 0.0654$ kg/mol) takes place at a temperature of 6.00×10^2 K, and the latent heat of sublimation is 1.99×10^6 J/kg. The pressure remains constant during the sublimation. Assume that the zinc vapor can be treated as a monatomic ideal gas and that the volume of solid zinc is negligible compared to the corresponding vapor. What is the change in the internal energy of the zinc when 1.50 kg of zinc sublimates?

Concept Questions and Answers What is sublimation and what is the latent heat of sublimation?

> *Answer* Sublimation is the process whereby a solid phase changes directly into a gas phase in response to the input of heat. The heat per kilogram needed to cause the phase change is called the latent heat of sublimation L_s. The heat Q needed to bring about the sublimation of a mass m of solid material is given by Equation 12.5 as $Q = mL_s$.

When a solid phase changes to a gas phase, does the volume of the material increase or decrease, and by how much?

> *Answer* For a given mass of material, gases generally have greater volumes than solids do, so the volume of the material increases. The increase in volume is $\Delta V = V_{gas} - V_{solid}$. Since the volume of the solid V_{solid} is negligibly small in comparison to the volume of the gas V_{gas}, we have $\Delta V = V_{gas}$. Using the ideal gas law as given in Equation 14.1, it follows that $V_{gas} = nRT/P$, so that $\Delta V = nRT/P$. In this result, n is the number of moles of material, R is the universal gas constant, and T is the Kelvin temperature.

As the material changes from a solid to a gas, does it do work on the environment or does the environment do work on it? How much work is involved?

> *Answer* To make room for itself, the expanding material must push against the environment and, in so doing, does work on the environment. Since the pressure remains constant, the work done by the material is given by Equation 15.2 as $W = P \Delta V$. Since $\Delta V = nRT/P$, the work becomes $W = P(nRT/P) = nRT$.

In this problem we begin with heat Q and realize that it is used for two purposes: First, it makes the solid change into a gas, which entails a change ΔU in the internal energy of the material, $\Delta U = U_{gas} - U_{solid}$. Second, it allows the expanding material to do work W on the environment. According to the conservation-of-energy principle, how is Q related to ΔU and W?

> *Answer* According to the conservation-of-energy principle, energy can neither be created nor destroyed, but can only be converted from one form to another (see Section 6.8). Therefore, part of the heat Q is used for ΔU and part for W, with the result that $Q = \Delta U + W$.

According to the first law of thermodynamics, how is Q related to ΔU and W?

> *Answer* As indicated in Equation 15.1, the first law of thermodynamics is $\Delta U = Q - W$. Rearranging this equation gives $Q = \Delta U + W$, which is identical to the result obtained from the conservation-of-energy principle.

Solution Using the facts that $Q = \Delta U + W$, $Q = mL_s$, and $W = nRT$, we have that

$$Q = \Delta U + W \quad \text{or} \quad mL_s = \Delta U + nRT$$

Solving for ΔU gives

$$\Delta U = mL_s - nRT$$

In this result, n is the number of moles of the ideal gas. According to the discussion in Section 14.1, the number of moles of gaseous zinc is the mass m of the sample divided by the mass per mole of zinc or $n = m/(0.0654 \text{ kg/mol})$. Therefore, we find

$$\Delta U = mL_s - nRT$$

$$= (1.50 \text{ kg})\left(1.99 \times 10^6 \frac{\text{J}}{\text{kg}}\right) - \left(\frac{1.50 \text{ kg}}{0.0654 \text{ kg/mol}}\right)\left(8.31 \frac{\text{J}}{\text{mol} \cdot \text{K}}\right)(6.00 \times 10^2 \text{ K})$$

$$= \boxed{2.87 \times 10^6 \text{ J}}$$

Heat engines can be used to perform work, as we have seen in this chapter. The concept of work, however, was first introduced in Chapter 6, along with the idea of kinetic energy and the work–energy theorem. The next example reviews some of the main features of heat engines, as well as kinetic energy and the work–energy theorem.

Concepts & Calculations Example 15
The Work–Energy Theorem

Each of two Carnot engines uses the same cold reservoir at a temperature of 275 K for its exhaust heat. Each engine receives 1450 J of input heat. The work from either of these engines is used to drive a pulley arrangement that uses a rope to accelerate a 125-kg crate from rest along a horizontal frictionless surface, as Figure 15.22 suggests. With engine 1 the crate attains a speed of 2.00 m/s, while with engine 2 it attains a speed of 3.00 m/s. Find the temperature of the hot reservoir for each engine.

Concept Questions and Answers With which engine is the change in the crate's kinetic energy greater?

Answer The change is greater with engine 2. Kinetic energy is $\text{KE} = \frac{1}{2}mv^2$, according to Equation 6.2, where m is the mass of the crate and v is its speed. The change in the kinetic energy is the final minus the initial value, or $\text{KE}_f - \text{KE}_0$. Since the crate starts from rest, it has zero initial kinetic energy. Thus, the change is equal to the final kinetic energy. Since engine 2 gives the crate the greater final speed, it causes the greater change in kinetic energy.

Which engine does more work?

Answer The work–energy theorem, as stated in Equation 6.3, indicates that the net work done on an object equals the change in the object's kinetic energy, or $W = \text{KE}_f - \text{KE}_0$. The net work is the work done by the net force. In Figure 15.22 the surface is horizontal, and the crate does not leave it. Therefore, the upward normal force that the surface applies to the crate must balance the downward weight of the crate. Furthermore, the surface is frictionless, so there is no friction force. The net force acting on the crate, then, consists of the single force due to the tension in the rope, which arises from the action of the engine. Thus, the work done by the engine is, in fact, the net work done on the crate. But we know that engine 2 causes the crate's kinetic energy to change by the greater amount, so that engine must do more work.

For which engine is the temperature of the hot reservoir greater?

Answer The temperature of the hot reservoir for engine 2 is greater. We know that engine 2 does more work, but each engine receives the same 1450 J of input heat. Therefore, engine 2 derives more work from the input heat. In other words, it is more efficient. But the efficiency of a Carnot engine depends only on the Kelvin temperatures of its hot and cold reservoirs. Since both engines use the same cold reservoir whose temperature is 275 K, only the temperatures of the hot reservoirs are different. Higher temperatures for the hot reservoir are associated with greater efficiencies, so the temperature of the hot reservoir for engine 2 is greater.

Solution According to Equation 15.11, the efficiency e of a heat engine is the magnitude of the work $|W|$ divided by the magnitude of the input heat $|Q_H|$, or $e = |W|/|Q_H|$. According to

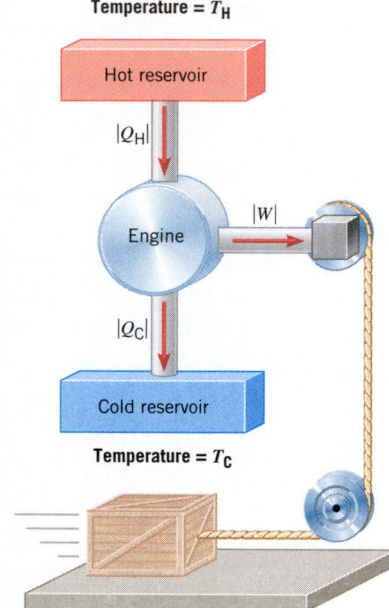

Temperature = T_H

Hot reservoir

$|Q_H|$

$|W|$

Engine

$|Q_C|$

Cold reservoir

Temperature = T_C

Figure 15.22 With the aid of pulleys and a rope, a Carnot engine provides the work that is used to accelerate the crate from rest along a horizontal frictionless surface. See Example 15.

Equation 15.15, the efficiency of a Carnot engine is $e_{Carnot} = 1 - T_C/T_H$, where T_C and T_H are, respectively, the Kelvin temperatures of the cold and hot reservoirs. Combining these two equations, we have

$$1 - \frac{T_C}{T_H} = \frac{|W|}{|Q_H|}$$

But $|W|$ is the magnitude of the net work done on the crate, and it equals the change in the crate's kinetic energy, or $|W| = KE_f - KE_0 = \frac{1}{2}mv^2$, according to Equations 6.2 and 6.3. With this substitution, the efficiency expression becomes

$$1 - \frac{T_C}{T_H} = \frac{\frac{1}{2}mv^2}{|Q_H|}$$

Solving for the temperature T_H, we find

$$T_H = \frac{T_C}{1 - \dfrac{mv^2}{2|Q_H|}}$$

As expected, the value of T_H for engine 2 is greater:

Engine 1
$$T_H = \frac{275 \text{ K}}{1 - \dfrac{(125 \text{ kg})(2.00 \text{ m/s})^2}{2(1450 \text{ J})}} = \boxed{332 \text{ K}}$$

Engine 2
$$T_H = \frac{275 \text{ K}}{1 - \dfrac{(125 \text{ kg})(3.00 \text{ m/s})^2}{2(1450 \text{ J})}} = \boxed{449 \text{ K}}$$

CONCEPT SUMMARY

If you need more help with a concept, use the Learning Aids noted next to the discussion or equation. Examples (**Ex.**) are in the text of this chapter. Go to **www.wiley.com/college/cutnell** for the following Learning Aids:

Interactive LearningWare (ILW) — Additional examples solved in a five-step interactive format.

Concept Simulations (CS) — Animated text figures or animations of important concepts.

Interactive Solutions (IS) — Models for certain types of problems in the chapter homework. The calculations are carried out interactively.

Topic	Discussion	Learning Aids
	15.1 THERMODYNAMIC SYSTEMS AND THEIR SURROUNDINGS A thermodynamic system is the collection of objects on which attention is being focused, and the surroundings are everything else in the environment. The state of a system is the physical condition of the system, as described by values for physical parameters, often pressure, volume, and temperature.	
Thermal equilibrium	**15.2 THE ZEROTH LAW OF THERMODYNAMICS** Two systems are in thermal equilibrium if there is no net flow of heat between them when they are brought into thermal contact.	
Temperature	Temperature is the indicator of thermal equilibrium in the sense that there is no net flow of heat between two systems in thermal contact that have the same temperature.	
Zeroth law of thermodynamics	The zeroth law of thermodynamics states that two systems individually in thermal equilibrium with a third system are in thermal equilibrium with each other.	
	15.3 THE FIRST LAW OF THERMODYNAMICS The first law of thermodynamics states that due to heat Q and work W, the internal energy of a system changes from its initial value of U_i to a final value of U_f according to	
First law of thermodynamics	$$\Delta U = U_f - U_i = Q - W \qquad (15.1)$$	**Ex. 1, 14**
Sign convention for Q and W	Q is positive when the system gains heat and negative when it loses heat. W is positive when work is done by the system and negative when work is done on the system.	
	The first law of thermodynamics is the conservation-of-energy principle applied to heat, work, and the change in the internal energy.	

Topic	Discussion	Learning Aids
Function of state	The internal energy is called a function of state because it depends only on the state of the system and not on the method by which the system came to be in a given state.	**Ex. 2**
Quasi-static process	**15.4 THERMAL PROCESSES** A thermal process is quasi-static when it occurs slowly enough that a uniform pressure and temperature exist throughout the system at all times.	
Isobaric process	An isobaric process is one that occurs at constant pressure. The work W done when a system changes at a constant pressure P from an initial volume V_i to a final volume V_f is	
Isobaric work	$$W = P\,\Delta V = P(V_f - V_i) \tag{15.2}$$	**Ex. 3**
Isochoric process	An isochoric process is one that takes place at constant volume, and no work is done in such a process.	
Isothermal process	An isothermal process is one that takes place at constant temperature.	
Adiabatic process	An adiabatic process is one that takes place without the transfer of heat.	
Work done as the area under a pressure–volume graph	The work done in any kind of quasi-static process is given by the area under the corresponding pressure-versus-volume graph.	**Ex. 4**
	15.5 THERMAL PROCESSES USING AN IDEAL GAS When n moles of an ideal gas change quasi-statically from an initial volume V_i to a final volume V_f at a constant Kelvin temperature T, the work done is	
Work done during an isothermal process	$$W = nRT \ln\left(\frac{V_f}{V_i}\right) \tag{15.3}$$	**Ex. 5** **ILW 15.1**
	When n moles of a monatomic ideal gas change quasi-statically and adiabatically from an initial temperature T_i to a final temperature T_f, the work done is	
Work done during an adiabatic process	$$W = \tfrac{3}{2}nR(T_i - T_f) \tag{15.4}$$ During an adiabatic process, and in addition to the ideal gas law, an ideal gas obeys the relation	
Adiabatic change in pressure and volume	$$P_i V_i^{\gamma} = P_f V_f^{\gamma} \tag{15.5}$$ where $\gamma = c_P/c_V$ is the ratio of the specific heat capacities at constant pressure and constant volume.	**IS 15.27**
	15.6 SPECIFIC HEAT CAPACITIES The molar specific heat capacity C of a substance determines how much heat Q is added or removed when the temperature of n moles of the substance changes by an amount ΔT:	
	$$Q = Cn\,\Delta T \tag{15.6}$$	**IS 15.97**
	For a monatomic ideal gas, the molar specific heat capacities at constant pressure and constant volume are, respectively,	
Specific heat capacities of a monatomic ideal gas	$$C_P = \tfrac{5}{2}R \tag{15.7}$$ $$C_V = \tfrac{3}{2}R \tag{15.8}$$	
	where R is the ideal gas constant. For a diatomic ideal gas at moderate temperatures that do not allow vibration to occur, these values are $C_P = \tfrac{7}{2}R$ and $C_V = \tfrac{5}{2}R$. For any type of ideal gas, the difference between C_P and C_V is	
	$$C_P - C_V = R \tag{15.10}$$	
The second law of thermodynamics (heat flow statement)	**15.7 THE SECOND LAW OF THERMODYNAMICS** The second law of thermodynamics can be stated in a number of equivalent forms. In terms of heat flow, the second law declares that heat flows spontaneously from a substance at a higher temperature to a substance at a lower temperature and does not flow spontaneously in the reverse direction.	
	15.8 HEAT ENGINES A heat engine produces work (magnitude $= \lvert W \rvert$) from input heat (magnitude $= \lvert Q_H \rvert$) that is extracted from a heat reservoir at a relatively high temperature. The engine rejects heat (magnitude $= \lvert Q_C \rvert$) into a reservoir at a relatively low temperature. The efficiency e of a heat engine is	
Efficiency of a heat engine	$$e = \frac{\text{Work done}}{\text{Input heat}} = \frac{\lvert W \rvert}{\lvert Q_H \rvert} \tag{15.11}$$	
	The conservation of energy requires that $\lvert Q_H \rvert$ must be equal to $\lvert W \rvert$ plus $\lvert Q_C \rvert$:	
Conservation of energy for a heat engine	$$\lvert Q_H \rvert = \lvert W \rvert + \lvert Q_C \rvert \tag{15.12}$$	**Ex. 6**

Topic	Discussion	Learning Aids
	By combining Equation 15.12 with Equation 15.11, the efficiency of a heat engine can also be written as	

$$e = 1 - \frac{|Q_C|}{|Q_H|} \qquad (15.13)$$

Reversible process

15.9 CARNOT'S PRINCIPLE AND THE CARNOT ENGINE A reversible process is one in which *both* the system and its environment can be returned to exactly the states they were in before the process occurred.

Carnot's principle

Carnot's principle is an alternative statement of the second law of thermodynamics. It states that no irreversible engine operating between two reservoirs at constant temperatures can have a greater efficiency than a reversible engine operating between the same temperatures. Furthermore, all reversible engines operating between the same temperatures have the same efficiency.

A Carnot engine

A Carnot engine is a reversible engine in which all input heat (magnitude $= |Q_H|$) originates from a hot reservoir at a single Kelvin temperature T_H and all rejected heat (magnitude $= |Q_C|$) goes into a cold reservoir at a single Kelvin temperature T_C. For a Carnot engine

$$\frac{|Q_C|}{|Q_H|} = \frac{T_C}{T_H} \qquad (15.14)$$

The efficiency e_{Carnot} of a Carnot engine is the maximum efficiency that an engine operating between two fixed temperatures can have:

Efficiency of a Carnot engine

$$e_{\text{Carnot}} = 1 - \frac{T_C}{T_H} \qquad (15.15)$$

Ex. 7, 8, 15
CS 15.1

15.10 REFRIGERATORS, AIR CONDITIONERS, AND HEAT PUMPS Refrigerators, air conditioners, and heat pumps are devices that utilize work (magnitude $= |W|$) to make heat (magnitude $= |Q_C|$) flow from a lower Kelvin temperature T_C to a higher Kelvin temperature T_H. In the process (the **Ex. 9** refrigeration process) they deposit heat (magnitude $= |Q_H|$) at the higher temperature. The principle of the conservation of energy requires that $|Q_H| = |W| + |Q_C|$.

Ex. 9
ILW 15.2

If the refrigeration process is ideal, in the sense that it occurs reversibly, the devices are called **Ex. 10** Carnot devices and the relation $|Q_C|/|Q_H| = T_C/T_H$ (Equation 15.14) holds.

Ex. 10

The coefficient of performance of a refrigerator or an air conditioner is

Coefficient of performance (refrigerator or air conditioner)

$$\frac{\text{Coefficient of}}{\text{performance}} = \frac{|Q_C|}{|W|} \qquad (15.16)$$

The coefficient of performance of a heat pump is

Coefficient of performance (heat pump)

$$\frac{\text{Coefficient of}}{\text{performance}} = \frac{|Q_H|}{|W|} \qquad (15.17)$$

15.11 ENTROPY The change in entropy ΔS for a process in which heat Q enters or leaves a system reversibly at a constant Kelvin temperature T is

Change in entropy

$$\Delta S = \left(\frac{Q}{T}\right)_R \qquad (15.18)$$ **Ex. 11, 13**

where the subscript R stands for "reversible."

The second law of thermodynamics (entropy statement)

The second law of thermodynamics can be stated in a number of equivalent forms. In terms of entropy, the second law states that the total entropy of the universe does not change when a reversible process occurs ($\Delta S_{\text{universe}} = 0$ J/K) and increases when an irreversible process occurs ($\Delta S_{\text{universe}} > 0$ J/K).

IS 15.79

Irreversible processes cause energy to be degraded in the sense that part of the energy becomes **Ex. 12** unavailable for the performance of work. The energy $W_{\text{unavailable}}$ that is unavailable for doing work because of an irreversible process is

Ex. 12

Unavailable work

$$W_{\text{unavailable}} = T_0 \Delta S_{\text{universe}} \qquad (15.19)$$

where $\Delta S_{\text{universe}}$ is the total entropy change of the universe and T_0 is the Kelvin temperature of the coldest reservoir into which heat can be rejected.

Entropy and disorder

Increased entropy is associated with a greater degree of disorder and decreased entropy with a lesser degree of disorder (more order).

The third law of thermodynamics

15.12 THE THIRD LAW OF THERMODYNAMICS The third law of thermodynamics states that it is not possible to lower the temperature of any system to absolute zero ($T = 0$ K) in a finite number of steps.

FOCUS ON CONCEPTS

Note to Instructors: The numbering of the questions shown here reflects the fact that they are only a representative subset of the total number that are available online. However, all of the questions are available for assignment via an online homework management program such as WileyPLUS or WebAssign.

Section 15.3   The First Law of Thermodynamics

1. The first law of thermodynamics states that the change ΔU in the internal energy of a system is given by $\Delta U = Q-W$, where Q is the heat and W is the work. Both Q and W can be positive or negative numbers. Q is a positive number if _____, and W is a positive number if _____.   **(a)** the system *loses* heat; work is done *by* the system   **(b)** the system *loses* heat; work is done *on* the system   **(c)** the system *gains* heat; work is done *by* the system   **(d)** the system *gains* heat; work is done *on* the system

Section 15.4   Thermal Processes

4. The drawing shows the expansion of three ideal gases. Rank the gases according to the work they do, largest to smallest.   **(a)** A, B, C   **(b)** A and B (a tie), C   **(c)** B and C (a tie), A   **(d)** B, C, A   **(e)** C, A, B

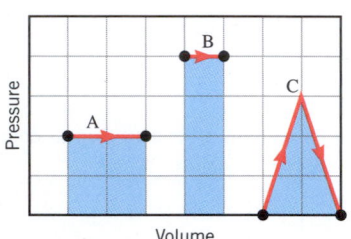

6. The pressure–volume graph shows three paths in which a gas expands from an initial state A to a final state B. The change $\Delta U_{A\rightarrow B}$ in internal energy is the same for each of the paths. Rank the paths according to the heat Q added to the gas, largest to smallest.   **(a)** 1, 2, 3   **(b)** 1, 3, 2   **(c)** 2, 1, 3   **(d)** 3, 1, 2   **(e)** 3, 2, 1

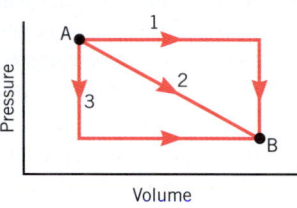

Section 15.5   Thermal Processes Using an Ideal Gas

8. An ideal monatomic gas expands isothermally from A to B, as the graph shows. What can be said about this process?   **(a)** The gas does no work.   **(b)** No heat enters or leaves the gas.   **(c)** The first law of thermodynamics does not apply to an isothermal process.   **(d)** The ideal gas law is not valid during an isothermal process.   **(e)** There is no change in the internal energy of the gas.

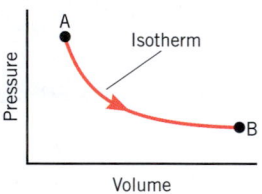

10. A monatomic ideal gas is thermally insulated, so no heat can flow between it and its surroundings. Is it possible for the temperature of the gas to rise?   **(a)** Yes. The temperature can rise if work is done *by* the gas.   **(b)** No. The only way that the temperature can rise is if heat is added to the gas.   **(c)** Yes. The temperature can rise if work is done *on* the gas.

Section 15.8   Heat Engines

13. A heat engine takes heat Q_H from a hot reservoir and uses part of this energy to perform work W. Assuming that Q_H cannot be changed, how can the efficiency of the engine be improved?   **(a)** Increase the work W; the heat Q_C rejected to the cold reservoir increases as a result.   **(b)** Increase the work W; the heat Q_C rejected to the cold reservoir remains unchanged.   **(c)** Increase the work W; the heat Q_C rejected to the cold reservoir decreases as a result.   **(d)** Decrease the work W; the heat Q_C rejected to the cold reservoir remains unchanged.   **(e)** Decrease the work W; the heat Q_C rejected to the cold reservoir decreases as a result.

Section 15.9   Carnot's Principle and the Carnot Engine

15. The three Carnot engines shown in the drawing operate with hot and cold reservoirs whose temperature differences are 100 K. Rank the efficiencies of the engines, largest to smallest.   **(a)** All engines have the same efficiency.   **(b)** A, B, C   **(c)** B, A, C   **(d)** C, B, A   **(e)** C, A, B

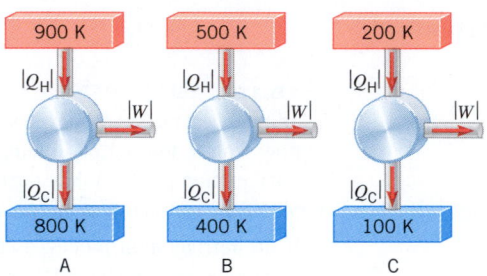

Section 15.10   Refrigerators, Air Conditioners, and Heat Pumps

17. A refrigerator operates for a certain time, and the work done by the electrical energy during this time is $W = 1000$ J. What can be said about the heat delivered to the room containing the refrigerator?   **(a)** The heat delivered to the room is less than 1000 J.   **(b)** The heat delivered to the room is equal to 1000 J.   **(c)** The heat delivered to the room is greater than 1000 J.

Section 15.11   Entropy

19. Heat is transferred from the sun to the earth via electromagnetic waves (see Chapter 24). Because of this transfer, the entropy of the sun _____, the entropy of the earth _____, and the entropy of the sun–earth system _____.   **(a)** increases, decreases, decreases   **(b)** decreases, increases, increases   **(c)** increases, increases, increases   **(d)** increases, decreases, increases   **(e)** decreases, increases, decreases

PROBLEMS

Note to Instructors: Most of the homework problems in this chapter are available for assignment via an online homework management program such as WileyPLUS or WebAssign, and those marked with the icon 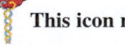 *are presented in WileyPLUS using a guided tutorial format that provides enhanced interactivity. See Preface for additional details.*

ssm   Solution is in the Student Solutions Manual.
www   Solution is available online at www.wiley.com/college/cutnell

    This icon represents a biomedical application.

Section 15.3   The First Law of Thermodynamics

1. In moving out of a dormitory at the end of the semester, a student does 1.6×10^4 J of work. In the process, his internal energy decreases by 4.2×10^4 J. Determine each of the following quantities (including the algebraic sign):   **(a)** W **(b)** ΔU **(c)** Q

2. In a game of football outdoors on a cold day, a player will begin to feel exhausted after using approximately 8.0×10^5 J of internal energy. **(a)** One player, dressed too lightly for the weather, has to leave the game after losing 6.8×10^5 J of heat. How much work has he done? **(b)** Another player, wearing clothes that offer better protection against heat loss, is able to remain in the game long enough to do 2.1×10^5 J of work. What is the magnitude of the heat that he has lost?

3. ssm A system does 164 J of work on its environment and gains 77 J of heat in the process. Find the change in the internal energy of **(a)** the system and **(b)** the environment.

4. **GO** A system does 4.8×10^4 J of work, and 7.6×10^4 J of heat flows into the system during the process. Find the change in the internal energy of the system.

5. ssm When one gallon of gasoline is burned in a car engine, 1.19×10^8 J of internal energy is released. Suppose that 1.00×10^8 J of this energy flows directly into the surroundings (engine block and exhaust system) in the form of heat. If 6.0×10^5 J of work is required to make the car go one mile, how many miles can the car travel on one gallon of gas?

6. Three moles of an ideal monatomic gas are at a temperature of 345 K. Then, 2438 J of heat is added to the gas, and 962 J of work is done on it. What is the final temperature of the gas?

***7.** In exercising, a weight lifter loses 0.150 kg of water through evaporation, the heat required to evaporate the water coming from the weight lifter's body. The work done in lifting weights is 1.40×10^5 J. **(a)** Assuming that the latent heat of vaporization of perspiration is 2.42×10^6 J/kg, find the change in the internal energy of the weight lifter. **(b)** Determine the minimum number of nutritional Calories of food (1 nutritional Calorie = 4186 J) that must be consumed to replace the loss of internal energy.

Section 15.4 Thermal Processes

8. A gas undergoes isochoric heating, during which it gains 5470 J of heat and attains a pressure of 3.45×10^5 Pa. Following this, it experiences an isobaric compression that is also adiabatic, in which its volume decreases by 6.84×10^{-3} m³. Find the total change in the internal energy of the gas for this two-step process. Be sure to include the algebraic sign (+ or −) of the total change in the internal energy.

9. ssm When a .22-caliber rifle is fired, the expanding gas from the burning gunpowder creates a pressure behind the bullet. This pressure causes the force that pushes the bullet through the barrel. The barrel has a length of 0.61 m and an opening whose radius is 2.8×10^{-3} m. A bullet (mass = 2.6×10^{-3} kg) has a speed of 370 m/s after passing through this barrel. Ignore friction and determine the average pressure of the expanding gas.

10. **GO** A system gains 2780 J of heat at a constant pressure of 1.26×10^5 Pa, and its internal energy increases by 3990 J. What is the change in the volume of the system, and is it an increase or a decrease?

11. A gas, while expanding under isobaric conditions, does 480 J of work. The pressure of the gas is 1.6×10^5 Pa, and its initial volume is 1.5×10^{-3} m³. What is the final volume of the gas?

12. The volume of a gas is changed along the curved line between A and B in the drawing. Do not assume that the curved line is an isotherm or that the gas is ideal. **(a)** Find the magnitude of the work for the process, and **(b)** determine whether the work is positive or negative.

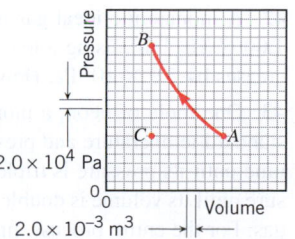

13. The pressure and volume of a gas are changed along the path ABCA. Using the data shown in the graph, determine the work done (including the algebraic sign) in each segment of the path: **(a)** A to B, **(b)** B to C, and **(c)** C to A.

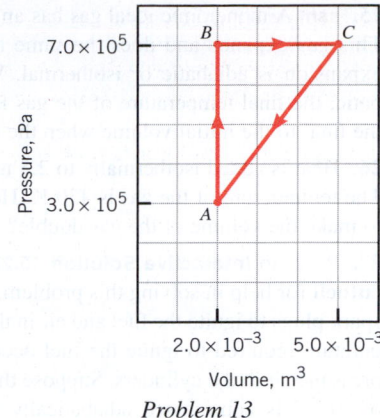

Problem 13

14. Refer to Multiple-Concept Example 3 to see how the concepts pertinent to this problem are used. The pressure of a gas remains constant while the temperature, volume, and internal energy of the gas increase by 53.0 C°, 1.40×10^{-3} m³, and 939 J, respectively. The mass of the gas is 24.0 g, and its specific heat capacity is 1080 J/(kg·C°). Determine the pressure.

15. ssm A system gains 1500 J of heat, while the internal energy of the system increases by 4500 J and the volume decreases by 0.010 m³. Assume that the pressure is constant and find its value.

***16.** A piece of aluminum has a volume of 1.4×10^{-3} m³. The coefficient of volume expansion for aluminum is $\beta = 69 \times 10^{-6}$ (C°)$^{-1}$. The temperature of this object is raised from 20 to 320 °C. How much work is done by the expanding aluminum if the air pressure is 1.01×10^5 Pa?

***17. ssm www** A monatomic ideal gas expands isobarically. Using the first law of thermodynamics, prove that the heat Q is positive, so that it is impossible for heat to flow out of the gas.

***18.** Refer to the drawing that accompanies Problem 91. When a system changes from A to B along the path shown on the pressure-versus-volume graph, it gains 2700 J of heat. What is the change in the internal energy of the system?

****19.** Water is heated in an open pan where the air pressure is one atmosphere. The water remains a liquid, which expands by a small amount as it is heated. Determine the ratio of the work done by the water to the heat absorbed by the water.

Section 15.5 Thermal Processes Using an Ideal Gas

20. Five moles of a monatomic ideal gas expand adiabatically, and its temperature decreases from 370 to 290 K. Determine **(a)** the work done (including the algebraic sign) by the gas, and **(b)** the change in its internal energy.

21. ssm Three moles of an ideal gas are compressed from 5.5×10^{-2} to 2.5×10^{-2} m³. During the compression, 6.1×10^3 J of work is done on the gas, and heat is removed to keep the temperature of the gas constant at all times. Find **(a)** ΔU, **(b)** Q, and **(c)** the temperature of the gas.

22. **GO** Three moles of neon expand isothermally to 0.250 from 0.100 m³. Into the gas flows 4.75×10^3 J of heat. Assuming that neon is an ideal gas, find its temperature.

23. The temperature of a monatomic ideal gas remains constant during a process in which 4700 J of heat flows out of the gas. How much work (including the proper + or − sign) is done?

24. The pressure of a monatomic ideal gas ($\gamma = \frac{5}{3}$) doubles during an adiabatic compression. What is the ratio of the final volume to the initial volume?

25. ssm A monatomic ideal gas has an initial temperature of 405 K. This gas expands and does the same amount of work whether the expansion is adiabatic or isothermal. When the expansion is adiabatic, the final temperature of the gas is 245 K. What is the ratio of the final to the initial volume when the expansion is isothermal?

26. Heat is added isothermally to 2.5 mol of a monatomic ideal gas. The temperature of the gas is 430 K. How much heat must be added to make the volume of the gas double?

*27. Refer to **Interactive Solution 15.27** at **www.wiley.com/college/ cutnell** for help in solving this problem. A diesel engine does not use spark plugs to ignite the fuel and air in the cylinders. Instead, the temperature required to ignite the fuel occurs because the pistons compress the air in the cylinders. Suppose that air at an initial temperature of 21 °C is compressed adiabatically to a temperature of 688 °C. Assume the air to be an ideal gas for which $\gamma = \frac{7}{5}$. Find the compression ratio, which is the ratio of the initial volume to the final volume.

*28. GO An ideal gas is taken through the three processes $(A \rightarrow B, B \rightarrow C,$ and $C \rightarrow A)$ shown in the drawing. In general, for each process the internal energy U of the gas can change because heat Q can be added to or removed from the gas and work W can be done by the gas or on the gas. For the three processes shown in the drawing, fill in the five missing entries in the following table.

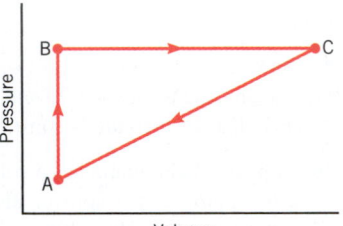

Process	ΔU	Q	W
$A \rightarrow B$	**(b)**	+561 J	**(a)**
$B \rightarrow C$	+4303 J	**(c)**	+3740 J
$C \rightarrow A$	**(d)**	**(e)**	+2867 J

*29. ssm The drawing refers to one mole of a monatomic ideal gas and shows a process that has four steps, two isobaric (A to B, C to D) and two isochoric (B to C, D to A). Complete the following table by calculating ΔU, W, and Q (including the algebraic signs) for each of the four steps.

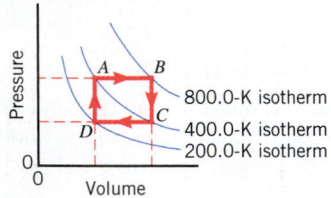

	ΔU	W	Q
A to B			
B to C			
C to D			
D to A			

*30. GO A monatomic ideal gas ($\gamma = \frac{5}{3}$) is contained within a perfectly insulated cylinder that is fitted with a movable piston. The initial pressure of the gas is 1.50×10^5 Pa. The piston is pushed so as to compress the gas, with the result that the Kelvin temperature doubles. What is the final pressure of the gas?

*31. The pressure and volume of an ideal monatomic gas change from A to B to C, as the drawing shows. The curved line between A and C is an isotherm. **(a)** Determine the total heat for the process and **(b)** state whether the flow of heat is into or out of the gas.

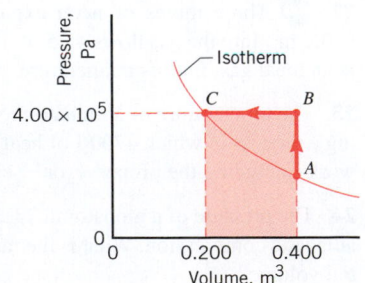

**32. Beginning with a pressure of 2.20×10^5 Pa and a volume of 6.34×10^{-3} m³, an ideal monatomic gas ($\gamma = \frac{5}{3}$) undergoes an adiabatic expansion such that its final pressure is 8.15×10^4 Pa. An alternative process leading to the same final state begins with an isochoric cooling to the final pressure, followed by an isobaric expansion to the final volume. How much more work does the gas do in the adiabatic process than in the alternative process?

33. ssm The drawing shows an adiabatically isolated cylinder that is divided initially into two identical parts by an adiabatic partition. Both sides contain one mole of a monatomic ideal gas ($\gamma = \frac{5}{3}$), with the initial temperature being 525 K on the left and 275 K on the right. The partition is then allowed to move slowly (i.e., quasi-statically) to the right, until the pressures on each side of the partition are the same. Find the final temperatures on the **(a) left and **(b)** right.

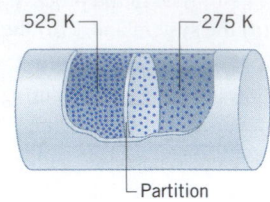

Section 15.6 Specific Heat Capacities

34. Three moles of a monatomic ideal gas are heated at a constant volume of 1.50 m³. The amount of heat added is 5.24×10^3 J. **(a)** What is the change in the temperature of the gas? **(b)** Find the change in its internal energy. **(c)** Determine the change in pressure.

35. ssm The temperature of 2.5 mol of a monatomic ideal gas is 350 K. The internal energy of this gas is doubled by the addition of heat. How much heat is needed when it is added at **(a)** constant volume and **(b)** constant pressure?

36. A monatomic ideal gas in a rigid container is heated from 217 K to 279 K by adding 8500 J of heat. How many moles of gas are there in the container?

37. ssm Heat is added to two identical samples of a monatomic ideal gas. In the first sample the heat is added while the volume of the gas is kept constant, and the heat causes the temperature to rise by 75 K. In the second sample, an identical amount of heat is added while the pressure (but not the volume) of the gas is kept constant. By how much does the temperature of this sample increase?

38. GO Under constant-volume conditions, 3500 J of heat is added to 1.6 moles of an ideal gas. As a result, the temperature of the gas increases by 75 K. How much heat would be required to cause the same temperature change under constant-pressure conditions? Do not assume anything about whether the gas is monatomic, diatomic, etc.

39. Heat Q is added to a monatomic ideal gas at constant pressure. As a result, the gas does work W. Find the ratio Q/W.

*40. A monatomic ideal gas expands at constant pressure. **(a)** What percentage of the heat being supplied to the gas is used to increase the internal energy of the gas? **(b)** What percentage is used for doing the work of expansion?

*41. Suppose that 31.4 J of heat is added to an ideal gas. The gas expands at a constant pressure of 1.40×10^4 Pa while changing its volume from 3.00×10^{-4} to 8.00×10^{-4} m³. The gas is not monatomic, so the relation $C_P = \frac{5}{2}R$ does not apply. **(a)** Determine the change in the internal energy of the gas. **(b)** Calculate its molar specific heat capacity C_P.

*42. A monatomic ideal gas is heated while at a constant volume of 1.00×10^{-3} m³, using a ten-watt heater. The pressure of the gas increases by 5.0×10^4 Pa. How long was the heater on?

**43. One mole of neon, a monatomic gas, starts out at conditions of standard temperature and pressure. The gas is heated at constant volume until its pressure is tripled, then further heated at constant pressure until its volume is doubled. Assume that neon behaves as an ideal gas. For the entire process, find the heat added to the gas.

Section 15.8 Heat Engines

44. Heat engines take input energy in the form of heat, use some of that energy to do work, and exhaust the remainder. Similarly, a person can be viewed as a heat engine that takes an input of internal energy, uses some of it to do work, and gives off the rest as heat. Suppose that a trained athlete can function as a heat engine with an efficiency of 0.11. **(a)** What is the magnitude of the internal energy that the athlete uses in order to do 5.1×10^4 J of work? **(b)** Determine the magnitude of the heat the athlete gives off.

45. ssm Multiple-Concept Example 6 deals with the concepts that are important in this problem. In doing 16 600 J of work, an engine rejects 9700 J of heat. What is the efficiency of the engine?

46. A lawnmower engine with an efficiency of 0.22 rejects 9900 J of heat every second. What is the magnitude of the work that the engine does in one second?

47. Due to a tune-up, the efficiency of an automobile engine increases by 5.0%. For an input heat of 1300 J, how much more work does the engine produce after the tune-up than before?

*48. **GO** A 52-kg mountain climber, starting from rest, climbs a vertical distance of 730 m. At the top, she is again at rest. In the process, her body generates 4.1×10^6 J of energy via metabolic processes. In fact, her body acts like a heat engine, the efficiency of which is given by Equation 15.11 as $e = |W|/|Q_H|$, where $|W|$ is the magnitude of the work she does and $|Q_H|$ is the magnitude of the input heat. Find her efficiency as a heat engine.

*49. ssm www Due to design changes, the efficiency of an engine increases from 0.23 to 0.42. For the same input heat $|Q_H|$, these changes increase the work done by the more efficient engine and reduce the amount of heat rejected to the cold reservoir. Find the ratio of the heat rejected to the cold reservoir for the improved engine to that for the original engine.

**50. An engine has an efficiency e_1. The engine takes input heat of magnitude $|Q_H|$ from a hot reservoir and delivers work of magnitude $|W_1|$. The heat rejected by this engine is used as input heat for a second engine, which has an efficiency e_2 and delivers work of magnitude $|W_2|$. The overall efficiency of this two-engine device is the magnitude of the total work delivered ($|W_1| + |W_2|$) divided by the magnitude $|Q_H|$ of the input heat. Find an expression for the overall efficiency e in terms of e_1 and e_2.

Section 15.9 Carnot's Principle and the Carnot Engine

51. A Carnot engine operates with an efficiency of 27.0% when the temperature of its cold reservoir is 275 K. Assuming that the temperature of the hot reservoir remains the same, what must be the temperature of the cold reservoir in order to increase the efficiency to 32.0%?

52. Five thousand joules of heat is put into a Carnot engine whose hot and cold reservoirs have temperatures of 500 and 200 K, respectively. How much heat is converted into work?

53. ssm A Carnot engine has an efficiency of 0.700, and the temperature of its cold reservoir is 378 K. **(a)** Determine the temperature of its hot reservoir. **(b)** If 5230 J of heat is rejected to the cold reservoir, what amount of heat is put into the engine?

54. A Carnot engine operates with a large hot reservoir and a much smaller cold reservoir. As a result, the temperature of the hot reservoir remains constant while the temperature of the cold reservoir slowly increases. This temperature change decreases the efficiency of the engine to 0.70 from 0.75. Find the ratio of the final temperature of the cold reservoir to its initial temperature.

55. An engine does 18 500 J of work and rejects 6550 J of heat into a cold reservoir whose temperature is 285 K. What would be the smallest possible temperature of the hot reservoir?

56. **GO** Carnot engine A has an efficiency of 0.60, and Carnot engine B has an efficiency of 0.80. Both engines utilize the same hot reservoir, which has a temperature of 650 K and delivers 1200 J of heat to each engine. Find the magnitude of the work produced by each engine and the temperatures of the cold reservoirs that they use.

57. Concept Simulation 15.1 at **www.wiley.com/college/cutnell** illustrates the concepts pertinent to this problem. A Carnot engine operates between temperatures of 650 and 350 K. To improve the efficiency of the engine, it is decided either to raise the temperature of the hot reservoir by 40 K or to lower the temperature of the cold reservoir by 40 K. Which change gives the greatest improvement? Justify your answer by calculating the efficiency in each case.

*58. The hot reservoir for a Carnot engine has a temperature of 890 K, while the cold reservoir has a temperature of 670 K. The heat input for this engine is 4800 J. The 670-K reservoir also serves as the hot reservoir for a second Carnot engine. This second engine uses the rejected heat of the first engine as input and extracts additional work from it. The rejected heat from the second engine goes into a reservoir that has a temperature of 420 K. Find the total work delivered by the two engines.

*59. ssm A power plant taps steam superheated by geothermal energy to 505 K (the temperature of the hot reservoir) and uses the steam to do work in turning the turbine of an electric generator. The steam is then converted back into water in a condenser at 323 K (the temperature of the cold reservoir), after which the water is pumped back down into the earth where it is heated again. The output power (work per unit time) of the plant is 84 000 kilowatts. Determine **(a)** the maximum efficiency at which this plant can operate and **(b)** the minimum amount of rejected heat that must be removed from the condenser every twenty-four hours.

*60. Suppose that the gasoline in a car engine burns at 631 °C, while the exhaust temperature (the temperature of the cold reservoir) is 139 °C and the outdoor temperature is 27 °C. Assume that the engine can be treated as a Carnot engine (a gross oversimplification). In an attempt to increase mileage performance, an inventor builds a second engine that functions between the exhaust and outdoor temperatures and uses the exhaust heat to produce additional work. Assume that the inventor's engine can also be treated as a Carnot engine. Determine the ratio of the total work produced by both engines to that produced by the first engine alone.

**61. ssm The drawing (not to scale) shows the way in which the pressure and volume change for an ideal gas that is used as the working substance in a Carnot engine. The gas begins at point a (pressure = P_a, volume = V_a) and expands isothermally at temperature T_H until point b (pressure = P_b, volume = V_b) is reached. During this expansion, the input heat of magnitude $|Q_H|$ enters the gas from the hot

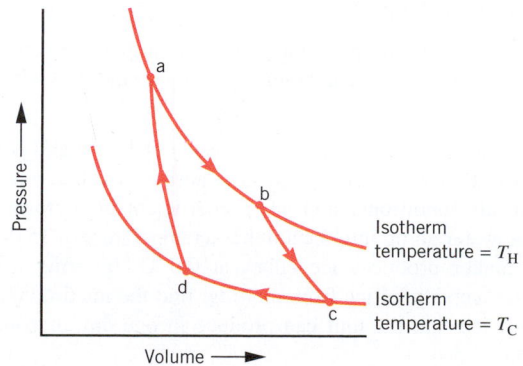

reservoir of the engine. Then, from point b to point c (pressure = P_c, volume = V_c), the gas expands adiabatically. Next, the gas is compressed isothermally at temperature T_C from point c to point d (pressure = P_d, volume = V_d). During this compression, heat of magnitude $|Q_C|$ is rejected to the cold reservoir of the engine. Finally, the gas is compressed adiabatically from point d to point a, where the gas is back in its initial state. The overall process a to b to c to d is called a Carnot cycle. Prove for this cycle that $|Q_C|/|Q_H| = T_C/T_H$.

**62. A nuclear-fueled electric power plant utilizes a so-called "boiling water reactor." In this type of reactor, nuclear energy causes water under pressure to boil at 285 °C (the temperature of the hot reservoir). After the steam does the work of turning the turbine of an electric generator, the steam is converted back into water in a condenser at 40 °C (the temperature of the cold reservoir). To keep the condenser at 40 °C, the rejected heat must be carried away by some means—for example, by water from a river. The plant operates at three-fourths of its Carnot efficiency, and the electrical output power of the plant is 1.2×10^9 watts. A river with a water flow rate of 1.0×10^5 kg/s is available to remove the rejected heat from the plant. Find the number of Celsius degrees by which the temperature of the river rises.

Section 15.10 Refrigerators, Air Conditioners, and Heat Pumps

63. ssm www The temperatures indoors and outdoors are 299 and 312 K, respectively. A Carnot air conditioner deposits 6.12×10^5 J of heat outdoors. How much heat is removed from the house?

64. GO The inside of a Carnot refrigerator is maintained at a temperature of 277 K, while the temperature in the kitchen is 299 K. Using 2500 J of work, how much heat can this refrigerator remove from its inside compartment?

65. A refrigerator operates between temperatures of 296 and 275 K. What would be its maximum coefficient of performance?

66. GO Two Carnot air conditioners, A and B, are removing heat from different rooms. The outside temperature is the same for both rooms, 309.0 K. The room serviced by unit A is kept at a temperature of 294.0 K, while the room serviced by unit B is kept at 301.0 K. The heat removed from either room is 4330 J. For both units, find the magnitude of the work required and the magnitude of the heat deposited outside.

67. A Carnot refrigerator is used in a kitchen in which the temperature is kept at 301 K. This refrigerator uses 241 J of work to remove 2561 J of heat from the food inside. What is the temperature inside the refrigerator?

68. A heat pump removes 2090 J of heat from the outdoors and delivers 3140 J of heat to the inside of a house. **(a)** How much work does the heat pump need? **(b)** What is the coefficient of performance of the heat pump?

69. A Carnot heat pump operates between an outdoor temperature of 265 K and an indoor temperature of 298 K. Find its coefficient of performance.

*70. The wattage of a commercial ice maker is 225 W and is the rate at which it does work. The ice maker operates just like a refrigerator or an air conditioner and has a coefficient of performance of 3.60. The water going into the unit has a temperature of 15.0 °C, and the ice maker produces ice cubes at 0.0 °C. Ignoring the work needed to keep stored ice from melting, find the maximum amount (in kg) of ice that the unit can produce in one day of continuous operation.

*71. ssm Review Conceptual Example 9 before attempting this problem. A window air conditioner has an average coefficient of performance of 2.0. This unit has been placed on the floor by the bed, in a futile attempt to cool the bedroom. During this attempt 7.6×10^4 J of heat is pulled in the front of the unit. The room is sealed and contains 3800 mol of air. Assuming that the molar specific heat capacity of the air is $C_V = \frac{5}{2}R$, determine the rise in temperature caused by operating the air conditioner in this manner.

*72. Interactive LearningWare 15.2 at www.wiley.com/college/cutnell explores one approach to problems such as this. Two kilograms of liquid water at 0 °C is put into the freezer compartment of a Carnot refrigerator. The temperature of the compartment is −15 °C, and the temperature of the kitchen is 27 °C. If the cost of electrical energy is ten cents per kilowatt · hour, how much does it cost to make two kilograms of ice at 0 °C?

*73. ssm www A Carnot refrigerator transfers heat from its inside (6.0 °C) to the room air outside (20.0 °C). **(a)** Find the coefficient of performance of the refrigerator. **(b)** Determine the magnitude of the minimum work needed to cool 5.00 kg of water from 20.0 to 6.0 °C when it is placed in the refrigerator.

**74. A Carnot engine uses hot and cold reservoirs that have temperatures of 1684 and 842 K, respectively. The input heat for this engine is $|Q_H|$. The work delivered by the engine is used to operate a Carnot heat pump. The pump removes heat from the 842-K reservoir and puts it into a hot reservoir at a temperature T'. The amount of heat removed from the 842-K reservoir is also $|Q_H|$. Find the temperature T'.

Section 15.11 Entropy

75. Consider three engines that each use 1650 J of heat from a hot reservoir (temperature = 550 K). These three engines reject heat to a cold reservoir (temperature = 330 K). Engine I rejects 1120 J of heat. Engine II rejects 990 J of heat. Engine III rejects 660 J of heat. One of the engines operates reversibly, and two operate irreversibly. However, of the two irreversible engines, one violates the second law of thermodynamics and could not exist. For each of the engines determine the total entropy change of the universe, which is the sum of the entropy changes of the hot and cold reservoirs. On the basis of your calculations, identify which engine operates reversibly, which operates irreversibly and could exist, and which operates irreversibly and could not exist.

76. Heat Q flows spontaneously from a reservoir at 394 K into a reservoir that has a lower temperature T. Because of the spontaneous flow, thirty percent of Q is rendered unavailable for work when a Carnot engine operates between the reservoir at temperature T and a reservoir at 248 K. Find the temperature T.

77. ssm Find the change in entropy of the H_2O molecules when **(a)** three kilograms of ice melts into water at 273 K and **(b)** three kilograms of water changes into steam at 373 K. **(c)** On the basis of the answers to parts (a) and (b), discuss which change creates more disorder in the collection of H_2O molecules.

78. On a cold day, 24 500 J of heat leaks out of a house. The inside temperature is 21 °C, and the outside temperature is −15 °C. What is the increase in the entropy of the universe that this heat loss produces?

*79. Refer to **Interactive Solution 15.79** at www.wiley.com/college/cutnell to review a method by which this problem can be solved. **(a)** After 6.00 kg of water at 85.0 °C is mixed in a perfect thermos with 3.00 kg of ice at 0.0 °C, the mixture is allowed to reach equilibrium. When heat is added to or removed from a solid or liquid of mass m and specific heat capacity c, the change in entropy can be shown to be $\Delta S = mc \ln(T_f/T_i)$, where T_i and T_f are the initial and final Kelvin temperatures. Using this expression and the change in entropy for melting, find the change in entropy that occurs. **(b)** Should the entropy of the universe increase or decrease as a result of the mixing process? Give your reasoning and state whether your answer in part (a) is consistent with your answer here.

*80. Heat flows from a reservoir at 373 K to a reservoir at

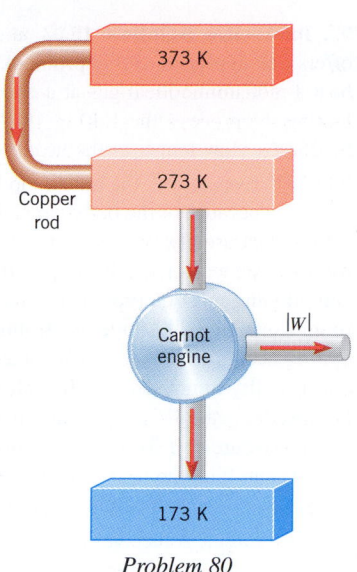

Problem 80

273 K through a 0.35-m copper rod with a cross-sectional area of $9.4 \times 10^{-4} \text{ m}^2$ (see the drawing). The heat then leaves the 273-K reservoir and enters a Carnot engine, which uses part of this heat to do work and rejects the remainder to a third reservoir at 173 K. How much of the heat leaving the 373-K reservoir is rendered unavailable for doing work in a period of 2.0 min?

*81. **GO** An irreversible engine operates between temperatures of 852 and 314 K. It absorbs 1285 J of heat from the hot reservoir and does 264 J of work. **(a)** What is the change $\Delta S_{universe}$ in the entropy of the universe associated with the operation of this engine? **(b)** If the engine were reversible, what would be the magnitude $|W|$ of the work it would have done, assuming that it operated between the same temperatures and absorbed the same heat as the irreversible engine? **(c)** Using the results of parts (a) and (b), find the difference between the work produced by the reversible and irreversible engines.

ADDITIONAL PROBLEMS

82. One-half mole of a monatomic ideal gas expands adiabatically and does 610 J of work. By how many kelvins does its temperature change? Specify whether the change is an increase or a decrease.

83. ssm One-half mole of a monatomic ideal gas absorbs 1200 J of heat while 2500 J of work is done by the gas. **(a)** What is the temperature change of the gas? **(b)** Is the change an increase or a decrease?

84. Multiple-Concept Example 6 deals with the same concepts as this problem does. What is the efficiency of a heat engine that uses an input heat of 5.6×10^4 J and rejects 1.8×10^4 J of heat?

85. A gas is contained in a chamber such as that in Figure 15.4. Suppose that the region outside the chamber is evacuated and the total mass of the block and the movable piston is 135 kg. When 2050 J of heat flows into the gas, the internal energy of the gas increases by 1730 J. What is the distance s through which the piston rises?

86. Engine 1 has an efficiency of 0.18 and requires 5500 J of input heat to perform a certain amount of work. Engine 2 has an efficiency of 0.26 and performs the same amount of work. How much input heat does the second engine require?

87. ssm A process occurs in which the entropy of a system increases by 125 J/K. During the process, the energy that becomes unavailable for doing work is zero. **(a)** Is this process reversible or irreversible? Give your reasoning. **(b)** Determine the change in the entropy of the surroundings.

88. See Multiple-Concept Example 10 to review the concepts that are important in this problem. The water in a deep underground well is used as the cold reservoir of a Carnot heat pump that maintains the temperature of a house at 301 K. To deposit 14 200 J of heat in the house, the heat pump requires 800 J of work. Determine the temperature of the well water.

89. A Carnot air conditioner maintains the temperature in a house at 297 K on a day when the temperature outside is 311 K. What is the coefficient of performance of the air conditioner?

90. A Carnot engine has an efficiency of 0.40. The Kelvin temperature of its hot reservoir is quadrupled, and the Kelvin temperature of its cold reservoir is doubled. What is the efficiency that results from these changes?

91. ssm **(a)** Using the data presented in the accompanying pressure-versus-volume graph, estimate the magnitude of the work done when the system changes from A to B to C along the path shown. **(b)** Determine whether the work is done by the system or on the system and, hence, whether the work is positive or negative.

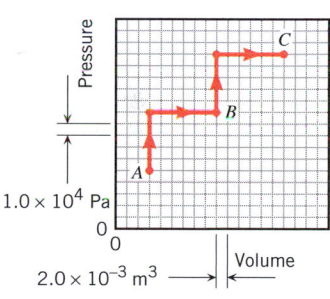

92. Refer to the drawing in Problem 12, where the curve between A and B is now an isotherm. An ideal gas begins at A and is changed along the horizontal line from A to C and then along the vertical line from C to B. **(a)** Find the heat for the process ACB and **(b)** determine whether it flows into or out of the gas.

*93. **ssm** Suppose a monatomic ideal gas is contained within a vertical cylinder that is fitted with a movable piston. The piston is frictionless and has a negligible mass. The area of the piston is 3.14×10^{-2} m^2, and the pressure outside the cylinder is 1.01×10^5 Pa. Heat (2093 J) is removed from the gas. Through what distance does the piston drop?

*94. An air conditioner keeps the inside of a house at a temperature of 19.0 °C when the outdoor temperature is 33.0 °C. Heat, leaking into the house at the rate of 10 500 joules per second, is removed by the air conditioner. Assuming that the air conditioner is a Carnot air conditioner, what is the work per second that must be done by the electrical energy in order to keep the inside temperature constant?

*95. A monatomic ideal gas expands from point *A* to point *B* along the path shown in the drawing. **(a)** Determine the work done by the gas. **(b)** The temperature of the gas at point *A* is 185 K. What is its temperature at point *B*? **(c)** How much heat has been added to or removed from the gas during the process?

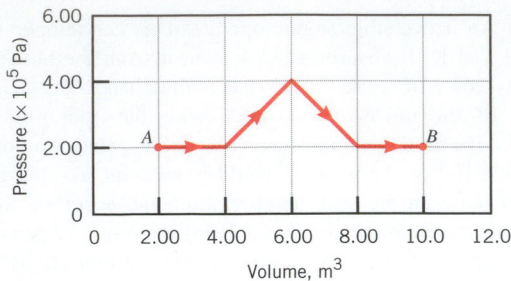

*96. The sun is a sphere with a radius of 6.96×10^8 m and an average surface temperature of 5800 K. Determine the amount by which the sun's thermal radiation increases the entropy of the entire universe each second. Assume that the sun is a perfect blackbody, and that the average temperature of the rest of the universe is 2.73 K. Do not consider the thermal radiation absorbed by the sun from the rest of the universe.

*97. **Interactive Solution 15.97** at **www.wiley.com/college/cutnell** offers one approach to this problem. A fifteen-watt heater is used to heat a monatomic ideal gas at a constant pressure of 7.60×10^5 Pa. During the process, the 1.40×10^{-3} m^3 volume of the gas increases by 25.0%. How long was the heater on?

98. Even at rest, the human body generates heat. The heat arises because of the body's metabolism—that is, the chemical reactions that are always occurring in the body to generate energy. In rooms designed for use by large groups, adequate ventilation or air conditioning must be provided to remove this heat. Consider a classroom containing 200 students. Assume that the metabolic rate of generating heat is 130 W for each student and that the heat accumulates during a fifty-minute lecture. In addition, assume that the air has a molar specific heat of $C_V = \frac{5}{2}R$ and that the room (volume = 1200 m^3, initial pressure = 1.01×10^5 Pa, and initial temperature = 21 °C) is sealed shut. If all the heat generated by the students were absorbed by the air, by how much would the air temperature rise during a lecture?

99. ssm Engine A receives three times more input heat, produces five times more work, and rejects two times more heat than engine B. Find the efficiency of **(a)** engine A and **(b)** engine B.

100. The work done by one mole of a monatomic ideal gas ($\gamma = \frac{5}{3}$) in expanding adiabatically is 825 J. The initial temperature and volume of the gas are 393 K and 0.100 m^3. Obtain **(a)** the final temperature and **(b)** the final volume of the gas.

WAVES AND SOUND

Foraging dolphins use echolocation to find their prey. The dolphins generate sound waves that are called ultrasonic waves because their frequencies range between 70 and 110 kHz, far above the highest frequency of 20 kHz that a healthy young person can hear. These waves reflect from a prey animal and return as an echo that the dolphins can hear. The nature, description, and a discussion of some uses of sound waves occupy most of this chapter. (© Flip Nicklin/Minden Pictures, Inc.)

16.1 THE NATURE OF WAVES

Water waves have two features common to all waves:

1. A wave is a traveling disturbance.
2. A wave carries energy from place to place.

In Figure 16.1 the wave created by the motorboat travels across the lake and disturbs the fisherman. However, *there is no bulk flow of water* outward from the motorboat. The wave is not a bulk movement of water such as a river, but, rather, a disturbance traveling on the surface of the lake. Part of the wave's energy in Figure 16.1 is transferred to the fisherman and his boat.

We will consider two basic types of waves, transverse and longitudinal. Figure 16.2 illustrates how a transverse wave can be generated using a Slinky, a remarkable toy in the form of a long, loosely coiled spring. If one end of the Slinky is jerked up and down, as in part *a*, an upward pulse is sent traveling toward the right. If the end is then jerked down and up, as in part *b*, a downward pulse is generated and also moves to the right. If the end is continually moved up and down in simple harmonic motion, an entire wave is produced. As part *c* illustrates, the wave consists of a series of alternating upward and downward sections that propagate to the right, disturbing the vertical position of the Slinky in the process. To focus attention on the disturbance, a colored dot is attached to the Slinky in part *c* of the drawing. As the wave advances, the dot is displaced up and down in simple harmonic motion. The motion of the dot occurs perpendicular, or transverse, to the

Figure 16.1 The wave created by the motorboat travels across the lake and disturbs the fisherman.

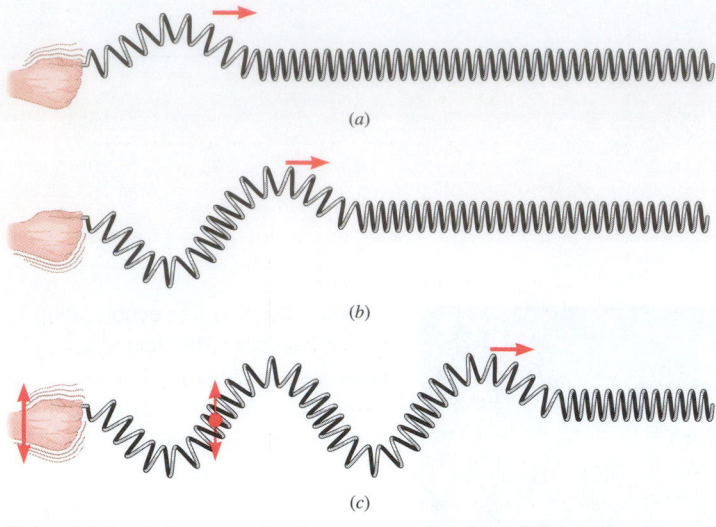

Figure 16.2 (*a*) An upward pulse moves to the right, followed by (*b*) a downward pulse. (*c*) When the end of the Slinky is moved up and down continuously, a transverse wave is produced.

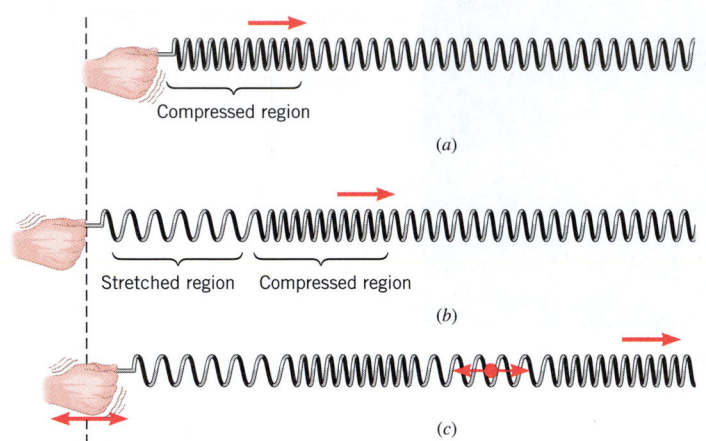

Figure 16.3 (*a*) A compressed region moves to the right, followed by (*b*) a stretched region. (*c*) When the end of the Slinky is moved back and forth continuously, a longitudinal wave is produced.

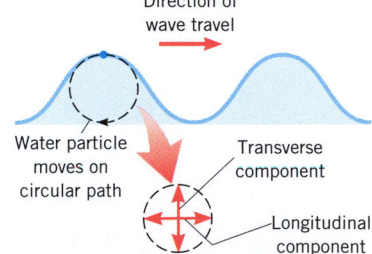

Figure 16.4 A water wave is neither transverse nor longitudinal, since water particles at the surface move clockwise on nearly circular paths as the wave moves from left to right.

direction in which the wave travels. Thus, *a transverse wave is one in which the disturbance occurs perpendicular to the direction of travel of the wave.* Radio waves, light waves, and microwaves are transverse waves. Transverse waves also travel on the strings of instruments such as guitars and banjos.

A longitudinal wave can also be generated with a Slinky, and Figure 16.3 demonstrates how. When one end of the Slinky is pushed forward along its length (i.e., longitudinally) and then pulled back to its starting point, as in part *a*, a region where the coils are squeezed together or compressed is sent traveling to the right. If the end is pulled backward and then pushed forward to its starting point, as in part *b*, a region where the coils are pulled apart or stretched is formed and also moves to the right. If the end is continually moved back and forth in simple harmonic motion, an entire wave is created. As part *c* shows, the wave consists of a series of alternating compressed and stretched regions that travel to the right and disturb the separation between adjacent coils. A colored dot is once again attached to the Slinky to emphasize the vibratory nature of the disturbance. In response to the wave, the dot moves back and forth in simple harmonic motion along the line of travel of the wave. Thus, *a longitudinal wave is one in which the disturbance occurs parallel to the line of travel of the wave.* A sound wave is a longitudinal wave.

Some waves are neither transverse nor longitudinal. For instance, in a water wave the motion of the water particles is not strictly perpendicular or strictly parallel to the line along which the wave travels. Instead, the motion includes both transverse and longitudinal components, since the water particles at the surface move on nearly circular paths, as Figure 16.4 indicates.

✓ **CHECK YOUR UNDERSTANDING**

(*The answers are given at the end of the book.*)

1. Considering the nature of a water wave (see Figure 16.4), which of the following statements correctly describes how a fishing float moves on the surface of a lake when a wave passes beneath it?   **(a)** It bobs up and down vertically.   **(b)** It moves back and forth horizontally.   **(c)** It moves in a vertical plane, exhibiting both motions described in (a) and (b) simultaneously.

2. Suppose that the longitudinal wave in Figure 16.3c moves to the right along the Slinky at a speed of 1 m/s. Does one coil of the Slinky move a distance of 1 mm to the right in a time of 1 ms?

16.2 PERIODIC WAVES

The transverse and longitudinal waves that we have been discussing are called *periodic waves* because they consist of cycles or patterns that are produced over and over again by the source. In Figures 16.2 and 16.3 the repetitive patterns occur as a result of the simple harmonic motion of the left end of the Slinky, so that every segment of the Slinky vibrates in simple harmonic motion. Sections 10.1 and 10.2 discuss the simple harmonic motion of an object on a spring and introduce the concepts of cycle, amplitude, period, and frequency. This same terminology is used to describe periodic waves, such as the sound waves we hear (discussed later in this chapter) and the light waves we see (discussed in Chapter 24).

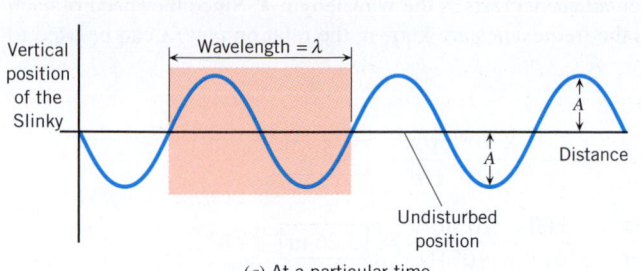

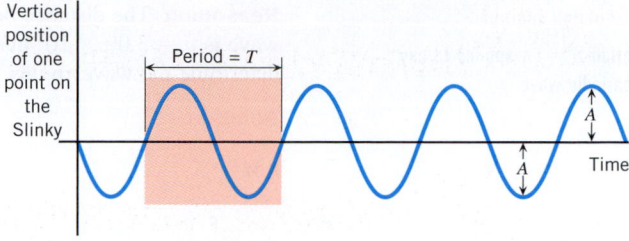

(a) At a particular time

(b) At a particular location

Figure 16.5 One cycle of the wave is shaded in color, and the amplitude of the wave is denoted as A.

Figure 16.5 uses a graphical representation of a transverse wave on a Slinky to review the terminology. One *cycle* of a wave is shaded in color in both parts of the drawing. A wave is a series of many cycles. In part *a* the vertical position of the Slinky is plotted on the vertical axis, and the corresponding distance along the length of the Slinky is plotted on the horizontal axis. Such a graph is equivalent to a photograph of the wave taken at one instant in time and shows the disturbance that exists at each point along the Slinky's length. As marked on this graph, the *amplitude A* is the maximum excursion of a particle of the medium (i.e., the Slinky) in which the wave exists from the particle's undisturbed position. The amplitude is the distance between a crest, or highest point on the wave pattern, and the undisturbed position; it is also the distance between a trough, or lowest point on the wave pattern, and the undisturbed position. The *wavelength* λ is the horizontal length of one cycle of the wave, as shown in Figure 16.5a. The wavelength is also the horizontal distance between two successive crests, two successive troughs, or any two successive equivalent points on the wave.

Part *b* of Figure 16.5 shows a graph in which time, rather than distance, is plotted on the horizontal axis. This graph is obtained by observing a single point on the Slinky. As the wave passes, the point under observation oscillates up and down in simple harmonic motion. As indicated on the graph, the *period T* is the time required for one complete up/down cycle, just as it is for an object vibrating on a spring. The period T is related to the *frequency f,* just as it is for any example of simple harmonic motion:

$$f = \frac{1}{T} \tag{10.5}$$

The period is commonly measured in seconds, and frequency is measured in cycles per second, or hertz (Hz). If, for instance, one cycle of a wave takes one-tenth of a second to pass an observer, then ten cycles pass the observer per second, as Equation 10.5 indicates [$f = 1/(0.1 \text{ s}) = 10$ cycles/s $= 10$ Hz].

A simple relation exists between the period, the wavelength, and the speed of any periodic wave, a relation that Figure 16.6 helps to introduce. Imagine waiting at a railroad crossing, while a freight train moves by at a constant speed v. The train consists of a long line of identical boxcars, each of which has a length λ and requires a time T to pass, so the speed is $v = \lambda/T$. This same equation applies for a wave and relates the speed of the wave to the wavelength λ and the period T. Since the frequency of a wave is $f = 1/T$, the expression for the speed is

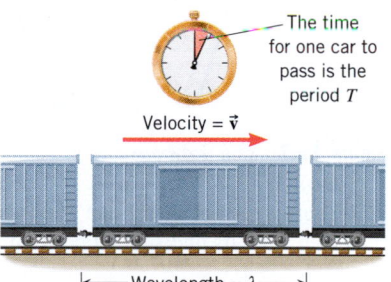

Figure 16.6 A train moving at a constant speed serves as an analogy for a traveling wave.

$$v = \frac{\lambda}{T} = f\lambda \tag{16.1}$$

The terminology just discussed and the fundamental relations $f = 1/T$ and $v = f\lambda$ apply to longitudinal as well as to transverse waves. Example 1 shows how the wavelength of a wave is determined by the wave speed and the frequency established by the source.

⬇ **Example 1** The Wavelengths of Radio Waves

AM and FM radio waves are transverse waves consisting of electric and magnetic disturbances traveling at a speed of 3.00×10^8 m/s. A station broadcasts an AM radio wave whose frequency is 1230×10^3 Hz (1230 kHz on the dial) and an FM radio wave whose frequency is 91.9×10^6 Hz (91.9 MHz on the dial). Find the distance between adjacent crests in each wave.

Reasoning The distance between adjacent crests is the wavelength λ. Since the speed of each wave is $v = 3.00 \times 10^8$ m/s and the frequencies are known, the relation $v = f\lambda$ can be used to determine the wavelengths.

Solution

AM
$$\lambda = \frac{v}{f} = \frac{3.00 \times 10^8 \text{ m/s}}{1230 \times 10^3 \text{ Hz}} = \boxed{244 \text{ m}}$$

FM
$$\lambda = \frac{v}{f} = \frac{3.00 \times 10^8 \text{ m/s}}{91.9 \times 10^6 \text{ Hz}} = \boxed{3.26 \text{ m}}$$

Notice that the wavelength of an AM radio wave is longer than two and one-half football fields!

✓ CHECK YOUR UNDERSTANDING

(The answer is given at the end of the book.)

3. A sound wave (a periodic longitudinal wave) from a loudspeaker travels from air into water. The frequency of the wave does not change, because the loudspeaker producing the sound determines the frequency. The speed of sound in air is 343 m/s, whereas the speed in fresh water is 1482 m/s. When the sound wave enters the water, does its wavelength increase, decrease, or remain the same?

16.3 THE SPEED OF A WAVE ON A STRING

The properties of the material* or medium through which a wave travels determine the speed of the wave. For example, Figure 16.7 shows a transverse wave on a string and draws attention to four string particles that have been drawn as colored dots. As the wave moves to the right, each particle is displaced, one after the other, from its undisturbed position. In the drawing, particles 1 and 2 have already been displaced upward, while particles 3 and 4 are not yet affected by the wave. Particle 3 will be next to move because the section of string immediately to its left (i.e., particle 2) will pull it upward.

Figure 16.7 leads us to conclude that the speed with which the wave moves to the right depends on how quickly one particle of the string is accelerated upward in response to the net pulling force exerted by its adjacent neighbors. In accord with Newton's second law, a stronger net force results in a greater acceleration, and, thus, a faster-moving wave. The ability of one particle to pull on its neighbors depends on how tightly the string is stretched—that is, on the tension (see

Figure 16.7 As a transverse wave moves to the right with speed v, each string particle is displaced, one after the other, from its undisturbed position.

Section 4.10 for a review of tension). The greater the tension, the greater the pulling force the particles exert on each other and the faster the wave travels, other things being equal. Along with the tension, a second factor influences the wave speed. According to Newton's second law, the inertia or mass of particle 3 in Figure 16.7 also affects how quickly it responds to the upward pull of particle 2. For a given net pulling force, a smaller mass has a greater acceleration than a larger mass. Therefore, other things being equal, a wave travels faster on a string whose particles have a small mass, or, as it turns out, on a string that has a small mass per unit length. The mass per unit length is called the *linear density* of the string. It is the mass m of the string divided by its length L, or m/L. Effects of the tension F and the mass per unit length are evident in the following expression for the speed v of a small-amplitude wave on a string:

$$v = \sqrt{\frac{F}{m/L}} \qquad (16.2)$$

The motion of transverse waves along a string is important in the operation of musical instruments, such as the guitar, the violin, and the piano. In these instruments, the strings are either plucked, bowed, or struck to produce transverse waves. Example 2 discusses the speed of the waves on the strings of a guitar.

---

*Electromagnetic waves (discussed in Chapter 24) can move through a vacuum, as well as through materials such as glass and water.

Example 2   Waves Traveling on Guitar Strings

Transverse waves travel on each string of an electric guitar after the string is plucked (see Figure 16.8). The length of each string between its two fixed ends is 0.628 m, and the mass is 0.208 g for the highest pitched E string and 3.32 g for the lowest pitched E string. Each string is under a tension of 226 N. Find the speeds of the waves on the two strings.

Reasoning The speed of a wave on a guitar string, as expressed by Equation 16.2, depends on the tension F in the string and its linear density m/L. Since the tension is the same for both strings, and smaller linear densities give rise to greater speeds, we expect the wave speed to be greatest on the string with the smallest linear density.

Solution The speeds of the waves are given by Equation 16.2 as

High-pitched E $v = \sqrt{\dfrac{F}{m/L}} = \sqrt{\dfrac{226\ \text{N}}{(0.208 \times 10^{-3}\ \text{kg})/(0.628\ \text{m})}} = \boxed{826\ \text{m/s}}$

Low-pitched E $v = \sqrt{\dfrac{F}{m/L}} = \sqrt{\dfrac{226\ \text{N}}{(3.32 \times 10^{-3}\ \text{kg})/(0.628\ \text{m})}} = \boxed{207\ \text{m/s}}$

Notice how fast the waves move: the speeds correspond to 1850 and 463 mi/h.

Conceptual Example 3 offers additional insight into the nature of a wave as a traveling disturbance.

Conceptual Example 3   Wave Speed Versus Particle Speed

As indicated in Figure 16.9, the speed of a transverse wave on a string is v_{wave}, and the speed at which a string particle moves is v_{particle}. Which of the following statements is correct? **(a)** The speeds v_{wave} and v_{particle} are identical. **(b)** The speeds v_{wave} and v_{particle} are different.

Reasoning A wave moves on a string at a speed v_{wave} that is determined by the properties of the string and has a constant value everywhere on the string at all times, assuming that these properties are the same everywhere on the string. Each particle on the string, however, moves in simple harmonic motion, assuming that the source generating the wave (e.g., the hand in Figure 16.2c) moves in simple harmonic motion. Each particle has a speed v_{particle} that is characteristic of simple harmonic motion.

Answer (a) is incorrect. The speed v_{wave} has a constant value at all times. In contrast, v_{particle} is not constant at all times, because it is the speed that characterizes simple harmonic motion and that speed varies as time passes. Thus, the two speeds are not identical.

Answer (b) is correct. The speed v_{wave} is determined by the tension F and the mass per unit length m/L of the string, according to $v_{\text{wave}} = \sqrt{\dfrac{F}{m/L}}$ (see Equation 16.2). The speed v_{particle} is characteristic of simple harmonic motion, according to $v_{\text{particle}} = A\omega \sin \omega t$ (Equation 10.7 without the minus sign, since we deal here only with speed, which is the magnitude of the velocity). The particle speed depends on the amplitude A and the angular frequency ω of the simple harmonic motion, as well as the time t; the speed is greatest when the particle is passing through the undisturbed position of the string, and it is zero when the particle has its maximum displacement. Thus, the two speeds are different, because v_{wave} depends on the properties of the string and v_{particle} depends on the properties of the source creating the wave.

Related Homework: *Problems 106, 107*

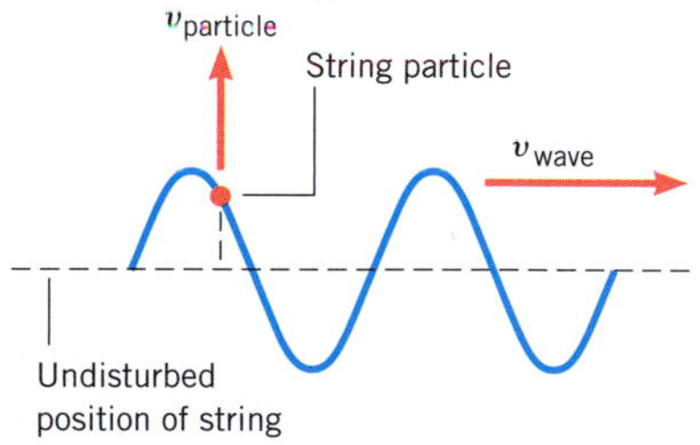

The physics of
waves on guitar strings.

Transverse vibration of the string

Figure 16.8 Plucking a guitar string generates transverse waves.

v_{particle}

String particle

v_{wave}

Undisturbed position of string

Figure 16.9 A transverse wave on a string is moving to the right with a constant speed v_{wave}. A string particle moves up and down in simple harmonic motion about the undisturbed position of the string. A string particle moves with a speed v_{particle}.

✓ **CHECK YOUR UNDERSTANDING**

(*The answers are given at the end of the book.*)

4. One end of each of two identical strings is attached to a wall. Each string is being pulled equally tight by someone at the other end. A transverse pulse is sent traveling along string A. A bit later an identical pulse is sent traveling along string B. What, if anything, can be done to make the pulse on string B catch up with and pass the pulse on string A?

5. In Section 4.10 the concept of a *massless* rope is discussed. Considering Equation 16.2, would it take any time for a transverse wave to travel the length of a truly massless rope?

Continued

6. A wire is strung tightly between two immovable posts. Review Section 12.4 and decide whether the speed of a transverse wave on this wire would increase, decrease, or remain the same when the temperature increases. Ignore any change in the mass per unit length of the wire.

7. Examine Conceptual Example 3 before addressing this question. A wave moves on a string with a constant velocity. Does this mean that the particles of the string always have zero acceleration?

8. A rope of mass *m* is hanging down from the ceiling. Nothing is attached to the loose end of the rope. As a transverse wave travels upward on the rope, does the speed of the wave increase, decrease, or remain the same?

9. String I and string II have the same length. However, the mass of string I is twice the mass of string II, and the tension in string I is eight times the tension in string II. A wave of the same amplitude and frequency travels on each of these strings. Which of the drawings correctly shows the waves, **(a)** A **(b)** B **(c)** C?

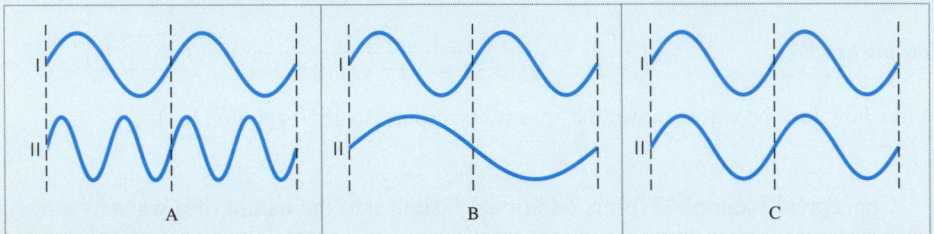

A   B   C

16.4   *THE MATHEMATICAL DESCRIPTION OF A WAVE

When a wave travels through a medium, it displaces the particles of the medium from their undisturbed positions. Suppose that a particle is located at a distance *x* from a coordinate origin. We would like to know the displacement *y* of this particle from its undisturbed position at any time *t* as the wave passes. For periodic waves that result from simple harmonic motion of the source, the expression for the displacement involves a sine or cosine, a fact that is not surprising. After all, in Chapter 10 simple harmonic motion is described using sinusoidal equations, and the graphs for a wave in Figure 16.5 look like a plot of displacement versus time for an object oscillating on a spring (see Figure 10.5).

Our approach will be to present the expression for the displacement and then show graphically that it gives a correct description. Equation 16.3 represents the displacement of a particle caused by a wave traveling in the $+x$ direction (to the right), with an amplitude A, frequency *f*, and wavelength λ. Equation 16.4 applies to a wave moving in the $-x$ direction (to the left).

Wave motion toward +x $$y = A \sin\left(2\pi f t - \frac{2\pi x}{\lambda}\right)$$ (16.3)

Wave motion toward −x $$y = A \sin\left(2\pi f t + \frac{2\pi x}{\lambda}\right)$$ (16.4)

These equations apply to transverse or longitudinal waves and assume that $y = 0$ m when $x = 0$ m and $t = 0$ s.

Consider a transverse wave moving in the $+x$ direction along a string. The term $(2\pi f t - 2\pi x/\lambda)$ in Equation 16.3 is called the *phase angle* of the wave. A string particle located at the origin ($x = 0$ m) exhibits simple harmonic motion with a phase angle of $2\pi f t$; that is, its displacement as a function of time is $y = A \sin(2\pi f t)$. A particle located at a distance *x* also exhibits simple harmonic motion, but its phase angle is

$$2\pi f t - \frac{2\pi x}{\lambda} = 2\pi f\left(t - \frac{x}{f\lambda}\right) = 2\pi f\left(t - \frac{x}{v}\right)$$

The quantity x/v is the time needed for the wave to travel the distance *x*. In other words, the simple harmonic motion that occurs at *x* is delayed by the time interval x/v compared to the motion at the origin.

Figure 16.10 shows the displacement *y* plotted as a function of position *x* along the string at a series of time intervals separated by one-fourth of the period *T* ($t = 0$ s, $\frac{1}{4}T, \frac{2}{4}T, \frac{3}{4}T, T$).

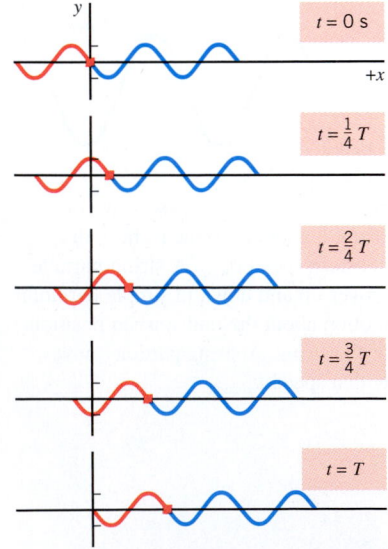

$t = 0$ s

$t = \frac{1}{4}T$

$t = \frac{2}{4}T$

$t = \frac{3}{4}T$

$t = T$

Figure 16.10 Equation 16.3 is plotted here at a series of times separated by one-fourth of the period *T*. The colored square in each graph marks the place on the wave that is located at $x = 0$ m when $t = 0$ s. As time passes, the wave moves to the right.

These graphs are constructed by substituting the corresponding value for *t* into Equation 16.3, remembering that $f = 1/T$, and then calculating *y* at a series of values for *x*. The graphs are like photographs taken at various times as the wave moves to the right. For reference, the colored square on each graph marks the place on the wave that is located at $x = 0$ m when $t = 0$ s. As time passes, the colored square moves to the right, along with the wave. In a similar manner, it can be shown that Equation 16.4 represents a wave moving in the $-x$ direction. Note that the phase angles $(2\pi ft - 2\pi x/\lambda)$ in Equation 16.3 and $(2\pi ft + 2\pi x/\lambda)$ in Equation 16.4 are measured in *radians*, not degrees. **Problem-solving insight:** ***When a calculator is used to evaluate the functions sin $(2\pi ft - 2\pi x/\lambda)$ or sin $(2\pi ft + 2\pi x/\lambda)$, it must be set to its radian mode.***

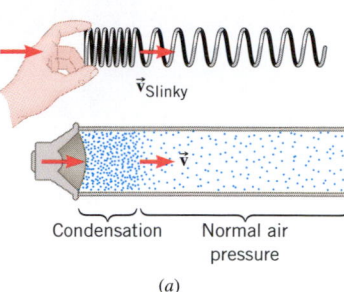

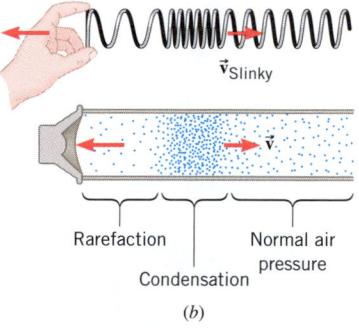

Condensation   Normal air pressure

(a)

Rarefaction   Normal air pressure

Condensation

(b)

Figure 16.11 (*a*) When the speaker diaphragm moves outward, it creates a condensation. (*b*) When the diaphragm moves inward, it creates a rarefaction. The condensation and rarefaction on the Slinky are included for comparison. In reality, the velocity of the wave on the Slinky $\vec{v}_{Slinky}$ is much smaller than the velocity of sound in air $\vec{v}$. For simplicity, the two waves are shown here to have the same velocity.

16.5 THE NATURE OF SOUND

LONGITUDINAL SOUND WAVES

Sound is a longitudinal wave that is created by a vibrating object, such as a guitar string, the human vocal cords, or the diaphragm of a loudspeaker. Moreover, sound can be created or transmitted only in a medium, such as a gas, liquid, or solid. As we will see, the particles of the medium must be present for the disturbance of the wave to move from place to place. Sound cannot exist in a vacuum.

The physics of a loudspeaker diaphragm. To see how sound waves are produced and why they are longitudinal, consider the vibrating diaphragm of a loudspeaker. When the diaphragm moves outward, it compresses the air directly in front of it, as in Figure 16.11*a*. This compression causes the air pressure to rise slightly. The region of increased pressure is called a ***condensation,*** and it travels away from the speaker at the speed of sound. The condensation is analogous to the compressed region of coils in a longitudinal wave on a Slinky, which is included in Figure 16.11*a* for comparison. After producing a condensation, the diaphragm reverses its motion and moves inward, as in part *b* of the drawing. The inward motion produces a region known as a ***rarefaction,*** where the air pressure is slightly less than normal. The rarefaction is similar to the stretched region of coils in a longitudinal Slinky wave. Following immediately behind the condensation, the rarefaction also travels away from the speaker at the speed of sound. Figure 16.12 further emphasizes the similarity between a sound wave and a longitudinal Slinky wave. As the wave passes, the colored dots attached both to the Slinky and to an air molecule execute simple harmonic motion about their undisturbed positions. The colored arrows on either side of the dots indicate that the simple harmonic motion occurs parallel to the line of travel. The drawing also shows that the wavelength λ is the distance between the centers of two successive condensations; λ is also the distance between the centers of two successive rarefactions.

Figure 16.13 illustrates a sound wave spreading out in space after being produced by a loudspeaker. When the condensations and rarefactions arrive at the ear, they force the eardrum to vibrate at the same frequency as the speaker diaphragm. The vibratory motion of the eardrum is interpreted by the brain as sound. It should be emphasized that sound is not a mass movement of air, like the wind. As the condensations and rarefactions of the sound wave travel outward from the vibrating diaphragm in Figure 16.13, the individual air molecules are not carried along with the wave. Rather, each molecule executes simple harmonic motion about a fixed location. In doing so, one molecule collides with its neighbor and passes the condensations and rarefactions forward. The neighbor, in turn, repeats the process.

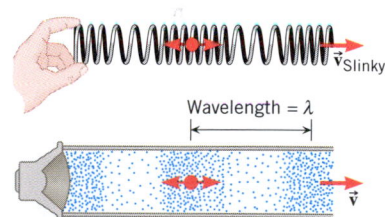

Wavelength = λ

Figure 16.12 Both the wave on the Slinky and the sound wave are longitudinal. The colored dots attached to the Slinky and to an air molecule vibrate back and forth parallel to the line of travel of the wave.

THE FREQUENCY OF A SOUND WAVE

Each cycle of a sound wave includes one condensation and one rarefaction, and the ***frequency*** is the number of cycles per second that passes by a given location. For example, if the diaphragm of a speaker vibrates back and forth in simple harmonic motion at a frequency of 1000 Hz, then 1000 condensations, each followed by a rarefaction, are generated every second, thus forming a sound wave whose frequency is also 1000 Hz. A sound with a single frequency is called a ***pure tone.*** Experiments

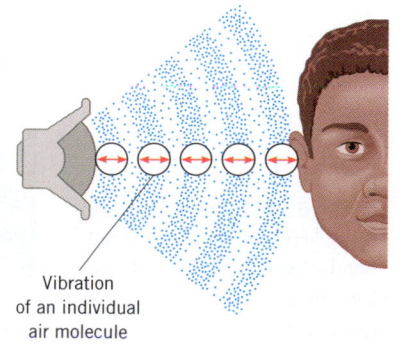

Vibration of an individual air molecule

Figure 16.13 Condensations and rarefactions travel from the speaker to the listener, but the individual air molecules do not move with the wave. A given molecule vibrates back and forth about a fixed location.

have shown that a healthy young person hears all sound frequencies from approximately 20 to 20 000 Hz (20 kHz). The ability to hear the high frequencies decreases with age, however, and a normal middle-aged adult hears frequencies only up to 12–14 kHz.

Pure tones are used in push-button telephones, such as the one shown in Figure 16.14. These phones simultaneously produce two pure tones when each button is pressed, a different pair of tones for each different button. The tones are transmitted electronically to the central telephone office, where they activate switching circuits that complete the call. For example, the drawing indicates that pressing the "5" button produces pure tones of 770 and 1336 Hz simultaneously, while the "9" button generates tones of 852 and 1477 Hz.

Sound can be generated whose frequency lies below 20 Hz or above 20 kHz, although humans normally do not hear it. Sound waves with frequencies below 20 Hz are said to be *infrasonic,* while those with frequencies above 20 kHz are referred to as *ultrasonic.* Some species of bats known as microbats use ultrasonic frequencies up to 120 kHz for locating prey and for navigating (Figure 16.15), while rhinoceroses use infrasonic frequencies as low as 5 Hz to call one another (Figure 16.16).

Frequency is an objective property of a sound wave because frequency can be measured with an electronic frequency counter. A listener's perception of frequency, however, is subjective. The brain interprets the frequency detected by the ear primarily in terms of the subjective quality called *pitch.* A pure tone with a large (high) frequency is interpreted as a high-pitched sound, while a pure tone with a small (low) frequency is interpreted as a low-pitched sound. For instance, a piccolo produces high-pitched sounds, and a tuba produces low-pitched sounds.

THE PRESSURE AMPLITUDE OF A SOUND WAVE

Figure 16.17 illustrates a pure-tone sound wave traveling in a tube. Attached to the tube is a series of gauges that indicate the pressure variations along the wave. The graph shows that the air pressure varies sinusoidally along the length of the tube. Although this graph has the appearance of a transverse wave, remember that the sound itself is a longitudinal wave. The graph also shows the *pressure amplitude* of the wave, which is the magnitude of the maximum change in pressure, measured relative to the undisturbed or atmospheric pressure. The pressure fluctuations in a sound wave are normally very small. For instance, in a typical conversation between two people the pressure amplitude is about 3×10^{-2} Pa, certainly a small amount compared with the atmospheric pressure of $1.01 \times 10^{+5}$ Pa. The ear is remarkable in being able to detect such small changes.

Loudness is an attribute of sound that depends primarily on the amplitude of the wave: the larger the amplitude, the louder the sound. The pressure amplitude is an objective property of a sound wave, since it can be measured. Loudness, on the other hand, is subjective. Each individual determines what is loud, depending on the acuteness of his or her hearing.

The physics of a push-button telephone.

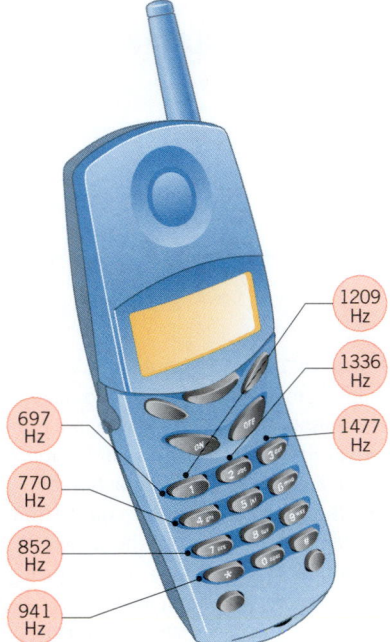

Figure 16.14 A push-button telephone and a schematic showing the two pure tones produced when each button is pressed.

✓ **CHECK YOUR UNDERSTANDING**

(*The answer is given at the end of the book.*)

10. In a traveling sound wave, are there any particles that are *always* at rest as the wave passes by?

Figure 16.15 Bats use ultrasonic sound waves for locating prey and for navigating. This bat has captured a centipede. (© Merlin D. Tuttle/BCI/Bat Conservation International, Inc.)

Figure 16.16 Rhinoceroses call to one another using infrasonic sound waves. (Ian Murphy/Stone/Getty Images)

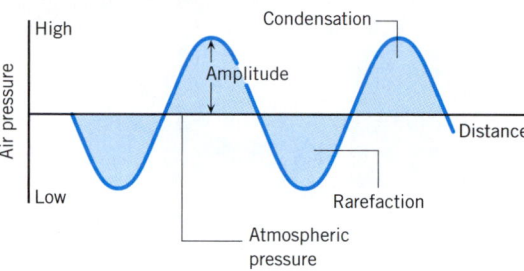

Figure 16.17 A sound wave is a series of alternating condensations and rarefactions. The graph shows that the condensations are regions of higher than normal air pressure, and the rarefactions are regions of lower than normal air pressure.

THE SPEED OF SOUND

16.6

GASES

Sound travels through gases, liquids, and solids at considerably different speeds, as Table 16.1 reveals. Near room temperature, the speed of sound in air is 343 m/s (767 mi/h) and is markedly greater in liquids and solids. For example, sound travels more than four times faster in water and more than seventeen times faster in steel than it does in air. In general, sound travels slowest in gases, faster in liquids, and fastest in solids.

Like the speed of a wave on a guitar string, the speed of sound depends on the properties of the medium. In a gas, it is only when molecules collide that the condensations and rarefactions of a sound wave can move from place to place. It is reasonable, then, to expect the speed of sound in a gas to have the same order of magnitude as the average molecular speed between collisions. For an ideal gas this average speed is the translational rms speed given by Equation 14.6: $v_{rms} = \sqrt{3kT/m}$, where T is the Kelvin temperature, m is the mass of a molecule, and k is Boltzmann's constant. Although the expression for v_{rms} overestimates the speed of sound, it does give the correct dependence on Kelvin temperature and particle mass. Careful analysis shows that the speed of sound in an ideal gas is given by

Ideal gas
$$v = \sqrt{\frac{\gamma kT}{m}} \qquad (16.5)$$

where $\gamma = c_P/c_V$ is the ratio of the specific heat capacity at constant pressure c_P to the specific heat capacity at constant volume c_V.

The factor γ is introduced in Section 15.5, where the adiabatic compression and expansion of an ideal gas are discussed. In Section 15.6 it is shown that γ has the value of $\gamma = \frac{5}{3}$ for ideal monatomic gases and a value of $\gamma = \frac{7}{5}$ for ideal diatomic gases. The value of γ appears in Equation 16.5 because the condensations and rarefactions of a sound wave are formed by adiabatic compressions and expansions of the gas. The regions that are compressed (the condensations) become slightly warmed, and the regions that are expanded (the rarefactions) become slightly cooled. However, no appreciable heat flows from a condensation to an adjacent rarefaction because the distance between the two (half a wavelength) is relatively large for most audible sound waves and a gas is a poor thermal conductor. Thus, the compression and expansion process is adiabatic. Example 4 illustrates the use of Equation 16.5.

Table 16.1 Speed of Sound in Gases, Liquids, and Solids

Substance	Speed (m/s)
Gases	
Air (0 °C)	331
Air (20 °C)	343
Carbon dioxide (0 °C)	259
Oxygen (0 °C)	316
Helium (0 °C)	965
Liquids	
Chloroform (20 °C)	1004
Ethyl alcohol (20 °C)	1162
Mercury (20 °C)	1450
Fresh water (20 °C)	1482
Seawater (20 °C)	1522
Solids	
Copper	5010
Glass (Pyrex)	5640
Lead	1960
Steel	5960

ANALYZING MULTIPLE-CONCEPT PROBLEMS

Example 4 An Ultrasonic Ruler

The physics of an ultrasonic ruler.

Figure 16.18 shows an ultrasonic ruler that is used to measure the distance to a target, such as a wall. To initiate the measurement, the ruler generates a pulse of ultrasonic sound that travels to the wall and, much like an echo, reflects from it. The reflected pulse returns to the ruler, which measures the time it takes for the round-trip. Using a preset value for the speed of

Continued

sound, the unit determines the distance to the wall and displays it on a digital read-out. Suppose that the round-trip travel time is 20.0 ms on a day when the air temperature is 32 °C. Assuming that air is an ideal diatomic gas ($\gamma = \frac{7}{5}$) and that the average molecular mass of air is 28.9 u, find the distance between the ultrasonic ruler and the wall.

Reasoning The distance between the ruler and the wall is equal to the speed of sound multiplied by the time it takes for the sound pulse to reach the wall. The speed v of sound can be determined from a knowledge of the air temperature T and the average mass m of an air molecule by using the relation $v = \sqrt{\gamma kT/m}$. The time can be deduced from the given data.

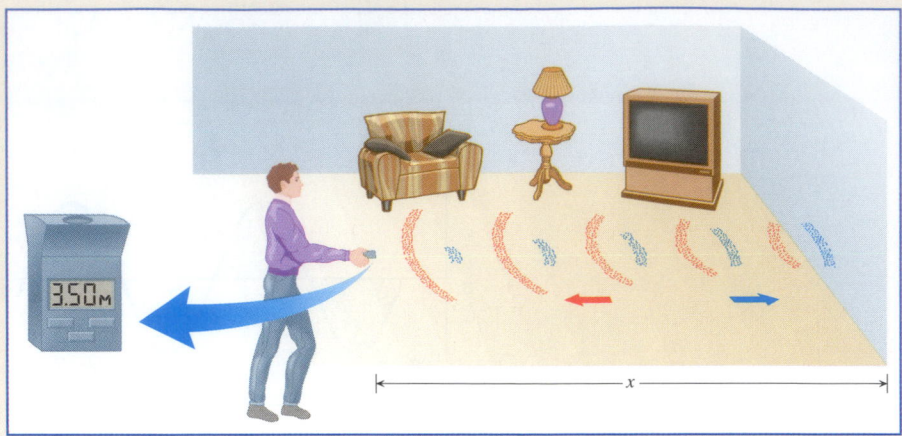

Figure 16.18 An ultrasonic ruler uses sound with a frequency greater than 20 kHz to measure the distance x to the wall. The blue arcs and blue arrow denote the outgoing sound wave, and the red arcs and red arrow denote the wave reflected from the wall.

Knowns and Unknowns The data for this problem are listed below:

Description	Symbol	Value	Comment
Round-trip time of sound	t_{RT}	20.0 ms	20.0 ms = 20.0×10^{-3} s
Air temperature	T_c	32 °C	
Ratio of specific heats for air	γ	$\frac{7}{5}$	
Average molecular mass of air	m	28.9 u	Must convert "u" to kilograms.
Unknown Variable			
Distance between ruler and wall	x	?	

Modeling the Problem

STEP 1 Kinematics Since sound moves at a constant speed, the distance x it travels is the product of its speed v and the time t, or $x = vt$. The time for the sound to reach the wall is one-half the round-trip time t_{RT}, so $t = \frac{1}{2}t_{RT}$. Thus, the distance to the wall is

$$x = v\left(\tfrac{1}{2}t_{RT}\right)$$

The round-trip time t_{RT} is known, but the speed of sound in air at 32 °C is not. We will find an expression for this speed in Step 2.

$$x = v\left(\tfrac{1}{2}t_{RT}\right) \quad (1)$$

?

STEP 2 Speed of Sound Since the air is assumed to be an ideal gas, the speed v of sound is related to the Kelvin temperature T and the average mass m of an air molecule by

$$v = \sqrt{\frac{\gamma kT}{m}} \quad (16.5)$$

where γ is the ratio of the specific heat capacity of air at constant pressure to that at constant volume (see Section 15.5), and k is Boltzmann's constant. The temperature in this expression must be the Kelvin temperature of the air, which is related to its Celsius temperature T_c by $T = T_c + 273.15$ (Equation 12.1). Thus, the speed of sound in air is

$$\boxed{v = \sqrt{\frac{\gamma k(T_c + 273.15)}{m}}}$$

$$x = v\left(\tfrac{1}{2}t_{RT}\right) \quad (1)$$

$$\boxed{v = \sqrt{\frac{\gamma k(T_c + 273.15)}{m}}}$$

This expression for v can be substituted into Equation 1, as shown on the right.

Solution Combining the results of the modeling steps, we have

STEP 1   STEP 2

$$x = v\left(\tfrac{1}{2}t_{RT}\right) = \sqrt{\frac{\gamma k(T_c + 273.15)}{m}}\left(\tfrac{1}{2}t_{RT}\right)$$

Since the average mass of an air molecule is given in atomic mass units (28.9 u), we must convert it to kilograms by using the conversion factor 1 u = 1.6605×10^{-27} kg (see Section 14.1). Thus,

$$m = (28.9\ \text{u})\left(\frac{1.6605 \times 10^{-27}\ \text{kg}}{1\ \text{u}}\right) = 4.80 \times 10^{-26}\ \text{kg}$$

The distance from the ultrasonic ruler to the wall is

$$
\begin{aligned}
x &= \sqrt{\frac{\gamma k(T_c + 273.15)}{m}}\left(\tfrac{1}{2}t_{RT}\right) \\
&= \sqrt{\frac{\tfrac{7}{5}(1.38 \times 10^{-23}\ \text{J/K})(32\ °\text{C} + 273.15)}{4.80 \times 10^{-26}\ \text{kg}}}\left[\tfrac{1}{2}(20.0 \times 10^{-3}\ \text{s})\right] = \boxed{3.50\ \text{m}}
\end{aligned}
$$

Related Homework: *Problems 48, 50*

Problem-solving insight
When using equation $v = \sqrt{\gamma kT/m}$ to calculate the speed of sound in an ideal gas, be sure to express the temperature T in kelvins and not in degrees Celsius or Fahrenheit.

Sonar (**so**und **na**vigation **r**anging) is a technique for determining water depth and locating underwater objects, such as reefs, submarines, and schools of fish. The core of a sonar unit consists of an ultrasonic transmitter and receiver mounted on the bottom of a ship. The transmitter emits a short pulse of ultrasonic sound, and at a later time the reflected pulse returns and is detected by the receiver. The water depth is determined from the electronically measured round-trip time of the pulse and a knowledge of the speed of sound in water; the depth registers automatically on an appropriate meter. Such a depth measurement is similar to the distance measurement discussed for the ultrasonic ruler in Example 4.

The physics of sonar.

Conceptual Example 5 illustrates how the speed of sound in air can be used to estimate the distance to a thunderstorm, using a handy rule of thumb.

 Conceptual Example 5 Lightning, Thunder, and a Rule of Thumb

In a thunderstorm, lightning and thunder occur nearly simultaneously. The light waves from the lightning travel at a speed of $v_{\text{light}} = 3.0 \times 10^8$ m/s, whereas the sound waves from the thunder travel at $v_{\text{sound}} = 343$ m/s. There is a rule of thumb for estimating how far away a storm is. After you see a lightning flash, count the seconds until you hear the thunder; divide the number of seconds by five to get the approximate distance (in miles) to the storm. In this rule, which of the two speeds plays a role? (**a**) Both v_{sound} and v_{light} (**b**) Only v_{sound} (**c**) Only v_{light}

Reasoning At a distance of one mile from a storm, the observer in Figure 16.19 detects either type of wave only after a time that is equal to the distance divided by the speed at which the wave travels. This fact will guide our analysis.

Answers (b) and (c) are incorrect. The rule involves the time that passes *between* seeing the lightning flash and hearing the thunder, not just the time at which either type of wave is detected. Therefore, both the speeds v_{light} and v_{sound} must play a role in the rule.

Answer (a) is correct. Light from the flash travels so rapidly that it reaches the observer almost instantaneously; its travel time for one mile (1.6×10^3 m) is only

$$t_{\text{light}} = \frac{1.6 \times 10^3\ \text{m}}{v_{\text{light}}} = \frac{1.6 \times 10^3\ \text{m}}{3.0 \times 10^8\ \text{m/s}} = 5 \times 10^{-6}\ \text{s}$$

In comparison, the sound of the thunder travels very slowly; its travel time for one mile is

$$t_{\text{sound}} = \frac{1.6 \times 10^3\ \text{m}}{v_{\text{sound}}} = \frac{1.6 \times 10^3\ \text{m}}{343\ \text{m/s}} = 5\ \text{s}$$

Since t_{light} is negligible compared to t_{sound}, the time between seeing the lightning flash and hearing the thunder is about 5 s for every mile of distance from the storm.

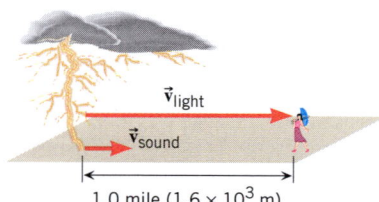

Figure 16.19 A lightning bolt from a thunderstorm generates a flash of light and sound (thunder) almost simultaneously. The speed of light is much greater than the speed of sound. Therefore, the light reaches the person first, followed later by the sound.

LIQUIDS

In a liquid, the speed of sound depends on the density ρ and the *adiabatic* bulk modulus B_{ad} of the liquid:

Liquid
$$v = \sqrt{\frac{B_{ad}}{\rho}}$$
(16.6)

The bulk modulus is introduced in Section 10.7 in a discussion of the volume deformation of liquids and solids. There it is tacitly assumed that the temperature remains constant while the volume of the material changes; that is, the compression or expansion is isothermal. However, the condensations and rarefactions in a sound wave occur under *adiabatic* rather than isothermal conditions. Thus, the adiabatic bulk modulus B_{ad} must be used when calculating the speed of sound in liquids. Values of B_{ad} will be provided as needed in this text.

Table 16.1 gives some data for the speed of sound in liquids. In seawater, for instance, the speed is 1522 m/s, which is more than four times as great as the speed in air. The speed of sound is an important parameter in the measurement of distance, as discussed for the ultrasonic ruler in Example 4. Accurate distance measurements using ultrasonic sound also play an important role in medicine, where the sound often travels through liquid-like materials in the body. A routine preoperative procedure in cataract surgery, for example, uses an ultrasonic probe called an A-scan to measure the length of the eyeball in front of the lens, the thickness of the lens, and the length of the eyeball between the lens and the retina (see Figure 16.20). The measurement is similar to that discussed in Example 4 and relies on the fact that the speed of sound in the material in front of and behind the lens of the eye is 1532 m/s, whereas that within the lens is 1641 m/s. In cataract surgery, the cataractous lens is removed and often replaced with an implanted artificial lens. Data provided by the A-scan facilitate the design of the lens implant (its size and the optical correction that it introduces).

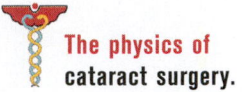

The physics of cataract surgery.

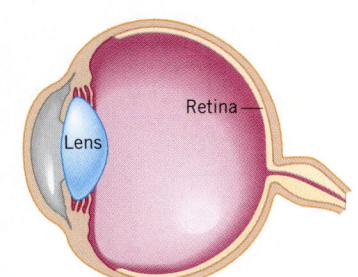

Figure 16.20 A cross-sectional view of the human eye.

SOLID BARS

When sound travels through a long, slender, solid bar, the speed of the sound depends on the properties of the medium according to

Long, slender, solid bar
$$v = \sqrt{\frac{Y}{\rho}}$$
(16.7)

where Y is Young's modulus (defined in Section 10.7) and ρ is the density.

✓ CHECK YOUR UNDERSTANDING

(The answers are given at the end of the book.)

11. Do you expect an echo to return to you more quickly on a hot day or a cold day, other things being equal?

12. Carbon monoxide (CO), hydrogen (H₂), and nitrogen (N₂) may be treated as ideal gases. Each has the same temperature and nearly the same value for the ratio of the specific heat capacities at constant pressure and constant volume. In which two of the three gases is the speed of sound approximately the same?

13. Jell-O starts out as a liquid and then sets to a gel. As the Jell-O sets and becomes more solid, does the speed of sound in this material increase, decrease, or remain the same?

16.7 SOUND INTENSITY

Sound waves carry energy that can be used to do work, like forcing the eardrum to vibrate. In an extreme case such as a sonic boom, the energy can be sufficient to cause damage to windows and buildings. The amount of energy transported per second by a sound wave is called the ***power*** of the wave and is measured in SI units of joules per second (J/s) or watts (W).

When a sound wave leaves a source, such as the loudspeaker in Figure 16.21, the power spreads out and passes through imaginary surfaces that have increasingly larger areas. For instance, the same sound power passes through the surfaces labeled 1 and 2 in the drawing. However, the power is spread out over a greater area in surface 2. We will bring together the ideas of sound power and the area through which the power passes and, in the process, will formulate the concept of sound intensity. The idea of wave intensity is not confined to sound waves. It will recur, for example, in Chapter 24 when we discuss another important type of waves, electromagnetic waves.

The *sound intensity I* is defined as the sound power P that passes perpendicularly through a surface divided by the area A of that surface:

$$I = \frac{P}{A} \tag{16.8}$$

The unit of sound intensity is power per unit area, or W/m^2. The next example illustrates how the sound intensity changes as the distance from a loudspeaker changes.

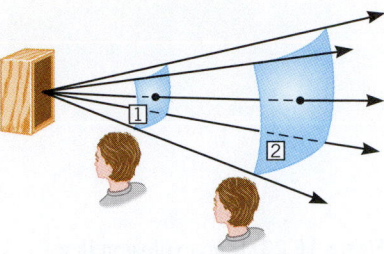

Figure 16.21 The power carried by a sound wave spreads out after leaving a source, such as a loudspeaker. Thus, the power passes perpendicularly through surface 1 and then through surface 2, which has the larger area.

 Example 6   Sound Intensities

In Figure 16.21, 12×10^{-5} W of sound power passes perpendicularly through the surfaces labeled 1 and 2. These surfaces have areas of $A_1 = 4.0$ m^2 and $A_2 = 12$ m^2. Determine the sound intensity at each surface and discuss why listener 2 hears a quieter sound than listener 1.

Reasoning   The sound intensity I is the sound power P passing perpendicularly through a surface divided by the area A of that surface. Since the same sound power passes through both surfaces and surface 2 has the greater area, the sound intensity is less at surface 2.

Solution   The sound intensity at each surface follows from Equation 16.8:

Surface 1 $\qquad I_1 = \dfrac{P}{A_1} = \dfrac{12 \times 10^{-5}\ \text{W}}{4.0\ \text{m}^2} = \boxed{3.0 \times 10^{-5}\ \text{W/m}^2}$

Surface 2 $\qquad I_2 = \dfrac{P}{A_2} = \dfrac{12 \times 10^{-5}\ \text{W}}{12\ \text{m}^2} = \boxed{1.0 \times 10^{-5}\ \text{W/m}^2}$

Problem-solving insight

Sound intensity I and sound power P are different concepts. They are related, however, since intensity equals power per unit area.

The sound intensity is less at the more distant surface, where the same power passes through a threefold greater area. The ear of a listener, with its fixed area, intercepts less power where the intensity, or power per unit area, is smaller. Thus, listener 2 intercepts less of the sound power than listener 1. With less power striking the ear, the sound is quieter.

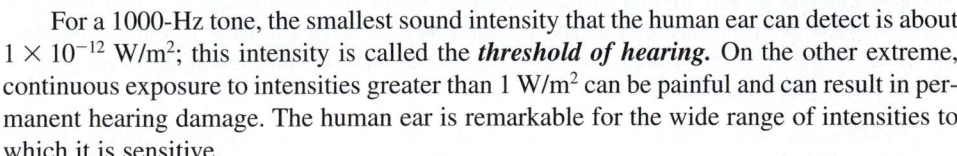

For a 1000-Hz tone, the smallest sound intensity that the human ear can detect is about 1×10^{-12} W/m^2; this intensity is called the *threshold of hearing.* On the other extreme, continuous exposure to intensities greater than 1 W/m^2 can be painful and can result in permanent hearing damage. The human ear is remarkable for the wide range of intensities to which it is sensitive.

If a source emits sound *uniformly in all directions,* the intensity depends on distance in a simple way. Figure 16.22 shows such a source at the center of an imaginary sphere (for clarity only a hemisphere is shown). The radius of the sphere is r. Since all the radiated sound power P passes through the spherical surface of area $A = 4\pi r^2$, the intensity at a distance r is

Spherically uniform radiation $\qquad I = \dfrac{P}{4\pi r^2} \tag{16.9}$

From this we see that the intensity of a source that radiates sound uniformly in all directions varies as $1/r^2$. For example, if the distance increases by a factor of two, the sound intensity decreases by a factor of $2^2 = 4$. Example 7 illustrates the effect of the $1/r^2$ dependence of intensity on distance.

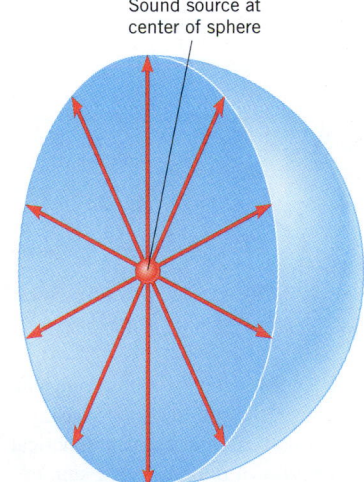

Figure 16.22 The sound source at the center of the sphere emits sound uniformly in all directions. In this drawing, only a hemisphere is shown for clarity.

 Example 7   Fireworks

During a fireworks display, a rocket explodes high in the air above the observers. Assume that the sound spreads out uniformly in all directions and that reflections from the ground can be

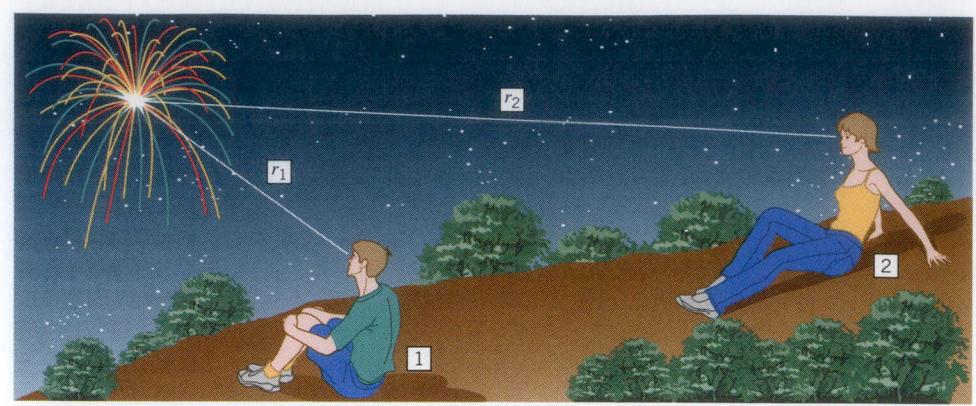

Figure 16.23 If an explosion in a fireworks display radiates sound uniformly in all directions, the intensity at any distance r is $I = P/(4\pi r^2)$, where P is the sound power of the explosion.

ignored. When the sound reaches listener 2 in Figure 16.23, who is $r_2 = 640$ m away from the explosion, the sound has an intensity of $I_2 = 0.10$ W/m². What is the sound intensity detected by listener 1, who is $r_1 = 160$ m away from the explosion?

Reasoning Listener 1 is four times closer to the explosion than listener 2. Therefore, the sound intensity detected by listener 1 is $4^2 = 16$ times greater than the sound intensity detected by listener 2.

Problem-solving insight

Equation 16.9 can be used only when the sound spreads out uniformly in all directions and there are no reflections of the sound waves.

Solution The ratio of the sound intensities can be found using Equation 16.9:

$$\frac{I_1}{I_2} = \frac{\dfrac{P}{4\pi r_1^2}}{\dfrac{P}{4\pi r_2^2}} = \frac{r_2^2}{r_1^2} = \frac{(640 \text{ m})^2}{(160 \text{ m})^2} = 16$$

As a result, $I_1 = (16)I_2 = (16)(0.10 \text{ W/m}^2) = \boxed{1.6 \text{ W/m}^2}$.

Equation 16.9 is valid only when no walls, ceilings, floors, etc. are present to reflect the sound and cause it to pass through the same surface more than once. Conceptual Example 8 demonstrates why this is so.

Conceptual Example 8   Reflected Sound and Sound Intensity

Suppose that the person singing in the shower in Figure 16.24 produces a sound power P. Sound reflects from the surrounding shower stall. At a distance r in front of the person, does the expression $I = P/(4\pi r^2)$ (Equation 16.9) **(a)** overestimate, **(b)** underestimate, or **(c)** give the correct total sound intensity?

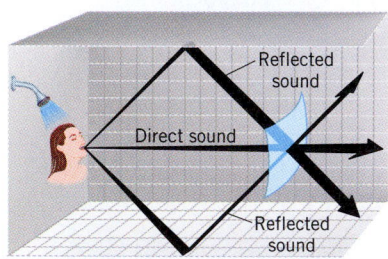

Figure 16.24 When someone sings in the shower, the sound power passing through part of an imaginary spherical surface (shown in blue) is the sum of the direct sound power and the reflected sound power.

Reasoning In arriving at Equation 16.9, it was assumed that the sound spreads out uniformly from the source and passes only once through the imaginary surface that surrounds it (see Figure 16.22). In Figure 16.24, only part of this imaginary surface (colored blue) is shown, but nonetheless, if Equation 16.9 is to apply, the same assumption must hold.

Answers (a) and (c) are incorrect. Equation 16.9 cannot overestimate the sound intensity, because it assumes that the sound passes through the imaginary surface only once and, hence, does not take into account the reflected sound within the shower stall. For the same reason, neither can Equation 16.9 give the correct sound intensity.

Answer (b) is correct. Figure 16.24 illustrates three paths by which the sound passes through the imaginary surface. The "direct" sound travels along a path from its source directly to the surface. It is the intensity of this sound that is given by $I = P/(4\pi r^2)$. The remaining paths are two of the many that characterize the sound reflected from the shower stall. The *total* sound power that passes through the surface is the sum of the direct and reflected powers. Thus the total sound intensity at a distance r from the source is greater than the intensity of the direct sound alone, so Equation 16.9 underestimates the sound intensity from the singing. People like to sing in the shower because their voices sound so much louder due to the enhanced intensity caused by the reflected sound.

Related Homework: *Problems 56, 108*

✓CHECK YOUR UNDERSTANDING

(*The answers are given at the end of the book.*)

14. Some animals rely on an acute sense of hearing for survival, and the visible parts of the ears on such animals are often relatively large. How does this anatomical feature help to increase the sensitivity of the animal's hearing for low-intensity sounds?

15. A source is emitting sound uniformly in all directions. There are no reflections anywhere. A *flat* surface faces the source. Is the sound intensity the same at all points on the surface?

16.8 DECIBELS

The *decibel* (dB) is a measurement unit used when comparing two sound intensities. The simplest method of comparison would be to compute the ratio of the intensities. For instance, we could compare $I = 8 \times 10^{-12}$ W/m^2 to $I_0 = 1 \times 10^{-12}$ W/m^2 by computing $I/I_0 = 8$ and stating that I is eight times as great as I_0. However, because of the way in which the human hearing mechanism responds to intensity, it is more appropriate to use a logarithmic scale for the comparison. For this purpose, the *intensity level β* (expressed in decibels) is defined as follows:

$$\beta = (10 \text{ dB}) \log\left(\frac{I}{I_0}\right) \qquad (16.10)$$

where "log" denotes the logarithm to the base ten. I_0 is the intensity of the reference level to which I is being compared and is sometimes the threshold of hearing; that is, $I_0 = 1.00 \times 10^{-12}$ W/m^2. With the aid of a calculator, the intensity level can be evaluated for the values of I and I_0 given above:

$$\beta = (10 \text{ dB}) \log\left(\frac{8 \times 10^{-12} \text{ W/m}^2}{1 \times 10^{-12} \text{ W/m}^2}\right) = (10 \text{ dB}) \log 8 = (10 \text{ dB})(0.9) = 9 \text{ dB}$$

This result indicates that I is 9 decibels greater than I_0. Although β is called the "intensity level," it is *not* an intensity and does *not* have intensity units of W/m^2. In fact, the decibel, like the radian, is dimensionless.

Notice that if both I and I_0 are at the threshold of hearing, then $I = I_0$, and the intensity level is 0 dB according to Equation 16.10:

$$\beta = (10 \text{ dB}) \log\left(\frac{I_0}{I_0}\right) = (10 \text{ dB}) \log 1 = 0 \text{ dB}$$

since $\log 1 = 0$. Thus, **an intensity level of zero decibels does not mean that the sound intensity I is zero; it means that $I = I_0$.**

Problem-solving insight

Intensity levels can be measured with a sound-level meter, such as the one in Figure 16.25. The intensity level β is displayed on its scale, assuming that the threshold of hearing is 0 dB. Table 16.2 lists the intensities I and the associated intensity levels β for some common sounds, using the threshold of hearing as the reference level.

Table 16.2 Typical Sound Intensities and Intensity Levels Relative to the Threshold of Hearing

	Intensity I (W/m^2)	Intensity Level β (dB)
Threshold of hearing	1.0×10^{-12}	0
Rustling leaves	1.0×10^{-11}	10
Whisper	1.0×10^{-10}	20
Normal conversation (1 meter)	3.2×10^{-6}	65
Inside car in city traffic	1.0×10^{-4}	80
Car without muffler	1.0×10^{-2}	100
Live rock concert	1.0	120
Threshold of pain	10	130

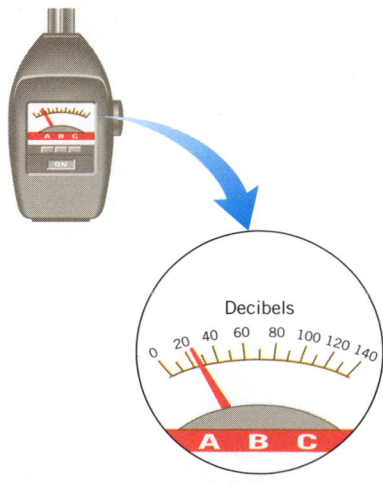

Figure 16.25 A sound-level meter and a close-up view of its decibel scale.

When a sound wave reaches a listener's ear, the sound is interpreted by the brain as loud or soft, depending on the intensity of the wave. Greater intensities give rise to louder sounds. However, the relation between intensity and loudness is not a simple proportionality, because doubling the intensity does *not* double the loudness, as we will now see.

Suppose you are sitting in front of a stereo system that is producing an intensity level of 90 dB. If the volume control on the amplifier is turned up slightly to produce a 91-dB level, you would just barely notice the change in loudness. *Hearing tests have revealed that a one-decibel (1-dB) change in the intensity level corresponds to approximately the smallest change in loudness that an average listener with normal hearing can detect.* Since 1 dB is the smallest perceivable increment in loudness, a change of 3 dB—say, from 90 to 93 dB—is still a rather small change in loudness. Example 9 determines the factor by which the sound intensity must be increased to achieve such a change.

Example 9 Comparing Sound Intensities

Audio system 1 produces an intensity level of $\beta_1 = 90.0$ dB, and system 2 produces an intensity level of $\beta_2 = 93.0$ dB. The corresponding intensities (in W/m^2) are I_1 and I_2. Determine the ratio I_2/I_1.

Reasoning Intensity levels are related to intensities by logarithms (see Equation 16.10), and it is a property of logarithms (see Appendix D) that $\log A - \log B = \log (A/B)$. Subtracting the two intensity levels and using this property, we find that

$$\beta_2 - \beta_1 = (10\text{ dB}) \log \left(\frac{I_2}{I_0}\right) - (10\text{ dB}) \log \left(\frac{I_1}{I_0}\right) = (10\text{ dB}) \log \left(\frac{I_2/I_0}{I_1/I_0}\right)$$

$$= (10\text{ dB}) \log \left(\frac{I_2}{I_1}\right)$$

Solution Using the result just obtained, we find

$$93.0\text{ dB} - 90.0\text{ dB} = (10\text{ dB}) \log \left(\frac{I_2}{I_1}\right)$$

$$0.30 = \log \left(\frac{I_2}{I_1}\right) \quad \text{or} \quad \frac{I_2}{I_1} = 10^{0.30} = \boxed{2.0}$$

Doubling the intensity changes the loudness by only a small amount (3 dB) and does not double it, so there is no simple proportionality between intensity and loudness.

200 watts

20 watts

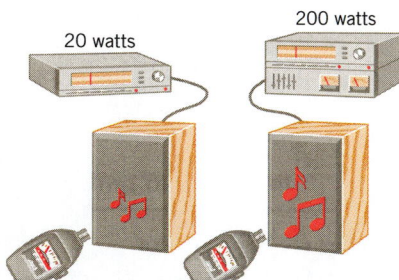

Figure 16.26 In spite of its tenfold greater power, the 200-watt audio system has only about double the loudness of the 20-watt system, when both are set for maximum volume.

To double the loudness of a sound, the intensity must be increased by more than a factor of two. *Experiment shows that if the intensity level increases by 10 dB, the new sound seems approximately twice as loud as the original sound.* For instance, a 70-dB intensity level sounds about twice as loud as a 60-dB level, and an 80-dB intensity level sounds about twice as loud as a 70-dB level. The factor by which the sound intensity must be increased to double the loudness can be determined as in Example 9:

$$\beta_2 - \beta_1 = 10.0\text{ dB} = (10\text{ dB}) \left[\log \left(\frac{I_2}{I_0}\right) - \log \left(\frac{I_1}{I_0}\right)\right]$$

Solving this equation reveals that $I_2/I_1 = 10.0$. Thus, increasing the sound intensity by a factor of ten will double the perceived loudness. Consequently, with both audio systems in Figure 16.26 set at maximum volume, the 200-watt system will sound only twice as loud as the much cheaper 20-watt system.

✓ CHECK YOUR UNDERSTANDING

(*The answers are given at the end of the book.*)

16. If two people talk simultaneously and each creates an intensity level of 65 dB at a certain point, does the total intensity level at this point equal 130 dB?

17. Two observation points are located at distances r_1 and r_2 from a source of sound. The sound spreads out uniformly from the source, and there are no reflecting surfaces in the environment. The sound intensity level at distance r_2 is 6 dB less than the level at distance r_1. **(a)** What is the ratio I_2/I_1 of the sound intensities at the two distances? **(b)** What is the ratio r_2/r_1 of the distances?

THE DOPPLER EFFECT

Have you ever heard an approaching fire truck and noticed the distinct change in the sound of the siren as the truck passes? The effect is similar to what you get when you put together the two syllables "eee" and "yow" to produce "eee-yow." While the truck approaches, the pitch of the siren is relatively high ("eee"), but as the truck passes and moves away, the pitch suddenly drops ("yow"). Something similar, but less familiar, occurs when an observer moves toward or away from a stationary source of sound. Such phenomena were first identified in 1842 by the Austrian physicist Christian Doppler (1803–1853) and are collectively referred to as the Doppler effect.

To explain why the Doppler effect occurs, we will bring together concepts that we have discussed previously—namely, the velocity of an object and the wavelength and frequency of a sound wave (Section 16.5). We will combine the effects of the velocities of the source and observer of the sound with the definitions of wavelength and frequency. In so doing, we will learn that the ***Doppler effect*** is the change in frequency or pitch of the sound detected by an observer because the sound source and the observer have different velocities with respect to the medium of sound propagation.

MOVING SOURCE

To see how the Doppler effect arises, consider the sound emitted by a siren on the stationary fire truck in Figure 16.27a. Like the truck, the air is assumed to be stationary with respect to the earth. Each solid blue arc in the drawing represents a condensation of the sound wave. Since the sound pattern is symmetrical, listeners standing in front of or behind the truck detect the same number of condensations per second and, consequently, hear the same frequency. Once the truck begins to move, the situation changes, as part b of the picture illustrates. Ahead of the truck, the condensations are now closer together, resulting in a decrease in the wavelength of the sound. This "bunching-up" occurs because the moving truck "gains ground" on a previously emitted condensation before emitting the next one. Since the condensations are closer together, the observer standing in front of the truck senses more of them arriving per second than she does when the truck is stationary. The increased rate of arrival corresponds to a greater sound frequency, which the observer hears as a higher pitch. Behind the moving truck, the condensations are farther apart than they are when the truck is stationary. This increase in the wavelength occurs because the truck pulls away from condensations emitted toward the rear. Consequently, fewer condensations per second arrive at the ear of an observer behind the truck, corresponding to a smaller sound frequency or lower pitch.

If the stationary siren in Figure 16.27a emits a condensation at the time $t = 0$ s, it will emit the next one at time T, where T is the period of the wave. The distance between these two condensations is the wavelength λ of the sound produced by the stationary source, as Figure 16.28a indicates. When the truck is moving with a speed v_s (the subscript "s" stands for the "source" of sound) toward a stationary observer, the siren also emits condensations at $t = 0$ s and at time T. However, prior to emitting the second condensation, the truck moves closer to the observer by a distance $v_s T$, as Figure 16.28b shows. As a result, the distance between successive condensations is no longer the wavelength λ created by the stationary siren, but, rather, a wavelength λ' that is shortened by the amount $v_s T$:

$$\lambda' = \lambda - v_s T$$

Let's denote the frequency perceived by the stationary observer as f_o, where the subscript "o" stands for "observer." According to Equation 16.1, f_o is equal to the speed of sound v divided by the shortened wavelength λ':

$$f_o = \frac{v}{\lambda'} = \frac{v}{\lambda - v_s T}$$

But for the stationary siren, we have $\lambda = v/f_s$ and $T = 1/f_s$, where f_s is the frequency of the sound emitted by the source (not the frequency f_o perceived by the observer). With the aid

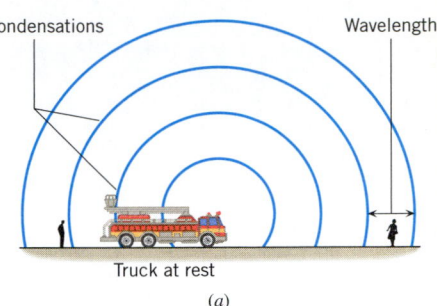

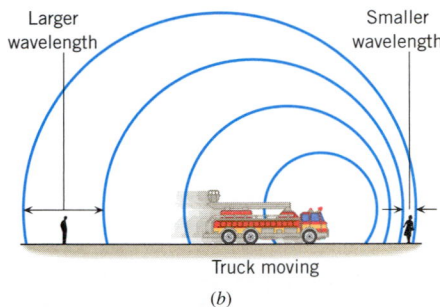

Figure 16.27 (a) When the truck is stationary, the wavelength of the sound is the same in front of and behind the truck. (b) When the truck is moving, the wavelength in front of the truck becomes smaller, while the wavelength behind the truck becomes larger.

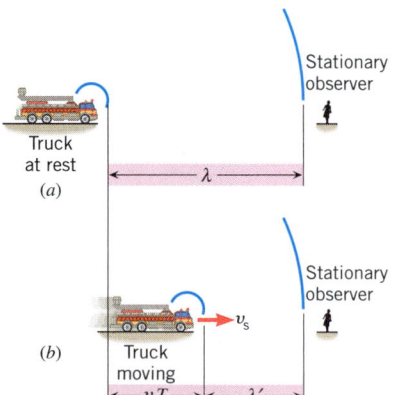

Figure 16.28 (a) When the fire truck is stationary, the distance between successive condensations is one wavelength λ. (b) When the truck moves with a speed v_s, the wavelength of the sound in front of the truck is shortened to λ'.

of these substitutions for λ and T, the expression for f_o can be arranged to give the following result:

Source moving
toward stationary
observer

$$f_o = f_s \left(\dfrac{1}{1 - \dfrac{v_s}{v}} \right) \qquad (16.11)$$

Since the term $1 - v_s/v$ is in the denominator in Equation 16.11 and is less than one, the frequency f_o heard by the observer is *greater* than the frequency f_s emitted by the source. The difference between these two frequencies, $f_o - f_s$, is called the **Doppler shift**, and its magnitude depends on the ratio of the speed of the source v_s to the speed of sound v.

When the siren moves away from, rather than toward, the observer, the wavelength λ' becomes *greater* than λ according to

$$\lambda' = \lambda + v_s T$$

Notice the presence of the "+" sign in this equation, in contrast to the "−" sign that appeared earlier. The same reasoning that led to Equation 16.11 can be used to obtain an expression for the observed frequency f_o:

Source moving
away from
stationary observer

$$f_o = f_s \left(\dfrac{1}{1 + \dfrac{v_s}{v}} \right) \qquad (16.12)$$

The denominator $1 + v_s/v$ in Equation 16.12 is greater than one, so the frequency f_o heard by the observer is *less* than the frequency f_s emitted by the source. The next example illustrates how large the Doppler shift is in a familiar situation.

 Example 10 The Sound of a Passing Train

A high-speed train is traveling at a speed of 44.7 m/s (100 mi/h) when the engineer sounds the 415-Hz warning horn. The speed of sound is 343 m/s. What are the frequency and wavelength of the sound, as perceived by a person standing at a crossing, when the train is **(a)** approaching and **(b)** leaving the crossing?

Reasoning When the train approaches, the person at the crossing hears a sound whose frequency is greater than 415 Hz because of the Doppler effect. As the train moves away, the person hears a frequency that is less than 415 Hz. We may use Equations 16.11 and 16.12, respectively, to determine these frequencies. In either case, the observed wavelength can be obtained according to Equation 16.1 as the speed of sound divided by the observed frequency.

Solution **(a)** When the train approaches, the observed frequency is

$$f_o = f_s \left(\dfrac{1}{1 - \dfrac{v_s}{v}} \right) = (415 \text{ Hz}) \left(\dfrac{1}{1 - \dfrac{44.7 \text{ m/s}}{343 \text{ m/s}}} \right) = \boxed{477 \text{ Hz}} \qquad (16.11)$$

The observed wavelength is

$$\lambda' = \dfrac{v}{f_o} = \dfrac{343 \text{ m/s}}{477 \text{ Hz}} = \boxed{0.719 \text{ m}} \qquad (16.1)$$

(b) When the train leaves the crossing, the observed frequency is

$$f_o = f_s \left(\dfrac{1}{1 + \dfrac{v_s}{v}} \right) = (415 \text{ Hz}) \left(\dfrac{1}{1 + \dfrac{44.7 \text{ m/s}}{343 \text{ m/s}}} \right) = \boxed{367 \text{ Hz}} \qquad (16.12)$$

In this case, the observed wavelength is

$$\lambda' = \dfrac{v}{f_o} = \dfrac{343 \text{ m/s}}{367 \text{ Hz}} = \boxed{0.935 \text{ m}} \qquad (16.1)$$

MOVING OBSERVER

Figure 16.29 shows how the Doppler effect arises when the sound source is stationary and the observer moves, again assuming that the air is stationary. The observer moves with a speed v_o ("o" stands for "observer") toward the stationary source and covers a distance $v_o t$ in a time t. During this time, the moving observer encounters all the condensations that he would if he were stationary, *plus an additional number*. The additional number of condensations encountered is the distance $v_o t$ divided by the distance λ between successive condensations, or $v_o t/\lambda$. Thus, the additional number of condensations encountered per second is v_o/λ. Since a stationary observer would hear a frequency f_s emitted by the source, the moving observer hears a greater frequency f_o given by

$$f_o = f_s + \frac{v_o}{\lambda} = f_s\left(1 + \frac{v_o}{f_s \lambda}\right)$$

Using the fact that $v = f_s \lambda$, where v is the speed of sound, we find that

Observer moving toward stationary source
$$f_o = f_s\left(1 + \frac{v_o}{v}\right) \qquad (16.13)$$

An observer moving *away from* a stationary source moves in the same direction as the sound wave and, as a result, intercepts *fewer* condensations per second than a stationary observer does. In this case, the moving observer hears a smaller frequency f_o that is given by

Observer moving away from stationary source
$$f_o = f_s\left(1 - \frac{v_o}{v}\right) \qquad (16.14)$$

It should be noted that the physical mechanism producing the Doppler effect is different when the source moves and the observer is stationary than when the observer moves and the source is stationary. When the source moves, as in Figure 16.28b, the wavelength of the sound perceived by the observer changes from λ to λ'. When the wavelength changes, the stationary observer hears a different frequency f_o than the frequency produced by the source. On the other hand, when the observer moves and the source is stationary, the *wavelength λ does not change* (see, for example, Figure 16.29). Instead, the moving observer intercepts a different number of wave condensations per second than does a stationary observer and therefore detects a different frequency f_o.

The next example illustrates how the observed frequency changes when the observer is accelerating.

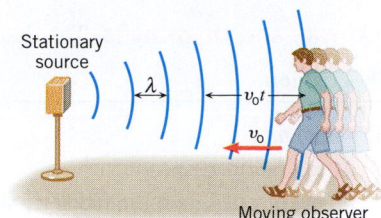

Figure 16.29 An observer moving with a speed v_o toward the stationary source intercepts more wave condensations per unit of time than does a stationary observer.

ANALYZING MULTIPLE-CONCEPT PROBLEMS

Example 11 An Accelerating Speedboat and the Doppler Effect

A speedboat, starting from rest, moves along a straight line away from a dock. The boat has a constant acceleration of $+3.00$ m/s^2 (see Figure 16.30). Attached to the dock is a siren that is producing a 755-Hz tone. If the air temperature is 20 °C, what is the frequency of the sound heard by a person on the boat when the boat's displacement from the dock is $+45.0$ m?

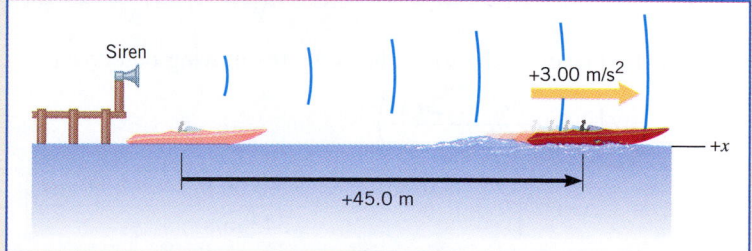

Figure 16.30 As the boat accelerates to the right, the person in it hears a sound frequency that is less than the frequency produced by the stationary siren.

Reasoning As the boat moves away from the dock, it is traveling in the same direction as the sound wave (see the drawing). Therefore, an observer on the moving boat intercepts fewer condensations and rarefactions per second than someone who is stationary. Consequently, the moving observer hears a frequency f_o that is smaller than the frequency emitted by the siren. The frequency f_o depends on the frequency of the siren and the speed of sound—both of which are known—and on the speed of the boat. The speed of the boat is the magnitude of its velocity, which can be determined by using one of the equations of kinematics.

Continued

Knowns and Unknowns The data for this problem are:

Description	Symbol	Value	Comment
Explicit Data			
Acceleration of speedboat	a_x	$+3.00$ m/s^2	
Frequency of sound produced by siren	f_s	755 Hz	Siren is stationary.
Displacement of speedboat	x	$+45.0$ m	
Air temperature	—	20 °C	
Implicit Data			
Initial velocity	$v_{0,o}$	0 m/s	Boat starts from rest.
Unknown Variable			
Frequency of sound heard by person on boat	f_o	?	Observer is moving.

Modeling the Problem

STEP 1 **The Doppler Effect** In this situation we have a stationary source (the siren) and an observer (the person in the boat) who is moving away from the siren. Therefore, the frequency f_o heard by the moving observer is given at the right by Equation 16.14. In this expression, f_s is the frequency of the sound emitted by the stationary siren, v_o is the speed of the moving observer (the speed of the boat), and v is the speed of sound. The frequency f_s is known, as is the speed of sound ($v = 343$ m/s at 20 °C; see Table 16.1). The speed v_o of the moving observer will be obtained in the next step.

$$f_o = f_s\left(1 - \frac{v_o}{v}\right) \quad (16.14)$$

STEP 2 **Kinematics** To determine the velocity of the observer, we note that the initial velocity $v_{0,o}$, acceleration a_x, and displacement x are known. Therefore, we turn to Equation 2.9 of the equations of kinematics, which relates these variables to the final velocity v_o of the moving observer by $v_o^2 = v_{0,o}^2 + 2a_x x$. Taking the square root of each side of this equation gives

$$v_o = \sqrt{v_{0,o}^2 + 2a_x x}$$

$$f_o = f_s\left(1 - \frac{v_o}{v}\right) \quad (16.14)$$

$$v_o = \sqrt{v_{0,o}^2 + 2a_x x}$$

In taking the square root, we have chosen the positive root because the boat is moving in the $+x$ direction (see the drawing). Since the velocity is positive, this expression also gives the magnitude of the velocity, which is the speed of the observer. The expression for v_o can be substituted into Equation 16.14, as indicated at the right. All the variables are known, so the frequency f_o heard by the moving observer can be found.

Solution Algebraically combining the results of the two steps, we have

STEP 1　　　STEP 2

$$f_o = f_s\left(1 - \frac{v_o}{v}\right) = f_s\left(1 - \frac{\sqrt{v_{0,o}^2 + 2a_x x}}{v}\right)$$

Thus, the frequency of the sound heard by the moving observer is

$$f_o = f_s\left(1 - \frac{\sqrt{v_{0,o}^2 + 2a_x x}}{v}\right)$$

$$= (755 \text{ Hz})\left[1 - \frac{\sqrt{(0 \text{ m/s})^2 + 2(+3.00 \text{ m/s}^2)(+45.0 \text{ m})}}{343 \text{ m/s}}\right] = \boxed{719 \text{ Hz}}$$

Related Homework: *Problems 81, 85, 86*

GENERAL CASE

It is possible for *both* the sound source and the observer to move with respect to the medium of sound propagation. If the medium is stationary, Equations 16.11–16.14 may be combined to give the observed frequency f_o as

(a)

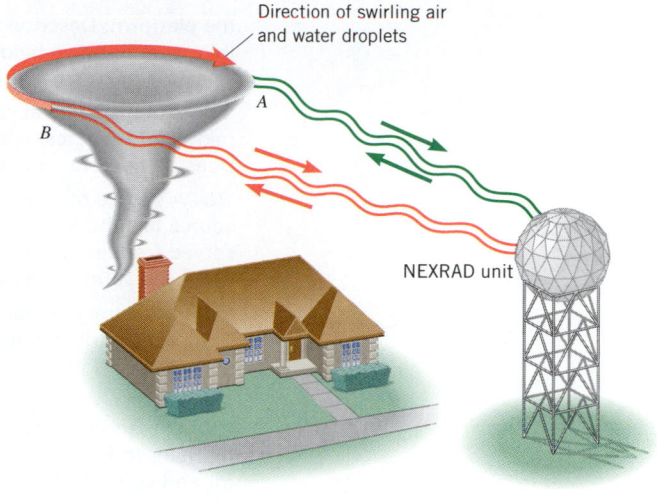

Direction of swirling air
and water droplets

A

B

NEXRAD unit

(b)

Figure 16.31 (a) A tornado is one of nature's most dangerous storms. (Steve Bronstein/The Image Bank/Getty Images) (b) The National Weather Service uses the NEXRAD system, which is based on Doppler-shifted radar, to identify the storms that are likely to spawn tornadoes.

Source and observer both moving

$$f_o = f_s \left(\frac{1 \pm \dfrac{v_o}{v}}{1 \mp \dfrac{v_s}{v}} \right)$$ (16.15)

In the numerator, the plus sign applies when the observer moves toward the source, and the minus sign applies when the observer moves away from the source. In the denominator, the minus sign is used when the source moves toward the observer, and the plus sign is used when the source moves away from the observer. The symbols v_o, v_s, and v denote numbers without an algebraic sign because the direction of travel has been taken into account by the plus and minus signs that appear directly in this equation.

NEXRAD

NEXRAD stands for **Nex**t Generation Weather **Rad**ar and is a nationwide system used by the National Weather Service to provide dramatically improved early warning of severe storms, such as the tornado in Figure 16.31. The system is based on radar waves, which are a type of electromagnetic wave (see Chapter 24) and, like sound waves, can exhibit the Doppler effect. The Doppler effect is at the heart of NEXRAD. As the drawing illustrates, a tornado is a swirling mass of air and water droplets. Radar pulses are sent out by a NEXRAD unit, whose protective covering is shaped like a soccer ball. The waves reflect from the water droplets and return to the unit, where the frequency is observed and compared to the outgoing frequency. For instance, droplets at point A in the drawing are moving toward the unit, and the radar waves reflected from them have their frequency Doppler-shifted to higher values. Droplets at point B, however, are moving away from the unit, and the frequency of the waves reflected from these droplets is Doppler-shifted to lower values. Computer processing of the Doppler frequency shifts leads to color-enhanced views on display screens (see Figure 16.32). These views reveal the direction and magnitude of the wind velocity and can identify, from distances up to 140 mi, the swirling air masses that are likely to spawn tornadoes. The equations that specify the Doppler frequency shifts are different from those given for sound waves by Equations 16.11–16.15. The reason for the difference is that radar waves propagate from one place to another by a different mechanism than that of sound waves (see Section 24.5).

**The physics of
Next Generation Weather Radar.**

Figure 16.32 This color-enhanced NEXRAD view of a tornado shows winds moving toward (green) and away from (red) a NEXRAD station, which is below and to the right of the figure. The white dot and arrow indicate the storm center and the direction of wind circulation. (Courtesy Kurt Hondl, National Severe Storms Laboratory, Norman, OK.)

✓ CHECK YOUR UNDERSTANDING

(The answers are given at the end of the book.)

18. At a swimming pool, a music fan up on a diving platform is listening to a radio. As the radio is playing a tone that has a constant frequency f_s, it is accidentally knocked off

Continued

the platform. Describe the Doppler effect heard by **(a)** the person on the platform and **(b)** a person down below in the water. In each case, state whether the observed frequency f_o is greater or smaller than f_s and describe how f_o changes (if it changes) as the radio falls.

19. When a car is at rest, its horn emits a frequency of 600 Hz. A person standing in the middle of the street with this car behind him hears the horn with a frequency of 580 Hz. Does he need to jump out of the way?

20. A source of sound produces the same frequency under water as it does in air. This source has the same velocity in air as it does under water. Consider the ratio f_o/f_s of the observed frequency f_o to the source frequency f_s. Is this ratio greater in air or under water when the source **(a)** approaches and **(b)** moves away from the observer?

21. Two cars, one behind the other, are traveling in the same direction at the same speed. Does either driver hear the other's horn at a frequency that is different from the frequency heard when both cars are at rest?

22. When a truck is stationary, its horn produces a frequency of 500 Hz. You are driving your car, and this truck is following behind. You hear its horn at a frequency of 520 Hz. **(a)** Refer to Equation 16.15 and decide which algebraic sign should be used in the numerator and which in the denominator. **(b)** Which driver, if either, is driving faster?

The physics of ultrasonic imaging.

16.10 APPLICATIONS OF SOUND IN MEDICINE

When ultrasonic waves are used in medicine for diagnostic purposes, high-frequency sound pulses are produced by a transmitter and directed into the body. As in sonar, reflections occur. They occur each time a pulse encounters a boundary between two tissues that have different densities or a boundary between a tissue and the adjacent fluid. By scanning ultrasonic waves across the body and detecting the echoes generated from various internal locations, it is possible to obtain an image, or sonogram, of the inner anatomy. Ultrasonic imaging is employed extensively in obstetrics to examine the developing fetus (Figure 16.33). The fetus, surrounded by the amniotic sac, can be distinguished from other anatomical features so that fetal size, position, and possible abnormalities can be detected. Ultrasound is also used in other medically related areas. For instance, tumors in the liver, kidney, brain, and pancreas can be detected with ultrasound. Yet another application involves monitoring the real-time movement of pulsating structures, such as heart valves ("echocardiography") and large blood vessels.

When ultrasound is used to form images of internal anatomical features or foreign objects in the body, the wavelength of the sound wave must be about the same size as, or smaller than, the object to be located. Therefore, high frequencies in the range from 1 to 15 MHz (1 MHz = 1 megahertz = 1×10^6 Hz) are the norm. For instance, the wavelength of 5-MHz ultrasound is $\lambda = v/f = 0.3$ mm, if a value of 1540 m/s is used for the speed of sound through tissue. A sound wave with a frequency higher than 5 MHz and a correspondingly shorter wavelength is required for locating objects smaller than 0.3 mm.

Figure 16.33 An ultrasonic scanner can be used to produce an image of the fetus as it develops in the uterus. The three-dimensional ultrasonic image on the right shows incredible detail in which the fetus appears to be sucking the thumb of its right hand. (*Left,* Jesse/ Photo Researchers, Inc.; *right,* GE Medical Systems/Photo Researchers, Inc.)

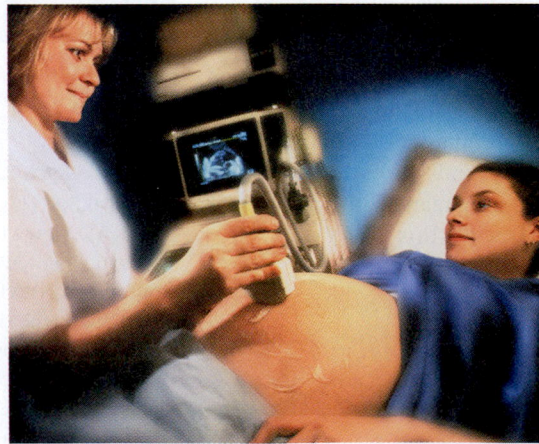
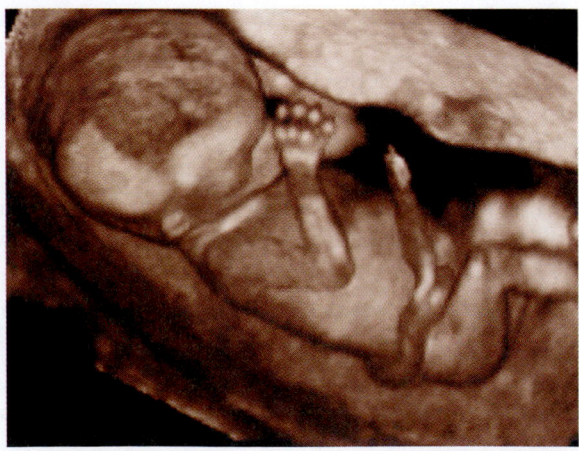

The physics of the cavitron ultrasonic surgical aspirator. Ultrasound also has applications other than imaging. Neurosurgeons use a device called a **c**avitron **u**ltrasonic **s**urgical **a**spirator (CUSA) to remove brain tumors once thought to be inoperable. Ultrasonic sound waves cause the slender tip of the CUSA probe (see Figure 16.34) to vibrate at approximately 23 kHz. The probe shatters any section of the tumor that it touches, and the fragments are flushed out of the brain with a saline solution. Because the tip of the probe is small, the surgeon can selectively remove small bits of malignant tissue without damaging the surrounding healthy tissue.

The physics of bloodless surgery with HIFU. Another application of ultrasound is in a new type of bloodless surgery, which can eliminate abnormal cells, such as those in benign hyperplasia of the prostate gland. This technique is known as HIFU (**h**igh-**i**ntensity **f**ocused **u**ltrasound). It is analogous to focusing the sun's electromagnetic waves by using a magnifying glass and producing a small region where the energy carried by the waves can cause localized heating. Ultrasonic waves can be used in a similar fashion. The waves enter directly through the skin and come into focus inside the body over a region that is sufficiently well defined to be surgically useful. Within this region the energy of the waves causes localized heating, leading to a temperature of about 56 °C (normal body temperature is 37 °C), which is sufficient to kill abnormal cells. The killed cells are eventually removed by the body's natural processes.

The physics of the Doppler flow meter. The Doppler flow meter is a particularly interesting medical application of the Doppler effect. This device measures the speed of blood flow, using transmitting and receiving elements that are placed directly on the skin, as in Figure 16.35. The transmitter emits a continuous sound whose frequency is typically about 5 MHz. When the sound is reflected from the red blood cells, its frequency is changed in a kind of Doppler effect because the cells are moving. The receiving element detects the reflected sound, and an electronic counter measures its frequency, which is Doppler-shifted relative to the transmitter frequency. From the change in frequency the speed of the blood flow can be determined. Typically, the change in frequency is around 600 Hz for flow speeds of about 0.1 m/s. The Doppler flow meter can be used to locate regions where blood vessels have narrowed, since greater flow speeds occur in the narrowed regions, according to the equation of continuity (see Section 11.8). In addition, the Doppler flow meter can be used to detect the motion of a fetal heart as early as 8–10 weeks after conception.

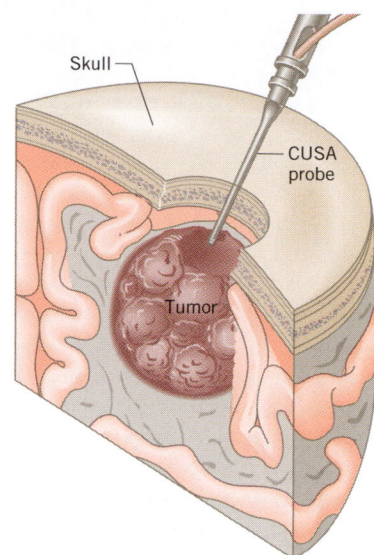

Figure 16.34 Neurosurgeons use a cavitron ultrasonic surgical aspirator (CUSA) to "cut out" brain tumors without adversely affecting the surrounding healthy tissue.

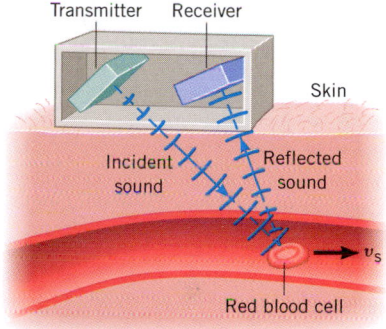

Figure 16.35 A Doppler flow meter measures the speed of red blood cells.

*THE SENSITIVITY OF THE HUMAN EAR

16.11

Although the ear is capable of detecting sound intensities as small as 1×10^{-12} W/m², it is *not* equally sensitive to all frequencies, as Figure 16.36 shows. This figure displays a series of graphs known as the *Fletcher–Munson curves,* after H. Fletcher and M. Munson, who first determined them in 1933. In these graphs the audible sound frequencies are plotted on the horizontal axis, and the sound intensity levels (in decibels) are plotted on the vertical axis. Each curve is a *constant loudness* curve because it shows the sound intensity level needed at each frequency to make the sound appear to have the same loudness. For example, the lowest (red) curve represents the threshold of hearing. It shows the intensity levels at which sounds of different frequencies just become audible. The graph indicates that the intensity level of a 100-Hz sound must be about 37 dB greater than the intensity level of a 1000-Hz sound to be at the threshold of hearing. Therefore, the ear is *less sensitive* to a 100-Hz sound than it is to a 1000-Hz sound. In general, Figure 16.36 reveals that the ear is most sensitive in the range of about 1–5 kHz, and becomes progressively less sensitive at higher and lower frequencies.

Each curve in Figure 16.36 represents a different loudness, and each is labeled according to its intensity level at 1000 Hz. For instance, the curve labeled "60" represents all sounds that have the same loudness as a 1000-Hz sound whose intensity level is 60 dB. These constant-loudness curves become flatter as the loudness increases, the relative flatness indicating that the ear is nearly equally sensitive to all frequencies when the sound is loud. Thus, when you listen to loud sounds, you hear the low frequencies, the middle frequencies, and the high frequencies about equally well. However, when you listen to quiet sounds, the high and low frequencies seem to be absent, because the ear is relatively insensitive to these frequencies under such conditions.

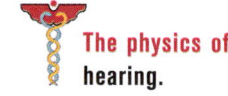

The physics of hearing.

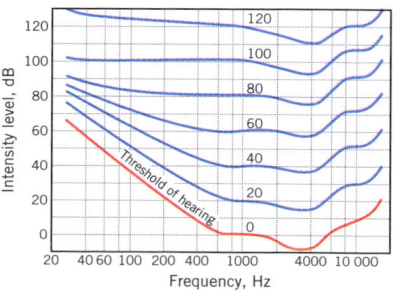

Figure 16.36 Each curve represents the intensity levels at which sounds of various frequencies have the same loudness. The curves are labeled by their intensity levels at 1000 Hz and are known as the Fletcher–Munson curves.

16.12

CONCEPTS & CALCULATIONS

One of the important concepts that we encountered in this chapter is a transverse wave. For instance, transverse waves travel along a guitar string when it is plucked or along a violin string when it is bowed. The next example reviews how the travel speed depends on the properties of the string and on the tension in it.

Concepts & Calculations Example 12
What Determines the Speed of a Wave on a String?

Figure 16.37 shows waves traveling on two strings. Each string is attached to a wall at one end and to a box that has a weight of 28.0 N at the other end. String 1 has a mass of 8.00 g and a length of 4.00 cm, and string 2 has a mass of 12.0 g and a length of 8.00 cm. Determine the speed of the wave on each string.

Concept Questions and Answers Is the tension the same in each string?

Answer Yes, the tension is the same. The two strings support the box, so the tension in each string is one-half the weight of the box. The fact that the strings have different masses and lengths does not affect the tension, which is determined only by the weight of the hanging box.

Is the speed of each wave the same?

Answer Not necessarily. The speed of a wave on a string depends on both the tension and the linear density, as Equation 16.2 indicates. The tension is the same in both strings, but if the linear densities of the strings are different, the speeds are different.

String 1 has a smaller mass and, hence, less inertia than string 2. Does this mean that the speed of the wave on string 1 is greater than the speed on string 2?

Answer Maybe yes, maybe no. The speed of a wave depends on the linear density of the string, which is its mass divided by its length. Depending on the lengths of the strings, string 1 could have a larger linear density and, hence, smaller speed, than string 2. The solution below illustrates this point.

Solution The speed of a wave on a string is given by Equation 16.2 as $v = \sqrt{F/(m/L)}$, where F is the tension and m/L is the mass per unit length, or linear density. Since both strings support the box, the tension in each is one-half the weight of the box, or $F = \frac{1}{2}(28.0 \text{ N}) = 14.0 \text{ N}$. The linear densities of the strings are

$$\frac{m_1}{L_1} = \frac{8.00 \text{ g}}{4.00 \text{ cm}} = 2.00 \text{ g/cm} = 0.200 \text{ kg/m}$$

$$\frac{m_2}{L_2} = \frac{12.0 \text{ g}}{8.00 \text{ cm}} = 1.50 \text{ g/cm} = 0.150 \text{ kg/m}$$

The speed of each wave is

$$v_1 = \sqrt{\frac{F}{m_1/L_1}} = \sqrt{\frac{14.0 \text{ N}}{0.200 \text{ kg/m}}} = \boxed{8.37 \text{ m/s}}$$

$$v_2 = \sqrt{\frac{F}{m_2/L_2}} = \sqrt{\frac{14.0 \text{ N}}{0.150 \text{ kg/m}}} = \boxed{9.66 \text{ m/s}}$$

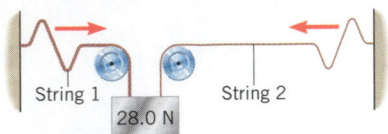

Figure 16.37 A wave travels on each of the two strings. The strings have different masses and lengths, and together support the 28.0-N box. Which is the faster wave?

The next example illustrates how the Doppler effect arises when an observer is moving away from or toward a stationary source of sound. In fact, we will see that it's possible for both situations to occur at the same time.

Concepts & Calculations Example 13
The Doppler Effect for a Moving Observer

A siren, mounted on a tower, emits a sound whose frequency is 2140 Hz. A person is driving a car away from the tower at a speed of 27.0 m/s. As Figure 16.38 illustrates, the sound reaches the person by two paths: the sound reflected from a building in front of the car, and the sound

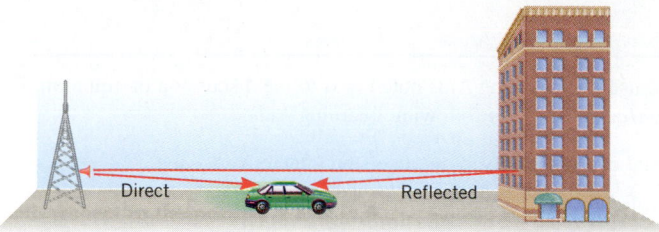

Figure 16.38 The sound from the siren reaches the car by a reflected path and a direct path. The direct and reflected sound waves, as well as the motion of the car, are assumed to lie along the same line. Because of the Doppler effect, the driver hears a different frequency for each sound.

coming directly from the siren. The speed of sound is 343 m/s. What frequency does the person hear for the **(a)** reflected and **(b)** direct sounds?

Concept Questions and Answers   One way that the Doppler effect can arise is that the wavelength of the sound changes. For either the direct or the reflected sound, does the wavelength change?

Answer   No. The wavelength changes only when the source of the sound is moving, as illustrated in Figure 16.28b. The siren is stationary, so the wavelength does not change.

Why does the driver hear a frequency for the reflected sound that is different from 2140 Hz, and is it greater than or smaller than 2140 Hz?

Answer   The car and the reflected sound are traveling in opposite directions, the car to the right and the reflected sound to the left. The driver intercepts more wave cycles per second than if the car were stationary. Consequently, the driver hears a frequency greater than 2140 Hz.

Why does the driver hear a frequency for the direct sound that is different from 2140 Hz, and is it greater than or smaller than 2140 Hz?

Answer   The car and the direct sound are traveling in the same direction. As the direct sound passes the car, the number of wave cycles per second intercepted by the driver is less than if the car were stationary. Thus, the driver hears a frequency that is less than 2140 Hz.

Solution   **(a)** For the reflected sound, the frequency f_o that the driver (the "observer") hears is equal to the frequency f_s of the waves emitted by the siren *plus* an additional number of cycles per second, because the car and the reflected sound are moving in opposite directions. The additional number of cycles per second is v_o/λ, where v_o is the speed of the car and λ is the wavelength of the sound (see the subsection "Moving Observer" in Section 16.9). According to Equation 16.1, the wavelength is equal to the speed of sound v divided by the frequency of the siren, $\lambda = v/f_s$. Thus, the frequency heard by the driver can be written as

$$f_o = f_s + \underbrace{\frac{v_o}{\lambda}}_{\substack{\text{Additional number of} \\ \text{cycles intercepted} \\ \text{per second}}} = f_s + \frac{v_o}{v/f_s}$$

$$= f_s\left(1 + \frac{v_o}{v}\right) = (2140 \text{ Hz})\left(1 + \frac{27.0 \text{ m/s}}{343 \text{ m/s}}\right) = \boxed{2310 \text{ Hz}}$$

(b) For the direct sound, the frequency f_o that the driver hears is equal to the frequency f_s of the waves emitted by the siren *minus* v_o/λ, because the car and direct sound are moving in the same direction:

$$f_o = f_s - \frac{v_o}{\lambda} = f_s - \frac{v_o}{v/f_s}$$

$$= f_s\left(1 - \frac{v_o}{v}\right) = (2140 \text{ Hz})\left(1 - \frac{27.0 \text{ m/s}}{343 \text{ m/s}}\right) = \boxed{1970 \text{ Hz}}$$

As expected, for the reflected wave, the driver hears a frequency greater than 2140 Hz, while for the direct sound he hears a frequency less than 2140 Hz.

CONCEPT SUMMARY

If you need more help with a concept, use the Learning Aids noted next to the discussion or equation. Examples (**Ex.**) are in the text of this chapter. Go to **www.wiley.com/college/cutnell** for the following Learning Aids:

Interactive LearningWare (ILW) — Additional examples solved in a five-step interactive format.

Concept Simulations (CS) — Animated text figures or animations of important concepts.

Interactive Solutions (IS) — Models for certain types of problems in the chapter homework. The calculations are carried out interactively.

Topic	Discussion	Learning Aids
Transverse wave Longitudinal wave	**16.1 THE NATURE OF WAVES** A wave is a traveling disturbance and carries energy from place to place. In a transverse wave, the disturbance occurs perpendicular to the direction of travel of the wave. In a longitudinal wave, the disturbance occurs parallel to the line along which the wave travels.	
Cycle Amplitude Wavelength Period Frequency	**16.2 PERIODIC WAVES** A periodic wave consists of cycles or patterns that are produced over and over again by the source of the wave. The amplitude of the wave is the maximum excursion of a particle of the medium from the particle's undisturbed position. The wavelength λ is the distance along the length of the wave between two successive equivalent points, such as two crests or two troughs. The period T is the time required for the wave to travel a distance of one wavelength. The frequency f (in hertz) is the number of wave cycles per second that passes an observer and is the reciprocal of the period (in seconds):	
Relation between frequency and period	$$f = \frac{1}{T} \qquad (10.5)$$	
Relation between speed, frequency, and wavelength	The speed v of a wave is related to its wavelength and frequency according to $$v = f\lambda \qquad (16.1)$$	**Ex. 1**
	16.3 THE SPEED OF A WAVE ON A STRING The speed of a wave depends on the properties of the medium in which the wave travels. For a transverse wave on a string that has a tension F and a mass per unit length m/L, the wave speed is	
Speed of a wave on a string	$$v = \sqrt{\frac{F}{m/L}} \qquad (16.2)$$	**Ex. 2, 3, 12** **CS 16.1** **IS 16.15**
Linear density	The mass per unit length is also called the linear density.	
	16.4 THE MATHEMATICAL DESCRIPTION OF A WAVE When a wave of amplitude A, frequency f, and wavelength λ moves in the $+x$ direction through a medium, the wave causes a displacement y of a particle at position x according to	
	$$y = A \sin\left(2\pi ft - \frac{2\pi x}{\lambda}\right) \qquad (16.3)$$	
	For a wave moving in the $-x$ direction, the expression is	
	$$y = A \sin\left(2\pi ft + \frac{2\pi x}{\lambda}\right) \qquad (16.4)$$	
Condensation Rarefaction	**16.5 THE NATURE OF SOUND** Sound is a longitudinal wave that can be created only in a medium; it cannot exist in a vacuum. Each cycle of a sound wave includes one condensation (a region of greater than normal pressure) and one rarefaction (a region of less than normal pressure).	
Pure tone Infrasonic frequency Ultrasonic frequency Pitch	A sound wave with a single frequency is called a pure tone. Frequencies less than 20 Hz are called infrasonic. Frequencies greater than 20 kHz are called ultrasonic. The brain interprets the frequency detected by the ear primarily in terms of the subjective quality known as pitch. A high-pitched sound is one with a large frequency (e.g., piccolo). A low-pitched sound is one with a small frequency (e.g., tuba).	
Pressure amplitude Loudness	The pressure amplitude of a sound wave is the magnitude of the maximum change in pressure, measured relative to the undisturbed pressure. The pressure amplitude is associated with the subjective quality of loudness. The larger the pressure amplitude, the louder the sound.	
	16.6 THE SPEED OF SOUND The speed of sound v depends on the properties of the medium. In an ideal gas, the speed of sound is	
Speed of sound in an ideal gas	$$v = \sqrt{\frac{\gamma kT}{m}} \qquad (16.5)$$	**Ex. 4, 5** **ILW 16.1**

Topic	Discussion	Learning Aids
	where $\gamma = c_P/c_V$ is the ratio of the specific heat capacities at constant pressure and constant volume, k is Boltzmann's constant, T is the Kelvin temperature, and m is the mass of a molecule of the gas. In a liquid, the speed of sound is	
Speed of sound in a liquid	$$v = \sqrt{\frac{B_{ad}}{\rho}} \qquad (16.6)$$	
	where B_{ad} is the adiabatic bulk modulus and ρ is the mass density. In a solid that has a Young's modulus of Y and the shape of a long slender bar, the speed of sound is	
Speed of sound in a solid bar	$$v = \sqrt{\frac{Y}{\rho}} \qquad (16.7)$$	
Intensity	**16.7 SOUND INTENSITY** The intensity I of a sound wave is the power P that passes perpendicularly through a surface divided by the area A of the surface	
	$$I = \frac{P}{A} \qquad (16.8)$$	Ex. 6 IS 16.61
Threshold of hearing	The SI unit for intensity is watts per square meter (W/m^2). The smallest sound intensity that the human ear can detect is known as the threshold of hearing and is about 1×10^{-12} W/m^2 for a 1-kHz sound. When a source radiates sound uniformly in all directions and no reflections are present, the intensity of the sound is inversely proportional to the square of the distance from the source, according to	
Spherically uniform radiation	$$I = \frac{P}{4\pi r^2} \qquad (16.9)$$	Ex. 7, 8
	16.8 DECIBELS The intensity level β (in decibels) is used to compare a sound intensity I to the sound intensity I_0 of a reference level:	
Intensity level in decibels	$$\beta = (10 \text{ dB}) \log\left(\frac{I}{I_0}\right) \qquad (16.10)$$	Ex. 9
	The decibel, like the radian, is dimensionless. An intensity level of zero decibels means that $I = I_0$. One decibel is approximately the smallest change in loudness that an average listener with healthy hearing can detect. An increase of ten decibels in the intensity level corresponds approximately to a doubling of the loudness of the sound.	
	16.9 THE DOPPLER EFFECT The Doppler effect is the change in frequency detected by an observer because the sound source and the observer have different velocities with respect to the medium of sound propagation. If the observer and source move with speeds v_o and v_s, respectively, and if the medium is stationary, the frequency f_o detected by the observer is	
The Doppler effect	$$f_o = f_s \left(\frac{1 \pm \dfrac{v_o}{v}}{1 \mp \dfrac{v_s}{v}} \right) \qquad (16.15)$$	Ex. 10, 11, 13 ILW 16.2 IS 16.83
	where f_s is the frequency of the sound emitted by the source and v is the speed of sound. In the numerator, the plus sign applies when the observer moves toward the source, and the minus sign applies when the observer moves away from the source. In the denominator, the minus sign is used when the source moves toward the observer, and the plus sign is used when the source moves away from the observer.	

FOCUS ON CONCEPTS

Note to Instructors: The numbering of the questions shown here reflects the fact that they are only a representative subset of the total number that are available online. However, all of the questions are available for assignment via an online homework management program such as WileyPLUS or WebAssign.

Section 16.1   The Nature of Waves

2. Domino toppling is an event in which a large number of dominoes are lined up close together and then allowed to topple, one after the other. The disturbance that propagates along the line of dominoes is _____. **(a)** partly transverse and partly longitudinal **(b)** transverse   **(c)** longitudinal

Section 16.2   Periodic Waves

3. A transverse wave on a string has an amplitude A. A tiny spot on the string is colored red. As one cycle of the wave passes by, what is the total distance traveled by the red spot? **(a)** A **(b)** $2A$ **(c)** $\frac{1}{2}A$ **(d)** $4A$ **(e)** $\frac{1}{4}A$

Section 16.3   The Speed of a Wave on a String

6. As a wave moves through a medium at a speed v, the particles of the medium move in simple harmonic motion about their undisturbed positions. The maximum speed of the simple harmonic motion is v_{max}. When the amplitude of the wave doubles, _____. (a) v doubles, but v_{max} remains the same   (b) v remains unchanged, but v_{max} doubles   (c) both v and v_{max} remain unchanged   (d) both v and v_{max} double

7. A rope is attached to a hook in the ceiling and is hanging straight down. The rope has a mass m, and nothing is attached to the free end of the rope. As a transverse wave travels down the rope from the top, _____. (a) the speed of the wave does not change   (b) the speed of the wave increases   (c) the speed of the wave decreases

Section 16.4   The Mathematical Description of a Wave

10. The equation that describes a transverse wave on a string is

$$y = (0.0120 \text{ m}) \sin[(483 \text{ rad/s})t - (3.00 \text{ rad/m})x]$$

where y is the displacement of a string particle and x is the position of the particle on the string. The wave is traveling in the $+x$ direction. What is the speed v of the wave?

Section 16.5   The Nature of Sound

11. As the amplitude of a sound wave in air decreases to zero, _____. (a) nothing happens to the condensations and rarefactions of the wave   (b) the condensations and rarefactions of the wave occupy more and more distance along the direction in which the wave is traveling   (c) the condensations of the wave disappear, but nothing happens to the rarefactions   (d) nothing happens to the condensations of the wave, but the rarefactions disappear   (e) both the condensations and the rarefactions of the wave disappear

Section 16.6   The Speed of Sound

12. An echo is sound that returns to you after being reflected from a distant surface (e.g., the side of a cliff). Assuming that the distances involved are the same, an echo under water and an echo in air return to you _____. (a) at different times, the echo under water returning more slowly   (b) at different times, the echo under water returning more quickly   (c) at the same time

13. A horn on a boat sounds a warning, and the sound penetrates the water. How does the frequency of the sound in the air compare to its frequency in the water? How does the wavelength in the air compare to the wavelength in the water?   (a) The frequency in the air is smaller than the frequency in the water, and the wavelength in the air is greater than the wavelength in the water.   (b) The frequency in the air is greater than the frequency in the water, and the wavelength in the air is smaller than the wavelength in the water.   (c) The frequency in the air is the same as the frequency in the water, and the wavelength in the air is the same as the wavelength in the water.   (d) The frequency in the air is the same as the frequency in the water, and the wavelength in the air is smaller than the wavelength in the water.   (e) The frequency in the air is the same as the frequency in the water, and the wavelength in the air is greater than the wavelength in the water.

Section 16.7   Sound Intensity

15. A source emits sound uniformly in all directions. There are no reflections of the sound. At a distance of 12 m from the source, the intensity of the sound is 5.0×10^{-3} W/m^2. What is the total sound power P emitted by the source?

Section 16.8   Decibels

17. A source emits sound uniformly in all directions. There are no reflections of the sound. At a distance r_1 from the source, the sound is 7.0 dB louder than it is at a distance r_2 from the source. What is the ratio r_1/r_2?

Section 16.9   The Doppler Effect

18. A red car and a blue car can move along the same straight one-lane road. Both cars can move only at one speed when they move (e.g., 60 mph). The driver of the red car sounds his horn. In which one of the following situations does the driver of the blue car hear the highest horn frequency?   (a) Both cars are moving at the same speed, and they are moving apart.   (b) Both cars are moving in the same direction at the same speed.   (c) Both cars are moving at the same speed, and they are moving toward each other.   (d) The red car is moving toward the blue car, which is stationary.   (e) The blue car is moving toward the red car, which is stationary.

19. What happens to the Doppler effect in air (i.e., the shift in frequency of a sound wave) as the temperature increases?   (a) It is greater at higher temperatures, but only in the case of a moving source and a stationary observer.   (b) It is greater at higher temperatures, but only in the case of a moving observer and a stationary source.   (c) It is greater at higher temperatures than at lower temperatures.   (d) It is less at higher temperatures than at lower temperatures.   (e) The Doppler effect does not change as the temperature increases.

PROBLEMS

Note to Instructors: Most of the homework problems in this chapter are available for assignment via an online homework management program such as WileyPLUS or WebAssign, and those marked with the icon **GO** *are presented in WileyPLUS using a guided tutorial format that provides enhanced interactivity. See Preface for additional details.*

ssm   Solution is in the Student Solutions Manual.
www   Solution is available online at www.wiley.com/college/cutnell

 This icon represents a biomedical application.

Section 16.1   The Nature of Waves,
Section 16.2   Periodic Waves

1. ssm Light is an electromagnetic wave and travels at a speed of 3.00×10^8 m/s. The human eye is most sensitive to yellow-green light, which has a wavelength of 5.45×10^{-7} m. What is the frequency of this light?

2. A car driving along a highway at a speed of 23 m/s strays onto the shoulder. Evenly spaced parallel grooves called "rumble strips" are carved into the pavement of the shoulder. Rolling over the rumble strips causes the car's wheels to oscillate up and down at a frequency of 82 Hz. How far apart are the centers of adjacent rumble-strip grooves?

3. A woman is standing in the ocean, and she notices that after a wave crest passes, five more crests pass in a time of 50.0 s. The distance between two successive crests is 32 m. Determine, if possible, the wave's   (a) period,   (b) frequency,   (c) wavelength,   (d) speed, and   (e) amplitude. If it is not possible to determine any of these quantities, then so state.

4. Tsunamis are fast-moving waves often generated by underwater earthquakes. In the deep ocean their amplitude is barely noticeable, but upon reaching shore, they can rise up to the astonishing height of a six-story building. One tsunami, generated off the Aleutian islands in Alaska, had a wavelength of 750 km and traveled a distance of 3700 km in 5.3 h. **(a)** What was the speed (in m/s) of the wave? For reference, the speed of a 747 jetliner is about 250 m/s. Find the wave's **(b)** frequency and **(c)** period.

5. **ssm** Suppose that the amplitude and frequency of the transverse wave in Figure 16.2c are, respectively, 1.3 cm and 5.0 Hz. Find the *total vertical distance* (in cm) through which the colored dot moves in 3.0 s.

6. A person lying on an air mattress in the ocean rises and falls through one complete cycle every five seconds. The crests of the wave causing the motion are 20.0 m apart. Determine **(a)** the frequency and **(b)** the speed of the wave.

7. Refer to the graphs that accompany Problem 26. From the data in these graphs, determine the speed of the wave.

*8. **GO** A jetskier is moving at 8.4 m/s in the direction in which the waves on a lake are moving. Each time he passes over a crest, he feels a bump. The bumping frequency is 1.2 Hz, and the crests are separated by 5.8 m. What is the wave speed?

*9. The speed of a transverse wave on a string is 450 m/s, and the wavelength is 0.18 m. The amplitude of the wave is 2.0 mm. How much time is required for a particle of the string to move through a total distance of 1.0 km?

*10. A 3.49-rad/s ($33\frac{1}{3}$ rpm) record has a 5.00-kHz tone cut in the groove. If the groove is located 0.100 m from the center of the record (see drawing), what is the wavelength in the groove?

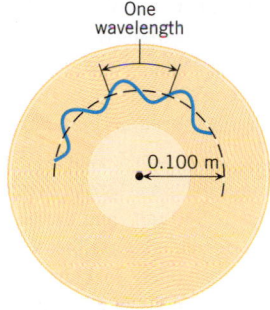

One wavelength

0.100 m

Problem 10

11. A water-skier is moving at a speed of 12.0 m/s. When she skis in the same direction as a traveling wave, she springs upward every 0.600 s because of the wave crests. When she skis in the direction opposite to the direction in which the wave moves, she springs upward every 0.500 s in response to the crests. The speed of the skier is greater than the speed of the wave. Determine **(a) the speed and **(b)** the wavelength of the wave.

Section 16.3 The Speed of a Wave on a String

12. A wire is stretched between two posts. Another wire is stretched between two posts that are twice as far apart. The tension in the wires is the same, and they have the same mass. A transverse wave travels on the shorter wire with a speed of 240 m/s. What would be the speed of the wave on the longer wire?

13. **ssm** Suppose that the linear density of the A string on a violin is 7.8×10^{-4} kg/m. A wave on the string has a frequency of 440 Hz and a wavelength of 65 cm. What is the tension in the string?

14. The mass of a string is 5.0×10^{-3} kg, and it is stretched so that the tension in it is 180 N. A transverse wave traveling on this string has a frequency of 260 Hz and a wavelength of 0.60 m. What is the length of the string?

15. Consult **Interactive Solution 16.15** at **www.wiley.com/college/ cutnell** in order to review a model for solving this problem. To measure the acceleration due to gravity on a distant planet, an astronaut hangs a 0.055-kg ball from the end of a wire. The wire has a length of 0.95 m and a linear density of 1.2×10^{-4} kg/m. Using electronic equipment, the astronaut measures the time for a transverse pulse to travel the length of the wire and obtains a value of 0.016 s. The mass of the wire is negligible compared to the mass of the ball. Determine the acceleration due to gravity.

16. Two wires are parallel, and one is directly above the other. Each has a length of 50.0 m and a mass per unit length of 0.020 kg/m. However, the tension in wire A is 6.00×10^2 N, and the tension in wire B is 3.00×10^2 N. Transverse wave pulses are generated simultaneously, one at the left end of wire A and one at the right end of wire B. The pulses travel toward each other. How much time does it take until the pulses pass each other?

17. The drawing shows two transverse waves traveling on strings. The linear density of each string is 0.065 kg/m. The tension is provided by a 26-N block that is hanging from the string. Find the speed of the wave in part **(a)** and in part **(b)** of the drawing.

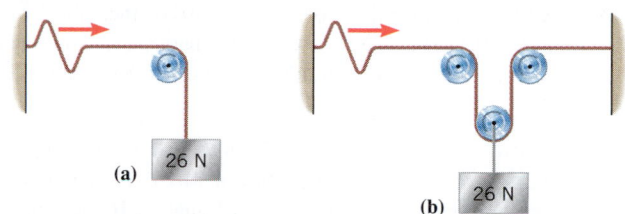

(a) 26 N

(b) 26 N

*18. **GO** The drawing shows a graph of two waves traveling to the right at the same speed. **(a)** Using the data in the drawing, determine the wavelength of each wave. **(b)** The speed of the waves is 12 m/s; calculate the frequency of each one. **(c)** What is the maximum speed for a particle attached to each wave?

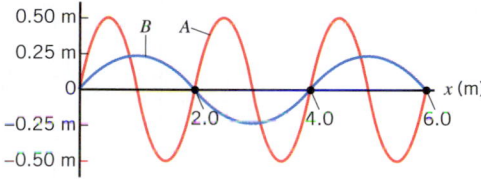

*19. **ssm** Two blocks are connected by a wire that has a mass per unit length of 8.50×10^{-4} kg/m. One block has a mass of 19.0 kg, and the other has a mass of 42.0 kg. These blocks are being pulled across a horizontal frictionless floor by a horizontal force $\vec{P}$ that is applied to the less massive block. A transverse wave travels on the wire between the blocks with a speed of 352 m/s (relative to the wire). The mass of the wire is negligible compared to the mass of the blocks. Find the magnitude of $\vec{P}$.

*20. A spider hangs from a strand of silk whose radius is 4.0×10^{-6} m. The density of the silk is 1300 kg/m³. When the spider moves, waves travel along the strand of silk at a speed of 280 m/s. Determine the mass of the spider.

*21. **GO** **(a)** A uniform rope of mass m and length L is hanging straight down from the ceiling. A small-amplitude transverse wave is sent up the rope from the bottom end. Derive an expression that gives the speed v of the wave on the rope in terms of the distance y above the bottom end of the rope and the magnitude g of the acceleration due to gravity. **(b)** Use the expression that you have derived to calculate the speeds at distances of 0.50 m and 2.0 m above the bottom end of the rope.

**22. A copper wire, whose cross-sectional area is 1.1×10^{-6} m², has a linear density of 9.8×10^{-3} kg/m and is strung between two walls. At the ambient temperature, a transverse wave travels with a speed of 46 m/s on this wire. The coefficient of linear expansion for copper is 17×10^{-6} (C°)⁻¹, and Young's modulus for copper is 1.1×10^{11} N/m². What will be the speed of the wave when the temperature is lowered by 14 C°? Ignore any change in the linear density caused by the change in temperature.

**23. The drawing shows a 15.0-kg ball being whirled in a circular path on the end of a string. The motion occurs on a frictionless, horizontal table. The angular speed of the ball is $\omega = 12.0$ rad/s. The string has a mass of 0.0230 kg. How much time does it take for a wave on the string to travel from the center of the circle to the ball?

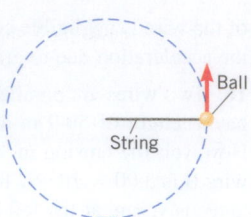

Section 16.4 The Mathematical Description of a Wave

(Note: The phase angles $(2\pi ft - 2\pi x/\lambda)$ and $(2\pi ft + 2\pi x/\lambda)$ are measured in radians, not degrees.)

24. A wave traveling along the x axis is described mathematically by the equation $y = 0.17 \sin(8.2\pi t + 0.54\pi x)$, where y is the displacement (in meters), t is in seconds, and x is in meters. What is the speed of the wave?

25. ssm A wave traveling in the $+x$ direction has an amplitude of 0.35 m, a speed of 5.2 m/s, and a frequency of 14 Hz. Write the equation of the wave in the form given by either Equation 16.3 or 16.4.

26. The drawing shows two graphs that represent a transverse wave on a string. The wave is moving in the $+x$ direction. Using the information contained in these graphs, write the mathematical expression (similar to Equation 16.3 or 16.4) for the wave.

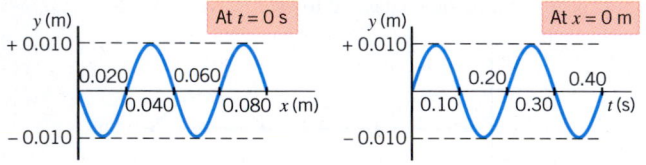

27. A wave has the following properties: amplitude = 0.37 m, period = 0.77 s, wave speed = 12 m/s. The wave is traveling in the $-x$ direction. What is the mathematical expression (similar to Equation 16.3 or 16.4) for the wave?

*28. GO A transverse wave is traveling on a string. The displacement y of a particle from its equilibrium position is given by $y = (0.021 \text{ m}) \sin(25t - 2.0x)$. Note that the phase angle $25t - 2.0x$ is in radians, t is in seconds, and x is in meters. The linear density of the string is 1.6×10^{-2} kg/m. What is the tension in the string?

*29. ssm The tension in a string is 15 N, and its linear density is 0.85 kg/m. A wave on the string travels toward the $-x$ direction; it has an amplitude of 3.6 cm and a frequency of 12 Hz. What are the **(a)** speed and **(b)** wavelength of the wave? **(c)** Write down a mathematical expression (like Equation 16.3 or 16.4) for the wave, substituting numbers for the variables A, f, and λ.

**30. A transverse wave on a string has an amplitude of 0.20 m and a frequency of 175 Hz. Consider the particle of the string at $x = 0$ m. It begins with a displacement of $y = 0$ m when $t = 0$ s, according to Equation 16.3 or 16.4. How much time passes between the first two instants when this particle has a displacement of $y = 0.10$ m?

Section 16.5 The Nature of Sound,
Section 16.6 The Speed of Sound

31. ssm Consult **Interactive LearningWare 16.1** at **www.wiley.com/college/cutnell** for insight into this problem. At what temperature is the speed of sound in helium (ideal gas, $\gamma = 1.67$, atomic mass = 4.003 u) the same as its speed in oxygen at 0 °C?

32. A bat emits a sound whose frequency is 91 kHz. The speed of sound in air at 20.0 °C is 343 m/s. However, the air temperature is 35 °C, so the speed of sound is not 343 m/s. Assume that air behaves like an ideal gas, and find the wavelength of the sound.

33. For research purposes a sonic buoy is tethered to the ocean floor and emits an infrasonic pulse of sound. The period of this sound is 71 ms. Determine the wavelength of the sound.

34. Have you ever listened for an approaching train by kneeling next to a railroad track and putting your ear to the rail? Young's modulus for steel is $Y = 2.0 \times 10^{11}$ N/m², and the density of steel is $\rho = 7860$ kg/m³. On a day when the temperature is 20 °C, how many times greater is the speed of sound in the rail than in the air?

35. ssm Argon (molecular mass = 39.9 u) is a monatomic gas. Assuming that it behaves like an ideal gas at 298 K ($\gamma = 1.67$), find **(a)** the rms speed of argon atoms and **(b)** the speed of sound in argon.

36. GO Suppose you are part of a team that is trying to break the sound barrier with a jet-powered car, which means that it must travel faster than the speed of sound in air. In the morning, the air temperature is 0 °C, and the speed of sound is 331 m/s. What speed must your car exceed if it is to break the sound barrier when the temperature has risen to 43 °C in the afternoon? Assume that air behaves like an ideal gas.

37. As the drawing illustrates, a siren can be made by blowing a jet of air through 20 equally spaced holes in a rotating disk. The time it takes for successive holes to move past the air jet is the period of the sound. The siren is to produce a 2200-Hz tone. What must be the angular speed ω (in rad/s) of the disk?

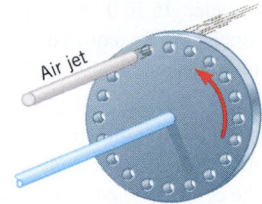

38. An observer stands 25 m behind a marksman practicing at a rifle range. The marksman fires the rifle horizontally, the speed of the bullets is 840 m/s, and the air temperature is 20 °C. How far does each bullet travel before the observer hears the report of the rifle? Assume that the bullets encounter no obstacles during this interval, and ignore both air resistance and the vertical component of the bullets' motion.

39. A sound wave is incident on a pool of fresh water (20 °C). The sound enters the water perpendicularly and travels a distance of 0.45 m before striking a 0.15-m-thick copper block lying on the bottom. The sound passes through the block, reflects from the bottom surface of the block, and returns to the top of the water along the same path. How much time elapses between when the sound enters and when it leaves the water?

40. GO An ultrasonic ruler, such as the one discussed in Example 4 in Section 16.6, displays the distance between the ruler and an object, such as a wall. The ruler sends out a pulse of ultrasonic sound and measures the time it takes for the pulse to reflect from the object and return. The ruler uses this time, along with a preset value for the speed of sound in air, to determine the distance. Suppose that you use this ruler under water, rather than in air. The actual distance from the ultrasonic ruler to an object is 25.0 m. The adiabatic bulk modulus and density of seawater are $B_{ad} = 2.37 \times 10^9$ Pa and $\rho = 1025$ kg/m³, respectively. Assume that the ruler uses a preset value of 343 m/s for the speed of sound in air. Determine the distance reading that the ruler displays.

41. An explosion occurs at the end of a pier. The sound reaches the other end of the pier by traveling through three media: air, fresh water, and a slender metal handrail. The speeds of sound in air, water, and the handrail are 343, 1482, and 5040 m/s, respectively. The sound travels a distance of 125 m in each medium. **(a)** Through which medium does the sound arrive first, second, and third? **(b)** After the first sound arrives, how much later do the second and third sounds arrive?

*42. As the drawing shows, one microphone is located at the origin, and a second microphone is located on the $+y$ axis. The microphones are separated by a distance of $D = 1.50$ m. A source of sound is located on the $+x$ axis, its distances from microphones 1 and 2 being L_1 and L_2, respectively. The speed of sound is 343 m/s. The sound reaches microphone 1 first, and then, 1.46 ms later, it reaches microphone 2. Find the distances L_1 and L_2.

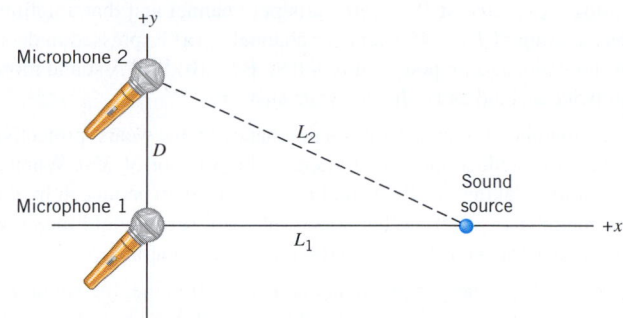

*43. **ssm** A hunter is standing on flat ground between two vertical cliffs that are directly opposite one another. He is closer to one cliff than to the other. He fires a gun and, after a while, hears three echoes. The second echo arrives 1.6 s after the first, and the third echo arrives 1.1 s after the second. Assuming that the speed of sound is 343 m/s and that there are no reflections of sound from the ground, find the distance between the cliffs.

*44. A monatomic ideal gas ($\gamma = 1.67$) is contained within a box whose volume is 2.5 m^3. The pressure of the gas is 3.5×10^5 Pa. The total mass of the gas is 2.3 kg. Find the speed of sound in the gas.

*45. **ssm** A long slender bar is made from an unknown material. The length of the bar is 0.83 m, its cross-sectional area is 1.3×10^{-4} m^2, and its mass is 2.1 kg. A sound wave travels from one end of the bar to the other end in 1.9×10^{-4} s. From which one of the materials listed in Table 10.1 is the bar most likely to be made?

*46. When an earthquake occurs, two types of sound waves are generated and travel through the earth. The primary, or P, wave has a speed of about 8.0 km/s and the secondary, or S, wave has a speed of about 4.5 km/s. A seismograph, located some distance away, records the arrival of the P wave and then, 78 s later, records the arrival of the S wave. Assuming that the waves travel in a straight line, how far is the seismograph from the earthquake?

*47. A sound wave travels twice as far in neon (Ne) as it does in krypton (Kr) in the same time interval. Both neon and krypton can be treated as monatomic ideal gases. The atomic mass of neon is 20.2 u, and the atomic mass of krypton is 83.8 u. The temperature of the krypton is 293 K. What is the temperature of the neon?

*48. Consult Multiple-Concept Example 4 in order to review a model for solving this type of problem. Suppose that you are standing by the side of a road in the Sahara desert where the temperature has reached a hot 56 °C (130 °F). A truck, traveling at a constant speed, passes by. After 4.00 s have elapsed, you use the ultrasonic ruler discussed in Example 4 to measure the distance to the truck. A sound pulse leaves the ultrasonic ruler and returns 0.120 s later. Assume that the average molecular mass of air is 28.9 u, air is an ideal diatomic gas ($\gamma = \frac{7}{5}$), and the truck moves a negligible distance in the time it takes for the sound pulse to reach it. Determine how fast the truck is moving.

**49. In a mixture of argon (atomic mass = 39.9 u) and neon (atomic mass = 20.2 u), the speed of sound is 363 m/s at 3.00×10^2 K. Assume that both monatomic gases behave as ideal gases. Find the percentage of the atoms that are argon and the percentage that are neon.

**50. Review Multiple-Concept Example 4 for background pertinent to this problem. A team of geophysicists is standing on the ground. Beneath their feet, at an unknown distance, is the ceiling of a cavern. The floor of the cavern is a distance h below this ceiling. To measure h, the team places microphones on the ground. At $t = 0$ s, a sound pulse is sent straight downward through the ground and into the cavern. When this pulse reaches the ceiling of the cavern, one part of it is reflected back toward the microphones, and a second part continues downward, eventually to be reflected from the cavern floor. The sound reflected from the cavern ceiling reaches the microphones at $t = 0.0245$ s, and the sound reflected from the cavern floor arrives at $t = 0.0437$ s. The cavern is presumed to be filled with air at a temperature of 9 °C. Assuming that air behaves like an ideal gas, what is the height h of the cavern?

51. **ssm www As a prank, someone drops a water-filled balloon out of a window. The balloon is released from rest at a height of 10.0 m above the ears of a man who is the target. Then, because of a guilty conscience, the prankster shouts a warning after the balloon is released. The warning will do no good, however, if shouted after the balloon reaches a certain point, even if the man could react infinitely quickly. Assuming that the air temperature is 20 °C and ignoring the effect of air resistance on the balloon, determine how far above the man's ears this point is.

Section 16.7 Sound Intensity

52. The average sound intensity inside a busy neighborhood restaurant is 3.2×10^{-5} W/m^2. How much energy goes into each ear (area = 2.1×10^{-3} m^2) during a one-hour meal?

53. **ssm www** At a distance of 3.8 m from a siren, the sound intensity is 3.6×10^{-2} W/m^2. Assuming that the siren radiates sound uniformly in all directions, find the total power radiated.

54. **GO** A source of sound is located at the center of two concentric spheres, parts of which are shown in the drawing. The source emits sound uniformly in all directions. On the spheres are drawn three small patches that may or may not have equal areas. However, the same sound power passes through each patch. The source produces 2.3 W of sound power, and the radii of the concentric spheres are $r_A = 0.60$ m and $r_B = 0.80$ m. (a) Determine the sound intensity at each of the three patches. (b) The sound power that passes through each of the patches is 1.8×10^{-3} W. Find the area of each patch.

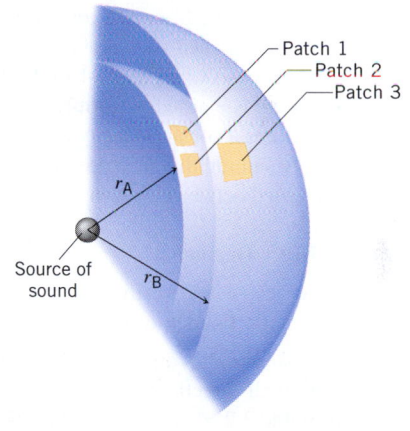

55. A rocket in a fireworks display explodes high in the air. The sound spreads out uniformly in all directions. The intensity of the sound is 2.0×10^{-6} W/m^2 at a distance of 120 m from the explosion. Find the distance from the source at which the intensity is 0.80×10^{-6} W/m^2.

56. Suppose that in Conceptual Example 8 (see Figure 16.24) the person is producing 1.1 mW of sound power. Some of the sound is reflected from the floor and ceiling. The intensity of this reflected sound at a distance of 3.0 m from the source is 4.4×10^{-6} W/m^2. What is the total sound intensity due to both the direct and reflected sounds, at this point?

57. ssm A loudspeaker has a circular opening with a radius of 0.0950 m. The electrical power needed to operate the speaker is 25.0 W. The average sound intensity at the opening is 17.5 W/m². What percentage of the electrical power is converted by the speaker into sound power?

58. A man stands at the midpoint between two speakers that are broadcasting an amplified static hiss uniformly in all directions. The speakers are 30.0 m apart and the total power of the sound coming from each speaker is 0.500 W. Find the total sound intensity that the man hears **(a)** when he is at his initial position halfway between the speakers, and **(b)** after he has walked 4.0 m directly toward one of the speakers.

****59.** Deep ultrasonic heating is used to promote healing of torn tendons. It is produced by applying ultrasonic sound over the affected area of the body. The sound transducer (generator) is circular with a radius of 1.8 cm, and it produces a sound intensity of 5.9×10^3 W/m². How much time is required for the transducer to emit 4800 J of sound energy?

****60.** **GO** Two sources of sound are located on the x axis, and each emits power uniformly in all directions. There are no reflections. One source is positioned at the origin and the other at $x = +123$ m. The source at the origin emits four times as much power as the other source. Where on the x axis are the two sounds equal in intensity? Note that there are two answers.

****61. ssm** Review **Interactive Solution 16.61** at **www.wiley.com/college/cutnell** for one approach to this problem. A dish of lasagna is being heated in a microwave oven. The effective area of the lasagna that is exposed to the microwaves is 2.2×10^{-2} m². The mass of the lasagna is 0.35 kg, and its specific heat capacity is 3200 J/(kg·C°). The temperature rises by 72 C° in 8.0 minutes. What is the intensity of the microwaves in the oven?

*****62.** A rocket, starting from rest, travels straight up with an acceleration of 58.0 m/s². When the rocket is at a height of 562 m, it produces sound that eventually reaches a ground-based monitoring station directly below. The sound is emitted uniformly in all directions. The monitoring station measures a sound intensity I. Later, the station measures an intensity $\frac{1}{3}I$. Assuming that the speed of sound is 343 m/s, find the time that has elapsed between the two measurements.

Section 16.8   Decibels

63. Humans can detect a difference in sound intensity levels of 1.0 dB. What is the ratio (louder to softer) of the sound intensities?

64. A woman stands a distance d from a loud motor that emits sound uniformly in all directions. The sound intensity at her position is an uncomfortable 3.2×10^{-3} W/m². At a position twice as far from the motor, what are **(a)** the sound intensity and **(b)** the sound intensity level relative to the threshold of hearing?

65. ssm When Gloria wears her hearing aid, the sound intensity level increases by 30.0 dB. By what factor does the sound intensity increase?

66. A recording engineer works in a soundproofed room that is 44.0 dB quieter than the outside. If the sound intensity in the room is 1.20×10^{-10} W/m², what is the intensity outside?

67. The sound intensity level at a rock concert is 115 dB, while that at a jazz fest is 95 dB. Determine the ratio of the sound intensity at the rock concert to the sound intensity at the jazz fest.

68. **GO** Using an intensity of 1×10^{-12} W/m² as a reference, the threshold of hearing for an average young person is 0 dB. Person 1 and person 2, who are not average, have thresholds of hearing that are $\beta_1 = -8.00$ dB and $\beta_2 = +12.0$ dB. What is the ratio I_1/I_2 of the

sound intensity I_1 when person 1 hears the sound at his own threshold of hearing compared to the sound intensity I_2 when person 2 hears the sound at his own threshold of hearing?

69. The equation $\beta = (10 \text{ dB}) \log (I/I_0)$, which defines the decibel, can be written in terms of power P (in watts) rather than intensity I (in watts/meter²). The form $\beta = (10 \text{ dB}) \log (P/P_0)$, can be used to compare two power levels in terms of decibels. Suppose that stereo amplifier A is rated at $P = 250$ watts per channel and that amplifier B has a rating of $P_0 = 45$ watts per channel. **(a)** Expressed in decibels, how much more powerful is A than B? **(b)** Will A sound more than twice as loud as B? Justify your answer.

****70.** A member of an aircraft maintenance crew wears protective earplugs that reduce the sound intensity by a factor of 350. When a jet aircraft is taking off, the sound intensity level experienced by the crew member is 88 dB. What sound intensity level would the crew member experience if he removed the protective earplugs?

****71. ssm** When one person shouts at a football game, the sound intensity level at the center of the field is 60.0 dB. When all the people shout together, the intensity level increases to 109 dB. Assuming that each person generates the same sound intensity at the center of the field, how many people are at the game?

****72.** Hearing damage may occur when a person is exposed to a sound intensity level of 90.0 dB (relative to the threshold of hearing) for a period of 9.0 hours. One particular eardrum has an area of 2.0×10^{-4} m². How much sound energy is incident on this eardrum during this time?

****73.** In a discussion person A is talking 1.5 dB louder than person B, and person C is talking 2.7 dB louder than person A. What is the ratio of the sound intensity of person C to the sound intensity of person B?

*****74.** A source emits sound uniformly in all directions. A radial line is drawn from this source. On this line, determine the positions of two points, 1.00 m apart, such that the intensity level at one point is 2.00 dB greater than the intensity level at the other.

*****75. ssm** Suppose that when a certain sound intensity level (in dB) triples, the sound intensity (in W/m²) also triples. Determine this sound intensity level.

Section 16.9   The Doppler Effect

76. A bird is flying directly toward a stationary bird-watcher and emits a frequency of 1250 Hz. The bird-watcher, however, hears a frequency of 1290 Hz. What is the speed of the bird, expressed as a percentage of the speed of sound?

77. ssm You are riding your bicycle directly away from a stationary source of sound and hear a frequency that is 1.0% lower than the emitted frequency. The speed of sound is 343 m/s. What is your speed?

78. Dolphins emit clicks of sound for communication and echolocation. A marine biologist is monitoring a dolphin swimming in seawater where the speed of sound is 1522 m/s. When the dolphin is swimming directly away at 8.0 m/s, the marine biologist measures the number of clicks occurring per second to be at a frequency of 2500 Hz. What is the difference (in Hz) between this frequency and the number of clicks per second actually emitted by the dolphin?

79. Interactive LearningWare 16.2 at **www.wiley.com/college/cutnell** provides some pertinent background for this problem. A convertible moves toward you and then passes you; all the while, its loudspeakers are producing a sound. The speed of the car is a constant 9.00 m/s, and the speed of sound is 343 m/s. What is the ratio of the frequency you hear while the car is approaching to the frequency you hear while the car is moving away?

80. You are flying in an ultra-light aircraft at a speed of 39 m/s. An eagle, whose speed is 18 m/s, is flying directly toward you. Each of the given speeds is relative to the ground. The eagle emits a shrill cry whose frequency is 3400 Hz. The speed of sound is 330 m/s. What frequency do you hear?

***81. ssm** Multiple-Concept Example 11 presents a model for solving this type of problem. A bungee jumper jumps from rest and screams with a frequency of 589 Hz. The air temperature is 20 °C. What is the frequency heard by the people on the ground below when she has fallen a distance of 11.0 m? Assume that the bungee cord has not yet taken effect, so she is in free-fall.

***82.** A car is parked 20.0 m directly south of a railroad crossing. A train is approaching the crossing from the west, headed directly east at a speed of 55.0 m/s. The train sounds a short blast of its 289-Hz horn when it reaches a point 20.0 m west of the crossing. What frequency does the car's driver hear when the horn blast reaches the car? The speed of sound in air is 343 m/s. (*Hint: Assume that only the component of the train's velocity that is directed toward the car affects the frequency heard by the driver.*)

***83.** Refer to **Interactive Solution 16.83** at **www.wiley.com/college/cutnell** for one approach to this type of problem. Two trucks travel at the same speed. They are far apart on adjacent lanes and approach each other essentially head-on. One driver hears the horn of the other truck at a frequency that is 1.14 times the frequency he hears when the trucks are stationary. The speed of sound is 343 m/s. At what speed is each truck moving?

***84.** **GO** The siren on an ambulance is emitting a sound whose frequency is 2450 Hz. The speed of sound is 343 m/s. **(a)** If the ambulance is stationary and you (the "observer") are sitting in a parked car, what are the wavelength and the frequency of the sound you hear? **(b)** Suppose that the ambulance is moving toward you at a speed of 26.8 m/s. Determine the wavelength and the frequency of the sound you hear. **(c)** If the ambulance is moving toward you at a speed of 26.8 m/s and you are moving toward it at a speed of 14.0 m/s, find the wavelength and frequency of the sound you hear.

***85.** Consult Multiple-Concept Example 11 in order to review a model for solving this type of problem. A car is accelerating while its horn is sounding. Just after the car passes a stationary person, the person hears a frequency of 966.0 Hz. Fourteen seconds later, the frequency heard by the person has decreased to 912.0 Hz. When the car is stationary, its horn emits a sound whose frequency is 1.00×10^3 Hz. The speed of sound is 343 m/s. What is the acceleration of the car?

***86.** **GO** Multiple-Concept Example 11 provides a model for solving this type of problem. A wireless transmitting microphone is mounted on a small platform that can roll down an incline, directly away from a loudspeaker that is mounted at the top of the incline. The loudspeaker broadcasts a tone that has a fixed frequency of 1.000×10^4 Hz, and the speed of sound is 343 m/s. At a time of 1.5 s following the release of the platform, the microphone detects a frequency of 9939 Hz. At a time of 3.5 s following the release of the platform, the microphone detects a frequency of 9857 Hz. What is the acceleration (assumed constant) of the platform?

***87. ssm** Two submarines are under water and approaching each other head-on. Sub A has a speed of 12 m/s and sub B has a speed of 8 m/s. Sub A sends out a 1550-Hz sonar wave that travels at a speed of 1522 m/s. **(a)** What is the frequency detected by sub B? **(b)** Part of the sonar wave is reflected from B and returns to A. What frequency does A detect for this reflected wave?

****88.** A microphone is attached to a spring that is suspended from the ceiling, as the drawing indicates. Directly below on the floor is a stationary 440-Hz source of sound. The microphone vibrates up and down in simple harmonic motion with a period of 2.0 s. The difference between the maximum and minimum sound frequencies detected by the microphone is 2.1 Hz. Ignoring any reflections of sound in the room and using 343 m/s for the speed of sound, determine the amplitude of the simple harmonic motion.

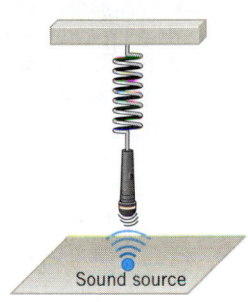

Sound source

ADDITIONAL PROBLEMS

WILEY PLUS

89. The bellow of a territorial bull hippopotamus has been measured at 115 dB above the threshold of hearing. What is the sound intensity?

90. The distance between a loudspeaker and the left ear of a listener is 2.70 m. **(a)** Calculate the time required for sound to travel this distance if the air temperature is 20 °C. **(b)** Assuming that the sound frequency is 523 Hz, how many wavelengths of sound are contained in this distance?

91. ssm At 20 °C the densities of fresh water and ethyl alcohol are, respectively, 998 and 789 kg/m³. Find the ratio of the adiabatic bulk modulus of fresh water to the adiabatic bulk modulus of ethyl alcohol at 20 °C.

92. The security alarm on a parked car goes off and produces a frequency of 960 Hz. The speed of sound is 343 m/s. As you drive toward this parked car, pass it, and drive away, you observe the frequency to change by 95 Hz. At what speed are you driving?

93. ssm www The middle C string on a piano is under a tension of 944 N. The period and wavelength of a wave on this string are 3.82 ms and 1.26 m, respectively. Find the linear density of the string.

94. Consider the freight train in Figure 16.6. Suppose that 15 boxcars pass by in a time of 12.0 s and each has a length of 14.0 m. **(a)** What is the frequency at which each boxcar passes? **(b)** What is the speed of the train?

95. ☤ **ssm** A middle-aged man typically has poorer hearing than a middle-aged woman. In one case a woman can just begin to hear a musical tone, while a man can just begin to hear the tone only when its intensity level is increased by 7.8 dB relative to the just-audible intensity level for the woman. What is the ratio of the sound intensity just detected by the man to the sound intensity just detected by the woman?

96. ☤ A typical adult ear has a surface area of 2.1×10^{-3} m². The sound intensity during a normal conversation is about 3.2×10^{-6} W/m² at the listener's ear. Assume that the sound strikes the surface of the ear perpendicularly. How much power is intercepted by the ear?

97. ssm In Figure 16.2*c* the hand moves the end of the Slinky up and down through two complete cycles in one second. The wave moves along the Slinky at a speed of 0.50 m/s. Find the distance between two adjacent crests on the wave.

98. The right-most key on a piano produces a sound wave that has a frequency of 4185.6 Hz. Assuming that the speed of sound in air is 343 m/s, find the corresponding wavelength.

99. ssm From a vantage point very close to the track at a stock car race, you hear the sound emitted by a moving car. You detect a frequency that is 0.86 times as small as the frequency emitted by the car when it is stationary. The speed of sound is 343 m/s. What is the speed of the car?

100. The speed of a sound in a container of hydrogen at 201 K is 1220 m/s. What would be the speed of sound if the temperature were raised to 405 K? Assume that hydrogen behaves like an ideal gas.

101. ssm A listener doubles his distance from a source that emits sound uniformly in all directions. By how many decibels does the sound intensity level change?

102. An amplified guitar has a sound intensity level that is 14 dB greater than the same unamplified sound. What is the ratio of the amplified intensity to the unamplified intensity?

* **103.** A portable radio is sitting at the edge of a balcony 5.1 m above the ground. The unit is emitting sound uniformly in all directions. By accident, it falls from rest off the balcony and continues to play on the way down. A gardener is working in a flower bed directly below the falling unit. From the instant the unit begins to fall, how much time is required for the sound intensity level heard by the gardener to increase by 10.0 dB?

* **104.** In Figure 16.3c the colored dot exhibits simple harmonic motion as the longitudinal wave passes. The wave has an amplitude of 5.4×10^{-3} m and a frequency of 4.0 Hz. Find the maximum acceleration of the dot.

* **105. ssm www** The drawing shows a frictionless incline and pulley. The two blocks are connected by a wire (mass per unit length = 0.0250 kg/m) and remain stationary. A transverse wave on the wire has a speed of

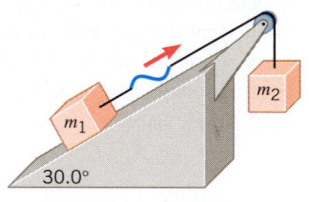

75.0 m/s. Neglecting the weight of the wire relative to the tension in the wire, find the masses m_1 and m_2 of the blocks.

* **106.** Review Conceptual Example 3 before starting this problem. The amplitude of a transverse wave on a string is 4.5 cm. The ratio of the maximum particle speed to the speed of the wave is 3.1. What is the wavelength (in cm) of the wave?

* **107.** Review Conceptual Example 3 before starting this problem. A horizontal wire is under a tension of 315 N and has a mass per unit length of 6.50×10^{-3} kg/m. A transverse wave with an amplitude of 2.50 mm and a frequency of 585 Hz is traveling on this wire. As the wave passes, a particle of the wire moves up and down in simple harmonic motion. Obtain (a) the speed of the wave and (b) the maximum speed with which the particle moves up and down.

* **108.** Review Conceptual Example 8 as background for this problem. A loudspeaker is generating sound in a room. At a certain point, the sound waves coming directly from the speaker (without reflecting from the walls) create an intensity level of 75.0 dB. The waves reflected from the walls create, by themselves, an intensity level of 72.0 dB at the same point. What is the total intensity level? (*Hint: The answer is not 147.0 dB.*)

** **109.** A jet is flying horizontally, as the drawing shows. When the plane is directly overhead at *B*, a person on the ground hears the sound coming from *A* in the drawing. The average temperature of the air is 20 °C. If the speed of the plane at *A* is 164 m/s, what is its speed at *B*, assuming that it has a constant acceleration?

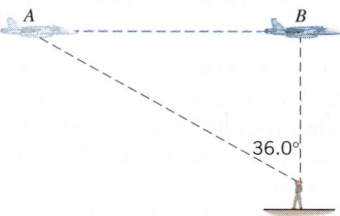

** **110.** The sonar unit on a boat is designed to measure the depth of fresh water ($\rho = 1.00 \times 10^3$ kg/m³, $B_{ad} = 2.20 \times 10^9$ Pa). When the boat moves into seawater ($\rho = 1025$ kg/m³, $B_{ad} = 2.37 \times 10^9$ Pa), the sonar unit is no longer calibrated properly. In seawater, the sonar unit indicates the water depth to be 10.0 m. What is the actual depth of the water?

CHAPTER 17

THE PRINCIPLE OF LINEAR SUPERPOSITION AND INTERFERENCE PHENOMENA

These musicians are playing alpenhorns, instruments used by mountain dwellers in Switzerland. Virtually all musical instruments produce their sound in a way that involves the principle of linear superposition, and the topics discussed in this chapter are related to this important principle.
(© Anneliese Villiger/zefa/Corbis)

17.1 THE PRINCIPLE OF LINEAR SUPERPOSITION

Often, two or more sound waves are present at the same place at the same time, as is the case with sound waves when everyone is talking at a party or when music plays from the speakers of a stereo system. To illustrate what happens when several waves pass simultaneously through the same region, let's consider Figures 17.1 and 17.2, which show two transverse pulses of equal heights moving toward each other along a Slinky. In Figure 17.1 both pulses are "up," while in Figure 17.2 one is "up" and the other is "down." Part *a* of each drawing shows the two pulses beginning to overlap. The pulses merge, and the Slinky assumes a shape that is *the sum of the shapes of the individual pulses.* Thus, when the two "up" pulses overlap completely, as in Figure 17.1*b*, the Slinky has a pulse height that is twice the height of an individual pulse. Likewise, when the "up" pulse and the "down" pulse overlap exactly, as in Figure 17.2*b*, they momentarily cancel, and the Slinky becomes straight. In either case, the two pulses move apart after overlapping, and the Slinky once again conforms to the shapes of the individual pulses, as in part *c* of both figures.

The adding together of individual pulses to form a resultant pulse is an example of a more general concept called the ***principle of linear superposition.***

THE PRINCIPLE OF LINEAR SUPERPOSITION

When two or more waves are present simultaneously at the same place, the resultant disturbance is the sum of the disturbances from the individual waves.

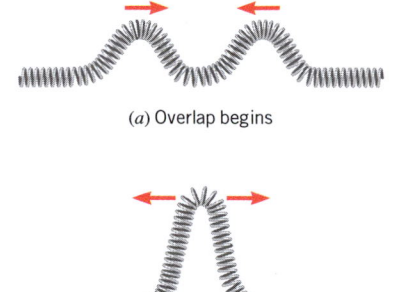

(a) Overlap begins

(b) Total overlap; the Slinky has twice the height of either pulse

(c) The receding pulses

Figure 17.1 Two transverse "up" pulses passing through each other.

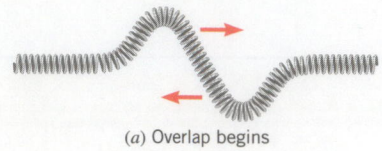

(a) Overlap begins

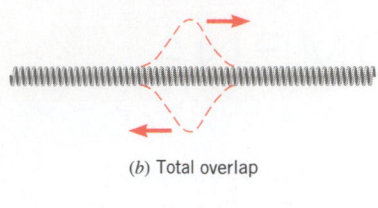

(b) Total overlap

(c) The receding pulses

Figure 17.2 Two transverse pulses, one "up" and one "down," passing through each other.

This principle can be applied to all types of waves, including sound waves, water waves, and electromagnetic waves such as light. It embodies one of the most important concepts in physics, and the remainder of this chapter deals with examples related to it.

✓ **CHECK YOUR UNDERSTANDING**

(The answers are given at the end of the book.)

1. The drawing shows a graph of two pulses traveling toward each other at $t = 0$ s. Each pulse has a constant speed of 1 cm/s. When $t = 2$ s, what is the height of the resultant pulse at **(a)** $x = 3$ cm and **(b)** $x = 4$ cm?

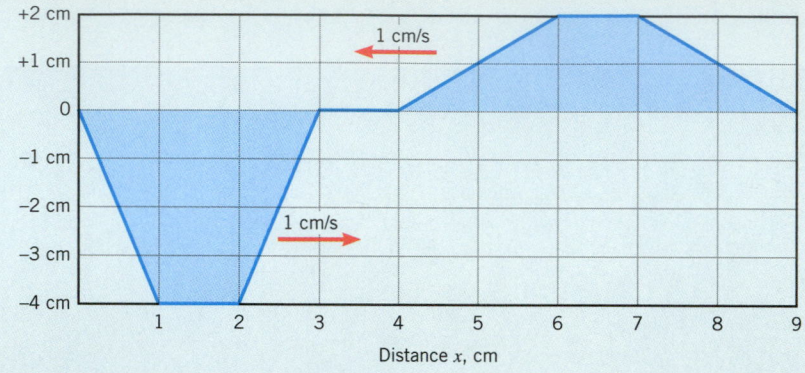

2. Does the principle of linear superposition imply that two sound waves, passing through the same place at the same time, always create a louder sound than is created by either wave alone?

17.2 CONSTRUCTIVE AND DESTRUCTIVE INTERFERENCE OF SOUND WAVES

Suppose that the sounds from two speakers overlap in the middle of a listening area, as in Figure 17.3, and that each speaker produces a sound wave of the same amplitude and frequency. For convenience, the wavelength of the sound is chosen to be $\lambda = 1$ m. In addition, assume that the diaphragms of the speakers vibrate in phase; that is, they move outward together and inward together. If the distance of each speaker from the overlap point is the same (3 m in the drawing), the condensations (C) of one wave always meet the condensations of the other when the waves come together; similarly, rarefactions (R) always meet rarefactions. According to the principle of linear superposition, the combined pattern is the sum of the individual patterns. As a result, the pressure fluctuations at the overlap point have twice the amplitude A that the individual waves have, and a listener at this spot hears a louder sound than the sound coming from either speaker alone. When two waves always meet condensation-to-condensation and rarefaction-to-rarefaction (or crest-to-crest and trough-to-trough), they are said to be *exactly in phase* and to exhibit *constructive interference.*

Figure 17.3 As a result of constructive interference between the two sound waves (amplitude = A), a loud sound (amplitude = 2A) is heard at an overlap point located equally distant from two in-phase speakers (C, condensation; R, rarefaction).

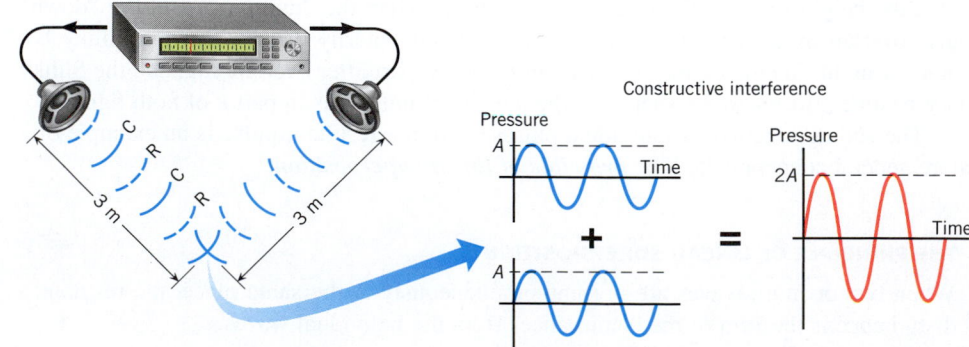

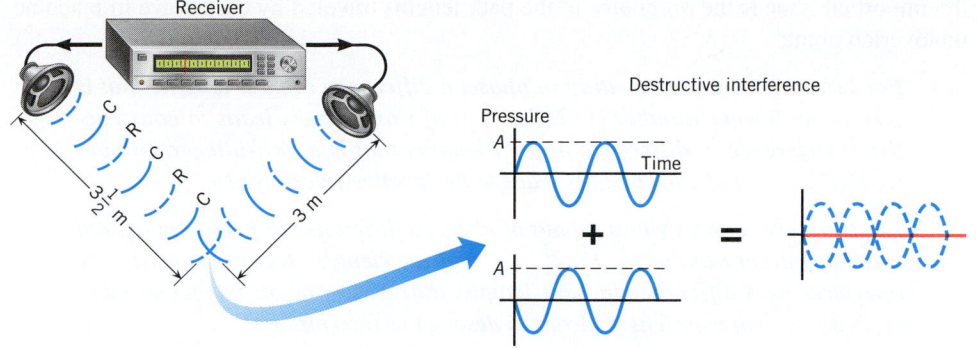

Figure 17.4 The speakers in this drawing vibrate in phase. However, the left speaker is one-half of a wavelength ($\frac{1}{2}$ m) farther from the overlap point than the right speaker. Because of destructive interference, no sound is heard at the overlap point (C, condensation; R, rarefaction).

Now consider what happens if one of the speakers is moved. The result is surprising. In Figure 17.4, the left speaker is moved away* from the overlap point by a distance equal to one-half of the wavelength, or 0.5 m. Therefore, at the overlap point, a condensation arriving from the left meets a rarefaction arriving from the right. Likewise, a rarefaction arriving from the left meets a condensation arriving from the right. According to the principle of linear superposition, the net effect is a mutual cancellation of the two waves. The condensations from one wave offset the rarefactions from the other, leaving only a *constant air pressure*. A constant air pressure, devoid of condensations and rarefactions, means that a listener detects no sound. When two waves always meet condensation-to-rarefaction (or crest-to-trough), they are said to be *exactly out of phase* and to exhibit *destructive interference.*

When two waves meet, they interfere constructively if they always meet exactly in phase and destructively if they always meet exactly out of phase. In either case, this means that the wave patterns do not shift relative to one another as time passes. Sources that produce waves in this fashion are called *coherent sources.*

Destructive interference is the basis of a useful technique for reducing the loudness of undesirable sounds. For instance, Figure 17.5 shows a pair of noise-canceling headphones. Small microphones are mounted inside the headphones and detect noise such as the engine noise that an airplane pilot would hear. The headphones also contain circuitry to process the electronic signals from the microphones and reproduce the noise in a form that is exactly out of phase compared to the original. This out-of-phase version is played back through the headphone speakers and, because of destructive interference, combines with the original noise to produce a quieter background.

If the left speaker in Figure 17.4 were moved away from the overlap point by *another* one-half wavelength ($3\frac{1}{2}$ m + $\frac{1}{2}$ m = 4 m), the two waves would again be in phase, and constructive interference would occur. The listener would hear a loud sound because the left wave travels one whole wavelength ($\lambda = 1$ m) farther than the right wave and, at the overlap point, condensation meets condensation and rarefaction meets rarefaction. In general,

The physics of noise-canceling headphones.

Figure 17.5 Noise-canceling headphones utilize destructive interference.

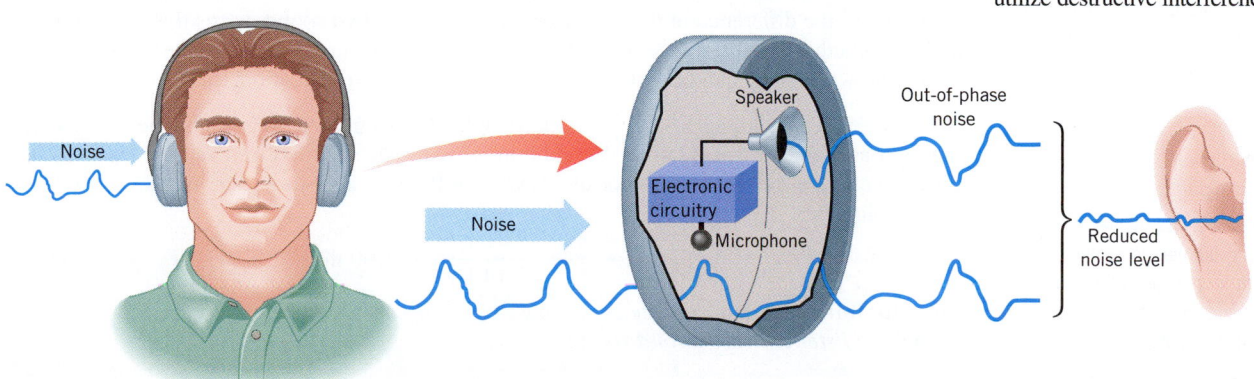

*When the left speaker is moved back, its sound intensity and, hence, its pressure amplitude decrease at the overlap point. In this chapter assume that the power delivered to the left speaker by the receiver is increased slightly to keep the amplitudes equal at the overlap point.

the important issue is the *difference* in the path lengths traveled by each wave in reaching the overlap point:

> *For two wave sources vibrating in phase, a difference in path lengths that is zero or an integer number (1, 2, 3, . . .) of wavelengths leads to constructive interference; a difference in path lengths that is a half-integer number ($\frac{1}{2}$, $1\frac{1}{2}$, $2\frac{1}{2}$, . . .) of wavelengths leads to destructive interference.*

> *For two wave sources vibrating out of phase, a difference in path lengths that is a half-integer number ($\frac{1}{2}$, $1\frac{1}{2}$, $2\frac{1}{2}$, . . .) of wavelengths leads to constructive interference; a difference in path lengths that is zero or an integer number (1, 2, 3, . . .) of wavelengths leads to destructive interference.*

Interference effects can also be detected if the two speakers are fixed in position and the listener moves about the room. Consider Figure 17.6, where the sound waves spread outward from each speaker, as indicated by the concentric circular arcs. Each solid arc represents the middle of a condensation, and each dashed arc represents the middle of a rarefaction. Where the two waves overlap, there are places of constructive interference and places of destructive interference. Constructive interference occurs at any spot where two condensations or two rarefactions intersect, and the drawing shows four such places as solid dots. A listener stationed at any one of these locations hears a loud sound. On the other hand, destructive interference occurs at any place where a condensation and a rarefaction intersect, such as the two open dots in the picture. A listener situated at a point of destructive interference hears no sound. At locations where neither constructive nor destructive interference occurs, the two waves partially reinforce or partially cancel, depending on the position relative to the speakers. Thus, it is possible for a listener to walk about the overlap region and hear marked variations in loudness.

The individual sound waves from the speakers in Figure 17.6 carry energy, and the energy delivered to the overlap region is the sum of the energies of the individual waves. This fact is consistent with the principle of conservation of energy, which we first encountered in Section 6.8. This principle states that energy can neither be created nor destroyed, but can only be converted from one form to another. One of the interesting consequences of interference is that the energy is redistributed, so there are places within the overlap region where the sound is loud and other places where there is no sound at all. Interference, so to speak, "robs Peter to pay Paul," but energy is always conserved in the process. Example 1 illustrates how to decide what a listener hears.

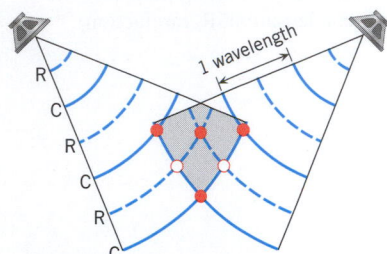

Figure 17.6 Two sound waves overlap in the shaded region. The solid lines denote the middle of the condensations (C), and the dashed lines denote the middle of the rarefactions (R). Constructive interference occurs at each solid dot (●) and destructive interference at each open dot (○).

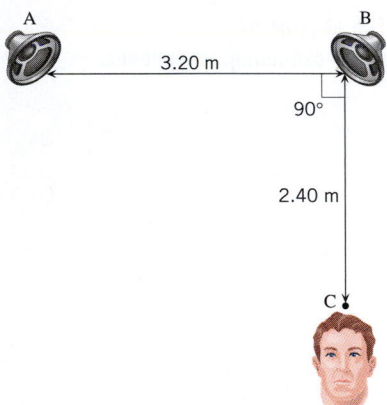

Figure 17.7 Example 1 discusses whether this setup leads to constructive or destructive interference at point C for 214-Hz sound waves.

Example 1 What Does a Listener Hear?

In Figure 17.7 two in-phase loudspeakers, A and B, are separated by 3.20 m. A listener is stationed at point C, which is 2.40 m in front of speaker B. The triangle ABC is a right triangle. Both speakers are playing identical 214-Hz tones, and the speed of sound is 343 m/s. Does the listener hear a loud sound or no sound?

Reasoning The listener will hear either a loud sound or no sound, depending on whether the interference occurring at point C is constructive or destructive. To determine which it is, we need to find the difference in the distances traveled by the two sound waves that reach point C and see whether the difference is an integer or half-integer number of wavelengths. In either event, the wavelength can be found from the relation $\lambda = v/f$ (Equation 16.1).

Solution Since the triangle ABC is a right triangle, the distance AC is given by the Pythagorean theorem as $\sqrt{(3.20 \text{ m})^2 + (2.40 \text{ m})^2} = 4.00$ m. The distance BC is given as 2.40 m. Thus, the difference in the travel distances for the waves is 4.00 m − 2.40 m = 1.60 m. The wavelength of the sound is

$$\lambda = \frac{v}{f} = \frac{343 \text{ m/s}}{214 \text{ Hz}} = 1.60 \text{ m} \qquad (16.1)$$

Since the difference in the distances is one wavelength, constructive interference occurs at point C, and the *listener hears a loud sound.*

Up to this point, we have been assuming that the speaker diaphragms vibrate synchronously, or in phase; that is, they move outward together and inward together. This may not be the case, however, and Conceptual Example 2 considers what happens then.

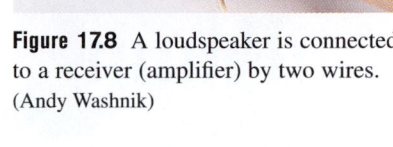

Conceptual Example 2   Out-of-Phase Speakers

The physics of wiring stereo speakers. To make a speaker operate, two wires must be connected between the speaker and the receiver (amplifier), as in Figure 17.8. To ensure that the diaphragms of two speakers vibrate in phase, it is necessary to make these connections in exactly the same way. If the wires for one speaker are not connected just as they are for the other speaker, the two diaphragms will vibrate out of phase. This means that whenever one diaphragm moves outward, the other moves inward, and vice versa. Suppose that in Figure 17.4 the connections are made so that the speaker diaphragms vibrate out of phase, everything else remaining the same. A listener at the overlap point would now hear **(a)** no sound because destructive interference occurs **(b)** a loud sound because constructive interference occurs.

Reasoning Since the speaker diaphragms in Figure 17.4 are now vibrating out of phase, one of them is moving exactly opposite to the way it was moving originally; let us assume that it is the left speaker. The effect of this change is that every condensation originating from the left speaker becomes a rarefaction, and every rarefaction becomes a condensation.

Answer (a) is incorrect. When the two speakers in Figure 17.4 are wired *in phase*, a condensation from one speaker always meets a rarefaction from the other at the overlap point, and destructive interference occurs. However, if one of the speakers were wired out of phase relative to the other, a condensation from one speaker would meet a condensation from the other, and destructive interference would *not* occur.

Answer (b) is correct. If the left speaker in Figure 17.4 were connected out of phase with respect to the right speaker, a condensation from the right speaker would meet a condensation (not a rarefaction) from the left speaker at the overlap point. Similarly, a rarefaction from the right speaker would meet a rarefaction from the left speaker. The result is *constructive* interference, and a loud sound would be heard.

Figure 17.8 A loudspeaker is connected to a receiver (amplifier) by two wires. (Andy Washnik)

The phenomena of constructive and destructive interference are exhibited by all types of waves, not just sound waves. We will encounter interference effects again in Chapter 27, in connection with light waves.

✓ **CHECK YOUR UNDERSTANDING**

(*The answers are given at the end of the book.*)

3. Suppose that you are sitting at the overlap point between the two speakers in Figure 17.4. Because of destructive interference, you hear no sound, even though both speakers are emitting identical sound waves. One of the speakers is suddenly shut off. Will you now hear a sound? **(a)** No. **(b)** Yes. **(c)** Yes, but only if you move a distance of one wavelength closer to the speaker that is still producing sound.

4. Starting at the overlap point in Figure 17.3, you walk along a straight path that is *perpendicular* to the line between the speakers and passes through the midpoint of that line. As you walk, the loudness of the sound **(a)** changes from loud to faint to loud **(b)** changes from faint to loud to faint **(c)** does not change.

5. Starting at the overlap point in Figure 17.3, you walk along a path that is *parallel* to the line between the speakers. As you walk, the loudness of the sound **(a)** changes from loud to faint to loud **(b)** changes from faint to loud to faint **(c)** does not change.

17.3   DIFFRACTION

Section 16.5 discusses the fact that sound is a pressure wave created by a vibrating object, such as a loudspeaker. The previous two sections of this chapter have examined what happens when two sound waves are present simultaneously at the same place; according to the principle of linear superposition, a resultant disturbance is formed from the sum of the individual waves. This principle reveals that overlapping sound waves exhibit interference effects, whereby the sound energy is redistributed within the overlap region.

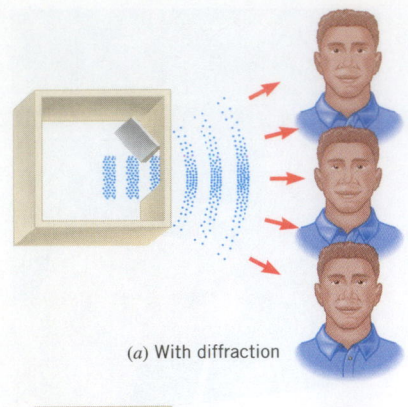

(a) With diffraction

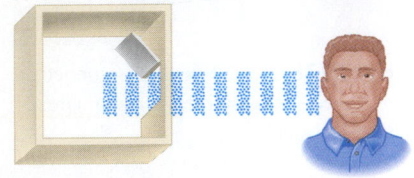

(b) Without diffraction

Figure 17.9 *(a)* The bending of a sound wave around the edges of the doorway is an example of diffraction. The source of the sound within the room is not shown. *(b)* If diffraction did not occur, the sound wave would not bend as it passed through the doorway.

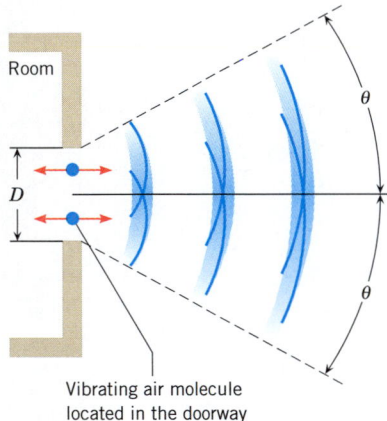

Vibrating air molecule located in the doorway

Figure 17.10 Each vibrating molecule of the air in the doorway generates a sound wave that expands outward and bends, or diffracts, around the edges of the doorway. Because of interference effects among the sound waves produced by all the molecules, the sound intensity is mostly confined to the region defined by the angle θ on either side of the doorway.

We will now use the principle of linear superposition to explore another interference effect, that of diffraction.

When a wave encounters an obstacle or the edges of an opening, it bends around them. For instance, a sound wave produced by a stereo system bends around the edges of an open doorway, as Figure 17.9*a* illustrates. If such bending did not occur, sound could be heard outside the room only at locations directly in front of the doorway, as part *b* of the drawing suggests. (It is assumed that no sound is transmitted directly through the walls.) The bending of a wave around an obstacle or the edges of an opening is called **diffraction.** All kinds of waves exhibit diffraction.

To demonstrate how the bending of waves arises, Figure 17.10 shows an expanded view of Figure 17.9*a*. When the sound wave reaches the doorway, the air in the doorway is set into longitudinal vibration. In effect, each molecule of the air in the doorway becomes a source of a sound wave in its own right, and, for purposes of illustration, the drawing shows two of the molecules. Each produces a sound wave that expands outward in three dimensions, much like a water wave does in two dimensions when a stone is dropped into a pond. The sound waves generated by all the molecules in the doorway must be added together to obtain the total sound wave at any location outside the room, in accord with the principle of linear superposition. However, even considering only the waves from the two molecules in the picture, it is clear that the expanding wave patterns reach locations off to either side of the doorway. The net effect is a "bending," or diffraction, of the sound around the edges of the opening. Further insight into the origin of diffraction can be obtained with the aid of Huygens' principle (see Section 27.5).

When the sound waves generated by every molecule in the doorway are added together, it is found that there are places where the intensity is a maximum and places where it is zero, in a fashion similar to that discussed in the previous section. Analysis shows that at a great distance from the doorway the intensity is a maximum directly opposite the center of the opening. As the distance to either side of the center increases, the intensity decreases and reaches zero, then rises again to a maximum, falls again to zero, rises back to a maximum, and so on. Only the maximum at the center is a strong one. The other maxima are weak and become progressively weaker at greater distances from the center. In Figure 17.10 the angle θ defines the location of the first minimum intensity point on either side of the center. Equation 17.1 gives θ in terms of the wavelength λ and the width D of the doorway and assumes that the doorway can be treated like a slit whose height is very large compared to its width:

Single slit—first minimum
$$\sin \theta = \frac{\lambda}{D} \qquad (17.1)$$

Waves also bend around the edges of openings other than single slits. Particularly important is the diffraction of sound by a circular opening, such as that in a loudspeaker. In this case, the angle θ is related to the wavelength λ and the diameter D of the opening by

Circular opening —first minimum
$$\sin \theta = 1.22 \frac{\lambda}{D} \qquad (17.2)$$

An important point to remember about Equations 17.1 and 17.2 is that the extent of the diffraction depends on the ratio of the wavelength to the size of the opening. If the ratio λ/D is small, then θ is small and little diffraction occurs. The waves are beamed in the forward direction as they leave an opening, much like the light from a flashlight. Such sound waves are said to have "narrow dispersion." Since high-frequency sound has a relatively small wavelength, it tends to have a narrow dispersion. On the other hand, for larger values of the ratio λ/D, the angle θ is larger. The waves spread out over a larger region and are said to have a "wide dispersion." Low-frequency sound, with its relatively large wavelength, typically has a wide dispersion.

In a stereo loudspeaker, a wide dispersion of the sound is desirable. Example 3 illustrates, however, that there are limitations to the dispersion that can be achieved, depending on the loudspeaker design.

ANALYZING MULTIPLE-CONCEPT PROBLEMS

Example 3 Designing a Loudspeaker for Wide Dispersion

A 1500-Hz sound and a 8500-Hz sound each emerges from a loudspeaker through a circular opening that has a diameter of 0.30 m (see Figure 17.11). Assuming that the speed of sound in air is 343 m/s, find the diffraction angle θ for each sound.

Reasoning The diffraction angle θ depends on the ratio of the wavelength λ of the sound to the diameter D of the opening, according to $\sin \theta = 1.22(\lambda/D)$ (Equation 17.2). The wavelength for each sound can be obtained from the given frequencies and the speed of sound.

Knowns and Unknowns The following data are available:

Description	Symbol	Value
Sound frequency	f	1500 Hz or 8500 Hz
Diameter of speaker opening	D	0.30 m
Speed of sound	v	343 m/s
Unknown Variable		
Diffraction angle	θ	?

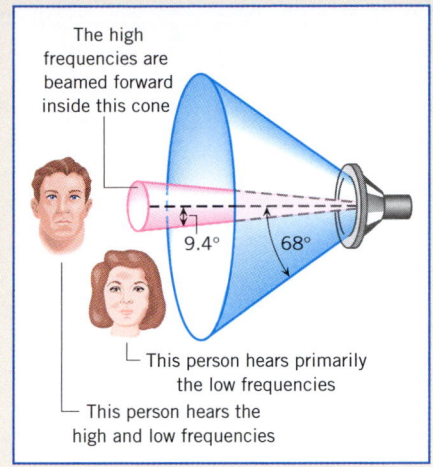

The high frequencies are beamed forward inside this cone

9.4°   68°

This person hears primarily the low frequencies

This person hears the high and low frequencies

Figure 17.11 Because the dispersion of high frequencies is less than the dispersion of low frequencies, you should be directly in front of the speaker to hear both the high and low frequencies equally well.

Modeling the Problem

STEP 1   The Diffraction Angle   Equation 17.2 indicates that the diffraction angle is related to the ratio of the wavelength λ of the sound to the diameter D of the opening by $\sin \theta = 1.22(\lambda/D)$. Solving this expression for θ gives Equation 1 at the right. A value for D is given. To determine value for λ, we turn to Step 2.

$$\theta = \sin^{-1}\left(1.22\frac{\lambda}{D}\right) \quad (1)$$

?

STEP 2   Wavelength   The wavelength λ is related to the frequency f and the speed v of the sound according to $v = f\lambda$ (Equation 16.1). Solving for λ gives

$$\lambda = \frac{v}{f}$$

which can be substituted into Equation 1 as shown at the right.

$$\theta = \sin^{-1}\left(1.22\frac{\lambda}{D}\right) \quad (1)$$

$$\lambda = \frac{v}{f}$$

Solution   Combining the results of each step algebraically, we find that

STEP 1   STEP 2

$$\theta = \sin^{-1}\left(1.22\frac{\lambda}{D}\right) = \sin^{-1}\left(1.22\frac{v/f}{D}\right)$$

Problem-solving insight: When a wave passes through an opening, the extent of diffraction is greater when the ratio λ/D is greater, where λ is the wavelength of the wave and D is the width or diameter of the opening.

Applying the above result to each of the sound frequencies shows that

1500-Hz sound $\theta = \sin^{-1}\left(1.22\frac{v}{fD}\right) = \sin^{-1}\left[1.22\frac{(343 \text{ m/s})}{(1500 \text{ Hz})(0.30 \text{ m})}\right] = \boxed{68°}$

8500-Hz sound $\theta = \sin^{-1}\left(1.22\frac{v}{fD}\right) = \sin^{-1}\left[1.22\frac{(343 \text{ m/s})}{(8500 \text{ Hz})(0.30 \text{ m})}\right] = \boxed{9.4°}$

The physics of tweeter loudspeakers. Figure 17.11 illustrates these results. With a 0.30-m opening, the dispersion of the higher-frequency sound is limited to only 9.4°. To increase the dispersion, a smaller opening is needed. It is for this reason that loudspeaker designers use a small-diameter speaker called a *tweeter* to generate the high-frequency sound (see Figure 17.12).

Related Homework: *Problems 13, 15, 16*

Tweeter

Figure 17.12 Small-diameter speakers, called tweeters, are used to produce high-frequency sound. The small diameter helps to promote a wider dispersion of the sound. (Sony Electronics, Inc./Sony USA)

As we have seen, diffraction is an interference effect, one in which some of the wave's energy is directed into regions that would otherwise not be accessible. Energy, of course, is conserved during this process, because energy is only redistributed during diffraction; no energy is created or destroyed.

✓ **CHECK YOUR UNDERSTANDING**

(*The answers are given at the end of the book.*)

6. At an open-air rock concert you are standing directly in front of a speaker. You hear the high-frequency sounds of a female vocalist as well as the low-frequency sounds of the rhythmic bass. As you walk to one side of the speaker, the sounds of the vocalist _____, and those of the rhythmic bass _____. **(a)** drop off noticeably; also drop off noticeably **(b)** drop off only slightly; drop off noticeably **(c)** drop off only slightly; also drop off only slightly **(d)** drop off noticeably; drop off only slightly

7. Refer to Example 1 in Section 16.2. Which type of radio wave, AM or FM, diffracts more readily around a given obstacle? **(a)** AM, because it has a greater wavelength **(b)** FM, because it has a greater wavelength **(c)** AM, because it has a greater frequency **(d)** FM, because it has a greater frequency

17.4 BEATS

In situations where waves with the *same frequency* overlap, we have seen how the principle of linear superposition leads to constructive and destructive interference and how it explains diffraction. We will see in this section that two overlapping waves with *slightly different frequencies* give rise to the phenomenon of beats. However, the principle of linear superposition again provides an explanation of what happens when the waves overlap.

A tuning fork has the property of producing a single-frequency sound wave when struck with a sharp blow. Figure 17.13 shows sound waves coming from two tuning forks placed side by side. The tuning forks in the drawing are identical, and each is designed to produce a 440-Hz tone. However, a small piece of putty has been attached to one fork, whose frequency is lowered to 438 Hz because of the added mass. When the forks are sounded simultaneously, the loudness of the resulting sound rises and falls periodically—faint, then loud, then faint, then loud, and so on. The periodic variations in loudness are called *beats* and result from the interference between two sound waves with slightly different frequencies.

For clarity, Figure 17.13 shows the condensations and rarefactions of the sound waves separately. In reality, however, the waves spread out and overlap. In accord with the principle of linear superposition, the ear detects the combined total of the two. Notice that there are places where the waves interfere constructively and places where they interfere destructively. When a region of constructive interference reaches the ear, a loud sound is heard. When a region of destructive interference arrives, the sound intensity drops to zero (assuming each of the waves has the same amplitude). The number of times per second that the loudness rises and falls is the **beat frequency** and is the **difference** between the two

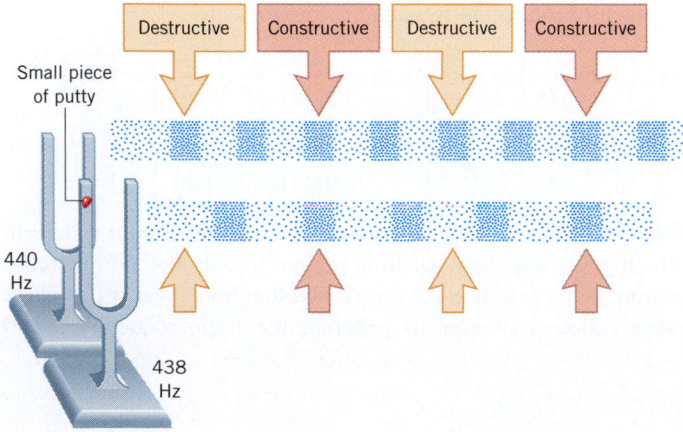

Figure 17.13 Two tuning forks have slightly different frequencies of 440 and 438 Hz. The phenomenon of beats occurs when the forks are sounded simultaneously. The sound waves are not drawn to scale.

sound frequencies. Thus, in the situation illustrated in Figure 17.13, an observer hears the sound loudness rise and fall at the rate of 2 times per second (440 Hz − 438 Hz).

Figure 17.14 helps to explain why the beat frequency is the difference between the two frequencies. The drawing displays graphical representations of the pressure patterns of a 10-Hz wave and a 12-Hz wave, along with the pressure pattern that results when the two overlap. These frequencies have been chosen for convenience, even though they lie below the audio range and are inaudible. Audible sound waves behave in exactly the same way. The top two drawings, in blue, show the pressure variations in a one-second interval of each wave. The third drawing, in red, shows the result of adding together the blue patterns according to the principle of linear superposition. Notice that the amplitude in the red drawing is not constant, as it is in the individual waves. Instead, the amplitude changes from a minimum to a maximum, back to a minimum, and so on. When such pressure variations reach the ear and occur in the audible frequency range, they produce a loud sound when the amplitude is a maximum and a faint sound when the amplitude is a minimum. Two loud–faint cycles, or beats, occur in the one-second interval shown in the drawing, corresponding to a beat frequency of 2 Hz. Thus, the beat frequency is the difference between the frequencies of the individual waves, or 12 Hz − 10 Hz = 2 Hz.

The physics of tuning a musical instrument. Musicians often tune their instruments by listening to a beat frequency. For instance, a guitar player plucks an out-of-tune string along with a tone that has the correct frequency. He then adjusts the string tension until the beats vanish, ensuring that the string is vibrating at the correct frequency.

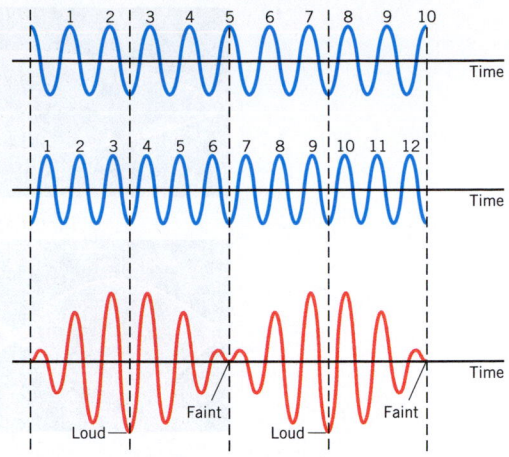

Figure 17.14 A 10-Hz and a 12-Hz sound wave, when added together, produce a wave with a beat frequency of 2 Hz. The drawings show the pressure patterns (in blue) of the individual waves and the pressure pattern (in red) that results when the two overlap. The time interval shown is one second.

✔ **CHECK YOUR UNDERSTANDING**

(The answers are given at the end of the book.)

8. Tuning fork A (frequency unknown) and tuning fork B (frequency = 384 Hz) together produce 6 beats in 2 seconds. When a small piece of putty is attached to tuning fork A, as in Figure 17.13, the beat frequency decreases. What is the frequency of tuning fork A before the putty is attached?   **(a)** 378 Hz   **(b)** 381 Hz   **(c)** 387 Hz   **(d)** 390 Hz

9. A tuning fork has a frequency of 440 Hz. The string of a violin and this tuning fork, when sounded together, produce a beat frequency of 1 Hz. From these two pieces of information alone, is it possible to determine the exact frequency of the violin string?   **(a)** Yes; the frequency of the violin string is 441 Hz.   **(b)** No, because the frequency of the violin string could be either 439 or 441 Hz.   **(c)** Yes; the frequency of the violin string is 439 Hz.

10. When the regions of constructive and destructive interference in Figure 17.13 move past a listener's ear, a beat frequency of 2 Hz is heard. Supposed that the tuning forks in the drawing are sounded under water and that the listener is also under water. The forks vibrate at 438 and 440 Hz, just as they do in air. However, sound travels four times faster in water than in air. The beat frequency heard by the underwater listener is   **(a)** 16 Hz   **(b)** 8 Hz   **(c)** 4 Hz   **(d)** 2 Hz.

17.5 TRANSVERSE STANDING WAVES

A standing wave is another interference effect that can occur when two waves overlap. Standing waves can arise with transverse waves, such as those on a guitar string, and also with longitudinal sound waves, such as those in a flute. In any case, the principle of linear superposition provides an explanation of the effect, just as it does for diffraction and beats.

Figure 17.15 shows some of the essential features of transverse standing waves. In this figure the left end of each string is vibrated back and forth, while the right end is attached to a wall. Regions of the string move so fast that they appear only as a blur in the photographs. Each of the patterns shown is called a *transverse standing wave pattern*. Notice that the patterns include special places called nodes and antinodes. The *nodes* are places that do not vibrate at all, and the *antinodes* are places where maximum vibration occurs. To the right of each photograph is a drawing that helps us to visualize the motion of the string as it vibrates in a standing wave pattern. These drawings freeze the shape of the

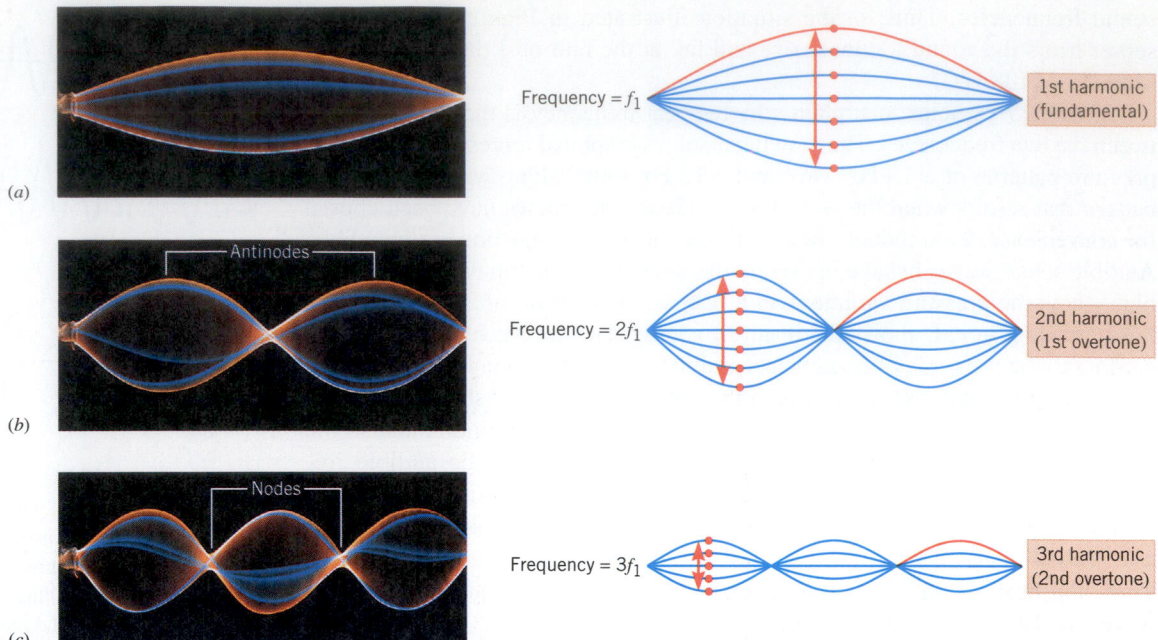

Frequency = f_1 — 1st harmonic (fundamental)

Frequency = $2f_1$ — 2nd harmonic (1st overtone)

Frequency = $3f_1$ — 3rd harmonic (2nd overtone)

Figure 17.15 Vibrating a string at certain unique frequencies sets up transverse standing wave patterns, such as the three shown in the photographs on the left. Each drawing on the right shows the various shapes that the string assumes at various times as it vibrates. The red dots attached to the strings focus attention on the maximum vibration that occurs at an antinode. In each of the drawings, one-half of a wave cycle is outlined in red. (Richard Megna/Fundamental Photographs)

string at various times and emphasize the maximum vibration that occurs at an antinode with the aid of a red dot attached to the string.

Each standing wave pattern is produced at a unique frequency of vibration. These frequencies form a series, the smallest frequency f_1 corresponding to the one-loop pattern and the larger frequencies being integer multiples of f_1, as Figure 17.15 indicates. Thus, if f_1 is 10 Hz, the frequency needed to establish the 2-loop pattern is $2f_1$ or 20 Hz, while the frequency needed to create the 3-loop pattern is $3f_1$ or 30 Hz, and so on. The frequencies in this series (f_1, $2f_1$, $3f_1$, etc.) are called *harmonics.* The lowest frequency f_1 is called the first harmonic, and the higher frequencies are designated as the second harmonic ($2f_1$), the third harmonic ($3f_1$), and so forth. The harmonic number (1st, 2nd, 3rd, etc.) corresponds to the number of loops in the standing wave pattern. The frequencies in this series are also referred to as the fundamental frequency, the first overtone, the second overtone, and so on. Thus, frequencies above the fundamental are *overtones* (see Figure 17.15).

Standing waves arise because identical waves travel on the string in *opposite directions* and combine in accord with the principle of linear superposition. A standing wave is said to be *standing* because it does not travel in one direction or the other, as do the individual waves that produce it. Figure 17.16 shows why there are waves traveling in both directions on the string. At the top of the picture, one-half of a wave cycle (the remainder of the wave is omitted for clarity) is moving toward the wall on the right. When the half-cycle reaches the wall, it causes the string to pull upward on the wall. Consistent with Newton's action–reaction law, the wall pulls downward on the string, and a downward-pointing half-cycle is sent back toward the left. Thus, the wave reflects from the wall. Upon arriving back at the point of origin, the wave reflects again, this time from the hand vibrating the string. For small vibration amplitudes, the hand is essentially fixed and behaves as the wall does in causing reflections. Repeated reflections at both ends of the string create a multitude of wave cycles traveling in both directions.

As each new cycle is formed by the vibrating hand, previous cycles that have reflected from the wall arrive and reflect again from the hand. Unless the timing is right, however, the new and the reflected cycles tend to offset one another, and a standing wave is not formed. Think about pushing someone on a swing and timing your pushes so that they reinforce one another. Such reinforcement in the case of the wave cycles leads to a large-amplitude standing wave. Suppose that the string has a length L and its left end is being vibrated at a frequency f_1. The time required to create a new wave cycle is the period T of the wave, where $T = 1/f_1$ (Equation 10.5). On the other hand, the time needed for a cycle to travel from the hand to the wall and back, a distance of $2L$, is $2L/v$, where v is the wave

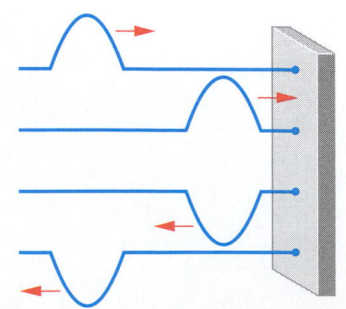

Figure 17.16 In reflecting from the wall, a forward-traveling half-cycle becomes a backward-traveling half-cycle that is inverted.

speed. Reinforcement between new and reflected cycles occurs if these two times are equal; that is, if $1/f_1 = 2L/v$. Thus, a standing wave is formed when the string is vibrated with a frequency of $f_1 = v/(2L)$.

Repeated reinforcement between newly created and reflected cycles causes a large-amplitude standing wave to develop on the string, *even when the hand itself vibrates with only a small amplitude.* Thus, the motion of the string is a resonance effect, analogous to that discussed in Section 10.6 for an object attached to a spring. The frequency f_1 at which resonance occurs is sometimes called a **natural frequency** of the string, similar to the frequency at which an object oscillates on a spring.

There is a difference between the resonance of the string and the resonance of a spring system, however. An object on a spring has only a single natural frequency, whereas the string has a *series* of natural frequencies. The series arises because a reflected wave cycle need not return to its point of origin in time to reinforce *every* newly created cycle. Reinforcement can occur, for instance, on *every other* new cycle, as it does if the string is vibrated at twice the frequency f_1, or $f_2 = 2f_1$. Likewise, if the vibration frequency is $f_3 = 3f_1$, reinforcement occurs on *every third* new cycle. Similar arguments apply for any frequency $f_n = nf_1$, where n is an integer. As a result, the series of natural frequencies that lead to standing waves on a string fixed at both ends is

String fixed at both ends $$f_n = n\left(\frac{v}{2L}\right) \qquad n = 1, 2, 3, 4, \ldots \qquad (17.3)$$

It is also possible to obtain Equation 17.3 in another way. In Figure 17.15, one-half of a wave cycle is outlined in red for each of the harmonics, to show that each loop in a standing wave pattern corresponds to one-half a wavelength. Since the two fixed ends of the string are nodes, the length L of the string must contain an integer number n of half-wavelengths: $L = n(\frac{1}{2}\lambda_n)$ or $\lambda_n = 2L/n$. Using this result for the wavelength in the relation $f_n\lambda_n = v$ shows that $f_n(2L/n) = v$, which can be rearranged to give Equation 17.3.

Standing waves on a string are important in the way many musical instruments produce sound. For instance, a guitar string is stretched between two supports and, when plucked, vibrates according to the series of natural frequencies given by Equation 17.3. The next two examples show how this series of frequencies governs the design of a guitar.

Problem-solving insight

The distance between two successive nodes (or between two successive antinodes) of a standing wave is equal to one-half of a wavelength.

ANALYZING MULTIPLE-CONCEPT PROBLEMS

Example 4 Playing a Guitar

The heaviest string on an electric guitar has a linear density of 5.28×10^{-3} kg/m and is stretched with a tension of 226 N. This string produces the musical note E when vibrating along its entire length in a standing wave at the fundamental frequency of 164.8 Hz. **(a)** Find the length L of the string between its two fixed ends (see Figure 17.17a). **(b)** A guitar player wants the string to vibrate at a fundamental frequency of 2×164.8 Hz $=$ 329.6 Hz, as it must if the musical note E is

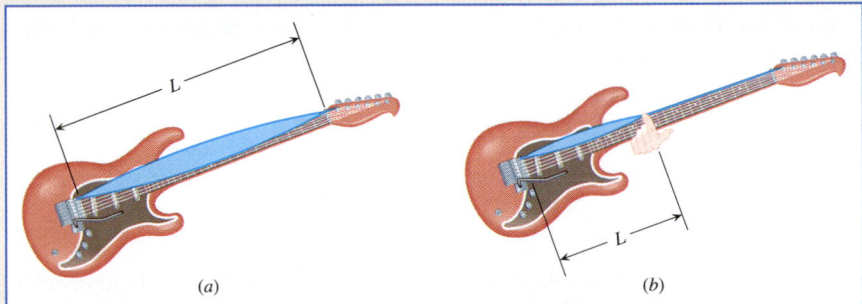

(a)   (b)

Figure 17.17 These drawings show the standing waves (in blue) that exist on a guitar string under different playing conditions.

to be sounded one octave higher in pitch. To accomplish this, he presses the string against the proper fret before plucking the string. Find the distance L between the fret and the bridge of the guitar (see Figure 17.17b).

Reasoning The series of natural frequencies (including the fundamental) for a string fixed at both ends is given by $f_n = nv/(2L)$ (Equation 17.3), where $n = 1, 2, 3$, etc. This equation can be solved for the length L. The speed v at which waves travel can be obtained from the tension and the linear density of the string. The fundamental frequencies that are given correspond to $n = 1$.

Knowns and Unknowns The given data are summarized as follows:

Description	Symbol	Value	Comment
Explicit Data			
Linear density of string	m/L	5.28×10^{-3} kg/m	
Tension in string	F	226 N	
Natural frequency at which string vibrates	f_n	164.8 Hz or 329.6 Hz	These are fundamental frequencies.
Implicit Data			
Integer variable in series of natural frequencies	n	1	Fundamental frequencies are given.
Unknown Variable			
Length	L	?	

Modeling the Problem

STEP 1 **Natural Frequencies** According to Equation 17.3, the natural frequencies for a string fixed at both ends are given by $f_n = nv/(2L)$, where n takes on the integer values 1, 2, 3, etc., v is the speed of the waves on the string, and L is the length between the two fixed ends. Solving this expression for L gives Equation 1 at the right. In this equation, only the speed v is unknown. We will obtain a value for it in Step 2.

$$L = \frac{nv}{2f_n} \qquad (1)$$

STEP 2 **Speed of the Waves on the String** The speed v of the waves traveling on the string is given by Equation 16.2 as

$$v = \sqrt{\frac{F}{m/L}}$$

where F is the tension and m/L is the linear density, both of which are given. The substitution of this expression for the speed into Equation 1 is shown at the right.

$$L = \frac{nv}{2f_n} \qquad (1)$$

$$v = \sqrt{\frac{F}{m/L}}$$

Solution Combining the results of each step algebraically, we find that

$$\overset{\text{STEP 1}}{\underset{}{L}} = \overset{}{\underset{}{\frac{nv}{2f_n}}} = \overset{\text{STEP 2}}{\underset{}{\frac{n}{2f_n}\sqrt{\frac{F}{m/L}}}}$$

It is given that the two natural frequencies at which the string vibrates are fundamental frequencies. Thus, $n = 1$ and $f_n = f_1$. We now determine the desired lengths in parts (a) and (b).

(a) $L = \dfrac{n}{2f_n}\sqrt{\dfrac{F}{m/L}} = \dfrac{1}{2(164.8 \text{ Hz})}\sqrt{\dfrac{226 \text{ N}}{5.28 \times 10^{-3} \text{ kg/m}}} = \boxed{0.628 \text{ m}}$

(b) $L = \dfrac{n}{2f_n}\sqrt{\dfrac{F}{m/L}} = \dfrac{1}{2(329.6 \text{ Hz})}\sqrt{\dfrac{226 \text{ N}}{5.28 \times 10^{-3} \text{ kg/m}}} = \boxed{0.314 \text{ m}}$

The length in part (b) is one-half the length in part (a) because the fundamental frequency in part (b) is twice the fundamental frequency in part (a).

Related Homework: *Problems 29, 32, 34*

🔻 **Conceptual Example 5** The Frets on a Guitar

**The physics of
the frets on a guitar.**

Figure 17.18 shows the frets on the neck of a guitar. They allow the player to produce a complete sequence of musical notes using a single string. Starting with the fret at the top of the neck, each successive fret shows where the player should press to get the next note in the sequence. Musicians call the sequence the chromatic scale, and every thirteenth note in it corresponds to one octave, or a doubling of the sound frequency. Which describes the spacing between the

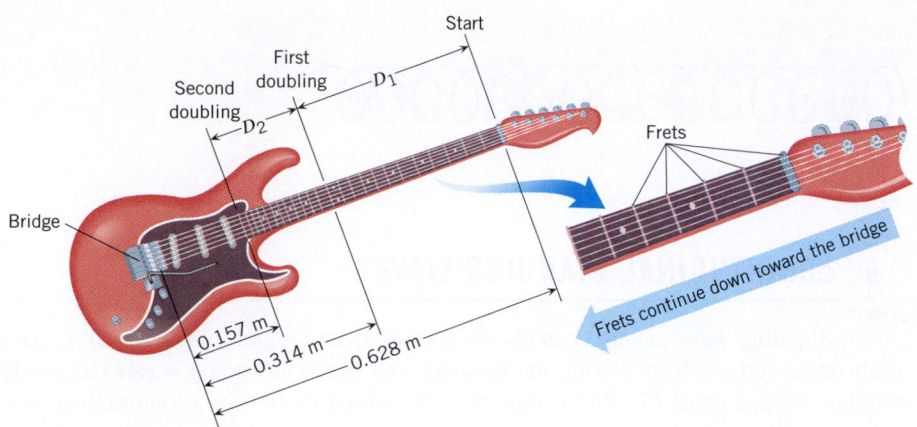

Figure 17.18 The spacing between the frets on the neck of a guitar changes going down the neck toward the bridge.

frets? It is **(a)** the same everywhere along the neck **(b)** greatest at the top of the neck and decreases with each additional fret further on down toward the bridge **(c)** smallest at the top of the neck and increases with each additional fret further on down toward the bridge.

Reasoning Our reasoning is based on the relation $f_1 = v/(2L)$ (Equation 17.3, with $n = 1$). The value of n is 1 because a string vibrates mainly at its fundamental frequency when plucked, as mentioned in Example 4. This equation shows that L, which is the length between a given fret and the bridge of the guitar, is inversely proportional to the fundamental frequency f_1, or $L = v/(2f_1)$. In Example 4 we found that the E string has a length of $L = 0.628$ m, corresponding to a fundamental frequency of $f_1 = 164.8$ Hz. We also found that the length between the bridge and the fret that must be pressed to double this frequency to 2×164.8 Hz = 329.6 Hz is one-half of 0.628 m, or $L = 0.314$ m. To understand the spacing between frets as one moves down the neck, consider the fret that must be pressed to double the frequency again, from 329.6 Hz to 659.2 Hz. The length between the bridge and this fret would be one-half of 0.314 m, or $L = 0.157$ m. Thus, the distances of the three frets that we have been discussing are 0.628 m, 0.314 m, and 0.157 m, as indicated in Figure 17.18. The distance D_1 between the first two of these frets is $D_1 = 0.628$ m $-$ 0.314 m = 0.314 m. Similarly, the distance between the second and third of these frets is $D_2 = 0.314$ m $-$ 0.157 m = 0.157 m.

Answers (a) and (c) are incorrect. The distances between the frets are shown in Figure 17.18. Clearly, the distances D_1 and D_2 are not equal, nor are they smaller at the top of the neck and greater further on down.

Answer (b) is correct. Figure 17.18 shows that D_1 is greater than D_2. Thus, the spacing between the frets is greatest at the top of the neck and decreases with each additional fret further on down.

Related Homework: *Problem 40*

✓ CHECK YOUR UNDERSTANDING

(The answers are given at the end of the book.)

11. A standing wave that corresponds to the fourth harmonic is set up on a string that is fixed at both ends. **(a)** How many loops are in this standing wave? **(b)** How many nodes (excluding the nodes at the ends of the string) does this standing wave have? **(c)** Is there a node or an antinode at the midpoint of the string? **(d)** If the frequency of this standing wave is 440 Hz, what is the frequency of the lowest-frequency standing wave that could be set up on this string?

12. The tension in a guitar string is doubled. By what factor does the frequency of the vibrating string change? **(a)** It increases by a factor of 2. **(b)** It increases by a factor of $\sqrt{2}$. **(c)** It decreases by a factor of 2. **(d)** It decreases by a factor of $\sqrt{2}$.

13. A string is vibrating back and forth as in Figure 17.15a. The tension in the string is decreased by a factor of four, with the frequency and length of the string remaining the same. A new standing wave pattern develops on the string. How many loops are in this new pattern? **(a)** 5 **(b)** 4 **(c)** 3 **(d)** 2

14. A rope is hanging vertically straight down. The top end is being vibrated back and forth, and a standing wave with many loops develops on the rope, analogous (but not identical) to a standing wave on a horizontal rope. The rope has mass. The separation between successive nodes is **(a)** everywhere the same along the rope **(b)** greater near the top of the rope than near the bottom **(c)** greater near the bottom of the rope than near the top.

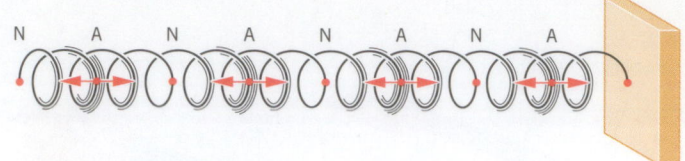

Figure 17.19 A longitudinal standing wave on a Slinky showing the displacement nodes (N) and antinodes (A).

17.6 LONGITUDINAL STANDING WAVES

Standing wave patterns can also be formed from longitudinal waves. For example, when sound reflects from a wall, the forward- and backward-going waves can produce a standing wave. Figure 17.19 illustrates the vibrational motion in a longitudinal standing wave on a Slinky. As in a transverse standing wave, there are nodes and antinodes. At the nodes the Slinky coils do not vibrate at all; that is, they have no displacement. At the antinodes the coils vibrate with maximum amplitude and, thus, have a maximum displacement. The red dots in Figure 17.19 indicate the lack of vibration at a node and the maximum vibration at an antinode. The vibration occurs along the line of travel of the individual waves, as is to be expected for longitudinal waves. In a standing wave of sound, the molecules or atoms of the medium behave as the red dots do.

Musical instruments in the wind family depend on longitudinal standing waves in producing sound. Since wind instruments (trumpet, flute, clarinet, pipe organ, etc.) are modified tubes or columns of air, it is useful to examine the standing waves that can be set up in such tubes. Figure 17.20 shows two cylindrical columns of air that are open at both ends. Sound waves, originating from a tuning fork, travel up and down within each tube, since they reflect from the ends of the tubes, even though the ends are open. If the frequency f of the tuning fork matches one of the natural frequencies of the air column, the downward- and upward-traveling waves combine to form a standing wave, and the sound of the tuning fork becomes markedly louder. To emphasize the longitudinal nature of the standing wave patterns, the left side of each pair of drawings in Figure 17.20 replaces the air in the tubes with Slinkies, on which the nodes and antinodes are indicated with red dots. As an additional aid in visualizing the standing waves, the right side of each pair of drawings shows blurred blue patterns within each tube. These patterns symbolize the amplitude of the vibrating air molecules at various locations. Wherever the pattern is widest, the amplitude of vibration is greatest (a displacement antinode), and wherever the pattern is narrowest there is no vibration (a displacement node).

To determine the natural frequencies of the air columns in Figure 17.20, notice that there is a displacement antinode at each end of the open tube because the air molecules there are free to move.* As in a transverse standing wave, the distance between two successive antinodes is one-half of a wavelength, so the length L of the tube must be an integer number n of half-wavelengths: $L = n(\frac{1}{2}\lambda_n)$ or $\lambda_n = 2L/n$. Using this wavelength in the relation $f_n = v/\lambda_n$ shows that the natural frequencies f_n of the tube are

Tube open at both ends
$$f_n = n\left(\frac{v}{2L}\right) \qquad n = 1, 2, 3, 4, \ldots \qquad (17.4)$$

At these frequencies, large-amplitude standing waves develop within the tube due to resonance. Examples 6 and 7 illustrate how Equation 17.4 is involved when a flute is played.

Figure 17.20 A pictorial representation of longitudinal standing waves on a Slinky (left side of each pair) and in a tube of air (right side of each pair) that is open at both ends (A, antinode; N, node).

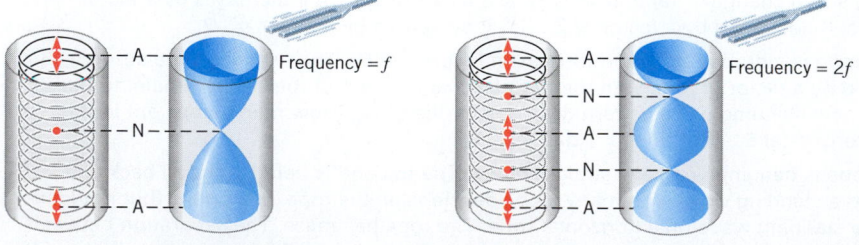

*In reality, the antinode does not occur exactly at the open end. However, if the tube's diameter is small compared to its length, little error is made in assuming that the antinode is located right at the end.

Example 6 Playing a Flute

The physics of a flute. When all the holes are closed on one type of flute, the lowest note it can sound is a middle C (fundamental frequency = 261.6 Hz). The air temperature is 293 K, and the speed of sound is 343 m/s. Assuming the flute is a cylindrical tube open at both ends, determine the distance L in Figure 17.21—that is, the distance from the mouthpiece to the end of the tube. (This distance is approximate, since the antinode does not occur exactly at the mouthpiece.)

Reasoning For a tube open at both ends, the series of natural frequencies (including the fundamental) is given by $f_n = nv/(2L)$ (Equation 17.4), where n takes on the integer values 1, 2, 3, etc. To obtain a value for L, we can solve this equation, since the given fundamental frequency corresponds to $n = 1$ and the speed v of sound is known.

Solution Solving Equation 17.4 for the length L, we obtain

$$L = \frac{nv}{2f_n} = \frac{(1)(343 \text{ m/s})}{2(261.6 \text{ Hz})} = \boxed{0.656 \text{ m}}$$

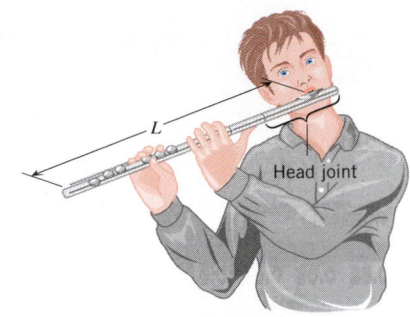

Figure 17.21 The length L of a flute between the mouthpiece and the end of the instrument determines the fundamental frequency of the lowest playable note.

ANALYZING MULTIPLE-CONCEPT PROBLEMS

Example 7 Tuning a Flute

A flautist is playing the flute discussed in Example 6, but now the temperature is 305 K instead of 293 K. Therefore, with the length calculated in Example 6, the note middle C does not have the proper fundamental frequency of 261.6 Hz. In other words, the flute is out of tune. To adjust the tuning, the flautist can alter the flute's length by changing the extent to which the head joint (see Figure 17.21) is inserted into the main stem of the instrument. To what length must the flute be adjusted to play the middle C at its proper frequency?

Reasoning As in Example 6, we will make use of the series of natural frequencies (including the fundamental) represented by $f_n = nv/(2L)$ (Equation 17.4), for a tube open at both ends. In this expression, n takes on the integer values 1, 2, 3, etc. and has the value of $n = 1$ for the given fundamental frequency. We can again solve Equation 17.4 for L, but now must deal with the fact that no value is given for the speed v of sound at the temperature of 305 K. To obtain the necessary value, we will assume that air behaves as an ideal gas and utilize the value given for v at 293 K to calculate a value at 305 K.

Knowns and Unknowns The given data are summarized as follows:

Description	Symbol	Value	Comment
Explicit Data			
Natural frequency for middle C	f_n	261.6 Hz	This is a fundamental frequency.
Speed of sound at 293 K	v_{293}	343 m/s	
Temperature at which flute is played	T_{305}	305 K	
Implicit Data			
Integer variable in series of natural frequencies	n	1	The fundamental frequency is given.
Unknown Variable			
Length	L	?	

Modeling the Problem

STEP 1 Natural Frequencies For a tube open at both ends, the natural frequencies are given by $f_n = nv_{305}/(2L)$, where n takes on the integer values 1, 2, 3, etc., v_{305} is the speed of sound at 305 K, and L is the length between the two open ends. Solving this expression for L gives Equation 1 at the right. Since the speed v_{305} is not given, we proceed to Step 2 in order to evaluate it.

$$L = \frac{nv_{305}}{2f_n} \qquad (1)$$

STEP 2 The Speed of Sound We assume that air behaves as an ideal gas. For an ideal gas, the speed v of sound is given by Equation 16.5 as $v = \sqrt{\gamma kT/m}$, where γ is the ratio of specific heat capacities at constant pressure and constant volume, k is Boltzmann's constant, T is the Kelvin temperature, and m is the average mass of the molecules and atoms

Continued

of which the air is composed. Applying this equation at the temperatures of 293 and 305 K, we have

$$v_{293} = \sqrt{\frac{\gamma k T_{293}}{m}} \quad \text{and} \quad v_{305} = \sqrt{\frac{\gamma k T_{305}}{m}}$$

Dividing these two expressions gives

$$\frac{v_{305}}{v_{293}} = \frac{\sqrt{\gamma k T_{305}/m}}{\sqrt{\gamma k T_{293}/m}} = \sqrt{\frac{T_{305}}{T_{293}}}$$

In this result, the speed v_{293} and the two temperatures are known, so that we may solve for the unknown speed v_{305}:

$$\boxed{v_{305} = v_{293}\sqrt{\frac{T_{305}}{T_{293}}}}$$

$$L = \frac{nv_{305}}{2f_n} \qquad (1)$$

$$\boxed{v_{305} = v_{293}\sqrt{\frac{T_{305}}{T_{293}}}}$$

This expression can be substituted into Equation 1, as shown at the right.

Solution Combining the results of each step algebraically, we find that

| STEP 1 | STEP 2 |

$$L = \frac{nv_{305}}{2f_n} = \frac{nv_{293}\sqrt{T_{305}/T_{293}}}{2f_n}$$

Since the given frequency is the fundamental frequency, it follows that in this result $n = 1$ and $f_n = f_1$. The length to which the flute must be adjusted is

$$L = \frac{nv_{293}\sqrt{T_{305}/T_{293}}}{2f_n} = \frac{(1)(343 \text{ m/s})\sqrt{(305 \text{ K})/(293 \text{ K})}}{2(261.6 \text{ Hz})} = \boxed{0.669 \text{ m}}$$

Comparing this result with that in Example 6, we see that to play in tune at the higher temperature, a flautist must lengthen the flute by 0.013 m.

Related Homework: *Problem 49*

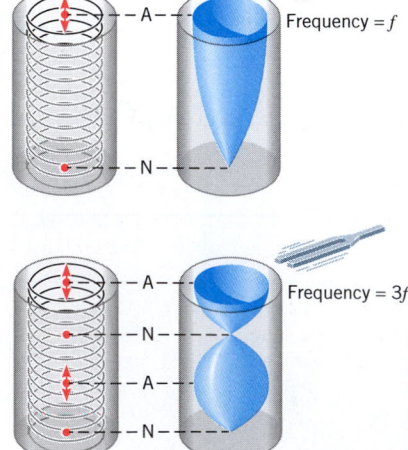

Figure 17.22 A pictorial representation of the longitudinal standing waves on a Slinky (left side of each pair) and in a tube of air (right side of each pair) that is open only at one end (A, antinode; N, node).

Standing waves can also exist in a tube with only one end open, as the patterns in Figure 17.22 indicate. Note the difference between these patterns and those in Figure 17.20. Here the standing waves have a displacement antinode at the open end and a displacement node at the closed end, where the air molecules are not free to move. Since the distance between a node and an adjacent antinode is one-fourth of a wavelength, the length L of the tube must be an odd number of quarter-wavelengths: $L = 1(\frac{1}{4}\lambda)$ and $L = 3(\frac{1}{4}\lambda)$ for the two standing wave patterns in Figure 17.22. In general, then, $L = n(\frac{1}{4}\lambda_n)$, where n is any odd integer ($n = 1, 3, 5, \ldots$). From this result it follows that $\lambda_n = 4L/n$, and the natural frequencies f_n can be obtained from the relation $f_n = v/\lambda_n$:

Tube open at only one end $$f_n = n\left(\frac{v}{4L}\right) \qquad n = 1, 3, 5, \ldots \qquad (17.5)$$

A tube open at only one end can develop standing waves only at the odd harmonic frequencies f_1, f_3, f_5, etc. In contrast, a tube open at both ends can develop standing waves at all harmonic frequencies f_1, f_2, f_3, etc. Moreover, the fundamental frequency f_1 of a tube open at only one end (Equation 17.5) is one-half that of a tube open at both ends (Equation 17.4). In other words, a tube open only at one end needs to be only one-half as long as a tube open at both ends in order to produce the *same* fundamental frequency.

Energy is also conserved when a standing wave is produced, either on a string or in a tube of air. The energy of the standing wave is the sum of the energies of the individual waves that comprise the standing wave. Once again, interference redistributes the energy of the individual waves to create locations of greatest energy (displacement antinodes) and locations of no energy (displacement nodes).

✓ **CHECK YOUR UNDERSTANDING**

(*The answers are given at the end of the book.*)

15. A cylindrical bottle, partially filled with water, is open at the top. When you blow across the top of the bottle a standing wave is set up inside it. Is there a node or an antinode **(a)** at the top of the bottle and **(b)** at the surface of the water? **(c)** If the standing wave is vibrating at its fundamental frequency, what is the distance between the top of the bottle and the surface of the water? Express your answer in terms of the wavelength λ of the standing wave. **(d)** If you take a sip from the bottle, is the fundamental frequency of the standing wave raised, lowered, or does it remain the same?

16. In Figure 17.20 both tubes are filled with air, in which the speed of sound is v_{air}. Suppose, instead, that the tube near the tuning fork labeled "Frequency = 2f" is filled not with air, but with another gas in which the speed of sound is v_{gas}. The frequency of each tuning fork remains unchanged. How should v_{gas} compare with v_{air} in order that the standing wave pattern in each tube has the same appearance? **(a)** $v_{gas} = \frac{1}{2} v_{air}$ **(b)** $v_{gas} = 2v_{air}$ **(c)** $v_{gas} = \frac{1}{4} v_{air}$ **(d)** $v_{gas} = 4v_{air}$

17. Standing waves can ruin the acoustics of a concert hall if there is excessive reflection of the sound waves that the performers generate. For example, suppose that a performer generates a 2093-Hz tone. If a large-amplitude standing wave is present, it is possible for a listener to move a distance of only 4.1 cm and hear the loudness of the tone change from loud to faint. What does the distance of 4.1 cm represent? **(a)** One wavelength of the sound **(b)** One-half the wavelength of the sound **(c)** One-fourth the wavelength of the sound

18. A wind instrument is brought into a warm house from the cold outdoors. What happens to the natural frequencies of the instrument? Neglect any change in the length of the instrument. **(a)** They increase. **(b)** They decrease. **(c)** They remain the same.

17.7 *COMPLEX SOUND WAVES

Musical instruments produce sound in a way that depends on standing waves. Examples 4 and 5 illustrate the role of transverse standing waves on the string of an electric guitar, while Examples 6 and 7 stress the role of longitudinal standing waves in the air column within a flute. In each example, sound is produced at the fundamental frequency of the instrument.

In general, however, a musical instrument does not produce just the fundamental frequency when it plays a note, but simultaneously generates a number of harmonics as well. Different instruments, such as a violin and a trumpet, generate harmonics to different extents, and the harmonics give the instruments their characteristic sound qualities or timbres. Suppose, for instance, that a violinist and a trumpet player both sound concert A, a note whose fundamental frequency is 440 Hz. Even though both instruments are playing the same note, most people can distinguish the sound of the violin from that of the trumpet. The instruments sound different because the relative amplitudes of the harmonics (880 Hz, 1320 Hz, etc.) that the instruments create are different.

The sound wave corresponding to a note produced by a musical instrument or a singer is called a ***complex sound wave*** because it consists of a mixture of the fundamental and harmonic frequencies. The pattern of pressure fluctuations in a complex wave can be obtained by using the principle of linear superposition, as Figure 17.23 indicates. This drawing shows a bar graph in which the heights of the bars give the relative amplitudes of the harmonics contained in a note such as a singer might produce. When the individual pressure patterns for each of the three harmonics are added together, they yield the complex pressure pattern shown at the top of the picture.*

In practice, a bar graph such as that in Figure 17.23 is determined with the aid of an electronic instrument known as a spectrum analyzer. When the note is produced, the complex

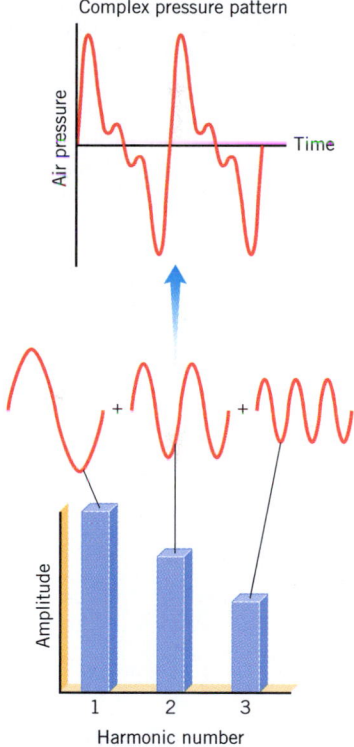

Figure 17.23 The topmost graph shows the pattern of pressure fluctuations such as a singer might produce. The pattern is the sum of the first three harmonics. The relative amplitudes of the harmonics correspond to the heights of the vertical bars in the bar graph.

The physics of a spectrum analyzer.

*In carrying out the addition, we assume that each individual pattern begins at zero at the origin when the time equals zero.

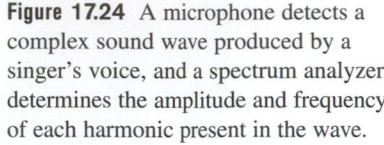

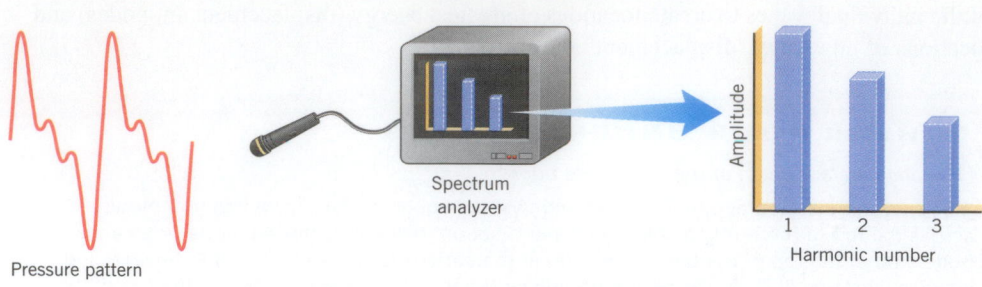

Pressure pattern

Spectrum analyzer

Amplitude

Harmonic number

Figure 17.24 A microphone detects a complex sound wave produced by a singer's voice, and a spectrum analyzer determines the amplitude and frequency of each harmonic present in the wave.

sound wave is detected by a microphone that converts the wave into an electrical signal. The electrical signal, in turn, is fed into the spectrum analyzer, as Figure 17.24 illustrates. The spectrum analyzer then determines the amplitude and frequency of each harmonic present in the complex wave and displays the results on its screen.

17.8 CONCEPTS & CALCULATIONS

Diffraction is the bending of a traveling wave around an obstacle or around the edges of an opening and is one of the consequences of the principle of linear superposition. As Equations 17.1 and 17.2 indicate, the extent of diffraction when a sound wave passes through an opening depends on the ratio of the wavelength of the sound to the width or diameter of the opening. Example 8 compares diffraction in two different media and reviews some of the fundamental properties of sound waves.

Concepts & Calculations Example 8
Diffraction in Two Different Media

A sound wave with a frequency of 15 kHz emerges through a circular opening that has a diameter of 0.20 m. Find the diffraction angle θ when the sound travels in air at a speed of 343 m/s and in water at a speed of 1482 m/s.

Concept Questions and Answers The diffraction angle for a circular opening is given by Equation 17.2 as $\sin \theta = 1.22 \lambda/D$, where λ is the wavelength of the sound and D is the diameter of the opening. How is the wavelength related to the frequency of the sound?

Answer According to Equation 16.1, the wavelength is given by $\lambda = v/f$, where v is the speed of sound and f is the frequency.

Is the wavelength of the sound in air greater than, smaller than, or equal to the wavelength in water?

Answer According to Equation 16.1, the wavelength is proportional to the speed v for a given value of the frequency f. Since sound travels at a slower speed in air than in water, the wavelength in air is smaller than the wavelength in water.

Is the diffraction angle in air greater than, smaller than, or equal to the diffraction angle in water?

Answer The extent of diffraction is determined by λ/D, the ratio of the wavelength to the diameter of the opening. Smaller ratios lead to less diffraction or smaller diffraction angles. The wavelength in air is smaller than in water, and the diameter of the opening is the same in both cases. Therefore, the ratio λ/D is smaller in air than in water, and the diffraction angle in air is smaller than the diffraction angle in water.

Solution Using $\sin \theta = 1.22 \lambda/D$ (Equation 17.2) and $\lambda = v/f$ (Equation 16.1), we have

$$\sin \theta = 1.22 \frac{\lambda}{D} = 1.22 \frac{v}{fD}$$

Applying this result for air and for water, we find

Air $\theta = \sin^{-1}\left(1.22\,\dfrac{v}{fD}\right) = \sin^{-1}\left[1.22\,\dfrac{(343\ \text{m/s})}{(15\ 000\ \text{Hz})(0.20\ \text{m})}\right] = \boxed{8.0°}$

Water $\theta = \sin^{-1}\left(1.22\,\dfrac{v}{fD}\right) = \sin^{-1}\left[1.22\,\dfrac{(1482\ \text{m/s})}{(15\ 000\ \text{Hz})(0.20\ \text{m})}\right] = \boxed{37°}$

As expected, the diffraction angle in air is smaller.

The next example deals with standing waves of sound in a gas. One of the factors that affect the formation of standing waves is the speed at which the individual waves travel. This example reviews how the speed of sound depends on the properties of the gas.

Concepts & Calculations Example 9
Standing Waves of Sound

Two tubes of gas are identical and are open at only one end. One tube contains neon (Ne) and the other krypton (Kr). Both are monatomic gases, have the same temperature, and may be assumed to be ideal gases. The fundamental frequency of the tube containing neon is 481 Hz. What is the fundamental frequency of the tube containing krypton?

Concept Questions and Answers For a gas-filled tube open at only one end, the fundamental frequency ($n = 1$) is given by Equation 17.5 as $f_1 = v/(4L)$, where v is the speed of sound and L is the length of the tube. How is the speed related to the properties of the gas?

Answer According to Equation 16.5, the speed is given by $v = \sqrt{\gamma kT/m}$, where γ is the ratio of the specific heat capacities at constant pressure and constant volume, k is Boltzmann's constant, T is the Kelvin temperature, and m is the mass of an atom of the gas.

All of the factors that affect the speed of sound are the same except the atomic mass. The periodic table located on the inside of the back cover gives the atomic masses of neon and krypton as 20.180 u and 83.80 u, respectively. Is the speed of sound in krypton greater than, smaller than, or equal to the speed of sound in neon?

Answer According to Equation 16.5 the speed of sound is $v = \sqrt{\gamma kT/m}$, so the speed is inversely proportional to the square root of the mass m of an atom. Thus, the speed is smaller when the mass is greater. Since krypton has the greater mass, the speed of sound in krypton is smaller than in neon.

Is the fundamental frequency of the tube containing krypton greater than, smaller than, or equal to the fundamental frequency of the tube containing neon?

Answer The fundamental frequency is given by Equation 17.5 as $f_1 = v/(4L)$. Since the speed of sound in krypton is smaller than in neon, the fundamental frequency of the krypton-filled tube is smaller than the fundamental frequency of the neon-filled tube.

Solution Using $f_1 = v/(4L)$ (Equation 17.5) and $v = \sqrt{\gamma kT/m}$ (Equation 16.5), we have

$$f_1 = \frac{v}{4L} = \frac{1}{4L}\sqrt{\frac{\gamma kT}{m}}$$

Applying this result to both tubes and taking the ratio of the frequencies, we obtain

$$\frac{f_{1,\,\text{Kr}}}{f_{1,\,\text{Ne}}} = \frac{\dfrac{1}{4L}\sqrt{\gamma kT/m_{\text{Kr}}}}{\dfrac{1}{4L}\sqrt{\gamma kT/m_{\text{Ne}}}} = \sqrt{\frac{m_{\text{Ne}}}{m_{\text{Kr}}}}$$

Solving for $f_{1,\,\text{Kr}}$ gives

$$f_{1,\,\text{Kr}} = f_{1,\,\text{Ne}}\sqrt{\frac{m_{\text{Ne}}}{m_{\text{Kr}}}} = (481\ \text{Hz})\sqrt{\frac{20.180\ \text{u}}{83.80\ \text{u}}} = \boxed{236\ \text{Hz}}$$

As expected, the fundamental frequency for the krypton-filled tube is smaller than the fundamental frequency for the neon-filled tube.

CONCEPT SUMMARY

If you need more help with a concept, use the Learning Aids noted next to the discussion or equation. Examples (**Ex.**) are in the text of this chapter. Go to **www.wiley.com/college/cutnell** for the following Learning Aids:

Interactive LearningWare (ILW) — Additional examples solved in a five-step interactive format.

Concept Simulations (CS) — Animated text figures or animations of important concepts.

Interactive Solutions (IS) — Models for certain types of problems in the chapter homework. The calculations are carried out interactively.

Topic	Discussion	Learning Aids
Principle of linear superposition	**17.1 THE PRINCIPLE OF LINEAR SUPERPOSITION** The principle of linear superposition states that when two or more waves are present simultaneously at the same place, the resultant disturbance is the sum of the disturbances from the individual waves.	**CS 17.1**
Constructive and destructive interference	**17.2 CONSTRUCTIVE AND DESTRUCTIVE INTERFERENCE OF SOUND WAVES** Constructive interference occurs at a point when two waves meet there crest-to-crest and trough-to-trough, thus reinforcing each other. Destructive interference occurs when the waves meet crest-to-trough and cancel each other.	
In-phase and out-of-phase waves	When waves meet crest-to-crest and trough-to-trough, they are exactly in phase. When they meet crest-to-trough, they are exactly out of phase.	
Conditions for constructive and destructive interference	For two wave sources vibrating in phase, a difference in path lengths that is zero or an integer number (1, 2, 3, . . .) of wavelengths leads to constructive interference; a difference in path lengths that is a half-integer number ($\frac{l}{2}$, $1\frac{l}{2}$, $2\frac{l}{2}$, . . .) of wavelengths leads to destructive interference.	**Ex. 1, 2** **IS 17.61**
	For two wave sources vibrating out of phase, a difference in path lengths that is a half-integer number ($\frac{l}{2}$, $1\frac{l}{2}$, $2\frac{l}{2}$, . . .) of wavelengths leads to constructive interference; a difference in path lengths that is zero or an integer number (1, 2, 3, . . .) of wavelengths leads to destructive interference.	
Diffraction	**17.3 DIFFRACTION** Diffraction is the bending of a wave around an obstacle or the edges of an opening. The angle through which the wave bends depends on the ratio of the wavelength λ of the wave to the width D of the opening; the greater the ratio λ/D, the greater the angle.	
	When a sound wave of wavelength λ passes through an opening, the first place where the intensity of the sound is a minimum relative to the center of the opening is specified by the angle θ. If the opening is a rectangular slit of width D, such as a doorway, the angle is	
Single slit–first minimum	$$\sin \theta = \frac{\lambda}{D} \qquad (17.1)$$	
	If the opening is a circular opening of diameter D, such as that in a loudspeaker, the angle is	
Circular opening–first minimum	$$\sin \theta = 1.22 \frac{\lambda}{D} \qquad (17.2)$$	**Ex. 3, 8**
Beats	**17.4 BEATS** Beats are the periodic variations in amplitude that arise from the linear superposition of two waves that have slightly different frequencies. When the waves are sound waves, the variations in amplitude cause the loudness to vary at the beat frequency, which is the difference between the frequencies of the waves.	**CS 17.2** **ILW 17.1**
Beat frequency		
Standing waves	**17.5 TRANSVERSE STANDING WAVES** A standing wave is the pattern of disturbance that results when oppositely traveling waves of the same frequency and amplitude pass through each other. A standing wave has places of minimum and maximum vibration called, respectively, nodes and antinodes.	
Nodes and antinodes		
Natural frequencies Harmonics	Under resonance conditions, standing waves can be established only at certain natural frequencies. The frequencies in this series (f_1, $2f_1$, $3f_1$, etc.) are called harmonics. The lowest frequency f_1 is called the first harmonic, the next frequency $2f_1$ is called the second harmonic, and so on.	
	For a string that is fixed at both ends and has a length L, the natural frequencies are	
String fixed at both ends	$$f_n = n\left(\frac{v}{2L}\right) \qquad n = 1, 2, 3, 4, \ldots \qquad (17.3)$$	**Ex. 4, 5** **ILW 17.2**
	where v is the speed of the wave on the string and n is a positive integer.	
	17.6 LONGITUDINAL STANDING WAVES For a gas in a cylindrical tube open at both ends, the natural frequencies of vibration are	
Tube open at both ends	$$f_n = n\left(\frac{v}{2L}\right) \qquad n = 1, 2, 3, 4, \ldots \qquad (17.4)$$	**Ex. 6, 7**
	where v is the speed of sound in the gas and L is the length of the tube.	

Topic	Discussion	Learning Aids
	For a gas in a cylindrical tube open at only one end, the natural frequencies of vibration are	
Tube open at only one end	$$f_n = n\left(\frac{v}{4L}\right) \qquad n = 1, 3, 5, 7, \ldots \qquad (17.5)$$	**Ex. 9** **IS 17.45**
Complex sound wave	**17.7 COMPLEX SOUND WAVES** A complex sound wave consists of a mixture of a fundamental frequency and overtone frequencies.	

FOCUS ON CONCEPTS

Note to Instructors: The numbering of the questions shown here reflects the fact that they are only a representative subset of the total number that are available online. However, all of the questions are available for assignment via an online homework management program such as WileyPLUS or WebAssign.

Section 17.1 The Principle of Linear Superposition

2. The drawing shows four moving pulses. Although shown as separated, the four pulses exactly overlap each other at the instant shown. Which combination of these pulses would produce a resultant pulse with the highest peak *and* the deepest valley at this instant? **(a)** 1 and 2 **(b)** 2, 3, and 4 **(c)** 2 and 3 **(d)** 1, 2, and 3 **(e)** 1 and 3

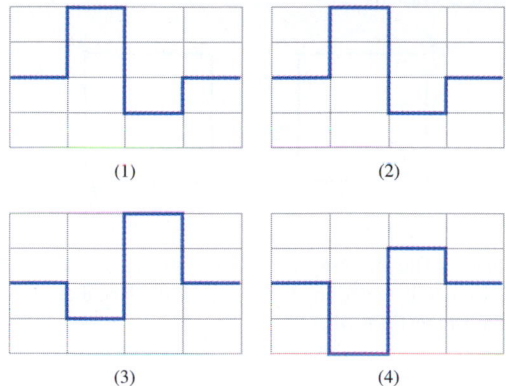

(1)   (2)

(3)   (4)

Section 17.2 Constructive and Destructive Interference of Sound Waves

3. Two cellists, one seated directly behind the other in an orchestra, play the same note for the conductor, who is directly in front of them. Because of the separation between the cellists, destructive interference occurs at the conductor. This separation is the smallest that produces destructive interference. Would this separation increase, decrease, or remain the same if the cellists produced a note with a higher frequency? **(a)** The separation between the cellists would remain the same. **(b)** The separation would decrease because the wavelength of the sound is greater. **(c)** The separation would decrease because the wavelength of the sound is smaller. **(d)** The separation would increase because the wavelength of the sound is greater. **(e)** The separation would increase because the wavelength of the sound is smaller.

Section 17.3 Diffraction

5. A loudspeaker is producing sound of a certain wavelength. Which combination of the wavelength λ (expressed as a multiple of λ_0) and the speaker's diameter D (expressed as a multiple of D_0) would exhibit the greatest amount of diffraction when the sound leaves the speaker and enters the room? **(a)** $\lambda = \lambda_0$, $D = D_0$ **(b)** $\lambda = 2\lambda_0$, $D = D_0$ **(c)** $\lambda = \lambda_0$, $D = 2D_0$ **(d)** $\lambda = 2\lambda_0$, $D = 2D_0$ **(e)** $\lambda = 3\lambda_0$, $D = 2D_0$

7. Sound of a given frequency leaves a loudspeaker and spreads out due to diffraction. The speaker is placed in a room that contains either air or helium. The speed of sound in helium is about three times as great as the speed of sound in air. In which room, if either, does the sound exhibit the greater diffraction when leaving the speaker? **(a)** The greater diffraction occurs in the air-filled room, because the wavelength of the sound is smaller in that room. **(b)** The greater diffraction occurs in the air-filled room, because the wavelength of the sound is greater in that room. **(c)** The diffraction is the same in both rooms. **(d)** The greater diffraction occurs in the helium-filled room, because the wavelength of the sound is smaller in that room. **(e)** The greater diffraction occurs in the helium-filled room, because the wavelength of the sound is greater in that room.

Section 17.4 Beats

8. Two musicians are comparing their trombones. The first produces a tone that is known to be 438 Hz. When the two trombones play together they produce 6 beats every 2 seconds. Which statement is true about the second trombone? **(a)** It is producing either a 432-Hz sound or a 444-Hz sound. **(b)** It is producing either a 436-Hz sound or a 440-Hz sound. **(c)** It is producing a 444-Hz sound, and could be producing no other sound frequency. **(d)** It is producing either a 435-Hz sound or a 441-Hz sound. **(e)** It is producing a 441-Hz sound and could be producing no other sound frequency.

Section 17.5 Transverse Standing Waves

11. Two transverse standing waves are shown in the drawing. The strings have the same tension and length, but the bottom string is more massive. Which standing wave, if either, is vibrating at the higher frequency? **(a)** The top standing wave has the higher frequency, because the traveling waves have a smaller speed due to the smaller mass of the string. 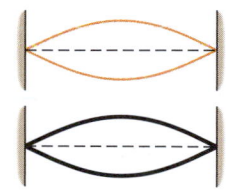 **(b)** The top standing wave has the higher frequency, because the traveling waves have a larger speed due to the smaller mass of the string. **(c)** Both standing waves have the same frequency, because the frequency of vibration does not depend on the mass of the string. **(d)** The bottom standing wave has the higher frequency, because the traveling waves have a smaller speed due to the larger mass of the string. **(e)** The bottom standing wave has the higher frequency, because the traveling waves have a larger speed due to the larger mass of the string.

12. A standing wave on a string fixed at both ends is vibrating at its fourth harmonic. If the length, tension, and linear density are kept constant, what can be said about the wavelength and frequency of the fifth harmonic relative to the fourth harmonic? **(a)** The wavelength of the fifth harmonic is longer, and its frequency is higher. **(b)** The wavelength of the fifth harmonic is longer, and its frequency is lower. **(c)** The wavelength of the fifth harmonic is shorter, and its frequency is higher. **(d)** The wavelength of the fifth harmonic is shorter, and its frequency is lower.

Section 17.6 Longitudinal Standing Waves

14. A longitudinal standing wave is established in a tube that is open at both ends (see the drawing). The length of the tube is 0.80 m. What is the wavelength of the waves that make up the standing wave? **(a)** 0.20 m **(b)** 0.40 m **(c)** 0.80 m **(d)** 1.20 m **(e)** 1.60 m

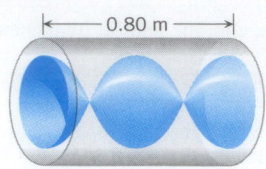

← 0.80 m →

16. A longitudinal standing wave is established in a tube open at only one end (see the drawing). The frequency of the standing wave is 660 Hz, and the speed of sound in air is 343 m/s. What is the length of the tube? **(a)** 0.13 m **(b)** 0.26 m **(c)** 0.39 m **(d)** 0.52 m **(e)** 0.65 m

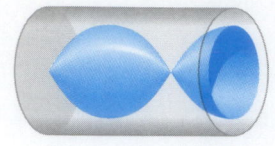

PROBLEMS

Note to Instructors: Most of the homework problems in this chapter are available for assignment via an online homework management program such as WileyPLUS or WebAssign, and those marked with the icon **GO** *are presented in* WileyPLUS *using a guided tutorial format that provides enhanced interactivity. See Preface for additional details.*

ssm   Solution is in the Student Solutions Manual.
www   Solution is available online at www.wiley.com/college/cutnell

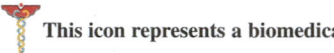 **This icon represents a biomedical application.**

Section 17.1   The Principle of Linear Superposition,
Section 17.2   Constructive and Destructive Interference
of Sound Waves

1. In Figure 17.7, suppose that the separation between speakers A and B is 5.00 m and the speakers are vibrating in phase. They are playing identical 125-Hz tones, and the speed of sound is 343 m/s. What is the largest possible distance between speaker B and the observer at C, such that he observes destructive interference?

2. Two speakers, one directly behind the other, are each generating a 245-Hz sound wave. What is the smallest separation distance between the speakers that will produce destructive interference at a listener standing in front of them? The speed of sound is 343 m/s.

3. ssm Concept Simulation 17.1 at **www.wiley.com/college/cutnell** illustrates the concept that is pertinent to this problem. The drawing graphs a string on which two pulses (half up and half down) are traveling at a constant speed of 1 cm/s at $t = 0$ s. Using the principle of linear superposition, draw the shape of the string's pulses at $t = 1$ s, 2 s, 3 s, and 4 s.

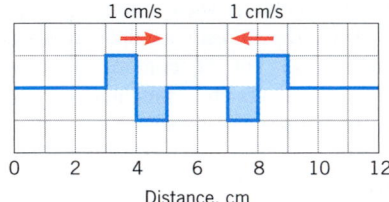

1 cm/s   1 cm/s

0   2   4   6   8   10   12
Distance, cm

4. Two waves are traveling in opposite directions on the same string. The displacements caused by the individual waves are given by $y_1 = (24.0 \text{ mm}) \sin(9.00\pi t - 1.25\pi x)$ and $y_2 = (35.0 \text{ mm}) \sin(2.88\pi t + 0.400\pi x)$. Note that the phase angles $(9.00\pi t - 1.25\pi x)$ and $(2.88\pi t + 0.400\pi x)$ are in radians, t is in seconds, and x is in meters. At $t = 4.00$ s, what is the net displacement (in mm) of the string at **(a)** $x = 2.16$ m and **(b)** $x = 2.56$ m? Be sure to include the algebraic sign (+ or −) with your answers.

5. ssm Two loudspeakers are vibrating in phase. They are set up as in Figure 17.7, and point C is located as shown there. The speed of sound is 343 m/s. The speakers play the same tone. What is the smallest frequency that will produce destructive interference at point C?

6. GO Both drawings show the same square, each of which has a side of length $L = 0.75$ m. An observer O is stationed at one corner of each square. Two loudspeakers are located at corners of the square, as in either drawing 1 or drawing 2. The speakers produce the same single-frequency tone in either drawing and are in phase. The speed

of sound is 343 m/s. Find the single smallest frequency that will produce both constructive interference in drawing 1 and destructive interference in drawing 2.

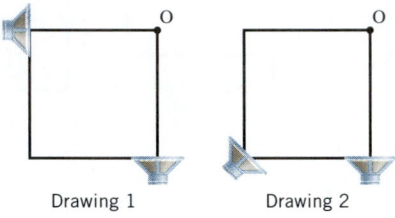

O   O

Drawing 1   Drawing 2

7. ssm www The drawing shows a loudspeaker A and point C, where a listener is positioned. A second loudspeaker B is located somewhere to the right of A. Both speakers vibrate in phase and are playing a 68.6-Hz tone. The speed of sound is 343 m/s. What is the closest to speaker A that speaker B can be located, so that the listener hears no sound?

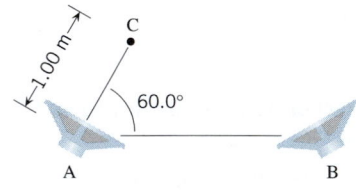

8. Suppose that the two speakers in Figure 17.7 are separated by 2.50 m and are vibrating exactly *out of phase* at a frequency of 429 Hz. The speed of sound is 343 m/s. Does the observer at C observe constructive or destructive interference when his distance from speaker B is **(a)** 1.15 m and **(b)** 2.00 m?

*9. The two speakers in the drawing are vibrating in phase, and a listener is standing at point P. Does constructive or destructive interference occur at P when the speakers produce sound waves whose frequency is **(a)** 1466 Hz and **(b)** 977 Hz? Justify your answers with appropriate calculations. Take the speed of sound to be 343 m/s.

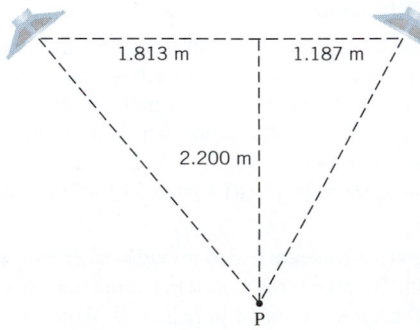

1.813 m   1.187 m

2.200 m

P

*10. **GO** A listener is standing in front of two speakers that are producing sound of the same frequency and amplitude, except that they are vibrating out of phase. Initially, the distance between the listener and each speaker is the same (see the drawing). As the listener moves sideways, the sound intensity gradually changes. When the distance x in the drawing is 0.92 m, the change reaches the maximum amount (either loud to soft, or soft to loud). Using the data shown in the drawing and 343 m/s for the speed of sound, determine the frequency of the sound coming from the speakers.

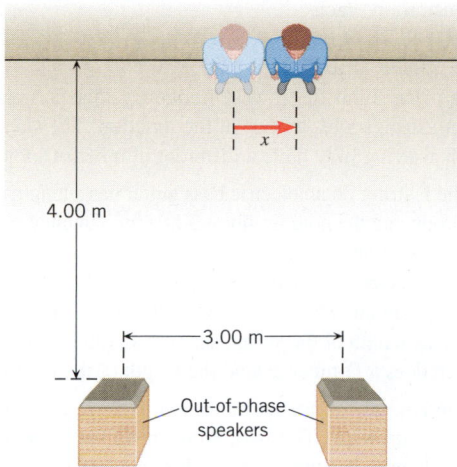

4.00 m

3.00 m

Out-of-phase speakers

**11. Speakers A and B are vibrating in phase. They are directly facing each other, are 7.80 m apart, and are each playing a 73.0-Hz tone. The speed of sound is 343 m/s. On the line between the speakers there are three points where constructive interference occurs. What are the distances of these three points from speaker A?

Section 17.3 Diffraction

12. Sound exits a diffraction horn loudspeaker through a rectangular opening like a small doorway. Such a loudspeaker is mounted outside on a pole. In winter, when the temperature is 273 K, the diffraction angle θ has a value of 15.0°. What is the diffraction angle for the same sound on a summer day when the temperature is 311 K?

13. **ssm** Consult Multiple-Concept Example 3 for background pertinent to this problem. A speaker has a diameter of 0.30 m. **(a)** Assuming that the speed of sound is 343 m/s, find the diffraction angle θ for a 2.0-kHz tone. **(b)** What speaker diameter D should be used to generate a 6.0-kHz tone whose diffraction angle is as wide as that for the 2.0-kHz tone in part (a)?

14. **GO** The following two lists give the diameters and sound frequencies for three loudspeakers. Pair each diameter with a frequency, so that the diffraction angle is the same for each of the speakers, and then find the common diffraction angle. Take the speed of sound to be 343 m/s.

Diameter, D	Frequency, f
0.050 m	6.0 kHz
0.10 m	4.0 kHz
0.15 m	12.0 kHz

15. Multiple-Concept Example 3 reviews the concepts that are important in this problem. The entrance to a large lecture room consists of two side-by-side doors, one hinged on the left and the other hinged on the right. Each door is 0.700 m wide. Sound of frequency 607 Hz is coming through the entrance from within the room. The speed of sound is 343 m/s. What is the diffraction angle θ of the sound after it passes through the doorway when **(a)** one door is open and **(b)** both doors are open?

16. **GO** For one approach to problems such as this, see Multiple-Concept Example 3. Sound emerges through a doorway, as in Figure 17.10. The width of the doorway is 77 cm, and the speed of sound is 343 m/s. Find the diffraction angle θ when the frequency of the sound is **(a)** 5.0 kHz and **(b)** 5.0×10^2 Hz.

*17. A 3.00-kHz tone is being produced by a speaker with a diameter of 0.175 m. The air temperature changes from 0 to 29 °C. Assuming air to be an ideal gas, find the *change* in the diffraction angle θ.

*18. A row of seats is parallel to a stage at a distance of 8.7 m from it. At the center and front of the stage is a diffraction horn loudspeaker. This speaker sends out its sound through an opening that is like a small doorway with a width D of 7.5 cm. The speaker is playing a tone that has a frequency of 1.0×10^4 Hz. The speed of sound is 343 m/s. What is the distance between two seats, located near the center of the row, at which the tone cannot be heard?

Section 17.4 Beats

19. **ssm** Two pure tones are sounded together. The drawing shows the pressure variations of the two sound waves, measured with respect to atmospheric pressure. What is the beat frequency?

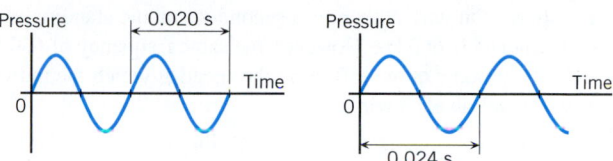

20. Two pianos each sound the same note simultaneously, but they are both out of tune. On a day when the speed of sound is 343 m/s, piano A produces a wavelength of 0.769 m, while piano B produces a wavelength of 0.776 m. How much time separates successive beats?

21. Two out-of-tune flutes play the same note. One produces a tone that has a frequency of 262 Hz, while the other produces a tone of 266 Hz. When a tuning fork is sounded together with the 262-Hz tone, a beat frequency of 1 Hz is produced. When the same tuning fork is sounded together with the 266-Hz tone, a beat frequency of 3 Hz is produced. What is the frequency of the tuning fork?

22. **ssm** When a guitar string is sounded along with a 440-Hz tuning fork, a beat frequency of 5 Hz is heard. When the same string is sounded along with a 436-Hz tuning fork, the beat frequency is 9 Hz. What is the frequency of the string?

23. In **Concept Simulation 17.2** at **www.wiley.com/college/cutnell** you can explore the concepts that are important in this problem. A 440.0-Hz tuning fork is sounded together with an out-of-tune guitar string, and a beat frequency of 3 Hz is heard. When the string is tightened, the frequency at which it vibrates increases, and the beat frequency is heard to decrease. What was the original frequency of the guitar string?

*24. **GO** Two cars have identical horns, each emitting a frequency of $f_s = 395$ Hz. One of the cars is moving with a speed of 12.0 m/s toward a bystander waiting at a corner, and the other car is parked. The speed of sound is 343 m/s. What is the beat frequency heard by the bystander?

*25. **GO** A sound wave is traveling in seawater, where the adiabatic bulk modulus and density are 2.31×10^9 Pa and 1025 kg/m³, respectively. The wavelength of the sound is 3.35 m. A tuning fork is struck under water and vibrates at 440.0 Hz. What would be the beat frequency heard by an underwater swimmer?

**26. Two carpenters are hammering at the same time, each at a different hammering frequency. The hammering frequency is the number of hammer blows per second. Every 4.6 s, both carpenters strike at the same instant, producing an effect very similar to a beat frequency. The first carpenter strikes a blow every 0.75 s. How many seconds elapse between the second carpenter's blows if the second carpenter hammers (a) more rapidly than the first carpenter, and (b) less rapidly than the first carpenter?

Section 17.5 Transverse Standing Waves

27. If the string in Figure 17.15c is vibrating at a frequency of 4.0 Hz and the distance between two successive nodes is 0.30 m, what is the speed of the waves on the string?

28. A string is fixed at both ends and is vibrating at 130 Hz, which is its third harmonic frequency. The linear density of the string is 5.6×10^{-3} kg/m, and it is under a tension of 3.3 N. Determine the length of the string.

29. ssm The approach to solving this problem is similar to that taken in Multiple-Concept Example 4. On a cello, the string with the largest linear density (1.56×10^{-2} kg/m) is the C string. This string produces a fundamental frequency of 65.4 Hz and has a length of 0.800 m between the two fixed ends. Find the tension in the string.

30. GO Two wires, each of length 1.2 m, are stretched between two fixed supports. On wire A there is a second-harmonic standing wave whose frequency is 660 Hz. However, the same frequency of 660 Hz is the third harmonic on wire B. Find the speed at which the individual waves travel on each wire.

31. ssm The A string on a string bass vibrates at a fundamental frequency of 55.0 Hz. If the string's tension were increased by a factor of four, what would be the new fundamental frequency?

32. Multiple-Concept Example 4 deals with the same concepts as this problem. A 41-cm length of wire has a mass of 6.0 g. It is stretched between two fixed supports and is under a tension of 160 N. What is the fundamental frequency of this wire?

33. A string has a linear density of 8.5×10^{-3} kg/m and is under a tension of 280 N. The string is 1.8 m
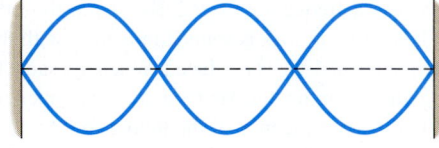
long, is fixed at both ends, and is vibrating in the standing wave pattern shown in the drawing. Determine the (a) speed, (b) wavelength, and (c) frequency of the traveling waves that make up the standing wave.

34. To review the concepts that play roles in this problem, consult Multiple-Concept Example 4. Sometimes, when the wind blows across a long wire, a low-frequency "moaning" sound is produced. This sound arises because a standing wave is set up on the wire, like a standing wave on a guitar string. Assume that a wire (linear density = 0.0140 kg/m) sustains a tension of 323 N because the wire is stretched between two poles that are 7.60 m apart. The lowest frequency that an average, healthy human ear can detect is 20.0 Hz. What is the lowest harmonic number n that could be responsible for the "moaning" sound?

*35. GO A copper block is suspended from a wire, as in part 1 of the drawing. A container of mercury is then raised up around the

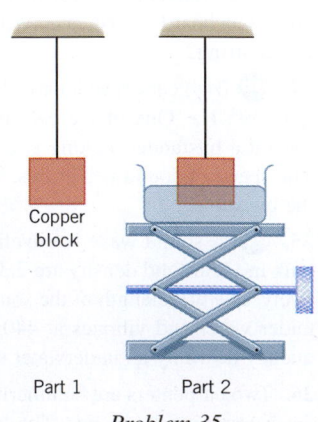
Part 1   Part 2

Problem 35

block, as in part 2, so that 50.0% of the block's volume is submerged in the mercury. The density of copper is 8890 kg/m³, and that of mercury is 13 600 kg/m³. Find the ratio of the fundamental frequency of the wire in part 2 to the fundamental frequency in part 1.

*36. Two strings have different lengths and linear densities, as the drawing shows. They are joined together and stretched so that the tension in each string is 190.0 N. The free ends of the joined string
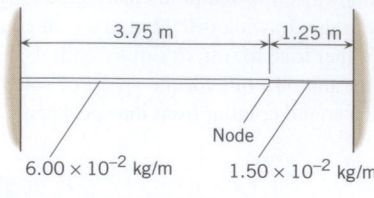
are fixed in place. Find the lowest frequency that permits standing waves in both strings with a node at the junction. The standing wave pattern in each string may have a different number of loops.

*37. ssm The E string on an electric bass guitar has a length of 0.628 m and, when producing the note E, vibrates at a fundamental frequency of 41.2 Hz. Players sometimes add to their instruments a device called a "D-tuner." This device allows the E string to be used to produce the note D, which has a fundamental frequency of 36.7 Hz. The D-tuner works by extending the length of the string, keeping all other factors the same. By how much does a D-tuner extend the length of the E string?

*38. GO Standing waves are set up on two strings fixed at each end, as shown in the drawing. The two strings have the same tension and mass per unit length, but they differ in length by 0.57 cm. The waves on the shorter string propagate with a speed of 41.8 m/s, and the fundamental frequency of the shorter string is 225 Hz. Determine the beat frequency produced by the two standing waves.

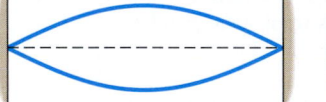 

39. ssm www The arrangement in the drawing shows a block (mass = 15.0 kg) that is held in position on a frictionless incline by a cord (length = 0.600 m). The mass per unit length of the cord is 1.20×10^{-2} kg/m, so the mass of the cord is negligible compared to the mass of the block. The cord is

being vibrated at a frequency of 165 Hz (vibration source not shown in the drawing). What are the values of the angle θ between 15.0° and 90.0° at which a standing wave exists on the cord?

40. Review Conceptual Example 5 before attempting this problem. As the drawing shows, the length of a guitar string is 0.628 m. The frets are numbered for convenience. A performer can play a musical scale on a single string because the spacing *between the frets* is designed according to the following rule: When the string is pushed against any fret j, the fundamental frequency of the shortened string is larger by a factor of the twelfth root of two ($\sqrt[12]{2}$) than it is when the string is pushed against the fret $j - 1$. Assuming that the tension in the string is the same for any note, find the spacing (a) between fret 1 and fret 0 and (b) between fret 7 and fret 6.

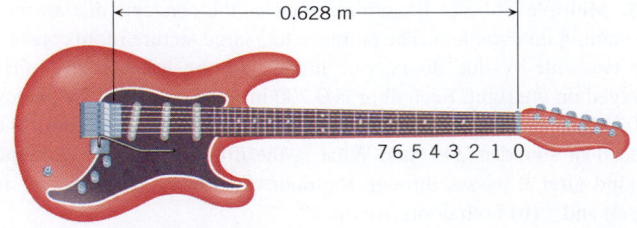

Section 17.6   Longitudinal Standing Waves,
Section 17.7   Complex Sound Waves

41. ssm A tube of air is open at only one end and has a length of 1.5 m. This tube sustains a standing wave at its third harmonic. What is the distance between one node and the adjacent antinode?

42.  Sound enters the ear, travels through the auditory canal, and reaches the eardrum. The auditory canal is approximately a tube open at only one end. The other end is closed by the eardrum. A typical length for the auditory canal in an adult is about 2.9 cm. The speed of sound is 343 m/s. What is the fundamental frequency of the canal? (Interestingly, the fundamental frequency is in the frequency range where human hearing is most sensitive.)

43. An organ pipe is open at both ends. It is producing sound at its third harmonic, the frequency of which is 262 Hz. The speed of sound is 343 m/s. What is the length of the pipe?

44. GO One method for measuring the speed of sound uses standing waves. A cylindrical tube is open at both ends, and one end admits sound from a tuning fork. A movable plunger is inserted into the other end at a distance L from the end of the tube where the tuning fork is. For a fixed frequency, the plunger is moved until the smallest value of L is measured that allows a standing wave to be formed. Suppose that the tuning fork produces a 485-Hz tone, and that the smallest value observed for L is 0.264 m. What is the speed of sound in the gas in the tube?

45. Refer to **Interactive Solution 17.45** at **www.wiley.com/college/ cutnell** to review a method by which this problem can be solved. The fundamental frequencies of two air columns are the same. Column A is open at both ends, while column B is open at only one end. The length of column A is 0.70 m. What is the length of column B?

46.  Divers working in underwater chambers at great depths must deal with the danger of nitrogen narcosis (the "bends"), in which nitrogen dissolves into the blood at toxic levels. One way to avoid this danger is for divers to breathe a mixture containing only helium and oxygen. Helium, however, has the effect of giving the voice a high-pitched quality, like that of Donald Duck's voice. To see why this occurs, assume for simplicity that the voice is generated by the vocal cords vibrating above a gas-filled cylindrical tube that is open only at one end. The quality of the voice depends on the harmonic frequencies generated by the tube; larger frequencies lead to higher-pitched voices. Consider two such tubes at 20 °C. One is filled with air, in which the speed of sound is 343 m/s. The other is filled with helium, in which the speed of sound is 1.00×10^3 m/s. To see the effect of helium on voice quality, calculate the ratio of the nth natural frequency of the helium-filled tube to the nth natural frequency of the air-filled tube.

47. ssm The fundamental frequency of a vibrating system is 400 Hz. For each of the following systems, give the three lowest frequencies (excluding the fundamental) at which standing waves can occur: **(a)** a string fixed at both ends, **(b)** a cylindrical pipe with both ends open, and   **(c)** a cylindrical pipe with only one end open.

*48. A thin 1.2-m aluminum rod sustains a longitudinal standing wave with vibration antinodes at each end of the rod. There are no other antinodes. The density and Young's modulus of aluminum are, respectively, 2700 kg/m³ and 6.9×10^{10} N/m². What is the frequency of the rod's vibration?

*49. Review Multiple-Concept Example 7 for background that is relevant to the kind of approach needed to solve this problem. Two ideal gases have the same temperature and the same value for γ (the ratio of the specific heat capacities at constant pressure and constant volume). A molecule of gas A has a mass of 7.31×10^{-26} kg, and a molecule of gas B has a mass of 1.06×10^{-25} kg. When gas A (speed of sound = 259 m/s) fills a tube that is open at both ends, the first overtone frequency of the tube is 386 Hz. Gas B fills another tube open at both ends, and this tube also has a first overtone frequency of 386 Hz. What is the length of the tube filled with gas B?

*50. GO ssm www A vertical tube is closed at one end and open to air at the other end. The air pressure is 1.01×10^5 Pa. The tube has a length of 0.75 m. Mercury (mass density = 13 600 kg/m³) is poured into it to shorten the effective length for standing waves. What is the absolute pressure at the bottom of the mercury column, when the fundamental frequency of the shortened, air-filled tube is equal to the third harmonic of the original tube?

*51. A person hums into the top of a well and finds that standing waves are established at frequencies of 42, 70.0, and 98 Hz. The frequency of 42 Hz is not necessarily the fundamental frequency. The speed of sound is 343 m/s. How deep is the well?

**52. A tube, open at only one end, is cut into two shorter (nonequal) lengths. The piece that is open at both ends has a fundamental frequency of 425 Hz, while the piece open only at one end has a fundamental frequency of 675 Hz. What is the fundamental frequency of the original tube?

ADDITIONAL PROBLEMS

53. The fundamental frequency of a string fixed at both ends is 256 Hz. How long does it take for a wave to travel the length of this string?

54. The range of human hearing is roughly from twenty hertz to twenty kilohertz. Based on these limits and a value of 343 m/s for the speed of sound, what are the lengths of the longest and shortest pipes (open at both ends and producing sound at their fundamental frequencies) that you expect to find in a pipe organ?

55. ssm Review Example 1 in the text. Speaker A is moved further to the left, while ABC remains a right triangle. What is the separation between the speakers when constructive interference occurs again at point C?

56. A pipe open only at one end has a fundamental frequency of 256 Hz. A second pipe, initially identical to the first pipe, is shortened by cutting off a portion of the open end. Now, when both pipes vibrate at their fundamental frequencies, a beat frequency of 12 Hz is heard. How many centimeters were cut off the end of the second pipe? The speed of sound is 343 m/s.

57. A tube is open only at one end. A certain harmonic produced by the tube has a frequency of 450 Hz. The next higher harmonic has a frequency of 750 Hz. The speed of sound in air is 343 m/s.   **(a)** What is the integer n that describes the harmonic whose frequency is 450 Hz? **(b)** What is the length of the tube?

58. The drawing graphs a string on which two rectangular pulses are traveling at a constant speed of 1 cm/s at time $t = 0$ s. Using the principle of linear superposition, draw the shape of the string's pulses at $t = 1$ s, 2 s, 3 s, and 4 s.

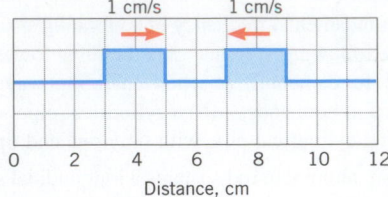

Distance, cm

59. ssm Suppose that the strings on a violin are stretched with the same tension and each has the same length between its two fixed ends. The musical notes and corresponding fundamental frequencies of two of these strings are G (196.0 Hz) and E (659.3 Hz). The linear density of the E string is 3.47×10^{-4} kg/m. What is the linear density of the G string?

* **60.** The drawing shows two strings that have the same length and linear density. The left end of each string is attached to a wall, while the right end passes over a pulley and is connected to objects of different weights (W_A and W_B). Different standing waves are set up on each string, but their frequencies are the same. If $W_A = 44$ N, what is W_B?

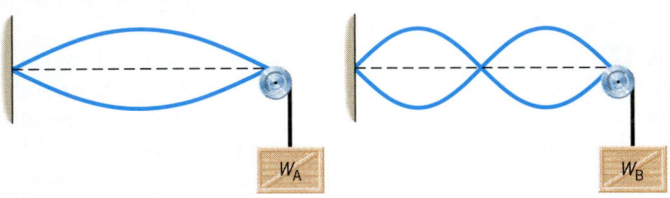

* **61.** Refer to **Interactive Solution 17.61** at **www.wiley.com/college/cutnell** to review a method by which this problem can be solved. Two loudspeakers on a concert stage are vibrating in phase. A listener is 50.5 m from the left speaker and 26.0 m from the right one. The listener can respond to all frequencies from 20 to 20 000 Hz, and the speed of sound is 343 m/s. What are the two lowest frequencies that can be heard loudly due to constructive interference?

** **62.** Two loudspeakers are mounted on a merry-go-round whose radius is 9.01 m. When stationary, the speakers both play a tone whose frequency is 100.0 Hz. As the drawing illustrates, they are situated at opposite ends of a diameter. The speed of sound is 343.00 m/s, and the merry-go-round revolves once every 20.0 s. What is the beat frequency that is detected by the listener when the merry-go-round is near the position shown?

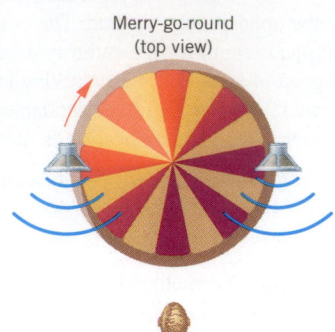

Merry-go-round (top view)

Listener

** **63. ssm www** The note that is three octaves above middle C is supposed to have a fundamental frequency of 2093 Hz. On a certain piano the steel wire that produces this note has a cross-sectional area of 7.85×10^{-7} m². The wire is stretched between two pegs. When the piano is tuned properly to produce the correct frequency at 25.0 °C, the wire is under a tension of 818.0 N. Suppose the temperature drops to 20.0 °C. In addition, as an approximation, assume that the wire is kept from contracting as the temperature drops. Consequently, the tension in the wire changes. What beat frequency is produced when this piano and another instrument (properly tuned) sound the note simultaneously?

CHAPTER 18

ELECTRIC FORCES AND ELECTRIC FIELDS

Electric charges are one of the fundamental building blocks of atoms. When these charges are removed from atoms, spectacular results can occur, such as the lightning in this photograph. Lightning is the flow of electric charge through the atmosphere. Electric charge and the force associated with it are topics discussed in this chapter. (© SUPERSTOCK)

18.1 THE ORIGIN OF ELECTRICITY

The electrical nature of matter is inherent in atomic structure. An atom consists of a small, relatively massive nucleus that contains particles called protons and neutrons. A proton has a mass of 1.673×10^{-27} kg, and a neutron has a slightly greater mass of 1.675×10^{-27} kg. Surrounding the nucleus is a diffuse cloud of orbiting particles called electrons, as Figure 18.1 suggests. An electron has a mass of 9.11×10^{-31} kg. Like mass, *electric charge* is an intrinsic property of protons and electrons, and only two types of charge have been discovered, positive and negative. A proton has a positive charge, and an electron has a negative charge. A neutron has no net electric charge.

Experiment reveals that the magnitude of the charge on the proton *exactly equals* the magnitude of the charge on the electron; the proton carries a charge $+e$, and the electron carries a charge $-e$. The SI unit for measuring the magnitude of an electric charge is the **coulomb*** (**C**), and e has been determined experimentally to have the value

$$e = 1.60 \times 10^{-19} \text{ C}$$

The symbol e represents only the magnitude of the charge on a proton or an electron and does not include the algebraic sign that indicates whether the charge is positive or negative. In nature, atoms are normally found with equal numbers of protons and electrons. Usually, then, an atom carries no net charge because the algebraic sum of the positive

*The definition of the coulomb depends on electric currents and magnetic fields, concepts that will be discussed later. Therefore, we postpone its definition until Section 21.7.

- electron
- proton
- neutron

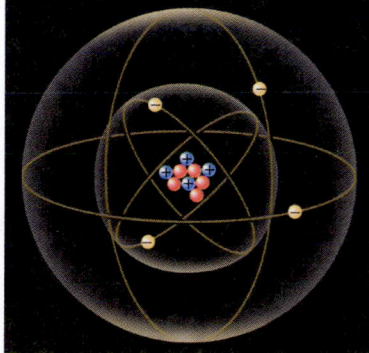

Figure 18.1 An atom contains a small, positively charged nucleus, about which the negatively charged electrons move. The closed-loop paths shown here are symbolic only. In reality, the electrons do not follow discrete paths, as Section 30.5 discusses.

charge of the nucleus and the negative charge of the electrons is zero. When an atom, or any object, carries no net charge, the object is said to be *electrically neutral.* The neutrons in the nucleus are electrically neutral particles.

Charges of larger magnitude than the charge on an electron or on a proton are built up on an object by adding or removing electrons. Thus, any charge of magnitude q is an integer multiple of e; that is, $q = Ne$, where N is an integer. Because any electric charge q occurs in integer multiples of elementary, indivisible charges of magnitude e, electric charge is said to be *quantized.* Example 1 emphasizes the quantized nature of electric charge.

Example 1 A Lot of Electrons

How many electrons are there in one coulomb of negative charge?

Reasoning The negative charge is due to the presence of excess electrons, since they carry negative charge. Because each electron has a charge whose magnitude is $e = 1.60 \times 10^{-19}$ C, the number of electrons is equal to the charge magnitude of one coulomb (1.00 C) divided by e.

Solution The number N of electrons is

$$N = \frac{1.00 \text{ C}}{e} = \frac{1.00 \text{ C}}{1.60 \times 10^{-19} \text{ C}} = \boxed{6.25 \times 10^{18}}$$

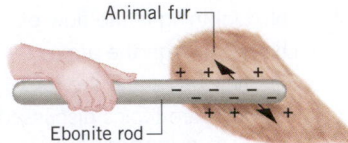

Figure 18.2 When an ebonite rod is rubbed against animal fur, electrons from atoms of the fur are transferred to the rod. This transfer gives the rod a negative charge ($-$) and leaves a positive charge ($+$) on the fur.

18.2 CHARGED OBJECTS AND THE ELECTRIC FORCE

Electricity has many useful applications that have come about because it is possible to transfer electric charge from one object to another. Usually electrons are transferred, and the body that gains electrons acquires an excess of negative charge. The body that loses electrons has an excess of positive charge. Such separation of charge occurs often when two unlike materials are rubbed together. For example, when an ebonite (hard, black rubber) rod is rubbed against animal fur, some of the electrons from atoms of the fur are transferred to the rod. The ebonite becomes negatively charged, and the fur becomes positively charged, as Figure 18.2 indicates. Similarly, if a glass rod is rubbed with a silk cloth, some of the electrons are removed from the atoms of the glass and deposited on the silk, leaving the silk negatively charged and the glass positively charged. There are many familiar examples of charge separation, as when you walk across a nylon rug or run a comb through dry hair. In each case, objects become "electrified" as surfaces rub against one another.

When an ebonite rod is rubbed with animal fur, the rubbing process serves only to separate electrons and protons already present in the materials. No electrons or protons are created or destroyed. Whenever an electron is transferred to the rod, a proton is left behind on the fur. Since the charges on the electron and proton have identical magnitudes but opposite signs, the algebraic sum of the two charges is zero, and the transfer does not change the net charge of the fur/rod system. If each material contains an equal number of protons and electrons to begin with, the net charge of the system is zero initially and remains zero at all times during the rubbing process.

Electric charges play a role in many situations other than rubbing two surfaces together. They are involved, for instance, in chemical reactions, electric circuits, and radioactive decay. A great number of experiments have verified that in any situation, the *law of conservation of electric charge* is obeyed.

LAW OF CONSERVATION OF ELECTRIC CHARGE

During any process, the net electric charge of an isolated system remains constant (is conserved).

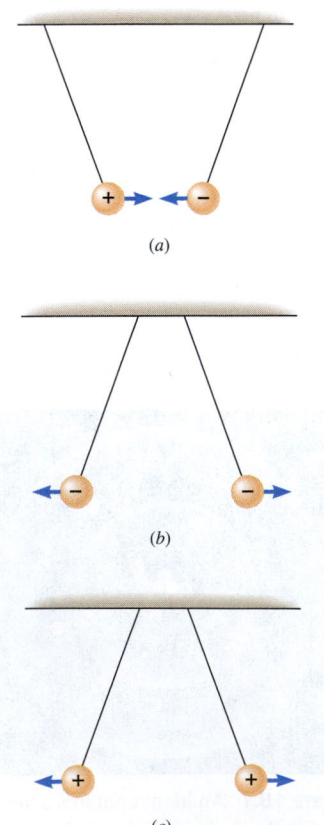

Figure 18.3 (*a*) A positive charge ($+$) and a negative charge ($-$) attract each other. (*b*) Two negative charges repel each other. (*c*) Two positive charges repel each other.

It is easy to demonstrate that two electrically charged objects exert a force on one another. Consider Figure 18.3*a*, which shows two small balls that have been *oppositely charged* and are light and free to move. The balls attract each other. On the other hand, balls with the *same* type of charge, either both positive or both negative, repel each other, as parts *b* and *c* of the drawing indicate. The behavior depicted in Figure 18.3 illustrates the following fundamental characteristic of electric charges:

> ### Like charges repel and unlike charges attract each other.

Like other forces that we have encountered, the ***electric force*** (also called the ***electrostatic force***) can alter the motion of an object. It can do so by contributing to the net external force $\Sigma\vec{\mathbf{F}}$ that acts on the object. Newton's second law, $\Sigma\vec{\mathbf{F}} = m\vec{\mathbf{a}}$, specifies the acceleration $\vec{\mathbf{a}}$ that arises because of the net external force. Any external electric force that acts on an object must be included when determining the net external force to be used in the second law.

A new technology based on the electric force may revolutionize the way books and other printed matter are made. This technology, called electronic ink, allows letters and graphics on a page to be changed instantly, much like the symbols displayed on a computer monitor. Figure 18.4*a* illustrates the essential features of electronic ink. It consists of millions of clear microcapsules, each having the diameter of a human hair and filled with a dark, inky liquid. Inside each microcapsule are several dozen extremely tiny white beads that carry a slightly negative charge. The microcapsules are sandwiched between two sheets, an opaque base layer and a transparent top layer, at which the reader looks. When a positive charge is applied to a small region of the base layer, as shown in part *b* of the drawing, the negatively charged white beads are drawn to it, leaving dark ink at the top layer. Thus, a viewer sees only the dark liquid. When a negative charge is applied to a

The physics of electronic ink.

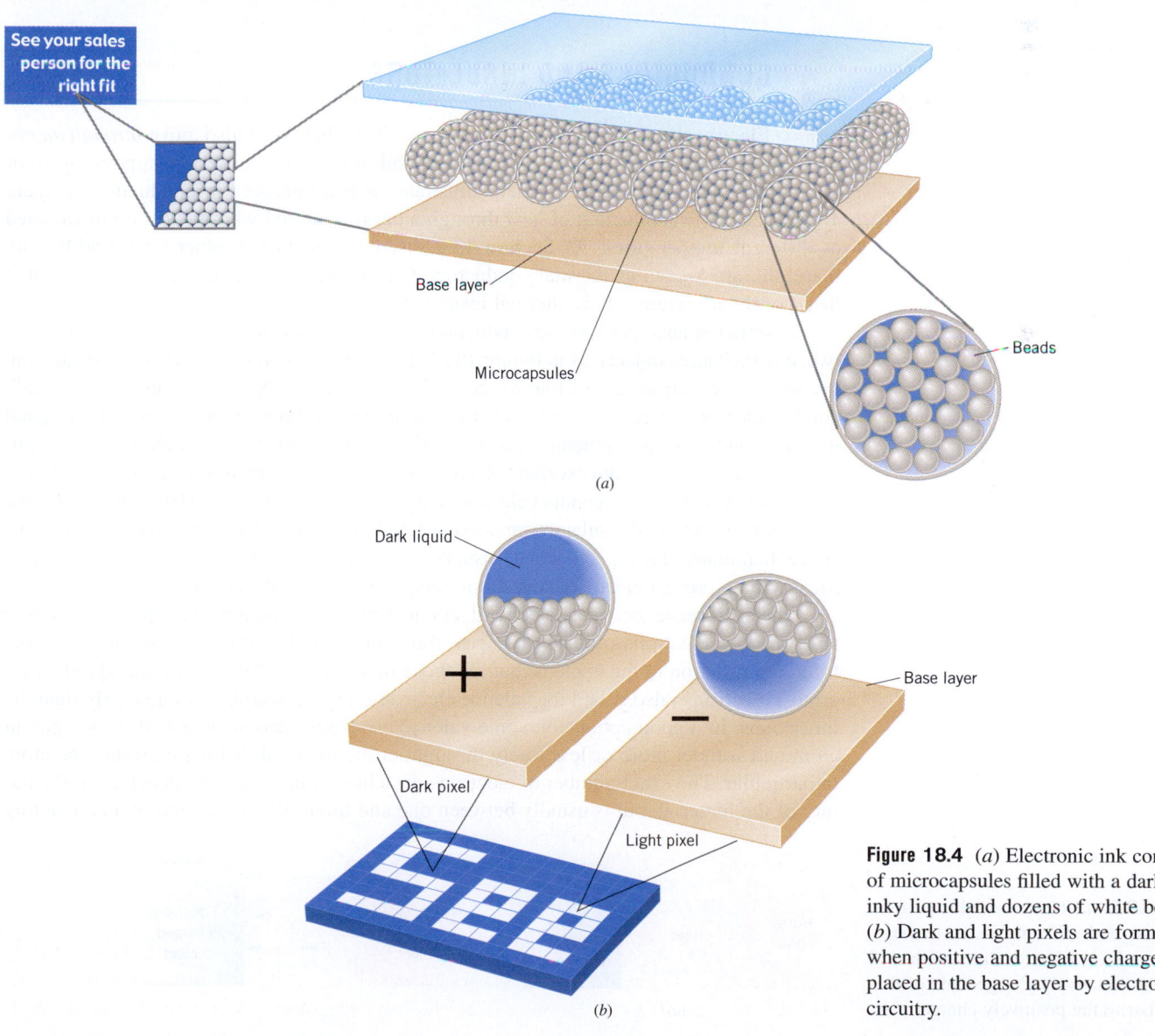

See your sales person for the right fit

Base layer

Microcapsules

Beads

(*a*)

Dark liquid

Base layer

Dark pixel

Light pixel

(*b*)

Figure 18.4 (*a*) Electronic ink consists of microcapsules filled with a dark, inky liquid and dozens of white beads. (*b*) Dark and light pixels are formed when positive and negative charges are placed in the base layer by electronic circuitry.

region of the base layer, the negatively charged white beads are repelled from it and are forced to the top of the microcapsules; now a viewer sees a white area due to the beads. Thus, electronic ink is based on the principle that like charges repel and unlike charges attract each other; a positive charge causes one color to appear, and a negative charge causes another color to appear. Each small region, whether dark or light, is known as a *pixel* (short for "picture element"). Computer chips provide the instructions to produce the negative and positive charges on the base layer of each pixel. Letters and graphics are produced by the patterns generated with the two colors.

✓ CHECK YOUR UNDERSTANDING

(The answers are given at the end of the book.)

1. An electrically neutral object acquires a net electric charge. Which one of the following statements concerning the mass of the object is true? **(a)** The mass does not change. **(b)** The mass increases if the charge is positive and decreases if it is negative. **(c)** The mass increases if the charge is negative and decreases if it is positive.

2. Object A and object B are each electrically neutral. Two million electrons are removed from A and placed on B. Expressed in coulombs, what is the resulting charge (algebraic sign and magnitude) on A and on B?

3. Object A has a charge of -1.6×10^{-13} C, and object B is electrically neutral. Two million electrons are removed from A and placed on B. Expressed in coulombs, what is the resulting charge (algebraic sign and magnitude) on A and on B?

18.3 CONDUCTORS AND INSULATORS

Electric charge can not only exist *on an object,* but it can also move *through an object.* However, materials differ vastly in their abilities to allow electric charge to move or be conducted through them. To help illustrate such differences in conductivity, Figure 18.5*a* recalls the conduction of heat through a bar of material whose ends are maintained at different temperatures. As Section 13.2 discusses, metals conduct heat readily and, therefore, are known as thermal conductors. On the other hand, substances that conduct heat poorly are referred to as thermal insulators.

A situation analogous to the conduction of heat arises when a metal bar is placed between two charged objects, as in Figure 18.5*b*. Electrons are conducted through the bar from the negatively charged object toward the positively charged object. Substances that readily conduct electric charge are called *electrical conductors.* Although there are exceptions, good thermal conductors are generally good electrical conductors. Metals such as copper, aluminum, silver, and gold are excellent electrical conductors and, therefore, are used in electrical wiring. Materials that conduct electric charge *poorly* are known as *electrical insulators.* In many cases, thermal insulators are also electrical insulators. Common electrical insulators are rubber, many plastics, and wood. Insulators, such as the rubber or plastic that coats electrical wiring, prevent electric charge from going where it is not wanted.

The difference between electrical conductors and insulators is related to atomic structure. As electrons orbit the nucleus, those in the outer orbits experience a weaker force of attraction to the nucleus than do those in the inner orbits. Consequently, the outermost electrons (also called the valence electrons) can be dislodged more easily than the inner ones. In a good conductor, some valence electrons become detached from a parent atom and wander more or less freely throughout the material, belonging to no one atom in particular. The exact number of electrons detached from each atom depends on the nature of the material, but is usually between one and three. When one end of a conducting

Figure 18.5 (*a*) Heat is conducted from the hotter end of the metal bar to the cooler end. (*b*) Electrons are conducted from the negatively charged end of the metal bar to the positively charged end.

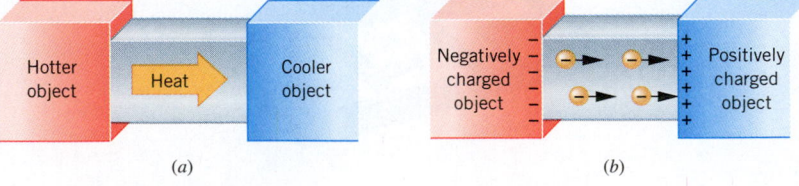

(*a*) (*b*)

bar is placed in contact with a negatively charged object and the other end in contact with a positively charged object, as in Figure 18.5*b*, the "free" electrons are able to move readily away from the negative end and toward the positive end. The ready movement of electrons is the hallmark of a good conductor. In an insulator the situation is different, for there are very few electrons free to move throughout the material. Virtually every electron remains bound to its parent atom. Without the "free" electrons, there is very little flow of charge when the material is placed between two oppositely charged bodies, so the material is an electrical insulator.

18.4 CHARGING BY CONTACT AND BY INDUCTION

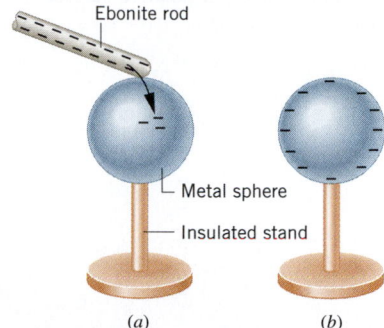

Figure 18.6 (*a*) Electrons are transferred by rubbing the negatively charged rod on the metal sphere. (*b*) When the rod is removed, the electrons distribute themselves over the surface of the sphere.

When a negatively charged ebonite rod is rubbed on a metal object, such as the sphere in Figure 18.6*a*, some of the excess electrons from the rod are transferred to the object. Once the electrons are on the metal sphere (where they can move readily) and the rod is removed, they repel one another and spread out over the sphere's surface. The insulated stand prevents them from flowing to the earth, where they could spread out even more. As shown in part *b* of the picture, the sphere is left with a negative charge distributed over its surface. In a similar manner, the sphere would be left with a positive charge after being rubbed with a positively charged rod. In this case, electrons from the sphere would be transferred to the rod. The process of giving one object a net electric charge by placing it in contact with another object that is already charged is known as ***charging by contact.***

It is also possible to charge a conductor in a way that does not involve contact. In Figure 18.7, a negatively charged rod is brought close to, *but does not touch,* a metal sphere. In the sphere, the free electrons closest to the rod move to the other side, as part *a* of the drawing indicates. As a result, the part of the sphere nearest the rod becomes positively charged and the part farthest away becomes negatively charged. These positively and negatively charged regions have been "induced" or "persuaded" to form because of the repulsive force between the negative rod and the free electrons in the sphere. If the rod were removed, the free electrons would return to their original places, and the charged regions would disappear.

Under most conditions the earth is a good electrical conductor. So when a metal wire is attached between the sphere and the ground, as in Figure 18.7*b*, some of the free electrons leave the sphere and distribute themselves over the much larger earth. If the grounding wire is then removed, followed by the ebonite rod, the sphere is left with a positive net charge, as part *c* of the picture shows. The process of giving one object a net electric charge *without* touching the object to a second charged object is called ***charging by induction.*** The process could also be used to give the sphere a negative net charge, if a positively charged rod were used. Then, electrons would be drawn up from the ground through the grounding wire and onto the sphere.

If the sphere in Figure 18.7 were made from an insulating material like plastic, instead of metal, the method of producing a net charge by induction would not work, because very little charge would flow through the insulating material and down the grounding wire. However, the electric force of the charged rod would have some effect on the insulating material. The electric force would cause the positive and negative charges in the molecules of the material to separate slightly, with the negative charges being "pushed" away from

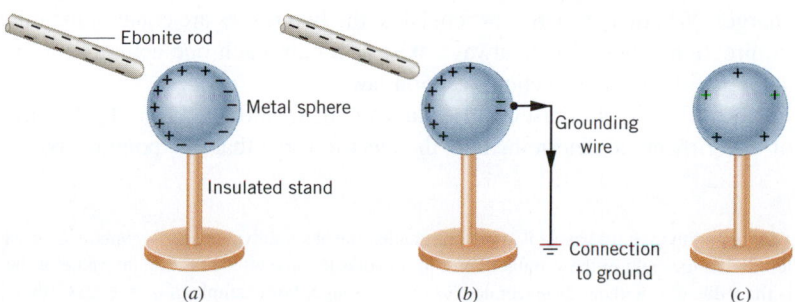

Figure 18.7 (*a*) When a charged rod is brought near the metal sphere without touching it, some of the positive and negative charges in the sphere are separated. (*b*) Some of the electrons leave the sphere through the grounding wire, with the result (*c*) that the sphere acquires a positive net charge.

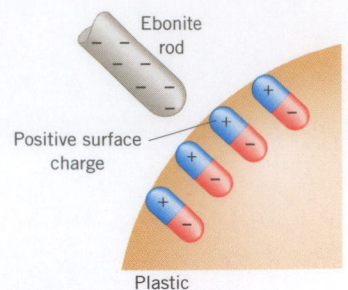

Figure 18.8 The negatively charged rod induces a slight positive surface charge on the plastic.

the negative rod, as Figure 18.8 illustrates. Although no net charge is created, the surface of the plastic does acquire a slight induced positive charge and is attracted to the negative rod. It is attracted in spite of the repulsive force between the negative rod and the negative charges in the plastic. This is because the negative charges in the plastic are further away from the rod than the positive charges are. For a similar reason, one piece of cloth can stick to another in the phenomenon known as "static cling," which occurs when an article of clothing has acquired an electric charge while being tumbled about in a clothes dryer.

✓ **CHECK YOUR UNDERSTANDING**

(*The answers are given at the end of the book.*)

4. Two metal spheres are identical. They are electrically neutral and are touching. An electrically charged ebonite rod is then brought near the spheres without touching them, as the drawing shows. After a while, with the rod held in place, the spheres are 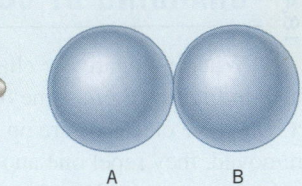 separated, and the rod is then removed. The following statements refer to the masses m_A and m_B of the spheres after they are separated and the rod is removed. Which one or more of the statements is true? **(a)** $m_A = m_B$ **(b)** $m_A > m_B$ if the rod is positive **(c)** $m_A < m_B$ if the rod is positive **(d)** $m_A > m_B$ if the rod is negative **(e)** $m_A < m_B$ if the rod is negative

5. Blow up a balloon, tie it shut, and rub it against your shirt a number of times, so that the balloon acquires a net electric charge. Now touch the balloon to the ceiling. When released, will the balloon remain stuck to the ceiling?

6. A rod made from insulating material carries a net charge (which may be positive or negative), whereas a copper sphere is electrically neutral. The rod is held close to the sphere but does not touch it. Which one of the following statements concerning the forces that the rod and sphere exert on each other is true? **(a)** The forces are always attractive. **(b)** The forces are always repulsive. **(c)** The forces are attractive when the rod is negative and repulsive when it is positive. **(d)** The forces are repulsive when the rod is negative and attractive when it is positive. **(e)** There are no forces.

COULOMB'S LAW

18.5

THE FORCE THAT POINT CHARGES EXERT ON EACH OTHER

The electrostatic force that stationary charged objects exert on each other depends on the amount of charge on the objects and the distance between them. Experiments reveal that the greater the charge and the closer together they are, the greater is the force. To set the stage for explaining these features in more detail, Figure 18.9 shows two charged bodies. These objects are so small, compared to the distance r between them, that they can be regarded as mathematical points. The "point charges" have magnitudes* $|q_1|$ and $|q_2|$. If the charges have *unlike* signs, as in part *a* of the picture, each object is *attracted* to the other by a force that is directed along the line between them; $+\vec{F}$ is the electric force exerted on object 1 by object 2 and $-\vec{F}$ is the electric force exerted on object 2 by object 1. If, as in part *b*, the charges have the *same* sign (both positive or both negative), each object is *repelled* from the other. The repulsive forces, like the attractive forces, act along the line between the charges. Whether attractive or repulsive, the two forces are equal in magnitude but opposite in direction. These forces always exist as a pair, each one acting on a different object, in accord with Newton's action–reaction law.

The French physicist Charles Augustin de Coulomb (1736–1806) carried out a number of experiments to determine how the electric force that one point charge applies to another

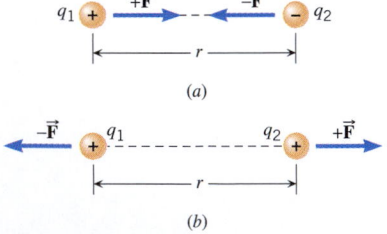

Figure 18.9 Each point charge exerts a force on the other. Regardless of whether the forces are (*a*) attractive or (*b*) repulsive, they are directed along the line between the charges and have equal magnitudes.

---

*The magnitude of a variable is sometimes called the absolute value and is symbolized by a vertical bar to the left and to the right of the variable. Thus, $|q|$ denotes the magnitude or absolute value of the variable q, which is the value of q without its algebraic plus or minus sign. For example, if $q = -2.0$ C, then $|q| = 2.0$ C.

depends on the amount of each charge and the separation between them. His result, now known as *Coulomb's law,* is stated below.

COULOMB'S LAW

The magnitude F of the electrostatic force exerted by one point charge q_1 on another point charge q_2 is directly proportional to the magnitudes $|q_1|$ and $|q_2|$ of the charges and inversely proportional to the square of the distance r between them:

$$F = k \frac{|q_1||q_2|}{r^2} \qquad (18.1)$$

where k is a proportionality constant: $k = 8.99 \times 10^9 \ \text{N} \cdot \text{m}^2/\text{C}^2$ in SI units. The electrostatic force is directed along the line joining the charges, and it is attractive if the charges have unlike signs and repulsive if the charges have like signs.

It is common practice to express k in terms of another constant ϵ_0, by writing $k = 1/(4\pi\epsilon_0)$; ϵ_0 is called the *permittivity of free space* and has a value that is given according to $\epsilon_0 = 1/(4\pi k) = 8.85 \times 10^{-12} \ \text{C}^2/(\text{N} \cdot \text{m}^2)$. Equation 18.1 gives the magnitude of the electrostatic force that each point charge exerts on the other. When using this equation, then, it is important to remember to substitute only the charge magnitudes (without algebraic signs) for $|q_1|$ and $|q_2|$, as Example 2 illustrates.

 Example 2   A Large Attractive Force

Two objects, whose charges are $+1.0$ and $-1.0 \ \text{C}$, are separated by 1.0 km. Compared to 1.0 km, the sizes of the objects are small. Find the magnitude of the attractive force that either charge exerts on the other.

Reasoning Considering that the sizes of the objects are small compared to the separation distance, we can treat the charges as point charges. Coulomb's law may then be used to find the magnitude of the attractive force, provided that only the *magnitudes of the charges* are used for the symbols $|q_1|$ and $|q_2|$ that appear in the law.

Solution The magnitude of the force is

$$F = k\frac{|q_1||q_2|}{r^2} = \frac{(8.99 \times 10^9 \ \text{N} \cdot \text{m}^2/\text{C}^2)(1.0 \ \text{C})(1.0 \ \text{C})}{(1.0 \times 10^3 \ \text{m})^2} = \boxed{9.0 \times 10^3 \ \text{N}} \quad (18.1)$$

The force calculated in Example 2 corresponds to about 2000 pounds and is so large because charges of $\pm 1.0 \ \text{C}$ are enormous. Such large charges are encountered only in the most severe conditions, as in a lightning bolt, where as much as 25 C can be transferred between the cloud and the ground. The typical charges produced in the laboratory are much smaller and are measured conveniently in microcoulombs (1 microcoulomb = $1 \ \mu\text{C} = 10^{-6} \ \text{C}$).

Coulomb's law has a form that is remarkably similar to Newton's law of gravitation ($F = Gm_1m_2/r^2$). The force in both laws depends on the inverse square ($1/r^2$) of the distance between the two objects and is directed along the line between them. In addition, the force is proportional to the product of an intrinsic property of each of the objects, the magnitudes of the charges $|q_1|$ and $|q_2|$ in Coulomb's law and the masses m_1 and m_2 in the gravitation law. However, there is a major difference between the two laws. The electrostatic force can be either repulsive or attractive, depending on whether or not the charges have the same sign; in contrast, the gravitational force is *always* an attractive force.

Section 5.5 discusses how the gravitational attraction between the earth and a satellite provides the centripetal force that keeps a satellite in orbit. Example 3 illustrates that the electrostatic force of attraction plays a similar role in a famous model of the atom created by the Danish physicist Niels Bohr (1885–1962).

ANALYZING MULTIPLE-CONCEPT PROBLEMS

Example 3 A Model of the Hydrogen Atom

In the Bohr model of the hydrogen atom, the electron (charge = $-e$) is in a circular orbit about the nuclear proton (charge = $+e$) at a radius of 5.29×10^{-11} m, as Figure 18.10 shows. The mass of the electron is 9.11×10^{-31} kg. Determine the speed of the electron.

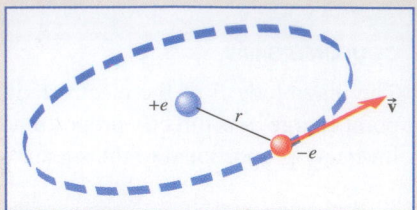

Figure 18.10 In the Bohr model of the hydrogen atom, the electron ($-e$) orbits the proton ($+e$) at a distance that is $r = 5.29 \times 10^{-11}$ m. The velocity of the electron is $\vec{v}$.

Reasoning Recall from Section 5.3 that a net force is required to keep an object such as an electron moving on a circular path. This net force is called the centripetal force and always points toward the center of the circle. The centripetal force has a magnitude given by $F_c = mv^2/r$, where m and v are, respectively, the mass and speed of the electron and r is the radius of the orbit. This equation can be solved for the speed of the electron. Since the mass and orbital radius are known, we can calculate the electron's speed provided that a value for the centripetal force can be found. For the electron in the hydrogen atom, the centripetal force is provided almost exclusively by the electrostatic force that the proton exerts on the electron. This attractive force points toward the center of the circle, and its magnitude is given by Coulomb's law. The electron is also pulled toward the proton by the gravitational force. However, the gravitational force is negligible in comparison to the electrostatic force.

Knowns and Unknowns The data for this problem are:

Description	Symbol	Value
Electron charge	$-e$	-1.60×10^{-19} C
Electron mass	m	9.11×10^{-31} kg
Proton charge	$+e$	$+1.60 \times 10^{-19}$ C
Radius of orbit	r	5.29×10^{-11} m
Unknown Variable		
Orbital speed of electron	v	?

Modeling the Problem

STEP 1 Centripetal Force An electron of mass m that moves with a constant speed v on a circular path of radius r experiences a net force, called the centripetal force. The magnitude F_c of this force is given by $F_c = mv^2/r$ (Equation 5.3). By solving this equation for the speed, we obtain Equation 1 at the right. The mass and radius in this expression are known. However, the magnitude of the centripetal force is not known, so we will evaluate it in Step 2.

$$v = \sqrt{\frac{rF_c}{m}} \qquad (1)$$
$$?$$

STEP 2 Coulomb's Law As the electron orbits the proton in the hydrogen atom, it is attracted to the proton by the electrostatic force. The magnitude F of the electrostatic force is given by Coulomb's law as $F = k|q_1||q_2|/r^2$ (Equation 18.1), where $|q_1|$ and $|q_2|$ are the magnitudes of the charges, r is the orbital radius, and $k = 8.99 \times 10^9$ N·m²/C². Since the centripetal force is provided almost entirely by the electrostatic force, it follows that $F_c = F$. Furthermore, $|q_1| = |-e|$ and $|q_2| = |+e|$. With these substitutions, Equation 18.1 becomes

$$F_c = k\frac{|-e||+e|}{r^2}$$

$$v = \sqrt{\frac{rF_c}{m}} \qquad (1)$$

$$F_c = k\frac{|-e||+e|}{r^2}$$

All the variables on the right side of this expression are known, so we substitute it into Equation 1, as indicated in the right column.

Solution Algebraically combining the results of the modeling steps, we have

STEP 1	STEP 2

$$v = \sqrt{\frac{rF_c}{m}} = \sqrt{\frac{r\left(k\dfrac{|-e||+e|}{r^2}\right)}{m}} = \sqrt{\frac{k|-e||+e|}{mr}}$$

The speed of the orbiting electron is

$$v = \sqrt{\frac{k|-e||+e|}{mr}}$$

$$= \sqrt{\frac{(8.99 \times 10^9 \text{ N} \cdot \text{m}^2/\text{C}^2)|-1.60 \times 10^{-19} \text{ C}||+1.60 \times 10^{-19} \text{ C}|}{(9.11 \times 10^{-31} \text{ kg})(5.29 \times 10^{-11} \text{ m})}} = \boxed{2.19 \times 10^6 \text{ m/s}}$$

Related Homework: *Problems 19, 23*

The physics of adhesion. Since the electrostatic force depends on the inverse square of the distance between the charges, it becomes larger for smaller distances, such as those involved when a strip of adhesive tape is stuck to a smooth surface. Electrons shift over the small distances between the tape and the surface. As a result, the materials become oppositely charged. Since the distance between the charges is relatively small, the electrostatic force of attraction is large enough to contribute to the adhesive bond. Figure 18.11 shows an image of the sticky surface of a piece of tape after it has been pulled off a metal surface. The image was obtained using an atomic-force microscope and reveals the tiny pits left behind when microscopic portions of the adhesive remain stuck to the metal because of the strong adhesive bonding forces.

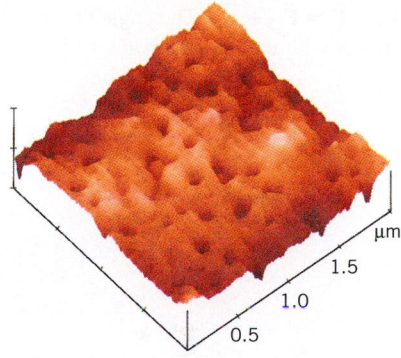

Figure 18.11 After a strip of tape has been pulled off a metal surface, there are tiny pits in the sticky surface of the tape, as this image shows. It was obtained using an atomic-force microscope. (Courtesy Louis Scudiero and J. Thomas Dickinson, Washington State University.)

THE FORCE ON A POINT CHARGE
DUE TO TWO OR MORE OTHER POINT CHARGES

Up to now, we have been discussing the electrostatic force on a point charge (magnitude $|q_1|$) due to another point charge (magnitude $|q_2|$). Suppose that a third point charge (magnitude $|q_3|$) is also present. What would be the net force on q_1 due to both q_2 and q_3? It is convenient to deal with such a problem in parts. First, find the magnitude and direction of the force exerted on q_1 by q_2 (ignoring q_3). Then, determine the force exerted on q_1 by q_3 (ignoring q_2). The *net force* on q_1 is the *vector sum* of these forces. Examples 4 and 5 illustrate this approach when the charges lie along a straight line and on a plane, respectively.

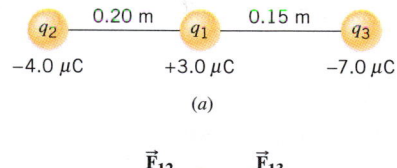

(a)

(b) Free-body diagram for q_1

Figure 18.12 (a) Three charges lying along the x axis. (b) The force exerted on q_1 by q_2 is $\vec{F}_{12}$, while the force exerted on q_1 by q_3 is $\vec{F}_{13}$.

 Example 4 Three Charges on a Line

Figure 18.12a shows three point charges that lie along the x axis in a vacuum. Determine the magnitude and direction of the net electrostatic force on q_1.

Reasoning Part b of the drawing shows a free-body diagram of the forces that act on q_1. Since q_1 and q_2 have opposite signs, they attract one another. Thus, the force exerted on q_1 by q_2 is $\vec{F}_{12}$, and it points to the left. Similarly, the force exerted on q_1 by q_3 is $\vec{F}_{13}$ and is also an attractive force. It points to the right in Figure 18.12b. The magnitudes of these forces can be obtained from Coulomb's law. The net force is the vector sum of $\vec{F}_{12}$ and $\vec{F}_{13}$.

Solution The magnitudes of the forces are

$$F_{12} = k \frac{|q_1||q_2|}{r_{12}^2} = \frac{(8.99 \times 10^9 \text{ N} \cdot \text{m}^2/\text{C}^2)(3.0 \times 10^{-6} \text{ C})(4.0 \times 10^{-6} \text{ C})}{(0.20 \text{ m})^2} = 2.7 \text{ N}$$

$$F_{13} = k \frac{|q_1||q_3|}{r_{13}^2} = \frac{(8.99 \times 10^9 \text{ N} \cdot \text{m}^2/\text{C}^2)(3.0 \times 10^{-6} \text{ C})(7.0 \times 10^{-6} \text{ C})}{(0.15 \text{ m})^2} = 8.4 \text{ N}$$

Since $\vec{F}_{12}$ points in the negative x direction, and $\vec{F}_{13}$ points in the positive x direction, the net force $\vec{F}$ is

$$\vec{F} = \vec{F}_{12} + \vec{F}_{13} = (-2.7 \text{ N}) + (8.4 \text{ N}) = \boxed{+5.7 \text{ N}}$$

The plus sign in the answer indicates that the net force points to the right in the drawing.

Problem-solving insight

When using Coulomb's law ($F = k |q_1||q_2|/r^2$), remember that the symbols $|q_1|$ and $|q_2|$ stand for the charge magnitudes. Do not substitute negative numbers for these symbols.

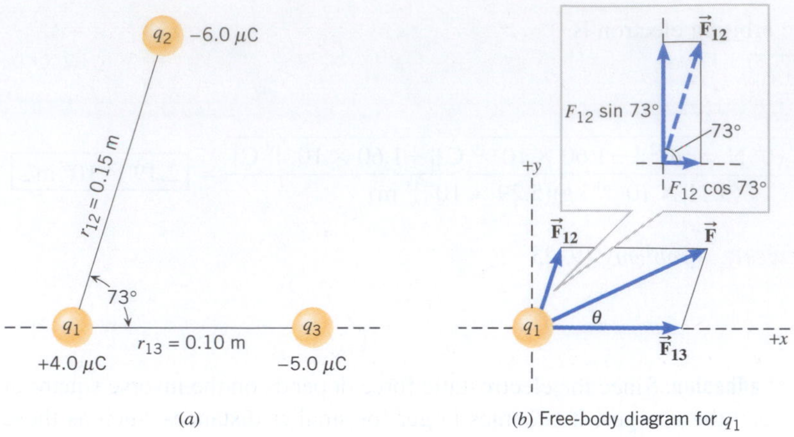

Figure 18.13 (*a*) Three charges lying in a plane. (*b*) The net force acting on q_1 is $\vec{F} = \vec{F}_{12} + \vec{F}_{13}$. The angle that $\vec{F}$ makes with the $+x$ axis is θ.

(*a*)

(*b*) Free-body diagram for q_1

Example 5 Three Charges in a Plane

Figure 18.13*a* shows three point charges that lie in the x, y plane in a vacuum. Find the magnitude and direction of the net electrostatic force on q_1.

Reasoning The force exerted on q_1 by q_2 is $\vec{F}_{12}$ and is an attractive force because the two charges have opposite signs. It points along the line between the charges. The force exerted on q_1 by q_3 is $\vec{F}_{13}$ and is also an attractive force. It points along the line between q_1 and q_3. Coulomb's law specifies the magnitudes of these forces. Since the forces point in different directions (see Figure 18.13*b*), we will use vector components to find the net force.

Solution The magnitudes of the forces are

$$F_{12} = k\frac{|q_1||q_2|}{r_{12}^2} = \frac{(8.99 \times 10^9\ \mathrm{N\cdot m^2/C^2})(4.0 \times 10^{-6}\ \mathrm{C})(6.0 \times 10^{-6}\ \mathrm{C})}{(0.15\ \mathrm{m})^2} = 9.6\ \mathrm{N}$$

$$F_{13} = k\frac{|q_1||q_3|}{r_{13}^2} = \frac{(8.99 \times 10^9\ \mathrm{N\cdot m^2/C^2})(4.0 \times 10^{-6}\ \mathrm{C})(5.0 \times 10^{-6}\ \mathrm{C})}{(0.10\ \mathrm{m})^2} = 18\ \mathrm{N}$$

The net force $\vec{F}$ is the vector sum of $\vec{F}_{12}$ and $\vec{F}_{13}$, as part *b* of the drawing shows. The components of $\vec{F}$ that lie in the x and y directions are $\vec{F}_x$ and $\vec{F}_y$, respectively. Our approach to finding $\vec{F}$ is the same as that used in Chapters 1 and 4. The forces $\vec{F}_{12}$ and $\vec{F}_{13}$ are resolved into x and y components. Then, the x components are combined to give $\vec{F}_x$, and the y components are combined to give $\vec{F}_y$. Once $\vec{F}_x$ and $\vec{F}_y$ are known, the magnitude and direction of $\vec{F}$ can be determined using trigonometry.

Problem-solving insight

The electrostatic force is a vector and has a direction as well as a magnitude. When adding electrostatic forces, take into account the directions of all forces, using vector components as needed.

Force	x component	y component
$\vec{F}_{12}$	$+(9.6\ \mathrm{N})\cos 73° = +2.8\ \mathrm{N}$	$+(9.6\ \mathrm{N})\sin 73° = +9.2\ \mathrm{N}$
$\vec{F}_{13}$	$+18\ \mathrm{N}$	$0\ \mathrm{N}$
$\vec{F}$	$\vec{F}_x = +21\ \mathrm{N}$	$\vec{F}_y = +9.2\ \mathrm{N}$

The magnitude F and the angle θ of the net force are

$$F = \sqrt{F_x^2 + F_y^2} = \sqrt{(21\ \mathrm{N})^2 + (9.2\ \mathrm{N})^2} = \boxed{23\ \mathrm{N}}$$

$$\theta = \tan^{-1}\left(\frac{F_y}{F_x}\right) = \tan^{-1}\left(\frac{9.2\ \mathrm{N}}{21\ \mathrm{N}}\right) = \boxed{24°}$$

✓ CHECK YOUR UNDERSTANDING

(*The answers are given at the end of the book.*)

7. Identical point charges are fixed to diagonally opposite corners of a square. Where does a third point charge experience the greater force?   (a) At the center of the square   (b) At one of the empty corners   (c) The question is unanswerable because the polarities of the charges are not given.

8. The drawing shows three point charges arranged in three different ways. The charges are $+q$, $-q$, and $-q$; each has the same magnitude, with one positive and the other two negative. In each of the arrangements the distance d is the same. Rank the arrangements in descending order (largest first) according to the magnitude of the net electrostatic force that acts on the positive charge.

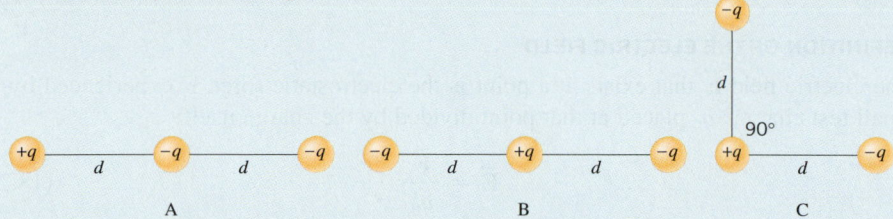

9. A proton and an electron are held in place on the x axis. The proton is at $x = -d$, while the electron is at $x = +d$. They are released simultaneously, and the only force that affects their motions significantly is the electrostatic force of attraction that each applies to the other. Which particle reaches the origin first?

10. A particle is attached to one end of a horizontal spring, and the other end of the spring is attached to a wall. When the particle is pushed so that the spring is compressed more and more, the particle experiences a greater and greater force from the spring. Similarly, a charged particle experiences a greater and greater force when pushed closer and closer to another particle that is fixed in position and has a charge of the same polarity. Considering this similarity, will the charged particle exhibit simple harmonic motion on being released, as will the particle on the spring?

THE ELECTRIC FIELD

DEFINITION

As we know, a charge can experience an electrostatic force due to the presence of other charges. For instance, the positive charge q_0 in Figure 18.14 experiences a force $\vec{F}$, which is the vector sum of the forces exerted by the charges on the rod and the two spheres. It is useful to think of q_0 as a **test charge** for determining the extent to which the surrounding charges generate a force. However, in using a test charge, we must be careful to select one with a very small magnitude, so that it does not alter the locations of the other charges. The next example illustrates how the concept of a test charge is applied.

Example 6 A Test Charge

The positive test charge shown in Figure 18.14 is $q_0 = +3.0 \times 10^{-8}$ C and experiences a force $F = 6.0 \times 10^{-8}$ N in the direction shown in the drawing. **(a)** Find the *force per coulomb* that the test charge experiences. **(b)** Using the result of part (a), predict the force that a charge of $+12 \times 10^{-8}$ C would experience if it replaced q_0.

Reasoning The charges in the environment apply a force $\vec{F}$ to the test charge q_0. The force per coulomb experienced by the test charge is $\vec{F}/q_0$. If q_0 is replaced by a new charge q, then the force on this new charge is the force per coulomb times q.

Solution (a) The force per coulomb of charge is

$$\frac{\vec{F}}{q_0} = \frac{6.0 \times 10^{-8}\,\text{N}}{3.0 \times 10^{-8}\,\text{C}} = \boxed{2.0\,\text{N/C}}$$

The direction of the force per coulomb is the same as the direction of $\vec{F}$ in Figure 18.14.

(b) The result from part (a) indicates that the surrounding charges can exert 2.0 newtons of force per coulomb of charge. Thus, a charge of $+12 \times 10^{-8}$ C would experience a force whose magnitude is

$$F = (2.0\,\text{N/C})(12 \times 10^{-8}\,\text{C}) = \boxed{24 \times 10^{-8}\,\text{N}}$$

The direction of this force would be the same as the direction of the force experienced by the test charge, since both have the same positive sign.

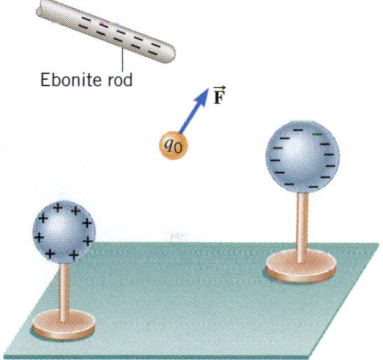

Figure 18.14 A positive charge q_0 experiences an electrostatic force $\vec{F}$ due to the surrounding charges on the ebonite rod and the two spheres.

The electric force per coulomb, $\vec{F}/q_0$, calculated in Example 6(a) is one illustration of an idea that is very important in the study of electricity. The idea is called the *electric field.* Equation 18.2 presents the definition of the electric field.

DEFINITION OF THE ELECTRIC FIELD

The electric field $\vec{E}$ that exists at a point is the electrostatic force $\vec{F}$ experienced by a small test charge* q_0 placed at that point divided by the charge itself:

$$\vec{E} = \frac{\vec{F}}{q_0} \qquad (18.2)$$

The electric field is a vector, and its direction is the same as the direction of the force $\vec{F}$ on a positive test charge.

SI Unit of Electric Field: newton per coulomb (N/C)

Equation 18.2 indicates that the unit for the electric field is that of force divided by charge, which is a newton/coulomb (N/C) in SI units.

   It is the surrounding charges that create an electric field at a given point. Any positive or negative charge placed at the point interacts with the field and, as a result, experiences a force, as the next example indicates.

 Example 7   An Electric Field Leads to a Force

In Figure 18.15 the charges on the two metal spheres and the ebonite rod create an electric field $\vec{E}$ at the spot indicated. This field has a magnitude of 2.0 N/C and is directed as in the drawing. Determine the force on a charge placed at that spot, if the charge has a value of **(a)** $q_0 = +18 \times 10^{-8}$ C and **(b)** $q_0 = -24 \times 10^{-8}$ C.

Reasoning The electric field at a given spot can exert a variety of forces, depending on the magnitude and sign of the charge placed there. The charge is assumed to be small enough that it does not alter the locations of the surrounding charges that create the field.

Solution **(a)** The magnitude of the force is the product of the magnitudes of q_0 and $\vec{E}$:

$$F = |q_0|\, E = (18 \times 10^{-8}\ \text{C})(2.0\ \text{N/C}) = \boxed{36 \times 10^{-8}\ \text{N}} \qquad (18.2)$$

Since q_0 is positive, the force points in the same direction as the electric field, as part *a* of the drawing indicates.

(b) In this case, the magnitude of the force is

$$F = |q_0|\, E = (24 \times 10^{-8}\ \text{C})(2.0\ \text{N/C}) = \boxed{48 \times 10^{-8}\ \text{N}} \qquad (18.2)$$

The force on the negative charge points in the direction *opposite* to the force on the positive charge—that is, opposite to the electric field (see part *b* of the drawing).

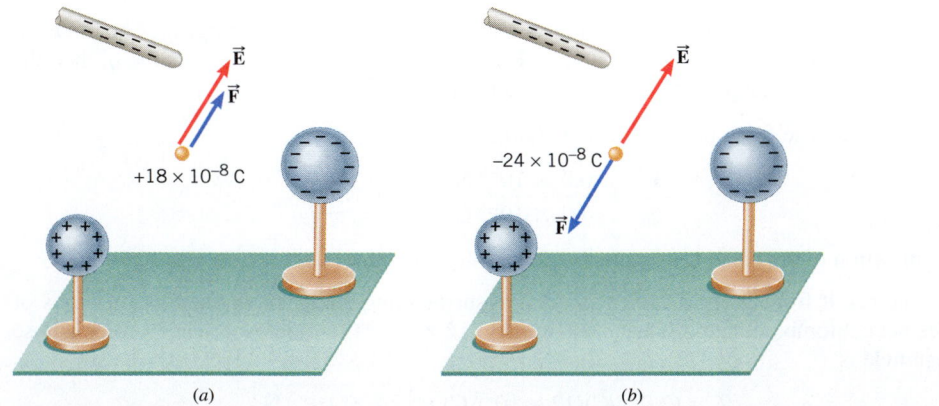

Figure 18.15 The electric field $\vec{E}$ that exists at a given spot can exert a variety of forces. The force exerted depends on the magnitude and sign of the charge placed at that spot. (*a*) The force on a positive charge points in the same direction as $\vec{E}$, while (*b*) the force on a negative charge points opposite to $\vec{E}$.

*As long as the test charge is small enough that it does not disturb the surrounding charges, it may be either positive or negative. Compared to a positive test charge, a negative test charge of the same magnitude experiences a force of the same magnitude that points in the opposite direction. However, the same electric field is given by Equation 18.2, in which $\vec{F}$ is replaced by $-\vec{F}$ and q_0 is replaced by $-q_0$.

At a particular point in space, each of the surrounding charges contributes to the net electric field that exists there. To determine the net field, it is necessary to obtain the various contributions separately and then find the vector sum of them all. Such an approach is an illustration of the principle of linear superposition, as applied to electric fields. (This principle is introduced in Section 17.1, in connection with waves.) Example 8 emphasizes the vector nature of the electric field, and Example 9 illustrates that a charged particle accelerates in an electric field.

 Example 8 Electric Fields Add as Vectors Do

Figure 18.16 shows two charged objects, A and B. Each contributes as follows to the net electric field at point P: $\vec{E}_A = 3.00$ N/C directed to the right, and $\vec{E}_B = 2.00$ N/C directed downward. Thus, $\vec{E}_A$ and $\vec{E}_B$ are perpendicular. What is the net electric field at P?

Reasoning The net electric field $\vec{E}$ is the vector sum of $\vec{E}_A$ and $\vec{E}_B$: $\vec{E} = \vec{E}_A + \vec{E}_B$. As illustrated in Figure 18.16, $\vec{E}_A$ and $\vec{E}_B$ are perpendicular, so $\vec{E}$ is the diagonal of the rectangle shown in the drawing. Thus, we can use the Pythagorean theorem to find the magnitude of $\vec{E}$ and trigonometry to find the directional angle θ.

Solution The magnitude of the net electric field is

$$E = \sqrt{E_A{}^2 + E_B{}^2} = \sqrt{(3.00 \text{ N/C})^2 + (2.00 \text{ N/C})^2} = \boxed{3.61 \text{ N/C}}$$

The direction of $\vec{E}$ is given by the angle θ in the drawing:

$$\theta = \tan^{-1}\left(\frac{E_B}{E_A}\right) = \tan^{-1}\left(\frac{2.00 \text{ N/C}}{3.00 \text{ N/C}}\right) = \boxed{33.7°}$$

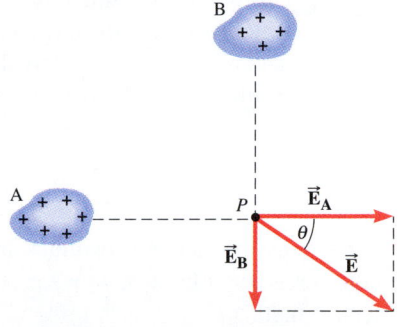

Figure 18.16 The electric field contributions $\vec{E}_A$ and $\vec{E}_B$, which come from the two charge distributions, are added vectorially to obtain the net field $\vec{E}$ at point P.

ANALYZING MULTIPLE-CONCEPT PROBLEMS

Example 9 A Proton Accelerating in an Electric Field

In a vacuum, a proton (charge $= +e$, mass $= 1.67 \times 10^{-27}$ kg) is moving parallel to a uniform electric field that is directed along the $+x$ axis (see Figure 18.17). The proton starts with a velocity of $+2.5 \times 10^4$ m/s and accelerates in the same direction as the electric field, which has a value of $+2.3 \times 10^3$ N/C. Find the velocity of the proton when its displacement is $+2.0$ mm from the starting point.

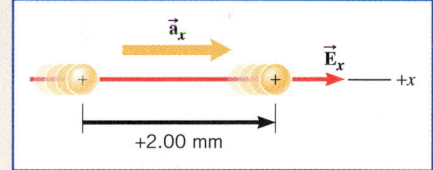

Figure 18.17 A proton, moving to the right, accelerates in the presence of the electric field $\vec{E}_x$.

Reasoning Since we know the initial velocity and displacement of the proton, we can determine its final velocity from an equation of kinematics, provided the proton's acceleration can be found. The acceleration is given by Newton's second law as the net force acting on the proton divided by its mass. The net force is the electrostatic force, since the proton is moving in an electric field. The electrostatic force depends on the proton's charge and the electric field, both of which are known.

Knowns and Unknowns The data for this problem are listed as follows:

Description	Symbol	Value	Comment
Proton charge	$+e$	$+1.60 \times 10^{-19}$ C	
Proton mass	m	1.67×10^{-27} kg	
Initial velocity of proton	v_{0x}	$+2.5 \times 10^4$ m/s	
Electric field	E_x	$+2.3 \times 10^3$ N/C	
Displacement of proton	x	$+2.0$ mm	2.0 mm $= 2.0 \times 10^{-3}$ m
Unknown Variable			
Final velocity of proton	v_x	?	

Continued

Modeling the Problem

STEP 1   Kinematics   To obtain the final velocity v_x of the proton we employ Equation 3.6a from the equations of kinematics: $v_x^2 = v_{0x}^2 + 2a_x x$. We have chosen this equation because two of the variables, the initial velocity v_{0x} and the displacement x, are known. Taking the square root of each side of this relation and choosing the $+$ sign, since the proton is moving in the $+x$ direction (see Figure 18.17), we arrive at Equation 1 in the right column. Although the acceleration a_x is not known, we will obtain an expression for it in Step 2.

$$v_x = +\sqrt{v_{0x}^2 + 2a_x x} \qquad (1)$$

STEP 2   Newton's Second Law   Newton's second law, as given in Equation 4.2a, states that the acceleration a_x of the proton is equal to the net force ΣF_x acting on it divided by the proton's mass m: $a_x = \Sigma F_x/m$. Only the electrostatic force F_x acts on the proton, so it is the net force. Setting $\Sigma F_x = F_x$ in Newton's second law gives

$$a_x = \frac{F_x}{m}$$

This expression can be substituted into Equation 1, as indicated at the right. The electrostatic force is not known, so we proceed to Step 3 to evaluate it using the concept of the electric field.

$$v_x = +\sqrt{v_{0x}^2 + 2a_x x} \qquad (1)$$

$$a_x = \frac{F_x}{m} \qquad (2)$$

STEP 3   The Electric Field   Since the proton is moving in a uniform electric field E_x, it experiences an electrostatic force F_x given by $F_x = q_0 E_x$ (Equation 18.2), where q_0 is the charge. Setting $q_0 = e$ for the proton, we have

$$F_x = eE_x$$

All the variables on the right side of this equation are known, so we substitute it into Equation 2, as shown in the right column.

$$v_x = +\sqrt{v_{0x}^2 + 2a_x x} \qquad (1)$$

$$a_x = \frac{F_x}{m} \qquad (2)$$

$$F_x = eE_x$$

Solution   Algebraically combining the results of the three steps, we have

$$\underset{\text{STEP 1}}{v_x =} + \sqrt{v_{0x}^2 + 2a_x x} = \underset{\text{STEP 2}}{+} \sqrt{v_{0x}^2 + 2\left(\frac{F_x}{m}\right)x} = \underset{\text{STEP 3}}{+}\sqrt{v_{0x}^2 + 2\left(\frac{eE_x}{m}\right)x}$$

The final velocity of the proton is

$$v_x = +\sqrt{v_{0x}^2 + 2\left(\frac{eE_x}{m}\right)x}$$

$$= +\sqrt{(2.5 \times 10^4 \text{ m/s})^2 + 2\left[\frac{(1.60 \times 10^{-19} \text{ C})(2.3 \times 10^3 \text{ N/C})}{1.67 \times 10^{-27} \text{ kg}}\right](2.0 \times 10^{-3} \text{ m})}$$

$$= \boxed{+3.9 \times 10^4 \text{ m/s}}$$

where the $+$ sign denotes that the final velocity points along the $+x$ axis.

Related Homework: *Problems 43, 46, 49*

POINT CHARGES

A more complete understanding of the electric field concept can be gained by considering the field created by a point charge, as in the following example.

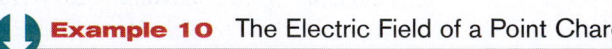

Example 10   The Electric Field of a Point Charge

There is an isolated point charge of $q = +15 \ \mu$C in a vacuum at the left in Figure 18.18a. Using a test charge of $q_0 = +0.80 \ \mu$C, determine the electric field at point P, which is 0.20 m away.

Reasoning Following the definition of the electric field, we place the test charge q_0 at point P, determine the force acting on the test charge, and then divide the force by the test charge.

Solution Coulomb's law (Equation 18.1), gives the magnitude of the force:

$$F = k\frac{|q_0||q|}{r^2} = \frac{(8.99 \times 10^9 \text{ N} \cdot \text{m}^2/\text{C}^2)(0.80 \times 10^{-6} \text{ C})(15 \times 10^{-6} \text{ C})}{(0.20 \text{ m})^2} = 2.7 \text{ N}$$

Equation 18.2 gives the magnitude of the electric field:

$$E = \frac{F}{|q_0|} = \frac{2.7 \text{ N}}{0.80 \times 10^{-6} \text{ C}} = \boxed{3.4 \times 10^6 \text{ N/C}}$$

The electric field $\vec{\mathbf{E}}$ points in the *same direction* as the force $\vec{\mathbf{F}}$ on the positive test charge. Since the test charge experiences a force of repulsion directed to the right, the electric field vector also points to the right, as Figure 18.18*b* shows.

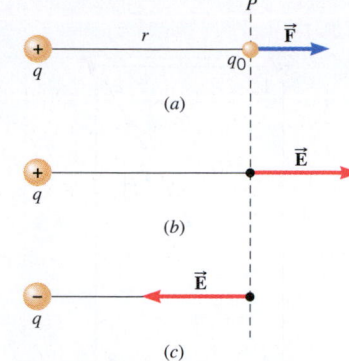

Figure 18.18 (*a*) At location P, a positive test charge q_0 experiences a repulsive force $\vec{\mathbf{F}}$ due to the positive point charge q. (*b*) At P, the electric field $\vec{\mathbf{E}}$ is directed to the right. (*c*) If the charge q were negative rather than positive, the electric field would have the same magnitude as in (*b*) but would point to the left.

The electric field produced by a point charge q can be obtained in general terms from Coulomb's law. First, note that the magnitude of the force exerted by the charge q on a test charge q_0 is $F = k|q||q_0|/r^2$. Then, divide this value by $|q_0|$ to obtain the magnitude of the field. Since $|q_0|$ is eliminated algebraically from the result, *the electric field does not depend on the test charge*:

Point charge q
$$E = \frac{k|q|}{r^2} \qquad (18.3)$$

As in Coulomb's law, the symbol $|q|$ denotes the magnitude of q in Equation 18.3, without regard to whether q is positive or negative. If q is positive, then $\vec{\mathbf{E}}$ is directed away from q, as in Figure 18.18*b*. On the other hand, if q is negative, then $\vec{\mathbf{E}}$ is directed toward q, since a negative charge attracts a positive test charge. For instance, Figure 18.18*c* shows the electric field that would exist at P if there were a charge of $-q$ instead of $+q$ at the left of the drawing. Example 11 reemphasizes the fact that all the surrounding charges make a contribution to the electric field that exists at a given place.

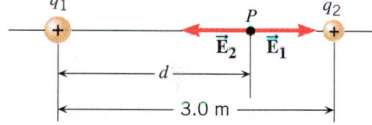

Figure 18.19 The two point charges q_1 and q_2 create electric fields $\vec{\mathbf{E}}_1$ and $\vec{\mathbf{E}}_2$ that cancel at a location P on the line between the charges.

Example 11   The Electric Fields from Separate Charges May Cancel

Two positive point charges, $q_1 = +16 \ \mu\text{C}$ and $q_2 = +4.0 \ \mu\text{C}$, are separated in a vacuum by a distance of 3.0 m, as Figure 18.19 illustrates. Find the spot on the line between the charges where the net electric field is zero.

Reasoning Between the charges the two field contributions have opposite directions, and the net electric field is zero at the place where the magnitude of $\vec{\mathbf{E}}_1$ equals the magnitude of $\vec{\mathbf{E}}_2$. However, since q_2 is smaller than q_1, this location must be *closer* to q_2, in order that the field of the smaller charge can balance the field of the larger charge. In the drawing, the cancellation spot is labeled P, and its distance from q_1 is d.

Solution At P, $E_1 = E_2$, and using the expression $E = k|q|/r^2$ (Equation 18.3), we have

$$\frac{k(16 \times 10^{-6} \text{ C})}{d^2} = \frac{k(4.0 \times 10^{-6} \text{ C})}{(3.0 \text{ m} - d)^2}$$

Rearranging this expression shows that $4.0(3.0 \text{ m} - d)^2 = d^2$. Taking the square root of each side of this equation reveals that

$$2.0(3.0 \text{ m} - d) = \pm d$$

The plus and minus signs on the right occur because either the positive or negative root can be taken. Therefore, there are two possible values for d: $+2.0$ m and $+6.0$ m. The value $+6.0$ m corresponds to a location off to the right of both charges, where the magnitudes of $\vec{\mathbf{E}}_1$ and $\vec{\mathbf{E}}_2$ are equal, but where the directions are the same. Thus, $\vec{\mathbf{E}}_1$ and $\vec{\mathbf{E}}_2$ do not cancel at this spot. The other value for d corresponds to the location shown in the drawing and is the zero-field location: $\boxed{d = +2.0 \text{ m}}$.

Problem-solving insight

Equation 18.3 gives only the magnitude of the electric field produced by a point charge. Therefore, do not use negative numbers for the symbol $|q|$ in this equation.

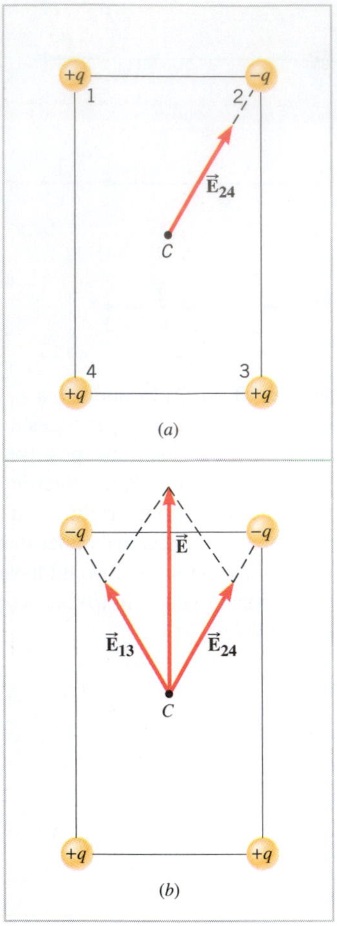

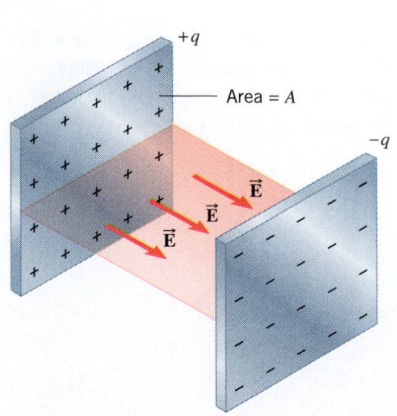

Figure 18.20 Charges of identical magnitude, but different signs, are placed at the corners of a rectangle. The charges give rise to different electric fields at the center *C* of the rectangle, depending on the signs of the charges.

When point charges are arranged in a symmetrical fashion, it is often possible to deduce useful information about the magnitude and direction of the electric field by taking advantage of the symmetry. Conceptual Example 12 illustrates the use of this technique.

Conceptual Example 12   Symmetry and the Electric Field

Four point charges all have the same magnitude, but they have different signs. These charges are fixed to the corners of a rectangle in two different ways, as Figure 18.20 shows. Consider the net electric field at the center *C* of the rectangle in each case. In which case, if either, is the net electric field greater? (**a**) It is greater in Figure 18.20*a*. (**b**) It is greater in Figure 18.20*b*. (**c**) The field has the same magnitude in both cases.

Reasoning The net electric field at *C* is the vector sum of the individual fields created there by the charges at each corner. Each of the individual fields has the same magnitude, since the charges all have the same magnitude and are equidistant from *C*. The directions of the individual fields are different, however. The field created by a positive charge points away from the charge, and the field created by a negative charge points toward the charge.

Answers (a) and (c) are incorrect. To see why these answers are incorrect, note that the charges on corners 2 and 4 are identical in both parts of Figure 18.20. Moreover, in Figure 18.20*a* the charges at corners 1 and 3 are both $+q$, so they contribute individual fields of the same magnitude at *C* that have *opposite* directions and, therefore, cancel. However, in Figure 18.20*b* the charges at corners 1 and 3 are $-q$ and $+q$, respectively. They contribute individual fields of the same magnitude at *C* that have the *same* directions and do not cancel, but combine to produce the field $\vec{E}_{13}$ shown in Figure 18.20*b*. The fact that this contribution to the net field at *C* is present in Figure 18.20*b* but not in Figure 18.20*a* means that the net fields in the two cases are different and that the net field in Figure 18.20*a* is less than (not greater than) the net field in Figure 18.20*b*.

Answer (b) is correct. To assess the net field at *C*, we need to consider the contribution from the charges at corners 2 and 4, which are $-q$ and $+q$, respectively, in both cases. This is just like the arrangement on corners 1 and 3, which was discussed previously. It leads to a contribution to the net field at *C* that is shown as $\vec{E}_{24}$ in both parts of the figure. In Figure 18.20*a* the net field at *C* is just $\vec{E}_{24}$, but in Figure 18.20*b* it is the vector sum of $\vec{E}_{13}$ and $\vec{E}_{24}$, which is clearly greater than either of these two values alone.

Related Homework: *Problem 41*

THE PARALLEL PLATE CAPACITOR

Equation 18.3, which gives the electric field of a point charge, is a very useful result. With the aid of integral calculus, this equation can be applied in a variety of situations where point charges are distributed over one or more surfaces. One such example that has considerable practical importance is the ***parallel plate capacitor.*** As Figure 18.21 shows, this device consists of two parallel metal plates, each with area *A*. A charge $+q$ is spread uniformly over one plate, while a charge $-q$ is spread uniformly over the other plate. In the region between the plates and away from the edges, the electric field points from the positive plate toward the negative plate and is perpendicular to both. It can be shown (see Example 16 in Section 18.9) that the electric field has a magnitude of

Parallel plate capacitor
$$E = \frac{q}{\epsilon_0 A} = \frac{\sigma}{\epsilon_0} \qquad (18.4)$$

where ϵ_0 is the permittivity of free space. In this expression the Greek symbol sigma (σ) denotes the charge per unit area ($\sigma = q/A$) and is sometimes called the ***charge density.*** Except in the region near the edges, the field has the same value at all places between the plates. The field does *not* depend on the distance from the charges, in distinct contrast to the field created by an isolated point charge.

Figure 18.21 A parallel plate capacitor.

✔ CHECK YOUR UNDERSTANDING

(The answers are given at the end of the book.)

11. There is an electric field at point *P*. A very small positive charge is placed at this point and experiences a force. Then the positive charge is replaced by a very small negative charge that has a magnitude different from that of the positive charge. Which one of the following statements is true concerning the forces that these charges experience at *P*? **(a)** They are identical. **(b)** They have the same magnitude but different directions. **(c)** They have different magnitudes but the same direction. **(d)** They have different magnitudes and different directions.

12. Suppose that in Figure 18.20*a* point charges −*q* are fixed in place at corners 1 and 3 and point charges +*q* are fixed in place at corners 2 and 4. What then would be the net electric field at the center *C* of the rectangle?

13. A positive point charge +*q* is fixed in position at the center of a square, as the drawing shows. A second point charge is fixed to corner B, C, or D. The net electric field that results at corner A is zero. **(a)** At which corner is the second charge located? **(b)** Is the second charge positive or negative? **(c)** Does the second charge have a greater, a smaller, or the same magnitude as the charge at the center?

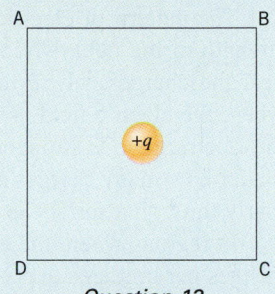

Question 13

14. A positive point charge is located to the left of a negative point charge. When both charges have the same magnitude, there is no place on the line passing through both charges where the net electric field due to the two charges is zero. Suppose, however, that the negative charge has a greater magnitude than the positive charge. On which part of the line, if any, is a place of zero net electric field now located? **(a)** To the left of the positive charge **(b)** Between the two charges **(c)** To the right of the negative charge **(d)** There is no zero place.

15. Three point charges are fixed to the corners of a square, one to a corner, in such a way that the net electric field at the empty corner is zero. Do these charges all have **(a)** the same sign and **(b)** the same magnitude (but, possibly, different signs)?

16. Consider two identical, thin, and nonconducting rods, A and B. On rod A, positive charge is spread evenly, so that there is the same amount of charge per unit length at every point. On rod B, positive charge is spread evenly over only the left half, and the same amount of negative charge is spread evenly over the right half. For each rod deduce the *direction* of the electric field at a point that is located directly above the midpoint of the rod.

ELECTRIC FIELD LINES

As we have seen, electric charges create an electric field in the space around them. It is useful to have a kind of "map" that gives the direction and strength of the field at various places. The great English physicist Michael Faraday (1791–1867) proposed an idea that provides such a "map"—the idea of ***electric field lines.*** Since the electric field is the electric force per unit charge, field lines are also called ***lines of force.***

To introduce the electric field line concept, Figure 18.22*a* shows a positive point charge +*q*. At the locations numbered 1–8, a positive test charge would experience a repulsive force, as the arrows in the drawing indicate. Therefore, the electric field created by the

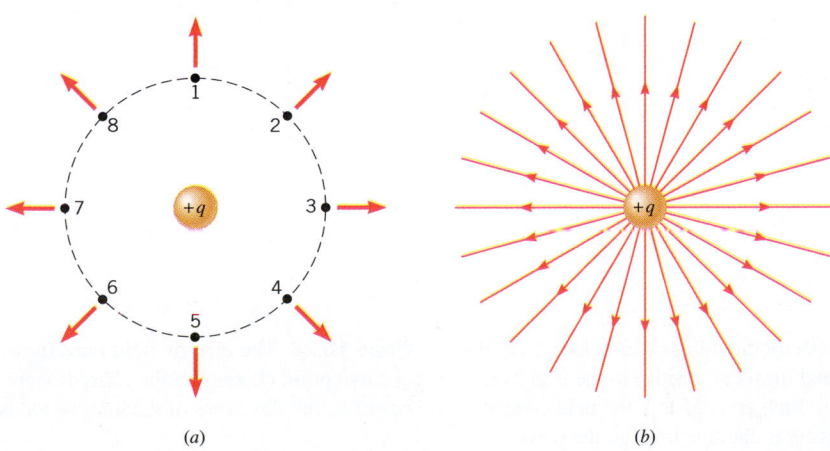

(a)   *(b)*

Figure 18.22 (*a*) At any of the eight marked spots around a positive point charge +*q*, a positive test charge would experience a repulsive force directed radially outward. (*b*) The electric field lines are directed radially outward from a positive point charge +*q*.

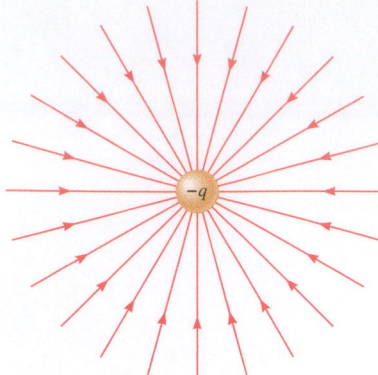

Figure 18.23 The electric field lines are directed radially inward toward a negative point charge $-q$.

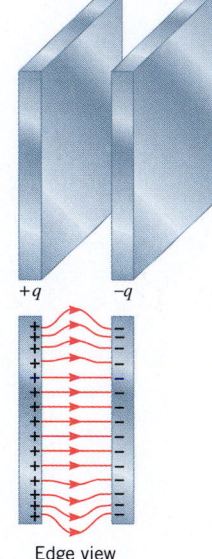

Edge view

Figure 18.24 In the central region of a parallel plate capacitor, the electric field lines are parallel and evenly spaced, indicating that the electric field there has the same magnitude and direction at all points.

charge $+q$ is directed radially outward. The electric field lines are lines drawn to show this direction, as part *b* of the drawing illustrates. They begin on the charge $+q$ and point radially *outward*. Figure 18.23 shows the field lines in the vicinity of a negative charge $-q$. In this case they are directed radially *inward* because the force on a positive test charge is one of attraction, indicating that the electric field points inward. In general, *electric field lines are always directed away from positive charges and toward negative charges.*

The electric field lines in Figures 18.22 and 18.23 are drawn in only two dimensions, as a matter of convenience. Field lines radiate from the charges in three dimensions, and an infinite number of lines could be drawn. However, for clarity only a small number are ever included in pictures. The number is chosen to be proportional to the magnitude of the charge; thus, five times as many lines would emerge from a $+5q$ charge as from a $+q$ charge.

The pattern of electric field lines also provides information about the magnitude or strength of the field. Notice that in Figures 18.22 and 18.23, the lines are closer together near the charges, where the electric field is stronger. At distances far from the charges, where the electric field is weaker, the lines are more spread out. It is true in general that the electric field is stronger in regions where the field lines are closer together. In fact, no matter how many charges are present, the number of lines per unit area passing perpendicularly through a surface is proportional to the magnitude of the electric field.

In regions where the electric field lines are equally spaced, there is the same number of lines per unit area everywhere, and the electric field has the same strength at all points. For example, Figure 18.24 shows that the field lines between the plates of a parallel plate capacitor are parallel and equally spaced, except near the edges where they bulge outward. The equally spaced, parallel lines indicate that the electric field has the same magnitude and direction at all points in the central region of the capacitor.

Often, electric field lines are curved, as in the case of an *electric dipole.* An electric dipole consists of two separated point charges that have the same magnitude but opposite signs. The electric field of a dipole is proportional to the product of the magnitude of one of the charges and the distance between the charges. This product is called the *dipole moment.* Many molecules, such as H_2O and HCl, have dipole moments. Figure 18.25 depicts the field lines in the vicinity of a dipole. For a curved field line, the electric field vector at a point is *tangent* to the line at that point (see points 1, 2, and 3 in the drawing). The pattern of the lines for the dipole indicates that the electric field is greatest in the region between and immediately surrounding the two charges, since the lines are closest together there.

Notice in Figure 18.25 that any given field line starts on the positive charge and ends on the negative charge. **Problem-solving insight:** *In general, electric field lines always begin*

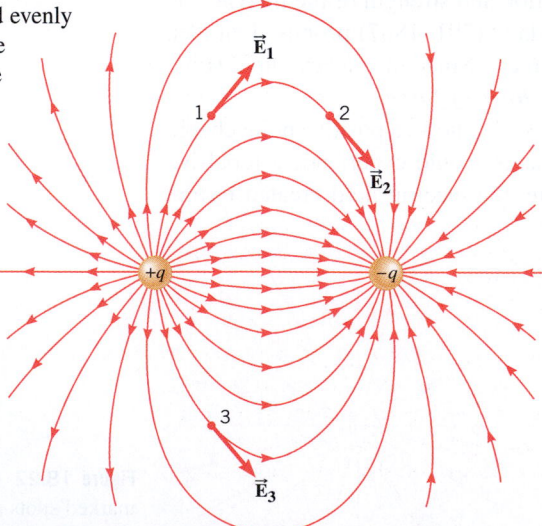

Figure 18.25 The electric field lines of an electric dipole are curved and extend from the positive to the negative charge. At any point, such as 1, 2, or 3, the field created by the dipole is tangent to the line through the point.

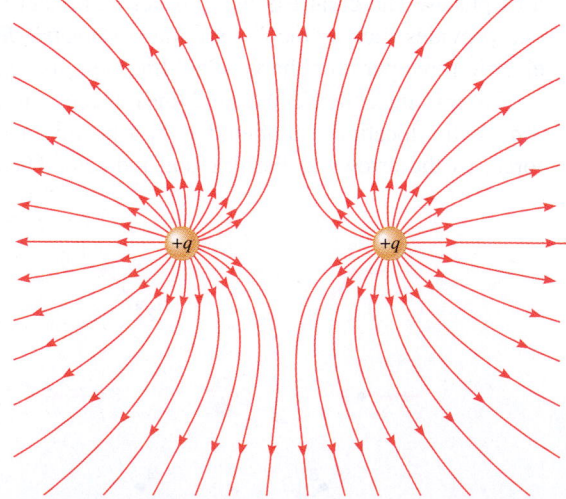

Figure 18.26 The electric field lines for two identical positive point charges. If the charges were both negative, the directions of the lines would be reversed.

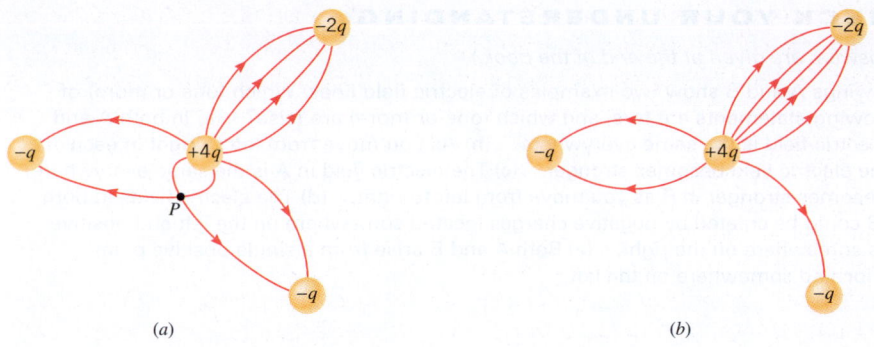

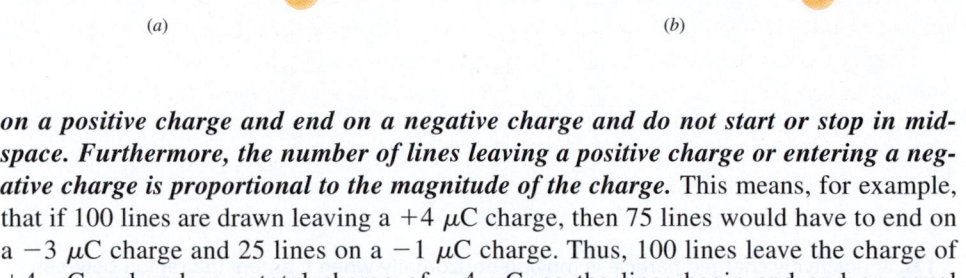

(a)   (b)   (c)

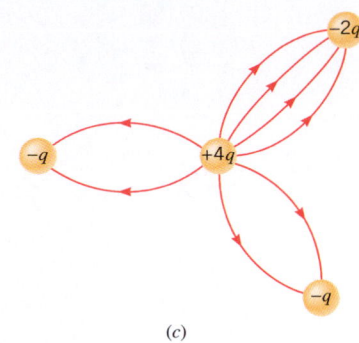

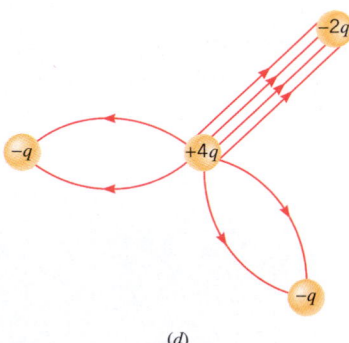

(d)

Figure 18.27 Only one of these drawings shows a representation of the electric field lines between the charges that could be correct. (See Conceptual Example 13.)

on a positive charge and end on a negative charge and do not start or stop in midspace. Furthermore, the number of lines leaving a positive charge or entering a negative charge is proportional to the magnitude of the charge. This means, for example, that if 100 lines are drawn leaving a $+4$ μC charge, then 75 lines would have to end on a -3 μC charge and 25 lines on a -1 μC charge. Thus, 100 lines leave the charge of $+4$ μC and end on a total charge of -4 μC, so the lines begin and end on equal amounts of total charge.

The electric field lines are also curved in the vicinity of two identical charges. Figure 18.26 shows the pattern associated with two positive point charges and reveals that there is an absence of lines in the region between the charges. The absence of lines indicates that the electric field is relatively weak between the charges.

Some of the important properties of electric field lines are considered in Conceptual Example 13.

Conceptual Example 13
Drawing Electric Field Lines

Figure 18.27 shows four possibilities for the electric field lines between three negative point charges ($-q$, $-q$, and $-2q$) and one positive point charge ($+4q$). Which of these is the only one of the four that could possibly show a correct representation of the field lines? **(a)** Figure 18.27a **(b)** Figure 18.27b **(c)** Figure 18.27c **(d)** Figure 18.27d

Reasoning Electric field lines begin on positive charges and end on negative charges. The tangent to a field line at a point gives the direction of the electric field at that point. Equally spaced parallel field lines indicate that the field has a constant value (magnitude and direction) in the corresponding region of space.

Answer (a) is incorrect. Field lines can never cross, as they do at point P in part a of the drawing. If two field lines were to intersect, there would be two electric fields at the point of intersection, one associated with each line. However, there can only be one value of the electric field at a given point.

Answer (b) is incorrect. The number of field lines that leave a positive charge or end on a negative charge is proportional to the magnitude of the charge. Since 8 lines leave the $+4q$ charge, one-half of them (or 4) must end on the $-2q$ charge, and one-fourth of them (or 2) must end on each of the $-q$ charges. Part b of the drawing incorrectly shows 5 lines ending on the $-2q$ charge and 1 line ending on one of the $-q$ charges.

Answer (d) is incorrect. Part d of the drawing is incorrect because the field lines between the $+4q$ charge and the $-2q$ charge are parallel and evenly spaced, which would indicate that the field everywhere in this region has a constant magnitude and direction. However, the field between the $+4q$ charge and the $-2q$ charge certainly is stronger in places close to either of the charges. The field lines should, therefore, have a curved nature, similar (but not identical) to the field lines that surround a dipole.

Answer (c) is correct. Part c of the drawing contains none of the errors discussed previously and, therefore, is the only drawing that could be correct.

Related Homework: *Problems 32, 63*

✔ **CHECK YOUR UNDERSTANDING**

(The answers are given at the end of the book.)

17. Drawings A and B show two examples of electric field lines. Which (one or more) of the following statements are true, and which (one or more) are false?   **(a)** In both A and B the electric field is the same everywhere.   **(b)** As you move from left to right in each case, the electric field becomes stronger.   **(c)** The electric field in A is the same everywhere, but it becomes stronger in B as you move from left to right.   **(d)** The electric fields in both A and B could be created by negative charges located somewhere on the left and positive charges somewhere on the right.   **(e)** Both A and B arise from a single positive point charge located somewhere on the left.

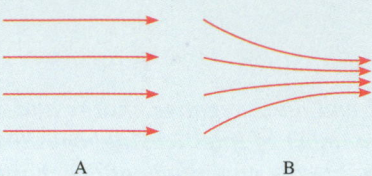

18. A positively charged particle is moving horizontally when it enters the region between the plates of a parallel plate capacitor, as the drawing illustrates. When the particle is within the capacitor, which of the following vectors, if any, are *parallel* to the electric field lines inside the capacitor?   **(a)** The particle's displacement   **(b)** Its velocity   **(c)** Its linear momentum   **(d)** Its acceleration

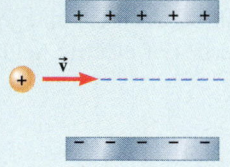

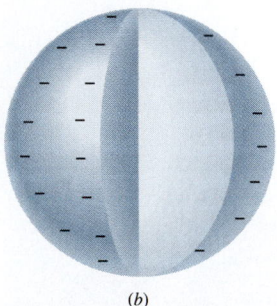

Figure 18.28 (*a*) Excess charge within a conductor (copper) moves quickly (*b*) to the surface.

18.8 THE ELECTRIC FIELD INSIDE A CONDUCTOR: SHIELDING

In conducting materials such as copper, electric charges move readily in response to the forces that electric fields exert. This property of conducting materials has a major effect on the electric field that can exist within and around them. Suppose that a piece of copper carries a number of excess electrons somewhere within it, as in Figure 18.28*a*. Each electron would experience a force of repulsion because of the electric field of its neighbors. And, since copper is a conductor, the excess electrons move readily in response to that force. In fact, as a consequence of the $1/r^2$ dependence on distance in Coulomb's law, they rush to the surface of the copper. Once static equilibrium is established with all of the excess charge on the surface, no further movement of charge occurs, as part *b* of the drawing indicates. Similarly, excess positive charge also moves to the surface of a conductor. In general, ***at equilibrium under electrostatic conditions, any excess charge resides on the surface of a conductor.***

Now consider the interior of the copper in Figure 18.28*b*. The interior is electrically neutral, although there are still free electrons that can move under the influence of an electric field. The absence of a net movement of these free electrons indicates that there is no net electric field present within the conductor. In fact, the excess charges arrange themselves on the conductor surface precisely in the manner needed to make the electric field zero within the material. Thus, ***at equilibrium under electrostatic conditions, the electric field is zero at any point within a conducting material.*** This fact has some fascinating implications.

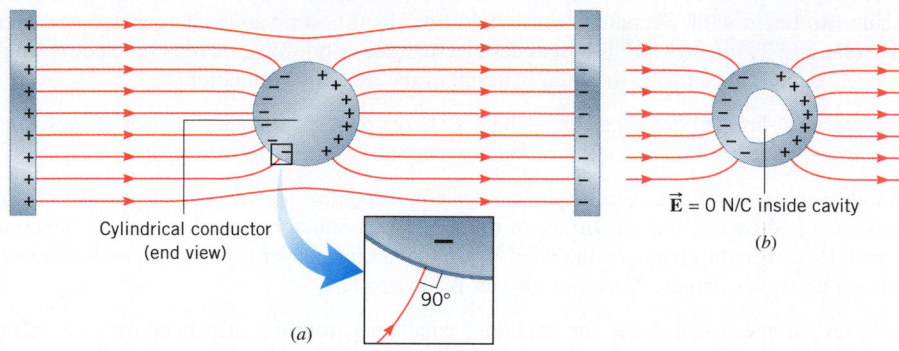

Figure 18.29 (*a*) A cylindrical conductor (shown as an end view) is placed between the oppositely charged plates of a capacitor. The electric field lines do not penetrate the conductor. The blowup shows that, just outside the conductor, the electric field lines are perpendicular to its surface. (*b*) The electric field is zero in a cavity within the conductor.

Figure 18.29*a* shows an uncharged, solid, cylindrical conductor at equilibrium in the central region of a parallel plate capacitor. Induced charges on the surface of the cylinder alter the electric field lines of the capacitor. Since an electric field cannot exist within the conductor under these conditions, the electric field lines do not penetrate the cylinder. Instead, they end or begin on the induced charges. Consequently, a test charge placed *inside* the conductor would feel no force due to the presence of the charges on the capacitor. In other words, *the conductor shields any charge within it from electric fields created outside the conductor.* The shielding results from the induced charges on the conductor surface.

Since the electric field is zero inside the conductor, nothing is disturbed if a cavity is cut from the interior of the material, as in part *b* of the drawing. Thus, the interior of the cavity is also shielded from external electric fields, a fact that has important applications, particularly for shielding electronic circuits. "Stray" electric fields are produced by various electrical appliances (e.g., hair dryers, blenders, and vacuum cleaners), and these fields can interfere with the operation of sensitive electronic circuits, such as those in stereo amplifiers, televisions, and computers. To eliminate such interference, circuits are often enclosed within metal boxes that provide shielding from external fields.

The blowup in Figure 18.29*a* shows another aspect of how conductors alter the electric field lines created by external charges. The lines are altered because *the electric field just outside the surface of a conductor is perpendicular to the surface at equilibrium under electrostatic conditions.* If the field were not perpendicular, there would be a component of the field parallel to the surface. Since the free electrons on the surface of the conductor can move, they would do so under the force exerted by that parallel component. In reality, however, no electron flow occurs at equilibrium. Therefore, there can be no parallel component, and the electric field is perpendicular to the surface.

The preceding discussion deals with features of the electric field within and around a conductor at equilibrium under electrostatic conditions. These features are related to the fact that conductors contain mobile free electrons and *do not apply to insulators,* which contain very few free electrons. Example 14 further explores the behavior of a conducting material in the presence of an electric field.

The physics of shielding electronic circuits.

 Conceptual Example 14 A Conductor in an Electric Field

A charge $+q$ is suspended at the center of a hollow, electrically neutral, spherical, metallic conductor, as Figure 18.30 illustrates. The table shows a number of possibilities for the charges that this suspended charge induces on the interior and exterior surfaces of the conductor. Which one of the possibilities is correct?

	Interior Surface	Exterior Surface
(a)	$-q$	0
(b)	$-\frac{1}{2}q$	$-\frac{1}{2}q$
(c)	$+q$	$-q$
(d)	$-q$	$+q$

Reasoning Three facts will guide our analysis. First: Since the suspended charge does not touch the conductor, the net charge on the conductor must remain zero, because it is electrically

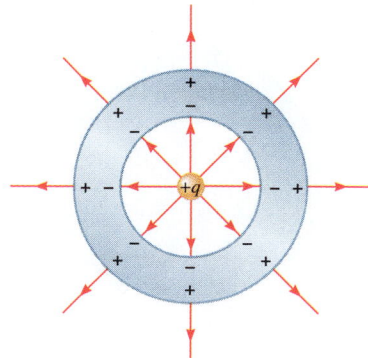

Figure 18.30 A positive charge $+q$ is suspended at the center of a hollow spherical conductor that is electrically neutral. Induced charges appear on the inner and outer surfaces of the conductor. The electric field within the conductor itself is zero.

neutral to begin with. Second: Electric field lines begin on positive charges and end on negative charges. Third: At equilibrium under electrostatic conditions, there is no electric field and, hence, no field lines inside the solid material of the metallic conductor.

Answers (a) and (b) are incorrect. The net charge on the conductor in each of these answers is $-q$, which cannot be, since the conductor's net charge must remain zero.

Answer (c) is incorrect. Electric field lines emanate from the suspended positive charge $+q$ and must end on the interior surface of the metallic conductor, since they do not penetrate the metal. However, the charge on the interior surface in this answer is positive, and field lines must end on negative charges. Thus, this answer is incorrect.

Answer (d) is correct. Since the field lines emanating from the suspended positive charge $+q$ terminate only on negative charges and do not penetrate the metal, there must be an induced *negative* charge on the interior surface. Furthermore, the lines begin and end on equal amounts of charge, so the magnitude of the total charge induced on the interior surface is the same as the magnitude of the suspended charge. Thus, the charge induced on the interior surface is $-q$. We know that the total net charge on the metallic conductor must remain at zero. Therefore, if a charge $-q$ is induced on the interior surface, there must also be a charge of $+q$ induced on the exterior surface, because excess charge cannot remain inside of the solid metal at equilibrium.

Related Homework: *Problem 65*

18.9 GAUSS' LAW

Section 18.6 discusses how a point charge creates an electric field in the space around the charge. There are also many situations in which an electric field is produced by charges that are spread out over a region, rather than by a single point charge. Such an extended collection of charges is called a "charge distribution." For example, the electric field within the parallel plate capacitor in Figure 18.21 is produced by positive charges spread uniformly over one plate and an equal number of negative charges spread over the other plate. As we will see, Gauss' law describes the relationship between a charge distribution and the electric field it produces. This law was formulated by the German mathematician and physicist Carl Friedrich Gauss (1777–1855).

In presenting Gauss' law, it will be necessary to introduce a new idea called *electric flux.* The idea of flux involves both the electric field and the surface through which it passes. By bringing together the electric field and the surface through which it passes, we will be able to define electric flux and then present Gauss' law.

We begin by developing a version of Gauss' law that applies only to a point charge, which we assume to be positive. The electric field lines for a positive point charge radiate outward in all directions from the charge, as Figure 18.22b indicates. The magnitude E of the electric field at a distance r from the charge is $E = kq/r^2$, according to Equation 18.3, in which we have replaced the symbol $|q|$ with the symbol q since we are assuming that the charge is positive. As mentioned in Section 18.5, the constant k can be expressed as $k = 1/(4\pi\epsilon_0)$, where ϵ_0 is the permittivity of free space. With this substitution, the magnitude of the electric field becomes $E = q/(4\pi\epsilon_0 r^2)$. We now place this point charge at the center of an imaginary spherical surface of radius r, as Figure 18.31 shows. Such a hypothetical closed surface is called a *Gaussian surface,* although in general it need not be spherical. The surface area A of a sphere is $A = 4\pi r^2$, and the magnitude of the electric field can be written in terms of this area as $E = q/(A\epsilon_0)$, or

Gauss' law for a point charge

$$\underbrace{EA}_{\substack{\text{Electric} \\ \text{flux, } \Phi_E}} = \frac{q}{\epsilon_0} \qquad (18.5)$$

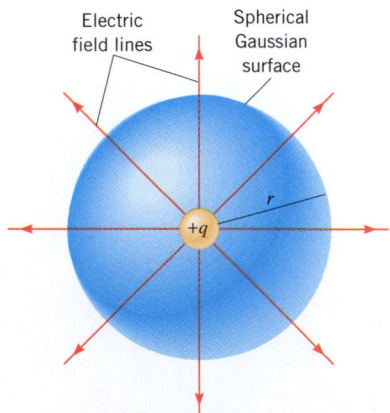

Figure 18.31 A positive point charge is located at the center of an imaginary spherical surface of radius r. Such a surface is one example of a Gaussian surface. Here the electric field is perpendicular to the surface and has the same magnitude everywhere on it.

Electric field lines

Spherical Gaussian surface

r

$+q$

The left side of Equation 18.5 is the product of the magnitude E of the electric field at any point on the Gaussian surface and the area A of the surface. In Gauss' law this product is especially important and is called the *electric flux,* Φ_E: $\Phi_E = EA$. (It will be necessary to modify this definition of flux when we consider the general case of a Gaussian surface with an arbitrary shape.)

Equation 18.5 is the result we have been seeking, for it is the form of Gauss' law that applies to a point charge. This result indicates that, aside from the constant ϵ_0, the electric flux Φ_E depends only on the charge q within the Gaussian surface and is independent of the radius r of the surface. We will now see how to generalize Equation 18.5 to account for distributions of charges and Gaussian surfaces with arbitrary shapes.

Figure 18.32 shows a charge distribution whose net charge is labeled Q. The charge distribution is surrounded by a Gaussian surface—that is, an imaginary closed surface. The surface can have *any arbitrary shape* (it need not be spherical), but *it must be closed* (an open surface would be like that of half an eggshell). The direction of the electric field is not necessarily perpendicular to the Gaussian surface. Furthermore, the magnitude of the electric field need not be constant on the surface but can vary from point to point.

To determine the electric flux through such a surface, we divide the surface into many tiny sections with areas ΔA_1, ΔA_2, and so on. Each section is so small that it is essentially flat and the electric field $\vec{E}$ is a constant (both in magnitude and direction) over it. For reference, a dashed line called the "normal" is drawn perpendicular to each section on the outside of the surface. To determine the electric flux for each of the sections, we use only the component of $\vec{E}$ that is perpendicular to the surface—that is, the component of the electric field that passes through the surface. From the drawing it can be seen that this component has a magnitude of $E \cos \phi$, where ϕ is the angle between the electric field and the normal. The electric flux through any one section is then $(E \cos \phi)\Delta A$. The electric flux Φ_E that passes through the entire Gaussian surface is the sum of all of these individual fluxes: $\Phi_E = (E_1 \cos \phi_1)\Delta A_1 + (E_2 \cos \phi_2)\Delta A_2 + \cdots$, or

$$\Phi_E = \Sigma(E \cos \phi)\Delta A \qquad (18.6)$$

where, as usual, the symbol Σ means "the sum of." Gauss' law relates the electric flux Φ_E to the net charge Q enclosed by the arbitrarily shaped Gaussian surface.

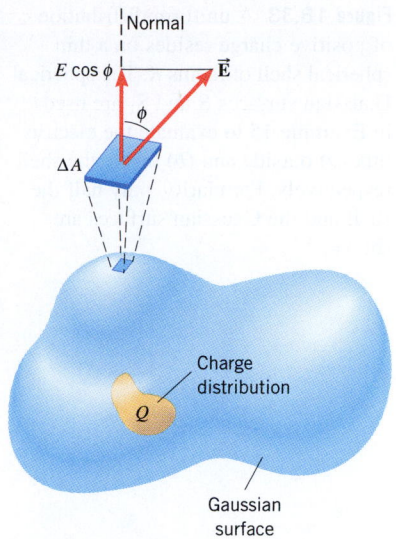

Figure 18.32 The charge distribution Q is surrounded by an arbitrarily shaped Gaussian surface. The electric flux Φ through any tiny segment of the surface is the product of $E \cos \phi$ and the area ΔA of the segment: $\Phi = (E \cos \phi)\Delta A$. The angle ϕ is the angle between the electric field and the normal to the surface.

GAUSS' LAW

The electric flux Φ_E through a Gaussian surface is equal to the net charge Q enclosed by the surface divided by ϵ_0, the permittivity of free space:

$$\underbrace{\Sigma(E \cos \phi)\Delta A}_{\text{Electric flux, }\Phi_E} = \frac{Q}{\epsilon_0} \qquad (18.7)$$

SI Unit of Electric Flux: $N \cdot m^2/C$

Although we arrived at Gauss' law by assuming the net charge Q was positive, Equation 18.7 also applies when Q is negative. In this case, the electric flux Φ_E is also negative. Gauss' law is often used to find the magnitude of the electric field produced by a distribution of charges. The law is most useful when the distribution is uniform and symmetrical. In the next two examples we will see how to apply Gauss' law in such situations.

Example 15 The Electric Field of a Charged Thin Spherical Shell

Figures 18.33*a* and *b* show a thin spherical shell of radius R (for clarity, only half of the shell is shown). A positive charge q is spread uniformly over the shell. Find the magnitude of the electric field at any point **(a)** outside the shell and **(b)** inside the shell.

Reasoning Because the charge is distributed uniformly over the spherical shell, the electric field is symmetrical. This means that the electric field is directed radially outward in all directions, and its magnitude is the same at all points that are equidistant from the shell. All such points lie on a sphere, so the symmetry is called *spherical symmetry*. With this symmetry in mind, we will use a spherical Gaussian surface to evaluate the electric flux Φ_E. We will then use Gauss' law to determine the magnitude of the electric field.

Figure 18.33 A uniform distribution of positive charge resides on a thin spherical shell of radius R. The spherical Gaussian surfaces S and S_1 are used in Example 15 to evaluate the electric flux (a) outside and (b) inside the shell, respectively. For clarity, only half the shell and the Gaussian surfaces are shown.

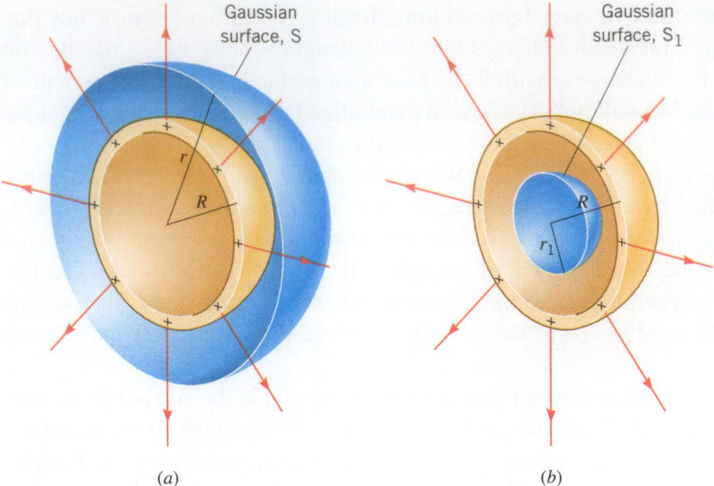

Solution (a) To find the magnitude of the electric field outside the charged shell, we evaluate the electric flux $\Phi_E = \Sigma(E \cos \phi)\Delta A$ by using a spherical Gaussian surface of radius r $(r > R)$ that is concentric with the shell. See the blue surface labeled S in Figure 18.33a. Since the electric field $\vec{E}$ is everywhere perpendicular to the Gaussian surface, $\phi = 0°$ and $\cos \phi = 1$. In addition, E has the same value at all points on the surface, since they are equidistant from the charged shell. Being constant over the surface, E can be factored outside the summation, with the result that

$$\Phi_E = \Sigma(E \cos 0°)\Delta A = E \underbrace{(\Sigma\Delta A)}_{\substack{\text{Area of} \\ \text{Gaussian} \\ \text{surface}}} = \underbrace{E (4\pi r^2)}_{\substack{\text{Surface area} \\ \text{of sphere}}}$$

The term $\Sigma\Delta A$ is just the sum of the tiny areas that make up the Gaussian surface. Since the area of a spherical surface is $4\pi r^2$, we have $\Sigma\Delta A = 4\pi r^2$. Setting the electric flux equal to Q/ϵ_0, as specified by Gauss' law, yields $E(4\pi r^2) = Q/\epsilon_0$. Since the only charge within the Gaussian surface is the charge q on the shell, it follows that the net charge within the Gaussian surface is $Q = q$. Thus, we can solve for E and find that

$$\boxed{E = \frac{q}{4\pi\epsilon_0 r^2} \qquad \text{(for } r > R)}$$

This is a surprising result, for it is the same as that for a point charge (see Equation 18.3 with $|q| = q$). Thus, the electric field outside a uniformly charged spherical shell is the same as if all the charge q were concentrated as a point charge at the center of the shell.

(b) To find the magnitude of the electric field inside the charged shell, we select a spherical Gaussian surface that lies inside the shell and is concentric with it. See the blue surface labeled S_1 in Figure 18.33b. Inside the charged shell, the electric field (if it exists) must also have spherical symmetry. Therefore, using reasoning like that in part (a), the electric flux through the Gaussian surface is $\Phi_E = \Sigma(E \cos \phi)\Delta A = E(4\pi r_1^2)$. In accord with Gauss' law, the electric flux must be equal to Q/ϵ_0, where Q is the net charge *inside* the Gaussian surface. But now $Q = 0$ C, since all the charge lies on the shell that is *outside* the surface S_1. Consequently, we have $E(4\pi r_1^2) = Q/\epsilon_0 = 0$, or

$$\boxed{E = 0 \text{ N/C} \qquad \text{(for } r < R)}$$

Gauss' law allows us to deduce that there is no electric field inside a uniform spherical shell of charge. An electric field exists only on the outside.

Example 16 The Electric Field Inside a Parallel Plate Capacitor

According to Equation 18.4, the electric field inside a parallel plate capacitor, and away from the edges, is constant and has a magnitude of $E = \sigma/\epsilon_0$, where σ is the charge density (the charge per unit area) on a plate. Use Gauss' law to obtain this result.

Reasoning Figure 18.34a shows the electric field inside a parallel plate capacitor. Because the positive and negative charges are distributed uniformly over the surfaces of the plates, symmetry requires that the electric field be perpendicular to the plates. We will take advantage of this symmetry by choosing our Gaussian surface to be a small cylinder whose axis is perpendicular to the plates (see part b of the figure). With this choice, we will be able to evaluate the electric flux and then, with the aid of Gauss' law, determine E.

Solution Figure 18.34b shows that we have placed our Gaussian cylinder so that its left end is inside the positive metal plate, and the right end is in the space between the plates. To determine the electric flux through this Gaussian surface, we evaluate the flux through each of the three parts—labeled 1, 2, and 3 in the drawing—that make up the total surface of the cylinder and then add up the fluxes.

Surface 1—the flat left end of the cylinder—is embedded inside the positive metal plate. As discussed in Section 18.8, the electric field is zero everywhere inside a conductor that is in equilibrium under electrostatic conditions. Since $E = 0$ N/C, the electric flux through this surface is also zero:

$$\Phi_1 = \Sigma(E \cos \phi)\Delta A = \Sigma[(0 \text{ N/C}) \cos \phi]\Delta A = 0$$

Surface 2—the curved wall of the cylinder—is everywhere parallel to the electric field between the plates, so that $\cos \phi = \cos 90° = 0$. Therefore, the electric flux through this surface is also zero:

$$\Phi_2 = \Sigma(E \cos \phi)\Delta A = \Sigma(E \cos 90°)\Delta A = 0$$

Surface 3—the flat right end of the cylinder—is perpendicular to the electric field between the plates, so $\cos \phi = \cos 0° = 1$. The electric field is constant over this surface, so E can be taken outside the summation in Equation 18.6. Noting that $\Sigma\Delta A = A$ is the area of surface 3, we find that the electric flux through this surface is

$$\Phi_3 = \Sigma(E \cos \phi)\Delta A = \Sigma(E \cos 0°)\Delta A = E(\Sigma\Delta A) = EA$$

The electric flux through the entire Gaussian cylinder is the sum of the three fluxes determined above:

$$\Phi_E = \Phi_1 + \Phi_2 + \Phi_3 = 0 + 0 + EA = EA$$

According to Gauss' law, we set the electric flux equal to Q/ϵ_0, where Q is the net charge *inside* the Gaussian cylinder: $EA = Q/\epsilon_0$. But Q/A is the charge per unit area, σ, on the plate. Therefore, we arrive at the value of the electric field inside a parallel plate capacitor: $\boxed{E = \sigma/\epsilon_0}$. The distance of the right end of the Gaussian cylinder from the positive plate does not appear in this result, indicating that the electric field is the same everywhere between the plates.

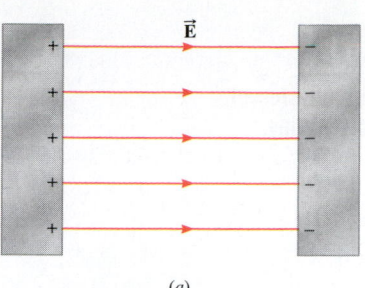

(a)

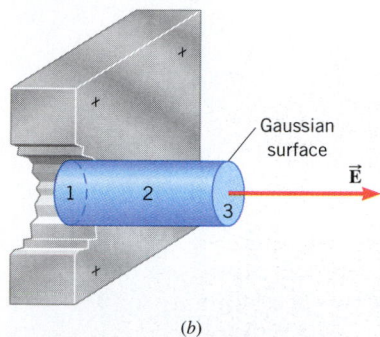

(b)

Figure 18.34 (a) A side view of a parallel plate capacitor, showing some of the electric field lines. (b) The Gaussian surface is a cylinder oriented so its axis is perpendicular to the positive plate and its left end is inside the plate.

✓ **CHECK YOUR UNDERSTANDING**

(*The answers are given at the end of the book.*)

19. A Gaussian surface contains a single charge within it, and as a result an electric flux passes through the surface. Suppose that the charge is then moved to another spot within the Gaussian surface. Does the flux through the surface change?

20. The drawing shows an arrangement of three charges. In parts *a* and *b* different Gaussian surfaces (both in blue) are shown. Through which surface, if either, does the greater electric flux pass?

21. The drawing shows three charges, labeled q_1, q_2, and q_3. A Gaussian surface (in blue) is drawn around q_1 and q_2. **(a)** Which charges determine the electric flux through the Gaussian surface? **(b)** Which charges produce the electric field that exists at the point *P*?

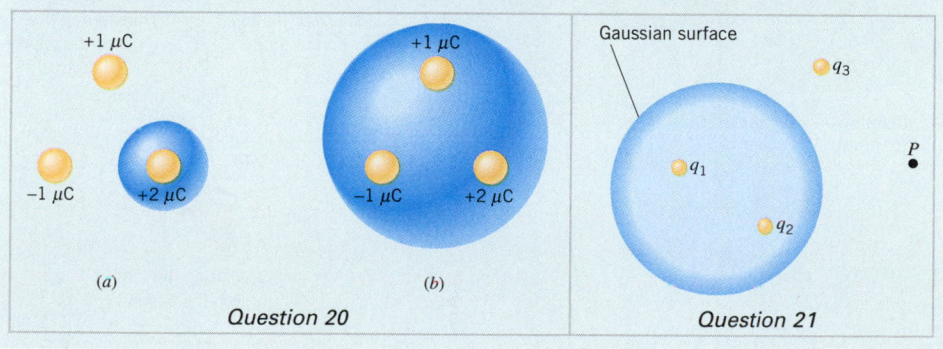

(a)

(b)

Question 20

Question 21

The physics of xerography.

*COPIERS AND COMPUTER PRINTERS

The electrostatic force that charged particles exert on one another plays the central role in an office copier. The copying process is called *xerography,* from the Greek *xeros* and *graphos,* meaning "dry writing." The heart of a copier is the xerographic drum, an aluminum cylinder coated with a layer of selenium (see Figure 18.35a). Aluminum is an excellent electrical conductor. Selenium, on the other hand, is a photoconductor: it is an insulator in the dark but becomes a conductor when exposed to light. Consequently, a positive charge deposited on the selenium surface will remain there, provided the selenium is kept in the dark. When the drum is exposed to light, however, electrons from the aluminum pass through the conducting selenium and neutralize the positive charge.

The photoconductive property of selenium is critical to the xerographic process, as Figure 18.35b illustrates. First, an electrode called a *corotron* gives the entire selenium surface a positive charge in the dark. Second, a series of lenses and mirrors focuses an image of a document onto the revolving drum. The dark and light areas of the document produce corresponding areas on the drum. The dark areas retain their positive charge, but the light areas become conducting and lose their positive charge, ending up neutralized. Thus, a positive-charge image of the document remains on the selenium surface. In the third step, a special dry black powder, called the *toner,* is given a negative charge and then spread onto the drum, where it adheres selectively to the positively charged areas. The fourth step involves transferring the toner onto a blank piece of paper. However, the attraction of the positive-charge image holds the toner to the drum. To transfer the toner, the paper is given a *greater positive charge* than that of the image, with the aid of another corotron. Last, the paper and adhering toner pass through heated pressure rollers, which melt the toner into the fibers of the paper and produce the finished copy.

The physics of a laser printer.

A laser printer is used with computers to provide high-quality copies of text and graphics. It is similar in operation to the xerographic machine, except that the information to be reproduced is not on paper. Instead, the information is transferred from the computer's

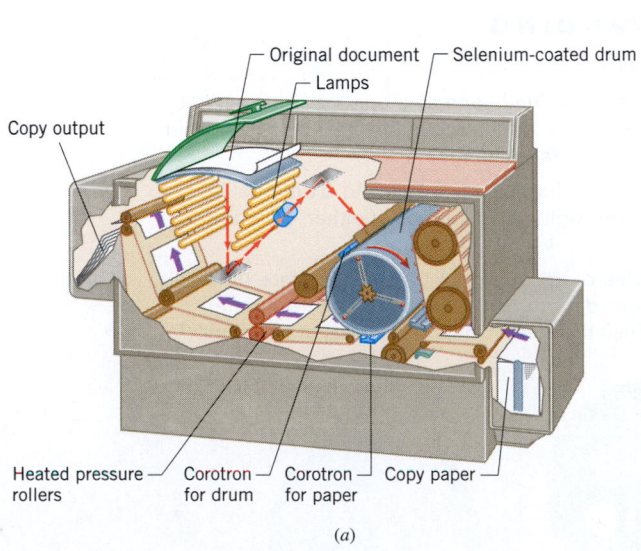

(a)

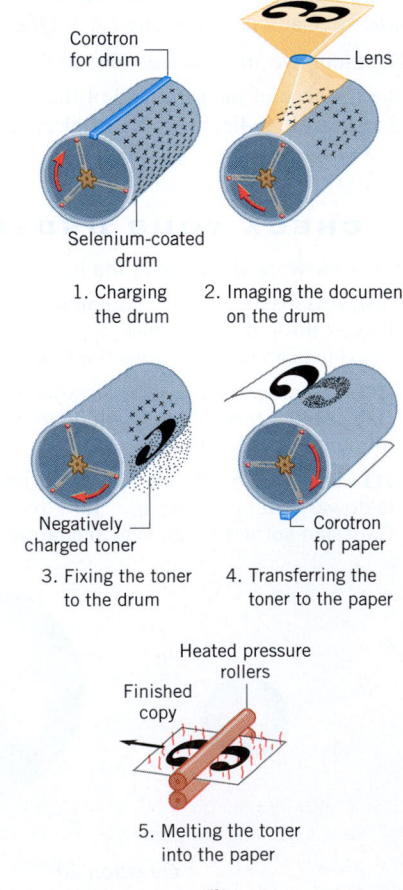

Corotron for drum — Lens

Selenium-coated drum

1. Charging the drum
2. Imaging the document on the drum

Negatively charged toner

Corotron for paper

3. Fixing the toner to the drum
4. Transferring the toner to the paper

Heated pressure rollers

Finished copy

5. Melting the toner into the paper

(b)

Figure 18.35 (a) This cutaway view shows the essential elements of a copying machine. (b) The five steps in the xerographic process.

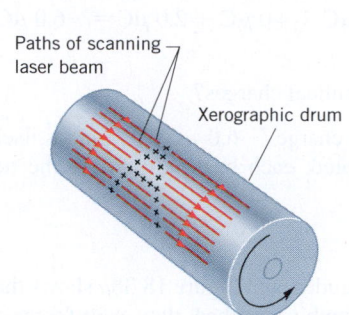

Figure 18.36 As the laser beam scans back and forth across the surface of the xerographic drum, a positive-charge image of the letter "A" is created.

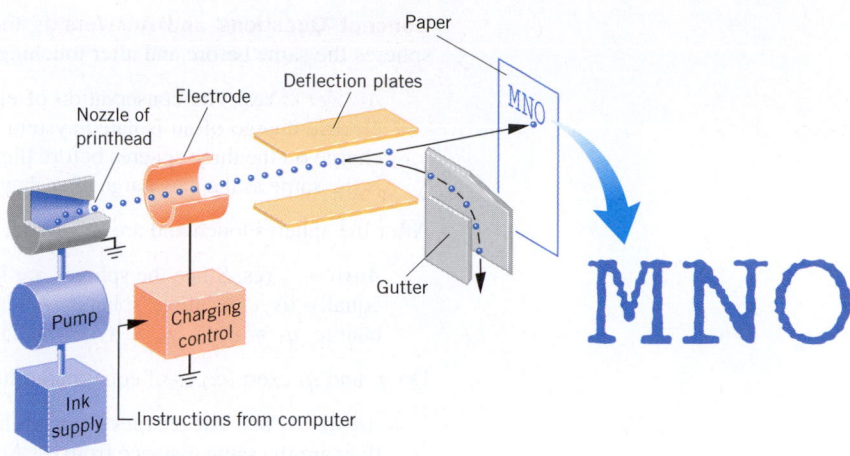

Figure 18.37 An inkjet printhead ejects a steady flow of ink droplets. Charged droplets are deflected into a gutter by the deflection plates, while uncharged droplets fly straight onto the paper. Letters formed by an inkjet printer look normal, except when greatly enlarged and the patterns from the drops become apparent.

memory to the printer, and laser light is used to copy it onto the selenium–aluminum drum. A laser beam, focused to a fine point, is scanned rapidly from side to side across the rotating drum, as Figure 18.36 indicates. While the light remains on, the positive charge on the drum is neutralized. As the laser beam moves, the computer turns the beam off at the right moments during each scan to produce the desired positive-charge image, which is the letter "A" in the picture.

An inkjet printer is another type of printer that uses electric charges in its operation. While shuttling back and forth across the paper, the inkjet printhead ejects a thin stream of ink. Figure 18.37 illustrates the elements of one type of printhead. The ink is forced out of a small nozzle and breaks up into extremely small droplets, with diameters that can be as small as 9×10^{-6} m. About 150 000 droplets leave the nozzle each second and travel with a speed of approximately 18 m/s toward the paper. During their flight, the droplets pass through two electrical components, an *electrode* and the *deflection plates* (a parallel plate capacitor). When the printhead moves over regions of the paper that are not to be inked, the charging control is turned on and an electric field is established between the printhead and the electrode. As the drops pass through the electric field, they acquire a net charge by the process of induction. The deflection plates divert the charged droplets into a gutter and thus prevent them from reaching the paper. Whenever ink is to be placed on the paper, the charging control, responding to instructions from the computer, turns off the electric field. The uncharged droplets fly straight through the deflection plates and strike the paper.

The physics of an inkjet printer.

CONCEPTS & CALCULATIONS

18.11

In this chapter we have studied electric forces and electric fields. We conclude now by presenting some examples that review important features of these concepts. The three-part format of the examples stresses the role of conceptual understanding in problem solving. First, the problem statement is given. Then, there is a concept question-and-answer section, followed by the solution section. The purpose of the concept question-and-answer section is to provide help in understanding the solution and to illustrate how a review of the concepts can help in anticipating some of the characteristics of the numerical answers.

Concepts & Calculations Example 17
The Vector Nature of Electric Forces

The charges on three identical metal spheres are $-12 \ \mu C$, $+4.0 \ \mu C$, and $+2.0 \ \mu C$. The spheres are brought together so they simultaneously touch each other. They are then separated and placed on the x and y axes, as in Figure 18.38a. What is the net force (magnitude and direction) exerted on the sphere at the origin? Treat the spheres as if they were particles.

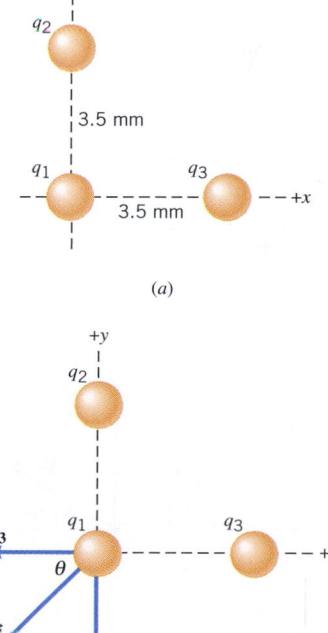

Figure 18.38 (a) Three equal charges lie on the x and y axes. (b) The net force exerted on q_1 by the other two charges is $\vec{F}$.

Concept Questions and Answers Is the net charge on the system comprising the three spheres the same before and after touching?

Answer Yes. The conservation of electric charge states that during any process the net electric charge of an isolated system remains constant (is conserved). Therefore, the net charge on the three spheres before they touch ($-12.0 \ \mu C + 4.0 \ \mu C + 2.0 \ \mu C = -6.0 \ \mu C$) is the same as the net charge after they touch.

After the spheres touch and are separated, do they have identical charges?

Answer Yes. Since the spheres are identical, the net charge ($-6.0 \ \mu C$) distributes itself equally over the three spheres. After they are separated, each has one-third of the net charge: $q_1 = q_2 = q_3 = \frac{1}{3}(-6.0 \ \mu C) = -2.0 \ \mu C$.

Do q_2 and q_3 exert forces of equal magnitude on q_1?

Answer Yes. The charges q_2 and q_3 have equal magnitudes, and Figure 18.38a shows that they are the same distance from q_1. According to Coulomb's law, then, they exert forces of equal magnitude on q_1.

Is the magnitude of the net force exerted on q_1 equal to $2F$, where F is the magnitude of the force that either q_2 or q_3 exerts on q_1?

Answer No. Although the two forces that act on q_1 have equal magnitudes, they have different directions. The forces are repulsive forces, since all of the charges in part *a* of the drawing are identical. Figure 18.38b shows the force $\vec{F}_{12}$ exerted on q_1 by q_2 and the force $\vec{F}_{13}$ exerted on q_1 by q_3. To obtain the net force $\vec{F}$, we must take these directions into account by using vector addition.

Solution The magnitude F_{12} of the force exerted on q_1 by q_2 is given by Coulomb's law, Equation 18.1, as

$$F_{12} = k\frac{|q_1||q_2|}{r_{12}^2} = \frac{(8.99 \times 10^9 \ \text{N}\cdot\text{m}^2/\text{C}^2)(2.0 \times 10^{-6} \ \text{C})(2.0 \times 10^{-6} \ \text{C})}{(3.5 \times 10^{-3} \ \text{m})^2}$$

$$= 2.9 \times 10^3 \ \text{N}$$

Note that we have used the magnitudes of q_1 and q_2 in Coulomb's law. As mentioned previously, the magnitude of the force $\vec{F}_{13}$ exerted on q_1 by q_3 has the same value as F_{12}, and so $F_{13} = 2.9 \times 10^3 \ \text{N}$. Since the forces $\vec{F}_{12}$ and $\vec{F}_{13}$ are perpendicular to each other (see Figure 18.38b), we may use the Pythagorean theorem to find the magnitude F of the net force:

$$F = \sqrt{F_{12}^2 + F_{13}^2} = \sqrt{(2.9 \times 10^3 \ \text{N})^2 + (2.9 \times 10^3 \ \text{N})^2} = \boxed{4.1 \times 10^3 \ \text{N}}$$

The angle θ that the net force makes with the $-x$ axis (see part *b* of the drawing) is

$$\theta = \tan^{-1}\left(\frac{F_{12}}{F_{13}}\right) = \tan^{-1}\left(\frac{2.9 \times 10^3 \ \text{N}}{2.9 \times 10^3 \ \text{N}}\right) = \boxed{45°}$$

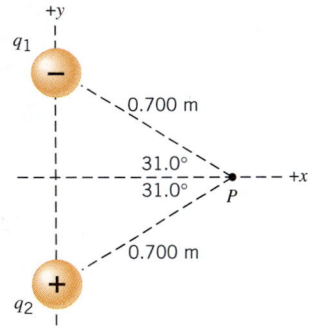

(a)

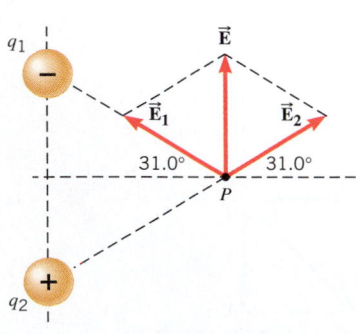

(b)

Figure 18.39 (a) Two charges q_1 and q_2 produce an electric field at the point P. (b) The electric fields $\vec{E}_1$ and $\vec{E}_2$ add to give the net electric field $\vec{E}$.

Concepts & Calculations Example 18

Becoming Familiar with Electric Fields

Two point charges are lying on the y axis in Figure 18.39a: $q_1 = -4.00 \ \mu C$ and $q_2 = +4.00 \ \mu C$. They are equidistant from the point P, which lies on the x axis. **(a)** What is the net electric field at P? **(b)** A small object of charge $q_0 = +8.00 \ \mu C$ and mass $m = 1.20$ g is placed at P. When it is released, what is its acceleration?

Concept Questions and Answers There is no charge at P in part (a). Is there an electric field at P?

Answer Yes. An electric field is produced by the charges q_1 and q_2, and it exists throughout the entire region that surrounds them. If a test charge were placed at this point, it would experience a force due to the electric field. The force would be the product of the charge and the electric field.

The charge q_1 produces an electric field at the point P. What is the direction of this field?

Answer The electric field created by a charge always points away from a positive charge and toward a negative charge. Since q_1 is negative, the electric field $\vec{E}_1$ points toward it (see Figure 18.39b).

What is the direction of the electric field produced by q_2 at P?

> *Answer* Since q_2 is positive, the electric field $\vec{E}_2$ that it produces points away from q_2, as shown in the drawing.

Is the magnitude of the net electric field equal to $E_1 + E_2$, where E_1 and E_2 are the magnitudes of the electric fields produced by q_1 and q_2?

> *Answer* No, because the electric fields have different directions. We must add the individual fields as vectors to obtain the net electric field. Only then can we determine its magnitude.

Solution (a) The magnitudes of the electric fields that q_1 and q_2 produce at P are given by Equation 18.3, for which the distances are specified in Figure 18.39a:

$$E_1 = \frac{k|q_1|}{r_1^2} = \frac{(8.99 \times 10^9 \text{ N·m}^2/\text{C}^2)(4.00 \times 10^{-6} \text{ C})}{(0.700 \text{ m})^2} = 7.34 \times 10^4 \text{ N/C}$$

$$E_2 = \frac{k|q_2|}{r_2^2} = \frac{(8.99 \times 10^9 \text{ N·m}^2/\text{C}^2)(4.00 \times 10^{-6} \text{ C})}{(0.700 \text{ m})^2} = 7.34 \times 10^4 \text{ N/C}$$

The x and y components of these fields and the total field $\vec{E}$ are given in the following table:

Electric field	x component	y component
$\vec{E}_1$	$-E_1 \cos 31.0° = -6.29 \times 10^4$ N/C	$+E_1 \sin 31.0° = +3.78 \times 10^4$ N/C
$\vec{E}_2$	$+E_2 \cos 31.0° = +6.29 \times 10^4$ N/C	$+E_2 \sin 31.0° = +3.78 \times 10^4$ N/C
$\vec{E}$	$\vec{E}_x = 0$ N/C	$\vec{E}_y = +7.56 \times 10^4$ N/C

The net electric field $\vec{E}$ has only a component along the $+y$ axis, so

$$\boxed{\vec{E} = 7.56 \times 10^4 \text{ N/C, directed along the } +y \text{ axis}}$$

(b) According to Newton's second law, Equation 4.1, the acceleration $\vec{a}$ of an object placed at this point is equal to the net force acting on it divided by its mass. The net force $\vec{F}$ is the product of the charge and the net electric field, $\vec{F} = q_0 \vec{E}$, as indicated by Equation 18.2. Thus, the acceleration is

$$\vec{a} = \frac{\vec{F}}{m} = \frac{q_0 \vec{E}}{m}$$

$$= \frac{(8.00 \times 10^{-6} \text{ C})(7.56 \times 10^4 \text{ N/C})}{1.20 \times 10^{-3} \text{ kg}} = \boxed{5.04 \times 10^2 \text{ m/s}^2, \text{ along the } +y \text{ axis}}$$

CONCEPT SUMMARY

If you need more help with a concept, use the Learning Aids noted next to the discussion or equation. Examples (**Ex.**) are in the text of this chapter. Go to **www.wiley.com/college/cutnell** for the following Learning Aids:

Interactive LearningWare (ILW) — Additional examples solved in a five-step interactive format.

Concept Simulations (CS) — Animated text figures or animations of important concepts.

Interactive Solutions (IS) — Models for certain types of problems in the chapter homework. The calculations are carried out interactively.

Topic	Discussion	Learning Aids
The coulomb (C)	**18.1 THE ORIGIN OF ELECTRICITY** There are two kinds of electric charge: positive and negative. The SI unit of electric charge is the coulomb (C). The magnitude of the charge on an electron or a proton is	
Magnitude of charge on electron or proton	$$e = 1.60 \times 10^{-19} \text{ C}$$ Since the symbol e denotes a magnitude, it has no algebraic sign. Thus, the electron carries a charge of $-e$, and the proton carries a charge of $+e$.	
Charge is quantized	The charge on any object, whether positive or negative, is quantized, in the sense that the charge consists of an integer number of protons or electrons.	**Ex. 1**
Law of conservation of electric charge	**18.2 CHARGED OBJECTS AND THE ELECTRIC FORCE** The law of conservation of electric charge states that the net electric charge of an isolated system remains constant during any process.	
Electric repulsion and attraction	Like charges repel and unlike charges attract each other.	

Topic	Discussion	Learning Aids
Conductor	**18.3 CONDUCTORS AND INSULATORS** An electrical conductor is a material, such as copper, that conducts electric charge readily.	
Insulator	An electrical insulator is a material, such as rubber, that conducts electric charge poorly.	
Charging by contact	**18.4 CHARGING BY CONTACT AND BY INDUCTION** Charging by contact is the process of giving one object a net electric charge by placing it in contact with an object that is already charged.	
Charging by induction	Charging by induction is the process of giving an object a net electric charge without touching it to a charged object.	
Point charge	**18.5 COULOMB'S LAW** A point charge is a charge that occupies so little space that it can be regarded as a mathematical point.	
	Coulomb's law gives the magnitude F of the electric force that two point charges q_1 and q_2 exert on each other:	**Ex. 2–5** **ILW 18.1** **CS 18.1**
Coulomb's law	$$F = k\frac{|q_1||q_2|}{r^2} \qquad (18.1)$$	**ILW 18.2** **Ex. 17** **IS 18.17**
	where $|q_1|$ and $|q_2|$ are the magnitudes of the charges and have no algebraic sign. The term k is a constant and has the value $k = 8.99 \times 10^9$ N·m²/C². The force specified by Equation 18.1 acts along the line between the two charges.	
Permittivity of free space	The permittivity of free space ϵ_0 is defined by the relation	
	$$k = \frac{1}{4\pi\epsilon_0}$$	
	18.6 THE ELECTRIC FIELD The electric field $\vec{E}$ at a given spot is a vector and is the electrostatic force $\vec{F}$ experienced by a very small test charge q_0 placed at that spot divided by the charge itself:	
Electric field	$$\vec{E} = \frac{\vec{F}}{q_0} \qquad (18.2)$$	**Ex. 6–9** **IS 18.73**
	The direction of the electric field is the same as the direction of the force on a positive test charge. The SI unit for the electric field is the newton per coulomb (N/C). The source of the electric field at any spot is the charged objects surrounding that spot.	
	The magnitude of the electric field created by a point charge q is	
Electric field of a point charge	$$E = \frac{k|q|}{r^2} \qquad (18.3)$$	**Ex. 10–12, 18** **ILW 18.3** **IS 18.47**
	where $|q|$ is the magnitude of the charge and has no algebraic sign and r is the distance from the charge. The electric field $\vec{E}$ points away from a positive charge and toward a negative charge.	
	For a parallel plate capacitor that has a charge per unit area of σ on each plate, the magnitude of the electric field between the plates is	
Electric field of a parallel plate capacitor	$$E = \frac{\sigma}{\epsilon_0} \qquad (18.4)$$	
Electric field lines	**18.7 ELECTRIC FIELD LINES** Electric field lines are lines that can be thought of as a "map," insofar as the lines provide information about the direction and strength of the electric field. The lines are directed away from positive charges and toward negative charges. The direction of the lines gives the	**Ex. 13**
Direction of electric field Strength of electric field	direction of the electric field, since the electric field vector at a point is tangent to the line at that point. The electric field is strongest in regions where the number of lines per unit area passing perpendicularly through a surface is the greatest—that is, where the lines are packed together most tightly.	**CS 18.2**
Excess charge carried by a conductor at equilibrium	**18.8 THE ELECTRIC FIELD INSIDE A CONDUCTOR: SHIELDING** Excess negative or positive charge resides on the surface of a conductor at equilibrium under electrostatic conditions. In such a situation, the electric field at any point within the conducting material is zero, and the electric field just outside the surface of the conductor is perpendicular to the surface.	**Ex. 14**
	18.9 GAUSS' LAW The electric flux Φ_E through a surface is related to the magnitude E of the electric field, the area A of the surface, and the angle ϕ that specifies the direction of the field relative to the normal to the surface:	
Electric flux	$$\Phi_E = \Sigma(E\cos\phi)\Delta A \qquad (18.6)$$	
	Gauss' law states that the electric flux through a closed surface (a Gaussian surface) is equal to the net charge Q enclosed by the surface divided by ϵ_0, the permittivity of free space:	
Gauss' law	$$\Phi_E = \Sigma(E\cos\phi)\Delta A = \frac{Q}{\epsilon_0} \qquad (18.7)$$	**Ex. 15, 16** **CS 18.3** **IS 18.61**

FOCUS ON CONCEPTS

Note to Instructors: The numbering of the questions shown here reflects the fact that they are only a representative subset of the total number that are available online. However, all of the questions are available for assignment via an online homework management program such as WileyPLUS or WebAssign.

Section 18.1   The Origin of Electricity

1. An object carries a charge of $-8.0\ \mu C$, while another carries a charge of $-2.0\ \mu C$. How many electrons must be transferred from the first object to the second so that both objects have the same charge?

Section 18.2   Charged Objects and the Electric Force

2. Each of three objects carries a charge. As the drawing shows, objects A and B attract each other, and objects C and A also attract each other. Which one of the following statements concerning objects B and C is true?   **(a)** They attract each other.   **(b)** They repel each other.   **(c)** They neither attract nor repel each other.   **(d)** This question cannot be answered without additional information.

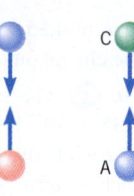

Section 18.4   Charging by Contact and by Induction

4. Each of two identical objects carries a net charge. The objects are made from conducting material. One object is attracted to a positively charged ebonite rod, and the other is repelled by the rod. After the objects are touched together, it is found that they are each repelled by the rod. What can be concluded about the initial charges on the objects?   **(a)** Initially both objects are positive, with both charges having the same magnitude.   **(b)** Initially both objects are negative, with both charges having the same magnitude.   **(c)** Initially one object is positive and one is negative, with the negative charge having a greater magnitude than the positive charge.   **(d)** Initially one object is positive and one is negative, with the positive charge having a greater magnitude than the negative charge.

5. Only one of three balls A, B, and C carries a net charge q. The balls are made from conducting material and are identical. One of the uncharged balls can become charged by touching it to the charged ball and then separating the two. This process of touching one ball to another and then separating the two balls can be repeated over and over again, with the result that the three balls can take on a variety of charges. Which one of the following distributions of charges could not possibly be achieved in this fashion, even if the process were repeated a large number of times?

(a) $q_A = \frac{1}{3}q$, $q_B = \frac{1}{3}q$, $q_C = \frac{1}{3}q$ **(c)** $q_A = \frac{1}{2}q$, $q_B = \frac{3}{8}q$, $q_C = \frac{1}{4}q$

(b) $q_A = \frac{1}{2}q$, $q_B = \frac{1}{4}q$, $q_C = \frac{1}{4}q$ **(d)** $q_A = \frac{3}{8}q$, $q_B = \frac{3}{8}q$, $q_C = \frac{1}{4}q$

Section 18.5   Coulomb's Law

8. Three point charges have equal magnitudes and are located on the same line. The separation d between A and B is the same as the separation between B and C. One of the charges is positive and two are negative, as the drawing shows. Consider the net electrostatic force that each charge experiences due to the other two charges. Rank the net forces in descending order (greatest first) according to magnitude.   **(a)** A, B, C   **(b)** B, C, A   **(c)** A, C, B   **(d)** C, A, B   **(e)** B, A, C

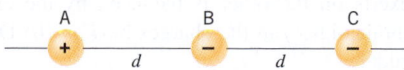

9. Three point charges have equal magnitudes and are fixed to the corners of an equilateral triangle. Two of the charges are positive and one is negative, as the drawing shows. At which one of the corners is the net force acting on the charge directed parallel to the x axis?   **(a)** A   **(b)** B   **(c)** C

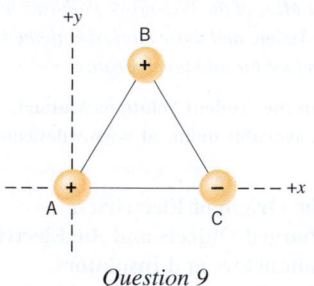

Question 9

Section 18.6   The Electric Field

12. A positive point charge q_1 creates an electric field of magnitude E_1 at a spot located at a distance r_1 from the charge. The charge is replaced by another positive point charge q_2, which creates a field of magnitude $E_2 = E_1$ at a distance of $r_2 = 2r_1$. How is q_2 related to q_1?   **(a)** $q_2 = 2q_1$ **(b)** $q_2 = \frac{1}{2}q_1$ **(c)** $q_2 = 4q_1$ **(d)** $q_2 = \frac{1}{4}q_1$ **(e)** $q_2 = \sqrt{2}q_1$

13. The drawing shows a positive and a negative point charge. The negative charge has the greater magnitude. Where on the line that passes through the charges is the one spot where the total electric field is zero?   **(a)** To the right of the negative charge   **(b)** To the left of the positive charge   **(c)** Between the charges, to the left of the midpoint   **(d)** Between the charges, to the right of the midpoint

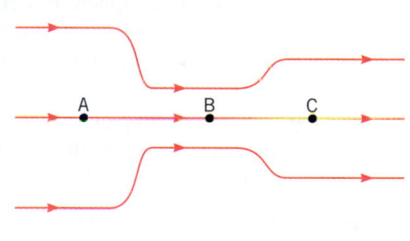

Section 18.7   Electric Field Lines

17. The drawing shows some electric field lines. For the points indicated, rank the magnitudes of the electric field in descending order (largest first).   **(a)** B, C, A   **(b)** B, A, C   **(c)** A, B, C   **(d)** A, C, B   **(e)** C, A, B

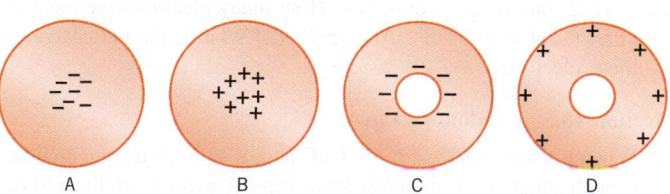

Section 18.8   The Electric Field Inside a Conductor: Shielding

18. The drawings show (in cross section) two solid spheres and two spherical shells. Each object is made from copper and has a net charge, as the plus and minus signs indicate. Which drawing correctly shows where the charges reside when they are in equilibrium?   **(a)** A   **(b)** B   **(c)** C   **(d)** D

Section 18.9   Gauss' Law

20. A cubical Gaussian surface surrounds two charges, $q_1 = +6.0 \times 10^{-12}\ C$ and $q_2 = -2.0 \times 10^{-12}\ C$, as the drawing shows. What is the electric flux passing through the surface?

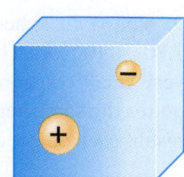

PROBLEMS

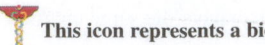

Problems that are not marked with a star are considered the easiest to solve. Problems that are marked with a single star (*) are more difficult, while those marked with a double star (**) are the most difficult.

Note to Instructors: *Most of the homework problems in this chapter are available for assignment via an online homework management program such as* WileyPLUS *or* WebAssign, *and those marked with the icon* **GO** *are presented in* WileyPLUS *using a guided tutorial format that provides enhanced interactivity. See Preface for additional details.*

ssm   Solution is in the Student Solutions Manual.
www   Solution is available online at www.wiley.com/college/cutnell

This icon represents a biomedical application.

Section 18.1   The Origin of Electricity,
Section 18.2   Charged Objects and the Electric Force,
Section 18.3   Conductors and Insulators,
Section 18.4   Charging by Contact and by Induction

1. ssm A metal sphere has a charge of $+8.0$ μC. What is the net charge after 6.0×10^{13} electrons have been placed on it?

2. Iron atoms have been detected in the sun's outer atmosphere, some with many of their electrons stripped away. What is the net electric charge (in coulombs) of an iron atom with 26 protons and 7 electrons? Be sure to include the algebraic sign $(+$ or $-)$ in your answer.

3. Four identical metallic objects carry the following charges: $+1.6$, $+6.2$, -4.8, and -9.4 μC. The objects are brought simultaneously into contact, so that each touches the others. Then they are separated. **(a)** What is the final charge on each object? **(b)** How many electrons (or protons) make up the final charge on each object?

4. GO A plate carries a charge of -3.0 μC, while a rod carries a charge of $+2.0$ μC. How many electrons must be transferred from the plate to the rod, so that both objects have the same charge?

5. ssm Consider three identical metal spheres, A, B, and C. Sphere A carries a charge of $+5q$. Sphere B carries a charge of $-q$. Sphere C carries no net charge. Spheres A and B are touched together and then separated. Sphere C is then touched to sphere A and separated from it. Last, sphere C is touched to sphere B and separated from it. **(a)** How much charge ends up on sphere C? What is the total charge on the three spheres **(b)** before they are allowed to touch each other and **(c)** after they have touched?

6. GO Four identical metal spheres have charges of $q_A = -8.0$ μC, $q_B = -2.0$ μC, $q_C = +5.0$ μC, and $q_D = +12.0$ μC. **(a)** Two of the spheres are brought together so they touch, and then they are separated. Which spheres are they, if the final charge on each one is $+5.0$ μC? **(b)** In a similar manner, which three spheres are brought together and then separated, if the final charge on each of the three is $+3.0$ μC? **(c)** The final charge on each of the three separated spheres in part (b) is $+3.0$ μC. How many electrons would have to be added to one of these spheres to make it electrically neutral?

* 7. Water has a mass per mole of 18.0 g/mol, and each water molecule (H_2O) has 10 electrons. **(a)** How many electrons are there in one liter (1.00×10^{-3} m^3) of water? **(b)** What is the net charge of all these electrons?

Section 18.5   Coulomb's Law

8. When point charges $q_1 = +8.4$ μC and $q_2 = +5.6$ μC are brought near each other, each experiences a repulsive force of magnitude 0.66 N. Determine the distance between the charges.

9. ssm Two spherical objects are separated by a distance that is 1.80×10^{-3} m. The objects are initially electrically neutral and are very small compared to the distance between them. Each object acquires the same negative charge due to the addition of electrons. As a result, each object experiences an electrostatic force that has a magni-

tude of 4.55×10^{-21} N. How many electrons did it take to produce the charge on one of the objects?

10. GO The masses of the earth and moon are 5.98×10^{24} and 7.35×10^{22} kg, respectively. Identical amounts of charge are placed on each body, such that the net force (gravitational plus electrical) on each is zero. What is the magnitude of the charge placed on each body?

11. ssm www Two very small spheres are initially neutral and separated by a distance of 0.50 m. Suppose that 3.0×10^{13} electrons are removed from one sphere and placed on the other. **(a)** What is the magnitude of the electrostatic force that acts on each sphere? **(b)** Is the force attractive or repulsive? Why?

12. Two point charges are fixed on the y axis: a negative point charge $q_1 = -25$ μC at $y_1 = +0.22$ m and a positive point charge q_2 at $y_2 = +0.34$ m. A third point charge $q = +8.4$ μC is fixed at the origin. The net electrostatic force exerted on the charge q by the other two charges has a magnitude of 27 N and points in the $+y$ direction. Determine the magnitude of q_2.

13. ssm Consult **Concept Simulation 18.1** at **www.wiley.com/college/cutnell** for insight into this problem. Three charges are fixed to an x, y coordinate system. A charge of $+18$ μC is on the y axis at $y = +3.0$ m. A charge of -12 μC is at the origin. Last, a charge of $+45$ μC is on the x axis at $x = +3.0$ m. Determine the magnitude and direction of the net electrostatic force on the charge at $x = +3.0$ m. Specify the direction relative to the $-x$ axis.

14. GO The drawings show three charges that have the same magnitude but different signs. In all cases the distance d between the charges is the same. The magnitude of the charges is $|q| = 8.6$ μC, and the distance between them is $d = 3.8$ mm. Determine the magnitude of the net force on charge 2 for each of the three drawings.

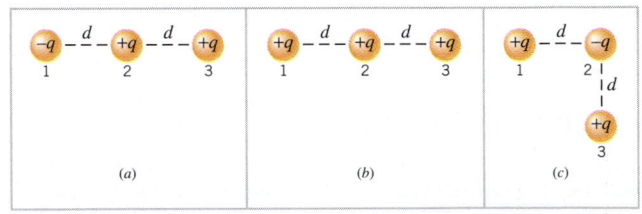

15. ssm Interactive LearningWare 18.1 at **www.wiley.com/college/cutnell** offers some perspective on this problem. Two tiny spheres have the same mass and carry charges of the same magnitude. The mass of each sphere is 2.0×10^{-6} kg. The gravitational force that each sphere exerts on the other is balanced by the electric force. **(a)** What algebraic signs can the charges have? **(b)** Determine the charge magnitude.

16. The force of repulsion that two like charges exert on each other is 3.5 N. What will the force be if the distance between the charges is increased to five times its original value?

17. Interactive Solution 18.17 at **www.wiley.com/college/cutnell** provides a model for solving this type of problem. Two small objects,

A and B, are fixed in place and separated by 3.00 cm in a vacuum. Object A has a charge of $+2.00$ μC, and object B has a charge of -2.00 μC. How many electrons must be removed from A and put onto B to make the electrostatic force that acts on each object an attractive force whose magnitude is 68.0 N?

18. A charge of -3.00 μC is fixed at the center of a compass. Two additional charges are fixed on the circle of the compass, which has a radius of 0.100 m. The charges on the circle are -4.00 μC at the position due north and $+5.00$ μC at the position due east. What are the magnitude and direction of the net electrostatic force acting on the charge at the center? Specify the direction relative to due east.

*** **19.** Multiple-Concept Example 3 provides some pertinent background for this problem. Suppose a single electron orbits about a nucleus containing two protons ($+2e$), as would be the case for a helium atom from which one of the two naturally occurring electrons is removed. The radius of the orbit is 2.65×10^{-11} m. Determine the magnitude of the electron's centripetal acceleration.

*** **20.** **GO** Three point charges have equal magnitudes, two being positive and one negative. These charges are fixed to the corners of an equilateral triangle, as the drawing shows. The magnitude of each of the charges is 5.0 μC, and the lengths of the sides of the triangle are 3.0 cm. Calculate the magnitude of the net force that each charge experiences.

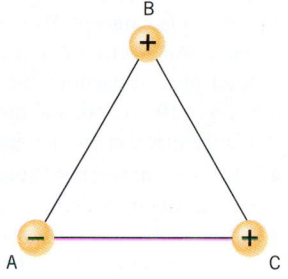

*** **21.** **Interactive LearningWare 18.2** at **www.wiley.com/college/cutnell** provides one approach to solving problems such as this one. The drawing shows three point charges fixed in place. The charge at the coordinate origin has a value of $q_1 = +8.00$ μC; the other two charges have identical magnitudes, but opposite signs: $q_2 = -5.00$ μC and $q_3 = +5.00$ μC. (a) Determine the net force (magnitude and direction) exerted on q_1 by the other two charges. (b) If q_1 had a mass of 1.50 g and it were free to move, what would be its acceleration?

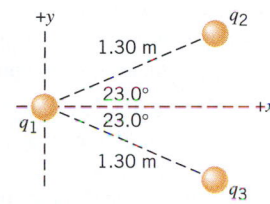

*** **22.** Four point charges have equal magnitudes. Three are positive, and one is negative, as the drawing shows. They are fixed in place on the same straight line, and adjacent charges are equally separated by a distance d. Consider the net electrostatic force acting on each charge. Calculate the ratio of the largest to the smallest net force.

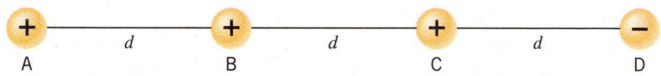

*** **23.** Multiple-Concept Example 3 illustrates several of the concepts that come into play in this problem. A single electron orbits a lithium nucleus that contains three protons ($+3e$). The radius of the orbit is 1.76×10^{-11} m. Determine the kinetic energy of the electron.

*** **24.** **GO** An electrically neutral model airplane is flying in a horizontal circle on a 3.0-m guideline, which is nearly parallel to the ground. The line breaks when the kinetic energy of the plane is 50.0 J. Reconsider the same situation, except that now there is a point charge of $+q$ on the plane and a point charge of $-q$ at the other end of the guideline. In this case, the line breaks when the kinetic energy of the plane is 51.8 J. Find the magnitude of the charges.

*** **25.** **ssm** In the rectangle in the drawing, a charge is to be placed at the empty corner to make the net force on the charge at corner A point

along the vertical direction. What charge (magnitude and algebraic sign) must be placed at the empty corner?

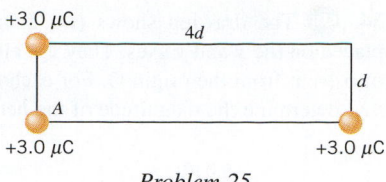

Problem 25

*** **26.** Two identical small insulating balls are suspended by separate 0.25-m threads that are attached to a common point on the ceiling. Each ball has a mass of 8.0×10^{-4} kg. Initially the balls are uncharged and hang straight down. They are then given identical positive charges and, as a result, spread apart with an angle of 36° between the threads. Determine (a) the charge on each ball and (b) the tension in the threads.

*** **27.** **ssm** A small spherical insulator of mass 8.00×10^{-2} kg and charge $+0.600$ μC is hung by a thin wire of negligible mass. A charge of -0.900 μC is held 0.150 m away from the sphere and directly to the right of it, so the wire makes an angle θ with the vertical (see the drawing). Find (a) the angle θ and (b) the tension in the wire.

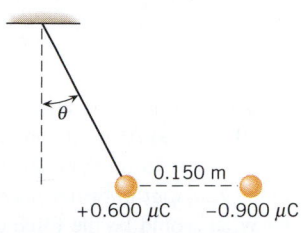

*** **28.** Two objects carry initial charges that are q_1 and q_2, respectively, where $|q_2| > |q_1|$. They are located 0.200 m apart and behave like point charges. They attract each other with a force that has a magnitude of 1.20 N. The objects are then brought into contact, so the net charge is shared equally, and then they are returned to their initial positions. Now it is found that the objects repel one another with a force whose magnitude is equal to the magnitude of the initial attractive force. What are the magnitudes of the initial charges on the objects?

Section 18.6 The Electric Field,
Section 18.7 Electric Field Lines,
Section 18.8 The Electric Field Inside a Conductor: Shielding

29. An electric field of 260 000 N/C points due west at a certain spot. What are the magnitude and direction of the force that acts on a charge of -7.0 μC at this spot?

30. **GO** Suppose you want to determine the electric field in a certain region of space. You have a small object of known charge and an instrument that measures the magnitude and direction of the force exerted on the object by the electric field. (a) The object has a charge of $+20.0$ μC and the instrument indicates that the electric force exerted on it is 40.0 μN, due east. What are the magnitude and direction of the electric field? (b) What are the magnitude and direction of the electric field if the object has a charge of -10.0 μC and the instrument indicates that the force is 20.0 μN, due west?

31. **ssm www** At a distance r_1 from a point charge, the magnitude of the electric field created by the charge is 248 N/C. At a distance r_2 from the charge, the field has a magnitude of 132 N/C. Find the ratio r_2/r_1.

32. Review Conceptual Example 13 as an aid in working this problem. Charges of $-4q$ are fixed to diagonally opposite corners of a square. A charge of $+5q$ is fixed to one of the remaining corners, and a charge of $+3q$ is fixed to the last corner. Assuming that ten electric field lines emerge from the $+5q$ charge, sketch the field lines in the vicinity of the four charges.

33. Four point charges have the same magnitude of 2.4×10^{-12} C and are fixed to the corners of a square that is 4.0 cm on a side. Three of the charges are positive and one is negative. Determine the magnitude of the net electric field that exists at the center of the square.

34. **GO** The drawing shows two situations in which charges are placed on the x and y axes. They are all located at the same distance of 6.1 cm from the origin O. For each of the situations in the drawing, determine the magnitude of the net electric field at the origin.

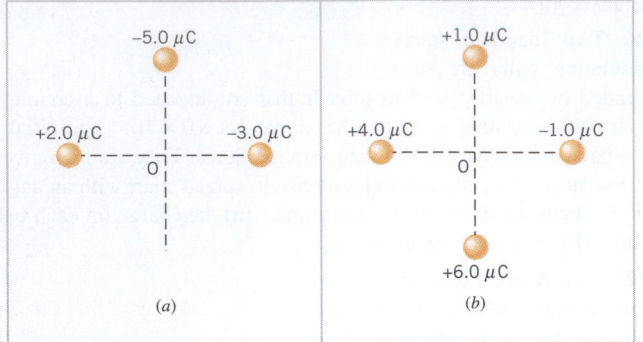

(a)

(b)

35. Two charges, -16 and $+4.0$ μC, are fixed in place and separated by 3.0 m. **(a)** At what spot along a line through the charges is the net electric field zero? Locate this spot relative to the positive charge. *(Hint: The spot does not necessarily lie between the two charges.)* **(b)** What would be the force on a charge of $+14$ μC placed at this spot?

36. The membrane surrounding a living cell consists of an inner and an outer wall that are separated by a small space. Assume that the membrane acts like a parallel plate capacitor in which the effective charge density on the inner and outer walls has a magnitude of 7.1×10^{-6} C/m^2. **(a)** What is the magnitude of the electric field within the cell membrane? **(b)** Find the magnitude of the electric force that would be exerted on a potassium ion (K$^+$; charge $= +e$) placed inside the membrane.

37. **ssm** Two charges are placed on the x axis. One of the charges ($q_1 = +8.5$ μC) is at $x_1 = +3.0$ cm and the other ($q_2 = -21$ μC) is at $x_1 = +9.0$ cm. Find the net electric field (magnitude and direction) at **(a)** $x = 0$ cm and **(b)** $x = +6.0$ cm.

38. Background pertinent to this problem is available in **Interactive LearningWare 18.3** at **www.wiley.com/college/cutnell**. A uniform electric field exists everywhere in the x, y plane. This electric field has a magnitude of 4500 N/C and is directed in the positive x direction. A point charge -8.0×10^{-9} C is placed at the origin. Determine the magnitude of the net electric field at **(a)** $x = -0.15$ m, **(b)** $x = +0.15$ m, and **(c)** $y = +0.15$ m.

39. **ssm** A small drop of water is suspended motionless in air by a uniform electric field that is directed upward and has a magnitude of 8480 N/C. The mass of the water drop is 3.50×10^{-9} kg. **(a)** Is the excess charge on the water drop positive or negative? Why? **(b)** How many excess electrons or protons reside on the drop?

40. **GO** A proton and an electron are moving due east in a constant electric field that also points due east. The electric field has a magnitude of 8.0×10^4 N/C. Determine the magnitude of the acceleration of the proton and the electron.

41. Review Conceptual Example 12 before attempting to work this problem. The magnitude of each of the charges in Figure 18.20 is 8.60×10^{-12} C. The lengths of the sides of the rectangles are 3.00 cm and 5.00 cm. Find the magnitude of the electric field at the center of the rectangle in Figures 18.20a and b.

42. Two charges are placed between the plates of a parallel plate capacitor. One charge is $+q_1$ and the other is $q_2 = +5.00$ μC. The charge per unit area on each of the plates has a magnitude of $\sigma = 1.30 \times 10^{-4}$ C/m^2. The magnitude of the force on q_1 due to q_2 equals the magnitude of the force on q_1 due to the electric field of the parallel plate capacitor. What is the distance r between the two charges?

*43. Multiple-Concept Example 9 illustrates the concepts employed in this problem. A small object, which has a charge $q = 7.5$ μC and mass $m = 9.0 \times 10^{-5}$ kg, is placed in a constant electric field. Starting from rest, the object accelerates to a speed of 2.0×10^3 m/s in a time of 0.96 s. Determine the magnitude of the electric field.

*44. A spring with an unstrained length of 0.074 m and a spring constant of 2.4 N/m hangs vertically downward from the ceiling. A uniform electric field directed vertically upward fills the region containing the spring. A sphere with a mass of 5.1×10^{-3} kg and a net charge of $+6.6$ μC is attached to the lower end of the spring. The spring is released slowly, until it reaches equilibrium. The equilibrium length of the spring is 0.059 m. What is the magnitude of the external electric field?

*45. **ssm www** At three corners of a rectangle (length $= 2d$, height $= d$), the following charges are located: $+q_1$ (upper left corner), $+q_2$ (lower right corner), and $-q$ (lower left corner). The net electric field at the (empty) upper right corner is zero. Find the magnitudes of q_1 and q_2. Express your answers in terms of q.

*46. Multiple-Concept Example 9 illustrates the concepts in this problem. An electron is released from rest at the negative plate of a parallel plate capacitor. The charge per unit area on each plate is $\sigma = 1.8 \times 10^{-7}$ C/m^2, and the plate separation is 1.5×10^{-2} m. How fast is the electron moving just before it reaches the positive plate?

*47. Review **Interactive Solution 18.47** at **www.wiley.com/college/cutnell** for help with this problem. The drawing shows two positive charges q_1 and q_2 fixed to a circle. At the center of the circle they produce a net electric field that is directed upward along the vertical axis. Determine the ratio $|q_2|/|q_1|$ of the charge magnitudes.

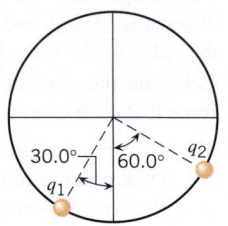

*48. **GO** The drawing shows a positive point charge $+q_1$, a second point charge q_2 that may be positive or negative, and a spot labeled P, all on the same straight line. The distance d between the two charges is the same as the distance between q_1 and the spot P. With q_2 present, the magnitude of the net electric field at P is twice what it is when q_1 is present alone. Given that $q_1 = +0.50$ μC, determine q_2 when it is **(a)** positive and **(b)** negative.

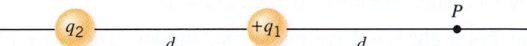

*49. In Multiple-Concept Example 9 you can see the concepts that are important in this problem. A particle of charge $+12$ μC and mass 3.8×10^{-5} kg is released from rest in a region where there is a constant electric field of $+480$ N/C. What is the displacement of the particle after a time of 1.6×10^{-2} s?

50. Two particles are in a uniform electric field that points in the $+x$ direction and has a magnitude of 2500 N/C. The mass and charge of particle 1 are $m_1 = 1.4 \times 10^{-5}$ kg and $q_1 = -7.0$ μC, while the corresponding values for particle 2 are $m_2 = 2.6 \times 10^{-5}$ kg and $q_2 = +18$ μC. Initially the particles are at rest. The particles are both located on the same electric field line but are separated from each other by a distance d. Particle 1 is located to the left of particle 2. When released, they accelerate but always remain at this same distance from each other. Find d.

51. **ssm** Two point charges of the same magnitude but opposite signs are fixed to either end of the base of an isosceles triangle, as the

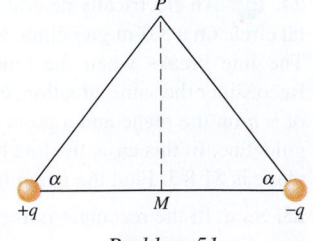

Problem 51

drawing shows. The electric field at the midpoint M between the charges has a magnitude E_M. The field directly above the midpoint at point P has a magnitude E_P. The ratio of these two field magnitudes is $E_M/E_P = 9.0$. Find the angle α in the drawing.

**52. An electrically charged point particle that has a mass of 1.50×10^{-6} kg is launched horizontally into a uniform electric field with a magnitude of 925 N/C. The particle's initial velocity is 8.80 m/s, directed eastward, and the external electric field is also directed eastward. A short time after being launched, the particle is 6.71×10^{-3} m below and 0.160 m east of its initial position. Determine the net charge carried by the particle, including the algebraic sign (+ or −).

**53. A small plastic ball with a mass of 6.50×10^{-3} kg and with a charge of $+0.150$ μC is suspended from an insulating thread and hangs between the plates of a capacitor (see the drawing). The ball is in equilibrium, with the thread making an angle of 30.0° with respect to the vertical. The area of each plate is 0.0150 m². What is the magnitude of the charge on each plate?

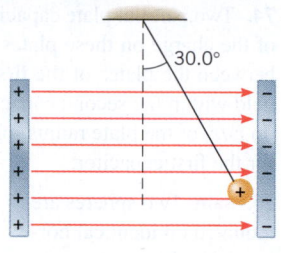

Section 18.9 Gauss' Law

54. A spherical surface completely surrounds a collection of charges. Find the electric flux through the surface if the collection consists of **(a)** a single $+3.5 \times 10^{-6}$ C charge, **(b)** a single -2.3×10^{-6} C charge, and **(c)** both of the charges in (a) and (b).

55. ssm The drawing shows an edge-on view of two planar surfaces that intersect and are mutually perpendicular. Surface 1 has an area of 1.7 m², while surface 2 has an area of 3.2 m². The electric field $\vec{E}$ in the drawing is uniform and has a magnitude of 250 N/C. Find the electric flux through **(a)** surface 1 and **(b)** surface 2.

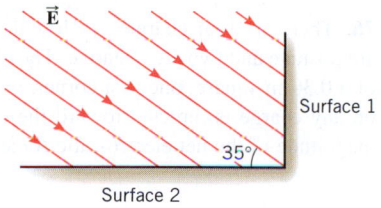

56. A circular surface with a radius of 0.057 m is exposed to a uniform external electric field of magnitude 1.44×10^4 N/C. The electric flux through the surface is 78 N·m²/C. What is the angle between the direction of the electric field and the normal to the surface?

57. A vertical wall (5.9 m $\times$ 2.5 m) in a house faces due east. A uniform electric field has a magnitude of 150 N/C. This field is parallel to the ground and points 35° north of east. What is the electric flux through the wall?

58. A charge Q is located inside a rectangular box. The electric flux through each of the six surfaces of the box is: $\Phi_1 = +1500$ N·m²/C, $\Phi_2 = +2200$ N·m²/C, $\Phi_3 = +4600$ N·m²/C, $\Phi_4 = -1800$ N·m²/C, $\Phi_5 = -3500$ N·m²/C, and $\Phi_6 = -5400$ N·m²/C. What is Q?

*59. ssm A cube is located with one corner situated at the origin of an x, y, z coordinate system. One of the cube's faces lies in the x, y plane, another in the y, z plane, and another in the x, z plane. In other words, the cube is in the first octant of the coordinate system. The edges of the cube are 0.20 m long. A uniform electric field is parallel to the x, y plane and points in the direction of the $+y$ axis. The magnitude of the field is 1500 N/C. **(a)** Find the electric flux through each of the six faces of the cube. **(b)** Add the six values obtained in part (a) to show that the electric flux through the cubical surface is zero, as Gauss' law predicts, since there is no net charge within the cube.

*60. GO Refer to **Concept Simulation 18.3** at **www.wiley.com/college/cutnell** for a perspective that is useful in solving this problem. Two spherical shells have a common center. A -1.6×10^{-6} C charge is spread uniformly over the inner shell, which has a radius of 0.050 m. A $+5.1 \times 10^{-6}$ C charge is spread uniformly over the outer shell, which has a radius of 0.15 m. Find the magnitude and direction of the electric field at a distance (measured from the common center) of **(a)** 0.20 m, **(b)** 0.10 m, and **(c)** 0.025 m.

*61. Interactive Solution 18.61 at **www.wiley.com/college/cutnell** offers help with this problem in an interactive environment. A solid nonconducting sphere has a positive charge q spread uniformly throughout its volume. The charge density or charge per unit volume, therefore, is $\dfrac{q}{\frac{4}{3}\pi R^3}$. Use Gauss' law to show that the electric field at a point within the sphere at a radius r has a magnitude of $\dfrac{qr}{4\pi\epsilon_0 R^3}$. *(Hint: For a Gaussian surface, use a sphere of radius r centered within the solid sphere. Note that the net charge within any volume is the charge density times the volume.)*

**62. A long, thin, straight wire of length L has a positive charge Q distributed uniformly along it. Use Gauss' law to show that the electric field created by this wire at a radial distance r has a magnitude of $E = \lambda/(2\pi\epsilon_0 r)$, where $\lambda = Q/L$. *(Hint: For a Gaussian surface, use a cylinder aligned with its axis along the wire and note that the cylinder has a flat surface at either end, as well as a curved surface.)*

ADDITIONAL PROBLEMS

63. Concept Simulation 18.2 at **www.wiley.com/college/cutnell** provides background concerning the electric field lines that are the focus of this problem. Review the important features of electric field lines discussed in Conceptual Example 13. Three point charges ($+q$, $+2q$, and $-3q$) are at the corners of an equilateral triangle. Sketch in six electric field lines between the three charges.

64. Two tiny conducting spheres are identical and carry charges of -20.0 μC and $+50.0$ μC. They are separated by a distance of 2.50 cm. **(a)** What is the magnitude of the force that each sphere experiences, and is the force attractive or repulsive? **(b)** The spheres are brought into contact and then separated to a distance of 2.50 cm. Determine the magnitude of the force that each sphere now experiences, and state whether the force is attractive or repulsive.

65. Conceptual Example 14 deals with the hollow spherical conductor in Figure 18.30. The conductor is initially electrically neutral, and then a charge $+q$ is placed at the center of the hollow space. Suppose the conductor initially has a net charge of $+2q$ instead of being neutral. What is the total charge on the interior and on the exterior surface when the $+q$ charge is placed at the center?

66. A tiny ball (mass = 0.012 kg) carries a charge of -18 μC. What electric field (magnitude and direction) is needed to cause the ball to float above the ground?

67. ssm A long, thin rod (length = 4.0 m) lies along the x axis, with its midpoint at the origin. In a vacuum, a +8.0 μC point charge is fixed to one end of the rod, and a −8.0 μC point charge is fixed to the other end. Everywhere in the x, y plane there is a constant external electric field (magnitude = 5.0 × 10³ N/C) that is perpendicular to the rod. With respect to the z axis, find the magnitude of the net torque applied to the rod.

68. In a vacuum, two particles have charges of q_1 and q_2, where q_1 = +3.5 μC. They are separated by a distance of 0.26 m, and particle 1 experiences an attractive force of 3.4 N. What is q_2 (magnitude and sign)?

69. ssm www Two particles, with identical positive charges and a separation of 2.60 × 10⁻² m, are released from rest. Immediately after the release, particle 1 has an acceleration $\vec{a}_1$ whose magnitude is 4.60 × 10³ m/s², while particle 2 has an acceleration $\vec{a}_2$ whose magnitude is 8.50 × 10³ m/s². Particle 1 has a mass of 6.00 × 10⁻⁶ kg. Find **(a)** the charge on each particle and **(b)** the mass of particle 2.

70. A charge of q = +7.50 μC is located in an electric field. The x and y components of the electric field are E_x = 6.00 × 10³ N/C and E_y = 8.00 × 10³ N/C, respectively. **(a)** What is the magnitude of the force on the charge? **(b)** Determine the angle that the force makes with the +x axis.

*__71.__ Two point charges are located along the x axis: q_1 = +6.0 μC at x_1 = +4.0 cm, and q_2 = +6.0 μC at x_2 = −4.0 cm. Two other charges are located on the y axis: q_3 = +3.0 μC at y_3 = +5.0 cm, and q_4 = −8.0 μC at y_4 = +7.0 cm. Find the net electric field (magnitude and direction) at the origin.

*__72.__ The drawing shows an equilateral triangle, each side of which has a

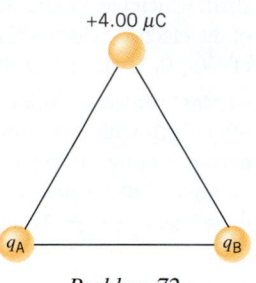

Problem 72

length of 2.00 cm. Point charges are fixed to each corner, as shown. The 4.00 μC charge experiences a net force due to the charges q_A and q_B. This net force points vertically downward and has a magnitude of 405 N. Determine the magnitudes and algebraic signs of the charges q_A and q_B.

*__73. Interactive Solution 18.73__ at **www.wiley.com/college/cutnell** provides a model for problems of this kind. A small object has a mass of 3.0 × 10⁻³ kg and a charge of −34 μC. It is placed at a certain spot where there is an electric field. When released, the object experiences an acceleration of 2.5 × 10³ m/s² in the direction of the +x axis. Determine the magnitude and direction of the electric field.

*__74.__ Two parallel plate capacitors have circular plates. The magnitude of the charge on these plates is the same. However, the electric field between the plates of the first capacitor is 2.2 × 10⁵ N/C, while the field within the second capacitor is 3.8 × 10⁵ N/C. Determine the ratio r_2/r_1 of the plate radius for the second capacitor to the plate radius for the first capacitor.

*__75. ssm__ Two spheres are mounted on identical horizontal springs and rest on a frictionless table, as in the drawing. When the spheres are uncharged, the spacing between them is 0.0500 m, and the springs are unstrained. When each sphere has a charge of +1.60 μC, the spacing doubles. Assuming that the spheres have a negligible diameter, determine the spring constant of the springs.

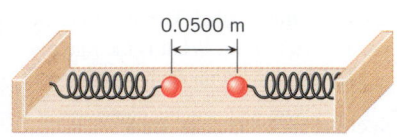

0.0500 m

**__76.__ There are four charges, each with a magnitude of 2.0 μC. Two are positive and two are negative. The charges are fixed to the corners of a 0.30-m square, one to a corner, in such a way that the net force on any charge is directed toward the center of the square. Find the magnitude of the net electrostatic force experienced by any charge.

CHAPTER 19

ELECTRIC POTENTIAL ENERGY AND THE ELECTRIC POTENTIAL

Berkeley, California, inventors Tom Burchill, Jonah Most, and Daniel Holtmann-Rice (with a portable solar panel on his back) show their "Solar Board," a three-person skateboard powered by the sun. Sunlight penetrates the solar cells of the panel and provides the energy that separates positive and negative charges in the materials from which the cells are made. Thus, each solar cell develops positive and negative terminals, much like the terminals of a battery and, in effect, converts solar energy into electric energy that powers the skateboard. Electric potential energy and the related concept of electric potential are the subjects of this chapter.

(© Lonny Shavelson/Zuma Press)

19.1 POTENTIAL ENERGY

In Chapter 18 we discussed the electrostatic force that two point charges exert on each other, the magnitude of which is $F = k|q_1||q_2|/r^2$. The form of this equation is similar to that for the gravitational force that two particles exert on each other, which is $F = Gm_1m_2/r^2$, according to Newton's law of universal gravitation (see Section 4.7). Both of these forces are conservative and, as Section 6.4 explains, a potential energy can be associated with a conservative force. Thus, an electric potential energy exists that is analogous to the gravitational potential energy. To set the stage for a discussion of the electric potential energy, let's review some of the important aspects of the gravitational counterpart.

Figure 19.1, which is essentially Figure 6.9, shows a basketball of mass m falling from point A to point B. The gravitational force, $m\vec{g}$, is the only force acting on the ball, where g is the magnitude of the acceleration due to gravity. As Section 6.3 discusses, the work W_{AB} done by the gravitational force when the ball falls from a height of h_A to a height of h_B is

$$W_{AB} = \underbrace{mgh_A}_{\substack{\text{Initial} \\ \text{gravitational} \\ \text{potential energy,} \\ \text{GPE}_A}} - \underbrace{mgh_B}_{\substack{\text{Final} \\ \text{gravitational} \\ \text{potential energy,} \\ \text{GPE}_B}} = \text{GPE}_A - \text{GPE}_B \qquad (6.4)$$

Recall that the quantity mgh is the gravitational potential energy* of the ball, $\text{GPE} = mgh$ (Equation 6.5), and represents the energy that the ball has by virtue of its position relative to the surface of the earth. Thus, the work done by the gravitational force equals the initial gravitational potential energy minus the final gravitational potential energy.

*The gravitational potential energy is now being denoted by GPE to distinguish it from the electric potential energy EPE.

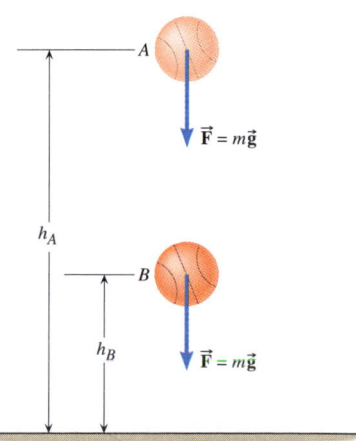

Figure 19.1 Gravity exerts a force, $\vec{F} = m\vec{g}$, on the basketball of mass m. Work is done by the gravitational force as the ball falls from A to B.

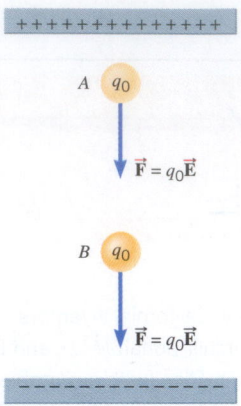

Figure 19.2 Because of the electric field $\vec{E}$, an electric force, $\vec{F} = q_0\vec{E}$, is exerted on a positive test charge $+q_0$. Work is done by the force as the charge moves from A to B.

Figure 19.2 clarifies the analogy between electric and gravitational potential energies. In this drawing a positive test charge $+q_0$ is situated at point A between two oppositely charged plates. Because of the charges on the plates, an electric field $\vec{E}$ exists in the region between the plates. Consequently, the test charge experiences an electric force, $\vec{F} = q_0\vec{E}$ (Equation 18.2), that is directed downward, toward the lower plate. (The gravitational force is being neglected here.) As the charge moves from A to B, work is done by this force, in a fashion analogous to the work done by the gravitational force in Figure 19.1. The work W_{AB} done by the electric force equals the difference between the electric potential energy EPE at A and the electric potential energy at B:

$$W_{AB} = \text{EPE}_A - \text{EPE}_B \tag{19.1}$$

This expression is similar to Equation 6.4. The path along which the test charge moves from A to B is of no consequence because the electric force is a conservative force, and so the work W_{AB} is the same for all paths (see Section 6.4).

19.2 THE ELECTRIC POTENTIAL DIFFERENCE

Since the electric force is $\vec{F} = q_0\vec{E}$, the work that it does as the charge moves from A to B in Figure 19.2 depends on the charge q_0. It is useful, therefore, to express this work on a per-unit-charge basis, by dividing both sides of Equation 19.1 by the charge:

$$\frac{W_{AB}}{q_0} = \frac{\text{EPE}_A}{q_0} - \frac{\text{EPE}_B}{q_0} \tag{19.2}$$

Notice that the right-hand side of this equation is the difference between two terms, each of which is an electric potential energy divided by the test charge, EPE/q_0. The quantity EPE/q_0 is the electric potential energy per unit charge and is an important concept in electricity. It is called the *electric potential* or, simply, the *potential* and is referred to with the symbol V, as in Equation 19.3.

DEFINITION OF ELECTRIC POTENTIAL

The electric potential V at a given point is the electric potential energy EPE of a small test charge q_0 situated at that point divided by the charge itself:

$$V = \frac{\text{EPE}}{q_0} \tag{19.3}$$

SI Unit of Electric Potential: joule/coulomb = volt (V)

The SI unit of electric potential is a joule per coulomb, a quantity known as a *volt*. The name honors Alessandro Volta (1745–1827), who invented the voltaic pile, the forerunner of the battery. In spite of the similarity in names, the electric potential energy EPE and the electric potential V are *not* the same. The electric potential energy, as its name implies, is an *energy* and, therefore, is measured in joules. In contrast, the electric potential is an *energy per unit charge* and is measured in joules per coulomb, or volts.

We can now relate the work W_{AB} done by the electric force when a charge q_0 moves from A to B to the potential difference $V_B - V_A$ between the points. Combining Equations 19.2 and 19.3, we have:

$$V_B - V_A = \frac{\text{EPE}_B}{q_0} - \frac{\text{EPE}_A}{q_0} = \frac{-W_{AB}}{q_0} \tag{19.4}$$

Often, the "delta" notation is used to express the difference (final value minus initial value) in potentials and in potential energies: $\Delta V = V_B - V_A$ and $\Delta(\text{EPE}) = \text{EPE}_B - \text{EPE}_A$. In terms of this notation, Equation 19.4 takes the following more compact form:

$$\Delta V = \frac{\Delta(\text{EPE})}{q_0} = \frac{-W_{AB}}{q_0} \tag{19.4}$$

Neither the potential V nor the potential energy EPE can be determined in an absolute sense, because only the *differences* ΔV and $\Delta(\text{EPE})$ are measurable in terms of the work W_{AB}. The gravitational potential energy has this same characteristic, since only the value at one height relative to the value at some reference height has any significance. Example 1 emphasizes the relative nature of the electric potential.

 Example 1 Work, Electric Potential Energy, and Electric Potential

In Figure 19.2, the work done by the electric force as the test charge ($q_0 = +2.0 \times 10^{-6}$ C) moves from A to B is $W_{AB} = +5.0 \times 10^{-5}$ J. **(a)** Find the value of the difference, $\Delta(\text{EPE}) = \text{EPE}_B - \text{EPE}_A$, in the electric potential energies of the charge between these points. **(b)** Determine the potential difference, $\Delta V = V_B - V_A$, between the points.

Reasoning The work done by the electric force when the charge travels from A to B is $W_{AB} = \text{EPE}_A - \text{EPE}_B$, according to Equation 19.1. Therefore, the difference in the electric potential energies (final value minus initial value) is $\Delta(\text{EPE}) = \text{EPE}_B - \text{EPE}_A = -W_{AB}$. The potential difference, $\Delta V = V_B - V_A$, is the difference in the electric potential energies divided by the charge q_0, according to Equation 19.4.

Solution **(a)** The difference in the electric potential energies of the charge between the points A and B is

$$\underbrace{\text{EPE}_B - \text{EPE}_A}_{= \Delta(\text{EPE})} = -W_{AB} = \boxed{-5.0 \times 10^{-5} \text{ J}} \qquad (19.1)$$

The fact that $\text{EPE}_B - \text{EPE}_A$ is negative means that the charge has a higher electric potential energy at A than at B.

(b) The potential difference ΔV between A and B is

$$\underbrace{V_B - V_A}_{= \Delta V} = \frac{\text{EPE}_B - \text{EPE}_A}{q_0} = \frac{-5.0 \times 10^{-5} \text{ J}}{2.0 \times 10^{-6} \text{ C}} = \boxed{-25 \text{ V}} \qquad (19.4)$$

The fact that $V_B - V_A$ is negative means that the electric potential is higher at A than at B.

This electric car stores energy from the charging station in its battery pack. The voltage of the battery pack and the magnitude of the charge sent through it by the charging station determine how much energy is stored. (© AP/Wide World Photos)

The potential difference between two points is measured in volts and, therefore, is often referred to as a "voltage." Everyone has heard of "voltage" because, as we will see in Chapter 20, it is frequently used in connection with everyday devices. For example, your TV requires a "voltage" of 120 V (which is applied between the two prongs of the plug on the power cord when it is inserted into an electrical wall outlet), and your cell phone and laptop computer use batteries that provide, for example, "voltages" of 1.5 V or 9 V (which exist between the two battery terminals).

In Figure 19.1 the speed of the basketball increases as it falls from A to B. Since point A has a greater gravitational potential energy than point B, we see that an object of mass m accelerates when it moves from a region of higher potential energy toward a region of lower potential energy. Likewise, the positive charge in Figure 19.2 accelerates as it moves from A to B because of the electric repulsion from the upper plate and the attraction to the lower plate. Since point A has a higher electric potential than point B, we conclude that *a positive charge accelerates from a region of higher electric potential toward a region of lower electric potential.* On the other hand, a negative charge placed between the plates in Figure 19.2 behaves in the opposite fashion, since the electric force acting on the negative charge is directed opposite to the electric force acting on the positive charge. *A negative charge accelerates from a region of lower potential toward a region of higher potential.* The next example illustrates the way positive and negative charges behave.

Problem-solving insight

Problem-solving insight

 Conceptual Example 2 How Positive and Negative Charges Accelerate

Three points, A, B, and C, are located along a horizontal line, as Figure 19.3 illustrates. A positive test charge is released from rest at A and accelerates toward B. Upon reaching B, the test charge continues to accelerate toward C. Assuming that only motion along the line is possible, what will a negative test charge do when it is released from rest at B? A negative test charge will **(a)** accelerate toward C, **(b)** remain stationary, **(c)** accelerate toward A.

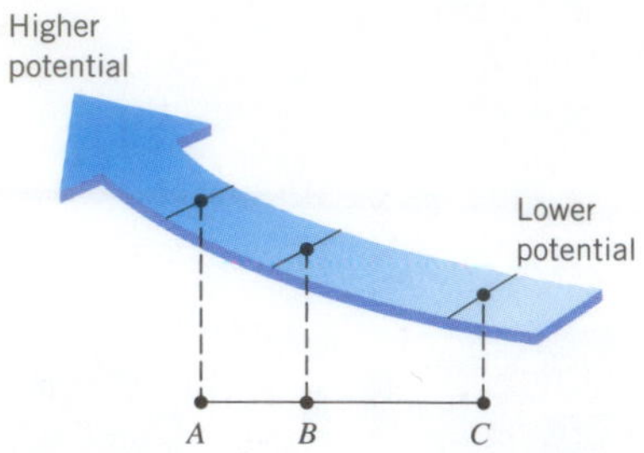

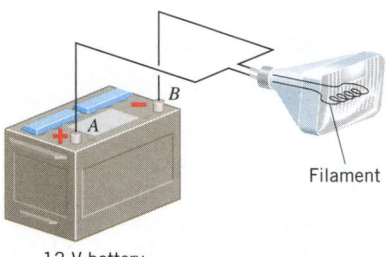

Higher
potential

Lower
potential

A   B   C

Figure 19.3 The electric potentials at points A, B, and C are different. Under the influence of these potentials, positive and negative charges accelerate in opposite directions.

Reasoning A positive charge accelerates from a region of higher potential toward a region of lower potential. A negative charge behaves in an opposite manner, because it accelerates from a region of lower potential toward a region of higher potential.

Answers (a) and (b) are incorrect. The positive charge accelerates from A to B and then from B to C. A negative charge placed at B also accelerates, but in a direction opposite to that of the positive charge. Therefore, a negative charge placed at B will not remain stationary, nor will it accelerate toward C.

Answer (c) is correct. Since the positive charge accelerates from A to B, the potential at A must exceed the potential at B. And since the positive test charge accelerates from B to C, the potential at B must exceed the potential at C. The potential at point B, then, must lie between the potential at points A and C, as Figure 19.3 illustrates. When the negative test charge is released from rest at B, it will accelerate toward the region of higher potential, so it will begin moving toward A.

B

+ A −

Filament

12-V battery

Figure 19.4 A headlight connected to a 12-V battery.

As a familiar application of electric potential energy and electric potential, Figure 19.4 shows a 12-V automobile battery with a headlight connected between its terminals. The positive terminal, point A, has a potential that is 12 V higher than the potential at the negative terminal, point B; in other words, $V_A - V_B = 12$ V. Positive charges would be repelled from the positive terminal and would travel through the wires and headlight toward the negative terminal.* As the charges pass through the headlight, virtually all their potential energy is converted into heat, which causes the filament to glow "white hot" and emit light. When the charges reach the negative terminal, they no longer have any potential energy. The battery then gives the charges an additional "shot" of potential energy by moving them to the higher-potential positive terminal, and the cycle is repeated. In raising the potential energy of the charges, the battery does work on them and draws from its reserve of chemical energy to do so. Example 3 illustrates the concepts of electric potential energy and electric potential as applied to a battery.

ANALYZING MULTIPLE-CONCEPT PROBLEMS

Example 3 Operating a Headlight

The wattage of the headlight in Figure 19.4 is 60.0 W. Determine the number of particles, each carrying a charge of 1.60×10^{-19} C (the magnitude of the charge on an electron), that pass between the terminals of the 12-V car battery when the headlight burns for one hour.

Reasoning The number of particles is the total charge that passes between the battery terminals in one hour divided by the magnitude of the charge on each particle. The total charge is the amount of charge needed to convey the energy used by the headlight in one hour. This energy is related to the wattage of the headlight, which specifies the power or rate at which energy is used, and the time the light is on.

Knowns and Unknowns The following table summarizes the data provided:

Description	Symbol	Value	Comment
Wattage of headlight	P	60.0 W	
Charge magnitude per particle	e	1.60×10^{-19} C	
Electric potential difference between battery terminals	$V_A - V_B$	12 V	See Figure 19.4.
Time headlight is on	t	3600 s	One hour.
Unknown Variable			
Number of particles	n	?	

*Historically, it was believed that positive charges flow in the wires of an electric circuit. Today, it is known that negative charges flow in wires from the negative toward the positive terminal. Here, however, we follow the customary practice of describing the flow of negative charges by specifying the opposite but equivalent flow of positive charges. This hypothetical flow of positive charges is called the "conventional electric current," as we will see in Section 20.1.

Modeling the Problem

STEP 1 **The Number of Particles** The number n of particles is the total charge q_0 that passes between the battery terminals in one hour divided by the magnitude e of the charge on each particle, as expressed by Equation 1 at the right. The value of e is given. To evaluate q_0, we proceed to Step 2.

$$n = \frac{q_0}{e} \quad (1)$$

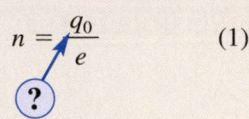

STEP 2 **The Total Charge Provided by the Battery** The battery must supply the total energy used by the headlight in one hour. The battery does this by supplying the charge q_0 to convey this energy. The energy is the difference between the electric potential energy EPE_A at terminal A and the electric potential energy EPE_B at terminal B of the battery (see Figure 19.4). According to Equation 19.4, this total energy is $\text{EPE}_A - \text{EPE}_B = q_0(V_A - V_B)$, where $V_A - V_B$ is the electric potential difference between the battery terminals. Solving this expression for q_0 gives

$$q_0 = \frac{\text{EPE}_A - \text{EPE}_B}{V_A - V_B}$$

which can be substituted into Equation 1, as shown at the right. As the data table shows, a value is given for $V_A - V_B$. In Step 3, we determine a value for $\text{EPE}_A - \text{EPE}_B$.

$$n = \frac{q_0}{e} \quad (1)$$

$$q_0 = \frac{\text{EPE}_A - \text{EPE}_B}{V_A - V_B} \quad (2)$$

?

STEP 3 **The Energy Used by the Headlight** The rate at which the headlight uses energy is the power P or wattage of the headlight. According to Equation 6.10b, the power is the total energy $\text{EPE}_A - \text{EPE}_B$ divided by the time t, so that $P = (\text{EPE}_A - \text{EPE}_B)/t$. Solving for the total energy gives

$$\text{EPE}_A - \text{EPE}_B = Pt$$

Since P and t are given, we substitute this result into Equation 2, as indicated at the right.

$$n = \frac{q_0}{e} \quad (1)$$

$$q_0 = \frac{\text{EPE}_A - \text{EPE}_B}{V_A - V_B} \quad (2)$$

$$\text{EPE}_A - \text{EPE}_B = Pt$$

Solution Combining the results of each step algebraically, we find that

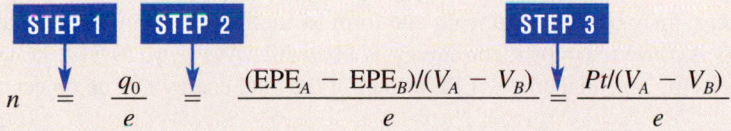

$$\overset{\text{STEP 1}}{n} = \overset{\text{STEP 2}}{\frac{q_0}{e}} = \frac{(\text{EPE}_A - \text{EPE}_B)/(V_A - V_B)}{e} = \overset{\text{STEP 3}}{\frac{Pt/(V_A - V_B)}{e}}$$

The number of particles that pass between the battery terminals in one hour is

$$n = \frac{Pt}{e(V_A - V_B)} = \frac{(60.0 \text{ W})(3600 \text{ s})}{(1.60 \times 10^{-19} \text{ C})(12 \text{ V})} = \boxed{1.1 \times 10^{23}}$$

Related Homework: *Problems 4, 5*

As used in connection with batteries, the volt is a familiar unit for measuring electric potential difference. The word "volt" also appears in another context, as part of a unit that is used to measure energy, particularly the energy of an atomic particle, such as an electron or a proton. This energy unit is called the *electron volt* (eV). **One electron volt is the magnitude of the amount by which the potential energy of an electron changes when the electron moves through a potential difference of one volt.** Since the magnitude of the change in potential energy is $|q_0 \Delta V| = |(-1.60 \times 10^{-19} \text{ C}) \times (1.00 \text{ V})| = 1.60 \times 10^{-19} \text{ J}$, it follows that

Problem-solving insight

$$1 \text{ eV} = 1.60 \times 10^{-19} \text{ J}$$

One million (10^{+6}) electron volts of energy is referred to as one MeV, and one billion (10^{+9}) electron volts of energy is one GeV, where the "G" stands for the prefix "giga" (pronounced "jig′a").

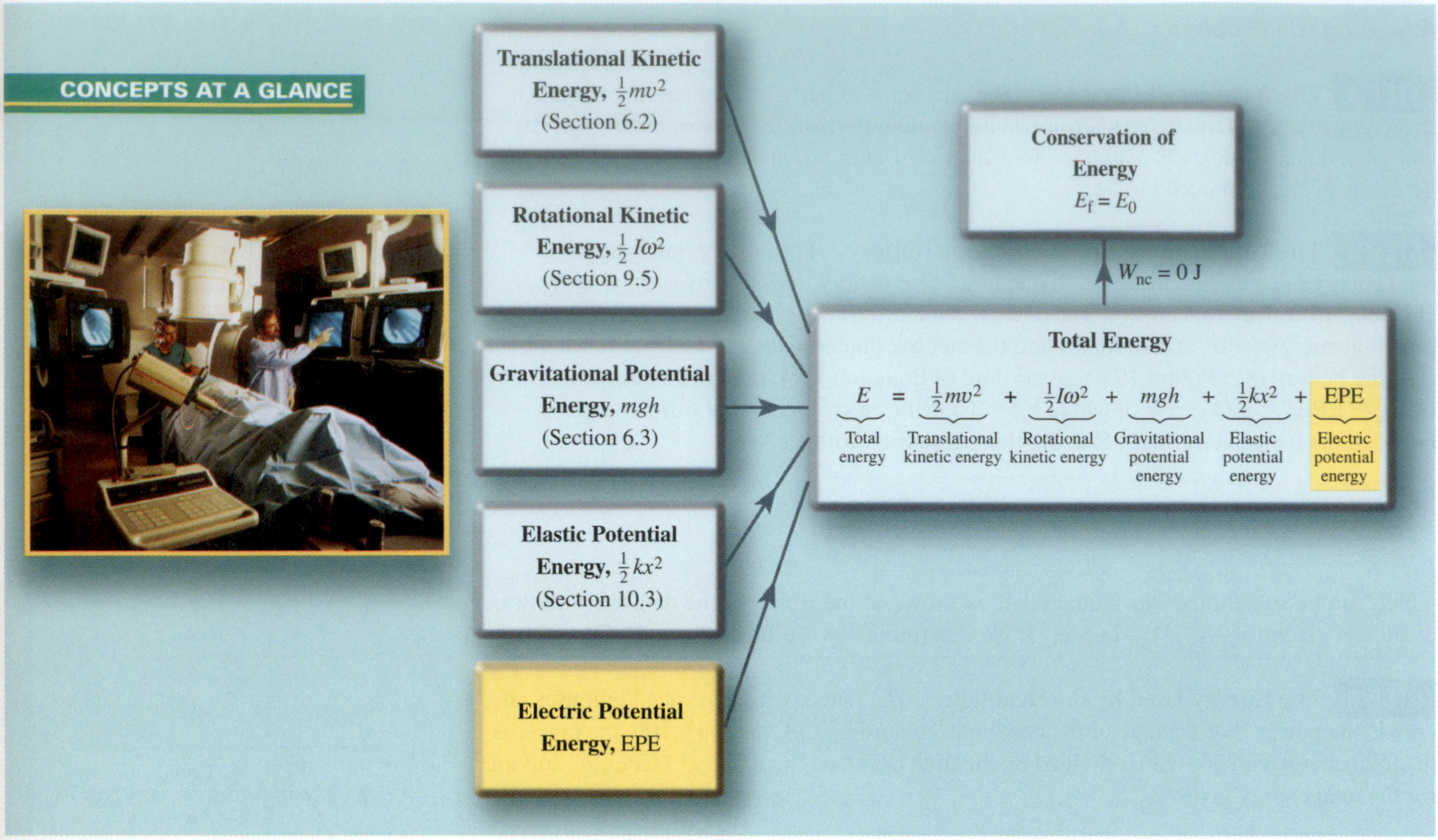

CONCEPTS AT A GLANCE

Translational Kinetic
Energy, $\frac{1}{2}mv^2$
(Section 6.2)

Rotational Kinetic
Energy, $\frac{1}{2}I\omega^2$
(Section 9.5)

Gravitational Potential
Energy, mgh
(Section 6.3)

Elastic Potential
Energy, $\frac{1}{2}kx^2$
(Section 10.3)

Electric Potential
Energy, EPE

Conservation of
Energy
$E_f = E_0$

$W_{nc} = 0$ J

Total Energy

$$E = \frac{1}{2}mv^2 + \frac{1}{2}I\omega^2 + mgh + \frac{1}{2}kx^2 + \boxed{EPE}$$

| Total energy | Translational kinetic energy | Rotational kinetic energy | Gravitational potential energy | Elastic potential energy | Electric potential energy |

Figure 19.5 CONCEPTS AT A GLANCE
Electric potential energy is another
type of energy. The patient in the
photograph is undergoing an X-ray
examination. To produce X-rays,
electrons are accelerated from rest
and allowed to collide with a target
material within the X-ray machine. As
each electron is accelerated, electric
potential energy is converted into
kinetic energy, and the total energy is
conserved during the process.
(Jonathan Nourok/Stone/Getty)

▶ **CONCEPTS AT A GLANCE** The Concepts-at-a-Glance chart in Figure 19.5 emphasizes
that the total energy of an object, which is the sum of its kinetic and potential energies, is
an important concept. Its significance lies in the fact that the total energy remains the same
(is conserved) during the object's motion, provided that nonconservative forces, such as friction, are either absent or do no net work. While the sum of the energies at each instant remains
constant, energy may be converted from one form to another; for example, gravitational potential energy is converted into kinetic energy as a ball falls. As Figure 19.5 illustrates, we now
include the electric potential energy EPE as part of the total energy that an object can have:

$$E = \frac{1}{2}mv^2 + \frac{1}{2}I\omega^2 + mgh + \frac{1}{2}kx^2 + EPE$$

| Total energy | Translational kinetic energy | Rotational kinetic energy | Gravitational potential energy | Elastic potential energy | Electric potential energy |

If the total energy is conserved as the object moves, then its final energy E_f is equal to its
initial energy E_0, or $E_f = E_0$. Example 4 illustrates how the conservation of energy is applied to a charge moving in an electric field. ◀

ANALYZING MULTIPLE-CONCEPT PROBLEMS

Example 4 The Conservation of Energy

A particle has a mass of 1.8×10^{-5} kg and a charge of
$+3.0 \times 10^{-5}$ C. It is released from rest at point A and accelerates until it reaches point B, as Figure 19.6a shows.
The particle moves on a horizontal straight line and does
not rotate. The only forces acting on the particle are the
gravitational force and an electrostatic force (neither is
shown in the drawing). The electric potential at A is 25 V
greater than that at B; in other words, $V_A - V_B = 25$ V.
What is the translational speed of the particle at point B?

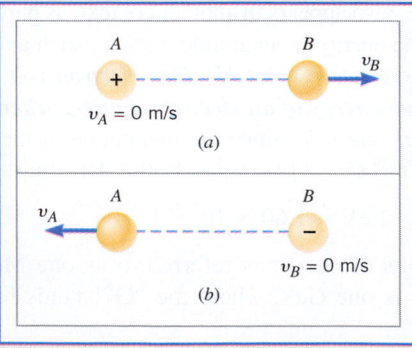

Figure 19.6 (a) A positive
charge starts from rest at
point A and accelerates toward
point B. (b) A negative charge
starts from rest at B and
accelerates toward A.

Reasoning The translational speed of the particle is related to the particle's translational kinetic energy, which forms one part of the total mechanical energy that the particle has. The total mechanical energy is conserved, because only the gravitational force and an electrostatic force, both of which are conservative forces, act on the particle (see Section 6.5). Thus, we will determine the speed at point B by utilizing the principle of conservation of mechanical energy.

Knowns and Unknowns We have the following data:

Description	Symbol	Value	Comment
Explicit Data			
Mass of particle	m	1.8×10^{-5} kg	
Electric charge of particle	q_0	$+3.0 \times 10^{-5}$ C	
Electric potential difference between points A and B	$V_A - V_B$	25 V	See Figure 19.6a.
Implicit Data			
Speed at point A	v_A	0 m/s	Particle released from rest.
Vertical height above ground	h	Remains constant	Particle travels horizontally.
Angular speed	ω	0 rad/s	Particle does not rotate during motion.
Elastic force	$F_{elastic}$	0 N	No elastic force acts on particle.
Unknown Variable			
Speed at point B	v_B	?	

Modeling the Problem

STEP 1 Conservation of Total Mechanical Energy   The particle's total mechanical energy E is

$$E = \underbrace{\tfrac{1}{2}mv^2}_{\substack{\text{Translational} \\ \text{kinetic energy}}} + \underbrace{\tfrac{1}{2}I\omega^2}_{\substack{\text{Rotational} \\ \text{kinetic energy}}} + \underbrace{mgh}_{\substack{\text{Gravitational} \\ \text{potential} \\ \text{energy}}} + \underbrace{\tfrac{1}{2}kx^2}_{\substack{\text{Elastic} \\ \text{potential} \\ \text{energy}}} + \underbrace{\text{EPE}}_{\substack{\text{Electric} \\ \text{potential} \\ \text{energy}}}$$

Since the particle does not rotate, the angular speed ω is always zero (see the data table) and since there is no elastic force (see the data table), we may omit the terms $\tfrac{1}{2}I\omega^2$ and $\tfrac{1}{2}kx^2$ from this expression. With this in mind, we express the fact that $E_B = E_A$ (energy is conserved) as follows:

$$\tfrac{1}{2}mv_B^2 + mgh_B + \text{EPE}_B = \tfrac{1}{2}mv_A^2 + mgh_A + \text{EPE}_A$$

This equation can be simplified further, since the particle travels horizontally, so that $h_B = h_A$ (see the data table), with the result that

$$\tfrac{1}{2}mv_B^2 + \text{EPE}_B = \tfrac{1}{2}mv_A^2 + \text{EPE}_A$$

Solving for v_B gives Equation 1 at the right. Values for v_A and m are available, and we turn to Step 2 in order to evaluate $\text{EPE}_A - \text{EPE}_B$.

$$v_B = \sqrt{v_A^2 + \frac{2(\text{EPE}_A - \text{EPE}_B)}{m}} \quad (1)$$

$\boxed{?}$

STEP 2 The Electric Potential Difference   According to Equation 19.4, the difference in electric potential energies $\text{EPE}_A - \text{EPE}_B$ is related to the electric potential difference $V_A - V_B$:

$$\boxed{\text{EPE}_A - \text{EPE}_B = q_0(V_A - V_B)}$$

$$v_B = \sqrt{v_A^2 + \frac{2(\text{EPE}_A - \text{EPE}_B)}{m}} \quad (1)$$

$$\boxed{\text{EPE}_A - \text{EPE}_B = q_0(V_A - V_B)}$$

The terms q_0 and $V_A - V_B$ are known, so we substitute this expression into Equation 1, as shown at the right.

Continued

Solution Combining the results of each step algebraically, we find that

STEP 1	STEP 2

$$v_B = \sqrt{v_A^2 + \frac{2(\text{EPE}_A - \text{EPE}_B)}{m}} = \sqrt{v_A^2 + \frac{2q_0(V_A - V_B)}{m}}$$

The speed of the particle at point B is

$$v_B = \sqrt{v_A^2 + \frac{2q_0(V_A - V_B)}{m}} = \sqrt{(0 \text{ m/s})^2 + \frac{2(+3.0 \times 10^{-5} \text{ C})(25 \text{ V})}{1.8 \times 10^{-5} \text{ kg}}} = \boxed{9.1 \text{ m/s}}$$

Note that if the particle had a negative charge of -3.0×10^{-5} C and were released from rest at point B, it would arrive at point A with the same speed of 9.1 m/s (see Figure 19.6b). This result can be obtained by returning to Modeling Step 1 and solving for v_A instead of v_B.

Related Homework: *Problems 2, 7, 8*

Problem-solving insight

A positive charge accelerates from a region of higher potential toward a region of lower potential. In contrast, a negative charge accelerates from a region of lower potential toward a region of higher potential.

✓ **CHECK YOUR UNDERSTANDING**

(The answers are given at the end of the book.)

1. An ion, starting from rest, accelerates from point *A* to point *B* due to a potential difference between the two points. Does the electric potential energy of the ion at point *B* depend on **(a)** the magnitude of its charge and **(b)** its mass? Does the speed of the ion at *B* depend on **(c)** the magnitude of its charge and **(d)** its mass?

2. The drawing shows three possibilities for the potentials at two points, *A* and *B*. In each case, the same positive charge is moved from *A* to *B*. In which case, if any, is the most work done on the positive charge by the electric force?

A	B	A	B	A	B
•	•	•	•	•	•
150 V	100 V	25 V	−25 V	−10 V	−60 V
Case 1		Case 2		Case 3	

3. A proton and an electron are released from rest at the midpoint between the plates of a charged parallel plate capacitor (see Chapter 18). Except for these particles, nothing else is between the plates. Ignore the attraction between the proton and the electron, and decide which particle strikes a capacitor plate first.

19.3 | **THE ELECTRIC POTENTIAL DIFFERENCE CREATED BY POINT CHARGES**

A positive point charge $+q$ creates an electric potential in a way that Figure 19.7 helps explain. This picture shows two locations A and B, at distances r_A and r_B from the charge. At any position between A and B an electrostatic force of repulsion $\vec{F}$ acts on a positive test charge $+q_0$. The magnitude of the force is given by Coulomb's law as $F = kq_0q/r^2$, where we assume for convenience that q_0 and q are positive, so that $|q_0| = q_0$ and $|q| = q$. When the test charge moves from A to B, work is done by this force. Since r varies between r_A and r_B, the force F also varies, and the work is not the product of the force and the distance between the points. (Recall from Section 6.1 that work is force times distance only if the force is constant.) However, the work W_{AB} can be found with the methods of integral calculus. The result is

$$W_{AB} = \frac{kqq_0}{r_A} - \frac{kqq_0}{r_B}$$

This result is valid whether q is positive or negative, and whether q_0 is positive or negative. The potential difference, $V_B - V_A$, between A and B can now be determined by substituting this expression for W_{AB} into Equation 19.4:

$$V_B - V_A = \frac{-W_{AB}}{q_0} = \frac{kq}{r_B} - \frac{kq}{r_A} \qquad (19.5)$$

As point B is located farther and farther from the charge q, r_B becomes larger and larger. In the limit that r_B is infinitely large, the term kq/r_B becomes zero, and it is

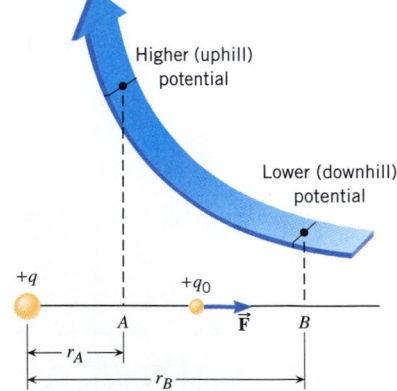

Figure 19.7 The positive test charge $+q_0$ experiences a repulsive force $\vec{F}$ due to the positive point charge $+q$. As a result, work is done by this force when the test charge moves from A to B. Consequently, the electric potential is higher (uphill) at A and lower (downhill) at B.

Higher (uphill) potential

Lower (downhill) potential

customary to set V_B equal to zero also. In this limit, Equation 19.5 becomes $V_A = kq/r_A$, and it is standard convention to omit the subscripts and write the potential in the following form:

Potential of a point charge
$$V = \frac{kq}{r} \qquad (19.6)$$

The symbol V in this equation does not refer to the potential in any absolute sense. Rather, $V = kq/r$ stands for the amount by which the potential at a distance r from a point charge differs from the potential at an infinite distance away. In other words, V refers to a potential difference with the arbitrary assumption that the potential at infinity is zero.

With the aid of Equation 19.6, we can describe the effect that a point charge q has on the surrounding space. When q is positive, the value of $V = kq/r$ is also positive, indicating that the positive charge has everywhere raised the potential above the zero reference value. Conversely, when q is negative, the potential V is also negative, indicating that the negative charge has everywhere decreased the potential below the zero reference value. The next example deals with these effects quantitatively.

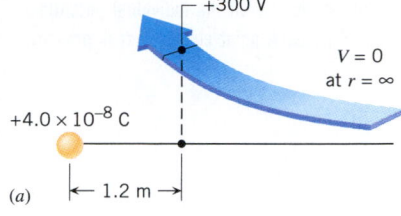

(a)

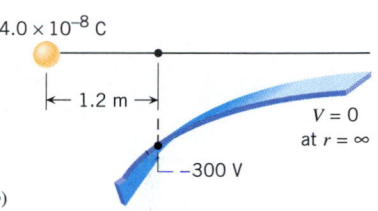

(b)

Figure 19.8 A point charge of 4.0×10^{-8} C alters the potential at a spot 1.2 m away. The potential is (a) increased by 300 V when the charge is positive and (b) decreased by 300 V when the charge is negative, relative to a zero reference potential at infinity.

 Example 5 The Potential of a Point Charge

Using a zero reference potential at infinity, determine the amount by which a point charge of 4.0×10^{-8} C alters the electric potential at a spot 1.2 m away when the charge is **(a)** positive and **(b)** negative.

Reasoning A point charge q alters the potential at every location in the surrounding space. In the expression $V = kq/r$, the effect of the charge in increasing or decreasing the potential is conveyed by the algebraic sign for the value of q.

Solution **(a)** Figure 19.8a shows the potential when the charge is positive:

$$V = \frac{kq}{r} = \frac{(8.99 \times 10^9 \ \text{N}\cdot\text{m}^2/\text{C}^2)(+4.0 \times 10^{-8} \ \text{C})}{1.2 \ \text{m}} = \boxed{+300 \ \text{V}} \qquad (19.6)$$

(b) Part b of the drawing illustrates the results when the charge is negative. A calculation similar to the one in part (a) shows that the potential is now negative: $\boxed{-300 \ \text{V}}$.

A single point charge raises or lowers the potential at a given location, depending on whether the charge is positive or negative. **When two or more charges are present, the potential due to all the charges is obtained by adding together the individual potentials,** as the next two examples show.

Problem-solving insight

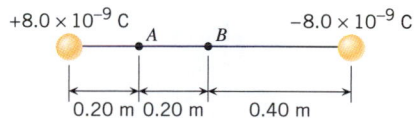 **Example 6** The Total Electric Potential

At locations A and B in Figure 19.9, find the total electric potential due to the two point charges.

Reasoning At each location, each charge contributes to the total electric potential. We obtain the individual contributions by using $V = kq/r$ and find the total potential by adding the individual contributions algebraically. The two charges have the same magnitude, but different signs. Thus, at A the total potential is positive because this spot is closer to the positive charge, whose effect dominates over that of the more distant negative charge. At B, midway between the charges, the total potential is zero, since the potential of one charge exactly offsets that of the other.

Figure 19.9 Both the positive and negative charges affect the electric potential at locations A and B.

Solution

Location	Contribution from + Charge		Contribution from − Charge	Total Potential
A	$\dfrac{(8.99 \times 10^9 \ \text{N}\cdot\text{m}^2/\text{C}^2)(+8.0 \times 10^{-9} \ \text{C})}{0.20 \ \text{m}}$	$+$	$\dfrac{(8.99 \times 10^9 \ \text{N}\cdot\text{m}^2/\text{C}^2)(-8.0 \times 10^{-9} \ \text{C})}{0.60 \ \text{m}} =$	$\boxed{+240 \ \text{V}}$
B	$\dfrac{(8.99 \times 10^9 \ \text{N}\cdot\text{m}^2/\text{C}^2)(+8.0 \times 10^{-9} \ \text{C})}{0.40 \ \text{m}}$	$+$	$\dfrac{(8.99 \times 10^9 \ \text{N}\cdot\text{m}^2/\text{C}^2)(-8.0 \times 10^{-9} \ \text{C})}{0.40 \ \text{m}} =$	$\boxed{0 \ \text{V}}$

Figure 19.10 Two point charges, one positive and one negative. The positive charge, $+2q$, has twice the magnitude of the negative charge, $-q$.

Problem-solving insight

At any location, the total electric potential is the algebraic sum of the individual potentials created by each point charge that is present.

Conceptual Example 7 Where Is the Potential Zero?

Two point charges are fixed in place, as in Figure 19.10. The positive charge is $+2q$ and has twice the magnitude of the negative charge, which is $-q$. On the line that passes through the charges, three spots are identified, P_1, P_2, and P_3. At which of these spots could the potential be equal to zero? **(a)** P_2 and P_3 **(b)** P_1 and P_3 **(c)** P_1 and P_2

Reasoning The total potential is the algebraic sum of the individual potentials created by each charge. It will be zero if the potential due to the positive charge is exactly offset by the potential due to the negative charge. The potential of a point charge is directly proportional to the charge and inversely proportional to the distance from the charge.

Answers (b) and (c) are incorrect. The total potential at P_1 cannot be zero. The positive charge has the larger magnitude and is closer to P_1 than is the negative charge. As a result, the potential of the positive charge at P_1 dominates over the potential of the negative charge, so the total potential cannot be zero.

Answer (a) is correct. Between the charges there is a location at which the individual potentials cancel each other. We saw a similar situation in Example 6, where the canceling occurred at the midpoint between the two charges that had equal magnitudes. Now the charges have unequal magnitudes, so the cancellation point does not occur at the midpoint. Instead, it occurs at the location P_2, which is closer to the charge with the smaller magnitude—namely, the negative charge. At P_2, since the potential of a point charge is inversely proportional to the distance from the charge, the effect of the smaller charge will be able to offset the effect of the more distant larger charge.

To the right of the negative charge there is also a location at which the individual potentials exactly cancel each other. All places on this section of the line are closer to the negative charge than to the positive charge. Therefore, there is a location P_3 in this region at which the potential of the smaller negative charge exactly cancels the potential of the more distant and larger positive charge.

Related Homework: *Problem 66*

In Example 6 we determined the total potential at a spot due to two point charges. In Example 8 we now extend this technique to find the total potential *energy* of three charges.

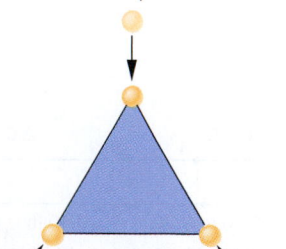

Figure 19.11 Three point charges are placed on the corners of an equilateral triangle. Example 8 illustrates how to determine the total electric potential energy of this group of charges.

Example 8 The Potential Energy of a Group of Charges

Three point charges initially are infinitely far apart. Then, as Figure 19.11 shows, they are brought together and placed at the corners of an equilateral triangle. Each side of the triangle has a length of 0.50 m. Determine the electric potential energy of the triangular group. In other words, determine the amount by which the electric potential energy of the group differs from that of the three charges in their initial, infinitely separated, locations.

Reasoning We will proceed in steps by adding charges to the triangle one at a time, and then determining the electric potential energy at each step. According to Equation 19.3, EPE $= q_0V$, the electric potential energy is the product of the charge and the electric potential at the spot where the charge is placed. The total electric potential energy of the triangular group is the sum of the energies of each step in assembling the group.

Solution The order in which the charges are put on the triangle does not matter; we begin with the charge of $+5.0$ μC. When this charge is placed at a corner of the triangle, it has no electric potential energy, according to EPE $= q_0V$. This is because the total potential V produced by the other two charges is zero at this corner, since they are infinitely far away. Once the charge is in place, the potential it creates at either empty corner ($r = 0.50$ m) is

$$V = \frac{kq}{r} = \frac{(8.99 \times 10^9 \text{ N} \cdot \text{m}^2/\text{C}^2)(+5.0 \times 10^{-6} \text{ C})}{0.50 \text{ m}} = +9.0 \times 10^4 \text{ V} \quad (19.6)$$

Therefore, when the $+6.0$ μC charge is placed at the second corner of the triangle, its electric potential energy is

$$\text{EPE} = qV = (+6.0 \times 10^{-6} \text{ C})(+9.0 \times 10^4 \text{ V}) = +0.54 \text{ J} \quad (19.3)$$

The electric potential at the remaining empty corner is the sum of the potentials due to the two charges that are already in place:

$$V = \frac{(8.99 \times 10^9 \text{ N} \cdot \text{m}^2/\text{C}^2)(+5.0 \times 10^{-6} \text{ C})}{0.50 \text{ m}}$$

$$+ \frac{(8.99 \times 10^9 \text{ N} \cdot \text{m}^2/\text{C}^2)(+6.0 \times 10^{-6} \text{ C})}{0.50 \text{ m}} = +2.0 \times 10^5 \text{ V}$$

When the third charge, $-2.0\ \mu\text{C}$, is placed at the remaining empty corner, its electric potential energy is

$$\text{EPE} = qV = (-2.0 \times 10^{-6} \text{ C})(+2.0 \times 10^5 \text{ V}) = -0.40 \text{ J} \qquad (19.3)$$

The total potential energy of the triangular group differs from that of the widely separated charges by an amount that is the sum of the potential energies calculated above:

$$\text{Total potential energy} = 0 \text{ J} + 0.54 \text{ J} - 0.40 \text{ J} = \boxed{+0.14 \text{ J}}$$

This energy originates in the work done to bring the charges together.

Problem-solving insight

Be careful to distinguish between the concepts of potential V and electric potential energy EPE. Potential is electric potential energy per unit charge: $V = \text{EPE}/q$.

✓ CHECK YOUR UNDERSTANDING

(The answers are given at the end of the book.)

4. The drawing shows four arrangements (A–D) of two point charges. In each arrangement consider the total electric potential that the charges produce at location P. Rank the arrangements (largest to smallest) according to the total potential. **(a)** B, C, A and D (a tie) **(b)** D, C, A, B **(c)** A and C (a tie), B, D **(d)** C, D, A, B

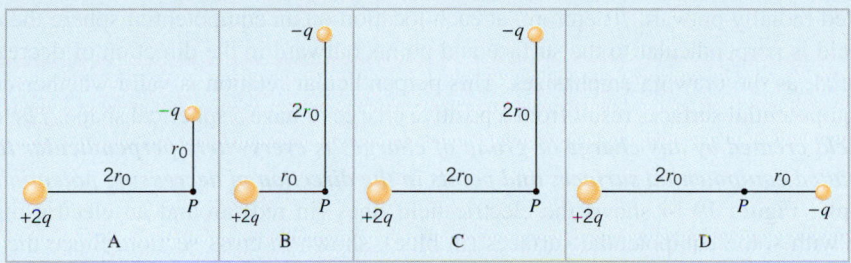

5. A positive point charge and a negative point charge have equal magnitudes. One charge is fixed to one corner of a square, and the other is fixed to another corner. On which corners should the charges be placed, so that the same potential exists at the empty corners? The charges should be placed at **(a)** adjacent corners, **(b)** diagonally opposite corners.

6. Three point charges have identical magnitudes, but two of the charges are positive and one is negative. These charges are fixed to the corners of a square, one to a corner. No matter how the charges are arranged, the potential at the empty corner is always **(a)** zero, **(b)** negative, **(c)** positive.

7. Consider a spot that is located midway between two identical point charges. Which one of the following statements concerning the electric field and the electric potential at this spot is true? **(a)** The electric field is zero, but the electric potential is not zero. **(b)** The electric field is not zero, but the electric potential is zero. **(c)** Both the electric field and the electric potential are zero. **(d)** Neither the electric field nor the electric potential is zero.

8. Four point charges have the same magnitude (but they may have different signs) and are placed at the corners of a square, as the drawing shows. What must be the sign (+ or −) of each charge so that *both* the electric field *and* the electric potential are zero at the center of the square? Assume that the potential has a zero value at infinity.

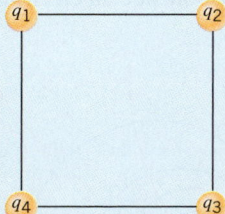

	q_1	q_2	q_3	q_4
(a)	−	−	−	−
(b)	+	+	−	−
(c)	+	+	+	+
(d)	+	−	+	−

9. An electric potential energy exists when two protons are separated by a certain distance. Does the electric potential energy increase, decrease, or remain the same **(a)** when both protons are replaced by electrons, and **(b)** when only one of the protons is replaced by an electron?

10. A proton is fixed in place. An electron is released from rest and allowed to collide with the proton. Then the roles of the proton and electron are reversed, and the same experiment is repeated. Which, if either, is traveling faster when the collision occurs, the proton or the electron?

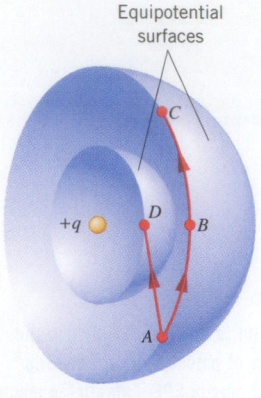

Equipotential
surfaces

Figure 19.12 The equipotential surfaces that surround the point charge $+q$ are spherical. The electric force does no work as a charge moves on a path that lies on an equipotential surface, such as the path *ABC*. However, work is done by the electric force when a charge moves between two equipotential surfaces, as along the path *AD*.

Problem-solving insight

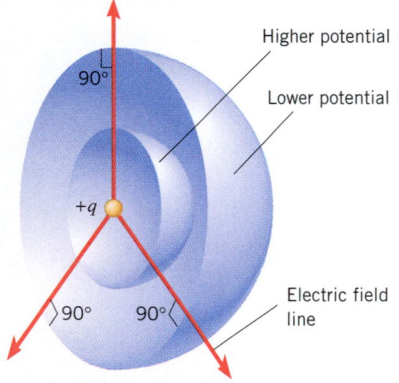

Higher potential

Lower potential

Electric field line

Figure 19.13 The radially directed electric field of a point charge is perpendicular to the spherical equipotential surfaces that surround the charge. The electric field points in the direction of *decreasing* potential.

Figure 19.14 A cross-sectional view of the equipotential surfaces (in blue) of an electric dipole. The surfaces are drawn to show that at every point they are perpendicular to the electric field lines (in red) of the dipole.

19.4 EQUIPOTENTIAL SURFACES AND THEIR RELATION TO THE ELECTRIC FIELD

An *equipotential surface* is a surface on which the electric potential is the same everywhere. The easiest equipotential surfaces to visualize are those that surround an isolated point charge. According to Equation 19.6, the potential at a distance r from a point charge q is $V = kq/r$. Thus, wherever r is the same, the potential is the same, and the equipotential surfaces are spherical surfaces centered on the charge. There are an infinite number of such surfaces, one for every value of r, and Figure 19.12 illustrates two of them. The larger the distance r, the smaller is the potential of the equipotential surface.

The net electric force does no work as a charge moves on an equipotential surface. This important characteristic arises because when an electric force does work W_{AB} as a charge moves from A to B, the potential changes according to $V_B - V_A = -W_{AB}/q_0$ (Equation 19.4). Since the potential remains the same everywhere on an equipotential surface, $V_A = V_B$, and we see that $W_{AB} = 0$ J. In Figure 19.12, for instance, the electric force does no work as a test charge moves along the circular arc *ABC*, which lies on an equipotential surface. In contrast, the electric force does work when a charge moves *between* equipotential surfaces, as from A to D in the picture.

The spherical equipotential surfaces that surround an isolated point charge illustrate another characteristic of all equipotential surfaces. Figure 19.13 shows two of the surfaces around a positive point charge, along with some electric field lines. The electric field lines give the direction of the electric field, and for a positive point charge the electric field is directed radially outward. Therefore, at each location on an equipotential sphere the electric field is perpendicular to the surface and points outward in the direction of decreasing potential, as the drawing emphasizes. This perpendicular relation is valid whether or not the equipotential surfaces result from a positive charge or have a spherical shape. ***The electric field created by any charge or group of charges is everywhere perpendicular to the associated equipotential surfaces and points in the direction of decreasing potential.*** For example, Figure 19.14 shows the electric field lines (in red) around an electric dipole, along with some equipotential surfaces (in blue), shown in cross section. Since the field lines are not simply radial, the equipotential surfaces are no longer spherical but, instead, have the shape necessary to be everywhere perpendicular to the field lines.

To see why an equipotential surface must be perpendicular to the electric field, consider Figure 19.15, which shows a hypothetical situation in which the perpendicular relation does *not* hold. If $\vec{\mathbf{E}}$ were not perpendicular to the equipotential surface, there would be a component of $\vec{\mathbf{E}}$ parallel to the surface. This field component would exert an electric force on a test charge placed on the surface. As the charge moved along the surface, work would be done by this component of the electric force. The work, according to Equation 19.4, would cause the potential to change, and, thus, the surface could not be an equipotential surface as assumed. The only way out of the dilemma is for the electric field to be perpendicular to the surface, so there is no component of the field parallel to the surface.

We have already encountered one equipotential surface. In Section 18.8, we found that the direction of the electric field just outside an electrical conductor is perpendicular to the

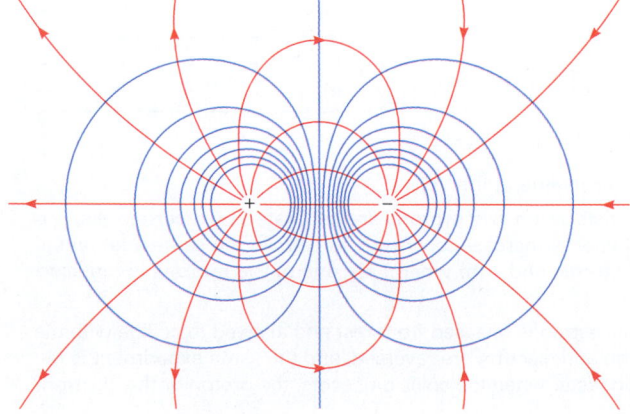

conductor's surface, when the conductor is at equilibrium under electrostatic conditions. Thus, the surface of any conductor is an equipotential surface under such conditions. In fact, since the electric field is zero everywhere inside a conductor whose charges are in equilibrium, the entire conductor can be regarded as an equipotential volume.

There is a quantitative relation between the electric field and the equipotential surfaces. One example that illustrates this relation is the parallel plate capacitor in Figure 19.16. As Section 18.6 discusses, the electric field $\vec{E}$ between the metal plates is perpendicular to them and is the same everywhere, ignoring fringe fields at the edges. To be perpendicular to the electric field, the equipotential surfaces must be planes that are parallel to the capacitor plates, which themselves are equipotential surfaces. The potential difference between the plates is given by Equation 19.4 as $\Delta V = V_B - V_A = -W_{AB}/q_0$, where A is a point on the positive plate and B is a point on the negative plate. The work done by the electric force as a positive test charge q_0 moves from A to B is $W_{AB} = F\Delta s$, where F refers to the electric force and Δs to the displacement along a line perpendicular to the plates. The force equals the product of the charge and the electric field E ($F = q_0 E$), so the work becomes $W_{AB} = F\Delta s = q_0 E \Delta s$. Therefore, the potential difference between the capacitor plates can be written in terms of the electric field as $\Delta V = -W_{AB}/q_0 = -q_0 E \Delta s/q_0$, or

$$E = -\frac{\Delta V}{\Delta s} \qquad (19.7a)$$

The quantity $\Delta V/\Delta s$ is referred to as the *potential gradient* and has units of volts per meter. In general, the relation $E = -\Delta V/\Delta s$ gives only the component of the electric field along the displacement Δs; it does not give the perpendicular component. The next example deals further with the equipotential surfaces between the plates of a capacitor.

 Example 9 The Electric Field and Potential Are Related

The plates of the capacitor in Figure 19.16 are separated by a distance of 0.032 m, and the potential difference between them is $\Delta V = V_B - V_A = -64$ V. Between the two labeled equipotential surfaces there is a potential difference of -3.0 V. Find the spacing between the two labeled surfaces.

Reasoning The electric field is $E = -\Delta V/\Delta s$. To find the spacing between the two labeled equipotential surfaces, we solve this equation for Δs, with $\Delta V = -3.0$ V and E equal to the electric field between the plates of the capacitor. A value for E can be obtained by using the values given for the distance and potential difference between the plates.

Solution The spacing between the labeled equipotential surfaces is given by

$$\Delta s = -\frac{\Delta V}{E} \qquad (1)$$

The electric field between the capacitor plates is

$$E = -\frac{\Delta V}{\Delta s} = -\frac{-64 \text{ V}}{0.032 \text{ m}} = 2.0 \times 10^3 \text{ V/m} \qquad (19.7a)$$

Substituting Equation 19.7a into Equation 1 gives

$$\Delta s = -\frac{\Delta V}{E} = -\frac{-3.0 \text{ V}}{2.0 \times 10^3 \text{ V/m}} = \boxed{1.5 \times 10^{-3} \text{ m}}$$

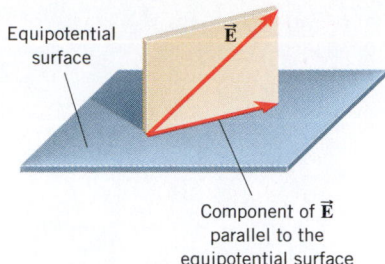

Figure 19.15 In this hypothetical situation, the electric field $\vec{E}$ is not perpendicular to the equipotential surface. As a result, there is a component of $\vec{E}$ parallel to the surface.

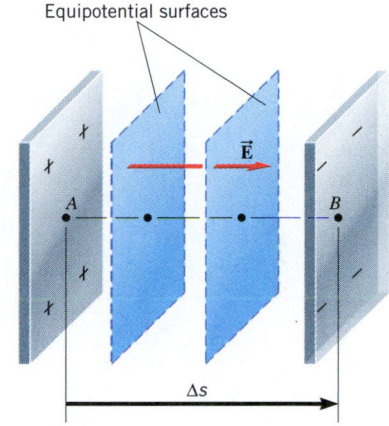

Figure 19.16 The metal plates of a parallel plate capacitor are equipotential surfaces. Two additional equipotential surfaces are shown between the plates. These two equipotential surfaces are parallel to the plates and are perpendicular to the electric field $\vec{E}$ between the plates.

Equation 19.7a gives the relationship between the electric field and the electric potential. It gives the component of the electric field along the displacement Δs in a region of space where the electric potential changes from place to place and applies to a wide variety of situations. When applied strictly to a parallel plate capacitor, however, this expression is often used in a slightly different form. In Figure 19.16, the metal plates of the capacitor are marked A (higher potential) and B (lower potential). Traditionally, in discussions of such a capacitor, the potential difference between the plates is referred to by using the symbol V to denote the amount by which the higher potential exceeds the lower potential ($V = V_A - V_B$). In this tradition, the symbol V is often referred to as simply the "voltage." For example, if the potential difference between the plates of a capacitor is 5 volts, it is common to say that the "voltage" of the capacitor is 5 volts. In addition, the displacement from plate A to plate B is expressed in terms of the separation d between the plates ($d = s_B - s_A$). With this nomenclature, Equation 19.7a becomes

$$E = -\frac{\Delta V}{\Delta s} = -\frac{V_B - V_A}{s_B - s_A} = \frac{V_A - V_B}{s_B - s_A} = \frac{V}{d} \qquad \text{(parallel plate capacitor)} \quad (19.7b)$$

✓ CHECK YOUR UNDERSTANDING

(The answers are given at the end of the book.)

11. The drawing shows a cross-sectional view of two spherical equipotential surfaces and two electric field lines that are perpendicular to these surfaces. When an electron moves from point A to point B (against the electric field), the electric force does $+3.2 \times 10^{-19}$ J of work. What are the electric potential differences **(a)** $V_B - V_A$, **(b)** $V_C - V_B$, and **(c)** $V_C - V_A$?

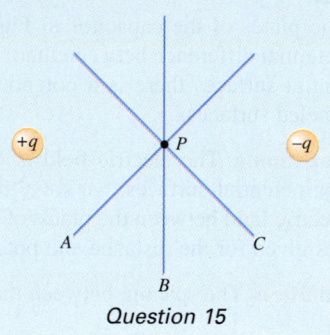

Electric field lines

Equipotential surfaces (Cross-sectional view)

Question 11

12. The electric potential is constant throughout a given region of space. Is the electric field zero or nonzero in this region?

13. In a region of space where the electric field is constant everywhere, as it is inside a parallel plate capacitor, is the potential constant everywhere? **(a)** Yes. **(b)** No, the potential is greatest at the positive plate. **(c)** No, the potential is greatest at the negative plate.

14. A positive test charge is placed in an electric field. In what direction should the charge be moved relative to the field, so that the charge experiences a constant electric potential? The charge should be moved **(a)** perpendicular to the electric field, **(b)** in the same direction as the electric field, **(c)** opposite to the direction of the electric field.

15. The location marked P in the drawing lies midway between the point charges $+q$ and $-q$. The blue lines labeled A, B, and C are edge-on views of three planes. Which of the planes is an equipotential surface? **(a)** A and C **(b)** A, B, and C **(c)** Only B **(d)** None of the planes is an equipotential surface.

16. Imagine that you are moving a positive test charge along the line between two identical point charges. With regard to the electric potential, is the midpoint on the line analogous to the top of a mountain or the bottom of a valley when the two point charges are **(a)** positive and **(b)** negative?

Question 15

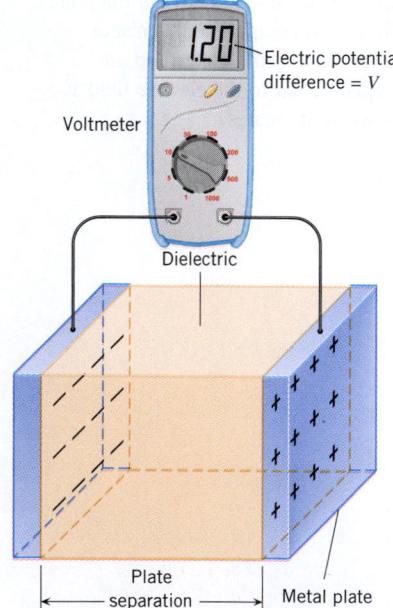

Voltmeter

Electric potential difference = V

Dielectric

Plate separation = d

Metal plate (area = A)

Figure 19.17 A parallel plate capacitor consists of two metal plates, one carrying a charge $+q$ and the other a charge $-q$. The potential of the positive plate exceeds that of the negative plate by an amount V. The region between the plates is filled with a dielectric.

19.5 CAPACITORS AND DIELECTRICS

THE CAPACITANCE OF A CAPACITOR

In Section 18.6 we saw that a parallel plate capacitor consists of two parallel metal plates placed near one another but not touching. This type of capacitor is only one among many. In general, a *capacitor* consists of two conductors of any shape placed near one another without touching. For a reason that will become clear later on, it is common practice to fill the region between the conductors or plates with an electrically insulating material called a *dielectric,* as Figure 19.17 illustrates.

A capacitor stores electric charge. Each capacitor plate carries a charge of the *same magnitude,* one positive and the other negative. Because of the charges, the electric potential of the positive plate exceeds that of the negative plate by an amount V, as Figure 19.17 indicates. Experiment shows that when the magnitude q of the charge on each plate is doubled, the magnitude V of the electric potential difference is also doubled, so q is proportional to V: $q \propto V$. Equation 19.8 expresses this proportionality with the aid of a proportionality constant C, which is the **capacitance** of the capacitor.

THE RELATION BETWEEN CHARGE AND POTENTIAL DIFFERENCE FOR A CAPACITOR

The magnitude q of the charge on each plate of a capacitor is directly proportional to the magnitude V of the potential difference between the plates:

$$q = CV \qquad (19.8)$$

where C is the capacitance.

SI Unit of Capacitance: coulomb/volt = farad (F)

Equation 19.8 shows that the SI unit of capacitance is the coulomb per volt (C/V). This unit is called the *farad* (F), named after the English scientist Michael Faraday (1791–1867). One farad is an enormous capacitance. Usually smaller amounts, such as a microfarad (1 μF = 10^{-6} F) or a picofarad (1 pF = 10^{-12} F), are used in electric circuits. The capacitance reflects the ability of the capacitor to store charge, in the sense that a larger capacitance C allows more charge q to be put onto the plates for a given value of the potential difference V.

The physics of random-access memory (RAM) chips. The ability of a capacitor to store charge lies at the heart of the random-access memory (RAM) chips used in computers, where information is stored in the form of the "ones" and "zeros" that comprise binary numbers. Figure 19.18 illustrates the role of a capacitor in a RAM chip. The capacitor is connected to a transistor switch, to which two lines are connected, an address line and a data line. A single RAM chip often contains millions of such transistor–capacitor units. The address line is used by the computer to locate a particular transistor–capacitor combination, and the data line carries the data to be stored. A pulse on the address line turns on the transistor switch. With the switch turned on, a pulse coming in on the data line can cause the capacitor to charge. A charged capacitor means that a "one" has been stored, whereas an uncharged capacitor means that a "zero" has been stored.

THE DIELECTRIC CONSTANT

If a dielectric is inserted between the plates of a capacitor, the capacitance can increase markedly because of the way in which the dielectric alters the electric field between the plates. Figure 19.19 shows how this effect comes about. In part *a*, the region between the charged plates is empty. The field lines point from the positive toward the negative plate. In part *b*, a dielectric has been inserted between the plates. Since the capacitor is not connected to anything, the charge on the plates remains constant as the dielectric is inserted. In many materials (e.g., water) the molecules possess permanent dipole moments, even though the molecules are electrically neutral. The dipole moment exists because one end of a molecule has a slight excess of negative charge while the other end has a slight excess of positive charge. When such molecules are placed between the charged plates of the capacitor, the negative ends are attracted to the positive plate and the positive ends are attracted to the negative plate. As a result, the dipolar molecules tend to orient themselves end to end, as in part *b*. Whether or not a molecule has a permanent dipole moment, the electric field can cause the electrons to shift position within a molecule, making one end slightly negative and the opposite end slightly positive. Because of the end-to-end orientation, the left surface of the dielectric becomes positively charged, and the right surface becomes negatively charged. The surface charges are shown in red in the picture.

Because of the surface charges on the dielectric, not all the electric field lines generated by the charges on the plates pass through the dielectric. As Figure 19.19c shows, some of the field lines end on the negative surface charges and begin again on the positive surface charges. Thus, the electric field inside the dielectric is less strong than the electric field inside the empty capacitor, assuming the charge on the plates remains constant. This reduction in the electric field is described by the **dielectric constant** κ, which is the ratio of the field magnitude E_0 without the dielectric to the field magnitude E inside the dielectric:

$$\kappa = \frac{E_0}{E} \qquad (19.9)$$

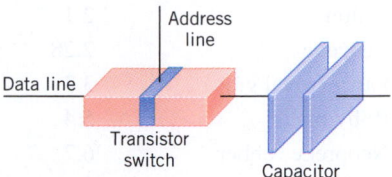

Figure 19.18 A transistor–capacitor combination is part of a RAM chip used in computer memories.

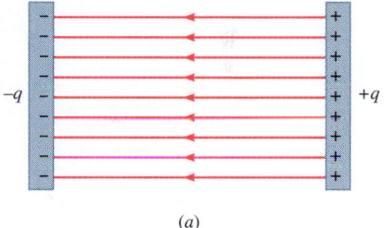

(a)

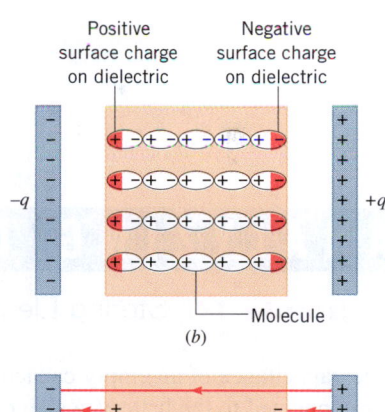

(b)

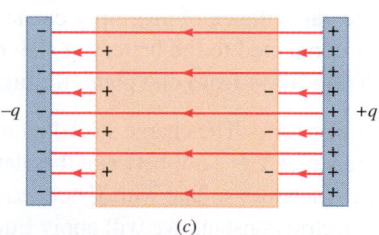

(c)

Figure 19.19 (a) The electric field lines inside an empty capacitor. (b) The electric field produced by the charges on the plates aligns the molecular dipoles within the dielectric end to end. The space between the dielectric and the plates is added for clarity. In reality, the dielectric fills the region between the plates. (c) The surface charges on the dielectric reduce the electric field inside the dielectric.

Table 19.1 Dielectric Constants of Some Common Substances[a]

Substance	Dielectric Constant, κ
Vacuum	1
Air	1.000 54
Teflon	2.1
Benzene	2.28
Paper (royal gray)	3.3
Ruby mica	5.4
Neoprene rubber	6.7
Methyl alcohol	33.6
Water	80.4

[a]Near room temperature.

Being a ratio of two field strengths, the dielectric constant is a number without units. Moreover, since the field $\vec{E}_0$ without the dielectric is greater than the field $\vec{E}$ inside the dielectric, the dielectric constant is greater than unity. The value of κ depends on the nature of the dielectric material, as Table 19.1 indicates.

THE CAPACITANCE OF A PARALLEL PLATE CAPACITOR

The capacitance of a capacitor is affected by the geometry of the plates and the dielectric constant of the material between them. For example, Figure 19.17 shows a parallel plate capacitor in which the area of each plate is A and the separation between the plates is d. The magnitude of the electric field inside the dielectric is given by Equation 19.7b as $E = V/d$, where V is the magnitude of the potential difference between the plates. If the charge on each plate is kept fixed, the electric field inside the dielectric is related to the electric field in the absence of the dielectric via Equation 19.9. Therefore,

$$E = \frac{E_0}{\kappa} = \frac{V}{d}$$

Since the electric field within an empty capacitor is $E_0 = q/(\epsilon_0 A)$ (see Equation 18.4), it follows that $q/(\kappa\epsilon_0 A) = V/d$, which can be solved for q to give

$$q = \left(\frac{\kappa\epsilon_0 A}{d}\right)V$$

A comparison of this expression with $q = CV$ (Equation 19.8) reveals that the capacitance C is

Parallel plate capacitor filled with a dielectric
$$C = \frac{\kappa\epsilon_0 A}{d} \tag{19.10}$$

Notice that only the geometry of the plates (A and d) and the dielectric constant κ affect the capacitance. With C_0 representing the capacitance of the empty capacitor ($\kappa = 1$), Equation 19.10 shows that $C = \kappa C_0$. In other words, the capacitance with the dielectric present is increased by a factor of κ over the capacitance without the dielectric. It can be shown that the relation $C = \kappa C_0$ applies to any capacitor, not just to a parallel plate capacitor. One reason, then, that capacitors are filled with dielectric materials is to increase the capacitance. Example 10 illustrates the effect that increasing the capacitance has on the charge stored by a capacitor.

ANALYZING MULTIPLE-CONCEPT PROBLEMS

Example 10 Storing Electric Charge

The capacitance of an empty capacitor is 1.2 μF. The capacitor is connected to a 12-V battery and charged up. With the capacitor connected to the battery, a slab of dielectric material is inserted between the plates. As a result, 2.6×10^{-5} C of *additional* charge flows from one plate, through the battery, and onto the other plate. What is the dielectric constant of the material?

Reasoning The charge stored by a capacitor is $q = CV$, according to Equation 19.8. The battery maintains a constant potential difference of $V = 12$ V between the plates of the capacitor while the dielectric is inserted. Inserting the dielectric causes the capacitance C to increase, so that with V held constant, the charge q must increase. Thus, additional charge flows onto the plates. To find the dielectric constant, we will apply Equation 19.8 to the capacitor filled with the dielectric material and then to the empty capacitor.

Knowns and Unknowns The following table summarizes the given data:

Description	Symbol	Value
Capacitance of empty capacitor	C_0	1.2 μF
Potential difference between capacitor plates	V	12 V
Additional charge that flows when dielectric is inserted	Δq	2.6×10^{-5} C
Unknown Variable		
Dielectric constant	κ	?

Modeling the Problem

STEP 1 **Dielectric Constant** According to Equations 19.8 and 19.10, the charge stored by the dielectric-filled capacitor is $q = \kappa C_0 V$, where κ is the dielectric constant, κC_0 is the capacitance of the capacitor with the dielectric inserted, C_0 is the capacitance of the empty capacitor, and V is the voltage provided by the battery. Solving for the dielectric constant gives Equation 1 at the right. In this expression, C_0 and V are known, and we will deal with the unknown quantity q in Step 2.

$$\kappa = \frac{q}{C_0 V} \quad (1)$$

STEP 2 **The Charge Stored by the Dielectric-Filled Capacitor** The charge q stored by the dielectric-filled capacitor is the charge q_0 stored by the empty capacitor plus the additional charge Δq that arises when the dielectric is inserted. Therefore, we have

$$q = q_0 + \Delta q$$

This result can be substituted into Equation 1, as indicated at the right. Δq is given, and a value for q_0 will be obtained in Step 3.

$$\kappa = \frac{q}{C_0 V} \quad (1)$$
$$q = q_0 + \Delta q \quad (2)$$

STEP 3 **The Charge Stored by the Empty Capacitor** According to Equation 19.8, the charge stored by the empty capacitor is

$$q_0 = C_0 V$$

The substitution of this expression into Equation 2 is shown at the right.

$$\kappa = \frac{q}{C_0 V} \quad (1)$$
$$q = q_0 + \Delta q \quad (2)$$
$$q_0 = C_0 V$$

Solution Combining the results of each step algebraically, we find that

$$\overset{\text{STEP 1}}{\kappa} = \overset{}{\frac{q}{C_0 V}} = \overset{\text{STEP 2}}{\frac{q_0 + \Delta q}{C_0 V}} = \overset{\text{STEP 3}}{\frac{C_0 V + \Delta q}{C_0 V}}$$

The dielectric constant can now be determined:

$$\kappa = \frac{C_0 V + \Delta q}{C_0 V} = 1 + \frac{\Delta q}{C_0 V} = 1 + \frac{2.6 \times 10^{-5}\ \text{C}}{(1.2 \times 10^{-6}\ \text{F})(12\ \text{V})} = \boxed{2.8}$$

Related Homework: *Problem 53*

In Example 10 the capacitor remains connected to the battery while the dielectric is inserted between the plates. The next example discusses what happens if the capacitor is disconnected from the battery before the dielectric is inserted.

Conceptual Example 11
The Effect of a Dielectric When a Capacitor Has a Constant Charge

An empty capacitor is connected to a battery and charged up. The capacitor is then disconnected from the battery, and a slab of dielectric material is inserted between the plates. Does the potential difference across the plates **(a)** increase, **(b)** remain the same, or **(c)** decrease?

Reasoning Our reasoning is guided by the following fact: Once the capacitor is disconnected from the battery, the charge on its plates remains constant, for there is no longer any way for charge to be added or removed. According to Equation 19.8, the magnitude q of the charge stored by the capacitor is $q = CV$, where C is its capacitance and V is the magnitude of the potential difference between the plates.

Answers (a) and (b) are incorrect. Placing a dielectric between the plates of a capacitor reduces the electric field in that region (see Figure 19.19c). The magnitude V of the potential difference is related to the magnitude E of the electric field by $V = Ed$, where d is the distance between the capacitor plates. Since E decreases when the dielectric is inserted and d is unchanged, the potential difference does not increase or remain the same.

Answer (c) is correct. Inserting the dielectric causes the capacitance C to increase. Since $q = CV$ and q is fixed, the potential difference V across the plates must decrease in order for q to remain unchanged. The amount by which the potential difference decreases from the value initially established by the battery depends on the dielectric constant of the slab.

Related Homework: *Problem 64*

Capacitors are used often in electronic devices, and Example 12 deals with one familiar application.

Example 12 A Computer Keyboard

Figure 19.20 In one kind of computer keyboard, each key, when pressed, changes the separation between the plates of a capacitor.

One kind of computer keyboard is based on the idea of capacitance. Each key is mounted on one end of a plunger, and the other end is attached to a movable metal plate (see Figure 19.20). The movable plate is separated from a fixed plate, the two plates forming a capacitor. When the key is pressed, the movable plate is pushed closer to the fixed plate, and the capacitance increases. Electronic circuitry enables the computer to detect the *change* in capacitance, thereby recognizing which key has been pressed. The separation of the plates is normally 5.00×10^{-3} m but decreases to 0.150×10^{-3} m when a key is pressed. The plate area is 9.50×10^{-5} m^2, and the capacitor is filled with a material whose dielectric constant is 3.50. Determine the change in capacitance that is detected by the computer.

Reasoning We can use Equation 19.10 directly to find the capacitance of the key, since the dielectric constant κ, the plate area A, and the plate separation d are known. We will use this relation twice, once to find the capacitance when the key is pressed and once when it is not pressed. The change in capacitance will be the difference between these two values.

Solution When the key is pressed, the capacitance is

$$C = \frac{\kappa \epsilon_0 A}{d} = \frac{(3.50)[8.85 \times 10^{-12}\ \text{C}^2/(\text{N} \cdot \text{m}^2)](9.50 \times 10^{-5}\ \text{m}^2)}{0.150 \times 10^{-3}\ \text{m}}$$

$$= 19.6 \times 10^{-12}\ \text{F} \quad (19.6\ \text{pF}) \tag{19.10}$$

A calculation similar to the one above reveals that when the key is *not* pressed, the capacitance has a value of 0.589×10^{-12} F (0.589 pF). The *change* in capacitance is an increase of $\boxed{19.0 \times 10^{-12}\ \text{F}\ (19.0\ \text{pF})}$. The *change* in the capacitance is greater with the dielectric present, which makes it easier for the circuitry within the computer to detect it.

ENERGY STORAGE IN A CAPACITOR

When a capacitor stores charge, it also stores energy. In charging up a capacitor, for example, a battery does work in transferring an increment of charge from one plate of the capacitor to the other plate. The work done is equal to the product of the charge increment and the potential difference between the plates. However, as each increment of charge is moved, the potential difference increases slightly, and a larger amount of work is needed to move the next increment. The total work W done in completely charging the capacitor is the product of the total charge q transferred and the average potential difference $\overline{V}$; $W = q\overline{V}$. Since the average potential difference is one-half the final potential V, or $\overline{V} = \frac{1}{2}V$, the total work done by the battery is $W = \frac{1}{2}qV$. This work does not disappear but is stored as electric potential energy in the capacitor, so that Energy $= \frac{1}{2}qV$. Equation 19.8 indicates that $q = CV$ or, equivalently, that $V = q/C$. We can see, then, that our expression for the energy can be cast into two additional equivalent forms by substituting for q or for V. Equations 19.11a–c summarize these results:

$$\text{Energy} = \tfrac{1}{2}qV \tag{19.11a}$$

$$\text{Energy} = \tfrac{1}{2}(CV)V = \tfrac{1}{2}CV^2 \tag{19.11b}$$

$$\text{Energy} = \tfrac{1}{2}q\left(\frac{q}{C}\right) = \frac{q^2}{2C} \tag{19.11c}$$

It is also possible to regard the energy as being stored in the electric field between the plates. The relation between energy and field strength can be obtained for a parallel plate capacitor by substituting $V = Ed$ (Equation 19.7b) and $C = \kappa \epsilon_0 A/d$ (Equation 19.10) into Equation 19.11b:

$$\text{Energy} = \tfrac{1}{2} CV^2 = \tfrac{1}{2} \left(\frac{\kappa \epsilon_0 A}{d} \right)(Ed)^2$$

Since the area A times the separation d is the volume between the plates, the energy per unit volume or **energy density** is

$$\text{Energy density} = \frac{\text{Energy}}{\text{Volume}} = \tfrac{1}{2} \kappa \epsilon_0 E^2 \qquad (19.12)$$

It can be shown that this expression is valid for any electric field strength, not just that between the plates of a capacitor.

The physics of an electronic flash attachment for a camera. The energy-storing capability of a capacitor is often put to good use in electronic circuits. For example, in an electronic flash attachment for a camera, energy from the battery pack is stored in a capacitor. The capacitor is then discharged between the electrodes of the flash tube, which converts the energy into light. Flash duration times range from 1/200 to 1/1 000 000 second or less, with the shortest flashes being used in high-speed photography (see Figure 19.21). Some flash attachments automatically control the flash duration by monitoring the light reflected from the photographic subject and quickly stopping or quenching the capacitor discharge when the reflected light reaches a predetermined level.

The physics of a defibrillator. During a heart attack, the heart produces a rapid, unregulated pattern of beats, a condition known as cardiac fibrillation. Cardiac fibrillation can often be stopped by sending a very fast discharge of electrical energy through the heart. For this purpose, emergency medical personnel use defibrillators, such as the one shown in Figure 19.22. A paddle is connected to each plate of a large capacitor, and the paddles are placed on the chest near the heart. The capacitor is charged to a potential difference of about a thousand volts. The capacitor is then discharged in a few thousandths of a second; the discharge current passes through a paddle, the heart, and the other paddle. Within a few seconds, the heart often returns to its normal beating pattern.

Figure 19.21 This time-lapse photo of a gymnast was obtained using a camera with an electronic flash attachment. The energy for each flash of light comes from the electrical energy stored in a capacitor. (© Bongarts/Getty Images)

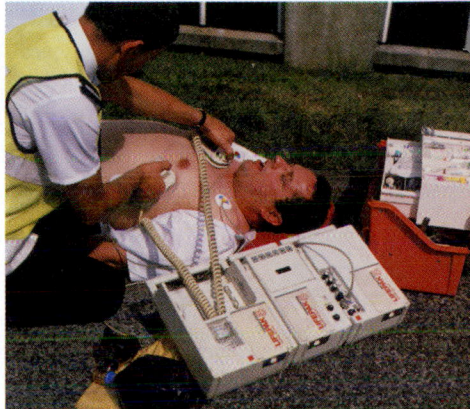

Figure 19.22 A portable defibrillator is being used in an attempt to revive this heart attack victim. A defibrillator uses the electrical energy stored in a capacitor to deliver a controlled electric current that can restore normal heart rhythm. (Adam Hart-Davis/SPL/Photo Researchers, Inc.)

✓ **CHECK YOUR UNDERSTANDING**

(*The answers are given at the end of the book.*)

17. An empty parallel plate capacitor is connected to a battery that maintains a constant potential difference between the plates. With the battery connected, a dielectric is then inserted between the plates. Do the following quantities decrease, remain the same, or increase when the dielectric is inserted? **(a)** The electric field between the plates **(b)** The capacitance **(c)** The charge on the plates **(d)** The energy stored by the capacitor

18. A parallel plate capacitor is charged up by a battery. The battery is then disconnected, but the charge remains on the plates. The plates are then pulled apart. Do the following quantities decrease, remain the same, or increase as the distance between the plates increases? **(a)** The capacitance of the capacitor **(b)** The potential difference between the plates **(c)** The electric field between the plates **(d)** The electric potential energy stored by the capacitor

*BIOMEDICAL APPLICATIONS OF ELECTRIC POTENTIAL DIFFERENCES

19.6

CONDUCTION OF ELECTRICAL SIGNALS IN NEURONS

The human nervous system is remarkable for its ability to transmit information in the form of electrical signals. These signals are carried by the nerves, and the concept of electric potential difference plays an important role in the process. For example, sensory information from our eyes and ears is carried to the brain by the optic nerves and auditory nerves, respectively. Other nerves transmit signals from the brain or spinal column to muscles, causing them to contract. Still other nerves carry signals within the brain.

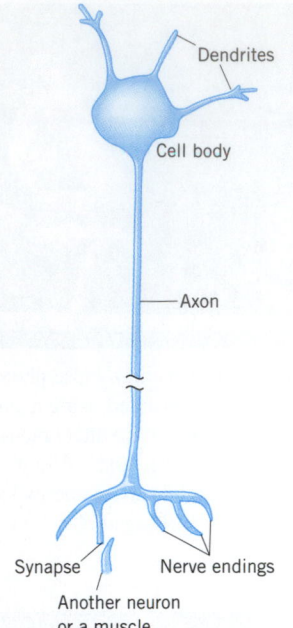

Figure 19.23 The anatomy of a typical neuron.

A nerve consists of a bundle of *axons,* and each axon is one part of a nerve cell, or *neuron.* As Figure 19.23 illustrates, a typical neuron consists of a cell body with numerous extensions, called *dendrites,* and a single axon. The dendrites convert stimuli, such as pressure or heat, into electrical signals that travel through the neuron. The axon sends the signal to the nerve endings, which transmit the signal across a gap (called a *synapse*) to the next neuron or to a muscle.

The fluid inside a cell, the intracellular fluid, is quite different from that outside the cell, the extracellular fluid. Both fluids contain concentrations of positive and negative ions. However, the extracellular fluid is rich in sodium (Na^+) and chlorine (Cl^-) ions, whereas the intracellular fluid is rich in potassium (K^+) ions and negatively charged proteins. These concentration differences between the fluids are extremely important to the life of the cell. If the cell membrane were freely permeable, the ions would diffuse across it until the concentrations on both sides were equal. (See Section 14.4 for a review of diffusion.) This does not happen, because a living cell has a selectively permeable membrane. Ions can enter or leave the cell only through membrane channels, and the permeability of the channels varies markedly from one ion to another. For example, it is much easier for K^+ ions to diffuse out of the cell than it is for Na^+ to enter the cell. As a result of selective membrane permeability, there is a small buildup of negative charges just on the inner side of the membrane and an equal amount of positive charges on the outer side (see Figure 19.24). The buildup of charge occurs very close to the membrane, so the membrane acts like a capacitor (see Problems 44 and 57). Elsewhere in the intracellular and extracellular fluids, there are equal numbers of positive and negative ions, so the fluids are overall electrically neutral. Such a separation of positive and negative charges gives rise to an electric potential difference across the membrane, called the *resting membrane potential.* In neurons, the resting membrane potential ranges from -40 to -90 mV, with a typical value of -70 mV. The minus sign indicates that the inner side of the membrane is negative relative to the outer side.

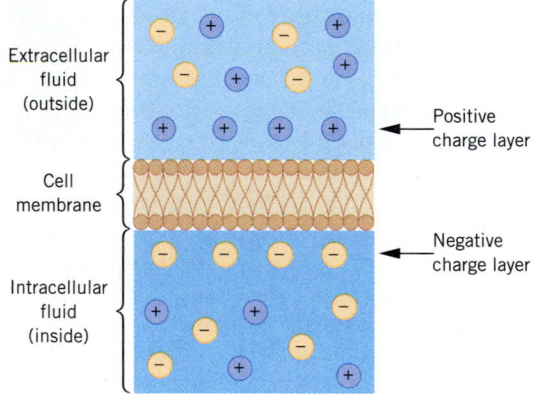

Figure 19.24 Positive and negative charge layers form on the outside and inside surfaces of a membrane during its resting state.

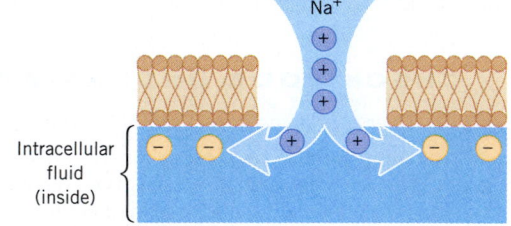

Figure 19.25 When a stimulus is applied to the cell, positive sodium ions (Na^+) rush into the cell, causing the interior surface of the membrane to become momentarily positive.

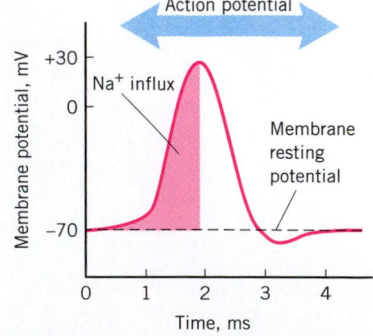

Figure 19.26 The action potential is caused by the rush of positive sodium ions into the cell and a subsequent return of the cell to its resting potential.

The physics of an action potential. A "resting" neuron is one that is not conducting an electrical signal. The *change* in the resting membrane potential is the key factor in the initiation and conduction of a signal. When a sufficiently strong stimulus is applied to a given point on the neuron, "gates" in the membrane open and sodium ions flood into the cell, as Figure 19.25 illustrates. The sodium ions are driven into the cell by attraction to the negative ions on the inner side of the membrane as well as by the relatively high concentration of sodium ions outside the cell. The large influx of Na^+ ions first neutralizes the negative ions on the interior of the membrane and then causes it to become positively charged. As a result, the membrane potential in this localized region goes from -70 mV, the resting potential, to about $+30$ mV in a very short time (see Figure 19.26). The sodium gates then close, and the cell membrane quickly returns to its normal resting potential. This change in potential, from -70 mV to $+30$ mV and back to -70 mV, is known as the *action potential.* The action potential lasts for a few milliseconds, and it is the electrical signal that propagates down the axon, typically at a speed of about 50 m/s, to the next neuron or to a muscle cell.

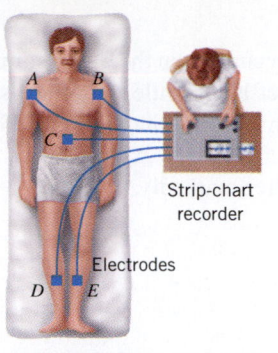

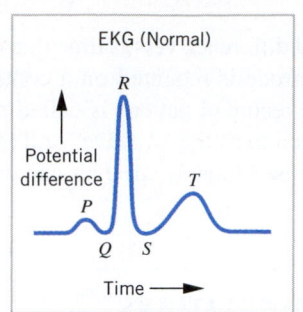

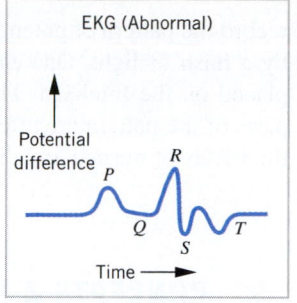

Figure 19.27 The potential differences generated by heart muscle activity provide the basis for electrocardiography. The normal and abnormal EKG patterns correspond to one heartbeat.

MEDICAL DIAGNOSTIC TECHNIQUES

Several important medical diagnostic techniques depend on the fact that the surface of the human body is *not* an equipotential surface. Between various points on the body there are small potential differences (approximately $30-500$ μV), which provide the basis for electrocardiography, electroencephalography, and electroretinography. The potential differences can be traced to the electrical characteristics of muscle cells and nerve cells. In carrying out their biological functions, these cells utilize positively charged sodium and potassium ions and negatively charged chlorine ions that exist within the cells and in the extracellular fluid. As a result of such charged particles, electric fields are generated that extend to the surface of the body and lead to the small potential differences.

Figure 19.27 shows electrodes placed on the body to measure potential differences in electrocardiography. The potential difference between two locations changes as the heart beats and forms a repetitive pattern. The recorded pattern of potential difference versus time is called an electrocardiogram (ECG or EKG), and its shape depends on which pair of points in the picture (*A* and *B*, *B* and *C*, etc.) is used to locate the electrodes. The figure also shows some EKGs and indicates the regions (*P*, *Q*, *R*, *S*, and *T*) associated with specific parts of the heart's beating cycle. The differences between the EKGs of normal and abnormal hearts provide physicians with a valuable diagnostic tool.

In electroencephalography the electrodes are placed at specific locations on the head, as Figure 19.28 indicates, and they record the potential differences that characterize brain behavior. The graph of potential difference versus time is known as an electroencephalogram (EEG). The various parts of the patterns in an EEG are often referred to as "waves" or "rhythms." The drawing shows an example of the main resting rhythm of the brain, the so-called alpha rhythm, and also illustrates the distinct differences that are found between the EEGs generated by healthy (normal) and diseased (abnormal) tissue.

The electrical characteristics of the retina of the eye lead to the potential differences measured in electroretinography. Figure 19.29 shows a typical electrode placement used to

The physics of electrocardiography.

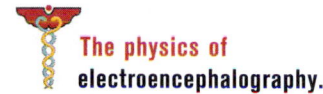

The physics of electroencephalography.

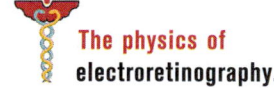

The physics of electroretinography.

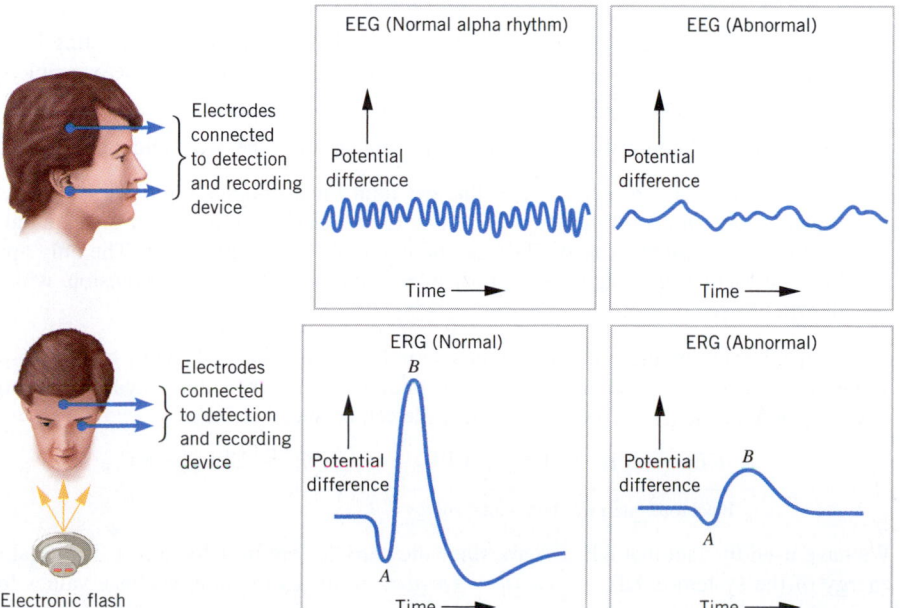

Figure 19.28 In electroencephalography the potential differences created by the electrical activity of the brain are used for diagnosing abnormal behavior.

Figure 19.29 The electrical activity of the retina of the eye generates the potential differences used in electroretinography.

record the pattern of potential difference versus time that occurs when the eye is stimulated by a flash of light. One electrode is mounted on a contact lens, while the other is often placed on the forehead. The recorded pattern is called an electroretinogram (ERG), and parts of the pattern are referred to as the "A wave" and the "B wave." As the graphs show, the ERGs of normal and diseased (abnormal) eyes can differ markedly.

19.7 CONCEPTS & CALCULATIONS

The conservation of energy (Chapter 6) and the conservation of linear momentum (Chapter 7) are two of the most broadly applicable principles in all of science. In this chapter, we have seen that electrically charged particles obey the conservation-of-energy principle, provided that the electric potential energy is taken into account. The behavior of electrically charged particles, however, must also be consistent with the conservation-of-momentum principle, as the first example in this section emphasizes.

Concepts & Calculations Example 13
Conservation Principles

Particle 1 has a mass of $m_1 = 3.6 \times 10^{-6}$ kg, while particle 2 has a mass of $m_2 = 6.2 \times 10^{-6}$ kg. Each has the same electric charge. These particles are initially held at rest, and the two-particle system has an initial electric potential energy of 0.150 J. Suddenly, the particles are released and fly apart because of the repulsive electric force that acts on each one (see Figure 19.30). The effects of the gravitational force are negligible, and no other forces act on the particles. At one instant following the release, the speed of particle 1 is measured to be $v_1 = 170$ m/s. What is the electric potential energy of the two-particle system at this instant?

Concept Questions and Answers What types of energy does the two-particle system have initially?

Answer Initially, the particles are at rest, so they have no kinetic energy. However, they do have electric potential energy. They also have gravitational potential energy, but it is negligible.

What types of energy does the two-particle system have at the instant illustrated in part *b* of the drawing?

Answer Since each particle is moving, each has kinetic energy. The two-particle system also has electric potential energy at this instant. Gravitational potential energy remains negligible.

Does the principle of conservation of energy apply?

Answer Yes. Since the gravitational force is negligible, the only force acting here is the conservative electric force. Nonconservative forces are absent. Thus, the principle of conservation of energy applies.

Does the principle of conservation of linear momentum apply to the two particles as they fly apart?

Answer Yes. This principle states that the total linear momentum of an isolated system remains constant. An isolated system is one for which the vector sum of the external forces acting on the system is zero. There are no external forces acting here. The only appreciable force acting on each of the two particles is the force of electric repulsion, which is an internal force.

Solution The conservation-of-energy principle indicates that the total energy of the two-particle system is the same at the later instant as it was initially. Considering that only kinetic energy and electric potential energy are present, the principle can be stated as follows:

$$\underbrace{\text{KE}_f + \text{EPE}_f}_{\text{Final total energy}} = \underbrace{\text{KE}_0 + \text{EPE}_0}_{\text{Initial total energy}} \quad \text{or} \quad \text{EPE}_f = \text{EPE}_0 - \text{KE}_f$$

We have used the fact that KE_0 is zero, since the particles are initially at rest. The final kinetic energy of the system is $\text{KE}_f = \frac{1}{2}m_1 v_{f1}^2 + \frac{1}{2}m_2 v_{f2}^2$. In this expression, we have values for both

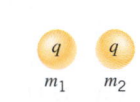

(a) Initial (at rest)

(b) Final

Figure 19.30 (a) Two particles have different masses, but the same electric charge q. They are initially held at rest. (b) At an instant following the release of the particles, they are flying apart due to the mutual force of electric repulsion.

masses and the speed v_{f1}. We can find the speed v_{f2} by using the principle of conservation of momentum:

$$\underbrace{m_1 v_{f1} + m_2 v_{f2}}_{\substack{\text{Final total} \\ \text{momentum}}} = \underbrace{m_1 v_{01} + m_2 v_{02}}_{\substack{\text{Initial total} \\ \text{momentum}}} \quad \text{or} \quad v_{f2} = -\frac{m_1}{m_2} v_{f1}$$

We have again used the fact that the particles are initially at rest, so that $v_{01} = v_{02} = 0$ m/s. Using this result for v_{f2}, we can now obtain the final electric potential energy from the conservation-of-energy equation:

$$
\begin{aligned}
\text{EPE}_f &= \text{EPE}_0 - \left[\tfrac{1}{2} m_1 v_{f1}^2 + \tfrac{1}{2} m_2 \left(-\frac{m_1}{m_2} v_{f1} \right)^2 \right] \\
&= \text{EPE}_0 - \tfrac{1}{2} m_1 \left(1 + \frac{m_1}{m_2} \right) v_{f1}^2 \\
&= 0.150 \text{ J} - \tfrac{1}{2}(3.6 \times 10^{-6} \text{ kg}) \left(1 + \frac{3.6 \times 10^{-6} \text{ kg}}{6.2 \times 10^{-6} \text{ kg}} \right)(170 \text{ m/s})^2 \\
&= \boxed{0.068 \text{ J}}
\end{aligned}
$$

Chapter 18 introduces the electric field, and the present chapter introduces the electric potential. These two concepts are central to the study of electricity, and it is important to distinguish between them, as the next example emphasizes.

Concepts & Calculations Example 14
Electric Field and Electric Potential

Two identical point charges ($+2.4 \times 10^{-9}$ C) are fixed in place, separated by 0.50 m (see Figure 19.31). Find the electric field and the electric potential at the midpoint of the line between the charges q_A and q_B.

Concept Questions and Answers The electric field is a vector and has a direction. At the midpoint, what are the directions of the individual electric-field contributions from q_A and q_B?

Answer Since both charges are positive, the individual electric-field contributions from q_A and q_B point away from each charge. Thus, at the midpoint they point in opposite directions.

Is the magnitude of the net electric field at the midpoint greater than, less than, or equal to zero?

Answer The magnitude of the net electric field is zero, because the electric field from q_A cancels the electric field from q_B. These charges have the same magnitude $|q|$, and the midpoint is the same distance r from each of them. According to $E = k|q|/r^2$ (Equation 18.3), then, each charge produces a field of the same strength. Since the individual electric-field contributions from q_A and q_B point in opposite directions at the midpoint, their vector sum is zero.

Is the total electric potential at the midpoint positive, negative, or zero?

Answer The total electric potential at the midpoint is the algebraic sum of the individual contributions from each charge. According to $V = kq/r$ (Equation 19.6), each contribution is positive, since each charge is positive. Thus, the total electric potential is also positive.

Does the total electric potential have a direction associated with it?

Answer No. The electric potential is a scalar quantity, not a vector quantity. Therefore, it has no direction associated with it.

Solution As discussed in the second concept question, the total electric field at the midpoint is $\boxed{E = 0 \text{ N/C}}$. Using $V = kq/r$ (Equation 19.6), we find that the total potential at the midpoint is

$$V_{\text{Total}} = \frac{kq_A}{r_A} + \frac{kq_B}{r_B}$$

Since $q_A = q_B = +2.4 \times 10^{-9}$ C and $r_A = r_B = 0.25$ m, we have

$$V = \frac{2kq_A}{r_A} = \frac{2(8.99 \times 10^9 \text{ N} \cdot \text{m}^2/\text{C}^2)(+2.4 \times 10^{-9} \text{ C})}{0.25 \text{ m}} = \boxed{170 \text{ V}}$$

Figure 19.31 Example 14 determines the electric field and the electric potential at the midpoint ($r_A = r_B$) between the identical charges ($q_A = q_B$).

CONCEPT SUMMARY

If you need more help with a concept, use the Learning Aids noted next to the discussion or equation. Examples (**Ex.**) are in the text of this chapter. Go to **www.wiley.com/college/cutnell** for the following Learning Aids:

Interactive LearningWare (ILW) — Additional examples solved in a five-step interactive format.

Concept Simulations (CS) — Animated text figures or animations of important concepts.

Interactive Solutions (IS) — Models for certain types of problems in the chapter homework. The calculations are carried out interactively.

Topic	Discussion	Learning Aids
Work and electric potential energy	**19.1 POTENTIAL ENERGY** When a positive test charge $+q_0$ moves from point A to point B in an electric field, work W_{AB} is done by the electric force. The work equals the electric potential energy (EPE) at A minus that at B: $$W_{AB} = \text{EPE}_A - \text{EPE}_B \qquad (19.1)$$	
Path independence	The electric force is a conservative force, so the path along which the test charge moves from A to B is of no consequence, for the work W_{AB} is the same for all paths.	
Electric potential	**19.2 THE ELECTRIC POTENTIAL DIFFERENCE** The electric potential V at a given point is the electric potential energy of a small test charge q_0 situated at that point divided by the charge itself: $$V = \frac{\text{EPE}}{q_0} \qquad (19.3)$$ The SI unit of electric potential is the joule per coulomb (J/C), or volt (V). The electric potential difference between two points A and B is	
Electric potential difference	$$V_B - V_A = \frac{\text{EPE}_B}{q_0} - \frac{\text{EPE}_A}{q_0} = \frac{-W_{AB}}{q_0} \qquad (19.4)$$ The potential difference between two points (or between two equipotential surfaces) is often called the "voltage."	**Ex. 1, 3** **ILW 19.1** **IS 19.5**
Acceleration of positive and negative charges	A positive charge accelerates from a region of higher potential toward a region of lower potential. Conversely, a negative charge accelerates from a region of lower potential toward a region of higher potential.	**Ex. 2**
Electron volt	An electron volt (eV) is a unit of energy. The relationship between electron volts and joules is $1 \text{ eV} = 1.60 \times 10^{-19}$ J. The total energy E of a system is the sum of its translational ($\frac{1}{2}mv^2$) and rotational ($\frac{1}{2}I\omega^2$) kinetic energies, gravitational potential energy (mgh), elastic potential energy ($\frac{1}{2}kx^2$), and electric potential energy (EPE):	
Total energy	$$E = \tfrac{1}{2}mv^2 + \tfrac{1}{2}I\omega^2 + mgh + \tfrac{1}{2}kx^2 + \text{EPE}$$	
Conservation of energy	If external nonconservative forces like friction do no net work, the total energy of the system is conserved. That is, the final total energy E_f is equal to the initial total energy E_0; $E_f = E_0$.	**Ex. 4, 13**
Electric potential of a point charge	**19.3 THE ELECTRIC POTENTIAL DIFFERENCE CREATED BY POINT CHARGES** The electric potential V at a distance r from a point charge q is $$V = \frac{kq}{r} \qquad (19.6)$$ where $k = 8.99 \times 10^9$ N·m²/C². This expression for V assumes that the electric potential is zero at an infinite distance away from the charge.	**Ex. 5** **CS 19.1** **IS 19.25**
Total electric potential	The total electric potential at a given location due to two or more charges is the algebraic sum of the potentials due to each charge.	**Ex. 6, 7, 14**
Total potential energy of a group of charges	The total potential energy of a group of charges is the amount by which the electric potential energy of the group differs from its initial value when the charges are infinitely far apart and far away. It is also equal to the work required to assemble the group, one charge at a time, starting with the charges infinitely far apart and far away.	**Ex. 8**
Equipotential surface	**19.4 EQUIPOTENTIAL SURFACES AND THEIR RELATION TO THE ELECTRIC FIELD** An equipotential surface is a surface on which the electric potential is the same everywhere. The electric force does no work as a charge moves on an equipotential surface, because the force is always perpendicular to the displacement of the charge.	

Topic	Discussion	Learning Aids
	The electric field created by any group of charges is everywhere perpendicular to the associated equipotential surfaces and points in the direction of decreasing potential.	
	The electric field is related to two equipotential surfaces by	
Relation between the electric field and the potential gradient	$$E = -\frac{\Delta V}{\Delta s} \qquad (19.7a)$$	**Ex. 9** **IS 19.37**
	where ΔV is the potential difference between the surfaces and Δs is the displacement. The term $\Delta V/\Delta s$ is called the potential gradient.	
A capacitor	**19.5 CAPACITORS AND DIELECTRICS** A capacitor is a device that stores charge and energy. It consists of two conductors or plates that are near one another, but not touching. The magnitude q of the charge on each plate is given by	
Relation between charge and potential difference	$$q = CV \qquad (19.8)$$	
	where V is the magnitude of the potential difference between the plates and C is the capacitance. The SI unit for capacitance is the coulomb per volt (C/V), or farad (F).	
	The insulating material included between the plates of a capacitor is called a dielectric. The dielectric constant κ of the material is defined as	
Dielectric constant	$$\kappa = \frac{E_0}{E} \qquad (19.9)$$	
	where E_0 and E are, respectively, the magnitudes of the electric fields between the plates without and with a dielectric, assuming the charge on the plates is kept fixed.	
	The capacitance of a parallel plate capacitor filled with a dielectric is	
Capacitance of a parallel plate capacitor	$$C = \frac{\kappa \epsilon_0 A}{d} \qquad (19.10)$$	**Ex. 10, 11, 12** **IS 19.53**
	where $\epsilon_0 = 8.85 \times 10^{-12}\ \mathrm{C^2/(N \cdot m^2)}$ is the permittivity of free space, A is the area of each plate, and d is the distance between the plates.	
	The electric potential energy stored in a capacitor is	
Energy stored in a capacitor	$$\text{Energy} = \tfrac{1}{2}qV = \tfrac{1}{2}CV^2 = q^2/(2C) \qquad (19.11a\text{–}c)$$	**ILW 19.2** **IS 19.45**
	The energy density is the energy stored per unit volume and is related to the magnitude E of the electric field as follows:	
Energy density	$$\text{Energy density} = \tfrac{1}{2}\kappa\epsilon_0 E^2 \qquad (19.12)$$	

FOCUS ON CONCEPTS

Note to Instructors: The numbering of the questions shown here reflects the fact that they are only a representative subset of the total number that are available online. However, all of the questions are available for assignment via an online homework management program such as WileyPLUS or WebAssign.

Section 19.2 The Electric Potential Difference

2. Two different charges, q_1 and q_2, are placed at two different locations, one charge at each location. The locations have the same electric potential V. Do the charges have the same electric potential energy? **(a)** Yes. If the electric potentials at the two locations are the same, the electric potential energies are also the same, regardless of the type (+ or −) and magnitude of the charges placed at these locations. **(b)** Yes, because electric potential and electric potential energy are just different names for the same concept. **(c)** No, because the electric potential V at a given location depends on the charge placed at that location, whereas the electric potential energy EPE does not. **(d)** No, because the electric potential energy EPE at a given location depends on the charge placed at that location as well as the electric potential V.

4. A proton is released from rest at point A in a constant electric field and accelerates to point B (see part a of the drawing). An electron is released from rest at point B and accelerates to point A (see part b of the drawing). How does the change in the proton's electric potential

energy compare with the change in the electron's electric potential energy? **(a)** The change in the proton's electric potential energy is the same as the change in the electron's electric potential energy. **(b)** The proton experiences a greater change in electric potential energy, since it has a greater charge magnitude. **(c)** The proton experiences a smaller change in electric potential energy, since it has a smaller charge magnitude. **(d)** The proton experiences a smaller change in electric potential energy, since it has a smaller speed at B than the electron has at A. This is due to the larger mass of the proton. **(e)** One cannot compare the change in electric potential energies because the proton and electron move in opposite directions.

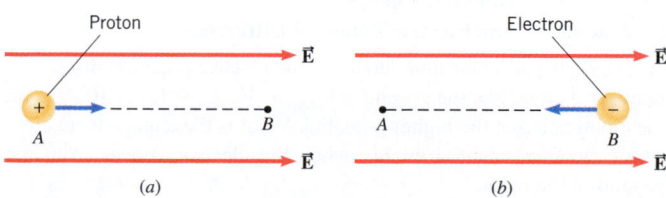

(a)　　　　　　　(b)

Section 19.3 The Electric Potential Difference Created by Point Charges

6. The drawing shows three arrangements of charged particles, all the same distance from the origin. Rank the arrangements, largest to smallest, according to the total electric potential V at the origin.

(a) A, B, C
(b) B, A, C
(c) B, C, A
(d) A, B and C (a tie)
(e) A and C (a tie), B

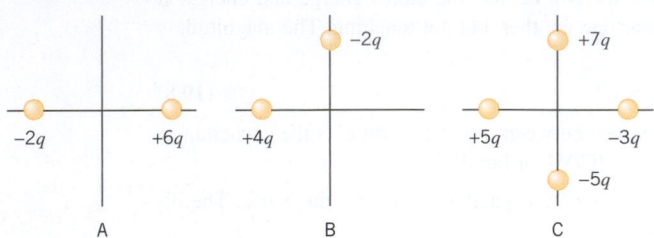

9. Four pairs of charged particles with identical separations are shown in the drawing. Rank the pairs according to their electric potential energy EPE, greatest (most positive) first.

(a) A and C (a tie), B and D (a tie)
(b) A, B, C, D
(c) C, B, D, A
(d) B, A, C and D (a tie)
(e) A and B (a tie), C, D

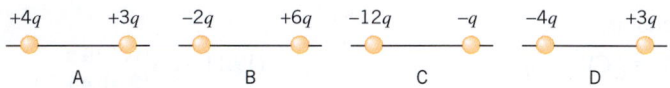

Section 19.4 Equipotential Surfaces and Their Relation to the Electric Field

11. The drawing shows edge-on views of three parallel plate capacitors with the same separation between the plates. The potential of each plate is indicated above it. Rank the capacitors as to the *magnitude* of the electric field inside them, largest to smallest.

(a) A, B, C
(b) A, C, B
(c) C, B, A
(d) C, A, B
(e) B, C, A

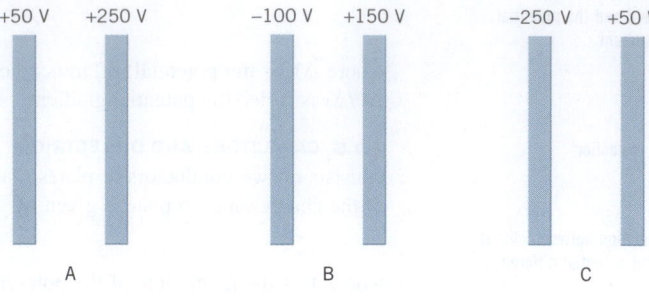

12. The drawing shows a plot of the electric potential V versus the displacement s. The plot consists of four segments. Rank the *magnitude* of the electric fields for the four segments, largest to smallest.

(a) D, C, B, A
(b) A and C (a tie), B and D (a tie)
(c) A, B, D, C
(d) B, D, C, A
(e) D, B, A and C (a tie)

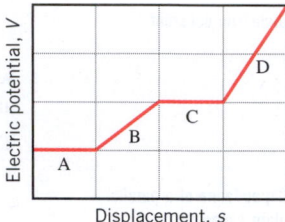

Section 19.5 Capacitors and Dielectrics

17. Which two or more of the following actions would increase the energy stored in a parallel plate capacitor when a constant potential difference is applied across the plates?

1. Increasing the area of the plates
2. Decreasing the area of the plates
3. Increasing the separation between the plates
4. Decreasing the separation between the plates
5. Inserting a dielectric between the plates

(a) 2, 4   (b) 2, 3, 5   (c) 1, 4, 5   (d) 1, 3

PROBLEMS

Note to Instructors: Most of the homework problems in this chapter are available for assignment via an online homework management program such as WileyPLUS or WebAssign, and those marked with the icon **GO** *are presented in WileyPLUS using a guided tutorial format that provides enhanced interactivity. See Preface for additional details.*

Note: All charges are assumed to be point charges unless specified otherwise.

ssm Solution is in the Student Solutions Manual.
www Solution is available online at www.wiley.com/college/cutnell

This icon represents a biomedical application.

Section 19.1 Potential Energy,
Section 19.2 The Electric Potential Difference

1. During a particular thunderstorm, the electric potential difference between a cloud and the ground is $V_{cloud} - V_{ground} = 1.3 \times 10^8$ V, with the cloud being at the higher potential. What is the change in an electron's electric potential energy when the electron moves from the ground to the cloud?

2. Multiple-Concept Example 4 provides useful background for this problem. Point A is at a potential of $+250$ V, and point B is at a potential of -150 V. An α-particle is a helium nucleus that contains two protons and two neutrons; the neutrons are electrically neutral. An α-particle starts from rest at A and accelerates toward B. When the α-particle arrives at B, what kinetic energy (in electron volts) does it have?

3. ssm The work done by an electric force in moving a charge from point A to point B is 2.70×10^{-3} J. The electric potential difference between the two points is $V_A - V_B = 50.0$ V. What is the charge?

4. Refer to Multiple-Concept Example 3 to review the concepts that are needed here. A cordless electric shaver uses energy at a rate of 4.0 W from a rechargeable 1.5-V battery. Each of the charged particles that the battery delivers to the shaver carries a charge that has a magnitude of 1.6×10^{-19} C. A fully charged battery allows the shaver to be used for its maximum operation time, during which 3.0×10^{22} of the charged particles pass between the terminals of the battery as the shaver operates. What is the shaver's maximum operation time?

5. Consult **Interactive Solution 19.5** at **www.wiley.com/college/cutnell** to review one method for modeling this problem. Multiple-Concept Example 3 employs some of the concepts that are needed here. An electric car accelerates for 7.0 s by drawing energy from its 290-V battery pack. During this time, 1200 C of charge passes through the battery pack. Find the minimum horsepower rating of the car.

6. GO An electron and a proton, starting from rest, are accelerated through an electric potential difference of the same magnitude. In the process, the electron acquires a speed v_e, while the proton acquires a speed v_p. Find the ratio v_e/v_p.

7. ssm Multiple-Concept Example 4 deals with the concepts that are important in this problem. As illustrated in Figure 19.6b, a negatively charged particle is released from rest at point B and accelerates until it reaches point A. The mass and charge of the particle are 4.0×10^{-6} kg and -2.0×10^{-5} C, respectively. Only the gravitational force and the electrostatic force act on the particle, which moves on a horizontal straight line without rotating. The electric potential at A is 36 V greater than that at B; in other words, $V_A - V_B = 36$ V. What is the translational speed of the particle at point A?

8. GO Review Multiple-Concept Example 4 to see the concepts that are pertinent here. In a television picture tube, electrons strike the screen after being accelerated from rest through a potential difference of 25 000 V. The speeds of the electrons are quite large, and for accurate calculations of the speeds, the effects of special relativity must be taken into account. Ignoring such effects, find the electron speed just before the electron strikes the screen.

***9. GO** During a lightning flash, there exists a potential difference of $V_{cloud} - V_{ground} = 1.2 \times 10^9$ V between a cloud and the ground. As a result, a charge of -25 C is transferred from the ground to the cloud. **(a)** How much work $W_{ground-cloud}$ is done on the charge by the electric force? **(b)** If the work done by the electric force were used to accelerate a 1100-kg automobile from rest, what would be its final speed? **(c)** If the work done by the electric force were converted into heat, how many kilograms of water at 0 °C could be heated to 100 °C?

***10.** A moving particle encounters an external electric field that decreases its kinetic energy from 9520 eV to 7060 eV as the particle moves from position A to position B. The electric potential at A is -55.0 V, and the electric potential at B is $+27.0$ V. Determine the charge of the particle. Include the algebraic sign (+ or −) with your answer.

***11. ssm www** The potential at location A is 452 V. A positively charged particle is released there from rest and arrives at location B with a speed v_B. The potential at location C is 791 V, and when released from rest from this spot, the particle arrives at B with twice the speed it previously had, or $2v_B$. Find the potential at B.

****12.** A particle is uncharged and is thrown vertically upward from ground level with a speed of 25.0 m/s. As a result, it attains a maximum height h. The particle is then given a positive charge $+q$ and

reaches the same maximum height h when thrown vertically upward with a speed of 30.0 m/s. The electric potential at the height h exceeds the electric potential at ground level. Finally, the particle is given a negative charge $-q$. Ignoring air resistance, determine the speed with which the negatively charged particle must be thrown vertically upward, so that it attains exactly the maximum height h. In all three situations, be sure to include the effect of gravity.

Section 19.3 The Electric Potential Difference Created by Point Charges

13. Two point charges, $+3.40$ μC and -6.10 μC, are separated by 1.20 m. What is the electric potential midway between them?

14. Point A is located 0.25 m away from a charge of -2.1×10^{-9} C. Point B is located 0.50 m away from the charge. What is the electric potential difference $V_B - V_A$ between these two points?

15. ssm Two charges A and B are fixed in place, at different distances from a certain spot. At this spot the potentials due to the two charges are equal. Charge A is 0.18 m from the spot, while charge B is 0.43 m from it. Find the ratio q_B/q_A of the charges.

16. GO The drawing shows a square, each side of which has a length of $L = 0.25$ m. On two corners of the square are fixed different positive charges, q_1 and q_2. Find the electric potential energy of a third charge $q_3 = -6.0 \times 10^{-9}$ C placed at corner A and then at corner B.

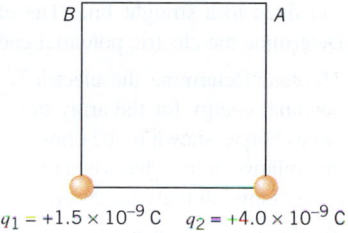

17. ssm www Two identical point charges are fixed to diagonally opposite corners of a square that is 0.500 m on a side. Each charge is $+3.0 \times 10^{-6}$ C. How much work is done by the electric force as one of the charges moves to an empty corner?

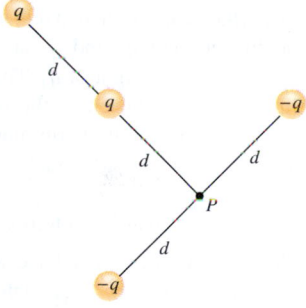

18. The drawing shows four point charges. The value of q is 2.0 μC, and the distance d is 0.96 m. Find the total potential at the location P. Assume that the potential of a point charge is zero at infinity.

Problem 18

19. The drawing shows six point charges arranged in a rectangle. The value of q is 9.0 μC, and the distance d is 0.13 m. Find the total electric potential at location P, which is at the center of the rectangle.

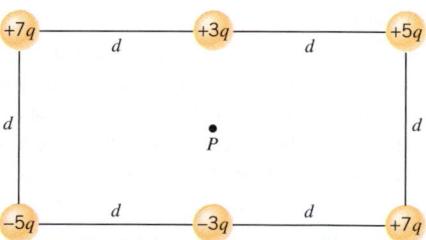

20. GO Charges of $-q$ and $+2q$ are fixed in place, with a distance of 2.00 m between them. A dashed line is drawn through the negative charge, perpendicular to the line between the charges. On the dashed line, at a distance L from the negative charge, there is at least one spot where the total potential is zero. Find L.

21. Identical $+1.8$ μC charges are fixed to adjacent corners of a square. What charge (magnitude and algebraic sign) should be fixed to one of the empty corners, so that the total electric potential at the remaining empty corner is 0 V?

22. Three point charges, -5.8×10^{-9} C, -9.0×10^{-9} C, and $+7.3 \times 10^{-9}$ C, are fixed at different positions on a circle. The total electric potential at the center of the circle is -2100 V. What is the radius of the circle?

*23. ssm A charge of -3.00 μC is fixed in place. From a horizontal distance of 0.0450 m, a particle of mass 7.20×10^{-3} kg and charge -8.00 μC is fired with an initial speed of 65.0 m/s directly toward the fixed charge. How far does the particle travel before its speed is zero?

*24. Two identical point charges ($q = +7.20 \times 10^{-6}$ C) are fixed at diagonally opposite corners of a square with sides of length 0.480 m. A test charge ($q_0 = -2.40 \times 10^{-8}$ C), with a mass of 6.60×10^{-8} kg, is released from rest at one of the empty corners of the square. Determine the speed of the test charge when it reaches the center of the square.

*25. Refer to **Interactive Solution 19.25** at **www.wiley.com/college/cutnell** to review one way in which this problem can be solved. Two protons are moving directly toward one another. When they are very far apart, their initial speeds are 3.00×10^6 m/s. What is the distance of closest approach?

*26. Four identical charges ($+2.0$ μC each) are brought from infinity and fixed to a straight line. The charges are located 0.40 m apart. Determine the electric potential energy of this group.

*27. ssm Determine the electric potential energy for the array of three charges shown in the drawing, relative to its value when the charges are infinitely far away.

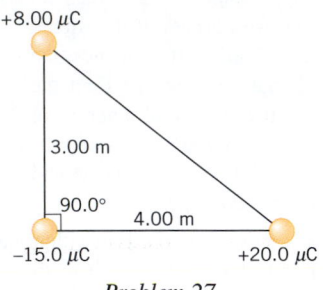

Problem 27

28. Charges q_1 and q_2 are fixed in place, q_2 being located at a distance d to the right of q_1. A third charge q_3 is then fixed to the line joining q_1 and q_2 at a distance d to the right of q_2. The third charge is chosen so the potential energy of the group is zero; that is, the potential energy has the same value as that of the three charges when they are widely separated. Determine the value for q_3, assuming that **(a) $q_1 = q_2 = q$ and **(b)** $q_1 = q$ and $q_2 = -q$. Express your answers in terms of q.

**29. Two particles each have a mass of 6.0×10^{-3} kg. One has a charge of $+5.0 \times 10^{-6}$ C, and the other has a charge of -5.0×10^{-6} C. They are initially held at rest at a distance of 0.80 m apart. Both are then released and accelerate toward each other. How fast is each particle moving when the separation between them is one-third its initial value?

**30. A positive charge of $+q_1$ is located 3.00 m to the left of a negative charge $-q_2$. The charges have different magnitudes. On the line through the charges, the net *electric field* is zero at a spot 1.00 m to the right of the negative charge. On this line there are also two spots where the potential is zero. Locate these two spots relative to the negative charge.

Section 19.4 Equipotential Surfaces and Their Relation to the Electric Field

31. Two equipotential surfaces surround a $+1.50 \times 10^{-8}$-C point charge. How far is the 190-V surface from the 75.0-V surface?

32. An equipotential surface that surrounds a point charge q has a potential of 490 V and an area of 1.1 m². Determine q.

33. ssm A spark plug in an automobile engine consists of two metal conductors that are separated by a distance of 0.75 mm. When an electric spark jumps between them, the magnitude of the electric field is 4.7×10^7 V/m. What is the magnitude of the potential difference ΔV between the conductors?

34. **GO** A positive point charge ($q = +7.2 \times 10^{-8}$ C) is surrounded by an equipotential surface A, which has a radius of $r_A = 1.8$ m. A positive test charge ($q_0 = +4.5 \times 10^{-11}$ C) moves from surface A to another equipotential surface B, which has a radius r_B. The work done as the test charge moves from surface A to surface B is $W_{AB} = -8.1 \times 10^{-9}$ J. Find r_B.

35. **ssm www** The inner and outer surfaces of a cell membrane carry a negative and a positive charge, respectively. Because of these charges, a potential difference of about 0.070 V exists across the membrane. The thickness of the cell membrane is 8.0×10^{-9} m. What is the magnitude of the electric field in the membrane?

36. The drawing shows a graph of a set of equipotential surfaces seen in cross section. Each is labeled according to its electric potential. A $+2.8 \times 10^{-7}$ C point charge is placed at position A. Find the work that is done on the point charge by the electric force when it is moved **(a)** from A to B, and **(b)** from A to C.

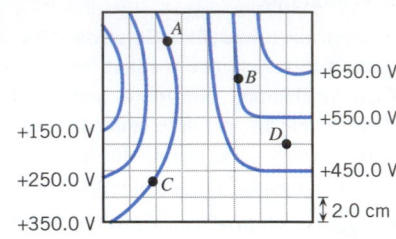

*37. Refer to **Interactive Solution 19.37** at **www.wiley.com/college/cutnell** to review a method by which this problem can be solved. An electric field has a constant value of 4.0×10^3 V/m and is directed downward. The field is the same everywhere. The potential at a point P within this region is 155 V. Find the potential at the following points: **(a)** 6.0×10^{-3} m directly above P, **(b)** 3.0×10^{-3} m directly below P, **(c)** 8.0×10^{-3} m directly to the right of P.

*38. **GO** An electron is released from rest at the negative plate of a parallel plate capacitor and accelerates to the positive plate (see the drawing). The plates are separated by a distance of 1.2 cm, and the electric field within the capacitor has a magnitude of 2.1×10^6 V/m. What is the kinetic energy of the electron just as it reaches the positive plate?

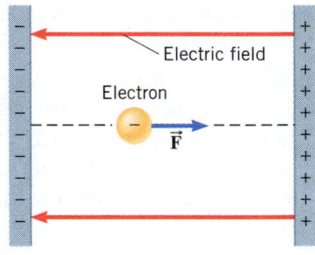

*39. ssm The drawing shows the potential at five points on a set of axes. Each of the four outer points is 6.0×10^{-3} m from the point at the origin. From the data shown, find the magnitude and direction of the electric field in the vicinity of the origin.

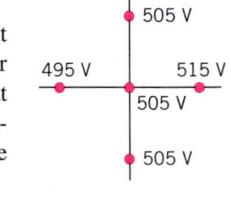

*40. **GO** Equipotential surface A has a potential of 5650 V, while equipotential surface B has a potential of 7850 V. A particle has a mass of 5.00×10^{-2} kg and a charge of $+4.00 \times 10^{-5}$ C. The particle has a speed of 2.00 m/s on surface A. A nonconservative outside force is applied to the particle, and it moves to surface B, arriving there with a speed of 3.00 m/s. How much work is done by the outside force in moving the particle from A to B?

*41. The drawing shows a uniform electric field that points in the negative y direction; the

Problem 41

magnitude of the field is 3600 N/C. Determine the electric potential difference **(a)** $V_B - V_A$ between points A and B, **(b)** $V_C - V_B$ between points B and C, and **(c)** $V_A - V_C$ between points C and A.

Section 19.5 Capacitors and Dielectrics

42. What voltage is required to store 7.2×10^{-5} C of charge on the plates of a 6.0-μF capacitor?

43. **ssm** The electric potential energy stored in the capacitor of a defibrillator is 73 J, and the capacitance is 120 μF. What is the potential difference that exists across the capacitor plates?

44. An axon is the relatively long tail-like part of a neuron, or nerve cell. The outer surface of the axon membrane (dielectric constant = 5, thickness = 1×10^{-8} m) is charged positively, and the inner portion is charged negatively. Thus, the membrane is a kind of capacitor. Assuming that the membrane acts like a parallel plate capacitor with a plate area of 5×10^{-6} m^2, what is its capacitance?

45. Refer to **Interactive Solution 19.45** at **www.wiley.com/college/cutnell** for one approach to this problem. The electronic flash attachment for a camera contains a capacitor for storing the energy used to produce the flash. In one such unit, the potential difference between the plates of an 850-μF capacitor is 280 V. **(a)** Determine the energy that is used to produce the flash in this unit. **(b)** Assuming that the flash lasts for 3.9×10^{-3} s, find the effective power or "wattage" of the flash.

46. **GO** Two capacitors have the same plate separation, but one has square plates and the other has circular plates. The square plates are a length L on each side, and the diameter of the circular plate is L. The capacitors have the same capacitance because they contain different dielectric materials. The dielectric constant of the material between the square plates has a value of $\kappa_{\text{square}} = 3.00$. What is the dielectric constant κ_{circle} of the material between the circular plates?

47. **ssm** A parallel plate capacitor has a capacitance of 7.0 μF when filled with a dielectric. The area of each plate is 1.5 m^2 and the separation between the plates is 1.0×10^{-5} m. What is the dielectric constant of the dielectric?

48. **GO** Two capacitors are identical, except that one is empty and the other is filled with a dielectric ($\kappa = 4.50$). The empty capacitor is connected to a 12.0-V battery. What must be the potential difference across the plates of the capacitor filled with a dielectric so that it stores the same amount of electrical energy as the empty capacitor?

49. A capacitor has a capacitance of 2.5×10^{-8} F. In the charging process, electrons are removed from one plate and placed on the other plate. When the potential difference between the plates is 450 V, how many electrons have been transferred?

50. **GO** Capacitor A and capacitor B both have the same voltage across their plates. However, the energy of capacitor A can melt m kilograms of ice at 0 °C, while the energy of capacitor B can boil away the same amount of water at 100 °C. The capacitance of capacitor A is 9.3 μF. What is the capacitance of capacitor B?

51. **ssm Interactive LearningWare 19.2** at **www.wiley.com/college/cutnell** reviews the concepts pertinent to this problem. What is the potential difference between the plates of a 3.3-F capacitor that stores sufficient energy to operate a 75-W light bulb for one minute?

52. A capacitor is constructed of two concentric conducting cylindrical shells. The radius of the inner cylindrical shell is 2.35×10^{-3} m, and that of the outer shell is 2.50×10^{-3} m. When the cylinders carry equal and opposite charges of magnitude 1.7×10^{-10} C, the electric field between the plates has an average magnitude of 4.2×10^4 V/m and is directed radially outward from the inner shell to the outer shell. Determine **(a)** the magnitude of the potential difference between the cylindrical shells and **(b)** the capacitance of this capacitor.

53. Refer to **Interactive Solution 19.53** at **www.wiley.com/college/cutnell** and Multiple-Concept Example 10 for help in solving this problem. An empty capacitor has a capacitance of 3.2 μF and is connected to a 12-V battery. A dielectric material ($\kappa = 4.5$) is inserted between the plates of this capacitor. What is the magnitude of the surface charge on the dielectric that is adjacent to either plate of the capacitor? (*Hint: The surface charge is equal to the difference in the charge on the plates with and without the dielectric.*)

54. An empty parallel plate capacitor is connected between the terminals of a 9.0-V battery and charged up. The capacitor is then disconnected from the battery, and the spacing between the capacitor plates is doubled. As a result of this change, what is the new voltage between the plates of the capacitor?

55. **ssm** The drawing shows a parallel plate capacitor. One-half of the region between the plates is filled with a material that has a dielectric constant κ_1. The other half is filled with a material that has a dielectric constant κ_2. The area of each plate is A, and the plate separation is d. The potential difference across the plates is V. Note especially that the charge stored by the capacitor is $q_1 + q_2 = CV$, where q_1 and q_2 are the charges on the area of the plates in contact with materials 1 and 2, respectively. Show that $C = \epsilon_0 A(\kappa_1 + \kappa_2)/(2d)$.

56. If the electric field inside a capacitor exceeds the *dielectric strength* of the dielectric between its plates, the dielectric will break down, discharging and ruining the capacitor. Thus, the dielectric strength is the maximum magnitude that the electric field can have without breakdown occurring. The dielectric strength of air is 3.0×10^6 V/m, and that of neoprene rubber is 1.2×10^7 V/m. A certain air-gap, parallel plate capacitor can store no more than 0.075 J of electrical energy before breaking down. How much energy can this capacitor store without breaking down after the gap between its plates is filled with neoprene rubber?

ADDITIONAL PROBLEMS

57. **ssm** The membrane that surrounds a certain type of living cell has a surface area of 5.0×10^{-9} m^2 and a thickness of 1.0×10^{-8} m. Assume that the membrane behaves like a parallel plate capacitor and has a dielectric constant of 5.0. **(a)** The potential on the outer surface of the membrane is $+60.0$ mV greater than that on the inside surface. How much charge resides on the outer surface?

(b) If the charge in part (a) is due to K$^+$ ions (charge $+e$), how many such ions are present on the outer surface?

58. A particle has a charge of $+1.5$ μC and moves from point A to point B, a distance of 0.20 m. The particle experiences a constant electric force, and its motion is along the line of action of the force.

The difference between the particle's electric potential energy at A and at B is $\text{EPE}_A - \text{EPE}_B = +9.0 \times 10^{-4}$ J. **(a)** Find the magnitude and direction of the electric force that acts on the particle. **(b)** Find the magnitude and direction of the electric field that the particle experiences.

59. An electron and a proton are initially very far apart (effectively an infinite distance apart). They are then brought together to form a hydrogen atom, in which the electron orbits the proton at an average distance of 5.29×10^{-11} m. What is $\text{EPE}_{\text{final}} - \text{EPE}_{\text{initial}}$, which is the change in the electric potential energy?

60. The drawing that accompanies Problem 36 shows a graph of a set of equipotential surfaces in cross section. The grid lines are 2.0 cm apart. Determine the magnitude and direction of the electric field at position D. Specify whether the electric field points toward the top or the bottom of the drawing.

61. **ssm** Suppose that the electric potential outside a living cell is higher than that inside the cell by 0.070 V. How much work is done by the electric force when a sodium ion (charge $= +e$) moves from the outside to the inside?

62. Location A is 3.00 m to the right of a point charge q. Location B lies on the same line and is 4.00 m to the right of the charge. The potential difference between the two locations is $V_B - V_A = 45.0$ V. What are the magnitude and sign of the charge?

63. A capacitor stores 5.3×10^{-5} C of charge when connected to a 6.0-V battery. How much charge does the capacitor store when connected to a 9.0-V battery?

64. Review Conceptual Example 11 before attempting this problem. An empty capacitor is connected to a 12.0-V battery and charged up. The capacitor is then disconnected from the battery, and a slab of dielectric material ($\kappa = 2.8$) is inserted between the plates. Find the amount by which the potential difference across the plates changes. Specify whether the change is an increase or a decrease.

65. The drawing shows the electric potential as a function of distance along the x axis. Determine the magnitude of the electric field in the region **(a)** A to B, **(b)** B to C, and **(c)** C to D.

Problem 65

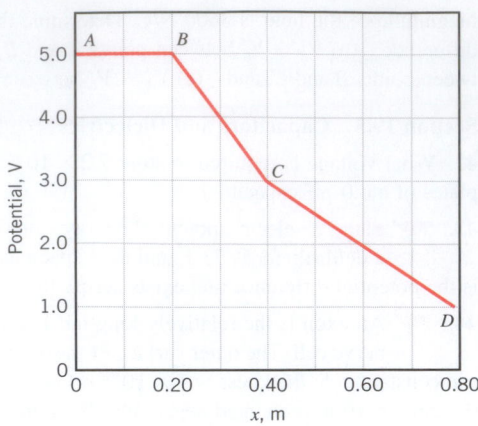

66. Review Conceptual Example 7 as background for this problem. A positive charge $+q_1$ is located to the left of a negative charge $-q_2$. On a line passing through the two charges, there are two places where the total potential is zero. The first place is between the charges and is 4.00 cm to the left of the negative charge. The second place is 7.00 cm to the right of the negative charge. **(a) What is the distance between the charges? **(b)** Find $|q_1|/|q_2|$, the ratio of the magnitudes of the charges.

67. **ssm www The potential difference between the plates of a capacitor is 175 V. Midway between the plates, a proton and an electron are released. The electron is released from rest. The proton is projected perpendicularly toward the negative plate with an initial speed. The proton strikes the negative plate at the same instant that the electron strikes the positive plate. Ignore the attraction between the two particles, and find the initial speed of the proton.

**68. One particle has a mass of 3.00×10^{-3} kg and a charge of $+8.00$ μC. A second particle has a mass of 6.00×10^{-3} kg and the same charge. The two particles are initially held in place and then released. The particles fly apart, and when the separation between them is 0.100 m, the speed of the 3.00×10^{-3}-kg particle is 125 m/s. Find the initial separation between the particles.

CHAPTER 20

ELECTRIC CIRCUITS

Without electric circuits this exuberant display of colorful lights from the hotels and casinos along the strip in Las Vegas, Nevada, would not be possible. Virtually every aspect of life in modern industrialized society utilizes or depends upon electric circuits in some way. (© Russell Kord/Alamy)

20.1 ELECTROMOTIVE FORCE AND CURRENT

Look around you. Chances are that there is an electrical device nearby—a radio, a hair dryer, a computer—something that uses electrical energy to operate. The energy needed to run an MP3 player, for instance, comes from batteries, as Figure 20.1 illustrates. The transfer of energy takes place via an electric circuit, in which the energy source (the battery pack) and the energy-consuming device (the MP3 player) are connected by conducting wires, through which electric charges move.

Within a battery, a chemical reaction occurs that transfers electrons from one terminal (leaving it positively charged) to another terminal (leaving it negatively charged). Figure 20.2 shows the two terminals of a car battery and a flashlight battery. The drawing also illustrates the symbol ($\dashv\vdash$) used to represent a battery in circuit drawings. Because of the positive and negative charges on the battery terminals, an electric potential difference

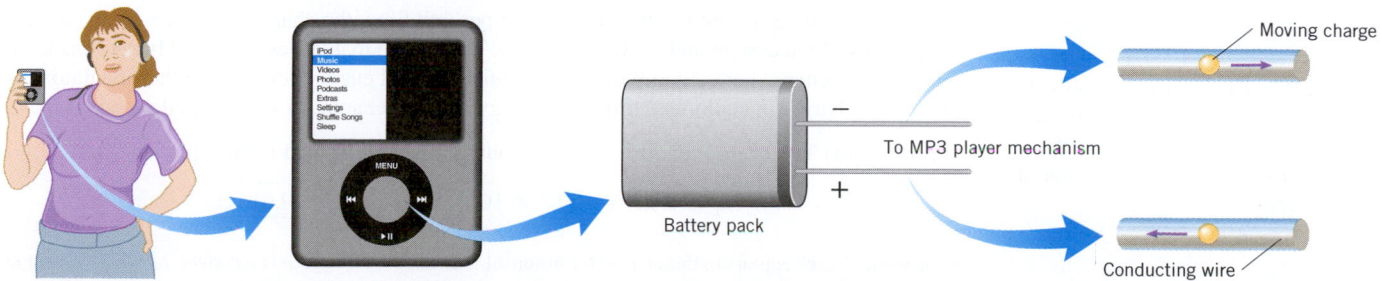

Figure 20.1 In an electric circuit, energy is transferred from a source (the battery pack) to a device (the MP3 player) by charges that move through a conducting wire.

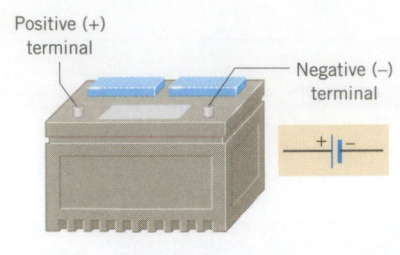

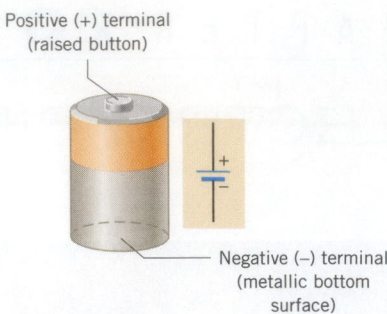

Figure 20.2 Typical batteries and the symbol ($\overset{+}{\rule{1.2em}{0.08em}}|\overset{-}{\rule{1.2em}{0.08em}}$) used to represent them in electric circuits.

exists between them. The maximum potential difference is called the *electromotive force** *(emf)* of the battery, for which the symbol $\mathscr{E}$ is used. In a typical car battery, the chemical reaction maintains the potential of the positive terminal at a maximum of 12 volts (12 joules/coulomb) higher than the potential of the negative terminal, so the emf is $\mathscr{E} = 12$ V. Thus, one coulomb of charge emerging from the battery and entering a circuit has at most 12 joules of energy. In a typical flashlight battery the emf is 1.5 V. In reality, the potential difference between the terminals of a battery is somewhat less than the maximum value indicated by the emf, for reasons that Section 20.9 discusses.

In a circuit such as the one shown in Figure 20.1, the battery creates an electric field within and parallel to the wire, directed from the positive toward the negative terminal. The electric field exerts a force on the free electrons in the wire, and they respond by moving. Figure 20.3 shows charges moving inside a wire and crossing an imaginary surface that is perpendicular to their motion. This flow of charge is known as an *electric current.* The electric current I is defined as the amount of charge per unit time that crosses the imaginary surface, as in Figure 20.3, in much the same sense that a river current is the amount of water per unit time that is flowing past a certain point. If the rate is constant, the current is

$$I = \frac{\Delta q}{\Delta t} \tag{20.1}$$

If the rate of flow is not constant, then Equation 20.1 gives the average current. Since the units for charge and time are the coulomb (C) and the second (s), the SI unit for current is a coulomb per second (C/s). One coulomb per second is referred to as an *ampere* (A), after the French mathematician André-Marie Ampère (1775–1836).

If the charges move around a circuit in the same direction at all times, the current is said to be *direct current (dc),* which is the kind produced by batteries. In contrast, the current is said to be *alternating current (ac)* when the charges move first one way and then the opposite way, changing direction from moment to moment. Many energy sources produce alternating current—for example, generators at power companies and microphones. Example 1 deals with direct current.

⬇ **Example 1** A Pocket Calculator

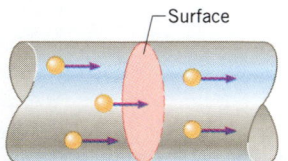

Figure 20.3 The electric current is the amount of charge per unit time that passes through an imaginary surface that is perpendicular to the motion of the charges.

The battery pack of a pocket calculator has a voltage[†] of 3.0 V and delivers a current of 0.17 mA. In one hour of operation, **(a)** how much charge flows in the circuit and **(b)** how much energy does the battery deliver to the calculator circuit?

Reasoning Since current is defined as charge per unit time, the charge that flows in one hour is the product of the current and the time (3600 s). The charge that leaves the 3.0-V battery pack has 3.0 joules of energy per coulomb of charge. Thus, the total energy delivered to the calculator circuit is the charge (in coulombs) times the energy per unit charge (in volts or joules/coulomb).

Solution **(a)** The charge that flows in one hour can be determined from Equation 20.1:

$$\Delta q = I(\Delta t) = (0.17 \times 10^{-3}\ \text{A})(3600\ \text{s}) = \boxed{0.61\ \text{C}}$$

---

*The word "force" appears in this context for historical reasons, even though it is incorrect. As we have seen in Section 19.2, electric potential is energy per unit charge, which is not force.

[†]The potential difference between two points, such as the terminals of a battery, is commonly called the voltage between the points.

(b) The energy delivered to the calculator circuit is

$$\text{Energy} = \text{Charge} \times \underbrace{\frac{\text{Energy}}{\text{Charge}}}_{\text{Battery voltage}} = (0.61\ \text{C})(3.0\ \text{V}) = \boxed{1.8\ \text{J}}$$

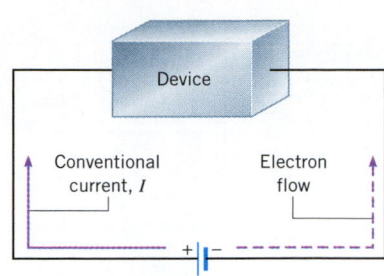

Figure 20.4 In a circuit, electrons actually flow through the metal wires. However, it is customary to use a conventional current I to describe the flow of charges.

Today, it is known that electrons flow in metal wires. Figure 20.4 shows the negative electrons emerging from the negative terminal of the battery and moving around the circuit toward the positive terminal. It is customary, however, *not* to use the flow of electrons when discussing circuits. Instead, a so-called **conventional current** is used, for reasons that date to the time when it was believed that positive charges moved through metal wires. Conventional current is the hypothetical flow of positive charges that would have the same effect in the circuit as the movement of negative charges that actually does occur. In Figure 20.4, negative electrons leave the negative terminal of the battery, pass through the device, and arrive at the positive terminal. The same effect would have been achieved if an equivalent amount of positive charge had left the positive terminal, passed through the device, and arrived at the negative terminal. Therefore, the drawing shows the conventional current originating from the positive terminal. A conventional current of hypothetical positive charges is consistent with our earlier use of a positive test charge for defining electric fields and potentials. The direction of conventional current is always from a point of higher potential toward a point of lower potential—that is, from the positive toward the negative terminal. In this text, the symbol I stands for conventional current.

20.2 OHM'S LAW

The current that a battery can push through a wire is analogous to the water flow that a pump can push through a pipe. Greater pump pressures lead to larger water flow rates, and, similarly, greater battery voltages lead to larger electric currents. In the simplest case, the current I is directly proportional to the voltage V; that is, $I \propto V$. Thus, a 12-V battery leads to twice as much current as a 6-V battery, when each is connected to the same circuit.

In a water pipe, the flow rate is not only determined by the pump pressure but is also affected by the length and diameter of the pipe. Longer and narrower pipes offer higher resistance to the moving water and lead to smaller flow rates for a given pump pressure. A similar situation exists in electric circuits, and to deal with it we introduce the concept of electrical resistance. Electrical resistance is defined in terms of two ideas that have already been discussed—the electric potential difference, or voltage (see Section 19.2), and the electric current (see Section 20.1).

The **resistance (R)** is defined as the ratio of the voltage V applied across a piece of material to the current I through the material, or $R = V/I$. When only a small current results from a large voltage, there is a high resistance to the moving charge. For many materials (e.g., metals), the ratio V/I is the same for a given piece of material over a wide range of voltages and currents. In such a case, the resistance is a constant. Then, the relation $R = V/I$ is referred to as **Ohm's law,** after the German physicist Georg Simon Ohm (1789–1854), who discovered it.

OHM'S LAW

The ratio V/I is a constant, where V is the voltage applied across a piece of material (such as a wire) and I is the current through the material:

$$\frac{V}{I} = R = \text{constant} \quad \text{or} \quad V = IR \tag{20.2}$$

R is the resistance of the piece of material.

SI Unit of Resistance: volt/ampere (V/A) = ohm (Ω)

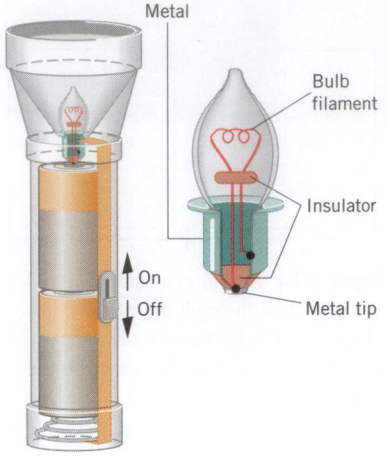

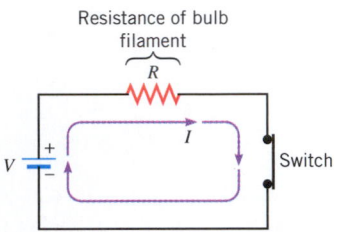

Figure 20.5 The circuit in this flashlight consists of a resistor (the filament of the light bulb) connected to a 3.0-V battery (two 1.5-V batteries).

The SI unit for resistance is a volt per ampere, which is called an *ohm* and is represented by the Greek capital letter omega (Ω). Ohm's law is not a fundamental law of nature like Newton's laws of motion. It is only a statement of the way certain materials behave in electric circuits.

To the extent that a wire or an electrical device offers resistance to the flow of charges, it is called a **resistor.** The resistance can have a wide range of values. The copper wires in a television set, for instance, have a very small resistance. On the other hand, commercial resistors can have resistances up to many kilohms (1 kΩ = 10^3 Ω) or megohms (1 MΩ = 10^6 Ω). Such resistors play an important role in electric circuits, where they are used to limit the amount of current and establish proper voltage levels.

In drawing electric circuits we follow the usual conventions: (1) a zigzag line ($-\!\!\bigwedge\!\!\bigwedge\!\!-$) represents a resistor and (2) a straight line (———) represents an ideal conducting wire, or one with a negligible resistance. Example 2 illustrates an application of Ohm's law to the circuit in a flashlight.

⬇ Example 2 A Flashlight

The filament in a light bulb is a resistor in the form of a thin piece of wire. The wire becomes hot enough to emit light because of the current in it. Figure 20.5 shows a flashlight that uses two 1.5-V batteries (effectively a single 3.0-V battery) to provide a current of 0.40 A in the filament. Determine the resistance of the glowing filament.

Reasoning The filament resistance is assumed to be the only resistance in the circuit. The potential difference applied across the filament is that of the 3.0-V battery. The resistance, given by Equation 20.2, is equal to this potential difference divided by the current.

Solution The resistance of the filament is

$$R = \frac{V}{I} = \frac{3.0 \text{ V}}{0.40 \text{ A}} = \boxed{7.5 \ \Omega} \qquad (20.2)$$

✓ CHECK YOUR UNDERSTANDING

(The answers are given at the end of the book.)

1. In circuit A the battery that supplies energy has twice as much voltage as the battery in circuit B. However, the current in circuit A is only one-half the current in circuit B. Circuit A presents _____ the resistance to the current that circuit B does. **(a)** twice **(b)** one-half **(c)** the same **(d)** four times **(e)** one-fourth

2. Two circuits present the same resistance to the current. In one circuit the battery causing the flow of current has a voltage of 9.0 V, and the current is 3.0 A. In the other circuit the battery has a voltage of 1.5 V. What is the current in this other circuit?

20.3 RESISTANCE AND RESISTIVITY

In a water pipe, the length and cross-sectional area of the pipe determine the resistance that the pipe offers to the flow of water. Longer pipes with smaller cross-sectional areas offer greater resistance. Analogous effects are found in the electrical case. For a wide range of materials, the resistance of a piece of material of length L and cross-sectional area A is

$$R = \rho \frac{L}{A} \qquad (20.3)$$

where ρ is a proportionality constant known as the **resistivity** of the material. It can be seen from Equation 20.3 that the unit for resistivity is the ohm·meter ($\Omega \cdot$m), and Table 20.1 lists values for various materials. All the conductors in the table are metals and have small resistivities. Insulators such as rubber have large resistivities. Materials like germanium and silicon have intermediate resistivity values and are, accordingly, called *semiconductors.*

Table 20.1 Resistivities[a] of Various Materials

Material	Resistivity ρ ($\Omega \cdot m$)	Material	Resistivity ρ ($\Omega \cdot m$)
Conductors		***Semiconductors***	
Aluminum	2.82×10^{-8}	Carbon	3.5×10^{-5}
Copper	1.72×10^{-8}	Germanium	0.5^{b}
Gold	2.44×10^{-8}	Silicon	20–2300^{b}
Iron	9.7×10^{-8}	***Insulators***	
Mercury	95.8×10^{-8}	Mica	10^{11}–10^{15}
Nichrome (alloy)	100×10^{-8}	Rubber (hard)	10^{13}–10^{16}
Silver	1.59×10^{-8}	Teflon	10^{16}
Tungsten	5.6×10^{-8}	Wood (maple)	3×10^{10}

[a]The values pertain to temperatures near 20 °C.
[b]Depending on purity.

Resistivity is an inherent property of a material, inherent in the same sense that density is an inherent property. Resistance, on the other hand, depends on both the resistivity and the geometry of the material. Thus, two wires can be made from copper, which has a resistivity of 1.72×10^{-8} $\Omega \cdot m$, but Equation 20.3 indicates that a short wire with a large cross-sectional area has a smaller resistance than does a long, thin wire. Wires that carry large currents, such as main power cables, are thick rather than thin so that the resistance of the wires is kept as small as possible. Similarly, electric tools that are to be used far away from wall sockets require thicker extension cords, as Example 3 illustrates.

The physics of electrical extension cords.

 Example 3 Longer Extension Cords

The instructions for an electric lawn mower suggest that a 20-gauge extension cord can be used for distances up to 35 m, but a thicker 16-gauge cord should be used for longer distances, to keep the resistance of the wire as small as possible. The cross-sectional area of 20-gauge wire is 5.2×10^{-7} m², while that of 16-gauge wire is 13×10^{-7} m². Determine the resistance of **(a)** 35 m of 20-gauge copper wire and **(b)** 75 m of 16-gauge copper wire.

Reasoning According to Equation 20.3, the resistance of a copper wire depends on the resistivity of copper and the length and cross-sectional area of the wire. The resistivity can be obtained from Table 20.1, while the length and cross-sectional area are given in the problem statement.

Solution According to Table 20.1 the resistivity of copper is 1.72×10^{-8} $\Omega \cdot m$. The resistance of the wires can be found using Equation 20.3:

20-gauge wire
$$R = \frac{\rho L}{A} = \frac{(1.72 \times 10^{-8}\ \Omega \cdot m)(35\ m)}{5.2 \times 10^{-7}\ m^2} = \boxed{1.2\ \Omega}$$

16-gauge wire
$$R = \frac{\rho L}{A} = \frac{(1.72 \times 10^{-8}\ \Omega \cdot m)(75\ m)}{13 \times 10^{-7}\ m^2} = \boxed{0.99\ \Omega}$$

Even though it is more than twice as long, the thicker 16-gauge wire has less resistance than the thinner 20-gauge wire. It is necessary to keep the resistance as low as possible to minimize heating of the wire, thereby reducing the possibility of a fire, as Conceptual Example 7 in Section 20.5 emphasizes.

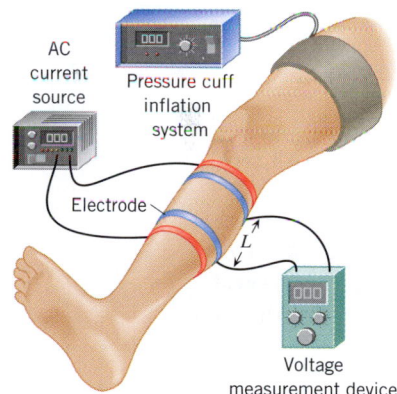

Figure 20.6 Using the technique of impedance plethysmography, the electrical resistance of the calf can be measured to diagnose deep venous thrombosis (blood clotting in the veins).

Equation 20.3 provides the basis for an important medical diagnostic technique known as impedance (or resistance) plethysmography. Figure 20.6 shows how the technique is applied to diagnose blood clotting in the veins (deep venous thrombosis) near the knee. A pressure cuff, like that used in blood pressure measurements, is placed around the midthigh, while electrodes are attached around the calf. The two outer electrodes are connected to a source that supplies a small amount of ac current. The two inner electrodes are separated by a distance L, and the voltage between them is measured. The voltage divided

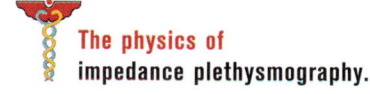 **The physics of impedance plethysmography.**

(a)

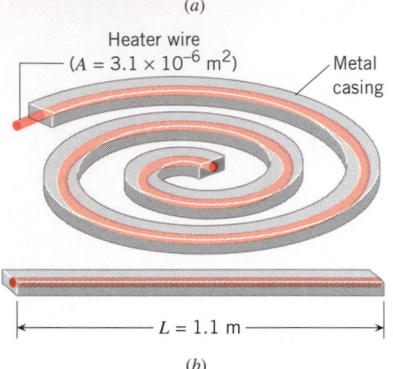

Heater wire
$(A = 3.1 \times 10^{-6} \text{ m}^2)$

Metal casing

$L = 1.1$ m

(b)

Figure 20.7 A heating element from an electric stove. (David Chasey/ PhotoDisc, Inc./Getty Images, Inc.)

by the current gives the resistance. The key to this technique is the fact that resistance can be related to the volume V_{calf} of the calf between the inner electrodes. The volume is the product of the length L and the cross-sectional area A of the calf, or $V_{calf} = LA$. Solving for A and substituting in Equation 20.3 shows that

$$R = \rho \frac{L}{A} = \rho \frac{L}{V_{calf}/L} = \rho \frac{L^2}{V_{calf}}$$

Thus, resistance is inversely proportional to volume, a fact that is exploited in diagnosing deep venous thrombosis. Blood flows from the heart into the calf through arteries in the leg and returns through the system of veins. The pressure cuff in Figure 20.6 is inflated to the point where it cuts off the venous flow but not the arterial flow. As a result, more blood enters than leaves the calf, the volume of the calf increases, and the electrical resistance decreases. When the cuff pressure is removed suddenly, the volume returns to a normal value, and so does the electrical resistance. With healthy (unclotted) veins, there is a rapid return to normal values. A slow return, however, reveals the presence of clotting.

The resistivity of a material depends on temperature. In metals, the resistivity increases with increasing temperature, whereas in semiconductors the reverse is true. For many materials and limited temperature ranges, it is possible to express the temperature dependence of the resistivity as follows:

$$\rho = \rho_0[1 + \alpha(T - T_0)] \qquad (20.4)$$

In this expression ρ and ρ_0 are the resistivities at temperatures T and T_0, respectively. The term α has the unit of reciprocal temperature and is the ***temperature coefficient of resistivity.*** When the resistivity increases with increasing temperature, α is positive, as it is for metals. When the resistivity decreases with increasing temperature, α is negative, as it is for the semiconductors carbon, germanium, and silicon. Since resistance is given by $R = \rho L/A$, both sides of Equation 20.4 can be multiplied by L/A to show that resistance depends on temperature according to

$$R = R_0[1 + \alpha(T - T_0)] \qquad (20.5)$$

The next example illustrates the role of the resistivity and its temperature coefficient in determining the electrical resistance of a piece of material.

The physics of a heating element on an electric stove.

Example 4 The Heating Element of an Electric Stove

Figure 20.7a shows a cherry-red heating element on an electric stove. The element contains a wire (length = 1.1 m, cross-sectional area = 3.1×10^{-6} m^2) through which electric charge flows. As Figure 20.7b shows, this wire is embedded within an electrically insulating material that is contained within a metal casing. The wire becomes hot in response to the flowing charge and heats the casing. The material of the wire has a resistivity of $\rho_0 = 6.8 \times 10^{-5}$ $\Omega \cdot$m at $T_0 = 320$ °C and a temperature coefficient of resistivity of $\alpha = 2.0 \times 10^{-3}$ (C°)$^{-1}$. Determine the resistance of the heater wire at an operating temperature of 420 °C.

Reasoning The expression $R = \rho L/A$ (Equation 20.3) can be used to find the resistance of the wire at 420 °C, once the resistivity ρ is determined at this temperature. Since the resistivity at 320 °C is given, Equation 20.4 can be employed to find the resistivity at 420 °C.

Solution At the operating temperature of 420 °C, the material of the wire has a resistivity of

$$\rho = \rho_0[1 + \alpha(T - T_0)]$$

$$\rho = (6.8 \times 10^{-5} \, \Omega \cdot \text{m})\{1 + [2.0 \times 10^{-3} \, (\text{C°})^{-1}](420 \, \text{°C} - 320 \, \text{°C})\} = 8.2 \times 10^{-5} \, \Omega \cdot \text{m}$$

This value of the resistivity can be used along with the given length and cross-sectional area to find the resistance of the heater wire:

$$R = \frac{\rho L}{A} = \frac{(8.2 \times 10^{-5} \, \Omega \cdot \text{m})(1.1 \, \text{m})}{3.1 \times 10^{-6} \, \text{m}^2} = \boxed{29 \, \Omega} \qquad (20.3)$$

There is an important class of materials whose resistivity suddenly goes to zero below a certain temperature T_c, which is called the *critical temperature* and is commonly a few degrees above absolute zero. Below this temperature, such materials are called *superconductors*. The name derives from the fact that with zero resistivity, these materials offer no resistance to electric current and are, therefore, perfect conductors. One of the remarkable properties of zero resistivity is that once a current is established in a superconducting ring, it continues indefinitely without the need for an emf. Currents have persisted in superconductors for many years without measurable decay. In contrast, the current in a nonsuperconducting material drops to zero almost immediately after the emf is removed.

Many metals become superconductors only at very low temperatures, such as aluminum ($T_c = 1.18$ K), tin ($T_c = 3.72$ K), lead ($T_c = 7.20$ K), and niobium ($T_c = 9.25$ K). Materials involving copper oxide complexes have been made that undergo the transition to the superconducting state at 175 K. Superconductors have many technological applications, including magnetic resonance imaging (Section 21.7), magnetic levitation of trains (Section 21.9), cheaper transmission of electric power, powerful (yet small) electric motors, and faster computer chips.

✓ CHECK YOUR UNDERSTANDING

(The answers are given at the end of the book.)

3. Two materials have different resistivities. Two wires of the same length are made, one from each of the materials. Is it possible for each wire to have the same resistance? **(a)** Yes, if the material with the greater resistivity is used for a thinner wire. **(b)** Yes, if the material with the greater resistivity is used for a thicker wire. **(c)** It is not possible.

4. How does the resistance of a copper wire change when both the length and diameter of the wire are doubled? **(a)** It decreases by a factor of two. **(b)** It increases by a factor of two. **(c)** It increases by a factor of four. **(d)** It decreases by a factor of four. **(e)** It does not change.

5. A resistor is connected between the terminals of a battery. This resistor is a wire, and the following five choices give possibilities for its length and radius in multiples of L_0 and r_0, respectively. For which one or more of the possibilities is the current in the resistor a minimum? **(a)** L_0 and r_0 **(b)** $\frac{1}{2}L_0$ and $\frac{1}{2}r_0$ **(c)** $2L_0$ and $2r_0$ **(d)** $2L_0$ and r_0 **(e)** $8L_0$ and $2r_0$

20.4 ELECTRIC POWER

One of the most important functions of the current in an electric circuit is to transfer energy from a source (such as a battery or a generator) to an electrical device (MP3 player, cell phone, etc.), as Figure 20.8 illustrates. Note that the positive (+) terminal of the battery is connected by a wire to the terminal labeled A on the device; likewise, the negative terminal (−) of the battery is connected to the B terminal. Thus, the battery maintains a constant potential difference between the terminals A and B, with A being at the higher potential. When an amount of positive charge Δq moves from the higher potential (A) to the lower potential (B), its electric potential energy decreases. In accordance with Equation 19.4,

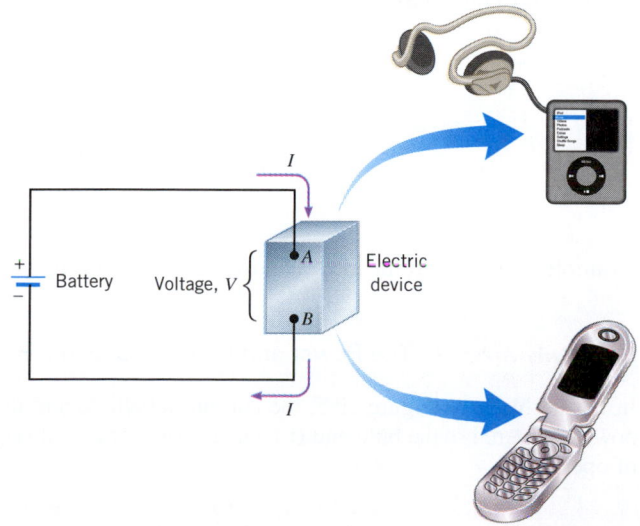

Figure 20.8 The current I in the circuit delivers energy to the electric device. The voltage between the terminals of the device is V.

Electric utility companies are planning for the replacement of power lines with high-temperature superconducting cables. Here, workmen are feeding a test cable through an existing passage at a power station. (Andrew Sacks/saxpix.com)

this decrease is $(\Delta q)V$, where V is the amount by which the electric potential at A exceeds that at B or, in other words, the voltage between the two points. Since the change in energy per unit time is the power P (Equation 6.10b), the electric power associated with this change in energy is

$$P = \frac{\text{Change in energy}}{\text{Time interval}} = \frac{(\Delta q)V}{\Delta t} = \underbrace{\frac{\Delta q}{\Delta t}}_{\text{Current, } I} V$$

The term $\Delta q/\Delta t$ is the charge per unit time, or the current I in the device, according to Equation 20.1. If follows, then, that the electric power is the product of the current and the voltage.

ELECTRIC POWER

When electric charge flows from point A to point B in a circuit, leading to a current I, and the voltage between the points is V, the electric power associated with this current and voltage is

$$P = IV \tag{20.6}$$

SI Unit of Power: watt (W)

Power is measured in watts, and Equation 20.6 indicates that the product of an ampere and a volt is equal to a watt.

When the charge moves through the device in Figure 20.8, the charge *loses* electric potential energy. The principle of conservation of energy tells us that the decrease in potential energy must be accompanied by a transfer of energy to some other form (or forms). In a cell phone, for example, the energy transferred appears as light energy (coming from the display screen), sound energy (emanating from the speaker), thermal energy (due to heating of the internal circuitry), and so on.

The charge in a circuit can also gain electrical energy. For example, when it moves through the battery in Figure 20.8, the charge goes from a lower to a higher potential, just the opposite of what happens in the electrical device. In this case, the charge *gains* electric potential energy. Consistent with the conservation of energy, this increase in potential energy must come from somewhere; in this case it comes from the chemical energy stored in the battery. Thus, the charge regains the energy it lost to the device, at the expense of the chemical energy of the battery.

Many electrical devices are essentially resistors that become hot when provided with sufficient electric power: toasters, irons, space heaters, heating elements on electric stoves, and incandescent light bulbs, to name a few. In such cases, it is convenient to have additional expressions that are equivalent to the power $P = IV$ but which include the resistance R explicitly. We can obtain two such equations by substituting $V = IR$, or equivalently $I = V/R$, into the relation $P = IV$:

$$P = IV \tag{20.6a}$$

$$P = I(IR) = I^2 R \tag{20.6b}$$

$$P = \left(\frac{V}{R}\right)V = \frac{V^2}{R} \tag{20.6c}$$

Example 5 deals with the electric power delivered to the bulb of a flashlight.

 Example 5   The Power and Energy Used in a Flashlight

In the flashlight in Figure 20.5, the current is 0.40 A, and the voltage is 3.0 V. Find **(a)** the power delivered to the bulb and **(b)** the electrical energy dissipated in the bulb in 5.5 minutes of operation.

Reasoning The electric power delivered to the bulb is the product of the current and voltage. Since power is energy per unit time, the energy delivered to the bulb is the product of the power and time.

Solution (a) The power is

$$P = IV = (0.40 \text{ A})(3.0 \text{ V}) = \boxed{1.2 \text{ W}} \qquad (20.6\text{a})$$

The "wattage" rating of this bulb would, therefore, be 1.2 W.

(b) The energy consumed in 5.5 minutes (330 s) follows from the definition of power as energy per unit time:

$$\text{Energy} = P \, \Delta t = (1.2 \text{ W})(330 \text{ s}) = \boxed{4.0 \times 10^2 \text{ J}}$$

Monthly electric bills specify the cost for the energy consumed during the month. Energy is the product of power and time, and electric companies compute your energy consumption by expressing power in kilowatts and time in hours. Therefore, a commonly used unit for energy is the *kilowatt · hour* (kWh). For instance, if you used an average power of 1440 watts (1.44 kW) for 30 days (720 h), your energy consumption would be (1.44 kW) (720 h) = 1040 kWh. At a cost of $0.12 per kWh, your monthly bill would be $125. As shown in Example 10 in Chapter 12, 1 kWh = 3.60×10^6 J of energy.

✓ **CHECK YOUR UNDERSTANDING**

(The answers are given at the end of the book.)

6. A toaster is designed to operate with a voltage of 120 V, and a clothes dryer is designed to operate with a voltage of 240 V. Based solely on this information, which appliance uses more power? **(a)** The toaster **(b)** The dryer **(c)** Insufficient information is given for an answer.

7. When an incandescent light bulb is turned on, a constant voltage is applied across the tungsten filament, which then becomes white hot. The temperature coefficient of resistivity for tungsten is a positive number. What happens to the power delivered to the bulb as the filament heats up? **(a)** It decreases. **(b)** It increases. **(c)** It remains constant.

8. The drawing shows a circuit that includes a bimetallic strip (made from brass and steel; see Section 12.4) with a resistance heater wire wrapped around it. When the switch is initially closed, a current appears in the circuit because charges flow through the heater wire (which becomes hot), the strip itself, the contact point, and the light bulb. The bulb glows in response. As long as the switch remains closed, does the bulb **(a)** continue to glow, **(b)** eventually turn off permanently, or **(c)** flash on and off?

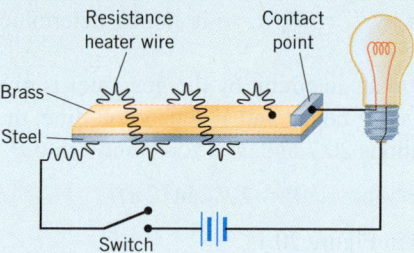

ALTERNATING CURRENT

20.5

Many electric circuits use batteries and involve direct current (dc). However, there are considerably more circuits that operate with alternating current (ac), in which the charge flow reverses direction periodically. The common generators that create ac electricity depend on magnetic forces for their operation and are discussed in Chapter 22. In an ac circuit, these generators serve the same purpose as a battery serves in a dc circuit; that is, they give energy to the moving electric charges. This section deals with ac circuits that contain only resistance.

Since the electrical outlets in a house provide alternating current, we all use ac circuits routinely. For example, the heating element of a toaster is essentially a thin wire of

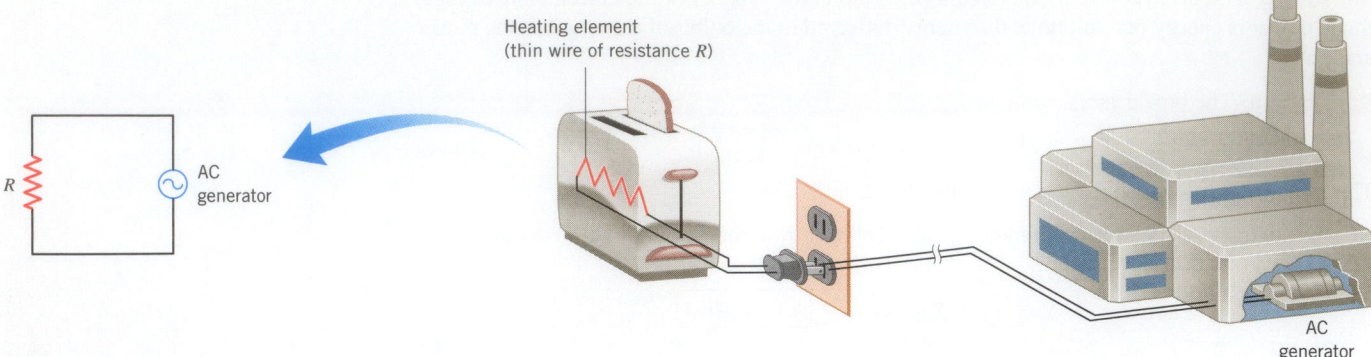

Heating element
(thin wire of resistance R)

R
AC generator

AC generator

Figure 20.9 This circuit consists of a toaster (resistance = R) and an ac generator at the electric power company.

resistance R and becomes red hot when electrical energy is dissipated in it. Figure 20.9 shows the ac circuit that is formed when a toaster is plugged into a wall socket. The circuit schematic in the picture introduces the symbol ⊝ that is used to represent the generator. In this case, the generator is located at the electric power company.

Figure 20.10 shows a graph that records the voltage V produced between the terminals of the ac generator in Figure 20.9 at each moment of time t. This is the most common type of ac voltage. It fluctuates sinusoidally between positive and negative values as a function of time:

$$V = V_0 \sin 2\pi f t \qquad (20.7)$$

where V_0 is the maximum or peak value of the voltage, and f is the frequency (in cycles/s or Hz) at which the voltage oscillates. The angle $2\pi f t$ in Equation 20.7 is expressed in radians, so a calculator must be set to its radian mode before the sine of this angle is evaluated. In the United States, the voltage at most home wall outlets has a *peak value* of approximately $V_0 = 170$ volts and oscillates with a frequency of $f = 60$ Hz. Thus, the period of each cycle is $\frac{1}{60}$ s, and the polarity of the generator terminals reverses twice during each cycle, as Figure 20.10 indicates.

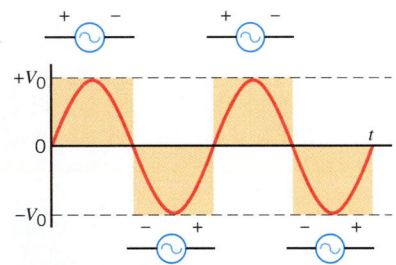

Figure 20.10 In the most common case, the ac voltage is a sinusoidal function of time. The relative polarity of the generator terminals during the positive and negative parts of the sine wave is indicated.

The current in an ac circuit also oscillates. In circuits that contain only resistance, the current reverses direction each time the polarity of the generator terminals reverses. Thus, the current in a circuit like that in Figure 20.9 would have a frequency of 60 Hz and would change direction twice during each cycle. Substituting $V = V_0 \sin 2\pi f t$ into $V = IR$ shows that the current can be represented as

$$I = \frac{V_0}{R} \sin 2\pi f t = I_0 \sin 2\pi f t \qquad (20.8)$$

The peak current is given by $I_0 = V_0/R$, so it can be determined if the peak voltage and the resistance are known.

The power delivered to an ac circuit by the generator is given by $P = IV$, just as it is in a dc circuit. However, since both I and V depend on time, the power fluctuates as time passes. Substituting Equations 20.7 and 20.8 for V and I into $P = IV$ gives

$$P = I_0 V_0 \sin^2 2\pi f t \qquad (20.9)$$

This expression is plotted in Figure 20.11.

Since the power fluctuates in an ac circuit, it is customary to consider the average power $\overline{P}$, which is one-half the peak power, as Figure 20.11 indicates:

$$\overline{P} = \tfrac{1}{2} I_0 V_0 \qquad (20.10)$$

On the basis of this expression, a kind of average current and average voltage can be introduced that are very useful when discussing ac circuits. A rearrangement of Equation 20.10 reveals that

$$\overline{P} = \left(\frac{I_0}{\sqrt{2}}\right)\left(\frac{V_0}{\sqrt{2}}\right) = I_{rms} V_{rms} \qquad (20.11)$$

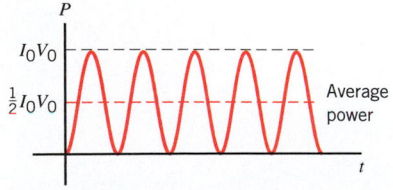

Figure 20.11 In an ac circuit, the power P delivered to a resistor oscillates between zero and a peak value of I_0V_0, where I_0 and V_0 are the peak current and voltage, respectively.

I_{rms} and V_{rms} are called the ***root-mean-square (rms)*** current and voltage, respectively, and may be calculated from the peak values by dividing them by $\sqrt{2}$:*

*This applies only for sinusoidal current and voltage.

$$I_{rms} = \frac{I_0}{\sqrt{2}} \qquad (20.12)$$

$$V_{rms} = \frac{V_0}{\sqrt{2}} \qquad (20.13)$$

Since the normal maximum ac voltage at a home wall socket in the United States is $V_0 = 170$ volts, the corresponding rms voltage is $V_{rms} = (170 \text{ volts})/\sqrt{2} = 120$ volts. Instructions for electrical devices usually specify this rms value. Similarly, when we specify an ac voltage or current in this text, it is an rms value, unless indicated otherwise. When we specify ac power, it is an average power, unless stated otherwise.

Except for dealing with average quantities, the relation $\overline{P} = I_{rms}V_{rms}$ has the same form as Equation 20.6a ($P = IV$). Moreover, Ohm's law can be written conveniently in terms of rms quantities:

$$V_{rms} = I_{rms}R \qquad (20.14)$$

Substituting Equation 20.14 into $\overline{P} = I_{rms}V_{rms}$ shows that the average power can be expressed in the following ways:

$$\overline{P} = I_{rms}V_{rms} \qquad (20.15a)$$

$$\overline{P} = I_{rms}^2 R \qquad (20.15b)$$

$$\overline{P} = \frac{V_{rms}^2}{R} \qquad (20.15c)$$

These expressions are completely analogous to $P = IV = I^2R = V^2/R$ for dc circuits. Example 6 deals with the average power in one familiar ac circuit.

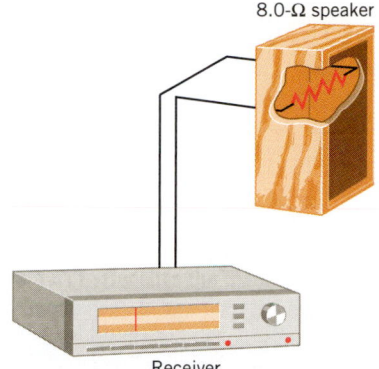

8.0-Ω speaker

Receiver

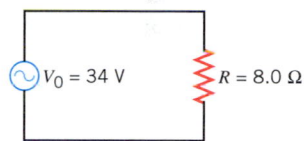

$V_0 = 34$ V $R = 8.0\ \Omega$

 Example 6   Electric Power Sent to a Loudspeaker

A stereo receiver applies a peak ac voltage of 34 V to a speaker. The speaker behaves approximately* as if it had a resistance of 8.0 Ω, as the circuit in Figure 20.12 indicates. Determine **(a)** the rms voltage, **(b)** the rms current, and **(c)** the average power for this circuit.

Figure 20.12 A receiver applies an ac voltage (peak value = 34 V) to an 8.0-Ω speaker.

Reasoning The rms voltage is, by definition, equal to the peak voltage divided by $\sqrt{2}$. Furthermore, we are assuming that the circuit contains only a resistor. Therefore, we can use Ohm's law (Equation 20.14) to calculate the rms current as the rms voltage divided by the resistance, and we can then determine the average power as the rms current times the rms voltage (Equation 20.15a).

Solution **(a)** The peak value of the voltage is $V_0 = 34$ V, so the corresponding rms value is

$$V_{rms} = \frac{V_0}{\sqrt{2}} = \frac{34 \text{ V}}{\sqrt{2}} = \boxed{24 \text{ V}} \qquad (20.13)$$

(b) The rms current can be obtained from Ohm's law:

$$I_{rms} = \frac{V_{rms}}{R} = \frac{24 \text{ V}}{8.0\ \Omega} = \boxed{3.0 \text{ A}} \qquad (20.14)$$

(c) The average power is

$$\overline{P} = I_{rms}V_{rms} = (3.0 \text{ A})(24 \text{ V}) = \boxed{72 \text{ W}} \qquad (20.15a)$$

Problem-solving insight

The rms values of the ac voltage and the ac current, V_{rms} and I_{rms}, respectively, are not the same as the peak values V_0 and I_0. The rms values are always smaller than the peak values by a factor of $\sqrt{2}$.

The electric power dissipated in a resistor causes the resistor to heat up. Excessive power can lead to a potential fire hazard, as Conceptual Example 7 discusses.

 Conceptual Example 7   Extension Cords and a Potential Fire Hazard

During the winter, many people use portable electric space heaters to keep warm. When the heater is located far from a 120-V wall receptacle, an extension cord is necessary (see Figure 20.13). To prevent fires, however, manufacturers sometimes caution about using extension cords. To minimize the risk of a fire, should a long extension cord used with a space heater be made from **(a)** larger-gauge or **(b)** smaller-gauge wire?

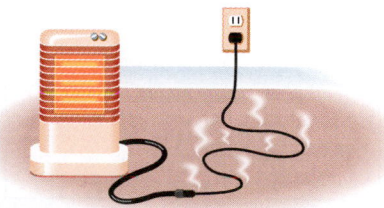

Figure 20.13 When an extension cord is used with a space heater, the cord must have a resistance that is sufficiently small to prevent overheating of the cord.

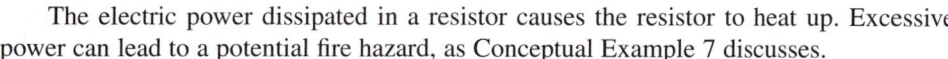

*Other factors besides resistance can affect the current and voltage in ac circuits; they are discussed in Chapter 23.

Reasoning An electric space heater contains a heater element that is a piece of wire of resistance R, which is heated to a high temperature. The heating occurs because of the power $I_{rms}^2 R$ dissipated in the heater element. A typical heater uses a relatively large current I_{rms} of about 12 A. On its way to the heater, this current passes through the wires of the extension cord. Since these additional wires offer resistance to the current, the extension cord can also heat up, just as the heater element does. This unwanted heating depends on the resistance of the wire in the extension cord and could lead to a fire. As Example 3 discusses, the larger-gauge wire is the one that has the smaller cross-sectional area. The cross-sectional area is important because it is one of the factors that determine the resistance of the wire in the extension cord.

Answer (a) is incorrect. To keep the heating of the extension cord to a safe level, the resistance of the wire must be kept small. Recall from Section 20.3 that the resistance of a wire depends inversely on its cross-sectional area. A larger-gauge wire has a smaller cross-sectional area and, therefore, has a greater resistance $R_{extension\ cord}$. This greater resistance would lead to more (not less) heating of the extension cord, because of the power $I_{rms}^2 R_{extension\ cord}$ dissipated in it.

Answer (b) is correct. Since the resistance of a wire depends inversely on its cross-sectional area, a smaller-gauge wire has a larger cross-sectional area and has a smaller resistance $R_{extension\ cord}$. This smaller resistance would lead to less heating of the extension cord due to the power $I_{rms}^2 R_{extension\ cord}$ dissipated in it, thus minimizing the risk of a fire.

Related Homework: *Problem 37*

✓ CHECK YOUR UNDERSTANDING

(The answers are given at the end of the book.)

9. The drawing shows a circuit in which a light bulb is connected to the household ac voltage via two switches, S_1 and S_2. This is the kind of wiring, for example, that allows you to turn on a carport light from either inside the house or out in the carport. Which one or more of the combinations of the switch positions will turn on the light?   **(a)** S_1 set to A and S_2 set to B   **(b)** S_1 set to B and S_2 set to B   **(c)** S_1 set to B and S_2 set to A   **(d)** S_1 set to A and S_2 set to A

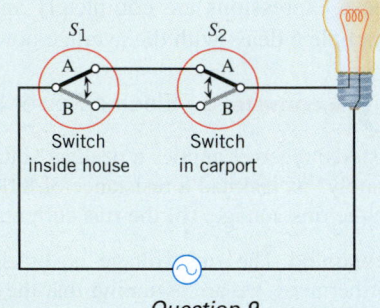

Question 9

10. Two light bulbs are designed for use with an ac voltage of 120 V and are rated at 75 W and 150 W. Which bulb, if either, has the greater filament resistance?

11. An ac circuit contains only a generator and a resistor. Which one of the following changes leads to the greatest average power being delivered to the circuit?   **(a)** Double the peak voltage of the generator and double the resistance.   **(b)** Double the resistance.   **(c)** Double the peak voltage of the generator.   **(d)** Double the peak voltage of the generator and reduce the resistance by a factor of two.

20.6 SERIES WIRING

Thus far, we have dealt with circuits that include only a single device, such as a light bulb or a loudspeaker. There are, however, many circuits in which more than one device is connected to a voltage source. This section introduces one method by which such connections may be made—namely, series wiring. *Series wiring means that the devices are connected in such a way that there is the same electric current through each device.* Figure 20.14 shows a circuit in which two different devices, represented by resistors R_1 and R_2, are connected in series with a battery. Note that if the current in one resistor is interrupted, the current in the other is too. This could occur, for example, if two light bulbs were connected in series and the filament of one bulb broke. Because of the series wiring, the voltage V supplied by the battery is divided between the two resistors. The drawing indicates that the portion of the voltage across R_1 is V_1, while the portion across R_2 is V_2, so $V = V_1 + V_2$. For the individual resistances, the definition of resistance indicates that $R_1 = V_1/I$ and $R_2 = V_2/I$, so that $V_1 = IR_1$ and $V_2 = IR_2$. Therefore, we have

$$V = V_1 + V_2 = IR_1 + IR_2 = I(R_1 + R_2) = IR_S$$

where R_S is called the *equivalent resistance* of the series circuit. Thus, two resistors in series are equivalent to a single resistor whose resistance is $R_S = R_1 + R_2$, in the sense that

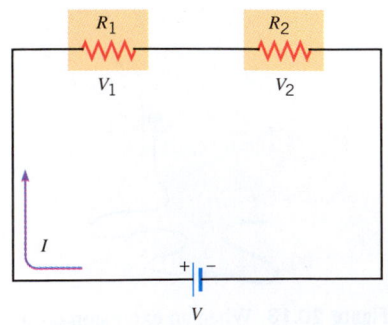

Figure 20.14 When two resistors are connected in series, the same current I is in both of them.

there is the same current through R_S as there is through the series combination of R_1 and R_2. This line of reasoning can be extended to any number of resistors in series if we note that, in general, *the voltage across all the resistors in series is the sum of the individual voltages across each resistor.* The result for the equivalent resistance is

Problem-solving insight

Series resistors $\qquad\qquad R_S = R_1 + R_2 + R_3 + \cdots \qquad\qquad$ (20.16)

Examples 8 and 9 illustrate the concept of equivalent resistance in series circuits.

 Example 8 A Series Circuit

Suppose that the resistances in Figure 20.14 are $R_1 = 47\ \Omega$ and $R_2 = 86\ \Omega$, and the battery voltage is 24 V. Determine the equivalent resistance of the two resistors and the current in the circuit.

Reasoning The two resistors are wired in series, since there is the same current through each one. The equivalent resistance R_S of a series circuit is the sum of the individual resistances, so $R_S = R_1 + R_2$. The current I can be obtained from Ohm's law as the voltage V divided by the equivalent resistance: $I = V/R_S$.

Solution The equivalent resistance is

$$R_S = R_1 + R_2 = 47\ \Omega + 86\ \Omega = \boxed{133\ \Omega} \qquad (20.16)$$

The current in the circuit is

$$I = \frac{V}{R_S} = \frac{24\ \text{V}}{133\ \Omega} = \boxed{0.18\ \text{A}} \qquad (20.2)$$

ANALYZING MULTIPLE-CONCEPT PROBLEMS

Example 9 Power Delivered to a Series Circuit

A 6.00-Ω resistor and a 3.00-Ω resistor are connected in series with a 12.0-V battery, as Figure 20.15 indicates. Assuming that the battery contributes no resistance to the circuit, find the power delivered to each of the resistors.

Reasoning The power P delivered to each resistor is the product of the current squared (I^2) and the corresponding resistance R, or $P = I^2R$. The resistances are known, and Ohm's law can be used to find the current. Ohm's law states that the current in the circuit (which is also the current through each

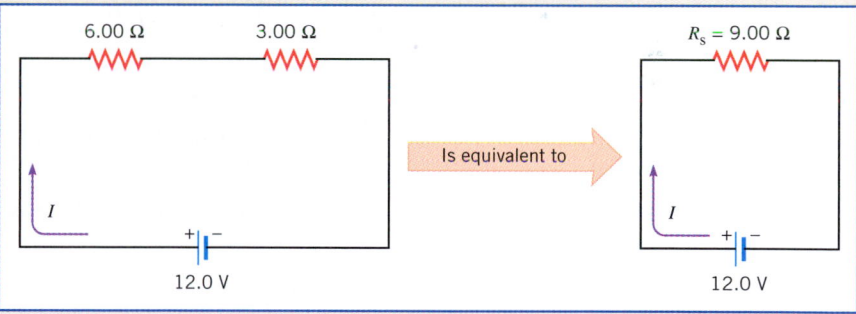

Figure 20.15 A 6.00-Ω and a 3.00-Ω resistor connected in series are equivalent to a single 9.00-Ω resistor.

of the resistors) equals the voltage V of the battery divided by the equivalent resistance R_S of the two resistors: $I = V/R_S$. Since the resistors are connected in series, we can obtain the equivalent resistance by adding the two resistances (see Figure 20.15).

Knowns and Unknowns The data for this problem are:

Description	Symbol	Value
Resistance of 6.00-Ω resistor	R_1	6.00 Ω
Resistance of 3.00-Ω resistor	R_2	3.00 Ω
Battery voltage	V	12.0 V
Unknown Variables		
Power delivered to 6.00-Ω resistor	P_1	?
Power delivered to 3.00-Ω resistor	P_2	?

Continued

Modeling the Problem

STEP 1   Power   The power P_1 delivered to the 6.00-Ω resistor is given by $P_1 = I^2 R_1$ (Equation 20.6b), where I is the current through the resistor and R_1 is the resistance. In this expression R_1 is a known quantity, and the current I will be determined in Step 2.

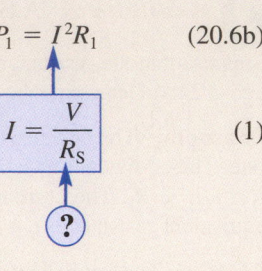

$$P_1 = I^2 R_1 \qquad (20.6b)$$

STEP 2   Ohm's Law   The current I in the circuit depends on the voltage V of the battery and the equivalent resistance R_S of the two resistors in series (see Figure 20.15). This dependence is given by Ohm's law (Equation 20.2) as

$$I = \frac{V}{R_S}$$

This result for the current can be substituted into Equation 20.6b, as indicated at the right. Note from the data table that the voltage is given. In Step 3 we will evaluate the equivalent resistance from the individual resistances R_1 and R_2.

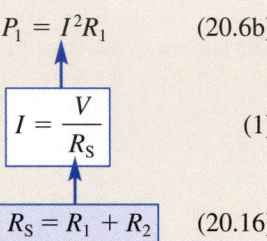

$$P_1 = I^2 R_1 \qquad (20.6b)$$

$$I = \frac{V}{R_S} \qquad (1)$$

STEP 3   Equivalent Resistance   Since the two resistors are wired in series, the equivalent resistance R_S is the sum of the two resistances (Equation 20.16):

$$R_S = R_1 + R_2$$

The resistances R_1 and R_2 are known. We substitute this expression for R_S into Equation 1, as shown in the right column.

$$P_1 = I^2 R_1 \qquad (20.6b)$$

$$I = \frac{V}{R_S} \qquad (1)$$

$$R_S = R_1 + R_2 \qquad (20.16)$$

Solution   Algebraically combining the results of the three steps, we have

STEP 1   STEP 2   STEP 3

$$P_1 \;\;\overset{\downarrow}{=}\;\; I^2 R_1 \;\;\overset{\downarrow}{=}\;\; \left(\frac{V}{R_S}\right)^2 R_1 \overset{\downarrow}{=} \left(\frac{V}{R_1 + R_2}\right)^2 R_1$$

The power delivered to the 6.00-Ω resistor is

$$P_1 = \left(\frac{V}{R_1 + R_2}\right)^2 R_1 = \left(\frac{12.0\ \text{V}}{6.00\ \Omega + 3.00\ \Omega}\right)^2 (6.00\ \Omega) = \boxed{10.7\ \text{W}}$$

In a similar fashion, it can be shown that the power delivered to the 3.00-Ω resistor is

$$P_2 = \left(\frac{V}{R_1 + R_2}\right)^2 R_2 = \left(\frac{12.0\ \text{V}}{6.00\ \Omega + 3.00\ \Omega}\right)^2 (3.00\ \Omega) = \boxed{5.3\ \text{W}}$$

Related Homework: *Problems 46, 116*

In Example 9 the total power sent to the two resistors is $P = 10.7\ \text{W} + 5.3\ \text{W} = 16.0\ \text{W}$. Alternatively, the total power could have been obtained by using the voltage across the two resistors (the battery voltage) and the equivalent resistance R_S:

$$P = \frac{V^2}{R_S} = \frac{(12.0\ \text{V})^2}{6.00\ \Omega + 3.00\ \Omega} = 16.0\ \text{W} \qquad (20.6c)$$

Problem-solving insight

In general, ***the total power delivered to any number of resistors in series is equal to the power delivered to the equivalent resistance.***

The physics of personal digital assistants.

Pressure-sensitive pads form the heart of computer input devices that function as personal digital assistants, or PDAs, and they offer an interesting application of series resistors. These devices are simple to use. You write directly on the pad with a plastic

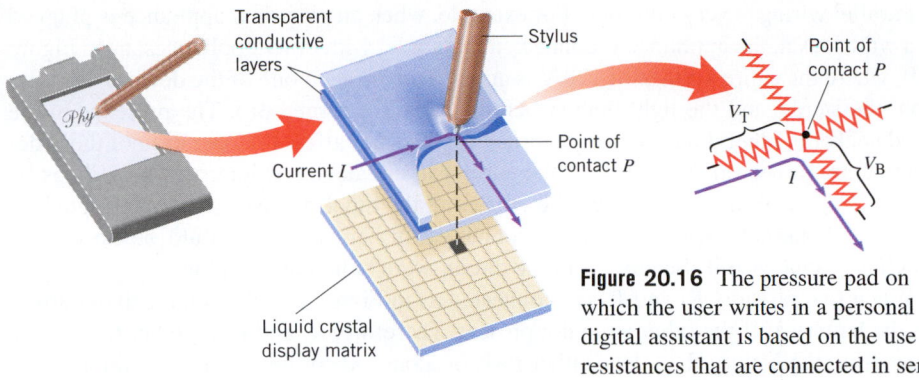

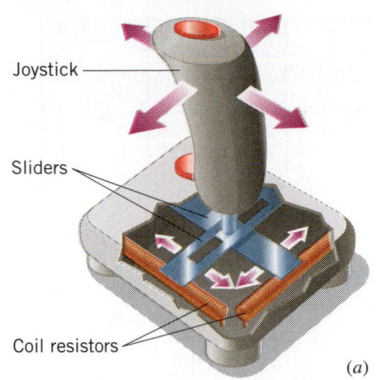

Figure 20.16 The pressure pad on which the user writes in a personal digital assistant is based on the use of resistances that are connected in series.

stylus that itself contains no electronics (see Figure 20.16). The writing appears as the stylus is moved, and recognition software interprets it as input for the built-in computer. The pad utilizes two transparent conductive layers that are separated by a small distance, except where pressure from the stylus brings them into contact (see point P in the drawing). Current I enters the positive side of the top layer, flows into the bottom layer through point P, and exits that layer through its negative side. Each layer provides resistance to the current, the amounts depending on where the point P is located. As the right side of the drawing indicates, the resistances from the layers are in series, since the same current exists in both of them. The voltage across the top-layer resistance is V_T, and the voltage across the bottom-layer resistance is V_B. In each case, the voltage is the current times the resistance. These two voltages are used to locate the point P and to activate (darken) one of the elements or pixels in a liquid crystal display matrix that lies beneath the transparent layers (see Section 24.6 for a discussion of liquid crystal displays). As the stylus is moved, the writing becomes visible as one element after another in the display matrix is activated.

The physics of a joystick. The joystick, found in computer games, also takes advantage of resistors connected in series. A joystick contains two straight coils of resistance wire that are oriented at 90° to each other (see Figure 20.17a). When the joystick is moved, it repositions the metallic slider on each of the coils. As part b of the drawing illustrates, each coil is connected across a 1.5-V battery.* Because one end of a coil is at 1.5 V and the other at 0 V, the voltage at the location of the slider is somewhere between these values; the voltage of the left slider in the drawing is labeled V_1 and the voltage of the right slider is V_2. The slider voltages are sent via wires to a computer, which translates them into positional data. In effect, the slider divides each resistance coil into two smaller resistance coils wired in series, and allows the voltage at the point where they are joined together to be detected.

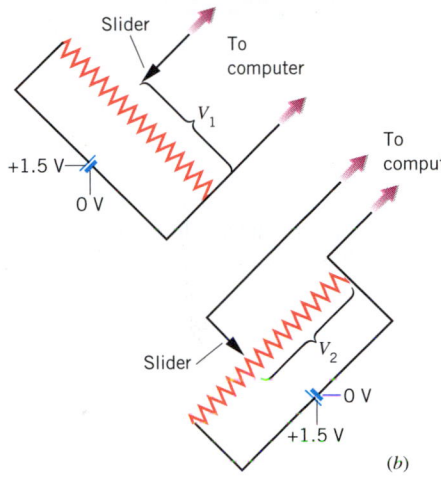

Figure 20.17 (a) A joystick uses two perpendicular movable sliders, and each makes contact with a coil resistor. (b) The sliders allow detection of the voltages V_1 and V_2, which a computer translates into positional data.

CHECK YOUR UNDERSTANDING

(*The answer is given at the end of the book.*)

12. The power rating of a 1000-W heater specifies the power that the heater uses when it is connected to an ac voltage of 120 V. What is the total power used by two of these heaters when they are connected in series with a single ac voltage of 120 V? **(a)** 4000 W **(b)** 3000 W **(c)** 2000 W **(d)** 1000 W **(e)** 500 W

20.7 **PARALLEL WIRING**

Parallel wiring is another method of connecting electrical devices. *Parallel wiring means that the devices are connected in such a way that the same voltage is applied across each device.* Figure 20.18 shows two resistors connected in parallel between the terminals of a battery. Part a of the picture is drawn so as to emphasize that the entire voltage of the battery is applied across each resistor. Actually, parallel connections are rarely drawn in this manner; instead they are drawn as in part b, where the dots indicate the points where the wires for the two branches are joined together. Parts a and b are equivalent representations of the same circuit.

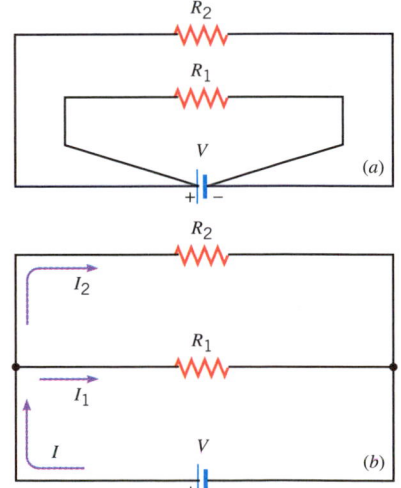

Figure 20.18 (a) When two resistors are connected in parallel, the same voltage V is applied across each resistor. (b) This drawing is equivalent to part a. I_1 and I_2 are the currents in R_1 and R_2.

*For clarity, two batteries are shown in Figure 20.17b, one associated with each resistance coil. In reality, both coils are connected across a single battery.

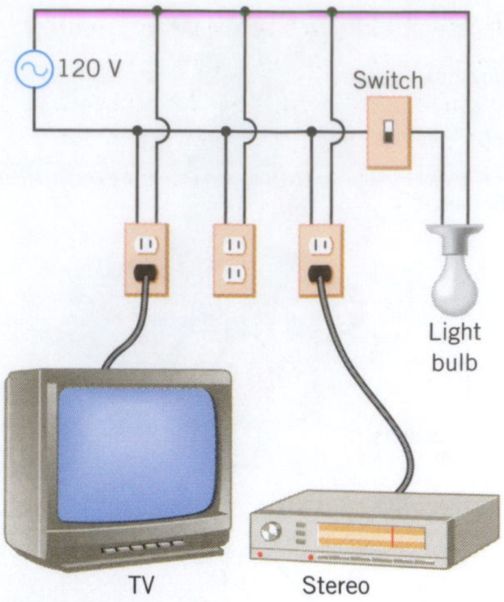

Figure 20.19 This drawing shows some of the parallel connections found in a typical home. Each wall socket provides 120 V to the appliance connected to it. In addition, 120 V is applied to the light bulb when the switch is turned on.

Parallel wiring is very common. For example, when an electrical appliance is plugged into a wall socket, the appliance is connected in parallel with other appliances, as in Figure 20.19, where the entire voltage of 120 V is applied across each one of the devices: the television, the stereo, and the light bulb (when the switch is turned on). The presence of the unused socket or other devices that are turned off does not affect the operation of those devices that are turned on. Moreover, if the current in one device is interrupted (perhaps by an opened switch or a broken wire), the current in the other devices is not interrupted. In contrast, if household appliances were connected in series, there would be no current through any appliance if the current in the circuit were halted at any point.

When two resistors R_1 and R_2 are connected as in Figure 20.18, each receives current from the battery as if the other were not present. Therefore, R_1 and R_2 together draw more current from the battery than does either resistor alone. According to the definition of resistance, $R = V/I$, a larger current implies a smaller resistance. Thus, the two parallel resistors behave as a single equivalent resistance that is *smaller* than either R_1 or R_2. Figure 20.20 returns to the water-flow analogy to provide additional insight into this important feature of parallel wiring. In part *a*, two sections of pipe that have the same length are connected in parallel with a pump. In part *b* these two sections have been replaced with a single pipe of the same length, whose cross-sectional area equals the combined cross-sectional areas of section 1 and section 2. The pump (analogous to a voltage source) can push more water per second (analogous to current) through the wider pipe in part *b* (analogous to a wider wire) than it can through *either* of the narrower pipes (analogous to narrower wires) in part *a*. In effect, the wider pipe offers less resistance to the flow of water than either of the narrower pipes offers individually.

As in a series circuit, it is possible to replace a parallel combination of resistors with an equivalent resistor that results in the same total current and power for a given voltage as the original combination. To determine the equivalent resistance for the two resistors in Figure 20.18*b*, note that the total current I from the battery is the sum of I_1 and I_2, where I_1 is the current in resistor R_1 and I_2 is the current in resistor R_2: $I = I_1 + I_2$. Since the same voltage V is applied across each resistor, the definition of resistance indicates that $I_1 = V/R_1$ and $I_2 = V/R_2$. Therefore,

$$I = I_1 + I_2 = \frac{V}{R_1} + \frac{V}{R_2} = V\left(\frac{1}{R_1} + \frac{1}{R_2}\right) = V\left(\frac{1}{R_P}\right)$$

where R_P is the equivalent resistance. Hence, when two resistors are connected in parallel, they are equivalent to a single resistor whose resistance R_P can be obtained from $1/R_P = 1/R_1 + 1/R_2$. In general, for any number of resistors wired in parallel, **the total current from the voltage source is the sum of the currents in the individual resistors.** Thus, a similar line of reasoning reveals that the equivalent resistance is

Problem-solving insight

Parallel resistors

$$\frac{1}{R_P} = \frac{1}{R_1} + \frac{1}{R_2} + \frac{1}{R_3} + \cdots \qquad (20.17)$$

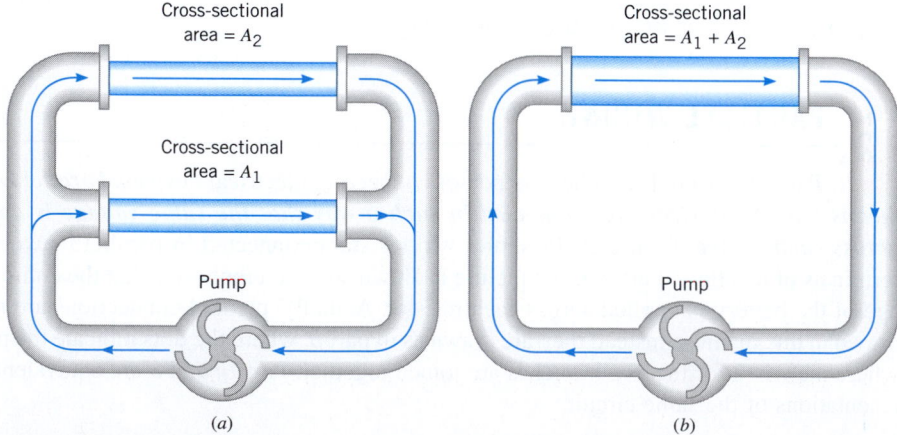

Figure 20.20 (*a*) Two equally long pipe sections, with cross-sectional areas A_1 and A_2, are connected in parallel to a water pump. (*b*) The two parallel pipe sections in part *a* are equivalent to a single pipe of the same length whose cross-sectional area is $A_1 + A_2$.

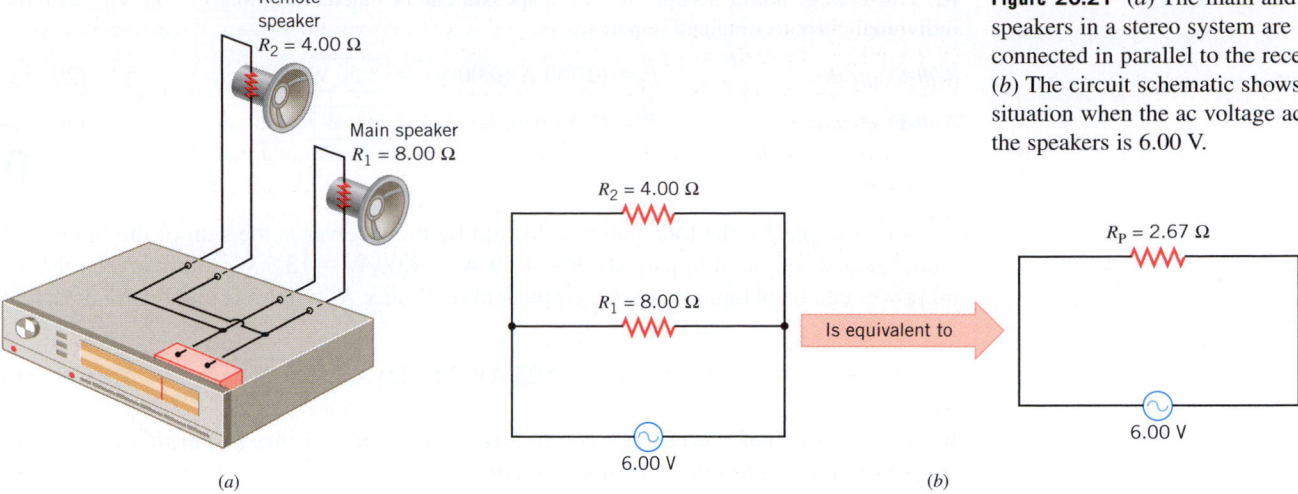

Figure 20.21 (*a*) The main and remote speakers in a stereo system are connected in parallel to the receiver. (*b*) The circuit schematic shows the situation when the ac voltage across the speakers is 6.00 V.

The next example deals with a parallel combination of resistors that occurs in a stereo system.

 Example 10 Main and Remote Stereo Speakers

Most receivers allow the user to connect "remote" speakers (to play music in another room, for instance) in addition to the main speakers. Figure 20.21 shows that the remote speaker and the main speaker for the right stereo channel are connected to the receiver in parallel (for clarity, the speakers for the left channel are not shown). At the instant represented in the picture, the ac voltage across the speakers is 6.00 V. The main-speaker resistance is 8.00 Ω, and the remote-speaker resistance is 4.00 Ω.* Determine **(a)** the equivalent resistance of the two speakers, **(b)** the total current supplied by the receiver, **(c)** the current in each speaker, and **(d)** the power dissipated in each speaker.

The physics of main and remote stereo speakers.

Reasoning The total current supplied to the two speakers by the receiver can be calculated as $I_{\text{rms}} = V_{\text{rms}}/R_P$, where R_P is the equivalent resistance of the two speakers in parallel and can be obtained from $1/R_P = 1/R_1 + 1/R_2$. The current in each speaker is different, however, since the speakers have different resistances. The average power delivered to a given speaker is the product of its current and voltage. In the parallel connection the same voltage is applied to each speaker.

Solution **(a)** According to Equation 20.17, the equivalent resistance of the two speakers is given by

$$\frac{1}{R_P} = \frac{1}{8.00\ \Omega} + \frac{1}{4.00\ \Omega} = \frac{3}{8.00\ \Omega} \quad \text{or} \quad R_P = \frac{8.00\ \Omega}{3} = \boxed{2.67\ \Omega}$$

This result is illustrated in part *b* of the drawing.

(b) Using the equivalent resistance in Ohm's law shows that the total current is

$$I_{\text{rms}} = \frac{V_{\text{rms}}}{R_P} = \frac{6.00\ \text{V}}{2.67\ \Omega} = \boxed{2.25\ \text{A}} \tag{20.14}$$

(c) Applying Ohm's law to each speaker gives the individual speaker currents:

8.00-Ω speaker
$$I_{\text{rms}} = \frac{V_{\text{rms}}}{R} = \frac{6.00\ \text{V}}{8.00\ \Omega} = \boxed{0.750\ \text{A}}$$

4.00-Ω speaker
$$I_{\text{rms}} = \frac{V_{\text{rms}}}{R} = \frac{6.00\ \text{V}}{4.00\ \Omega} = \boxed{1.50\ \text{A}}$$

The sum of these currents is equal to the total current obtained in part (b).

Problem-solving insight

The equivalent resistance R_P of a number of resistors in parallel has a reciprocal given by $R_P^{-1} = R_1^{-1} + R_2^{-1} + R_3^{-1} + \ldots$, where R_1, R_2, and R_3 are the individual resistances. After adding together the reciprocals R_1^{-1}, R_2^{-1}, and R_3^{-1}, do not forget to take the reciprocal of the result to find R_P.

---

*In reality, frequency-dependent characteristics (see Chapter 23) play a role in the operation of a loudspeaker. We assume here, however, that the frequency of the sound is low enough that the speakers behave as pure resistances.

(d) The average power dissipated in each speaker can be calculated using $\overline{P} = I_{rms}V_{rms}$ with the individual currents obtained in part (c):

8.00-Ω speaker $\overline{P} = (0.750 \text{ A})(6.00 \text{ V}) = \boxed{4.50 \text{ W}}$ (20.15a)

4.00-Ω speaker $\overline{P} = (1.50 \text{ A})(6.00 \text{ V}) = \boxed{9.00 \text{ W}}$ (20.15a)

In Example 10, the total power delivered by the receiver is the sum of the individual values that were found in part (d), $\overline{P} = 4.50 \text{ W} + 9.00 \text{ W} = 13.5 \text{ W}$. Alternatively, the total power can be obtained from the equivalent resistance $R_P = 2.67 \text{ Ω}$ and the total current in part (b):

$$\overline{P} = I_{rms}^2 R_P = (2.25 \text{ A})^2(2.67 \text{ Ω}) = \boxed{13.5 \text{ W}}$$ (20.15b)

Problem-solving insight

In general, *the total power delivered to any number of resistors in parallel is equal to the power delivered to the equivalent resistor.*

In a parallel combination of resistances, it is the *smallest* resistance that has the largest impact in determining the equivalent resistance. In fact, if one resistance approaches zero, then according to Equation 20.17, the equivalent resistance also approaches zero. In such a case, the near-zero resistance is said to *short out* the other resistances by providing a near-zero resistance path for the current to follow as a shortcut around the other resistances.

An interesting application of parallel wiring occurs in a three-way light bulb, as Conceptual Example 11 discusses.

Conceptual Example 11   A Three-Way Light Bulb and Parallel Wiring

The physics of a three-way light bulb.

Three-way light bulbs are popular because they can provide three levels of illumination (e.g., 50 W, 100 W, and 150 W) using a 120-V socket. The socket, however, must be equipped with a special three-way switch that enables one to select the illumination level. This switch does not select different voltages, because a three-way bulb uses a single voltage of 120 V. Within the bulb are two separate filaments. When the bulb is producing its highest illumination level and one of the filaments burns out (i.e., vaporizes), the bulb shines at one of the other illumination levels (either the lowest or the intermediate one). When the bulb is set to its highest level of illumination, how are the two filaments connected, **(a)** in parallel or **(b)** in series?

Reasoning In a series connection, the filaments would be connected in such a way that there is the same current through each one. The current would enter one filament and then leave that filament and enter into the other one. In a parallel connection, the same voltage would be applied across each filament, but the current through each would, in general, be different, the two currents existing independently of one another.

Answer (b) is incorrect. If the filaments were wired in series and one of them burned out, no current would pass through the bulb and none of the illumination levels would be available, contrary to what is observed. Therefore, the filaments are not wired in series.

Answer (a) is correct. Since the filaments are not in series, they must be in parallel, as Figure 20.22 helps to explain. The power dissipated in a resistance R is $\overline{P} = V_{rms}^2/R$, according to Equation 20.15c. With a single value of 120 V for the voltage V_{rms}, three different power ratings for the bulb can be obtained only if three different values for the resistance R are available. In a 50-W/100-W/150-W bulb, for example, one resistance R_{50} is provided by the 50-W filament, and the second resistance R_{100} comes from the 100-W filament. The third resistance R_{150} is the parallel combination of the other two and can be obtained from $1/R_{150} = 1/R_{50} + 1/R_{100}$. Figure 20.22 illustrates a simplified version of how the three-way switch operates in such a bulb. The first position of the switch closes contact A and leaves contact B open, energizing only the 50-W filament. The second position closes contact B and leaves contact A open, energizing only the 100-W filament. The third position closes both contacts A and B, so that both filaments light up to give the highest level of illumination.

Related Homework: *Problem 54*

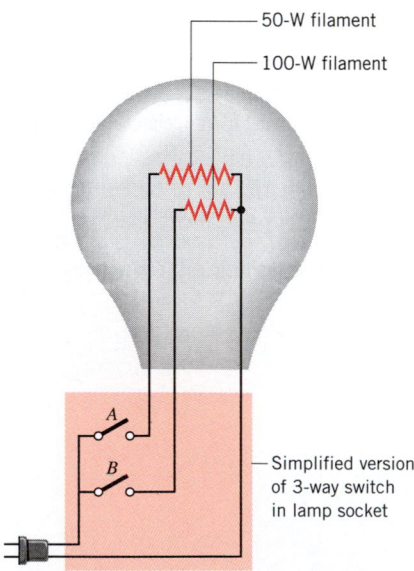

50-W filament

100-W filament

A

B

Simplified version of 3-way switch in lamp socket

Figure 20.22 A three-way light bulb uses two connected filaments. The filaments can be turned on one at a time or both together in parallel.

✓ **CHECK YOUR UNDERSTANDING**

(The answers are given at the end of the book.)

13. A car has two headlights, and their power is derived from the car's battery. The filament in one burns out, but the other headlight stays on. Are the headlights connected in series or in parallel?

14. Two identical light bulbs are connected to identical batteries in two different ways. In method A the bulbs are connected in parallel, and the parallel combination is connected between the one battery's terminals. In method B the bulbs are connected in series, and the series combination is connected between the other battery's terminals. What is the ratio of the power supplied by the battery in method A to the power supplied in method B?
(a) $\frac{1}{4}$ **(b)** 4 **(c)** $\frac{1}{2}$ **(d)** 2 **(e)** 1

20.8 CIRCUITS WIRED PARTIALLY IN SERIES AND PARTIALLY IN PARALLEL

Often an electric circuit is wired partially in series and partially in parallel. The key to determining the current, voltage, and power in such a case is to deal with the circuit in parts, with the resistances in each part being either in series or in parallel with each other. Example 12 shows how such an analysis is carried out.

⬇ **Example 12**   A Four-Resistor Circuit

Figure 20.23a shows a circuit composed of a 24-V battery and four resistors, whose resistances are 110, 180, 220, and 250 Ω. Find **(a)** the total current supplied by the battery and **(b)** the voltage between points A and B in the circuit.

Reasoning The total current that is supplied by the battery can be obtained from Ohm's law, $I = V/R$, where R is the equivalent resistance of the four resistors. The equivalent resistance can be calculated by dealing with the circuit in parts. The voltage V_{AB} between the two points A and B is also given by Ohm's law, $V_{AB} = IR_{AB}$, where I is the current and R_{AB} is the equivalent resistance between the two points.

Solution (a) The 220-Ω resistor and the 250-Ω resistor are in series, so they are equivalent to a single resistor whose resistance is 220 Ω + 250 Ω = 470 Ω (see Figure 20.23a). The 470-Ω resistor is in parallel with the 180-Ω resistor. Their equivalent resistance can be obtained from Equation 20.17:

$$\frac{1}{R_{AB}} = \frac{1}{470\ \Omega} + \frac{1}{180\ \Omega} = 0.0077\ \Omega^{-1}$$

$$R_{AB} = \frac{1}{0.0077\ \Omega^{-1}} = 130\ \Omega$$

The circuit is now equivalent to a circuit containing a 110-Ω resistor in series with a 130-Ω resistor (see Figure 20.23b). This combination behaves like a single resistor whose resistance is $R = 110\ \Omega + 130\ \Omega = 240\ \Omega$ (see Figure 20.23c). The total current from the battery is, then,

$$I = \frac{V}{R} = \frac{24\ \text{V}}{240\ \Omega} = \boxed{0.10\ \text{A}}$$

(b) The current $I = 0.10$ A passes through the resistance between points A and B. Therefore, Ohm's law indicates that the voltage across the 130-Ω resistor between points A and B is

$$V_{AB} = IR_{AB} = (0.10\ \text{A})(130\ \Omega) = \boxed{13\ \text{V}}$$

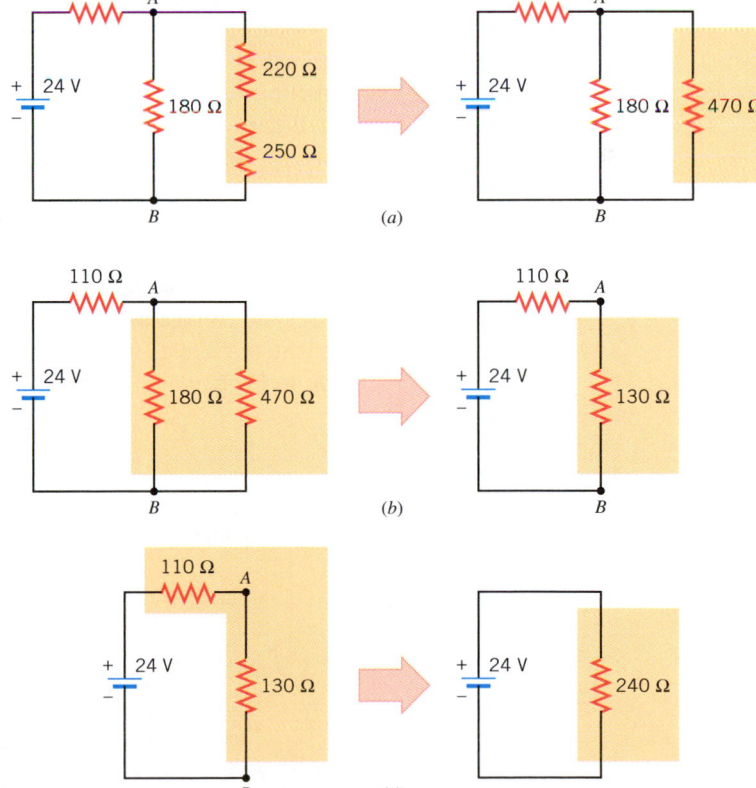

Figure 20.23 The circuits shown in this picture are equivalent.

✓ **CHECK YOUR UNDERSTANDING**

(The answers are given at the end of the book.)

15. In one of the circuits in the drawing, none of the resistors is in series or in parallel with any of the other resistors. Which circuit is it?

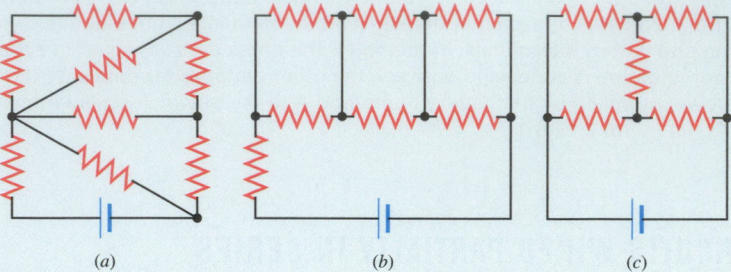

(a)   (b)   (c)

16. You have three resistors, each of which has a resistance R. By connecting all three together in various ways, which one or more of the following resistance values can you obtain? **(a)** $3R$ **(b)** $\frac{3}{2}R$ **(c)** R **(d)** $\frac{2}{3}R$ **(e)** $\frac{1}{3}R$

17. You have four resistors, each of which has a resistance R. It is possible to connect these four together so that the equivalent resistance of the combination is also R. How many ways can you do it? There is more than one way.

20.9 INTERNAL RESISTANCE

So far, the circuits we have considered include batteries or generators that contribute only their emfs to a circuit. In reality, however, such devices also add some resistance. This resistance is called the ***internal resistance*** of the battery or generator because it is located inside the device. In a battery, the internal resistance is due to the chemicals within the battery. In a generator, the internal resistance is the resistance of wires and other components within the generator.

Figure 20.24 presents a schematic representation of the internal resistance r of a battery. The drawing emphasizes that when an external resistance R is connected to the battery, the resistance is connected *in series* with the internal resistance. The internal resistance of a functioning battery is typically small (several thousandths of an ohm for a new car battery). Nevertheless, the effect of the internal resistance may not be negligible. Example 13 illustrates that when current is drawn from a battery, the internal resistance causes the voltage between the terminals to drop below the maximum value specified by the battery's emf. The actual voltage between the terminals of a battery is known as the ***terminal voltage.***

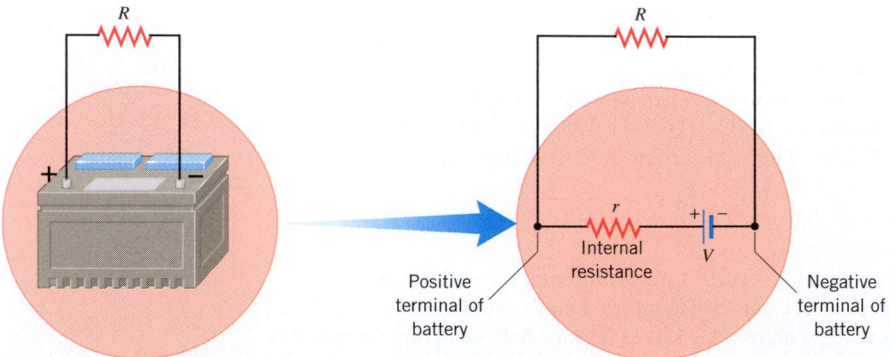

Figure 20.24 When an external resistance R is connected between the terminals of a battery, the resistance is connected in series with the internal resistance r of the battery.

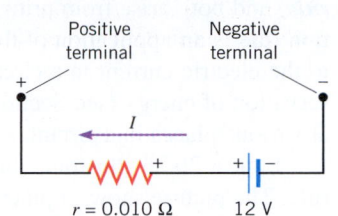

To car's electrical system
(ignition, lights, radio, etc.)

Positive terminal   Negative terminal

$r = 0.010\ \Omega$ 12 V

Figure 20.25 A car battery whose emf is 12 V and whose internal resistance is r.

 Example 13 The Terminal Voltage of a Battery

Figure 20.25 shows a car battery whose emf is 12.0 V and whose internal resistance is 0.010 Ω. This resistance is relatively large because the battery is old and the terminals are corroded. What is the terminal voltage when the current I drawn from the battery is **(a)** 10.0 A and **(b)** 100.0 A?

The physics of automobile batteries.

Reasoning The voltage between the terminals is not the entire 12.0-V emf, because part of the emf is needed to make the current go through the internal resistance. The amount of voltage needed can be determined from Ohm's law as the current I through the battery times the internal resistance r. For larger currents, a larger amount of voltage is needed, leaving less of the emf between the terminals.

Solution **(a)** The voltage needed to make a current of $I = 10.0$ A go through an internal resistance of $r = 0.010\ \Omega$ is

$$V = Ir = (10.0\ \text{A})(0.010\ \Omega) = 0.10\ \text{V}$$

To find the terminal voltage, remember that the direction of conventional current is always from a higher toward a lower potential. To emphasize this fact in the drawing, plus and minus signs have been included at the right and left ends, respectively, of the resistance r. The terminal voltage can be calculated by starting at the negative terminal and keeping track of how the voltage increases and decreases as we move toward the positive terminal. The voltage rises by 12.0 V due to the battery's emf. However, the voltage drops by 0.10 V because of the potential difference across the internal resistance. Therefore, the terminal voltage is $12.0\ \text{V} - 0.10\ \text{V} = \boxed{11.9\ \text{V}}$.

(b) When the current through the battery is 100.0 A, the voltage needed to make the current go through the internal resistance is

$$V = Ir = (100.0\ \text{A})(0.010\ \Omega) = 1.0\ \text{V}$$

The terminal voltage now decreases to $12.0\ \text{V} - 1.0\ \text{V} = \boxed{11.0\ \text{V}}$.

Example 13 indicates that the terminal voltage of a battery is smaller when the current drawn from the battery is larger, an effect that any car owner can demonstrate. Turn the headlights on before starting your car, so that the current through the battery is about 10 A, as in part (a) of Example 13. Then start the car. The starter motor draws a large amount of additional current from the battery, momentarily increasing the total current by an appreciable amount. Consequently, the terminal voltage of the battery decreases, causing the headlights to become dimmer.

20.10 KIRCHHOFF'S RULES

Electric circuits that contain a number of resistors can often be analyzed by combining individual groups of resistors in series and parallel, as Section 20.8 discusses. However, there are many circuits in which no two resistors are in series or in parallel. To deal with such circuits it is necessary to employ methods other than the series–parallel

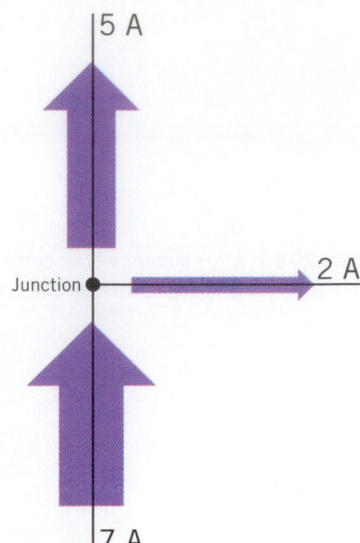

Figure 20.26 A junction is a point in a circuit where a number of wires are connected together. If 7 A of current is directed into the junction, then a total of 7 A (5 A + 2 A) of current must be directed out of the junction.

method. One alternative is to take advantage of Kirchhoff's rules, named after their developer Gustav Kirchhoff (1824–1887). There are two rules, the *junction rule* and the *loop rule,* and both arise from principles and ideas that we have encountered earlier. The junction rule is an application of the law of conservation of electric charge (see Section 18.2) to the electric current in a circuit. The loop rule is an application of the principle of conservation of energy (see Section 6.8) to the electric potential (see Section 19.2) that exists at various places in a circuit.

Figure 20.26 illustrates in greater detail the basic idea behind Kirchhoff's junction rule. The picture shows a junction where several wires are connected together. As Section 18.2 discusses, electric charge is conserved. Therefore, since there is no accumulation of charges at the junction itself, the total charge per second flowing into the junction must equal the total charge per second flowing out of it. In other words, *the junction rule states that the total current directed into a junction must equal the total current directed out of the junction,* or 7 A = 5 A + 2 A for the specific case shown in the picture.

To help explain Kirchhoff's loop rule, Figure 20.27 shows a circuit in which a 12-V battery is connected to a series combination of a 5-Ω and a 1-Ω resistor. The plus and minus signs associated with each resistor remind us that, outside a battery, conventional current is directed from a higher toward a lower potential. From left to right, there is a potential drop of 10 V across the first resistor and another drop of 2 V across the second resistor. Keeping in mind that potential is the electric potential energy per unit charge, let us follow a positive test charge clockwise* around the circuit. Starting at the negative terminal of the battery, we see that the test charge gains potential energy because of the 12-V rise in potential due to the battery. The test charge then loses potential energy because of the 10-V and 2-V drops in potential across the resistors, ultimately arriving back at the negative terminal. In traversing the closed-circuit loop, the test charge is like a skier gaining gravitational potential energy in going up a hill on a chair lift and then losing it to friction in coming down and stopping. When the skier returns to the starting point, the gain equals the loss, so there is no net change in potential energy. Similarly, when the test charge arrives back at its starting point, there is no net change in electric potential energy, the gains matching the losses. *The loop rule expresses this example of energy conservation in terms of the electric potential and states that for a closed-circuit loop, the total of all the potential drops is the same as the total of all the potential rises,* or 10 V + 2 V = 12 V for the specific case in Figure 20.27.

Kirchhoff's rules can be applied to any circuit, even when the resistors are not in series or in parallel. The two rules are summarized below, and Examples 14 and 15 illustrate how to use them.

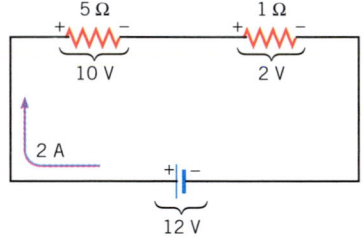

Figure 20.27 Following a positive test charge clockwise around the circuit, we see that the total voltage drop of 10 V + 2 V across the two resistors equals the voltage rise of 12 V due to the battery. The plus and minus signs on the resistors emphasize that, outside a battery, conventional current is directed from a higher potential (+) toward a lower potential (−).

KIRCHHOFF'S RULES

Junction rule. The sum of the magnitudes of the currents directed into a junction equals the sum of the magnitudes of the currents directed out of the junction.

Loop rule. Around any closed-circuit loop, the sum of the potential drops equals the sum of the potential rises.

⬇ **Example 14** Using Kirchhoff's Loop Rule

Figure 20.28 shows a circuit that contains two batteries and two resistors. Determine the current *I* in the circuit.

Reasoning The first step is to draw the current, which we have chosen to be clockwise around the circuit. The choice of direction is *arbitrary,* and if it is incorrect, *I* will turn out to be negative.

The second step is to mark the resistors with plus and minus signs, which serve as an aid in identifying the potential drops and rises for Kirchhoff's loop rule. **Problem-solving insight:** *Remember that, outside a battery, conventional current is always directed from a higher*

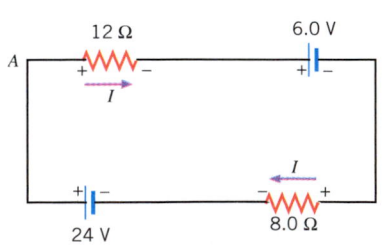

Figure 20.28 A single-loop circuit that contains two batteries and two resistors.

*The choice of the clockwise direction is arbitrary.

potential (+) toward a lower potential (−). Thus, we **must** mark the resistors as indicated in Figure 20.28, to be consistent with the clockwise direction chosen for the current.

We may now apply Kirchhoff's loop rule to the circuit, starting at corner *A*, proceeding clockwise around the loop, and identifying the potential drops and rises as we go. The potential across each resistor is given by Ohm's law as $V = IR$. The clockwise direction is arbitrary, and the same result is obtained with a counterclockwise path.

Solution Starting at corner *A* and moving clockwise around the loop, there is

1. A potential drop (+ to −) of $IR = I(12\ \Omega)$ across the 12-Ω resistor

2. A potential drop (+ to −) of 6.0 V across the 6.0-V battery

3. A potential drop (+ to −) of $IR = I(8.0\ \Omega)$ across the 8.0-Ω resistor

4. A potential rise (− to +) of 24 V across the 24-V battery

Setting the sum of the potential drops equal to the sum of the potential rises, as required by Kirchhoff's loop rule, gives

$$\underbrace{I(12\ \Omega) + 6.0\ \text{V} + I(8.0\ \Omega)}_{\text{Potential drops}} = \underbrace{24\ \text{V}}_{\text{Potential rises}}$$

Solving this equation for the current yields $\boxed{I = 0.90\ \text{A}}$. The current is a positive number, indicating that our initial choice for the direction (clockwise) of the current was correct.

Example 15 The Electrical System of a Car

In a car, the headlights are connected to the battery and would discharge the battery if it were not for the alternator, which is run by the engine. Figure 20.29 indicates how the car battery, headlights, and alternator are connected. The circuit includes an internal resistance of 0.0100 Ω for the car battery and its leads and a resistance of 1.20 Ω for the headlights. For the sake of simplicity, the alternator is approximated as an additional 14.00-V battery with an internal resistance of 0.100 Ω. Determine the currents through the car battery (I_B), the headlights (I_H), and the alternator (I_A).

The physics of an automobile electrical system.

Reasoning The drawing shows the directions chosen arbitrarily for the currents I_B, I_H, and I_A. If any direction is incorrect, the analysis will reveal a negative value for the corresponding current.

Next, we mark the resistors with the plus and minus signs that serve as an aid in identifying the potential drops and rises for the loop rule, recalling that, outside a battery, conventional current is always directed from a higher potential (+) toward a lower potential (−). Thus, given the directions selected for I_B, I_H, and I_A, the plus and minus signs *must* be those indicated in Figure 20.29. Kirchhoff's junction and loop rules can now be used.

Problem-solving insight

In problems involving Kirchhoff's rules, it is always helpful to mark the resistors with plus and minus signs to keep track of the potential rises and drops in the circuit.

Solution The junction rule can be applied to junction *B* or junction *E*. In either case, the same equation results:

Junction rule applied at B

$$\underbrace{I_A + I_B}_{\substack{\text{Into} \\ \text{junction}}} = \underbrace{I_H}_{\substack{\text{Out of} \\ \text{junction}}}$$

In applying the loop rule to the lower loop *BEFA*, we start at point *B*, move clockwise around the loop, and identify the potential drops and rises. There is a potential rise (− to +) of

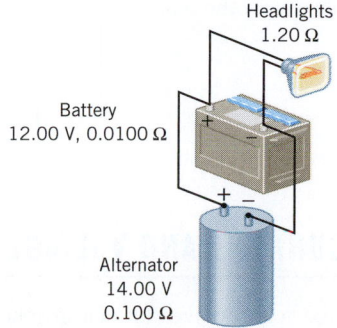

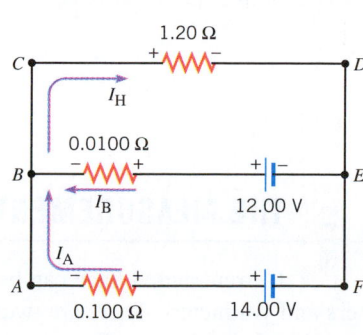

Figure 20.29 A circuit showing the headlight(s), battery, and alternator of a car.

I_B (0.0100 Ω) across the 0.0100-Ω resistor and then a drop (+ to −) of 12.00 V due to the car battery. Continuing around the loop, we find a 14.00-V rise (− to +) across the alternator, followed by a potential drop (+ to −) of I_A (0.100 Ω) across the 0.100-Ω resistor. Setting the sum of the potential drops equal to the sum of the potential rises gives the following result:

Loop rule: BEFA $\underbrace{I_A(0.100\ \Omega) + 12.00\ \text{V}}_{\text{Potential drops}} = \underbrace{I_B(0.0100\ \Omega) + 14.00\ \text{V}}_{\text{Potential rises}}$

Since there are three unknown variables in this problem, I_B, I_H, and I_A, a third equation is needed for a solution. To obtain the third equation, we apply the loop rule to the upper loop *CDEB,* choosing a clockwise path for convenience. The result is

Loop rule: CDEB $\underbrace{I_B(0.0100\ \Omega) + I_H(1.20\ \Omega)}_{\text{Potential drops}} = \underbrace{12.00\ \text{V}}_{\text{Potential rises}}$

These three equations can be solved simultaneously to show that

$$\boxed{I_B = -9.0\ \text{A},\ I_H = 10.1\ \text{A},\ I_A = 19.1\ \text{A}}$$

The negative answer for I_B indicates that the current through the battery is not directed from right to left, as drawn in Figure 20.29. Instead, the 9.0-A current is directed from left to right, opposite to the way current would be directed if the alternator were not connected. It is the left-to-right current created by the alternator that keeps the battery charged.

Note that we can check our results by applying Kirchhoff's loop rule to the outer loop *ABCDEF.* If our results are correct, then the sum of the potential drops around this loop will be equal to the sum of the potential rises.

REASONING STRATEGY

Applying Kirchhoff's Rules

1. Draw the current in each branch of the circuit. Choose any direction. If your choice is incorrect, the value obtained for the current will turn out to be a negative number.

2. Mark each resistor with a plus sign at one end and a minus sign at the other end, in a way that is consistent with your choice for the current direction in Step 1. Outside a battery, conventional current is always directed from a higher potential (the end marked +) toward a lower potential (the end marked −).

3. Apply the junction rule and the loop rule to the circuit, obtaining in the process as many independent equations as there are unknown variables.

4. Solve the equations obtained in Step 3 simultaneously for the unknown variables. (See Appendix C.3.)

✓ CHECK YOUR UNDERSTANDING

(The answer is given at the end of the book.)

18. The drawing shows a circuit containing three resistors and three batteries. In preparation for applying Kirchhoff's rules, the currents in each resistor have been drawn. For these currents, write down the equations that result from applying Kirchhoff's junction rule and loop rule. Apply the loop rule to loops *ABCD* and *BEFC.*

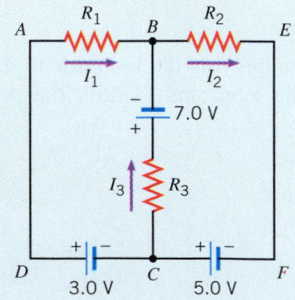

20.11 THE MEASUREMENT OF CURRENT AND VOLTAGE

Current and voltage can be measured with devices known, respectively, as ammeters and voltmeters. There are two types of such devices: those that use digital electronics and those that do not. The essential feature of nondigital devices is the dc *galvanometer.*

As Figure 20.30a illustrates, a galvanometer consists of a magnet, a coil of wire, a spring, a pointer, and a calibrated scale. The coil is mounted so that it can rotate, which causes the pointer to move in relation to the scale. The coil rotates in response to the torque applied by the magnet when there is a current in the coil (see Section 21.6). The coil stops rotating when this torque is balanced by the torque of the spring.

Two characteristics of a galvanometer are important when it is used as part of a measurement device. First, the amount of dc current that causes full-scale deflection of the pointer indicates the sensitivity of the galvanometer. For instance, Figure 20.30a shows an instrument that deflects full scale when the current in the coil is 0.10 mA. The second important characteristic is the resistance R_C of the wire in the coil. Figure 20.30b shows how a galvanometer with a coil resistance of R_C is represented in a circuit diagram.

The physics of an ammeter. Since an **ammeter** is an instrument that measures current, it must be inserted in the circuit so the current passes directly through it, as Figure 20.31 shows. (This is true for both digital and nondigital ammeters; Figure 20.31 shows a digital instrument.)

A nondigital ammeter includes a galvanometer and one or more *shunt resistors,* which are connected in parallel with the galvanometer and provide a bypass for current in excess of the galvanometer's full-scale limit. The bypass allows the ammeter to be used to measure a current exceeding the full-scale limit. In Figure 20.32, for instance, a current of 60.0 mA enters terminal A of an ammeter (nondigital), which uses a galvanometer with a full-scale current of 0.100 mA. The shunt resistor R can be selected so that 0.100 mA of current enters the galvanometer, while 59.9 mA bypasses it. In such a case, the measurement scale on the ammeter would be labeled 0 to 60.0 mA. To determine the shunt resistance, it is necessary to know the coil resistance R_C (see Problem 87).

When an ammeter is inserted into a circuit, the equivalent resistance of the ammeter adds to the circuit resistance. Any increase in circuit resistance causes a reduction in current, and this is a problem, because an ammeter should only measure the current, not change it. Therefore, an *ideal* ammeter would have zero resistance. In practice, a good ammeter is designed with an equivalent resistance small enough so there is only a negligible reduction of the current in the circuit when the ammeter is inserted.

The physics of a voltmeter. A **voltmeter** is an instrument that measures the voltage between two points, A and B, in a circuit. Figure 20.33 shows that the voltmeter must be connected between the points and is *not* inserted into the circuit as an ammeter is. (This is true for both digital and nondigital voltmeters; Figure 20.33 shows a digital instrument).

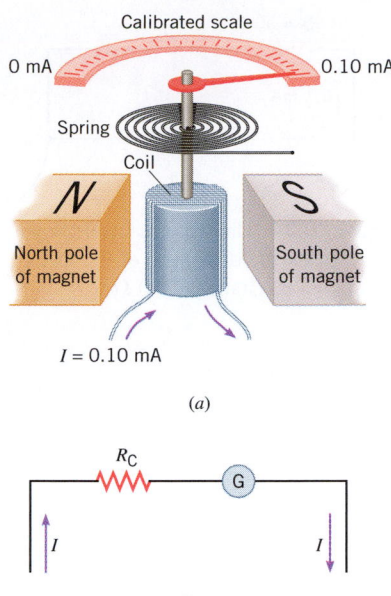

Figure 20.30 (a) A dc galvanometer. The coil of wire and pointer rotate when there is a current in the wire. (b) A galvanometer with a coil resistance of R_C is represented in a circuit diagram as shown here.

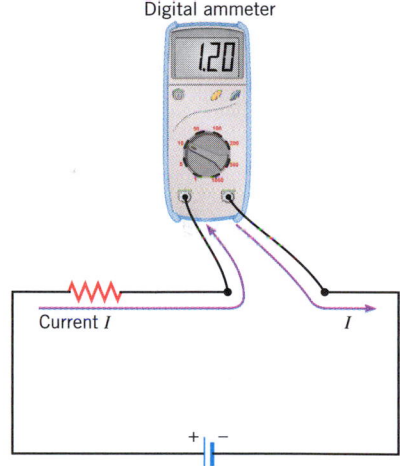

Figure 20.31 An ammeter must be inserted into a circuit so that the current passes directly through it.

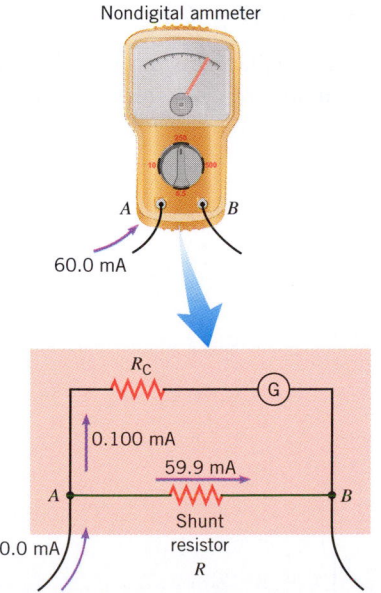

Figure 20.32 If a galvanometer with a full-scale limit of 0.100 mA is to be used to measure a current of 60.0 mA, a shunt resistance R must be used, so the excess current of 59.9 mA can detour around the galvanometer coil.

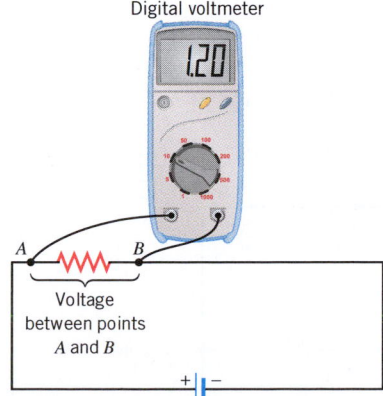

Figure 20.33 To measure the voltage between two points A and B in a circuit, a voltmeter is connected between the points.

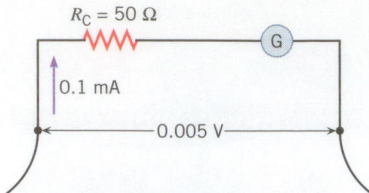

Figure 20.34 The galvanometer shown has a full-scale deflection of 0.1 mA and a coil resistance of 50 Ω.

A nondigital voltmeter includes a galvanometer whose scale is calibrated in volts. Suppose, for instance, that the galvanometer in Figure 20.34 has a full-scale current of 0.1 mA and a coil resistance of 50 Ω. Under full-scale conditions, the voltage across the coil would, therefore, be $V = IR_C = (0.1 \times 10^{-3} \, \text{A})(50 \, \Omega) = 0.005 \, \text{V}$. Thus, this galvanometer could be used to register voltages in the range $0-0.005$ V. A nondigital voltmeter, then, is a galvanometer used in this fashion, along with some provision for adjusting the range of voltages to be measured. To adjust the range, an additional resistance R is connected in series with the coil resistance R_C (see Problem 88).

Ideally, the voltage registered by a voltmeter should be the same as the voltage that exists when the voltmeter is not connected. However, a voltmeter takes some current from a circuit and, thus, alters the circuit voltage to some extent. An *ideal* voltmeter would have infinite resistance and would draw away only an infinitesimal amount of current. In reality, a good voltmeter is designed with a resistance that is large enough so the unit does not appreciably alter the voltage in the circuit to which it is connected.

✓ **CHECK YOUR UNDERSTANDING**

(The answers are given at the end of the book.)

19. An ideal ammeter has _____ resistance, whereas an ideal voltmeter has _____ resistance. **(a)** zero, zero **(b)** infinite, infinite **(c)** zero, infinite **(d)** infinite, zero

20. What would happen to the current in a circuit if a voltmeter, inadvertently mistaken for an ammeter, were inserted into the circuit? The current would **(a)** increase markedly **(b)** decrease markedly **(c)** remain the same.

20.12 CAPACITORS IN SERIES AND IN PARALLEL

Figure 20.35 shows two different capacitors connected in parallel to a battery. Since the capacitors are in parallel, they have the same voltage V across their plates. However, the capacitors *contain different amounts of charge*. The charge stored by a capacitor is $q = CV$ (Equation 19.8), so $q_1 = C_1V$ and $q_2 = C_2V$.

As with resistors, it is always possible to replace a parallel combination of capacitors with an *equivalent capacitor* that stores the same charge and energy for a given voltage as the combination does. To determine the equivalent capacitance C_P, note that the total charge q stored by the two capacitors is

$$q = q_1 + q_2 = C_1V + C_2V = (C_1 + C_2)V = C_PV$$

This result indicates that two capacitors in parallel can be replaced by an equivalent capacitor whose capacitance is $C_P = C_1 + C_2$. For any number of capacitors in parallel, the equivalent capacitance is

Parallel capacitors $\qquad\qquad C_P = C_1 + C_2 + C_3 + \cdots$ \hfill (20.18)

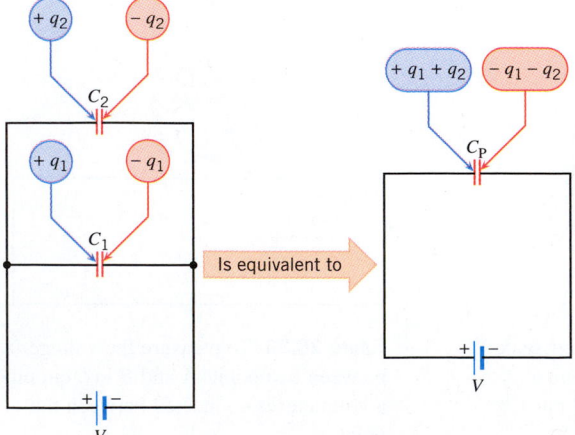

Figure 20.35 In a parallel combination of capacitances C_1 and C_2, the voltage V across each capacitor is the same, but the charges q_1 and q_2 on each capacitor are different.

Capacitances in parallel simply add together to give an equivalent capacitance. This behavior contrasts with that of resistors in parallel, which combine as reciprocals, according to Equation 20.17. The reason for this difference is that the charge q on a capacitor is directly proportional to the capacitance C ($q = CV$), whereas the current I in a resistor is inversely proportional to the resistance R ($I = V/R$).

The equivalent capacitor not only stores the same amount of charge as the parallel combination of capacitors, but also stores the same amount of energy. For instance, the energy stored in a single capacitor is $\frac{1}{2}CV^2$ (Equation 19.11b), so the total energy stored by two capacitors in parallel is

$$\text{Total energy} = \tfrac{1}{2}C_1V^2 + \tfrac{1}{2}C_2V^2 = \tfrac{1}{2}(C_1 + C_2)V^2 = \tfrac{1}{2}C_PV^2$$

which is equal to the energy stored in the equivalent capacitor C_P.

When capacitors are connected in series, the equivalent capacitance is different from when they are in parallel. As an example, Figure 20.36 shows two capacitors in series and reveals the following important fact. ***All capacitors in series, regardless of their capacitances, contain charges of the same magnitude, $+q$ and $-q$, on their plates.*** The battery places a charge of $+q$ on plate a of capacitor C_1, and this charge induces a charge of $+q$ to depart from the opposite plate a', leaving behind a charge $-q$. The $+q$ charge that leaves plate a' is deposited on plate b of capacitor C_2 (since these two plates are connected by a wire), where it induces a $+q$ charge to move away from the opposite plate b', leaving behind a charge of $-q$. Thus, all capacitors in series contain charges of the same magnitude on their plates. Note the difference between charging capacitors in parallel and in series. When charging parallel capacitors, the battery moves a charge q that is the sum of the charges moved for each of the capacitors: $q = q_1 + q_2 + q_3 + \cdots$. In contrast, when charging a series combination of n capacitors, the battery only moves a charge q, not nq, because the charge q passes by induction from one capacitor directly to the next one in line.

Problem-solving insight

The equivalent capacitance C_S for the series connection in Figure 20.36 can be determined by observing that the battery voltage V is shared by the two capacitors. The drawing indicates that the voltages across C_1 and C_2 are V_1 and V_2, so that $V = V_1 + V_2$. Since the voltages across the capacitors are $V_1 = q/C_1$ and $V_2 = q/C_2$, we find that

$$V = V_1 + V_2 = \frac{q}{C_1} + \frac{q}{C_2} = q\left(\frac{1}{C_1} + \frac{1}{C_2}\right) = q\left(\frac{1}{C_S}\right)$$

Thus, two capacitors in series can be replaced by an equivalent capacitor whose capacitance C_S can be obtained from $1/C_S = 1/C_1 + 1/C_2$. For any number of capacitors connected in series the equivalent capacitance is given by

Series capacitors
$$\frac{1}{C_S} = \frac{1}{C_1} + \frac{1}{C_2} + \frac{1}{C_3} + \cdots \tag{20.19}$$

Equation 20.19 indicates that capacitances in series combine as reciprocals and do not simply add together as resistors in series do. It is left as an exercise (Problem 109) to show that the equivalent series capacitance stores the same electrostatic energy as the sum of the energies of the individual capacitors.

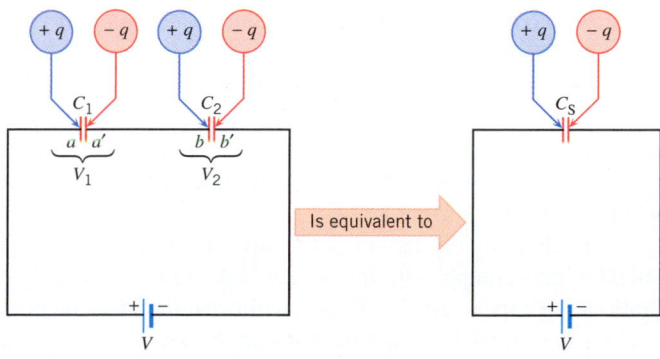

Figure 20.36 In a series combination of capacitances C_1 and C_2, the same amount of charge q is on the plates of each capacitor, but the voltages V_1 and V_2 across each capacitor are different.

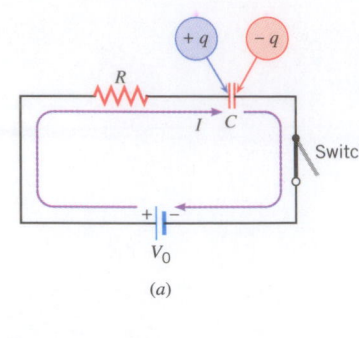

(a)

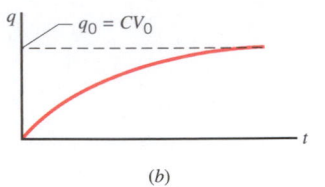

(b)

Figure 20.37 Charging a capacitor.

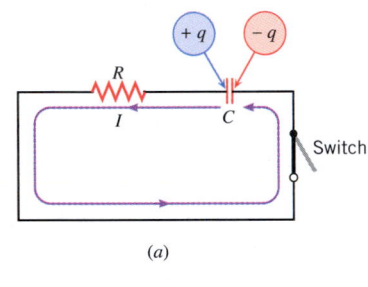

(a)

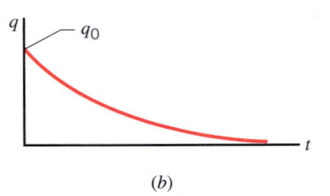

(b)

Figure 20.38 Discharging a capacitor.

The physics of heart pacemakers.

The physics of windshield wipers.

It is possible to simplify circuits containing a number of capacitors in the same general fashion as that outlined for resistors in Example 12 and Figure 20.23. The capacitors in a parallel grouping can be combined according to Equation 20.18, and those in a series grouping can be combined according to Equation 20.19.

20.13 RC CIRCUITS

Many electric circuits contain both resistors and capacitors. Figure 20.37 illustrates an example of a resistor–capacitor circuit, or *RC* circuit. Part *a* of the drawing shows the circuit at a time *t* after the switch has been closed and the battery has begun to charge up the capacitor plates. The charge on the plates builds up gradually to its equilibrium value of $q_0 = CV_0$, where V_0 is the voltage of the battery. Assuming that the capacitor is uncharged at time $t = 0$ s when the switch is closed, it can be shown that the magnitude q of the charge on the plates at time t is

Capacitor charging $\qquad\qquad q = q_0[1 - e^{-t/(RC)}]$ (20.20)

where the exponential e has the value of 2.718. . . . Part *b* of the drawing shows a graph of this expression, which indicates that the charge is $q = 0$ C when $t = 0$ s and increases gradually toward the equilibrium value of $q_0 = CV_0$. The voltage V across the capacitor at any time can be obtained from Equation 20.20 by dividing the charges q and q_0 by the capacitance C, since $V = q/C$ and $V_0 = q_0/C$.

The term RC in the exponent in Equation 20.20 is called the *time constant τ* of the circuit:

$$\tau = RC \qquad (20.21)$$

The time constant is measured in seconds; verification of the fact that an ohm times a farad is equivalent to a second is left as an exercise (see Check Your Understanding Question 21). The time constant is the amount of time required for the capacitor to accumulate 63.2% of its equilibrium charge, as can be seen by substituting $t = \tau = RC$ in Equation 20.20; $q_0(1 - e^{-1}) = q_0(0.632)$. The charge approaches its equilibrium value rapidly when the time constant is small and slowly when the time constant is large.

Figure 20.38*a* shows a circuit at a time *t* after a switch is closed to allow a charged capacitor to begin discharging. There is no battery in this circuit, so the charge $+q$ on the left plate of the capacitor can flow counterclockwise through the resistor and neutralize the charge $-q$ on the right plate. Assuming that the capacitor has a charge q_0 at time $t = 0$ s when the switch is closed, it can be shown that

Capacitor discharging $\qquad\qquad q = q_0 e^{-t/(RC)}$ (20.22)

where q is the amount of charge remaining on either plate at time t. The graph of this expression in part *b* of the drawing shows that the charge begins at q_0 when $t = 0$ s and decreases gradually toward zero. Smaller values of the time constant RC lead to a more rapid discharge. Equation 20.22 indicates that when $t = \tau = RC$, the magnitude of the charge remaining on each plate is $q_0 e^{-1} = q_0(0.368)$. Therefore, the time constant is also the amount of time required for a charged capacitor to *lose* 63.2% of its charge.

The charging/discharging of a capacitor has many applications. Heart pacemakers, for instance, incorporate *RC* circuits to control the timing of voltage pulses that are delivered to a malfunctioning heart to regulate its beating cycle. The pulses are delivered by electrodes attached externally to the chest or located internally near the heart when the pacemaker is implanted surgically (see Figure 20.39). A voltage pulse is delivered when the capacitor discharges to a preset level, following which the capacitor is recharged rapidly and the cycle repeats. The value of the time constant *RC* controls the pulsing rate, which is about once per second.

The charging/discharging of a capacitor is also used in automobiles that have windshield wipers equipped for intermittent operation during a light drizzle. In this mode of operation, the wipers remain off for a while and then turn on briefly. The timing of the on–off cycle is determined by the time constant of a resistor–capacitor combination.

✓ CHECK YOUR UNDERSTANDING

(The answers are given at the end of the book.)

21. The time constant τ of a series *RC* circuit is measured in seconds (s) and is given by $\tau = RC$, where the resistance *R* is measured in ohms (Ω) and the capacitance *C* is measured in farads (F). Verify that an ohm times a farad is equivalent to a second.

22. The drawings show two different resistor–capacitor arrangements. The time constant for arrangement A is 0.20 s. What is the time constant for arrangement B?
(a) 0.050 s **(b)** 0.10 s **(c)** 0.20 s **(d)** 0.40 s **(e)** 0.80 s

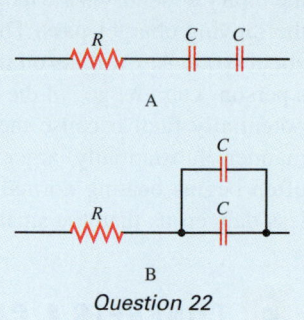

Question 22

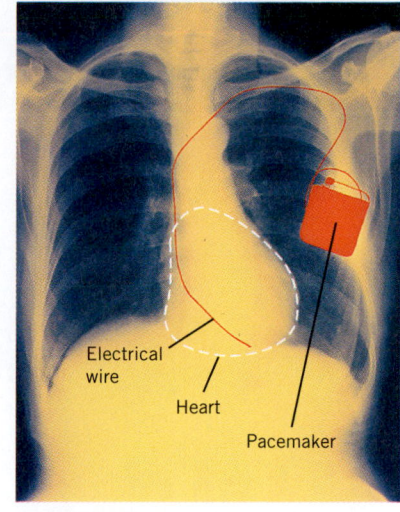

Figure 20.39 This X-ray photograph shows a heart pacemaker that has been implanted surgically. (RNHRD NHS Trust/Stone/Getty Images, Inc.)

Electrical wire

Heart

Pacemaker

20.14

SAFETY AND THE PHYSIOLOGICAL EFFECTS OF CURRENT

Electric circuits, although very useful, can also be hazardous. To reduce the danger inherent in using circuits, proper *electrical grounding* is necessary. The next two figures help to illustrate what electrical grounding means and how it is achieved.

Figure 20.40*a* shows a clothes dryer connected to a wall socket via a two-prong plug. The dryer is operating normally; that is, the wires inside are insulated from the metal casing of the dryer, so no charge flows through the casing itself. Notice that one terminal of the ac generator is customarily connected to ground ($\perp$) by the electric power company. Part *b* of the drawing shows the hazardous result that occurs if a wire comes loose and contacts the metal casing. A person touching it receives a shock, since electric charge flows through the casing, the person's body, and the ground on the way back to the generator.

Figure 20.41 shows the same appliance connected to a wall socket via a three-prong plug that provides safe electrical grounding. The third prong connects the metal casing directly to a copper rod driven into the ground or to a copper water pipe that is in the ground. This arrangement protects against electrical shock in the event that a broken wire touches the metal casing. In this event, charge would flow through the casing, through the third prong, and into the ground, returning eventually to the generator. No charge would flow through the person's body, because the copper rod provides much less electrical resistance than does the body.

The physics of
safe electrical grounding.

Figure 20.40 (*a*) A normally operating clothes dryer that is connected to a wall socket via a two-prong plug. (*b*) An internal wire accidentally touches the metal casing, and a person who touches the casing receives an electrical shock.

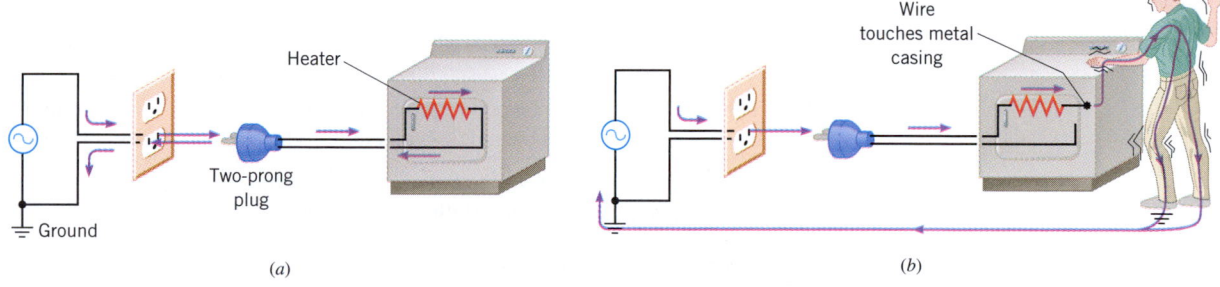

Heater

Two-prong plug

Ground

(a)

Wire touches metal casing

(b)

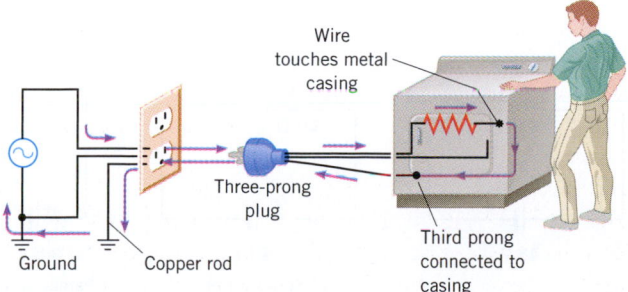

Wire touches metal casing

Three-prong plug

Ground   Copper rod

Third prong connected to casing

Figure 20.41 A safely connected dryer. If the dryer malfunctions, a person touching it receives no shock, since electric charge flows through the third prong and into the ground via a copper rod, rather than through the person's body.

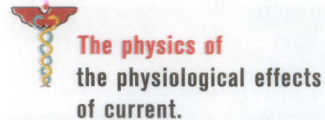

The physics of the physiological effects of current.

Serious and sometimes fatal injuries can result from electrical shock. The severity of the injury depends on the magnitude of the current and the parts of the body through which the moving charges pass. The amount of current that causes a mild tingling sensation is about 0.001 A. Currents on the order of 0.01–0.02 A can lead to muscle spasms, in which a person "can't let go" of the object causing the shock. Currents of approximately 0.2 A are potentially fatal because they can make the heart fibrillate, or beat in an uncontrolled manner. Substantially larger currents stop the heart completely. However, since the heart often begins beating normally again after the current ceases, the larger currents can be less dangerous than the smaller currents that cause fibrillation.

20.15 ## CONCEPTS & CALCULATIONS

Series and parallel wiring are two common ways in which devices, such as light bulbs, can be connected to a circuit. The next example reviews the concepts of voltage, current, resistance, and power in the context of these two types of circuits.

Concepts & Calculations Example 16
The Brightness of Light Bulbs Wired in Series and in Parallel

A circuit contains a 48-V battery and a single light bulb whose resistance is 240 Ω. A second, identical, light bulb can be wired either in series or in parallel with the first one (see Figure 20.42). Determine the power delivered to a single bulb when the circuit contains **(a)** only one bulb, **(b)** two bulbs in series, and **(c)** two bulbs in parallel. Assume that the battery has no internal resistance.

Concept Questions and Answers How is the power P that is delivered to a light bulb related to the bulb's resistance R and the voltage V across it?

Answer Equation 20.6c gives the answer as $P = V^2/R$.

When there is only one light bulb in the circuit, what is the voltage V?

Answer Since there is only one bulb in the circuit, V is the voltage of the battery, $V = 48$ V.

The more power delivered to a light bulb, the brighter it is. When two bulbs are wired in series, does the brightness of each bulb increase, decrease, or remain the same relative to the brightness when only a single bulb is in the circuit?

Answer Since the bulbs are identical and are wired in series, each receives one-half the battery voltage V. The power delivered to each bulb, then, is

$$P = \frac{(\frac{1}{2}V)^2}{R} = \frac{1}{4}\left(\frac{V^2}{R}\right)$$

This result shows that the power delivered to each bulb in the series circuit is only one-fourth the power delivered to the single-bulb circuit. Thus, the brightness of each bulb decreases.

When two bulbs are wired in parallel, does the brightness of each bulb increase, decrease, or remain the same relative to the brightness when only a single bulb is in the circuit?

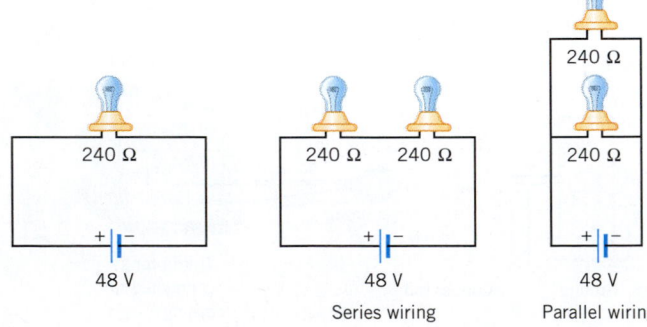

Figure 20.42 When a second identical light bulb is connected to the circuit on the left, it can be connected either in series or in parallel with the original bulb.

Answer Since the bulbs are wired in parallel, each receives the full battery voltage V. Thus, the power delivered to each bulb remains the same as when only one bulb is in the circuit, so the brightness of each bulb does not change.

Solution **(a)** When only one bulb is in the circuit, the power that the bulb receives is

$$P = \frac{V^2}{R} = \frac{(48\ \text{V})^2}{240\ \Omega} = \boxed{9.6\ \text{W}} \qquad (20.6\text{c})$$

(b) When the two (identical) light bulbs are wired in series, each receives one-half the battery voltage V. According to Equation 20.6c, the power P delivered to each bulb of resistance R is

$$P = \frac{(\frac{1}{2}V)^2}{R} = \frac{[\frac{1}{2}(48\ \text{V})]^2}{240\ \Omega} = \boxed{2.4\ \text{W}}$$

As expected, the power delivered to each bulb is only one-fourth the power delivered when there is only one bulb in the circuit.

(c) When the two bulbs are wired in parallel, the voltage across each is the same as the voltage of the battery. Therefore, the power delivered to each bulb is given by

$$P = \frac{V^2}{R} = \frac{(48\ \text{V})^2}{240\ \Omega} = \boxed{9.6\ \text{W}} \qquad (20.6\text{c})$$

As expected, the power delivered, and hence the brightness, does not change relative to that in the single-bulb circuit.

Kirchhoff's junction rule and loop rule are important tools for analyzing the currents and voltages in complex circuits. The rules are easy to use, once some of the subtleties are understood. The next example explores these subtleties in a two-loop circuit.

Concepts & Calculations Example 17
Using Kirchhoff's Rules

For the circuit in Figure 20.43, use Kirchhoff's junction rule and loop rule to find the currents through the three resistors.

Concept Questions and Answers Notice that there are two loops, labeled 1 and 2, in this circuit. Does it matter that there is no battery in loop 1, but only two resistors?

Answer No, it doesn't matter. A loop can have any number of batteries, including none at all.

The currents through the three resistors are labeled as I_1, I_2 and I_3. Does it matter which direction, left-to-right or right-to-left, has been chosen for each current?

Answer No. If we initially select the wrong direction for a current, it does not matter. The value obtained for that particular current will turn out to be a negative number, indicating that the actual current is in the opposite direction.

When we place the + and − signs at the ends of each resistor, does it matter which end is + and which is −?

Answer Yes, it does. Once the direction of the current has been selected, the + and − signs must be chosen so that the current goes from the + end toward the − end of the resistor. Notice that this is the case for each of the three resistors in Figure 20.43.

When we evaluate the potential drops and rises around a closed loop, does it matter which direction, clockwise or counterclockwise, is chosen for the evaluation?

Answer No, the direction is arbitrary. If we choose a clockwise direction, for example, we will have a certain number of potential drops and rises. If we choose a counterclockwise direction, all the drops become rises, and vice versa. Since we always set the potential drops equal to the potential rises, it does not matter which direction is picked for evaluating them. In the Solution that follows, we will use a counterclockwise direction for loop 1 and a clockwise direction for loop 2.

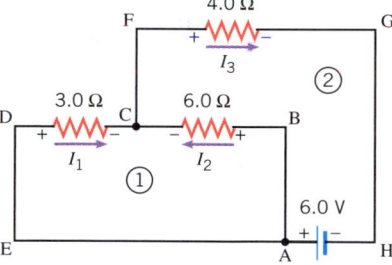

Figure 20.43 The three currents in this circuit are evaluated using Kirchhoff's loop and junction rules.

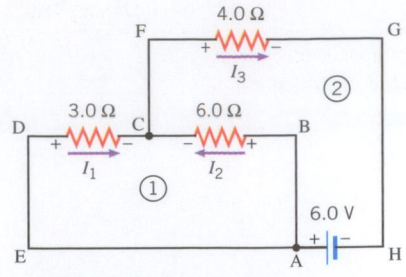

Figure 20.43 (Repeated) The three currents in this circuit are evaluated using Kirchhoff's loop and junction rules.

Solution Let's begin by expressing the current I_3 in terms of I_1 and I_2. In Figure 20.43 (which is repeated here in the margin for convenience), notice that I_1 and I_2 flow into junction C and that I_3 flows out of it. Kirchhoff's junction rule states that

$$\underbrace{I_1 + I_2}_{\substack{\text{Currents} \\ \text{into junction}}} = \underbrace{I_3}_{\substack{\text{Currents} \\ \text{out of junction}}} \qquad (20.23)$$

We now apply the loop rule to loop 1 (ABCDEA), starting at point A and proceeding counterclockwise:

$$\underbrace{I_2(6.0\ \Omega)}_{\text{Potential drops}} = \underbrace{I_1(3.0\ \Omega)}_{\text{Potential rises}} \qquad (20.24)$$

The two equations above contain three unknown variables, so we now apply the loop rule to loop 2 to obtain another equation that contains these variables. Starting at point A, we will move clockwise around loop 2 (ABCFGHA):

$$\underbrace{I_2(6.0\ \Omega) + I_3(4.0\ \Omega)}_{\text{Potential drops}} = \underbrace{6.0\ \text{V}}_{\text{Potential rises}} \qquad (20.25)$$

Substituting Equation 20.23 for I_3 into Equation 20.25 gives

$$I_2(6.0\ \Omega) + (I_1 + I_2)(4.0\ \Omega) = 6.0\ \text{V} \qquad (20.26)$$

Solving Equation 20.24 for I_2 gives $I_2 = \frac{1}{2}I_1$. Substituting this result into Equation 20.26, we obtain

$$\frac{1}{2}I_1\ (6.0\ \Omega) + (I_1 + \frac{1}{2}I_1)(4.0\ \Omega) = 6.0\ \text{V}$$

Solving this result for I_1 yields $I_1 = \boxed{\frac{2}{3}\ \text{A}}$. Substituting this result back into Equation 20.24 and solving for I_2 gives $I_2 = \boxed{\frac{1}{3}\ \text{A}}$. Equation 20.23 indicates, then, that $I_3 = I_1 + I_2 = \boxed{1.0\ \text{A}}$.

CONCEPT SUMMARY

If you need more help with a concept, use the Learning Aids noted next to the discussion or equation. Examples (**Ex.**) are in the text of this chapter. Go to **www.wiley.com/college/cutnell** for the following Learning Aids:

Interactive LearningWare (ILW) — Additional examples solved in a five-step interactive format.

Concept Simulations (CS) — Animated text figures or animations of important concepts.

Interactive Solutions (IS) — Models for certain types of problems in the chapter homework. The calculations are carried out interactively.

Topic	Discussion	Learning Aids
Electromotive force	**20.1 ELECTROMOTIVE FORCE AND CURRENT** There must be at least one source or generator of electrical energy in an electric circuit. The electromotive force (emf) of a generator, such as a battery, is the maximum potential difference (in volts) that exists between the terminals of the generator.	
	The rate of flow of charge is called the electric current. If the rate is constant, the current I is given by	
Current	$$I = \frac{\Delta q}{\Delta t} \qquad (20.1)$$	**Ex. 1**
The ampere	where Δq is the magnitude of the charge crossing a surface in a time Δt, the surface being perpendicular to the motion of the charge. The SI unit for current is the coulomb per second (C/s), which is referred to as an ampere (A).	
Direct current	When the charges flow only in one direction around a circuit, the current is called direct current (dc). When the direction of charge flow changes from moment to moment, the current is known as alternating current (ac).	
Alternating current		
Conventional current	Conventional current is the hypothetical flow of positive charges that would have the same effect in a circuit as the movement of negative charges that actually does occur.	
Resistance	**20.2 OHM'S LAW** The definition of electrical resistance R is $R = V/I$, where V (in volts) is the voltage applied across a piece of material and I (in amperes) is the current through the material. Resistance is measured in volts per ampere, a unit called an ohm (Ω).	
The ohm		

Topic	Discussion	Learning Aids

If the ratio of the voltage to the current is constant for all values of voltage and current, the resistance is constant. In this event, the definition of resistance becomes Ohm's law, as follows:

Ohm's law

$$\frac{V}{I} = R = \text{constant} \quad \text{or} \quad V = IR \qquad (20.2)$$

Ex. 2
CS 20.1

20.3 RESISTANCE AND RESISTIVITY The resistance of a piece of material of length L and cross-sectional area A is

Resistance in terms of length and area

$$R = \rho \frac{L}{A} \qquad (20.3)$$ **Ex. 3**

Resistivity

where ρ is the resistivity of the material.

The resistivity of a material depends on the temperature. For many materials and limited temperature ranges, the temperature dependence is given by

Temperature dependence of resistivity

$$\rho = \rho_0[1 + \alpha(T - T_0)] \qquad (20.4)$$ **Ex. 4**

Temperature coefficient of resistivity

where ρ and ρ_0 are the resistivities at temperatures T and T_0, respectively, and α is the temperature coefficient of resistivity.

The temperature dependence of the resistance R is given by

Temperature dependence of resistance

$$R = R_0[1 + \alpha(T - T_0)] \qquad (20.5)$$

where R and R_0 are the resistances at temperatures T and T_0, respectively.

20.4 ELECTRIC POWER When electric charge flows from point A to point B in a circuit, leading to a current I, and the voltage between the points is V, the electric power associated with this current and voltage is

Electric power

$$P = IV \qquad (20.6a)$$ **Ex. 5**

For a resistor, Ohm's law applies, and it follows that the power delivered to the resistor is also given by either of the following two equations:

$$P = I^2 R \qquad (20.6b)$$

Power delivered to a resistor

$$P = \frac{V^2}{R} \qquad (20.6c)$$ **Ex. 16**

20.5 ALTERNATING CURRENT The alternating voltage between the terminals of an ac generator can be represented by

Alternating voltage

$$V = V_0 \sin 2\pi f t \qquad (20.7)$$

where V_0 is the peak value of the voltage, t is the time, and f is the frequency (in Hertz) at which the voltage oscillates.

Correspondingly, in a circuit containing only resistance, the ac current is

Alternating current

$$I = I_0 \sin 2\pi f t \qquad (20.8)$$

where I_0 is the peak value of the current and is related to the peak voltage via $I_0 = V_0/R$.

For sinusoidal current and voltage, the root mean square (rms) current and voltage are related to the peak values according to the following equations:

Root mean square current

$$I_{\text{rms}} = \frac{I_0}{\sqrt{2}} \qquad (20.12)$$

Ex. 6, 7

Root mean square voltage

$$V_{\text{rms}} = \frac{V_0}{\sqrt{2}} \qquad (20.13)$$

The power in an ac circuit is the product of the current and the voltage and oscillates in time. The average power is

Average ac power

$$\overline{P} = I_{\text{rms}} V_{\text{rms}} \qquad (20.15a)$$

For a resistor, Ohm's law applies, so that $V_{\text{rms}} = I_{\text{rms}} R$ and the average power delivered to the resistor is also given by the following two equations:

Average ac power delivered to a resistor

$$\overline{P} = I_{\text{rms}}^2 R \qquad (20.15b)$$

$$\overline{P} = \frac{V_{\text{rms}}^2}{R} \qquad (20.15c)$$

Topic	Discussion	Learning Aids
	20.6 SERIES WIRING When devices are connected in series, there is the same current through each device. The equivalent resistance R_S of a series combination of resistances (R_1, R_2, R_3, etc.) is	
Equivalent series resistance	$$R_S = R_1 + R_2 + R_3 + \cdots \qquad (20.16)$$	**IS 20.119** **Ex. 8, 9**
	The power delivered to the equivalent resistance is equal to the total power delivered to any number of resistors in series.	
	20.7 PARALLEL WIRING When devices are connected in parallel, the same voltage is applied across each device. In general, devices wired in parallel carry different currents. The reciprocal of the equivalent resistance R_P of a parallel combination of resistances (R_1, R_2, R_3, etc.) is	
Equivalent parallel resistance	$$\frac{1}{R_P} = \frac{1}{R_1} + \frac{1}{R_2} + \frac{1}{R_3} + \cdots \qquad (20.17)$$	**IS 20.57** **Ex. 10, 11**
	The power delivered to the equivalent resistance is equal to the total power delivered to any number of resistors in parallel.	
	20.8 CIRCUITS WIRED PARTIALLY IN SERIES AND PARTIALLY IN PARALLEL Sometimes, one section of a circuit is wired in series, while another is wired in parallel. In such cases the circuit can be analyzed in parts, according to the respective series and parallel equivalent resistances of the various sections.	**Ex. 12** **CS 20.2, 20.5**
Internal resistance Terminal voltage	**20.9 INTERNAL RESISTANCE** The internal resistance of a battery or generator is the resistance within the battery or generator. The terminal voltage is the voltage between the terminals of a battery or generator and is equal to the emf only when there is no current through the device. When there is a current I, the internal resistance r causes the terminal voltage to be less than the emf by an amount Ir.	**Ex. 13**
The junction rule	**20.10 KIRCHHOFF'S RULES** Kirchhoff's junction rule states that the sum of the magnitudes of the currents directed into a junction equals the sum of the magnitudes of the currents directed out of the junction.	**Ex. 14, 15, 17** **ILW 20.1**
The loop rule	Kirchhoff's loop rule states that, around any closed-circuit loop, the sum of the potential drops equals the sum of the potential rises. The Reasoning Strategy given at the end of Section 20.10 explains how these two rules are applied to analyze any circuit.	**CS 20.3**
Galvanometer Ammeter Voltmeter	**20.11 THE MEASUREMENT OF CURRENT AND VOLTAGE** A galvanometer is a device that responds to electric current and is used in nondigital ammeters and voltmeters. An ammeter is an instrument that measures current and must be inserted into a circuit in such a way that the current passes directly through the ammeter. A voltmeter is an instrument for measuring the voltage between two points in a circuit. A voltmeter must be connected between the two points and is not inserted into a circuit as an ammeter is.	
	20.12 CAPACITORS IN SERIES AND IN PARALLEL The equivalent capacitance C_P for a parallel combination of capacitances (C_1, C_2, C_3, etc.) is	
Equivalent parallel capacitance	$$C_P = C_1 + C_2 + C_3 + \cdots \qquad (20.18)$$	
	In general, each capacitor in a parallel combination carries a different amount of charge. The equivalent capacitor carries the same total charge and stores the same total energy as the parallel combination.	
	The reciprocal of the equivalent capacitance C_S for a series combination (C_1, C_2, C_3, etc.) of capacitances is	
Equivalent series capacitance	$$\frac{1}{C_S} = \frac{1}{C_1} + \frac{1}{C_2} + \frac{1}{C_3} + \cdots \qquad (20.19)$$	
	The equivalent capacitor carries the same amount of charge as *any one* of the capacitors in the combination and stores the same total energy as the series combination.	
	20.13 *RC* CIRCUITS The charging or discharging of a capacitor in a dc series circuit (resistance R, capacitance C) does not occur instantaneously. The charge on a capacitor builds up gradually, as described by the following equation:	
Charging a capacitor	$$q = q_0[1 - e^{-t/(RC)}] \qquad (20.20)$$	**IS 20.107**

Topic	Discussion	Learning Aids
	where q is the charge on the capacitor at time t and q_0 is the equilibrium value of the charge. The time constant τ of the circuit is	
Time constant	$\tau = RC$	(20.21) **CS 20.4**
	The discharging of a capacitor through a resistor is described as follows:	
Discharging a capacitor	$q = q_0 e^{-t/(RC)}$	(20.22)
	where q_0 is the charge on the capacitor at time $t = 0$ s.	

FOCUS ON CONCEPTS

Note to Instructors: *The numbering of the questions shown here reflects the fact that they are only a representative subset of the total number that are available online. However, all of the questions are available for assignment via an online homework management program such as* WileyPLUS *or* WebAssign.

Section 20.1 Electromotive Force and Current

1. In 2.0 s, 1.9×10^{19} electrons pass a certain point in a wire. What is the current I in the wire?

Section 20.2 Ohm's Law

2. Which one of the following graphs correctly represents Ohm's law, where V is the voltage and I is the current? **(a)** A **(b)** B **(c)** C **(d)** D

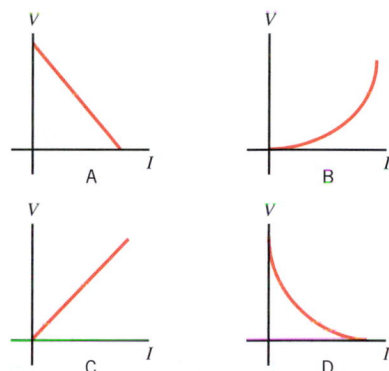

Section 20.3 Resistance and Resistivity

3. Two wires are made from the same material. One wire has a resistance of 0.10 Ω. The other wire is twice as long as the first wire and has a radius that is half as much. What is the resistance of the second wire? **(a)** 0.40 Ω **(b)** 0.20 Ω **(c)** 0.10 Ω **(d)** 0.050 Ω **(e)** 0.80 Ω

Section 20.4 Electric Power

5. A single resistor is connected across the terminals of a battery. Which one or more of the following changes in voltage and current leaves unchanged the electric power dissipated in the resistor?

 (A) Doubling the voltage and reducing the current by a factor of two

 (B) Doubling the voltage and increasing the resistance by a factor of four

 (C) Doubling the current and reducing the resistance by a factor of four

(a) A, B, C **(b)** A, B **(c)** B, C **(d)** A **(e)** B

Section 20.5 Alternating Current

7. The average power dissipated in a 47-Ω resistor is 2.0 W. What is the peak value I_0 of the ac current in the resistor?

Section 20.6 Series Wiring

8. For the circuit shown in the drawing, what is the voltage V_1 across resistance R_1?

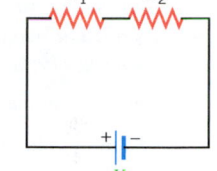

(a) $V_1 = \left(\dfrac{R_1}{R_2}\right)V$ **(d)** $V_1 = \left(\dfrac{R_1}{R_1 + R_2}\right)V$

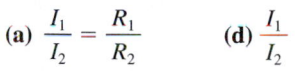

(b) $V_1 = \left(\dfrac{R_2}{R_1}\right)V$ **(e)** $V_1 = \left(\dfrac{R_1 + R_2}{R_1}\right)V$

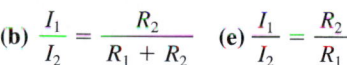

(c) $V_1 = V$

Section 20.7 Parallel Wiring

10. For the circuit shown in the drawing, what is the ratio of the current I_1 in resistance R_1 to the current I_2 in resistance R_2?

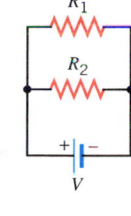

(a) $\dfrac{I_1}{I_2} = \dfrac{R_1}{R_2}$ **(d)** $\dfrac{I_1}{I_2} = 1$

(b) $\dfrac{I_1}{I_2} = \dfrac{R_2}{R_1 + R_2}$ **(e)** $\dfrac{I_1}{I_2} = \dfrac{R_2}{R_1}$

(c) $\dfrac{I_1}{I_2} = \dfrac{R_1}{R_1 + R_2}$

Section 20.8 Circuits Wired Partially in Series and Partially in Parallel

12. In the following three arrangements each resistor has the same resistance R. Rank the equivalent resistances of the arrangements in descending order (largest first). **(a)** A, B, C **(b)** B, A, C **(c)** B, C, A **(d)** A, C, B **(e)** C, B, A

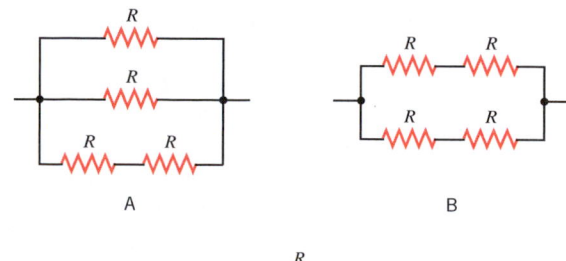

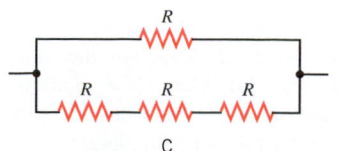

Section 20.9 Internal Resistance

13. A battery has an emf of V and an internal resistance of r. What resistance R, when connected across the terminals of this battery, will cause the terminal voltage of the battery to be $\frac{1}{2}V$?　**(a)** $R = \frac{1}{2}r$　**(b)** $R = 2r$　**(c)** $R = 4r$　**(d)** $R = r$　**(e)** $R = \frac{1}{4}r$

Section 20.10 Kirchhoff's Rules

15. When applying Kirchhoff's rules, one of the essential steps is to mark each resistor with plus and minus signs to label how the potential changes from one end of the resistor to the other. The circuit in the drawing contains four resistors, each marked with the associated plus and

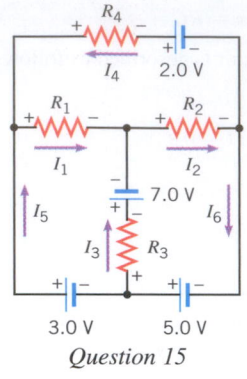

3.0 V　　　5.0 V

Question 15

minus signs. However, one resistor is marked incorrectly. Which one is it?　**(a)** R_1　**(b)** R_2　**(c)** R_3　**(d)** R_4

Section 20.12 Capacitors in Series and in Parallel

18. Three capacitors are identical, each having a capacitance C. Two of them are connected in series. Then, this series combination is connected in parallel with the third capacitor. What is the equivalent capacitance of the entire connection?　**(a)** $\frac{1}{2}C$　**(b)** $\frac{1}{3}C$　**(c)** $3C$　**(d)** $\frac{2}{3}C$　**(e)** $\frac{3}{2}C$

Section 20.13 RC Circuits

20. The time constant of an RC circuit is 2.6 s. How much time t is required for the capacitor (uncharged initially) to gain one-half of its full equilibrium charge?

PROBLEMS

Note to Instructors: Most of the homework problems in this chapter are available for assignment via an online homework management program such as WileyPLUS *or* WebAssign, *and those marked with the icon* **GO** *are presented in* WileyPLUS *using a guided tutorial format that provides enhanced interactivity. See Preface for additional details.*

Note: For problems that involve ac conditions, the current and voltage are rms values and the power is an average value, unless indicated otherwise.

ssm　Solution is in the Student Solutions Manual.
www　Solution is available online at www.wiley.com/college/cutnell

　This icon represents a biomedical application.

Section 20.1 Electromotive Force and Current,
Section 20.2 Ohm's Law

1. A defibrillator is used during a heart attack to restore the heart to its normal beating pattern (see Section 19.5). A defibrillator passes 18 A of current through the torso of a person in 2.0 ms. **(a)** How much charge moves during this time?　**(b)** How many electrons pass through the wires connected to the patient?

2. An especially violent lightning bolt has an average current of 1.26×10^3 A lasting 0.138 s. How much charge is delivered to the ground by the lightning bolt?

3. ssm A fax machine uses 0.110 A of current in its normal mode of operation, but only 0.067 A in the standby mode. The machine uses a potential difference of 120 V. In one minute　**(a)** how much more charge passes through the machine in the normal mode than in the standby mode, and　**(b)** how much more energy is used?

4. Suppose that the resistance between the walls of a biological cell is 5.0×10^9 Ω.　**(a)** What is the current when the potential difference between the walls is 75 mV?　**(b)** If the current is composed of Na$^+$ ions ($q = +e$), how many such ions flow in 0.50 s?

5. The heating element of a clothes dryer has a resistance of 11 Ω and is connected across a 240-V electrical outlet. What is the current in the heating element?

*6. GO The resistance of a bagel toaster is 14 Ω. To prepare a bagel, the toaster is operated for one minute from a 120-V outlet. How much energy is delivered to the toaster?

*7. A resistor is connected across the terminals of a 9.0-V battery, which delivers 1.1×10^5 J of energy to the resistor in six hours. What is the resistance of the resistor?

*8. A car battery has a rating of 220 ampere·hours (A·h). This rating is one indication of the *total charge* that the battery can provide to a circuit before failing.　**(a)** What is the total charge (in coulombs) that this battery can provide?　**(b)** Determine the maximum current that the battery can provide for 38 minutes.

9. ssm A beam of protons is moving toward a target in a particle accelerator. This beam constitutes a current whose value is 0.50 μA. **(a) How many protons strike the target in 15 s?　**(b)** Each proton has a kinetic energy of 4.9×10^{-12} J. Suppose the target is a 15-gram block of aluminum, and all the kinetic energy of the protons goes into heating it up. What is the change in temperature of the block that results from the 15-s bombardment of protons?

Section 20.3 Resistance and Resistivity

10. GO The resistance and the magnitude of the current depend on the path that the current takes. The drawing shows three situations in which the current takes different paths through a piece of material. Each of the rectangular pieces is made from a material whose resistivity is $\rho = 1.50 \times 10^{-2}$ $\Omega \cdot$m, and the unit of length in the drawing is $L_0 = 5.00$ cm. Each piece of material is connected to a 3.00-V battery. Find　**(a)** the resistance and　**(b)** the current in each case.

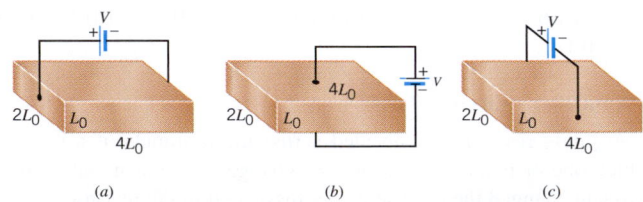

(a)　　　　　　(b)　　　　　　(c)

11. Two wires are identical, except that one is aluminum and one is copper. The aluminum wire has a resistance of 0.20 Ω. What is the resistance of the copper wire?

12. A cylindrical copper cable carries a current of 1200 A. There is a potential difference of 1.6×10^{-2} V between two points on the cable that are 0.24 m apart. What is the radius of the cable?

13. In Section 12.3 it was mentioned that temperatures are often measured with electrical resistance thermometers made of platinum wire. Suppose that the resistance of a platinum resistance thermometer is 125 Ω when its temperature is 20.0 °C. The wire is then immersed in boiling chlorine, and the resistance drops to 99.6 Ω. The temperature coefficient of resistivity of platinum is $\alpha = 3.72 \times 10^{-3}$ (C°)$^{-1}$. What is the temperature of the boiling chlorine?

14. High-voltage power lines are a familiar sight throughout the country. The aluminum wire used for some of these lines has a cross-sectional area of 4.9×10^{-4} m^2. What is the resistance of ten kilometers of this wire?

15. ssm www Two wires have the same length and the same resistance. One is made from aluminum and the other from copper. Obtain the ratio of the cross-sectional area of the aluminum wire to that of the copper wire.

16. A large spool in an electrician's workshop has 75 m of insulation-coated wire coiled around it. When the electrician connects a battery to the ends of the spooled wire, the resulting current is 2.4 A. Some weeks later, after cutting off various lengths of wire for use in repairs, the electrician finds that the spooled wire carries a 3.1-A current when the same battery is connected to it. What is the length of wire remaining on the spool?

* **17.** The temperature coefficient of resistivity for the metal gold is 0.0034 (C°)$^{-1}$, and for tungsten it is 0.0045 (C°)$^{-1}$. The resistance of a gold wire increases by 7.0% due to an increase in temperature. For the same increase in temperature, what is the percentage increase in the resistance of a tungsten wire?

* **18.** A toaster uses a Nichrome heating wire. When the toaster is turned on at 20 °C, the initial current is 1.50 A. A few seconds later, the toaster warms up and the current now has a value of 1.30 A. The average temperature coefficient of resistivity for Nichrome wire is 4.5×10^{-4} (C°)$^{-1}$. What is the temperature of the heating wire?

* **19. ssm www** A wire has a resistance of 21.0 Ω. It is melted down, and from the same volume of metal a new wire is made that is three times longer than the original wire. What is the resistance of the new wire?

* **20.** The filament in an incandescent light bulb is made from tungsten. The light bulb is plugged into a 120-V outlet and draws a current of 1.24 A. If the radius of the tungsten wire is 0.0030 mm, how long must the wire be?

** **21.** An aluminum wire is hung between two towers and has a length of 175 m. A current of 125 A exists in the wire, and the potential difference between the ends of the wire is 0.300 V. The density of aluminum is 2700 kg/m^3. Find the mass of the wire.

Section 20.4   Electric Power

22. The heating element in an iron has a resistance of 24 Ω. The iron is plugged into a 120-V outlet. What is the power delivered to the iron?

23. A cigarette lighter in a car is a resistor that, when activated, is connected across the 12-V battery. Suppose that a lighter uses 33 W of power. Find   **(a)** the resistance of the lighter and   **(b)** the current that the battery delivers to the lighter.

24. A blow-dryer and a vacuum cleaner each operate with a voltage of 120 V. The current rating of the blow-dryer is 11 A, and that of the vacuum cleaner is 4.0 A. Determine the power consumed by   **(a)** the blow-dryer and   **(b)** the vacuum cleaner.   **(c)** Determine the ratio of the energy used by the blow-dryer in 15 minutes to the energy used by the vacuum cleaner in one-half hour.

25. There are approximately 110 million households that use TVs in the United States. Each TV uses, on average, 75 W of power and is turned on for 6.0 hours a day. If electrical energy costs $0.12 per kWh, how much money is spent every day in keeping 110 million TVs turned on?

26. **GO** Each of the four circuits in the drawing consists of a single resistor whose resistance is either R or $2R$, and a single battery whose voltage is either V or $2V$. The unit of voltage in each circuit is $V = 12.0$ V and the unit of resistance is $R = 6.00$ Ω. Determine **(a)** the power supplied to each resistor and   **(b)** the current delivered to each resistor.

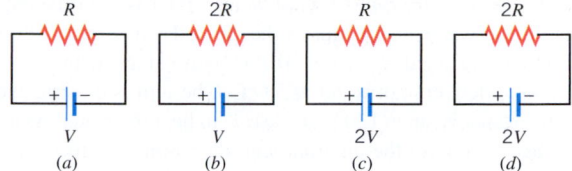

(a)   (b)   (c)   (d)

27. ssm In doing a load of clothes, a clothes dryer uses 16 A of current at 240 V for 45 min. A personal computer, in contrast, uses 2.7 A of current at 120 V. With the energy used by the clothes dryer, how long (in hours) could you use this computer to "surf" the Internet?

* **28.** A piece of Nichrome wire has a radius of 6.5×10^{-4} m. It is used in a laboratory to make a heater that uses 4.00×10^2 W of power when connected to a voltage source of 120 V. Ignoring the effect of temperature on resistance, estimate the necessary length of wire.

* **29.** An electric heater used to boil small amounts of water consists of a 15-Ω coil that is immersed directly in the water. It operates from a 120-V socket. How much time is required for this heater to raise the temperature of 0.50 kg of water from 13 °C to the normal boiling point?

* **30.** The rear window of a van is coated with a layer of ice at 0 °C. The density of ice is 917 kg/m^3. The driver of the van turns on the rear-window defroster, which operates at 12 V and 23 A. The defroster directly heats an area of 0.52 m^2 of the rear window. What is the maximum thickness of ice coating this area that the defroster can melt in 3.0 minutes?

* **31. ssm** Tungsten has a temperature coefficient of resistivity of 0.0045 (C°)$^{-1}$. A tungsten wire is connected to a source of constant voltage via a switch. At the instant the switch is closed, the temperature of the wire is 28 °C, and the initial power delivered to the wire is P_0. At what wire temperature will the power that is delivered to the wire be decreased to $\frac{1}{2}P_0$?

Section 20.5   Alternating Current

32. According to Equation 20.7, an ac voltage V is given as a function of time t by $V = V_0 \sin 2\pi f t$, where V_0 is the peak voltage and f is the frequency (in hertz). For a frequency of 60.0 Hz, what is the smallest value of the time at which the voltage equals one-half of the peak value?

33. The rms current in a copy machine is 6.50 A, and the resistance of the machine is 18.6 Ω. What are   **(a)** the average power and   **(b)** the peak power delivered to the machine?

34. The current in a circuit is ac and has a peak value of 2.50 A. Determine the rms current.

35. ssm The average power used by a stereo speaker is 55 W. Assuming that the speaker can be treated as a 4.0-Ω resistance, find the peak value of the ac voltage applied to the speaker.

36. A 550-W space heater is designed for operation in Germany, where household electrical outlets supply 230 V (rms) service. What is the power output of the heater when plugged into a 120-V (rms) electrical outlet in a house in the United States? Ignore the effects of temperature on the heater's resistance.

37. Review Conceptual Example 7 as an aid in solving this problem. A portable electric heater uses 18 A of current. The manufacturer recommends that an extension cord attached to the heater receive no more than 2.0 W of power per meter of length. What is the smallest radius of copper wire that can be used in the extension cord? *(Note: An extension cord contains two wires.)*

*38. **GO** The *recovery time* of a hot water heater is the time required to heat all the water in the unit to the desired temperature. Suppose that a 52-gal (1.00 gal $= 3.79 \times 10^{-3}$ m^3) unit starts with cold water at 11 °C and delivers hot water at 53 °C. The unit is electric and utilizes a resistance heater (120 V ac, 3.0 Ω) to heat the water. Assuming that no heat is lost to the environment, determine the recovery time (in hours) of the unit.

*39. **ssm** A light bulb is connected to a 120.0-V wall socket. The current in the bulb depends on the time t according to the relation $I = (0.707$ A$) \sin [(314$ Hz$)t]$.   **(a)** What is the frequency of the alternating current?   **(b)** Determine the resistance of the bulb's filament. **(c)** What is the average power delivered to the light bulb?

**40. To save on heating costs, the owner of a greenhouse keeps 660 kg of water around in barrels. During a winter day, the water is heated by the sun to 10.0 °C. During the night the water freezes into ice at 0.0 °C in nine hours. What is the minimum ampere rating of an electric heating system (240 V) that would provide the same heating effect as the water does?

Section 20.6   Series Wiring

41. **ssm** The current in a series circuit is 15.0 A. When an additional 8.00-Ω resistor is inserted in series, the current drops to 12.0 A. What is the resistance in the original circuit?

42. A 60.0-W lamp is placed in series with a resistor and a 120.0-V source. If the voltage across the lamp is 25 V, what is the resistance R of the resistor?

43. **ssm** Three resistors, 25, 45, and 75 Ω, are connected in series, and a 0.51-A current passes through them. What are   **(a)** the equivalent resistance and   **(b)** the potential difference across the three resistors?

44. An 86-Ω resistor and a 67-Ω resistor are connected in series across a battery. The voltage across the 86-Ω resistor is 27 V. What is the voltage across the 67-Ω resistor?

45. The current in a 47-Ω resistor is 0.12 A. This resistor is in series with a 28-Ω resistor, and the series combination is connected across a battery. What is the battery voltage?

*46. Multiple-Concept Example 9 reviews the concepts that are important to this problem. A light bulb is wired in series with a 144-Ω resistor, and they are connected across a 120.0-V source. The power delivered to the light bulb is 23.4 W. What is the resistance of the light bulb? Note that there are two possible answers.

*47. **ssm** An extension cord is used with an electric weed trimmer that has a resistance of 15.0 Ω. The extension cord is made of copper wire that has a cross-sectional area of 1.3×10^{-6} m^2. The combined length of the two wires in the extension cord is 92 m.   **(a)** Determine the resistance of the extension cord.   **(b)** The extension cord is plugged into a 120-V socket. What voltage is applied to the trimmer itself?

*48. One heater uses 340 W of power when connected by itself to a battery. Another heater uses 240 W of power when connected by itself to the same battery. How much total power do the heaters use when they are both connected in series across the battery?

49. Two resistances, R_1 and R_2, are connected in series across a 12-V battery. The current increases by 0.20 A when R_2 is removed, leaving R_1 connected across the battery. However, the current increases by just 0.10 A when R_1 is removed, leaving R_2 connected across the battery. Find   **(a) R_1 and   **(b)** R_2.

Section 20.7   Parallel Wiring

50. What resistance must be placed in parallel with a 155-Ω resistor to make the equivalent resistance 115 Ω?

51. **ssm** A 16-Ω loudspeaker and an 8.0-Ω loudspeaker are connected in parallel across the terminals of an amplifier. Assuming that the speakers behave as resistors, determine the equivalent resistance of the two speakers.

52. **GO** The drawing shows three different resistors in two different circuits. The battery has a voltage of $V = 24.0$ V, and the resistors have values of $R_1 = 50.0$ Ω, $R_2 = 25.0$ Ω, and $R_3 = 10.0$ Ω. **(a)** For the circuit on the left, determine the current through and the voltage across each resistor.   **(b)** Repeat part (a) for the circuit on the right.

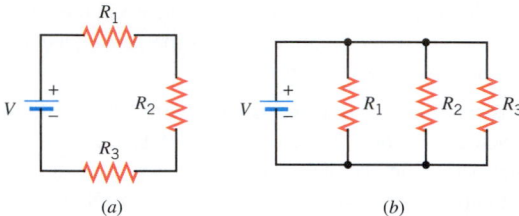

(a)   (b)

53. The drawing shows a circuit that contains a battery, two resistors, and a switch. What is the equivalent resistance of the circuit when the switch is   **(a)** open and   **(b)** closed? What is the total power delivered to the resistors when the switch is   **(c)** open and **(d)** closed?

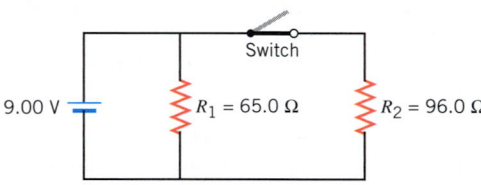

54. For the three-way bulb (50 W, 100 W, 150 W) discussed in Conceptual Example 11, find the resistance of each of the two filaments. Assume that the wattage ratings are not limited by significant figures, and ignore any heating effects on the resistances.

55. **ssm** Two resistors, 42.0 and 64.0 Ω, are connected in parallel. The current through the 64.0-Ω resistor is 3.00 A.   **(a)** Determine the current in the other resistor.   **(b)** What is the total power supplied to the two resistors?

56. Two identical resistors are connected in parallel across a 25-V battery, which supplies them with a total power of 9.6 W. While the battery is still connected, one of the resistors is heated so that its resistance doubles. The resistance of the other resistor remains unchanged. Find   **(a)** the initial resistance of each resistor and   **(b)** the total power delivered to the resistors after one resistor has been heated.

57. Help with problems of this kind is available in **Interactive Solution 20.57** at **www.wiley.com/college/cutnell**. A coffee cup heater and a lamp are connected in parallel to the same 120-V outlet. Together, they use a total of 111 W of power. The resistance of the heater is 4.0×10^2 Ω. Find the resistance of the lamp.

*58. **GO** The drawing shows two circuits, and the same battery is used in each. The two resistances R_A in circuit A are the same, and the two resistances R_B in circuit B are the same. Knowing that the same total power is delivered in each circuit, find the ratio R_B/R_A for the circuits in the drawing.

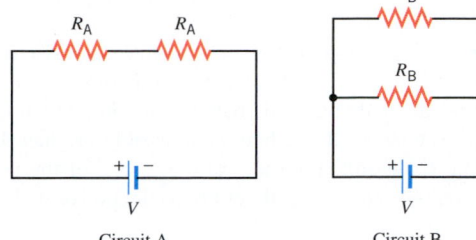

Circuit A   Circuit B

*59. **ssm** The total current delivered to a number of devices connected in parallel is the sum of the individual currents in each device. Circuit breakers are resettable automatic switches that protect against a dangerously large total current by "opening" to stop the current at a specified safe value. A 1650-W toaster, a 1090-W iron, and a 1250-W microwave oven are turned on in a kitchen. As the drawing shows, they are all connected through a 20-A circuit breaker (which has negligible resistance) to an ac voltage of 120 V. **(a)** Find the equivalent resistance of the three devices. **(b)** Obtain the total current delivered by the source and determine whether the breaker will "open" to prevent an accident.

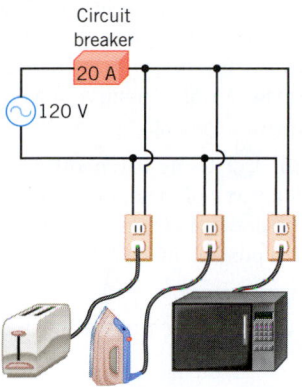

*60. A cylindrical aluminum pipe of length 1.50 m has an inner radius of 2.00×10^{-3} m and an outer radius of 3.00×10^{-3} m. The interior of the pipe is completely filled with copper. What is the resistance of this unit? *(Hint: Imagine that the pipe is connected between the terminals of a battery and decide whether the aluminum and copper parts of the pipe are in series or in parallel.)*

**61. The rear window defogger of a car consists of thirteen thin wires (resistivity = 88.0×10^{-8} $\Omega \cdot$m) embedded in the glass. The wires are connected in parallel to the 12.0-V battery, and each has a length of 1.30 m. The defogger can melt 2.10×10^{-2} kg of ice at 0 °C into water at 0 °C in two minutes. Assume that all the power delivered to the wires is used immediately to melt the ice. Find the cross-sectional area of each wire.

Section 20.8   Circuits Wired Partially in Series and Partially in Parallel

62. **GO** The circuit in the drawing contains three identical resistors. Each resistor has a value of 10.0 Ω. Determine the equivalent resistance between the points a and b, b and c, and a and c.

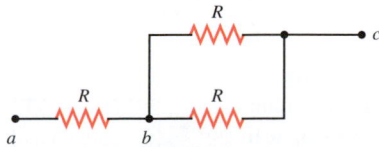

63. **ssm** A 14-Ω coffee maker and a 16-Ω frying pan are connected in series across a 120-V source of voltage. A 23-Ω bread maker is also connected across the 120-V source and is in parallel with the series combination. Find the total current supplied by the source of voltage.

64. Find the equivalent resistance between points A and B in the drawing.

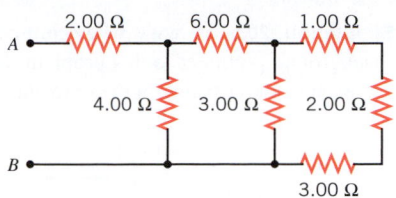

65. **ssm** Determine the equivalent resistance between the points A and B for the group of resistors in the drawing.

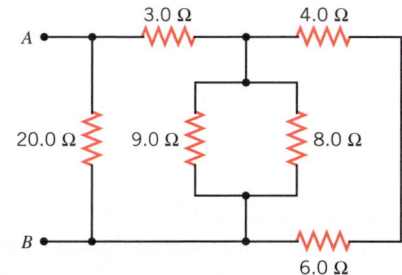

66. Find the equivalent resistance between the points A and B in the drawing.

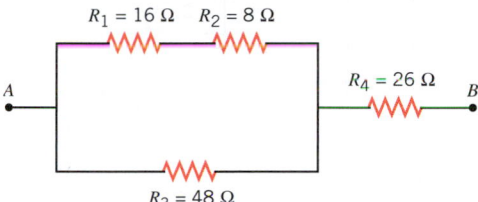

67. A 60.0-Ω resistor is connected in parallel with a 120.0-Ω resistor. This parallel group is connected in series with a 20.0-Ω resistor. The total combination is connected across a 15.0-V battery. Find **(a)** the current and **(b)** the power delivered to the 120.0-Ω resistor.

*68. **GO** Each resistor in the three circuits in the drawing has the same resistance R, and the batteries have the same voltage V. The values for R and V are 9.0 Ω and 6.0 V, respectively. Determine the total power delivered by the battery in each of the three circuits.

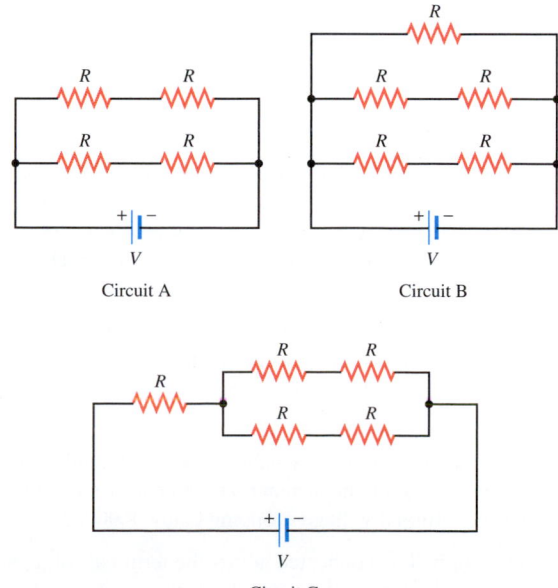

Circuit A   Circuit B

Circuit C

*69. Eight different values of resistance can be obtained by connecting together three resistors (1.00, 2.00, and 3.00 Ω) in all possible ways. What are the values?

*70. **Concept Simulation 20.5** at **www.wiley.com/college/cutnell** provides some background pertinent to this problem. Determine the power supplied to each of the resistors in the drawing.

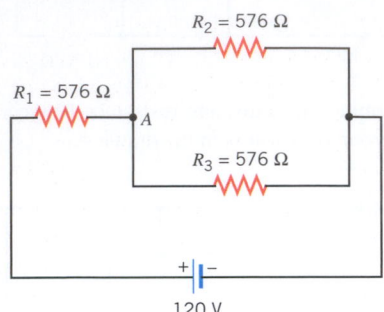

120 V

*71. **ssm www** The current in the 8.00-Ω resistor in the drawing is 0.500 A. Find the current in (a) the 20.0-Ω resistor and in (b) the 9.00-Ω resistor.

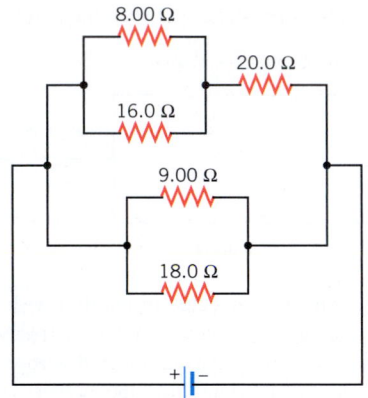

**72. The circuit shown in the drawing is constructed with six identical resistors and an ideal battery. When the resistor R_4 is removed from the circuit, the current in the battery decreases by 1.9 A. Determine the resistance of each resistor.

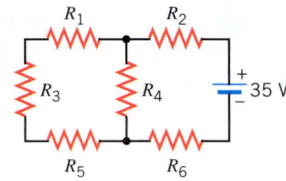

Section 20.9 Internal Resistance

73. **ssm** A battery has an internal resistance of 0.50 Ω. A number of identical light bulbs, each with a resistance of 15 Ω, are connected in parallel across the battery terminals. The terminal voltage of the battery is observed to be one-half the emf of the battery. How many bulbs are connected?

74. A new "D" battery has an emf of 1.5 V. When a wire of negligible resistance is connected between the terminals of the battery, a current of 28 A is produced. Find the internal resistance of the battery.

75. A battery has an internal resistance of 0.012 Ω and an emf of 9.00 V. What is the maximum current that can be drawn from the battery without the terminal voltage dropping below 8.90 V?

76. When a light bulb is connected across the terminals of a battery, the battery delivers 24 W of power to the bulb. A voltage of 11.8 V

exists between the terminals of the battery, which has an internal resistance of 0.10 Ω. What is the emf of the battery?

*77. **GO** A battery delivering a current of 55.0 A to a circuit has a terminal voltage of 23.4 V. The electric power being dissipated by the internal resistance of the battery is 34.0 W. Find the emf of the battery.

*78. When a "dry-cell" flashlight battery with an internal resistance of 0.33 Ω is connected to a 1.50-Ω light bulb, the bulb shines dimly. However, when a lead-acid "wet-cell" battery with an internal resistance of 0.050 Ω is connected, the bulb is noticeably brighter. Both batteries have the same emf. Find the ratio $P_{\text{wet}}/P_{\text{dry}}$ of the power delivered to the bulb by the wet-cell battery to the power delivered by the dry-cell battery.

Section 20.10 Kirchhoff's Rules

79. **ssm** Consider the circuit in the drawing. Determine (a) the magnitude of the current in the circuit and (b) the magnitude of the voltage between the points labeled A and B. (c) State which point, A or B, is at the higher potential.

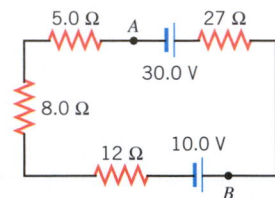

80. **GO** Using Kirchhoff's loop rule, find the value of the current I in part c of the drawing, where $R = 5.0 \ \Omega$. (Note: Parts a and b of the drawing are used in the online tutorial help that is provided for this problem in the *WileyPLUS* homework management program.)

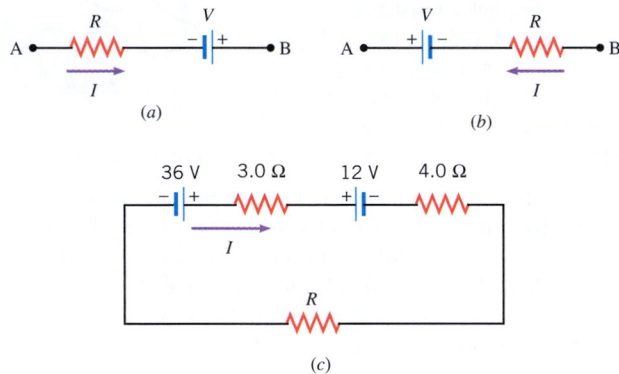

81. For the circuit shown in the drawing, find the current I through the 2.00-Ω resistor and the voltage V of the battery to the left of this resistor.

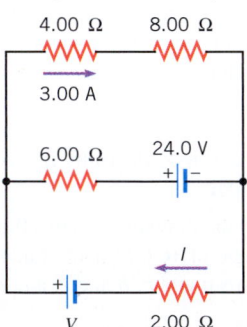

82. The drawing shows a portion of a larger circuit. Current flows left to right in each resistor. What is the current in the resistor R?

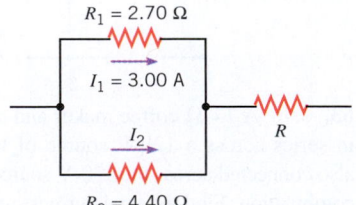

83. Determine the current (magnitude and direction) in the 8.0- and 2.0-Ω resistors in the drawing.

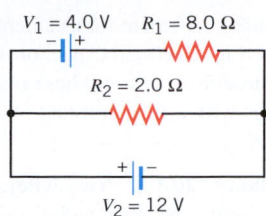

*84. Concept Simulation 20.3 at **www.wiley.com/college/cutnell** allows you to verify your answer for this problem. Find the current in the 4.00-Ω resistor in the drawing. Specify the direction of the current.

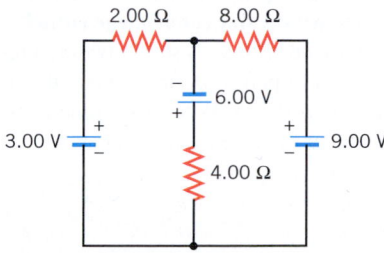

*85. ssm Determine the voltage across the 5.0-Ω resistor in the drawing. Which end of the resistor is at the higher potential?

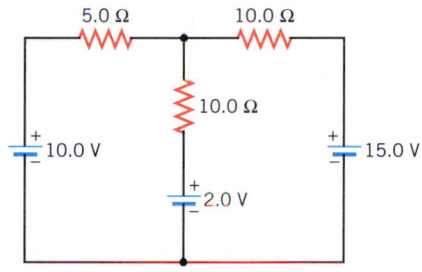

86. None of the resistors in the circuit shown in the drawing is connected in series or in parallel with one another. Find **(a) the current I_5 and the resistances **(b)** R_2 and **(c)** R_3.

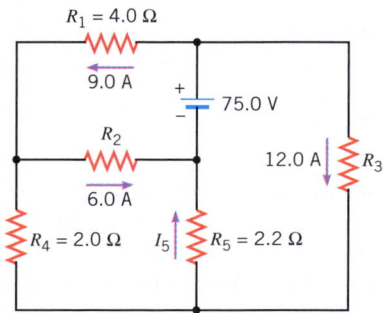

Section 20.11 The Measurement of Current and Voltage

87. ssm A galvanometer has a full-scale current of 0.100 mA and a coil resistance of 50.0 Ω. This instrument is used with a shunt resistor to form a nondigital ammeter that will register full scale for a current of 60.0 mA. Determine the resistance of the shunt resistor.

88. The coil of wire in a galvanometer has a resistance of $R_C = 60.0\,\Omega$. The galvanometer exhibits a full-scale deflection when the current through it is 0.400 mA. A resistor is connected in series with this combination so as to produce a nondigital voltmeter. The voltmeter is to have a full-scale deflection when it measures a potential difference of 10.0 V. What is the resistance of this resistor?

89. Nondigital voltmeter A has an equivalent resistance of $2.40 \times 10^5\,\Omega$ and a full-scale voltage of 50.0 V. Nondigital voltmeter B, using the same galvanometer as voltmeter A, has an equivalent resistance of $1.44 \times 10^5\,\Omega$. What is its full-scale voltage?

90. A galvanometer with a coil resistance of 12.0 Ω and a full-scale current of 0.150 mA is used with a shunt resistor to make a nondigital ammeter. The ammeter registers a maximum current of 4.00 mA. Find the equivalent resistance of the ammeter.

*91. ssm Two scales on a nondigital voltmeter measure voltages up to 20.0 and 30.0 V, respectively. The resistance connected in series with the galvanometer is 1680 Ω for the 20.0-V scale and 2930 Ω for the 30.0-V scale. Determine the coil resistance and the full-scale current of the galvanometer that is used in the voltmeter.

92. In measuring a voltage, a voltmeter uses some current from the circuit. Consequently, the voltage measured is only an approximation to the voltage present when the voltmeter is not connected. Consider a circuit consisting of two 1550-Ω resistors connected in series across a 60.0-V battery. **(a) Find the voltage across one of the resistors. **(b)** A nondigital voltmeter has a full-scale voltage of 60.0 V and uses a galvanometer with a full-scale deflection of 5.00 mA. Determine the voltage that this voltmeter registers when it is connected across the resistor used in part (a).

Section 20.12 Capacitors in Series and in Parallel

93. Two capacitors are connected in parallel across the terminals of a battery. One has a capacitance of 2.0 μF and the other a capacitance of 4.0 μF. These two capacitors together store 5.4×10^{-5} C of charge. What is the voltage of the battery?

94. Three capacitors are connected in series. The equivalent capacitance of this combination is 3.00 μF. Two of the individual capacitances are 6.00 μF and 9.00 μF. What is the third capacitance (in μF)?

95. ssm Three capacitors have identical geometries. One is filled with a material whose dielectric constant is 2.50. Another is filled with a material whose dielectric constant is 4.00. The third capacitor is filled with a material whose dielectric constant κ is such that this single capacitor has the same capacitance as the series combination of the other two. Determine κ.

96. GO Two capacitors are connected to a battery. The battery voltage is $V = 60.0$ V, and the capacitances are $C_1 = 2.00\,\mu$F and $C_2 = 4.00\,\mu$F. Determine the total energy stored by the two capacitors when they are wired **(a)** in parallel and **(b)** in series.

97. Determine the equivalent capacitance between A and B for the group of capacitors in the drawing.

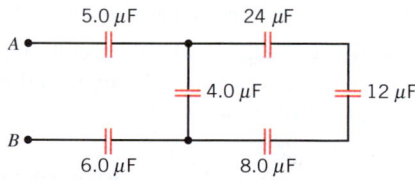

98. You have three capacitors: $C_1 = 67\,\mu$F, $C_2 = 45\,\mu$F, and $C_3 = 33\,\mu$F. Determine the maximum equivalent capacitance you can obtain by connecting two of the capacitors in parallel and then connecting the parallel combination in series with the remaining capacitor.

99. A 2.00-μF and a 4.00-μF capacitor are connected to a 60.0-V battery. What is the total charge supplied to the capacitors when they are wired **(a)** in parallel and **(b)** in series with each other?

*100. A 3.00-μF and a 5.00-μF capacitor are connected in series across a 30.0-V battery. A 7.00-μF capacitor is then connected in parallel across the 3.00-μF capacitor. Determine the voltage across the 7.00-μF capacitor.

*101. **ssm** A sheet of gold foil (negligible thickness) is placed between the plates of a capacitor and has the same area as each of the plates. The foil is parallel to the plates, at a position one-third of the way from one to the other. Before the foil is inserted, the capacitance is C_0. What is the capacitance after the foil is in place? Express your answer in terms of C_0.

**102. The drawing shows two capacitors that are fully charged ($C_1 = 2.00$ μF, $q_1 = 6.00$ μC; $C_2 = 8.00$ μF, $q_2 = 12.0$ μC). The switch is closed, and charge flows until equilibrium is reestablished (i.e., until both capacitors have the same voltage across their plates). Find the resulting voltage across either capacitor.

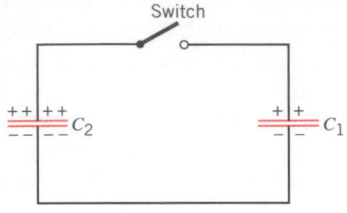

Section 20.13 RC Circuits

103. **ssm** In a heart pacemaker, a pulse is delivered to the heart 81 times per minute. The capacitor that controls this pulsing rate discharges through a resistance of 1.8×10^6 Ω. One pulse is delivered every time the fully charged capacitor loses 63.2% of its original charge. What is the capacitance of the capacitor?

104. The circuit in the drawing contains two resistors and two capacitors that are connected to a battery via a switch. When the switch is

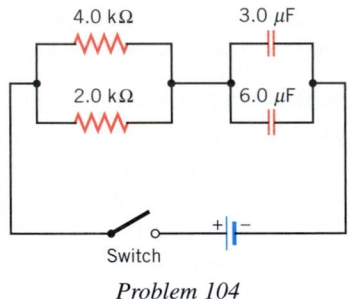

Problem 104

closed, the capacitors begin to charge up. What is the time constant for the charging process?

105. An electronic flash attachment for a camera produces a flash by using the energy stored in a 150-μF capacitor. Between flashes, the capacitor recharges through a resistor whose resistance is chosen so the capacitor recharges with a time constant of 3.0 s. Determine the value of the resistance.

*106. **Concept Simulation 20.4** at **www.wiley.com/college/cutnell** provides background for this problem and gives you the opportunity to verify your answer graphically. How many time constants must elapse before a capacitor in a series RC circuit is charged to 80.0% of its equilibrium charge?

*107. For one approach to problems like this one, consult **Interactive Solution 20.107** at **www.wiley.com/college/cutnell.** Four identical capacitors are connected with a resistor in two different ways. When they are connected as in part a of the drawing, the time constant to charge up this circuit is 0.72 s. What is the time constant when they are connected with the same resistor, as in part b?

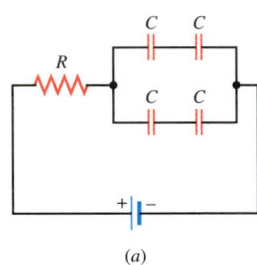

(a)

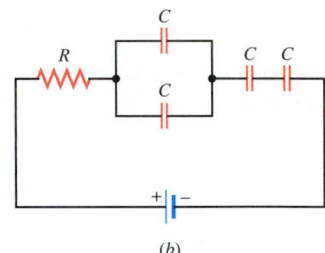

(b)

ADDITIONAL PROBLEMS

108. A portable CD player operates with a voltage of 4.5 V, and its power usage is 0.11 W. Find the current in the player.

109. **ssm** Suppose that two capacitors (C_1 and C_2) are connected in series. Show that the sum of the energies stored in these capacitors is equal to the energy stored in the equivalent capacitor. [*Hint: The energy stored in a capacitor can be expressed as $q^2/(2C)$.*]

110. A battery charger is connected to a dead battery and delivers a current of 6.0 A for 5.0 hours, keeping the voltage across the battery terminals at 12 V in the process. How much energy is delivered to the battery?

111. **ssm** In the Arctic, electric socks are useful. A pair of socks uses a 9.0-V battery pack for each sock. A current of 0.11 A is drawn from each battery pack by wire woven into the socks. Find the resistance of the wire in one sock.

112. **Interactive LearningWare 20.1** at **www.wiley.com/college/cutnell** provides background for this problem. Find the magnitude and direction of the current in the 2.0-Ω resistor in the drawing.

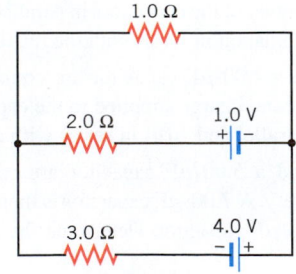

113. A coil of wire has a resistance of 38.0 Ω at 25 °C and 43.7 Ω at 55 °C. What is the temperature coefficient of resistivity?

114. The circuit in the drawing shows two resistors, a capacitor, and a battery. When the capacitor is fully charged, what is the magnitude q of the charge on one of its plates?

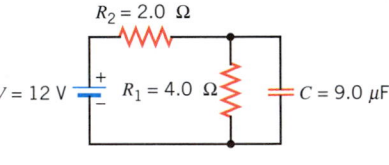

115. The coil of a galvanometer has a resistance of 20.0 Ω, and its meter deflects full scale when a current of 6.20 mA passes through it. To make the galvanometer into a nondigital ammeter, a 24.8-mΩ shunt resistor is added to it. What is the maximum current that this ammeter can read?

116. Multiple-Concept Example 9 discusses the physics principles used in this problem. Three resistors, 2.0, 4.0, and 6.0 Ω, are connected in series across a 24-V battery. Find the power delivered to each resistor.

*117. **ssm** The circuit in the drawing contains five identical resistors. The 45-V battery delivers 58 W of power to the circuit. What is the resistance R of each resistor?

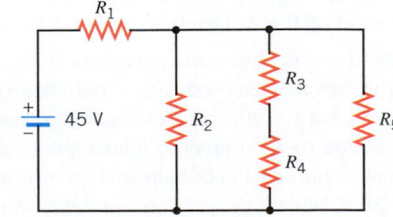

* **118.** A resistor (resistance = R) is connected first in parallel and then in series with a 2.00-Ω resistor. A battery delivers five times as much current to the parallel combination as it does to the series combination. Determine the two possible values for R.

* **119.** ssm Interactive Solution 20.119 at **www.wiley.com/college/ cutnell** provides one approach to problems like this one. Three resistors are connected in series across a battery. The value of each resistance and its maximum power rating are as follows: 2.0 Ω and 4.0 W, 12.0 Ω and 10.0 W, and 3.0 Ω and 5.0 W. (a) What is the greatest voltage that the battery can have without one of the resistors burning up? (b) How much power does the battery deliver to the circuit in (a)?

* **120.** A 75.0-Ω and a 45.0-Ω resistor are connected in parallel. When this combination is connected across a battery, the current delivered by the battery is 0.294 A. When the 45.0-Ω resistor is disconnected, the current from the battery drops to 0.116 A. Determine (a) the emf and (b) the internal resistance of the battery.

* **121.** ssm www Two wires have the same cross-sectional area and are joined end to end to form a single wire. One is tungsten, which has a temperature coefficient of resistivity of $\alpha = 0.0045$ (C°)$^{-1}$. The

other is carbon, for which $\alpha = -0.0005$ (C°)$^{-1}$. The total resistance of the composite wire is the sum of the resistances of the pieces. The total resistance of the composite *does not change with temperature.* What is the ratio of the lengths of the tungsten and carbon sections? Ignore any changes in length due to thermal expansion.

* **122.** Two cylindrical rods, one copper and the other iron, are identical in lengths and cross-sectional areas. They are joined end to end to form one long rod. A 12-V battery is connected across the free ends of the copper–iron rod. What is the voltage between the ends of the copper rod?

* **123.** ssm A 7.0-μF and a 3.0-μF capacitor are connected in series across a 24-V battery. What voltage is required to charge a parallel combination of the two capacitors to the same total energy?

** **124.** A digital thermometer employs a thermistor as the temperature-sensing element. A thermistor is a kind of semiconductor and has a large negative temperature coefficient of resistivity α. Suppose that $\alpha = -0.060$ (C°)$^{-1}$ for the thermistor in a digital thermometer used to measure the temperature of a patient. The resistance of the thermistor decreases to 85% of its value at the normal body temperature of 37.0 °C. What is the patient's temperature?

MAGNETIC FORCES AND MAGNETIC FIELDS

This beautiful display of light in the sky is known as the northern lights (aurora borealis). It occurs when electrons, streaming from the sun, become trapped by the earth's magnetic field. The electrons collide with molecules in the upper atmosphere, and the result is the production of light. Magnetic forces and magnetic fields are the subjects of this chapter. (United States Air Force photo by Senior Airman Joshua Strang)

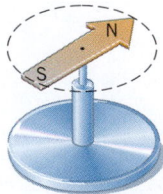

Figure 21.1 The needle of a compass is a permanent magnet that has a north magnetic pole (N) at one end and a south magnetic pole (S) at the other.

21.1 MAGNETIC FIELDS

Permanent magnets have long been used in navigational compasses. As Figure 21.1 illustrates, the compass needle is a permanent magnet supported so it can rotate freely in a plane. When the compass is placed on a horizontal surface, the needle rotates until one end points approximately to the north. The end of the needle that points north is labeled the ***north magnetic pole;*** the opposite end is the ***south magnetic pole.***

Magnets can exert forces on each other. Figure 21.2 shows that the magnetic forces between north and south poles have the property that

like poles repel each other, and unlike poles attract.

This behavior is similar to that of like and unlike electric charges. However, there is a significant difference between magnetic poles and electric charges. It is possible to separate positive from negative electric charges and produce isolated charges of either kind. In contrast, no one has found a magnetic monopole (an isolated north or south pole). Any attempt to separate north and south poles by cutting a bar magnet in half fails, because each piece becomes a smaller magnet with its own north and south poles.

Surrounding a magnet, there is a ***magnetic field.*** The magnetic field is analogous to the electric field that exists in the space around electric charges. Like the electric field, the magnetic field has both a magnitude and a direction. We postpone a discussion of the magnitude until Section 21.2, concentrating our attention here on the direction. ***The direction of the magnetic field at any point in space is the direction indicated by the north pole of a small compass needle placed at that point.*** In Figure 21.3 the compass needle is symbolized by an arrow, with the head of the arrow representing the north pole. The drawing shows how compasses can be used to map out the magnetic field in the space around a bar magnet. Since like poles repel and unlike poles attract, the needle of each compass becomes aligned relative to the magnet in the manner shown in the picture. The compass needles provide a visual picture of the magnetic field that the bar magnet creates.

To help visualize the electric field, we introduced electric field lines in Section 18.7. In a similar fashion, it is possible to draw magnetic field lines, and Figure 21.4a illustrates some of the lines around a bar magnet. The lines appear to originate from the north pole and end on the south pole; they do not start or stop in midspace. A visual image of the magnetic field lines can be created by sprinkling finely ground iron filings on a piece of paper that covers the magnet. Iron filings in a magnetic field behave like tiny compasses and align themselves along the field lines, as the photo in Figure 21.4b shows.

As is the case with electric field lines, the magnetic field at any point is tangent to the magnetic field line at that point. Furthermore, the strength of the field is proportional to the number of lines per unit area that passes through a surface oriented perpendicular to the lines. Thus, the magnetic field is stronger in regions where the field lines are relatively close together and weaker where they are relatively far apart. For instance, in Figure 21.4a the lines are closest together near the north and south poles, reflecting the fact that the strength of the field is greatest in these regions. Away from the poles, the magnetic field becomes weaker. Notice in part c of the drawing that the field lines in the gap between the poles of the horseshoe magnet are nearly parallel and equally spaced, indicating that the magnetic field there is approximately constant.

Although the north pole of a compass needle points northward, it does not point exactly at the north geographic pole. The north geographic pole is that point where the earth's axis of rotation crosses the surface in the northern hemisphere (see Figure 21.5). Measurements of the magnetic field surrounding the earth show that the earth behaves magnetically almost as if it were a bar magnet.* As the drawing illustrates, the orientation of this fictitious bar magnet defines a magnetic axis for the earth. The location at which the magnetic axis crosses the surface in the northern hemisphere is known as the north magnetic pole. It is so named because it is the location toward which the north end of a compass needle points. Since unlike poles attract, the south pole of the earth's fictitious bar magnet lies beneath the north magnetic pole, as Figure 21.5 indicates.

The north magnetic pole does not coincide with the north geographic pole but, instead, lies at a latitude of nearly 80°, just northwest of Ellef Ringnes Island in extreme northern Canada. It is interesting to note that the position of the north magnetic pole is not fixed, but moves over the years. Pointing as it does at the north magnetic pole, a compass needle deviates from the north geographic pole. The angle that a compass needle deviates is called the *angle of declination*. For New York City, the present angle of declination is about 13° west, meaning that a compass needle points 13° west of geographic north.

Figure 21.5 shows that the earth's magnetic field lines are not parallel to the surface at all points. For instance, near the north magnetic pole the field lines are almost perpendicular

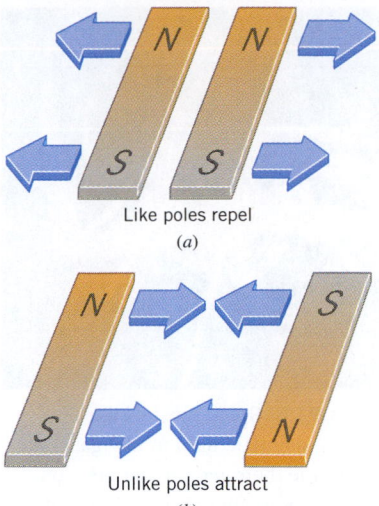

Figure 21.2 Bar magnets have a north magnetic pole at one end and a south magnetic pole at the other end. (a) Like poles repel each other, and (b) unlike poles attract.

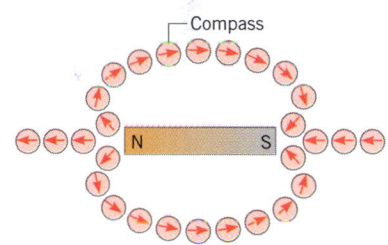

Figure 21.3 At any location in the vicinity of a magnet, the north pole (the arrowhead in this drawing) of a small compass needle points in the direction of the magnetic field at that location.

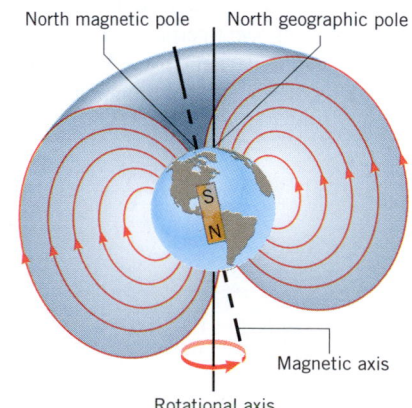

Figure 21.5 The earth behaves magnetically almost as if a bar magnet were located near its center. The axis of this fictitious bar magnet does not coincide with the earth's rotational axis; the two axes are currently about 11.5° apart.

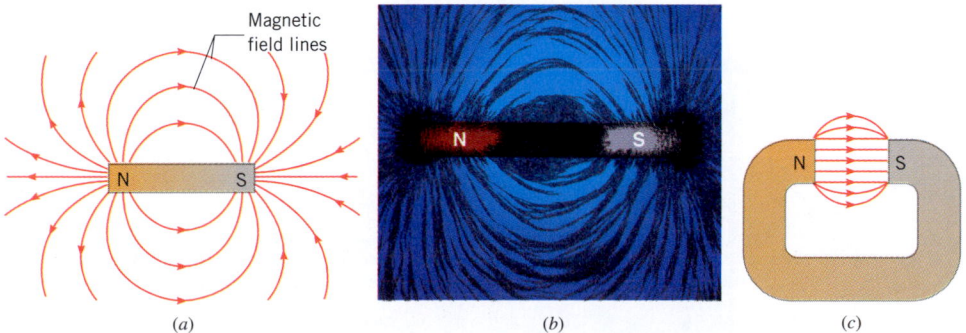

Figure 21.4 (a) The magnetic field lines and (b) the pattern of iron filings (dark, curved regions) in the vicinity of a bar magnet. (© Yoav Levy/Phototake) (c) The magnetic field lines in the gap of a horseshoe magnet.

*At present it is not known with certainty what causes the earth's magnetic field. The magnetic field seems to originate from electric currents that in turn arise from electric charges circulating within the liquid outer region of the earth's core. Section 21.7 discusses how a current produces a magnetic field.

Figure 21.6 Spiny lobsters use the earth's magnetic field to navigate and determine their geographic position. (Courtesy Kenneth Lohmann, University of North Carolina at Chapel Hill)

to the surface of the earth. The angle that the magnetic field makes with respect to the surface at any point is known as the *angle of dip*.

The physics of navigation in animals. Some animals can sense the earth's magnetic field and use it for navigational purposes. Until recently (2004), the only examples of this ability were in vertebrates, or animals that have a backbone, such as migratory birds. Now, however, researchers have found that the spiny lobster (see Figure 21.6), which is an invertebrate, can also use the earth's magnetic field to navigate and can determine its geographic location in a way similar to that of a person using the Global Positioning System (see Section 5.5). This ability may be related to the presence in the lobsters of the mineral magnetite, a magnetic material used for compass needles.

21.2 THE FORCE THAT A MAGNETIC FIELD EXERTS ON A MOVING CHARGE

When a charge is placed in an electric field, it experiences an electric force, as Section 18.6 discusses. When a charge is placed in a magnetic field, it also experiences a force, provided that certain conditions are met, as we will see. The *magnetic force,* like all the forces we have studied (e.g., the gravitational, elastic, and electric forces), may contribute to the net force that causes an object to accelerate. Thus, when present, the magnetic force must be included in Newton's second law.

The following two conditions must be met for a charge to experience a magnetic force when placed in a magnetic field:

1. The charge must be moving, because no magnetic force acts on a stationary charge.

2. The velocity of the moving charge must have a component that is perpendicular to the direction of the magnetic field.

Figure 21.7 (a) No magnetic force acts on a charge moving with a velocity $\vec{v}$ that is parallel or antiparallel to a magnetic field $\vec{B}$. (b) The charge experiences a maximum force $\vec{F}_{max}$ when the charge moves perpendicular to the field. (c) If the charge travels at an angle θ with respect to $\vec{B}$, only the velocity component perpendicular to $\vec{B}$ gives rise to a magnetic force $\vec{F}$, which is smaller than $\vec{F}_{max}$. This component is $v \sin \theta$.

To examine the second condition more closely, consider Figure 21.7, which shows a positive test charge $+q_0$ moving with a velocity $\vec{v}$ through a magnetic field. In the drawing, the magnetic field vector is labeled by the symbol $\vec{B}$. The field is produced by an arrangement of magnets not shown in the drawing and is assumed to be constant in both magnitude and direction. If the charge moves *parallel or antiparallel* to the field, as in part *a* of the drawing, the charge experiences *no magnetic force*. If, on the other hand, the charge moves *perpendicular* to the field, as in part *b*, the charge experiences the *maximum possible force* $\vec{F}_{max}$. In general, if a charge moves at an angle θ* with respect to the field (see part *c* of the drawing), only the velocity component $v \sin \theta$, which is perpendicular to the field, gives rise to a magnetic force. This force $\vec{F}$ is smaller than the maximum possible force. The component of the velocity that is parallel to the magnetic field yields no force.

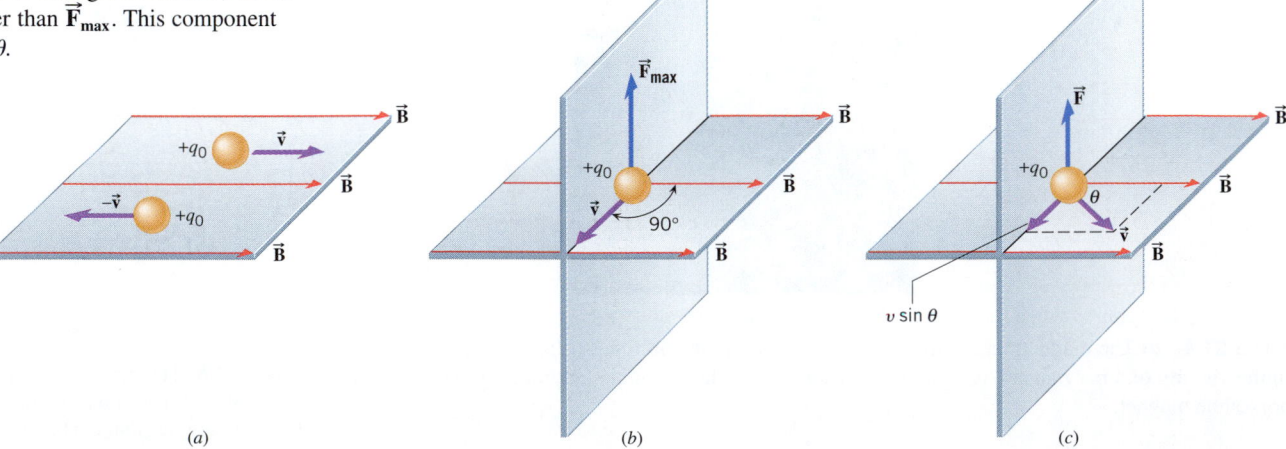

(a)   (b)   (c)

*The angle θ between the velocity of the charge and the magnetic field is chosen so that it lies in the range $0 \leq \theta \leq 180°$.

Figure 21.7 shows that the direction of the magnetic force $\vec{F}$ is perpendicular to both the velocity $\vec{v}$ and the magnetic field $\vec{B}$; in other words, $\vec{F}$ is perpendicular to the plane defined by $\vec{v}$ and $\vec{B}$. As an aid in remembering the direction of the force, it is convenient to use *Right-Hand Rule No. 1 (RHR-1)*, as Figure 21.8 illustrates:

> ***Right-Hand Rule No. 1.*** Extend the right hand so the fingers point along the direction of the magnetic field $\vec{B}$ and the thumb points along the velocity $\vec{v}$ of the charge. The palm of the hand then faces in the direction of the magnetic force $\vec{F}$ that acts on a positive charge.

It is as if the open palm of the right hand pushes on the positive charge in the direction of the magnetic force. **Problem-solving insight:** *If the moving charge is negative instead of positive, the direction of the magnetic force is opposite to that predicted by RHR-1.* Thus, there is an easy method for finding the force on a moving negative charge. First, assume that the charge is positive and use RHR-1 to find the direction of the force. Then, reverse this direction to find the direction of the force acting on the negative charge.

We will now use what we know about the magnetic force to define the magnetic field, in a procedure that is analogous to that used in Section 18.6 to define the electric field. Recall that the electric field at any point in space is the force per unit charge that acts on a test charge q_0 placed at that point. In other words, to determine the electric field $\vec{E}$, we divide the electrostatic force $\vec{F}$ by the charge q_0: $\vec{E} = \vec{F}/q_0$. In the magnetic case, however, the test charge is moving, and the force depends not only on the charge q_0, but also on the velocity component $v \sin \theta$ that is perpendicular to the magnetic field. Therefore, to determine the magnitude of the magnetic field, we divide the magnitude of the magnetic force by the magnitude $|q_0|$ of the charge and also by $v \sin \theta$, according to the following definition:

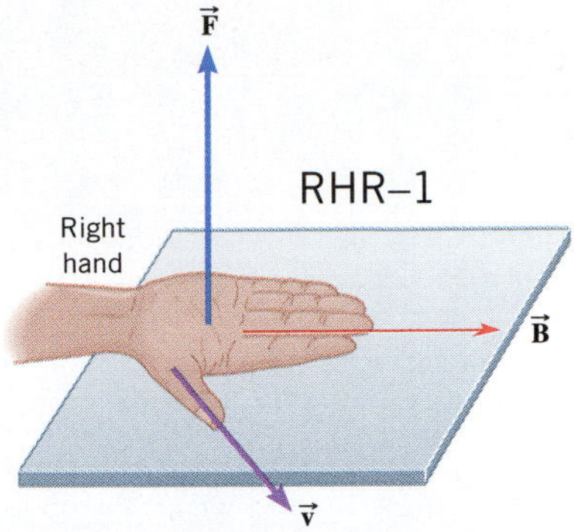

RHR–1

Figure 21.8 Right-Hand Rule No. 1 is illustrated. When the right hand is oriented so the fingers point along the magnetic field $\vec{B}$ and the thumb points along the velocity $\vec{v}$ of a positively charged particle, the palm faces in the direction of the magnetic force $\vec{F}$ applied to the particle.

DEFINITION OF THE MAGNETIC FIELD

The magnitude B of the magnetic field at any point in space is defined as

$$B = \frac{F}{|q_0|(v \sin \theta)} \tag{21.1}$$

where F is the magnitude of the magnetic force on a test charge, $|q_0|$ is the magnitude of the test charge, and v is the magnitude of the charge's velocity, which makes an angle θ $(0 \le \theta \le 180°)$ with the direction of the magnetic field. The magnetic field $\vec{B}$ is a vector, and its direction can be determined by using a small compass needle.

SI Unit of Magnetic Field: $\dfrac{\text{newton} \cdot \text{second}}{\text{coulomb} \cdot \text{meter}} = 1$ tesla (T)

The unit of magnetic field strength that follows from Equation 21.1 is the $\text{N} \cdot \text{s}/(\text{C} \cdot \text{m})$. This unit is called the *tesla* (T), a tribute to the Croatian-born American engineer Nikola Tesla (1856–1943). Thus, one tesla is the strength of the magnetic field in which a unit test charge, traveling perpendicular to the magnetic field at a speed of one meter per second, experiences a force of one newton. Because a coulomb per second is an ampere (1 C/s = 1 A, see Section 20.1), the tesla is often written as $1 \text{ T} = 1 \text{ N}/(\text{A} \cdot \text{m})$.

In many situations the magnetic field has a value that is considerably less than one tesla. For example, the strength of the magnetic field near the earth's surface is approximately 10^{-4} T. In such circumstances, a magnetic field unit called the *gauss* (G) is sometimes used. Although not an SI unit, the gauss is a convenient size for many applications involving magnetic fields. The relation between the gauss and the tesla is

$$1 \text{ gauss} = 10^{-4} \text{ tesla}$$

Example 1 deals with the magnetic force exerted on a moving proton and on a moving electron.

Example 1 Magnetic Forces on Charged Particles

A proton in a particle accelerator has a speed of 5.0×10^6 m/s. The proton encounters a magnetic field whose magnitude is 0.40 T and whose direction makes an angle of $\theta = 30.0°$ with respect to the proton's velocity (see Figure 21.7c). Find **(a)** the magnitude and direction of the magnetic force on the proton and **(b)** the acceleration of the proton. **(c)** What would be the force and acceleration if the particle were an electron instead of a proton?

Reasoning For both the proton and the electron, the magnitude of the magnetic force is given by Equation 21.1. The magnetic forces that act on these particles have opposite directions, however, because the charges have opposite signs. In either case, the acceleration is given by Newton's second law, which applies to the magnetic force just as it does to any force. In using the second law, we must take into account the fact that the masses of the proton and the electron are different.

Solution **(a)** The positive charge on a proton is 1.60×10^{-19} C, and according to Equation 21.1, the magnitude of the magnetic force is $F = |q_0|vB \sin \theta$. Therefore,

$$F = (1.60 \times 10^{-19} \text{ C})(5.0 \times 10^6 \text{ m/s})(0.40 \text{ T})(\sin 30.0°) = \boxed{1.6 \times 10^{-13} \text{ N}}$$

The direction of the magnetic force is given by RHR-1 and is directed **_upward_** in Figure 21.7c, with the magnetic field pointing to the right.

(b) The magnitude a of the proton's acceleration follows directly from Newton's second law as the magnitude of the net force divided by the mass m_p of the proton. Since the only force acting on the proton is the magnetic force F, it is the net force. Thus,

$$a = \frac{F}{m_p} = \frac{1.6 \times 10^{-13} \text{ N}}{1.67 \times 10^{-27} \text{ kg}} = \boxed{9.6 \times 10^{13} \text{ m/s}^2} \tag{4.1}$$

The direction of the acceleration is the same as the direction of the net force (the magnetic force).

(c) The magnitude of the magnetic force on the electron is the same as that on the proton, since both have the same velocity and charge magnitude. However, the direction of the force on the electron is opposite to that on the proton, or **_downward_** in Figure 21.7c, since the electron charge is negative. Furthermore, the electron has a smaller mass m_e and, therefore, experiences a significantly greater acceleration:

$$a = \frac{F}{m_e} = \frac{1.6 \times 10^{-13} \text{ N}}{9.11 \times 10^{-31} \text{ kg}} = \boxed{1.8 \times 10^{17} \text{ m/s}^2}$$

The direction of this acceleration is downward in Figure 21.7c.

✓ CHECK YOUR UNDERSTANDING

(The answers are given at the end of the book.)

1. Suppose that you accidentally use your left hand, instead of your right hand, to determine the direction of the magnetic force that acts on a positive charge moving in a magnetic field. Do you get the correct answer? **(a)** Yes, because either hand can be used **(b)** No, because the direction you get will be perpendicular to the correct direction **(c)** No, because the direction you get will be opposite to the correct direction

2. Two particles, having the same charge but different velocities, are moving in a constant magnetic field (see the drawing, where the velocity vectors are drawn to scale). Which particle, if either, experiences the greater magnetic force? **(a)** Particle 1 experiences the greater force, because it is moving perpendicular to the magnetic field. **(b)** Particle 2 experiences the greater force, because it has the greater speed. **(c)** Particle 2 experiences the greater force, because a component of its velocity is parallel to the magnetic field. **(d)** Both particles experience the same magnetic force, because the component of each velocity that is perpendicular to the magnetic field is the same.

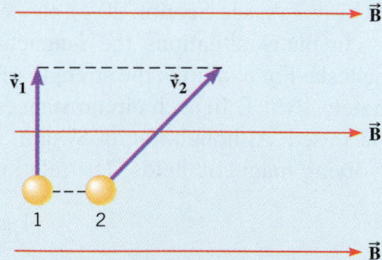

3. A charged particle, passing through a certain region of space, has a velocity whose magnitude and direction remain constant. **(a)** If it is known that the external magnetic field is zero everywhere in this region, can you conclude that the external electric field is also zero? **(b)** If it is known that the external electric field is zero everywhere, can you conclude that the external magnetic field is also zero?

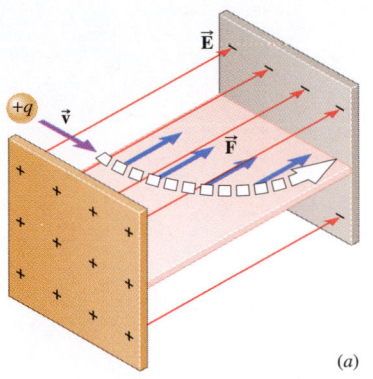

(a)

21.3 THE MOTION OF A CHARGED PARTICLE IN A MAGNETIC FIELD

COMPARING PARTICLE MOTION IN ELECTRIC AND MAGNETIC FIELDS

The motion of a charged particle in an electric field is noticeably different from the motion in a magnetic field. For example, Figure 21.9a shows a positive charge moving between the plates of a parallel plate capacitor. Initially, the charge is moving perpendicular to the direction of the electric field. Since the direction of the electric force on a positive charge is in the same direction as the electric field, the particle is deflected sideways. Part b of the drawing shows the same particle traveling initially at right angles to a magnetic field. An application of RHR-1 shows that when the charge enters the field, the charge is deflected upward (not sideways) by the magnetic force. As the charge moves upward, the direction of the magnetic force changes, always remaining perpendicular to both the magnetic field and the velocity. Conceptual Example 2 focuses on the difference in how electric and magnetic fields apply forces to a moving charge.

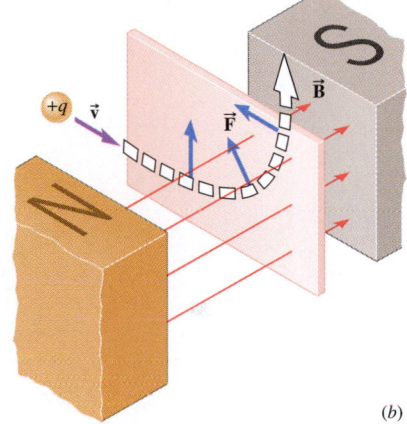

(b)

Figure 21.9 (a) The electric force $\vec{F}$ that acts on a positive charge is parallel to the electric field $\vec{E}$. (b) The magnetic force $\vec{F}$ is perpendicular to both the magnetic field $\vec{B}$ and the velocity $\vec{v}$.

Conceptual Example 2 A Velocity Selector

The physics of a velocity selector. A velocity selector is a device for measuring the velocity of a charged particle. The device operates by applying electric and magnetic forces to the particle in such a way that these forces balance. Figure 21.10a shows a particle with a positive charge $+q$ and a velocity $\vec{v}$, which is perpendicular to a constant magnetic field* $\vec{B}$. Figure 21.10b illustrates a velocity selector, which is a cylindrical tube that is located within the magnetic field. Inside the tube is a parallel plate capacitor that produces an electric field $\vec{E}$ (not shown) that is perpendicular to the magnetic field. The charged particle enters the left end of the tube, moving perpendicular to the magnetic field. If the strengths of $\vec{E}$ and $\vec{B}$ are adjusted properly, the electric and magnetic forces acting on the particle will cancel each other. With no net force acting on the particle, the velocity remains unchanged, according to Newton's first law. As a result, the particle moves in a straight line at a constant speed and exits at the right end of the tube. The magnitude of the velocity that is "selected" can be determined from a knowledge of the strengths of the electric and magnetic fields. Particles with velocities different from the one "selected" are deflected and do not exit at the right end of the tube.

How should the electric field $\vec{E}$ be directed so that the force it applies to the particle can balance the magnetic force produced by $\vec{B}$? The electric field should be directed: **(a)** in the same direction as the magnetic field; **(b)** in a direction opposite to that of the magnetic field; **(c)** from the upper plate of the parallel plate capacitor toward the lower plate; **(d)** from the lower plate of the parallel plate capacitor toward the upper plate.

Reasoning If the electric and magnetic forces are to cancel each other, they must have opposite directions. The direction of the magnetic force can be found by applying Right-Hand Rule No. 1 (RHR-1) to the moving charged particle. This rule reveals that the magnetic force acting

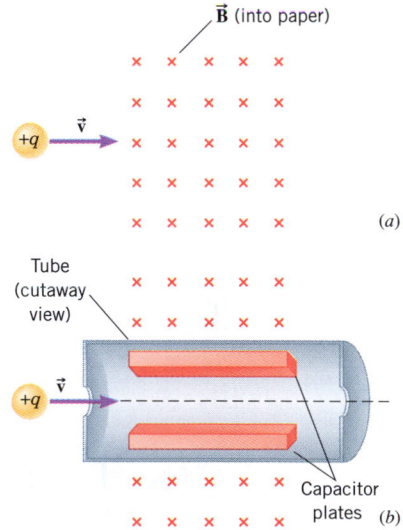

Figure 21.10 (a) A particle with a positive charge q and velocity $\vec{v}$ moves perpendicularly into a magnetic field $\vec{B}$. (b) A velocity selector is a tube in which an electric field (not shown) is perpendicular to a magnetic field, and the field magnitudes are adjusted so that the electric and magnetic forces acting on the particle balance.

*In many instances it is convenient to orient the magnetic field $\vec{B}$ so its direction is perpendicular to the page. In these cases it is customary to use a dot to symbolize the magnetic field pointing *out of the page* (toward the reader); this dot symbolizes the tip of the arrow representing the $\vec{B}$ vector. A region in which a magnetic field is directed *into the page* is drawn as a series of crosses that indicate the tail feathers of the arrows representing the $\vec{B}$ vectors. Therefore, regions in which a magnetic field is directed out of the page or into the page are drawn as shown below:

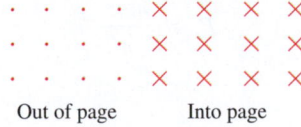

Out of page   Into page

on the positively charged particle in Figure 21.10*a* is directed upward, toward the top of the page. Since the particle is positively charged, the direction of the electric force is the same as the direction of the electric field produced by the capacitor plates.

Answers (a), (b), and (d) are incorrect. If a charged particle moves in the same direction as the magnetic field or in the opposite direction, as in answers (a) and (b), the particle does not experience a magnetic force. For the velocity selector to work, the particle must experience both a magnetic and an electric force. Answer (d) is not correct because the electric force on the positively charged particle would also be upward, which is the same direction as the magnetic force.

Answer (c) is correct. Since the magnetic force is directed upward, the electric force must be directed downward. The force applied to a positive charge by an electric field has the same direction as the field itself, so the electric field must point downward, from the upper plate of the capacitor toward the lower plate. As a result, the upper plate must be positively charged.

Related Homework: *Problems 24, 26*

We have seen that a charged particle traveling in a magnetic field experiences a magnetic force that is always perpendicular to the field. In contrast, the force applied by an electric field is always parallel (or antiparallel) to the field direction. Because of the difference in the way that electric and magnetic fields exert forces, the work done on a charged particle by each field is different, as we now discuss.

THE WORK DONE ON A CHARGED PARTICLE MOVING THROUGH ELECTRIC AND MAGNETIC FIELDS

In Figure 21.9*a* an electric field applies a force to a positively charged particle, and, consequently, the path of the particle bends in the direction of the force. Because there is a component of the particle's displacement in the direction of the electric force, the force does work on the particle, according to Equation 6.1. This work increases the kinetic energy and, hence, the speed of the particle, in accord with the work–energy theorem (see Section 6.2). In contrast, the magnetic force in Figure 21.9*b* always acts in a direction that is perpendicular to the motion of the charge. Consequently, the displacement of the moving charge never has a component in the direction of the magnetic force. As a result, *the magnetic force cannot do work and change the kinetic energy of the charged particle* in Figure 21.9*b*. Thus, the speed of the particle *does not* change, although the force does alter the direction of the motion.

THE CIRCULAR TRAJECTORY

To describe the motion of a charged particle in a constant magnetic field more completely, let's discuss the special case in which the velocity of the particle is perpendicular to a uniform magnetic field. As Figure 21.11 illustrates, the magnetic force serves to move the particle in a circular path. To understand why, consider two points on the circumference labeled 1 and 2. When the positively charged particle is at point 1, the magnetic force $\vec{F}$ is perpendicular to the velocity $\vec{v}$ and points directly upward in the drawing. This force causes the trajectory to bend upward. When the particle reaches point 2, the magnetic force still remains perpendicular to the velocity but is now directed to the left in the drawing. **Problem-solving insight:** *The magnetic force always remains perpendicular to the velocity and is directed toward the center of the circular path.*

To find the radius of the path in Figure 21.11, we use the concept of centripetal force from Section 5.3. The centripetal force is the net force, directed toward the center of the circle, that is needed to keep a particle moving along a circular path. The magnitude F_c of this force depends on the speed v and mass m of the particle, as well as the radius r of the circle:

$$F_c = \frac{mv^2}{r} \tag{5.3}$$

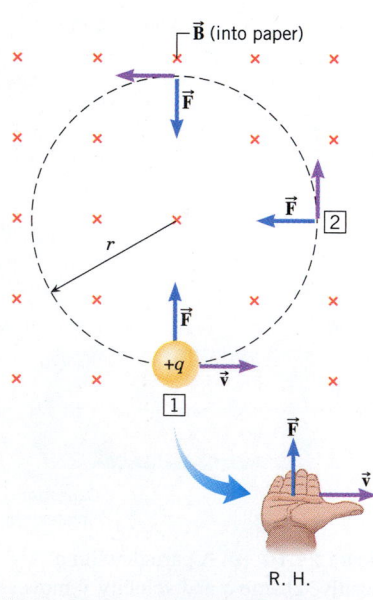

R. H.

Figure 21.11 A positively charged particle is moving perpendicular to a constant magnetic field. The magnetic force $\vec{F}$ causes the particle to move on a circular path (R.H. = right hand).

In the present situation, the magnetic force furnishes the centripetal force. Being perpendicular to the velocity, the magnetic force does no work in keeping the charge $+q$ on the circular path. According to Equation 21.1, the magnitude of the magnetic force is given by $|q|vB \sin 90°$, so $|q|vB = mv^2/r$ or

$$r = \frac{mv}{|q|B} \tag{21.2}$$

This result shows that the radius of the circle is inversely proportional to the magnitude of the magnetic field, with stronger fields producing "tighter" circular paths. Example 3 illustrates an application of Equation 21.2.

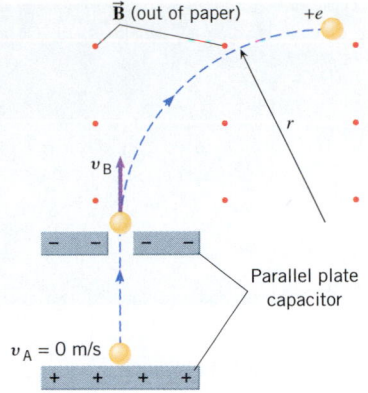

Figure 21.12 A proton, starting from rest at the positive plate of the capacitor, accelerates toward the negative plate. After leaving the capacitor, the proton enters a magnetic field, where it moves on a circular path of radius r.

Example 3
The Motion of a Proton

A proton is released from rest at point A, which is located next to the positive plate of a parallel plate capacitor (see Figure 21.12). The proton then accelerates toward the negative plate, leaving the capacitor at point B through a small hole in the plate. The electric potential of the positive plate is 2100 V greater than that of the negative plate, so $V_A - V_B = 2100$ V. Once outside the capacitor, the proton travels at a constant velocity until it enters a region of constant magnetic field of magnitude 0.10 T. The velocity is perpendicular to the magnetic field, which is directed out of the page in Figure 21.12. Find **(a)** the speed v_B of the proton when it leaves the negative plate of the capacitor, and **(b)** the radius r of the circular path on which the proton moves in the magnetic field.

Reasoning The only force that acts on the proton (charge = $+e$) while it is between the capacitor plates is the conservative electric force. Thus, we can use the conservation of energy to find the speed of the proton when it leaves the negative plate. The total energy of the proton is the sum of its kinetic energy, $\frac{1}{2}mv^2$, and its electric potential energy, EPE. Following Example 4 in Chapter 19, we set the total energy at point B equal to the total energy at point A:

$$\underbrace{\tfrac{1}{2}mv_B{}^2 + \text{EPE}_B}_{\text{Total energy at } B} = \underbrace{\tfrac{1}{2}mv_A{}^2 + \text{EPE}_A}_{\text{Total energy at } A}$$

We note that $v_A = 0$ m/s, since the proton starts from rest, and use Equation 19.4 to set $\text{EPE}_A - \text{EPE}_B = e(V_A - V_B)$. Then the conservation of energy reduces to $\frac{1}{2}mv_B{}^2 = e(V_A - V_B)$. Solving for v_B gives $v_B = \sqrt{2e(V_A - V_B)/m}$. The proton enters the magnetic field with this speed and moves on a circular path with a radius that is given by Equation 21.2.

Solution **(a)** The speed of the proton is

$$v_B = \sqrt{\frac{2e(V_A - V_B)}{m}} = \sqrt{\frac{2(1.60 \times 10^{-19}\text{ C})(2100\text{ V})}{1.67 \times 10^{-27}\text{ kg}}} = \boxed{6.3 \times 10^5 \text{ m/s}}$$

(b) When the proton moves in the magnetic field, the radius of the circular path is

$$r = \frac{mv_B}{eB} = \frac{(1.67 \times 10^{-27}\text{ kg})(6.3 \times 10^5\text{ m/s})}{(1.60 \times 10^{-19}\text{ C})(0.10\text{ T})} = \boxed{6.6 \times 10^{-2} \text{ m}} \tag{21.2}$$

One of the prominent areas in physics today is the study of elementary particles, which are the basic building blocks from which all matter is constructed. Important information about an elementary particle can be obtained from its motion in a magnetic field, with the aid of a device known as a bubble chamber. A bubble chamber contains a superheated liquid such as hydrogen, which will boil and form bubbles readily. When an electrically charged particle passes through the chamber, a thin track of bubbles is left in its wake. This track can be photographed to show how a magnetic field affects the particle motion. Conceptual Example 4 illustrates how physicists deduce information from such photographs.

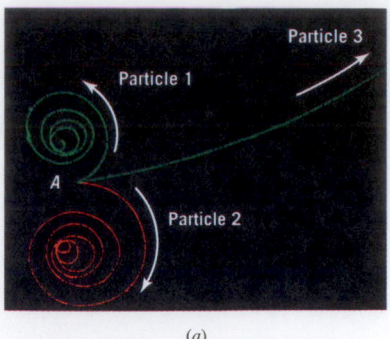

(a)

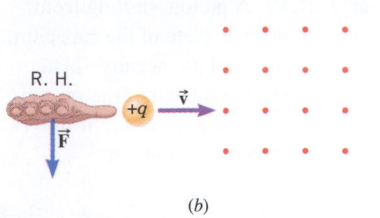

(b)

Figure 21.13 (a) A photograph of tracks in a bubble chamber. A magnetic field is directed out of the paper. At point A a gamma ray (not visible) spontaneously transforms into two charged particles, and a third charged particle is knocked out of a hydrogen atom in the chamber. (Lawrence Berkeley Laboratory/Photo Researchers, Inc.) (b) In accord with RHR-1, the magnetic field applies a downward force to a positively charged particle that moves to the right.

Conceptual Example 4   Particle Tracks in a Bubble Chamber

Figure 21.13a shows the bubble-chamber tracks resulting from an event that begins at point A. At this point a gamma ray (emitted by certain radioactive substances), traveling in from the left, spontaneously transforms into two charged particles. There is no track from the gamma ray itself. These particles move away from point A, producing the two spiral tracks. A third charged particle is knocked out of a hydrogen atom and moves forward, producing the long track with the slight upward curvature. Each of the three particles has the same mass and carries a charge of the same magnitude. A uniform magnetic field is directed out of the paper toward you. What is the sign (+ or −) of the charge carried by each particle?

	Particle 1	Particle 2	Particle 3
(a)	−	−	+
(b)	−	+	−
(c)	+	−	−
(d)	+	−	+

Reasoning Figure 21.13b shows a positively charged particle traveling with a velocity $\vec{v}$ that is perpendicular to a magnetic field. The field is directed out of the paper, just like it is in part a of the drawing. RHR-1 indicates that the magnetic force points downward. This magnetic force provides the centripetal force that causes a particle to move on a circular path (see Section 5.3). The centripetal force is directed toward the center of the circular path. Thus, in Figure 21.13a a positive charge would move on a downward-curving track, and a negative charge would move on an upward-curving track.

Answers (a), (c), and (d) are incorrect. Since particles 1 and 3 move on upward-curving tracks, they are negatively charged, not positively charged.

Answer (b) is correct. A downward-curving track in the photograph indicates a positive charge, while an upward-curving track indicates a negative charge. Thus, particles 1 and 3 carry a negative charge. They are, in fact, electrons (e⁻). In contrast, particle 2 carries a positive charge. It is called a positron (e⁺), an elementary particle that has the same mass as an electron but an opposite charge.

Related Homework: *Check Your Understanding 5, 6, Problem 28*

✓ CHECK YOUR UNDERSTANDING

(The answers are given at the end of the book.)

4. Suppose that the positive charge in Figure 21.9a were launched from the negative plate toward the positive plate, in a direction *opposite* to the electric field $\vec{E}$. A sufficiently strong electric field would prevent the charge from striking the positive plate. Suppose that the positive charge in Figure 21.9b were launched from the south pole toward the north pole, in a direction *opposite* to the magnetic field $\vec{B}$. Would a sufficiently strong magnetic field prevent the charge from reaching the north pole?   **(a)** Yes   **(b)** No, because a magnetic field cannot exert a force on a charged particle that is moving antiparallel to the field   **(c)** No, because the magnetic force would cause the charge to move faster as it moved toward the north pole

5. Review Conceptual Example 4 and **Concept Simulation 21.1** at **www.wiley.com/college/cutnell** as background for this question. Three particles move through a constant magnetic field and follow the paths shown in the drawing. Determine whether each particle is positively (+) charged, negatively (−) charged, or neutral.

	Particle 1	Particle 2	Particle 3
(a)	neutral	+	neutral
(b)	−	neutral	+
(c)	−	−	−
(d)	+	neutral	−
(e)	+	+	+

6. Suppose that the three particles in Figure 21.13a have identical charge magnitudes and masses. Which particle has the greatest speed? Refer to Conceptual Example 4 as needed.

7. A positive charge moves along a circular path under the influence of a magnetic field. The magnetic field is perpendicular to the plane of the circle, as in Figure 21.11. If the velocity of the particle is reversed at some point along the path, will the particle retrace its path? **(a)** Yes **(b)** No, because it will move around a *different* circle in a counterclockwise direction

8. Refer to Figure 21.11. Assume that the particle in the picture is a proton. If an electron is projected at point 1 with the same velocity $\vec{v}$, it will not follow exactly the same path as the proton, unless the magnetic field is adjusted in the following manner: the magnitude of the magnetic field must be _____, and the direction of the magnetic field must be _____. **(a)** the same, reversed **(b)** increased, the same **(c)** reduced, reversed

9. The drawing shows a top view of four interconnected chambers. A *negative* charge is fired into chamber 1. By turning on separate magnetic fields in each chamber, the charge can be made to exit from chamber 4, as shown. How should the magnetic field in each chamber be directed: out of the page or into the page?

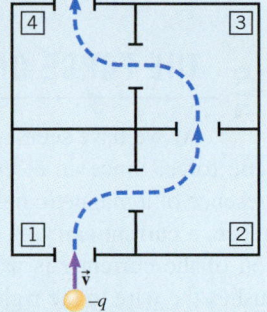

	Chamber 1	Chamber 2	Chamber 3	Chamber 4
(a)	out of	into	out of	into
(b)	into	out of	out of	into
(c)	out of	into	into	out of
(d)	into	out of	into	out of

10. The drawing shows a particle carrying a positive charge $+q$ at the coordinate origin, as well as a target located in the third quadrant. A uniform magnetic field is directed perpendicularly into the plane of the paper. The charge can be projected in the plane of the paper only, along the positive or negative *x* or *y* axis. There are four possible directions ($+x$, $-x$, $+y$, $-y$) for the initial velocity of the particle. The particle can be made to hit the target for only two of the four directions. Which two directions are they? **(a)** $+y$, $-y$ **(b)** $-y$, $+x$ **(c)** $-x$, $+y$ **(d)** $+x$, $-x$

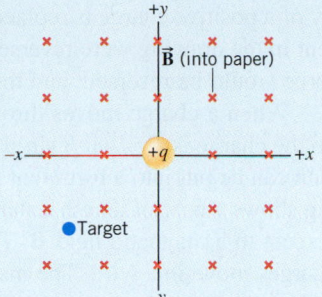

21.4 THE MASS SPECTROMETER

Physicists use mass spectrometers for determining the relative masses and abundances of isotopes.* Chemists use these instruments to help identify unknown molecules produced in chemical reactions. Mass spectrometers are also used during surgery, where they give the anesthesiologist information on the gases, including the anesthetic, in the patient's lungs.

In the type of mass spectrometer illustrated in Figure 21.14, the atoms or molecules are first vaporized and then ionized by the ion source. The ionization process removes one electron from the particle, leaving it with a net positive charge of $+e$. The positive ions are then accelerated through the potential difference V, which is applied between the ion source and the metal plate. With a speed v, the ions pass through a hole in the plate and enter a region of constant magnetic field $\vec{B}$, where they are deflected in semicircular paths. Only those ions following a path with the proper radius r strike the detector, which records the number of ions arriving per second.

The mass m of the detected ions can be expressed in terms of r, B, and v by recalling that the radius of the path followed by a particle of charge $+e$ is $r = mv/(eB)$ (Equation 21.2). In addition, the Reasoning section in Example 3 shows that the ion speed v can be expressed in terms of the potential difference V as $v = \sqrt{2eV/m}$. This expression for v is the same as that used in Example 3, except that, for convenience, we have replaced the potential difference, $V_A - V_B$, by the symbol V. Eliminating v from these two equations algebraically and solving for the mass gives

$$m = \left(\frac{er^2}{2V}\right)B^2$$

*Isotopes are atoms that have the same atomic number but different atomic masses due to the presence of different numbers of neutrons in the nucleus. They are discussed in Section 31.1.

The physics of a mass spectrometer.

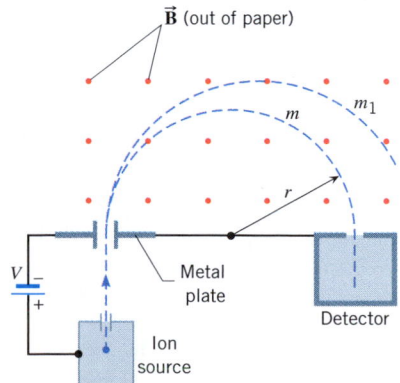

Figure 21.14 In this mass spectrometer the dashed lines are the paths traveled by ions of different masses. Ions with mass m follow the path of radius r and enter the detector. Ions with the larger mass m_1 follow the outer path and miss the detector.

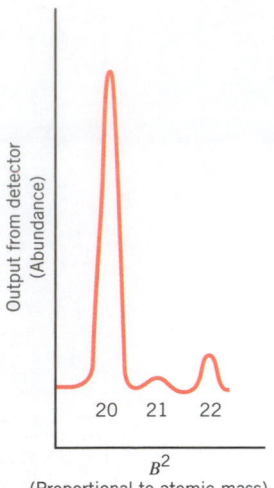

Figure 21.15 The mass spectrum (not to scale) of naturally occurring neon, showing three isotopes whose atomic mass numbers are 20, 21, and 22. The larger the peak, the more abundant the isotope.

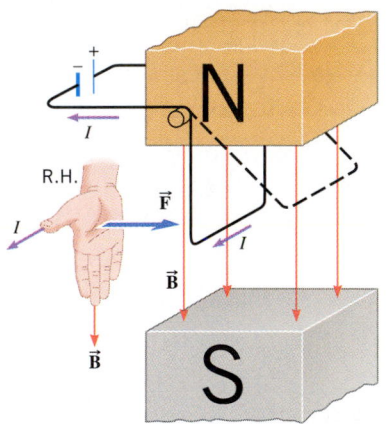

Figure 21.16 The wire carries a current I, and the bottom segment of the wire is oriented perpendicular to a magnetic field $\vec{\mathbf{B}}$. A magnetic force deflects the wire to the right.

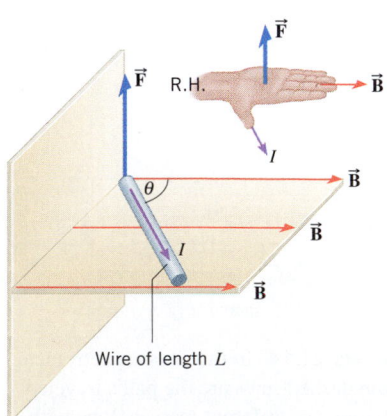

Figure 21.17 The current I in the wire, oriented at an angle θ with respect to a magnetic field $\vec{\mathbf{B}}$, is acted upon by a magnetic force $\vec{\mathbf{F}}$.

This result shows that the mass of each ion reaching the detector is proportional to B^2. Experimentally changing the value of B and keeping the term in the parentheses constant will allow ions of different masses to enter the detector. A plot of the detector output as a function of B^2 then gives an indication of what masses are present and the abundance of each mass.

Figure 21.15 shows a record obtained by a mass spectrometer for naturally occurring neon gas. The results show that the element neon has three isotopes whose atomic mass numbers are 20, 21, and 22. These isotopes occur because neon atoms exist with different numbers of neutrons in the nucleus. Notice that the isotopes have different abundances, with neon-20 being the most abundant.

21.5 THE FORCE ON A CURRENT IN A MAGNETIC FIELD

As we have seen, a charge moving through a magnetic field can experience a magnetic force. Since an electric current is a collection of moving charges, a current in the presence of a magnetic field can also experience a magnetic force. In Figure 21.16, for instance, a current-carrying wire is placed between the poles of a magnet. When the direction of the current I is as shown, the moving charges experience a magnetic force that pushes the wire to the right in the drawing. The direction of the force is determined in the usual manner by using RHR-1, with the minor modification that the direction of the velocity of a positive charge is replaced by the direction of the conventional current I. If the current in the drawing were reversed by switching the leads to the battery, the direction of the force would be reversed, and the wire would be pushed to the left.

When a charge moves through a magnetic field, the magnitude of the force that acts on the charge is $F = |q|vB \sin \theta$ (Equation 21.1). With the aid of Figure 21.17, this expression can be put into a form that is more suitable for use with an electric current. The drawing shows a wire of length L that carries a current I. The wire is oriented at an angle θ with respect to a magnetic field $\vec{\mathbf{B}}$. This picture is similar to Figure 21.7c, except that now the charges move in a wire. The magnetic force exerted on this length of wire is the net force acting on the total amount of charge moving in the wire. Suppose that an amount of conventional positive charge Δq travels the length of the wire in a time interval Δt. The magnitude of the magnetic force on this amount of charge is given by Equation 21.1 as $F = (\Delta q)vB \sin \theta$. Multiplying and dividing the right side of this equation by Δt, we find that

$$F = \underbrace{\left(\frac{\Delta q}{\Delta t}\right)}_{I} \underbrace{(v \, \Delta t)}_{L} B \sin \theta$$

The term $\Delta q/\Delta t$ is the current I in the wire (see Equation 20.1), and the term $v \, \Delta t$ is the length L of the wire. With these two substitutions, the expression for the magnetic force exerted on a current-carrying wire becomes

Magnetic force on a current-carrying wire of length L $\qquad F = ILB \sin \theta \qquad\qquad$ (21.3)

As in the case of a single charge traveling in a magnetic field, the magnetic force on a current-carrying wire is a maximum when the wire is oriented perpendicular to the field ($\theta = 90°$) and vanishes when the current is parallel or antiparallel to the field ($\theta = 0°$ or $180°$). The direction of the magnetic force is given by RHR-1, as Figure 21.17 indicates.

The physics of a loudspeaker. Most loudspeakers operate on the principle that a magnetic field exerts a force on a current-carrying wire. Figure 21.18a shows a speaker design that consists of three basic parts: a cone, a voice coil, and a permanent magnet. The cone is mounted so it can vibrate back and forth. When vibrating, it pushes and pulls on the air in front of it, thereby creating sound waves. Attached to the apex of the cone is the voice coil, which is a hollow cylinder around which coils of wire are wound. The voice coil is slipped over one of the poles of the stationary permanent magnet (the north pole in the drawing) and can move freely. The two ends of the voice-coil wire are connected to the speaker terminals on the back panel of a receiver.

The receiver acts as an ac generator, sending an alternating current to the voice coil. The alternating current interacts with the magnetic field to generate an alternating force

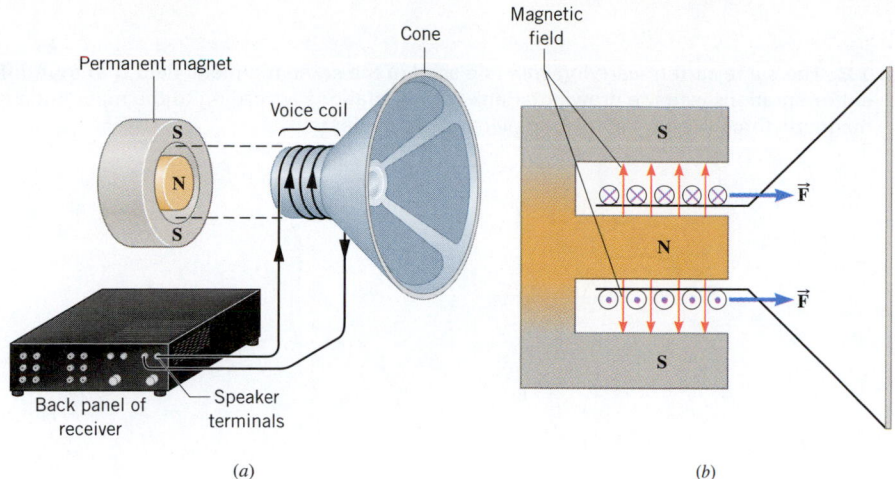

Figure 21.18 (*a*) An "exploded" view of one type of speaker design, which shows a cone, a voice coil, and a permanent magnet. (*b*) Because of the current in the voice coil (shown as ⊗ and ⊙), the magnetic field causes a force $\vec{\mathbf{F}}$ to be exerted on the voice coil and cone.

that pushes and pulls on the voice coil and the attached cone. To see how the magnetic force arises, consider Figure 21.18*b*, which is a cross-sectional view of the voice coil and the magnet. In the cross-sectional view, the current is directed into the page in the upper half of the voice coil (⊗⊗⊗) and out of the page in the lower half (⊙⊙⊙). In both cases the magnetic field is perpendicular to the current, so the maximum possible force is exerted on the wire. An application of RHR-1 to both the upper and lower halves of the voice coil shows that the magnetic force $\vec{\mathbf{F}}$ in the drawing is directed to the right, causing the cone to accelerate in that direction. One-half of a cycle later when the current is reversed, the direction of the magnetic force is also reversed, and the cone accelerates to the left. If, for example, the alternating current from the receiver has a frequency of 1000 Hz, the alternating magnetic force causes the cone to vibrate back and forth at the same frequency, and a 1000-Hz sound wave is produced. Thus, it is the magnetic force on a current-carrying wire that is responsible for converting an electrical signal into a sound wave. In Example 5 a typical force and acceleration in a loudspeaker are determined.

Problem-solving insight

Whenever the current in a wire reverses direction, the force exerted on the wire by a given magnetic field also reverses direction.

 Example 5 The Force and Acceleration in a Loudspeaker

The voice coil of a speaker has a diameter of $d = 0.025$ m, contains 55 turns of wire, and is placed in a 0.10-T magnetic field. The current in the voice coil is 2.0 A. (**a**) Determine the magnetic force that acts on the coil and cone. (**b**) The voice coil and cone have a combined mass of 0.020 kg. Find their acceleration.

Reasoning The magnetic force that acts on the current-carrying voice coil is given by Equation 21.3 as $F = ILB \sin \theta$. The effective length L of the wire in the voice coil is very nearly the number of turns N times the circumference (πd) of one turn: $L = N\pi d$. The acceleration of the voice coil and cone is given by Newton's second law as the magnetic force divided by the combined mass.

Solution (**a**) Since the magnetic field acts perpendicular to all parts of the wire, $\theta = 90°$ and the force on the voice coil is

$$F = ILB \sin \theta = I(N\pi d)B \sin \theta = (2.0 \text{ A})[55\pi(0.025 \text{ m})](0.10 \text{ T})\sin 90° = \boxed{0.86 \text{ N}} \quad (21.3)$$

(**b**) According to Newton's second law, the acceleration of the voice coil and cone is

$$a = \frac{F}{m} = \frac{0.86 \text{ N}}{0.020 \text{ kg}} = \boxed{43 \text{ m/s}^2} \quad (4.1)$$

This acceleration is more than four times the acceleration due to gravity.

✓ **CHECK YOUR UNDERSTANDING**

(*The answers are given at the end of the book.*)

11. Refer to Figure 21.16. (**a**) What happens to the direction of the magnetic force if the current is reversed? (**b**) What happens to the direction of the force if *both* the current *and* the magnetic poles are reversed?

Continued

12. The same current-carrying wire is placed in the same magnetic field $\vec{B}$ in four different orientations (see the drawing). Rank the orientations according to the magnitude of the magnetic force exerted on the wire, largest to smallest.

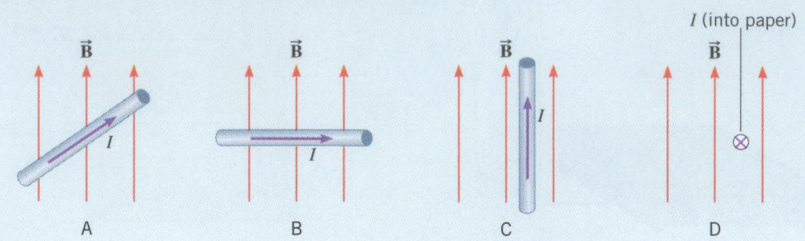

THE TORQUE ON A CURRENT-CARRYING COIL

21.6

We have seen that a current-carrying wire can experience a force when placed in a magnetic field. If a loop of wire is suspended properly in a magnetic field, the magnetic force produces a torque that tends to rotate the loop. This torque is responsible for the operation of a widely used type of electric motor.

Figure 21.19a shows a rectangular loop of wire attached to a vertical shaft. The shaft is mounted so that it is free to rotate in a uniform magnetic field. When there is a current in the loop, the loop rotates because magnetic forces act on the two vertical sides, labeled 1 and 2 in the drawing. Part b shows a top view of the loop and the magnetic forces $\vec{F}$ and $-\vec{F}$ that act on the two sides. These two forces have the same magnitude, but an application of RHR-1 shows that they point in opposite directions, so the loop experiences no net force. The loop does, however, experience a net torque that tends to rotate it in a clockwise fashion about the vertical shaft. Figure 21.20a shows that the torque is maximum when the normal to the plane of the loop is perpendicular to the field. In contrast, part b shows that the torque is zero when the normal is parallel to the field. **When a current-carrying loop is placed in a magnetic field, the loop tends to rotate such that its normal becomes aligned with the magnetic field.** In this respect, a current loop behaves like a magnet (e.g., a compass needle) suspended in a magnetic field, since a magnet also rotates to align itself with the magnetic field.

It is possible to determine the magnitude of the torque on the loop. From Equation 21.3 the magnetic force on each vertical side has a magnitude of $F = ILB \sin 90°$, where L is the length of side 1 or side 2, and $\theta = 90°$ because the current I always remains perpendicular to the magnetic field as the loop rotates. As Section 9.1 discusses, the torque produced by a force is the product of the magnitude of the force and the lever arm. In Figure 21.19b the lever arm is the perpendicular distance from the line of action of the force to the shaft. This distance is given by $(w/2) \sin \phi$, where w is the width of the loop,

Figure 21.19 (a) A current-carrying loop of wire, which can rotate about a vertical shaft, is situated in a magnetic field. (b) A top view of the loop. The current in side 1 is directed out of the page ($\odot$), while the current in side 2 is directed into the page ($\otimes$). The current in side 1 experiences a force $\vec{F}$ that is opposite to the force exerted on side 2. The two forces produce a clockwise torque about the shaft.

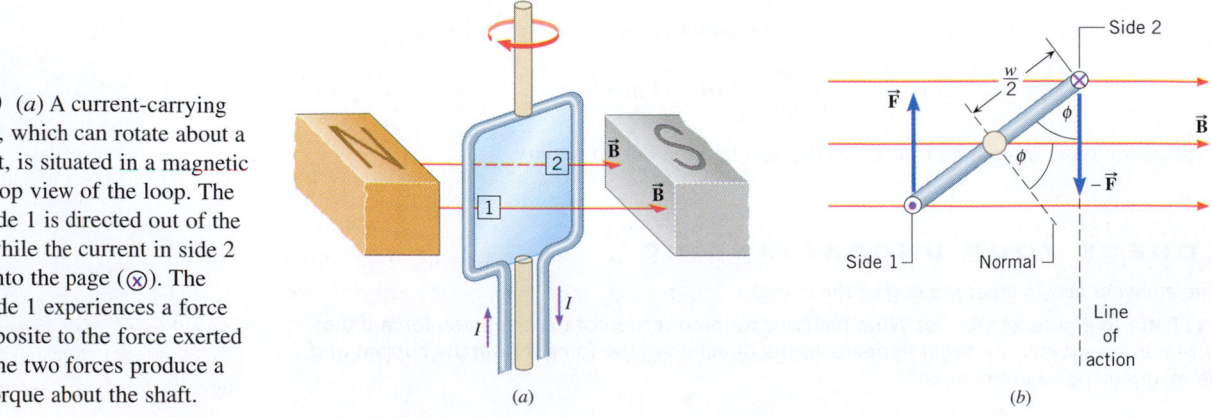

and ϕ is the angle between the normal to the plane of the loop and the direction of the magnetic field. The net torque is the sum of the torques on the two sides, so

$$\text{Net torque} = \tau = ILB\left(\tfrac{1}{2}w \sin \phi\right) + ILB\left(\tfrac{1}{2}w \sin \phi\right) = IAB \sin \phi$$

In this result the product Lw has been replaced by the area A of the loop. If the wire is wrapped so as to form a coil containing N loops, each of area A, the force on each side is N times larger, and the torque becomes proportionally greater:

$$\tau = NIAB \sin \phi \qquad (21.4)$$

Equation 21.4 has been derived for a rectangular coil, but it is valid for any shape of flat coil, such as a circular coil. The torque depends on the geometric properties of the coil and the current in it through the quantity NIA. This quantity is known as the ***magnetic moment*** of the coil, and its unit is ampere · meter² ($A \cdot m^2$). The greater the magnetic moment of a current-carrying coil, the greater is the torque that the coil experiences when placed in a magnetic field. Example 6 discusses the torque that a magnetic field applies to such a coil.

 Example 6 The Torque Exerted on a Current-Carrying Coil

A coil of wire has an area of 2.0×10^{-4} m², consists of 100 loops or turns, and contains a current of 0.045 A. The coil is placed in a uniform magnetic field of magnitude 0.15 T. **(a)** Determine the magnetic moment of the coil. **(b)** Find the maximum torque that the magnetic field can exert on the coil.

Reasoning and Solution **(a)** The magnetic moment of the coil is

$$\text{Magnetic moment} = NIA = (100)(0.045 \text{ A})(2.0 \times 10^{-4} \text{ m}^2) = \boxed{9.0 \times 10^{-4} \text{ A} \cdot \text{m}^2}$$

(b) According to Equation 21.4, the torque is the product of the magnetic moment NIA and $B \sin \phi$. However, the maximum torque occurs when $\phi = 90°$, so

$$\tau = \underbrace{(NIA)}_{\substack{\text{Magnetic} \\ \text{moment}}}(B \sin 90°) = (9.0 \times 10^{-4} \text{ A} \cdot \text{m}^2)(0.15 \text{ T}) = \boxed{1.4 \times 10^{-4} \text{ N} \cdot \text{m}}$$

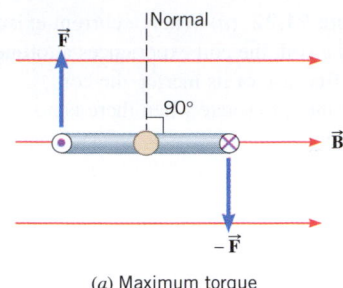

(a) Maximum torque

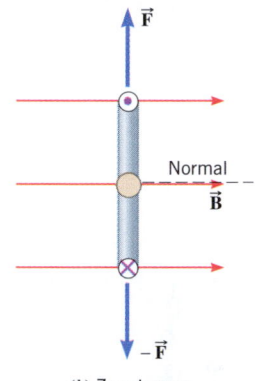

(b) Zero torque

Figure 21.20 (*a*) Maximum torque occurs when the normal to the plane of the loop is perpendicular to the magnetic field. (*b*) The torque is zero when the normal is parallel to the field.

The physics of a direct-current electric motor. The electric motor is found in many devices, such as CD players, automobiles, washing machines, and air conditioners. Figure 21.21 shows that a direct-current (dc) motor consists of a coil of wire placed in a magnetic field and free to rotate about a vertical shaft. The coil of wire contains many turns and is wrapped around an iron cylinder that rotates with the coil, although these features have been omitted to simplify the drawing. The coil and iron cylinder assembly is known as the armature. Each end of the wire coil is attached to a metallic half-ring. Rubbing against each of the half-rings is a graphite contact called a brush. While the half-rings rotate with the coil, the graphite brushes remain stationary. The two half-rings and the associated brushes are referred to as a split-ring commutator (see below).

The operation of a motor can be understood by considering Figure 21.22. In part *a* the current from the battery enters the coil through the left brush and half-ring, goes around the coil, and then leaves through the right half-ring and brush. Consistent with RHR-1, the directions of the magnetic forces $\vec{F}$ and $-\vec{F}$ on the two sides of the coil are as shown in the drawing. These forces produce the torque that turns the coil. Eventually, the coil reaches the position shown in part *b* of the drawing. In this position the half-rings momentarily lose electrical contact with the brushes, so that there is no current in the coil and no applied torque. However, like any moving object, the rotating coil does not stop immediately, for its inertia carries it onward. When the half-rings reestablish contact with the brushes, there again is a current in the coil, and a magnetic torque again rotates the coil in the same direction. The split-ring commutator ensures that the current is always in the proper direction to yield a torque that produces a continuous rotation of the coil.

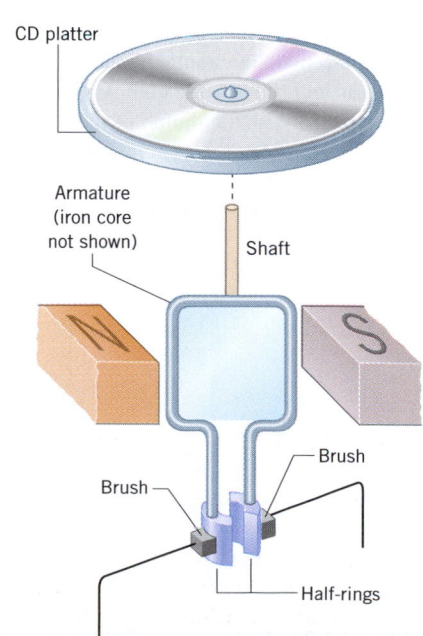

Figure 21.21 The basic components of a dc motor. A CD platter is shown as it might be attached to the motor.

Figure 21.22 (*a*) When a current exists in the coil, the coil experiences a torque. (*b*) Because of its inertia, the coil continues to rotate when there is no current.

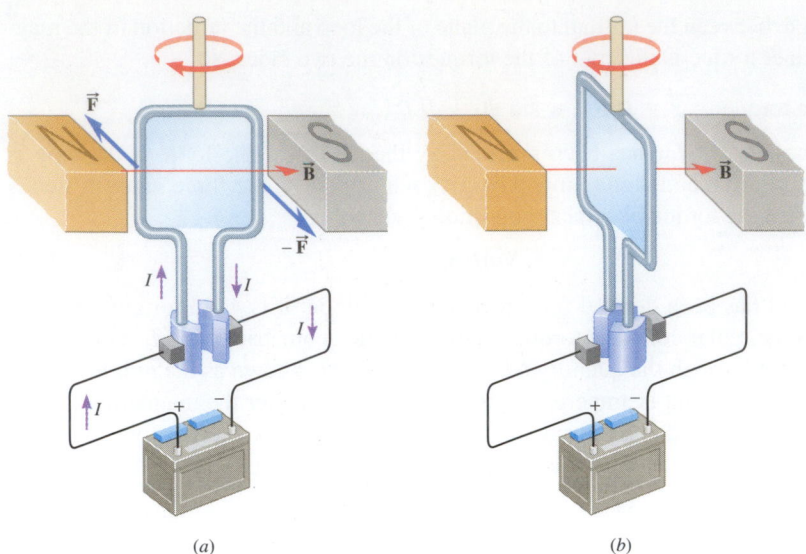

(a)   (b)

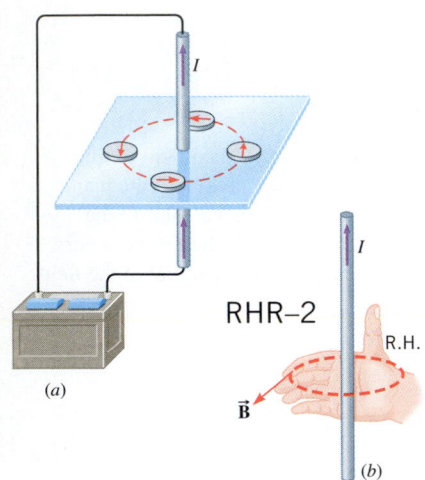

(a)

RHR-2
R.H.

$\vec{B}$

(b)

Figure 21.23 (*a*) A very long, straight, current-carrying wire produces magnetic field lines that are circular about the wire, as indicated by the compass needles. (*b*) With the thumb of the right hand (R.H.) along the current *I*, the curled fingers point in the direction of the magnetic field, according to RHR-2.

MAGNETIC FIELDS PRODUCED BY CURRENTS

We have seen that a current-carrying wire can experience a magnetic force when placed in a magnetic field that is produced by an external source, such as a permanent magnet. *A current-carrying wire also produces a magnetic field of its own,* as we will see in this section. Hans Christian Oersted (1777–1851) first discovered this effect in 1820 when he observed that a current-carrying wire influences the orientation of a nearby compass needle. The compass needle aligns itself with the net magnetic field produced by the current and the magnetic field of the earth. Oersted's discovery, which linked the motion of electric charges with the creation of a magnetic field, marked the beginning of an important discipline called *electromagnetism.*

A LONG, STRAIGHT WIRE

Figure 21.23*a* illustrates Oersted's discovery with a very long, straight wire. When a current is present, the compass needles point in a circular pattern about the wire. The pattern indicates that the magnetic field lines produced by the current are circles centered on the wire. If the direction of the current is reversed, the needles also reverse their directions, indicating that the direction of the magnetic field has reversed. The direction of the field can be obtained by using *Right-Hand Rule No. 2 (RHR-2),* as part *b* of the drawing indicates:

> *Right-Hand Rule No. 2.* Curl the fingers of the right hand into the shape of a half-circle. Point the thumb in the direction of the conventional current *I*, and the tips of the fingers will point in the direction of the magnetic field $\vec{B}$.

Experimentally, it is found that the magnitude *B* of the magnetic field produced by an infinitely long, straight wire is directly proportional to the current *I* and inversely proportional to the radial distance *r* from the wire: $B \propto I/r$. As usual, this proportionality is converted into an equation by introducing a proportionality constant, which, in this instance, is written as $\mu_0/(2\pi)$. Thus, the magnitude of the magnetic field is

Infinitely long, straight wire
$$B = \frac{\mu_0 I}{2\pi r} \qquad (21.5)$$

The constant μ_0 is known as the *permeability of free space,* and its value is $\mu_0 = 4\pi \times 10^{-7}$ T·m/A. The magnetic field becomes stronger nearer the wire, where *r* is smaller. Therefore, the field lines near the wire are closer together than those located farther away, where the field is weaker. Figure 21.24 shows the pattern of field lines.

The magnetic field that surrounds a current-carrying wire can exert a force on a moving charge, as the next example illustrates.

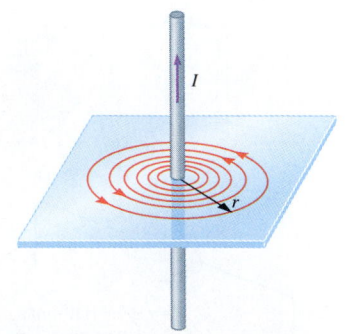

Figure 21.24 The magnetic field becomes stronger as the radial distance *r* decreases, so the field lines are closer together near the wire.

ANALYZING MULTIPLE-CONCEPT PROBLEMS

Example 7 A Current Exerts a Magnetic Force on a Moving Charge

Figure 21.25 shows a very long, straight wire carrying a current of 3.0 A. A particle has a charge of $+6.5 \times 10^{-6}$ C and is moving parallel to the wire at a distance of 0.050 m. The speed of the particle is 280 m/s. Determine the magnitude and direction of the magnetic force exerted on the charged particle by the current in the wire.

Reasoning The current generates a magnetic field in the space around the wire. The charged particle moves in the presence of this field and, therefore, can experience a magnetic force. The magnitude of this force is given by Equation 21.1, and the direction can be determined by applying RHR-1 (see Section 21.2). Note in Figure 21.25 that the magnetic field $\vec{\mathbf{B}}$ produced by the current lies in a plane that is perpendicular to both the wire and the velocity $\vec{\mathbf{v}}$ of the particle. Thus, the angle between $\vec{\mathbf{B}}$ and $\vec{\mathbf{v}}$ is $\theta = 90.0°$.

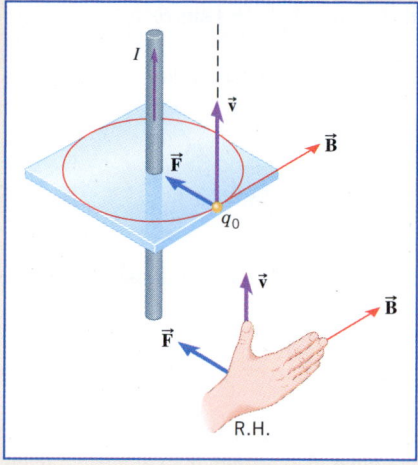

Figure 21.25 The positive charge q_0 moves with a velocity $\vec{\mathbf{v}}$ and experiences a magnetic force $\vec{\mathbf{F}}$ because of the magnetic field $\vec{\mathbf{B}}$ produced by the current in the wire.

Knowns and Unknowns The following list summarizes the data that are given:

Description	Symbol	Value	Comment
Explicit Data			
Current in wire	I	3.0 A	
Electric charge of particle	q_0	$+6.5 \times 10^{-6}$ C	
Distance of particle from wire	r	0.050 m	Particle moves parallel to wire; see Figure 21.25.
Speed of particle	v	280 m/s	
Implicit Data			
Directional angle of particle velocity with respect to magnetic field	θ	90.0°	Particle moves parallel to wire; see **Reasoning**.
Unknown Variable			
Magnitude of magnetic force exerted on particle	F	?	

Modeling the Problem

STEP 1 Magnetic Force on the Particle The magnitude F of the magnetic force acting on the charged particle is given at the right by Equation 21.1, where $|q_0|$ is the magnitude of the charge, v is the particle speed, B is the magnitude of the magnetic field produced by the wire, and θ is the angle between the particle velocity and the magnetic field. Values are given for $|q_0|$, v, and θ. The value of B, however, is unknown, and we determine it in Step 2.

$$F = |q_0|vB \sin \theta \quad (21.1)$$

$?$

STEP 2 Magnetic Field Produced by the Wire The magnitude B of the magnetic field produced by a current I in an infinitely long, straight wire is given by Equation 21.5:

$$B = \frac{\mu_0 I}{2\pi r} \quad (21.5)$$

$$F = |q_0|vB \sin \theta \quad (21.1)$$

$$B = \frac{\mu_0 I}{2\pi r} \quad (21.5)$$

where μ_0 is the permeability of free space and r is the distance from the wire. This expression can be substituted into Equation 21.1, as shown at the right.

Continued

Solution Combining the results of each step algebraically, we find that

STEP 1   **STEP 2**

$$F = |q_0|vB \sin\theta = |q_0|v\left(\frac{\mu_0 I}{2\pi r}\right)\sin\theta$$

The magnitude of the magnetic force on the charged particle is

$$F = |q_0|v\left(\frac{\mu_0 I}{2\pi r}\right)\sin\theta$$

$$= (6.5 \times 10^{-6}\,\text{C})(280\,\text{m/s})\,\frac{(4\pi \times 10^{-7}\,\text{T·m/A})(3.0\,\text{A})}{2\pi(0.050\,\text{m})}\,\sin 90.0° = \boxed{2.2 \times 10^{-8}\,\text{N}}$$

The direction of the magnetic force is predicted by RHR-1 and, as shown in Figure 21.25, is radially inward toward the wire.

Related Homework: *Problem 76*

We have now seen that an electric current can create a magnetic field of its own. Earlier, we have also seen that an electric current can experience a force created by another magnetic field. Therefore, the magnetic field that one current creates can exert a force on another nearby current. Examples 8 and 9 deal with this magnetic interaction between currents.

ANALYZING MULTIPLE-CONCEPT PROBLEMS

Example 8 Two Current-Carrying Wires Exert Magnetic Forces on One Another

Figure 21.26 shows two parallel, straight wires that are very long. The wires are separated by a distance of 0.065 m and carry currents of $I_1 = 15$ A and $I_2 = 7.0$ A. Find the magnitude and direction of the force that the magnetic field of wire 1 applies to a 1.5-m section of wire 2 when the currents have **(a)** opposite and **(b)** the same directions.

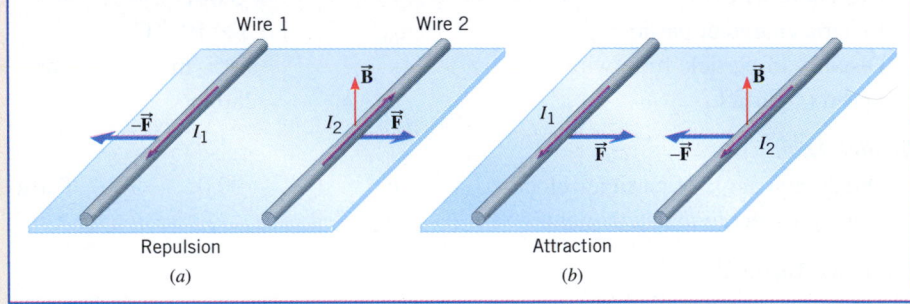

Figure 21.26 (*a*) Two long, parallel wires carrying currents I_1 and I_2 in opposite directions repel each other. (*b*) The wires attract each other when the currents are in the same direction.

Reasoning The current I_2 is situated in the magnetic field produced by the current I_1. Therefore, a length L of wire 2 experiences a magnetic force due to this field. The magnitude of the force is given by Equation 21.3 and the direction by RHR-1. Note in Figure 21.26 that the direction of the magnetic field $\vec{B}$ produced by the current I_1 is upward at the location of wire 2, as part *a* of the figure shows. The direction can be obtained using RHR-2 (thumb of right hand along I_1, curled fingers point upward at wire 2 and indicate the direction of $\vec{B}$). Thus, $\vec{B}$ is perpendicular to wire 2, and it follows that the angle between $\vec{B}$ and I_2 is $\theta = 90.0°$.

Knowns and Unknowns The following data are available:

Description	Symbol	Value	Comment
Explicit Data			
Separation between wires	r	0.065 m	Wires are parallel; see Figure 21.26.
Current in wire 1	I_1	15 A	
Current in wire 2	I_2	7.0 A	
Length of section of wire 2	L	1.5 m	
Implicit Data			
Directional angle of I_2 with respect to magnetic field produced by I_1	θ	90.0°	Wires are parallel; see *Reasoning*.
Unknown Variable			
Magnitude of magnetic force exerted on the section of wire 2	F	?	

Modeling the Problem

STEP 1 **Magnetic Force on the Section of Wire 2** The magnitude F of the magnetic force acting on the section of wire 2 is given at the right by Equation 21.3. In this expression I_2 is the current in wire 2, L is the length of the section, B is the magnitude of the magnetic field produced by I_1, and θ is the angle between I_2 and the magnetic field. We have values for I_2, L, and θ. To determine B, we turn to Step 2.

$$F = I_2 L B \sin \theta \qquad (21.3)$$

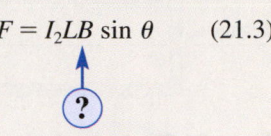

STEP 2 **Magnetic Field Produced by the Current in Wire 1** The magnitude B of the magnetic field produced by the current I_1 is given by Equation 21.5:

$$B = \frac{\mu_0 I_1}{2\pi r} \qquad (21.5)$$

where μ_0 is the permeability of free space and r is the distance from the wire. This expression can be substituted into Equation 21.3, as shown at the right.

$$F = I_2 L B \sin \theta \qquad (21.3)$$

$$B = \frac{\mu_0 I_1}{2\pi r} \qquad (21.5)$$

Solution Combining the results of each step algebraically, we find that

STEP 1   STEP 2

$$F = I_2 L B \sin \theta = I_2 L \left(\frac{\mu_0 I_1}{2\pi r} \right) \sin \theta$$

(a) The magnitude of the magnetic force on the 1.5-m section of wire 2 is

$$F = I_2 L \left(\frac{\mu_0 I_1}{2\pi r} \right) \sin \theta = (7.0 \text{ A})(1.5 \text{ m}) \frac{(4\pi \times 10^{-7} \text{ T} \cdot \text{m/A})(15 \text{ A})}{2\pi(0.065 \text{ m})} \sin 90.0°$$

$$= \boxed{4.8 \times 10^{-4} \text{ N}}$$

The direction of the magnetic force is away from wire 1, as Figure 21.26a indicates. The direction is found by using RHR-1 (fingers of the right hand extend upward along $\vec{B}$, thumb points along I_2, palm pushes in the direction of the force $\vec{F}$).

In a like manner, the current in wire 2 also creates a magnetic field that produces a force on wire 1. Reasoning similar to that above shows that a 1.5-m section of wire 1 is repelled from wire 2 with a force that also has a magnitude of 4.8×10^{-4} N. Thus, each wire generates a force on the other, and if the currents are in *opposite* directions, the wires *repel* each other. The fact that the two wires exert oppositely directed forces of equal magnitude on each other is consistent with Newton's third law, the action–reaction law.*

(b) If the current in wire 2 is reversed, as Figure 21.26b shows, wire 2 is attracted to wire 1 because the direction of the magnetic force is reversed. However, the magnitude of the force is the same as that calculated in part a. Likewise, wire 1 is attracted to wire 2. Two parallel wires carrying currents in the *same* direction *attract* each other.

Related Homework: *Problem 80*

---

*In this example the currents in the wires and the distance between them are known; therefore, the magnetic force that one wire exerts on the other can be calculated. If, instead, the force and the distance were known and the wires carried the same current, that current could be calculated. This is, in fact, the procedure used to define the ampere, which is the unit for electric current. This procedure is used because force and distance are quantities that can be measured with a high degree of precision. One ampere is the amount of electric current in each of two long, parallel wires that gives rise to a magnetic force per unit length of 2×10^{-7} N/m on each wire when the wires are separated by one meter. With the ampere defined in terms of force and distance, the coulomb is defined as the quantity of electric charge that passes a given point in one second when the current is one ampere or one coulomb per second.

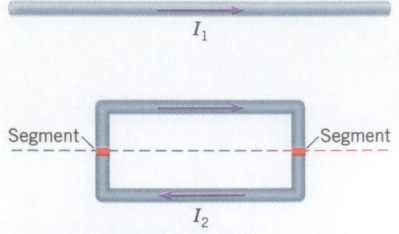

Figure 21.27 A very long, straight wire carries a current I_1, and a rectangular coil carries a current I_2. The dashed line is parallel to the wire and locates a small segment on each short side of the coil.

Conceptual Example 9 The Net Force That a Current-Carrying Wire Exerts on a Current-Carrying Coil

Figure 21.27 shows a very long, straight wire carrying a current I_1 and a rectangular coil carrying a current I_2. The wire and the coil lie in the same plane, with the wire parallel to the long sides of the rectangle. The coil is **(a)** attracted to the wire, **(b)** repelled from the wire, **(c)** neither attracted to nor repelled from the wire.

Reasoning The current in the straight wire exerts a force on each of the four sides of the coil. The net force acting on the coil is the vector sum of these four forces. To determine whether the net force is attractive, repulsive, or neither, we need to consider the directions and magnitudes of the individual forces. For each side of the rectangular coil we will first use RHR-2 to determine the direction of the magnetic field produced by the long, straight wire. Then, we will employ RHR-1 to find the direction of the magnetic force exerted on each side.

It should be noted that the magnetic forces that act on the two short sides of the rectangular coil play no role. To see why, consider a small segment of each of the short sides, located at the same distance from the straight wire, as indicated by the dashed line in Figure 21.27. Each of these segments experiences the same magnetic field from the current I_1. RHR-2 shows that this field is directed downward into the plane of the paper, so that it is perpendicular to the current I_2 in each segment. However, the directions of I_2 in the segments are opposite. As a result, RHR-1 reveals that the magnetic field from the straight wire applies a magnetic force to one segment that is opposite to the magnetic force applied to the other segment. Thus, the forces on the two short sides of the coil cancel.

Answers (b) and (c) are incorrect. There is indeed a net magnetic force that acts on the rectangular coil, but it is not a force that repels the coil from the straight wire. To see why, consider the following explanation.

Answer (a) is correct. In the long side of the coil near the wire, the current I_2 has the same direction as the current I_1, and we have just seen in Example 8 that two such currents *attract* each other. In the other long side of the coil the current I_2 has a direction opposite to that of I_1 and, according to Example 8, they repel one another. However, the attractive force is stronger than the repulsive force because the magnetic field produced by the current I_1 is stronger at shorter distances than at greater distances. Consequently, the rectangular coil is attracted to the long, straight wire.

A LOOP OF WIRE

If a current-carrying wire is bent into a circular loop, the magnetic field lines around the loop have the pattern shown in Figure 21.28a. At the *center* of a loop of radius R, the magnetic field is perpendicular to the plane of the loop and has the value $B = \mu_0 I/(2R)$, where I is the current in the loop. Often, the loop consists of N turns of wire that are wound sufficiently close together that they form a flat coil with a single radius. In this case, the magnetic fields of the individual turns add together to give a net field that is N times greater than the field of a single loop. For such a coil the magnetic field at the center is

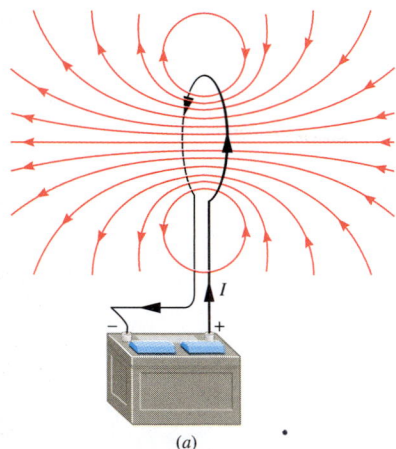

(a)

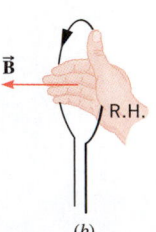

(b)

Figure 21.28 (a) The magnetic field lines in the vicinity of a current-carrying circular loop. (b) The direction of the magnetic field at the center of the loop is given by RHR-2.

Center of a circular loop
$$B = N\frac{\mu_0 I}{2R} \qquad (21.6)$$

The direction of the magnetic field at the center of the loop can be determined with the help of RHR-2. If the thumb of the right hand is pointed in the direction of the current and the curled fingers are placed at the center of the loop, as in Figure 21.28b, the fingers indicate that the magnetic field points from right to left.

Example 10 shows how the magnetic fields produced by the current in a loop of wire and the current in a long, straight wire combine to form a net magnetic field.

Example 10 Finding the Net Magnetic Field

A long, straight wire carries a current of $I_1 = 8.0$ A. As Figure 21.29a illustrates, a circular loop of wire lies immediately to the right of the straight wire. The loop has a radius of $R = 0.030$ m and carries a current of $I_2 = 2.0$ A. Assuming that the thickness of each wire is negligible, find the magnitude and direction of the net magnetic field at the center C of the loop.

Reasoning The net magnetic field at the point C is the sum of two contributions: (1) the field $\vec{B}_1$ produced by the long, straight wire, and (2) the field $\vec{B}_2$ produced by the circular loop. An application of RHR-2 shows that at point C the field $\vec{B}_1$ is directed upward, perpendicular to the plane containing the straight wire and the loop (see part b of the drawing). Similarly, RHR-2 shows that the magnetic field $\vec{B}_2$ is directed downward, opposite to the direction of $\vec{B}_1$.

Solution If we choose the upward direction in Figure 21.29b as positive, the net magnetic field at point C is

$$B = \underbrace{\frac{\mu_0 I_1}{2\pi r}}_{\substack{\text{Long, straight} \\ \text{wire}}} - \underbrace{\frac{\mu_0 I_2}{2R}}_{\substack{\text{Center of a} \\ \text{circular loop}}} = \frac{\mu_0}{2}\left(\frac{I_1}{\pi r} - \frac{I_2}{R}\right)$$

$$B = \frac{(4\pi \times 10^{-7}\ \text{T} \cdot \text{m/A})}{2}\left[\frac{8.0\ \text{A}}{\pi(0.030\ \text{m})} - \frac{2.0\ \text{A}}{0.030\ \text{m}}\right] = \boxed{1.1 \times 10^{-5}\ \text{T}}$$

Problem-solving insight: Do not confuse the formula for the magnetic field produced at the center of a circular loop with that of a very long, straight wire. The formulas are similar, differing only by a factor of π in the denominator.

The net field is positive, so it is directed upward, perpendicular to the plane.

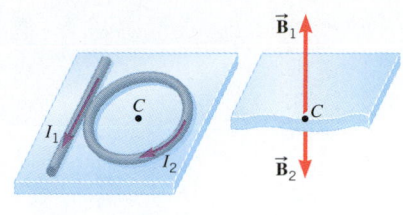

Figure 21.29 (*a*) A long, straight wire carrying a current I_1 lies next to a circular loop that carries a current I_2. (*b*) The magnetic fields at the center C of the loop produced by the straight wire ($\vec{B}_1$) and the loop ($\vec{B}_2$).

A comparison of the magnetic field lines around the current loop in Figure 21.28a with those in the vicinity of the short bar magnet in Figure 21.30a shows that the two patterns are similar. Not only are the patterns similar, but the loop itself behaves as a bar magnet with a "north pole" on one side and a "south pole" on the other side. To emphasize that the loop may be imagined to be a bar magnet, Figure 21.30b includes a "phantom" bar magnet at the center of the loop. The side of the loop that acts like a north pole can be determined with the aid of RHR-2: curl the fingers of the right hand into the shape of a half-circle, point the thumb along the current I, and place the curled fingers at the center of the loop. The fingers not only point in the direction of $\vec{B}$, but they also point toward the north pole of the loop.

Because a current-carrying loop acts like a bar magnet, two adjacent loops can be either attracted to or repelled from each other, depending on the relative directions of the currents. Figure 21.31 includes a "phantom" magnet for each loop and shows that the loops are attracted to each other when the currents are in the same direction and repelled from each other when the currents are in opposite directions. This behavior is analogous to that of the two long, straight wires discussed in Example 8 (see Figure 21.26).

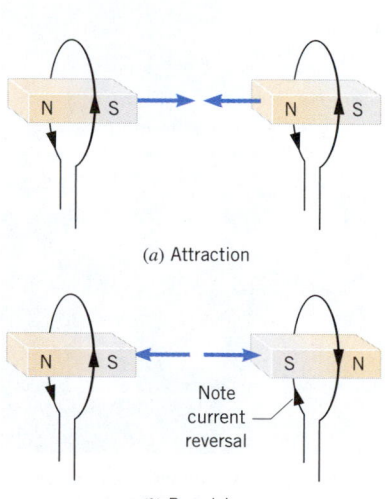

(*a*) Attraction

(*b*) Repulsion

Figure 21.31 (*a*) The two current loops attract each other if the directions of the currents are the same and (*b*) repel each other if the directions are opposite. The "phantom" magnets help explain the attraction and repulsion.

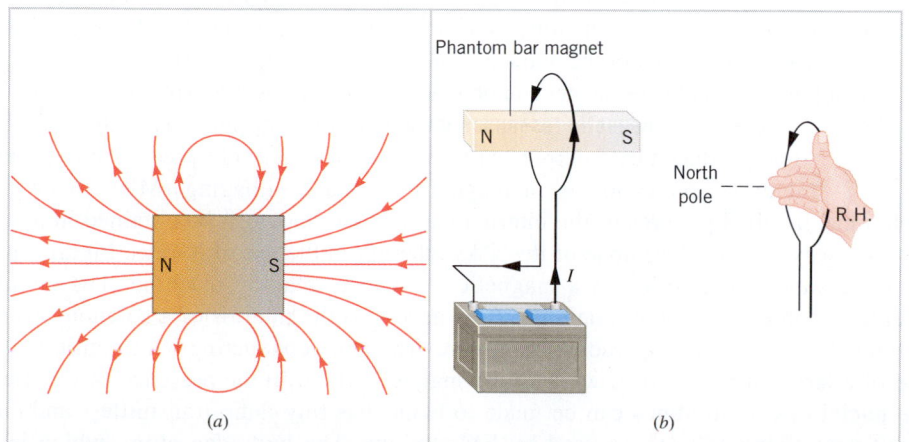

Phantom bar magnet

North pole

R.H.

(*a*)   (*b*)

Figure 21.30 (*a*) The field lines around the bar magnet resemble those around the loop in Figure 21.28a. (*b*) The current loop can be imagined to be a "phantom" bar magnet with a north pole and a south pole.

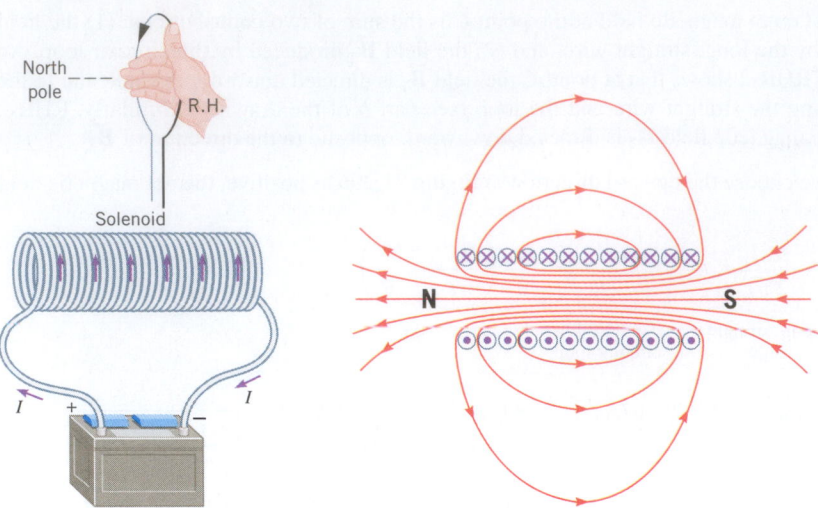

Figure 21.32 A solenoid and a cross-sectional view of it, showing the magnetic field lines and the north and south poles.

A SOLENOID

A solenoid is a long coil of wire in the shape of a helix (see Figure 21.32). If the wire is wound so the turns are packed close to each other and the solenoid is long compared to its diameter, the magnetic field lines have the appearance shown in the drawing. Notice that the field inside the solenoid and away from its ends is nearly constant in magnitude and directed parallel to the axis. The direction of the field inside the solenoid is given by RHR-2, just as it is for a circular current loop. The magnitude of the magnetic field in the interior of a long solenoid is

Interior of a long solenoid

$$B = \mu_0 n I \tag{21.7}$$

where n is the number of turns per unit length of the solenoid and I is the current. If, for example, the solenoid contains 100 turns and has a length of 0.05 m, the number of turns per unit length is $n = (100 \text{ turns})/(0.05 \text{ m}) = 2000 \text{ turns/m}$. The magnetic field outside the solenoid is not constant and is much weaker than the interior field. In fact, the magnetic field outside is nearly zero if the length of the solenoid is much greater than its diameter.

As with a single loop of wire, a solenoid can also be imagined to be a bar magnet, for the solenoid is just an array of connected current loops. And, as with a circular current loop, the location of the north pole can be determined with RHR-2. Figure 21.32 shows that the left end of the solenoid acts as a north pole, and the right end behaves as a south pole. Solenoids are often referred to as *electromagnets,* and they have several advantages over permanent magnets. For one thing, the strength of the magnetic field can be altered by changing the current and/or the number of turns per unit length. Furthermore, the north and south poles of an electromagnet can be readily switched by reversing the current.

The physics of magnetic resonance imaging (MRI). Applications of the magnetic field produced by a current-carrying solenoid are widespread. An important medical application is in magnetic resonance imaging (MRI). With this technique, detailed pictures of the internal parts of the body can be obtained in a non-invasive way that involves none of the risks inherent in the use of X-rays. Figure 21.33 shows a patient in position in a magnetic resonance imaging machine. The circular opening visible in the photograph provides access to the interior of a solenoid, which is typically made from superconducting wire. The superconducting wire facilitates the use of a large current to produce a strong magnetic field. In the presence of this field, the nuclei of certain atoms can be made to behave as tiny radio transmitters and emit radio waves similar to those used by FM stations. The hydrogen atom, which is so prevalent in the human body, can be made to behave in this fashion. The strength of the magnetic field determines where a given collection of hydrogen atoms will "broadcast" on an imaginary FM dial. With a magnetic field that has a slightly different strength at

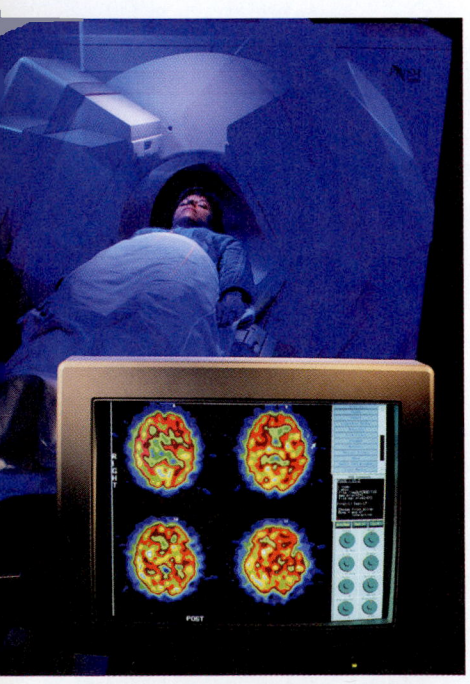

Figure 21.33 A magnetic resonance imaging (MRI) machine. The circular opening in which the patient is positioned provides access to the interior of a solenoid. MRI scans of the brain can be seen on the monitor. (David Mendelsohn/Masterfile)

different places, it is possible to associate the location on this imaginary FM dial with a physical location within the body. Computer processing of these locations produces the magnetic resonance image. When hydrogen atoms are used in this way, the image is essentially a map showing their distribution within the body. Remarkably detailed images can now be obtained, such as the one in Figure 21.34. They provide doctors with a powerful diagnostic tool that complements those from X-ray and other techniques. Surgeons can now perform operations more accurately by stepping inside specially designed MRI scanners along with the patient and seeing live images of the area into which they are cutting.

The physics of television screens and computer display monitors. Some television sets and computer display monitors use electromagnets (solenoids) to produce images by exerting magnetic forces on moving electrons. An evacuated glass tube, called a cathode-ray tube (CRT), contains an electron gun that sends a narrow beam of high-speed electrons toward the screen of the tube, as Figure 21.35a illustrates. The inner surface of the screen is covered with a phosphor coating, and when the electrons strike it, they generate a spot of visible light. This spot is called a pixel (a contraction of "picture element").

To create a black-and-white picture, the electron beam is scanned rapidly from left to right across the screen. As the beam makes each horizontal scan, the number of electrons per second striking the screen is changed by electronics controlling the electron gun, making the scan line brighter in some places and darker in others. When the beam reaches the right side of the screen, it is turned off and returned to the left side slightly below where it started (see part b of the figure). The beam is then scanned across the next line, and so on. In current TV sets, a complete picture consists of 525 scan lines (or 625 in Europe) and is formed in $\frac{1}{30}$ of a second. High-definition TV sets have about 1100 scan lines, giving a much sharper, more detailed picture.

The electron beam is deflected by a pair of electromagnets placed around the neck of the tube, between the electron gun and the screen. One electromagnet is responsible for producing the horizontal deflection of the beam and the other for the vertical deflection. For clarity, Figure 21.35a shows the net magnetic field generated by the electromagnets at one instant, and not the electromagnets themselves. The electric current in the electromagnets produces a net magnetic field that exerts a force on the moving electrons, causing their trajectories to bend and reach different points on the screen. Changing the current changes the field, so the electrons can be deflected to any point on the screen.

A color TV operates with three electron guns instead of one. Also, the single phosphor of a black-and-white TV is replaced by a large number of three-dot clusters of phosphors that glow red, green, and blue when struck by an electron beam, as indicated in Figure 21.35c. Each red, green, and blue color in a cluster is produced when electrons from one of the three guns strike the corresponding phosphor dot. The three dots are so close together that, from a normal viewing distance, they cannot be separately distinguished. Red, green, and blue are primary colors, so virtually all other colors can be created by varying the intensities of the three beams focused on a cluster.

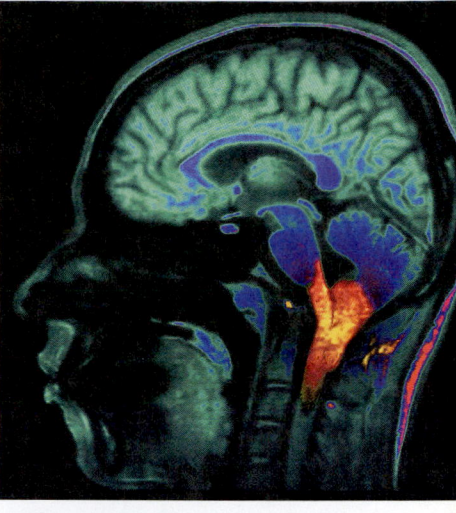

Figure 21.34 Magnetic resonance imaging provides one way to diagnose brain disorders. This image shows the brain of a patient with an Arnold-Chiari deformity. In this congenital anomaly, an abnormally elongated cerebellum and medulla oblongata (the red-orange-yellow region just to the right of the top of the spine) protrude down into the spinal canal. (© ISM/Phototake)

Figure 21.35 (a) A cathode-ray tube contains an electron gun, a magnetic field for deflecting the electron beam, and a phosphor-coated screen. A CRT color TV actually uses three guns, although only one is shown here for clarity. (b) The image is formed by scanning the electron beam across the screen. (c) The red, green, and blue phosphors of a color TV.

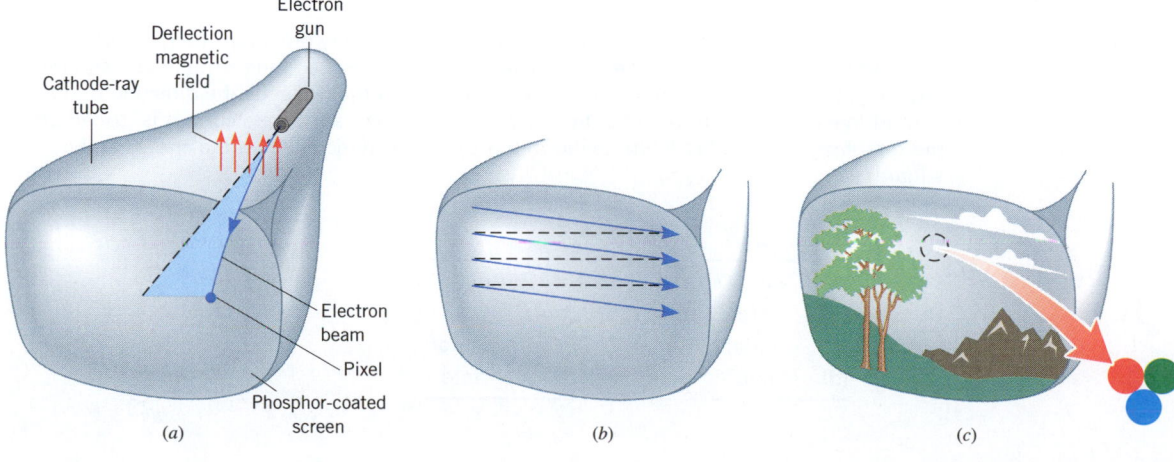

Cathode-ray tube

Deflection magnetic field

Electron gun

Electron beam

Pixel

Phosphor-coated screen

(a)

(b)

(c)

✔ **CHECK YOUR UNDERSTANDING**

(The answers are given at the end of the book.)

13. The drawing shows a conducting wire wound into a helical shape. The helix acts like a spring and expands back toward its original shape after the coils are squeezed together and released. The bottom end of the wire just barely touches the mercury (a good electrical conductor) in the cup. After the switch is closed, current in the circuit causes the light bulb to glow. Does the bulb **(a)** repeatedly turn on and off like a turn signal on a car, **(b)** glow continually, or **(c)** glow briefly and then go out?

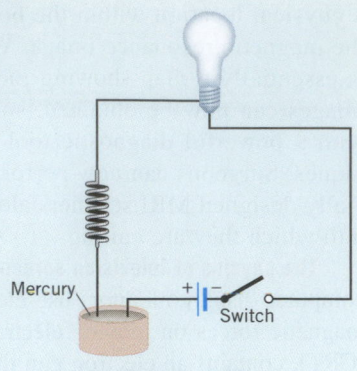
Question 13

14. For each electromagnet at the left in the drawing, will it be attracted to or repelled from the permanent magnet immediately to its right?

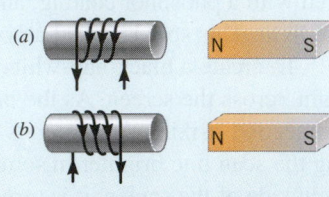

15. For each electromagnet at the left in the drawing, will it be attracted to or repelled from the electromagnet immediately to its right?

16. Refer to Figure 21.5. If the earth's magnetism is assumed to originate from a large circular loop of current within the earth, then the plane of the current loop must be oriented _____ to the earth's magnetic axis, and the direction of the current around the loop (when looking down on the loop from the north magnetic pole) is _____. **(a)** parallel, clockwise **(b)** parallel, counterclockwise **(c)** perpendicular, clockwise **(d)** perpendicular, counterclockwise

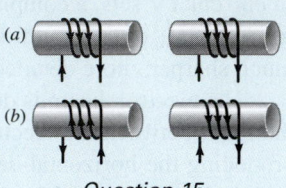
Question 14

17. The drawing shows an end-on view of three parallel wires that are perpendicular to the plane of the paper. In two of the wires the current is directed into the paper, while in the remaining wire the current is directed out of the paper. The two outermost wires are held rigidly in place. Which way will the middle wire move? **(a)** To the left **(b)** To the right **(c)** It will not move at all.

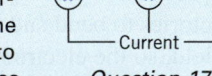
Question 15

18. In Figure 21.26, assume that the current I_1 is larger than the current I_2. In parts *a* and *b*, decide whether there are places at which the total magnetic field is zero. State whether these places are located to the left of both wires, between the wires, or to the right of both wires.

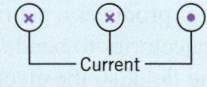
Question 17

19. Each of the four drawings shows the same three concentric loops of wire. The currents in the loops have the same magnitude *I* and have the directions shown. Rank the magnitude of the net magnetic field produced at the center of each of the four drawings, largest to smallest.

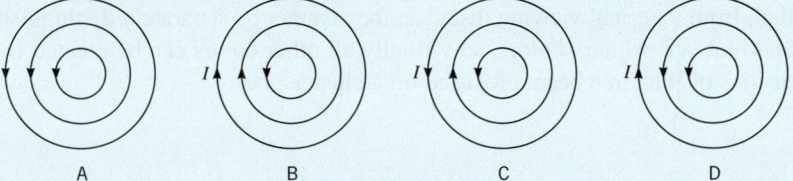

20. There are four wires viewed end-on in the drawing. They are long, straight, and perpendicular to the plane of the paper. Their cross sections lie at the corners of a square. The magnitude of the current in each wire is the same. What must be the direction of the current in each wire (into or out of the page), so that when *any* single current is turned off, the total magnetic field at *P* (the center of the square) is directed toward a corner of the square?

	Wire 1	Wire 2	Wire 3	Wire 4
(a)	out of	out of	into	out of
(b)	into	out of	into	into
(c)	out of	into	out of	out of
(d)	into	into	into	into

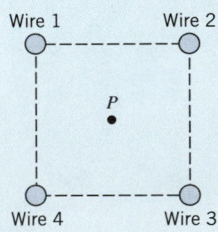

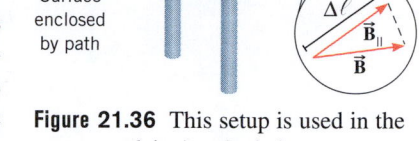

21.8 AMPÈRE'S LAW

We have seen that an electric current creates a magnetic field. However, the magnitude and direction of the field at any point in space depends on the specific geometry of the wire carrying the current. For instance, distinctly different magnetic fields surround a long, straight wire, a circular loop of wire, and a solenoid. Although different, each of these fields can be obtained from a general law known as *Ampère's law,* which is valid for a wire of any geometrical shape. Ampère's law specifies the relationship between an electric current and the magnetic field that it creates.

To see how Ampère's law is stated, consider Figure 21.36, which shows two wires carrying currents I_1 and I_2. In general, there may be any number of currents. Around the wires we construct an arbitrarily shaped but closed path. This path encloses a surface and is constructed from a large number of short segments, each of length $\Delta\ell$. Ampère's law deals with the product of $\Delta\ell$ and $B_\parallel$ for each segment, where $B_\parallel$ is the component of the magnetic field that is *parallel* to $\Delta\ell$ (see the blow-up view in the drawing). For magnetic fields that do not change as time passes, the law states that the sum of all the $B_\parallel \Delta\ell$ terms is proportional to the net current I passing through the surface enclosed by the path. For the specific example in Figure 21.36, we see that $I = I_1 + I_2$. Ampère's law is stated in equation form as follows:

Figure 21.36 This setup is used in the text to explain Ampère's law.

AMPÈRE'S LAW FOR STATIC MAGNETIC FIELDS

For any current geometry that produces a magnetic field that does not change in time,

$$\Sigma B_\parallel \Delta\ell = \mu_0 I \qquad (21.8)$$

where $\Delta\ell$ is a small segment of length along a closed path of arbitrary shape around the current, $B_\parallel$ is the component of the magnetic field parallel to $\Delta\ell$, I is the net current passing through the surface bounded by the path, and μ_0 is the permeability of free space. The symbol Σ indicates that the sum of all $B_\parallel \Delta\ell$ terms must be taken around the closed path.

To illustrate the use of Ampère's law, we apply it in Example 11 to the special case of the current in a long, straight wire and show that it leads to the proper expression for the magnetic field.

Example 11   An Infinitely Long, Straight, Current-Carrying Wire

Use Ampère's law to obtain the magnetic field produced by the current in an infinitely long, straight wire.

Reasoning Figure 21.23a shows that compass needles point in a circular pattern around the wire, so we know that the magnetic field lines are circular. Therefore, it is convenient to use a circular path of radius r when applying Ampère's law, as Figure 21.37 indicates.

Solution Along the circular path in Figure 21.37, the magnetic field is everywhere parallel to $\Delta\ell$ and has a constant magnitude, since each point is at the same distance from the wire. Thus, $B_\parallel = B$ and, according to Ampère's law, we have

$$\Sigma B_\parallel \Delta\ell = B\,(\Sigma \Delta\ell) = \mu_0 I$$

However, $\Sigma\Delta\ell$ is just the circumference $2\pi r$ of the circle, so Ampère's law reduces to

$$B\,(\Sigma \Delta\ell) = B\,(2\pi r) = \mu_0 I$$

Dividing both sides by $2\pi r$ shows that $\boxed{B = \mu_0 I/(2\pi r)}$, as given earlier in Equation 21.5.

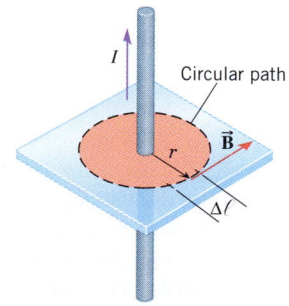

Figure 21.37 Example 11 uses Ampère's law to find the magnetic field in the vicinity of this long, straight, current-carrying wire.

MAGNETIC MATERIALS

FERROMAGNETISM

The similarity between the magnetic field lines in the neighborhood of a bar magnet and those around a current loop suggests that the magnetism in each case arises from a common cause. The field that surrounds the loop is created by the charges moving in the wire. The magnetic field around a bar magnet is also due to the motion of charges, but the motion is not that of a bulk current through the magnetic material. Instead, the motion responsible for the magnetism is that of the electrons within the atoms of the material.

The magnetism produced by electrons within an atom can arise from two motions. First, each electron orbiting the nucleus behaves like an atomic-sized loop of current that generates a small magnetic field, similar to the field created by the current loop in Figure 21.28*a*. Second, each electron possesses a spin that also gives rise to a magnetic field. The net magnetic field created by the electrons within an atom is due to the combined fields created by their orbital and spin motions.

In most substances the magnetism produced at the atomic level tends to cancel out, with the result that the substance is nonmagnetic overall. However, there are some materials, known as *ferromagnetic materials,* in which the cancellation does not occur for groups of approximately 10^{16}–10^{19} neighboring atoms, because they have electron spins that are naturally aligned parallel to each other. This alignment results from a special type of quantum mechanical* interaction between the spins. The result of the interaction is a small but highly magnetized region of about 0.01 to 0.1 mm in size, depending on the nature of the material; this region is called a *magnetic domain.* Each domain behaves as a small magnet with its own north and south poles. Common ferromagnetic materials are iron, nickel, cobalt, chromium dioxide, and alnico (an *al*uminum–*ni*ckel–*co*balt alloy).

INDUCED MAGNETISM

Often the magnetic domains in a ferromagnetic material are arranged randomly, as Figure 21.38*a* illustrates for a piece of iron. In such a situation, the magnetic fields of the domains cancel each other, so the iron displays little, if any, overall magnetism. However, an unmagnetized piece of iron can be magnetized by placing it in an external magnetic field provided by a permanent magnet or an electromagnet. The external magnetic field penetrates the unmagnetized iron and *induces* (or brings about) a state of magnetism in the iron by causing two effects. Those domains whose magnetism is parallel or nearly parallel to the external magnetic field grow in size at the expense of other domains that are not so oriented. In Figure 21.38*b* the growing domains are colored gold. In addition, the magnetic alignment of some domains may rotate and become more oriented in the direction of the external field. The resulting preferred alignment of the domains gives the iron an overall magnetism, so the iron behaves like a magnet with associated north and south poles. In some types of ferromagnetic materials, such as the chromium dioxide used in cassette tapes, the domains remain aligned for the most part when the external magnetic field is removed, and the material thus becomes permanently magnetized.

Figure 21.38 (*a*) Each magnetic domain is a highly magnetized region that behaves like a small magnet (represented by an arrow whose head indicates a north pole). An unmagnetized piece of iron consists of many domains that are randomly aligned. The size of each domain is exaggerated for clarity. (*b*) The external magnetic field of the permanent magnet causes those domains that are parallel or nearly parallel to the field to grow in size (shown in gold color).

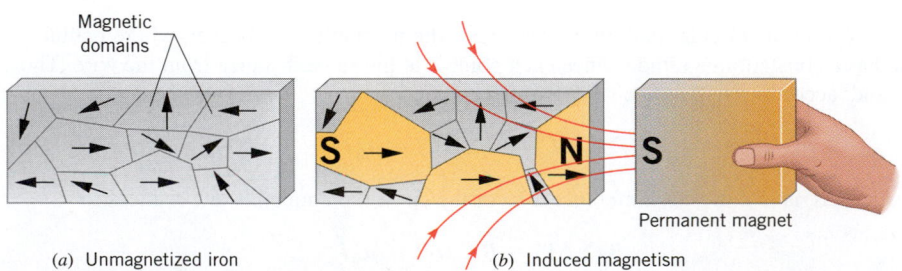

*The branch of physics called quantum mechanics is mentioned in Section 29.5. A detailed discussion of it is beyond the scope of this book.

The magnetism induced in a ferromagnetic material can be surprisingly large, even in the presence of a weak external field. For instance, it is not unusual for the induced magnetic field to be a hundred to a thousand times stronger than the external field that causes the alignment. For this reason, high-field electromagnets are constructed by wrapping the current-carrying wire around a solid core made from iron or other ferromagnetic material.

Induced magnetism explains why a permanent magnet sticks to a refrigerator door and why an electromagnet can pick up scrap iron at a junkyard. Notice in Figure 21.38*b* that there is a north pole at the end of the iron that is closest to the south pole of the permanent magnet. The net result is that the two opposite poles give rise to an attraction between the iron and the permanent magnet. Conversely, the north pole of the permanent magnet would also attract the piece of iron by inducing a south pole in the nearest side of the iron. In non-ferromagnetic materials, such as aluminum and copper, the formation of magnetic domains does not occur, so magnetism cannot be induced into these substances. Consequently, magnets do not stick to aluminum cans.

The physics of detecting fingerprints. The attraction that induced magnetism creates between a permanent magnet and a ferromagnetic material is used in crime-scene investigations, where powder is dusted onto surfaces to make fingerprints visible. Magnetic fingerprint powder allows investigators to recover evidence from surfaces like the neoprene glove in Figure 21.39, which are very difficult to examine without damaging the prints in the process. Brushing excess conventional powder away ruins the delicate ridges of the pattern that allow the print to be identified reliably. Magnetic fingerprint powder, however, consists of tiny iron flakes coated with an organic material that allows them to stick to the greasy residue in the print. A permanent magnet eliminates the need for brushing away excess powder by creating induced magnetism in the iron and pulling away the powder not stuck directly to the greasy residue.

Figure 21.39 Magnetic fingerprint powder has been used to reveal fingerprints on this neoprene glove. The powder consists of tiny iron flakes coated with an organic material that enables them to stick to the greasy residue in the print. Because of induced magnetism, a permanent magnet can be used to remove (without brushing and possible damage to the print) any excess powder that would obscure the evidence. (James King-Holmes/SPL./Photo Researchers, Inc.)

MAGNETIC TAPE RECORDING

The physics of magnetic tape recording. The process of magnetic tape recording uses induced magnetism, as Figure 21.40 illustrates. The weak electrical signal from a microphone is routed to an amplifier where it is amplified. The current from the output of the amplifier is then sent to the recording head, which is a coil of wire wrapped around an iron core. The iron core has the approximate shape of a horseshoe with a small gap between the two ends. The ferromagnetic iron substantially enhances the magnetic field produced by the current in the wire.

When there is a current in the coil, the recording head becomes an electromagnet with a north pole at one end and a south pole at the other end. The magnetic field lines pass through the iron core and cross the gap. Within the gap, the lines are directed from the north pole to the south pole. Some of the field lines in the gap "bow outward," as Figure 21.40 indicates; the bowed region of the magnetic field is called the *fringe field*. The fringe field penetrates the magnetic coating on the tape and induces magnetism in the coating. This induced magnetism is retained when the tape leaves the vicinity of the recording head and, thus, provides a means for storing audio information. Audio information is stored, because at any instant in time the way in which the tape is magnetized depends on the amount and direction of current in the recording head. The current, in turn, depends on the sound picked up by the microphone, so that changes in the sound that occur from moment to moment are preserved as changes in the tape's induced magnetism.

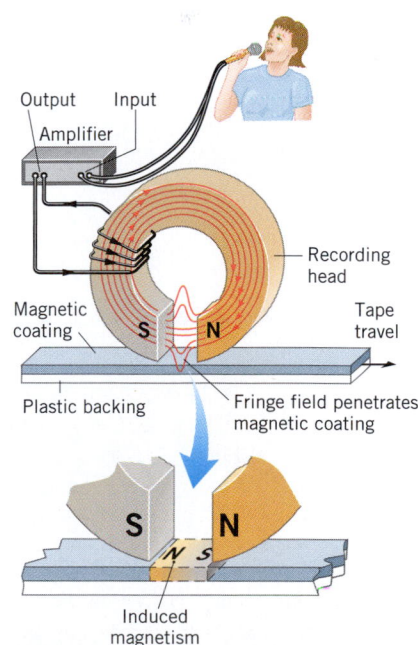

Figure 21.40 The magnetic fringe field of the recording head penetrates the magnetic coating on the tape and magnetizes it.

MAGLEV TRAINS

The physics of a magnetically levitated train. A magnetically levitated train—maglev, for short—uses forces that arise from induced magnetism to levitate or float above a guideway. Since it rides a few centimeters above the guideway, a maglev does not need wheels. Freed from friction with the guideway, the train can achieve significantly greater speeds than do conventional trains. For example, the Transrapid maglev in Figure 21.41 has achieved speeds of 110 m/s (250 mph).

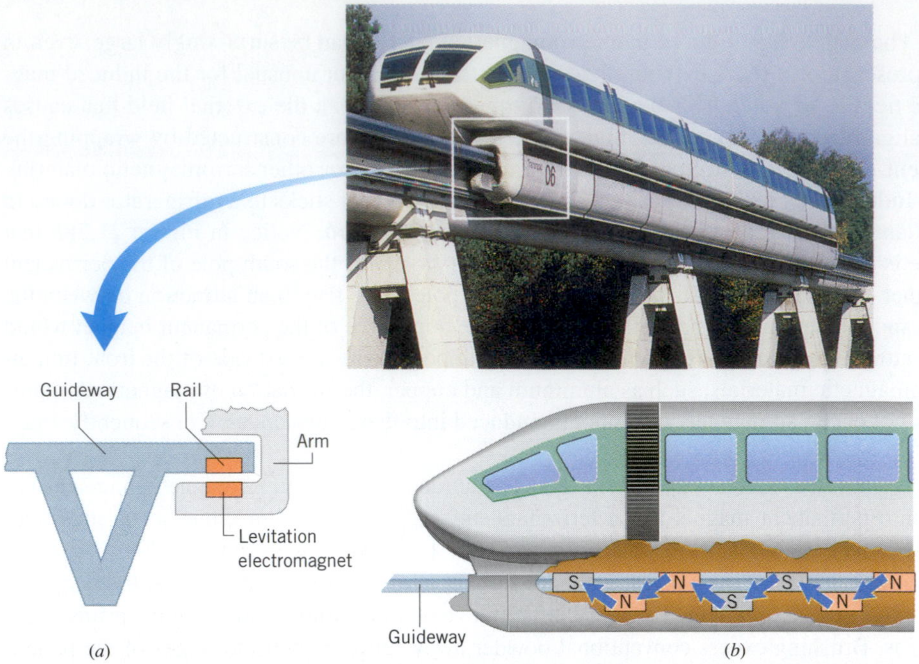

Figure 21.41 (*a*) The Transrapid maglev (a German train) has achieved speeds of 110 m/s (250 mph). The levitation electromagnets are drawn up toward the rail in the guideway, levitating the train. (Courtesy Deutsche Bundesbahn/Transrapid) (*b*) The magnetic propulsion system.

Figure 21.41*a* shows that the Transrapid maglev achieves levitation with electromagnets mounted on arms that extend around and under the guideway. When a current is sent to an electromagnet, the resulting magnetic field creates induced magnetism in a rail mounted in the guideway. The upward attractive force from the induced magnetism is balanced by the weight of the train, so the train moves without touching the rail or the guideway.

Magnetic levitation only lifts the train and does not move it forward. Figure 21.41*b* illustrates how magnetic propulsion is achieved. In addition to the levitation electromagnets, propulsion electromagnets are also placed along the guideway. By controlling the direction of the currents in the train and guideway electromagnets, it is possible to create an unlike pole in the guideway just ahead of each electromagnet on the train and a like pole just behind. Each electromagnet on the train is thus both pulled and pushed forward by electromagnets in the guideway. By adjusting the timing of the like and unlike poles in the guideway, the speed of the train can be adjusted. Reversing the poles in the guideway electromagnets serves to brake the train.

✓ CHECK YOUR UNDERSTANDING

(*The answers are given at the end of the book.*)

21. In a TV commercial that advertises a soda pop, a strong electromagnet picks up a delivery truck carrying cans of the soft drink. The picture switches to the interior of the truck, where cans are seen to fly upward and stick to the roof just beneath the electromagnet. Are these cans made entirely of aluminum?

22. Suppose that you have two bars. Bar 1 is a permanent magnet and bar 2 is not a magnet, but is made from a ferromagnetic material like iron. A third bar (bar 3), which is known to be a permanent magnet, is brought close to one end of bar 1 and then to one end of bar 2. Which one of the following statements is true?   **(a)** Bars 1 and 3 will either be attracted to or repelled from each other, while bars 2 and 3 will always be repelled from each other.   **(b)** Bars 1 and 3 will either be attracted to or repelled from each other, while bars 2 and 3 will always be attracted to each other.   **(c)** Bars 1 and 3 will always be repelled from each other, while bars 2 and 3 will either be attracted to or repelled from each other.   **(d)** Bars 1 and 3 will always be attracted to each other, while bars 2 and 3 will either be attracted to or repelled from each other.

21.10 CONCEPTS & CALCULATIONS

Both magnetic fields and electric fields can apply forces to an electric charge. However, there are distinct differences in the way the two types of fields apply their forces. This chapter discusses the magnetic field, and Chapter 18 deals with the electric field. The following example serves to review how the two types of fields behave.

Concepts & Calculations Example 12 Magnetic and Electric Fields

Figure 21.42 shows a particle that carries a charge of $q_0 = -2.80 \times 10^{-6}$ C. It is moving along the $+y$ axis at a speed of $v = 4.80 \times 10^6$ m/s. A magnetic field of magnitude 3.35×10^{-5} T is directed along the $+z$ axis, and an electric field of magnitude 123 N/C points along the $-x$ axis. Determine the magnitude and direction of the net force that acts on the particle.

Concept Questions and Answers What is the net force?

Answer The net force is the vector sum of all the forces acting on the particle, which in this case are the magnetic force and the electric force.

How do you determine the direction of the magnetic force acting on the negative charge?

Answer First, remember that the direction for either a positive or negative charge is perpendicular to the plane formed by the magnetic field vector $\vec{B}$ and the velocity vector $\vec{v}$. Then, apply Right-Hand Rule No. 1 (RHR-1) to find the direction of the force as if the charge were positive. Finally, reverse the direction indicated by RHR-1 to determine the direction of the force on the negative charge.

How do you determine the direction of the electric force acting on the negative charge?

Answer The direction for either a positive or negative charge is along the line of the electric field. The electric force on a positive charge is in the same direction as the electric field, while the force on a negative charge is in the direction opposite to the electric field.

Does the fact that the charge is moving affect the values of the magnetic and electric forces?

Answer The fact that the charge is moving affects the value of the magnetic force, for without motion, there is no magnetic force. The electric force, in contrast, has the same value whether or not the charge is moving.

Solution An application of RHR-1 (see Figure 21.8) reveals that if the charge were positive, the magnetic force would point along the $+x$ axis. The charge is negative, however, so the magnetic force points along the $-x$ axis. According to Equation 21.1, the magnitude of the magnetic force is $F_{magnetic} = |q_0|vB \sin \theta$, where $|q_0|$ is the magnitude of the charge. Since the velocity and the magnetic field are perpendicular, $\theta = 90°$, and we have $F_{magnetic} = |q_0|vB$.

The electric field points along the $-x$ axis and would apply a force in that same direction if the charge were positive. The charge is negative, however, so the electric force points in the opposite direction, or along the $+x$ axis. According to Equation 18.2, the magnitude of the electric force is $F_{electric} = |q_0|E$.

Using plus and minus signs to take into account the different directions of the magnetic and electric forces, we find that the net force is

$$\Sigma F = +F_{electric} - F_{magnetic} = +|q_0|E - |q_0|vB$$
$$= +|-2.80 \times 10^{-6} \text{ C}|(123 \text{ N/C}) - |-2.80 \times 10^{-6} \text{ C}|(4.80 \times 10^6 \text{ m/s})(3.35 \times 10^{-5} \text{ T})$$
$$= \boxed{-1.06 \times 10^{-4} \text{ N}}$$

The fact that the answer is negative means that the net force points along the $-x$ axis.

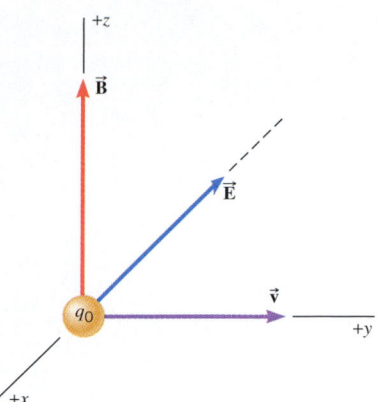
Figure 21.42 The net force acting on the moving charge in this drawing is calculated in Example 12.

Problem-solving insight
The direction of the magnetic force exerted on a negative charge is opposite to that exerted on a positive charge, assuming both charges are moving in the same direction in the same magnetic field.

A number of forces can act simultaneously on a rigid object, and some of them can produce torques, as Chapter 9 discusses. If, as a result of the forces and torques, the object has no acceleration of any kind, it is in equilibrium. Section 9.2 discusses the equilibrium of rigid objects. The next example illustrates that the magnetic force can be one of the forces keeping an object in equilibrium.

 Concepts & Calculations Example 13 Equilibrium

A 125-turn rectangular coil of wire is hung from one arm of a balance, as Figure 21.43 shows. With the magnetic field $\vec{B}$ turned off, a mass M is added to the pan on the other arm to balance the weight of the coil. When the constant 0.200-T magnetic field is turned on and there is a current of 8.50 A in the coil, how much additional mass m must be added to regain the balance?

Concept Questions and Answers In a balanced, or equilibrium, condition the device has no angular acceleration. What does this imply about the net torque acting on the device?

Answer According to Newton's second law as applied to rotational motion (Equation 9.7), when the angular acceleration is zero, the net torque is zero. Counterclockwise torques must balance clockwise torques for an object that is in equilibrium.

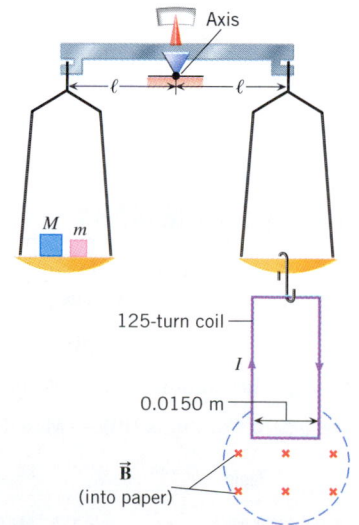
Figure 21.43 This setup is the focus of Example 13.

What is a torque?

Answer According to Equation 9.1, the torque produced by a force is the magnitude of the force times the lever arm of the force. The lever arm is the perpendicular distance between the line of action of the force and the axis of rotation. Counterclockwise torques are positive, and clockwise torques are negative.

In calculating the torques acting on an object in equilibrium, where do you locate the axis of rotation?

Answer The location of the axis is arbitrary because, if there is no angular acceleration, the torques must add up to zero with respect to any axis whatsoever. Choose an axis that simplifies your calculations.

Solution In a condition of equilibrium the sum of the torques acting on the device is zero. For the purpose of calculating torques, we use an axis that is perpendicular to the page at the knife-edge supporting the balance arms (see the drawing). The forces that can produce torques are the weights of the pans, the masses in the left pan, the coil hanging from the right pan, and the magnetic force on the coil. The weight of each pan is W_{pan}, and the weight of the coil is W_{coil}. The weights of the masses M and m are Mg and mg, where g is the magnitude of the acceleration due to gravity. The lever arm for each of these forces is ℓ (see the drawing). When the magnetic field is turned off and the mass m is not present, the sum of the torques is

Field off $$\Sigma\tau = \underbrace{(+W_{pan}\ell + Mg\ell)}_{\text{Left pan}} + \underbrace{(-W_{pan}\ell - W_{coil}\ell)}_{\text{Right pan}} = Mg\ell - W_{coil}\ell = 0$$

When the field is turned on, a magnetic force is applied to the 0.0150-m segment of the coil that is in the field. Since the magnetic field points into the page and the current is from right to left, an application of RHR-1 reveals that the magnetic force points downward. According to Equation 21.3, the magnitude of the magnetic force is $F = ILB \sin \theta$, where I is the current and L is the length of the wire in the field. Since the current and the field are perpendicular, $\theta = 90°$, and we have $F = ILB$. The sum of the torques with the field turned on and the mass m added to the left pan is

Field on $$\Sigma\tau = \underbrace{(+W_{pan}\ell + Mg\ell + mg\ell)}_{\text{Left pan}} + \underbrace{(-W_{pan}\ell - W_{coil}\ell - ILB\ell)}_{\text{Right pan}}$$

$$= Mg\ell + mg\ell - W_{coil}\ell - ILB\ell = 0$$

From the field-off equation, it follows that $Mg\ell - W_{coil}\ell = 0$. With this substitution, the field-on equation simplifies to the following:

$$mg\ell - ILB\ell = 0$$

Solving this expression for the additional mass m gives

$$m = \frac{ILB}{g} = \frac{(8.50 \text{ A})[125(0.0150 \text{ m})](0.200 \text{ T})}{9.80 \text{ m/s}^2} = \boxed{0.325 \text{ kg}}$$

In this calculation we have used the fact that the coil has 125 turns, so the length of the wire in the field is 125 (0.0150 m).

CONCEPT SUMMARY

If you need more help with a concept, use the Learning Aids noted next to the discussion or equation. Examples (**Ex.**) are in the text of this chapter. Go to **www.wiley.com/college/cutnell** for the following Learning Aids:

Interactive LearningWare (ILW) — Additional examples solved in a five-step interactive format.

Concept Simulations (CS) — Animated text figures or animations of important concepts.

Interactive Solutions (IS) — Models for certain types of problems in the chapter homework. The calculations are carried out interactively.

Topic	Discussion	Learning Aids
North and south poles	**21.1 MAGNETIC FIELDS** A magnet has a north pole and a south pole. The north pole is the end that points toward the north magnetic pole of the earth when the magnet is freely suspended. Like magnetic poles repel each other, and unlike poles attract each other.	

Topic	Discussion	Learning Aids
Direction of the magnetic field	A magnetic field exists in the space around a magnet. The magnetic field is a vector whose direction at any point is the direction indicated by the north pole of a small compass needle placed at that point.	
Magnetic field lines	As an aid in visualizing the magnetic field, magnetic field lines are drawn in the vicinity of a magnet. The lines appear to originate from the north pole and end on the south pole. The magnetic field at any point in space is tangent to the magnetic field line at the point. Furthermore, the strength of the magnetic field is proportional to the number of lines per unit area that passes through a surface oriented perpendicular to the lines.	
Magnetic force (direction)	**21.2 THE FORCE THAT A MAGNETIC FIELD EXERTS ON A MOVING CHARGE** The direction of the magnetic force acting on a charge moving with a velocity $\vec{v}$ in a magnetic field $\vec{B}$ is perpendicular to both $\vec{v}$ and $\vec{B}$. For a positive charge the direction can be determined with the aid of Right-Hand Rule No. 1 (see below). The magnetic force on a moving negative charge is opposite to the force on a moving positive charge.	
Right-Hand Rule No. 1	Extend the right hand so the fingers point along the direction of the magnetic field $\vec{B}$ and the thumb points along the velocity $\vec{v}$ of the charge. The palm of the hand then faces in the direction of the magnetic force $\vec{F}$ that acts on a positive charge.	
	The magnitude B of the magnetic field at any point in space is defined as	
Magnetic field (magnitude)	$$B = \frac{F}{\lvert q_0 \rvert v \sin \theta} \qquad (21.1)$$	**Ex. 1**
The tesla and the gauss	where F is the magnitude of the magnetic force on a test charge, $\lvert q_0 \rvert$ is the magnitude of the test charge, and v is the magnitude of the charge's velocity, which makes an angle θ with the direction of the magnetic field. The SI unit for the magnetic field is the tesla (T). Another, smaller unit for the magnetic field is the gauss; 1 gauss $= 10^{-4}$ tesla. The gauss is not an SI unit.	
	21.3 THE MOTION OF A CHARGED PARTICLE IN A MAGNETIC FIELD When a charged particle moves in a region that contains both magnetic and electric fields, the net force on the particle is the vector sum of the magnetic and electric forces.	**Ex. 2, 12**
A constant magnetic force does no work	A magnetic force does no work on a particle, because the direction of the force is always perpendicular to the motion of the particle. Being unable to do work, the magnetic force cannot change the kinetic energy, and hence the speed, of the particle; however, the magnetic force does change the direction in which the particle moves.	
	When a particle of charge q (magnitude $= \lvert q \rvert$) and mass m moves with speed v perpendicular to a uniform magnetic field of magnitude B, the magnetic force causes the particle to move on a circular path of radius	
Radius of circular path	$$r = \frac{mv}{\lvert q \rvert B} \qquad (21.2)$$	**CS 21.1** **Ex. 3, 4** **ILW 21.1** **IS 21.23**
	21.4 THE MASS SPECTROMETER The mass spectrometer is an instrument for measuring the abundance of ionized atoms or molecules that have different masses. The atoms or molecules are ionized $(+e)$, accelerated to a speed v by a potential difference V, and sent into a uniform magnetic field of magnitude B. The magnetic field causes the particles (each with a mass m) to move on a circular path of radius r. The relation between m and B is	
Relation between mass and magnetic field	$$m = \left(\frac{er^2}{2V} \right) B^2$$	
	21.5 THE FORCE ON A CURRENT IN A MAGNETIC FIELD An electric current, being composed of moving charges, can experience a magnetic force when placed in a magnetic field of magnitude B. For a straight wire that has a length L and carries a current I, the magnetic force has a magnitude of	
Force on a current (magnitude)	$$F = ILB \sin \theta \qquad (21.3)$$	**Ex. 5, 13** **ILW 21.2** **IS 21.39**
Force on a current (direction)	where θ is the angle between the directions of the current and the magnetic field. The direction of the force is perpendicular to both the current and the magnetic field and is given by Right-Hand Rule No. 1.	
	21.6 THE TORQUE ON A CURRENT-CARRYING COIL Magnetic forces can exert a torque on a current-carrying loop of wire and thus cause the loop to rotate. When a current I exists in a coil of wire with N turns, each of area A, in the presence of a magnetic field of magnitude B, the coil experiences a net torque of magnitude	
Torque on a current-carrying coil	$$\tau = NIAB \sin \phi \qquad (21.4)$$	**Ex. 6** **IS 21.51**
	where ϕ is the angle between the direction of the magnetic field and the normal to the plane of the coil.	

Topic	Discussion	Learning Aids
Magnetic moment	The quantity *NIA* is known as the magnetic moment of the coil.	

21.7 MAGNETIC FIELDS PRODUCED BY CURRENTS An electric current produces a magnetic field, with different current geometries giving rise to different field patterns. For an infinitely long, straight wire, the magnetic field lines are circles centered on the wire, and their direction is given by Right-Hand Rule No. 2 (see below). The magnitude of the magnetic field at a radial distance *r* from the wire is

Long, straight wire	$$B = \frac{\mu_0 I}{2\pi r} \qquad (21.5)$$	Ex. 7, 8, 9 ILW 21.3 CS 21.2
Permeability of free space	where *I* is the current in the wire and μ_0 is a constant known as the permeability of free space ($\mu_0 = 4\pi \times 10^{-7}$ T·m/A).	
Right-Hand Rule No. 2	Curl the fingers of the right hand into the shape of a half-circle. Point the thumb in the direction of the conventional current *I*, and the tips of the fingers will point in the direction of the magnetic field $\vec{B}$.	

The magnitude of the magnetic field at the center of a flat circular loop consisting of *N* turns, each of radius *R*, is

Center of a circular loop	$$B = N\frac{\mu_0 I}{2R} \qquad (21.6)$$	CS 21.3 Ex. 10 IS 21.61

The loop has associated with it a north pole on one side and a south pole on the other side. The side of the loop that behaves like a north pole can be predicted by using Right-Hand Rule No. 2.

A solenoid is a coil of wire wound in the shape of a helix. Inside a long solenoid the magnetic field is nearly constant and has a magnitude of

Interior of a solenoid	$$B = \mu_0 n I \qquad (21.7)$$	CS 21.4

where *n* is the number of turns per unit length of the solenoid. One end of the solenoid behaves like a north pole, and the other end like a south pole. The end that is the north pole can be predicted by using Right-Hand Rule No. 2.

21.8 AMPÈRE'S LAW Ampère's law specifies the relationship between a current and its associated magnetic field. For any current geometry that produces a magnetic field that does not change in time, Ampère's law states that

Ampère's law	$$\Sigma B_{\parallel} \Delta \ell = \mu_0 I \qquad (21.8)$$	Ex. 11 IS 21.70

where $\Delta \ell$ is a small segment of length along a closed path of arbitrary shape around the current, $B_{\parallel}$ is the component of the magnetic field parallel to $\Delta \ell$, *I* is the net current passing through the surface bounded by the path, and μ_0 is the permeability of free space. The symbol Σ indicates that the sum of all $B_{\parallel} \Delta \ell$ terms must be taken around the closed path.

Ferromagnetic materials	**21.9 MAGNETIC MATERIALS** Ferromagnetic materials, such as iron, are made up of tiny regions called magnetic domains, each of which behaves as a small magnet. In an unmagnetized ferromagnetic material, the domains are randomly aligned. In a permanent magnet, many of the domains are aligned, and a high degree of magnetism results. An unmagnetized ferromagnetic material can be induced into becoming magnetized by placing it in an external magnetic field.

FOCUS ON CONCEPTS

Note to Instructors: The numbering of the questions shown here reflects the fact that they are only a representative subset of the total number that are available online. However, all of the questions are available for assignment via an online homework management program such as WileyPLUS or WebAssign.

Section 21.2 The Force That a Magnetic Field Exerts on a Moving Charge

2. At a location near the equator, the earth's magnetic field is horizontal and points north. An electron is moving vertically upward from the ground. What is the direction of the magnetic force that acts on the electron? **(a)** North **(b)** East **(c)** South **(d)** West **(e)** The magnetic force is zero.

3. The drawing shows four situations in which a positively charged particle is moving with a velocity $\vec{v}$ through a magnetic field $\vec{B}$. In each case, the magnetic field is directed out of the page toward you,

and the velocity is directed to the right. In only one of these drawings is the magnetic force $\vec{F}$ physically reasonable. Which drawing is it? **(a)** 1 **(b)** 2 **(c)** 3 **(d)** 4

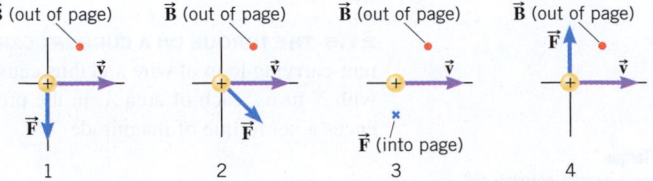

Section 21.3   The Motion of a Charged Particle in a Magnetic Field

6. Three particles are moving perpendicular to a uniform magnetic field and travel on circular paths (see the drawing). The particles have the same mass and speed. List the particles in order of their charge magnitude, largest to smallest.   **(a)** 3, 2, 1   **(b)** 3, 1, 2   **(c)** 2, 3, 1   **(d)** 1, 3, 2   **(e)** 1, 2, 3

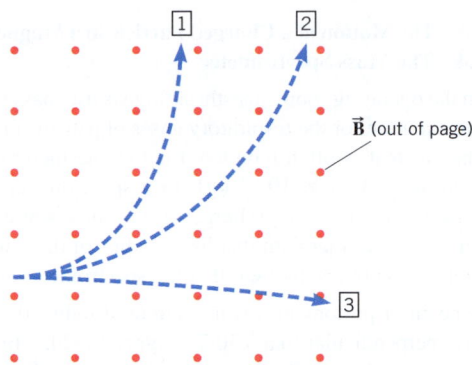

8. The drawing shows the circular paths of an electron and a proton. These particles have the same charge magnitudes, but the proton is more massive. They travel at the same speed in a uniform magnetic field $\vec{B}$, which is directed into the page everywhere. Which particle follows the larger circle, and does it travel clockwise or counterclockwise?

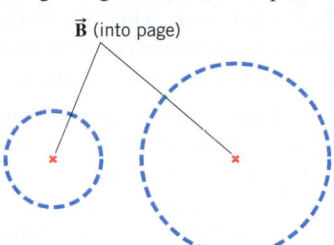

$\vec{B}$ (into page)

	Particle	Direction of Travel
(a)	electron	clockwise
(b)	electron	counterclockwise
(c)	proton	clockwise
(d)	proton	counterclockwise

Section 21.5   The Force on a Current in a Magnetic Field

12. Four views of a horseshoe magnet and a current-carrying wire are shown in the drawing. The wire is perpendicular to the page, and the current is directed out of the page toward you. In which one or more of these situations does the magnetic force on the current point due north?   **(a)** 1 and 2   **(b)** 3 and 4   **(c)** 2   **(d)** 3   **(e)** 1

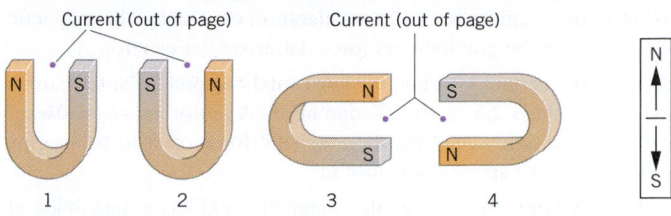
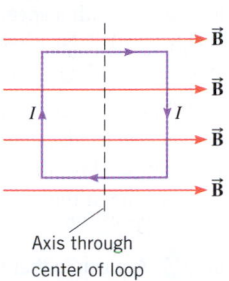

Section 21.6   The Torque on a Current-Carrying Coil

13. A square, current-carrying loop is placed in a uniform magnetic field $\vec{B}$ with the plane of the loop parallel to the magnetic field (see the drawing). The dashed line is the axis of rotation. The magnetic field exerts _____.   **(a)** a net force and a net torque on the loop   **(b)** a net force, but not a net torque, on the loop   **(c)** a net torque, but not a net force, on the loop   **(d)** neither a net force nor a net torque on the loop

Axis through center of loop

Section 21.7   Magnetic Fields Produced by Currents

16. The drawing shows four situations in which two very long wires are carrying the same current, although the directions of the currents may be different. The point P in the drawings is equidistant from each wire. Which one (or more) of these situations gives rise to a zero net magnetic field at P?   **(a)** 2 and 4   **(b)** Only 1   **(c)** Only 2   **(d)** 2 and 3   **(e)** 3 and 4

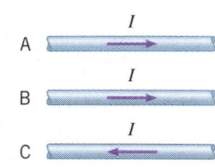

17. Three long, straight wires are carrying currents that have the same magnitude. In C the current is opposite to the current in A and B. The wires are equally spaced. Each wire experiences a net force due to the other two wires. Which wire experiences a net force with the greatest magnitude?   **(a)** A   **(b)** B   **(c)** C   **(d)** All three wires experience a net force that has the same magnitude.

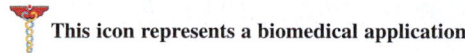

PROBLEMS

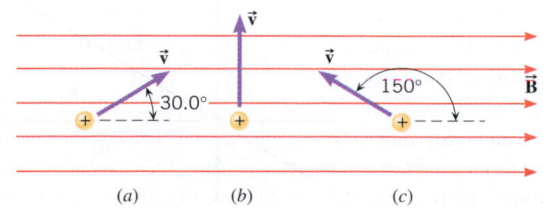 WILEY PLUS

Note to Instructors: Most of the homework problems in this chapter are available for assignment via an online homework management program such as WileyPLUS or WebAssign, and those marked with the icon **GO** *are presented in WileyPLUS using a guided tutorial format that provides enhanced interactivity. See Preface for additional details.*

ssm   Solution is in the Student Solutions Manual.
www   Solution is available online at www.wiley.com/college/cutnell

This icon represents a biomedical application.

Section 21.1   Magnetic Fields,
Section 21.2   The Force That a Magnetic Field Exerts on a Moving Charge

1.   ssm A particle with a charge of $+8.4 \ \mu\text{C}$ and a speed of 45 m/s enters a uniform magnetic field whose magnitude is 0.30 T. For each of the three cases in the drawing, find the magnitude and direction of the magnetic force on the particle.

2. In New England, the horizontal component of the earth's magnetic field has a magnitude of 1.6×10^{-5} T. An electron is shot vertically straight up from the ground with a speed of 2.1×10^6 m/s. What is the magnitude of the acceleration caused by the magnetic force? Ignore the gravitational force acting on the electron.

3. ssm At a certain location, the horizontal component of the earth's magnetic field is 2.5×10^{-5} T, due north. A proton moves eastward with just the right speed for the magnetic force on it to balance its weight. Find the speed of the proton.

4. In a certain region, the earth's magnetic field has a magnitude of 5.4×10^{-5} T and is directed north at an angle of 58° below the horizontal. An electrically charged bullet is fired north and 11° above the horizontal, with a speed of 670 m/s. The magnetic force on the bullet is 2.8×10^{-10} N, directed due east. Determine the bullet's electric charge, including its algebraic sign (+ or −).

5. When a charged particle moves at an angle of 25° with respect to a magnetic field, it experiences a magnetic force of magnitude F. At what angle (less than 90°) with respect to this field will this particle, moving at the same speed, experience a magnetic force of magnitude $2F$?

6. GO A particle that has an 8.2-μC charge moves with a velocity of magnitude 5.0×10^5 m/s along the $+x$ axis. It experiences no magnetic force, although there is a magnetic field present. The maximum possible magnetic force that the charge could experience has a magnitude of 0.48 N. Find the magnitude and direction of the magnetic field. Note that there are two possible answers for the direction of the field.

7. A charge is moving perpendicular to a magnetic field and experiences a force whose magnitude is 2.7×10^{-3} N. If this same charge were to move at the same speed and the angle between its velocity and the same magnetic field were 38°, what would be the magnitude of the magnetic force that the charge would experience?

8. Two charged particles move in the same direction with respect to the same magnetic field. Particle 1 travels three times faster than particle 2. However, each particle experiences a magnetic force of the same magnitude. Find the ratio $|q_1|/|q_2|$ of the magnitudes of the charges.

*9. The drawing shows a parallel plate capacitor that is moving with a speed of 32 m/s through a 3.6-T magnetic field. The velocity $\vec{v}$ is perpendicular to the magnetic field. The electric field within the capacitor has a value of 170 N/C, and each plate has an area of 7.5×10^{-4} m². What is the magnetic force (magnitude and direction) exerted on the positive plate of the capacitor?

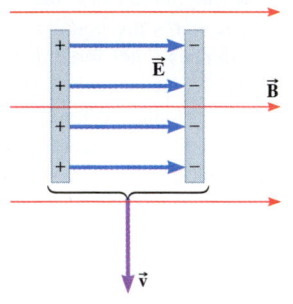

*10. GO A particle has a charge of $q = +5.60$ μC and is located at the coordinate origin. As the drawing shows, an electric field of $E_x = +245$ N/C exists along the $+x$ axis. A magnetic field also exists, and its x and y components are $B_x = +1.80$ T and $B_y = +1.40$ T. Calculate the force (magnitude and direction) exerted on the particle by each of the three fields when it is **(a)** stationary, **(b)** moving along the $+x$ axis at a speed of 375 m/s, and **(c)** moving along the $+z$ axis at a speed of 375 m/s.

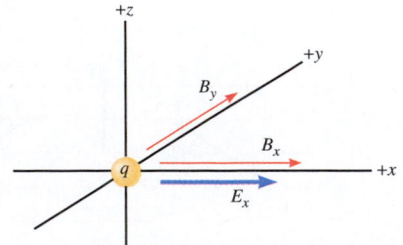

*11. ssm www The electrons in the beam of a television tube have a kinetic energy of 2.40×10^{-15} J. Initially, the electrons move horizontally from west to east. The vertical component of the earth's magnetic field points down, toward the surface of the earth, and has a magnitude of 2.00×10^{-5} T. **(a)** In what direction are the electrons deflected by this field component? **(b)** What is the acceleration of an electron in part (a)?

Section 21.3 The Motion of a Charged Particle in a Magnetic Field, Section 21.4 The Mass Spectrometer

12. In the operating room, anesthesiologists use mass spectrometers to monitor the respiratory gases of patients undergoing surgery. One gas that is often monitored is the anesthetic isoflurane (molecular mass = 3.06×10^{-25} kg). In a spectrometer, a singly ionized molecule of isoflurane (charge = $+e$) moves at a speed of 7.2×10^3 m/s on a circular path that has a radius of 0.10 m. What is the magnitude of the magnetic field that the spectrometer uses?

13. ssm A beam of protons moves in a circle of radius 0.25 m. The protons move perpendicular to a 0.30-T magnetic field. **(a)** What is the speed of each proton? **(b)** Determine the magnitude of the centripetal force that acts on each proton.

14. GO A charged particle with a charge-to-mass ratio of $|q|/m = 5.7 \times 10^8$ C/kg travels on a circular path that is perpendicular to a magnetic field whose magnitude is 0.72 T. How much time does it take for the particle to complete one revolution?

15. A charged particle enters a uniform magnetic field and follows the circular path shown in the drawing. **(a)** Is the particle positively or negatively charged? Why? **(b)** The particle's speed is 140 m/s, the magnitude of the magnetic field is 0.48 T, and the radius of the path is 960 m. Determine the mass of the particle, given that its charge has a magnitude of 8.2×10^{-4} C.

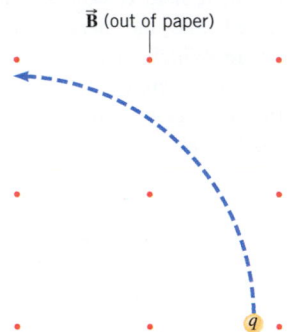

$\vec{B}$ (out of paper)

16. GO A proton is projected perpendicularly into a magnetic field that has a magnitude of 0.50 T. The field is then adjusted so that an electron will follow a circular path of the same radius when it is projected perpendicularly into the field with the same velocity that the proton had. What is the magnitude of the field used for the electron?

17. ssm The solar wind is a thin, hot gas given off by the sun. Charged particles in this gas enter the magnetic field of the earth and can experience a magnetic force. Suppose that a charged particle traveling with a speed of 9.0×10^6 m/s encounters the earth's magnetic field at an altitude where the field has a magnitude of 1.2×10^{-7} T. Assuming that the particle's velocity is perpendicular to the magnetic field, find the radius of the circular path on which the particle would move if it were **(a)** an electron and **(b)** a proton.

18. When beryllium-7 ions ($m = 11.65 \times 10^{-27}$ kg) pass through a mass spectrometer, a uniform magnetic field of 0.283 T curves their path directly to the center of the detector (see Figure 21.14). For the same accelerating potential difference, what magnetic field should be used to send beryllium-10 ions ($m = 16.63 \times 10^{-27}$ kg) to the same location in the detector? Both types of ions are singly ionized ($q = +e$).

19. Two isotopes of carbon, carbon-12 and carbon-13, have masses of 19.93×10^{-27} kg and 21.59×10^{-27} kg, respectively. These two isotopes are singly ionized ($+e$) and each is given a speed of 6.667×10^5 m/s. The ions then enter the bending region of a mass spectrometer where the magnetic field is 0.8500 T. Determine the spatial separation between the two isotopes after they have traveled through a half-circle.

20. **GO** Particle 1 and particle 2 have masses of $m_1 = 2.3 \times 10^{-8}$ kg and $m_2 = 5.9 \times 10^{-8}$ kg, but they carry the same charge q. The two particles accelerate from rest through the same electric potential difference V and enter the same magnetic field, which has a magnitude B. The particles travel perpendicular to the magnetic field on circular paths. The radius of the circular path for particle 1 is $r_1 = 12$ cm. What is the radius (in cm) of the circular path for particle 2?

21. An α-particle has a charge of $+2e$ and a mass of 6.64×10^{-27} kg. It is accelerated from rest through a potential difference that has a value of 1.20×10^6 V and then enters a uniform magnetic field whose magnitude is 2.20 T. The α-particle moves perpendicular to the magnetic field at all times. What is   **(a)** the speed of the α-particle, **(b)** the magnitude of the magnetic force on it, and   **(c)** the radius of its circular path?

*22. The ion source in a mass spectrometer produces both singly and doubly ionized species, X^+ and X^{2+}. The difference in mass between these species is too small to be detected. Both species are accelerated through the same electric potential difference, and both experience the same magnetic field, which causes them to move on circular paths. The radius of the path for the species X^+ is r_1, while the radius for species X^{2+} is r_2. Find the ratio r_1/r_2 of the radii.

*23. ssm Consult **Interactive Solution 21.23** at www.wiley.com/college/cutnell to review a model for solving this problem. A proton with a speed of 3.5×10^6 m/s is shot into a region between two plates that are separated by a distance of 0.23 m. As the drawing shows, a magnetic field exists between the plates, and it is perpendicular to the velocity of the proton. What must be the magnitude of the magnetic field so the proton just misses colliding with the opposite plate?

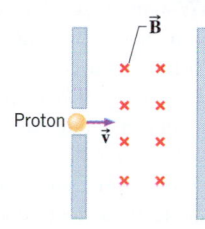

*24. Review Conceptual Example 2 as an aid in understanding this problem. A velocity selector has an electric field of magnitude 2470 N/C, directed vertically upward, and a horizontal magnetic field that is directed south. Charged particles, traveling east at a speed of 6.50×10^3 m/s, enter the velocity selector and are able to pass completely through without being deflected. When a different particle with an electric charge of $+4.00 \times 10^{-12}$ C enters the velocity selector traveling east, the net force (due to the electric and magnetic fields) acting on it is 1.90×10^{-9} N, pointing directly upward. What is the speed of this particle?

*25. ssm A particle of charge $+7.3\ \mu$C and mass 3.8×10^{-8} kg is traveling perpendicular to a 1.6-T magnetic field, as the drawing shows. The speed of the particle is 44 m/s.   **(a)** What is the value of the angle θ, such that the particle's subsequent path will intersect the y axis at the greatest possible value of y?   **(b)** Determine this value of y.

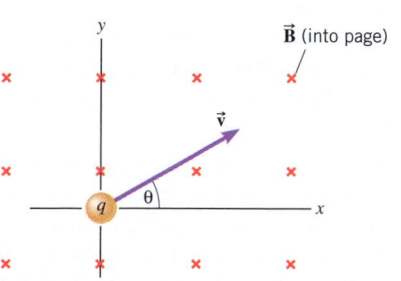

*26. Review Conceptual Example 2 as background for this problem. A charged particle moves through a velocity selector at a constant speed in a straight line. The electric field of the velocity selector is 3.80×10^3 N/C, while the magnetic field is 0.360 T. When the electric field is turned off, the charged particle travels on a circular path whose radius is 4.30 cm. Find the charge-to-mass ratio of the particle.

*27. A positively charged particle of mass 7.2×10^{-8} kg is traveling due east with a speed of 85 m/s and enters a 0.31-T uniform magnetic field. The particle moves through one-quarter of a circle in a time of 2.2×10^{-3} s, at which time it leaves the field heading due south. All during the motion the particle moves perpendicular to the magnetic field.   **(a)** What is the magnitude of the magnetic force acting on the particle?   **(b)** Determine the magnitude of its charge.

*28. Conceptual Example 4 provides background pertinent to this problem. An electron has a kinetic energy of 2.0×10^{-17} J. It moves on a circular path that is perpendicular to a uniform magnetic field of magnitude 5.3×10^{-5} T. Determine the radius of the path.

**29. Refer to Check Your Understanding Question 10 before starting this problem. Suppose that the target discussed there is located at the coordinates $x = -0.10$ m and $y = -0.10$ m. In addition, suppose that the particle is a proton and the magnetic field has a magnitude of 0.010 T. The speed at which the particle is projected is the same for either of the two paths leading to the target. Find the speed.

Section 21.5   The Force on a Current in a Magnetic Field

30. At New York City, the earth's magnetic field has a vertical component of 5.2×10^{-5} T that points downward (perpendicular to the ground) and a horizontal component of 1.8×10^{-5} T that points toward geographic north (parallel to the ground). What is the magnitude and direction of the magnetic force on a 6.0-m long, straight wire that carries a current of 28 A perpendicularly into the ground?

31. ssm A 45-m length of wire is stretched horizontally between two vertical posts. The wire carries a current of 75 A and experiences a magnetic force of 0.15 N. Find the magnitude of the earth's magnetic field at the location of the wire, assuming the field makes an angle of $60.0°$ with respect to the wire.

32. A horizontal wire of length 0.53 m, carrying a current of 7.5 A, is placed in a uniform external magnetic field. When the wire is horizontal, it experiences no magnetic force. When the wire is tilted upward at an angle of $19°$, it experiences a magnetic force of 4.4×10^{-3} N. Determine the magnitude of the external magnetic field.

33. A square coil of wire containing a single turn is placed in a uniform 0.25-T magnetic field, as the drawing shows. Each side has a length of 0.32 m, and the current in the coil is 12 A. Determine the magnitude of the magnetic force on each of the four sides.

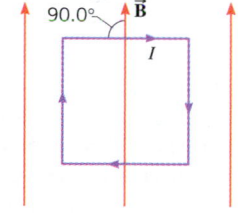

34. **GO** The drawing shows a wire composed of three segments, AB, BC, and CD. There is a current of $I = 2.8$ A in the wire. There is also a magnetic field $\vec{B}$ (magnitude $= 0.26$ T) that is the same everywhere and points in the direction of the $+z$ axis. The lengths of the wire segments are $L_{AB} = 1.1$ m, $L_{BC} = 0.55$ m, and $L_{CD} = 0.55$ m. Find the magnitude of the force that acts on each segment.

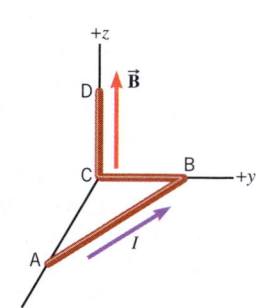

Problem 34

35. ssm A wire carries a current of 0.66 A. This wire makes an angle of $58°$ with respect to a magnetic field of magnitude 4.7×10^{-5} T. The wire experiences a magnetic force of magnitude 7.1×10^{-5} N. What is the length of the wire?

36. Two insulated wires, each 2.40 m long, are taped together to form a two-wire unit that is 2.40 m long. One wire carries a current of 7.00 A; the other carries a smaller current I in the opposite direction. The two-wire unit is placed at an angle of $65.0°$ relative to a magnetic field whose magnitude is 0.360 T. The magnitude of the net magnetic force experienced by the two-wire unit is 3.13 N. What is the current I?

37. The x, y, and z components of a magnetic field are $B_x = 0.10$ T, $B_y = 0.15$ T, and $B_z = 0.17$ T. A 25-cm wire is oriented along the z axis and carries a current of 4.3 A. What is the magnitude of the magnetic force that acts on this wire?

*38. GO A horizontal wire is hung from the ceiling of a room by two massless strings. The wire has a length of 0.20 m and a mass of 0.080 kg. A uniform magnetic field of magnitude 0.070 T is directed from the ceiling to the floor. When a current of $I = 42$ A exists in the wire, the wire swings upward and, at equilibrium, makes an angle ϕ with respect to the vertical, as the drawing shows. Find   (a) the angle ϕ and   (b) the tension in each of the two strings.

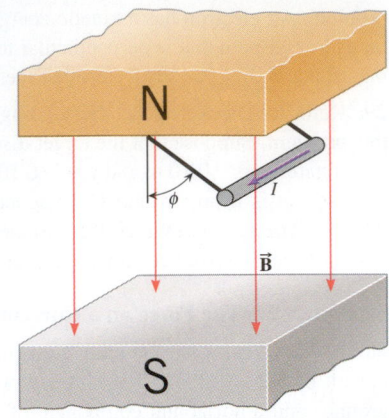

*39. ssm Consult **Interactive Solution 21.39** at **www.wiley.com/ college/cutnell** to explore a model for solving this problem. The drawing shows a thin, uniform rod that has a length of 0.45 m and a mass of 0.094 kg. This rod lies in the plane of the paper and is attached to the floor by a hinge at point P. A uniform magnetic field of 0.36 T is directed perpendicularly into the plane of the paper. There is a current $I = 4.1$ A in the rod, which does not rotate clockwise or counterclockwise. Find the angle θ. (*Hint: The magnetic force may be taken to act at the center of gravity.*)

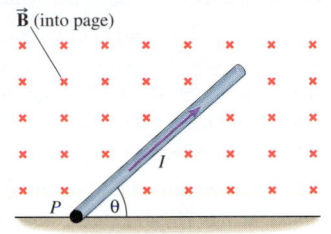

*40. A copper rod of length 0.85 m is lying on a frictionless table (see the drawing). Each end of the rod is attached to a fixed wire by an unstretched spring that has a spring constant of $k = 75$ N/m. A magnetic field with a strength of 0.16 T is oriented perpendicular to the surface of the table.   (a) What must be the direction of the current in the copper rod that causes the springs to stretch?   (b) If the current is 12 A, by how much does each spring stretch?

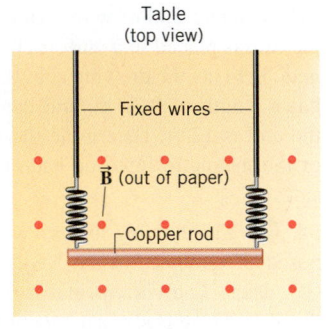

Table (top view)

Fixed wires

$\vec{B}$ (out of paper)

Copper rod

41. The two conducting rails in the drawing are tilted upward so they each make an angle of 30.0° with respect to the ground. The vertical magnetic field has a magnitude of 0.050 T. The 0.20-kg aluminum rod (length = 1.6 m) slides *without friction* down the rails at a constant velocity. How much current flows through the rod?

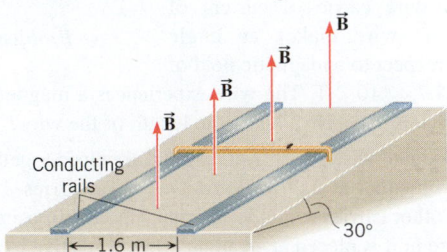

Conducting rails

←1.6 m→

30°

Section 21.6   The Torque on a Current-Carrying Coil

42. Two coils have the same number of circular turns and carry the same current. Each rotates in a magnetic field as in Figure 21.19. Coil 1 has a radius of 5.0 cm and rotates in a 0.18-T field. Coil 2 rotates in a 0.42-T field. Each coil experiences the same maximum torque. What is the radius (in cm) of coil 2?

43. The maximum torque experienced by a coil in a 0.75-T magnetic field is 8.4×10^{-4} N·m. The coil is circular and consists of only one turn. The current in the coil is 3.7 A. What is the length of the wire from which the coil is made?

44. Two circular coils of current-carrying wire have the same magnetic moment. The first coil has a radius of 0.088 m, has 140 turns, and carries a current of 4.2 A. The second coil has 170 turns and carries a current of 9.5 A. What is the radius of the second coil?

45. A wire has a length of 7.00×10^{-2} m and is used to make a circular coil of one turn. There is a current of 4.30 A in the wire. In the presence of a 2.50-T magnetic field, what is the maximum torque that this coil can experience?

46. A coil carries a current and experiences a torque due to a magnetic field. The value of the torque is 80.0% of the maximum possible torque.   (a) What is the smallest angle between the magnetic field and the normal to the plane of the coil?   (b) Make a drawing that shows how this coil would be oriented relative to the magnetic field. Be sure to include the angle in the drawing.

47. ssm www The rectangular loop in the drawing consists of 75 turns and carries a current of $I = 4.4$ A. A 1.8-T magnetic field is directed along the $+y$ axis. The loop is free to rotate about the z axis.   (a) Determine the magnitude of the net torque exerted on the loop and   (b) state whether the 35° angle will increase or decrease.

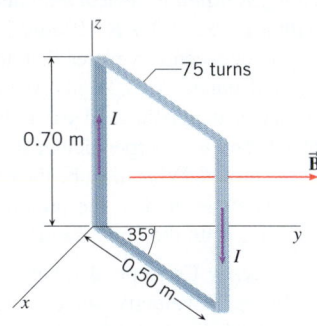

75 turns

0.70 m

35°

0.50 m

$\vec{B}$

48. GO You have a wire of length $L = 1.00$ m from which to make the square coil of a dc motor. The current in the coil is $I = 1.7$ A, and the magnetic field of the motor has a magnitude of $B = 0.34$ T. Find the maximum torque exerted on the coil when the wire is used to make a single-turn square coil and a two-turn square coil.

49. Two pieces of the same wire have the same length. From one piece, a square coil containing a single loop is made. From the other, a circular coil containing a single loop is made. The coils carry different currents. When placed in the same magnetic field with the same orientation, they experience the same torque. What is the ratio $I_{\text{square}}/I_{\text{circle}}$ of the current in the square coil to the current in the circular coil?

*50. A square coil and a rectangular coil are each made from the same length of wire. Each contains a single turn. The long sides of the rectangle are twice as long as the short sides. Find the ratio $\tau_{\text{square}}/\tau_{\text{rectangle}}$ of the maximum torques that these coils experience in the same magnetic field when they contain the same current.

*51. Consult **Interactive Solution 21.51** at **www.wiley.com/college/ cutnell** to see how this problem can be solved. The coil in Figure 21.20a contains 410 turns and has an area per turn of 3.1×10^{-3} m^2. The magnetic field is 0.23 T, and the current in the coil is 0.26 A. A brake shoe is pressed perpendicularly against the shaft to keep the coil from turning. The coefficient of static friction between the shaft and the brake shoe is 0.76. The radius of the shaft is 0.012 m. What is the magnitude of the minimum normal force that the brake shoe exerts on the shaft?

**52. A charge of 4.0×10^{-6} C is placed on a small conducting sphere that is located at the end of a thin insulating rod whose length is 0.20 m. The rod rotates with an angular speed of $\omega = 150$ rad/s about an axis that passes perpendicularly through its other end. Find the magnetic moment of the rotating charge. (*Hint: The charge travels around a circle in a time equal to the period of the motion.*)

Section 21.7 Magnetic Fields Produced by Currents

53. ssm Suppose in Figure 21.26a that $I_1 = I_2 = 25$ A and that the separation between the wires is 0.016 m. By applying an external magnetic field (created by a source other than the wires) it is possible to cancel the mutual repulsion of the wires. This external field must point along the vertical direction.   (a) Does the external field point up or down? Explain.   (b) What is the magnitude of the external field?

54. GO A long solenoid has 1400 turns per meter of length, and it carries a current of 3.5 A. A small circular coil of wire is placed inside the solenoid with the normal to the coil oriented at an angle of 90.0° with respect to the axis of the solenoid. The coil consists of 50 turns, has an area of 1.2×10^{-3} m², and carries a current of 0.50 A. Find the torque exerted on the coil.

55. ssm The magnetic field produced by the solenoid in a magnetic resonance imaging (MRI) system designed for measurements on whole human bodies has a field strength of 7.0 T, and the current in the solenoid is 2.0×10^2 A. What is the number of turns per meter of length of the solenoid? Note that the solenoid used to produce the magnetic field in this type of system has a length that is not very long compared to its diameter. Because of this and other design considerations, your answer will be only an approximation.

56. What must be the radius of a circular loop of wire so the magnetic field at its center is 1.8×10^{-4} T when the loop carries a current of 12 A?

57. ssm Two circular loops of wire, each containing a single turn, have the same radius of 4.0 cm and a common center. The planes of the loops are perpendicular. Each carries a current of 1.7 A. What is the magnitude of the net magnetic field at the common center?

58. The drawing shows four insulated wires overlapping one another, forming a square with 0.050-m sides. All four wires are much longer than the sides of the square. The net magnetic field at the center of the square is 61 μT. Calculate the current I.

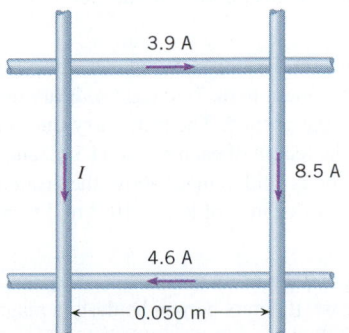

59. Two long, straight wires are separated by 0.120 m. The wires carry currents of 8.0 A in opposite directions, as the drawing indicates. Find the magnitude of the net magnetic field at the points labeled   (a) A and   (b) B.

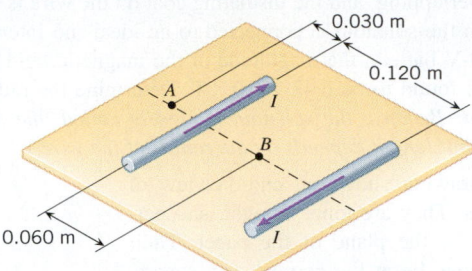

60. GO A very long, straight wire carries a current of 0.12 A. This wire is tangent to a single-turn, circular wire loop that also carries a current. The directions of the currents are such that the net magnetic field at the center of the loop is zero. Both wires are insulated and have diameters that can be neglected. How much current is there in the loop?

*61. Review **Interactive Solution 21.61** at **www.wiley.com/college/cutnell** for one approach to this problem. Two circular coils are concentric and lie in the same plane. The inner coil contains 140 turns of wire, has a radius of 0.015 m, and carries a current of 7.2 A. The outer coil contains 180 turns and has a radius of 0.023 m. What must be the magnitude and direction (relative to the current in the inner coil) of the current in the outer coil, so that the net magnetic field at the common center of the two coils is zero?

*62. GO The drawing shows two perpendicular, long, straight wires, both of which lie in the plane of the paper. The current in each of the wires is $I = 5.6$ A. Find the magnitudes of the net magnetic fields at points A and B.

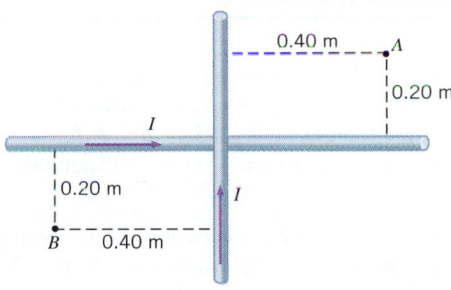

*63. Two infinitely long, straight wires are parallel and separated by a distance of one meter. They carry currents in the same direction. Wire 1 carries four times the current that wire 2 carries. On a line drawn perpendicular to both wires, locate the spot (relative to wire 1) where the net magnetic field is zero. Assume that wire 1 lies to the left of wire 2 and note that there are three regions to consider on this line: to the left of wire 1, between wire 1 and wire 2, and to the right of wire 2.

*64. A small compass is held horizontally, the center of its needle a distance of 0.280 m directly north of a long wire that is perpendicular to the earth's surface. When there is no current in the wire, the compass needle points due north, which is the direction of the horizontal component of the earth's magnetic field at that location. This component is parallel to the earth's surface. When the current in the wire is 25.0 A, the needle points 23.0° east of north.   (a) Does the current in the wire flow toward or away from the earth's surface?   (b) What is the magnitude of the horizontal component of the earth's magnetic field at the location of the compass?

*65. ssm A piece of copper wire has a resistance per unit length of 5.90×10^{-3} Ω/m. The wire is wound into a thin, flat coil of many turns that has a radius of 0.140 m. The ends of the wire are connected to a 12.0-V battery. Find the magnetic field strength at the center of the coil.

**66. A solenoid is formed by winding 25.0 m of insulated silver wire around a hollow cylinder. The turns are wound as closely as possible without overlapping, and the insulating coat on the wire is negligibly thin. When the solenoid is connected to an ideal (no internal resistance) 3.00-V battery, the magnitude of the magnetic field inside the solenoid is found to be 6.48×10^{-3} T. Determine the radius of the wire. (*Hint: Because the solenoid is closely coiled, the number of turns per unit length depends on the radius of the wire.*)

**67. The drawing shows an end-on view of three wires. They are long, straight, and perpendicular to the plane of the paper. Their cross sections lie at the corners of a square. The currents in wires 1 and 2 are $I_1 = I_2 = I$ and are directed into the paper. What is the direction of the current in wire 3, and what is the ratio I_3/I, so that the net magnetic field at the empty corner is zero?

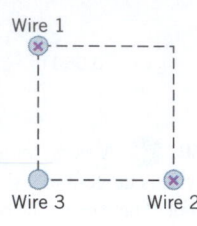

Section 21.8 Ampère's Law

68. The wire in Figure 21.37 carries a current of 12 A. Suppose that a second long, straight wire is placed right next to this wire. The current in the second wire is 28 A. Use Ampère's law to find the magnitude of the magnetic field at a distance of $r = 0.72$ m from the wires when the currents are (**a**) in the same direction and (**b**) in opposite directions.

69. ssm Suppose that a uniform magnetic field is everywhere perpendicular to this page. The field points directly upward toward you. A circular path is drawn on the page. Use Ampère's law to show that there can be no net current passing through the circular surface.

*70. Refer to **Interactive Solution 21.70** at www.wiley.com/college/cutnell for help with problems like this one. A very long, hollow cylinder is formed by rolling up a thin sheet of copper. Electric charges flow along the copper sheet parallel to the axis of the cylinder. The arrangement is, in effect, a hollow tube of current I. Use Ampère's law to show that the magnetic field (**a**) is $\mu_0 I/(2\pi r)$ outside the cylinder at a distance r from the axis and (**b**) is zero at any point within the hollow interior of the cylinder. (*Hint: For closed paths, use circles perpendicular to and centered on the axis of the cylinder.*)

**71. A long, cylindrical conductor is solid throughout and has a radius R. Electric charges flow parallel to the axis of the cylinder and pass uniformly through the entire cross section. The arrangement is, in effect, a solid tube of current I_0. The current per unit cross-sectional area (i.e., the current density) is $I_0/(\pi R^2)$. Use Ampère's law to show that the magnetic field inside the conductor at a distance r from the axis is $\mu_0 I_0 r/(2\pi R^2)$. (*Hint: For a closed path, use a circle of radius r perpendicular to and centered on the axis. Note that the current through any surface is the area of the surface times the current density.*)

ADDITIONAL PROBLEMS

72. A charge of -8.3 μC is traveling at a speed of 7.4×10^6 m/s in a region of space where there is a magnetic field. The angle between the velocity of the charge and the field is 52°. A force of magnitude 5.4×10^{-3} N acts on the charge. What is the magnitude of the magnetic field?

73. A long, straight wire carries a current of 48 A. The magnetic field produced by this current at a certain point is 8.0×10^{-5} T. How far is the point from the wire?

74. An electron is moving through a magnetic field whose magnitude is 8.70×10^{-4} T. The electron experiences only a magnetic force and has an acceleration of magnitude 3.50×10^{14} m/s². At a certain instant, it has a speed of 6.80×10^6 m/s. Determine the angle θ (less than 90°) between the electron's velocity and the magnetic field.

75. ssm The 1200-turn coil in a dc motor has an area per turn of 1.1×10^{-2} m². The design for the motor specifies that the magnitude of the maximum torque is 5.8 N·m when the coil is placed in a 0.20-T magnetic field. What is the current in the coil?

76. Multiple-Concept Example 7 discusses how problems like this one can be solved. A $+6.00$-μC charge is moving with a speed of 7.50×10^4 m/s parallel to a very long, straight wire. The wire is 5.00 cm from the charge and carries a current of 67.0 A in a direction opposite to that of the moving charge. Find the magnitude and direction of the force on the charge.

77. ssm In a lightning bolt, 15 C of charge flows during a time of 1.5×10^{-3} s. Assuming that the lightning bolt can be represented as a long, straight line of current, what is the magnitude of the magnetic field at a distance of 25 m from the bolt?

78. The triangular loop of wire shown in the drawing carries a current of $I = 4.70$ A. A uniform magnetic field is directed parallel to side AB of the triangle and has a magnitude of 1.80 T. (**a**) Find the magnitude and direction of the magnetic force exerted on each side of the triangle. (**b**) Determine the magnitude of the net force exerted on the triangle.

Problem 78

79. A magnetic field has a magnitude of 1.2×10^{-3} T, and an electric field has a magnitude of 4.6×10^3 N/C. Both fields point in the same direction. A positive 1.8-μC charge moves at a speed of 3.1×10^6 m/s in a direction that is perpendicular to both fields. Determine the magnitude of the net force that acts on the charge.

80. Multiple-Concept Example 8 reviews the concepts from this chapter that are pertinent here. Two rigid rods are oriented parallel to each other and to the ground. The rods carry the same current in the same direction. The length of each rod is 0.85 m, and the mass of each is 0.073 kg. One rod is held in place above the ground, while the other floats beneath it at a distance of 8.2×10^{-3} m. Determine the current in the rods.

*81. ssm www A particle of mass 6.0×10^{-8} kg and charge $+7.2$ μC is traveling due east. It enters perpendicularly a magnetic field whose magnitude is 3.0 T. After entering the field, the particle completes one-half of a circle and exits the field traveling due west. How much time does the particle spend traveling in the magnetic field?

*82. Two parallel rods are each 0.50 m in length. They are attached at their centers to either end of a spring (spring constant = 150 N/m) that is initially neither stretched nor compressed. When 950 A of current is in each rod in the same direction, the spring is observed to be compressed by 2.0 cm. Treat the rods as long, straight wires and find the separation between them when the current is present.

*83. One component of a magnetic field has a magnitude of 0.048 T and points along the $+x$ axis, while the other component has a magnitude of 0.065 T and points along the $-y$ axis. A particle carrying a charge of $+2.0 \times 10^{-5}$ C is moving along the $+z$ axis at a speed of 4.2×10^3 m/s. **(a)** Find the magnitude of the net magnetic force that acts on the particle. **(b)** Determine the angle that the net force makes with respect to the $+x$ axis.

**84. The drawing shows two long, straight wires that are suspended from a ceiling. The mass per unit length of each wire is 0.050 kg/m. Each of the four strings suspending the wires has a length of 1.2 m. When the wires carry identical currents in opposite directions, the angle between the strings holding the two wires is 15°. What is the current in each wire?

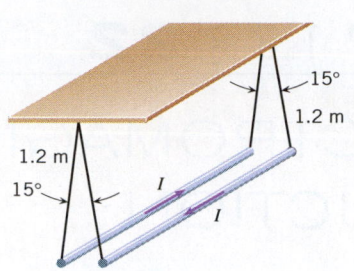

Problem 84

**85. ssm In the model of the hydrogen atom created by Niels Bohr, the electron moves around the proton at a speed of 2.2×10^6 m/s in a circle of radius 5.3×10^{-11} m. Considering the orbiting electron to be a small current loop, determine the magnetic moment associated with this motion. *(Hint: The electron travels around the circle in a time equal to the period of the motion.)*

CHAPTER 22

ELECTROMAGNETIC INDUCTION

Alicia Keys, like most singers, uses a microphone when she performs. Many microphones operate using magnetism and a phenomenon known as electromagnetic induction. (© Al Pereira/WireImage/Getty Images)

22.1 INDUCED EMF AND INDUCED CURRENT

There are a number of ways a magnetic field can be used to generate an electric current, and Figure 22.1 illustrates one of them. This drawing shows a bar magnet and a helical coil of wire to which an ammeter is connected. When there is no *relative* motion between the magnet and the coil, as in part *a* of the drawing, the ammeter reads zero, indicating that no current exists. However, when the magnet moves toward the coil, as in part *b*, a current *I* appears. As the magnet approaches, the magnetic field $\vec{\mathbf{B}}$ that it creates at the location of the coil becomes stronger and stronger, and it is this *changing* field that produces the current. When the magnet moves away from the coil, as in part *c*, a current is also produced, but with a reversed direction. Now the magnetic field at the coil becomes weaker as the magnet moves away. Once again it is the *changing* field that generates the current.

A current would also be created in Figure 22.1 if the magnet were held stationary and the coil were moved, because the magnetic field at the coil would be changing as the coil approached or receded from the magnet. Only relative motion between the magnet and the coil is needed to generate a current; it does not matter which one moves.

The current in the coil is called an ***induced current*** because it is brought about (or "induced") by a changing magnetic field. Since a source of emf (electromotive force) is always needed to produce a current, the coil itself behaves as if it were a source of emf. This emf is known as an ***induced emf.*** Thus, a changing magnetic field induces an emf in the coil, and the emf leads to an induced current.

Induced emf and induced current are frequently used in the cruise control devices found in many cars. Figure 22.2 illustrates how a cruise control device operates. Usually two magnets are mounted on opposite sides of the vehicle's drive shaft, with a stationary

The physics of
an automobile cruise control device.

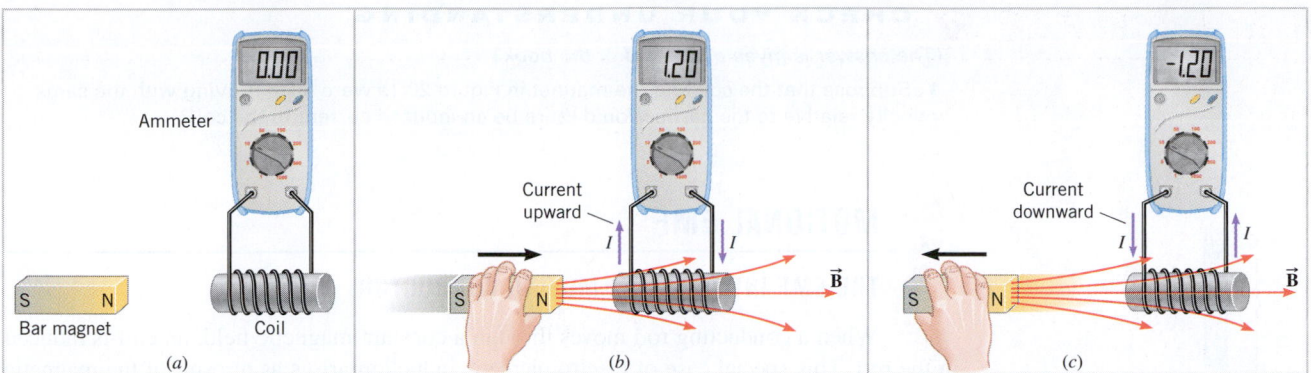

(a)   (b)   (c)

Figure 22.1 (a) When there is no relative motion between the coil of wire and the bar magnet, there is no current in the coil. (b) A current is created in the coil when the magnet moves toward the coil. (c) A current also exists when the magnet moves away from the coil, but the direction of the current is opposite to that in (b).

sensing coil positioned nearby. As the shaft turns, the magnets pass by the coil and cause an induced emf and current to appear in it. A microprocessor (the "brain" of a computer) counts the pulses of current and, with the aid of its internal clock and a knowledge of the shaft's radius, determines the rotational speed of the drive shaft. The rotational speed, in turn, is related to the car's speed. Thus, once the driver sets the desired cruising speed with the speed control switch (mounted near the steering wheel), the microprocessor can compare it with the measured speed. To the extent that the selected cruising speed and the measured speed differ, a signal is sent to a servo, or control, mechanism, which causes the throttle/fuel injector to send more or less fuel to the engine. The car speeds up or slows down accordingly, until the desired cruising speed is reached.

Figure 22.3 shows another way to induce an emf and a current in a coil. An emf can be induced by *changing the area* of a coil in a constant magnetic field. Here the shape of the coil is being distorted to reduce the area. As long as the area is changing, an induced emf and current exist; they vanish when the area is no longer changing. If the distorted coil is returned to its original shape, thereby increasing the area, an oppositely directed current is generated while the area is changing.

In each of the previous examples, both an emf and a current are induced in the coil because the coil is part of a complete, or closed, circuit. If the circuit were open—perhaps because of an open switch—there would be no induced current. However, an emf would still be induced in the coil, whether the current exists or not.

Changing a magnetic field and changing the area of a coil are methods that can be used to create an induced emf. The phenomenon of producing an induced emf with the aid of a magnetic field is called ***electromagnetic induction.*** The next section discusses yet another method by which an induced emf can be created.

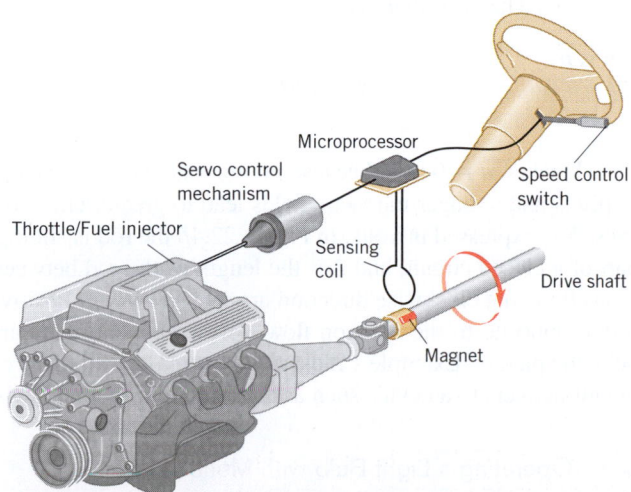

Figure 22.2 Induced emf lies at the heart of an automobile cruise control device. The emf is induced in a sensing coil by magnets attached to the rotating drive shaft.

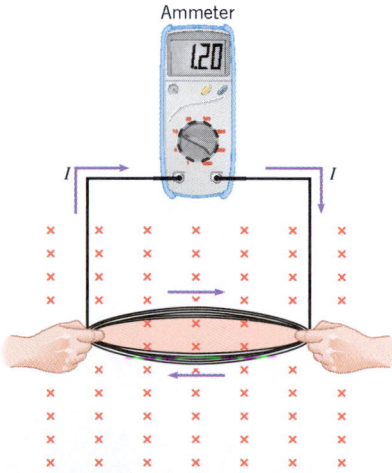

Figure 22.3 While the area of the coil is changing, an induced emf and current are generated.

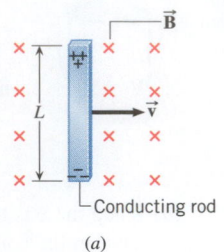

(a)

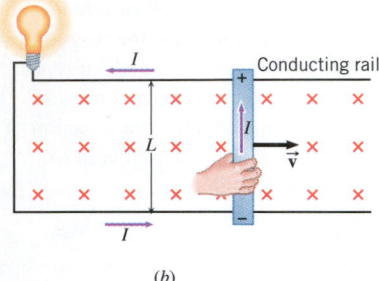

(b)

Figure 22.4 (*a*) When a conducting rod moves at right angles to a constant magnetic field, the magnetic force causes opposite charges to appear at the ends of the rod, giving rise to an induced emf. (*b*) The induced emf causes an induced current *I* to appear in the circuit.

22.2 MOTIONAL EMF

THE EMF INDUCED IN A MOVING CONDUCTOR

When a conducting rod moves through a constant magnetic field, an emf is induced in the rod. This special case of electromagnetic induction arises as a result of the magnetic force that acts on a moving charge (see Section 21.2). Consider the metal rod of length *L* moving to the right in Figure 22.4*a*. The velocity $\vec{v}$ of the rod is constant and is perpendicular to a uniform magnetic field $\vec{B}$. Each charge *q* within the rod also moves with a velocity $\vec{v}$ and experiences a magnetic force of magnitude $F = |q|vB$, according to Equation 21.1. By using RHR-1, it can be seen that the mobile, free electrons are driven to the bottom of the rod, leaving behind an equal amount of positive charge at the top. (Remember to reverse the direction of the force that RHR-1 predicts, since the electrons have a negative charge. See Section 21.2.) The positive and negative charges accumulate until the attractive electric force that they exert on each other becomes equal in magnitude to the magnetic force. When the two forces balance, equilibrium is reached and no further charge separation occurs.

The separated charges on the ends of the moving conductor give rise to an induced emf, called a *motional emf* because it originates from the motion of charges through a magnetic field. The emf exists as long as the rod moves. If the rod is brought to a halt, the magnetic force vanishes, with the result that the attractive electric force reunites the positive and negative charges and the emf disappears. The emf of the moving rod is analogous to the emf between the terminals of a battery. However, the emf of a battery is produced by chemical reactions, whereas the motional emf is created by the agent that moves the rod through the magnetic field (like the hand in Figure 22.4*b*.)

The fact that the electric and magnetic forces balance at equilibrium in Figure 22.4*a* can be used to determine the magnitude of the motional emf $\mathscr{E}$. Acccording to Equation 18.2, the magnitude of the electric force acting on the positive charge *q* at the top of the rod is *Eq*, where *E* is the magnitude of the electric field due to the separated charges. And according to Equation 19.7a (without the minus sign), the electric field magnitude is given by the voltage between the ends of the rod (the emf $\mathscr{E}$) divided by the length *L* of the rod. Thus, the electric force is $Eq = (\mathscr{E}/L)q$. Since we are dealing now with a positive charge, the magnetic force is $B|q|(v \sin 90°) = Bqv$, according to Equation 21.1, because the charge *q* moves perpendicular to the magnetic field. Since these two forces balance, it follows that $(\mathscr{E}/L)q = Bqv$. The emf, then, is

Motional emf when $\vec{v}, \vec{B}$,
and L are mutually　　　　　　　　$$\mathscr{E} = vBL \qquad\qquad (22.1)$$
perpendicular

As expected, $\mathscr{E} = 0$ V when $v = 0$ m/s, because no motional emf is developed in a stationary rod. Greater speeds and stronger magnetic fields lead to greater emfs for a given length *L*. As with batteries, $\mathscr{E}$ is expressed in volts. In Figure 22.4*b* the rod is sliding on conducting rails that form part of a closed circuit, and *L* is the length of the rod between the rails. Due to the emf, electrons flow in a clockwise direction around the circuit. Positive charge would flow in the direction opposite to the electron flow, so the conventional current *I* is drawn counterclockwise in the picture. Example 1 illustrates how to determine the electrical energy that the motional emf delivers to a device such as the light bulb in the drawing.

🔔 **Example 1**　　Operating a Light Bulb with Motional Emf

Suppose that the rod in Figure 22.4*b* is moving at a speed of 5.0 m/s in a direction perpendicular to a 0.80-T magnetic field. The rod has a length of 1.6 m and a negligible electrical resistance. The rails also have negligible resistance. The light bulb, however, has a resistance of 96 Ω.

Find **(a)** the emf produced by the rod, **(b)** the current induced in the circuit, **(c)** the electric power delivered to the bulb, and **(d)** the energy used by the bulb in 60.0 s.

Reasoning The moving rod acts like an imaginary battery and supplies a motional emf of vBL to the circuit. The induced current can be determined from Ohm's law as the motional emf divided by the resistance of the bulb. The electric power delivered to the bulb is the product of the induced current and the potential difference across the bulb (which, in this case, is the motional emf). The energy used is the product of the power and the time.

Solution **(a)** The motional emf is given by Equation 22.1 as

$$\mathscr{E} = vBL = (5.0 \text{ m/s})(0.80 \text{ T})(1.6 \text{ m}) = \boxed{6.4 \text{ V}}$$

(b) According to Ohm's law, the induced current is equal to the motional emf divided by the resistance of the circuit:

$$I = \frac{\mathscr{E}}{R} = \frac{6.4 \text{ V}}{96 \ \Omega} = \boxed{0.067 \text{ A}} \qquad (20.2)$$

(c) The electric power P delivered to the light bulb is the product of the current I and the potential difference across the bulb:

$$P = I\mathscr{E} = (0.067 \text{ A})(6.4 \text{ V}) = \boxed{0.43 \text{ W}} \qquad (20.6a)$$

(d) Since power is energy per unit time, the energy E used in 60.0 s is the product of the power and the time:

$$E = Pt = (0.43 \text{ W})(60.0 \text{ s}) = \boxed{26 \text{ J}} \qquad (6.10b)$$

MOTIONAL EMF AND ELECTRICAL ENERGY

Motional emf arises because a magnetic force acts on the charges in a conductor that is moving through a magnetic field. Whenever this emf causes a current, a second magnetic force enters the picture. In Figure 22.4b, for instance, the second force arises because the current I in the rod is perpendicular to the magnetic field. The current, and hence the rod, experiences a magnetic force $\vec{\mathbf{F}}$ whose magnitude is given by Equation 21.3 as $F = ILB \sin 90°$. Using the values of I, L, and B given in Example 1, we see that $F = (0.067 \text{ A})(1.6 \text{ m})(0.80 \text{ T}) = 0.086 \text{ N}$. The direction of $\vec{\mathbf{F}}$ is specified by RHR-1 and is *opposite* to the velocity $\vec{\mathbf{v}}$ of the rod, and thus points to the left (see Figure 22.5). By itself, $\vec{\mathbf{F}}$ would *slow down* the rod, and here lies the crux of the matter. To keep the rod moving to the right with a constant velocity, a counterbalancing force must be applied to the rod by an external agent, such as the hand in the picture. This force is labeled $\vec{\mathbf{F}}_{\text{hand}}$ in the drawing. The counterbalancing force must have a magnitude of 0.086 N and must be directed opposite to the magnetic force $\vec{\mathbf{F}}$. If the counterbalancing force were removed, the rod would decelerate under the influence of $\vec{\mathbf{F}}$ and eventually come to rest. During the deceleration, the motional emf would decrease and the light bulb would eventually go out.

We can now answer an important question—Who or what provides the 26 J of electrical energy that the light bulb in Example 1 uses in 60.0 seconds? The provider is the external agent that applies the 0.086-N counterbalancing force needed to keep the rod moving. This agent does work, and Example 2 shows that the work done is equal to the electrical energy used by the bulb.

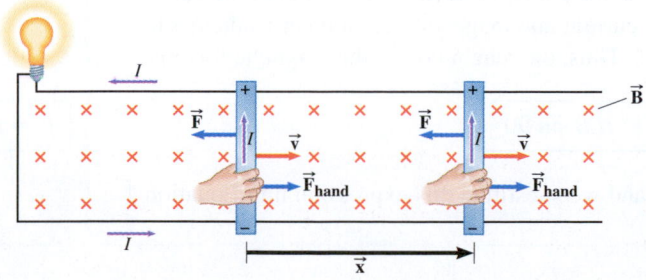

Figure 22.5 A magnetic force $\vec{\mathbf{F}}$ is exerted on the current I in the moving rod and is directed opposite to the rod's velocity $\vec{\mathbf{v}}$. Since the force $\vec{\mathbf{F}}_{\text{hand}}$ counterbalances the magnetic force $\vec{\mathbf{F}}$, the rod moves to the right at a constant velocity.

ANALYZING MULTIPLE-CONCEPT PROBLEMS

Example 2 The Work Needed to Keep the Light Bulb Burning

As we saw in Example 1, an induced current of 0.067 A exists in the circuit due to the moving rod. As Figure 22.5 shows, the hand provides a force $\vec{F}_{hand}$ that keeps the rod moving to the right. Determine the work done by this force in a time of 60.0 s. Assume, as in Example 1, that the magnetic field has a magnitude of 0.80 T and that the rod has a length of 1.6 m and moves at a constant speed of 5.0 m/s.

Reasoning According to the discussion in Section 6.1, the work done by the hand is equal to the product of (1) the magnitude F_{hand} of the force exerted by the hand, (2) the magnitude x of the rod's displacement, and (3) the cosine of the angle between the force and the displacement. Since the rod moves to the right at a constant speed, it has no acceleration and is, therefore, in equilibrium (see Section 4.11). Thus, the force exerted by the hand must be equal in magnitude, but opposite in direction, to the magnetic force $\vec{F}$ exerted on the rod, since they are the only two forces acting on the rod along the direction of the motion. We will determine the magnitude F of the magnetic force by using Equation 21.3, and this will enable us to find F_{hand}. Since the rod moves at a constant speed, the magnitude x of its displacement is the product of its speed and the time of travel.

Knowns and Unknowns The data for this problem are:

Description	Symbol	Value
Current	I	0.067 A
Length of rod	L	1.6 m
Speed of rod	v	5.0 m/s
Time during which rod moves	t	60.0 s
Magnitude of magnetic field	B	0.80 T
Unknown Variable		
Work done by hand	W	?

Modeling the Problem

STEP 1 **Work** The work done by the hand in Figure 22.5 is given by $W = F_{hand}x \cos \theta'$ (Equation 6.1). In this equation, F_{hand} is the magnitude of the force that the hand exerts on the rod, x is the magnitude of the rod's displacement, and θ' is the angle between the force and the displacement. Since the force and displacement point in the same direction, $\theta' = 0°$, so

$$W = F_{hand}x \cos 0°$$

Two forces act on the rod; the force $\vec{F}_{hand}$, which points to the right, and the magnetic force $\vec{F}$, which points to the left. Since the rod moves at a constant velocity, the magnitudes of these two forces are equal, so that $F_{hand} = F$. Substituting this relation into the expression for the work gives Equation 1 at the right. At this point, neither F nor x is known, and they will be evaluated in Steps 2 and 3, respectively.

$$W = Fx \cos 0° \qquad (1)$$

STEP 2 **Magnetic Force Exerted on a Current-Carrying Rod** We have seen in Section 21.5 that a rod of length L that carries a current I in a magnetic field of magnitude B experiences a magnetic force of magnitude F. The magnitude of the force is given by $F = ILB \sin \theta$ (Equation 21.3), where θ is the angle between the direction of the current and the magnetic field. In this case, the current and magnetic field are perpendicular to each other (see Figure 22.5), so $\theta = 90°$. Thus, the magnitude of the magnetic force is

$$\boxed{F = ILB \sin 90°}$$

The quantities, I, L, and B are known, and we substitute this expression into Equation 1 at the right.

$$W = Fx \cos 0° \qquad (1)$$

$$\boxed{F = ILB \sin 90°}$$

STEP 3 **Kinematics** Since the rod is moving at a constant speed, the distance x it travels is the product of its speed v and the time t:

$$x = vt$$

The variables v and t are known. We can also substitute this relation into Equation 1, as shown in the right column.

$$W = Fx \cos 0° \quad \text{(1)}$$

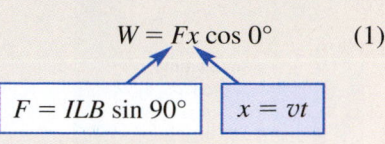

$$F = ILB \sin 90° \qquad x = vt$$

Solution Algebraically combining the results of the three steps, we have

STEP 1   **STEP 2**   **STEP 3**

$$W = Fx \cos 0° = (ILB \sin 90°) \, x \cos 0° = (ILB \sin 90°) \, (vt) \cos 0°$$

The work done by the force of the hand is

$$W = (ILB \sin 90°)(vt) \cos 0°$$

$$= (0.067 \text{ A})(1.6 \text{ m})(0.80 \text{ T})(\sin 90°)(5.0 \text{ m/s})(60.0 \text{ s})(\cos 0°) = \boxed{26 \text{ J}}$$

The 26 J of work done on the rod by the hand is mechanical energy and is the same as the 26 J of energy consumed by the light bulb (see Example 1). Hence, the moving rod and the magnetic force convert mechanical energy into electrical energy, much as a battery converts chemical energy into electrical energy.

Related Homework: *Problem 7*

It is important to realize that the direction of the current in Figure 22.5 is consistent with the principle of conservation of energy. Consider what would happen if the direction of the current were reversed, as in Figure 22.6. With the direction of the current reversed, the direction of the magnetic force $\vec{F}$ would also be reversed and would point in the direction of the velocity $\vec{v}$ of the rod. As a result, the force would cause the rod to accelerate rather than decelerate. The rod would accelerate without the need for an external force (like that provided by the hand in Figure 22.5) and would create a motional emf that supplies energy to the light bulb. Thus, this hypothetical generator would produce energy out of nothing, since there is no external agent. Such a device cannot exist because it violates the principle of conservation of energy, which states that energy cannot be created or destroyed, but can only be converted from one form to another. Therefore, the current cannot be directed clockwise around the circuit, as in Figure 22.6. In situations such as the one in Examples 1 and 2, when a motional emf leads to an induced current, a magnetic force always appears that opposes the motion, in accord with the principle of conservation of energy. Conceptual Example 3 deals further with the important issue of energy conservation.

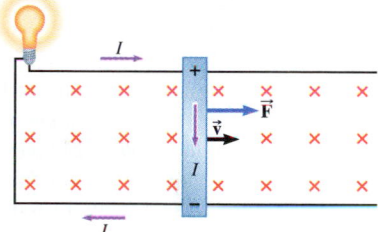

Figure 22.6 The current cannot be directed clockwise in this circuit, because the magnetic force $\vec{F}$ exerted on the rod would then be in the same direction as the velocity $\vec{v}$. The rod would accelerate to the right and create energy on its own, violating the principle of conservation of energy.

Conceptual Example 3 Conservation of Energy

Figure 22.7*a* illustrates a conducting rod that is free to slide down between two vertical copper tracks. There is no kinetic friction between the rod and the tracks, although the rod maintains electrical contact with the tracks during its fall. A constant magnetic field $\vec{B}$ is directed perpendicular to the motion of the rod, as the drawing shows. Because there is no friction, the only force that acts on the rod is its weight $\vec{W}$, so the rod falls with an acceleration equal to the acceleration due to gravity, which has a magnitude of $g = 9.8 \text{ m/s}^2$. Suppose that a resistance R is connected between the tops of the tracks, as in part *b* of the drawing. Is the magnitude of the acceleration with which the rod now falls **(a)** equal to g, **(b)** greater than g, or **(c)** less than g?

Reasoning As the rod falls perpendicular to the magnetic field, a motional emf is induced between its ends. This emf is induced whether or not the resistance R is attached between the tracks. However, when R is present, a complete circuit is formed, and the emf produces an

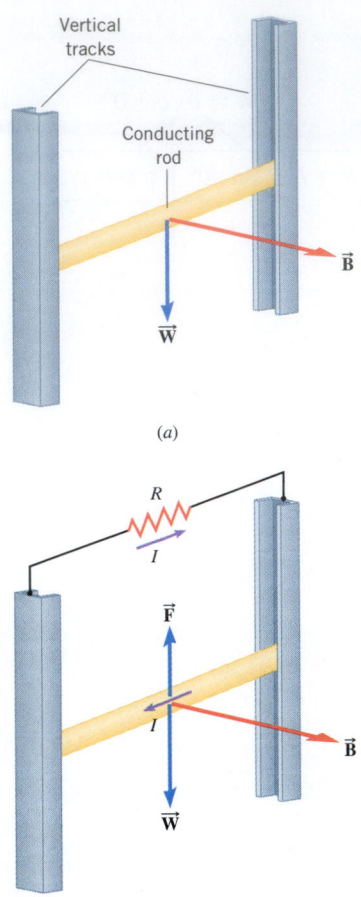

Figure 22.7 (a) Because there is no kinetic friction between the falling rod and the tracks, the only force acting on the rod is its weight $\vec{\mathbf{W}}$. (b) When an induced current I exists in the circuit, a magnetic force $\vec{\mathbf{F}}$ also acts on the rod.

induced current I that is perpendicular to the field. The direction of this current is such that the rod experiences an upward magnetic force $\vec{\mathbf{F}}$, opposite to the weight of the rod (see part b of the drawing and use RHR-1). The net force acting on the rod is $\vec{\mathbf{W}} + \vec{\mathbf{F}}$, which is *less* than the weight, since $\vec{\mathbf{F}}$ points upward and the weight $\vec{\mathbf{W}}$ points downward. In accord with Newton's second law of motion, the downward acceleration is proportional to the net force.

Answers (a) and (b) are incorrect. Since the net downward force on the rod in Figure 22.7b is less than the rod's weight and since the downward acceleration is proportional to the net force, the rod cannot have an acceleration whose magnitude is equal to or greater than g.

Answer (c) is correct. Since the net downward force on the rod when R is present is less than the rod's weight and since the downward acceleration is proportional to the net force, the downward acceleration has a magnitude less than g. Thus, the rod gains speed as it falls but does so less rapidly than if R were not present. As the speed of the rod in Figure 22.7b increases during the descent, the magnetic force increases, until the time comes when its magnitude equals the magnitude of the rod's weight. When this occurs, the net force and the rod's acceleration will be zero. From this moment on, the rod will fall at a constant velocity. In any event, the rod always has a smaller speed than does a freely falling rod (i.e., R is absent) at the same place. The speed is smaller because only part of the gravitational potential energy (GPE) is being converted into kinetic energy (KE) as the rod falls, with part also being dissipated as heat in the resistance R. In fact, when the rod eventually attains a constant velocity, none of the GPE is converted into KE and all of it is dissipated as heat.

Related Homework: *Problem 10*

✓ CHECK YOUR UNDERSTANDING

(The answers are given at the end of the book.)

2. Consider the induced emf being generated in Figure 22.4. Suppose that the length of the rod is reduced by a factor of four. For the induced emf to be the same, what should be done? **(a)** Without changing the speed of the rod, increase the strength of the magnetic field by a factor of four. **(b)** Without changing the magnetic field, increase the speed of the rod by a factor of four. **(c)** Increase both the speed of the rod and the strength of the magnetic field by a factor of two. **(d)** All of the previous three methods may be used.

3. In the discussion concerning Figure 22.5, we saw that a force of 0.086 N from an external agent was required to keep the rod moving at a constant velocity. Suppose that friction is absent and that the light bulb is suddenly removed from its socket while the rod is moving. How much force does the external agent then need to apply to the rod to keep it moving at a constant velocity? **(a)** 0 N **(b)** Greater than 0 N but less than 0.086 N **(c)** More than 0.086 N **(d)** 0.086 N

4. Eddy currents are electric currents that can arise in a piece of metal when it moves through a region where the magnetic field is not the same everywhere. The drawing shows, for example, a metal sheet moving to the right at a velocity $\vec{\mathbf{v}}$ and a magnetic field $\vec{\mathbf{B}}$ that is directed perpendicular to the sheet. At the instant represented, the field only extends over the left half of the sheet. An emf is induced that leads to the eddy current indicated. Such eddy currents cause the velocity of the moving sheet to decrease and are used in various devices as a brake to damp out unwanted motion. Does the eddy current in the drawing circulate **(a)** counterclockwise or **(b)** clockwise?

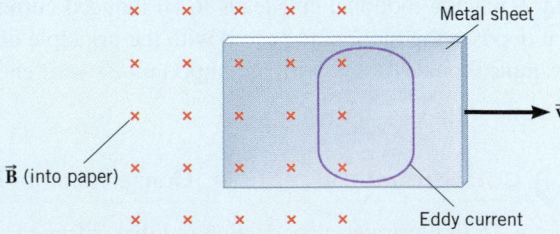

MAGNETIC FLUX

MOTIONAL EMF AND MAGNETIC FLUX

Motional emf, as well as any other type of induced emf, can be described in terms of a concept called *magnetic flux*. Magnetic flux is analogous to electric flux, which deals with the electric field and the surface through which it passes (see Section 18.9 and Figure 18.32). Magnetic flux is defined in a similar way by bringing together the magnetic field and the surface through which it passes.

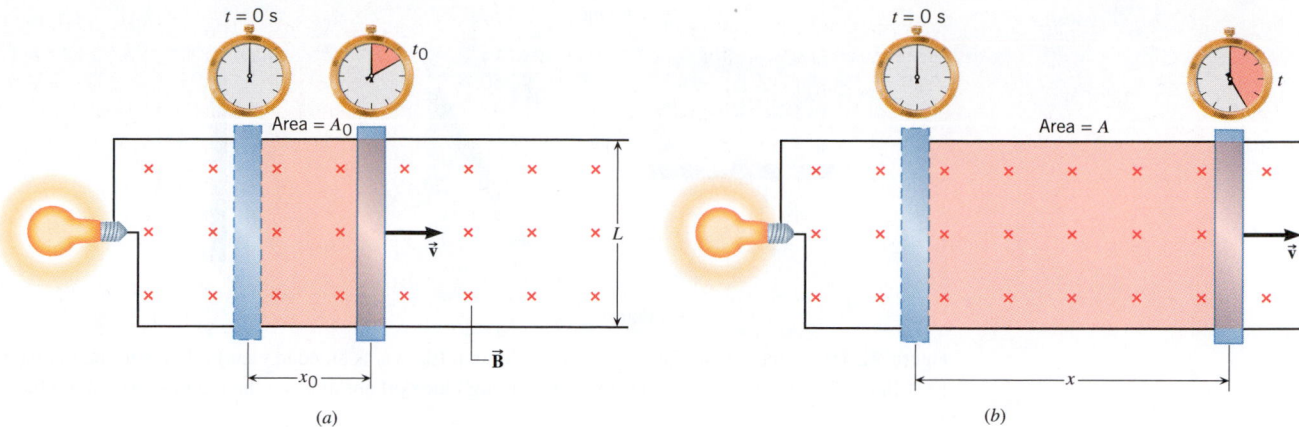

(a) (b)

Figure 22.8 (a) In a time t_0, the moving rod sweeps out an area $A_0 = x_0 L$. (b) The area swept out in a time t is $A = xL$. In both parts of the figure the areas are shaded in color.

We can see how the motional emf is related to the magnetic flux with the aid of Figure 22.8, which shows the rod used to derive Equation 22.1 ($\mathscr{E} = vBL$). In this drawing the rod moves through a magnetic field beginning at time $t = 0$ s. In part a the rod has moved a distance x_0 to the right at time t_0, whereas in part b it has moved a greater distance x at a later time t. The speed v of the rod is the distance traveled divided by the elapsed time: $v = (x - x_0)/(t - t_0)$. Substituting this expression for v into $\mathscr{E} = vBL$ gives

$$\mathscr{E} = \left(\frac{x - x_0}{t - t_0}\right) BL = \left(\frac{xL - x_0 L}{t - t_0}\right) B$$

As the drawing indicates, the term $x_0 L$ is the area A_0 swept out by the rod in moving a distance x_0, while xL is the area A swept out in moving a distance x. Thus, the emf becomes

$$\mathscr{E} = \left(\frac{A - A_0}{t - t_0}\right) B = \frac{(BA) - (BA)_0}{t - t_0}$$

The product BA of the magnetic field strength and the area appears in the numerator of this expression. This product is called **_magnetic flux_** and is represented by the symbol Φ (Greek capital letter phi); thus $\Phi = BA$. The magnitude of the induced emf is the *change* in flux $\Delta\Phi = \Phi - \Phi_0$ divided by the time interval $\Delta t = t - t_0$ during which the change occurs:

$$\mathscr{E} = \frac{\Phi - \Phi_0}{t - t_0} = \frac{\Delta\Phi}{\Delta t}$$

In other words, the induced emf equals the time rate of change of the magnetic flux.

You will almost always see the previous equation written with a minus sign—namely, $\mathscr{E} = -\Delta\Phi/\Delta t$. The minus sign is introduced for the following reason: The direction of the current induced in the circuit is such that the magnetic force $\vec{\mathbf{F}}$ acts on the rod to *oppose* its motion, thereby tending to slow down the rod (see Figure 22.5). The minus sign ensures that the polarity of the induced emf sends the induced current in the proper direction so as to give rise to this opposing magnetic force.* This issue of the polarity of the induced emf will be discussed further in Section 22.5.

The advantage of writing the induced emf as $\mathscr{E} = -\Delta\Phi/\Delta t$ is that this relation is far more general than our present discussion suggests. In Section 22.4 we will see that $\mathscr{E} = -\Delta\Phi/\Delta t$ can be applied to *all possible ways of generating induced emfs*.

A GENERAL EXPRESSION FOR MAGNETIC FLUX

In Figure 22.8 the direction of the magnetic field $\vec{\mathbf{B}}$ is perpendicular to the surface swept out by the moving rod. In general, however, $\vec{\mathbf{B}}$ may not be perpendicular to the surface. For instance, in Figure 22.9 the direction perpendicular to the surface is indicated by the normal to the surface, but the magnetic field is inclined at an angle ϕ with respect to this direction. In such a case the flux is computed using only the component of the field

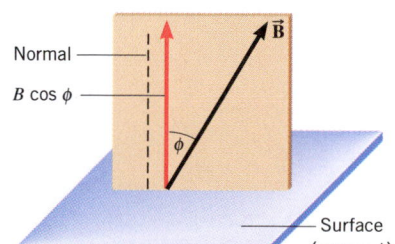

Figure 22.9 When computing the magnetic flux, the component of the magnetic field that is perpendicular to the surface must be used; this component is $B \cos \phi$.

*A detailed mathematical discussion of why the minus sign arises is beyond the scope of this book.

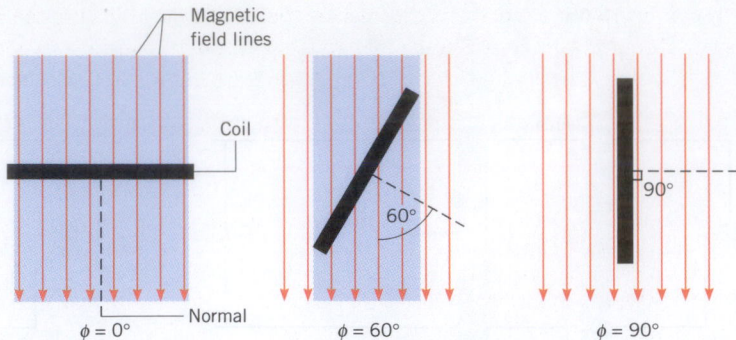

Figure 22.10 Three orientations of a rectangular coil (drawn as an edge view) relative to the magnetic field lines. The magnetic field lines that pass through the coil are those in the regions shaded in blue.

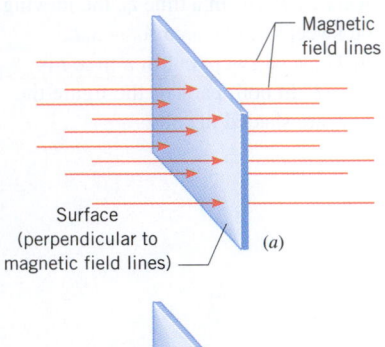

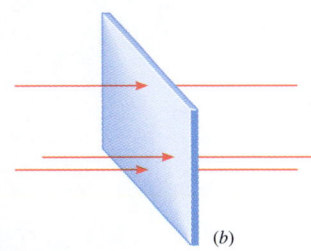

Figure 22.11 The magnitude of the magnetic field in (*a*) is three times as great as that in (*b*) because the number of magnetic field lines crossing the surfaces is in the ratio of 3:1.

Problem-solving insight

The magnetic flux Φ is determined by more than just the magnitude B of the magnetic field and the area A. It also depends on the angle ϕ (see Figure 22.9 and Equation 22.2).

that is perpendicular to the surface, $B \cos \phi$. The general expression for magnetic flux is

$$\Phi = (B \cos \phi)A = BA \cos \phi \qquad (22.2)$$

If either the magnitude B of the magnetic field or the angle ϕ is not constant over the surface, (i.e., if they are not the same at each point on the surface), an average value for the product $B \cos \phi$ must be used to compute the flux. Equation 22.2 shows that the unit of magnetic flux is the tesla · meter^2 (T · m^2). This unit is called a *weber* (Wb), after the German physicist Wilhelm Weber (1804–1891): 1 Wb = 1 T · m^2. Example 4 illustrates how to determine the magnetic flux for three different orientations of the surface of a coil relative to the magnetic field.

Example 4 Magnetic Flux

A rectangular coil of wire is situated in a constant magnetic field whose magnitude is 0.50 T. The coil has an area of 2.0 m^2. Determine the magnetic flux for the three orientations, $\phi = 0°$, 60.0°, and 90.0°, shown in Figure 22.10.

Reasoning The magnetic flux Φ is defined as $\Phi = BA \cos \phi$, where B is the magnitude of the magnetic field, A is the area of the surface through which the magnetic field passes, and ϕ is the angle between the magnetic field and the normal to the surface.

Solution The magnetic flux for the three cases is:

$\phi = 0°$	$\Phi = (0.50 \text{ T})(2.0 \text{ m}^2) \cos 0° = \boxed{1.0 \text{ Wb}}$
$\phi = 60.0°$	$\Phi = (0.50 \text{ T})(2.0 \text{ m}^2) \cos 60.0° = \boxed{0.50 \text{ Wb}}$
$\phi = 90.0°$	$\Phi = (0.50 \text{ T})(2.0 \text{ m}^2) \cos 90.0° = \boxed{0 \text{ Wb}}$

GRAPHICAL INTERPRETATION OF MAGNETIC FLUX

It is possible to interpret the magnetic flux graphically because the magnitude of the magnetic field $\vec{B}$ is proportional to the number of field lines per unit area that pass through a surface perpendicular to the lines (see Section 21.1). For instance, the magnitude of $\vec{B}$ in Figure 22.11*a* is three times larger than it is in part *b* of the drawing, since the number of field lines drawn through the identical surfaces is in the ratio of 3:1. Because Φ is directly proportional to B for a given area, the flux in part *a* is also three times larger than the flux in part *b*. Therefore, *the magnetic flux is proportional to the number of field lines that pass through a surface.*

The graphical interpretation of flux also applies when the surface is oriented at an angle with respect to $\vec{B}$. For example, as the coil in Figure 22.10 is rotated from $\phi = 0°$ to 60° to 90°, the number of magnetic field lines passing through the surface (see the field lines in the regions shaded in blue) changes in the ratio of 8:4:0 or 2:1:0. The results of Example 4 show that the flux in the three orientations changes by the same ratio. Because the magnetic flux is proportional to the number of field lines passing through a surface, we often use phrases such as "the flux that passes through a surface bounded by a loop of wire."

✓ CHECK YOUR UNDERSTANDING

(The answers are given at the end of the book.)

5. A magnetic field has the same direction and the same magnitude B everywhere. A circular area A is bounded by a loop of wire. Which of the following statements is true concerning the magnitude of the magnetic flux that passes through this area?   **(a)** It is zero. **(b)** It is BA.   **(c)** Its maximum possible value is BA.   **(d)** Its minimum possible value is BA.

6. Suppose that a magnetic field is constant everywhere on a flat 1.0-m² surface and that the magnetic flux through this surface is 2.0 Wb. From these data, which one of the following pieces of information can be determined about the magnetic field?   **(a)** The magnitude of the field   **(b)** The magnitude of the component of the field that is perpendicular to the surface   **(c)** The magnitude of the component of the field that is parallel to the surface

22.4 FARADAY'S LAW OF ELECTROMAGNETIC INDUCTION

Two scientists are given credit for the discovery of electromagnetic induction: the Englishman Michael Faraday (1791–1867) and the American Joseph Henry (1797–1878). Since Faraday investigated electromagnetic induction in more detail and published his findings first, the law that describes the phenomenon bears his name.

Faraday discovered that whenever there is a *change in flux* through a loop of wire, an emf is induced in the loop. In this context, the word "change" refers to a change as time passes. A flux that is constant in time creates no emf. Faraday's law of electromagnetic induction is expressed by bringing together the idea of magnetic flux and the time interval during which it changes. In fact, Faraday found that the magnitude of the induced emf is equal to the time rate of change of the magnetic flux. This is consistent with the relation we obtained in Section 22.3 for the specific case of motional emf: $\mathcal{E} = -\Delta\Phi/\Delta t$.

Often the magnetic flux passes through a coil of wire containing more than one loop (or turn). If the coil consists of N loops, and if the same flux passes through each loop, it is found experimentally that the total induced emf is N times that induced in a single loop. An analogous situation occurs in a flashlight when two 1.5-V batteries are stacked in series on top of one another to give a total emf of 3.0 volts. For the general case of N loops, the total induced emf is described by Faraday's law of electromagnetic induction in the following manner:

FARADAY'S LAW OF ELECTROMAGNETIC INDUCTION

The average emf $\mathcal{E}$ induced in a coil of N loops is

$$\mathcal{E} = -N\left(\frac{\Phi - \Phi_0}{t - t_0}\right) = -N\frac{\Delta\Phi}{\Delta t} \qquad (22.3)$$

where $\Delta\Phi$ is the change in magnetic flux through one loop and Δt is the time interval during which the change occurs. The term $\Delta\Phi/\Delta t$ is the average time rate of change of the flux that passes through one loop.

SI Unit of Induced Emf: volt (V)

Faraday's law states that an emf is generated if the magnetic flux changes for any reason. Since the flux is given by Equation 22.2 as $\Phi = BA \cos \phi$, it depends on three factors, B, A, and ϕ, any of which may change. Example 5 considers a change in B.

Example 5   The Emf Induced by a Changing Magnetic Field

A coil of wire consists of 20 turns, or loops, each with an area of 1.5×10^{-3} m². A magnetic field is perpendicular to the surface of each loop at all times, so that $\phi = \phi_0 = 0°$. At time $t_0 = 0$ s, the magnitude of the field at the location of the coil is $B_0 = 0.050$ T. At a later time

$t = 0.10$ s, the magnitude of the field at the coil has increased to $B = 0.060$ T. **(a)** Find the average emf induced in the coil during this time. **(b)** What would be the value of the average induced emf if the magnitude of the magnetic field decreased from 0.060 T to 0.050 T in 0.10 s?

Reasoning To find the induced emf, we use Faraday's law of electromagnetic induction (Equation 22.3), combining it with the definition of magnetic flux from Equation 22.2. We note that only the magnitude of the magnetic field changes in time. All other factors remain constant.

Solution **(a)** Since $\phi = \phi_0$, the induced emf is

$$\mathcal{E} = -N\left(\frac{\Phi - \Phi_0}{t - t_0}\right) = -N\left(\frac{BA \cos\phi - B_0 A \cos\phi}{t - t_0}\right)$$

$$= -NA\cos\phi\left(\frac{B - B_0}{t - t_0}\right)$$

$$\mathcal{E} = -(20)(1.5 \times 10^{-3}\text{ m}^2)(\cos 0°)\left(\frac{0.060\text{ T} - 0.050\text{ T}}{0.10\text{ s} - 0\text{ s}}\right) = \boxed{-3.0 \times 10^{-3}\text{ V}}$$

(b) The calculation here is similar to that in part (a), except the initial and final values of B are interchanged. This interchange reverses the sign of the emf, so $\boxed{\mathcal{E} = +3.0 \times 10^{-3}\text{ V}}$. Because the algebraic sign or polarity of the emf is reversed, the direction of the induced current would be opposite to that in part (a).

The next example demonstrates that an emf can be created when a coil is rotated in a magnetic field.

Example 6 The Emf Induced in a Rotating Coil

A flat coil of wire has an area of 0.020 m² and consists of 50 turns. At $t_0 = 0$ s the coil is oriented so the normal to its surface has the same direction ($\phi_0 = 0°$) as a constant magnetic field of magnitude 0.18 T. The coil is then rotated through an angle of $\phi = 30.0°$ in a time of 0.10 s (see Figure 22.10). **(a)** Determine the average induced emf. **(b)** What would be the induced emf if the coil were returned to its initial orientation in the same time of 0.10 s?

Reasoning As in Example 5 we can determine the induced emf by using Faraday's law of electromagnetic induction, along with the definition of magnetic flux. In the present case, however, only ϕ (the angle between the normal to the surface of the coil and the magnetic field) changes in time. All other factors remain constant.

Solution **(a)** Faraday's law yields

$$\mathcal{E} = -N\left(\frac{\Phi - \Phi_0}{t - t_0}\right) = -N\left(\frac{BA\cos\phi - BA\cos\phi_0}{t - t_0}\right)$$

$$= -NBA\left(\frac{\cos\phi - \cos\phi_0}{t - t_0}\right)$$

$$\mathcal{E} = -(50)(0.18\text{ T})(0.020\text{ m}^2)\left(\frac{\cos 30.0° - \cos 0°}{0.10\text{ s} - 0\text{ s}}\right) = \boxed{+0.24\text{ V}}$$

(b) When the coil is rotated back to its initial orientation in a time of 0.10 s, the initial and final values of ϕ are interchanged. As a result, the induced emf has the same magnitude but opposite polarity, so $\boxed{\mathcal{E} = -0.24\text{ V}}$.

One application of Faraday's law that is found in the home is a safety device called a ground fault interrupter. This device protects against electrical shock from an appliance, such as a clothes dryer. It plugs directly into a wall socket, as in Figure 22.12 or, in new home construction, replaces the socket entirely. The interrupter consists of a circuit breaker that can be triggered to stop the current to the dryer, depending on whether an induced voltage appears across a sensing coil. This coil is wrapped around an iron ring, through which the current-carrying wires pass. In the drawing, the current going to the dryer is shown in red, and the returning current is shown in green. Each of the currents creates a

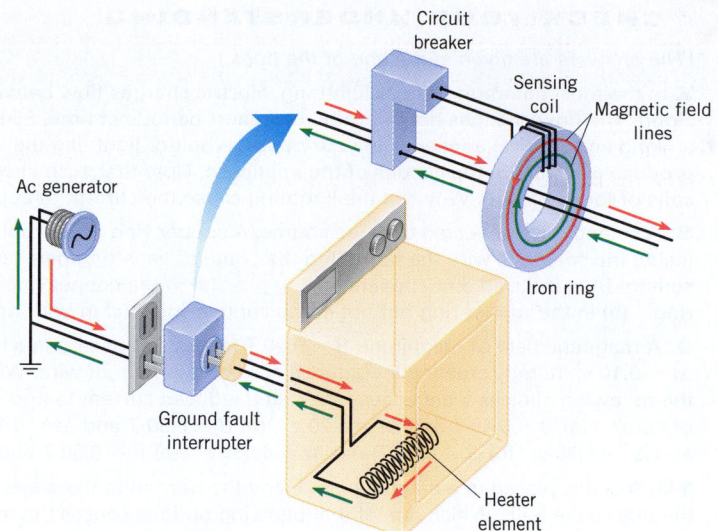

Figure 22.12 The clothes dryer is connected to the wall socket through a ground fault interrupter. The dryer is operating normally.

magnetic field that encircles the corresponding wire, according to RHR-2 (see Section 21.7). However, the field lines have opposite directions since the currents have opposite directions. As the drawing shows, the iron ring guides the field lines through the sensing coil. Since the current is ac, the fields from the red and green current are changing, but the red and green field lines always have opposite directions and the opposing fields cancel at all times. As a result, the net flux through the coil remains zero, and no induced emf appears in the coil. Thus, when the dryer operates normally, the circuit breaker is not triggered and does not shut down the current. The picture changes if the dryer malfunctions, as when a wire inside the unit breaks and accidentally contacts the metal case. When someone touches the case, some of the current begins to pass through the person's body and into the ground, returning to the ac generator *without using the return wire that passes through the ground fault interrupter*. Under this condition, the net magnetic field through the sensing coil is no longer zero and changes with time, since the current is ac. The changing flux causes an induced voltage to appear in the sensing coil, which triggers the circuit breaker to stop the current. Ground fault interrupters work very fast (in less than a millisecond) and turn off the current before it reaches a dangerous level.

Conceptual Example 7 discusses another application of electromagnetic induction—namely, how a stove can cook food without getting hot.

 Conceptual Example 7 An Induction Stove

Figure 22.13 shows two pots of water that were placed on an induction stove at the same time. There are two interesting features in this drawing. First, the stove itself is cool to the touch. Second, the water in the ferromagnetic metal pot is boiling while the water in the glass pot is not. How can such a "cool" stove boil water, and why isn't the water in the glass pot boiling?

Reasoning and Solution The key to this puzzle is related to the fact that one pot is made from a ferromagnetic metal and one from glass. We know that metals are good conductors, while glass is an insulator. Perhaps the stove causes electricity to flow directly in the metal pot. This is exactly what happens. The stove is called an *induction stove* because it operates by using electromagnetic induction. Just beneath the cooking surface is a metal coil that carries an ac current (frequency about 25 kHz). This current produces an alternating magnetic field that extends outward to the location of the metal pot. As the changing field crosses the pot's bottom surface, an emf is induced in it. Because the pot is metallic, an induced current is generated by the induced emf. The metal has a finite resistance to the induced current, however, and heats up as energy is dissipated in this resistance. The fact that the metal is ferromagnetic is important. Ferromagnetic materials contain magnetic domains (see Section 21.9), and the boundaries between them move extremely rapidly in response to the external magnetic field, thus enhancing the induction effect. A normal aluminum cooking pot, in contrast, is not ferromagnetic, so this enhancement is absent and such cookware is not used with induction stoves. An emf is also induced in the glass pot and the cooking surface of the stove. However, these materials are insulators, so very little induced current exists within them. Thus, they do not heat up very much and remain cool to the touch.

The physics of an induction stove.

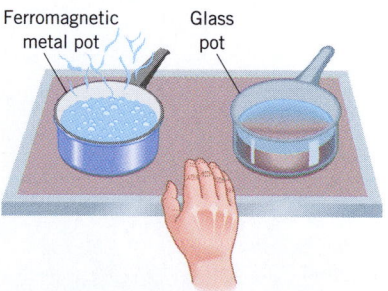

Figure 22.13 The water in the ferromagnetic metal pot is boiling—yet the water in the glass pot is not boiling, and the stove top is cool to the touch. The stove operates in this way by using electromagnetic induction.

✔ **CHECK YOUR UNDERSTANDING**

(The answers are given at the end of the book.)

7. In the most common form of lightning, electric charges flow between the ground and a cloud. The flow changes dramatically over short periods of time. Even without directly striking an electrical appliance in your house, a bolt of lightning that strikes nearby can produce a current in the circuits of the appliance. Note that such circuits typically contain coils or loops of wire. Why can the lightning cause the current to appear?

8. A solenoid is connected to an ac source. A copper ring and a rubber ring are placed inside the solenoid, with the normal to the plane of each ring parallel to the axis of the solenoid. An induced emf appears _____. **(a)** in the copper ring but not in the rubber ring **(b)** in the rubber ring but not in the copper ring **(c)** in both rings

9. A magnetic field of magnitude $B = 0.20$ T is reduced to zero in a time interval of $\Delta t = 0.10$ s, thereby creating an induced current in a loop of wire. Which one or more of the following choices would cause the same induced current to appear in the same loop of wire? **(a)** $B = 0.40$ T and $\Delta t = 0.20$ s **(b)** $B = 0.30$ T and $\Delta t = 0.10$ s **(c)** $B = 0.30$ T and $\Delta t = 0.30$ s **(d)** $B = 0.10$ T and $\Delta t = 0.050$ s **(e)** $B = 0.50$ T and $\Delta t = 0.40$ s

10. A coil is placed in a magnetic field, and the normal to the plane of the coil remains parallel to the field. Which one of the following options causes the magnitude of the average emf induced in the coil to be as large as possible? **(a)** The magnitude of the field is small, and its rate of change is large. **(b)** The magnitude of the field is large, and its rate of change is small. **(c)** The magnitude of the field is large, and it does not change.

22.5 LENZ'S LAW

An induced emf drives current around a circuit just as the emf of a battery does. With a battery, conventional current is directed out of the positive terminal, through the attached device, and into the negative terminal. The same is true for an induced emf, although the locations of the positive and negative terminals are generally not as obvious. Therefore, a method is needed for determining the polarity or algebraic sign of the induced emf, so the terminals can be identified. As we discuss this method, it will be helpful to keep in mind that the net magnetic field penetrating a coil of wire results from two contributions. One is the original magnetic field that produces the changing flux that leads to the induced emf. The other arises because of the induced current, which, like any current, creates its own magnetic field. The field created by the induced current is called the *induced magnetic field.*

To determine the polarity of the induced emf, we will use a method based on a discovery made by the Russian physicist Heinrich Lenz (1804–1865). This discovery is known as *Lenz's law.*

LENZ'S LAW

The induced emf resulting from a changing magnetic flux has a polarity that leads to an induced current whose direction is such that the induced magnetic field opposes the original flux change.

Lenz's law is best illustrated with examples. Each will be worked out according to the following reasoning strategy:

REASONING STRATEGY

Determining the Polarity of the Induced Emf

1. Determine whether the magnetic flux that penetrates a coil is increasing or decreasing.

2. Find what the direction of the induced magnetic field must be so that it can *oppose the change in flux* by adding to or subtracting from the original field.

3. Having found the direction of the induced magnetic field, use RHR-2 (see Section 21.7) to determine the direction of the induced current. Then the polarity of the induced emf can be assigned because conventional current is directed out of the positive terminal, through the external circuit, and into the negative terminal.

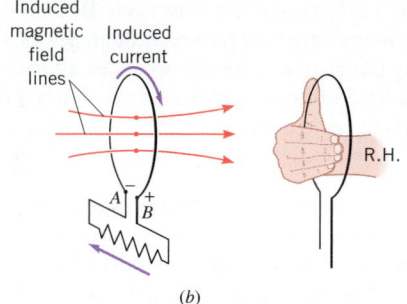

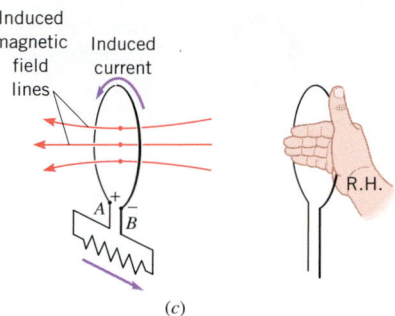

(a)   (b)   (c)

Conceptual Example 8   The Emf Produced by a Moving Magnet

Figure 22.14a shows a permanent magnet approaching a loop of wire. The external circuit attached to the loop consists of the resistance R, which could represent the filament in a light bulb, for instance. In Figure 22.14a, what is the polarity of the induced emf? In other words, **(a)** is point A positive and point B negative or **(b)** is point A negative and point B positive?

Reasoning We will apply Lenz's law, the essence of which is that the change in magnetic flux must be opposed by the induced magnetic field. The flux through the loop is increasing, since the magnitude of the magnetic field at the loop is increasing as the magnet approaches. To oppose the increase in the flux, the direction of the induced magnetic field must be opposite to the field of the bar magnet. Thus, since the field of the bar magnet passes through the loop from left to right in part a of the drawing, the induced field must pass through the loop from right to left. An induced current creates this induced field, and from the direction of this current we will be able to decide the polarity of the induced emf.

Answer (b) is incorrect. If point A were negative and point B were positive, as in Figure 22.14b, the induced current in the loop would be as shown in that part of the drawing, because conventional current exits from the positive terminal and returns through the external circuit (the resistance R) to the negative terminal. Application of RHR-2 reveals that this induced current would lead to an induced field that passes through the loop from left to right, not right to left as needed to oppose the flux change.

Answer (a) is correct. Figure 22.14c shows the situation with point A positive and point B negative and the induced current that results. An application of RHR-2 reveals that the induced field produced by this current indeed passes through the loop from right to left, as needed. We conclude, therefore, that the polarity shown in Figure 22.14c is correct.

In Conceptual Example 8 the direction of the induced magnetic field is opposite to the direction of the external field of the bar magnet. **Problem-solving insight: *The induced field is not always opposite to the external field, however, because Lenz's law requires only that it must oppose the change in the flux that generates the emf.*** Conceptual Example 9 illustrates this point.

Conceptual Example 9   The Emf Produced by a Moving Copper Ring

In Figure 22.15 there is a constant magnetic field in a rectangular region of space. This field is directed perpendicularly into the page. Outside this region there is no magnetic field. A copper ring slides through the region, from position 1 to position 5. Since the field is zero outside the rectangular region, no flux passes through the ring in positions 1 and 5, there is no change in the flux through the ring, and there is no induced emf or current in the ring. Which one of the following options correctly describes the induced current in the ring as it passes through positions 2, 3, and 4? **(a)** I_2 is clockwise, I_3 is counterclockwise, I_4 is counterclockwise. **(b)** I_2 is counterclockwise, I_3 is clockwise, I_4 is clockwise. **(c)** I_2 is clockwise, $I_3 = 0$ A, I_4 is counterclockwise. **(d)** I_2 is counterclockwise, $I_3 = 0$ A, I_4 is clockwise.

Reasoning Lenz's law will guide us. It requires that the induced magnetic field oppose the change in flux. Sometimes this means that the induced field is opposite to the external magnetic field, as in Example 8. However, we will see that the induced field sometimes has the same direction as the external field in order to oppose the flux change.

Figure 22.14 (a) As the magnet moves to the right, the magnetic flux through the loop increases. The external circuit attached to the loop has a resistance R. (b) One possibility and (c) another possibility for the direction of the induced current. See Conceptual Example 8.

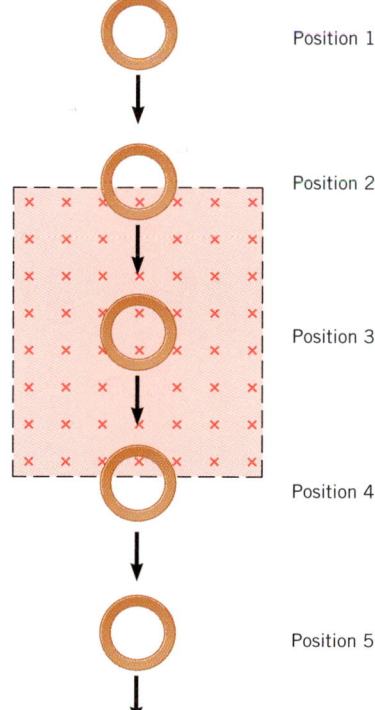

Figure 22.15 A constant magnetic field is directed into the page in the shaded rectangular region. Conceptual Example 9 discusses what happens to the induced emf and current in a copper ring that slides through the region from position 1 to position 5.

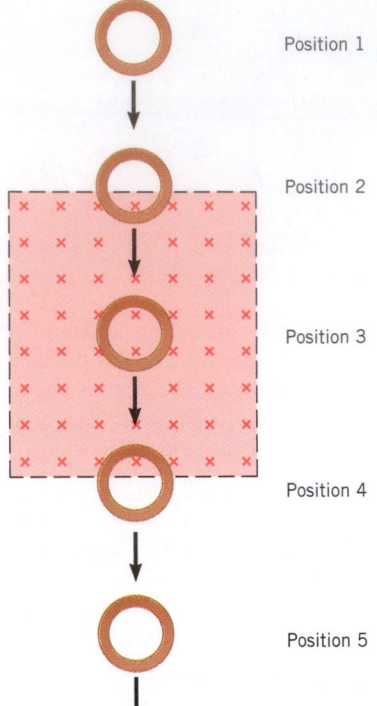

Position 1

Position 2

Position 3

Position 4

Position 5

Figure 22.15 (Repeated) A constant magnetic field is directed into the page in the shaded rectangular region. Conceptual Example 9 discusses what happens to the induced emf and current in a copper ring that slides through the region from position 1 to position 5.

Answers (a) and (b) are incorrect. Both of these answers specify that there is an induced current I_3 in the ring as it passes through position 3, contrary to fact. The external field within the rectangular region certainly produces a flux through the ring. (See Figure 22.15, which is repeated here in the margin for convenience.) However, the external field is constant, so the flux through the ring does not change as the ring moves. In order for an induced emf to exist and to cause an induced current, the flux must change.

Answer (c) is incorrect. As the ring moves out of the field region in position 4, the flux through the ring decreases, so there is an induced emf and an induced current. Lenz's law requires that the induced current must lead to an induced magnetic field that opposes this flux decrease. To oppose the decrease, the induced field must point in the same direction as the external field. To create an induced field pointing into the page, the induced current I_4 must be clockwise (use RHR-2), not counterclockwise as this answer specifies.

Answer (d) is correct. In position 2 the flux increases and, according to Lenz's law, the induced current must create an induced magnetic field that opposes the increase. To oppose the increase the induced field must point opposite to the external field and, therefore, must point out of the page. RHR-2 indicates that the induced current must be counterclockwise, as this answer states. In position 4 the flux through the ring decreases, and the induced magnetic field must oppose the decrease by pointing in the same direction as the external field—namely, into the page. RHR-2 reveals that the induced current must be clockwise, as this answer indicates. In position 3 the flux through the loop is not changing, so there is no induced emf and current, as this answer specifies.

Related Homework: *Problem 35*

Lenz's law should not be thought of as an independent law, because it is a consequence of the law of conservation of energy. The connection between energy conservation and induced emf has already been discussed in Section 22.2 for the specific case of motional emf. However, the connection is valid for any type of induced emf. In fact, the polarity of the induced emf, as specified by Lenz's law, ensures that energy is conserved.

✓ CHECK YOUR UNDERSTANDING

(The answers are given at the end of the book.)

11. In Figure 22.3 a coil of wire is being stretched. What would be the direction of the induced current if the direction of the external magnetic field in the figure were reversed? **(a)** Clockwise **(b)** Counterclockwise

12. A circular loop of wire is lying flat on a horizontal table, and you are looking down at it. An external magnetic field has a constant direction that is perpendicular to the table, and there is an induced clockwise current in the loop. Is the external field directed upward toward you or downward away from you, and is its magnitude increasing or decreasing? Note that there are two possible answers.

13. When the switch in the drawing is closed, the current in the coil increases to its equilibrium value. While the current is increasing there is an induced current in the metal ring. The ring is free to move. What happens to the ring? **(a)** It does not move. **(b)** It jumps downward. **(c)** It jumps upward.

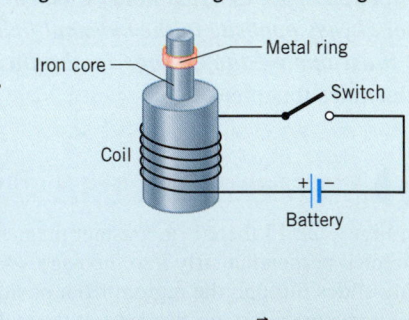

14. A conducting rod is free to slide along a pair of conducting rails, in a region where a uniform and constant (in time) magnetic field is directed into the plane of the paper, as the drawing illustrates. Initially the rod is at rest. There is no friction between the rails and the rod. What happens to the rod after the switch is closed? If any induced emf develops, be sure to account for its effect. **(a)** The rod accelerates to the right, its velocity increasing without limit. **(b)** The rod does not move. **(c)** The rod accelerates to the right for a while and then slows down and comes to a halt. **(d)** The rod accelerates to the right and eventually reaches a constant velocity at which it continues to move.

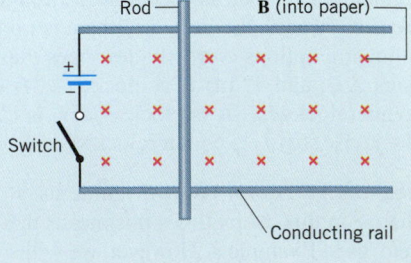

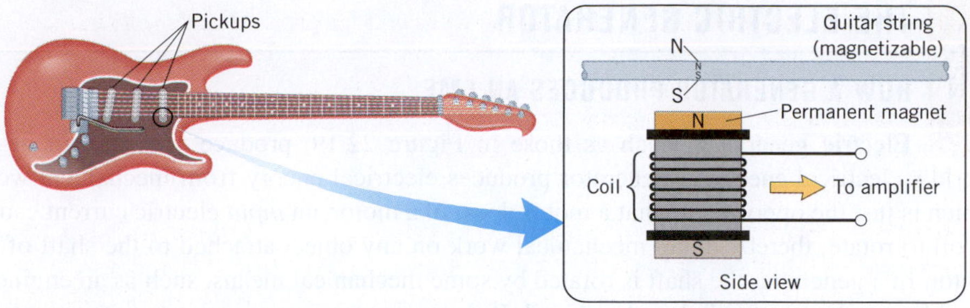

Pickups

Guitar string (magnetizable)

N

S

N — Permanent magnet

Coil

To amplifier

S

Side view

Figure 22.16 When the string of an electric guitar vibrates, an emf is induced in the coil of the pickup. The two ends of the coil are connected to the input of an amplifier.

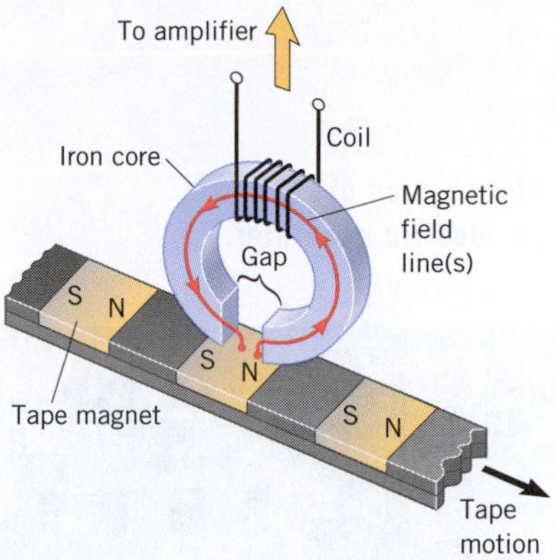

To amplifier

Iron core

Coil

Gap

Magnetic field line(s)

Tape magnet

S N

S N

S N

Tape motion

Figure 22.17 The playback head of a tape deck. As each tape magnet goes by the gap, some magnetic field lines pass through the core and coil. The changing flux in the coil creates an induced emf. The gap width has been exaggerated.

*APPLICATIONS OF ELECTROMAGNETIC INDUCTION TO THE REPRODUCTION OF SOUND

22.6

Electromagnetic induction plays an important role in the technology used for the reproduction of sound. Virtually all electric guitars, for example, use electromagnetic pickups in which an induced emf is generated in a coil of wire by a vibrating string. Each pickup is located below the strings, as Figure 22.16 illustrates, and each is sensitive to different harmonics that the strings produce. Each string is made from a magnetizable metal, and the pickup consists of a coil of wire within which a permanent magnet is located. The magnetic field of the magnet penetrates the guitar string, causing it to become magnetized with north and south poles. When the magnetized string is plucked, it oscillates, thereby changing the magnetic flux that passes through the coil. The changing flux induces an emf in the coil, and the polarity of this emf alternates with the vibratory motion of the string. A string vibrating at 440 Hz, for example, induces a 440-Hz ac emf in the coil. This signal, after being amplified, is sent to loudspeakers, which produce a 440-Hz sound wave (concert A).

The physics of the electric guitar pickup.

The playback head of a tape deck uses a moving tape to generate an emf in a coil of wire. Figure 22.17 shows a section of magnetized tape in which a series of "tape magnets" have been created in the magnetic layer of the tape during the recording process (see Section 21.9). The tape moves beneath the playback head, which consists of a coil of wire wrapped around an iron core. The iron core has the approximate shape of a horseshoe with a small gap between the two ends. Some of the field lines of the tape magnet under the gap are routed through the highly magnetizable iron core, and hence through the coil, as they proceed from the north pole to the south pole. Consequently, the flux through the coil changes as the tape moves past the gap. The change in flux leads to an ac emf, which is amplified and sent to the speakers, where the original sound is reproduced.

The physics of the playback head of a tape deck.

The physics of a moving-coil and a moving-magnet microphone. There are a number of types of microphones, and Figure 22.18 illustrates the one known as a moving-coil microphone. When a sound wave strikes the diaphragm of the microphone, the diaphragm vibrates back and forth, and the attached coil moves along with it. Nearby is a stationary magnet. As the coil alternately approaches and recedes from the magnet, the flux through the coil changes. Consequently, an ac emf is induced in the coil. This electrical signal is sent to an amplifier and then to the speakers. In a moving-magnet microphone, the magnet is attached to the diaphragm and moves relative to a stationary coil.

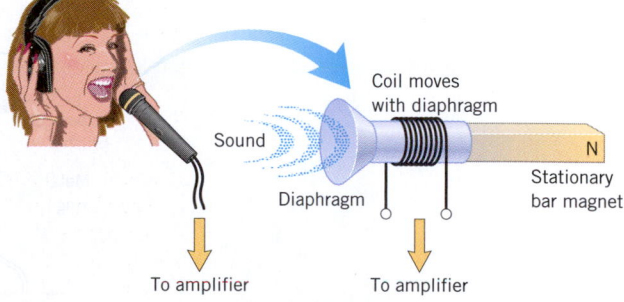

Coil moves with diaphragm

Sound

N

Stationary bar magnet

Diaphragm

To amplifier

To amplifier

Figure 22.18 A moving-coil microphone.

✓ **CHECK YOUR UNDERSTANDING**

(*The answer is given at the end of the book.*)

15. The string of an electric guitar vibrates in a standing wave pattern that consists of nodes and antinodes. (Section 17.5 discusses standing waves.) Where should an electromagnetic pickup be located in the standing wave pattern to produce a maximum emf? **(a)** At a node **(b)** At an antinode

THE ELECTRIC GENERATOR

HOW A GENERATOR PRODUCES AN EMF

**The physics of
an electric generator.**

Figure 22.19 Electric generators such as these supply electric power by producing an induced emf according to Faraday's law of electromagnetic induction. (Michael Melford/The Image Bank/Getty Images)

Electric generators, such as those in Figure 22.19, produce virtually all of the world's electrical energy. A generator produces electrical energy from mechanical work, which is just the opposite of what a motor does. In a motor, an *input* electric current causes a coil to rotate, thereby doing mechanical work on any object attached to the shaft of the motor. In a generator, the shaft is rotated by some mechanical means, such as an engine or a turbine, and an emf is induced in a coil. If the generator is connected to an external circuit, an electric current is the *output* of the generator.

In its simplest form, an ac generator consists of a coil of wire that is rotated in a uniform magnetic field, as Figure 22.20a indicates. Although not shown in the picture, the wire is usually wound around an iron core. As in an electric motor, the coil/core combination is called the *armature*. Each end of the wire forming the coil is connected to the external circuit by means of a metal ring that rotates with the coil. Each ring slides against a stationary carbon brush, to which the external circuit (the lamp in the drawing) is connected.

To see how current is produced by the generator, consider the two vertical sides of the coil in Figure 22.20b. Since each is moving in a magnetic field $\vec{\mathbf{B}}$, the magnetic force exerted on the charges in the wire causes them to flow, thus creating a current. With the aid of RHR-1 (fingers of extended right hand point along $\vec{\mathbf{B}}$, thumb along the velocity $\vec{\mathbf{v}}$, palm pushes in the direction of the force on a positive charge), it can be seen that the direction of the current is from bottom to top in the left side and from top to bottom in the right side. Thus, charge flows around the loop. The upper and lower segments of the loop are also moving. However, these segments can be ignored because the magnetic force on the charges within them points toward the sides of the wire and not along the length.

The magnitude of the motional emf developed in a conductor moving through a magnetic field is given by Equation 22.1. To apply this expression to the left side of the coil, whose length is L (see Figure 22.20c), we need to use the velocity component $v_\perp$ that is perpendicular to $\vec{\mathbf{B}}$. Letting θ be the angle between $\vec{\mathbf{v}}$ and $\vec{\mathbf{B}}$ (see Figure 22.20b), it follows that $v_\perp = v \sin \theta$, and, with the aid of Equation 22.1, the emf can be written as

$$\mathscr{E} = BLv_\perp = BLv \sin \theta$$

The emf induced in the right side has the same magnitude as that in the left side. Since the emfs from both sides drive current in the same direction around the loop, the emf for

Figure 22.20 (*a*) This electric generator consists of a coil (only one loop is shown) of wire that is rotated in a magnetic field $\vec{\mathbf{B}}$ by some mechanical means. (*b*) The current *I* arises because of the magnetic force exerted on the charges in the moving wire. (*c*) The dimensions of the coil.

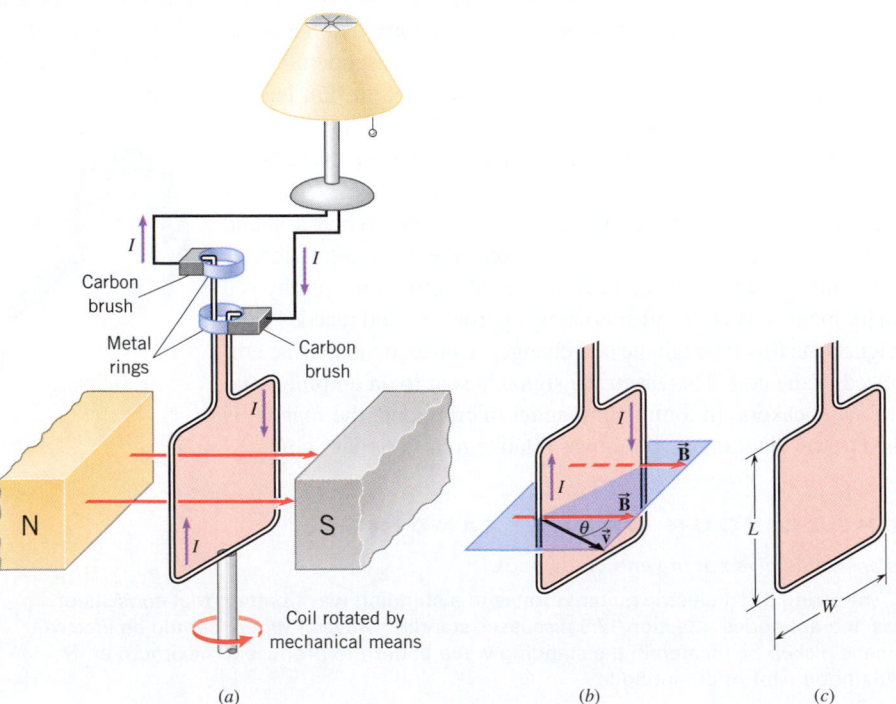

the complete loop is $\mathcal{E} = 2BLv \sin\theta$. If the coil consists of N loops, the net emf is N times as great as that of one loop, so

$$\mathcal{E} = N(2BLv \sin\theta)$$

It is convenient to express the variables v and θ in terms of the angular speed ω at which the coil rotates. Equation 8.2 shows that the angle θ is the product of the angular speed and the time, $\theta = \omega t$, if it is assumed that $\theta = 0$ rad when $t = 0$ s. Furthermore, any point on each vertical side moves on a circular path of radius $r = W/2$, where W is the width of the coil (see Figure 22.20c). Thus, the tangential speed v of each side is related to the angular speed ω via Equation 8.9 as $v = r\omega = (W/2)\omega$. Substituting these expressions for θ and v in the previous equation for $\mathcal{E}$, and recognizing that the product LW is the area A of the coil, we can write the induced emf as

Emf induced in a rotating planar coil
$$\mathcal{E} = NAB\omega \sin\omega t = \mathcal{E}_0 \sin\omega t \qquad \text{where } \omega = 2\pi f \qquad (22.4)$$

In this result, the angular speed ω is in radians per second and is related to the frequency f [in cycles per second or hertz (Hz)] according to $\omega = 2\pi f$ (Equation 10.6).

Although Equation 22.4 was derived for a rectangular coil, the result is valid for any planar shape of area A (e.g., circular) and shows that the emf varies sinusoidally with time. The peak, or maximum, emf $\mathcal{E}_0$ occurs when $\sin\omega t = 1$ and has the value $\mathcal{E}_0 = NAB\omega$. Figure 22.21 shows a plot of Equation 22.4 and reveals that the emf changes polarity as the coil rotates. This changing polarity is exactly the same as that discussed for an ac voltage in Section 20.5 and illustrated in Figure 20.10. If the external circuit connected to the generator is a closed circuit, an alternating current results that changes direction at the same frequency f as the emf changes polarity. Therefore, this electric generator is also called an *alternating current (ac) generator*. The next two examples show how Equation 22.4 is applied.

Squeezing this hand-held generator produces electric power for portable devices such as cell phones, CD players, and pagers. (Mario Tama/ Getty Images News and Sport Services)

 Example 10 An AC Generator

In Figure 22.20 the coil of the ac generator rotates at a frequency of $f = 60.0$ Hz and develops an emf of 120 V (rms; see Section 20.5). The coil has an area of $A = 3.0 \times 10^{-3}$ m^2 and consists of $N = 500$ turns. Find the magnitude of the magnetic field in which the coil rotates.

Reasoning The magnetic field can be found from the relation $\mathcal{E}_0 = NAB\omega$. However, in using this equation we must remember that $\mathcal{E}_0$ is the peak emf, whereas the given value of 120 V is not a peak value but an rms value. The peak emf is related to the rms emf by $\mathcal{E}_0 = \sqrt{2}\mathcal{E}_{rms}$, according to Equation 20.13.

Solution Solving $\mathcal{E}_0 = NAB\omega$ for B and using the facts that $\mathcal{E}_0 = \sqrt{2}\mathcal{E}_{rms}$ and $\omega = 2\pi f$, we find that the magnitude of the magnetic field is

$$B = \frac{\mathcal{E}_0}{NA\omega} = \frac{\sqrt{2}\mathcal{E}_{rms}}{NA2\pi f} = \frac{\sqrt{2}\,(120\text{ V})}{(500)(3.0 \times 10^{-3}\text{ m}^2)2\pi(60.0\text{ Hz})} = \boxed{0.30\text{ T}}$$

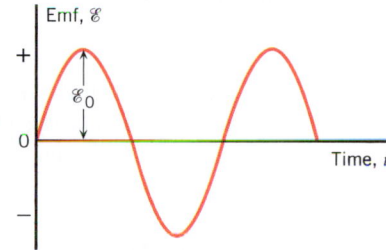

Figure 22.21 An ac generator produces this alternating emf $\mathcal{E}$ according to $\mathcal{E} = \mathcal{E}_0 \sin\omega t$.

Problem-solving insight

In the equation $\mathcal{E}_0 = NAB\omega$, remember that the angular frequency ω must be in rad/s and is related to the frequency f (in Hz) according to $\omega = 2\pi f$ (Equation 10.6).

ANALYZING MULTIPLE-CONCEPT PROBLEMS

Example 11 A Bike Generator

A bicyclist is traveling at night, and a generator mounted on the bike powers a headlight. A small rubber wheel on the shaft of the generator presses against the bike tire and turns the coil of the generator at an angular speed that is 44 times as great as the angular speed of the tire itself. The tire has a radius of 0.33 m. The coil consists of 75 turns, has an area of 2.6×10^{-3} m^2, and rotates in a 0.10-T magnetic field. When the peak emf being generated is 6.0 V, what is the linear speed of the bike?

The physics of a bike generator.

Continued

Reasoning Since the tires are rolling, the linear speed v of the bike is related to the angular speed ω_{tire} of its tires by $v = r\omega_{\text{tire}}$ (see Section 8.6), where r is the radius of a tire. We are given that the angular speed ω_{coil} of the coil is 44 times as great as that of the tire. Thus, $\omega_{\text{tire}} = \frac{1}{44}\omega_{\text{coil}}$ and the linear speed of the bike can be related to the angular speed of the coil. Furthermore, according to the discussion in this section on electric generators, the angular speed of the coil is related to the peak emf developed by the rotating coil, the number of turns in the coil, the area of the coil, and the magnetic field, all of which are known.

Knowns and Unknowns The data for this problem are:

Description	Symbol	Value	Comment
Radius of tire	r	0.33 m	
Number of turns in coil	N	75	
Angular speed of coil	ω_{coil}	$44\omega_{\text{tire}}$	Angular speed of coil is 44 times as great as that of tire.
Area of coil	A	2.6×10^{-3} m^2	
Magnitude of magnetic field	B	0.10 T	
Peak emf produced by generator	$\mathcal{E}_0$	6.0 V	
Unknown Variable			
Linear speed of bike	v	?	

Modeling the Problem

STEP 1 **Rolling Motion** When a tire rolls without slipping on the ground, the linear speed v of the tire (the speed at which its axle is moving forward) is related to the angular speed ω_{tire} of the tire about the axle. This relationship is given by

$$v = r\omega_{\text{tire}} \qquad (8.12)$$

where r is the radius of the tire. We are given that the angular speed ω_{coil} of the coil is 44 times as great as the angular speed of the tire, so $\omega_{\text{coil}} = 44\omega_{\text{tire}}$. Solving this equation for ω_{tire} and substituting the result into $v = r\omega_{\text{tire}}$, we obtain Equation 1 at the right. The radius of the tire is known, but the angular speed of the coil is not; we will evaluate it in Step 2.

$$v = r\left(\tfrac{1}{44}\right)\omega_{\text{coil}} \qquad (1)$$

$?$

STEP 2 **Peak Emf Induced in a Rotating Planar Coil** A generator produces an emf when a coil of wire rotates in a magnetic field. According to Equation 22.4, the peak emf $\mathcal{E}_0$ is given by $\mathcal{E}_0 = NAB\omega_{\text{coil}}$, where N is the number of turns in the coil, A is the area of the coil, B is the magnitude of the magnetic field, and ω_{coil} is the angular speed of the rotating coil. Solving this relation for ω_{coil} gives

$$\omega_{\text{coil}} = \frac{\mathcal{E}_0}{NAB}$$

$$v = r\left(\tfrac{1}{44}\right)\omega_{\text{coil}} \qquad (1)$$

$$\omega_{\text{coil}} = \frac{\mathcal{E}_0}{NAB}$$

Note from the data table that all the variables on the right side of this equation are known. We substitute this result into Equation 1, as indicated at the right.

Solution Algebraically combining the results of the two steps, we have

STEP 1   **STEP 2**

$$v = r\left(\tfrac{1}{44}\right)\omega_{\text{coil}} = r\left(\tfrac{1}{44}\right)\left(\frac{\mathcal{E}}{NAB}\right)$$

The linear speed of the bicycle is

$$v = r\left(\tfrac{1}{44}\right)\left(\frac{\mathcal{E}_0}{NAB}\right) = (0.33\text{ m})\left(\tfrac{1}{44}\right)\left[\frac{6.0\text{ V}}{(75)(2.6 \times 10^{-3}\text{ m}^2)(0.10\text{ T})}\right] = \boxed{2.3\text{ m/s}}$$

Related Homework: *Problem 47*

THE ELECTRICAL ENERGY DELIVERED BY A GENERATOR AND THE COUNTERTORQUE

Some power-generating stations burn fossil fuel (coal, gas, or oil) to heat water and produce pressurized steam for turning the blades of a turbine whose shaft is linked to the generator. Others use nuclear fuel or falling water as a source of energy. As the turbine rotates, the generator coil also rotates and mechanical work is transformed into electrical energy.

The devices to which the generator supplies electricity are known collectively as the "load," because they place a burden or load on the generator by taking electrical energy from it. If all the devices are switched off, the generator runs under a no-load condition, because there is no current in the external circuit and the generator does not supply electrical energy. Then, work needs to be done on the turbine only to overcome friction and other mechanical losses within the generator itself, and fuel consumption is at a minimum.

Figure 22.22 illustrates a situation in which a load is connected to a generator. Because there is now a current $I = I_1 + I_2$ in the coil of the generator and the coil is situated in a magnetic field, the current experiences a magnetic force $\vec{F}$. Figure 22.23 shows the magnetic force acting on the left side of the coil, with the direction of $\vec{F}$ given by RHR-1. A force of equal magnitude but opposite direction also acts on the right side of the coil, although this force is not shown in the drawing. The magnetic force $\vec{F}$ gives rise to a *countertorque* that opposes the rotational motion. The greater the current drawn from the generator, the greater the countertorque, and the harder it is for the turbine to turn the coil. To compensate for this countertorque and keep the coil rotating at a constant angular speed, work must be done by the turbine, which means that more fuel must be burned. This is another example of the law of conservation of energy, since the electrical energy consumed by the load must ultimately come from the energy source used to drive the turbine.

THE BACK EMF GENERATED BY AN ELECTRIC MOTOR

A generator converts mechanical work into electrical energy; in contrast, an electric motor converts electrical energy into mechanical work. Both devices are similar and consist of a coil of wire that rotates in a magnetic field. In fact, as the armature of a motor rotates, the magnetic flux passing through the coil changes and an emf is induced in the coil. Thus, when a motor is operating, two sources of emf are present: (1) the applied emf V that provides current to drive the motor (e.g., from a 120-V outlet), and (2) the emf $\mathcal{E}$ induced by the generator-like action of the rotating coil. The circuit diagram in Figure 22.24 shows these two emfs.

Consistent with Lenz's law, the induced emf $\mathcal{E}$ acts to oppose the applied emf V and is called the **back emf** or the **counter emf** of the motor. The greater the speed of the motor, the greater is the flux change through the coil, and the greater is the back emf. Because V and $\mathcal{E}$ have opposite polarities, the net emf in the circuit is $V - \mathcal{E}$. In Figure 22.24, R is the resistance of the wire in the coil, and the current I drawn by the motor is determined from Ohm's law as the net emf divided by the resistance:

$$I = \frac{V - \mathcal{E}}{R} \tag{22.5}$$

The next example uses this result to illustrate that the current in a motor depends on both the applied emf V and the back emf $\mathcal{E}$.

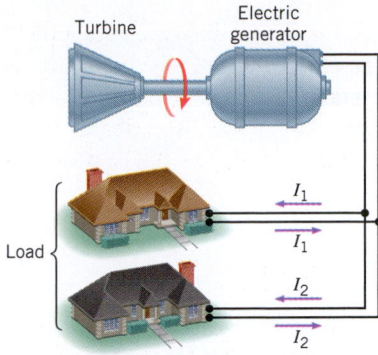

Figure 22.22 The generator supplies a total current of $I = I_1 + I_2$ to the load.

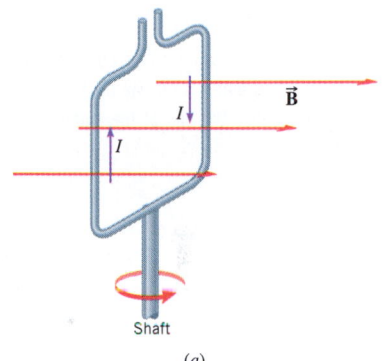

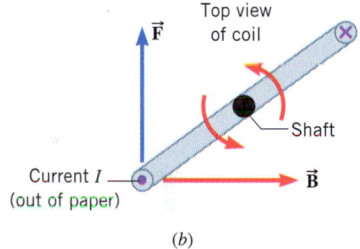

Figure 22.23 (*a*) A current I exists in the rotating coil of the generator. (*b*) A top view of the coil, showing the magnetic force $\vec{F}$ exerted on the left side of the coil.

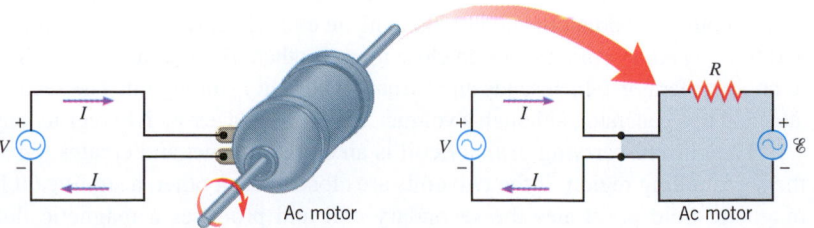

Figure 22.24 The applied emf V supplies the current I to drive the motor. The circuit on the right shows V along with the electrical equivalent of the motor, including the resistance R of its coil and the back emf $\mathcal{E}$.

The physics of operating a motor.

Example 12 Operating a Motor

The coil of an ac motor has a resistance of $R = 4.1 \ \Omega$. The motor is plugged into an outlet where $V = 120.0$ volts (rms), and the coil develops a back emf of $\mathcal{E} = 118.0$ volts (rms) when rotating at normal constant speed. The motor is turning a wheel. Find **(a)** the current when the motor first starts up and **(b)** the current when the motor is operating at normal speed.

Reasoning Once normal operating speed is attained, the motor need only work to compensate for frictional losses. But in bringing the wheel up to speed from rest, the motor must also do work to increase the wheel's rotational kinetic energy. Thus, bringing the wheel up to speed requires more work, and hence more current, than maintaining the normal operating speed. We expect our answers to parts (a) and (b) to reflect this fact.

Solution **(a)** When the motor just starts up, the coil is not rotating, so there is no back emf induced in the coil and $\mathcal{E} = 0$ V. The start-up current drawn by the motor is

$$I = \frac{V - \mathcal{E}}{R} = \frac{120 \text{ V} - 0 \text{ V}}{4.1 \ \Omega} = \boxed{29 \text{ A}} \tag{22.5}$$

(b) At normal speed, the motor develops a back emf of $\mathcal{E} = 118.0$ volts, so the current is

$$I = \frac{V - \mathcal{E}}{R} = \frac{120.0 \text{ V} - 118.0 \text{ V}}{4.1 \ \Omega} = \boxed{0.49 \text{ A}}$$

Problem-solving insight

The current in an electric motor depends on both the applied emf V and any back emf $\mathcal{E}$ developed because the coil of the motor is rotating.

Example 12 illustrates that when a motor is just starting, there is little back emf, and, consequently, a relatively large current exists in the coil. As the motor speeds up, the back emf increases until it reaches a maximum value when the motor is rotating at normal speed. The back emf becomes almost equal to the applied emf, and the current is reduced to a relatively small value, which is sufficient to provide the torque on the coil needed to overcome frictional and other losses in the motor and to drive the load (e.g., a fan).

✔ **CHECK YOUR UNDERSTANDING**

(*The answers are given at the end of the book.*)

16. In a car, the generator-like action of the alternator occurs while the engine is running and keeps the battery fully charged. The headlights would discharge an old and failing battery quickly if it were not for the alternator. Why does the engine of a parked car run more quietly with the headlights off than with them on when the battery is in bad shape?

17. You have a fixed length of wire and need to design a generator that will produce the greatest peak emf for a given frequency and magnetic field strength. You should use **(a)** a one-turn square coil, **(b)** a two-turn square coil, **(c)** either a one- or a two-turn square coil because both give the same peak emf for a given frequency and magnetic field strength.

18. An electric motor in a hair dryer is running at its normal constant operating speed and, thus, is drawing a relatively small current, as in part (b) of Example 12. The wire in the coil of the motor has some resistance. What happens to the temperature of the coil if the shaft of the motor is prevented from turning, so that the back emf is suddenly reduced to zero? **(a)** Nothing. **(b)** The temperature decreases. **(c)** The temperature increases (the coil could even burn up).

22.8 MUTUAL INDUCTANCE AND SELF-INDUCTANCE

MUTUAL INDUCTANCE

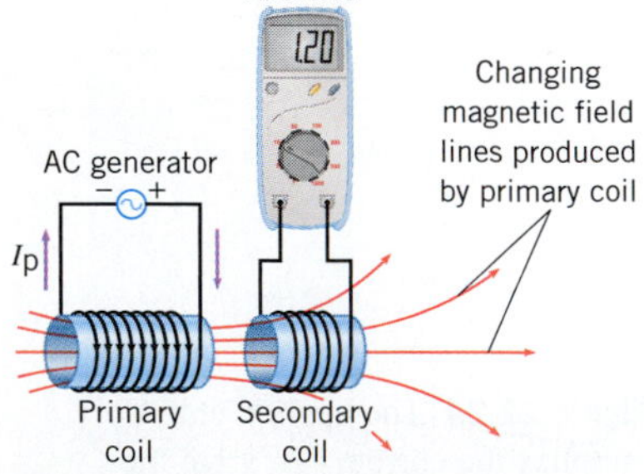

We have seen that an emf can be induced in a coil by keeping the coil stationary and moving a magnet nearby, or by moving the coil near a stationary magnet. Figure 22.25 illustrates another important method of inducing an emf. Here, two coils of wire, the *primary coil* and the *secondary coil*, are placed close to each other. The primary coil is the one connected to an ac generator, which sends an alternating current I_p through it. The secondary coil is not attached to a generator, although a voltmeter is connected across it to register any induced emf.

The current-carrying primary coil is an electromagnet and creates a magnetic field in the surrounding region. If the two coils are close to each other, a significant fraction of this magnetic field penetrates the secondary coil and produces a magnetic flux. The flux is

Figure 22.25 An alternating current I_p in the primary coil creates an alternating magnetic field. This changing field induces an emf in the secondary coil.

changing, since the current in the primary coil and its associated magnetic field are changing. Because of the change in flux, an emf is induced in the secondary coil.

The effect in which a changing current in one circuit induces an emf in another circuit is called **mutual induction.** According to Faraday's law of electromagnetic induction, the average emf $\mathcal{E}_s$ induced in the secondary coil is proportional to the change in flux $\Delta\Phi_s$ passing through it. However, $\Delta\Phi_s$ is produced by the change in current ΔI_p in the primary coil. Therefore, it is convenient to recast Faraday's law into a form that relates $\mathcal{E}_s$ to ΔI_p. To see how this recasting is accomplished, note that the net magnetic flux passing through the secondary coil is $N_s\Phi_s$, where N_s is the number of loops in the secondary coil and Φ_s is the flux through one loop (assumed to be the same for all loops). The net flux is proportional to the magnetic field, which, in turn, is proportional to the current I_p in the primary coil. Thus, we can write $N_s\Phi_s \propto I_p$. This proportionality can be converted into an equation in the usual manner by introducing a proportionality constant M, known as the **mutual inductance:**

$$N_s\Phi_s = MI_p \quad \text{or} \quad M = \frac{N_s\Phi_s}{I_p} \tag{22.6}$$

Substituting this equation into Faraday's law, we find that

$$\mathcal{E}_s = -N_s\frac{\Delta\Phi_s}{\Delta t} = -\frac{\Delta(N_s\Phi_s)}{\Delta t} = -\frac{\Delta(MI_p)}{\Delta t} = -M\frac{\Delta I_p}{\Delta t}$$

Emf due to mutual induction

$$\mathcal{E}_s = -M\frac{\Delta I_p}{\Delta t} \tag{22.7}$$

Writing Faraday's law in this manner makes it clear that the average emf $\mathcal{E}_s$ induced in the secondary coil is due to the change in the current ΔI_p in the primary coil.

Equation 22.7 shows that the measurement unit for the mutual inductance M is V·s/A, which is called a henry (H) in honor of Joseph Henry: 1 V·s/A = 1 H. The mutual inductance depends on the geometry of the coils and the nature of any ferromagnetic core material that is present. Although M can be calculated for some highly symmetrical arrangements, it is usually measured experimentally. In most situations, values of M are less than 1 H and are often on the order of millihenries (1 mH = 1×10^{-3} H) or microhenries (1 μH = 1×10^{-6} H).

A new technique that shows promise for the treatment of psychiatric disorders such as depression is based on mutual induction. This technique is called transcranial magnetic stimulation (TMS) and is a type of indirect and gentler electric shock therapy. In traditional electric shock therapy, electric current is delivered directly through the skull and penetrates the brain, disrupting its electrical circuitry and in the process alleviating the symptoms of the psychiatric disorder. The treatment is not gentle and requires an anesthetic, because relatively large electric currents must be used to penetrate the skull. In contrast, TMS produces its electric current by using a time-varying magnetic field. A primary coil is held over the part of the brain to be treated (see Figure 22.26), and a time-varying current is applied to

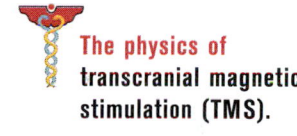

The physics of transcranial magnetic stimulation (TMS).

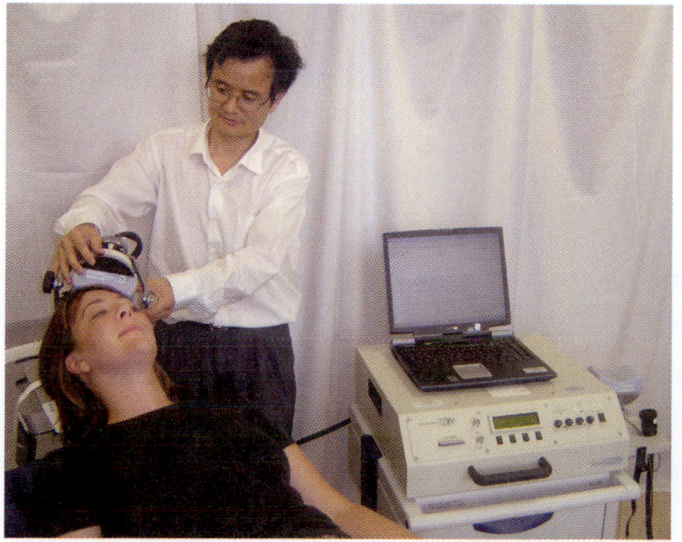

Figure 22.26 In the technique of transcranial magnetic stimulation (TMS), a time-varying electric current is applied to a primary coil, which is held over a region of the brain, as this photograph illustrates. The time-varying magnetic field produced by the coil penetrates the brain and creates an induced emf within it. This induced emf leads to an induced current that disrupts the electric circuits of the brain, thereby relieving some of the symptoms of psychiatric disorders such as depression. (Courtesy Mark S. George, Xingbao Li, and Chris Molnar, Medical University of South Carolina)

this coil. The arrangement is analogous to that in Figure 22.25, except that the brain and the electrically conductive pathways within it take the place of the secondary coil. The magnetic field produced by the primary coil penetrates the brain and, since the field is changing in time, it induces an emf in the brain. This induced emf causes an electric current to flow in the conductive brain tissue, with therapeutic results similar to those of conventional electric shock treatment. The current delivered to the brain, however, is much smaller than the current in the conventional treatment, so that patients receive TMS treatments without anesthetic and without severe after-effects such as headaches and memory loss. TMS remains in the experimental stage, however, and the optimal protocol for applying the technique has not yet been determined.

SELF-INDUCTANCE

In all the examples of induced emfs presented so far, the magnetic field has been produced by an external source, such as a permanent magnet or an electromagnet. However, the magnetic field need not arise from an external source. An emf can be induced in a current-carrying coil by a change in the magnetic field that the current itself produces. For instance, Figure 22.27 shows a coil connected to an ac generator. The alternating current creates an alternating magnetic field that, in turn, creates a changing flux through the coil. The change in flux induces an emf in the coil, in accord with Faraday's law. The effect in which a changing current in a circuit induces an emf in the same circuit is referred to as *self-induction.*

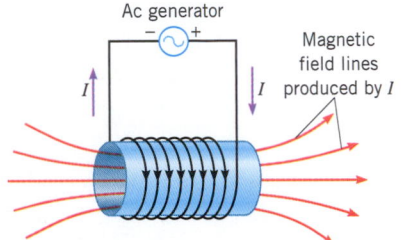

Figure 22.27 The alternating current in the coil generates an alternating magnetic field that induces an emf in the coil.

When dealing with self-induction, as with mutual induction, it is customary to recast Faraday's law into a form in which the induced emf is proportional to the change in current in the coil rather than to the change in flux. If Φ is the magnetic flux that passes through one turn of the coil, then $N\Phi$ is the net flux through a coil of N turns. Since Φ is proportional to the magnetic field, and the magnetic field is proportional to the current I, it follows that $N\Phi \propto I$. By inserting a constant L, called the *self-inductance,* or simply the *inductance,* of the coil, we can convert this proportionality into Equation 22.8:

$$N\Phi = LI \quad \text{or} \quad L = \frac{N\Phi}{I} \tag{22.8}$$

Faraday's law of induction now gives the average induced emf as

$$\mathscr{E} = -N\frac{\Delta\Phi}{\Delta t} = -\frac{\Delta(N\Phi)}{\Delta t} = -\frac{\Delta(LI)}{\Delta t} = -L\frac{\Delta I}{\Delta t}$$

Emf due to self-induction

$$\mathscr{E} = -L\frac{\Delta I}{\Delta t} \tag{22.9}$$

Like mutual inductance, L is measured in henries. The magnitude of L depends on the geometry of the coil and on the core material. Wrapping the coil around a ferromagnetic (iron) core substantially increases the magnetic flux—and therefore the inductance—relative to that for an air core. Because of their self-inductance, coils are known as *inductors* and are widely used in electronics. Inductors come in all sizes, typically in the range between millihenries and microhenries. Example 13 shows how to determine the emf induced in a solenoid from knowledge of its self-inductance and the rate at which the current in the solenoid changes.

ANALYZING MULTIPLE-CONCEPT PROBLEMS

Example 13 The Emf Induced in a Long Solenoid

A long solenoid of length 8.0×10^{-2} m and cross-sectional area 5.0×10^{-5} m² contains 6500 turns per meter of length. Determine the emf induced in the solenoid when the current in the solenoid changes from 0 to 1.5 A during the time interval from 0 to 0.20 s.

Reasoning According to the relation $\mathscr{E} = -L(\Delta I/\Delta t)$, the emf $\mathscr{E}$ induced in the solenoid depends on the self-inductance L of the solenoid and the rate $\Delta I/\Delta t$ at which the solenoidal current changes. The variables ΔI and Δt are known, and we can evaluate the self-inductance by using the definition, $L = N\Phi/I$, where N is the number of turns in the coil, Φ is the magnetic flux that passes through one turn, and I is the current. The magnetic flux Φ is given by $\Phi = BA \cos \phi$ (Equation 22.2), and the magnetic field inside a solenoid has the value of $B = \mu_0 n I$ (Equation 21.7), where n is the number of turns per unit length. These relations will enable us to evaluate the emf induced in the solenoid.

Knowns and Unknowns The data for this problem are listed below:

Description	Symbol	Value
Length of solenoid	ℓ	8.0×10^{-2} m
Cross-sectional area of solenoid	A	5.0×10^{-5} m^2
Number of turns per unit length	n	6500 turns/meter
Change in current	ΔI	1.5 A
Change in time	Δt	0.20 s
Unknown Variable		
Emf induced in solenoid	$\mathscr{E}$	?

Modeling the Problem

STEP 1 **Emf Due to Self-Induction** The emf $\mathscr{E}$ produced by self-induction is given at the right by Faraday's law of induction in the form expressed in Equation 22.9, where L is the self-inductance of the solenoid, ΔI is the change in current, and Δt is the change in time. Both ΔI and Δt are known (see the data table), and L will be evaluated in the next step.

$$\mathscr{E} = -L\frac{\Delta I}{\Delta t} \qquad (22.9)$$

$$?$$

STEP 2 **Self-Inductance** The self-inductance L is defined by $L = N\Phi/I$ (Equation 22.8), where N is the number of turns, Φ is the magnetic flux that passes through one turn, and I is the current in the coil. The number of turns is equal to the number n of turns per meter of length times the length ℓ of the solenoid, or $N = n\ell$. Substituting this expression for N into $L = N\Phi/I$ gives

$$L = \frac{n\ell\Phi}{I}$$

This result for the self-inductance can be substituted into Equation 22.9, as indicated at the right. Note from the data table that n and ℓ are known. The magnetic flux Φ is not known and, along with the current I, will be dealt with in Step 3.

$$\mathscr{E} = -L\frac{\Delta I}{\Delta t} \qquad (22.9)$$

$$L = \frac{n\ell\Phi}{I} \qquad (1)$$

$$?$$

STEP 3 **Magnetic Flux** According to the discussion in Section 22.3, the magnetic flux Φ is given by $\Phi = BA \cos \phi$ (Equation 22.2), where B is the magnitude of the magnetic field that penetrates the surface of area A, and ϕ is the angle between the magnetic field and the normal to the surface. In the case of a solenoid, the interior magnetic field is directed perpendicular to the plane of the loops (see Section 21.7), so $\phi = 0°$. Thus, the magnetic flux is $\Phi = BA \cos 0°$. According to the discussion in Section 21.7, the magnitude of the magnetic field inside a long solenoid is given by $B = \mu_0 n I$ (Equation 21.7). Substituting this expression for B into the equation for the flux yields

$$\Phi = (\mu_0 n I) A \cos 0°$$

$$\mathscr{E} = -L\frac{\Delta I}{\Delta t} \qquad (22.9)$$

$$L = \frac{n\ell\Phi}{I} \qquad (1)$$

$$\Phi = (\mu_0 n I) A \cos 0°$$

All the variables on the right side of this equation, except the current, are known. We now substitute this expression into Equation 1, as indicated in the right column. Note that the current I appears in both the numerator and denominator in the expression for the self-inductance L, so I will be algebraically eliminated from the final result.

Continued

Solution Algebraically combining the results of the three steps, we have

STEP 1	STEP 2		STEP 3

$$\mathscr{E} = -L\frac{\Delta I}{\Delta t} = -\frac{n\ell\Phi}{I}\left(\frac{\Delta I}{\Delta t}\right) = -\frac{n\ell(\mu_0 nI\, A\cos 0°)}{I}\left(\frac{\Delta I}{\Delta t}\right) = \underbrace{-\mu_0 n^2 \ell A}_{\substack{\text{Self-}\\\text{inductance}}}\left(\frac{\Delta I}{\Delta t}\right)$$

The self-inductance of the solenoid is $L = \mu_0 n^2 \ell A$, and the emf induced in the solenoid is

$$\mathscr{E} = -\mu_0 n^2 \ell A\left(\frac{\Delta I}{\Delta t}\right)$$

$$= -(4\pi \times 10^{-7}\ \text{T}\cdot\text{m/A})(6500\ \text{turns/m})^2(8.0 \times 10^{-2}\ \text{m})(5.0 \times 10^{-5}\ \text{m}^2)\left(\frac{1.5\ \text{A}}{0.20\ \text{s}}\right)$$

$$= \boxed{-1.6 \times 10^{-3}\ \text{V}}$$

Related Homework: *Problems 55, 56, 58*

THE ENERGY STORED IN AN INDUCTOR

An inductor, like a capacitor, can store energy. This stored energy arises because a generator does work to establish a current in an inductor. Suppose that an inductor is connected to a generator whose terminal voltage can be varied continuously from zero to some final value. As the voltage is increased, the current I in the circuit rises continuously from zero to its final value. While the current is rising, an induced emf $\mathscr{E} = -L(\Delta I/\Delta t)$ appears across the inductor. Conforming to Lenz's law, the polarity of the induced emf $\mathscr{E}$ is opposite to the polarity of the generator voltage, so as to oppose the increase in the current. Thus, the generator must do work to push the charges through the inductor against this induced emf. The increment of work ΔW done by the generator in moving a small amount of charge ΔQ through the inductor is $\Delta W = -(\Delta Q)\mathscr{E} = -(\Delta Q)[-L(\Delta I/\Delta t)]$, according to Equation 19.4. Since $\Delta Q/\Delta t$ is the current I, the work done is

$$\Delta W = LI(\Delta I)$$

In this expression ΔW represents the work done by the generator to increase the current in the inductor by an amount ΔI. To determine the total work W done while the current is changed from zero to its final value, all the small increments ΔW must be added together. This summation is left as an exercise (see Problem 80). The result is $W = \frac{1}{2}LI^2$, where I represents the final current in the inductor. This work is stored as energy in the inductor, so that

Energy stored in an inductor $$\text{Energy} = \tfrac{1}{2}LI^2 \qquad (22.10)$$

It is possible to regard the energy in an inductor as being stored in its magnetic field. For the special case of a long solenoid, we found in Example 13 that the self-inductance is $L = \mu_0 n^2 A\ell$, where n is the number of turns per unit length, A is the cross-sectional area, and ℓ is the length of the solenoid. As a result, the energy stored in a long solenoid is

$$\text{Energy} = \tfrac{1}{2}LI^2 = \tfrac{1}{2}\mu_0 n^2 A\ell I^2$$

Since $B = \mu_0 nI$ at the interior of a long solenoid (Equation 21.7), this energy can be expressed as

$$\text{Energy} = \frac{1}{2\mu_0} B^2 A\ell$$

The term $A\ell$ is the volume inside the solenoid where the magnetic field exists, so the energy per unit volume, or ***energy density***, is

$$\text{Energy density} = \frac{\text{Energy}}{\text{Volume}} = \frac{1}{2\mu_0} B^2 \qquad (22.11)$$

Although this result was obtained for the special case of a long solenoid, it is quite general and is valid for any point where a magnetic field exists in air or vacuum or in a nonmagnetic material. Thus, energy can be stored in a magnetic field, just as it can in an electric field.

22.9 TRANSFORMERS

One of the most important applications of mutual induction and self-induction takes place in a transformer. A *transformer* is a device for increasing or decreasing an ac voltage. For instance, whenever a cordless device (e.g., a cell phone) is plugged into a wall recepta-cle to recharge the batteries, a transformer plays a role in reducing the 120-V ac voltage to a much smaller value. Typically, between 3 and 9 V are needed to energize batteries. In another example, a picture tube in a television set needs about 15 000 V to accelerate the electron beam, and a transformer is used to obtain this high voltage from the 120 V at a wall socket.

Figure 22.28 shows a drawing of a transformer. The transformer consists of an iron core on which two coils are wound: a primary coil with N_p turns and a secondary coil with N_s turns. The primary coil is the one connected to the ac generator. For the moment, sup-pose that the switch in the secondary circuit is open, so there is no current in this circuit.

The alternating current in the primary coil establishes a changing magnetic field in the iron core. Because iron is easily magnetized, it greatly enhances the magnetic field rela-tive to that in an air core and guides the field lines to the secondary coil. In a well-designed core, nearly all the magnetic flux Φ that passes through each turn of the primary also goes through each turn of the secondary. Since the magnetic field is changing, the flux through the primary and secondary coils is also changing, and consequently an emf is induced in both coils. In the secondary coil the induced emf $\mathcal{E}_s$ arises from mutual induction and is given by Faraday's law as

The physics of transformers.

$$\mathcal{E}_s = -N_s \frac{\Delta\Phi}{\Delta t}$$

In the primary coil the induced emf $\mathcal{E}_p$ is due to self-induction and is specified by Fara-day's law as

$$\mathcal{E}_p = -N_p \frac{\Delta\Phi}{\Delta t}$$

The term $\Delta\Phi/\Delta t$ is the same in both of these equations, since the same flux penetrates each turn of both coils. Dividing the two equations shows that

$$\frac{\mathcal{E}_s}{\mathcal{E}_p} = \frac{N_s}{N_p}$$

In a high-quality transformer the resistances of the coils are negligible, so the magnitudes of the emfs, $\mathcal{E}_s$ and $\mathcal{E}_p$, are nearly equal to the terminal voltages, V_s and V_p, across the coils (see Section 20.9 for a discussion of terminal voltage). The relation $\mathcal{E}_s/\mathcal{E}_p = N_s/N_p$ is called the *transformer equation* and is usually written in terms of the terminal voltages:

Transformer equation

$$\frac{V_s}{V_p} = \frac{N_s}{N_p} \tag{22.12}$$

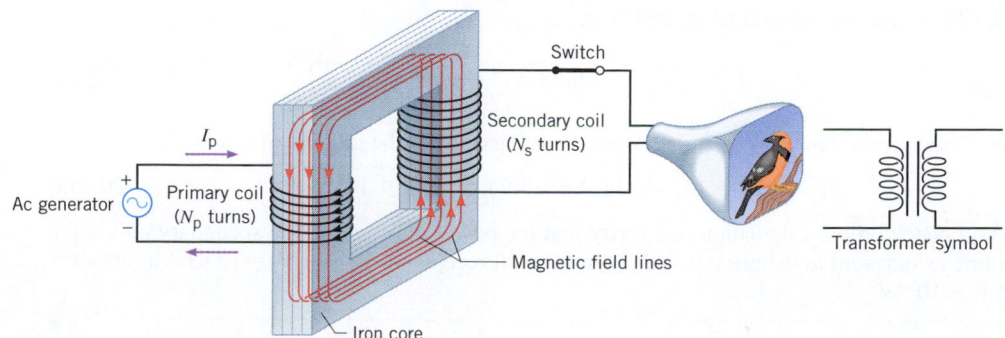

Figure 22.28 A transformer consists of a primary coil and a secondary coil, both wound on an iron core. The changing magnetic flux produced by the current in the primary coil induces an emf in the secondary coil. At the far right is the symbol for a transformer.

Power distribution stations use high-voltage transformers similar to these to step up or step down voltages. (Lester Lefkowitz/Photographer's Choice/ Getty Images)

Problem-solving insight

According to the transformer equation, if N_s is greater than N_p, the secondary (output) voltage is greater than the primary (input) voltage. In this case we have a *step-up* transformer. On the other hand, if N_s is less than N_p, the secondary voltage is less than the primary voltage, and we have a *step-down* transformer. The ratio N_s/N_p is referred to as the *turns ratio* of the transformer. A turns ratio of 8/1 (often written as 8 : 1) means, for example, that the secondary coil has eight times more turns than the primary coil. Conversely, a turns ratio of 1 : 8 implies that the secondary has one-eighth as many turns as the primary.

A transformer operates with ac electricity and not with dc. A steady direct current in the primary coil produces a flux that does not change in time, and thus no emf is induced in the secondary coil. The ease with which transformers can change voltages from one value to another is a principal reason why ac is preferred over dc.

With the switch in the secondary circuit of Figure 22.28 closed, a current I_s exists in the circuit and electrical energy is fed to the TV tube. This energy comes from the ac generator connected to the primary coil. Although the secondary voltage V_s may be larger or smaller than the primary voltage V_p, energy is not being created or destroyed by the transformer. Energy conservation requires that the energy delivered to the secondary coil must be the same as the energy delivered to the primary coil, provided no energy is dissipated in heating these coils or is otherwise lost. In a well-designed transformer, less than 1% of the input energy is lost in the form of heat. Noting that power is energy per unit time, and assuming 100% energy transfer, the average power $\overline{P}_p$ delivered to the primary coil is equal to the average power $\overline{P}_s$ delivered to the secondary coil: $\overline{P}_p = \overline{P}_s$. However, $\overline{P} = IV$ (Equation 20.15a), so $I_p V_p = I_s V_s$, or

$$\frac{I_s}{I_p} = \frac{V_p}{V_s} = \frac{N_p}{N_s} \tag{22.13}$$

Observe that V_s/V_p is equal to the turns ratio N_s/N_p, while I_s/I_p is equal to the inverse turns ratio N_p/N_s. **Consequently, a transformer that steps up the voltage simultaneously steps down the current, and a transformer that steps down the voltage steps up the current.** However, the power is neither stepped up nor stepped down, since $\overline{P}_p = \overline{P}_s$. Example 14 emphasizes this fact.

 Example 14 A Step-Down Transformer

A step-down transformer inside a stereo receiver has 330 turns in the primary coil and 25 turns in the secondary coil. The plug connects the primary coil to a 120-V wall socket, and there is a current of 0.83 A in the primary coil while the receiver is turned on. Connected to the secondary coil are the transistor circuits of the receiver. Find **(a)** the voltage across the secondary coil, **(b)** the current in the secondary coil, and **(c)** the average electric power delivered to the transistor circuits.

Reasoning The transformer equation, Equation 22.12, states that the secondary voltage V_s is equal to the product of the primary voltage V_p and the turns ratio N_s/N_p. On the other hand, Equation 22.13 indicates that the secondary current I_s is equal to the product of the primary current I_p and the inverse turns ratio N_p/N_s. The average power delivered to the transistor circuits is the product of the secondary current and the secondary voltage.

Solution **(a)** The voltage across the secondary coil can be found from the transformer equation:

$$V_s = V_p \frac{N_s}{N_p} = (120 \text{ V})\left(\frac{25}{330}\right) = \boxed{9.1 \text{ V}} \tag{22.12}$$

(b) The current in the secondary coil is

$$I_s = I_p \frac{N_p}{N_s} = (0.83 \text{ A})\left(\frac{330}{25}\right) = \boxed{11 \text{ A}} \tag{22.13}$$

(c) The average power $\overline{P}_s$ delivered to the secondary coil is the product of I_s and V_s:

$$\overline{P}_s = I_s V_s = (11 \text{ A})(9.1 \text{ V}) = \boxed{1.0 \times 10^2 \text{ W}} \tag{20.15a}$$

As a check on our calculation, we verify that the power delivered to the secondary coil is the same as that sent to the primary coil from the wall receptacle: $\overline{P}_p = I_p V_p = (0.83 \text{ A})(120 \text{ V}) = 1.0 \times 10^2 \text{ W}$.

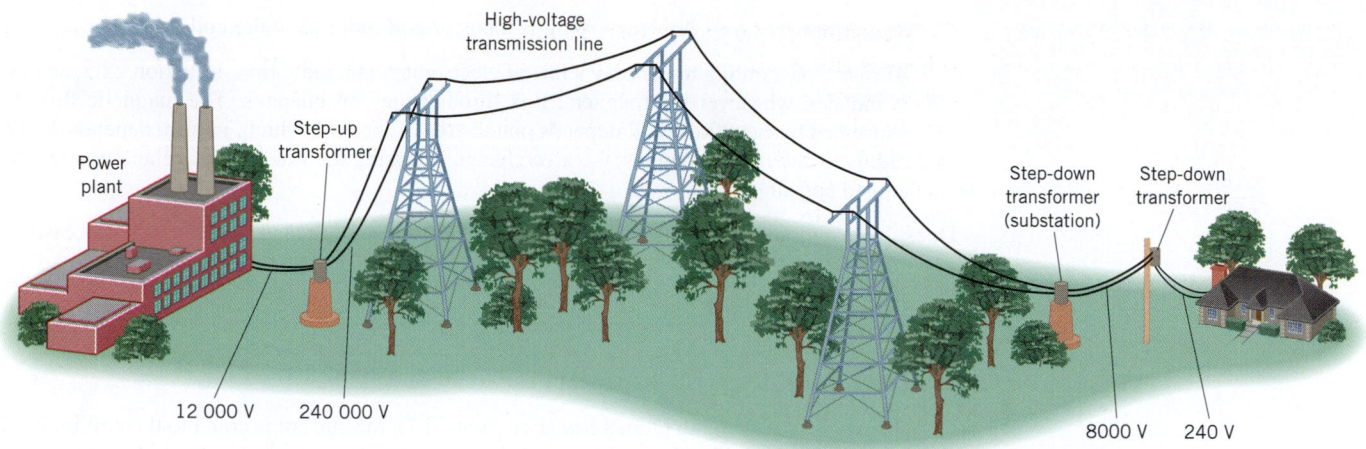

Figure 22.29 Transformers play a key role in the transmission of electric power.

Transformers play an important role in the transmission of power between electrical generating plants and the communities they serve. Whenever electricity is transmitted, there is always some loss of power in the transmission lines themselves due to resistive heating. Since the resistance of the wires is proportional to their length, the longer the wires the greater is the power loss. Power companies reduce this loss by using transformers that step up the voltage to high levels while reducing the current. A smaller current means less power loss, since $P = I^2R$, where R is the resistance of the transmission wires (see Problem 67). Figure 22.29 shows one possible way of transmitting power. The power plant produces a voltage of 12 000 V. This voltage is then raised to 240 000 V by a 20:1 step-up transformer. The high-voltage power is sent over the long-distance transmission line. Upon arrival at the city, the voltage is reduced to about 8000 V at a substation using a 1:30 step-down transformer. However, before any domestic use, the voltage is further reduced to 240 V (or possibly 120 V) by another step-down transformer that is often mounted on a utility pole. The power is then distributed to consumers.

✓ CHECK YOUR UNDERSTANDING

(The answers are given at the end of the book.)

19. A transformer changes the 120-V voltage at a wall socket to 12 000 V. The current delivered by the wall socket is **(a)** stepped up by a factor of 100, **(b)** stepped down by a factor of 100, **(c)** neither stepped up nor stepped down.

20. A transformer that stepped up the voltage and the current simultaneously would **(a)** produce less power at the secondary coil than was supplied at the primary coil, **(b)** produce more power at the secondary coil than was supplied at the primary coil, **(c)** produce the same amount of power at the secondary coil that was supplied at the primary coil, **(d)** violate the law of conservation of energy. Choose one or more.

CONCEPTS & CALCULATIONS

In this chapter we have seen that there are three ways to create an induced emf in a coil: by changing the magnitude of a magnetic field, by changing the direction of the field relative to the coil, and by changing the area of the coil. We have already explored the first two methods, and Example 15 now illustrates the third method. In the process, it provides a review of Faraday's law of electromagnetic induction.

Concepts & Calculations Example 15
The Emf Produced by a Changing Area

A circular coil of radius 0.11 m contains a single turn and is located in a constant magnetic field of magnitude 0.27 T. The magnetic field has the same direction as the normal to the plane of the coil. The radius increases to 0.30 m in a time of 0.080 s. **(a)** Determine the magnitude of the emf induced in the coil. **(b)** The coil has a resistance of 0.70 Ω. Find the magnitude of the induced current.

Concept Questions and Answers Why is there an emf induced in the coil?

Answer According to Faraday's law of electromagnetic induction, Equation 22.3, an emf is induced whenever the magnetic flux through the coil changes. The magnetic flux, as expressed by Equation 22.2, depends on the area of the coil, which, in turn, depends on the radius. If the radius changes, the area changes, causing the flux to change and an induced emf to appear.

Does the magnitude of the induced emf depend on whether the area is increasing or decreasing?

Answer No. The magnitude of the induced emf depends only on the magnitude of the rate $\Delta\Phi/\Delta t$ at which the flux changes. It does not matter whether the flux is increasing or decreasing.

What determines the amount of current induced in the coil?

Answer According to Ohm's law (Equation 20.2), the current is equal to the emf induced in the coil divided by the resistance of the coil. Therefore, the larger the induced emf and the smaller the resistance, the larger will be the induced current.

If the coil is cut so it is no longer one continuous piece, are there an induced emf and induced current?

Answer An induced emf is generated in the coil regardless of whether the coil is whole or cut. However, an induced current exists only if the coil is continuous. Cutting the coil is like opening a switch. There is no longer a continuous path for the current to follow, so the current stops.

Solution **(a)** The average induced emf is given by Faraday's law of electromagnetic induction.

$$\mathcal{E} = -N\frac{\Delta\Phi}{\Delta t} = -N\left(\frac{\Phi - \Phi_0}{t - t_0}\right) \tag{22.3}$$

where Φ and Φ_0 are the final and initial fluxes, respectively. The definition of magnetic flux is $\Phi = BA\cos\phi$, according to Equation 22.2, where $\phi = 0°$ because the field has the same direction as the normal. The area of a circle is $A = \pi r^2$. With these expressions for Φ and A, Faraday's law becomes

$$\mathcal{E} = -N\left(\frac{BA\cos\phi - BA_0\cos\phi}{t - t_0}\right) = -NB(\cos\phi)\left(\frac{\pi r^2 - \pi r_0^2}{t - t_0}\right)$$

$$= -(1)(0.27\text{ T})(\cos 0°)\left[\frac{\pi(0.30\text{ m})^2 - \pi(0.11\text{ m})^2}{0.080\text{ s}}\right] = -0.83\text{ V}$$

Thus, the magnitude of the induced emf is $\boxed{0.83\text{ V}}$.

(b) The magnitude of the induced current I is equal to the magnitude $|\mathcal{E}|$ of the induced emf divided by the resistance R of the coil:

$$I = \frac{|\mathcal{E}|}{R} = \frac{0.83\text{ V}}{0.70\ \Omega} = \boxed{1.2\text{ A}} \tag{20.2}$$

One of the most important applications of Faraday's law of electromagnetic induction is the electric generator, because it is the source of virtually all the electrical energy that we use. A generator produces an emf as a rotating coil changes its orientation relative to a fixed magnetic field. Example 16 reviews the characteristics of this type of induced emf.

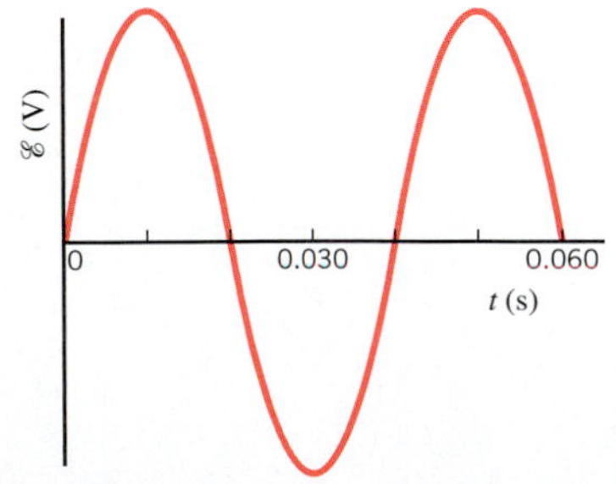

Figure 22.30 A plot of the emf produced by a generator as a function of time.

Concepts & Calculations Example 16
The Emf Produced by a Generator

The graph in Figure 22.30 shows the emf produced by a generator as a function of time t. The coil of the generator has an area of $A = 0.15\text{ m}^2$ and consists of $N = 10$ turns. The coil rotates in a magnetic field of magnitude 0.27 T. **(a)** Determine the period of the motion. **(b)** What is the angular frequency of the rotating coil? **(c)** Find the value of the emf when $t = \frac{1}{4}T$, where T denotes the period of the coil motion. **(d)** What is the emf when $t = 0.025$ s?

Concept Questions and Answers Can the period of the rotating coil be determined from the graph?

> *Answer* Yes. The period is the time for the coil to rotate through one revolution, or cycle. During this time, the emf is positive for one half of the cycle and negative for the other half of the cycle.

The emf produced by a generator depends on its angular frequency. How is the angular frequency of the rotating coil related to the period?

> *Answer* According to Equation 10.6, the angular frequency ω is related to the period T by $\omega = 2\pi/T$.

Starting at $t = 0$ s, how much time is required for the generator to produce its peak emf? Express the answer in terms of the period T of the motion (e.g., $t = \frac{1}{10}T$).

> *Answer* An examination of the graph shows that the emf reaches its maximum or peak value one-quarter of the way through a cycle. Since the time to complete one cycle is the period T, the time to reach the peak emf is $t = \frac{1}{4}T$.

How often does the polarity of the emf change in one cycle?

> *Answer* The graph shows that in every cycle the emf has an interval of positive polarity followed by an interval of negative polarity. In other words, the polarity changes from positive to negative and then from negative to positive during each cycle. Thus, the polarity changes twice during each cycle.

Solution (a) The period is the time for the generator to make one complete cycle. From the graph it can be seen that this time is $T = \boxed{0.040 \text{ s}}$.

(b) The angular frequency ω is related to the period by $\omega = 2\pi/T$, so

$$\omega = \frac{2\pi}{T} = \frac{2\pi}{0.040 \text{ s}} = \boxed{160 \text{ rad/s}}$$

(c) When $t = \frac{1}{4}T$, the emf has reached its peak value. According to Equation 22.4, the peak emf is

$$\mathcal{E}_0 = NAB\omega = (10)(0.15 \text{ m}^2)(0.27 \text{ T})(160 \text{ rad/s}) = \boxed{65 \text{ V}}$$

Problem-solving insight

In evaluating sin ωt a calculator must be set to the radian mode (not the degree mode) because ω is expressed in units of rad/s.

(d) The emf produced by the generator as a function of time is given by Equation 22.4 as

$$\mathcal{E} = \mathcal{E}_0 \sin \omega t = (65 \text{ V}) \sin [(160 \text{ rad/s})(0.025 \text{ s})] = \boxed{-49 \text{ V}}$$

CONCEPT SUMMARY

If you need more help with a concept, use the Learning Aids noted next to the discussion or equation. Examples (**Ex.**) are in the text of this chapter. Go to **www.wiley.com/college/cutnell** for the following Learning Aids:

Interactive LearningWare (ILW) — Additional examples solved in a five-step interactive format.

Concept Simulations (CS) — Animated text figures or animations of important concepts.

Interactive Solutions (IS) — Models for certain types of problems in the chapter homework. The calculations are carried out interactively.

Topic	Discussion	Learning Aids
Electromagnetic induction	**22.1 INDUCED EMF AND INDUCED CURRENT** Electromagnetic induction is the phenomenon in which an emf is induced in a piece of wire or a coil of wire with the aid of a magnetic field. The emf is called an induced emf, and any current that results from the emf is called an induced current.	
	22.2 MOTIONAL EMF An emf $\mathcal{E}$ is induced in a conducting rod of length L when the rod moves with a speed v in a magnetic field of magnitude B, according to	
Motional emf	$$\mathcal{E} = vBL \qquad (22.1)$$	**Ex. 1** **ILW 22.1**
	Equation 22.1 applies when the velocity of the rod, the length of the rod, and the magnetic field are mutually perpendicular.	

Topic	Discussion	Learning Aids

Conservation of energy

When the motional emf is used to operate an electrical device, such as a light bulb, the energy delivered to the device originates in the work done to move the rod, and the law of conservation of energy applies. **Ex. 2, 3**

22.3 MAGNETIC FLUX The magnetic flux Φ that passes through a surface is

Magnetic flux

$$\Phi = BA \cos \phi \qquad (22.2) \quad \textbf{Ex. 4}$$

where B is the magnitude of the magnetic field, A is the area of the surface, and ϕ is the angle between the field and the normal to the surface.

The magnetic flux is proportional to the number of magnetic field lines that pass through the surface.

22.4 FARADAY'S LAW OF ELECTROMAGNETIC INDUCTION Faraday's law of electromagnetic induction states that the average emf $\mathscr{E}$ induced in a coil of N loops is

Faraday's Law

$$\mathscr{E} = -N\left(\frac{\Phi - \Phi_0}{t - t_0}\right) = -N\frac{\Delta\Phi}{\Delta t} \qquad (22.3)$$

Ex. 5–7
CS 22.1
ILW 22.2
IS 22.77

where $\Delta\Phi$ is the change in magnetic flux through one loop and Δt is the time interval during which the change occurs. Motional emf is a special case of induced emf.

22.5 LENZ'S LAW Lenz's law provides a way to determine the polarity of an induced emf. Lenz's **Ex. 8, 9** law is stated as follows: The induced emf resulting from a changing magnetic flux has a polarity **IS 22.37** that leads to an induced current whose direction is such that the induced magnetic field opposes the original flux change. This statement is a consequence of the law of conservation of energy.

Lenz's law

22.7 THE ELECTRIC GENERATOR In its simplest form, an electric generator consists of a coil of N loops that rotates in a uniform magnetic field $\vec{\mathbf{B}}$. The emf produced by this generator is

Emf of a generator

$$\mathscr{E} = NAB\omega \sin \omega t = \mathscr{E}_0 \sin \omega t \qquad (22.4)$$

Ex. 10, 11
ILW 22.3
IS 22.43

where A is the area of the coil, ω is the angular speed (in rad/s) of the coil, and $\mathscr{E}_0 = NAB\omega$ is the peak emf. The angular speed in rad/s is related to the frequency f in cycles/s, or Hz, according to $\omega = 2\pi f$.

When an electric motor is running, it exhibits a generator-like behavior by producing an induced emf, called the back emf. The current I needed to keep the motor running at a constant speed is

Back emf of a motor

$$I = \frac{V - \mathscr{E}}{R} \qquad (22.5) \quad \textbf{Ex. 12}$$

where V is the emf applied to the motor by an external source, $\mathscr{E}$ is the back emf, and R is the resistance of the motor coil.

22.8 MUTUAL INDUCTANCE AND SELF-INDUCTANCE Mutual induction is the effect in which a changing current in the primary coil induces an emf in the secondary coil. The average emf $\mathscr{E}_s$ induced in the secondary coil by a change in current ΔI_p in the primary coil is

Emf due to mutual induction

$$\mathscr{E}_s = -M\frac{\Delta I_p}{\Delta t} \qquad (22.7)$$

Mutual inductance

where Δt is the time interval during which the change occurs. The constant M is the mutual inductance between the two coils and is measured in henries (H).

Self-induction is the effect in which a change in current ΔI in a coil induces an average emf $\mathscr{E}$ in the same coil, according to

Emf due to self-induction

$$\mathscr{E} = -L\frac{\Delta I}{\Delta t} \qquad (22.9) \quad \textbf{Ex. 13}$$

Self-inductance

The constant L is the self-inductance, or inductance, of the coil and is measured in henries.

To establish a current I in an inductor, work must be done by an external agent. This work is stored as energy in the inductor, the amount of energy being

Energy stored in an inductor

$$\text{Energy} = \tfrac{1}{2}LI^2 \qquad (22.10)$$

The energy stored in an inductor can be regarded as being stored in its magnetic field. At any point in air or vacuum or in a nonmagnetic material where a magnetic field $\vec{\mathbf{B}}$ exists, the energy density, or the energy stored per unit volume, is

Energy density of a magnetic field

$$\text{Energy density} = \frac{1}{2\mu_0}B^2 \qquad (22.11)$$

Topic	Discussion	Learning Aids
	22.9 TRANSFORMERS A transformer consists of a primary coil of N_p turns and a secondary coil of N_s turns. If the resistances of the coils are negligible, the voltage V_p across the primary coil and the voltage V_s across the secondary coil are related according to the transformer equation:	
Transformer equation	$$\frac{V_s}{V_p} = \frac{N_s}{N_p} \qquad (22.12)$$	**Ex. 14** **IS 22.61**
Turns ratio	where the ratio N_s/N_p is called the turns ratio of the transformer.	
	A transformer functions with ac electricity, not with dc. If the transformer is 100% efficient in transferring power from the primary to the secondary coil, the ratio of the secondary current I_s to the primary current I_p is	
	$$\frac{I_s}{I_p} = \frac{N_p}{N_s} \qquad (22.13)$$	

FOCUS ON CONCEPTS

Note to Instructors: The numbering of the questions shown here reflects the fact that they are only a representative subset of the total number that are available online. However, all of the questions are available for assignment via an online homework management program such as WileyPLUS or WebAssign.

Section 22.2 Motional Emf

2. You have three light bulbs; bulb A has a resistance of 240 Ω, bulb B has a resistance of 192 Ω, and bulb C has a resistance of 144 Ω. Each of these bulbs is used for the same amount of time in a setup like the one in the drawing. In each case the speed of the rod and the magnetic field strength are the same. Rank the setups in descending order, according to how much work the hand in the drawing must do (largest amount of work first). **(a)** A, B, C **(b)** A, C, B **(c)** B, C, A **(d)** B, A, C **(e)** C, B, A

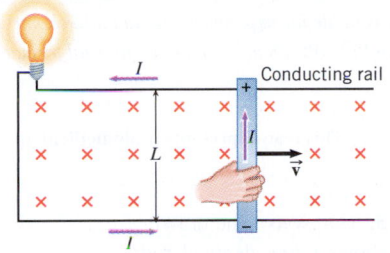

Section 22.3 Magnetic Flux

4. The drawing shows a cube. The dashed lines in the drawing are perpendicular to faces 1, 2, and 3 of the cube. Magnetic fields are oriented with respect to these faces as shown, and each of the three fields $\vec{B}_1$, $\vec{B}_2$, and $\vec{B}_3$ has the same magnitude. Note that $\vec{B}_2$ is parallel to face 2 of the cube. Rank the magnetic fluxes that pass through the faces 1, 2, and 3 of the cube in decreasing order (largest first). **(a)** Φ_1, Φ_2, Φ_3 **(b)** Φ_1, Φ_3, Φ_2 **(c)** Φ_2, Φ_1, Φ_3 **(d)** Φ_2, Φ_3, Φ_1 **(e)** Φ_3, Φ_2, Φ_1

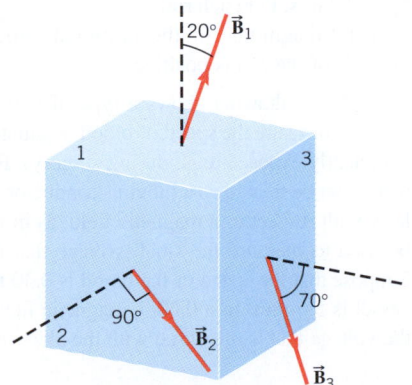

Section 22.4 Faraday's Law of Electromagnetic Induction

7. The drawing shows three flat coils, one square and two rectangular, that are each being pushed into a region where there is a uniform magnetic field directed into the page. Outside of this region the magnetic field is zero. In each case the magnetic field within the region has the same magnitude, and the coil is being pushed at the same velocity $\vec{v}$. Each coil begins with one side just at the edge of the field region. Consider the magnitude of the average emf induced as each coil is pushed from the starting position shown in the drawing until the coil is just completely within the field region. Rank the magnitudes of the average emfs in descending order (largest first). **(a)** $\mathscr{E}_A, \mathscr{E}_B, \mathscr{E}_C$ **(b)** $\mathscr{E}_A, \mathscr{E}_C, \mathscr{E}_B$ **(c)** $\mathscr{E}_B, \mathscr{E}_A$ and $\mathscr{E}_C$ (a tie) **(d)** $\mathscr{E}_C, \mathscr{E}_A$ and $\mathscr{E}_B$ (a tie)

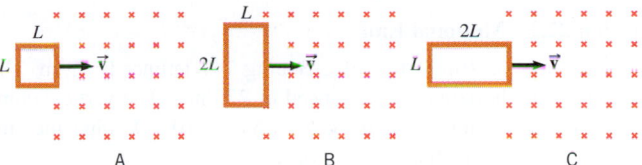

8. A long, vertical, straight wire carries a current I. The wire is perpendicular to the plane of a circular metal loop and passes through the center of the loop (see the drawing). The loop is allowed to fall and maintains its orientation with respect to the straight wire while doing so. In what direction does the current induced in the loop flow? **(a)** There is no induced current. **(b)** It is flowing around the loop from A to B to C to A. **(c)** It is flowing around the loop from C to B to A to C.

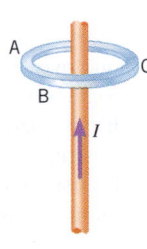

Section 22.5 Lenz's Law

9. The drawing shows a top view of two circular coils of conducting wire lying on a flat surface. The centers of the coils coincide. In the larger coil there are a switch and a battery. The smaller coil contains no switch and no battery. Describe the induced current that appears in the smaller coil when the switch in the larger coil is closed. **(a)** It flows counterclockwise forever after the switch is closed. **(b)** It flows clockwise forever after the switch is closed. **(c)** It flows counterclockwise, but only for a short period just after the switch is closed. **(d)** It flows clockwise, but only for a short period just after the switch is closed.

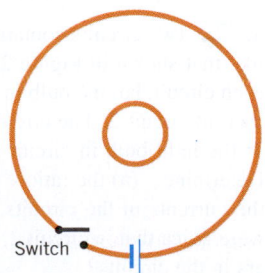

Section 22.7   The Electric Generator

10. You have a fixed length of conducting wire. From it you can construct a single-turn flat coil that has the shape of a square, a circle, or a rectangle with the long side twice the length of the short side. Each can be used with the same magnetic field to produce a generator that operates at the same frequency. Rank the peak emfs $\mathscr{E}_0$ of the three generators in descending order (largest first).   **(a)** $\mathscr{E}_{0,\,\text{square}}$, $\mathscr{E}_{0,\,\text{circle}}$, $\mathscr{E}_{0,\,\text{rectangle}}$ **(b)** $\mathscr{E}_{0,\,\text{circle}}$, $\mathscr{E}_{0,\,\text{square}}$, $\mathscr{E}_{0,\,\text{rectangle}}$ **(c)** $\mathscr{E}_{0,\,\text{square}}$, $\mathscr{E}_{0,\,\text{rectangle}}$, $\mathscr{E}_{0,\,\text{circle}}$ **(d)** $\mathscr{E}_{0,\,\text{rectangle}}$, $\mathscr{E}_{0,\,\text{square}}$, $\mathscr{E}_{0,\,\text{circle}}$ **(e)** $\mathscr{E}_{0,\,\text{rectangle}}$, $\mathscr{E}_{0,\,\text{circle}}$, $\mathscr{E}_{0,\,\text{square}}$

12. An electric motor is plugged into a standard wall socket and is running at normal speed. Suddenly, some dirt prevents the shaft of the motor from turning quite so rapidly. What happens to the back emf of the motor, and what happens to the current that the motor draws from the wall socket?   **(a)** The back emf increases, and the current drawn from the socket decreases.   **(b)** The back emf increases, and the current drawn from the socket increases.   **(c)** The back emf decreases, and the current drawn from the socket decreases.   **(d)** The back emf decreases, and the current drawn from the socket increases.

Section 22.8   Mutual Inductance and Self-Inductance

14. Inductor 1 stores the same amount of energy as inductor 2, although it has only one-half the inductance of inductor 2. What is the ratio I_1/I_2 of the currents in the two inductors?   **(a)** 2.000   **(b)** 1.414   **(c)** 4.000   **(d)** 0.500   **(e)** 0.250

Section 22.9   Transformers

18. The primary coil of a step-up transformer is connected across the terminals of a standard wall socket, and resistor 1 with a resistance R_1 is connected across the secondary coil. The current in the resistor is then measured. Next, resistor 2 with a resistance R_2 is connected directly across the terminals of the wall socket (without the transformer). The current in this resistor is also measured and found to be the same as the current in resistor 1. How does the resistance R_2 compare to the resistance R_1?   **(a)** The resistance R_2 is less than the resistance R_1.   **(b)** The resistance R_2 is greater than the resistance R_1.   **(c)** The resistance R_2 is the same as the resistance R_1.

PROBLEMS

Note to Instructors: Most of the homework problems in this chapter are available for assignment via an online homework management program such as WileyPLUS or WebAssign, and those marked with the icon  *are presented in WileyPLUS using a guided tutorial format that provides enhanced interactivity. See Preface for additional details.*

ssm   Solution is in the Student Solutions Manual.
www   Solution is available online at www.wiley.com/college/cutnell

This icon represents a biomedical application.

Section 22.2   Motional Emf

1. The wingspan (tip to tip) of a Boeing 747 jetliner is 59 m. The plane is flying horizontally at a speed of 220 m/s. The vertical component of the earth's magnetic field is 5.0×10^{-6} T. Find the emf induced between the plane's wing tips.

2. A 0.80-m aluminum bar is held with its length parallel to the east–west direction and dropped from a bridge. Just before the bar hits the river below, its speed is 22 m/s, and the emf induced across its length is 6.5×10^{-4} V. Assuming the horizontal component of the earth's magnetic field at the location of the bar points directly north, **(a)** determine the magnitude of the horizontal component of the earth's magnetic field, and **(b)** state whether the east end or the west end of the bar is positive.

3. In 1996, NASA performed an experiment called the Tethered Satellite experiment. In this experiment a 2.0×10^4-m length of wire was let out by the space shuttle *Atlantis* to generate a motional emf. The shuttle had an orbital speed of 7.6×10^3 m/s, and the magnitude of the earth's magnetic field at the location of the wire was 5.1×10^{-5} T. If the wire had moved perpendicular to the earth's magnetic field, what would have been the motional emf generated between the ends of the wire?

4. GO Two circuits contain an emf produced by a moving metal rod, like that shown in Figure 22.4b. The speed of the rod is the same in each circuit, but the bulb in circuit 1 has one-half the resistance of the bulb in circuit 2. The circuits are otherwise identical. The resistance of the light bulb in circuit 1 is 55 Ω, and that in circuit 2 is 110 Ω. Determine **(a)** the ratio $\mathscr{E}_1/\mathscr{E}_2$ of the emfs and **(b)** the ratio I_1/I_2 of the currents in the circuits.   **(c)** If the speed of the rod in circuit 1 were twice that in circuit 2, what would be the ratio P_1/P_2 of the powers in the circuits?

5. **ssm www** The drawing shows three identical rods (A, B, and C) moving in different planes. A constant magnetic field of magnitude 0.45 T is directed along the +y axis. The length of each rod is $L = 1.3$ m, and the speeds are the same, $v_A = v_B = v_C = 2.7$ m/s. For each rod, find the magnitude of the motional emf, and indicate which end (1 or 2) of the rod is positive.

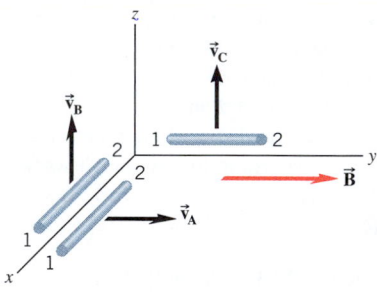

6. The drawing shows a type of flow meter that can be used to measure the speed of blood in situations when a blood vessel is sufficiently exposed (e.g., during surgery). Blood is conductive enough that it can be treated as a moving conductor. When it flows perpendicularly with respect to a magnetic field, as in the drawing, electrodes can be used to measure the small voltage that develops across the vessel. Suppose that the speed of the blood is 0.30 m/s and the diameter of the vessel is 5.6 mm. In a 0.60-T magnetic field what is the magnitude of the voltage that is measured with the electrodes in the drawing?

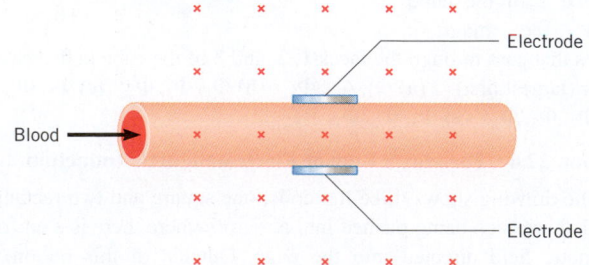

*7. Multiple-Concept Example 2 discusses the concepts that are used in this problem. Suppose that the magnetic field in Figure 22.5 has a magnitude of 1.2 T, the rod has a length of 0.90 m, and the hand keeps the rod moving to the right at a constant speed of 3.5 m/s. If the current in the circuit is 0.040 A, what is the average power being delivered to the circuit by the hand?

*8. Refer to the drawing that accompanies Check Your Understanding Question 14. Suppose that the voltage of the battery in the circuit is 3.0 V, the magnitude of the magnetic field (directed perpendicularly into the plane of the paper) is 0.60 T, and the length of the rod between the rails is 0.20 m. Assuming that the rails are very long and have negligible resistance, find the maximum speed attained by the rod after the switch is closed.

*9. ssm Suppose that the light bulb in Figure 22.4b is a 60.0-W bulb with a resistance of 240 Ω. The magnetic field has a magnitude of 0.40 T, and the length of the rod is 0.60 m. The only resistance in the circuit is that due to the bulb. What is the shortest distance along the rails that the rod would have to slide for the bulb to remain lit for one-half second?

**10. Review Conceptual Example 3 and Figure 22.7b. A conducting rod slides down between two frictionless vertical copper tracks at a constant speed of 4.0 m/s perpendicular to a 0.50-T magnetic field. The resistance of the rod and tracks is negligible. The rod maintains electrical contact with the tracks at all times and has a length of 1.3 m. A 0.75-Ω resistor is attached between the tops of the tracks. (a) What is the mass of the rod? (b) Find the change in the gravitational potential energy that occurs in a time of 0.20 s. (c) Find the electrical energy dissipated in the resistor in 0.20 s.

Section 22.3 Magnetic Flux

For problems in this set, assume that the magnetic flux is a positive quantity.

11. A magnetic field has a magnitude of 0.078 T and is uniform over a circular surface whose radius is 0.10 m. The field is oriented at an angle of $\phi = 25°$ with respect to the normal to the surface. What is the magnetic flux through the surface?

12. Two flat surfaces are exposed to a uniform, horizontal magnetic field of magnitude 0.47 T. When viewed edge-on, the first surface is tilted at an angle of 12° from the horizontal, and a net magnetic flux of 8.4×10^{-3} Wb passes through it. The same net magnetic flux passes through the second surface. (a) Determine the area of the first surface. (b) Find the smallest possible value for the area of the second surface.

13. ssm A standard door into a house rotates about a vertical axis through one side, as defined by the door's hinges. A uniform magnetic field is parallel to the ground and perpendicular to this axis. Through what angle must the door rotate so that the magnetic flux that passes through it decreases from its maximum value to one-third of its maximum value?

14. GO A loop of wire has the shape shown in the drawing. The top part of the wire is bent into a semicircle of radius $r = 0.20$ m. The normal to the plane of the loop is parallel to a constant magnetic field ($\phi = 0°$) of magnitude 0.75 T. What is the change $\Delta\Phi$ in the magnetic flux that passes through the loop when, starting with the position shown in the drawing, the semicircle is rotated through half a revolution?

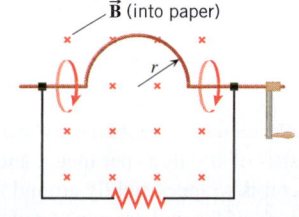

15. The drawing shows two surfaces that have the same area. A uniform magnetic field $\vec{B}$ fills the space occupied by these surfaces and is oriented parallel to the yz plane as shown. Find the ratio Φ_{xz}/Φ_{xy} of the magnetic fluxes that pass through the surfaces.

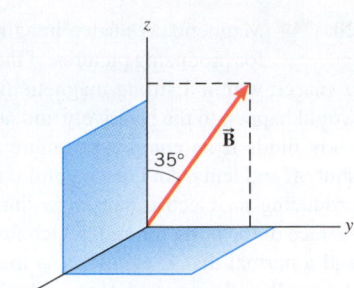

*16. A long, narrow, rectangular loop of wire is moving toward the bottom of the page with a speed of 0.020 m/s (see the drawing). The loop is leaving a region in which a 2.4-T magnetic field exists; the magnetic field outside this region is zero. During a time of 2.0 s, what is the magnitude of the *change* in the magnetic flux?

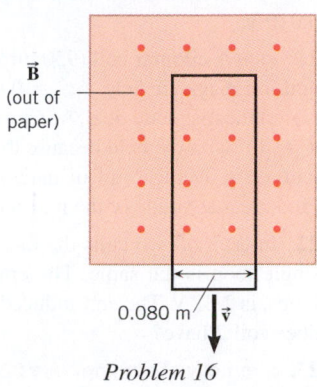

Problem 16

*17. ssm www A five-sided object, whose dimensions are shown in the drawing, is placed in a uniform magnetic field. The magnetic field has a magnitude of 0.25 T and points along the positive y direction. Determine the magnetic flux through each of the five sides.

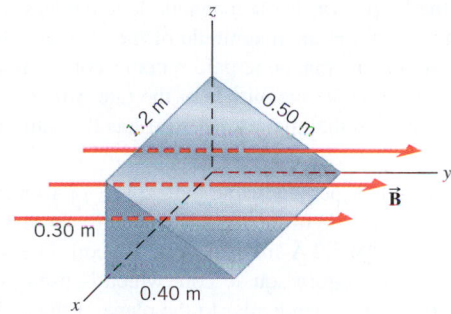

Section 22.4 Faraday's Law of Electromagnetic Induction

18. GO A magnetic field passes through a stationary wire loop, and its magnitude changes in time according to the graph in the drawing. The direction of the field remains constant, however. There are three equal time intervals indicated in the graph: 0–3.0 s, 3.0–6.0 s, and 6.0–9.0 s. The loop consists of 50 turns of wire and has an area of 0.15 m². The magnetic field is oriented parallel to the normal to the loop. For purposes of this problem, this means that $\phi = 0°$ in Equation 22.2. (a) For each interval, determine the induced emf. (b) The wire has a resistance of 0.50 Ω. Determine the induced current for the first and third intervals.

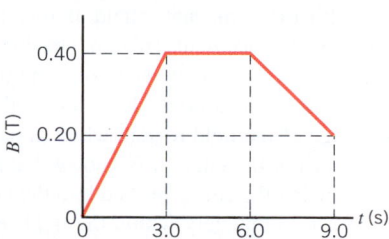

19. A planar coil of wire has a single turn. The normal to this coil is parallel to a uniform and constant (in time) magnetic field of 1.7 T. An emf that has a magnitude of 2.6 V is induced in this coil because the coil's area A is shrinking. What is the magnitude of $\Delta A/\Delta t$, which is the rate (in m²/s) at which the area changes?

20. Magnetic resonance imaging (MRI) is a medical technique for producing pictures of the interior of the body. The patient is placed within a strong magnetic field. One safety concern is what would happen to the positively and negatively charged particles in the body fluids if an equipment failure caused the magnetic field to be shut off suddenly. An induced emf could cause these particles to flow, producing an electric current within the body. Suppose the largest surface of the body through which flux passes has an area of 0.032 m^2 and a normal that is parallel to a magnetic field of 1.5 T. Determine the smallest time period during which the field can be allowed to vanish if the magnitude of the average induced emf is to be kept less than 0.010 V.

21. ssm A circular coil (950 turns, radius = 0.060 m) is rotating in a uniform magnetic field. At $t = 0$ s, the normal to the coil is perpendicular to the magnetic field. At $t = 0.010$ s, the normal makes an angle of $\phi = 45°$ with the field because the coil has made one-eighth of a revolution. An average emf of magnitude 0.065 V is induced in the coil. Find the magnitude of the magnetic field at the location of the coil.

22. In each of two coils the rate of change of the magnetic flux in a single loop is the same. The emf induced in coil 1, which has 184 loops, is 2.82 V. The emf induced in coil 2 is 4.23 V. How many loops does coil 2 have?

23. Interactive LearningWare 22.2 at **www.wiley.com/college/cutnell** reviews the fundamental approach in problems such as this. A constant magnetic field passes through a single rectangular loop whose dimensions are 0.35 m × 0.55 m. The magnetic field has a magnitude of 2.1 T and is inclined at an angle of 65° with respect to the normal to the plane of the loop. **(a)** If the magnetic field decreases to zero in a time of 0.45 s, what is the magnitude of the average emf induced in the loop? **(b)** If the magnetic field remains constant at its initial value of 2.1 T, what is the magnitude of the rate $\Delta A/\Delta t$ at which the area should change so that the average emf has the same magnitude as in part (a)?

24. A magnetic field is perpendicular to the plane of a single-turn circular coil. The magnitude of the field is changing, so that an emf of 0.80 V and a current of 3.2 A are induced in the coil. The wire is then re-formed into a single-turn square coil, which is used in the same magnetic field (again perpendicular to the plane of the coil and with a magnitude changing at the same rate). What emf and current are induced in the square coil?

*25. Parts a and b of the drawing show the same uniform and constant (in time) magnetic field $\vec{B}$ directed perpendicularly into the paper over a rectangular region. Outside this region, there is no field. Also shown is a rectangular coil (one turn), which lies in the plane of the paper. In part a the long side of the coil (length = L) is just at the edge of the field region, while in part b the short side (width = W) is just at the edge. It is known that $L/W = 3.0$. In both parts of the drawing the coil is pushed into the field with the same velocity $\vec{v}$ until it is completely within the field region. The magnitude of the average emf induced in the coil in part a is 0.15 V. What is its magnitude in part b?

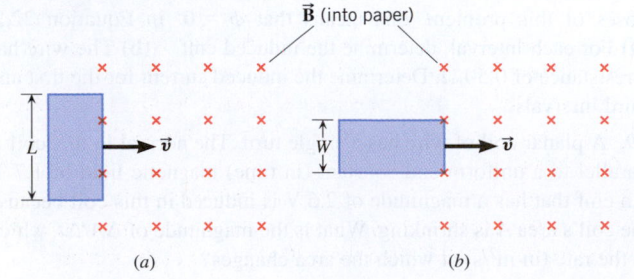

(a)   (b)

*26. **GO** A flat coil of wire has an area A, N turns, and a resistance R. It is situated in a magnetic field, such that the normal to the coil is parallel to the magnetic field. The coil is then rotated through an angle of 90°, so that the normal becomes perpendicular to the magnetic field. The coil has an area of 1.5×10^{-3} m^2, 50 turns, and a resistance of 140 Ω. During the time while it is rotating, a charge of 8.5×10^{-5} C flows in the coil. What is the magnitude of the magnetic field?

*27. ssm www A piece of copper wire is formed into a single circular loop of radius 12 cm. A magnetic field is oriented parallel to the normal to the loop, and it increases from 0 to 0.60 T in a time of 0.45 s. The wire has a resistance per unit length of 3.3×10^{-2} Ω/m. What is the average electrical energy dissipated in the resistance of the wire?

*28. A flat circular coil with 105 turns, a radius of 4.00×10^{-2} m, and a resistance of 0.480 Ω is exposed to an external magnetic field that is directed perpendicular to the plane of the coil. The magnitude of the external magnetic field is changing at a rate of $\Delta B/\Delta t = 0.783$ T/s, thereby inducing a current in the coil. Find the magnitude of the magnetic field at the center of the coil that is produced by the induced current.

*29. The drawing shows a copper wire (negligible resistance) bent into a circular shape with a radius of 0.50 m. The radial section BC is fixed in place, while the copper bar AC sweeps around at an angular speed of 15 rad/s. The bar makes electrical contact with the wire at all times. The wire and the bar have negligible resistance. A uniform magnetic field exists everywhere, is perpendicular to the plane of the circle, and has a magnitude of 3.8×10^{-3} T. Find the magnitude of the current induced in the loop ABC.

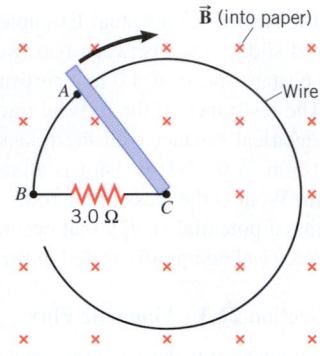

*30. **GO** The drawing shows a coil of copper wire that consists of two semicircles joined by straight sections of wire. In part a the coil is lying flat on a horizontal surface. The dashed line also lies in the plane of the horizontal surface. Starting from the orientation in part a, the smaller semicircle rotates at an angular frequency ω about the dashed line, until its plane becomes perpendicular to the horizontal surface, as shown in part b. A uniform magnetic field is constant in time and is directed upward, perpendicular to the horizontal surface. The field completely fills the region occupied by the coil in either part of the drawing. The magnitude of the magnetic field is 0.35 T. The resistance of the coil is 0.025 Ω, and the smaller semicircle has a radius of 0.20 m. The angular frequency at which the small semicircle rotates is 1.5 rad/s. Determine the average current I, if any, induced in the coil as the coil changes shape from that in part a of the drawing to that in part b. Be sure to include an explicit plus or minus sign along with your answer.

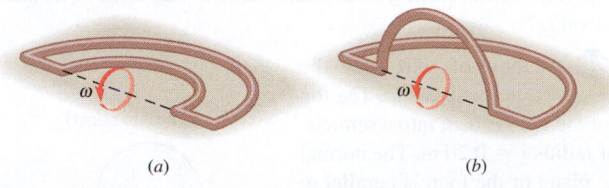

(a)   (b)

**31. ssm A solenoid has a cross-sectional area of 6.0×10^{-4} m^2, consists of 400 turns per meter, and carries a current of 0.40 A. A 10-turn coil is wrapped tightly around the circumference of the solenoid. The ends of the coil are connected to a 1.5-Ω resistor. Suddenly, a switch is opened, and the current in the solenoid dies to zero in a time of 0.050 s. Find the average current induced in the coil.

Section 22.5 Lenz's Law

32. The plane of a flat, circular loop of wire is horizontal. An external magnetic field is directed perpendicular to the plane of the loop. The magnitude of the external magnetic field is increasing with time. Because of this increasing magnetic field, an induced current is flowing clockwise in the loop, as viewed from above. What is the direction of the external magnetic field? Justify your conclusion.

33. ssm In Figure 22.1, suppose that the north and south poles of the magnet were interchanged. Determine the direction of the current through the ammeter in parts *b* and *c* of the picture (left to right or right to left). Give your rationale.

34. GO The drawing shows a straight wire carrying a current *I*. Above the wire is a rectangular loop that contains a resistor *R*. If the current *I* is decreasing in time, what is the direction of the induced current through the resistor *R*—left-to-right or right-to-left?

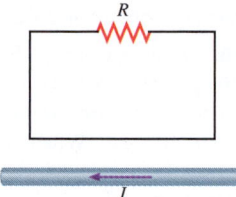

35. ssm Review Conceptual Example 9 as an aid in understanding this problem. A long, straight wire lies on a table and carries a current *I*. As the drawing shows, a small circular loop of wire is pushed across the top of the table from position 1 to position 2. Determine the direction of the induced current, clockwise or counterclockwise, as the loop moves past **(a)** position 1 and **(b)** position 2. Justify your answers.

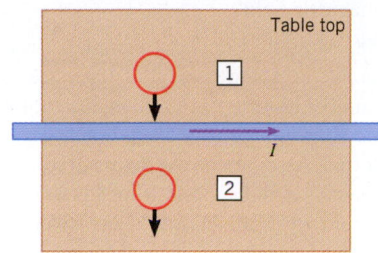

36. The drawing shows that a uniform magnetic field is directed perpendicularly into the plane of the paper and fills the entire region to the left of the *y* axis. There is no magnetic field to the right of the *y* axis. A rigid right triangle *ABC* is made of copper wire. The triangle rotates counterclockwise about the origin at point *C*. What is the direction (clockwise or counterclockwise) of the induced current when the triangle is crossing **(a)** the +*y* axis, **(b)** the −*x* axis, **(c)** the −*y* axis, and **(d)** the +*x* axis? For each case, justify your answer.

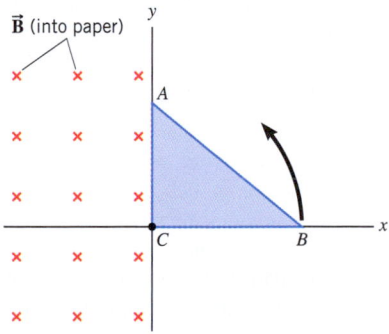

*37. Consult **Interactive Solution 22.37** at **www.wiley.com/college/cutnell** for one approach to this problem. A circular loop of wire rests on a table. A long, straight wire lies on this loop, directly over its center, as the drawing illustrates. The current *I* in the straight wire is decreasing. In what direction is the induced current, if any, in the loop? Give your reasoning.

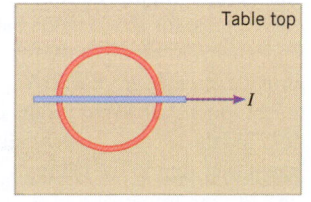

*38. Indicate the direction of the electric field between the plates of the parallel plate capacitor shown in the drawing if the magnetic field is decreasing in time. Give your reasoning.

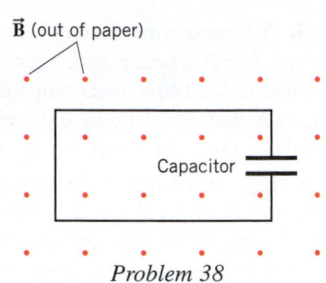

Problem 38

39. A wire loop is suspended from a string that is attached to point *P* in the drawing. When released, the loop swings downward, from left to right, through a uniform magnetic field, with the plane of the loop remaining perpendicular to the plane of the paper at all times. **(a) Determine the direction of the current induced in the loop as it swings past the locations labeled I and II. Specify the direction of the current in terms of the points *x*, *y*, and *z* on the loop (e.g., *x* → *y* → *z* or *z* → *y* → *x*). The points *x*, *y*, and *z* lie behind the plane of the paper. **(b)** What is the direction of the induced current at the locations II and I when the loop swings back, from right to left? Provide reasons for your answers.

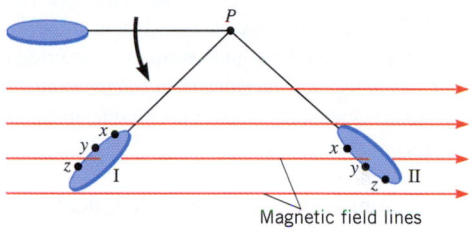

Section 22.7 The Electric Generator

40. A 120.0-V motor draws a current of 7.00 A when running at normal speed. The resistance of the armature wire is 0.720 Ω. **(a)** Determine the back emf generated by the motor. **(b)** What is the current at the instant when the motor is just turned on and has not begun to rotate? **(c)** What series resistance must be added to limit the starting current to 15.0 A?

41. ssm www The drawing shows a plot of the output emf of a generator as a function of time *t*. The coil of this device has a cross-sectional area per turn of 0.020 m² and contains 150 turns. Find **(a)** the frequency *f* of the generator in hertz, **(b)** the angular speed *ω* in rad/s, and **(c)** the magnitude of the magnetic field.

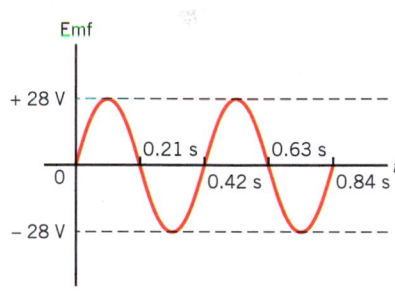

42. When its coil rotates at a frequency of 280 Hz, a certain generator has a peak emf of 75 V. **(a)** What is the peak emf of the generator when its coil rotates at a frequency of 45 Hz? **(b)** Determine the frequency of the coil's rotation when the peak emf of the generator is 180 V.

43. Interactive Solution 22.43 at **www.wiley.com/college/cutnell** provides one model for solving this problem. The maximum strength of the earth's magnetic field is about 6.9×10^{-5} T near the south magnetic pole. In principle, this field could be used with a rotating coil to generate 60.0-Hz ac electricity. What is the minimum number of turns (area per turn = 0.022 m²) that the coil must have to produce an rms voltage of 120 V?

44. A vacuum cleaner is plugged into a 120.0-V socket and uses 3.0 A of current in normal operation when the back emf generated by the electric motor is 72.0 V. Find the coil resistance of the motor.

*45. A generator uses a coil that has 100 turns and a 0.50-T magnetic field. The frequency of this generator is 60.0 Hz, and its emf has an rms value of 120 V. Assuming that each turn of the coil is a square (an approximation), determine the length of the wire from which the coil is made.

*46. **GO** At its normal operating speed, an electric fan motor draws only 15.0% of the current it draws when it just begins to turn the fan blade. The fan is plugged into a 120.0-V socket. What back emf does the motor generate at its normal operating speed?

*47. **ssm** Consult Multiple-Concept Example 11 for background material relating to this problem. A small rubber wheel on the shaft of a bicycle generator presses against the bike tire and turns the coil of the generator at an angular speed that is 38 times as great as the angular speed of the tire itself. Each tire has a radius of 0.300 m. The coil consists of 125 turns, has an area of 3.86×10^{-3} m², and rotates in a 0.0900-T magnetic field. The bicycle starts from rest and has an acceleration of +0.550 m/s². What is the peak emf produced by the generator at the end of 5.10 s?

**48. A 60.0-Hz generator delivers an average power of 75 W to a single light bulb. When an induced current exists in the rotating coil of a generator, a torque—called a countertorque—is exerted on the coil. Determine the maximum countertorque in the generator coil. (*Hint: The peak current, peak emf, and maximum countertorque all occur at the same instant.*)

Section 22.8 Mutual Inductance and Self-Inductance

49. ssm The earth's magnetic field, like any magnetic field, stores energy. The maximum strength of the earth's field is about 7.0×10^{-5} T. Find the maximum magnetic energy stored in the space above a city if the space occupies an area of 5.0×10^{8} m² and has a height of 1500 m.

50. The current through a 3.2-mH inductor varies with time according to the graph shown in the drawing. What is the average induced emf during the time intervals **(a)** 0–2.0 ms, **(b)** 2.0–5.0 ms, and **(c)** 5.0–9.0 ms?

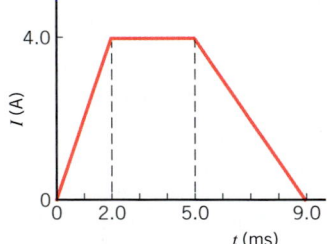

51. Two coils of wire are placed close together. Initially, a current of 2.5 A exists in one of the coils, but there is no current in the other. The current is then switched off in a time of 3.7×10^{-2} s. During this time, the average emf induced in the other coil is 1.7 V. What is the mutual inductance of the two-coil system?

52. **GO** A constant current of $I = 15$ A exists in a solenoid whose inductance is $L = 3.1$ H. The current is then reduced to zero in a certain amount of time. **(a)** If the current goes from 15 to 0 A in a time of 75 ms, what is the emf induced in the solenoid? **(b)** How much electrical energy is stored in the solenoid? **(c)** At what rate must the electrical energy be removed from the solenoid when the current is reduced to 0 A in a time of 75 ms? Note that the rate at which energy is removed is the power.

53. Mutual induction can be used as the basis for a metal detector. A typical setup uses two large coils that are parallel to each other and have a common axis. Because of mutual induction, the ac generator connected to the primary coil causes an emf of 0.46 V to be induced in the secondary coil. When someone without metal objects walks through the coils, the mutual inductance and, thus, the induced emf do not change much. But when a person carrying a handgun walks through, the mutual inductance increases. The change in emf can be used to trigger an alarm. If the mutual inductance increases by a factor of three, find the new value of the induced emf.

54. **GO** During a 72-ms interval, a change in the current in a primary coil occurs. This change leads to the appearance of a 6.0-mA current in a nearby secondary coil. The secondary coil is part of a circuit in which the resistance is 12 Ω. The mutual inductance between the two coils is 3.2 mH. What is the change in the primary current?

55. ssm Multiple-Concept Example 13 reviews some of the principles used in this problem. Suppose you wish to make a solenoid whose self-inductance is 1.4 mH. The inductor is to have a cross-sectional area of 1.2×10^{-3} m² and a length of 0.052 m. How many turns of wire are needed?

*56. Multiple-Concept Example 13 provides useful background for this problem. A 5.40×10^{-5} H solenoid is constructed by wrapping 65 turns of wire around a cylinder with a cross-sectional area of 9.0×10^{-4} m². When the solenoid is shortened by squeezing the turns closer together, the inductance increases to 8.60×10^{-5} H. Determine the change in the length of the solenoid.

*57. A magnetic field has a magnitude of 12 T. What is the magnitude of an electric field that stores the same energy per unit volume as this magnetic field?

*58. Multiple-Concept Example 13 reviews the concepts used in this problem. A long solenoid (cross-sectional area $= 1.0 \times 10^{-6}$ m², number of turns per unit length $= 2400$ turns/m) is bent into a circular shape so it looks like a donut. This wire-wound donut is called a toroid. Assume that the diameter of the solenoid is small compared to the radius of the toroid, which is 0.050 m. Find the emf induced in the toroid when the current decreases to 1.1 A from 2.5 A in a time of 0.15 s.

59. Coil 1 is a flat circular coil that has N_1 turns and a radius R_1. At its center is a much smaller flat, circular coil that has N_2 turns and radius R_2. The planes of the coils are parallel. Assume that coil 2 is so small that the magnetic field due to coil 1 has nearly the same value at all points covered by the area of coil 2. Determine an expression for the mutual inductance between these two coils in terms of μ_0, N_1, R_1, N_2, and R_2.

Section 22.9 Transformers

60. In some places, insect "zappers," with their blue lights, are a familiar sight on a summer's night. These devices use a high voltage to electrocute insects. One such device uses an ac voltage of 4320 V, which is obtained from a standard 120.0-V outlet by means of a transformer. If the primary coil has 21 turns, how many turns are in the secondary coil?

61. Interactive Solution 22.61 at **www.wiley.com/college/cutnell** offers one approach to problems such as this one. The secondary coil of a step-up transformer provides the voltage that operates an electrostatic air filter. The turns ratio of the transformer is 50:1. The primary coil is plugged into a standard 120-V outlet. The current in the secondary coil is 1.7×10^{-3} A. Find the power consumed by the air filter.

62. **GO** The rechargeable batteries for a laptop computer need a much smaller voltage than what a wall socket provides. Therefore, a transformer is plugged into the wall socket and produces the necessary voltage for charging the batteries. The batteries are rated at 9.0 V, and a current of 225 mA is used to charge them. The wall socket provides a voltage of 120 V. **(a)** Determine the turns ratio of the transformer. **(b)** What is the current coming from the wall socket? **(c)** Find the average power delivered by the wall socket and the average power sent to the batteries.

63. ssm A step-down transformer (turns ratio = 1:8) is used with an electric train to reduce the voltage from the wall receptacle to a value needed to operate the train. When the train is running, the current in the secondary coil is 1.6 A. What is the current in the primary coil?

64. A transformer consisting of two coils wrapped around an iron core is connected to a generator and a resistor, as shown in the drawing. There are 11 turns in the primary coil and 18 turns in the secondary coil.

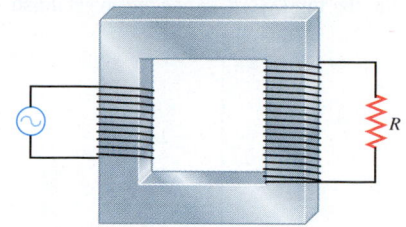

The peak voltage across the resistor is 67 V. What is the peak emf of the generator?

65. The resistances of the primary and secondary coils of a transformer are 56 and 14 Ω, respectively. Both coils are made from lengths of the same copper wire. The circular turns of each coil have the same diameter. Find the turns ratio N_s/N_p.

***66.** Suppose there are two transformers between your house and the high-voltage transmission line that distributes the power. In addition, assume that your house is the only one using electric power. At a substation the primary coil of a step-down transformer (turns ratio = 1:29) receives the voltage from the high-voltage transmission line. Because of your usage, a current of 48 mA exists in the primary coil of this transformer. The secondary coil is connected to the primary of another

step-down transformer (turns ratio = 1:32) somewhere near your house, perhaps up on a telephone pole. The secondary coil of this transformer delivers a 240-V emf to your house. How much power is your house using? Remember that the current and voltage given in this problem are rms values.

***67. ssm** A generating station is producing 1.2×10^6 W of power that is to be sent to a small town located 7.0 km away. Each of the two wires that comprise the transmission line has a resistance per kilometer of 5.0×10^{-2} Ω/km. **(a)** Find the power used to heat the wires if the power is transmitted at 1200 V. **(b)** A 100:1 step-up transformer is used to raise the voltage before the power is transmitted. How much power is now used to heat the wires?

***68.** **GO** In a television set the power needed to operate the picture tube comes from the secondary of a transformer. The primary of the transformer is connected to a 120-V receptacle on a wall. The picture tube of the television set uses 91 W, and there is 5.5 mA of current in the secondary coil of the transformer to which the tube is connected. Find the turns ratio N_s/N_p of the transformer.

****69.** A generator is connected across the primary coil (N_p turns) of a transformer, while a resistance R_2 is connected across the secondary coil (N_s turns). This circuit is equivalent to a circuit in which a single resistance R_1 is connected directly across the generator, without the transformer. Show that $R_1 = (N_p/N_s)^2 R_2$, by starting with Ohm's law as applied to the secondary coil.

ADDITIONAL PROBLEMS

70. The magnetic flux that passes through one turn of a 12-turn coil of wire changes to 4.0 from 9.0 Wb in a time of 0.050 s. The average induced current in the coil is 230 A. What is the resistance of the wire?

71. ssm A generator has a square coil consisting of 248 turns. The coil rotates at 79.1 rad/s in a 0.170-T magnetic field. The peak output of the generator is 75.0 V. What is the length of one side of the coil?

72. A rectangular loop of wire with sides 0.20 and 0.35 m lies in a plane perpendicular to a constant magnetic field (see part *a* of the drawing). The magnetic field has a magnitude of 0.65 T and is directed parallel to the normal of the loop's surface. In a time of 0.18 s, one-half of the loop is then folded back onto the other half, as indicated in part *b* of the drawing. Determine the magnitude of the average emf induced in the loop.

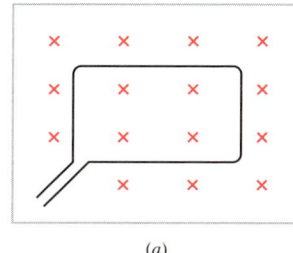

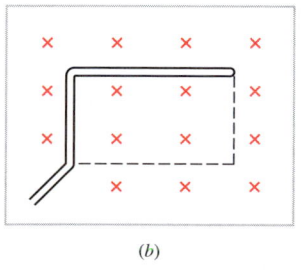

(a)   (b)

73. ssm The drawing depicts a copper loop lying flat on a table (not shown) and connected to a battery via a closed switch. The current I in the loop generates the magnetic field lines shown in the drawing. The switch is then opened and the current goes to zero. There are also two smaller conducting loops A and B lying flat on the table, but not connected to batteries. Determine the direction of the induced

current in **(a)** loop A and **(b)** loop B. Specify the direction of each induced current to be clockwise or counterclockwise when viewed from above the table. Provide a reason for each answer.

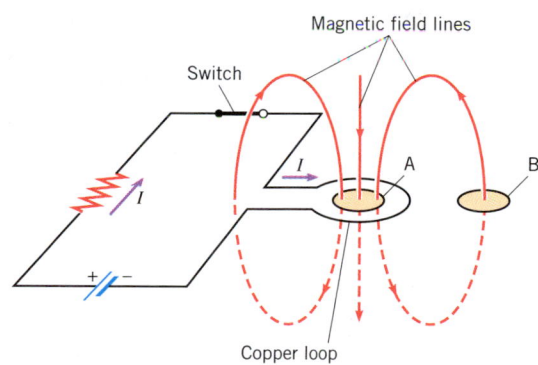

74. The batteries in a portable CD player are recharged by a unit that plugs into a wall socket. Inside the unit is a step-down transformer with a turns ratio of 1:13. The wall socket provides 120 V. What voltage does the secondary coil of the transformer provide?

75. ssm The coil within an ac generator has an area per turn of 1.2×10^{-2} m² and consists of 500 turns. The coil is situated in a 0.13-T magnetic field and is rotating at an angular speed of 34 rad/s. What is the emf induced in the coil at the instant when the normal to the loop makes an angle of 27° with respect to the direction of the magnetic field?

***76.** A 3.0-μF capacitor has a voltage of 35 V between its plates. What must be the current in a 5.0-mH inductor so that the energy stored in the inductor equals the energy stored in the capacitor?

*77. **Interactive Solution 22.77** at www.wiley.com/college/cutnell offers some help for this problem. A copper rod is sliding on two conducting rails that make an angle of 19° with respect to each other, as in the

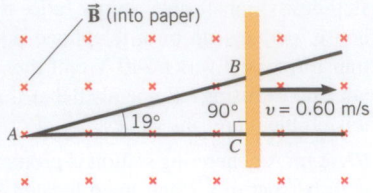

drawing. The rod is moving to the right with a constant speed of 0.60 m/s. A 0.38-T uniform magnetic field is perpendicular to the plane of the paper. Determine the magnitude of the average emf induced in the triangle ABC during the 6.0-s period after the rod has passed point A.

*78. The drawing shows a bar magnet falling through a metal ring. In part a the ring is solid all the way around, but in part b it has been cut through. **(a)** Explain why the motion of the magnet in part a is retarded when the magnet is above the ring and below the ring as well. Draw

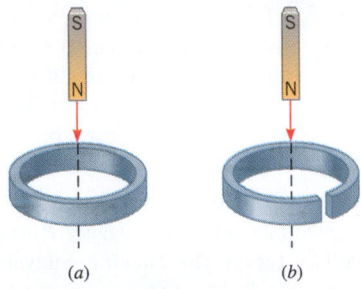

any induced currents that appear in the ring. **(b)** Explain why the motion of the magnet is unaffected by the ring in part b.

*79. **ssm** A magnetic field is passing through a loop of wire whose area is 0.018 m². The direction of the magnetic field is parallel to the normal to the loop, and the magnitude of the field is increasing at the rate of 0.20 T/s. **(a)** Determine the magnitude of the emf induced in the loop. **(b)** Suppose that the area of the loop can be enlarged or shrunk. If the magnetic field is increasing as in part (a), at what rate (in m²/s) should the area be changed at the instant when $B = 1.8$ T if the induced emf is to be zero? Explain whether the area is to be enlarged or shrunk.

*80. The purpose of this problem is to show that the work W needed to establish a final current I_f in an inductor is $W = \frac{1}{2}LI_f^2$ (Equation 22.10). In Section 22.8 we saw that the amount of work ΔW needed to change the current through an inductor by an amount ΔI is $\Delta W = LI(\Delta I)$, where L is the inductance. The drawing shows a graph

of LI versus I. Notice that $LI(\Delta I)$ is the area of the shaded vertical rectangle whose height is LI and whose width is ΔI. Use this fact to show that the total work W needed to establish a current I_f is $W = \frac{1}{2}LI_f^2$.

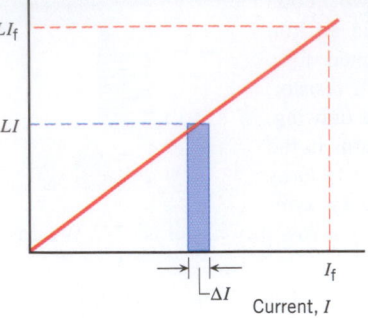

**81. Two 0.68-m-long conducting rods are rotating at the same speed in opposite directions, and both are perpendicular to a 4.7-T magnetic field. As the drawing shows, the ends of these rods come to within 1.0 mm of each other as they rotate. Moreover, the fixed ends about which the rods are rotating are connected by a wire, so these ends are at the same electric potential. If a potential difference of 4.5×10^3 V is required to cause a 1.0-mm spark in air, what is the angular speed (in rad/s) of the rods when a spark jumps across the gap?

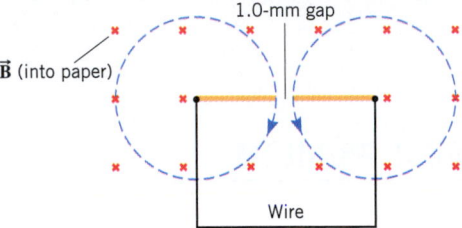

82. A motor is designed to operate on 117 V and draws a current of 12.2 A when it first starts up. At its normal operating speed, the motor draws a current of 2.30 A. Obtain **(a) the resistance of the armature coil, **(b)** the back emf developed at normal speed, and **(c)** the current drawn by the motor at one-third of the normal speed.

CHAPTER 23

ALTERNATING CURRENT CIRCUITS

A view of the interior of the air traffic control tower at Los Angeles International Airport (LAX). All of the electronic devices in the tower, as well as the lights in the city, use alternating current (ac), the subject of this chapter. (© Chad Slattery/Stone/Getty Images)

23.1 CAPACITORS AND CAPACITIVE REACTANCE

Our experience with capacitors so far has been in dc circuits. As we have seen in Section 20.13, charge flows in a dc circuit only for the brief period after the battery voltage is applied across the capacitor. In other words, charge flows only while the capacitor is charging up. After the capacitor becomes fully charged, no more charge leaves the battery. However, suppose that the battery connections to the fully charged capacitor were suddenly reversed. Then charge would flow again, but in the reverse direction, until the battery recharges the capacitor according to the new connections. What happens in an ac circuit is similar. The polarity of the voltage applied to the capacitor continually switches back and forth, and, in response, charges flow first one way around the circuit and then the other way. This flow of charge, surging back and forth, constitutes an alternating current. Thus, charge flows continuously in an ac circuit containing a capacitor.

To help set the stage for the present discussion, recall that, for a purely resistive ac circuit, the rms voltage V_{rms} across the resistor is related to the rms current I_{rms} through it by $V_{rms} = I_{rms}R$ (Equation 20.14). The resistance R has the same value for any frequency of the ac voltage or current. Figure 23.1 emphasizes this fact by showing that a graph of resistance versus frequency is a horizontal straight line.

For the rms voltage across a capacitor the following expression applies, which is analogous to $V_{rms} = I_{rms}R$:

$$V_{rms} = I_{rms}X_C \tag{23.1}$$

The term X_C appears in place of the resistance R and is called the **capacitive reactance.** The capacitive reactance, like resistance, is measured in *ohms* and determines how much rms current exists in a capacitor in response to a given rms voltage across the capacitor. It is found experimentally that the capacitive reactance X_C is inversely proportional to both the frequency f and the capacitance C, according to the following equation:

$$X_C = \frac{1}{2\pi f C} \tag{23.2}$$

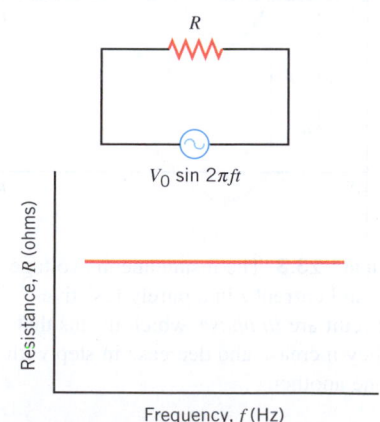

Figure 23.1 The resistance in a purely resistive circuit has the same value at all frequencies. The maximum emf of the generator is V_0.

719

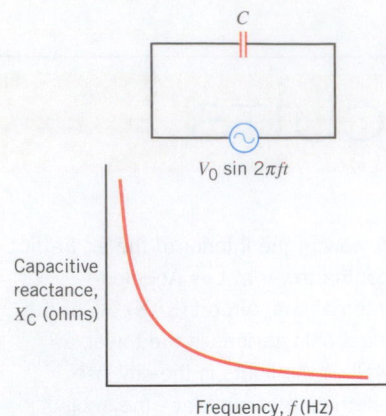

$V_0 \sin 2\pi f t$

Capacitive reactance, X_C (ohms)

Frequency, f (Hz)

Figure 23.2 The capacitive reactance X_C is inversely proportional to the frequency f according to $X_C = 1/(2\pi f C)$.

For a fixed value of the capacitance C, Figure 23.2 gives a plot of X_C versus frequency, according to Equation 23.2. A comparison of this drawing with Figure 23.1 reveals that a capacitor and a resistor behave differently. As the frequency becomes very large, Figure 23.2 shows that X_C approaches zero, signifying that a capacitor offers only a negligibly small opposition to the alternating current. In contrast, in the limit of zero frequency (i.e., direct current), X_C becomes infinitely large, and a capacitor provides so much opposition to the motion of charges that there is no current.

Example 1 illustrates the use of Equation 23.2 and also demonstrates how frequency and capacitance determine the amount of current in an ac circuit.

Example 1 A Capacitor in an AC Circuit

For the circuit in Figure 23.2, the capacitance of the capacitor is 1.50 μF, and the rms voltage of the generator is 25.0 V. What is the rms current in the circuit when the frequency of the generator is **(a)** 1.00×10^2 Hz and **(b)** 5.00×10^3 Hz?

Reasoning The current can be found from $I_{rms} = V_{rms}/X_C$, once the capacitive reactance X_C is determined. The values for the capacitive reactance will reflect the fact that the capacitor provides more opposition to the current when the frequency is smaller.

Solution **(a)** At a frequency of 1.00×10^2 Hz, we find

$$X_C = \frac{1}{2\pi f C} = \frac{1}{2\pi(1.00 \times 10^2 \text{ Hz})(1.50 \times 10^{-6} \text{ F})} = 1060 \ \Omega \quad (23.2)$$

$$I_{rms} = \frac{V_{rms}}{X_C} = \frac{25.0 \text{ V}}{1060 \ \Omega} = \boxed{0.0236 \text{ A}} \quad (23.1)$$

(b) When the frequency is 5.00×10^3 Hz, the calculations are similar:

$$X_C = \frac{1}{2\pi f C} = \frac{1}{2\pi(5.00 \times 10^3 \text{ Hz})(1.50 \times 10^{-6} \text{ F})} = 21.2 \ \Omega \quad (23.2)$$

$$I_{rms} = \frac{V_{rms}}{X_C} = \frac{25.0 \text{ V}}{21.2 \ \Omega} = \boxed{1.18 \text{ A}} \quad (23.1)$$

Problem-solving insight

The capacitive reactance X_C is inversely proportional to the frequency f of the voltage, so if the frequency increases by a factor of 50, as it does in Example 1, the capacitive reactance decreases by a factor of 50.

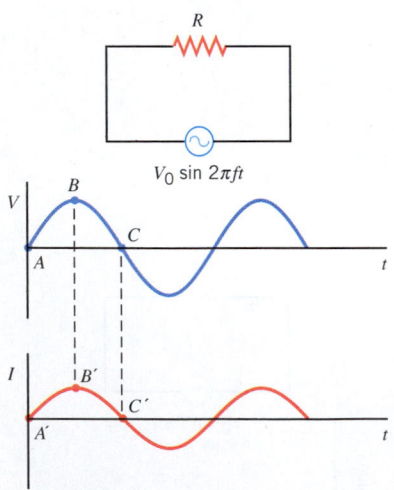

$V_0 \sin 2\pi f t$

Figure 23.3 The instantaneous voltage V and current I in a purely resistive circuit are *in phase*, which means that they increase and decrease in step with one another.

We now consider the behavior of the instantaneous (not rms) voltage and current. For comparison, Figure 23.3 shows graphs of voltage and current versus time in a resistive circuit. These graphs indicate that when only resistance is present, the voltage and current are proportional to each other at every moment. For example, when the voltage increases from A to B on the graph, the current follows along in step, increasing from A' to B' during the same time interval. Likewise, when the voltage decreases from B to C, the current decreases from B' to C'. For this reason, the current in a resistance R is said to be *in phase* with the voltage across the resistance.

For a capacitor, this in-phase relation between instantaneous voltage and current does *not* exist. Figure 23.4 shows graphs of the ac voltage and current versus time for a circuit that contains only a capacitor. As the voltage increases from A to B, the charge on the capacitor increases and reaches its full value at B. The current, however, is not the same thing as the charge. The current is the rate of flow of charge and has a maximum positive value at the start of the charging process at A'. It is a maximum because there is no charge on the capacitor at the start and hence no capacitor voltage to oppose the generator voltage. When the capacitor is fully charged at B, the capacitor voltage has a magnitude equal to that of the generator and completely opposes the generator voltage. The result is that the current decreases to zero at B'. While the capacitor voltage decreases from B to C, the charges flow out of the capacitor in a direction opposite to that of the charging current, as indicated by the negative current from B' to C'. Thus, voltage and current are not in phase but are, in fact, one-quarter wave cycle out of step, or out of phase. More specifically, assuming that the voltage fluctuates as $V_0 \sin 2\pi f t$, the current varies as $I_0 \sin (2\pi f t + \pi/2) = I_0 \cos 2\pi f t$. Since $\pi/2$ radians correspond to 90° and since the current reaches its maximum value *before* the voltage does, it is said that *the current in a capacitor leads the voltage across the capacitor by a phase angle of 90°*.

The fact that the current and voltage for a capacitor are 90° out of phase has an important consequence from the point of view of electric power, since power is the product of current and voltage. For the time interval between points A and B (or A' and B') in Figure 23.4, both current and voltage are positive. Therefore, the instantaneous power is also positive, meaning that the generator is delivering energy to the capacitor. However, during the period between B and C (or B' and C'), the current is negative while the voltage remains positive, and the power, which is the product of the two, is negative. During this period, the capacitor is returning energy to the generator. Thus, the power alternates between positive and negative values for equal periods of time. In other words, the capacitor alternately absorbs and releases energy. **Problem-solving insight:** *Consequently, the average power (and, hence, the average energy) used by a capacitor in an ac circuit is zero.*

It will prove useful later on to use a model for the voltage and current when analyzing ac circuits. In this model, voltage and current are represented by rotating arrows, often called *phasors,* whose lengths correspond to the maximum voltage V_0 and maximum current I_0, as Figure 23.5 indicates. These phasors rotate counterclockwise at a frequency f. For a resistor, the phasors are co-linear as they rotate (see part *a* of the drawing) because voltage and current are in phase. For a capacitor (see part *b*), the phasors remain perpendicular while rotating because the phase angle between the current and the voltage is 90°. Since current leads voltage for a capacitor, the current phasor is ahead of the voltage phasor in the direction of rotation.

Note from the two circuit drawings in Figure 23.5 that the instantaneous voltage across the resistor or the capacitor is $V_0 \sin 2\pi f t$. We can find this instantaneous value of the voltage directly from the phasor diagram. Imagine that the voltage phasor V_0 in this diagram represents the hypotenuse of a right triangle. Then, the *vertical component* of the phasor would be $V_0 \sin 2\pi f t$. In a similar manner, the instantaneous current can be found as the vertical component of the current phasor.

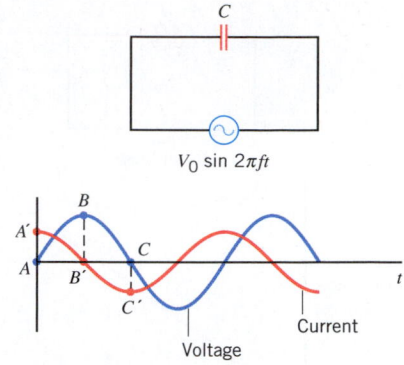

Figure 23.4 In a circuit containing only a capacitor, the instantaneous voltage and current are not in phase. Instead, the current *leads* the voltage by one-quarter of a cycle or by a phase angle of 90°.

✓ CHECK YOUR UNDERSTANDING

(The answer is given at the end of the book.)

1. One circuit contains only an ac generator and a resistor, and the rms current in this circuit is I_R. Another circuit contains only an ac generator and a capacitor, and the rms current in this circuit is I_C. The maximum, or peak, voltage of the generator is the same in both circuits and does not change. If the frequency of each generator is tripled, by what factor does the ratio I_R/I_C change? Specify whether the change is an increase or a decrease.

23.2 INDUCTORS AND INDUCTIVE REACTANCE

As Section 22.8 discusses, an inductor is usually a coil of wire, and the basis of its operation is Faraday's law of electromagnetic induction. According to Faraday's law, an inductor develops a voltage that opposes a change in the current. This voltage V is given by $V = -L(\Delta I / \Delta t)$ (see Equation 22.9*), where $\Delta I / \Delta t$ is the rate at which the current changes and L is the inductance of the inductor. In an ac circuit the current is always changing, and Faraday's law can be used to show that the rms voltage across an inductor is

$$V_{rms} = I_{rms} X_L \qquad (23.3)$$

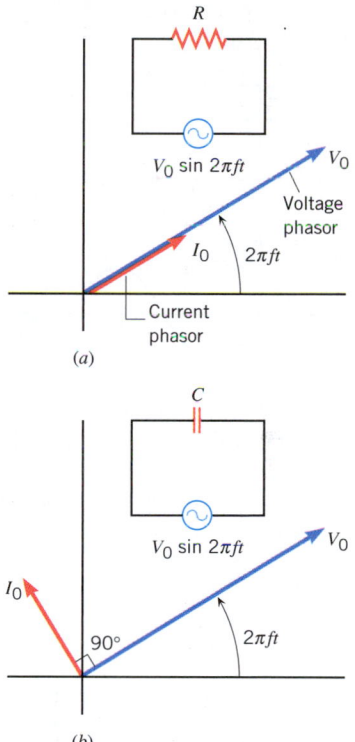

Figure 23.5 These rotating-arrow models represent the voltage and the current in ac circuits that contain (*a*) only a resistor and (*b*) only a capacitor.

Equation 23.3 is analogous to $V_{rms} = I_{rms} R$, with the term X_L appearing in place of the resistance R and being called the *inductive reactance.* The inductive reactance is measured in ohms and determines how much rms current exists in an inductor for a given rms voltage across the inductor. It is found experimentally that the inductive reactance X_L is directly proportional to the frequency f and the inductance L, as indicated by the following equation:

$$X_L = 2\pi f L \qquad (23.4)$$

*When an inductor is used in a circuit, the notation is simplified if we designate the potential difference across the inductor as the voltage V, rather than the emf $\mathscr{E}$.

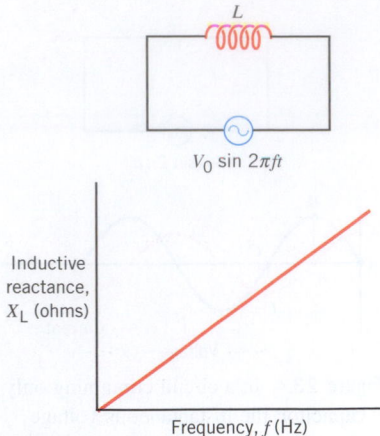

Figure 23.6 In an ac circuit the inductive reactance X_L is directly proportional to the frequency f, according to $X_L = 2\pi f L$.

This relation indicates that the larger the inductance, the larger is the inductive reactance. Note that the inductive reactance is directly proportional to the frequency ($X_L \propto f$), whereas the capacitive reactance is inversely proportional to the frequency ($X_C \propto 1/f$).

Figure 23.6 shows a graph of the inductive reactance versus frequency for a fixed value of the inductance, according to Equation 23.4. As the frequency becomes very large, X_L also becomes very large. In such a situation, an inductor provides a large opposition to the alternating current. In the limit of zero frequency (i.e., direct current), X_L becomes zero, indicating that an inductor does not oppose direct current at all. The next example demonstrates the effect of inductive reactance on the current in an ac circuit.

⬇ **Example 2** An Inductor in an AC Circuit

The circuit in Figure 23.6 contains a 3.60-mH inductor. The rms voltage of the generator is 25.0 V. Find the rms current in the circuit when the generator frequency is **(a)** 1.00×10^2 Hz and **(b)** 5.00×10^3 Hz.

Reasoning The current can be calculated from $I_{rms} = V_{rms}/X_L$, provided the inductive reactance is obtained first. The inductor offers more opposition to the changing current when the frequency is larger, and the values for the inductive reactance will reflect this fact.

Solution **(a)** At a frequency of 1.00×10^2 Hz, we find

$$X_L = 2\pi f L = 2\pi(1.00 \times 10^2 \text{ Hz})(3.60 \times 10^{-3} \text{ H}) = 2.26 \ \Omega \qquad (23.4)$$

$$I_{rms} = \frac{V_{rms}}{X_L} = \frac{25.0 \text{ V}}{2.26 \ \Omega} = \boxed{11.1 \text{ A}} \qquad (23.3)$$

(b) The calculation is similar when the frequency is 5.00×10^3 Hz:

$$X_L = 2\pi f L = 2\pi(5.00 \times 10^3 \text{ Hz})(3.60 \times 10^{-3} \text{ H}) = 113 \ \Omega \qquad (23.4)$$

$$I_{rms} = \frac{V_{rms}}{X_L} = \frac{25.0 \text{ V}}{113 \ \Omega} = \boxed{0.221 \text{ A}} \qquad (23.3)$$

Problem-solving insight: The inductive reactance X_L is directly proportional to the frequency f of the voltage. If the frequency increases by a factor of 50, for example, the inductive reactance also increases by a factor of 50.

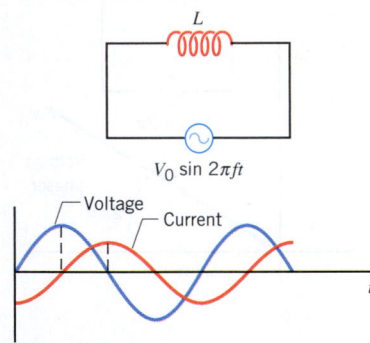

Figure 23.7 The instantaneous voltage and current in a circuit containing only an inductor are not in phase. The current *lags behind* the voltage by one-quarter of a cycle or by a phase angle of 90°.

By virtue of its inductive reactance, an inductor affects the amount of current in an ac circuit. The inductor also influences the current in another way, as Figure 23.7 shows. This figure displays graphs of voltage and current versus time for a circuit containing only an inductor. At a maximum or minimum on the current graph, the current does not change much with time, so the voltage generated by the inductor to oppose a change in the current is zero. At the points on the current graph where the current is zero, the graph is at its steepest, and the current has the largest rate of increase or decrease. Correspondingly, the voltage generated by the inductor to oppose a change in the current has the largest positive or negative value. Thus, current and voltage are not in phase but are one-quarter of a wave cycle out of phase. If the voltage varies as $V_0 \sin 2\pi f t$, the current fluctuates as $I_0 \sin (2\pi f t - \pi/2) = -I_0 \cos 2\pi f t$. The current reaches its maximum *after* the voltage does, and it is said that *the current in an inductor lags behind the voltage by a phase angle of 90°* ($\pi/2$ radians). In a purely capacitive circuit, in contrast, the current leads the voltage by 90° (see Figure 23.4).

In an inductor the 90° phase difference between current and voltage leads to the same result for average power that it does in a capacitor. An inductor alternately absorbs and releases energy for equal periods of time. **Problem-solving insight:** *Thus, the average power (and, hence, the average energy) used by an inductor in an ac circuit is zero.*

As an alternative to the graphs in Figure 23.7, Figure 23.8 uses phasors to describe the instantaneous voltage and current in a circuit containing only an inductor. The voltage and current phasors remain perpendicular as they rotate, because there is a 90° phase angle between them. The current phasor lags behind the voltage phasor, relative to the direction of rotation, in contrast to the equivalent picture in Figure 23.5b for a capacitor. Once again, the instantaneous values are given by the vertical components of the phasors.

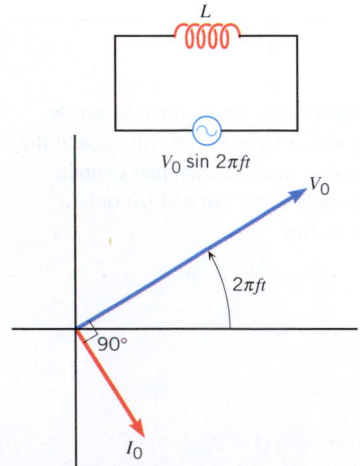

Figure 23.8 This phasor model represents the voltage and current in a circuit that contains only an inductor.

✓ **CHECK YOUR UNDERSTANDING**

(The answer is given at the end of the book.)

2. The drawing shows three ac circuits: one contains a resistor, one a capacitor, and one an inductor. The frequency of each ac generator is reduced to one-half its initial value. Which circuit experiences **(a)** the *greatest increase* in current and **(b)** the *least change* in current?

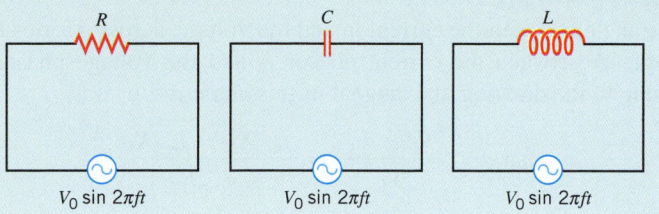

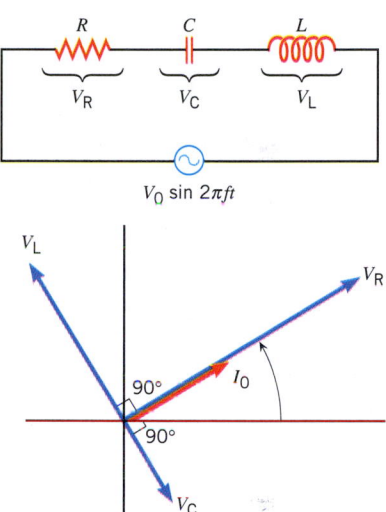

Figure 23.9 A series RCL circuit contains a resistor, a capacitor, and an inductor.

CIRCUITS CONTAINING RESISTANCE, CAPACITANCE, AND INDUCTANCE

23.3

Capacitors and inductors can be combined along with resistors in a single circuit. The simplest combination contains a resistor, a capacitor, and an inductor in series, as Figure 23.9 shows. In a series RCL circuit the total opposition to the flow of charge is called the ***impedance*** of the circuit and comes partially from (1) the resistance R, (2) the capacitive reactance X_C, and (3) the inductive reactance X_L. It is tempting to follow the analogy of a series combination of resistors and calculate the impedance by simply adding together R, X_C, and X_L. However, such a procedure is not correct. Instead, the phasors shown in Figure 23.10 must be used. The lengths of the voltage phasors in this drawing represent the maximum voltages V_R, V_C, and V_L across the resistor, the capacitor, and the inductor, respectively. The current is the same for each device, since the circuit is wired in series. The length of the current phasor represents the maximum current I_0. Notice that the drawing shows the current phasor to be (1) in phase with the voltage phasor for the resistor, (2) ahead of the voltage phasor for the capacitor by 90°, and (3) behind the voltage phasor for the inductor by 90°. These three facts are consistent with our earlier discussion in Sections 23.1 and 23.2.

The basis for dealing with the voltage phasors in Figure 23.10 is Kirchhoff's loop rule. In an ac circuit this rule applies to the *instantaneous* voltages across each circuit component and the generator. Therefore, it is necessary to take into account the fact that these voltages do not have the same phase; that is, the phasors V_R, V_C, and V_L point in different directions in the drawing. Kirchhoff's loop rule indicates that the phasors add together to give the total voltage V_0 that is supplied to the circuit by the generator. The addition, however, must be like a vector addition, to take into account the different directions of the phasors. Since V_L and V_C point in opposite directions, they combine to give a resultant phasor of $V_L - V_C$, as Figure 23.11 shows. In this drawing the resultant $V_L - V_C$ is perpendicular to V_R and may be combined with it to give the total voltage V_0. Using the Pythagorean theorem, we find

$$V_0^2 = V_R^2 + (V_L - V_C)^2$$

In this equation each of the symbols stands for a maximum voltage and when divided by $\sqrt{2}$ gives the corresponding rms voltage. Therefore, it is possible to divide both sides of the equation by $(\sqrt{2})^2$ and obtain a result for $V_{rms} = V_0/\sqrt{2}$. This result has exactly the same form as that above, but involves the rms voltages $V_{R\text{-}rms}$, $V_{C\text{-}rms}$, and $V_{L\text{-}rms}$. However, to avoid such awkward symbols, we simply interpret V_R, V_C, and V_L as rms quantities in the following expression:

$$V_{rms}^2 = V_R^2 + (V_L - V_C)^2 \tag{23.5}$$

The last step in determining the impedance of the circuit is to remember that $V_R = I_{rms}R$, $V_C = I_{rms}X_C$, and $V_L = I_{rms}X_L$. With these substitutions Equation 23.5 can be written as

$$V_{rms} = I_{rms} \sqrt{R^2 + (X_L - X_C)^2}$$

Therefore, for the entire RCL circuit, it follows that

$$V_{rms} = I_{rms}Z \tag{23.6}$$

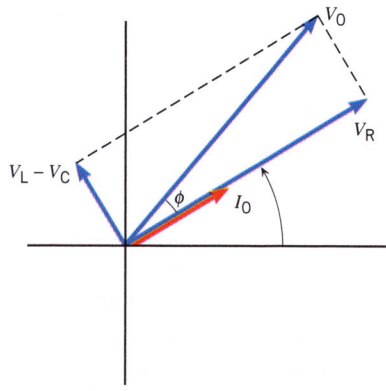

Figure 23.10 The three voltage phasors (V_R, V_C, and V_L) and the current phasor (I_0) for a series RCL circuit.

Figure 23.11 This simplified version of Figure 23.10 results when the phasors V_L and V_C, which point in opposite directions, are combined to give a resultant of $V_L - V_C$.

where the impedance Z of the circuit is defined as

Series RCL combination
$$Z = \sqrt{R^2 + (X_L - X_C)^2} \qquad (23.7)$$

The impedance of the circuit, like R, X_C, and X_L, is measured in ohms. In Equation 23.7, $X_L = 2\pi f L$ and $X_C = 1/(2\pi f C)$.

The phase angle between the current in and the voltage across a series RCL combination is the angle ϕ between the current phasor I_0 and the voltage phasor V_0 in Figure 23.11. According to the drawing, the tangent of this angle is

$$\tan \phi = \frac{V_L - V_C}{V_R} = \frac{I_{rms}X_L - I_{rms}X_C}{I_{rms}R}$$

Series RCL combination
$$\tan \phi = \frac{X_L - X_C}{R} \qquad (23.8)$$

The phase angle ϕ is important because it has a major effect on the average power $\overline{P}$ delivered to the circuit. Remember that, on the average, only the resistance consumes power; that is, $\overline{P} = I_{rms}^2 R$ (Equation 20.15b). According to Figure 23.11, $\cos \phi = V_R/V_0 = (I_{rms}R)/(I_{rms}Z) = R/Z$, so that $R = Z \cos \phi$. Therefore,

$$\overline{P} = I_{rms}^2 Z \cos \phi = I_{rms}(I_{rms}Z) \cos \phi$$
$$\overline{P} = I_{rms}V_{rms} \cos \phi \qquad (23.9)$$

where $V_{rms} = I_{rms}Z$ is the rms voltage of the generator, according to Equation 23.6. The term $\cos \phi$ is called the **power factor** of the circuit. As a check on the validity of Equation 23.9, note that if no resistance is present, $R = 0\ \Omega$, and $\cos \phi = R/Z = 0$. Consequently, $\overline{P} = I_{rms}V_{rms} \cos \phi = 0$, a result that is expected since, on the average, neither a capacitor nor an inductor consumes energy. Conversely, if only resistance is present, $Z = \sqrt{R^2 + (X_L - X_C)^2} = R$, and $\cos \phi = R/Z = 1$. In this case, $\overline{P} = I_{rms}V_{rms} \cos \phi = I_{rms}V_{rms}$, which is the expression for the average power delivered to a resistor. Examples 3 and 4 deal with the current, voltages, and power for a series RCL circuit.

ANALYZING MULTIPLE-CONCEPT PROBLEMS

Example 3   Current in a Series RCL Circuit

A series RCL circuit contains a 148-Ω resistor, a 1.50-μF capacitor, and a 35.7-mH inductor. The generator has a frequency of 512 Hz and an rms voltage of 35.0 V. Determine the rms current in the circuit.

Reasoning The rms current in the circuit is equal to the rms voltage of the generator divided by the impedance of the circuit, according to Equation 23.6. The impedance of the circuit can be found from the resistance of the resistor and the reactances of the capacitor and the inductor via Equation 23.7.

Knowns and Unknowns The following table summarizes the data that we have:

Description	Symbol	Value	Comment
Resistance of resistor	R	148 Ω	
Capacitance of capacitor	C	1.50 μF	$1\ \mu F = 10^{-6}\ F$
Inductance of inductor	L	35.7 mH	$1\ mH = 10^{-3}\ H$
Frequency of generator	f	512 Hz	
Rms voltage of generator	V_{rms}	35.0 V	
Unknown Variable			
Rms current in circuit	I_{rms}	?	

Modeling the Problem

STEP 1 Current According to Equation 23.6, the rms voltage V_{rms} of the generator, the rms current I_{rms}, and the impedance Z of the circuit are related according to

$$V_{rms} = I_{rms}Z$$

Solving for the current gives Equation 1 at the right, in which V_{rms} is known. The impedance is unknown, but it can be dealt with as in Step 2.

$$I_{rms} = \frac{V_{rms}}{Z} \quad (1)$$

$$\uparrow$$

$$?$$

STEP 2 Impedance For a series RCL circuit, the impedance Z is related to the resistance R, the inductive reactance X_L, and the capacitive reactance X_C, according to

$$Z = \sqrt{R^2 + (X_L - X_C)^2} \quad (23.7)$$

As indicated at the right, this expression can be substituted into Equation 1. The resistance is given, and we turn to Step 3 to deal with the reactances.

$$I_{rms} = \frac{V_{rms}}{Z} \quad (1)$$

$$\uparrow$$

$$Z = \sqrt{R^2 + (X_L - X_C)^2} \quad (23.7)$$

$$\underbrace{\qquad}$$

$$\uparrow$$

$$?$$

STEP 3 Inductive and Capacitive Reactances The inductive reactance X_L and the capacitive reactance X_C are given by Equation 23.4 and Equation 23.2 as

$$X_L = 2\pi f L \quad \text{and} \quad X_C = \frac{1}{2\pi f C}$$

where L is the inductance, C is the capacitance, and f is the frequency. Using these two expressions, we find that

$$X_L - X_C = 2\pi f L - \frac{1}{2\pi f C}$$

This result can now be substituted into Equation 23.7, as shown at the right.

$$I_{rms} = \frac{V_{rms}}{Z} \quad (1)$$

$$\uparrow$$

$$Z = \sqrt{R^2 + (X_L - X_C)^2} \quad (23.7)$$

$$\underbrace{\qquad}$$

$$\uparrow$$

$$X_L - X_C = 2\pi f L - \frac{1}{2\pi f C}$$

Solution Combining the results of each step algebraically, we find that

$$\underset{\text{STEP 1}}{I_{rms}} = \underset{\text{STEP 2}}{\frac{V_{rms}}{Z}} = \frac{V_{rms}}{\sqrt{R^2 + (X_L - X_C)^2}} = \underset{\text{STEP 3}}{\frac{V_{rms}}{\sqrt{R^2 + \left(2\pi f L - \frac{1}{2\pi f C}\right)^2}}}$$

The rms current in the circuit is

$$I_{rms} = \frac{V_{rms}}{\sqrt{R^2 + \left(2\pi f L - \frac{1}{2\pi f C}\right)^2}}$$

$$= \frac{35.0 \text{ V}}{\sqrt{(148 \ \Omega)^2 + \left[2\pi(512 \text{ Hz})(35.7 \times 10^{-3} \text{ H}) - \frac{1}{2\pi(512 \text{ Hz})(1.50 \times 10^{-6} \text{ F})}\right]^2}}$$

$$= \boxed{0.201 \text{ A}}$$

Related Homework: *Problems 21, 23*

Example 4 Voltages and Power in a Series RCL Circuit

For the series RCL circuit discussed in Example 3, the resistance, capacitance, and inductance are $R = 148 \ \Omega$, $C = 1.50 \ \mu\text{F}$, and $L = 35.7$ mH, respectively. The generator has a frequency of 512 Hz and an rms voltage of 35.0 V. In Example 3, it is found that the rms current in the circuit is $I_{rms} = 0.201$ A. Find **(a)** the rms voltage across each circuit element and **(b)** the average electric power delivered by the generator.

Reasoning The rms voltages across each circuit element can be determined from $V_R = I_{rms}R$, $V_C = I_{rms}X_C$, and $V_L = I_{rms}X_L$. In these expressions, the rms current is known. The resistance R is given, and the capacitive reactance X_C and inductive reactance X_L can be determined from Equations 23.2 and 23.4. The average power delivered to the circuit by the generator is specified by Equation 23.9.

Solution (a) The individual reactances are

$$X_C = \frac{1}{2\pi fC} = \frac{1}{2\pi(512 \text{ Hz})(1.50 \times 10^{-6} \text{ F})} = 207\ \Omega \qquad (23.2)$$

$$X_L = 2\pi fL = 2\pi(512 \text{ Hz})(35.7 \times 10^{-3} \text{ H}) = 115\ \Omega \qquad (23.4)$$

The rms voltages across each circuit element are

$$V_R = I_{rms}R = (0.201 \text{ A})(148\ \Omega) = \boxed{29.7 \text{ V}} \qquad (20.14)$$

$$V_C = I_{rms}X_C = (0.201 \text{ A})(207\ \Omega) = \boxed{41.6 \text{ V}} \qquad (23.1)$$

$$V_L = I_{rms}X_L = (0.201 \text{ A})(115\ \Omega) = \boxed{23.1 \text{ V}} \qquad (23.3)$$

Observe that these three voltages do not add up to give the generator's rms voltage of 35.0 V. Instead, the rms voltages satisfy Equation 23.5. It is the sum of the *instantaneous* voltages across R, C, and L that equals the generator's *instantaneous* voltage, according to Kirchhoff's loop rule. *The rms voltages do not satisfy the loop rule.*

(b) The average power delivered by the generator is $\overline{P} = I_{rms}V_{rms}\cos\phi$ (Equation 23.9). Therefore, a value for the phase angle ϕ is needed and can be obtained from Equation 23.8 as follows:

$$\tan\phi = \frac{X_L - X_C}{R} \quad \text{or} \quad \phi = \tan^{-1}\left(\frac{X_L - X_C}{R}\right) = \tan^{-1}\left(\frac{115\ \Omega - 207\ \Omega}{148\ \Omega}\right) = -32°$$

The phase angle is negative since the circuit is more capacitive than inductive (X_C is greater than X_L), and the current leads the voltage. The average power delivered by the generator is

$$\overline{P} = I_{rms}V_{rms}\cos\phi = (0.201 \text{ A})(35.0 \text{ V})\cos(-32°) = \boxed{6.0 \text{ W}}$$

This amount of power is delivered only to the resistor, since neither the capacitor nor the inductor uses power, on average.

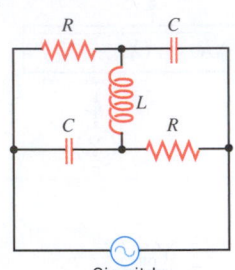

Circuit I

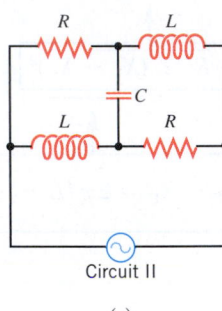

Circuit II

(a)

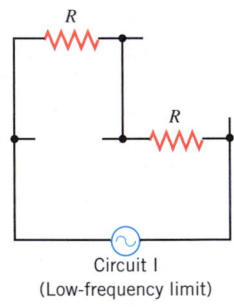

Circuit I
(Low-frequency limit)

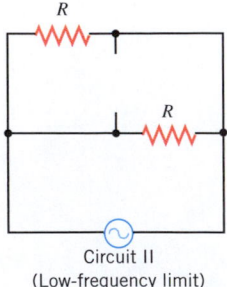

Circuit II
(Low-frequency limit)

(b)

Figure 23.12 (a) These circuits are discussed in the limit of very small or low frequency in Conceptual Example 5. (b) For a frequency very near zero, the circuits in part a behave as if they were as shown here.

In addition to the series RCL circuit, there are many different ways to connect resistors, capacitors, and inductors. In analyzing these additional possibilities, it helps to keep in mind the behavior of capacitors and inductors at the extreme limits of the frequency. When the frequency approaches zero (i.e., dc conditions), the reactance of a capacitor becomes so large that no charge can flow through the capacitor. It is as if the capacitor were cut out of the circuit, leaving an open gap in the connecting wire. In the limit of zero frequency the reactance of an inductor is vanishingly small. The inductor offers no opposition to a dc current. It is as if the inductor were replaced with a wire of zero resistance. In the limit of very large frequency, the behaviors of a capacitor and an inductor are reversed. The capacitor has a very small reactance and offers little opposition to the current, as if it were replaced by a wire with zero resistance. In contrast, the inductor has a very large reactance when the frequency is very large. The inductor offers so much opposition to the current that it might as well be cut out of the circuit, leaving an open gap in the connecting wire. Conceptual Example 5 illustrates how to gain insight into more complicated circuits using the limiting behaviors of capacitors and inductors.

 Conceptual Example 5 The Limiting Behavior of Capacitors and Inductors

Figure 23.12a shows two circuits. The rms voltage of the generator is the same in each case. The values of the resistance R, the capacitance C, and the inductance L in these circuits are the same. The frequency of the ac generator is very nearly zero. In which circuit does the generator supply more rms current, (a) circuit I or (b) circuit II?

Reasoning According to Equation 23.6, the rms current is given by $I_{rms} = V_{rms}/Z$. However, the impedance Z cannot be obtained from Equation 23.7, since the circuits in Figure 23.12a are not series RCL circuits. Since V_{rms} is the same in each case, the greater current is delivered to the circuit with the smaller impedance Z. In the limit of very small frequencies, the capacitors have very large impedances and, thus, allow very little current to flow through them. In essence, the capacitors behave as if they were cut out of the circuit, leaving gaps in the connecting wires. On the other hand, in the limit of very small frequencies, the inductors have very small impedances and behave as if they were replaced by wires with zero resistance. Figure 23.12*b* shows the circuits as they would appear according to these changes.

Answer (a) is incorrect. According to Figure 23.12*b*, circuit I behaves as if it contained only two identical resistors wired in series, with a total impedance of $Z = R + R = 2R$. In contrast, circuit II behaves as if it contained two identical resistors wired in parallel, in which case the total impedance is given by $1/Z = 1/R + 1/R = 2/R$, or $Z = R/2$. Circuit I contains the greater impedance, so the generator supplies less, not more, rms current to that circuit.

Answer (b) is correct. The rms current I_{rms} in a circuit is given by $I_{rms} = V_{rms}/Z$. Since V_{rms} is the same for both circuits and circuit II has the smaller impedance [see the explanation for why Answer (a) is incorrect], its generator supplies the greater rms current.

Related Homework: *Problem 46*

The impedance of an ac circuit contains important information about the resistance, capacitance, and inductance in the circuit. As an example of a very complex circuit, consider the human body. It contains muscle, which is a relatively good conductor of electricity due to its high water content, and also fat, which is a relatively poor conductor due to its low water content. The impedance that the body offers to ac electricity is referred to as bioelectrical impedance and is largely determined by resistance and capacitive reactance. Capacitance enters the picture because cell membranes can act like tiny capacitors. Bioelectric impedance analysis provides the basis for the determination of body-fat percentage by the body-fat scales (see Figure 23.13) that are widely available for home use. When you stand barefoot on such a scale, electrodes beneath your feet send a weak ac current (approximately 800 μA, 50 kHz) through your lower body in order to measure your body's impedance. The scale also measures your weight. A built-in computer combines the impedance and weight with information you provide about height, age, and sex to determine the percentage of fat in your body to an accuracy of about 5%. For men (age 20 to 39) a percentage of 8 to 19% is considered average, whereas the corresponding range of values for women of similar ages is 21 to 33%.

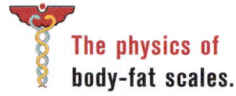

The physics of body-fat scales.

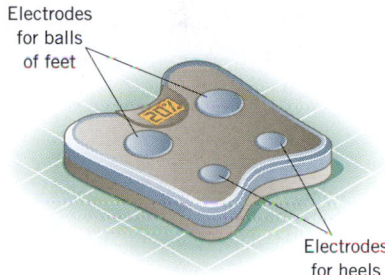

Electrodes for balls of feet

Electrodes for heels

Figure 23.13 Bathroom scales are now widely available that can provide estimates of your body-fat percentage. When you stand barefoot on the scale, electrodes beneath your feet send a small ac current through your lower body that allows the body's electrical impedance to be measured. This impedance is correlated with the percentage of fat in the body.

The physics of transcutaneous electrical nerve stimulation (TENS). Weak ac electricity with a much lower frequency than that used to measure bioelectrical impedance is used in a technique called transcutaneous electrical nerve stimulation (TENS). TENS is the most commonly used form of electroanalgesia in pain-management situations and, in its conventional form, uses an ac frequency between 40 and 150 Hz. Ac current is passed between two electrodes attached to the body and inhibits the transmission of pain-related nerve impulses. The technique is thought to work by affecting the "gates" in a nerve cell that control the passage of sodium ions into and out of the cell (see Section 19.6). Figure 23.14 shows TENS being applied during assessment of pain control following suspected damage to the radial nerve in the forearm.

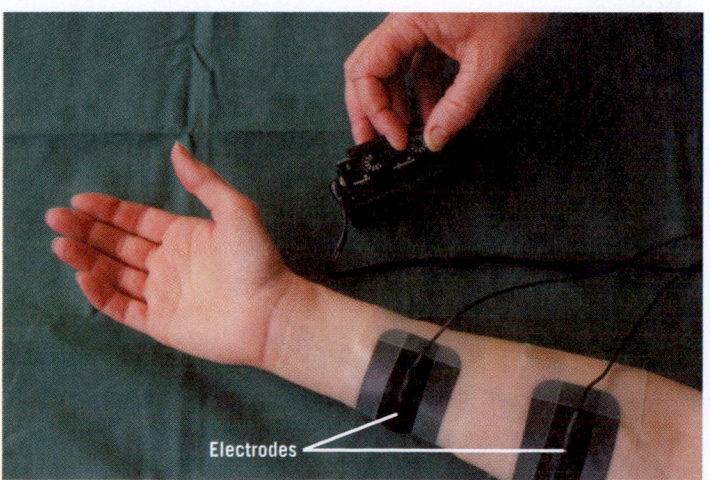

Electrodes

Figure 23.14 Transcutaneous electrical nerve stimulation (TENS) is shown here being applied to the forearm, in an assessment of pain control following suspected damage to the radial nerve. (Martin Dohrn/SPL/Photo Researchers, Inc.)

✔ CHECK YOUR UNDERSTANDING

(The answers are given at the end of the book.)

3. A long wire of finite resistance is connected to an ac generator. The wire is then wound into a coil of many loops and reconnected to the generator. Is the current in the circuit with the coil greater than, less than, or the same as the current in the circuit with the uncoiled wire?

4. A light bulb and a parallel plate capacitor (containing a dielectric material between the plates) are connected in series to the 60-Hz ac voltage present at a wall outlet. When the dielectric material is removed from the space between the plates, does the brightness of the bulb increase, decrease, or remain the same?

5. An air-core inductor is connected in series with a light bulb, and this circuit is plugged into an ac electrical outlet. When a piece of iron is inserted inside the inductor, does the brightness of the bulb increase, decrease, or remain the same?

6. Consider the circuit in Figure 23.9. With the capacitor and the inductor present, a certain amount of current is in the circuit. Then the capacitor and the inductor are removed, so that only the resistor remains connected to the generator. Is it possible that, under a certain condition, the current in the simplified circuit has the same rms value as in the original circuit? **(a)** No **(b)** Yes, when $X_L = R$ **(c)** Yes, when $X_C = R$ **(d)** Yes, when $X_C = X_L$

7. Review Conceptual Example 5 as an aid in understanding this question. An inductor and a capacitor are connected in parallel across the terminals of an ac generator. Does the current from the generator decrease, remain the same, or increase as the frequency becomes **(a)** very large and **(b)** very small?

8. Review Conceptual Example 5 as an aid in understanding this question. For which of the two circuits discussed there does the generator deliver more current when the frequency is very large? **(a)** Circuit I **(b)** Circuit II

23.4 RESONANCE IN ELECTRIC CIRCUITS

The behavior of the current and voltage in a series RCL circuit can give rise to a condition of *resonance.* Resonance occurs when the frequency of a vibrating force exactly matches a natural (resonant) frequency of the object to which the force is applied. When resonance occurs, the force can transmit a large amount of energy to the object, leading to a large-amplitude vibratory motion. We have already encountered several examples of resonance. First, resonance can occur when a vibrating force is applied to an object of mass m that is attached to a spring whose spring constant is k (Section 10.6). In this case there is one natural frequency f_0, whose value is $f_0 = [1/(2\pi)]\sqrt{k/m}$. Second, resonance occurs when standing waves are set up on a string (Section 17.5) or in a tube of air (Section 17.6). The string and tube of air have many natural frequencies, one for each allowed standing wave. As we will now see, a condition of resonance can also be established in a series RCL circuit. In this case there is only one natural frequency, and the vibrating force is provided by the oscillating electric field that is related to the voltage of the generator.

Figure 23.15 helps us to understand why there is a resonant frequency for an ac circuit. This drawing presents an analogy between the electrical case (ignoring resistance)

Figure 23.15 The oscillation of an object on a spring is analogous to the oscillation of the electric and magnetic fields that occur, respectively, in a capacitor and in an inductor. PE, potential energy; KE, kinetic energy.

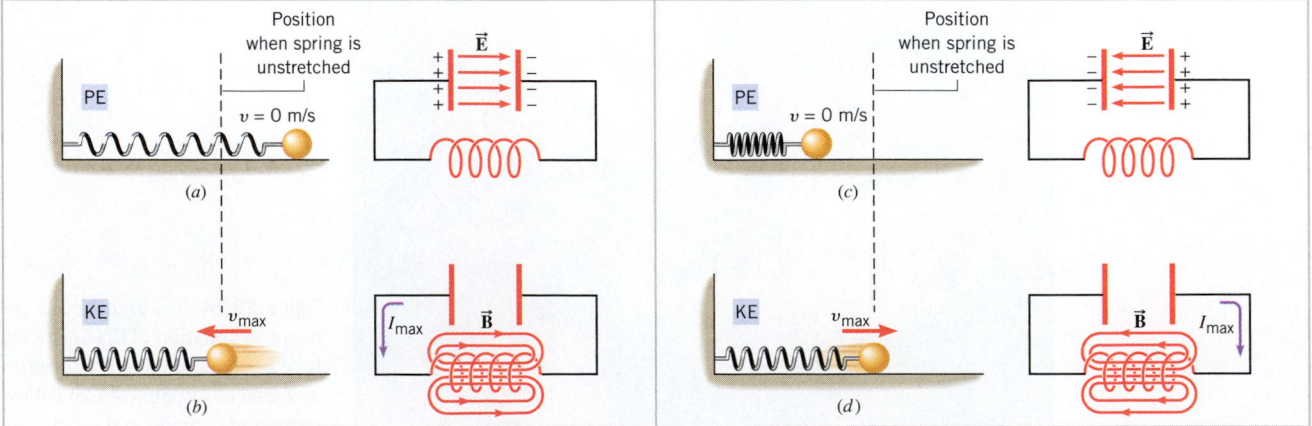

and the mechanical case of an object attached to a horizontal spring (ignoring friction). Part *a* shows a fully stretched spring that has just been released, and the initial speed *v* of the object is zero. All the energy is stored in the form of elastic potential energy. When the object begins to move, it gradually loses potential energy and picks up kinetic energy. In part *b*, the object moves with speed v_{max} and maximum kinetic energy through the position where the spring is unstretched (zero potential energy). Because of its inertia, the moving object coasts through this position and eventually comes to a halt in part *c* when the spring is fully compressed and all kinetic energy has been converted back into elastic potential energy. Part *d* of the picture is like part *b*, except that the direction of motion is reversed. The resonant frequency f_0 of the object on the spring is the natural frequency at which the object vibrates and is given as $f_0 = [1/(2\pi)]\sqrt{k/m}$ according to Equations 10.6 and 10.11. In this expression, *m* is the mass of the object, and *k* is the spring constant.

In the electrical case, Figure 23.15*a* begins with a fully charged capacitor that has just been connected to an inductor. At this instant the energy is stored in the electric field between the capacitor plates. As the capacitor discharges, the electric field $\vec{E}$ between the plates decreases, while a magnetic field $\vec{B}$ builds up around the inductor because of the increasing current in the circuit. The maximum current and the maximum magnetic field exist at the instant when the capacitor is completely discharged, as in part *b* of the figure. Energy is now stored entirely in the magnetic field of the inductor. The voltage induced in the inductor keeps the charges flowing until the capacitor again becomes fully charged, but now with reversed polarity, as in part *c*. Once again, the energy is stored in the electric field between the plates, and no energy resides in the magnetic field of the inductor. Part *d* of the cycle repeats part *b*, but with reversed directions of current and magnetic field. Thus, an ac circuit can have a resonant frequency because there is a natural tendency for energy to shuttle back and forth between the electric field of the capacitor and the magnetic field of the inductor.

To determine the resonant frequency at which energy shuttles back and forth between the capacitor and the inductor, we note that the current in a series RCL circuit is $I_{rms} = V_{rms}/Z$ (Equation 23.6). In this expression Z is the impedance of the circuit and is given by $Z = \sqrt{R^2 + (X_L - X_C)^2}$ (Equation 23.7). As Figure 23.16 illustrates, the rms current is a maximum when the impedance is a minimum, assuming a given generator voltage. The minimum impedance of $Z = R$ occurs when the frequency is f_0, such that $X_L = X_C$ or $2\pi f_0 L = 1/(2\pi f_0 C)$. This result can be solved for f_0, which is the resonant frequency:

$$f_0 = \frac{1}{2\pi\sqrt{LC}} \qquad (23.10)$$

The resonant frequency is determined by the inductance and the capacitance, but not the resistance.

The effect of resistance on electrical resonance is to make the "sharpness" of the circuit response less pronounced, as Figure 23.17 indicates. When the resistance is small, the current-versus-frequency graph falls off suddenly on either side of the maximum current. When the resistance is large, the falloff is more gradual, and there is less current at the maximum.

The following example deals with one application of resonance in electric circuits. In this example the focus is on the oscillation of energy between a capacitor and an inductor. Once a capacitor/inductor combination is energized, the energy will oscillate indefinitely as in Figure 23.15, provided there is some provision to replace any dissipative losses that occur because of resistance. Circuits that include this type of provision are called oscillator circuits.

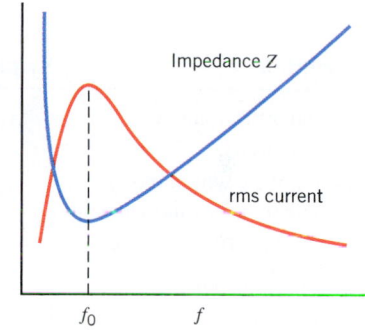

Figure 23.16 In a series RCL circuit the impedance is a minimum, and the current is a maximum, when the frequency *f* equals the resonant frequency f_0 of the circuit.

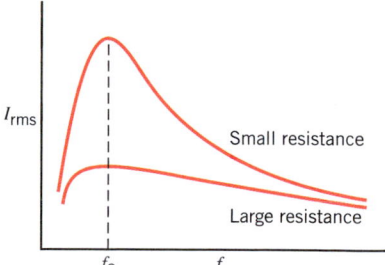

Figure 23.17 The effect of resistance on the current in a series RCL circuit.

ANALYZING MULTIPLE-CONCEPT PROBLEMS

Example 6 A Heterodyne Metal Detector

Figure 23.18 shows a heterodyne metal detector being used. As Figure 23.19 illustrates, this device utilizes two capacitor/inductor oscillator circuits, A and B. Each produces its own resonant

The physics of a heterodyne metal detector.

Continued

frequency, $f_{0A} = 1/(2\pi\sqrt{L_A C})$ and $f_{0B} = 1/(2\pi\sqrt{L_B C})$. Any difference between these frequencies is detected through earphones as a beat frequency $|f_{0B} - f_{0A}|$. In the absence of any nearby metal object, the inductances L_A and L_B are the same, and f_{0A} and f_{0B} are identical. There is no beat frequency. When inductor B (the search coil) comes near a piece of metal, the inductance L_B decreases, the corresponding oscillator frequency f_{0B} increases, and a beat frequency is heard. Suppose that initially each inductor is adjusted so that $L_B = L_A$, and each oscillator has a resonant frequency of 855.5 kHz. Assuming that the inductance of search coil B decreases by 1.000% due to a nearby piece of metal, determine the beat frequency heard through the earphones.

Reasoning The beat frequency is $|f_{0B} - f_{0A}|$, and to find it we need to determine the effect that the 1.000% decrease in the inductance L_B has on the resonant frequency f_{0B}. We will do this by a direct application of $f_{0B} = 1/(2\pi\sqrt{L_B C})$ (Equation 23.10).

Knowns and Unknowns The following table summarizes the available data:

Description	Symbol	Value	Comment
Resonant frequency of circuit A	f_{0A}	855.5 kHz	Does not change.
Amount by which metal object causes inductance L_B to decrease	—	1.000%	
Unknown Variable			
Beat frequency	$\|f_{0B} - f_{0A}\|$	?	

Figure 23.18 A heterodyne metal detector can be used to locate buried metal objects. Researchers (like the one in this photograph) also use metal detectors to locate meteorite fragments, because many of the fragments contain large amounts of metals such as iron or nickel. (David Parker/Photo Researchers, Inc.)

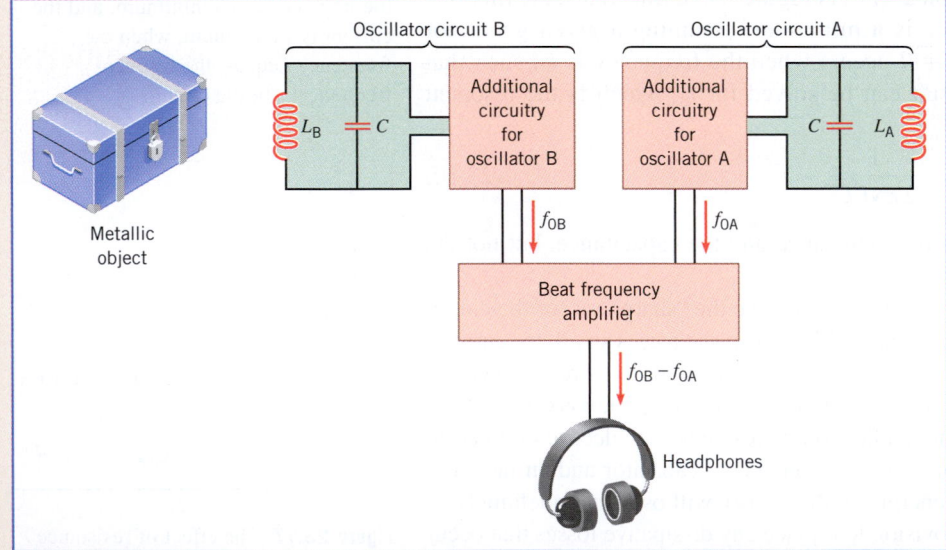

Figure 23.19 A heterodyne metal detector uses two electrical oscillators, A and B, in its operation. When the resonant frequency of oscillator B is changed due to the proximity of a metallic object, a beat frequency, whose value is $|f_{0B} - f_{0A}|$, is heard in the headphones.

Modeling the Problem

STEP 1 **Beat Frequency** The beat frequency is the magnitude of the difference between the two resonant frequencies, as indicated by Equation 1 at the right. The vertical bars, as usual, indicate that the magnitude or absolute value of the enclosed quantity is needed. The frequency f_{0A} is given, but the frequency f_{0B} is determined by the presence of the metal object and is unknown. We deal with it in Step 2.

$$\text{Beat frequency} = |f_{0B} - f_{0A}| \quad (1)$$

STEP 2 **Effect of Metal Object on Frequency f_{0B}** According to Equation 23.10, the resonant frequency of circuit B is

$$f_{0B} = \frac{1}{2\pi\sqrt{L_B C}} \qquad (23.10)$$

Initially, each inductor is adjusted so that $L_B = L_A$. Since the metal object causes the inductance of coil B to decrease by 1.000%, the new value of L_B becomes $L_B = 0.99000 L_A$. Substituting this expression into Equation 23.10 gives

$$f_{0B} = \frac{1}{2\pi\sqrt{L_B C}} = \frac{1}{2\pi\sqrt{0.99000 L_A C}}$$

Rearranging this result, we find that

$$f_{0B} = \frac{1}{2\pi\sqrt{0.99000 L_A C}} = \left(\frac{1}{\sqrt{0.99000}}\right)\frac{1}{2\pi\sqrt{L_A C}}$$

Finally, we recognize that $f_{0A} = 1/(2\pi\sqrt{L_A C})$ and can then express the resonant frequency of circuit B in the presence of the metal object as follows:

$$\boxed{f_{0B} = \left(\frac{1}{\sqrt{0.99000}}\right)f_{0A}}$$

This result can now be substituted into Equation 1, as shown at the right.

$$\text{Beat frequency} = |f_{0B} - f_{0A}| \qquad (1)$$

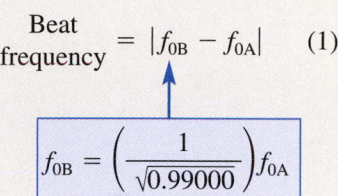

Solution Combining the results of each step algebraically, we find that

STEP 1	STEP 2

$$\text{Beat frequency} = \overset{\text{STEP 1}}{|f_{0B} - f_{0A}|} = \left|\overset{\text{STEP 2}}{\left(\frac{1}{\sqrt{0.99000}}\right)f_{0A}} - f_{0A}\right|$$

The beat frequency, then, is

$$\text{Beat frequency} = \left|\left(\frac{1}{\sqrt{0.99000}}\right)f_{0A} - f_{0A}\right| = \left|\left(\frac{1}{\sqrt{0.99000}}\right) - 1\right|f_{0A}$$

$$= \left|\left(\frac{1}{\sqrt{0.99000}}\right) - 1\right|(855.5 \text{ kHz}) = \boxed{4.3 \text{ kHz}}$$

Related Homework: *Problem 38*

✓ **CHECK YOUR UNDERSTANDING**

(*The answers are given at the end of the book.*)

9. The resistance in a series RCL circuit is doubled.   **(a)** Does the resonant frequency increase, decrease, or remain the same?   **(b)** Does the maximum current in the circuit increase, decrease, or remain the same?

10. In a series RCL circuit at resonance, does the current lead or lag behind the voltage across the generator, or is the current in phase with the voltage?

11. Is it possible for two series RCL circuits to have the same resonant frequencies and yet have   **(a)** different R values and   **(b)** different C and L values?

12. Suppose the generator connected to a series RCL circuit has a frequency that is greater than the resonant frequency of the circuit. Suppose, in addition, that it is necessary to match the resonant frequency of the circuit to the frequency of the generator. To accomplish this, should you add a second capacitor   **(a)** in series or   **(b)** in parallel with the one already present?

Figure 23.20 In a typical audio system, diodes are used in the power supply to create a dc voltage from the ac voltage present at the wall socket. This dc voltage is necessary so the transistors in the amplifier can perform their task of enlarging the small ac voltages originating in the compact disc player, etc.

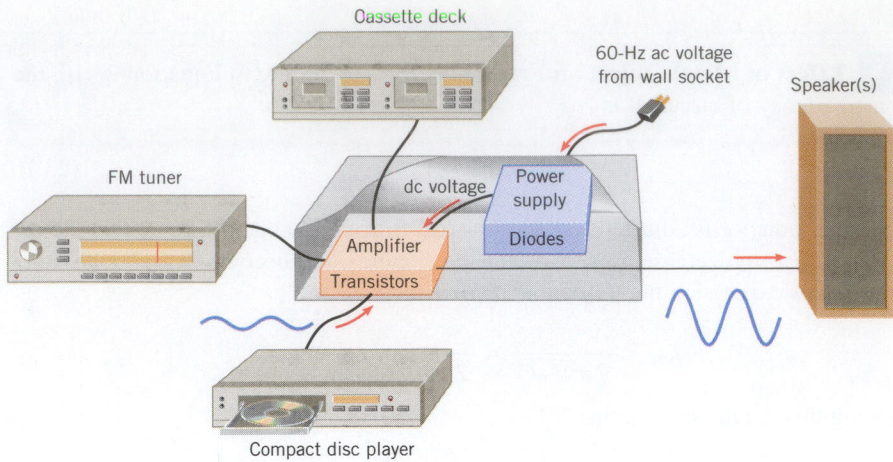

23.5 SEMICONDUCTOR DEVICES

Semiconductor devices such as diodes and transistors are widely used in modern electronics, and Figure 23.20 illustrates one application. The drawing shows an audio system in which small ac voltages (originating in a compact disc player, an FM tuner, or a cassette deck) are amplified so they can drive the speaker(s). The electric circuits that accomplish the amplification do so with the aid of a dc voltage provided by the power supply. In portable units the power supply is simply a battery. In nonportable units, however, the power supply is a separate electric circuit containing diodes, along with other elements. As we will see, the diodes convert the 60-Hz ac voltage present at a wall outlet into the dc voltage needed by the amplifier, which, in turn, performs its job of amplification with the aid of transistors.

n-TYPE AND *p*-TYPE SEMICONDUCTORS

The materials used in diodes and transistors are semiconductors, such as silicon and germanium. However, they are not pure materials because small amounts of "impurity" atoms (about one part in a million) have been added to them to change their conductive properties. For instance, Figure 23.21*a* shows an array of atoms that symbolizes the crystal structure in pure silicon. Each silicon atom has four outer-shell* electrons, and each electron participates with electrons from neighboring atoms in forming the bonds that hold the crystal together. Since they participate in forming bonds, these electrons generally do not move throughout the crystal. Consequently, pure silicon and germanium are not good conductors of electricity. It is possible, however, to increase their conductivities by adding tiny amounts of impurity atoms, such as phosphorus or arsenic, whose atoms have five outer-shell electrons. For example, when a phosphorus atom replaces a silicon atom in the crystal, only four of the five outer-shell electrons of phosphorus fit into the crystal structure. The extra fifth electron does not fit in and is relatively free to diffuse throughout the crystal, as part *b* of the drawing suggests. A semiconductor containing small amounts of phosphorus can therefore be envisioned as containing immobile, positively charged phosphorus atoms and a pool of electrons that are free to wander throughout the material. These mobile electrons allow the semiconductor to conduct electricity.

The process of adding impurity atoms is called *doping*. A semiconductor doped with an impurity that contributes mobile electrons is called an ***n*-type semiconductor,** since the mobile charge carriers have a **n**egative charge. Note that an *n*-type semiconductor is overall electrically neutral, since it contains equal numbers of positive and negative charges.

It is also possible to dope a silicon crystal with an impurity whose atoms have only three outer-shell electrons (e.g., boron or gallium). Because of the missing fourth electron, there is a "hole" in the lattice structure at the boron atom, as Figure 23.21*c* illustrates.

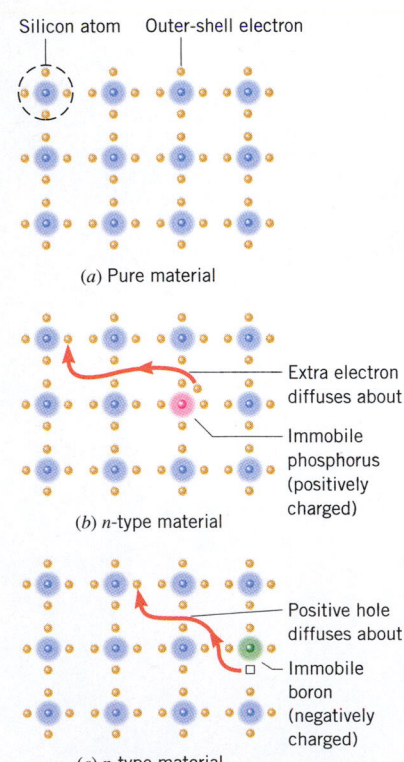

(a) Pure material

Extra electron diffuses about

Immobile phosphorus (positively charged)

(b) *n*-type material

Positive hole diffuses about

Immobile boron (negatively charged)

(c) *p*-type material

Silicon atom   Outer-shell electron

Figure 23.21 A silicon crystal that is (*a*) undoped, or pure, (*b*) doped with phosphorus to produce an *n*-type material, and (*c*) doped with boron to produce a *p*-type material.

*Section 30.6 discusses the electronic structure of the atom in terms of "shells."

An electron from a neighboring silicon atom can move into this hole, in which event the region around the boron atom, having acquired the electron, becomes negatively charged. Of course, when a nearby electron does move, it leaves behind a hole. This hole is positively charged, since it results from the removal of an electron from the vicinity of a neutral silicon atom. The vast majority of atoms in the lattice are silicon, so the hole is almost always next to another silicon atom. Consequently, an electron from one of these adjacent atoms can move into the hole, with the result that the hole moves to yet another location. In this fashion, a positively charged hole can wander through the crystal. This type of semiconductor can, therefore, be viewed as containing immobile, negatively charged boron atoms and an equal number of positively charged, mobile holes. Because of the mobile holes, the semiconductor can conduct electricity. In this case the charge carriers are positive. A semiconductor doped with an impurity that introduces mobile **positive** holes is called a ***p-type semiconductor.***

THE SEMICONDUCTOR DIODE

The physics of a semiconductor diode. A ***p-n junction diode*** is a device that is formed from a *p*-type semiconductor and an *n*-type semiconductor. The *p-n* junction between the two materials is of fundamental importance to the operation of diodes and transistors. Figure 23.22 shows separate *p*-type and *n*-type semiconductors, each electrically neutral. Figure 23.23*a* shows them joined together to form a diode. Mobile electrons from the *n*-type semiconductor and mobile holes from the *p*-type semiconductor flow across the junction and combine. This process leaves the *n*-type material with a positive charge layer and the *p*-type material with a negative charge layer, as part *b* of the drawing indicates. The positive and negative charge layers on the two sides of the junction set up an electric field $\vec{E}$, much like the field in a parallel plate capacitor. This electric field tends to prevent any further movement of charge across the junction, and all charge flow quickly stops.

Suppose now that a battery is connected across the *p-n* junction, as in Figure 23.24*a*, where the negative terminal of the battery is attached to the *n*-material, and the positive terminal is attached to the *p*-material. In this situation the junction is said to be in a condition of ***forward bias,*** and, as a result, there is a current in the circuit. The negative terminal of the battery repels the mobile electrons in the *n*-type material, and they move toward the junction. Likewise, the positive terminal repels the positive holes in the *p*-type material, and they also move toward the junction. At the junction the electrons fill the holes. In the meantime, the negative terminal provides a fresh supply of electrons to the *n*-material, and the positive terminal pulls off electrons from the *p*-material, forming new holes in the process. Consequently, a continual flow of charge, and hence a current, is maintained.

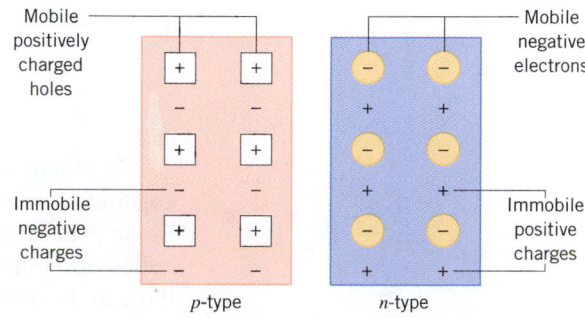

Figure 23.22 A *p*-type semiconductor and an *n*-type semiconductor.

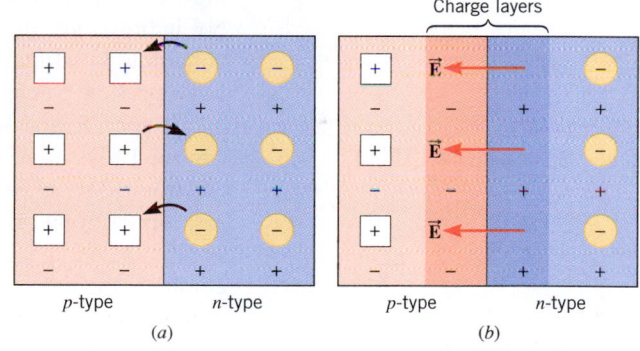

Figure 23.23 At the junction between *n* and *p* materials, (*a*) mobile electrons and holes combine and (*b*) create positive and negative charge layers. The electric field produced by the charge layers is $\vec{E}$.

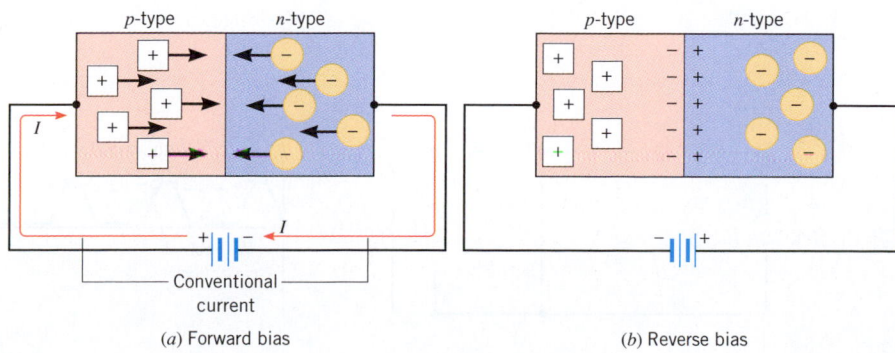

(*a*) Forward bias   (*b*) Reverse bias

Figure 23.24 (*a*) There is an appreciable current through the diode when the diode is forward-biased. (*b*) Under a reverse-bias condition, there is almost no current through the diode.

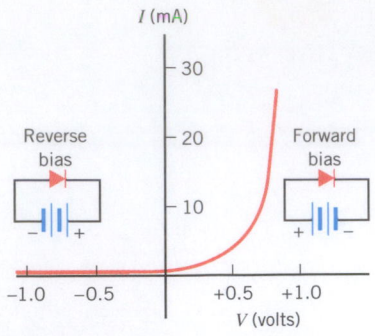

Figure 23.25 The current-versus-voltage characteristics of a typical *p-n* junction diode.

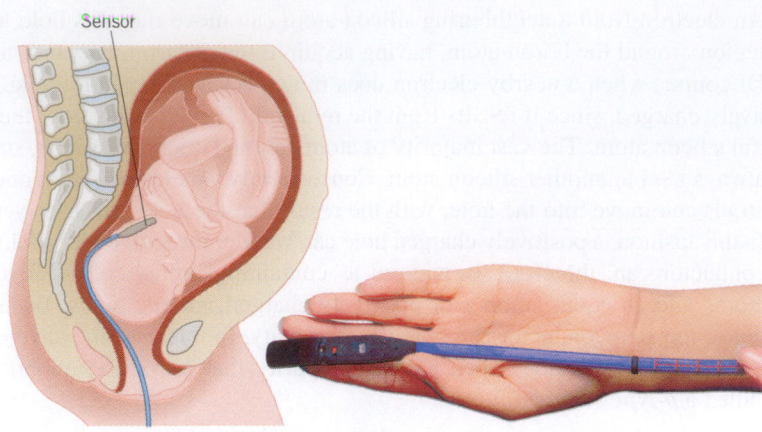

Figure 23.26 A fetal oxygen monitor uses a sensor that contains LEDs to measure the level of oxygen in the fetus's blood. (Reprinted by permission of Nellcor Puritan Bennett, Inc., Pleasanton, California)

In Figure 23.24*b* the battery polarity has been reversed, and the *p-n* junction is in a condition known as ***reverse bias.*** The battery forces electrons in the *n*-material and holes in the *p*-material away from the junction. As a result, the potential across the junction builds up until it opposes the battery potential, and very little current can be sustained through the diode. The diode, then, is a unidirectional device, for it allows current to pass in only one direction.

The graph in Figure 23.25 shows the dependence of the current on the magnitude and polarity of the voltage applied across a *p-n* junction diode. The exact values of the current depend on the nature of the semiconductor and the extent of the doping. Also shown in the drawing is the symbol used for a diode (►──). The direction of the arrowhead in the symbol indicates the direction of the conventional current in the diode under a forward-bias condition. In a forward-bias condition, the side of the symbol that contains the arrowhead has a positive potential relative to the other side.

The physics of light-emitting diodes (LEDs).

A special kind of diode is called an **LED,** which stands for **light-emitting diode.** You can see LEDs in the form of small bright red, green, or yellow lights that appear on most electronic devices, such as computers, TV sets, and stereo systems. These diodes, like others, carry current in only one direction. Imagine a forward-biased diode, like that shown in Figure 23.24*a*, in which a current exists. An LED emits light whenever electrons and holes combine, the light coming from the *p-n* junction. Commercial LEDs are often made from gallium, suitably doped with arsenic and phosphorus atoms.

The physics of a fetal oxygen monitor.

A **fetal oxygen monitor** uses LEDs to measure the level of oxygen in a fetus's blood. A sensor is inserted into the mother's uterus and positioned against the cheek of the fetus, as indicated in Figure 23.26. Two LEDs are located within the sensor, and each shines light of a different wavelength (or color) into the fetal tissue. The light is reflected by the oxygen-carrying red blood cells and is detected by an adjacent photodetector. Light from one of the LEDs is used to measure the level of oxyhemoglobin in the blood, and light from the other LED is used to measure the level of deoxyhemoglobin. From a comparison of these two levels, the oxygen saturation in the blood is determined.

Figure 23.27 A half-wave rectifier circuit, together with a capacitor and a transformer (not shown), constitutes a dc power supply because the rectifier converts an ac voltage into a dc voltage.

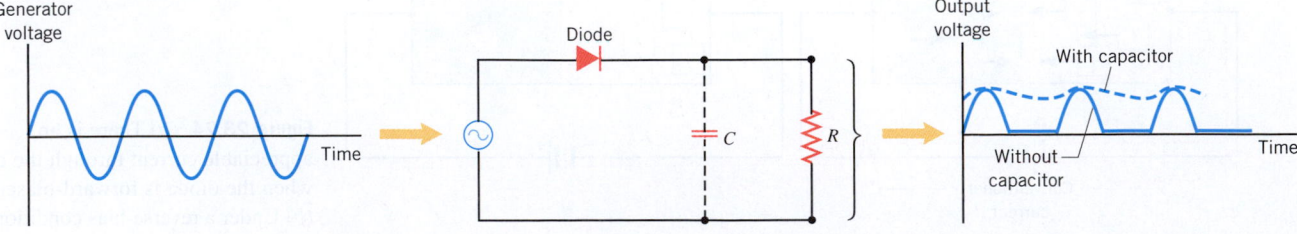

Because diodes are unidirectional devices, they are commonly used in ***rectifier circuits,*** which convert an ac voltage into a dc voltage. For instance, Figure 23.27 shows a circuit in which charges flow through the resistance *R* only while the ac generator biases the diode in the forward direction. Since current occurs only during one-half of every generator voltage cycle, the circuit is called a *half-wave rectifier*. A plot of the output voltage across the resistor reveals that only the positive halves of each cycle are present. If a capacitor is added in parallel with the resistor, as indicated in the drawing, the capacitor charges up and keeps the voltage from dropping to zero between each positive half-cycle. It is also possible to construct *full-wave rectifier circuits*, in which both halves of every cycle of the generator voltage drive current through the load resistor in the same direction.

When a circuit such as the one in Figure 23.27 includes a capacitor and also a transformer to establish the desired voltage level, the circuit is called a *power supply*. In the audio system in Figure 23.20, the power supply receives the 60-Hz ac voltage from a wall socket and produces a dc output voltage that is used for the transistors within the amplifier. Power supplies using diodes are also found in virtually all electronic appliances, such as televisions and microwave ovens.

The physics of rectifier circuits.

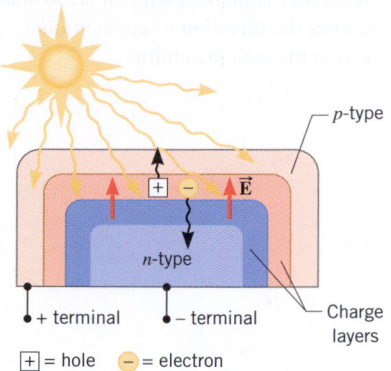

Figure 23.28 A solar cell formed from a *p-n* junction. When sunlight strikes it, the solar cell acts like a battery, with + and − terminals.

SOLAR CELLS

The physics of solar cells. Solar cells use *p-n* junctions to convert sunlight directly into electricity, as Figure 23.28 illustrates. The solar cell in this drawing consists of a *p*-type semiconductor surrounding an *n*-type semiconductor. As discussed earlier, charge layers form at the junction between the two types of semiconductors, leading to an electric field $\vec{E}$ pointing from the *n*-type toward the *p*-type layer. The outer covering of *p*-type material is so thin that sunlight penetrates into the charge layers and ionizes some of the atoms there. In the process of ionization, the energy of the sunlight causes a negative electron to be ejected from the atom, leaving behind a positive hole. As the drawing indicates, the electric field in the charge layers causes the electron and the hole to move away from the junction. The electron moves into the *n*-type material, and the hole moves into the *p*-type material. As a result, the sunlight causes the solar cell to develop negative and positive terminals, much like the terminals of a battery. The current that a single solar cell can provide is small, so applications of solar cells often use many of them mounted to form large panels, as Figure 23.29 illustrates.

Figure 23.29 This car is propelled entirely by electrical energy, generated by solar cells that cover its top surface. (© AP/Wide World Photos)

TRANSISTORS

The physics of transistors. A number of different kinds of transistors are in use today. One type is the ***bipolar-junction transistor,*** which consists of two *p-n* junctions formed by three layers of doped semiconductors. As Figure 23.30 indicates, there are *pnp* and *npn* transistors. In either case, the middle region is made very thin compared to the outer regions.

A transistor is useful because it can be used in circuits that amplify a smaller voltage into a larger one. A transistor plays the same kind of role in an amplifier circuit that a valve does when it controls the flow of water through a pipe. A small change in the valve setting produces a large change in the amount of water per second that flows through the pipe. Similarly, a small change in the voltage input to a transistor produces a large change in the output from the transistor.

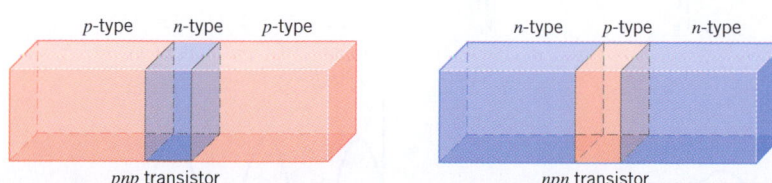

Figure 23.30 There are two kinds of bipolar-junction transistors, *pnp* and *npn*.

Figure 23.31 A *pnp* transistor, along with its bias voltages V_E and V_C. On the symbol for the *pnp* transistor at the right, the emitter is marked with an arrow that denotes the direction of conventional current through the emitter.

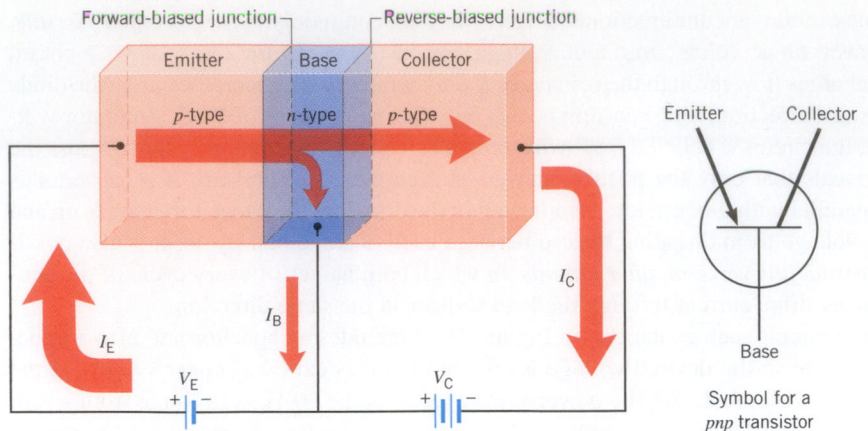

Figure 23.31 shows a *pnp* transistor connected to two batteries, labeled V_E and V_C. The voltages V_E and V_C are applied in such a way that the *p-n* junction on the left has a forward bias, while the *p-n* junction on the right has a reverse bias. Moreover, the voltage V_C is usually much larger than V_E for a reason to be discussed shortly. The drawing also shows the standard symbol and nomenclature for the three sections of the *pnp* transistor—namely, the *emitter,* the *base,* and the *collector.* The arrow in the symbol points in the direction of the conventional current through the emitter.

The positive terminal of V_E pushes the mobile positive holes in the *p*-type material of the emitter toward the emitter/base junction. Since this junction has a forward bias, the holes enter the base region readily. Once in the base region, the holes come under the strong influence of V_C and are attracted to its negative terminal. Since the base is so thin (about 10^{-6} m or so), approximately 98% of the holes are drawn through the base and into the collector. The remaining 2% of the holes combine with free electrons in the base region, thereby giving rise to a small base current I_B. As the drawing shows, the moving holes in the emitter and collector constitute currents that are labeled I_E and I_C, respectively. From Kirchhoff's junction rule it follows that $I_C = I_E - I_B$.

Because the base current I_B is small, the collector current is determined primarily by current from the emitter ($I_C = I_E - I_B \approx I_E$). This means that a change in I_E will cause a change in I_C of nearly the same amount. Furthermore, a substantial change in I_E can be caused by only a small change in the forward-bias voltage V_E. To see that this is the case, look back at Figure 23.25 and notice how steep the current-versus-voltage curve is for a *p-n* junction: small changes in the forward-bias voltage give rise to large changes in the current.

With the help of Figure 23.32 we can now appreciate what was meant by the earlier statement that a small change in the voltage input to a transistor leads to a large change in

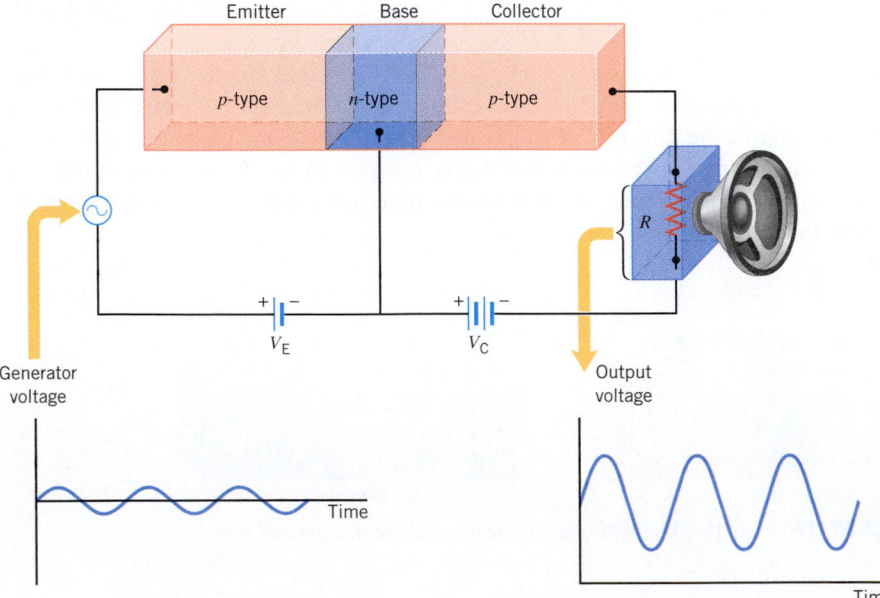

Figure 23.32 The basic *pnp* transistor amplifier in this drawing amplifies a small generator voltage to produce an enlarged voltage across the resistance *R*.

the output. This picture shows an ac generator connected in series with the battery V_E and a resistance R connected in series with the collector. The generator voltage could originate from many sources, such as an electric guitar pickup or a compact disc player, while the resistance R could represent a loudspeaker. The generator introduces small voltage changes in the forward bias across the emitter/base junction and, thus, causes large corresponding changes in the current leaving the collector and passing through the resistance R. As a result, the output voltage across R is an enlarged or amplified version of the input voltage of the generator. The operation of an *npn* transistor is similar to that of a *pnp* transistor. The main difference is that the bias voltages and current directions are reversed.

It is important to realize that the increased power available at the output of a transistor amplifier does *not* come from the transistor itself. Rather, it comes from the power provided by the voltage source V_C. The transistor, acting like an automatic valve, merely allows the small, weak signals from the input generator to control the power taken from the source V_C and delivered to the resistance R.

Today it is possible to combine arrays of billions of transistors, diodes, resistors, and capacitors on a tiny chip of silicon that usually measures less than a centimeter on a side. These arrays are called *integrated circuits* (ICs) and can be designed to perform almost any desired electronic function. Integrated circuits, such as the type in Figure 23.33, have revolutionized the electronics industry and lie at the heart of computers, cellular phones, digital watches, and programmable appliances.

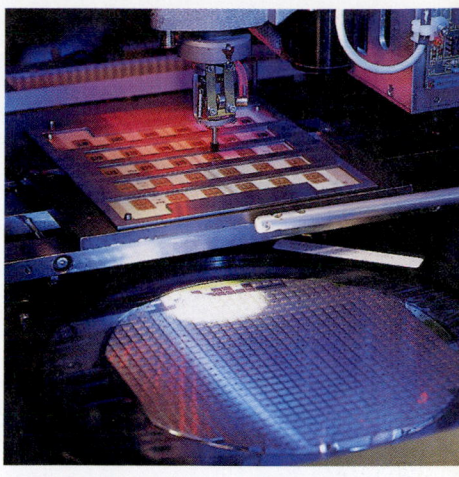

Figure 23.33 Integrated circuit (IC) chips are manufactured on wafers of semiconductor material. Shown here is one wafer containing many chips. Some of the so-called smart cards in which the chips are used are also shown. (Courtesy ORGA Card Systems, Inc.)

✓ **CHECK YOUR UNDERSTANDING**

(*The answer is given at the end of the book.*)

13. The drawing shows a full-wave rectifier circuit, in which the direction of the current through the load resistor R is the same for both positive and negative halves of the generator's voltage cycle. What is the direction of the current through the resistor (left to right, or right to left) when **(a)** the top of the generator is positive and the bottom is negative and **(b)** the top of the generator is negative and the bottom is positive?

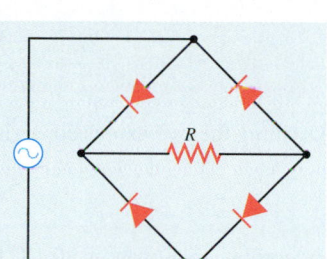

CONCEPTS & CALCULATIONS

A capacitor is one of the important elements found in ac circuits. As we have seen in this chapter, the capacitance of a capacitor influences the amount of current in a circuit. The capacitance, in turn, is determined by the geometry of the capacitor and the material that fills the space between its plates, as Section 19.5 discusses. When capacitors are connected together, the equivalent capacitance depends on the nature of the connection—for example, whether it is a series or a parallel connection, as Section 20.12 discusses. The next example provides a review of these issues concerning capacitors.

⬇ **Concepts & Calculations Example 7**   Capacitors in Ac Circuits

Two parallel plate capacitors are filled with the same dielectric material and have the same plate area. However, the plate separation of capacitor 1 is twice that of capacitor 2. When capacitor 1 is connected across the terminals of an ac generator, the generator delivers an rms current of 0.60 A. What is the current delivered by the generator when both capacitors are connected in parallel across its terminals? In both cases the generator produces the same frequency and voltage.

Concept Questions and Answers   Which of the two capacitors has the greater capacitance?

Answer   The capacitance of a capacitor is given by Equation 19.10 as $C = \kappa\epsilon_0 A/d$, where κ is the dielectric constant of the material between the plates, ϵ_0 is the permittivity of free space, A is the area of each plate, and d is the separation between the plates. Since the dielectric constant and the plate area are the same for each capacitor, the capacitance is inversely proportional to the plate separation. Therefore, capacitor 2, with the smaller plate separation, has the greater capacitance.

Is the equivalent capacitance of the parallel combination greater or smaller than the capacitance of capacitor 1?

Answer According to Equation 20.18, the equivalent capacitance of the two capacitors in parallel is $C_P = C_1 + C_2$, where C_1 and C_2 are the individual capacitances. Therefore, the value for C_P is greater than the value for C_1.

Is the capacitive reactance for C_P greater or smaller than for C_1?

Answer The capacitive reactance is given by $X_C = 1/(2\pi fC)$, according to Equation 23.2, where f is the frequency. For a given frequency, the reactance is inversely proportional to the capacitance. Since the capacitance C_P is greater than C_1, the corresponding reactance is smaller.

When both capacitors are connected in parallel across the terminals of the generator, is the current from the generator greater or smaller than when capacitor 1 is connected alone?

Answer According to Equation 23.1, the current is given by $I_{rms} = V_{rms}/X_C$, where V_{rms} is the rms voltage of the generator and X_C is the capacitive reactance. For a given voltage, the current is inversely proportional to the reactance. Since the reactance in the parallel case is smaller than for C_1 alone, the current in the parallel case is greater.

Solution Using Equation 23.1 to express the current as $I_{rms} = V_{rms}/X_C$ and Equation 23.2 to express the reactance as $X_C = 1/(2\pi fC)$, we find that the current is

$$I_{rms} = \frac{V_{rms}}{X_C} = \frac{V_{rms}}{1/(2\pi fC)} = V_{rms}2\pi fC$$

Applying this result to the case where C_1 is connected alone to the generator and to the case where the two capacitors are connected in parallel, we obtain

$$\underbrace{I_{1,\,rms} = V_{rms}2\pi fC_1}_{C_1 \text{ alone}} \quad \text{and} \quad \underbrace{I_{P,\,rms} = V_{rms}2\pi fC_P}_{C_1 \text{ and } C_2 \text{ in parallel}}$$

Dividing the two expressions gives

$$\frac{I_{P,\,rms}}{I_{1,\,rms}} = \frac{V_{rms}2\pi fC_P}{V_{rms}2\pi fC_1} = \frac{C_P}{C_1}$$

According to Equation 20.18, the effective capacitance of the two capacitors in parallel is $C_P = C_1 + C_2$, so that the result for the current ratio becomes

$$\frac{I_{P,\,rms}}{I_{1,\,rms}} = \frac{C_1 + C_2}{C_1} = 1 + \frac{C_2}{C_1}$$

Since the capacitance of a capacitor is given by Equation 19.10 as $C = \kappa\epsilon_0 A/d$, we find that

$$\frac{I_{P,\,rms}}{I_{1,\,rms}} = 1 + \frac{\kappa\epsilon_0 A/d_2}{\kappa\epsilon_0 A/d_1} = 1 + \frac{d_1}{d_2}$$

We know that $d_1 = 2d_2$, so that the current in the parallel case is

$$I_{P,\,rms} = I_{1,\,rms}\left(1 + \frac{d_1}{d_2}\right) = (0.60 \text{ A})\left(1 + \frac{2d_2}{d_2}\right) = \boxed{1.8 \text{ A}}$$

As expected, the current in the parallel case is greater.

In ac circuits that contain capacitance, inductance, and resistance, it is only the resistance that, on average, consumes power. The average power delivered to a capacitor or an inductor is zero. However, the presence of a capacitor and/or an inductor does influence the rms current in the circuit. When the current changes for any reason, the power consumed by a resistor also changes, as Example 8 illustrates.

🔵 **Concepts & Calculations Example 8** Only Resistance Consumes Power

An ac generator has a frequency of 1200 Hz and a constant rms voltage. When a 470-Ω resistor is connected between the terminals of the generator, an average power of 0.25 W is consumed by the resistor. Then, a 0.080-H inductor is connected in series with the resistor, and the combination is connected between the generator terminals. What is the average power consumed in the series combination?

Concept Questions and Answers In which case does the generator deliver a greater rms current?

Answer When only the resistor is present, the current is given by Equation 20.14 as $I_{rms} = V_{rms}/R$, where V_{rms} is the rms voltage of the generator. When the resistor and the inductor are connected in series, the current is given by $I_{rms} = V_{rms}/Z$, according to Equation 23.6. In this expression Z is the impedance and is given by Equation 23.7 as $Z = \sqrt{R^2 + X_L^2}$, where X_L is the inductive reactance and is given by Equation 23.4 as $X_L = 2\pi fL$. Since Z is greater than R, the current is greater when only the resistance is present.

In which case is a greater average power consumed in the circuit?

Answer Only the resistor in an ac circuit consumes power on average, the amount of the average power being $\overline{P} = I_{rms}^2 R$, according to Equation 20.15b. Since the resistance R is the same in both cases, a greater average power is consumed when the current I_{rms} is greater. Since the current is greater when only the resistor is present, more power is consumed in that case.

Solution When only the resistor is present, the average power is $\overline{P} = I_{rms}^2 R$, according to Equation 20.15b, and the current is given by $I_{rms} = V_{rms}/R$, according to Equation 20.14. Therefore, we find in this case that

Resistor only
$$\overline{P} = I_{rms}^2 R = \left(\frac{V_{rms}}{R}\right)^2 R = \frac{V_{rms}^2}{R}$$

When the inductor is also present, the average power is still $\overline{P} = I_{rms}^2 R$, but the current is now given by $I_{rms} = V_{rms}/Z$, according to Equation 23.6, in which Z is the impedance. In this case, then, we have

Resistor and inductor
$$\overline{P} = I_{rms}^2 R = \left(\frac{V_{rms}}{Z}\right)^2 R = \frac{V_{rms}^2 R}{Z^2}$$

Dividing this result by the analogous result for the resistor-only case, we obtain

$$\frac{\overline{P}_{\text{Resistor and inductor}}}{\overline{P}_{\text{Resistor only}}} = \frac{V_{rms}^2 R/Z^2}{V_{rms}^2/R} = \frac{R^2}{Z^2}$$

In this expression Z is the impedance and is given by Equation 23.7 as $Z = \sqrt{R^2 + X_L^2}$, where X_L is the inductive reactance and is given by Equation 23.4 as $X_L = 2\pi fL$. With these substitutions, the ratio of the powers becomes

$$\frac{\overline{P}_{\text{Resistor and inductor}}}{\overline{P}_{\text{Resistor only}}} = \frac{R^2}{R^2 + (2\pi fL)^2}$$

The power consumed in the series combination, then, is

$$\overline{P}_{\text{Resistor and inductor}} = \overline{P}_{\text{Resistor only}} \left[\frac{R^2}{R^2 + (2\pi fL)^2}\right]$$

$$= (0.25 \text{ W}) \frac{(470 \text{ }\Omega)^2}{(470 \text{ }\Omega)^2 + [2\pi(1200 \text{ Hz})(0.080 \text{ H})]^2} = \boxed{0.094 \text{ W}}$$

As expected, more power is consumed in the resistor-only case.

CONCEPT SUMMARY

If you need more help with a concept, use the Learning Aids noted next to the discussion or equation. Examples (**Ex.**) are in the text of this chapter. Go to **www.wiley.com/college/cutnell** for the following Learning Aids:

Interactive LearningWare (ILW) — Additional examples solved in a five-step interactive format.

Concept Simulations (CS) — Animated text figures or animations of important concepts.

Interactive Solutions (IS) — Models for certain types of problems in the chapter homework. The calculations are carried out interactively.

Topic	Discussion	Learning Aids
	23.1 CAPACITORS AND CAPACITIVE REACTANCE In an ac circuit the rms voltage V_{rms} across a capacitor is related to the rms current I_{rms} by	
Relation between voltage and current	$$V_{rms} = I_{rms} X_C \qquad (23.1)$$	

Topic	Discussion	Learning Aids

where X_C is the capacitive reactance. The capacitive reactance is measured in ohms (Ω) and is given by

Capacitive reactance

$$X_C = \frac{1}{2\pi fC} \qquad (23.2)$$

Ex. 1, 7
CS 23.1

where f is the frequency and C is the capacitance.

Phase angle between current and voltage

The ac current in a capacitor leads the voltage across the capacitor by a phase angle of 90° or $\pi/2$ radians. As a result, a capacitor consumes no power, on average.

Voltage and current phasors

The phasor model is useful for analyzing the voltage and current in an ac circuit. In this model, the voltage and current are represented by rotating arrows, called phasors.

The length of the voltage phasor represents the maximum voltage V_0, and the length of the current phasor represents the maximum current I_0. The phasors rotate in a counterclockwise direction at a frequency f. Since the current leads the voltage by 90° in a capacitor, the current phasor is ahead of the voltage phasor by 90° in the direction of rotation.

The instantaneous values of the voltage and current are equal to the vertical components of the corresponding phasors.

23.2 INDUCTORS AND INDUCTIVE REACTANCE In an ac circuit the rms voltage V_{rms} across an inductor is related to the rms current I_{rms} by

Relation between voltage and current

$$V_{rms} = I_{rms} X_L \qquad (23.3)$$

where X_L is the inductive reactance. The inductive reactance is measured in ohms (Ω) and is given by

Inductive reactance

$$X_L = 2\pi fL \qquad (23.4)$$

Ex. 2
CS 23.1

where f is the frequency and L is the inductance.

Phase angle between current and voltage

The ac current in an inductor lags behind the voltage across the inductor by a phase angle of 90° or $\pi/2$ radians. Consequently, an inductor, like a capacitor, consumes no power, on average.

Voltage and current phasors

The voltage and current phasors in a circuit containing only an inductor also rotate in a counterclockwise direction at a frequency f. However, since the current lags the voltage by 90° in an inductor, the current phasor is behind the voltage phasor by 90° in the direction of rotation.

The instantaneous values of the voltage and current are equal to the vertical components of the corresponding phasors.

23.3 CIRCUITS CONTAINING RESISTANCE, CAPACITANCE, AND INDUCTANCE When a resistor, a capacitor, and an inductor are connected in series, the rms voltage across the combination is related to the rms current according to

Relation between voltage and current

$$V_{rms} = I_{rms} Z \qquad (23.6)$$

where Z is the impedance of the combination. The impedance is measured in ohms (Ω) and is given by

Impedance

$$Z = \sqrt{R^2 + (X_L - X_C)^2} \qquad (23.7)$$

IS 23.27
CS 23.2

where R is the resistance, and X_L and X_C are, respectively, the inductive and capacitive reactances.

The tangent of the phase angle ϕ between current and voltage in a series RCL circuit is

Phase angle between current and voltage

$$\tan\phi = \frac{X_L - X_C}{R} \qquad (23.8)$$

Only the resistor in the RCL combination consumes power, on average. The average power $\overline{P}$ consumed in the circuit is

Average power consumed

$$\overline{P} = I_{rms}V_{rms}\cos\phi \qquad (23.9)$$

Ex. 3, 4, 5, 8

Power factor

where $\cos\phi$ is called the power factor of the circuit.

23.4 RESONANCE IN ELECTRIC CIRCUITS A series RCL circuit has a resonant frequency f_0 that is given by

Resonant frequency

$$f_0 = \frac{1}{2\pi\sqrt{LC}} \qquad (23.10)$$

Ex. 6

where L is the inductance and C is the capacitance. At resonance, the impedance of the circuit has a minimum value equal to the resistance R, and the rms current has a maximum value.

Topic	Discussion	Learning Aids
n-type semiconductor	**23.5 SEMICONDUCTOR DEVICES** In an *n*-type semiconductor, mobile negative electrons carry the current. An *n*-type material is produced by doping a semiconductor such as silicon with a small amount of impurity atoms such as phosphorus.	
p-type semiconductor	In a *p*-type semiconductor, mobile positive "holes" in the crystal structure carry the current. A *p*-type material is produced by doping a semiconductor with a small amount of impurity atoms such as boron.	
	These two types of semiconductors are used in *p-n* junction diodes, light-emitting diodes, and solar cells, and in *pnp* and *npn* bipolar junction transistors.	

FOCUS ON CONCEPTS

Note to Instructors: The numbering of the questions shown here reflects the fact that they are only a representative subset of the total number that are available online. However, all of the questions are available for assignment via an online homework management program such as WileyPLUS or WebAssign.

Section 23.1   Capacitors and Capacitive Reactance

1. A circuit contains an ac generator and a resistor. What happens to the average power dissipated in the resistor when the frequency is doubled and the rms voltage is tripled?   **(a)** Nothing happens, because the average power does not depend on either the frequency or the rms voltage.   **(b)** The average power doubles because it is proportional to the frequency.   **(c)** The average power triples because it is proportional to the rms voltage.   **(d)** The average power increases by a factor of $3^2 = 9$ because it is proportional to the square of the rms voltage.   **(e)** The average power increases by a factor of $2 \times 3 = 6$ because it is proportional to the product of the frequency and the rms voltage.

Section 23.2   Inductors and Inductive Reactance

4. What happens to the capacitive reactance X_C and the inductive reactance X_L if the frequency of the ac voltage is doubled?   **(a)** X_C increases by a factor of 2, and X_L decreases by a factor of 2.   **(b)** X_C and X_L both increase by a factor of 2.   **(c)** X_C and X_L do not change.   **(d)** X_C and X_L both decrease by a factor of 2.   **(e)** X_C decreases by a factor of 2, and X_L increases by a factor of 2.

8. Each of the four phasor diagrams represents a different circuit. V_0 and I_0 represent, respectively, the maximum voltage of the generator and the current in the circuit. Which circuit contains only a resistor?   **(a)** A   **(b)** B   **(c)** C   **(d)** D

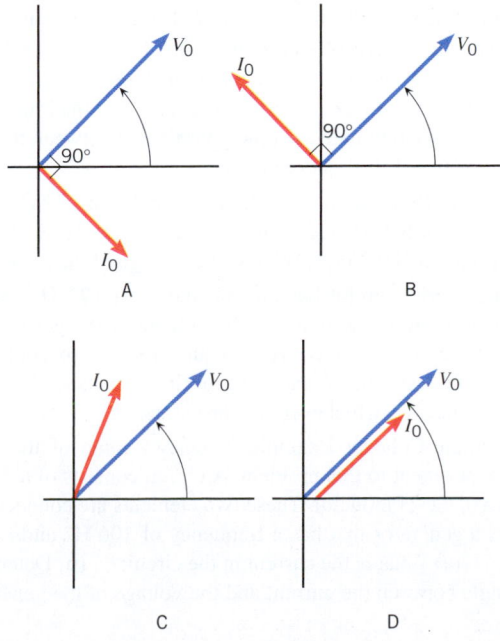

Section 23.3   Circuits Containing Resistance, Capacitance, and Inductance

11. The table shows the rms voltage V_C across the capacitor and the rms voltage V_L across the inductor for three series RCL circuits. In which circuit does the rms voltage across the entire RCL combination lead the current through the combination?   **(a)** Circuit 1   **(b)** Circuit 2   **(c)** Circuit 3   **(d)** The total rms voltage across the RCL combination does not lead the current in any of the circuits.

Circuit	V_C	V_L
1	50 V	100 V
2	100 V	50 V
3	50 V	50 V

15. A capacitor and an inductor are connected to an ac generator in two ways: in series and in parallel (see the drawing). At low frequencies, which circuit has the greater current?

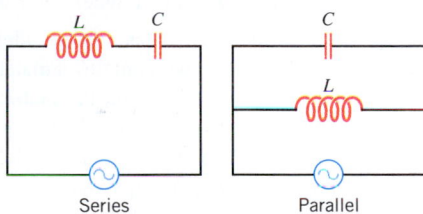

(a) The series circuit, because the impedance of the circuit is small due to the small reactances of both the inductor and the capacitor.   **(b)** The series circuit, because the impedance of the circuit is large due to the large reactances of both the inductor and the capacitor.   **(c)** The parallel circuit, because the impedance of the circuit is small due to the large reactance of the inductor.   **(d)** The parallel circuit, because the impedance of the circuit is large due to the large reactance of the inductor.   **(e)** The parallel circuit, because the impedance of the circuit is small due to the small reactance of the inductor.

Section 23.4   Resonance in Electric Circuits

18. In an RCL circuit a second capacitor is added in parallel to the capacitor already present. Does the resonant frequency of the circuit increase, decrease, or remain the same?   **(a)** The resonant frequency increases, because it depends inversely on the square root of the capacitance, and the equivalent capacitance decreases when a second capacitor is added in parallel.   **(b)** The resonant frequency increases, because it is directly proportional to the capacitance, and the equivalent capacitance increases when a second capacitor is added in parallel.   **(c)** The resonant frequency decreases, because it is directly proportional to the capacitance, and the equivalent capacitance decreases when a second capacitor is added in parallel.   **(d)** The resonant frequency decreases, because it depends inversely on the square root of the capacitance, and the equivalent capacitance increases when a second capacitor is added in parallel.   **(e)** The resonant frequency remains the same.

PROBLEMS

Note to Instructors: Most of the homework problems in this chapter are available for assignment via an online homework management program such as WileyPLUS or WebAssign, and those marked with the icon 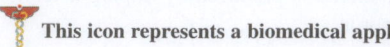 are presented in WileyPLUS using a guided tutorial format that provides enhanced interactivity. See Preface for additional details.

Note: For problems in this set, the ac current and voltage are rms values, and the power is an average value, unless indicated otherwise.

ssm Solution is in the Student Solutions Manual.
www Solution is available online at www.wiley.com/college/cutnell

This icon represents a biomedical application.

Section 23.1 Capacitors and Capacitive Reactance

1. ssm A capacitor is attached to a 5.00-Hz generator. The instantaneous current is observed to reach a maximum value at a certain time. What is the least amount of time that passes before the instantaneous voltage across the capacitor reaches its maximum value?

2. GO Two identical capacitors are connected in parallel to an ac generator that has a frequency of 610 Hz and produces a voltage of 24 V. The current in the circuit is 0.16 A. What is the capacitance of each capacitor?

3. ssm A capacitor is connected across the terminals of an ac generator that has a frequency of 440 Hz and supplies a voltage of 24 V. When a second capacitor is connected in parallel with the first one, the current from the generator increases by 0.18 A. Find the capacitance of the second capacitor.

4. A 63.0-μF capacitor is connected to a generator operating at a low frequency. The rms voltage of the generator is 4.00 V and is constant. A fuse in series with the capacitor has negligible resistance and will burn out when the rms current reaches 15.0 A. As the generator frequency is increased, at what frequency will the fuse burn out?

5. The reactance of a capacitor is 68 Ω when the ac frequency is 460 Hz. What is the reactance when the frequency is 870 Hz?

*6. GO Two parallel plate capacitors are identical, except that one of them is empty and the other contains a material with a dielectric constant of 4.2 in the space between the plates. The empty capacitor is connected between the terminals of an ac generator that has a fixed frequency and rms voltage. The generator delivers a current of 0.22 A. What current does the generator deliver after the other capacitor is connected in parallel with the first one?

*7. A capacitor is connected across an ac generator whose frequency is 750 Hz and whose *peak* output voltage is 140 V. The rms current in the circuit is 3.0 A. **(a)** What is the capacitance of the capacitor? **(b)** What is the magnitude of the *maximum* charge on one plate of the capacitor?

**8. A capacitor (capacitance C_1) is connected across the terminals of an ac generator. Without changing the voltage or frequency of the generator, a second capacitor (capacitance C_2) is added in series with the first one. As a result, the current delivered by the generator decreases by a factor of three. Suppose that the second capacitor had been added in parallel with the first one, instead of in series. By what factor would the current delivered by the generator have increased?

Section 23.2 Inductors and Inductive Reactance

9. ssm At what frequency (in Hz) are the reactances of a 52-mH inductor and a 76-μF capacitor equal?

10. An inductor is to be connected to the terminals of a generator (rms voltage = 15.0 V) so that the resulting rms current will be 0.610 A. Determine the required inductive reactance.

11. The current in an inductor is 0.20 A, and the frequency is 750 Hz. If the inductance is 0.080 H, what is the voltage across the inductor?

12. GO An ac generator has a frequency of 2.2 kHz and a voltage of 240 V. An inductance L_1 = 6.0 mH is connected across its terminals.

Then a second inductance L_2 = 9.0 mH is connected in parallel with L_1. Find the current that the generator delivers to L_1 and to the parallel combination.

13. ssm www A 40.0-μF capacitor is connected across a 60.0-Hz generator. An inductor is then connected in parallel with the capacitor. What is the value of the inductance if the rms currents in the inductor and capacitor are equal?

14. GO An ac generator has a frequency of 7.5 kHz and a voltage of 39 V. When an inductor is connected between the terminals of this generator, the current in the inductor is 42 mA. What is the inductance of the inductor?

*15. A 30.0-mH inductor has a reactance of 2.10 kΩ. **(a)** What is the frequency of the ac current that passes through the inductor? **(b)** What is the capacitance of a capacitor that has the same reactance at this frequency? The frequency is tripled, so that the reactances of the inductor and capacitor are no longer equal. What are the new reactances of **(c)** the inductor and **(d)** the capacitor?

**16. Two inductors are connected in parallel across the terminals of a generator. One has an inductance of L_1 = 0.030 H, and the other has an inductance of L_2 = 0.060 H. A single inductor, with an inductance L, is connected across the terminals of a second generator that has the same frequency and voltage as the first one. The current delivered by the second generator is equal to the *total* current delivered by the first generator. Find L.

Section 23.3 Circuits Containing Resistance, Capacitance, and Inductance

17. In a series circuit, a generator (1350 Hz, 15.0 V) is connected to a 16.0-Ω resistor, a 4.10-μF capacitor, and a 5.30-mH inductor. Find the voltage across each circuit element.

18. When a resistor is connected across the terminals of an ac generator (112 V) that has a fixed frequency, there is a current of 0.500 A in the resistor. When an inductor is connected across the terminals of this same generator, there is a current of 0.400 A in the inductor. When both the resistor and the inductor are connected in series between the terminals of this generator, what are **(a)** the impedance of the series combination and **(b)** the phase angle between the current and the voltage of the generator?

19. ssm A series RCL circuit includes a resistance of 275 Ω, an inductive reactance of 648 Ω, and a capacitive reactance of 415 Ω. The current in the circuit is 0.233 A. What is the voltage of the generator?

20. GO An ac series circuit has an impedance of 192 Ω, and the phase angle between the current and the voltage of the generator is $\phi = -75°$. The circuit contains a resistor and either a capacitor or an inductor. Find the resistance R and the capacitive reactance X_C or the inductive reactance X_L, whichever is appropriate.

21. ssm Multiple-Concept Example 3 reviews some of the basic ideas that are pertinent to this problem. A circuit consists of a 215-Ω resistor and a 0.200-H inductor. These two elements are connected in series across a generator that has a frequency of 106 Hz and a voltage of 234 V. **(a)** What is the current in the circuit? **(b)** Determine the phase angle between the current and the voltage of the generator.

22. A circuit consists of a resistor in series with an inductor and an ac generator that supplies a voltage of 115 V. The inductive reactance is 52.0 Ω, and the current in the circuit is 1.75 A. Find the average power delivered to the circuit.

23. Multiple-Concept Example 3 reviews some of the concepts needed for this problem. An ac generator has a frequency of 4.80 kHz and produces a current of 0.0400 A in a series circuit that contains only a 232-Ω resistor and a 0.250-μF capacitor. Obtain   **(a)** the voltage of the generator and   **(b)** the phase angle between the current and the voltage across the resistor/capacitor combination.

* **24.** **GO** Part *a* of the drawing shows a resistor and a charged capacitor wired in series. When the switch is closed, the capacitor discharges as charge moves from one plate to the other. Part *b* shows the amount *q* of charge remaining on each plate of the capacitor as a function of time. In part *c* of the drawing, the switch has been removed and an ac generator has been inserted into the circuit. The circuit elements in the drawing have the following values: $R = 18\ \Omega$, $V_{rms} = 24$ V for the generator, and $f = 380$ Hz for the generator. The time constant for the circuit in part *a* is $\tau = 3.0 \times 10^{-4}$ s. What is the rms current in the circuit in part *c*?

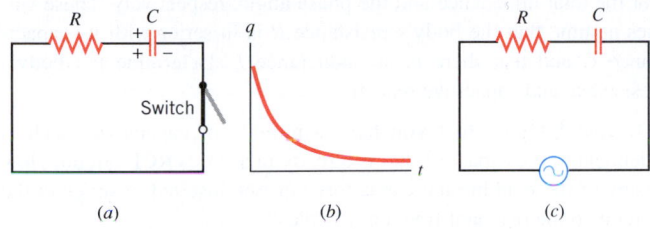

(a)   (b)   (c)

* **25.** **ssm www** A circuit consists of an 85-Ω resistor in series with a 4.0-μF capacitor, and the two are connected between the terminals of an ac generator. The voltage of the generator is fixed. At what frequency is the current in the circuit one-half the value that exists when the frequency is very large?

* **26.** A series circuit contains only a resistor and an inductor. The voltage V of the generator is fixed. If $R = 16\ \Omega$ and $L = 4.0$ mH, find the frequency at which the current is one-half its value at zero frequency.

* **27.** Refer to **Interactive Solution 23.27** at **www.wiley.com/college/cutnell** for help with problems like this one. A series RCL circuit contains only a capacitor ($C = 6.60\ \mu$F), an inductor ($L = 7.20$ mH), and a generator (*peak* voltage = 32.0 V, frequency = 1.50×10^3 Hz). When $t = 0$ s, the instantaneous value of the voltage is zero, and it rises to a maximum one-quarter of a period later.   **(a)** Find the *instantaneous* value of the voltage across the capacitor/inductor combination when $t = 1.20 \times 10^{-4}$ s.   **(b)** What is the *instantaneous* value of the current when $t = 1.20 \times 10^{-4}$ s? *(Hint: The instantaneous values of the voltage and current are, respectively, the vertical components of the voltage and current phasors.)*

* **28.** An 84.0-mH inductor and a 5.80-μF capacitor are connected in series with a generator whose frequency is 375 Hz. The rms voltage across the capacitor is 2.20 V. Determine the rms voltage across the inductor.

** **29.** When a resistor is connected by itself to an ac generator, the average power delivered to the resistor is 1.000 W. When a capacitor is added in series with the resistor, the power delivered is 0.500 W. When an inductor is added in series with the resistor (without the capacitor), the power delivered is 0.250 W. Determine the power delivered when both the capacitor and the inductor are added in series with the resistor.

Section 23.4   Resonance in Electric Circuits

30. A *tank circuit* in a radio transmitter is a series RCL circuit connected to an antenna. The antenna broadcasts radio signals at the resonant frequency of the tank circuit. Suppose that a certain tank circuit in a shortwave radio transmitter has a fixed capacitance of 1.8×10^{-11} F and a variable inductance. If the antenna is intended to broadcast radio signals ranging in frequency from 4.0 MHz to 9.0 MHz, find the   **(a)** minimum and   **(b)** maximum inductance of the tank circuit.

31. ssm A series RCL circuit has a resonant frequency of 690 kHz. If the value of the capacitance is 2.0×10^{-9} F, what is the value of the inductance?

32. GO The resistor in a series RCL circuit has a resistance of 92 Ω, while the voltage of the generator is 3.0 V. At resonance, what is the average power delivered to the circuit?

33. ssm A 10.0-Ω resistor, a 12.0-μF capacitor, and a 17.0-mH inductor are connected in series with a 155-V generator.   **(a)** At what frequency is the current a maximum?   **(b)** What is the maximum value of the rms current?

34. GO The resonant frequency of an RCL circuit is 1.3 kHz, and the value of the inductance is 7.0 mH. What is the resonant frequency (in kHz) when the value of the inductance is 1.5 mH?

35. A series RCL circuit is at resonance and contains a variable resistor that is set to 175 Ω. The power delivered to the circuit is 2.6 W. Assuming that the voltage remains constant, how much power is delivered when the variable resistor is set to 562 Ω?

36. GO The capacitance in a series RCL circuit is $C_1 = 2.60\ \mu$F, and the corresponding resonant frequency is $f_{01} = 7.30$ kHz. The generator frequency is 5.60 kHz. What is the value of the capacitance C_2 that should be added to the circuit so that the circuit will have a resonant frequency that matches the generator frequency?

37. A series RCL circuit has a resonant frequency of 1500 Hz. When operating at a frequency other than 1500 Hz, the circuit has a capacitive reactance of 5.0 Ω and an inductive reactance of 30.0 Ω. What are the values of   **(a)** L and   **(b)** C?

* **38.** Consult Multiple-Concept Example 6 for background on how to approach this problem. In the absence of a nearby metal object, the two inductances (L_A and L_B) in a heterodyne metal detector are the same, and the resonant frequencies of the two oscillator circuits have the same value of 630 kHz. When the search coil (inductor B) is brought near a buried metal object, a beat frequency of 7.3 kHz is heard. By what percentage does the buried object reduce the inductance of the search coil?

* **39. ssm www** A series RCL circuit contains a 5.10-μF capacitor and a generator whose voltage is 11.0 V. At a resonant frequency of 1.30 kHz the power delivered to the circuit is 25.0 W. Find the values of   **(a)** the inductance and   **(b)** the resistance.   **(c)** Calculate the power factor when the generator frequency is 2.31 kHz.

* **40. GO** A charged capacitor and an inductor are connected as shown in the drawing (this circuit is the same as that in Figure 23.15a). There is no resistance in the circuit. As Section 23.4 discusses, the electrical energy initially present in the charged capacitor then oscillates back and forth between the inductor and the capacitor. The initial charge on the capacitor has a magnitude of $q = 2.90\ \mu$C. The capacitance is $C = 3.60\ \mu$F, and the inductance is $L = 75.0$ mH.   **(a)** What is the electrical energy stored initially by the charged capacitor?   **(b)** Find the maximum current in the inductor.

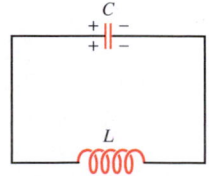

**41. In a series RCL circuit the dissipated power drops by a factor of two when the frequency of the generator is changed from the resonant frequency to a nonresonant frequency. The peak voltage is held constant while this change is made. Determine the power factor of the circuit at the nonresonant frequency.

**42. A 108-Ω resistor, a 0.200-μF capacitor, and a 5.42-mH inductor are connected in series to a generator whose voltage is 26.0 V. The current in the circuit is 0.141 A. Because of the shape of the current-versus-frequency graph (see Figure 23.16), there are two possible values for the frequency that correspond to this current. Obtain these two values.

ADDITIONAL PROBLEMS

43. ssm What voltage is needed to create a current of 35 mA in a circuit containing only a 0.86-μF capacitor, when the frequency is 3.4 kHz?

44. A 2700-Ω resistor and a 1.1-μF capacitor are connected in series across a generator (60.0 Hz, 120 V). Determine the power delivered to the circuit.

45. Suppose that the inductance is zero ($L = 0$ H) in the series RCL circuit shown in Figure 23.10. The rms voltages across the generator and the resistor are 45 and 24 V, respectively. What is the rms voltage across the capacitor?

46. Review Conceptual Example 5 and Figure 23.12. Find the ratio of the current in circuit I to the current in circuit II in the *high-frequency limit* for the same generator voltage.

47. ssm www An 8.2-mH inductor is connected to an ac generator (10.0 V rms, 620 Hz). Determine the *peak value* of the current supplied by the generator.

48. Two ac generators supply the same voltage. However, the first generator has a frequency of 1.5 kHz, and the second has a frequency of 6.0 kHz. When an inductor is connected across the terminals of the first generator, the current delivered is 0.30 A. How much current is delivered when this inductor is connected across the terminals of the second generator?

49. A light bulb has a resistance of 240 Ω. It is connected to a standard wall socket (120 V, 60.0 Hz). **(a)** Determine the current in the bulb. **(b)** Determine the current in the bulb after a 10.0-μF capacitor is added in series in the circuit. **(c)** It is possible to return the current in the bulb to the value calculated in part (a) by adding an inductor in series with the bulb and the capacitor. What is the value of the inductance of this inductor?

*50. In one measurement of the body's bioelectric impedance, values of $Z = 4.50 \times 10^2$ Ω and $\phi = -9.80°$ are obtained for the total impedance and the phase angle, respectively. These values assume that the body's resistance R is in series with its capacitance C and that there is no inductance L. Determine the body's resistance and capacitive reactance.

*51. ssm Suppose that you have a number of capacitors. Each is identical to the capacitor that is already in a series RCL circuit. How many of these additional capacitors must be inserted in series in the circuit so the resonant frequency triples?

**52. A generator is connected to a resistor and a 0.032-H inductor in series. The rms voltage across the generator is 8.0 V. When the generator frequency is set to 130 Hz, the rms voltage across the inductor is 2.6 V. Determine the resistance of the resistor in this circuit.

CHAPTER 24

ELECTROMAGNETIC WAVES

Each of the brilliant colors in this parade float corresponds to a different wavelength in the visible region of the spectrum of electromagnetic waves. As we will see in this chapter, the visible wavelengths comprise only a small part of the total spectrum. (© Ken Levine/Allsport/Getty Images)

24.1 THE NATURE OF ELECTROMAGNETIC WAVES

In Section 13.3 we saw that energy is transported to us from the sun via a class of waves known as electromagnetic waves. This class includes the familiar visible, ultraviolet, and infrared waves. In Sections 18.6, 21.1, and 21.2 we studied the concepts of electric and magnetic fields. It was the great Scottish physicist James Clerk Maxwell (1831–1879) who showed that these two fields fluctuating together can form a propagating *electromagnetic wave.* We will now bring together our knowledge of electric and magnetic fields in order to understand this important type of wave.

Figure 24.1 illustrates one way to create an electromagnetic wave. The setup consists of two straight metal wires that are connected to the terminals of an ac generator and serve as an antenna. The potential difference between the terminals changes sinusoidally with time t and has a period T. Part a shows the instant $t = 0$ s, when there is no charge at the ends of either wire. Since there is no charge, there is no electric field at the point P just to the right of the antenna. As time passes, the top wire becomes positively charged and the bottom wire negatively charged. One-quarter of a cycle later ($t = \frac{1}{4}T$), the charges have attained their maximum values, as part b of the drawing indicates. The corresponding electric field $\vec{E}$ at point P is represented by the red arrow and has increased to its maximum strength in the downward direction.* Part b also shows that the electric field created at earlier times (see the black arrow in the picture) has not disappeared but has moved to the right. Here lies the crux of the matter: At distant points, the electric field of the charges is not felt immediately. Instead, the field is created first near the wires and then, like the effect of a pebble dropped into a pond, moves outward as a wave in all directions. Only the field moving to the right is shown in the picture for the sake of clarity.

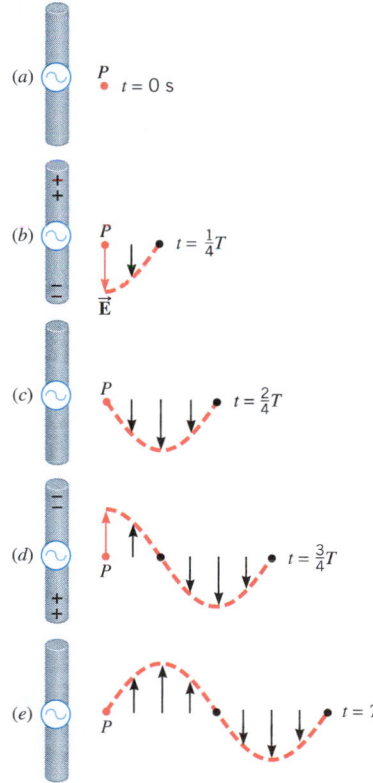

Figure 24.1 In each part of the drawing, the red arrow represents the electric field $\vec{E}$ produced at point P by the oscillating charges on the antenna at the indicated time. The black arrows represent the electric fields created at earlier times. For simplicity, only the fields propagating to the right are shown.

---

*The direction of the electric field can be obtained by imagining a positive test charge at P and determining the direction in which it would be pushed because of the charges on the wires.

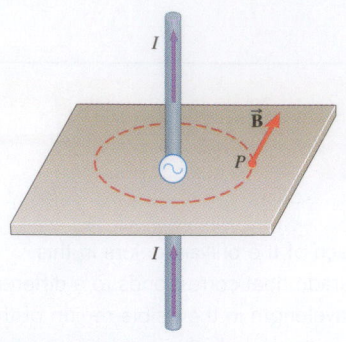

Figure 24.2 The oscillating current I in the antenna wires creates a magnetic field $\vec{B}$ at point P that is tangent to a circle centered on the wires. The field is directed as shown when the current is upward and is directed in the opposite direction when the current is downward.

Parts c–e of Figure 24.1 show the creation of the electric field at point P (red arrow) at later times during the generator cycle. In each part, the fields produced earlier in the sequence (black arrows) continue propagating toward the right. Part d shows the charges on the wires when the polarity of the generator has reversed, so the top wire is negative and the bottom wire is positive. As a result, the electric field at P has reversed its direction and points upward. In part e of the sequence, a complete sine wave has been drawn through the tips of the electric field vectors to emphasize that the field changes sinusoidally.

Along with the electric field in Figure 24.1, a magnetic field $\vec{B}$ is also created, because the charges flowing in the antenna constitute an electric current, which produces a magnetic field. Figure 24.2 illustrates the field direction at point P at the instant when the current in the antenna wire is upward. With the aid of Right-Hand Rule No. 2 (thumb of the right hand points along the current I, fingers curl in the direction of $\vec{B}$), the magnetic field at P can be seen to point into the page. As the oscillating current changes, the magnetic field changes accordingly. The magnetic fields created at earlier times propagate outward as a wave, just as the electric fields do.

Notice that the magnetic field in Figure 24.2 is perpendicular to the page, whereas the electric field in Figure 24.1 lies in the plane of the page. Thus, the electric and magnetic fields created by the antenna are mutually perpendicular and remain so as they move outward. Moreover, both fields are perpendicular to the direction of travel. These perpendicular electric and magnetic fields, moving together, constitute an electromagnetic wave.

The electric and magnetic fields in Figures 24.1 and 24.2 decrease to zero rapidly with increasing distance from the antenna. Therefore, they exist mainly near the antenna and together are called the ***near field.*** Electric and magnetic fields do form a wave at large distances from the antenna, however. These fields arise from an effect that is different from the one that produces the near field and are referred to as the ***radiation field.*** Faraday's law of induction provides part of the basis for the radiation field. As Section 22.4 discusses, this law describes the emf or potential difference produced by a changing magnetic field. And, as Section 19.4 explains, a potential difference can be related to an electric field. Thus, a changing magnetic field produces an electric field. Maxwell predicted that the reverse effect also occurs—namely, that a changing electric field produces a magnetic field. The radiation field arises because the changing magnetic field creates an electric field that fluctuates in time and the changing electric field creates the magnetic field.

Figure 24.3 shows the electromagnetic wave of the radiation field far from the antenna. The picture shows only the part of the wave traveling along the $+x$ axis. The parts traveling in the other directions have been omitted for clarity. It should be clear from the drawing that ***an electromagnetic wave is a transverse wave*** because the electric and magnetic fields are both perpendicular to the direction in which the wave travels. Moreover, an electromagnetic wave, unlike a wave on a string or a sound wave, does not require a medium in which to propagate. ***Electromagnetic waves can travel through a vacuum or a material substance,*** since electric and magnetic fields can exist in either one.

Electromagnetic waves can be produced in situations that do not involve a wire antenna. In general, any electric charge that is accelerating emits an electromagnetic wave, whether the charge is inside a wire or not. In an alternating current, an electron oscillates in simple harmonic motion along the length of the wire and is one example of an accelerating charge.

All electromagnetic waves move through a vacuum at the same speed, and the symbol c is used to denote its value. This speed is called the ***speed of light in a vacuum*** and

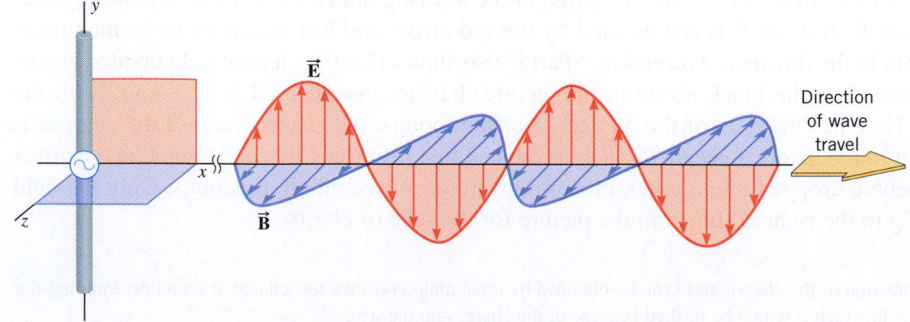

Figure 24.3 This picture shows the wave of the radiation field far from the antenna. Observe that $\vec{E}$ and $\vec{B}$ are perpendicular to each other, and both are perpendicular to the direction of travel.

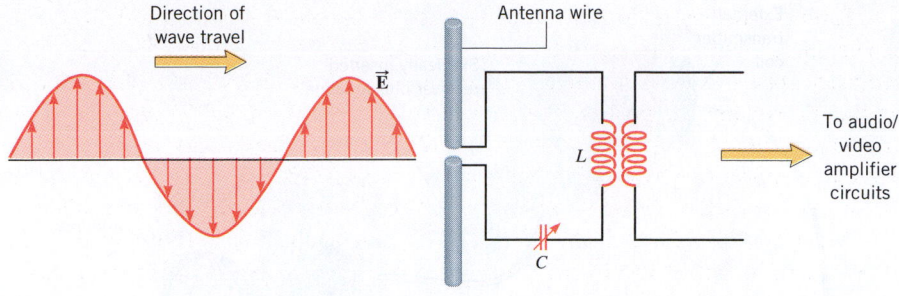

Figure 24.4 A radio wave can be detected with a receiving antenna wire that is parallel to the electric field of the wave. The magnetic field of the radio wave has been omitted for simplicity.

is $c = 3.00 \times 10^8$ m/s. In air, electromagnetic waves travel at nearly the same speed as they do in a vacuum, but, in general, they move through a substance such as glass at a speed that is substantially less than c.

The frequency of an electromagnetic wave is determined by the oscillation frequency of the electric charges at the source of the wave. In Figures 24.1–24.3 the wave frequency would equal the frequency of the ac generator. Suppose, for example, that the antenna is broadcasting electromagnetic waves known as radio waves. The frequencies of AM radio waves lie between 545 and 1605 kHz, these numbers corresponding to the limits of the AM broadcast band on the radio dial. The frequencies of FM radio waves lie between 88 and 108 MHz on the dial. Television channels 2–6, on the other hand, utilize electromagnetic waves with frequencies between 54 and 88 MHz, and channels 7–13 use frequencies between 174 and 216 MHz.

Radio and television reception involves a process that is the reverse of that outlined earlier for the creation of electromagnetic waves. When broadcasted waves reach a receiving antenna, they interact with the electric charges in the antenna wires. Either the electric field or the magnetic field of the waves can be used. To take full advantage of the electric field, the wires of the receiving antenna must be parallel to the electric field, as Figure 24.4 indicates. The electric field acts on the electrons in the wire, forcing them to oscillate back and forth along the length of the wire. Consequently, an ac current exists in the antenna and the circuit connected to it. The variable-capacitor C (——⊬——) and the inductor L in the circuit provide one way to select the frequency of the desired electromagnetic wave. By adjusting the value of the capacitance, it is possible to adjust the corresponding resonant frequency f_0 of the circuit [$f_0 = 1/(2\pi\sqrt{LC})$, Equation 23.10] to match the frequency of the wave. Under the condition of resonance, there will be a maximum oscillating current in the inductor. Because of mutual inductance, this current creates a maximum voltage in the second coil in the drawing, and this voltage can then be amplified and processed by the remaining radio or television circuitry.

To detect the magnetic field of a broadcasted radio wave, a receiving antenna in the form of a loop can be used, as Figure 24.5 shows. For best reception, the normal to the plane of the wire loop must be oriented parallel to the magnetic field. Then, as the wave sweeps by, the magnetic field penetrates the loop, and the changing magnetic flux induces a voltage and a current in the loop, in accord with Faraday's law. Once again, the resonant frequency of a capacitor/inductor combination can be adjusted to match the frequency of the desired electromagnetic wave. Both straight wire and loop antennas can be seen on the ship in Figure 24.6.

The physics of radio and television reception.

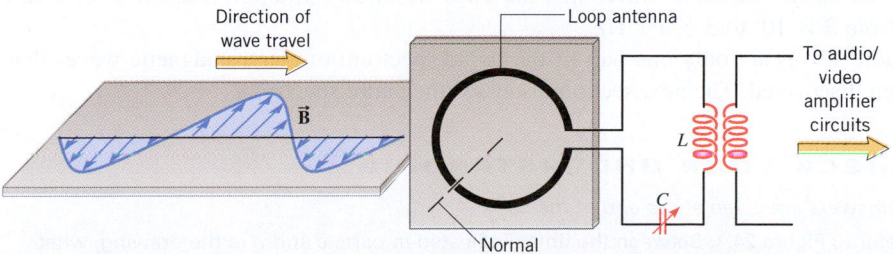

Figure 24.5 With a receiving antenna in the form of a loop, the magnetic field of a broadcasted radio wave can be detected. The normal to the plane of the loop should be parallel to the magnetic field for best reception. For clarity, the electric field of the radio wave has been omitted.

Figure 24.6 This ship uses both straight and loop antennas to communicate with other vessels and on-shore stations. (Harvey Lloyd)

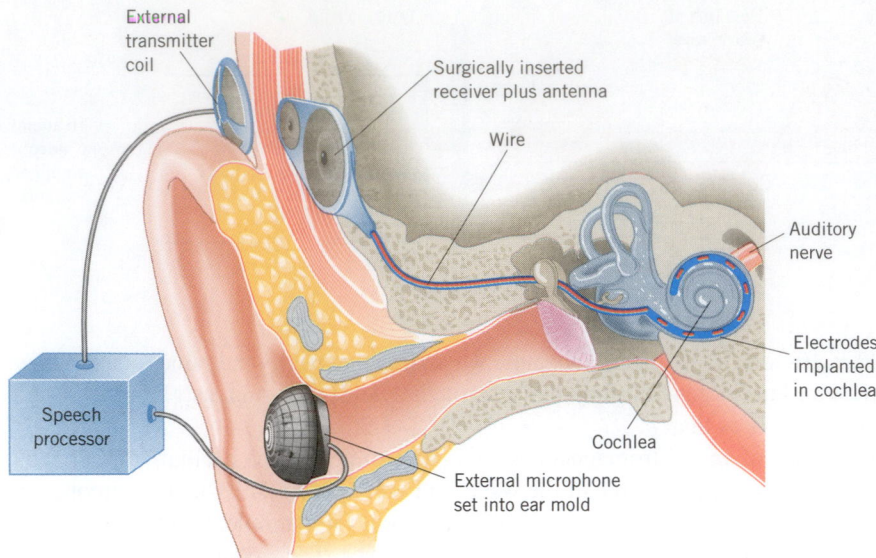

Figure 24.7 Hearing-impaired people can sometimes recover part of their hearing with the help of a cochlear implant. Broadcasting and receiving electromagnetic waves play central roles in this device.

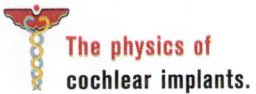

The physics of
cochlear implants.

The physics of
wireless capsule endoscopy.

Figure 24.8 This wireless capsule endoscope is designed to be swallowed. As it passes through a patient's intestines, it broadcasts video images of the interior of the intestines. (Courtesy Given Imaging, Ltd.)

Cochlear implants use the broadcasting and receiving of radio waves to provide assistance to hearing-impaired people who have auditory nerves that are at least partially intact. These implants utilize radio waves to bypass the damaged part of the hearing mechanism and access the auditory nerve directly, as Figure 24.7 illustrates. An external microphone (often set into an ear mold) detects sound waves and sends a corresponding electrical signal to a speech processor small enough to be carried in a pocket. The speech processor encodes these signals into a radio wave, which is broadcast from an external transmitter coil placed over the site of a miniature receiver (and its receiving antenna) that has been surgically inserted beneath the skin. The receiver acts much like a radio. It detects the broadcasted wave and from the encoded audio information produces electrical signals that represent the sound wave. These signals are sent along a wire to electrodes that are implanted in the cochlea of the inner ear. The electrodes stimulate the auditory nerves that feed directly between structures within the cochlea and the brain. To the extent that the nerves are intact, a person can learn to recognize sounds.

The broadcasting and receiving of radio waves are also now being used in the practice of endoscopy. In this medical diagnostic technique a device called an endoscope is used to peer inside the body. For example, to examine the interior of the colon for signs of cancer, a conventional endoscope (known as a colonoscope) is inserted through the rectum. (See Section 26.3.) The wireless capsule endoscope shown in Figure 24.8 bypasses this invasive procedure completely. With a size of about 11×26 mm, this capsule can be swallowed and carried through the gastrointestinal tract by the involuntary contractions of the walls of the intestines (peristalsis). The capsule is self-contained and uses no external wires. A marvel of miniaturization, it contains a radio transmitter and its associated antenna, batteries, a white-light-emitting diode (see Section 23.5) for illumination, and an optical system to capture the digital images. As the capsule moves through the intestine, the transmitter broadcasts the images to an array of small receiving antennas attached to the patient's body. These receiving antennas also are used to determine the position of the capsule within the body. The radio waves that are used lie in the ultrahigh-frequency, or UHF, band, from 3×10^8 to 3×10^9 Hz.

Radio waves are only one part of the broad spectrum of electromagnetic waves that has been discovered. The next section discusses the entire spectrum.

✓ **CHECK YOUR UNDERSTANDING**

(The answers are given at the end of the book.)

1. Refer to Figure 24.1. Between the times indicated in parts *c* and *d* in the drawing, what is the direction of the magnetic field at the point *P* for the electromagnetic wave being generated? Is it directed **(a)** upward along the length of the wire, **(b)** downward along the length of the wire, **(c)** into the plane of the paper, or **(d)** out of the plane of the paper?

2. A transmitting antenna is located at the origin of an *x, y, z* axis system and broadcasts an electromagnetic wave whose electric field oscillates along the *y* axis. The wave travels along the +*x* axis. Three possible wire loops are available for use with an LC-tuned circuit to detect this wave: **(a)** a loop that lies in the *xy* plane, **(b)** a loop that lies in the *xz* plane, and **(c)** a loop that lies in the *yz* plane. Which one of the loops will detect the wave?

3. Why does the peak value of the emf induced in a loop antenna (see Figure 24.5) depend on the frequency of the electromagnetic wave?

24.2 THE ELECTROMAGNETIC SPECTRUM

An electromagnetic wave, like any periodic wave, has a frequency f and a wavelength λ that are related to the speed v of the wave by $v = f\lambda$ (Equation 16.1). For electromagnetic waves traveling through a vacuum or, to a good approximation, through air, the speed is $v = c$, so $c = f\lambda$.

As Figure 24.9 shows, electromagnetic waves exist with an enormous range of frequencies, from values less than 10^4 Hz to greater than 10^{24} Hz. Since all these waves travel through a vacuum at the same speed of $c = 3.00 \times 10^8$ m/s, Equation 16.1 can be used to find the correspondingly wide range of wavelengths that the picture also displays. The ordered series of electromagnetic wave frequencies or wavelengths in Figure 24.9 is called the **electromagnetic spectrum.** Historically, regions of the spectrum have been given names such as radio waves and infrared waves. Although the boundary between adjacent regions is shown as a sharp line in the drawing, the boundary is not so well defined in practice, and the regions often overlap.

Beginning on the left in Figure 24.9, we find radio waves. Lower-frequency radio waves are generally produced by electrical oscillator circuits, while higher-frequency radio waves (called microwaves) are usually generated using electron tubes called klystrons. Infrared radiation, sometimes loosely called heat waves, originates with the vibration and rotation of molecules within a material. Visible light is emitted by hot objects, such as the sun, a burning log, or the filament of an incandescent light bulb, when the temperature is high enough to excite the electrons within an atom. Ultraviolet frequencies can be produced from the discharge of an electric arc. X-rays are produced by the sudden deceleration of high-speed electrons. And, finally, gamma rays are radiation from nuclear decay.

Astronomers use the different regions of the electromagnetic spectrum to gather information about distant celestial objects. Figure 24.10, for example, shows four views of the Crab Nebula, each in a different region of the spectrum. The Crab Nebula is located 6.0×10^{16} km away from the earth and is the remnant of a star that underwent a supernova explosion in 1054 AD.

The physics of astronomy and the electromagnetic spectrum.

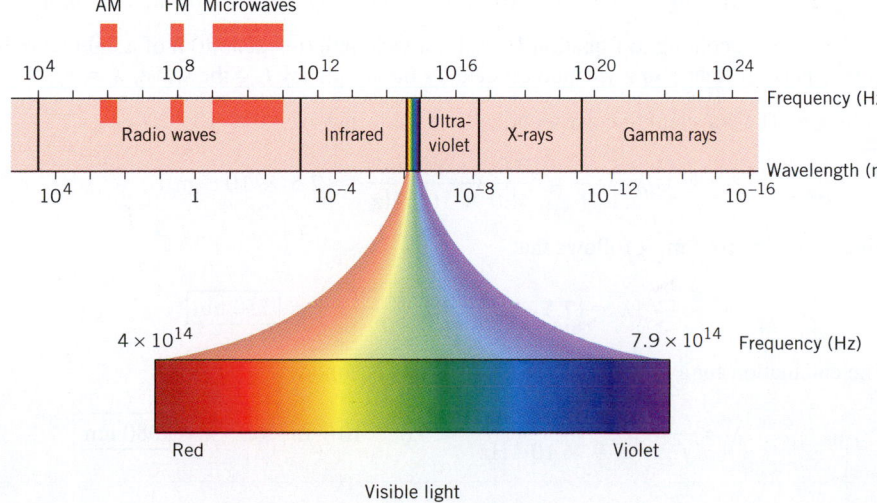

Figure 24.9 The electromagnetic spectrum.

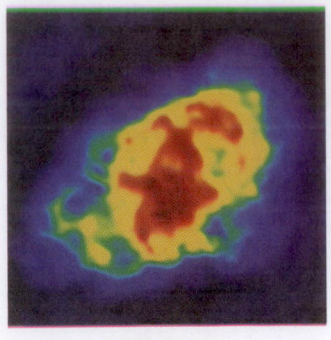

(a) Radio wave

(b) Infrared

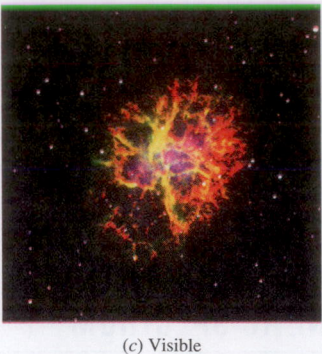

(c) Visible

(d) X-ray

Figure 24.10 Four views of the Crab Nebula. Each view is in a different region of the electromagnetic spectrum, as indicated. [(a) NRAO/AUI/NSF/Science Photo Library/Photo Researchers; (b) photo by Ken Chambers of UH/IFA—photo provided courtesy of Keck Observatory; (c) Mount Stromlo and Siding Spring Observatories/Photo Researchers, Inc.; (d) Courtesy NASA.]

**The physics of
a pyroelectric ear thermometer.**

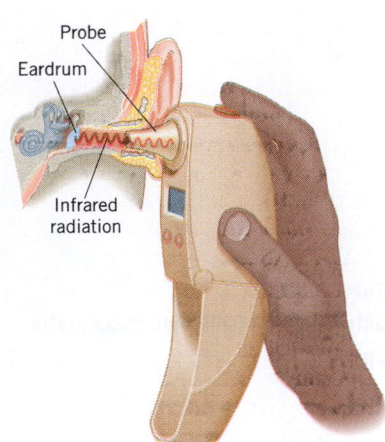

Probe

Eardrum

Infrared
radiation

Figure 24.11 A pyroelectric thermometer measures body temperature by determining the amount of infrared radiation emitted by the eardrum and surrounding tissue.

The human body, like any object, radiates infrared radiation, and the amount emitted depends on the temperature of the body. Although infrared radiation cannot be seen by the human eye, it can be detected by sensors. An ear thermometer, like the pyroelectric thermometer shown in Figure 24.11, measures the body's temperature by determining the amount of infrared radiation that emanates from the eardrum and surrounding tissue. The ear is one of the best places for this measurement because it is close to the hypothalamus, an area at the bottom of the brain that controls body temperature. The ear is also not cooled or warmed by eating, drinking, or breathing. When the probe of the thermometer is inserted into the ear canal, infrared radiation travels down the barrel of the probe and strikes the sensor. The absorption of infrared radiation warms the sensor, and, as a result, its electrical conductivity changes. The change in electrical conductivity is measured by an electronic circuit. The output from the circuit is sent to a microprocessor, which calculates the body temperature and displays the result on a digital readout.

Of all the frequency ranges in the electromagnetic spectrum, the most familiar is that of visible light, although it is the most narrow (see Figure 24.9). Only waves with frequencies between about 4.0×10^{14} Hz and 7.9×10^{14} Hz are perceived by the human eye as visible light. Usually visible light is discussed in terms of wavelengths (in vacuum) rather than frequencies. As Example 1 indicates, the wavelengths of visible light are extremely small and, therefore, are normally expressed in *nanometers* (nm); 1 nm = 10^{-9} m. An obsolete (non-SI) unit occasionally used for wavelengths is the *angstrom* (Å); 1 Å = 10^{-10} m.

Example 1 The Wavelengths of Visible Light

Find the range in wavelengths (in vacuum) for visible light in the frequency range between 4.0×10^{14} Hz (red light) and 7.9×10^{14} Hz (violet light). Express the answers in nanometers.

Reasoning According to Equation 16.1, the wavelength (in vacuum) λ of a light wave is equal to the speed of light c in a vacuum divided by the frequency f of the wave, $\lambda = c/f$.

Solution The wavelength corresponding to a frequency of 4.0×10^{14} Hz is

$$\lambda = \frac{c}{f} = \frac{3.00 \times 10^8 \text{ m/s}}{4.0 \times 10^{14} \text{ Hz}} = 7.5 \times 10^{-7} \text{ m}$$

Since 1 nm = 10^{-9} m, it follows that

$$\lambda = (7.5 \times 10^{-7} \text{ m})\left(\frac{1 \text{ nm}}{10^{-9} \text{ m}}\right) = \boxed{750 \text{ nm}}$$

The calculation for a frequency of 7.9×10^{14} Hz is similar:

$$\lambda = \frac{c}{f} = \frac{3.00 \times 10^8 \text{ m/s}}{7.9 \times 10^{14} \text{ Hz}} = 3.8 \times 10^{-7} \text{ m} \quad \text{or} \quad \lambda = \boxed{380 \text{ nm}}$$

The eye/brain recognizes light of different wavelengths as different colors. A wavelength of 750 nm (in vacuum) is approximately the longest wavelength of red light, whereas 380 nm (in vacuum) is approximately the shortest wavelength of violet light. Between these limits are found the other familiar colors, as Figure 24.9 indicates.

The association between color and wavelength in the visible part of the electromagnetic spectrum is well known. The wavelength also plays a central role in governing the behavior and use of electromagnetic waves in all regions of the spectrum. For instance, Conceptual Example 2 considers the influence of the wavelength on diffraction.

 Conceptual Example 2 The Diffraction of AM and FM Radio Waves

As we have discussed in Section 17.3, diffraction is the ability of a wave to bend around an obstacle or around the edges of an opening. Based on that discussion, which type of radio wave would you expect to bend more readily around an obstacle such as a building, **(a)** an AM radio wave or **(b)** an FM radio wave?

The physics of AM and FM radio reception.

Reasoning Section 17.3 points out that, other things being equal, sound waves exhibit diffraction to a greater extent when the wavelength is longer than when it is shorter. Based on this information, we expect that longer-wavelength electromagnetic waves will bend more readily around obstacles than will shorter-wavelength waves.

Answer (b) is incorrect. Figure 24.9 shows that FM radio waves have considerably shorter wavelengths than do AM waves. Therefore, FM radio waves exhibit less diffraction than AM waves do and bend less readily around obstacles.

Answer (a) is correct. Since AM radio waves have greater wavelengths than FM waves do (see Figure 24.9), they exhibit greater diffraction and bend more readily around obstacles than FM waves do.

The picture of light as a wave is supported by experiments that will be discussed in Chapter 27. However, there are also experiments indicating that light can behave as if it were composed of discrete particles rather than waves. These experiments will be discussed in Chapter 29. Wave theories and particle theories of light have been around for hundreds of years, and it is now widely accepted that light, as well as other electromagnetic radiation, exhibits a dual nature. Either wave-like or particle-like behavior can be observed, depending on the kind of experiment being performed.

24.3 THE SPEED OF LIGHT

At a speed of 3.00×10^8 m/s, light travels from the earth to the moon in a little over a second, so the time required for light to travel between two places on earth is very short. Therefore, the earliest attempts at measuring the speed of light had only limited success. One of the first accurate measurements employed a rotating mirror, and Figure 24.12 shows a simplified version of the setup. It was used first by the French scientist Jean Foucault (1819–1868) and later in a more refined version by the American physicist Albert Michelson (1852–1931). If the angular speed of the rotating eight-sided mirror in Figure 24.12 is adjusted correctly, light reflected from one side travels to the fixed mirror, reflects, and can be detected after reflecting from another side that has rotated into place at just the right time. The minimum angular speed must be such that one side of the mirror rotates one-eighth of a revolution during the time it takes for the light to make the round trip between the mirrors. For one of his experiments, Michelson placed his fixed mirror and rotating mirror on Mt. San Antonio and Mt. Wilson in California, a distance of 35 km apart. From a value of the minimum angular speed in such experiments, he obtained the value of $c = (2.997\,96 \pm 0.000\,04) \times 10^8$ m/s in 1926.

Today, the speed of light has been determined with such high accuracy that it is used to define the meter. As discussed in Section 1.2, the speed of light is now *defined* to be

Speed of light in a vacuum

$$c = 299\,792\,458 \text{ m/s}$$

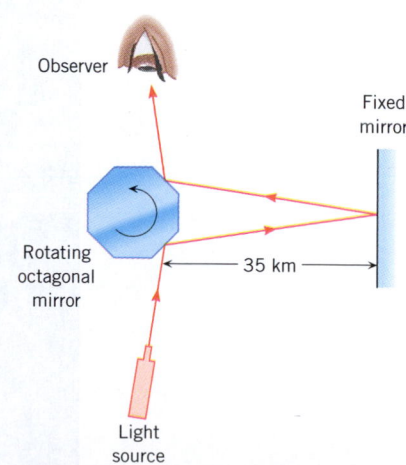

Figure 24.12 Between 1878 and 1931, Michelson used a rotating eight-sided mirror to measure the speed of light. This is a simplified version of the setup.

However, a value of 3.00×10^8 m/s is adequate for most calculations. The second is defined in terms of a cesium clock, and the meter is then defined as the distance light travels in a vacuum during a time of $1/(299\,792\,458)$ s. Although the speed of light in a vacuum is large, it is finite, so it takes a finite amount of time for light to travel from one place to another. The travel time is especially long for light traveling between astronomical objects, as Conceptual Example 3 discusses.

Conceptual Example 3 Looking Back in Time

A supernova is a violent explosion that occurs at the death of certain stars. For a few days after the explosion, the intensity of the emitted light can become a billion times greater than that of our own sun. After several years, however, the intensity usually returns to zero. Supernovae are relatively rare events in the universe, for only six have been observed in our galaxy within the past 400 years. One of them was recorded in 1987. It occurred in a neighboring galaxy, approximately 1.66×10^{21} m away. Figure 24.13 shows a photograph of the sky (*a*) before and (*b*) a few hours after the explosion. Astronomers say that viewing an event like the supernova is like looking back in time. Which one of the following statements correctly describes what we see when we view such events? (**a**) The nearer the event is to the earth, the further back in time we are looking. (**b**) The farther the event is from the earth, the further back in time we are looking.

Reasoning The light from the supernova traveled to earth at a speed of $c = 3.00 \times 10^8$ m/s. The time t required for the light to travel the distance d between the event and the earth is $t = d/c$ and is proportional to the distance.

Answer (a) is incorrect. Since the time required for the light to travel the distance between the event and the earth is proportional to the distance, the light from near-earth events reaches us sooner rather than later. Therefore, the nearer the event is to the earth, the less into the past it allows us to see, contrary to what this answer implies.

Answer (b) is correct. The travel time for light from the supernova is

$$t = \frac{d}{c} = \frac{1.66 \times 10^{21} \text{ m}}{3.00 \times 10^8 \text{ m/s}} = 5.53 \times 10^{12} \text{ s}$$

This corresponds to 175 000 years, so when astronomers saw the explosion in 1987, they were actually seeing the light that left the supernova 175 000 years earlier. In other words, they were looking back in time. Greater values for the distance d mean greater values for the time t.

Related Homework: *Problem 19*

Figure 24.13 A view of the sky (*a*) before and (*b*) a few hours after the 1987 supernova. (Courtesy Anglo-Australian Telescope Board)

(*a*) (*b*)

In 1865, Maxwell determined theoretically that electromagnetic waves propagate through a vacuum at a speed given by

$$c = \frac{1}{\sqrt{\epsilon_0 \mu_0}} \qquad (24.1)$$

where $\epsilon_0 = 8.85 \times 10^{-12}$ C^2/(N·m^2) is the (electric) permittivity of free space and $\mu_0 = 4\pi \times 10^{-7}$ T·m/A is the (magnetic) permeability of free space. Originally ϵ_0 was introduced in Section 18.5 as an alternative way of writing the proportionality constant k in Coulomb's law [$k = 1/(4\pi\epsilon_0)$] and, hence, plays a basic role in determining the strengths of the electric fields created by point charges. The role of μ_0 is similar for magnetic fields; it was introduced in Section 21.7 as part of a proportionality constant in the expression for the magnetic field created by the current in a long, straight wire. Substituting the values for ϵ_0 and μ_0 into Equation 24.1 shows that

$$c = \frac{1}{\sqrt{[8.85 \times 10^{-12}\,\text{C}^2/(\text{N·m}^2)](4\pi \times 10^{-7}\,\text{T·m/A})}} = 3.00 \times 10^8 \text{ m/s}$$

The experimental and theoretical values for c agree. Maxwell's success in predicting c provided a basis for inferring that light behaves as a wave consisting of oscillating electric and magnetic fields.

✔ **CHECK YOUR UNDERSTANDING**

(The answer is given at the end of the book.)

4. The frequency of electromagnetic wave A is twice that of electromagnetic wave B. For these two waves, what is the ratio λ_A/λ_B of the wavelengths in a vacuum? **(a)** $\lambda_A/\lambda_B = 2$, because wave A has twice the speed that wave B has. **(b)** $\lambda_A/\lambda_B = 2$, because wave A has one-half the speed that wave B has. **(c)** $\lambda_A/\lambda_B = \frac{1}{2}$, because wave A has one-half the speed that wave B has. **(d)** $\lambda_A/\lambda_B = \frac{1}{2}$, because wave A has twice the speed that wave B has. **(e)** $\lambda_A/\lambda_B = \frac{1}{2}$, because both waves have the same speed.

24.4 THE ENERGY CARRIED BY ELECTROMAGNETIC WAVES

Electromagnetic waves, like water waves or sound waves, carry energy. The energy is carried by the electric and magnetic fields that comprise the wave. In a microwave oven, for example, microwaves penetrate food and deliver their energy to it, as Figure 24.14 illustrates. The electric field of the microwaves is largely responsible for delivering the energy, and water molecules in the food absorb it. The absorption occurs because each water molecule has a permanent dipole moment; that is, one end of a molecule has a slight positive charge, and the other end has a negative charge of equal magnitude. As a result, the positive and negative ends of different molecules can form a bond. However, the electric field of the microwaves exerts forces on the positive and negative ends of a molecule, causing it to spin. Because the field is oscillating rapidly—about 2.4×10^9 times a second—the water molecules are kept spinning at a high rate. In the process, the energy of the microwaves is used to break bonds between neighboring water molecules and ultimately is converted into internal energy. As the internal energy increases, the temperature of the water increases, and the food cooks.

The energy carried by electromagnetic waves in the infrared and visible regions of the spectrum plays the key role in the greenhouse effect that is a contributing factor to global warming. The infrared waves from the sun are largely prevented from reaching the earth's surface by carbon dioxide and water in the atmosphere, which reflect them back into space. The visible waves do reach the earth's surface, however, and the energy they carry heats the earth. Heat also flows to the surface from the interior of the earth. The heated surface in turn radiates infrared waves outward, which, if they could, would carry their energy into space. However, the atmospheric carbon dioxide and water reflect these infrared waves back toward the earth, just as they reflect the infrared waves from the sun. Thus, their energy is trapped, and the earth becomes warmer, like plants in a greenhouse. In a greenhouse, however, energy is trapped mainly for a different reason—namely, the lack of effective convection currents to carry warm air past the cold glass walls.

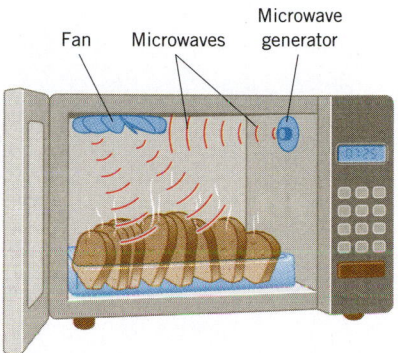

Figure 24.14 A microwave oven. The rotating fan blades reflect the microwaves to all parts of the oven.

The physics of a microwave oven.

The physics of the greenhouse effect.

A measure of the energy stored in the electric field $\vec{E}$ of an electromagnetic wave, such as a microwave, is provided by the electric energy density. As we saw in Section 19.5, this density is the electric energy per unit volume of space in which the electric field exists:

$$\begin{array}{c} \text{Electric energy} \\ \text{density} \end{array} = \frac{\text{Electric energy}}{\text{Volume}} = \frac{1}{2} \kappa \epsilon_0 E^2 = \frac{1}{2} \epsilon_0 E^2 \qquad (19.12)$$

In this equation, the dielectric constant κ has been set equal to unity, since we are dealing with an electric field in a vacuum (or in air). From Section 22.8, the analogous expression for the magnetic energy density is

$$\begin{array}{c} \text{Magnetic energy} \\ \text{density} \end{array} = \frac{\text{Magnetic energy}}{\text{Volume}} = \frac{1}{2\mu_0} B^2 \qquad (22.11)$$

The *total energy density* u of an electromagnetic wave in a vacuum is the sum of these two energy densities:

$$u = \frac{\text{Total energy}}{\text{Volume}} = \frac{1}{2} \epsilon_0 E^2 + \frac{1}{2\mu_0} B^2 \qquad (24.2a)$$

In an electromagnetic wave propagating through a vacuum or air, the electric field and the magnetic field carry equal amounts of energy per unit volume of space. Since $\frac{1}{2}\epsilon_0 E^2 = \frac{1}{2}(B^2/\mu_0)$, it is possible to rewrite Equation 24.2a for the total energy density in two additional, but equivalent, forms:

$$u = \epsilon_0 E^2 \qquad (24.2b)$$

$$u = \frac{1}{\mu_0} B^2 \qquad (24.2c)$$

The fact that the two energy densities are equal implies that the electric and magnetic fields are related. To see how, we set the electric energy density equal to the magnetic energy density and obtain

$$\frac{1}{2} \epsilon_0 E^2 = \frac{1}{2\mu_0} B^2 \quad \text{or} \quad E^2 = \frac{1}{\epsilon_0 \mu_0} B^2$$

However, according to Equation 24.1, $c = 1/\sqrt{\epsilon_0 \mu_0}$, so it follows that $E^2 = c^2 B^2$. Taking the square root of both sides of this result shows that the relation between the magnitudes of the electric and magnetic fields in an electromagnetic wave is

$$E = cB \qquad (24.3)$$

In an electromagnetic wave, the electric and magnetic fields fluctuate sinusoidally in time, so Equations 24.2a–c give the energy density of the wave at any instant in time. If an average value $\bar{u}$ for the total energy density is desired, average values are needed for E^2 and B^2. In Section 20.5 we faced a similar situation for alternating currents and voltages and introduced rms (root mean square) quantities. Using an analogous procedure here, we find that the rms values for the electric and magnetic fields, E_{rms} and B_{rms}, are related to the maximum values of these fields, E_0 and B_0, by

$$E_{rms} = \frac{1}{\sqrt{2}} E_0 \quad \text{and} \quad B_{rms} = \frac{1}{\sqrt{2}} B_0$$

Equations 24.2a–c can now be interpreted as giving the average energy density $\bar{u}$, provided the symbols E and B are interpreted to mean the rms values given above. The average density of the sunlight reaching the earth is determined in the next example.

⬇ Example 4   The Average Energy Density of Sunlight

Sunlight enters the top of the earth's atmosphere with an electric field whose rms value is $E_{rms} = 720$ N/C. Find **(a)** the average total energy density of this electromagnetic wave and **(b)** the rms value of the sunlight's magnetic field.

Reasoning The average total energy density $\bar{u}$ of the sunlight can be obtained from Equation 24.2b, provided the rms value is used for the electric field. Since the magnitudes of the magnetic and electric fields are related according to Equation 24.3, the rms value of the magnetic field is $B_{rms} = E_{rms}/c$.

Solution (a) According to Equation 24.2b, the average total energy density is

$$\overline{u} = \epsilon_0 E_{rms}^2 = [8.85 \times 10^{-12} \text{ C}^2/(\text{N} \cdot \text{m}^2)](720 \text{ N/C})^2 = \boxed{4.6 \times 10^{-6} \text{ J/m}^3}$$

(b) Using Equation 24.3, we find that the rms magnetic field is

$$B_{rms} = \frac{E_{rms}}{c} = \frac{720 \text{ N/C}}{3.0 \times 10^8 \text{ m/s}} = \boxed{2.4 \times 10^{-6} \text{ T}}$$

As an electromagnetic wave moves through space, it carries energy from one region to another. This energy transport is characterized by the **intensity** of the wave. We have encountered the concept of intensity before, in connection with sound waves in Section 16.7. The sound intensity is the sound power that passes perpendicularly through a surface divided by the area of the surface. The intensity of an electromagnetic wave is defined similarly. For an electromagnetic wave, the intensity is the electromagnetic power divided by the area of the surface.

Using this definition of intensity, we can show that the electromagnetic intensity S is related to the energy density u. According to Equation 16.8 the intensity is the power P that passes perpendicularly through a surface divided by the area A of that surface, or $S = P/A$. Furthermore, the power is equal to the total energy passing through the surface divided by the elapsed time t (Equation 6.10b), so that $P = $ (Total energy)/t. Combining these two relations gives

$$S = \frac{P}{A} = \frac{\text{Total energy}}{tA}$$

Now, consider Figure 24.15, which shows an electromagnetic wave traveling in a vacuum along the x axis. In a time t the wave travels the distance ct, passing through the surface of area A. Consequently, the volume of space through which the wave passes is ctA. The total (electric and magnetic) energy in this volume is

$$\text{Total energy} = (\text{Total energy density}) \times \text{Volume} = u(ctA)$$

Using this result in the expression for the intensity, we obtain

$$S = \frac{\text{Total energy}}{tA} = \frac{uctA}{tA} = cu \tag{24.4}$$

Thus, the intensity and the energy density are related by the speed of light, c. Substituting Equations 24.2a–c, one at a time, into Equation 24.4 shows that the intensity of an electromagnetic wave depends on the electric and magnetic fields according to the following equivalent relations:

$$S = cu = \frac{1}{2}c\epsilon_0 E^2 + \frac{c}{2\mu_0}B^2 \tag{24.5a}$$

$$S = c\epsilon_0 E^2 \tag{24.5b}$$

$$S = \frac{c}{\mu_0}B^2 \tag{24.5c}$$

If the rms values for the electric and magnetic fields are used in Equations 24.5a–c, the intensity becomes an average intensity, $\overline{S}$, as Example 5 illustrates.

> **Problem-solving insight**
> The concepts of power and intensity are similar, but they are not the same. Intensity is the power that passes perpendicularly through a surface divided by the area of the surface.

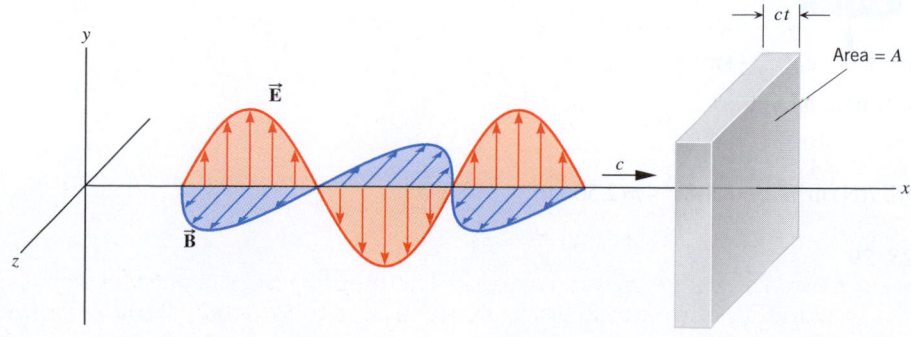

Figure 24.15 In a time t, an electromagnetic wave moves a distance ct along the x axis and passes through a surface of area A.

ANALYZING MULTIPLE-CONCEPT PROBLEMS

Example 5 Power and Intensity

Figure 24.16 shows a tiny source that is emitting light uniformly in all directions. At a distance of 2.50 m from the source, the rms electric field strength of the light is 19.0 N/C. Assuming that the light does not reflect from anything in the environment, determine the average power of the light emitted by the source.

Reasoning Recall from Section 16.7 that the power crossing a surface perpendicularly is equal to the intensity at the surface times the area of the surface (see Equation 16.8). Since the source emits light uniformly in all directions, the light intensity is the same at all points on the imaginary spherical surface in Figure 24.16. Moreover, the light crosses this surface perpendicularly. Equation 24.5b relates the average light intensity at the surface to its rms electric field strength (which is known), and the area of the surface can be found from a knowledge of its radius.

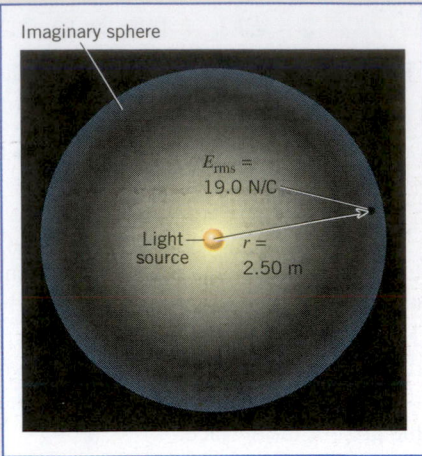

Figure 24.16 At a distance of 2.50 m from the light source, the rms electric field of the light has a value of 19.0 N/C.

Knowns and Unknowns The following data are available:

Description	Symbol	Value
Rms electric field strength 2.50 m from light source	E_{rms}	19.0 N/C
Distance from light source	r	2.50 m
Unknown Variable		
Average power emitted by light source	$\overline{P}$	?

Modeling the Problem

STEP 1 **Average Intensity** According to the discussion in Section 16.7, the average power $\overline{P}$ that passes perpendicularly through the imaginary spherical surface is equal to the average intensity $\overline{S}$ times the area A of the surface, or $\overline{P} = \overline{S}A$. The area of a spherical surface is $A = 4\pi r^2$, where r is the radius of the sphere. Thus, the average power can be written as in Equation 1 at the right. The radius is known, but the average light intensity is not, so we turn to Step 2 to evaluate it.

$$\overline{P} = \overline{S}(4\pi r^2) \qquad (1)$$

$$\textcircled{?}$$

STEP 2 **Average Intensity and Electric Field** The average intensity $\overline{S}$ of the light passing through the imaginary spherical surface is related to the known rms electric field strength E_{rms} at the surface by Equation 24.5b:

$$\overline{S} = c\epsilon_0 E_{rms}^2$$

where c is the speed of light in a vacuum and ϵ_0 is the permittivity of free space. We can substitute this expression into Equation 1, as indicated at the right.

$$\overline{P} = \overline{S}(4\pi r^2) \qquad (1)$$

$$\overline{S} = c\epsilon_0 E_{rms}^2$$

Solution Algebraically combining the results of each step, we have

STEP 1	STEP 2

$$\overline{P} = \overline{S}(4\pi r^2) = c\epsilon_0 E_{rms}^2(4\pi r^2)$$

The average power emitted by the light source is

$$\overline{P} = c\epsilon_0 E_{rms}^2(4\pi r^2)$$

$$= (3.00 \times 10^8 \text{ m/s})[8.85 \times 10^{-12} \text{ C}^2/(\text{N} \cdot \text{m}^2)](19.0 \text{ N/C})^2 \, 4\pi(2.50 \text{ m})^2 = \boxed{75.3 \text{ W}}$$

Related Homework: *Problems 26, 28, 29*

✓ **CHECK YOUR UNDERSTANDING**

(*The answers are given at the end of the book.*)

5. If both the electric and magnetic fields of an electromagnetic wave double in magnitude, how does the intensity of the wave change? The intensity   **(a)** decreases by a factor of four   **(b)** decreases by a factor of two   **(c)** increases by a factor of two   **(d)** increases by a factor of four   **(e)** increases by a factor of eight.

6. Suppose that the electric field of an electromagnetic wave decreases in magnitude. Does the magnitude of the magnetic field   **(a)** increase,   **(b)** decrease, or   **(c)** remain the same?

24.5 THE DOPPLER EFFECT AND ELECTROMAGNETIC WAVES

Section 16.9 presents a discussion of the Doppler effect that sound waves exhibit when either the source of a sound wave, the observer of the wave, or both are moving with respect to the medium of propagation (e.g., air). This effect is one in which the observed sound frequency is greater or smaller than the frequency emitted by the source. A different Doppler effect arises when the source moves than when the observer moves.

Electromagnetic waves also can exhibit a Doppler effect, but it differs from that for sound waves for two reasons. First, sound waves require a medium such as air in which to propagate. In the Doppler effect for sound, it is the motion (of the source, the observer, and the waves themselves) relative to this medium that is important. In the Doppler effect for electromagnetic waves, motion relative to a medium plays no role, because the waves do not require a medium in which to propagate. They can travel in a vacuum. Second, in the equations for the Doppler effect in Section 16.9, the speed of sound plays an important role, and it depends on the reference frame relative to which it is measured. For example, the speed of sound with respect to moving air is different than it is with respect to stationary air. As we will see in Section 28.2, electromagnetic waves behave in a different way. The speed at which they travel has the same value, whether it is measured relative to a stationary observer or relative to one moving at a constant velocity. For these two reasons, the same Doppler effect arises for electromagnetic waves when either the source or the observer of the waves moves; only the relative motion of the source and the observer with respect to one another is important.

When electromagnetic waves and the source and the observer of the waves all travel along the same line in a vacuum (or in air, to a good degree of approximation), the single equation that specifies the Doppler effect is

$$f_o = f_s \left(1 \pm \frac{v_{rel}}{c} \right) \qquad \text{if } v_{rel} \ll c \qquad (24.6)$$

In this expression, f_o is the observed frequency, and f_s is the frequency emitted by the source. The symbol v_{rel} stands for the speed of the source and the observer relative to one another, and c is the speed of light in a vacuum. Equation 24.6 applies only if v_{rel} is very small compared to c—that is, if $v_{rel} \ll c$. It is essential to realize that v_{rel} is the ***relative*** speed of the source and the observer. Thus, if the source is moving due east at a speed of 28 m/s with respect to the earth, while the observer is moving due east at a speed of 22 m/s, the value for v_{rel} is 28 m/s − 22 m/s = 6 m/s. Because v_{rel} is the relative ***speed,*** it has no algebraic sign. The direction of the relative motion is taken into account by choosing the plus or minus sign in Equation 24.6. The plus sign is used when the source and the observer come together, and the minus sign is used when they move apart. For example, suppose the source is traveling at a velocity of 28 m/s due east. The observer, traveling ahead of the source, is moving at a slower velocity of 22 m/s due east. The source and observer are coming together (the source is catching up with the observer), and the plus sign is used in Equation 24.6. Suppose, on the other hand, that the velocity of the observer is 34 m/s due east (greater than the source velocity of 28 m/s due east). Now the source and observer are moving apart (the observer is pulling away from the source) and the minus sign is used in Equation 24.6. Example 6 illustrates one familiar use of the Doppler effect for electromagnetic waves.

Problem-solving insight

The Doppler effect for electromagnetic waves depends on the speed v_{rel} of the observer and the source of the waves relative to one another. In general, do not use the speed of the observer or of the source with respect to the ground in Equation 24.6.

ANALYZING MULTIPLE-CONCEPT PROBLEMS

Example 6 Radar Guns and the Doppler Effect

Police use radar guns and the Doppler effect to catch speeders. Figure 24.17 illustrates a moving car approaching a stationary police car. A radar gun emits an electromagnetic wave that reflects from the oncoming car. The reflected wave returns to the police car with a frequency (measured by on-board equipment) that is different from the emitted frequency. One such radar gun emits a wave whose frequency is 8.0×10^9 Hz. When the speed of the car is 39 m/s and the approach is essentially head-on, what is the *difference* between the frequency of the wave returning to the police car and that emitted by the radar gun?

The physics of radar speed traps.

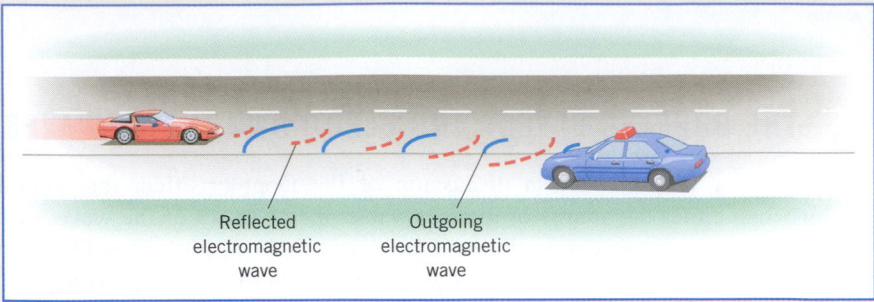

Reflected electromagnetic wave

Outgoing electromagnetic wave

Figure 24.17 A radar gun in the police car emits an electromagnetic wave that reflects from the moving car. The frequencies of the emitted and reflected waves are different because of the Doppler effect.

Reasoning As the car moves toward the wave emitted by the radar gun, the car intercepts a greater number of wave crests per second than it would if it were stationary. Thus, the car "observes" a wave frequency f_o that is greater than the frequency f_s emitted by the radar gun. The car then reflects the wave back toward the police car. In effect, the car becomes a moving source of radar waves that are emitted with a frequency f_o. Since the car is moving toward the police car, the reflected crests arrive at the police car with a frequency f'_o that is even greater than f_o. Thus, there are two frequency changes due to the Doppler effect, one associated with f_o and the other with f'_o. We will employ Equation 24.6 to determine these frequencies.

Knowns and Unknowns The data for this problem are listed below:

Description	Symbol	Value	Comment
Frequency emitted by radar gun	f_s	8.0×10^9 Hz	
Relative speed between moving car and police car	v_{rel}	39 m/s	The relative speed is just that of the moving car, since the police car is stationary.
Unknown Variable			
Frequency difference	$f'_o - f_s$	?	Subscript "o" indicates that f'_o is the frequency detected by the "observer" (the police car).

Modeling the Problem

STEP 1 Frequency Difference The desired frequency difference is $f'_o - f_s$, where f_s is the frequency emitted by the radar gun and is known; f'_o is the frequency of the wave returning to the police car after being reflected from the moving car and is not known. However, f'_o and the frequency f_o observed by the moving car are related by Equation 24.6:

$$f'_o = f_o\left(1 + \frac{v_{rel}}{c}\right)$$

where v_{rel} is the relative speed between the moving car and the stationary police car. We have chosen the + sign in Equation 24.6, since the moving car and the police car are coming together. With this expression for f'_o, the desired frequency difference can be written as shown in Equation 1 at the right. The relative speed v_{rel} and the frequency f_s are known; the frequency f_o will be determined in Step 2.

$$f'_o - f_s = f_o\left(1 + \frac{v_{rel}}{c}\right) - f_s \quad (1)$$

(?)

STEP 2 **Frequency of Wave "Observed" by Moving Car** The radar gun sends out a wave whose frequency is f_s. As the moving car approaches this wave, the car "observes" a frequency f_o that is greater than f_s. The relation between these frequencies is given by Equation 24.6:

$$f_o = f_s\left(1 + \frac{v_{rel}}{c}\right)$$

Again, the + sign has been chosen in Equation 24.6, since the moving car and the police car are coming together. All the variables on the right side of this equation are known, and we substitute this expression for f_o into Equation 1, as indicated at the right.

$$f'_o - f_s = f_o\left(1 + \frac{v_{rel}}{c}\right) - f_s \quad (1)$$

$$f_o = f_s\left(1 + \frac{v_{rel}}{c}\right)$$

Solution Algebraically combining the results of the two steps above, we obtain

$$\overset{\text{STEP 1}}{\overbrace{\hspace{3cm}}} \quad \overset{\text{STEP 2}}{\overbrace{\hspace{3cm}}}$$

$$f'_o - f_s = f_o\left(1 + \frac{v_{rel}}{c}\right) - f_s = \underbrace{f_s\left(1 + \frac{v_{rel}}{c}\right)}_{f_o}\left(1 + \frac{v_{rel}}{c}\right) - f_s$$

Multiplying and combining terms in this equation yields

$$f'_o - f_s = f_s\left[1 + \frac{2v_{rel}}{c} + \left(\frac{v_{rel}}{c}\right)^2\right] - f_s = f_s\left(\frac{v_{rel}}{c}\right)\left(2 + \frac{v_{rel}}{c}\right)$$

The value of v_{rel}/c is (39 m/s)/(3.0 × 10^8 m/s) = 13 × 10^{-8}. This is an extremely small number when compared to the number 2, so the term $(2 + v_{rel}/c)$ is very nearly equal to 2. Thus, we find that the difference between the frequency of the wave returning to the police car and that emitted by the radar gun is

$$f'_o - f_s = f_s\left(\frac{v_{rel}}{c}\right)(2) = (8.0 \times 10^9 \text{ Hz})\left(\frac{39 \text{ m/s}}{3.0 \times 10^8 \text{ m/s}}\right)(2) = \boxed{2.1 \times 10^3 \text{ Hz}}$$

Related Homework: *Problems 36, 54*

The Doppler effect of electromagnetic waves provides a powerful tool for astronomers. For instance, Example 10 in Chapter 5 discusses how astronomers have identified a supermassive black hole at the center of galaxy M87 by using the Hubble Space Telescope. They focused the telescope on regions to either side of the center of the galaxy (see Figure 5.15). From the light emitted by these two regions, they were able to use the Doppler effect to determine that one side is moving away from the earth, while the other side is moving toward the earth. In other words, the galaxy is rotating. The speeds of recession and approach enabled astronomers to determine the rotational speed of the galaxy, and Example 10 in Chapter 5 shows how the value for this speed leads to the identification of the black hole. Astronomers routinely study the Doppler effect of the light that reaches the earth from distant parts of the universe. From such studies, they have determined the speeds at which distant light-emitting objects are receding from the earth.

The physics of
astronomy and the Doppler effect.

✓ **CHECK YOUR UNDERSTANDING**

(The answers are given at the end of the book.)

7. An astronomer measures the Doppler change in frequency for the light reaching the earth from a distant star. From this measurement, can the astronomer tell whether the star is moving away from the earth or the earth is moving away from the star?

Continued

8. The drawing shows three situations—A, B, and C—in which an observer and a source of electromagnetic waves are moving along the same line. In each case the source emits a wave of the same frequency. The arrows in each situation denote velocity vectors relative to the ground and have magnitudes of either v or $2v$. Rank the magnitudes of the frequencies of the observed waves in descending order (largest first).

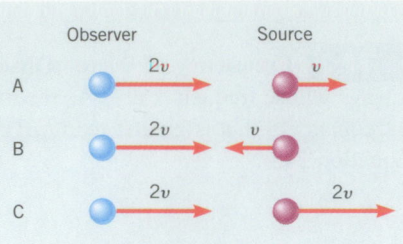

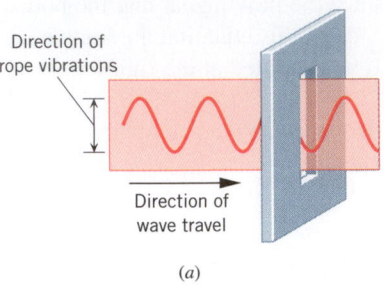

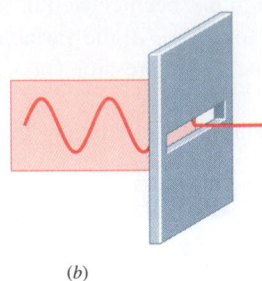

Figure 24.18 A transverse wave is linearly polarized when its vibrations always occur along one direction. (*a*) A linearly polarized wave on a rope can pass through a slit that is parallel to the direction of the rope vibrations, but (*b*) cannot pass through a slit that is perpendicular to the vibrations.

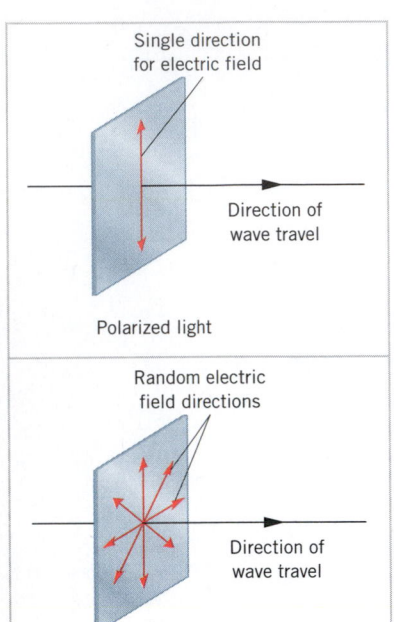

Figure 24.19 In polarized light, the electric field of the electromagnetic wave fluctuates along a single direction. Unpolarized light consists of short bursts of electromagnetic waves emitted by many different atoms. The electric field directions of these bursts are perpendicular to the direction of wave travel but are distributed randomly about it.

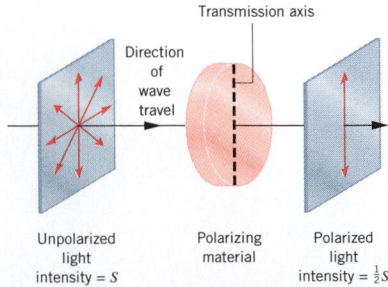

Figure 24.20 With the aid of a piece of polarizing material, polarized light may be produced from unpolarized light. The transmission axis of the material is the direction of polarization of the light that passes through the material.

POLARIZATION

24.6

POLARIZED ELECTROMAGNETIC WAVES

One of the essential features of electromagnetic waves is that they are transverse waves, and because of this feature they can be polarized. Figure 24.18 illustrates the idea of polarization by showing a transverse wave as it travels along a rope toward a slit. The wave is said to be ***linearly polarized,*** which means that its vibrations always occur along one direction. This direction is called the direction of polarization. In part *a* of the picture, the direction of polarization is vertical, parallel to the slit. Consequently, the wave passes through easily. However, when the slit is turned perpendicular to the direction of polarization, as in part *b*, the wave cannot pass, because the slit prevents the rope from oscillating. For longitudinal waves, such as sound waves, the notion of polarization has no meaning. In a longitudinal wave the direction of vibration is along the direction of travel, and the orientation of the slit would have no effect on the wave.

In an electromagnetic wave such as the one in Figure 24.3, the electric field oscillates along the y axis. Similarly, the magnetic field oscillates along the z axis. Therefore, the wave is linearly polarized, with the direction of polarization taken arbitrarily to be the direction along which the electric field oscillates. If the wave is a radio wave generated by a straight-wire antenna, the direction of polarization is determined by the orientation of the antenna. In comparison, the visible light given off by an incandescent light bulb consists of electromagnetic waves that are completely unpolarized. In this case the waves are emitted by a large number of atoms in the hot filament of the bulb. When an electron in an atom oscillates, the atom behaves as a miniature antenna that broadcasts light for brief periods of time, about 10^{-8} seconds. However, the directions of these atomic antennas change randomly as a result of collisions. Unpolarized light, then, consists of many individual waves, emitted in short bursts by many "atomic antennas," each with its own direction of polarization. Figure 24.19 compares polarized and unpolarized light. In the unpolarized case, the arrows shown around the direction of wave travel symbolize the random directions of polarization of the individual waves that comprise the light.

Linearly polarized light can be produced from unpolarized light with the aid of certain materials. One commercially available material goes under the name of Polaroid. Such materials allow only the component of the electric field along one direction to pass through, while absorbing the field component perpendicular to this direction. As Figure 24.20 indicates, the direction of polarization that a polarizing material allows through is called the ***transmission axis.*** No matter how this axis is oriented, the average intensity of

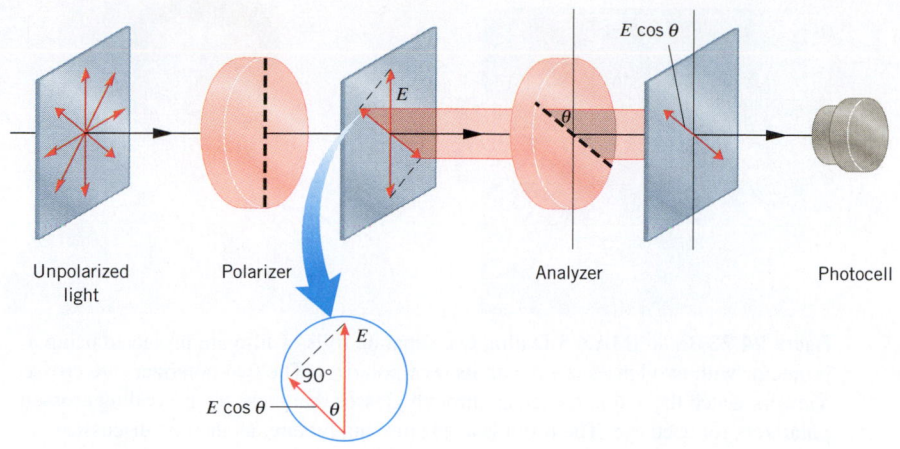

Figure 24.21 Two sheets of polarizing material, called the polarizer and the analyzer, may be used to adjust the polarization direction and intensity of the light reaching the photocell. This can be done by changing the angle θ between the transmission axes of the polarizer and analyzer.

the transmitted polarized light is one-half the average intensity of the incident unpolarized light. The reason for this is that the unpolarized light contains all polarization directions to an equal extent. Moreover, the electric field for each direction can be resolved into components perpendicular and parallel to the transmission axis, with the result that the average components perpendicular and parallel to the axis are equal. As a result, the polarizing material absorbs as much of the electric (and magnetic) field strength as it transmits.

MALUS' LAW

Once polarized light has been produced with a piece of polarizing material, it is possible to use a second piece to change the polarization direction and simultaneously adjust the intensity of the light. Figure 24.21 shows how. As in this picture, the first piece of polarizing material is called the **polarizer** and the second piece is referred to as the **analyzer.** The transmission axis of the analyzer is oriented at an angle θ relative to the transmission axis of the polarizer. If the electric field strength of the polarized light incident on the analyzer is E, the field strength passing through is the component parallel to the transmission axis, or $E \cos \theta$. According to Equation 24.5b, the intensity is proportional to the square of the electric field strength. Consequently, the average intensity of polarized light passing through the analyzer is proportional to $\cos^2\theta$. Thus, both the polarization direction and the intensity of the light can be adjusted by rotating the transmission axis of the analyzer relative to that of the polarizer. The average intensity $\overline{S}$ of the light leaving the analyzer, then, is

Malus' law
$$\overline{S} = \overline{S}_0 \cos^2 \theta \qquad (24.7)$$

where $\overline{S}_0$ is the average intensity of the light entering the analyzer. Equation 24.7 is sometimes called **Malus' law,** for it was discovered by the French engineer Étienne Louis Malus (1775–1812). Example 7 illustrates the use of Malus' law.

 Example 7 Using Polarizers and Analyzers

What value of θ should be used in Figure 24.21 so that the average intensity of the polarized light reaching the photocell will be one-tenth the average intensity of the unpolarized light?

Reasoning Both the polarizer and the analyzer reduce the intensity of the light. The polarizer reduces the intensity by a factor of one-half, as discussed earlier. Therefore, if the average intensity of the unpolarized light is $\overline{I}$, the average intensity of the polarized light leaving the polarizer and striking the analyzer is $\overline{S}_0 = \overline{I}/2$. The angle θ must now be selected so that the average intensity of the light leaving the analyzer will be $\overline{S} = \overline{I}/10$. Malus' law provides the solution.

Solution Using $\overline{S}_0 = \overline{I}/2$ and $\overline{S} = \overline{I}/10$ in Malus' law, we find

$$\tfrac{1}{10}\overline{I} = \tfrac{1}{2}\overline{I} \cos^2 \theta$$

$$\tfrac{1}{5} = \cos^2 \theta \quad \text{or} \quad \theta = \cos^{-1}\left(\frac{1}{\sqrt{5}}\right) = \boxed{63.4^\circ}$$

Problem-solving insight

Remember that when unpolarized light strikes a polarizer, only one-half of the incident light is transmitted, the other half being absorbed by the polarizer.

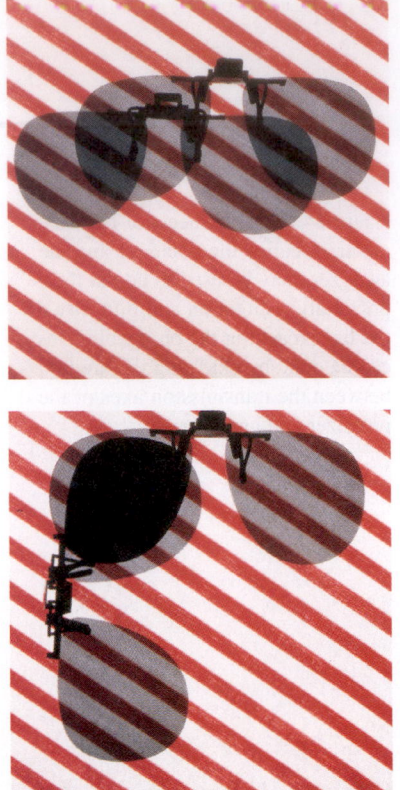

Figure 24.22 When Polaroid sunglasses are uncrossed (top photograph), the transmitted light is dimmed due to the extra thickness of tinted plastic. However, when they are crossed (bottom photograph), the intensity of the transmitted light is reduced to zero because of the effects of polarization. (Diane Schiumo/Fundamental Photographs)

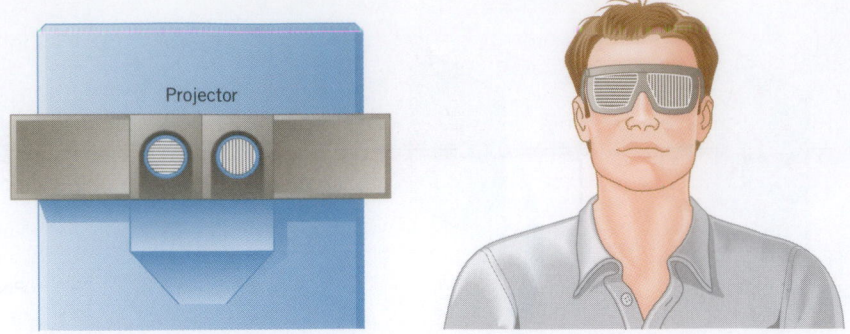

Figure 24.23 In an IMAX 3-D film, two separate rolls of film are projected using a projector with two lenses, each with its own polarizer. The two polarizers are crossed. Viewers watch the action on-screen through glasses that have corresponding crossed polarizers for each eye. The result is a 3-D moving picture, as the text discusses.

When $\theta = 90°$ in Figure 24.21, the polarizer and analyzer are said to be *crossed,* and no light is transmitted by the polarizer/analyzer combination. As an illustration of this effect, Figure 24.22 shows two pairs of Polaroid sunglasses in uncrossed and crossed configurations.

The physics of **IMAX 3-D films.** An exciting application of crossed polarizers is used in viewing IMAX 3-D movies. These movies are recorded on two separate rolls of film, using a camera that provides images from the two different perspectives that correspond to what is observed by human eyes and allow us to see in three dimensions. The camera has two apertures or openings located at roughly the spacing between our eyes. The films are projected using a projector with two lenses, as Figure 24.23 indicates. Each lens has its own polarizer, and the two polarizers are crossed (see the drawing). In one type of theater, viewers watch the action on-screen using glasses with corresponding polarizers for the left and right eyes, as the drawing shows. Because of the crossed polarizers the left eye sees only the image from the left lens of the projector, and the right eye sees only the image from the right lens. Since the two images have the approximate perspectives that the left and right eyes would see in reality, the brain combines the images to produce a realistic 3-D effect.

Conceptual Example 8 illustrates an interesting result that occurs when a piece of polarizing material is inserted between a crossed polarizer and analyzer.

Conceptual Example 8
How Can a Crossed Polarizer and Analyzer Transmit Light?

As explained earlier, no light reaches the photocell in Figure 24.21 when the polarizer and the analyzer are crossed. Suppose that a third piece of polarizing material is inserted between the polarizer and analyzer, as in Figure 24.24*a*. With the insert in place, will light reach the photocell when **(a)** $\theta = 0°$, **(b)** $\theta = 90°$, or **(c)** θ is between 0 and 90°?

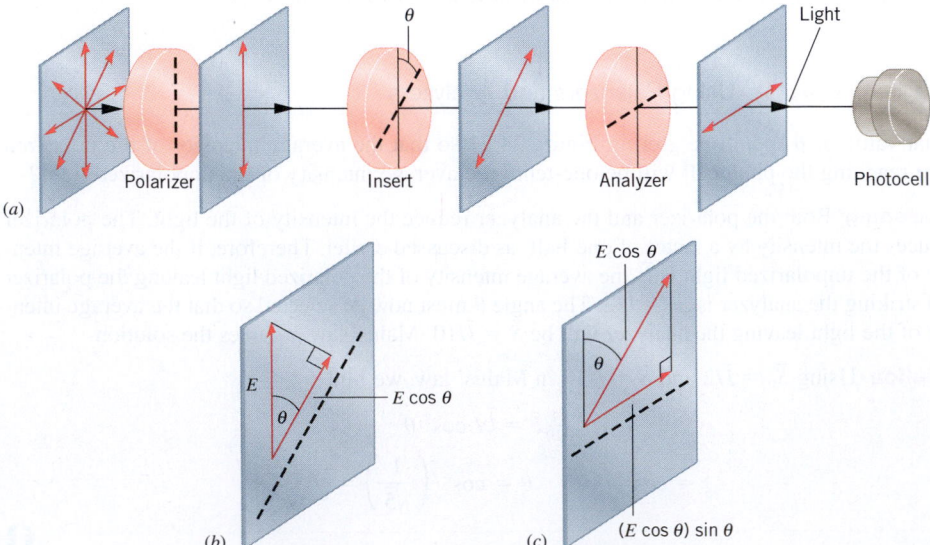

Figure 24.24 (*a*) Light reaches the photocell when a piece of polarizing material is inserted between a crossed polarizer and analyzer. (*b*) The electric-field component parallel to the insert's transmission axis is $E \cos \theta$. (*c*) Light incident on the analyzer has a component $(E \cos \theta) \sin \theta$ parallel to its transmission axis.

Reasoning If any light is to pass through the analyzer, it must have an electric field component parallel to the transmission axis of the analyzer. Thus, without the insert in Figure 24.24a, no light reaches the photocell, because the analyzer and polarizer are crossed, which means that the electric field of the light reaching the analyzer has no component parallel to the analyzer's transmission axis. We need to consider, then, whether the presence of the insert leads to an electric field component parallel to the analyzer's transmission axis.

Answers (a) and (b) are incorrect. With $\theta = 0°$, the polarizer and the insert have parallel transmission axes, so the light leaving the polarizer passes through the insert unaffected. It reaches the analyzer with its electric field perpendicular to the analyzer's transmission axis and is, thus, prevented from reaching the photocell. With $\theta = 90°$, the polarizer and the insert are crossed, so no light leaves the insert to reach the analyzer and the photocell.

Answer (c) is correct. Parts b and c of Figure 24.24 show that, with the insert present, the light reaching the analyzer has an electric field component that is parallel to the analyzer's transmission axis when θ is between 0 and 90°. In part b the electric field E of the light leaving the polarizer makes an angle θ with respect to the transmission axis of the insert and has a component $E \cos \theta$ with respect to that axis. This component passes through the insert. In part c the field $(E \cos \theta)$ incident on the analyzer has a component parallel to the transmission axis of the analyzer—namely, $(E \cos \theta) \sin \theta$. This component passes through the analyzer and reaches the photocell.

Related Homework: *Problems 43, 58*

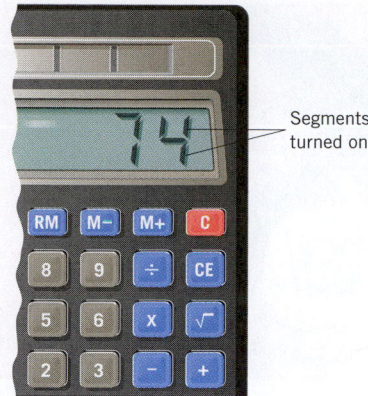

Figure 24.25 Liquid crystal displays (LCDs) use liquid crystal segments to form the numbers.

An application of a crossed polarizer/analyzer combination occurs in one kind of liquid crystal display (LCD). LCDs are widely used in pocket calculators and cell phones. The display usually consists of blackened numbers and letters set against a light gray background. As Figure 24.25 indicates, each number or letter is formed from a combination of liquid crystal segments that have been turned on and appear black. The liquid crystal part of an LCD segment consists of the liquid crystal material sandwiched between two transparent electrodes, as in Figure 24.26. When a voltage is applied between the electrodes, the liquid crystal is said to be "on." Part a of the picture shows that linearly polarized incident light passes through the "on" material without having its direction of polarization affected. When the voltage is removed, as in part b, the liquid crystal is said to be "off" and now rotates the direction of polarization by 90°. A complete LCD segment also includes a crossed polarizer/analyzer combination, as Figure 24.27 illustrates. The polarizer, analyzer, electrodes, and liquid crystal material are packaged as a single unit. The polarizer produces polarized light from incident unpolarized light. With the display segment

The physics of a liquid crystal display.

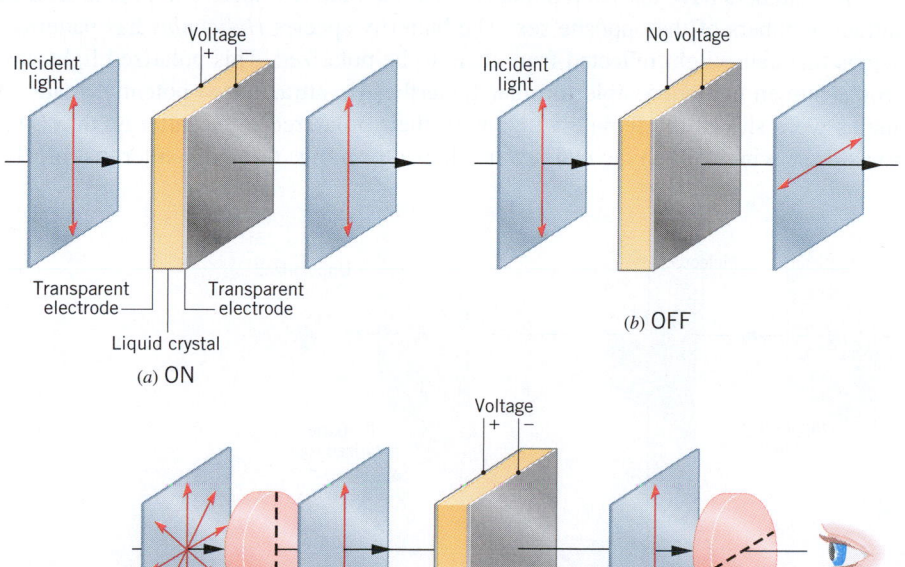

(a) ON   (b) OFF

ON

Figure 24.26 A liquid crystal in its (a) "on" state and (b) "off" state.

Figure 24.27 An LCD incorporates a crossed polarizer/analyzer combination. When the LCD segment is turned on (voltage applied), no light is transmitted through the analyzer, and the observer sees a black segment.

Figure 24.28 Want to see a photograph of yourself? Just take a picture with a properly equipped cell phone and look at the LCD display. (© Mango Productions/Corbis)

The physics of
Polaroid sunglasses.

The physics of
butterflies and polarized light.

turned on, as in Figure 24.27, the polarized light emerges from the liquid crystal only to be absorbed by the analyzer, since the light is polarized perpendicular to the transmission axis of the analyzer. Since no light emerges from the analyzer, an observer sees a black segment against a light gray background, as in Figure 24.25. On the other hand, the segment is turned off when the voltage is removed, in which case the liquid crystal rotates the direction of polarization by 90° to coincide with the axis of the analyzer. The light now passes through the analyzer and enters the eye of the observer. However, the light coming from the segment has been designed to have the same color and shade (light gray) as the background of the display, so the segment becomes indistinguishable from the background.

Color LCD display screens and computer monitors are popular because they occupy less space and weigh less than traditional cathode-ray tube (CRT) units do. An LCD display screen, such as the one in Figure 24.28, uses thousands of LCD segments arranged like the squares on graph paper. To produce color, three segments are grouped together to form a tiny picture element (or "pixel"). Color filters are used to enable one segment in the pixel to produce red light, one to produce green, and one to produce blue. The eye blends the colors from each pixel into a composite color. By varying the intensity of the red, green, and blue colors, the pixel can generate an entire spectrum of colors.

THE OCCURRENCE OF POLARIZED LIGHT IN NATURE

Polaroid is a familiar material because of its widespread use in sunglasses. Such sunglasses are designed so that the transmission axis of the Polaroid is oriented vertically when the glasses are worn in the usual fashion. Thus, the glasses prevent any light that is polarized horizontally from reaching the eye. Light from the sun is unpolarized, but a considerable amount of horizontally polarized sunlight originates by reflection from horizontal surfaces such as that of a lake. Section 26.4 discusses this effect. Polaroid sunglasses reduce glare by preventing the horizontally polarized reflected light from reaching the eyes.

Polarized sunlight also originates from the scattering of light by molecules in the atmosphere. Figure 24.29 shows light being scattered by a single atmospheric molecule. The electric fields in the unpolarized sunlight cause the electrons in the molecule to vibrate perpendicular to the direction in which the light is traveling. The electrons, in turn, reradiate the electromagnetic waves in different directions, as the drawing illustrates. The light radiated straight ahead in direction *A* is unpolarized, just like the incident light; but light radiated perpendicular to the incident light in direction *C* is polarized. Light radiated in the intermediate direction *B* is partially polarized.

Researchers have discovered that at least one butterfly species uses polarized light to attract members of the opposite sex. The butterfly species *Heliconius* has patterns on its wings that cause light reflected from them to be polarized. This polarized light, invisible to the human eye but visible to other butterflies, is attractive to potential mates. When males were shown the female wings with their polarized light patterns, they swarmed toward the wings. When the males were shown the wings through a filter that blocked out

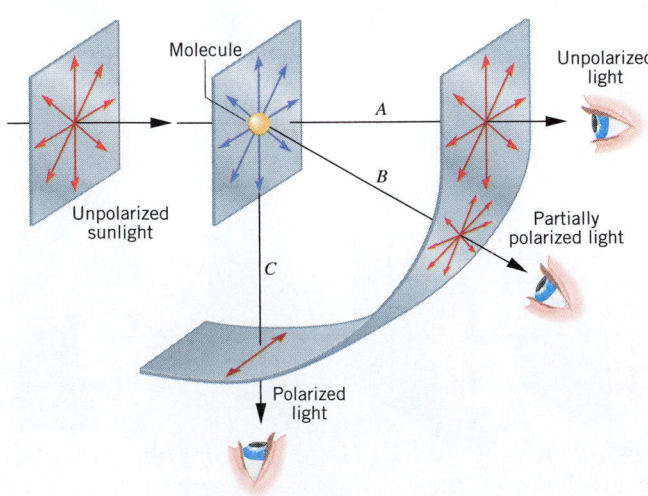

Figure 24.29 In the process of being scattered from atmospheric molecules, unpolarized light from the sun becomes partially polarized.

the polarization effects, they largely ignored the wings. Figure 24.30 shows the polarized light reflected from the wings of the *Heliconius cydno* butterfly. The left wing is shown as it normally appears. The light reflected from the white pattern is highly polarized. The right wing is shown as it appears when viewed through a polarizing filter whose transmission axis is crossed with respect to the direction in which the reflected light is polarized. The white patterns in the right wing are black when viewed through the filter, a clear indication that the light reflected from them is indeed polarized. There is also experimental evidence that some bird species use polarized light as a navigational aid.

✓ CHECK YOUR UNDERSTANDING

(The answers are given at the end of the book.)

9. Malus' law applies to the setup in Figure 24.21, which shows the analyzer rotated through an angle θ and the polarizer held fixed. Does Malus' law apply when the analyzer is held fixed and the polarizer is rotated?

10. In Example 7, we saw that when the angle between the polarizer and analyzer is 63.4°, the average intensity of the transmitted light drops to one-tenth of the average intensity of the incident unpolarized light. What happens to the light intensity that is not transmitted?

11. The drawing shows two sheets of polarizing material. The transmission axis of one is vertical, and that of the other makes an angle of 45° with the vertical. Unpolarized light shines on this arrangement first from the left and then from the right. From which direction does at least some light pass through both sheets? **(a)** From the left **(b)** From the right **(c)** From either direction **(d)** From neither direction What is the answer when the light is horizontally polarized? What is the answer when the light is vertically polarized?

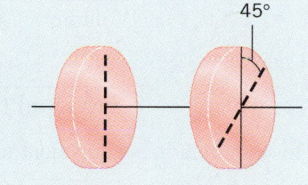

12. You are sitting upright on the beach near a lake on a sunny day, wearing Polaroid sunglasses. When you lie down on your side, facing the lake, the sunglasses don't work as well as they do while you are sitting upright. Why not?

Figure 24.30 A *Heliconius cydno* butterfly. The left wing is shown as it appears normally, and the right wing as it appears when viewed through a polarizing filter. The light reflected from the white patterns is polarized, and these patterns in the right wing are black because the transmission axis of the filter is crossed with respect to the polarization direction of the reflected light. [Courtesy Alison Sweeney, Duke University. Image from *Nature* 423: 31–32 (May 1, 2003). Reproduced with permission.]

24.7 CONCEPTS & CALCULATIONS

One of the central ideas of this chapter is that electromagnetic waves carry energy. Two concepts are used to describe this energy—the wave's intensity and its energy density. The next example reviews these important ideas.

Concepts & Calculations Example 9
↓ Intensity and Energy Density of a Wave

Figure 24.31 shows the popular dish antenna that receives digital TV signals from a satellite. The average intensity of the electromagnetic wave that carries a particular TV program is $\overline{S} = 7.5 \times 10^{-14}$ W/m², and the circular aperture of the antenna has a radius of $r = 15$ cm. **(a)** Determine the electromagnetic energy delivered to the dish during a one-hour program. **(b)** What is the average energy density of the electromagnetic wave?

Concept Questions and Answers How is the average power passing through the circular aperture of the antenna related to the average intensity of the TV signal?

> *Answer* According to Equation 16.8, the average intensity $\overline{S}$ of a wave is equal to the average power $\overline{P}$ that passes perpendicularly through a surface divided by the area A of the surface. Thus, $\overline{S} = \overline{P}/A$, and the average power is $\overline{P} = \overline{S}A$, where $A = \pi r^2$.

How much energy does the antenna receive in a time t?

> *Answer* Since the average power is the energy per unit time according to Equation 6.10b, the energy received by the antenna is the product of the average power $\overline{P}$ and the time t, or Energy $= \overline{P}t$.

What is the average energy density, or average energy per unit volume, of the electromagnetic wave?

> *Answer* The energy delivered to the dish is carried by the electromagnetic wave. The wave passes perpendicularly through an imaginary surface that has an area A that matches

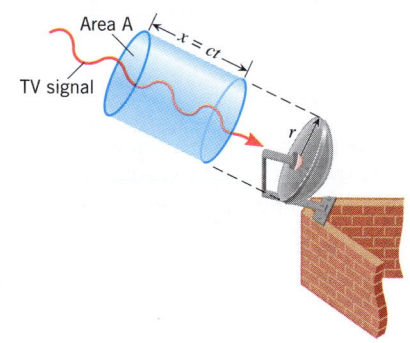

Figure 24.31 A dish antenna picks up the TV signal from a satellite. The signal extends over the dish's entire aperture (radius r).

the circular area of the dish's aperture. In a time t, all the energy that passes through the surface is contained, therefore, in a cylinder of length x (see Figure 24.31). Since the energy is being carried by an electromagnetic wave, it travels at the speed of light c. Thus, the length of the cylinder is $x = ct$. The energy density u of the wave, or energy per unit volume, is the energy contained within the cylinder divided by its volume.

Solution (a) According to Equation 6.10b, the energy received by the antenna during a time t is equal to the average power multiplied by the time, Energy $= \overline{P}t$. On the other hand, Equation 16.8 gives the average power as the product of the average intensity and the area, $\overline{P} = \overline{S}A$, so that

$$\text{Energy} = \underbrace{\overline{S}A}_{\text{Average power}} t = \overline{S}(\pi r^2)t$$

$$= (7.5 \times 10^{-14} \text{ W/m}^2)\pi(0.15 \text{ m})^2(3600 \text{ s}) = \boxed{1.9 \times 10^{-11} \text{ J}}$$

(b) The energy density is equal to the energy divided by the volume of the cylinder in Figure 24.31. The volume of the cylinder is equal to the cross-sectional area πr^2 times its length ct.

$$u = \frac{\text{Energy}}{\text{Volume}} = \frac{\text{Energy}}{(\pi r^2)(ct)}$$

$$= \frac{1.9 \times 10^{-11} \text{ J}}{\pi(0.15 \text{ m})^2(3.00 \times 10^8 \text{ m/s})(3600 \text{ s})} = \boxed{2.5 \times 10^{-22} \text{ J/m}^3}$$

This calculation is equivalent to using Equation 24.4, which gives the energy density as

$$u = \frac{\text{Energy}}{\text{Volume}} = \frac{\overline{S}}{c} = \frac{7.5 \times 10^{-14} \text{ W/m}^2}{3.00 \times 10^8 \text{ m/s}} = 2.5 \times 10^{-22} \text{ J/m}^3$$

We have seen how the intensities of completely polarized or completely unpolarized light beams can change as they pass through a polarizer. But what about light that is partially polarized and partially unpolarized? Can the concepts that we discussed in Section 24.6 be applied to such light? The answer is "yes," and Example 10 illustrates how.

Concepts & Calculations Example 10
Partially Polarized and Partially Unpolarized Light

The light beam in Figure 24.32 passes through a polarizer whose transmission axis makes an angle ϕ with the vertical. The beam is partially polarized and partially unpolarized, and the average intensity $\overline{S}_0$ of the incident light is the sum of the average intensity $\overline{S}_{0, \text{polar}}$ of the polarized light and the average intensity $\overline{S}_{0, \text{unpolar}}$ of the unpolarized light; $\overline{S}_0 = \overline{S}_{0, \text{polar}} + \overline{S}_{0, \text{unpolar}}$. The intensity $\overline{S}$ of the transmitted light is also the sum of two parts: $\overline{S} = \overline{S}_{\text{polar}} + \overline{S}_{\text{unpolar}}$. As the polarizer is rotated clockwise, the intensity of the transmitted light has a minimum value of $\overline{S} = 2.0 \text{ W/m}^2$ when $\phi = 20.0°$ and has a maximum value of $\overline{S} = 8.0 \text{ W/m}^2$ when the angle is $\phi = \phi_{\text{max}}$. (a) What is the intensity $\overline{S}_{0, \text{unpolar}}$ of the incident light that is unpolarized? (b) What is the intensity $\overline{S}_{0, \text{polar}}$ of the incident light that is polarized?

Concept Questions and Answers How is $\overline{S}_{\text{unpolar}}$ related to $\overline{S}_{0, \text{unpolar}}$?

Answer When unpolarized light strikes a polarizer, one half is absorbed and the other half is transmitted, so that $\overline{S}_{\text{unpolar}} = \frac{1}{2}\overline{S}_{0, \text{unpolar}}$. This is true for any angle ϕ of the transmission axis.

How is $\overline{S}_{\text{polar}}$ related to $\overline{S}_{0, \text{polar}}$?

Answer When polarized light strikes a polarizer, the transmitted intensity is given by Malus' law (Equation 24.7), $\overline{S}_{\text{polar}} = \overline{S}_{0, \text{polar}} \cos^2 \theta$, where θ is the angle between the direction of polarization of the incident light and the transmission axis. Note that θ is not the same as ϕ because the direction of polarization of the incident beam is not given and may not be in the vertical direction.

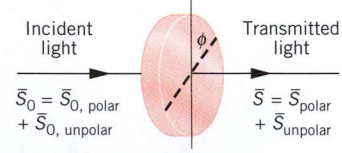

Figure 24.32 Light that is partially polarized and partially unpolarized is incident on a sheet of polarizing material.

Incident light
$\overline{S}_0 = \overline{S}_{0, \text{polar}} + \overline{S}_{0, \text{unpolar}}$

Transmitted light
$\overline{S} = \overline{S}_{\text{polar}} + \overline{S}_{\text{unpolar}}$

The minimum transmitted intensity is 2.0 W/m². Why isn't it 0 W/m²?

Answer When the polarized part of the incident beam passes through the polarizer, the intensity of its transmitted portion changes as the angle ϕ changes, ranging between some maximum value and 0 W/m², in accord with Malus' law. However, when the unpolarized part of the beam passes through the polarizer, its transmitted intensity does not change as the angle ϕ changes. Thus, the minimum intensity of the transmitted beam is 2.0 W/m², rather than 0 W/m², because of the unpolarized light.

At what angle ϕ_{max} is the intensity of the light leaving the polarizing material a maximum?

Answer When the polarized part of the beam passes through the polarizer, the intensity of its transmitted portion is zero when $\phi = 20.0°$. For this angle, the direction of polarization of the polarized light is perpendicular to the transmission axis of the polarizer. The intensity rises to a maximum when the polarizer is rotated by 90.0° relative to this position, at which orientation the direction of polarization is aligned with the transmission axis. Thus, the intensity is a maximum when $\phi = \phi_{max} = 20.0° + 90.0° = 110.0°$.

Solution (a) The average intensity $\overline{S}$ of the transmitted light is (see Figure 24.32) $\overline{S} = \overline{S}_{polar} + \overline{S}_{unpolar}$. We are given that the minimum intensity is $\overline{S} = 2.0$ W/m² when $\phi = 20.0°$, and we know that $\overline{S}_{polar} = 0$ W/m² for this angle. Thus, $\overline{S}_{unpolar} = \overline{S} - \overline{S}_{polar} = 2.0$ W/m². The intensity $\overline{S}_{0, unpolar}$ of the unpolarized incident light is twice this amount because the polarizer absorbs one-half the incident light:

$$\overline{S}_{0, unpolar} = 2\overline{S}_{unpolar} = 2(2.0 \text{ W/m}^2) = \boxed{4.0 \text{ W/m}^2}$$

(b) When $\phi = \phi_{max} = 110°$, the intensity of the transmitted beam is at its maximum value of $\overline{S} = 8.0$ W/m². Writing the transmitted intensity as the sum of two parts, we have

$$\overline{S} = 8.0 \text{ W/m}^2 = \underbrace{\overline{S}_{0, polar} \cos^2 \theta}_{\overline{S}_{polar}} + \underbrace{2.0 \text{ W/m}^2}_{\overline{S}_{unpolar}}$$

The angle θ is the angle between the direction of polarization of the incident polarized light and the transmission axis of the polarizer. Since $\overline{S}_{polar}$ is a maximum, we know that $\theta = 0°$. Solving this equation for $\overline{S}_{0, polar}$ yields

$$\overline{S}_{0, polar} = 8.0 \text{ W/m}^2 - 2.0 \text{ W/m}^2 = \boxed{6.0 \text{ W/m}^2}$$

CONCEPT SUMMARY

If you need more help with a concept, use the Learning Aids noted next to the discussion or equation. Examples (**Ex.**) are in the text of this chapter. Go to **www.wiley.com/college/cutnell** for the following Learning Aids:

Interactive LearningWare (ILW) — Additional examples solved in a five-step interactive format.

Concept Simulations (CS) — Animated text figures or animations of important concepts.

Interactive Solutions (IS) — Models for certain types of problems in the chapter homework. The calculations are carried out interactively.

Topic	Discussion	Learning Aids
Electromagnetic wave	**24.1 THE NATURE OF ELECTROMAGNETIC WAVES** An electromagnetic wave consists of mutually perpendicular and oscillating electric and magnetic fields. The wave is a transverse wave, since the fields are perpendicular to the direction in which the wave travels. Electromagnetic waves can travel through a vacuum or a material substance. All electromagnetic waves travel through a vacuum at the same speed, which is known as the speed of light c ($c = 3.00 \times 10^8$ m/s).	
Speed of light		
Relation between frequency, wavelength, and speed of light in a vacuum	**24.2 THE ELECTROMAGNETIC SPECTRUM** The frequency f and wavelength λ of an electromagnetic wave in a vacuum are related to its speed c through the relation $$c = f\lambda$$	Ex. 1, 2
Electromagnetic spectrum	The series of electromagnetic waves, arranged in order of their frequencies or wavelengths, is called the electromagnetic spectrum. In increasing order of frequency (decreasing order of wavelength), the spectrum includes radio waves, infrared radiation, visible light, ultraviolet	

Topic	Discussion	Learning Aids

| | radiation, X-rays, and gamma rays. Visible light has frequencies between about 4.0×10^{14} and 7.9×10^{14} Hz. The human eye and brain perceive different frequencies or wavelengths as different colors. | |

24.3 THE SPEED OF LIGHT James Clerk Maxwell showed that the speed of light in a vacuum is

Speed of light

$$c = \frac{1}{\sqrt{\epsilon_0 \mu_0}} \qquad (24.1) \quad \text{Ex. 3}$$

where ϵ_0 is the (electric) permittivity of free space and μ_0 is the (magnetic) permeability of free space.

24.4 THE ENERGY CARRIED BY ELECTROMAGNETIC WAVES The total energy density u of an electromagnetic wave is the total energy per unit volume of the wave and, in a vacuum, is given by

Total energy density

$$u = \frac{1}{2} \epsilon_0 E^2 + \frac{1}{2\mu_0} B^2 \qquad (24.2a)$$

where E and B, respectively, are the magnitudes of the electric and magnetic fields of the wave. Since the electric and magnetic parts of the total energy density are equal, the following two equations are equivalent to Equation 24.2a:

$$u = \epsilon_0 E^2 \qquad (24.2b)$$

$$u = \frac{1}{\mu_0} B^2 \qquad (24.2c)$$

In a vacuum, E and B are related according to

Relation between electric and magnetic fields

$$E = cB \qquad (24.3)$$

Equations 24.2a–c can be used to determine the average total energy density, if the rms average values E_{rms} and B_{rms} are used in place of the symbols E and B. The rms values are related to the peak values E_0 and B_0 in the usual way:

Ex. 4

Root mean square fields

$$E_{\text{rms}} = \frac{1}{\sqrt{2}} E_0 \quad \text{and} \quad B_{\text{rms}} = \frac{1}{\sqrt{2}} B_0$$

Intensity The intensity of an electromagnetic wave is the power that the wave carries perpendicularly through a surface divided by the area of the surface. In a vacuum, the intensity S is related to the total energy density u according to

Ex. 5

IS 24.31

Relation between intensity and total energy density

$$S = cu \qquad (24.4) \quad \text{Ex. 9}$$

24.5 THE DOPPLER EFFECT AND ELECTROMAGNETIC WAVES When electromagnetic waves and the source and observer of the waves all travel along the same line in a vacuum, the Doppler effect is given by

Doppler effect

$$f_o = f_s \left(1 \pm \frac{v_{\text{rel}}}{c} \right) \qquad \text{if } v_{\text{rel}} \ll c \qquad (24.6) \quad \begin{array}{l}\text{Ex. 6} \\ \text{IS 24.35}\end{array}$$

where f_o and f_s are, respectively, the observed and emitted wave frequencies and v_{rel} is the relative speed of the source and the observer. The plus sign is used when the source and the observer come together, and the minus sign is used when they move apart.

Linearly polarized electromagnetic wave

24.6 POLARIZATION A linearly polarized electromagnetic wave is one in which the oscillation of the electric field occurs only along one direction, which is taken to be the direction of polarization. The magnetic field also oscillates along only one direction, which is perpendicular to the electric field direction. In an unpolarized wave such as the light from an incandescent bulb, the direction of polarization does not remain fixed, but fluctuates randomly in time.

Unpolarized electromagnetic wave

Transmission axis

Polarizing materials allow only the component of the wave's electric field along one direction (and the associated magnetic field component) to pass through them. The preferred transmission direction for the electric field is called the transmission axis of the material.

When unpolarized light is incident on a piece of polarizing material, the transmitted polarized light has an average intensity that is one-half the average intensity of the incident light.

Topic	Discussion	Learning Aids
Polarizer and analyzer	When two pieces of polarizing material are used one after the other, the first is called the polarizer, and the second is referred to as the analyzer. If the average intensity of polarized light falling on the analyzer is $\overline{S}_0$, the average intensity $\overline{S}$ of the light leaving the analyzer is given by Malus' law as	
Malus' law	$$\overline{S} = \overline{S}_0 \cos^2 \theta \qquad (24.7)$$	**Ex. 7, 8, 10** **ILW 24.1**
Crossed polarizer and analyzer	where θ is the angle between the transmission axes of the polarizer and analyzer. When $\theta = 90°$, the polarizer and the analyzer are said to be "crossed," and no light passes through the analyzer.	

FOCUS ON CONCEPTS

Note to Instructors: The numbering of the questions shown here reflects the fact that they are only a representative subset of the total number that are available online. However, all of the questions are available for assignment via an online homework management program such as WileyPLUS or WebAssign.

Section 24.1 The Nature of Electromagnetic Waves

1. The drawing shows an x, y, z coordinate system. A circular loop of wire lies in the z, x plane and, when used with an LC-tuned circuit, detects an electromagnetic wave. Which one of the following statements is correct? **(a)** The wave travels along the x axis, and its electric field oscillates along the y axis. **(b)** The wave travels along the z axis, and its electric field oscillates along the x axis. **(c)** The wave travels along the z axis, and its electric field oscillates along the y axis. **(d)** The wave travels along the y axis, and its electric field oscillates along the x axis. **(e)** The wave travels along the y axis, and its electric field oscillates along the z axis.

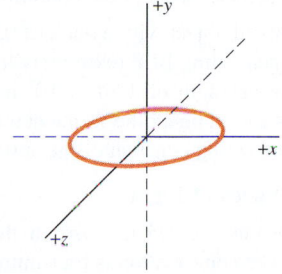

Section 24.2 The Electromagnetic Spectrum

2. An electromagnetic wave travels in a vacuum. The wavelength of the wave is tripled. How is this accomplished? **(a)** By tripling the frequency of the wave **(b)** By tripling the speed of the wave **(c)** By reducing the frequency of the wave by a factor of three **(d)** By reducing the speed of the wave by a factor of three **(e)** By tripling the magnitudes of the electric and magnetic fields that comprise the wave

Section 24.4 The Energy Carried by Electromagnetic Waves

3. An electromagnetic wave is traveling in a vacuum. The magnitudes of the electric and magnetic fields of the wave are _____, and the electric and magnetic energies carried by the wave are _____. **(a)** equal, proportional (but not equal) to each other **(b)** proportional (but not equal) to each other, equal **(c)** equal, equal **(d)** proportional (but not equal) to each other, unequal

Section 24.5 The Doppler Effect and Electromagnetic Waves

6. The drawing shows four situations—A, B, C, and D—in which an observer and a source of electromagnetic waves can move along the same line. In each case the source emits a wave of the same frequency, and in each case only the source or the observer is moving. The arrow in each situation denotes the velocity vector, which has the same magnitude in each situation. When there is no arrow, the observer or the source is stationary. Rank the frequencies of the observed electromagnetic waves in descending order (largest first) according to magnitude. **(a)** A and B (a tie), C and D (a tie) **(b)** C and D (a tie), A and B (a tie) **(c)** A and D (a tie), B and C (a tie) **(d)** B and D (a tie), A and C (a tie) **(e)** B and C (a tie), A and D (a tie)

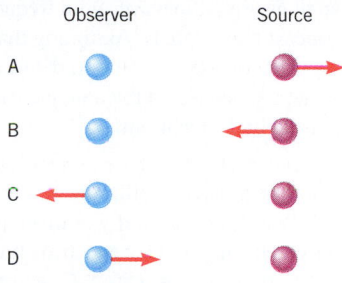

Section 24.6 Polarization

10. The drawing shows two sheets of polarizing material. Polarizer 1 has its transmission axis aligned vertically, and polarizer 2 has its transmission axis aligned at an angle of $45°$ with respect to the vertical. Light that is completely polarized along the vertical direction is incident either from the far left or from the far right. In either case, the average intensity of the incident light is the same. Which one of the following statements is true concerning the average intensity of the light that is transmitted by the pair of sheets? **(a)** When the light is incident from either the left or the right, the transmitted intensity is one-half the incident intensity. **(b)** When the light is incident from either the left or the right, the transmitted intensity is one-fourth the incident intensity. **(c)** When the light is incident from the left, the transmitted intensity is one-half the incident intensity; when the light is incident from the right, the transmitted intensity is zero. **(d)** When the light is incident from the left, the transmitted intensity is one-fourth the incident intensity; when the light is incident from the right, the transmitted intensity is one-half the incident intensity. **(e)** When the light is incident from the left, the transmitted intensity is one-half the incident intensity; when the light is incident from the right, the transmitted intensity is one-fourth the incident intensity.

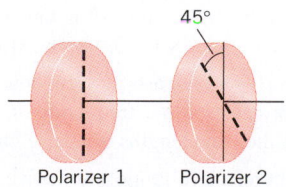

Polarizer 1   Polarizer 2

PROBLEMS

Note to Instructors: Most of the homework problems in this chapter are available for assignment via an online homework management program such as WileyPLUS or WebAssign, and those marked with the icon 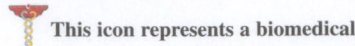 *are presented in WileyPLUS using a guided tutorial format that provides enhanced interactivity. See Preface for additional details.*

ssm   Solution is in the Student Solutions Manual.
www   Solution is available online at www.wiley.com/college/cutnell

This icon represents a biomedical application.

Section 24.1   The Nature of Electromagnetic Waves

1. During a flare-up from a sunspot, X-rays (electromagnetic waves) are emitted. If the distance between the sun and the earth is 1.50×10^{11} m, how long (in minutes) does it take for the X-rays to reach the earth?

2. The team monitoring a space probe exploring the outer solar system finds that radio transmissions from the probe take 2.53 hours to reach earth. How distant (in meters) is the probe?

3. ssm In astronomy, distances are often expressed in light-years. One light-year is the distance traveled by light in one year. The distance to Alpha Centauri, the closest star other than our own sun that can be seen by the naked eye, is 4.3 light-years. Express this distance in meters.

4. GO FM radio stations use radio waves with frequencies from 88.0 to 108 MHz to broadcast their signals. Assuming that the inductance in Figure 24.4 has a value of 6.00×10^{-7} H, determine the range of capacitance values that are needed so the antenna can pick up all the radio waves broadcasted by FM stations.

*5. ssm Equation 16.3, $y = A \sin(2\pi ft - 2\pi x/\lambda)$, gives the mathematical representation of a wave oscillating in the y direction and traveling in the positive x direction. Let y in this equation equal the electric field of an electromagnetic wave traveling in a vacuum. The maximum electric field is $A = 156$ N/C, and the frequency is $f = 1.50 \times 10^8$ Hz. Plot a graph of the electric field strength versus position, using for x the following values: 0, 0.50, 1.00, 1.50, and 2.00 m. Plot this graph for   **(a)** a time $t = 0$ s and   **(b)** a time t that is one-fourth of the wave's period.

**6. A flat coil of wire is used with an LC-tuned circuit as a receiving antenna. The coil has a radius of 0.25 m and consists of 450 turns. The transmitted radio wave has a frequency of 1.2 MHz. The magnetic field of the wave is parallel to the normal to the coil and has a maximum value of 2.0×10^{-13} T. Using Faraday's law of electromagnetic induction and the fact that the magnetic field changes from zero to its maximum value in one-quarter of a wave period, find the magnitude of the average emf induced in the antenna during this time.

Section 24.2   The Electromagnetic Spectrum

7. Obtain the wavelengths in vacuum for   **(a)** blue light whose frequency is 6.34×10^{14} Hz, and   **(b)** orange light whose frequency is 4.95×10^{14} Hz. Express your answers in nanometers (1 nm = 10^{-9} m).

8. In a certain UHF radio wave, the shortest distance between positions at which the electric and magnetic fields are zero is 0.34 m. Determine the frequency of this UHF radio wave.

9. ssm www At one time television sets used "rabbit-ears" antennas. Such an antenna consists of a pair of metal rods. The length of each rod can be adjusted to be one-quarter of a wavelength of an electromagnetic wave whose frequency is 60.0 MHz. How long is each rod?

10. TV channel 3 (VHF) broadcasts at a frequency of 63.0 MHz. TV channel 23 (UHF) broadcasts at a frequency of 527 MHz. Find the ratio (VHF/UHF) of the wavelengths for these channels.

11. The human eye is most sensitive to light with a frequency of about 5.5×10^{14} Hz, which is in the yellow-green region of the electromagnetic spectrum. How many wavelengths of this light can fit across the width of your thumb, a distance of about 2.0 cm?

12. GO A certain type of laser emits light that has a frequency of 5.2×10^{14} Hz. The light, however, occurs as a series of short pulses, each lasting for a time of 2.7×10^{-11} s.   **(a)** How many wavelengths are there in one pulse?   **(b)** The light enters a pool of water. The frequency of the light remains the same, but the speed of the light slows down to 2.3×10^8 m/s. How many wavelengths are there now in one pulse?

13. Two radio waves are used in the operation of a cellular telephone. To receive a call, the phone detects the wave emitted at one frequency by the transmitter station or base unit. To send your message to the base unit, your phone emits its own wave at a different frequency. The difference between these two frequencies is fixed for all channels of cell phone operation. Suppose that the wavelength of the wave emitted by the base unit is 0.34339 m and the wavelength of the wave emitted by the phone is 0.36205 m. Using a value of 2.9979×10^8 m/s for the speed of light, determine the difference between the two frequencies used in the operation of a cell phone.

*14. A positively charged object with a mass of 0.115 kg oscillates at the end of a spring, generating ELF (extremely low frequency) radio waves that have a wavelength of 4.80×10^7 m. The frequency of these radio waves is the same as the frequency at which the object oscillates. What is the spring constant of the spring?

Section 24.3   The Speed of Light

15. ssm Two astronauts are 1.5 m apart in their spaceship. One speaks to the other. The conversation is transmitted to earth via electromagnetic waves. The time it takes for sound waves to travel at 343 m/s through the air between the astronauts equals the time it takes for the electromagnetic waves to travel to the earth. How far away from the earth is the spaceship?

16. GO Figure 24.12 illustrates Michelson's setup for measuring the speed of light with the mirrors placed on Mt. San Antonio and Mt. Wilson in California, which are 35 km apart. Using a value of 3.00×10^8 m/s for the speed of light, find the minimum angular speed (in rev/s) for the rotating mirror.

17. A lidar (laser radar) gun is an alternative to the standard radar gun that uses the Doppler effect to catch speeders. A lidar gun uses an infrared laser and emits a precisely timed series of pulses of infrared electromagnetic waves. The time for each pulse to travel to the speeding vehicle and return to the gun is measured. In one situation a lidar gun in a stationary police car observes a difference of 1.27×10^{-7} s in round-trip travel times for two pulses that are emitted 0.450 s apart. Assuming that the speeding vehicle is approaching the police car essentially head-on, determine the speed of the vehicle.

18. A laptop computer communicates with a router wirelessly, by means of radio signals. The router is connected by cable directly to the Internet. The laptop is 8.1 m from the router, and is downloading text and images from the Internet at an average rate of 260 Mbps, or 260 megabits per second. (A *bit*, or *binary digit*, is the smallest unit of digital information.) On average, how many bits are downloaded to the laptop in the time it takes the wireless signal to travel from the router to the laptop?

19. Review Conceptual Example 3 for information pertinent to this problem. When we look at the star Polaris (the North Star), we are seeing it as it was 680 years ago. How far away from us (in meters) is Polaris?

*20. **GO** A mirror faces a cliff located some distance away. Mounted on the cliff is a second mirror, directly opposite the first mirror and facing toward it. A gun is fired very close to the first mirror. The speed of sound is 343 m/s. How many times does the flash of the gunshot travel the round-trip distance between the mirrors before the echo of the gunshot is heard?

*21. A communications satellite is in a synchronous orbit that is 3.6×10^7 m directly above the equator. The satellite is located midway between Quito, Equador, and Belém, Brazil, two cities almost on the equator that are separated by a distance of 3.5×10^6 m. Find the time it takes for a telephone call to go by way of satellite between these cities. Ignore the curvature of the earth.

Section 24.4 The Energy Carried by Electromagnetic Waves

22. A laser emits a narrow beam of light. The radius of the beam is 1.0×10^{-3} m, and the power is 1.2×10^{-3} W. What is the intensity of the laser beam?

23. An industrial laser is used to burn a hole through a piece of metal. The average intensity of the light is $\overline{S} = 1.23 \times 10^9$ W/m². What is the rms value of **(a)** the electric field and **(b)** the magnetic field in the electromagnetic wave emitted by the laser?

24. **GO** The magnitude of the electric field of an electromagnetic wave increases from 315 to 945 N/C. **(a)** Determine the wave intensities for the two values of the electric field. **(b)** What is the magnitude of the magnetic field associated with each electric field? **(c)** Determine the wave intensity for each value of the magnetic field.

25. **ssm** A future space station in orbit about the earth is being powered by an electromagnetic beam from the earth. The beam has a cross-sectional area of 135 m² and transmits an average power of 1.20×10^4 W. What are the rms values of the **(a)** electric and **(b)** magnetic fields?

26. Multiple-Concept Example 5 discusses the principles used in this problem. A neodymium-glass laser emits short pulses of high-intensity electromagnetic waves. The electric field of such a wave has an rms value of $E_{rms} = 2.0 \times 10^9$ N/C. Find the average power of each pulse that passes through a 1.6×10^{-5}-m² surface that is perpendicular to the laser beam.

27. **ssm** The average intensity of sunlight at the top of the earth's atmosphere is 1390 W/m². What is the maximum energy that a 25-m $\times$ 45-m solar panel could collect in one hour in this sunlight?

28. **GO** Consult Multiple-Concept Example 5 to review the concepts on which this problem depends. A light bulb emits light uniformly in all directions. The average emitted power is 150.0 W. At a distance of 5.00 m from the bulb, determine **(a)** the average intensity of the light, **(b)** the rms value of the electric field, and **(c)** the peak value of the electric field.

29. Multiple-Concept Example 5 provides some pertinent background for this problem. The mean distance between earth and the sun is 1.50×10^{11} m. The average intensity of solar radiation incident on the upper atmosphere of the earth is 1390 W/m². Assuming that the sun emits radiation uniformly in all directions, determine the total power radiated by the sun.

30. **GO** A stationary particle of charge $q = 2.6 \times 10^{-8}$ C is placed in a laser beam (an electromagnetic wave) whose intensity is

2.5×10^3 W/m². Determine the magnitude of the **(a)** electric and **(b)** magnetic forces exerted on the charge. If the charge is moving at a speed of 3.7×10^4 m/s perpendicular to the magnetic field of the electromagnetic wave, find the magnitudes of the **(c)** electric and **(d)** magnetic forces exerted on the particle.

*31. **Interactive Solution 24.31** at **www.wiley.com/college/cutnell** provides one model for problems like this one. The drawing shows an edge-on view of the solar panels on a communications satellite. The dashed line specifies the normal to the panels. Sunlight strikes the panels at an angle θ with respect to the normal. If the solar power impinging on the panels is 2600 W when $\theta = 65°$, what is it when $\theta = 25°$?

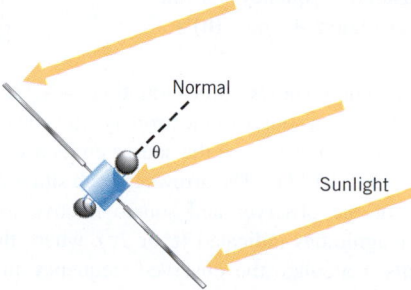

*32. The electromagnetic wave that delivers a cellular phone call to a car has a magnetic field with an rms value of 1.5×10^{-10} T. The wave passes perpendicularly through an open window, the area of which is 0.20 m². How much energy does this wave carry through the window during a 45-s phone call?

*33. What fraction of the power radiated by the sun is intercepted by the planet Mercury? The radius of Mercury is 2.44×10^6 m, and its mean distance from the sun is 5.79×10^{10} m. Assume that the sun radiates uniformly in all directions.

**34. A gamma-ray telescope intercepts a pulse of gamma radiation from a *magnetar*, a type of star with a spectacularly large magnetic field. The pulse lasts 0.24 s and delivers 8.4×10^{-6} J of energy perpendicularly to the 75-m² surface area of the telescope's detector. The magnetar is thought to be 4.5×10^{20} m (about 50 000 light-years) from earth, and to have a radius of 9.0×10^3 m. Find the magnitude of the rms magnetic field of the gamma-ray pulse at the surface of the magnetar, assuming that the pulse radiates uniformly outward in all directions.

Section 24.5 The Doppler Effect and Electromagnetic Waves

35. Review **Interactive Solution 24.35** at **www.wiley.com/college/cutnell** to see one model for solving this problem. A distant galaxy emits light that has a wavelength of 434.1 nm. On earth, the wavelength of this light is measured to be 438.6 nm. **(a)** Decide whether this galaxy is approaching or receding from the earth. Give your reasoning. **(b)** Find the speed of the galaxy relative to the earth.

36. Multiple-Concept Example 6 explores the concepts that are involved in this problem. Suppose that the police car in that example is moving to the right at 27 m/s, while the speeder is coming up from behind at a speed of 39 m/s, both speeds being with respect to the ground. Assume that the electromagnetic wave emitted by the radar gun has a frequency of 8.0×10^9 Hz. Find the difference between the frequency of the wave that returns to the police car after reflecting from the speeder's car and the original frequency emitted by the police car.

*37. ssm A distant galaxy is simultaneously rotating and receding from the earth. As the drawing shows, the galactic center is receding from the earth at a relative speed of $u_G = 1.6 \times 10^6$ m/s. Relative to the center, the tangential speed is $v_T = 0.4 \times 10^6$ m/s for locations A and B, which are equidistant from the center. When the frequencies of the light coming from regions A and B are measured on earth, they are not the same and each is different from the emitted frequency of 6.200×10^{14} Hz. Find the measured frequency for the light from **(a)** region A and **(b)** region B.

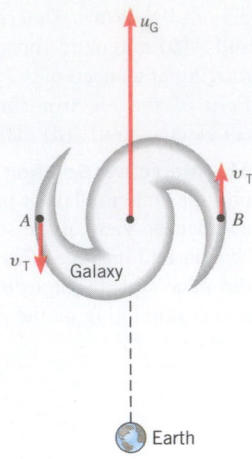

*38. GO The drawing shows three situations—A, B, and C—in which an observer and a source of electromagnetic waves are moving along the same line. In each case the source emits a wave that has a frequency of 4.57×10^{14} Hz. The arrows in each situation denote velocity vectors of the observer and source relative to the ground and have the magnitudes indicated (v or $2v$), where the speed v is 1.50×10^6 m/s. Calculate the observed frequency in each of the three cases.

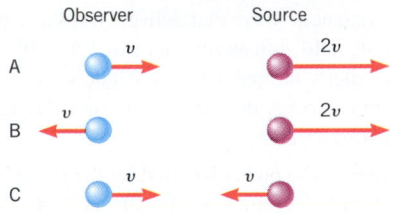

Section 24.6 Polarization

39. Unpolarized light whose intensity is 1.10 W/m^2 is incident on the polarizer in Figure 24.21. **(a)** What is the intensity of the light leaving the polarizer? **(b)** If the analyzer is set at an angle of $\theta = 75°$ with respect to the polarizer, what is the intensity of the light that reaches the photocell?

40. GO The drawing shows light incident on a polarizer whose transmission axis is parallel to the z axis. The polarizer is rotated clockwise through an angle α. The average intensity of the incident light is 7.0 W/m^2. Determine the average intensity of the transmitted light for each of the six cases shown in the table.

	Intensity of Transmitted Light	
Incident Light	$\alpha = 0°$	$\alpha = 35°$
(a) Unpolarized		
(b) Polarized parallel to z axis		
(c) Polarized parallel to y axis		

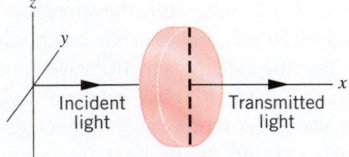

41. ssm In the polarizer/analyzer combination in Figure 24.21, 90.0% of the light intensity falling on the analyzer is absorbed. Determine the angle between the transmission axes of the polarizer and the analyzer.

42. The average intensity of light emerging from a polarizing sheet is 0.764 W/m^2, and the average intensity of the horizontally polarized light incident on the sheet is 0.883 W/m^2. Determine the angle that the transmission axis of the polarizing sheet makes with the horizontal.

43. Review Conceptual Example 8 before solving this problem. Suppose that unpolarized light of intensity 150 W/m^2 falls on the polarizer in Figure 24.24a, and the angle θ in the drawing is $30.0°$. What is the light intensity reaching the photocell?

44. GO The drawing shows three polarizer/analyzer pairs. The incident light beam for each pair is unpolarized and has the same average intensity of 48 W/m^2. Find the average intensity of the transmitted beam for each of the three cases (A, B, and C) shown in the drawing.

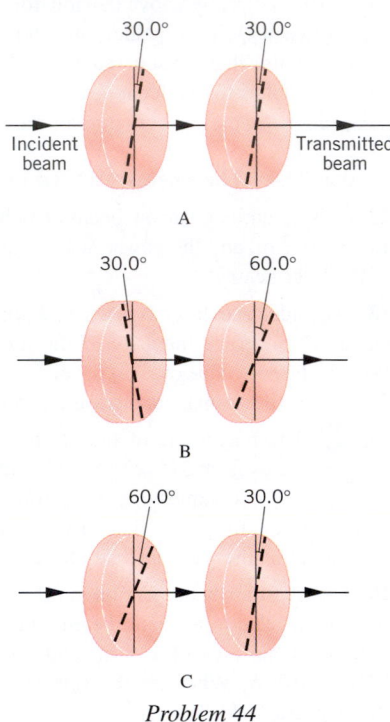

Problem 44

45. In experiment 1, unpolarized light falls on the polarizer in Figure 24.21. The angle of the analyzer is $\theta = 60.0°$. In experiment 2, the unpolarized light is replaced by light of the same intensity, but the light is polarized along the direction of the polarizer's transmission axis. By how many *additional* degrees must the analyzer be rotated so that the light falling on the photocell will have the same intensity as it did in experiment 1? Explain whether θ is increased or decreased by this additional number of degrees.

*46. A beam of polarized light with an average intensity of 15 W/m^2 is sent through a polarizer. The transmission axis makes an angle of $25°$ with respect to the direction of polarization. Determine the rms value of the electric field of the transmitted beam.

*47. ssm www More than one analyzer can be used in a setup like the one in Figure 24.21, each analyzer following the previous one. Suppose that the transmission axis of the first analyzer is rotated $27°$ relative to the transmission axis of the polarizer, and that the transmission axis of each additional analyzer is rotated $27°$ relative to the transmission axis of the previous one. What is the minimum number of analyzers needed for the light reaching the photocell to have an intensity that is reduced by at least a factor of 100 relative to the intensity of the light striking the first analyzer?

*48. GO The drawing shows four sheets of polarizing material, each with its transmission axis oriented differently. Light that is polarized in the vertical direction is incident from the left and has an average intensity of 27 W/m^2. Determine the average intensity of the light that emerges on the right in the drawing when sheet A alone is removed, when sheet B alone is removed, when sheet C alone is removed, and when sheet D alone is removed.

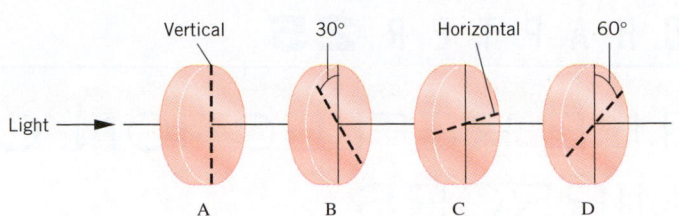

ADDITIONAL PROBLEMS

49. A truck driver is broadcasting at a frequency of 26.965 MHz with a CB (citizen's band) radio. Determine the wavelength of the electromagnetic wave being used. The speed of light is $c = 2.9979 \times 10^8$ m/s.

50. The maximum strength of the magnetic field in an electromagnetic wave is 3.3×10^{-6} T. What is the maximum strength of the wave's electric field?

51. ssm www An AM station is broadcasting a radio wave whose frequency is 1400 kHz. The value of the capacitance in Figure 24.4 is 8.4×10^{-11} F. What must be the value of the inductance in order that this station can be tuned in by the radio?

52. Magnetic resonance imaging, or MRI (see Section 21.7), and positron emission tomography, or PET scanning (see Section 32.6), are two medical diagnostic techniques. Both employ electromagnetic waves. For these waves, find the ratio of the MRI wavelength (frequency = 6.38×10^7 Hz) to the PET scanning wavelength (frequency = 1.23×10^{20} Hz).

53. ssm The microwave radiation left over from the Big Bang explosion of the universe has an average energy density of 4×10^{-14} J/m³. What is the rms value of the electric field of this radiation?

54. Multiple-Concept Example 6 reviews the concepts that play a role in this problem. A speeder is pulling directly away and increasing his distance from a police car that is moving at 25 m/s with respect to the ground. The radar gun in the police car emits an electromagnetic wave with a frequency of 7.0×10^9 Hz. The wave reflects from the speeder's car and returns to the police car, where its frequency is measured to be 320 Hz less than the emitted frequency. Find the speeder's speed with respect to the ground.

55. ssm For one approach to this problem, consult **Interactive LearningWare 24.1** at **www.wiley.com/college/cutnell**. For each of the three sheets of polarizing material shown in the drawing, the orientation of the transmission axis is labeled relative to the vertical. The incident beam of light is unpolarized and has an intensity of 1260.0 W/m². What is the intensity of the beam transmitted through the three sheets when $\theta_1 = 19.0°$, $\theta_2 = 55.0°$, and $\theta_3 = 100.0°$?

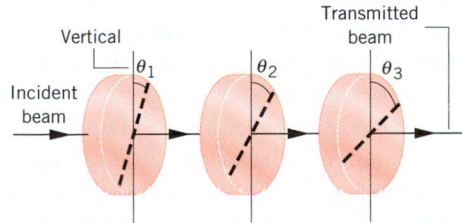

*56. An electromagnetic wave strikes a 1.30-cm² section of wall perpendicularly. The rms value of the wave's magnetic field is determined to be 6.80×10^{-4} T. How long does it take for the wave to deliver 1850 J of energy to the wall?

*57. In a traveling electromagnetic wave, the electric field is represented mathematically as

$$E = E_0 \sin\left[(1.5 \times 10^{10} \text{ s}^{-1})t - (5.0 \times 10^1 \text{ m}^{-1})x\right]$$

where E_0 is the maximum field strength. **(a)** What is the frequency of the wave? **(b)** This wave and the wave that results from its reflection can form a standing wave, in a way similar to that in which standing waves can arise on a string (see Section 17.5). What is the separation between adjacent nodes in the standing wave?

*58. Before attempting this problem, review Conceptual Example 8. The intensity of the light that reaches the photocell in Figure 24.24a is 110 W/m², when $\theta = 23°$. What would be the intensity reaching the photocell if the *analyzer* were removed from the setup, everything else remaining the same?

*59. ssm www The power radiated by the sun is 3.9×10^{26} W. The earth orbits the sun in a nearly circular orbit of radius 1.5×10^{11} m. The earth's axis of rotation is tilted by 27° relative to the plane of the orbit (see the drawing), so sunlight does not strike the equator perpendicularly. What power strikes a 0.75-m² patch of flat land at the equator at point Q?

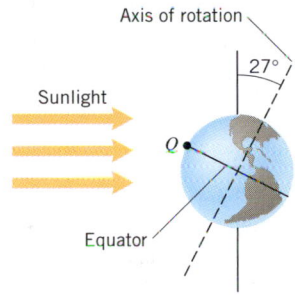

*60. A heat lamp emits infrared radiation whose rms electric field is $E_{\text{rms}} = 2800$ N/C. **(a)** What is the average intensity of the radiation? **(b)** The radiation is focused on a person's leg over a circular area of radius 4.0 cm. What is the average power delivered to the leg? **(c)** The portion of the leg being radiated has a mass of 0.28 kg and a specific heat capacity of 3500 J/(kg·C°). How long does it take to raise its temperature by 2.0 C°? Assume that there is no other heat transfer into or out of the portion of the leg being heated.

*61. An argon-ion laser produces a cylindrical beam of light whose average power is 0.750 W. How much energy is contained in a 2.50-m length of the beam?

**62. Suppose that the light falling on the polarizer in Figure 24.21 is partially polarized (average intensity = $\bar{S}_P$) and partially unpolarized (average intensity = $\bar{S}_U$). The total incident intensity is $\bar{S}_P + \bar{S}_U$, and the percentage polarization is $100 \, \bar{S}_P/(\bar{S}_P + \bar{S}_U)$. When the polarizer is rotated in such a situation, the intensity reaching the photocell varies between a minimum value of $\bar{S}_{\text{min}}$ and a maximum value of $\bar{S}_{\text{max}}$. Show that the percentage polarization can be expressed as $100 \, (\bar{S}_{\text{max}} - \bar{S}_{\text{min}})/(\bar{S}_{\text{max}} + \bar{S}_{\text{min}})$.

CHAPTER 25

THE REFLECTION OF LIGHT: MIRRORS

The reflection of light from the plane surface of the water acts to double the presence of this tiger. This chapter discusses the images formed by the reflection of light from plane and spherical mirrors. (© Tim Fitzharris/Minden Pictures, Inc.)

25.1

WAVE FRONTS AND RAYS

Mirrors are usually close at hand. It is difficult, for example, to put on makeup, shave, or drive a car without them. We see images in mirrors because some of the light that strikes them is reflected into our eyes. To discuss reflection, it is necessary to introduce the concepts of a wave front and a ray of light, and we can do so by taking advantage of the familiar topic of sound waves (see Chapter 16). Both sound and light are kinds of waves. Sound is a pressure wave, whereas light is electromagnetic in nature. However, the ideas of a wave front and a ray apply to both.

Consider a small spherical object whose surface is pulsating in simple harmonic motion. A sound wave is emitted that moves spherically outward from the object at a constant speed. To represent this wave, we draw surfaces through all points of the wave that are in the same phase of motion. These surfaces of constant phase are called *wave fronts.* Figure 25.1 shows a hemispherical view of the wave fronts. In this view they appear as concentric spherical shells about the vibrating object. If the wave fronts are drawn through the condensations, or crests, of the sound wave, as they are in the picture, the distance between adjacent wave fronts equals the wavelength λ. The radial lines pointing outward from the source and perpendicular to the wave fronts are called *rays.* The rays point in the direction of the velocity of the wave.

Figure 25.2*a* shows small sections of two adjacent spherical wave fronts. At large distances from the source, the wave fronts become less and less curved and approach the shape of flat surfaces, as in part *b* of the drawing. Waves whose wave fronts are flat surfaces (i.e., planes) are known as *plane waves* and are important in understanding the properties of mirrors and lenses. Since rays are perpendicular to the wave fronts, the rays for a plane wave are parallel to each other.

Figure 25.1 A hemispherical view of a sound wave emitted by a pulsating sphere. The wave fronts are drawn through the condensations of the wave, so the distance between two successive wave fronts is the wavelength λ. The rays are perpendicular to the wave fronts and point in the direction of the velocity of the wave.

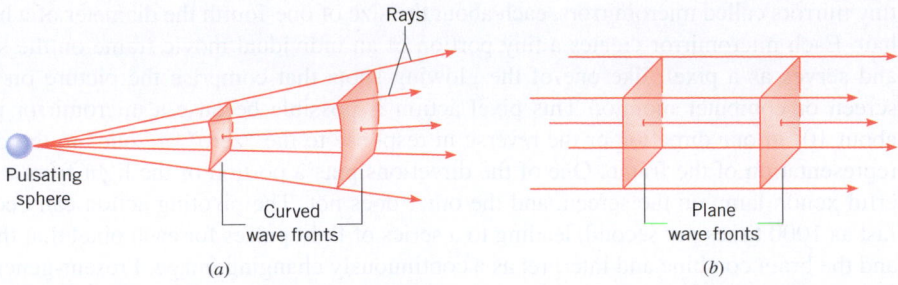

Figure 25.2 (*a*) Portions of two spherical wave fronts are shown. The rays are perpendicular to the wave fronts and diverge. (*b*) For a plane wave, the wave fronts are flat surfaces, and the rays are parallel to each other.

The concepts of wave fronts and rays can also be used to describe light waves. For light waves, the ray concept is particularly convenient when showing the path taken by the light. We will make frequent use of light rays, which can be regarded essentially as narrow beams of light much like those that lasers produce.

25.2 THE REFLECTION OF LIGHT

Most objects reflect a certain portion of the light falling on them. Suppose that a ray of light is incident on a flat, shiny surface, such as the mirror in Figure 25.3. As the drawing shows, the *angle of incidence* θ_i is the angle that the incident ray makes with respect to the normal, which is a line drawn perpendicular to the surface at the point of incidence. The *angle of reflection* θ_r is the angle that the reflected ray makes with the normal. The **law of reflection** describes the behavior of the incident and reflected rays.

LAW OF REFLECTION

The incident ray, the reflected ray, and the normal to the surface all lie in the same plane, and the angle of reflection θ_r equals the angle of incidence θ_i:

$$\theta_r = \theta_i$$

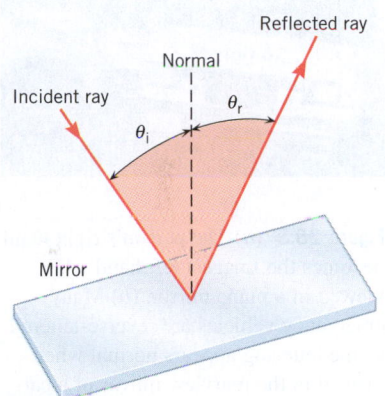

Figure 25.3 The angle of reflection θ_r equals the angle of incidence θ_i. These angles are measured with respect to the normal, which is a line drawn perpendicular to the surface of the mirror at the point of incidence.

When parallel light rays strike a smooth, plane surface, such as the ones in Figure 25.4*a*, the reflected rays are parallel to each other. This type of reflection is one example of what is known as *specular reflection* and is important in determining the properties of mirrors. Most surfaces, however, are not perfectly smooth, because they contain irregularities the sizes of which are equal to or greater than the wavelength of the light. The law of reflection applies to each ray, but the irregular surface reflects the light rays in various directions, as part *b* of the drawing suggests. This type of reflection is known as *diffuse reflection*. Common surfaces that give rise to diffuse reflection are most papers, wood, nonpolished metals, and walls covered with a "flat" (nongloss) paint.

A revolution in digital technology is occurring in the movie industry, where digital techniques are now being used to produce films. Until recently, films have been viewed primarily by using projectors that shine light directly through a strip of film containing the images. Now, however, projectors are available that allow a movie produced using digital techniques to be viewed completely without film by using a digital representation (zeros and ones) of the images. One form of digital projector depends on the law of reflection and

The physics of digital movie projectors and micromirrors.

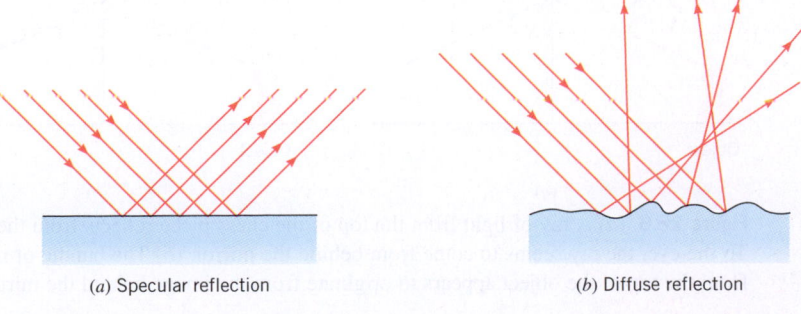

(a) Specular reflection   (b) Diffuse reflection

Figure 25.4 (*a*) The drawing shows specular reflection from a polished plane surface, such as a mirror. The reflected rays are parallel to each other. (*b*) A rough surface reflects the light rays in all directions; this type of reflection is known as diffuse reflection.

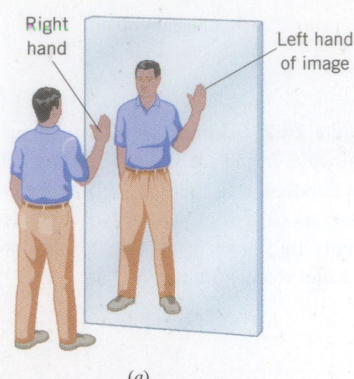

Right hand

Left hand of image

(a)

(b)

Figure 25.5 (a) The person's right hand becomes the image's left hand when viewed in a plane mirror. (b) Many emergency vehicles are reverse-lettered so the lettering appears normal when viewed in the rearview mirror of a car. (Zoran Milich/Masterfile)

tiny mirrors called micromirrors, each about the size of one-fourth the diameter of a human hair. Each micromirror creates a tiny portion of an individual movie frame on the screen and serves as a pixel, like one of the glowing spots that comprise the picture on a TV screen or computer monitor. This pixel action is possible because a micromirror pivots about 10° in one direction or the reverse in response to the "zero" or "one" in the digital representation of the frame. One of the directions puts a portion of the light from a powerful xenon lamp on the screen, and the other does not. The pivoting action can occur as fast as 1000 times per second, leading to a series of light pulses for each pixel that the eye and the brain combine and interpret as a continuously changing image. Present-generation digital micromirror projectors use up to several million micromirrors to reproduce each of the three primary colors (red, green, and blue) that comprise a color image.

25.3 THE FORMATION OF IMAGES BY A PLANE MIRROR

When you look into a plane (flat) mirror, you see an image of yourself that has three properties:

1. The image is upright.
2. The image is the same size as you are.
3. The image is located as far behind the mirror as you are in front of it.

As Figure 25.5a illustrates, the image of yourself in the mirror is also reversed right to left and left to right. If you wave your *right* hand, it is the *left* hand of the image that waves back. Similarly, letters and words held up to a mirror are reversed. Ambulances and other emergency vehicles are often lettered in reverse, as in Figure 25.5b, so that the letters will appear normal when seen in the rearview mirror of a car.

To illustrate why an image appears to originate from behind a plane mirror, Figure 25.6a shows a light ray leaving the top of an object. This ray reflects from the mirror (angle of reflection equals angle of incidence) and enters the eye. To the eye, it appears that the ray originates from behind the mirror, somewhere back along the dashed line. Actually, rays going in all directions leave each point on the object, but only a small bundle of such rays is intercepted by the eye. Part b of the figure shows a bundle of two rays leaving the top of the object. All the rays that leave a given point on the object, no matter what angle θ they have when they strike the mirror, appear to originate from a corresponding point on the image behind the mirror (see the dashed lines in part b). For each point on the object, there is a single corresponding point on the image, and it is this fact that makes the image in a plane mirror a sharp and undistorted one.

Although rays of light *seem* to come from the image, it is evident from Figure 25.6b that they do not originate from behind the plane mirror where the image appears to be. Because none of the light rays actually emanate from the image, it is called a ***virtual image.*** In this text the parts of the light rays that appear to come from a virtual image are represented by dashed lines. *Curved* mirrors, on the other hand, can produce images from

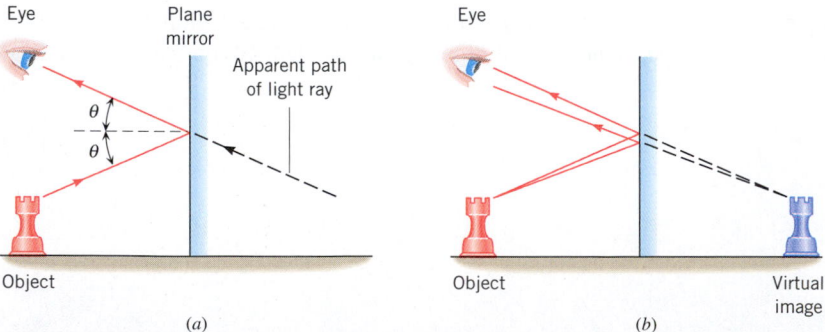

Figure 25.6 (a) A ray of light from the top of the chess piece reflects from the mirror. To the eye, the ray seems to come from behind the mirror. (b) The bundle of rays from the top of the object appears to originate from the image behind the mirror.

which all the light rays actually do emanate. Such images are known as *real images* and are discussed later.

With the aid of the law of reflection, it is possible to show that the image is located as far behind a plane mirror as the object is in front of it. In Figure 25.7 the object distance is d_o and the image distance is d_i. A ray of light leaves the base of the object, strikes the mirror at an angle of incidence θ, and is reflected at the same angle. To the eye, this ray appears to come from the base of the image. For the angles β_1 and β_2 in the drawing it follows that $\theta + \beta_1 = 90°$ and $\alpha + \beta_2 = 90°$. But the angle α is equal to the angle of reflection θ, since the two are opposite angles formed by intersecting lines. Therefore, $\beta_1 = \beta_2$. As a result, triangles *ABC* and *DBC* are identical (congruent) because they share a common side *BC* and have equal angles ($\beta_1 = \beta_2$) at the top and equal angles (90°) at the base. Thus, the magnitude of the object distance d_o equals the magnitude of the image distance d_i.

By starting with a light ray from the top of the object, rather than the bottom, we can use the same line of reasoning to show also that the height of the image equals the height of the object.

Conceptual Examples 1 and 2 discuss some interesting features of plane mirrors.

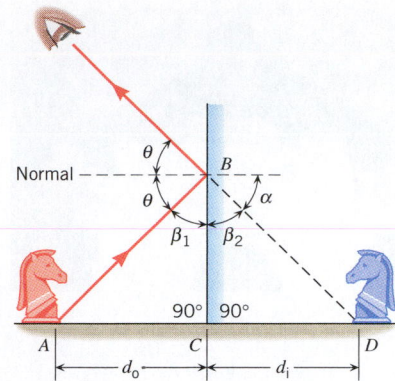

Figure 25.7 This drawing illustrates the geometry used with a plane mirror to show that the image distance d_i equals the object distance d_o.

Conceptual Example 1   Full-Length Versus Half-Length Mirrors

In Figure 25.8 a woman is standing in front of a plane mirror. Is the minimum mirror height that is necessary for her to see her full image **(a)** equal to her height, or **(b)** equal to one-half her height?

Reasoning The woman sees her image because light emanating from her body is reflected by the mirror (labeled *ABCD* in Figure 25.8) and enters her eyes. Consider, for example, a ray of light from her foot *F*. This ray strikes the mirror at *B* and enters her eyes at *E*. According to the law of reflection, the angles of incidence and reflection are both θ. This law will allow us to deduce how the height of the mirror is related to her own height.

Answer (a) is incorrect. The mirror in Figure 25.8 is the same height as the woman. Any light from her foot that strikes the mirror below *B* is reflected toward a point on her body that is below her eyes. Since light striking the mirror below *B* does not enter her eyes, the part of the mirror between *B* and *A* may be removed. Thus, the necessary minimum height of the mirror is not equal to the woman's height.

Answer (b) is correct. As discussed above, the section *AB* of the mirror is not necessary in order for the woman to see her full image. The section *BC* of the mirror that produces the image is one-half the woman's height between *F* and *E*. This follows because the right triangles *FBM* and *EBM* are identical. They are identical because they share a common side *BM* and have two angles, θ and 90°, that are the same. The blowup in Figure 25.8 illustrates a similar line of reasoning, starting with a ray from the woman's head at *H*. This ray is reflected from the mirror at *P* and enters her eyes. The top mirror section *PD* may be removed without disturbing this reflection. The necessary section *CP* is one-half the woman's height between her head at *H* and her eyes at *E*. We find, then, that only the sections *BC* and *CP* are needed for the woman to see her full length. The height of section *BC* plus section *CP* is exactly one-half the woman's height. The conclusions here are valid regardless of how far the person stands from the mirror. Thus, to view one's full length in a mirror, only a half-length mirror is needed.

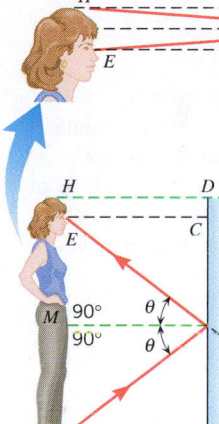

Figure 25.8 A woman stands in front of a plane mirror and sees her full image.

Related Homework: *Problems 6, 7*

Conceptual Example 2   Multiple Reflections

A person is sitting in front of two mirrors that intersect at a 90° angle. As Figure 25.9*a* illustrates, the person sees three images of herself. (The person herself is only partially visible at the bottom of the photo.) These images arise because rays of light emanate from her body, reflect from the mirrors, and enter her eyes. Consider the light that enters her eyes and appears to come from each of the three images identified in Figure 25.9*b*. The table shows three possibilities for the number of reflections that the light undergoes before entering her eyes. Which possibility is correct?

(a)

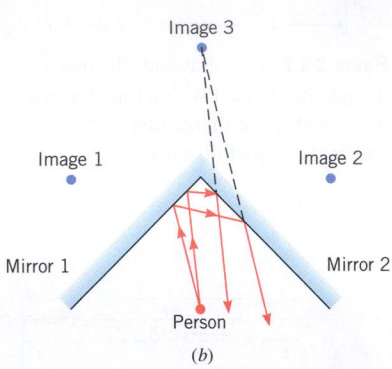

(b)

Figure 25.9 (a) These two perpendicular plane mirrors produce three images of the person (not completely visible) sitting in front of them. (Andy Washnik) (b) A "double" reflection, one from each mirror, can also occur.

	Number of Reflections		
Possibility	Image 1	Image 2	Image 3
(a)	2	2	3
(b)	3	3	3
(c)	1	1	2

Reasoning Images of the woman are formed when light emanating from her body enters her eyes after being reflected by one, or both, mirrors. For each reflection, the angle of the light reflected from a mirror is equal to the angle of the light incident on the mirror (law of reflection). We will see that there are three ways that light can reach her eyes from the two mirrors.

Answers (a) and (b) are incorrect. Figure 25.9b represents a top view of the person in front of the two mirrors. It is a straightforward matter to understand two of the images that she sees. These are the images that are normally seen when one sits in front of a mirror. Sitting in front of mirror 1, she sees image 1, which is located as far behind that mirror as she is in front of it. She also sees image 2 behind mirror 2, at a distance that matches her distance in front of that mirror. Each of these images arises from light emanating from her body and reflecting only once from a single mirror. Therefore, each ray of light does not reflect two or three times before entering her eyes.

Answer (c) is correct. As discussed above, images 1 and 2 arise, respectively, from *single* reflections from mirrors 1 and 2. The third image arises when light undergoes two reflections in sequence, first from one mirror and then from the other. When such a double reflection occurs, an additional image becomes possible. Figure 25.9b shows two rays of light that strike mirror 1. Each one, according to the law of reflection, has an angle of reflection that equals the angle of incidence. The rays then strike mirror 2, where they again are reflected according to the law of reflection. When the outgoing rays are extended backward (see the dashed lines in the drawing), they intersect and appear to originate from image 3. Thus, the third image arises when an incident ray of light is reflected twice, once from each mirror, before entering her eyes.

Related Homework: *Problem 2*

✓ CHECK YOUR UNDERSTANDING

(*The answers are given at the end of the book.*)

1. The drawing shows a light ray undergoing multiple reflections from a mirrored corridor. The walls of the corridor are either parallel or perpendicular to one another. If the initial angle of incidence is 35°, what is the angle of reflection when the ray makes its last reflection?

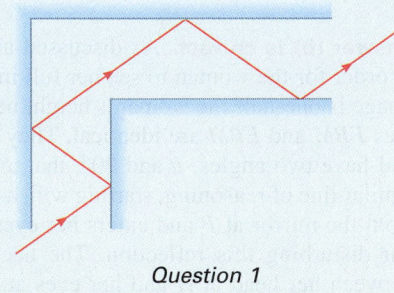

Question 1

2. A sign painted on a store window is reversed when viewed from inside the store. If a person inside the store views the reversed sign in a plane mirror, does the sign appear as it would when viewed from outside the store? (Try it by writing some letters on a transparent sheet of paper and then holding the back side of the paper up to a mirror.)

3. If a clock is held in front of a mirror, its image is reversed left to right. From the point of view of a person looking into the mirror, does the image of the second hand rotate in the reverse (counterclockwise) direction?

25.4 SPHERICAL MIRRORS

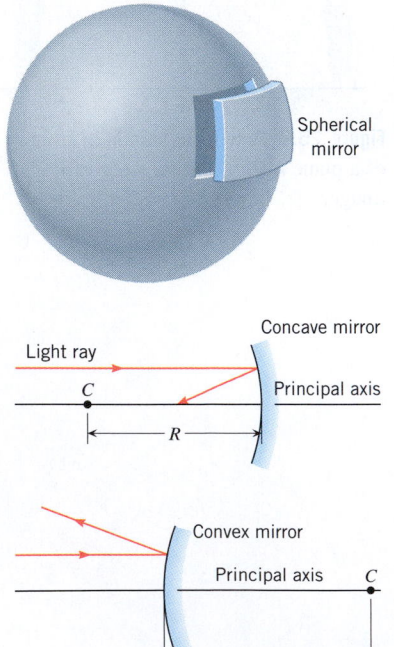

Figure 25.10 A spherical mirror has the shape of a segment of a spherical surface. The center of curvature is point C and the radius is R. For a concave mirror, the reflecting surface is the inner one; for a convex mirror it is the outer one.

The most common type of curved mirror is a spherical mirror. As Figure 25.10 shows, a spherical mirror has the shape of a section from the surface of a sphere. If the inside surface of the mirror is polished, it is a **concave mirror.** If the outside surface is polished, it is a **convex mirror.** The drawing shows both types of mirrors, with a light ray reflecting from the polished surface. The law of reflection applies, just as it does for a plane mirror. For either type of spherical mirror, the normal is drawn perpendicular to the mirror at the

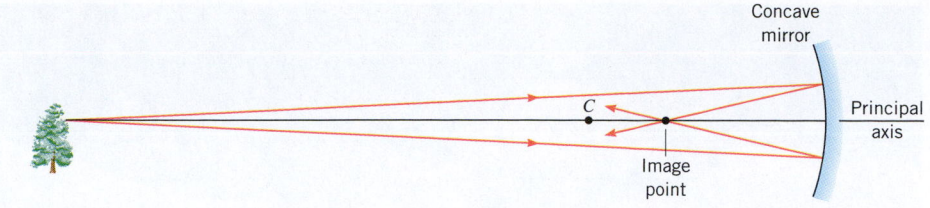

Figure 25.11 A point on the tree lies on the principal axis of the concave mirror. Rays from this point that are near the principal axis are reflected from the mirror and cross the axis at the image point.

point of incidence. For each type, the center of curvature is located at point *C*, and the radius of curvature is *R*. The **principal axis** of the mirror is a straight line drawn through *C* and the midpoint of the mirror.

Figure 25.11 shows a tree in front of a concave mirror. A point on this tree lies on the principal axis of the mirror and is beyond the center of curvature *C*. Light rays emanate from this point and reflect from the mirror, consistent with the law of reflection. If the rays are near the principal axis, they cross it at a common point after reflection. This point is called the *image point*. The rays continue to diverge from the image point as if there were an object there. Since light rays actually come from the image point, the image is a real image.

If the tree in Figure 25.11 is infinitely far from the mirror, the rays are parallel to each other and to the principal axis as they approach the mirror. Figure 25.12 shows rays near and parallel to the principal axis, as they reflect from the mirror and pass through an image point. In this special case the image point is referred to as the **focal point F** of the mirror. Therefore, an object infinitely far away on the principal axis gives rise to an image at the focal point of the mirror. The distance between the focal point and the middle of the mirror is the **focal length f** of the mirror.

We can show that the focal point *F* lies halfway between the center of curvature *C* and the middle of a concave mirror. In Figure 25.13, a light ray parallel to the principal axis strikes the mirror at point *A*. The line *CA* is the radius of the mirror and, therefore, is the normal to the spherical surface at the point of incidence. The ray reflects from the mirror such that the angle of reflection θ equals the angle of incidence. Furthermore, the angle *ACF* is also θ because the radial line *CA* is a transversal of two parallel lines. Since two of its angles are equal, the colored triangle *CAF* is an isosceles triangle; thus, sides *CF* and *FA* are equal. However, when the incoming ray lies close to the principal axis, the angle of incidence θ is small, and the distance *FA* does not differ appreciably from the distance *FB*. Therefore, in the limit that θ is small, *CF* = *FA* = *FB*, and so the focal point *F* lies halfway between the center of curvature and the mirror. In other words, the focal length *f* is one-half of the radius *R*:

Focal length of a concave mirror

$$f = \tfrac{1}{2}R \qquad (25.1)$$

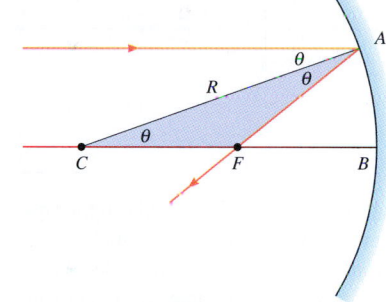

Figure 25.12 Light rays near and parallel to the principal axis are reflected from a concave mirror and converge at the focal point *F*. The focal length *f* is the distance between *F* and the mirror.

Figure 25.13 This drawing is used to show that the focal point *F* of a concave mirror is halfway between the center of curvature *C* and the mirror at point *B*.

Rays that lie close to the principal axis are known as **paraxial rays,** * and Equation 25.1 is valid only for such rays. Rays that are far from the principal axis do not converge to a single point after reflection from the mirror, as Figure 25.14 shows. The result is a blurred image. The fact that a spherical mirror does not bring all rays parallel to the principal axis to a single image point is known as **spherical aberration.** Spherical aberration can be minimized by using a mirror whose height is small compared to the radius of curvature.

The physics of capturing solar energy with mirrors; automobile headlights. A sharp image point can be obtained with a large mirror, if the mirror is parabolic in shape instead of spherical. The shape of a parabolic mirror is such that all light rays parallel to the principal axis, regardless of their distance from the axis, are reflected through a single image point. However, parabolic mirrors are costly to manufacture and are used where the sharpest images are required, as in research-quality telescopes. Parabolic mirrors are also used in one method of collecting solar energy for commercial purposes. Figure 25.15 shows a long row of concave parabolic mirrors that reflect the sun's rays to the focal point. Located at the focal point and running the length of the row is an oil-filled pipe. The focused rays of the sun heat the oil. In a solar-thermal electric plant, the heat from many such rows is used to generate steam. The steam, in turn, drives a turbine connected to an electric generator. Another application of parabolic mirrors is in automobile headlights. Here, however, the situation is reversed from the operation of a solar collector. In a headlight, a high-intensity light bulb is placed at the focal point of the mirror, and light emerges parallel to the principal axis.

*Paraxial rays are close to the principal axis but not necessarily parallel to it.

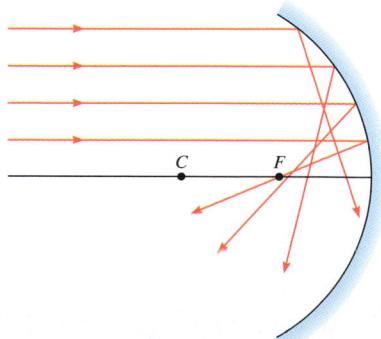

Figure 25.14 Rays that are farthest from the principal axis have the greatest angle of incidence and miss the focal point *F* after reflection from the mirror.

Figure 25.15 This long row of parabolic mirrors focuses the sun's rays to heat an oil-filled pipe located at the focal point of each mirror. It is one of many that are used by a solar-thermal electric plant in the Mojave Desert. (© Jim West/Alamy)

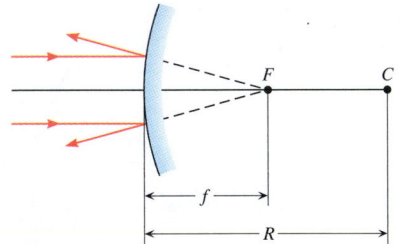

Figure 25.16 When paraxial light rays that are parallel to the principal axis strike a convex mirror, the reflected rays appear to originate from the focal point *F*. The radius of curvature is *R* and the focal length is *f*.

A convex mirror also has a focal point, and Figure 25.16 illustrates its meaning. In this picture, parallel rays are incident on a convex mirror. Clearly, the rays diverge after being reflected. If the incident parallel rays are paraxial, the reflected rays seem to come from a single point *F* behind the mirror. This point is the focal point of the convex mirror, and its distance from the midpoint of the mirror is the *focal length f*. The focal length of a convex mirror is also one-half of the radius of curvature, just as it is for a concave mirror. However, we assign the focal length of a convex mirror a negative value because it will be convenient later on:

Focal length of
a convex mirror

$$f = -\tfrac{1}{2}R \qquad (25.2)$$

Spherical aberration is a problem with convex mirrors, just as it is with concave mirrors. Rays that emanate from a single point on an object but are far from the principal axis do not appear to originate from a single image point after reflection from the mirror. As with a concave mirror, the result is a blurred image.

✓ CHECK YOUR UNDERSTANDING

(The answers are given at the end of the book.)

4. A section of a sphere has a radius of curvature of 0.60 m, and both the inside and outside surfaces have a mirror-like polish. What are the focal lengths of the inside and outside surfaces?

5. The photograph shows an experimental device at Sandia National Laboratories in New Mexico. This device is a mirror that focuses sunlight to heat sodium to a boil, which then heats helium gas in an engine. The engine does the work of driving a generator to produce electricity. The sodium unit and the engine are labeled in the photo. **(a)** What kind of mirror, concave or convex, is being used? **(b)** Where is the sodium unit located relative to the mirror? Express your answer in terms of the focal length of the mirror.

(Courtesy Sandia National Laboratories)

Question 5

6. Refer to Figure 25.14 and the related discussion about spherical aberration. To bring the top ray closer to the focal point *F* after reflection, describe how you would change the shape of the mirror. Would you open it up to produce a more gently curving shape or bring the top and bottom edges closer to the principal axis?

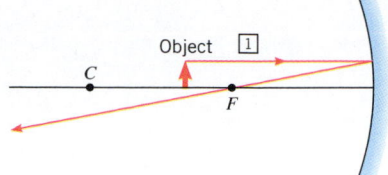

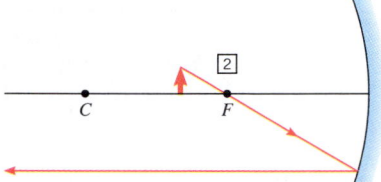

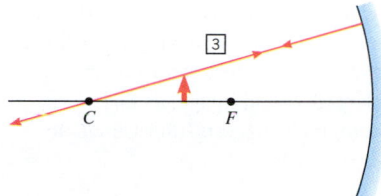

THE FORMATION OF IMAGES BY SPHERICAL MIRRORS

25.5

As we have seen, some of the light rays emitted from an object in front of a mirror strike the mirror, reflect from it, and form an image. We can analyze the image produced by either concave or convex mirrors by using a graphical method called *ray tracing.* This method is based on the law of reflection and the notion that a spherical mirror has a center of curvature C and a focal point F. Ray tracing enables us to find the location of the image, as well as its size, by taking advantage of the following fact: paraxial rays leave from a point on the object and intersect at a corresponding point on the image after reflection.

CONCAVE MIRRORS

Three specific paraxial rays are especially convenient to use in the ray-tracing method. Figure 25.17 shows an object in front of a concave mirror, and these three rays leave from a point on the top of the object. The rays are labeled 1, 2, and 3, and when tracing their paths, we use the following reasoning strategy.

REASONING STRATEGY

Ray Tracing for a Concave Mirror

Ray 1. This ray is initially parallel to the principal axis and, therefore, passes through the focal point F after reflection from the mirror.

Ray 2. This ray initially passes through the focal point F and is reflected parallel to the principal axis. Ray 2 is analogous to ray 1 except that the reflected, rather than the incident, ray is parallel to the principal axis.

Ray 3. This ray travels along a line that passes through the center of curvature C and follows a radius of the spherical mirror; as a result, the ray strikes the mirror perpendicularly and reflects back on itself.

Figure 25.17 The rays labeled 1, 2, and 3 are useful in locating the image of an object placed in front of a concave spherical mirror. The object is represented as a vertical arrow.

If rays 1, 2, and 3 are superimposed on a scale drawing, they converge at a point on the top of the image, as can be seen in Figure 25.18a.* Although three rays have been used here to locate the image, only two are really needed; the third ray is usually drawn to serve as a check. In a similar fashion, rays from all other points on the object locate corresponding points on the image, and the mirror forms a complete image of the object. If you were to place your eye as shown in the drawing, you would see an image that is *larger* and *inverted* relative to the object. The image is *real* because the light rays actually pass through the image.

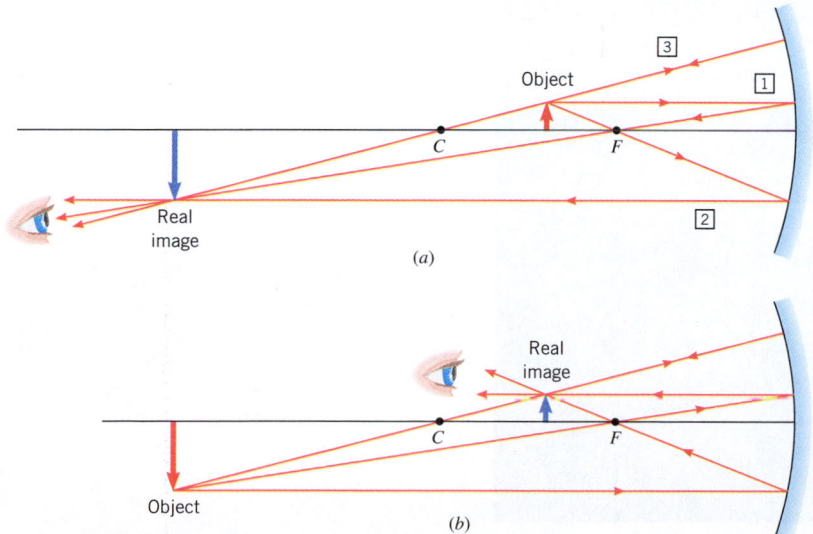

Figure 25.18 (*a*) When an object is placed between the focal point F and the center of curvature C of a concave mirror, a real image is formed. The image is enlarged and inverted relative to the object. (*b*) When the object is located beyond the center of curvature C, a real image is created that is reduced in size and inverted relative to the object.

*In the drawings that follow, we assume that the rays are paraxial, although the distance between the rays and the principal axis is often exaggerated for clarity.

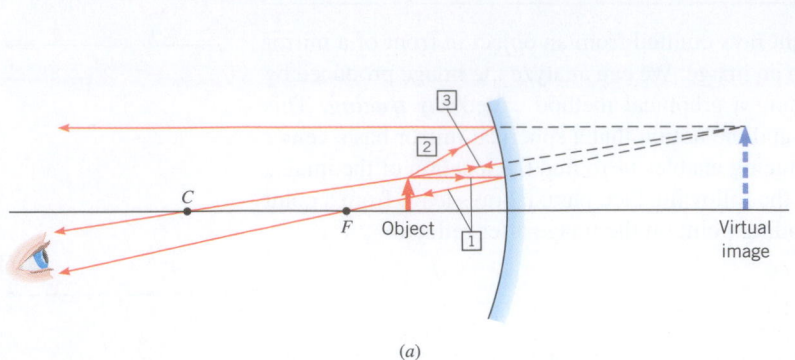

(a)   (b)

Figure 25.19 (a) When an object is located between a concave mirror and its focal point F, an enlarged, upright, and virtual image is produced. (b) A makeup mirror (or shaving mirror) is a concave mirror that can form an enlarged virtual image, as this photograph suggests. (Stockdisc/Getty Images, Inc.)

If the locations of the object and image in Figure 25.18a are interchanged, the situation in part b of the drawing results. The three rays in part b are the same as those in part a, except that the directions are reversed. **Problem-solving insight:** These drawings illustrate the ***principle of reversibility,*** which states that ***if the direction of a light ray is reversed, the light retraces its original path.*** This principle is quite general and is not restricted to reflection from mirrors. The image is *real,* and it is *smaller* and *inverted* relative to the object.

When the object is placed between the focal point F and a concave mirror, as in Figure 25.19a, three rays can again be drawn to find the image. Now, however, ray 2 does not go through the focal point on its way to the mirror, since the object is inside the focal point. When projected backward, though, ray 2 appears to come from the focal point. Therefore, after reflection, ray 2 is directed parallel to the principal axis. In this case the three reflected rays diverge from each other and do not converge to a common point. However, when projected behind the mirror, the three rays appear to come from a common point; thus, a *virtual* image is formed. This virtual image is *larger* than the object and *upright*. **The physics of makeup and shaving mirrors.** Makeup and shaving mirrors are concave mirrors. When you place your face between the mirror and its focal point, you see an enlarged virtual image of yourself, as part b of the drawing shows.

The physics of a head-up display for automobiles. Concave mirrors are also used in one method for displaying the speed of a car. The method presents a digital readout (e.g., "55 mph") that the driver sees when looking directly through the windshield, as in Figure 25.20a. The advantage of the method, which is called a head-up display (HUD), is that the

Figure 25.20 (a) A head-up display (HUD) presents the driver with a digital readout of the car's speed in the field of view seen through the windshield. (Copyright GM Corporation. All rights reserved.) (b) One version of a HUD uses a concave mirror. (See text for explanation.)

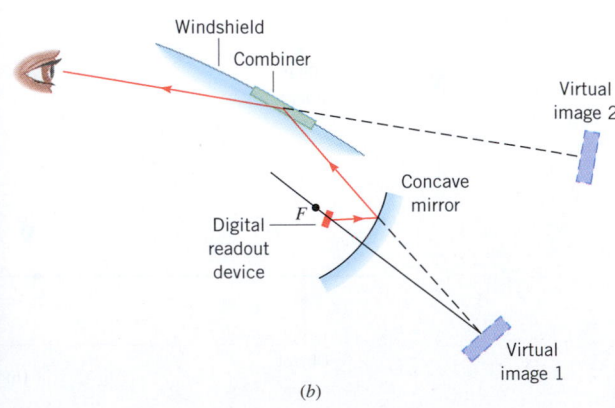

(a)   (b)

driver does not need to take his or her eyes off the road to monitor the speed. Figure 25.20*b* shows how a HUD works. Located below the windshield is a readout device that displays the speed in digital form. This device is located in front of a concave mirror but within its focal point. The arrangement is similar to the one in Figure 25.19*a* and produces a virtual, upright, and enlarged image of the speed readout (see virtual image 1 in Figure 25.20*b*). Light rays that appear to come from this image strike the windshield at a place where a so-called "combiner" is located. The purpose of the combiner is to combine the digital readout information with the field of view that the driver sees through the windshield. The combiner is virtually undetectable by the driver because it allows all colors except one to pass through it unaffected. The one exception is the color produced by the digital readout device. For this color, the combiner behaves as a plane mirror and reflects the light that appears to originate from image 1. Thus, the combiner produces image 2, which is what the driver sees. The location of image 2 is out above the front bumper. The driver can then read the speed with eyes focused just as they are to see the road.

CONVEX MIRRORS

The ray-tracing procedure for determining the location and size of an image in a convex mirror is similar to that for a concave mirror. The same three rays are used. However, the focal point and center of curvature of a convex mirror lie behind the mirror, not in front of it. Figure 25.21*a* shows the rays. When tracing their paths, we use the following reasoning strategy, which takes into account these locations of the focal point and center of curvature.

REASONING STRATEGY

Ray Tracing for a Convex Mirror

Ray 1. This ray is initially parallel to the principal axis and, therefore, appears to originate from the focal point *F* after reflection from the mirror.

Ray 2. This ray heads toward *F*, emerging parallel to the principal axis after reflection. Ray 2 is analogous to ray 1, except that the reflected, rather than the incident, ray is parallel to the principal axis.

Ray 3. This ray travels toward the center of curvature *C*; as a result, the ray strikes the mirror perpendicularly and reflects back on itself.

Figure 25.21 (*a*) An object placed in front of a convex mirror always produces a virtual image behind the mirror. The virtual image is reduced in size and upright. (*b*) The sun shield on this pilot's helmet acts as a convex mirror and reflects an image of his plane. (Chad Slattery/Stone/Getty Images)

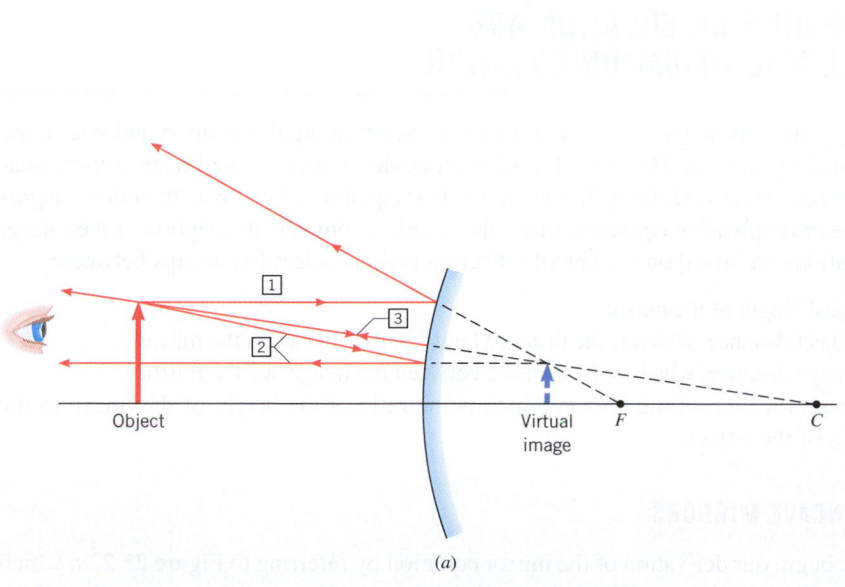

(*a*)

(*b*)

The three rays in Figure 25.21*a* appear to come from a point on a *virtual* image that is behind the mirror. The virtual image is *diminished in size* and *upright,* relative to the object. A convex mirror *always* forms a virtual image of the object, no matter where in front of the mirror the object is placed. Figure 25.21*b* shows an example of such an image.

Because of its shape, a convex mirror gives a wider field of view than do other types of mirrors. Therefore, they are often used in stores for security purposes. A mirror with a wide field of view is also needed to give a driver a good rear view. Thus, the outside mirror on the passenger side is often a convex mirror. Printed on such a mirror is usually the warning "VEHICLES IN MIRROR ARE CLOSER THAN THEY APPEAR." The reason for the warning is that, as in Figure 25.21*a*, the virtual image is reduced in size and therefore looks smaller, just as a distant object would appear in a plane mirror. An unwary driver, thinking that the side-view mirror is a plane mirror, might incorrectly deduce from the small size of the image that the car behind is far enough away to ignore.

The physics of
passenger-side automobile mirrors.

✓ CHECK YOUR UNDERSTANDING

(The answers are given at the end of the book.)

7. Concept Simulation 25.3 at www.wiley.com/college/cutnell allows you to explore the concepts to which this question relates. Is it possible to use a convex mirror to produce an image that is larger than the object?

8. (a) When you look at the back side of a shiny teaspoon held at arm's length, do you see yourself upright or upside down?   **(b)** When you look at the other side of the spoon, do you see yourself upright or upside down? Assume in both cases that the distance between you and the spoon is greater than the focal length of the spoon.

9. (a) Can the image formed by a concave mirror ever be projected directly onto a screen without the help of other mirrors or lenses? If so, specify where the object should be placed relative to the mirror.   **(b)** Repeat part (a) assuming that the mirror is convex.

10. Suppose that you stand in front of a spherical mirror (concave or convex). Is it possible for your image to be   **(a)** real and upright   **(b)** virtual and inverted?

11. An object is placed between the focal point and the center of curvature of a concave mirror. The object is then moved closer to the mirror, but still remains between the focal point and the center of curvature. Do the magnitudes of   **(a)** the image distance and **(b)** the image height become larger or smaller?

12. When you see the image of yourself formed by a mirror, it is because (1) light rays actually coming from a real image enter your eyes or (2) light rays appearing to come from a virtual image enter your eyes. If light rays from the image do not enter your eyes, you do not see yourself. Are there any places on the principal axis where you cannot see yourself when you are standing in front of a mirror that is   **(a)** convex   **(b)** concave? If so, where are these places? Assume that you have only the one mirror to use.

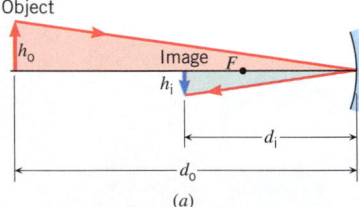

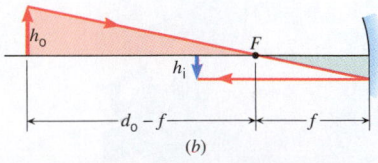

Figure 25.22 These diagrams are used to derive the mirror equation and the magnification equation. (*a*) The two colored triangles are similar triangles. (*b*) If the ray is close to the principal axis, the two colored regions are almost similar triangles.

25.6 THE MIRROR EQUATION AND THE MAGNIFICATION EQUATION

Ray diagrams drawn to scale are useful for determining the location and size of the image formed by a mirror. However, for an accurate description of the image, a more analytical technique is needed. We will now derive two equations, known as the ***mirror equation*** and the ***magnification equation,*** that will provide a complete description of the image. These equations are based on the law of reflection and provide relationships between:

f = the focal length of the mirror
d_o = the object distance, which is the distance between the object and the mirror
d_i = the image distance, which is the distance between the image and the mirror
m = the magnification of the mirror, which is the ratio of the height of the image to the height of the object.

CONCAVE MIRRORS

We begin our derivation of the mirror equation by referring to Figure 25.22*a*, which shows a ray leaving the top of the object and striking a concave mirror at the point where the principal axis intersects the mirror. Since the principal axis is perpendicular to the mirror, it is also the normal at this point of incidence. Therefore, the ray reflects at an equal angle and passes through the image. The two colored triangles are similar triangles because they have equal angles, so

$$\frac{h_o}{-h_i} = \frac{d_o}{d_i}$$

where h_o is the height of the object and h_i is the height of the image. The minus sign appears on the left in this equation because the image is inverted in Figure 25.22a. In part *b* another ray leaves the top of the object, this one passing through the focal point *F*, reflecting parallel to the principal axis, and then passing through the image. Provided the ray remains close to the axis, the two colored areas can be considered to be similar triangles, with the result that

$$\frac{h_o}{-h_i} = \frac{d_o - f}{f}$$

Setting the two equations above equal to each other yields $d_o/d_i = (d_o - f)/f$. Rearranging this result gives the ***mirror equation:***

Mirror equation
$$\frac{1}{d_o} + \frac{1}{d_i} = \frac{1}{f} \qquad (25.3)$$

We have derived this equation for a real image formed in front of a concave mirror. In this case, the image distance is a positive quantity, as are the object distance and the focal length. However, we have seen in the last section that a concave mirror can also form a virtual image, if the object is located between the focal point and the mirror. Equation 25.3 can also be applied to such a situation, provided that we adopt the convention that d_i is negative for an image behind the mirror, as it is for a virtual image.

In deriving the magnification equation, we remember that the ***magnification*** *m* of a mirror is the ratio of the image height to the object height: $m = h_i/h_o$. If the image height is less than the object height, the magnitude of *m* is less than one, and if the image is larger than the object, the magnitude of *m* is greater than one. We have already shown that $h_o/(-h_i) = d_o/d_i$, so it follows that

Magnification equation
$$m = \frac{\text{Image height, } h_i}{\text{Object height, } h_o} = -\frac{d_i}{d_o} \qquad (25.4)$$

As Examples 3 and 4 show, the value of *m* is negative if the image is inverted and positive if the image is upright.

The reflections of these three friends in this fun-house mirror are distorted because the mirror is curved and not flat. It is not a spherical mirror, however. (Gail Mooney/Masterfile)

 Example 3   A Real Image Formed by a Concave Mirror

A 2.0-cm-high object is placed 7.10 cm from a concave mirror whose radius of curvature is 10.20 cm. Find **(a)** the location of the image and **(b)** its size.

Reasoning Since $f = \frac{1}{2}R = \frac{1}{2}(10.20 \text{ cm}) = 5.10 \text{ cm}$, the object is located between the focal point *F* and the center of curvature *C* of the mirror, as in Figure 25.18a. Based on this figure, we expect that the image is real and that, relative to the object, it is farther away from the mirror, inverted, and larger.

Solution **(a)** With $d_o = 7.10$ cm and $f = 5.10$ cm, the mirror equation (Equation 25.3) can be used to find the image distance:

$$\frac{1}{d_i} = \frac{1}{f} - \frac{1}{d_o} = \frac{1}{5.10 \text{ cm}} - \frac{1}{7.10 \text{ cm}} = 0.055 \text{ cm}^{-1} \quad \text{or} \quad \boxed{d_i = 18 \text{ cm}}$$

In this calculation, *f* and d_o are positive numbers, indicating that the focal point and the object are in front of the mirror. The positive answer for d_i means that the image is also in front of the mirror, and the reflected rays actually pass through the image, as Figure 25.18a shows. In other words, the positive value for d_i indicates that the image is a real image.

Problem-solving insight

According to the mirror equation, the image distance d_i has a reciprocal given by $d_i^{-1} = f^{-1} - d_o^{-1}$. After combining the reciprocals f^{-1} and d_o^{-1}, do not forget to take the reciprocal of the result to find d_i.

(**b**) According to the magnification equation (Equation 25.4), the image height h_i is related to the object height h_o and the magnification m by $h_i = mh_o$, where $m = -d_i/d_o$. Thus, we find that

$$h_i = -\left(\frac{d_i}{d_o}\right)h_o = -\left(\frac{18 \text{ cm}}{7.10 \text{ cm}}\right)(2.0 \text{ cm}) = \boxed{-5.1 \text{ cm}}$$

The negative value for h_i indicates that the image is inverted with respect to the object, as in Figure 25.18a.

 Example 4   A Virtual Image Formed by a Concave Mirror

An object is placed 6.00 cm in front of a concave mirror that has a 10.0-cm focal length. (**a**) Determine the location of the image. (**b**) The object is 1.2 cm high. Find the image height.

Reasoning The object is located between the focal point and the mirror, as in Figure 25.19a. The setup is analogous to a person using a makeup or shaving mirror. Therefore, we expect that the image is virtual and that, relative to the object, it is upright and larger.

Solution (**a**) Using the mirror equation with $d_o = 6.00$ cm and $f = 10.0$ cm, we have

$$\frac{1}{d_i} = \frac{1}{f} - \frac{1}{d_o} = \frac{1}{10.0 \text{ cm}} - \frac{1}{6.00 \text{ cm}} = -0.067 \text{ cm}^{-1} \quad \text{or} \quad \boxed{d_i = -15 \text{ cm}}$$

The answer for d_i is negative, indicating that the image is *behind* the mirror. Thus, as expected, the image is virtual.

(**b**) The image height h_i can be found from the magnification equation, which indicates that $h_i = mh_o$, where h_o is the object height and $m = -d_i/d_o$. It follows, then, that

$$h_i = -\left(\frac{d_i}{d_o}\right)h_o = -\left(\frac{-15 \text{ cm}}{6.00 \text{ cm}}\right)(1.2 \text{ cm}) = \boxed{3.0 \text{ cm}}$$

The image is larger than the object, and the positive value for h_i indicates that the image is upright (see Figure 25.19a).

CONVEX MIRRORS

The mirror equation and the magnification equation can also be used with convex mirrors, provided the focal length f is taken to be a *negative number,* as indicated explicitly in Equation 25.2. One way to remember this is to recall that the focal point of a convex mirror lies *behind* the mirror. Example 5 deals with a convex mirror.

 Example 5   A Virtual Image Formed by a Convex Mirror

A convex mirror is used to reflect light from an object placed 66 cm in front of the mirror. The focal length of the mirror is $f = -46$ cm (note the minus sign). Find (**a**) the location of the image and (**b**) the magnification.

Reasoning We have seen that a convex mirror always forms a virtual image, as in Figure 25.21a, where the image is upright and smaller than the object. These characteristics should also be indicated by the results of our analysis here.

Solution (**a**) With $d_o = 66$ cm and $f = -46$ cm, the mirror equation gives

$$\frac{1}{d_i} = \frac{1}{f} - \frac{1}{d_o} = \frac{1}{-46 \text{ cm}} - \frac{1}{66 \text{ cm}} = -0.037 \text{ cm}^{-1} \quad \text{or} \quad \boxed{d_i = -27 \text{ cm}}$$

The negative sign for d_i indicates that the image is behind the mirror and, therefore, is a virtual image.

(**b**) According to the magnification equation, the magnification is

$$m = -\frac{d_i}{d_o} = -\frac{(-27 \text{ cm})}{66 \text{ cm}} = \boxed{0.41}$$

The image is smaller (m is less than one) and upright (m is positive) with respect to the object.

Convex mirrors, like plane (flat) mirrors, always produce virtual images behind the mirror. However, the virtual image in a convex mirror is closer to the mirror than it would be if the mirror were planar, as Example 6 illustrates.

Example 6   A Convex Versus a Plane Mirror

An object is placed 9.00 cm in front of a mirror. The image is 3.00 cm closer to the mirror when the mirror is convex than when it is planar (see Figure 25.23). Find the focal length of the convex mirror.

Reasoning For a plane mirror, the image and the object are the same distance on either side of the mirror. Thus, the image would be 9.00 cm behind a plane mirror. If the image in a convex mirror is 3.00 cm closer than this, the image must be located 6.00 cm behind the convex mirror. In other words, when the object distance is $d_o = 9.00$ cm, the image distance for the convex mirror is $d_i = -6.00$ cm (negative because the image is virtual). The mirror equation can be used to find the focal length of the mirror.

Solution According to the mirror equation, the reciprocal of the focal length is

$$\frac{1}{f} = \frac{1}{d_o} + \frac{1}{d_i} = \frac{1}{9.00 \text{ cm}} + \frac{1}{-6.00 \text{ cm}} = -0.056 \text{ cm}^{-1} \quad \text{or} \quad \boxed{f = -18 \text{ cm}}$$

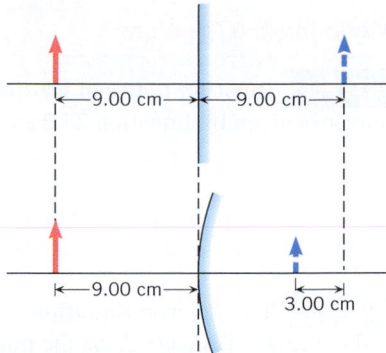

Figure 25.23 The object distance (9.00 cm) is the same for the plane mirror (top part of drawing) as for the convex mirror (bottom part of drawing). However, as discussed in Example 6, the image formed by the convex mirror is 3.00 cm closer to the mirror.

Contact lenses are worn to correct vision problems. Optometrists take advantage of the mirror equation and the magnification equation in providing lenses that fit the patient's eyes properly, as the next example illustrates.

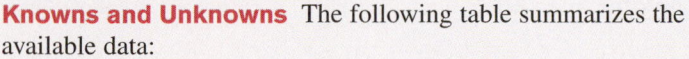

ANALYZING MULTIPLE-CONCEPT PROBLEMS

Example 7   Measuring the Curvature of the Cornea of the Eye

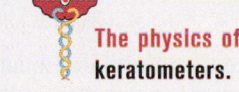

The physics of keratometers.

A contact lens rests against the cornea of the eye. Figure 25.24 shows an optometrist using a keratometer to measure the radius of curvature of the cornea, thereby ensuring that the prescribed lenses fit accurately. In the keratometer, light from an illuminated object reflects from the corneal surface, which acts like a convex mirror and forms an upright virtual image that is smaller than the object (see Figure 25.21a). With the object placed 9.0 cm in front of the cornea, the magnification of the corneal surface is measured to be 0.046. Determine the radius of the cornea.

Reasoning The radius of a convex mirror can be determined from the mirror's focal length, since the two are related. The focal length is related to the distances of the object and its image from the mirror via the mirror equation. The magnification of the mirror is also related to the object and image distances according to the magnification equation. By using the mirror equation and the magnification equation, we will be able to determine the focal length and, hence, the radius.

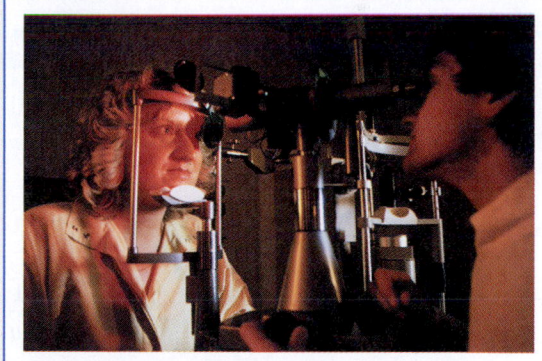

Figure 25.24 An optometrist is using a keratometer to measure the radius of curvature of the cornea of the eye, which is the surface against which a contact lens rests. (Andrew McLenaghan/SPL/Photo Researchers, Inc.).

Knowns and Unknowns The following table summarizes the available data:

Description	Symbol	Value	Comment
Object distance	d_o	9.0 cm	Distance of object from cornea.
Magnification of corneal surface	m	0.046	Cornea acts like a convex mirror and forms a virtual image.
Unknown Variable			
Radius of cornea	R	?	

Continued

Modeling the Problem

STEP 1 **Relation between Radius and Focal Length** The focal length f of a convex mirror is given by Equation 25.2 as

$$f = -\tfrac{1}{2}R$$

where R is the radius of the spherical surface. Solving this expression for the radius gives Equation 1 at the right. In Step 2, we determine the unknown focal length.

$$R = -2f \tag{1}$$

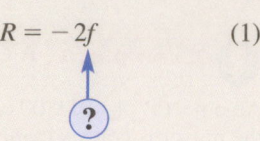

STEP 2 **The Mirror Equation** The focal length is related to the object distance d_o and the image distance d_i via the mirror equation, which specifies that

$$\frac{1}{f} = \frac{1}{d_o} + \frac{1}{d_i} \tag{25.3}$$

Solving this equation for f gives

$$f = \left(\frac{1}{d_o} + \frac{1}{d_i} \right)^{-1}$$

which can be substituted into Equation 1 as shown in the right column. A value for d_o is given in the data table, and we turn to Step 3 to determine a value for d_i.

$$R = -2f \tag{1}$$

$$f = \left(\frac{1}{d_o} + \frac{1}{d_i} \right)^{-1} \tag{2}$$

STEP 3 **The Magnification Equation** According to the magnification equation, the magnification m is given by

$$m = -\frac{d_i}{d_o} \tag{25.4}$$

Solving for d_i, we obtain

$$d_i = -md_o$$

and can substitute the result into Equation 2, as shown at the right.

$$R = -2f \tag{1}$$

$$f = \left(\frac{1}{d_o} + \frac{1}{d_i} \right)^{-1} \tag{2}$$

$$d_i = -md_o$$

Solution Combining the results of each step algebraically, we find that

| STEP 1 | STEP 2 | | STEP 3 |

$$R = -2f = -2\left(\frac{1}{d_o} + \frac{1}{d_i} \right)^{-1} = -2\left[\frac{1}{d_o} + \frac{1}{(-md_o)} \right]^{-1}$$

Rearranging terms gives

$$R = -2\left[\frac{1}{d_o} + \frac{1}{(-md_o)} \right]^{-1} = -2\left[\frac{1}{d_o}\left(1 - \frac{1}{m}\right) \right]^{-1} = 2\left[\frac{1}{d_o}\left(\frac{1-m}{m}\right) \right]^{-1} = \frac{2d_o m}{1-m}$$

Thus, the radius is

$$R = \frac{2d_o m}{1-m} = \frac{2(9.0 \text{ cm})(0.046)}{1 - 0.046} = \boxed{0.87 \text{ cm}}$$

Related Homework: *Problems 31, 44*

The following Reasoning Strategy summarizes the sign conventions that are used with the mirror equation and the magnification equation. These conventions apply to both concave and convex mirrors.

REASONING STRATEGY

Summary of Sign Conventions for Spherical Mirrors

Focal length

f is + for a concave mirror.

f is − for a convex mirror.

Object distance

d_o is + if the object is in front of the mirror (real object).

d_o is − if the object is behind the mirror (virtual object).*

Image distance

d_i is + if the image is in front of the mirror (real image).

d_i is − if the image is behind the mirror (virtual image).

Magnification

m is + for an image that is upright with respect to the object.

m is − for an image that is inverted with respect to the object.

✓ CHECK YOUR UNDERSTANDING

(The answers are given at the end of the book.)

13. An object is placed in front of a spherical mirror, and the magnification of the system is $m = -6$. What does this number tell you about the image? (Select one or more of the following choices.) **(a)** The image is larger than the object. **(b)** The image is smaller than the object. **(c)** The image is upright relative to the object. **(d)** The image is inverted relative to the object. **(e)** The image is a real image. **(f)** The image is a virtual image.

14. Concept Simulation 25.3 at **www.wiley.com/college/cutnell** reviews the concepts that are important in this question. Plane mirrors and convex mirrors form virtual images. With a plane mirror, the image may be infinitely far behind the mirror, depending on where the object is located in front of the mirror. For an object in front of a single convex mirror, what is the greatest distance behind the mirror at which the image can be found?

25.7 CONCEPTS & CALCULATIONS

Relative to the object in front of a spherical mirror, the image can differ in a number of respects. The image can be real (in front of the mirror) or virtual (behind the mirror). It can be larger or smaller than the object, and it can be upright or inverted. As you solve problems dealing with spherical mirrors, keep these image characteristics in mind. They can help guide you to the correct answer, as Examples 8 and 9 illustrate.

Concepts & Calculations Example 8

 Finding the Focal Length

An object is located 7.0 cm in front of a mirror. The virtual image is located 4.5 cm away from the mirror and is smaller than the object. Find the focal length of the mirror.

Concept Questions and Answers Based solely on the fact that the image is virtual, is the mirror concave or convex, or is either type possible?

>*Answer* Either type is possible. A concave mirror can form a virtual image if the object is within the focal point of the mirror, as Figure 25.19a illustrates. A convex mirror always forms a virtual image, as Figure 25.21a shows.

---

*Sometimes optical systems use two (or more) mirrors, and the image formed by the first mirror serves as the object for the second mirror. Occasionally, such an object falls *behind* the second mirror. In this case the object distance is negative, and the object is said to be a virtual object.

The image is smaller than the object, as well as virtual. Do these characteristics together indicate a concave or convex mirror, or do they indicate either type?

Answer They indicate a convex mirror. A concave mirror can produce an image that is smaller than the object if the object is located beyond the center of curvature of the mirror, as in Figure 25.18*b*. However, the image in Figure 25.18*b* is real, not virtual. A convex mirror, in contrast, always produces an image that is virtual and smaller than the object, as Figure 25.21*a* illustrates.

Is the focal length positive or negative?

Answer The focal length is negative because the mirror is convex. A concave mirror has a positive focal length.

Solution The virtual image is located behind the mirror and, therefore, has a negative image distance, $d_i = -4.5$ cm. Using this value together with the object distance of $d_o = 7.0$ cm, we can apply the mirror equation to find the focal length:

$$\frac{1}{f} = \frac{1}{d_o} + \frac{1}{d_i} = \frac{1}{7.0 \text{ cm}} + \frac{1}{-4.5 \text{ cm}} = -0.079 \text{ cm}^{-1} \quad \text{or} \quad \boxed{f = -13 \text{ cm}}$$

As expected, the focal length is negative.

 Concepts & Calculations Example 9 Two Choices

The radius of curvature of a mirror is 24 cm. A diamond ring is placed in front of this mirror. The image is twice the size of the ring. Find the object distance of the ring.

Concept Questions and Answers Is the mirror concave or convex, or is either type possible?

Answer A convex mirror always forms an image that is smaller than the object, as Figure 25.21*a* shows. Therefore, the mirror must be concave.

How many places are there in front of a concave mirror where the ring can be placed and produce an image that is twice the size of the object?

Answer There are two places. Figure 25.18*a* illustrates that one of the places is between the center of curvature and the focal point. The enlarged image is real and inverted. Figure 25.19*a* shows that another possibility is between the focal point and the mirror, in which case the enlarged image is virtual and upright.

What are the possible values for the magnification of the image of the ring?

Answer Since the image is inverted in Figure 25.18*a*, the magnification for this possibility is $m = -2$. In Figure 25.19*a*, however, the image is upright, so the magnification is $m = +2$ for this option. In either case, the image is twice the size of the ring.

Solution According to the mirror equation and the magnification equation, we have

$$\underbrace{\frac{1}{d_o} + \frac{1}{d_i} = \frac{1}{f}}_{\text{Mirror equation}} \quad \text{and} \quad \underbrace{m = -\frac{d_i}{d_o}}_{\text{Magnification equation}}$$

We can solve the magnification equation for the image distance and obtain $d_i = -md_o$. Substituting this expression for d_i into the mirror equation gives

$$\frac{1}{d_o} + \frac{1}{(-md_o)} = \frac{1}{f} \quad \text{or} \quad d_o = \frac{f(m-1)}{m}$$

Applying this result with the two magnifications (and noting that $f = \frac{1}{2}R = 12$ cm), we obtain

$$m = -2 \qquad d_o = \frac{f(m-1)}{m} = \frac{(12 \text{ cm})(-2-1)}{-2} = \boxed{18 \text{ cm}}$$

$$m = +2 \qquad d_o = \frac{f(m-1)}{m} = \frac{(12 \text{ cm})(+2-1)}{+2} = \boxed{6.0 \text{ cm}}$$

CONCEPT SUMMARY

If you need more help with a concept, use the Learning Aids noted next to the discussion or equation. Examples (**Ex.**) are in the text of this chapter. Go to **www.wiley.com/college/cutnell** for the following Learning Aids:

Interactive LearningWare (ILW) — Additional examples solved in a five-step interactive format.

Concept Simulations (CS) — Animated text figures or animations of important concepts.

Interactive Solutions (IS) — Models for certain types of problems in the chapter homework. The calculations are carried out interactively.

Topic	Discussion	Learning Aids
Wave fronts Plane waves	**25.1 WAVE FRONTS AND RAYS** Wave fronts are surfaces on which all points of a wave are in the same phase of motion. Waves whose wave fronts are flat surfaces are known as plane waves.	
Rays	Rays are lines that are perpendicular to the wave fronts and point in the direction of the velocity of the wave.	

Law of reflection	**25.2 THE REFLECTION OF LIGHT** When light reflects from a smooth surface, the reflected light obeys the law of reflection:
	
	  a. The incident ray, the reflected ray, and the normal to the surface all lie in the same plane.
	  b. The angle of reflection θ_r equals the angle of incidence θ_i; $\theta_r = \theta_i$.	

Virtual image	**25.3 THE FORMATION OF IMAGES BY A PLANE MIRROR** A virtual image is one from which all the rays of light do not actually come, but only appear to do so.	
Real image	A real image is one from which all the rays of light actually do emanate.	
Plane mirror	A plane mirror forms an upright, virtual image that is located as far behind the mirror as the object is in front of it. In addition, the heights of the image and the object are equal.	**CS 25.1** **Ex. 1, 2**
Concave and convex mirrors	**25.4 SPHERICAL MIRRORS** A spherical mirror has the shape of a section from the surface of a sphere. If the inside surface of the mirror is polished, it is a concave mirror. If the outside surface is polished, it is a convex mirror.	
Principal axis Paraxial rays	The principal axis of a mirror is a straight line drawn through the center of curvature and the middle of the mirror's surface. Rays that are close to the principal axis are known as paraxial rays. Paraxial rays are not necessarily parallel to the principal axis.	
Radius of curvature	The radius of curvature R of a mirror is the distance from the center of curvature to the mirror.	
Focal point of a concave mirror	The focal point of a concave spherical mirror is a point on the principal axis, in front of the mirror. Incident paraxial rays that are parallel to the principal axis converge to the focal point after being reflected from the concave mirror.	
Focal point of a convex mirror	The focal point of a convex spherical mirror is a point on the principal axis, behind the mirror. For a convex mirror, incident paraxial rays that are parallel to the principal axis diverge after reflecting from the mirror. These rays seem to originate from the focal point.	
Spherical aberration	The fact that a spherical mirror does not bring all rays parallel to the principal axis to a single image point after reflection is known as spherical aberration.	
Focal length	The focal length f of a mirror is the distance along the principal axis between the focal point and the mirror. The focal length and the radius of curvature R are related by	
Concave mirror	$$f = \tfrac{1}{2}R \qquad\qquad (25.1)$$	
Convex mirror	$$f = -\tfrac{1}{2}R \qquad\qquad (25.2)$$	
Ray tracing	**25.5 THE FORMATION OF IMAGES BY SPHERICAL MIRRORS** The image produced by a mirror can be located by a graphical method known as ray tracing.	
	For a concave mirror, the following paraxial rays are especially useful for ray tracing (see Figure 25.17):	**CS 25.2**
Three rays for a concave mirror	**Ray 1.** This ray leaves the object traveling parallel to the principal axis. The ray reflects from the mirror and passes through the focal point. **Ray 2.** This ray leaves the object and passes through the focal point. The ray reflects from the mirror and travels parallel to the principal axis. **Ray 3.** This ray leaves the object and travels along a line that passes through the center of curvature. The ray strikes the mirror perpendicularly and reflects back on itself.	

Topic	Discussion	Learning Aids
	For a convex mirror, these paraxial rays are useful for ray tracing (see Figure 25.21a):	CS 25.3
Three rays for a convex mirror	**Ray 1.** This ray leaves the object traveling parallel to the principal axis. After reflection from the mirror, the ray appears to originate from the focal point of the mirror.	
	Ray 2. This ray leaves the object and heads toward the focal point. After reflection, the ray travels parallel to the principal axis.	
	Ray 3. This ray leaves the object and travels toward the center of curvature. The ray strikes the mirror perpendicularly and reflects back on itself.	

25.6 THE MIRROR EQUATION AND THE MAGNIFICATION EQUATION   The mirror equation specifies the relation between the object distance d_o, the image distance d_i, and the focal length f of the mirror: **Ex. 3, 4**

Mirror equation

$$\frac{1}{d_o} + \frac{1}{d_i} = \frac{1}{f}$$ (25.3)   **ILW 25.1**

The mirror equation can be used with either concave or convex mirrors.   **Ex. 5–9**

The magnification m of a mirror is the ratio of the image height h_i to the object height h_o: **IS 25.21**

Magnification

$$m = \frac{h_i}{h_o}$$ **IS 25.45**

The magnification is also related to d_i and d_o by the magnification equation:

Magnification equation

$$m = -\frac{d_i}{d_o}$$ (25.4)

The algebraic sign conventions for the variables appearing in these equations are summarized in the Reasoning Strategy at the end of Section 25.6.

FOCUS ON CONCEPTS

Note to Instructors: The numbering of the questions shown here reflects the fact that they are only a representative subset of the total number that are available online. However, all of the questions are available for assignment via an online homework management program such as WileyPLUS or WebAssign.

Section 25.1   Wave Fronts and Rays

2. A ray is _____.   **(a)** always parallel to other rays   **(b)** parallel to the velocity of the wave   **(c)** perpendicular to the velocity of the wave   **(d)** parallel to the wave fronts

Section 25.2   The Reflection of Light

4. The drawing shows a top view of an object located to the right of a mirror. A single ray of light is shown leaving the object. After reflection from the mirror, through which location, A, B, C, or D, does the ray pass?   **(a)** A   **(b)** B   **(c)** C   **(d)** D

Question 4

Section 25.3   The Formation of Images by a Plane Mirror

6. A friend is standing 2 m in front of a plane mirror. You are standing 3 m directly behind your friend. What is the distance between you and the image of your friend?   **(a)** 2 m   **(b)** 3 m   **(c)** 5 m   **(d)** 7 m   **(e)** 10 m

7. You hold the words **TOP DOG** in front of a plane mirror. What does the image of these words look like?   **(a)** ꓷОТ ꓨОꓷ   **(b)** ꓨОꓷ ТОꓑ   **(c)** **TOP DOG**   **(d)** **DOG TOP**   **(e)** ꓕОꓑ ꓷОꓨ

Section 25.4   Spherical Mirrors

8. Rays of light coming from the sun (a very distant object) are near and parallel to the principal axis of a concave mirror. After reflecting from the mirror, where will the rays cross each other at a single point? The rays _____.   **(a)** will not cross each other after reflecting from a concave mirror   **(b)** will cross at the point where the principal axis intersects the mirror   **(c)** will cross at the center of curvature   **(d)** will cross at the focal point   **(e)** will cross at a point beyond the center of curvature

Section 25.5   The Formation of Images by Spherical Mirrors

12. Which one of the following statements concerning spherical mirrors is correct?   **(a)** Only a convex mirror can produce an enlarged image.   **(b)** Both concave and convex mirrors can produce an enlarged image.   **(c)** Only a concave mirror can produce an enlarged image, provided the object distance is less than the radius of curvature.   **(d)** Only a concave mirror can produce an enlarged image, provided the object distance is greater than the radius of curvature.

13. Suppose that you hold up a small convex mirror in front of your face. Which answer describes the image of your face?   **(a)** Virtual, inverted   **(b)** Virtual, upright   **(c)** Virtual, enlarged   **(d)** Real, inverted   **(e)** Real, reduced in size

Section 25.6   The Mirror Equation and the Magnification Equation

14. An object is placed at a known distance in front of a mirror whose focal length is also known. You apply the mirror equation and find that the image distance is a negative number. This result tells you that _____.   **(a)** the image is larger than the object   **(b)** the image is smaller than the object   (c) the image is inverted relative to the object   (d) the image is real   (e) the image is virtual

15. An object is situated at a known distance in front of a convex mirror whose focal length is also known. A friend of yours does a calcu-

lation that shows that the magnification is -2. After some thought, you conclude correctly that _____.   (a) your friend's answer is correct   (b) the magnification should be $+2$ (c) the magnification should be $+\frac{1}{2}$ (d) the magnification should be $-\frac{1}{2}$

PROBLEMS

Note to Instructors: Most of the homework problems in this chapter are available for assignment via an online homework management program such as WileyPLUS or WebAssign, and those marked with the icon **GO** *are presented in WileyPLUS using a guided tutorial format that provides enhanced interactivity. See Preface for additional details.*

ssm Solution is in the Student Solutions Manual.
www Solution is available online at www.wiley.com/college/cutnell

This icon represents a biomedical application.

Section 25.2 The Reflection of Light,
Section 25.3 The Formation of Images by a Plane Mirror

1. ssm www Two diverging light rays, originating from the same point, have an angle of 10° between them. After the rays reflect from a plane mirror, what is the angle between them? Construct one possible ray diagram that supports your answer.

2. Review Conceptual Example 2. Suppose that in Figure 25.9b the two perpendicular plane mirrors are represented by the $-x$ and $-y$ axes of an x, y coordinate system. An object is in front of these mirrors at a point whose coordinates are $x = -2.0$ m and $y = -1.0$ m. Find the coordinates that locate each of the three images.

3. You are trying to photograph a bird sitting on a tree branch, but a tall hedge is blocking your view. However, as the drawing shows, a plane mirror reflects light from the bird into your camera. For what distance must you set the focus of the camera lens in order to snap a sharp picture of the bird's image?

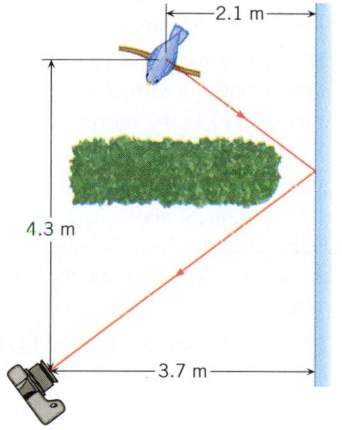

Problem 3

4. GO Suppose that you are walking perpendicularly with a velocity of $+0.90$ m/s toward a stationary plane mirror. What is the velocity of your image relative to you? The direction in which you walk is the positive direction.

5. ssm Two plane mirrors are separated by 120°, as the drawing illustrates. If a ray strikes mirror M_1 at a 65° angle of incidence, at what angle θ does it leave mirror M_2?

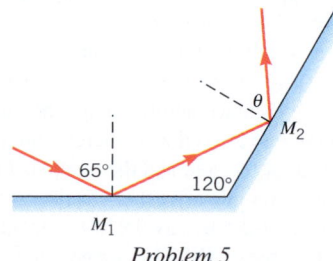

Problem 5

6. Review Conceptual Example 1 before attempting this problem. A person whose eyes are 1.70 m above the floor stands in front of a plane mirror. The top of her head is 0.12 m above her eyes.   **(a)** What is the height of the shortest mirror in which she can see her entire image?   **(b)** How far above the floor should the bottom edge of the mirror be placed?

7. Review Conceptual Example 1 as an aid in understanding this problem. The drawings show two arrows, A and B, that are located in

front of a plane mirror. A person at point P is viewing the image of each arrow. Which images can be seen in their entirety? Determine your answers by drawing a ray from the head and foot of each arrow that reflects from the mirror according to the law of reflection and reaches point P. Only if *both* rays reach point P after reflection can the image of that arrow be seen in its entirety.

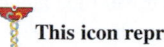

*8. GO A small mirror is attached to a vertical wall, and it hangs a distance of 1.80 m above the floor. The mirror is facing due east, and a ray of sunlight strikes the mirror early in the morning and then again later in the morning. The incident and reflected rays lie in a plane that is perpendicular to both the wall and the floor. Early in the morning, the reflected ray strikes the floor at a distance of 3.86 m from the base of the wall. Later on in the morning, the ray is observed to strike the floor at a distance of 1.26 m from the wall. The earth rotates at a rate of 15.0° per hour. How much time (in hours) has elapsed between the two observations?

*9. A ray of light strikes a plane mirror at a 45° angle of incidence. The mirror is then rotated by 15° into the position shown in red in the drawing, while the incident ray is kept fixed.   **(a)** Through what angle ϕ does the reflected ray rotate?   **(b)** What is the answer to part (a) if the angle of incidence is 60° instead of 45°?

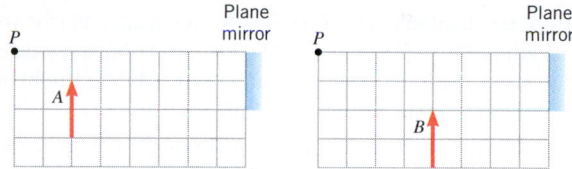

*10. In an experiment designed to measure the speed of light, a laser is aimed at a mirror that is 50.0 km due north. A detector is placed 117 m due east of the laser. The mirror is to be aligned so that light from the laser reflects into the detector.   **(a)** When properly aligned, what angle should the normal to the surface of the mirror make with due south?   **(b)** Suppose the mirror is misaligned, so that the actual angle between the normal to the surface and due south is too large by 0.004°. By how many meters (due east) will the reflected ray miss the detector?

*11. GO You walk at an angle of $\theta = 50.0°$ toward a plane mirror, as in the drawing. Your walking velocity has a magnitude of 0.90 m/s. What is the velocity of your image relative to you (magnitude and direction)?

Problem 11

**12. A lamp is twice as far in front of a plane mirror as a person is. Light from the lamp reaches the person via two paths, reflected and direct. It strikes the mirror at a 30.0° angle of incidence and reflects from it before reaching the person. The total time for the light to travel this path includes the time to travel to the mirror and the time to travel from the mirror to the person. The light also travels directly to the person without reflecting. Find the ratio of the total travel time along the reflected path to the travel time along the direct path.

Section 25.4 Spherical Mirrors,
Section 25.5 The Formation of Images by Spherical Mirrors

13. When an object is located very far away from a convex mirror, the image of the object is 18 cm behind the mirror. Using a ray diagram drawn to scale, determine where the image is located when the object is placed 9.0 cm in front of the mirror. Note that the mirror must be drawn to scale also. In your drawing, assume that the height of the object is 3.0 cm.

14. An object is placed 11 cm in front of a concave mirror whose focal length is 18 cm. The object is 3.0 cm tall. Using a ray diagram drawn to scale, measure **(a)** the location and **(b)** the height of the image. The mirror must be drawn to scale.

15. ssm The radius of curvature of a convex mirror is 1.00×10^2 cm. An object that is 10.0 cm high is placed 25.0 cm in front of this mirror. Using a ray diagram drawn to scale, measure **(a)** the location and **(b)** the height of the image. The mirror must be drawn to scale.

16. An object is placed in front of a convex mirror. Draw the convex mirror (radius of curvature = 15 cm) to scale, and place the object 25 cm in front of it. Make the object height 4 cm. Using a ray diagram, locate the image and measure its height. Now move the object closer to the mirror, so the object distance is 5 cm. Again, locate its image using a ray diagram. As the object moves closer to the mirror, **(a)** does the magnitude of the image distance become larger or smaller, and **(b)** does the magnitude of the image height become larger or smaller? **(c)** What is the ratio of the image height when the object distance is 5 cm to its height when the object distance is 25 cm? Give your answer to one significant figure.

17. Concept Simulation 25.3 at www.wiley.com/college/cutnell illustrates the concepts pertinent to this problem. A convex mirror has a focal length of −40.0 cm. A 12.0-cm-tall object is located 40.0 cm in front of this mirror. Using a ray diagram drawn to scale, determine the **(a)** location and **(b)** size of the image. Note that the mirror must be drawn to scale.

*18. A plane mirror and a concave mirror ($f = 8.0$ cm) are facing each other and are separated by a distance of 20.0 cm. An object is placed between the mirrors and is 10.0 cm from each mirror. Consider the light from the object that reflects first from the plane mirror and then from the concave mirror. Using a ray diagram drawn to scale, find the location of the image that this light produces in the concave mirror. Specify this distance relative to the concave mirror.

Section 25.6 The Mirror Equation
and the Magnification Equation

19. ssm The image produced by a concave mirror is located 26 cm in front of the mirror. The focal length of the mirror is 12 cm. How far in front of the mirror is the object located?

20. An object that is 25 cm in front of a convex mirror has an image located 17 cm behind the mirror. How far behind the mirror is the image located when the object is 19 cm in front of the mirror?

21. Refer to **Interactive Solution 25.21** at www.wiley.com/college/cutnell for help with problems like this one. A concave mirror ($R = 56.0$ cm) is used to project a transparent slide onto a wall. The slide is located at a distance of 31.0 cm from the mirror, and a small flashlight shines light through the slide and onto the mirror. The setup is similar to that in Figure 25.18a. **(a)** How far from the wall should the mirror be located? **(b)** The height of the object on the slide is 0.95 cm. What is the height of the image? **(c)** How should the slide be oriented, so that the picture on the wall looks normal?

22. GO A small statue has a height of 3.5 cm and is placed in front of a concave mirror. The image of the statue is inverted, 1.5 cm tall, and located 13 cm in front of the mirror. Find the focal length of the mirror.

23. A mirror produces an image that is located 34.0 cm behind the mirror when the object is located 7.50 cm in front of the mirror. What is the focal length of the mirror, and is the mirror concave or convex?

24. GO ssm A concave mirror has a focal length of 42 cm. The image formed by this mirror is 97 cm in front of the mirror. What is the object distance?

25. The outside mirror on the passenger side of a car is convex and has a focal length of −7.0 m. Relative to this mirror, a truck traveling in the rear has an object distance of 11 m. Find **(a)** the image distance of the truck and **(b)** the magnification of the mirror.

26. GO A convex mirror has a focal length of −27.0 cm. Find the magnification produced by the mirror when the object distance is 9.0 cm and 18.0 cm.

27. ssm www When viewed in a spherical mirror, the image of a setting sun is a virtual image. The image lies 12.0 cm behind the mirror. **(a)** Is the mirror concave or convex? Why? **(b)** What is the radius of curvature of the mirror?

28. A drop of water on a countertop reflects light from a flower held 3.0 cm directly above it. The flower's diameter is 2.0 cm, and the diameter of the flower's image is 0.10 cm. What is the focal length of the water drop, assuming that it may be treated as a convex spherical mirror?

*29. ssm An object is located 14.0 cm in front of a convex mirror, the image being 7.00 cm behind the mirror. A second object, twice as tall as the first one, is placed in front of the mirror, but at a different location. The image of this second object has the same height as the other image. How far in front of the mirror is the second object located?

*30. GO A tall tree is growing across a river from you. You would like to know the distance between yourself and the tree, as well as its height, but are unable to make the measurements directly. However, by using a mirror to form an image of the tree and then measuring the image distance and the image height, you can calculate the distance to the tree as well as its height. Suppose that this mirror produces an image of the sun, and the image is located 0.9000 m from the mirror. The same mirror is then used to produce an image of the tree. The image of the tree is 0.9100 m from the mirror. **(a)** How far away is the tree? **(b)** The image height of the tree has a magnitude of 0.12 m. How tall is the tree?

*31. Multiple-Concept Example 7 reviews the concepts that are needed to solve this problem. In Figure 25.20b the head-up display is designed so that the distance between the digital readout device and virtual image 1 is 2.00 m. The magnification of virtual image 1 is 4.00. Find the focal length of the concave mirror. (*Hint: Remember that the image distance for virtual image 1 is a negative quantity.*)

* **32.**  A candle is placed 15.0 cm in front of a convex mirror. When the convex mirror is replaced with a plane mirror, the image moves 7.0 cm farther away from the mirror. Find the focal length of the convex mirror.

* **33.** A spherical mirror is polished on both sides. When the convex side is used as a mirror, the magnification is +1/4. What is the magnification when the concave side is used as a mirror, the object remaining the same distance from the mirror?

** **34.** A man holds a double-sided spherical mirror so that he is looking directly into its convex surface, 45 cm from his face. The magnification of the image of his face is +0.20. What will be the

image distance when he reverses the mirror (looking into its concave surface), maintaining the same distance between the mirror and his face? Be sure to include the algebraic sign (+ or −) with your answer.

** **35.** A spacecraft is in a circular orbit about the moon, 1.22×10^5 m above its surface. The speed of the spacecraft is 1620 m/s, and the radius of the moon is 1.74×10^6 m. If the moon were a smooth, reflective sphere, **(a)** how far below the moon's surface would the image of the spacecraft appear, and **(b)** what would be the apparent speed of the spacecraft's image? (*Hint: Both the spacecraft and its image have the same angular speed about the center of the moon.*)

ADDITIONAL PROBLEMS

WILEY ⊕
PLUS

36. Concept Simulation 25.2 at **www.wiley.com/college/cutnell** illustrates the concepts pertinent to this problem. A 2.0-cm-high object is situated 15.0 cm in front of a concave mirror that has a radius of curvature of 10.0 cm. Using a ray diagram drawn to scale, measure **(a)** the location and **(b)** the height of the image. The mirror must be drawn to scale.

37. ssm The image behind a convex mirror (radius of curvature = 68 cm) is located 22 cm from the mirror. **(a)** Where is the object located and **(b)** what is the magnification of the mirror? Determine whether the image is **(c)** upright or inverted and **(d)** larger or smaller than the object.

38. A concave mirror ($f = 45$ cm) produces an image whose distance from the mirror is one-third the object distance. Determine **(a)** the object distance and **(b)** the (positive) image distance.

39. The drawing shows a laser beam shining on a plane mirror that is perpendicular to the floor. The beam's angle of incidence is 33.0°. The beam emerges from the laser at a point that is 1.10 m from the mirror and 1.80 m above the floor. After reflection, how far from the base of the mirror does the beam strike the floor?

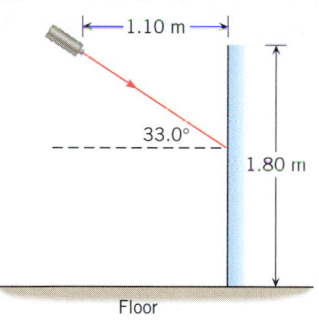

40. A concave mirror has a focal length of 12 cm. This mirror forms an image located 36 cm in front of the mirror. What is the magnification of the mirror?

41. ssm www A small postage stamp is placed in front of a concave mirror (radius = R) so that the image distance equals the object distance. **(a)** In terms of R, what is the object distance? **(b)** What is the magnification of the mirror? **(c)** State whether the image is upright or inverted relative to the object. Draw a ray diagram to guide your thinking.

* **42.** Two plane mirrors are facing each other. They are parallel, 3.00 cm apart, and 17.0 cm in length, as the drawing indicates. A laser beam is directed at the top mirror from the left edge of the bottom mirror.

What is the smallest angle of incidence with respect to the top mirror, such that the laser beam **(a)** hits only one of the mirrors and **(b)** hits each mirror only once?

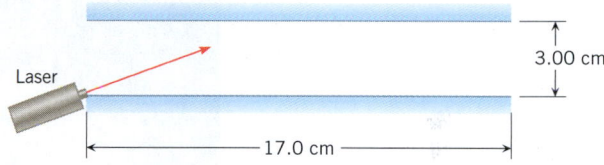

* **43. ssm www** Identical objects are located at the same distance from two spherical mirrors, A and B. The magnifications produced by the mirrors are $m_A = 4.0$ and $m_B = 2.0$. Find the ratio f_A/f_B of the focal lengths of the mirrors.

* **44.**  Consult Multiple-Concept Example 7 to see a model for solving this type of problem. A concave makeup mirror is designed so the virtual image it produces is twice the size of the object when the distance between the object and the mirror is 14 cm. What is the radius of curvature of the mirror?

* **45.** Consult **Interactive Solution 25.45** at **www.wiley.com/college/cutnell** for insight into this problem. An object is placed in front of a convex mirror, and the size of the image is one-fourth that of the object. What is the ratio d_o/f of the object distance to the focal length of the mirror?

** **46.** A concave mirror has a focal length of 30.0 cm. The distance between an object and its image is 45.0 cm. Find the object and image distances, assuming that **(a)** the object lies beyond the center of curvature and **(b)** the object lies within the focal point.

** **47. ssm** The drawing shows a top view of a square room. One wall is missing, and the wall on the right is a mirror. From point P in the center of the open side, a laser is pointed at the mirrored wall. At what angle of incidence must the light strike the right-hand wall so that, after being reflected, the light hits the left corner of the back wall?

THE REFRACTION OF LIGHT: LENSES AND OPTICAL INSTRUMENTS

The sparkle of properly cut diamonds is one of the reasons that they are treasured as gemstones. They sparkle because of the refraction of light. Refraction is the change in the direction of the light when it moves from one medium into another and is one of the central themes of this chapter. (© Matthias Kulka/zefa/Corbis)

Table 26.1 Index of Refraction[a] for Various Substances

Substance	Index of Refraction, n
Solids at 20 °C	
Diamond	2.419
Glass, crown	1.523
Ice (0 °C)	1.309
Sodium chloride	1.544
Quartz	
Crystalline	1.544
Fused	1.458
Liquids at 20 °C	
Benzene	1.501
Carbon disulfide	1.632
Carbon tetrachloride	1.461
Ethyl alcohol	1.362
Water	1.333
Gases at 0 °C, 1 atm	
Air	1.000 293
Carbon dioxide	1.000 45
Oxygen, O_2	1.000 271
Hydrogen, H_2	1.000 139

[a]Measured with light whose wavelength in a vacuum is 589 nm.

26.1 THE INDEX OF REFRACTION

As Section 24.3 discusses, light travels through a vacuum at a speed of $c = 3.00 \times 10^8$ m/s. It can also travel through many materials, such as air, water, and glass. Atoms in the material absorb, reemit, and scatter the light, however. Therefore, light travels through the material at a speed that is less than c, the actual speed depending on the nature of the material. In general, we will see that the change in speed as a ray of light goes from one material to another causes the ray to deviate from its incident direction. This change in direction is called *refraction.* To describe the extent to which the speed of light in a material medium differs from that in a vacuum, we use a parameter called the *index of refraction* (or *refractive index*). The index of refraction is an important parameter because it appears in Snell's law of refraction, which will be discussed in the next section. This law is the basis of all the phenomena discussed in this chapter.

DEFINITION OF THE INDEX OF REFRACTION

The index of refraction n of a material is the ratio of the speed c of light in a vacuum to the speed v of light in the material:

$$n = \frac{\text{Speed of light in a vacuum}}{\text{Speed of light in the material}} = \frac{c}{v} \qquad (26.1)$$

Table 26.1 lists the refractive indices for some common substances. The values of n are greater than unity because the speed of light in a material medium is less than it is in a vacuum. For example, the index of refraction for diamond is $n = 2.419$, so the speed of light in diamond is $v = c/n = (3.00 \times 10^8 \text{ m/s})/2.419 = 1.24 \times 10^8 \text{ m/s}$. In contrast, the index of refraction for air (and also for other gases) is so close to unity that $n_{air} = 1$ for most purposes. The index of refraction depends slightly on the wavelength of the light, and the values in Table 26.1 correspond to a wavelength of $\lambda = 589$ nm in a vacuum.

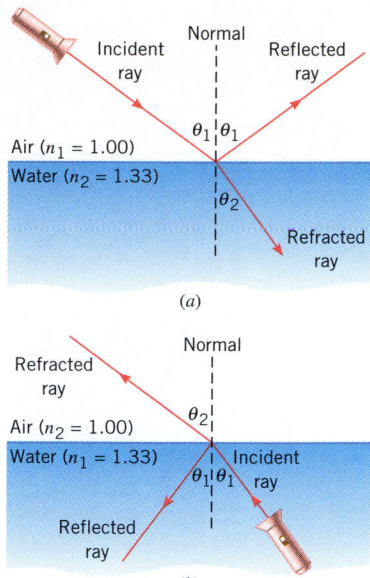

SNELL'S LAW AND THE REFRACTION OF LIGHT

26.2

SNELL'S LAW

When light strikes the interface between two transparent materials, such as air and water, the light generally divides into two parts, as Figure 26.1a illustrates. Part of the light is reflected, with the angle of reflection equaling the angle of incidence. The remainder is transmitted across the interface. If the incident ray does not strike the interface at normal incidence, the transmitted ray has a different direction than the incident ray. When a ray enters the second material and changes direction, it is said to be refracted and behaves in one of the following two ways:

1. When light travels from a medium where the refractive index is smaller into a medium where it is larger, the refracted ray is bent toward the normal, as in Figure 26.1a.

2. When light travels from a medium where the refractive index is larger into a medium where it is smaller, the refracted ray is bent away from the normal, as in Figure 26.1b.

These two possibilities illustrate that both the incident and refracted rays obey the principle of reversibility. Thus, the directions of the rays in part a of the drawing can be reversed to give the situation depicted in part b. In part b the reflected ray lies in the water rather than in the air.

In both parts of Figure 26.1 the angles of incidence, refraction, and reflection are measured relative to the normal. Note that the index of refraction of air is labeled n_1 in part a, whereas it is n_2 in part b, because *we label all variables associated with the incident (and reflected) ray with subscript 1 and all variables associated with the refracted ray with subscript 2.*

The angle of refraction θ_2 depends on the angle of incidence θ_1 and on the indices of refraction, n_2 and n_1, of the two media. The relation between these quantities is known as *Snell's law of refraction,* after the Dutch mathematician Willebrord Snell (1591–1626), who discovered it experimentally. At the end of this section is a proof of Snell's law.

Figure 26.1 (a) When a ray of light is directed from air into water, part of the light is reflected at the interface and the remainder is refracted into the water. The refracted ray is bent *toward* the normal ($\theta_2 < \theta_1$). (b) When a ray of light is directed from water into air, the refracted ray in air is bent *away* from the normal ($\theta_2 > \theta_1$).

SNELL'S LAW OF REFRACTION

When light travels from a material with refractive index n_1 into a material with refractive index n_2, the refracted ray, the incident ray, and the normal to the interface between the materials all lie in the same plane. The angle of refraction θ_2 is related to the angle of incidence θ_1 by

$$n_1 \sin \theta_1 = n_2 \sin \theta_2 \qquad (26.2)$$

Example 1 illustrates the use of Snell's law.

Example 1 Determining the Angle of Refraction

A light ray strikes an air/water surface at an angle of 46° with respect to the normal. The refractive index for water is 1.33. Find the angle of refraction when the direction of the ray is **(a)** from air to water and **(b)** from water to air.

Reasoning Snell's law of refraction applies to both part (a) and part (b). However, in part (a) the incident ray is in air, whereas in part (b) it is in water. We keep track of this difference by always labeling the variables associated with the incident ray with a subscript 1 and the variables associated with the refracted ray with a subscript 2.

Solution (a) The incident ray is in air, so $\theta_1 = 46°$ and $n_1 = 1.00$. The refracted ray is in water, so $n_2 = 1.33$. Snell's law can be used to find the angle of refraction θ_2:

$$\sin \theta_2 = \frac{n_1 \sin \theta_1}{n_2} = \frac{(1.00) \sin 46°}{1.33} = 0.54 \qquad (26.2)$$

$$\theta_2 = \sin^{-1}(0.54) = \boxed{33°}$$

Since θ_2 is less than θ_1, the refracted ray is bent *toward* the normal, as Figure 26.1a shows.

(b) With the incident ray in water, we find that

$$\sin \theta_2 = \frac{n_1 \sin \theta_1}{n_2} = \frac{(1.33) \sin 46°}{1.00} = 0.96$$

$$\theta_2 = \sin^{-1}(0.96) = \boxed{74°}$$

Since θ_2 is greater than θ_1, the refracted ray is bent *away* from the normal, as in Figure 26.1b.

Problem-solving insight

The angle of incidence θ_1 and the angle of refraction θ_2 that appear in Snell's law are measured with respect to the normal to the surface, and not with respect to the surface itself.

We have seen that reflection and refraction of light waves occur simultaneously at the interface between two transparent materials. It is important to keep in mind that light waves are composed of electric and magnetic fields, which carry energy. The principle of conservation of energy (see Chapter 6) indicates that the energy reflected plus the energy refracted must add up to equal the energy carried by the incident light, provided that none of the energy is absorbed by the materials. The percentage of incident energy that appears as reflected versus refracted light depends on the angle of incidence and the refractive indices of the materials on either side of the interface. For instance, when light travels from air toward water at perpendicular incidence, most of its energy is refracted and little is reflected. But when the angle of incidence is nearly 90° and the light barely grazes the water surface, most of its energy is reflected, with only a small amount refracted into the water. On a rainy night, you probably have experienced the annoying glare that results when light from an oncoming car just grazes the wet road. Under such conditions, most of the light energy reflects into your eyes.

The simultaneous reflection and refraction of light have applications in a number of devices. For instance, interior rearview mirrors in cars often have adjustment levers. One position of the lever is for day driving, while the other is for night driving and reduces glare from the headlights of the car behind. As Figure 26.2a indicates, this kind of mirror is a glass wedge with a back side that is silvered and highly reflecting. Part b of the picture shows the day setting. Light from the car behind follows path ABCD in reaching the driver's eye. At points A and C, where the light strikes the air–glass surface, there are both reflected and refracted rays. The reflected rays are drawn as thin lines, the thinness denoting that only a small percentage (about 10%) of the light during the day is reflected at A and C. The weak reflected rays at A and C do not reach the driver's eye. In contrast, almost all the light reaching the silvered back surface at B is reflected toward the driver. Since most of the light follows path ABCD, the driver sees a bright image of the car behind. During the night, the adjustment lever can be used to rotate the top of the mirror away from the driver (see part c of the drawing). Now, most of the light from the headlights behind

The physics of rearview mirrors.

Figure 26.2 A car's interior rearview mirror with a day–night adjustment lever.

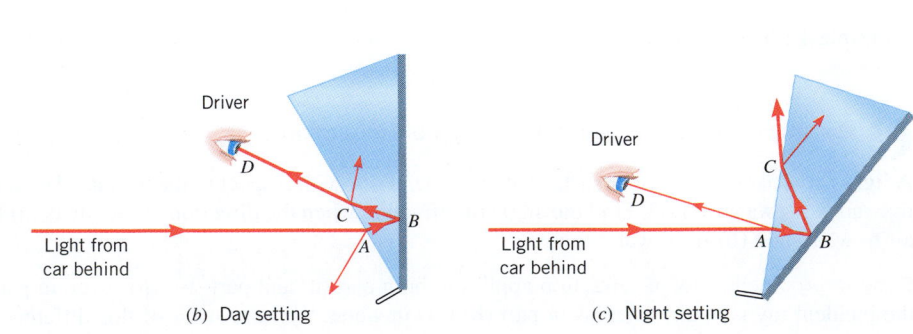

(a)   (b) Day setting   (c) Night setting

follows path *ABC* and does not reach the driver. Only the light that is weakly reflected from the front surface along path *AD* is seen, and the result is significantly less glare.

APPARENT DEPTH

One interesting consequence of refraction is that an object lying under water appears to be closer to the surface than it actually is. Example 2 sets the stage for explaining why, by showing what must be done to shine a light on such an object.

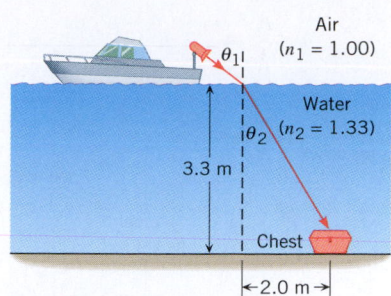

Figure 26.3 The beam from the searchlight is refracted when it enters the water.

Example 2 Finding a Sunken Chest

A searchlight on a yacht is being used at night to illuminate a sunken chest, as in Figure 26.3. At what angle of incidence θ_1 should the light be aimed?

Reasoning The angle of incidence θ_1 must be such that, after refraction, the light strikes the chest. The angle of incidence can be obtained from Snell's law, once the angle of refraction θ_2 is determined. This angle can be found using the data in Figure 26.3 and trigonometry. The light travels from a region of lower into a region of higher refractive index, so the light is bent toward the normal and we expect θ_1 to be greater than θ_2.

Solution From the data in the drawing it follows that $\tan \theta_2 = (2.0 \text{ m})/(3.3 \text{ m})$, so $\theta_2 = 31°$. With $n_1 = 1.00$ for air and $n_2 = 1.33$ for water, Snell's law gives

$$\sin \theta_1 = \frac{n_2 \sin \theta_2}{n_1} = \frac{(1.33) \sin 31°}{1.00} = 0.69$$

$$\theta_1 = \sin^{-1}(0.69) = \boxed{44°}$$

As expected, θ_1 is greater than θ_2.

Problem-solving insight

Remember that the refractive indices are written as n_1 for the medium in which the incident light travels and n_2 for the medium in which the refracted light travels.

When the sunken chest in Example 2 is viewed from the boat (Figure 26.4*a*), light rays from the chest pass upward through the water, refract away from the normal when they enter the air, and then travel to the observer. This picture is similar to Figure 26.3, except that the direction of the rays is reversed and the searchlight is replaced by an observer. When the rays entering the air are extended back into the water (see the dashed lines), they indicate that the observer sees a virtual image of the chest at an *apparent depth* that is less than the actual depth. The image is virtual because light rays do not actually pass through it. For the situation shown in Figure 26.4*a*, it is difficult to determine the apparent depth. A much simpler case is shown in part *b*, where the observer is *directly above* the submerged object, and the apparent depth d' is related to the actual depth d by

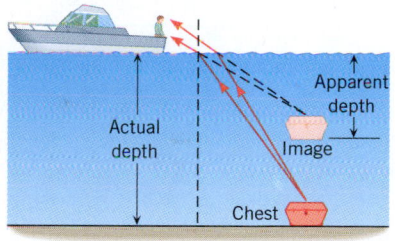

(*a*)

*Apparent depth,
observer directly
above object*

$$d' = d \left(\frac{n_2}{n_1} \right) \tag{26.3}$$

In this result, n_1 is the refractive index of the medium associated with the incident ray (the medium in which the object is located), and n_2 refers to the medium associated with the refracted ray (the medium in which the observer is situated). The proof of Equation 26.3 is the focus of Problem 19 at the end of the chapter. Example 3 illustrates that the effect of apparent depth is quite noticeable in water.

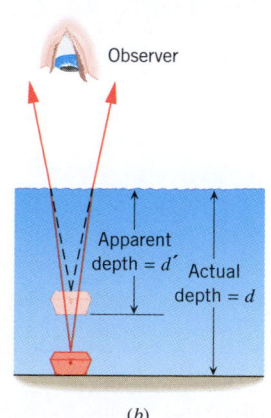

(*b*)

Figure 26.4 (*a*) Because light from the chest is refracted away from the normal when the light enters the air, the apparent depth of the image is less than the actual depth. (*b*) The observer is viewing the submerged object from directly overhead.

Example 3 The Apparent Depth of a Swimming Pool

A swimmer is treading water (with her head above the water) at the surface of a pool 3.00 m deep. She sees a coin on the bottom directly below. How deep does the coin appear to be?

Reasoning Equation 26.3 may be used to find the apparent depth, provided we remember that the light rays travel from the coin to the swimmer. Therefore, the incident ray is coming from the coin under the water ($n_1 = 1.33$), while the refracted ray is in the air ($n_2 = 1.00$).

Solution The apparent depth d' of the coin is

$$d' = d \left(\frac{n_2}{n_1} \right) = (3.00 \text{ m}) \left(\frac{1.00}{1.33} \right) = \boxed{2.26 \text{ m}} \tag{26.3}$$

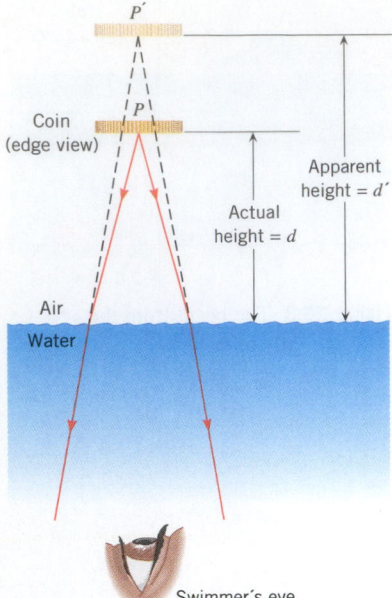

Figure 26.5 Rays from point P on a coin in the air above the water refract toward the normal as they enter the water. An underwater swimmer perceives the rays as originating from a point P' that is farther above the surface than the actual point P.

In Example 3, a person sees a coin on the bottom of a pool at an apparent depth that is less than the actual depth. Conceptual Example 4 considers the reverse situation—namely, a person looking from under the water at a coin in the air.

Conceptual Example 4   On the Inside Looking Out

A swimmer is under water and looking up at the surface. Someone holds a coin in the air, directly above the swimmer's eyes. To the swimmer, the coin appears to be at a certain height above the water. Is the apparent height of the coin (**a**) greater than, (**b**) less than, or (**c**) the same as its actual height?

Reasoning Figure 26.5 shows two rays of light leaving a point P on the coin. When the rays enter the water, they are refracted toward the normal because water has a larger index of refraction than air has. By extending the refracted rays backward (see the dashed lines in the drawing), we find that they appear to originate from a point P' on a virtual image, which is what the swimmer sees.

Answers (b) and (c) are incorrect. These answers are incorrect because the point P' in Figure 26.5 is located at a height that is greater than, not less than or the same as, the actual height of the coin.

Answer (a) is correct. The point P' in Figure 26.5 is on a virtual image that is located at an apparent height d' that is greater than the actual height d. Equation 26.3 [$d' = d(n_2/n_1)$] reveals the same result, because n_1 represents the medium (air) associated with the incident ray and n_2 represents the medium (water) associated with the refracted ray. Since n_2 for water is greater than n_1 for air, the ratio n_2/n_1 is greater than one and d' is larger than d. This situation is the opposite of that in Figure 26.4b, where an object beneath the water appears to a person above the water to be closer to the surface than it actually is.

Related Homework: *Problems 23, 108*

THE DISPLACEMENT OF LIGHT BY A TRANSPARENT SLAB OF MATERIAL

A windowpane is an example of a transparent slab of material. It consists of a plate of glass with parallel surfaces. When a ray of light passes through the glass, the emergent ray is parallel to the incident ray but displaced from it, as Figure 26.6 shows. This result can be verified by applying Snell's law to each of the two glass surfaces, with the result that $n_1 \sin \theta_1 = n_2 \sin \theta_2 = n_3 \sin \theta_3$. Since air surrounds the glass, $n_1 = n_3$, and it follows that $\sin \theta_1 = \sin \theta_3$. Therefore, $\theta_1 = \theta_3$, and the emergent and incident rays are parallel. However, as the drawing shows, the emergent ray is displaced laterally relative to the incident ray. The extent of the displacement depends on the angle of incidence, the thickness of the slab, and the refractive index of the slab.

DERIVATION OF SNELL'S LAW

Snell's law can be derived by considering what happens to the wave fronts when the light passes from one medium into another. Figure 26.7a shows light propagating from

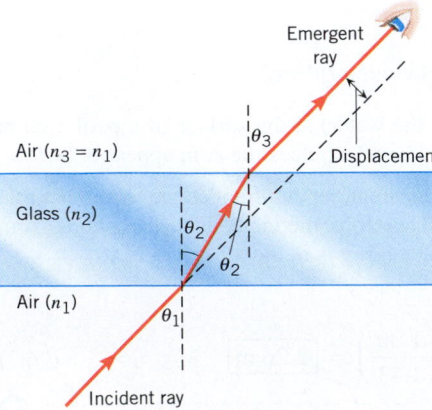

Figure 26.6 When a ray of light passes through a pane of glass that has parallel surfaces and is surrounded by air, the emergent ray is parallel to the incident ray ($\theta_3 = \theta_1$) but is displaced from it.

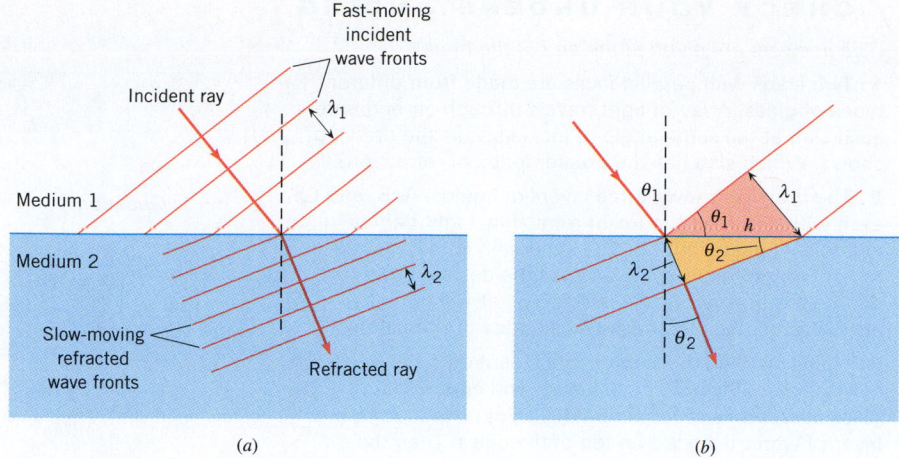

Figure 26.7 (*a*) The wave fronts are refracted as the light passes from medium 1 into medium 2. (*b*) An enlarged view of the incident and refracted wave fronts at the surface.

medium 1, where the speed is relatively large, into medium 2, where the speed is smaller; therefore, n_1 is less than n_2. The plane wave fronts in this picture are drawn perpendicular to the incident and refracted rays. Since the part of each wave front that penetrates medium 2 slows down, the wave fronts in medium 2 are rotated clockwise relative to those in medium 1. Correspondingly, the refracted ray in medium 2 is bent toward the normal, as the drawing shows.

Although the incident and refracted waves have different speeds, *they have the same frequency f.* The fact that the frequency does not change can be understood in terms of the atomic mechanism underlying the generation of the refracted wave. When the electromagnetic wave strikes the surface, the oscillating electric field forces the electrons in the molecules of medium 2 to oscillate at the same frequency as the wave. The accelerating electrons behave like atomic antennas that radiate "extra" electromagnetic waves, which combine with the original wave. The net electromagnetic wave within medium 2 is a superposition of the original wave plus the extra radiated waves, and it is this superposition that constitutes the refracted wave. Since the extra waves are radiated at the same frequency as the original wave, the refracted wave also has the same frequency as the original wave.

The distance between successive wave fronts in Figure 26.7*a* has been chosen to be the wavelength λ. Since the frequencies are the same in both media but the speeds are different, it follows from Equation 16.1 that the wavelengths are different: $\lambda_1 = v_1/f$ and $\lambda_2 = v_2/f$. Since v_1 is assumed to be larger than v_2, λ_1 is larger than λ_2, and the wave fronts are farther apart in medium 1.

Figure 26.7*b* shows an enlarged view of the incident and refracted wave fronts at the surface. The angles θ_1 and θ_2 within the colored right triangles are, respectively, the angles of incidence and refraction. In addition, the triangles share the same hypotenuse h. Therefore,

$$\sin \theta_1 = \frac{\lambda_1}{h} = \frac{v_1/f}{h} = \frac{v_1}{hf}$$

and

$$\sin \theta_2 = \frac{\lambda_2}{h} = \frac{v_2/f}{h} = \frac{v_2}{hf}$$

Combining these two equations into a single equation by eliminating the common term hf gives

$$\frac{\sin \theta_1}{v_1} = \frac{\sin \theta_2}{v_2}$$

By multiplying each side of this result by c, the speed of light in a vacuum, and recognizing that the ratio c/v is the index of refraction n, we arrive at Snell's law of refraction: $n_1 \sin \theta_1 = n_2 \sin \theta_2$.

✓ CHECK YOUR UNDERSTANDING

(The answers are given at the end of the book.)

1. Two slabs with parallel faces are made from different types of glass. A ray of light travels through air and enters each slab at the same angle of incidence, as the drawing shows. Which slab has the greater index of refraction?

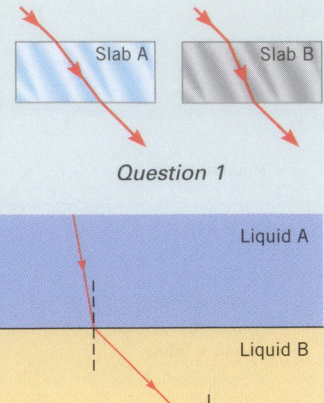

Question 1

2. The drawing shows three layers of liquids, A, B, and C, each with a different index of refraction. Light begins in liquid A, passes into B, and eventually into C, as the ray of light in the drawing shows. The dashed lines denote the normals to the interfaces between the layers. Which liquid has the smallest index of refraction?

3. Light traveling through air is incident on a flat piece of glass at a 35° angle of incidence and enters the glass at an angle of refraction θ_{glass}. Suppose that a layer of water is added on top of the glass. Then the light travels through air and is incident on the water at the 35° angle of incidence. Does the light enter the glass at the same angle of refraction θ_{glass} as it did when the water was not present?

Question 2

4. Two identical containers, one filled with water ($n = 1.33$) and the other with benzene ($n = 1.50$) are viewed from directly above. Which container (if either) appears to have a greater depth of fluid?

5. When an observer peers over the edge of a deep, empty, metal bowl on a kitchen table, he does not see the entire bottom surface. Therefore, a small object lying on the bottom is hidden from view, but the object can be seen when the bowl is filled with liquid A. When the bowl is filled with liquid B, however, the object remains hidden from view. Which liquid has the greater index of refraction?

6. A man is fishing from a dock, using a bow and arrow. To strike a fish that he sees beneath the water, should he aim **(a)** somewhat above the fish, **(b)** directly at the fish, or **(c)** somewhat below the fish?

7. A man is fishing from a dock. He is using a laser gun that emits an intense beam of light. To strike a fish that he sees beneath the water, should he aim **(a)** somewhat above the fish, **(b)** directly at the fish, or **(c)** somewhat below the fish?

8. Two rays of light converge to a point on a screen. A thick plate of glass with parallel surfaces is placed in the path of this converging light, with the parallel surfaces parallel to the screen. Will the point of convergence **(a)** move away from the glass plate, **(b)** move toward the glass plate, or **(c)** remain on the screen?

26.3 TOTAL INTERNAL REFLECTION

When light passes from a medium of larger refractive index into one of smaller refractive index—for example, from water to air—the refracted ray bends *away* from the normal, as in Figure 26.8a. As the angle of incidence increases, the angle of refraction also increases. When the angle of incidence reaches a certain value, called the *critical angle* θ_c, the angle of refraction is 90°. Then the refracted ray points along the surface; part *b* illustrates what happens at the critical angle. When the angle of incidence exceeds the critical angle, as in part *c* of the drawing, there is no refracted light. All the incident light is reflected back into the medium from which it came, a phenomenon called *total internal reflection*. Total internal reflection occurs only when light travels from a higher-index

Figure 26.8 (*a*) When light travels from a higher-index medium (water) into a lower-index medium (air), the refracted ray is bent away from the normal. (*b*) When the angle of incidence is equal to the critical angle θ_c, the angle of refraction is 90°. (*c*) If θ_1 is greater than θ_c, there is no refracted ray, and total internal reflection occurs.

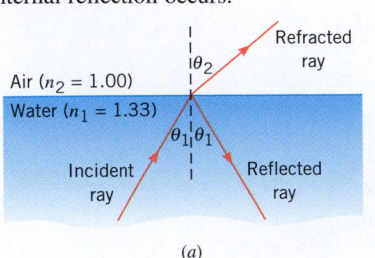

(a)

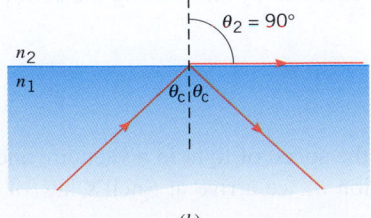

(b)

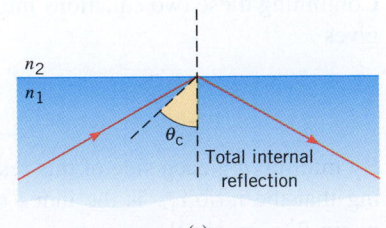

(c)

medium toward a lower-index medium. It does not occur when light propagates in the reverse direction—for example, from air to water.

An expression for the critical angle θ_c can be obtained from Snell's law by setting $\theta_1 = \theta_c$ and $\theta_2 = 90°$ (see Figure 26.8b):

$$\sin \theta_c = \frac{n_2 \sin 90°}{n_1}$$

Critical angle $\qquad\qquad \sin \theta_c = \dfrac{n_2}{n_1} \qquad (n_1 > n_2) \qquad\qquad$ (26.4)

For instance, the critical angle for light traveling from water ($n_1 = 1.33$) to air ($n_2 = 1.00$) is $\theta_c = \sin^{-1}(1.00/1.33) = 48.8°$. For incident angles greater than $48.8°$, Snell's law predicts that $\sin \theta_2$ is greater than unity, a value that is not possible. Thus, light rays with incident angles exceeding $48.8°$ yield no refracted light, and the light is totally reflected back into the water, as Figure 26.8c indicates. Then, the air–water interface acts like a mirror. Figure 26.9, for example, shows the mirror-like ability of the interface to form a reflected image of a floating reptile whose head just barely breaks the surface of the water. The body of the reptile hangs downward, and light from it that strikes the surface at angles exceeding the critical angle is reflected to form the image in the upper part of the photograph.

The next example illustrates how the critical angle changes when the indices of refraction change.

Figure 26.9 This underwater photograph shows a reptile floating with its head just breaking the surface and its body hanging downward. Light from the body that strikes the air–water interface at angles greater than the critical angle is reflected, and the interface acts like a mirror to form the image in the upper part of the photograph. (© Joel Sartore/ National Geographic Society)

 Example 5   Total Internal Reflection

A beam of light is propagating through diamond ($n_1 = 2.42$) and strikes a diamond–air interface at an angle of incidence of $28°$. **(a)** Will part of the beam enter the air ($n_2 = 1.00$) or will the beam be totally reflected at the interface? **(b)** Repeat part (a), assuming that the diamond is surrounded by water ($n_2 = 1.33$) instead of air.

Reasoning Total internal reflection occurs only when the beam of light has an angle of incidence that is greater than the critical angle θ_c. The critical angle is different in parts (a) and (b), since it depends on the ratio n_2/n_1 of the refractive indices of the incident (n_1) and refracting (n_2) media.

Solution **(a)** The critical angle θ_c for total internal reflection at the diamond–air interface is given by Equation 26.4 as

$$\theta_c = \sin^{-1}\left(\frac{n_2}{n_1}\right) = \sin^{-1}\left(\frac{1.00}{2.42}\right) = 24.4°$$

Because the angle of incidence of $28°$ is greater than the critical angle, there is no refraction, and the light is totally reflected back into the diamond.

(b) If water, rather than air, surrounds the diamond, the critical angle for total internal reflection becomes larger:

$$\theta_c = \sin^{-1}\left(\frac{n_2}{n_1}\right) = \sin^{-1}\left(\frac{1.33}{2.42}\right) = 33.3°$$

Now a beam of light that has an angle of incidence of $28°$ (less than the critical angle of $33.3°$) at the diamond–water interface is refracted into the water.

The critical angle plays an important role in why a diamond sparkles, as Conceptual Example 6 discusses.

 Conceptual Example 6   The Sparkle of a Diamond

A diamond gemstone is famous for its sparkle in air because the light coming from it glitters as the diamond is moved about. The sparkle is related to the total internal reflection of light that occurs within the diamond. What happens to the sparkle when the diamond is placed under water? **(a)** Nothing happens, for the water has no effect on total internal reflection. **(b)** The water reduces the sparkle markedly by making the total internal reflection less likely to occur.

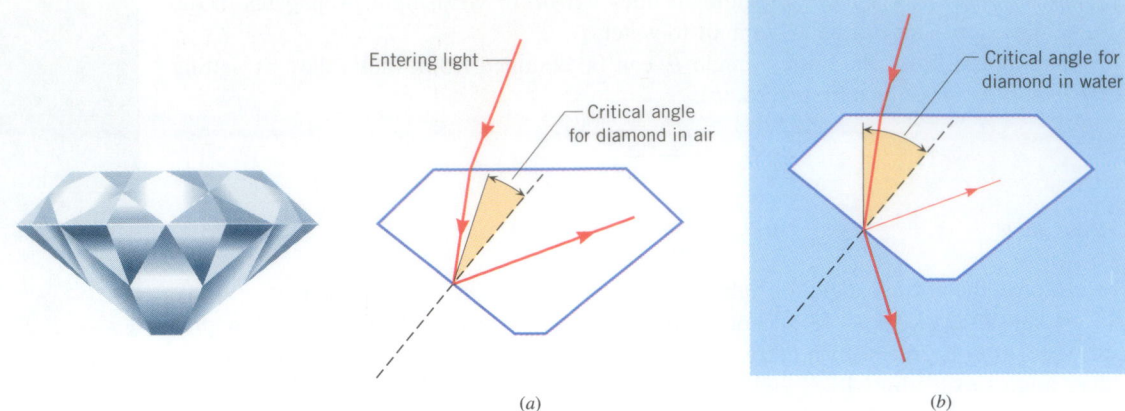

Entering light

Critical angle
for diamond in air

Critical angle for
diamond in water

(a)   (b)

Figure 26.10 (a) Near the bottom of the diamond, light is totally internally reflected, because the incident angle exceeds the critical angle for diamond and air. (b) When the diamond is in water, the same light is partially reflected and partially refracted, since the incident angle is less than the critical angle for diamond and water.

The physics of
why a diamond sparkles.

Reasoning When a diamond is held in a certain way in air, the intensity of the light coming from within it is greatly enhanced. Figure 26.10 helps to explain that this enhancement or sparkle is related to total internal reflection. Part a of the drawing shows a ray of light striking a lower facet of the diamond at an angle of incidence that exceeds the critical angle for a diamond–air interface. As a result, this ray undergoes total internal reflection back into the diamond and eventually exits the top surface. Since diamond has a relatively small critical angle in air, many of the rays striking a lower facet behave in this fashion and create the diamond's sparkle. Part (a) of Example 5 shows that the critical angle is 24.4° and is so small because the index of refraction of diamond ($n = 2.42$) is large compared to that of air ($n = 1.00$).

Answer (a) is incorrect. The water does indeed have an effect on the total internal reflection that occurs. This is because the critical angle depends on the index of refraction of the water as well as that of the diamond (see Equation 26.4).

Answer (b) is correct. Figure 26.10b illustrates what happens to the same ray of light within the diamond when the diamond is surrounded by water. Because water has a larger index of refraction than air does, the critical angle for the diamond–water interface is no longer 24.4° but increases to 33.3°, as part (b) of Example 5 shows. Therefore, this particular ray is no longer totally internally reflected. As Figure 26.10b indicates, only some of the light is now reflected back into the diamond, while the remainder escapes into the water. Consequently, less light exits from the top of the diamond, causing it to lose much of its sparkle.

Figure 26.11 Total internal reflection at a glass–air interface can be used to turn a ray of light through an angle of (a) 90° or (b) 180°. (c) Two prisms, each reflecting the light twice by total internal reflection, are sometimes used in binoculars to produce a lateral displacement of a light ray.

Many optical instruments, such as binoculars, periscopes, and telescopes, use glass prisms and total internal reflection to turn a beam of light through 90° or 180°. Figure 26.11a shows a light ray entering a 45°–45°–90° glass prism ($n_1 = 1.5$) and striking the hypotenuse of the prism at an angle of incidence of $\theta_1 = 45°$. The critical angle for a glass–air interface is $\theta_c = \sin^{-1}(n_2/n_1) = \sin^{-1}(1.0/1.5) = 42°$. Since the angle of incidence is greater than the critical angle, the light is totally reflected at the hypotenuse and is directed vertically upward in the drawing, having been turned through an angle of 90°.

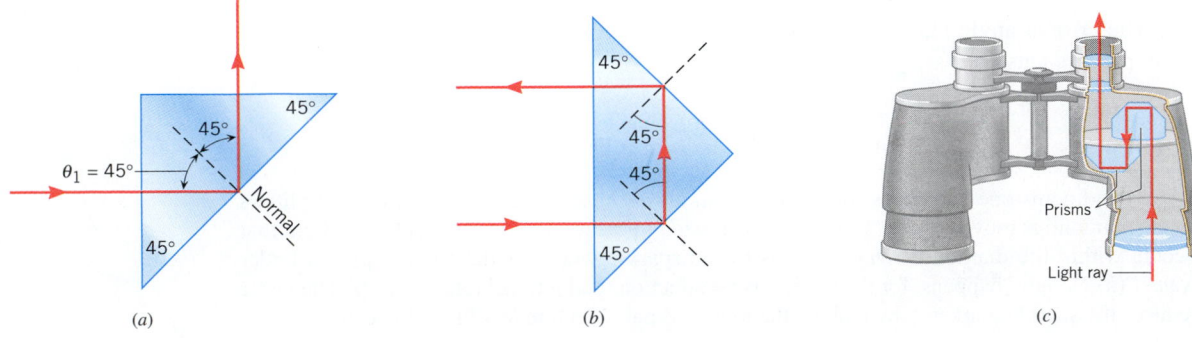

$\theta_1 = 45°$

45°

45°

45°

Normal

(a)

45°

45°

45°

45°

(b)

Prisms

Light ray

(c)

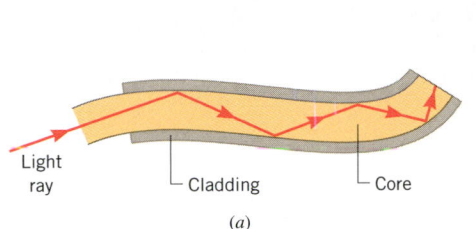

(a)

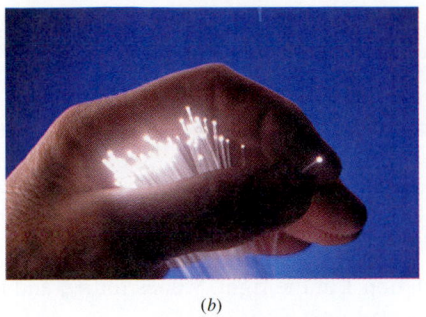

(b)

Figure 26.12 (*a*) Light can travel with little loss in a curved optical fiber, because the light is totally reflected whenever it strikes the core–cladding interface and because the absorption of light by the core itself is small. (*b*) Light being transmitted by a bundle of optical fibers. (PhotoDisc, Inc./Getty Images)

Part *b* of the picture shows how the same prism can turn the beam through 180° when total internal reflection occurs twice. Prisms can also be used in tandem to produce a lateral displacement of a light ray, while leaving its initial direction unaltered. Figure 26.11*c* illustrates such an application in binoculars.

An important application of total internal reflection occurs in fiber optics, where hair-thin threads of glass or plastic, called optical fibers, "pipe" light from one place to another. Figure 26.12*a* shows that an optical fiber consists of a cylindrical inner *core* that carries the light and an outer concentric shell, the *cladding*. The core is made from transparent glass or plastic that has a relatively high index of refraction. The cladding is also made of glass, but of a type that has a relatively low index of refraction. Light enters one end of the core, strikes the core/cladding interface at an angle of incidence greater than the critical angle, and, therefore, is reflected back into the core. Light thus travels inside the optical fiber along a zigzag path. In a well-designed fiber, little light is lost as a result of absorption by the core, so light can travel many kilometers before its intensity diminishes appreciably. Optical fibers are often bundled together to produce cables. Because the fibers themselves are so thin, the cables are relatively small and flexible and can fit into places inaccessible to larger metal wires. Example 7 deals with the light entering and traveling in an optical fiber.

The physics of fiber optics.

ANALYZING MULTIPLE-CONCEPT PROBLEMS

Example 7 An Optical Fiber

Figure 26.13 shows an optical fiber that consists of a core made of flint glass ($n_{\text{flint}} = 1.667$) surrounded by a cladding made of crown glass ($n_{\text{crown}} = 1.523$). A ray of light in air enters the fiber at an angle θ_1 with respect to the normal. What is θ_1 if this light also strikes the core–cladding interface at an angle that just barely exceeds the critical angle?

Reasoning The angle of incidence θ_1 is related to the angle of refraction θ_2 (see Figure 26.13) by Snell's law, where θ_2 is part of the right triangle in the drawing. The critical angle θ_c for the core–cladding interface is also part of the same right triangle, so that $\theta_2 = 90° - \theta_c$. The critical angle can be determined from a knowledge of the indices of refraction of the core and the cladding.

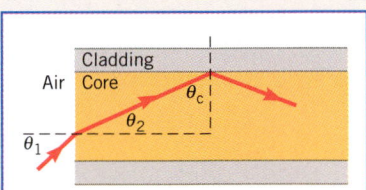

Figure 26.13 A ray of light enters the left end of an optical fiber and strikes the core–cladding interface at an angle that just barely exceeds the critical angle θ_c.

Knowns and Unknowns The data used in this problem are:

Description	Symbol	Value	Comment
Index of refraction of core material (flint glass)	n_{flint}	1.667	
Index of refraction of cladding material (crown glass)	n_{crown}	1.523	
Index of refraction of air	n_{air}	1.000	See Table 26.1.
Unknown Variable			
Angle of incidence of light ray entering optical fiber	θ_1	?	

Continued

Modeling the Problem

STEP 1 **Snell's Law of Refraction**   The ray of light, initially traveling in air, strikes the left end of the optical fiber at an angle of incidence labeled θ_1 in Figure 26.13. When the light enters the flint-glass core, its angle of refraction is θ_2. Snell's law of refraction gives the relation between these angles as

$$n_{air} \sin \theta_1 = n_{flint} \sin \theta_2 \qquad (26.2)$$

Solving this equation for θ_1 yields Equation 1 at the right. Values for n_{air} and n_{flint} are known. The angle θ_2 will be evaluated in the next step.

$$\theta_1 = \sin^{-1}\left(\frac{n_{flint} \sin \theta_2}{n_{air}}\right) \qquad (1)$$

?

STEP 2 **The Critical Angle**   We know that the light ray inside the core strikes the core–cladding interface at an angle that just barely exceeds the critical angle θ_c. When the angle of incidence exceeds the critical angle, all the light is reflected back into the core. From the right triangle in Figure 26.13, we see that the critical angle is related to θ_2 by $\theta_2 = 90° - \theta_c$. The critical angle depends on the indices of refraction of the core (flint glass) and cladding (crown glass) according to Equation 26.4:

$$\sin \theta_c = \frac{n_{crown}}{n_{flint}} \quad \text{or} \quad \theta_c = \sin^{-1}\left(\frac{n_{crown}}{n_{flint}}\right)$$

Substituting this expression for θ_c into $\theta_2 = 90° - \theta_c$ gives

$$\boxed{\theta_2 = 90° - \sin^{-1}\left(\frac{n_{crown}}{n_{flint}}\right)}$$

This result for θ_2 can be substituted into Equation 1, as indicated at the right.

$$\theta_1 = \sin^{-1}\left(\frac{n_{flint} \sin \theta_2}{n_{air}}\right) \qquad (1)$$

$$\boxed{\theta_2 = 90° - \sin^{-1}\left(\frac{n_{crown}}{n_{flint}}\right)} \qquad (2)$$

Solution   Combining the results of Steps 1 and 2 algebraically to produce a single equation gives a rather cumbersome result. Hence, we follow the simpler procedure of evaluating Equation 2 numerically and then substituting the result into Equation 1:

$$\theta_2 = 90° - \sin^{-1}\left(\frac{n_{crown}}{n_{flint}}\right) = 90° - \underbrace{\sin^{-1}\left(\frac{1.523}{1.667}\right)}_{66.01°} = 23.99° \qquad (2)$$

$$\theta_1 = \sin^{-1}\left(\frac{n_{flint} \sin \theta_2}{n_{air}}\right) = \sin^{-1}\left(\frac{1.667 \sin 23.99°}{1.000}\right) = \boxed{42.67°}$$

Related Homework: *Problem 35*

The physics of
endoscopy.

Optical fiber cables are the medium of choice for high-quality telecommunications because the cables are relatively immune to external electrical interference and because a light beam can carry information through an optical fiber just as electricity carries information through copper wires. The information-carrying capacity of light, however, is thousands of times greater than that of electricity. A laser beam traveling through a single optical fiber can carry tens of thousands of telephone conversations and several TV programs simultaneously.

In the field of medicine, optical fiber cables have had extraordinary impact. In the practice of endoscopy, for instance, a device called an endoscope is used to peer inside the body. Figure 26.14 shows a bronchoscope being used, which is a kind of endoscope that is inserted through the nose or mouth, down the bronchial tubes, and into the lungs. It consists of two optical fiber cables. One provides light to illuminate interior body parts, while the other sends back an image for viewing. A bronchoscope greatly simplifies the diagnosis of pulmonary disease. Tissue samples can even be collected with some bronchoscopes. A colonoscope is another kind of endoscope, and its design is similar to that of the bronchoscope. It is inserted through the rectum and used to examine the interior of the colon

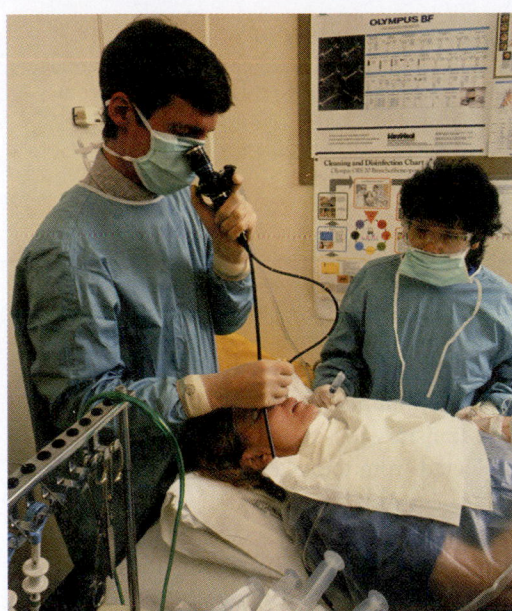

Figure 26.14 A doctor is using an endoscope to collect samples of tissue and fluid from the lung of a patient who has a history of asthma and allergies. Subsequently, the samples are examined under a microscope to obtain a diagnosis. (© James King-Holmes/SPL/Photo Researchers, Inc.)

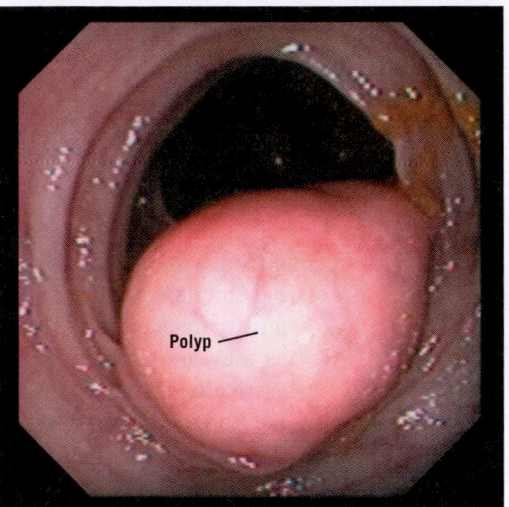

Figure 26.15 A colonoscope reveals a benign (noncancerous) polyp attached to the wall of a patient's colon (large intestine). This is a lipoma, a tumor arising from fatty tissue. Polyps that turn cancerous or grow large enough to obstruct the colon are surgically removed. (© David M. Martin/SPL/Photo Researchers, Inc.)

(see Figure 26.15). The colonoscope currently offers the best hope for diagnosing colon cancer in its early stages, when it can be treated.

The use of optical fibers has also revolutionized surgical techniques. In arthroscopic surgery, a small surgical instrument, several millimeters in diameter, is mounted at the end of an optical fiber cable. The surgeon can insert the instrument and cable into a joint, such as the knee, with only a tiny incision and minimal damage to the surrounding tissue (see Figure 26.16). Consequently, recovery from the procedure is relatively rapid compared to recovery from traditional surgical techniques.

The physics of arthroscopic surgery.

✓ CHECK YOUR UNDERSTANDING

(The answers are given at the end of the book.)

9. The drawing shows a 30°–60°–90° prism and two light rays, A and B, which both strike the prism perpendicularly. The prism is surrounded by an unknown liquid, which is the same in both parts of the drawing. When ray A reaches the hypotenuse in the drawing, it is totally internally reflected. Which one of the following statements applies to ray B when it reaches the hypotenuse? **(a)** It may or may not be totally internally reflected, depending on what the surrounding liquid is. **(b)** It is not totally internally reflected, no matter what the surrounding liquid is. **(c)** It is totally internally reflected, no matter what the surrounding liquid is.

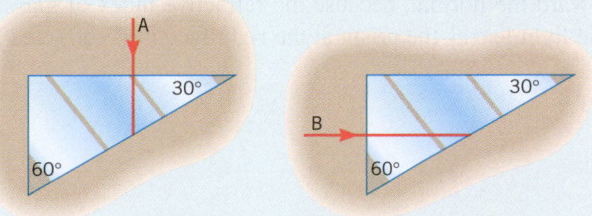

10. A shallow swimming pool has a constant depth. A point source of light is located in the middle of the bottom of this pool and emits light in all directions. However, no light exits the surface of the water except through a relatively small circular area that is centered on and directly above the light source. Why does the light exit the water through such a limited area?

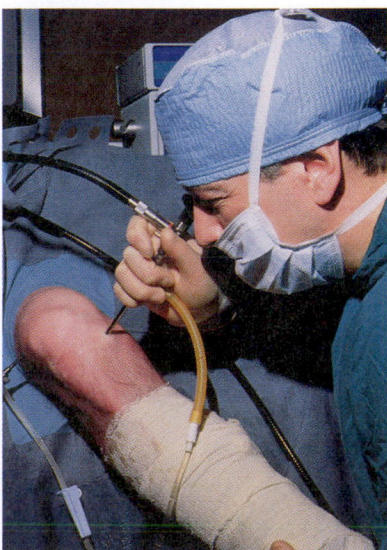

Figure 26.16 Optical fibers have made arthroscopic surgery possible, such as the repair of the damaged knee shown in this photograph. (© Margaret Rose Orthopaedic Hospital/Photo Researchers, Inc.)

Continued

11. Refer to Figure 26.6. Note that the ray within the glass slab is traveling from a medium with a larger refractive index toward a medium with a smaller refractive index. Is it possible, for θ_1 less than 90°, that the ray within the glass will experience total internal reflection at the glass–air interface?

26.4 POLARIZATION AND THE REFLECTION AND REFRACTION OF LIGHT

For incident angles other than 0°, unpolarized light becomes partially polarized in reflecting from a nonmetallic surface, such as water. To demonstrate this fact, rotate a pair of Polaroid sunglasses in the sunlight reflected from a lake. You will see that the light intensity transmitted through the glasses is a minimum when the glasses are oriented as they are normally worn. Since the transmission axis of the glasses is aligned vertically, it follows that the light reflected from the lake is partially polarized in the horizontal direction.

There is one special angle of incidence at which the reflected light is completely polarized parallel to the surface, the refracted ray being only partially polarized. This angle is called the **Brewster angle** θ_B. Figure 26.17 summarizes what happens when unpolarized light strikes a nonmetallic surface at the Brewster angle. The value of θ_B is given by **Brewster's law,** in which n_1 and n_2 are, respectively, the refractive indices of the materials in which the incident and refracted rays propagate:

Brewster's law $$\tan \theta_B = \frac{n_2}{n_1} \qquad (26.5)$$

This relation is named after the Scotsman David Brewster (1781–1868), who discovered it. Figure 26.17 also indicates that the reflected and refracted rays are perpendicular to each other when light strikes the surface at the Brewster angle (see Problem 41 at the end of the chapter).

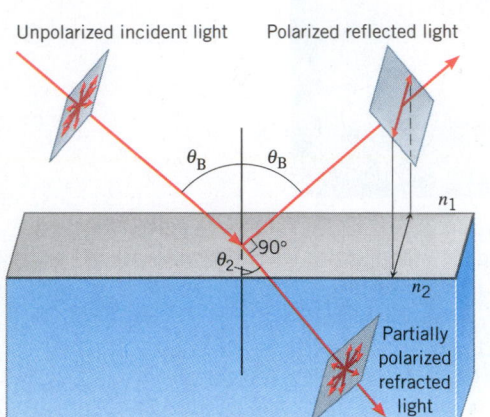

Figure 26.17 When unpolarized light is incident on a nonmetallic surface at the Brewster angle θ_B, the reflected light is 100% polarized in a direction parallel to the surface. The angle between the reflected and refracted rays is 90°.

✓ CHECK YOUR UNDERSTANDING

(The answer is given at the end of the book.)

12. You are sitting by the shore of a lake on a sunny and windless day. When are your Polaroid sunglasses most effective in reducing the glare of the sunlight reflected from the lake surface? When the angle of incidence of the sunlight on the lake is _____. **(a)** almost 90° because the sun is low in the sky **(b)** 0° because the sun is directly overhead **(c)** somewhere between 90° and 0°

26.5 THE DISPERSION OF LIGHT: PRISMS AND RAINBOWS

Figure 26.18*a* shows a ray of monochromatic light passing through a glass prism surrounded by air. When the light enters the prism at the left face, the refracted ray is bent toward the normal, because the refractive index of glass is greater than that of air. When the light leaves the prism at the right face, it is refracted away from the normal. Thus, the

Figure 26.18 (*a*) A ray of light is refracted as it passes through the prism. The prism is surrounded by air. (*b*) Two different colors are refracted by different amounts. For clarity, the amount of refraction has been exaggerated. (*c*) Sunlight is dispersed into its color components by this prism. (© David Parker/Photo Researchers, Inc.)

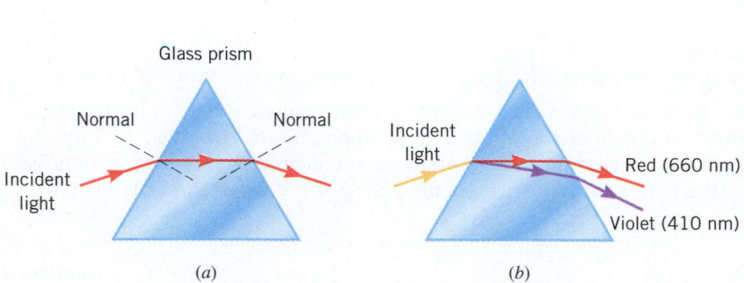
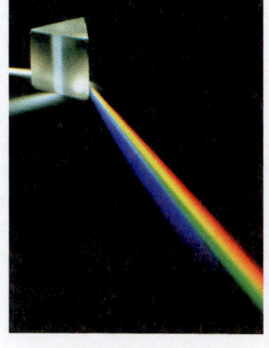

net effect of the prism is to change the direction of the ray, causing it to bend downward upon entering the prism, and downward again upon leaving. Because the refractive index of the glass depends on wavelength (see Table 26.2), rays corresponding to different colors are bent by different amounts by the prism and depart traveling in different directions. The greater the index of refraction for a given color, the greater the bending, and part *b* of the drawing shows the refractions for the colors red and violet, which are at opposite ends of the visible spectrum. If a beam of sunlight, which contains all colors, is sent through the prism, the sunlight is separated into a spectrum of colors, as part *c* shows. The spreading of light into its color components is called ***dispersion.***

In Figure 26.18*a* the ray of light is refracted twice by a glass prism surrounded by air. Conceptual Example 8 explores what happens to the light when the prism is surrounded by materials other than air.

Conceptual Example 8
The Refraction of Light Depends on Two Refractive Indices

In Figure 26.18*a* the glass prism is surrounded by air and bends the ray of light downward. It is also possible for the prism to bend the ray upward, as in Figure 26.19*a*, or to not bend the ray at all, as in part *b* of the drawing. How can the situations illustrated in Figure 26.19 arise?

Reasoning and Solution Snell's law of refraction includes the refractive indices of *both* materials on either side of an interface. With this in mind, we note that the ray bends upward, or away from the normal, as it enters the prism in Figure 26.19*a*. A ray bends away from the normal when it travels from a medium with a larger refractive index into a medium with a smaller refractive index. When the ray leaves the prism, it again bends upward, which is toward the normal at the point of exit. A ray bends toward the normal when traveling from a smaller toward a larger refractive index. Thus, *the situation in Figure 26.19a could arise if the prism were immersed in a fluid, such as carbon disulfide, that has a larger refractive index than does glass* (see Table 26.1).

We have seen in Figures 26.18*a* and 26.19*a* that a glass prism can bend a ray of light either downward or upward, depending on whether the surrounding fluid has a smaller or larger index of refraction than the glass. It seems logical to conclude, then, that *a prism will not bend a ray at all, neither up nor down, if the surrounding fluid has the same index of refraction as the glass*—a condition known as *index matching.* This is exactly what is happening in Figure 26.19*b*, where the ray proceeds straight through the prism as if the prism were not even there. If the index of refraction of the surrounding fluid equals that of the glass prism, then $n_1 = n_2$, and Snell's law ($n_1 \sin \theta_1 = n_2 \sin \theta_2$) reduces to $\sin \theta_1 = \sin \theta_2$. Therefore, the angle of refraction equals the angle of incidence, and no bending of the light occurs.

Related Homework: *Problem 46*

The physics of rainbows. Another example of dispersion occurs in rainbows, in which refraction by water droplets gives rise to the colors. You can often see a rainbow just as a storm is leaving, if you look at the departing rain with the sun at your back. When light from the sun enters a spherical raindrop, as in Figure 26.20, light of each color is refracted or bent by an amount that depends on the refractive index of water for that wavelength. After reflection from the back surface of the droplet, the different colors are again refracted as they reenter the air. Although each droplet disperses the light into its full spectrum of colors, the observer in Figure 26.21*a* sees only one color of light coming from any given droplet, since only one color travels in the right direction to reach the observer's eyes. However, all colors are visible in a rainbow (see Figure 26.21*b*) because each color originates from different droplets at different angles of elevation.

Table 26.2 Indices of Refraction n of Crown Glass at Various Wavelengths

Color[a]	Vacuum Wavelength (nm)	Index of Refraction, n
Red	660	1.520
Orange	610	1.522
Yellow	580	1.523
Green	550	1.526
Blue	470	1.531
Violet	410	1.538

[a]Approximate

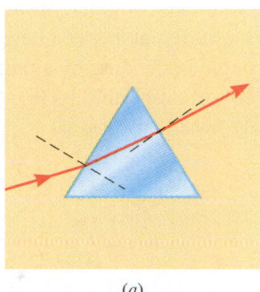

(a)

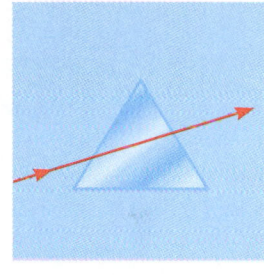

(b)

Figure 26.19 A ray of light passes through identical prisms, each surrounded by a different fluid. The ray of light is (*a*) refracted upward and (*b*) not refracted at all.

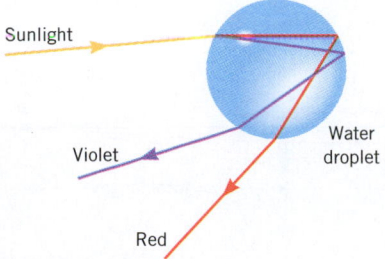

Figure 26.20 When sunlight emerges from a water droplet, the light is dispersed into its constituent colors, of which only two are shown.

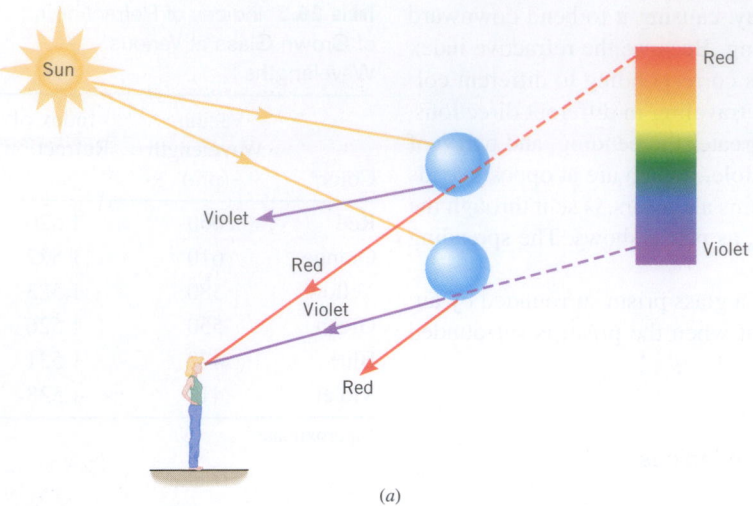

Figure 26.21 (a) The different colors seen in a rainbow originate from water droplets at different angles of elevation. (b) A couple is standing by a rainbow at the foot of a waterfall. (© Eric Bean/ The Image Bank/Getty Images)

✓ CHECK YOUR UNDERSTANDING

(*The answers are given at the end of the book.*)

13. Two blocks, made from the same transparent material, are immersed in different liquids. A ray of light strikes each block at the same angle of incidence. From the drawing, determine which liquid, A or B, has the greater index of refraction.

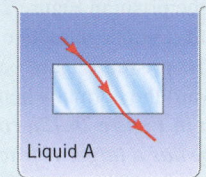

Liquid A

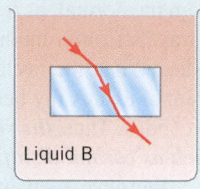

Liquid B

14. A beam of violet-colored light is propagating in crown glass. When the light reaches the boundary between the glass and the surrounding air, the beam is totally reflected back into the glass. What happens if the light is red and has the same angle of incidence θ_1 at the glass–air interface as does the violet-colored light? **(a)** Depending on the value for θ_1, red light may not be totally reflected, and some of it may be refracted into the air. **(b)** No matter what the value for θ_1, the red light behaves exactly the same as the violet-colored light. (*Hint: Refer to Table 26.2 and review Section 26.3.*)

26.6 LENSES

The lenses used in optical instruments, such as eyeglasses, cameras, and telescopes, are made from transparent materials that refract light. They refract the light in such a way that an image of the source of the light is formed. Figure 26.22a shows a crude lens formed from two glass prisms. Suppose that an object centered on the principal axis is infinitely far from the lens so the rays from the object are parallel to the principal axis. In passing through the prisms, these rays are bent toward the axis because of refraction. Unfortunately, the rays do not all cross the axis at the same place, and, therefore, such a crude lens gives rise to a blurred image of the object.

A better lens can be constructed from a single piece of transparent material with properly curved surfaces, often spherical, as in part b of the drawing. With this improved lens, rays that are near the principal axis (paraxial rays) and parallel to it converge to a single point on the axis after emerging from the lens. This point is called the *focal point F* of the lens. Thus, an object located infinitely far away on the principal axis leads to an

Figure 26.22 (a) These two prisms cause rays of light that are parallel to the principal axis to change direction and cross the axis at different points. (b) With a converging lens, paraxial rays that are parallel to the principal axis converge to the focal point F after passing through the lens.

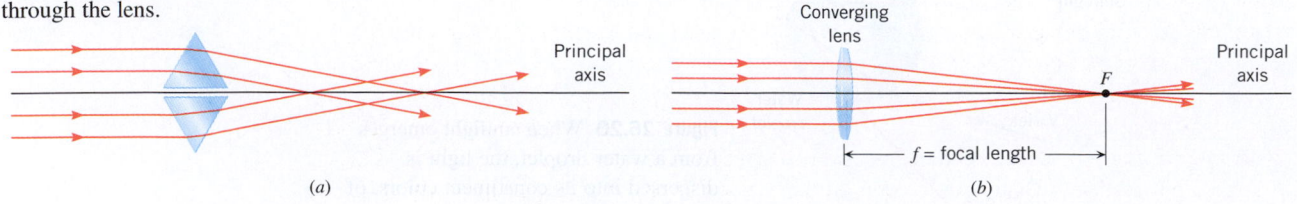

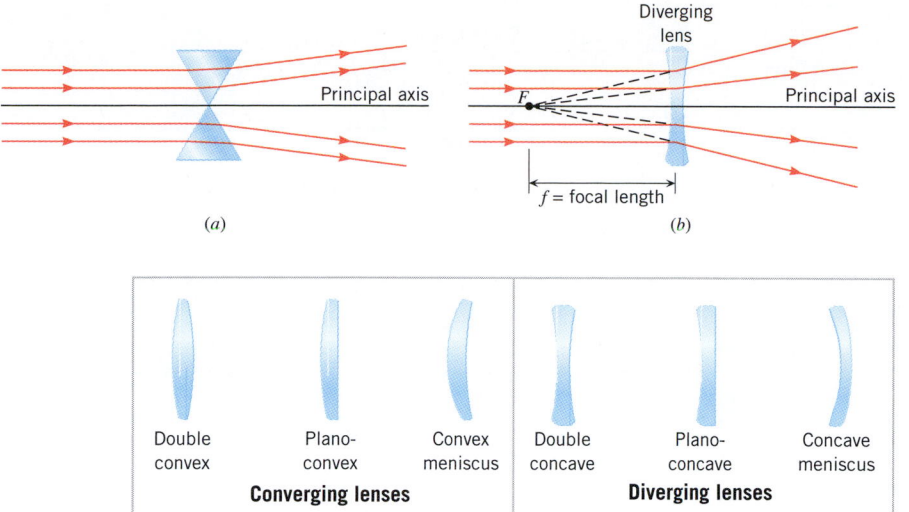

Figure 26.23 (*a*) These two prisms cause parallel rays to diverge. (*b*) With a diverging lens, paraxial rays that are parallel to the principal axis appear to originate from the focal point *F* after passing through the lens.

Double convex	Plano-convex	Convex meniscus	Double concave	Plano-concave	Concave meniscus
	Converging lenses			**Diverging lenses**	

Figure 26.24 Converging and diverging lenses come in a variety of shapes.

image at the focal point of the lens. The distance between the focal point and the lens is the *focal length f.* In what follows, we assume the lens is so thin compared to *f* that it makes no difference whether *f* is measured between the focal point and either surface of the lens or the center of the lens. The type of lens in Figure 26.22*b* is known as a **converging lens** because it causes incident parallel rays to converge at the focal point.

Another type of lens found in optical instruments is a **diverging lens,** which causes incident parallel rays to diverge after exiting the lens. Two prisms can also be used to form a crude diverging lens, as in Figure 26.23*a*. In a properly designed diverging lens, such as the one in part *b* of the picture, paraxial rays that are parallel to the principal axis appear to originate from a single point on the axis after passing through the lens. This point is the focal point *F*, and its distance *f* from the lens is the focal length. Again, we assume that the lens is thin compared to the focal length.

Converging and diverging lenses come in a variety of shapes, as Figure 26.24 illustrates. Observe that converging lenses are thicker at the center than at the edges, whereas diverging lenses are thinner at the center.

✓ CHECK YOUR UNDERSTANDING

(*The answers are given at the end of the book.*)

15. A beacon in a lighthouse is to produce a parallel beam of light. The beacon consists of a light source and a converging lens. Should the light source be placed **(a)** between the focal point and the lens, **(b)** at the focal point of the lens, or **(c)** beyond the focal point? (*Hint: Refer to Section 25.5 and review the principle of reversibility.*)

16. Review Conceptual Example 8 as an aid in answering this question. Is it possible for a lens to behave as a converging lens when surrounded by air but to behave as a diverging lens when surrounded by another medium?

26.7 THE FORMATION OF IMAGES BY LENSES

RAY DIAGRAMS AND RAY TRACING

Each point on an object emits light rays in all directions, and when some of these rays pass through a lens, they form an image. As with mirrors, ray diagrams can be drawn to determine the location and size of the image. Lenses differ from mirrors, however, in that light can pass through a lens from left to right or from right to left. Therefore, when constructing ray diagrams, begin by locating a focal point *F* on *each side of the lens;* each point lies on the principal axis at the same distance *f* from the lens. The lens is assumed to be thin, in that its thickness is small compared with the focal length and the distances of the object and the image from the lens. For convenience, it is also assumed that the object is located to the left of the lens and is oriented perpendicular to the principal axis. There are three paraxial rays that leave a point on the top of the object and are especially helpful

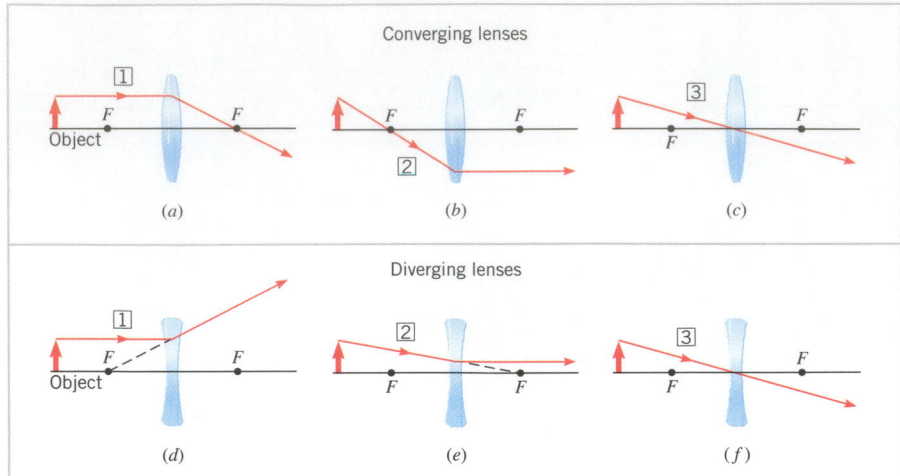

Figure 26.25 The rays shown here are useful in determining the nature of the images formed by converging and diverging lenses.

in drawing ray diagrams. They are labeled 1, 2, and 3 in Figure 26.25. When tracing their paths, we use the following reasoning strategy.

REASONING STRATEGY

Ray Tracing for Converging and Diverging Lenses

Converging Lens	Diverging Lens
Ray 1	
This ray initially travels parallel to the principal axis. In passing through a converging lens, the ray is refracted toward the axis and travels through the focal point on the right side of the lens, as Figure 26.25a shows.	This ray initially travels parallel to the principal axis. In passing through a diverging lens, the ray is refracted away from the axis, and *appears* to have originated from the focal point on the left of the lens. The dashed line in Figure 26.25d represents the apparent path of the ray.
Ray 2	
This ray first passes through the focal point on the left and then is refracted by the lens in such a way that it leaves traveling parallel to the axis, as in Figure 26.25b.	This ray leaves the object and moves toward the focal point on the right of the lens. Before reaching the focal point, however, the ray is refracted by the lens so as to exit parallel to the axis. See Figure 26.25e, where the dashed line indicates the ray's path in the absence of the lens.
Ray 3*	
This ray travels directly through the center of the thin lens without any appreciable bending, as in Figure 26.25c.	This ray travels directly through the center of the thin lens without any appreciable bending, as in Figure 26.25f.

*Ray 3 does not bend as it proceeds through the lens because the left and right surfaces of each type of lens are nearly parallel at the center. Thus, in either case, the lens behaves as a transparent slab. As Figure 26.6 shows, the rays incident on and exiting from a slab travel in the same direction with only a lateral displacement. If the lens is sufficiently thin, the displacement is negligibly small.

IMAGE FORMATION BY A CONVERGING LENS

Figure 26.26a illustrates the formation of a real image by a converging lens. Here the object is located at a distance from the lens that is greater than twice the focal length (beyond the point labeled 2F). To locate the image, any two of the three special rays can be drawn from the tip of the object, although all three are shown in the drawing. The point

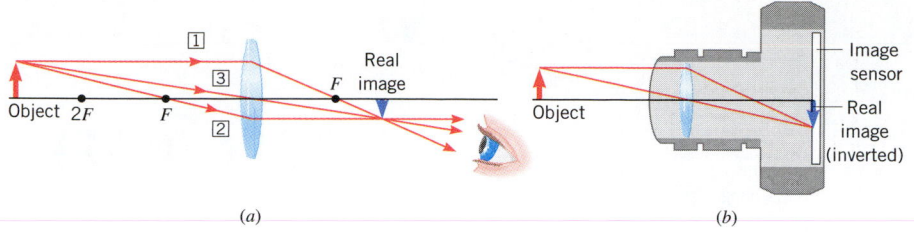

Figure 26.26 (*a*) When the object is placed to the left of the point labeled 2*F*, a real, inverted, and smaller image is formed. (*b*) The arrangement in part *a* is like that used in a camera.

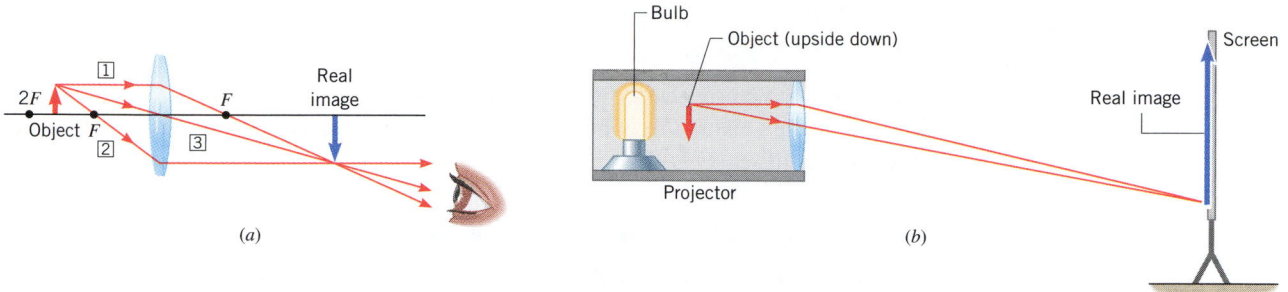

Figure 26.27 (*a*) When the object is placed between 2*F* and *F*, the image is real, inverted, and larger than the object. (*b*) This arrangement is found in projectors.

on the right side of the lens where these rays intersect locates the tip of the image. The ray diagram indicates that the image is real, inverted, and smaller than the object. This optical arrangement is similar to that used in a camera, where an image sensor* or a piece of film records the image (see part *b* of the drawing).

The physics of a camera.

When the object is placed between 2*F* and *F*, as in Figure 26.27*a*, the image is still real and inverted; however, the image is now larger than the object. This optical system is used in a slide or film projector in which a small piece of film is the object and the enlarged image falls on a screen. However, to obtain an image that is right-side up, the film must be placed in the projector upside down.

The physics of a slide or film projector.

When the object is located between the focal point and the lens, as in Figure 26.28, the rays diverge after leaving the lens. To a person viewing the diverging rays, they appear to come from an image behind (to the left of) the lens. Because none of the rays actually come from the image, it is a virtual image. The ray diagram shows that the virtual image is upright and enlarged. A magnifying glass uses this arrangement, as can be seen in part *b* of the drawing.

IMAGE FORMATION BY A DIVERGING LENS

We have seen that a converging lens can form a real image or a virtual image, depending on where the object is located with respect to the lens. In contrast, regardless of

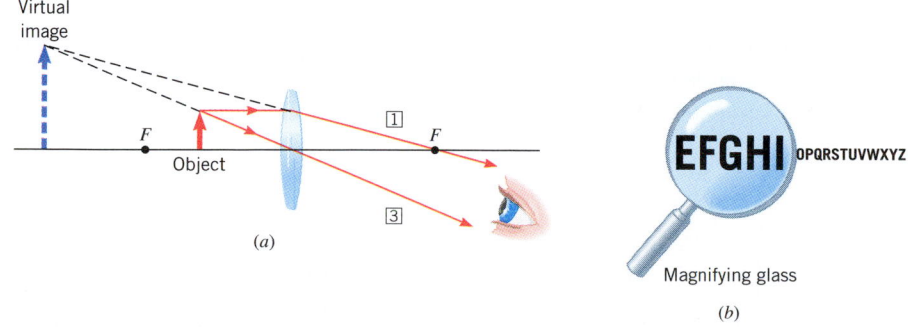

Figure 26.28 (*a*) When an object is placed within the focal point *F* of a converging lens, an upright, enlarged, and virtual image is created. (*b*) Such an image is seen when looking through a magnifying glass.

*One type of image sensor used in today's digital cameras utilizes a charge-coupled device (CCD). See Section 29.3 for a discussion of CCDs.

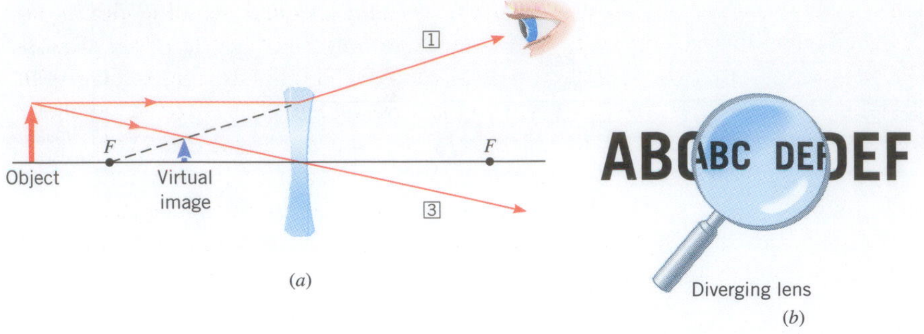

Figure 26.29 (*a*) A diverging lens always forms a virtual image of a real object. The image is upright and smaller relative to the object. (*b*) The image seen through a diverging lens.

the position of a real object, a diverging lens always forms a virtual image that is on the same side of the lens as the object and is upright and smaller relative to the object, as Figure 26.29 illustrates.

✓ **CHECK YOUR UNDERSTANDING**

(*The answer is given at the end of the book.*)

17. A converging lens is used to produce a real image, as in Figure 26.27*a*. A piece of black tape is then placed over the upper half of the lens. Which one of the following statements is true concerning the image that results with the tape in place? **(a)** The image is of the entire object, although its brightness is reduced since fewer rays produce it. **(b)** The image is of the object's lower half only, but its brightness is not reduced. **(c)** The image is of the object's upper half only, but its brightness is not reduced.

26.8 THE THIN-LENS EQUATION AND THE MAGNIFICATION EQUATION

When an object is placed in front of a spherical mirror, we can determine the location, size, and nature of its image by using the technique of ray tracing or the mirror and magnification equations. Both options are based on the law of reflection. The mirror and magnification equations relate the distances d_o and d_i of the object and image from the mirror to the focal length f and magnification m. For an object placed in front of a lens, Snell's law of refraction leads to the technique of ray tracing and to equations that are identical to the mirror and magnification equations. Thus, lenses work because of the refraction of light, whereas mirrors work because of the reflection of light, a distinction between the two devices that is important to keep in mind.

The equations that result from applying Snell's law to lenses are referred to as the thin-lens equation and the magnification equation:

Thin-lens equation
$$\frac{1}{d_o} + \frac{1}{d_i} = \frac{1}{f} \tag{26.6}$$

Magnification equation
$$m = \frac{\text{Image height}}{\text{Object height}} = \frac{h_i}{h_o} = -\frac{d_i}{d_o} \tag{26.7}$$

Figure 26.30 defines the symbols in these expressions with the aid of a thin converging lens, but the expressions also apply to a diverging lens, if it is thin. The derivations of these equations are presented at the end of this section.

Figure 26.30 The drawing shows the focal length f, the object distance d_o, and the image distance d_i for a converging lens. The object and image heights are, respectively, h_o and h_i.

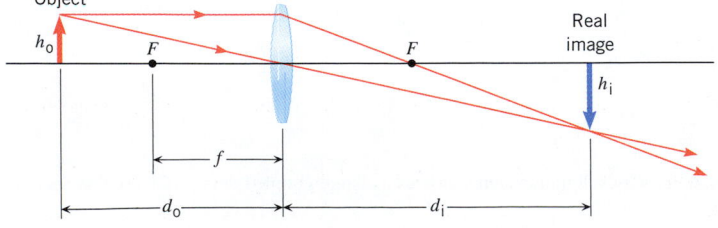

Certain sign conventions accompany the use of the thin-lens and magnification equations, and the conventions are similar to those used with mirrors in Section 25.6. The issue of real versus virtual images, however, is slightly different with lenses than with mirrors. With a mirror, a real image is formed on the *same side* of the mirror as the object (see Figure 25.18), in which case the image distance d_i is a positive number. With a lens, a positive value for d_i also means the image is real. However, starting with an actual object, a real image is formed on the side of the lens *opposite to* the object (see Figure 26.30). The *sign conventions* listed in the following Reasoning Strategy apply to light rays traveling from left to right from a real object.

REASONING STRATEGY

Summary of Sign Conventions for Lenses

Focal length

f is + for a converging lens.

f is − for a diverging lens.

Object distance

d_o is + if the object is to the left of the lens (real object), as is usual.

d_o is − if the object is to the right of the lens (virtual object).*

Image distance

d_i is + for an image (real) formed to the right of the lens by a real object.

d_i is − for an image (virtual) formed to the left of the lens by a real object.

Magnification

m is + for an image that is upright with respect to the object.

m is − for an image that is inverted with respect to the object.

---

* This situation arises in systems containing more than one lens, where the image formed by the first lens becomes the object for the second lens. In such a case, the object of the second lens may lie to the right of that lens, in which event d_o is assigned a negative value and the object is called a virtual object.

Examples 9 and 10 illustrate the use of the thin-lens and magnification equations.

 Example 9   The Real Image Formed by a Camera Lens

A person 1.70 m tall is standing 2.50 m in front of a digital camera. The camera uses a converging lens whose focal length is 0.0500 m. **(a)** Find the image distance (the distance between the lens and the image sensor) and determine whether the image is real or virtual. **(b)** Find the magnification and the height of the image on the image sensor.

Reasoning This optical arrangement is similar to that in Figure 26.26*a*, where the object distance is greater than twice the focal length of the lens. Therefore, we expect the image to be real, inverted, and smaller than the object.

Solution **(a)** To find the image distance d_i we use the thin-lens equation with $d_o = 2.50$ m and $f = 0.0500$ m:

$$\frac{1}{d_i} = \frac{1}{f} - \frac{1}{d_o} = \frac{1}{0.0500 \text{ m}} - \frac{1}{2.50 \text{ m}} = 19.6 \text{ m}^{-1} \quad \text{or} \quad \boxed{d_i = 0.0510 \text{ m}}$$

Since the image distance is a positive number, a $\boxed{\text{real image}}$ is formed on the image sensor.

(b) The magnification follows from the magnification equation:

$$m = -\frac{d_i}{d_o} = -\frac{0.0510 \text{ m}}{2.50 \text{ m}} = \boxed{-0.0204}$$

The image is 0.0204 times as large as the object, and it is inverted since m is negative. Since the object height is $h_o = 1.70$ m, the image height is

$$h_i = mh_o = (-0.0204)(1.70 \text{ m}) = \boxed{-0.0347 \text{ m}}$$

Problem-solving insight

In the thin-lens equation, the reciprocal of the image distance d_i is given by $d_i^{-1} = f^{-1} - d_o^{-1}$, where f is the focal length and d_o is the object distance. After combining the reciprocals f^{-1} and d_o^{-1}, do not forget to take the reciprocal of the result to find d_i.

Example 10 The Virtual Image Formed by a Diverging Lens

An object is placed 7.10 cm to the left of a diverging lens whose focal length is $f = -5.08$ cm (a diverging lens has a negative focal length). **(a)** Find the image distance and determine whether the image is real or virtual. **(b)** Obtain the magnification.

Reasoning This situation is similar to that in Figure 26.29a. The ray diagram shows that the image is virtual, erect, and smaller than the object.

Solution **(a)** The thin-lens equation can be used to find the image distance d_i:

$$\frac{1}{d_i} = \frac{1}{f} - \frac{1}{d_o} = \frac{1}{-5.08 \text{ cm}} - \frac{1}{7.10 \text{ cm}}$$

$$= -0.338 \text{ cm}^{-1} \quad \text{or} \quad \boxed{d_i = -2.96 \text{ cm}}$$

The image distance is negative, indicating that the image is $\boxed{\text{virtual}}$ and located to the left of the lens.

(b) Since d_i and d_o are known, the magnification equation shows that

$$m = -\frac{d_i}{d_o} = -\frac{-2.96 \text{ cm}}{7.10 \text{ cm}} = \boxed{0.417}$$

The image is upright (m is $+$) and smaller ($m < 1$) than the object.

The thin-lens and magnification equations can be derived by considering rays 1 and 3 in Figure 26.31a. Ray 1 is shown separately in part b of the drawing, where the angle θ is the same in each of the two colored triangles. Thus, $\tan \theta$ is the same for each triangle:

$$\tan \theta = \frac{h_o}{f} = \frac{-h_i}{d_i - f}$$

A minus sign has been inserted in the numerator of the ratio $h_i/(d_i - f)$ for the following reason. The angle θ in Figure 26.31b is assumed to be positive. Since the image is inverted

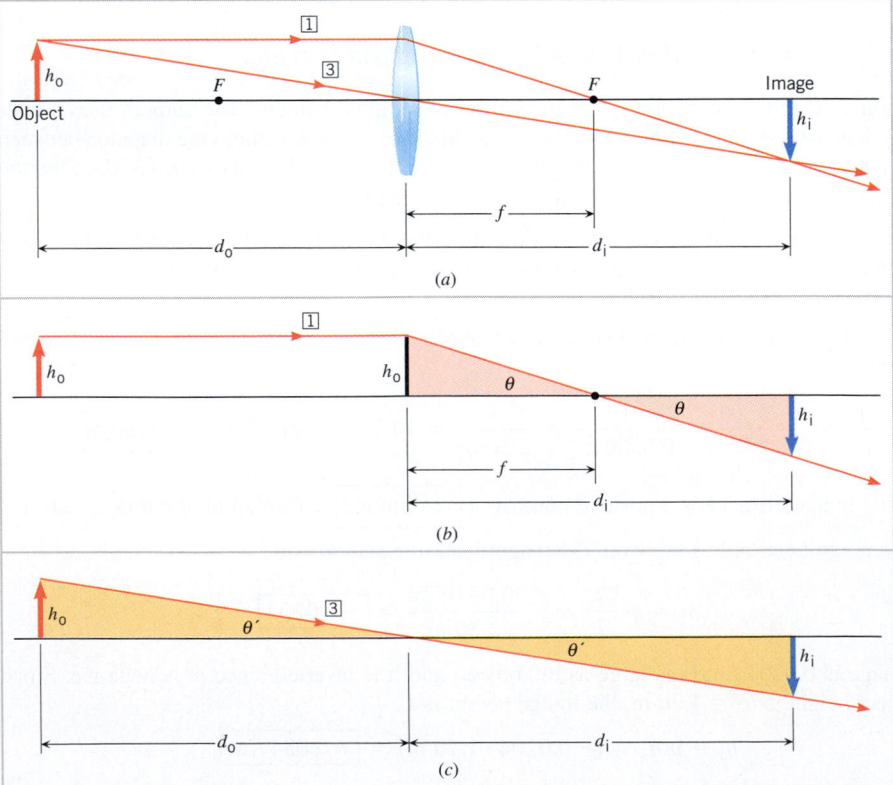

Figure 26.31 These ray diagrams are used for deriving the thin-lens and magnification equations.

relative to the object, the image height h_i is a negative number. The insertion of the minus sign ensures that the term $-h_i/(d_i - f)$, and hence tan θ, is a positive quantity.

Ray 3 is shown separately in part c of the drawing, where the angles θ' are the same. Therefore,

$$\tan \theta' = \frac{h_o}{d_o} = \frac{-h_i}{d_i}$$

A minus sign has been inserted in the numerator of the term h_i/d_i for the same reason that a minus sign was inserted earlier—namely, to ensure that tan θ' is a positive quantity. The first equation gives $h_i/h_o = -(d_i - f)/f$, while the second equation yields $h_i/h_o = -d_i/d_o$. Equating these two expressions for h_i/h_o and rearranging the result produces the thin-lens equation, $1/d_o + 1/d_i = 1/f$. The magnification equation follows directly from the equation $h_i/h_o = -d_i/d_o$, if we recognize that h_i/h_o is the magnification m of the lens.

✓ CHECK YOUR UNDERSTANDING

(The answers are given at the end of the book.)

18. A spherical mirror and a lens are immersed in water. Compared to the way they work in air, which one do you expect will be more affected by the water?

19. An object is located at a distance d_o in front of a lens. The lens has a focal length f and produces an upright image that is twice as tall as the object. What kind of lens is it, and what is the object distance? Express your answer as a fraction or multiple of the focal length.

20. In an old movie a photographic film negative is introduced as evidence in a trial. The negative shows an image of a house that no longer exists. The verdict depends on knowing exactly how far above the ground a window ledge was (the object height h_o). The distance between the ground and the ledge on the negative (the image height h_i) can be measured. What additional information is needed to calculate h_o?   **(a)** Nothing else is needed.   **(b)** Just the object distance d_o, which is the distance between the house and the camera lens.   **(c)** Just the focal length f of the lens.   **(d)** Both d_o and f are needed.

26.9   LENSES IN COMBINATION

Many optical instruments, such as microscopes and telescopes, use a number of lenses together to produce an image. Among other things, a multiple-lens system can produce an image that is magnified more than is possible with a single lens. For instance, Figure 26.32a shows a two-lens system used in a microscope. The first lens, the lens closest to the object, is referred to as the *objective*. The second lens is known as the *eyepiece* (or *ocular*). The object is placed just outside the focal point F_o of the objective. The image formed by the objective—called the "first image" in the drawing—is real, inverted, and enlarged compared to the object. This first image then serves as the object for the eyepiece. Since the first image falls between the eyepiece and its focal point F_e, the eyepiece forms an enlarged, virtual, final image, which is what the observer sees.

The location of the final image in a multiple-lens system can be determined by applying the thin-lens equation to each lens separately. The key point to remember in such situations is that **the image produced by one lens serves as the object for the next lens,** as the next example illustrates.

Problem-solving insight

Example 11   A Microscope—Two Lenses in Combination

The objective and eyepiece of the compound microscope in Figure 26.32 are both converging lenses and have focal lengths of $f_o = 15.0$ mm and $f_e = 25.5$ mm. A distance of 61.0 mm separates the lenses. The microscope is being used to examine an object placed $d_{o1} = 24.1$ mm in front of the objective. Find the final image distance.

Reasoning The thin-lens equation can be used to locate the final image produced by the eyepiece. We know the focal length of the eyepiece, but to determine the final image distance from the thin-lens equation we also need to know the object distance, which is not given. To obtain this distance, we recall that the image produced by one lens (the objective) is the object for the next lens (the eyepiece). We can use the thin-lens equation to locate the image produced by the

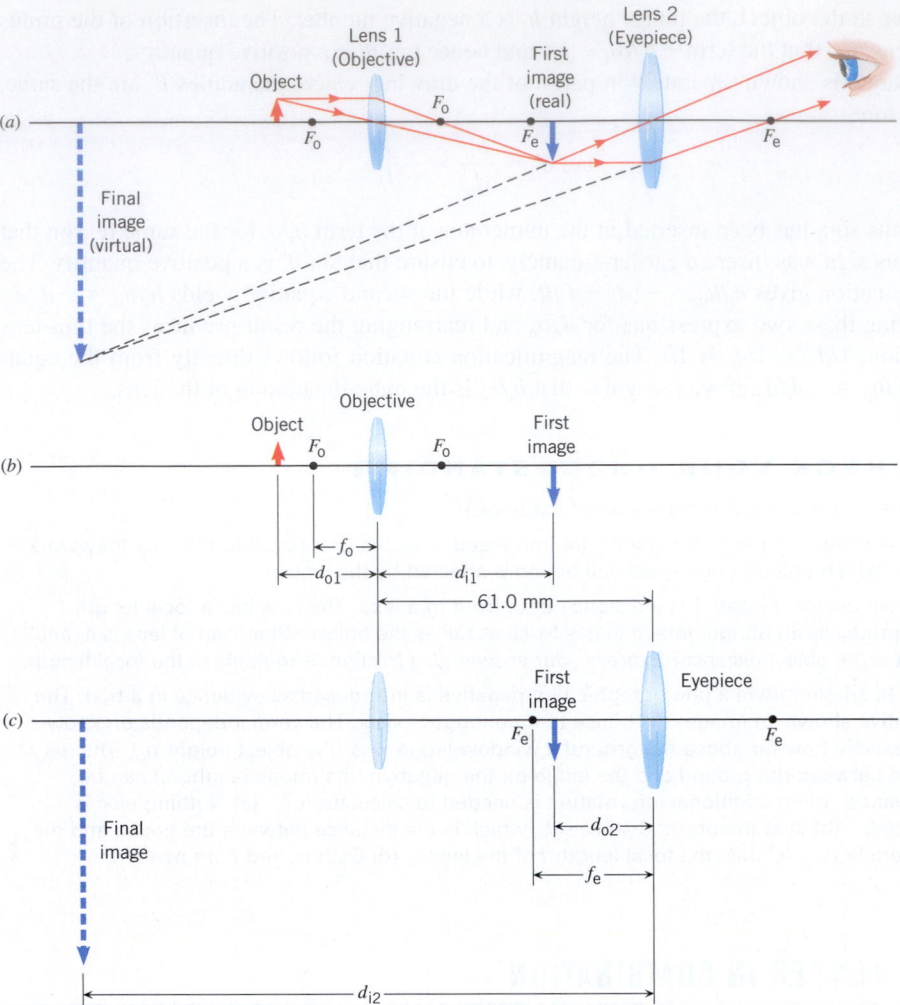

Figure 26.32 (a) This two-lens system can be used as a compound microscope to produce a virtual, enlarged, and inverted final image. (b) The objective forms the first image and (c) the eyepiece forms the final image.

objective, since the focal length and the object distance for this lens are given. The location of this image relative to the eyepiece will tell us the object distance for the eyepiece.

Solution The final image distance relative to the eyepiece is d_{i2}, and we can determine it by using the thin-lens equation:

$$\frac{1}{d_{i2}} = \frac{1}{f_e} - \frac{1}{d_{o2}}$$

The focal length f_e of the eyepiece is known, but to obtain a value for the object distance d_{o2} we must locate the first image produced by the objective. The first image distance d_{i1} (see Figure 26.32b) can be determined using the thin-lens equation with $d_{o1} = 24.1$ mm and $f_o = 15.0$ mm.

$$\frac{1}{d_{i1}} = \frac{1}{f_o} - \frac{1}{d_{o1}} = \frac{1}{15.0 \text{ mm}} - \frac{1}{24.1 \text{ mm}}$$

$$= 0.0252 \text{ mm}^{-1} \quad \text{or} \quad d_{i1} = 39.7 \text{ mm}$$

The first image now becomes the object for the eyepiece (see part c of the drawing). Since the distance between the lenses is 61.0 mm, the object distance for the eyepiece is $d_{o2} = 61.0 \text{ mm} - d_{i1} = 61.0 \text{ mm} - 39.7 \text{ mm} = 21.3 \text{ mm}$. Noting that the focal length of the eyepiece is $f_e = 25.5$ mm, we can find the final image distance with the thin-lens equation:

$$\frac{1}{d_{i2}} = \frac{1}{f_e} - \frac{1}{d_{o2}} = \frac{1}{25.5 \text{ mm}} - \frac{1}{21.3 \text{ mm}}$$

$$= -0.0077 \text{ mm}^{-1} \quad \text{or} \quad \boxed{d_{i2} = -130 \text{ mm}}$$

The fact that d_{i2} is negative indicates that the final image is virtual. It lies to the left of the eyepiece, as the drawing shows.

The overall magnification m of a two-lens system is the product of the magnifications m_1 and m_2 of the individual lenses, or $m = m_1 \times m_2$. Suppose, for example, that the image of lens 1 is magnified by a factor of 5 relative to the original object. As we know, the image of lens 1 serves as the object for lens 2. Suppose, in addition, that lens 2 magnifies this object further by a factor of 8. The final image of the two-lens system, then, would be $5 \times 8 = 40$ times as large as the original object. In other words, $m = m_1 \times m_2$.

Problem-solving insight

26.10

THE HUMAN EYE

ANATOMY

Without doubt, the human eye is the most remarkable of all optical devices. Figure 26.33 shows some of its main anatomical features. The eyeball is approximately spherical with a diameter of about 25 mm. Light enters the eye through a transparent membrane (the *cornea*). This membrane covers a clear liquid region (the *aqueous humor*), behind which are a diaphragm (the *iris*), the *lens*, a region filled with a jelly-like substance (the *vitreous humor*), and, finally, the *retina*. The retina is the light-sensitive part of the eye, consisting of millions of structures called *rods* and *cones*. When stimulated by light, these structures send electrical impulses via the *optic nerve* to the brain, which interprets the image detected by the retina.

The iris is the colored portion of the eye and controls the amount of light reaching the retina. The iris acts as a controller because it is a muscular diaphragm with a variable opening at its center, through which the light passes. The opening is called the *pupil*. The diameter of the pupil varies from about 2 to 7 mm, decreasing in bright light and increasing (dilating) in dim light.

Of prime importance to the operation of the eye is the fact that the lens is flexible, and its shape can be altered by the action of the *ciliary muscle*. The lens is connected to the ciliary muscle by the *suspensory ligaments* (see the drawing). We will see shortly how the shape-changing ability of the lens affects the focusing ability of the eye.

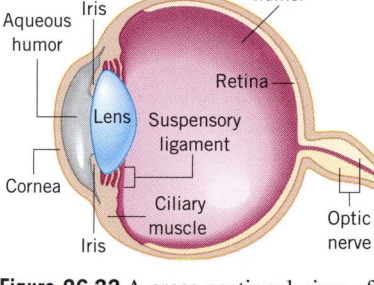

Figure 26.33 A cross-sectional view of the human eye.

OPTICS

Optically, the eye and the camera are similar; both have a lens system and a diaphragm with a variable opening or aperture at its center. Moreover, the retina of the eye and the image sensor in a camera serve similar functions, for both record the image formed by the lens system. In the eye, the image formed on the retina is real, inverted, and smaller than the object, just as it is in a camera. Although the image on the retina is inverted, it is interpreted by the brain as being right-side up.

For clear vision, the eye must refract the incoming light rays, so as to form a sharp image on the retina. In reaching the retina, the light travels through five different media, each with a different index of refraction n: air ($n = 1.00$), the cornea ($n = 1.38$), the aqueous humor ($n = 1.33$), the lens ($n = 1.40$, on the average), and the vitreous humor ($n = 1.34$). Each time light passes from one medium into another, it is refracted at the boundary. The greatest amount of refraction, about 70% or so, occurs at the air/cornea boundary. According to Snell's law, the large refraction at this interface occurs primarily because the refractive index of air ($n = 1.00$) is so different from that of the cornea ($n = 1.38$). The refraction at all the other boundaries is relatively small because the indices of refraction on either side of these boundaries are nearly equal. The lens itself contributes only about 20–25% of the total refraction, since the surrounding aqueous and vitreous humors have indices of refraction that are nearly the same as that of the lens.

Even though the lens contributes only a quarter of the total refraction or less, its function is an important one. The eye has a fixed image distance; that is, the distance between the lens and the retina is constant. Therefore, the only way that objects located at different distances can produce sharp images on the retina is for the focal length of the lens to be adjustable. It is the ciliary muscle that adjusts the focal length. When the eye looks at a very distant object, the ciliary muscle is not tensed. The lens has its least curvature and, consequently, its longest focal length. Under this condition the eye is said to be "fully

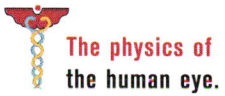

The physics of the human eye.

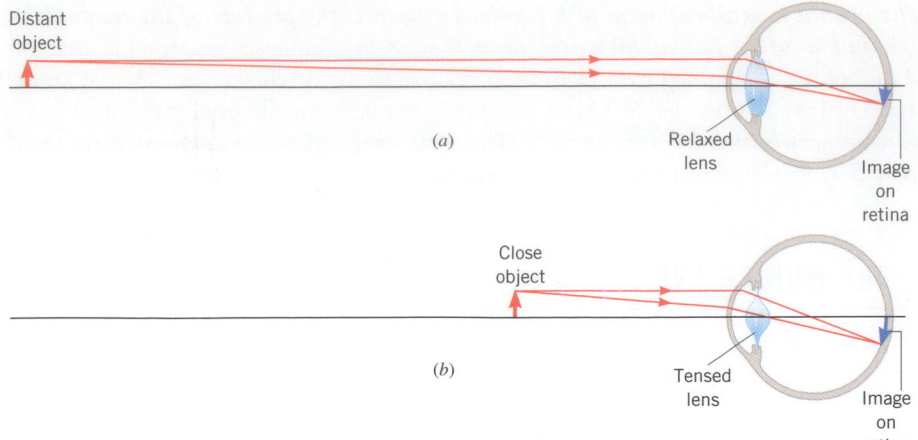

Figure 26.34 (*a*) When fully relaxed, the lens of the eye has its longest focal length, and an image of a very distant object is formed on the retina. (*b*) When the ciliary muscle is tensed, the lens has a shorter focal length. Consequently, an image of a closer object is formed on the retina.

relaxed," and the rays form a sharp image on the retina, as in Figure 26.34*a*. When the object moves closer to the eye, the ciliary muscle tenses automatically, thereby increasing the curvature of the lens, shortening the focal length, and permitting a sharp image to form again on the retina (Figure 26.34*b*). When a sharp image of an object is formed on the retina, we say the eye is "focused" on the object. The process in which the lens changes its focal length to focus on objects at different distances is called *accommodation.*

When you hold a book too close, the print is blurred because the lens cannot adjust enough to bring the book into focus. The point nearest the eye at which an object can be placed and still produce a sharp image on the retina is called the *near point* of the eye. The ciliary muscle is fully tensed when an object is placed at the near point. For people in their early twenties with normal vision, the near point is located about 25 cm from the eye. It increases to about 50 cm at age 40 and to roughly 500 cm at age 60. Since most reading material is held at a distance of 25–45 cm from the eye, older adults typically need eyeglasses to overcome the loss of accommodation. The *far point* of the eye is the location of the farthest object on which the fully relaxed eye can focus. A person with normal eyesight can see objects very far away, such as the planets and stars, and thus has a far point located nearly at infinity.

NEARSIGHTEDNESS

The physics of nearsightedness.

A person who is *nearsighted (myopic)* can focus on nearby objects but cannot clearly see objects far away. For such a person, the far point of the eye is not at infinity and may even be as close to the eye as three or four meters. When a nearsighted eye tries to focus on a distant object, the eye is fully relaxed, like a normal eye. However, the nearsighted eye has a focal length that is shorter than it should be, so rays from the distant object form a sharp image in front of the retina, as Figure 26.35*a* shows, and blurred vision results.

The nearsighted eye can be corrected with glasses or contacts that use *diverging* lenses, as Figure 26.35*b* suggests. The rays from the object diverge after leaving the eyeglass lens. Therefore, when they are subsequently refracted toward the principal axis by the eye, a sharp image is formed farther back and falls on the retina. Since the relaxed (but nearsighted) eye can focus on an object at the eye's far point—but not on objects farther away—the diverging lens is designed to transform a very distant object into an image located at the far point. Part *c* of the drawing shows this transformation, and the next example illustrates how to determine the focal length of the diverging lens that accomplishes it.

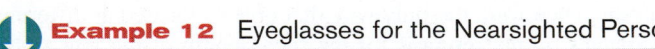

Example 12 Eyeglasses for the Nearsighted Person

A nearsighted person has a far point located only 521 cm from the eye. Assuming that eyeglasses are to be worn 2 cm in front of the eye, find the focal length needed for the diverging lenses of the glasses so the person can see distant objects.

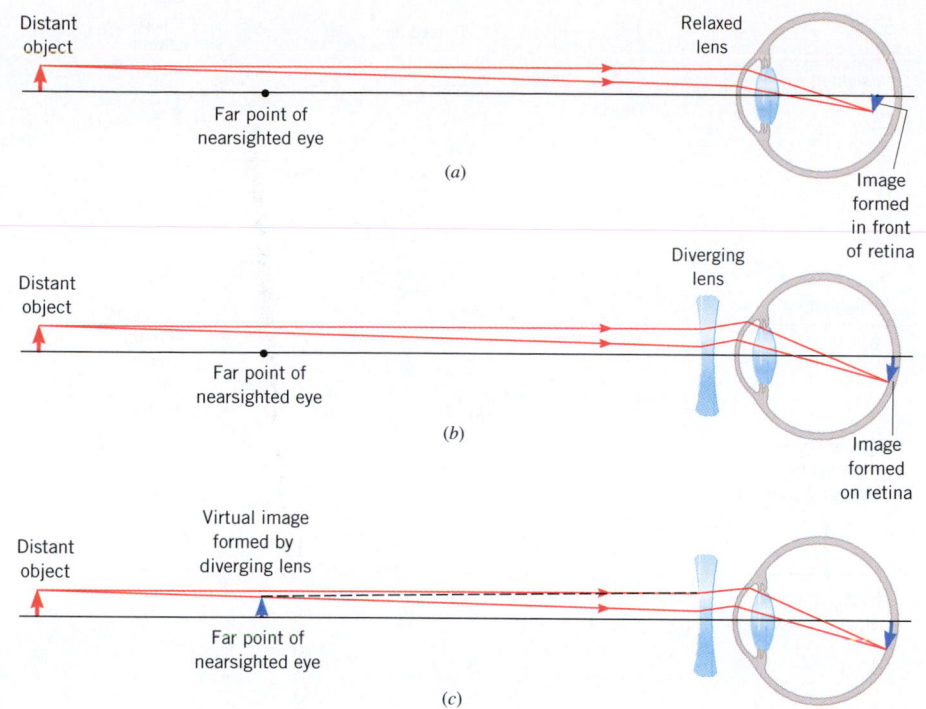

Figure 26.35 (a) When a nearsighted person views a distant object, the image is formed in front of the retina. The result is blurred vision. (b) With a diverging lens in front of the eye, the image is moved onto the retina and clear vision results. (c) The diverging lens is designed to form a virtual image at the far point of the nearsighted eye.

Reasoning In Figure 26.35c the far point is 521 cm away from the eye. Since the glasses are worn 2 cm from the eye, the far point is 519 cm to the left of the diverging lens. The image distance, then, is −519 cm, the negative sign indicating that the image is a virtual image formed to the left of the lens. The object is assumed to be infinitely far from the diverging lens. The thin-lens equation can be used to find the focal length of the eyeglasses. We expect the focal length to be negative, since the lens is a diverging lens.

Problem-solving insight

Eyeglasses are worn about 2 cm from the eyes. Be sure, if necessary, to take this 2 cm into account when determining the object and image distances (d_o and d_i) that are used in the thin-lens equation.

Solution With $d_i = -519$ cm and $d_o = \infty$, the focal length can be found as follows:

$$\frac{1}{f} = \frac{1}{d_o} + \frac{1}{d_i} = \frac{1}{\infty} + \frac{1}{-519 \text{ cm}} \quad \text{or} \quad \boxed{f = -519 \text{ cm}} \qquad (26.6)$$

The value for f is negative, as expected for a diverging lens.

FARSIGHTEDNESS

The physics of farsightedness.

A *farsighted (hyperopic)* person can usually see distant objects clearly, but cannot focus on those nearby. Whereas the near point of a young and normal eye is located about 25 cm from the eye, the near point of a farsighted eye may be considerably farther away than that, perhaps as far as several hundred centimeters. When a farsighted eye tries to focus on a book held closer than the near point, it accommodates and shortens its focal length as much as it can. However, even at its shortest, the focal length is longer than it should be. Therefore, the light rays from the book would form a sharp image behind the retina if they could do so, as Figure 26.36a suggests. In reality, no light passes through the retina, but a blurred image does form on it.

Figure 26.36b shows that farsightedness can be corrected by placing a *converging* lens in front of the eye. The lens refracts the light rays more toward the principal axis before they enter the eye. Consequently, when the rays are refracted even more by the eye, they converge to form an image on the retina. Part c of the figure illustrates what the eye sees when it looks through the converging lens. The lens is designed so that the eye perceives the light to be coming from a virtual image located at the near point. Example 13 shows how the focal length of the converging lens is determined to correct for farsightedness.

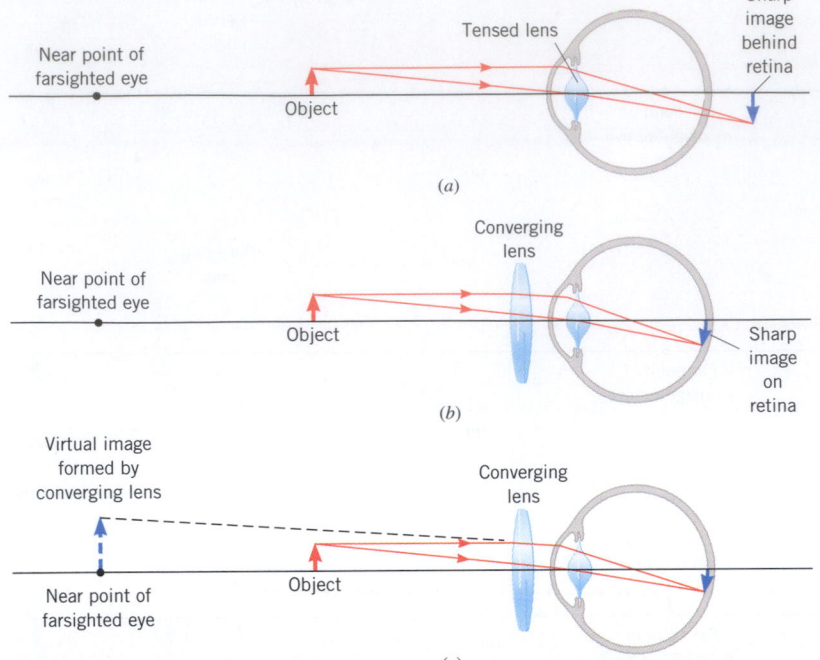

Figure 26.36 (a) When a farsighted person views an object located inside the near point, a sharp image would be formed behind the retina if light could pass through it. Only a blurred image forms on the retina. (b) With a converging lens in front of the eye, the sharp image is moved onto the retina and clear vision results. (c) The converging lens is designed to form a virtual image at the near point of the farsighted eye.

⬇ **Example 13** Contact Lenses for the Farsighted Person

A farsighted person has a near point located 210 cm from the eyes. Obtain the focal length of the converging lenses in a pair of contacts that can be used to read a book held 25.0 cm from the eyes.

Reasoning A contact lens is placed directly against the eye. Thus, the object distance, which is the distance from the book to the lens, is 25.0 cm. The lens forms an image of the book at the near point of the eye, so the image distance is -210 cm. The minus sign indicates that the image is a virtual image formed to the left of the lens, as in Figure 26.36c. The focal length can be obtained from the thin-lens equation.

Solution With $d_o = 25.0$ cm and $d_i = -210$ cm, the focal length can be determined from the thin-lens equation as follows:

$$\frac{1}{f} = \frac{1}{d_o} + \frac{1}{d_i} = \frac{1}{25.0 \text{ cm}} + \frac{1}{-210 \text{ cm}} = 0.0352 \text{ cm}^{-1} \quad \text{or} \quad \boxed{f = 28.4 \text{ cm}}$$

THE REFRACTIVE POWER OF A LENS—THE DIOPTER

The extent to which rays of light are refracted by a lens depends on its focal length. However, optometrists who prescribe correctional lenses and opticians who make the lenses do not specify the focal length directly in prescriptions. Instead, they use the concept of *refractive power* to describe the extent to which a lens refracts light:

$$\frac{\text{Refractive power}}{\text{of a lens (in diopters)}} = \frac{1}{f \text{ (in meters)}} \tag{26.8}$$

The refractive power is measured in units of *diopters*. One diopter is 1 m^{-1}.

Equation 26.8 shows that a converging lens has a refractive power of 1 diopter if it focuses parallel light rays to a focal point 1 m beyond the lens. If a lens refracts parallel rays even more and converges them to a focal point only 0.25 m beyond the lens, the lens has four times more refractive power, or 4 diopters. Since a converging lens has a positive focal length and a diverging lens has a negative focal length, the refractive power of a converging lens is positive and that of a diverging lens is negative. Thus, the eyeglasses in Example 12 would be described in an optometrist's prescription in the following way: Refractive power = $1/(-5.19 \text{ m}) = -0.193$ diopters. The contact lenses in Example 13 would be described in a similar fashion: Refractive power = $1/(0.284 \text{ m}) = 3.52$ diopters.

✓ **CHECK YOUR UNDERSTANDING**

(The answers are given at the end of the book.)

21. Two people who wear glasses are camping. One is nearsighted, and the other is far-sighted. Whose glasses may be useful in starting a fire by concentrating the sun's rays into a small region at the focal point of the lens used in the glasses?

22. Suppose that a person with a near point of 26 cm is standing in front of a plane mirror. How close can he stand to the mirror and still see himself in focus?

23. ⚕ To a swimmer under water, objects look blurred. However, goggles that keep the water away from the eyes allow the swimmer to see objects in sharp focus. Why?

24. When glasses use diverging lenses to correct for nearsightedness or converging lenses to correct for farsightedness, the eyes of the person wearing the glasses lie between the lenses and their focal points. When you look at the eyes of this person, they do not appear to have their normal size. Which one of the following describes what you see? **(a)** The converging lenses make the eyes appear smaller, and the diverging lenses make the eyes appear larger. **(b)** The converging lenses make the eyes appear larger, and the diverging lenses make the eyes appear smaller. **(c)** Both types of lenses make the eyes appear larger. **(d)** Both types of lenses make the eyes appear smaller.

26.11 ANGULAR MAGNIFICATION AND THE MAGNIFYING GLASS

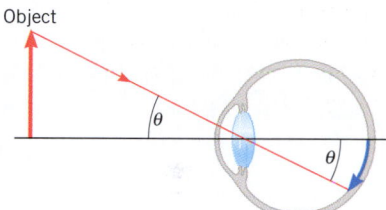

Figure 26.37 The angle θ is the angular size of both the image and the object.

If you hold a penny at arm's length, the penny looks larger than the moon. The reason is that the penny, being so close, forms a larger image on the retina of the eye than does the more distant moon. The brain interprets the larger image of the penny as arising from a larger object. The size of the image on the retina determines how large an object appears to be. However, the size of the image on the retina is difficult to measure. Alternatively, the angle θ subtended by the image can be used as an indication of the image size. Figure 26.37 shows this alternative, which has the advantage that θ is also the angle subtended by the object and, hence, can be measured more easily. The angle θ is called the ***angular size*** of both the image and the object. The larger the angular size, the larger the image on the retina, and the larger the object appears to be.

According to Equation 8.1, the angle θ (measured in radians) is the length of the circular arc that is subtended by the angle divided by the radius of the arc, as Figure 26.38a indicates. Part b of the drawing shows the situation for an object of height h_o viewed at a distance d_o from the eye. When θ is small, h_o is approximately equal to the arc length and d_o is nearly equal to the radius, so that

$$\theta \ (\text{in radians}) = \text{Angular size} \approx \frac{h_o}{d_o}$$

This approximation is good to within one percent for angles of 9° or smaller. In the next example the angular size of a penny is compared with that of the moon.

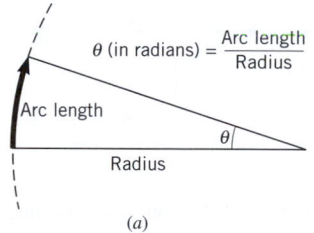

(a)

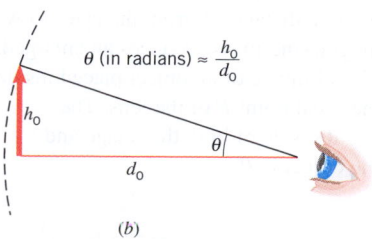

(b)

⬇ **Example 14** A Penny and the Moon

Compare the angular size of a penny (diameter = h_o = 1.9 cm) held at arm's length (d_o = 71 cm) with the angular size of the moon (diameter = h_o = 3.5 × 10⁶ m, and d_o = 3.9 × 10⁸ m).

Reasoning The angular size θ of an object is given approximately by its height h_o divided by its distance d_o from the eye, $\theta \approx h_o/d_o$, provided that the angle involved is less than roughly 9°; this approximation applies here. The "heights" of the penny and the moon are their diameters.

Solution The angular sizes of the penny and moon are

Penny $\qquad \theta \approx \dfrac{h_o}{d_o} = \dfrac{1.9 \ \text{cm}}{71 \ \text{cm}} = \boxed{0.027 \ \text{rad} \ (1.5°)}$

Moon $\qquad \theta \approx \dfrac{h_o}{d_o} = \dfrac{3.5 \times 10^6 \ \text{m}}{3.9 \times 10^8 \ \text{m}} = \boxed{0.0090 \ \text{rad} \ (0.52°)}$

The penny thus appears to be about three times as large as the moon.

Figure 26.38 (a) The angle θ, measured in radians, is the arc length divided by the radius. (b) For small angles (less than 9°), θ in radians is approximately equal to h_o/d_o, where h_o and d_o are the object height and distance.

An optical instrument, such as a magnifying glass, allows us to view small or distant objects because it produces a larger image on the retina than would be possible otherwise. In other words, an optical instrument magnifies the angular size of the object. The *angular magnification* (or *magnifying power*) M is the angular size θ' of the final image produced by the instrument divided by a reference angular size θ. The reference angular size is the angular size of the object when seen without the instrument.

$$\text{\textit{Angular}}\atop\text{\textit{magnification}} \qquad M = \dfrac{\begin{array}{c}\text{Angular size of}\\\text{final image produced}\\\text{by optical instrument}\end{array}}{\begin{array}{c}\text{Reference angular size}\\\text{of object seen without}\\\text{optical instrument}\end{array}} = \dfrac{\theta'}{\theta} \qquad (26.9)$$

The physics of a magnifying glass.

A magnifying glass is the simplest device that provides angular magnification. In this case, the reference angular size θ is chosen to be the angular size of the object when placed at the near point of the eye and seen without the magnifying glass. Since an object cannot be brought closer than the near point and still produce a sharp image on the retina, θ represents the largest angular size obtainable without the magnifying glass. Figure 26.39a indicates that the reference angular size is $\theta \approx h_o/N$, where N is the distance from the eye to the near point. To compute θ', recall from Section 26.7 and Figure 26.28 that a magnifying glass is usually a single converging lens, with the object located inside the focal point. In this situation, Figure 26.39b indicates that the lens produces a virtual image that is enlarged and upright with respect to the object. Assuming that the eye is next to the magnifying glass, the angular size θ' seen by the eye is $\theta' \approx h_o/d_o$, where d_o is the object distance. The angular magnification is

$$M = \frac{\theta'}{\theta} \approx \frac{h_o/d_o}{h_o/N} = \frac{N}{d_o}$$

According to the thin-lens equation, d_o is related to the image distance d_i and the focal length f of the lens by

$$\frac{1}{d_o} = \frac{1}{f} - \frac{1}{d_i}$$

Substituting this expression for $1/d_o$ into the previous expression for M leads to the following result:

Angular magnification of a magnifying glass
$$M = \frac{\theta'}{\theta} \approx \left(\frac{1}{f} - \frac{1}{d_i}\right)N \qquad (26.10)$$

Two special cases of this result are of interest, depending on whether the image is located as close to the eye as possible or as far away as possible. To be seen clearly, the closest the image can be relative to the eye is at the near point, or $d_i = -N$. The minus sign indicates that the image lies to the left of the lens and is virtual. In this event, Equation 26.10 becomes $M \approx (N/f) + 1$. The farthest the image can be from the eye is at infinity $(d_i = -\infty)$; this occurs when the object is placed at the focal point of the lens. When the image is at infinity, Equation 26.10 simplifies to $M \approx N/f$. Clearly, the angular magnification is greater when the image is at the near point of the eye rather than at infinity. In either case, however, the greatest magnification is achieved by using a magnifying glass with the shortest possible focal length. Example 15 illustrates how to determine the angular magnification of a magnifying glass that is used in these two ways.

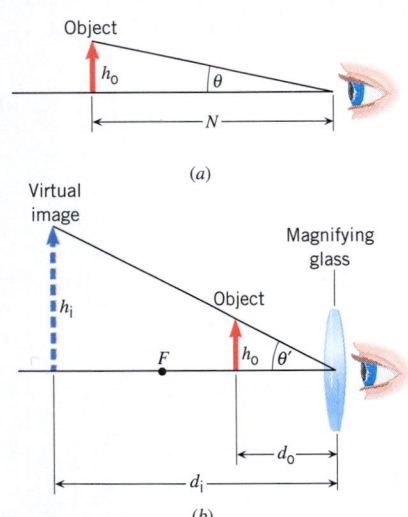

Figure 26.39 (*a*) Without a magnifying glass, the largest angular size θ occurs when the object is placed at the near point, a distance N from the eye. (*b*) A magnifying glass produces an enlarged, virtual image of an object placed inside the focal point F of the lens. The angular size of both the image and the object is θ'.

 Example 15 Examining a Diamond with a Magnifying Glass

A jeweler, whose near point is 40.0 cm from his eye and whose far point is at infinity, is using a small magnifying glass (called a loupe) to examine a diamond. The lens of the magnifying glass has a focal length of 5.00 cm, and the image of the gem is −185 cm from the lens. The image distance is negative because the image is virtual and is formed on the same side of the lens as the object. **(a)** Determine the angular magnification of the magnifying glass. **(b)** Where should the image be located so the jeweler's eye is fully relaxed and has the least strain? What is the angular magnification under this "least strain" condition?

Reasoning The angular magnification of the magnifying glass can be determined from Equation 26.10. In part (a) the image distance is -185 cm. In part (b) the ciliary muscle of the jeweler's eye is fully relaxed, so the image must be located infinitely far from the eye, at its far point, as Section 26.10 discusses.

Solution (a) With $f = 5.00$ cm, $d_i = -185$ cm, and $N = 40.0$ cm, the angular magnification is

$$M = \left(\frac{1}{f} - \frac{1}{d_i}\right)N = \left(\frac{1}{5.00 \text{ cm}} - \frac{1}{-185 \text{ cm}}\right)(40.0 \text{ cm}) = \boxed{8.22}$$

(b) With $f = 5.00$ cm, $d_i = -\infty$, and $N = 40.0$ cm, the angular magnification is

$$M = \left(\frac{1}{f} - \frac{1}{d_i}\right)N = \left(\frac{1}{5.00 \text{ cm}} - \frac{1}{-\infty}\right)(40.0 \text{ cm}) = \boxed{8.00}$$

Jewelers often prefer to minimize eyestrain when viewing objects, even though it means a slight reduction in angular magnification.

An employee of the department for counterfeit money in Munich, Germany, uses a magnifying glass to examine a 200-euro note. (© Peter Kneffel/dpa/Landov)

✓ CHECK YOUR UNDERSTANDING

(The answers are given at the end of the book.)

25. A bird-watcher sees the following three raptors in the air at the distances indicated: a kestrel (wing span = 0.58 m at a distance of 21 m), a bald eagle (wing span = 2.29 m at a distance of 95 m), and a red-tailed hawk (wing span = 1.27 m at a distance of 41 m). Rank the raptors in descending order (largest first) according to the angular size seen by the bird-watcher.

26. Who benefits more from using a magnifying glass, a person whose near point is located at a distance away from the eyes of **(a)** 75 cm or **(b)** 25 cm?

27. A person who has a near point of 25.0 cm is looking with unaided eyes at an object that is located at the near point. The object has an angular size of 0.012 rad. Then, holding a magnifying glass ($f = 10.0$ cm) next to her eye, she views the image of this object, the image being located at her near point. What is the angular size of the image?

26.12 THE COMPOUND MICROSCOPE

The physics of the compound microscope.

To increase the angular magnification beyond that possible with a magnifying glass, an additional converging lens can be included to "premagnify" the object before the magnifying glass comes into play. The result is an optical instrument known as the *compound microscope* (Figure 26.40). As discussed in Section 26.9, the magnifying glass is called the eyepiece, and the additional lens is called the objective.

The angular magnification of the compound microscope is $M = \theta'/\theta$ (Equation 26.9), where θ' is the angular size of the final image and θ is the reference angular size. As with the magnifying glass in Figure 26.39, the reference angular size is determined by the height h_o of the object when the object is located at the near point of the unaided eye: $\theta \approx h_o/N$, where N is the distance between the eye and the near point. Assuming that the object is placed just outside the focal point F_o of the objective (see Figure 26.32a) and that the final image is very far from the eyepiece (i.e., near infinity; see Figure 26.32c), it can be shown that

Angular magnification of a compound microscope

$$M \approx -\frac{(L - f_e)N}{f_o f_e} \qquad (L > f_o + f_e) \tag{26.11}$$

In Equation 26.11, f_o and f_e are, respectively, the focal lengths of the objective and the eyepiece. The angular magnification is greatest when f_o and f_e are as small as possible (since they are in the denominator in Equation 26.11) and when the distance L between the lenses is as large as possible. Furthermore, L must be greater than the sum of f_o and f_e for this equation to be valid. Example 16 deals with the angular magnification of a compound microscope.

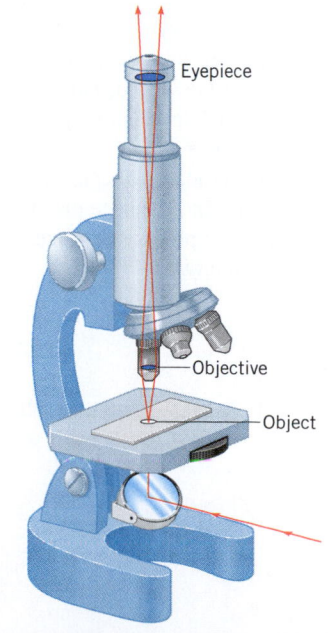

Figure 26.40 A compound microscope.

(Labels: Eyepiece, Objective, Object)

⬇ **Example 16** The Angular Magnification of a Compound Microscope

The focal length of the objective of a compound microscope is $f_o = 0.40$ cm, and the focal length of the eyepiece is $f_e = 3.0$ cm. The two lenses are separated by a distance of $L = 20.0$ cm. A person with a near point distance of $N = 25$ cm is using the microscope. (a) Determine the angular magnification of the microscope. (b) Compare the answer in part (a) with the largest angular magnification obtainable by using the eyepiece alone as a magnifying glass.

Reasoning The angular magnification of the compound microscope can be obtained directly from Equation 26.11, since all the variables are known. When the eyepiece is used alone as a magnifying glass, as in Figure 26.39b, the largest angular magnification occurs when the image seen through the eyepiece is as close as possible to the eye. The image in this case is at the near point, and according to Equation 26.10, the angular magnification is $M \approx (N/f_e) + 1$.

Solution (a) The angular magnification of the compound microscope is

$$M \approx -\frac{(L - f_e)N}{f_o f_e} = -\frac{(20.0 \text{ cm} - 3.0 \text{ cm})(25 \text{ cm})}{(0.40 \text{ cm})(3.0 \text{ cm})} = \boxed{-350}$$

The minus sign indicates that the final image is inverted relative to the initial object.

(b) The maximum angular magnification of the eyepiece by itself is

$$M \approx \frac{N}{f_e} + 1 = \frac{25 \text{ cm}}{3.0 \text{ cm}} + 1 = \boxed{9.3}$$

The effect of the objective is to increase the angular magnification of the compound microscope by a factor of $350/9.3 = 38$ compared to the angular magnification of a magnifying glass.

⬆

26.13 THE TELESCOPE

The physics of the telescope.

A telescope is an instrument for magnifying distant objects, such as stars and planets. Like a microscope, a telescope consists of an objective and an eyepiece (also called the ocular). When the objective is a lens, as is the case in this section, the telescope is referred to as a *refracting* telescope, since lenses utilize the refraction of light.*

Usually the object being viewed is far away, so the light rays entering the telescope are nearly parallel, and the "first image" is formed just beyond the focal point F_o of the objective, as Figure 26.41a illustrates. The first image is real and inverted. Unlike the first image in the compound microscope, however, this image is *smaller* than the object. If, as in part b of the drawing, the telescope is constructed so the first image lies just inside the focal point F_e of the eyepiece, the eyepiece acts like a magnifying glass. It forms a final image that is greatly enlarged, virtual, and located near infinity. This final image can then be viewed with a fully relaxed eye.

The angular magnification M of a telescope, like that of a magnifying glass or a microscope, is the angular size θ' subtended by the final image of the telescope divided by the reference angular size θ of the object. For an astronomical object, such as a planet, it is convenient to use as a reference the angular size of the object seen in the sky with the unaided eye. Since the object is far away, the angular size seen by the unaided eye is nearly the same

Figure 26.41 (a) An astronomical telescope is used to view distant objects. (Note the "break" in the principal axis, between the object and the objective.) The objective produces a real, inverted first image. (b) The eyepiece magnifies the first image to produce the final image near infinity.

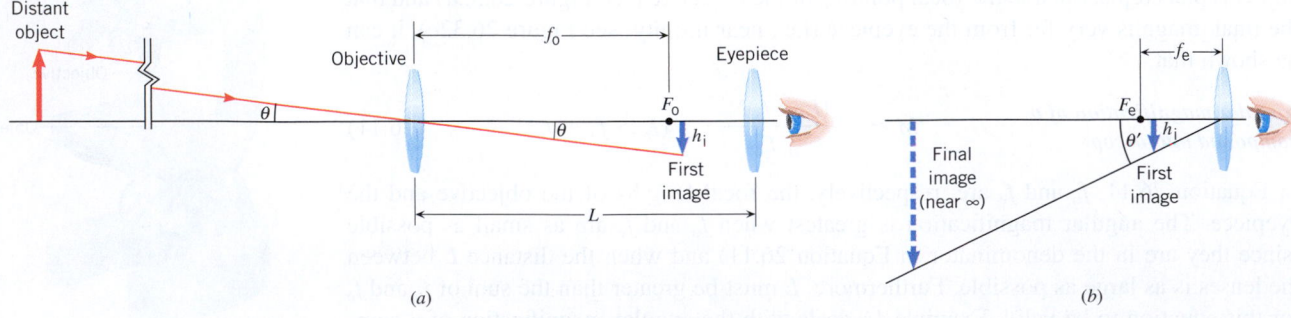

*Another type of telescope utilizes a mirror instead of a lens for the objective and is called a *reflecting* telescope.

as the angle θ subtended at the objective of the telescope in Figure 26.41a. Moreover, θ is also the angle subtended by the first image, so $\theta \approx -h_i/f_o$, where h_i is the height of the first image and f_o is the focal length of the objective. A minus sign has been inserted into this equation because the first image is inverted relative to the object and the image height h_i is a negative number. The insertion of the minus sign ensures that the term $-h_i/f_o$, and hence θ, is a positive quantity. To obtain an expression for θ', refer to Figure 26.41b and note that the first image is located very near the focal point F_e of the eyepiece, which has a focal length f_e. Therefore, $\theta' \approx h_i/f_e$. The angular magnification of the telescope is approximately

Angular magnification of an astronomical telescope
$$M = \frac{\theta'}{\theta} \approx \frac{h_i/f_e}{-h_i/f_o} \approx -\frac{f_o}{f_e} \qquad (26.12)$$

The angular magnification is determined by the ratio of the focal length of the objective to the focal length of the eyepiece. For large angular magnifications, the objective should have a long focal length and the eyepiece a short one. Some of the design features of a telescope are the topic of the next example.

Example 17 The Angular Magnification of an Astronomical Telescope

The telescope shown in Figure 26.42 has the following specifications: f_o = 985 mm and f_e = 5.00 mm. From these data, find **(a)** the angular magnification of the telescope and **(b)** the approximate length of the telescope.

Reasoning The angular magnification of the telescope follows directly from Equation 26.12, since the focal lengths of the objective and eyepiece are known. We can find the length of the telescope by noting that it is approximately equal to the distance L between the objective and eyepiece. Figure 26.41 shows that the first image is located just beyond the focal point F_o of the objective and just inside the focal point F_e of the eyepiece. These two focal points are, therefore, very close together, so the distance L is approximately the sum of the two focal lengths: $L \approx f_o + f_e$.

Solution **(a)** The angular magnification is approximately

$$M \approx -\frac{f_o}{f_e} = -\frac{985 \text{ mm}}{5.00 \text{ mm}} = \boxed{-197} \qquad (26.12)$$

(b) The approximate length of the telescope is

$$L \approx f_o + f_e = 985 \text{ mm} + 5.00 \text{ mm} = \boxed{990 \text{ mm}}$$

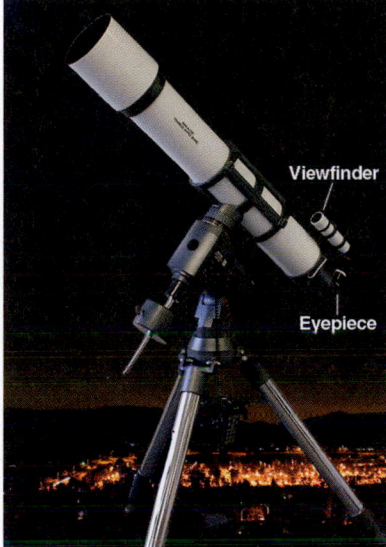

Figure 26.42 An astronomical telescope. The viewfinder is a separate small telescope with low magnification and serves as an aid in locating the object. Once the object has been found, the viewer looks through the eyepiece to obtain the full magnification of the telescope. (© Tony Freeman/PhotoEdit)

✓ CHECK YOUR UNDERSTANDING

(The answers are given at the end of the book.)

28. In the construction of a telescope, one of two lenses is to be used as the objective and one as the eyepiece. The focal lengths of the lenses are **(a)** 3 cm and **(b)** 45 cm. Which lens should be used as the objective?

29. Two refracting telescopes have identical eyepieces, although one telescope is twice as long as the other. Which telescope has the greater angular magnification?

30. A well-designed optical instrument is composed of two converging lenses separated by 14 cm. The focal lengths of the lenses are 0.60 and 4.5 cm. Is the instrument a microscope or a telescope?

31. It is often thought that virtual images are somehow less important than real images. To show that this is not true, identify which of the following instruments normally produce final images that are virtual: **(a)** a projector, **(b)** a camera, **(c)** a magnifying glass, **(d)** eyeglasses, **(e)** a compound microscope, and **(f)** an astronomical telescope.

26.14 LENS ABERRATIONS

Rather than forming a sharp image, a single lens typically forms an image that is slightly out of focus. This lack of sharpness arises because the rays originating from a single point on the object are not focused to a single point on the image. As a result, each point

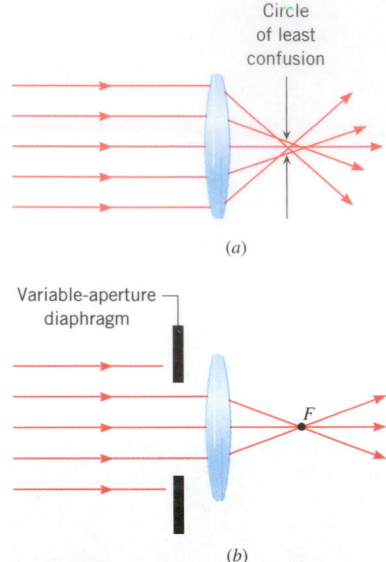

Figure 26.43 (*a*) In a converging lens, spherical aberration prevents light rays parallel to the principal axis from converging to a common point. (*b*) Spherical aberration can be reduced by allowing only rays near the principal axis to pass through the lens. The refracted rays now converge more nearly to a single focal point *F*.

on the image becomes a small blur. The lack of point-to-point correspondence between object and image is called an *aberration*.

One common type of aberration is **spherical aberration,** and it occurs with converging and diverging lenses made with spherical surfaces. Figure 26.43*a* shows how spherical aberration arises with a converging lens. Ideally, all rays traveling parallel to the principal axis are refracted so they cross the axis at the same point after passing through the lens. However, rays far from the principal axis are refracted more by the lens than are those closer in. Consequently, the outer rays cross the axis closer to the lens than do the inner rays, so a lens with spherical aberration does not have a unique focal point. Instead, as the drawing suggests, there is a location along the principal axis where the light converges to the smallest cross-sectional area. This area is circular and is known as the *circle of least confusion.* The circle of least confusion is where the most satisfactory image can be formed by the lens.

Spherical aberration can be reduced substantially by using a variable-aperture diaphragm to allow only those rays close to the principal axis to pass through the lens. Figure 26.43*b* indicates that a reasonably sharp focal point can be achieved by this method, although less light now passes through the lens. Lenses with parabolic surfaces are also used to reduce this type of aberration, but they are difficult and expensive to make.

Chromatic aberration also causes blurred images. It arises because the index of refraction of the material from which the lens is made varies with wavelength. Section 26.5 discusses how this variation leads to the phenomenon of dispersion, in which different colors refract by different amounts. Figure 26.44*a* shows sunlight incident on a converging lens, in which the light spreads into its color spectrum because of dispersion. For clarity, however, the picture shows only the colors at the opposite ends of the visible spectrum—red and violet. Violet is refracted more than red, so the violet ray crosses the principal axis closer to the lens than does the red ray. Thus, the focal length of the lens is shorter for violet than for red, with intermediate values of the focal length corresponding to the colors in between. As a result of chromatic aberration, an undesirable color fringe surrounds the image.

Chromatic aberration can be greatly reduced by using a compound lens, such as the combination of a converging lens and a diverging lens shown in Figure 26.44*b*. Each lens is made from a different type of glass. With this lens combination the red and violet rays almost come to a common focus and, thus, chromatic aberration is reduced. A lens combination designed to reduce chromatic aberration is called an *achromatic lens* (from the Greek "achromatos," meaning "without color"). All high-quality cameras use achromatic lenses.

✔ **CHECK YOUR UNDERSTANDING**

(*The answer is given at the end of the book.*)

32. Why does chromatic aberration occur in lenses but not in mirrors?

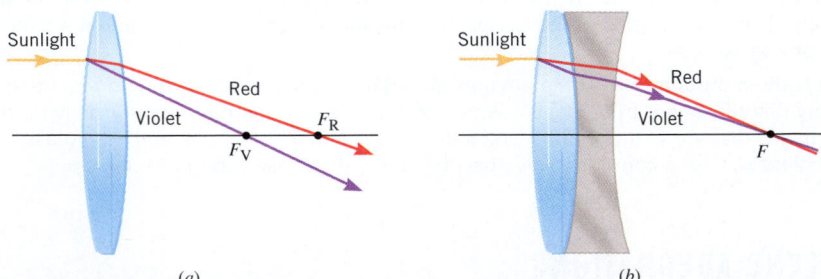

Figure 26.44 (*a*) Chromatic aberration arises when different colors are focused at different points along the principal axis: F_V = focal point for violet light, F_R = focal point for red light. (*b*) A converging and a diverging lens in tandem can be designed to bring different colors more nearly to the same focal point *F*.

CONCEPTS & CALCULATIONS

One important phenomenon discussed in this chapter is how a ray of light is refracted when it goes from one medium into another. Example 18 reviews some of the important aspects of refraction, including Snell's law, the concept of a critical angle, and the notion of index matching.

Concepts & Calculations Example 18
Refraction

A ray of light is incident on a glass–water interface at the critical angle θ_c, as Figure 26.45 illustrates. The reflected light then passes through a liquid (immiscible with water) and into air. The indices of refraction for the four substances are given in the drawing. Determine the angle of refraction θ_5 for the ray as it passes into the air.

Concept Questions and Answers What determines the critical angle when the ray strikes the glass–water interface?

Answer According to Equation 26.4, the critical angle θ_c is determined by the indices of refraction n_{glass} and n_{water}: $\theta_c = \sin^{-1}(n_{water}/n_{glass})$.

When light is incident at the glass–water interface at the critical angle, what is the angle of refraction, and how is the angle of reflection θ_1 related to the critical angle?

Answer When light is incident at the critical angle, the angle of refraction is 90°, so the refracted ray skims along the surface. (For simplicity, this ray is not shown in the drawing.) We know from our work with mirrors that the angle of reflection is equal to the angle of incidence. Therefore, the angle of reflection is equal to the critical angle, so $\theta_1 = \theta_c$.

When the reflected ray strikes the glass–liquid interface, how is the angle of refraction θ_3 related to the angle of incidence θ_2? Note that the two materials have the same indices of refraction.

Answer According to Snell's law, Equation 26.2, the angle of refraction θ_3 is related to the angle of incidence θ_2 by $\sin \theta_3 = (n_{glass} \sin \theta_2)/n_{liquid}$. Since the two indices of refraction are equal, a condition known as "index matching," $\sin \theta_3 = \sin \theta_2$, and the two angles are equal. Therefore, the ray crosses the glass–liquid interface without being refracted.

When the ray passes from the liquid into the air, is the ray refracted?

Answer Yes. Because the two materials have different indices of refraction, the angle of refraction θ_5 is different from the angle of incidence θ_4. The angle of refraction can be found from Snell's law.

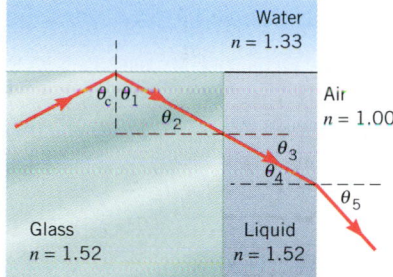

Figure 26.45 A ray of light strikes the glass–water interface at the critical angle θ_c. The reflected ray passes through a liquid and then into air.

Solution Let us follow the ray of light as it progresses from left to right in Figure 26.45. It strikes the glass–water interface at the critical angle of

$$\theta_c = \sin^{-1}\left(\frac{n_{water}}{n_{glass}}\right) = \sin^{-1}\left(\frac{1.33}{1.52}\right) = 61.0°$$

The angle of reflection θ_1 from the glass–water interface is the same as the critical angle, so $\theta_1 = 61.0°$.

The ray then strikes the glass–liquid interface with an angle of incidence of $\theta_2 = 90.0° - 61.0° = 29.0°$. Note that, as usual, the angle of incidence is measured relative to the normal. At the glass–liquid interface, the angle of refraction θ_3 is the same as the angle of incidence θ_2, since the indices of refraction of the two materials are the same ($n_{glass} = n_{liquid} = 1.52$). Thus, $\theta_3 = \theta_2 = 29.0°$.

At the liquid–air interface we use Snell's law, Equation 26.2, to determine the angle θ_5 at which the refracted ray enters the air: $n_{liquid} \sin \theta_4 = n_{air} \sin \theta_5$. Note from the drawing that the angle of incidence θ_4 is the same as θ_3, so $\theta_4 = 29.0°$. The angle of refraction is

$$\theta_5 = \sin^{-1}\left(\frac{1.52 \sin 29.0°}{1.00}\right) = \boxed{47.5°}$$

One of the most important uses of refraction is in lenses, the behavior of which is governed by the thin-lens and magnification equations. Example 19 discusses how these equations are applied to a two-lens system and reviews the all-important sign conventions that must be followed (see the Reasoning Strategy in Section 26.8). In particular, the example shows how to account for a virtual object, in which the object lies to the right (rather than to the left) of a lens.

Concepts & Calculations Example 19
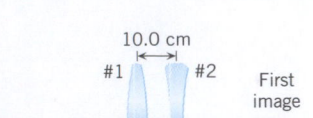
A Two-Lens System

In Figure 26.46 a converging lens ($f_1 = +20.0$ cm) and a diverging lens ($f_2 = -15.0$ cm) are separated by a distance of 10.0 cm. An object with a height of $h_{o1} = 5.00$ mm is placed at a distance of $d_{o1} = 45.0$ cm to the left of the first (converging) lens. What are (a) the image distance d_{i1} and (b) the height h_{i1} of the image produced by the first lens? (c) What is the object distance for the second (diverging) lens? Find (d) the image distance d_{i2} and (e) the height h_{i2} of the image produced by the second lens.

Figure 26.46 The image produced by the first lens, the "first image," falls to the right of the second lens, so the image becomes a virtual object for the second lens.

Concept Questions and Answers Is the image produced by the first (converging) lens real or virtual?

Answer A converging lens can form either a real or a virtual image, depending on where the object is located relative to the focal point. If the object is to the left of the focal point, as it is in this example, the image is real and falls to the right of the lens. (Had the object been located between the focal point and the lens, the image would have been a virtual image located to the left of the converging lens.)

As far as the second lens is concerned, what role does the image produced by the first lens play?

Answer The image produced by the first lens acts as the object for the second lens.

Note in Figure 26.46 that the image produced by the first lens is called the "first image," and it falls to the right of the second lens. This image acts as the object for the second lens. Normally, however, an object would lie to the left of a lens. How do we take into account that this object lies to the right of the diverging lens?

Answer According to the Reasoning Strategy in Section 26.8, an object that is located to the right of a lens is called a virtual object and is assigned a negative object distance.

How do we find the location of the image produced by the second lens when its object is a virtual object?

Answer Once the object distance for the second lens has been assigned a negative number, we can use the thin-lens equation in the usual manner to find the image distance.

Solution (a) The distance d_{i1} of the image from the first lens can be found from the thin-lens equation, Equation 26.6. The focal length, $f_1 = +20.0$ cm, is positive because the lens is a converging lens, and the object distance, $d_{o1} = +45.0$ cm, is positive because it lies to the left of the lens:

$$\frac{1}{d_{i1}} = \frac{1}{f_1} - \frac{1}{d_{o1}} = \frac{1}{20.0 \text{ cm}} - \frac{1}{45.0 \text{ cm}} = 0.0278 \text{ cm}^{-1} \quad \text{or} \quad d_{i1} = \boxed{36.0 \text{ cm}}$$

This image distance is positive, indicating that the image is real.

(b) The height h_{i1} of the image produced by the first lens can be obtained from the magnification equation, Equation 26.7:

$$h_{i1} = h_{o1}\left(-\frac{d_{i1}}{d_{o1}}\right) = (5.00 \text{ mm})\left(-\frac{36.0 \text{ cm}}{45.0 \text{ cm}}\right) = \boxed{-4.00 \text{ mm}}$$

The minus sign indicates that the first image is inverted with respect to the object (see Figure 26.46).

(c) The first image falls 36.0 cm to the right of the first lens. However, the second lens is located 10.0 cm to the right of the first lens, and therefore the first image is located $36.0 \text{ cm} - 10.0 \text{ cm} = 26.0$ cm to the right of the second lens. This image acts as the object for the second lens. Since the object for the second lens lies to the right of it, the object is a virtual object and is assigned a negative number: $d_{o2} = \boxed{-26.0 \text{ cm}}$.

(d) The distance d_{i2} of the image from the second lens can be found from the thin-lens equation. The focal length, $f_2 = -15.0$ cm, is negative because the lens is a diverging lens, and the object distance, $d_{o2} = -26.0$ cm, is negative because it lies to the right of the lens:

$$\frac{1}{d_{i2}} = \frac{1}{f_2} - \frac{1}{d_{o2}} = \frac{1}{-15.0 \text{ cm}} - \frac{1}{-26.0 \text{ cm}} = -0.0282 \text{ cm}^{-1}$$

or

$$d_{i2} = \boxed{-35.5 \text{ cm}}$$

The negative sign for d_{i2} means that the image is formed to the left of the diverging lens and, hence, is a virtual image.

(e) The image height h_{i1} produced by the first lens becomes the object height h_{o2} for the second lens: $h_{o2} = h_{i1} = -4.00$ mm. The height h_{i2} of the image produced by the second lens follows from the magnification equation:

$$h_{i2} = h_{o2}\left(-\frac{d_{i2}}{d_{o2}}\right) = (-4.00 \text{ mm})\left(-\frac{-35.5 \text{ cm}}{-26.0 \text{ cm}}\right) = \boxed{5.46 \text{ mm}}$$

CONCEPT SUMMARY

If you need more help with a concept, use the Learning Aids noted next to the discussion or equation. Examples (**Ex.**) are in the text of this chapter. Go to **www.wiley.com/college/cutnell** for the following Learning Aids:

Interactive LearningWare (ILW) — Additional examples solved in a five-step interactive format.

Concept Simulations (CS) — Animated text figures or animations of important concepts.

Interactive Solutions (IS) — Models for certain types of problems in the chapter homework. The calculations are carried out interactively.

Topic	Discussion	Learning Aids
Refraction	**26.1 THE INDEX OF REFRACTION** The change in speed as a ray of light goes from one material to another causes the ray to deviate from its incident direction. This change in direction is called refraction. The index of refraction n of a material is the ratio of the speed c of light in a vacuum to the speed v of light in the material:	
Index of refraction	$$n = \frac{c}{v} \qquad (26.1)$$	**IS 26.7**
	The values for n are greater than unity, because the speed of light in a material medium is less than it is in a vacuum.	
	26.2 SNELL'S LAW AND THE REFRACTION OF LIGHT The refraction that occurs at the interface between two materials obeys Snell's law of refraction. This law states that (1) the refracted ray, the incident ray, and the normal to the interface all lie in the same plane, and (2) the angle of refraction θ_2 is related to the angle of incidence θ_1 according to	
Snell's law of refraction	$$n_1 \sin\theta_1 = n_2 \sin\theta_2 \qquad (26.2)$$	**Ex. 1**
	where n_1 and n_2 are the indices of refraction of the incident and refracting media, respectively. The angles are measured relative to the normal.	
	Because of refraction, a submerged object has an apparent depth that is different from its actual depth. If the observer is directly above (or below) the object, the apparent depth (or height) d' is related to the actual depth (or height) d according to	**Ex. 2**
Apparent depth	$$d' = d\left(\frac{n_2}{n_1}\right) \qquad (26.3)$$	**Ex. 3, 4**
	where n_1 and n_2 are the refractive indices of the materials (the media) in which the object and the observer, respectively, are located.	
	26.3 TOTAL INTERNAL REFLECTION When light passes from a material with a larger refractive index n_1 into a material with a smaller refractive index n_2, the refracted ray is bent away from the	

Topic	Discussion	Learning Aids

normal. If the incident ray is at the critical angle θ_c, the angle of refraction is 90°. The critical angle is determined from Snell's law and is given by

Critical angle

$$\sin \theta_c = \frac{n_2}{n_1} \qquad (n_1 > n_2) \tag{26.4}$$

Total internal reflection

When the angle of incidence exceeds the critical angle, all the incident light is reflected back into the material from which it came, a phenomenon known as total internal reflection.

26.4 POLARIZATION AND THE REFLECTION AND REFRACTION OF LIGHT When light is incident on a nonmetallic surface at the Brewster angle θ_B, the reflected light is completely polarized parallel to the surface. The Brewster angle is given by

Brewster angle

Brewster's law

$$\tan \theta_B = \frac{n_2}{n_1} \tag{26.5}$$

where n_1 and n_2 are the refractive indices of the incident and refracting media, respectively. When light is incident at the Brewster angle, the reflected and refracted rays are perpendicular to each other.

26.5 THE DISPERSION OF LIGHT: PRISMS AND RAINBOWS A glass prism can spread a beam of sunlight into a spectrum of colors because the index of refraction of the glass depends on the wavelength of the light. Thus, a prism bends the refracted rays corresponding to different colors by different amounts. The spreading of light into its color components is known as dispersion. The dispersion of light by water droplets in the air leads to the formation of rainbows.

Dispersion

Index matching

A prism will not bend a light ray at all, neither up nor down, if the surrounding fluid has the same refractive index as the glass, a condition known as index matching.

26.6 LENSES 26.7 THE FORMATION OF IMAGES BY LENSES Converging lenses and diverging lenses depend on the phenomenon of refraction in forming an image. With a converging lens, paraxial rays that are parallel to the principal axis are focused to a point on the axis by the lens. This point is called the focal point of the lens, and its distance from the lens is the focal length f. Paraxial light rays that are parallel to the principal axis of a diverging lens appear to originate from its focal point after passing through the lens. The distance of this point from the lens is the focal length f. The image produced by a converging or a diverging lens can be located via a technique known as ray tracing, which utilizes the three rays outlined in the Reasoning Strategy given in Section 26.7.

Focal point and focal length of a converging lens

Focal point and focal length of a diverging lens

Ray tracing

Image formed by a converging lens

The nature of the image formed by a converging lens depends on where the object is situated relative to the lens. When the object is located at a distance from the lens that is greater than twice the focal length, the image is real, inverted, and smaller than the object. When the object is located at a distance from the lens that is between the focal length and twice the focal length, the image is real, inverted, and larger than the object. When the object is located within the focal length, the image is virtual, upright, and larger than the object.

Image formed by a diverging lens

Regardless of the position of a real object, a diverging lens always produces an image that is virtual, upright, and smaller than the object.

26.8 THE THIN-LENS EQUATION AND THE MAGNIFICATION EQUATION The thin-lens equation can be used with either converging or diverging lenses that are thin, and it relates the object distance d_o, the image distance d_i, and the focal length f of the lens:

Thin-lens equation

$$\frac{1}{d_o} + \frac{1}{d_i} = \frac{1}{f} \tag{26.6}$$

Magnification

The magnification m of a lens is the ratio of the image height h_i to the object height h_o and is also related to d_o and d_i by the magnification equation:

Magnification equation

$$m = \frac{h_i}{h_o} = -\frac{d_i}{d_o} \tag{26.7}$$

The algebraic sign conventions for the variables appearing in the thin-lens and magnification equations are summarized in the Reasoning Strategy given in Section 26.8.

26.9 LENSES IN COMBINATION When two or more lenses are used in combination, the image produced by one lens serves as the object for the next lens.

Topic	Discussion	Learning Aids
	26.10 THE HUMAN EYE In the human eye, a real, inverted image is formed on a light-sensitive surface, called the retina. Accommodation is the process by which the focal length of the eye is automatically adjusted, so that objects at different distances produce sharp images on the retina. The near point of the eye is the point nearest the eye at which an object can be placed and still have a sharp image produced on the retina. The far point of the eye is the location of the farthest object on which the fully relaxed eye can focus. For a young and normal eye, the near point is located 25 cm from the eye, and the far point is located at infinity.	

Accommodation

Near point
Far point

Nearsightedness

A nearsighted (myopic) eye is one that can focus on nearby objects, but not on distant objects. Nearsightedness can be corrected with eyeglasses or contacts made from diverging lenses. A far- **Ex. 12, 13**

Farsightedness

sighted (hyperopic) eye can see distant objects clearly, but not objects close up. Farsightedness can **IS 26.75** be corrected with converging lenses.

The refractive power of a lens is measured in diopters and is given by

Refractive power

$$\text{Refractive power (in diopters)} = \frac{1}{f \text{ (in meters)}} \tag{26.8}$$

where f is the focal length of the lens and must be expressed in meters. A converging lens has a positive refractive power, and a diverging lens has a negative refractive power.

26.11 ANGULAR MAGNIFICATION AND THE MAGNIFYING GLASS The angular size of an object is the angle that it subtends at the eye of the viewer. For small angles, the angular size θ in radians is

Angular size

$$\theta \text{ (in radians)} \approx \frac{h_o}{d_o}$$

Ex. 14

where h_o is the height of the object and d_o is the object distance. The angular magnification M of an optical instrument is the angular size θ' of the final image produced by the instrument divided by the reference angular size θ of the object, which is that seen without the instrument:

Angular magnification

$$M = \frac{\theta'}{\theta} \tag{26.9}$$

A magnifying glass is usually a single converging lens that forms an enlarged, upright, and virtual image of an object placed at or inside the focal point of the lens. For a magnifying glass held close to the eye, the angular magnification M is approximately

Angular magnification
of a magnifying glass

$$M \approx \left(\frac{1}{f} - \frac{1}{d_i} \right) N \tag{26.10}$$ **Ex. 15**

where f is the focal length of the lens, d_i is the image distance, and N is the distance of the viewer's near point from the eye.

26.12 THE COMPOUND MICROSCOPE A compound microscope usually consists of two lenses, an objective and an eyepiece. The final image is enlarged, inverted, and virtual. The angular magnification M of such a microscope is approximately

Angular magnification
of a compound microscope

$$M \approx -\frac{(L - f_e)N}{f_o f_e} \qquad (L > f_o + f_e) \tag{26.11}$$ **Ex. 16**

where f_o and f_e are, respectively, the focal lengths of the objective and eyepiece, L is the distance between the two lenses, and N is the distance of the viewer's near point from his or her eye.

26.13 THE TELESCOPE An astronomical telescope magnifies distant objects with the aid of an objective and an eyepiece, and it produces a final image that is inverted and virtual. The angular magnification M of a telescope is approximately

Angular magnification of an
astronomical telescope

$$M \approx -\frac{f_o}{f_e} \tag{26.12}$$ **Ex. 17**

where f_o and f_e are, respectively, the focal lengths of the objective and the eyepiece.

26.14 LENS ABERRATIONS Lens aberrations limit the formation of perfectly focused or sharp images by optical instruments. Spherical aberration occurs because rays that pass through the outer edge of a lens with spherical surfaces are not focused at the same point as rays that pass through near the center of the lens.

Spherical aberration

Chromatic aberration

Chromatic aberration arises because a lens focuses different colors at slightly different points.

FOCUS ON CONCEPTS

Note to Instructors: The numbering of the questions shown here reflects the fact that they are only a representative subset of the total number that are available online. However, all of the questions are available for assignment via an online homework management program such as WileyPLUS or WebAssign.

Section 26.2   Snell's Law and the Refraction of Light

1. The drawings show two examples in which a ray of light is refracted at the interface between two liquids. In each example the incident ray is in liquid A and strikes the interface at the same angle of incidence. In one case the ray is refracted into liquid B, and in the other it is refracted into liquid C. The dashed lines denote the normals to the interfaces. Rank the indices of refraction of the three liquids in descending order (largest first). **(a)** n_A, n_B, n_C **(b)** n_A, n_C, n_B **(c)** n_C, n_A, n_B **(d)** n_B, n_A, n_C **(e)** n_C, n_B, n_A

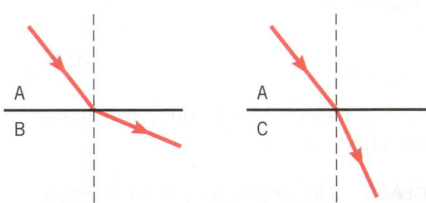

5. A coin is resting on the bottom of an empty container. The container is then filled to the brim three times, each time with a different liquid. An observer (in air) is directly above the coin and looks down at it. With liquid A in the container, the apparent depth of the coin is 7 cm, with liquid B it is 6 cm, and with liquid C it is 5 cm. Rank the indices of refraction of the liquids in descending order (largest first). **(a)** n_A, n_B, n_C **(b)** n_A, n_C, n_B **(c)** n_C, n_A, n_B **(d)** n_C, n_B, n_A **(e)** n_B, n_A, n_C

Section 26.3   Total Internal Reflection

6. The refractive index of material A is greater than the refractive index of material B. A ray of light is incident on the interface between these two materials in a number of ways, as the drawings illustrate. The dashed lines denote the normals to the interfaces. Which one of the drawings shows a situation that is *not* possible? **(a)** Drawing 1 **(b)** Drawing 2 **(c)** Drawing 3 **(d)** Drawing 4

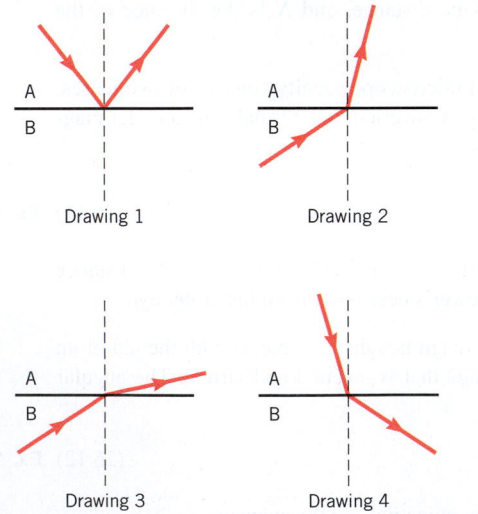

7. The drawing shows a rectangular block of glass ($n = 1.52$) surrounded by air. A ray of light starts out within the glass and travels toward point A, where some or all of it is reflected toward point B. At which points does some of the light escape the glass? **(a)** Only at point A **(b)** Only at point B **(c)** At both points A and B **(d)** At neither point A nor point B

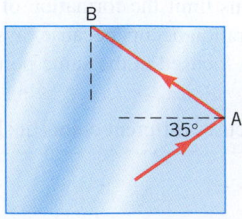

Section 26.4   Polarization and the Reflection and Refraction of Light

8. A diamond ($n = 2.42$) is lying on a table. At what angle of incidence θ is the light that is reflected from one of the facets of the diamond completely polarized?

Section 26.5   The Dispersion of Light: Prisms and Rainbows

9. The indices of refraction for red, green, and violet light in glass are $n_{red} = 1.520$, $n_{green} = 1.526$, and $n_{violet} = 1.538$. When a ray of light passes through a transparent slab of glass, the emergent ray is parallel to the incident ray, but can be displaced relative to it. For light passing through a glass slab that is surrounded by air, which color is displaced the most? **(a)** All colors are displaced equally. **(b)** Red **(c)** Green **(d)** Violet

Section 26.7   The Formation of Images by Lenses

11. An object is situated to the left of a lens. A ray of light from the object is close to and parallel to the principal axis of the lens. The ray passes through the lens. Which one of the following statements is true? **(a)** The ray crosses the principal axis at a distance from the lens equal to twice the focal length, no matter whether the lens is converging or diverging. **(b)** The ray passes through the lens without changing direction, no matter whether the lens is converging or diverging. **(c)** The ray passes through a focal point of the lens, no matter whether the lens is converging or diverging. **(d)** The ray passes through a focal point of the lens only if the lens is a diverging lens. **(e)** The ray passes through a focal point of the lens only if the lens is a converging lens.

12. What type of single lens produces a virtual image that is inverted with respect to the object? **(a)** Both a converging and a diverging lens can produce such an image. **(b)** Neither a converging nor a diverging lens produces such an image. **(c)** A converging lens **(d)** A diverging lens

Section 26.9   Lenses in Combination

15. Two converging lenses have the same focal length of 5.00 cm. They have a common principal axis and are separated by 21.0 cm. An object is located 10.0 cm to the left of the left-hand lens. What is the image distance (relative to the lens on the right) of the final image produced by this two-lens system?

Section 26.10   The Human Eye

17. Here are a number of statements concerning the refractive power of lenses.

    A. A positive refractive power means that a lens always creates an image that is larger than the object.

    B. Two lenses with the same refractive power have the same focal lengths.

    C. A lens with a positive refractive power is a converging lens, whereas a lens with a negative refractive power is a diverging lens.

    D. Two lenses with different refractive powers can have the same focal length.

    E. The fact that lens A has twice the refractive power of lens B means that the focal length of lens A is twice that of lens B.

Which of these statements are false? **(a)** A, B, C **(b)** C, D, E **(c)** A, D, E **(d)** B, C, E **(e)** B, C, D

Section 26.11 Angular Magnification and the Magnifying Glass

18. The table lists the angular sizes in radians and distances from the eye for three objects, A, B, and C. In each case the angular size is small.

Object	Angular Size (in Radians)	Distance of Object from Eye
A	θ	d_o
B	2θ	$2d_o$
C	θ	$2d_o$

Rank the heights of these objects in descending order (largest first).
(a) B, C, A **(b)** B, A, C **(c)** A, B, C **(d)** A, C, B **(e)** C, A, B

Section 26.13 The Telescope

19. An astronomical telescope has an angular magnification of –125 when used properly. What would the angular magnification M be if the objective and the eyepiece were interchanged?

Section 26.14 Lens Aberrations

20. Which one of the five choices below best completes the following statement? The fact that the refractive index depends on the wavelength of light is the cause of _____. **(a)** dispersion **(b)** chromatic aberration **(c)** spherical aberration **(d)** dispersion and chromatic aberration **(e)** spherical aberration and chromatic aberration

PROBLEMS

Note to Instructors: Most of the homework problems in this chapter are available for assignment via an online homework management program such as WileyPLUS or WebAssign, and those marked with the icon **GO** *are presented in WileyPLUS using a guided tutorial format that provides enhanced interactivity. See Preface for additional details.*

Unless specified otherwise, use the values given in Table 26.1 for the refractive indices.

ssm Solution is in the Student Solutions Manual.
www Solution is available online at www.wiley.com/college/cutnell

 This icon represents a biomedical application.

Section 26.1 The Index of Refraction

1. ssm Light travels at a speed of 2.201×10^8 m/s in a certain substance. What substance from Table 26.1 could this be? For the speed of light in a vacuum use 2.998×10^8 m/s; show your calculations.

2. In an ultra-low-temperature experiment, a collection of sodium atoms enter a special state called a *Bose-Einstein condensate* in which the index of refraction is 1.57×10^7. What is the speed of light in this condensate?

3. The refractive indices of materials A and B have a ratio of $n_A/n_B = 1.33$. The speed of light in material A is 1.25×10^8 m/s. What is the speed of light in material B?

4. The frequency of a light wave is the same when the light travels in ethyl alcohol or in carbon disulfide. Find the ratio of the wavelength of the light in ethyl alcohol to that in carbon disulfide.

5. ssm www A plate glass window ($n = 1.5$) has a thickness of 4.0×10^{-3} m. How long does it take light to pass perpendicularly through the plate?

6. Light has a wavelength of 340.0 nm and a frequency of 5.403×10^{14} Hz when traveling through a certain substance. What substance from Table 26.1 could this be? Show your calculations.

*7. Interactive Solution 26.7 at **www.wiley.com/college/cutnell** offers one model for problems like this one. In a certain time, light travels 6.20 km in a vacuum. During the same time, light travels only 3.40 km in a liquid. What is the refractive index of the liquid?

*8. A flat sheet of ice has a thickness of 2.0 cm. It is on top of a flat sheet of crystalline quartz that has a thickness of 1.1 cm. Light strikes the ice perpendicularly and travels through it and then through the quartz. In the time it takes the light to travel through the two sheets, how far (in centimeters) would it have traveled in a vacuum?

Section 26.2 Snell's Law and the Refraction of Light

9. ssm A light ray in air is incident on a water surface at a 43° angle of incidence. Find the angles of **(a)** reflection and **(b)** refraction.

10. **GO** The drawing shows four different situations in which a light ray is traveling from one medium into another. In some of the cases, the refraction is not shown correctly. For cases (a), (b), and (c), the angle of incidence is 55°; for case (d), the angle of incidence is 0°.

Determine the angle of refraction in each case. If the drawing shows the refraction incorrectly, explain why it is incorrect.

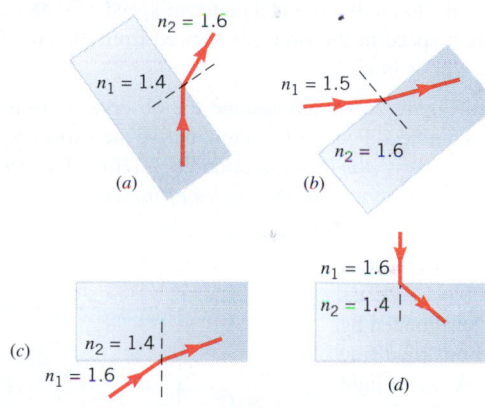

11. A ray of light impinges from air onto a block of ice ($n = 1.309$) at a 60.0° angle of incidence. Assuming that this angle remains the same, find the difference $\theta_{2,\,ice} - \theta_{2,\,water}$ in the angles of refraction when the ice turns to water ($n = 1.333$).

12. A spotlight on a boat is 2.5 m above the water, and the light strikes the water at a point that is 8.0 m horizontally displaced from the spotlight (see the drawing). The depth of the water is 4.0 m. Determine the distance d, which locates the point where the light strikes the bottom.

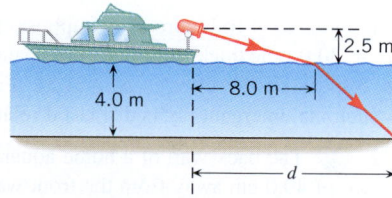

13. ssm The drawing shows a coin resting on the bottom of a beaker filled with an unknown liquid. A ray of light from the coin travels to the surface of the liquid and is refracted as it enters into the air. A person sees the ray as it skims just above the surface of the liquid. How fast is the light traveling in the liquid?

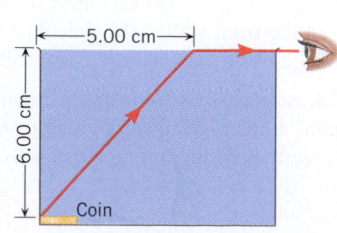

14. A person working on the transmission of a car accidentally drops a bolt into a tray of oil. The oil is 5.00 cm deep. The bolt appears to be 3.40 cm beneath the surface of the oil, when viewed from directly above. What is the index of refraction of the oil?

15. A scuba diver, submerged under water, looks up and sees sunlight at an angle of 28.0° from the vertical. At what angle, measured from the vertical, does this sunlight strike the surface of the water?

16. **GO** The drawing shows a ray of light traveling through three materials whose surfaces are parallel to each other. The refracted rays (but not the reflected rays) are shown as the light passes through each material. A ray of light strikes the *a*–*b* interface at a 50.0° angle of incidence. The index of refraction of material *a* is $n_a = 1.20$.

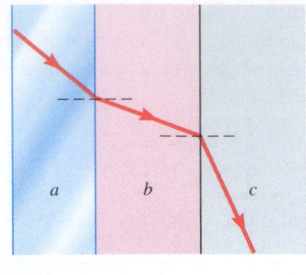

The angles of refraction in materials *b* and *c* are, respectively, 45.0° and 56.7°. Find the indices of refraction in these two media.

17. Light in a vacuum is incident on a transparent glass slab. The angle of incidence is 35.0°. The slab is then immersed in a pool of liquid. When the angle of incidence for the light striking the slab is 20.3°, the angle of refraction for the light entering the slab is the same as when the slab was in a vacuum. What is the index of refraction of the liquid?

*****18.** A stone held just beneath the surface of a swimming pool is released and sinks to the bottom at a constant speed of 0.48 m/s. What is the apparent speed of the stone, as viewed from directly above by an observer who is in air?

*****19.** Refer to Figure 26.4*b* and assume the observer is nearly above the submerged object. For this situation, derive the expression for the apparent depth: $d' = d(n_2/n_1)$, Equation 26.3. (*Hint: Use Snell's law of refraction and the fact that the angles of incidence and refraction are small, so $\tan \theta \approx \sin \theta$.*)

*****20.** The drawing shows a rectangular block of glass ($n = 1.52$) surrounded by liquid carbon disulfide ($n = 1.63$). A ray of light is incident on the glass at point A with a 30.0° angle of incidence. At what angle of refraction does the ray leave the glass at point B?

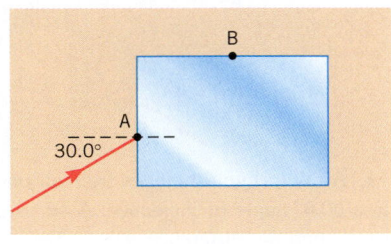

*****21.** **ssm www** In Figure 26.6, suppose that the angle of incidence is $\theta_1 = 30.0°$, the thickness of the glass pane is 6.00 mm, and the refractive index of the glass is $n_2 = 1.52$. Find the amount (in mm) by which the emergent ray is displaced relative to the incident ray.

*****22.** **GO** The back wall of a home aquarium is a mirror that is a distance of 40.0 cm away from the front wall. The walls of the tank are negligibly thin. A fish, swimming midway between the front and back walls, is being viewed by a person looking through the front wall. The index of refraction of air is $n_{air} = 1.000$ and that of water is $n_{water} = 1.333$. **(a)** Calculate the apparent distance between the fish and the front wall. **(b)** Calculate the apparent distance between the image of the fish and the front wall.

*****23.** **ssm** Review Conceptual Example 4 as background for this problem. A man in a boat is looking straight down at a fish in the water directly beneath him. The fish is looking straight up at the man. They

are equidistant from the air–water interface. To the man, the fish appears to be 2.0 m beneath his eyes. To the fish, how far above its eyes does the man appear to be?

******24.** A small logo is embedded in a thick block of crown glass ($n = 1.52$), 3.20 cm beneath the top surface of the glass. The block is put under water, so there is 1.50 cm of water above the top surface of the block. The logo is viewed from directly above by an observer in air. How far beneath the top surface of the water does the logo appear to be?

******25.** A beaker has a height of 30.0 cm. The lower half of the beaker is filled with water, and the upper half is filled with oil ($n = 1.48$). To a person looking down into the beaker from above, what is the apparent depth of the bottom?

Section 26.3 Total Internal Reflection

26. A glass is half-full of water, with a layer of vegetable oil ($n = 1.47$) floating on top. A ray of light traveling downward through the oil is incident on the water at an angle of 71.4°. Determine the critical angle for the oil–water interface and decide whether the ray will penetrate into the water.

27. **ssm** One method of determining the refractive index of a transparent solid is to measure the critical angle when the solid is in air. If θ_c is found to be 40.5°, what is the index of refraction of the solid?

28. A point source of light is submerged 2.2 m below the surface of a lake and emits rays in all directions. On the surface of the lake, directly above the source, the area illuminated is a circle. What is the maximum radius that this circle could have?

29. **Interactive Solution 26.29** at **www.wiley.com/college/cutnell** provides one model for solving problems such as this. A glass block ($n = 1.56$) is immersed in a liquid. A ray of light within the glass hits a glass–liquid surface at a 75.0° angle of incidence. Some of the light enters the liquid. What is the smallest possible refractive index for the liquid?

30. **GO** The drawing shows three materials, *a*, *b*, and *c*. A ray of light strikes the *a*–*b* interface at an angle that just barely exceeds its critical angle of 40.0°. The reflected ray then strikes the *a*–*c* interface at an angle of incidence that just barely exceeds its critical angle (which is not 40.0°). The index of refraction of material *a* is $n_a = 1.80$. Find the indices of refraction for the two other materials.

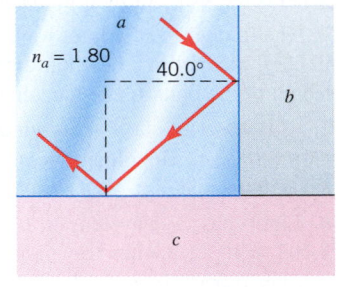

31. The drawing shows a crown glass slab with a rectangular cross section. As illustrated, a laser beam strikes the upper surface at an angle of 60.0°. After reflecting from the upper surface, the beam reflects from the side and bottom surfaces. **(a)** If the glass is surrounded by air, determine where part of the beam first exits the glass, at point *A*, *B*, or *C*. **(b)** Repeat part (a), assuming that the glass is surrounded by water instead of air.

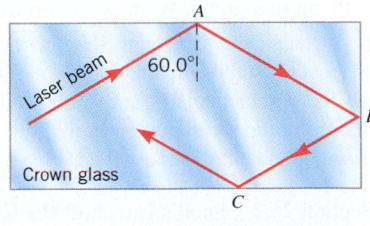

32. **GO** The drawing shows three layers of different materials, with air above and below the layers. The interfaces between the layers are parallel. The index of refraction of each layer is given in the drawing. Identical rays of light are sent into the layers, and light zigzags through each layer,

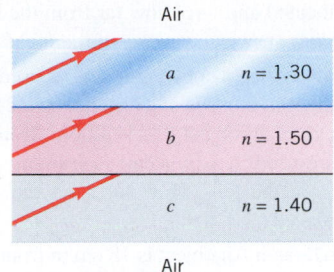

reflecting from the top and bottom surfaces. The index of refraction for air is $n_{air} = 1.00$. For each layer, the ray of light has an angle of incidence of 75.0°. For the cases in which total internal refection is possible from either the top or bottom surface of a layer, determine the amount by which the angle of incidence exceeds the critical angle.

*33. The drawing shows a ray of light traveling from point A to point B, a distance of 4.60 m in a material that has an index of refraction n_1. At point B, the light encounters a different substance whose index of refraction is $n_2 = 1.63$. The light strikes the interface at the critical angle of $\theta_c = 48.1°$. How much time does it take for the light to travel from A to B?

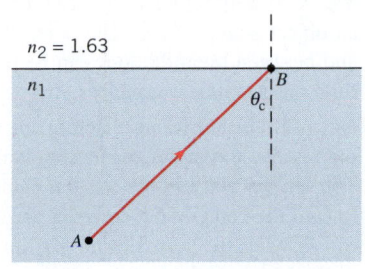

*34. A layer of liquid B floats on liquid A. A ray of light begins in liquid A and undergoes total internal reflection at the interface between the liquids when the angle of incidence exceeds 36.5°. When liquid B is replaced with liquid C, total internal reflection occurs for angles of incidence greater than 47.0°. Find the ratio n_B/n_C of the refractive indices of liquids B and C.

*35. **ssm** Multiple-Concept Example 7 and **Interactive LearningWare 26.1** at **www.wiley.com/college/cutnell** provide helpful background for this problem. The drawing shows a crystalline quartz slab with a rectangular cross section. A ray of light strikes the slab at an incident angle of $\theta_1 = 34°$, enters the quartz, and travels to point P. This slab is surrounded by a fluid with a

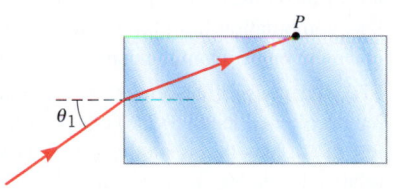

refractive index n. What is the maximum value of n for which total internal reflection occurs at point P?

Section 26.4 Polarization and the Reflection and Refraction of Light

36. For light that originates within a liquid and strikes the liquid–air interface, the critical angle is 39°. What is Brewster's angle for this light?

37. **ssm** Light is reflected from a glass coffee table. When the angle of incidence is 56.7°, the reflected light is completely polarized parallel to the surface of the glass. What is the index of refraction of the glass?

38. A laser is mounted in air, at a distance of 0.476 m above the edge of a large, horizontal pane of crown glass, as shown in the drawing. The laser is aimed at the glass in such a way that the reflected

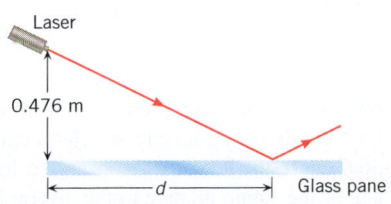

beam is 100% polarized. Determine the distance d between the edge of the pane and the point at which the laser beam is reflected.

39. **ssm www** When light strikes the surface between two materials from above, the Brewster angle is 65.0°. What is the Brewster angle when the light encounters the same surface from below?

40. Light is incident from air onto the surface of a liquid. The angle of incidence is 53.0°, and the angle of refraction is 34.0°. At what angle of incidence would the reflected light be 100% polarized?

*41. In Section 26.4 it is mentioned that the reflected and refracted rays are perpendicular to each other when light strikes the surface at the Brewster angle. This is equivalent to saying that the angle of reflection plus the angle of refraction is 90°. Using Snell's law and Brewster's law, prove that the angle of reflection plus the angle of refraction is 90°.

*42. When red light in a vacuum is incident at the Brewster angle on a certain type of glass, the angle of refraction is 29.9°. What are **(a)** the Brewster angle and **(b)** the index of refraction of the glass?

Section 26.5 The Dispersion of Light: Prisms and Rainbows

43. A ray of sunlight is passing from diamond into crown glass; the angle of incidence is 35.00°. The indices of refraction for the blue and red components of the ray are: blue ($n_{diamond} = 2.444$, $n_{crown\ glass} = 1.531$), and red ($n_{diamond} = 2.410$, $n_{crown\ glass} = 1.520$). Determine the angle between the refracted blue and red rays in the crown glass.

44. Red light ($n = 1.520$) and violet light ($n = 1.538$) traveling in air are incident on a slab of crown glass. Both colors enter the glass at the same angle of refraction. The red light has an angle of incidence of 30.00°. What is the angle of incidence of the violet light?

45. **ssm** A beam of sunlight encounters a plate of crown glass at a 45.00° angle of incidence. Using the data in Table 26.2, find the angle between the violet ray and the red ray in the glass.

46. Refer to Conceptual Example 8 as an aid in understanding this problem. The drawing shows a ray of light traveling through a gas ($n = 1.00$), a solid ($n = 1.55$), and a liquid ($n = 1.55$). At what angle θ does the light enter the liquid?

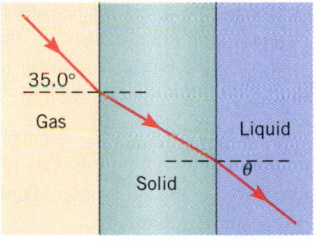

*47. **ssm** This problem relates to Figure 26.18, which illustrates the dispersion of light by a prism. The prism is made from glass, and its cross section is an equilateral triangle. The indices of refraction for the red and violet light are 1.662 and 1.698, respectively. The angle of incidence for both the red and the violet light is 60.0°. Find the angles of refraction at which the red and violet rays emerge into the air from the prism.

*48. The drawing shows a horizontal ray of white light incident perpendicularly on the vertical face of a prism (crown glass). The light enters the prism, and part of it undergoes refraction at the slanted face and emerges into the surrounding material. The rest of it is totally internally reflected and exits through the horizontal base of the prism. The colors of light that emerge from the slanted face may be chosen by altering the index of refraction of the material surrounding the prism. Find the required index of refraction of the surrounding material so that **(a)** only red light and **(b)** all colors except violet emerge from the slanted face. (See Table 26.2.)

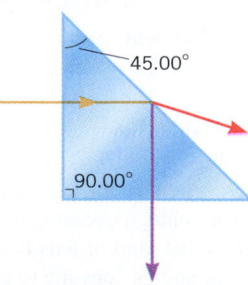

Section 26.6 Lenses, Section 26.7 The Formation of Images by Lenses, Section 26.8 The Thin-Lens Equation and the Magnification Equation

(Note: When drawing ray diagrams, be sure that the object height h_o is much smaller than the focal length f of the lens or mirror.)

49. ssm An object is located 30.0 cm to the left of a converging lens whose focal length is 50.0 cm. **(a)** Draw a ray diagram to scale and from it determine the image distance and the magnification. **(b)** Use the thin-lens and magnification equations to verify your answers to part (a).

50. The owner of a van installs a rear-window lens that has a focal length of −0.300 m. When the owner looks out through the lens at a person standing directly behind the van, the person appears to be just 0.240 m from the back of the van, and appears to be 0.34 m tall. **(a)** How far from the van is the person actually standing, and **(b)** how tall is the person?

51. A converging lens ($f = 12.0$ cm) is held 8.00 cm in front of a newspaper that has a print size with a height of 2.00 mm. Find **(a)** the image distance (in cm) and **(b)** the height (in mm) of the magnified print.

52. A tourist takes a picture of a mountain 14 km away using a camera that has a lens with a focal length of 50 mm. She then takes a second picture when she is only 5.0 km away. What is the ratio of the height of the mountain's image on the camera's image sensor for the second picture to its height on the image sensor for the first picture?

53. A slide projector has a converging lens whose focal length is 105.00 mm. **(a)** How far (in meters) from the lens must the screen be located if a slide is placed 108.00 mm from the lens? **(b)** If the slide measures 24.0 mm × 36.0 mm, what are the dimensions (in mm) of its image?

54. GO To focus a camera on objects at different distances, the converging lens is moved toward or away from the image sensor, so a sharp image always falls on the sensor. A camera with a telephoto lens ($f = 200.0$ mm) is to be focused on an object located first at a distance of 3.5 m and then at 50.0 m. Over what distance must the lens be movable?

55. A diverging lens has a focal length of −25 cm. **(a)** Find the image distance when an object is placed 38 cm from the lens. **(b)** Is the image real or virtual?

56. GO An object is placed to the left of a lens, and a real image is formed to the right of the lens. The image is inverted relative to the object and is one-half the size of the object. The distance between the object and the image is 90.0 cm. **(a)** How far from the lens is the object? **(b)** What is the focal length of the lens?

57. A camper is trying to start a fire by focusing sunlight onto a piece of paper. The diameter of the sun is 1.39×10^9 m, and its mean distance from the earth is 1.50×10^{11} m. The camper is using a converging lens whose focal length is 10.0 cm. **(a)** What is the area of the sun's image on the paper? **(b)** If 0.530 W of sunlight passes through the lens, what is the intensity of the sunlight at the paper?

*58. Concept Simulation 26.4 at **www.wiley.com/college/cutnell** provides the option of exploring the ray diagram that applies to this problem. The distance between an object and its image formed by a diverging lens is 49.0 cm. The focal length of the lens is −233.0 cm. Find **(a)** the image distance and **(b)** the object distance.

*59. ssm An office copier uses a lens to place an image of a document onto a rotating drum. The copy is made from this image. **(a)** What kind of lens is used, converging or diverging? If the document and its copy are to have the same size, but are inverted with respect to one another, **(b)** how far from the document is the lens

located and **(c)** how far from the lens is the image located? Express your answers in terms of the focal length f of the lens.

*60. When a converging lens is used in a camera (as in Figure 26.26b), the film must be at a distance of 0.210 m from the lens to record an image of an object that is 4.00 m from the lens. The same lens and film are used in a projector (see Figure 26.27b), with the screen 0.500 m from the lens. How far from the projector lens should the film be placed?

*61. ssm An object is 18 cm in front of a diverging lens that has a focal length of −12 cm. How far in front of the lens should the object be placed so that the size of its image is reduced by a factor of 2.0?

*62. An object is in front of a converging lens ($f = 0.30$ m). The magnification of the lens is $m = 4.0$. **(a)** Relative to the lens, in what direction should the object be moved so that the magnification changes to $m = −4.0$? **(b)** Through what distance should the object be moved?

**63. A converging lens ($f = 25.0$ cm) is used to project an image of an object onto a screen. The object and the screen are 125 cm apart, and between them the lens can be placed at either of two locations. Find the two object distances.

64. A filmmaker wants to achieve an interesting visual effect by filming a scene through a converging lens with a focal length of 50.0 m. The lens is placed between the camera and a horse, which canters toward the camera at a constant speed of 7.0 m/s. The camera starts rolling when the horse is 40.0 m from the lens. Find the average speed of the image of the horse **(a) during the first 2.0 s after the camera starts rolling and **(b)** during the following 2.0 s.

Section 26.9 Lenses in Combination

65. Two identical diverging lenses are separated by 16 cm. The focal length of each lens is −8.0 cm. An object is located 4.0 cm to the left of the lens that is on the left. Determine the final image distance relative to the lens on the right.

66. GO A converging lens ($f_1 = 24.0$ cm) is located 56.0 cm to the left of a diverging lens ($f_2 = −28.0$ cm). An object is placed to the left of the converging lens, and the final image produced by the two-lens combination lies 20.7 cm to the left of the diverging lens. How far is the object from the converging lens?

67. Two converging lenses are separated by 24.00 cm. The focal length of each lens is 12.00 cm. An object is placed 36.00 cm to the left of the lens that is on the left. Determine the final image distance relative to the lens on the right.

68. GO Two systems are formed from a converging lens and a diverging lens, as shown in parts a and b of the drawing. (The point labeled "$F_{\text{converging}}$" is the focal point of the converging lens.) An object is placed inside the focal point of lens 1 at a distance of 10.00 cm to the left of lens 1. The focal lengths of the converging and diverging lenses are 15.00 and −20.0 cm, respectively. The distance between the lenses is 50.0 cm. Determine the final image distance for each system, measured with respect to lens 2.

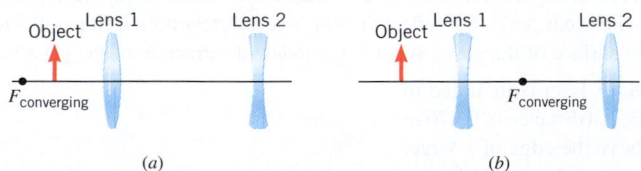

(a)   *(b)*

69. ssm A converging lens ($f = 12.0$ cm) is located 30.0 cm to the left of a diverging lens ($f = −6.00$ cm). A postage stamp is placed 36.0 cm to the left of the converging lens. **(a)** Locate the final image of the stamp relative to the diverging lens. **(b)** Find the overall magnification. **(c)** Is the final image real or virtual? With respect to the original object, is the final image   **(d)** upright or inverted, and is it   **(e)** larger or smaller?

70. Interactive LearningWare 26.2 at **www.wiley.com/college/cutnell** offers a review of the concepts that play roles in this problem. A diverging lens ($f = -10.0$ cm) is located 20.0 cm to the left of a converging lens ($f = 30.0$ cm). A 3.00-cm-tall object stands to the left of the diverging lens, exactly at its focal point.   **(a)** Determine the distance of the final image relative to the converging lens.   **(b)** What is the height of the final image (including the proper algebraic sign)?

*__71. ssm__ An object is placed 20.0 cm to the left of a diverging lens ($f = -8.00$ cm). A concave mirror ($f = 12.0$ cm) is placed 30.0 cm to the right of the lens.   **(a)** Find the final image distance, measured relative to the mirror.   **(b)** Is the final image real or virtual?   **(c)** Is the final image upright or inverted with respect to the original object?

*__72.__ Visitors at a science museum are invited to sit in a chair to the right of a full-length diverging lens ($f_1 = -3.00$ m) and observe a friend sitting in a second chair, 2.00 m to the left of the lens. The visitor then presses a button and a converging lens ($f_2 = +4.00$ m) rises from the floor to a position 1.60 m to the right of the diverging lens, allowing the visitor to view the friend through both lenses at once. Find   **(a)** the magnification of the friend when viewed through the diverging lens only and   **(b)** the overall magnification of the friend when viewed through both lenses. Be sure to include the algebraic signs ($+$ or $-$) with your answers.

**__73.__ A coin is located 20.0 cm to the left of a converging lens ($f = 16.0$ cm). A second, identical lens is placed to the right of the first lens in such a way that the image formed by the combination will have the same size and orientation as the original coin. Find the separation between the lenses.

Section 26.10   The Human Eye

74. A student is reading material written on a blackboard. Her contact lenses have a refractive power of 57.50 diopters; the lens-to-retina distance is 1.750 cm.   **(a)** How far (in meters) is the blackboard from her eyes?   **(b)** If the material written on the blackboard is 5.00 cm high, what is the size of the image on her retina?

75. **Interactive Solution 26.75** at **www.wiley.com/college/cutnell** illustrates one approach to solving problems such as this one. A farsighted person has a near point that is 67.0 cm from her eyes. She wears eyeglasses that are designed to enable her to read a newspaper held at a distance of 25.0 cm from her eyes. Find the focal length of the eyeglasses, assuming that they are worn   **(a)** 2.2 cm from the eyes and   **(b)** 3.3 cm from the eyes.

76. A nearsighted patient's far point is 0.690 m from her eyes. She is able to see distant objects in focus when wearing glasses with a refractive power of -1.50 diopters. What is the distance between her eyes and the glasses?

77. ssm Your friend has a near point of 138 cm, and she wears contact lenses that have a focal length of 35.1 cm. How close can she hold a magazine and still read it clearly?

78. **GO** A farsighted woman breaks her current eyeglasses and is using an old pair whose refractive power is 1.660 diopters. Since these eyeglasses do not completely correct her vision, she must hold a newspaper 42.00 cm from her eyes in order to read it. She wears the eyeglasses 2.00 cm from her eyes. How far is her near point from her eyes?

79. ssm An optometrist prescribes contact lenses that have a focal length of 55.0 cm.   **(a)** Are the lenses converging or diverging, and   **(b)** is the person who wears them nearsighted or farsighted?   **(c)** Where is the unaided near point of the person located,

if the lenses are designed so that objects no closer than 35.0 cm can be seen clearly?

*__80.__ A farsighted man uses eyeglasses with a refractive power of 3.80 diopters. Wearing the glasses 0.025 m from his eyes, he is able to read books held no closer than 0.280 m from his eyes. He would like a prescription for contact lenses to serve the same purpose. What is the correct contact lens prescription, in diopters?

__81.__ The far point of a nearsighted person is 6.0 m from her eyes, and she wears contacts that enable her to see distant objects clearly. A tree is 18.0 m away and 2.0 m high.   **(a) When she looks through the contacts at the tree, what is its image distance?   **(b)** How high is the image formed by the contacts?

__82.__ Bill is farsighted and has a near point located 125 cm from his eyes. Anne is also farsighted, but her near point is 75.0 cm from her eyes. Both have glasses that correct their vision to a normal near point (25.0 cm from the eyes), and both wear the glasses 2.0 cm from the eyes. Relative to the eyes, what is the closest object that can be seen clearly   **(a) by Anne when she wears Bill's glasses and   **(b)** by Bill when he wears Anne's glasses?

Section 26.11   Angular Magnification and the Magnifying Glass

83. ssm A quarter (diameter $= 2.4$ cm) is held at arm's length (70.0 cm). The sun has a diameter of 1.39×10^9 m and is 1.50×10^{11} m from the earth. What is the ratio of the angular size of the quarter to that of the sun?

84. A jeweler whose near point is 72 cm from his eye uses a magnifying glass as in Figure 26.39b to examine a watch. The watch is held 4.0 cm from the magnifying glass. Find the angular magnification of the magnifying glass.

85. An engraver uses a magnifying glass ($f = 9.50$ cm) to examine some work, as in Figure 26.39b. The image he sees is located 25.0 cm from his eye, which is his near point.   **(a)** What is the distance between the work and the magnifying glass?   **(b)** What is the angular magnification of the magnifying glass?

86. A dentist is examining a dental filling in a patient's tooth. The diameter of the filling is 2.4 mm, and the dentist's near point is 17.0 cm. To get a better look at the filling, the dentist dons safety goggles fitted with magnifying glasses ($f = 6.0$ cm). Find the greatest possible angular size (in radians) of the patient's filling when viewed by the dentist, both   **(a)** without and   **(b)** with the magnifying glasses.

87. An object has an angular size of 0.0150 rad when placed at the near point (21.0 cm) of an eye. When the eye views this object using a magnifying glass, the largest possible angular size of the image is 0.0380 rad. What is the focal length of the magnifying glass?

*__88.__ A stamp collector is viewing a stamp with a magnifying glass held next to her eye. Her near point is 25 cm from her eye.   **(a)** What is the refractive power of a magnifying glass that has an angular magnification of 6.0 when the image of the stamp is located at the near point?   **(b)** What is the angular magnification when the image of the stamp is 45 cm from the eye?

**__89. ssm__ A farsighted person can read printing as close as 25.0 cm when she wears contacts that have a focal length of 45.4 cm. One day, she forgets her contacts and uses a magnifying glass, as in Figure 26.39b. Its maximum angular magnification is 7.50 for a young person with a normal near point of 25.0 cm. What is the maximum angular magnification that the magnifying glass can provide for her?

Section 26.12   The Compound Microscope

90. An anatomist is viewing heart muscle cells with a microscope that has two selectable objectives with refracting powers of 100 and

300 diopters. When he uses the 100-diopter objective, the image of a cell subtends an angle of 3×10^{-3} rad with the eye. What angle is subtended when he uses the 300-diopter objective?

91. ssm A compound microscope has a barrel whose length is 16.0 cm and an eyepiece whose focal length is 1.4 cm. The viewer has a near point located 25 cm from his eyes. What focal length must the objective have so that the angular magnification of the microscope will be -320?

92. An insect subtends an angle of only 4.0×10^{-3} rad at the unaided eye when placed at the near point. What is the angular size (magnitude only) when the insect is viewed through a microscope whose angular magnification has a magnitude of 160?

93. A microscope for viewing blood cells has an objective with a focal length of 0.50 cm and an eyepiece with a focal length of 2.5 cm. The distance between the objective and eyepiece is 14.0 cm. If a blood cell subtends an angle of 2.1×10^{-5} rad when viewed with the naked eye at a near point of 25.0 cm, what angle (magnitude only) does it subtend when viewed through the microscope?

*94. In a compound microscope, the objective has a focal length of 0.60 cm, while the eyepiece has a focal length of 2.0 cm. The separation between the objective and the eyepiece is $L = 12.0$ cm. Another microscope that has the same angular magnification can be constructed by interchanging the two lenses, provided that the distance between the lenses is adjusted to a value L'. Find L'.

*95. In a compound microscope, the focal length of the objective is 3.50 cm and that of the eyepiece is 6.50 cm. The distance between the lenses is 26.0 cm. **(a)** What is the angular magnification of the microscope if the person using it has a near point of 35.0 cm? **(b)** If, as usual, the first image lies just inside the focal point of the eyepiece (see Figure 26.32), how far is the object from the objective? **(c)** What is the magnification (not the angular magnification) of the objective?

Section 26.13 The Telescope

96. A stargazer has an astronomical telescope with an objective whose focal length is 180 cm and an eyepiece whose focal length is 1.20 cm. He wants to increase the angular magnification of a galaxy under view by replacing the telescope's eyepiece. Once the eyepiece is replaced, the barrel of the telescope must be adjusted to bring the

galaxy back into focus. If the barrel can only be shortened by 0.50 cm from its current length, what is the best angular magnification the stargazer will be able to achieve?

97. ssm Mars subtends an angle of 8.0×10^{-5} rad at the unaided eye. An astronomical telescope has an eyepiece with a focal length of 0.032 m. When Mars is viewed using this telescope, it subtends an angle of 2.8×10^{-3} rad. Find the focal length of the telescope's objective lens.

98. An astronomical telescope has an angular magnification of -132. Its objective has a refractive power of 1.50 diopters. What is the refractive power of its eyepiece?

99. ssm An astronomical telescope has an angular magnification of -184 and uses an objective with a focal length of 48.0 cm. What is the focal length of the eyepiece?

100. An amateur astronomer decides to build a telescope from a discarded pair of eyeglasses. One of the lenses has a refractive power of 11 diopters, and the other has a refractive power of 1.3 diopters. **(a)** Which lens should be the objective? **(b)** How far apart should the lenses be separated? **(c)** What is the angular magnification of the telescope?

*101. A refracting telescope has an angular magnification of -83.00. The length of the barrel is 1.500 m. What are the focal lengths of **(a)** the objective and **(b)** the eyepiece?

*102. **GO** The lengths of three telescopes are $L_A = 455$ mm, $L_B = 615$ mm, and $L_C = 824$ mm. The focal length of the eyepiece for each telescope is 3.00 mm. Find the angular magnification of each telescope.

*103. The telescope at Yerkes Observatory in Wisconsin has an objective whose focal length is 19.4 m. Its eyepiece has a focal length of 10.0 cm. **(a)** What is the angular magnification of the telescope? **(b)** If the telescope is used to look at a lunar crater whose diameter is 1500 m, what is the size of the first image, assuming that the surface of the moon is 3.77×10^8 m from the surface of the earth? **(c)** How close does the crater appear to be when seen through the telescope?

**104. The angular magnification of a telescope is 32 800 times as large when you look through the correct end of the telescope as when you look through the wrong end. What is the angular magnification of the telescope?

ADDITIONAL PROBLEMS

105. ssm An object is located 9.0 cm in front of a converging lens ($f = 6.0$ cm). Using an accurately drawn ray diagram, determine where the image is located.

106. The near point of a naked eye is 25 cm. When placed at the near point and viewed by the naked eye, a tiny object would have an angular size of 5.2×10^{-5} rad. When viewed through a compound microscope, however, it has an angular size of -8.8×10^{-3} rad. (The minus sign indicates that the image produced by the microscope is inverted.) The objective of the microscope has a focal length of 2.6 cm, and the distance between the objective and the eyepiece is 16 cm. Find the focal length of the eyepiece.

107. ssm Concept Simulation 26.1 at **www.wiley.com/college/cutnell** illustrates the concepts that are pertinent to this problem. A ray of light is traveling in glass and strikes a glass–liquid interface. The angle of incidence is 58.0°, and the index of refraction of glass is $n = 1.50$. **(a)** What must be the index of refraction of the liquid so that the direction of the light entering the liquid is not changed?

(b) What is the largest index of refraction that the liquid can have, so that none of the light is transmitted into the liquid and all of it is reflected back into the glass?

108. As an aid in understanding this problem, refer to Conceptual Example 4. A swimmer, who is looking up from under the water, sees a diving board directly above at an apparent height of 4.0 m above the water. What is the actual height of the diving board?

109. ssm A person has far points of 5.0 m from the right eye and 6.5 m from the left eye. Write a prescription for the refractive power of each corrective contact lens.

110. Amber ($n = 1.546$) is a transparent brown-yellow fossil resin. An insect, trapped and preserved within the amber, appears to be 2.5 cm beneath the surface when viewed directly from above. How far below the surface is the insect actually located?

111. ssm A macroscopic (or macro) lens for a camera is usually a converging lens of normal focal length built into a lens barrel that can be adjusted to provide the additional distance needed between the lens and the image sensor when focusing at very close range. Suppose that a macro lens ($f = 50.0$ mm) has a maximum distance of 275 mm between the lens and the image sensor. How close can the object be located in front of the lens?

112. Concept Simulation 26.3 at **www.wiley.com/college/cutnell** reviews the concepts that play a role in this problem. A converging lens has a focal length of 88.00 cm. An object 13.0 cm tall is located 155.0 cm in front of this lens. **(a)** What is the image distance? **(b)** Is the image real or virtual? **(c)** What is the image height? Be sure to include the proper algebraic sign.

113. ssm A beam of light is traveling in air and strikes a material. The angles of incidence and refraction are 63.0° and 47.0°, respectively. Obtain the speed of light in the material.

114. The near point of a naked eye is 32 cm. When an object is placed at the near point and viewed by the naked eye, it has an angular size of 0.060 rad. A magnifying glass has a focal length of 16 cm, and is held next to the eye. The enlarged image that is seen is located 64 cm from the magnifying glass. Determine the angular size of the image.

115. A nearsighted person cannot read a sign that is more than 5.2 m from his eyes. To deal with this problem, he wears contact lenses that do not correct his vision completely, but do allow him to read signs located up to distances of 12.0 m from his eyes. What is the focal length of the contacts?

116. Violet light and red light travel through air and strike a block of plastic at the same angle of incidence. The angle of refraction is 30.400° for the violet light and 31.200° for the red light. The index of refraction for violet light in plastic is greater than that for red light by 0.0400. Delaying any rounding off of calculations until the very end, find the index of refraction for violet light in plastic.

117. Consult **Interactive Solution 26.117** at **www.wiley.com/college/cutnell** to review the concepts on which this problem depends. A camera is supplied with two interchangeable lenses, whose focal lengths are 35.0 and 150.0 mm. A woman whose height is 1.60 m stands 9.00 m in front of the camera. What is the height (including sign) of her image on the image sensor, as produced by **(a)** the 35.0-mm lens and **(b)** the 150.0-mm lens?

118. (a) For a diverging lens ($f = -20.0$ cm), construct a ray diagram to scale and find the image distance for an object that is 20.0 cm from the lens. **(b)** Determine the magnification of the lens from the diagram.

***119. ssm www** The moon's diameter is 3.48×10^6 m, and its mean distance from the earth is 3.85×10^8 m. The moon is being photographed by a camera whose lens has a focal length of 50.0 mm. **(a)** Find the diameter of the moon's image on the slide film. **(b)** When the slide is projected onto a screen that is 15.0 m from the lens of the projector ($f = 110.0$ mm), what is the diameter of the moon's image on the screen?

***120.** A person using a magnifying glass as in Figure 26.39b observes that for clear vision its maximum angular magnification is 1.25 times as large as its minimum angular magnification. Assuming that the person has a near point located 25 cm from her eye, what is the focal length of the magnifying glass?

***121. ssm** At age forty, a certain man requires contact lenses ($f = 65.0$ cm) to read a book held 25.0 cm from his eyes. At age forty-five, while wearing these contacts he must now hold a book 29.0 cm from his eyes. **(a)** By what distance has his near point *changed*? **(b)** What focal-length lenses does he require at age forty-five to read a book at 25.0 cm?

***122.** An object is placed in front of a converging lens in such a position that the lens ($f = 12.0$ cm) creates a real image located 21.0 cm from the lens. Then, with the object remaining in place, the lens is replaced with another converging lens ($f = 16.0$ cm). A new, real image is formed. What is the image distance of this new image?

****123. ssm** An astronomical telescope is being used to examine a relatively close object that is only 114.00 m away from the objective of the telescope. The objective and eyepiece have focal lengths of 1.500 and 0.070 m, respectively. Noting that the expression $M \approx -f_o/f_e$ is no longer applicable because the object is so close, use the thin-lens and magnification equations to find the angular magnification of this telescope. *(Hint: See Figure 26.41 and note that the focal points F_o and F_e are so close together that the distance between them may be ignored.)*

****124.** The equation

$$\frac{1}{d_o} + \frac{1}{d_i} = \frac{1}{f}$$

is called the *Gaussian form of the thin-lens equation. The drawing shows the variables d_o, d_i, and f. The drawing also shows the distances x and x', which are, respectively, the distance from the object to the focal point on the left of the lens and the distance from the focal point on the right of the lens to the image. An equivalent form of the thin-lens equation, involving x, x', and f, is called the *Newtonian* form. Show that the Newtonian form of the thin-lens equation can be written as $xx' = f^2$.

****125.** The contacts worn by a farsighted person allow her to see objects clearly that are as close as 25.0 cm, even though her uncorrected near point is 79.0 cm from her eyes. When she is looking at a poster, the contacts form an image of the poster at a distance of 217 cm from her eyes. **(a)** How far away is the poster actually located? **(b)** If the poster is 0.350 m tall, how tall is the image formed by the contacts?

CHAPTER 27

INTERFERENCE AND THE WAVE NATURE OF LIGHT

When two or more light waves exist simultaneously at the same place, the phenomenon of interference occurs, which is the topic of this chapter. Among the effects that can arise because of wave interference are the multicolors often seen in a thin film, such as those displayed in this fantastic soap bubble. (© Alberto Paredes/Age Fotostock America, Inc.)

27.1 THE PRINCIPLE OF LINEAR SUPERPOSITION

Chapter 17 examines what happens when several sound waves are present at the same place at the same time. The pressure disturbance that results is governed by the *principle of linear superposition, which states that the resultant disturbance is the sum of the disturbances from the individual waves.* Light is also a wave, an electromagnetic wave, and it too obeys the superposition principle. When two or more light waves pass through a given point, their electric fields combine according to the principle of linear superposition and produce a resultant electric field. According to Equation 24.5b, the square of the electric field strength is proportional to the intensity of the light, which, in turn, is related to its brightness. Thus, interference can and does alter the brightness of light, just as it affects the loudness of sound.

Figure 27.1 illustrates what happens when two identical waves (same wavelength λ and same amplitude) arrive at the point P in phase—that is, crest-to-crest and trough-to-trough. According to the principle of linear superposition, the waves reinforce each other and *constructive interference* occurs. The resulting total wave at P has an amplitude that is twice the amplitude of either individual wave, and, in the case of light waves, the brightness at P is greater than that due to either wave alone. The waves start out in phase and are in phase at P because the distances ℓ_1 and ℓ_2 between this spot and the sources of the waves differ by one wavelength λ. In Figure 27.1, these distances are $\ell_1 = 2\frac{1}{4}$ wavelengths and $\ell_2 = 3\frac{1}{4}$ wavelengths. In general, when the waves start out in phase, constructive interference will result at P whenever the distances are the same or differ by any integer number of wavelengths—in other words, assuming that ℓ_2 is the larger distance, whenever $\ell_2 - \ell_1 = m\lambda$, where $m = 0, 1, 2, 3, \ldots$.

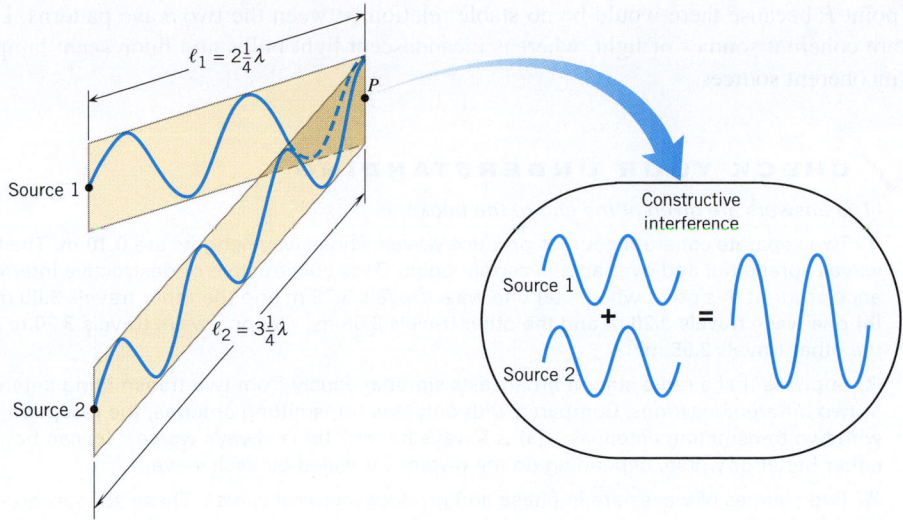

Figure 27.1 The waves emitted by source 1 and source 2 start out in phase and arrive at point P in phase, leading to constructive interference at that point.

Figure 27.2 shows what occurs when two identical waves arrive at the point P out of phase with one another, or crest-to-trough. Now the waves mutually cancel, according to the principle of linear superposition, and **destructive interference** results. With light waves this would mean that there is no brightness. The waves begin with the same phase but are out of phase at P because the distances through which they travel in reaching this spot differ by one-half of a wavelength ($\ell_1 = 2\frac{3}{4}\lambda$ and $\ell_2 = 3\frac{1}{4}\lambda$ in the drawing). In general, for waves that start out in phase, destructive interference will take place at P whenever the distances differ by any odd integer number of half-wavelengths—that is, whenever $\ell_2 - \ell_1 = \frac{1}{2}\lambda, \frac{3}{2}\lambda, \frac{5}{2}\lambda, \ldots$, where ℓ_2 is the larger distance. This is equivalent to $\ell_2 - \ell_1 = (m + \frac{1}{2})\lambda$, where $m = 0, 1, 2, 3, \ldots$.

Examples illustrating the application of the principle of linear superposition to explain the interference of light waves can be found throughout this chapter. For relatively straight-forward examples that deal with two sources of sound waves and the resulting constructive or destructive interference, see Examples 1 and 2 in Chapter 17 and Example 11 in this chapter.

If constructive or destructive interference is to continue occurring at a point, the sources of the waves must be **coherent sources.** Two sources are coherent if the waves they emit maintain a constant phase relation. Effectively, this means that the waves do not shift relative to one another as time passes. For instance, suppose that the wave pattern of source 1 in Figure 27.2 shifted forward or backward by random amounts at random moments. Then, on average, neither constructive nor destructive interference would be observed at

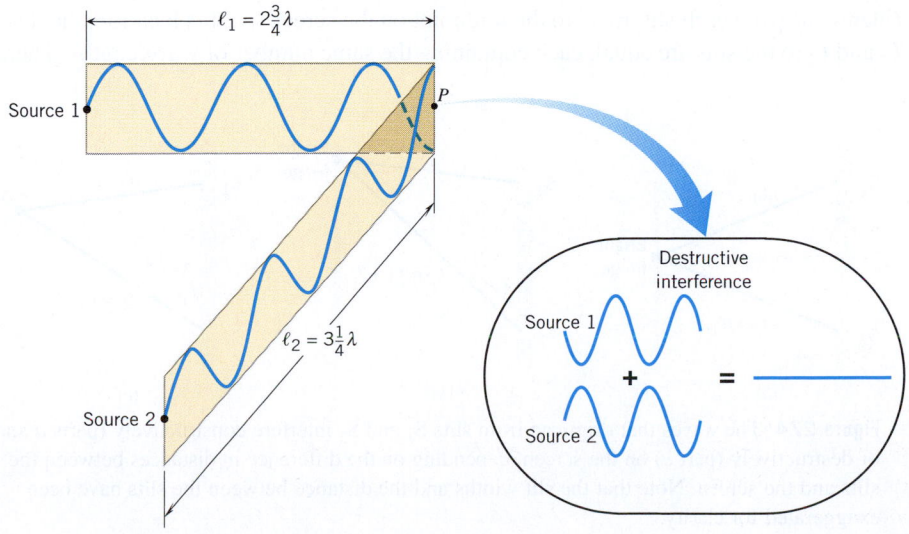

Figure 27.2 The waves emitted by the two sources have the same phase to begin with, but they arrive at point P out of phase. As a result, destructive interference occurs at P.

point P because there would be no stable relation between the two wave patterns. Lasers are coherent sources of light, whereas incandescent light bulbs and fluorescent lamps are incoherent sources.

✓ CHECK YOUR UNDERSTANDING

(The answers are given at the end of the book.)

1. Two separate coherent sources produce waves whose wavelengths are 0.10 m. The two waves spread out and overlap at a certain point. Does constructive or destructive interference occur at this point when **(a)** one wave travels 3.20 m and the other travels 3.00 m, **(b)** one wave travels 3.20 m and the other travels 3.05 m, **(c)** one wave travels 3.20 m and the other travels 2.95 m?

2. Suppose that a radio station broadcasts simultaneously from two transmitting antennas at *two different locations*. Compared with only one transmitting antenna, the reception with two transmitting antennas **(a)** is always better **(b)** is always worse **(c)** can be either better or worse, depending on the distance traveled by each wave.

3. Two sources of waves are in phase and produce identical waves. These sources are mounted at the corners of a square and broadcast waves uniformly in all directions. At the center of the square, will the waves always produce constructive interference no matter which two corners of the square are occupied by the sources?

27.2 YOUNG'S DOUBLE-SLIT EXPERIMENT

In 1801 the English scientist Thomas Young (1773–1829) performed a historic experiment that demonstrated the wave nature of light by showing that two overlapping light waves can interfere with each other. His experiment was particularly important because he was also able to determine the wavelength of the light from his measurements, the first such determination of this important property. Figure 27.3 shows one arrangement of Young's experiment, in which light of a single wavelength (monochromatic light) passes through a single narrow slit and falls on two closely spaced, narrow slits S_1 and S_2. These two slits act as coherent sources of light waves that interfere constructively and destructively at different points on the screen to produce a pattern of alternating bright and dark fringes. The purpose of the single slit is to ensure that only light from one direction falls on the double slit. Without it, light coming from different points on the light source would strike the double slit from different directions and cause the pattern on the screen to be washed out. The slits S_1 and S_2 act as coherent sources of light waves because the light from each originates from the same primary source—namely, the single slit.

To help explain the origin of the bright and dark fringes, Figure 27.4 presents three top views of the double slit and the screen. Part *a* illustrates how a bright fringe arises directly opposite the midpoint between the two slits. In this part of the drawing the waves (identical) from each slit travel to the midpoint on the screen. At this location, the distances ℓ_1 and ℓ_2 to the slits are equal, each containing the same number of wavelengths. Therefore,

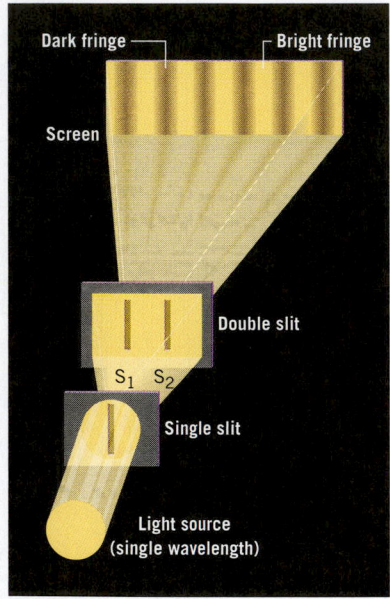

Figure 27.3 In Young's double-slit experiment, two slits S_1 and S_2 act as coherent sources of light. Light waves from these slits interfere constructively and destructively on the screen to produce, respectively, the bright and dark fringes. The slit widths and the distance between the slits have been exaggerated for clarity.

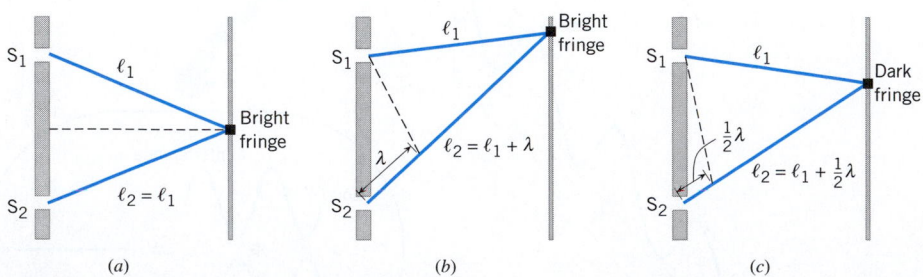

Figure 27.4 The waves that originate from slits S_1 and S_2 interfere constructively (parts *a* and *b*) or destructively (part *c*) on the screen, depending on the difference in distances between the slits and the screen. Note that the slit widths and the distance between the slits have been exaggerated for clarity.

constructive interference results, leading to the bright fringe. Part *b* indicates that constructive interference produces another bright fringe on one side of the midpoint when the distance ℓ_2 is larger than ℓ_1 by exactly one wavelength. A bright fringe also occurs symmetrically on the other side of the midpoint when the distance ℓ_1 exceeds ℓ_2 by one wavelength; for clarity, however, this additional bright fringe is not shown. Constructive interference produces other bright fringes (also not shown) on both sides of the middle wherever the difference between ℓ_1 and ℓ_2 is an integer number of wavelengths: λ, 2λ, 3λ, and so on. Part *c* shows how the first dark fringe arises. Here the distance ℓ_2 is larger than ℓ_1 by exactly one-half a wavelength, so the waves interfere destructively, giving rise to the dark fringe. Destructive interference creates other dark fringes on both sides of the center wherever the difference between ℓ_1 and ℓ_2 equals an odd integer number of half-wavelengths: $1(\frac{\lambda}{2})$, $3(\frac{\lambda}{2})$, $5(\frac{\lambda}{2})$, and so on.

The brightness of the fringes in Young's experiment varies, as the photograph in Figure 27.5 shows. Below the photograph is a graph to suggest the way in which the intensity varies for the fringe pattern. The central fringe is labeled with a zero, and the other bright fringes are numbered in ascending order on either side of the center. It can be seen that the central fringe has the greatest intensity. To either side of the center, the intensities of the other fringes decrease symmetrically in a way that depends on how small the slit widths are relative to the wavelength of the light.

The position of the fringes observed on the screen in Young's experiment can be calculated with the aid of Figure 27.6. If the screen is located far away compared with the separation *d* of the slits, then the lines labeled ℓ_1 and ℓ_2 in part *a* are nearly parallel. Being nearly parallel, these lines make approximately equal angles θ with the horizontal. The distances ℓ_1 and ℓ_2 differ by an amount $\Delta\ell$, which is the length of the short side of the colored triangle in part *b* of the drawing. Since the triangle is a right triangle, it follows that $\Delta\ell = d \sin \theta$. Constructive interference occurs when the distances differ by an integer number *m* of wavelengths λ, or $\Delta\ell = d \sin \theta = m\lambda$. Therefore, the angle θ for the interference maxima can be determined from the following expression:

Bright fringes of a double slit
$$\sin \theta = m \frac{\lambda}{d} \qquad m = 0, 1, 2, 3, \ldots \qquad (27.1)$$

The value of *m* specifies the *order* of the fringe. Thus, $m = 2$ identifies the "second-order" bright fringe. Part *c* of Figure 27.6 stresses that the angle θ given by Equation 27.1 locates bright fringes on either side of the midpoint between the slits. A similar line of reasoning leads to the conclusion that the dark fringes, which lie between the bright fringes, are located according to

Dark fringes of a double slit
$$\sin \theta = (m + \tfrac{1}{2}) \frac{\lambda}{d} \qquad m = 0, 1, 2, 3, \ldots \qquad (27.2)$$

Example 1 illustrates how to determine the distance of a higher-order bright fringe from the central bright fringe with the aid of Equation 27.1.

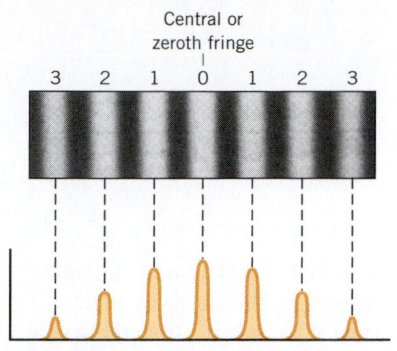

Figure 27.5 The results of Young's double-slit experiment, showing a photograph of the bright and dark fringes formed on the screen and a graph of the light intensity. The central or zeroth fringe is the brightest fringe (greatest intensity). (From Michel Cagnet, et al., *Atlas of Optical Phenomena*, Springer-Verlag, Berlin, 1962)

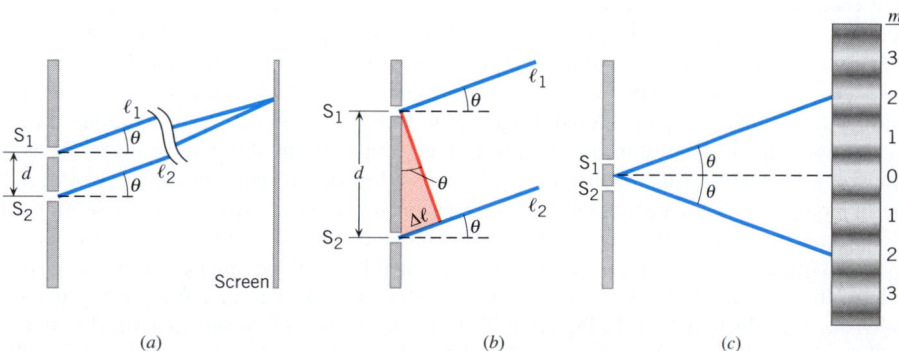

Figure 27.6 (*a*) Rays from slits S_1 and S_2, which make approximately the same angle θ with the horizontal, strike a distant screen at the same spot. (*b*) The difference in the path lengths of the two rays is $\Delta\ell = d \sin \theta$. (*c*) The angle θ is the angle at which a bright fringe ($m = 2$, here) occurs on either side of the central bright fringe ($m = 0$).

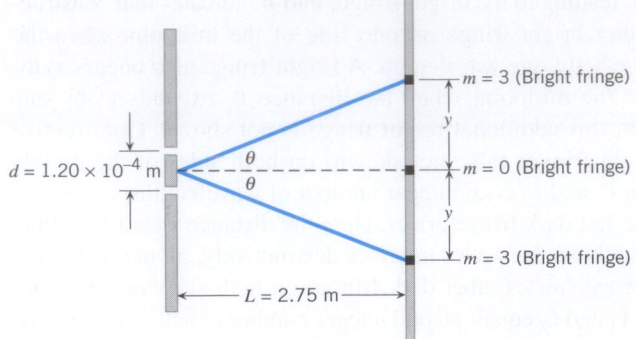

Figure 27.7 The third-order bright fringe ($m = 3$) is observed on the screen at a distance y from the central bright fringe ($m = 0$).

⬇ **Example 1** Young's Double-Slit Experiment

Red light ($\lambda = 664$ nm in vacuum) is used in Young's experiment with the slits separated by a distance $d = 1.20 \times 10^{-4}$ m. The screen in Figure 27.7 is located at a distance of $L = 2.75$ m from the slits. Find the distance y on the screen between the central bright fringe and the third-order bright fringe.

Reasoning This problem can be solved by first using Equation 27.1 to determine the value of θ that locates the third-order ($m = 3$) bright fringe. Then trigonometry can be used to obtain the distance y.

Solution According to Equation 27.1, we find

$$\theta = \sin^{-1}\left(\frac{m\lambda}{d}\right) = \sin^{-1}\left[\frac{3(664 \times 10^{-9} \text{ m})}{1.20 \times 10^{-4} \text{ m}}\right] = 0.951°$$

According to Figure 27.7, the distance y can be calculated from $\tan \theta = y/L$:

$$y = L \tan \theta = (2.75 \text{ m}) \tan 0.951° = \boxed{0.0456 \text{ m}}$$

In the preceding version of Young's experiment, monochromatic light has been used. Light that contains a mixture of wavelengths can also be used. Conceptual Example 2 deals with some of the interesting features of the resulting interference pattern.

⬇ **Conceptual Example 2** White Light and Young's Experiment

Figure 27.8 shows a photograph that illustrates the kind of interference fringes that can result when white light, which is a mixture of all colors, is used in Young's experiment. Except for the central fringe, which is white, the bright fringes are a rainbow of colors. Why does Young's experiment separate white light into its constituent colors? In any group of colored fringes, such as the two singled out in Figure 27.8, why is red farther out from the central fringe than green is? And finally, why is the central fringe white rather than colored?

Reasoning and Solution To understand how the color separation arises, we need to remember that each color corresponds to a different wavelength λ and that constructive and destructive interference depend on the wavelength. According to Equation 27.1 ($\sin \theta = m\lambda/d$), there is a different angle that locates a bright fringe for each value of λ, and thus for each color. These different angles lead to the separation of colors on the observation screen. In fact, on either side of the central fringe, there is one group of colored fringes for $m = 1$ and another for each additional value of m.

Now, consider what it means that, within any single group of colored fringes, red is farther out from the central fringe than green is. It means that, in the equation $\sin \theta = m\lambda/d$, red light has a larger angle θ than green light does. Does this make sense? Yes, because red has the larger wavelength (see Table 26.2, where $\lambda_{\text{red}} = 660$ nm and $\lambda_{\text{green}} = 550$ nm).

In Figure 27.8, the central fringe is distinguished from all the other colored fringes by being white. In Equation 27.1, the central fringe is different from the other fringes because it is the only one for which $m = 0$. In Equation 27.1, a value of $m = 0$ means that $\sin \theta = m\lambda/d = 0$, which reveals that $\theta = 0°$, no matter what the wavelength λ is. In other words, all wavelengths have a zeroth-order bright fringe located at the same place on the screen, so that all colors strike the screen there and mix together to produce the white central fringe.

Related Homework: *Problem 8*

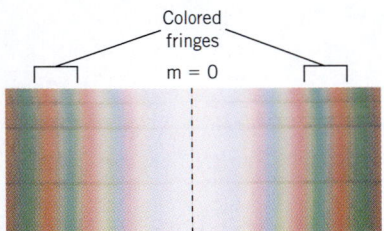

Figure 27.8 This photograph shows the results observed on the screen in one version of Young's experiment in which white light (a mixture of all colors) is used. (© Andy Washnik)

Historically, Young's experiment provided strong evidence that light has a wave-like character. If light behaved only as a stream of "tiny particles," as others believed at the time,* then the two slits would deliver the light energy into only two bright fringes located directly opposite the slits on the screen. Instead, Young's experiment shows that wave interference redistributes the energy from the two slits into many bright fringes.

✓ CHECK YOUR UNDERSTANDING

(The answers are given at the end of the book.)

4. Replace the slits S_1 and S_2 in Figure 27.3 with identical in-phase loudspeakers and use the same ac electrical signal to drive them. The two sound waves produced will then be identical, and you will have the audio equivalent of Young's double-slit experiment. In terms of loudness and softness, what would you hear as you walk along the screen, starting from the center and going to either end? **(a)** Loud, then soft, then loud, then soft, etc., with the loud sounds *decreasing in intensity* as you walk away from the center **(b)** Loud, then soft, then loud, then soft, etc., with the loud sounds *increasing in intensity* as you walk away from the center **(c)** Soft, then loud, then soft, then loud, etc., with the loud sounds *decreasing in intensity* as you walk away from the center **(d)** Soft, then loud, then soft, then loud, etc., with the loud sounds *increasing in intensity* as you walk away from the center

5. The drawing shows two double slits that have slit separations of d_1 and d_2. Light whose wavelength is either λ_1 or λ_2 passes through the slits. For comparison, the wavelengths are also illustrated in the drawing. For which combination of slit separation and wavelength would the pattern of bright and dark fringes on the observation screen be **(a)** the most spread out and **(b)** the least spread out?

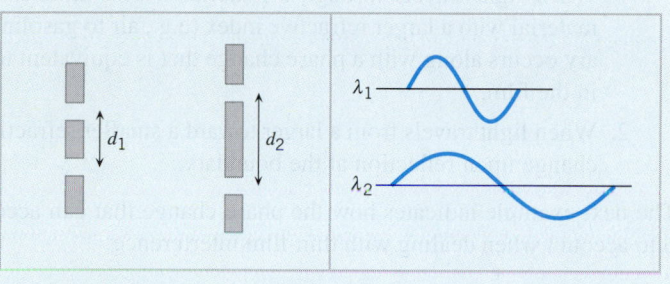

6. Suppose the light waves coming from *both* slits in a Young's double-slit experiment had their phases shifted by an amount equivalent to a half-wavelength. **(a)** Would the pattern be the same or would the positions of the light and dark fringes be interchanged? **(b)** Would the pattern be the same or would the positions of the light and dark fringes be interchanged if the light coming from *only one* of the slits had its phase shifted by an amount equivalent to a half-wavelength?

7. In Young's double-slit experiment, is it possible to see interference fringes when the wavelength of the light is greater than the distance between the slits?

27.3 THIN-FILM INTERFERENCE

Young's double-slit experiment is one example of interference between light waves. Interference also occurs in more common circumstances. For instance, Figure 27.9 shows a thin film such as gasoline floating on water. To begin with, let us assume that the film has a constant thickness. Consider what happens when monochromatic light (a single wavelength) strikes the film nearly perpendicularly. At the top surface of the film reflection occurs and produces the light wave represented by ray 1. However, refraction also occurs, and some light enters the film. Part of this light reflects from the bottom surface of the film and passes back up through the film, eventually reentering the air. Thus, a second light wave, which is represented by ray 2, also exists. Moreover, this wave, having traversed the film twice, has traveled farther than wave 1. Because of the extra travel distance, there can be interference between the two waves. If constructive interference occurs, an observer whose eyes detect the superposition of waves 1 and 2 would see a uniformly bright film. If destructive interference occurs, an observer would see a uniformly dark film.

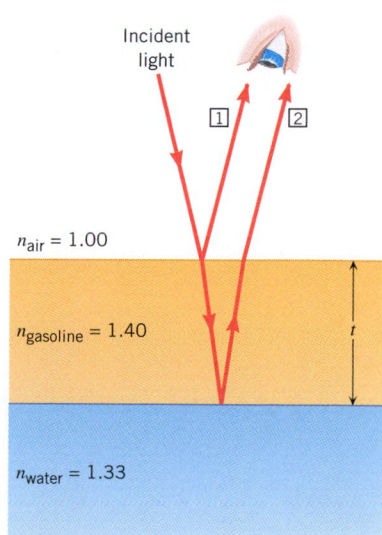

Figure 27.9 Because of reflection and refraction, two light waves, represented by rays 1 and 2, enter the eye when light shines on a thin film of gasoline floating on a thick layer of water.

---

*It is now known that the particle, or corpuscular, theory of light, which Isaac Newton promoted, does indeed explain some experiments that the wave theory cannot explain. Today, light is regarded as having both particle and wave characteristics. Chapter 29 discusses this dual nature of light.

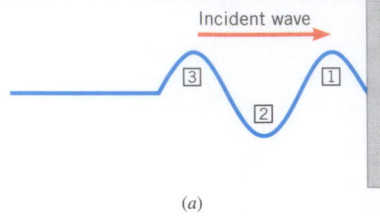

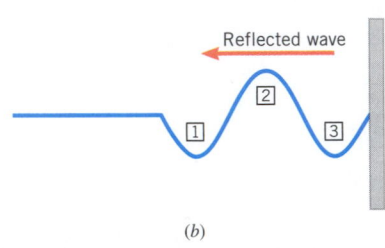

Figure 27.10 When a wave on a string reflects from a wall, the wave undergoes a phase change. Thus, an upward-pointing half-cycle of the wave becomes, after reflection, a downward-pointing half-cycle, and vice versa, as the numbered labels in the drawing indicate.

In Figure 27.9 the difference in path lengths between waves 1 and 2 occurs inside the thin film. Therefore, *the wavelength that is important for thin-film interference is the wavelength within the film, not the wavelength in vacuum.* The wavelength within the film can be calculated from the wavelength in vacuum by using the index of refraction n for the film. With the aid of Equations 26.1 and 16.1, it can be shown that $n = c/v = (c/f)/(v/f) = \lambda_{\text{vacuum}}/\lambda_{\text{film}}$. In other words,

$$\lambda_{\text{film}} = \frac{\lambda_{\text{vacuum}}}{n} \qquad (27.3)$$

In explaining the interference that can occur in Figure 27.9, we need to add one other important part to the story. Whenever waves reflect at a boundary, it is possible for them to change phase. Figure 27.10, for example, shows that a wave on a string is inverted when it reflects from the end that is tied to a wall (see also Figure 17.16). This inversion is equivalent to a half-cycle of the wave, as if the wave had traveled an additional distance of one-half of a wavelength. In contrast, a phase change does not occur when a wave on a string reflects from the end of a string that is hanging free. When light waves undergo reflection, similar phase changes occur as follows:

1. When light travels through a material with a smaller refractive index toward a material with a larger refractive index (e.g., air to gasoline), reflection at the boundary occurs along with a phase change that is equivalent to one-half of a wavelength in the film.

2. When light travels from a larger toward a smaller refractive index, there is no phase change upon reflection at the boundary.

The next example indicates how the phase change that can accompany reflection is taken into account when dealing with thin-film interference.

Example 3   A Colored Thin Film of Gasoline

A thin film of gasoline floats on a puddle of water. Sunlight falls almost perpendicularly on the film and reflects into your eyes. Although sunlight is white since it contains all colors, the film looks yellow because destructive interference eliminates the color of blue ($\lambda_{\text{vacuum}} = 469$ nm) from the reflected light. The refractive indices of the blue light in gasoline and in water are 1.40 and 1.33, respectively. Determine the minimum nonzero thickness t of the film.

Reasoning To solve this problem, we must express the condition for destructive interference in terms of the film thickness t and the wavelength λ_{film} in the gasoline film. We must also take into account any phase changes that occur upon reflection.

In Figure 27.9, the phase change for wave 1 is equivalent to one-half of a wavelength, since this light travels from a smaller refractive index ($n_{\text{air}} = 1.00$) toward a larger refractive index ($n_{\text{gasoline}} = 1.40$). In contrast, there is no phase change when wave 2 reflects from the bottom surface of the film, since this light travels from a material with a larger refractive index ($n_{\text{gasoline}} = 1.40$) toward a material with a smaller one ($n_{\text{water}} = 1.33$). The net phase change between waves 1 and 2 due to reflection is, thus, equivalent to one-half of a wavelength, $\frac{1}{2}\lambda_{\text{film}}$. This half-wavelength must be combined with the extra travel distance for wave 2, to determine the condition for destructive interference. For destructive interference, the combined total must be an odd integer number of half-wavelengths. Since wave 2 travels back and forth through the film and since light strikes the film nearly perpendicularly, the extra travel distance is twice the film thickness, or $2t$. Thus, the condition for destructive interference is

$$\underbrace{2t}_{\substack{\text{Extra distance} \\ \text{traveled by} \\ \text{wave 2}}} + \underbrace{\tfrac{1}{2}\lambda_{\text{film}}}_{\substack{\text{Half-wavelength} \\ \text{net phase change} \\ \text{due to reflection}}} = \underbrace{\tfrac{1}{2}\lambda_{\text{film}}, \tfrac{3}{2}\lambda_{\text{film}}, \tfrac{5}{2}\lambda_{\text{film}}, \ldots}_{\substack{\text{Condition for} \\ \text{destructive interference}}}$$

After subtracting the term $\frac{1}{2}\lambda_{\text{film}}$ from the left-hand side of this equation and from each term on the right-hand side, we can solve for the thickness t of the film that yields destructive interference:

$$t = \frac{m\lambda_{\text{film}}}{2} \qquad m = 0, 1, 2, 3, \ldots$$

Solution In order to calculate t, we need to know the wavelength of the blue light in the film. Equation 27.3, with $n = 1.40$, gives this wavelength as

$$\lambda_{film} = \frac{\lambda_{vacuum}}{n} = \frac{469 \text{ nm}}{1.40} = 335 \text{ nm}$$

Problem-solving insight

When analyzing thin-film interference effects, remember to use the wavelength of the light in the film (λ_{film}) instead of the wavelength in a vacuum (λ_{vacuum}).

With this value for λ_{film} and $m = 1$, our result for t gives the minimum nonzero film thickness for which the blue color is missing in the reflected light as follows:

$$t = \frac{m\lambda_{film}}{2} = \frac{(1)(335 \text{ nm})}{2} = \boxed{168 \text{ nm}}$$

In Example 3 a half-wavelength net phase change occurs due to the reflections at the upper and lower surfaces of the thin film. Depending on the refractive indices of the materials above and below the film, it is also possible that these reflections yield a zero net phase change. See Example 12 in Section 27.10 for a situation in which this occurs.

The thin film in Example 3 has the same yellow color everywhere. In nature, such a uniformly colored thin film would be unusual, and the next example deals with a more realistic situation.

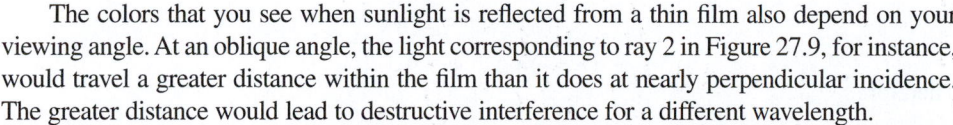

Conceptual Example 4 Multicolored Thin Films

Under natural conditions, thin films, like gasoline on water or like the soap bubble in Figure 27.11, have a multicolored appearance that often changes while you are watching them. Why are such films multicolored, and what can be inferred from the fact that the colors change in time?

Reasoning and Solution In Example 3 we have seen that a thin film can appear yellow if destructive interference removes blue light from the reflected sunlight. The thickness of the film is the key. If the thickness were different, so that destructive interference removed green light from the reflected sunlight, the film would appear magenta. Constructive interference can also cause certain colors to appear brighter than others in the reflected light and thereby give the film a colored appearance. The colors that are enhanced by constructive interference, like those removed by destructive interference, depend on the thickness of the film. Thus, we conclude that *the different colors in a thin film of gasoline on water or in a soap bubble arise because the thickness is different in different places on the film. Moreover, the fact that the colors change as you watch them indicates that the thickness is changing.* A number of factors can cause the thickness to change, including air currents, temperature fluctuations, and the pull of gravity, which tends to make a vertical film sag, leading to thicker regions at the bottom than at the top.

Problem-solving insight

Related Homework: *Problem 17*

The colors that you see when sunlight is reflected from a thin film also depend on your viewing angle. At an oblique angle, the light corresponding to ray 2 in Figure 27.9, for instance, would travel a greater distance within the film than it does at nearly perpendicular incidence. The greater distance would lead to destructive interference for a different wavelength.

Thin-film interference can be beneficial in optical instruments. For example, some cameras contain six or more lenses. Reflections from all the lens surfaces can reduce

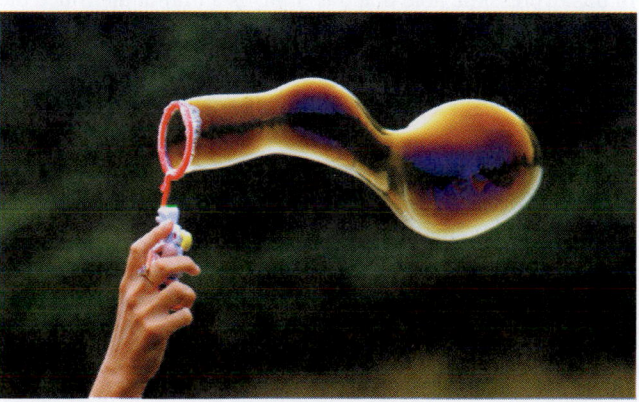

Figure 27.11 A soap bubble is multicolored when viewed in sunlight because of the effects of thin-film interference. (© Raymond Forbes/Age Fotostock America, Inc.)

considerably the amount of light directly reaching the film. In addition, multiple reflections from the lenses often reach the film indirectly and degrade the quality of the image. To minimize such unwanted reflections, high-quality lenses are often covered with a thin nonreflective coating of magnesium fluoride ($n = 1.38$). The thickness of the coating is usually chosen to ensure that destructive interference eliminates the reflection of green light, which is in the middle of the visible spectrum. It should be pointed out that the absence of any reflected light does not mean that it has been destroyed by the nonreflective coating. Rather, the "missing" light has been transmitted into the coating and the lens.

Another interesting illustration of thin-film interference is the air wedge. As Figure 27.12a shows, an air wedge is formed when two flat plates of glass are separated along one side, perhaps by a thin sheet of paper. The thickness of this film of air varies between zero, where the plates touch, and the thickness of the paper. When monochromatic light reflects from this arrangement, alternate bright and dark fringes are formed by constructive and destructive interference, as the drawing indicates and Example 5 discusses.

Example 5 An Air Wedge

(a) Assuming that green light ($\lambda_{vacuum} = 552$ nm) strikes the glass plates nearly perpendicularly in Figure 27.12, determine the number of bright fringes that occur between the place where the plates touch and the edge of the sheet of paper (thickness $= 4.10 \times 10^{-5}$ m). **(b)** Explain why there is a dark fringe where the plates touch.

Reasoning A bright fringe occurs wherever there is constructive interference, as determined by any phase changes due to reflection and by the thickness of the air wedge. There is no phase change upon reflection for wave 1, since this light travels from a larger (glass) toward a smaller (air) refractive index. In contrast, there is a half-wavelength phase change for wave 2, since the ordering of the refractive indices is reversed at the lower air/glass boundary where reflection occurs. The net phase change due to reflection for waves 1 and 2, then, is equivalent to a half-wavelength. Now we combine any extra distance traveled by ray 2 with this half-wavelength and determine the condition for the constructive interference that creates the bright fringes. Constructive interference occurs whenever the *combination* yields an integer number of wavelengths. At nearly perpendicular incidence, the extra travel distance for wave 2 is approximately twice the thickness t of the wedge at any point, so the condition for constructive interference is

$$\underbrace{2t}_{\substack{\text{Extra distance}\\\text{traveled by}\\\text{wave 2}}} + \underbrace{\tfrac{1}{2}\lambda_{film}}_{\substack{\text{Half-wavelength}\\\text{net phase change}\\\text{due to reflection}}} = \underbrace{\lambda_{film}, 2\lambda_{film}, 3\lambda_{film}, \ldots}_{\substack{\text{Condition for}\\\text{constructive interference}}}$$

Subtracting the term $\tfrac{1}{2}\lambda_{film}$ from the left-hand side of this equation and from each term on the right-hand side yields

$$2t = \underbrace{\tfrac{1}{2}\lambda_{film}, \tfrac{3}{2}\lambda_{film}, \tfrac{5}{2}\lambda_{film}, \ldots}_{(m + \frac{1}{2})\lambda_{film} \quad m = 0, 1, 2, 3, \ldots}$$

Therefore,

$$t = \frac{(m + \frac{1}{2})\lambda_{film}}{2} \qquad m = 0, 1, 2, 3, \ldots$$

In this expression, note that the "film" is a film of air. Since the refractive index of air is nearly one, λ_{film} is virtually the same as that in a vacuum, so $\lambda_{film} = 552$ nm.

Solution **(a)** When t equals the thickness of the paper holding the plates apart, the corresponding value of m can be obtained from the equation above:

$$m = \frac{2t}{\lambda_{film}} - \frac{1}{2} = \frac{2(4.10 \times 10^{-5}\ \text{m})}{552 \times 10^{-9}\ \text{m}} - \frac{1}{2} = 148$$

Since the first bright fringe occurs when $m = 0$, the number of bright fringes is $m + 1 = \boxed{149}$.

(b) Where the plates touch, there is a dark fringe because of destructive interference between the light waves represented by rays 1 and 2. Destructive interference occurs because the thickness of the wedge is zero here and the only difference between the rays is the half-wavelength phase change due to reflection from the lower plate.

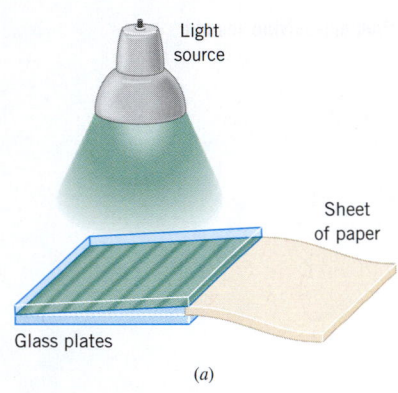

Light source

Sheet of paper

Glass plates

(a)

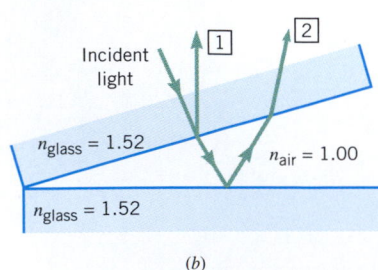

Incident light

$\boxed{1}$ $\boxed{2}$

$n_{glass} = 1.52$

$n_{air} = 1.00$

$n_{glass} = 1.52$

(b)

Figure 27.12 (a) The wedge of air formed between two flat glass plates causes an interference pattern of alternating dark and bright fringes to appear in reflected light. (b) A side view of the glass plates and the air wedge.

Another type of air wedge can also be used to determine the degree to which the surface of a lens or mirror is spherical. When an accurate spherical surface is put in contact with an optically flat plate, as in Figure 27.13a, the circular interference fringes shown in part b of the figure can be observed. The circular fringes are called *Newton's rings*. They arise in the same way that the straight fringes arise in Figure 27.12a.

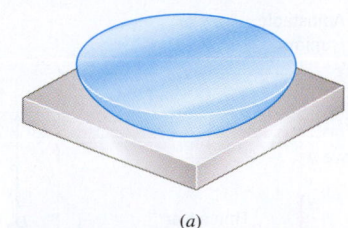

(a)

✓ CHECK YOUR UNDERSTANDING

(*The answers are given at the end of the book.*)

8. A camera lens is covered with a nonreflective coating that eliminates the reflection of perpendicularly incident green light. Recalling Snell's law of refraction (see Section 26.2), would you expect the reflected green light to be eliminated if it were incident on the nonreflective coating at an angle of 45° rather than perpendicularly?   **(a)** No, because the distance traveled by the light in the film is less than twice the film thickness.   **(b)** No, because the distance traveled by the light in the film is greater than twice the film thickness.   **(c)** Yes, the green light will still be eliminated.

9. Two pieces of the same glass are covered with thin films of different materials. In reflected sunlight, however, the films have different colors. Why?   **(a)** The films could have the same thickness, but different refractive indices.   **(b)** The films could have different thicknesses, but the same refractive indices.   **(c)** Both of the preceding answers could be correct.

10. A transparent coating is deposited on a glass plate and has a refractive index that is *larger than that of the glass.* For a certain wavelength within the coating, the thickness of the coating is a quarter-wavelength. Does the coating enhance or reduce the reflection of the light corresponding to this wavelength?

11. The drawings show three situations—A, B, and C—in which light reflects almost perpendicularly from the top and bottom surfaces of a thin film, with the indices of refraction as shown.   **(a)** For which situation(s) will there be a net phase shift (due to reflection) between waves 1 and 2 that is equivalent to either zero wavelengths or one wavelength (λ_{film}), where λ_{film} is the wavelength of the light in the film?   **(b)** For which

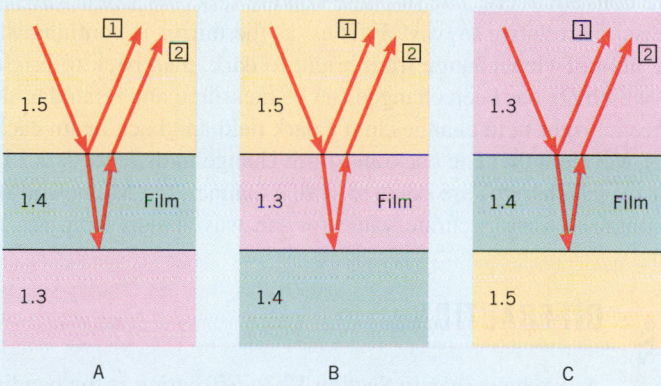

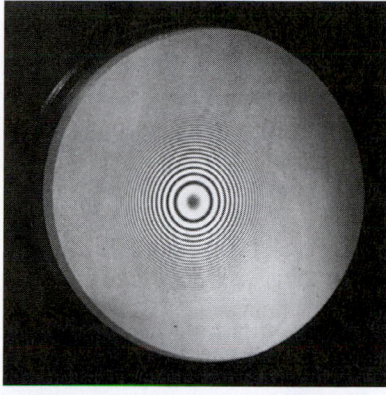

(b)

Figure 27.13 (a) The air wedge between a convex spherical glass surface and an optically flat plate leads to (b) a pattern of circular interference fringes that are known as Newton's rings. (Courtesy Bausch & Lomb)

situation(s) will the film appear dark when the thickness of the film is equal to $\frac{1}{2}\lambda_{film}$?

12. When sunlight reflects from a thin film of soapy water, the film appears multicolored, in part because destructive interference removes different wavelengths from the light reflected at different places, depending on the thickness of the film. What happens as the film becomes thinner and thinner?   **(a)** Nothing happens, and the film remains multicolored.   **(b)** The film looks brighter and brighter in reflected light, appearing totally white just before it breaks.   **(c)** The film looks darker and darker in reflected light, appearing black just before it breaks.

13. Two thin films are floating on water ($n = 1.33$). The films have refractive indices of $n_1 = 1.20$ and $n_2 = 1.45$. Suppose that the thickness of each film approaches zero. In reflected light, film 1 will look ____ and film 2 will look ____.   **(a)** bright, bright   **(b)** bright, dark   **(c)** dark, bright   **(d)** dark, dark

THE MICHELSON INTERFEROMETER

27.4

An interferometer is an apparatus that can be used to measure the wavelength of light by utilizing interference between two light waves. One particularly famous interferometer was developed by Albert A. Michelson (1852–1931). The Michelson interferometer uses reflection to set up conditions where two light waves interfere.

The physics of the Michelson interferometer.

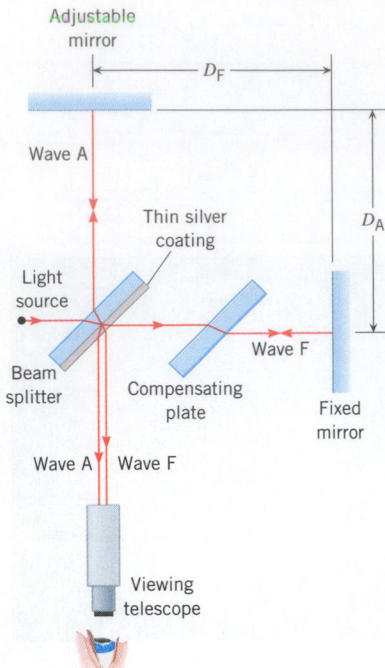

Figure 27.14 A schematic drawing of a Michelson interferometer.

Figure 27.14 presents a schematic drawing of the instrument. Waves emitted by the monochromatic light source strike a *beam splitter*, so called because it splits the beam of light into two parts. The beam splitter is a glass plate, the far side of which is coated with a thin layer of silver that reflects part of the beam upward as wave A in the drawing. The coating is so thin, however, that it also allows the remainder of the beam to pass directly through as wave F. Wave A strikes an adjustable mirror and reflects back on itself. It again crosses the beam splitter and then enters the viewing telescope. Wave F strikes a fixed mirror and returns, to be partly reflected into the viewing telescope by the beam splitter. Note that wave A passes through the glass plate of the beam splitter three times in reaching the viewing scope, while wave F passes through it only once. The compensating plate in the path of wave F has the same thickness as the beam splitter plate and ensures that wave F also passes three times through the same thickness of glass on the way to the viewing scope. Thus, an observer who views the superposition of waves A and F through the telescope sees constructive or destructive interference, depending only on the difference in path lengths D_A and D_F traveled by the two waves.

Now suppose that the mirrors are perpendicular to each other, the beam splitter makes a 45° angle with each, and the distances D_A and D_F are equal. Waves A and F travel the same distance, and the field of view in the telescope is uniformly bright due to constructive interference. However, if the adjustable mirror is moved away from the telescope by a distance of $\frac{1}{4}\lambda$, wave A travels back and forth by an amount that is twice this value, leading to an extra distance of $\frac{1}{2}\lambda$. Then, the waves are out of phase when they reach the viewing scope, destructive interference occurs, and the viewer sees a dark field. If the adjustable mirror is moved farther, full brightness returns as soon as the waves are in phase and interfere constructively. The in-phase condition occurs when wave A travels a total extra distance of λ relative to wave F. Thus, as the mirror is continuously moved, the viewer sees the field of view change from bright to dark, then back to bright, and so on. The amount by which D_A has been changed can be measured and related to the wavelength of the light, since a bright field changes into a dark field and back again each time D_A is changed by a half-wavelength. (The back-and-forth change in distance is λ.) If a sufficiently large number of wavelengths are counted in this manner, the Michelson interferometer can be used to obtain a very accurate value for the wavelength from the measured changes in D_A.

27.5 DIFFRACTION

As we have seen in Section 17.3, *diffraction* is the bending of waves around obstacles or the edges of an opening. In Figure 27.15, sound waves are leaving a room through an open doorway. Because the exiting sound waves bend, or diffract, around the edges of the opening, a listener outside the room can hear the sound even when standing around the corner from the doorway.

Diffraction is an interference effect, and the Dutch scientist Christian Huygens (1629–1695) developed a principle that is useful in explaining why diffraction arises. *Huygens' principle* describes how a wave front that exists at one instant gives rise to the wave front that exists later on. This principle states that:

> *Every point on a wave front acts as a source of tiny wavelets that move forward with the same speed as the wave; the wave front at a later instant is the surface that is tangent to the wavelets.*

We begin by using Huygens' principle to explain the diffraction of sound waves in Figure 27.15. The drawing shows the top view of a plane wave front of sound approaching a doorway and identifies five points on the wave front just as it is leaving the opening. According to Huygens' principle, each of these points acts as a source of wavelets, which are shown as red circular arcs at some moment after they are emitted. The tangent to the wavelets from points 2, 3, and 4 indicates that in front of the doorway the wave front is flat and moving straight ahead. At the edges, however, points 1 and 5 are the last points that produce wavelets. Huygens' principle suggests that in conforming to the curved shape of the wavelets near the edges, the new wave front moves into regions that it would not reach otherwise. The sound wave, then, bends or diffracts around the edges of the doorway.

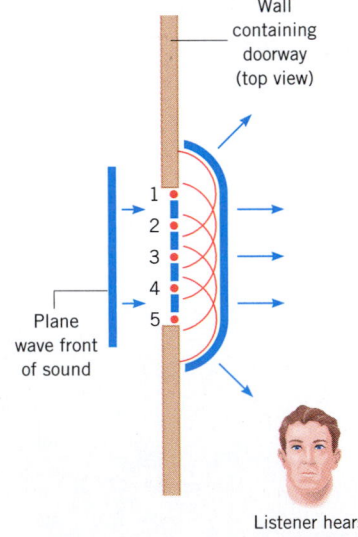

Figure 27.15 Sound bends, or diffracts, around the edges of a doorway, so even a person who is not standing directly in front of the opening can hear the sound. The five red points within the doorway act as sources and emit the five Huygens wavelets shown in red.

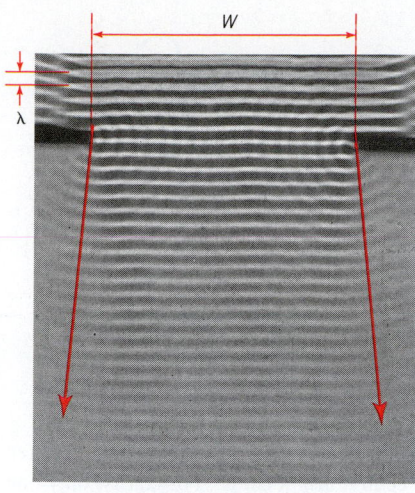

W

λ

(a) Smaller value for λ/W,
less diffraction

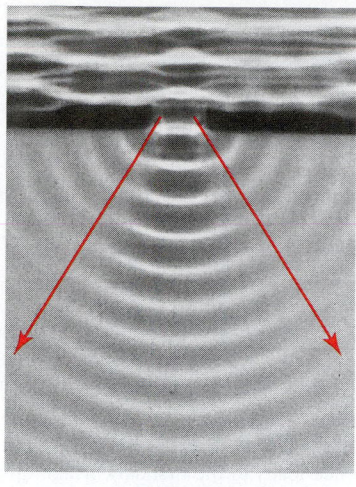

(b) Larger value for λ/W,
more diffraction

Figure 27.16 These photographs show water waves (horizontal lines) approaching an opening whose width W is greater in (*a*) than in (*b*). In addition, the wavelength λ of the waves is smaller in (*a*) than in (*b*). Therefore, the ratio λ/W increases from (*a*) to (*b*) and so does the extent of the diffraction, as the red arrows indicate. (Courtesy Education Development Center)

Huygens' principle applies not just to sound waves, but to all kinds of waves. For instance, light has a wave-like nature and, consequently, exhibits diffraction. Therefore, you may ask, "Since I can hear around the edges of a doorway, why can't I also see around them?" As a matter of fact, light waves do bend around the edges of a doorway. However, the degree of bending is extremely small, so the diffraction of light is not enough to allow you to see around the corner.

As we will learn, the extent to which a wave bends around the edges of an opening is determined by the ratio λ/W, where λ is the wavelength of the wave and W is the width of the opening. The photographs in Figure 27.16 illustrate the effect of this ratio on the diffraction of water waves. The degree to which the waves are diffracted or bent is indicated by the two red arrows in each photograph. In part *a*, the ratio λ/W is small because the wavelength (as indicated by the distance between the wave fronts) is small relative to the width of the opening. The wave fronts move through the opening with little bending or diffraction into the regions around the edges. In part *b*, the wavelength is larger and the width of the opening is smaller. As a result, the ratio λ/W is larger, and the degree of bending becomes more pronounced, with the wave fronts penetrating more into the regions around the edges of the opening.

Based on the pictures in Figure 27.16, we might expect that light waves of wavelength λ will bend or diffract appreciably when they pass through an opening whose width W is small enough to make the ratio λ/W sufficiently large. This is indeed the case, as Figure 27.17 illustrates. In this picture, it is assumed that parallel rays (or plane wave fronts) of light fall

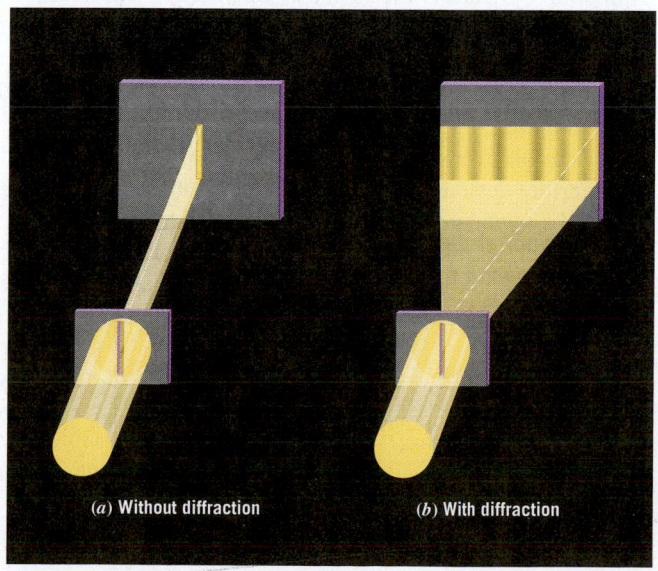

(a) Without diffraction   (b) With diffraction

Figure 27.17 (*a*) If light were to pass through a very narrow slit *without* being diffracted, only the region on the screen directly opposite the slit would be illuminated. (*b*) Diffraction causes the light to bend around the edges of the slit into regions it would not otherwise reach, forming a pattern of alternating bright and dark fringes on the screen. The slit width has been exaggerated for clarity.

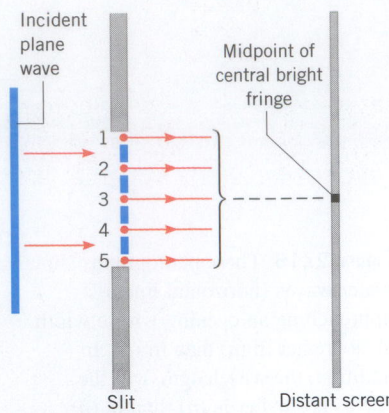

Figure 27.18 A plane wave front is incident on a single slit. This top view of the slit shows five sources of Huygens wavelets. The wavelets travel toward the midpoint of the central bright fringe on the screen, as the red rays indicate. The screen is very far from the slit.

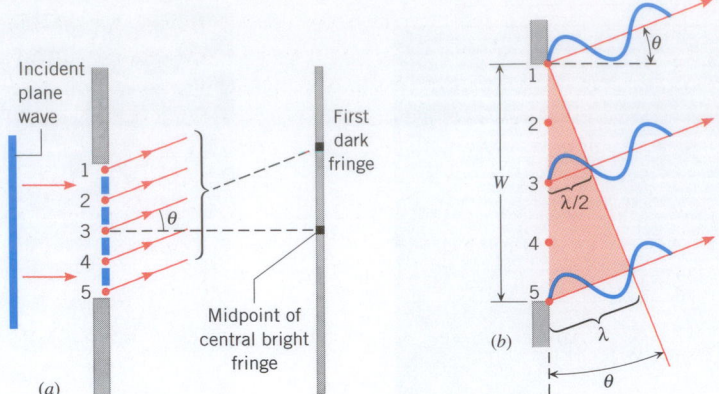

Figure 27.19 These drawings pertain to single-slit diffraction and show how destructive interference leads to the first dark fringe on either side of the central bright fringe. For clarity, only one of the dark fringes is shown. The screen is very far from the slit.

on a very narrow slit and illuminate a viewing screen that is located far from the slit. Part *a* of the drawing shows what would happen if light were *not* diffracted: it would pass through the slit without bending around the edges and would produce an image of the slit on the screen. Part *b* shows what actually happens. The light diffracts around the edges of the slit and brightens regions on the screen that are not directly opposite the slit. The diffraction pattern on the screen consists of a bright central band, accompanied by a series of narrower faint fringes that are parallel to the slit itself.

To help explain how the pattern of diffraction fringes arises, Figure 27.18 shows a top view of a plane wave front approaching the slit and singles out five sources of Huygens wavelets. Consider how the light from these five sources reaches the midpoint on the screen. To simplify things, the screen is assumed to be so far from the slit that the rays from each Huygens source are nearly parallel.* Then, all the wavelets travel virtually the same distance to the midpoint, arriving there in phase. As a result, constructive interference creates a bright central fringe on the screen, directly opposite the slit.

The wavelets emitted by the Huygens sources in the slit can also interfere destructively on the screen, as Figure 27.19 illustrates. Part *a* shows light rays directed from each source toward the first dark fringe. The angle θ gives the position of this dark fringe relative to the line between the midpoint of the slit and the midpoint of the central bright fringe. Since the screen is very far from the slit, the rays from each Huygens source are nearly parallel and are oriented at nearly the same angle θ, as in part *b* of the drawing. The wavelet from source 1 travels the shortest distance to the screen, while the wavelet from source 5 travels the farthest. Destructive interference creates the first dark fringe when the extra distance traveled by the wavelet from source 5 is exactly one wavelength, as the colored right triangle in the drawing indicates. Under this condition, the extra distance traveled by the wavelet from source 3 at the center of the slit is exactly one-half of a wavelength.

Therefore, wavelets from sources 1 and 3 in Figure 27.19*b* are exactly out of phase and interfere destructively when they reach the screen. Similarly, a wavelet that originates slightly below source 1 cancels a wavelet that originates the same distance below source 3. Thus, each wavelet from the upper half of the slit cancels a corresponding wavelet from the lower half, and no light reaches the screen. As can be seen from the colored right triangle, the angle θ locating the first dark fringe is given by $\sin \theta = \lambda/W$, where W is the width of the slit.

Figure 27.20 shows the condition that leads to destructive interference at the second dark fringe on either side of the midpoint on the screen. In reaching the screen, the light

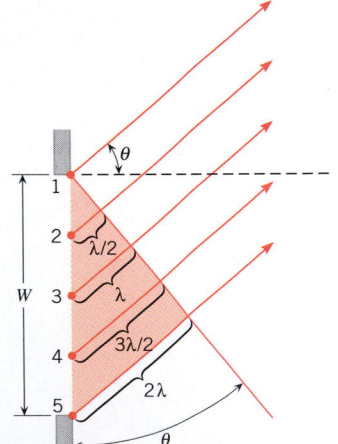

Figure 27.20 In a single-slit diffraction pattern, multiple dark fringes occur on either side of the central bright fringe. This drawing shows how destructive interference creates the second dark fringe on a very distant screen.

*When the rays are parallel, the diffraction is called Fraunhofer diffraction in tribute to the German optician Joseph von Fraunhofer (1787–1826). When the rays are not parallel, the diffraction is referred to as Fresnel diffraction, named for the French physicist Augustin Jean Fresnel (1788–1827).

from source 5 now travels a distance of two wavelengths farther than the light from source 1. Under this condition, the wavelet from source 5 travels one wavelength farther than the wavelet from source 3, and the wavelet from source 3 travels one wavelength farther than the wavelet from source 1. Therefore, each half of the slit can be treated as the entire slit was in the previous paragraph; all the wavelets from the top half interfere destructively with each other, and all the wavelets from the bottom half do likewise. As a result, no light from either half reaches the screen, and another dark fringe occurs. The colored triangle in the drawing shows that this second dark fringe occurs when $\sin \theta = 2\lambda/W$. Similar arguments hold for the third- and higher-order dark fringes, with the general result being

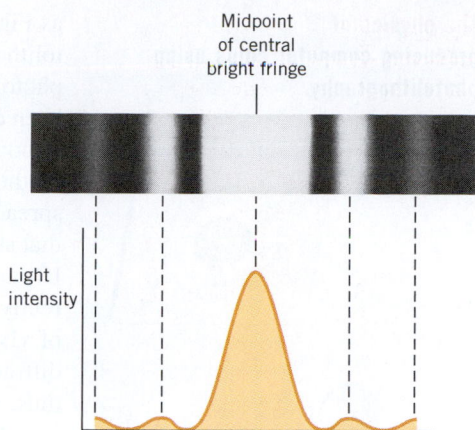

Dark fringes for single-slit diffraction

$$\sin \theta = m\frac{\lambda}{W} \qquad m = 1, 2, 3, \ldots \qquad (27.4)$$

Between each pair of dark fringes there is a bright fringe due to constructive interference. The brightness of the fringes is related to the light intensity, just as loudness is related to sound intensity. The intensity of the light at any location on the screen is the amount of light energy per second per unit area that strikes the screen there. Figure 27.21 gives a graph of the light intensity, along with a photograph of a single-slit diffraction pattern. The central bright fringe, which is approximately four times as wide as the other bright fringes, has by far the greatest intensity.

The width of the central fringe provides some indication of the extent of the diffraction, as Example 6 illustrates.

Figure 27.21 The photograph shows a single-slit diffraction pattern, with a bright and wide central fringe. The higher-order bright fringes are much less intense than the central fringe, as the graph indicates. (From Michel Cagnet et al., *Atlas of Optical Phenomena*, Springer-Verlag, Berlin, 1962)

⬇ **Example 6** Single-Slit Diffraction

Light passes through a slit and shines on a flat screen that is located $L = 0.40$ m away (see Figure 27.22). The wavelength of the light in a vacuum is $\lambda = 410$ nm. The distance between the midpoint of the central bright fringe and the first dark fringe is y. Determine the width $2y$ of the central bright fringe when the width of the slit is **(a)** $W = 5.0 \times 10^{-6}$ m and **(b)** $W = 2.5 \times 10^{-6}$ m.

Reasoning The width of the central bright fringe is determined by two factors. One is the angle θ that locates the first dark fringe on either side of the midpoint. The other is the distance L between the screen and the slit. Larger values for θ and L lead to a wider central bright fringe. Larger values of θ mean greater diffraction and occur when the ratio λ/W is larger. Thus, we expect the width of the central bright fringe to be greater when the slit width W is smaller.

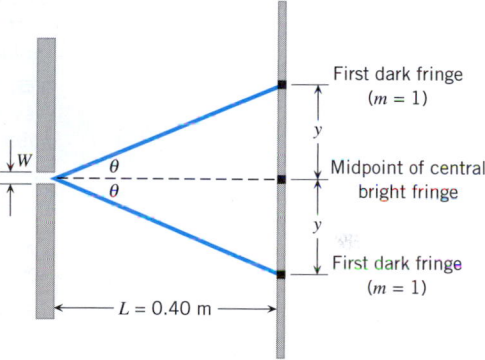

Figure 27.22 The distance $2y$ is the width of the central bright fringe.

Solution **(a)** The angle θ in Equation 27.4 locates the first dark fringe when $m = 1$: $\sin \theta = (1)\lambda/W$. Therefore,

$$\theta = \sin^{-1}\left(\frac{\lambda}{W}\right) = \sin^{-1}\left(\frac{410 \times 10^{-9}\text{ m}}{5.0 \times 10^{-6}\text{ m}}\right) = 4.7°$$

According to Figure 27.22, $\tan \theta = y/L$, so the width $2y$ of the central bright fringe is

$$2y = 2L\tan\theta = 2(0.40\text{ m})\tan 4.7° = \boxed{0.066\text{ m}}$$

(b) Repeating the calculation in part (a) with $W = 2.5 \times 10^{-6}$ m shows that $\boxed{2y = 0.13\text{ m}}$. As expected for a given wavelength, the width $2y$ of the central maximum in the diffraction pattern is greater when the width of the slit is smaller.

In the production of computer chips, it is important to minimize the effects of diffraction. Such chips are very small and yet contain enormous numbers of electronic components,

The physics of
producing computer chips using
photolithography.

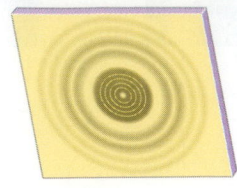

Opaque disk

Light

Figure 27.23 The diffraction pattern formed by an opaque disk consists of a small bright spot in the center of the dark shadow, circular bright fringes within the shadow, and concentric bright and dark fringes surrounding the shadow.

as Figure 23.33 illustrates. Such miniaturization is achieved using the techniques of photolithography. The patterns on the chip are created first on a "mask," which is similar to a photographic slide. Light is then directed through the mask onto silicon wafers that have been coated with a photosensitive material. The light-activated parts of the coating can be removed chemically, to leave the ultrathin lines that correspond to the miniature patterns on the chip. As the light passes through the narrow slit-like patterns on the mask, the light spreads out due to diffraction. If excessive diffraction occurs, the light spreads out so much that sharp patterns are not formed on the photosensitive material coating the silicon wafer. Ultraminiaturization of the patterns requires the absolute minimum of diffraction, and currently this is achieved by using ultraviolet light, which has a wavelength shorter than that of visible light. The shorter the wavelength λ, the smaller the ratio λ/W, and the less the diffraction. The wavelengths of X-rays are much shorter than those of ultraviolet light and, thus, will reduce diffraction even more, allowing further miniaturization.

Another example of diffraction can be seen when light from a point source falls on an opaque disk, such as a coin (Figure 27.23). The effects of diffraction modify the dark shadow cast by the disk in several ways. First, the light waves diffracted around the circular edge of the disk interfere constructively at the center of the shadow to produce a small bright spot. There are also circular bright fringes in the shadow area. In addition, the boundary between the circular shadow and the lighted screen is not sharply defined but consists of concentric bright and dark fringes. The various fringes are analogous to those produced by a single slit and are due to interference between Huygens wavelets that originate from different points near the edge of the disk.

✓ **CHECK YOUR UNDERSTANDING**

(The answers are given at the end of the book.)

14. A diffraction pattern is produced on a viewing screen by using a single slit with blue light. Does the pattern broaden or contract (become narrower) **(a)** when the blue light is replaced by red light **(b)** when the slit width is increased?

15. A sound wave has a much greater wavelength than does a light wave. When the two waves pass through a doorway, which one, if either, diffracts to a greater extent? **(a)** The sound wave **(b)** The light wave **(c)** Both waves diffract by the same amount.

27.6 RESOLVING POWER

Figure 27.24 shows three photographs of an automobile's headlights taken at progressively greater distances from the camera. In parts *a* and *b*, the two separate headlights can be seen clearly. In part *c*, however, the car is so far away that the headlights are barely distinguishable and appear almost as a single light. The *resolving power* of an optical instrument, such as a camera, is its ability to distinguish between two closely spaced objects. If a camera with a higher resolving power had taken these pictures, the photograph in part *c* would have shown two distinct and separate headlights. Any instrument used for viewing objects that are close together must have a high resolving power. This is true, for example, for a telescope used to view distant stars or for a microscope used to view tiny organisms. We will now see that diffraction occurs when light passes through the circular, or nearly circular, openings that admit light into cameras, telescopes, microscopes, and human eyes. The resulting diffraction pattern places a limit on the resolving power of these instruments.

Figure 27.24 These automobile headlights were photographed at various distances from the camera, closest in part (*a*) and farthest in part (*c*). In part (*c*), the headlights are so far away that they are barely distinguishable. (© Truax/The Image Finders)

(*a*)

(*b*)

(*c*)

Figure 27.25 shows the diffraction pattern created by a small circular opening when the viewing screen is far from the opening. The pattern consists of a central bright circular region, surrounded by alternating bright and dark circular fringes. These fringes are analogous to the rectangular fringes that a single slit produces. The angle θ in the picture locates the first circular dark fringe relative to the center of the central bright region and is given by

$$\sin \theta = 1.22 \frac{\lambda}{D} \qquad (27.5)$$

where λ is the wavelength of the light and D is the diameter of the opening. This expression is similar to Equation 27.4 for a slit ($\sin \theta = \lambda/W$, when $m = 1$) and is valid when the distance to the screen is much larger than the diameter D.

An optical instrument with the ability to resolve two closely spaced objects can produce images of them that can be identified separately. Think about the images on the image sensor when light from two widely separated point objects passes through the circular aperture of a camera. As Figure 27.26 illustrates, each image is a circular diffraction pattern, but the two patterns do not overlap and are completely resolved. On the other hand, if the objects are sufficiently close together, the intensity patterns created by the diffraction overlap, as Figure 27.27a suggests. In fact, if the overlap is extensive, it may no longer be possible to distinguish the patterns separately. In such a case, the picture from a camera would show a single blurred object instead of two separate objects. In Figure 27.27b the diffraction patterns overlap, but not enough to prevent us from seeing that two objects are present. Ultimately, then, diffraction limits the ability of an optical instrument to produce distinguishable images of objects that are close together.

It is useful to have a criterion for judging whether two closely spaced objects will be resolved by an optical instrument. Figure 27.27a presents the **Rayleigh criterion** for resolution, first proposed by Lord Rayleigh (1842–1919):

> *Two point objects are just resolved when the first dark fringe in the diffraction pattern of one falls directly on the central bright fringe in the diffraction pattern of the other.*

The minimum angle θ_{min} between the two objects in the drawing is the angle given by Equation 27.5. If θ_{min} is small (less than about 10°) and is expressed in radians, $\sin \theta_{min} \approx \theta_{min}$. Then, Equation 27.5 can be rewritten as

$$\theta_{min} \approx 1.22 \frac{\lambda}{D} \qquad (\theta_{min} \text{ in radians}) \qquad (27.6)$$

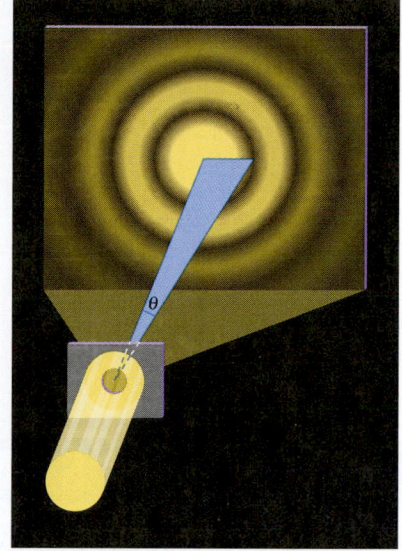

Figure 27.25 When light passes through a small circular opening, a circular diffraction pattern is formed on a screen. The angle θ locates the first dark fringe relative to the central bright region. The intensities of the bright fringes and the diameter of the opening have been exaggerated for clarity.

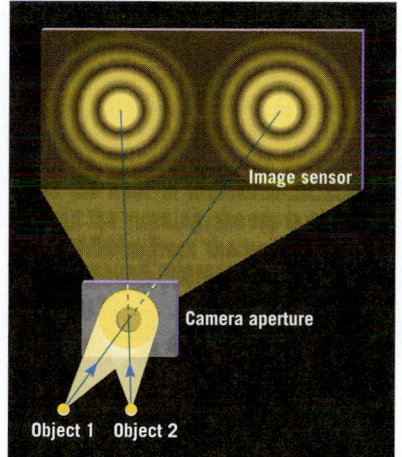

Figure 27.26 When light from two point objects passes through the circular aperture of a camera, two circular diffraction patterns are formed as images on the image sensor. The images here are completely separated or resolved because the objects are widely separated.

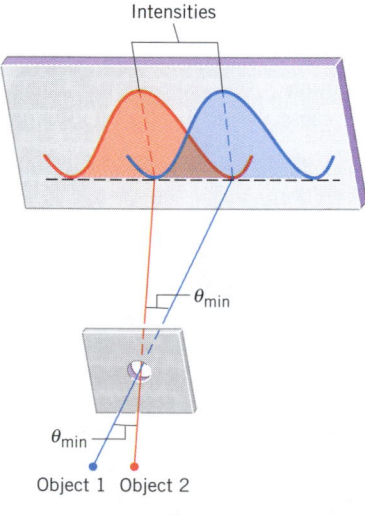

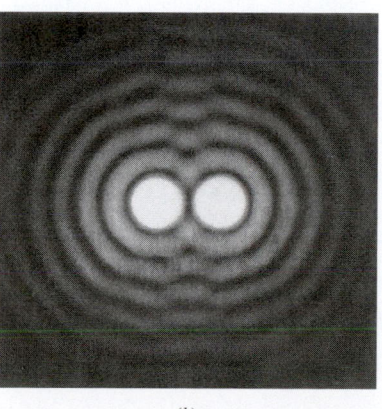

(a)

(b)

Figure 27.27 (a) According to the Rayleigh criterion, two point objects are just resolved when the first dark fringe (zero intensity) of one image falls on the central bright fringe (maximum intensity) of the other image. (b) This photograph shows two overlapping but still resolvable diffraction patterns. (From Michel Cagnet et al., *Atlas of Optical Phenomena,* Springer-Verlag, Berlin, 1962)

For a given wavelength λ and aperture diameter D, this result specifies the smallest angle that two point objects can subtend at the aperture and still be resolved. According to Equation 27.6, optical instruments designed to resolve closely spaced objects (small values of θ_{min}) must utilize the smallest possible wavelength and the largest possible aperture diameter. For example, when short-wavelength ultraviolet light is collected by its large 2.4-m-diameter mirror, the Hubble Space Telescope is capable of resolving two closely spaced stars that have an angular separation of about $\theta_{min} = 1 \times 10^{-7}$ rad. This angle is equivalent to resolving two objects only 1 cm apart when they are 1×10^5 m (about 62 miles) from the telescope. Example 7 deals with the resolving power of the human eye.

ANALYZING MULTIPLE-CONCEPT PROBLEMS

Example 7 The Human Eye Versus the Eagle's Eye

(a) A hang glider is flying at an altitude of 120 m. Green light (wavelength = 555 nm in vacuum) enters the pilot's eye through a pupil that has a diameter of 2.5 mm, Determine how far apart two point objects must be on the ground if the pilot is to have any hope of distinguishing between them (see Figure 27.28). (b) An eagle's eye has a pupil with a diameter of 6.2 mm. Repeat part (a) for an eagle flying at the same altitude as the glider.

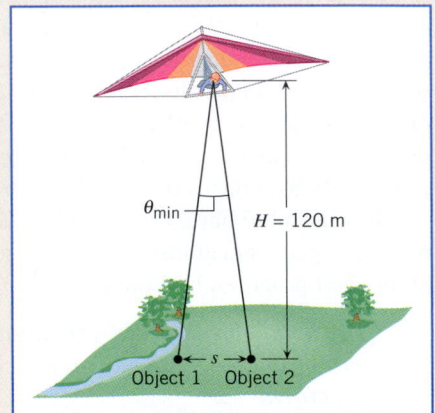

Reasoning A greater distance s of separation between the objects on the ground makes it easier for the eye of the observer (the pilot or the eagle; see Figure 27.28) to resolve them as separate objects. This is because the angle that the two objects subtend at the pupil of the eye is greater when the separation distance is greater. This angle must be at least as large as the angle θ_{min} specified by the Rayleigh criterion for resolution. In applying this criterion, we will use the concept of the radian to express the angle as an arc length (approximately the separation distance) divided by a radius (the altitude), as discussed in Section 8.1.

Figure 27.28 The Rayleigh criterion can be used to estimate the smallest distance s that can separate two objects on the ground, if a person on a hang glider is to be able to distinguish between them.

Knowns and Unknowns The following table summarizes the data that are given:

Description	Symbol	Value	Comment
Altitude	H	120 m	Same for pilot and eagle.
Wavelength of light in vacuum	λ	555 nm	1 nm = 10^{-9} m
Diameter of pupil of eye	D	2.5 mm or 6.2 mm	Smaller value is for pilot; larger value is for eagle.
Unknown Variable			
Separation between objects on ground	s	?	

Modeling the Problem

STEP 1 Radian Measure The angle that the two objects subtend at the pupil of the eye must be at least as large as the angle θ_{min} specified by the Rayleigh criterion for resolution. Using radian measure as discussed in Section 8.1, we refer to Figure 27.28 and express this angle as

$$\theta_{min} \approx \frac{s}{H}$$

This is an approximate application of Equation 8.1, which states that an angle in radians is the arc length divided by the radius. Here, the arc length is approximately the separation distance s, assuming that the altitude H is much greater than s. Solving for s gives Equation 1 at the right. In this result, the altitude is known, and we proceed to Step 2 to evaluate the angle θ_{min}.

$$s \approx \theta_{min} H \qquad (1)$$

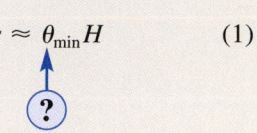

STEP 2 **The Rayleigh Criterion** The Rayleigh criterion specifies the angle θ_{min} in radians as

$$\boxed{\theta_{min} \approx 1.22 \frac{\lambda}{D}} \qquad (27.6)$$

$$s \approx \theta_{min} H \qquad (1)$$

$$\boxed{\theta_{min} \approx 1.22 \frac{\lambda}{D}} \qquad (27.6)$$

where λ is the wavelength of the light in vacuum* and D is the diameter of the pupil of the eye. The substitution of this expression into Equation 1 is shown at the right.

Solution Combining the results of each step algebraically, we find that

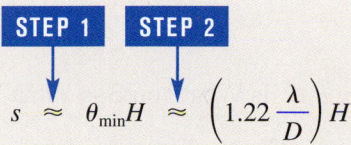

$$s \approx \theta_{min} H \approx \left(1.22 \frac{\lambda}{D}\right) H$$

The separation distance between the objects on the ground can now be obtained.
(a) For the pilot to have any hope of distinguishing between the objects, the separation distance must be at least

Problem-solving insight
The minimum angle θ_{min} between two objects that are just resolved must be expressed in radians, not degrees, when using Equation 27.6 ($\theta_{min} \approx 1.22\ \lambda/D$).

$$s \approx \left(1.22 \frac{\lambda}{D}\right) H = 1.22 \left(\frac{555 \times 10^{-9}\ \text{m}}{2.5 \times 10^{-3}\ \text{m}}\right)(120\ \text{m}) = \boxed{0.033\ \text{m}}$$

(b) For the eagle, we find that

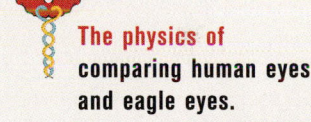

The physics of comparing human eyes and eagle eyes.

$$s \approx \left(1.22 \frac{\lambda}{D}\right) H = 1.22 \left(\frac{555 \times 10^{-9}\ \text{m}}{6.2 \times 10^{-3}\ \text{m}}\right)(120\ \text{m}) = \boxed{0.013\ \text{m}}$$

Since the pupil of the eagle's eye is larger than that of a human eye, diffraction creates less of a limitation for the eagle; the two objects can be closer together and still be resolved by the eagle's eye.

Related Homework: *Problems 33, 41*

Many optical instruments have a resolving power exceeding that of the human eye. The typical camera does, for instance. Conceptual Example 8 compares the abilities of the human eye and a camera to resolve two closely spaced objects.

*In applying the Rayleigh criterion, we use the given wavelength in vacuum because it is nearly identical to the wavelength in air. We use the wavelength in air or vacuum, even though the diffraction occurs within the eye, where the index of refraction is about $n = 1.36$ and the wavelength is $\lambda_{eye} = \lambda_{vacuum}/n$, according to Equation 27.3. The reason is that in entering the eye, the light is refracted according to Snell's law (Section 26.2), which also includes an effect due to the index of refraction. If the angle of incidence is small, the effect of n in Snell's law cancels the effect of n in Equation 27.3, to a good degree of approximation.

Figure 27.29 This person is about to take a photograph of a famous painting by Georges Seurat, who developed the technique of using tiny dots of color to construct his images. Conceptual Example 8 discusses what the person sees when the photograph is printed. (Georges Seurat, *A Sunday on La Grande Jatte.* © 2008. All rights reserved. Reproduced with permission of The Art Institute of Chicago)

Conceptual Example 8   Is What You See What You Get?

The French postimpressionist artist Georges Seurat developed a painting technique in which dots of color are placed close together on the canvas. From sufficiently far away the individual dots are not distinguishable, and the images in the picture take on a more normal appearance. Figure 27.29 shows a person in a museum looking at one of Seurat's paintings. Suppose that the person stands close to the painting, then backs up until the dots just become indistinguishable to his eyes and takes a picture from this position. The light enters his eyes through pupils that have diameters of 2.5 mm and enters the digital camera through an aperture, or opening, with a diameter of 25 mm. He then goes home and prints an enlarged photograph of the painting. Can he see the individual dots in the photograph? **(a)** No, because if his eye cannot see the dots at the museum, the camera is also unable to record the individual dots. **(b)** Yes, because the camera gathers light through a much larger aperture than does the eye. **(c)** Yes, because, unlike the eye, a photograph taken by a camera is not limited by the effects of diffraction.

Reasoning   To answer this question, we turn to the Rayleigh criterion for resolving two point objects (such as the dots)—namely, $\theta_{min} \approx 1.22\,\lambda/D$. Here θ_{min} is the minimum angle that exists between light rays from two adjacent dots as the rays pass through the aperture (see Figure 27.27), λ is the wavelength of the light, and D is the diameter of the aperture. A larger value of D implies a smaller value for θ_{min}, which, in turn, means that the instrument has a greater resolving power.

Answer (a) is incorrect.   Diffraction limits the ability of any instrument to see two closely spaced objects as distinct. This ability depends on the diameter of the aperture through which the light enters the instrument. Since the eye and the camera have apertures of different diameters, the camera may record the individual dots in the painting even though the eye does not see them as distinct.

Answer (c) is incorrect.   The effects of diffraction limit the resolving power of both the eye and the camera.

Answer (b) is correct.   For the eye and the camera, the aperture diameters are $D_{eye} = 2.5$ mm and $D_{camera} = 25$ mm, so the diameter for the camera is ten times larger than that for the eye. Thus, at the distance at which the eye loses its ability to resolve the individual dots in the painting, the camera can still easily resolve them. As discussed in the footnote to Example 7, we can ignore the effect on the wavelength of the index of refraction of the material from which the eye is made.

Related Homework: *Check Your Understanding Question 18, Problem 38*

✔ CHECK YOUR UNDERSTANDING

(The answers are given at the end of the book.)

16. Suppose that the pupil of your eye were elliptical instead of circular in shape, with the long axis of the ellipse oriented in the vertical direction. Would the resolving power of your eye be the same in the horizontal and vertical directions and, if not, in which direction would it be the greatest? The resolving power would **(a)** be the same in both directions **(b)** be greater in the horizontal direction **(c)** be greater in the vertical direction.

17. Suppose that you were designing an eye and could select the size of the pupil and the wavelengths of the electromagnetic waves to which the eye is sensitive. As far as the limitation created by diffraction is concerned, rank the following design choices in order of decreasing resolving power (greatest first): **(a)** Large pupil and ultraviolet wavelengths **(b)** Small pupil and infrared wavelengths **(c)** Small pupil and ultraviolet wavelengths

18. Review Conceptual Example 8 before answering this question. A person is viewing one of Seurat's paintings that consists of dots of color. She is so close to the painting that the dots are distinguishable. Without moving, she squints, thus reducing the size of the opening in her eyes. Does squinting make the painting take on a more normal appearance?

19. On many cameras one can select the *f*-number setting, or *f*-stop. The *f*-number gives the ratio of the focal length of the camera lens to the diameter of the aperture through which light enters the camera. If you want to resolve two closely spaced objects in a picture, should you use a small or a large *f*-number setting?

27.7 THE DIFFRACTION GRATING

Diffraction patterns of bright and dark fringes occur when monochromatic light passes through a single or double slit. Fringe patterns also result when light passes through more than two slits, and an arrangement consisting of a large number of parallel, closely spaced slits called a ***diffraction grating*** has proved very useful. Gratings with as many as 40 000 slits per centimeter can be made, depending on the production method. In one method a diamond-tipped cutting tool is used to inscribe closely spaced parallel lines on a glass plate, the spaces between the lines serving as the slits. In fact, the number of slits per centimeter is often quoted as the number of lines per centimeter.

The physics of a diffraction grating.

Figure 27.30 illustrates how light travels to a distant viewing screen from each of five slits in a grating and forms the central bright fringe and the first-order bright fringes on either side. Higher-order bright fringes are also formed but are not shown in the drawing. Each bright fringe is located by an angle θ relative to the central fringe. These bright fringes are sometimes called the *principal fringes* or *principal maxima*, since they are places where the light intensity is a maximum. The term "principal" distinguishes them from other, much less bright, fringes that are referred to as *secondary fringes* or *secondary maxima*.

Constructive interference creates the principal fringes. To show how, we assume the screen is far from the grating, so that the rays remain nearly parallel while the light travels toward the screen. In reaching the place on the screen where the first-order maximum is

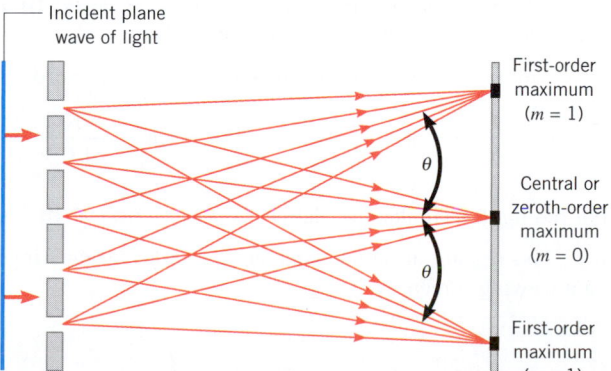

Figure 27.30 When light passes through a diffraction grating, a central bright fringe ($m = 0$) and higher-order bright fringes ($m = 1, 2, \ldots$) form when the light falls on a distant viewing screen.

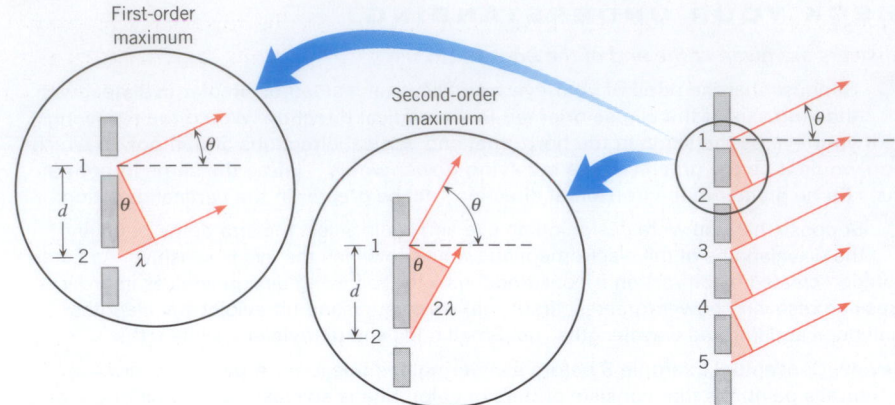

Figure 27.31 The conditions shown here lead to the first- and second-order intensity maxima in the diffraction pattern produced by the diffraction grating on the right.

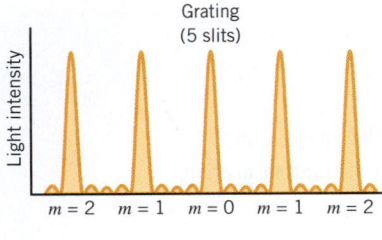

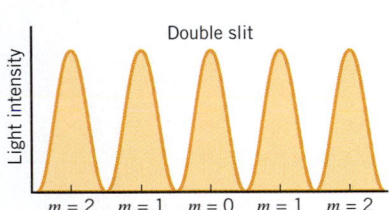

Figure 27.32 The bright fringes produced by a diffraction grating are much narrower than those produced by a double slit. Note the three small secondary bright fringes between the principal bright fringes of the grating.

located, light from slit 2 travels a distance of one wavelength farther than light from slit 1, as in Figure 27.31. Similarly, light from slit 3 travels one wavelength farther than light from slit 2, and so forth, as emphasized by the four colored right triangles on the right-hand side of the drawing. For the first-order maximum, the enlarged view of one of these right triangles shows that constructive interference occurs when $\sin \theta = \lambda/d$, where d is the separation between the slits. The second-order maximum forms when the extra distance traveled by light from adjacent slits is two wavelengths, so that $\sin \theta = 2\lambda/d$. The general result is

Principal maxima of *** a diffraction grating***	$\sin \theta = m \dfrac{\lambda}{d} \qquad m = 0, 1, 2, 3, \ldots$	(27.7)

The separation d between the slits can be calculated from the number of slits per centimeter of grating; for instance, a grating with 2500 slits per centimeter has a slit separation of $d = (1/2500)$ cm $= 4.0 \times 10^{-4}$ cm. Equation 27.7 is identical to Equation 27.1 for the double slit. A grating, however, produces bright fringes that are much *narrower* or *sharper* than those from a double slit, as the intensity patterns in Figure 27.32 reveal. Note also from the drawing that between the principal maxima of a diffraction grating there are secondary maxima with much smaller intensities. For a large number of slits, these secondary maxima become very small.

The next example illustrates the ability of a grating to separate the components in a mixture of colors.

Example 9 Separating Colors with a Diffraction Grating

A mixture of violet light ($\lambda = 410$ nm in vacuum) and red light ($\lambda = 660$ nm in vacuum) falls on a grating that contains 1.0×10^4 lines/cm. For each wavelength, find the angle θ that locates the first-order maxima.

Reasoning Before Equation 27.7 can be used here, a value for the separation d between the slits is needed: $d = 1/(1.0 \times 10^4$ lines/cm$) = 1.0 \times 10^{-4}$ cm, or 1.0×10^{-6} m. For violet light, the angle θ_{violet} for the first-order maxima ($m = 1$) is given by $\sin \theta_{\text{violet}} = m\lambda_{\text{violet}}/d$, with an analogous equation applying for the red light.

Solution For violet light, the angle locating the first-order maxima is

$$\theta_{\text{violet}} = \sin^{-1} \frac{\lambda_{\text{violet}}}{d} = \sin^{-1} \left(\frac{410 \times 10^{-9} \text{ m}}{1.0 \times 10^{-6} \text{ m}} \right) = \boxed{24°}$$

For red light, a similar calculation with $\lambda_{\text{red}} = 660 \times 10^{-9}$ m shows that $\boxed{\theta_{\text{red}} = 41°}$. Because θ_{violet} and θ_{red} are different, separate first-order bright fringes are seen for violet and red light on a viewing screen.

If the light in Example 9 had been sunlight, the angles for the first-order maxima would cover all values in the range between 24° and 41°, since sunlight contains all colors or wavelengths between violet and red. Consequently, a rainbow-like dispersion of the colors would be observed to either side of the central fringe on a screen, as can be seen in Figure 27.33. This drawing shows that the spectrum of colors associated with the $m = 2$ order is completely separate from the spectrum of the $m = 1$ order. For higher orders, however, the spectra from adjacent orders may overlap (see Problems 51 and 61). The central maximum ($m = 0$) is white because all the colors overlap there.

The physics of a grating spectroscope. An instrument designed to measure the angles at which the principal maxima of a grating occur is called a grating spectroscope. With a measured value of the angle, calculations such as those in Example 9 can be turned around to provide the corresponding value of the wavelength. As we will discuss in Chapter 30, the atoms in a hot gas emit discrete wavelengths, and determining the values of these wavelengths is one important technique used to identify the atoms. Figure 27.34 shows the principle of a grating spectroscope. The slit that admits light from the source (e.g., a hot gas) is located at the focal point of the collimating lens, so the light rays striking the grating are parallel. The telescope is used to detect the bright fringes and, hence, to measure the angle θ.

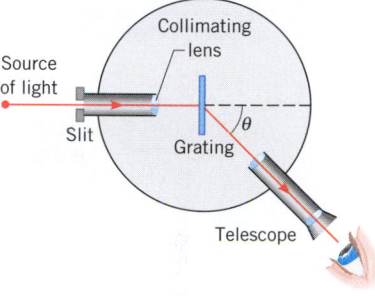

Figure 27.33 When sunlight falls on a diffraction grating, a rainbow of colors is produced at each principal maximum ($m = 1, 2, \ldots$). The central maximum ($m = 0$), however, is white but is not shown in the drawing.

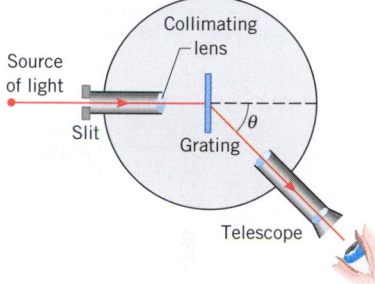

Figure 27.34 A grating spectroscope.

✓ CHECK YOUR UNDERSTANDING

(*The answers are given at the end of the book.*)

20. The drawing shows a top view of a diffraction grating and the mth-order principal maxima that are obtained with red and blue light. Red light has the longer wavelength. **(a)** Which principal maximum is associated with blue light, the one farther from or the one closer to the central maximum? **(b)** If the number of slits per centimeter in the grating were increased, would these two principal maxima move away from the central maximum, move toward the central maximum, or remain in the same place?

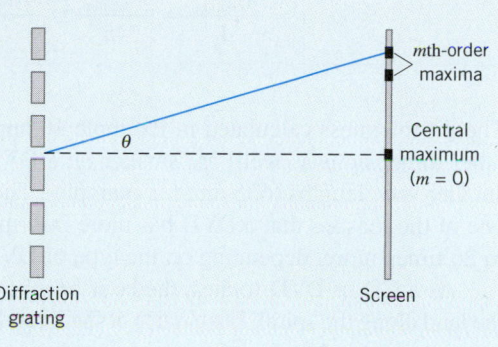

21. What would happen to the distance between the bright fringes produced by a diffraction grating if the entire interference apparatus (light source, grating, and screen) were immersed in water?

27.8

*COMPACT DISCS, DIGITAL VIDEO DISCS, AND THE USE OF INTERFERENCE

The compact disc (CD) and the digital video disc (DVD) have revolutionized how text, graphics, music, and movies are stored for use in computers, stereo sound systems, and televisions. The operation of these discs uses interference effects in some interesting ways. Each disc contains a spiral track that holds the information, which is detected using a laser beam that reflects from the bottom of the disc, as Figure 27.35 illustrates. The information is encoded in the form of raised areas on the bottom of the disc. Under a microscope these raised areas appear as "pits" when viewed from the *top* or labeled side of the disc. They are separated by flat areas called "land." The pits and land are covered with a transparent plastic coating, which has been omitted from the drawing for simplicity.

The physics of retrieving information from compact discs and digital video discs. As the disc rotates, the laser beam reflects off the bottom and into a detector. The reflected light intensity fluctuates as the pits and land areas pass by, and the fluctuations convey the information as a series of binary numbers (zeros and ones). To make the fluctuations easier to detect, the pit thickness t (see Figure 27.35) is chosen with destructive interference in mind. As the laser beam overlaps the edges of a pit, part of the beam is reflected from the

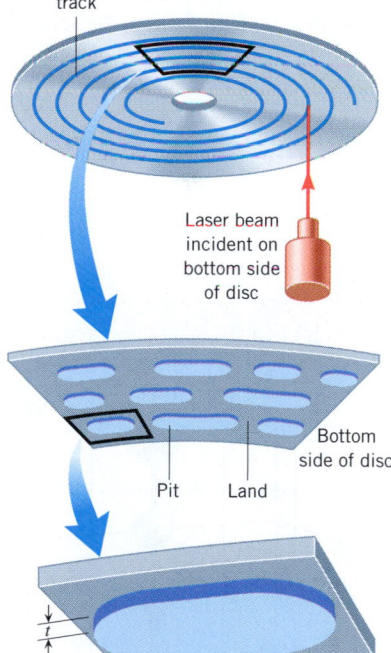

Figure 27.35 The bottom surface of a compact disc (CD) or digital video disc (DVD) carries information in the form of raised areas ("pits") and flat areas ("land") along a spiral track. A CD or DVD is played by using a laser beam that strikes the bottom surface and reflects from it.

raised pit surface and part from the land. The part that reflects from the land travels an additional distance of $2t$. The thickness of the pits is chosen so that $2t$ is one-half of a wavelength of the laser beam in the plastic coating. With this choice, destructive interference occurs when the two parts of the reflected beam combine. As a result, there is markedly less reflected intensity when the laser beam passes over a pit edge than when it passes over other places on the surface. Thus, the fluctuations in reflected light that occur while the disc rotates are large enough to detect because of the effects of destructive interference. Example 10 determines the theoretical thickness of the pits on a compact disc. In reality, a value slightly less than that obtained in the example is used for technical reasons that are not pertinent here.

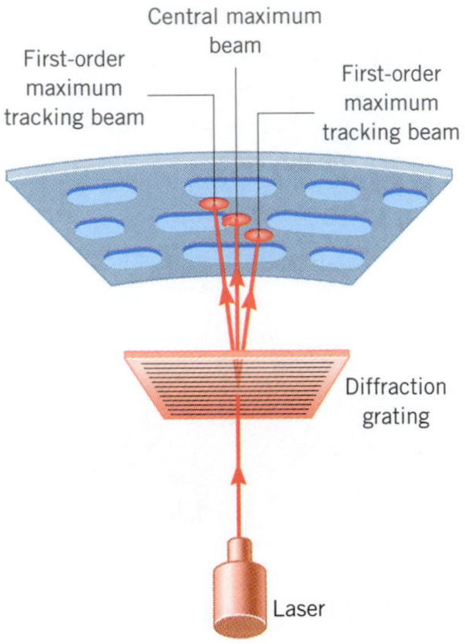

Example 10 Pit Thickness on a Compact Disc

The laser in a CD player has a wavelength of 780 nm in a vacuum. The plastic coating over the pits has an index of refraction of $n = 1.5$. Find the thickness of the pits on a CD.

Reasoning As we have discussed, the thickness t is chosen so that $2t = \frac{1}{2}\lambda_{\text{coating}}$ in order to achieve destructive interference. Equation 27.3 gives the wavelength in the plastic coating as $\lambda_{\text{coating}} = \lambda_{\text{vacuum}}/n$.

Solution The thickness of the pits is

$$t = \frac{\lambda_{\text{coating}}}{4} = \frac{\lambda_{\text{vacuum}}}{4n} = \frac{780 \times 10^{-9}\text{ m}}{4(1.5)} = \boxed{1.3 \times 10^{-7}\text{ m}}$$

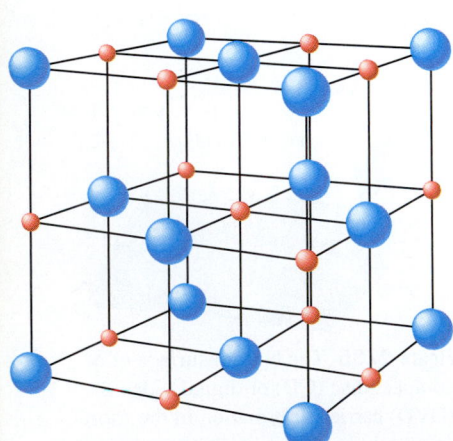

Figure 27.36 A three-beam tracking method has been used in CD players to ensure that the laser follows the spiral track correctly. The three beams are derived from a single laser beam with the aid of a diffraction grating.

The pit thickness calculated in Example 10 applies only to a CD. The pit thickness (and other dimensions as well) are smaller on a DVD because the lasers used for DVDs have smaller wavelengths (635 nm, for example). The fact that the pit dimensions are smaller is one of the reasons that a DVD has more information storage capacity than a CD—from 7 to 26 times more, depending on the type of DVD.

As a CD or DVD rotates, the laser beam must accurately follow or track the pits and the land along the spiral. **The physics of the three-beam tracking method for compact discs.** One method that has been used to ensure accurate tracking for CDs is the three-beam method, in which a diffraction grating is the key element, as Figure 27.36 shows. Before the laser beam strikes the CD, the beam passes through a grating that produces a central maximum and two first-order maxima, one on either side. As the picture indicates, the central maximum beam falls directly on the spiral track. This beam reflects into a detector, and the reflected light intensity fluctuates as the pits and land areas pass by, the fluctuations conveying the information. The two first-order maxima beams are called *tracking beams*. They hit the CD between the arms of the spiral and also reflect into detectors of their own. Under perfect conditions, the intensities of the two reflected tracking beams do not fluctuate, since they originate from the smooth surface between the arms of the spiral where there are no pits. As a result, each tracking-beam detector puts out the same constant electrical signal. However, if the tracking drifts to either side, the reflected intensity of each tracking beam changes because of the pits. In response, the tracking-beam detectors produce different electrical signals. The difference between the signals is used in a "feedback" circuit to correct for the drift and put the three beams back into their proper positions.

27.9 X-RAY DIFFRACTION

The physics of X-ray diffraction. Not all diffraction gratings are commercially made. Nature also creates diffraction gratings, although these gratings do not look like an array of closely spaced slits. Instead, nature's gratings are the arrays of regularly spaced atoms that exist in crystalline solids. For example, Figure 27.37 shows the structure of a crystal of ordinary salt (NaCl). Typically, the atoms in a crystalline solid are separated by distances of about 1.0×10^{-10} m, so we might expect a crystalline array of atoms to act like a grating with roughly this "slit" spacing for electromagnetic waves of the appropriate wavelength. Assuming that $\sin \theta = 0.5$ and that $m = 1$ in Equation 27.7, then $0.5 = \lambda/d$.

Figure 27.37 In this drawing of the crystalline structure of sodium chloride, the small red spheres represent positive sodium ions, and the large blue spheres represent negative chloride ions.

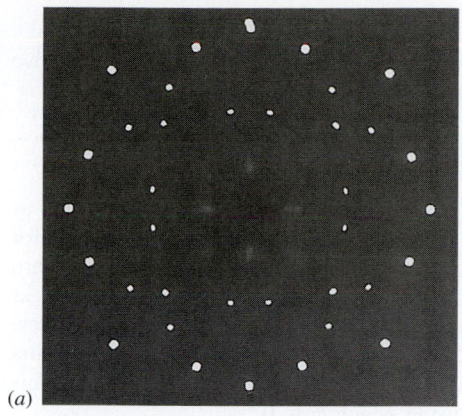
(a)

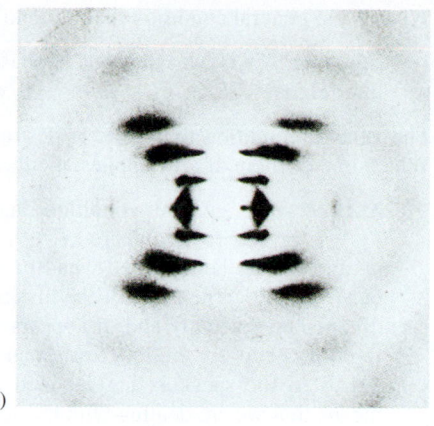
(b)

Figure 27.38 The X-ray diffraction patterns from (*a*) crystalline NaCl and (*b*) DNA. The image of DNA was obtained by Rosalind Franklin in 1953, the year in which Watson and Crick discovered DNA's structure. (*a*, Courtesy Edwin Jones, University of South Carolina; *b*, © Omikron/Photo Researchers, Inc.)

A value of $d = 1.0 \times 10^{-10}$ m in this equation gives a wavelength of $\lambda = 0.5 \times 10^{-10}$ m. This wavelength is much shorter than that of visible light and falls in the X-ray region of the electromagnetic spectrum. (See Figure 24.9.)

A diffraction pattern does indeed result when X-rays are directed onto a crystalline material, as Figure 27.38a illustrates for a crystal of NaCl. The pattern consists of a complicated arrangement of spots because a crystal has a complex three-dimensional structure. It is from patterns such as this that the spacing between atoms and the nature of the crystal structure can be determined. X-ray diffraction has also been applied with great success toward understanding the structure of biologically important molecules, such as proteins and nucleic acids. One of the most famous results was the discovery in 1953 by James Watson and Francis Crick that the structure of the nucleic acid DNA is a double helix. X-ray diffraction patterns such as that in Figure 27.38b played the pivotal role in their research.

27.10

CONCEPTS & CALCULATIONS

The ability to exhibit interference effects is a fundamental characteristic of any kind of wave. Our understanding of these effects depends on the principle of linear superposition, which we first encountered in Chapter 17. Only by means of this principle can we understand the constructive and destructive interference of light waves that lie at the heart of every topic in this chapter. Therefore, it is fitting that we review the essence of the principle in the next example.

Concepts & Calculations Example 11
The Principle of Linear Superposition

A square is 3.5 m on a side, and point A is the midpoint of one of its sides. On the side opposite this spot, two in-phase loudspeakers are located at adjacent corners, as Figure 27.39 indicates. Standing at point A, you hear a loud sound because constructive interference occurs between the identical sound waves coming from the speakers. As you walk along the side of the square toward either empty corner, the loudness diminishes gradually but does not entirely disappear until you reach either empty corner, where you hear no sound at all. Thus, at each empty corner destructive interference occurs. Find the wavelength of the sound waves.

Concept Questions and Answers Why does constructive interference occur at point A?

Answer The loudspeakers are in-phase sources of identical sound waves. The waves coming from one speaker travel a distance ℓ_1 in reaching point A and the waves from the second speaker travel a distance ℓ_2. The condition that leads to constructive interference is $\ell_2 - \ell_1 = m\lambda$, where λ is the wavelength of the waves and $m = 0, 1, 2, 3, \ldots$. Thus, the two distances are the same or differ by an integer number of wavelengths. Point A is the midpoint of a side of the square, so that ℓ_1 and ℓ_2 are the same, and constructive interference occurs.

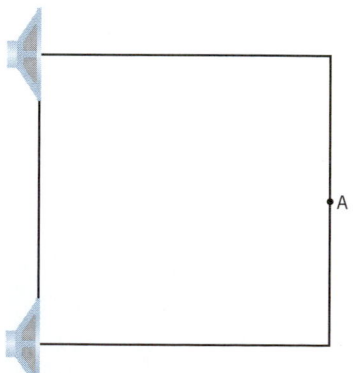

Figure 27.39 Two loudspeakers are located at the corners of a square and produce identical sound waves. As Example 11 discusses, constructive interference occurs at point A, which is at the midpoint of the opposite side of the square, and destructive interference occurs at either empty corner.

What is the general condition that leads to destructive interference?

Answer   The waves start out in phase, so the general condition that leads to destructive interference is $\ell_2 - \ell_1 = (m + \frac{1}{2})\lambda$, where $m = 0, 1, 2, 3, \ldots$.

The general condition that leads to destructive interference entails a number of possibilities. Which one of them, if any, applies at either empty corner of the square?

Answer   The general condition that leads to destructive interference is written as $\ell_2 - \ell_1 = (m + \frac{1}{2})\lambda$, where $m = 0, 1, 2, 3, \ldots$. There are many possible values for m, and we are being asked what the specific value is. As you walk from point A toward either empty corner, one of the distances decreases and the other increases; the loudness diminishes gradually and disappears at either empty corner. Therefore, destructive interference occurs for the first time at an empty corner, when the difference between the distances to the speakers has attained the smallest possible value of $\ell_2 - \ell_1 = \frac{1}{2}\lambda$. This means that we are dealing with the specific case in which $m = 0$.

Solution   Consider the destructive interference that occurs at either empty corner. Using L to denote the length of a side of the square and taking advantage of the Pythagorean theorem, we have

$$\ell_1 = L \quad \text{and} \quad \ell_2 = \sqrt{L^2 + L^2} = \sqrt{2}\, L$$

The specific condition for the destructive interference at the empty corner is

$$\ell_2 - \ell_1 = \sqrt{2}\, L - L = \frac{1}{2}\lambda$$

Solving for the wavelength of the waves gives

$$\lambda = 2L\,(\sqrt{2} - 1) = 2(3.5 \text{ m})(\sqrt{2} - 1) = \boxed{2.9 \text{ m}}$$

The next example deals with thin-film interference and serves as a review of the factors that must be considered in such cases.

Concepts & Calculations Example 12
A Soap Film

A soap film ($n = 1.33$) is 375 nm thick and coats a flat piece of glass ($n = 1.52$). Thus, air is on one side of the film and glass is on the other side, as Figure 27.40 illustrates. Sunlight, whose wavelengths (in vacuum) extend from 380 to 750 nm, travels through the air and strikes the film nearly perpendicularly. For which wavelength(s) in this range does constructive interference cause the film to look bright in reflected light?

Concept Questions and Answers   What, if any, phase change occurs when light, traveling in air, reflects from the air–film interface?

Answer   In air the index of refraction is $n = 1.00$, whereas in the film it is $n = 1.33$. A phase change occurs whenever light travels through a material with a smaller refractive index toward a material with a larger refractive index and reflects from the boundary between the two. The phase change is equivalent to $\frac{1}{2}\lambda_{\text{film}}$, where λ_{film} is the wavelength in the film.

What, if any, phase change occurs when light, traveling in the film, reflects from the film–glass interface?

Answer   In the film the index of refraction is $n = 1.33$, whereas in the glass it is $n = 1.52$. Thus, the light is again traveling through a material with a smaller value of n toward a material with a greater value of n, and reflection at the film–glass interface is accompanied by a phase change that is equivalent to $\frac{1}{2}\lambda_{\text{film}}$.

Is the wavelength of the light in the film greater than, smaller than, or equal to the wavelength in a vacuum?

Answer   The wavelength of the light in the film (refractive index $= n$) is given by Equation 27.3 as $\lambda_{\text{film}} = \lambda_{\text{vacuum}}/n$. Since the refractive index of the film is $n = 1.33$, the wavelength in the film is less than the wavelength in a vacuum.

Incident
light

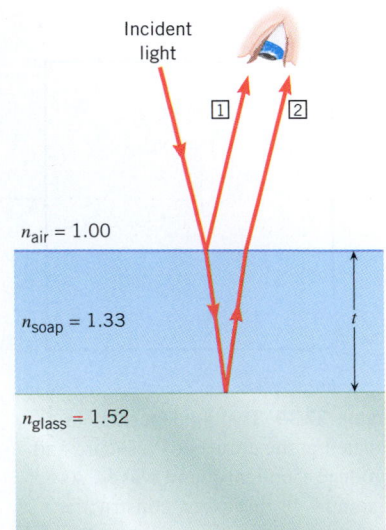

$n_{\text{air}} = 1.00$

$n_{\text{soap}} = 1.33$

$n_{\text{glass}} = 1.52$

Figure 27.40 The constructive interference that occurs with this thin soap film is analyzed in Example 12.

Solution Figure 27.40 shows the soap film and the two rays of light that represent the interfering light waves. At nearly perpendicular incidence, ray 2 travels a distance of $2t$ farther than ray 1, where t is the thickness of the film. In addition, as discussed in the Concept Questions, ray 2 experiences a phase shift of $\frac{1}{2}\lambda_{film}$ upon reflection at the bottom film surface, while ray 1 experiences the same phase shift at the upper film surface. Therefore, there is no net phase change for the two reflected rays, and only the extra travel distance determines the type of interference that occurs. For constructive interference the extra travel distance must be an integer number of wavelengths in the film:

$$\underbrace{2t}_{\substack{\text{Extra distance} \\ \text{traveled by} \\ \text{ray 2}}} + \underbrace{0}_{\substack{\text{Zero net phase} \\ \text{change due to} \\ \text{reflection}}} = \underbrace{\lambda_{film}, 2\lambda_{film}, 3\lambda_{film}, \ldots}_{\substack{\text{Condition for} \\ \text{constructive interference}}}$$

This result is equivalent to

$$2t = m\lambda_{film} \qquad m = 1, 2, 3, \ldots$$

The wavelength in the film is $\lambda_{film} = \lambda_{vacuum}/n$ (Equation 27.3), so the condition for constructive interference becomes

$$2t = m\lambda_{film} = m\frac{\lambda_{vacuum}}{n} \qquad m = 1, 2, 3, \ldots$$

Solving for the vacuum wavelength gives

$$\lambda_{vacuum} = \frac{2nt}{m} \qquad m = 1, 2, 3, \ldots$$

For the first four values of m and the given values for n and t, we find that

$$m = 1 \qquad \lambda_{vacuum} = \frac{2nt}{m} = \frac{2(1.33)(375 \text{ nm})}{1} = 998 \text{ nm}$$

$$m = 2 \qquad \lambda_{vacuum} = \frac{2nt}{m} = \frac{2(1.33)(375 \text{ nm})}{2} = 499 \text{ nm}$$

$$m = 3 \qquad \lambda_{vacuum} = \frac{2nt}{m} = \frac{2(1.33)(375 \text{ nm})}{3} = 333 \text{ nm}$$

$$m = 4 \qquad \lambda_{vacuum} = \frac{2nt}{m} = \frac{2(1.33)(375 \text{ nm})}{4} = 249 \text{ nm}$$

The range of visible wavelengths (in vacuum) extends from 380 to 750 nm. Therefore, the only visible wavelength in which the film appears bright due to constructive interference is $\boxed{499 \text{ nm}}$.

CONCEPT SUMMARY

If you need more help with a concept, use the Learning Aids noted next to the discussion or equation. Examples (**Ex.**) are in the text of this chapter. Go to **www.wiley.com/college/cutnell** for the following Learning Aids:

Interactive LearningWare (ILW) — Additional examples solved in a five-step interactive format.

Concept Simulations (CS) — Animated text figures or animations of important concepts.

Interactive Solutions (IS) — Models for certain types of problems in the chapter homework. The calculations are carried out interactively.

Topic	Discussion	Learning Aids
Principle of linear superposition	**27.1 THE PRINCIPLE OF LINEAR SUPERPOSITION** The principle of linear superposition states that when two or more waves are present simultaneously in the same region of space, the resultant disturbance is the sum of the disturbances from the individual waves.	**Ex. 11**
Constructive interference	Constructive interference occurs at a point when two waves meet there crest-to-crest and trough-to-trough, thus reinforcing each other. When two waves that start out in phase and have traveled	

Topic	Discussion	Learning Aids

some distance meet at a point, constructive interference occurs whenever the travel distances are the same or differ by any integer number of wavelengths: $\ell_2 - \ell_1 = m\lambda$, where ℓ_1 and ℓ_2 are the distances traveled by the waves, and $m = 0, 1, 2, 3, \ldots$.

Destructive interference

Destructive interference occurs at a point when two waves meet there crest-to-trough, thus mutually canceling each other. When two waves that start out in phase and have traveled some distance meet at a point, destructive interference occurs whenever the travel distances differ by any odd integer number of half-wavelengths: $\ell_2 - \ell_1 = (m + \frac{1}{2})\lambda$, where ℓ_1 and ℓ_2 are the distances traveled by the waves, and $m = 0, 1, 2, 3, \ldots$.

Coherent sources

Two sources are coherent if the waves they emit maintain a constant phase relation. In other words, the waves do not shift relative to one another as time passes. If constructive and destructive interference are to be observed, coherent sources are necessary.

Double-slit experiment

27.2 YOUNG'S DOUBLE-SLIT EXPERIMENT In Young's double-slit experiment, light passes through a pair of closely spaced narrow slits and produces a pattern of alternating bright and dark fringes on a viewing screen. The fringes arise because of constructive and destructive interference. The angle θ that locates the mth-order bright fringe is given by

Bright fringes of a double slit

$$\sin \theta = \frac{m\lambda}{d} \qquad m = 0, 1, 2, 3, \ldots \qquad (27.1)$$

Ex. 1, 2
ILW 27.1
IS 27.9

where λ is the wavelength of the light, and d is the spacing between the slits. The angle that locates the mth dark fringe is given by

Dark fringes of a double slit

$$\sin \theta = \frac{(m + \frac{1}{2})\lambda}{d} \qquad m = 0, 1, 2, 3, \ldots \qquad (27.2)$$

27.3 THIN-FILM INTERFERENCE Constructive and destructive interference of light waves can occur with thin films of transparent materials. The interference occurs between light waves that reflect from the top and bottom surfaces of the film.

One important factor in thin-film interference is the thickness of a film relative to the wavelength of the light within the film. The wavelength λ_{film} within a film is

Wavelength within a film

$$\lambda_{\text{film}} = \frac{\lambda_{\text{vacuum}}}{n} \qquad (27.3)$$

Ex. 3, 4, 5, 12
ILW 27.2
IS 27.20

where λ_{vacuum} is the wavelength in a vacuum, and n is the index of refraction of the film.

Phase change due to reflection

A second important factor is the phase change that can occur when light reflects at each surface of the film:

1. When light travels through a material with a smaller index of refraction toward a material with a larger index of refraction, reflection at the boundary occurs along with a phase change that is equivalent to one-half a wavelength in the film.

2. When light travels through a material with a larger index of refraction toward a material with a smaller index of refraction, there is no phase change upon reflection at the boundary.

Michelson interferometer

27.4 THE MICHELSON INTERFEROMETER An interferometer is an instrument that can be used to measure the wavelength of light by employing interference between two light waves. The Michelson interferometer splits the light into two beams. One beam travels to a fixed mirror, reflects from it, and returns. The other beam travels to a movable mirror, reflects from it, and returns. When the two returning beams are combined, interference is observed, the amount of which depends on the travel distances.

Diffraction
Huygens' principle

27.5 DIFFRACTION Diffraction is the bending of waves around obstacles or around the edges of an opening. Diffraction is an interference effect that can be explained with the aid of Huygens' principle. This principle states that every point on a wave front acts as a source of tiny wavelets that move forward with the same speed as the wave; the wave front at a later instant is the surface that is tangent to the wavelets.

CS 27.1

When light passes through a single narrow slit and falls on a viewing screen, a pattern of bright and dark fringes is formed because of the superposition of Huygens wavelets. The angle θ that specifies the mth dark fringe on either side of the central bright fringe is given by

Dark fringes for single-slit diffraction

$$\sin \theta = m\frac{\lambda}{W} \qquad m = 1, 2, 3, \ldots \qquad (27.4)$$

Ex. 6
ILW 27.3

where λ is the wavelength of the light and W is the width of the slit.

Topic	Discussion	Learning Aids
Resolving power	**27.6 RESOLVING POWER** The resolving power of an optical instrument is the ability of the instrument to distinguish between two closely spaced objects. Resolving power is limited by the diffraction that occurs when light waves enter an instrument, often through a circular opening.	
Rayleigh criterion	The Rayleigh criterion specifies that two point objects are just resolved when the first dark fringe in the diffraction pattern of one falls directly on the central bright fringe in the diffraction pattern of the other. According to this specification, the minimum angle (in radians) that two point objects can subtend at a circular aperture of diameter D and still be resolved as separate objects is	
Minimum angle for resolving two point objects	$$\theta_{min} \approx 1.22 \frac{\lambda}{D} \qquad (\theta_{min} \text{ in radians}) \qquad (27.6)$$ where λ is the wavelength of the light.	**Ex. 7, 8** **IS 27.41**
Diffraction grating	**27.7 THE DIFFRACTION GRATING** A diffraction grating consists of a large number of parallel, closely spaced slits. When light passes through a diffraction grating and falls on a viewing screen, the light forms a pattern of bright and dark fringes. The bright fringes are referred to as principal maxima and are found at an angle θ such that	
Principal maxima of a diffraction grating	$$\sin \theta = m\frac{\lambda}{d} \qquad m = 0, 1, 2, 3, \ldots \qquad (27.7)$$ where λ is the wavelength of the light and d is the separation between two adjacent slits.	**Ex. 9**
	27.8 COMPACT DISCS, DIGITAL VIDEO DISCS, AND THE USE OF INTERFERENCE Compact discs and digital video discs depend on interference for their operation.	**Ex. 10**
X-ray diffraction	**27.9 X-RAY DIFFRACTION** A diffraction pattern forms when X-rays are directed onto a crystalline material. The pattern arises because the regularly spaced atoms in a crystal act like a diffraction grating. Because the spacing is extremely small, on the order of 1×10^{-10} m, the wavelength of the electromagnetic waves must also be very small—hence, the use of X-rays. The crystal structure of a material can be determined from its X-ray diffraction pattern.	

FOCUS ON CONCEPTS

Note to Instructors: The numbering of the questions shown here reflects the fact that they are only a representative subset of the total number that are available online. However, all of the questions are available for assignment via an online homework management program such as WileyPLUS or WebAssign.

Section 27.1   The Principle of Linear Superposition

1. The two loudspeakers in the drawing are producing identical sound waves. The waves spread out and overlap at the point P. What is the difference $\ell_2 - \ell_1$ in the two path lengths if point P is at the third sound intensity minimum from the central sound intensity maximum? Express this difference in terms of the wavelength λ of the sound.   **(a)** $\frac{1}{2}\lambda$ **(b)** λ **(c)** $\frac{3}{2}\lambda$ **(d)** 3λ **(e)** $\frac{5}{2}\lambda$

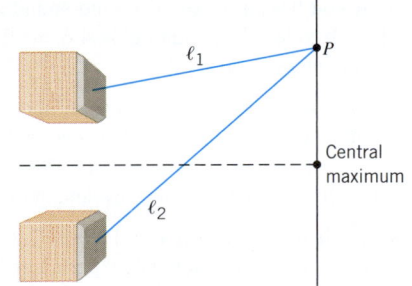

Section 27.2   Young's Double-Slit Experiment

2. In a certain Young's double-slit experiment, a diffraction pattern is formed on a distant screen. The angle that locates a given bright fringe is small, so that the approximation $\sin \theta \approx \theta$ is valid. Assuming that θ remains small, by what factor does it change if the wavelength λ is doubled and the slit separation d is doubled?   **(a)** The angle does not change.   **(b)** The angle

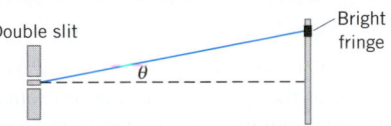

increases by a factor of 2.   **(c)** The angle increases by a factor of 4.   **(d)** The angle decreases by a factor of 2.   **(e)** The angle decreases by a factor of 4.

Section 27.3   Thin-Film Interference

6. Light of wavelength 600 nm in vacuum is incident nearly perpendicularly on a thin film whose index of refraction is 1.5. The light travels from the top surface of the film to the bottom surface, reflects from the bottom surface, and returns to the top surface, as the drawing indicates. How far has the light traveled inside the film? Express your answer in terms of the wavelength λ_{film} of the light within the film.   **(a)** $2\lambda_{film}$ **(b)** $3\lambda_{film}$ **(c)** $4\lambda_{film}$ **(d)** $6\lambda_{film}$ **(e)** $12\lambda_{film}$

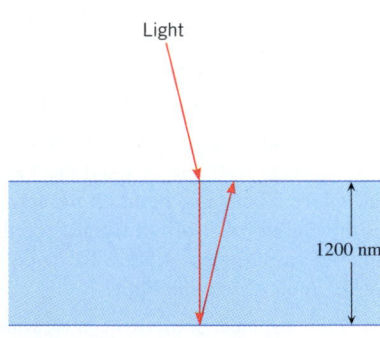

8. Light is incident perpendicularly on four transparent films of different thickness. The thickness of each film is given in the drawings in terms of the wavelength λ_{film} of the light within the film. The index of refraction of each film is 1.5, and each is surrounded by air. Which film (or films) will appear bright due to constructive interference when

viewed from the top surface, upon which the light is incident?
(a) 1, 2, 3, 4   **(b)** 2, 3   **(c)** 3   **(d)** 3, 4   **(e)** 4

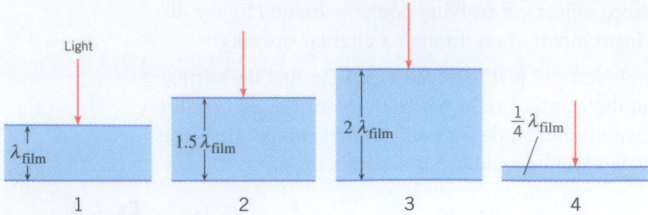

Light

λ_{film} 1   1.5 λ_{film} 2   2 λ_{film} 3 $\frac{1}{4} \lambda_{\text{film}}$ 4

Section 27.5   Diffraction

12. Light passes through a single slit. If the width of the slit is reduced, what happens to the width of the central bright fringe?   **(a)** The width of the central bright fringe does not change, because it depends only on the wavelength of the light and not on the width of the slit.   **(b)** The central bright fringe becomes wider, because the angle that locates the first dark fringe on either side of the central bright fringe becomes smaller.   **(c)** The central bright fringe becomes wider, because the angle that locates the first dark fringe on either side of the central bright fringe becomes larger.   **(d)** The central bright fringe becomes narrower, because the angle that locates the first dark fringe on either side of the central bright fringe becomes larger.   **(e)** The central bright fringe becomes narrower, because the angle that locates the first dark fringe on either side of the central bright fringe becomes smaller.

13. Light of wavelength λ passes through a single slit of width W and forms a diffraction pattern on a viewing screen. If this light is then replaced by light of wavelength 2λ, the original diffraction pattern is exactly reproduced if the width of the slit _____.   **(a)** is changed to $\frac{1}{4}W$ **(b)** is changed to $\frac{1}{2}W$ **(c)** is changed to $2W$ **(d)** is changed to $4W$ **(e)** remains the same—no change is necessary

Section 27.6   Resolving Power

15. Suppose that you are using a microscope to view two closely spaced cells. For a given lens diameter, which color of light would you use to achieve the best possible resolving power?   **(a)** Red   **(b)** Yellow   **(c)** Green   **(d)** Blue   **(e)** All the colors give the same resolving power.

Section 27.7   The Diffraction Grating

18. A diffraction grating is illuminated with yellow light. The diffraction pattern seen on a viewing screen consists of three yellow bright fringes, one at the central maximum ($\theta = 0°$) and one on either side of it at $\theta = \pm 50°$. Then the grating is simultaneously illuminated with red light. Where a red and a yellow fringe overlap, an orange fringe is produced. The new pattern consists of _____.   **(a)** only red fringes at $0°$ and $\pm 50°$ **(b)** only yellow fringes at $0°$ and $\pm 50°$ **(c)** only orange fringes at $0°$ and $\pm 50°$ **(d)** an orange fringe at $0°$, yellow fringes at $\pm 50°$, and red fringes farther out   **(e)** an orange fringe at $0°$, yellow fringes at $\pm 50°$, and red fringes closer in

PROBLEMS

Note to Instructors: Most of the homework problems in this chapter are available for assignment via an online homework management program such as WileyPLUS or WebAssign, and those marked with the icon **GO** *are presented in WileyPLUS using a guided tutorial format that provides enhanced interactivity. See Preface for additional details.*

ssm   Solution is in the Student Solutions Manual.
www   Solution is available online at www.wiley.com/college/cutnell

 This icon represents a biomedical application.

Section 27.1   The Principle of Linear Superposition,
Section 27.2   Young's Double-Slit Experiment

1. **ssm** The transmitting antenna for a radio station is 7.00 km from your house. The frequency of the electromagnetic wave broadcast by this station is 536 kHz. The station builds a second transmitting antenna that broadcasts an identical electromagnetic wave in phase with the original one. The new antenna is 8.12 km from your house. Does interference occur at the receiving antenna of your radio? If so, is it constructive or destructive? Show your calculations.

2. In a Young's double-slit experiment, two rays of monochromatic light emerge from the slits and meet at a point on a distant screen, as in Figure 27.6a. The point on the screen where these two rays meet is the eighth-order bright fringe. The difference in the distances that the two rays travel is 4.57×10^{-6} m. What is the wavelength (in nm) of the monochromatic light?

3. Two in-phase sources of waves are separated by a distance of 4.00 m. These sources produce identical waves that have a wavelength of 5.00 m. On the line between them, there are two places at which the same type of interference occurs.   **(a)** Is it constructive or destructive interference, and   **(b)** where are the places located?

4. **GO** Point A is the midpoint of one of the sides of a square. On the side opposite this spot, two in-phase loudspeakers are located at adjacent corners, as in Figure 27.39. Standing at point A you hear a loud sound because of constructive interference between the identical sound waves coming from the speakers. As you walk

along the side of the square toward either empty corner, the loudness diminishes gradually to nothing and then increases again until you hear a maximally loud sound at the corner. If the length of each side of the square is 4.6 m, find the wavelength of the sound waves.

5. In a Young's double-slit experiment, the seventh dark fringe is located 0.025 m to the side of the central bright fringe on a flat screen, which is 1.1 m away from the slits. The separation between the slits is 1.4×10^{-4} m. What is the wavelength of the light being used?

6. **GO** In a setup like that in Figure 27.7, a wavelength of 625 nm is used in a Young's double-slit experiment. The separation between the slits is $d = 1.4 \times 10^{-5}$ m. The total width of the screen is 0.20 m. In one version of the setup, the separation between the double slit and the screen is $L_A = 0.35$ m, whereas in another version it is $L_B = 0.50$ m. On one side of the central bright fringe, how many bright fringes lie on the screen in the two versions of the setup? Do not include the central bright fringe in your counting.

7. **Interactive LearningWare 27.1** at www.wiley.com/college/cutnell explores the approach taken in problems such as this one. Two parallel slits are illuminated by light composed of two wavelengths, one of which is 645 nm. On a viewing screen, the light whose wavelength is known produces its third dark fringe at the same place where the light whose wavelength is unknown produces its fourth-order bright fringe. The fringes are counted relative to the central or zeroth-order bright fringe. What is the unknown wavelength?

*8. Review Conceptual Example 2 before attempting this problem. Two slits are 0.158 mm apart. A mixture of red light (wavelength = 665 nm) and yellow-green light (wavelength = 565 nm) falls on the slits. A flat observation screen is located 2.24 m away. What is the distance on the screen between the third-order red fringe and the third-order yellow-green fringe?

*9. Refer to **Interactive Solution 27.9** at **www.wiley.com/college/cutnell** for help in solving this problem. In a Young's double-slit experiment the separation y between the second-order bright fringe and the central bright fringe on a flat screen is 0.0180 m when the light has a wavelength of 425 nm. Assume that the angles that locate the fringes on the screen are small enough so that $\sin \theta \approx \tan \theta$. Find the separation y when the light has a wavelength of 585 nm.

**10. In Young's experiment a mixture of orange light (611 nm) and blue light (471 nm) shines on the double slit. The centers of the first-order bright blue fringes lie at the outer edges of a screen that is located 0.500 m away from the slits. However, the first-order bright orange fringes fall off the screen. By how much and in which direction (toward or away from the slits) should the screen be moved so that the centers of the first-order bright orange fringes will just appear on the screen? It may be assumed that θ is small, so that $\sin \theta \approx \tan \theta$.

11. **ssm www A sheet that is made of plastic ($n = 1.60$) covers *one slit* of a double slit (see the drawing). When the double slit is illuminated by monochromatic light ($\lambda_{vacuum} = 586$ nm), the center of the screen appears dark rather than bright. What is the minimum thickness of the plastic?

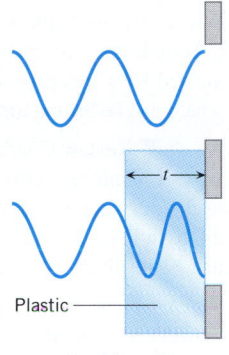

Problem 11

Section 27.3 Thin-Film Interference

12. A nonreflective coating of magnesium fluoride ($n = 1.38$) covers the glass ($n = 1.52$) of a camera lens. Assuming that the coating prevents reflection of yellow-green light (wavelength in vacuum = 565 nm), determine the minimum nonzero thickness that the coating can have.

13. **ssm** Light of wavelength 691 nm (in vacuum) is incident perpendicularly on a soap film ($n = 1.33$) suspended in air. What are the two smallest nonzero film thicknesses (in nm) for which the reflected light undergoes constructive interference?

14. When monochromatic light shines perpendicularly on a soap film ($n = 1.33$) with air on each side, the second smallest nonzero film thickness for which destructive interference of reflected light is observed is 296 nm. What is the vacuum wavelength of the light in nm?

15. **Interactive LearningWare 27.2** at **www.wiley.com/college/cutnell** provides some pertinent background for this problem. A transparent film ($n = 1.43$) is deposited on a glass plate ($n = 1.52$) to form a nonreflecting coating. The film has a thickness that is 1.07×10^{-7} m. What is the longest possible wavelength (in vacuum) of light for which this film has been designed?

16. **GO** A soap film ($n = 1.33$) is 465 nm thick and lies on a glass plate ($n = 1.52$). Sunlight, whose wavelengths (in vacuum) extend from 380 to 750 nm, travels through the air and strikes the film perpendicularly. For which wavelength(s) in this range does destructive interference cause the film to look dark in reflected light?

17. Review Conceptual Example 4 before beginning this problem. A soap film with different thicknesses at different places has an unknown refractive index n and air on both sides. In reflected light it looks multicolored. One region looks yellow because destructive interference has removed blue ($\lambda_{vacuum} = 469$ nm) from the reflected light, while another looks magenta because destructive interference

has removed green ($\lambda_{vacuum} = 555$ nm). In these regions the film has the minimum nonzero thickness t required for the destructive interference to occur. Find the ratio $t_{magenta}/t_{yellow}$.

*18. **GO** A beam of light is sent directly down onto a glass plate ($n = 1.5$) and a plastic plate ($n = 1.2$) that form a thin wedge of air (see the drawing). An observer looking down through the glass plate sees the fringe pattern shown in the lower part of the drawing, with the dark fringes at the ends A and B. The wavelength of the light is 520 nm. Using the fringe pattern shown in the drawing, determine the thickness of the air wedge at B.

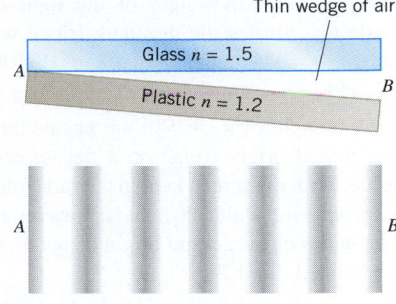

*19. **ssm www** Orange light ($\lambda_{vacuum} = 611$ nm) shines on a soap film ($n = 1.33$) that has air on either side of it. The light strikes the film perpendicularly. What is the minimum thickness of the film for which constructive interference causes it to look bright in reflected light?

*20. Consult **Interactive Solution 27.20** at **www.wiley.com/college/cutnell** to review a model for solving this problem. A film of oil lies on wet pavement. The refractive index of the oil exceeds that of the water. The film has the minimum nonzero thickness such that it appears dark due to destructive interference when viewed in red light (wavelength = 640.0 nm in vacuum). Assuming that the visible spectrum extends from 380 to 750 nm, for which visible wavelength(s) (in vacuum) will the film appear bright due to constructive interference?

**21. A piece of curved glass has a radius of curvature of $r = 10.0$ m and is used to form Newton's rings, as in Figure 27.13. Not counting the dark spot at the center of the pattern, there are one hundred dark fringes, the last one being at the outer edge of the curved piece of glass. The light being used has a wavelength of 654 nm in vacuum. What is the radius R of the outermost dark ring in the pattern? *(Hint: Note that r is much greater than R, and you may assume that $\tan \theta = \theta$ for small angles, where θ must be expressed in radians.)*

**22. A circular drop of oil lies on a smooth, horizontal surface. The drop is thickest in the center and tapers to zero thickness at the edge. When illuminated from above by blue light ($\lambda = 455$ nm), 56 concentric bright rings are visible, including a bright fringe at the edge of the drop. In addition, there is a bright spot in the center of the drop. When the drop is illuminated from above by red light ($\lambda = 637$ nm), a bright spot again appears at the center, along with a different number of bright rings. Ignoring the bright spot, how many bright rings appear in red light? Assume that the index of refraction of the oil is the same for both wavelengths.

Section 27.5 Diffraction

23. (a) As Section 17.3 discusses, high-frequency sound waves exhibit less diffraction than low-frequency sound waves do. However, even high-frequency sound waves exhibit much more diffraction under normal circumstances than do light waves that pass through the same opening. The highest frequency that a healthy ear can typically hear is 2.0×10^4 Hz. Assume that a sound wave with this frequency travels at 343 m/s and passes through a doorway that has a width of 0.91 m. Determine the angle that locates the first minimum to either side of the central maximum in the diffraction pattern for the sound. This minimum is equivalent to the first dark fringe in a single-slit diffraction pattern for light. (b) Suppose that yellow light (wavelength = 580 nm in vacuum) passes through a doorway and that the first dark fringe in its diffraction pattern is located at the angle determined in part (a). How wide would this hypothetical doorway have to be?

24. A single slit has a width of 2.1×10^{-6} m and is used to form a diffraction pattern. Find the angle that locates the second dark fringe when the wavelength of the light is **(a)** 430 nm and **(b)** 660 nm.

25. ssm A diffraction pattern forms when light passes through a single slit. The wavelength of the light is 675 nm. Determine the angle that locates the first dark fringe when the width of the slit is **(a)** 1.8×10^{-4} m and **(b)** 1.8×10^{-6} m.

26. GO A slit has a width of $W_1 = 2.3 \times 10^{-6}$ m. When light with a wavelength of $\lambda_1 = 510$ nm passes through this slit, the width of the central bright fringe on a flat observation screen has a certain value. With the screen kept in the same place, this slit is replaced with a second slit (width W_2), and a wavelength of $\lambda_2 = 740$ nm is used. The width of the central bright fringe on the screen is observed to be unchanged. Find W_2.

27. ssm www A flat screen is located 0.60 m away from a single slit. Light with a wavelength of 510 nm (in vacuum) shines through the slit and produces a diffraction pattern. The width of the central bright fringe on the screen is 0.050 m. What is the width of the slit?

28. GO Light shines through a single slit whose width is 5.6×10^{-4} m. A diffraction pattern is formed on a flat screen located 4.0 m away. The distance between the middle of the central bright fringe and the first dark fringe is 3.5 mm. What is the wavelength of the light?

***29.** The width of a slit is 2.0×10^{-5} m. Light with a wavelength of 480 nm passes through this slit and falls on a screen that is located 0.50 m away. In the diffraction pattern, find the width of the bright fringe that is next to the central bright fringe.

***30.** Light waves with two different wavelengths, 632 nm and 474 nm, pass simultaneously through a single slit whose width is 7.15×10^{-5} m and strike a screen 1.20 m from the slit. Two diffraction patterns are formed on the screen. What is the distance (in cm) between the common center of the diffraction patterns and the first occurrence of a dark fringe from one pattern falling on top of a dark fringe from the other pattern?

***31. ssm** In a single-slit diffraction pattern on a flat screen, the central bright fringe is 1.2 cm wide when the slit width is 3.2×10^{-5} m. When the slit is replaced by a second slit, the wavelength of the light and the distance to the screen remaining unchanged, the central bright fringe broadens to a width of 1.9 cm. What is the width of the second slit? It may be assumed that θ is so small that $\sin \theta \approx \tan \theta$.

****32.** In a single-slit diffraction pattern, the central fringe is 450 times as wide as the slit. The screen is 18 000 times farther from the slit than the slit is wide. What is the ratio λ/W, where λ is the wavelength of the light shining through the slit and W is the width of the slit? Assume that the angle that locates a dark fringe on the screen is small, so that $\sin \theta \approx \tan \theta$.

Section 27.6 Resolving Power

33. Multiple-Concept Example 7 reviews the concepts that are important in this problem. You are looking down at the earth from inside a jetliner flying at an altitude of 8690 m. The pupil of your eye has a diameter of 2.00 mm. Determine how far apart two cars must be on the ground if you are to have any hope of distinguishing between them in **(a)** red light (wavelength = 665 nm in vacuum) and **(b)** violet light (wavelength = 405 nm in vacuum).

34. Two stars are 3.7×10^{11} m apart and are equally distant from the earth. A telescope has an objective lens with a diameter of 1.02 m and just detects these stars as separate objects. Assume that light of wavelength 550 nm is being observed. Also assume that diffraction effects, rather than atmospheric turbulence, limit the resolving power of the telescope. Find the maximum distance that these stars could be from the earth.

35. ssm Late one night on a highway, a car speeds by you and fades into the distance. Under these conditions the pupils of your eyes have diameters of about 7.0 mm. The taillights of this car are separated by a distance of 1.2 m and emit red light (wavelength = 660 nm in vacuum). How far away from you is this car when its taillights appear to merge into a single spot of light because of the effects of diffraction?

36. GO An inkjet printer uses tiny dots of red, green, and blue ink to produce an image. Assume that the dot separation on the printed page is the same for all colors. At normal viewing distances, the eye does not resolve the individual dots, regardless of color, so that the image has a normal look. The wavelengths for red, green, and blue are $\lambda_{red} = 660$ nm, $\lambda_{green} = 550$ nm, and $\lambda_{blue} = 470$ nm. The diameter of the pupil through which light enters the eye is 2.0 mm. For a viewing distance of 0.40 m, what is the maximum allowable dot separation?

37. A large group of football fans comes to the game with colored cards that spell out the name of their team when held up simultaneously. Most of the cards are colored blue ($\lambda_{vacuum} = 480$ nm). When displayed, the average distance between neighboring cards is 5.0 cm. If the cards are to blur together into solid blocks of color when viewed by a spectator at the other end of the stadium (160 m away), what must be the maximum diameter (in mm) of the spectator's pupils?

38. Review Conceptual Example 8 as background for this problem. In addition to the data given there, assume that the dots in the painting are separated by 1.5 mm and that the wavelength of the light is $\lambda_{vacuum} = 550$ nm. Find the distance at which the dots can just be resolved by **(a)** the eye and **(b)** the camera.

39. Astronomers have discovered a planetary system orbiting the star Upsilon Andromedae, which is at a distance of 4.2×10^{17} m from the earth. One planet is believed to be located at a distance of 1.2×10^{11} m from the star. Using visible light with a vacuum wavelength of 550 nm, what is the minimum necessary aperture diameter that a telescope must have so that it can resolve the planet and the star?

***40. GO** A spotlight sends red light (wavelength = 694.3 nm) to the moon. At the surface of the moon, which is 3.77×10^8 m away, the light strikes a reflector left there by astronauts. The reflected light returns to the earth, where it is detected. When it leaves the spotlight, the circular beam of light has a diameter of about 0.20 m, and diffraction causes the beam to spread as the light travels to the moon. In effect, the first circular dark fringe in the diffraction pattern defines the size of the central bright spot on the moon. Determine the diameter (not the radius) of the central bright spot on the moon.

***41. ssm www** Consult Multiple-Concept Example 7 to see a model for solving this kind of problem. Refer to **Interactive Solution 27.41** at **www.wiley.com/college/cutnell** to review a method by which this problem can be solved. You are using a microscope to examine a blood sample. Recall from Section 26.12 that the sample should be placed just outside the focal point of the objective lens of the microscope. **(a)** If the specimen is being illuminated with light of wavelength λ and the diameter of the objective equals its focal length, determine the closest distance between two blood cells that can just be resolved. Express your answer in terms of λ. **(b)** Based on your answer to (a), should you use light with a longer wavelength or a shorter wavelength if you wish to resolve two blood cells that are even closer together?

****42.** Two concentric circles of light emit light whose wavelength is 555 nm. The larger circle has a radius of 4.0 cm, and the smaller circle has a radius of 1.0 cm. When taking a picture of these lighted circles, a camera admits light through an aperture whose diameter is 12.5 mm. What is the maximum distance at which the camera can **(a)** distinguish one circle from the other and **(b)** reveal that the inner circle is a circle of light rather than a solid disk of light?

Section 27.7   The Diffraction Grating,
Section 27.8   Compact Discs, Digital Video Discs,
and the Use of Interference

43. ssm The diffraction gratings discussed in the text are transmission gratings because light *passes through* them. There are also gratings in which the light *reflects from* the grating to form a pattern of fringes. Equation 27.7 also applies to a reflection grating with straight parallel lines when the incident light shines perpendicularly on the grating. The surface of a compact disc (CD) has a multicolored appearance because it acts like a reflection grating and spreads sunlight into its colors. The arms of the spiral track on the CD are separated by 1.1×10^{-6} m. Using Equation 27.7, estimate the angle of the first-order maximum for a wavelength of   **(a)** 660 nm (red) and **(b)** 410 nm (violet).

44. The light shining on a diffraction grating has a wavelength of 495 nm (in vacuum). The grating produces a second-order bright fringe whose position is defined by an angle of 9.34°. How many lines per centimeter does the grating have?

45. ssm The wavelength of the laser beam used in a compact disc player is 780 nm. Suppose that a diffraction grating produces first-order tracking beams that are 1.2 mm apart at a distance of 3.0 mm from the grating. Estimate the spacing between the slits of the grating.

46. GO Two diffraction gratings, A and B, are located at the same distance from the observation screens. Light with the same wavelength λ is used for each. The separation between adjacent principal maxima for grating A is 2.7 cm, and for grating B it is 3.2 cm. Grating A has 2000 lines per meter. How many lines per meter does grating B have? (*Hint: The diffraction angles are small enough that the approximation $\sin \theta \approx \tan \theta$ can be used.*)

47. For a wavelength of 420 nm, a diffraction grating produces a bright fringe at an angle of 26°. For an unknown wavelength, the same grating produces a bright fringe at an angle of 41°. In both cases the bright fringes are of the same order m. What is the unknown wavelength?

48. A diffraction grating is 1.50 cm wide and contains 2400 lines. When used with light of a certain wavelength, a third-order maximum is formed at an angle of 18.0°. What is the wavelength (in nm)?

*49. A diffraction grating has 2604 lines per centimeter, and it produces a principal maximum at $\theta = 30.0°$. The grating is used with light that contains all wavelengths between 410 and 660 nm. What is (are) the wavelength(s) of the incident light that could have produced this maximum?

*50. GO Light of wavelength 410 nm (in vacuum) is incident on a diffraction grating that has a slit separation of 1.2×10^{-5} m. The distance between the grating and the viewing screen is 0.15 m. A diffraction pattern is produced on the screen that consists of a central bright fringe and higher-order bright fringes (see the drawing).   **(a)** Determine the distance y from the central bright fringe to the second-order bright fringe. (*Hint: The diffraction angles are small enough that the approximation $\tan \theta \approx \sin \theta$ can be used.*)   **(b)** If the entire apparatus is submerged in water ($n_{water} = 1.33$), what is the distance y?

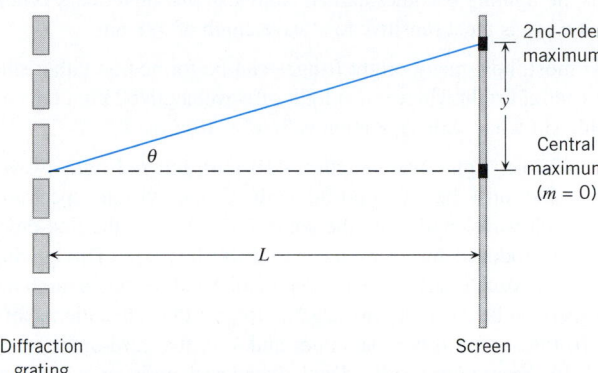

*51. ssm The same diffraction grating is used with two different wavelengths of light, λ_A and λ_B. The fourth-order principal maximum of light A exactly overlaps the third-order principal maximum of light B. Find the ratio λ_A/λ_B.

**52. The distance between adjacent slits of a certain diffraction grating is 1.250×10^{-5} m. The grating is illuminated by monochromatic light with a wavelength of 656.0 nm, and is then heated so that its temperature increases by 100.0 C°. Determine the change in the angle of the seventh-order principal maximum that occurs as a result of the thermal expansion of the grating. The coefficient of linear expansion for the diffraction grating is 1.30×10^{-4} $(C°)^{-1}$. Be sure to include the proper algebraic sign with your answer: $+$ if the angle increases, $-$ if the angle decreases.

**53. There are 5620 lines per centimeter in a grating that is used with light whose wavelength is 471 nm. A flat observation screen is located at a distance of 0.750 m from the grating. What is the minimum width that the screen must have so the *centers* of all the principal maxima formed on either side of the central maximum fall on the screen?

ADDITIONAL PROBLEMS

54. A tank of gasoline ($n = 1.40$) is open to the air ($n = 1.00$). A thin film of liquid floats on the gasoline and has a refractive index that is between 1.00 and 1.40. Light that has a wavelength of 625 nm (in vacuum) shines perpendicularly down through the air onto this film, and in this light the film looks bright due to constructive interference. The thickness of the film is 242 nm and is the minimum nonzero thickness for which constructive interference can occur. What is the refractive index of the film?

55. ssm In a Young's double-slit experiment, the wavelength of the light used is 520 nm (in vacuum), and the separation between the slits is 1.4×10^{-6} m. Determine the angle that locates **(a)** the dark fringe for which $m = 0$,   **(b)** the bright fringe for which $m = 1$,   **(c)** the dark fringe for which $m = 1$, and   **(d)** the bright fringe for which $m = 2$.

56. The dark fringe for $m = 0$ in a Young's double-slit experiment is located at an angle of $\theta = 15°$. What is the angle that locates the dark fringe for $m = 1$?

57. ssm A flat observation screen is placed at a distance of 4.5 m from a pair of slits. The separation on the screen between the central bright fringe and the first-order bright fringe is 0.037 m. The light illuminating the slits has a wavelength of 490 nm. Determine the slit separation.

58. Light that has a wavelength of 668 nm passes through a slit 6.73×10^{-6} m wide and falls on a screen that is 1.85 m away. What is the distance on the screen from the center of the central bright fringe to the third dark fringe on either side?

59. A hunter who is a bit of a braggart claims that from a distance of 1.6 km he can selectively shoot either of two squirrels who are sitting ten centimeters apart on the same branch of a tree. What's more, he claims that he can do this without the aid of a telescopic sight on his rifle. **(a)** Determine the diameter of the pupils of his eyes that would be required for him to be able to resolve the squirrels as separate objects. In this calculation use a wavelength of 498 nm (in vacuum) for the light. **(b)** State whether his claim is reasonable, and provide a reason for your answer. In evaluating his claim, consider that the human eye automatically adjusts the diameter of its pupil over a typical range of 2 to 8 mm, the larger values coming into play as the lighting becomes darker. Note also that under dark conditions, the eye is most sensitive to a wavelength of 498 nm.

*60. At most, how many bright fringes can be formed on either side of the central bright fringe when light of wavelength 625 nm falls on a double slit whose slit separation is 3.76×10^{-6} m?

*61. **ssm** Violet light (wavelength = 410 nm) and red light (wavelength = 660 nm) lie at opposite ends of the visible spectrum. **(a)** For each wavelength, find the angle θ that locates the first-order maximum produced by a grating with 3300 lines/cm. This grating converts a mixture of all colors between violet and red into a rainbow-like dispersion between the two angles. Repeat the calculation above for **(b)** the second-order maximum and **(c)** the third-order maximum. **(d)** From your results, decide whether there is an overlap between any of the "rainbows" and, if so, specify which orders overlap.

*62. **GO** The pupil of an eagle's eye has a diameter of 6.0 mm. Two field mice are separated by 0.010 m. From a distance of 176 m, the eagle sees them as one unresolved object and dives toward them at a speed of 17 m/s. Assume that the eagle's eye detects light that has a wavelength of 550 nm in vacuum. How much time passes until the eagle sees the mice as separate objects?

*63. How many dark fringes will be produced on either side of the central maximum if light (λ = 651 nm) is incident on a single slit that is 5.47×10^{-6} m wide?

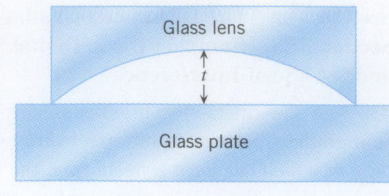

Glass lens

t

Glass plate

*64. The drawing shows a cross section of a plano-concave lens resting on a flat glass plate. (A plano-concave lens has one surface that is a plane and the other that is concave spherical.) The thickness t is 1.37×10^{-5} m. The lens is illuminated with monochromatic light (λ_{vacuum} = 550 nm), and a series of concentric bright and dark rings is formed, much like Newton's rings. How many bright rings are there? *(Hint: The cross section shown in the drawing reveals that a kind of air wedge exists between the place where the two pieces of glass touch and the top of the curved surface where the distance t is marked.)*

65. Two gratings A and B have slit separations d_A and d_B, respectively. They are used with the same light and the same observation screen. When grating A is replaced with grating B, it is observed that the first-order maximum of A is exactly replaced by the second-order maximum of B. **(a) Determine the ratio d_B/d_A of the spacings between the slits of the gratings. **(b)** Find the next two principal maxima of grating A and the principal maxima of B that exactly replace them when the gratings are switched. Identify these maxima by their order numbers.

66. **Interactive LearningWare 27.2 at **www.wiley.com/college/cutnell** reviews the concepts that are important in this problem. A uniform layer of water (n = 1.33) lies on a glass plate (n = 1.52). Light shines perpendicularly on the layer. Because of constructive interference, the layer looks maximally bright when the wavelength of the light is 432 nm in vacuum and *also* when it is 648 nm in vacuum. **(a)** Obtain the minimum thickness of the film. **(b)** Assuming that the film has the minimum thickness and that the visible spectrum extends from 380 to 750 nm, determine the visible wavelength(s) (in vacuum) for which the film appears completely dark.

CHAPTER **28**

SPECIAL RELATIVITY

This elephant has been fitted with a Global Positioning System (GPS) tracking collar. The collar will allow scientists to track the elephant's movements because the GPS technology can locate objects on the earth with remarkable accuracy. The accuracy results, in part, because the system incorporates Einstein's theory of special relativity. (© SIAN BROWN/Newhouse News Service/ Landov LLC)

28.1 EVENTS AND INERTIAL REFERENCE FRAMES

In the theory of special relativity, an **event**, such as the launching of the space shuttle in Figure 28.1, is a physical "happening" that occurs at a certain place and time. In this drawing two observers are watching the lift-off, one standing on the earth and one in an airplane that is flying at a constant velocity relative to the earth. To record the event, each observer uses a **reference frame** that consists of a set of *x, y, z* axes (called a *coordinate system*) and a clock. The coordinate systems are used to establish where the event occurs, and the clocks to specify when. Each observer is at rest relative to his own reference frame. However, the earth-based observer and the airborne observer are moving relative to each other and so, also, are their respective reference frames.

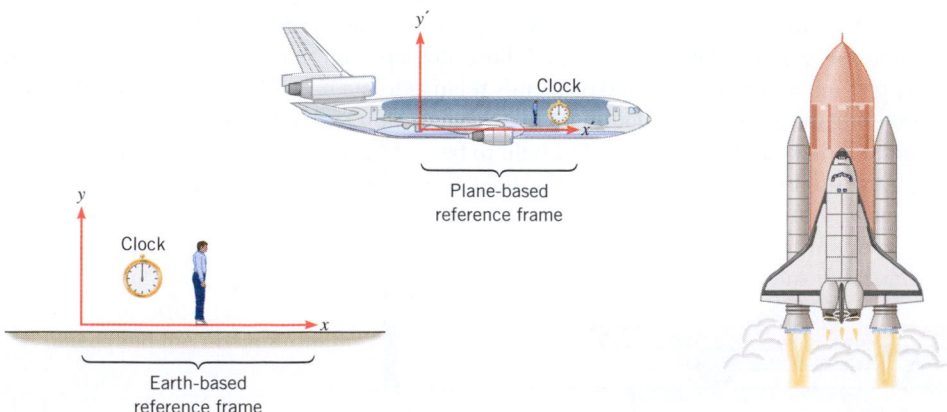

Figure 28.1 Using an earth-based reference frame, an observer standing on the earth records the location and time of an event (the space shuttle lift-off). Likewise, an observer in the airplane uses a plane-based reference frame to describe the event.

875

The theory of special relativity deals with a "special" kind of reference frame, called an ***inertial reference frame.*** As Section 4.2 discusses, an inertial reference frame is one in which Newton's law of inertia is valid. That is, if the net force acting on a body is zero, the body either remains at rest or moves at a constant velocity. In other words, the acceleration of such a body is zero when measured in an inertial reference frame. Rotating and otherwise accelerating reference frames are not inertial reference frames. The earth-based reference frame in Figure 28.1 is not quite an inertial frame because it is subjected to centripetal accelerations as the earth spins on its axis and revolves around the sun. In most situations, however, the effects of these accelerations are small, and we can neglect them. To the extent that the earth-based reference frame is an inertial frame, so is the plane-based reference frame, because the plane moves at a constant velocity relative to the earth. The next section discusses why inertial reference frames are important in relativity.

28.2 THE POSTULATES OF SPECIAL RELATIVITY

Einstein built his theory of special relativity on two fundamental assumptions or postulates about the way nature behaves.

THE POSTULATES OF SPECIAL RELATIVITY

1. ***The Relativity Postulate.*** The laws of physics are the same in every inertial reference frame.
2. ***The Speed-of-Light Postulate.*** The speed of light in a vacuum, measured in any inertial reference frame, always has the same value of *c,* no matter how fast the source of light and the observer are moving relative to each other.

It is not too difficult to accept the relativity postulate. For instance, in Figure 28.1 each observer, using his own inertial reference frame, can make measurements on the motion of the space shuttle. The relativity postulate asserts that both observers find their data to be consistent with Newton's laws of motion. Similarly, both observers find that the behavior of the electronics on the space shuttle is described by the laws of electromagnetism. According to the relativity postulate, ***any inertial reference frame is as good as any other for expressing the laws of physics.*** As far as inertial reference frames are concerned, nature does not play favorites.

Since the laws of physics are the same in all inertial reference frames, there is no experiment that can distinguish between an inertial frame that is at rest and one that is moving at a constant velocity. When you are seated on the aircraft in Figure 28.1, for instance, it is just as valid to say that you are at rest and the earth is moving as it is to say the converse. It is not possible to single out one particular inertial reference frame as being at "absolute rest." Consequently, it is meaningless to talk about the "absolute velocity" of an object—that is, its velocity measured relative to a reference frame at "absolute rest." Thus, the earth moves relative to the sun, which itself moves relative to the center of our galaxy. And the galaxy moves relative to other galaxies, and so on. According to Einstein, only the relative velocity between objects, not their absolute velocities, can be measured and is physically meaningful.

Whereas the relativity postulate seems plausible, the speed-of-light postulate defies common sense. For instance, Figure 28.2 illustrates a person standing on the bed of a truck that is moving at a constant speed of 15 m/s relative to the ground. Now, suppose that you are standing on the ground and the person on the truck shines a flashlight at you. The person on the truck observes the speed of light to be *c*. What do you measure for the speed of light? You might guess that the speed of light would be *c* + 15 m/s. However, this guess

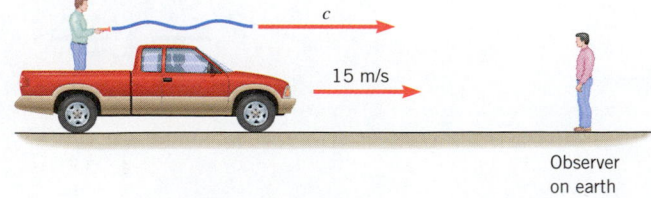

Figure 28.2 Both the person on the truck and the observer on the earth measure the speed of the light to be *c*, regardless of the speed of the truck.

is inconsistent with the speed-of-light postulate, which states that all observers in inertial reference frames measure the speed of light to be c—nothing more, nothing less. Therefore, you must also measure the speed of light to be c, the same as that measured by the person on the truck. According to the speed-of-light postulate, the fact that the flashlight is moving has no influence whatsoever on the speed of the light approaching you. This property of light, although surprising, has been verified many times by experiment.

Since waves, such as water waves and sound waves, require a medium through which to propagate, it was natural for scientists before Einstein to assume that light did too. This hypothetical medium was called the *luminiferous ether* and was assumed to fill all of space. Furthermore, it was believed that light traveled at the speed c only when measured with respect to the ether. According to this view, an observer moving relative to the ether would measure a speed for light that was smaller or greater than c, depending on whether the observer moved with or against the light, respectively. During the years 1883–1887, however, the American scientists A. A. Michelson and E. W. Morley carried out a series of famous experiments whose results were not consistent with the ether theory. Their results indicated that the speed of light is indeed the same in all inertial reference frames and does not depend on the motion of the observer. These experiments, and others, led eventually to the demise of the ether theory and the acceptance of the theory of special relativity.

▶ CONCEPTS AT A GLANCE The remainder of this chapter reexamines, from the viewpoint of special relativity, a number of fundamental concepts that have been discussed in earlier chapters from the viewpoint of classical physics. These concepts are time, length, momentum, kinetic energy, and the addition of velocities. We will see that each is modified by special relativity in a way that depends on the speed v of a moving object relative to the speed c of light in a vacuum. Figure 28.3 illustrates that when the object moves slowly [v is much smaller than c ($v \ll c$)], the modification is negligibly small, and the classical version of each concept provides an accurate description of reality. However, when the object moves so rapidly that v is an appreciable fraction of the speed of light [v is approximately equal to c ($v \approx c$)], the effects of special relativity must be considered. The gold panel in Figure 28.3 lists the various equations that convey the modifications

Figure 28.3 CONCEPTS AT A GLANCE This chart emphasizes that it is the speed v of a moving object, as compared to the speed c of light in a vacuum, that determines whether the effects of special relativity are measurably important. Albert Einstein (1879–1955), the author of the theory of special relativity, is one of the most famous scientists of the twentieth century. (© Hulton Archive/Getty Images)

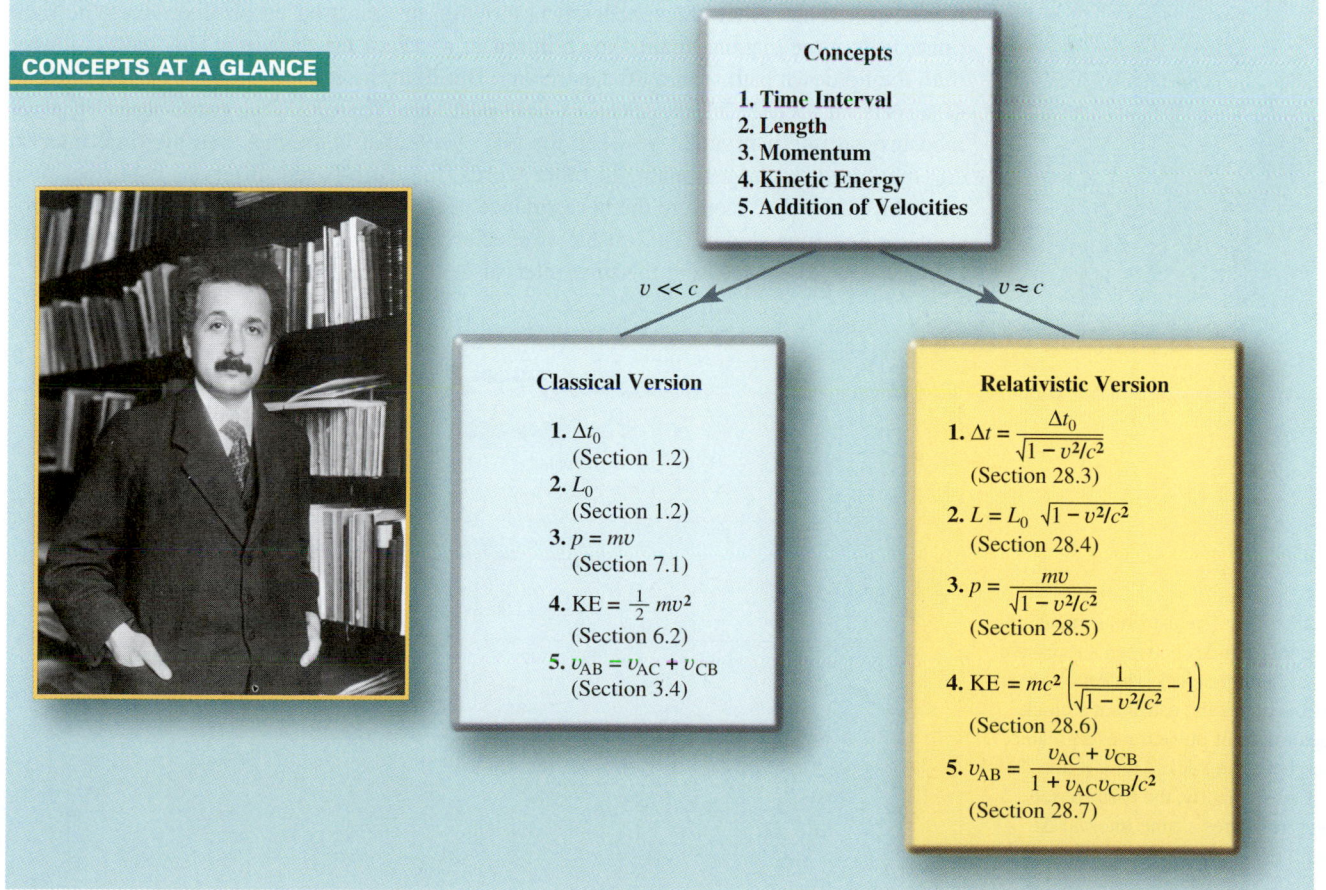

CONCEPTS AT A GLANCE

Concepts

1. Time Interval
2. Length
3. Momentum
4. Kinetic Energy
5. Addition of Velocities

$v \ll c$ $v \approx c$

Classical Version

1. Δt_0
   (Section 1.2)
2. L_0
   (Section 1.2)
3. $p = mv$
   (Section 7.1)
4. $KE = \frac{1}{2}mv^2$
   (Section 6.2)
5. $v_{AB} = v_{AC} + v_{CB}$
   (Section 3.4)

Relativistic Version

1. $\Delta t = \dfrac{\Delta t_0}{\sqrt{1 - v^2/c^2}}$
   (Section 28.3)
2. $L = L_0\sqrt{1 - v^2/c^2}$
   (Section 28.4)
3. $p = \dfrac{mv}{\sqrt{1 - v^2/c^2}}$
   (Section 28.5)
4. $KE = mc^2\left(\dfrac{1}{\sqrt{1 - v^2/c^2}} - 1\right)$
   (Section 28.6)
5. $v_{AB} = \dfrac{v_{AC} + v_{CB}}{1 + v_{AC}v_{CB}/c^2}$
   (Section 28.7)

imposed by special relativity. Each of these equations will be discussed in later sections of this chapter. ◄

It is important to realize that the modifications imposed by special relativity do not imply that the classical concepts of time, length, momentum, kinetic energy, and the addition of velocities, as developed by Newton and others, are wrong. They are just limited to speeds that are very small compared to the speed of light. In contrast, the relativistic view of the concepts applies to all speeds between zero and the speed of light.

28.3 THE RELATIVITY OF TIME: TIME DILATION

TIME DILATION

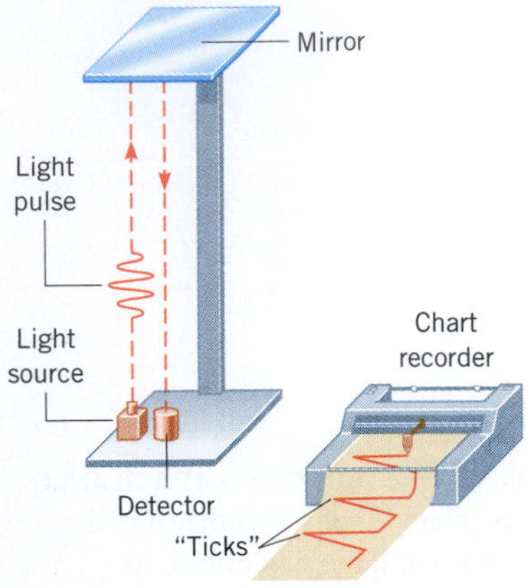

Common experience indicates that time passes just as quickly for a person standing on the ground as it does for an astronaut in a spacecraft. In contrast, special relativity reveals that the person on the ground measures time passing more slowly for the astronaut than for herself. We can see how this curious effect arises with the help of the clock illustrated in Figure 28.4, which uses a pulse of light to mark time. A short pulse of light is emitted by a light source, reflects from a mirror, and then strikes a detector that is situated next to the source. Each time a pulse reaches the detector, a "tick" registers on the chart recorder, another short pulse of light is emitted, and the cycle repeats. Thus, the time interval between successive "ticks" is marked by a beginning event (the firing of the light source) and an ending event (the pulse striking the detector). The source and detector are so close to each other that the two events can be considered to occur at the same location.

Suppose two identical clocks are built. One is kept on earth, and the other is placed aboard a spacecraft that travels at a constant velocity relative to the earth. The astronaut is at rest with respect to the clock on the spacecraft and, therefore, sees the light pulse move along the up/down path shown in Figure 28.5a. According to the astronaut, the time interval Δt_0 required for the light to follow this path is the distance $2D$ divided by the speed of light c; $\Delta t_0 = 2D/c$. To the astronaut, Δt_0 is the time interval between the "ticks" of the spacecraft clock—that is, the time interval between the beginning and ending events of the clock. An earth-based observer, however, does *not* measure Δt_0 as the time interval between these two events. Since the spacecraft is moving, the earth-based observer sees the light pulse follow the diagonal path shown in red in part *b* of the drawing. This path is longer than the up/down path seen by the astronaut. But light travels at the *same speed c* for both observers, in accord with the speed-of-light postulate. Therefore, the earth-based observer measures a time interval Δt between the two events that is *greater* than the time interval Δt_0 measured by the astronaut. In other words, the earth-based observer, using her own earth-based clock to measure the performance of the astronaut's clock, finds that the astronaut's clock runs slowly. This result of special relativity is known as *time dilation*. (To *dilate* means to expand, and the time interval Δt is "expanded" relative to Δt_0.)

Figure 28.4 A light clock.

Figure 28.5 (*a*) The astronaut measures the time interval Δt_0 between successive "ticks" of his light clock. (*b*) An observer on earth watches the astronaut's clock and sees the light pulse travel a greater distance between "ticks" than it does in part *a*. Consequently, the earth-based observer measures a time interval Δt between "ticks" that is greater than Δt_0.

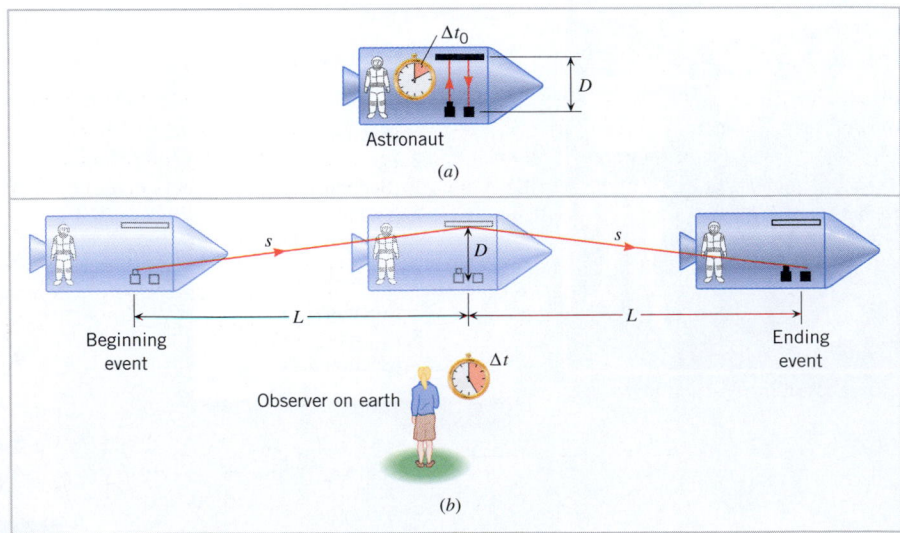

The time interval Δt that the earth-based observer measures in Figure 28.5*b* can be determined as follows. While the light pulse travels from the source to the detector, the space-craft moves a distance $2L = v\,\Delta t$ to the right, where v is the speed of the spacecraft relative to the earth. From the drawing it can be seen that the light pulse travels a total diagonal distance of $2s$ during the time interval Δt. Applying the Pythagorean theorem, we find that

$$2s = 2\sqrt{D^2 + L^2} = 2\sqrt{D^2 + \left(\frac{v\,\Delta t}{2}\right)^2}$$

But the distance $2s$ is also equal to the speed of light times the time interval Δt, so that $2s = c\,\Delta t$. Therefore,

$$c\,\Delta t = 2\sqrt{D^2 + \left(\frac{v\,\Delta t}{2}\right)^2}$$

Squaring this result and solving for Δt gives

$$\Delta t = \frac{2D}{c}\,\frac{1}{\sqrt{1 - \dfrac{v^2}{c^2}}}$$

However, $2D/c = \Delta t_0$, where Δt_0 is the time interval between successive "ticks" of the spacecraft's clock as measured by the astronaut. With this substitution, the equation for Δt can be expressed as

***Time
dilation***
$$\Delta t = \frac{\Delta t_0}{\sqrt{1 - \dfrac{v^2}{c^2}}} \tag{28.1}$$

The symbols in this formula are defined as follows:

Δt_0 = proper time interval, which is the interval between two events as measured by an observer who is at rest with respect to the events and who views them as occurring *at the same place*

Δt = dilated time interval, which is the interval measured by an observer who is in motion with respect to the events and who views them as occurring at *different places*

v = relative speed between the two observers

c = speed of light in a vacuum

For a speed v that is less than c, the term $\sqrt{1 - v^2/c^2}$ in Equation 28.1 is less than 1, and the dilated time interval Δt is greater than Δt_0. Example 1 illustrates the extent of this time dilation effect.

 Example 1   Time Dilation

The spacecraft in Figure 28.5 is moving past the earth at a constant speed v that is 0.92 times the speed of light. Thus, $v = (0.92)(3.0 \times 10^8 \text{ m/s})$, which is often written as $v = 0.92c$. The astronaut measures the time interval between successive "ticks" of the spacecraft clock to be $\Delta t_0 = 1.0$ s. What is the time interval Δt that an earth observer measures between "ticks" of the astronaut's clock?

Reasoning Since the clock on the spacecraft is moving relative to the earth, the earth-based observer measures a greater time interval Δt between "ticks" than does the astronaut, who is at rest relative to the clock. The dilated time interval Δt can be determined from the time dilation relation, Equation 28.1.

Solution The dilated time interval is

$$\Delta t = \frac{\Delta t_0}{\sqrt{1 - \dfrac{v^2}{c^2}}} = \frac{1.0 \text{ s}}{\sqrt{1 - \left(\dfrac{0.92c}{c}\right)^2}} = \boxed{2.6 \text{ s}}$$

From the point of view of the earth-based observer, the astronaut is using a clock that is running slowly, because the earth-based observer measures a time between "ticks" that is longer (2.6 s) than what the astronaut measures (1.0 s).

The physics of
the Global Positioning System
and special relativity.

Present-day spacecrafts fly nowhere near as fast as the craft in Example 1. Yet circumstances exist in which time dilation can create appreciable errors if not accounted for. The Global Positioning System (GPS), for instance, uses highly accurate and stable atomic clocks on board each of 24 satellites orbiting the earth at speeds of about 4000 m/s. These clocks make it possible to measure the time it takes for electromagnetic waves to travel from a satellite to a ground-based GPS receiver. From the speed of light and the times measured for signals from three or more of the satellites, it is possible to locate the position of the receiver (see Section 5.5). The stability of the clocks must be better than one part in 10^{13} to ensure the positional accuracy demanded of the GPS. Using Equation 28.1 and the speed of the GPS satellites, we can calculate the difference between the dilated time interval and the proper time interval as a fraction of the proper time interval and compare the result to the stability of the GPS clocks:

$$\frac{\Delta t - \Delta t_0}{\Delta t_0} = \frac{1}{\sqrt{1 - v^2/c^2}} - 1 = \frac{1}{\sqrt{1 - (4000 \text{ m/s})^2/(3.00 \times 10^8 \text{ m/s})^2}} - 1$$

$$= \frac{1}{1.1 \times 10^{10}}$$

This result is approximately one thousand times greater than the GPS-clock stability of one part in 10^{13}. Thus, if not taken into account, time dilation would cause an error in the measured position of the earth-based GPS receiver roughly equivalent to the error caused by a thousand-fold degradation in the stability of the atomic clocks.

PROPER TIME INTERVAL

In Figure 28.5 both the astronaut and the person standing on the earth are measuring the time interval between a beginning event (the firing of the light source) and an ending event (the light pulse striking the detector). For the astronaut, who is at rest with respect to the light clock, the two events occur at the same location. (Remember, we are assuming that the light source and detector are so close together that they are considered to be at the same place.) Being at rest with respect to a clock is the usual or "proper" situation, so the time interval Δt_0 measured by the astronaut is called the **proper time interval.** In general, the proper time interval Δt_0 between two events is the time interval measured by an observer who is at rest relative to the events and sees them at the *same location* in space. On the other hand, the earth-based observer does not see the two events occurring at the same location in space, since the spacecraft is in motion. The time interval Δt that the earth-based observer measures is, therefore, not a proper time interval in the sense that we have defined it.

To understand situations involving time dilation, it is essential to distinguish between Δt_0 and Δt. It is helpful if one first identifies the two events that define the time interval. These may be something other than the firing of a light source and the light pulse striking a detector. Then determine the reference frame in which the two events occur at the same place. An observer at rest in this reference frame measures the proper time interval Δt_0.

SPACE TRAVEL

One of the intriguing aspects of time dilation occurs in conjunction with space travel. Since enormous distances are involved, travel to even the closest star outside our solar system would take a long time. However, as the following example shows, the travel time can be considerably less for the passengers than one might guess.

Shown here in orbit is astronaut David A. Wolf as he works on the International Space Station during a session of extravehicular activity. (Courtesy NASA)

Example 2   Space Travel

Alpha Centauri, a nearby star in our galaxy, is 4.3 light-years away. This means that, as measured by a person on earth, it would take light 4.3 years to reach this star. If a rocket leaves for Alpha Centauri and travels at a speed of $v = 0.95c$ relative to the earth, by how much will the passengers have aged, according to their own clock, when they reach their destination? Assume that the earth and Alpha Centauri are stationary with respect to one another.

Reasoning The two events in this problem are the departure from earth and the arrival at Alpha Centauri. At departure, earth is just outside the spaceship. Upon arrival at the destination, Alpha Centauri is just outside. Therefore, relative to the passengers, the two events occur at the same place—namely, just outside the spaceship. Thus, the passengers measure the proper time interval Δt_0 on their clock, and it is this interval that we must find. For a person left behind on earth, the events occur at *different places,* so such a person measures the dilated time interval Δt rather than the proper time interval. To find Δt we note that the time to travel a given distance is inversely proportional to the speed. Since it takes 4.3 years to traverse the distance between earth and Alpha Centauri at the speed of light, it would take even longer at the slower speed of $v = 0.95c$. Thus, a person on earth measures the dilated time interval to be $\Delta t = (4.3 \text{ years})/0.95 = 4.5$ years. This value can be used with the time-dilation equation to find the proper time interval Δt_0.

Solution Using the time-dilation equation, we find that the proper time interval by which the passengers judge their own aging is

$$\Delta t_0 = \Delta t \sqrt{1 - \frac{v^2}{c^2}} = (4.5 \text{ years}) \sqrt{1 - \left(\frac{0.95c}{c}\right)^2} = \boxed{1.4 \text{ years}}$$

Thus, the people aboard the rocket will have aged by only 1.4 years when they reach Alpha Centauri, and not the 4.5 years an earthbound observer has calculated.

Problem-solving insight

In dealing with time dilation, decide which interval is the proper time interval as follows: (1) Identify the two events that define the interval. (2) Determine the reference frame in which the events occur at the same place; an observer at rest in this frame measures the proper time interval Δt_0.

The physics of
space travel and special relativity.

VERIFICATION OF TIME DILATION

A striking confirmation of time dilation was achieved in 1971 by an experiment carried out by J. C. Hafele and R. E. Keating.* They transported very precise cesium-beam atomic clocks around the world on commercial jets. Since the speed of a jet plane is considerably less than c, the time-dilation effect is extremely small. However, the atomic clocks were accurate to about $\pm 10^{-9}$ s, so the effect could be measured. The clocks were in the air for 45 hours, and their times were compared to reference atomic clocks kept on earth. The experimental results revealed that, within experimental error, the readings on the clocks on the planes were different from those on earth by an amount that agreed with the prediction of relativity.

The behavior of subatomic particles called *muons* provides additional confirmation of time dilation. These particles are created high in the atmosphere, at altitudes of about 10 000 m. When at rest, muons exist only for about 2.2×10^{-6} s before disintegrating. With such a short lifetime, these particles could never make it down to the earth's surface, even traveling at nearly the speed of light. However, *a large number of muons do reach the earth.* The only way they can do so is to live longer because of time dilation, as Example 3 illustrates.

Example 3   The Lifetime of a Muon

The average lifetime of a muon at rest is 2.2×10^{-6} s. A muon created in the upper atmosphere, thousands of meters above sea level, travels toward the earth at a speed of $v = 0.998c$. Find, on the average, **(a)** how long a muon lives according to an observer on earth, and **(b)** how far the muon travels before disintegrating.

Reasoning The two events of interest are the generation and subsequent disintegration of the muon. When the muon is at rest, these events occur at the same place, so the muon's average (at rest) lifetime of 2.2×10^{-6} s is a proper time interval Δt_0. When the muon moves at a speed $v = 0.998c$ relative to the earth, an observer on the earth measures a dilated lifetime Δt that is given by Equation 28.1. The average distance x traveled by a muon, as measured by an earth observer, is equal to the muon's speed times the dilated time interval.

*J. C. Hafele and R. E. Keating, "Around-the-World Atomic Clocks: Observed Relativistic Time Gains," *Science,* Vol. 177, July 14, 1972, p. 168.

Solution (a) The observer on earth measures a dilated lifetime. Using the time-dilation equation, we find that

$$\Delta t = \frac{\Delta t_0}{\sqrt{1 - \dfrac{v^2}{c^2}}} = \frac{2.2 \times 10^{-6} \text{ s}}{\sqrt{1 - \left(\dfrac{0.998c}{c}\right)^2}} = \boxed{35 \times 10^{-6} \text{ s}} \tag{28.1}$$

Problem-solving insight
The proper time interval Δt_0 is always shorter than the dilated time interval Δt.

(b) The distance traveled by the muon before it disintegrates is

$$x = v \, \Delta t = (0.998)(3.00 \times 10^8 \text{ m/s})(35 \times 10^{-6} \text{ s}) = \boxed{1.0 \times 10^4 \text{ m}}$$

Thus, the dilated, or extended, lifetime provides sufficient time for the muon to reach the surface of the earth. If its lifetime were only 2.2×10^{-6} s, a muon would travel only 660 m before disintegrating and could never reach the earth.

✓ CHECK YOUR UNDERSTANDING

(*The answers are given at the end of the book.*)

1. Which one of the following statements concerning the dilated time interval is false? **(a)** It is always greater than the proper time interval. **(b)** It depends on the relative speed between the observers who measure the proper and dilated time intervals. **(c)** It depends on the speed of light in a vacuum. **(d)** It is measured by an observer who is at rest with respect to the events that define the time interval.

2. A baseball player at home plate hits a pop fly straight up (the beginning event) that is caught by the catcher at home plate (the ending event). Which one or more of the following observers record(s) the proper time interval between the two events? **(a)** A spectator sitting in the stands **(b)** A spectator watching the game at home on TV **(c)** The third baseman running in to cover the play

3. A playground carousel is a circular platform that can rotate about an axis perpendicular to the plane of the platform at its center. An observer is looking down at this platform from an inertial reference frame directly above the rotational axis. Three clocks are attached to the platform. Clock A is attached directly to the axis. Clock B is attached to a point midway between the axis and the outer edge of the platform. Clock C is attached to the outer edge of the platform. Rank the clocks according to how slow the observer finds them to be running (slowest first).

28.4 THE RELATIVITY OF LENGTH: LENGTH CONTRACTION

Because of time dilation, observers moving at a constant velocity relative to each other measure different time intervals between two events. For instance, Example 2 in the previous section illustrates that a trip from earth to Alpha Centauri at a speed of $v = 0.95c$ takes 4.5 years according to a clock on earth, but only 1.4 years according to a clock in the rocket. These two times differ by the factor $\sqrt{1 - v^2/c^2}$. Since the times for the trip are different, one might ask whether the observers measure different distances between earth and Alpha Centauri. The answer, according to special relativity, is yes. After all, both the earth-based observer and the rocket passenger agree that the relative speed between the rocket and earth is $v = 0.95c$. Since speed is distance divided by time and the time is different for the two observers, it follows that the distances must also be different, if the relative speed is to be the same for both individuals. Thus, the earth observer determines the distance to Alpha Centauri to be $L_0 = v \, \Delta t = (0.95c)(4.5 \text{ years}) = 4.3$ light-years. On the other hand, a passenger aboard the rocket finds the distance is only $L = v \, \Delta t_0 = (0.95c)(1.4 \text{ years}) = 1.3$ light-years. The passenger, measuring the shorter time, also measures the shorter distance. This shortening of the distance between two points is one example of a phenomenon known as **length contraction.**

The relation between the distances measured by two observers in relative motion at a constant velocity can be obtained with the aid of Figure 28.6. Part *a* of the drawing shows the situation from the point of view of the earth-based observer. This person measures the time of the trip to be Δt, the distance to be L_0, and the relative speed of the rocket to be

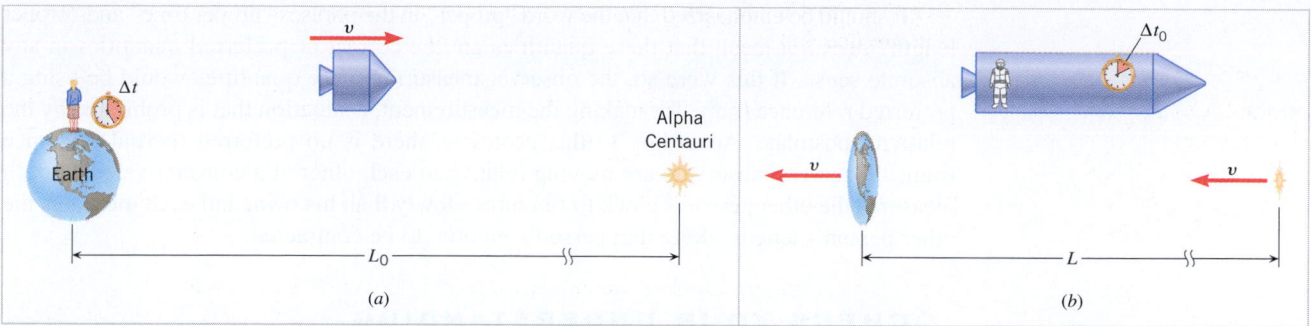

(a)   (b)

Figure 28.6 (a) As measured by an observer on the earth, the distance to Alpha Centauri is L_0, and the time required to make the trip is Δt. (b) According to the passenger on the spacecraft, the earth and Alpha Centauri move with speed v relative to the craft. The passenger measures the distance and time of the trip to be L and Δt_0, respectively, both quantities being less than those in part a.

$v = L_0/\Delta t$. Part b of the drawing presents the point of view of the passenger, for whom the rocket is at rest, and the earth and Alpha Centauri appear to move by at a speed v. The passenger determines the distance of the trip to be L, the time to be Δt_0, and the relative speed to be $v = L/\Delta t_0$. Since the relative speed computed by the passenger equals that computed by the earth-based observer, it follows that $v = L/\Delta t_0 = L_0/\Delta t$. Using this result and the time-dilation equation, Equation 28.1, we obtain the following relation between L and L_0:

Length contraction

$$L = L_0 \sqrt{1 - \frac{v^2}{c^2}} \qquad (28.2)$$

The length L_0 is called the **proper length;** it is the length (or distance) between two points *as measured by an observer at rest with respect to them.* Since v is less than c, the term $\sqrt{1 - v^2/c^2}$ is less than 1, and L is less than L_0. It is important to note that this length contraction occurs only along the direction of the motion. Those dimensions that are perpendicular to the motion are not shortened, as the next example discusses.

 Example 4 The Contraction of a Spacecraft

An astronaut, using a meter stick that is at rest relative to a cylindrical spacecraft, measures the length and diameter of the spacecraft to be 82 m and 21 m, respectively. The spacecraft moves with a constant speed of $v = 0.95c$ relative to the earth, as in Figure 28.6. What are the dimensions of the spacecraft, as measured by an observer on earth?

Reasoning The length of 82 m is a proper length L_0, since it is measured using a meter stick that is at rest relative to the spacecraft. The length L measured by the observer on earth can be determined from the length-contraction formula, Equation 28.2. On the other hand, the diameter of the spacecraft is perpendicular to the motion, so the earth observer does not measure any change in the diameter.

Solution The length L of the spacecraft, as measured by the observer on earth, is

$$L = L_0 \sqrt{1 - \frac{v^2}{c^2}} = (82 \text{ m})\sqrt{1 - \left(\frac{0.95c}{c}\right)^2} = \boxed{26 \text{ m}}$$

Problem-solving insight

The proper length L_0 is always larger than the contracted length L.

Both the astronaut and the observer on earth measure the same value for the diameter of the spacecraft: Diameter = $\boxed{21 \text{ m}}$. Figure 28.6a shows the size of the spacecraft as measured by the earth observer, and part b shows the size as measured by the astronaut.

When dealing with relativistic effects we need to distinguish carefully between the criteria for the proper time interval and the proper length. The proper time interval Δt_0 between two events is the time interval measured by an observer who is at rest relative to the events and sees them occurring at the *same place.* All other moving inertial observers will measure a larger value for this time interval. The proper length L_0 of an object is the length measured by an observer who is *at rest* with respect to the object. All other moving inertial observers will measure a shorter value for this length. The observer who measures the proper time interval may not be the same one who measures the proper length. For instance, Figure 28.6 shows that the astronaut measures the proper time interval Δt_0 for the trip between earth and Alpha Centauri, whereas the earth-based observer measures the proper length (or distance) L_0 for the trip.

It should be emphasized that the word "proper" in the phrases "proper time" and "proper length" does *not* mean that these quantities are the correct or preferred quantities in any absolute sense. If this were so, the observer measuring these quantities would be using a preferred reference frame for making the measurement, a situation that is prohibited by the relativity postulate. According to this postulate, there is no preferred inertial reference frame. When two observers are moving relative to each other at a constant velocity, each measures the other person's clock to run more slowly than his own, and each measures the other person's length, along that person's motion, to be contracted.

✓ **CHECK YOUR UNDERSTANDING**

(*The answers are given at the end of the book.*)

4. If the speed c of light in a vacuum were infinitely large instead of 3.0×10^8 m/s, would the effects of time dilation and length contraction be observable?

5. Suppose that you are standing at a railroad crossing, watching a train go by. Both you and a passenger in the train are looking at a clock on the train. Who measures the proper time interval, and who measures the proper length of a train car? **(a)** You measure the proper time interval, and the passenger measures the proper length. **(b)** You measure both the proper time interval and the proper length. **(c)** The passenger measures both the proper time interval and the proper length. **(d)** You measure the proper length, and the passenger measures the proper time interval.

6. Which of the following quantities will two observers always measure to be the *same*, regardless of the relative velocity between the observers? **(a)** The time interval between two events **(b)** The length of an object **(c)** The speed of light in a vacuum **(d)** The relative speed between the observers

7. The drawing shows an object that has the shape of a square when it is at rest in inertial reference frame R. When the object moves relative to this reference frame, the object's velocity vector is in the plane of the square and is parallel to the diagonal AC. Since the speed of the motion is an appreciable fraction of the speed of light in a vacuum, noticeable length contraction occurs. Does an observer in reference frame R see the object as a square? *(Hint: Consider what happens to each of the diagonals AC and BD.)*

28.5 RELATIVISTIC MOMENTUM

Thus far we have discussed how time intervals and distances between two events are measured by observers moving at a constant velocity relative to each other. Special relativity also alters our ideas about momentum and energy.

Recall from Section 7.2 that when two or more objects interact, the principle of conservation of linear momentum applies if the system of objects is isolated. This principle states that the total linear momentum of an isolated system remains constant at all times. (An isolated system is one in which the sum of the external forces acting on the objects is zero.) The conservation of linear momentum is a law of physics and, in accord with the relativity postulate, is valid in all inertial reference frames. That is, when the total linear momentum is conserved in one inertial reference frame, it is conserved in all inertial reference frames.

As an example of momentum conservation, suppose that several people are watching two billiard balls collide on a frictionless pool table. One person is standing next to the pool table, and the other is moving past the table with a constant velocity. Since the two balls constitute an isolated system, the relativity postulate requires that both observers must find the total linear momentum of the two-ball system to be the same before, during, and after the collision. For this kind of situation, Section 7.1 defines the classical linear momentum $\vec{p}$ of an object to be the product of its mass m and velocity $\vec{v}$. As a result, the magnitude of the classical momentum is $p = mv$. As long as the speed of an object is considerably smaller than the speed of light, this definition is adequate. However, when the speed approaches the speed of light, an analysis of the collision shows that the total linear momentum is not conserved in all inertial reference frames if one defines linear momentum simply as the product of mass and velocity. In order to preserve the conservation of linear momentum, it is necessary to modify this definition. The theory of special relativity reveals that the magnitude of the *relativistic momentum* must be defined as in Equation 28.3:

Magnitude of the
relativistic momentum

$$p = \frac{mv}{\sqrt{1 - \frac{v^2}{c^2}}}$$

(28.3)

The total relativistic momentum of an isolated system is conserved in all inertial reference frames.

From Equation 28.3, we can see that the magnitudes of the relativistic and nonrelativistic momenta differ by the same factor of $\sqrt{1 - v^2/c^2}$ that occurs in the time-dilation and length-contraction equations. Since this factor is always less than 1 and occurs in the denominator in Equation 28.3, the relativistic momentum is always larger than the nonrelativistic momentum. To illustrate how the two quantities differ as the speed v increases, Figure 28.7 shows a plot of the ratio of the momentum magnitudes (relativistic/nonrelativistic) as a function of v. According to Equation 28.3, this ratio is just $1/\sqrt{1 - v^2/c^2}$. The graph shows that for speeds attained by ordinary objects, such as cars and planes, the relativistic and nonrelativistic momenta are almost equal because their ratio is nearly 1. Thus, at speeds much less than the speed of light, either the nonrelativistic momentum or the relativistic momentum can be used to describe collisions. On the other hand, when the speed of the object becomes comparable to the speed of light, the relativistic momentum becomes significantly greater than the nonrelativistic momentum and must be used. Example 5 deals with the relativistic momentum of an electron traveling close to the speed of light.

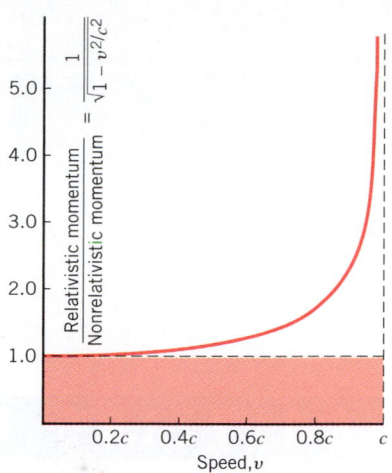

Figure 28.7 This graph shows how the ratio of the magnitude of the relativistic momentum to the magnitude of the nonrelativistic momentum increases as the speed of an object approaches the speed of light.

Example 5 The Relativistic Momentum of a High-Speed Electron

The particle accelerator at Stanford University (Figure 28.8) is 3 km long and accelerates electrons to a speed of 0.999 999 999 7c, which is very nearly equal to the speed of light. Find the magnitude of the relativistic momentum of an electron that emerges from the accelerator, and compare it with the nonrelativistic value.

Reasoning and Solution The magnitude of the electron's relativistic momentum can be obtained from Equation 28.3 if we recall that the mass of an electron is $m = 9.11 \times 10^{-31}$ kg:

$$p = \frac{mv}{\sqrt{1 - \frac{v^2}{c^2}}} = \frac{(9.11 \times 10^{-31} \text{ kg})(0.999\ 999\ 999\ 7c)}{\sqrt{1 - \frac{(0.999\ 999\ 999\ 7c)^2}{c^2}}} = \boxed{1 \times 10^{-17} \text{ kg} \cdot \text{m/s}}$$

This value for the magnitude of the momentum agrees with the value measured experimentally. The relativistic momentum is greater than the nonrelativistic momentum by a factor of

$$\frac{1}{\sqrt{1 - \frac{v^2}{c^2}}} = \frac{1}{\sqrt{1 - \frac{(0.999\ 999\ 999\ 7c)^2}{c^2}}} = \boxed{4 \times 10^4}$$

Particle accelerator

THE EQUIVALENCE OF MASS AND ENERGY

THE TOTAL ENERGY OF AN OBJECT

One of the most astonishing results of special relativity is that mass and energy are equivalent, in the sense that a gain or loss of mass can be regarded equally well as a gain or loss of energy. Consider, for example, an object of mass m traveling at a speed v. Einstein showed that the *total energy* E of the moving object is related to its mass and speed by the following relation:

Total energy
of an object

$$E = \frac{mc^2}{\sqrt{1 - \frac{v^2}{c^2}}}$$

(28.4)

Figure 28.8 The Stanford 3-km linear accelerator accelerates electrons almost to the speed of light. (© Bill Marsh/ Photo Researchers, Inc.)

To gain some understanding of Equation 28.4, consider the special case in which the object is at rest. When $v = 0$ m/s, the total energy is called the *rest energy E_0*, and Equation 28.4 reduces to Einstein's now-famous equation:

Rest energy of an object
$$E_0 = mc^2 \qquad (28.5)$$

The rest energy represents the energy equivalent of the mass of an object at rest. As Example 6 shows, even a small mass is equivalent to an enormous amount of energy.

ANALYZING MULTIPLE-CONCEPT PROBLEMS

Example 6 The Energy Equivalent of a Golf Ball

A 0.046-kg golf ball is lying on the green, as Figure 28.9 illustrates. If the rest energy of this ball were used to operate a 75-W light bulb, how long would the bulb remain lit?

Reasoning The average power delivered to the light bulb is 75 W, which means that it uses 75 J of energy per second. Therefore, the time that the bulb would remain lit is equal to the total energy used by the light bulb divided by the energy per second (i.e., the average power) delivered to it. This energy comes from the rest energy of the golf ball, which is equal to its mass times the speed of light squared.

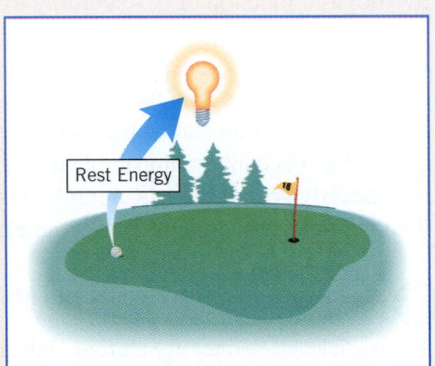

Figure 28.9 The rest energy of a golf ball is sufficient to keep a 75-W light bulb burning for an incredibly long time (see Example 6).

Knowns and Unknowns The data for this problem are:

Description	Symbol	Value
Mass of golf ball	m	0.046 kg
Average power delivered to light bulb	$\overline{P}$	75 W
Unknown Variable		
Time that light bulb would remain lit	t	?

Modeling the Problem

STEP 1 Power The average power $\overline{P}$ is equal to the energy delivered to the light bulb divided by the time t (see Section 6.7 and Equation 6.10b), or $\overline{P} = $ Energy/t. In this case the energy comes from the rest energy E_0 of the golf ball, so $\overline{P} = E_0/t$. Solving for the time gives Equation 1 at the right. The average power is known, and the rest energy will be evaluated in Step 2.

$$t = \frac{E_0}{\overline{P}} \qquad (1)$$
$?$

STEP 2 Rest Energy The rest energy E_0 is the total energy of the golf ball as it rests on the green. If the golf ball's mass is m, then its rest energy is

$$\boxed{E_0 = mc^2} \qquad (28.5)$$

where c is the speed of light in a vacuum. Both m and c are known, so we substitute this expression for the rest energy into Equation 1, as indicated at the right.

$$t = \frac{E_0}{\overline{P}} \qquad (1)$$
$$\boxed{E_0 = mc^2} \qquad (28.5)$$

Solution Algebraically combining the results of each step, we have

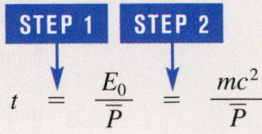

$$t = \frac{E_0}{\overline{P}} = \frac{mc^2}{\overline{P}}$$

The time that the light bulb would remain lit is

$$t = \frac{mc^2}{\overline{P}} = \frac{(0.046 \text{ kg})(3.0 \times 10^8 \text{ m/s})^2}{75 \text{ W}} = \boxed{5.5 \times 10^{13} \text{ s}}$$

Expressed in years (1 yr $= 3.2 \times 10^7$ s), this time is equivalent to

$$(5.5 \times 10^{13} \text{ s})\left(\frac{1 \text{ yr}}{3.2 \times 10^7 \text{ s}}\right) = 1.7 \times 10^6 \text{ yr} \quad \text{or} \quad 1.7 \text{ million years!}$$

Related Homework: *Problems 30, 31*

When an object is accelerated from rest to a speed v, the object acquires kinetic energy in addition to its rest energy. The total energy E is the sum of the rest energy E_0 and the kinetic energy KE, or $E = E_0 + \text{KE}$. Therefore, the kinetic energy is the difference between the object's total energy and its rest energy. Using Equations 28.4 and 28.5, we can write the kinetic energy as

Kinetic energy
of an object

$$\text{KE} = E - E_0 = mc^2\left(\frac{1}{\sqrt{1 - \dfrac{v^2}{c^2}}} - 1\right) \tag{28.6}$$

This equation is the relativistically correct expression for the kinetic energy of an object of mass m moving at speed v.

Equation 28.6 looks nothing like the kinetic energy expression introduced in Section 6.2—namely, $\text{KE} = \frac{1}{2}mv^2$ (Equation 6.2). However, for speeds much less than the speed of light ($v \ll c$), the relativistic equation for the kinetic energy reduces to $\text{KE} = \frac{1}{2}mv^2$, as can be seen by using the binomial expansion* to represent the square root term in Equation 28.6:

$$\frac{1}{\sqrt{1 - \dfrac{v^2}{c^2}}} = 1 + \tfrac{1}{2}\left(\frac{v^2}{c^2}\right) + \tfrac{3}{8}\left(\frac{v^2}{c^2}\right)^2 + \cdots$$

Suppose that v is much smaller than c—say, $v = 0.01c$. The second term in the expansion has the value $\frac{1}{2}(v^2/c^2) = 5.0 \times 10^{-5}$, while the third term has the much smaller value $\frac{3}{8}(v^2/c^2)^2 = 3.8 \times 10^{-9}$. The additional terms are smaller still, so if $v \ll c$, we can neglect the third and additional terms in comparison with the first and second terms. Substituting the first two terms into Equation 28.6 gives

$$\text{KE} \approx mc^2\left(1 + \tfrac{1}{2}\frac{v^2}{c^2} - 1\right) = \tfrac{1}{2}mv^2$$

which is the familiar form for the kinetic energy. However, Equation 28.6 gives the correct kinetic energy for all speeds and must be used for speeds near the speed of light, as in Example 7.

 Example 7 A High-Speed Electron

An electron ($m = 9.109 \times 10^{-31}$ kg) is accelerated from rest to a speed of $v = 0.9995c$ in a particle accelerator. Determine the electron's **(a)** rest energy, **(b)** total energy, and **(c)** kinetic energy in millions of electron volts or MeV.

Reasoning and Solution **(a)** The electron's rest energy is

$$E_0 = mc^2 = (9.109 \times 10^{-31} \text{ kg})(2.998 \times 10^8 \text{ m/s})^2 = 8.187 \times 10^{-14} \text{ J} \tag{28.5}$$

---

*The binomial expansion states that $(1 - x)^n = 1 - nx + n(n - 1)x^2/2 + \cdots$. In our case, $x = v^2/c^2$ and $n = -1/2$.

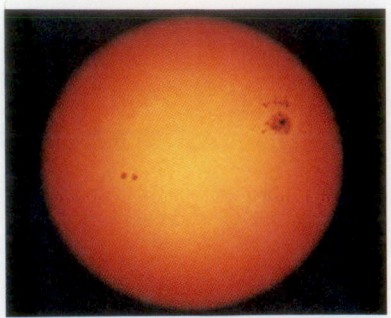

Visible light image

X-ray image

Figure 28.10 The sun emits electromagnetic energy over a broad portion of the electromagnetic spectrum. These photographs were obtained using that energy in the indicated regions of the spectrum. (*Top,* © Mark Marten/ NASA/Photo Researchers, Inc.; *bottom,* © Dr. Leon Golub/Photo Researchers, Inc.)

Since 1 eV = 1.602×10^{-19} J, the electron's rest energy is

$$(8.187 \times 10^{-14} \text{ J})\left(\frac{1 \text{ eV}}{1.602 \times 10^{-19} \text{ J}}\right) = \boxed{5.11 \times 10^5 \text{ eV} \quad \text{or} \quad 0.511 \text{ MeV}}$$

(b) The total energy of an electron traveling at a speed of $v = 0.9995c$ is

$$E = \frac{mc^2}{\sqrt{1 - \dfrac{v^2}{c^2}}} = \frac{(9.109 \times 10^{-31} \text{ kg})(2.998 \times 10^8 \text{ m/s})^2}{\sqrt{1 - \left(\dfrac{0.9995c}{c}\right)^2}} \qquad (28.4)$$

$$= \boxed{2.59 \times 10^{-12} \text{ J} \quad \text{or} \quad 16.2 \text{ MeV}}$$

(c) The kinetic energy is the difference between the total energy and the rest energy:

$$\text{KE} = E - E_0 = 2.59 \times 10^{-12} \text{ J} - 8.2 \times 10^{-14} \text{ J} \qquad (28.6)$$

$$= \boxed{2.51 \times 10^{-12} \text{ J} \quad \text{or} \quad 15.7 \text{ MeV}}$$

For comparison, if the kinetic energy of the electron had been calculated from $\frac{1}{2}mv^2$, a value of only 0.26 MeV would have been obtained.

Since mass and energy are equivalent, any change in one is accompanied by a corresponding change in the other. For instance, life on earth is dependent on electromagnetic energy (light) from the sun. Because this energy is leaving the sun (see Figure 28.10), there is a decrease in the sun's mass. Example 8 illustrates how to determine this decrease.

Example 8 The Sun Is Losing Mass

The sun radiates electromagnetic energy at the rate of 3.92×10^{26} W. **(a)** What is the change in the sun's mass during each second that it is radiating energy? **(b)** The mass of the sun is 1.99×10^{30} kg. What fraction of the sun's mass is lost during a human lifetime of 75 years?

Reasoning Since 1 W = 1 J/s, the amount of electromagnetic energy radiated during each second is 3.92×10^{26} J. Thus, during each second, the sun's rest energy decreases by this amount. The change ΔE_0 in the sun's rest energy is related to the change Δm in its mass by $\Delta E_0 = (\Delta m)c^2$, according to Equation 28.5.

Solution **(a)** For each second that the sun radiates energy, the change in its mass is

$$\Delta m = \frac{\Delta E_0}{c^2} = \frac{3.92 \times 10^{26} \text{ J}}{(3.00 \times 10^8 \text{ m/s})^2} = \boxed{4.36 \times 10^9 \text{ kg}}$$

Over 4 billion kilograms of mass are lost by the sun during each second.

(b) The amount of mass lost by the sun in 75 years is

$$\Delta m = (4.36 \times 10^9 \text{ kg/s})\left(\frac{3.16 \times 10^7 \text{ s}}{1 \text{ year}}\right)(75 \text{ years}) = 1.0 \times 10^{19} \text{ kg}$$

Although this is an enormous amount of mass, it represents only a tiny fraction of the sun's total mass:

$$\frac{\Delta m}{m_{\text{sun}}} = \frac{1.0 \times 10^{19} \text{ kg}}{1.99 \times 10^{30} \text{ kg}} = \boxed{5.0 \times 10^{-12}}$$

Any change ΔE_0 in the rest energy of a system causes a change in the mass of the system according to $\Delta E_0 = (\Delta m)c^2$. It does not matter whether the change in energy is due to a change in electromagnetic energy, potential energy, thermal energy, or so on. Although any change in energy gives rise to a change in mass, in most instances the change in mass is too small to be detected. For instance, when 4186 J of heat is used to raise the temperature of 1 kg of water by 1 C°, the mass changes by only $\Delta m = \Delta E_0/c^2 = (4186 \text{ J})/(3.00 \times 10^8 \text{ m/s})^2 = 4.7 \times 10^{-14}$ kg. Conceptual Example 9 illustrates further how a change in the energy of an object leads to an equivalent change in its mass.

⬇ Conceptual Example 9 | When Is a Massless Spring Not Massless?

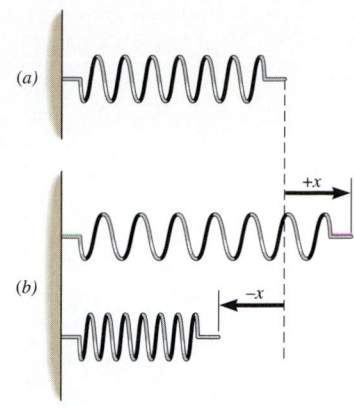

Figure 28.11*a* shows a top view of a *massless* spring on a horizontal table. Initially the spring is unstrained. Then the spring is either stretched or compressed by an amount *x* from its unstrained length, as Figure 28.11*b* illustrates. What is the mass of the spring in Figure 28.11*b*? **(a)** It is greater than zero by an amount that is larger when the spring is stretched. **(b)** It is greater than zero by an amount that is larger when the spring is compressed. **(c)** It is greater than zero by an amount that is the same when the spring is stretched or compressed. **(d)** It remains zero.

Reasoning When a spring is stretched or compressed, its elastic potential energy changes. As discussed in Section 10.3, the elastic potential energy of an ideal spring is equal to $\frac{1}{2}kx^2$, where *k* is the spring constant and *x* is the amount of stretch or compression. Consistent with the theory of special relativity, any change in the total energy of a system, including a change in the elastic potential energy, is equivalent to a change in the mass of the system.

Answers (a), (b), and (d) are incorrect. In being stretched or compressed by the same amount *x*, the spring acquires the same amount of elastic potential energy ($\frac{1}{2}kx^2$). Therefore, according to special relativity, the spring acquires the same mass regardless of whether it is stretched or compressed, so these answers must be incorrect.

Answer (c) is correct. The spring acquires elastic potential energy in being stretched or compressed. Special relativity indicates that this additional energy is equivalent to additional mass. Since the amount of stretch or compression is the same, the potential energy is the same in either case, and so is the additional mass.

Related Homework: *Problem 26*

Figure 28.11 (*a*) This spring is unstrained and assumed to have no mass. (*b*) When the spring is either stretched or compressed by an amount *x*, it gains elastic potential energy and, hence, mass.

It is also possible to transform matter itself into other forms of energy. For example, the positron (see Section 31.4) has the same mass as an electron but an opposite electrical charge. If these two particles of matter collide, they are completely annihilated, and a burst of high-energy electromagnetic waves is produced. Thus, matter is transformed into electromagnetic waves, the energy of the electromagnetic waves being equal to the total energies of the two colliding particles. The medical diagnostic technique known as positron emission tomography or PET scanning depends on the electromagnetic energy produced when a positron and an electron are annihilated (see Section 32.6).

The transformation of electromagnetic waves into matter also happens. In one experiment, an extremely high-energy electromagnetic wave, called a gamma ray (see Section 31.4), passes close to the nucleus of an atom. If the gamma ray has sufficient energy, it can create an electron and a positron. The gamma ray disappears, and the two particles of matter appear in its place. Except for picking up some momentum, the nearby nucleus remains unchanged. The process in which the gamma ray is transformed into the two particles is known as *pair production*.

THE RELATION BETWEEN TOTAL ENERGY AND MOMENTUM

It is possible to derive a useful relation between the total relativistic energy *E* and the relativistic momentum *p*. We begin by rearranging Equation 28.3 for the momentum, to obtain

$$\frac{m}{\sqrt{1 - v^2/c^2}} = \frac{p}{v}$$

With this substitution, Equation 28.4 for the total energy becomes

$$E = \frac{mc^2}{\sqrt{1 - v^2/c^2}} = \frac{pc^2}{v} \quad \text{or} \quad \frac{v}{c} = \frac{pc}{E}$$

Using this expression to replace *v*/*c* in Equation 28.4 gives

$$E = \frac{mc^2}{\sqrt{1 - v^2/c^2}} = \frac{mc^2}{\sqrt{1 - p^2c^2/E^2}} \quad \text{or} \quad E^2 = \frac{m^2c^4}{1 - p^2c^2/E^2}$$

Solving this expression for E^2 shows that

$$E^2 = p^2c^2 + m^2c^4 \tag{28.7}$$

THE SPEED OF LIGHT IN A VACUUM IS THE ULTIMATE SPEED

One of the important consequences of the theory of special relativity is that objects with mass cannot reach the speed of light in a vacuum. Thus, the speed of light in a vacuum represents the ultimate speed. To see that this speed limitation is a consequence of special relativity, consider Equation 28.6, which gives the kinetic energy of a moving object. As v approaches the speed of light c, the $\sqrt{1 - v^2/c^2}$ term in the denominator approaches zero. Hence, the kinetic energy becomes infinitely large. However, the work–energy theorem (Section 6.2) tells us that an infinite amount of work would have to be done to give the object an infinite kinetic energy. Since an infinite amount of work is not available, we are left with the conclusion that objects with mass cannot attain the speed of light c.

✓ CHECK YOUR UNDERSTANDING

(*The answers are given at the end of the book.*)

8. Consider the same cup of coffee sitting on the same table in the following four situations: **(a)** The coffee is hot (95°C), and the table is at sea level. **(b)** The coffee is cold (60°C), and the table is at sea level. **(c)** The coffee is hot (95°C), and the table is on a mountain top. **(d)** The coffee is cold (60°C), and the table is on a mountain top. In which situation does the cup of coffee have the greatest mass, and in which the smallest mass?

9. A system consists of two positive charges. Is the total mass of the system greater when the two charges are **(a)** separated by a finite distance or **(b)** infinitely far apart?

10. A parallel plate capacitor is initially uncharged. Then it is fully charged up by removing electrons from one plate and placing them on the other plate. Is the mass of the capacitor greater when it is **(a)** uncharged or **(b)** fully charged?

11. It takes work to accelerate a particle from rest to a given speed close to the speed of light in a vacuum. For which particle is less work required, **(a)** an electron or **(b)** a proton?

28.7 THE RELATIVISTIC ADDITION OF VELOCITIES

The velocity of an object relative to an observer plays a central role in special relativity, and to determine this velocity, it is sometimes necessary to add two or more velocities together. We first encountered relative velocity in Section 3.4, so we will begin by reviewing some of the ideas presented there. Figure 28.12 illustrates a truck moving at a constant velocity of $v_{TG} = +15$ m/s relative to an observer standing on the ground, where the plus sign denotes a direction to the right. Suppose someone on the truck throws a baseball toward the observer at a velocity of $v_{BT} = +8.0$ m/s relative to the truck. We might conclude that the observer on the ground would see the ball approaching at a velocity of $v_{BG} = v_{BT} + v_{TG} = 8.0$ m/s $+ 15$ m/s $= +23$ m/s. These symbols are similar to those used in Section 3.4 and have the following meaning:

$$v_{\boxed{BG}} = \text{velocity of the } \boxed{\text{Baseball}} \text{ relative to the } \boxed{\text{Ground}} = +23 \text{ m/s}$$

$$v_{\boxed{BT}} = \text{velocity of the } \boxed{\text{Baseball}} \text{ relative to the } \boxed{\text{Truck}} = +8.0 \text{ m/s}$$

$$v_{\boxed{TG}} = \text{velocity of the } \boxed{\text{Truck}} \text{ relative to the } \boxed{\text{Ground}} = +15.0 \text{ m/s}$$

Although the result that $v_{BG} = +23$ m/s seems reasonable, careful measurements would show that it is not quite right. According to special relativity, the equation $v_{BG} = v_{BT} + v_{TG}$ is not valid for the following reason. If the velocity of the truck had a

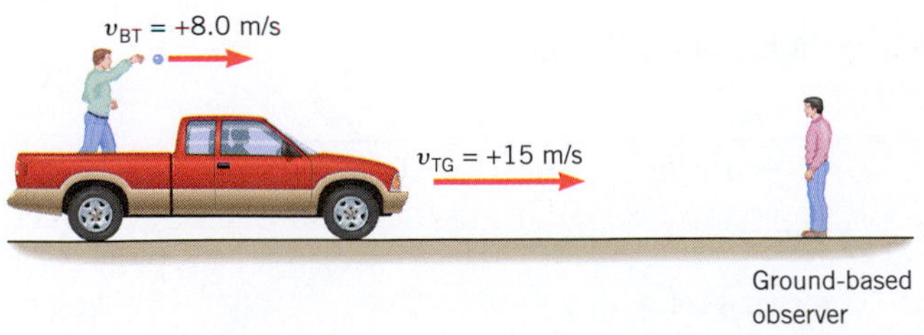

Figure 28.12 The truck is approaching the ground-based observer at a relative velocity of $v_{TG} = +15$ m/s. The velocity of the baseball relative to the truck is $v_{BT} = +8.0$ m/s.

$v_{BT} = +8.0$ m/s

$v_{TG} = +15$ m/s

Ground-based observer

magnitude sufficiently close to the speed of light in a vacuum, the equation would predict that the observer on the earth could see the baseball moving faster than the speed of light. This is not possible, since no object with a finite mass can move faster than the speed of light in a vacuum.

For the case in which the truck and ball are moving along the same straight line, the theory of special relativity reveals that the velocities are related according to

$$v_{BG} = \frac{v_{BT} + v_{TG}}{1 + \dfrac{v_{BT}v_{TG}}{c^2}}$$

The subscripts in this equation have been chosen for the specific situation shown in Figure 28.12. For the general situation, the relative velocities are related by the *velocity-addition formula:*

Velocity addition
$$v_{AB} = \frac{v_{AC} + v_{CB}}{1 + \dfrac{v_{AC}v_{CB}}{c^2}} \qquad (28.8)$$

where all the velocities are assumed to be constant and the symbols have the following meanings:

$$v_{\boxed{AB}} = \text{velocity of } \boxed{\text{object A}} \text{ relative to } \boxed{\text{object B}}$$

$$v_{\boxed{AC}} = \text{velocity of } \boxed{\text{object A}} \text{ relative to } \boxed{\text{object C}}$$

$$v_{\boxed{CB}} = \text{velocity of } \boxed{\text{object C}} \text{ relative to } \boxed{\text{object B}}$$

The ordering of the subscripts in Equation 28.8 follows the discussion in Section 3.4. For motion along a straight line, the velocities can have either positive or negative values, depending on whether they are directed along the positive or negative direction. Furthermore, switching the order of the subscripts changes the sign of the velocity, so, for example, $v_{BA} = -v_{AB}$ (see Example 12 in Chapter 3).

Equation 28.8 differs from the nonrelativistic formula ($v_{AB} = v_{AC} + v_{CB}$) by the presence of the $v_{AC}v_{CB}/c^2$ term in the denominator. This term arises because of the effects of time dilation and length contraction that occur in special relativity. When v_{AC} and v_{CB} are small compared to c, the $v_{AC}v_{CB}/c^2$ term is small compared to 1, so the velocity-addition formula reduces to $v_{AB} \approx v_{AC} + v_{CB}$. However, when either v_{AC} or v_{CB} is comparable to c, the results can be quite different, as Example 10 illustrates.

 Example 10   The Relativistic Addition of Velocities

Imagine a hypothetical situation in which the truck in Figure 28.12 is moving relative to the ground with a velocity of $v_{TG} = +0.8c$. A person riding on the truck throws a baseball at a velocity relative to the truck of $v_{BT} = +0.5c$. What is the velocity v_{BG} of the baseball relative to a person standing on the ground?

Reasoning The observer standing on the ground does *not* see the baseball approaching at $v_{BG} = 0.5c + 0.8c = 1.3c$. This cannot be, because the speed of the ball would then exceed the speed of light in a vacuum. The velocity-addition formula gives the correct velocity, which has a magnitude less than the speed of light.

Solution The ground-based observer sees the ball approaching with a velocity of

$$v_{BG} = \frac{v_{BT} + v_{TG}}{1 + \dfrac{v_{BT}v_{TG}}{c^2}} = \frac{0.5c + 0.8c}{1 + \dfrac{(0.5c)(0.8c)}{c^2}} = \boxed{0.93c} \qquad (28.8)$$

Example 10 discusses how the speed of a baseball is viewed by observers in different inertial reference frames. The next example deals with a similar situation, except that the baseball is replaced by the light of a laser beam.

Figure 28.13 An intergalactic cruiser, closing in on a hostile spacecraft, fires a beam of laser light.

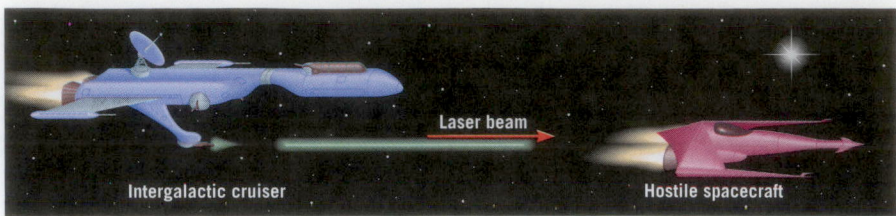

Laser beam

Intergalactic cruiser

Hostile spacecraft

Figure 28.13 An intergalactic cruiser, closing in on a hostile spacecraft, fires a beam of laser light.

Conceptual Example 11   The Speed of a Laser Beam

Figure 28.13 shows an intergalactic cruiser approaching a hostile spacecraft. Both vehicles move at a constant velocity. The velocity of the cruiser relative to the spacecraft is $v_{CS} = +0.7c$, the direction to the right being the positive direction. The cruiser fires a beam of laser light at the hostile renegades. The velocity of the laser beam relative to the cruiser is $v_{LC} = +c$. Which one of the following statements correctly describes the velocity v_{LS} of the laser beam relative to the renegades' spacecraft and the velocity v at which the renegades see the laser beam move away from the cruiser? (a) $v_{LS} = +0.7c$ and $v = +c$ (b) $v_{LS} = +0.3c$ and $v = +c$ (c) $v_{LS} = +c$ and $v = +0.7c$ (d) $v_{LS} = +c$ and $v = +0.3c$

Reasoning Since both vehicles move at a constant velocity, each constitutes an inertial reference frame. According to the speed-of-light postulate, *all* observers in inertial reference frames measure the speed of light in a vacuum to be c.

Answers (a) and (b) are incorrect. Since the renegades' spacecraft constitutes an inertial reference frame, the velocity of the laser beam relative to it can only have a value of $v_{LS} = +c$, according to the speed-of-light postulate.

Answer (c) is incorrect. The velocity at which the renegades see the laser beam move away from the cruiser cannot be $v = +0.7c$, because they see the cruiser moving at a velocity of $+0.7c$ and the laser beam moving at a velocity of only $+c$ (not $+1.4c$).

Answer (d) is correct. The renegades see the cruiser approach them at a relative velocity of $v_{CS} = +0.7c$ and see the laser beam approach them at a relative velocity of $v_{LS} = +c$. Both these velocities are measured relative to the *same* inertial reference frame—namely, that of their own spacecraft. Therefore, the renegades see the laser beam move away from the cruiser at a velocity that is the difference between these two velocities, or $+c - (+0.7c) = +0.3c$. The velocity-addition formula, Equation 28.8, does not apply here because both velocities are measured relative to the *same* inertial reference frame. Equation 28.8 is used only when the velocities are measured relative to different inertial reference frames.

Related Homework: *Problem 46*

It is a straightforward matter to show that the velocity-addition formula is consistent with the speed-of-light postulate. Consider Figure 28.14, which shows a person riding on a truck and holding a flashlight. The velocity of the light relative to the person on the truck is $v_{LT} = +c$. The velocity v_{LG} of the light relative to the observer standing on the ground is given by the velocity-addition formula as

$$v_{LG} = \frac{v_{LT} + v_{TG}}{1 + \dfrac{v_{LT}v_{TG}}{c^2}} = \frac{c + v_{TG}}{1 + \dfrac{cv_{TG}}{c^2}} = \frac{(c + v_{TG})c}{(c + v_{TG})} = c$$

Thus, the velocity-addition formula indicates that the observer on the ground and the person on the truck both measure the speed of light to be c, independent of the relative velocity v_{TG} between them. This is exactly what the speed-of-light postulate states.

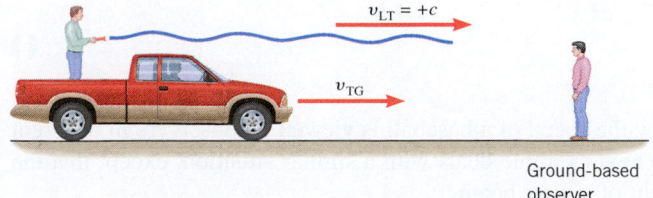

$v_{LT} = +c$

v_{TG}

Ground-based observer

Figure 28.14 The speed of the light emitted by the flashlight is c relative to both the truck and the observer on the ground.

✓ **CHECK YOUR UNDERSTANDING**

(*The answer is given at the end of the book.*)

12. Car A and car B are both traveling due east on a straight section of an interstate highway. The speed of car A relative to the ground is 10 m/s faster than the speed of car B relative to the ground. According to special relativity, is the speed of car A relative to car B **(a)** 10 m/s, **(b)** less than 10 m/s, or **(c)** greater than 10 m/s?

28.8   CONCEPTS & CALCULATIONS

There are many astonishing consequences of special relativity, two of which are time dilation and length contraction. Example 12 reviews these important concepts in the context of a golf game in a hypothetical world where the speed of light is a little faster than that of a golf cart.

Concepts & Calculations Example 12

⬇ Golf and Special Relativity

Imagine playing golf in a world where the speed of light is only $c = 3.40$ m/s. Golfer A drives a ball down a flat horizontal fairway for a distance that he measures as 75.0 m. Golfer B, riding in a cart, happens to pass by just as the ball is hit (see Figure 28.15). Golfer A stands at the tee and watches while golfer B moves down the fairway toward the ball at a constant speed of 2.80 m/s. **(a)** How far is the ball hit according to a measurement made by golfer B? **(b)** According to each golfer, how much time does it take for golfer B to reach the ball?

Concept Questions and Answers Who measures the proper length of the drive, and who measures the contracted length?

> *Answer*   Consider two coordinate systems, one attached to the earth and the other to the golf cart. The proper length L_0 is the distance between two points as measured by an observer *who is at rest with respect to them*. Since golfer A is standing at the tee and is at rest relative to the earth, golfer A measures the proper length of the drive, which is $L_0 = 75.0$ m. Golfer B is not at rest with respect to the two points, however, and measures a contracted length for the drive that is less than 75.0 m.

Who measures the proper time interval, and who measures the dilated time interval?

> *Answer*   The proper time interval Δt_0 is the time interval measured by an observer who is at rest in his or her coordinate system and *who views the beginning and ending events as occurring at the same place*. The beginning event is when the ball is hit, and the ending event is when golfer B arrives at the ball. When the ball is hit, it is alongside the origin of golfer B's coordinate system. When golfer B arrives at the ball down the fairway, it is again alongside the origin of his coordinate system. Thus, golfer B measures the proper time interval. Golfer A, who does not see the beginning and ending events occurring at the same place, measures a dilated, or longer, time interval.

Solution **(a)** Golfer A, being at rest with respect to the beginning and ending points, measures the proper length of the drive, so $L_0 = 75.0$ m. Golfer B, who is moving, measures a contracted length L given by Equation 28.2:

$$L = L_0 \sqrt{1 - \frac{v^2}{c^2}} = (75.0 \text{ m}) \sqrt{1 - \frac{(2.80 \text{ m/s})^2}{(3.40 \text{ m/s})^2}} = \boxed{42.5 \text{ m}}$$

Thus, the moving golfer measures the length of the drive to be a shortened 42.5 m rather than the 75.0 m measured by the stationary golfer.

(b) According to golfer B, the time interval Δt_0 it takes to reach the ball is equal to the contracted length L that he measures divided by the speed of the ground with respect to him. The speed of the ground with respect to golfer B is the same as the speed v of the cart with respect to the ground. Thus, we find

$$\Delta t_0 = \frac{L}{v} = \frac{42.5 \text{ m}}{2.80 \text{ m/s}} = \boxed{15.2 \text{ s}}$$

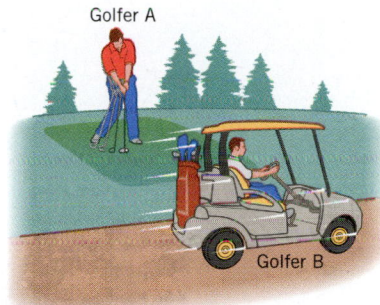

Golfer A

Golfer B

Figure 28.15 Just as golfer A hits the ball, golfer B passes by in a golf cart. According to special relativity, each golfer measures a different distance for how far the ball is hit.

Golfer A, standing at the tee, measures a dilated time interval Δt, which is related to the proper time interval by Equation 28.1:

$$\Delta t = \frac{\Delta t_0}{\sqrt{1 - \dfrac{v^2}{c^2}}} = \frac{15.2 \text{ s}}{\sqrt{1 - \dfrac{(2.80 \text{ m/s})^2}{(3.40 \text{ m/s})^2}}} = \boxed{26.8 \text{ s}}$$

In summary, golfer A measures the proper length (75.0 m) and a dilated time interval (26.8 s), and golfer B measures a shortened length (42.5 m) and a proper time interval (15.2 s).

Other important consequences of special relativity are the equivalence of mass and energy, and the dependence of kinetic energy on the total energy and on the rest energy. Example 13 compares these properties for three different particles.

Concepts & Calculations Example 13

Mass and Energy

The rest energy E_0 and the total energy E of three particles, expressed in terms of a basic amount of energy $E' = 5.98 \times 10^{-10}$ J, are listed in the table below. The speeds of these particles are large, in some cases approaching the speed of light. For each particle, determine its **(a)** mass and **(b)** kinetic energy.

Particle	Rest Energy	Total Energy
a	E'	$2E'$
b	E'	$4E'$
c	$5E'$	$6E'$

Concept Questions and Answers Given the rest energies specified in the table, what is the ranking (largest first) of the masses of the particles?

Answer The rest energy is the energy that an object has when its speed is zero. According to special relativity, the rest energy E_0 and the mass m are equivalent. The relation between the two is given by Equation 28.5 as $E_0 = mc^2$, where c is the speed of light in a vacuum. Thus, the rest energy is directly proportional to the mass. From the table it can be seen that particles a and b have identical rest energies, so they have identical masses. Particle c has the greatest rest energy, so it has the greatest mass. The ranking of the masses, largest first, is c, then a and b (a tie).

Is the kinetic energy KE given by the expression $\text{KE} = \frac{1}{2}mv^2$, and what is the ranking (largest first) of the kinetic energies of the particles?

Answer No, because the expression $\text{KE} = \frac{1}{2}mv^2$ applies only when the speed of the object is much, much less than the speed of light. According to special relativity, the kinetic energy is the difference between the total energy E and the rest energy E_0, so $\text{KE} = E - E_0$. Therefore, we can examine the table and determine the kinetic energy of each particle in terms of E'. The kinetic energies of particles a, b, and c are, respectively, $2E' - E' = E'$, $4E' - E' = 3E'$, and $6E' - 5E' = E'$. The ranking of the kinetic energies, largest first, is b, then a and c (a tie).

Solution **(a)** The mass of particle a can be determined from its rest energy $E_0 = mc^2$. Since $E_0 = E'$ (see the table), its mass is

$$m_a = \frac{E'}{c^2} = \frac{5.98 \times 10^{-10} \text{ J}}{(3.00 \times 10^8 \text{ m/s})^2} = \boxed{6.64 \times 10^{-27} \text{ kg}}$$

In a similar manner, we find that the masses of particles b and c are

$$\boxed{m_b = 6.64 \times 10^{-27} \text{ kg}} \quad \text{and} \quad \boxed{m_c = 33.2 \times 10^{-27} \text{ kg}}$$

As expected, the ranking is $m_c > m_a = m_b$.

(b) According to Equation 28.6, the kinetic energy KE of a particle is equal to its total energy E minus its rest energy E_0; $KE = E - E_0$. For particle a, its total energy is $E = 2E'$ and its rest energy is $E_0 = E'$, so its kinetic energy is

$$KE_a = 2E' - E' = E' = \boxed{5.98 \times 10^{-10} \text{ J}}$$

The kinetic energies of particles b and c can be determined in a similar fashion:

$$\boxed{KE_b = 17.9 \times 10^{-10} \text{ J}} \quad \text{and} \quad \boxed{KE_c = 5.98 \times 10^{-10} \text{ J}}$$

As anticipated, the ranking is $KE_b > KE_a = KE_c$.

CONCEPT SUMMARY

If you need more help with a concept, use the Learning Aids noted next to the discussion or equation. Examples (**Ex.**) are in the text of this chapter. Go to **www.wiley.com/college/cutnell** for the following Learning Aids:

Interactive LearningWare (ILW) — Additional examples solved in a five-step interactive format.

Concept Simulations (CS) — Animated text figures or animations of important concepts.

Interactive Solutions (IS) — Models for certain types of problems in the chapter homework. The calculations are carried out interactively.

Topic	Discussion	Learning Aids
Reference frame	**28.1 EVENTS AND INERTIAL REFERENCE FRAMES** An event is a physical "happening" that occurs at a certain place and time. To record the event an observer uses a reference frame that consists of a coordinate system and a clock. Different observers may use different reference frames.	
Inertial reference frame	The theory of special relativity deals with inertial reference frames. An inertial reference frame is one in which Newton's law of inertia is valid. Accelerating reference frames are not inertial reference frames.	
Relativity postulate Speed-of-light postulate	**28.2 THE POSTULATES OF SPECIAL RELATIVITY** The theory of special relativity is based on two postulates. The relativity postulate states that the laws of physics are the same in every inertial reference frame. The speed-of-light postulate says that the speed of light in a vacuum, measured in any inertial reference frame, always has the same value of c, no matter how fast the source of the light and the observer are moving relative to each other.	
Proper time interval Dilated time interval	**28.3 THE RELATIVITY OF TIME: TIME DILATION** The proper time interval Δt_0 between two events is the time interval measured by an observer who is at rest relative to the events and views them occurring at the same place. An observer who is in motion with respect to the events and who views them as occurring at different places measures a dilated time interval Δt. The dilated time interval is greater than the proper time interval, according to the time-dilation equation:	CS 28.1
Time-dilation equation	$$\Delta t = \frac{\Delta t_0}{\sqrt{1 - \dfrac{v^2}{c^2}}} \tag{28.1}$$ In this expression, v is the relative speed between the observer who measures Δt_0 and the observer who measures Δt.	Ex. 1–3 IS 28.5
Proper length Contracted length	**28.4 THE RELATIVITY OF LENGTH: LENGTH CONTRACTION** The proper length L_0 between two points is the length measured by an observer who is at rest relative to the points. An observer moving with a relative speed v parallel to the line between the two points does not measure the proper length. Instead, such an observer measures a contracted length L given by the length-contraction formula:	
Length-contraction formula	$$L = L_0 \sqrt{1 - \frac{v^2}{c^2}} \tag{28.2}$$ Length contraction occurs only along the direction of the motion. Those dimensions that are perpendicular to the motion are not shortened. The observer who measures the proper length may not be the observer who measures the proper time interval.	Ex. 4, 12 ILW 28.1 IS 28.13

Topic	Discussion	Learning Aids

28.5 RELATIVISTIC MOMENTUM An object of mass m, moving with speed v, has a relativistic momentum whose magnitude p is given by

Relativistic momentum

$$p = \frac{mv}{\sqrt{1 - \dfrac{v^2}{c^2}}}$$ (28.3) **Ex. 5**

28.6 THE EQUIVALENCE OF MASS AND ENERGY Energy and mass are equivalent. The total energy E of an object of mass m, moving at speed v, is

Total energy

$$E = \frac{mc^2}{\sqrt{1 - \dfrac{v^2}{c^2}}}$$ (28.4)

The rest energy E_0 is the total energy of an object at rest ($v = 0$ m/s):

Rest energy

$$E_0 = mc^2$$ (28.5) **Ex. 6, 8, 9**

An object's total energy is the sum of its rest energy and its kinetic energy KE, or $E = E_0 + \text{KE}$. Therefore, the kinetic energy is

Kinetic energy

$$\text{KE} = E - E_0 = mc^2 \left(\frac{1}{\sqrt{1 - \dfrac{v^2}{c^2}}} - 1 \right)$$ (28.6) **Ex. 7, 13**

The relativistic total energy and momentum are related according to

Relation between total energy and momentum

$$E^2 = p^2 c^2 + m^2 c^4$$ (28.7)

The ultimate speed Objects with mass cannot attain the speed of light c, which is the ultimate speed for such objects.

28.7 THE RELATIVISTIC ADDITION OF VELOCITIES According to special relativity, the velocity-addition formula specifies how the relative velocities of moving objects are related. For objects that move along the same straight line, this formula is

Velocity-addition formula

$$v_{AB} = \frac{v_{AC} + v_{CB}}{1 + \dfrac{v_{AC} v_{CB}}{c^2}}$$ (28.8) **Ex. 10, 11**

where v_{AB} is the velocity of object A relative to object B, v_{AC} is the velocity of object A relative to object C, and v_{CB} is the velocity of object C relative to object B. The velocities can have positive or negative values, depending on whether they are directed along the positive or negative direction. Furthermore, switching the order of the subscripts changes the sign of the velocity, so that, for example, $v_{BA} = -v_{AB}$.

FOCUS ON CONCEPTS

Note to Instructors: The numbering of the questions shown here reflects the fact that they are only a representative subset of the total number that are available online. However, all of the questions are available for assignment via an online homework management program such as WileyPLUS or WebAssign.

Section 28.1   Events and Inertial Reference Frames

1. Consider a person along with a frame of reference in each of the following situations. In which one or more of the following situations is the frame of reference an inertial frame of reference?   **(a)** The person is oscillating in simple harmonic motion at the end of a bungee cord.   **(b)** The person is in a car going around a circular curve at a constant speed.   **(c)** The person is in a plane that is landing on an aircraft carrier.   **(d)** The person is in the space shuttle during lift-off.   **(e)** None of the above.

Section 28.3   The Relativity of Time: Time Dilation

2. On a highway there is a flashing light to mark the start of a section of the road where work is being done. Who measures the proper time between two flashes of light?   **(a)** A worker standing still on the road

(b) A driver in a car approaching at a constant velocity   **(c)** Both the worker and the driver   **(d)** Neither the worker nor the driver

Section 28.4   The Relativity of Length: Length Contraction

4. Two spacecrafts A and B are moving relative to each other at a constant velocity. Observers in spacecraft A see spacecraft B. Likewise, observers in spacecraft B see spacecraft A. Who sees the proper length of either spacecraft?   **(a)** Observers in spacecraft A see the proper length of spacecraft B.   **(b)** Observers in spacecraft B see the proper length of spacecraft A.   **(c)** Observers in both spacecrafts see the proper length of the other spacecraft.   **(d)** Observers in neither spacecraft see the proper length of the other spacecraft.

6. In a baseball game the batter hits the ball into center field and takes off for first base. The catcher can only stand and watch. Assume that the batter runs at a constant velocity. Who measures the proper time it takes for the runner to reach first base, and who measures the proper length between home plate and first base? **(a)** The catcher measures the proper time, and the runner measures the proper length. **(b)** The runner measures the proper time, and the catcher measures the proper length. **(c)** The catcher measures both the proper time and the proper length. **(d)** The runner measures both the proper time and the proper length.

7. To which one or more of the following situations do the time-dilation and length-contraction equations apply? **(a)** With respect to an inertial frame, two observers have different constant accelerations. **(b)** With respect to an inertial frame, two observers have the same constant acceleration. **(c)** With respect to an inertial frame, two observers are moving with different constant velocities. **(d)** With respect to an inertial frame, one observer has a constant velocity, and another observer has a constant acceleration. **(e)** All of the above.

Section 28.5 Relativistic Momentum

10. Which one of the following statements about linear momentum is true (p = magnitude of the momentum, m = mass, and v = speed)? **(a)** When the magnitude p of the momentum is defined as $p = \dfrac{mv}{\sqrt{1 - v^2/c^2}}$, the linear momentum of an isolated system is conserved only if the speeds of the various parts of the system are very high. **(b)** When the magnitude p of the momentum is defined as $p = mv$, the linear momentum of an isolated system is conserved only if the speeds of the various parts of the system are very high. **(c)** When the magnitude p of the momentum is defined as $p = \dfrac{mv}{\sqrt{1 - v^2/c^2}}$, the linear momentum of an isolated system is conserved no matter what the speeds of the various parts of the system are. **(d)** When the magnitude p of the momentum is defined as $p = mv$, the linear momentum of an isolated system is conserved no matter what the speeds of the various parts of the system are.

11. Which of the following two expressions for the magnitude p of the linear momentum can be used when the speed v of an object of mass m is very small compared to the speed of light c in a vacuum?

A. $p = mv$

B. $p = \dfrac{mv}{\sqrt{1 - \dfrac{v^2}{c^2}}}$

(a) Only A   **(b)** Only B   **(c)** Neither A nor B   **(d)** Both A and B

Section 28.6 The Equivalence of Mass and Energy

13. Consider the following three possibilities for a glass of water at rest on a kitchen counter. The temperature of the water is 0 °C. Rank the mass of the water in descending order (largest first).

A. The water is half liquid and half ice.

B. The water is all liquid.

C. The water is all ice.

(a) C, A, B   **(b)** B, A, C   **(c)** A, C, B   **(d)** B, C, A   **(e)** C, B, A

15. An object has a kinetic energy KE and a potential energy PE. It also has a rest energy E_0. Which one of the following is the correct way to express the object's total energy E? **(a)** $E = $ KE + PE **(b)** $E = E_0 + $ KE **(c)** $E = E_0 + $ KE + PE **(d)** $E = E_0 + $ KE − PE

17. The kinetic energy of an object of mass m is equal to its rest energy. What is the magnitude p of the object's momentum? **(a)** $p = \sqrt{3}mc$ **(b)** $p = 2mc$ **(c)** $p = 4mc$ **(d)** $p = \sqrt{2}mc$ **(e)** $p = 3mc$

Section 28.7 The Relativistic Addition of Velocities

18. Two spaceships are traveling in the same direction. With respect to an inertial frame of reference, spaceship A has a speed of $0.900c$. With respect to the same inertial frame, spaceship B has a speed of $0.500c$. Find the speed v_{AB} of spaceship A relative to spaceship B.

PROBLEMS

Note to Instructors: *Most of the homework problems in this chapter are available for assignment via an online homework management program such as WileyPLUS or WebAssign, and those marked with the icon* **GO** *are presented in WileyPLUS using a guided tutorial format that provides enhanced interactivity. See Preface for additional details.*

Before doing any calculations involving time dilation or length contraction, it is useful to identify which observer measures the proper time interval Δt_0 or the proper length L_0.

ssm Solution is in the Student Solutions Manual.
www Solution is available online at www.wiley.com/college/cutnell

 This icon represents a biomedical application.

Section 28.3 The Relativity of Time: Time Dilation

1. ssm A law enforcement officer in an intergalactic "police car" turns on a red flashing light and sees it generate a flash every 1.5 s. A person on earth measures that the time between flashes is 2.5 s. How fast is the "police car" moving relative to the earth?

2. A particle known as a pion lives for a short time before breaking apart into other particles. Suppose that a pion is moving at a speed of $0.990c$, and an observer who is stationary in a laboratory measures the pion's lifetime to be 3.5×10^{-8} s. **(a)** What is the lifetime according to a hypothetical person who is riding along with the pion? **(b)** According to this hypothetical person, how far does the laboratory move before the pion breaks apart?

3. ssm A Klingon spacecraft has a speed of $0.75c$ with respect to the earth. The Klingons measure 37.0 h for the time interval between two events on the earth. What value for the time interval would they measure if their ship had a speed of $0.94c$ with respect to the earth?

4. GO Suppose that you are traveling on board a spacecraft that is moving with respect to the earth at a speed of $0.975c$. You are breathing at a rate of 8.0 breaths per minute. As monitored on earth, what is your breathing rate?

*5. Interactive Solution 28.5 at **www.wiley.com/college/cutnell** illustrates one way to model this problem. A 6.00-kg object oscillates back and forth at the end of a spring whose spring constant is 76.0 N/m. An observer is traveling at a speed of 1.90×10^8 m/s relative to the fixed end of the spring. What does this observer measure for the period of oscillation?

*6. A spaceship travels at a constant speed from earth to a planet orbiting another star. When the spacecraft arrives, 12 years have elapsed on earth, and 9.2 years have elapsed on board the ship. How far away (in meters) is the planet, according to observers on earth?

**7. As observed on earth, a certain type of bacterium is known to double in number every 24.0 hours. Two cultures of these bacteria are prepared, each consisting initially of one bacterium. One culture is left on earth and the other placed on a rocket that travels at a speed of 0.866c relative to the earth. At a time when the earthbound culture has grown to 256 bacteria, how many bacteria are in the culture on the rocket, according to an earth-based observer?

Section 28.4   The Relativity of Length: Length Contraction

8. A tourist is walking at a speed of 1.3 m/s along a 9.0-km path that follows an old canal. If the speed of light in a vacuum were 3.0 m/s, how long would the path be, according to the tourist?

9. ssm www How fast must a meter stick be moving if its length is observed to shrink to one-half of a meter?

10. Interactive LearningWare 28.1 at **www.wiley.com/college/ cutnell** reviews the concepts that play roles in this problem. The distance from earth to the center of our galaxy is about 23 000 ly (1 ly = 1 light-year = 9.47×10^{15} m), as measured by an earth-based observer. A spaceship is to make this journey at a speed of 0.9990c. According to a clock on board the spaceship, how long will it take to make the trip? Express your answer in years (1 yr = 3.16×10^7 s).

11. ssm A UFO streaks across the sky at a speed of 0.90c relative to the earth. A person on earth determines the length of the UFO to be 230 m along the direction of its motion. What length does the person measure for the UFO when it lands?

12. GO A Martian leaves Mars in a spaceship that is heading to Venus. On the way, the spaceship passes earth with a speed $v = 0.80c$ relative to it. Assume that the three planets do not move relative to each other during the trip. The distance between Mars and Venus is 1.20×10^{11} m, as measured by a person on earth. **(a)** What does the Martian measure for the distance between Mars and Venus? **(b)** What is the time of the trip (in seconds) as measured by the Martian?

13. Interactive Solution 28.13 at **www.wiley/com/college/cutnell** illustrates one approach to solving this problem. A space traveler moving at a speed of 0.70c with respect to the earth makes a trip to a distant star that is stationary relative to the earth. He measures the length of this trip to be 6.5 light-years. What would be the length of this same trip (in light-years) as measured by a traveler moving at a speed of 0.90c with respect to the earth?

14. An unstable high-energy particle is created in the laboratory, and it moves at a speed of 0.990c. Relative to a stationary reference frame fixed to the laboratory, the particle travels a distance of 1.05×10^{-3} m before disintegrating. What are **(a)** the proper distance and **(b)** the distance measured by a hypothetical person traveling with the particle? Determine the particle's **(c)** proper lifetime and **(d)** its dilated lifetime.

*15. As the drawing shows, a carpenter on a space station has constructed a 30.0° ramp. A rocket moves past the space station with a relative speed of 0.730c in a direction parallel to side x. What does a person aboard the rocket measure for the angle of the ramp?

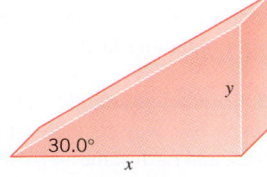

**16. An object is made of glass and has the shape of a cube 0.11 m on a side, according to an observer at rest relative to it. However, an observer moving at high speed parallel to one of the object's edges and knowing that the object's mass is 3.2 kg determines its density to be 7800 kg/m³, which is much greater than the density of glass. What is the moving observer's speed (in units of c) relative to the cube?

17. GO Twins who are 19.0 years of age leave the earth and travel to a distant planet 12.0 light-years away. Assume that the planet and earth are at rest with respect to each other. The twins depart at the same time on different spaceships. One twin travels at a speed of 0.900c, and the other twin travels at 0.500c. **(a) According to the theory of special relativity, what is the difference between their ages when they meet again at the earliest possible time? **(b)** Which twin is older?

Section 28.5   Relativistic Momentum

18. The speed of an ion in a particle accelerator is doubled from 0.460c to 0.920c. The initial relativistic momentum of the ion is 5.08×10^{-17} kg·m/s. Determine **(a)** the mass and **(b)** the final relativistic momentum of the ion.

19. A jetliner has a mass of 1.2×10^5 kg and flies at a speed of 140 m/s. **(a)** Find the magnitude of its momentum. **(b)** If the speed of light in a vacuum had the hypothetical value of 170 m/s, what would be the magnitude of the jetliner's momentum?

20. GO Three particles are listed in the table. The mass and speed of each particle are given as multiples of the variables m and v, which have the values $m = 1.20 \times 10^{-8}$ kg and $v = 0.200c$. The speed of light in a vacuum is $c = 3.00 \times 10^8$ m/s. Determine the momentum for each particle according to special relativity.

Particle	Mass	Speed
a	m	v
b	$\frac{1}{2}m$	$2v$
c	$\frac{1}{4}m$	$4v$

21. ssm A woman is 1.6 m tall and has a mass of 55 kg. She moves past an observer with the direction of the motion parallel to her height. The observer measures her relativistic momentum to have a magnitude of 2.0×10^{10} kg·m/s. What does the observer measure for her height?

22. GO A spacecraft has a nonrelativistic (or classical) momentum whose magnitude is 1.3×10^{13} kg·m/s. The spacecraft moves at such a speed that the pilot measures the proper time interval between two events to be one-half the dilated time interval. Find the relativistic momentum of the spacecraft.

*23. ssm Starting from rest, two skaters push off against each other on smooth level ice, where friction is negligible. One is a woman and one is a man. The woman moves away with a velocity of +2.5 m/s relative to the ice. The mass of the woman is 54 kg, and the mass of the man is 88 kg. Assuming that the speed of light is 3.0 m/s, so that the relativistic momentum must be used, find the recoil velocity of the man relative to the ice. (*Hint: This problem is similar to Example 6 in Chapter 7.*)

Section 28.6   The Equivalence of Mass and Energy

24. An electron and a positron have masses of 9.11×10^{-31} kg. They collide and both vanish, with only electromagnetic radiation appearing after the collision. If each particle is moving at a speed of 0.20c relative to the laboratory before the collision, determine the energy of the electromagnetic radiation.

25. ssm Determine the ratio of the relativistic kinetic energy to the nonrelativistic kinetic energy ($\frac{1}{2}mv^2$) when a particle has a speed of **(a)** $1.00 \times 10^{-3}c$ and **(b)** 0.970c.

26. Review Conceptual Example 9 for background pertinent to this problem. Suppose that the speed of light in a vacuum were one million times smaller than its actual value, so that $c = 3.00 \times 10^2$ m/s.

The spring constant of a spring is 850 N/m. Determine how far you would have to compress the spring from its equilibrium length in order to increase its mass by 0.010 g.

27. Suppose that one gallon of gasoline produces 1.1×10^8 J of energy, and this energy is sufficient to operate a car for twenty miles. An aspirin tablet has a mass of 325 mg. If the aspirin could be converted completely into thermal energy, how many miles could the car go on a single tablet?

28. **GO** Two kilograms of water are changed **(a)** from ice at 0 °C into liquid water at 0 °C and **(b)** from liquid water at 100 °C into steam at 100 °C. For each situation, determine the change in mass of the water.

29. **ssm** How much work must be done on an electron to accelerate it from rest to a speed of $0.990c$?

30. Multiple-Concept Example 6 reviews the principles that play a role in this problem. A nuclear power reactor generates 3.0×10^9 W of power. In one year, what is the change in the mass of the nuclear fuel due to the energy being taken from the reactor?

*31. Multiple-Concept Example 6 explores the approach taken in problems such as this one. Quasars are believed to be the nuclei of galaxies in the early stages of their formation. Suppose that a quasar radiates electromagnetic energy at the rate of 1.0×10^{41} W. At what rate (in kg/s) is the quasar losing mass as a result of this radiation?

*32. **GO** The table gives the total energy and the rest energy for three objects in terms of an energy increment ϵ. For each object, determine the speed as a multiple of the speed c of light in a vacuum.

Object	Total Energy (E)	Rest Energy (E_0)
A	2.00ϵ	ϵ
B	3.00ϵ	ϵ
C	3.00ϵ	2.00ϵ

*33. An object has a total energy of 5.0×10^{15} J and a kinetic energy of 2.0×10^{15} J. What is the magnitude of the object's relativistic momentum?

Section 28.7 The Relativistic Addition of Velocities

34. Galaxy A is moving away from us with a speed of $0.75c$ relative to the earth. Galaxy B is moving away from us in the opposite direction with a relative speed of $0.55c$. Assume that the earth and the galaxies are moving at constant velocities, so they are inertial reference frames. How fast is galaxy A moving according to an observer in galaxy B?

35. **ssm** A spacecraft approaching the earth launches an exploration vehicle. After the launch, an observer on earth sees the spacecraft approaching at a speed of $0.50c$ and the exploration vehicle approaching at a speed of $0.70c$. What is the speed of the exploration vehicle relative to the spaceship?

36. **GO** You are driving down a two-lane country road, and a truck in the opposite lane is traveling toward you. Suppose that the speed of light in a vacuum is $c = 65$ m/s. Determine the speed of the truck relative to you when **(a)** your speed is 25 m/s and the truck's speed is 35 m/s and **(b)** your speed is 5.0 m/s and the truck's speed is 55 m/s. The speeds given in parts (a) and (b) are relative to the ground.

37. The spaceship *Enterprise 1* is moving directly away from earth at a velocity that an earth-based observer measures to be $+0.65c$. A sister ship, *Enterprise 2*, is ahead of *Enterprise 1* and is also moving directly away from earth along the same line. The velocity of *Enterprise 2* relative to *Enterprise 1* is $+0.31c$. What is the velocity of *Enterprise 2*, as measured by the earth-based observer?

*38. A person on earth notices a rocket approaching from the right at a speed of $0.75c$ and another rocket approaching from the left at $0.65c$. What is the relative speed between the two rockets, as measured by a passenger on one of them?

*39. **ssm www** The crew of a rocket that is moving away from the earth launches an escape pod, which they measure to be 45 m long. The pod is launched toward the earth with a speed of $0.55c$ relative to the rocket. After the launch, the rocket's speed relative to the earth is $0.75c$. What is the length of the escape pod as determined by an observer on earth?

*40. **GO** Two identical spaceships are under construction. The constructed length of each spaceship is 1.50 km. After being launched, spaceship A moves away from earth at a constant velocity (speed is $0.850c$) with respect to the earth. Spaceship B follows in the same direction at a different constant velocity (speed is $0.500c$) with respect to the earth. Determine the length that a passenger on one spaceship measures for the other spaceship.

41. Two atomic particles approach each other in a head-on collision. Each particle has a mass of 2.16×10^{-25} kg. The speed of each particle is 2.10×10^8 m/s when measured by an observer standing in the laboratory. **(a) What is the speed of one particle as seen by the other particle? **(b)** Determine the relativistic momentum of one particle, as it would be observed by the other.

ADDITIONAL PROBLEMS

42. Radium is a radioactive element whose nucleus emits an α particle (a helium nucleus) with a kinetic energy of about 7.8×10^{-13} J (4.9 MeV). To what amount of mass is this energy equivalent?

43. **ssm www** Two spaceships A and B are exploring a new planet. Relative to this planet, spaceship A has a speed of $0.60c$, and spaceship B has a speed of $0.80c$. What is the ratio D_A/D_B of the values for the planet's diameter that each spaceship measures in a direction that is parallel to its motion?

44. What is the magnitude of the relativistic momentum of a proton with a relativistic total energy of 2.7×10^{-10} J?

45. Suppose that you are planning a trip in which a spacecraft is to travel at a constant velocity for exactly six months, as measured by a clock on board the spacecraft, and then return home at the same speed. Upon your return, the people on earth will have advanced exactly one hundred years into the future. According to special relativity, how fast must you travel? Express your answer to five significant figures as a multiple of c—for example, 0.955 85c.

*46. Refer to Conceptual Example 11 as an aid in solving this problem. An intergalactic cruiser has two types of guns: a photon cannon that fires a beam of laser light and an ion gun that shoots ions at a velocity of 0.950c relative to the cruiser. The cruiser closes in on an alien spacecraft at a velocity of 0.800c relative to this spacecraft. The captain fires both types of guns. At what velocity do the aliens see **(a)** the laser light and **(b)** the ions approach them? At what velocity do the aliens see **(c)** the laser light and **(d)** the ions move away from the cruiser?

*47. An unstable particle is at rest and suddenly breaks up into two fragments. No external forces act on the particle or its fragments.

One of the fragments has a velocity of +0.800c and a mass of 1.67×10^{-27} kg, and the other has a mass of 5.01×10^{-27} kg. What is the velocity of the more massive fragment? *(Hint: This problem is similar to Example 6 in Chapter 7.)*

*48. How close would two stationary electrons have to be positioned so that their total mass is twice what it is when the electrons are very far apart?

49. ssm www A rectangle has the dimensions of 3.0 m × 2.0 m when viewed by someone at rest with respect to it. When you move past the rectangle along one of its sides, the rectangle looks like a square. What dimensions do you observe when you move at the same speed along the adjacent side of the rectangle?

CHAPTER **29**

PARTICLES AND WAVES

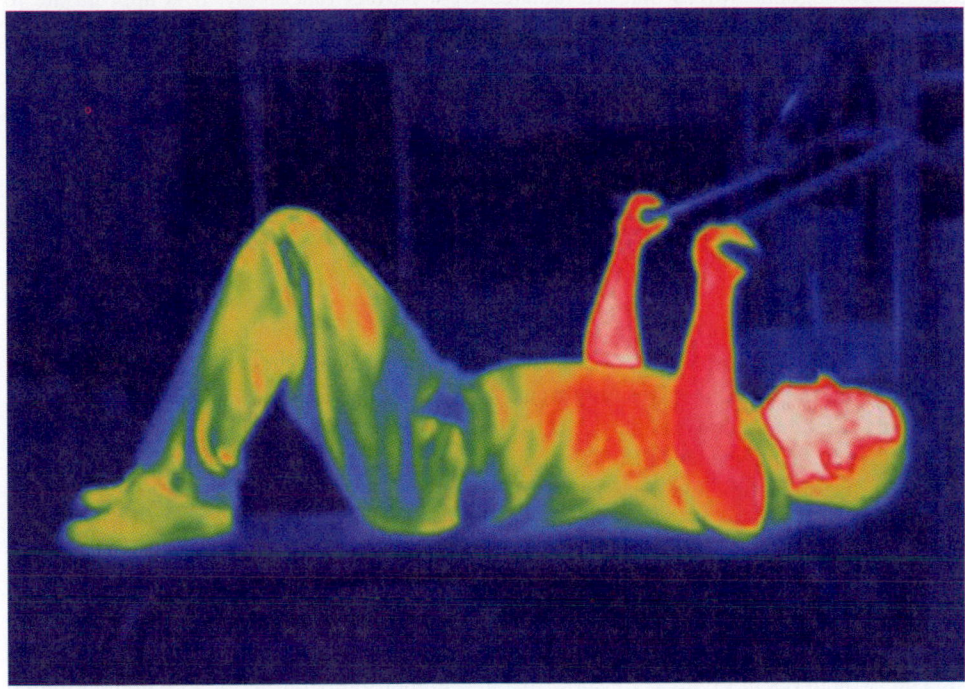

This thermogram shows a bodybuilder lying on a bench press. Thermography is a technique for visualizing temperatures by recording the infrared electromagnetic waves emitted by different parts of the body. The waves are detected electronically and displayed with different colors representing different temperatures from hot (white) to cold (blue). Electromagnetic waves are composed of particle-like entities called photons, one of the subjects of this chapter. (© Ted Kinsman/Photo Researchers, Inc.)

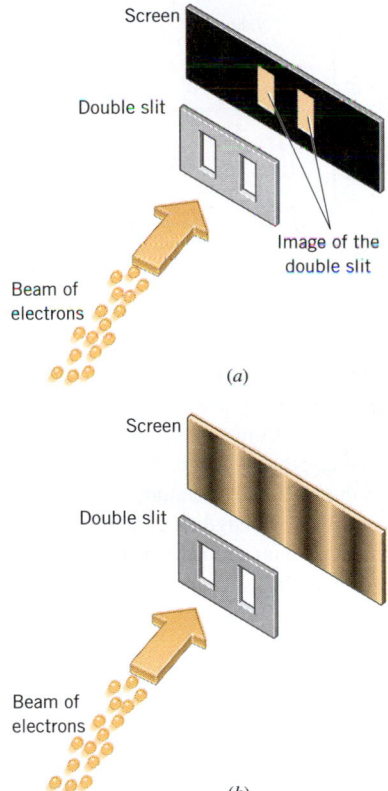

Figure 29.1 (*a*) If electrons behaved as discrete particles with no wave properties, they would pass through one or the other of the two slits and strike the screen, causing it to glow and produce exact images of the slits. (*b*) In reality, the screen reveals a pattern of bright and dark fringes, similar to the pattern produced when a beam of light is used and interference occurs between the light waves coming from each slit.

29.1 THE WAVE–PARTICLE DUALITY

The ability to exhibit interference effects is an essential characteristic of waves. For instance, Section 27.2 discusses Young's famous experiment in which light passes through two closely spaced slits and produces a pattern of bright and dark fringes on a screen (see Figure 27.3). The fringe pattern is a direct indication that interference is occurring between the light waves coming from each slit.

One of the most incredible discoveries of twentieth-century physics is that particles can also behave like waves and exhibit interference effects. For instance, Figure 29.1 shows a version of Young's experiment performed by directing *a beam of electrons* onto a double slit. In this experiment, the screen is like a television screen and glows wherever an electron strikes it. Part *a* of the drawing indicates the pattern that would be seen on the screen if each electron, behaving strictly as a particle, were to pass through one slit or the other and strike the screen. The pattern would consist of an image of each slit. Part *b* shows the pattern actually observed, which consists of bright and dark fringes, reminiscent of what is obtained when light waves pass through a double slit. The fringe pattern indicates that the electrons are exhibiting the interference effects associated with waves.

But how can electrons behave like waves in the experiment shown in Figure 29.1*b*? And what kind of waves are they? The answers to these profound questions will be discussed later in this chapter. For the moment, we intend only to emphasize that the concept of an electron as a tiny discrete particle of matter does not account for the fact that the electron can behave as a wave in some circumstances. In other words, the electron exhibits a dual nature, with both particle-like characteristics and wave-like characteristics.

Here is another interesting question: If a particle can exhibit wave-like properties, can waves exhibit particle-like behavior? As the next three sections reveal, the answer is yes. In fact, experiments that demonstrated the particle-like behavior of waves were performed near the beginning of the twentieth century, before the experiments that demonstrated the wave-like properties of the electrons. Scientists now accept the *wave–particle duality* as an essential part of nature:

> *Waves can exhibit particle-like characteristics, and particles can exhibit wave-like characteristics.*

Section 29.2 begins the remarkable story of the wave–particle duality by discussing the electromagnetic waves that are radiated by a perfect blackbody. It is appropriate to begin with blackbody radiation, because it provided the first link in the chain of experimental evidence leading to our present understanding of the wave–particle duality.

29.2 BLACKBODY RADIATION AND PLANCK'S CONSTANT

All bodies, no matter how hot or cold, continuously radiate electromagnetic waves. For instance, we see the glow of very hot objects because they emit electromagnetic waves in the visible region of the spectrum. Our sun, which has a surface temperature of about 6000 K, appears yellow, while the cooler star Betelgeuse has a red-orange appearance due to its lower surface temperature of 2900 K. However, at relatively low temperatures, cooler objects emit visible light waves only weakly and, as a result, do not appear to be glowing. Certainly the human body, at only 310 K, does not emit enough visible light to be seen in the dark with the unaided eye. But the body does emit electromagnetic waves in the infrared region of the spectrum, and these can be detected with infrared-sensitive devices.

At a given temperature, the intensities of the electromagnetic waves emitted by an object vary from wavelength to wavelength throughout the visible, infrared, and other regions of the spectrum. Figure 29.2 illustrates how the intensity per unit wavelength depends on wavelength for a perfect blackbody emitter. As Section 13.3 discusses, a perfect blackbody at a constant temperature absorbs and reemits all the electromagnetic radiation that falls on it. The two curves in Figure 29.2 show that at a higher temperature the maximum emitted intensity per unit wavelength increases and shifts to shorter wavelengths, toward the visible region of the spectrum. In accounting for the shape of these curves, the German physicist Max Planck (1858–1947) took the first step toward our present understanding of the wave–particle duality.

In 1900 Planck calculated the blackbody radiation curves, using a model that represents a blackbody as a large number of atomic oscillators, each of which emits and absorbs electromagnetic waves. To obtain agreement between the theoretical and experimental curves, Planck assumed that the energy E of an atomic oscillator could have only the discrete values of $E = 0$, hf, $2hf$, $3hf$, and so on. In other words, he assumed that

$$E = nhf \qquad n = 0, 1, 2, 3, \ldots \qquad (29.1)$$

where n is either zero or a positive integer, f is the frequency of vibration (in hertz), and h is a constant now called *Planck's constant.** It has been determined experimentally that Planck's constant has a value of

$$h = 6.626\,068\,96 \times 10^{-34}\ \text{J} \cdot \text{s}$$

The radical feature of Planck's assumption was that the energy of an atomic oscillator could have only discrete values (hf, $2hf$, $3hf$, etc.), with energies in between these values being forbidden. Whenever the energy of a system can have only certain definite values, and nothing in between, the energy is said to be *quantized*. This quantization of the energy was unexpected on the basis of the traditional physics of the time. However, it was soon realized that energy quantization had wide-ranging and valid implications.

Conservation of energy requires that the energy carried off by the radiated electromagnetic waves must equal the energy lost by the atomic oscillators in Planck's model. Suppose, for example, that an oscillator with an energy of $3hf$ emits an electromagnetic wave. According to Equation 29.1, the next smallest allowed value for the energy of the

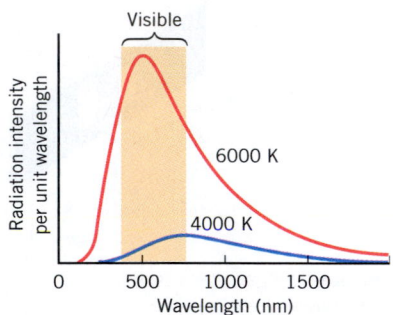

Figure 29.2 The electromagnetic radiation emitted by a perfect blackbody has an intensity per unit wavelength that varies from wavelength to wavelength, as each curve indicates. At the higher temperature, the intensity per unit wavelength is greater, and the maximum occurs at a shorter wavelength.

---

*It is now known that the energy of a harmonic oscillator is $E = (n + \frac{1}{2})hf$; the extra term of $\frac{1}{2}$ is unimportant to the present discussion.

CONCEPTS AT A GLANCE

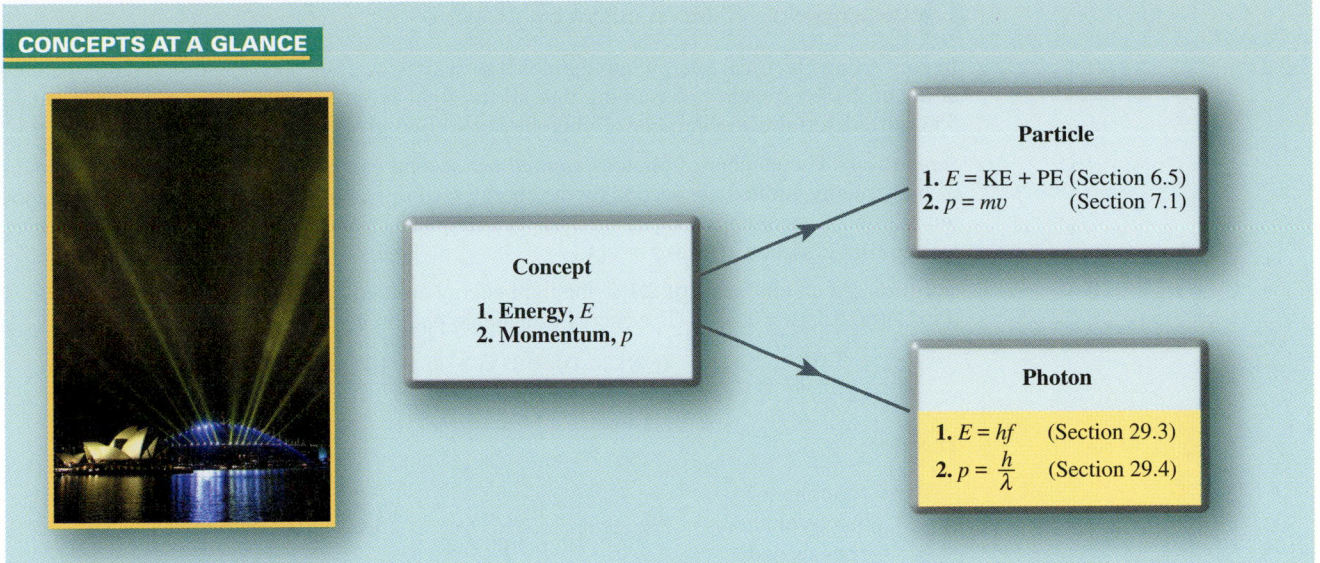

Figure 29.3 CONCEPTS AT A GLANCE
A moving particle has energy E and momentum p. An electromagnetic wave is composed of particle-like entities called photons, each of which also has energy and momentum. Although the spotlight beams in the photograph look like continuous beams of light, each is composed of discrete photons. (© Cameron Spencer/Getty Images News and Sport Services)

oscillator is $2hf$. In such a case, the energy carried off by the electromagnetic wave would have the value of hf, equaling the amount of energy lost by the oscillator. Thus, Planck's model for blackbody radiation sets the stage for the idea that electromagnetic energy occurs as a collection of discrete amounts, or packets, of energy, with the energy of a packet being equal to hf. Einstein made the proposal that light consists of such energy packets.

29.3 PHOTONS AND THE PHOTOELECTRIC EFFECT

▶ **CONCEPTS AT A GLANCE**   The total energy E and the linear momentum $\vec{p}$ are fundamental concepts in physics. We have seen in Chapters 6 and 7 how they apply to moving particles, such as electrons and protons. The total energy of a (nonrelativistic) particle is the sum of its kinetic energy (KE) and potential energy (PE), or $E = KE + PE$. The magnitude p of the particle's momentum is the product of its mass m and speed v, or $p = mv$. These particle concepts are listed in the upper-right portion of the Concepts-at-a-Glance chart in Figure 29.3. We will now discuss the fact that electromagnetic waves are composed of particle-like entities called **photons**, and the lower-right portion of the chart shows that the ideas of energy and momentum also apply to them. However, as we will see, the equations defining photon energy ($E = hf$) and momentum ($p = h/\lambda$) are different from those for a particle, as the chart indicates. ◀

Experimental evidence that light consists of photons comes from a phenomenon called the **photoelectric effect**, in which electrons are emitted from a metal surface when light shines on it. Figure 29.4 illustrates the effect. The electrons are emitted if the light being used has a sufficiently high frequency. The ejected electrons move toward a positive electrode called the *collector* and cause a current to register on the ammeter. Because the electrons are ejected with the aid of light, they are called **photoelectrons**. As will be discussed shortly, a number of features of the photoelectric effect could not be explained solely with the ideas of classical physics.

In 1905 Einstein presented an explanation of the photoelectric effect that took advantage of Planck's work concerning blackbody radiation. It was primarily for his theory of the photoelectric effect that he was awarded the Nobel Prize in physics in 1921. In his photoelectric theory, Einstein proposed that light of frequency f could be regarded as a collection of discrete packets of energy (photons), each packet containing an amount of energy E given by

Energy of a photon
$$E = hf \qquad (29.2)$$

where h is Planck's constant. The light energy given off by a light bulb, for instance, is carried by photons. The brighter the bulb, the greater is the number of photons emitted per second. Example 1 estimates the number of photons emitted per second by a typical light bulb.

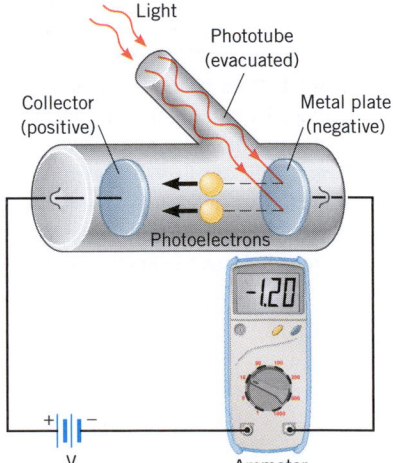

Figure 29.4 In the photoelectric effect, light with a sufficiently high frequency ejects electrons from a metal surface. These photoelectrons, as they are called, are drawn to the positive collector, thus producing a current.

Example 1 Photons from a Light Bulb

In converting electrical energy into light energy, a sixty-watt incandescent light bulb operates at about 2.1% efficiency. Assuming that all the light is green light (vacuum wavelength = 555 nm), determine the number of photons per second given off by the bulb.

Reasoning The number of photons emitted per second can be found by dividing the amount of light energy emitted per second by the energy E of one photon. The energy of a single photon is $E = hf$, according to Equation 29.2. The frequency f of the photon is related to its wavelength λ by Equation 16.1 as $f = c/\lambda$.

Solution At an efficiency of 2.1%, the light energy emitted per second by a sixty-watt bulb is $(0.021)(60.0 \text{ J/s}) = 1.3 \text{ J/s}$. The energy of a single photon is

$$E = hf = \frac{hc}{\lambda} = \frac{(6.63 \times 10^{-34}\text{ J}\cdot\text{s})(3.00 \times 10^8\text{ m/s})}{555 \times 10^{-9}\text{ m}} = 3.58 \times 10^{-19}\text{ J}$$

Therefore,

$$\text{Number of photons emitted per second} = \frac{1.3\text{ J/s}}{3.58 \times 10^{-19}\text{ J/photon}} = \boxed{3.6 \times 10^{18}\text{ photons/s}}$$

According to Einstein, when light shines on a metal, a photon can give up its energy to an electron in the metal. If the photon has enough energy to do the work of removing the electron from the metal, the electron can be ejected. The work required depends on how strongly the electron is held. For the *least strongly* held electrons, the necessary work has a minimum value W_0 and is called the **work function** of the metal. If a photon has energy in excess of the work needed to remove an electron, the excess appears as kinetic energy of the ejected electron. Thus, the least strongly held electrons are ejected with the maximum kinetic energy KE_{max}. Einstein applied the conservation-of-energy principle and proposed the following relation to describe the photoelectric effect:

$$\underbrace{hf}_{\substack{\text{Photon}\\\text{energy}}} = \underbrace{\text{KE}_{max}}_{\substack{\text{Maximum}\\\text{kinetic energy}\\\text{of ejected}\\\text{electron}}} + \underbrace{W_0}_{\substack{\text{Minimum}\\\text{work needed to}\\\text{eject electron}}} \tag{29.3}$$

According to this equation, $\text{KE}_{max} = hf - W_0$, which is plotted in Figure 29.5, with KE_{max} along the y axis and f along the x axis. The graph is a straight line that crosses the x axis at $f = f_0$. At this frequency, the electron departs from the metal with no kinetic energy ($\text{KE}_{max} = 0$ J). According to Equation 29.3, when $\text{KE}_{max} = 0$ J the energy hf_0 of the incident photon is equal to the work function W_0 of the metal: $hf_0 = W_0$.

The photon concept provides an explanation for a number of features of the photoelectric experiment that are difficult to explain without photons. It is observed, for instance, that only light with a frequency above a certain minimum value f_0 will eject electrons. If the frequency is below this value, no electrons are ejected, regardless of how intense the light is. Example 2 determines the minimum frequency value for a silver surface.

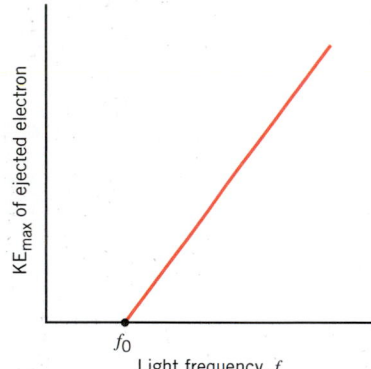

Figure 29.5 Photons can eject electrons from a metal when the light frequency is above a minimum value f_0. For frequencies above this value, ejected electrons have a maximum kinetic energy KE_{max} that is linearly related to the frequency, as the graph shows.

Example 2 The Photoelectric Effect for a Silver Surface

The work function for a silver surface is $W_0 = 4.73$ eV. Find the minimum frequency that light must have to eject electrons from this surface.

Reasoning The minimum frequency f_0 is that frequency at which the photon energy equals the work function W_0 of the metal, so the electron is ejected with zero kinetic energy. Since 1 eV = 1.60×10^{-19} J, the work function expressed in joules is

$$W_0 = (4.73 \text{ eV})\left(\frac{1.60 \times 10^{-19}\text{ J}}{1\text{ eV}}\right) = 7.57 \times 10^{-19}\text{ J}$$

Using Equation 29.3, we find

$$hf_0 = \underbrace{\text{KE}_{max}}_{= 0\text{ J}} + W_0 \quad \text{or} \quad f_0 = \frac{W_0}{h}$$

Problem-solving insight

The work function of a metal is the minimum energy needed to eject an electron from the metal. An electron that has received this minimum energy has no kinetic energy once outside the metal.

Solution The minimum frequency f_0 is

$$f_0 = \frac{W_0}{h} = \frac{7.57 \times 10^{-19} \text{ J}}{6.63 \times 10^{-34} \text{ J} \cdot \text{s}} = \boxed{1.14 \times 10^{15} \text{ Hz}}$$

Photons with frequencies less than f_0 do not have enough energy to eject electrons from a silver surface. Since $\lambda_0 = c/f_0$, the wavelength of this light is $\lambda_0 = 263$ nm, which is in the ultraviolet region of the electromagnetic spectrum.

In Example 2 the electrons are ejected with no kinetic energy, because the light shining on the silver surface has the minimum possible frequency that will eject them. When the frequency of the light exceeds this minimum value, the electrons that are ejected do have kinetic energy. The next example deals with such a situation.

ANALYZING MULTIPLE-CONCEPT PROBLEMS

Example 3 The Maximum Speed of Ejected Photoelectrons

Light with a wavelength of 95 nm shines on a selenium surface, which has a work function of 5.9 eV. The ejected electrons have some kinetic energy. Determine the maximum speed with which electrons are ejected.

Reasoning The maximum speed of the ejected electrons is related to their maximum kinetic energy. Conservation of energy dictates that this maximum kinetic energy is related to the work function of the surface and the energy of the incident photons. The work function is given. The energy of the photons can be obtained from the frequency of the light, which is related to the wavelength.

Knowns and Unknowns We have the following data:

Description	Symbol	Value	Comment
Wavelength of light	λ	95 nm	1 nm $= 10^{-9}$ m
Work function of selenium surface	W_0	5.9 eV	Will be converted to joules.
Unknown Variable			
Maximum speed of photoelectrons	v_{max}	?	

Modeling the Problem

STEP 1 **Kinetic Energy and Speed** The maximum kinetic energy KE_{max} of the ejected electrons is $KE_{max} = \frac{1}{2}mv_{max}^2$, where m is the mass of an electron. Solving for the maximum speed v_{max} gives Equation 1 at the right. The mass of the electron is $m = 9.11 \times 10^{-31}$ kg (see inside of front cover). The maximum kinetic energy is unknown, but we will evaluate it in Step 2.

$$v_{max} = \sqrt{\frac{2KE_{max}}{m}} \quad (1)$$

$\boxed{?}$

STEP 2 **Conservation of Energy** According to the principle of conservation of energy, as expressed by Equation 29.3, we have

$$\underbrace{hf}_{\substack{\text{Photon} \\ \text{energy}}} = \underbrace{KE_{max}}_{\substack{\text{Maximum} \\ \text{kinetic energy} \\ \text{of ejected} \\ \text{electron}}} + \underbrace{W_0}_{\substack{\text{Minimum} \\ \text{work needed to} \\ \text{eject electron}}}$$

where f is the frequency of the light. Solving for KE_{max} gives

$$\boxed{KE_{max} = hf - W_0} \quad (2)$$

which can be substituted into Equation 1, as shown at the right. In this expression the work function W_0 is known, and we will deal with the unknown frequency f in Step 3.

$$v_{max} = \sqrt{\frac{2KE_{max}}{m}} \quad (1)$$

$$\boxed{KE_{max} = hf - W_0} \quad (2)$$

$\boxed{?}$

Continued

STEP 3 **Relationship between Frequency and Wavelength** The frequency and wavelength of the light are related to the speed of light c according to $f\lambda = c$ (Equation 16.1). Solving for the frequency gives

$$f = \frac{c}{\lambda}$$

which we substitute into Equation 2, as shown at the right.

$$v_{max} = \sqrt{\frac{2KE_{max}}{m}} \qquad (1)$$

$$KE_{max} = hf - W_0 \qquad (2)$$

$$f = \frac{c}{\lambda}$$

Solution Combining the results of each step algebraically, we find that

| STEP 1 | STEP 2 | STEP 3 |

$$v_{max} = \sqrt{\frac{2KE_{max}}{m}} = \sqrt{\frac{2(hf - W_0)}{m}} = \sqrt{\frac{2\left(h\dfrac{c}{\lambda} - W_0\right)}{m}}$$

Thus, the maximum speed of the photoelectrons is

$$v_{max} = \sqrt{\frac{2\left(h\dfrac{c}{\lambda} - W_0\right)}{m}}$$

$$v_{max} = \sqrt{\frac{2\left[(6.63 \times 10^{-34}\ \text{J}\cdot\text{s})\dfrac{(3.00 \times 10^8\ \text{m/s})}{(95 \times 10^{-9}\ \text{m})} - (5.9\ \text{eV})\dfrac{(1.60 \times 10^{-19}\ \text{J})}{(1\ \text{eV})}\right]}{9.11 \times 10^{-31}\ \text{kg}}}$$

$$= \boxed{1.6 \times 10^6\ \text{m/s}}$$

Note that in this calculation we have converted the value of the work function from electron volts to joules.

Related Homework: *Problem 11*

Another significant feature of the photoelectric effect is that the maximum kinetic energy of the ejected electrons remains the same when the intensity of the light increases, provided the light frequency remains the same. As the light intensity increases, more photons per second strike the metal, and consequently more electrons per second are ejected. However, since the frequency is the same for each photon, the energy of each photon is also the same. Thus, the ejected electrons always have the same maximum kinetic energy.

Whereas the photon model of light explains the photoelectric effect satisfactorily, the electromagnetic wave model of light does not. Certainly, it is possible to imagine that the electric field of an electromagnetic wave would cause electrons in the metal to oscillate and tear free from the surface when the amplitude of oscillation becomes large enough. However, were this the case, higher-intensity light would eject electrons with a greater maximum kinetic energy, a fact that experiment does not confirm. Moreover, in the electromagnetic wave model, a relatively long time would be required with low-intensity light before the electrons would build up a sufficiently large oscillation amplitude to tear free. Instead, experiment shows that even the weakest light intensity causes electrons to be ejected almost instantaneously, provided the frequency of the light is above the minimum value f_0. The failure of the electromagnetic wave model to explain the photoelectric effect does not mean that the wave model should be abandoned. However, we must recognize that the wave model does not account for all the characteristics of light. The photon model also makes an important contribution to our understanding of the way light behaves when it interacts with matter.

(a)   (b)

Figure 29.6 (a) Digital cameras like this one use an array of charge-coupled devices instead of film to capture an image. (© Travelpix Ltd/Getty Images) (b) Images taken by a digital camera can be easily downloaded to a computer and sent to your friends via the Internet. (© Ross Woodhall/Taxi/Getty Images)

Because a photon has energy, the photon can eject an electron from a metal surface when it interacts with the electron. However, a photon is different from a normal particle. A normal particle has a mass and can travel at speeds up to, but not equal to, the speed of light. A photon, on the other hand, travels at the speed of light in a vacuum and does not exist as an object at rest. The energy of a photon is entirely kinetic in nature, because it has no rest energy and no mass. To show that a photon has no mass, we rewrite Equation 28.4 for the total energy E as

$$E\sqrt{1 - \frac{v^2}{c^2}} = mc^2$$

The term $\sqrt{1 - (v^2/c^2)}$ is zero because a photon travels at the speed of light, $v = c$. Since the energy E of the photon is finite, the left side of the equation above is zero. Thus, the right side must also be zero, so $m = 0$ kg and the photon has no mass.

The physics of charge-coupled devices and digital cameras. One of the most exciting and useful applications of the photoelectric effect is the charge-coupled device (CCD). An array of these devices is used instead of film in digital cameras (see Figure 29.6) to capture images in the form of many small groups of electrons. CCD arrays are also used in digital camcorders and electronic scanners, and they provide the method of choice with which astronomers capture those spectacular images of the planets and the stars. For use with visible light, a CCD array consists of a sandwich of semiconducting silicon, insulating silicon dioxide, and a number of electrodes, as Figure 29.7 shows. The array is divided into many small sections, or pixels, sixteen of which are shown in the drawing. Each pixel captures a small part of a picture. Digital cameras can have up to eight million pixels, depending on price. The greater the number of pixels, the better is the resolution of the photograph. The blow-up in Figure 29.7 shows a single pixel. Incident photons of visible light strike the silicon and generate electrons via the photoelectric effect. The range of energies of the visible photons is such that approximately one electron is released when a photon interacts with a silicon atom. The electrons do not escape from the silicon, but are trapped within a pixel because of a positive voltage applied to the electrodes beneath the insulating layer. Thus, the number of electrons that are released and trapped is proportional to the number of photons striking the pixel. In this fashion, each pixel in the CCD array accumulates an accurate representation of the light intensity at that point on the image. Color information is provided using red, green, or blue filters or a system of prisms to separate the colors. Astronomers use CCD arrays not only in the visible region of the electromagnetic spectrum but in other regions as well.

In addition to trapping the photoelectrons, the electrodes beneath the pixels are used to read out the electron representation of the picture. By changing the positive voltages applied to the electrodes, it is possible to cause all of the electrons trapped in one row of pixels to be transferred to the adjacent row. In this fashion, for instance, row 1 in Figure 29.7 is transferred into row 2, row 2 into row 3, and row 3 into the bottom row, which serves a special purpose. The bottom row functions as a horizontal shift register, from

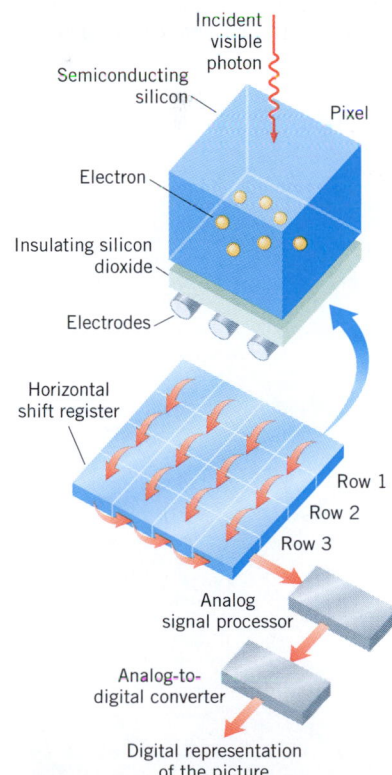

Figure 29.7 A CCD array can be used to capture photographic images using the photoelectric effect.

which the contents of each pixel can be shifted to the right, one at a time, and read into an analog signal processor. This processor senses the varying number of electrons in each pixel in the shift register as a kind of wave that has a fluctuating amplitude. After another shift in rows, the information in the next row is read out, and so forth. The output of the analog signal processor is sent to an analog-to-digital converter, which produces a digital representation of the image in terms of the zeros and ones that computers recognize.

Another application of the photoelectric effect depends on the fact that the moving photoelectrons in Figure 29.4 constitute a current—a current that changes as the intensity of the light changes. All automatic garage door openers have a safety feature that prevents the door from closing when it encounters an obstruction (person, vehicle, etc.). As Figure 29.8 illustrates, a sending unit transmits an invisible (infrared) beam across the opening of the door. The beam is detected by a receiving unit that contains a photodiode. A photodiode is a type of *p-n* junction diode (see Section 23.5). When infrared photons strike the photodiode, electrons bound to the atoms absorb the photons and become liberated. These liberated, mobile electrons cause the current in the photodiode to increase. When a person walks through the beam, the light is momentarily blocked from reaching the receiving unit, and the current in the photodiode decreases. The change in current is sensed by electronic circuitry that immediately stops the downward motion of the door and then causes it to rise up.

Figure 29.9a shows a portion of the Eagle Nebula, a giant star-forming region some 7000 light-years from earth. The photo was taken by the Hubble Space Telescope and reveals clouds of molecular gas and dust, in which there is dramatic evidence of the energy carried by photons. These clouds extend more than a light-year from base to tip and are the birthplace of stars. A star begins to form within a cloud when the gravitational force pulls together sufficient gas to create a high-density "ball." When the gaseous ball becomes sufficiently dense, thermonuclear fusion (see Section 32.5) occurs at its core, and the star begins to shine. The newly born stars are buried within the cloud and cannot be seen from earth. However, the process of photoevaporation allows astronomers to see many of the high-density regions where stars are being formed. Photoevaporation is the process in which high-energy, ultraviolet (UV) photons from hot stars outside the cloud heat it up, much like microwave photons heat food in a microwave oven. Figure 29.9a shows streamers of gas photoevaporating from the cloud as it is illuminated by stars located beyond the photograph's upper edge. As photoevaporation proceeds, globules of gas that are denser than their surroundings are exposed. The globules are known as *evaporating gaseous globules* (EGGs), and they are slightly larger than our solar system. The drawing in part *b* of Figure 29.9 shows that the EGGs shade the gas and dust behind them from the UV photons, creating the many finger-like protrusions seen on the surface of the cloud. Astronomers believe that some of these EGGs contain young stars within them.

The physics of a safety feature of garage door openers.

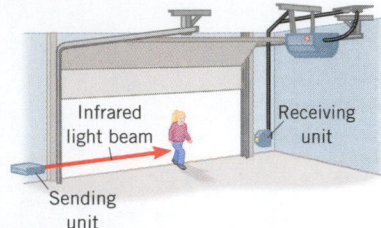

Figure 29.8 When an obstruction prevents the infrared light beam from reaching the receiving unit, the current in the receiving unit drops. This drop in current is detected by an electronic circuit that stops the downward movement of the garage door and then causes it to rise.

The physics of photoevaporation and star formation.

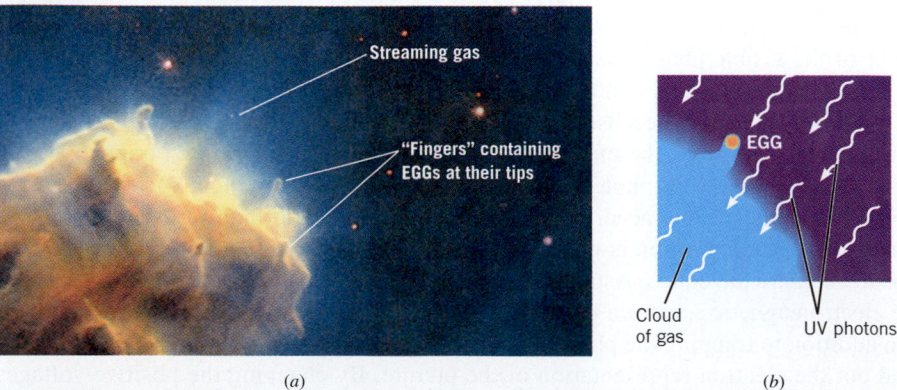

Figure 29.9 (*a*) Photoevaporation produces finger-like projections on the surface of the gas clouds in the Eagle Nebula. At the fingertips are high-density evaporating gaseous globules (EGGs). (Courtesy NASA) (*b*) This drawing illustrates the photoevaporation that is occurring in the photograph in part (*a*).

✓ **CHECK YOUR UNDERSTANDING**

(The answers are given at the end of the book.)

1. The photons emitted by a source of light do *not* all have the same energy. Is the source monochromatic? (A monochromatic light source emits light that has a single wavelength.)

2. Which colored light bulb—red, orange, yellow, green, or violet—emits photons with **(a)** the least energy and **(b)** the greatest energy? (See Example 1 in Chapter 24.)

3. Does a photon emitted by a higher-wattage red light bulb have more energy than a photon emitted by a lower-wattage red bulb?

4. Radiation of a given wavelength causes electrons to be emitted from the surface of metal 1 but not from the surface of metal 2. Why could this be?   **(a)** Metal 1 has a greater work function than metal 2 has.   **(b)** Metal 1 has a smaller work function than metal 2 has.   **(c)** The energy of a photon striking metal 1 is greater than the energy of a photon striking metal 2.

5. In the photoelectric effect, electrons are ejected from the surface of a metal when light shines on it. Which one or more of the following would lead to an increase in the maximum kinetic energy of the ejected electrons?   **(a)** Increasing the frequency of the incident light   **(b)** Increasing the number of photons per second striking the surface   **(c)** Using photons whose frequency f_0 is less than W_0/h, where W_0 is the work function of the metal and h is Planck's constant   **(d)** Selecting a metal that has a greater work function

6. In the photoelectric effect, suppose that the intensity of the light is increased, while the frequency of the light is kept constant. The frequency is greater than the minimum frequency f_0. State whether each of the following will increase, decrease, or remain the same: **(a)** The current in the phototube   **(b)** The number of electrons emitted per second from the metal surface   **(c)** The maximum kinetic energy that an electron could have   **(d)** The maximum momentum that an electron could have

THE MOMENTUM OF A PHOTON AND THE COMPTON EFFECT

29.4

Although Einstein presented his photon model for the photoelectric effect in 1905, it was not until 1923 that the photon concept began to achieve widespread acceptance. It was then that the American physicist Arthur H. Compton (1892–1962) used the photon model to explain his research on the scattering of X-rays by the electrons in graphite. X-rays are high-frequency electromagnetic waves and, like light, they are composed of photons.

Figure 29.10 illustrates what happens when an X-ray photon strikes an electron in a piece of graphite. Like two billiard balls colliding on a pool table, the X-ray photon scatters in one direction after the collision, and the electron recoils in another direction. Compton observed that the scattered photon has a frequency f' that is smaller than the frequency f of the incident photon, indicating that the photon loses energy during the collision. In addition, he found that the difference between the two frequencies depends on the angle θ at which the scattered photon leaves the collision. The phenomenon in which an X-ray photon is scattered from an electron, with the scattered photon having a smaller frequency than the incident photon, is called the *Compton effect.*

In Section 7.3, collisions between two objects are analyzed using the fact that the total kinetic energy and the total linear momentum of the objects are the same before and after an elastic collision. Similar analysis can be applied to the collision between a photon and an electron. The electron is assumed to be initially at rest and essentially free—that is, not bound to the atoms of the material. According to the principle of conservation of energy,

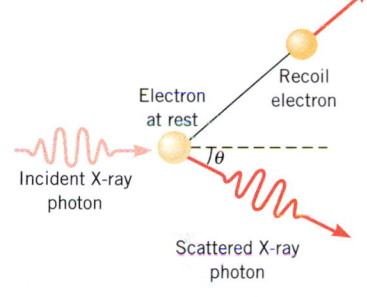

Figure 29.10 In an experiment performed by Arthur H. Compton, an X-ray photon collides with a stationary electron. The scattered photon and the recoil electron depart from the collision in different directions.

$$\underbrace{hf}_{\substack{\text{Energy of}\\\text{incident}\\\text{photon}}} = \underbrace{hf'}_{\substack{\text{Energy of}\\\text{scattered}\\\text{photon}}} + \underbrace{\text{KE}}_{\substack{\text{Kinetic energy}\\\text{of recoil}\\\text{electron}}} \qquad (29.4)$$

where the relation $E = hf$ has been used for the photon energies. It follows, then, that $hf' = hf - \text{KE}$, which shows that the energy and corresponding frequency f' of the scattered photon are less than the energy and frequency of the incident photon, just as Compton observed. Since $\lambda' = c/f'$ (Equation 16.1), the wavelength of the scattered X-rays is larger than that of the incident X-rays.

For an initially stationary electron, conservation of total linear momentum requires that

$$\underset{\substack{\text{Momentum of}\\\text{incident photon}}}{} = \underset{\substack{\text{Momentum of}\\\text{scattered photon}}}{} + \underset{\substack{\text{Momentum of}\\\text{recoil electron}}}{} \qquad (29.5)$$

To find an expression for the magnitude p of the photon's momentum, we use Equations 28.3 and 28.4. According to these equations, the momentum p and the total energy E of any particle are

$$p = \frac{mv}{\sqrt{1 - (v^2/c^2)}} \qquad (28.3)$$

$$E = \frac{mc^2}{\sqrt{1 - (v^2/c^2)}} \qquad (28.4)$$

Rearranging Equation 28.4 to show that $\dfrac{m}{\sqrt{1 - (v^2/c^2)}} = \dfrac{E}{c^2}$ and substituting this result into Equation 28.3 gives

$$p = \frac{mv}{\sqrt{1 - (v^2/c^2)}} = \frac{Ev}{c^2}$$

A photon travels at the speed of light, so that we have $v = c$, and the momentum of a photon is

$$p = \frac{Ev}{c^2} = \frac{Ec}{c^2} = \frac{E}{c}$$

This result only applies to a photon and does not apply to a particle with mass, because such a particle cannot travel at the speed of light. We also know that the energy of a photon is related to its frequency f according to $E = hf$ (Equation 29.2) and that the speed c of a photon is related to its frequency and wavelength λ according to $c = f\lambda$ (Equation 16.1). With these substitutions, our expression for the momentum of a photon becomes

$$p = \frac{E}{c} = \frac{hf}{f\lambda} = \frac{h}{\lambda} \qquad (29.6)$$

Using Equations 29.4, 29.5, and 29.6, Compton showed that the difference between the wavelength λ' of the scattered photon and the wavelength λ of the incident photon is related to the scattering angle θ by

$$\lambda' - \lambda = \frac{h}{mc}(1 - \cos\theta) \qquad (29.7)$$

In this equation m is the mass of the electron. The quantity $h/(mc)$ is called the **Compton wavelength of the electron** and has the value $h/(mc) = 2.43 \times 10^{-12}$ m. Since $\cos\theta$ varies between $+1$ and -1, the shift $\lambda' - \lambda$ in the wavelength can vary between zero and $2h/(mc)$, depending on the value of θ, a fact observed by Compton.

The photoelectric effect and the Compton effect provide compelling evidence that light can exhibit particle-like characteristics attributable to energy packets called photons. But what about the interference phenomena discussed in Chapter 27, such as Young's double-slit experiment and single-slit diffraction, which demonstrate that light behaves as a wave? Does light have two distinct personalities, in which it behaves like a stream of particles in some experiments and like a wave in others? The answer is yes, for physicists now believe that this wave–particle duality is an inherent property of light. Light is a far more interesting (and complex) phenomenon than just a stream of particles or an electromagnetic wave.

In the Compton effect the electron recoils because it gains some of the photon's momentum. In principle, then, the momentum that photons have can be used to make other objects move. Conceptual Example 4 considers a propulsion system for an interstellar spaceship that is based on the momentum of a photon.

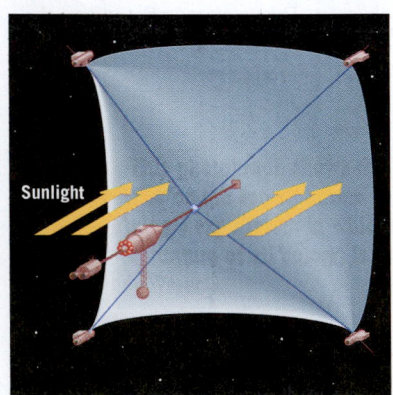

Figure 29.11 A solar sail provides the propulsion for this interstellar spaceship.

The physics of
solar sails and spaceship propulsion.

⬇ Conceptual Example 4 Solar Sails and the Propulsion of Spaceships

One propulsion method that is currently being studied for interstellar travel uses a large sail. The intent is that sunlight striking the sail will create a force that pushes the ship away from the sun (Figure 29.11), much as the wind propels a sailboat. To get the greatest possible force, the surface of the sail facing the sun **(a)** should be shiny like a mirror, **(b)** should be black, or **(c)** could be either shiny or black, since the same force will be created for either type of surface.

Reasoning In Conceptual Example 3 in Chapter 7, we found that hailstones striking the roof of a car exert a force on it because the collision changes their momentum. Photons also have momentum, so, like the hailstones, they can apply a force to the sail.

As in Chapter 7, we will be guided by the impulse–momentum theorem (Equation 7.4) in assessing the force. This theorem states that when a net force acts on an object, the impulse of the net force is equal to the change in momentum of the object. Greater impulses lead to greater forces for a given time interval. Thus, when a photon collides with the sail, the photon's momentum changes because of the force that the sail applies to the photon. Newton's action–reaction law (Section 4.5) indicates that the photon simultaneously applies a force of equal magnitude, but opposite direction, to the sail. It is this reaction force that propels the spaceship, and it will be greater when the momentum change experienced by the photon is greater. The surface of the sail facing the sun, then, should be such that it causes the largest possible momentum change for the impinging photons.

Answers (b) and (c) are incorrect. In Conceptual Example 3 in Chapter 7 we further examined whether hailstones or raindrops exert the greater force when they strike the roof of a car. The difference is that hailstones, being hard objects, bounce off the roof, while raindrops splatter and do not bounce very much. We concluded that hailstones, because of the bounce, experience a greater change in momentum than raindrops do, so the roof exerts a greater force on the hailstones. By Newton's action–reaction law, the car roof, then, experiences a greater force from the hailstones than from the raindrops. We saw in Section 13.3 that radiation is reflected from a shiny mirror-like surface and is absorbed by a black surface. Therefore, by analogy with raindrops that stick to the car roof, the sail experiences a smaller force when the surface facing the sun is black.

Answer (a) is correct. The sun's radiation reflects from a shiny mirror-like surface and is absorbed by a black surface. Now, consider a photon that strikes the sail perpendicularly. When a photon reflects from a mirror-like surface, the photon's momentum changes from its value in the forward direction to a value of the same magnitude in the reverse direction. This is a greater change than the one that occurs when the photon is absorbed by a black surface. In the latter case, the momentum changes only from its value in the forward direction to a value of zero. Consequently, the surface of the sail facing the sun should be shiny in order to produce the greatest possible propulsion force. A shiny surface causes the photons to bounce like hailstones on the roof of a car and, in doing so, to apply a greater force to the sail.

Related Homework: *Problem 24.*

✓ CHECK YOUR UNDERSTANDING

(The answers are given at the end of the book.)

7. In the Compton effect, an incident X-ray photon of wavelength λ is scattered by an electron, and the scattered photon has a wavelength of λ'. Suppose that the incident photon is scattered by a proton instead of an electron. For a given scattering angle θ, is the change $\lambda' - \lambda$ in the wavelength of the photon scattered by the proton greater than, less than, or the same as the wavelength of the photon scattered by the electron? *(Hint: Use Equation 29.7 for a proton instead of an electron.)*

8. In a Compton scattering experiment, an incident X-ray photon is traveling along the $+x$ direction. An electron, initially at rest, is struck by the photon and is accelerated straight ahead in the same direction as the incident X-ray photon. Which way does the scattered photon move? **(a)** Along the $+y$ direction **(b)** Along the $-y$ direction **(c)** Along the $-x$ direction *(Hint: Use the principle of conservation of momentum to guide your reasoning.)*

9. The speed of a particle is much less than the speed of light. Thus, when the particle's speed doubles, the particle's momentum doubles, and its kinetic energy becomes four times greater. However, when the momentum of a photon doubles, its energy becomes **(a)** two times greater **(b)** four times greater **(c)** one-half as much **(d)** one-fourth as much.

10. Review Conceptual Example 4 as background for this question. The photograph shows a device called a radiometer. The four regular panels are black on one side and shiny like a mirror on the other side. In bright light, the panel arrangement spins around in a direction from the black side of a panel toward the shiny side. Do photon collisions with both sides of the panels cause the observed spinning?

(Courtesy Sargent-Welch Scientific Company)

Question 10

CONCEPTS AT A GLANCE

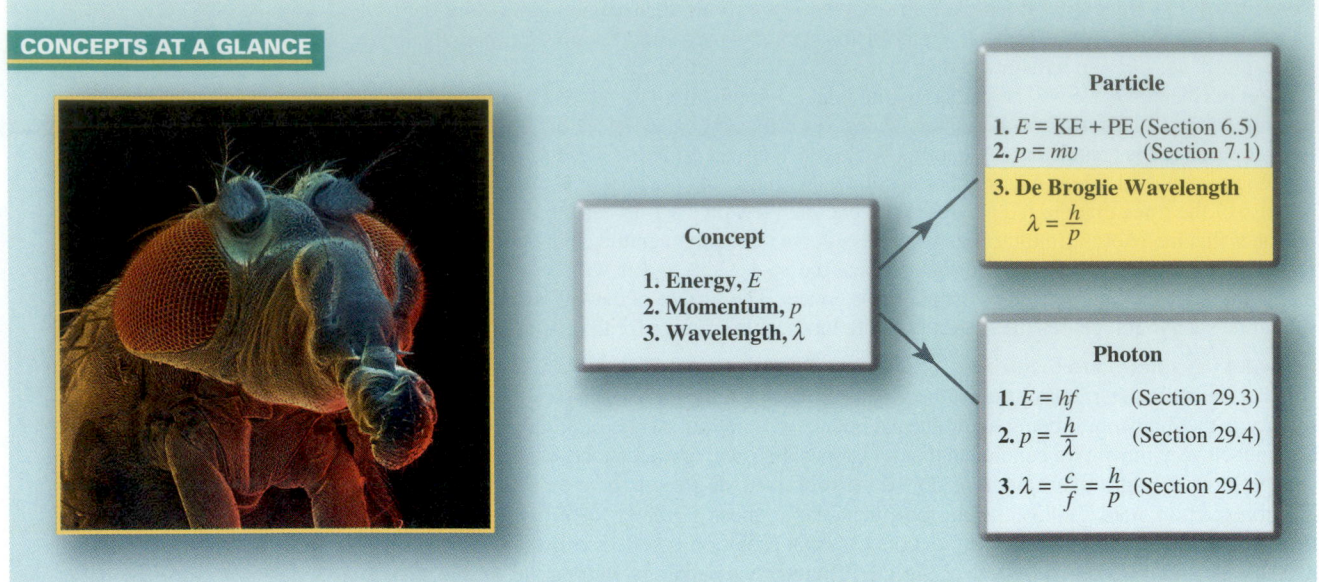

Concept

1. Energy, E
2. Momentum, p
3. Wavelength, λ

Particle

1. $E = KE + PE$ (Section 6.5)
2. $p = mv$ (Section 7.1)
3. **De Broglie Wavelength**
   $\lambda = \dfrac{h}{p}$

Photon

1. $E = hf$ (Section 29.3)
2. $p = \dfrac{h}{\lambda}$ (Section 29.4)
3. $\lambda = \dfrac{c}{f} = \dfrac{h}{p}$ (Section 29.4)

Figure 29.12 CONCEPTS AT A GLANCE Both a moving particle and a wave possess energy, momentum, and a wavelength. The photograph shows a highly magnified view of a *Drosophila* fruit fly, made with a scanning electron microscope. This microscope uses electrons instead of light. The resolution of the fine detail is exceptional because the wavelength of an electron can be made much smaller than that of visible light. (© David Scharf/SPL/Photo Researchers, Inc.)

(a)

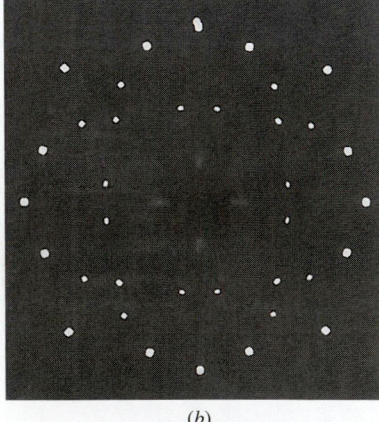

(b)

Figure 29.13 (*a*) The neutron diffraction pattern and (*b*) the X-ray diffraction pattern for a crystal of sodium chloride (NaCl). (*a*, From Wollan, Shull, and Marney, *Phys. Rev.* 73 (5): 527, 1948; *b*, Courtesy Edwin Jones, University of South Carolina)

29.5 THE DE BROGLIE WAVELENGTH AND THE WAVE NATURE OF MATTER

▶ CONCEPTS AT A GLANCE As a graduate student in 1923, Louis de Broglie (1892–1987) made the astounding suggestion that since light waves could exhibit particle-like behavior, particles of matter should exhibit wave-like behavior. De Broglie proposed that all moving matter has a wavelength associated with it, just as a wave does. The Concepts-at-a-Glance chart in Figure 29.12, which is a continuation of the chart in Figure 29.3, shows that the notions of energy, momentum, and wavelength are applicable to particles as well as to waves. ◀

De Broglie made the explicit proposal that the wavelength λ of a particle is given by the same relation (Equation 29.6) that applies to a photon:

De Broglie wavelength
$$\lambda = \frac{h}{p} \qquad (29.8)$$

where h is Planck's constant and p is the magnitude of the relativistic momentum of the particle. Today, λ is known as the ***de Broglie wavelength*** of the particle.

Confirmation of de Broglie's suggestion came in 1927 from the experiments of the American physicists Clinton J. Davisson (1881–1958) and Lester H. Germer (1896–1971) and, independently, those of the English physicist George P. Thomson (1892–1975). Davisson and Germer directed a beam of electrons onto a crystal of nickel and observed that the electrons exhibited a diffraction behavior, analogous to that seen when X-rays are diffracted by a crystal (see Section 27.9 for a discussion of X-ray diffraction). The wavelength of the electrons revealed by the diffraction pattern matched that predicted by de Broglie's hypothesis, $\lambda = h/p$. More recently, Young's double-slit experiment has been performed with electrons and reveals the effects of wave interference illustrated in Figure 29.1.

Particles other than electrons can also exhibit wave-like properties. For instance, neutrons are sometimes used in diffraction studies of crystal structure. Figure 29.13 compares the neutron diffraction pattern and the X-ray diffraction pattern caused by a crystal of rock salt (NaCl).

Although all moving particles have a de Broglie wavelength, the effects of this wavelength are observable only for particles whose masses are very small, on the order of the mass of an electron or a neutron, for instance. Example 5 illustrates why.

Example 5 The De Broglie Wavelength of an Electron and of a Baseball

Determine the de Broglie wavelength for **(a)** an electron (mass = 9.1×10^{-31} kg) moving at a speed of 6.0×10^6 m/s and **(b)** a baseball (mass = 0.15 kg) moving at a speed of 13 m/s.

Reasoning In each case, the de Broglie wavelength is given by Equation 29.8 as Planck's constant divided by the magnitude of the momentum. Since the speeds are small compared to the speed of light, we can ignore relativistic effects and express the magnitude of the momentum as the product of the mass and the speed.

Solution **(a)** Since the magnitude p of the momentum is the product of the mass m of the particle and its speed v, we have $p = mv$. Using this expression in Equation 29.8 for the de Broglie wavelength, we obtain

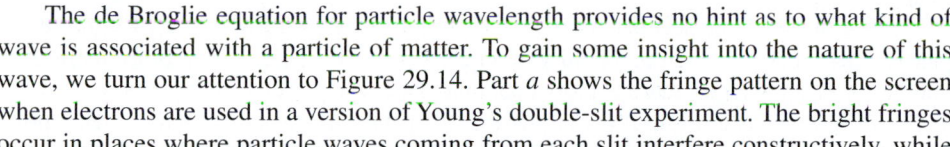

$$\lambda = \frac{h}{p} = \frac{h}{mv} = \frac{6.63 \times 10^{-34}\ \text{J}\cdot\text{s}}{(9.1 \times 10^{-31}\ \text{kg})(6.0 \times 10^6\ \text{m/s})} = \boxed{1.2 \times 10^{-10}\ \text{m}}$$

A de Broglie wavelength of 1.2×10^{-10} m is about the size of the interatomic spacing in a solid, such as the nickel crystal used by Davisson and Germer, and, therefore, leads to the observed diffraction effects.

(b) A calculation similar to that in part (a) shows that the de Broglie wavelength of the baseball is $\boxed{\lambda = 3.3 \times 10^{-34}\ \text{m}}$. This wavelength is incredibly small, even by comparison with the size of an atom (10^{-10} m) or a nucleus (10^{-14} m). Thus, the ratio λ/W of this wavelength to the width W of an ordinary opening, such as a window, is so small that the diffraction of a baseball passing through the window cannot be observed.

The de Broglie equation for particle wavelength provides no hint as to what kind of wave is associated with a particle of matter. To gain some insight into the nature of this wave, we turn our attention to Figure 29.14. Part *a* shows the fringe pattern on the screen when electrons are used in a version of Young's double-slit experiment. The bright fringes occur in places where particle waves coming from each slit interfere constructively, while the dark fringes occur in places where the waves interfere destructively.

When an electron passes through the double-slit arrangement and strikes a spot on the screen, the screen glows at that spot, and parts *b*, *c*, and *d* of Figure 29.14 illustrate how the spots accumulate in time. As more and more electrons strike the screen, the spots eventually form the fringe pattern that is evident in part *d*. Bright fringes occur where there is a high probability of electrons striking the screen, and dark fringes occur where there is a low probability. Here lies the key to understanding particle waves. *Particle waves are waves of probability,* waves whose magnitude at a point in space gives an indication of the probability that the particle will be found at that point. At the place where the screen is located, the pattern of probabilities conveyed by the particle waves causes the fringe pattern to emerge. The fact that no fringe pattern is apparent in part *b* of the figure does not mean that there are no probability waves present; it just means that too few electrons have struck the screen for the pattern to be recognizable.

The pattern of probabilities that leads to the fringes in Figure 29.14 is analogous to the pattern of light intensities that is responsible for the fringes in Young's original experiment with light waves (see Figure 27.3). Section 24.4 discusses the fact that the intensity of the light is proportional to either the square of the electric field strength or the square of the magnetic field strength of the wave. In an analogous fashion in the case of particle waves, the probability is proportional to the square of the magnitude Ψ (Greek letter Psi) of the wave. Ψ is referred to as the *wave function* of the particle.

In 1925 the Austrian physicist Erwin Schrödinger (1887–1961) and the German physicist Werner Heisenberg (1901–1976) independently developed theoretical frameworks for determining the wave function. In so doing, they established a new branch of physics called *quantum mechanics.* The word "quantum" refers to the fact that in the world of the atom, where particle waves must be considered, the particle energy is quantized, so only certain energies are allowed. To understand the structure of the atom and the phenomena related to it, quantum mechanics is essential, and the Schrödinger equation for calculating the wave function is now widely used. In the next chapter, we will explore the structure of the atom based on the ideas of quantum mechanics.

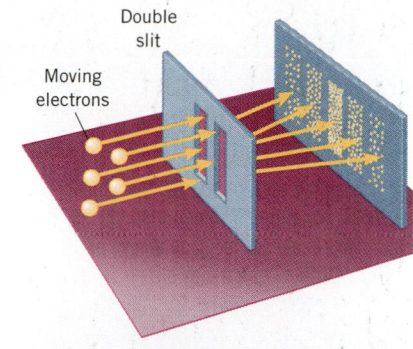

(a)

(b) After 100 electrons

(c) After 3000 electrons

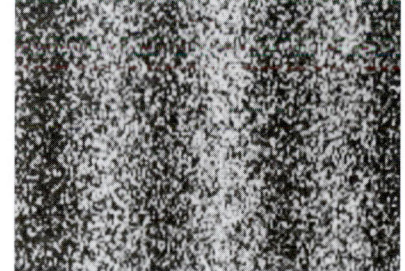

(d) After 70 000 electrons

Figure 29.14 In this electron version of Young's double-slit experiment, the characteristic fringe pattern becomes recognizable only after a sufficient number of electrons have struck the screen. (From A. Tonomura, J. Endo, T. Matsuda, T. Kawasaki, and H. Ezawa *Am. J. Phys.* 57 (2): 117, Feb. 1989.)

11. A stone is dropped from the top of a building. As the stone falls, does its de Broglie wavelength increase, decrease, or remain the same?

12. An electron and a neutron have different masses. Is it possible, according to Equation 29.8, that they can have the same de Broglie wavelength? **(a)** Yes, provided the magnitudes of their momenta are different. **(b)** Yes, provided their speeds are different. **(c)** No; two particles with different masses always have different de Broglie wavelengths.

13. In Figure 29.1*b*, replace the electrons with protons that have the same speed. With the aid of Equation 27.1 for the bright fringes in Young's double-slit experiment and Equation 29.8, decide whether the angular separation between the fringes would increase, decrease, or remain the same, relative to the angular separation produced by the electrons.

14. A beam of electrons passes through a single slit, and a beam of protons passes through a second, but identical, slit. The electrons and the protons have the same speed. Which one of the following correctly describes the beam that experiences the greatest amount of diffraction? **(a)** The electrons, because they have the smaller momentum and, hence, the smaller de Broglie wavelength **(b)** The electrons, because they have the smaller momentum and, hence, the larger de Broglie wavelength **(c)** The protons, because they have the smaller momentum and, hence, the smaller de Broglie wavelength **(d)** The protons, because they have the larger momentum and, hence, the smaller de Broglie wavelength **(e)** Both beams experience the same amount of diffraction, because the electrons and protons have the same de Broglie wavelength.

29.6 THE HEISENBERG UNCERTAINTY PRINCIPLE

As the previous section discusses, the bright fringes in Figure 29.14 indicate the places where there is a high probability of an electron striking the screen. Since there are a number of bright fringes, there is more than one place where each electron has some probability of hitting. As a result, it is not possible to specify in advance exactly where on the screen an individual electron will hit. All we can do is speak of the probability that the electron may end up in a number of different places. No longer is it possible to say, as Newton's laws would suggest, that a single electron, fired through the double slit, will travel directly forward in a straight line and strike the screen. This simple model just does not apply when a particle as small as an electron passes through a pair of closely spaced narrow slits. Because the wave nature of particles is important in such circumstances, we lose the ability to predict with 100% certainty the path that a single particle will follow. Instead, only the average behavior of large numbers of particles is predictable, and the behavior of any individual particle is uncertain.

To see more clearly into the nature of the uncertainty, consider electrons passing through a single slit, as in Figure 29.15. After a sufficient number of electrons strike the screen, a diffraction pattern emerges. The electron diffraction pattern consists of alternating bright and dark fringes and is analogous to the pattern for light waves shown in Figure 27.21. Figure 29.15 shows the slit and locates the first dark fringe on either side of the

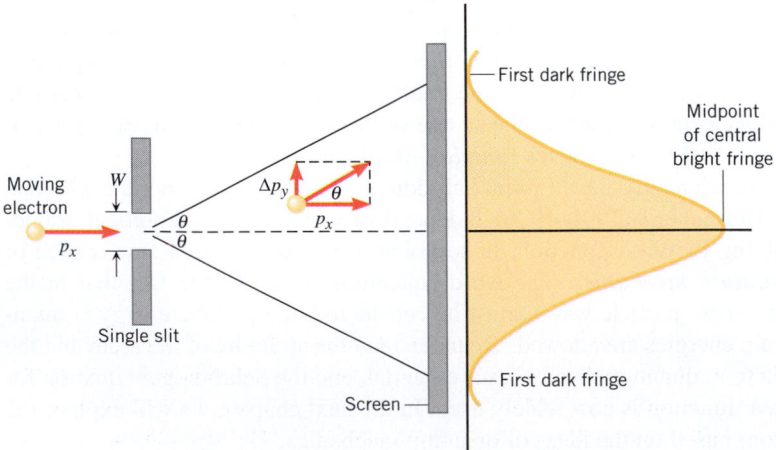

Figure 29.15 When a sufficient number of electrons pass through a single slit and strike the screen, a diffraction pattern of bright and dark fringes emerges. (Only the central bright fringe is shown.) This pattern is due to the wave nature of the electrons and is analogous to that produced by light waves.

central bright fringe. The central fringe is bright because electrons strike the screen over the entire region between the dark fringes. If the electrons striking the screen outside the central bright fringe are neglected, the extent to which the electrons are diffracted is given by the angle θ in the drawing. To reach locations within the central bright fringe, some electrons must have acquired momentum in the y direction, despite the fact that they enter the slit traveling along the x direction and have no momentum in the y direction to start with. The figure illustrates that the y component of the momentum may be as large as Δp_y. The notation Δp_y indicates the difference between the maximum value of the y component of the momentum after the electron passes through the slit and its value of zero before the electron passes through the slit. Δp_y represents the *uncertainty* in the y component of the momentum in that the y component may have any value from zero to Δp_y.

It is possible to relate Δp_y to the width W of the slit. To do this, we assume that Equation 27.4, which applies to light waves, also applies to particle waves whose de Broglie wavelength is λ. This equation, $\sin \theta = \lambda/W$, specifies the angle θ that locates the first dark fringe. If θ is small, then $\sin \theta \approx \tan \theta$. Moreover, Figure 29.15 indicates that $\tan \theta = \Delta p_y/p_x$, where p_x is the x component of the momentum of the electron. Therefore, $\Delta p_y/p_x \approx \lambda/W$. However, $p_x = h/\lambda$ according to de Broglie's equation, so that

$$\frac{\Delta p_y}{p_x} = \frac{\Delta p_y}{h/\lambda} \approx \frac{\lambda}{W}$$

As a result,

$$\Delta p_y \approx \frac{h}{W} \tag{29.9}$$

which indicates that a smaller slit width leads to a larger uncertainty in the y component of the electron's momentum.

It was Heisenberg who first suggested that the uncertainty Δp_y in the y component of the momentum is related to the uncertainty in the y position of the electron as the electron passes through the slit. To get a feel for this relationship, let's assume that the center of the slit is at $y = 0$ m. Because the width of the slit is W, the electron is somewhere within $\pm\frac{1}{2}W$ from the center of the slit. Thus, we take the uncertainty in the y position of the electron to be $\Delta y = \frac{1}{2}W$, so that $W = 2\,\Delta y$. Substituting this relation into Equation 29.9 shows that $\Delta p_y \approx h/(2\,\Delta y)$ or $(\Delta p_y)(\Delta y) \approx \frac{1}{2}h$. The result of Heisenberg's more complete analysis is given below in Equation 29.10 and is known as the ***Heisenberg uncertainty principle.*** Note that the Heisenberg principle is a general principle with wide applicability. It does not just apply to the case of single-slit diffraction, which we have used here for the sake of convenience.

THE HEISENBERG UNCERTAINTY PRINCIPLE

Momentum and position

$$(\Delta p_y)(\Delta y) \geq \frac{h}{4\pi} \tag{29.10}$$

Δy = uncertainty in a particle's position along the y direction
Δp_y = uncertainty in the y component of the linear momentum of the particle

Energy and time

$$(\Delta E)(\Delta t) \geq \frac{h}{4\pi} \tag{29.11}$$

ΔE = uncertainty in the energy of a particle when the particle is in a certain state
Δt = time interval during which the particle is in the state

The Heisenberg uncertainty principle places limits on the accuracy with which the momentum and position of a particle can be specified simultaneously. These limits are not just limits due to faulty measuring techniques. They are fundamental limits imposed by nature, and there is no way to circumvent them. Equation 29.10 indicates that Δp_y and Δy

cannot both be arbitrarily small at the same time. If one is small, then the other must be large, so that their product equals or exceeds (≥) Planck's constant divided by 4π. For example, if the position of a particle is known exactly, so that Δy is zero, then Δp_y is an infinitely large number, and the momentum of the particle is completely uncertain. Conversely, if we assume that Δp_y is zero, then Δy is an infinitely large number, and the position of the particle is completely uncertain. In other words, the Heisenberg uncertainty principle states that it is impossible to specify precisely both the momentum and position of a particle at the same time.

There is also an uncertainty principle that deals with energy and time, as expressed by Equation 29.11. The product of the uncertainty ΔE in the energy of a particle and the time interval Δt during which the particle remains in a given energy state is greater than or equal to Planck's constant divided by 4π. Therefore, the shorter the lifetime of a particle in a given state, the greater is the uncertainty in the energy of that state.

Example 6 shows that the uncertainty principle has significant consequences for the motion of tiny particles such as electrons but has little effect on the motion of macroscopic objects, even those with as little mass as a Ping-Pong ball.

 Example 6 The Heisenberg Uncertainty Principle

Assume that the position of an object is known so precisely that the uncertainty in the position is only $\Delta y = 1.5 \times 10^{-11}$ m. **(a)** Determine the minimum uncertainty in the momentum of the object. Find the corresponding minimum uncertainty in the speed of the object in the case when the object is **(b)** an electron (mass = 9.1×10^{-31} kg) and **(c)** a Ping-Pong ball (mass = 2.2×10^{-3} kg).

Problem-solving insight

The Heisenberg uncertainty principle states that the product of Δp_y and Δy is greater than or equal to $h/4\pi$. For a given value of Δp_y or Δy, the minimum uncertainty in the other term occurs when the product is equal to $h/4\pi$.

Reasoning The minimum uncertainty Δp_y in the y component of the momentum is given by the Heisenberg uncertainty principle as $\Delta p_y = h/(4\pi \Delta y)$, where Δy is the uncertainty in the position of the object along the y direction. Both the electron and the Ping-Pong ball have the same uncertainty in their momenta because they have the same uncertainty in their positions. However, these objects have very different masses. As a result, we will find that the uncertainty in the speeds of these objects is very different.

Solution **(a)** The minimum uncertainty in the y component of the momentum is

$$\Delta p_y = \frac{h}{4\pi \Delta y} = \frac{6.63 \times 10^{-34} \text{ J}\cdot\text{s}}{4\pi(1.5 \times 10^{-11} \text{ m})} = \boxed{3.5 \times 10^{-24} \text{ kg}\cdot\text{m/s}} \qquad (29.10)$$

(b) Since $\Delta p_y = m \Delta v_y$, the minimum uncertainty in the speed of the electron is

$$\Delta v_y = \frac{\Delta p_y}{m} = \frac{3.5 \times 10^{-24} \text{ kg}\cdot\text{m/s}}{9.1 \times 10^{-31} \text{ kg}} = \boxed{3.8 \times 10^6 \text{ m/s}}$$

Thus, the small uncertainty in the y position of the electron gives rise to a large uncertainty in the speed of the electron.

(c) The uncertainty in the speed of the Ping-Pong ball is

$$\Delta v_y = \frac{\Delta p_y}{m} = \frac{3.5 \times 10^{-24} \text{ kg}\cdot\text{m/s}}{2.2 \times 10^{-3} \text{ kg}} = \boxed{1.6 \times 10^{-21} \text{ m/s}}$$

Because the mass of the Ping-Pong ball is relatively large, the small uncertainty in its y position gives rise to an uncertainty in its speed that is much smaller than that for the electron. Thus, in contrast to the electron, we can know simultaneously where the ball is and how fast it is moving, to a very high degree of certainty.

Example 6 emphasizes how the uncertainty principle imposes different uncertainties on the speeds of an electron (small mass) and a Ping-Pong ball (large mass). For objects like the ball, which have relatively large masses, the uncertainties in position and speed are so small that they have no effect on our ability to determine simultaneously where such objects are and how fast they are moving. The uncertainties calculated in Example 6 depend on more than just the mass, however. They also depend on Planck's constant, which is a very small number. It is interesting to speculate about what life would be like if Planck's constant were much larger than 6.63×10^{-34} J · s. Conceptual Example 7 deals with just such speculation.

Conceptual Example 7 What If Planck's Constant Were Large?

Suppose that you are target shooting at a stationary target. A bullet leaving the barrel of a gun is analogous to an electron passing through the single slit in Figure 29.15. With this analogy in mind and assuming that the magnitude of the bullet's momentum is not abnormally large, what would target shooting be like if Planck's constant had a relatively large value instead of its extremely small value of 6.63×10^{-34} J·s? **(a)** It would be more accurate because there would be less uncertainty in where the bullet hits the target. **(b)** It would be less accurate because there would be greater uncertainty in where the bullet hits the target. **(c)** There would be no difference.

Reasoning Let's assume that the bullet is moving down the barrel of the gun in the $+x$ direction and that the target lies on the x axis. When it exits the barrel, the bullet—like the electron passing through a single slit—acquires a momentum component that is perpendicular (in the y direction) to the barrel. This happens even though inside the barrel the bullet travels only along the x direction and has no momentum component in the y direction. Analogous to the discussion surrounding Figure 29.15, the y component of the momentum may be as large as Δp_y, where Δp_y indicates the difference between the maximum value of the y component of the momentum after the bullet leaves the barrel and its value of zero while the bullet is in the barrel. Δp_y is related to Planck's constant h and the diameter W of the barrel opening via the relation $\Delta p_y \approx h/W$ (Equation 29.9). Since we are now postulating that Planck's constant is large, Δp_y is also large.

Answers (a) and (c) are incorrect. Target shooting becomes more accurate if Planck's constant becomes smaller, not larger. Here's the reason: Inside the barrel the bullet is moving in the $+x$ direction. However, upon exiting the barrel, the bullet acquires a momentum component Δp_y in the y direction and begins to deviate from its original path by moving in the y direction. According to $\Delta p_y \approx h/W$ (Equation 29.9), the smaller the value of h, the smaller is Δp_y. If, in the extreme limit, Planck's constant were zero, Δp_y would also be zero, and the bullet would move only in the $+x$ direction and, thus, would hit the target.

Answer (b) is correct. If the bullet, after leaving the barrel, had only a momentum component that was parallel to the barrel, the bullet would strike the target. However, upon leaving the barrel, the bullet also acquires a momentum component Δp_y that is perpendicular to the barrel. The relation $\Delta p_y \approx h/W$ (Equation 29.9) shows that the larger the value of Planck's constant h, the larger is the value of Δp_y. Since this momentum component is perpendicular to the barrel itself, the bullet can strike at locations other than the target. Thus, target shooting would be less accurate if Planck's constant had a relatively large value.

 ## CONCEPTS & CALCULATIONS

In the photoelectric effect, electrons can be emitted from a metal surface when light shines on it, as we have seen. Einstein explained this phenomenon in terms of photons and the conservation of energy. The same basic explanation applies when the light has a single wavelength or when it consists of a mixture of wavelengths, as sunlight does. Our earlier discussion in Section 29.3 deals with a single wavelength. Example 8, in contrast, considers a mixture of wavelengths and reviews the basic concepts that come into play in this important phenomenon.

Concepts & Calculations Example 8

Sunlight and the Photoelectric Effect

Sunlight, whose visible wavelengths range from 380 to 750 nm, is incident on a sodium surface. The work function is $W_0 = 2.28$ eV. Find the maximum kinetic energy KE_{max} (in joules) of the photoelectrons emitted from the surface and the range of wavelengths that will cause photoelectrons to be emitted.

Concept Questions and Answers Will electrons with the greatest value of KE_{max} be emitted when the incident photons have a relatively greater or a relatively smaller amount of energy?

Answer When a photon ejects an electron from the sodium, the photon energy is used for two purposes. Some of it is used to do the work of ejecting the electron, and what remains is carried away as kinetic energy of the emitted electron. The minimum work to eject an electron is equal to the work function W_0, and in this case the electron carries away a maximum amount KE_{max} of kinetic energy. According to the principle of conservation of energy, the energy of an incident photon is equal to the sum of W_0 and KE_{max}. Thus, for a given value of W_0, greater values of KE_{max} will occur when the incident photon has a greater energy to begin with.

In the range of visible wavelengths, which wavelength corresponds to incident photons that carry the greatest energy?

Answer According to Equation 29.2, the energy of a photon is related to the frequency f and Planck's constant h according to $E = hf$. However, the frequency is given by Equation 16.1 as $f = c/\lambda$, where c is the speed of light and λ is the wavelength. Therefore, we have $E = hc/\lambda$, so the smallest value of the wavelength corresponds to the greatest energy.

What is the smallest value of KE_{max} with which an electron can be ejected from the sodium?

Answer The smallest value of KE_{max} with which an electron can be ejected from the sodium is 0 J. This occurs when an incident photon carries an energy that exactly equals the work function W_0, with no energy left over to be carried away by the electron as kinetic energy.

Problem-solving insight

Values for the work function are usually given in electron volts (eV). Be sure to convert the value into joules before combining it with other quantities that are specified in joules.

Solution The value given for the work function is in electron volts, so we first convert it into joules by using the fact that $1 \text{ eV} = 1.60 \times 10^{-19}$ J:

$$W_0 = (2.28 \text{ eV}) \frac{1.60 \times 10^{-19} \text{ J}}{1 \text{ eV}} = 3.65 \times 10^{-19} \text{ J}$$

The incident photons that have the greatest energy are those with the smallest wavelength or $\lambda = 380$ nm. Then, according to Equation 29.3 and Equation 16.1 ($f = c/\lambda$), the greatest value for KE_{max} is

$$\text{KE}_{\text{max}} = hf - W_0 = \frac{hc}{\lambda} - W_0$$

$$= \frac{(6.63 \times 10^{-34} \text{ J} \cdot \text{s})(3.00 \times 10^8 \text{ m/s})}{380 \times 10^{-9} \text{ m}} - 3.65 \times 10^{-19} \text{ J} = \boxed{1.58 \times 10^{-19} \text{ J}}$$

As the wavelength increases, the energy of the incident photons decreases until it equals the work function, at which point electrons are ejected with zero kinetic energy. Thus, we have

$$\text{KE}_{\text{max}} = 0 \text{ J} = \frac{hc}{\lambda} - W_0$$

$$\lambda = \frac{hc}{W_0} = \frac{(6.63 \times 10^{-34} \text{ J} \cdot \text{s})(3.00 \times 10^8 \text{ m/s})}{3.65 \times 10^{-19} \text{ J}} = 5.45 \times 10^{-7} \text{ m} = 545 \text{ nm}$$

The range of wavelengths over which sunlight ejects electrons from the sodium surface is, then, $\boxed{380 \text{ to } 545 \text{ nm}}$.

Kinetic energy and momentum, respectively, are discussed in Chapters 6 and 7. In the present chapter we have seen that moving particles of matter not only possess kinetic energy and momentum but are also characterized by a wavelength that is known as the de Broglie wavelength. We conclude with an example that reviews kinetic energy and momentum and focuses on the relation between these fundamental physical quantities and the de Broglie wavelength.

Concepts & Calculations Example 9
The De Broglie Wavelength

An electron and a proton have the same kinetic energy and are moving at speeds much less than the speed of light. Determine the ratio of the de Broglie wavelength of the electron to that of the proton.

Concept Questions and Answers How is the de Broglie wavelength λ related to the magnitude p of the momentum?

Answer According to Equation 29.8, the de Broglie wavelength is given by $\lambda = h/p$, where h is Planck's constant. The greater the momentum, the smaller is the de Broglie wavelength, and vice versa.

How is the magnitude of the momentum related to the kinetic energy of a particle of mass m that is moving at a speed that is much less than the speed of light?

Answer According to Equation 7.2 the magnitude of the momentum is $p = mv$, where v is the speed. Equation 6.2 gives the kinetic energy as $\text{KE} = \frac{1}{2}mv^2$, which can be solved to show that $v = \sqrt{2(\text{KE})/m}$. Substituting this result into Equation 7.2 shows that $p = m\sqrt{2(\text{KE})/m} = \sqrt{2m(\text{KE})}$.

Which has the greater de Broglie wavelength, the electron or the proton?

Answer According to $\lambda = h/p$, the particle with the greater wavelength is the one with the smaller momentum. However, $p = \sqrt{2m(\text{KE})}$ indicates that, for a given kinetic energy, the particle with the smaller momentum is the one with the smaller mass. The masses of the electron and proton are $m_{\text{electron}} = 9.11 \times 10^{-31}$ kg and $m_{\text{proton}} = 1.67 \times 10^{-27}$ kg. Thus, the electron, with its smaller mass, has the greater de Broglie wavelength.

Solution Using Equation 29.8 for the de Broglie wavelength and the fact that the magnitude of the momentum is related to the kinetic energy by $p = \sqrt{2m(\text{KE})}$, we have

$$\lambda = \frac{h}{p} = \frac{h}{\sqrt{2m(\text{KE})}}$$

Applying this result to the electron and the proton gives

$$\frac{\lambda_{\text{electron}}}{\lambda_{\text{proton}}} = \frac{h/\sqrt{2m_{\text{electron}}(\text{KE})}}{h/\sqrt{2m_{\text{proton}}(\text{KE})}} = \sqrt{\frac{m_{\text{proton}}}{m_{\text{electron}}}} = \sqrt{\frac{1.67 \times 10^{-27} \text{ kg}}{9.11 \times 10^{-31} \text{ kg}}} = \boxed{42.8}$$

As expected, the wavelength for the electron is greater than that for the proton.

CONCEPT SUMMARY

If you need more help with a concept, use the Learning Aids noted next to the discussion or equation. Examples (**Ex.**) are in the text of this chapter. Go to **www.wiley.com/college/cutnell** for the following Learning Aids:

Interactive LearningWare (ILW) — Additional examples solved in a five-step interactive format.

Concept Simulations (CS) — Animated text figures or animations of important concepts.

Interactive Solutions (IS) — Models for certain types of problems in the chapter homework. The calculations are carried out interactively.

Topic	Discussion	Learning Aids
	29.1 THE WAVE–PARTICLE DUALITY, 29.2 BLACKBODY RADIATION AND PLANCK'S CONSTANT	
Wave–particle duality	The wave–particle duality refers to the fact that a wave can exhibit particle-like characteristics and a particle can exhibit wave-like characteristics.	
Perfect blackbody	At a constant temperature, a perfect blackbody absorbs and reemits all the electromagnetic radiation that falls on it. Max Planck calculated the emitted radiation intensity per unit wavelength as a function of wavelength. In his theory, Planck assumed that a blackbody consists of atomic oscillators that can have only discrete, or quantized, energies. Planck's quantized energies are given by	
Energies of atomic oscillators	$$E = nhf \qquad n = 0, 1, 2, 3, \ldots \qquad (29.1)$$ where h is Planck's constant (6.63×10^{-34} J · s) and f is the vibration frequency of an oscillator.	
	29.3 PHOTONS AND THE PHOTOELECTRIC EFFECT All electromagnetic radiation consists of photons, which are packets of energy. The energy of a photon is	
Energy of a photon	$$E = hf \qquad (29.2)$$ where h is Planck's constant and f is the frequency of the photon. A photon has no mass and always travels at the speed of light c in a vacuum.	**Ex. 1** **ILW 29.1**
Photoelectric effect Work function	The photoelectric effect is the phenomenon in which light shining on a metal surface causes electrons to be ejected from the surface. The work function W_0 of a metal is the minimum work that	

Topic	Discussion	Learning Aids

must be done to eject an electron from the metal. In accordance with the conservation of energy, the electrons ejected from a metal have a maximum kinetic energy KE_{max} that is related to the energy hf of the incident photon and the work function of the metal by

Conservation of energy and the photoelectric effect

$$hf = KE_{max} + W_0 \qquad (29.3)$$

Ex. 2, 3, 8
CS 29.1

29.4 THE MOMENTUM OF A PHOTON AND THE COMPTON EFFECT The magnitude p of a photon's momentum is

Magnitude of a photon's momentum

$$p = \frac{h}{\lambda} \qquad (29.6)$$

IS 29.19

where h is Planck's constant and λ is the wavelength of the photon.

Compton effect

The Compton effect is the scattering of a photon by an electron in a material, the scattered photon having a smaller frequency and, hence, a smaller energy than the incident photon. Part of the photon's energy and momentum are transferred to the recoiling electron. The difference between the wavelength λ' of the scattered photon and the wavelength λ of the incident photon is related to the scattering angle θ by

Wavelength difference in the Compton effect

$$\lambda' - \lambda = \frac{h}{mc}(1 - \cos\theta) \qquad (29.7)$$

Ex. 4

Compton wavelength of the electron

where m is the mass of the electron. The quantity $h/(mc)$ is known as the Compton wavelength of the electron.

29.5 THE DE BROGLIE WAVELENGTH AND THE WAVE NATURE OF MATTER The de Broglie wavelength λ of a particle is

De Broglie wavelength

$$\lambda = \frac{h}{p} \qquad (29.8)$$

Ex. 5, 9
IS 29.33

where h is Planck's constant and p is the magnitude of the relativistic momentum of the particle. Because of its wavelength, a particle can exhibit wave-like characteristics. The wave associated with a particle is a wave of probability.

29.6 THE HEISENBERG UNCERTAINTY PRINCIPLE The Heisenberg uncertainty principle places limits on our knowledge about the behavior of a particle. The uncertainty principle indicates that

Uncertainty principle–momentum and position

$$(\Delta p_y)(\Delta y) \geq \frac{h}{4\pi} \qquad (29.10)$$

Ex. 6, 7

where Δy is the uncertainty in the particle's position along the y direction, and Δp_y is the uncertainty in the y component of the linear momentum of the particle.

The uncertainty principle also states that

Uncertainty principle–energy and time

$$(\Delta E)(\Delta t) \geq \frac{h}{4\pi} \qquad (29.11)$$

where ΔE is the uncertainty in the energy of a particle when it is in a certain state, and Δt is the time interval during which the particle is in the state.

FOCUS ON CONCEPTS

Note to Instructors: The numbering of the questions shown here reflects the fact that they are only a representative subset of the total number that are available online. However, all of the questions are available for assignment via an online homework management program such as WileyPLUS or WebAssign.

Section 29.2 Blackbody Radiation and Planck's Constant

1. An astronomer is measuring the electromagnetic radiation emitted by two stars, which are both assumed to be perfect blackbody emitters. For each star she makes a plot of the radiation intensity per unit wavelength as a function of wavelength. She notices that the curve for star A has a maximum that occurs at a shorter wavelength than does the curve for star B. What can she conclude about the surface temperatures of the two stars? **(a)** Star A has the greater surface temperature. **(b)** Star B has the greater surface temperature. **(c)** Both stars, being perfect blackbody emitters, have the same surface temperature. **(d)** There is not enough information to draw a conclusion about the temperatures.

Section 29.3 Photons and the Photoelectric Effect

2. Photons are generated by a microwave oven in a kitchen and by an X-ray machine at a dentist's office. Which type of photon has the greater frequency, the greater energy, and the greater wavelength?

	Greater Frequency	Greater Energy	Greater Wavelength
(a)	X-ray	X-ray	X-ray
(b)	X-ray	X-ray	Microwave
(c)	Microwave	X-ray	Microwave
(d)	Microwave	Microwave	X-ray
(e)	X-ray	Microwave	Microwave

5. A laser emits a beam of light whose photons all have the same frequency. When the beam strikes the surface of a metal, photoelectrons are ejected from the surface. What happens if the laser emits twice the number of photons per second? **(a)** The photoelectrons are ejected from the surface with twice the maximum kinetic energy. **(b)** The photoelectrons are ejected from the surface with the same maximum kinetic energy. **(c)** The number of photoelectrons ejected per second from the surface doubles. **(d)** Both b and c happen. **(e)** Both a and c happen.

6. The surface of a metal plate is illuminated with light of a certain frequency. Which of the following conditions determine whether or not photoelectrons are ejected from the metal?

1. The number of photons per second emitted by the light source
2. The length of time that the light is turned on
3. The thermal conductivity of the metal
4. The surface area of the metal illuminated by the light
5. The type of metal from which the plate is made

(a) 1 and 2   **(b)** 5   **(c)** 3 and 5   **(d)** 4   **(e)** 2 and 3

Section 29.4   The Momentum of a Photon and the Compton Effect

9. Does a photon, like a moving particle such as an electron, have a momentum? **(a)** No, because a photon is a wave, and a wave does not have a momentum. **(b)** No, because a photon has no mass, and mass is necessary in order to have a momentum. **(c)** No, because a photon, always traveling at the speed of light in a vacuum, would have an infinite momentum. **(d)** Yes, and the magnitude p of the photon's momentum is related to its wavelength λ by $p = h/\lambda$, where h is Planck's constant. **(e)** Yes, and the magnitude p of the photon's momentum is related to its wavelength λ by $p = h\lambda$, where h is Planck's constant.

Section 29.5   The De Broglie Wavelength and the Wave Nature of Matter

13. Two particles, A and B, have the same mass, but particle A has a charge of $+q$ and B has a charge of $+2q$. The particles are accelerated from rest through the same potential difference. Which one has the longer de Broglie wavelength at the end of the acceleration? **(a)** Particle A, because it has the greater momentum, and, hence, the longer de Broglie wavelength **(b)** Particle B, because it has the greater momentum, and, hence, the longer de Broglie wavelength **(c)** Particle A, because it has the smaller momentum, and, hence, the longer de Broglie wavelength **(d)** Particle B, because it has the smaller momentum, and, hence, the longer de Broglie wavelength **(e)** Both particles have the same de Broglie wavelength.

Section 29.6   The Heisenberg Uncertainty Principle

16. Suppose that the momentum of an electron is measured with complete accuracy (i.e., the uncertainty in its momentum is zero). The uncertainty in a simultaneous measurement of the electron's position _____. **(a)** is also zero **(b)** is infinitely large **(c)** is some finite value between zero and infinity **(d)** cannot be measured, because one cannot measure the position and momentum of a particle, such as an electron, simultaneously

PROBLEMS

Note to Instructors: *Most of the homework problems in this chapter are available for assignment via an online homework management program such as* WileyPLUS *or* WebAssign, *and those marked with the icon* **GO** *are presented in* WileyPLUS *using a guided tutorial format that provides enhanced interactivity. See Preface for additional details.*

In working these problems, ignore relativistic effects unless instructed otherwise, and assume that wavelengths referred to are in a vaccum unless otherwise specified.

ssm   Solution is in the Student Solutions Manual.
www   Solution is available online at www.wiley.com/college/cutnell

⚕ **This icon represents a biomedical application.**

Section 29.3   Photons and the Photoelectric Effect

1. An FM radio station broadcasts at a frequency of 98.1 MHz. The power radiated from the antenna is 5.0×10^4 W. How many photons per second does the antenna emit?

2. The dissociation energy of a molecule is the energy required to break the molecule apart into its separate atoms. The dissociation energy for the cyanogen molecule is 1.22×10^{-18} J. Suppose that this energy is provided by a single photon. Determine the **(a)** wavelength and **(b)** frequency of the photon. **(c)** In what region of the electromagnetic spectrum (see Figure 24.9) does this photon lie?

3. **ssm** Ultraviolet light with a frequency of 3.00×10^{15} Hz strikes a metal surface and ejects electrons that have a maximum kinetic energy of 6.1 eV. What is the work function (in eV) of the metal?

4. **GO** The work function of a metal surface is 4.80×10^{-19} J. The maximum speed of the electrons emitted from the surface is $v_A = 7.30 \times 10^5$ m/s when the wavelength of the light is λ_A. However, a maximum speed of $v_B = 5.00 \times 10^5$ m/s is observed when the wavelength is λ_B. Find the wavelengths λ_A and λ_B.

5. **ssm** Ultraviolet light is responsible for sun tanning. Find the wavelength (in nm) of an ultraviolet photon whose energy is 6.4×10^{-19} J.

6. The maximum wavelength that an electromagnetic wave can have and still eject electrons from a metal surface is 485 nm. What is the work function W_0 of this metal? Express your answer in electron volts.

7. A magnesium surface has a work function of 3.68 eV. Electromagnetic waves with a wavelength of 215 nm strike the surface and eject electrons. Find the maximum kinetic energy of the ejected electrons. Express your answer in electron volts.

8. **GO** Light is shining perpendicularly on the surface of the earth with an intensity of 680 W/m². Assuming that all the photons in the light have the same wavelength (in vacuum) of 730 nm, determine the number of photons per second per square meter that reach the earth.

*9. **ssm** Consult **Interactive LearningWare 29.1** at **www.wiley.com/college/cutnell** for background material relating to this problem. An owl has good night vision because its eyes can detect a light intensity as small as 5.0×10^{-13} W/m². What is the minimum number of photons per second that an owl eye can detect if its pupil has a diameter of 8.5 mm and the light has a wavelength of 510 nm?

*10. When light with a wavelength of 221 nm is incident on a certain metal surface, electrons are ejected with a maximum kinetic energy of 3.28×10^{-19} J. Determine the wavelength (in nm) of light that should be used to double the maximum kinetic energy of the electrons ejected from this surface.

*11. Multiple-Concept Example 3 reviews the concepts necessary to solve this problem. Light is incident on the surface of metallic sodium, whose work function is 2.3 eV. The maximum speed of the photoelectrons emitted by the surface is 1.2×10^6 m/s. What is the wavelength of the light?

*12. **GO** A glass plate has a mass of 0.50 kg and a specific heat capacity of 840 J/(kg·C°). The wavelength of infrared light is 6.0×10^{-5} m, while the wavelength of blue light is 4.7×10^{-7} m. Find the number of infrared photons and the number of blue photons needed to raise the temperature of the glass plate by 2.0 C°, assuming that all the photons are absorbed by the glass.

13. ssm www A laser emits 1.30×10^{18} photons per second in a beam of light that has a diameter of 2.00 mm and a wavelength of 514.5 nm. Determine **(a)** the average electric field strength and **(b)** the average magnetic field strength for the electromagnetic wave that constitutes the beam.

14. (a) How many photons (wavelength = 620 nm) must be absorbed to melt a 2.0-kg block of ice at 0 °C into water at 0 °C? **(b)** On the average, how many H_2O molecules does one photon convert from the ice phase to the water phase?

Section 29.4 The Momentum of a Photon and the Compton Effect

15. ssm In a Compton scattering experiment, the incident X-rays have a wavelength of 0.2685 nm, and the scattered X-rays have a wavelength of 0.2703 nm. Through what angle θ in Figure 29.10 are the X-rays scattered?

16. A photon of red light (wavelength = 720 nm) and a Ping-Pong ball (mass = 2.2×10^{-3} kg) have the same momentum. At what speed is the ball moving?

17. A light source emits a beam of photons, each of which has a momentum of 2.3×10^{-29} kg · m/s. **(a)** What is the frequency of the photons? **(b)** To what region of the electromagnetic spectrum do the photons belong? Consult Figure 24.9 if necessary.

18. GO In the Compton effect, momentum conservation applies, so the total momentum of the photon and the electron is the same before and after the scattering occurs. Suppose that in Figure 29.10 the incident photon moves in the +x direction and the scattered photon emerges at an angle of $\theta = 90.0°$, which is in the −y direction. The incident photon has a wavelength of 9.00×10^{-12} m. Find the x and y components of the momentum of the scattered electron.

19. Refer to **Interactive Solution 29.19** at **www.wiley.com/college/cutnell** for help in solving this problem. An incident X-ray photon of wavelength 0.2750 nm is scattered from an electron that is initially at rest. The photon is scattered at an angle of $\theta = 180.0°$ in Figure 29.10 and has a wavelength of 0.2825 nm. Use the conservation of linear momentum to find the momentum gained by the electron.

20. A sample is bombarded by incident X-rays, and free electrons in the sample scatter some of the X-rays at an angle of $\theta = 122.0°$ with respect to the incident X-rays (see Figure 29.10). The scattered X-rays have a momentum whose magnitude is 1.856×10^{-24} kg · m/s. Determine the wavelength (in nm) of the incident X-rays. (For accuracy, use $h = 6.626 \times 10^{-34}$ J · s, $c = 2.998 \times 10^8$ m/s, and $m = 9.109 \times 10^{-31}$ kg for the mass of an electron.)

*21. **ssm www** What is the maximum amount by which the wavelength of an incident photon could change when it undergoes Compton scattering from a nitrogen molecule (N_2)?

*22. **GO** A photon of wavelength 0.45000 nm strikes a free electron that is initially at rest. The photon is scattered straight backward. What is the speed of the recoil electron after the collision?

*23. The X-rays detected at a scattering angle of $\theta = 163°$ in Figure 29.10 have a wavelength of 0.1867 nm. Find **(a)** the wavelength of an incident photon, **(b)** the energy of an incident photon, **(c)** the energy of a scattered photon, and **(d)** the kinetic energy of the recoil electron. (For accuracy, use $h = 6.626 \times 10^{-34}$ J · s and $c = 2.998 \times 10^8$ m/s.)

24. Some scientists have suggested that spacecraft with sails of the kind described in Conceptual Example 4 could be propelled by lasers. Suppose that such a sail is constructed of a highly reflective material thin enough so that one square meter of the sail has a mass of just 3.0×10^{-3} kg. The sail is to be propelled by an ultraviolet laser beam (wavelength = 225 nm) that will strike its surface perpendicularly. **(a)** Use the impulse–momentum theorem (Section 7.1) to determine the number of photons per second that must strike each square meter of the sail in order to cause an acceleration of 9.8×10^{-6} m/s², which is one million times smaller than the gravitational acceleration at the earth's surface. Assume that no other forces act on the sail, and that all the incident photons are reflected. **(b)** Determine the intensity (power per unit area) that the laser beam must have when it strikes the sail.

Section 29.5 The De Broglie Wavelength and the Wave Nature of Matter

25. ssm The de Broglie wavelength of a proton in a particle accelerator is 1.30×10^{-14} m. Determine the kinetic energy (in joules) of the proton.

26. GO In a Young's double-slit experiment that uses electrons, the angle that locates the first-order bright fringes is $\theta_A = 1.6 \times 10^{-4}$ degrees when the magnitude of the electron momentum is $p_A = 1.2 \times 10^{-22}$ kg · m/s. With the same double slit, what momentum magnitude p_B is necessary so that an angle of $\theta_B = 4.0 \times 10^{-4}$ degrees locates the first-order bright fringes?

27. The interatomic spacing in a crystal of table salt is 0.282 nm. This crystal is being studied in a neutron diffraction experiment, similar to the one that produced the photograph in Figure 29.13a. How fast must a neutron (mass = 1.67×10^{-27} kg) be moving to have a de Broglie wavelength of 0.282 nm?

28. A particle has a speed of 1.2×10^6 m/s. Its de Broglie wavelength is 8.4×10^{-14} m. What is the mass of the particle?

29. Recall from Section 14.3 that the average kinetic energy of an atom in a monatomic ideal gas is given by $\overline{KE} = \frac{3}{2}kT$, where $k = 1.38 \times 10^{-23}$ J/K and T is the Kelvin temperature of the gas. Determine the de Broglie wavelength of a helium atom (mass = 6.65×10^{-27} kg) that has the average kinetic energy at room temperature (293 K).

30. Find the de Broglie wavelength of an electron with a speed of $0.88c$. Take relativistic effects into account.

*31. **ssm** A particle has a de Broglie wavelength of 2.7×10^{-10} m. Then its kinetic energy doubles. What is the particle's new de Broglie wavelength, assuming that relativistic effects can be ignored?

*32. **GO** From a cliff that is 9.5 m above a lake, a young woman (mass = 41 kg) jumps from rest, straight down into the water. At the instant she strikes the water, what is her de Broglie wavelength?

*33. Consult **Interactive Solution 29.33** at **www.wiley.com/college/cutnell** to explore a model for solving this problem. In a television picture tube the electrons are accelerated from rest through a potential difference V. Just before an electron strikes the screen, its de Broglie wavelength is 0.900×10^{-11} m. What is the potential difference?

*34. **GO** Particle A is at rest, and particle B collides head-on with it. The collision is completely inelastic, so the two particles stick together after the collision and move off with a common velocity. The masses of the particles are different, and no external forces act on them. The de Broglie wavelength of particle B before the collision is 2.0×10^{-34} m. What is the de Broglie wavelength of the object that moves off after the collision?

*35. **ssm** An electron, starting from rest, accelerates through a potential difference of 418 V. What is the final de Broglie wavelength of the electron, assuming that its final speed is much less than the speed of light?

**36. The kinetic energy of a particle is equal to the energy of a photon. The particle moves at 5.0% of the speed of light. Find the ratio of the photon wavelength to the de Broglie wavelength of the particle.

Section 29.6 The Heisenberg Uncertainty Principle

37. ssm www In the lungs there are tiny sacs of air, which are called alveoli. An oxygen molecule (mass = 5.3×10^{-26} kg) is trapped within a sac, and the uncertainty in its position is 0.12 mm. What is the minimum uncertainty in the velocity of this oxygen molecule?

38. GO Particles pass through a single slit of width 0.200 mm (see Figure 29.15). The de Broglie wavelength of each particle is 633 nm. After the particles pass through the slit, they spread out over a range of angles. Assume that the uncertainty in the position of the particles is one-half the width of the slit, and use the Heisenberg uncertainty principle to determine the minimum range of angles.

39. ssm www Suppose that an electron is trapped within a small region and the uncertainty in its position is 3.0×10^{-15} m. What is the minimum uncertainty in the electron's momentum?

40. An object is moving along a straight line, and the uncertainty in its position is 2.5 m. **(a)** Find the minimum uncertainty in the momentum of the object. Find the minimum uncertainty in the object's velocity, assuming that the object is **(b)** a golf ball (mass = 0.045 kg) and **(c)** an electron.

*41. The minimum uncertainty Δy in the position y of a particle is equal to its de Broglie wavelength. Determine the minimum uncertainty in the speed of the particle, where this minimum uncertainty Δv_y is expressed as a percentage of the particle's speed v_y $\left(\text{Percentage} = \dfrac{\Delta v_y}{v_y} \times 100\%\right)$. Assume that relativistic effects can be ignored.

*42. A subatomic particle created in an experiment exists in a certain state for a time of $\Delta t = 7.4 \times 10^{-20}$ s before decaying into other particles. Apply both the Heisenberg uncertainty principle and the equivalence of energy and mass (see Section 28.6) to determine the minimum uncertainty involved in measuring the mass of this short-lived particle.

ADDITIONAL PROBLEMS

43. ssm As Section 17.3 discusses, sound waves diffract, or bend, around the edges of a doorway. Larger wavelengths diffract more than smaller wavelengths. **(a)** The speed of sound is 343 m/s. With what speed would a 55.0-kg person have to move through a doorway to diffract to the same extent as a 128-Hz bass tone? **(b)** At the speed calculated in part (a), how long (in years) would it take the person to move a distance of one meter?

44. Two sources produce electromagnetic waves. Source B produces a wavelength that is three times the wavelength produced by source A. Each photon from source A has an energy of 2.1×10^{-18} J. What is the energy of a photon from source B?

45. A bacterium (mass = 2×10^{-15} kg) in the blood is moving at 0.33 m/s. What is the de Broglie wavelength of this bacterium?

46. What are **(a)** the wavelength of a 5.0-eV photon and **(b)** the de Broglie wavelength of a 5.0-eV electron?

47. Radiation of a certain wavelength causes electrons with a maximum kinetic energy of 0.68 eV to be ejected from a metal whose work function is 2.75 eV. What will be the maximum kinetic energy

(in eV) with which this same radiation ejects electrons from another metal whose work function is 2.17 eV?

48. How fast does a proton have to be moving in order to have the same de Broglie wavelength as an electron that is moving with a speed of 4.50×10^6 m/s?

*49. The width of the central bright fringe in a diffraction pattern on a screen is identical when either electrons or red light (vacuum wavelength = 661 nm) pass through a single slit. The distance between the screen and the slit is the same in each case and is large compared to the slit width. How fast are the electrons moving?

*50. A proton is located at a distance of 0.420 m from a point charge of + 8.30 μC. The repulsive electric force moves the proton until it is at a distance of 1.58 m from the charge. Suppose that the electric potential energy lost by the system is carried off by a photon that is emitted during the process. What is its wavelength?

**51. An X-ray photon is scattered at an angle of $\theta = 180.0°$ from an electron that is initially at rest. After scattering, the electron has a speed of 4.67×10^6 m/s. Find the wavelength of the incident X-ray photon.

CHAPTER 30

THE NATURE OF THE ATOM

This photograph shows a 3D CAT scan of a human heart. The right atrium is shown in purple and the right ventricle in teal (greenish blue). The pulmonary artery is shown in light blue and the aorta in orange. CAT scanning is an important noninvasive medical diagnostic technique that utilizes X-rays to provide images of the interior of the human body. The production of X-rays is related to the structure of the atom, and that structure is the main topic of this chapter. (© Living Art Enterprises, LLC/Photo Researchers, Inc.)

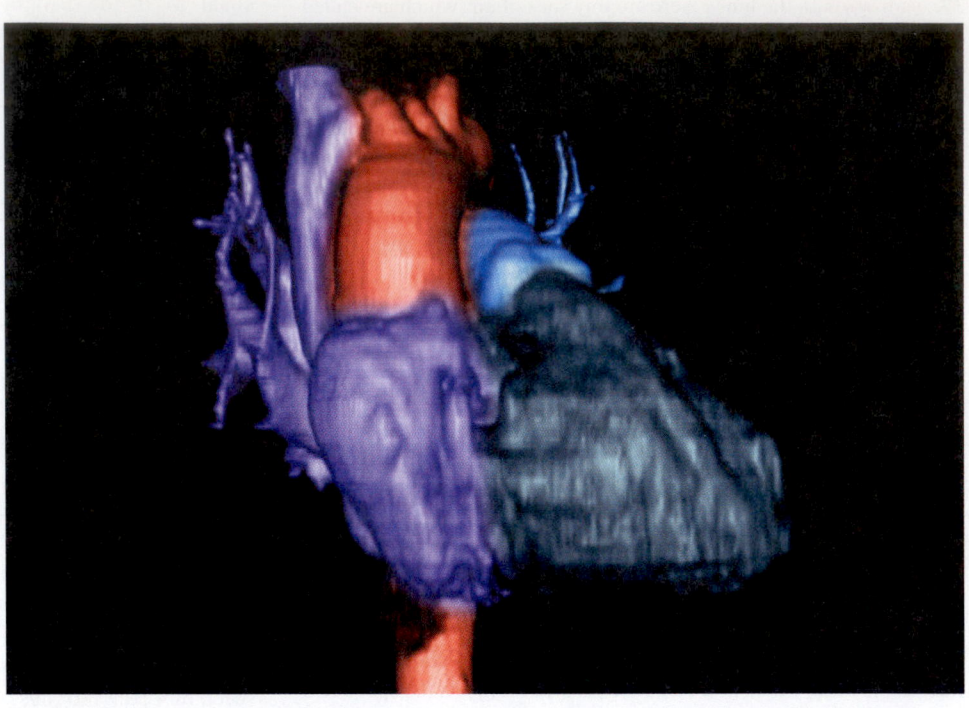

Negative electron

Positive nucleus

Figure 30.1 In the nuclear atom a small positively charged nucleus is surrounded at relatively large distances by a number of electrons.

30.1 RUTHERFORD SCATTERING AND THE NUCLEAR ATOM

An atom contains a small, positively charged nucleus (radius $\approx 10^{-15}$ m), which is surrounded at relatively large distances (radius $\approx 10^{-10}$ m) by a number of electrons, as Figure 30.1 illustrates. In the natural state, an atom is electrically neutral because the nucleus contains a number of protons (each with a charge of $+e$) that equals the number of electrons (each with a charge of $-e$). This model of the atom is universally accepted now and is referred to as the **nuclear atom**.

The nuclear atom is a relatively recent idea. In the early part of the twentieth century a widely accepted model, developed by the English physicist Joseph J. Thomson (1856–1940), pictured the atom very differently. In Thomson's view there was no nucleus at the center of an atom. Instead, the positive charge was assumed to be spread throughout the atom, forming a kind of paste or pudding, in which the negative electrons were suspended like plums.

The "plum-pudding" model was discredited in 1911 when the New Zealand physicist Ernest Rutherford (1871–1937) published experimental results that the model could not explain. As Figure 30.2 indicates, Rutherford and his co-workers directed a beam of alpha particles (α particles) at a thin metal foil made of gold. Alpha particles are positively charged particles (the nuclei of helium atoms, although this was not recognized at the time) emitted by some radioactive materials. If the plum-pudding model were correct, the α particles would be expected to pass nearly straight through the foil. After all, there is nothing in this model to deflect the relatively massive α particles, since the electrons have a comparatively small mass and the positive charge is spread out in a "diluted" pudding. Using a zinc sulfide screen, which flashed briefly when struck by an α particle, Rutherford and co-workers were able to determine that not all the α particles passed straight through the foil. Instead, some were deflected at large angles, even backward. Rutherford himself said,

"It was almost as incredible as if you had fired a fifteen-inch shell at a piece of tissue and it came back and hit you." Rutherford concluded that the positive charge, instead of being distributed thinly and uniformly throughout the atom, was concentrated in a small region called the nucleus.

But how could the electrons in a nuclear atom remain separated from the positively charged nucleus? If the electrons were stationary, they would be pulled inward by the attractive electric force of the nuclear charge. Therefore, the electrons must be moving around the nucleus in some fashion, like planets in orbit around the sun. In fact, the nuclear model of the atom is sometimes referred to as the "planetary" model. The dimensions of the atom, however, are such that it contains a larger fraction of empty space than our solar system does, as Conceptual Example 1 discusses.

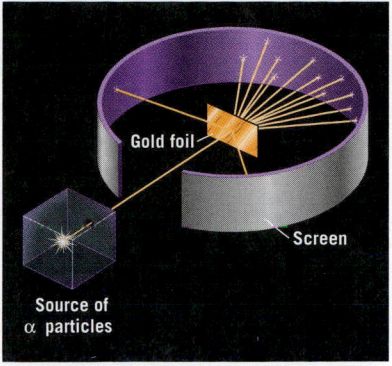

Figure 30.2 A Rutherford scattering experiment in which α particles are scattered by a thin gold foil. The entire apparatus is located within a vacuum chamber (not shown).

Conceptual Example 1   Are Atoms Mostly Empty Space?

In the planetary model of the atom, the radius of the nucleus ($\approx 1 \times 10^{-15}$ m) is analogous to the radius of the sun ($\approx 7 \times 10^8$ m). The electrons orbit the nucleus at a radial distance ($\approx 1 \times 10^{-10}$ m) that is analogous to the radial distance ($\approx 1.5 \times 10^{11}$ m) at which the earth orbits the sun. Suppose that the dimensions of the sun and the earth's orbit had the same proportions as those of an atomic nucleus and an electron's orbit. What then would be true about the distance between the earth and the sun? **(a)** It would be much greater than it actually is. **(b)** It would be much smaller than it actually is. **(c)** It would be roughly the same as it actually is.

Reasoning The radius of an electron orbit is one hundred thousand times larger than the radius of the nucleus: $(1 \times 10^{-10}$ m$)/(1 \times 10^{-15}$ m$) = 10^5$. Using this factor with the radius of the sun will reveal the correct answer.

Answers (b) and (c) are incorrect. Suppose that the earth's orbital radius about the sun were indeed 10^5 times the sun's radius. The distance between the earth and the sun, then, would be $10^5 \times (7 \times 10^8$ m$) = 7 \times 10^{13}$ m, which is neither smaller than nor roughly the same as the actual distance.

Answer (a) is correct. If the earth's orbital radius about the sun were 10^5 times the sun's radius, the distance between the earth and the sun would be $10^5 \times (7 \times 10^8$ m$) = 7 \times 10^{13}$ m, which is more than four hundred times greater than the actual orbital radius of 1.5×10^{11} m. In fact, the earth would be more than ten times farther from the sun than is Pluto, which has an orbital radius of about 6×10^{12} m. An atom, then, contains a much greater fraction of empty space than does our solar system.

Related Homework: *Problem 4*

Although the planetary model of the atom is easy to visualize, it too is fraught with difficulties. For instance, an electron moving on a curved path has a centripetal acceleration, as Section 5.2 discusses. And when an electron is accelerating, it radiates electromagnetic waves, as Section 24.1 discusses. These waves carry away energy. With their energy constantly being depleted, the electrons would spiral inward and eventually collapse into the nucleus. Since matter is stable, such a collapse does not occur. Thus, the planetary model, although providing a more realistic picture of the atom than the "plum-pudding" model, must be telling only part of the story. The full story of atomic structure is fascinating, and the next section describes another aspect of it.

30.2 LINE SPECTRA

We have seen in Sections 13.3 and 29.2 that all objects emit electromagnetic waves, and we will see in Section 30.3 how this radiation arises. For a solid object, such as the hot filament of a light bulb, these waves have a continuous range of wavelengths, some of which are in the visible region of the spectrum. The continuous range of wavelengths is characteristic of the entire collection of atoms that make up the solid. In contrast, individual atoms, free of the strong interactions that are present in a solid, emit only certain specific wavelengths rather than a continuous range. These wavelengths are characteristic

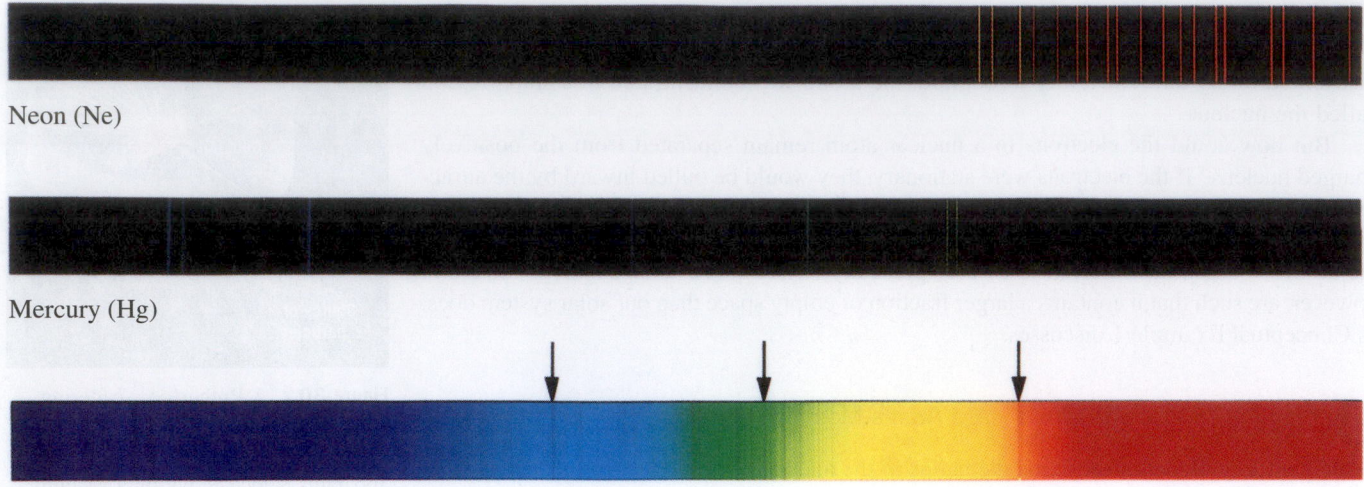

Neon (Ne)

Mercury (Hg)

Solar absorption spectrum (Fraunhofer lines)

Figure 30.3 The line spectra for neon and mercury, along with the continuous spectrum of the sun. The dark lines in the sun's spectrum are called Fraunhofer lines, three of which are marked by arrows. (Courtesy Bausch & Lomb)

The physics of

neon signs and mercury vapor street lamps.

of the atom and provide important clues about its structure. To study the behavior of individual atoms, low-pressure gases are used in which the atoms are relatively far apart.

A low-pressure gas in a sealed tube can be made to emit electromagnetic waves by applying a sufficiently large potential difference between two electrodes located within the tube. With a grating spectroscope like that in Figure 27.34, the individual wavelengths emitted by the gas can be separated and identified as a series of bright fringes. The series of fringes is called a *line spectrum* because each bright fringe appears as a thin rectangle (a "line") resulting from the large number of parallel, closely spaced slits in the grating of the spectroscope. Figure 30.3 shows the visible parts of the line spectra for neon and mercury. The specific visible wavelengths emitted by neon and mercury give neon signs and mercury vapor street lamps their characteristic colors.

The simplest line spectrum is that of atomic hydrogen, and much effort has been devoted to understanding the pattern of wavelengths that it contains. Figure 30.4 illustrates in schematic form some of the groups or series of lines in the spectrum of atomic hydrogen. The group of lines in the visible region is known as the *Balmer series,* in recognition of Johann J. Balmer (1825–1898), a Swiss schoolteacher who found an empirical equation that gave the values for the observed wavelengths. This equation is given next, along with similar equations that apply to the *Lyman series* at shorter wavelengths and the *Paschen series* at longer wavelengths, which are also shown in the drawing:

Lyman series
$$\frac{1}{\lambda} = R\left(\frac{1}{1^2} - \frac{1}{n^2}\right) \qquad n = 2, 3, 4, \ldots \qquad (30.1)$$

Balmer series
$$\frac{1}{\lambda} = R\left(\frac{1}{2^2} - \frac{1}{n^2}\right) \qquad n = 3, 4, 5, \ldots \qquad (30.2)$$

Paschen series
$$\frac{1}{\lambda} = R\left(\frac{1}{3^2} - \frac{1}{n^2}\right) \qquad n = 4, 5, 6, \ldots \qquad (30.3)$$

In these equations, the constant term R has the value of $R = 1.097 \times 10^7 \text{ m}^{-1}$ and is called the *Rydberg constant.* An essential feature of each group of lines is that there are

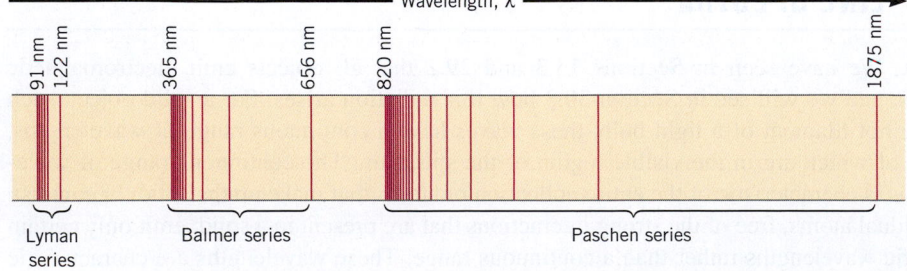

Figure 30.4 Line spectrum of atomic hydrogen. Only the Balmer series lies in the visible region of the electromagnetic spectrum.

long and short wavelength limits, with the lines being increasingly crowded toward the short wavelength limit. Figure 30.4 also gives these limits for each series, and Example 2 determines them for the Balmer series.

 Example 2   The Balmer Series

Find **(a)** the longest and **(b)** the shortest wavelengths of the Balmer series.

Reasoning Each wavelength in the series corresponds to one value for the integer n in Equation 30.2. Longer wavelengths are associated with smaller values of n. The longest wavelength occurs when n has its smallest value of $n = 3$. The shortest wavelength arises when n has a very large value, so that $1/n^2$ is essentially zero.

Solution **(a)** With $n = 3$, Equation 30.2 reveals that for the longest wavelength

$$\frac{1}{\lambda} = R\left(\frac{1}{2^2} - \frac{1}{n^2}\right) = (1.097 \times 10^7 \text{ m}^{-1})\left(\frac{1}{2^2} - \frac{1}{3^2}\right) = 1.524 \times 10^6 \text{ m}^{-1}$$

$$\text{or} \quad \boxed{\lambda = 656 \text{ nm}}$$

(b) With $1/n^2 = 0$, Equation 30.2 reveals that for the shortest wavelength

$$\frac{1}{\lambda} = (1.097 \times 10^7 \text{ m}^{-1})\left(\frac{1}{2^2} - 0\right) = 2.743 \times 10^6 \text{ m}^{-1} \quad \text{or} \quad \boxed{\lambda = 365 \text{ nm}}$$

Equations 30.1–30.3 are useful because they reproduce the wavelengths that hydrogen atoms radiate. However, these equations are empirical and provide no insight as to *why* certain wavelengths are radiated and others are not. It was the great Danish physicist, Niels Bohr (1885–1962), who provided the first model of the atom that predicted the discrete wavelengths emitted by atomic hydrogen. Bohr's model started us on the way toward understanding how the structure of the atom restricts the radiated wavelengths to certain values. In 1922 Bohr received the Nobel Prize in physics for his accomplishment.

30.3 THE BOHR MODEL OF THE HYDROGEN ATOM

In 1913 Bohr presented a model that led to equations such as Balmer's for the wavelengths that the hydrogen atom radiates. Bohr's theory begins with Rutherford's picture of an atom as a nucleus surrounded by electrons moving in circular orbits. In his theory, Bohr made a number of assumptions and combined the new quantum ideas of Planck and Einstein with the traditional description of a particle in uniform circular motion.

Adopting Planck's idea of quantized energy levels (see Section 29.2), Bohr hypothesized that in a hydrogen atom there can be only certain values of the total energy (electron kinetic energy plus potential energy). These allowed energy levels correspond to different orbits for the electron as it moves around the nucleus, the larger orbits being associated with larger total energies. Figure 30.5 illustrates two of the orbits. In addition, Bohr assumed that an electron in one of these orbits *does not* radiate electromagnetic waves. For this reason, the orbits are called **stationary orbits** or **stationary states**. Bohr recognized that radiationless orbits violated the laws of physics, as they were then known. But the assumption of such orbits was necessary, because the traditional laws indicated that an electron radiates electromagnetic waves as it accelerates around a circular path, and the loss of the energy carried by the waves would lead to the collapse of the orbit.

To incorporate Einstein's photon concept (see Section 29.3), Bohr theorized that a photon is emitted only when the electron *changes* orbits from a larger one with a higher energy to a smaller one with a lower energy, as Figure 30.5 indicates. How do electrons get into the higher-energy orbits in the first place? They get there by picking up energy when atoms collide, which happens more often when a gas is heated, or by acquiring energy when a high voltage is applied to a gas.

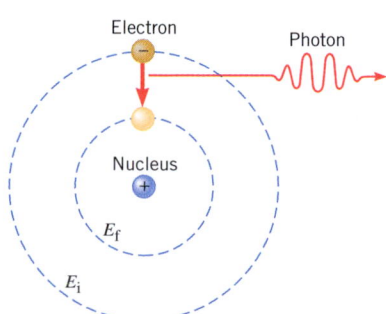

Figure 30.5 In the Bohr model, a photon is emitted when the electron drops from a larger, higher-energy orbit (energy $= E_i$) to a smaller, lower-energy orbit (energy $= E_f$).

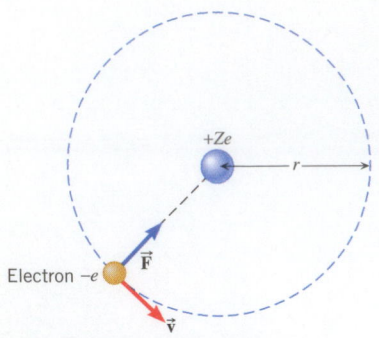

Figure 30.6 In the Bohr model, the electron is in uniform circular motion around the nucleus. The centripetal force $\vec{F}$ is the electrostatic force of attraction that the positive nuclear charge exerts on the electron.

When an electron in an initial orbit with a larger energy E_i changes to a final orbit with a smaller energy E_f, the emitted photon has an energy of $E_i - E_f$, consistent with the law of conservation of energy. But according to Einstein, the energy of a photon is hf, where f is its frequency and h is Planck's constant. As a result, we find that

$$E_i - E_f = hf \qquad (30.4)$$

Since the frequency of an electromagnetic wave is related to the wavelength by $f = c/\lambda$, Bohr could use Equation 30.4 to determine the wavelengths radiated by a hydrogen atom. First, however, he had to derive expressions for the energies E_i and E_f.

THE ENERGIES AND RADII OF THE BOHR ORBITS

For an electron of mass m and speed v in an orbit of radius r (see Figure 30.6), the total energy is the kinetic energy (KE $= \frac{1}{2}mv^2$) of the electron plus the electric potential energy EPE. The potential energy is the product of the charge $(-e)$ on the electron and the electric potential produced by the positive nuclear charge, in accord with Equation 19.3. We assume that the nucleus contains Z protons,* for a total nuclear charge of $+Ze$. The electric potential at a distance r from a point charge of $+Ze$ is given as $+kZe/r$ by Equation 19.6, where the constant k is $k = 8.988 \times 10^9$ N·m²/C². The electric potential energy is, then, EPE $= (-e)(+kZe/r)$. Consequently, the total energy E of the atom is

$$E = \text{KE} + \text{EPE}$$
$$= \tfrac{1}{2}mv^2 - \frac{kZe^2}{r} \qquad (30.5)$$

But a centripetal force of magnitude mv^2/r (Equation 5.3) acts on a particle in uniform circular motion. As Figure 30.6 indicates, the centripetal force is provided by the electrostatic force of attraction $\vec{F}$ that the protons in the nucleus exert on the electron. According to Coulomb's law (Equation 18.1), the magnitude of the electrostatic force is $F = kZe^2/r^2$. Therefore, $mv^2/r = kZe^2/r^2$, or

$$mv^2 = \frac{kZe^2}{r} \qquad (30.6)$$

We can use this relation to eliminate the term mv^2 from Equation 30.5, with the result that

$$E = \frac{1}{2}\left(\frac{kZe^2}{r}\right) - \frac{kZe^2}{r} = -\frac{kZe^2}{2r} \qquad (30.7)$$

The total energy of the atom is negative because the negative electric potential energy is larger in magnitude than the positive kinetic energy.

A value for the radius r is needed, if Equation 30.7 is to be useful. To determine r, Bohr made an assumption about the orbital angular momentum of the electron. The magnitude L of the angular momentum is given by Equation 9.10 as $L = I\omega$, where $I = mr^2$ is the moment of inertia of the electron moving on its circular path and $\omega = v/r$ is the angular speed of the electron in radians per second. Thus, the angular momentum is $L = (mr^2)(v/r) = mvr$. Bohr conjectured that the angular momentum can assume only certain discrete values; in other words, L is quantized. He postulated that the allowed values are integer multiples of Planck's constant divided by 2π:

$$L_n = mv_n r_n = n\frac{h}{2\pi} \qquad n = 1, 2, 3, \ldots \qquad (30.8)$$

Solving this equation for v_n and substituting the result into Equation 30.6 lead to the following expression for the radius r_n of the nth Bohr orbit:

$$r_n = \left(\frac{h^2}{4\pi^2 mke^2}\right)\frac{n^2}{Z} \qquad n = 1, 2, 3, \ldots \qquad (30.9)$$

* For hydrogen, $Z = 1$, but we also wish to consider situations in which Z is greater than 1.

With $h = 6.626 \times 10^{-34}$ J·s, $m = 9.109 \times 10^{-31}$ kg, $k = 8.988 \times 10^{9}$ N·m²/C², and $e = 1.602 \times 10^{-19}$ C, this expression reveals that

Radii for Bohr orbits (in meters)
$$r_n = (5.29 \times 10^{-11} \text{ m}) \frac{n^2}{Z} \qquad n = 1, 2, 3, \ldots \tag{30.10}$$

Therefore, in the hydrogen atom ($Z = 1$) the smallest Bohr orbit ($n = 1$) has a radius of $r_1 = 5.29 \times 10^{-11}$ m. This particular value is called the **Bohr radius.** Figure 30.7 shows the first three Bohr orbits for the hydrogen atom.

Equation 30.9 for the radius of a Bohr orbit can be substituted into Equation 30.7 to show that the corresponding total energy for the nth orbit is

$$E_n = -\left(\frac{2\pi^2 m k^2 e^4}{h^2}\right)\frac{Z^2}{n^2} \qquad n = 1, 2, 3, \ldots \tag{30.11}$$

Substituting values for h, m, k, and e into this expression yields

Bohr energy levels in joules
$$E_n = -(2.18 \times 10^{-18} \text{ J})\frac{Z^2}{n^2} \qquad n = 1, 2, 3, \ldots \tag{30.12}$$

Often, atomic energies are expressed in units of electron volts rather than joules. Since 1.60×10^{-19} J $= 1$ eV, Equation 30.12 can be rewritten as

Bohr energy levels in electron volts
$$E_n = -(13.6 \text{ eV})\frac{Z^2}{n^2} \qquad n = 1, 2, 3, \ldots \tag{30.13}$$

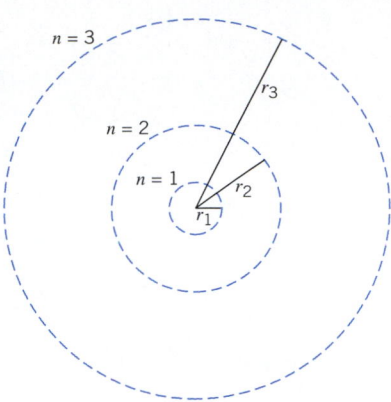

Figure 30.7 The first Bohr orbit in the hydrogen atom has a radius $r_1 = 5.29 \times 10^{-11}$ m. The second and third Bohr orbits have radii $r_2 = 4r_1$ and $r_3 = 9r_1$, respectively.

ENERGY LEVEL DIAGRAMS

It is useful to represent the energy values given by Equation 30.13 on an *energy level diagram,* as in Figure 30.8. In this diagram, which applies to the hydrogen atom ($Z = 1$), the highest energy level corresponds to $n = \infty$ in Equation 30.13 and has an energy of 0 eV. This is the energy of the atom when the electron is completely removed ($r = \infty$) from the nucleus and is at rest. In contrast, the lowest energy level corresponds to $n = 1$ and has a value of -13.6 eV. The lowest energy level is called the **ground state,** to distinguish it from the higher levels, which are called **excited states.** Observe how the energies of the excited states come closer and closer together as n increases.

The electron in a hydrogen atom at room temperature spends most of its time in the ground state. To raise the electron from the ground state ($n = 1$) to the highest possible excited state ($n = \infty$), 13.6 eV of energy must be supplied. Supplying this amount of energy removes the electron from the atom, producing the positive hydrogen ion H⁺. This is the minimum energy needed to remove the electron and is called the **ionization energy.** Thus, the Bohr model predicts that the ionization energy of atomic hydrogen is 13.6 eV, in excellent agreement with the experimental value. In Example 3 the Bohr model is applied to doubly ionized lithium.

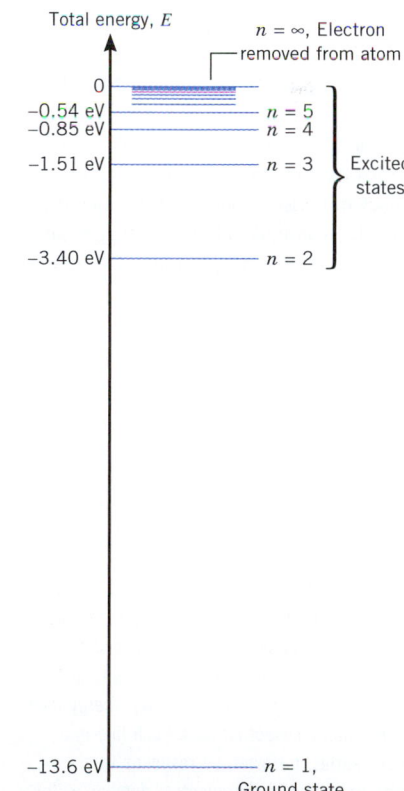

Figure 30.8 Energy level diagram for the hydrogen atom.

🔱 **Example 3** The Ionization Energy of Li²⁺

The Bohr model does not apply when more than one electron orbits the nucleus because it does not account for the electrostatic force that one electron exerts on another. For instance, an electrically neutral lithium atom (Li) contains three electrons in orbit around a nucleus that includes three protons ($Z = 3$), and Bohr's analysis is not applicable. However, the Bohr model can be used for the doubly charged positive ion of lithium (Li²⁺) that results when two electrons are removed from the neutral atom, leaving only one electron to orbit the nucleus. Obtain the ionization energy that is needed to remove the remaining electron from Li²⁺.

Reasoning The lithium ion Li²⁺ contains three times the positive nuclear charge that the hydrogen atom contains. Therefore, the orbiting electron is attracted more strongly to the nucleus in Li²⁺ than to the nucleus in the hydrogen atom. As a result, we expect that more energy is required to ionize Li²⁺ than the 13.6 eV required for atomic hydrogen.

Solution The Bohr energy levels for Li^{2+} are obtained from Equation 30.13 with $Z = 3$; $E_n = -(13.6 \text{ eV})(3^2/n^2)$. Therefore, the ground state ($n = 1$) energy is

$$E_1 = -(13.6 \text{ eV})\frac{3^2}{1^2} = -122 \text{ eV}$$

Removing the electron from Li^{2+} requires 122 eV of energy: $\boxed{\text{Ionization energy} = 122 \text{ eV}}$. This value for the ionization energy agrees well with the experimental value of 122.4 eV and, as expected, is greater than the 13.6 eV required for atomic hydrogen.

THE LINE SPECTRA OF THE HYDROGEN ATOM

To predict the wavelengths in the line spectrum of the hydrogen atom, Bohr combined his ideas about atoms (electron orbits are stationary orbits and the angular momentum of an electron is quantized) with Einstein's idea of the photon. As applied by Bohr, the photon concept is inherent in Equation 30.4, $E_i - E_f = hf$, which states that the frequency f of the photon is proportional to the difference between two energy levels of the hydrogen atom. If we substitute Equation 30.11 for the total energies E_i and E_f into Equation 30.4 and recall from Equation 16.1 that $f = c/\lambda$, we obtain the following result:

$$\frac{1}{\lambda} = \frac{2\pi^2 m k^2 e^4}{h^3 c}(Z^2)\left(\frac{1}{n_f^2} - \frac{1}{n_i^2}\right) \quad (30.14)$$

$$n_i, n_f = 1, 2, 3, \ldots \quad \text{and} \quad n_i > n_f$$

Using known values for h, m, k, e, and c, we find that $2\pi^2 m k^2 e^4/(h^3 c) = 1.097 \times 10^7 \text{ m}^{-1}$, in agreement with the Rydberg constant R that appears in Equations 30.1–30.3. The agreement between the theoretical and experimental values of the Rydberg constant was a major accomplishment of Bohr's theory.

With $Z = 1$ and $n_f = 1$, Equation 30.14 reproduces Equation 30.1 for the Lyman series. Thus, Bohr's model shows that the Lyman series of lines occurs when electrons make transitions from higher energy levels with $n_i = 2, 3, 4, \ldots$ to the first energy level where $n_f = 1$. Figure 30.9 shows these transitions. Notice that when an electron makes a transition from $n_i = 2$ to $n_f = 1$, the longest wavelength photon in the Lyman series is emitted, since the energy change is the smallest possible. When an electron makes a transition from the highest level where $n_i = \infty$ to the lowest level where $n_f = 1$, the shortest wavelength is emitted, since the energy change is the largest possible. Since the higher energy levels are increasingly close together, the lines in the series become more and more crowded toward the short wavelength limit, as can be seen in Figure 30.4. Figure 30.9 also shows the energy level transitions for the Balmer series, where $n_i = 3, 4, 5, \ldots$, and $n_f = 2$. In the Paschen series (see Figure 30.4) $n_i = 4, 5, 6, \ldots$, and $n_f = 3$. The next example deals further with the line spectrum of the hydrogen atom.

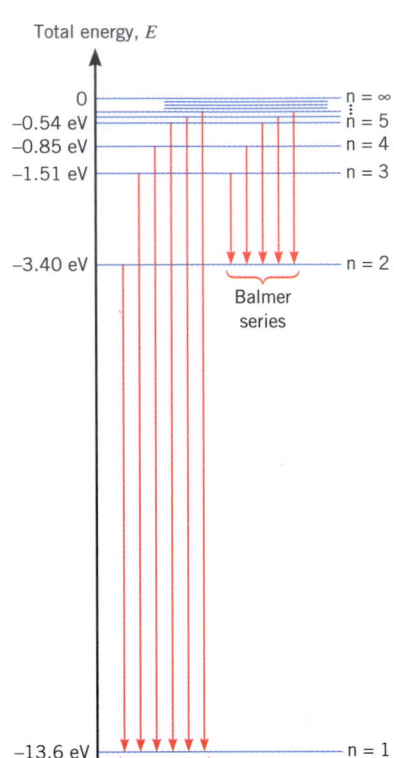

Total energy, E

$n = \infty$
-0.54 eV — $n = 5$
-0.85 eV — $n = 4$
-1.51 eV — $n = 3$

-3.40 eV — $n = 2$

Balmer series

-13.6 eV — $n = 1$

Lyman series

Figure 30.9 The Lyman and Balmer series of lines in the hydrogen atom spectrum correspond to transitions that the electron makes between higher and lower energy levels, as indicated here.

 Example 4 The Brackett Series for Atomic Hydrogen

In the line spectrum of atomic hydrogen there is also a group of lines known as the Brackett series. These lines are produced when electrons, excited to high energy levels, make transitions to the $n = 4$ level. Determine **(a)** the longest wavelength in this series and **(b)** the wavelength that corresponds to the transition from $n_i = 6$ to $n_f = 4$. **(c)** Refer to Figure 24.9 and identify the spectral region in which these lines are found.

Reasoning The longest wavelength corresponds to the transition that has the smallest energy change, which is between the $n_i = 5$ and $n_f = 4$ levels in Figure 30.8. The wavelength for this transition, as well as that for the transition from $n_i = 6$ to $n_f = 4$, can be obtained from Equation 30.14.

Solution **(a)** Using Equation 30.14 with $Z = 1$, $n_i = 5$, and $n_f = 4$, we find that

$$\frac{1}{\lambda} = (1.097 \times 10^7 \text{ m}^{-1})(1^2)\left(\frac{1}{4^2} - \frac{1}{5^2}\right) = 2.468 \times 10^5 \text{ m}^{-1} \quad \text{or} \quad \boxed{\lambda = 4051 \text{ nm}}$$

Problem-solving insight

In the line spectrum of atomic hydrogen, all lines in a given series (e.g., the Brackett series) are identified by a single value of the quantum number n_f for the *lower energy level into which an electron falls.* Each line in a given series, however, corresponds to a different value of the quantum number n_i for the *higher energy level where an electron originates.*

(b) The calculation here is similar to that in part (a):

$$\frac{1}{\lambda} = (1.097 \times 10^7 \text{ m}^{-1})(1^2)\left(\frac{1}{4^2} - \frac{1}{6^2}\right) = 3.809 \times 10^5 \text{ m}^{-1} \quad \text{or} \quad \boxed{\lambda = 2625 \text{ nm}}$$

(c) According to Figure 24.9, these lines lie in the $\boxed{\text{infrared region}}$ of the spectrum.

The various lines in the hydrogen atom spectrum are produced when electrons change from higher to lower energy levels and photons are emitted. Consequently, the spectral lines are called *emission lines.* Electrons can also make transitions in the reverse direction, from lower to higher levels, in a process known as *absorption.* In this case, an atom absorbs a photon that has precisely the energy needed to produce the transition. Thus, if photons with a continuous range of wavelengths pass through a gas and then are analyzed with a grating spectroscope, a series of dark *absorption lines* appears in the continuous spectrum. The dark lines indicate the wavelengths that have been removed by the absorption process. Such absorption lines can be seen in Figure 30.3 in the spectrum of the sun, where they are called Fraunhofer lines, after their discoverer. They are due to atoms, located in the outer and cooler layers of the sun, that absorb radiation coming from the interior. The interior portion of the sun is too hot for individual atoms to retain their structures and, therefore, the interior emits a continuous spectrum of wavelengths.

The physics of absorption lines in the sun's spectrum.

The Bohr model provides a great deal of insight into atomic structure. However, this model is now known to be oversimplified and has been superseded by a more detailed picture provided by quantum mechanics and the Schrödinger equation (see Section 30.5).

✓ **CHECK YOUR UNDERSTANDING**

(The answers are given at the end of the book.)

1. Which one of the following statements is true? **(a)** An atom is less easily ionized when its outermost electron is in an excited state than when it is in the ground state. **(b)** An atom is more easily ionized when its outermost electron is in an excited state than when it is in the ground state. **(c)** The energy state (excited state or ground state) of the outermost electron in an atom has nothing to do with how easily the atom can be ionized.

2. An electron in the hydrogen atom is in the $n = 4$ energy level. When this electron makes a transition to a lower energy level, is the wavelength of the photon emitted in **(a)** the Lyman series only, **(b)** the Balmer series only, **(c)** the Paschen series only, or **(d)** could it be in the Lyman, the Balmer, or the Paschen series?

3. A tube contains atomic hydrogen, and nearly all of the electrons in the atoms are in the ground state or $n = 1$ energy level. Electromagnetic radiation with a continuous spectrum of wavelengths (including those in the Lyman, Balmer, and Paschen series) enters one end of the tube and leaves the other end. The exiting radiation is found to contain strong absorption lines. To which one or more of the series do the wavelengths of these absorption lines correspond? Assume that once an electron absorbs a photon and jumps to a higher energy level, it does not absorb yet another photon and jump to an even higher energy level.

30.4

DE BROGLIE'S EXPLANATION OF BOHR'S ASSUMPTION ABOUT ANGULAR MOMENTUM

Of all the assumptions Bohr made in his model of the hydrogen atom, perhaps the most puzzling is the assumption about the angular momentum of the electron $[L_n = mv_nr_n = nh/(2\pi); n = 1, 2, 3, \ldots]$. Why should the angular momentum have only those values that are integer multiples of Planck's constant divided by 2π? In 1923, ten years after Bohr's work, de Broglie pointed out that his own theory for the wavelength of a moving particle could provide an answer to this question.

In de Broglie's way of thinking, the electron in its circular Bohr orbit must be pictured as a particle wave. And like waves traveling on a string, particle waves can lead to standing waves under resonant conditions. Section 17.5 discusses these conditions for a string. Standing waves form when the total distance traveled by a wave down the string and back is one wavelength, two wavelengths, or any integer number of wavelengths. The total

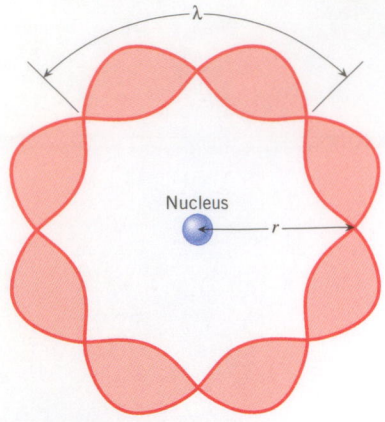

Figure 30.10 De Broglie suggested standing particle waves as an explanation for Bohr's angular momentum assumption. Here, a standing particle wave is illustrated on a Bohr orbit where four de Broglie wavelengths fit into the circumference of the orbit.

distance around a Bohr orbit of radius r is the circumference of the orbit or $2\pi r$. By the same reasoning, then, the condition for standing particle waves for the electron in a Bohr orbit would be

$$2\pi r = n\lambda \qquad n = 1, 2, 3, \ldots$$

where n is the number of whole wavelengths that fit into the circumference of the circle. But according to Equation 29.8 the de Broglie wavelength of the electron is $\lambda = h/p$, where p is the magnitude of the electron's momentum. If the speed of the electron is much less than the speed of light, the momentum is $p = mv$, and the condition for standing particle waves becomes $2\pi r = nh/(mv)$. A rearrangement of this result gives

$$mvr = n\frac{h}{2\pi} \qquad n = 1, 2, 3, \ldots$$

which is just what Bohr assumed for the angular momentum of the electron. As an example, Figure 30.10 illustrates the standing particle wave on a Bohr orbit for which $2\pi r = 4\lambda$.

De Broglie's explanation of Bohr's assumption about angular momentum emphasizes an important fact—namely, that particle waves play a central role in the structure of the atom. Moreover, the theoretical framework of quantum mechanics includes the Schrödinger equation for determining the wave function Ψ (Greek letter Psi) that represents a particle wave. The next section deals with the picture that quantum mechanics gives for atomic structure, a picture that supersedes the Bohr model. In any case, the Bohr expression for the energy levels (Equation 30.11) can be applied when a single electron orbits the nucleus, whereas the theoretical framework of quantum mechanics can be applied, in principle, to atoms that contain an arbitrary number of electrons.

30.5 THE QUANTUM MECHANICAL PICTURE OF THE HYDROGEN ATOM

The picture of the hydrogen atom that quantum mechanics and the Schrödinger equation provide differs in a number of ways from the Bohr model. The Bohr model uses a single integer number n to identify the various electron orbits and the associated energies. Because this number can have only discrete values, rather than a continuous range of values, n is called a **quantum number**. In contrast, quantum mechanics reveals that four different quantum numbers are needed to describe each state of the hydrogen atom. These four are described below.

1. **The principal quantum number n.** As in the Bohr model, this number determines the total energy of the atom and can have only integer values: $n = 1, 2, 3, \ldots$. In fact, the Schrödinger equation predicts* that the energy of the hydrogen atom is identical to the energy obtained from the Bohr model: $E_n = -(13.6 \text{ eV}) Z^2/n^2$.

2. **The orbital quantum number ℓ.** This number determines the angular momentum of the electron due to its orbital motion. The values that ℓ can have depend on the value of n, and only the following integers are allowed:

$$\ell = 0, 1, 2, \ldots, (n - 1)$$

For instance, if $n = 1$, the orbital quantum number can have only the value $\ell = 0$, but if $n = 4$, the values $\ell = 0, 1, 2$, and 3 are possible. The magnitude L of the angular momentum of the electron is

$$L = \sqrt{\ell(\ell + 1)}\,\frac{h}{2\pi} \tag{30.15}$$

3. **The magnetic quantum number m_ℓ.** The word "magnetic" is used here because an externally applied magnetic field influences the energy of the atom, and this

*This prediction requires that small relativistic effects and small interactions within the atom be ignored, and assumes that the hydrogen atom is not located in an external magnetic field.

Table 30.1 Quantum Numbers for the Hydrogen Atom

Name	Symbol	Allowed Values
Principal quantum number	n	1, 2, 3, . . .
Orbital quantum number	ℓ	0, 1, 2, . . . , $(n-1)$
Magnetic quantum number	m_ℓ	$-\ell, \ldots, -2, -1, 0, +1, +2, \ldots, +\ell$
Spin quantum number	m_s	$-\frac{1}{2}, +\frac{1}{2}$

quantum number is used in describing the effect. Since the effect was discovered by the Dutch physicist Pieter Zeeman (1865–1943), it is known as the *Zeeman effect*. When there is no external magnetic field, m_ℓ plays no role in determining the energy. In either event, the magnetic quantum number determines the component of the angular momentum along a specific direction, which is called the z direction by convention. The values that m_ℓ can have depend on the value of ℓ, with only the following positive and negative integers being permitted:

$$m_\ell = -\ell, \ldots, -2, -1, 0, +1, +2, \ldots, +\ell$$

For example, if the orbital quantum number is $\ell = 2$, then the magnetic quantum number can have the values $m_\ell = -2, -1, 0, +1,$ and $+2$. The component L_z of the angular momentum in the z direction is

$$L_z = m_\ell \frac{h}{2\pi} \tag{30.16}$$

4. **The spin quantum number m_s.** This number is needed because the electron has an intrinsic property called "spin angular momentum." Loosely speaking, we can view the electron as spinning while it orbits the nucleus, analogous to the way the earth spins as it orbits the sun. There are two possible values for the spin quantum number of the electron:

$$m_s = +\tfrac{1}{2} \quad \text{or} \quad m_s = -\tfrac{1}{2}$$

Sometimes the phrases "spin up" and "spin down" are used to refer to the directions of the spin angular momentum associated with the values for m_s.

Table 30.1 summarizes the four quantum numbers that are needed to describe each state of the hydrogen atom. One set of values for n, ℓ, m_ℓ, and m_s corresponds to one state. As the principal quantum number n increases, the number of possible combinations of the four quantum numbers rises rapidly, as Example 5 illustrates.

 Example 5   Quantum Mechanical States of the Hydrogen Atom

Determine the number of possible states for the hydrogen atom when the principal quantum number is **(a)** $n = 1$ and **(b)** $n = 2$.

Reasoning Each different combination of the four quantum numbers summarized in Table 30.1 corresponds to a different state. We begin with the value for n and find the allowed values for ℓ. Then, for each ℓ value we find the possibilities for m_ℓ. Finally, m_s may be $+\frac{1}{2}$ or $-\frac{1}{2}$ for each group of values for n, ℓ, and m_ℓ.

Solution **(a)** The diagram below shows the possibilities for ℓ, m_ℓ, and m_s when $n = 1$:

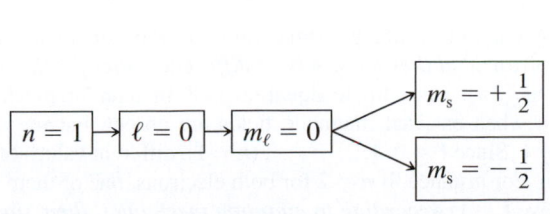

	State		
n	ℓ	m_ℓ	m_s
1	0	0	$+\frac{1}{2}$
1	0	0	$-\frac{1}{2}$

Thus, there are two different states for the hydrogen atom. In the absence of an external magnetic field, these two states have the same energy, since they have the same value of n.

(b) When $n = 2$, there are eight possible combinations for the values of n, ℓ, m_ℓ, and m_s, as the diagram below indicates:

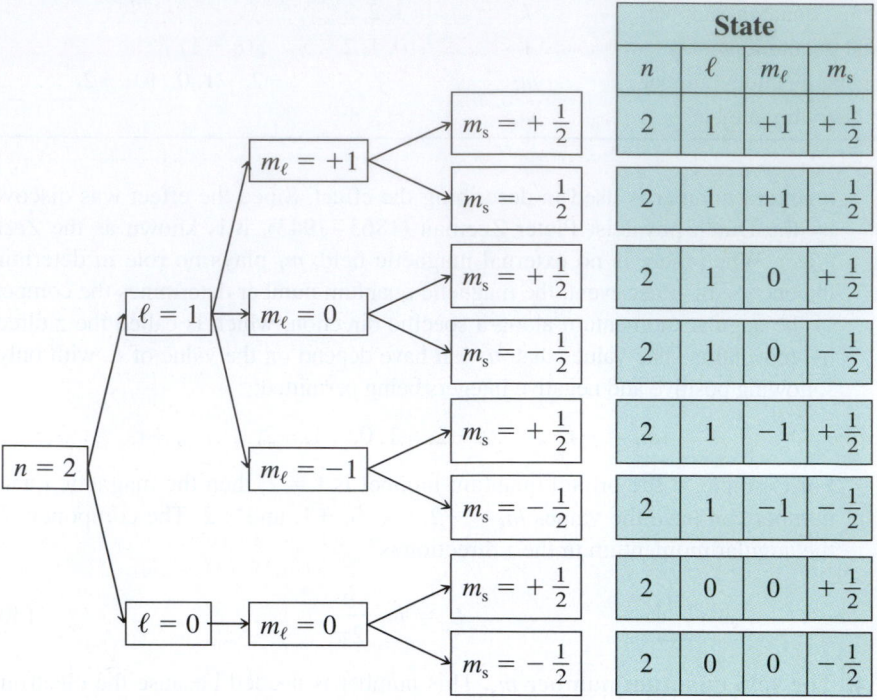

State			
n	ℓ	m_ℓ	m_s
2	1	+1	$+\frac{1}{2}$
2	1	+1	$-\frac{1}{2}$
2	1	0	$+\frac{1}{2}$
2	1	0	$-\frac{1}{2}$
2	1	−1	$+\frac{1}{2}$
2	1	−1	$-\frac{1}{2}$
2	0	0	$+\frac{1}{2}$
2	0	0	$-\frac{1}{2}$

With the same value of $n = 2$, all eight states have the same energy when there is no external magnetic field.

Quantum mechanics provides a more accurate picture of atomic structure than does the Bohr model. It is important to realize that the two pictures differ substantially, as Conceptual Example 6 illustrates.

Conceptual Example 6 The Bohr Model Versus Quantum Mechanics

Consider two hydrogen atoms. There are no external magnetic fields present, and the electron in each atom has the same energy. According to the Bohr model and to quantum mechanics, is it possible for the electrons in these atoms **(a)** to have zero orbital angular momentum and **(b)** to have different orbital angular momenta?

Reasoning and Solution **(a)** In both the Bohr model and quantum mechanics, the energy is proportional to $1/n^2$, according to Equation 30.13, where n is the principal quantum number. Moreover, the value of n may be $n = 1, 2, 3, \ldots$, and may not be zero. *In the Bohr model, the fact that n may not be zero means that it is not possible for the orbital angular momentum to be zero* because the angular momentum is proportional to n, according to Equation 30.8. In the quantum mechanical picture the magnitude of the orbital angular momentum is proportional to $\sqrt{\ell(\ell + 1)}$, as given by Equation 30.15. Here, ℓ is the orbital quantum number and may take on the values $\ell = 0, 1, 2, \ldots, (n - 1)$. We note that ℓ [and therefore $\sqrt{\ell(\ell + 1)}$] may be zero, no matter what the value for n is. Consequently, *the orbital angular momentum may be zero according to quantum mechanics,* in contrast to the case for the Bohr model.

(b) If the electrons have the same energy, they have the same value for the principal quantum number n. *In the Bohr model, this means that they cannot have different values for the orbital angular momentum L_n,* since $L_n = nh/(2\pi)$, according to Equation 30.8. In quantum mechanics, the energy is also determined by n when external magnetic fields are absent, but the orbital angular momentum is determined by ℓ. Since $\ell = 0, 1, 2, \ldots, (n - 1)$, different values of ℓ are compatible with the same value of n. For instance, if $n = 2$ for both electrons, one of them could have $\ell = 0$, while the other could have $\ell = 1$. *According to quantum mechanics, then, the electrons could have different orbital angular momenta, even though they have the same energy.*

The following table summarizes the discussion from parts (a) and (b):

	Bohr Model	Quantum Mechanics
(a) For a given n, can the angular momentum ever be zero?	No	Yes
(b) For a given n, can the angular momentum have different values?	No	Yes

Related Homework: *Problem 29*

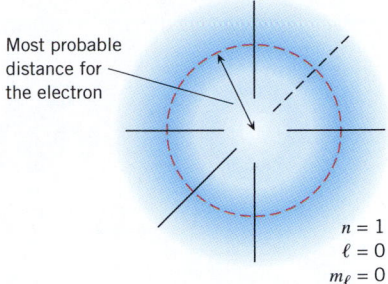

Figure 30.11 The electron probability cloud for the ground state ($n = 1$, $\ell = 0$, $m_\ell = 0$) of the hydrogen atom.

According to the Bohr model, the nth orbit is a circle of radius r_n, and every time the position of the electron in this orbit is measured, the electron is found exactly at a distance r_n away from the nucleus. This simplistic picture is now known to be incorrect, and the quantum mechanical picture of the atom has replaced it. Suppose that the electron is in a quantum mechanical state for which $n = 1$, and we imagine making a number of measurements of the electron's position with respect to the nucleus. We would find that its position is uncertain, in the sense that there is a probability of finding the electron sometimes very near the nucleus, sometimes very far from the nucleus, and sometimes at intermediate locations. The probability is determined by the wave function Ψ, as Section 29.5 discusses. We can make a three-dimensional picture of our findings by marking a dot at each location where the electron is found. More dots occur at places where the probability of finding the electron is higher, and after a sufficient number of measurements, a picture of the quantum mechanical state emerges. Figure 30.11 shows the spatial distribution for an electron in a state for which $n = 1$, $\ell = 0$, and $m_\ell = 0$. This picture is constructed from so many measurements that the individual dots are no longer visible but have merged to form a kind of probability "cloud" whose density changes gradually from place to place. The dense regions indicate places where the probability of finding the electron is higher, and the less dense regions indicate places where the probability is lower. Also indicated in Figure 30.11 is the radius where quantum mechanics predicts the greatest probability per unit radial distance of finding the electron in the $n = 1$ state. This radius matches exactly the radius of 5.29×10^{-11} m found for the first Bohr orbit.

For a principal quantum number of $n = 2$, the probability clouds are different than for $n = 1$. In fact, more than one cloud shape is possible because with $n = 2$ the orbital quantum number can be either $\ell = 0$ or $\ell = 1$. Although the value of ℓ does not affect the energy of the hydrogen atom, the value does have a significant effect on the shape of the probability clouds. Figure 30.12*a* shows the cloud for $n = 2$, $\ell = 0$, and $m_\ell = 0$. Part *b* of the drawing shows that when $n = 2$, $\ell = 1$, and $m_\ell = 0$, the cloud has a two-lobe shape with the nucleus at the center between the lobes. For larger values of n, the probability clouds become increasingly complex and are spread out over larger volumes of space.

The probability cloud picture of the electron in a hydrogen atom is very different from the well-defined orbit of the Bohr model. The fundamental reason for this difference is to be found in the Heisenberg uncertainty principle, as Conceptual Example 7 discusses.

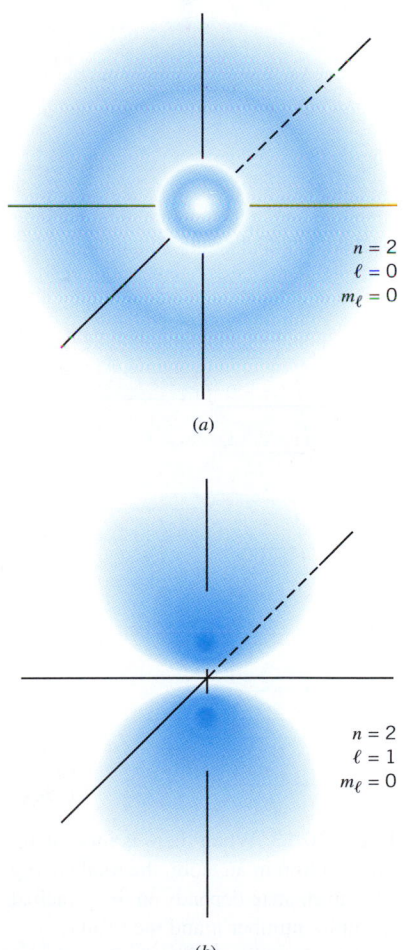

Figure 30.12 The electron probability clouds for the hydrogen atom when (a) $n = 2$, $\ell = 0$, $m_\ell = 0$ and (b) $n = 2$, $\ell = 1$, $m_\ell = 0$.

Conceptual Example 7 The Uncertainty Principle and the Hydrogen Atom

In the Bohr model of the hydrogen atom, the electron in the ground state ($n = 1$) is in an orbit that has a radius of exactly 5.29×10^{-11} m, so that the uncertainty Δy in its radial position is $\Delta y = 0$ m. According to the Heisenberg uncertainty principle, what does the fact that there is no uncertainty in the electron's radial position imply about the electron's radial speed? The uncertainty principle implies **(a)** nothing about the radial speed, **(b)** that the radial speed has only a small uncertainty, **(c)** that the radial speed has an infinitely large uncertainty.

Reasoning We need to obtain an expression for the uncertainty in the radial speed of the electron to use as a guide. As stated in Equation 29.10, the Heisenberg principle is $(\Delta p_y)(\Delta y) \geq h/(4\pi)$. In the present context, Δy is the uncertainty in the electron's radial position, so that Δp_y is the uncertainty in the electron's radial momentum. According to Equation 7.2, however, the magnitude of the momentum is $p_y = mv_y$, where m is the electron's mass and

v_y is the electron's radial speed. As a result, the uncertainty in the momentum is $\Delta p_y = \Delta(mv_y) = m\,\Delta v_y$. With this substitution for Δp_y, the Heisenberg principle becomes $(m\,\Delta v_y)(\Delta y) \geq h/(4\pi)$, which can be rearranged to show that

$$\Delta v_y \geq \frac{h}{m(\Delta y)(4\pi)}$$

Answers (a) and (b) are incorrect. Our result for Δv_y shows that the Heisenberg principle does indeed imply something about the uncertainty in the electron's radial speed. Since $\Delta y = 0$ m in the Bohr model, our result for Δv_y shows that the uncertainty in the speed is infinitely large (Δy is in the denominator on the right). Therefore, answers (a) and (b) cannot be correct.

Answer (c) is correct. Our expression for Δv_y shows that, in fact, the uncertainty in the radial speed is infinitely large ($\Delta y = 0$ m in the Bohr model and Δy is in the denominator on the right in the expression above). Such a large uncertainty in the radial speed means that the electron may be moving very rapidly in the radial direction and, therefore, would not remain in its Bohr orbit. Quantum mechanics, with its probability-cloud picture of atomic structure, correctly represents the positional and motional uncertainty that the Heisenberg principle reveals. The Bohr model does not correctly represent this aspect of reality at the atomic level.

✓ CHECK YOUR UNDERSTANDING

(The answers are given at the end of the book.)

4. In the Bohr model for the hydrogen atom, the closer the electron is to the nucleus, the smaller is the total energy of the electron. Is this also true in the quantum mechanical picture of the hydrogen atom?

5. In the quantum mechanical picture of the hydrogen atom, the orbital angular momentum of the electron may be zero in any of the possible energy states. For which energy state *must* the orbital angular momentum be zero?

6. Consider two different hydrogen atoms. The electron in each atom is in a different excited state, so that each electron has a different total energy. Is it possible for the electrons to have the same orbital angular momentum L, according to **(a)** the Bohr model and **(b)** quantum mechanics?

7. The magnitude of the orbital angular momentum of the electron in a hydrogen atom is observed to increase. According to **(a)** the Bohr model and **(b)** quantum mechanics, does this necessarily mean that the total energy of the electron also increases?

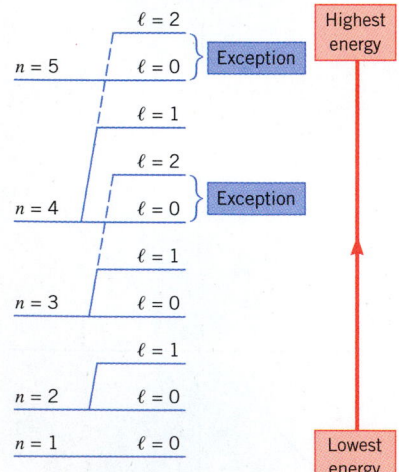

Figure 30.13 When there is more than one electron in an atom, the total energy of a given state depends on the principal quantum number n and the orbital quantum number ℓ. The energy increases with increasing n (with some exceptions) and, for a fixed n, with increasing ℓ. For clarity, levels for $n = 6$ and higher are not shown.

30.6 THE PAULI EXCLUSION PRINCIPLE AND THE PERIODIC TABLE OF THE ELEMENTS

Except for hydrogen, all electrically neutral atoms contain more than one electron, with the number given by the atomic number Z of the element. In addition to being attracted by the nucleus, the electrons repel each other. This repulsion contributes to the total energy of a multiple-electron atom. As a result, the one-electron energy expression for hydrogen, $E_n = -(13.6 \text{ eV})\, Z^2/n^2$, does not apply to other neutral atoms. However, the simplest approach for dealing with a multiple-electron atom still uses the four quantum numbers n, ℓ, m_ℓ, and m_s.

Detailed quantum mechanical calculations reveal that the energy level of each state of a multiple-electron atom depends on both the principal quantum number n and the orbital quantum number ℓ. Figure 30.13 illustrates that the energy generally increases as n increases, but there are exceptions, as the drawing indicates. Furthermore, for a given n, the energy also increases as ℓ increases.

In a multiple-electron atom, all electrons with the same value of n are said to be in the same *shell*. Electrons with $n = 1$ are in a single shell (sometimes called the K shell), electrons with $n = 2$ are in another shell (the L shell), those with $n = 3$ are in a third shell (the M shell), and so on. Those electrons with the same values for both n and ℓ are often referred to as being in the same *subshell*. The $n = 1$ shell consists of a single $\ell = 0$ subshell. The $n = 2$ shell has two subshells, one with $\ell = 0$ and one with $\ell = 1$. Similarly, the $n = 3$ shell has three subshells, one with $\ell = 0$, one with $\ell = 1$, and one with $\ell = 2$.

In the hydrogen atom near room temperature, the electron spends most of its time in the lowest energy level, or ground state—namely, in the $n = 1$ shell. Similarly, when an atom contains more than one electron and is near room temperature, the electrons spend most of their time in the lowest energy levels possible. The lowest energy state for an atom is called the **ground state.** However, when a multiple-electron atom is in its ground state, not every electron is in the $n = 1$ shell in general because the electrons obey a principle discovered by the Austrian physicist Wolfgang Pauli (1900–1958).

THE PAULI EXCLUSION PRINCIPLE

No two electrons in an atom can have the same set of values for the four quantum numbers n, ℓ, m_ℓ, and m_s.

Suppose two electrons in an atom have three quantum numbers that are identical: $n = 3$, $m_\ell = 1$, and $m_s = -\frac{1}{2}$. According to the exclusion principle, it is not possible for each to have $\ell = 2$, for example, since each would then have the same four quantum numbers. Each electron must have a different value for ℓ (for instance, $\ell = 1$ and $\ell = 2$) and, consequently, would be in a different subshell. With the aid of the Pauli exclusion principle, we can determine which energy levels are occupied by the electrons in an atom in its ground state, as the next example demonstrates.

 Example 8 Ground States of Atoms

Determine which of the energy levels in Figure 30.13 are occupied by the electrons in the ground state of hydrogen (1 electron), helium (2 electrons), lithium (3 electrons), beryllium (4 electrons), and boron (5 electrons).

Reasoning In the ground state of an atom the electrons are in the lowest available energy levels. Consistent with the Pauli exclusion principle, they fill those levels "from the bottom up"— that is, from the lowest to the highest energy.

Solution As the colored dot (●) in Figure 30.14 indicates, the electron in the hydrogen atom (H) is in the $n = 1$, $\ell = 0$ subshell, which has the lowest possible energy. A second electron is present in the helium atom (He), and both electrons can have the quantum numbers $n = 1$, $\ell = 0$, and $m_\ell = 0$. However, in accord with the Pauli exclusion principle, each electron must have a different spin quantum number, $m_s = +\frac{1}{2}$ for one electron and $m_s = -\frac{1}{2}$ for the other. Thus, the drawing shows both electrons in the lowest energy level.

The third electron that is present in the lithium atom (Li) would violate the exclusion principle if it were also in the $n = 1$, $\ell = 0$ subshell, no matter what the value for m_s is. Thus, the $n = 1$, $\ell = 0$ subshell is filled when occupied by two electrons. With this level filled, the $n = 2$, $\ell = 0$ subshell becomes the next lowest energy level available and is where the third electron of lithium is found (see Figure 30.14). In the beryllium atom (Be), the fourth electron is in the $n = 2$, $\ell = 0$ subshell, along with the third electron. This is possible, since the third and fourth electrons can have different values for m_s.

With the first four electrons in place as just discussed, the fifth electron in the boron atom (B) cannot fit into the $n = 1$, $\ell = 0$ or the $n = 2$, $\ell = 0$ subshell without violating the exclusion principle. Therefore, the fifth electron is found in the $n = 2$, $\ell = 1$ subshell, which is the next available energy level with the lowest energy, as Figure 30.14 indicates. For this electron, m_ℓ can be -1, 0, or $+1$, and m_s can be $+\frac{1}{2}$ or $-\frac{1}{2}$ in each case. However, in the absence of an external magnetic field, all six of these possibilities correspond to the same energy.

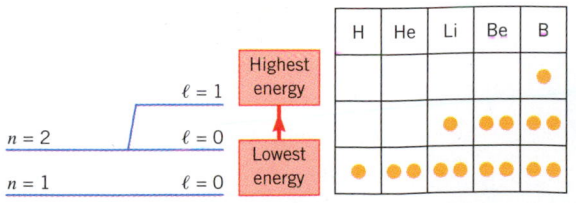

Figure 30.14 The electrons (●) in the ground state of an atom fill the available energy levels "from the bottom up"— that is, from the lowest to the highest energy, consistent with the Pauli exclusion principle. The ranking of the energy levels in this figure is meant to apply for a given atom only.

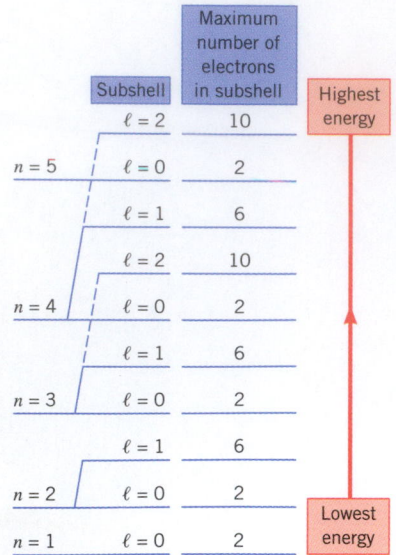

Figure 30.15 The maximum number of electrons that the ℓth subshell can hold is $2(2\ell + 1)$.

Table 30.2 The Convention of Letters Used to Refer to the Orbital Quantum Number

Orbital Quantum Number ℓ	Letter
0	s
1	p
2	d
3	f
4	g
5	h

Because of the Pauli exclusion principle, there is a maximum number of electrons that can fit into an energy level or subshell. Example 8 shows that the $n = 1$, $\ell = 0$ subshell can hold at most two electrons. The $n = 2$, $\ell = 1$ subshell, however, can hold six electrons because with $\ell = 1$, there are three possibilities for m_ℓ (−1, 0, and +1), and for each of these, the value of m_s can be $+\frac{1}{2}$ or $-\frac{1}{2}$. In general, m_ℓ can have the values 0, ±1, ±2, . . . , ±ℓ, for $2\ell + 1$ possibilities. Since each of these can be combined with two possibilities for m_s, the total number of different combinations for m_ℓ and m_s is $2(2\ell + 1)$. This, then, is the maximum number of electrons the ℓth subshell can hold, as Figure 30.15 summarizes.

For historical reasons, there is a widely used convention in which each subshell of an atom is referred to by a letter rather than by the value of its orbital quantum number ℓ. For instance, an $\ell = 0$ subshell is called an s subshell. An $\ell = 1$ subshell and an $\ell = 2$ subshell are known as p and d subshells, respectively. The higher values of $\ell = 3$, 4, and so on, are referred to as f, g, and so on, in alphabetical sequence, as Table 30.2 indicates.

This convention of letters is used in a shorthand notation that is convenient for indicating simultaneously the principal quantum number n, the orbital quantum number ℓ, and the number of electrons in the n, ℓ subshell. An example of this notation follows:

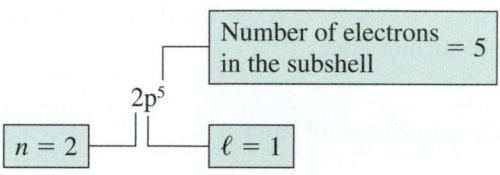

With this notation, the arrangement, or configuration, of the electrons in an atom can be specified efficiently. For instance, in Example 8, we found that the electron configuration for boron has two electrons in the $n = 1$, $\ell = 0$ subshell, two in the $n = 2$, $\ell = 0$ subshell, and one in the $n = 2$, $\ell = 1$ subshell. In shorthand notation this arrangement is expressed as $1s^2\, 2s^2\, 2p^1$. Table 30.3 gives the ground-state electron configurations written in this fashion for elements containing up to thirteen electrons. The first five entries are those worked out in Example 8.

Each entry in the periodic table of the elements often includes the ground-state electronic configuration, as Figure 30.16 illustrates for argon. To save space, only the configuration of the outermost electrons and unfilled subshells is specified, using the shorthand notation just discussed. Originally the periodic table was developed by

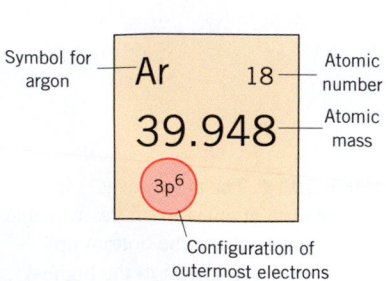

Figure 30.16 The entries in the periodic table of the elements often include the ground-state configuration of the outermost electrons.

Table 30.3 Ground-State Electronic Configurations of Atoms

Element	Number of Electrons	Configuration of the Electrons
Hydrogen (H)	1	$1s^1$
Helium (He)	2	$1s^2$
Lithium (Li)	3	$1s^2\, 2s^1$
Beryllium (Be)	4	$1s^2\, 2s^2$
Boron (B)	5	$1s^2\, 2s^2\, 2p^1$
Carbon (C)	6	$1s^2\, 2s^2\, 2p^2$
Nitrogen (N)	7	$1s^2\, 2s^2\, 2p^3$
Oxygen (O)	8	$1s^2\, 2s^2\, 2p^4$
Fluorine (F)	9	$1s^2\, 2s^2\, 2p^5$
Neon (Ne)	10	$1s^2\, 2s^2\, 2p^6$
Sodium (Na)	11	$1s^2\, 2s^2\, 2p^6\, 3s^1$
Magnesium (Mg)	12	$1s^2\, 2s^2\, 2p^6\, 3s^2$
Aluminum (Al)	13	$1s^2\, 2s^2\, 2p^6\, 3s^2\, 3p^1$

the Russian chemist Dmitri Mendeleev (1834–1907) on the basis that certain groups of elements exhibit similar chemical properties. There are eight of these groups, plus the transition elements in the middle of the table, which include the lanthanide series and the actinide series. The similar chemical properties within a group can be explained on the basis of the configurations of the outer electrons of the elements in the group. Thus, quantum mechanics and the Pauli exclusion principle offer an explanation for the chemical behavior of the atoms. The full periodic table can be found on the inside of the back cover.

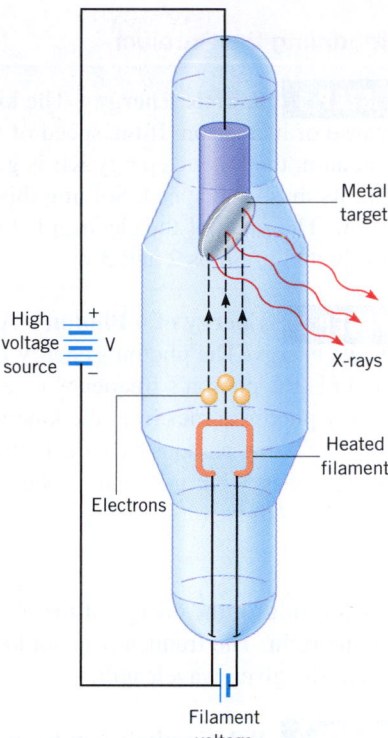

Figure 30.17 In an X-ray tube, electrons are emitted by a heated filament, accelerate through a large potential difference V, and strike a metal target. The X-rays originate when the electrons interact with the target.

✓ **CHECK YOUR UNDERSTANDING**

(*The answers are given at the end of the book.*)

8. Using the convention of letters to refer to the orbital quantum number, write down the ground-state configuration of the electrons in krypton ($Z = 36$).

9. Can a 5g subshell contain **(a)** 22 electrons? **(b)** 17 electrons?

10. An electronic configuration for manganese ($Z = 25$) is written as $1s^2\, 2s^2\, 2p^6\, 3s^2\, 3p^6\, 4s^2\, 3d^4\, 4p^1$. Does this configuration represent **(a)** the ground state or **(b)** an excited state?

X-RAYS

The physics of X-rays. X-rays were discovered by Wilhelm K. Roentgen (1845–1923), a Dutch physicist who performed much of his work in Germany. X-rays can be produced when electrons, accelerated through a large potential difference, collide with a metal target made from molybdenum or platinum, for example. The target is contained within an evacuated glass tube, as Figure 30.17 shows. Example 9 discusses the relationship between the wavelength of the emitted X-rays and the speed of the impinging electrons.

ANALYZING MULTIPLE-CONCEPT PROBLEMS

Example 9 X-Rays and Electrons

The highest-energy X-rays produced by an X-ray tube have a wavelength of 1.20×10^{-10} m. What is the speed of the electrons in Figure 30.17 just before they strike the metal target? Assume that the speed is much less than the speed of light in a vacuum.

Reasoning We can find the electron's speed from a knowledge of its kinetic energy, since the two are related. Moreover, it is the electron's kinetic energy that determines the energy of any photon. The energy of any photon, on the other hand, is directly proportional to its frequency, as discussed in Section 29.3. But we know from our study of waves (see Section 16.2) that the frequency of a wave is inversely proportional to its wavelength. Thus, we will be able to find the speed of an impinging electron from the given X-ray wavelength.

Knowns and Unknowns The following table summarizes what we know and what we seek:

Description	Symbol	Value
Wavelength of highest-energy X-ray photon	λ	1.20×10^{-10} m
Unknown Variable		
Speed of electron just before it strikes target	v	?

Continued

Modeling the Problem

STEP 1 **Kinetic Energy** The kinetic energy of an electron is the energy it has because of its motion. If the speed of the electron is much less than the speed of light in a vacuum, the kinetic energy KE is given by Equation 6.2 as $KE = \frac{1}{2}mv^2$, where m and v are its mass and speed. Solving this expression for the speed gives Equation 1 at the right. The mass of the electron is known but its kinetic energy is not, so this will be evaluated in Steps 2 and 3.

$$v = \sqrt{\frac{2(KE)}{m}} \quad (1)$$

?

STEP 2 **Energy of a Photon** An X-ray photon is a discrete packet of electromagnetic-wave energy. The photon's energy E is given by $E = hf$, where h is Planck's constant and f is the photon's frequency (see Equation 29.2). The energy needed to produce an X-ray photon comes from the kinetic energy of an electron striking the target. We know that the X-ray photons have the highest possible energy. This means that all of an electron's kinetic energy KE goes into producing an X-ray photon, so $KE = E$. Substituting $KE = E$ into $E = hf$ gives

$$\boxed{KE = hf} \quad (2)$$

This result for the energy of the photon can be substituted into Equation 1, as indicated at the right. The frequency is not known, but Step 3 discusses how it can be obtained from the given wavelength.

$$v = \sqrt{\frac{2(KE)}{m}} \quad (1)$$

$$\boxed{KE = hf} \quad (2)$$

?

STEP 3 **Relation between Frequency and Wavelength** Electromagnetic waves such as X-rays travel at the speed c of light in a vacuum. According to Equation 16.1, this speed is related to the frequency f and the wavelength λ by $c = f\lambda$. Solving for the frequency gives

$$\boxed{f = \frac{c}{\lambda}}$$

All the variables on the right side of this equation are known, so we substitute it into Equation 2 for the kinetic energy, as shown in the right column.

$$v = \sqrt{\frac{2(KE)}{m}} \quad (1)$$

$$\boxed{KE = hf} \quad (2)$$

$$\boxed{f = \frac{c}{\lambda}}$$

Solution Algebraically combining the results of the three steps, we have

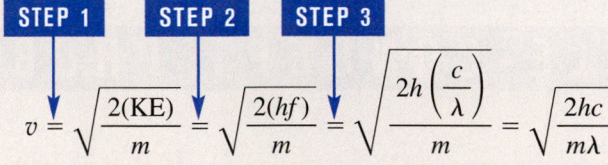

$$v = \sqrt{\frac{2(KE)}{m}} = \sqrt{\frac{2(hf)}{m}} = \sqrt{\frac{2h\left(\dfrac{c}{\lambda}\right)}{m}} = \sqrt{\frac{2hc}{m\lambda}}$$

The speed of an electron just before it strikes the metal target is

$$v = \sqrt{\frac{2hc}{m\lambda}} = \sqrt{\frac{2(6.63 \times 10^{-34}\,\text{J} \cdot \text{s})(3.00 \times 10^{8}\,\text{m/s})}{(9.11 \times 10^{-31}\,\text{kg})(1.20 \times 10^{-10}\,\text{m})}} = \boxed{6.03 \times 10^{7}\,\text{m/s}}$$

Related Homework: *Problem 42*

A plot of X-ray intensity per unit wavelength versus the wavelength looks similar to Figure 30.18 and consists of sharp peaks or lines superimposed on a broad continuous spectrum. The sharp peaks are called characteristic lines or ***characteristic X-rays*** because they are characteristic of the target material. The broad continuous spectrum is referred to as ***Bremsstrahlung*** (German for "braking radiation") and is emitted when the electrons decelerate or "brake" upon hitting the target.

In Figure 30.18 the characteristic lines are marked K_α and K_β because they involve the $n = 1$ or K shell of a metal atom. If an electron with enough energy strikes the target, one of the K-shell electrons can be knocked out. An electron in one of the outer shells can then fall into the K shell, and an X-ray photon is emitted in the process. The K_α line arises when an electron in the $n = 2$ level falls into the vacancy that the impinging electron has created in the $n = 1$ level. Similarly, the K_β line arises when an electron in the $n = 3$ level falls to the $n = 1$ level. Example 10 shows that a large potential difference is needed to operate an X-ray tube so that the electrons impinging on the metal target will have sufficient energy to generate the characteristic X-rays. Example 11 determines an estimate for the K_α wavelength of platinum.

 Example 10 Operating an X-Ray Tube

Strictly speaking, the Bohr model does not apply to multiple-electron atoms, but it can be used to make estimates. Use the Bohr model to estimate the minimum energy that an incoming electron must have to knock a K-shell electron entirely out of an atom in a platinum ($Z = 78$) target in an X-ray tube.

Reasoning According to the Bohr model, the energy of a K-shell electron is given by Equation 30.13, $E_n = -(13.6 \text{ eV}) Z^2/n^2$, with $n = 1$. When striking a platinum target, an incoming electron must have at least enough energy to raise the K-shell electron from this low energy level up to the 0-eV level that corresponds to a very large distance from the nucleus. Only then will the incoming electron knock the K-shell electron entirely out of a target atom.

Solution The energy of the Bohr $n = 1$ level is

$$E_1 = -(13.6 \text{ eV}) \frac{Z^2}{n^2} = -(13.6 \text{ eV}) \frac{(78 - 1)^2}{1^2} = -8.1 \times 10^4 \text{ eV}$$

In this calculation we have used $78 - 1$ rather than 78 for the value of Z. In so doing, we account approximately for the fact that each of the two K-shell electrons applies a repulsive force to the other. This repulsive force tends to balance the attractive force of one nuclear proton. In effect, one electron shields the other from the force of that proton. Therefore, to raise the K-shell electron up to the 0-eV level, the minimum energy for an incoming electron is $\boxed{8.1 \times 10^4 \text{ eV}}$. One electron volt is the kinetic energy acquired when an electron accelerates from rest through a potential difference of one volt. Thus, a potential difference of 81 000 V must be applied to the X-ray tube.

 Example 11 The K_α Characteristic X-Ray for Platinum

Use the Bohr model to estimate the wavelength of the K_α line in the X-ray spectrum of platinum ($Z = 78$).

Reasoning This example is very similar to Example 4, which deals with the emission line spectrum of the hydrogen atom. As in that example, we use Equation 30.14, this time with the initial value of n being $n_i = 2$ and the final value being $n_f = 1$. As in Example 10, a value of 77 rather than 78 is used for Z to account approximately for the shielding effect of the single K-shell electron in canceling out the attraction of one nuclear proton.

Solution Using Equation 30.14, we find that

$$\frac{1}{\lambda} = (1.097 \times 10^7 \text{ m}^{-1})(78 - 1)^2 \left(\frac{1}{1^2} - \frac{1}{2^2} \right) = 4.9 \times 10^{10} \text{ m}^{-1} \quad \text{or}$$

$$\boxed{\lambda = 2.0 \times 10^{-11} \text{ m}}$$

This answer is close to an experimental value of 1.9×10^{-11} m.

Another interesting feature of the X-ray spectrum in Figure 30.18 is the sharp cutoff that occurs at a wavelength of λ_0 on the short-wavelength side of the Bremsstrahlung. This cutoff wavelength is independent of the target material but depends on the energy of the impinging electrons. An impinging electron cannot give up any more than all of its kinetic energy when decelerated by the metal target in an X-ray tube. Thus, at most, an emitted X-ray photon can have an energy equal to the kinetic energy KE of the electron and a frequency given by Equation 29.2 as $f = (\text{KE})/h$, where h is Planck's constant. But the kinetic energy acquired by an electron (charge magnitude $= e$) in accelerating from rest through

Figure 30.18 When a molybdenum target is bombarded with electrons that have been accelerated from rest through a potential difference of 45 000 V, this X-ray spectrum is produced. The vertical axis is not to scale.

Problem-solving insight

Equation 30.13 for the Bohr energy levels [$E_n = -(13.6 \text{ eV})Z^2/n^2, n = 1$] can be used in rough calculations of the energy levels involved in the production of K_α X-rays. In this equation, however, the atomic number Z must be reduced by one, to account approximately for the shielding of one K-shell electron by the other K-shell electron.

Radiologist Steven Sirr demonstrates for violinist Desiree Ruhstrat the X-ray procedure called computed tomography (CT) on her violin. CT scans can reveal defects, information about how instruments are made, and distinctive "fingerprints" that can aid in identifying lost or stolen instruments. (© AP/Wide World Photos)

a potential difference V is e times V, according to earlier discussions in Section 19.2; V is the potential difference applied across the X-ray tube (see Figure 30.17). Thus, the maximum photon frequency is $f_0 = (eV)/h$. Since $f_0 = c/\lambda_0$, a maximum frequency corresponds to a minimum wavelength, which is the cutoff wavelength λ_0:

$$\lambda_0 = \frac{hc}{eV} \tag{30.17}$$

Figure 30.18, for instance, assumes a potential difference of 45 000 V, which corresponds to a cutoff wavelength of

$$\lambda_0 = \frac{(6.63 \times 10^{-34} \text{ J·s})(3.00 \times 10^8 \text{ m/s})}{(1.60 \times 10^{-19} \text{ C})(45\,000 \text{ V})} = 2.8 \times 10^{-11} \text{ m}$$

The medical profession began using X-rays for diagnostic purposes almost immediately after their discovery. When a conventional X-ray is obtained, the patient is typically positioned in front of a piece of photographic film, and a single burst of radiation is directed through the patient and onto the film. Since the dense structure of bone absorbs X-rays much more than soft tissue does, a shadow-like picture is recorded on the film. As useful as such pictures are, they have an inherent limitation. The image on the film is a superposition of all the "shadows" that result as the radiation passes through one layer of body material after another. Interpreting which part of a conventional X-ray corresponds to which layer of body material is very difficult.

The physics of CAT scanning.

The technique known as CAT scanning, or CT scanning, has greatly extended the ability of X-rays to provide images from specific locations within the body. The acronym CAT stands for **c**omputerized **a**xial **t**omography or **c**omputer-**a**ssisted **t**omography, and the shorter version CT stands for **c**omputerized **t**omography. In this technique a series of X-ray images are obtained as indicated in Figure 30.19. A number of X-ray beams form a "fanned out" array of radiation and pass simultaneously through the patient. Each of the beams is detected on the other side by a detector, which records the beam intensity. The various intensities are different, depending on the nature of the body material through which the beams have passed. The feature of CAT scanning that leads to dramatic improvements over the conventional technique is that the X-ray source can be rotated to different orientations, so that the fanned-out array of beams can be sent through the patient from various directions. Figure 30.19a singles out two directions for illustration. In reality many different orientations are used, and the intensity of each beam in the array is recorded as a function of orientation. The way in which the intensity of a beam changes from one orientation to another is used as input to a computer. The computer then constructs a highly resolved image of the cross-sectional slice of body material through which the fan of radiation has passed. In effect, the CAT scanning technique makes it possible to take an X-ray picture of a cross-sectional "slice" that is perpendicular to the body's long axis. In fact, the word "axial" in the phrase "computerized axial tomography" refers to the body's long axis. Figure 30.20 shows three-dimensional CAT scans of parts of the human anatomy.

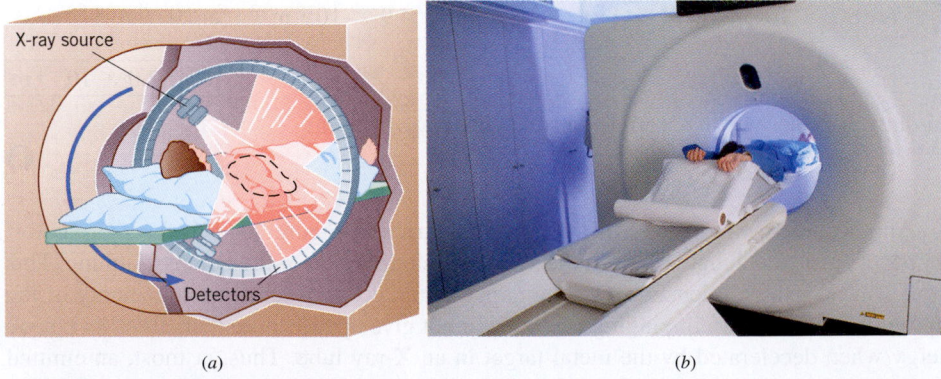

(a) (b)

Figure 30.19 (a) In CAT scanning, a "fanned-out" array of X-ray beams is sent through the patient from different orientations. (b) A patient in a CAT scanner. (© Laurent/American Hospital of Paris/Photo Researchers, Inc.)

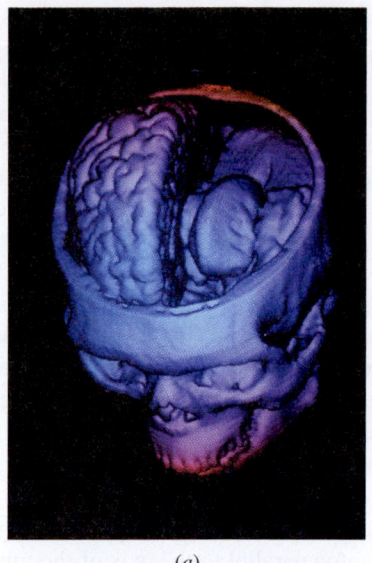 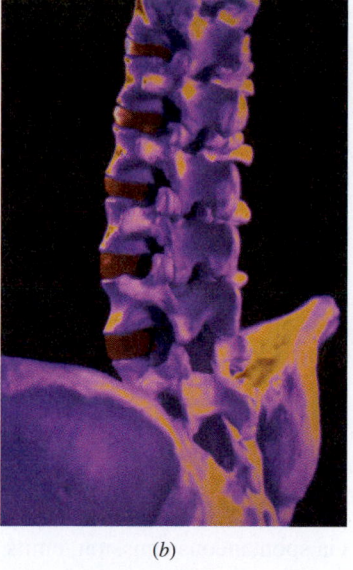

(a) (b)

Figure 30.20 These three-dimensional CAT scans show an image of (a) a skull and brain and (b) the pelvis and part of the spine, including the intervertebral discs. (a, © GJLP/Phototake; b, © Collection CNRI/Phototake)

✓ **CHECK YOUR UNDERSTANDING**

(The answers are given at the end of the book.)

11. X-ray tube A and X-ray tube B use the same voltage to accelerate the electrons. However, tube A uses a copper target, whereas tube B uses a silver target. Which one of the following statements is true? **(a)** The cutoff wavelength is greater for tube A. **(b)** The cutoff wavelength is greater for tube B. **(c)** Both tubes have the same cutoff wavelengths.

12. Is it possible to adjust the electric potential V used to operate an X-ray tube so that Bremsstrahlung X-rays are created but characteristic X-rays are not created? **(a)** Yes, if V is small enough. **(b)** Yes, if V is large enough. **(c)** No, regardless of the value of V.

13. Which one of the following statements is true? **(a)** The K_α wavelength can be smaller than the cutoff wavelength λ_0, assuming that both are produced by the same X-ray tube. **(b)** The K_α wavelength is produced when an electron undergoes a transition from the $n = 1$ energy level to the $n = 2$ energy level. **(c)** The K_β wavelength is always smaller than the K_α wavelength for a given metal target.

30.8 THE LASER

The physics of the laser. The laser is one of the most useful inventions of the twentieth century. Today, there are many types of lasers, and most of them work in a way that depends directly on the quantum mechanical structure of the atom.

When an electron makes a transition from a higher energy state to a lower energy state, a photon is emitted. The emission process can be one of two types, spontaneous or stimulated. In *spontaneous emission* (see Figure 30.21a), the photon is emitted spontaneously, in a random direction, without external provocation. In *stimulated emission* (see Figure 30.21b), an incoming photon induces, or stimulates, the electron to change energy levels. To produce stimulated emission, however, the incoming photon must have an energy that exactly matches the difference between the energies of the two levels—namely, $E_i - E_f$. Stimulated emission is similar to a resonant process, in which the incoming photon "jiggles" the electron at just the frequency to which it is particularly sensitive and causes the change between energy levels. This frequency is given by Equation 30.4 as $f = (E_i - E_f)/h$. The operation of lasers depends on stimulated emission.

Stimulated emission has three important features. First, one photon goes in and two photons come out (see Figure 30.21b). In this sense, the process amplifies the number of photons. In fact, this is the origin of the word "laser," which is an acronym for **l**ight **a**mplification by the **s**timulated **e**mission of **r**adiation. Second, the emitted photon travels in the same direction as the incoming photon. Third, the emitted photon is exactly in step with or has the same phase as the incoming photon. In other words, the two electromagnetic waves that these two photons represent are coherent (see Section 17.2) and are locked in

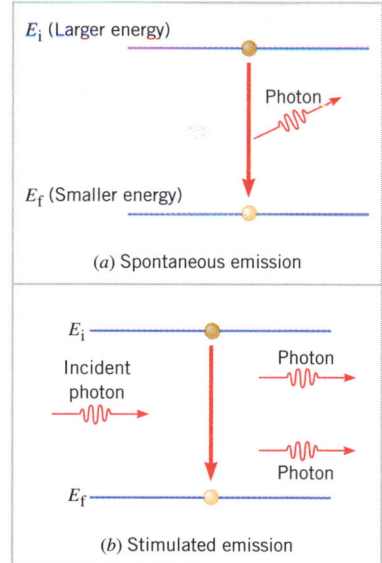

Figure 30.21 (a) Spontaneous emission of a photon occurs when the electron (●) makes an unprovoked transition from a higher to a lower energy level, the photon departing in a random direction. (b) Stimulated emission of a photon occurs when an incoming photon with the correct energy induces an electron to change energy levels, the emitted photon traveling in the same direction as the incoming photon.

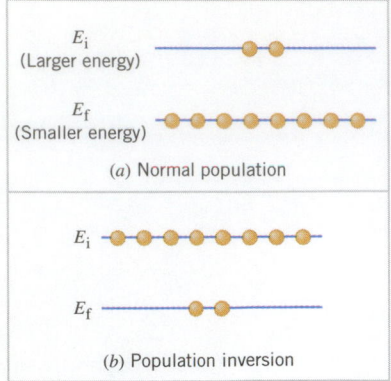

E_i
(Larger energy)

E_f
(Smaller energy)

(a) Normal population

E_i

E_f

(b) Population inversion

Figure 30.22 (a) In a normal situation at room temperature, most of the electrons in atoms are found in a lower or ground-state energy level. (b) If an external energy source is provided to excite electrons into a higher energy level, a population inversion can be created in which more electrons are in the higher level than in the lower level.

step with one another. In contrast, two photons emitted by the filament of an incandescent light bulb are emitted independently. They are not coherent, since one does not stimulate the emission of the other.

Although stimulated emission plays a pivotal role in a laser, other factors are also important. For instance, an external source of energy must be available to excite electrons into higher energy levels. The energy can be provided in a number of ways, including intense flashes of ordinary light and high-voltage discharges. If sufficient energy is delivered to the atoms, more electrons will be excited to a higher energy level than remain in a lower level, a condition known as a **population inversion.** Figure 30.22 compares a normal energy level population with a population inversion. The population inversions used in lasers involve a higher energy state that is **metastable,** in the sense that electrons remain in it for a much longer period of time than they do in an ordinary excited state (10^{-3} s versus 10^{-8} s, for example). The requirement of a metastable higher energy state is essential, so that there is more time to enhance the population inversion.

Figure 30.23 shows the widely used helium/neon laser. To sustain the necessary population inversion, a high voltage is discharged across a low-pressure mixture of 15% helium and 85% neon contained in a glass tube. The laser process begins when an atom, via spontaneous emission, emits a photon parallel to the axis of the tube. This photon, via stimulated emission, causes another atom to emit two photons parallel to the tube axis. These two photons, in turn, stimulate two more atoms, yielding four photons. Four yield eight, and so on, in a kind of avalanche. To ensure that more and more photons are created by stimulated emission, both ends of the tube are silvered to form mirrors that reflect the photons back and forth through the helium/neon mixture. One end is only partially silvered, however, so that some of the photons can escape from the tube to form the laser beam. When the stimulated emission involves only a single pair of energy levels, the output beam has a single frequency or wavelength and is said to be monochromatic.

A laser beam is also exceptionally narrow. The width is determined by the size of the opening through which the beam exits, and very little spreading-out occurs, except that due to diffraction around the edges of the opening. A laser beam does not spread much because any photons emitted at an angle with respect to the tube axis are quickly reflected out the sides of the tube by the silvered ends (see Figure 30.23). These ends are carefully arranged to be perpendicular to the tube axis. Since all the power in a laser beam can be confined to a narrow region, the intensity, or power per unit area, can be quite large.

Figure 30.24 shows the pertinent energy levels for a helium/neon laser. By coincidence, helium and neon have nearly identical metastable higher energy states, respectively

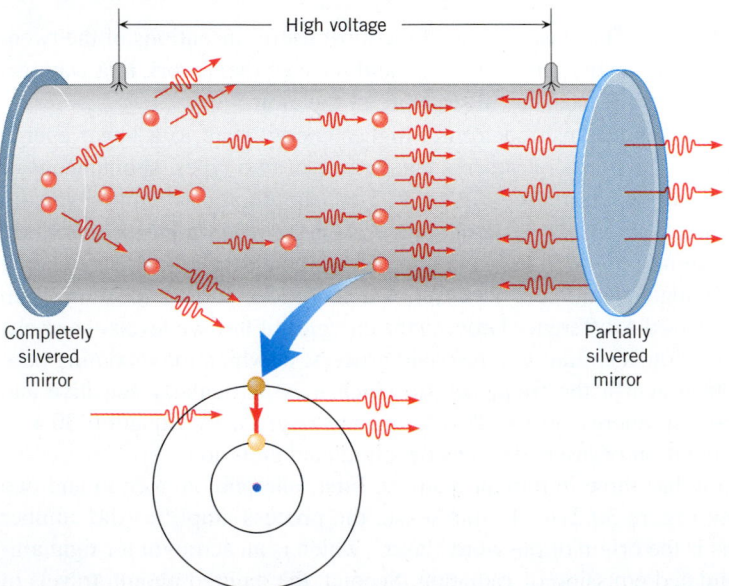

High voltage

Completely
silvered
mirror

Partially
silvered
mirror

Figure 30.23 A schematic drawing of a helium/neon laser. The blow-up shows the stimulated emission that occurs when an electron in a neon atom is induced to change from a higher to a lower energy level.

located 20.61 and 20.66 eV above the ground state. The high-voltage discharge across the gaseous mixture excites electrons in helium atoms to the 20.61-eV state. Then, when an excited helium atom collides inelastically with a neon atom, the 20.61 eV of energy is given to an electron in the neon atom, along with 0.05 eV of kinetic energy from the moving atoms. As a result, the electron in the neon atom is raised to the 20.66-eV state. In this fashion, a population inversion is sustained in the neon, relative to an energy level that is 18.70 eV above the ground state. In producing the laser beam, stimulated emission causes electrons in neon to drop from the 20.66-eV level to the 18.70-eV level. The energy change of 1.96 eV corresponds to a wavelength of 633 nm, which is in the red region of the visible spectrum.

The helium/neon laser is not the only kind of laser. There are many different types, including the ruby laser, the argon-ion laser, the carbon dioxide laser, the gallium arsenide solid-state laser, and chemical dye lasers. Depending on the type and whether the laser operates continuously or in pulses, the available beam power ranges from milliwatts to megawatts. Since lasers provide coherent monochromatic electromagnetic radiation that can be confined to an intense narrow beam, they are useful in a wide variety of situations. Today they are used to reproduce music in compact disc players, to weld parts of automobile frames together, to transmit telephone conversations and other forms of communication over long distances, to study molecular structure, and to measure distances accurately in surveying. For example, Figure 30.25 shows a three-dimensional map of the Martian topography that was obtained by the Mars Orbiter Laser Altimeter (MOLA) on the Mars Global Surveyor spacecraft. The map was constructed from 27 million height measurements, each made by sending laser pulses to the Martian surface and measuring their return times. The large *Hellas Planitia* impact basin (dark blue) is at the lower left and is 1800 km wide. At the upper right edge of the image is *Elysium Mons* (red, surrounded by a small band of yellow), a large volcano.

Many other uses have been found since the laser was invented in 1960, and the next section discusses some of them in the field of medicine.

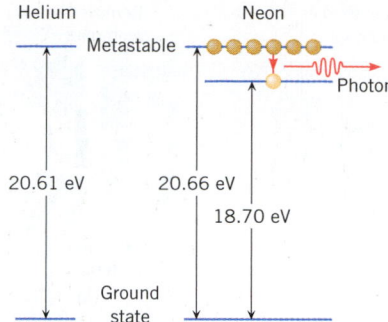

Figure 30.24 These energy levels are involved in the operation of a helium/neon laser.

**The physics of
a laser altimeter.**

✔ **CHECK YOUR UNDERSTANDING**

(*The answers are given at the end of the book.*)

14. A certain laser is designed to operate continuously. Which one of the following statements is false? **(a)** The population inversion used in this laser involves a higher energy state and a lower energy state. **(b)** The population inversion used in this laser involves a metastable higher energy state. **(c)** The laser needs an external source of energy to operate. **(d)** The external energy source that the laser uses can be disconnected once the population inversion is established.

15. Laser A produces green light. Laser B produces red light. Which laser utilizes energy levels that have a larger energy difference between them? **(a)** Laser A **(b)** Laser B **(c)** The energy difference between the levels is the same for both lasers.

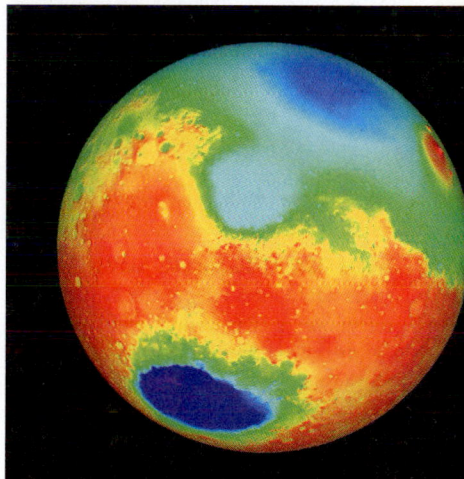

Figure 30.25 A three-dimensional map of the topography of Mars. The elevation is color-coded from white (highest) through red, yellow, green, blue, and purple (lowest). (NASA/Photo Researchers, Inc.)

30.9 *MEDICAL APPLICATIONS OF THE LASER

One of the medical areas in which the laser has had a substantial impact is in ophthalmology, which deals with the structure, function, and diseases of the eye. Section 26.10 discusses the human eye and the use of contact lenses and eyeglasses to correct nearsightedness and farsightedness. In these conditions, the eye cannot refract light properly and produces blurred images on the retina.

A laser-based procedure known as PRK (**p**hoto**r**efractive **k**eratectomy) offers an alternative treatment for nearsightedness and farsightedness that does not rely on lenses. It involves the use of a laser to remove small amounts of tissue from the cornea of the eye (see Figure 26.33) and thereby change its curvature. As Section 26.10 points out, light enters the eye through the cornea, and it is at the air/cornea boundary that most of the refraction of the light occurs. Therefore, changing the curvature of that boundary can correct deficiencies in the way the eye refracts light, thus causing the image to be focused onto the retina where it belongs. Ideally, the cornea is dome-shaped. If the dome is too steep,

**The physics of
PRK eye surgery.**

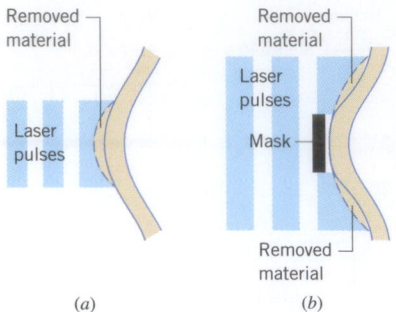

Figure 30.26 (*a*) To correct for myopia (nearsightedness) using the PRK procedure, a laser vaporizes tissue (dashed line) on the center of the cornea, thereby flattening it. (*b*) To correct for hyperopia (farsightedness), a laser vaporizes tissue on the peripheral region of the cornea, thereby steepening its contour.

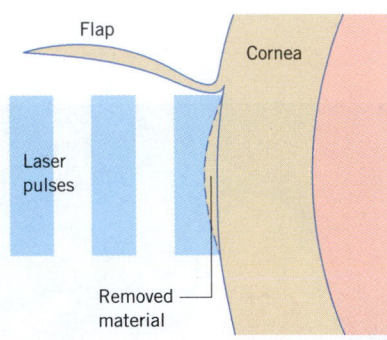

Figure 30.27 To correct for myopia (nearsightedness) using the LASIK technique, a laser vaporizes tissue (dashed line) on the cornea, thereby flattening it.

however, the rays of light are focused in front of the retina and nearsightedness results. As Figure 30.26*a* shows, the laser light removes tissue from the center of the cornea, thereby flattening it and increasing the eye's effective focal length. On the other hand, if the shape of the cornea is too flat, light rays would come to a focus behind the retina if they could, and farsightedness occurs. As part *b* of the drawing illustrates, the center of the cornea is now masked and the laser is used to remove peripheral tissue. This steepens the shape of the cornea, thereby shortening the eye's effective focal length and allowing rays to be focused on the retina.

The physics of LASIK eye surgery. The LASIK (laser-assisted in situ keratomileusis) procedure uses a motor-powered blade known as a microkeratome to partially detach a thin flap (about 0.2 mm thick) in the front of the cornea (see Figure 30.27). The flap is pulled back and the laser beam then remodels the corneal tissue underneath by vaporizing cells. Afterward, the flap is folded back into place, with no stitches being required. The laser light in the PRK and LASIK techniques is pulsed and comes from an ultraviolet excimer laser that produces a wavelength of 193 nm. The cornea absorbs this wavelength extremely well, so that weak pulses can be used, leading to highly precise and controllable removal of corneal tissue. Typically, 0.1 to 0.5 μm of tissue is removed by each pulse without damaging adjacent layers.

The physics of removing port-wine stains. Another medical application of the laser is in the treatment of congenital capillary malformations known as port-wine stains, which affect 0.3% of children at birth. These birthmarks are usually found on the head and neck, as Figure 30.28*a* illustrates. Preferred treatment for port-wine stains now utilizes a pulsed dye laser. Figure 30.28*b* shows an example of an excellent result after irradiation with laser light of wavelength 585 nm, in the form of pulses lasting 0.45 ms and occurring every 3 s. The laser beam was focused to a spot 5 mm in diameter. The light is absorbed by oxyhemoglobin in the malformed capillaries, which are destroyed in the process without damage to adjacent normal tissue. Eventually the destroyed capillaries are replaced by normal blood vessels, which causes the port-wine stain to fade.

The physics of photodynamic therapy for cancer. In the treatment of cancer, the laser is being used along with light-activated drugs in photodynamic therapy. The procedure involves administering the drug intravenously, so that the tumor can absorb it from the bloodstream, the advantage being that the drug is then located right near the cancer cells. When the drug is activated by laser light, a chemical reaction ensues that disintegrates the cancer cells and the small blood vessels that feed them. In Figure 30.29 a patient is being treated for cancer of the esophagus. An endoscope that uses optical fibers is inserted down the patient's throat to guide the red laser light to the tumor site and activate the drug. Photodynamic therapy works best with small tumors in their early stages.

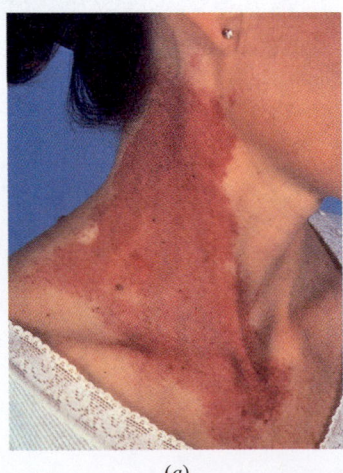
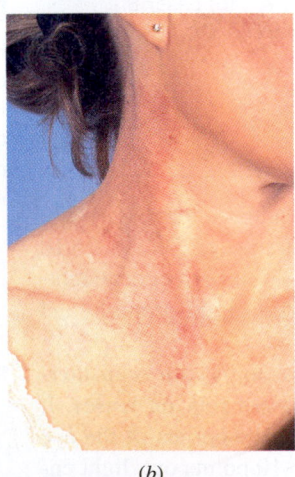

Figure 30.28 A patient with a port-wine stain (*a*) before and (*b*) after treatment using a pulsed dye laser (wavelength = 585 nm). [From E. Tan and C. Vinciullo, Pulsed Dye Laser Treatment of Port Wine Stains, *Med J. Australia*, 164 (6):333, 1996. © 1996. Reproduced with permission.]

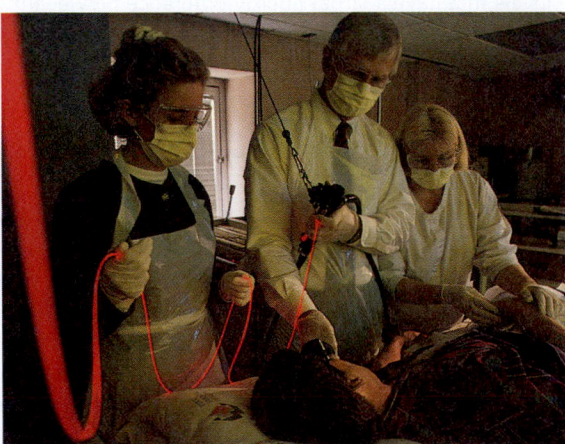

Figure 30.29 Photodynamic therapy to treat cancer of the esophagus is being administered to this patient. Red laser light is routed to the tumor site with an endoscope that incorporates optical fibers. (© Fritz Hoffmann/The Image Works)

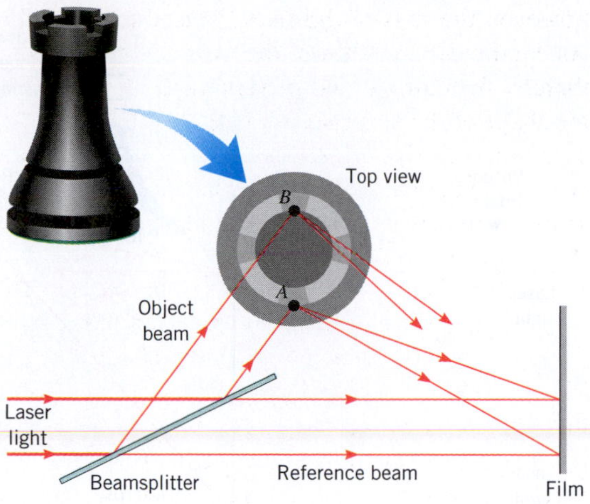

Figure 30.30 An arrangement used to produce a hologram.

30.10 *HOLOGRAPHY

One of the most familiar applications of lasers is in holography, which is a process for producing three-dimensional images. The information used to produce a holographic image is captured on photographic film, which is referred to as a hologram. Figure 30.30 illustrates how a hologram is made. Laser light strikes a half-silvered mirror, or beamsplitter, which reflects part and transmits part of the light. In the drawing, the reflected part is called the *object beam* because it illuminates the object (a chess piece). The transmitted part is called the *reference beam*. The object beam reflects from the chess piece at points such as A and B and, together with the reference beam, falls on the film. One of the main characteristics of laser light is that it is coherent. Thus, the light from the two beams has a stable phase relationship, like the light from the two slits in Young's double-slit experiment (see Section 27.2). Because of the stable phase relationship and because the two beams travel different distances, an interference pattern is formed on the film. This pattern is the hologram and, although much more complex, is analogous to the pattern of bright and dark fringes formed in the double-slit experiment.

Figure 30.31 shows in greater detail how a holographic interference pattern arises. This drawing considers only the reference beam and the light (wavelength $=\lambda$) coming from point A on the chess piece. As we know from Section 27.1, constructive interference between the two light waves leads to a bright fringe; it occurs when the waves, in reaching the film, travel distances that differ by an integer number m of wavelengths. In the drawing, ℓ_m is the distance between point A and the place on the film where the mth-order bright fringe occurs. ℓ_0 is the perpendicular distance that the reference beam would travel from point A to the $m = 0$ bright fringe. In addition, r_m is the distance along the film that locates the bright fringe. In terms of these distances, we know that

$$\ell_m - \ell_0 = m\lambda \qquad \text{(condition for constructive interference)}$$

$$\ell_0^2 + r_m^2 = \ell_m^2 \qquad \text{(Pythagorean theorem)}$$

The first equation indicates that $\ell_m = m\lambda + \ell_0$, which can be substituted into the second equation. The result can be rearranged to show that

$$r_m^2 = m\lambda(m\lambda + 2\ell_0)$$

Since ℓ_0 is typically much larger than λ (for instance, $\ell_0 \approx 10^{-1}$ m and $\lambda \approx 10^{-6}$ m), it follows that $r_m \approx \sqrt{m\lambda 2\ell_0}$. In other words, r_m is roughly proportional to $\sqrt{m}$. Therefore, the fringes are farther apart near the top of the film than they are near the bottom. For example, for the $m = 1$ and $m = 2$ fringes, we have $r_2 - r_1 \propto \sqrt{2} - \sqrt{1} = 0.41$, whereas for the $m = 2$ and $m = 3$ fringes, we have $r_3 - r_2 \propto \sqrt{3} - \sqrt{2} = 0.32$.

In addition to the fringe pattern just discussed, the total interference pattern on the hologram includes interference effects that are related to light coming from point B and other locations on the object in Figure 30.30. The total pattern is very complicated. Nevertheless, the fringe pattern for point A alone is sufficient to illustrate the fact that a

The physics of holography.

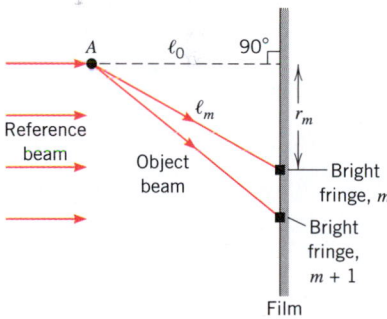

Figure 30.31 This drawing helps to explain how the interference pattern arises on the film when light from point A (see Figure 30.30) and light from the reference beam combine there.

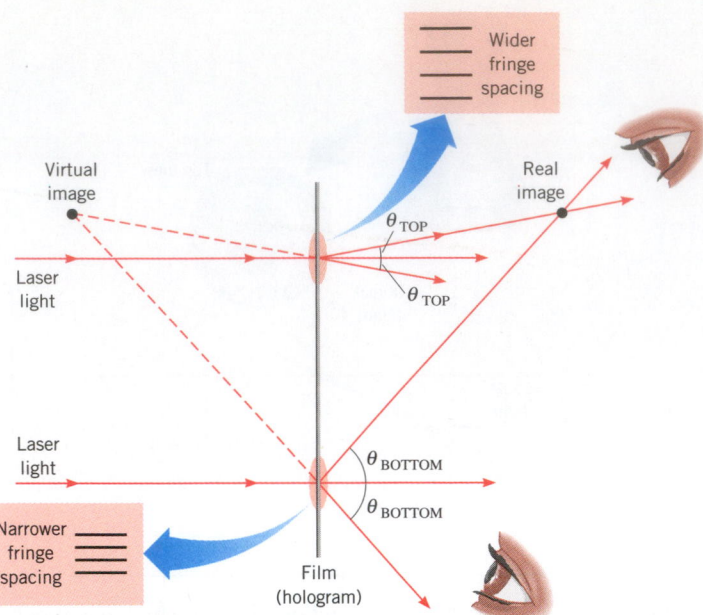

Figure 30.32 When the laser light used to produce a hologram is shone through it, both a real and a virtual image of the object are produced.

hologram can be used to produce both a virtual image and a real image of the object, as we will now see.

To produce the holographic images, the laser light is directed through the interference pattern on the film, as in Figure 30.32 The pattern can be thought of as a kind of diffraction grating, with the bright fringes analogous to the spaces between the slits of the grating. Section 27.7 discusses how light passing through a grating is split into higher-order bright fringes that are symmetrically located on either side of a central bright fringe. Figure 30.32 shows the three rays that correspond to the central and first-order bright fringes, as they originate from a spot near the top and a spot near the bottom of the film. The angle θ, which locates the first-order fringes relative to the central fringe, is given by Equation 27.7 (with $m = 1$) as $\sin \theta = \lambda/d$, where d is the separation between the slits of the grating. When the slit separation is greater, as it is near the top of the film, the angle is smaller. When the slit separation is smaller, as it is near the bottom, the angle is larger. Thus, Figure 30.32 has been drawn with θ_{TOP} smaller than θ_{BOTTOM}. Of the three rays emerging from the film at the top and the three at the bottom, we use the uppermost one in each case to locate the real image of point A on the chess piece. The real image is located where these two rays intersect, when extended to the right. To locate the virtual image, we use the lower ray in each of the three-ray bundles at the top and bottom of the film. When projected to the left, these rays appear to be originating from the spot where the projections intersect—that is, from the virtual image.

A holographic image differs greatly from a photographic image. The most obvious difference is that a hologram provides a three-dimensional image, whereas photographs are two-dimensional. The three-dimensional nature of the holographic image is inherent in the interference pattern formed on the film. In Figure 30.30 part of this pattern arises because the light emitted by point A travels different distances in reaching different spots on the film than does the light in the reference beam. The same can be said about the light emitted from point B and other places as well. As a result, the total interference pattern contains information about how much farther from the film the various points on the object are, and because of this information holographic images are three-dimensional. Furthermore, as Figure 30.33 illustrates, it is possible to "walk around" a holographic image and view it from different angles, as you would the original object.

A vast difference exists between the methods used to produce holograms and photographs. As Section 26.7 discusses, a camera uses a converging lens to produce a photograph. The lens focuses the light rays originating from a point on the object to a corresponding point on the film. To produce a hologram, lenses are not used in this way, and a point on the object *does not correspond* to a single point on the film. In Figure 30.31, light from point A diverges on its way to the film, and there is no lens to make it converge to a single

 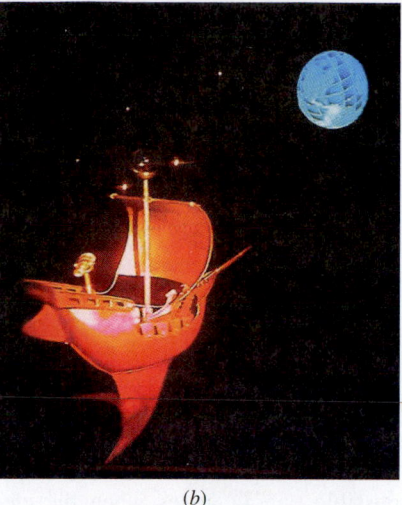

(a)　　　　　　　　　　　　(b)

Figure 30.33 A hologram of a galleon spaceship approaching earth. Two views of the same hologram are shown: (a) looking at the hologram perpendicularly and (b) looking at an angle to the right of the perpendicular. (© Holographics North, Inc., and Sprint. Reproduced with permission of Sprint Corporation)

corresponding point. The light falls over the entire exposed region of the film and contributes everywhere to the interference pattern that is formed. Thus, a hologram may be cut into smaller pieces, and each piece will contain some information about the light originating from point A. For this reason, each smaller piece can be used to produce a three-dimensional image of the object. In contrast, it is not possible to reconstruct the entire image in a photograph from only a small piece of the original film.

The holograms discussed here are typically viewed with the aid of the laser light used to produce them. There are also other kinds of holograms. Credit cards, for example, use holograms for identification purposes. This kind of hologram is called a rainbow hologram and is designed to be viewed in white light that is reflected from it. Other applications of holography include head-up displays for instrument panels in high-performance fighter planes, laser scanners at checkout counters, computerized data storage and retrieval systems, and methods for high-precision biomedical measurements.

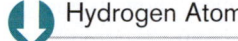

CONCEPTS & CALCULATIONS

The Bohr model of the hydrogen atom introduces a number of important features that characterize the quantum picture of atomic structure. Among them are the concepts of quantized energy levels and the photon emission that occurs when an electron makes a transition from a higher to a lower energy state. Example 12 deals with these ideas.

Concepts & Calculations Example 12
Hydrogen Atom Energy Levels

A hydrogen atom ($Z = 1$) is in the third excited state. It makes a transition to a different state, and a photon is either absorbed or emitted. Determine the quantum number n_f of the final state and the energy of the photon when the photon is **(a)** emitted with the shortest possible wavelength, **(b)** emitted with the longest possible wavelength, and **(c)** absorbed with the longest possible wavelength.

Concept Questions and Answers What is the quantum number of the third excited state?

Answer The lowest-energy state is the ground state, and its quantum number is $n = 1$. The first excited state corresponds to $n = 2$, the second excited state to $n = 3$, and the third excited state to $n = 4$.

When an atom emits a photon, is the final quantum number n_f of the atom greater than or less than the initial quantum number n_i?

Answer Since the photon carries energy away, the final energy of the atom is less than its initial energy. Lower energies correspond to lower quantum numbers (see Figure 30.8). Therefore, the final quantum number is less than the initial quantum number.

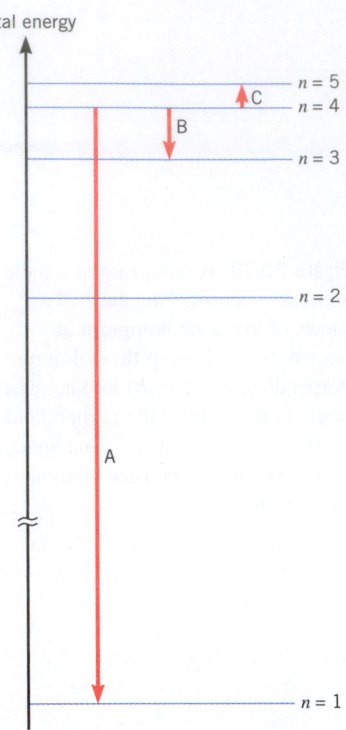

Total energy

$n = 5$
$n = 4$
$n = 3$

$n = 2$

$n = 1$

Figure 30.34 A photon is emitted when an electron in the $n = 4$ state jumps to either the $n = 1$ or the $n = 3$ state. An electron makes a transition from the $n = 4$ state to the $n = 5$ state when a photon of the proper energy is absorbed.

When an atom absorbs a photon, is the final quantum number n_f of the atom greater than or less than the initial quantum number n_i?

Answer When an atom absorbs a photon, the atom gains the energy of the photon, so the final energy of the atom is greater than its initial energy. Greater energies correspond to higher quantum numbers, so the final quantum number is greater than the initial quantum number.

How is the wavelength of a photon related to its energy?

Answer The energy E of a photon is given by Equation 29.2 as $E = hf$, where h is Planck's constant and f is the photon's frequency. But the frequency and wavelength λ are related by $f = c/\lambda$, according to Equation 16.1, where c is the speed of light. Combining these relations gives $E = hc/\lambda$, or $\lambda = hc/E$. Thus, we see that the wavelength is inversely proportional to the energy.

Solution (a) When a photon is emitted with the shortest possible wavelength, it has the largest possible energy, since the wavelength is inversely proportional to the energy. The largest possible energy arises when the electron jumps from the initial state ($n_i = 4$) to the ground state ($n_f = 1$), as shown by transition A in Figure 30.34. Therefore, the quantum number of the final state is $\boxed{n_f = 1}$. The energy E of the photon is the difference between the energies of the two states, so $E = E_4 - E_1$. The energy of the nth state is given by Equation 30.13 as $E_n = -(13.6\ \text{eV})\,Z^2/n^2$, so the energy of the photon is

$$E = (-13.6\ \text{eV})(1)^2 \left(\frac{1}{4^2} - \frac{1}{1^2} \right) = \boxed{12.8\ \text{eV}}$$

(b) When a photon is emitted with the longest possible wavelength, it has the smallest possible energy. The smallest possible energy arises when the electron jumps from the initial state ($n_i = 4$) to the next lower state ($n_f = 3$), as shown by transition B in Figure 30.34. Therefore, the quantum number of the final state is $\boxed{n_f = 3}$. The energy E of the photon is the difference between the energies of the two states, so $E = E_4 - E_3$:

$$E = (-13.6\ \text{eV})(1)^2 \left(\frac{1}{4^2} - \frac{1}{3^2} \right) = \boxed{0.661\ \text{eV}}$$

(c) When a photon is absorbed by the hydrogen atom, the electron jumps to a higher energy state. The photon has the longest possible wavelength when its energy is the smallest. The smallest possible energy change in the hydrogen atom arises when the electron jumps from the initial state ($n_i = 4$) to the next higher state ($n_f = 5$), as shown by transition C in the drawing. Therefore, the quantum number of the final state is $\boxed{n_f = 5}$. The energy E of the photon is the difference between the energies of the two states, so $E = E_5 - E_4$:

$$E = (-13.6\ \text{eV})(1)^2 \left(\frac{1}{5^2} - \frac{1}{4^2} \right) = \boxed{0.306\ \text{eV}}$$

The next example reviews the physics of how a K_α X-ray is produced, how its energy is related to the ionization energies of the target atom, and how to determine the minimum voltage needed to produce it in an X-ray tube.

Concepts & Calculations Example 13
The Production of K_α X-Rays

The K-shell and L-shell ionization energies of a metal are 8979 eV and 951 eV, respectively. (a) Assuming that there is a vacancy in the L shell, what must be the minimum voltage across an X-ray tube with a target made from this metal to produce K_α X-ray photons? (b) Determine the wavelength of a K_α photon.

Concept Questions and Answers How is the K_α photon produced and how much energy does it have?

Answer A K_α photon is produced when an electron in a metal atom jumps from the higher-energy $n = 2$ state to the lower-energy $n = 1$ state (see transition A in Figure 30.35). The energy E of the photon is the difference between the energies of these two states: $E = E_2 - E_1$.

What must be the minimum voltage across the X-ray tube to produce a K_α photon?

Answer   It takes energy to produce a K_α photon, and this energy comes from the electrons striking the target in the X-ray tube. According to Equation 19.3, the energy possessed by each electron when it arrives at the target is eV, where e is the magnitude of the electron's charge and V is the potential difference or voltage across the tube. To cause the target to emit a K_α X-ray photon, the impinging electron must have enough energy to create a vacancy in the K shell. The vacancy is created because the impinging electron provides the energy to move a K-shell electron into a higher energy level or to remove it entirely from the atom. In the present case, with a vacancy in the L shell, the impinging electron can create the K-shell vacancy if it has the energy needed to elevate a K-shell electron to the L shell, the energy needed being $E_2 - E_1$ (see Figure 30.35). This, then, is the minimum energy that the impinging electron must have. Correspondingly, the minimum voltage across the tube must be such that $eV_{min} = E_2 - E_1$, or $V_{min} = (E_2 - E_1)/e$.

What is meant by the phrases "K-shell ionization energy" and "L-shell ionization energy"?

Answer   The K-shell ionization energy is the energy needed to remove an electron in the K shell ($n_i = 1$) completely from the atom (see transition B in Figure 30.35). The removed electron is assumed to have no kinetic energy and is infinitely far away ($n_f = \infty$), so that it has no electric potential energy. Thus, the total energy of the removed electron is zero. Similarly, the L-shell ionization energy is the energy needed to remove an electron in the L shell ($n_i = 2$) completely from the atom (see transition C in Figure 30.35).

What does the difference between the K-shell and L-shell ionization energies represent?

Answer   As transitions B and C in Figure 30.35 suggest, the difference between the two ionization energies is equal to the difference between the total energy E_2 of the L shell and the total energy E_1 of the K shell.

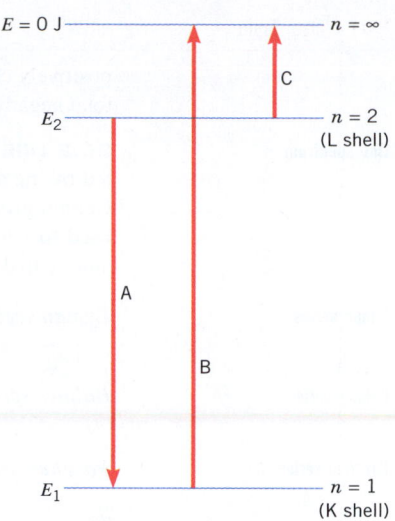

Figure 30.35 A K_α X-ray photon is emitted when an electron jumps from the $n = 2$ state to the $n = 1$ state (transition A). The K-shell ionization energy is the energy required to raise an electron from the $n = 1$ to the $n = \infty$ state, where it has zero total energy (transition B). Similarly, the L-shell ionization energy is the energy needed for the transition from $n = 2$ to $n = \infty$ (transition C). For clarity, the energy levels are not drawn to scale.

Solution   (a) The minimum voltage across the X-ray tube is $V_{min} = (E_2 - E_1)/e$. But the difference $E_2 - E_1$ in energies between the $n = 2$ and $n = 1$ states is equal to the difference in energies between the K-shell and L-shell ionization energies, or 8979 eV − 951 eV = 8028 eV. Thus, the minimum voltage is

$$V_{min} = \frac{E_2 - E_1}{e} = \frac{(8028 \text{ eV})\left(\dfrac{1.602 \times 10^{-19} \text{ J}}{1 \text{ eV}}\right)}{1.602 \times 10^{-19} \text{ C}} = \boxed{8028 \text{ V}}$$

(b) The wavelength of the K_α photon is $\lambda = c/f$, where f is its frequency. The frequency is related to the energy $E_2 - E_1$ of the photon by Equation 29.2, $f = (E_2 - E_1)/h$, where h is Planck's constant. Combining these relations gives

$$\lambda = \frac{c}{f} = \frac{c}{(E_2 - E_1)/h} = \frac{hc}{E_2 - E_1}$$

$$= \frac{(6.626 \times 10^{-34} \text{ J} \cdot \text{s})(2.998 \times 10^8 \text{ m/s})}{(8028 \text{ eV})\left(\dfrac{1.602 \times 10^{-19} \text{ J}}{1 \text{ eV}}\right)} = \boxed{1.545 \times 10^{-10} \text{ m}}$$

CONCEPT SUMMARY

If you need more help with a concept, use the Learning Aids noted next to the discussion or equation. Examples (**Ex.**) are in the text of this chapter. Go to **www.wiley.com/college/cutnell** for the following Learning Aids:

Interactive LearningWare (ILW) — Additional examples solved in a five-step interactive format.

Concept Simulations (CS) — Animated text figures or animations of important concepts.

Interactive Solutions (IS) — Models for certain types of problems in the chapter homework. The calculations are carried out interactively.

Topic	Discussion	Learning Aids
Nuclear atom	**30.1 RUTHERFORD SCATTERING AND THE NUCLEAR ATOM** The idea of a nuclear atom originated in 1911 as a result of experiments by Ernest Rutherford in which α particles were scattered by a thin metal foil. The phrase "nuclear atom" refers to the fact that an atom consists of a small,	

Topic	Discussion	Learning Aids

positively charged nucleus surrounded at relatively large distances by a number of electrons, whose total negative charge equals the positive nuclear charge when the atom is electrically neutral. **Ex. 1**

Line spectrum

30.2 LINE SPECTRA A line spectrum is a series of discrete electromagnetic wavelengths emitted by the atoms of a low-pressure gas that is subjected to a sufficiently high potential difference. Certain groups of discrete wavelengths are referred to as "series." The following equations can be used to determine the wavelengths in three of the series that are found in the line spectrum of atomic hydrogen:

Lyman series

Lyman series
$$\frac{1}{\lambda} = R\left(\frac{1}{1^2} - \frac{1}{n^2}\right) \qquad n = 2, 3, 4, \ldots \tag{30.1}$$

Balmer series

Balmer series
$$\frac{1}{\lambda} = R\left(\frac{1}{2^2} - \frac{1}{n^2}\right) \qquad n = 3, 4, 5, \ldots \tag{30.2}$$ **Ex. 2**

Paschen series

Paschen series
$$\frac{1}{\lambda} = R\left(\frac{1}{3^2} - \frac{1}{n^2}\right) \qquad n = 4, 5, 6, \ldots \tag{30.3}$$

The constant term R is called the Rydberg constant and has the value $R = 1.097 \times 10^7 \text{ m}^{-1}$.

30.3 THE BOHR MODEL OF THE HYDROGEN ATOM The Bohr model applies to atoms or ions that have only a single electron orbiting a nucleus containing Z protons. This model assumes that

Stationary orbit

the electron exists in circular orbits that are called stationary orbits because the electron does not radiate electromagnetic waves while in them. According to this model, a photon is emitted only when an electron changes from an orbit with a higher energy E_i to an orbit with a lower energy E_f. The orbital energies and the photon frequency f are related according to

Photon frequency

$$E_i - E_f = hf \tag{30.4}$$

where h is Planck's constant. The model also assumes that the magnitude L_n of the orbital angular momentum of the electron can only have the following discrete values:

Angular momentum

$$L_n = n\frac{h}{2\pi} \qquad n = 1, 2, 3, \ldots \tag{30.8}$$

With these assumptions, it can be shown that the nth Bohr orbit has a radius r_n and is associated with a total energy E_n of

Orbital radius

$$r_n = (5.29 \times 10^{-11} \text{ m})\frac{n^2}{Z} \qquad n = 1, 2, 3, \ldots \tag{30.10}$$

Total orbital energy

$$E_n = -(13.6 \text{ eV})\frac{Z^2}{n^2} \qquad n = 1, 2, 3, \ldots \tag{30.13}$$ **Ex. 12**

Ionization energy

The ionization energy is the minimum energy needed to remove an electron completely from an atom. The Bohr model predicts that the wavelengths comprising the line spectrum emitted by a hydrogen atom can be calculated from **Ex. 3**

ILW 30.1

Line spectrum wavelengths

$$\frac{1}{\lambda} = RZ^2\left(\frac{1}{n_f^2} - \frac{1}{n_i^2}\right) \tag{30.14}$$ **Ex. 4**

$$n_i, n_f = 1, 2, 3, \ldots \quad \text{and} \quad n_i > n_f$$ **CS 30.1**

30.4 DE BROGLIE'S EXPLANATION OF BOHR'S ASSUMPTION ABOUT ANGULAR MOMENTUM Louis de Broglie proposed that the electron in a circular Bohr orbit should be considered as a particle wave and that standing particle waves around the orbit offer an explanation of the angular momentum assumption in the Bohr model.

30.5 THE QUANTUM MECHANICAL PICTURE OF THE HYDROGEN ATOM Quantum mechanics describes the hydrogen atom in terms of the following four quantum numbers:

Principal quantum number
Orbital quantum number
Magnetic quantum number

(1) The principal quantum number n, which can have the integer values $n = 1, 2, 3, \ldots$ **Ex. 5, 6**
(2) The orbital quantum number ℓ, which can have the integer values $\ell = 0, 1, 2, \ldots, (n-1)$
(3) The magnetic quantum number m_ℓ, which can have the positive and negative values **IS 30.57** $m_\ell = -\ell, \ldots, -2, -1, 0, +1, +2, \ldots, +\ell$

Spin quantum number

(4) The spin quantum number m_s, which, for an electron, can be either $m_s = +\frac{1}{2}$ or $m_s = -\frac{1}{2}$

According to quantum mechanics, an electron does not reside in a circular orbit but, rather, has some probability of being found at various distances from the nucleus. **Ex. 7**

30.6 THE PAULI EXCLUSION PRINCIPLE AND THE PERIODIC TABLE OF THE ELEMENTS The Pauli

Pauli exclusion principle

exclusion principle states that no two electrons in an atom can have the same set of values for the four quantum numbers n, ℓ, m_ℓ, and m_s. This principle determines the way in which the electrons

Topic	Discussion	Learning Aids
	in multiple-electron atoms are distributed into shells (defined by the value of n) and subshells (defined by the value of ℓ).	Ex. 8
Convention of letters	The following notation is used to refer to the orbital quantum numbers: s denotes $\ell = 0$, p denotes $\ell = 1$, d denotes $\ell = 2$, f denotes $\ell = 3$, g denotes $\ell = 4$, h denotes $\ell = 5$, and so on.	
	The arrangement of the periodic table of the elements is related to the Pauli exclusion principle.	
X-rays	**30.7 X-RAYS** X-rays are electromagnetic waves emitted when high-energy electrons strike a metal target contained within an evacuated glass tube. The emitted X-ray spectrum of wavelengths	Ex. 9
Characteristic X-rays	consists of sharp "peaks" or "lines," called characteristic X-rays, superimposed on a broad continu-	Ex. 10
Bremsstrahlung	ous range of wavelengths called Bremsstrahlung. The K_α characteristic X-ray is emitted when an electron in the $n = 2$ level (L shell) drops into a vacancy in the $n = 1$ level (K shell). Similarly,	Ex. 11, 13
	the K_β characteristic X-ray is emitted when an electron in the $n = 3$ level (M shell) drops into a	ILW 30.2
Cutoff wavelength	vacancy in the $n = 1$ level (K shell). The minimum wavelength, or cutoff wavelength λ_0, of the Bremsstrahlung is determined by the kinetic energy of the electrons striking the target in the X-ray	IS 30.39
	tube, according to	

$$\lambda_0 = \frac{hc}{eV} \qquad (30.17)$$

where h is Planck's constant, c is the speed of light in a vacuum, e is the magnitude of the charge on an electron, and V is the potential difference applied across the X-ray tube.

Topic	Discussion	Learning Aids
	30.8 THE LASER, 30.9 MEDICAL APPLICATIONS OF THE LASER A laser is a device that gener-	
Stimulated emission	ates electromagnetic waves via a process known as stimulated emission. In this process, one photon stimulates the production of another photon by causing an electron in an atom to fall from a higher energy level to a lower energy level. The emitted photon travels in the same direction as the photon causing the stimulation. Because of this mechanism of photon production, the electromagnetic waves generated by a laser are coherent and may be confined to a very narrow beam.	
Spontaneous emission	Stimulated emission contrasts with the process known as spontaneous emission, in which an electron in an atom also falls from a higher to a lower energy level, but does so spontaneously, in a random direction, without any external provocation.	

FOCUS ON CONCEPTS

Note to Instructors: The numbering of the questions shown here reflects the fact that they are only a representative subset of the total number that are available online. However, all of the questions are available for assignment via an online homework management program such as WileyPLUS or WebAssign.

Section 30.3   The Bohr Model of the Hydrogen Atom

1. Consider applying the Bohr model to a neutral helium atom ($Z = 2$). The model takes into account a number of factors. Which one of the following does it not take into account?   (a) The quantization of the orbital angular momentum of an electron   (b) The centripetal acceleration of an electron   (c) The electric potential energy of an electron   (d) The electrostatic repulsion between electrons   (e) The electrostatic attraction between the nucleus and an electron

3. According to the Bohr model, what determines the shortest wavelength in a given series of wavelengths emitted by the atom? (a) The quantum number n_i that identifies the higher energy level from which the electron falls into a lower energy level   (b) The quantum number n_f that identifies the lower energy level into which the electron falls from a higher energy level   (c) The ratio n_f/n_i, where n_f is the quantum number that identifies the lower energy level into which the electron falls and n_i is the quantum number that identifies the higher level from which the electron falls   (d) The sum $n_f + n_i$ of two quantum numbers, where n_f identifies the lower energy level into which the electron falls and n_i identifies the higher level from which the electron falls   (e) The difference $n_f - n_i$ of

two quantum numbers, where n_f identifies the lower energy level into which the electron falls and n_i identifies the higher level from which the electron falls

Section 30.5   The Quantum Mechanical Picture of the Hydrogen Atom

6. According to quantum mechanics, only one of the following combinations of the principal quantum number n and the orbital quantum number ℓ is possible for the electron in a hydrogen atom. Which combination is it?   (a) $n = 3, \ell = 3$ (b) $n = 2, \ell = 3$ (c) $n = 1, \ell = 2$ (d) $n = 0, \ell = 0$ (e) $n = 3, \ell = 1$

8. Which one of the following statements is false?   (a) The orbits in the Bohr model have precise sizes, whereas in the quantum mechanical picture of the hydrogen atom they do not.   (b) In the absence of external magnetic fields, both the Bohr model and quantum mechanics predict the same total energy for the electron in the hydrogen atom.   (c) The spin angular momentum of the electron plays a role in both the Bohr model and the quantum mechanical picture of the hydrogen atom.   (d) The magnitude of the orbital angular momentum cannot be zero in the Bohr model, but it can be zero in the quantum mechanical picture of the hydrogen atom.

Section 30.6 The Pauli Exclusion Principle and the Periodic Table of the Elements

10. Each of the following answers indicates the quantum mechanical states of two electrons, A and B. Which pair of states could *not* describe two of the electrons in a multiple-electron atom?

(a)

	n	ℓ	m_ℓ	m_s
A	4	1	+1	$-\frac{1}{2}$
B	3	1	-1	$-\frac{1}{2}$

(b)

	n	ℓ	m_ℓ	m_s
A	3	2	-1	$-\frac{1}{2}$
B	3	1	-1	$+\frac{1}{2}$

(c)

	n	ℓ	m_ℓ	m_s
A	2	0	0	$-\frac{1}{2}$
B	2	1	+1	$+\frac{1}{2}$

(d)

	n	ℓ	m_ℓ	m_s
A	5	3	+1	$-\frac{1}{2}$
B	4	1	0	$+\frac{1}{2}$

(e)

	n	ℓ	m_ℓ	m_s
A	3	2	-2	$+\frac{1}{2}$
B	3	2	-2	$+\frac{1}{2}$

11. Consider the 5f and 6h subshells in a multiple-electron atom. Which of these subshells can contain 19 electrons? **(a)** Only the 6h subshell **(b)** Only the 5f subshell **(c)** Both subshells **(d)** Neither subshell

Section 30.7 X-Rays

14. Silver ($Z = 47$), copper ($Z = 29$), and platinum ($Z = 78$) can be used as the target in an X-ray tube. Rank in descending order (largest first) the energies needed for impinging electrons to knock a K-shell electron completely out of an atom in each of these targets. **(a)** Silver, copper, platinum **(b)** Silver, platinum, copper **(c)** Platinum, silver, copper **(d)** Platinum, copper, silver **(e)** Copper, silver, platinum

16. The voltage applied across an X-ray tube is doubled. What happens to the cutoff wavelength in the spectrum of wavelengths emitted by the tube's metal target? **(a)** It also doubles. **(b)** It decreases by a factor of two. **(c)** It increases by a factor of four. **(d)** It decreases by a factor of four. **(e)** Nothing happens to it.

Section 30.8 The Laser

17. Consider two energy levels that characterize the atoms of a material used in a laser. A population inversion between these two levels _____. **(a)** has the lower energy level more populated than it normally is and the higher energy level less populated than it normally is **(b)** is the same thing as a metastable state **(c)** requires no external source of energy to be sustained **(d)** has the higher energy level more populated than it normally is and the lower energy level less populated than it normally is

PROBLEMS

Note to Instructors: Most of the homework problems in this chapter are available for assignment via an online homework management program such as WileyPLUS or WebAssign, and those marked with the icon **GO** are presented in WileyPLUS using a guided tutorial format that provides enhanced interactivity. See Preface for additional details.

In working these problems, ignore relativistic effects.

ssm Solution is in the Student Solutions Manual.
www Solution is available online at www.wiley.com/college/cutnell

⚕ This icon represents a biomedical application.

Section 30.1 Rutherford Scattering and the Nuclear Atom

1. ssm The nucleus of the hydrogen atom has a radius of about 1×10^{-15} m. The electron is normally at a distance of about 5.3×10^{-11} m from the nucleus. Assuming that the hydrogen atom is a sphere with a radius of 5.3×10^{-11} m, find **(a)** the volume of the atom, **(b)** the volume of the nucleus, and **(c)** the percentage of the volume of the atom that is occupied by the nucleus.

2. In a Rutherford scattering experiment, each atom in a thin gold foil can be considered a target with a circular cross section. Alpha particles are fired at this target, with the nucleus as the bull's-eye. The ratio of the cross-sectional area of the gold nucleus to the cross-sectional area of the atom is equal to 2.6×10^{-7}. The radius of the gold atom is 1.4×10^{-11} m. What is the radius of the gold nucleus?

3. In a Rutherford scattering experiment a target nucleus has a diameter of 1.4×10^{-14} m. The incoming α particle has a mass of 6.64×10^{-27} kg. What is the kinetic energy of an α particle that has a de Broglie wavelength equal to the diameter of the target nucleus? Ignore relativistic effects.

4. Review Conceptual Example 1 and use the information therein as an aid in working this problem. Suppose that you're building a scale model of the hydrogen atom, and the nucleus is represented by a ball of radius 3.2 cm (somewhat smaller than a baseball). How many miles away (1 mi = 1.61×10^5 cm) should the electron be placed?

*5. ssm There are Z protons in the nucleus of an atom, where Z is the atomic number of the element. An α particle carries a charge of $+2e$. In a scattering experiment, an α particle, heading directly toward a nucleus in a metal foil, will come to a halt when all the particle's kinetic energy is converted to electric potential energy. In such a situation, how close will an α particle with a kinetic energy of 5.0×10^{-13} J come to a gold nucleus ($Z = 79$)?

*6. GO The nucleus of a copper atom contains 29 protons and has a radius of 4.8×10^{-15} m. How much work (in electron volts) is done by the electric force as a proton is brought from infinity, where it is at rest, to the "surface" of a copper nucleus?

Section 30.2 Line Spectra,
Section 30.3 The Bohr Model of the Hydrogen Atom

7. ssm www In the line spectrum of atomic hydrogen there is also a group of lines known as the Pfund series. These lines are produced when electrons, excited to high energy levels, make transitions to the $n = 5$ level. Determine **(a)** the longest wavelength and **(b)** the shortest wavelength in this series. **(c)** Refer to Figure 24.9 and identify the region of the electromagnetic spectrum in which these lines are found.

8. A singly ionized helium atom (He^+) has only one electron in orbit about the nucleus. What is the radius of the ion when it is in the second excited state?

9. Using the Bohr model, determine the ratio of the energy of the nth orbit of a triply ionized beryllium atom (Be^{3+}, $Z = 4$) to the energy of the nth orbit of a hydrogen atom (H).

10. (a) What is the minimum energy (in electron volts) that is required to remove the electron from the ground state of a singly ionized helium atom (He^+, $Z = 2$)?   (b) What is the ionization energy for He^+?

11. ssm Find the energy (in joules) of the photon that is emitted when the electron in a hydrogen atom undergoes a transition from the $n = 7$ energy level to produce a line in the Paschen series.

12. GO (a) What is the ionization energy of a hydrogen atom that is in the $n = 4$ excited state?   (b) For a hydrogen atom, determine the ratio of the ionization energy for the $n = 4$ excited state to the ionization energy for the ground state.

13. A hydrogen atom is in the ground state. It absorbs energy and makes a transition to the $n = 3$ excited state. The atom returns to the ground state by emitting two photons. What are their wavelengths?

14. Consider the Bohr energy expression (Equation 30.13) as it applies to singly ionized helium He^+ ($Z = 2$) and doubly ionized lithium Li^{2+} ($Z = 3$). This expression predicts equal electron energies for these two species for certain values of the quantum number n (the quantum number is different for each species). For quantum numbers less than or equal to 9, what are the lowest three energies (in electron volts) for which the helium energy level is equal to the lithium energy level?

15. In the hydrogen atom the radius of orbit B is sixteen times greater than the radius of orbit A. The total energy of the electron in orbit A is -3.40 eV. What is the total energy of the electron in orbit B?

*16. GO A sodium atom ($Z = 11$) contains 11 protons in its nucleus. Strictly speaking, the Bohr model does not apply, because the neutral atom contains 11 electrons instead of a single electron. However, we can apply the model to the outermost electron as an approximation, provided that we use an effective value $Z_{effective}$ rather than 11 for the number of protons in the nucleus.   (a) The ionization energy for the outermost electron in a sodium atom is 5.1 eV. Use the Bohr model with $Z = Z_{effective}$ to calculate a value for $Z_{effective}$.   (b) Using $Z = 11$ and $Z = Z_{effective}$, determine the corresponding two values for the radius of the outermost Bohr orbit.

*17. ssm A wavelength of 410.2 nm is emitted by the hydrogen atoms in a high-voltage discharge tube. What are the initial and final values of the quantum number n for the energy level transition that produces this wavelength?

*18. The energy of the $n = 2$ Bohr orbit is -30.6 eV for an unidentified ionized atom in which only one electron moves about the nucleus. What is the radius of the $n = 5$ orbit for this species?

*19. ssm The Bohr model can be applied to singly ionized helium He^+ ($Z = 2$). Using this model, consider the series of lines that is produced when the electron makes a transition from higher energy levels into the $n_f = 4$ level. Some of the lines in this series lie in the visible region of the spectrum (380–750 nm). What are the values of n_i for the energy levels from which the electron makes the transitions corresponding to these lines?

*20. Doubly ionized lithium Li^{2+} ($Z = 3$) and triply ionized beryllium Be^{3+} ($Z = 4$) each emit a line spectrum. For a certain series of lines in the lithium spectrum, the shortest wavelength is 40.5 nm. For the same series of lines in the beryllium spectrum, what is the shortest wavelength?

**21. A certain species of ionized atoms produces an emission line spectrum according to the Bohr model, but the number of protons Z in the nucleus is unknown. A group of lines in the spectrum forms a series in which the shortest wavelength is 22.79 nm and the longest wavelength is 41.02 nm. Find the next-to-the-longest wavelength in the series of lines.

**22. In the Bohr model of hydrogen, the electron moves in a circular orbit around the nucleus. Determine the angular speed of the electron, in revolutions per second, when it is in   (a) the ground state and   (b) the first excited state.

Section 30.5   The Quantum Mechanical Picture of the Hydrogen Atom

23. The orbital quantum number for the electron in a hydrogen atom is $\ell = 5$. What is the smallest possible value (algebraically) for the total energy of this electron? Give your answer in electron volts.

24. GO The table lists quantum numbers for five states of the hydrogen atom. Which (if any) of them are not possible? For those that are not possible, explain why.

	n	ℓ	m_ℓ
(a)	3	3	0
(b)	2	1	-1
(c)	4	2	3
(d)	5	-3	2
(e)	4	0	0

25. A hydrogen atom is in its second excited state. Determine, according to quantum mechanics,   (a) the total energy (in eV) of the atom,   (b) the magnitude of the maximum angular momentum the electron can have in this state, and   (c) the maximum value that the z component L_z of the angular momentum can have.

26. The maximum value for the magnetic quantum number in state A is $m_\ell = 2$, while in state B it is $m_\ell = 1$. What is the ratio L_A/L_B of the magnitudes of the orbital angular momenta of an electron in these two states?

27. ssm www The principal quantum number for an electron in an atom is $n = 6$, and the magnetic quantum number is $m_\ell = 2$. What possible values for the orbital quantum number ℓ could this electron have?

*28. The electron in a certain hydrogen atom has an angular momentum of 8.948×10^{-34} J·s. What is the largest possible magnitude for the z component of the angular momentum of this electron? For accuracy, use $h = 6.626 \times 10^{-34}$ J·s.

*29. ssm www Review Conceptual Example 6 as background for this problem. For the hydrogen atom, the Bohr model and quantum mechanics both give the same value for the energy of the nth state. However, they do not give the same value for the orbital angular momentum L.   (a) For $n = 1$, determine the values of L [in units of $h/(2\pi)$] predicted by the Bohr model and quantum mechanics.   (b) Repeat part (a) for $n = 3$, noting that quantum mechanics permits more than one value of ℓ when the electron is in the $n = 3$ state.

**30. GO An electron is in the $n = 5$ state. What is the smallest possible value for the angle between the z component of the orbital angular momentum and the orbital angular momentum?

Section 30.6   The Pauli Exclusion Principle and the Periodic Table of the Elements

31. Two of the three electrons in a lithium atom have quantum numbers of $n = 1$, $\ell = 0$, $m_\ell = 0$, $m_s = +\frac{1}{2}$ and $n = 1$, $\ell = 0$, $m_\ell = 0$, $m_s = -\frac{1}{2}$. What quantum numbers can the third electron have if the atom is in   (a) its ground state and   (b) its first excited state?

32. Referring to Figure 30.15 for the order in which the subshells fill and following the style used in Table 30.3, determine the ground-state electronic configuration for cadmium Cd ($Z = 48$).

33. Figure 30.15 was constructed using the Pauli exclusion principle and indicates that the $n = 1$ shell holds 2 electrons, the $n = 2$ shell holds 8 electrons, and the $n = 3$ shell holds 18 electrons. These numbers can be obtained by adding the numbers given in the figure for the subshells contained within a given shell. How many electrons can be put into the $n = 5$ shell, which is only partly shown in the figure?

34. **GO** Which of the following subshell configurations are not allowed? For those that are not allowed, give the reason why.   (a) $3s^1$ (b) $2d^2$ (c) $3s^4$ (d) $4p^8$ (e) $5f^{12}$

35. Write down the fourteen sets of the four quantum numbers that correspond to the electrons in a completely filled 4f subshell.

*36. In the ground state, the outermost shell ($n = 1$) of helium (He) is filled with electrons, as is the outermost shell ($n = 2$) of neon (Ne). The full outermost shells of these two elements distinguish them as the first two so-called "noble gases." Suppose that the spin quantum number m_s had *three* possible values, rather than two. If that were the case, which elements would be   (a) the first and   (b) the second noble gases? Assume that the possible values for the other three quantum numbers are unchanged, and that the Pauli exclusion principle still applies.

Section 30.7   X-Rays

37. **ssm** Molybdenum has an atomic number of $Z = 42$. Using the Bohr model, estimate the wavelength of the K_α X-ray.

38. When a certain element is bombarded with high-energy electrons, K_α X-rays that have an energy of 9890 eV are emitted. Determine the atomic number Z of the element, and identify the element. Use the Bohr model as necessary.

39. **Interactive Solution 30.39** at **www.wiley.com/college/cutnell** provides one model for solving problems such as this one. An X-ray tube is being operated at a potential difference of 52.0 kV. What is the Bremsstrahlung wavelength that corresponds to 35.0% of the kinetic energy with which an electron collides with the metal target in the tube?

40. **GO** The voltage across an X-ray tube is 35.0 kV. Suppose that the molybdenum ($Z = 42$) target in the X-ray tube is replaced by a silver ($Z = 47$) target. Determine   (a) the tube's cutoff wavelength and   (b) the wavelengths of the K_α X-ray photons emitted by the molybdenum and silver targets.

41. The K_β characteristic X-ray line for tungsten has a wavelength of 1.84×10^{-11} m. What is the difference in energy between the two energy levels that give rise to this line? Express the answer in (a) joules and   (b) electron volts.

*42. Multiple-Concept Example 9 reviews the concepts that are important in this problem. An electron, traveling at a speed of 6.00×10^7 m/s, strikes the target of an X-ray tube. Upon impact, the electron decelerates to one-quarter of its original speed, emitting an X-ray in the process. What is the wavelength of the X-ray photon?

*43. **ssm** An X-ray tube contains a silver ($Z = 47$) target. The high voltage in this tube is increased from zero. Using the Bohr model, find the value of the voltage at which the K_α X-ray just appears in the X-ray spectrum.

*44. **GO** The Bohr model, although not strictly applicable, can be used to estimate the minimum energy E_{min} that an incoming electron must have in an X-ray tube in order to knock a K-shell electron entirely out of an atom in the metal target. The K_α X-ray wavelength of metal A is 2.0 times the K_α X-ray wavelength of metal B. What is the ratio of $E_{min, A}$ for metal A to $E_{min, B}$ for metal B?

Section 30.8   The Laser

45. **ssm www** A laser is used in eye surgery to weld a detached retina back into place. The wavelength of the laser beam is 514 nm, and the power is 1.5 W. During surgery, the laser beam is turned on for 0.050 s. During this time, how many photons are emitted by the laser?

46. **GO** A pulsed laser emits light in a series of short pulses, each having a duration of 25.0 ms. The average power of each pulse is 5.00 mW, and the wavelength of the light is 633 nm. Find the number of photons in each pulse.

47. The dye laser used in the treatment of the port-wine stain in Figure 30.28 (see Section 30.9) has a wavelength of 585 nm. A carbon dioxide laser produces a wavelength of 1.06×10^{-5} m. What is the minimum number of photons that the carbon dioxide laser must produce to deliver at least as much or more energy to a target as does a single photon from the dye laser?

48. **GO** The drawing shows three energy levels of a laser that are involved in the lasing action. These levels are analogous to the levels in the Ne atoms of a He-Ne laser. The E_2 level is a metastable level, and the E_0 level is the ground state. The difference between the energy levels of the laser is shown in the drawing.   (a) What energy (in eV per electron) must an external source provide to start the lasing action?   (b) What is the wavelength of the laser light?   (c) In what region of the electromagnetic spectrum (see Figure 24.9) does the laser light lie?

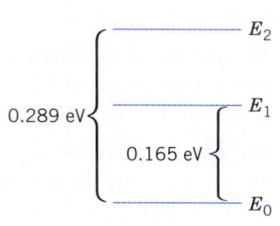

49. A laser peripheral iridotomy is a procedure for treating an eye condition known as narrow-angle glaucoma, in which pressure buildup in the eye can lead to loss of vision. A neodymium YAG laser (wavelength = 1064 nm) is used in the procedure to punch a tiny hole in the peripheral iris, thereby relieving the pressure buildup. In one application the laser delivers 4.1×10^{-3} J of energy to the iris in creating the hole. How many photons does the laser deliver?

*50. Fusion is the process by which the sun produces energy. One experimental technique for creating controlled fusion utilizes a solid-state laser that emits a wavelength of 1060 nm and can produce a power of 1.0×10^{14} W for a pulse duration of 1.1×10^{-11} s. In contrast, the helium/neon laser used in a bar-code scanner at the checkout counter emits a wavelength of 633 nm and produces a power of about 1.0×10^{-3} W. How long (in days) would the helium/neon laser have to operate to produce the same number of photons that the solid-state laser produces in 1.1×10^{-11} s?

ADDITIONAL PROBLEMS

51. In the style shown in Table 30.3, write down the ground-state electronic configuration for arsenic As ($Z = 33$). Refer to Figure 30.15 for the order in which the subshells fill.

52. **Concept Simulation 30.1** at **www.wiley.com/college/cutnell** reviews the concepts on which the solution to this problem depends. The electron in a hydrogen atom is in the first excited state, when the electron acquires an additional 2.86 eV of energy. What is the quantum number n of the state into which the electron moves?

53. ssm When an electron makes a transition between energy levels of an atom, there are no restrictions on the initial and final values of the principal quantum number n. According to quantum mechanics, however, there is a rule that restricts the initial and final values of the orbital quantum number ℓ. This rule is called a *selection rule* and states that $\Delta\ell = \pm 1$. In other words, when an electron makes a transition between energy levels, the value of ℓ can only increase or decrease by one. The value of ℓ may not remain the same nor increase or decrease by more than one. According to this rule, which of the following energy level transitions are allowed? **(a)** $2s \rightarrow 1s$ **(b)** $2p \rightarrow 1s$ **(c)** $4p \rightarrow 2p$ **(d)** $4s \rightarrow 2p$ **(e)** $3d \rightarrow 3s$

54. Interactive LearningWare 30.2 at **www.wiley.com/college/cutnell** reviews the concepts that are pertinent to this problem. By using the Bohr model, decide which element is likely to emit a K_{α} X-ray with a wavelength of 4.5×10^{-9} m.

55. ssm For a doubly ionized lithium atom Li^{2+} ($Z = 3$), what is the principal quantum number of the state in which the electron has the same total energy as a ground-state electron has in the hydrogen atom?

56. In the X-ray spectrum of niobium ($Z = 41$), a K_{α} peak is observed at a wavelength of 7.462×10^{-11} m. **(a)** Determine the magnitude of the difference between the observed wavelength of the K_{α} X-ray for niobium and that predicted by the Bohr model. **(b)** Express the magnitude of this difference as a percentage of the observed wavelength.

*57. Interactive Solution 30.57 at **www.wiley.com/college/cutnell** offers one approach to problems of this type. For an electron in a hydrogen atom, the z component of the angular momentum has a *maximum* value of $L_z = 4.22 \times 10^{-34}$ J · s. Find the three smallest possible values (algebraically) for the total energy (in electron volts) that this atom could have.

*58. Interactive LearningWare 30.1 at **www.wiley.com/college/cutnell** reviews the concepts that play roles in this problem. A hydro-

gen atom emits a photon that has momentum with a magnitude of 5.452×10^{-27} kg · m/s. This photon is emitted because the electron in the atom falls from a higher energy level into the $n = 1$ level. What is the quantum number of the level from which the electron falls? Use a value of 6.626×10^{-34} J · s for Planck's constant.

*59. ssm For atomic hydrogen, the Paschen series of lines occurs when $n_f = 3$, whereas the Brackett series occurs when $n_f = 4$ in Equation 30.14. Using this equation, show that the ranges of wavelengths in these two series overlap.

*60. Consider a particle of mass m that can exist only between $x = 0$ m and $x = +L$ on the x axis. We could say that this particle is confined to a "box" of length L. In this situation, imagine the standing de Broglie waves that can fit into the box. For example, the drawing shows the first three possibilities. Note in this picture that there are either one, two, or three half-wavelengths that fit into the distance L. Use Equation 29.8 for

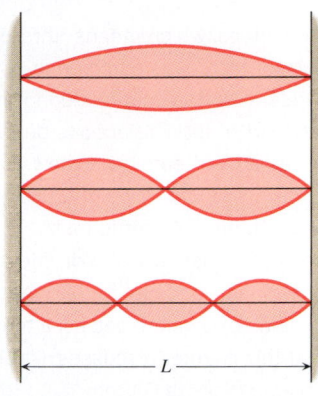

the de Broglie wavelength of a particle and derive an expression for the allowed energies (only kinetic energy) that the particle can have. This expression involves m, L, Planck's constant, and a quantum number n that can have only the values $1, 2, 3, \ldots$.

61. ssm www **(a) Derive an expression for the speed of the electron in the nth Bohr orbit, in terms of Z, n, and the constants k, e, and h. For the hydrogen atom, determine the speed in **(b)** the $n = 1$ orbit and **(c)** the $n = 2$ orbit. **(d)** Generally, when speeds are less than one-tenth the speed of light, the effects of special relativity can be ignored. Are the speeds found in (b) and (c) consistent with ignoring relativistic effects in the Bohr model?

NUCLEAR PHYSICS AND RADIOACTIVITY

Triceratops, which means "three-horned face," was common in western North America during the last part of the Cretaceous period. It was a plant eater and grew to 9 meters long and weighed up to 6 U.S. tons (5.4 metric tons). Paleontologists use the disintegration of radioactive nuclei to date the fossil remains of extinct species, a method that this chapter discusses. (© Louie Psihoyos/Science Faction)

31.1 NUCLEAR STRUCTURE

Atoms consist of electrons in orbit about a central nucleus. As we have seen in Chapter 30, the electron orbits are quantum mechanical in nature and have interesting characteristics. Little has been said about the nucleus, however. Since the nucleus is interesting in its own right, we now consider it in greater detail.

The nucleus of an atom consists of neutrons and protons, collectively referred to as *nucleons.* The *neutron,* discovered in 1932 by the English physicist James Chadwick (1891–1974), carries no electric charge and has a mass slightly larger than that of a proton (see Table 31.1).

The number of protons in the nucleus is different in different elements and is given by the *atomic number Z.* In an electrically neutral atom, the number of nuclear protons equals the number of electrons in orbit around the nucleus. The number of neutrons in the

Table 31.1 Properties of Select Particles

Particle	Electric Charge (C)	Mass	
		Kilograms (kg)	Atomic Mass Units (u)
Electron	-1.60×10^{-19}	$9.109\,382 \times 10^{-31}$	$5.485\,799 \times 10^{-4}$
Proton	$+1.60 \times 10^{-19}$	$1.672\,622 \times 10^{-27}$	$1.007\,276$
Neutron	0	$1.674\,927 \times 10^{-27}$	$1.008\,665$
Hydrogen atom	0	$1.673\,534 \times 10^{-27}$	$1.007\,825$

nucleus is N. The total number of protons and neutrons is referred to as the ***atomic mass number*** A because the total nuclear mass is *approximately* equal to A times the mass of a single nucleon:

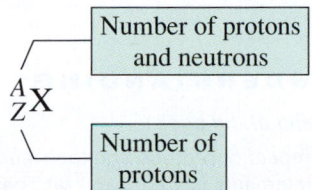

$$A = Z + N \qquad (31.1)$$

Number of protons and neutrons (atomic mass number or nucleon number) = Number of protons (atomic number) + Number of neutrons

Sometimes A is also called the ***nucleon number.*** A shorthand notation is often used to specify Z and A along with the chemical symbol for the element. For instance, the nuclei of all naturally occurring aluminum atoms have $A = 27$, and the atomic number for aluminum is $Z = 13$. In shorthand notation, then, the aluminum nucleus is specified as $^{27}_{13}\text{Al}$. The number of neutrons in an aluminum nucleus is $N = A - Z = 14$. In general, for an element whose chemical symbol is X, the symbol for the nucleus is

$$^A_Z\text{X}$$

Number of protons and neutrons

Number of protons

For a proton the symbol is ^1_1H, since the proton is the nucleus of a hydrogen atom. A neutron is denoted by ^1_0n. In the case of an electron we use $^{\,0}_{-1}\text{e}$, where $A = 0$ because an electron is not composed of protons or neutrons and $Z = -1$ because the electron has a negative charge.

Nuclei that contain the same number of protons, but a different number of neutrons, are known as ***isotopes.*** Carbon, for example, occurs in nature in two stable forms. In most carbon atoms (98.90%), the nucleus is the $^{12}_6\text{C}$ isotope and consists of six protons and six neutrons. A small fraction (1.10%), however, contain nuclei that have six protons and seven neutrons—namely, the $^{13}_6\text{C}$ isotope. The percentages given here are the natural abundances of the isotopes. The atomic masses in the periodic table are average atomic masses, taking into account the abundances of the various isotopes.

The protons and neutrons in the nucleus are clustered together to form an approximately spherical region, as Figure 31.1 illustrates. Experiment shows that the radius r of the nucleus depends on the atomic mass number A and is given approximately in meters by

$$r \approx (1.2 \times 10^{-15}\text{ m})A^{1/3} \qquad (31.2)$$

The radius of the aluminum nucleus ($A = 27$), for example, is $r \approx (1.2 \times 10^{-15}\text{ m})27^{1/3} = 3.6 \times 10^{-15}\text{ m}$. Equation 31.2 leads to an important conclusion concerning the nuclear density of different atoms, as Conceptual Example 1 discusses.

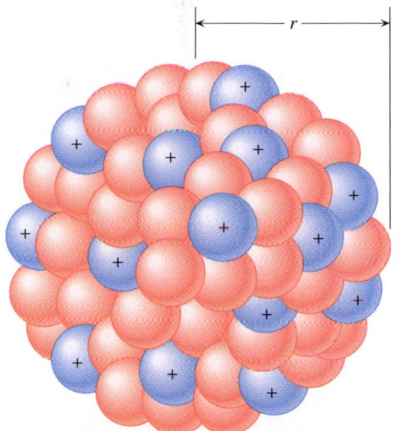

Figure 31.1 The nucleus is approximately spherical (radius $= r$) and contains protons ($\oplus$) clustered closely together with neutrons ($\bullet$).

 Conceptual Example 1 Nuclear Density

It is well known that lead and oxygen contain different atoms and that the density of solid lead is much greater than the density of gaseous oxygen. Using the definition of density along with Equation 31.2, decide whether the density of the *nucleus* in a lead atom is **(a)** greater than, **(b)** approximately equal to, or **(c)** less than the density of the *nucleus* in an oxygen atom.

Reasoning The density ρ of an object, such as the nucleus, is defined as its mass M divided by its volume V: $\rho = M/V$ (Equation 11.1). The mass of a nucleus is approximately equal to the number A of nucleons in the nucleus times the mass m of a single nucleon, since the masses of a proton and a neutron are nearly the same. Thus, we have that $M \approx Am$, where A is greater for lead than for oxygen, but m is the same for both. The nucleus is approximately spherical with a radius r, so its volume V is given by $V = \frac{4}{3}\pi r^3$. The radius, however, depends on the number A of nucleons through the relation $r \approx (1.2 \times 10^{-15}\text{ m})A^{1/3}$ (Equation 31.2). Therefore, we can write the density of a nucleus as follows:

$$\rho = \frac{M}{V} \approx \frac{Am}{\frac{4}{3}\pi r^3} = \frac{Am}{\frac{4}{3}\pi[(1.2 \times 10^{-15}\text{ m})\,A^{1/3}]^3} \approx \frac{m}{\frac{4}{3}\pi(1.2 \times 10^{-15}\text{ m})^3}$$

Note that the nucleon number A has been eliminated algebraically from this result, as a direct consequence of Equation 31.2.

Answers (a) and (c) are incorrect. The result obtained in the Reasoning section for the nuclear density ρ depends only on numerical factors and the value of m, which is the mass of a single nucleon no matter where the nucleon is located. The nuclear density does not depend on the nuclear number A. Thus, the nuclear density of lead, which is the ratio of its mass to its volume, is neither greater than nor less than the nuclear density of oxygen.

Answer (b) is correct. The result obtained in the Reasoning section for the nuclear density ρ indicates that the density of the nucleus in a lead atom is approximately the same as it is in an oxygen atom. In general, because of Equation 31.2, the *nuclear* density has nearly the same value in all atoms. The difference in densities between solid lead and gaseous oxygen, however, arises mainly because of the difference in how closely the atoms are packed together in the solid and gaseous phases.

Related Homework: *Problem 9*

✓ CHECK YOUR UNDERSTANDING

(*The answers are given at the end of the book.*)

1. Two nuclei differ in their numbers of protons and their numbers of neutrons. Which one or more of the following statements is/are true?   **(a)** They are different isotopes of the same element.   **(b)** They have the same electric charge.   **(c)** They could have the same radii.   **(d)** They have approximately the same nuclear density.

2. A material is known to be an isotope of lead, although the particular isotope is not known. From such limited information, which of the following quantities can you specify? **(a)** Its atomic number   **(b)** Its neutron number   **(c)** Its atomic mass number

3. Two nuclei have different nucleon numbers A_1 and A_2. Are the two nuclei necessarily isotopes of the same element?

4. Can two nuclei have the same radius, even though they contain different numbers of protons and different numbers of neutrons?

31.2 THE STRONG NUCLEAR FORCE AND THE STABILITY OF THE NUCLEUS

Two positive charges that are as close together as they are in a nucleus repel one another with a very strong electrostatic force. What, then, keeps the nucleus from flying apart? Clearly, some kind of attractive force must hold the nucleus together, since many kinds of naturally occurring atoms contain stable nuclei. The gravitational force of attraction between nucleons is too weak to counteract the repulsive electric force, so a different type of force must hold the nucleus together. This force is the *strong nuclear force* and is one of only three fundamental forces that have been discovered, fundamental in the sense that all forces in nature can be explained in terms of these three. The gravitational force is also one of these forces, as is the electroweak force (see Section 31.5).

Many features of the strong nuclear force are well known. For example, it is almost independent of electric charge. At a given separation distance, nearly the same nuclear force of attraction exists between two protons, between two neutrons, or between a proton and a neutron. The range of action of the strong nuclear force is extremely short, with the force of attraction being very strong when two nucleons are as close as 10^{-15} m and essentially zero at larger distances. In contrast, the electric force between two protons decreases to zero only gradually as the separation distance increases to large values and, therefore, has a relatively long range of action.

The limited range of action of the strong nuclear force plays an important role in the stability of the nucleus. For a nucleus to be stable, the electrostatic repulsion between the protons must be balanced by the attraction between the nucleons due to the strong nuclear force. But one proton repels all other protons within the nucleus, since the electrostatic force has such a long range of action. In contrast, a proton or a neutron attracts only its nearest neighbors via the strong nuclear force. As the number Z of protons in the nucleus increases under these conditions, the number N of neutrons has to increase even more, if stability is to be maintained. Figure 31.2 shows a plot of N versus Z for naturally occurring elements that have stable nuclei. For reference, the plot also includes the straight line that represents the condition $N = Z$. With few exceptions, the points representing stable nuclei fall above this reference line, reflecting the fact that the number of neutrons becomes greater than the number of protons as the atomic number Z increases.

As more and more protons occur in a nucleus, there comes a point when a balance of repulsive and attractive forces cannot be achieved by an increased number of neutrons. Eventually, the limited range of action of the strong nuclear force prevents extra neutrons from balancing the long-range electric repulsion of extra protons. The stable nucleus with the largest number of protons ($Z = 83$) is that of bismuth, $^{209}_{83}$Bi, which contains 126 neutrons. All nuclei with more than 83 protons (e.g., uranium, $Z = 92$) are unstable and spontaneously break apart or rearrange their internal structures as time passes. This spontaneous disintegration or rearrangement of internal structure is called *radioactivity,* first discovered in 1896 by the French physicist Henri Becquerel (1852–1908). Section 31.4 discusses radioactivity in greater detail.

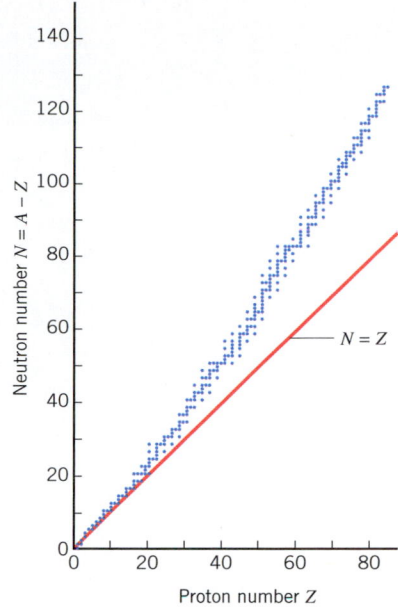

Figure 31.2 With few exceptions, the naturally occurring stable nuclei have a number N of neutrons that equals or exceeds the number Z of protons. Each dot in this plot represents a stable nucleus.

<table>
<tr><td>31.3</td><td>

THE MASS DEFECT OF THE NUCLEUS AND NUCLEAR BINDING ENERGY

</td></tr>
</table>

Because of the strong nuclear force, the nucleons in a stable nucleus are held tightly together. Therefore, energy is required to separate a stable nucleus into its constituent protons and neutrons, as Figure 31.3 illustrates. The more stable the nucleus is, the greater is the amount of energy needed to break it apart. The required energy is called the *binding energy* of the nucleus.

Two ideas that we have studied previously come into play as we discuss the binding energy of a nucleus. These are the rest energy of an object (Section 28.6) and mass (Section 4.2). In Einstein's theory of special relativity, energy and mass are equivalent; in fact, the rest energy E_0 and the mass m are related via $E_0 = mc^2$ (Equation 28.5), where c is the speed of light in a vacuum. Therefore, a change ΔE_0 in the rest energy of the system is equivalent to a change Δm in the mass of the system, according to $\Delta E_0 = (\Delta m)c^2$. We see, then, that the binding energy used in Figure 31.3 to disassemble the nucleus appears as extra mass of the separated and stationary nucleons. In other words, the sum of the individual masses of the separated protons and neutrons is greater by an amount Δm than the mass of the stable nucleus. The difference in mass Δm is known as the *mass defect* of the nucleus.

As Example 2 shows, the binding energy of a nucleus can be determined from the mass defect according to Equation 31.3:

$$\text{Binding energy} = (\text{Mass defect})c^2 = (\Delta m)c^2 \qquad (31.3)$$

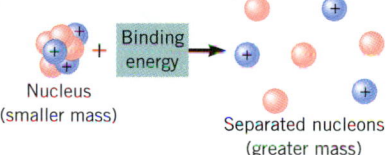

Figure 31.3 Energy, called the binding energy, must be supplied to break the nucleus apart into its constituent protons and neutrons. Each of the separated nucleons is at rest and out of the range of the forces of the other nucleons.

Example 2   The Binding Energy of the Helium Nucleus

The most abundant isotope of helium has a $^4_2$He nucleus whose mass is 6.6447×10^{-27} kg. For this nucleus, find **(a)** the mass defect and **(b)** the binding energy.

Reasoning The symbol $^4_2$He indicates that the helium nucleus contains $Z = 2$ protons and $N = 4 - 2 = 2$ neutrons. To obtain the mass defect Δm, we first determine the sum of the individual masses of the separated protons and neutrons. Then we subtract from this sum the mass of the $^4_2$He nucleus. Finally, we use Equation 31.3 to calculate the binding energy from the value for Δm.

Solution **(a)** Using data from Table 31.1, we find that the sum of the individual masses of the nucleons is

$$\underbrace{2(1.6726 \times 10^{-27} \text{ kg})}_{\text{Two protons}} + \underbrace{2(1.6749 \times 10^{-27} \text{ kg})}_{\text{Two neutrons}} = 6.6950 \times 10^{-27} \text{ kg}$$

This value is greater than the mass of the intact $^4_2$He nucleus, and the mass defect is

$$\Delta m = 6.6950 \times 10^{-27} \text{ kg} - 6.6447 \times 10^{-27} \text{ kg} = \boxed{0.0503 \times 10^{-27} \text{ kg}}$$

(b) According to Equation 31.3, the binding energy is

$$\text{Binding energy} = (\Delta m)c^2 = (0.0503 \times 10^{-27} \text{ kg})(3.00 \times 10^8 \text{ m/s})^2 = 4.53 \times 10^{-12} \text{ J}$$

Usually, binding energies are expressed in energy units of electron volts instead of joules $(1 \text{ eV} = 1.60 \times 10^{-19} \text{ J})$:

$$\text{Binding energy} = (4.53 \times 10^{-12} \text{ J})\left(\frac{1 \text{ eV}}{1.60 \times 10^{-19} \text{ J}}\right) = 2.83 \times 10^7 \text{ eV} = \boxed{28.3 \text{ MeV}}$$

In this result, one million electron volts is denoted by the unit MeV. The value of 28.3 MeV is more than two million times greater than the energy required to remove an orbital electron from an atom.

In calculations such as that in Example 2, it is customary to use the ***atomic mass unit*** (u) instead of the kilogram. As introduced in Section 14.1, the atomic mass unit is one-twelfth of the mass of a $^{12}_{6}\text{C}$ atom of carbon. In terms of this unit, the mass of a $^{12}_{6}\text{C}$ atom is exactly 12 u. Table 31.1 also gives the masses of the electron, the proton, and the neutron in atomic mass units. For future calculations, the energy equivalent of one atomic mass unit can be determined by observing that the mass of a proton is 1.6726×10^{-27} kg or 1.0073 u, so that

$$1 \text{ u} = (1 \text{ u})\left(\frac{1.6726 \times 10^{-27} \text{ kg}}{1.0073 \text{ u}}\right) = 1.6605 \times 10^{-27} \text{ kg}$$

and

$$\Delta E_0 = (\Delta m)c^2 = (1.6605 \times 10^{-27} \text{ kg})(2.9979 \times 10^8 \text{ m/s})^2 = 1.4924 \times 10^{-10} \text{ J}$$

In electron volts, therefore, one atomic mass unit is equivalent to

$$1 \text{ u} = (1.4924 \times 10^{-10} \text{ J})\left(\frac{1 \text{ eV}}{1.6022 \times 10^{-19} \text{ J}}\right) = 9.315 \times 10^8 \text{ eV} = 931.5 \text{ MeV}$$

Data tables for isotopes, such as the table in Appendix F, give masses in atomic mass units. Typically, however, the given masses are not nuclear masses. They are *atomic masses*—that is, the masses of neutral atoms, including the mass of the orbital electrons. Example 3 deals again with the $^{4}_{2}\text{He}$ nucleus and shows how to take into account the effect of the orbital electrons when using such data to determine binding energies.

 Example 3 The Binding Energy of the Helium Nucleus, Revisited

The atomic mass of helium $^{4}_{2}\text{He}$ is 4.0026 u, and the atomic mass of hydrogen $^{1}_{1}\text{H}$ is 1.0078 u. Using atomic mass units instead of kilograms, obtain the binding energy of the $^{4}_{2}\text{He}$ nucleus.

Reasoning To determine the binding energy, we calculate the mass defect in atomic mass units and then use the fact that one atomic mass unit is equivalent to 931.5 MeV of energy. The mass of 4.0026 u for $^{4}_{2}\text{He}$ *includes the mass of the two electrons in the neutral helium atom*. To calculate the mass defect, we must subtract 4.0026 u from the sum of the individual masses of the nucleons, including the mass of the electrons. As Figure 31.4 illustrates, the electron mass will be included if the masses of two hydrogen atoms are used in the calculation instead of the masses of two protons. The mass of a $^{1}_{1}\text{H}$ hydrogen atom is given in Table 31.1 as 1.0078 u, and the mass of a neutron as 1.0087 u.

Figure 31.4 Data tables usually give the mass of the neutral atom (including the orbital electrons) rather than the mass of the nucleus. When data from such tables are used to determine the mass defect of a nucleus, the mass of the orbital electrons must be taken into account, as this drawing illustrates for the $^{4}_{2}\text{He}$ isotope of helium. See Example 3.

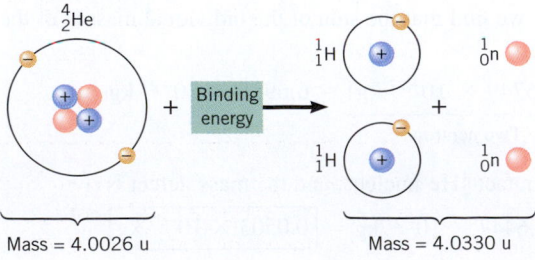

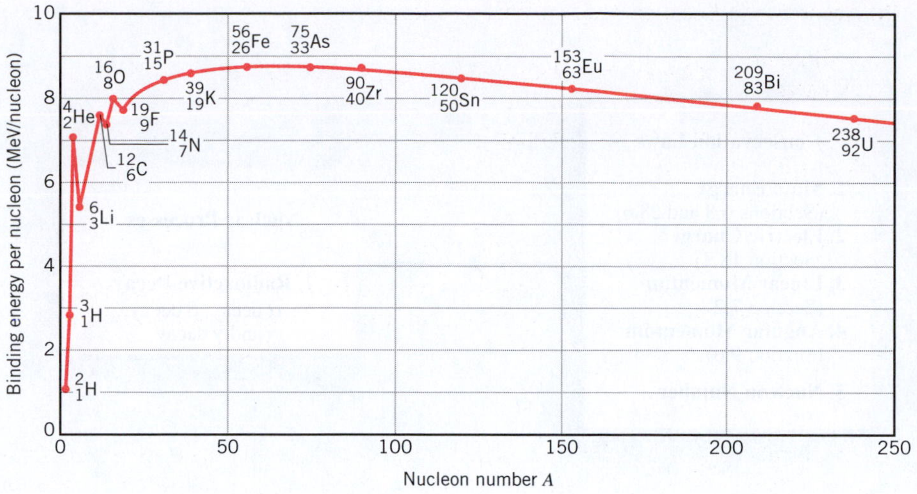

Figure 31.5 A plot of binding energy per nucleon versus the nucleon number A.

Solution The sum of the individual masses is

$$\underbrace{2(1.0078 \text{ u})}_{\substack{\text{Two hydrogen} \\ \text{atoms}}} + \underbrace{2(1.0087 \text{ u})}_{\text{Two neutrons}} = 4.0330 \text{ u}$$

The mass defect is $\Delta m = 4.0330 \text{ u} - 4.0026 \text{ u} = 0.0304 \text{ u}$. Since 1 u is equivalent to 931.5 MeV, the binding energy is $\boxed{\text{Binding energy} = 28.3 \text{ MeV}}$, which matches the result obtained in Example 2.

To see how the nuclear binding energy varies from nucleus to nucleus, it is necessary to compare the binding energy for each nucleus on a per-nucleon basis. The graph in Figure 31.5 shows a plot in which the binding energy divided by the nucleon number A is plotted against the nucleon number itself. In the graph, the peak for the ^4_2He isotope of helium indicates that the ^4_2He nucleus is particularly stable. The binding energy per nucleon increases rapidly for nuclei with small masses and reaches a maximum of approximately 8.7 MeV/nucleon for a nucleon number of about $A = 60$. For greater nucleon numbers, the binding energy per nucleon decreases gradually. Eventually, the binding energy per nucleon decreases enough so there is insufficient binding energy to hold the nucleus together. Nuclei more massive than the $^{209}_{83}\text{Bi}$ nucleus of bismuth are unstable and hence radioactive.

✔ **CHECK YOUR UNDERSTANDING**

(The answers are given at the end of the book.)

5. Using Figure 31.5, rank the following nuclei in ascending order according to the binding energy per nucleon (smallest first):   **(a)** Phosphorus $^{31}_{15}\text{P}$ **(b)** Cobalt $^{59}_{27}\text{Co}$ **(c)** Tungsten $^{184}_{74}\text{W}$ **(d)** Thorium $^{232}_{90}\text{Th}$

6. The following table gives values for the mass defect Δm for four hypothetical nuclei, A, B, C, and D. Which statement is true regarding the stability of these nuclei?   **(a)** Nucleus D is the most stable, and A is the least stable.   **(b)** Nucleus C is stable, whereas A, B, and D are not.   **(c)** Nucleus A is the most stable, and D is not stable.   **(d)** Nuclei A and B are stable, but B is more stable than A.

	A	B	C	D
Mass defect, Δm	$+6.0 \times 10^{-29}$ kg	$+2.0 \times 10^{-29}$ kg	0 kg	-6.0×10^{-29} kg

31.4 RADIOACTIVITY

When an unstable or radioactive nucleus disintegrates spontaneously, certain kinds of particles and/or high-energy photons are released. These particles and photons are collectively called "rays." Three kinds of rays are produced by naturally occurring

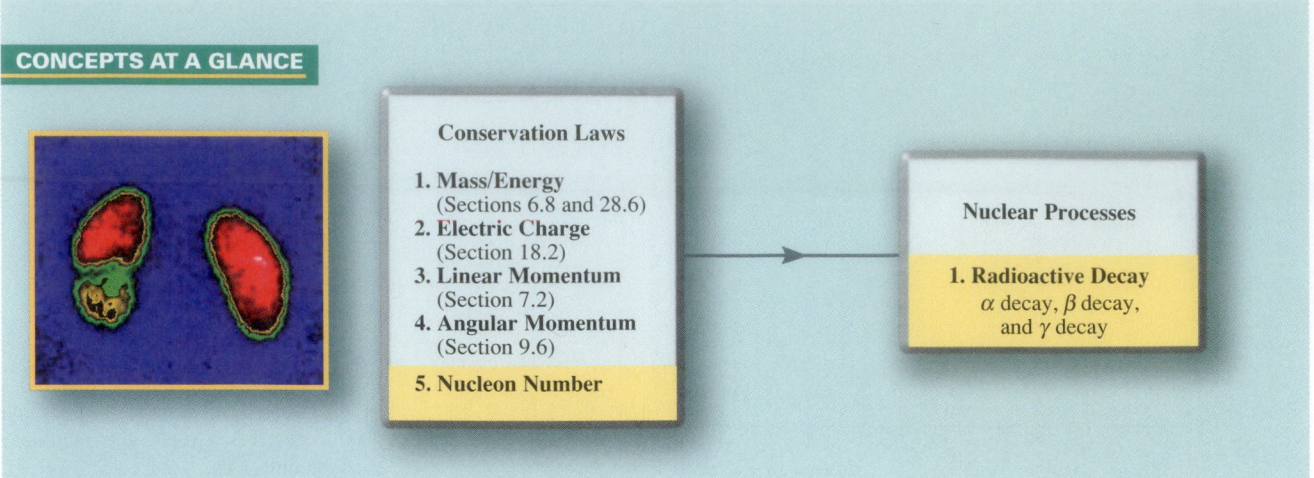

CONCEPTS AT A GLANCE

Conservation Laws

1. **Mass/Energy**
   (Sections 6.8 and 28.6)
2. **Electric Charge**
   (Section 18.2)
3. **Linear Momentum**
   (Section 7.2)
4. **Angular Momentum**
   (Section 9.6)
5. **Nucleon Number**

Nuclear Processes

1. **Radioactive Decay**
   α decay, β decay,
   and γ decay

Figure 31.6 CONCEPTS AT A GLANCE
The conservation laws listed at the left side of this chart are obeyed when a nucleus undergoes radioactive decay. The three types of naturally occurring decay are α decay, β decay, and γ decay. Nuclear medicine uses radioactive decay to produce scans of organs. This photograph shows a nuclear scan of two kidneys, the one on the left displaying an invasive cancer. (ISM/Phototake)

radioactivity: **α rays, β rays,** and **γ rays.** They are named according to the first three letters of the Greek alphabet, alpha (α), beta (β), and gamma (γ), to indicate the extent of their ability to penetrate matter. α rays are the least penetrating, being blocked by a thin (≈ 0.01 mm) sheet of lead, whereas β rays penetrate lead to a much greater distance (≈ 0.1 mm). γ rays are the most penetrating and can pass through an appreciable thickness (≈ 100 mm) of lead.

▶ **CONCEPTS AT A GLANCE** The nuclear disintegration process that produces α, β, and γ rays must obey the laws of physics that we have studied in previous chapters. As the Concepts-at-a-Glance chart in Figure 31.6 reminds us, these laws are called conservation laws because each of them deals with a property (such as mass/energy, electric charge, linear momentum, and angular momentum) that is conserved or does not change during a process. To the first four conservation laws in Figure 31.6, we now add a fifth, the conservation of nucleon number. In all radioactive decay processes it has been observed that the number of nucleons (protons plus neutrons) present before the decay is equal to the number of nucleons after the decay. Therefore, the number of nucleons is conserved during a nuclear disintegration. As applied to the disintegration of a nucleus, the conservation laws require that the energy, electric charge, linear momentum, angular momentum, and nucleon number that a nucleus possesses must remain unchanged when it disintegrates into nuclear fragments and accompanying α, β, or γ rays. ◀

The three types of radioactivity that occur naturally can be observed in a relatively simple experiment. A piece of radioactive material is placed at the bottom of a narrow hole in a lead cylinder. The cylinder is located within an evacuated chamber, as Figure 31.7 illustrates. A magnetic field is directed perpendicular to the plane of the paper, and a photographic plate is positioned to the right of the hole. Three spots appear on the developed plate, which are associated with the radioactivity of the nuclei in the material. Since moving particles are deflected by a magnetic field only when they are electrically charged, this experiment reveals that two types of radioactivity (α and β rays, as it turns out) consist of charged particles, whereas the third type (γ rays) does not.

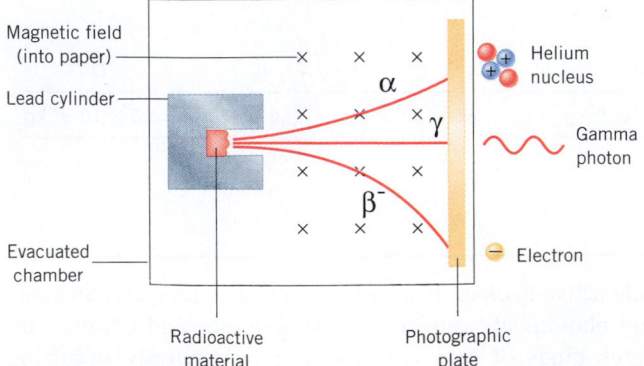

Magnetic field (into paper)

Lead cylinder

α

γ

β^-

Helium nucleus

Gamma photon

Electron

Evacuated chamber

Radioactive material

Photographic plate

Figure 31.7 α and β rays are deflected by a magnetic field and, therefore, consist of moving charged particles. γ rays are not deflected by a magnetic field and, consequently, must be uncharged.

α DECAY

When a nucleus disintegrates and produces α rays, it is said to undergo **α decay.** Experimental evidence shows that α rays consist of positively charged particles, each one being the $_2^4$He nucleus of helium. Thus, an α particle has a charge of $+2e$ and a nucleon number of $A = 4$. Since the grouping of 2 protons and 2 neutrons in a $_2^4$He nucleus is particularly stable, as we have seen in connection with Figure 31.5, it is not surprising that an α particle can be ejected as a unit from a more massive unstable nucleus.

Figure 31.8 shows the disintegration process for one example of α decay:

$$_{92}^{238}\text{U} \longrightarrow \, _{90}^{234}\text{Th} + \, _2^4\text{He}$$

Parent	Daughter	α particle
nucleus	nucleus	(helium
(uranium)	(thorium)	nucleus)

The original nucleus is referred to as the *parent nucleus* (P), and the nucleus remaining after disintegration is called the *daughter nucleus* (D). Upon emission of an α particle, the uranium $_{92}^{238}$U parent is converted into the $_{90}^{234}$Th daughter, which is an isotope of thorium. The parent and daughter nuclei are different, so α decay converts one element into another, a process known as **transmutation.**

Electric charge is conserved during α decay. In Figure 31.8, for instance, 90 of the 92 protons in the uranium nucleus end up in the thorium nucleus, and the remaining 2 protons are carried off by the α particle. The total number of 92, however, is the same before and after disintegration. α decay also conserves the number of nucleons, because the number is the same before (238) and after (234 + 4) disintegration. Consistent with the conservation of electric charge and nucleon number, the general form for α decay is

α decay

$$_Z^A\text{P} \longrightarrow \, _{Z-2}^{A-4}\text{D} + \, _2^4\text{He}$$

Parent	Daughter	α particle
nucleus	nucleus	(helium nucleus)

When a nucleus releases an α particle, the nucleus also releases energy. In fact, the energy released by radioactive decay is responsible, in part, for keeping the interior of the earth hot and, in some places, even molten. The following example shows how the conservation of mass/energy can be used to determine the amount of energy released in α decay.

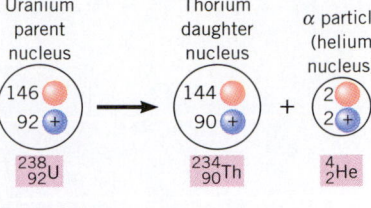

Figure 31.8 α decay occurs when an unstable parent nucleus emits an α particle and in the process is converted into a different, or daughter, nucleus.

 Example 4 α Decay and the Release of Energy

The atomic mass of uranium $_{92}^{238}$U is 238.0508 u, that of thorium $_{90}^{234}$Th is 234.0436 u, and that of an α particle $_2^4$He is 4.0026 u. Determine the energy released when α decay converts $_{92}^{238}$U into $_{90}^{234}$Th.

Reasoning Since energy is released during the decay, the combined mass of the $_{90}^{234}$Th daughter nucleus and the α particle is less than the mass of the $_{92}^{238}$U parent nucleus. The difference in mass is equivalent to the energy released. We will determine the difference in mass in atomic mass units and then use the fact that 1 u is equivalent to 931.5 MeV.

Solution The decay and the masses are shown below:

$$_{92}^{238}\text{U} \longrightarrow \, _{90}^{234}\text{Th} + \, _2^4\text{He}$$

238.0508 u	234.0436 u	4.0026 u
	238.0462 u	

The decrease in mass, or mass defect for the decay process, is 238.0508 u − 238.0462 u = 0.0046 u. As usual, the masses are atomic masses and include the mass of the orbital electrons. But this causes no error here because the same total number of electrons is included for $_{92}^{238}$U, on the one hand, and for $_{90}^{234}$Th plus $_2^4$He, on the other. Since 1 u is equivalent to 931.5 MeV, the released energy is $\boxed{4.3 \text{ MeV}}$.

When α decay occurs as in Example 4, the energy released appears as kinetic energy of the recoiling $_{90}^{234}$Th nucleus and the α particle, except for a small portion carried away as a γ ray. Conceptual Example 5 discusses how the $^{234}_{90}$Th nucleus and the α particle share in the released energy.

Conceptual Example 5 How Energy Is Shared During the α Decay of $^{238}_{92}$U

In Example 4, the energy released by the α decay of $^{238}_{92}$U is found to be 4.3 MeV. Since this energy is carried away as kinetic energy of the recoiling $^{234}_{90}$Th nucleus and the α particle, it follows that $KE_{Th} + KE_\alpha = 4.3$ MeV. However, KE_{Th} and KE_α are not equal. Which particle carries away more kinetic energy, the $^{234}_{90}$Th nucleus or the α particle?

Reasoning and Solution Kinetic energy depends on the mass m and speed v of a particle, since $KE = \frac{1}{2}mv^2$ (Equation 6.2). The $^{234}_{90}$Th nucleus has a much greater mass than the α particle, and since the kinetic energy is proportional to the mass, it is tempting to conclude that the $^{234}_{90}$Th nucleus has the greater kinetic energy. This conclusion is not correct, however, since it does not take into account the fact that the $^{234}_{90}$Th nucleus and the α particle have different speeds after the decay. In fact, we expect the thorium nucleus to recoil with the smaller speed precisely *because* it has the greater mass. The decaying $^{238}_{92}$U is like a father and his young daughter on ice skates, pushing off against one another. The more massive father recoils with much less speed than the daughter. We can use the principle of conservation of linear momentum to verify our expectation.

As Section 7.2 discusses, the conservation principle states that the total linear momentum of an isolated system remains constant. An isolated system is one for which the vector sum of the external forces acting on the system is zero, and the decaying $^{238}_{92}$U nucleus fits this description. It is stationary initially, and since momentum is mass times velocity, its initial momentum is zero. In its final form, the system consists of the $^{234}_{90}$Th nucleus and the α particle and has a final total momentum of $m_{Th}v_{Th} + m_\alpha v_\alpha$. According to momentum conservation, the initial and final values of the total momentum of the system must be the same, so that $m_{Th}v_{Th} + m_\alpha v_\alpha = 0$. Solving this equation for the velocity of the thorium nucleus, we find that $v_{Th} = -m_\alpha v_\alpha/m_{Th}$. Since m_{Th} is much greater than m_α, we can see that the speed of the thorium nucleus is less than the speed of the α particle. Moreover, the kinetic energy depends on the square of the speed and only the first power of the mass. As a result of its much greater speed, *the α particle has the greater kinetic energy.*

Related Homework: *Problem 61*

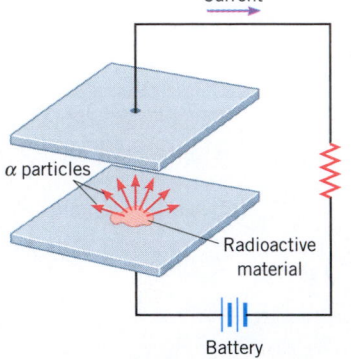

Figure 31.9 A smoke detector.

The physics of
radioactivity and smoke detectors.

One widely used application of α decay is in smoke detectors. Figure 31.9 illustrates how a smoke detector operates. Two small and parallel metal plates are separated by a distance of about one centimeter. A tiny amount of radioactive material at the center of one of the plates emits α particles, which collide with air molecules. During the collisions, the air molecules are ionized to form positive and negative ions. The voltage from a battery causes one plate to be positive and the other negative, so that each plate attracts ions of opposite charge. As a result there is a current in the circuit attached to the plates. The presence of smoke particles between the plates reduces the current, since the ions that collide with a smoke particle are usually neutralized. The drop in current that smoke particles cause is used to trigger an alarm.

β DECAY

The β rays in Figure 31.7 are deflected by the magnetic field in a direction opposite to that of the positively charged α rays. Consequently, these β rays, which are the most common kind, consist of negatively charged particles or β^- particles. Experiment shows that β^- particles are electrons. As an illustration of β^- decay, consider the thorium $^{234}_{90}$Th nucleus, which decays by emitting a β^- particle, as in Figure 31.10:

$$^{234}_{90}\text{Th} \longrightarrow {}^{234}_{91}\text{Pa} + {}^{0}_{-1}\text{e}$$

Parent nucleus (thorium)	Daughter nucleus (protactinium)	β^- particle (electron)

Figure 31.10 β decay occurs when a neutron in an unstable parent nucleus decays into a proton and an electron, the electron being emitted as the β^- particle. In the process, the parent nucleus is transformed into the daughter nucleus.

β^- decay, like α decay, causes a transmutation of one element into another. In this case, thorium $^{234}_{90}$Th is converted into protactinium $^{234}_{91}$Pa. The law of conservation of charge is obeyed, since the net number of positive charges is the same before (90) and after (91 − 1) the β^- emission. The law of conservation of nucleon number is obeyed, since the nucleon number remains at $A = 234$. The general form for β^- decay is

β^- decay
$$^A_Z P \longrightarrow ^A_{Z+1} D + ^0_{-1} e$$

Parent nucleus   Daughter nucleus β^- particle (electron)

The electron emitted in β^- decay does *not* actually exist within the parent nucleus and is *not* one of the orbital electrons. Instead, the electron is created when a neutron decays into a proton and an electron; when this occurs, the proton number of the parent nucleus increases from Z to $Z + 1$ and the nucleon number remains unchanged. The newly created electron is usually fast-moving and escapes from the atom, leaving behind a positively charged atom.

Example 6 illustrates that energy is released during β^- decay, just as it is during α decay, and that the conservation of mass/energy applies.

 Example 6 β^- Decay and the Release of Energy

The atomic mass of thorium $^{234}_{90}$Th is 234.043 59 u, and the atomic mass of protactinium $^{234}_{91}$Pa is 234.043 30 u. Find the energy released when β^- decay changes $^{234}_{90}$Th into $^{234}_{91}$Pa.

Reasoning To find the energy released, we follow the usual procedure of determining how much the mass has decreased because of the decay and then calculating the equivalent energy.

Solution The decay and the masses are shown below:

$$^{234}_{90}\text{Th} \longrightarrow ^{234}_{91}\text{Pa} + ^0_{-1}e$$
234.043 59 u   234.043 30 u

When the $^{234}_{90}$Th nucleus of a thorium atom is converted into a $^{234}_{91}$Pa nucleus, the number of orbital electrons remains the same, so the resulting protactinium atom is missing one orbital electron. However, the given mass includes all 91 electrons of a neutral protactinium atom. In effect, then, the value of 234.043 30 u for $^{234}_{91}$Pa already includes the mass of the β^- particle. The mass decrease that accompanies the β^- decay is

$$234.043\ 59\ \text{u} - 234.043\ 30\ \text{u} = 0.000\ 29\ \text{u}$$

The equivalent energy (1 u = 931.5 MeV) is $\boxed{0.27\ \text{MeV}}$. This is the maximum kinetic energy that the emitted electron can have.

Problem-solving insight

In β^- decay, be careful not to include the mass of the electron ($^0_{-1}$e) twice. As discussed here for the daughter atom ($^{234}_{91}$Pa), the atomic mass already includes the mass of the emitted electron.

A second kind of β decay sometimes occurs.* In this process the particle emitted by the nucleus is a *positron* rather than an electron. A positron, also called a β^+ particle, has the same mass as an electron but carries a charge of $+e$ instead of $-e$. The disintegration process for β^+ decay is

β^+ decay
$$^A_Z P \longrightarrow ^A_{Z-1} D + ^0_1 e$$

Parent nucleus   Daughter nucleus β^+ particle (positron)

The emitted positron does *not* exist within the nucleus but, rather, is created when a nuclear proton is transformed into a neutron. In the process, the proton number of the parent nucleus decreases from Z to $Z−1$, and the nucleon number remains the same. As with β^- decay, the laws of conservation of charge and nucleon number are obeyed, and there is a transmutation of one element into another.

*A third kind of β decay also occurs in which a nucleus pulls in, or captures, one of the orbital electrons from outside the nucleus. The process is called *electron capture*, or **K capture**, since the electron normally comes from the innermost, or K, shell.

γ DECAY

The nucleus, like the orbital electrons, exists only in discrete energy states or levels. When a nucleus changes from an excited energy state (denoted by an asterisk *) to a lower energy state, a photon is emitted. The process is similar to the one discussed in Section 30.3 for the photon emission that leads to the hydrogen atom line spectrum. With nuclear energy levels, however, the photon has a much greater energy and is called a γ ray. The γ decay process is written as follows:

γ *decay*
$$\underset{\substack{\text{Excited} \\ \text{energy state}}}{{}^{A}_{Z}\text{P*}} \longrightarrow \underset{\substack{\text{Lower} \\ \text{energy state}}}{{}^{A}_{Z}\text{P}} + \underset{\substack{\gamma\,\text{ray}}}{\gamma}$$

γ decay does *not* cause a transmutation of one element into another. In the next example the wavelength of one particular γ-ray photon is determined.

ANALYZING MULTIPLE-CONCEPT PROBLEMS

Example 7 The Wavelength of a Photon Emitted During γ Decay

What is the wavelength (in vacuum) of the 0.186-MeV γ-ray photon emitted by radium ${}^{226}_{88}\text{Ra}$?

Reasoning The wavelength of the photon is related to the speed of light and the frequency of the photon. The frequency is not given, but it can be obtained from the 0.186-MeV energy of the photon. The photon is emitted with this energy when the nucleus changes from one energy state to a lower energy state. The energy is the difference ΔE between the two nuclear energy levels, in a way very similar to that discussed in Section 30.3 for the energy levels of the electron in the hydrogen atom. In that section, we saw that the energy difference ΔE is related to the frequency f and Planck's constant h, so that we will be able to obtain the frequency from the given energy value.

Knowns and Unknowns The following table summarizes the available data:

Description	Symbol	Value	Comment
Energy of γ-ray photon	ΔE	0.186 MeV	Will be converted into joules.
Unknown Variable			
Wavelength of γ-ray photon	λ	?	

Problem-solving insight

The energy ΔE of a γ-ray photon, like that of photons in other regions of the electromagnetic spectrum (visible, infrared, microwave, etc.), is equal to the product of Planck's constant h and the frequency f of the photon: $\Delta E = hf$.

Modeling the Problem

STEP 1 **The Relation of Wavelength to Frequency** The photon wavelength λ is related to the photon frequency f and the speed c of light in a vacuum according to Equation 16.1, as shown at the right. We have no value for the frequency, so we turn to Step 2 to evaluate it.

$$\lambda = \frac{c}{f} \qquad (16.1)$$

$$\boxed{?}$$

STEP 2 **Photon Frequency and Photon Energy** Section 30.3 discusses the fact that the photon emitted when the electron in a hydrogen atom changes from a higher to a lower energy level has an energy ΔE, which is the difference between the energy levels. A similar situation exists here when the nucleus changes from a higher to a lower energy level. The γ-ray photon that is emitted has an energy ΔE given by $\Delta E = hf$ (Equation 30.4). Solving for the frequency, we obtain

$$\boxed{f = \frac{\Delta E}{h}}$$

which we can substitute into Equation 16.1, as indicated at the right.

$$\lambda = \frac{c}{f} \qquad (16.1)$$

$$\boxed{f = \frac{\Delta E}{h}}$$

Solution Combining the results of each step algebraically, we find that

STEP 1   STEP 2

$$\lambda \;\; \overset{\downarrow}{=} \;\; \frac{c}{f} \;\; \overset{\downarrow}{=} \;\; \frac{c}{\dfrac{\Delta E}{h}}$$

The wavelength of the γ-ray photon is

$$\lambda = \frac{hc}{\Delta E} = \frac{(6.63 \times 10^{-34} \text{ J} \cdot \text{s})(3.00 \times 10^{8} \text{ m/s})}{(0.186 \times 10^{6} \text{ eV})\left(\dfrac{1.60 \times 10^{-19} \text{ J}}{1 \text{ eV}}\right)} = \boxed{6.68 \times 10^{-12} \text{ m}}$$

Note that we have converted the value of $\Delta E = 0.186 \times 10^{6}$ eV into joules by using the fact that 1 eV $= 1.60 \times 10^{-19}$ J.

Related Homework: *Problem 26*

MEDICAL APPLICATIONS OF RADIOACTIVITY

Gamma Knife radiosurgery is becoming a very promising medical procedure for treating certain problems of the brain, including benign and cancerous tumors as well as blood vessel malformations. The procedure, which involves no knife at all, uses powerful, highly focused beams of γ rays aimed at the tumor or malformation. The γ rays are emitted by a radioactive cobalt-60 source. As Figure 31.11*a* illustrates, the patient wears a protective metal helmet that is perforated with many small holes. Part *b* of the figure shows that the holes focus the γ rays to a single tiny target within the brain. The target tissue thus receives a very intense dose of radiation and is destroyed, while the surrounding healthy tissue is undamaged. Gamma Knife surgery is a noninvasive, painless, and bloodless procedure that is often performed under local anesthesia. Hospital stays are 70 to 90% shorter than with conventional surgery, and patients often return to work within a few days.

**The physics of
Gamma Knife radiosurgery.**

An exercise thallium heart scan is a test that uses radioactive thallium to produce images of the heart muscle. When combined with an exercise test, such as walking on a treadmill, the thallium scan helps identify regions of the heart that do not receive enough blood. The scan is especially useful in diagnosing the presence of blockages in the coronary arteries, which supply oxygen-rich blood to the heart muscle. During the test, a small amount of thallium is injected into a vein while the patient walks on a treadmill. The thallium attaches to the red blood cells and is carried throughout the body. The thallium enters the heart muscle by way of the coronary arteries and collects in heart-muscle cells that come into contact with the blood. The thallium isotope used, $^{201}_{81}$Tl, emits γ rays, which a special camera records. Since the thallium reaches those regions of the heart that have an adequate blood supply, lesser amounts show up in areas where the blood flow has been

**The physics of
an exercise thallium heart scan.**

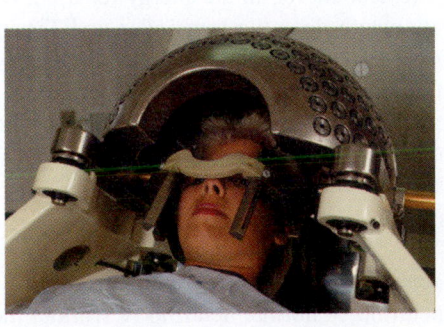

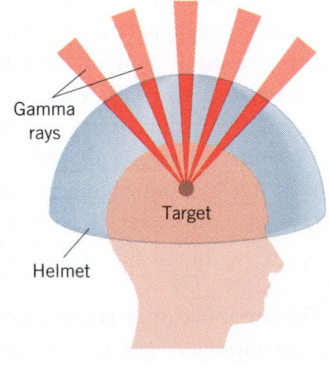

Gamma rays

Target

Helmet

(a)

(b)

Figure 31.11 (*a*) In Gamma Knife radiosurgery, a protective metal helmet containing many small holes is placed over the patient's head. (© Custom Medical Stock Photo) (*b*) The holes focus the beams of γ rays to a tiny target within the brain.

reduced due to arterial blockages (see Figure 31.12). A second set of images is taken several hours later, while the patient is resting. These images help differentiate between regions of the heart that temporarily do not receive enough blood (the blood flow returns to normal after the exercise) and regions that are permanently damaged due to, for example, a previous heart attack (the blood flow does not return to normal).

The use of radioactive isotopes to deliver radiation to specific targets in the body is an important medical technique. In treating cancer, for example, the method of delivery should ideally apply a high dose of radiation to a malignant tumor in order to kill it, while applying only a small (non-damaging) dose to healthy surrounding tissue. Brachytherapy implants offer such a delivery method. In this type of treatment, radioactive isotopes are formed into small seeds and implanted directly in the tumor according to a predesigned pattern. The energy and type of radiation emitted by the isotopes can be exploited to optimize a treatment design and minimize damage to healthy tissue. Seeds containing iridium $^{192}_{77}\text{Ir}$ are used to treat many cancers, and seeds containing iodine $^{125}_{53}\text{I}$ and palladium $^{103}_{46}\text{Pd}$ are used for prostate cancer. Research has also indicated that brachytherapy implants may have an important role to play in the treatment of atherosclerosis, in which blood vessels become blocked with plaque. Such blockages are often treated using the technique of balloon angioplasty. With the aid of a catheter inserted into an occluded coronary artery, a balloon is inflated to open the artery and place a stent (a metallic mesh that provides support for the arterial wall) at the site of the blockage. Sometimes the arterial wall is damaged in this process, and as it heals, the artery often becomes blocked again. Brachytherapy implants (using iridium $^{192}_{77}\text{Ir}$ or phosphorus $^{32}_{15}\text{P}$, for instance) have been found to inhibit repeat blockages following angioplasty.

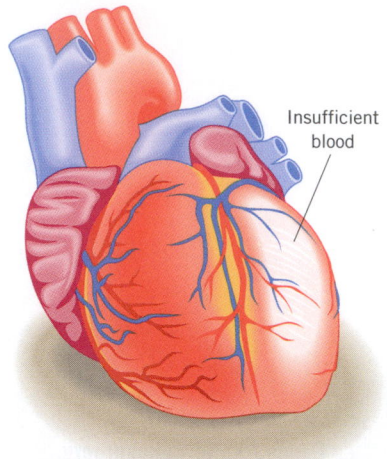

The physics of brachytherapy implants.

Figure 31.12 An exercise thallium heart scan indicates regions of the heart that receive insufficient blood during exercise.

✓ CHECK YOUR UNDERSTANDING

(*The answers are given at the end of the book.*)

7. Polonium $^{216}_{84}\text{Po}$ undergoes α decay to produce a daughter nucleus that itself undergoes β^- decay. Which one of the following nuclei is the one that ultimately results? **(a)** $^{211}_{82}\text{Pb}$ **(b)** $^{211}_{81}\text{Tl}$ **(c)** $^{212}_{81}\text{Tl}$ **(d)** $^{212}_{83}\text{Bi}$ **(e)** $^{213}_{82}\text{Pb}$

8. Uranium $^{238}_{92}\text{U}$ decays into thorium $^{234}_{90}\text{Th}$ by means of α decay, as Example 4 in the text discusses. Another possibility is that the $^{238}_{92}\text{U}$ nucleus just emits a single proton instead of an α particle. This hypothetical decay scheme is shown below, along with the pertinent atomic masses:

$$\begin{array}{ccc} ^{238}_{92}\text{U} & \longrightarrow & ^{237}_{91}\text{Pa} & + & ^{1}_{1}\text{H} \\ \text{Uranium} & & \text{Protactinium} & & \text{Proton} \\ 238.050\,78\text{ u} & & 237.051\,14\text{ u} & & 1.007\,83\text{ u} \end{array}$$

For a decay to be possible, it must bring the parent nucleus toward a more stable state by allowing the release of energy. Compare the total mass of the products of this hypothetical decay with the mass of $^{238}_{92}\text{U}$ and decide whether the emission of a single proton is possible for $^{238}_{92}\text{U}$.

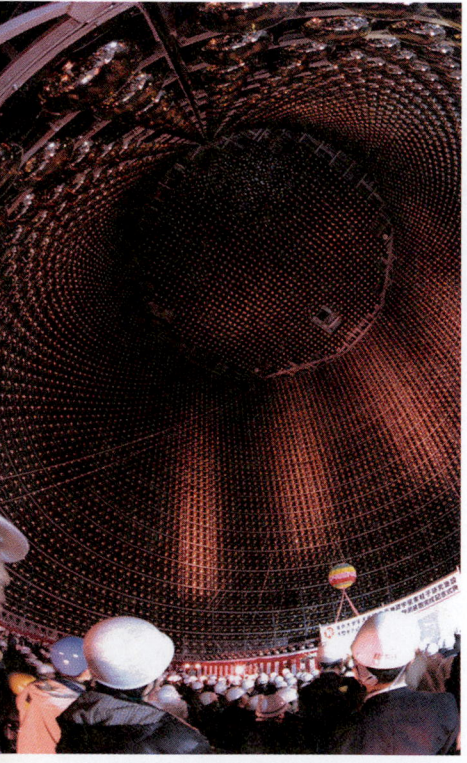

Figure 31.13 The Super-Kamiokande neutrino detector in Japan consists of a steel cylindrical tank containing 12.5 million gallons of ultrapure water. Its inner wall is lined with 11 000 photomultiplier tubes. (© Kyodo News International)

31.5 | THE NEUTRINO

When a β particle is emitted by a radioactive nucleus, energy is simultaneously released, as Example 6 illustrates. Experimentally, however, it is found that most β particles do not have enough kinetic energy to account for all the energy released. If a β particle carries away only part of the energy, where does the remainder go? The question puzzled physicists until 1930, when Wolfgang Pauli proposed that part of the energy is carried away by another particle that is emitted along with the β particle. This additional particle is called the **_neutrino_**, and its existence was verified experimentally in 1956. The Greek letter nu (ν) is used to symbolize the neutrino. For instance, the β^- decay of thorium $^{234}_{90}\text{Th}$ (see Section 31.4) is more correctly written as

$$^{234}_{90}\text{Th} \longrightarrow ^{234}_{91}\text{Pa} + {}^{0}_{-1}\text{e} + \overline{\nu}$$

The bar above the ν is included because the neutrino emitted in this particular decay process is an antimatter neutrino, or antineutrino. A normal neutrino (ν without the bar) is emitted when β^+ decay occurs.

The neutrino has zero electric charge and is extremely difficult to detect because it interacts very weakly with matter. For example, the average neutrino can penetrate one light-year of lead (about 9.5×10^{15} m) without interacting with it. Thus, even though trillions of neutrinos pass through our bodies every second, they have no effect. Although difficult, it is possible to detect neutrinos. Figure 31.13 shows the Super-Kamiokande neutrino detector in Japan. It is located 915 m underground and consists of a steel cylindrical tank, ten stories tall, whose inner wall is lined with 11 000 photomultiplier tubes (see Section 31.9). The tank is filled with 12.5 million gallons of ultrapure water. Neutrinos colliding with the water molecules produce light patterns that the photomultiplier tubes detect.

One of the major scientific questions of our time is whether neutrinos have mass. The question is important because neutrinos are so plentiful in the universe. Even a very small mass could account for a significant portion of the mass in the universe and, possibly, could have an effect on the formation of galaxies. In 1998 the Super-Kamiokande detector yielded the first strong, but indirect, evidence that neutrinos do indeed have a small mass. (The mass of the electron neutrino is less than 0.0004% of the mass of an electron.) This finding implies that neutrinos travel at less than the speed of light. If the neutrino's mass were zero, like that of a photon, it would travel at the speed of light.

The emission of neutrinos and β particles involves a force called the *weak nuclear force* because it is much weaker than the strong nuclear force. It is now known that the weak nuclear force and the electromagnetic force are two different manifestations of a single, more fundamental force, the **electroweak force.** The theory for the electroweak force was developed by Sheldon Glashow (1932–   ), Abdus Salam (1926–1996), and Steven Weinberg (1933–   ), who shared a Nobel Prize for their achievement in 1979. The electroweak force, the gravitational force, and the strong nuclear force are the three fundamental forces in nature.

31.6 RADIOACTIVE DECAY AND ACTIVITY

The question of which radioactive nucleus in a group of nuclei disintegrates at a given instant is decided like the winning numbers in a state lottery: individual disintegrations occur randomly. As time passes, the number N of parent nuclei decreases, as Figure 31.14 shows. This graph of N versus time indicates that the decrease occurs in a smooth fashion, with N approaching zero after enough time has passed. To help describe the graph, it is useful to define the **half-life** $T_{1/2}$ of a radioactive isotope as the time required for one-half of the nuclei present to disintegrate. For example, radium $^{226}_{88}$Ra has a half-life of 1600 years, because it takes this amount of time for one-half of a given quantity of this isotope to disintegrate. In another 1600 years, one-half of the remaining radium atoms will have disintegrated, leaving only one-fourth of the original number intact. In Figure 31.14, the number of nuclei present at time $t = 0$ s is $N = N_0$, and the number present at $t = T_{1/2}$ is $N = \frac{1}{2}N_0$. The number present at $t = 2T_{1/2}$ is $N = \frac{1}{4}N_0$, and so on. The value of the half-life depends on the nature of the radioactive nucleus. Values ranging from a fraction of a second to billions of years have been found (see Table 31.2).

The physics of radioactive radon gas in houses. Radon $^{222}_{86}$Rn is a naturally occurring radioactive gas produced when radium $^{226}_{88}$Ra undergoes α decay. There is a nationwide concern about radon as a health hazard because radon in the soil is gaseous and can enter the basement of homes through cracks in the foundation. (It should be noted, however, that the mechanism of indoor radon entry is not well understood and that entry via foundation cracks is likely only part of the story.) Once inside, the concentration of radon can rise markedly, depending on the type of housing construction and the concentration of radon in the surrounding soil. Radon gas decays into daughter nuclei that are also radioactive. The radioactive nuclei can attach to dust and smoke particles that can be inhaled, and they remain in the lungs to release tissue-damaging radiation. Prolonged exposure to high levels of radon can lead to lung cancer. Since radon gas concentrations can be measured with inexpensive monitoring devices, it is recommended that all homes be tested for radon. Example 8 deals with the half-life of radon $^{222}_{86}$Rn.

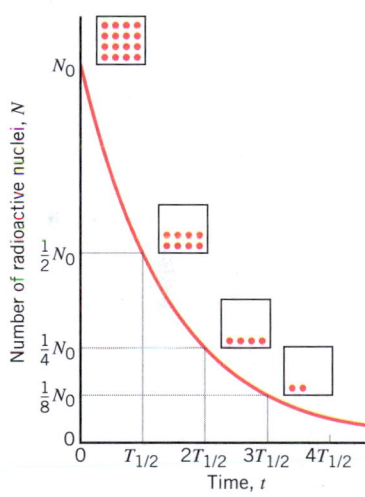

Figure 31.14 The half-life $T_{1/2}$ of a radioactive decay is the time in which one-half of the radioactive nuclei disintegrate.

Table 31.2 Some Half-Lives for Radioactive Decay

Isotope		Half-Life
Polonium	$^{214}_{84}$Po	1.64×10^{-4} s
Krypton	$^{89}_{36}$Kr	3.16 min
Radon	$^{222}_{86}$Rn	3.83 d
Strontium	$^{90}_{38}$Sr	29.1 yr
Radium	$^{226}_{88}$Ra	1.6×10^3 yr
Carbon	$^{14}_{6}$C	5.73×10^3 yr
Uranium	$^{238}_{92}$U	4.47×10^9 yr
Indium	$^{115}_{49}$In	4.41×10^{14} yr

Example 8 The Radioactive Decay of Radon Gas

Suppose that 3.0×10^7 radon atoms are trapped in a basement at the time the basement is sealed against further entry of the gas. The half-life of radon is 3.83 days. How many radon atoms remain after 31 days?

Reasoning During each half-life, the number of radon atoms is reduced by a factor of two. Thus, we determine the number of half-lives there are in a period of 31 days. At the end of each half-life we reduce the number of radon atoms present at the beginning of that half-life by a factor of two.

Solution In a period of 31 days there are (31 days)/(3.83 days) = 8.1 half-lives. In 8 half-lives the number of radon atoms is reduced by a factor of $2^8 = 256$. Ignoring the difference between 8 and 8.1 half-lives, we find that the number of atoms remaining after 31 days is $(3.0 \times 10^7)/256 = \boxed{1.2 \times 10^5}$.

The *activity* of a radioactive sample is the number of disintegrations per second that occur. Each time a disintegration occurs, the number N of radioactive nuclei decreases. As a result, the activity can be obtained by dividing ΔN, the change in the number of nuclei, by Δt, the time interval during which the change takes place; the average activity over the time interval Δt is the magnitude of $\Delta N/\Delta t$, or $|\Delta N/\Delta t|$. Since the decay of any individual nucleus is completely random, the number of disintegrations per second that occurs in a sample is proportional to the number of radioactive nuclei present, so that

$$\frac{\Delta N}{\Delta t} = -\lambda N \tag{31.4}$$

where λ is a proportionality constant referred to as the *decay constant.* The minus sign is present in this equation because each disintegration decreases the number N of nuclei originally present.

The SI unit for activity is the *becquerel* (Bq), named after Antoine Becquerel (1852–1908). One becquerel equals one disintegration per second. Activity is also measured in terms of a unit called the *curie* (Ci), in honor of Marie (1867–1934) and Pierre (1859–1906) Curie, the discoverers of radium and polonium. Historically, the curie was chosen as a unit because it is roughly the activity of one gram of pure radium. In terms of becquerels,

$$1 \text{ Ci} = 3.70 \times 10^{10} \text{ Bq}$$

The activity of the radium put into the dial of a watch to make it glow in the dark is about 4×10^4 Bq, and the activity used in radiation therapy for cancer is approximately a billion times greater, or 4×10^{13} Bq.

The mathematical expression for the graph of N versus t shown in Figure 31.14 can be obtained from Equation 31.4 with the aid of calculus. The result for the number N of radioactive nuclei present at time t is

$$N = N_0 e^{-\lambda t} \tag{31.5}$$

assuming that the number present at $t = 0$ s is N_0. The exponential e has the value $e = 2.718 \ldots$, and many calculators provide the value of e^x. We can relate the half-life $T_{1/2}$ of a radioactive nucleus to its decay constant λ in the following manner. By substituting $N = \frac{1}{2}N_0$ and $t = T_{1/2}$ into Equation 31.5, we find that $\frac{1}{2} = e^{-\lambda T_{1/2}}$. Taking the natural logarithm of both sides of this equation reveals that $\ln 2 = \lambda T_{1/2}$ or

$$T_{1/2} = \frac{\ln 2}{\lambda} = \frac{0.693}{\lambda} \tag{31.6}$$

The following example illustrates the use of Equations 31.5 and 31.6.

 Example 9 The Activity of Radon $^{222}_{86}$Rn

As in Example 8, suppose that there are 3.0×10^7 radon atoms ($T_{1/2} = 3.83$ days or 3.31×10^5 s) trapped in a basement. **(a)** How many radon atoms remain after 31 days? Find the activity **(b)** just after the basement is sealed against further entry of radon and **(c)** 31 days later.

Reasoning The number N of radon atoms remaining after a time t is given by $N = N_0 e^{-\lambda t}$, where $N_0 = 3.0 \times 10^7$ is the original number of atoms when $t = 0$ s and λ is the decay constant. The decay constant is related to the half-life $T_{1/2}$ of the radon atoms by $\lambda = 0.693/T_{1/2}$. The activity can be obtained from Equation 31.4, $\Delta N/\Delta t = -\lambda N$.

Solution (a) The decay constant is

$$\lambda = \frac{0.693}{T_{1/2}} = \frac{0.693}{3.83 \text{ days}} = 0.181 \text{ days}^{-1} \qquad (31.6)$$

and the number N of radon atoms remaining after 31 days is

$$N = N_0 e^{-\lambda t} = (3.0 \times 10^7)e^{-(0.181 \text{ days}^{-1})(31 \text{ days})} = \boxed{1.1 \times 10^5} \qquad (31.5)$$

This value is slightly less than that found in Example 8 because there we ignored the difference between 8.0 and 8.1 half-lives.

(b) The activity can be obtained from Equation 31.4, provided the decay constant is expressed in reciprocal seconds:

$$\lambda = \frac{0.693}{T_{1/2}} = \frac{0.693}{3.31 \times 10^5 \text{ s}} = 2.09 \times 10^{-6} \text{ s}^{-1} \qquad (31.6)$$

Thus, the number of disintegrations per second is

$$\frac{\Delta N}{\Delta t} = -\lambda N = -(2.09 \times 10^{-6} \text{ s}^{-1})(3.0 \times 10^7) = -63 \text{ disintegrations/s} \qquad (31.4)$$

The activity is the magnitude of $\Delta N/\Delta t$, so initially $\boxed{\text{Activity} = 63 \text{ Bq}}$.

(c) From part (a), the number of radioactive nuclei remaining at the end of 31 days is $N = 1.1 \times 10^5$, and reasoning similar to that in part (b) reveals that $\boxed{\text{Activity} = 0.23 \text{ Bq}}$.

✓ CHECK YOUR UNDERSTANDING

(The answers are given at the end of the book.)

9. The thallium $^{208}_{81}\text{Tl}$ nucleus is radioactive, with a half-life of 3.053 min. At a given instant, the activity of a certain sample of thallium is 2400 Bq. Using the concept of a half-life, and without doing any written calculations, determine whether the activity 9 minutes later is (a) a little less than $\frac{1}{8}(2400 \text{ Bq}) = 300 \text{ Bq}$, (b) a little more than $\frac{1}{8}(2400 \text{ Bq}) = 300 \text{ Bq}$, (c) a little less than $\frac{1}{3}(2400 \text{ Bq}) = 800 \text{ Bq}$, or (d) a little more than $\frac{1}{3}(2400 \text{ Bq}) = 800 \text{ Bq}$.

10. The half-life of indium $^{115}_{49}\text{In}$ is 4.41×10^{14} yr. Thus, one-half of the nuclei in a sample of this isotope will decay in this time, which is very long. Is it possible for any single nucleus in the sample to decay after only one second?

11. Is it possible for two different samples of the same radioactive element to have different activities?

31.7 RADIOACTIVE DATING

The physics of radioactive dating. One important application of radioactivity is the determination of the age of archaeological or geological samples. If an object contains radioactive nuclei when it is formed, then the decay of these nuclei marks the passage of time like a clock, half of the nuclei disintegrating during each half-life. If the half-life is known, a measurement of the number of nuclei present today relative to the number present initially can give the age of the sample. According to Equation 31.4, the activity of a sample is proportional to the number of radioactive nuclei, so one way to obtain the age is to compare present activity with initial activity. A more accurate way is to determine the present number of radioactive nuclei with the aid of a mass spectrometer.

The present activity of a sample can be measured, but how is it possible to know what the original activity was, perhaps thousands of years ago? Radioactive dating methods entail certain assumptions that make it possible to estimate the original activity. For instance, the

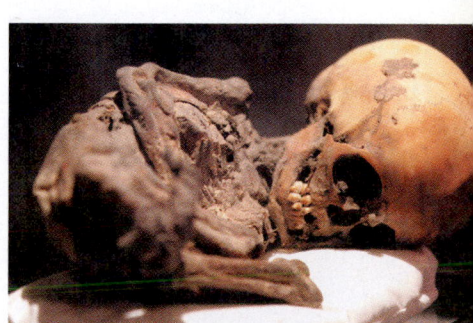

These mummified remains were found in a cave in Mexico in 2002. They are thought to be about 2300 years old. Radioactive dating is one of the techniques used to determine the age of such remains. (© David Aguilar/Reuters/Landov LLC)

radiocarbon technique utilizes the $^{14}_{6}C$ isotope of carbon, which undergoes β^- decay with a half-life of 5730 yr. This isotope is present in the earth's atmosphere at an equilibrium concentration of about one atom for every 8.3×10^{11} atoms of normal carbon $^{12}_{6}C$. It is often assumed* that this value has remained constant over the years because $^{14}_{6}C$ is created when cosmic rays interact with the earth's upper atmosphere, a production method that offsets the loss via β^- decay. Moreover, nearly all living organisms ingest the equilibrium concentration of $^{14}_{6}C$. However, once an organism dies, metabolism no longer sustains the input of $^{14}_{6}C$, and β^- decay causes half of the $^{14}_{6}C$ nuclei to disintegrate every 5730 years. Example 10 illustrates how to determine the $^{14}_{6}C$ activity of one gram of carbon in a living organism.

Example 10
$^{14}_{6}C$ Activity Per Gram of Carbon in a Living Organism

(a) Determine the number of carbon $^{14}_{6}C$ atoms present for every gram of carbon $^{12}_{6}C$ in a living organism. Find (b) the decay constant and (c) the activity of this sample.

Reasoning The total number of carbon $^{12}_{6}C$ atoms in one gram of carbon $^{12}_{6}C$ is equal to the corresponding number of moles times Avogadro's number (see Section 14.1). Since there is only one $^{14}_{6}C$ atom for every 8.3×10^{11} atoms of $^{12}_{6}C$, the number of $^{14}_{6}C$ atoms is equal to the total number of $^{12}_{6}C$ atoms divided by 8.3×10^{11}. The decay constant λ for $^{14}_{6}C$ is $\lambda = 0.693/T_{1/2}$, where $T_{1/2}$ is the half-life. The activity is equal to the magnitude of $\Delta N/\Delta t$, which is equal to the decay constant times the number of $^{14}_{6}C$ atoms present, according to Equation 31.4.

Solution (a) One gram of carbon $^{12}_{6}C$ (atomic mass = 12 u) is equivalent to 1.0/12 mol. Since Avogadro's number is 6.02×10^{23} atoms/mol and since there is one $^{14}_{6}C$ atom for every 8.3×10^{11} atoms of $^{12}_{6}C$, the number of $^{14}_{6}C$ atoms is

$$\begin{matrix} \text{Number of } ^{14}_{6}C \\ \text{atoms for every } 1.0 \\ \text{gram of carbon } ^{12}_{6}C \end{matrix} = \left(\frac{1.0}{12} \text{ mol} \right) \left(6.02 \times 10^{23} \frac{\text{atoms}}{\text{mol}} \right) \left(\frac{1}{8.3 \times 10^{11}} \right)$$

$$= \boxed{6.0 \times 10^{10} \text{ atoms}}$$

(b) Since the half-life of $^{14}_{6}C$ is 5730 yr (1.81×10^{11} s), the decay constant is

$$\lambda = \frac{0.693}{T_{1/2}} = \frac{0.693}{1.81 \times 10^{11} \text{ s}} = \boxed{3.83 \times 10^{-12} \text{ s}^{-1}} \qquad (31.6)$$

(c) Equation 31.4 indicates that $\Delta N/\Delta t = -\lambda N$, so the magnitude of $\Delta N/\Delta t$ is λN.

$$\begin{matrix} \text{Activity of } ^{14}_{6}C \text{ for} \\ \text{every } 1.0 \text{ gram of} \\ \text{carbon } ^{12}_{6}C \text{ in a} \\ \text{living organism} \end{matrix} = \lambda N = (3.83 \times 10^{-12} \text{ s}^{-1})(6.0 \times 10^{10} \text{ atoms}) = \boxed{0.23 \text{ Bq}}$$

An organism that lived thousands of years ago presumably had an activity of about 0.23 Bq per gram of carbon. When the organism died, the activity began decreasing. From a sample of the remains, the current activity per gram of carbon can be measured and compared to the value of 0.23 Bq to determine the time that has transpired since death. This procedure is illustrated in Example 11.

---

*The assumption that the $^{14}_{6}C$ concentration has always been at its present equilibrium value has been evaluated by comparing $^{14}_{6}C$ ages with ages determined by counting tree rings. More recently, ages determined using the radioactive decay of uranium $^{238}_{92}U$ have been used for comparison. These comparisons indicate that the equilibrium value of the $^{14}_{6}C$ concentration has indeed remained constant for the past 1000 years. However, from there back about 30 000 years, it appears that the $^{14}_{6}C$ concentration in the atmosphere was larger than its present value by up to 40%. As a first approximation we ignore such discrepancies.

ANALYZING MULTIPLE-CONCEPT PROBLEMS

Example 11 The Ice Man

On September 19, 1991, German tourists on a walking trip in the Italian Alps found a Stone-Age traveler, later dubbed the Ice Man, whose body had become trapped in a glacier. Figure 31.15 shows the well-preserved remains, which were dated using the radiocarbon method. Material found with the body had a $^{14}_6C$ activity of about 0.121 Bq per gram of carbon. Find the age of the Ice Man's remains.

Reasoning In the radiocarbon method, the number of radioactive nuclei remaining at a given instant is related to the number present initially, the time that has passed since the Ice Man died, and the decay constant for $^{14}_6C$. Thus, to determine the age of the remains, we will need information about the number of nuclei present when the body was discovered and the number present initially, which can be related to the activity of the material found with the body and the initial activity. To determine the age, we will also need the decay constant, which can be obtained from the half-life of $^{14}_6C$.

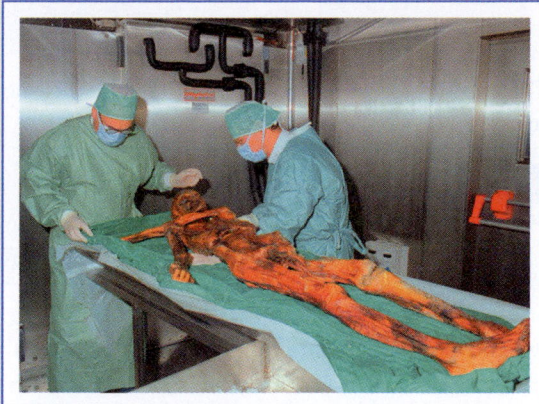

Figure 31.15 The two scientists in this photograph are studying the Ice Man or "Oetzi," as he is also called. His frozen remains and various artifacts were discovered in the ice of a glacier in the Italian Alps in 1991. Radiocarbon dating has revealed his age. (© AFP/Getty Images News)

Knowns and Unknowns We have the following data:

Description	Symbol	Value	Comment
Explicit Data			
Activity of material found with body	A	0.121 Bq	This is the activity per gram of carbon.
Implicit Data			
Half-life of $^{14}_6C$	$T_{1/2}$	5730 yr	The radiocarbon dating method is specified.
Initial activity of material found with body	A_0	0.23 Bq	This activity is assumed for one gram of carbon in a living organism.
Unknown Variable			
Age of Ice Man's remains	t	?	

Modeling the Problem

STEP 1 Radioactive Decay   The number N of radioactive nuclei present at a time t is

$$N = N_0 e^{-\lambda t} \qquad (31.5)$$

where N_0 is the number present initially at $t = 0$ s and λ is the decay constant for $^{14}_6C$. Rearranging terms gives

$$\frac{N}{N_0} = e^{-\lambda t}$$

Taking the natural logarithm of both sides of this result, we find that

$$\ln\left(\frac{N}{N_0}\right) = -\lambda t$$

Solving for t shows that the age of the Ice Man's remains is given by Equation 1 at the right. To use this result, we need information about the ratio N/N_0 and λ. We deal with N/N_0 in Step 2 and with λ in Step 3.

$$t = -\left(\frac{1}{\lambda}\right)\ln\left(\frac{N}{N_0}\right) \qquad (1)$$

$$\boxed{?} \qquad \boxed{?}$$

STEP 2 Activity   The activity A is the number of disintegrations per second, or $\left|\dfrac{\Delta N}{\Delta t}\right|$, where ΔN is the number of disintegrations that occur in the time interval Δt. Noting that $\dfrac{\Delta N}{\Delta t} = -\lambda N$ according to Equation 31.4, we find for the activity that

$$A = \left|\frac{\Delta N}{\Delta t}\right| = |-\lambda N| = \lambda N$$

Continued

Using this expression, we have that

$$\frac{N}{N_0} = \frac{\lambda N}{\lambda N_0} = \frac{A}{A_0}$$

The substitution of this result into Equation 1 is shown at the right. We turn now to Step 3, in order to evaluate the decay constant λ.

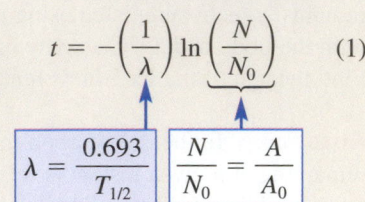

$$t = -\left(\frac{1}{\lambda}\right)\ln\left(\frac{N}{N_0}\right) \quad (1)$$

$$\frac{N}{N_0} = \frac{A}{A_0}$$

STEP 3 **Decay Constant** The decay constant is related to the half-life $T_{1/2}$ according to

$$\lambda = \frac{0.693}{T_{1/2}} \quad (31.6)$$

which we substitute into Equation 1, as shown at the right.

$$t = -\left(\frac{1}{\lambda}\right)\ln\left(\frac{N}{N_0}\right) \quad (1)$$

$$\lambda = \frac{0.693}{T_{1/2}} \qquad \frac{N}{N_0} = \frac{A}{A_0}$$

Solution Combining the results of each step algebraically, we find that

STEP 1 **STEP 2** **STEP 3**

$$t = -\left(\frac{1}{\lambda}\right)\ln\left(\frac{N}{N_0}\right) = -\left(\frac{1}{\lambda}\right)\ln\left(\frac{A}{A_0}\right) = -\left(\frac{1}{0.693/T_{1/2}}\right)\ln\left(\frac{A}{A_0}\right)$$

This result reveals that the age of the Ice Man's remains is

$$t = -\left(\frac{T_{1/2}}{0.693}\right)\ln\left(\frac{A}{A_0}\right) = -\left(\frac{5730 \text{ yr}}{0.693}\right)\ln\left(\frac{0.121 \text{ Bq}}{0.23 \text{ Bq}}\right) = \boxed{5300 \text{ yr}}$$

Note that this solution implies for the activity that

$$A = A_0 e^{-\lambda t}$$

This can be seen by combining the result from Step 2 ($N/N_0 = A/A_0$) with Equation 31.5 ($N = N_0 e^{-\lambda t}$).

Related Homework: *Problems 46, 58*

Radiocarbon dating is not the only radioactive dating method. For example, other methods utilize uranium $^{238}_{92}\text{U}$, potassium $^{40}_{19}\text{K}$, and lead $^{210}_{82}\text{Pb}$. For such methods to be useful, the half-life of the radioactive species must be neither too short nor too long relative to the age of the sample to be dated, as Conceptual Example 12 discusses.

Conceptual Example 12 Dating a Bottle of Wine

A bottle of red wine is thought to have been sealed about 5 years ago. The wine contains a number of different atoms, including carbon, oxygen, and hydrogen. Each of these has a radioactive isotope. The radioactive isotope of carbon is the familiar $^{14}_{6}\text{C}$, with a half-life of 5730 yr. The radioactive isotope of oxygen is $^{15}_{8}\text{O}$ and has a half-life of 122.2 s. The radioactive isotope of hydrogen, called tritium, is $^{3}_{1}\text{H}$; its half-life is 12.33 yr. The activity of each of these isotopes is known at the time the bottle was sealed. However, only one of the isotopes is useful for determining the age of the wine accurately from a measurement of its current activity. Which is it? **(a)** $^{14}_{6}\text{C}$ **(b)** $^{15}_{8}\text{O}$ **(c)** $^{3}_{1}\text{H}$

Reasoning To find the age of the wine, it is necessary to determine the ratio of the current activity A to the initial activity A_0 (see Example 11). If the age of the sample is very small relative to the half-life of the nuclei, relatively few of the nuclei have decayed during the wine's life, and the measured activity changes little from its initial value ($A \approx A_0$). To obtain an accurate age from such a small change would require prohibitively precise measurements. On the other hand, if the age of the sample is many times greater than the half-life of the nuclei, virtually all of the nuclei would have decayed, and the current activity would be so small ($A \approx 0$) that it would be virtually impossible to detect.

Answer (a) is incorrect. The expected age of the wine is about 5 years. This period is only a tiny fraction of the 5730-yr half-life of $^{14}_{6}\text{C}$. As a result, relatively few of the $^{14}_{6}\text{C}$ nuclei would decay during the wine's life, and the current activity would be nearly the same as the initial activity ($A \approx A_0$), thus requiring prohibitively precise measurements.

Answer (b) is incorrect. The $^{15}_{8}O$ isotope is not very useful either, because of its relatively short half-life of 122.2 s. During a 5-year period, so many half-lives of 122.2 s would occur that the current activity would be vanishingly small ($A \approx 0$) and undetectable.

Answer (c) is correct. The only remaining option is the $^{3}_{1}H$ isotope. The expected age of 5 yr is long enough relative to the half-life of 12.33 yr that a measurable change in activity will occur, but not so long that the current activity will have completely vanished for all practical purposes.

Related Homework: *Problem 49*

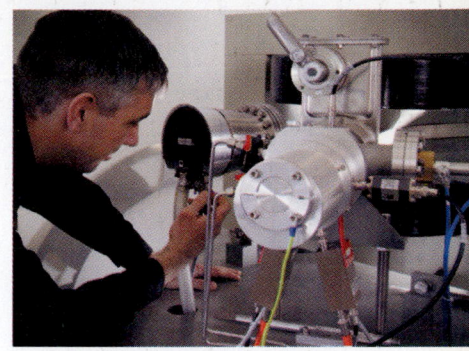

In the radioactive dating technique that uses the carbon $^{14}_{6}C$ isotope, the relatively few $^{14}_{6}C$ atoms can be detected by measuring their activity, as we have seen. It is also possible to use an accelerator mass spectrometer, such as the one in this photograph, to detect these atoms more accurately. (© James King Holmes/SPL/Photo Researchers, Inc.)

✓ **CHECK YOUR UNDERSTANDING**

(The answers are given at the end of the book.)

12. To which one or more of the following objects, each about 1000 yr old, can the radiocarbon dating technique *not* be applied? **(a)** A wooden box **(b)** A gold statue **(c)** Some well-preserved animal fur

13. Suppose there were a greater number of carbon $^{14}_{6}C$ atoms in a plant living 5000 yr ago than is currently believed. When the seeds of this plant are tested using radiocarbon dating, is the age obtained too small or too large?

14. Review Conceptual Example 12 as an aid in answering this question. Tritium is an isotope of hydrogen and undergoes β^- decay with a half-life of 12.33 yr. Like carbon $^{14}_{6}C$, tritium is produced in the atmosphere because of cosmic rays and can be used in a radioactive dating technique. Can tritium dating be used to determine a reliable date for a sample that is about 700 yr old?

31.8 RADIOACTIVE DECAY SERIES

When an unstable parent nucleus decays, the resulting daughter nucleus is sometimes also unstable. If so, the daughter then decays and produces its own daughter, and so on, until a completely stable nucleus is produced. This sequential decay of one nucleus after another is called a *radioactive decay series.* Examples 4–6 discuss the first two steps of a series that begins with uranium $^{238}_{92}U$:

$$\text{Uranium} \qquad \text{Thorium}$$
$$^{238}_{92}U \longrightarrow \ ^{234}_{90}Th + \ ^{4}_{2}He$$
$$\longrightarrow \ ^{234}_{91}Pa + \ ^{0}_{-1}e$$
$$\text{Protactinium}$$

Furthermore, Examples 8 and 9 deal with radon $^{222}_{86}Rn$, which is formed down the line in the $^{238}_{92}U$ radioactive decay series. Figure 31.16 shows the entire series. At several points, branches occur because more than one kind of decay is possible for an intermediate species. Ultimately, however, the series ends with lead $^{206}_{82}Pb$, which is stable.

The $^{238}_{92}U$ series and other such series are the only sources of some of the radioactive elements found in nature. Radium $^{226}_{88}Ra$, for instance, has a half-life of 1600 yr, which is short enough that all the $^{226}_{88}Ra$ created when the earth was formed billions of years ago has now disappeared. The $^{238}_{92}U$ series provides a continuing supply of $^{226}_{88}Ra$, however.

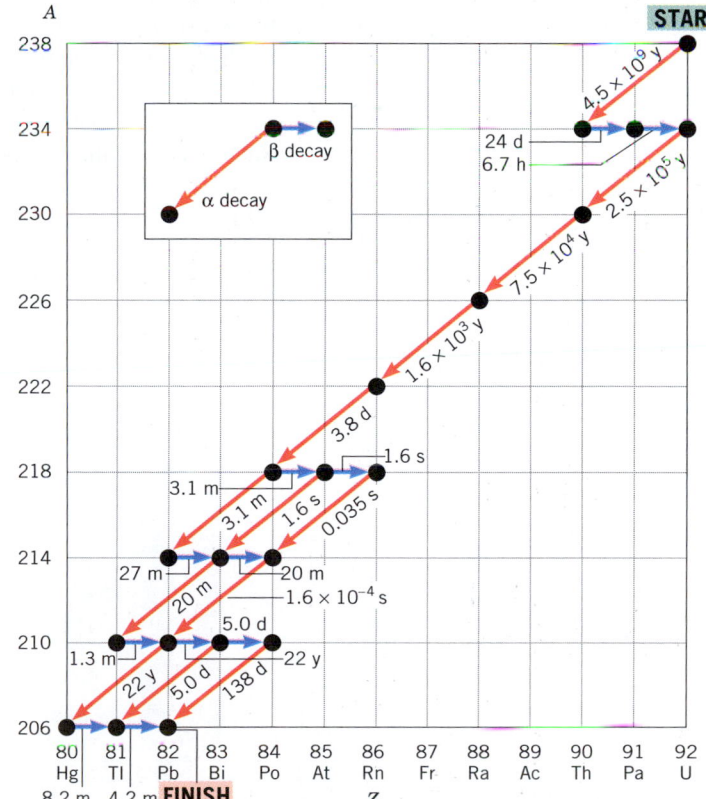

Figure 31.16 The radioactive decay series that begins with uranium $^{238}_{92}U$ and ends with lead $^{206}_{82}Pb$. Half-lives are given in seconds (s), minutes (m), hours (h), days (d), or years (y). The inset in the upper left identifies the type of decay that each nucleus undergoes.

✓ **CHECK YOUR UNDERSTANDING**

(The answer is given at the end of the book.)

15. Because of radioactive decay, one element can be transmuted into another. Thus, a container of uranium $^{238}_{92}U$ ultimately becomes a container of lead $^{206}_{82}Pb$, as Figure 31.16 indicates. Roughly, how long does it take for $^{238}_{92}U$ to transmute entirely into $^{206}_{82}Pb$? **(a)** Several decades **(b)** Several centuries **(c)** Thousands of years **(d)** Millions of years **(e)** Billions of years

31.9 RADIATION DETECTORS

The physics of radiation detectors.

There are a number of devices that can be used to detect the particles and photons (γ rays) emitted when a radioactive nucleus decays. Such devices detect the ionization that these particles and photons cause as they pass through matter.

The most familiar detector is the *Geiger counter,* which Figure 31.17 illustrates. The Geiger counter consists of a gas-filled metal cylinder. The α, β, or γ rays enter the cylinder through a thin window at one end. γ rays can also penetrate directly through the metal. A wire electrode runs along the center of the tube and is kept at a high positive voltage (1000–3000 V) relative to the outer cylinder. When a high-energy particle or photon enters the cylinder, it collides with and ionizes a gas molecule. The electron produced from the gas molecule accelerates toward the positive wire, ionizing other molecules in its path. Additional electrons are formed, and an avalanche of electrons rushes toward the wire, leading to a pulse of current through the resistor *R*. This pulse can be counted or made to produce a "click" in a loudspeaker. The number of counts or clicks is related to the number of disintegrations that produced the particles or photons.

The *scintillation counter* is another important radiation detector. As Figure 31.18 indicates, this device consists of a scintillator mounted on a photomultiplier tube. Often the scintillator is a crystal (e.g., cesium iodide) containing a small amount of impurity (thallium), but plastic, liquid, and gaseous scintillators are also used. In response to ionizing radiation, the scintillator emits a flash of visible light. The photons of the flash then strike the photocathode of the photomultiplier tube. The photocathode is made of a material that emits electrons because of the photoelectric effect. These photoelectrons are then attracted to a special electrode kept at a voltage of about +100 V relative to the photocathode. The electrode is coated with a substance that emits several additional electrons for every electron striking it. The additional electrons are attracted to a second similar electrode (voltage = +200 V), where they generate even more electrons. Commercial photomultiplier tubes contain as many as 15 of these special electrodes, so photoelectrons resulting from the light flash of the scintillator lead to a cascade of electrons and a pulse of current. As in a Geiger tube, the current pulses can be counted.

Ionizing radiation can also be detected with several types of *semiconductor detectors.* Such devices utilize *n*- and *p*-type materials (see Section 23.5), and their operation depends

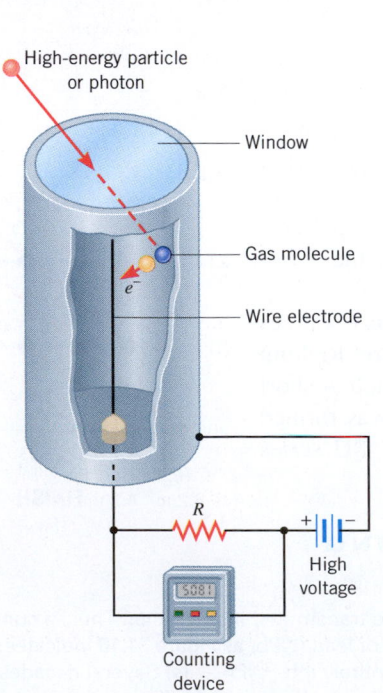

Figure 31.17 A Geiger counter.

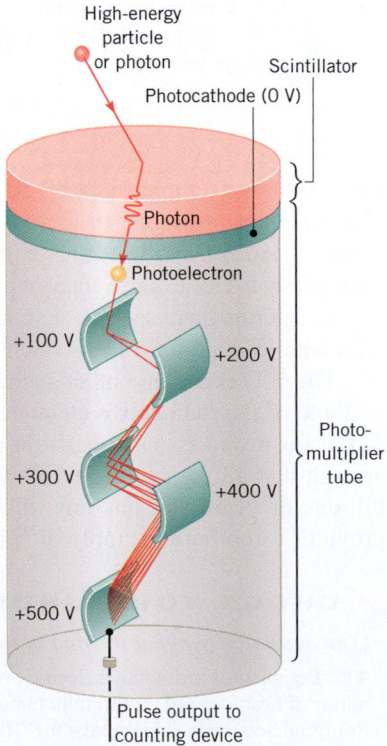

Figure 31.18 A scintillation counter.

on the electrons and holes formed in the materials as a result of the radiation. One of the main advantages of semiconductor detectors is their ability to discriminate between two particles with only slightly different energies.

A number of instruments provide a pictorial representation of the path that high-energy particles follow after they are emitted from unstable nuclei. In a *cloud chamber,* a gas is cooled just to the point at which it will condense into droplets, provided nucleating agents are available on which the droplets can form. When a high-energy particle, such as an α particle or a β particle, passes through the gas, the ions it leaves behind serve as nucleating agents, and droplets form along the path of the particle. A *bubble chamber* works in a similar fashion, except that it contains a liquid that is just at the point of boiling. Tiny bubbles form along the trail of a high-energy particle passing through the liquid. Paths revealed in a cloud or bubble chamber can be photographed to provide a permanent record of the event. Figure 31.19 shows a photograph of tracks in a bubble chamber. A *photographic emulsion* also can be used directly to produce a record of the path taken by a particle of ionizing radiation. Ions formed as the particle passes through the emulsion cause silver to be deposited along the track when the emulsion is developed.

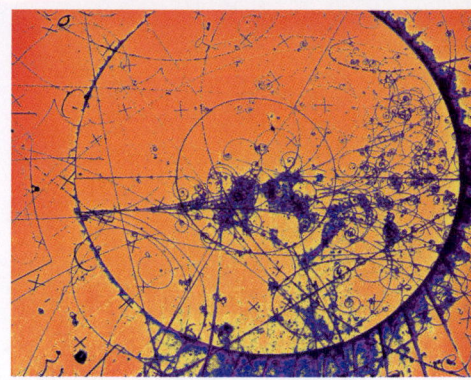

Figure 31.19 A photograph showing particle tracks in a bubble chamber. (© CERN/Photo Researchers, Inc.)

31.10 CONCEPTS & CALCULATIONS

Radioactive decay obeys the conservation laws of physics, and we have studied five of these laws, as the Concepts-at-a-Glance chart in Figure 31.6 illustrates. Two of them are particularly important in understanding the types of radioactivity that occur and the nuclear changes that accompany them. These are the conservation of electric charge and the conservation of nucleon number. Example 13 emphasizes their importance and reviews how they are applied.

Concepts & Calculations Example 13
Electric Charge and Nucleon Number

Thorium $^{228}_{90}$Th produces a daughter nucleus that is radioactive. The daughter, in turn, produces its own radioactive daughter, and so on. This process continues until bismuth $^{212}_{83}$Bi is reached. What are the total number N_α of α particles and the total number N_β of β^- particles that are generated in this series of radioactive decays?

Concept Questions and Answers How many of the 90 protons in the thorium nucleus are carried off by the α particles?

Answer Each α particle is a helium $^4_2$He nucleus and carries off two protons. Therefore, the total number of protons carried off by the α particles is $N_\alpha(2)$.

How many protons are left behind when the β^- particles are emitted?

Answer Each β^- particle is an electron $_{-1}^{0}$e and is emitted when a neutron in the nucleus decays into a proton and an electron. Therefore, the total number of protons left behind by the β^- particles is N_β.

How many of the 228 nucleons in the thorium nucleus are carried off by the α particles?

Answer Each α particle is a helium $^4_2$He nucleus and carries off four nucleons. Therefore, the total number of nucleons carried off by the α particles is $N_\alpha(4)$.

Does the departure of a β^- particle alter the number of nucleons?

Answer No. Each β^- particle is an electron $_{-1}^{0}$e and is emitted when a neutron in the nucleus decays into a proton and an electron. In effect, a neutron is replaced by a proton. But since each is a nucleon, the number of nucleons is not changed.

Solution The overall decay process can be written as

$$^{228}_{90}\text{Th} \longrightarrow {}^{212}_{83}\text{Bi} + N_\alpha({}^4_2\text{He}) + N_\beta({}^0_{-1}\text{e})$$

Since electric charge must be conserved, we know that the charge on the left side of this reaction must be equal to the total charge on the right side, the result being

$$90 = 83 + N_\alpha(2) + N_\beta(-1)$$

Since the nucleon number must be conserved also, we know that the 228 nucleons on the left must equal the total number of nucleons on the right, so that

$$228 = 212 + N_\alpha(4) + N_\beta(0)$$

Solving this result for the number of α particles gives $\boxed{N_\alpha = 4}$. Substituting $N_\alpha = 4$ into the conservation-of-charge equation, we find that the number of β^- particles is $\boxed{N_\beta = 1}$.

Energy in the form of heat is needed to raise the temperature of an object, as Section 12.7 discusses. One source of the heat can be radioactive decay, as we see in the next example.

 Concepts & Calculations Example 14   The Energy from Radioactive Decay

A one-gram sample of thorium $^{228}_{90}$Th contains 2.64×10^{21} atoms and undergoes α decay with a half-life of 1.913 yr (1.677×10^4 h). Each disintegration releases an energy of 5.52 MeV (8.83×10^{-13} J). Assuming that all of the energy is used to heat a 3.8-kg sample of water, find the change in temperature of the water that occurs in one hour.

Concept Questions and Answers   How much heat Q is needed to raise the temperature of a mass m of water by ΔT degrees?

> *Answer*   According to Equation 12.4, the heat needed is $Q = cm\,\Delta T$, where the specific heat capacity of water is $c = 4186$ J/(kg $\cdot$ C°).

The energy released by each disintegration is E. What is the total energy E_{Total} released by a number n of disintegrations?

> *Answer*   The total energy released is just the number of disintegrations times the energy for each one, or $E_{\text{Total}} = nE$.

What is the number n of disintegrations that occur during a time t?

> *Answer*   The number of radioactive nuclei remaining after a time t is given by Equations 31.5 and 31.6 as
>
> $$N = N_0 e^{-0.693t/T_{1/2}}$$
>
> where N_0 is the number of radioactive nuclei present at $t = 0$ s and $T_{1/2}$ is the half-life for the decay. Therefore, the number n of disintegrations that occur during the time t is $N_0 - N$ or
>
> $$n = N_0 - N = N_0(1 - e^{-0.693t/T_{1/2}}) \tag{31.7}$$

Solution   According to Equation 12.4, the heat needed to change the temperature of the water is $Q = cm\,\Delta T$. This heat is provided by the total energy released in one hour of radioactive decay, or $E_{\text{Total}} = nE$. Setting $Q = E_{\text{Total}}$, we obtain $nE = cm\,\Delta T$, which can be solved for the change in temperature ΔT:

$$\Delta T = \frac{nE}{cm}$$

Using Equation 31.7 for the number of disintegrations n that occur in the time t, we obtain

$$\Delta T = \frac{N_0(1 - e^{-0.693t/T_{1/2}})E}{cm}$$

$$= \frac{(2.64 \times 10^{21})[1 - e^{-0.693(1.000\text{ h})/(1.677 \times 10^4\text{ h})}](8.83 \times 10^{-13}\text{ J})}{[4186\text{ J/(kg}\cdot\text{C}°)](3.8\text{ kg})} = \boxed{6.1\text{ C}°}$$

CONCEPT SUMMARY

If you need more help with a concept, use the Learning Aids noted next to the discussion or equation. Examples (**Ex.**) are in the text of this chapter. Go to **www.wiley.com/college/cutnell** for the following Learning Aids:

Interactive LearningWare (ILW) — Additional examples solved in a five-step interactive format.

Concept Simulations (CS) — Animated text figures or animations of important concepts.

Interactive Solutions (IS) — Models for certain types of problems in the chapter homework. The calculations are carried out interactively.

Topic	Discussion	Learning Aids
Nucleons	**31.1 NUCLEAR STRUCTURE** The nucleus of an atom consists of protons and neutrons, which are collectively referred to as nucleons. A neutron is an elecrically neutral particle whose mass is	

Topic	Discussion	Learning Aids
Atomic number	slightly larger than that of the proton. The atomic number Z is the number of protons in the nucleus. The atomic mass number A (or nucleon number) is the total number of protons and neutrons in the nucleus:	
Atomic mass number (or nucleon number)	$$A = Z + N \qquad (31.1)$$ where N is the number of neutrons. For an element whose chemical symbol is X, the symbol for the nucleus is ^A_ZX.	
Isotopes	Nuclei that contain the same number of protons, but a different number of neutrons, are called isotopes.	
Radius of a nucleus	The approximate radius (in meters) of a nucleus is given by $$r \approx (1.2 \times 10^{-15} \text{ m}) A^{1/3} \qquad (31.2)$$	**Ex. 1**
Strong nuclear force	**31.2 THE STRONG NUCLEAR FORCE AND THE STABILITY OF THE NUCLEUS** The strong nuclear force is the force of attraction between nucleons (protons and neutrons) and is one of the three fundamental forces of nature. This force balances the electrostatic force of repulsion between protons and holds the nucleus together. The strong nuclear force has a very short range of action and is almost independent of electric charge.	
	31.3 THE MASS DEFECT OF THE NUCLEUS AND NUCLEAR BINDING ENERGY The binding energy of a nucleus is the energy required to separate the nucleus into its constituent protons and neutrons. The binding energy is equal to	
Binding energy	$$\text{Binding energy} = (\Delta m)c^2 \qquad (31.3)$$	**Ex. 2, 3**
Mass defect	where Δm is the mass defect of the nucleus and c is the speed of light in a vacuum. The mass defect is the amount by which the sum of the individual masses of the protons and neutrons exceeds the mass of the intact nucleus.	
Atomic mass unit	When specifying nuclear masses, it is customary to use the atomic mass unit (u). One atomic mass unit has a mass of 1.6605×10^{-27} kg and is equivalent to an energy of 931.5 MeV.	
Radioactivity	**31.4 RADIOACTIVITY** Unstable nuclei spontaneously decay by breaking apart or rearranging their internal structure in a process called radioactivity. Naturally occurring radioactivity produces α, β, and γ rays. α rays consist of positively charged particles, each particle being the $^4_2$He nucleus of helium. The general form for α decay is	
α decay	$$\underbrace{^A_Z\text{P}}_{\substack{\text{Parent} \\ \text{nucleus}}} \longrightarrow \underbrace{^{A-4}_{Z-2}\text{D}}_{\substack{\text{Daughter} \\ \text{nucleus}}} + \underbrace{^4_2\text{He}}_{\substack{\alpha \text{ particle} \\ \text{(helium nucleus)}}}$$	**Ex. 4, 5** **IS 31.27**
β^- decay	The most common kind of β ray consists of negatively charged particles, or β^- particles, which are electrons. The general form for β^- decay is $$\underbrace{^A_Z\text{P}}_{\substack{\text{Parent} \\ \text{nucleus}}} \longrightarrow \underbrace{^A_{Z+1}\text{D}}_{\substack{\text{Daughter} \\ \text{nucleus}}} + \underbrace{^{\ 0}_{-1}\text{e}}_{\substack{\beta^- \text{ particle} \\ \text{(electron)}}}$$	**Ex. 6, 13**
Positron	β^+ decay produces another kind of β ray, which consists of positively charged particles, or β^+ particles. A β^+ particle, also called a positron, has the same mass as an electron, but carries a charge of $+e$ instead of $-e$.	
Transmutation	If a radioactive parent nucleus disintegrates into a daughter nucleus that has a different atomic number, as occurs in α and β decay, one element has been converted into another element, the conversion being referred to as a transmutation.	
γ decay	γ rays are high-energy photons emitted by a radioactive nucleus. The general form for γ decay is $$\underbrace{^A_Z\text{P*}}_{\substack{\text{Excited} \\ \text{energy state}}} \longrightarrow \underbrace{^A_Z\text{P}}_{\substack{\text{Lower} \\ \text{energy state}}} + \underbrace{\gamma}_{\substack{\gamma \text{ ray}}}$$	**Ex. 7** **ILW 31.1**
	γ decay does not cause a transmutation of one element into another.	
Neutrino	**31.5 THE NEUTRINO** The neutrino is an electrically neutral particle that is emitted along with β particles and has a mass that is much, much smaller than the mass of an electron.	
Half-life Activity	**31.6 RADIOACTIVE DECAY AND ACTIVITY** The half-life of a radioactive isotope is the time required for one-half of the nuclei present to disintegrate or decay. The activity is the number of disintegrations per second that occur. Activity is the magnitude of $\Delta N/\Delta t$, where ΔN is the change in the number N of radioactive nuclei and Δt is the time interval during which the change occurs.	**Ex. 8**

Topic	Discussion	Learning Aids
	In other words, activity is $\lvert \Delta N/\Delta t \rvert$. The SI unit for activity is the becquerel (Bq), one becquerel being one disintegration per second. Activity is sometimes also measured in a unit called the curie (Ci); 1 Ci = 3.70×10^{10} Bq.	
	Radioactive decay obeys the following relation	
	$$\frac{\Delta N}{\Delta t} = -\lambda N$$ (31.4)	**IS 31.41**
Decay constant	where λ is the decay constant. This equation can be solved by the methods of integral calculus to show that	
Number of nuclei remaining as a function of time	$$N = N_0 e^{-\lambda t}$$ (31.5)	**CS 31.1** **Ex. 14**
	where N_0 is the original number of nuclei. The decay constant λ is related to the half-life $T_{1/2}$ according to	
Decay constant and half-life	$$\lambda = \frac{0.693}{T_{1/2}}$$ (31.6)	**Ex. 9**
	31.7 RADIOACTIVE DATING If an object contained radioactive nuclei when it was formed, then the decay of these nuclei can be used to determine the age of the object. One way to obtain the age is to relate the present activity A of an object to its initial activity A_0:	**Ex. 10**
Activity as a function of time	$$A = A_0 e^{-\lambda t}$$	**Ex. 11**
	where λ is the decay constant and t is the age of the object. For radiocarbon dating that uses the $^{14}_{6}C$ isotope of carbon, the initial activity is often assumed to be $A_0 = 0.23$ Bq.	**Ex. 12**
Radioactive decay series	**31.8 RADIOACTIVE DECAY SERIES** The sequential decay of one nucleus after another is called a radioactive decay series. A decay series starts with a radioactive nucleus and ends with a completely stable nucleus. Figure 31.16 illustrates one such series that begins with uranium $^{238}_{92}U$ and ends with lead $^{206}_{82}Pb$.	
Radiation detectors	**31.9 RADIATION DETECTORS** A number of devices are used to detect α and β particles as well as γ rays. These include the Geiger counter, the scintillation counter, semiconductor detectors, cloud and bubble chambers, and photographic emulsions.	

FOCUS ON CONCEPTS

Note to Instructors: The numbering of the questions shown here reflects the fact that they are only a representative subset of the total number that are available online. However, all of the questions are available for assignment via an online homework management program such as WileyPLUS or WebAssign.

Section 31.1 Nuclear Structure

1. An indium (In) nucleus contains 49 protons and 66 neutrons. Which one of the following symbols describes this nucleus? **(a)** $^{49}_{115}In$ **(b)** $^{49}_{66}In$ **(c)** $^{115}_{66}In$ **(d)** $^{66}_{49}In$ **(e)** $^{115}_{49}In$

2. The notation for a particular nucleus is $^{85}_{37}Rb$. In an electrically neutral atom, how many electrons are in orbit about this nucleus? **(a)** 37 + 85 = 122 **(b)** 85 **(c)** 37 **(d)** 85 − 37 = 48 **(e)** The number of electrons cannot be determined from the notation.

Section 31.3 The Mass Defect of the Nucleus and Nuclear Binding Energy

6. Suppose that we lived in a hypothetical world in which the mass of each proton and each neutron were exactly 1 u. In this world, the atomic mass of copper $^{63}_{29}Cu$ is 62.5 u. What would be the mass defect for this nucleus? **(a)** 63 u **(b)** 29 u **(c)** 63 u − 29 u = 34 u **(d)** 0.5 u **(e)** 63 u + 29 u = 92 u

Section 31.4 Radioactivity

7. Which one or more of the three decay processes (α, β^-, or γ) results in a new element? **(a)** α and β^- **(b)** Only α **(c)** Only β^- **(d)** β^- and γ **(e)** Only γ

9. A nucleus can undergo α, β^-, or γ decay. For each type of decay, is the radius of the daughter nucleus greater than, less than, or about the same as the radius of the parent nucleus?

	α Decay	β^- Decay	γ Decay
(a)	Greater than	Greater than	About the same as
(b)	Greater than	Less than	About the same as
(c)	Less than	About the same as	Greater than
(d)	Less than	About the same as	About the same as
(e)	About the same as	Less than	Less than

Section 31.6 Radioactive Decay and Activity

13. Two samples contain different radioactive isotopes. Is it possible for these samples to have the same activity? **(a)** Yes, if they have the same number of nuclei but different half-lives. **(b)** Yes, if they have different numbers of nuclei and different half-lives. **(c)** Yes, if they have different numbers of nuclei but the same half-lives. **(d)** No, because they can have different half-lives. **(e)** No, because they can have different numbers of nuclei.

16. The drawing shows the activities of three radioactive samples. Rank the samples according to half-life, largest first. **(a)** 2, 3, 1 **(b)** 1, 2, 3 **(c)** 3, 2, 1 **(d)** 1, 3, 2 **(e)** 3, 1, 2

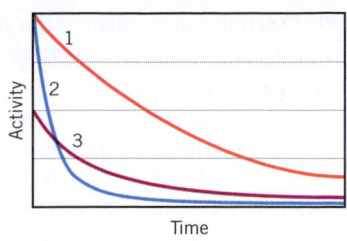

Section 31.7   Radioactive Dating

18. The bones from an animal found at an archaeological dig have a $^{14}_{6}$C activity of 0.10 Bq per gram of carbon. The half-life of the radioactive isotope $^{14}_{6}$C is 5730 yr. Which one of the following best describes the age of the bones?   **(a)** It is less than 2000 years.   **(b)** It is between 2000 and 3000 years.   **(c)** It is between 3000 and 4000 years.   **(d)** It is between 4000 and 5000 years.   **(e)** It is more than 5000 years.

PROBLEMS

Note to Instructors: Most of the homework problems in this chapter are available for assignment via an online homework management program such as WileyPLUS *or* WebAssign, *and those marked with the icon* **GO** *are presented in* WileyPLUS *using a guided tutorial format that provides enhanced interactivity. See Preface for additional details.*

The data given for atomic masses in these problems include the mass of the electrons orbiting the nucleus of the electrically neutral atom.

ssm   Solution is in the Student Solutions Manual.
www   Solution is available online at www.wiley.com/college/cutnell

 This icon represents a biomedical application.

Section 31.1   Nuclear Structure,
Section 31.2   The Strong Nuclear Force and the Stability of the Nucleus

1. ssm For $^{208}_{82}$Pb find   **(a)** the net electrical charge of the nucleus, **(b)** the number of neutrons,   **(c)** the number of nucleons,   **(d)** the approximate radius of the nucleus, and   **(e)** the nuclear density.

2. In electrically neutral atoms, how many   **(a)** protons are in the uranium $^{238}_{92}$U nucleus,   **(b)** neutrons are in the mercury $^{202}_{80}$Hg nucleus, and   **(c)** electrons are in orbit about the niobium $^{93}_{41}$Nb nucleus?

3. In each of the following cases, what element does the symbol X represent and how many neutrons are in the nucleus? Use the periodic table on the inside of the back cover as needed.   **(a)** $^{195}_{78}$X **(b)** $^{32}_{16}$X   **(c)** $^{63}_{29}$X   **(d)** $^{11}_{5}$X   **(e)** $^{239}_{94}$X

4. GO In a nucleus, each proton experiences a repulsive electrostatic force from each of the other protons. In a nucleus of gold $^{197}_{79}$Au, what is the magnitude of the least possible electrostatic force of repulsion that one proton can exert on another?

5. ssm www What is the radius of a nucleus of titanium $^{48}_{22}$Ti?

6. GO The largest stable nucleus has a nucleon number of 209, and the smallest has a nucleon number of 1. If each nucleus is assumed to be a sphere, what is the ratio (largest/smallest) of the surface areas of these spheres?

* **7.** An unknown nucleus contains 70 neutrons and has twice the volume of the nickel $^{60}_{28}$Ni nucleus. Identify the unknown nucleus in the form $^{A}_{Z}$X. Use the periodic table on the inside of the back cover as needed.

* **8.** The ratio r_X/r_T of the radius of an unknown nucleus $^{A}_{Z}$X to a tritium nucleus $^{3}_{1}$T is $\dfrac{r_X}{r_T} = 1.10$. Both nuclei contain the same number of neutrons. Identify the unknown nucleus in the form $^{A}_{Z}$X. Use the periodic table on the inside of the back cover as needed.

** **9. ssm www** Refer to Conceptual Example 1 for a discussion of nuclear densities. A neutron star is composed of neutrons and has a density that is approximately the same as that of a nucleus. What is the radius of a neutron star whose mass is 0.40 times the mass of the sun?

** **10.** Suppose that you could pack neutrons (mass $= 1.67 \times 10^{-27}$ kg) inside a tennis ball (radius $= 0.032$ m) in the same way as neutrons and protons are packed together in the nucleus of an atom. **(a)** Approximately how many neutrons would fit inside the tennis

ball?   **(b)** A small object is placed 2.0 m from the center of the neutron-packed tennis ball, and the tennis ball exerts a gravitational force on it. When the object is released, what is the magnitude of the acceleration that it experiences? Ignore the gravitational force exerted on the object by the earth.

Section 31.3   The Mass Defect of the Nucleus and Nuclear Binding Energy *(Note: The atomic mass for hydrogen $^{1}_{1}$H is 1.007 825 u; this includes the mass of one electron.)*

11. ssm Find the binding energy (in MeV) for lithium $^{7}_{3}$Li (atomic mass $= 7.016\ 003$ u).

12. Determine the mass defect (in atomic mass units) for   **(a)** helium $^{3}_{2}$He, which has an atomic mass of 3.016 030 u, and   **(b)** the isotope of hydrogen known as tritium $^{3}_{1}$T, which has an atomic mass of 3.016 050 u.   **(c)** On the basis of your answers to parts (a) and (b), state which nucleus requires more energy to disassemble it into its separate and stationary constituent nucleons. Give your reasoning.

13. Mercury $^{202}_{80}$Hg has an atomic mass of 201.970 617 u. Obtain the binding energy *per nucleon* (in MeV/nucleon).

14. A 245-kg boulder is dropped into a mine shaft that is 3.0×10^3 m deep. During the boulder's fall, the system consisting of the earth and the boulder loses a certain amount of gravitational potential energy. It would take an equal amount of energy to "free" the boulder from the shaft by raising it back to the top, so this can be considered the system's binding energy.   **(a)** Determine the binding energy (in joules) of the earth–boulder system.   **(b)** How much mass does the earth–boulder system lose when the boulder falls to the bottom of the shaft?

15. Use the plot of binding energy per nucleon in Figure 31.5 to determine the mass defect for the oxygen $^{16}_{8}$O nucleus. Express your answer in kilograms.

* **16. GO** A copper penny has a mass of 3.0 g. Determine the energy (in MeV) that would be required to break all the copper nuclei into their constituent protons and neutrons. Ignore the energy that binds the electrons to the nucleus and the energy that binds one atom to another in the structure of the metal. For simplicity, assume that all the copper nuclei are $^{63}_{29}$Cu (atomic mass $= 62.939\ 598$ u).

* **17. ssm** Two isotopes of a certain element have binding energies that differ by 5.03 MeV. The isotope with the larger binding energy contains one more neutron than the other isotope. Find the difference in atomic mass between the two isotopes.

*18. (a) Energy is required to separate a nucleus into its constituent nucleons, as Figure 31.3 indicates; this energy is the *total* binding energy of the nucleus. In a similar way one can speak of the energy that binds a single nucleon to the remainder of the nucleus. For example, separating nitrogen $^{14}_{7}N$ into nitrogen $^{13}_{7}N$ and a neutron takes energy equal to the binding energy of the neutron, as shown below:

$$^{14}_{7}N + \text{Energy} \longrightarrow ^{13}_{7}N + ^{1}_{0}n$$

Find the energy (in MeV) that binds the neutron to the $^{14}_{7}N$ nucleus by considering the mass of $^{13}_{7}N$ (atomic mass = 13.005 738 u) and the mass of $^{1}_{0}n$ (atomic mass = 1.008 665 u), as compared to the mass of $^{14}_{7}N$ (atomic mass = 14.003 074 u).   (b) Similarly, one can speak of the energy that binds a single proton to the $^{14}_{7}N$ nucleus:

$$^{14}_{7}N + \text{Energy} \longrightarrow ^{13}_{6}C + ^{1}_{1}H$$

Following the procedure outlined in part (a), determine the energy (in MeV) that binds the proton (atomic mass = 1.007 825 u) to the $^{14}_{7}N$ nucleus. The atomic mass of carbon $^{13}_{6}C$ is 13.003 355 u.   (c) Which nucleon is more tightly bound, the neutron or the proton?

Section 31.4 Radioactivity

19. **ssm www** Write the β^+ decay process for each of the following nuclei, being careful to include Z and A and the proper chemical symbol for each daughter nucleus: (a) $^{18}_{9}F$ (b) $^{15}_{8}O$

20. Osmium $^{191}_{76}Os$ (atomic mass = 190.960 920 u) is converted into iridium $^{191}_{77}Ir$ (atomic mass = 190.960 584 u) via β^- decay. What is the energy (in MeV) released in this process?

21. **ssm** Write the β^- decay process for $^{35}_{16}S$, including the chemical symbol and values for Z and A.

22. **GO** The β^- decay of phosphorus $^{32}_{15}P$ (atomic mass = 31.973 907 u) produces a daughter nucleus that is sulfur $^{32}_{16}S$ (atomic mass = 31.972 070 u), a β^- particle, and an antineutrino. The kinetic energy of the β^- particle is 0.90 MeV. Find the maximum possible energy (in MeV) that the antineutrino could carry away.

23. Find the energy (in MeV) released when α decay converts radium $^{226}_{88}Ra$ (atomic mass = 226.025 40 u) into radon $^{222}_{86}Rn$ (atomic mass = 222.017 57 u). The atomic mass of an α particle is 4.002 603 u.

24. Lead $^{207}_{82}Pb$ is a stable daughter nucleus that can result from either an α decay or a β^- decay. Write the decay processes, including the chemical symbols and values for Z and A of the parent nuclei, for (a) the α decay and   (b) the β^- decay.

25. Complete the following decay processes by stating what the symbol X represents (X = α, β^-, β^+, or γ):

(a) $^{211}_{82}Pb \rightarrow ^{211}_{83}Bi + X$ (c) $^{231}_{90}Th^* \rightarrow ^{231}_{90}Th + X$

(b) $^{11}_{6}C \rightarrow ^{11}_{5}B + X$ (d) $^{210}_{84}Po \rightarrow ^{206}_{82}Pb + X$

26. Multiple-Concept Example 7 reviews the concepts needed to solve this problem. When uranium $^{235}_{92}U$ decays, it emits (among other things) a γ ray that has a wavelength of 1.14×10^{-11} m. Determine the energy (in MeV) of this γ ray.

*27. Refer to **Interactive Solution 31.27** at **www.wiley.com/college/ cutnell** to review a model for solving this type of problem. Polonium $^{210}_{84}Po$ (atomic mass = 209.982 848 u) undergoes α decay. Assuming that all the released energy is in the form of kinetic energy of the α particle (atomic mass = 4.002 603 u) and ignoring the recoil of the daughter nucleus (lead $^{206}_{82}Pb$, 205.974 440 u), find the speed of the α particle. Ignore relativistic effects.

*28. **GO** Radon $^{220}_{86}Rn$ produces a daughter nucleus that is radioactive. The daughter, in turn, produces its own radioactive daughter, and so on. This process continues until lead $^{208}_{82}Pb$ is reached. What are the total number N_α of α particles and the total number N_β of β^- particles that are generated in this series of radioactive decays?

*29. Determine the symbol $^{A}_{Z}X$ for the parent nucleus whose α decay produces the same daughter as the β^- decay of thallium $^{208}_{81}Tl$.

30. Interactive LearningWare 31.1 at **www.wiley.com/college/ cutnell** reviews the concepts that are involved in this problem. An isotope of beryllium (atomic mass = 7.017 u) emits a γ ray and recoils with a speed of 2.19×10^4 m/s. Assuming that the beryllium nucleus is stationary to begin with, find the wavelength of the γ ray.

31. ssm Find the energy (in MeV) released when β^+ decay converts sodium $^{22}_{11}Na$ (atomic mass = 21.994 434 u) into neon $^{22}_{10}Ne$ (atomic mass = 21.991 383 u). Notice that the atomic mass for $^{22}_{11}Na$ includes the mass of 11 electrons, whereas the atomic mass for $^{22}_{10}Ne$ includes the mass of only 10 electrons.

Section 31.6 Radioactive Decay and Activity

32. The half-lives in two different samples, A and B, of radioactive nuclei are related according to $T_{1/2,B} = \frac{1}{2}T_{1/2,A}$. In a certain period the number of radioactive nuclei in sample A decreases to one-fourth the number present initially. In this same period the number of radioactive nuclei in sample B decreases to a fraction f of the number present initially. Find f.

33. **ssm** How many half-lives are required for the number of radioactive nuclei to decrease to one-millionth of the initial number?

34. The isotope $^{224}_{88}Ra$ of radium has a decay constant of 2.19×10^{-6} s^{-1}. What is the half-life (in days) of this isotope?

35. ⚕ Strontium $^{90}_{38}Sr$ has a half-life of 29.1 yr. It is chemically similar to calcium, enters the body through the food chain, and collects in the bones. Consequently, $^{90}_{38}Sr$ is a particularly serious health hazard. How long (in years) will it take for 99.9900% of the $^{90}_{38}Sr$ released in a nuclear reactor accident to disappear?

36. **GO** Two radioactive waste products from nuclear reactors are strontium $^{90}_{38}Sr$ ($T_{1/2} = 29.1$ yr) and cesium $^{134}_{55}Cs$ ($T_{1/2} = 2.06$ yr). These two species are present initially in a ratio of $N_{0,Sr}/N_{0,Cs} = 7.80 \times 10^{-3}$. What is the ratio N_{Sr}/N_{Cs} fifteen years later?

37. **ssm** The number of radioactive nuclei present at the start of an experiment is 4.60×10^{15}. The number present twenty days later is 8.14×10^{14}. What is the half-life (in days) of the nuclei?

38. ⚕ Iodine $^{131}_{53}I$ is used in diagnostic and therapeutic techniques in the treatment of thyroid disorders. This isotope has a half-life of 8.04 days. What percentage of an initial sample of $^{131}_{53}I$ remains after 30.0 days?

39. **ssm** If the activity of a radioactive substance is initially 398 disintegrations/min and two days later it is 285 disintegrations/min, what is the activity four days later still, or six days after the start? Give your answer in disintegrations/min.

*40. One day, a cell phone company sends a text message to each of its 5800 subscribers, announcing that they have been automatically enrolled as contestants in a promotional lottery modeled on nuclear decay. On the first day, 10% of the 5800 contestants are notified by text message that they have been randomly eliminated from the lottery. The other 90% of the contestants automatically advance to the next round. On each of the following days, 10% of the remaining contestants are randomly eliminated, until fewer than 10 contestants remain. Determine   (a) the activity (number of contestants eliminated per day) on the second day of the lottery,   (b) the decay constant (in d^{-1}) of the lottery, and   (c) the half-life (in d) of the lottery.

*41. Refer to **Interactive Solution 31.41** at **www.wiley.com/college/ cutnell** for one approach to solving this problem. To see why one curie of activity was chosen to be 3.7×10^{10} Bq, determine the activity (in disintegrations per second) of one gram of radium $^{226}_{88}Ra$ ($T_{1/2} = 1.6 \times 10^3$ yr).

***42.** **GO** A one-gram sample of radium $^{224}_{88}$Ra (atomic mass = 224.020 186 u, $T_{1/2}$ = 3.66 days) contains 2.69×10^{21} nuclei and undergoes α decay to produce radon $^{220}_{86}$Rn (atomic mass = 220.011 368 u). The atomic mass of an α particle is 4.002 603 u. The latent heat of fusion for water is 33.5×10^4 J/kg. With the energy released in 3.66 days, how many kilograms of ice could be melted at 0°C?

***43.** A sample of ore containing radioactive strontium $^{90}_{38}$Sr has an activity of 6.0×10^5 Bq. The atomic mass of strontium is 89.908 u, and its half-life is 29.1 yr. How many grams of strontium are in the sample?

***44.** **GO** Outside the nucleus, the neutron itself is radioactive and decays into a proton, an electron, and an antineutrino. The half-life of a neutron (mass = 1.675×10^{-27} kg) outside the nucleus is 10.4 min. On average, over what distance (in meters) would a beam of 5.00-eV neutrons travel before the number of neutrons decreased to 75.0% of its initial value?

***45.** **ssm** Two radioactive nuclei A and B are present in equal numbers to begin with. Three days later, there are three times as many A nuclei as there are B nuclei. The half-life of species B is 1.50 days. Find the half-life of species A.

Section 31.7   Radioactive Dating

46. Review Multiple-Concept Example 11 for help in approaching this problem. An archaeological specimen containing 9.2 g of carbon has an activity of 1.6 Bq. How old (in years) is the specimen?

***47.** **ssm** The practical limit to ages that can be determined by radiocarbon dating is about 41 000 yr. In a 41 000-yr-old sample, what percentage of the original $^{14}_{6}$C atoms remains?

48. **GO** The half-life for the α decay of uranium $^{238}_{92}$U is 4.47×10^9 yr. Determine the age (in years) of a rock specimen that contains 60.0% of its original number of $^{238}_{92}$U atoms.

49. Review Conceptual Example 12 before starting to solve this problem. The number of unstable nuclei remaining after a time $t = 5.00$ yr is N, and the number present initially is N_0. Find the ratio N/N_0 for   **(a)** $^{14}_{6}$C (half-life = 5730 yr),   **(b)** $^{15}_{8}$O (half-life = 122.2 s; use $t = 1.00$ h, since otherwise the answer is out of the range of your calculator), and   **(c)** $^{3}_{1}$H (half-life = 12.33 yr). Verify that your answers are consistent with the reasoning in Conceptual Example 12.

50. When a sample from a meteorite is analyzed, it is determined that 93.8% of the original mass of a certain radioactive isotope is still present. Based on this finding, the age of the meteorite is calculated to be 4.51×10^9 yr. What is the half-life (in yr) of the isotope used to date the meteorite?

***51.** **ssm** When any radioactive dating method is used, experimental error in the measurement of the sample's activity leads to error in the estimated age. In an application of the radiocarbon dating technique to certain fossils, an activity of 0.100 Bq per gram of carbon is measured to within an accuracy of $\pm 10.0\%$. Find the age of the fossils and the maximum error (in years) in the value obtained. Assume that there is no error in the 5730-year half-life of $^{14}_{6}$C nor in the value of 0.23 Bq per gram of carbon in a living organism.

****52.** **(a)** A sample is being dated by the radiocarbon technique. If the sample were uncontaminated, its activity would be 0.011 Bq per gram of carbon. Find the true age (in years) of the sample.   **(b)** Suppose the sample is contaminated, so that only 98.0% of its carbon is ancient carbon. The remaining 2.0% is fresh carbon, in the sense that the $^{14}_{6}$C it contains has not had any time to decay. Assuming that the lab technician is unaware of the contamination, what apparent age (in years) would be determined for the sample?

ADDITIONAL PROBLEMS

53. For lead $^{206}_{82}$Pb (atomic mass = 205.974 440 u) obtain   **(a)** the mass defect in atomic mass units,   **(b)** the binding energy (in MeV), and   **(c)** the binding energy per nucleon (in MeV/nucleon).

54. By what factor does the nucleon number of a nucleus have to increase in order for the nuclear radius to double?

55. A device used in radiation therapy for cancer contains 0.50 g of cobalt $^{60}_{27}$Co (59.933 819 u). The half-life of $^{60}_{27}$Co is 5.27 yr. Determine the activity of the radioactive material.

56. In the form $^{A}_{Z}$X, identify the daughter nucleus that results when   **(a)** plutonium $^{242}_{94}$Pu undergoes α decay,   **(b)** sodium $^{24}_{11}$Na undergoes β^- decay, and   **(c)** nitrogen $^{13}_{7}$N undergoes β^+ decay.

57. Find the energy released when lead $^{211}_{82}$Pb (atomic mass = 210.988 735 u) undergoes β^- decay to become bismuth $^{211}_{83}$Bi (atomic mass = 210.987 255 u).

58. Multiple-Concept Example 11 reviews most of the concepts that are needed to solve this problem. Material found with a mummy in the arid highlands of southern Peru has a $^{14}_{6}$C activity per gram of carbon that is 78.5% of the activity present initially. How long ago (in years) did this individual die?

***59.** **ssm www** The photomultiplier tube in a commercial scintillation counter contains 15 of the special electrodes, or dynodes. Each dynode produces 3 electrons for every electron that strikes it. One photoelectron strikes the first dynode. What is the maximum number of electrons that strike the 15th dynode?

***60.** The isotope $^{198}_{79}$Au (atomic mass = 197.968 u) of gold has a half-life of 2.69 days and is used in cancer therapy. What mass (in grams) of this isotope is required to produce an activity of 315 Ci?

***61.** Review Conceptual Example 5 as background for this problem. The α decay of uranium $^{238}_{92}$U produces thorium $^{234}_{90}$Th (atomic mass = 234.0436 u). In Example 4, the energy released in this decay is determined to be 4.3 MeV. Determine how much of this energy is carried away by the recoiling $^{234}_{90}$Th daughter nucleus and how much by the α particle (atomic mass = 4.002 603 u). Assume that the energy of each particle is kinetic energy, and ignore the small amount of energy carried away by the γ ray that is also emitted. In addition, ignore relativistic effects.

***62.** In a radioactive decay series similar to that shown in Fig. 31.16, thorium $^{228}_{90}$Th (atomic mass = 228.028 715 u) undergoes four successive α decays, producing a daughter nucleus.   **(a)** Determine the symbol $^{A}_{Z}$X for the nucleus produced by four successive α decays of $^{228}_{90}$Th.   **(b)** What is the total amount of energy (in MeV) released in this series of α decays? The mass of the daughter nucleus can be obtained by using the result of part (a) and consulting Appendix F at the back of the book. The mass of a single α particle is 4.002 603 u.

CHAPTER 32

IONIZING RADIATION, NUCLEAR ENERGY, AND ELEMENTARY PARTICLES

Nuclear fission is the process in which the nucleus of a larger atom is split into smaller fragments, with the release of a relatively large amount of energy. This process is the basis for commercial energy production via nuclear reactors. The controlled-fission chain reaction that occurs in these reactors is monitored by technicians in control rooms such as the one shown here. Nuclear fission and chain reactions are among the topics discussed in this chapter. (© Mark Romine/Agefotostock America, Inc.)

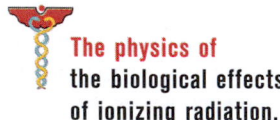

The physics of the biological effects of ionizing radiation.

32.1 BIOLOGICAL EFFECTS OF IONIZING RADIATION

Ionizing radiation consists of photons and/or moving particles that have sufficient energy to knock an electron out of an atom or molecule, thus forming an ion. The photons usually lie in the ultraviolet, X-ray, or γ-ray regions of the electromagnetic spectrum (see Figure 24.9), whereas the moving particles can be the α and β particles emitted during radioactive decay. An energy of roughly 1 to 35 eV is needed to ionize an atom or molecule, and the particles and γ rays emitted during nuclear disintegration often have energies of several million eV. Therefore, a single α particle, β particle, or γ ray can ionize thousands of molecules.

Nuclear radiation is potentially harmful to humans because the ionization it produces can significantly alter the structure of molecules within a living cell. The alterations can lead to the death of the cell and even of the organism itself. Despite the potential hazards, however, ionizing radiation is used in medicine for diagnostic and therapeutic purposes, such as locating bone fractures and treating cancer. The hazards can be minimized only if the fundamentals of radiation exposure, including dose units and the biological effects of radiation, are understood.

Exposure is a measure of the ionization produced in air by X-rays or γ rays, and it is defined in the following manner: A beam of X-rays or γ rays is sent through a mass m of dry air at standard temperature and pressure (STP: 0 °C, 1 atm pressure). In passing through the air, the beam produces positive ions whose total charge is q. Exposure is defined as the total charge per unit mass of air: exposure $= q/m$. The SI unit for exposure is coulombs per kilogram (C/kg). However, the first radiation unit to be defined was the *roentgen* (R), and it is still used today. With q expressed in coulombs (C) and m in kilograms (kg), the exposure in roentgens is given by

$$\text{Exposure (in roentgens)} = \left(\frac{1}{2.58 \times 10^{-4}}\right)\frac{q}{m} \qquad (32.1)$$

Thus, when X-rays or γ rays produce an exposure of one roentgen, $q = 2.58 \times 10^{-4}$ C of positive charge are produced in $m = 1$ kg of dry air:

$$1\ \text{R} = 2.58 \times 10^{-4}\ \text{C/kg} \qquad \text{(dry air, at STP)}$$

Since the concept of exposure is defined in terms of the ionizing abilities of X-rays and γ rays in air, it does not specify the effect of radiation on living tissue. For biological purposes, the ***absorbed dose*** is a more suitable quantity because it is the energy absorbed from the radiation per unit mass of absorbing material:

$$\text{Absorbed dose} = \frac{\text{Energy absorbed}}{\text{Mass of absorbing material}} \qquad (32.2)$$

The SI unit of absorbed dose is the *gray* (Gy), which is the unit of energy divided by the unit of mass: 1 Gy = 1 J/kg. Equation 32.2 is applicable to all types of radiation and absorbing media. Another unit is often used for absorbed dose—namely, the *rad* (sometimes abbreviated rd). The word "rad" is an acronym for **r**adiation **a**bsorbed **d**ose. The rad and the gray are related by

$$1\ \text{rad} = 0.01\ \text{gray}$$

Example 1 deals with the gray and the rad as units for the absorbed dose.

ANALYZING MULTIPLE-CONCEPT PROBLEMS

Example 1 Absorbed Dose of γ Rays

Figure 32.1 shows γ rays being absorbed by water. What is the absorbed dose (in rads) of γ rays that will heat the water from 20.0 to 50.0 °C?

Reasoning When γ rays are absorbed by the water, they cause it to heat up. The absorbed dose of γ rays is the energy (heat) absorbed by the water divided by its mass. According to the discussion in Section 12.7, the heat that must be absorbed by the water in order for its temperature to increase by a given amount depends on the mass and specific heat capacity of the water. We will use the concept of specific heat capacity to evaluate the absorbed dose of γ rays.

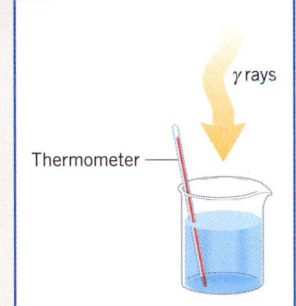

Figure 32.1 When the water absorbs the γ rays, its temperature rises.

Knowns and Unknowns The following table summarizes the given information:

Description	Symbol	Value
Initial temperature of water	T_0	20.0 °C
Final temperature of water	T	50.0 °C
Unknown Variable		
Absorbed dose of γ rays (in rads)	Absorbed dose	?

Modeling the Problem

STEP 1 Absorbed Dose   The absorbed dose of γ rays is the energy (heat) Q absorbed by the water divided by its mass m (see Equation 32.2), as indicated in the right column. Neither Q nor m is known. Both variables will be dealt with in Step 2.

$$\text{Absorbed dose} = \frac{Q}{m} \qquad (32.2)$$
$$\boxed{?}$$

STEP 2 Heat Needed to Increase the Temperature of the Water   The heat Q that is needed to increase the temperature of a mass m of water by an amount ΔT is $Q = cm\,\Delta T$ (Equation 12.4), where c is the specific heat capacity of water. The change in temperature

Continued

ΔT is equal to the higher temperature T minus the lower temperature T_0, or $\Delta T = T - T_0$. Thus, the heat can be expressed as

$$Q = cm\,\Delta T = cm(T - T_0) \qquad (12.4)$$

$$\text{Absorbed dose} = \frac{Q}{m} \qquad (32.2)$$

$$Q = cm(T - T_0) \qquad (12.4)$$

This expression for the heat absorbed by the water can be substituted into Equation 32.2 for the absorbed dose, as indicated in the right column. Note that the mass m of the water appears in both the numerator and denominator, so it can be eliminated algebraically.

Solution Algebraically combining the results of the two modeling steps gives

STEP 1   STEP 2

$$\text{Absorbed dose} = \frac{Q}{m} = \frac{cm(T - T_0)}{m} = c(T - T_0)$$

Taking the specific heat capacity c of water from Table 12.2, we find that the absorbed dose of γ rays [expressed in grays (Gy)] is

$$\text{Absorbed dose} = c(T - T_0) = [4186 \text{ J/(kg} \cdot \text{C}°)](50.0\ °\text{C} - 20.0\ °\text{C}) = 1.26 \times 10^5 \text{ Gy}$$

The problem asks that the absorbed dose be expressed in rads, rather than in grays. To this end, we note that 1 rad = 0.01 Gy, so

$$\text{Absorbed dose} = (1.26 \times 10^5 \text{ Gy})\left(\frac{1 \text{ rad}}{0.01 \text{ Gy}}\right) = \boxed{1.26 \times 10^7 \text{ rad}}$$

Related Homework: *Problems 7, 10, 50*

The amount of biological damage produced by ionizing radiation is different for different kinds of radiation. For instance, a 1-rad dose of neutrons is far more likely to produce eye cataracts than a 1-rad dose of X-rays. To compare the damage caused by different types of radiation, the ***relative biological effectiveness*** (RBE) is used.* The relative biological effectiveness of a particular type of radiation is the ratio of the dose of 200-keV X-rays needed to produce a certain biological effect to the dose of the radiation needed to produce the same biological effect:

$$\begin{array}{l}\text{Relative biological} \\ \text{effectiveness (RBE)}\end{array} = \dfrac{\begin{array}{c}\text{The dose of 200-keV X-rays that} \\ \text{produces a certain biological effect}\end{array}}{\begin{array}{c}\text{The dose of radiation that} \\ \text{produces the same biological effect}\end{array}} \qquad (32.3)$$

The RBE depends on the nature of the ionizing radiation and its energy, as well as on the type of tissue being irradiated. Table 32.1 lists some typical RBE values for different kinds of radiation, assuming that an "average" biological tissue is being irradiated. A value of RBE = 1 for γ rays and β^- particles indicates that they produce the same biological damage as do 200-keV X-rays. The larger RBE values for protons, α particles, and neutrons indicate that they cause substantially more damage. The RBE is often used in conjunction with the absorbed dose to reflect the damage-producing character of the radiation. The product of the absorbed dose in rads (not in grays) and the RBE is the ***biologically equivalent dose:***

Table 32.1 Relative Biological Effectiveness (RBE) for Various Types of Radiation

Type of Radiation	RBE
200-keV X-rays	1
γ rays	1
β^- particles (electrons)	1
Protons	10
α particles	10–20
Neutrons	
Slow	2
Fast	10

$$\begin{array}{l}\text{Biologically equivalent dose} \\ \text{(in rems)}\end{array} = \begin{array}{l}\text{Absorbed dose} \\ \text{(in rads)}\end{array} \times \text{RBE} \qquad (32.4)$$

The unit for the biologically equivalent dose is the *rem,* short for **r**oentgen **e**quivalent, **m**an. Example 2 illustrates the use of the biologically equivalent dose.

*The RBE is sometimes called the *quality factor* (QF).

🔵 **Example 2** Comparing Absorbed Doses of γ Rays and Neutrons

A biological tissue is irradiated with γ rays that have an RBE of 0.70. The absorbed dose of γ rays is 850 rad. The tissue is then exposed to neutrons whose RBE is 3.5. The biologically equivalent dose of the neutrons is the same as that of the γ rays. What is the absorbed dose of neutrons?

Reasoning The biologically equivalent doses of the neutrons and the γ rays are the same. Therefore, the tissue damage produced in each case is the same. However, the RBE of the neutrons is larger than the RBE of the γ rays by a factor of 3.5/0.70 = 5.0. Consequently, we will find that the absorbed dose of the neutrons is only one-fifth as great as that of the γ rays.

Solution According to Equation 32.4, the biologically equivalent dose is the product of the absorbed dose (in rads) and the RBE; it is the same for the γ rays and the neutrons. Therefore, we have

$$\text{Biologically equivalent dose} = (\text{Absorbed dose})_{\gamma\,\text{rays}}\,\text{RBE}_{\gamma\,\text{rays}} = (\text{Absorbed dose})_{\text{neutrons}}\,\text{RBE}_{\text{neutrons}}$$

Solving for the absorbed dose of the neutrons gives

$$(\text{Absorbed dose})_{\text{neutrons}} = (\text{Absorbed dose})_{\gamma\,\text{rays}}\left(\frac{\text{RBE}_{\gamma\,\text{rays}}}{\text{RBE}_{\text{neutrons}}}\right) = (850\ \text{rad})\left(\frac{0.70}{3.5}\right) = \boxed{170\ \text{rad}}$$

🔵

Everyone is continually exposed to background radiation from natural sources, such as cosmic rays (high-energy particles that come from outside the solar system), radioactive materials in the environment, radioactive nuclei (primarily carbon $^{14}_{6}$C and potassium $^{40}_{19}$K) within our own bodies, and radon. Table 32.2 lists the average biologically equivalent doses received from these sources by a person in the United States. According to this table, radon is a major contributor to the natural background radiation. Radon is an odorless radioactive gas and poses a health hazard because, when inhaled, it can damage the lungs and cause cancer. Radon is found in soil and rocks and enters houses via a mechanism that is not well understood. One possibility for indoor radon entry is through cracks and crevices in the foundation. The amount of radon in the soil varies greatly throughout the country, with some localities having significant amounts and others having virtually none. Accordingly, the dose that any individual receives can vary widely from the average value of 200 mrem/yr given in Table 32.2 (1 mrem = 10^{-3} rem). In many houses, the entry of radon can be reduced significantly by sealing the foundation against entry of the gas and providing good ventilation so it does not accumulate.

To the natural background of radiation, a significant amount of human-made radiation has been added, mostly from medical/dental diagnostic X-rays. Table 32.2 indicates an average total dose of 360 mrem/yr from all sources.

The effects of radiation on humans can be grouped into two categories, according to the time span between initial exposure and the appearance of physiological symptoms: (1) short-term or acute effects that appear within a matter of minutes, days, or weeks, and (2) long-term or latent effects that appear years, decades, or even generations later.

Radiation sickness is the general term applied to the acute effects of radiation. Depending on the severity of the dose, a person with radiation sickness can exhibit nausea, vomiting, fever, diarrhea, and loss of hair. Ultimately, death can occur. The severity of radiation sickness is related to the dose received, and in the following discussion the biologically equivalent doses quoted are whole-body, single doses. A dose less than 50 rem causes no short-term ill effects. A dose between 50 and 300 rem brings on radiation sickness, the severity increasing with increasing dosage. A whole-body dose in the range of 400–500 rem is classified as an LD_{50} dose, meaning that it is a lethal dose (LD) for about 50% of the people so exposed; death occurs within a few months. Whole-body doses greater than 600 rem result in death for almost all individuals.

Long-term or latent effects of radiation may appear as a result of high-level brief exposure or low-level exposure over a long period of time. Some long-term effects are hair loss, eye cataracts, and various kinds of cancer. In addition, genetic defects caused by mutated genes may be passed on from one generation to the next.

Because of the hazards of radiation, the federal government has established dose limits. The permissible dose for an individual is defined as the dose, accumulated over a long

Table 32.2 Average Biologically Equivalent Doses of Radiation Received by a U. S. Resident[a]

Source of Radiation	Biologically Equivalent Dose (mrem/yr)[b]
Natural background radiation	
Cosmic rays	28
Radioactive earth and air	28
Internal radioactive nuclei	39
Inhaled radon	≈200
Human-made radiation	
Consumer products	10
Medical/dental diagnostics	39
Nuclear medicine	14
Rounded total:	360

[a] National Council on Radiation Protection and Measurements, Report No. 93, "Ionizing Radiation Exposure of the Population of the United States," 1987. The data shown are under review, and it is anticipated that an updated report will be published late in 2008.
[b] 1 mrem = 10^{-3} rem.

period of time or resulting from a single exposure, that carries negligible probability of a severe health hazard. Federal standards (1991) state that an individual in the general population should not receive more than 500 mrem of human-made radiation each year, *exclusive* of medical sources. A person exposed to radiation in the workplace (e.g., a radiation therapist) should not receive more than 5 rem per year from work-related sources.

✓ **CHECK YOUR UNDERSTANDING**

(*The answers are given at the end of the book.*)

1. Two different types of radiation have the same RBE. Is it possible for these two types of radiation to deliver different biologically equivalent doses of radiation to a given tissue sample?

2. The damage-producing character of a given type of ionizing radiation depends on **(a)** only the RBE of the radiation, **(b)** only the absorbed dose of the radiation, **(c)** both the RBE and the absorbed dose of the radiation.

3. A person faces the possibility of receiving the following absorbed doses of ionizing radiation: 20 rad of γ rays (RBE = 1), 5 rad of neutrons (RBE = 10), and 2 rad of α particles (RBE = 20). Rank the amount of biological damage that these possibilities will cause in decreasing order (greatest damage first).

32.2 INDUCED NUCLEAR REACTIONS

Section 31.4 discusses how a radioactive parent nucleus disintegrates spontaneously into a daughter nucleus. It is also possible to bring about, or induce, the disintegration of a stable nucleus by striking it with another nucleus, an atomic or subatomic particle, or a γ-ray photon. A **nuclear reaction** is said to occur whenever an incident nucleus, particle, or photon causes a change to occur in a target nucleus.

In 1919, Ernest Rutherford observed that when an α particle strikes a nitrogen nucleus, an oxygen nucleus and a proton are produced. This nuclear reaction is written as

$$\underbrace{{}^{4}_{2}\text{He}}_{\substack{\text{Incident} \\ \alpha \text{ particle}}} + \underbrace{{}^{14}_{7}\text{N}}_{\substack{\text{Nitrogen} \\ \text{(target)}}} \longrightarrow \underbrace{{}^{17}_{8}\text{O}}_{\text{Oxygen}} + \underbrace{{}^{1}_{1}\text{H}}_{\text{Proton, } p}$$

Because the incident α particle induces the transmutation of nitrogen into oxygen, this reaction is an example of an **induced nuclear transmutation.**

Nuclear reactions are often written in a shorthand form. For example, the reaction above is designated by ${}^{14}_{7}\text{N}$ (α, p) ${}^{17}_{8}\text{O}$. The first and last symbols represent the initial and final nuclei, respectively. The symbols within the parentheses denote the incident α particle (on the left) and the small emitted particle or proton p (on the right). Some other induced nuclear transmutations are listed below, together with the equivalent shorthand notations:

Nuclear Reaction	Notation
${}^{1}_{0}\text{n} + {}^{10}_{5}\text{B} \rightarrow {}^{7}_{3}\text{Li} + {}^{4}_{2}\text{He}$	${}^{10}_{5}\text{B}$ (n, α) ${}^{7}_{3}\text{Li}$
$\gamma + {}^{25}_{12}\text{Mg} \rightarrow {}^{24}_{11}\text{Na} + {}^{1}_{1}\text{H}$	${}^{25}_{12}\text{Mg}$ (γ, p) ${}^{24}_{11}\text{Na}$
${}^{1}_{1}\text{H} + {}^{13}_{6}\text{C} \rightarrow {}^{14}_{7}\text{N} + \gamma$	${}^{13}_{6}\text{C}$ (p, γ) ${}^{14}_{7}\text{N}$

▶ **CONCEPTS AT A GLANCE** Induced nuclear reactions, like the radioactive decay process discussed in Section 31.4, obey the conservation laws of physics. We have discussed these laws in previous chapters, as the Concepts-at-a-Glance chart in Figure 32.2 indicates. This chart is similar to the one shown earlier in Figure 31.6 and emphasizes the important role that the conservation laws play in all nuclear processes. In particular, both the total electric charge of the nucleons and the total number of nucleons are conserved during an induced nuclear reaction. The fact that these quantities are conserved makes it possible to identify the nucleus produced in a reaction, as the next example illustrates. ◀

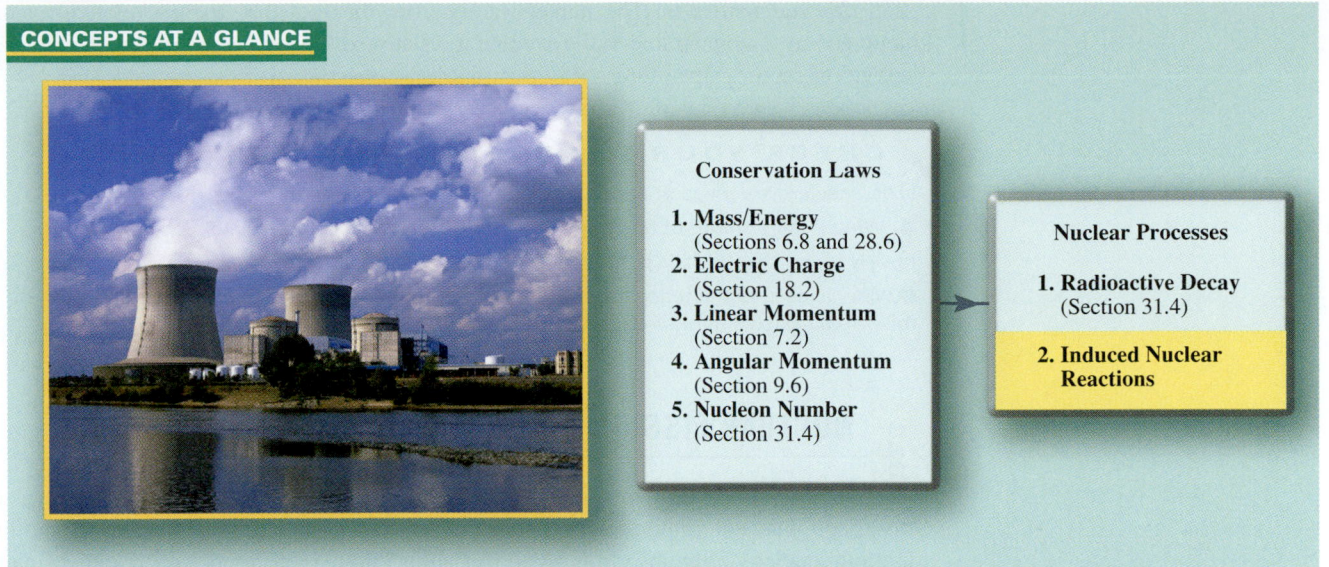

Conservation Laws

1. **Mass/Energy**
   (Sections 6.8 and 28.6)
2. **Electric Charge**
   (Section 18.2)
3. **Linear Momentum**
   (Section 7.2)
4. **Angular Momentum**
   (Section 9.6)
5. **Nucleon Number**
   (Section 31.4)

Nuclear Processes

1. **Radioactive Decay**
   (Section 31.4)

2. **Induced Nuclear Reactions**

Figure 32.2 CONCEPTS AT A GLANCE In any nuclear process, such as radioactive decay or an induced nuclear reaction, the conservation laws of physics are obeyed. Such processes provide the basis for the commercial generation of power, as in the nuclear power plant shown in the photograph. (© Steve Allen/Agefotostock)

Example 3   An Induced Nuclear Transmutation

An α particle strikes an aluminum $^{27}_{13}\text{Al}$ nucleus. As a result, an unknown nucleus $^{A}_{Z}\text{X}$ and a neutron $^{1}_{0}\text{n}$ are produced:

$$^{4}_{2}\text{He} + {}^{27}_{13}\text{Al} \longrightarrow {}^{A}_{Z}\text{X} + {}^{1}_{0}\text{n}$$

Identify the nucleus produced, including its atomic number Z (the number of protons) and its atomic mass number A (the number of nucleons).

Reasoning The total electric charge of the nucleons is conserved, so that we can set the total number of protons before the reaction equal to the total number after the reaction. The total number of nucleons is also conserved, so that we can set the total number before the reaction equal to the total number after the reaction. These two conserved quantities will allow us to identify the nucleus $^{A}_{Z}\text{X}$.

Solution The conservation of total electric charge and total number of nucleons leads to the equations listed below:

Conserved Quantity	Before Reaction		After Reaction
Total electric charge (number of protons)	$2 + 13$	$=$	$Z + 0$
Total number of nucleons	$4 + 27$	$=$	$A + 1$

Solving these equations for Z and A gives $Z = 15$ and $A = 30$. Since $Z = 15$ identifies the element as phosphorus (P), the nucleus produced is $\boxed{^{30}_{15}\text{P}}$.

Induced nuclear transmutations can be used to produce isotopes that are not found naturally. In 1934, Enrico Fermi suggested a method for producing elements with a higher atomic number than uranium ($Z = 92$). These elements—neptunium ($Z = 93$), plutonium ($Z = 94$), americium ($Z = 95$), and so on—are known as *transuranium elements,* and none occurs naturally. They are created in a nuclear reaction between a suitably chosen lighter element and a small incident particle, usually a neutron or an α particle. For example, Figure 32.3 shows a reaction that produces plutonium from uranium. A neutron is captured by a uranium $^{238}_{92}\text{U}$ nucleus, producing $^{239}_{92}\text{U}$ and a γ ray. The $^{239}_{92}\text{U}$ nucleus is radioactive and decays with a half-life of 23.5 min into neptunium $^{239}_{93}\text{Np}$. Neptunium is also radioactive and disintegrates with a half-life of 2.4 days into plutonium $^{239}_{94}\text{Pu}$. Plutonium is the final product and has a half-life of 24 100 yr.

The neutrons that participate in nuclear reactions can have kinetic energies that cover a wide range. In particular, those that have a kinetic energy of about 0.04 eV or less are

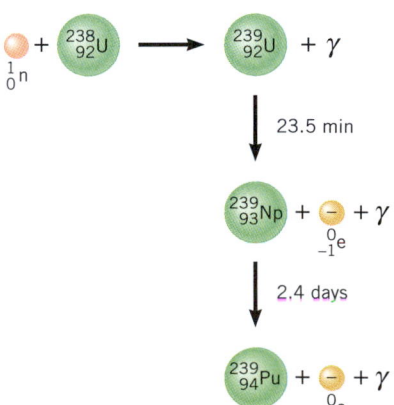

Figure 32.3 An induced nuclear reaction is shown in which $^{238}_{92}\text{U}$ is transmuted into the transuranium element plutonium $^{239}_{94}\text{Pu}$.

called ***thermal neutrons.*** The name derives from the fact that such a relatively small kinetic energy is comparable to the average translational kinetic energy of a molecule in an ideal gas at room temperature.

✓ CHECK YOUR UNDERSTANDING

(The answers are given at the end of the book.)

4. Which one or more of the following nuclear reactions could possibly occur?

(a) $^{15}_{7}\text{N}\,(\alpha, \gamma)\,^{18}_{9}\text{F}$ **(b)** $^{14}_{6}\text{C}\,(p, n)\,^{15}_{8}\text{O}$ **(c)** $^{14}_{7}\text{N}\,(p, \gamma)\,^{15}_{8}\text{O}$ **(d)** $^{15}_{8}\text{O}\,(n, p)\,^{13}_{6}\text{C}$

5. Why is each of the following reactions *not* allowed?   **(a)** $^{60}_{28}\text{Ni}\,(\alpha, p)\,^{62}_{29}\text{Cu}$
(b) $^{27}_{13}\text{Al}\,(n, n)\,^{28}_{13}\text{Al}$ **(c)** $^{39}_{19}\text{K}\,(p, \alpha)\,^{36}_{17}\text{Cl}$

32.3 NUCLEAR FISSION

In 1939 four German scientists, Otto Hahn, Lise Meitner, Fritz Strassmann, and Otto Frisch, made an important discovery that ushered in the atomic age. They found that a uranium nucleus, after absorbing a neutron, splits into two fragments, each with a smaller mass than the original nucleus. The splitting of a massive nucleus into two less massive fragments is known as ***nuclear fission.***

Figure 32.4 shows a fission reaction in which a uranium $^{235}_{92}\text{U}$ nucleus is split into barium $^{141}_{56}\text{Ba}$ and krypton $^{92}_{36}\text{Kr}$ nuclei. The reaction begins when $^{235}_{92}\text{U}$ absorbs a slowly moving neutron, creating a "compound nucleus," $^{236}_{92}\text{U}$. The compound nucleus disintegrates quickly into $^{141}_{56}\text{Ba}$, $^{92}_{36}\text{Kr}$, and three neutrons according to the following reaction:

$$^{1}_{0}\text{n} + \,^{235}_{92}\text{U} \longrightarrow \underbrace{^{236}_{92}\text{U}}_{\substack{\text{Compound} \\ \text{nucleus} \\ \text{(unstable)}}} \longrightarrow \underbrace{^{141}_{56}\text{Ba}}_{\text{Barium}} + \underbrace{^{92}_{36}\text{Kr}}_{\text{Krypton}} + \underbrace{3\,^{1}_{0}\text{n}}_{\text{3 neutrons}}$$

This reaction is only one of the many possible reactions that can occur when uranium fissions. For example, another reaction is

$$^{1}_{0}\text{n} + \,^{235}_{92}\text{U} \longrightarrow \underbrace{^{236}_{92}\text{U}}_{\substack{\text{Compound} \\ \text{nucleus} \\ \text{(unstable)}}} \longrightarrow \underbrace{^{140}_{54}\text{Xe}}_{\text{Xenon}} + \underbrace{^{94}_{38}\text{Sr}}_{\text{Strontium}} + \underbrace{2\,^{1}_{0}\text{n}}_{\text{2 neutrons}}$$

Some reactions produce as many as 5 neutrons; however, the average number produced per fission is 2.5.

When a neutron collides with and is absorbed by a uranium nucleus, the uranium nucleus begins to vibrate and becomes distorted. The vibration continues until the distortion

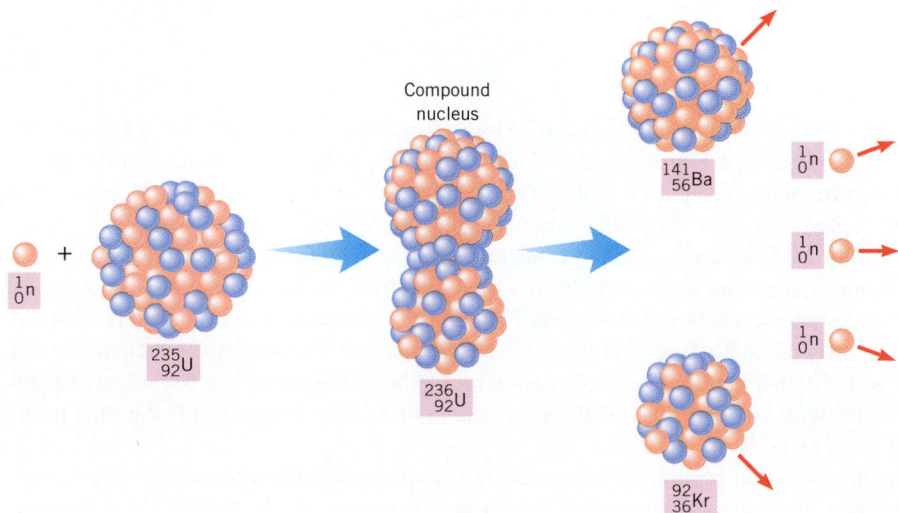

Figure 32.4 A slowly moving neutron causes the uranium nucleus $^{235}_{92}\text{U}$ to fission into barium $^{141}_{56}\text{Ba}$, krypton $^{92}_{36}\text{Kr}$, and three neutrons.

becomes so severe that the attractive strong nuclear force can no longer balance the electrostatic repulsion between the nuclear protons. At this point, the nucleus bursts apart into fragments, which carry off energy, primarily in the form of kinetic energy. The energy carried off by the fragments is enormous and was stored in the original nucleus mainly in the form of electric potential energy. An average of roughly 200 MeV of energy is released per fission. This energy is approximately 10^8 times greater than the energy released per molecule in an ordinary chemical reaction, such as the combustion of gasoline or coal. Example 4 demonstrates how to estimate the energy released during the fission of a nucleus.

 Example 4 The Energy Released During Nuclear Fission

Estimate the amount of energy released when a massive nucleus ($A = 240$) fissions.

Reasoning Figure 31.5 shows that the binding energy of a nucleus with $A = 240$ is about 7.6 MeV per nucleon. We assume that this nucleus fissions into two fragments, each with $A \approx 120$. According to Figure 31.5, the binding energy of the fragments increases to about 8.5 MeV per nucleon. Consequently, when a massive nucleus fissions, there is a release of about 8.5 MeV − 7.6 MeV = 0.9 MeV of energy per nucleon.

Solution Since there are 240 nucleons involved in the fission process, the total energy released per fission is approximately (0.9 MeV/nucleon)(240 nucleons) $\approx$ $\boxed{200 \text{ MeV}}$.

Virtually all naturally occurring uranium is composed of two isotopes. These isotopes and their natural abundances are $^{238}_{92}U$ (99.275%) and $^{235}_{92}U$ (0.720%). Although $^{238}_{92}U$ is by far the most abundant isotope, the probability that it will capture a neutron and fission is very small. For this reason, $^{238}_{92}U$ is not the isotope of choice for generating nuclear energy. In contrast, the isotope $^{235}_{92}U$ readily captures a neutron and fissions, *provided the neutron is a thermal neutron* (kinetic energy ≈ 0.04 eV or less). The probability of a thermal neutron causing $^{235}_{92}U$ to fission is about 500 times greater than the probability for a neutron whose energy is relatively high—say, 1 MeV. Thermal neutrons can also be used to fission other nuclei, such as plutonium $^{239}_{94}Pu$. Conceptual Example 5 deals with one of the reasons why thermal neutrons are useful for inducing nuclear fission.

 Conceptual Example 5 Neutrons Versus Protons or Alpha Particles

A thermal neutron has a relatively small amount of kinetic energy but, nevertheless, can penetrate a nucleus. To penetrate the same nucleus, would a proton or an α particle need **(a)** the same small amount of kinetic energy as the neutron needs, **(b)** a much larger amount of kinetic energy than the neutron needs, or **(c)** much less kinetic energy than the neutron needs?

Reasoning To penetrate a nucleus, a particle such as a neutron, a proton, or an α particle must have enough kinetic energy to do the work of overcoming any repulsive force that it encounters. A repulsive force can arise because protons in the nucleus are electrically charged. Since a neutron is electrically neutral, however, it encounters no electrostatic force of repulsion as it approaches the nuclear protons, and, hence, needs relatively little energy to reach the nucleus.

Answers (a) and (c) are incorrect. A proton and an α particle each carry a positive charge, so that each would encounter an electrostatic force of repulsion as it approached the nuclear protons, a force that a thermal neutron does not encounter. These answers ignore the additional kinetic energy that a proton or an α particle would need to overcome the repulsion.

Answer (b) is correct. A proton and an α particle, each being positively charged, would each require much more kinetic energy than a neutron does, in order to overcome the electrostatic force of repulsion from the nuclear protons. It is true that each would also experience the attractive strong nuclear force from the nuclear protons and neutrons. However, this force has an extremely short range of action and, therefore, would came into play only after an impinging particle reached the target nucleus. In comparison, the electrostatic force has a long range of action and is encountered throughout the entire journey to the target.

Related Homework: *Problem 45*

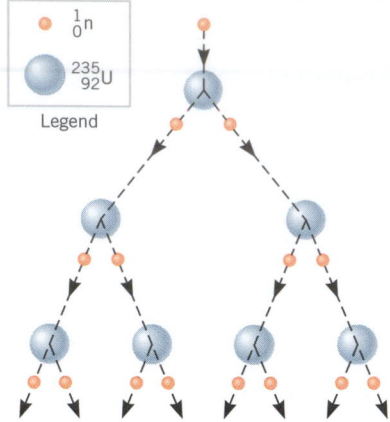

Figure 32.5 A chain reaction. For clarity, it is assumed that each fission generates two neutrons (2.5 neutrons are actually liberated on the average). The fission fragments are not shown.

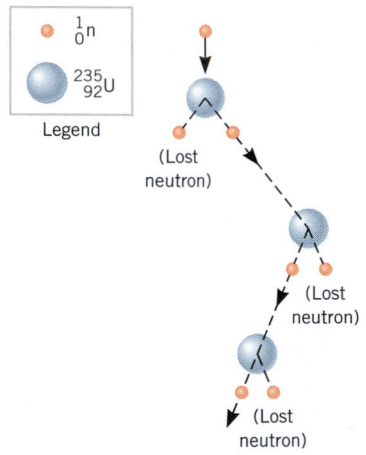

Figure 32.6 In a controlled chain reaction, only one neutron, on average, from each fission event causes another nucleus to fission. The "lost neutron" is absorbed by a material (not shown) that does not fission. As a result, energy is released at a steady or controlled rate.

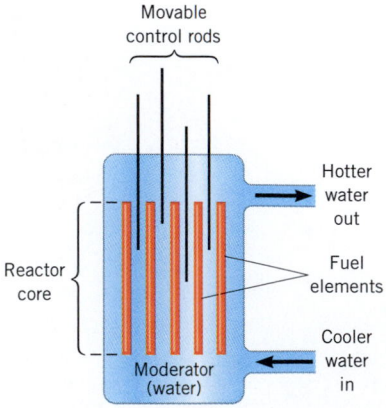

Figure 32.7 A nuclear reactor consists of fuel elements, control rods, and a moderator (in this case, water).

The fact that the uranium fission reaction releases 2.5 neutrons, on the average, makes it possible for a self-sustaining series of fissions to occur. As Figure 32.5 illustrates, each neutron released can initiate another fission event, resulting in the emission of still more neutrons, followed by more fissions, and so on. A ***chain reaction*** is a series of nuclear fissions whereby some of the neutrons produced by each fission cause additional fissions. During an uncontrolled chain reaction, it would not be unusual for the number of fissions to increase a thousandfold within a few millionths of a second. With an average energy of about 200 MeV being released per fission, an uncontrolled chain reaction can generate an incredible amount of energy in a very short time, as happens in an atomic bomb (which is actually a *nuclear* bomb).

By using a material that can absorb neutrons without fissioning, it is possible to limit the number of neutrons in the environment of the fissile nuclei. In this way, a condition can be established whereby each fission event contributes, on average, only *one neutron* that fissions another nucleus (see Figure 32.6). Thus, the chain reaction and the rate of energy production are controlled. The controlled-fission chain reaction is the principle behind nuclear reactors used in the commercial generation of electric power.

✓ **CHECK YOUR UNDERSTANDING**

(*The answers are given at the end of the book.*)

6. When the nucleus of a certain element absorbs a thermal neutron, fission usually occurs, with the production of nuclear fragments and a number N of neutrons. However, in a collection of atoms of this element, a small fraction of the thermal neutrons absorbed by the nuclei does not lead to fission. For which one of the following values of N can a self-sustaining chain reaction *not* be produced using this element?   **(a)** $N = 4$ **(b)** $N = 3$ **(c)** $N = 2$ **(d)** $N = 1$

7. Thermal neutrons, thermal protons, and thermal electrons all have the same kinetic energy of about 0.04 eV. Rank the speeds of these particles in descending order (greatest speed first).

8. Would a release of energy accompany the fission of a nucleus of nucleon number 25 into two fragments of about equal mass? Consult Figure 31.5 as needed.

32.4 **NUCLEAR REACTORS**

The physics of nuclear reactors. A nuclear reactor is a type of furnace in which energy is generated by a controlled-fission chain reaction. The first nuclear reactor was built by Enrico Fermi in 1942, on the floor of a squash court under the west stands of Stagg Field at the University of Chicago. Today, there are a number of kinds and sizes of reactors, and many have the same three basic components: fuel elements, a neutron moderator, and control rods. Figure 32.7 illustrates these components.

The ***fuel elements*** contain the fissile fuel and, for example, may be thin rods about 1 cm in diameter. In a large power reactor there may be thousands of fuel elements placed close together, and the entire region of fuel elements is known as the ***reactor core***. Uranium $^{235}_{92}\text{U}$ is a common reactor fuel. Since the natural abundance of this isotope is only about 0.7%, there are special uranium-enrichment plants to increase the percentage. Most commercial reactors use uranium in which the amount of $^{235}_{92}\text{U}$ has been enriched to about 3%.

Whereas neutrons with energies of about 0.04 eV (or less) readily fission $^{235}_{92}\text{U}$, the neutrons released during the fission process have significantly greater energies of several MeV or so. Consequently, a nuclear reactor must contain some type of material that will decrease or moderate the speed of such energetic neutrons so they can readily fission additional $^{235}_{92}\text{U}$ nuclei. The material that slows down the neutrons is called a ***moderator.*** One commonly used moderator is water. When an energetic neutron leaves a fuel element, the neutron enters the surrounding water and collides with water molecules. With each collision, the neutron loses an appreciable fraction of its energy and slows down. Once slowed down to thermal energy by the moderator, a process that takes less than 10^{-3} s, the neutron is capable of initiating a fission event upon reentering a fuel element.

If the output power from a reactor is to remain constant, only one neutron from each fission event must trigger a new fission, as Figure 32.6 illustrates. When each fission leads

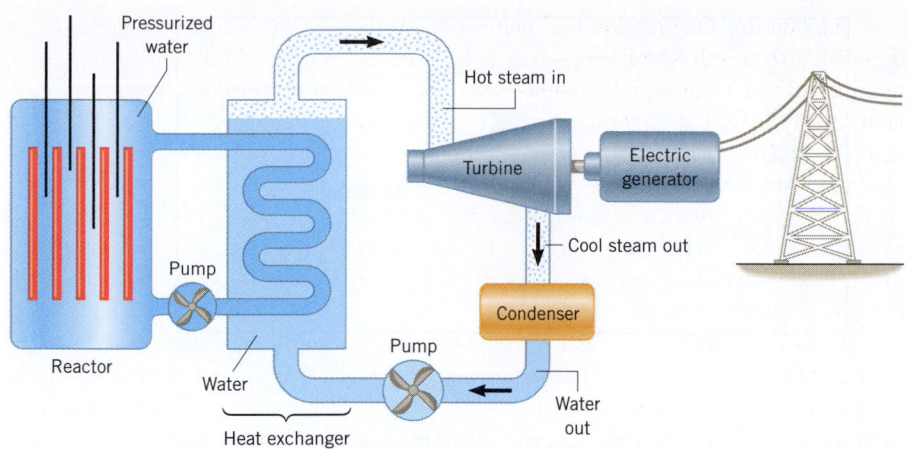

Figure 32.8 Diagram of a nuclear power plant that uses a pressurized water reactor.

to one additional fission—no more or no less—the reactor is said to be *critical*. A reactor normally operates in a critical condition, because then it produces a steady output of energy. The reactor is *subcritical* when, on average, the neutrons from each fission trigger *less than one* subsequent fission. In a subcritical reactor, the chain reaction is not self-sustaining and eventually dies out. When the neutrons from each fission trigger *more than one* additional fission, the reactor is *supercritical*. During a supercritical condition, the energy released by a reactor increases. If left unchecked, the increasing energy can lead to a partial or total meltdown of the reactor core, with the possible release of radioactive material into the environment.

Clearly, a control mechanism is needed to keep the reactor in its normal, or critical, state. This control is accomplished by a number of **control rods** that can be moved into and out of the reactor core (see Figure 32.7). The control rods contain an element, such as boron or cadmium, that readily absorbs neutrons without fissioning. If the reactor becomes supercritical, the control rods are automatically moved farther into the core to absorb the excess neutrons causing the condition. In response, the reactor returns to its critical state. Conversely, if the reactor becomes subcritical, the control rods are partially withdrawn from the core. Fewer neutrons are absorbed, more neutrons are available for fission, and the reactor again returns to its critical state.

Figure 32.8 illustrates a pressurized water reactor. In such a reactor, the heat generated within the fuel rods is carried away by water that surrounds the rods. To remove as much heat as possible, the water temperature is allowed to rise to a high value (about 300 °C). To prevent boiling, which occurs at 100 °C at 1 atmosphere of pressure, the water is pressurized in excess of 150 atmospheres. The hot water is pumped through a heat exchanger, where heat is transferred to water flowing in a second, closed system. The heat transferred to the second system produces steam that drives a turbine. The turbine is coupled to an electric generator, whose output electric power is delivered to consumers via high-voltage transmission lines. After exiting the turbine, the steam is condensed back into water that is returned to the heat exchanger.

32.5 NUCLEAR FUSION

In Example 4 in Section 32.3, the plot of binding energy per nucleon in Figure 31.5 is used to estimate the amount of energy released in the fission process. As summarized in Figure 32.9, the massive nuclei at the right end of the curve have a binding energy of about 7.6 MeV per nucleon. The less massive fission fragments are near the center of the curve and have a binding energy of approximately 8.5 MeV per nucleon. The energy released per nucleon by fission is the difference between these two values, or about 0.9 MeV per nucleon.

A glance at the far left end of the diagram in Figure 32.9 suggests another means of generating energy. Two nuclei with very low mass and relatively small binding energies per nucleon could be combined or "fused" into a single, more massive nucleus that has a greater binding energy per nucleon. This process is called **nuclear fusion**. A substantial amount of energy can be released during a fusion reaction, as Example 6 shows.

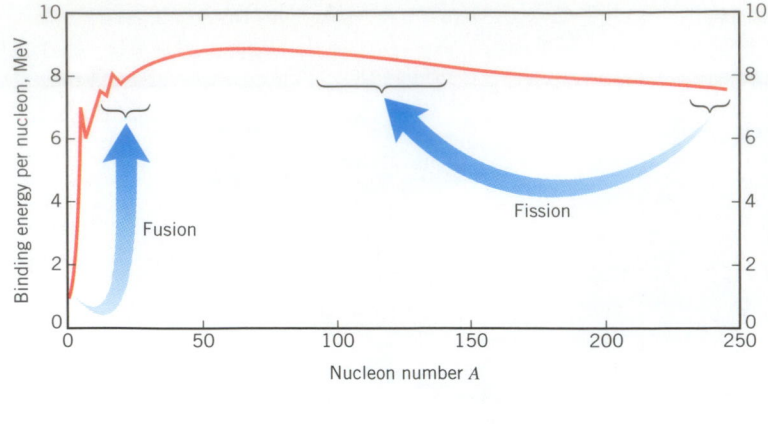

Figure 32.9 When fission occurs, a massive nucleus divides into two fragments whose binding energy per nucleon is greater than that of the original nucleus. When fusion occurs, two low-mass nuclei combine to form a more massive nucleus whose binding energy per nucleon is greater than that of the original nuclei.

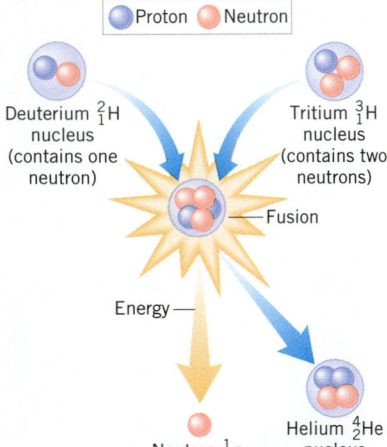

Figure 32.10 Deuterium and tritium are fused together to form a helium nucleus ($_2^4$He). The result is the release of an enormous amount of energy, mainly carried by a single high-energy neutron ($_0^1$n).

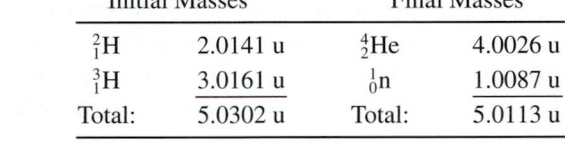 **Example 6** The Energy Released During Nuclear Fusion

Two isotopes of hydrogen, $_1^2$H (deuterium, D) and $_1^3$H (tritium, T), fuse to form $_2^4$He and a neutron according to the following reaction:

$$_1^2\text{H} + _1^3\text{H} \longrightarrow _2^4\text{He} + _0^1\text{n}$$

Determine the energy that is released by this particular fusion reaction, which is illustrated in Figure 32.10.

Reasoning Energy is released, so the total mass of the final nuclei is less than the total mass of the initial nuclei. To determine the energy released, we find the amount (in atomic mass units u) by which the total mass has decreased. Then, we use the fact that 1 u is equivalent to 931.5 MeV of energy, as determined in Section 31.3. This approach is the same as that used in Section 31.4 for radioactive decay.

Solution The masses of the initial and final nuclei in this reaction, as well as the mass of the neutron, are

Initial Masses		Final Masses	
$_1^2$H	2.0141 u	$_2^4$He	4.0026 u
$_1^3$H	3.0161 u	$_0^1$n	1.0087 u
Total:	5.0302 u	Total:	5.0113 u

The decrease in mass, or the mass defect, is $\Delta m = 5.0302$ u $-$ 5.0113 u $= 0.0189$ u. Since 1 u is equivalent to 931.5 MeV, the energy released is $\boxed{17.6 \text{ MeV}}$.

The deuterium nucleus contains 2 nucleons, and the tritium nucleus contains 3. Thus, there are 5 nucleons that participate in the fusion, so the energy released per nucleon is about 3.5 MeV. This energy per nucleon is greater than the energy released in a fission process ($\approx$0.9 MeV per nucleon).

Because fusion reactions release so much energy, there is considerable interest in fusion reactors, although to date no commercial units have been constructed. The difficulties in building a fusion reactor arise mainly because the two low-mass nuclei must be brought sufficiently near each other so that the short-range strong nuclear force can pull them together, leading to fusion. However, each nucleus has a positive charge and repels the other electrically. For the nuclei to get sufficiently close in the presence of the repulsive electric force, they must have large kinetic energies, and hence large temperatures, to start with. For example, a temperature of around a hundred million °C is needed to start the deuterium–tritium reaction discussed in Example 6.

Reactions that require such extremely high temperatures are called ***thermonuclear reactions.*** The most important thermonuclear reactions occur in stars, such as our own sun. The energy radiated by the sun comes from deep within its core, where the temperature is high enough to initiate the fusion process. One group of reactions thought to occur in the sun is the *proton–proton* cycle, which is a series of reactions whereby six protons form a helium nucleus, two positrons, two γ rays, two protons, and two neutrinos. The energy released by the proton–proton cycle is about 25 MeV (see Problem 38).

Human-made fusion reactions have been carried out in a fusion-type nuclear bomb—commonly called a hydrogen bomb. In a hydrogen bomb, the fusion reaction is ignited by a fission bomb using uranium or plutonium. The temperature produced by the fission bomb is sufficiently high to initiate a thermonuclear reaction where, for example, hydrogen isotopes are fused into helium, releasing even more energy. For fusion to be useful as a commercial energy source, the energy must be released in a steady, controlled manner—unlike the energy in a bomb. To date, scientists have not succeeded in constructing a fusion device that produces more energy on a continual basis than is expended in operating the device. A fusion device uses a high temperature to start a reaction, and under such a condition, all the atoms are completely ionized to form a *plasma* (a gas composed of charged particles, like $_1^2H^+$ and e^-). The problem is to confine the hot plasma for a long enough time so that collisions among the ions can lead to fusion.

One ingenious method of confining the plasma is called *magnetic confinement* because it uses a magnetic field to contain and compress the charges in the plasma. Charges moving in the magnetic field are subject to magnetic forces. As the forces increase, the associated pressure builds, and the temperature rises. The gas becomes a superheated plasma, ultimately fusing when the pressure and temperature are high enough.

The physics of nuclear fusion using magnetic confinement.

Another type of confinement scheme, known as *inertial confinement,* is also being developed. Tiny, solid pellets of fuel are dropped into a container. As each pellet reaches the center of the container, a number of high-intensity laser beams strike the pellet simultaneously. The heating causes the exterior of the pellet to vaporize almost instantaneously. However, the inertia of the vaporized atoms keeps them from expanding outward as fast as the vapor is being formed. As a result, high pressures, high densities, and high temperatures are achieved at the center of the pellet, thus causing fusion. A variant of inertial confinement fusion, called "Z pinch," is under development at Sandia National Laboratories in New Mexico. This device would also implode tiny fuel pellets, but without using lasers. Instead, scientists are using a cylindrical array of fine tungsten wires that are connected to a gigantic capacitor. When the capacitor is discharged, a huge current is sent through the wires. The heated wires vaporize almost instantly, generating a hot gas of ions, or plasma. The plasma is driven inward upon itself by the huge magnetic field produced by the current. The compressed plasma becomes superhot and generates a gigantic X-ray pulse. This pulse, it is hoped, would implode the solid fuel pellets to temperatures and pressures at which fusion would occur.

The physics of nuclear fusion using inertial confinement.

When compared to fission, fusion has some attractive features as an energy source. As we have seen in Example 6, fusion yields more energy per nucleon of fuel than fission does. Moreover, one type of fuel, $_1^2H$ (deuterium), is found in the waters of the oceans and is plentiful, cheap, and relatively easy to separate from the common $_1^1H$ isotope of hydrogen. Fissile materials like naturally occurring uranium $_{92}^{235}U$ are much less available, and supplies could be depleted within a century or two. However, the commercial use of fusion to provide cheap energy remains in the future.

✓ CHECK YOUR UNDERSTANDING

(The answers are given at the end of the book.)

9. Which one or more of the following statements correctly describe differences between fission and fusion?   **(a)** Fission involves the combining of low-mass nuclei to form a more massive nucleus, whereas fusion involves the splitting of a massive nucleus into less massive fragments.   **(b)** Fission involves the splitting of a massive nucleus into less massive fragments, whereas fusion involves the combining of low-mass nuclei to form a more massive nucleus.   **(c)** More energy per nucleon is released when a fission event occurs than when a fusion event occurs.   **(d)** Less energy per nucleon is released when a fission event occurs than when a fusion event occurs.

10. Which one or more of the following statements concerning fission and fusion are true? **(a)** Both fission and fusion reactions are characterized by a mass defect.   **(b)** Both fission and fusion reactions always obey the conservation laws of physics.   **(c)** Both fission and fusion take advantage of the fact that the binding energy per nucleon varies with the nucleon number of the nucleus.

11. Would the fusion of two nuclei, each with a nucleon number of 60, release energy? Consult Figure 32.9 as needed.

ELEMENTARY PARTICLES

32.6

SETTING THE STAGE

By 1932 the electron, the proton, and the neutron had been discovered and were thought to be nature's three *elementary particles,* in the sense that they were the basic building blocks from which all matter is constructed. Experimental evidence obtained since then, however, shows that several hundred additional particles exist, and scientists no longer believe that the proton and the neutron are elementary particles.

Most of these new particles have masses greater than the electron's mass, and many are more massive than protons or neutrons. Most of the new particles are unstable and decay in times between about 10^{-6} and 10^{-23} s.

Often, new particles are produced by accelerating protons or electrons to high energies and letting them collide with a target nucleus. For example, Figure 32.11 shows a collision between an energetic proton and a stationary proton. If the incoming proton has sufficient energy, the collision produces an entirely new particle, the *neutral pion* (π^0). The π^0 particle lives for only about 0.8×10^{-16} s before it decays into two γ-ray photons. Since the pion did not exist before the collision, it was created from part of the incident proton's energy. Because a new particle such as the neutral pion is often created from energy, it is customary to report the mass of the particle in terms of its equivalent *rest energy* (see Equation 28.5). Often, energy units of MeV are used. For instance, detailed analyses of experiments reveal that the mass of the π^0 particle is equivalent to a rest energy of 135.0 MeV. For comparison, the more massive proton has a rest energy of 938.3 MeV. Analyses of experiments also provide the electric charge and other properties of particles created in high-energy collisions. In the limited space available here, it is not possible to describe all the new particles that have been found. However, we will highlight some of the more significant discoveries.

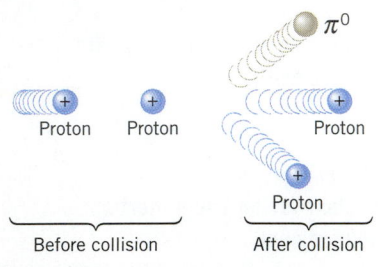

Figure 32.11 When an energetic proton collides with a stationary proton, a neutral pion (π^0) is produced. Part of the energy of the incident proton goes into creating the pion.

NEUTRINOS

In 1930, Wolfgang Pauli suggested that a particle called the *neutrino* (now known as the "electron neutrino") should accompany the β decay of a radioactive nucleus. As Section 31.5 discusses, the neutrino has no electric charge, has a very small mass (a tiny fraction of the mass of an electron), and travels at speeds approaching (but less than) the speed of light. Neutrinos were finally discovered in 1956. Today, neutrinos are created in abundance in nuclear reactors and particle accelerators and are thought to be plentiful in the universe.

POSITRONS AND ANTIPARTICLES

The year 1932 saw the discovery of the *positron* (a contraction for "positive electron"). The positron has the same mass as the electron but carries an opposite charge of $+e$. A collision between a positron and an electron is likely to annihilate both particles, converting them into electromagnetic energy in the form of γ rays. For this reason, positrons never coexist with ordinary matter for any appreciable length of time. The mutual annihilation of a positron and an electron lies at the heart of an important medical diagnostic technique, as Conceptual Example 7 discusses.

The physics of PET scanning.

 Conceptual Example 7 Positron Emission Tomography

Certain radioactive isotopes decay by positron emission—for example, oxygen $^{15}_{8}\text{O}$. In the medical diagnostic technique known as PET scanning (positron emission tomography), such isotopes are injected into the body, where they collect at specific sites. A positron ($^{0}_{1}\text{e}$) emitted by the decaying isotope immediately encounters an electron ($^{0}_{-1}\text{e}$) in the body tissue, and the resulting mutual annihilation produces two γ-ray photons ($^{0}_{1}\text{e} + ^{0}_{-1}\text{e} \rightarrow \gamma + \gamma$), which are detected by devices mounted on a ring around the patient. As Figure 32.12*a* shows, the two photons strike oppositely positioned detectors and, in so doing, reveal the line on which the annihilation occurred. Such information leads to a computer-generated image that can be useful in diagnosing

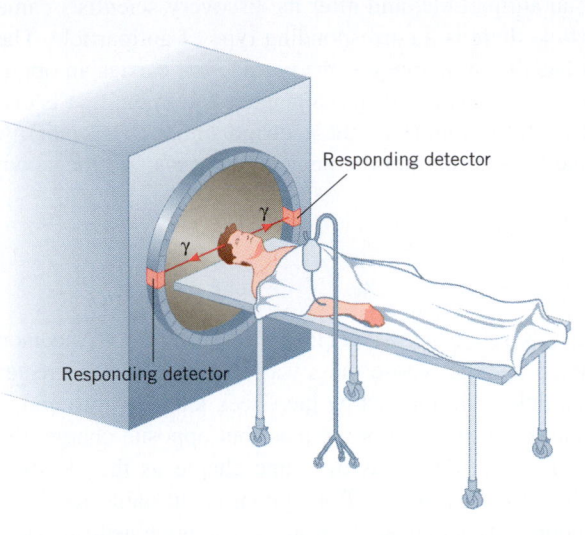

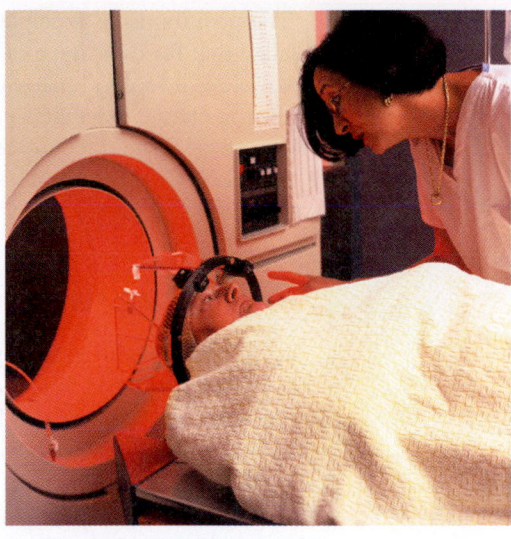

(a)   (b)

abnormalities at the site where the radioactive isotope collects (see Figure 32.13). Which conservation principle accounts for the fact that the photons strike oppositely positioned detectors, the principle of conservation of **(a)** linear momentum or **(b)** energy?

Reasoning Momentum is a vector concept and, therefore, has a direction. The momentum-conservation principle states that the total linear momentum of an isolated system remains constant (see Section 7.2). An isolated system is one on which no net external force acts. Energy is not a vector concept and has no direction associated with it. The energy-conservation principle states that energy can neither be created nor destroyed, but can only be converted from one form to another (see Section 6.8).

Answer (b) is incorrect. Energy is not a vector and has no direction associated with it. Therefore, it cannot, by itself, reveal the directional line along which the γ-ray photons were emitted.

Answer (a) is correct. The positron and the electron constitute an isolated system, so that momentum conservation applies. They do exert electrostatic forces on one another, since they carry electric charges. However, these forces are internal, not external, forces and cannot change the total linear momentum of the two-particle system. The total photon momentum, then, must equal the total momentum of the positron and the electron, which is nearly zero, to the extent that these particles have much less momentum than the photons do. With a total linear momentum of zero, the momentum vectors of the photons must point in opposite directions. Thus, the two photons depart from the annihilation site heading toward oppositely located detectors.

Related Homework: *Problem 44*

Figure 32.12 (*a*) In positron emission tomography, or PET scanning, a radioactive isotope is injected into the body. The isotope decays by emitting a positron, which annihilates an electron in the body tissue, producing two γ-ray photons. These photons strike detectors mounted on opposite sides of a ring that surrounds the patient. (*b*) A nurse is talking to a patient about to undergo a PET scan of the brain. (© CC Studio/ SPL/Photo Researchers, Inc.)

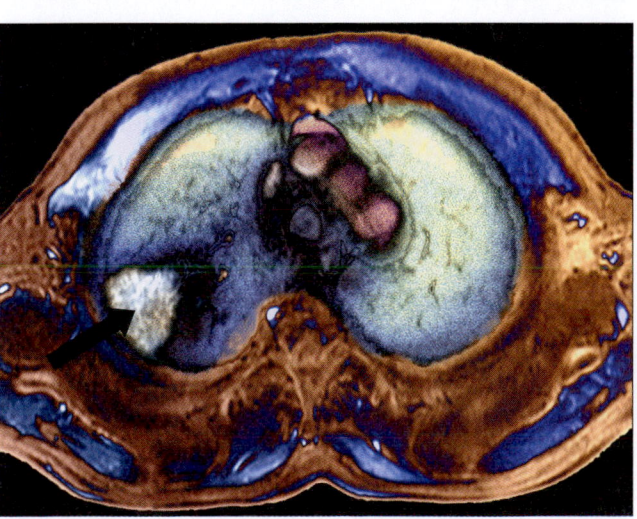

Figure 32.13 Positron emission tomography (PET) provides an important medical diagnostic technique. This PET scan of the chest, axial section, shows cancer in a lung (see arrow). (© ISM/Phototake)

The positron is an example of an antiparticle, and after its discovery scientists came to realize that for every type of particle there is a corresponding type of antiparticle. The antiparticle is a form of matter that has the same mass as the particle but carries an opposite electric charge (e.g., the electron–positron pair) or has a magnetic moment that is oriented in an opposite direction relative to the spin (e.g., the neutrino–antineutrino pair). A few electrically neutral particles, like the photon and the neutral pion (π^0), are their own antiparticles.

MUONS AND PIONS

In 1937, the American physicists S. H. Neddermeyer (1907–1988) and C. D. Anderson (1905–1991) discovered a new charged particle whose mass was about 207 times greater than the mass of the electron. The particle is designated by the Greek letter μ (mu) and is known as a *muon*. There are two muons that have the same mass but opposite charge: the particle μ^- and its antiparticle μ^+. The μ^- muon has the same charge as the electron, whereas the μ^+ muon has the same charge as the positron. Both muons are unstable, and have a lifetime of 2.2×10^{-6} s. The μ^- muon decays into an electron (β^-), a muon neutrino (ν_μ), and an electron antineutrino ($\bar{\nu}_e$), according to the following reaction:

$$\mu^- \longrightarrow \beta^- + \nu_\mu + \bar{\nu}_e$$

The μ^+ muon decays into a positron (β^+), a muon antineutrino ($\bar{\nu}_\mu$), and an electron neutrino (ν_e):

$$\mu^+ \longrightarrow \beta^+ + \bar{\nu}_\mu + \nu_e$$

Muons interact with protons and neutrons via the weak nuclear force (see Section 31.5).

The Japanese physicist Hideki Yukawa (1907–1981) predicted in 1935 that *pions* exist, but they were not discovered until 1947. Pions come in three varieties: one that is positively charged, the negatively charged antiparticle with the same mass, and the neutral pion, mentioned earlier, which is its own antiparticle. The symbols for these pions are, respectively, π^+, π^-, and π^0. The charged pions are unstable and have a lifetime of 2.6×10^{-8} s. The decay of a charged pion almost always produces a muon:

$$\pi^- \longrightarrow \mu^- + \bar{\nu}_\mu$$
$$\pi^+ \longrightarrow \mu^+ + \nu_\mu$$

As mentioned earlier, the neutral pion π^0 is also unstable and decays into two γ-ray photons, the lifetime being 0.8×10^{-16} s. The pions are of great interest because, unlike the muons, the pions interact with protons and neutrons via the strong nuclear force.

CLASSIFICATION OF PARTICLES

It is useful to group the known particles into three families—the bosons, the leptons, and the hadrons—as Table 32.3 summarizes. The ***boson family*** is composed of particles (all are types of bosons) that play central roles in nature's three fundamental forces (see Section 4.6). The photon is associated with the electromagnetic force, which is one manifestation of the electroweak force. The W^-, W^+, and Z^0 particles are associated with the weak nuclear force, which is the other manifestation of the electroweak force. The gluons are associated with the strong nuclear force, whereas the graviton is thought to be associated with the gravitational force.

The ***lepton family*** consists of particles that interact by means of the *weak nuclear force*. Leptons can also exert gravitational and (if the leptons are charged) electromagnetic forces on other particles. The four better-known leptons are the electron, the muon, the electron neutrino ν_e, and the muon neutrino ν_μ. Table 32.3 lists these particles together with their antiparticles. Two other leptons have also been discovered, the tau particle (τ) and its neutrino (ν_τ), bringing the number of particles in the lepton family to six.

Table 32.3 Some Particles and Their Properties

Family	Particle	Particle Symbol	Antiparticle Symbol	Rest Energy (MeV)	Lifetime (s)
Boson[a]	Photon	γ	Self[b]	0	Stable
	$W^{\pm}$	W^-	W^+	8.04×10^4	3×10^{-25}
	Z^0	Z^0	Self[b]	9.12×10^4	3×10^{-25}
	Gluons[c]	g	—	0	—
	Graviton[c]	—	—	0	—
Lepton	Electron	e^- or β^-	e^+ or β^+	0.511	Stable
	Muon	μ^-	μ^+	105.7	2.2×10^{-6}
	Tau	τ^-	τ^+	1777	2.9×10^{-13}
	Electron neutrino	ν_e	$\bar{\nu}_e$	$<2 \times 10^{-6}$	Stable
	Muon neutrino	ν_μ	$\bar{\nu}_\mu$	<0.19	Stable
	Tau neutrino	ν_τ	$\bar{\nu}_\tau$	<18.2	Stable
Hadron					
Mesons					
	Pion	π^+	π^-	139.6	2.6×10^{-8}
		π^0	Self[b]	135.0	8.4×10^{-17}
	Kaon	K^+	K^-	493.7	1.2×10^{-8}
		K^0_S	$\bar{K}^0_S$	497.7	8.9×10^{-11}
		K^0_L	$\bar{K}^0_L$	497.7	5.2×10^{-8}
	Eta	η^0	Self[b]	547.3	$<10^{-18}$
Baryons					
	Proton	p	$\bar{p}$	938.3	Stable
	Neutron	n	$\bar{n}$	939.6	886
	Lambda	Λ^0	$\bar{\Lambda}^0$	1116	2.6×10^{-10}
	Sigma	Σ^+	$\bar{\Sigma}^-$	1189	8.0×10^{-11}
		Σ^0	$\bar{\Sigma}^0$	1193	7.4×10^{-20}
		Σ^-	$\bar{\Sigma}^+$	1197	1.5×10^{-10}
	Omega	Ω^-	Ω^+	1672	8.2×10^{-11}

[a] The particles in this family are types of bosons that are associated with (or mediate) nature's fundamental forces.
[b] The particle is its own antiparticle.
[c] Free gluons and the graviton have not been observed experimentally.

The **hadron family** contains the particles that interact by means of *both the strong nuclear force and the weak nuclear force.* Hadrons can also interact by gravitational and electromagnetic forces, but at short distances ($\leq 10^{-15}$ m) the strong nuclear force dominates. Among the hadrons are the proton, the neutron, and the pions. As Table 32.3 indicates, most hadrons are short-lived. The hadrons are subdivided into two groups, the **mesons** and the **baryons,** for a reason that will be discussed in connection with the idea of quarks.

QUARKS

As more and more hadrons were discovered, it became clear that they were not all elementary particles. The suggestion was made that the hadrons are made up of smaller, more elementary particles called **quarks.** In 1963, a quark theory was advanced independently by M. Gell-Mann (1929–) and G. Zweig (1937–). The theory proposed that there are three quarks and three corresponding antiquarks, and that hadrons are constructed from combinations of these. Thus, the quarks are elevated to the status of elementary particles for the hadron family. The particles in the lepton family are considered to be elementary, and as such they are not composed of quarks.

Table 32.4 Quarks and Antiquarks

| Name | Quarks | | Antiquarks | |
	Symbol	Charge	Symbol	Charge
Up	u	$+\frac{2}{3}e$	$\bar{u}$	$-\frac{2}{3}e$
Down	d	$-\frac{1}{3}e$	$\bar{d}$	$+\frac{1}{3}e$
Strange	s	$-\frac{1}{3}e$	$\bar{s}$	$+\frac{1}{3}e$
Charmed	c	$+\frac{2}{3}e$	$\bar{c}$	$-\frac{2}{3}e$
Top	t	$+\frac{2}{3}e$	$\bar{t}$	$-\frac{2}{3}e$
Bottom	b	$-\frac{1}{3}e$	$\bar{b}$	$+\frac{1}{3}e$

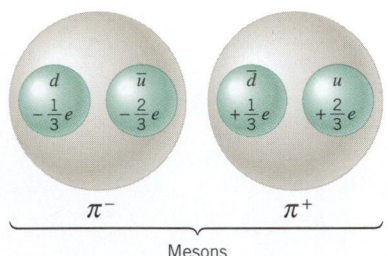

Mesons

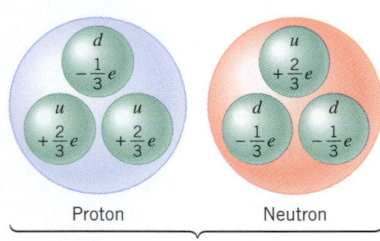

Baryons

Figure 32.14 According to the original quark model of hadrons, all mesons consist of a quark and an antiquark, whereas baryons contain three quarks.

The three quarks were named *up* (*u*), *down* (*d*), and *strange* (*s*), and were assumed to have, respectively, fractional charges of $+\frac{2}{3}e$, $-\frac{1}{3}e$, and $-\frac{1}{3}e$. In other words, a quark possesses a charge magnitude smaller than that of an electron, which has a charge of $-e$. Table 32.4 lists the symbols and electric charges of these quarks and the corresponding antiquarks. Experimentally, quarks should be recognizable by their fractional charges, but in spite of an extensive search for them, free quarks have never been found.

According to the original quark theory, the mesons are different from the baryons, because each meson consists of only two quarks—a quark and an antiquark—whereas a baryon contains three quarks. For instance, the π^- pion (a meson) is composed of a d quark and a $\bar{u}$ antiquark, $\pi^- = d + \bar{u}$, as Figure 32.14 shows. These two quarks combine to give the π^- pion a net charge of $-e$. Similarly, the π^+ pion is a combination of the $\bar{d}$ and u quarks, $\pi^+ = \bar{d} + u$. In contrast, protons and neutrons, being baryons, consist of three quarks. A proton contains the combination $d + u + u$, and a neutron contains the combination $d + d + u$ (see Figure 32.14). These groups of three quarks give the correct charges for the proton and neutron.

The original quark model was extremely successful in predicting not only the correct charges for the hadrons, but other properties as well. However, in 1974 a new particle, the J/ψ meson, was discovered. This meson has a rest energy of 3097 MeV, much larger than the rest energies of other known mesons. The existence of the J/ψ meson could be explained only if a new quark–antiquark pair existed; this new quark was called *charmed* (*c*). With the discovery of more and more particles, it has been necessary to postulate a fifth and a sixth quark; their names are *top* (*t*) and *bottom* (*b*), although some scientists prefer to call these quarks *truth* and *beauty*. Today, there is firm evidence for all six quarks, each with its corresponding antiquark. All of the hundreds of known hadrons can be accounted for in terms of these six quarks and their antiquarks.

In addition to electric charge, quarks also have other properties. For example, each quark possesses a characteristic called **color**, for which there are three possibilities: blue, green, or red. The corresponding possibilities for the antiquarks are antiblue, antigreen, and antired. The use of the term "color" and the specific choices of blue, green, and red are arbitrary, for the visible colors of the electromagnetic spectrum have nothing to do with quark properties. The quark property of color, however, is important, because it brings the quark model into agreement with the Pauli exclusion principle and enables the model to account for experimental observations that are otherwise difficult to explain.

THE STANDARD MODEL

The various elementary particles that have been discovered can interact via one or more of the following four forces: the gravitational force, the strong nuclear force, the weak nuclear force, and the electromagnetic force. In particle physics, the phrase *"the standard model"* refers to the currently accepted explanation for the strong nuclear force, the weak nuclear force, and the electromagnetic force. In this model, the strong nuclear force between quarks is described in terms of the concept of color, and the theory is referred to as quantum chromodynamics. According to the standard model, the weak nuclear force

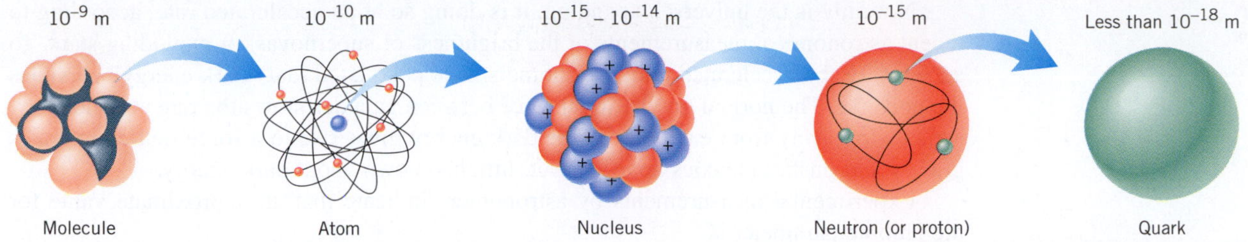

10^{-9} m	10^{-10} m	$10^{-15} - 10^{-14}$ m	10^{-15} m	Less than 10^{-18} m
Molecule	Atom	Nucleus	Neutron (or proton)	Quark

Figure 32.15 The current view of how matter is composed of basic units, starting with a molecule and ending with a quark. The approximate sizes of each unit are also listed.

and the electromagnetic force are separate manifestations of a single, even more fundamental, force, referred to as the electroweak force, as we have seen in Section 31.5.

In the standard model, our understanding of the building blocks of matter follows the hierarchical pattern illustrated in Figure 32.15. Molecules, such as water (H_2O) and glucose ($C_6H_{12}O_6$), are composed of atoms. Each atom consists of a nucleus that is surrounded by a cloud of electrons. The nucleus, in turn, is made up of protons and neutrons, which are composed of quarks.

✓ **CHECK YOUR UNDERSTANDING**

(*The answers are given at the end of the book.*)

12. Of the following particles, which ones are *not* composed of quarks and antiquarks? **(a)** A proton **(b)** An electron **(c)** A neutron **(d)** A neutrino

13. The sigma-minus particle Σ^- has a charge of $-e$. Which one of the following quark combinations corresponds to this particle? **(a)** *dds* **(b)** *ssu* **(c)** *uus*

32.7 COSMOLOGY

Cosmology is the study of the structure and evolution of the universe. In this study, both the very large and the very small aspects of the universe are important. Astronomers, for example, study stars located at enormous distances from the earth, up to billions of light-years away. In contrast, particle physicists focus their efforts on the very small elementary particles (10^{-18} m or smaller) that comprise matter. The synergy between the work of astronomers and that of particle physicists has led to significant advances in our understanding of the universe. Central to that understanding is the belief that the universe is expanding, and we begin by discussing the evidence that justifies this belief.

THE EXPANDING UNIVERSE AND THE BIG BANG

The idea that the universe is expanding originated with the astronomer Edwin P. Hubble (1889–1953). He found that light reaching the earth from distant galaxies is Doppler-shifted toward greater wavelengths—that is, toward the red end of the visible spectrum. As Section 24.5 discusses, this type of Doppler shift results when the observer and the source of the light are moving away from each other. The speed at which a galaxy is receding from the earth can be determined from the measured Doppler shift in wavelength. Hubble found that a galaxy located at a distance d from the earth recedes from the earth at a speed v given by

Hubble's law
$$v = Hd \tag{32.5}$$

where H is a constant known as the Hubble parameter. In other words, the recession speed is proportional to the distance d, so that more distant galaxies are moving away from the earth at greater speeds. Equation 32.5 is referred to as Hubble's law.

Hubble's picture of an expanding universe does not mean that the earth is at the center of the expansion. In fact, there is no literal center. Imagine a loaf of raisin bread expanding as it bakes. Each raisin moves away from every other raisin, without any single one acting as a center for the expansion. Galaxies in the universe behave in a similar fashion. Observers in other galaxies would see distant galaxies moving away, just as we do.

The physics of an expanding universe.

**The physics of
"dark energy."**

Not only is the universe expanding, it is doing so at an accelerated rate, according to recent astronomical measurements of the brightness of supernovas, or exploding stars. To account for the accelerated rate, astronomers have postulated that "dark energy" pervades the universe. The normal gravitational force between galaxies slows the rate at which they are moving away from each other. The dark energy gives rise to a force that counteracts gravity and pushes galaxies apart. As yet, little is known about dark energy.

Experimental measurements by astronomers indicate that an approximate value for the Hubble parameter is

$$H = 0.022 \frac{\text{m}}{\text{s} \cdot \text{light-year}}$$

The value for the Hubble parameter is believed to be accurate within 10%. Scientists are very interested in obtaining an accurate value for H because it can be related to an age for the universe, as the next example illustrates.

Example 8   An Age for the Universe

Determine an estimate of the age of the universe using Hubble's law.

Reasoning   Consider a galaxy currently located at a distance d from the earth. According to Hubble's law, this galaxy is moving away from us at a speed of $v = Hd$. At an earlier time, therefore, this galaxy must have been closer. We can imagine, in fact, that in the remote past the separation distance was relatively small and that the universe originated at such a time. To estimate the age of the universe, we calculate the time it has taken the galaxy to recede to its present position. For this purpose, time is simply distance divided by speed, or $t = d/v$.

Solution   Using Hubble's law and the fact that a distance of 1 light-year is 9.46×10^{15} m, we estimate the age of the universe to be

$$t = \frac{d}{v} = \frac{d}{Hd} = \frac{1}{H}$$

$$t = \frac{1}{0.022 \dfrac{\text{m}}{\text{s} \cdot \text{light-year}}} = \frac{1}{\left(0.022 \dfrac{\text{m}}{\text{s} \cdot \text{light-year}}\right)\left(\dfrac{1 \text{ light-year}}{9.46 \times 10^{15} \text{ m}}\right)}$$

$$= 4.3 \times 10^{17} \text{ s} \quad \text{or} \quad \boxed{1.4 \times 10^{10} \text{ yr}}$$

The idea presented in Example 8, that our galaxy and other galaxies in the universe were very close together at some earlier instant in time, lies at the heart of the **_Big Bang theory._** This theory postulates that the universe had a definite beginning in a cataclysmic event, sometimes called the primeval fireball. Dramatic evidence supporting the theory was discovered in 1965 by Arno A. Penzias (1933–) and Robert W. Wilson (1936–). Using a radio telescope, they discovered that the earth is being bathed in weak electromagnetic waves in the microwave region of the spectrum (wavelength = 7.35 cm, see Figure 24.9). They observed that the intensity of these waves is the same, no matter where in the sky they pointed their telescope, and concluded that the waves originated outside of our galaxy. This microwave background radiation, as it is called, represents radiation left over from the Big Bang and is a kind of cosmic afterglow. Subsequent measurements have confirmed the research of Penzias and Wilson and have shown that the microwave radiation is consistent with a perfect blackbody (see Sections 13.3 and 29.2) radiating at a temperature of 2.7 K, in agreement with theoretical analysis of the Big Bang. In 1978, Penzias and Wilson received a Nobel Prize for their discovery.

THE STANDARD MODEL FOR THE EVOLUTION OF THE UNIVERSE

Based on the recent experimental and theoretical research in particle physics, scientists have proposed an evolutionary sequence of events following the Big Bang. This sequence is known as the **_standard cosmological model_** and is illustrated in Figure 32.16.

Immediately following the Big Bang, the temperature of the universe was incredibly high, about 10^{32} K. During this initial period, the three fundamental forces (the gravitational force, the strong nuclear force, and the electroweak force) all behaved as a single

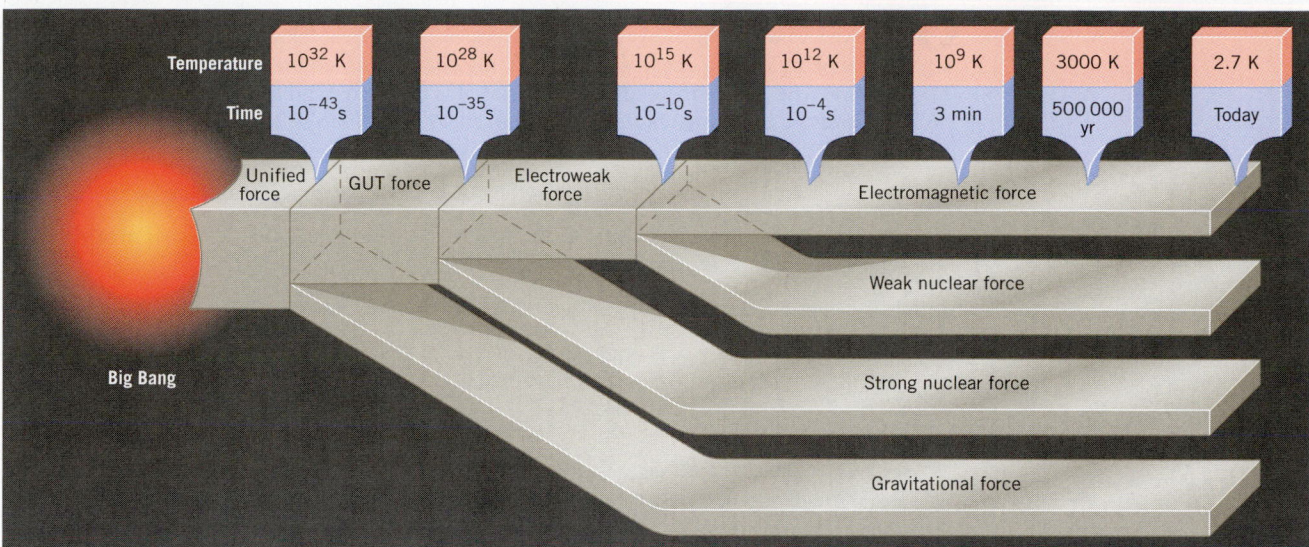

Figure 32.16 According to the standard cosmological model, the universe has evolved as illustrated here. In this model, the universe is presumed to have originated with a cataclysmic event known as the Big Bang. The times shown are those following this event.

unified force. Very quickly, in about 10^{-43} s, the gravitational force took on a separate identity all its own, as Figure 32.16 indicates. Meanwhile, the strong nuclear force and the electroweak force continued to act as a single force, which is sometimes referred to as the GUT force. GUT stands for the grand unified theory that presumably would explain such a force. Slightly later, at about 10^{-35} s after the Big Bang, the GUT force separated into the strong nuclear force and the electroweak force, the universe expanding and cooling somewhat to a temperature of roughly 10^{28} K (see Figure 32.16). From this point on, the strong nuclear force behaved as we know it today, while the electroweak force maintained its identity. In this scenario, note that the weak nuclear force and the electromagnetic force have not yet manifested themselves as separate entities. The disappearance of the electroweak force and the appearance of the weak nuclear force and the electromagnetic force eventually occurred at approximately 10^{-10} s after the Big Bang, when the temperature of the expanding universe had cooled to about 10^{15} K.

From the Big Bang up until the strong nuclear force separated from the GUT force at a time of 10^{-35} s, all particles of matter were similar, and there was no distinction between quarks and leptons. After this time, quarks and leptons became distinguishable. Eventually the quarks and antiquarks formed hadrons, such as protons and neutrons and their antiparticles. By a time of 10^{-4} s after the Big Bang, however, the temperature had cooled to approximately 10^{12} K, and the hadrons had mostly disappeared. Protons and neutrons survived only as a very small fraction of the total number of particles, the majority of which were leptons such as electrons, positrons, and neutrinos. Like most of the hadrons before them, most of the electrons and positrons eventually disappeared. However, they did leave behind a relatively small number of electrons to join the small number of protons and neutrons at a time of about 3 min following the Big Bang. At this time the temperature of the expanding universe had decreased to about 10^9 K, and small nuclei such as that of helium began forming. Later, when the universe was about 500 000 years old and the temperature had dropped to near 3000 K, hydrogen and helium atoms began forming. As the temperature decreased further, stars and galaxies formed, and today we find a temperature of 2.7 K characterizing the cosmic background radiation of the universe.

CONCEPTS & CALCULATIONS

32.8

When considering the biological effects of ionizing radiation, the concept of the biologically equivalent dose is especially important. Its importance lies in the fact that the biologically equivalent dose incorporates both the amount of energy per unit mass that is absorbed and the effectiveness of a particular type of radiation in producing a certain biological effect. Example 9 examines this concept and also reviews the notions of power (Section 6.7) and intensity (Section 16.7) of a wave.

⬇ **Concepts & Calculations Example 9** The Biologically Equivalent Dose

A patient is being given a chest X-ray. The X-ray beam is turned on for 0.20 s, and its intensity is 0.40 W/m². The area of the chest being exposed is 0.072 m², and the radiation is absorbed by 3.6 kg of tissue. The relative biological effectiveness (RBE) of the X-rays for this tissue is 1.1. Determine the biologically equivalent dose received by the patient.

Concept Questions and Answers How is the power of the beam related to the beam intensity?

> *Answer* The power P is equal to the intensity I times the cross-sectional area A of the beam, as indicated by Equation 16.8 ($P = IA$).

How is the energy absorbed by the tissue related to the power of the beam?

> *Answer* The energy E is equal to the product of the power P delivered by the beam and the time t of exposure, according to Equation 6.10b ($E = Pt$).

What is the absorbed dose?

> *Answer* The absorbed dose is defined by Equation 32.2 as the energy E absorbed by the tissue divided by the tissue's mass m, or Absorbed dose = E/m. The unit of the absorbed dose is the gray (Gy).

How is the biologically equivalent dose related to the absorbed dose?

> *Answer* The biologically equivalent dose is equal to the product of the absorbed dose and the relative biological effectiveness (RBE), as indicated by Equation 32.4. However, the absorbed dose must be expressed in rad units, not in gray units. The relation between the two units is 1 rad = 0.01 Gy.

Solution The biologically equivalent dose is given by Equation 32.4 as

$$\text{Biologically equivalent dose} = [\text{Absorbed dose (in rads)}](\text{RBE})$$

At this point we need to convert the absorbed dose from Gy units to rad units:

$$\text{Absorbed dose (in rads)} = [\text{Absorbed dose (in Gy)}]\left(\frac{1 \text{ rad}}{0.01 \text{ Gy}}\right)$$

The biologically equivalent dose then becomes

$$\text{Biologically equivalent dose} = [\text{Absorbed dose (in Gy)}]\left(\frac{1 \text{ rad}}{0.01 \text{ Gy}}\right)(\text{RBE})$$

Equation 32.2 indicates that the absorbed dose (in Gy) is equal to the energy E divided by the mass m of the tissue, so that we have

$$\text{Biologically equivalent dose} = \left(\frac{E}{m}\right)\left(\frac{1 \text{ rad}}{0.01 \text{ Gy}}\right)(\text{RBE})$$

From the answers to the first and second Concept Questions, we can conclude that the absorbed energy E is related to the intensity I, the cross-sectional area A, and the time duration t of the beam via $E = Pt = IAt$. Substituting this result into the equation for the biologically equivalent dose gives

$$\begin{aligned}
\text{Biologically equivalent dose} &= \left(\frac{IAt}{m}\right)\left(\frac{1 \text{ rad}}{0.01 \text{ Gy}}\right)(\text{RBE}) \\
&= \frac{(0.40 \text{ W/m}^2)(0.072 \text{ m}^2)(0.20 \text{ s})}{3.6 \text{ kg}}\left(\frac{1 \text{ rad}}{0.01 \text{ Gy}}\right)(1.1) = \boxed{0.18 \text{ rem}}
\end{aligned}$$

⬆

The next example illustrates the decay of a particle into two photons, so matter is completely converted into electromagnetic waves. The example discusses how to find the energy, frequency, and wavelength of each photon. It also provides an opportunity to review the principle of conservation of energy and the principle of conservation of linear momentum.

⬇ **Concepts & Calculations Example 10** The Decay of the π^0 Meson

The π^0 meson is a particle that has a rest energy of 135.0 MeV (see Table 32.3). It lives for a very short time and then decays into two γ-ray photons: $\pi^0 \rightarrow \gamma + \gamma$. Suppose that one of the γ-ray photons travels along the $+x$ axis. If the π^0 is at rest when it decays, find **(a)** the energy (in MeV), **(b)** the frequency and wavelength, and **(c)** the momentum of each γ-ray photon.

Concept Questions and Answers How is the energy E of each γ-ray photon related to the rest energy E_0 of the π^0 particle?

Answer According to the principle of conservation of energy, the sum of the energies of the two γ-ray photons must equal the rest energy E_0 of the π^0 particle, since that particle is at rest. Therefore, the energy E of each photon is $E = \frac{1}{2}E_0$.

How can the frequency and wavelength of a photon be determined from its energy?

Answer A photon is a packet of electromagnetic energy, and each contains an amount of energy given by Equation 29.2 as $E = hf$, where h is Planck's constant and f is the frequency. Equation 16.1 indicates that the wavelength λ of the photon is inversely proportional to its frequency, $\lambda = c/f$, where c is the speed of light.

How is the total linear momentum of the photons related to the momentum of the π^0 particle, and what is the momentum of each photon?

Answer The principle of conservation of linear momentum states that the total linear momentum of an isolated system is constant (see Section 7.2). Since the π^0 particle is at rest, the total linear momentum before the decay is zero. Since the π^0 particle is an isolated system, momentum is conserved, and the total linear momentum of the two γ-ray photons must also be zero. The magnitude p of the momentum of a photon is given by Equation 29.6 as $p = h/\lambda$. Thus, if one photon travels along the $+x$ axis with a momentum of $+h/\lambda$, the other must travel along the $-x$ axis with a momentum of $-h/\lambda$, ensuring that the total momentum is zero.

Solution (a) Since the sum of the energies of the two γ-ray photons must be equal to the rest energy E_0 of the π^0 particle, the energy of each photon is

$$E = \tfrac{1}{2}E_0 = \tfrac{1}{2}(135.0 \text{ MeV}) = \boxed{67.5 \text{ MeV}}$$

(b) The frequency f of each photon is given by Equation 29.2 as its energy (in units of joules) divided by Planck's constant. Using the fact that $1 \text{ eV} = 1.60 \times 10^{-19}$ J, we find that

$$f = \frac{E}{h} = \frac{(67.5 \times 10^6 \text{ eV})\left(\dfrac{1.60 \times 10^{-19} \text{ J}}{1 \text{ eV}}\right)}{6.63 \times 10^{-34} \text{ J} \cdot \text{s}} = \boxed{1.63 \times 10^{22} \text{ Hz}}$$

The wavelength of each photon can be found directly from Equation 16.1:

$$\lambda = \frac{c}{f} = \frac{3.00 \times 10^8 \text{ m/s}}{1.63 \times 10^{22} \text{ Hz}} = \boxed{1.84 \times 10^{-14} \text{ m}}$$

(c) The magnitude of each photon's momentum is related to its wavelength by Equation 29.6:

$$p = \frac{h}{\lambda} = \frac{6.63 \times 10^{-34} \text{ J} \cdot \text{s}}{1.84 \times 10^{-14} \text{ m}} = 3.60 \times 10^{-20} \text{ kg} \cdot \text{m/s}$$

Since the photons move along the x axis in opposite directions, their momenta are $\boxed{+3.60 \times 10^{-20} \text{ kg} \cdot \text{m/s}}$ and $\boxed{-3.60 \times 10^{-20} \text{ kg} \cdot \text{m/s}}$.

CONCEPT SUMMARY

If you need more help with a concept, use the Learning Aids noted next to the discussion or equation. Examples (**Ex.**) are in the text of this chapter. Go to **www.wiley.com/college/cutnell** for the following Learning Aids:

Interactive LearningWare (ILW) — Additional examples solved in a five-step interactive format.

Concept Simulations (CS) — Animated text figures or animations of important concepts.

Interactive Solutions (IS) — Models for certain types of problems in the chapter homework. The calculations are carried out interactively.

Topic	Discussion	Learning Aids
Ionizing radiation	**32.1 BIOLOGICAL EFFECTS OF IONIZING RADIATION** Ionizing radiation consists of photons and/or moving particles that have enough energy to ionize an atom or molecule. Exposure is a measure of the ionization produced in air by X-rays or γ rays. When a beam of X-rays or γ rays is sent through a mass m of dry air (0 °C, 1 atm pressure) and produces positive ions whose total charge is q, the exposure in coulombs per kilogram (C/kg) is	
Exposure (in C/kg)	Exposure (in coulombs per kilogram) $= \dfrac{q}{m}$	

Topic	Discussion	Learning Aids
	With q in coulombs (C) and m in kilograms (kg), the exposure in roentgens is	
Exposure (in roentgens)	$$\text{Exposure (in roentgens)} = \left(\frac{1}{2.58 \times 10^{-4}} \right) \frac{q}{m} \qquad (32.1)$$	
	The absorbed dose is the amount of energy absorbed from the radiation per unit mass of absorbing material:	
Absorbed dose	$$\text{Absorbed dose} = \frac{\text{Energy absorbed}}{\text{Mass of absorbing material}} \qquad (32.2)$$	**Ex. 1**
	The SI unit of absorbed dose is the gray (Gy); 1 Gy = 1 J/kg. However, the rad is another unit that is often used: 1 rad = 0.01 Gy.	
	The amount of biological damage produced by ionizing radiation is different for different types of radiation. The relative biological effectiveness (RBE) is the absorbed dose of 200-keV X-rays required to produce a certain biological effect divided by the dose of a given type of radiation that produces the same biological effect.	
Relative biological effectiveness	$$RBE = \frac{\begin{array}{c}\text{The dose of 200-keV X-rays that}\\\text{produces a certain biological effect}\end{array}}{\begin{array}{c}\text{The dose of radiation that}\\\text{produces the same biological effect}\end{array}} \qquad (32.3)$$	
	The biologically equivalent dose (in rems) is the product of the absorbed dose (in rads) and the RBE:	
Biologically equivalent dose	$$\begin{array}{c}\text{Biologically equivalent dose}\\\text{(in rems)}\end{array} = \begin{array}{c}\text{Absorbed dose}\\\text{(in rads)}\end{array} \times RBE \qquad (32.4)$$	**Ex. 2, 9**
Induced nuclear reaction **Induced nuclear transmutation**	**32.2 INDUCED NUCLEAR REACTIONS** An induced nuclear reaction occurs whenever a target nucleus is struck by an incident nucleus, an atomic or subatomic particle, or a γ-ray photon and undergoes a change as a result. An induced nuclear transmutation is a reaction in which the target nucleus is changed into a nucleus of a new element.	
Nuclear reactions obey conservation laws	All nuclear reactions (induced or spontaneous) obey the conservation laws of physics as they relate to mass/energy, electric charge, linear momentum, angular momentum, and nucleon number.	**Ex. 3**
Shorthand notation for nuclear reactions	Nuclear reactions are often written in a shorthand form, such as $^{14}_{7}N \, (\alpha, p) \, ^{17}_{8}O$. The first and last symbols $^{14}_{7}N$ and $^{17}_{8}O$ denote, respectively, the initial and final nuclei. The symbols within the parentheses denote the incident α particle (on the left) and the small emitted particle or proton p (on the right).	
Thermal neutron	A thermal neutron is one that has a kinetic energy of about 0.04 eV.	
Nuclear fission **Chain reaction** **Controlled chain reaction**	**32.3 NUCLEAR FISSION** Nuclear fission occurs when a massive nucleus splits into two less massive fragments. Fission can be induced by the absorption of a thermal neutron. When a massive nucleus fissions, energy is released because the binding energy per nucleon is greater for the fragments than for the original nucleus. Neutrons are also released during nuclear fission. These neutrons can, in turn, induce other nuclei to fission and lead to a process known as a chain reaction. A chain reaction is said to be controlled if each fission event contributes, on average, only one neutron that fissions another nucleus.	**Ex. 4, 5** **ILW 32.1** **CS 32.1**
Nuclear reactor **Fuel elements** **Neutron moderator** **Control rods** **Critical state** **Subcritical state** **Supercritical state**	**32.4 NUCLEAR REACTORS** A nuclear reactor is a device that generates energy by a controlled chain reaction. Many reactors in use today have the same three basic components: fuel elements, a neutron moderator, and control rods. The fuel elements contain the fissile fuel, and the entire region of fuel elements is known as the reactor core. The neutron moderator is a material (water, for example) that slows down the neutrons released in a fission event to thermal energies so they can initiate additional fission events. Control rods contain material that readily absorbs neutrons without fissioning. They are used to keep the reactor in its normal, or critical, state, in which each fission event leads to one additional fission, no more, no less. The reactor is subcritical when, on average, the neutrons from each fission trigger less than one subsequent fission. The reactor is supercritical when, on average, the neutrons from each fission trigger more than one additional fission.	
Nuclear fusion **Magnetic confinement** **Inertial confinement**	**32.5 NUCLEAR FUSION** In a fusion process, two nuclei with smaller masses combine to form a single nucleus with a larger mass. Energy is released by fusion when the binding energy per nucleon is greater for the larger nucleus than for the smaller nuclei. Fusion reactions are said to be thermonuclear because they require extremely high temperatures to proceed. Current studies of nuclear fusion utilize either magnetic confinement or inertial confinement to contain the fusing nuclei at the high temperatures that are necessary.	**Ex. 6** **IS 32.31**

Topic	Discussion	Learning Aids
Boson family Lepton family Hadron family	**32.6 ELEMENTARY PARTICLES** Subatomic particles are divided into three families: the boson family (which includes the photon), the lepton family (which includes the electron), and the hadron family (which includes the proton and the neutron).	**Ex. 7, 10**
Elementary particles	Elementary particles are the basic building blocks of matter. All members of the boson and lepton families are elementary particles.	
Quark theory	The quark theory proposes that the hadrons are not elementary particles but are composed of elementary particles called quarks. Currently, the hundreds of hadrons can be accounted for in terms of six quarks (up, down, strange, charmed, top, and bottom) and their antiquarks.	
The standard model	The standard model consists of two parts: (1) the currently accepted explanation for the strong nuclear force in terms of the quark concept of "color" and (2) the theory of the electroweak interaction.	
Cosmology	**32.7 COSMOLOGY** Cosmology is the study of the structure and evolution of the universe. Our universe is expanding. The speed v at which a distant galaxy recedes from the earth is given by Hubble's law:	
Hubble's law	$$v = Hd \qquad (32.5)$$	**Ex. 8**
Hubble parameter	where $H = 0.022$ m/(s·light-year) is called the Hubble parameter and d is the distance of the galaxy from the earth.	
Big Bang theory	The Big Bang theory postulates that the universe had a definite beginning in a cataclysmic event, sometimes called the primeval fireball. The radiation left over from this event is in the microwave region of the electromagnetic spectrum, and it is consistent with a perfect blackbody radiating at a temperature of 2.7 K, in agreement with theoretical analysis of the Big Bang.	
Standard cosmological model	The standard cosmological model for the evolution of the universe is summarized in Figure 32.16.	

FOCUS ON CONCEPTS

Note to Instructors: *The numbering of the questions shown here reflects the fact that they are only a representative subset of the total number that are available online. However, all of the questions are available for assignment via an online homework management program such as WileyPLUS or WebAssign.*

Section 32.1 Biological Effects of Ionizing Radiation

1. Biologically equivalent doses are specified in units called_____.
(a) rads **(b)** grays **(c)** rems **(d)** J/kg **(e)** roentgens

Section 32.2 Induced Nuclear Reactions

4. Determine the unknown nuclear species $^A_Z X$ in the following nuclear reaction:

$$^A_Z X + {}^{14}_7 N \longrightarrow {}^{14}_6 C + {}^1_1 H$$

(a) $^2_1 H$ **(b)** $^1_0 n$ **(c)** γ ray **(d)** $^4_2 He$

Section 32.3 Nuclear Fission

6. The fission of $^{235}_{92} U$ can occur via many different reactions. In general, they can be written as follows:

$$^1_0 n + {}^{235}_{92} U \longrightarrow {}^{A_X}_{Z_X} X + {}^{A_Y}_{Z_Y} Y + \eta {}^1_0 n$$

where X and Y refer to the identities of the fission fragments and η is the number of neutrons produced. Which one or more of the following statements are true?

A. The compound nucleus that is formed to initiate the fission process is the same, no matter what X and Y refer to.

B. The greater the number η of neutrons produced by the reaction, the smaller is the sum of the nucleon numbers A_X and A_Y.

C. The sum of the atomic numbers Z_X and Z_Y is the same, no matter what X and Y refer to.

(a) A **(b)** A and B **(c)** A and C **(d)** B and C **(e)** A, B, and C

Section 32.5 Nuclear Fusion

9. In each of the following three nuclear fusion reactions, the masses of the nuclei are given beneath each nucleus. Rank the energy produced by each reaction in descending order (greatest first).

Reaction I $\qquad \underset{2.0141\ u}{^2_1 H} + \underset{2.0141\ u}{^2_1 H} \longrightarrow \underset{3.0160\ u}{^3_2 He} + \underset{1.0087\ u}{^1_0 n}$

Reaction II $\qquad \underset{3.0160\ u}{^3_2 He} + \underset{3.0160\ u}{^3_2 He} \longrightarrow \underset{4.0026\ u}{^4_2 He} + \underset{1.0078\ u}{^1_1 H} + \underset{1.0078\ u}{^1_1 H}$

Reaction III $\qquad \underset{15.0001\ u}{^{15}_7 N} + \underset{1.0078\ u}{^1_1 H} \longrightarrow \underset{12.0000\ u}{^{12}_6 C} + \underset{4.0026\ u}{^4_2 He}$

(a) I, II, III **(b)** I, III, II **(c)** II, III, I **(d)** II, I, III **(e)** III, II, I

Section 32.6 Elementary Particles

10. The drawings show four possibilities for hadrons in the quark theory. In each of the possibilities, the symbols for the quarks are shown together with the corresponding electric charges. Note that e stands for the magnitude of the charge on an electron. Which one shows the quark structure for an antiproton? **(a)** A **(b)** B **(c)** C **(d)** D

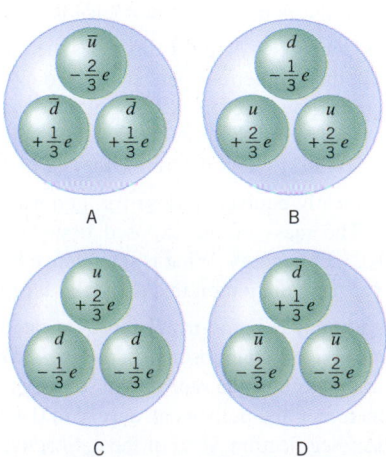

Section 32.7 Cosmology

11. Which one of the following statements is *not* accepted as a part of the current picture of the universe? **(a)** The universe is expanding. **(b)** Early in the history of the universe, there was only one fundamental force. **(c)** The universe began with a cataclysmic event known as the Big Bang. **(d)** The weak electromagnetic radiation in the microwave region of the spectrum that bathes the earth provides evidence of the Big Bang theory. **(e)** Hubble's law indicates that a galaxy located at a distance d from the earth is moving away from the earth at a speed that is inversely proportional to d.

PROBLEMS

Note to Instructors: Most of the homework problems in this chapter are available for assignment via an online homework management program such as WileyPLUS or WebAssign, and those marked with the icon  *are presented in WileyPLUS using a guided tutorial format that provides enhanced interactivity. See Preface for additional details.*

ssm Solution is in the Student Solutions Manual.
www Solution is available online at www.wiley.com/college/cutnell

🍷 **This icon represents a biomedical application.**

Section 32.1 Biological Effects of Ionizing Radiation

1. ssm www Neutrons (RBE = 2.0) and α particles have the same biologically equivalent dose. However, the absorbed dose of the neutrons is six times the absorbed dose of the α particles. What is the RBE for the α particles?

2. Over a full course of treatment, two different tumors are to receive the same absorbed dose of therapeutic radiation. The smaller of the tumors (mass = 0.12 kg) absorbs a total of 1.7 J of energy. **(a)** Determine the absorbed dose, in Gy. **(b)** What is the total energy absorbed by the larger of the tumors (mass = 0.15 kg)?

3. A person receives a single whole-body dose of α particles. The absorbed dose is 38 rad, and the RBE of the α particles is 12. **(a)** Determine the biologically equivalent dose received by this person. **(b)** With this dose, which one of the following would you expect to happen: no short-term ill effects, the onset of radiation sickness, a 50% chance of dying, or almost certain death?

4. Over a year's time, a person receives a biologically equivalent dose of 24 mrem (millirems) from cosmic rays, which consist primarily of high-energy protons bombarding earth's atmosphere from space. The relative biological effectiveness of protons is 10. **(a)** What is the person's absorbed dose in rads? **(b)** The person absorbs 1.9×10^{-3} J of energy from cosmic rays in a year. What is the person's mass?

5. 🍷 **ssm** A beam of particles is directed at a 0.015-kg tumor. There are 1.6×10^{10} particles per second reaching the tumor, and the energy of each particle is 4.0 MeV. The RBE for the radiation is 14. Find the biologically equivalent dose given to the tumor in 25 s.

6. GO Someone stands near a radioactive source and receives doses of the following types of radiation: γ rays (20 mrad, RBE = 1), electrons (30 mrad, RBE = 1), protons (5 mrad, RBE = 10), and slow neutrons (5 mrad, RBE = 2). Rank the types of radiation, highest first, as to which produces the largest biologically equivalent dose.

7. 🍷 Multiple-Concept Example 1 discusses the concepts that are relevant to this problem. A person undergoing radiation treatment for a cancerous growth receives an absorbed dose of 2.1 Gy. All the radiation is absorbed by the growth. If the growth has a specific heat capacity of 4200 J/(kg·C°), determine the rise in its temperature.

8. GO The biologically equivalent dose for a typical chest X-ray is 2.5×10^{-2} rem. The mass of the exposed tissue is 21 kg, and it absorbs 6.2×10^{-3} J of energy. What is the relative biological effectiveness (RBE) for the radiation on this particular type of tissue?

***9.** 🍷 A 2.0-kg tumor is being irradiated by a radioactive source. The tumor receives an absorbed dose of 12 Gy in a time of 850 s. Each disintegration of the radioactive source produces a particle that enters the tumor and delivers an energy of 0.40 MeV. What is the activity $\Delta N/\Delta t$ (see Section 31.6) of the radioactive source?

***10.** Multiple-Concept Example 1 discusses some of the physics principles that are used to solve this problem. What absorbed dose (in rads) of γ rays is required to change a block of ice at $0.0\,°C$ into steam at $100.0\,°C$?

***11. ssm** A beam of nuclei is used for cancer therapy. Each nucleus has an energy of 130 MeV, and the relative biological effectiveness (RBE) of this type of radiation is 16. The beam is directed onto a 0.17-kg tumor, which receives a biologically equivalent dose of 180 rem. How many nuclei are in the beam?

Section 32.2 Induced Nuclear Reactions

12. Identify the unknown species ${}^{A}_{Z}X$ in the following nuclear reaction: ${}^{22}_{11}\text{Na}\,(d, \alpha)\,{}^{A}_{Z}X$. Here, d stands for the deuterium isotope ${}^{2}_{1}H$ of hydrogen.

13. ssm www Write the equation for the reaction ${}^{17}_{8}\text{O}\,(\gamma, \alpha n)\,{}^{12}_{6}\text{C}$. The notation "$\alpha n$" means that an α particle and a neutron are produced by the reaction.

14. GO For each of the nuclear reactions listed below, determine the unknown particle ${}^{A}_{Z}X$. Use the periodic table on the inside of the back cover as needed.

(a) ${}^{A}_{Z}X + {}^{14}_{7}\text{N} \longrightarrow {}^{1}_{1}\text{H} + {}^{17}_{8}\text{O}$ **(c)** ${}^{1}_{1}\text{H} + {}^{27}_{13}\text{Al} \longrightarrow {}^{A}_{Z}X + {}^{1}_{0}\text{n}$

(b) ${}^{15}_{7}\text{N} + {}^{A}_{Z}X \longrightarrow {}^{12}_{6}\text{C} + {}^{4}_{2}\text{He}$ **(d)** ${}^{7}_{3}\text{Li} + {}^{1}_{1}\text{H} \longrightarrow {}^{4}_{2}\text{He} + {}^{A}_{Z}X$

15. A neutron causes ${}^{232}_{90}\text{Th}$ to change according to the reaction

$${}^{1}_{0}\text{n} + {}^{232}_{90}\text{Th} \longrightarrow {}^{A}_{Z}X + \gamma$$

(a) Identify the unknown nucleus ${}^{A}_{Z}X$, giving its atomic mass number A, its atomic number Z, and the symbol X for the element. **(b)** The ${}^{A}_{Z}X$ nucleus subsequently undergoes β^{-} decay, and its daughter does too. Identify the final nucleus, giving its atomic mass number, atomic number, and name.

16. Complete the following nuclear reactions, assuming that the unknown quantity signified by the question mark is a single entity: **(a)** ${}^{34}_{18}\text{Ar}\,(n, \alpha)?$ **(b)** ${}^{82}_{34}\text{Se}\,(?, n)\,{}^{82}_{35}\text{Br}$ **(c)** ${}^{58}_{28}\text{Ni}\,({}^{40}_{18}\text{Ar}, ?)\,{}^{57}_{27}\text{Co}$ **(d)** $?(\gamma, \alpha){}^{16}_{8}\text{O}$

17. Write the reactions below in the shorthand form discussed in the text. Note in reaction **(c)** that ${}^{2}_{1}H$ is a deuteron d.

(a) ${}^{1}_{0}\text{n} + {}^{14}_{7}\text{N} \longrightarrow {}^{14}_{6}\text{C} + {}^{1}_{1}\text{H}$

(b) ${}^{1}_{0}\text{n} + {}^{238}_{92}\text{U} \longrightarrow {}^{239}_{92}\text{U} + \gamma$

(c) ${}^{1}_{0}\text{n} + {}^{24}_{12}\text{Mg} \longrightarrow {}^{23}_{11}\text{Na} + {}^{2}_{1}\text{H}$

***18.** During a nuclear reaction, an unknown particle is absorbed by a copper ${}^{63}_{29}\text{Cu}$ nucleus, and the reaction products are ${}^{62}_{29}\text{Cu}$, a neutron, and a proton. What are the name, atomic number, and nucleon number of the nucleus formed temporarily when the copper ${}^{63}_{29}\text{Cu}$ nucleus absorbs the unknown particle?

***19. ssm** Consider the induced nuclear reaction $^2_1\text{H} + ^{14}_7\text{N} \rightarrow$ $^{12}_6\text{C} + ^4_2\text{He}$. The atomic masses are ^2_1H (2.014 102 u), $^{14}_7\text{N}$ (14.003 074 u), $^{12}_6\text{C}$ (12.000 000 u), and ^4_2He (4.002 603 u). Determine the energy (in MeV) released by the reaction.

Section 32.3 Nuclear Fission, Section 32.4 Nuclear Reactors

20. $^{235}_{92}\text{U}$ absorbs a thermal neutron and fissions into rubidium $^{93}_{37}\text{Rb}$ and cesium $^{141}_{55}\text{Cs}$. What nucleons are produced by the fission, and how many are there?

21. ssm When a $^{235}_{92}\text{U}$ (235.043 924 u) nucleus fissions, about 200 MeV of energy is released. What is the ratio of this energy to the rest energy of the uranium nucleus?

22. How many neutrons are produced when ^{235}U fissions in the following way? $^1_0\text{n} + ^{235}_{92}\text{U} \rightarrow ^{132}_{50}\text{Sn} + ^{101}_{42}\text{Mo} + \text{neutrons}$

23. Interactive LearningWare 32.1 at **www.wiley.com/college/ cutnell** reviews the concepts that lie at the heart of this problem. What energy (in MeV) is released by the following fission reaction?

$$^1_0\text{n} + ^{235}_{92}\text{U} \longrightarrow ^{140}_{54}\text{Xe} + ^{94}_{38}\text{Sr} + 2^1_0\text{n}$$
$$1.009 \text{ u} \quad 235.044 \text{ u} \qquad 139.922 \text{ u} \quad 93.915 \text{ u} \quad 2(1.009 \text{ u})$$

24. The energy released by each fission within the core of a nuclear reactor is 2.0×10^2 MeV. The number of fissions occurring each second is 2.0×10^{19}. Determine the power (in watts) that the reactor generates.

25. Uranium $^{235}_{92}\text{U}$ fissions into two fragments plus three neutrons: $^1_0\text{n} + ^{235}_{92}\text{U} \rightarrow (2 \text{ fragments}) + 3^1_0\text{n}$. The mass of a neutron is 1.008 665 u and the mass of $^{235}_{92}\text{U}$ is 235.043 924 u. If 225.0 MeV of energy is released, what is the total mass of the two fragments?

***26. GO** The energy consumed in one year in the United States is about 9.3×10^{19} J. With each $^{235}_{92}\text{U}$ fission, about 2.0×10^2 MeV of energy is released. How many kilograms of $^{235}_{92}\text{U}$ would be needed to generate this energy if all the nuclei fissioned?

***27. ssm** (a) If each fission reaction of a $^{235}_{92}\text{U}$ nucleus releases about 2.0×10^2 MeV of energy, determine the energy (in joules) released by the complete fissioning of 1.0 gram of $^{235}_{92}\text{U}$. (b) How many grams of $^{235}_{92}\text{U}$ would be consumed in one year to supply the energy needs of a household that uses 30.0 kWh of energy per day, on the average?

***28. GO** The water that cools a reactor core enters the reactor at 216 °C and leaves at 287 °C. (The water is pressurized, so it does not turn to steam.) The core is generating 5.6×10^9 W of power. Assume that the specific heat capacity of water is 4420 J/(kg·C°) over the temperature range stated above, and find the mass of water that passes through the core each second.

****29. ssm www** A nuclear power plant is 25% efficient, meaning that only 25% of the power it generates goes into producing electricity. The remaining 75% is wasted as heat. The plant generates 8.0×10^8 watts of electric power. If each fission releases 2.0×10^2 MeV of energy, how many kilograms of $^{235}_{92}\text{U}$ are fissioned per year?

****30.** When a nuclear reactor is in a critical state, the neutrons released in each fission trigger an average of exactly one additional fission. If the average number of additional fissions triggered rises above one, the reactor enters a supercritical state in which the fission rate and the power output grow very rapidly. A reactor in a critical state has a power output of 25 kW. The reactor then enters a supercritical state in which each fission triggers an average of 1.01 additional fissions. The average time for the neutrons released by one generation of fissions to trigger the next generation of fissions is 1.2×10^{-8} s. How much time elapses before the power output from a single generation of fissions grows to 3300 MW (roughly the normal output of a commercial reactor)?

Section 32.5 Nuclear Fusion

31. Interactive Solution 32.31 at **www.wiley.com/college/cutnell** illustrates one way to approach problems of this type. The fusion of two deuterium nuclei (^2_1H, mass = 2.0141 u) can yield a helium nucleus (^3_2He, mass = 3.0160 u) and a neutron (^1_0n, mass = 1.0087 u). What is the energy (in MeV) released in this reaction?

32. GO In one type of fusion reaction a proton fuses with a neutron to form a deuterium nucleus:

$$^1_1\text{H} + ^1_0\text{n} \longrightarrow ^2_1\text{H} + \gamma$$

The masses are ^1_1H (1.0078 u), ^1_0n (1.0087 u), and ^2_1H (2.0141 u). The γ-ray photon is massless. How much energy (in MeV) is released by this reaction?

33. Two deuterium atoms (^2_1H) react to produce tritium (^3_1H) and hydrogen (^1_1H) according to the following reaction:

$$^2_1\text{H} + ^2_1\text{H} \longrightarrow ^3_1\text{H} + ^1_1\text{H}$$
$$2.014\ 102 \text{ u} \quad 2.014\ 102 \text{ u} \quad 3.016\ 050 \text{ u} \quad 1.007\ 825 \text{ u}$$

What is the energy (in MeV) released by this deuterium–deuterium reaction?

34. Tritium (^3_1H) is a rare isotope of hydrogen that can be produced by the following fusion reaction:

$$^A_Z\text{X} + ^A_1\text{Y} \longrightarrow ^3_1\text{H} + \gamma$$
$$1.0087 \text{ u} \quad 2.0141 \text{ u} \qquad 3.0161 \text{ u}$$

(a) Determine the atomic mass number A, the atomic number Z, and the names X and Y of the unknown particles. (b) Using the masses given in the reaction, determine how much energy (in MeV) is released by this reaction.

***35. ssm** Imagine that your car is powered by a fusion engine in which the following reaction occurs: $3^2_1\text{H} \rightarrow ^4_2\text{He} + ^1_1\text{H} + ^1_0\text{n}$. The masses are ^2_1H (2.0141 u), ^4_2He (4.0026 u), ^1_1H (1.0078 u), and ^1_0n (1.0087 u). The engine uses 6.1×10^{-6} kg of deuterium ^2_1H fuel. If one gallon of gasoline produces 2.1×10^9 J of energy, how many gallons of gasoline would have to be burned to equal the energy released by all the deuterium fuel?

***36. GO** In Example 6 it was determined that 17.6 MeV of energy is released when the following fusion reaction occurs:

$$^2_1\text{H} + ^3_1\text{H} \longrightarrow ^4_2\text{He} + ^1_0\text{n}$$
$$2.0141 \text{ u} \quad 3.0161 \text{ u} \qquad 4.0026 \text{ u} \quad 1.0087 \text{ u}$$

Ignore relativistic effects and determine the kinetic energies of the neutron and the α particles.

***37.** Deuterium (^2_1H) is an attractive fuel for fusion reactions because it is abundant in the oceans, where about 0.015% of the hydrogen atoms in the water (H_2O) are deuterium atoms. (a) How many deuterium atoms are there in one kilogram of water? (b) If each deuterium nucleus produces about 7.2 MeV in a fusion reaction, how many kilograms of water would be needed to supply the energy needs of the United States for one year, estimated to be 1.1×10^{20} J?

****38.** The proton–proton cycle thought to occur in the sun consists of the following sequence of reactions:

$$(1) \ ^1_1\text{H} + ^1_1\text{H} \longrightarrow ^2_1\text{H} + ^0_1\text{e} + \nu$$
$$(2) \ ^1_1\text{H} + ^2_1\text{H} \longrightarrow ^3_2\text{He} + \gamma$$
$$(3) \ ^3_2\text{He} + ^3_2\text{He} \longrightarrow ^4_2\text{He} + ^1_1\text{H} + ^1_1\text{H}$$

In these reactions ^0_1e is a positron (mass = 0.000 549 u), ν is a neutrino (mass ≈ 0 u), and γ is a gamma ray photon (mass = 0 u). Note that reaction (3) uses two ^3_2He nuclei, which are formed by *two* reactions of type (1) and *two* reactions of type (2). Verify that the proton–proton cycle generates about 25 MeV of energy. The atomic masses are ^1_1H (1.007 825 u), ^2_1H (2.014 102 u), ^3_2He (3.016 030 u), and ^4_2He (4.002 603 u). Be sure to account for the fact that there are two electrons in two hydrogen atoms, whereas there is one electron in a single deuterium (^2_1H) atom. The mass of one electron is 0.000 549 u.

Section 32.6   Elementary Particles

39. The main decay mode for the negative pion is $\pi^- \rightarrow \mu^- + \bar{\nu}_\mu$. Find the energy (in MeV) released in this decay. Consult Table 32.3 for rest energies and assume that the rest energy for $\bar{\nu}_\mu$ is ≈ 0 MeV.

40. The lambda particle Λ^0 has an electric charge of zero. It is a baryon and, hence, is composed of three quarks. They are all different. One of these quarks is the up quark u, and there are no antiquarks present. Make a list of the three possibilities for the quarks contained in Λ^0. (Other information is needed to decide which one of these possibilities is the actual Λ^0 particle.)

41. A collision between two protons produces three new particles: $p + p \rightarrow p + \pi^+ + \Lambda^0 + K^0$. The rest energies of the new particles are π^+ (139.6 MeV), Λ^0 (1116 MeV), and K^0 (497.7 MeV). Note that one proton disappears during the reaction. How much of the protons' incident energy (in MeV) is transformed into matter during this reaction?

42. A neutral pion π^0 (rest energy = 135.0 MeV) produced in a high-energy particle experiment moves at a speed of 0.780 c. After a very short time, it decays into two γ-ray photons. One of the γ-ray photons has an energy of 192 MeV. What is the energy (in MeV) of the second γ-ray photon? Take relativistic effects into account.

43. ssm www Suppose a neutrino is created and has an energy of 35 MeV. **(a)** Assuming the neutrino, like the photon, has no mass and travels at the speed of light, find the momentum of the neutrino. **(b)** Determine the de Broglie wavelength of the neutrino.

***44.**  Review Conceptual Example 7 as background for this problem. An electron and its antiparticle annihilate each other, producing two γ-ray photons. The kinetic energies of the particles are negligible. Determine the magnitude of the momentum of each photon.

***45.** Review Conceptual Example 5 as background for this problem. An energetic proton is fired at a stationary proton. For the reaction to produce new particles, the two protons must approach each other to within a distance of about 8.0×10^{-15} m. The moving proton must have a sufficient speed to overcome the repulsive Coulomb force. What must be the minimum initial kinetic energy (in MeV) of the proton?

ADDITIONAL PROBLEMS

46. A film badge worn by a radiologist indicates that she has received an absorbed dose of 2.5×10^{-5} Gy. The mass of the radiologist is 65 kg. How much energy has she absorbed?

47. ssm The K^- particle has a charge of $-e$ and contains one quark and one antiquark. **(a)** Which quarks can the particle *not* contain? **(b)** Which antiquarks can the particle *not* contain?

48. What is the atomic number Z, the atomic mass number A, and the element X in the reaction $^{10}_{5}B(\alpha, p)^{A}_{Z}X$?

49. Neutrons released by a fission reaction must be slowed by collisions with the moderator nuclei before the neutrons can cause further fissions. Suppose a 1.5-MeV neutron leaves each collision with 65% of its incident energy. How many collisions are required to reduce the neutron's energy to at least 0.040 eV, which is the energy of a thermal neutron?

***50.** Multiple-Concept Example 1 uses an approach similar to that needed in this problem, except here the temperature remains constant while a phase change occurs. A sample of liquid water at 100 °C and 1 atm pressure boils into steam at 100 °C because it is irradiated with a large dose of ionizing radiation. What is the absorbed dose of the radiation in rads?

***51. ssm** One proposed fusion reaction combines lithium $^{6}_{3}$Li (6.015 u) with deuterium $^{2}_{1}$H (2.014 u) to give helium $^{4}_{2}$He (4.003 u): $^{2}_{1}$H + $^{6}_{3}$Li $\rightarrow$ 2^{4}_{2}He. How many kilograms of lithium would be needed to supply the energy needs of one household for a year, estimated to be 3.8×10^{10} J?

***52.** When 1.0 kg of coal is burned, approximately 3.0×10^7 J of energy is released. If the energy released during each $^{235}_{92}$U fission is 2.0×10^2 MeV, how many kilograms of coal must be burned to produce the same energy as 1.0 kg of $^{235}_{92}$U?

***53.** Suppose that the $^{239}_{94}$Pu nucleus fissions into two fragments whose mass ratio is 0.32 : 0.68. With the aid of Figure 32.9, estimate the energy (in MeV) released during this fission.

****54.** One kilogram of dry air at STP conditions is exposed to 1.0 R of X-rays. One roentgen is defined by Equation 32.1. An equivalent definition can be based on the fact that an exposure of one roentgen deposits 8.3×10^{-3} J of energy per kilogram of dry air. Using the two definitions and assuming that all ions produced are singly charged, determine the average energy (in eV) needed to produce one ion in air.

POWERS OF TEN AND SCIENTIFIC NOTATION

In science, very large and very small decimal numbers are conveniently expressed in terms of powers of ten, some of which are listed below:

$$10^3 = 10 \times 10 \times 10 = 1000 \qquad 10^{-3} = \frac{1}{10 \times 10 \times 10}$$
$$= 0.001$$

$$10^2 = 10 \times 10 = 100 \qquad 10^{-2} = \frac{1}{10 \times 10} = 0.01$$

$$10^1 = 10 \qquad 10^{-1} = \frac{1}{10} = 0.1$$

$$10^0 = 1$$

Using powers of ten, we can write the radius of the earth in the following way, for example:

$$\text{Earth radius} = 6\,380\,000 \text{ m} = 6.38 \times 10^6 \text{ m}$$

The factor of ten raised to the sixth power is ten multiplied by itself six times, or one million, so the earth's radius is 6.38 million meters. Alternatively, the factor of ten raised to the sixth power indicates that the decimal point in the term 6.38 is to be moved six places *to the right* to obtain the radius as a number without powers of ten.

For numbers less than one, negative powers of ten are used. For instance, the Bohr radius of the hydrogen atom is

$$\text{Bohr radius} = 0.000\,000\,000\,0529 \text{ m} = 5.29 \times 10^{-11} \text{ m}$$

The factor of ten raised to the minus eleventh power indicates that the decimal point in the term 5.29 is to be moved eleven places *to the left* to obtain the radius as a number without powers of ten. Numbers expressed with the aid of powers of ten are said to be in *scientific notation.*

Calculations that involve the multiplication and division of powers of ten are carried out as in the following examples:

$$(2.0 \times 10^6)(3.5 \times 10^3) = (2.0 \times 3.5) \times 10^{6+3} = 7.0 \times 10^9$$

$$\frac{9.0 \times 10^7}{2.0 \times 10^4} = \left(\frac{9.0}{2.0}\right) \times 10^7 \times 10^{-4}$$

$$= \left(\frac{9.0}{2.0}\right) \times 10^{7-4} = 4.5 \times 10^3$$

The general rules for such calculations are

$$\frac{1}{10^n} = 10^{-n} \tag{A-1}$$

$$10^n \times 10^m = 10^{n+m} \qquad \text{(Exponents added)} \tag{A-2}$$

$$\frac{10^n}{10^m} = 10^{n-m} \qquad \text{(Exponents subtracted)} \tag{A-3}$$

where n and m are any positive or negative number.

Scientific notation is convenient because of the ease with which it can be used in calculations. Moreover, scientific notation provides a convenient way to express the significant figures in a number, as Appendix B discusses.

SIGNIFICANT FIGURES

The number of *significant figures* in a number is the number of digits whose values are known with certainty. For instance, a person's height is measured to be 1.78 m, with the measurement error being in the third decimal place. All three digits are known with certainty, so that the number contains three significant figures. If a zero is given as the last digit to the right of the decimal point, the zero is presumed to be significant. Thus, the number 1.780 m contains four significant figures. As another example, consider a distance of 1500 m. This number contains only two significant figures, the one and the five. The zeros immediately to the left of the unexpressed decimal point are not counted as significant figures. However, zeros located between significant figures are significant, so a distance of 1502 m contains four significant figures.

Scientific notation is particularly convenient from the point of view of significant figures. Suppose it is known that a certain distance is fifteen hundred meters, to four significant figures. Writing the number as 1500 m presents a problem because it implies that only two significant figures are known. In contrast, the scientific notation of 1.500×10^3 m has the advantage of indicating that the distance is known to four significant figures.

When two or more numbers are used in a calculation, the number of significant figures in the answer is limited by the number of significant figures in the original data. For instance, a rectangular garden with sides of 9.8 m and 17.1 m has an area of (9.8 m)(17.1 m). A calculator gives 167.58 m² for this product. However, one of the original lengths is known only to two significant figures, so the final answer is limited to only two significant figures and should be rounded off to 170 m². In general, *when numbers are multiplied or divided, the number of significant figures in the final answer equals the smallest number of significant figures in any of the original factors.*

The number of significant figures in the answer to an addition or a subtraction is also limited by the original data. Consider the total distance along a biker's trail that consists of three segments with the distances shown as follows:

	2.5 km
	11 km
	5.26 km
Total	18.76 km

The distance of 11 km contains no significant figures to the right of the decimal point. Therefore, neither does the sum of the three distances, and the total distance should not be reported as 18.76 km. Instead, the answer is rounded off to 19 km. In general, *when numbers are added or subtracted, the last significant figure in the answer occurs in the last column (counting from left to right) containing a number that results from a combination of digits that are all significant.* In the answer of 18.76 km, the eight is the sum of $2 + 1 + 5$, each digit being significant. However, the seven is the sum of $5 + 0 + 2$, and the zero is not significant, since it comes from the 11-km distance, which contains no significant figures to the right of the decimal point.

APPENDIX C

ALGEBRA

C.1 PROPORTIONS AND EQUATIONS

Physics deals with physical variables and the relations between them. Typically, variables are represented by the letters of the English and Greek alphabets. Sometimes, the relation between variables is expressed as a proportion or inverse proportion. Other times, however, it is more convenient or necessary to express the relation by means of an equation, which is governed by the rules of algebra.

If two variables are ***directly proportional*** and one of them doubles, then the other variable also doubles. Similarly, if one variable is reduced to one-half its original value, then the other is also reduced to one-half its original value. In general, if x is directly proportional to y, then increasing or decreasing one variable by a given factor causes the other variable to change in the same way by the same factor. This kind of relation is expressed as $x \propto y$, where the symbol $\propto$ means "is proportional to."

Since the proportional variables x and y always increase and decrease by the same factor, the ratio of x to y must have a constant value, or $x/y = k$, where k is a constant, independent of the values for x and y. Consequently, a proportionality such as $x \propto y$ can also be expressed in the form of an equation: $x = ky$. The constant k is referred to as a ***proportionality constant.***

If two variables are ***inversely proportional*** and one of them increases by a given factor, then the other decreases by the same factor. An inverse proportion is written as $x \propto 1/y$. This kind of proportionality is equivalent to the following equation: $xy = k$, where k is a proportionality constant, independent of x and y.

C.2 SOLVING EQUATIONS

Some of the variables in an equation typically have known values, and some do not. It is often necessary to solve the equation so that a variable whose value is unknown is expressed in terms of the known quantities. ***In the process of solving an equation, it is permissible to manipulate the equation in any way, as long as a change made on one side of the equals sign is also made on the other side.*** For example, consider the equation $v = v_0 + at$. Suppose values for v, v_0, and a are available, and the value of t is required. To solve the equation for t, we begin by subtracting v_0 from *both* sides:

$$
\begin{aligned}
v &= v_0 + at \\
-v_0 &= -v_0 \\
\hline
v - v_0 &= at
\end{aligned}
$$

Next, we divide both sides of $v - v_0 = at$ by the quantity a:

$$\frac{v - v_0}{a} = \frac{at}{a} = (1)t$$

On the right side, the a in the numerator divided by the a in the denominator equals one, so that

$$t = \frac{v - v_0}{a}$$

It is always possible to check the correctness of the algebraic manipulations performed in solving an equation by substituting the answer back into the original equation. In the previous example, we substitute the answer for t into $v = v_0 + at$:

$$v = v_0 + a\left(\frac{v - v_0}{a}\right) = v_0 + (v - v_0) = v$$

The result $v = v$ implies that our algebraic manipulations were done correctly.

Algebraic manipulations other than addition, subtraction, multiplication, and division may play a role in solving an equation. The same basic rule applies, however: Whatever is done to the left side of an equation must also be done to the right side. As another example, suppose it is necessary to express v_0 in terms of v, a, and x, where $v^2 = v_0^2 + 2ax$. By subtracting $2ax$ from both sides, we isolate v_0^2 on the right:

$$
\begin{aligned}
v^2 &= v_0^2 + 2ax \\
-2ax &= -2ax \\
\hline
v^2 - 2ax &= v_0^2
\end{aligned}
$$

To solve for v_0, we take the positive and negative square root of *both* sides of $v^2 - 2ax = v_0^2$:

$$v_0 = \pm \sqrt{v^2 - 2ax}$$

C.3 SIMULTANEOUS EQUATIONS

When more than one variable in a single equation is unknown, additional equations are needed if solutions are to be found for all of the unknown quantities. Thus, the equation $3x + 2y = 7$ cannot be solved by itself to give unique values for both x and y. However, if x and y also (i.e., simultaneously) obey the equation $x - 3y = 6$, then both unknowns can be found.

There are a number of methods by which such simultaneous equations can be solved. One method is to solve one equation for x in terms of y and substitute the result into the other equation to obtain an expression containing only the single unknown variable y. The equation $x - 3y = 6$, for instance, can be solved for x by adding $3y$ to each side, with the result that $x = 6 + 3y$. The substitution of this expression for x into the equation $3x + 2y = 7$ is shown below:

$$
\begin{aligned}
3x + 2y &= 7 \\
3(6 + 3y) + 2y &= 7 \\
18 + 9y + 2y &= 7
\end{aligned}
$$

We find, then, that $18 + 11y = 7$, a result that can be solved for y:

$$
\begin{aligned}
18 + 11y &= 7 \\
-18 &\quad -18 \\
\hline
11y &= -11
\end{aligned}
$$

Dividing both sides of this result by 11 shows that $y = -1$. The value of $y = -1$ can be substituted in either of the original equations to obtain a value for x:

$$
\begin{aligned}
x - 3y &= 6 \\
x - 3(-1) &= 6 \\
x + 3 &= 6 \\
-3 &\quad -3 \\
\hline
x &= 3
\end{aligned}
$$

C.4 THE QUADRATIC FORMULA

Equations occur in physics that include the square of a variable. Such equations are said to be *quadratic* in that variable, and often can be put into the following form:

$$ax^2 + bx + c = 0 \qquad \text{(C-1)}$$

where a, b, and c are constants independent of x. This equation can be solved to give the **quadratic formula,** which is

$$x = \frac{-b \pm \sqrt{b^2 - 4ac}}{2a} \qquad \text{(C-2)}$$

The $\pm$ in the quadratic formula indicates that there are two solutions. For instance, if $2x^2 - 5x + 3 = 0$, then $a = 2$, $b = -5$, and $c = 3$. The quadratic formula gives the two solutions as follows:

Solution 1:
Plus sign

$$x = \frac{-b + \sqrt{b^2 - 4ac}}{2a}$$
$$= \frac{-(-5) + \sqrt{(-5)^2 - 4(2)(3)}}{2(2)}$$
$$= \frac{+5 + \sqrt{1}}{4} = \frac{3}{2}$$

Solution 2:
Minus sign

$$x = \frac{-b - \sqrt{b^2 - 4ac}}{2a}$$
$$= \frac{-(-5) - \sqrt{(-5)^2 - 4(2)(3)}}{2(2)}$$
$$= \frac{+5 - \sqrt{1}}{4} = 1$$

APPENDIX D

EXPONENTS AND LOGARITHMS

Appendix A discusses powers of ten, such as 10^3, which means ten multiplied by itself three times, or $10 \times 10 \times 10$. The three is referred to as an **exponent.** The use of exponents extends beyond powers of ten. In general, the term y^n means the factor y is multiplied by itself n times. For example, y^2, or y squared, is familiar and means $y \times y$. Similarly, y^5 means $y \times y \times y \times y \times y$.

The rules that govern algebraic manipulations of exponents are the same as those given in Appendix A (see Equations A-1, A-2, and A-3) for powers of ten:

$$\frac{1}{y^n} = y^{-n} \qquad \text{(D-1)}$$

$$y^n y^m = y^{n+m} \qquad \text{(Exponents added)} \qquad \text{(D-2)}$$

$$\frac{y^n}{y^m} = y^{n-m} \qquad \text{(Exponents subtracted)} \qquad \text{(D-3)}$$

To the three rules above we add two more that are useful. One of these is

$$y^n z^n = (yz)^n \qquad \text{(D-4)}$$

The following example helps to clarify the reasoning behind this rule:

$$3^2 5^2 = (3 \times 3)(5 \times 5) = (3 \times 5)(3 \times 5) = (3 \times 5)^2$$

The other additional rule is

$$(y^n)^m = y^{nm} \qquad \text{(Exponents multiplied)} \qquad \text{(D-5)}$$

To see why this rule applies, consider the following example:

$$(5^2)^3 = (5^2)(5^2)(5^2) = 5^{2+2+2} = 5^{2\times3}$$

Roots, such as a square root or a cube root, can be represented with fractional exponents. For instance,

$$\sqrt{y} = y^{1/2} \quad \text{and} \quad \sqrt[3]{y} = y^{1/3}$$

In general, the nth root of y is given by

$$\sqrt[n]{y} = y^{1/n} \qquad \text{(D-6)}$$

The rationale for Equation D-6 can be explained using the fact that $(y^n)^m = y^{nm}$. For instance, the fifth root of y is the number that, when

multiplied by itself five times, gives back y. As shown below, the term $y^{1/5}$ satisfies this definition:

$$(y^{1/5})(y^{1/5})(y^{1/5})(y^{1/5})(y^{1/5}) = (y^{1/5})^5 = y^{(1/5)\times5} = y$$

Logarithms are closely related to exponents. To see the connection between the two, note that it is possible to express any number y as another number B raised to the exponent x. In other words,

$$y = B^x \qquad \text{(D-7)}$$

The exponent x is called the **logarithm** of the number y. The number B is called the **base number.** One of two choices for the base number is usually used. If $B = 10$, the logarithm is known as the *common logarithm,* for which the notation "log" applies:

Common logarithm $\qquad y = 10^x \quad \text{or} \quad x = \log y \qquad \text{(D-8)}$

If $B = e = 2.718 \ldots$, the logarithm is referred to as the *natural logarithm,* and the notation "ln" is used:

Natural logarithm $\qquad y = e^z \quad \text{or} \quad z = \ln y \qquad \text{(D-9)}$

The two kinds of logarithms are related by

$$\ln y = 2.3026 \log y \qquad \text{(D-10)}$$

Both kinds of logarithms are often given on calculators.

The logarithm of the product or quotient of two numbers A and C can be obtained from the logarithms of the individual numbers according to the rules below. These rules are illustrated here for natural logarithms, but they are the same for any kind of logarithm.

$$\ln (AC) = \ln A + \ln C \qquad \text{(D-11)}$$

$$\ln \left(\frac{A}{C} \right) = \ln A - \ln C \qquad \text{(D-12)}$$

Thus, the logarithm of the product of two numbers is the sum of the individual logarithms, and the logarithm of the quotient of two numbers is the difference between the individual logarithms. Another useful rule concerns the logarithm of a number A raised to an exponent n:

$$\ln A^n = n \ln A \qquad \text{(D-13)}$$

Rules D-11, D-12, and D-13 can be derived from the definition of the logarithm and the rules governing exponents.

APPENDIX E

GEOMETRY AND TRIGONOMETRY

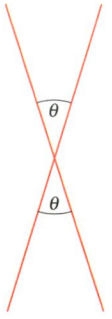 E.1   GEOMETRY

ANGLES

Two angles are equal if

1. They are vertical angles (see Figure E1).
2. Their sides are parallel (see Figure E2).

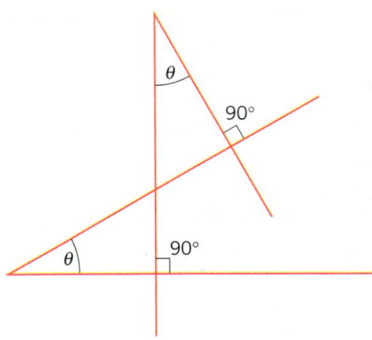

Figure E1   **Figure E2**

3. Their sides are mutually perpendicular (see Figure E3).

Figure E3

TRIANGLES

1. The *sum of the angles* of any triangle is 180° (see Figure E4).

$\alpha + \beta + \gamma = 180°$

Figure E4

2. A *right triangle* has one angle that is 90°.
3. An *isosceles triangle* has two sides that are equal.
4. An *equilateral triangle* has three sides that are equal. Each angle of an equilateral triangle is 60°.
5. Two triangles are *similar* if two of their angles are equal (see Figure E5). The corresponding sides of similar triangles are proportional to each other:

$$\frac{a_1}{a_2} = \frac{b_1}{b_2} = \frac{c_1}{c_2}$$

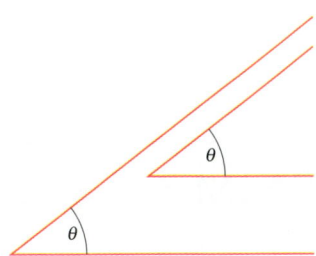

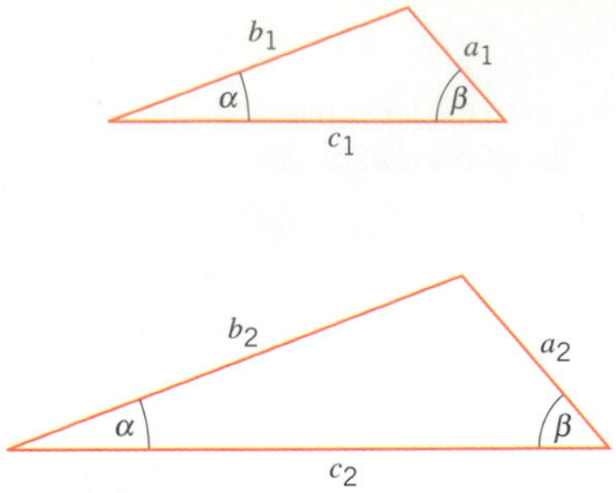

Figure E5

6. Two similar triangles are *congruent* if they can be placed on top of one another to make an exact fit.

CIRCUMFERENCES, AREAS, AND VOLUMES OF SOME COMMON SHAPES

1. Triangle of base b and altitude h (see Figure E6):

$$\text{Area} = \tfrac{1}{2}bh$$

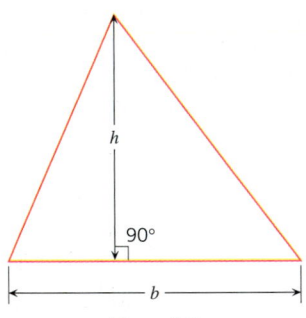

Figure E6

2. Circle of radius r:

$$\text{Circumference} = 2\pi r$$
$$\text{Area} = \pi r^2$$

3. Sphere of radius r:

$$\text{Surface area} = 4\pi r^2$$
$$\text{Volume} = \tfrac{4}{3}\pi r^3$$

4. Right circular cylinder of radius r and height h (see Figure E7):

$$\text{Surface area} = 2\pi r^2 + 2\pi rh$$
$$\text{Volume} = \pi r^2 h$$

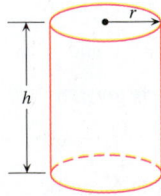

Figure E7

E.2 TRIGONOMETRY

BASIC TRIGONOMETRIC FUNCTIONS

1. For a right triangle, the sine, cosine, and tangent of an angle θ are defined as follows (see Figure E8):

$$\sin \theta = \frac{\text{Side opposite } \theta}{\text{Hypotenuse}} = \frac{h_o}{h}$$

$$\cos \theta = \frac{\text{Side adjacent to } \theta}{\text{Hypotenuse}} = \frac{h_a}{h}$$

$$\tan \theta = \frac{\text{Side opposite } \theta}{\text{Side adjacent to } \theta} = \frac{h_o}{h_a}$$

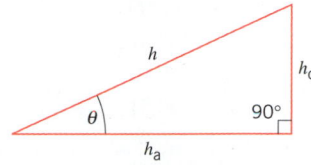

Figure E8

2. The secant (sec θ), cosecant (csc θ), and cotangent (cot θ) of an angle θ are defined as follows:

$$\sec \theta = \frac{1}{\cos \theta} \qquad \csc \theta = \frac{1}{\sin \theta} \qquad \cot \theta = \frac{1}{\tan \theta}$$

TRIANGLES AND TRIGONOMETRY

1. The **Pythagorean theorem** states that the square of the hypotenuse of a right triangle is equal to the sum of the squares of the other two sides (see Figure E8):

$$h^2 = h_o^2 + h_a^2$$

2. The *law of cosines* and the *law of sines* apply to any triangle, not just a right triangle, and they relate the angles and the lengths of the sides (see Figure E9):

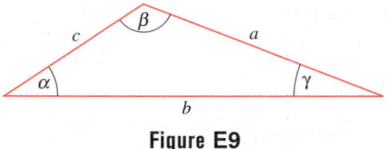

Figure E9

Law of cosines $\qquad c^2 = a^2 + b^2 - 2ab \cos \gamma$

Law of sines $\qquad \dfrac{a}{\sin \alpha} = \dfrac{b}{\sin \beta} = \dfrac{c}{\sin \gamma}$

OTHER TRIGONOMETRIC IDENTITIES

1. $\sin (-\theta) = -\sin \theta$
2. $\cos (-\theta) = \cos \theta$
3. $\tan (-\theta) = -\tan \theta$
4. $(\sin \theta) / (\cos \theta) = \tan \theta$
5. $\sin^2 \theta + \cos^2 \theta = 1$
6. $\sin (\alpha \pm \beta) = \sin \alpha \cos \beta \pm \cos \alpha \sin \beta$

If $\alpha = 90°$, $\sin (90° \pm \beta) = \cos \beta$

If $\alpha = \beta$, $\sin 2\beta = 2 \sin \beta \cos \beta$

7. $\cos (\alpha \pm \beta) = \cos \alpha \cos \beta \mp \sin \alpha \sin \beta$

If $\alpha = 90°$, $\cos (90° \pm \beta) = \mp \sin \beta$

If $\alpha = \beta$, $\cos 2\beta = \cos^2 \beta - \sin^2 \beta = 1 - 2 \sin^2 \beta$

APPENDIX F

SELECTED ISOTOPES[a]

Atomic No. Z	Element	Symbol	Atomic Mass No. A	Atomic Mass u	% Abundance, or Decay Mode If Radioactive	Half-Life (if Radioactive)
0	(Neutron)	n	1	1.008 665	β^-	10.37 min
1	Hydrogen	H	1	1.007 825	99.985	
	Deuterium	D	2	2.014 102	0.015	
	Tritium	T	3	3.016 050	β^-	12.33 yr
2	Helium	He	3	3.016 030	0.000 138	
			4	4.002 603	≈ 100	
3	Lithium	Li	6	6.015 121	7.5	
			7	7.016 003	92.5	
4	Beryllium	Be	7	7.016 928	EC, γ	53.29 days
			9	9.012 182	100	
5	Boron	B	10	10.012 937	19.9	
			11	11.009 305	80.1	

[a]Data for atomic masses are taken from *Handbook of Chemistry and Physics*, 66th ed., CRC Press, Boca Raton, FL. The masses are those for the neutral atom, including the Z electrons. Data for percent abundance, decay mode, and half-life are taken from E. Browne and R. Firestone, *Table of Radioactive Isotopes*, V. Shirley, Ed., Wiley, New York, 1986. α = alpha particle emission, β^- = negative beta emission, β^+ = positron emission, γ = γ-ray emission, EC = electron capture.

APPENDIX F Selected Isotopes *(continued)*

Atomic No. Z	Element	Symbol	Atomic Mass No. A	Atomic Mass u	% Abundance, or Decay Mode If Radioactive	Half-Life (if Radioactive)
6	Carbon	C	11	11.011 432	β^+, EC	20.39 min
			12	12.000 000	98.90	
			13	13.003 355	1.10	
			14	14.003 241	β^-	5730 yr
7	Nitrogen	N	13	13.005 738	β^+, EC	9.965 min
			14	14.003 074	99.634	
			15	15.000 108	0.366	
8	Oxygen	O	15	15.003 065	β^+, EC	122.2 s
			16	15.994 915	99.762	
			18	17.999 160	0.200	
9	Fluorine	F	18	18.000 937	EC, β^+	1.8295 h
			19	18.998 403	100	
10	Neon	Ne	20	19.992 435	90.51	
			22	21.991 383	9.22	
11	Sodium	Na	22	21.994 434	β^+, EC, γ	2.602 yr
			23	22.989 767	100	
			24	23.990 961	β^-, γ	14.659 h
12	Magnesium	Mg	24	23.985 042	78.99	
13	Aluminum	Al	27	26.981 539	100	
14	Silicon	Si	28	27.976 927	92.23	
			31	30.975 362	β^-, γ	2.622 h
15	Phosphorus	P	31	30.973 762	100	
			32	31.973 907	β^-	14.282 days
16	Sulfur	S	32	31.972 070	95.02	
			35	34.969 031	β^-	87.51 days
17	Chlorine	Cl	35	34.968 852	75.77	
			37	36.965 903	24.23	
18	Argon	Ar	40	39.962 384	99.600	
19	Potassium	K	39	38.963 707	93.2581	
			40	39.963 999	β^-, EC, γ	1.277×10^9 yr
20	Calcium	Ca	40	39.962 591	96.941	
21	Scandium	Sc	45	44.955 910	100	
22	Titanium	Ti	48	47.947 947	73.8	
23	Vanadium	V	51	50.943 962	99.750	
24	Chromium	Cr	52	51.940 509	83.789	
25	Manganese	Mn	55	54.938 047	100	
26	Iron	Fe	56	55.934 939	91.72	
27	Cobalt	Co	59	58.933 198	100	
			60	59.933 819	β^-, γ	5.271 yr
28	Nickel	Ni	58	57.935 346	68.27	
			60	59.930 788	26.10	
29	Copper	Cu	63	62.939 598	69.17	
			65	64.927 793	30.83	
30	Zinc	Zn	64	63.929 145	48.6	
			66	65.926 034	27.9	
31	Gallium	Ga	69	68.925 580	60.1	
32	Germanium	Ge	72	71.922 079	27.4	
			74	73.921 177	36.5	
33	Arsenic	As	75	74.921 594	100	
34	Selenium	Se	80	79.916 520	49.7	
35	Bromine	Br	79	78.918 336	50.69	
36	Krypton	Kr	84	83.911 507	57.0	
			89	88.917 640	β^-, γ	3.16 min
			92	91.926 270	β^-, γ	1.840 s
37	Rubidium	Rb	85	84.911 794	72.165	

APPENDIX F Selected Isotopes *(continued)*

Atomic No. Z	Element	Symbol	Atomic Mass No. A	Atomic Mass u	% Abundance, or Decay Mode If Radioactive	Half-Life (if Radioactive)
38	Strontium	Sr	86	85.909 267	9.86	
			88	87.905 619	82.58	
			90	89.907 738	β^-	29.1 yr
			94	93.915 367	β^-, γ	1.235 s
39	Yttrium	Y	89	88.905 849	100	
40	Zirconium	Zr	90	89.904 703	51.45	
41	Niobium	Nb	93	92.906 377	100	
42	Molybdenum	Mo	98	97.905 406	24.13	
43	Technetium	Tc	98	97.907 215	β^-, γ	4.2×10^6 yr
44	Ruthenium	Ru	102	101.904 348	31.6	
45	Rhodium	Rh	103	102.905 500	100	
46	Palladium	Pd	106	105.903 478	27.33	
47	Silver	Ag	107	106.905 092	51.839	
			109	108.904 757	48.161	
48	Cadmium	Cd	114	113.903 357	28.73	
49	Indium	In	115	114.903 880	95.7; β^-	4.41×10^{14} yr
50	Tin	Sn	120	119.902 200	32.59	
51	Antimony	Sb	121	120.903 821	57.3	
52	Tellurium	Te	130	129.906 229	38.8; β^-	2.5×10^{21} yr
53	Iodine	I	127	126.904 473	100	
			131	130.906 114	β^-, γ	8.040 days
54	Xenon	Xe	132	131.904 144	26.9	
			136	135.907 214	8.9	
			140	139.921 620	β^-, γ	13.6 s
55	Cesium	Cs	133	132.905 429	100	
			134	133.906 696	$\beta^-,$ EC, γ	2.062 yr
56	Barium	Ba	137	136.905 812	11.23	
			138	137.905 232	71.70	
			141	140.914 363	β^-, γ	18.27 min
57	Lanthanum	La	139	138.906 346	99.91	
58	Cerium	Ce	140	139.905 433	88.48	
59	Praseodymium	Pr	141	140.907 647	100	
60	Neodymium	Nd	142	141.907 719	27.13	
61	Promethium	Pm	145	144.912 743	EC, α, γ	17.7 yr
62	Samarium	Sm	152	151.919 729	26.7	
63	Europium	Eu	153	152.921 225	52.2	
64	Gadolinium	Gd	158	157.924 099	24.84	
65	Terbium	Tb	159	158.925 342	100	
66	Dysprosium	Dy	164	163.929 171	28.2	
67	Holmium	Ho	165	164.930 319	100	
68	Erbium	Er	166	165.930 290	33.6	
69	Thulium	Tm	169	168.934 212	100	
70	Ytterbium	Yb	174	173.938 859	31.8	
71	Lutetium	Lu	175	174.940 770	97.41	
72	Hafnium	Hf	180	179.946 545	35.100	
73	Tantalum	Ta	181	180.947 992	99.988	
74	Tungsten (wolfram)	W	184	183.950 928	30.67	
75	Rhenium	Re	187	186.955 744	62.60; β^-	4.6×10^{10} yr
76	Osmium	Os	191	190.960 920	β^-, γ	15.4 days
			192	191.961 467	41.0	
77	Iridium	Ir	191	190.960 584	37.3	
			193	192.962 917	62.7	
78	Platinum	Pt	195	194.964 766	33.8	

APPENDIX F Selected Isotopes (continued)

Atomic No. Z	Element	Symbol	Atomic Mass No. A	Atomic Mass u	% Abundance, or Decay Mode If Radioactive	Half-Life (if Radioactive)
79	Gold	Au	197	196.966 543	100	
			198	197.968 217	β^-, γ	2.6935 days
80	Mercury	Hg	202	201.970 617	29.80	
81	Thallium	Tl	205	204.974 401	70.476	
			208	207.981 988	β^-, γ	3.053 min
82	Lead	Pb	206	205.974 440	24.1	
			207	206.975 872	22.1	
			208	207.976 627	52.4	
			210	209.984 163	α, β^-, γ	22.3 yr
			211	210.988 735	β^-, γ	36.1 min
			212	211.991 871	β^-, γ	10.64 h
			214	213.999 798	β^-, γ	26.8 min
83	Bismuth	Bi	209	208.980 374	100	
			211	210.987 255	α, β^-, γ	2.14 min
			212	211.991 255	β^-, α, γ	1.0092 h
84	Polonium	Po	210	209.982 848	α, γ	138.376 days
			212	211.988 842	α, γ	45.1 s
			214	213.995 176	α, γ	163.69 μs
			216	216.001 889	α, γ	150 ms
85	Astatine	At	218	218.008 684	α, β^-	1.6 s
86	Radon	Rn	220	220.011 368	α, γ	55.6 s
			222	222.017 570	α, γ	3.825 days
87	Francium	Fr	223	223.019 733	α, β^-, γ	21.8 min
88	Radium	Ra	224	224.020 186	α, γ	3.66 days
			226	226.025 402	α, γ	1.6×10^3 yr
			228	228.031 064	β^-, γ	5.75 yr
89	Actinium	Ac	227	227.027 750	α, β^-, γ	21.77 yr
			228	228.031 015	β^-, γ	6.13 h
90	Thorium	Th	228	228.028 715	α, γ	1.913 yr
			231	231.036 298	β^-, γ	1.0633 days
			232	232.038 054	100; α, γ	1.405×10^{10}yr
			234	234.043 593	β^-, γ	24.10 days
91	Protactinium	Pa	231	231.035 880	α, γ	3.276×10^4 yr
			234	234.043 303	β^-, γ	6.70 h
			237	237.051 140	β^-, γ	8.7 min
92	Uranium	U	232	232.037 130	α, γ	68.9 yr
			233	233.039 628	α, γ	1.592×10^5 yr
			235	235.043 924	0.7200; α, γ	7.037×10^8 yr
			236	236.045 562	α, γ	2.342×10^7 yr
			238	238.050 784	99.2745; α, γ	4.468×10^9 yr
			239	239.054 289	β^-, γ	23.47 min
93	Neptunium	Np	239	239.052 933	β^-, γ	2.355 days
94	Plutonium	Pu	239	239.052 157	α, γ	2.411×10^4 yr
			242	242.058 737	α, γ	3.763×10^5 yr
95	Americium	Am	243	243.061 375	α, γ	7.380×10^3 yr
96	Curium	Cm	245	245.065 483	α, γ	8.5×10^3 yr
97	Berkelium	Bk	247	247.070 300	α, γ	1.38×10^3 yr
98	Californium	Cf	249	249.074 844	α, γ	350.6 yr
99	Einsteinium	Es	254	254.088 019	α, γ, β^-	275.7 days
100	Fermium	Fm	253	253.085 173	EC, α, γ	3.00 days
101	Mendelevium	Md	255	255.091 081	EC, α	27 min
102	Nobelium	No	255	255.093 260	EC, α	3.1 min
103	Lawrencium	Lr	257	257.099 480	α, EC	646 ms
104	Rutherfordium	Rf	261	261.108 690	α	1.08 min
105	Dubnium	Db	262	262.113 760	α	34 s

ANSWERS TO CHECK YOUR UNDERSTANDING

CHAPTER 1

CYU 1: (a) Yes.
  (b) No.
CYU 2: No.
CYU 3: a, b, c, f
CYU 4: No.
CYU 5: b, d
CYU 6: (a) 11 m
  (b) 5 m
CYU 7: No.
CYU 8: Yes.
CYU 9: (a) The magnitude of $\vec{B}$ is equal to the magnitude of $\vec{A}$.
  (b) The direction of $\vec{B}$ is opposite to the direction of $\vec{A}$.
CYU 10: Vector $\vec{A}$ is perpendicular to vector $\vec{B}$.
CYU 11: Vector $\vec{A}$ points in the same direction as vector $\vec{B}$.
CYU 12: $\vec{A}$ and $\vec{D}$
CYU 13: (a) A_x is $-$ and A_y is $+$
  (b) B_x is $+$ and B_y is $-$
  (c) R_x is $+$ and R_y is $+$
CYU 14: No.
CYU 15: Yes.
CYU 16: (a) $A_x = 0$ units and $A_y = +12$ units
  (b) $A_x = -12$ units and $A_y = 0$ units
  (c) $A_x = 0$ units and $A_y = -12$ units
  (d) $A_x = +12$ units and $A_y = 0$ units
CYU 17: No.
CYU 18: a

CHAPTER 2

CYU 1: 0 m
CYU 2: scalar quantity
CYU 3: No.
CYU 4: a
CYU 5: average velocity = 2.7 m/s due east, average speed = 8.0 m/s
CYU 6: Yes.
CYU 7: No.
CYU 8: c
CYU 9: No.
CYU 10: No.
CYU 11: the rifle with the short barrel
CYU 12: $1.73\,v$
CYU 13: a
CYU 14: b
CYU 15: b
CYU 16: b

CHAPTER 3

CYU 1: b
CYU 2: c
CYU 3: a and c
CYU 4: (a) Yes; when the object is at its highest point.
  (b) No.
CYU 5: No.
CYU 6: b
CYU 7: (a) when the ball is at its highest point in the trajectory
  (b) at the initial and final positions of the motion
CYU 8: Yes.
CYU 9: Both bullets reach the ground at the same time.
CYU 10: (a) The displacement is greater for the stone thrown horizontally.
  (b) The impact speed is greater for the stone thrown horizontally.
  (c) The time of flight is the same for both stones.
CYU 11: No.
CYU 12: Ball A has the greater launch speed.
CYU 13: (a) $+70$ m/s
  (b) $+30$ m/s
  (c) $+40$ m/s
  (d) -60 m/s
CYU 14: No.
CYU 15: The two times are the same.
CYU 16: (a) The range toward the front is the same as the range toward the rear.
  (b) The range toward the front is greater than the range toward the rear.
CYU 17: swimmer A

CHAPTER 4

CYU 1: b
CYU 2: c
CYU 3: d
CYU 4: No, because two or more forces can cancel each other, leading to a net force of zero.
CYU 5: c
CYU 6: a and d
CYU 7: b
CYU 8: Yes, because the ratio of the two weights depends only on the masses of the objects, which are the same on the earth and on Mars.
CYU 9: a
CYU 10: d
CYU 11: No.
CYU 12: a
CYU 13: b
CYU 14: c
CYU 15: To pull, because the upward component of the pulling force reduces the normal force and, therefore, also reduces the force of kinetic friction acting on the sled.
CYU 16: 43°
CYU 17: c, a, b
CYU 18: a
CYU 19: a
CYU 20: b
CYU 21: d
CYU 22: No, because there must always be a vertical (upward) component of the tension force in the rope to balance the weight of the crate.
CYU 23: c
CYU 24: No, because the transfer described does not change the total mass being pulled by the engine.
CYU 25: a

CHAPTER 5

CYU 1: (a) The velocity is due south and the acceleration is due west.
  (b) The velocity is due west and the acceleration is due north.
CYU 2: Yes, if you are going around a curve.
CYU 3: the person at the equator
CYU 4: a and b
CYU 5: *AB* or *DE, CD, BC*
CYU 6: (a) $4r$
  (b) $4r$
CYU 7: No.
CYU 8: the same
CYU 9: edge of the turntable
CYU 10: car B
CYU 11: less than
CYU 12: (a) less than
  (b) equal to
CYU 13: (a) Yes.
  (b) Yes.
  (c) Yes.
  (d) Yes.
CYU 14: vertical

CHAPTER 6

CYU 1: b
CYU 2: d
CYU 3: d

CYU 4: a
CYU 5: No.
CYU 6: c
CYU 7: false
CYU 8: c
CYU 9: a
CYU 10: a, b, and c
CYU 11: b
CYU 12: d
CYU 13: b and d
CYU 14: e
CYU 15: c
CYU 16: a
CYU 17: No.

CHAPTER 7
CYU 1: No.
CYU 2: The total linear momentum is approximately zero because of the random directions and random speeds of the moving people.
CYU 3: (a) Yes.
        (b) No.
CYU 4: (a) No.
        (b) Yes.
CYU 5: b
CYU 6: (a) No.
        (b) The impulse of the thrust is equal in magnitude and opposite in direction to the impulse of the force due to air resistance.
CYU 7: (a) No.
        (b) No.
CYU 8: equal to
CYU 9: Yes.
CYU 10: a
CYU 11: decrease
CYU 12: (a) No.
        (b) decrease
CYU 13: b, c, d
CYU 14: the cannonball
CYU 15: d
CYU 16: No. It is the total kinetic energy of the *system* that is the same before and after the collision.
CYU 17: c
CYU 18: nearer the heavier end
CYU 19: (a) zero
        (b) Yes, opposite to the motion of the sunbather.
CYU 20: a

CHAPTER 8
CYU 1: Both axes lie in the plane of the paper. One passes through point A and is parallel to the line BC. The other passes through point A and the midpoint of the line BC.
CYU 2: B, C, A
CYU 3: No. The instantaneous angular speed of each blade is the same, but the blades are rotating in opposite directions.
CYU 4: c

CYU 5: 1.0 rev/s
CYU 6: b
CYU 7: Case A
CYU 8: a
CYU 9: at the north pole or at the south pole
CYU 10: c
CYU 11: 0.30 m
CYU 12: d
CYU 13: c
CYU 14: b
CYU 15: 8.0 m/s^2
CYU 16: a
CYU 17: Among the many possible answers are the motions of a Frisbee through the air, the earth in its orbit, a twirling baton that has been thrown into the air, the blades on a moving lawn mower cutting the grass, and an ice skater performing a quadruple jump.

CHAPTER 9
CYU 1: 0°, 45°, 90°
CYU 2: greater torque
CYU 3: (a) Yes, if the lever arm is very small.
        (b) Yes, if the lever arm is very large.
CYU 4: the box at the far right
CYU 5: (a) C
        (b) A
        (c) B
CYU 6: Additional forces are necessary.
CYU 7: a
CYU 8: b
CYU 9: Bob
CYU 10: A, B, C
CYU 11: axis B
CYU 12: (a) remains the same
        (b) remains the same
CYU 13: (a) remains the same
        (b) increases
CYU 14: (a) Both have the same translational speed.
        (b) Both have the same translational speed.
CYU 15: axis B
CYU 16: solid sphere, solid cylinder, spherical shell, hoop
CYU 17: (a) decreases
        (b) remains the same
CYU 18: decrease
CYU 19: greater than
CYU 20: No.

CHAPTER 10
CYU 1: No, because the force of gravity acting on the ball is constant, unlike the restoring force of simple harmonic motion.
CYU 2: Both boxes experience the same net force due to the springs.

CYU 3: 180 N/m
CYU 4: The spring stretches more when attached to the wall.
CYU 5: object II
CYU 6: at the position $x = 0$ m
CYU 7: The particle can cover the greater distance in the same time because at larger amplitudes the maximum speed is greater
CYU 8: The same amount of energy is stored in both cases, since the elastic potential energy is proportional to the square of the displacement x.
CYU 9: b, c, a
CYU 10: The amplitude is unchanged. The frequency and maximum speed each decrease by a factor of $\sqrt{2}$.
CYU 11: a
CYU 12: the simple-pendulum clock, because its period depends on the acceleration due to gravity
CYU 13: Use a shoe and the shoe laces to make a simple pendulum whose period is related to the magnitude g of the acceleration due to gravity (see Equations 10.5 and 10.16). Measure the period of your pendulum and calculate g.
CYU 14: Yes, because for small angles the period of each person's motion is the same.
CYU 15: Yes, because the frequency depends on the mass of the car and its occupants (see Equations 10.6 and 10.11).
CYU 16: $v = \dfrac{d}{2\pi}\sqrt{\dfrac{k}{m}}$
CYU 17: b
CYU 18: The rod with the square cross section is longer.
CYU 19: No, because the value of B given in Table 10.3 applies to solid aluminum, not to a can that is mostly empty space.
CYU 20: No, because pressure involves a force that acts *perpendicular* to an area. In Equation 10.18 for shear deformation, the force acts parallel, not perpendicular, to the area A (see Figure 10.31).
CYU 21: Face B experiences the largest stress, and face C experiences the smallest stress.

CHAPTER 11
CYU 1: (a) outward
        (b) inward
CYU 2: b
CYU 3: a
CYU 4: (a) increase
        (b) decrease
        (c) remain constant

CYU 5: A noticeable amount of water will remain in the tank.

CYU 6: b

CYU 7: Yes; see Equation 11.4, in which P_2 is the pressure at his wrist and P_1 is the pressure above the water.

CYU 8: c

CYU 9: Both beams experience the same buoyant force.

CYU 10: (a) The readings are the same.
(b) The final reading is greater than the initial reading.

CYU 11: No. You float because the weight of the water you displace equals your weight. Each weight is proportional to g, so its value makes no difference.

CYU 12: No. F_B depends only on the weight of the water she displaces, which doesn't change.

CYU 13: b

CYU 14: No.

CYU 15: d

CYU 16: c

CYU 17: e

CYU 18: c

CYU 19: c

CYU 20: b

CYU 21: a

CYU 22: c

CHAPTER 12

CYU 1: 178 °X

CYU 2: (a) No. (b) Yes. (c) No.

CYU 3: It decreases (see Equations 10.5 and 10.16).

CYU 4: With equal values for α, concrete and steel expand (contract) by the same amount as the temperature increases (decreases), thus minimizing problems with thermal stress.

CYU 5: The bottom is bowed outward, because it acts like a bimetallic strip.

CYU 6: b and d

CYU 7: cooled

CYU 8: No. When the temperature changes, the change in volume of the cavity within the glass would exactly compensate for the change in volume of the mercury, which would never rise or fall in the capillary tube of the thermometer.

CYU 9: Less than. The buoyant force is equal to the weight of the displaced water (see Section 11.6, Archimedes' principle), which is proportional to the water's density. Here, warmer water has a smaller density than cooler water does (see Figure 12.20).

CYU 10: a and b

CYU 11: the object with the smaller mass

CYU 12: c, b, d, a

CYU 13: Because heat is released when the water freezes at 0 °C (consistent with the latent heat of fusion of water), and this heat warms the blossoms.

CYU 14: c, a, b

CYU 15: No, because at sea level water boils at a higher temperature and the stove may not generate enough heat.

CYU 16: Because water in an open pot boils at 100 °C, thus preventing the temperature from rising further, whereas under the elevated pressure in the autoclave water has a boiling point above 100 °C.

CYU 17: Boiling water has a vapor pressure of one atmosphere, and the cool water in the sealed jar has a lower vapor pressure. The excess external pressure crates a net force pushing on the lid, making it hard to unscrew.

CYU 18: Under pressure in the sealed bottle, the soda has a freezing point lower than normal (see Figure 12.35b). The outside temperature is not cold enough to freeze it. When the bottle is opened, the pressure on the liquid decreases to one atmosphere, and the freezing point rises to its normal value. The liquid is now cold enough to freeze.

CYU 19: As the water vapor is removed, more forms in an attempt to reestablish equilibrium between liquid and vapor. When the pumping is rapid, the required latent heat is supplied mostly by the remaining liquid, which cools and eventually freezes.

CYU 20: 100%

CYU 21: Yes. The dew points on the two nights could be different, Tuesday's being higher than Monday's due to a greater partial pressure of water vapor in the air on Tuesday than on Monday.

CYU 22: The air above the swimming pool probably has a greater partial pressure of water vapor (due to inefficient humidity control) and, therefore, a higher dew point than that in the other room. Evidently, the temperature at the inner window-surfaces is below the dew point of the room with the swimming pool but above the dew point in the other room.

CHAPTER 13

CYU 1: a

CYU 2: b

CYU 3: the house with the snow on the roof

CYU 4: b

CYU 5: c

CYU 6: c

CYU 7: hollow, air-filled strands

CYU 8: c

CYU 9: b

CYU 10: forced convection

CYU 11: strip B

CYU 12: a

CYU 13: d

CYU 14: b

CYU 15: e

CHAPTER 14

CYU 1: Both have the same number of molecules, but oxygen has the greater mass.

CYU 2: In general, the number of molecules would be different. But they could be the same, if the molecular masses of the two types of molecules happen to be the same.

CYU 3: 66.4%

CYU 4: The ideal gas law gives the pressure as $P = nRT/V$, where T and V are constant. The fan reduces n in the house and increases it in the attic, so pressure decreases in the house and increases in the attic. The fan has a harder job pushing air out against the higher attic pressure.

CYU 5: The ideal gas law gives the gas pressure as $P = nRT/V$, where V and n are constant. As T increases, the pressure increases and could cause the can to burst.

CYU 6: The ideal gas law gives the gas pressure as $P = nRT/V$, where V and n are constant. As T increases, the pressure increases.

CYU 7: The ideal gas law gives the gas pressure as $P = nRT/V$, where T and n are constant. As V decreases due to the incoming tide, the pressure increases, and your ears pop inward, as if you were climbing down a mountain.

CYU 8: The ideal gas law gives the gas volume as $V = nRT/P$, where T and n are constant. As the pressure P decreases during the balloon's ascent, the volume increases. The balloon would overinflate if not underinflated to start with.

CYU 9: Boyle's law gives the final pressure in the bottle after the cork is pressed in: $P_f = P_i(V_i/V_f)$, where V_i/V_f is the volume of air above the wine before the cork is pressed in divided by the volume after the cork is pressed in. This ratio is much larger for the full bottle than for the half-full bottle, creating a pressure large enough to push the cork out.

CYU 10: Xenon has the greatest and argon the smallest temperature.

CYU 11: less than, which follows directly from the impulse–momentum theorem

CYU 12: No. The average kinetic energy is proportional to the Kelvin, not the Celsius, temperature.

CYU 13: It remains unchanged.

CYU 14: argon

CYU 15: $v_{rms,\,new}/v_{rms,\,inital} = 0.707$

CYU 16: L must be small and there must be many alveoli so that the total effective area A is large.

CYU 17: c, a, b

CHAPTER 15

CYU 1: d

CYU 2: b

CYU 3: c

CYU 4: $A \rightarrow B$: $Q = +$ and $W = +$
$B \rightarrow C$: $\Delta U = +$ and $W = 0$

CYU 5: a

CYU 6: b

CYU 7: a

CYU 8: c

CYU 9: c

CYU 10: c

CYU 11: b

CYU 12: No, because Carnot's principle only states that a reversible engine operating between two temperatures is more efficient than an irreversible engine operating between the *same temperatures.*

CYU 13: d

CYU 14: d

CYU 15: b

CYU 16: c

CYU 17: a

CYU 18: c

CYU 19: b

CYU 20: a

CYU 21: d

CYU 22: c and d

CYU 23: the popcorn that results from the kernels; a salad after it has been tossed; a messy apartment

CYU 24: b

CYU 25: c

CHAPTER 16

CYU 1: c

CYU 2: No. The coil moves back and forth in simple harmonic motion.

CYU 3: The wavelength increases.

CYU 4: The person pulling on string B should pull harder to increase the tension in the string.

CYU 5: In Equation 16.2, the speed would be infinitely large if m were zero, so it would take no time at all.

CYU 6: decrease

CYU 7: No, because the particles exhibit simple harmonic motion, in which the acceleration is not always zero.

CYU 8: increase

CYU 9: a

CYU 10: No, because each particle executes simple harmonic motion as the wave passes by.

CYU 11: hot day

CYU 12: CO and N_2

CYU 13: increase

CYU 14: Large outer ears intercept and direct more sound power into the auditory system than smaller ones do.

CYU 15: No, because not all points on the surface are at the same distance from the source.

CYU 16: No, because it is the intensities I_1 and I_2 that add to give to a total intensity I_{total}. The intensity levels β_1 and β_2 do not add to give a total intensity level β_{total}.

CYU 17: (a) 1/4
(b) 2

CYU 18: (a) f_o is smaller than f_s, and f_o decreases during the fall.
(b) f_o is greater than f_s, and f_o increases during the fall.

CYU 19: No, because the observed frequency is less than the source frequency, so the car is moving away from him.

CYU 20: (a) greater in air
(b) greater under water

CYU 21: No, because there is no relative motion of the cars.

CYU 22: (a) minus sign in both places
(b) the truck driver

CHAPTER 17

CYU 1: (a) -3 cm
(b) -2 cm

CYU 2: No, because if the two sound waves have the same amplitude and frequency, they might cancel in a way analogous to that illustrated in Figure 17.2*b* and no sound will be heard.

CYU 3: b

CYU 4: c

CYU 5: a

CYU 6: d

CYU 7: a

CYU 8: c

CYU 9: b

CYU 10: d

CYU 11: (a) 4
(b) 3
(c) node
(d) 110 Hz

CYU 12: b

CYU 13: d

CYU 14: b

CYU 15: (a) antinode
(b) node
(c) $\frac{1}{4}\lambda$
(d) lowered

CYU 16: b

CYU 17: c ($\frac{1}{4}\lambda$ is the distance between an antinode and an adjacent node.)

CYU 18: a

CHAPTER 18

CYU 1: c

CYU 2: $+3.2 \times 10^{-13}$ C on object A and -3.2×10^{-13} C on object B

CYU 3: $+1.6 \times 10^{-13}$ C on object A and -3.2×10^{-13} C on object B

CYU 4: b and e

CYU 5: Yes, because the charge on the balloon will induce a slight charge of opposite polarity in the surface of the ceiling, analogous to that in Figure 18.8.

CYU 6: a

CYU 7: b

CYU 8: C, A, B

CYU 9: the electron, because, being less massive, it has the greater acceleration

CYU 10: No, because the force of the spring changes direction when the spring is stretched compared to when it is compressed, while the electrostatic force does not have this characteristic.

CYU 11: d

CYU 12: 0 N/C

CYU 13: (a) corner C
(b) negative
(c) greater

CYU 14: a

CYU 15: (a) No.
(b) No.

CYU 16: For rod A, the field points perpendicularly away from the rod. For rod B, it points parallel to the rod and is directed from the positive toward the negative half.

CYU 17: (a) false
(b) false
(c) true
(d) false
(e) false

CYU 18: d

CYU 19: The flux does not change, as long as the charge remains within the Gaussian surface.

CYU 20: The same flux passes through each, since each encloses the same net charge.

CYU 21: (a) q_1 and q_2
(b) q_1, q_2, and q_3

CHAPTER 19

CYU 1: (a) Yes.
(b) No.

(c) Yes.

(d) Yes.

CYU 2: The work is the same in all three cases (see Equation 19.4).

CYU 3: The electron arrives at a plate first.

CYU 4: a

CYU 5: b

CYU 6: c

CYU 7: a

CYU 8: d

CYU 9: (a) remains the same

(b) decreases

CYU 10: the electron

CYU 11: (a) +2.0 V

(b) 0 V

(c) +2.0 V

CYU 12: The electric field is zero.

CYU 13: b

CYU 14: a

CYU 15: c

CYU 16: (a) bottom of a valley

(b) top of a mountain

CYU 17: (a) decreases

(b) increases

(c) increases

(d) increases

CYU 18: (a) decreases

(b) increases

(c) remains the same

(d) increases

CHAPTER 20

CYU 1: d

CYU 2: 0.50 A

CYU 3: b

CYU 4: a

CYU 5: b, d, and e

CYU 6: c (A value for the current is also needed.)

CYU 7: a

CYU 8: c

CYU 9: b and d

CYU 10: The 75-W bulb. See Equation 20.15c.

CYU 11: d

CYU 12: e

CYU 13: in parallel

CYU 14: b

CYU 15: c

CYU 16: a, b, d, and e

CYU 17: There are two ways. One is to form two groups of two parallel resistors and then connect the groups in series. The other is to form two groups of two series resistors and then connect the groups in parallel.

CYU 18: Junction rule: $I_1 + I_3 = I_2$

Loop rule, loop *ABCD:*

$3.0 \text{ V} + 7.0 \text{ V} + I_3R_3 = I_1R_1$

Loop rule, loop *BEFC:*

$5.0 \text{ V} = I_3R_3 + 7.0 \text{ V} + I_2R_2$

CYU 19: c

CYU 20: b

CYU 21: ohm × farad

= (volt/ampere)(coulomb/volt)

= coulomb/ampere

= coulomb/(coulomb/second)

= second

CYU 22: e

CHAPTER 21

CYU 1: c

CYU 2: d

CYU 3: (a) Yes.

(b) No, because the particle could move either parallel or anti-parallel to the magnetic field.

CYU 4: b

CYU 5: d

CYU 6: particle 3

CYU 7: b

CYU 8: c

CYU 9: b

CYU 10: c

CYU 11: (a) The direction of the magnetic force reverses.

(b) The direction of the magnetic force does not change.

CYU 12: B and D (a tie), A, C

CYU 13: a

CYU 14: (a) repelled

(b) repelled

CYU 15: (a) attracted

(b) repelled

CYU 16: c

CYU 17: a

CYU 18: Part *a*: There is a point to the right of both wires where the total magnetic field is zero.

Part *b*: There is a point between the wires where the total magnetic field is zero. This point is closer to the wire carrying the current I_2.

CYU 19: A, D, C, B

CYU 20: d

CYU 21: No, because aluminum is a non-ferromagnetic material.

CYU 22: b

CHAPTER 22

CYU 1: No. With both the magnet and coil moving at the same velocity with respect to the earth, there is no relative motion between the magnet and the coil, which is needed for there to be an induced current in the coil.

CYU 2: d

CYU 3: a

CYU 4: b

CYU 5: c

CYU 6: b

CYU 7: A lightning bolt is a large electric current that changes in time and, thus, produces a magnetic field that also changes in time. When

this changing field passes through a coil or loop of wire in an appliance, it can, via Faraday's law, create an induced emf, which can lead to an induced current.

CYU 8: c

CYU 9: a and d

CYU 10: a

CYU 11: b

CYU 12: Answer 1: downward and decreasing

Answer 2: upward and increasing

CYU 13: c

CYU 14: d

CYU 15: b

CYU 16: With the headlights off, the engine does not need to do the work of keeping the battery charged.

CYU 17: a

CYU 18: c

CYU 19: b

CYU 20: b and d

CHAPTER 23

CYU 1: The ratio decreases by a factor of 3.

CYU 2: (a) the circuit containing the inductor

(b) the circuit containing the resistor

CYU 3: less than

CYU 4: decreases

CYU 5: decreases

CYU 6: d

CYU 7: (a) increases

(b) increases

CYU 8: a

CYU 9: (a) remains the same

(b) decreases

CYU 10: in phase (see Equation 23.8, in which $X_L = X_C$)

CYU 11: (a) Yes.

(b) Yes.

CYU 12: a

CYU 13: (a) left to right

(b) left to right

CHAPTER 24

CYU 1: d

CYU 2: a

CYU 3: because, according to Faraday's law of electromagnetic induction, the emf depends on how rapidly the magnetic field of the wave is changing and this is determined by the frequency of the wave

CYU 4: e

CYU 5: d

CYU 6: b

CYU 7: No. The same Doppler change results when the star moves away from the earth and when the earth moves away from the star. Only the relative motion between the star and the earth can be detected.

CYU 8: B, A, C

CYU 9: Yes.

CYU 10: The light intensity that is not transmitted is absorbed by the polarizer and the analyzer. The polarizer absorbs one-half of the incident intensity, and the analyzer absorbs four-tenths of the incident intensity.

CYU 11: unpolarized: c
horizontally polarized: b
vertically polarized: c

CYU 12: because the transmission axis of the Polaroid material is nearly horizontal, in the same direction as the polarized light reflected from the lake

CHAPTER 25

CYU 1: 55°
CYU 2: Yes.
CYU 3: Yes.
CYU 4: $f_{inside} = +0.30$ m, $f_{outside} = -0.30$ m
CYU 5: (a) concave
(b) The sodium unit and engine are located at the focal point of the mirror.
CYU 6: Open the surface up to produce a more gently curving shape.
CYU 7: No.
CYU 8: (a) upright
(b) upside down
CYU 9: (a) Yes, provided the object distance is greater than the focal length of the mirror.
(b) It is not possible for a convex mirror to project an image directly onto a screen.
CYU 10: (a) No.
(b) No.
CYU 11: (a) The magnitude of the image distance becomes larger.
(b) The magnitude of the image height becomes larger.
CYU 12: (a) No. You can see yourself anywhere on the principal axis.
(b) You cannot see yourself when you are between the center of curvature and the focal point of the mirror because your image is behind you.
CYU 13: A, D, and E
CYU 14: The image will never be located beyond the focal point (behind the mirror).

CHAPTER 26

CYU 1: slab B
CYU 2: liquid B
CYU 3: Yes. To see why, apply Snell's law at the air–water interface and at the water–glass interface.
CYU 4: the one filled with water
CYU 5: liquid A
CYU 6: c
CYU 7: b

CYU 8: a
CYU 9: c
CYU 10: The critical angle for a water–air interface is 48.8° (see Equation 26.4). Any light emitted at an angle greater than 48.8° with respect to the vertical is incident on the surface at an angle exceeding the critical angle. It is totally internally reflected and doesn't exit the water.
CYU 11: No. To see why, apply Snell's law at both surfaces of the glass slab and use Equation 26.4.
CYU 12: c (They are most effective when the angle of incidence is the Brewster angle and the reflected light is 100% polarized.)
CYU 13: liquid A
CYU 14: a (Since $n = 1.520$ for red light and $n = 1.538$ for violet-colored light, the critical angle for total internal reflection is greater for red than for violet-colored light.)
CYU 15: b
CYU 16: Yes.
CYU 17: a
CYU 18: the lens
CYU 19: converging lens, $d_o = \frac{1}{2}f$
CYU 20: d
CYU 21: the glasses of the farsighted person, since they use converging lenses
CYU 22: 13 cm
CYU 23: Light normally passes from air ($n = 1.00$) into the cornea ($n = 1.38$), at which time most of the eye's refraction of the light occurs. If water ($n = 1.33$) replaces air, the similarity of the index of refraction of water to that of the cornea reduces the eye's normal refraction and causes blurred vision. Goggles preserve the air–cornea boundary.
CYU 24: b
CYU 25: hawk, kestrel, eagle
CYU 26: a
CYU 27: 0.042 rad
CYU 28: b
CYU 29: the longer telescope
CYU 30: microscope
CYU 31: c, d, e, f
CYU 32: because chromatic aberration is related to the refraction of light and not to the reflection of light

CHAPTER 27

CYU 1: (a) constructive
(b) destructive
(c) destructive
CYU 2: c
CYU 3: Yes.
CYU 4: a

CYU 5: (a) d_1 and λ_2
(b) d_2 and λ_1
CYU 6: (a) The pattern would be the same.
(b) The positions of the light and dark fringes would be interchanged.
CYU 7: No, because θ in Equations 27.1 and 27.2 approaches 90° as λ becomes larger and larger.
CYU 8: b
CYU 9: c
CYU 10: enhances
CYU 11: (a) A and C
(b) B
CYU 12: c
CYU 13: b
CYU 14: (a) broadens
(b) contracts
CYU 15: a
CYU 16: c
CYU 17: a, c, b
CYU 18: Yes.
CYU 19: small f-number setting
CYU 20: (a) the maximum that is closer to the central maximum
(b) away from the central maximum
CYU 21: The distance between the bright fringes would decrease.

CHAPTER 28

CYU 1: d
CYU 2: a, b
CYU 3: C, B, A
CYU 4: No, because the term v^2/c^2 in Equations 28.1 and 28.2 would then be zero.
CYU 5: c
CYU 6: c, d
CYU 7: No, because the two diagonals are perpendicular, so that diagonal AC is contracted, whereas diagonal BD is not contracted.
CYU 8: greatest mass: c, smallest mass: b
CYU 9: a, because then they have more electric potential energy (see Example 8 in Chapter 19)
CYU 10: b, because the fully charged capacitor stores electric potential energy (see Section 19.5)
CYU 11: a. The work is the change in kinetic energy, which is proportional to the mass (see the work–energy theorem in Section 6.2). The electron has the smaller mass.
CYU 12: c

CHAPTER 29

CYU 1: No.
CYU 2: (a) red
(b) violet
CYU 3: No.
CYU 4: b
CYU 5: a

CYU 6: (a) increases
(b) increases
(c) remains the same
(d) remains the same

CYU 7: less than

CYU 8: c

CYU 9: a

CYU 10: No. Photon collisions would cause spinning in a direction from the shiny side of a panel toward the black side.

CYU 11: decreases

CYU 12: b

CYU 13: decreases

CYU 14: b

CHAPTER 30

CYU 1: b

CYU 2: d

CYU 3: The absorption lines belong only to the Lyman series, since very few electrons are present with $n = 2$ or $n = 3$.

CYU 4: No, because the location of the electron in a given quantum mechanical energy state is uncertain.

CYU 5: when the electron is in the $n = 1$ state, because then the only possible value for the orbital quantum number is $\ell = 0$

CYU 6: (a) No, because the Bohr model uses the same quantum number n for the total energy and the orbital angular momentum (see Equations 30.13 and 30.8).
(b) Yes, because quantum mechanics uses the quantum number n for the total energy but the quantum number ℓ for the orbital angular momentum.

CYU 7: (a) Yes, because the Bohr model uses the same quantum number n for the orbital angular momentum and the total energy (see Equations 30.8 and 30.13).
(b) No, because quantum mechanics uses the quantum number ℓ for the orbital angular momentum but the quantum number n for the total energy.

CYU 8: $1s^2\, 2s^2\, 2p^6\, 3s^2\, 3p^6\, 4s^2\, 3d^{10}\, 4p^6$

CYU 9: (a) No.
(b) Yes.

CYU 10: b

CYU 11: c

CYU 12: a

CYU 13: c

CYU 14: d

CYU 15: a

CHAPTER 31

CYU 1: c, d

CYU 2: a

CYU 3: No, because they could have different numbers of protons (different atomic numbers).

CYU 4: Yes, because the total number A of nucleons could be the same, and it is the value of A that determines the radius.

CYU 5: c

CYU 6: d, c, a, b

CYU 7: d

CYU 8: It is not possible, because the total mass of the decay products is greater than the mass of the parent nucleus, $^{238}_{92}\text{U}$, indicating that energy would not be released.

CYU 9: b

CYU 10: Yes, because the decay of any single nucleus occurs randomly and can happen at any moment.

CYU 11: Yes.

CYU 12: b, because the gold statue does not contain carbon atoms

CYU 13: too small

CYU 14: No, because in 700 years the activity of a sample would have decreased to an immeasurably small fraction of its initial value.

CYU 15: e

CHAPTER 32

CYU 1: Yes, if the absorbed dose of the radiation is different for each type of radiation (see Equation 32.4).

CYU 2: c

CYU 3: neutrons, α particles, γ rays

CYU 4: c

CYU 5: (a) because it violates the conservation of nucleon number
(b) because it violates the conservation of nucleon number
(c) because it violates the conservation of electric charge

CYU 6: d

CYU 7: electrons, protons, neutrons

CYU 8: No, because the binding energy per nucleon is greater for the original nucleus than for the two fragments, as indicated in Figure 31.5.

CYU 9: b and d

CYU 10: a, b, and c

CYU 11: No, because the binding energy per nucleon is greater for the original nuclei than for the nucleus resulting from the fusion, as indicated in Figure 32.9.

CYU 12: b and d

CYU 13: a

ANSWERS
TO ODD-NUMBERED PROBLEMS

CHAPTER 1

1. 124 m^2
3. 10 159 m
5. 0.75 m^2/s
7. 2.0 magnums
9. (a), (d), and (e) are dimensionally correct.
11. $\dfrac{[M]}{[T]^2}$
13. 25.9° south of west
15. 0.25 m
17. 0.487 nm
19. 35.3°
21. 1.2 × 10^2 m
23. (a) 5.70 × 10^2 newtons
    (b) 33.6° south of west
25. (a) $F = 551$ newtons, $\theta = 36.1°$ north of west
    (b) $F = 551$ newtons, $\theta = 36.1°$ south of west
27. smallest magnitude: $\vec{F}_1 + \vec{F}_3 =$ 10.0 newtons, due east; largest magnitude: $\vec{F}_3 + \vec{F}_4 =$ 70.0 newtons, due west
29. (a) 1200 m
    (b) 26° south of east
31. (a) 5600 newtons
    (b) along the dashed line
33. (a) 142 newtons, 67° south of east
    (b) 142 newtons, 67° north of west
35. (a) 15.8 m/s
    (b) 6.37 m/s
37. (a) 46 paces
    (b) 88 paces
39. 222 m, 55.8°
41. Vectors $\vec{A}$ and $\vec{C}$ are equal (each has a magnitude of 100.0 m, and each is oriented at an angle of 36.9° above the +x axis).
43. (a) 10.4 units
    (b) 12.0 units
45. 7.1 m, 9.9° north of east
47. 30.2 m, 10.2°
49. 0.90 km, 56° north of west
51. (a) 2.7 km
    (b) 6.0 × 10^1 degrees, north of east
53. 6.88 km, 26.9°
55. (a) 178 units
    (b) 164 units
57. 5.5 km
59. $\vec{C}$ has the largest x component. $\vec{B}$ has the largest y component.
61. (a) 147 km
    (b) 47.9 km
63. (a) 9.4 ft
    (b) 69°

65. (a) 25.0°
    (b) 34.8 newtons
67. (a) 78 newtons
    (b) 34°
69. x component: −288 units
    y component: +156 units

CHAPTER 2

1. (a) 12.4 km
   (b) 8.8 km, due east
3. 5 × 10^4 yr
5. (a) +8.0 m/s
   (b) −8.0 m/s
   (c) +20.0 m/s
7. 109 m
9. 52 m
11. 7.2 × 10^3 m
13. 2.1 s
15. (a) 0 m/s^2
    (b) −14 m/s^2
17. (a) 3.0 × 10^2 days
    (b) +1.04 × 10^{-4} m/s^2
19. (a) 18 m/s
    (b) 6.0 m/s
    (c) 6.0 m/s
    (d) 18 m/s
21. 4.5 m/s^2
23. 8.0 m/s (Cycle A was initially traveling faster.)
25. 3.1 m/s^2 directed southward
27. (a) 1.7 × 10^2 cm/s^2
    (b) 0.15 s
29. (a) 1.5 m/s^2
    (b) 1.5 m/s^2
    (c) 76 m
31. 0.74 m/s
33. (a) 2.2 m/s
    (b) 4.4 m/s
    (c) 0.021 m/s^2
35. 52.8 m
37. 96.9 m/s
39. +25.7 m
41. 14 s
43. (a) −9.80 m/s^2
    (b) 5.7 m
45. 6.12 s
47. $d_1 = 0.018$ m, $d_2 = 0.071$ m, $d_3 = 0.16$ m
49. 1.7 s
51. (a) −7.9 m/s
    (b) 3.2 m
53. 1.1 s
55. −5.06 m
57. 0.767 m/s
59. 10.6 m

61. 0.40 s
63. −11 m/s
65. segment A: 1.9 m/s^2, segment B: 0 m/s^2, segment C: 3.3 m/s^2
67. (a) segments A and C: positive, segment B: negative, segment D: zero
    (b) segment A: +6.3 km/h, segment B: −3.8 km/h, segment C: +0.63 km/h, segment D: 0 km/h
69. −8.3 km/h^2
71. (a) 6.6 s
    (b) 5.3 m/s
73. 44.1 m/s
75. −1.0 m/s^2
77. 1.6 × 10^{-2} s
79. 2.8 s
81. (a) 2.67 × 10^4 m
    (b) 6.74 m/s, due north
83. 2.0 × 10^1 m
85. 11.1 s
87. (a) 13 m/s
    (b) 0.93 m/s^2

CHAPTER 3

1. 2.8 m
3. $v_x = 11$ m/s, $v_y = 13$ m/s
5. $x = 75.3$ km, $y = 143$ km
7. 8.8 × 10^2 m
9. (a) 2.47 m/s^2
   (b) 2.24 m/s^2
11. (a) 1.35 km, 21° north of west
    (b) 0.540 km/h, 21° north of west
13. 14.6 s
15. 5.4 m/s
17. 1.7 s
19. 4.90 m
21. (a) 1.78 s
    (b) 20.8 m/s
23. (a) 239 m/s, 57.1° above the horizontal
    (b) 239 m/s, 57.1° above the horizontal
25. 1130 m/s, 37.7° above the +x axis
27. 30.0 m
29. 0.844 m
31. 14.1 m/s
33. 5.2 m
35. 33.2 m
37. (a) 1380 m
    (b) 66.0° below the ground
39. 14.7 m/s
41. 56 m
43. 14.9 m
45. 5.79 m/s
47. 0.141° and 89.860°
49. $D = 850$ m, $H = 31$ m
51. $\theta_1 = 28.1°$ and $\theta_2 = 67.7°$

53. (a) 2.0×10^3 s
   (b) 1.8×10^3 m
55. 4.5 m/s
57. (a) 41 m/s, due east
   (b) 41 m/s, due west
59. 6.3 m/s, 18° north of east
61. 2.3 m/s
63. 5.2 m/s, 52° west of south
65. 3.05 m/s, 14.8° north of west
67. 8.6 m/s
69. (a) 6.0×10^1 m
   (b) 290 m
71. (a) 1.1 s
   (b) 1.3 s
73. 5.17 s
75. 42°
77. 21.9 m/s, 40.0° above the horizontal
79. 8.79 m/s, 81.5°

CHAPTER 4

1. 93 N
3. (a) $+6$ N
   (b) -24 N
   (c) -9.0 N
5. 130 N
7. 3560 N
9. (a) 3.6 N
   (b) 0.40 N
11. 1.83 m/s², directed to the left
13. 10.3 m/s², directed above the horizontal at 21.9°
15. 1.39×10^6 N
17. 0.78 m, 21° south of east
19. 18.4 N, 68° north of east
21. (a) weight $= 1.13 \times 10^3$ N, mass $= 115$ kg
   (b) weight $= 0$ N, mass $= 115$ kg
23. (a) 5.1×10^{-6} N
   (b) 5.1×10^{-6} N
25. (a) 10.5 m/s²
   (b) 1.07
27. 0.223 m/s²
29. (a) 1.04×10^3 N
   (b) 1.04×10^3 N
   (c) 2.45 m/s²
   (d) 1.74×10^{-22} m/s²
31. 1.76×10^{24} kg
33. 4.7 kg
35. 178
37. $x = +0.414 L$
39. 39 N
41. (a) 0.97
   (b) 0.82
43. (a) 1.6×10^4 N
   (b) 4.3×10^3 N
45. 0.444
47. (a) 390 N
   (b) 7.7 m/s, direction is toward second base
49. 4.4 s
51. 1.65 m/s², 34.6° above the x axis
53. 1.00×10^2 N, 53.1° south of east
55. 11.6 N

57. (a) 1400 N
   (b) 2400 N
59. 9.70 N
61. 62 N
63. 1.9×10^2 N
65. 0.141
67. 406 N
69. 0.29
71. 18.0 m/s², 56.3° above the $+x$ axis
73. 1730 N, due west
75. (a) 1.3 N
   (b) 6.5 N
77. 6.6 m/s
79. 160 N
81. 2730 N
83. (a) 4.25 m/s²
   (b) 1080 N
85. 0.265 m
87. 820 N
89. (a) $\Delta T_A = 0$ N, $\Delta T_B = -4.7$ N,
      $\Delta T_C = 0$ N
   (b) $\Delta T_A = 0$ N, $\Delta T_B = 0$ N,
      $\Delta T_C = +4.7$ N
91. (a) 13.7 N
   (b) 1.37 m/s²
93. (a) 0.60 m/s²
   (b) left string: 104 N,
      right string: 230 N
95. 1.8×10^{-7} N
97. 7.3×10^2 N
99. (a) 447 N
   (b) 241 N
101. 29 400 N
103. 4290 N
105. (a) 3.75 m/s²
   (b) 2.4×10^2 N
107. (a) 3.56 m/s²
   (b) 281 N
109. 286 N
111. 8.7 s
113. 33 s
115. 16.3 N
117. 1.2 s
119. 68°

CHAPTER 5

1. 160 s
3. 1.6 m
5. 0.79 m/s²
7. 3600
9. 332 m
11. 10 600 rev/min
13. 606 N
15. 0.68 m/s
17. 0.31 m
19. (a) 88 N
   (b) 181 N
21. 28°
23. (a) 3510 N
   (b) 14.9 m/s
25. (a) 19 m/s
   (b) 23 m/s
27. 39°

29. 184 m
31. satellite A: 7690 m/s
   satellite B: 7500 m/s
33. 1.33×10^4 m/s
35. 1/27
37. 12 m/s
39. 2.45×10^4 N
41. 14.0 m/s
43. 17 m/s
45. Twelve o'clock: 14 N
   Six o'clock: 18 N
47. 8.48 m/s
49. (a) 1.2×10^4 N
   (b) 1.7×10^4 N
51. 6.9 m/s²
53. 33 m/s
55. (a) 1.70×10^3 N
   (b) 1.66×10^3 N
57. (a) 3.0×10^5 m/s²
   (b) 3.1×10^4 g
59. 3500 N
61. speed $= 19$ m/s: 23 N
   speed $= 38$ m/s: 77 N

CHAPTER 6

1. 1.20×10^4 J
3. (a) 4.1×10^5 J
   (b) -4.1×10^5 J
5. 42.8°
7. (a) More net work is done during the dive.
   (b) 6.8×10^7 J
9. (a) 1.8×10^3 J
   (b) -1.2×10^3 J
11. 45 N
13. 3.2×10^3 J
15. 39 m/s
17. 2.5×10^7 J
19. (a) 3.1×10^3 J
   (b) 2.2×10^2 J
21. 9×10^3 m/s
23. 18%
25. 1.4×10^{11} J
27. 5.4×10^2 J
29. 444 J
31. (a) -3.0×10^4 J
   (b) The resistive force is not a conservative force.
33. (a) 27 J
   (b) 36 J
   (c) 8.8 J
   (d) The change in gravitational potential energy is -27 J $= -W$, where W is the work done by the weight.
35. (a) -1086 J
   (b) The skater is 2.01 m below the starting point.
37. At $h = 20.0$ m: KE $= 0$ J, PE $= 392$ J, and $E = 392$ J
   At $h = 10.0$ m: KE $= 196$ J, PE $= 196$ J, and $E = 392$ J
   At $h = 0$ m: KE $= 392$ J, PE $= 0$ J, and $E = 392$ J

39. (a) 28.3 m/s
  (b) 28.3 m/s
  (c) 28.3 m/s
41. (a) 52.2 J
  (b) 48.8 m/s
43. 3.29 m/s
45. 6.33 m
47. 18 m
49. 0.33 m
51. 1.0×10^3 J
53. -4.51×10^4 J
55. -1.21×10^6 J
57. (a) -270 J
  (b) 140 N
59. 2450 N
61. 13.5 m
63. 3.6×10^6 J
65. (a) 3.3×10^4 W
  (b) 5.1×10^4 W
67. (a) 1.0×10^4 W
  (b) 13 hp
69. 6.7×10^2 N
71. (a) Bow 1 requires more work.
  (b) 25 J
73. (a) 93 J
  (b) No work is done on the skater.
  (c) 2.3 m/s
75. 7.07 m/s
77. 6.6 m/s
79. 2.2×10^3 J
81. 256 N
83. (a) 2.8 J
  (b) 35 N
85. 1.7 m/s
87. 3.40×10^2 N

CHAPTER 7

1. 9.6 ms
3. 1.8 N, downward
5. -8.7 kg·m/s
7. $+5.1 \times 10^7$ kg·m/s
9. (a) $+2.2 \times 10^3$ N
  (b) $+4.4 \times 10^3$ N
11. 6.7 m
13. 3.7 N·s
15. $+344$ N
17. 84 kg
19. 96 kg
21. 4500 m/s, in the same direction that the
  rocket had before the explosion
23. $m_1 = 1.00$ kg, $m_2 = 1.00$ kg
25. 0.707
27. $+547$ m/s
29. (a) -0.432 m/s
  (b) $+1.82$ m/s
31. 7.4%
33. (a) 5.00-kg ball: -0.400 m/s, 7.50-kg
    ball: $+1.60$ m/s
  (b) both balls: $+0.800$ m/s
35. $+9.3$ m/s
37. $+9.09$ m/s
39. (a) $+8.9$ m/s

(b) -3.6×10^4 N·s
  (c) 5.9 m
41. (a) 73.0°
  (b) 4.28 m/s
43. 2.175×10^{-3}
45. (a) 5.56 m/s
  (b) 1.50-kg ball: -2.83 m/s,
    4.60-kg ball: $+2.73$ m/s
  (c) 1.50-kg ball: 0.409 m,
    4.60-kg ball: 0.380 m
47. 8 bounces
49. (a) $+1.0$ m/s
  (b) $+1.0$ m/s
  (c) equal to
51. (a) -1.05 m/s
  (b) $+2.53$ m/s
53. (a) -1.5 m/s
  (b) $+1.1$ m/s
55. (a) 4.89 m/s
  (b) 1.22 m
57. -1.5×10^{-4} m/s
59. 9.5%
61. 1.5×10^{-10} m/s
63. (a) -0.14 m/s
  (b) -7.1×10^{-3} m/s
65. 0.097 m

CHAPTER 8

1. 63.7 grad
3. (a) $+7.3 \times 10^{-5}$ rad/s
  (b) $+2.0 \times 10^{-7}$ rad/s
5. 21 rad
7. (a) $+0.75$ rad/s^2
  (b) -0.75 rad/s^2
  (c) $+1.0$ rad/s^2
  (d) -2.0 rad/s^2
9. 8.0 s
11. 128 s
13. 1200 s
15. 6.05 m
17. (a) 2.00×10^{-2} s
  (b) 4.00×10^{-2} s
19. 25 rev
21. (a) 4.00×10^1 rad
  (b) 15.0 rad/s
23. (a) 54.0 rad/s
  (b) 486 rad
25. (a) 10.0 s
  (b) -2.00 rad/s^2
27. $+267$ rad
29. 1.95×10^4 rad
31. 2.1 rev
33. 7.37 s
35. (a) 7.50 rad/s
  (b) -1.73×10^{-3} rad/s^2 (The angular
    velocity is decreasing.)
37. 0.18 m
39. 22 rev/s
41. (a) 1.25 m/s
  (b) 7.98 rev/s
43. (a) 3.61 rad/s
  (b) 6.53 rad/s^2

45. 0.87
47. (a) 0.583 m/s^2
  (b) 31.0°
49. 0.577
51. (a) 2.4×10^5 m/s
  (b) 5.3×10^{20} N
53. 8.71 rad/s^2
55. 693 rad
57. (a) -1.4 rad/s^2
  (b) $+33$ rad
59. (a) 8.33 rad/s, counterclockwise
  (b) 14.7 rad/s, clockwise
61. 974 rev
63. 2 rev
65. (a) 2.5 m/s^2
  (b) 3.1 m/s^2
67. 157.3 rad/s
69. 0.62 m
71. 4.63 m/s
73. 12.5 s
75. 1.43×10^{-1} m/s

CHAPTER 9

1. 1.70×10^3 N·m
3. 2.1×10^2 N
5. (a) 27 N·m
  (b) 34°
7. 43.7°
9. (a) FL
  (b) FL
  (c) FL
11. 1.03 m
13. (a) 2590 N
  (b) 2010 N
15. 1200 N, to the left
17. 24 m/s
19. $\vec{T} = 56.4$ N, down
  $\vec{F} = 70.6$ N, up
21. $V = 170$ N, $P = 270$ N, $H = 210$ N
23. 37.6°
25. (a) 2260 N
  (b) horizontal component: 1450 N,
    vertical component: 1450 N
27. (a) 1.21×10^3 N
  (b) 1.01×10^3 N, downward
29. 51.4 N
31. 8.0×10^{-4} N·m
33. 0.027 kg·m^2
35. hoop: 0.20 N·m, disk: 0.10 N·m
37. 0.50 rad/s^2
39. 0.060 kg·m^2
41. 2.0 s
43. (a) system A: 229 kg·m^2,
    system B: 321 kg·m^2
  (b) system A: -1270 N·m,
    system B: 0 N·m
  (c) system A: -27.7 rad/s,
    system B: 0 rad/s
45. 0.78 N
47. 22.0 kg
49. 6.1×10^5 rev/min
51. (a) object 1: 12.0 m/s, object 2: 9.00 m/s,
    object 3: 18.0 m/s

(b) 1.08×10^3 J
   (c) 60.0 kg·m^2
   (d) 1.08×10^3 J
53. 2/7
55. 3/4
57. 1.3 m/s
59. 4.4 kg·m^2
61. 1.83 rad/s
63. 0.26 rad/s
65. 8%
67. 0.573 m
69. (a) 13 500 N·m
   (b) 132 000 N·m
71. 8.2 rad/s^2
73. (a) 5.94 rad/s^2
   (b) 44.0 N
75. (a) 7.40×10^2 N, downward
   (b) 0.851 m
77. 34 m/s (for each module)
79. 1.7 m

CHAPTER 10

1. 237 N
3. 650 N/m
5. 0.012 m
7. 1.4 kg
9. 0.240 m
11. 2.29×10^{-3} m
13. (a) 0.407 m
   (b) 397 N
15. 3.5×10^4 N/m
17. (a) -9.84 N
   (b) 10.5 rad/s
   (c) 1.26 m/s
   (d) 13.2 m/s^2
19. (a) 0.450 m
   (b) 3.31 rad/s
   (c) 1.49 m/s
21. 4.3 kg
23. 140 N/m
25. (a) -1.84×10^2 J
   (b) $+1.84 \times 10^2$ J
   (c) 0 J
27. (a) 58.8 N/m
   (b) 11.4 rad/s
29.

h (meters)	KE	PE (gravity)	PE (elastic)	E
0	0 J	0 J	8.76 J	8.76 J
0.200	1.00 J	3.92 J	3.84 J	8.76 J
0.400	0 J	7.84 J	0.92 J	8.76 J

31. 14 m/s
33. 7.18×10^{-2} m
35. 0.50 m/s
37. 24.2 rad/s
39. 0.44 m/s
41. 2.37×10^3 N/m
43. 0.40 s
45. (a) 3.5 rad/s
   (b) 2.0×10^{-2} J
   (c) 0.41 m/s
47. (a) 1.64 s
   (b) 1.64 s

49. 0.54 s
51. 5.2×10^{-4} m
53. 260 m
55. 2.9×10^{-2} m
57. 3.7×10^{-5} m
59. 1.6×10^5 N
61. 2.3×10^{-6} m
63. (a) 1.8×10^{-7} m
   (b) 1.0×10^{-6} m
65. 1.2×10^{11} N/m^2
67. (a) 2.5×10^{-4}
   (b) 7.5×10^{-5} m
69. 4.6×10^{-4}
71. 12 m
73. 2.1×10^{-5} m
75. 6.0 rad/s
77. 61 kg
79. $+0.50$ m
81. (a) amplitude = 3.59×10^{-2} m, frequency = 4.24 Hz
   (b) amplitude = 5.08×10^{-2} m, frequency = 4.24 Hz
83. (a) 2.66 Hz
   (b) 0.0350 m
85. 1.25 m/s (11.2-kg block), 0.645 m/s (21.7-kg block)
87. 33.4 m/s
89. -4.4×10^{-5}
91. (a) 0.25 s
   (b) 0.75 s

CHAPTER 11

1. 8750 N. The bed should not be purchased.
3. 317 m^2
5. 3.91×10^{-6} m^3
7. 1.9 gal
9. 4240 s
11. 1.1×10^3 N
13. 24
15. 32 N
17. 2.4×10^3 Pa
19. 0.750 m
21. 7.0×10^5 Pa
23. 2.9×10^4 Pa
25. 0.50 m
27. (a) 2.45×10^5 Pa
   (b) 1.73×10^5 Pa
29. 31.3 rad/s
31. 2.3×10^8 N
33. 3.8×10^5 N
35. (a) 93.0 N
   (b) 94.9 N
37. 108 N
39. 8.50×10^5 N·m
41. 4.89 m
43. 250 kg/m^3
45. 390 kg/m^3
47. 7.9×10^{-4} m^3
49. 6.3×10^{-3} kg
51. 7.6×10^{-2} m
53. 1120 N
55. 4.5×10^{-5} kg/s

57. (a) 0.18 m
   (b) 0.14 m
59. 0.816
61. (a) 150 Pa
   (b) The pressure inside the roof is greater than the pressure on the outside. Therefore, there is a net outward force on the roof that can blow it outward if the wind speed is sufficiently high.
63. 1.92×10^5 N
65. 96 Pa
67. 3.0×10^5 Pa
69. (a) 14 m/s
   (b) 0.98 m^3/s
71. (a) 32.8 m/s
   (b) 54.9 m
73. 33 m/s
75. 7.78 m/s
77. (a) 1.01×10^5 Pa
   (b) 1.19×10^5 Pa
79. 1.19
81. 1.7 m
83. 2.25
85. (a) 1.26×10^5 Pa
   (b) 19.4 m
87. 59 N
89. 8.3×10^3 lb
91. 7.0×10^{-2} m
93. 10.3 m
95. (a) 1.6×10^{-4} m^3/s
   (b) 2.0×10^1 m/s
97. 0.20 m
99. 0.74 m
101. 78.4 gal/min
103. 1.41×10^5 N, downward

CHAPTER 12

1. -459.67 °F
3. (a) 102 °C (day), -173 °C (night)
   (b) 215 °F (day), -2.80×10^2 °F (night)
5. (a) -196 °C
   (b) -321 °F
7. -164 °C
9. $T_R = T_F + 459.67$
11. 1500 m
13. (a) The radius will be larger.
   (b) 0.0017
15. 1.7×10^{-5} (C°)$^{-1}$
17. 5.8 m
19. 49 °C
21. 2.0027 s
23. 41 °C
25. 26 °C
27. 18 N
29. 2.5×10^{-7} m^3
31. 230 C°
33. 7.3×10^{-6} m^3
35. 0.33 gal
37. (a) The apparent weight will be larger after the sphere cools.
   (b) 18 N
39. 45 atm

41. 6.9

43. 43.0 °C

45. 19 °C

47. 21.03 °C

49. 940 °C

51. 650 W

53. 1.2×10^{-2} kg

55. 4.0×10^{5} J

57. 3.9×10^{5} J

59. (a) 4.52×10^{6} J
   (b) 5.36×10^{6} J

61. 9.49×10^{-3} kg

63. 0.42 kg

65. 64 °C

67. 0.223

69. 1.9×10^{4} J/kg

71. 3.50×10^{2} m/s

73. 0.237 kg

75. 5.5

77. 87%

79. 2.8×10^{5} J

81. 39%

83. 28%

85. 4.9×10^{-2} m

87. 3.9×10^{-3} kg

89. 110 C°

91. 3.1×10^{-3} m^3

93. 33%

95. 2.6×10^{-3} kg

97. 4.4×10^{3} N

99. 0.016 C°

101. 1.1×10^{3} N

CHAPTER 13

1. 1.5 C°

3. 8.0×10^{2} J/s

5. 2.0×10^{-3} m

7. 17

9. 287 °C

11. 85 J

13. (a) 21 °C
   (b) 18 °C

15. 103.3 °C

17. (a) 101.2 °C
   (b) 110.6 °C

19. (a) 2.0
   (b) 0.61

21. 14.5 d

23. (a) 6.3 J/s
   (b) 4.8 J/s

25. 1.2×10^{4} s

27. 532 K

29. 320 K

31. 0.39 kg

33. 12

35. 12 J

37. 5800 K

39. 0.70

41. −15 °C

43. 4.5

CHAPTER 14

1. 1.07×10^{-22} kg

3. (a) 294.307 u

(b) 4.887×10^{-25} kg

5. 1.00×10^{-2} g

7. (a) 2.3×10^{3} mol
   (b) 1.4×10^{27}

9. 42.4 mol

11. 1.1 g

13. (a) 201 mol
   (b) 1.21×10^{5} Pa

15. 67.0 m^3

17. 12

19. 925 K

21. 39

23. (a) 3.3×10^{2} K
   (b) 2.8×10^{5} Pa

25. 5.9×10^{4} g

27. 0.93 mol/m^3

29. 0.090 %

31. 308 K

33. 750 K

35. 3.9×10^{5} J

37. 327 m/s

39. 1.6×10^{-15} kg

41. 343 m/s

43. (a) −120 N (assuming the bullets travel
   in the + direction)
   (b) 120 N
   (c) 4.0×10^{5} Pa

45. 0.14 kg/m^3

47. 1.34×10^{-7} kg

49. (a) 2.1 s
   (b) 1.6×10^{-5} s
   (c) because the diffusional path is a zigzag
   path and not a straight-line path

51. (a) The answer is a derivation.
   (b) 31 s

53. 304 K

55. 2.2 kg/m^3

57. (a) 46.3 m^2/s^2
   (b) 40.1 m^2/s^2

59. 8.1 g

61. 2820 m

63. 7.23×10^{-20} J

CHAPTER 15

1. (a) $+1.6 \times 10^{4}$ J
   (b) -4.2×10^{4} J
   (c) -2.6×10^{4} J

3. (a) −87 J
   (b) +87 J

5. 32 miles

7. (a) -5.03×10^{5} J
   (b) 1.20×10^{2} nutritional calories

9. 1.2×10^{7} Pa

11. 4.5×10^{-3} m^3

13. (a) 0 J
   (b) $+2.1 \times 10^{3}$ J
   (c) -1.5×10^{3} J

15. 3.0×10^{5} Pa

17. The answer is a proof.

19. 4.99×10^{-6}

21. (a) 0 J
   (b) -6.1×10^{3} J
   (c) 310 K

23. −4700 J

25. 1.81

27. 19.3

29. A to B: $\Delta U = 4990$ J, $W = 3320$ J,
   $Q = 8310$ J
   B to C: $\Delta U = -4990$ J, $W = 0$ J,
   $Q = -4990$ J
   C to D: $\Delta U = -2490$ J, $W = -1660$ J,
   $Q = -4150$ J
   D to A: $\Delta U = 2490$ J, $W = 0$ J,
   $Q = 2490$ J

31. (a) -8.00×10^{4} J
   (b) Heat flows out of the gas.

33. (a) 477 K
   (b) 323 K

35. (a) 1.1×10^{4} J
   (b) 1.8×10^{4} J

37. 45 K

39. 5/2

41. (a) 24.4 J
   (b) 37.3 J/(mol·K)

43. 2.38×10^{4} J

45. 0.631

47. 65 J

49. 0.75

51. 256 K

53. (a) 1260 K
   (b) 1.74×10^{4} J

55. 1090 K

57. lowering the temperature of the cold
   reservoir

59. (a) 0.360
   (b) 1.3×10^{13} J

61. The answer is a proof.

63. 5.86×10^{5} J

65. 13

67. 275 K

69. 9.03

71. 0.48 K

73. (a) 2.0×10^{1}
   (b) 1.5×10^{4} J

75. engine I: +0.4 J/K (irreversible, could
   exist)
   engine II: 0 J/K (reversible)
   engine III: −1.0 J/K (irreversible, could
   not exist)

77. (a) 3.68×10^{3} J/K
   (b) 1.82×10^{4} J/K
   (c) The vaporization process creates
   more disorder.

79. (a) $+8.0 \times 10^{2}$ J/K
   (b) The entropy of the universe increases.

81. (a) +1.74 J/K
   (b) 811 J
   (c) 546 J

83. (a) -2.1×10^{2} K
   (b) decrease

85. 0.24 m

87. (a) reversible
   (b) −125 J/K

89. 21

91. (a) $+3.0 \times 10^{3}$ J
   (b) Work is done by the system.

93. 0.264 m
95. (a) 2.00×10^6 J
    (b) 925 K
    (c) 4.40×10^6 J
97. 44.3 s
99. (a) 5/9
    (b) 1/3

CHAPTER 16

1. 5.50×10^{14} Hz
3. (a) 10.0 s
    (b) 0.100 Hz
    (c) 32 m
    (d) 3.2 m/s
    (e) It is not possible to determine the amplitude.
5. 78 cm
7. 0.20 m/s
9. 5.0×10^1 s
11. (a) 1.09 m/s
    (b) 6.55 m
13. 64 N
15. 7.7 m/s^2
17. (a) 2.0×10^1 m/s
    (b) 1.4×10^1 m/s
19. 153 N
21. (a) $v = \sqrt{yg}$
    (b) 2.2 m/s ($y = 0.50$ m), 4.4 m/s ($y = 2.0$ m)
23. 3.26×10^{-3} s
25. $y = (0.35 \text{ m}) \sin[(88 \text{ rad/s})t - (17 \text{ m}^{-1})x]$
27. $y = (0.37 \text{ m}) \sin[(8.2 \text{ rad/s})t + (0.68 \text{ m}^{-1})x]$
29. (a) 4.2 m/s
    (b) 0.35 m
    (c) $y = (3.6 \times 10^{-2} \text{ m}) \sin[(75 \text{ rad/s})t + (18 \text{ m}^{-1})x]$
31. 28.8 K
33. 110 m
35. (a) 431 m/s
    (b) 322 m/s
37. 690 rad/s
39. 6.7×10^{-4} s
41. (a) first in metal, second in water, third in air
    (b) Second sound arrives 0.059 s later, and third sound arrives 0.339 s later.
43. 650 m
45. tungsten
47. 283 K
49. 57% argon, 43% neon
51. 0.404 m
53. 6.5 W
55. 190 m
57. 1.98%
59. 8.0×10^2 s
61. 7.6×10^3 W/m^2
63. 1.3
65. 1000
67. 1.0×10^2
69. (a) 7.4 dB
    (b) No, since it requires an increase of 10 dB to double the loudness.

71. 79 400
73. 2.6
75. 2.39 dB
77. 3.4 m/s
79. 1.054
81. 615 Hz
83. 22 m/s
85. 1.5 m/s^2
87. (a) 1570 Hz
    (b) 1590 Hz
89. 0.316 W/m^2
91. 2.06
93. 8.68×10^{-3} kg/m
95. 6.0
97. 0.25 m
99. 56 m/s
101. -6.0 dB
103. 0.84 s
105. $m_1 = 28.7$ kg, $m_2 = 14.3$ kg
107. (a) 2.20×10^2 m/s
    (b) 9.19 m/s
109. 239 m/s

CHAPTER 17

1. 8.42 m
3. The answer is a series of drawings.
5. 107 Hz
7. 3.89 m
9. (a) destructive interference
    (b) constructive interference
11. 3.90 m, 1.55 m, 6.25 m
13. (a) 44°
    (b) 0.10 m
15. (a) 53.8°
    (b) 23.8°
17. 3.7°
19. 8 Hz
21. 263 Hz
23. 437 Hz
25. 8 Hz
27. 2.4 m/s
29. 171 N
31. 1.10×10^2 Hz
33. (a) 180 m/s
    (b) 1.2 m
    (c) 150 Hz
35. 0.485
37. 0.077 m
39. 20.8° and 53.1°
41. 0.50 m
43. 1.96 m
45. 0.35 m
47. (a) $f_2 = 800$ Hz, $f_3 = 1200$ Hz, $f_4 = 1600$ Hz
    (b) $f_2 = 800$ Hz, $f_3 = 1200$ Hz, $f_4 = 1600$ Hz
    (c) $f_3 = 1200$ Hz, $f_5 = 2000$ Hz, $f_7 = 2800$ Hz
49. 0.557 m
51. 6.1 m
53. 1.95×10^{-3} s
55. 5.06 m
57. (a) 3

    (b) 0.57 m
59. 3.93×10^{-3} kg/m
61. 28 Hz and 42 Hz
63. 12 Hz

CHAPTER 18

1. -1.6 μC
3. (a) -1.6 μC
    (b) 1.0×10^{13}
5. (a) $+1.5$ q
    (b) $+4$ q
    (c) $+4$ q
7. (a) 3.35×10^{26} electrons
    (b) -5.36×10^7 C
9. 8
11. (a) 0.83 N
    (b) attractive
13. 0.38 N, 49° below the $-x$ axis
15. (a) both positive or both negative
    (b) 1.7×10^{-16} C
17. 3.8×10^{12}
19. 7.19×10^{23} m/s^2
21. (a) $+0.166$ N (directed along the $+y$ axis)
    (b) $+111$ m/s^2 (directed along the $+y$ axis)
23. 1.96×10^{-17} J
25. -3.3×10^{-6} C
27. (a) 15.4°
    (b) 0.813 N
29. 1.8 N due east
31. 1.37
33. 54 N/C
35. (a) 3.0 m from the positive charge (not between the charges)
    (b) 0 N
37. (a) -6.2×10^7 N/C (directed along the $-x$ axis)
    (b) $+2.9 \times 10^8$ N/C (directed along the $+x$ axis)
39. (a) positive
    (b) 2.53×10^7 protons
41. 3.11×10^2 N/C
43. 2.5×10^4 N/C
45. $|q_1| = 0.716$ q, $|q_2| = 0.0895$ q
47. 0.577
49. $+1.9 \times 10^{-2}$ m
51. 61°
53. 3.25×10^{-8} C
55. (a) 350 N·m^2/C
    (b) 460 N·m^2/C
57. 1.8×10^3 N·m^2/C
59. (a) The flux through the face in x, z plane at $y = 0$ m is -6.0×10^1 N·m^2/C. The flux through the face parallel to the x, z plane at $y = 0.20$ m is $+6.0 \times 10^1$ N·m^2/C. The flux through each of the remaining four faces is zero.
    (b) 0 N·m^2/C
61. The answer is a proof.
63. The answer is a drawing.
65. $-q$ on the interior surface and $+3$ q on the exterior surface
67. 0.16 N·m

69. (a) 4.56×10^{-8} C
(b) 3.25×10^{-6} kg
71. 3.9×10^6 N/C, in the $+y$ direction
73. 2.2×10^5 N/C, in the $-x$ direction
75. 92.0 N/m

CHAPTER 19
1. -2.1×10^{-11} J
3. 5.40×10^{-5} C
5. 67 hp
7. 19 m/s
9. (a) 3.0×10^{10} J
(b) 7.4×10^3 m/s
(c) 7.2×10^4 kg
11. 339 V
13. -4.05×10^4 V
15. 2.4
17. -4.7×10^{-2} J
19. $+7.8 \times 10^6$ V
21. -3.1×10^{-6} C
23. 0.0342 m
25. 1.53×10^{-14} m
27. -0.746 J
29. Each particle is moving at a speed of 9.7 m/s.
31. 1.1 m
33. 3.5×10^4 V
35. 8.8×10^6 V/m
37. (a) 179 V
(b) 143 V
(c) 155 V
39. 1.7×10^3 V/m, to the left
41. (a) 0 V
(b) $+290$ V
(c) -290 V
43. 1.1×10^3 V
45. (a) 33 J
(b) 8500 W
47. 5.3
49. 7.0×10^{13}
51. 52 V
53. 1.3×10^{-4} C
55. The answer is a proof.
57. (a) 1.3×10^{-12} C
(b) 8.1×10^6
59. -4.35×10^{-18} J
61. 1.1×10^{-20} J
63. 8.0×10^{-5} C
65. (a) 0 V/m
(b) 1.0×10^1 V/m
(c) 5.0 V/m
67. 2.77×10^6 m/s

CHAPTER 20
1. (a) 3.6×10^{-2} C
(b) 2.3×10^{17}
3. (a) 2.6 C
(b) 310 J
5. 22 A
7. 16 Ω
9. (a) 4.7×10^{13}
(b) 17 C°
11. 0.12 Ω
13. $-34.6°$ C

15. 1.64
17. 9.3%
19. 189 Ω
21. 9.7×10^2 kg
23. (a) 4.4 Ω
(b) 2.8 A
25. \$5.9 $\times 10^6$
27. 8.9 h
29. 190 s
31. 250° C
33. (a) 786 W
(b) 1572 W
35. 21 V
37. 1.3×10^{-3} m
39. (a) 50.0 Hz
(b) 2.40×10^2 Ω
(c) 60.0 W
41. 32 Ω
43. (a) 145 Ω
(b) 74 V
45. 9.0 V
47. (a) 1.2 Ω
(b) 110 V
49. (a) 35 Ω
(b) 5.0×10^1 Ω
51. 5.3 Ω
53. (a) 65.0 Ω
(b) 38.8 Ω
(c) 1.25 W
(d) 2.09 W
55. (a) 4.57 A
(b) 1450 W
57. 190 Ω
59. (a) 3.6 Ω
(b) 33 A. The breaker will open.
61. 3.58×10^{-8} m^2
63. 9.2 A
65. 4.6 Ω
67. (a) 8.33×10^{-2} A
(b) 0.833 W
69. 6.00 Ω, 0.545 Ω, 3.67 Ω, 2.75 Ω, 2.20 Ω, 1.50 Ω, 1.33 Ω, 0.833 Ω
71. (a) 0.750 A
(b) 2.11 A
73. 30
75. 8.3 A
77. 24.0 V
79. (a) 0.38 A
(b) 2.0×10^1 V
(c) B
81. $I = 5.00$ A, $V = 46.0$ V
83. 6.0 A (left to right) in the 2.0-Ω resistor, 2.0 A (left to right) in the 8.0-Ω resistor
85. 0.75 V. Left end is at higher potential.
87. 0.0835 Ω
89. 30.0 V
91. 820 Ω, 8.00×10^{-3} A
93. 9.0 V
95. 1.54
97. 2.0 μF
99. (a) 3.60×10^{-4} C
(b) 8.00×10^{-5} C
101. C_0

103. 4.1×10^{-7} F
105. 2.0×10^4 Ω
107. 0.29 s
109. The answer is a proof.
111. 82 Ω
113. 0.0050 (C°)$^{-1}$
115. 5.01 A
117. 25 Ω
119. (a) 15.5 V
(b) 14.2 W
121. $L_{\text{tungsten}}/L_{\text{carbon}} = 70$
123. 11 V

CHAPTER 21
1. (a) 5.7×10^{-5} N, into the paper
(b) 1.1×10^{-4} N, into the paper
(c) 5.7×10^{-5} N, into the paper
3. 4.1×10^{-3} m/s
5. 58°
7. 1.7×10^{-3} N
9. 1.3×10^{-10} N, directed out of the page
11. (a) due south
(b) 2.55×10^{14} m/s^2
13. (a) 7.2×10^6 m/s
(b) 3.5×10^{-13} N
15. (a) negatively charged
(b) 2.7×10^{-3} kg
17. (a) 4.3×10^2 m
(b) 7.8×10^5 m
19. 1.63×10^{-2} m
21. (a) 1.08×10^7 m/s
(b) 7.60×10^{-12} N
(c) 0.102 m
23. 0.16 T
25. (a) 0°
(b) 0.29 m
27. (a) 4.4×10^{-3} N
(b) 1.7×10^{-4} C
29. 9.6×10^4 m/s
31. 5.1×10^{-5} T
33. top side: 0.96 N, bottom side: 0.96 N, each of the other two sides: 0 N
35. 2.7 m
37. 0.19 N
39. 44°
41. 14 A
43. 0.062 m
45. 4.19×10^{-3} N·m
47. (a) 170 N·m
(b) The angle will increase.
49. 1.27
51. 8.3 N
53. (a) down
(b) 3.1×10^{-4} T
55. 2.8×10^4 turns/m
57. 3.8×10^{-5} T
59. (a) 4.3×10^{-5} T
(b) 5.3×10^{-5} T
61. 8.6 A. The current in the outer coil must have an opposite direction to the current in the inner coil.
63. 0.800 m, to the right of wire 1
65. 1.04×10^{-2} T

67. The current in wire 3 is directed out of the plane of the paper. $I_3/I = 2$
69. The answer is a proof.
71. The answer is a proof.
73. 0.12 m
75. 2.2 A
77. 8.0×10^{-5} T
79. 1.1×10^{-2} N
81. 8.7×10^{-3} s
83. (a) 6.8×10^{-3} N
  (b) 36°
85. 9.3×10^{-24} A·m²

CHAPTER 22
1. 0.065 V
3. 7800 V
5. rod A: emf = 0 V; rod B: emf = 1.6 V, with end 2 being positive;
  rod C: emf = 0 V
7. 0.15 W
9. 250 m
11. 2.2×10^{-3} Wb
13. 70.5°
15. 0.70
17. both triangular ends: 0 Wb;
  bottom surface: 0 Wb;
  1.2 m × 0.30 m surface: 0.090 Wb;
  1.2 m × 0.50 m surface: 0.090 Wb
19. 1.5 m²/s
21. 8.6×10^{-5} T
23. (a) 0.38 V
  (b) 0.43 m²/s
25. 0.050 V
27. 6.6×10^{-2} J
29. 2.4×10^{-3} A
31. 1.6×10^{-5} A
33. Figure 22.1*b*: right to left
  Figure 22.1*c*: left to right
35. (a) clockwise
  (b) clockwise
37. There is no induced current.
39. (a) location I: $x \rightarrow y \rightarrow z$
    location II: $z \rightarrow y \rightarrow x$
  (b) location I: $z \rightarrow y \rightarrow x$
    location II: $x \rightarrow y \rightarrow z$
41. (a) 2.4 Hz
  (b) 15 rad/s
  (c) 0.62 T
43. 3.0×10^5
45. 38 m
47. 15.4 V
49. 1.5×10^9 J
51. 2.5×10^{-2} H
53. 1.4 V
55. 220
57. 3.6×10^9 N/C
59. $M = \mu_0 \pi N_1 N_2 R_2^2/(2R_1)$
61. 1.0×10^1 W
63. 0.20 A
65. 0.25
67. (a) 7.0×10^5 W
  (b) 7.0×10^1 W
69. The answer is a proof.

71. 0.150 m
73. (a) clockwise
  (b) counterclockwise
75. 12 V
77. 0.14 V
79. (a) 3.6×10^{-3} V
  (b) 2.0×10^{-3} m²/s. Area must be shrunk.
81. 2100 rad/s

CHAPTER 23
1. 5.00×10^{-2} s
3. 2.7×10^{-6} F
5. 36 Ω
7. (a) 6.4×10^{-6} F
  (b) 9.0×10^{-4} C
9. 8.0×10^1 Hz
11. 75 V
13. 176 mH
15. (a) 1.11×10^4 Hz
  (b) 6.83×10^{-9} F
  (c) 6.30×10^3 Ω
  (d) 7.00×10^2 Ω
17. resistor: 10.5 V, capacitor: 19.0 V, inductor: 29.6 V
19. 83.9 V
21. (a) 0.925 A
  (b) 31.8°
23. (a) 10.7 V
  (b) −29.8°
25. 270 Hz
27. (a) 29.0 V
  (b) −0.263 A
29. 0.651 W
31. 2.7×10^{-5} H
33. (a) 352 Hz
  (b) 15.5 A
35. 0.81 W
37. (a) 1.3×10^{-3} H
  (b) 8.7×10^{-6} F
39. (a) 2.94×10^{-3} H
  (b) 4.84 Ω
  (c) 0.163
41. 0.707
43. 1.9 V
45. 38 V
47. 0.44 A
49. (a) 0.50 A
  (b) 0.34 A
  (c) Yes. 0.704 H
51. 8

CHAPTER 24
1. 8.33 min
3. 4.1×10^{16} m
5. The answers are in graphical form.
7. (a) 473 nm
  (b) 606 nm
9. 1.25 m
11. 3.7×10^4
13. 4.500×10^7 Hz
15. 1.3×10^6 m
17. 42.3 m/s
19. 6.4×10^{18} m

21. 0.24 s
23. (a) 6.81×10^5 N/C
  (b) 2.27×10^{-3} T
25. (a) 183 N/C
  (b) 6.10×10^{-7} T
27. 5.6×10^9 J
29. 3.93×10^{26} W
31. 5600 W
33. 4.44×10^{-10}
35. (a) receding
  (b) 3.1×10^6 m/s
37. (a) 6.175×10^{14} Hz
  (b) 6.159×10^{14} Hz
39. (a) 0.55 W/m²
  (b) 3.7×10^{-2} W/m²
41. 71.6°
43. 14 W/m²
45. 9.3°. The angle θ is increased.
47. 20
49. 11.118 m
51. 1.5×10^{-4} H
53. 0.07 N/C
55. 206 W/m²
57. (a) 2.4×10^9 Hz
  (b) 0.063 m
59. 920 W
61. 6.25×10^{-9} J

CHAPTER 25
1. 10°
3. 7.2 m
5. 55°
7. arrow A
9. (a) 30°
  (b) 30°
11. 1.2 m/s, in the $-x$ direction
13. The image is located 6.0 cm behind the mirror.
15. (a) The image is located 16.7 cm behind the mirror.
  (b) 6.67 cm
17. (a) The image is located 20.0 cm behind the mirror.
  (b) 6.0 cm
19. 22 cm
21. (a) 290 cm
  (b) −8.9 cm
  (c) upside down
23. 9.62 cm, concave
25. (a) −4.3 m
  (b) 0.39
27. (a) convex
  (b) 24.0 cm
29. 42.0 cm
31. 0.533 m
33. $-\frac{1}{2}$
35. (a) 1.07×10^5 m
  (b) 1420 m/s
37. (a) +62 cm
  (b) +0.35
  (c) upright
  (d) smaller
39. 1.67 m

41. (a) R
　　(b) -1
　　(c) inverted
43. 0.67
45. -3
47. 33.7°

CHAPTER 26

1. ethyl alcohol
3. 1.66×10^8 m/s
5. 2.0×10^{-11} s
7. 1.82
9. (a) 43°
　　(b) 31°
11. 0.9°
13. 1.92×10^8 m/s
15. 38.7°
17. 1.65
19. The answer is a derivation.
21. 1.19 mm
23. 2.7 m
25. 21.4 cm
27. 1.54
29. 1.51
31. (a) B
　　(b) A
33. 3.36×10^{-8} s
35. 1.35
37. 1.52
39. 25.0°
41. The answer is a proof.
43. 0.86°
45. 0.35°
47. red ray: 52.7°, violet ray: 56.2°
49. (a) $d_i = -75$ cm, $m = +2.5$
　　(b) $d_i = -75.0$ cm, $m = +2.50$
51. (a) -24 cm
　　(b) 6.0 mm
53. (a) 3.78 m
　　(b) width $= 8.40 \times 10^2$ mm,
　　　　height $= 1.26 \times 10^3$ mm
55. (a) -15 cm
　　(b) virtual
57. (a) 6.74×10^{-7} m^2
　　(b) 7.86×10^5 W/m^2
59. (a) converging lens
　　(b) $2f$
　　(c) $2f$
61. 48 cm
63. $+35$ cm and $+90.5$ cm
65. 5.6 cm to the left of the right-hand lens
67. -12 cm
69. (a) 4.00 cm to the left of the diverging
　　　　lens
　　(b) -0.167
　　(c) virtual
　　(d) inverted
　　(e) smaller
71. (a) 18.1 cm
　　(b) real
　　(c) inverted
　　　　160.0 cm

75. (a) 35.2 cm
　　(b) 32.9 cm
77. 28.0 cm
79. (a) converging
　　(b) farsighted
　　(c) 96.3 cm
81. (a) -4.5 m
　　(b) 0.50 m
83. 3.7
85. (a) 6.88 cm
　　(b) 3.63
87. 13.7 cm
89. 15.4
91. 0.81 cm
93. 4.8×10^{-3} rad
95. (a) -30.0
　　(b) 4.27 cm
　　(c) -4.57
97. 1.1 m
99. 0.261 cm
101. (a) 1.482 m
　　(b) 0.018 m
103. (a) -194
　　(b) -7.8×10^{-5} m
　　(c) 1.94×10^6 m
105. $d_i = 18$ cm
107. (a) 1.50
　　(b) 1.27
109. right eye: -0.20 diopters
　　left eye: -0.15 diopters
111. 61.1 mm
113. 2.46×10^8 m/s
115. -9.2 m
117. (a) -0.00625 m
　　(b) -0.0271 m
119. (a) 4.52×10^{-4} m
　　(b) 6.12×10^{-2} m
121. (a) 11.8 cm
　　(b) 47.8 cm
123. -31
125. (a) 31.3 cm
　　(b) 2.43 m

CHAPTER 27

1. Constructive interference occurs.
3. (a) Destructive interference occurs.
　　(b) 3.25 m and 0.75 m from one of the
　　　　sources
5. 4.9×10^{-7} m
7. 403 nm
9. 0.0248 m
11. 487 nm
13. 1.30×10^2 nm and 3.90×10^2 nm
15. 6.12×10^{-7} m
17. 1.18
19. 115 nm
21. 0.0256 m
23. (a) 1.1°
　　(b) 3.0×10^{-5} m
25. (a) 0.21°
　　(b) 22°
27. 1.2×10^{-5} m

29. 0.012 m
31. 2.0×10^{-5} m
33. (a) 3.53 m
　　(b) 2.15 m
35. 1.0×10^4 m
37. 1.9 mm
39. 2.3 m
41. (a) 1.22λ
　　(b) shorter wavelength
43. (a) 37°
　　(b) 22°
45. 4.0×10^{-6} m
47. 630 nm
49. 640 nm and 480 nm
51. $\frac{3}{4}$
53. 1.95 m
55. (a) 11°
　　(b) 22°
　　(c) 34°
　　(d) 48°
57. 6.0×10^{-5} m
59. (a) 9.7×10^{-3} m
　　(b) The hunter's claim is not reasonable.
61. (a) violet light: $\theta = 7.9°$
　　　　red light: $\theta = 13°$
　　(b) violet light: $\theta = 16°$
　　　　red light: $\theta = 26°$
　　(c) violet light: $\theta = 24°$
　　　　red light: $\theta = 41°$
　　(d) The second and third orders overlap.
63. 8
65. (a) 2
　　(b) $m_B = 4$, $m_A = 2$ and $m_B = 6$, $m_A = 3$

CHAPTER 28

1. 2.4×10^8 m/s
3. 72 h
5. 2.28 s
7. 16
9. 2.60×10^8 m/s
11. 530 m
13. 4.0 light-years
15. 40.2°
17. (a) 4.3 yr
　　(b) the twin traveling at $0.500\,c$
19. (a) 1.7×10^7 kg·m/s
　　(b) 3.0×10^7 kg·m/s
21. 1.0 m
23. -2.0 m/s
25. (a) 1.0
　　(b) 6.6
27. 5.3×10^6 mi
29. 5.0×10^{-13} J
31. 1.1×10^{24} kg/s
33. 1.3×10^7 kg·m/s
35. $0.31\,c$
37. $+0.80\,c$
39. 42 m
41. (a) 2.82×10^8 m/s
　　(b) 1.8×10^{-16} kg·m/s
43. 1.3
45. $0.999\ 95\,c$

47. $-0.406\,c$
49. 3.0 m × 1.3 m

CHAPTER 29
1. 7.7×10^{29} photons/s
3. 6.3 eV
5. 310 nm
7. 2.10 eV
9. 73 photons/s
11. 1.9×10^{-7} m
13. (a) 7760 N/C
    (b) 2.59×10^{-5} T
15. 75°
17. (a) 1.0×10^{13} Hz
    (b) infrared
19. 4.755×10^{-24} kg·m/s
21. 9.50×10^{-17} m
23. (a) 0.1819 nm
    (b) 1.092×10^{-15} J
    (c) 1.064×10^{-15} J
    (d) 2.8×10^{-17} J
25. 7.77×10^{-13} J
27. 1.41×10^{3} m/s
29. 7.38×10^{-11} m
31. 1.9×10^{-10} m
33. 1.86×10^{4} V
35. 6.01×10^{-11} m
37. 8.3×10^{-6} m/s
39. 1.8×10^{-20} kg·m/s
41. 8.0%
43. (a) 4.50×10^{-36} m/s
    (b) 7.05×10^{27} years
45. 1×10^{-18} m
47. 1.26 eV
49. 1.10×10^{3} m/s
51. 3.09×10^{-10} m

CHAPTER 30
1. (a) 6.2×10^{-31} m^3
    (b) 4×10^{-45} m^3
    (c) 7×10^{-13}%
3. 1.7×10^{-13} J
5. 7.3×10^{-14} m
7. (a) 7458 nm
    (b) 2279 nm
    (c) infrared region
9. 16
11. 1.98×10^{-19} J
13. 6.56×10^{-7} m and 1.22×10^{-7} m
15. −0.213 eV
17. $n_i = 6$ and $n_f = 2$
19. $6 \le n_i \le 19$
21. 30.39 nm
23. −0.378 eV
25. (a) −1.51 eV
    (b) 2.58×10^{-34} J·s
    (c) 2.11×10^{-34} J·s
27. 2, 3, 4, and 5
29. (a) Bohr model: $L = h/(2\pi)$,
        quantum mechanics: $L = 0$ J·s
    (b) Bohr model: $L = 3h/(2\pi)$,
        quantum mechanics: $L = 0$ J·s,
        $L = \sqrt{2}h/(2\pi)$, and $L = \sqrt{6}h/(2\pi)$

31. (a)

n	ℓ	m_ℓ	m_s
2	0	0	1/2
2	0	0	−1/2

(b)

n	ℓ	m_ℓ	m_s
2	1	1	1/2
2	1	1	−1/2
2	1	0	1/2
2	1	0	−1/2
2	1	−1	1/2
2	1	−1	−1/2

33. 50

35.

n	ℓ	m_ℓ	m_s
4	3	3	1/2
4	3	3	−1/2
4	3	2	1/2
4	3	2	−1/2
4	3	1	1/2
4	3	1	−1/2
4	3	0	1/2
4	3	0	−1/2
4	3	−1	1/2
4	3	−1	−1/2
4	3	−2	1/2
4	3	−2	−1/2
4	3	−3	1/2
4	3	−3	−1/2

37. 7.230×10^{-11} m
39. 6.83×10^{-11} m
41. (a) 1.08×10^{-14} J
    (b) 6.75×10^{4} eV
43. 21 600 V
45. 1.9×10^{17}
47. 19
49. 2.2×10^{16}
51. $1s^2\ 2s^2\ 2p^6\ 3s^2\ 3p^6\ 4s^2\ 3d^{10}\ 4p^3$
53. (a) not allowed
    (b) allowed
    (c) not allowed
    (d) allowed
    (e) not allowed
55. 3
57. −0.544 eV, −0.378 eV, −0.278 eV
59. The answer is a proof.
61. (a) $v_n = 2\pi ke^2 Z/(nh)$
    (b) 2.19×10^{6} m/s
    (c) 1.09×10^{6} m/s
    (d) Yes.

CHAPTER 31
1. (a) $+1.31 \times 10^{-17}$ C
    (b) 126
    (c) 208
    (d) 7.1×10^{-15} m
    (e) 2.3×10^{17} kg/m^3

3. (a) X = Pt (platinum), 117 neutrons
    (b) X = S (sulfur), 16 neutrons
    (c) X = Cu (copper), 34 neutrons
    (d) X = B (boron), 6 neutrons
    (e) X = Pu (plutonium), 145 neutrons
5. 4.4×10^{-15} m
7. $^{120}_{50}$Sn
9. 9.4×10^{3} m
11. 39.25 MeV
13. 7.90 MeV/nucleon
15. 2.28×10^{-28} kg
17. 1.003 27 u
19. (a) $^{18}_{9}\text{F} \rightarrow \,^{18}_{8}\text{O} + \,^{0}_{+1}e$
    (b) $^{15}_{8}\text{O} \rightarrow \,^{15}_{7}\text{N} + \,^{0}_{+1}e$
21. $^{35}_{16}\text{S} \rightarrow \,^{35}_{17}\text{Cl} + \,^{0}_{-1}e$
23. 4.87 MeV
25. (a) X = β^-
    (b) X = β^+
    (c) X = γ ray
    (d) X = α particle
27. 1.61×10^{7} m/s
29. $^{212}_{84}$Po
31. 1.82 MeV
33. 19.9
35. 387 yr
37. 8.00 days
39. 146 disintegrations/min
41. 3.7×10^{10} disintegrations/s
43. 1.2×10^{-7} g
45. 7.23 days
47. 0.70%
49. (a) 0.999
    (b) 1.36×10^{-9}
    (c) 0.755
51. age of fossils = 6900 yr, maximum error in the age of the fossils = 900 yr
53. (a) 1.741 670 u
    (b) 1622 MeV
    (c) 7.87 MeV/nucleon
55. 2.1×10^{13} Bq
57. 1.38 MeV
59. 4 782 969 electrons
61. energy carried away by $^{234}_{90}$Th daughter nucleus = 0.072 MeV, energy carried away by α particle = 4.2 MeV

CHAPTER 32
1. 12
3. (a) 460 rem
    (b) 50% chance of dying
5. 2.4×10^{4} rem
7. 5.0×10^{-4} C°
9. 4.4×10^{11} s^{-1}
11. 9.2×10^{8}
13. $\gamma + \,^{17}_{8}\text{O} \rightarrow \,^{12}_{6}\text{C} + \,^{4}_{2}\text{He} + \,^{1}_{0}n$
15. (a) $A = 233$, $Z = 90$, thorium $^{233}_{90}$Th
    (b) $A = 233$, $Z = 92$, uranium $^{233}_{92}$U
17. (a) $^{14}_{7}\text{N}\ (n, p)\ ^{14}_{6}\text{C}$, where n denotes $^{1}_{0}n$ and p denotes $^{1}_{1}$H
    (b) $^{238}_{92}\text{U}\ (n, \gamma)\ ^{239}_{92}\text{U}$
    (c) $^{24}_{12}\text{Mg}\ (n, d)\ ^{23}_{11}\text{Na}$, where d denotes $^{2}_{1}$H
19. 13.6 MeV

21. 9.0×10^{-4}
23. 184 MeV
25. 232.7851 u
27. (a) 8.2×10^{10} J
    (b) 0.48 g
29. 1200 kg
31. 3.3 MeV
33. 4.03 MeV

35. 1.0 gal
37. (a) 1.0×10^{22}
    (b) 9.6×10^9 kg
39. 33.9 MeV
41. 815 MeV
43. (a) 1.9×10^{-20} kg·m/s
    (b) 3.5×10^{-14} m
45. 0.18 MeV

47. (a) The κ^- particle does not contain u, c, or t quarks.
    (b) The κ^- particle does not contain $\bar{d}$, $\bar{s}$, or $\bar{b}$ antiquarks.
49. 41
51. 1.1×10^{-4} kg
53. 160 MeV

INDEX

APPENDIX

USING THE TEXT IN PREPARATION FOR THE AP PHYSICS B EXAM

You want to do well on the AP Physics B exam. That's one of the primary reasons you signed up for this course—to show, through your exam performance, that you understand physics at the college level. To prepare for the AP exam, you will use every resource available to you. Yes, some of these can be valuable, especially your AP Physics class, and practice tests from the College Board. But don't underestimate the usefulness of your textbook. The eighth edition of Cutnell and Johnson has been designed with the AP Physics B exam specifically in mind.

The AP Physics B exam is written by high school and college physics teachers to test the physics reasoning and problem-solving skills that are expected in college physics courses. The members of the test development committee put together what is possibly the most authentic standardized test you will ever take—meaning that if you know physics well, you will do well on the exam. Thus, studying for the exam is what you do every day in your AP Physics class. Pay attention in class, do the assigned problems in the textbook, *read* the textbook. Use the time spent in class, reading the text, and on homework problems as "study" time. Because the exam does such a good job of finding out what you know about physics, everything you can do to improve your physics understanding will simultaneously improve your likelihood of success on the AP exam.

KNOW THE LEVEL OF THE EXAM

It helps to have a clear understanding of the unique nature of the AP Physics exam. Most important, the difficulty of the exam is beyond what you have experienced on most tests in high school. Though you may be used to scoring 90% or above to earn an A, the top score of 5 on the AP exam is earned by garnering only 60–65% of the available points. That means you are not expected to do every part of every problem perfectly.

Your teacher will (of course) assign some problems from the text, and you may choose to look at other problems on your own. Each problem is rated with stars: no-stars, one-star, or two-stars. The AP exam will include questions at the level of the no-star and one-star problems.

This no-star problem could be similar to an AP multiple-choice question, or one part of an AP free-response problem. [p. 189]

21. Multiple-Concept Example 4 and **Interactive LearningWare 6.2** at **www.wiley.com/college/cutnell** review the concepts that are important in this problem. A 5.0×10^4-kg space probe is traveling at a speed of 11 000 m/s through deep space. Retrorockets are fired along the line of motion to reduce the probe's speed. The retrorockets generate a force of 4.0×10^5 N over a distance of 2500 km. What is the final speed of the probe?

This one-star problem could be a partial or entire AP free-response problem.

*22. The concepts in this problem are similar to those in Multiple-Concept Example 4, except that the force doing the work in this problem is the tension in the cable. A rescue helicopter lifts a 79-kg person straight up by means of a cable. The person has an upward acceleration of 0.70 m/s^2 and is lifted from rest through a distance of 11 m. **(a)** What is the tension in the cable? How much work is done by **(b)** the tension in the cable and **(c)** the person's weight? **(d)** Use the work–energy theorem and find the final speed of the person.

This level of difficulty should inform your study habits. Don't expect to be able to solve a one-star problem perfectly by yourself as soon as you look at it. Eventually, on average, you will get more than halfway through most problems. Use your initial foray at a textbook problem as a starting point for collaboration with classmates and your teacher.

The exam doesn't test how well you KNOW physics, but rather how well you can DO physics. AP exam problems will not be simply textbook questions with the numbers changed. Your collaboration after your initial attempt at a problem like the ones above will develop and reinforce problem-solving skills and strategies. Therefore, while you should not panic because you didn't get all the way through a problem by yourself, you nevertheless need to figure out how to finish. That's where the additional features of the text besides the end-of-chapter problems will be most useful.

KNOW STRUCTURE OF THE AP EXAM

The AP exam consists of two parts. Part I is 70 multiple-choice questions, to be answered in 90 minutes. Part II is 6–8 free-response problems, also to be answered in 90 minutes.

You are not allowed to use a calculator on the multiple-choice portion of the test. That's okay— you won't NEED a calculator, because the multiple-choice questions aren't testing your number crunching ability. For that matter, the free-response questions are not about how well you can use a calculator, even though you have access to one. Physics isn't about numbers. It's about understanding the natural world. And both types of questions will focus on your understanding, not on how well you can do arithmetic.

Though you will *practice* both multiple-choice and free-response questions, the physics skills necessary for each part of the exam are the same. Rather than "studying" for one or the other question structure, prepare instead for the categories of question you might have to answer.

PAY ATTENTION IN LABORATORY

One of the free-response questions, or at least a couple of PARTS of a question, will ask about an experiment. You may be asked to devise and describe experimental methods, or to analyze data that were collected in an experiment. Of course, the very best way to prepare for these questions is to take your own class's lab seriously. Don't just do the lab to get it done, figure out what you're doing and why.

When you are asked to describe an experimental procedure, you should use plain and direct language. Assume that your reader knows basic laboratory protocols. It's unnecessary to say "be sure to wear goggles!" or "record the data in a notebook." You're not writing a lab manual with step-by-step, detailed instructions. Rather, you need to say merely what will be measured, and HOW and with what device the measurement will be made. Try to answer any experimental procedure question in no more than three sentences.

Some *Check Your Understanding* questions ask you to design an experimental procedure, just like the AP exam will. [p. 300]

13. Concept Simulation 10.2 at **www.wiley.com/college/cutnell** deals with the concept on which this question is based. Suppose you were kidnapped and held prisoner by space invaders in a completely isolated room, with nothing but a watch and a pair of shoes (including shoe laces of known length). How could you determine whether you were on earth or on the moon?

KNOW YOUR EQUATIONS

The free-response portion of the AP exam comes with an equation sheet showing the fundamental physics relationships which you should use to approach the problems. The multiple-choice part of the exam, though, does not include an equation sheet. Thus, you are expected to know—I suggest this means *memorize*—the equations on the sheet.

Memorizing an equation doesn't merely mean being able to spit it back by rote. You must know what each variable means, under what circumstances the equation is valid, and be able to state in words the relationship represented by the equation. Throughout the text, purple boxes highlight most of the critical equations to know.

The purple boxes tell you everything you need to know about the equations that you must learn for the AP exam.

IMPULSE–MOMENTUM THEOREM

When a net average force $\Sigma \vec{\mathbf{F}}$ acts on an object during a time interval Δt, the impulse of this force is equal to the change in momentum of the object:

$$\underbrace{(\Sigma \vec{\mathbf{F}}) \, \Delta t}_{\text{Impulse}} = \underbrace{m \vec{\mathbf{v}}_{\mathbf{f}}}_{\substack{\text{Final} \\ \text{momentum}}} - \underbrace{m \vec{\mathbf{v}}_{\mathbf{0}}}_{\substack{\text{Initial} \\ \text{momentum}}} \qquad (7.4)$$

Impulse = Change in momentum

DIAGRAMS AND GRAPHS

- **Diagrams and graphs with special meaning**

Some types of graphs and diagrams show up regularly on the AP exam. For example, you can expect to see motion graphs, pV diagrams for thermal processes, electric and magnetic field lines, and more. When a kind of diagram can be sure to show up on the AP exam, the text usually will devote an entire section to the analysis.

2.7

GRAPHICAL ANALYSIS OF VELOCITY AND ACCELERATION

15.4

THERMAL PROCESSES

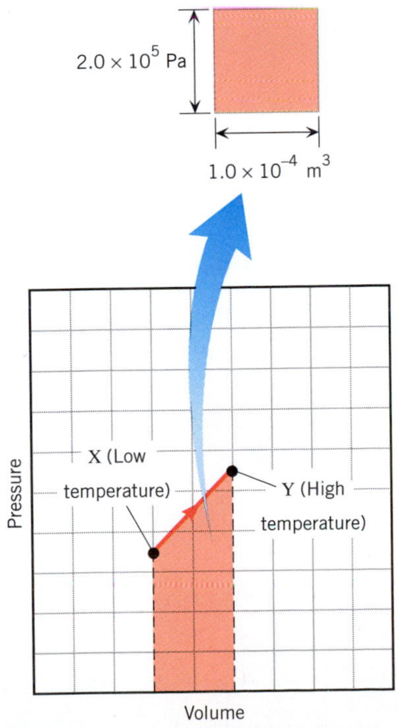

You're sure to see motion graphs and thermodynamic pressure-volume diagrams, so they each get an entire section of a chapter. [p. 48, 441]

2.0×10^5 Pa

1.0×10^{-4} m^3

Pressure

X (Low temperature)

Y (High temperature)

Volume

Figure 15.7 The colored area gives the work done by the gas for the process from X to Y.

- **Three things you can do with a graph**

Many AP problems test your ability to interpret physics concepts graphically. There are—usually—*three things* you can do with a graph. If you know your fundamental relationships, you can use your physics intuition to decide which of the three things is correct in each case. This textbook takes special care to demonstrate pictorially when a relationship can be represented and analyzed on a graph. The sidebar notes in every chapter show possible graphs that you might deal with on the AP exam.

Thing 1: Read off the axis of a graph

This may sound obvious, but students often forget under the pressure of the AP exam: When the quantity you're looking for is one of the axes, just read off that axis!

In this force-time graph, the average force can be read off the vertical axis, while a time interval is read off the horizontal axis. [p. 195]

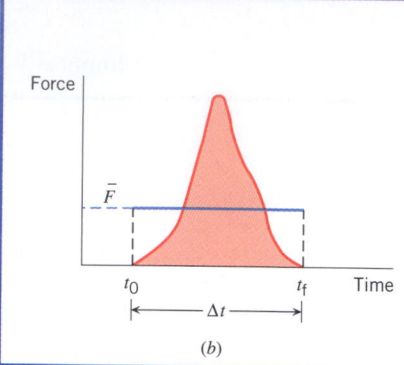

Figure 7.1 (*a*) The collision time between a bat and a ball is very short, often less than a millisecond, but the force can be quite large. (*b*) When the bat strikes the ball, the magnitude of the force exerted on the ball rises to a maximum value and then returns to zero when the ball leaves the bat. The time interval during which the force acts is Δt, and the magnitude of the average force is $\bar{F}$. (*a*, Chuck Savage/Corbis Images)

Thing 2: Take the slope of a graph

When the quantity you're looking for involves the vertical axis of a graph divided by the horizontal axis, take the slope.

The slope of a position-time graph is velocity because

$$v = \frac{\Delta x}{\Delta t}. \text{ [p. 48]}$$

Figure 2.18 A graph of position vs. time for an object moving with a constant velocity of $v = \Delta x/\Delta t = +4$ m/s.

Thing 3: Take the area under a graph

When the quantity you're looking for involves the vertical axis of a graph *multiplied* by the horizontal axis, find the area between the graph and the horizontal axis.

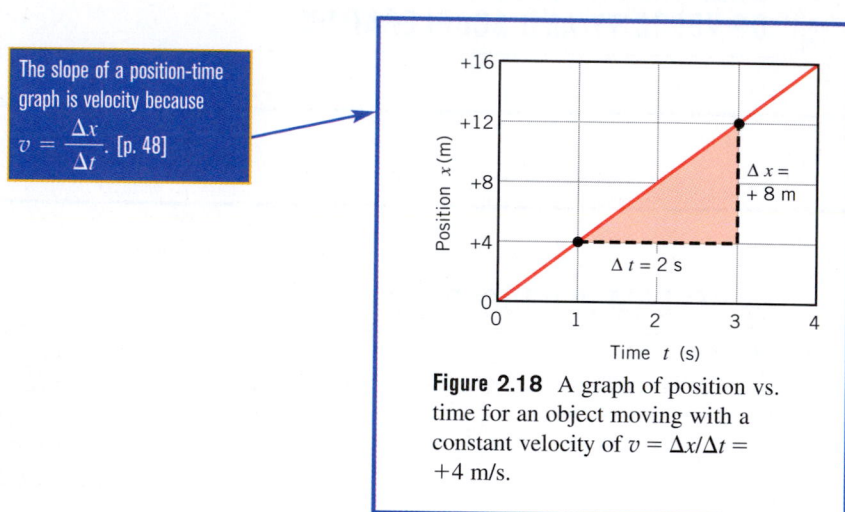

6.9 **WORK DONE BY A VARIABLE FORCE**

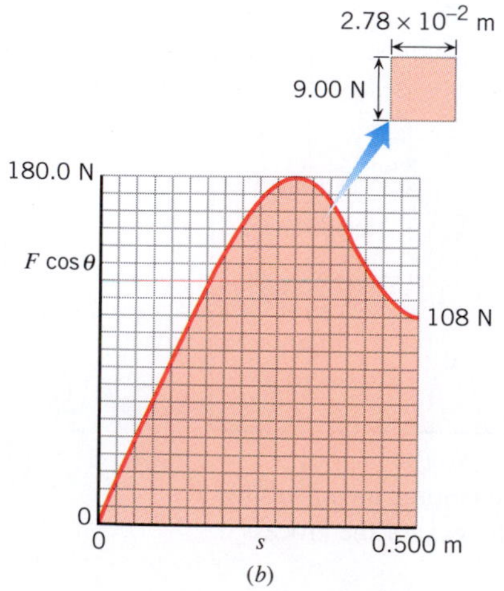

The area under a force-position graph is the work done by that force, because work = $F \cdot \Delta$x. [p. 182]

Figure 6.21 (*a*) A compound bow. (*b*) A plot of $F \cos \theta$ versus s as the bowstring is drawn back.

KNOW HOW TO SOLVE THE DIFFERENT TYPES OF QUESTIONS YOU'LL BE ASKED

While every single test item doesn't necessarily fit into a neat category, it is nevertheless possible to identify commonalities among questions. It might surprise you to learn that most AP exam questions do NOT require extensive calculation—and, since a calculator is forbidden on part I, the multiple-choice questions never demand lots of math. Here are a few types of problems you'll see again and again.

> The blue "Check Your Understanding" features provide AP-level conceptual questions. [from p. 204]

• One common type of question tests conceptual understanding. To prepare for free-response items that require a verbal response, try some of the "Check Your Understanding" questions at the end of chapter sections, like problems 5–8 on page 169 about work and energy. At right is a "Check Your Understanding" question about momentum:

> ✔ **CHECK YOUR UNDERSTANDING**
>
> (*The answers are given at the end of the book.*)
>
> **11.** You are a passenger on a jetliner that is flying at a constant velocity. You get up from your seat and walk toward the front of the plane. Because of this action, your forward momentum increases. Does the forward momentum of the plane itself decrease, remain the same, or increase?

When you answer this one, don't just say "decrease." Give a brief written response: "Momentum of the system consisting of me and the plane is conserved. Therefore, a gain in my momentum must be offset by a loss in the plane's momentum." An answer of one to two sentences, referencing a fundamental physics principle or relationship, is sufficient.

> "Focus on Concepts" questions in each chapter provide some of the best AP multiple-choice prep available anywhere. [from p. 216]

At the end of each chapter, right before the set of problems, is a section called "Focus on Concepts." This section includes conceptual questions posed as multiple-choice items.

To use these questions most effectively, start by answering within a minute or two—just as you would on the real AP exam. Then, check your answer with a classmate or your teacher. If you're wrong, write out a complete verbal explanation, just like you did with the "Check Your Understanding" questions. For starters, you might look at page 216 and the variety of conceptual multiple-choice items there.

> **6.** A particle is moving along the $+x$ axis, and the graph shows its momentum p as a function of time t. Use the impulse–momentum theorem to rank (largest to smallest) the three regions according to the *magnitude* of the impulse applied to the particle. **(a)** A, B, C **(b)** A, C, B **(c)** A and C (a tie), B **(d)** C, A, B **(e)** B, A, C

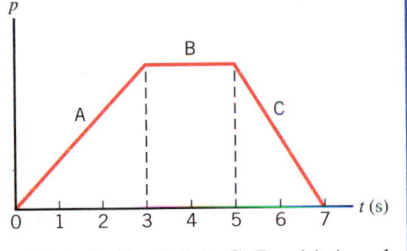

• A second common type of question doesn't ask just for a calculated answer, but rather for how that answer would change given a change in the situation surrounding the problem. To prepare for these questions, start with the no-star end-of-chapter problems that you're doing as homework.

> **40.** The skateboarder in the drawing starts down the left side of the ramp with an initial speed of 5.4 m/s. If nonconservative forces, such as kinetic friction and air resistance, are negligible, what would be the height h of the highest point reached by the skateboarder on the right side of the ramp?

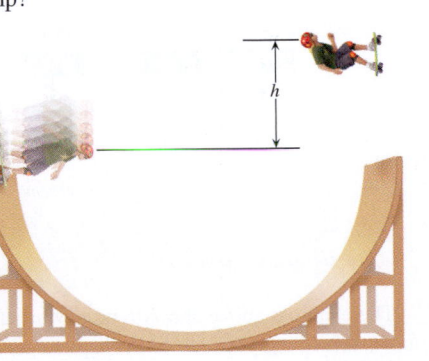

> For every no-star problem you do, ask how the answer would change if one of the given quantities were doubled. [from p. 190]

Say you do this problem as a homework assignment and find that the skateboarder goes 1.5 m above the ramp. Prepare for the AP exam by continuing the problem: How high would she go if her initial speed were doubled? Answer quickly!

It would take an awfully long time to redo the problem from scratch with 10.8 m/s as the initial speed. On the multiple-choice section, you only have about a minute per problem; on the free-response, you'd probably only have a few minutes to tackle this part. What to do?

You might think the height would certainly be doubled as well, but that's not correct. Rather, look at your solution. Try leaving the initial speed as a variable v rather than plugging in 5.6 m/s. You'll see that when you use the conservation of energy equation, the height depends on v *squared*. So doubling the speed *quadruples* the height!

Okay, now what if her mass were doubled?

You can take pretty much any no-star end-of-chapter problem and ask yourself, what if the (pick any given value) were doubled? This simple question will get you in the habit of solving in variables and recognizing within the timeframe of the test how those variables can change. Try problem 15 in chapter 6 about the speed of an arrow out of a bow. See what happens when you double the arrow's mass. Then, try doubling the distance that the bowstring stretches. How does the arrow's speed change (or not change)?

DON'T BE INTIMIDATED BY FREE-RESPONSE PROBLEMS

Often it is the cumulative nature of the AP exam that flummoxes a student. You will get comfortable approaching a problem that focuses on ONE equation or relevant concept, as you will see on the multiple-choice questions. However, free-response problems tend to combine two or three relevant concepts into a single problem. You might have to use BOTH conservation of energy AND conservation of momentum; you might see the gravitational force on a planet also act as a centripetal force to keep the planet in its orbit.

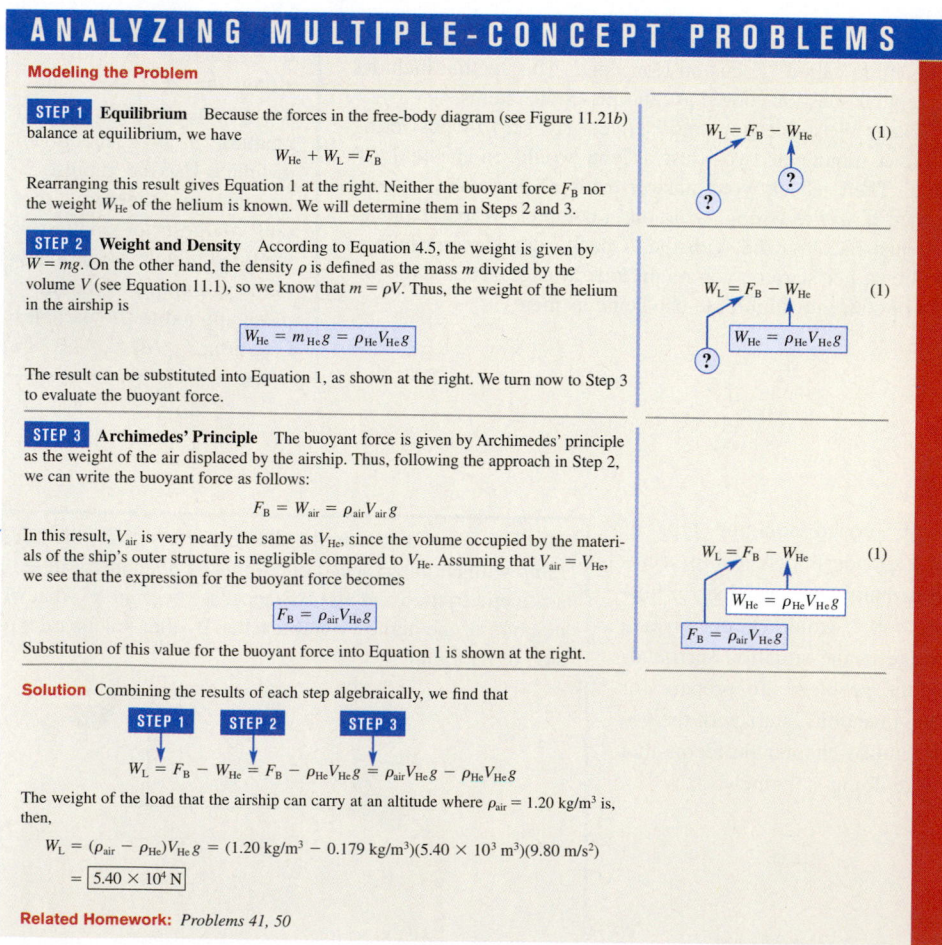

ANALYZING MULTIPLE-CONCEPT PROBLEMS

Modeling the Problem

STEP 1 **Equilibrium** Because the forces in the free-body diagram (see Figure 11.21b) balance at equilibrium, we have

$$W_{He} + W_L = F_B$$

Rearranging this result gives Equation 1 at the right. Neither the buoyant force F_B nor the weight W_{He} of the helium is known. We will determine them in Steps 2 and 3.

$$W_L = F_B - W_{He} \quad (1)$$

STEP 2 **Weight and Density** According to Equation 4.5, the weight is given by $W = mg$. On the other hand, the density ρ is defined as the mass m divided by the volume V (see Equation 11.1), so we know that $m = \rho V$. Thus, the weight of the helium in the airship is

$$W_{He} = m_{He}g = \rho_{He}V_{He}g$$

The result can be substituted into Equation 1, as shown at the right. We turn now to Step 3 to evaluate the buoyant force.

$$W_L = F_B - W_{He} \quad (1)$$
$$W_{He} = \rho_{He}V_{He}g$$

STEP 3 **Archimedes' Principle** The buoyant force is given by Archimedes' principle as the weight of the air displaced by the airship. Thus, following the approach in Step 2, we can write the buoyant force as follows:

$$F_B = W_{air} = \rho_{air}V_{air}g$$

In this result, V_{air} is very nearly the same as V_{He}, since the volume occupied by the materials of the ship's outer structure is negligible compared to V_{He}. Assuming that $V_{air} = V_{He}$, we see that the expression for the buoyant force becomes

$$F_B = \rho_{air}V_{He}g$$

Substitution of this value for the buoyant force into Equation 1 is shown at the right.

$$W_L = F_B - W_{He} \quad (1)$$
$$W_{He} = \rho_{He}V_{He}g$$
$$F_B = \rho_{air}V_{He}g$$

Solution Combining the results of each step algebraically, we find that

STEP 1   STEP 2   STEP 3

$$W_L = F_B - W_{He} = F_B - \rho_{He}V_{He}g = \rho_{air}V_{He}g - \rho_{He}V_{He}g$$

The weight of the load that the airship can carry at an altitude where $\rho_{air} = 1.20$ kg/m³ is, then,

$$W_L = (\rho_{air} - \rho_{He})V_{He}g = (1.20 \text{ kg/m}^3 - 0.179 \text{ kg/m}^3)(5.40 \times 10^3 \text{ m}^3)(9.80 \text{ m/s}^2)$$
$$= \boxed{5.40 \times 10^4 \text{ N}}$$

Related Homework: Problems 41, 50

The text extensively models the cumulative, "multiple-concept" problems that you will see throughout the AP free-response sections. [p. 332]

The example above shows a problem that requires you to analyze forces on an object in equilibrium—something you learned in chapter 4—AND to use Archimedes' principle, covered in Chapter 11. Note the form of the solution. The textbook describes each step separately but in detail, and then shows

2op: 1, 6, 9b, 15, 17

Today's Hero

Admiral Robert Peary became the first man to reach the North Pole on April 6, 1909. Peary, a civil engineer by trade, had earlier explored the northern tip of Greenland with his then-pregnant wife, who gave birth to their child while still on the island. During his expeditions to Greenland, he discovered and named Independence Bay, explored the northeastern coast, and named the island's northern-most point, Cape Morris Jesup. As a result of his explorations, Peary proved that Greenland is an island rather than a continent.

the relevant mathematics alongside. Look at the "related homework" questions. Try these (problems 41 and 50 in chapter 9). Pay especially close attention to those at the one-star difficulty level.

A free-response question will often "talk you through" the reasoning by breaking the solution into multiple, lettered parts. Thus, in the example above, step one might be "part (a)," step two "part (b)," and so on. Focus on one part at a time. Do the same thing when solving problems from the text: describe each step in turn. Don't try to answer the entire problem in one minute.

> **Get partial credit–always try the ENTIRE problem**
>
> Remember, you are only expected to get 60–65% of the points to earn a 5! So, don't give up on an entire free-response problem just because you can't do part (b). If you can do (a) and (c), you're still in line for a 5. When you can't figure out one part, make up an answer and move on to the next part. Communicate with the reader that you DO know how to solve part of the problem. You'll receive partial credit appropriate to the understanding you've shown.

THE CALCULATOR IS A CONVENIENCE, NOT A CRUTCH

You are allowed a calculator on the free-response section. But, you should not be spending much time at all pounding buttons! Solve every problem on paper, explaining the relevant principles, using the relevant equations, and writing down all substitutions, before you finally use the calculator to get a final numerical answer. An appropriate problem presentation that minimizes calculator use until the very end is modeled in every example problem.

This equation can be solved for v_f, the common velocity of the two cars after the collision:

$$v_f = \frac{m_1 v_{01} + m_2 v_{02}}{m_1 + m_2}$$

$$= \frac{(65 \times 10^3 \text{ kg})(0.80 \text{ m/s}) + (92 \times 10^3 \text{ kg})(1.3 \text{ m/s})}{(65 \times 10^3 \text{ kg} + 92 \times 10^3 \text{ kg})} = \boxed{+1.1 \text{ m/s}}$$

> When solving a problem, write out equations and substitutions clearly *before* touching your calculator. Note that numbers are rarely inserted until the final steps of an example problem. [p. 202]

THE EQUATION SHEET IS ALSO A CONVENIENCE, NOT A CRUTCH

On the free-response portion of the test, you'll be given an equation sheet. In fact, it might be useful to get yourself a copy of the equation sheet via **collegeboard.com** or from your teacher. But, it is important that you use the equation sheet correctly. Experienced AP Physics students can offer two "do's" and one "don't" about the equation sheet:

- *DO* **use the equation sheet as a study guide.**

The sheet is organized by topic. So, after you've studied each chapter in the text, compare your copy of the equation sheet with the equations listed in the "Concept Summary" at the end of the chapter. Be able to spit back, in your own words, what each equation means and when it's valid. If you have a test in your class tomorrow, these concept summaries, in conjunction with the equation sheet, are the fundamental issues you need to review. When you're stumped on a homework problem, use the equation sheet to remind you of the principles you have available to work with.

> At the end of each chapter, the Concept Summary lists these important relationships and what they mean in an easy-to-read, at-a-glance format. It's often useful to read over these briefly to remind yourself of what physics tools you have available–this is the same way you'll use the equation sheet on the AP exam. [p. 562]

CONCEPT SUMMARY

Point charge   **18.5 COULOMB'S LAW**   A point charge is a charge that occupies so little space that it can be regarded as a mathematical point.

Coulomb's law gives the magnitude F of the electric force that two point charges q_1 and q_2 exert on each other:

Coulomb's law

$$F = k \frac{|q_1||q_2|}{r^2} \tag{18.1}$$

where $|q_1|$ and $|q_2|$ are the magnitudes of the charges and have no algebraic sign. The term k is a constant and has the value $k = 8.99 \times 10^9 \text{ N} \cdot \text{m}^2/\text{C}^2$. The force specified by Equation 18.1 acts along the line between the two charges.

Permittivity of free space   The permittivity of free space ϵ_0 is defined by the relation

$$k = \frac{1}{4\pi\epsilon_0}$$

Recognize which equations in the text might be stated using slightly unorthodox notation on the AP exam: for example, on the official AP equation sheet, Coulomb's law is expressed as $F = \dfrac{1}{4\pi\epsilon_0} \dfrac{q_1 q_2}{r^2}$, which is slightly different from what you'll see in chapter 18 (where the Coulomb's law constant k is used). Know about these minor differences ahead of time, so they don't throw you.

It is a most useful exercise to mark up your equation sheet. Write down the kinds of problems in which an equation is useful. Define any confusing variables. Use the text to create *your* "concept summary," written in your own words. Of course, you cannot take a marked-up equation sheet into the AP exam—you have to use the one provided with the exam. But the process of making and using your own annotated sheet throughout the year will familiarize you with the material you are expected to know.

- *DO* **look at the equation sheet during the exam to double-check the form of an equation you intend to use.**

Imagine that part of a problem gives information about an object moving in a straight line. You know the displacement, the (constant) acceleration, and the final velocity; the problem asks for the object's initial velocity. What's your thought process?

First, you categorize the problem as one-dimensional kinematics. Your problem-solving strategy is, therefore (as discussed in Chapter 2.4) is to choose the kinematics equation that includes the three known variables, allowing you to solve for the unknown.

But wait . . . even though you are pretty confident about what these equations say, you might not be 100% sure of the precise form: Is it $v_f^2 = v_0^2$ PLUS $2ax$, or is it MINUS $2ax$? Granted, you should know this, because a multiple-choice problem may require you to make a simple calculation without the equation sheet. On the free-response section, though, better a quick glance to be sure than to make a preventable error. A quick glance shows a PLUS sign. Solve and move on.

2.4 EQUATIONS OF KINEMATICS FOR CONSTANT ACCELERATION   2.5 APPLICATIONS OF THE EQUATIONS OF KINEMATICS   The equations of kinematics apply when an object moves with a constant acceleration along a straight line. These equations relate the displacement $x - x_0$, the acceleration a, the final velocity v, the initial velocity v_0, and the elapsed time $t - t_0$. Assuming that $x_0 = 0$ m at $t_0 = 0$ s, the equations of kinematics are

Ex. 5–9, 18

$$v = v_0 + at \tag{2.4}$$

CS 2.1, 2.2

$$x = \tfrac{1}{2}(v_0 + v)t \tag{2.7}$$

$$x = v_0 t + \tfrac{1}{2}at^2 \tag{2.8}$$

ILW 2.1, 2.2

$$v^2 = v_0^2 + 2ax \tag{2.9}$$

IS 2.36, 2.84

Know your equations. But if you have any doubts, look the equation up just to be sure you've got it right. [p. 52]

- *DON'T* **go fishing for equations that might possibly be useful.**

It is neither good physics nor a useful test-taking strategy to hunt through the equation sheet without knowing already what you're looking for. It is laudable, as discussed above, to double-check whether the force of a spring is $F = kx$ or $F = \tfrac{1}{2}kx^2$. You knew before you searched that you needed the force of a spring equation, and you remembered generally what that equation looked like. It would be a major hindrance both to your physics understanding and to your test score if you instead looked all over the equation sheet for any old equation with an F in it.

PRACTICE SOLVING PROBLEMS UNDER TIME PRESSURE

The free-response questions, on average, must be solved in just about one minute per point. Of course, smart test-taking strategy can help you use that time wisely; for example, if you can pick an easy question or two to knock out quickly, then you have more time for the toughies. Nevertheless, you may find yourself pressed for time on the free-response section.

So, practice solving problems in limited time. Sure, at the beginning of your course, as you're still learning how to deal with the unique nature of physics problem solving, you might feel the need to spend an hour or two on a just a couple of homework problems. Fine. As the course progresses, try removing the luxury of unlimited time from your daily work. (You probably don't want to do that much homework anyway, right?)

Use the time frame of the free-response exam to determine your approach to homework as the year progresses. Spend about 5–10 minutes on a no-star question, or about 10–15 minutes on a one-star question. Then, STOP. You probably won't be finished, but you should have gotten somewhere.

Go talk to a friend or your teacher before you finish the problem. The AP exam structure does not allow you to stare at a problem without getting somewhere, and you won't have any magical insights after 15 minutes that you didn't have in the first 5. See how far you got on your own, and think to yourself how you could get further next time. Eventually, you'll find that you can answer many homework questions fully, by yourself, and in the time allotted. That's when you'll know you're ready for the AP exam.

> **5. ssm** A car is hauling a 92-kg trailer, to which it is connected by a spring. The spring constant is 2300 N/m. The car accelerates with an acceleration of 0.30 m/s^2. By how much does the spring stretch?
>
> * **20. GO** Objects of equal mass are oscillating up and down in simple harmonic motion on two different vertical springs. The spring constant of spring 1 is 174 N/m. The motion of the object on spring 1 has twice the amplitude as the motion of the object on spring 2. The magnitude of the maximum velocity is the same in each case. Find the spring constant of spring 2.

Spend about 5 or 10 minutes on this no-star problem before either finishing or seeking help. [p. 313]

Spend about 10 or 15 minutes on this one-star problem before either finishing or seeking help. [p. 314]

KNOW HOW THE FREE-RESPONSE IS GRADED

Every summer, 150–200 high school and college physics teachers descend on Fort Collins, Colorado, to grade the AP Physics free-response tests—including yours. For each problem, these "readers" develop a rubric that awards points not just for the correct answers but also for correct solution methods and correct statements of physics principles. *You can earn substantial partial credit on every part of every free-response problem.*

So show your work clearly. Communicate with the reader. Use words as well as mathematics in presenting your solution. The more you can tell the reader about your approach to the problem, the more likely the reader will be able to award credit for your understanding.

Whenever the textbook presents an example problem, effective communication is modeled. Notice how the solution is explicit about what equations and principles are used; what simplifying assumptions are made; what values are plugged into the final equation; and WHY those values are used. Though you may not use quite so many words when you have only a few minutes to write your answer, your solution still should cover all these issues in order to earn as much credit as possible.

> **Solution** Figure 6.11b shows the gymnast moving upward. The initial and final heights are $h_0 = 1.20$ m and $h_f = 4.80$ m, respectively. The initial speed is v_0 and the final speed is $v_f = 0$ m/s, since the gymnast comes to a momentary halt at the highest point. Since $v_f = 0$ m/s, the final kinetic energy is $KE_f = 0$ J, and the work–energy theorem becomes $W = KE_f - KE_0 = -KE_0$. The work W is that due to gravity, so this theorem reduces to $W_{gravity} = mg(h_0 - h_f) = -\frac{1}{2}mv_0^2$. Solving for v_0 gives
>
> $$v_0 = \sqrt{-2g(h_0 - h_f)} = \sqrt{-2(9.80 \text{ m/s}^2)(1.20 \text{ m} - 4.80 \text{ m})} = \boxed{8.40 \text{ m/s}}$$

The problem may have merely asked you to determine the initial speed of a gymnast, but notice that a proper solution requires more than just "8.40 m/s." [p. 170]

So, how much work do you *need* to show? Take a look at the "analyzing multiple concepts problems" feature. The right-hand column of these boxes always shows a well-crafted solution without a lot of writing. Don't be afraid to use words or annotations in your answer, as is always modeled in the features. The better presented your work, the greater the likelihood of partial credit even if your answer is wrong.

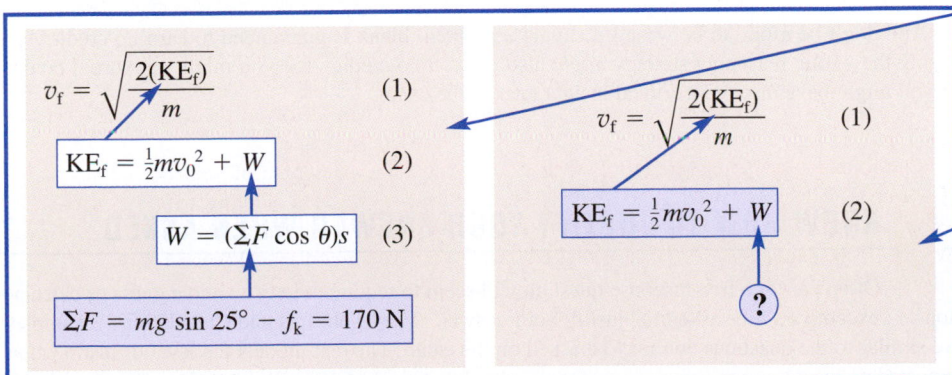

The text shows you how to present a mathematical solution in a form that can be followed by an AP reader. Even if, say, step 2 is incorrect, or if you plug in wrong values, you'll still earn substantial partial credit for a correct approach. [p. 168, p. 166]

Use words wherever possible to annotate your solution. Make it crystal clear to the reader what you're doing. [p. 166]

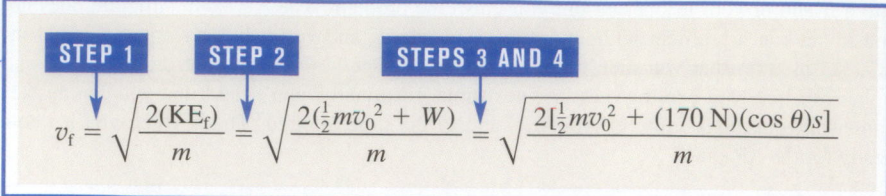

$$v_{\mathrm{f}} = \sqrt{\frac{2(KE_{\mathrm{f}})}{m}} = \sqrt{\frac{2(\frac{1}{2}mv_0^2 + W)}{m}} = \sqrt{\frac{2[\frac{1}{2}mv_0^2 + (170 \text{ N})(\cos\theta)s]}{m}}$$

Remember, YOU are not invited to read and grade the free-response questions. What you write on the exam is all the reader sees about your solution. After an in-class test, you might be able to go up to your teacher and say, "Hey, let me tell you what I meant on problem 2, because you might not be sure based on what I wrote down." Not so on the AP exam . . . you must leave no doubt in the reader's mind that you know what you're doing.

CATEGORIZE PROBLEMS AND HAVE A STRATEGY FOR EACH CATEGORY

Sometimes you might stare at a problem and simply have no idea where even to start. Hopefully this is a rarer occurrence as the school year progresses. But, it *will* happen to you. Prepare a technique that might jump-start your pencil into doing something useful: when you're uncertain, try categorizing the problem. Figure out which topic is being covered. If you're not 100% sure of the topic at hand, just guess—writing *something* is always better than writing nothing.

When you have a type of problem identified, you can use a reasoning strategy you're familiar with. Pink boxes with clear Reasoning Strategies are spread throughout the text. Apply a Reasoning Strategy to your difficult problem. You will likely find that merely the process of writing will get you going in the right direction.

Use the Reasoning Strategies, presented in the text for many physics principles, to get started. Reasoning Strategies can be a useful crutch even when you're a bit confused. [p. 203]

REASONING STRATEGY

Applying the Principle of Conservation of Linear Momentum

1. Decide which objects are included in the system.

2. Relative to the system that you have chosen, identify the internal forces and the external forces.

3. Verify that the system is isolated. In other words, verify that the sum of the external forces applied to the system is zero. Only if this sum is zero can the conservation principle be applied. If the sum of the average external forces is not zero, consider a different system for analysis.

4. Set the total final momentum of the isolated system equal to the total initial momentum. Remember that linear momentum is a vector. If necessary, apply the conservation principle separately to the various vector components.

And don't be afraid to be wrong. Leaving a problem blank is guaranteed to earn no credit. Applying the wrong reasoning strategy *might* also get you no credit—but you might get partial credit, or you might have inspired a critical insight later in the exam.

KNOW HOW TO JUSTIFY YOUR ANSWER WHEN ASKED

Often part of a free-response question will seem to require merely a simple guess or calculation . . . except you'll be asked to "justify your answer." The "Concepts and Calculations" examples are similar to the questions you may be asked on the exam. The text models the level of justification expected on the AP exam.

⬇ Concepts & Calculations Example 10   A Scalar and a Vector

Two joggers, Jim and Tom, are both running at a speed of 4.00 m/s. Jim has a mass of 90.0 kg, and Tom has a mass of 55.0 kg. Find the kinetic energy and momentum of the two-jogger system when **(a)** Jim and Tom are both running due north (Figure 7.18a) and **(b)** Jim is running due north and Tom is running due south (Figure 7.18b).

Concept Questions and Answers   Does the total kinetic energy have a smaller value in case (a) or (b), or is it the same in both cases?

Answer   Everything is the same in both cases, except that both joggers are running due north in (a), but one is running due north and one is running due south in (b). Kinetic energy is a scalar quantity and does not depend on directional differences such as these. Therefore, the kinetic energy of the two-jogger system is the same in both cases.

> This "Concept Questions and Answers" feature poses a simple-sounding question that requires a couple of sentences of explanation, just like you'll see on the AP exam. [p. 213]

Two to three sentences, sometimes (but not always) including an equation or a brief calculation, are sufficient. Be forewarned not to go overboard to impress the readers. For one thing, long answers waste time that could be better spent elsewhere on the exam. For another, you can't get extra credit on a part, no matter how impressive your answer may be. And finally, you can in fact LOSE credit that you already earned if you accidentally contradict yourself, or if you say both the right answer and the wrong answer. There's no point in hedging your bets, because readers are instructed to penalize incorrect statements even in the presence of correct ones. Just write your couple of sentences, and stop while you're ahead.

CYU 13:   Use a shoe and the shoe laces to make a simple pendulum whose period is related to the magnitude g of the acceleration due to gravity (see Equations 10.5 and 10.16). Measure the period of your pendulum and calculate g.

> The answer in the back of the book models the straightforward, concise response that will serve you best. [p. A-10]

USE THE RESOURCES PROVIDED BY THE COLLEGE BOARD AND OTHER AVAILABLE STUDY AIDS

Your teacher can get access to all sorts of previously administered exam materials. Every free-response test is released to teachers, as are the multiple-choice sections from some years. By looking at some examples of real AP questions (and by looking through this appendix), you can get a good feel for which style of questions in this text most closely mimic the actual AP Physics exam.

Attempt some of these authentic AP questions under test conditions. Try 3 questions in 45 minutes or so. Then, grade the problems using the official rubric written by the readers. (Your teacher can provide these rubrics for you—ask!) You'll see what you're good at—and what you need to improve.

If you need more practice problems in AP format, get the indispensable Student Study Guide for Advanced Placement Physics B. This book, written by a former member of the AP Physics Test Development Committee, provides you with all sorts of practice questions. You can use this book as review after a chapter, or as review for the actual AP exam in May.

PRACTICE, PRACTICE, PRACTICE

The resources listed above are certainly useful, but you can prepare for the AP exam even with just this textbook. Try making a short practice test. Pick 12 "Focus on Concepts" questions at random from the chapters listed in the preface that you have covered in class. Take 15–20 minutes to answer these. Then, pick three one-star problems at random from the listed chapters, to be answered in 45 minutes. Check your answers with the book, with a friend who took the same "practice test," or with your teacher. Then do the same "test" again, except do it perfectly, explaining your answers to the multiple-choice questions you missed the first time. The process of discovering, admitting, and correcting your mistakes is critical to your success in AP Physics.

9. Force $\vec{F}_1$ acts on a particle and does work W_1. Force $\vec{F}_2$ acts simultaneously on the particle and does work W_2. The speed of the particle does not change. Which one of the following must be true? **(a)** W_1 is zero and W_2 is positive **(b)** $W_1 = -W_2$ **(c)** W_1 is positive and W_2 is zero **(d)** W_1 is positive and W_2 is positive

2. Six runners have the mass (in multiples of m_0), speed (in multiples of v_0), and direction of travel that are indicated in the table. Which two runners have identical momenta? **(a)** B and C **(b)** A and C **(c)** C and D **(d)** A and E **(e)** D and F

Runner	Mass	Speed	Direction of Travel
A	$\frac{1}{2}m_0$	v_0	Due north
B	m_0	v_0	Due east
C	m_0	$2v_0$	Due south
D	$2m_0$	v_0	Due west
E	m_0	$\frac{1}{2}v_0$	Due north
F	$2m_0$	$2v_0$	Due west

6. Ball 1 is thrown into the air and it follows the trajectory for projectile motion shown in the drawing. At the instant that ball 1 is at the top of its trajectory, ball 2 is dropped from rest from the same height. Which ball reaches the ground first? **(a)** Ball 1 reaches the ground first, since it is moving at the top of the trajectory, while ball 2 is dropped from rest. **(b)** Ball 2 reaches the ground first, because it has the shorter distance to travel. **(c)** Both balls reach the ground at the same time. **(d)** There is not enough information to tell which ball reaches the ground first.

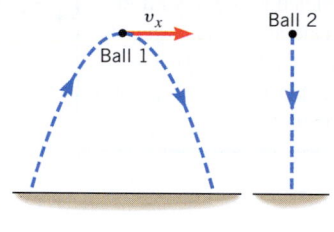

Just copy and paste some "Focus on Concepts" questions from each chapter, and you have a multiple-choice practice test–or a three-question quiz like this one here. [p. 187, 216, 81]

This text provides you all the guidance you'll need to ace the AP Physics exam. You just have to use it actively. All the features in the world can't prepare you for the exam without your intensive participation. Now that you've read this appendix, you know which parts and features of the book are most useful in your studying. It merely remains for you to follow the advice herein, to engage the problems and questions, and to correct your mistakes along the way. Learning physics is not merely about reading a text but is a participatory process.

FALL PHYSICS EXAM:

☐ MULTIPLE CHOICE & PROBLEMS
   (35-40)   (6)

- KINEMATICS, VECTORS, FORCES

- LOOK OVER OLD TESTS

- POSS. WA REVIEW

- NO RAMPS & FRICTION

- COMMONLY MISSED PROBLEMS, #3 FORCES TEST

- T